THERMAL ENGINEERING

(SI UNITS)

[BE (Mech.), B.Tech (Mech.), UPSC (Engineering Services), AMIE, M.Tech. etc.)

By

Prof. D.K. Chavan
Mechanical Engineering Department,
Marathwada Mitra Mandal's
College of Engineering
(M.M.C.O.E.) Pune–52
Ex. Assistant Professor
Mechanical Engineering Department,
M.I.T., Pune–38

Prof. G.K. Pathak
Sr. Faculty Member
Mechanical Engineering Department,
Maharashtra Institute of Technology
M.I.T., Pune–38

STANDARD BOOK HOUSE

Unit of: **RAJSONS PUBLICATIONS PVT. LTD.**

1705-A, Nai Sarak, P.B. No. 1074, Delhi - 110006 Ph.: +91-(011)-23265506 • Fax: +91-(011)-25555295
Show Room: 4262/3, First Lane, G-Floor, Gali Punjabian, Ansari Road, Darya Ganj,
New Delhi-110002 Ph.: +91-(011)-43551085, +91-(011)-43551185
E-mail: sbh10@hotmail.com • www.standardbookhouse.com

Published by:
RAJINDER KUMAR JAIN
Standard Book House
Unit of: Rajsons Publications Pvt. Ltd.
1705-A, Nai Sarak, Delhi - 110006
Post Box: 1074
Ph.: +91-(011)-23265506 Fax: +91-(011)-25555295

Showroom:
4262/3, First Lane, G-Floor, Gali Punjabian,
Ansari Road, Darya Ganj,
New Delhi-110002
Ph.: +91-(011)-43551085, +91-(011)-43551185
E-mail: sbh10@hotmail.com
Web: www.standardbookhouse.com

Second Edition: 2013

Price: **₹ 825.00**

ISBN: 978-81-89401-20-7

Typeset by:
C.S.M.S. Computers, Delhi.

Printed by:
Radha Press, Kailsh Nagar, Delh-110031

DEDICATED

TO

Late Shri Kishanrao L Chavan

Late Smt. Umabai Kishanrao Chavan

Late Shri Ramkrishna I Zad

Late Shri Jitendra R Zad

Preface

It gives us great pleasure to present a book on 'Thermal Engineering', for the students of Mechanical Engineering from first year to final year engineering, Post graduate student.

In this book we will study basics of mechanical engineering, steam generators, fuels and combustion, compressors, steam nozzles, turbines, I.C. engines and their systems, gas turbines, jet propulsion, refrigeration and air-conditioning, non-conventional power sources.

Also introduction to metal cutting processes, metal joining, sheet metal working, machine design, elements and introduction to Mechatronics are also included to increase the usefulness of the book for first-year students for various universities all over the country.

Every care has been taken to present the matter in precise and simple language. Self explanatory figures are included to enable the students to draw the same in the exams very easily.

From a long teaching experience it has been observed that students commit mistakes in the units, as such, for every step in the solution of the problems, units have been made clear.

To develop more confidence in students appropriate questions from various universities are given at the end of each chapter. In the entire book S.I. system of units is used.

All the necessary care has been taken to avoid mistakes/misprints in the book. However it is quite likely that some mistakes/misprints might have been passed unnoticed. Such mistakes and misprints of the book if brought to notice will be gratefully acknowledged.

Any suggestions to improve the usefulness of the book will be gratefully accepted.

We express our sincere thanks to,

(i) Prof. Dr. A.N. Gaikwad, Prof. S.N. Satpati of MMCOE, Pune–52
(ii) Prof. Raghubir Singh, Principal, Wingss Aeronautical College, Warje, Pune.
(iii) Prof. D. S. Pavaskar, Nishant Prakashan, Pune.
(iv) Prof. Molkeri Chandrashekhar M., H.O.D., Mechanical Engineering Department, Rural Engineering College, Bhalki, Bidar, Karnataka.
(v) Prof. Pujari G. K., Hyderabad.
(vi) Prof. N. K. Joshi, H.O.D., Mechanical Department, Tolani Maritime Institute Talegaon, Pune.
(vii) Prof. V. C. Tapali, P.D.A. College of Engineering, Gulbarga, Kamataka

We also express our sincere thanks to entire staff of Standard Book House, New Delhi for publishing the book in the very short span of time.

Authors

August, 2013

Contents

PART 1

CHAPTER 21. STEAM CONDENSERS 831–856

CHAPTER 22. IMPROVED RANKINE CYCLE 857–953

CHAPTER 35. JET PROPULSION AND ROCKET PROPULSION 1242–1295

Symbols Used in this Book

A	=	Area
a	=	Acceleration
C	=	Specific heat, velocity
°C	=	Degree centigrade
C_p	=	Specific heat at constant pressure
C_v	=	Specific heat at constant volume
F	=	Force
g	=	Acceleration due to gravity
g_c	=	Constant of proportionality in Newton's law
h	=	Specific enthalpy
H	=	Enthalpy
J	=	Mechanical equivalent of heat or joules equivalent
K	=	Degree kelvin
kg	=	Kilograms
KE	=	Kinetic Energy
L	=	Length
m	=	Metre
M	=	Molecular weight
n	=	Number of moles, Polytropic index
N	=	Revolutions per minute (RPM)
P	=	Pressure
PE	=	Potential energy
q, Q	=	Heat tranfer
R	=	Specific gas constant
$\bar{R}$	=	Universal gas constant
sS	=	Entropy
t	=	Temperature celsius
T	=	Absolute temperature
u, U	=	Internal energy
v, V	=	Volume
w, W	=	Work
x	=	Dryness fraction
Z	=	Elevation above datum

GREEK SYMBOLS

γ	=	Specific heat ratio
Δ	=	Change in a property
δ	=	Small change in a property
η	=	Efficiency
θ	=	Temperature
ρ	=	Density

Introduction

SOURCES OF ENERGY

All energy eventually has common origin from the sun. Energy, especially electric energy is important for day-to-day living and the industry. Energy from various sources—like fossil fuels, coal, water etc. is converted into electrical energy and then distributed wherever needed.

Thermal

1. About 65% of power production is through thermal.
2. Production cost is less.
3. Stock of coal is limited and quality of coal is also poor.
4. Plants also have less production capacity.

Gas

1. Cost of gas is increasing in other countries, so costlier projects.
2. Countries supplying gas, have increased the cost of gas in recent days.
3. Supply of gas is also not continuous.

Hydro

1. Because of environmental and rehabilitation problems, projects are getting delayed.
2. Time required for construction and starting up is about 1.5 times than the schedule.

Nuclear

1. Less than 3% of total power share
2. Continuous supply of Uranium is necessary.
3. Thorium based energy plants are cheaper, but because of technical problems it requires more time.

POWER GENERATION SCENARIO IN THE FOLLOWING COUNTRIES

1. India

(a) Out of total energy requirement 3.5% is through Nuclear, upto 2050 MW it has to be increased to 2.5%.
(b) In India less capacity about 220 MW plants are about 17.
(c) Uranium deficity is the main cause.
(d) To develop technology it requires time.
(e) Tremendous Stocks of Thorium, but technological problems, otherwise use can produce energy for a very long period of time.

2. America

(a) About 21% of total power requirements is provided through Nuclear energy source.
(b) In 32 states nearly 100 Nuclear power plants.

3. France

(a) About 77% of total power requirement is produced through Nuclear energy.
(b) For power generation, there are 19 plants in the country.

4. China

(a) More share is through coal and hydro.
(b) 10 Nuclear power plants are active.
(c) On 10 work is under progress and 31 other are sanctioned.

Challenges

(a) About half of our population resides in dark.
(b) Supplying electricity to all parts of country, it requires more production of electricity.
(c) Coal is depleting continuously stocks are available for about 100 more years.
(d) Energy generation technology is to be developed, otherwise it will effect economy of our country.

PART–1

CHAPTER 1

Basic Concepts of Thermodynamics and Zeroth Law

1.1 INTRODUCTION

In order to understand any science, everyone must get familiar with its unique terminology. This chapter is dedicated for introduction to the basic terms and concepts involved in thermodynamics. Along with the introduction to various concepts and definitions of different terms we shall also discuss the basic system of units which will be followed in the numerical part of this chapter.

1.2 THERMODYNAMIC SYSTEMS

Sometime we have to visit a general physician for certain health related problem. At that time, the physician examines our body for the accurate diagnosis of the illness. But, when we consult an ophthalmologist, he concen-

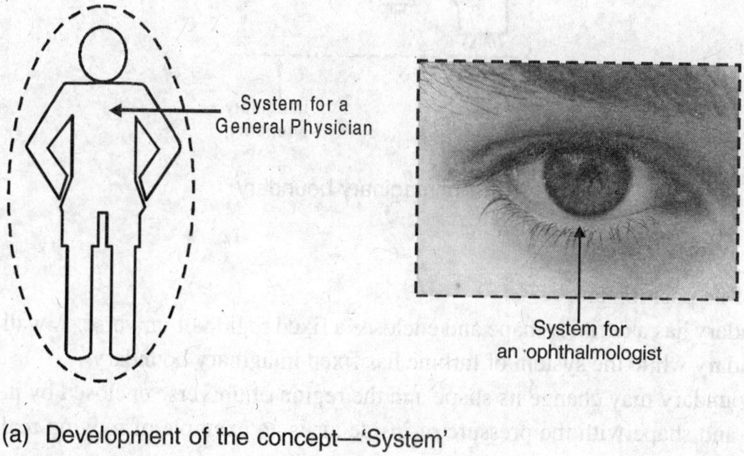

(a) Development of the concept—'System'

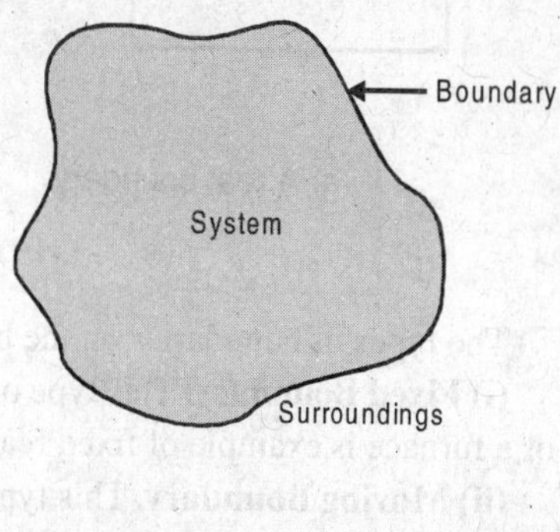

(b) Thermodynamic system

Fig. 1.1

trates only a part of our body i.e., Eye. Thus, every doctor decides the portion of body upon which concentration is made, according to his scope of diagnosis and treatment.

Above mentioned fact is elaborated in view of thermodynamic studies and is specified with the terms like System, Surrounding and Boundary.

A Thermodynamic System is defined as *quantity of matter or region of space upon which attention is concentrated in the analysis of the problem.*

Thermodynamic system may be defined as any part of the *Universe* which we want to study.

In order to fix the limits of a system, there must be a partition around the system. This partition is known as the *boundary of the system.*

Everything external to the boundary is called the *Surrounding.* The system along with surrounding forms an *Universe.* [Ref. Fig. 1.1(b)].

1.2.1 Types of Boundary

The working of a thermodynamic system depends on the energy interactions between the system and surrounding, which takes place across the boundary. Hence we must start our studies with the type of boundaries and its properties.

The types of boundaries on the basis of its existence are:

(i) Real Boundary. Consider an oven used for baking the food products. The wall of the oven physically encloses the heating coils and the food products. Thus it forms a real boundary. The cylinder-piston in IC engine provides a real boundary during expansion or compression of inside gases. [Refer Fig. 1.2 (a)].

(ii) Imaginary Boundary. When we wish to study the working of a tap of water, we separate the valve system from rest of the thing using an imaginary boundary. In thermodynamics the systems like nozzles, turbines, carburettor of vehicle can be defined with a concept of such an imaginary boundary. [Refer Fig. 1.2 (b)].

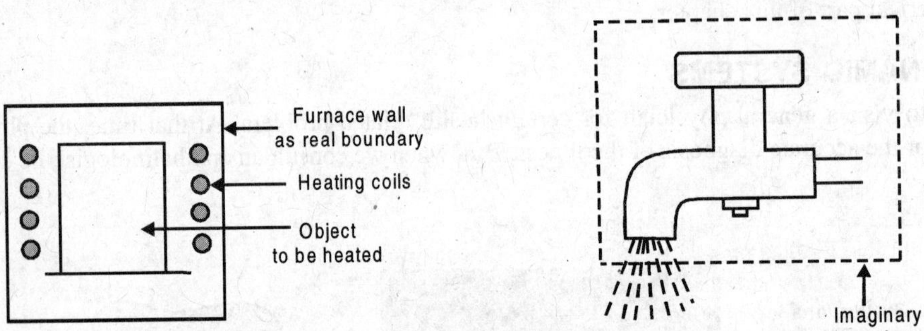

(a) A real boundary (b) An imaginary boundary

Fig. 1.2

The types of boundaries on the basis of its position are:

(i) Fixed Boundary. This type of boundary has a definite shape and encloses a fixed region of universe. A wall of a furnace is example of fixed real boundary while the system of turbine has fixed imaginary boundary.

(ii) Moving Boundary. This type of boundary may change its shape and the region of universe enclosed by it. A wall of balloon which changes its size and shape with the pressure of inside air is an example of moving real boundary.

1.2.2 Types of System

The types of systems may be viewed in a tree form as shown in Fig. 1.3

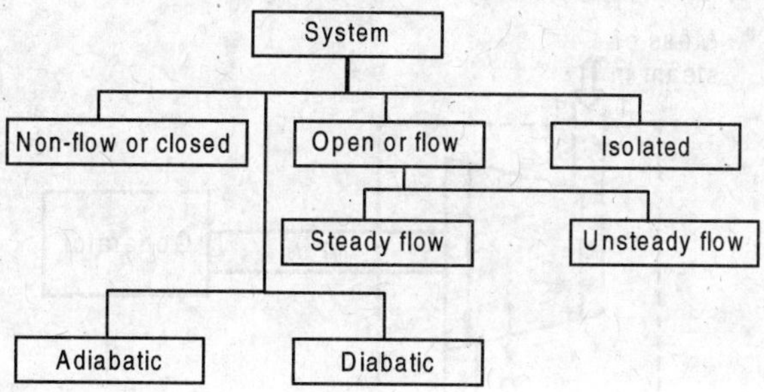

Fig. 1.3

(a) Closed or Non-flow System. It is fixed mass i.e., there is no mass transfer across the boundary. But there may be energy transfer into and out of the system as shown in Fig. 1.4 (a).

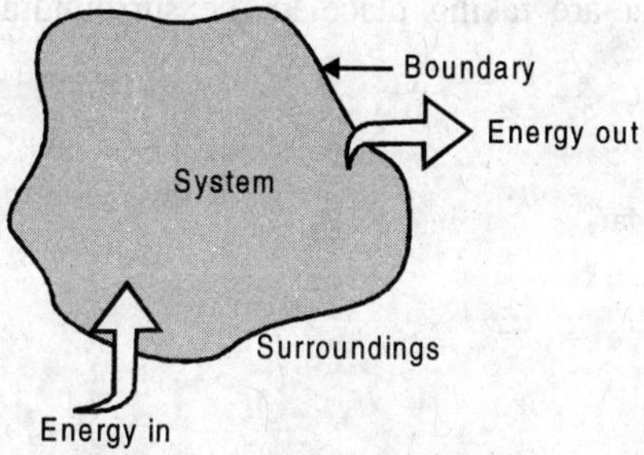

(a) A closed system

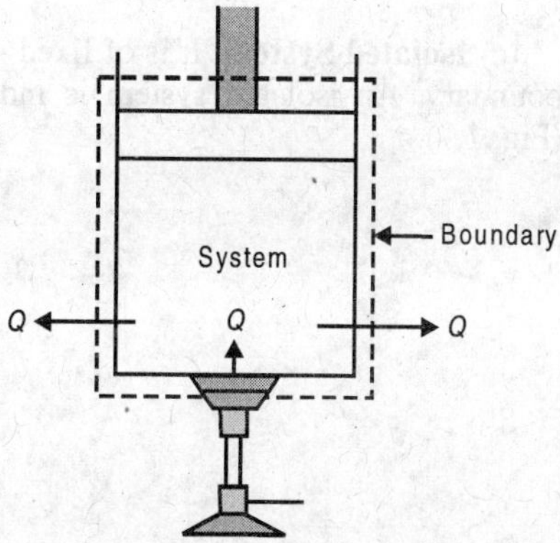

(b) Example of a closed system

Fig. 1.4

As an example, consider a system consisting of a gas, in a cylinder fitted with a piston as shown in Fig. 1.4 (b). If heat is supplied to the gas, the gas expands i.e., the volume of the gas increases so work will be done by the gas on the piston. Thus heat and work will cross the system boundary. But since no mass crosses the boundary, this system is called a *closed system.*

(b) Open System or Flow System. It is one in which both mass and energy can enter and leave the system as shown in Fig. 1.5. Most of the engineering devices are open systems.

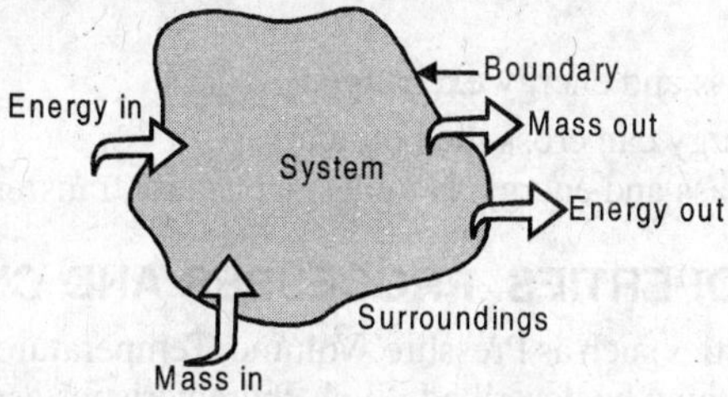

Fig. 1.5 *An open system*

For example, consider a portion of steam plant as shown in Fig. 1.6. Mass of steam enters at section (1) and after expanding into the turbine, leaves at section (2). In this case work is produced when the turbine is driven. So, as shown, the mass, work and heat are crossing the system boundary.

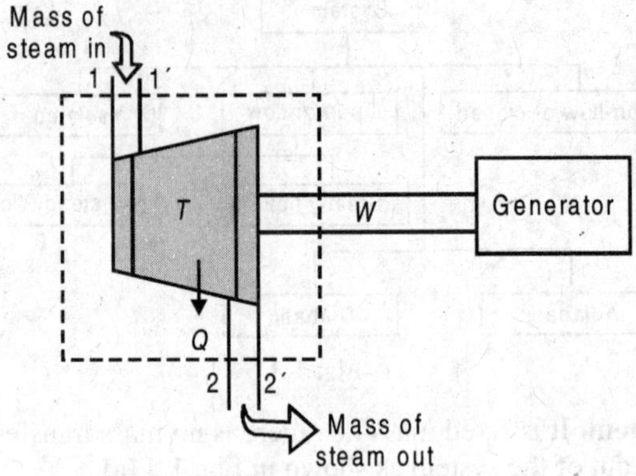

Fig. 1.6 *Example of an open system*

(c) Isolated System. It is of fixed mass and energy and there is no mass or energy transfer across the system boundary. An isolated system is independent of all the changes that are taking place in the surroundings. (Fig. 1.7).

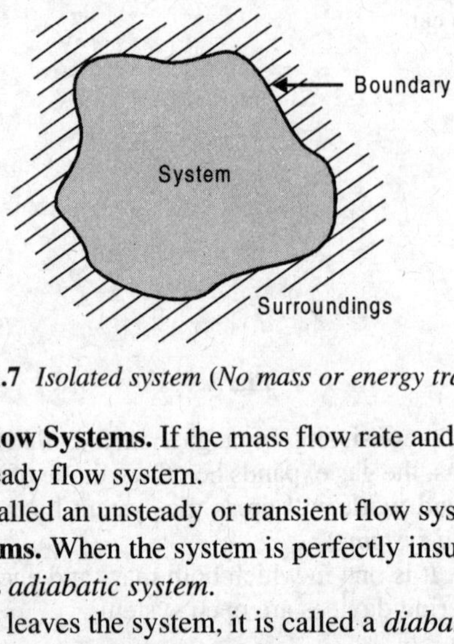

Fig. 1.7 *Isolated system (No mass or energy transfer)*

(d) Steady Flow and Unsteady Flow Systems. If the mass flow rate and energy flow rate at inlet and outlet are same, then the system is known as steady flow system.

If the rates are varying, then it is called an unsteady or transient flow system.

(e) Adiabatic and Diabatic Systems. When the system is perfectly insulated, so that neither heat enters nor it leaves the system, then it is known as *adiabatic system.*

When heat energy, either enters or leaves the system, it is called a *diabatic system.* It is not insulated.

Points to Remember :

- Closed system is of fixed mass and energy exchange.
- In open system mass and energy can cross the boundary.
- Isolated system is of fixed mass and energy. No energy or mass transfer occurs.

1.3 THERMODYNAMIC PROPERTIES, PROCESSES AND CYCLES

Every system has certain characteristics such as Pressure, Volume, Temperature, Density, Internal Energy, Enthalpy etc. by which its physical condition may be described. Such characteristics are called as *Properties* of the system.

Properties may be of two types:

1. Intensive Properties. They are independent of total mass in the system e.g., Pressure, Temperature, Density.

2. Extensive Properties. These are dependent on the total mass in the system. When the mass of the system changes, the values of the properties will change proportionately, e.g., Volume, Weight, Energy, Enthalpy, Entropy.

Properties may also be classified as —

1. Primary properties (P, V, T)
2. Secondary or derived properties (Internal energy u, Enthalpy h and Entropy s)

The properties of unit mass are known as *Specific properties* for example specific volume, specific internal energy etc. Thus all the specific properties indicate the values for unit mass and they are independent of total mass, so *all the specific properties are intensive properties.*

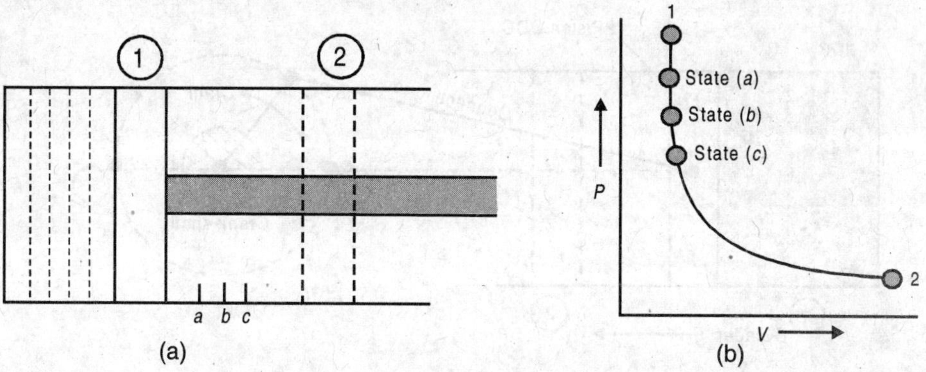

Fig. 1.8

When all the properties of a system have definite values (such as P, V, T) then the system is said to exist at a *Definite state* and the properties of the state are said to be *State functions or point functions.* Any operation in which one or more properties of the system changes, is called *Change of state,* and the series of states (and substates like a, b, c, etc. as shown in Fig. 1.9) passed during a change of state from state (1) to state (2) is called the *path* of change of state. The properties of the states and substances are called as *Path functions.*

When the path is completely specified, then the change of state is called a *Process.*

A *Process* is defined as the transformation of the system from one fixed state to another fixed state.

When any one of the properties changes, the working substance or system is said to have undergone a process.

One way of classifying a thermodynamic process is—Reversible process and Irreversible process.

(A) Reversible Process. The system will undergo a reversible process in three different conditions:

(i) Thermally Reversible Process. When the system has uniform temperature throughout the process and is in equilibrium with the surroundings, then the system is said to be in thermal equilibrium. In other words temperature gradient in the system does not exist.

Let us consider an example of working substance undergoing a process, during which it receives heat from the source. This is shown by 1–2 in Fig. 1.9.

When the working substance retraces its path in the reverse direction, then the process is called *Reversible process.* In the forward direction consider a point A where the temperature is T. At that moment the temperature of the source will be $(T + \Delta T)$, so that the heat transfer takes place from source to working substance.

In the reverse direction, the working substance will have to reject heat. At point A the working substance at a temperature of T has to transfer heat to source at $(T + \Delta T)$, that is not possible and the process cannot be retraced back and the process will not be reversible. To have the process reversible, the temperature gradient must be zero *i.e.*, $\Delta T = 0$. In a thermally reversible process, therefore, temperature gradient does not exists between the surroundings (source) and the system.

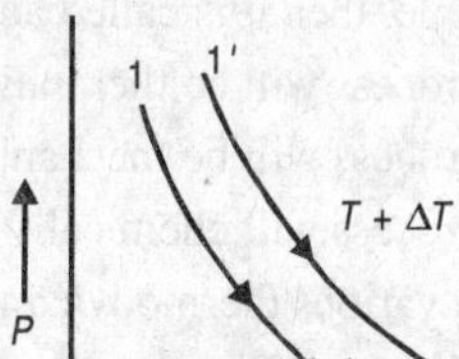

Fig. 1.9

(ii) Mechanically Reversible Process. A system is said to be mechanically reversible, when there are no mechanical losses either internally or externally.

Let us consider a simple piston-cylinder arrangement as shown in Fig 1.10. Position 1 represents Inner Dead Centre (IDC) and position 2 represents Outer Dead Centre position (ODC). Let space between the piston and cylinder cover be filled with a gas at a pressure of P, volume V and temperature T. And the pressure on the other side of piston be atmospheric P_{atm}. When the piston is released, then because of pressure potential, the piston starts moving towards RHS direction — till ODC and after that piston moves towards LHS direction while the crank is rotating.

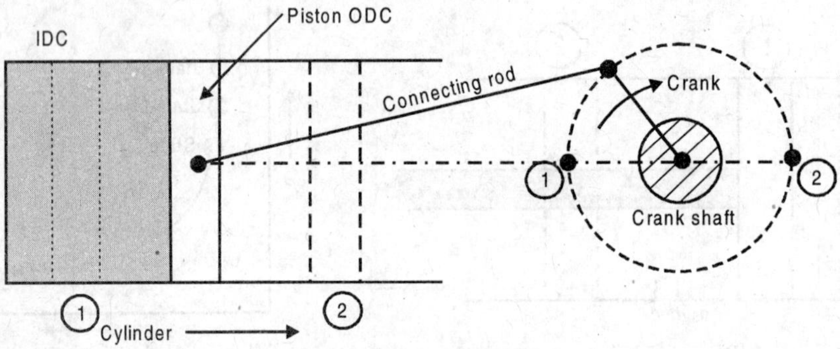

Fig. 1.10

If there are no mechanical losses the piston will reach to its original position. This type of process is called as *Mechanically Reversible Process.*

In actual practice this does not exist. Mechanical losses are mainly *frictional losses.* If the process is mechanically reversible, then the friction between the moving parts must be zero.

We can have other example as a tennis ball released from a certain height, then it will start bouncing ultimately it comes to rest. If there would not have been any mechanical losses, at the point of contact, then it would have been bouncing continuously. In practice this does not happen.

(iii) Chemically Reversible Process. The process is called chemically reversible process, when there are no chemical reactions during the process.

A system will be in *thermodynamic equilibrium*, when it is in thermal equilibrium, mechanical equilibrium, and chemical equilibrium simultaneously.

A system in thermodynamic equilibrium is incapable of any spontaneous change and it is in complete balance with its surroundings.

(B) Irreversible Process. When the process is not thermally reversible, mechanically reversible and chemically reversible, then it is called an *Irreversible process.*

A process will be thermally irreversible, when the heat transfer with finite temperature difference takes place.

A process will be mechanically irreversible, when the friction exists during the process.

A process will chemically irreversible, when chemical reactions take place during the process.

The various thermodynamic processes (Refer Figs. 1.11 a–c) which are used in engineering practice are:

1. Constant Pressure Process or Isobaric Process. A process during which pressure remains constant is called a constant pressure process. The law for the process is ($P = C$) and is represented on the PV-diagram by means of a horizontal line 1–2.

2. Constant Volume Process or Isochoric Process. A process during which volume remains constant is called a constant volume process. The law for the process is $V = C$ and is represented by means of a vertical line 3–4 on the PV-diagram.

3. Constant Temperature or Isothermal Process. A process during which the temperature remains constant is called a constant temperature on isothermal process. The law for the process is $PV = C$ and is represented by the curve 5–6 on the *PV*-diagram. Also the law of the process may be given as $T = C$.

4. Adiabatic or Isentropic Process. A process during which heat exchange with the surroundings is zero, then it is called an *Adiabatic process*. A reversible adiabatic process is called as an Isentropic process, during which entropy remains constant ($S = C$). The law for the adiabatic process is $PV^{\gamma} = C$. Irreversible adiabatic process is called a *Throttling process*, during which enthalpy remains constant. It is given by law h = constant.

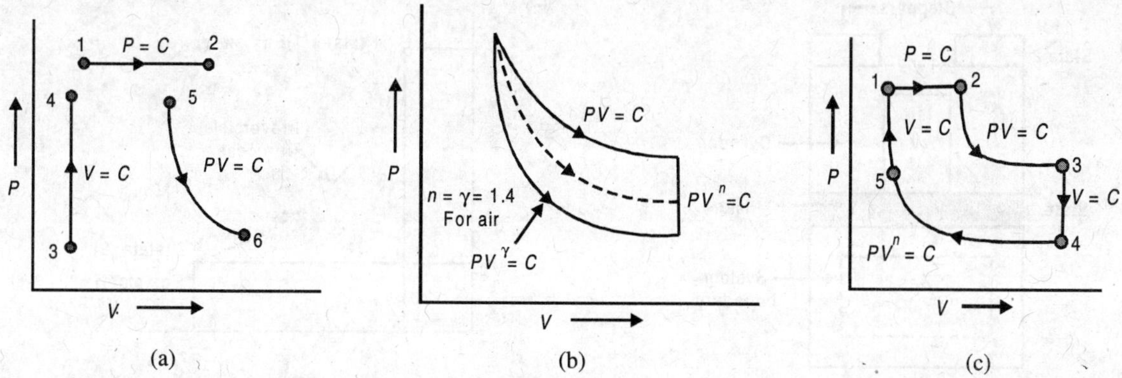

Fig. 1.11

5. General Law Process or Polytropic Process. The law for the process is $PV^n = C$. In actual practice neither we get isothermal process, nor we get adiabatic process, but what we get is the process in between these two and is known as polytropic process [Refer Fig 1.11(b)].

Thermodynamic Cycle. It is defined as the series of state changes such that the final state is identical with the initial state.

or

When the system undergoes the series of processes such that the end states are same, the system is said to have undergone a thermodynamic cycle.

or

Series of thermodynamic processes, the end states of which are identical, is called a *thermodynamic cycle.*

In a thermodynamic cycle, chemical composition of the working fluid during the process does not change. Thus all the properties in the initial and final states remain unchanged.

For example, water that circulates through the steam power plant. There is a change of phase during the process, but the end states are same.

Points to Remember :

- Intensive property is independent of mass
- Extensive property depends on mass.

1.4 QUASI-STATIC PROCESS

Quasi means nearly or almost. So quasi-static process means nearly static process or nearly stationary process, or a process which proceeds with extreme slowness.

Quasi-static process proceeds from one equilibrium state to another equilibrium state till the end of the process i.e., all the states passed during the process are all equilibrium states.

In each state the process deviates from the equilibrium state by a very small amount and immediately attains its equilibrium.

Let us consider a system consisting of a gas in a cylinder fitted with a piston as shown in Fig. 1.12 (a). Let *W* be the weight kept on the piston which just balances the upward force exerted by the gas. The system is initially in the equilibrium state. During initial condition, let the properties be P_I V_1 and t_1. If the weight is removed then there will be an unbalanced force between the system and the surroundings and the piston will move up till it hits the stops. Then again the system comes to another equilibrium state described by the properties P_2, V_2, t_2 but the intermediate states passed through by the system are non-equilibrium states. In Fig. 1.12 (b) points (1) and (2) are initial and final equilibrium states and since the system is passing through the non-equilibrium states, it is joined by dotted line. Such a process is called as *Irreversible process.*

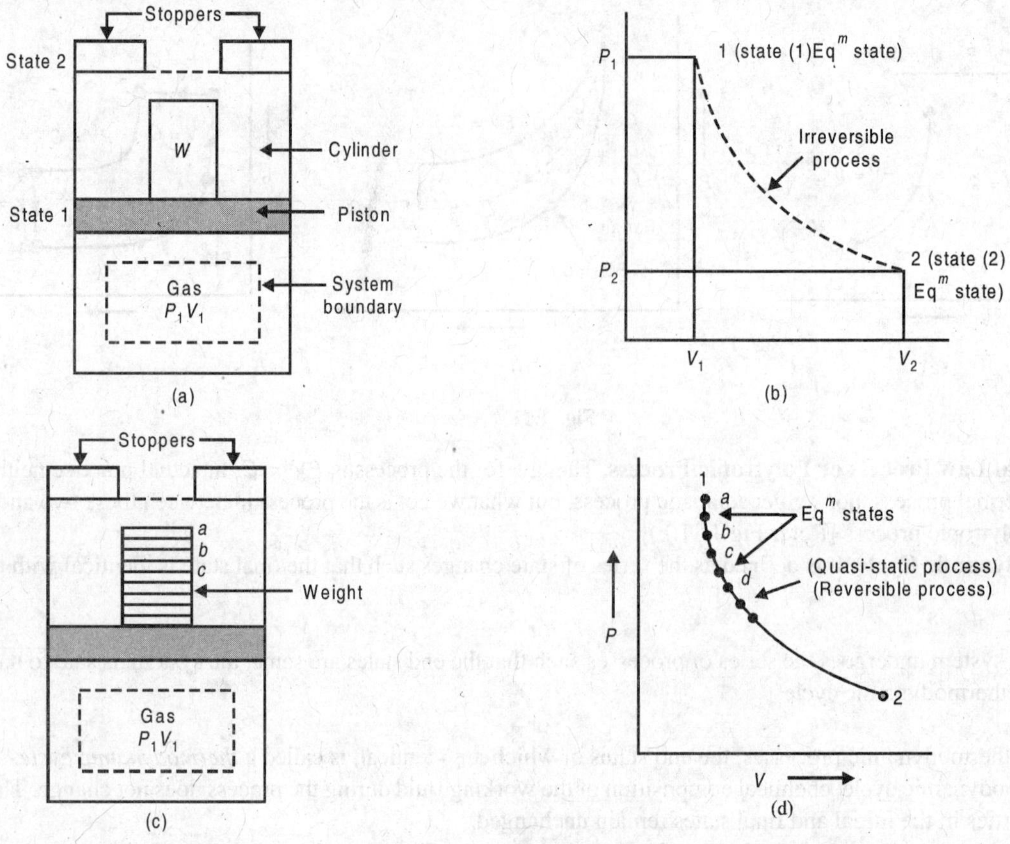

Fig. 1.12

Now if the single weight on the piston is made up of many very small pieces of weights as shown in Fig. 1.12 (c) and these weights are removed one by one, then the departure of the state of the system from the thermodynamic equilibrium state will be very small. So, every state passed through by the system will be an equilibrium state.

A Quasi-static process is also called as *Reversible process* and is represented on the *PV*-diagram by a continuous line as shown in Fig. 1.12 (d).

In other words suppose when one weight say *a* is removed then the system will deviate from its equilibrium by a very small amount and immediately comes to an equilibrium state *a* (as shown in *PV*-diagram Fig. 1.12 (d) and when weight *b* is removed, the system again deviates by a small amount and immediately comes to another equilibrium state *b*. Thus when these small weights are removed one by one the system proceeds from one equilibrium state to another equilibrium state till end, such a process which passes through all the equilibrium states is known as quasistatic process.

1.5 WORKING SUBSTANCE

"It is a fluid, in which energy can be stored and from which energy can be removed".

or

"It is a fluid which is used in thermodynamic systems, for converting one form of energy into the other".

The working substance may be a liquid, vapour or a gas. Generally air and water are used as working substances, as they are freely available. However NH_3 and Freon-12 are also used depending upon specific applications.

1.6 UNITS AND DIMENSIONS

Nowadays generally SI (System International) system of units is used. The basic units in this system are:

System. Basic Units

Sr. No.	*Quantity*	*Unit*	*Symbols of Unit*
1.	Length	Metre	m
2.	Mass	Kilogram	kg
3.	Time	Seconds	s
4.	Temperature	Celsius/Kelvin	°C/K

The dimensions of all other quantities are derived from these basic units which are:

Sr. No.	*Quantity*	*Unit*	*Symbol of Unit*	*Alternative Unit*
1.	Force	Newton	N	—
2.	Energy (Work or Heat)	Joule	J	Nm
3.	Power	Watt	W	J/sec
4.	Pressure	Pascal	Pa	N/m^2
5.	Volume	Cubic metres	m^3	Cu-m

A proper understanding of units of Force, Pressure, Temperature, Work, Heat and Power is necessary for solving numerical problems in Thermal Engineering.

1. Force. The unit of force in SI units is Newtons (N). A force of 1N produces an acceleration of 1 m/sec² when applied to a mass of 1 kg.

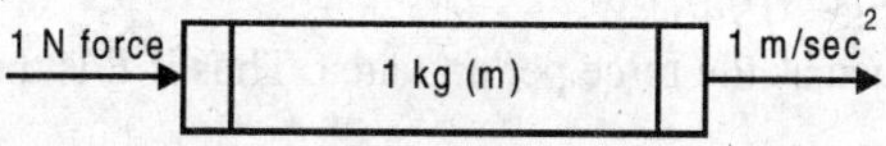

Fig. 1.13

We know that from Newton's Second law of motion $F \propto ma$

or
$$F = \frac{ma}{g_c}, \text{ where } g_c = \text{constant of proportionality.}$$

As 1 N force produces an acceleration of 1 m/sec² in 1 kg mass, we have,

$$1 \text{ Newton} = \frac{1}{g_c} \times 1 \text{ kg} \times 1 \text{ m/sec}^2$$

$$\therefore \quad g_c = 1 \text{ in SI units and units will be}$$

$$g_c = \frac{1 \text{ kg-m}}{\text{N}-\text{sec}^2}$$

When $g_c = 1$, the product of mass and acceleration becomes the magnitude of force.

If the mass of 1 kg is allowed to fall freely under the action of standard gravitational force, it is accelerated at the rate of 9.806 m/sec^2 (9.81 m/sec^2) and we have,

$$\text{Force} = 1 \text{ kg} \times 9.81 \text{ m/sec}^2 = 9.81 \text{ N}$$

$\therefore$ It follows that the weight of 1 kg mass equals 9.81 N and since the weight is the force,

$$\mathbf{1\ kgf = 9.81\ N}$$

2. Weight. Weight of a body is the force exerted on its mass due to gravity.

3. Density. Density is defined as mass per unit volume of the substance. It is denoted by ρ

$$\therefore \quad \rho = \frac{\text{mass}}{\text{volume}} = \frac{\text{kg}}{\text{m}^3}$$

Since unit of volume is m^3 and mass is kg.

4. Work (mechanical). When a force F acts on a body and causes displacement through a distance dl in the direction of force, then the work is said to be done. This work is equal to the product of force and distance moved.

$$W = F \times dl$$

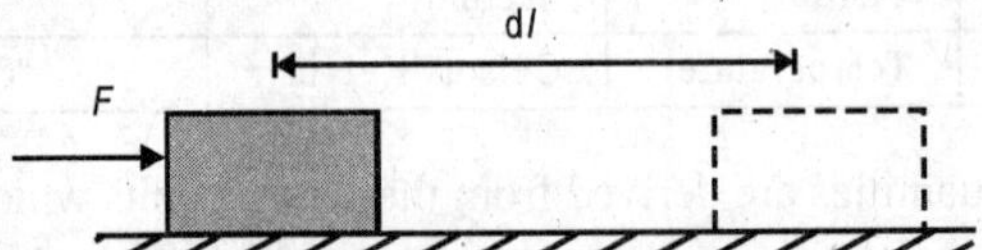

Fig. 1.14

If force F is in Newton and distance dl is in metre, then the resultant unit of W.D. will be Nm and 1 Nm = 1 joule.

5. Power. It is defined as the rate of doing work.

$$\text{or} \quad \text{Power} = \frac{\text{W.D.}}{\text{Time taken}}$$

If the unit of W.D. is in joules and the time taken is in seconds, then the unit of Power is J/sec.
And the rate of doing work of 1 J/sec is called watt (W)

$$\therefore \quad \text{Power} = \frac{\text{J}}{\text{sec}} \text{ or W}$$

Earlier there used to be another unit of Power known as Horse Power (H.P.)

$$1 \text{ HP} = 746 \text{ watts} = 0.746 \text{ kW}$$

6. Pressure (P). Pressure is defined as the force per unit area. Thus if F is the force applied to an area A then,

$P = \dfrac{F}{A}$ and if F is in newton and A is in m^2 then the unit will be $\dfrac{\text{N}}{\text{m}^2}$.

This unit of pressure is called as pascal (Pa).

1.7 MECHANICAL AND THERMODYNAMIC WORK

1.7.1 Mechanical Work

$$\text{W.D.} = F \times dl$$

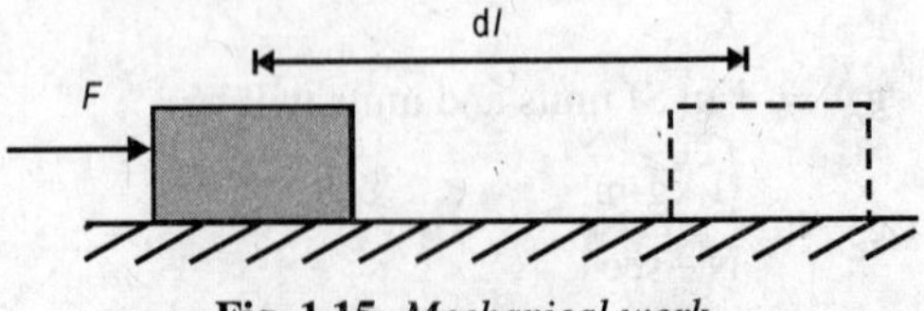

Fig. 1.15 *Mechanical work*

When a force F acts on a body and causes a displacement through a distance in the direction of force, then the work is said to be done and this work is equal to the product of force and distance moved.

i.e., $$\text{Work done} = F \times dl$$

If F is in N, and dl is in m then the resultant unit will be Nm or Joule.

1.7.2 Thermodynamic Work

"It is an interaction between the system and the surroundings and the work is said to be done by the system on surroundings, if the complete external effect is lifting up of the body".

Let us consider a battery and motor system as shown in Fig. 1.16. The motor is driving a fan. The system is doing work on the surroundings.

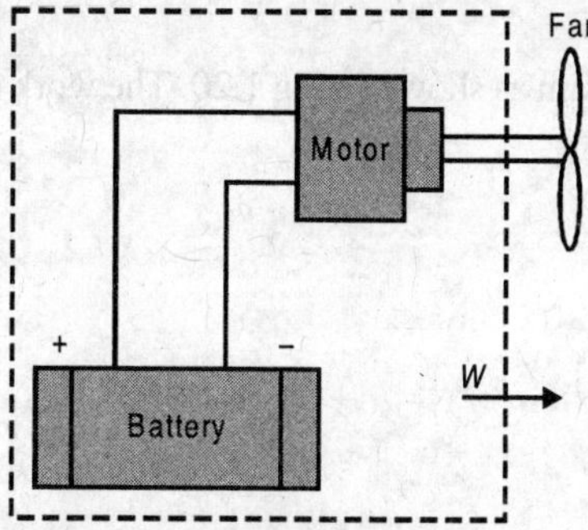

Fig. 1.16

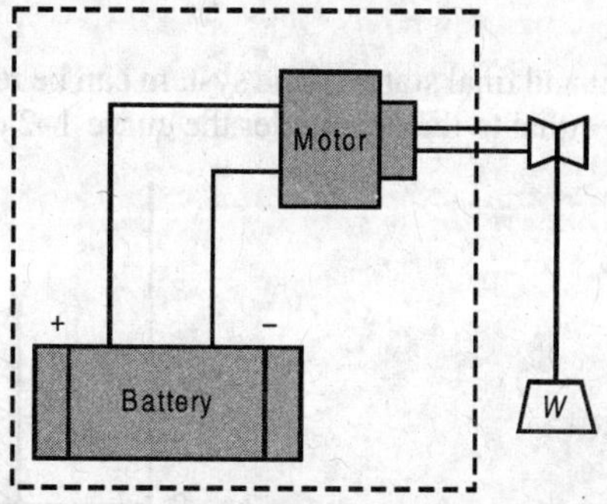

Fig. 1.17

When the fan is replaced by a pulley and weight as shown in Fig. 1.17, then the weight may be raised as the pulley is driven by the motor. Then the complete external effect is raising up the body, so the system is doing thermodynamic work.

When the work is done by the system, then it is taken as + ve, and when the work is done on the system, it is taken as –ve. (Refer Fig. 1.18)

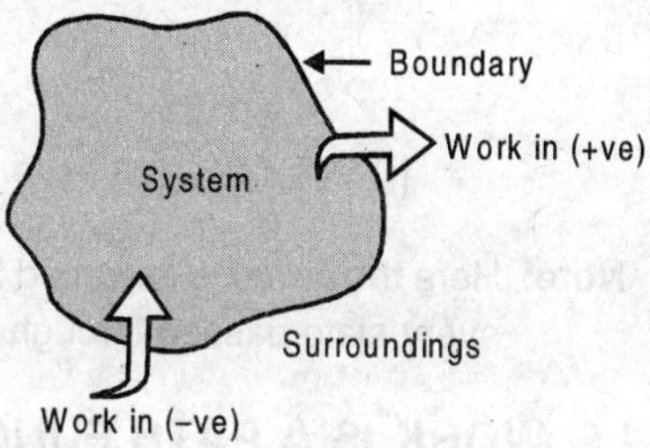

Fig. 1.18

1.8 *PdV*-WORK

Consider a system consisting of a gas in a cylinder fitted with a piston as shown in Fig. 1.19. During the initial condition of the piston i.e., when the piston is at position (1), the pressure inside the cylinder is P_1 and volume is V_1. Let the gas expands as the piston moves to position (2). Then the pressure falls to P_2 and volume will increase to V_2.

Now, consider a small movement of the piston dl during which pressure P is assumed to be constant. If a is the cross sectional area in m^2, then the force acting on the piston is given by

$$F = P \times a$$

And the small amount of work done by the gas on the piston in causing the displacement

dl = force × displacement.

$$\delta W = F \times dl = P \times a \times dl \qquad (\because F = P \times a)$$

$$\delta W = PdV,$$

where $dV = a \times dl$ = small amount of displacement volume.

When the piston moves from position (1) to position (2) volume will change from V_1 to V_2 then the total amount of work done is given by,

$$W_{1-2} = \int_{V_1}^{V_2} PdV$$

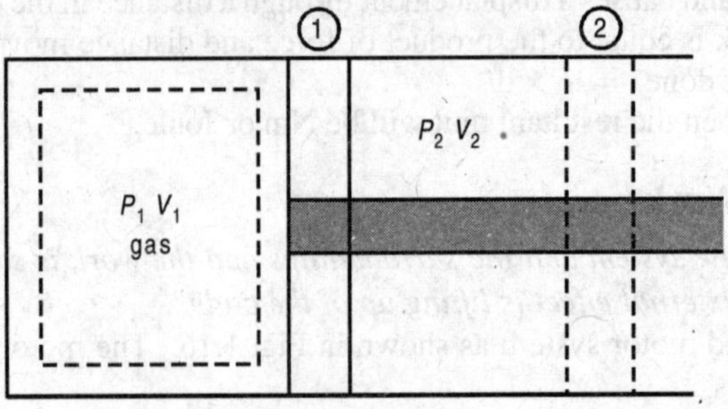

Fig. 1.19

The initial and final states of the system can be represented on *PV*-diagram as shown in Fig 1.20. The work done W_{1-2} will be equal to the area under the curve 1–2 on the *PV*-diagram.

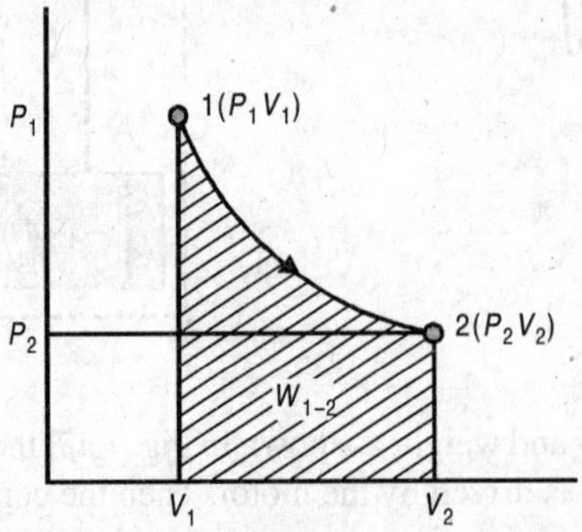

Fig. 1.20

Note: Here the piston is assumed to move very very slowly so that path 1–2 is Quasi-static (i.e., nearly static) and every state passed through is an equilibrium state.

1.9 WORK IS A PATH FUNCTION AND PROPERTIES ARE POINT FUNCTIONS

With reference to Fig. 1.21 it is possible to take a system from state (1) to state (2) by following many quasi-static paths such as *A*, *B*, *C* etc.

Now if we consider the path 1–*A*–2, then the work done during the process 1–*A*–2 will be equal to the area under the curve 1–*A*–2 and if we consider path 1–*B*–2, then the work done during the path 1–*B*–2 will be equal to the area under the curve 1–*B*–2.

So, the work done during the process directly depends upon the path and not on the end states. So, work is called Path function and δW is an inexact or imperfect differential.

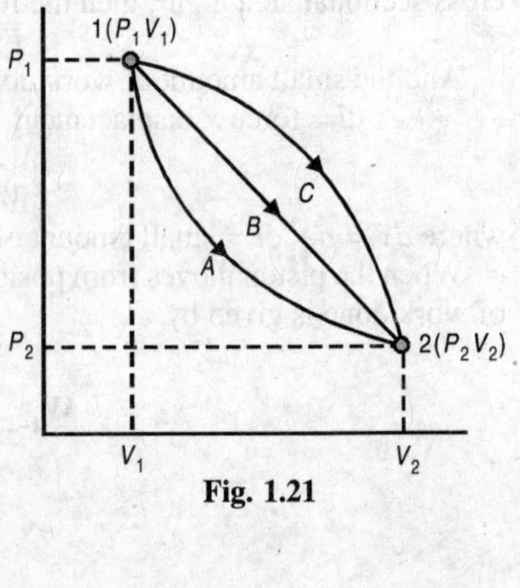

Fig. 1.21

i.e. $$\int_1^2 \delta W \neq W_2 - W_1$$

But $$\int_1^2 \delta W = W_{1-2} \text{ or } {}_1W_2$$

Thermodynamic properties are point functions, since for a given state,

02there is a definite value for each property. During change of states, the values of the properties will change and this change in values of the properties does not depend on path, but it depends upon end states and hence all the properties are point functions.

The differentials of point functions are exact or perfect differentials and the integration is simply, e.g.,

$$\int_1^2 dV = V_2 - V_1$$

Operator δ is used to denote inexact differentials, operator *d* is used to denote exact differentials.

1.10 EQUATIONS FOR WORK DONE IN VARIOUS PROCESSES

1.10.1 *PdV* Work in Various Quasi-static Processes

1. Constant pressure process (Isobaric process). Here pressure is constant.

We know that Work Done is given by,

$$W_{1-2} = \int_{V_1}^{V_2} PdV = P\int_{V_1}^{V_2} dV$$

$$= P(V_2 - V_1) \text{ Nm, if } P \text{ is N/m}^2 \times V \text{ is m}^3$$

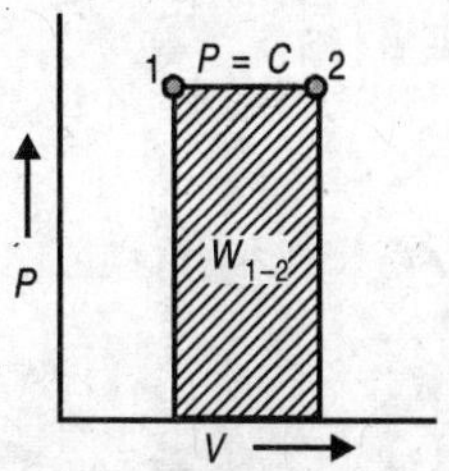

Fig. 1.22

2. Constant volume process. (Isochoric process)

$$W_{1-2} = \int PdV$$
$$= 0$$

During this process $V = C$ hence W.D. $= 0$

Since derivative of a constant is zero.

From *PV*-diagram area under the line 1–2 is zero.

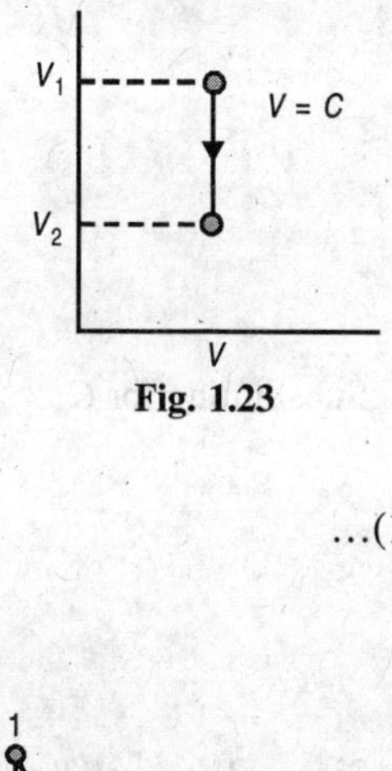

Fig. 1.23

3. Isothermal process. We know that work done in a non-flow system

$$W_{1-2} = \int_1^2 P.dV \qquad ...(1)$$

The law for the process is, $PV = C$

i.e. $PV = P_1V_1 = C$

$$\therefore \quad P = \frac{P_1 V_1}{V}$$

Substituting this in Eq. (1) we get,

$$\therefore \quad W_{1-2} = \int_1^2 \frac{P_1 V_1}{V} \cdot dV = P_1 V_1 \int_1^2 \frac{dV}{V}$$

$$\mathbf{W_{1-2} = P_1V_1 \ln\frac{V_2}{V_1}}$$

$$\mathbf{W_{1-2} = P_1V_1 \ln\frac{P_1}{P_2}}$$

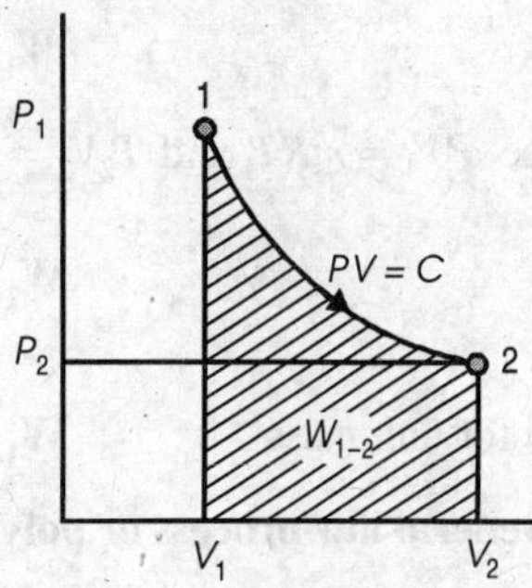

Fig. 1.24

4. Adiabatic process: The law for the process is $PV^{\gamma} = C$ where γ is an adiabatic index. We know that, the work done in non-flow system,

$$W_{1-2} = \int_1^2 PdV \qquad \ldots(2)$$

But for a polytropic process,

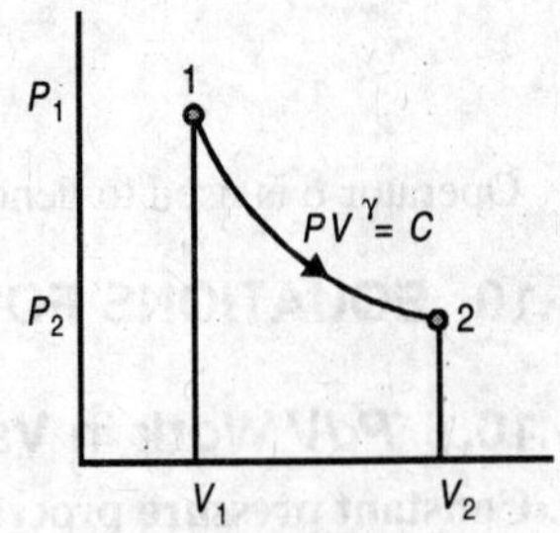

Fig. 1.25

$$PV^{\gamma} = P_1V_1^{\gamma} = P_2V_2^{\gamma} = C$$

$$\therefore \quad P = \frac{C}{V^{\gamma}}$$

Substituting in Eq. (2), we get,

$$W_{1-2} = \int_1^2 \frac{C}{V^{\gamma}} dV$$

$$W_{1-2} = \int_1^2 C V^{-\gamma} dV$$

$$= C\left[\frac{V^{-\gamma+1}}{-\gamma+1}\right]_1^2$$

$$= C\left[\frac{V_2^{-\gamma+1} - V_1^{-\gamma+1}}{-\gamma+1}\right] = \frac{CV_2^{-\gamma+1} - CV_1^{-\gamma+1}}{-\gamma+1}$$

$$= \frac{CV_2^{-\gamma+1}}{-\gamma+1} - \frac{CV_1^{-\gamma+1}}{-\gamma+1}$$

Substituting for C

$$= \frac{P_2V_2^{\gamma} \cdot V_2^{-\gamma+1}}{-\gamma+1} - \frac{P_1V_1^{\gamma} \cdot V_1^{-\gamma+1}}{-\gamma+1}$$

$$= \frac{P_2V_2^{\gamma-\gamma+1} - P_1V_1^{\gamma-\gamma+1}}{1-\gamma}$$

$$W_{1-2} = \frac{P_2V_2 - P_1V_1}{1-\gamma} \text{ or } \frac{P_1V_1 - P_2V_2}{\gamma-1} \qquad \ldots(3)$$

Since $P_1V_1 = mRT_1$ and $P_2V_2 = mRT_2$

$$\mathbf{W_{1-2}} = \frac{\mathbf{mR(T_2-T_1)}}{\mathbf{1-r}}$$

And for unit mass

$$\mathbf{W_{1-2}} = \frac{\mathbf{R(T_2-T_1)}}{\mathbf{1-r}} \qquad \ldots(4)$$

5. General law process or polytropic process. The law of the process is $PV^n = C$ where n is polytropic index.

The work done for this process, $\int P \cdot dV$ can be obtained by substituting γ by n.

$$\therefore \quad \mathbf{W_{1-2}} = \frac{\mathbf{P_1\ V_1 - P_2\ V_2}}{\mathbf{n-1}}$$

1.11 HEAT

"Heat is a form of energy, which crosses the system boundary due to the temperature difference between the system and the surroundings".

When the two bodies, one hot and the other cold are kept in contact with each other, then the hot body loses heat and becomes colder and the cold body gains heat and becomes hotter and this process continues till the thermal equilibrium is reached. At this stage the two bodies will be at the same temperature.

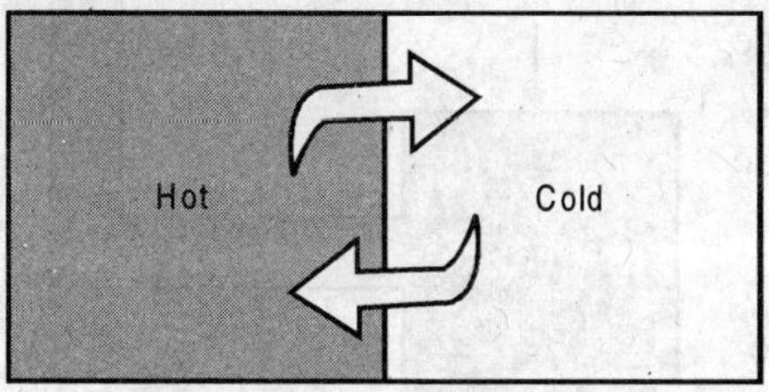

Fig. 1.26

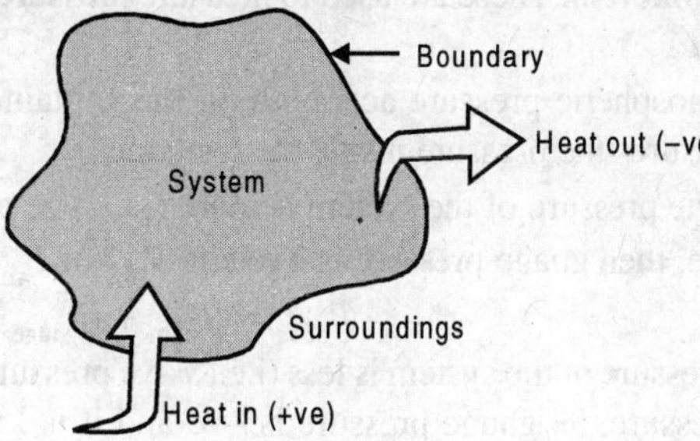

Fig. 1.27

Like work, heat is a path function and we know that the differentials of path functions are imperfect differentials. If Q is the heat transfer, then the magnitude of heat transfer during the process 1–2 is given by,

$$\int_1^2 \delta Q = Q_{1-2} \text{ or } Q_2$$

$$\int_1^2 \delta Q \neq Q_2 - Q_1$$

Note: When heat flows into the system then it is taken as +ve and when heat flows out of the system then it is taken as –ve.

1.12 PRESSURE

It is defined as force per unit area i.e.,

$$P = \frac{F}{A}$$

$$F = \text{Force in newton (N)}$$

$$A = \text{Area is a m}^2$$

Then the unit of pressure will be N/m^2, which is the basic unit of pressure, in SI units. This is also sometimes called as Pascal (Pa). Since this unit is very small, when compared to many engineering values, the units like, KPa, MPa, bar are used.

$$1 \text{ bar} = 10^5 \text{ N/m}^2 = 100 \text{ kN/m}^2 = 100 \text{ kPa}$$

Pressures are also measured in mm, or cm, of Hg or H_2O column. The pressure exerted by the atmosphere is known as atmospheric pressure and is denoted by 1 atm.

In various units, the value of 1 atm pressure are given by,

$$1 \text{ atm} = 760 \text{ mm of Hg} = 101325 \text{ N/m}^2 = 1.01325 \text{ bar.}$$
$$= 10.33 \text{ m of } H_2O = 1.033 \text{ kg/cm}^2$$

Another useful unit of pressure is 1 Torr and this is equal to pressure exerted due to 1 mm of Hg = 133.34 N/m^2.

1.13 PRESSURE MEASUREMENT

Pressure is measured by using,

1. Barometer. This measures the atmospheric pressure.

2. Bourdon Pressure Gauge. This is used to measure pressure inside the pipes/containers etc.

3. Manometers. These are used to measure pressure inside the pipes/vessels etc.

The atmospheric pressure acts both on the container and the gauge which measures the pressure inside the container.

When the pressure of the system is more (i.e., +ve pressure) than the atmosphere, then guage pressure is +ve and P_{abs} or P_{real} is given by,

$$P_{abs} = P_{atm} + P_{gauge}$$

If the pressure of the system is less (i.e., *s* –ve pressure) than the atmospheric pressure, its gauge pressure is –ve and it is known as vacuum and,

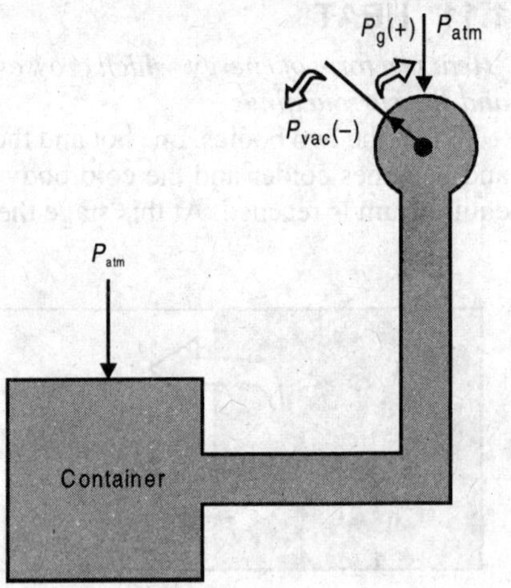

Fig. 1.28

$$P_{abs} = P_{atm} - P_{vac}$$

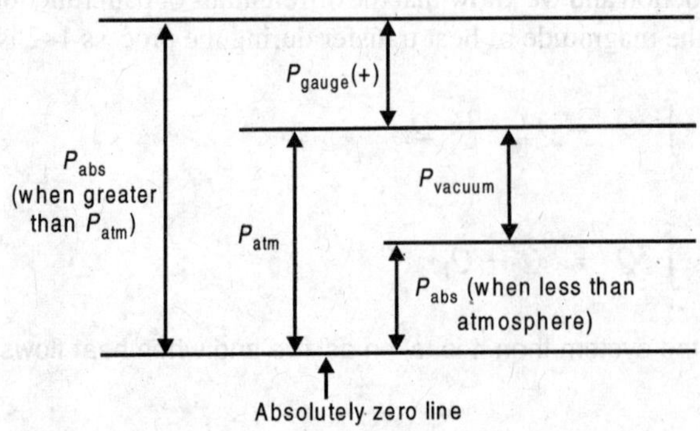

Fig. 1.29

Figure 1.29 shows the relationship between absolute and gauge pressure.

1.14 PRESSURE EXERTED DUE TO A COLUMN OF FLUID

Figure 1.30 shows a fluid of density ρ contained in container to a depth *z* or *h*. Now consider an elemental vertical cylinder of cross sectional area *a*.

Then we know that,

$$\text{Density} = \frac{m}{v}$$

$$\therefore \quad \text{Mass} = \text{density} \times \text{volume}$$

$$= \rho \times a \times Z$$

But

$$\text{Force} = \text{mass} \times \text{acceleration}$$

$$= \rho \,.\, a \,.\, Z \times g$$

But pressure exerted due to a column of fluid at a depth *Z*

$$= \frac{\text{Force}}{\text{Area}} = \frac{\rho \cdot a \cdot Z \cdot g}{a}$$

$$\mathbf{P} = \rho g Z$$

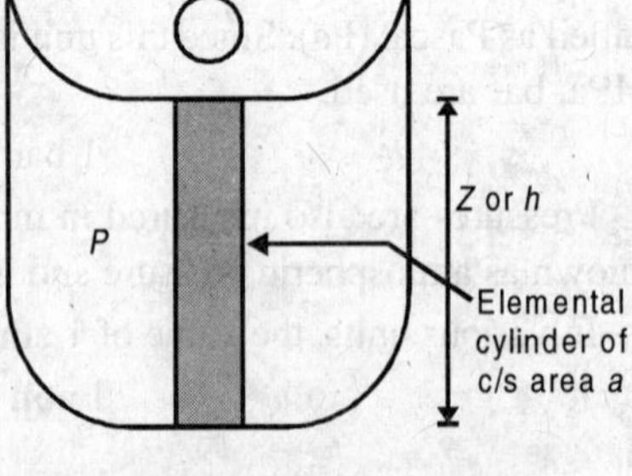

Fig. 1.30

Note:

Units $$P = \frac{kg}{m^3} \times \frac{m}{sec^2} \times m$$

We can write this as, $$= kg \times \frac{m}{sec^2} \times \frac{1}{m^2} = [\text{mass} \times \text{acceleration}] \times \frac{1}{m^2}$$

$$P = \frac{Force}{m^2} = \frac{N}{m^2} \text{ resultant unit.}$$

1.15 BAROMETER

It is used to measure atmospheric pressure. Consider Hg in the container as shown in Fig. 1.31 (a) and (b). First of all, let a glass tube open at both the ends is immersed in the Hg as shown in Fig. 1.31 (a). Then the atmospheric pressure acting on the surface of Hg level is same as shown.

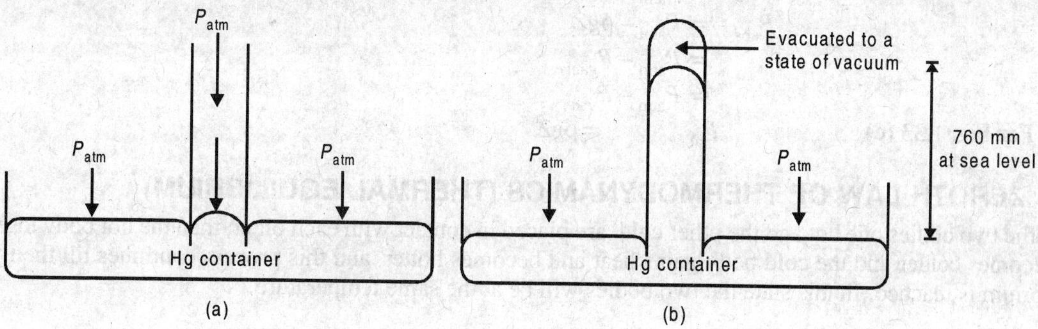

Fig. 1.31 *Barometer.*

Now if the glass tube with one end sealed and evacuated to a state of vacuum is immersed as shown in Fig. 1.31 (b) then, due to atmospheric pressure acting on the surface of Hg, the Hg will rise in the glass tube and this will be equal to 760 mm at sea level.

1.16 BOURDON PRESSURE GAUGE

This is the most common type of pressure gauge. This is used to measure the pressure inside the pipes, containers, vessels etc.

The basic element of this gauge is the tube *A*. It is elliptical in cross section and is bent into an arc of circle as shown in Fig. 1.32. End *B* of this tube is sealed, whereas open end *C* is connected to the connecting union *D* through which it can be mounted over the pipes or containers of whose pressures are to be measured.

If the pressure of the system is more than the atmosphere (i.e., +ve pressure) then the tube will tend to curl out. Conversely if the pressure of the system is less than atmosphere (i.e., –ve pressure) then the tube will tend to curl in. The change in pressure will therefore appear as the movement of the end *B*. This end *B* is connected to the quadrant gear *F* through the link *E*. The quadrant gear *F* engages with die small gear *G*, on to which a pointer *H* is attached.

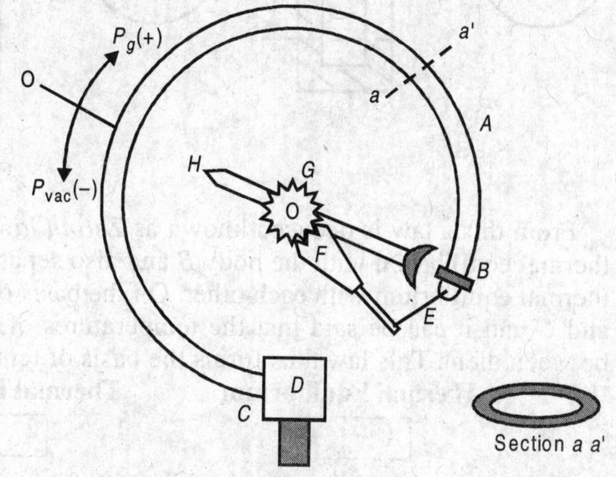

Fig. 1.32

Thus any movement due to change in pressure of the end *B* will be transmitted to the quadrant gear which will rotate the gear *G* and hence the pointer. The pointer rotates on the calibrated scale and thus gives the pressure readings directly.

1.17 MANOMETERS

These are used to measure the pressure inside the pipes, containers. It consists of a bent glass tube or a U-type glass tube which contains a liquid generally H_2O or Hg. One end of the manometer is connected to the container or pipe of whose pressure measurement is to be done and the other end is generally open to the atmosphere. The U-tube with one end closed and evacuated to state of vacuum indicates the absolute pressure directly as shown in Fig. 1.33(c).

Balancing the pressure values about the line 1–1', we get for Fig. 1.33 (a).

$$P_{abs} = P_{abs} + \rho g Z = P_{abs} + P_{gauge}$$

For Fig. 1.33 (b) $\quad P_{abs} + \rho g Z = P_{atm}$

i.e.,

$$P_{abs} = P_{atm} - \rho g Z$$
$$= P_{atm} - P_{gauge}$$
$$= P_{atm} - P_{vac}$$

For Fig. 1.33 (c) $\quad P_{abs} = \rho g Z$

1.18 ZEROTH LAW OF THERMODYNAMICS (THERMAL EQUILIBRIUM)

When the two bodies one hot and the other cold, are placed in contact with each other, then the hot body loses heat and becomes colder and the cold body gains heat and becomes hotter, and this process continues till the thermal equilibrium is reached. In this state the two bodies will be at the same temperature.

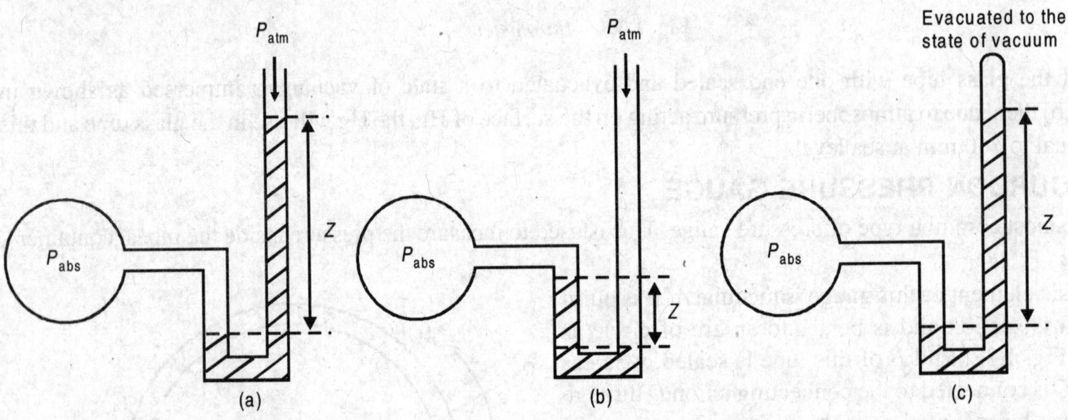

Fig. 1.33

From this a law is deduced known as *Zeroth law of thermodynamics.* It states that, "When a body *A* is in thermal equilibrium with the body *B* and also separately with the body *C*, then the bodies *B* and *C* will be in thermal equilibrium with each other. On the basis of this law, it is possible to compare the temperatures of *B* and *C* and it can be said that the temperatures of *B* and *C* are equal without actually making any contact between them. This law thus forms the basis of temperature measurement: by comparison method.

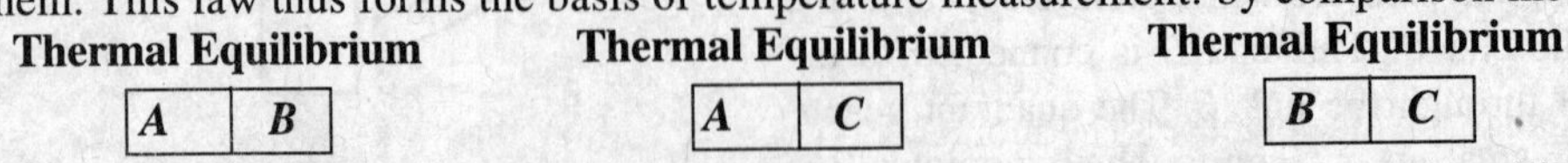

If *A* is thermal equilibrium separately with *B* and *C* then *B* and *C* will also be in thermal equilibrium.

1.19 THERMOMETRIC PROPERTY AND THERMOMETERS

Temperature. It is property of the system or of a body which distinguishes a hot body from cold one.

Thermometers. In order to obtain, a quantitative measure of temperature, a reference body is used, and a certain physical characteristic of this body which changes with the temperature is selected. The selected characteristic is called the *thermometric property* and the reference body which is used in the determination of temperature is called *thermometer.*

A very common thermometer, consists of a small amount of Hg in an evacuated capillary tube. In this case the change in length of Hg is taken as the measure of temperature.

There are five different kinds of thermometers, each of them have their own thermometric properties viz.

Thermometer	*Thermometric property*	*Thermometric substance*
1. Constant Volume gas Thermometer	Volume	Gas
2. Constant Pressure gas Thermometer	Pressure	Gas
3. Electrical resistance Thermometer	Resistance	Wire
4. Thermo couple	Thermal emf	Wire
5. Hg in glass Thermometer	Length	Hg

1.20 PRINCIPLE OF TEMPERATURE MEASUREMENT

Let x be the value of any thermometric property as given in the table and θ be the temperature corresponding to x.

When the temperature changes, the values of the thermometric properties change proportionately. Therefore the value of temperature is proportional to the values of thermometric properties.

So we have, $$\theta \propto x$$

i.e., $$\theta = ax \qquad \ldots(1)$$

For Eqs (1) and (2) $$\theta_1 = ax_1$$

$$\theta_1 = ax_2 \qquad \ldots(2)$$

where a is the constant of proportionality.

Dividing we get, $$\frac{\theta_1}{\theta_1} = \frac{x_1}{x_2} \qquad \ldots(3)$$

Thus the ratio of temperature is equal to the ratio of values of properties, such as length of Hg, volume, pressure etc.

Now, to construct a scale we need a reference point. The internationally accepted reference point is the *Triple point of water,* which is a temperature, at which all the 3-phases of water i.e., ice, water and vapour, exists in equilibrium. And the value of triple point temperature is 273.16 K (or 0.01°C) at a pressure of 0.006 bar, where *K* stands for degree Kelvin.

∴ Equation (1) becomes,

$$273.16 = ax_t \Rightarrow \frac{273.16}{x_t} = a$$

where x_t is the value of the property at triple point.

Now susbtitute this value of a in Eq. (1)

i.e. $$\theta = ax \text{ we get, } \theta = \frac{273.16}{x_t}\cdot x$$

$$\boldsymbol{\theta = 273.16 \times \frac{x}{x_t}}$$

Thus for any change in value of x (i.e., Thermometric property) the temperature θ can be found out. Thus from this equation. we can construct a scale and calibrate it to give us direct temperature readings. Here it is to be noted that x_t is the value of the property at triple point and is constant.

1.21 SCALE OF TEMPERATURE

A scale of temperature is based on some reference or fixed point to which an arbitrary value of temperature is assigned. Thus on a thermometer with some fixed or reference point, any temperature has a unique numerical value w.r.t. the fixed reference point.

1.21.1 Scales of Temperature

(1) (a) Celsius scale, (b) Fahrenheit scale.
(2) Absolute Temperature scale.
(3) Ideal Gas temperature scale.

1. (a) Celsius Scale. This scale is named after Anders the Celsius, who devised this scale. There are two fixed points on this scale. One is ice point and the other is the steam point. These two points are numbered as 0 and 100 on Celsius scale. The interval between them is divided into 100 equal parts and each division, is known as one degree Celsius and degree Celsius denoted as °C.

(b) Fahrenheit Scale. On this scale the ice point is numbered as 32 and steam point is numbered as 212. The interval between them is divided into 180 equal parts and each division, is known as 1 °F.

The temperature readings on one scale can be converted into a reading on the other scale by the following formula.

$$\frac{C}{100} = \frac{F-32}{180}$$

$$\therefore \quad C = \frac{100}{180}(F-32) = \frac{5}{9}(F-32)$$

and

$$F = \frac{9}{5}(C)+32$$

2. Absolute Temperature Scale or Kelvin Temperature Scale or Thermodynamic Temperature Scale. This is an internationally accepted scale. In this scale the temperature is measured from absolute zero temperature.

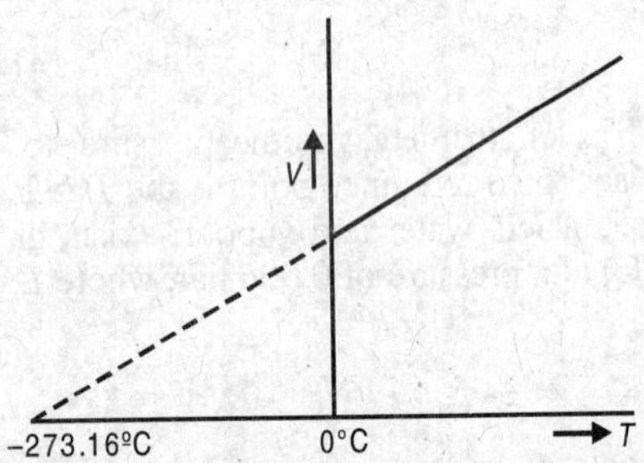

Fig. 1.34

Absolute zero is the temperature at which all the vibratory, translatory and rotational motion of the molecules of a gas, is supposed to cease. A gas on cooling contracts in volume. In case of perfect gases, 'Charle's found that decreases in volume per degree centigrade decrease in temperature is 1/273. 16th of its volume at 0°C and pressure remaining constant. Thus the volume of gas will be zero at temperature –273.16°C. This temperature 273.16°C below or behind 0ºC (or –273.16°C) is called the *absolute zero temperature.*

In this scale there is one more reference point taken, the Triple point of water. This is a temperature at which all the three phases of water exists in equilibrium, and the value of triple point temperature is 0.01°C or 273.16 K at a pressure of 0.006 bar. Here it is to be noted that, the ice point temperature on this scale is 273.15 K, and steam point temperature is 373.15 K and therefore to convert the temperature given in °C to K, just add 273.15.

K = °C + 273.15 or in general, K = °C + 273. Here it is to be noted that the value of Absolute zero on this scale is –273.16°C.

3. Ideal Gas-temperature Scale. This is similar to absolute temperature scale. This scale is constructed by using constant volume gas thermometer.

In a constant volume gas thermometer, gas will be used as the working substance and as all the gases behave as ideal gases at low pressure, thus scale is known as Ideal gas temperature scale. The value of absolute zero on this scale is same as that given by Absolute temperature scale i.e., – 273.16ºC.

1.22 MICROSCOPIC AND MACROSCOPIC POINT OF VIEW

There are two points of views.

1. Microscopic or Statistical Thermodynamics.
2. Macroscopic or Classical Thermodynamics, from which the behaviour of matter or working of a system can be studied.

From microscopic point of view, it is considered that, the matter is not continuous, but it is made up of a large number of identical particles called molecules. For example, consider a gas in a cylinder as a system. Then this gas is made up of number of molecules. Each molecule of a gas, at a given instant, has certain position, energy, etc., and for each molecule these changes very rapidly because of the collisions (strikings). The behaviour of the gas is described by summing up the behaviours of each molecule. And for summing up the behaviour of molecules statistical methods are employed and hence it is also called as *statistical thermodynamics.*

In macroscopic point of view a certain quantity of matter is considered, and the events that are taking place at the molecular level are not taken into account. For example, let us consider a system of an IC engine consisting of a charge in the engine cylinder. At any instant the system has certain volume depending upon the position of the piston, this volume is easily measurable.

Another quantity to describe the system is the pressure of the gas inside the cylinder. A pressure gauge can be used to measure the same. Similarly temperature, chemical composition etc. may be described. Thus, in macroscopic point of view the system will be described by large scale properties.

Though the approach based on microscopic point of view seems to be different, from that based on macroscopic point of view, but there exists a relationship between them. The relationship between macroscopic and microscopic point of view lies in the fact that the macroscopic properties are in fact the average properties of a large number of microscopic characteristics.

Hence, when both the methods are applied to a particular system, they give the same result.

Points to Remember:

- ❒ The microscopic point of view postulates the existence of molecules, their movement and collisions and hence it is constantly being changed. Whereas the macroscopic properties can be measured and their existence can be felt by our senses.

1.23 ENERGY IN TRANSIT

When energy flows from one system (or one body) to another system (other body), then it is referred as energy in transit.

In general heat and work are called energies in transit. The energy transfer is due to some energy potential e.g.,

(i) Mechanical energy or work energy are transferred due to difference in pressure potential.

(ii) Electrical energy is transferred due to difference in voltage potential.

(iii) Heat energy is transferred due to difference in temperature potential.

LIST OF FORMULAE

Note: (1) Unit of work in SI units is Nm or joules

(2) Unit of pressure in SI units is N/m^2 or Pa

(3) 1 bar $= 10^5\ N/m^2 = 100\ kN/m^2 = 100$ kPa

(4) 1 atm $= 760$ mm of Hg

$= 10.33$ m of water

$= 101325\ N/m^2 = 1.01325$ bar

$= 1.033\ kg/cm^2$

SOLVED EXAMPLES

Example 1.1 *A fluid in a cylinder is at a pressure of 700 kN/m². It is expanded at constant pressure from a volume of 0.28 m³ to a volume of 1.68 m³. Determine the work done.*

Data: $P = 700\ kN/m^2 = 700 \times 10^3\ N/m^2$

Expanded at constant pressure,

$V_1 = 0.28\ m^3\ V_2 = 1.68\ m^3$

Solution

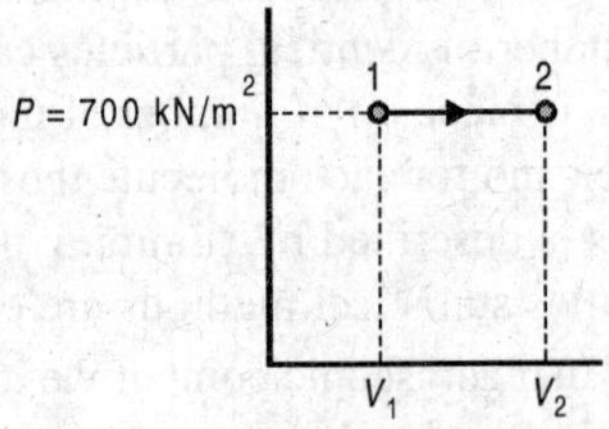

Fig. Ex. 1.1

We know that, work done $= \int_{V_1}^{V_2} PdV$

i.e., Work done $= W_{1-2} = P(V_2 - V_1)$ since $(P = C)$

Note : Units $\frac{N}{m^2} \times m^3 = Nm$

$$= 700 \times 10^3 (1.68 - 0.28)\ Nms$$
$$= 700 \times 10^3 \times 1.4\ Nm \text{ or Joules}$$
$$= 980000\ J = 9.8 \times 10^3\ J$$
$$\mathbf{W_{1-2} = 980\ kJ = 0.98\ MJ}$$

Example 1.2 *0.112 m³ of gas has a pressure of 138 kN/m². It is compressed to 690 kN/m², according to the law $PV^{1.4} = C$. Determine the new volume of the gas.*

Data: The gas is compressed according to the $PV^{1.4} = C$

$P_1 = 138\ kN/m^2$ $\quad P_2 = 690\ kN/m^2$

$V_1 = 0.112\ m^3$ $\quad V_2 = ?$

Solution

We know that polytropic process is given by $PV^n = C$

i.e., $$PV^n = P_1 V_1^n = P_2 V_2^n = C$$

$\therefore$ $$P_1 V_1^{1.4} = P_2 V_2^{1.4}$$

$$\frac{P_1}{P_2} = \left[\frac{V_2}{V_1}\right]^{1.4}$$

or $$\frac{V_2}{V_1} = \left[\frac{P_1}{P_2}\right]^{\frac{1}{1.4}}$$

or $$V_2 = V_1 \times \left[\frac{P_1}{P_2}\right]^{\frac{1}{1.4}}$$

$$= 0.112\ m^3 \times \left[\frac{138\ kN/m^2}{690\ kN/m^2}\right]^{1/1.4}$$

$$= 0.112\ m^3 \times (0.2)^{1/1.4} = 0.112 \times 0.3168$$

$$\mathbf{V_2 = 0.03555\ m^3}$$

Example 1.3 *0.014 m^3 gas at a pressure of 2070 kN/m^2 expands to a pressure of 207 kN/m^2, according to the law $PV^{1.35} = C$. Determine the work done by the gas during the expansion.*

Data: Gas expands according to the law $PV^{1.35} = C$

$V_1 = 0.14\ m^3$ $\quad P_1 = 2070\ kN/m^2$

$P_2 = 207\ kN/m^2$ $\quad$ Work done = ?

Solution

The work done during a polytropic process is given by,

$$W_{1-2} = \frac{P_1V_1 - P_2V_2}{n-1} = \frac{P_2V_2 - P_1V_1}{1-n}$$

Since V_2 is not given, to determine V_2,

We know that since it is a polytropic process,

$$P_1V_1^n = P_2V_2^n$$

$$\frac{P_1}{P_2} = \left[\frac{V_2}{V_1}\right]^n$$

$$V_2 = V_1 \times \left[\frac{P_1}{P_2}\right]^{\frac{1}{n}} = 0.014\ \text{m}^3 \left[\frac{2070\ \text{kN/m}^2}{207\ \text{kN/m}^2}\right]^{\frac{1}{1.35}}$$

$$= 0.014 \times 5.5048\ \text{m}^3$$

$$\mathbf{V_2 = 0.0771\ m^3}$$

$$\text{Work done} = W_{1-2} = \frac{P_1V_1 - P_2V_2}{n-1}$$

$$= \frac{2070\times10^3\ \text{N/m}^2 \times 0.014\ \text{m}^3 - 207\times10^3\ \text{N/m}^2 \times 0.077\text{m}^3}{1.35-1}$$

$$= \frac{28980 - 15959.7}{0.35} = \frac{13020.3}{0.35} = 37200.857\ \text{Nm}$$

$$= 37.2\times10^3\ \text{Nm or Joules}$$

$$\mathbf{W_{1-2} = 37.2\ kJ}$$

Example 1.4 *A gas is compressed hyperbolically (Isothermally) from a pressure and volume of 100 kN/m^2 and 0.056 m^3 respectively to a volume of 0.007 m^3. Determine the final pressure and the work done on the gas.*

Data: Since the gas is compressed Isothermally, PV =C i.e. $P_1V_1 = P_2V_2$

$P_1 = 100\ kN/m^2 = 100 \times 10^3\ N/m^2$

$P_2 = ?$

Work done = ?

$V_1 = 0.056\ m^3$ $\quad V_2 = 0.007\ m^3$

Solution

Since it is an Isothermal process $P_1V_1 = P_2V_2$

$$P_2 = \frac{P_1V_1}{V_2}$$

$$P_2 = \frac{100\times1000\ \text{N/m}^2 \times 0.056\ \text{m}^3}{0.007\ \text{m}^3}$$

$$P_2 = \frac{5600}{0.007}$$

$$P_2 = 8\times10^5 \text{ N/m}^2$$

$$\mathbf{P_2 = 8\times10^2 kN/m^2 = 800\ kN/m^2}$$

We know that, Work done $= P_1V_1 \ln \dfrac{V_2}{V_1}$

$$= 100\times1000\times0.056\times\ln\left(\frac{0.007}{0.056}\right)$$

$$= 100 \times 1000 \times 0.056 \times -2.079$$

$$W_{1-2} = -11644.973 \text{ Nm} = -11.65 \text{ kNm or kJ}$$

∴ The work done on the gas = 11.65 kJ

Example 1.5 *A non -flow reversible, quasi static process occurs for which pressure P = [V^2 + 8/V] bar. Determine the work done if volume changes from 1 m^3 to 3 m^3.*

Data: $P = [V_2 + 8/V]\ bar = [V^2 + 8/V]\times 10^5\ N/m^2$

$V_1 = 1 m^3,\ V_2 = 3m^2$, work done = ?

Solution

We know that, work done in a closed system or displacement work done

$$W_{1-2} = \int PdV$$

$$W_{1-2} = \int_{V_1=1}^{V_2=3} PdV, \text{ subtitute for } P,$$

$$= \int_1^3 \left[V^2 + \frac{8}{V}\right]\times10^5 dV = 10^5 \int_1^3 \left[V^2 + \frac{8}{V}\right] dV$$

$$= 10^5 \left[\frac{V^3}{3} + 8\ln V\right]_1^3 = 10^5\left[\frac{3^3 - 1^3}{3} + 8\ln\frac{3}{1}\right]$$

$$= 10^5\left[\frac{27-1}{3} + 8\times1.0986\right] = 10^5\left[\frac{26}{3} + 8.789\right]$$

$$\mathbf{W_{1-2} = 1745556.5\ Nm = 17.455\times10^5\ Nm\ or\ Joules}$$

Example 1.6 *In a certain thermodynamic process of an ideal gas, the volume changes from 0.2 m^3 to 0.5 m^3 while pressure change occurs according to the law P = 1500 $\left[\dfrac{V}{100} + 1\right]$. Where P is in N/m^2 and V is in m^3, then find out the work done by the gas in kJ.*

Data: $V_1 = 0.2\,m^3$; $V_2 = 0.5\,m^3$ and $P = 1500\left[\dfrac{V}{100} + 1\right] N/m^2$

Solution

We know that, work done $= W_{1-2} = \int PdV$; substitute for P

$$= \int_{0.2}^{0.5} 1500\left[\frac{V}{100}+1\right] dV = 1500\times\int_{0.2}^{0.5}\left[\frac{V}{100}+1\right] dV$$

$$= 1500\times\left[\frac{V^2}{2\times100}+V\right]_{0.2}^{0.5} = 1500\times\left[\frac{0.5^2-0.2^2}{200}+(0.5-0.2)\right]$$

$$= 1500\times(0.00105+0.3) = 451.575 \text{ Nm}$$

$$\mathbf{W_{1-2} = 0.4515 \text{ kJ}}$$

Example 1.7 *Show that for Van-der-Waals equation,* $\left[P+\frac{a}{V^2}\right](V-b) = mRT$

The work done is given by, $mRT \log_e \frac{V_2-b}{V_1-b} - a\left[\frac{1}{V_1}-\frac{1}{V_2}\right]$, *where* V_1 *and* V_2 *are initial and final volumes.*

Solution

From the given equation i.e.,

$$\left[P+\frac{a}{V^2}\right](V-b) = mRT$$

$$P = \frac{mRT}{V-b}-\frac{a}{V^2}$$

$$\therefore \quad \text{Work done} = W_{1-2} = \int PdV = \int_1^2\left[\frac{mRT}{V-b}-\frac{a}{V^2}\right]dV$$

$$= mRT\int_1^2\frac{1}{V-b}dV - \int_1^2\frac{a}{V^2}dV$$

$$\mathbf{W_{1-2} = mRT \log_e \frac{V_2-b}{V_1-b} - a\left[\frac{1}{V_1}-\frac{1}{V_2}\right]}$$

Example 1.8 *A gas in a cylinder and piston arrangement comprise the system. Gas expands frictionlessly from 1.5* m^3 *to 2* m^3 *while receiving 2 Nm of work from a Paddle wheel. Pressure of gas remains constant at 6* N/m^2*. Determine the net work done by the system.*

Solution

Work done by the gas on the piston,

W_{gas} = Force × Displacement = $F \times dl = P \times A \times dl = P \times dV$

The total work done by the gas, when the volume changes from V_1 to V_2 and the pressure remaining constant = $\int PdV$

i.e.,
$$W_{gas} = \int_1^2 PdV$$

$$= P\int_{1.5}^2 dV \quad \text{since } P \text{ is constant}$$

$$= 6[V]_{1.5}^2$$

$$= 6\,[2-1.5] = 3 \text{ Nm}$$

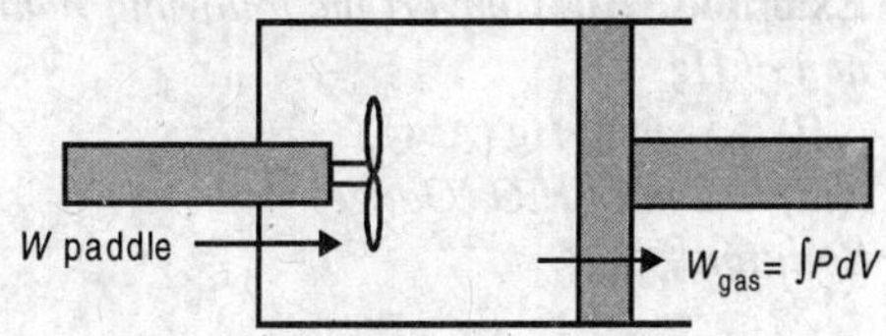

Fig. Ex. 1.8

But gas receives 2 Nm of work from the paddle wheel,

∴ **Net work done by the system = 3–2 = 1 Nm**

Example 1.9 *Convert the following readings of pressure to kPa assuming that the barometer reads 760 mm of Hg*

(a) 80 cm of Hg *(b) 40 cm of Hg vacuum*

(c) 1.5 mt of water *(d) 5.2 bar*

Solution

Let us assume ρ of Hg = 13596 kg/m³

and g = 9.806 m/sec², if the values are not given

Let Z_o = Barometric pressure = 760 mm of Hg

∴ Pressure exerted due to 760 mm of Hg column in the barometer,

$$P_{atm} = \rho gh \text{ or } \rho g Z_o = 13596 \times 9.806 \times \frac{760}{1000}$$

$$= 101325.01 \text{ N/m}^2 \text{ or Pa} = 101.325 \text{ kN/m}^2 \text{ or kPa}$$

(a) Pressure of 80 cm of Hg

For 76 cm = 101.325 kPa

∴ 80 cm = ?

$$\mathbf{\frac{80}{76} \times 101.325 = 106.6578 \text{ kPa}}$$

or

We know that, pressure exerted due to a column of fluid = ρ.g.Z

$$= 13596 \times 9.806 \times 0.8 = 106657.8 \text{ N/m}^2 = 106.657 \text{ kPa}$$

(b) 40 cm of Hg vacuum

In this case, $P_{abs} = P_{atm} - P_{vac}$ = 76 cm – 40 cm = 36 cm

Since for 76 cm = 101.325 kPa

∴ 36 cm = ?

$$\frac{36}{76} \times 101.325 = 47.996 \text{ kPa}$$

(c) Pressure due to 1.5 mt of H_2O column on kPa

$$= \rho_w \cdot g \cdot Z_w = 1000 \times 9.806 \times 1.5$$

$$= \mathbf{14709 \text{ N/m}^2 \text{ or Pa} = 14709 \text{ kPa}}$$

(d) 5.2 bar = 5.2 × 10⁵ N/m² or Pa

$$= \mathbf{5.2 \times 10^2 \text{ kPa} = 520 \text{ kPa}}$$

Note: Pressure of 1 atm. equals 760 mm of Hg or 101325 N/m². These are standard values. To get this valve of 101325 N/m², we have to use r of Hg = 13596 kg/m³ and g = 9.806 m/sec².
If we use r of Hg = 13.6 × 10³ kg/m³ and g = 9.81 m/sec² then we get P_{atm} = 101396 N/m².

Example 1.10 *Convert the following readings of pressure into kPa (Abs) assuming barometric pressure of 750 mm of Hg.*

(i) 55 cm of Hg (Abs) *(ii) 3.3 Atm (Gauge)*

(iii) 3.2 m of H_2O (Gauge) *(iv) 4.6 bar (Abs)*

Solution

For $$P_{atm} = \rho_{Hg} \cdot g \cdot Z_o = \frac{13596 \times 9.81 \times 0.75}{1000}$$

$$= 100.32 \text{ kPa}$$

(i) 55 cm of Hg (Abs)

$$P_{abs} = 13596 \times 9.81 \times 0.55 = 73357 \text{ N/m}^2$$
$$= 73.35 \text{ kPa}$$

(ii) 3.3 atm (Gauge)

Since 1 atm = 101325 N/m²

∴ 3.3 atm = 3.3 × 101325 N/m²

$$\therefore \quad P_{abs} = P_{atm} + P_{gauge} = 100.032 + \frac{3.3 \times 101325}{1000}$$
$$= 434.405 \text{ kPa}$$

(iii) 3.2 m of H_2O (Gauge)

$$P_{abs} = P_{atm} + P_{gauge} = 100.032 + \frac{100 \times 9.81 \times 3.2}{1000}$$
$$= 131.42 \text{ kPa}$$

(iv) 4.6 bar (Abs)

Since 1 bar = 10^5 N/m² = 10^2 kN/m²

∴ 4.6 bar = 4.6 × 100 kPa

= **460 kPa**

Example 1.11 *A manometer shows a reading of 50 cm of water. If atmospheric pressure is 763 mm of Hg. Find absolute pressure in kPa. Take ρ of Hg = 13.6 × 10³ kg/m³, g = 9.81 m/sec².*

Solution

$$P_{abs} = P_{atm} + P_{gauge}$$

$$P_{atm} = \rho \cdot g \cdot Z_o = 13.6 \times 10^3 \frac{\text{kg}}{\text{m}^3} \times 9.81 \frac{\text{m}}{\text{sec}^2} \times 0.765 \text{ m}$$
$$= 102063.24 \text{ N/m}^2 \text{ or Pa}$$
$$= 102.063 \text{ kPa}$$

and $P_{manometer}$ or $P_{gauge} = \rho_w \cdot g \cdot Z_w = 1000 \times 9.81 \times 0.50 = 4905 \text{ kPa}$

$$= 4.905 \text{ kPa}$$

$$P_{abs} = P_{atm} + P_{gauge}$$

$$P_{abs} = 102.063 + 4.905 = 106.968 \text{ kPa}$$

Example 1.12 *A vacuum gauge mounted on the condenser reads 70 cm of Hg. If atmospheric pressure is 101.325 kPa. Find absolute pressure in kPa.*

Solution

We know that $P_{abs} = P_{atm} - P_{vac}$

To find $P_{vac} = \rho \cdot g \cdot Z = 13.6 \times 10^3 \times 9.81 \times 0.70$

$$= 93391.2 \text{ Pa} = 93.3912 \text{ kPa}$$

$$\mathbf{P_{abs} = 101.325 - 93.3912}$$
$$\mathbf{= 7.9338 \text{ kPa}}$$

Example 1.13 *A vacuum gauge connected to the tank reads 30 kPa at a location where a barometric reading is 755 mm of Hg. Determine the absolute pressure in the tank. Take density of Hg = 13590 kg/m³.*

Solution

$$P_{vac} = 30 \text{ kPa}$$
$$Z_0 = 755 \text{ mm of Hg}$$
$$P_{abs} = ?$$

$$P_{abs} = P_{atm} - P_{vac} \quad \ldots(1)$$

To find P_{atm}

$$P_{atm} = \rho_{Hg} \cdot g \cdot Z_o = 13590 \times 9.81 \times 0.755$$

$$P_{atm} = 100655.01 \text{ N/m}^2$$

$$P_{atm} = 100.655 \text{ kPa} \quad \ldots(2)$$

∴ From (1)

$$P_{abs} = P_{atm} - P_{vac} = 100.655 - 30$$

$$\mathbf{P_{abs} = 70.655 \text{ kPa}}$$

Example 1.14 *The pressure of air in a pipe line measured with Hg manometer having one arm open to the atmosphere. The level in the open arm is 562 mm higher than the arm connected to the air pipe. Calculate the air pressure. The barometer reads 760 mm of Hg, the acceleration due to gravity is 9.81 m/sec² and the density of Hg is 13596 kg /m³.*

Data: $z = 0.560$ m $Z_o = 0.760$ m
$g = 9.81$ m/sec² $\rho_{Hg} = 13596$ kg/m²

Solution

Note: Since the plane 1–1′ is horizontal plane, and is in the same continuous static mass of liquid, the pressure heads at the marked (*) points in figure will be equal.

Therefore the pressure along a horizontal line 1–1' is same. Balancing the pressure values about the line 1–1' we get,

$$P_{abs} = P_{atm} + \rho \cdot g \cdot Z \quad \ldots(1)$$

But

$$P_{atm} = \rho \times g \times Z_o$$

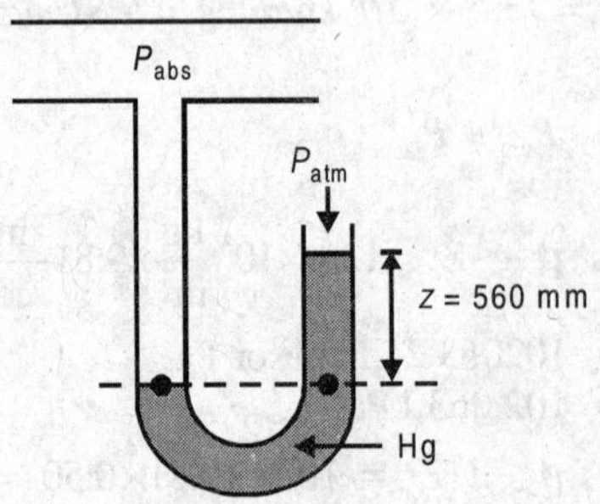

Fig. Ex. 1.14

$$= 13640 \times 9.79 \times 0.761 \quad \ldots(2)$$

$$= 101620.59 \text{ N/m}^2$$

$$= 101.621 \text{ kN/m}^2$$

The pressure exerted due to the column of Hg of height $Z = \rho \times g \times Z$

$$= 13640 \times 9.79 \times 0.562 = 75047 \text{ N/m}^2 \quad \ldots(3)$$

$$= 75.047 \text{ kN/m}^2$$

∴ From Eqs. (1), (2) and (3)

$$\mathbf{P_{abs} = 101.621 + 75.047 = 176.668 \text{ kN/m}^2}$$

Example 1.15 *A U-tube manometer contains a liquid having a density of 800 kg/m³. When the manometer is connected to a gas pipe, the level in the open arm is 30 cm higher than the level in the column connected to the gas pipe. Find the absolute pressure of the gas in bar, if the barometric pressure is 75 cm of Hg.*

Solution

We have, Density of liquid (ρ_{liq}) = 800 kg/m³

Z = 30 cm

$\therefore$ We have, gauge pressure $= \rho g Z = 800 \times 9.81 \times 0.3$

$= 2354.4 \text{ N/m}^2$

$= 0.023544 \text{ bar}$

Now $P_{atm} = \rho \cdot g \cdot Z = 75 \text{ cm of Hg}$

$= 13.596 \times 10^3 \times 9.81 \times 0.75$

$= 99991.78 \text{ N/m}^2$

$= 0.9999 \text{ bar}$

Also, $P_{abs} = P_{gauge} + P_{atm} = 0.023544 + 0.9999$

$\mathbf{P_{abs} = 1.234618 \text{ bar}}$

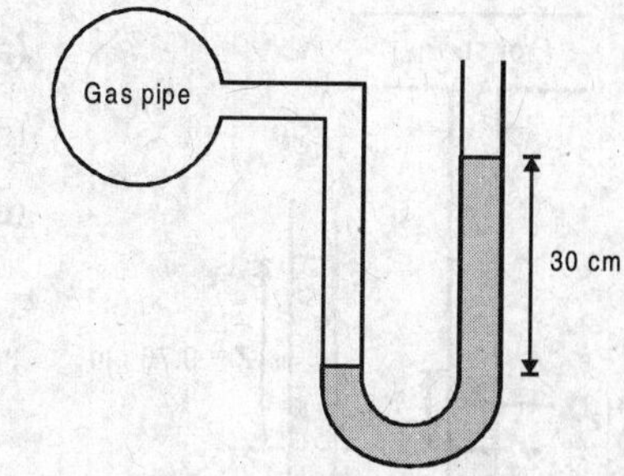

Fig. Ex. 1.15

Example 1.16. A U-tube manometer is connected to a gas pipe. The level of the liquid in the manometer arm open to the atm. is 17 cm lower than the level of liquid in the arm connected to the gas pipe. The liquid in the manometer has specific gravity of 0.8. Find the abs. pressure of the gas if the barometer reads760 mm of Hg. Take $\rho_{Hg} = 13596$ and $g = 9.806$.

Data: $Z = 0.17 \text{ m}$ $\quad P_{abs} = ?$

$Z_0 = 0.76$ $\quad Sp.gr = 0.8$

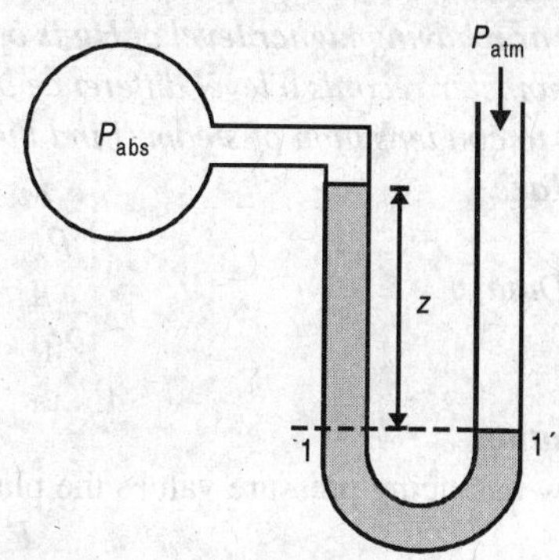

Fig. Ex. 1.16

Solution

Note: Specific gravity = $\dfrac{\rho \text{ of given liquid}}{\rho \text{ of water}}$

$\therefore$ ρ of given liquid $= 0.8 \times 10^3 \text{ kg/m}^2$

Balancing the pressure values about the plane 1–1'

$$P_{abs} + \rho g Z = P_{atm}$$

$$P_{abs} = P_{atm} - \rho \cdot g \cdot Z$$

(a) $$P_{atm} = \rho \cdot g \cdot Z_o \quad \ldots(1)$$

$$= 13596 \text{ kg/m}^3 \times 9.806 \text{ m/sec}^2 \times 0.7 \text{ m}$$

$$= 101325 \text{ N/m}^2$$

and $$\rho g Z = 0.8 \times 10^3 \times 9.806 \times 0.17 = 1333.616 \text{ N/m}^2$$

$\therefore$ $$P_{abs} = 101325 - 1333.616 = 99991.384 \text{ N/m}^2$$

$$\mathbf{P_{abs} = 0.999913 \text{ bar}}$$

Example 1.17. *A U-tube manometer with one arm open to atmosphere is used to measure the pressure of steam flowing through a pipe. Hg is used in the manometer. The height of the column open to the atm. is 9.75 cm greater than the Hg level in the arm connected to the steam pipe. Steam condenses in the manometer on the steam side.*

The column of condensate is 3.4 cm. The atmospheric pressure is 76 cm of Hg. Find the absolute pressure of steam. Take,

$g = 9.806 \text{ m/sec}^2$ *and*

$\rho_{Hg} = 13596 \text{ kg/m}^3$

$P_{abs} = P \text{ of steam} = ?$

Data: $P_{abs} = ?$

$Z = 9.75 \text{ cm}$

Solution

Balancing the pressure values about the Plane 1–1'

$$P_{abs} + \rho_w \cdot g \cdot Z_w = P_{atm} + \rho g Z$$

Now to find,

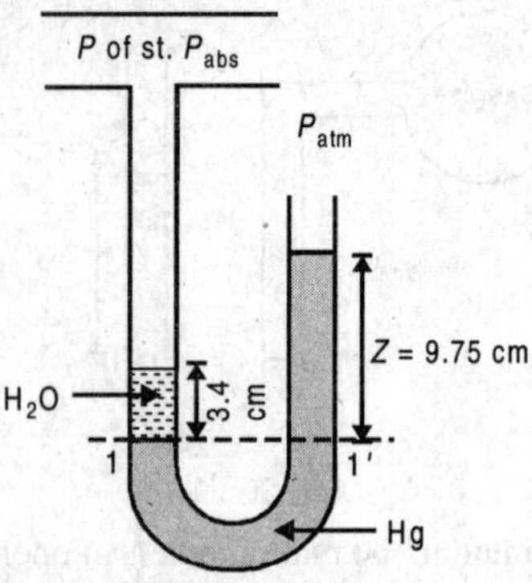

Fig. Ex. 1.17

(a) $\rho_w \cdot g \cdot Z_w$ = $1000 \times 9.806 \times 0.034 = 333.404 \text{ N/m}^2$ ...(2)

(b) $P_{atm} = \rho \cdot g \cdot Z_o$ = $13596 \times 9.806 \times 0.76 = 101325 \text{ N/m}^2$ ...(3)

(c) $\rho \cdot g \cdot Z$ = $13596 \times 9.806 \times 0.0975$

$= 12998.932 \text{ N/m}^2$ ...(4)

∴ Now substituting the values of Eqs (2), (3) and (4) in Eq. (1) we get,

$$P_{abs} + 333.404 = 101325 + 12998.932$$

$$P_{abs} = 114323.93 - 333.404 = 113990.53 \text{ N/m}^2$$

$$\mathbf{P_{abs} = 1.139905 \text{ bar}}$$

Example 1.18 *A Hg manometer is used to measure the pressure of steam in the mains. (i.e., main steam pipe). Its one arm having higher level of Hg is open to atmosphere at 98 kPa. The other arm is connected to steam pipe. The manometer records a level difference of 60 cm of Hg. On the lower level of Hg, 4 cm of water column accumulates due to condensation of steam. Find the absolute pressure of steam in bar.*

Take $g = 9.7 \text{ m/sec}^2$ *and* $\rho_{Hg} = 13.69 \text{ gm/cc}.$

$\rho_w = 1000 \text{ kg/m}^3$

Data: $g = 9.7 \text{ m/sec}^2$

$\rho_{Hg} = 13.69 \text{ gm/cc} = 13.69 \times 10^3 \text{ kg/m}^3.$

$\rho_{water} = 1000 \text{ kg/m}^3$

Solution

Now balancing pressure values the plane 1–1′

$$P_{abs} + \rho_w g Z_w = P_{atm} + \rho \cdot g \cdot Z \quad \text{...(1)}$$

Now to find,

(a) $\rho_w g Z_w = 1000 \times 9.7 \times 0.04$

$= 388 \text{ N/m}^2$

$= 0.338 \text{ kN/m}^2$ or kPa ...(2)

(b) $\rho \cdot g \cdot Z = 13.69 \times 10^3 \times 9.7 \times 0.6$

$= 79675.8 \text{ N/m}^2 = 79.6758 \text{ kPa}$

∴ From (1) $P_{abs} = P_{atm} + \rho g Z - \rho_w g Z_w$

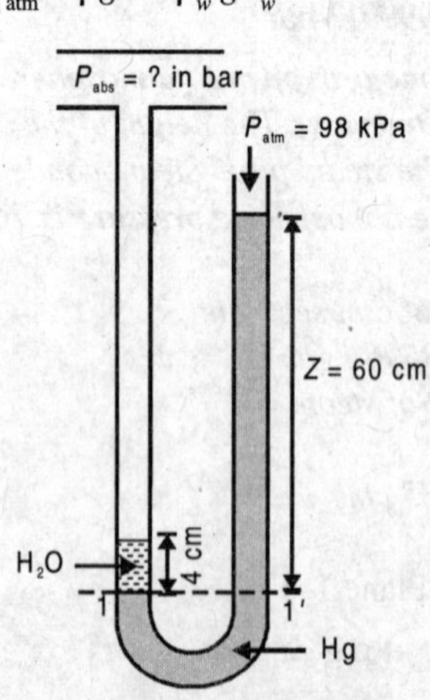

Fig. Ex. 1.18

$$= 98 + 79.68 - 0.388$$
$$= 177.68 - 0.388$$
$$= 177.292 \text{ kPa}$$

Since $1 \text{ bar} = 10^5 \text{ N/m}^2 = 100 \text{ kPa}$

$$\mathbf{P_{abs} = 1.77292 \text{ bar}}$$

Example 1.19 *A U-tube manometer with one arm open to atmosphere is used to measure the pressure of steam flowing through a pipe. Mercury is used as a manometric fluid. The height of the column of mercury in the open end is 10 cm greater than the mercury level in the arm connected to the steam pipe. Steam condenses in the manometer on the steam side. The column of condensate is 3.5 cm in height. The atmospheric pressure is 760 mm mercury. What is the absolute pressure of steam in the pipe in bar ?*

Data: $Z_o = 0.76$ m $\quad P_{abs}$ or $P_{steam} = ?$

$g = 9.806 \quad \rho_{Hg} = 13596$

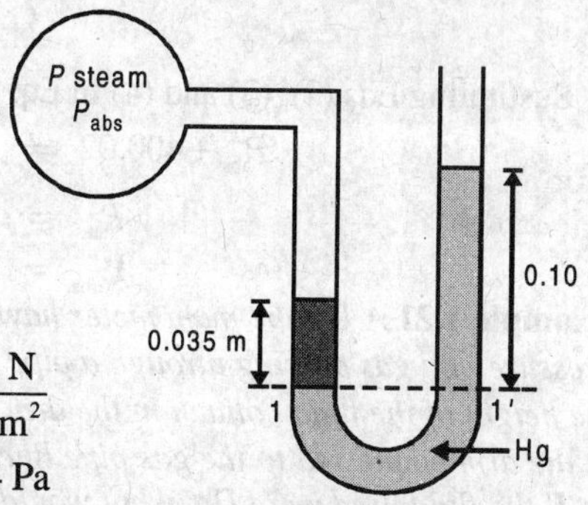

Fig. Ex. 1.19

Solution

$$P_{abs} + \rho_w g Z_w = P_{atm} + \rho_{Hg}\, g\, Z_{Hg}$$

$$P_{atm} = \rho_{Hg}\, g\, Z_o$$

$$= 13596 \times 9.806 \times 0.76 = 101325 \frac{\text{N}}{\text{m}^2}$$

$$\rho_w g Z_w = 1000 \times 9.806 \times 0.035 = 333.404 \text{ Pa}$$

$$\rho_{Hg}\, g\, Z_{Hg} = 13596 \times 9.806 \times 0.10 = 12998.93 \text{ Pa}$$

∴ From (1) $\quad P_{abs} = P_{atm} + \rho_{Hg} g Z_{Hg} - \rho_w g Z_w$

$$= 101325 + 12998.9 - 333.404$$

[From (1), (2), (3) and (4)] $\quad = 113990.52 \text{ Pa} = 1.139 \text{ bar}$

$$\mathbf{P_{abs} = 1.14 \text{ bar}}$$

Example 1.20 *A U-tube manometer with one arm open to atmosphere is used to measure the pressure of kerosene vapour passing in a pipe. Hg level in the open arm is 10 cm greater than that in the arm connected to the kerosene pipe. There is a condensation of kerosene vapour on the Hg column of 5.1 cm height. If the atmospheric pressure is 75.5 cm of Hg. Find the absolute pressure of vapour in kPa. Assume specific gravity of kerosene as 0.8 and that of Hg as 13.6.*

Solution

Balancing the pressure values @ the plane 1–1'

$$P_{\text{abs of Kero. vap}} + \rho_{kero}\, g Z_{kero} = P_{atm} + \rho_{Hg}\, g Z_{Hg} \quad ...(1)$$

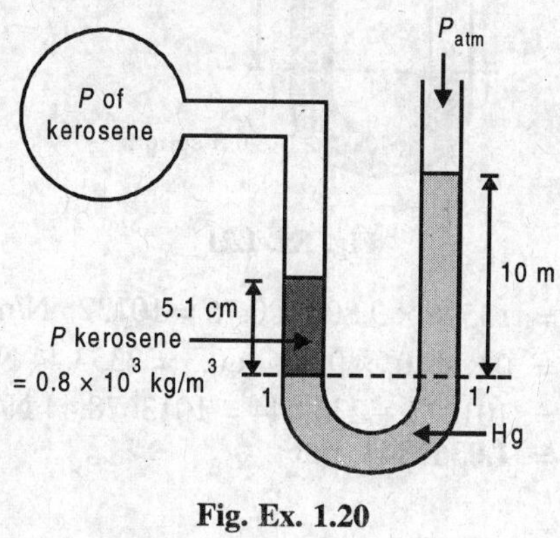

Fig. Ex. 1.20

Now to find

(a) $$\rho_{kero} \cdot g \cdot Z_{kero.} = 800 \times 9.81 \times 0.051 \text{ m}$$

$$= 400.02 \text{ N/m}^2 \quad ...(2)$$

(b) $$P_{atm} = \rho_{Hg}\, g Z_o$$

$$= 13600 \times 9.81 \times 0.755 \text{ m}$$

$$= 100726.08 \text{ N/m}^2 \quad ...(3)$$

(c) $$\rho_{Hg}\, g Z_{Hg} = 13600 \times 9.81 \times 0.1 \text{ m}$$

$$= 13341.6 \text{ N/m}^2 \quad ...(4)$$

Sustituting Eqs (2), (3) and (4) in Eq. (1) we get,

$$P_{abs} + 400.02 = 100729.08 + 13341.6$$

$$P_{abs} = 113670.68 \text{ N/m}^2$$

$$\mathbf{P_{abs} = 113.67068 \text{ kP}}$$

Example 1.21 *A U-tube manometer having equal arms with one arm open to atmosphere is used to measure the pressure of a gas passing through a pipe line. The fluid used in the manometer has a specific gravity of 0.8, while the height of the fluid column in the arm open to the atmosphere is 30 cm greater than the height of fluid column in the arm connected to the gas pipe line. Determine, the absolute pressure of the gas in pipe line.*

If the fluid used were Hg, what would be the difference of height of Hg-column in the 2-arms.

Take $\rho_{Hg} = 13596$ *and* $g = 9.806$ *m/sec*^{2}

Solution

r of fluid used in the manometer = 0.8×1000 kg/m^3

$$P_{abs} = P_{atm} + rgZ$$

Let the barometer reads 760 cm of Hg

$$P_{atm} = rgZ_o$$

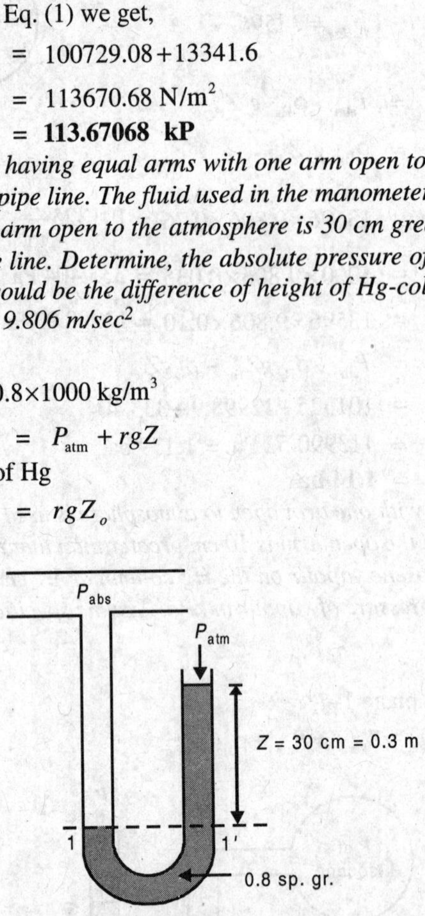

Fig. Ex. 1.21

$$= 13596 \times 9.806 \times 0.76 = 101325 \text{ N/m}^2$$

and $$rgZ = 0.8 \times 10^3 \times 9.806 \times 0.3 = 2353.44 \text{ N/m}^2$$

$\therefore$ $$P_{abs} = 101325 + 2353.44 = 1013678.44 \text{ N/m}^2$$

$$\mathbf{P_{abs} = 1.0367844 \text{ bar}}$$

If Hg is used, then to find Z,

(i.e., to produce the same absolute pressure, we have to find the equivalent column of Hg)

$$P_{abs} = P_{atm} + rgZ$$

$$103678.44 = 101325 + 13596 \times 9.806 \times Z$$

$$Z = \frac{103678.44 - 101325}{13596 \times 9.806} = 0.01765 \text{ mt of Hg}$$

$$\mathbf{Z = 1.765 \text{ cm of Hg}}$$

Example 1.22 *A vertical composite liquid column with its upper end exposed to atmospheric pressure, comprise of 45 cms of Hg (sp. gr. 13.6); 65 cm of water and 80 cm of oil (sp. gr. 0.8). Calculate the pressure in bar,*

(i) At the bottom of the column.

(ii) At the inter surface of oil and water.

(iii) At the inter surface of water and Hg.

Assume atmospheric pressure as 100 kPa and g = 9.81 m/sec²

Solution

Since it is given P_{atm} = 100 kPa i.e., equal to 1 bar

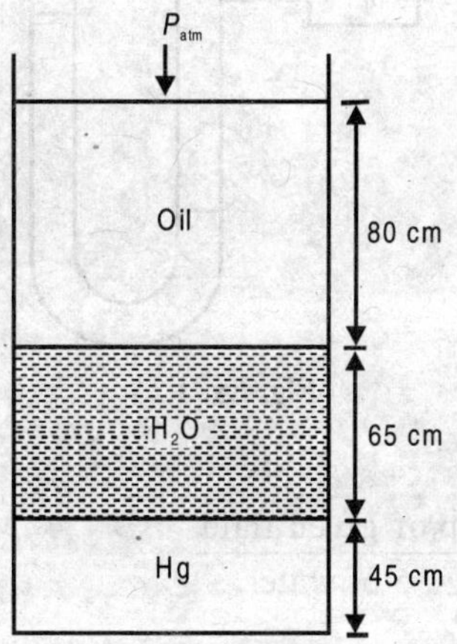

Fig. Ex. 1.22

(Since 10^5 N/m² = 1 bar = 100 kPa)

(i)

$$P_{\text{at bottom}} = P_{atm} + P_{oil} + P_{water} + P_{Hg}$$

$$= 1 + \rho_{oil} g Z_{oil} + \rho_w g Z_w + \rho g Z$$

$$= 1 + \frac{0.8 \times 1000 \times 9.81 \times 0.8}{10^5} + \frac{1000 \times 9.81 \times 0.65}{10^5} + \frac{13.6 \times 10^3 \times 9.81 \times 0.45}{10^5}$$

Note: If r is in kg/m³, g is in m/sec², Z is in m, then the resultant unit will be N/m².

∴ To get in bar divide by 10^5

(i)

$$P_{\text{at bottom}} = 1 + 0.062784 + 0.063765 + 0.6$$

$$\mathbf{P_{\text{at bottom}} = 1.72692 \text{ bar}}$$

(ii) Pressure at the interface of oil and water

$$= 1 + 0.062784 = 1.062784 \text{ bar}$$

(iii) Pressure at the inter surface of waterand Hg = 1 + 0.062784 + 0.063765

= 1.1265 bar

Example 1.23 *An air tank has three manometers connected to it. The fluids in them are oil (Specific Gravity = 0.8) water and mercury (Specific Gravity = 13.6). If the absolute pressure in the tank is 1.2 bar and the barometer reads 760 mm of Hg. Estimate the height of fluid in each of manometer.*

Data: Fluids in the manometers are oil (Sp. Gr. = 0.8)

Water and mercury (Sp. Gr. = 13.6)

$$P_{abs} = 1.2 \text{ bar}$$

$$\text{Barometric reading} = 760 \text{ mm of Hg}$$

Height of fluid in each of manometer = ?

Solution

We know that,

$$P_{abs} = P_{atm} + P_{gauge}$$

$$P_{atm} = \rho_{Hg} g Z_o = 13.6\times10^3 \times 9.81\times0.76 = 101396.16 \text{ N}$$

$$P_{abs} = 1.2 \text{ bar} = 1.2\times10^5 \text{ N/m}^2$$

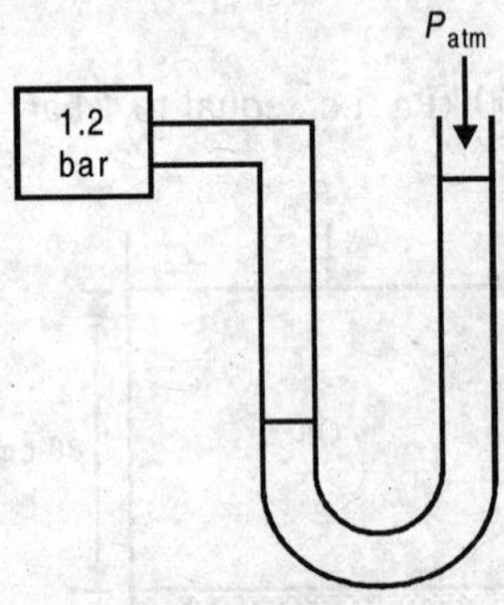

Fig. Ex. 1.23

Case I: For Oil Manometer

(a)
$$\text{Sp. Gravity} = \frac{\rho \text{ of given fluid}}{\rho \text{ of water}}$$

$$0.8 = \frac{\rho_{oil}}{1000}$$

$$\therefore \quad \rho_{oil} = 800 \text{ kg/m}^3$$

$$\therefore \quad P_{gauge} = \rho_{oil} g Z_{oil} = 800\times9.81\times Z_{oil}$$

Subtituting these values in Eq. (1) we get,

$$1.2\times10^5 = 101396.6 + 800\times9.81\times Z_{oil}$$

$$\mathbf{Z_{oil} = 2.37 \text{ m}}$$

Case II: For Water Manometer

$$P_{abs} = P_{atm} + \rho_w g Z_w$$

$$1.2\times10^5 = 101396.6 + 1000\times9.81\times Z_w$$

$$\mathbf{Z_w = 1.89 \text{ m}}$$

Case III: For Hg Manometer

$$P_{abs} = P_{atm} + \rho_{Hg} g Z_{Hg}$$

$$1.2\times10^5 = 101396.6 + 13.6\times10^3 \times 9.81\times Z_{Hg}$$

$$\mathbf{Z_{Hg} = 0.139 \text{ Hg}}$$

Example 1.24 *A basic barometer can be used as an altitude measuring device in aeroplanes. The ground control reports a barometric reading of 753 mm of Hg, while the Pilot's reading is 690 mm of Hg.*

Estimate the altitude of the plane from the ground level if the average air density is 1.25 kg/m³ and g = 9.8 m/sec².

Solution

Barometer reading at ground level $Z_{ground} = 0.753$ m

$$\text{Pilot's Barometric reading } Z_{plane} = 0.690 \text{ m}$$

$$\text{Altitude} = ?$$

$$\text{Pressure at the ground level} = \rho g Z_{ground}$$

$$= 13600 \times 9.8 \times 0.753 = 100359.84 \text{ N/m}^2 \quad ...(1)$$

$$\text{Pressure at plane level} = \rho g Z_{plane}$$

$$= 13600 \times 9.8 \times 0.690$$

$$= 91963.2 \text{ N/m}^2 \quad ...(2)$$

$\therefore$ Change of pressure at the ground level and that of plane level,

$$\Delta P = 100359.84 - 91963.2$$

$$= 8396.64 \text{ N/m}^2$$

$\therefore$ This is the difference of pressure, now to find altitude for this much difference of pressure,

$$\Delta P = \rho_{air} g Z_{altitude}$$

$$8396.64 = 1.25 \times 9.8 \times Z_{altitude}$$

$$\mathbf{Z_{altitude} = 684.44 \text{ m}}$$

Example 1.25 *A manometer connected between 2 pipes as shown in Fig. Ex. 1.25. Take,*

$\rho_{water} = 1000$ kg/m³ $\qquad \rho_{Hg} = 13590$ kg/m³

Pressure at A = 400 kPa $\qquad g = 9.8$ m/sec².

Find Pressure at B = ?

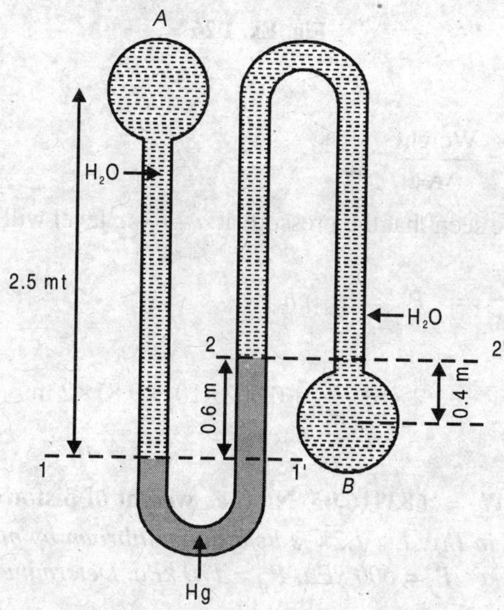

Fig. Ex. 1.25

Solution

Note. If there are more than one U-turns, then start from point and as we go downwards take it as positive, and as we go upwards take it as negative. Then equate this pressure equation to pressure at the second point.

So, writing the pressure equation,

$$P_A + P_w\, g\, Z_w - \rho_{Hg}\, g\, Z_{Hg} + \rho_w\, g\, Z_w = P_B$$

$$400 \times 1000 + 1000 \times 9.8 \times 2.5 - 13590 \times 9.8 \times 0.6 + 1000 \times 9.8 \times 0.4 = P_B$$

$$400000 + 24500 - 79409.2 + 3920 = P_B$$

$$\mathbf{P_B = 348510.8\ Pascals}$$

$$\mathbf{P_B = 348.5108\ kPa}$$

Example 1.26 *The cylinder and tubing shown in Fig. Ex. 1.26 contain oil of density 0.902×10^3 kg/m³ for gauge reading of 2 bar. What is the total weight of piston and slab placed on it?*

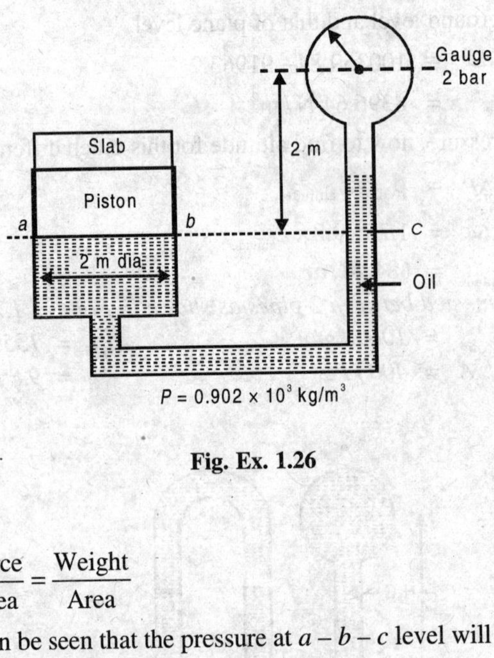

Fig. Ex. 1.26

Solution

$$\text{We know that, Pressure} = \frac{\text{Force}}{\text{Area}} = \frac{\text{Weight}}{\text{Area}}$$

And from Fig. Ex. 1.26 it can be seen that the pressure at $a - b - c$ level will be same.

$$\therefore \quad \frac{W}{A} = P_{gauge} + \rho_{oil}\, g h_{oil}$$

$$\therefore \quad \frac{W}{\frac{\pi}{4}(2)^2} = 2 \times 10^5 \frac{\text{N}}{\text{m}^2} + 0.902 \times 10^3 \times 9.81 \times 2\ \text{m}$$

$$\mathbf{W = 683916.05\ N} \quad \textbf{(i.e., weight of piston and slab)}$$

Example 1.27 *The piston shown in Fig. Ex. 1.27 is held in equilibrium by pressure of gas flowing through the pipe. The piston has a mass of 21 kg. $P_I = 600$ kPa; $P_{II} = 170$ kPa. Determine the pressure in the gas P_{III}.*

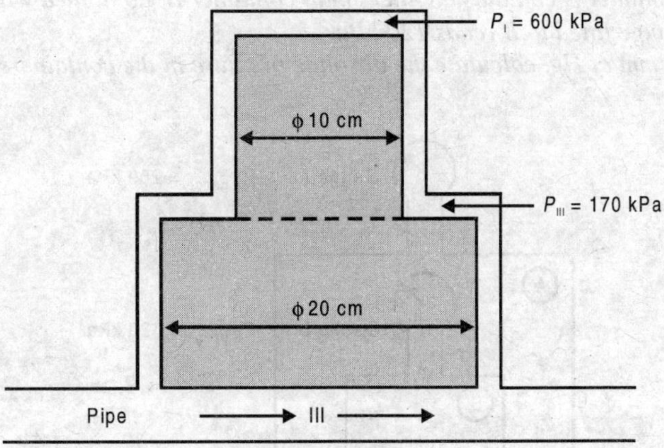

Fig. Ex. 1.27

Solution

Total down ward force acting will be:

(i) Due to self weight. (ii) Pressure P_I acting on 10 cm dia. piston. (iii) Pressure P_{II} acting on 20 cm dia piston. And this is balanced by pressure P_{III}.

$$\therefore \quad \text{Force due to mass} = 21 \times 9.81 = 206.01 \text{ N} = 0.20601 \text{ kN}$$

$$\text{Then force I acting on 10 cm dia piston} = \text{Pr}.\times\text{Area} = 600\times\frac{\pi}{4}(0.1)^2 = 4.712389 \text{ kN}$$

$$\text{And force II acting on 20 cm dia piston} = 170\times\left(\frac{\pi}{4}(0.2^2-0.1^2)\right) = 4.0055306 \text{ kN}$$

$$\therefore \quad \text{Total downward load/force} = 0.20601+4.712389+4.005306 = 8.9239296 \text{ kN}$$

This is balanced by P_{III}

$$\therefore \quad 8.9239296 = P_{III}\times\text{Area} = P_{III}\times\frac{\pi}{4}(0.2)^2 = P_{III}\times 0.0314$$

$$\mathbf{P_{III} = 284.0575 \text{ kPa}}$$

Example 1.28 *The basic barometer can be used to measure the height of a building. If the barometric readings at top and bottom of a building are 730 and 755 mm of Hg respectively: Determine the height of the building. Assume an average air density of 1.18 kg/m³.*

Solution

The difference between pressure at top and bottom is equal to the air head of height equal to that of the building.

$$\therefore \quad P_1 - P_2 = \rho_{air}gh$$

$$(0.755-0.73)\times 13600\times 9.81 = 1.18\times 9.81\times h$$

$$h = \frac{13600\times 0.025}{1.18}$$

$$\mathbf{h = 288.14 \text{ m}}$$

Example 1.29 *A large container A contains another small container B, both fitted with gauges. The gauge fitted to A reads 200 kPa and gauge fitted to B reads 120 kPa.*

If barometer reads 750 mm of Hg, calculate the absolute pressure in the containers A and B.

Solution

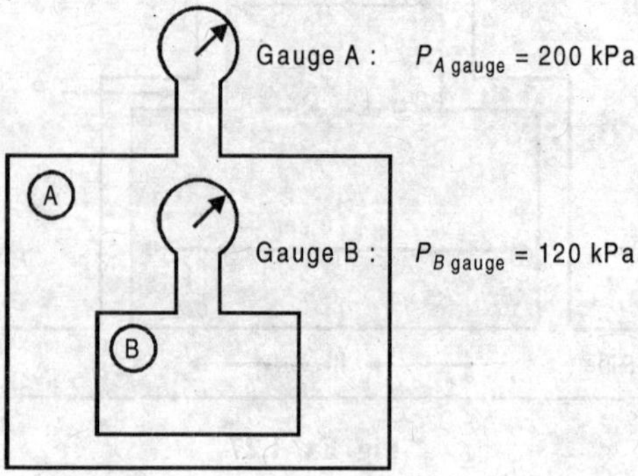

Fig. Ex. 1.29

Barometer reading h_{baro} = 750 mm of Hg = Z_o

$$P_{atm} = \rho \cdot g \cdot h_{baro}$$
$$= 13596 \times 9.806 \times 0.750$$
$$= 100.03 \text{ kPa}$$
$$P_{A\,abs} = P_{A\,gauge} + P_{atm}$$
$$= 200 + 100.03$$
$$= 300.03 \text{ kPa}$$

Note that gauge *B* reads pressure relative to its surrounding presure i.e., gauge *B* is surrounded by pressure of 300.03 kPa.

$$P_{B\,abs} = P_{B\,gauge} + P_{atms} \text{ of } B$$
$$= 120 + 300.03$$
$$\mathbf{P_{B\,abs} = 420.03\ kPa}$$

Example 1.30 *Convert the following temperatures into degree Fahrenheit.*

(i) 40 °C (ii) – 20 °C

Solution

(i) Using the relation
$$\frac{C}{100} = \frac{F-32}{180}$$
$$\frac{40}{100} = \frac{F-32}{180}$$
$$F-32 = \frac{40\times180}{100} = 72$$
$$\mathbf{F = 72+32 = 104\,°F}$$

(ii)
$$\frac{-20}{100} = \frac{F-32}{180}$$
$$F-32 = \frac{-20\times180}{100} = -36$$
$$\mathbf{F = 32-36 = -4\,°F}$$

Example 1.31 *Find the temperature which has the same value on both the Centigrade and Fahrenheit scales.*
Solution

Using the relation $$\frac{C}{100} = \frac{F-32}{180}$$

Putting C = F, $$\therefore \quad \frac{C}{100} = \frac{C-32}{180}$$

$$\therefore \quad 180\,C = 100\,C - 3200$$

$$\therefore \quad C = -40$$

$$\therefore \quad \mathbf{-40\,°C = -40\,°F}$$

Example 1.32 *A temperature scale of a certain thermometer is given by the relation $t = a \log_e P + b$, where a and b constants and P is the thermometric property of the fluid in the thermometer. If at the ice point and steam point, thermometric properties are found to be 1.5 and 7.5 respectively. What would be the temperature corresponding to the thermometric property 3.5 on Celsius scale.*
Solution

$$t = a \log_e P + b$$

$$100 = a \log_e 7.5 + b$$

$$0 = a \log_e 1.5 + b$$

Solving theses two equations we get,

$\therefore$ $a = 62.1336$ and $b = -25.168$

$\therefore$ For $P = 3.5$, $t = 62.1336 \log_e 3.5 - 25.168$

$$\mathbf{t = 52.670676\,°C}$$

Example 1.33 *The temperature scale of a certain thermometer is given by $t = a \ln P + b$, when a and b are constants and P is the thermometric property. The temperature of ice point and steam point are 100 and 300 respectively on this scale. Experiment gives values of thermometric properties of $P_1 = 1.86$ and $P_2 = 6.81$ at ice point and steam point respectively. Evaluate the temperature corresponding to a reading of thermometric property at P = 2.5.*
Solution

$$t = a \ln P + b \quad ...(1)$$

$$300 = a \ln 6.8 + b$$

$$100 = a \ln 1.86 + b \quad ...(2)$$

On subatracting we get, $$200 = a \ln\left(\frac{6.81}{1.86}\right)$$

$$a = \frac{200}{1.3} = 154 \quad ...(3)$$

Substituting this value of *a* in (2) we get,

$$100 = 154 \ln 1.86 + b$$

$$\therefore \quad b = 100 - 154 \ln 1.86 = 4.43$$

$\therefore$ For $P = 2.5$,

$$t = 154 \ln 2.5 + 4.43$$

$$\mathbf{t = 145.4°}$$

Example 1.34 *The pressure in a constant volume gas thermometer is measured as 32 mm of Hg above the atmospheric pressure at triple point of water, determine the temperature in °C, when pressure is 76 mm of Hg above atmospheric pressure. Barometric pressure is 752 mm of Hg.*

Solution

Note: In a constant volume gas thermometer, volume remains constant but as the temperature changes, pressure change occurs.

$$\frac{\theta_1}{\theta_2} = \left(\frac{P_2}{P_1}\right) \text{constant volume} \quad ...(1)$$

Given that $\theta_1 = 273.16$ K (triple point of water)

$$P_1 = P_{abs} + P_{gauge} = 752 + 32$$

$$P_1 = 784 \text{ mm of Hg}$$

and $P_2 = P_{atm} + P_{gauge} = 752 + 76 = 828$ mm of Hg

Substituting these in Eq. (1) we get

$$\frac{273.16}{\theta_2} = \frac{784}{828}$$

$$\theta_2 = 288.49 \text{ K}$$

$$\boldsymbol{\theta_2 = 15.49°C}$$

Example 1.35 *The temperature T on a thermometric scale is defined in terms of a property P by the relation, T = a ln P + b where a and b are constants.*

The temperatures of the ice point and steam points are assigned the numbers 32 and 212 respectively. Experiment gives values of P of 1.86 and 6.81 at the ice point and steam point respectively. Evaluate the temperature corresponding to a reading of P = 2.5 on the thermometer.

Solution

We have $T = a \ln P + b$

Substituting the values for ice point and steam point,

$$32 = a \ln 1.86 + b$$

and $212 = a \ln 6.81 + b$

Solving we get $a = 138.7,\ b = -54.07$

∴ Equation becomes $T = 138.7 \ln P - 54.07$

Now for $P = 2.5$, $T = 138.7 \ln 2.5 - 54.07 = \mathbf{73}$ ■■■

THEORY QUESTIONS

1.1. What is Thermodynamics?

1.2. Define Thermodynamic system? Explain its types with suitable examples.

1.3. Discuss : (1) System, (2) Surrounding, (3) Boundary, (4) Universe.

1.4. What do you understand by thermodynamic system ? How these are classified ? Explain each with suitable sketch, or

Explain

1. Open system
2. Closed system
3. Isolated system

and state the examples of each.

1.5. Distinguish between: Open system and closed system.

1.6. State with reasoning whether the following systems are closed, open or isolated.

1. IC engine
2. Refrigerator
3. Pressure cooker.

1.7. Explain Thermodynamic properties, Processes and cycles.

1.8. Discuss :

1. State
2. Property

1.9. Explain : Thermodynamic properties.

1.10. Compare: Intensive and extensive properties.

1.11. What do you understand by state function and path function ?

1.12. Compare state function and path function.

1.13. Discuss – Unit extensive property is an intensive property.

1.14. Define the terms: Process.

1.15. Explain – 1. Cycle; 2. Process.

1.16. Define property and classify it with proper examples.

1.17. What is meant by Working Substance? Explain with an example.

1.18. Write a short note on Units and Dimensions.

1.19. What is Mechanical and Thermodynamic Work ? Explain.

1.20. What do you mean by thermodynamic work and mechanical work ?

1.21. What is *PdV*-work ? or Work done at the moving boundary ? or Displacement work.

1.22. Define heat and work and explain why neither is a property.

1.23. Prove that Work is a path function and Properties are point functions.

1.24. Define heat and work and explain why neither is a property.

1.25. 'Property is an exact differential'. Explain.

1.26. Derive the equations for work done in various quasi-static processes.

1.27. Write a note on Heat Energy.

1.28. Define the terms : Work and heat.

1.29. Explain the similarities and differences between heat and work.

1.30. Define the property – Pressure. State the various units used in practice.

1.31. How Pressure measurement is done ?

1.32. Write a note on Pressure exerted due to a column of fluid.

1.33. Write a short note on Barometer.

1.34. Briefly explain why you cannot use a mercury barometer to measure pressure in a gravity free environment.

1.35. Write a note on Bourdon Pressure gauge.

1.36. Explain – Bourdon's Pressure gauge with sketch.

1.37. Write a note on Manometer.

1.38. Define Thermal Equilibrium. State and explain the Zeroth law of Thermodynamics.

1.39. State and explain zeroth law of thermodynamics and explain how it leads to the concept of temperature.

1.40. What is meant by thermometric property. Enlist the temperature measuring devices.

1.41. Discuss various thermodynamic properties used for temperature measurement.

1.42. What is the principle of Temperature Measurement ?

1.43. What are the different Scales of Temperature?

1.44. Write a note on Microscopic and Macroscopic point of view.

1.45. What do you understand by microscopic and macroscopic approach of thermodynamic study ? Explain.

1.46. What is the difference between the classical and statistical approaches to Thermodynamics.

1.47. Write a note on Energy in transit.

1.48. Explain the term — Energy in transit.

1.49. What is a Quasi–Static Process? Explain.

1.50. What do you mean by Quasi-static process.

1.51. Explain why a non-quasi equilibrium compression process requires a larger work input than the corresponding quasi equilibrium one, and, why a non-quasi equilibrium expansion process delivers less work than the corresponding quasi-equilibrium one.

PROBLEMS FOR PRACTICE

1.1. The pressure of a fluid of specific gravity 0.8 flowing in a horizontal pipe line is determined with a simple mercury U-tube manometer. The level of Hg surface in the right limb which is open to the atmosphere is 90 mm above the centre of the pipes. The level of Hg in the left limb which is connected to the pipe is 60 mm below the centre of pipe. Determine absolute pressure of liquid in the pipe is N/m^2. Assume standard P_{atm} to be equal to 10 m of water. (*Ans.* 117641.52 N/m^2)

1.2. An inverted U-tube is connected across two pipes *A* and *B* carrying fluid as shown in Fig. P1.2. Calculate the pressure difference between the centre lines of the pipes.

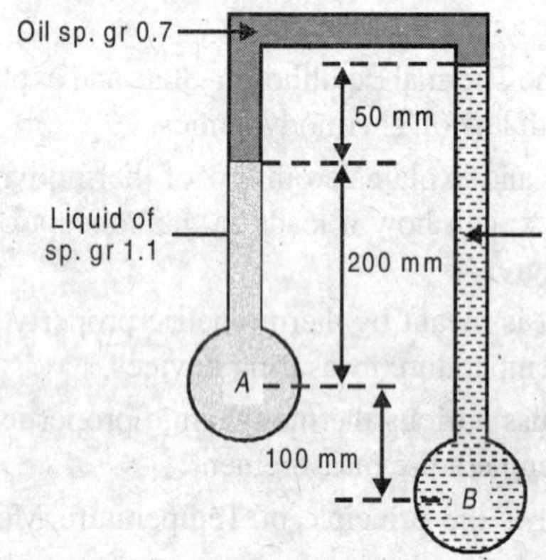

Fig. P1.2

1.3. In the arrangement shown in Fig. P1.3 P_3. Find the difference of pressure in the pipe *A* and *B*. (*Ans.* $P_A - P_B$ = 29135.7 N/m^2)

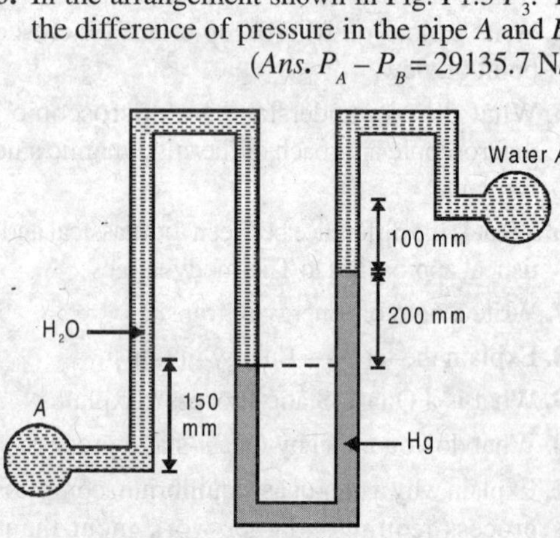

Fig. P1.3

1.4. A pressure gauge is attached to closed tank which contains some gas. The gauge reads 1.3 bar when barometer reads 75 cm of Hg. If the barometer changes to 72 cm of Hg, what will be the reading on the gauge for the same absolute pressure.

1.5. A U-tube manometer with one arm open to atmosphere is used to measure the pressure of the steam flowing through a pipe using mercury as manometric fluid. The level of the Hg column in the arm connected to the pipe is 125 mm below the level of Hg in the open arm. The column of condensate collected is 35 mm. Taking g = 9.81 m/s^2, ρ_{Hg} = 13560 kg/m^3 and atm. pressure as 760 mm of Hg, estimate the absolute pressure of steam in the pipe in bar.

1.6. Both a Bourdon gauge and manometer are attached to a gas tank. If the reading on the pressure gauge is 80 kPa, determine the distance between the two fluid levels of the manometer if the fluid is (i) Hg having sp. gravity of 13.6, (ii) water.

1.7. A U-tube manometer with one arm open to atmosphere is used to measure the pressure of steam flowing through a pipe. The height of Hg column open to atmosphere is 15 cm greater than the Hg level in the arm connected to steam pipe. The column of condensate is 5 cm. The barometer reading is 70 cm of Hg. Find the absolute pressure of steam in kPa.

1.8. A water pump is required to supply water at 900 lit/min into a tank at a height of 20 m. Find the power required to drive the pump in kW.

1.9. Estimate the height of the atmospheric column to produce a pressure of 1 bar at the earth surface. Assume that the pressure *P* in Pascal and sp. volume *V* in m^3/kg are related according to the law $PV^{1.39} = 2.25 \times 10^5$ and g = 9.8 m/s^2.

1.10. An air tank has three manometers connected to it. The fluids in them are oil (sp. gr. = 0.85), water and Hg. If the absolute pressure in the tank is 1.24 bar and the barometric height is 75 cm of Hg, estimate the height of fluid in each manometer.

1.11. A gas expands in a piston—cylinder device from 0.01 to 0.03 m^3, the process being described by $P = aV^{-1} + b$ where *P* is in bar and *V* in m^3. If a = 0.06 bar/m^3 and work done by the gas on the piston is 10.6 kg, evaluate *b*.

1.12. In a piston cylinder arrangement the pressure is inversely proportional to the square of the volume. The initial pressure is 10 bar in the cylinder and initial volume is 0.1 m^3. The volume is now changed so that the final pressure is 2 bar. Find the work done in kJ.

IMPORTANT POINTS TO BE REMEMBERED

1. *Thermodynamics* is a branch of science which deals with the transfomation of one kind of energy into other kind, particularly heat energy into mechanical energy.
2. *System is any* part of the universe which we want to study specifically. A system may be closed, open system or isolated system.
3. *System* may also be homogeneous or heterogeneous.
4. *Phase* of matter is state of matter. It may be solid, liquid or gaseous.
5. *The property* of the system is the descriptive characterstic.

   The prpoerty may be *intensive* or *extensive*

   Intensive property does not depend on the *mass* of system extensive property depends on the mss of the system.

   Property may also be primary property and does not depend on the mass of the system P, Q, T are the primary properties.

   Secondary property is a derived property.
6. *State* of the system is the condition.
7. *Weeking substance* is a fluid in the thermodynamic system which converts one kind of energy into another.
8. *Thermodynamic process* is defined as the transformation from one state to another state.
9. *Process* is reversible process or irreversible process.
10. *Reversible process* is that process which be retraced back. This is

    (i) Mechanical reversible process and

    (ii) Thermally reversible process and

    (iii) Chemical reversible process.

    These three process is combined as Thermodynamically reversible process.
11. *Thermodynamic cycle* is the series of processes the end states which is same.
12. *Quasi* means nearly or almost. Quasi static process means nerly static process or nearly stationary process or a process which process with extreme slowness.
13. *Zeroth law of thermodynamics* states that when body A is in thermal equilibrium with the body B and also separately with the body C, then the bodies B and C will be in thermal equilibrium with each other. This law forms the basis of temperature measurement by comparison method.
14. *Pressure* is defined as the force per unit area N/m^2, bar = $\dfrac{10^5 \times \text{N}}{\text{m}^2}$
15. *Mechanical work* is the product of force and displacement.

    Work done = Force × Diplacement
16. *PdV* work is the work done when both there is a relation between pressure and volume change during the process. Total work done = $W = \int_1^2 PdV$.
17. Work is a path function and properties and point function.
18. Work $= P(V_2 - V_1)$ for constant pressure process

    = 0 for constant volume process

$= P_1 V_1 \log e \dfrac{V_2}{V_1}$ for constant isothermal process

$= \dfrac{P_1V_1 - P_2V_2}{\gamma - 1}$ for constatnt Rev-adiabatic process

$= \dfrac{P_1V_1 - P_2V_2}{n - 1}$ for constant General law process

19. *Heat.* Heat is a form of energy which crosses the system boundary due to the temperature difference between the system and the surroundings. It is generally denoted by Q.
20. *Heat Q* is a path function.
21. *Barometer* is used to measure atmospheric pressure.

$$1 \text{ atm} = 760 \text{ mm Hg column} = 10.33 \text{ m of water} = 1.01325 \text{ bar}$$

CHAPTER 2

First Law of Thermodynamics

2.1 INTRODUCTION

Energy is inherent in all matters. Energy may appear in many different forms. Conversion can be made from one form of energy to another. We are unable to define the general term energy in a simple way, but we can define with precision the various forms in which it appears.

One of the consequences of Einstein's theory of Relativity is that mass may be converted into energy and energy into mass, the relation being given by the famous equation,

$$\text{Energy } (E) = m \text{ (mass)} \times C^2 \qquad \text{(where } C - \text{velocity of light)}$$

Energy is a scalar quantity, not a vector quantity.

Except the nuclear reaction, (where mass is converted into energy) total energy of the universe is constant. For this matter, First Law of Thermodynamics can be expressed as—

Energy can neither be created nor destroyed (except in nuclear reactions). This is the law of conservation of energy.

2.2 JOULE'S EXPERIMENT

Joule carried out experiments in 1843, which led to the formulation of the First Law of Thermodynamics. He took a known quantity of water in a rigid vessel, which was insulated adiabatically from the surroundings. The vessel was fitted with a paddle wheel and a thermometer as shown in Fig. 2.1.

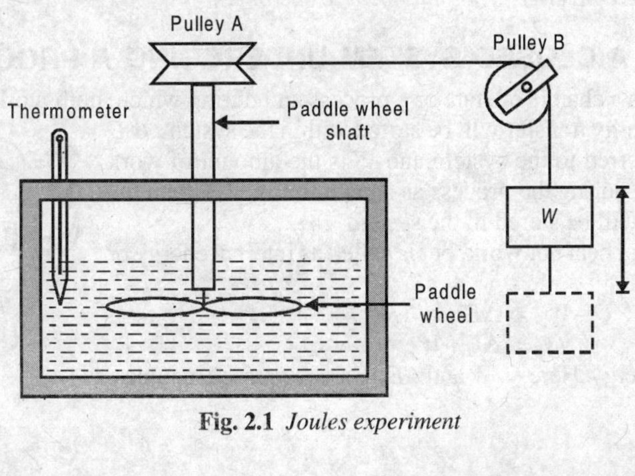

Fig. 2.1 *Joules experiment*

Now let a certain amount of work W_{1-2} be done upon the closed system by the paddle wheel. The quantity of work can be measured by the fall of weight W which drives the paddle wheel.

Let t_1 be the initial temperature of water before work transfer and after work transfer let the temperature rise to t_2. So, the system is changing its state from state (1) to state (2) and the process 1–2 undergone by the system is shown in Fig. 2.2.

Now let the insulation be removed. Then the system will interact by heat transfer (i.e., heat will be dissipated from the system to the surroundings) till the system comes to original temperature t_1. The amount of heat transfer Q_{2-1} from the system during the process 2–1 can be estimated: ($Q_{2-1} = mC_p\,\Delta T$).

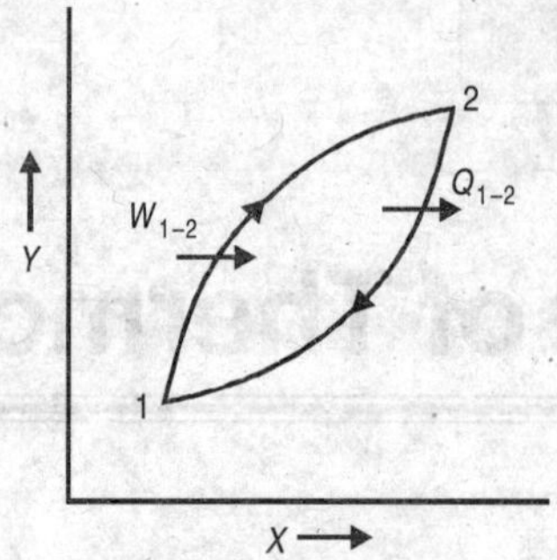

Fig. 2.2

The system thus undergoes a cycle, which consists of a definite amount of work input W_{1-2} to the system, followed by a heat transfer Q_{2-1} from the system.

Joule, repeated this experiment for different weights and different heights and in each case he found that, net work transfer during a cycle is proportional to net heat transfer,

i.e.,
$$\Sigma\delta W \propto \Sigma\delta Q$$

Or in other words *when a closed system undergoes any cyclic process, the cyclic integral of work is proportional to the cyclic integral of heat*. This is known as the first law of thermodynamics for a closed system undergoing a cycle.

i.e.,
$$\oint \delta W \propto \oint \delta Q$$
$$\oint \delta W = J\oint \delta Q. \text{ Since in SI units J} = 1$$
$$\oint \delta W = \oint \delta Q$$

Thus, the I-Law states the general principle of conservation of energy i.e., *Energy can neither be created nor be destroyed, but energy can be converted from one form to the other.*

2.3 FIRST LAW FOR A CLOSED SYSTEM UNDERGOING A PROCESS

If a closed system undergoes a change of state or a process and during which, both work transfer and heat transfer are involved, then the net energy transfer will be stored within the system. If Q is the amount of heat transferred to the system and W is the amount of work transferred from the system, during the process as shown in Fig. 2.3, then the net energy transfer ($Q - W$) will be stored in the system.

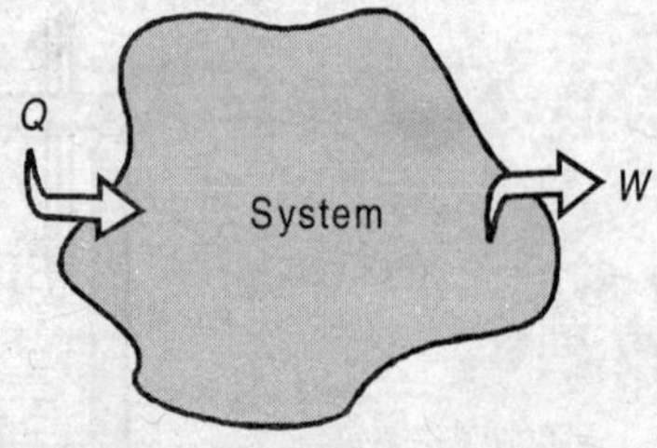

Fig. 2.3

Energy in storage is neither heat nor work, but is called as Internal energy or simply energy of the system.

$\therefore$
$$Q - W = \Delta E$$
or
$$Q = \Delta E + W$$

where ΔE is the change in energy. Here Q, W and ΔE all are expressed in joules.

2.4 ENERGY — A PROPERTY OF THE SYSTEM

Consider a system which changes its state from state (1) to state (2) by following the path A, and returns from state (2) to state (1) by following the path B as shown in Fig. 2.4. So the system undergoes a cycle. Now writing the I-law for the path A,

$$Q_A - W_A = \Delta E_A \quad \ldots(1)$$

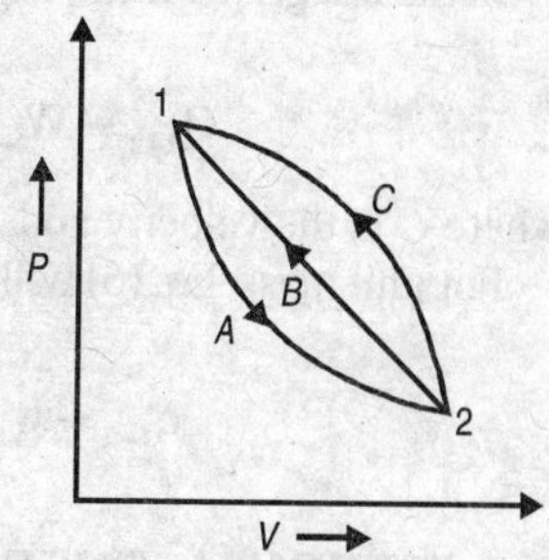

Fig. 2.4

and for path B.

$$Q_B - W_B = \Delta E_B \quad \ldots(2)$$

The processes 1–A–2 and 2–B–1 together will constitute a cycle for which,

$$\oint \delta W = \oint \delta Q$$

i.e., Total work transfer during cycle = Total heat transfer during the cycle.

$$W_A + W_B = Q_A + Q_B$$

Rewriting $\quad W_B - Q_B = Q_A - W_A \quad \ldots(3)$

$\therefore \quad -(Q_B - W_B) = Q_A - W_A \quad \ldots(4)$

$$-\Delta E_B = \Delta E_A$$

i.e., $\quad \Delta E_A = -\Delta E_B \quad \ldots(5)$

Similarly if the system returns from point (2) to state point (1) by following the path C instead of path B then,

$$\Delta E_A = -\Delta E_C \quad \ldots(6)$$

From Eqs (4) and (5),

$$\mathbf{\Delta E_B = \Delta E_C}$$

Thus the change in energy for the path B and C are same. Hence change in energy does not depend upon path, so it depends on end states. Hence, it is a point function and since properties are point functions, *Energy is a property of the system.*

2.5 DIFFERENT FORMS OF STORED ENERGIES

The total energy E is made up of Kinetic Energy (K.E.), Potential Energy (P.E.) and Internal Energy (I.E.).

$$\therefore \quad E = \text{K.E.} + \text{P.E.} + \text{I.E.}$$

Kinetic Energy is due to the motion of the fluid in the system. Potential Energy is the energy due to the gravitational force.

A part of the total energy which is stored in the molecular and atomic structure is known as *Internal energy* and is denoted by U.

In the absence of Kinetic Energy and Potential Energy i.e., when

$$\text{KE} = 0 \text{ and } \text{PE} = 0$$

$$E = U, \text{ and the I-Law for process}$$

$$Q = \Delta E + W$$

becomes $\quad Q = \Delta U + W$

In the differential form, Eqs (1) and (2) become

$$\delta Q = dE + \delta W$$

i.e., $\quad \delta Q = dE + PdV \,(\text{as } \delta W = PdV)$

Also, $\quad \delta Q = dU + \delta W \,(\text{when } E = U)$

i.e., $\quad \delta Q = dU + PdV$

When KE and PE are considered, then the I-Law for a process will be

$$\delta Q - \delta W = dU + d(\text{KE}) + d(\text{PE})$$

In the integrated form, Eq. (5) becomes

$$Q_{1-2} - W_{1-2} = U_2 - U_1 + \frac{m\left(C_2^2 - C_1^2\right)}{2} + mg\left(Z_2 - Z_1\right)$$

where C is the velocity and Z is the height.

For unit mass, Eq. (6) will be,

$$q_{1-2} - w_{1-2} = u_2 - u_1 + \frac{\left(C_2^2 - C_1^2\right)}{2} + g\left(Z_2 - Z_1\right)$$

2.6 INTERNAL ENERGY (U)

It is the energy stored in the molecular structure due to heat and work interactions.

or

In case of gases we know that the gas is made up of a number of molecules, which are moving continuously. I.E. is the energy which arises from the motion of these molecules.

If the temperature of the gas is increased, the molecular activity increases, therefore the I.E. also increases. Thus I.E. is a function of temperature and its value can be increased or decreased by adding or removing heat to or from the system.

Due to practical difficulties it is very very difficult to determine the absolute value of Internal energy. Fortunately, in most of the thermodynamic applications change in I.E. is used, since changes in the states of a system is considered. Change in I.E. is denoted by ΔU and $\Delta U = U_2 - U_1$.

For unit mass, $\Delta u = u_2 - u_1$

2.7 ENTHALPY (H)

It is an extensive property, since its value depends on mass. Enthalpy of a substance is given by the sum of Internal energy and Pressure–Volume product.

i.e., $\mathbf{H = U + PV}$ and specific enthalpy, $\mathbf{h = u + Pv}$

2.8 SPECIFIC HEAT AT CONSTANT PRESSURE AND SPECIFIC HEAT AT CONSTANT VOLUME

Specific heat of a gas is defined as the amount of heat required to rise the temperature of unit mass of a gas through one degree.

Let
C = General specific heat
Q = Heat transfer (J)
ΔT = Change in temperature K
m = mass (kg)

Then from the definition, $$C = \frac{Q}{m \cdot \Delta T} \left[\text{units will be } \frac{\text{J}}{\text{kg-K}}\right] \quad \text{...(1)}$$

or $$Q = m \cdot C \cdot \Delta T \quad \text{...(2)}$$

In the differential form, $$\delta Q = m \cdot C \cdot dT \quad \text{...(3)}$$

and for unit mass, $$\delta q = C \cdot dT \quad \text{...(4)}$$

or $$C = \frac{\delta q}{dT} \quad \text{...(5)}$$

There are 2-types of specific heats,

(1) Specific heat at constant volume C_v

(2) Specific heat at constant pressure C_p

1. Specific Heat at Constant Volume C_v

It is the amount of heat required to rise the temperature of unit mass of a gas through one degree when the volume is constant.

If a unit mass of a gas is taken in a closed vessel and is heated, the volume of the gas remains constant, but the temperature increases. As the volume remains constant, there is no external work done by the gas and as temperature of the gas increases, there is increase in internal energy of the gas.

Therefore heat supplied to the gas is completely utilised in increasing the I.E. of the gas,

and $$C_V = \left(\frac{\delta q}{dT}\right)_V = \left(\frac{du}{dT}\right)_V$$

Therefore C_v is also defined as, *the rate of change of specifitc internal energy with respect to temperature when the volume is kept constant.*

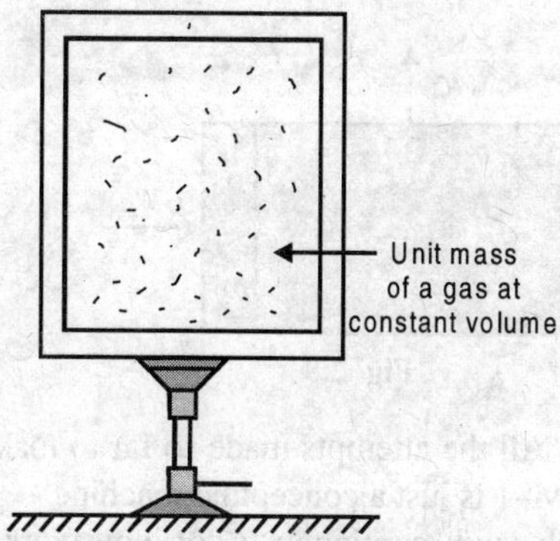

Fig. 2.5

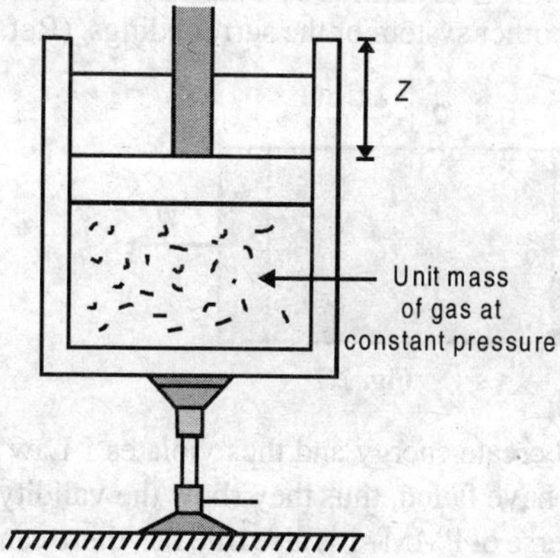

Fig. 2.6

2. Specific Heat at Constant Pressure C_p

It is the amount of heat required to rise the temperature of unit mass of a gas through one degree when the pressure is kept constant.

Consider a unit mass of a gas in a cylinder fitted with a frictionless piston as shown in the Fig. 2.6. When the gas is heated, the piston moves up, maintaining the same pressure. But the volume and temperature of the gas increases during heating. As there is increase in volume, there is external work done by the gas and as there is increase in temperature, there is increase in I.E.

Thus heat supplied when the pressure is constant, is utilised for two purposes,

(a) To do some external work.

(b) To increase the I.E. of the gas.

Whereas in case of constant volume heating, the heat supplied is completely utilised for increasing the I.E.

∴ Specific heat at constant pressure is greater than specific heat at constant volume.

i.e., $$C_P > C_V \quad \text{and,} \quad C_P = \left[\frac{\delta q}{dT}\right]_P$$

Also $$C_P = \left[\frac{dh}{dT}\right]_P$$

Therefore, C_p is also defined as, *the rate of change of specific enthalpy with respect to temperature when the pressure is kept constant.*

2.9 ADIABATIC INDEX

It is the ratio of specific heat at constant pressure to the specific heat at constant volume and is given by,

$$K \text{ or } \gamma = \frac{C_p}{C_v}$$

Note: For air, $C_p = 1.005 \frac{\text{kJ}}{\text{kg}-\text{K}}$ and $C_v = 0.718 \frac{\text{kJ}}{\text{kg}-\text{K}}$ and $\gamma = 1.4$

2.10 PERPETUAL MOTION MACHINE OF FIRST KIND [PMM – 1]

First law states the general principle of conservation of energy i.e., energy can neither be created nor be destroyed but it can be transformed from one form to the other.

But PMM–1 is defined as a machine which will produce continuous work output without receiving any energy from any other system or the surroundings. (Ref. Fig. 2.7).

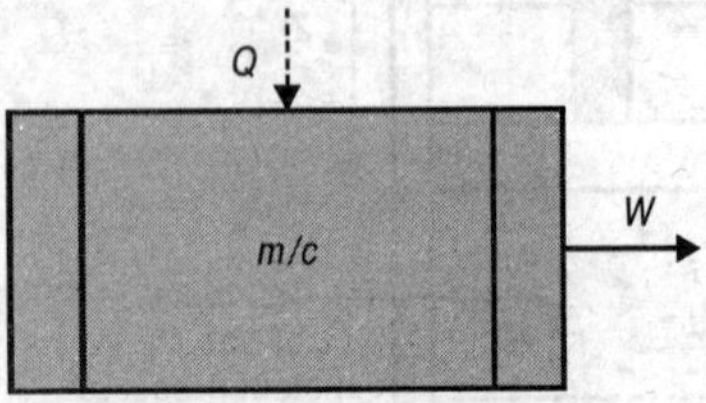

Fig. 2.7

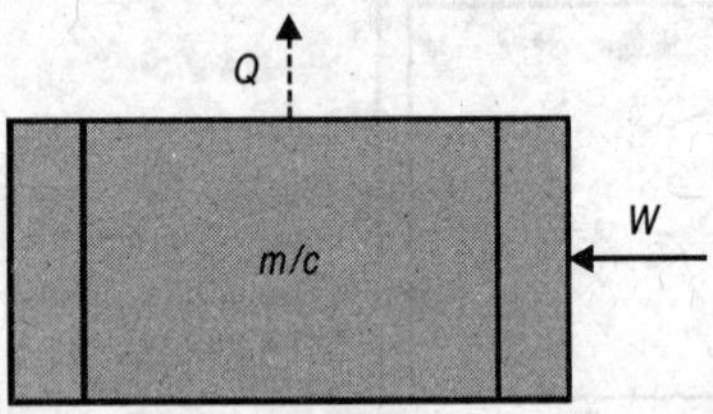

Fig. 2.8

It will create energy and thus violates I-Law of thermodynamics. All the attempts made so far to make such a machine have failed, thus they show the validity of I–law. Thus PMM–1 is just a conceptual machine.

Converse of PMM–1 is also true i.e., there can be no machine which would continuously consume work, without producing some other form of energy (Fig. 2.8).

2.11 FLOW PROCESS; CONTROL VOLUME; CONTROL SURFACE

Flow Process. Processes performed in the open systems are called as Flow-processes.

Figure 2.9 shows a portion of steam power plant. High pressure steam enters the steam turbines at Section (I), expands and produces work output and then it leaves the turbine at section (2).

For analysing the expansion process, 2-methods are used.

In the first method certain mass of the fluid Marked *A* is considered and this is suppose to flow through the turbine.

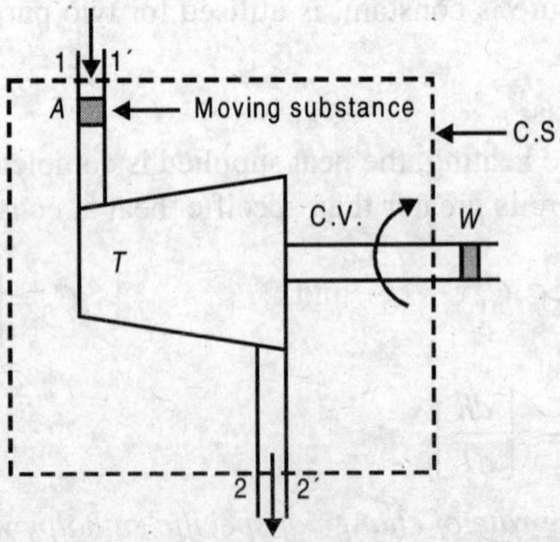

Fig. 2.9

In the second method, certain fixed volume of the system known as *control volume (CV)* is considered. The moving substance flows through this control volume. The surface of the control volume is known as *control surface (CS)*.

2.12 FLOW WORK OR FLOW ENERGY

In case of flow processes certain amount of work or energy is required to push the fluid into and out of the system. This work energy is known as *flow work or flow energy.*

Consider a flow process as shown in Fig. 2.10.

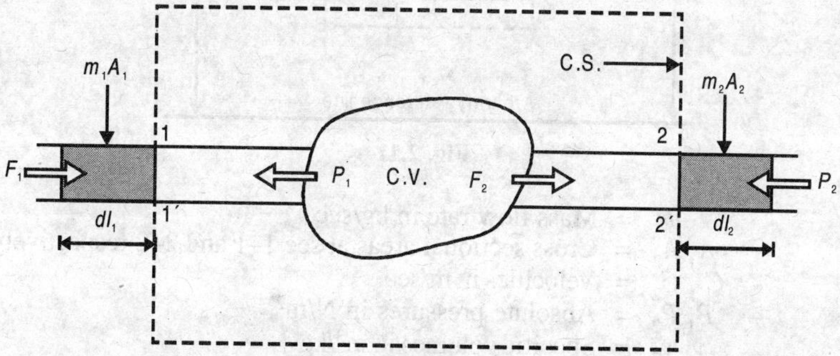

Fig. 2.10

Let F_1 be the force which forces the mass m_1 of cross sectional area A_1 into the system against the pressure P_1 of the system. Now consider a small amount of work done on the system in causing the displacement dl_1 for the mass m_1.

Thus the small amount of work done on the system in causing the displacement dl_1,

$$\delta W = -F_1 \times dl_1$$

Note : –ve sign, since the work is done on the system.

But $$F_1 = P_1 \times A_1$$

$\therefore$ $$\delta W = -P_1 \times A_1 \times dl_1$$

Since $$A_1 \times dl_1 = dV_1 = \text{small amount of displacement volume.}$$

$\therefore$ $$\delta W = -P_1 \times dV_1$$

$\therefore$ Total flow work at Sec. (1) $= -P_1 V_1$

Similarly work done by the system to force the fluid out of the system at Sec. $(2) = P_2V_2$

$\therefore$ **Net flow work** $= P_2 V_2 - P_1 V_1$

and **for unit mass** $= P_2 v_2 - P_1 v_1$

Note: Since the flow work is entirely expressed in terms of properties of the system, the net flow work depends on the end states and it is a *Property*. Whereas the other forms of work are path functions so in problems involving flow processes, flow work is to be calculated separately.

2.13 CONDITIONS OF STEADY FLOW SYSTEM

A steady flow process should satisfy the following conditions:

(i) The mass flow rate into and out of the system are equal and do not vary with time i.e., mass in the system does not change.

(If the mass flow rate at the inlet is 10 kg/sec., then the mass flow rate at the exit will be 10 kg/sec and it does not vary with time i.e., if we note the mass flow rate say, when we come to the college and while going home also it will be same.)

(ii) The energy of the fluid at the entrance and at the exit are same and do not vary with time.

(iii) The rates of heat and work transfer into and out of the system do not vary with time.

2.14 STEADY FLOW ENERGY EQUATION (S.F.E.E.)

Steady flow energy equation is obtained by applying the first law of thermodynamics to a steady flow system.

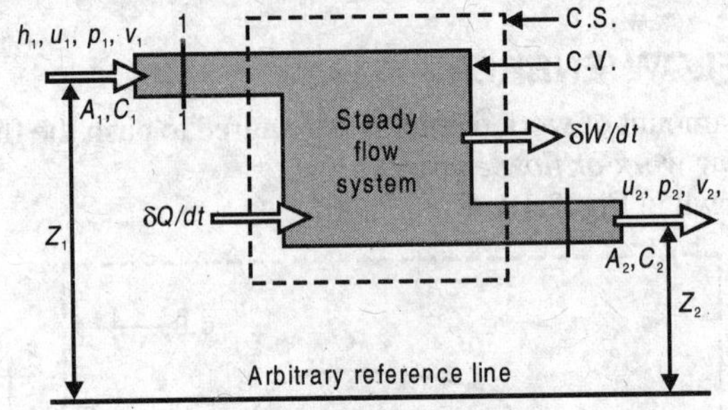

Fig. 2.11

Let,

$\dot{m}$ = Mass flow rate in kg/sec

$A_1\ A_2$ = Cross sectional areas at sec 1–1 and 2–2 respectively in m^2

$C_1\ C_2$ = Velocities in m/sec

$P_1\ P_2$ = Absolute pressures in N/m^2

$v_1\ v_2$ = Specific volumes in m^3/kg

$u_1\ u_2$ = Specific internal energies in J/kg

$h_1\ h_2$ = Specific enthalpies in J/kg

$Z_1\ Z_2$ = Elevations above arbitrary reference datum line in m

Q = Net heat transfer in Joules

$\frac{\delta Q}{dt}$ = Net rate of heat transfer in J/sec

$\frac{\delta Q}{dm}$ = q = Net rate of heat transfer in J/kg

W = Net work transfer in joules

$\frac{\delta W}{dt}$ = Net rate of work transfer in J/sec

$\frac{\delta W}{dm}$ = w = Net rate of work transfer in J/kg

t = time in seconds

Figure 2.11 shows a steady flow system. Fluid entes at sec. 1–1 and leaves at sec. 2–2. Since it is a steady flow system. According to 2nd condition.

Total energy at the entrance = Total energy at the exit

∴ We can write,

$$\dot{m}\begin{bmatrix}\text{Energy carried}\\ \text{into the system}\end{bmatrix}+\dot{m}\begin{bmatrix}\text{Specific enthalpy}\\ \text{of entering fluid}\end{bmatrix}+\begin{bmatrix}\text{Net rate of heat}\\ \text{transfer into}\\ \text{system}\end{bmatrix}$$

$$=\ \dot{m}\begin{bmatrix}\text{Energy carried}\\ \text{out of the system}\end{bmatrix}+\dot{m}\begin{bmatrix}\text{Specific enthalpy}\\ \text{of the fluid}\\ \text{at the exit}\end{bmatrix}+\begin{bmatrix}\text{Net rate of work}\\ \text{transfer by the}\\ \text{system}\end{bmatrix}$$

i.e. , $\dot{m}[\text{KE}_1 + \text{PE}_1] + \dot{m}(h_1) + \frac{\delta Q}{dt} = \dot{m}[\text{KE}_2 + \text{PE}_2] + \dot{m}(h_2) + \frac{\delta W}{dt}$

i.e., $\dot{m}\left[\frac{C_1^2}{2} + g\ Z_1\right] + \dot{m}(h_1) + \frac{\delta Q}{dt} = \dot{m}\left[\frac{C_2^2}{2} + g\ Z_2\right] + \dot{m}(h_2) + \frac{\delta W}{dt}$

Note: m of KE and PE will be unity, since we are deriving the S.F.E.E. for unit mass.

$$\therefore \quad \dot{\mathbf{m}}\left[\frac{\mathbf{C}_1^2}{2} + \mathbf{g}\ \mathbf{Z}_1 + \mathbf{h}_1\right] + \frac{\delta \mathbf{Q}}{\mathbf{dt}} = \dot{\mathbf{m}}\left[\frac{\mathbf{C}_2^2}{2} + \mathbf{g}\ \mathbf{Z}_2 + \mathbf{h}_2\right] + \frac{\delta \mathbf{W}}{\mathbf{dt}} \quad \ldots(1)$$

This is steady flow energy equation on *time basis.*

Also, since $h_1 = u_1 + P_1v_1$ and $h_2 = u_2 + P_2v_2$

$\therefore$ Equation (1) becomes,

$$\therefore \quad \dot{m}\left[\frac{C_1^2}{2} + g\ Z_1 + u_1 + P_1v_1\right] + \frac{\delta Q}{dt} = \dot{m}\left[\frac{C_2^2}{2} + g\ Z_2 + u_2 + P_2v_2\right] + \frac{\delta W}{dt}$$

$$\text{or} \quad \dot{\mathbf{m}}[\mathbf{KE}_1 + \mathbf{PE}_1 + \mathbf{IE}_1 + \mathbf{FE}_1] + \frac{\delta \mathbf{W}}{\mathbf{dt}} = \dot{\mathbf{m}}[\mathbf{KE}_2 + \mathbf{PE}_2 + \mathbf{IE}_2 + \mathbf{FE}_2] + \frac{\delta \mathbf{W}}{\mathbf{dt}} \quad \ldots(2)$$

2.15 STEADY FLOW ENERGY EQUATION ON MASS BASIS

For deriving this, we have to consider $\dot{m}$ = 1 kg/sec and all other quantities will be for per kg mass such as $\delta W/dm$ and $\delta Q/dm$.

$\therefore$ Equation (1) becomes,

$$\left[\frac{C_1^2}{2} + gZ_1 + h_1\right] + \frac{\delta Q}{dm} = \left[\frac{C_2^2}{2} + gZ_2 + h_2\right] + \frac{\delta W}{dm}$$

$$\therefore \quad \left[\frac{\mathbf{C}_1^2}{2} + \mathbf{gZ}_1 + \mathbf{h}_1\right] + \mathbf{q} = \left[\frac{\mathbf{C}_2^2}{2} + \mathbf{gZ}_2 + \mathbf{h}_2\right] + \mathbf{w}$$

This is the SFEE on mass basis.

2.16 IN A STEADY FLOW SYSTEM W.D.= $-\int v.dP$ WHEN CHANGES IN KE AND PE ARE NEGLECTED

We know that, SFEE on mass basis. From Eq. (3)

$$\left[\frac{C_1^2}{2} + gZ_1 + h_1\right] + q = \left[\frac{C_2^2}{2} + gZ_2 + h_2\right] + w \quad \ldots(1)$$

$$\text{i.e.,} \quad \left[\frac{C_1^2}{2} + gZ_1 + u_1 + P_1v_1\right] + q = \left[\frac{C_2^2}{2} + gZ_2 + u_2 + P_2v_2\right] + w$$

$$\therefore \quad q = u_2 - u_1 + P_2v_2 - P_1v_1 + \frac{C_2^2 - C_1^2}{2} + g(Z_2 - Z_1) + w$$

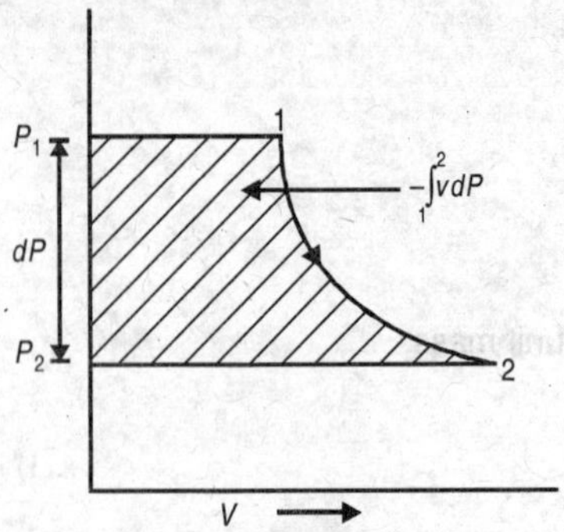

Fig. 2.12

$$\therefore \quad q = \Delta u + \Delta(Pv) + \Delta(\text{KE}) + d(\text{PE}) + w$$

In differential form, $\delta q = du + d(Pv) + d(\text{KE}) + d(\text{PE}) + \delta w$ ...(2)

But we also know that I Law for a closed system undergoing a process is,

$$Q - W = \Delta E = \Delta U$$

For unit mass, $q - w = \Delta u$

i.e., $q = \Delta u + w$

In the differential form, $\delta q = du + \delta w$

i.e., $\delta q = du + Pdv$ ...(3)

Assuming that δq *is same for closed system and for the steady flow system, equating Eqs* (2) *and* (3)

$$du + Pdv = du + d(Pv) + d(\text{KE}) + d(\text{PE}) + \delta w \quad ...(4)$$

i.e., $du + Pdv = du + Pdv + vdP + d(\text{KE}) + d(\text{PE}) + \delta w$

$\therefore$ We can write, $-v\,dP = \delta w + d(\text{KE}) + d(\text{PE})$

$\therefore$ For total mass, $-\int vdP = w + \Delta(\text{KE}) + \Delta(\text{PE})$ ...(5)

When change in KE and PE are neglected,

$$-\int \mathbf{vdP} = \mathbf{w}$$

This work done is given by the area behind the curve 1–2 as shown in Fig. 2.12.

2.17 SIGNIFICANCE OF $\int Pdv$ IN CASE OF STEADY FLOW PROCESS AND NON-FLOW PROCESS

Upto Eq. (4) derivation is same as in the above article.

From Eq. (4), $du + Pdv = du + d(Pv) + d(\text{KE}) + d(\text{PE}) + \delta w$

$$Pdv = d(Pv) + d(\text{KE}) + d(\text{PE}) + \delta w$$

$\therefore$ For total mass, $\int_1^2 \mathbf{Pdv} = \mathbf{\Delta(Pv) + \Delta(KE) + \Delta(PE) + w}$

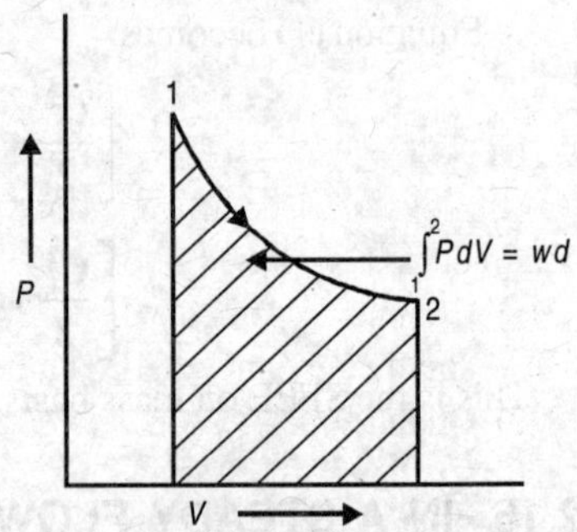

Fig. 2.13

For a non-flow process (i.e. for a closed system),

$\int Pdv$ =work done and is given by the area under the curve 1–2 as shown in Fig. 2.13.

2.18 PROOFS

2.18.1 $-\int \mathbf{vdP} = \mathbf{Q} - \mathbf{\Delta H}$

We know that, for a steady flow system,

$$Q - W = \Delta H + \Delta\text{PE} + \Delta\text{KE} \quad ...(1)$$

Consider the case when the changes in PE and KE are neglected,

$$Q - W = \Delta H$$

$$\therefore \quad W = Q - \Delta H \quad ...(2)$$

Also we know that, in a reversible steady flow process,

$$\text{Work done} = -\int v\,dP \text{ when changes in KE and PE are neglected.} \quad ...(3)$$

Hence from (2) and (3) we can write,

$$\therefore \quad -\int \mathbf{v\,dP} = \mathbf{Q - \Delta H}$$

2.18.2 $\int \mathbf{P dv = Q - \Delta U}$

We know that, for a non-flow process,

$$Q - W = \Delta U$$

$$W = Q - \Delta U \quad ...(1)$$

And for a Non-flow process, $W = \int P dv$ ...(2)

$\therefore$ From (1) and (2) $\quad \int \mathbf{Pdv} = \mathbf{Q - \Delta U}$

2.18.3 $-\int v\,\mathbf{dP} - \int \mathbf{Pdv} = -\Delta \mathbf{P}v$

We know that, $\quad -\int vdP = w + \Delta\text{KE} + \Delta\text{PE}$

Also we know that, significance of $\int P\,dv$ in case of steady flow process and non-flow process,

$$\int P\,dv = w + \Delta(\text{PE}) + \Delta(\text{KE}) + \Delta Pv \quad ...(3)$$

$\therefore$ Substracting Eq. (2) from (1) we get,

$$-\int \mathbf{vdP} - \int \mathbf{Pdv} = -\Delta\mathbf{Pv}$$

2.19 APPLICATIONS OF ENERGY EQUATIONS

(a) Boiler. It is a steam generator.

Characteristics of the boiler are:

(1) Change in velocity of the fluid at the entrance and exit is very small, and hence ΔKE can be neglected, i.e., $\Delta\text{KE} = 0$

(2) Change in elevation between the inlet and exit point is very small so change in PE may be neglected, i.e., $\Delta\text{PE} = 0$

(3) Since no work is done in a boiler $w = 0$

$\therefore$ From SFEE on mass basis,

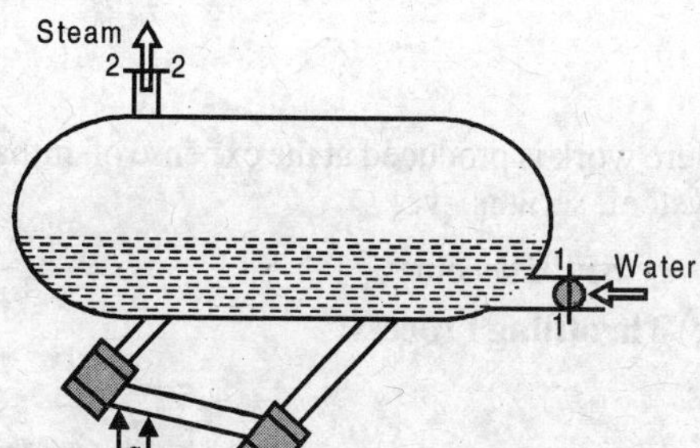

Fig. 2.14

$$\left[\frac{C_1^2}{2} + gZ_1 + h_1\right] + q = \left[\frac{C_2^2}{2} + gZ_2 + h_2\right] + w$$

i.e., $$q = \left[\frac{(C_2^2 - C_1^2)}{2} + g(Z_2 - Z_1) + (h_2 - h_1)\right] + w \quad ...(1)$$

On substituting the above conditions Eq. (1) becomes,

$$q = [0 + 0 + (h_2 - h_1)] + 0$$

i.e., $$\mathbf{q = h_2 - h_{1ss}}$$

(b) Turbine (Engine)/Compressor (or pump)

The turbines and engines are power producing devices whereas Compressors, Pumps, Blowers etc. are power consuming devices.

Charcteristic of a turbine are,

(i) Since it is insulated, negligible heat transfer.

$\therefore \quad q = 0$

(ii) Change in velocity of the fluid at the entrance and at the exit is negligible.

$\therefore \quad \Delta KE = 0$

(iii) Since change in elevation between inlet and exit point is negligible,

$\Delta PE = 0$

$\therefore$ From SFEE i.e., from Eq. (1),

$$q = \left[\frac{(C_2^2 - C_1^2)}{2} + g(Z_2 - Z_1) + (h_2 - h_1)\right] + w$$

$\therefore$ The equation reduces to, $\quad 0 = [0 + 0 + (h_2 - h_1)] + w$

or $\quad \mathbf{w = h_1 - h_2}$

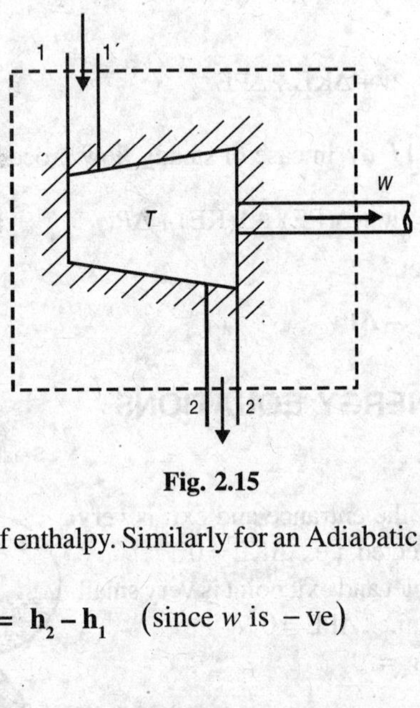

Fig. 2.15

Here work is produced at the expense of enthalpy. Similarly for an Adiabatic Pump/Compressor, work is done on the system, so w is –ve.

$\therefore$ SFEE becomes, $\quad \mathbf{w = h_2 - h_1} \quad$ (since w is $-$ ve)

(c) Throttling Process:

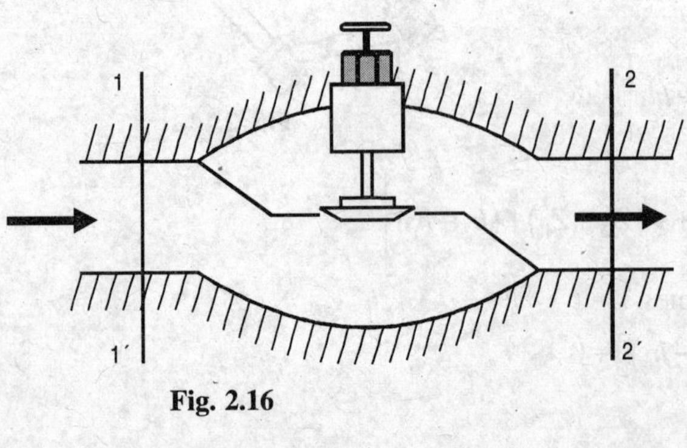

Fig. 2.16

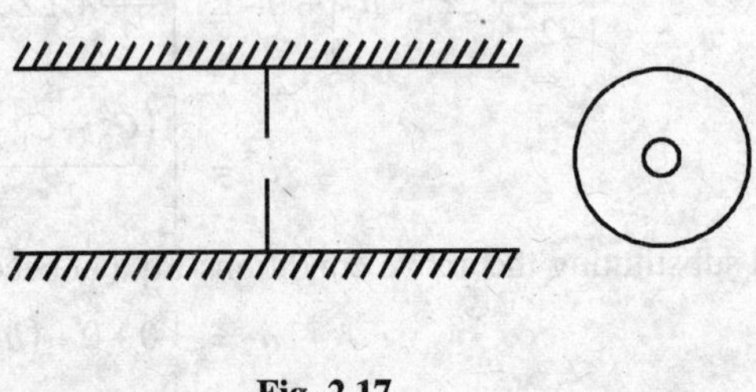

Fig. 2.17

When a fluid flows through a partially opened valve or an orifice then there is an appreciable pressure drop and the flow is said to be throttled.

Characteristics are,

(i) $q = 0$, (ii) $w = 0$, (iii) $\Delta PE = 0$

Substituting these in SFEE we get

$$q = \left[\frac{C_2^2 - C_1^2}{2} + g(Z_2 - Z_1) + (h_2 - h_1)\right] + w$$

$$0 = \left[\frac{(C_2^2 - C_1^2)}{2} + 0 + (h_2 - h_1)\right] + 0$$

i.e.,
$$\frac{\mathbf{C_1^2}}{\mathbf{2}} + \mathbf{h_1} = \frac{\mathbf{C_2^2}}{\mathbf{2}} + \mathbf{h_2}$$

(d) Nozzle. It is a device which is used to increase the KE of flowing fluids

(i) $q = 0$

(ii) $w = 0$

(iii) $\Delta PE = 0$

$\therefore$ Substituting these in SFEE we get,

$$q = \left[\frac{C_2^2 - C_1^2}{2} + g(Z_2 - Z_1) + (h_2 - h_1)\right] + w$$

$$0 = \left[\frac{C_2^2 - C_1^2}{2} + 0 + (h_2 - h_1)\right] + 0$$

i.e.,
$$\frac{\mathbf{C_1^2}}{\mathbf{2}} + \mathbf{h_1} = \frac{\mathbf{C_2^2}}{\mathbf{2}} + \mathbf{h_2}$$

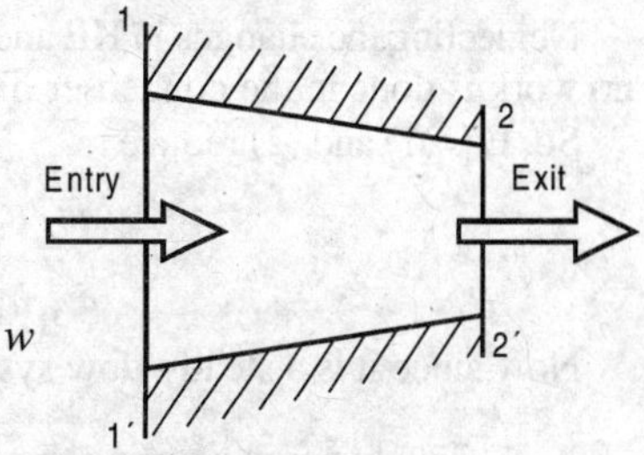

Fig. 2.18

(e) Condenser or to show in a steady flow process with multiple streams of fluid without mixing.

$$\frac{\dot{m}_1}{\dot{m}_3} = \frac{h_4 - h_3}{h_1 - h_2}$$

Figure 2.19 shows a surface condenser. In this case, cooling water enters at Section 1– 1′ passes through the tubes and leaves the condenser at Section 2 – 2′.

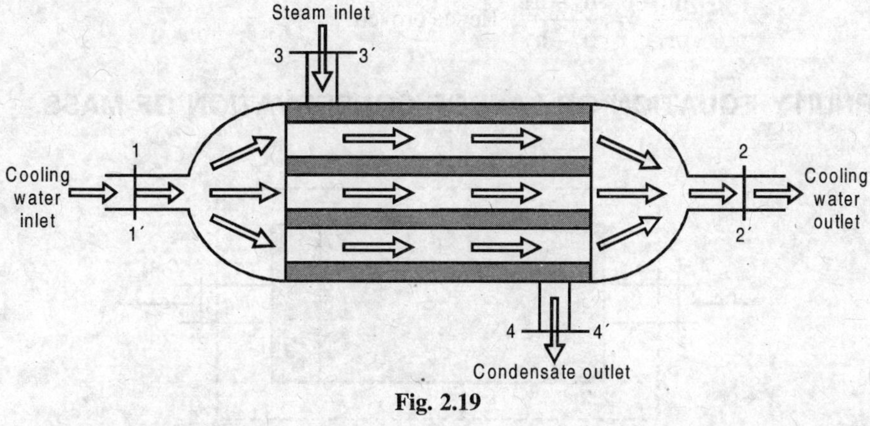

Fig. 2.19

Steam to be condensed enters the condenser at Section 3–3 and when it comes in contact with the cold surface of water tubes, it gets condensed and the condensate leaves the condenser at Section 4–4′.

So, in this case working fluids are water and steam i.e., two fluids flow through the system and since they do not mix with each other, it may be considered as a steady flow system with multiple streams of fluid without mixing.

Let, $\dot{m}_1\ \dot{m}_2$ — Mass flow rates of cooling water at Secs. 1–1′ and 2– 2′ in kg/sec

$\dot{m}_3\ \dot{m}_4$ — Mass flow rates of steam and condenstate Secs. 3–3′ and 4–4′ respectively in kg/sec.

$h_1\ h_2$ — Specific enthalpies of cooling water at Secs. 1–1′ asnd 2–2′ respectively in J/kg.

$h_3\ h_4$ — Specific enthalpies of steam and condensate at Secs. 3–3′ and 4–4′ respectively in J/kg.

Since it is steady flow system, writing SFEE for the coolant.

$$\dot{m}_1\left[\frac{C_1^2}{2}+gZ_1+h_1\right]+\frac{\delta Q}{dt} = \dot{m}_2\left[\frac{C_2^2}{2}+gZ_2+h_2\right]+\frac{\delta W}{dt} \quad \ldots(1)$$

And for steam,

$$\dot{m}_3\left[\frac{C_3^2}{2}+gZ_3+h_3\right]+\frac{\delta Q}{dt} = \dot{m}_4\left[\frac{C_4^2}{2}+gZ_4+h_4\right]+\frac{\delta W}{dt} \quad \ldots(2)$$

Neglecting the changes in KE and PE, since they are very small and as the condenser is insulated $\delta Q/dt = 0$ and as no work is done in the condenser $\delta W/dt = 0$.

So, Eqs (1) and (2) reduce to,

$$\dot{m}_1\ h_1 = \dot{m}_2\ h_2 \quad \ldots(3)$$

$$\dot{m}_3\ h_3 = \dot{m}_4\ h_4 \quad \ldots(4)$$

Now since it is a steady flow system,

$$\left[\begin{array}{l}\text{Total energy entering the}\\ \text{condenser with steam}\\ \text{and coolant}\end{array}\right] = \left[\begin{array}{l}\text{Total energy leaving the}\\ \text{condenser with coolant}\\ \text{and condensate}\end{array}\right]$$

$$\therefore \quad \dot{m}_1\ h_1+\dot{m}_3\ h_3 = \dot{m}_2\ h_2+\dot{m}_4\ h_4$$

But for a steady flow system

$$\dot{m}_1 = \dot{m}_2 \text{ and } \dot{m}_3 = \dot{m}_4$$

$$\therefore \quad \dot{m}_1\left(h_1-h_2\right) = \dot{m}_3\left(h_4-h_3\right)$$

$$\therefore \quad \frac{\dot{\mathbf{m}}_1}{\dot{\mathbf{m}}_3} = \frac{\mathbf{h}_4-\mathbf{h}_3}{\mathbf{h}_1-\mathbf{h}_2}. \text{ Hence proved.}$$

2.20 CONTINUITY EQUATION OR LAW OF CONSERVATION OF MASS

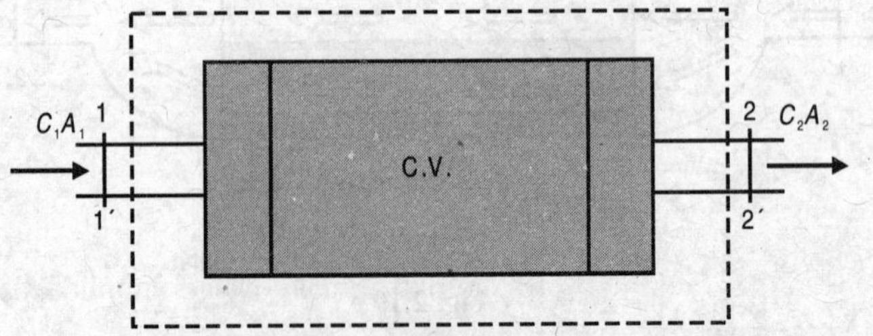

Fig. 2.20

Continuity equation is based on the principle of conservation of mass

Let $\rho_1 \rho_2$ — Densities of the fluid at sec 1–1′ and 2– 2′ respectively in kg/m²

$A_1 A_2$ — Cross sectional areas at sec 1–1′ and 2– 2′ respectively in m²

$C_1 C_2$ — Velocities of the fluid at sec 1–1′ and 2– 2′ respectively in m/sec

Then the volume of the fluid entering at sec 1–1′ per unit time $= A_1 \times C_1$

$\left(\because A_1 \text{ is in m}^2 \text{ and } C_1 \text{ is in m/sec then resultant unit will be m}^3\text{/sec}\right)$

∴ Mass of the fluid entering at sec 1–1′ per unit time,

$$= \text{Density} \times \text{Volume}$$

$$= \rho_1 \times A_1 \times C_1$$

Simlarly mass of fluid leaving at sec 2–2′ per unit $= \rho_1 A_2 C_2$

Since it is steady flow system,

Mass flow rate at sec 1–1′ = Mass flow rate at sec 2–2′

$$\therefore \quad \rho_1 A_1 C_1 = \rho_2 A_2 C_2$$

Since

$$\rho = \frac{1}{v}$$

$$\therefore \quad \frac{A_1C_1}{v_1} = \frac{A_2C_2}{v_2}$$

or

$$\rho AC = \text{constant}$$

$$\frac{AC}{v} = \text{constant}$$

These are the continuity equations.

2.21 LIMITATIONS OF FIRST LAW OF THERMODYNAMICS

We have seen that first law of thermodynamics is simply a law of conservation of energy and has its own limitations. These limitations are:

1. The law considers all forms of energies equivalent i.e., the first law of thermodynamics is a law of energy equivalence. This law is necessary condition so far as the account of energy balance is concerned with the possibility of transformation of one kind of energy to another.
2. The first law does not consider the direction of energy transformation.
3. The first law of thermodynamics does not consider the grade of the energy or energy reservoirs. It, assumes that all energy reservoirs are identical.

The second law of thermodynamics relates to the direction of energy exchange processes.

One of the statements of the second law of thermodynamics is Clausius statement and states that no cyclic process is prossible whose result is the flow of heat out of a heat reservoir at one temperature and the flow of an equal quantity of heat into a second reservoir at a higher temperature, without external work. Or, heat can not flow from one body at a lower temperature to the another body at higher temperature without any external work.

This principle is used in a refrigerator. For refrigerator external power is supplied to the compressor.

Without proof or discussion we here define the third law of thermodynamics.

The third law states that the entropy of a pure substance is zero at absolute zero temperature.

■■■

SOLVED EXAMPLES

Example 2.1 *During a cycle consisting of 4-processes, the heat transfers are 60 kJ, – 8 kJ, – 34 kJ and 6 kJ. Determine the net work for the cycle.*

Solution

We know that from the first law for a closed system undergoing a cycle,

Total heat transfer in a cycle = Total work transfer in a cycle

i.e.,
$$\oint \delta Q = \oint \delta W$$

As given in the problem, the cycle consists of 4-processes and heat transfers for the processes are given.

$$\therefore \quad \oint \delta Q = 60\,\text{kJ} - 8\,\text{kJ} - 34\,\text{kJ} + 6\,\text{kJ}$$

$$\oint \delta Q = 24\,\text{kJ} = \oint \delta W$$

$\therefore$ **Net work transfer in a cycle = 24 kJ**

Example 2.2 *In a non-flow reversible process 300 kJ of heat leaves the system consisting of a gas. The Internal Energy of the gas remains the same. Calculate the work done.*

Data:

$Q = -300$ kJ *(Since heat flows out of the system)*

U = *Constant*

$\therefore \quad \Delta U = 0$

$W = ?$

Solution

We know that first law for a closed system undergoing a process $Q - W = \Delta U$. As given in problem $Q - W = 0$, since $\Delta U = 0$.

$$\mathbf{Q = W = -300\ kJ}$$

i.e., 300 kJ of work is done on the gas.

Example 2.3 *In a non-flow reversible process, the pressure and volume are related by*

$$P = \left[\frac{3.5}{V} + 3\right] bar$$

where P is in bar and V is in m³. During the process the volume changes from 1.5 m³ to 4.5 m³ and heat added is 2000 kJ. Determine change in Internal Energy (I.E.)?

Data:

$$P = \left[\frac{3.5}{V} + 3\right] \times 10^2\ kPa$$

$V_1 = 1.5\ m^3$

$V_2 = 4.5\ m^3$

$Q = 2000\ kJ$

$\Delta U = ?$

Solution

We know that first law for a closed system undergoing a process,

$$Q - W = \Delta U$$

Q is given, ΔU is to be found

$\therefore$ To find W,

Also we know that,

Work done $= \int P\,dV$ in closed system

Substituting
$$P = \left[\frac{3.5}{V} + 3\right] \times 10^2\ \text{kPa}$$

We get Work done $= \int \left[\frac{3.5}{V} + 3\right] \times 10^2 \; dV$

$$= 100 \int_{1.5}^{4.5} \left(\frac{3.5}{V} + 3\right) dV \text{ kJ}$$

$$= 100\left[(3.5 \ln V + 3V)\right]_{1.5}^{4.5}$$

$$= 100\left[3.5 \ln \frac{4.5}{1.5} + 3(4.5 - 1.5)\right]$$

$$= 100\left[3.5 \ln 3 + 3 \times (3)\right]$$

$$= 100[3.845 + 9] = 100[12.845]$$

Work done $= 1284.5143$ kJ

$\therefore \quad Q - W = \Delta U$

$2000 - 1284.5143 = \Delta U$

$\therefore \quad$ **ΔU = 715.4857**

Example 2.4 *In a boiler water enters with an enthalpy of 35 kJ/kg and steam leaves with the enthalpy of 705 kJ/kg. Find the heat transfer per kg of steam. The change in KE and PE may be neglected.*

Solution

We know that, SFEE on mass basis,

$$\left[\frac{C_1^2}{2} + gZ_1 + h_1\right] + q = \left[\frac{C_2^2}{2} + gZ_2 + h_2\right] + w$$

$$(0 + 0 + h_1) + q = (0 + 0 + h_2) + w$$

$$h_1 + q = h_2 + w$$

But in a boiler since no work is done $w = 0$

$\therefore \quad q = h_2 - h_1$

$q = 705 - 35$

$\therefore \quad$ **q = 670 kJ/kg**

Example 2.5 *A piston and cylinder machine contains a fluid system which passes through a complete cycle of 4-processes. During a cycle the sum of all heat transfers is – 170 kJ. The system completes 100 cycles/min. Complete the following table showing the method for each item and calculate the net rate of work output in kW.*

Process	*Q (kJ/min)*	*W (kJ/min)*	*4E (kJ/min)*
a–b	*0*	*2170*	—
b–c	*21,000*	*0*	—
c–d	*– 2100*	—	*– 36600*
d–a	—	—	—

Solution

Process a–b

$$Q = \Delta E + W$$

$$0 = \Delta E + 2170$$

ΔE = –2170 kJ/min

Process b–c

$$Q = \Delta E + W$$
$$21000 = \Delta E + 0$$
$$\therefore \quad \mathbf{\Delta E = 21000\ kJ/min}$$

Process c–d

$$Q = \Delta E + W$$
$$-2100 = -36600 + W$$
$$\mathbf{W = 34500\ kJ/min}$$

Process d–a

It is given that, sum of all heat transfers/cycle

i.e., $$\oint Q = -170 \text{ kJ}$$

Secondly, the system completes 100 cycles/min

$$\therefore \quad \text{Total heat transfer/min} = -17000 \text{ kJ/min}$$
$$Q_{a-b} + Q_{b-c} + Q_{c-d} + Q_{d-a} = -17000 \text{ kJ/min}$$
$$0 + 21000 - 21000 + Q_{d-a} = -1700 \text{ kJ/min}$$
$$\therefore \quad \mathbf{Q_{d-a} = -35900\ kJ/min}$$

While studying energy a property of the system, we have studied that change in energy for the process $1 - a - 2 = \Delta E_a$

And change in energy for the process,

$$2 - b - 1 = \Delta E_b \text{ (– ve)}$$

$\therefore$ Net change in energy for the cycle

$$= \Delta E_a - \Delta E_b = 0$$

$\therefore$ We can write cyclic integral of any property is zero or $\oint dE = 0$

So, $$\Delta E_{a-b} + \Delta E_{b-c} + \Delta E_{c-d} + \Delta E_{d-a} = 0$$
$$\therefore \quad -2170 + 21000 - 36600 + \Delta E_{d-a} = 0$$
$$\therefore \quad \mathbf{\Delta E_{d-a} = 17770\ kJ/min}$$

We know that, $$Q_{d-a} = \Delta E_{d-a} + W_{d-a}$$
$$W_{d-a} = Q_{d-a} - \Delta E_{d-a} = -35900 - 17770$$
$$\mathbf{W_{d-a} = -53670\ kJ/min}$$

Lastly rate of work output since

$$\oint \delta Q = \oint \delta W$$
$$= -17000 \text{ kJ/min}$$
$$\oint \delta \mathbf{W} = \textbf{Net rate of work = – 283.33 kW or kJ/sec}$$

Example 2.6 *A certain cycle consists of 3 processes. The energy transfer in each process are tabulated below.*

Process	*Q (kJ/kg)*	*w (kJ/kg)*	*Δu (kJ/kg)*
1 – 2	*50*	—	*20*
2 – 3	*– 30*	*– 40*	—
3 – 1	—	—	*– 30*

If the network done l kg of fluid is + 30 kJ/kg complete the table. Also find the power developed, if the mass of fluid in the cycle is 0.1 kg and the system completes 10 such cycles per sec.

Solution Since all the equalities are given for unit mass,

For process 1–2

$$q = \Delta u + w$$
$$50 = 20 + w$$
$$\therefore \quad \mathbf{w = 30\ kJ/kg}$$

For process 2–3

$$q = \Delta u + w$$
$$-30 = \Delta u - 40$$
$$\mathbf{\Delta u = 10\ kJ/kg}$$

For process 3–1

It is given that, net w.d./kg = – 30 kJ /kg

$$\therefore \quad W_{1-2} + w_{2-3} + w_{3-1} = 30\ \text{kJ/kg}$$
$$30 - 40 + w_{3-1} = 30$$
$$30 - 40 + w_{3-1} = 0$$
$$\therefore \quad \mathbf{W_{3-1} = 40\ kJ/kg}$$
$$\therefore \quad q = \Delta u + w$$
$$q = -30 + 40$$
$$\mathbf{q = 10\ kJ/kg}$$

It is given that w.d/kg = 30 kJ/kg and mass of the fluid flowing = 0.1 kg/cycle and system completes 10 cycle per sec.

$$\therefore \quad \dot{m} = 0.1\ \text{kg/cycle} \times 10\ \text{cycle/sec}$$
$$\therefore \quad \dot{m} = 1\ \text{kg/sec}$$
$$\therefore \quad \text{We know that w.d./kg} = 30\ \text{kJ/kg}$$
$$\therefore \quad \text{Rate of w.d. or power} = 30\ \text{kJ/kg} \times 1\ \text{kg/sec} = 30\ \text{kJ/sec} = 30\ \text{kW}$$

Remember:

$$\frac{\delta w}{dt} = \dot{m}\frac{\delta w}{dm}$$

Unit:

$$\frac{\text{kJ}}{\text{sec}} = \frac{\text{kg}}{\text{sec}} \times \frac{\text{kJ}}{\text{kg}}$$
$$\frac{\text{kJ}}{\text{sec}} = \frac{\text{kJ}}{\text{sec}}$$

Example 2.7 *The I.E. of a certain substance is given by the following equation,*

$$u = 3.56Pv + 84$$

where u is in kJ/kg, P is in kPa, v is in m^3/kg.

A system composed of 3 kg of this substance expands from an initial pressure of 500 kPa and a volume of 0.22 m^3 to a final pressure of 100 kPa in a process in which pressure and volume are related by $PV^{1.2} = C$.

(i) If the expansion is quasi-static, find Q, ΔU and W for the process.

(ii) In another process the same system expands according to the same pressure-volume relation as in case (i) and from the same initial state and same final state as in case (i), but the heat transfer for this case is 30 kJ. Find the work transfer for this process.

(iii) Explain the difference in work transfer in case (i) and (ii).

Data : u is in kJ/kg

P is in kPa

$$v = m^3/kg$$
$$m = 3kg$$
$$P_1 = 500\ kPa$$
$$V_1 = 0.22m^3$$
$$P_2 = 100\ kPa$$
$$PV^{12} = C$$
$$Q = ?\ \Delta U = ?\ W = ?$$

Solution

It is given that $u = 3.56\,Pv + 84$

Change in IE for unit mass

$$\Delta u = u_2 - u_1$$
$$= 3.56\,P_2V_2 + 84 - 3.56\,P_1V_1 - 84$$
$$\Delta u = 3.56\,(P_2V_2 - P_1V_1)$$
$$\therefore \quad \text{Total change in IE} = \Delta U = 3.56\left(P_2V_1 - P_1V_1\right)$$

Note: V is the total volume for 3 kg mass amd it is an extensive property.

$\therefore$ It depends on mass and P is an intensive property.

Since in Eq. (1) V_2 is not given,

$$\therefore \quad P_1 V_1^{1.2} = P_2 V_2^{1.2}$$

$$\therefore \quad V_2 = V_1\left(\frac{P_1}{P_2}\right)^{\frac{1}{1.2}} = 0.22\left(\frac{500}{100}\right)^{\frac{1}{1.2}}$$

$$V_2 = 0.22 \times 3.83 = 0.845\ \text{m}^3$$

Now from (1)

$$\Delta U = 3.56\left[100 \times 0.845 - 500 \times 0.22\right]\ \text{kJ}$$
$$= 356\,[1 \times 0.845 - 5 \times 0.22]\ \text{kJ}$$
$$\mathbf{\Delta U = -91\ kJ}$$

For a quasi-static process, W.D. $= \int PdV$

Since it is a polytropic process,

$$W_{1-2} = \frac{P_2V_2 - P_1V_1}{1-n}$$

$$W_{1-2} = \frac{100 \times 0.845 - 500 \times 0.222}{1 - 1.2}$$

$$\mathbf{W_{1-2} = 127.5\ kJ}$$

$$\therefore \quad Q = \Delta U + W$$
$$Q = -91 + 127.5$$
$$\mathbf{Q = 36.5\ kJ}$$

Case (ii) Here $Q = 30$ kJ

Since the end states are same as in case (i)

$$W = Q - \Delta U = 30 - (-91)$$
$$\mathbf{W = 121\ kJ}$$

(iii) The work in case (ii) is not equal to $\int PdV$. So the process is not quasi-static.

Example 2.8 *The I.E. of a certain substance is given by the equation U = 3.5 PV + 80. Where U is in kJ, pressure P is in kPa, and volume V is in m³. A system composed of 5 kg of this substance expands from an initial pressure of 5 bar and volume of 0.22 m³ to a final pressure of 1 bar in a process in which pressure and volume are related by $PV^{1.2} = C$.*

Find for quasi-static process,

(1) Heat transfer, (2)Total change in I.E. (3) Non-flow work

Data: *U in kJ ;* *P in kPa*

V in m³;

$$m = 5\ kg$$
$$P_1 = 5\ bar = 500\ kPa$$
$$V_1 = 0.22\ m^3\ ;$$

$$P_2 = 1\ bar = 100\ kPa$$

$$P_1 V_1^{1.2} = P_2 V_2^{1.2} = C$$

$$\therefore \quad V_2 = V_1 \times \left(\frac{P_1}{P_2}\right)^{\frac{1}{1.2}}$$

$$V_2 = 0.22 \times (5)^{1/1.2} = 0.84\ \text{m}^3$$

Solution

Now ΔU i.e., change in I.E. will be,

$$\Delta U = U_2 - U_1 = 3.5(P_2V_2 - P_1V_1)$$

$$= 3.5[100 \times 0.84 - 500 \times 0.22]$$

$$\mathbf{\Delta U = -91\ kJ}$$

$$W_{1-2} = \frac{P_1V_1 - P_2V_2}{n-1} = \frac{500 \times 0.22 - 100 \times 0.84}{0.2}$$

$$\mathbf{W_{1-2} = 130\ kJ}$$

$$Q = \Delta U + W = -91 + 130$$

$$\mathbf{Q = 39\ kJ}$$

Example 2.9 *A fluid is confined in a cylinder by a spring loaded frictionless piston, so that the pressure in the fluid is a linear function of the volume, P = a + bV. The internal energy of the fluid is given by the following equation U = 34 + 3.15 PV.*

Where U is in kJ, P in kPa, V in m³. If the fluid changes from an initial state of 170 kPa, 0.03 m³ to a final state of 400 kPa, 0.06 m³ with no work other than that done on the piston. Find the direction and magnitude of the work and heat transfer.

Data: $P = a + bV$; $U = 34 + 3.15\ PV$

$P_1 = 170\ kPa$; $V_1 = 0.03\ m^3$

$P_2 = 400\ kPa$; $V_2 = 0.06\ m^3$

$W = ?$ $Q = ?$

Solution

The change in I.E. of the fluid during the process is given as,

$$U_2 - U_1 = 3.15\,(P_2V_2 - P_1V_1)$$

$$\left[\textbf{Note: } \text{Units} \frac{\text{kN}}{\text{m}^2} \times \text{m}^3 = \text{kNm} = \text{kJ}\right]$$

$$U_2 - U_1 = 3.15\,[400 \times 0.06 - 1.70 \times 0.03]$$

$$= 100 \times 3.15\,[4 \times 0.06 - 1.70 \times 0.03]$$

$$U_2 - U_1 = 315 \times 0.189$$

$$\mathbf{U_2 - U_1 = 59.5\ kJ}$$

Now

$$P = a + bV$$

$$170 = a + b \times 0.03$$

$$400 = a + b \times 0.06$$

From these two equations on solving we get,

$$a = -60\ \text{kN/m}^2$$

$$b = 7667\ \text{kN/m}^5$$

We know that, W.D. during the process $W_{1-2} = \int PdV$

$$= \int_{V_1}^{V_2} (a+bV)dV$$

$$= a(V_2 - V_1) + b\frac{(V_2^2 - V_1^2)}{2}$$

$$= (V_2 - V_1)\left[a + \frac{b}{2}(V_1 + V_2)\right]$$

$$= 0.03 \text{ m}^3 \left[-60\frac{\text{kN}}{\text{m}^2} + \frac{7667}{2}\frac{\text{kN}}{\text{m}^5} \times 0.09 \text{ m}^3\right]$$

$$\mathbf{W_{1-2} = 8.55 \ kJ}$$

Since W_{1-2} is positive, work is done by the system,

∴ Heat transfer involved is given by

$$Q = \Delta U + W$$

$$Q_{1-2} = U_2 - U_1 + W_{1-2} = 59.5 + 8.55$$

$$Q_{1-2} = 68.05 \text{ kJ}$$

Since Q_{1-2} is positive, heat flows into the system during the process.

Example 2.10 *The expression for the I.E. of a certain closed system is given by U = M + N (PV) where P is the pressure, V is the volume and M and N are constants. The system now undergoes a reversible process where Q = 0. Show that the relation between P and V is given by, $PV^K = C$ where $K = \frac{N+1}{N}$.*

Solution

As given in the problem,

$$U = M + N(P.V) \text{ differentiating we get,} \quad ...(1)$$

$$dU = O + N.P.dV + N.V.dP$$

We know that from first law for the process,

$$Q - W = \Delta U$$

$$Q = \Delta U + W$$

or $$\delta Q = dU + \delta W$$

$$\delta Q = dU + P.dV$$

Since $$Q = 0, 0 = dU + P.dV$$

Given, $$dU = -P.dV \quad ...(2)$$

Equating (1) and (2)

$$-P.dV = N.P.dV + N.V\,dP$$

$$O = N.P.dV + N.V\,dP + P.dV$$

$$= (N+1)P.dV + N.V.dP$$

Dividing throughout by N, we get

$$O = \frac{N+1}{N}P.dV + V \cdot dP$$

i.e. $$-V.dP = \frac{N+1}{N}P.dV$$

$$-VdP = K.P.dV \text{ as given } \frac{N+1}{N} = K$$

$$-\frac{dP}{P} = K.\frac{dV}{V}$$

Integrating, $$\ln\frac{1}{P} = \ln V^{K}$$

or $$\mathbf{PV^{K} = C}$$

where $$K = \frac{N+1}{N}$$

Example 2.11 *Following data refers to a steady flow process.*

	At entrance	*At exit*
Enthalpy (kJ/kg)	*4000*	*4100*
Velocity (m/sec)	*50*	*20*
Height (m)	*50*	*10*

Mass flow rate $\dot{m}$ *= 1 kg/sec*
Heat transfer rate = 200 kJ/sec to the system.
Determine the power capacity of the system and state whether it is a power producing system or otherwise.

Solution

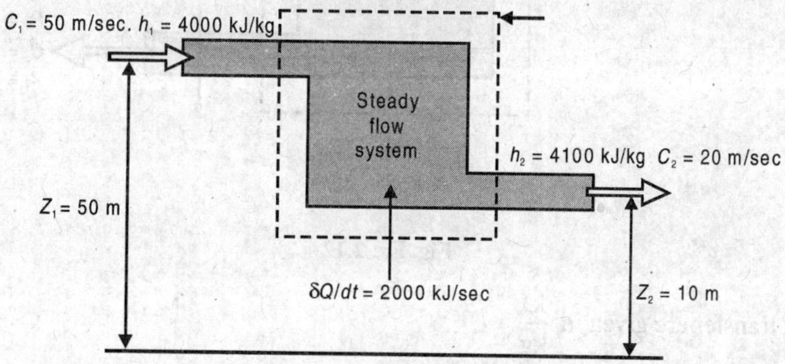

Fig. Ex. 2.11

$$\dot{m} = 1\ \text{kg/sec}$$

Since heat transfer is given in kJ/sec, we will have to use SFEE on time basis.

$$\dot{m}\left[\frac{C_1^2}{2} + gZ_1 + h_1\right] + \frac{\delta Q}{dt} = \dot{m}\left[\frac{C_2^2}{2} + gZ_2 + h_2\right] + \frac{\delta W}{dt}$$

Since power is to found,

$$\frac{\delta W}{dt} = \dot{m}\left[\frac{C_1^2 - C_2^2}{2} + g(Z_1 - Z_2) + (h_1 - h_2)\right] + \frac{\delta Q}{dt}$$

$$\frac{\delta W}{dt} = \frac{1\ \text{kg}}{\text{sec}}\left[\frac{50^2 - 20^2}{2\times1000}\ \frac{\text{kJ}}{\text{kg}} + \frac{9.81\,(40)}{1000}\ \frac{\text{kJ}}{\text{kg}} + (4000 - 4100)\right] + 200\frac{\text{kJ}}{\text{sec}}$$

$$\frac{\delta W}{dt} = \frac{1\ \text{kg}}{\text{sec}}\left[1.05\frac{\text{kJ}}{\text{kg}} + 0.3924\ \frac{\text{kJ}}{\text{kg}} + -100\ \frac{\text{kJ}}{\text{kg}}\right] + 200\ \frac{\text{kJ}}{\text{sec}}$$

$$\frac{\delta W}{dt} = 1.05\frac{\text{kJ}}{\text{sec}} + 0.3924\frac{\text{kJ}}{\text{sec}} - 100\frac{\text{kJ}}{\text{sec}} + 200\frac{\text{kJ}}{\text{sec}}$$

$$\mathbf{\frac{\delta W}{dt} = 101.4484\ kJ/sec\ or\ kW}$$

Since $\frac{\delta W}{dt}$ is + ve, it is a power producing system.

Example 2.12 *In a steady flow system 135 kJ/kg of work is done by the system. The specific volume of the fluid, pressure and velocity at the inlet are 0.37 m³/kg, 600 kPa, 16 m/sec. The inlet is 32 m above the floor level. The discharge conditions are 0.62 m³/kg, 100 kPa and 270 m/sec. The total heat loss between inlet and discharge is 9 kJ/kg. In flowing through this system, does the specific I.E. increase or decrease and by how much? The outlet is at flow level.*

Solution

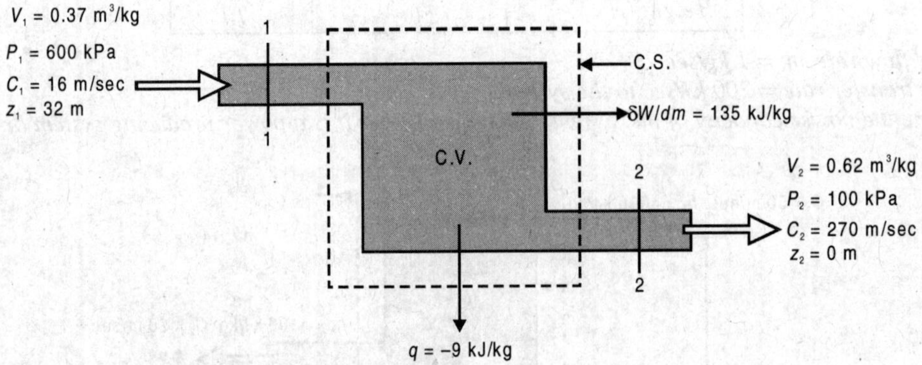

Fig. Ex. 2.12

Since work and heat transfer are given in $\frac{\text{kJ}}{\text{kg}}$

Using SFEE on mass basis,

$$\left[\frac{C_1^2}{2} + gZ_1 + h_1\right] + q = \left[\frac{C_2^2}{2} + gZ_2 + h_2\right] + w$$

Since pressure, specific volumes are given and change in Sp. I.E., is to be found, substitute $h_1 = u_1 + P_1v_1$ and $h_2 = u_2 + P_2v_2$ we get,

$$\left[\frac{C_1^2}{2} + gZ_1 + u_1 + P_1v_1\right] + q = \left[\frac{C_2^2}{2} + gZ_2 + u_2 + P_2v_2\right] + w$$

$$\therefore \qquad u_1 - u_2 = \frac{C_2^2 - C_1^2}{2} + g(Z_2 - Z_1) + (P_2v_2 - P_1v_1) + w - q$$

$$u_1 - u_2 = \frac{270^2 - 16^2}{2 \times 1000}\frac{\text{kJ}}{\text{kg}} + \frac{9.81(-32)}{1000}\frac{\text{kJ}}{\text{kg}}$$

$$+(100 \times .62 - 600 \times 0.37)\frac{\text{kJ}}{\text{kg}} + 135\frac{\text{kJ}}{\text{kg}} - \left(-9\frac{\text{kJ}}{\text{kg}}\right)$$

$$u_1 - u_2 = 36.45 - 0.314 - 160 + 135 + 9$$
$$u_1 - u_2 = 36.45 - 20.136 \text{ kJ/kg}$$
$$u_1 - u_2 = 20.136 \text{ kJ/kg}$$
$$\therefore \quad \mathbf{u_2 - u_1 = -20.136 \text{ kJ/kg}}$$

$\therefore$ Specific I.E. decreases by 20.236 kJ/kg

Example 2.13 *In a steady flow system, a fluid flows at the rate of 5 kg/sec. It enters at a pressure of 620 kPa, velocity of 300 m/sec. I. E., 2100 kJ/kg and specific volume 0.37 m³/kg.*

It leaves the system at a pressure of 130 kPa, a velocity of 150 m/sec; I.E. 1500 kJ/kg and specific volume 1.2 m³/kg. During its flow through system there is a heat loss of 30 kJ/kg. Determine the power capacity of the system in kW. State whether it is from or to the system. Neglect change in P.E.

Solution

Since heat transfer is given in $\frac{\text{kJ}}{\text{kg}}$, using SFEE on mass basis.

Note : Since *P*, *u*, *v* are given, writing the equation in terms of these quantities.

$$\left[\frac{C_1^2}{2} + gZ_1 + u_1 + P_1v_1\right] + \frac{\delta Q}{dm} = \left[\frac{C_2^2}{2} + gZ_2 + u_2 + P_2v_2\right] + \frac{\delta W}{dm}$$

Since power capacity is to be found out,

$$\frac{\delta W}{dm} = \left[\frac{C_1^2 - C_2^2}{2} + g(Z_1 - Z_2) + (u_1 - u_2) + (P_1v_1 - P_2v_2)\right] + \frac{\delta Q}{dm}$$

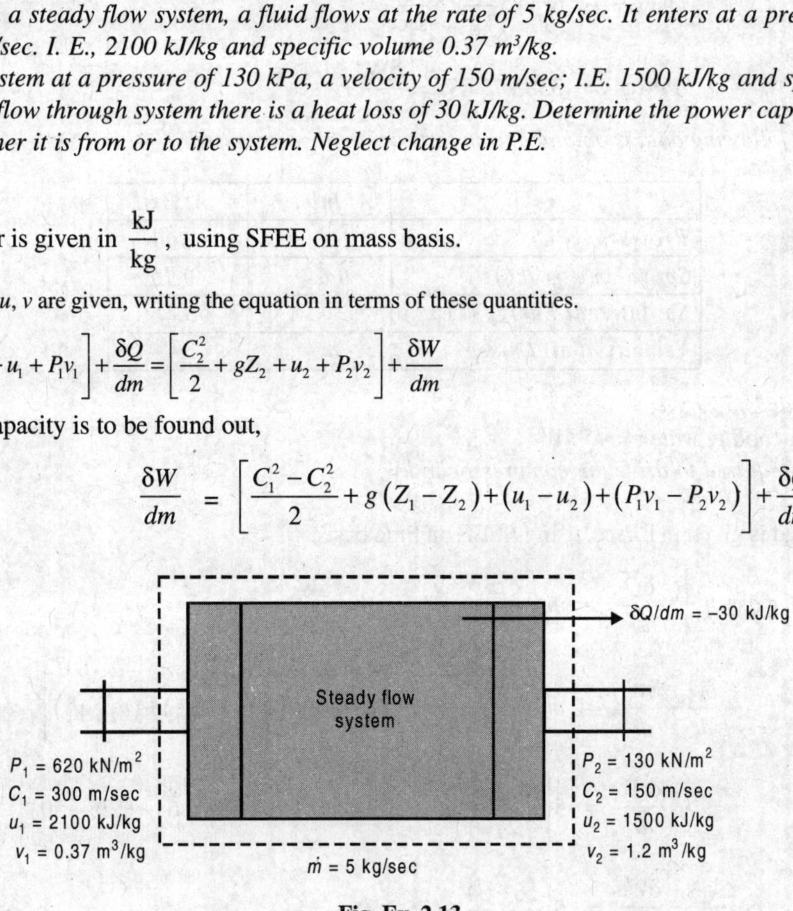

Fig. Ex. 2.13

Neglecting ΔPE as given in the problem,

$$\frac{\delta W}{dm} = \left[\frac{C_1^2 - C_2^2}{2} + (u_1 - u_2) + (P_1v_1 - P_2v_2)\right] + \frac{\delta Q}{dm}$$

$$\frac{\delta W}{dm} = \frac{300^2 - 150^2}{2 \times 1000}\frac{\text{kJ}}{\text{kg}} + (2100 - 1500) + (620 \times 0.37 - 130 \times 1.2) - 30\frac{\text{kJ}}{\text{kg}}$$

$$\frac{\delta W}{dm} = \left[\frac{67500}{2000} + 600 + 73\right]\frac{\text{kJ}}{\text{kg}} - 30\frac{\text{kJ}}{\text{kg}} = 33.75 + 600 + 73 - 30$$

$$\frac{\delta \mathbf{W}}{\mathbf{dm}} = \mathbf{676.75\ kJ/kg}$$

Since power is required to be expressed in kW,

Using $$\frac{\delta W}{dt} = \dot{m}\frac{\delta W}{dm}$$

Note : Mass flow rate = 5 kg/sec is given

$\therefore$ Power = $676.75 \frac{\text{kJ}}{\text{kg}} \times 5 \frac{\text{kg}}{\text{sec}}$

Power = 3383.75 kW = $\frac{\delta \mathbf{W}}{\mathbf{dt}}$

Example 2.14 *The following data is obtained from tests on an air compressor.*

	At Inlet	At Exit
Pressure (kPa)	100	500
Sp. volume (m^3/kg)	0.6	0.15
Sp. Internal Energy (kJ/kg)	50	125
Velocity of air (m/sec)	8	4

Rate of flow of air is 5 kg/sec.
Heat rejected to cooling water = 45 kW
Find the power required to drive the compressor in kW.

Solution

Since heat is rejected is given in kJ/sec, using SFEE on time basis,

$$\dot{m}\left[\frac{C_1^2}{2} + gZ_1 + h_1\right] + \frac{\delta Q}{dt} = \dot{m}\left[\frac{C_2^2}{2} + gZ_2 + h_2\right] + \frac{\delta W}{dt}$$

$$\therefore \quad \frac{\delta Q}{dt} - \frac{\delta W}{dt} = \dot{m}\left[\frac{C_2^2 - C_1^2}{2} + g(Z_2 - Z_1) + (P_2v_2 - P_1v_1) + (u_2 - u_1)\right]$$

$$-45 - \frac{\delta W}{dt} = 5\left[\frac{4^2 - 8^2}{2\times 1000} + 0 + (500\times 0.15 - 100\times 0.6) + (125 - 50)\right]$$

$$-45 - \frac{\delta W}{dt} = 5\left[\frac{16 - 64}{2000} + (75 - 60) + 75\right]$$

$$-45 - \frac{\delta W}{dt} = 5[-0.024 + 15 + 75]$$

$$\Rightarrow \quad \frac{\delta \mathbf{W}}{\mathbf{dm}} = \mathbf{-494.88\ kW}$$

$\therefore$ Power required to drive the compressor = 494.88 kW.

Example 2.15 *Steam enters a steam turbine with a velocity of 40 m/sec, enthalpy 2500 kJ/kg and leaves with a velocity of 90 m/sec and enthalpy 2030 kJ/kg. Heat losses from the turbine insulation to the surroundings are 250 kJ/min, and the steam flow rate is 5000 kg/hr. Neglect the change of P.E. Find the power developed by the turbine in kW.*

Solution

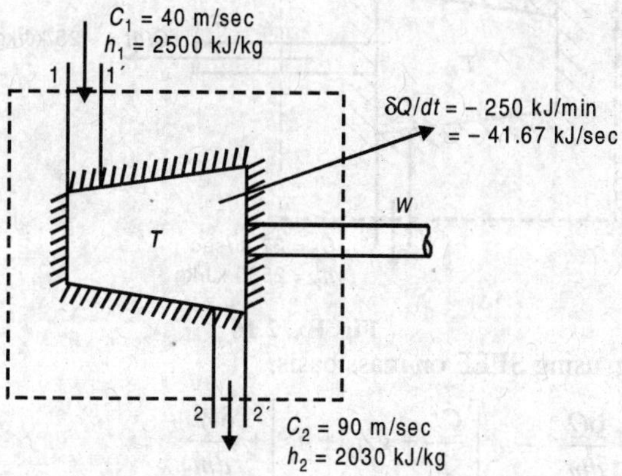

Fig. Ex. 2.15

$$\text{Steam flow rate} = 5000 \text{ kg/hr}$$

$$= \frac{5000}{3600} \text{ kg/sec} = 1.389 \text{ kg/sec}$$

$$\dot{m}\left[\frac{C_1^2}{2} + gZ_1 + h_1\right] + \frac{\delta Q}{dt} = \dot{m}\left[\frac{C_2^2}{2} + gZ_2 + h_2\right] + \frac{\delta W}{dt}$$

$$\dot{m}\left[\frac{C_1^2 - C_2^2}{2} + (h_1 - h_2)\right] + \frac{\delta Q}{dt} = \frac{\delta W}{dt}$$

$$1.389 \frac{\text{kg}}{\text{sec}}\left[\frac{40^2 - 90^2}{2\times1000}\frac{\text{kJ}}{\text{kg}} + (2500 - 2030)\frac{\text{kJ}}{\text{kg}}\right] - 4.167\frac{\text{kJ}}{\text{sec}} = \frac{\delta W}{dt}$$

$$1.389\left[-3.25\frac{\text{kJ}}{\text{kg}} + (470)\frac{\text{kJ}}{\text{kg}}\right] - 4.167\frac{\text{kJ}}{\text{sec}} = \frac{\delta W}{dt}$$

$$1.389\frac{\text{kg}}{\text{sec}}\left[466.75\frac{\text{kJ}}{\text{kg}}\right] - 4.167\frac{\text{kJ}}{\text{sec}} = \frac{\delta W}{dt}$$

$$\frac{\delta \mathbf{W}}{\mathbf{dt}} = \mathbf{644.149 \text{ kW}}$$

Example 2.16 *Steam enters a steam turbine with a velocity of 16 m/sec and specific enthalpy 2990 kJ/kg. The steam leaves the turbine with a velocity of 37 m/sec and specific enthalpy of 2530 kJ/kg. The heat lost to the surroudings is 25 kJ/kg. The steam flow rate is 3,60,000 kg/hr. Determine the work output from the turbine in kW.*

Solution

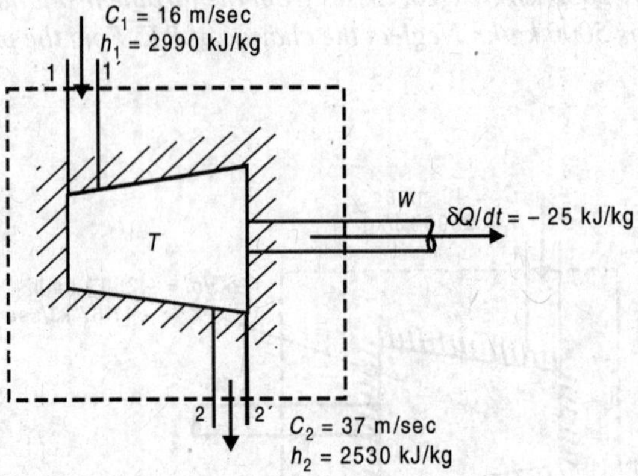

Fig. Ex. 2.16

Since heat lost is given in kJ/kg, using SFEE on mass basis,

$$\left[\frac{C_1^2}{2}+gZ_1+h_1\right]+\frac{\delta Q}{dm} = \left[\frac{C_2^2}{2}+gZ_2+h_2\right]+\frac{\delta W}{dm}$$

Since nothing is mentioned about elevation, neglecting ΔPE.

$$\therefore \quad \frac{C_1^2}{2}+h_1+\frac{\delta Q}{dm} = \frac{C_2^2}{2}+h_2+\frac{\delta W}{dm}$$

$$\therefore \quad \frac{\delta W}{dm} = \frac{\left(C_1^2-C_2^2\right)}{2}+\left(h_1-h_2\right)+\frac{\delta Q}{dm}$$

$$\frac{\delta W}{dm} = \frac{16^2-37^2}{2\times1000}\,\frac{\text{kJ}}{\text{kg}}+(2990-2530)\,\frac{\text{kJ}}{\text{kg}}-25\,\frac{\text{kJ}}{\text{kg}}$$

$$\frac{\delta W}{dm} = \frac{-1113}{2000}+460-25$$

$$\mathbf{\frac{\delta W}{dm} = 434.443\ \frac{kJ}{kg}}$$

To express power in kW, using

$$\frac{\delta W}{dt} = \dot{m}\frac{\delta W}{dm}$$

It is given that mass flow rate = 360000 kg/hr = 100 kg/sec.

$$\therefore \quad \text{Power} = 434.443\,\frac{\text{kJ}}{\text{kg}}\times100\,\frac{\text{kg}}{\text{sec}}$$

$$\textbf{Power} = \mathbf{434443\ kW} = \mathbf{\frac{\delta W}{dt}}$$

Example 2.17 *The gas turbine of a turbojet engine receives a steady flow of gases at pressure of 7.2 bar temperature of 850°C and velocity of 160 m/sec. It discharges the gases at a pressure of 1.15 bar, temperature 450°C and a velocity of 250 m/sec. Determine the external work output of the turbine in 5 kJ/kg of gas flow. The process may be assumed to be adiabatic and C_p for gas as 1.04 kJ/kg K.*

Data: $P_1 = 7.2\times10^2\ kPa$ $\quad W = ?$ i.e. $\dfrac{\delta W}{dm} = ?$

$Cp_{gases} = 1.04\ kJ/kg\text{-}K$

$T_1 = 850°C$ $\quad C_1 = 160\ m/sec$

Solution

Insulation (Adiabtic Turbine) $\therefore Q = 0$

$P_2 = 1.15\times10^2$ kPa

$T_2 = 450°$ C

$C_2 = 250$ m/sec

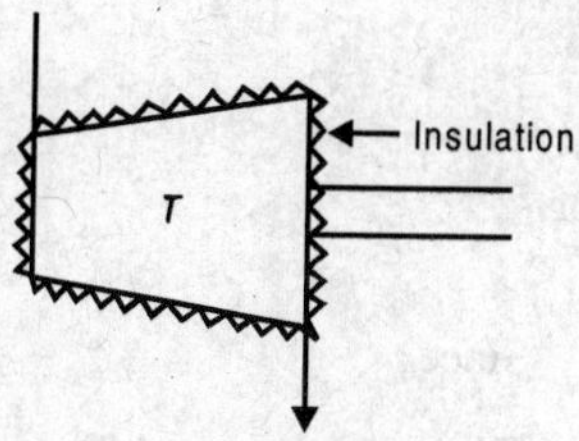

Fig. Ex. 2.17

We know that steady Flow Energy equation on mass basis is given by,

$$\left[\frac{C_1^2}{2} + gZ_1 + h_1\right] + \frac{\delta Q}{dm} = \left[\frac{C_2^2}{2} + gZ_2 + h_2\right] + \frac{\delta W}{dm} \quad ...(1)$$

Since Adiabtic Expansion process $\dfrac{\delta Q}{dm} = 0$

Since they have not mentioned anything about elevation $gZ_1 = 0, gZ_2 = 0$. Then Eq. (1) becomes,

$$\left[\frac{C_1^2}{2} + h_1\right] = \left[\frac{C_2^2}{2} + h_2\right] + \frac{\delta W}{dm}$$

i.e.,

$$\frac{\left(C_1^2 - C_2^2\right)}{2} + (h_1 - h_2) = \frac{\delta W}{dm} \quad ...(2)$$

Since

$$h = C_P T$$

$$\therefore \quad \frac{\delta W}{dm} = (h_1 - h_2) + \frac{\left(C_1^2 - C_2^2\right)}{2}$$

$$= C_P(850 - 450) + \frac{160^2 - 250^2}{2\times1000}$$

$$= 1.04(850 - 450) + \frac{160^2 - 250^2}{2\times1000} \quad 416 + (-18.45)$$

$$\mathbf{\frac{\delta W}{dm} = 397.55\ kJ/kg}$$

Example 2.18 *The power output of an adiabatic steam turbine is 5 MW, and the inlet and exit conditions are as under,*

	Inlet	*A Exit*
Pressure	*2 MPa*	*0.15 bar*
Temperature (°C)	*400*	*Dryness 0.9*
Velocity (m/sec)	*50*	*180*
Elevation m	*10*	*6*

Determine the work done per unit of mass of the steam flowing through turbine and mass flow rate of the steam.

Solution

We know that, since work done per unit of mass of steam is to be found, using SFEE on mass basis.

$$\left[\frac{C_1^2}{2} + gZ_1 + h_1\right] + q = \left[\frac{C_2^2}{2} + gZ_2 + h_2\right] + w$$

or $$w = (h_1 - h_2) + \frac{(C_1^2 - C_2^2)}{2} + g(Z_1 - Z_2)$$

Since $$h_1 = 3248.7 \text{ kJ/kg}$$

[**Note:** From steam tables coresponding to and 20 bar (2 MPa)].

$$h_2 = h_f + x \cdot h_{fg}.$$ Since the steam is wet at 0.15 bar from steam tables

and $$h_2 = 226 + 0.9 \times 2373.2 = 2361.88 \text{ kJ/kg}$$

$$w = (3248.7 - 2361.88) + \frac{50^2 - 180^2}{2000} + \frac{4 \times 9.81}{1000}$$

$$w = 871.91 \text{ kJ/kg} = \frac{\delta W}{dm}$$

Also, we know that, $$\frac{\delta W}{dt} = \dot{m}\frac{\delta W}{dm}$$

$$5000\frac{\text{kJ}}{\text{sec}} = \dot{m} \times 871.91 \text{ kJ/kg}$$

$\therefore$ $$\dot{\mathbf{m}} = \mathbf{5.734 \text{ kg/sec}}$$

Example 2.19 *A turbine operating on air has inlet conditions as10 bar, 750 K and 200 m/s. While exit conditions are 1.25 bar and 40 m/sec. The mass flow rate of air is 1000 kg/hr. The flow of air is assumed to be reversible adiabatic (isentropic). Calculate,*

(i) The temperature of air at exit.

(ii) The power output of turbine.

Assume C_p = 1.053 kJ/kg K and k = 1.375 adiabatic index

Solution

$$\dot{m} = 1000 \text{ kg/hr} = 0.2777 \text{ kg/sec}$$

To find T_2, since the expansion process is adiabatic,

We know that, $$\frac{T_2}{T_1} = \left(\frac{P_2}{P_1}\right)^{\frac{k-1}{k}}$$

$\Rightarrow$ $$T_2 = T_1 \times \left(\frac{P_2}{P_1}\right)^{\frac{k-1}{k}} = 750 \times \left(\frac{1.25 \times 100}{10 \times 100}\right)^{\frac{1.375-1}{1.375}}$$

$$\mathbf{T_2 = 425.367 \text{ K}}$$

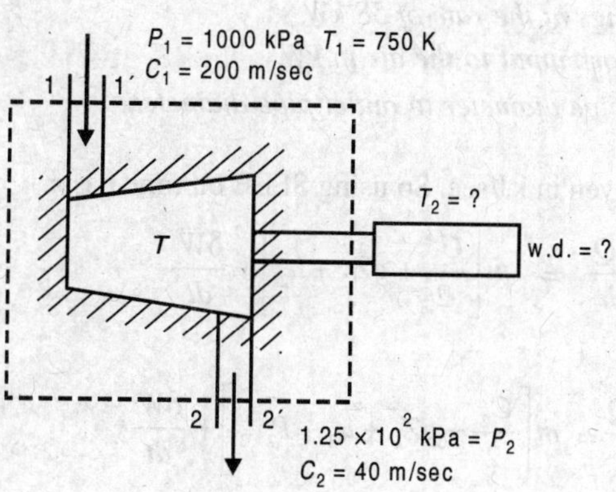

Fig. Ex. 2.19

We also know that, the SFEE is,

$$\dot{m}\left[\frac{C_1^2}{2}+gZ_1+h_1\right]+\frac{\delta Q}{dt} = \dot{m}\left[\frac{C_2^2}{2}+gZ_2+h_2\right]+\frac{\delta W}{dt}$$

Neglecting the ΔPE , and $\frac{\delta Q}{dt}=0$ since the turbine is adiabatic.

$$\dot{m}\left[\frac{C_1^2}{2}+h_1\right] = \dot{m}\left[\frac{C_2^2}{2}+h_2\right]+\frac{\delta W}{dt}$$

$$-\frac{\delta W}{dt} = \dot{m}\left[\frac{\left(C_1^2-C_2^2\right)}{2}+h_2-h_1\right] \qquad ...(1)$$

$$h_2-h_1 = \Delta h = m\times C_p\,.\,\Delta T$$

$$= 1\times 1.053\times(425.367-750) \qquad [\because m = 1 \text{ for unit mass}]$$

$$(h_2-h_1) = -341.83855 \text{ kJ} \qquad ...(2)$$

Substituting (2) in (1) we get,

$$-\frac{\delta W}{dt} = 0.277\,\frac{\text{kg}}{\text{sec}}\left[\frac{40^2-200^2}{2\times 1000}+(-341.838)\right]$$

$$-\frac{\delta W}{dt} = 0.277[-19.2-341.838]=0.277[-361.038]$$

$$= -100.007 \text{ kW}$$

$$\boldsymbol{\frac{\delta W}{dt} = +100.007 \text{ kW.}}$$

Example 2.20 *Air flows steadily at the rate of 0.5 kg/sec, through the air compressor, entering at 7 m/sec velocity, 100 kPa pressure, 0.95 m³/kg volume and leaving at 5 m/sec, 700 kPa, 0.19 m³/kg. The I.E. of the air leaving is 90 kJ/kg greater than that of air entering.*

Heat rejected to the surroundings at the rate of 58 kW.

(a) Calculate the rate of work input to the air in kW.

(b) Find the ratio of inlet pipe diameter to outlet pipe diameter.

Solution

Heat lost to the surroundings is given in kJ/sec. So using SFEE on time basis.

$$\dot{m}\left[\frac{C_1^2}{2}+gZ_1+h_1\right]+\frac{\delta Q}{dt} = \dot{m}\left[\frac{C_2^2}{2}+gZ_2+h_2\right]+\frac{\delta W}{dt}$$

Since, P, u, and v are given.

$$\dot{m}\left[\frac{C_1^2}{2}+gZ_1+u_1+P_1v_1\right]+\frac{\delta Q}{dt} = \dot{m}\left[\frac{C_2^2}{2}+gZ_2+u_2+P_2v_2\right]+\frac{\delta W}{dt}$$

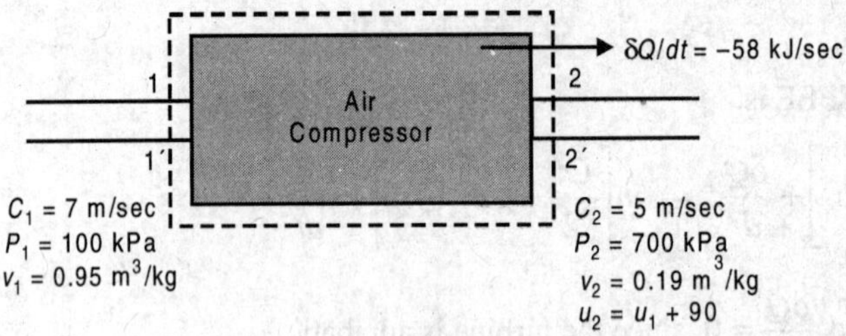

Fig. Ex. 2.20

Since power is to be found,

$$\frac{\delta W}{dt} = \dot{m}\left[\frac{C_2^2-C_1^2}{2}+g(Z_2-Z_1)+(P_2v_2-P_1v_1)+(u_2-u_1)\right]+\frac{\delta Q}{dt}$$

$$= 0.5\,\frac{\text{kg}}{\text{sec}}\left[\frac{5^2-7^2}{2\times1000}+0+(700\times0.19-100\times0.95)\frac{\text{kJ}}{\text{kg}}+90\,\frac{\text{kJ}}{\text{kg}}\right]-58\,\frac{\text{kJ}}{\text{sec}}$$

$$= 0.5\,\frac{\text{kg}}{\text{sec}}\left[(-0.012+38+90)\frac{\text{kJ}}{\text{kg}}\right]-58\,\frac{\text{kJ}}{\text{sec}}$$

$$= 0.5\,\frac{\text{kg}}{\text{sec}}\left[127.988\,\frac{\text{kJ}}{\text{kg}}\right]-58\,\frac{\text{kJ}}{\text{sec}}$$

$$\frac{\delta W}{dt} = -63.994\,\frac{\text{kJ}}{\text{sec}}-58\,\frac{\text{kJ}}{\text{sec}}$$

$$\boldsymbol{\frac{\delta W}{dt} = -121.994\ \text{kW}}$$

(b) From continuity equation, $\dot{m} = \dfrac{A_1C_1}{v_1}=\dfrac{A_2C_2}{v_2}$

$$\therefore\quad \frac{A_1}{A_2} = \frac{v_1}{v_2}\cdot\frac{C_2}{C_1}=\frac{0.95}{0.19}\times\frac{5}{7}$$

$$\boldsymbol{\frac{A_1}{A_2} = 3.57}$$

Also we know that,
$$\frac{A_1}{A_2} = \frac{\frac{\pi}{4}\cdot d_1^2}{\frac{\pi}{4}\cdot d_2^2} = 3.57$$

$$\therefore \quad \frac{d_1}{d_2} = \sqrt{3.57}$$

$$\frac{\mathbf{d_1}}{\mathbf{d_2}} = \mathbf{1.89}$$

Example 2.21 *At the inlet to a certain nozzle, the enthalpy of the fluid passing is 3000 kJ/kg and the velocity is 60 m/sec. At the discharge end the enthalpy is 2757 kJ/kg. The nozzle is horizontal. The heat loss during the flow is negligible.*

(1) Find the velocity at the exit.

(2) If the inlet area is 0.1 m² and specific volume is 0.187 m³/kg, find the mass flow rate.

(3) If the specific volume at the outlet is 0.498 m³/kg.

Find the area at the exit of the nozzle.

Solution

Since the nozzle is horizontal, $\Delta PE = 0$

Since the heat loss during the flow is negligible, $q = 0$

Since no work is done in the nozle, $w = 0$

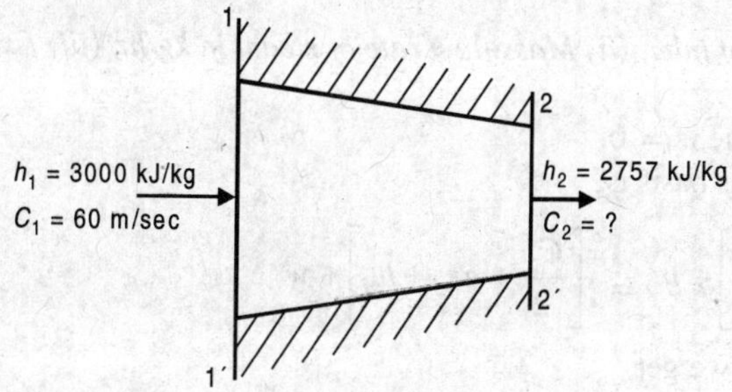

Fig. Ex. 2.21

We know that SFEE on mass basis is,

$$\left[\frac{C_1^2}{2} + gZ_1 + h_1\right] + q = \left[\frac{C_2^2}{2} + gZ_2 + h_2\right] + w \qquad \ldots(1)$$

or
$$q - w = \frac{C_2^2 - C_1^2}{2} + g(Z_2 - Z_1) + h_2 - h_1$$

On substituting the given characteristics, Eq. (1) reduces to,

$$h_1 + \frac{C_1^2}{2} = h_2 + \frac{C_2^2}{2}$$

$$3000\frac{\text{kJ}}{\text{kg}} + \frac{60^2}{2\times1000} = 2757\frac{\text{kJ}}{\text{kg}} + \frac{C_2^2}{2\times1000}$$

$$3000 + 1.8 = 2757 + \frac{C_2^2}{2000}$$

$$244.8 = \frac{C_2^2}{2000}$$

$\therefore$ $$C_2^2 = 489600$$

$\therefore$ $$\mathbf{C_2 = 699.71 \text{ m/sec}}$$

(ii) Mass flow rate $$\dot{m} = \frac{A_1 C_1}{v_1} = \frac{A_2 C_2}{v_2}$$

$$= \frac{0.1 \text{ m}^2 \times 60 \text{ m/sec}}{0.187}$$

$$\mathbf{\dot{m} = 32.08 \text{ kg/sec}}$$

(iii) Area at the exit $$\dot{m} = \frac{A_1 C_1}{v_1} = \frac{A_2 C_2}{v_2}$$

$$A_2 = \frac{m v_2}{C_2} = \frac{32.08 \times 0.498}{699.71}$$

$\therefore$ $$\mathbf{A_2 = 0.0228 \text{ m}^3}$$

Example 2.22 *A nozzle is a device used to increase velocity of steam flowing through it. At inlet the specific enthalpy of steam is 3000 kJ/kg. Specific volume 0.187 m³/kg and area 0.1 m² and at outlet sp. enthalpy 2762 kJ/kg, specific volume 0.498 m³/kg. There is no loss of heat, nozzle is horizontal and stream enters with the velocity of 60 m/sec.*

Find: (i) Velocity of steam of inlet, (ii) Mass flow rate of steam in kg/hr, (iii) Exit area of nozzle.

Solution

For nozzle since no work is done $w = 0$.

We know that SFEE on mass basis is,

$$\left[\frac{C_1^2}{2} + gz_1 + h_1\right] + q = \left[\frac{C_2^2}{2} + gz_2 + h_2\right] + w$$

On substituting characteristic, we get,

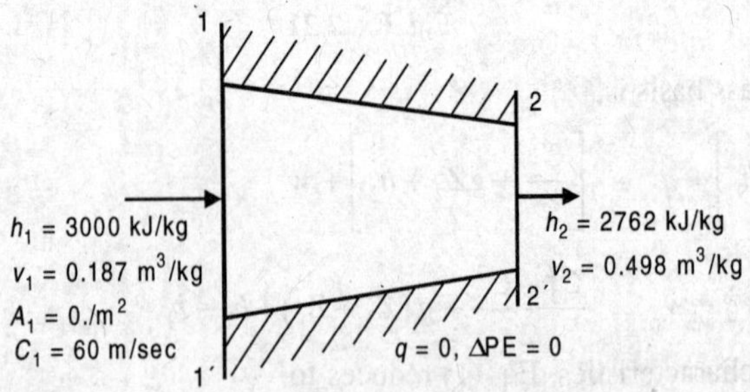

Fig. Ex. 2.22

$$h_1 + \frac{C_1^2}{2} = h_2 + \frac{C_2^2}{2}$$

$$3000 + \frac{60^2}{2 \times 1000} = 2762 + \frac{C_2^2}{2 \times 100}$$

$\Rightarrow$ $$\mathbf{C_2 = 692.53 \text{ m/sec}}$$

(ii) Mass flow rate $\dot{m} = \frac{A_1 C_1}{v_1} = \frac{A_2 C_2}{v_2} = \frac{0.1\times 60}{0.187}$

$$\dot{\mathbf{m}} = \mathbf{32.856\ kg/sec}$$

(iii) Area at the exit, $\dot{m} = \frac{A_1 C_1}{v_1} = \frac{A_2 C_2}{v_2}$

$$A_2 = \frac{m \cdot v_2}{C_2} = \frac{32.856\times 0.4948}{692.53}$$

$$\mathbf{A_2 = 0.0231\ m^2}$$

Example 2.23 *Air enters a nozzle with a velocity of 40 m/sec. The decrease in the enthalpy in the nozzle is 180000 J/kg. Determine the exit velocity. Assume nozzle to be adiabatic.*

Solution

We know that, SFEE on mass basis is,

$$\left[\frac{C_1^2}{2} + gZ_1 + h_1\right] + q = \left[\frac{C_2^2}{2} + gZ_2 + h_2\right] + w \quad ...(1)$$

Since for a nozzle characteristics are,

$$\Delta PE = 0,\ w = 0,\ q = 0 \text{ since adiabatic}$$

Equation (1) reduces to,

$$\frac{C_1^2}{2} + h_1 = \frac{C_2^2}{2} + h_2$$

$$\frac{C_1^2}{2} + (h_1 - h_2) = \frac{C_2^2}{2}$$

$$\frac{(40)^2}{2\times 1000} + 180000 = \frac{C_2^2}{2\times 1000}$$

$$\therefore \quad \mathbf{C_2 = 601.33\ m/sec}$$

Example 2.24 *Air is compressed continuously by an air compressor according to the law $PV^{1.3} = C$. The pressure at the inlet is 1 bar, and at the outlet is 5 bar. The volume of the air changes from 3 m³/kg to 0.8 m³/kg and the velocity changes 25 m/sec at inlet to 130 m/sec at the delivery. The delivery connection is 12 mt above the inlet. What is the shaft power of the compressor. Is it a power absorbing or power producing system.*

Solution

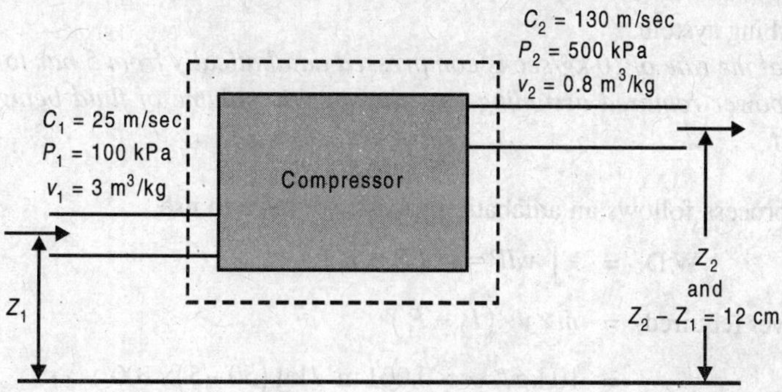

Fig. Ex. 2.24

Note. If the steady flow process or system follows. $PV^n = C$ then follow this method to solve the problem.

i.e., use $$-\int v\,dP = w + \Delta KE + \Delta PE$$

We know that,

	Isothermal ($Pv = C$) (For unit mass)	Polytropic ($Pv^n = C$)
(i) W.D. in closed system = $\int P dv$	$P_1 v_1 \ln_e \frac{v_2}{v_1}$	$\frac{P_2 v_2 - P_1 v_1}{1-n}$
(ii) W.D. for open system = $-\int v dP$	Steady flow system, $P_1 v_1 \ln_e \frac{v_2}{v_1}$ or $\frac{P_1}{P_2}$	$\frac{n}{n-1} \times P_2 v_2 - P_1 v_1$

For Polytropic Process,

$$-\int_1^2 v dP = \frac{n}{n-1} \times P_1 v_1 - P_2 v_2$$

$$= \frac{1.3}{1.3-1} \times (100 \times 3 - 500 \times 0.8) \frac{\text{kJ}}{\text{kg}}$$

$$= -433.333 \text{ kJ/kg}$$

and, $$\Delta KE = \frac{1}{2}\left(\frac{C_2^2 - C_1^2}{1000}\right) = \frac{130^2 - 25^2}{2 \times 1000}$$

and $$\Delta KE = 8.1375 \frac{\text{kJ}}{\text{kg}}$$

$$\Delta PE = g(Z_2 - Z_1) = \frac{9.81 \times 12}{1000} 0.11772 \frac{\text{kJ}}{\text{kg}}$$

Substituting these in Eq. (1) we get,

$$\therefore \quad -\int v dP = w + \Delta KE + \Delta PE$$

$$-433.333 = w + 8.1375 + 0.11772$$

$$\therefore \quad \mathbf{w = -441.588 \text{ kJ/kg}}$$

$$= \text{Shaft work}$$

Since w is (–) negative work is done on the system.

∴ It is a power absorbing system.

Example 2.25 *A fluid at the rate of 10 kg/sec is compressed adiabatically from 5 bar to 50 bar in a steady flow process. Calculate the power required assuming that the specific volume of fluid being as 0.001 m³/kg which remains almost constant.*

Solution

Since the compression process follows an adiabatic process, we have to use,

$$\text{W.D.} = -\int v dP = v.(P_2 - P_1)$$

$$\therefore \quad \text{Power required} = \dot{m} \times v \times (P_2 - P_1)$$

$$= 10 \text{ kg/sec} \times 0.001 \text{ m}^3/\text{kg}\,(50-5) \times 100$$

Power required = 45 kW

Example 2.26 *A blower handles 2 kg/sec of air at 20°C and consumes a power of 30 kW. The inlet and outlet velocities are 100 in/sec and 150 m/sec respectively. Estimate the exit air temperature assuming adiabatic conditions.*

Take $R \text{ for air} = 0.287 \text{ kJ/kg-K}$

$$\gamma = \frac{C_p}{C_v} = 1.4$$

and $C_p = 1.005 \text{ kJ/kg-K}$

Solution

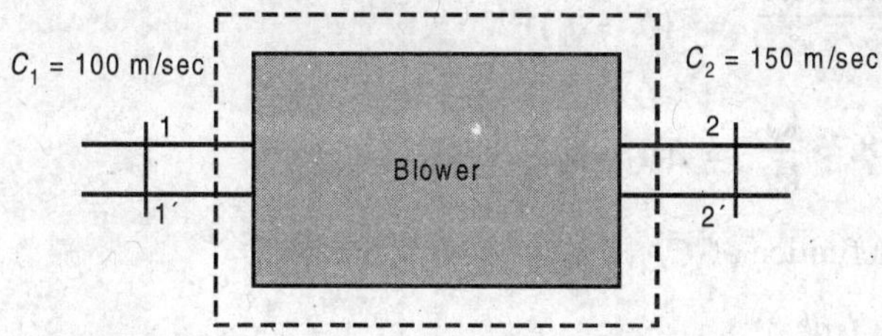

Fig. Ex. 2.26

$$\dot{m} = 2 \text{ kg/sec}$$

$$\frac{\delta W}{dt} = -30 \text{ kW} = -30 \text{ kJ/sec.}$$

$$t_2 = ? \; t_1 = 20°\text{C}$$

Since $\frac{\delta W}{dt}$ is given using the SFEE on time basis.

$$\dot{m}\left[\frac{C_1^2}{2} + gZ_1 + h_1\right] + \frac{\delta Q}{dt} = \dot{m}\left[\frac{C_2^2}{2} + gZ_2 + h_2\right] + \frac{\delta W}{dt}$$

$$\frac{\delta W}{dt} = -\dot{m}\left[\frac{C_2^2 - C_1^2}{2} + g(Z_2 - Z_1) + (h_2 - h_1)\right] + \frac{\delta Q}{dt}$$

Since adiabatic conditions are given, $\frac{\delta Q}{dt} = 0$

Neglecting the change in PE, we have,

$$\frac{\delta W}{dt} = -\dot{m}\left[\frac{C_2^2 - C_1^2}{2} + (h_2 - h_1)\right] \quad ...(1)$$

$$\frac{\delta W}{dt} = -\dot{m}\left[\frac{C_2^2 - C_1^2}{2} + (h_2 - h_1)\right]$$

$$-30\frac{\text{kJ}}{\text{sec}} = -2\frac{\text{kg}}{\text{sec}}\left[\frac{150^2 - 100^2}{2 \times 1000} + (h_2 - h_1)\right]$$

$$-30\frac{\text{kJ}}{\text{sec}} = -2\frac{\text{kg}}{\text{sec}}\left[\frac{12500}{2000} + (h_2 - h_1)\right]$$

$$-30\,\frac{\text{kJ}}{\text{sec}} = -2\,\frac{\text{kg}}{\text{sec}}\left[6.25\,\frac{\text{kJ}}{\text{kg}} + (h_2 - h_1)\right]$$

$$-30\,\frac{\text{kJ}}{\text{sec}} = -12.5\,\frac{\text{kJ}}{\text{sec}} - 2\,\frac{\text{kg}}{\text{sec}} \times (h_2 - h_1)$$

$$-30\frac{\text{kJ}}{\text{sec}} + 12.5\frac{\text{kJ}}{\text{sec}} = -2\frac{\text{kg}}{\text{sec}} \times (h_2 - h_1)$$

$$\frac{-17.5\ \text{kJ/sec}}{2\ \text{kg/sec}} = (h_2 - h_1)$$

$$\mathbf{8.75\,\frac{kJ}{kg} = (h_2 - h_1)}$$

We also know from the definition of C_P,

$$\left(\frac{dh}{dT}\right)_P = C_P$$

or $$dh = C_P \,.\, dT \quad (m = 1 \text{ for unit mass})$$

or $$\text{Change enthalpy } \Delta h = C_P\,\Delta T$$

$$h_2 - h_1 = C_P\,(t_2 - t_1)$$

$$8.75 = 1.005\,(t_2 - 20)$$

$$\mathbf{t_2 = 28.75°C}$$

Example 2.27 *In a water cooled compressor 0.6 kg of air is compressed/sec. Power required to run the compressor is 40 kW. Heat lost to the cooling water is 30% of input and 10% of input is lost in bearings and other frictional effects. Air enters the compressor of 1 bar and 30°C. If the changes in PE and KE are neglected, estimate the exit air temperature. Take C_p = 1.005 kJ/kg-K.*

Data : $\dot{m}$ = 0.6 kg/sec. $\qquad \dfrac{\delta W}{dt} = 40$ kW

$P_1 = 1$ bar $= 10^5$ N/m² $\qquad T_1 = 30°C$

$\Delta PE = 0,\ \Delta KE = 0 \qquad T_2 = ?$

Solution

We know that, $$\Delta H = m\,C_P\,(T_2 - T_1)$$

$$= 0.6 \times 1.005\,(T_2 - 30)$$

$$\Delta H = 0.603\,(T_2 - 30) \qquad ...(1)$$

Net heat lossses, $$= 40\% \text{ of input}$$

$$= \frac{40}{100} \times 40 \text{ kW}$$

$$\frac{\delta Q}{dt} = 0.4 \times 40 = 16 \text{ kW.}$$

SFEE is given by,

$$\dot{m}\left[\frac{C_1^2}{2}+gZ_1+h_1\right]+\frac{\delta Q}{dt} = \dot{m}\left[\frac{C_2^2}{2}+gZ_2+h_2\right]+\frac{\delta W}{dt}$$

Since $\Delta KE = 0$ and $\Delta PE = 0$

$$\dot{m}\,h_1+\frac{\delta Q}{dt} = \dot{m}\,h_2+\frac{\delta W}{dt}$$

$$\therefore \quad \frac{\delta W}{dt}-\frac{\delta Q}{dt} = \dot{m}\left(h_1-h_2\right)=\Delta H \qquad ...(2)$$

From (1) and (2)

$$40-16 = 0.603\,(T_2-30)$$

$$\mathbf{T_2 = 70^\circ\,C}$$

Example 2.28 *A room is fitted with 2 fans each consuming 0.2 kW power. There are three lamps in the room which are provided to heat the room each consuming 200 W. Ventilation air enters the room with the enthalpy of 85 kJ/kg and leaves the room with the enthalpy of 60 kJ/kg. The rate of air flow is 100 kg/hr. There are five persons in the room and heat generated by each person is 600 kJ/hr. Determine the rate at which the heat is to be removed by a room cooler, so that steady state is maintained in the room.*

Solution

Air flow rate = 100 kg/hr

Heat generated by each person = 600 kJ/hr = 0.1666 kJ/sec

∴ For 5 persons = 0.8333 kJ/sec

Since work input is given in kW using SFEE on time basis.

$$\dot{m}\left[\frac{C_1^2}{2}+gZ_1+h_1\right]+\frac{\delta Q}{dt} = \dot{m}\left[\frac{C_1^2}{2}+gZ_2+h_2\right]+\frac{\delta W}{dt}$$

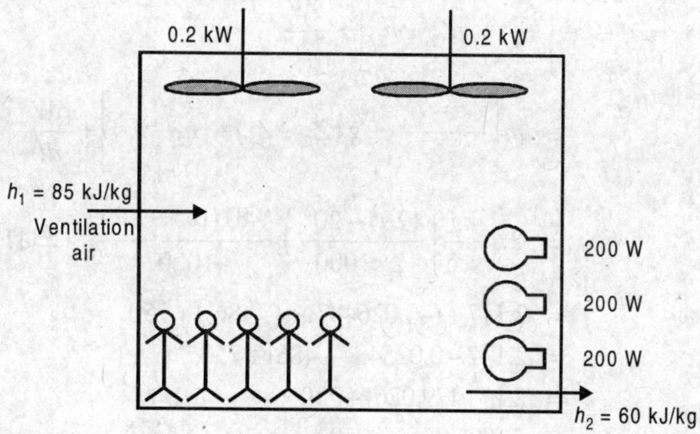

Fig. Ex. 2.28

Neglecting ΔKE and ΔPE.

$$\dot{m}\,[h_1-h_2]+\frac{\delta Q}{dt} = \frac{\delta W}{dt}$$

i.e., $\dot{m}\left[h_1 - h_2\right] + \left[\begin{pmatrix}\text{heat added due}\\ \text{to lamps}\end{pmatrix} + \begin{pmatrix}\text{heat generated}\\ \text{by 5 persons}\end{pmatrix} - \begin{pmatrix}\text{heat to be removed}\\ \text{by the cooler}\end{pmatrix}\right]$

$$= \text{Work input to the fans (which should be } -\text{ve)}$$

$$\therefore \frac{100}{3600}\frac{\text{kJ}}{\text{sec}}\left[85\frac{\text{kJ}}{\text{kg}} - 60\frac{\text{kJ}}{\text{kg}}\right] + \left[0.6\frac{\text{kJ}}{\text{sec}} + 0.833\frac{\text{kJ}}{\text{sec}} - X\right] = -0.4\frac{\text{kJ}}{\text{sec}}$$

$$0.6944 + [0.6 + 0.833 - X] = -0.4$$

$$0.6944 + 0.6 + 0.833 + 0.4$$

$\therefore$ **X = 2.574 kJ/sec. Heat to be removed by cooler**

Example 2.29 *In a steady flow device, the inlet and outlet conditions are given below. Determine the heat loss/ gain by the system.*

Property	*Inlet*	*Outlet*
Pressure (bar)	*10*	*0.15*
Specific volume (m^3/kg)	*0.206*	*8.93*
Specific enthalpy (kJ/kg)	*2827*	*2341*
Velocity (m/s)	*20*	*120*
Elevation m	*3.2*	*0.5*

The fluid flow rate through the device is 2.1 kg/s. The work output of the device is 750 kW.

Data: $Q = ?$

$$\frac{\delta W}{dt} = +750 \text{ kW}$$

Solution

Since work output is given in $\frac{\delta W}{dt}$, using SFEE on time basis,

$$\dot{m}\left[\frac{C_1^2}{2} + gZ_1 + h_1\right] + \frac{\delta Q}{dt} = \dot{m}\left[\frac{C_2^2}{2} + gZ_2 + h_2\right] + \frac{\delta W}{dt}$$

$$= \dot{m}\left[\frac{C_2^2 - C_1^2}{2} + g(Z_2 - Z_1) + (h_2 - h_1)\right] + \frac{\delta W}{dt}$$

$$= 2.1\frac{\text{kg}}{\text{sec}}\left[\frac{120^2 - 20^2}{2\times1000} + \frac{9.81(0.5-3.2)}{1000} + (2341 - 2827)\right] + 750$$

$$= 2.1\,[7 + (-0.026487) + (-486)] + 750$$

$$= 2.1\,[7 - 0.025487 - 486] + 750$$

$$= 2.1\,[-479.026] + 750$$

$$= -1005.9556 + 750$$

$$\frac{\boldsymbol{\delta Q}}{\textbf{dt}} = \textbf{-255.95562 kW}$$

Note: –ve sign implies that heat will flow out of system.

Example 2.30 *Air at a temperature of 15°C passes through a heat exchanger at a velocity of 30 m/c where its temperature is raised to 800°C. It then enters a turbine with the same velocity of 30 m/s and expands until the temperature falls to 650°C. On leaving the turbine, the air is taken at a velocity of 60 m/s to a nozzle where it expands until the temperature has fallen to 500° C. If the air flow rate is 2 kg/s calculate,*

(a) *Rate of heat transfer to the air in the heat exchanger.*

(b) *Power output from the turbine, assuming no heat loss, and*

(c) *Velocity of air at exit from nozzle, assuming no heat loss, take the specific enthalpy of air as $h = C_p T$, where $C_p = 1.005$ kJ/kg K, and T = Temperature.*

Solution

For heat exchanger:

$$q_{1-2} + \dot{m}\left(h_1 + \frac{C_1^2}{2} + gz_1\right) = w_{1-2} + \dot{m}\left(h_2 + \frac{C_2^2}{2} + gz_2\right)$$

Then

$$q_{1-2} = \dot{m}(h_2 - h_1) \; (\because C_1 = C_2) = \dot{m}\, C_P (T_2 - T_1)$$

$$q_{1-2} = 2 \times 1.005 \times (800 - 15)$$

$$\mathbf{q_{1-2} = 1577.85 \text{ kJ/s}}$$

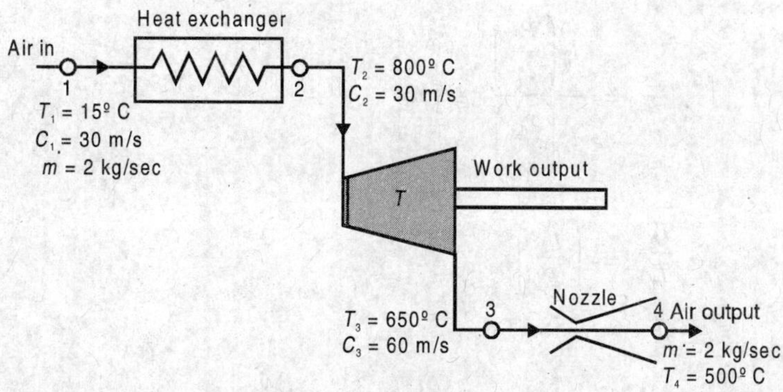

Fig. Ex. 2.30

For the turbine:

$$q_{2-3} + \dot{m}\left(h_2 + \frac{C_2^2}{2} + gz_2\right) = w_{2-3} + \dot{m}\left(h_3 + \frac{C_3^2}{2} + gz_3\right)$$

$\therefore$

$$w_{2-3} = \dot{m}\left[(h_2 - h_3) + \frac{C_2^2 - C_3^2}{2}\right] \qquad (\because q_{2-3} = 0)$$

$$= \dot{m}\left[C_P (T_2 - T_3) + \frac{C_2^2 - C_3^2}{2}\right]$$

$$= 2\left[1.005(800 - 650) + \left(\frac{30^2 - 60^2}{2 \times 1000}\right)\right]$$

$$\mathbf{W_{2-3} = 298.8 \text{ kW}}$$

For nozzle:

$$q_{3-4} + \dot{m}\left[h_3 + \frac{C_3^2}{2} + gz_3\right] = w_{3-4} + \dot{m}\left[h_4 + \frac{C_4^2}{2} + gz_4\right]$$

$$C_4^2 - C_3^2 = 2\times1000\times C_P(T_3 - T_4) \quad (\because q_{3-4} = 0 \text{ and } w_{3-4} = 0)$$

$$\therefore \quad C_4^2 - C_3^2 = 2\times1000\times C_P(T_3 - T_4) = 2\times1000\times1.005(650-500) = 301500$$

$$\therefore \quad \mathbf{C_4 = 552.36 \text{ m/s}}$$

Example 2.31 *Helium gas is expanded polytropically in a turbine, from 4 bar, 300°C to 1 bar such that final volume is 2.5 times the volume at inlet. Velocity of gas at exit is 50 m/s. What is the mass flow rate of gas required to produce 1 MW turbine output? How much is the heat transfer during the process ? Also determine exit area of turbine.*

Assume specific heat of helium = 5.193 kJ/kg K at constant pressure.

Solution

We know that

$$R = \frac{G}{M} = \frac{8.314}{4} = 2.0785$$

$$v_1 = \frac{RT_1}{P_1} = \frac{2.0785\times573}{400} = 2.9774\ \frac{\text{m}^3}{\text{kg}}$$

$$v_2 = 2.5\ v_1 = 7.4436 \text{ m}^3/\text{kg}$$

For Polytropic Process,

$$P_1v_1^n = P_2v_2^n$$

$$\therefore \quad n = \frac{l_n\left(\dfrac{P_2}{P_1}\right)}{l_n\left(\dfrac{v_1}{v_2}\right)} = 1.513$$

and

$$\frac{T_2}{T_1} = \left(\frac{P_2}{P_1}\right)^{\frac{n-1}{n}}$$

$$\therefore \quad T_2 = 358.11 \text{ K}$$

$$\text{Specific work} = -\int_1^2 v\,dP = \frac{n}{n-1}RT_1\left[1 - \frac{P_2v_2}{P_1v_1}\right]$$

$$= \frac{1.513}{0.513}\times2.0785\times573\times\left[1 - \frac{1\times2.5}{4}\right] = 1317.2 \text{ kJ/kg}$$

$$\Delta\text{KE} = \frac{1}{2}C_2^2 = 1250\ \frac{\text{J}}{\text{kg}} \text{ or } 1.25 \text{ kJ/kg}$$

$$q - w_s = \Delta h + \Delta\text{KE}$$

and

$$-\int v\,dP = w_s + \Delta\text{KE} \quad \text{as} \quad \Delta\text{PE} = 0$$

$$\therefore \quad \mathbf{W_s = 1317.2 - 1.25 = 1315.95 \text{ kJ/kg}}$$

and

$$q = 1315.95 + CP(T_2 - T_1) + 1.25 = 1315.95 + 5.193\,(358.11 - 573) + 1.25$$

$$\mathbf{q = 201.28 \text{ kJ/kg}}$$

Then

$$\dot{\mathbf{m}} = \frac{\mathbf{100}}{\mathbf{W_s}}\ \mathbf{kJ/s = 0.7599\ kg/s}$$

$$\therefore \quad \textbf{Exit area } \mathbf{A_2} = \frac{\dot{\mathbf{m}}\ \mathbf{v_2}}{\mathbf{C_2}} = \mathbf{0.1131\ m^2}$$

■■■

THEORY QUESTIONS

2.1. Write a note on 'Joule's Experiment' ?

2.2. State and explain the first law of thermodynamics for a closed system undergoing a cyclic change.

2.3. Derive an equation for energy transfer for a closed system.

2.4. Explain the First Law of Thermodynamics and hence show their cyclic integral of δQ and cyclic integral of δW are equal for a closed system.

2.5. State and explain the first law of thermodynamics for a closed system undergoing a process.

2.6. Prove that 'Internal Energy' is a property of a system.

2.7. Write a note on different forms of stored energies.

2.8. Define the term 'Enthalpy' (H).

2.9. Write a note on'Internal Energy' (U).

2.10. Distinguish between heat and internal energy.

2.11. Define specific heat at constant pressure and at constant volume for a perfect gas.

2.12. Define Adiabatic Index (g).

2.13. What is Perpetual Motion Machine of first kind (PMM-1) ?

2.14. Define a flow process and explain the terms Control Volume and control surface.

2.15. Explain the term flow work and discuss how it differs from the work done by a closed system.

2.16. Write a note on Flow work or Flow energy.

2.17. How is 'Steady flow system characterized'?

2.18. What are the conditions for a steady flow process?

2.19. State the conditions of First Law of Thermodynamics for a Steady Flow Processes.

2.20. Explain the essentials of non-flow and flow systems. Give suitable examples of each. What are advantages of steady flow system over non-flow system.

2.21. Derive the Steady Flow Energy Equation (SFEE).

2.22. Apply first law of thermodynamics to steady flow system and derive the equation of energy.

2.23. Derive Steady Flow Energy Equation on mass basis.

2.24. Explain the significance of $-\int v\,dP$

2.25. Explain the significance of $\int P dv$ in case of steady flow process and non-flow process.

2.26. Prove that, $-\int vdP = Q - \Delta H$.

2.27. Prove that, $\int Pdv = Q - \Delta U$.

2.28. Obtain the energy equations for the following open systems:

(a) Boiler (b) Turbine

(c) Compressor (d) Throttling process

(e) Nozzles (f) Condenser

2.29. Derive Continuity equation.

PROBLEMS FOR PRACTICE

2.1. A centrifugal air compressor delivers 15 kg of air/min. Air enters at I bar, I0 m/sec, 0.5 m³/kg. It is discharged at 7 bar, 80 m/sec and 0.15 m³/kg. During the process, its enthalpy increases by 160 kJ/kg and the air rejects 720 kJ/min of heat to the surroundings. Find,

(i) Power required to run the compressor.

(ii) Ratio of inlet to outlet pipe diameter. Neglect ΔPE.

2.2. For a steady flow system, following particulars are noted,

	Inlet	Exit
1. Pressure (bar)	1	5
2. Vol (m³/kg)	2	0.58
3. Velocity (m/sec)	20	30
4. Elevation (m)	0	2

If the mass flow rate of the working fluid is 1 kg/sec, find the power of the system. Is it a power absorbing or power producing system?

Assume that the steady flow system follows $Pv^{1.3} = C$.

2.3. One kg of working substance undergoes a reversible constant pressure process at 1.2 bar during which its volume changes from 1 m³ to 1.8 m³ and temperature changes from 50°C to 37°C. The sp. heat of a substance at constant pressure is given by $C_p = \left(1.1 + \frac{40}{t+30}\right)$ kg/kg°C where t is in °C. Find out

(a) Heat supplied (b) Work done

(c) Change in I.E. (d) Change in enthalpy

2.4. A fluid is contained in a cylinder piston arrangement. The piston is spring loaded and frictionless such that the pressure of the fluid varies with its volume according to the law $P = a + bV$. The internal energy of the fluid is given by $V = 30 + 3.2\ PV$ where U is in kJ. P in kPa, V in m³. The fluid undergoes a change of state from 150 kPa and 0.025 m³ to 450 kPa and 0.05 m³. Assuming no work other than that on the piston determine the direction and magnitude of work and heat transfer.

2.5. What at the rate of 10 kg/sec is compressed adiabatically from 5 bar to 50 bar in a steady flow process. Calculate the power required assuming that sp. volume of water being as 0.001 m³/kg which remains almost constant.

2.6. Steam enters a turbine steadily at 10 mPa and 550°C with a velocity of 60 m/s and leaves at 25 kPa with a dryness fraction of 0.95. A heat loss of 30 kJ/kg occurs during the process. The inlet area of the turbine is 150 cm² and exit area is 1400 cm². Determine : (1) mass flow rate of steam. (2) exit velocity (3) power output.

2.7. A closed system receives 200 kg of heat at constant volume. The system then undergoes a constant pressure process during which 150 kg of heat is rejected and 50 kJ of work is done on the system. Calculate how much of work is required to be done if the system is to be restored to its original state by means of an adiabatic process. If the initial internal energy is 50 kJ, determine the values of internal energy at the end of the constant volume and constant pressure processes.

2.8. In a reversible steady flow process 225 kW of heat is rejected per kg of the fluid passing through the system. During the process, pressure and volume change from 6 bar and 0.075 m³/kg to 2 bar and 0.2 m³/kg. Assuming that flow process follows the law PV = constant and ignoring changes in K.E. and P.E., determine the change in enthalpy of the fluid during the process.

2.9. Air flow steadily at the rate of 0.4 kg/sec through an air compressor, entering at 6 m/sec with pressure of 1 bar and sp. volume of 0.85 m³/kg and leaving at 4.5 m/sec with a pressure of 6.9 bar and sp. volume of 0.16 m³/kg. The sp. internal energy of the air leaving is 88 kJ/kg greater than that of the air entering, cooling water in the jacket surrounding the cylinder absorbs heat from the air at the rate of 59 kg/sec. Calculate the power required to drive the compressor and the inlet and outlet pipe areas.

2.10. In the turbine of a gas turbine unit the gases flow through the turbine at 17 kg/sec and the power developed by the turbine is 14 MW. The sp. enthalpies of the gases at inlet and outlet are 1200 kJ/kg and 360 kJ/kg respectively. Calculate the rate at which heat is rejected from the turbine. Find also the area of the inlet pipe given that sp. volume of the gases at inlet is 0.5 m³/kg.

2.11. A steam turbine operates under steady flow conditions, receiving steam and leaving the steam at the following rate.

	Inlet	Outlet
Pressure	15 bar	130 kPa
I.E. (kJ/kg)	2500	1500
Velocity (m/sec)	300	200
Elevation (m)	3	0
Volume (m³)	0.15	1.2 m³ /kg

2.12. A closed system undergoes a cycle 1–2–3–1. If $Q_{12} = 30$, $Q_{23} = 10$, $W_{12} = 5$, $W_{31} = 25$ and $\Delta E_{31} = 15$, determine W_{23} and ΔE_{23}.

2.13. The following data is available for a test on air compressor for an air flow rate of 6 kg/sec.

	Inlet	Outlet
Pressure	95 kPa	7.5 bar
Sp. Internal Energy (kJ/kg)	40	150
Sp. Volume (m³/kg)	0.95	0.162
Velocity of air (m/sec)	6	1

Estimate: (1) Power required to drive the compressor in kW. (2) Ratio of inlet pipe diameter to outlet pipe diameter.

2.14. A non-flow system undergoes a frictionless process according to the law $P = \frac{4.5}{v} + 2$ where P is in bar, V is in m³/kg. During this process volume changes from 0.12 m³/kg to 0.04 m³/kg and temperature increases by 133°C. The change in I.E. of fluid is given by $du = C_v\, dT$ where $C_v = 0.71$ kJ/kg K and dT is temperature change. Find out (1) Heat transfer, (2) Change in enthalpy. Assume fluid mass of 5 kg.

IMPORTANT POINTS TO BE REMEMBERED

1. *Joule experiment* : When a closed system undergoes any cyclic process, the cyclic integral of work is proportional to the cyclic integral of heat.
$$\Sigma \partial W = \Sigma \partial Q$$
2. *First law of thermodynamics* states that energy can neither be created nor be destroyed but energy can be converted from one form to the other. It is a law of conservation of energy.
3. *For closed system* $Q = \Delta E + W$ or $\Delta U + W$.
4. Stored energies are (a) Kinetic energy (b) Potential energy and (c) Internal energy.
5. Internal energy (U) is the energy stored in the molecular structure due to heat and work interactions.
6. *Kinetic energy KE* (K) is due to motion of the system $\frac{C^2}{2}$.
7. *Potential energy PE* (P) is the energy due to its position (greavitational force).
8. *Enthalpy H.* It is an extensive property and is a sum of internal energy (u) and P-V product. It is denoted by H or h. $H = U + PV$ or $h = u + p.v.$
9. *Specific heat* (C) is the amount of heat required to increase the temprature of unit mass of working substance (gas) through one degree (KI/kg. K) $Q = mC\,(t_2 - t_1)$.
10. *Specific heat at constant volume* (Cv) is the amount of heat required to raise the temperature of unit mass of a gass through one degree when the volume is constant.
11. *Specific heat at constant pressure* (Cp) is the amount of heat required to raise the temperature of unit mass of a gas through one degree when the pressure is constant.
12. *Adiabatic index* γ is the ratio of Cp.
13. *Perpetual motion machine of first kind (PMM-I)* is defined as a machine which will produce continuous work output without receiving any energy from any other system or the surroundings.
14. *Flow process.* Processes performed in the open systems are called as flow processes.
15. *Flow work or flow energy* is the work energy in the flow process required to push the fluid into and out of the system. (W_f).
16. *Steady flow system* is the system in which mass and energy flow rate is same at inlet and outlet of the system. These rates do not vary with time.
17. *Stead flow energy equation* is give by
$$Q = \Delta K + \Delta P + \Delta U + \Delta W_f + W_{sf}$$
18. *Work done* in non-flow process is $\int_1^2 P \cdot dv$ and is area unable the p-V diagram.
19. *Work done* in steady flow process is $-\int Vdp$ if ΔK and $\Delta P = 0$. It is the area behind the p-V diagram
20. *Continuity equation or law of conservation of mass is*
$$\rho AV = \text{constant } \rho \rightarrow \text{Density, } A\text{-area, } V\text{-velocity}$$
or
$$\frac{AV}{v} = \text{constant } v \text{ is the specific volume.}$$
21. First law limitations are
    (i) It is a law of energy equivalence.
    (ii) Does not consider direction of energy transformation.
    (iii) Does not consider grade of the energy or energy reservoirs.

CHAPTER 3

Second Law of Thermodynamics

3.1 INTRODUCTION

The first law of thermodynamics deals with the conversion of energy. However it does not specify the direction of the process. The second law of thermodynamics is about the direction, quality and quantity.

3.2 LIMITATIONS OF FIRST LAW OF THERMODYNAMICS

First-law states that, when a closed system undergoes any cyclic process, then the cyclic integral of work is equal to the cyclic integral of heat

i.e. $$\oint \delta W = \oint \delta Q$$

Thus the law merely states that work transfer during a cycle is equal to the heat transfer and does not place any restriction on the direction of heat and the work transfer. It does not specify whether the process is possible in a particular direction or not at all. According to this law it can be assumed that the energy transfer can take place in either direction, since it does not specify the direction of energy transfer. But by practical experience it is observed that, even if a proposed cycle satisfies the first law, it doesn't ensure that the cycle will occur actually. This resulted in the formulation of *Second Law of Thermodynamics*. Thus a cycle will proceed only if both I and II laws of thermodynamics are satisfied.

Thus the second law involves the fact that processes will proceed in a certain direction but not in the opposite direction, for example,

(i) Consider the Joule's experiment in which the fall of weight W rotates the paddle wheel and increases the temperature of water. Here work is converted into heat. However reverse of this process is not possible i.e., by heating water, paddle wheel will not rotate and lift the weight. This shows that the system can operate in a cyclic process only when work is done on the system and heat is rejected out of the system. But it cannot work in a cyclic process when both work and heat transfer are positive even though such a process will not violate the first law of thermodynamics.

(ii) Consider hot tea in a cup, here heat flows from tea to the cooler surroundings. Once it is cooled, it can never be heated by the addition of heat from the cooler surroundings, without doing any work on it.

(iii) A cup of ice cream when kept in atmosphere absorbs heat and melts. But it will not solidify giving back heat without doing any work on it.

(iv) When a vehicle is stopped by applying mechanical brakes, the brakes get hot. The I.E. of the brake increases by an amount equal to the decrease in K.E. of the vehicle. According to the first law, the hot brake on cooling, should give back its increase in I.E. to the wheels of the vehicle, causing it to regain its speed. This never happens in practice. From the above examples it is obvious that, even if the first law is satisfied the processes will proceed in a particular direction while they are impossible in the opposite direction. Thus the first law of thermodynamics is necessary but not sufficient condition for the processes to take place.

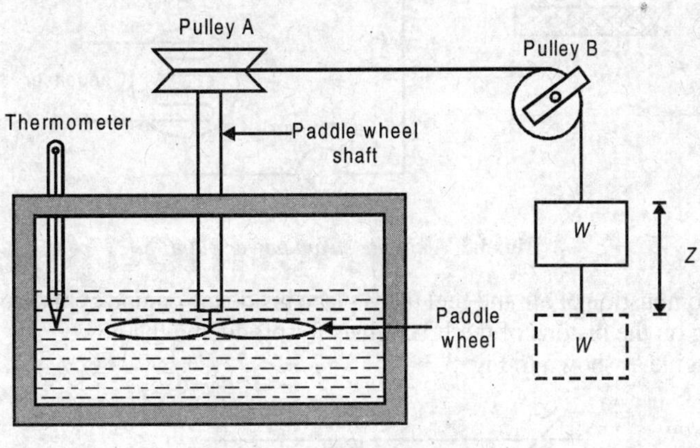

Fig. 3.1. *Joule's experiments*

It has also been practically found that all forms of energies are not equally convertible into work and the first law is silent about the extent of conversion of energy. It is necessary to study *Second Law of Thermodynamics* in this regard.

Points to Remember

- First Law is silent about the direction.
- It gives no indication about extent of conversion of energy.

3.3 HEAT ENGINE

Heat engine converts heat energy into mechanical energy. Heat engine works on thermodynamic cycle like Otto, Diesel, Rankine in which there is a net heat transfer to the system and a net work transfer from the system.

As we have studied, heat engines are broadly classified into two types,

(i) External combustion engines.

(ii) Internal combustion engines.

Figure 3.2 shows a simple steam power plant. In this power plant an amount of heat Q_1 is supplied from the furnace (Also called as high temperature reservoir) to the water in the boiler drum. An amount of heat Q_2 is rejected to the low temperature reservoir, like a coolant in the condenser and in doing so, an amount of work W will be produced.

The thermal efficiency η_{Th} of the heat engine is defined as,

$$\eta_{Th} = \frac{\text{Energy output}}{\text{Energy input}} \quad \text{or} \quad \eta_{Th} = \frac{\text{Energy sought for}}{\text{Energy that costs}}$$

$$\eta_{Th} = \frac{\text{Work output}}{\text{Heat supplied}} \quad \text{or} \quad \eta_{Th} = \frac{W}{Q_1} = \frac{Q_1 - Q_2}{Q_2}$$

$$\eta_{Th} = 1 - \frac{Q_2}{Q_1}$$

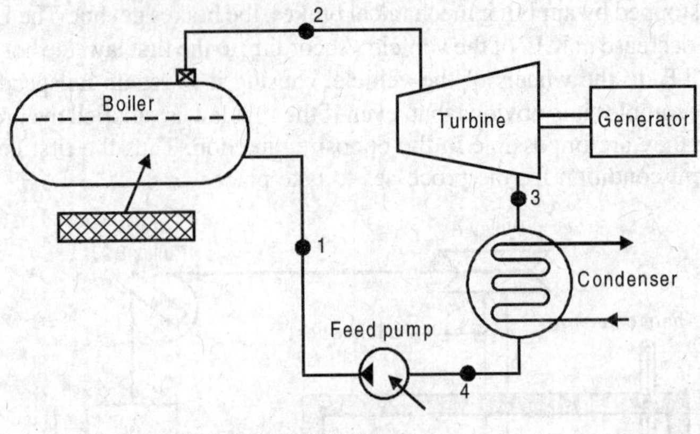

Fig. 3.2. *A simple steam power plant*

In case of IC engines combustion of air and fuel takes place inside the engine cylinder. The products of combustion will be directly acting on the pistons of the IC engines for producing the power:

Now consider an IC engine as shown in Fig. 3.3.

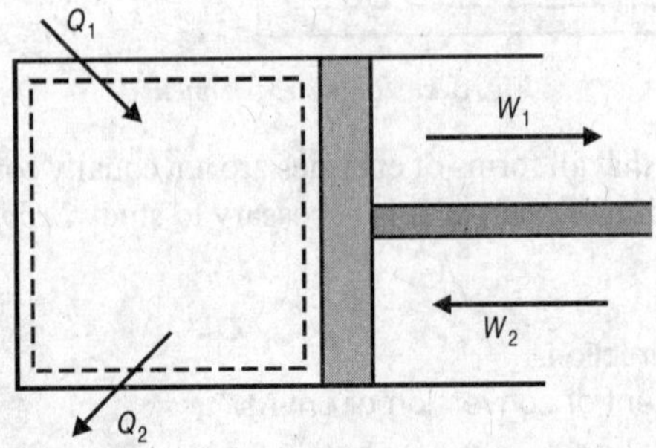

Fig. 3.3

Let Q_1 is the heat transferred to the system
Q_2 is the heat transferred from the system
W_1 is the work done by the system
W_2 is the work done on the system

Then, net heat transfer in a cycle $= Q_{net} = Q_1 - Q_2$ and

Net work transfer in a cycle $= W_{net} = W_1 - W_2$.

According to the first law of thermodynamics,

$$\oint \delta W = \oint \delta Q$$

or

$$W_{net} = Q_{net}$$

∴ We can write,

$$\mathbf{W_1 - W_2 = Q_1 - Q_2}$$

3.4 HEAT RESERVOIRS

"Heat reservoir is a body of infinite heat capacity, which is capable of absorbing or rejecting any quantity of heat without suffering any change in any of its thermodynamic properties".

A heat reservoir which supplies an amount of heat Q_1 to the heat engine operating in heat engine cycle is called the High Temperature Reservoir (HTR) or Source.

The heat reservoir to which an amount of heat Q_2 is rejected by the heat engine is called the Low Temperature Reservoir (LTR) or Sink.

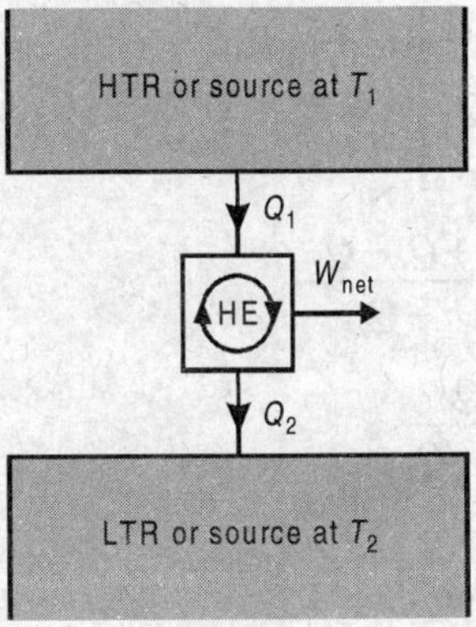

Fig. 3.4

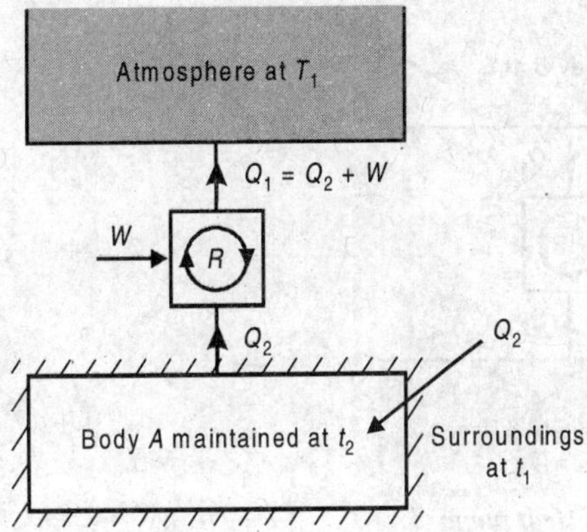

Fig. 3.5

3.5 REFRIGERATOR

"It is a device, which produces and maintains the temperature below atmospheric temperature" or

"It is a device which operating in cycle, maintains an enclosed space at a temperature lower than the temperature of the surroundings".

Let a body *A* is to be maintained at a temperature t_2, which is lower than the atmospheric temperature t_1. Even though the body, *A* is insulated, there will be a heat leakage Q_2 into the body because of the temperature difference between the body and the surroundings. In order to maintain the body *A* at constant temperature t_2, same amount of heat is to be removed at the same rate, at which it is leaking into the body.

A refrigerator is a device which operating in a cycle, removes heat Q_2 from the body a t_2 and transfers this heat to the atmosphere at t_1 by consuming an amount of work *W*.

In the refrigeration cycle, body *A*, Q_2 and *W* are of prime importance. Just like the efficiency of heat engine cycle, here a parameter called *co-efficient of performance* is important. The copies defined by,

$$[\text{COP}]_{\text{ref.}} = \frac{\text{Desired effect}}{\text{Work output}} = \frac{Q_2}{W} = \frac{Q_2}{Q_1 - Q_2} \quad \ldots(1)$$

3.6 HEAT PUMP

It is just opposite to the refrigerator i.e., "It is a device operating in a cycle, maintains a body say *B* at a temperature higher than the temperature of the surroundings".

Because of the temperature difference between the body and the surroundings there will be a heat leakage Q_1 from the body. The body will be maintained at constant temperature t_1, if heat is supplied into the body at the same rate at which heat leaking out of the body.

A heat pump is a device which operating in a cycle extracts heat from the atmosphere and transfers an amount of heat Q_1 to the body *B* and maintains the body *B* at constant temperature t_1.

In a heat pump, body *B*, Q_1 and *W* are of prime importance.

$$[\text{COP}]_{\text{HP}} = \frac{\text{Desired effect}}{\text{Work output}} = \frac{Q_1}{W}$$

$$= \frac{Q_1}{Q_1 - Q_2}$$

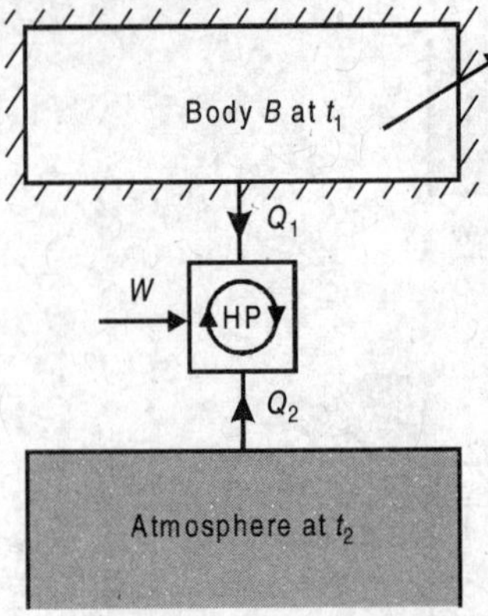

Fig. 3.6. *Heat pump*

From Eqs (1) and (2) we can write,

$$[COP]_{HP} = [COP]_{ref} + 1$$

i.e.

$$\frac{Q_1}{Q_1 - Q_2} = \frac{Q_2}{Q_1 - Q_2} + 1$$

$$\frac{Q_1}{Q_1 - Q_2} = \frac{Q_2 + Q_1 - Q_2}{Q_1 - Q_2}$$

$$\frac{Q_1}{Q_1 - Q_2} = \frac{Q_1}{Q_1 - Q_2}$$

LHS = RHS

$$\mathbf{[COP]_{HP} = [COP]_{ref} + 1}$$

3.7 STATEMENTS OF SECOND LAW OF THERMODYNAMICS

There are two important statements of second law. They are,

(i) Kelvin Planck Statement, (ii) Clausius Statement.

(i) Kelvin Planck's Statement

"It is impossible for a heat engine to produce net work in a cycle, if it exchanges heat with a single reservoir".

or

"Heat engine cannot produce work output, if it exchanges heat with a single reservoir".

We have seen that, $\eta_{Th} = \frac{\text{Energy output}}{\text{Energy input}} = \frac{W}{Q_1}$

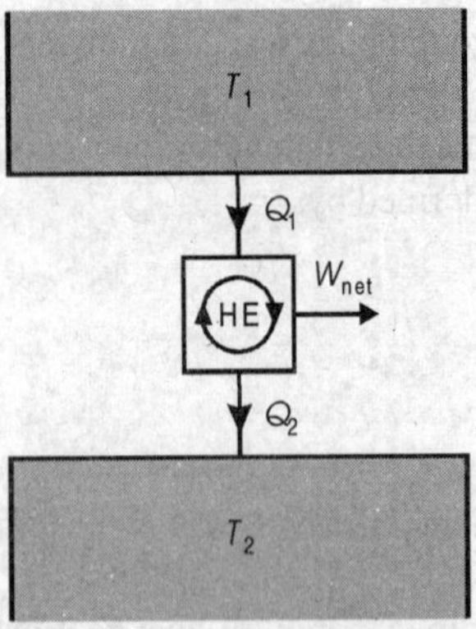

Fig. 3.7. *Heat Engine producing W_{net} by Exchanging Heat with 2 Reservoirs*

Fig. 3.8. *PMM-2*

By practical experience, it is noted that total heat supplied cannot be converted into useful work W i.e., $Q_1 \neq W$. A heat engine can never be 100% efficient. But $W < Q_1$, therefore there has to be heat rejection i.e., $Q_2 > 0$.

Therefore for a heat engine to produce network in a cycle, it has to exchange heat with two reservoirs. (Ref. Fig. 3.7)

But if, $Q_2 = 0$, then $Q_1 = W$ or $\eta_{Th} = 100\%$, the heat engine will produce network in a complete cycle by exchanging heat with only one reservoir, thus it violates Kelvin Planck's statement. Such a heat engine is called a *Perpetual Motion Machine of the second kind. (PMM-2)* Ref. Fig. 3.8. A PMM-2 is impossible and it is just a conceptual engine. All the attempts made so far to make such a machine have failed. Thus they show the validity of Kelvin Planck's, statement.

(ii) Clausius Statement

We know that, heat always flow from a hot body to a cold body. The reverse process never occurs by itself.

Clausius statement is given as, "It is impossible to construct a device, which operating in a cycle will produce no effect other than transfer of heat from a low temperature body to a high temperature body". Or

"Heat cannot flow by itself from a cold body to a hot body, in order to achieve this some work must be expended".

For example, Refrigerator—in this case heat is removed from the cold body A and transferred to the atmosphere but by consuming work input W.

3.8 EQUIVALENCE OF KELVIN PLANCK AND CLAUSIUS STATEMENTS

We have studied, the Kelvin Plank and Clausius statement as follow.

(i) Kelvin Planck's Statement. "Far a heat engine to produce W_{net} it has to exchange heat with two reservoirs".

(ii) Clausius Statement. "Heat cannot flow by itself from a cold body to a hot body. In order to achieve this some work must be expended".

At first sight these two statements will appear to be two different statements and are unconnected. But it can be easily shown that, these are the two parallel statements of second law and are equivalent in all respects.

The equivalence of the two statements will be proved if we prove that violation of one statement results into the violation of the other.

(a) Violation of Clausius statement leads into the violation of Kelvin Planck's statement. Consider a cyclic heat pump P, which transfers heat from a LTR at t_2 to HTR at t_1, without consuming work input (i.e., $W = 0$), thus violating Clausius statement.

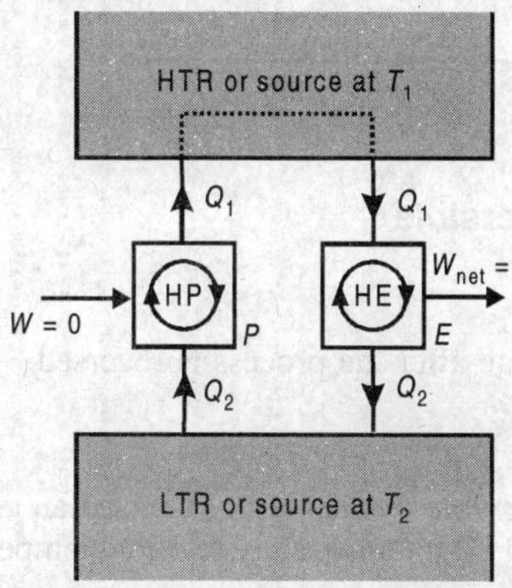

Fig. 3.9. *Violation of Clausius Statement*

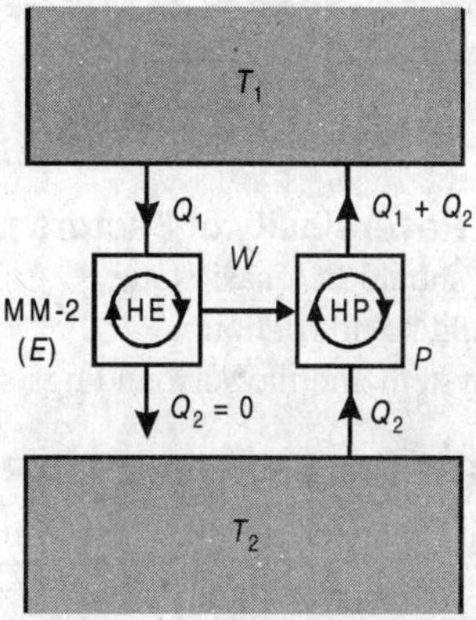

Fig. 3.10. *Violation of Kelvin–Planck Statement*

Now, let us assume a cyclic heat engine E which also operates between the same two reservoirs at t_1 and t_2 respectively. The rate of working of the heat engine is such that it draws an amount of heat Q_1 from HTR equal to that discharged by the heat pump. Then the HTR may be eliminated and heat Q_1 discharged by the heat pump may be directly fed to the heat engine. So, the heat pump P and heat engine E acting together will form a heat engine, operating in cycles and produce net work by exchanging heat with one reservoir. This violates the Kelvin Planck's statement.

(b) Violation of Kelvin Planck Statement Leads into the Violation of Clausius Statement. Let us consider a PMM-2 (E), which produces W_{net} in a cycle by exchanging heat with only one reservoir at t_1 and thus violates Kelvin Planck's statement (ref. Fig. 3.10).

Now, let us assume a cycle heat pump P extracting heat Q_2 from LTR at t_2 and supplying heat to HTR at t_1 by consuming work W equal to that PMM-2 supplies in a complete cycle. So, E and P together will form a heat pump and producing the complete effect of transferring heat from LTR to HTR, without any external aid.

Points to Remember

⇨ At first sight the two statements appear to be different but these two are the two parallel statements and they are equivalent in all respects.

3.9 REVERSIBILITY AND IRREVERSIBILITY

Quasi—means nearly or almost. So, quasi-static process means nearly static process or a process which proceeds with extreme slowness. A quasi-static process proceeds from one equilibrium state to another equilibrium state till the end of process. All the states passed through by the system, during the process are all equilibrium states.

A quasi-static process is represented on the *PV*-diagram by a continuous curve and if we carry out the process in the reverse direction, then it is possible to retrace the same path. Hence it is known as a *Reversible process.*

Whereas in case of *Irreversible process* only end states are equilibrium states and all the intermediate states are non-equilibrium states and is represented by means of a dotted curve as shown in Fig. 3.11.

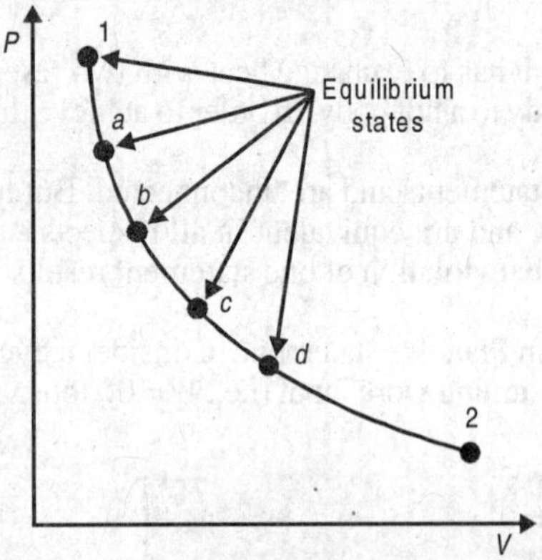

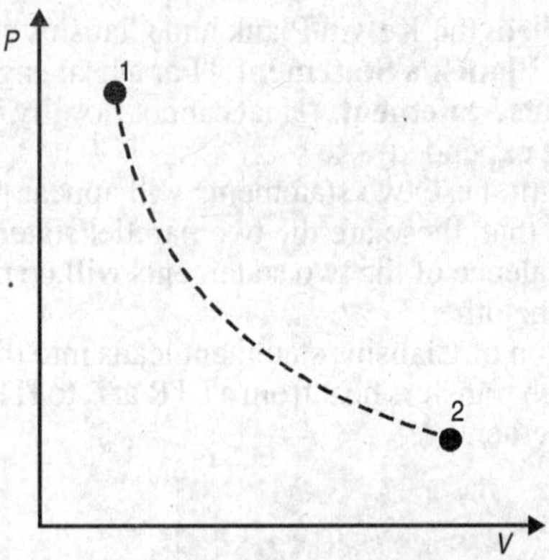

Fig. 3.11

Conditions of Reversibility or Factors Making the Process Irreversible

(i) A process should be quasi-static

(ii) There should be no friction

(iii) Both the system and the surroundings should restore their initial state after the process is reversed.

3.10 CARNOT ENGINE AND CARNOT CYCLE

Carnot, a French Engineer, was the first to introduce the concept of reversible cycle. Carnot devised an engine, which is named after him as Carnot engine. Carnot engine works between high temperature and low temperature reservoirs as shown in Fig. 3.12.

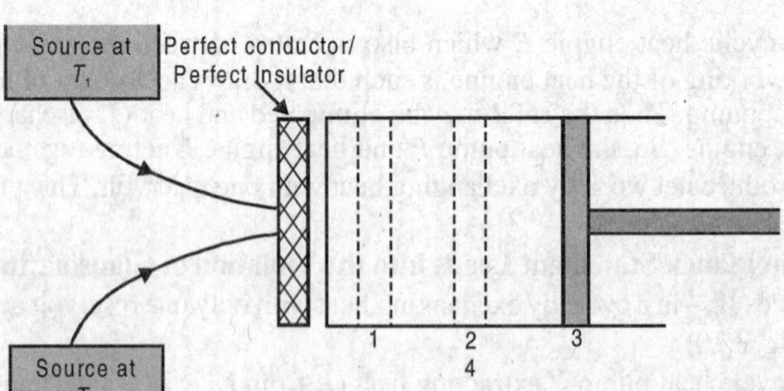

Fig. 3.12. *Carnot's engine*

Carnot engine works on carnot cycle. It consists of an alternate series of two reversible isothermal and two reversible adiabatic processes. Since each process is reversible, the Carnot cycle as a whole is reversible.

Carnot cycle (Figs 3.12 and 3.13) is independent of nature of working fluid. It can work with any working substance like gas, vapour or any other working substance. Let us assume that the working fluid of the engine is a gas.

Figure 3.12 shows the proposed reversible engine working between source at T_1 and sink at T_2. It consists of a cylinder, which has a piston working in it without friction. The walls of the cylinder and piston are assumed to be perfect insulators of heat. The cylinder head is so arranged that it can be a perfect conductor and perfect insulator alternatively.

During the beginning of the stroke the cylinder head is made perfect conductor and the heat source at T_1 is brought in contact with the gas in the cylinder. The gas expands isothermally at T_1 from state point (1) to state point (2). During the expansion process, heat supplied Q_1 is used in the expansion of the gas.

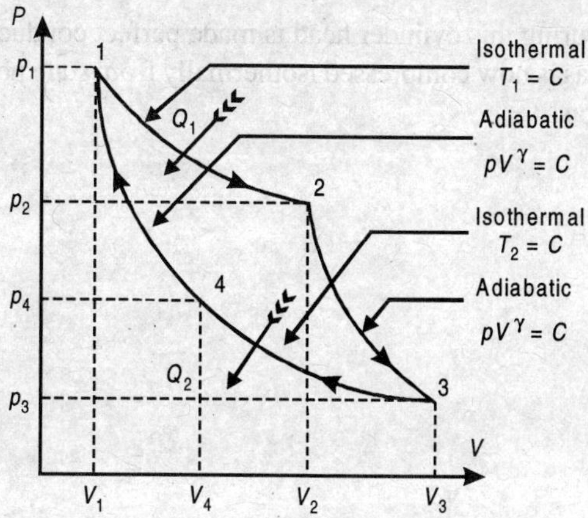

Fig. 3.13

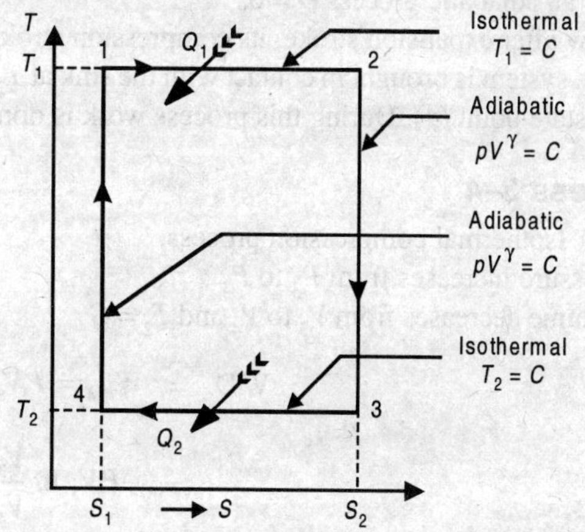

Fig. 3.14

Process 1–2

It is a Isothermal expansion process.

Law for the process is $PV = C$

Pressure falls from P_1 to P_2

Volume increases from V_1 to V_2 and, $T = C$.

$$\text{Work done} = \int_1^2 PdV$$

$$W_{1-2} = P_1V_1 \ln \frac{V_2}{V_1}$$

Since, $$P_1V_1 = mRT_1$$

$$W_{1-2} = mRT_1 \ln \frac{V_2}{V_1}$$

and for an Isothermal process, $$Q_1 = W_{1-2} = mRT_1 \ln \frac{V_2}{V_1} \quad \ldots(1)$$

Note. For derivation refer Ideal gases work done during a Isothermal process.

When the state point (2) is reached, the cylinder head is made perfect insulator and the gas in the cylinder is allowed to expand till the end of the stroke by following $PV^\gamma = C$.

Process 2–3

It is an adiabatic expansion process.

Pressure falls from P_2 to P_3
Volume increases from V_2 to V_3
Temperature falls from T_1 to T_2

$$\text{Work done} = \int PdV$$

$$W_{2-3} = \frac{P_2V_2 - P_3V_3}{\gamma - 1} = \frac{mR(T_1 - T_2)}{\gamma - 1}$$

For an adiabatic process $Q = 0$.

Now after expansion stroke, its compression stroke starts. During this cylinder head is made perfect conductor and the system is brought in contact with the sink at T_2 and the gas is now compressed isothermally from state point (3) to state point (4). During this process work is done on the gas.

Process 3–4

It is an Isothermal compression process.

Pressure increases from P_3 to P_4
Volume decreases from V_3 to V_4 and $T_2 = C$

$$\text{W.D.} = W_{3-4} = P_3V_3 \ln \frac{V_3}{V_4}$$

i.e.

$$= -P_3V_3 \ln \frac{V_3}{V_4}$$

$$W_{3-4} = -mRT_2 \ln \frac{V_3}{V_4}$$

– ve sign implies work is done on the gas during compression.

For isothermal process $\quad Q = W$

$$\therefore \quad Q_{3-4} = Q_2 = -mRT_2 \ln \frac{V_3}{V_4} \quad \ldots(2)$$

Again the cylinder head is made perfect insulator and the gas is compressed adiabatically to state point (1).

Process 4–1

It is an adiabatic compression process.

Pressure increases from P_4 to P_1
Volume decreases from V_4 to V_1
Temperature increases from T_2 to T_1

$$\text{W.D.} = \frac{(P_4V_4 - P_1V_1)}{\gamma - 1} = \frac{-(P_1V_1 - P_4V_4)}{\gamma - 1}$$

and $Q = 0$ for an adiabatic process.

We know that,

$$\eta_{Th} = \frac{\text{Work done}}{\text{Heat applied}} = \frac{W}{Q_1} = \frac{Q_1 - Q_2}{Q_1}$$

$\therefore$ From (1) and (2)

$$\eta_{Th} = \frac{mRT_1 \ln \frac{V_2}{V_1} - mRT_2 \ln \frac{V_3}{V_4}}{mRT_1 \ln \frac{V_2}{V_1}}$$

Note that the ratio $$\frac{V_2}{V_1} = \frac{V_3}{V_4}$$

$$\therefore \quad \eta_{Th} = \frac{mR \ln \frac{V_2}{V_1}(T_1 - T_2)}{mR \ln \frac{V_2}{V_1} \cdot T_1}$$

$$\boldsymbol{\eta_{Th} \text{ or } \eta_{Carnot} = \frac{T_1 - T_2}{T_1}}$$

Note. Carnot cycle is a reversible cycle. If the processes are carried out in anticlockwise direction then it will work as heat pump.

For all the four process, the cycle will be completed when the isothermal expansion and isothermal compression ratios are same. The proof for this is given below.

From the diagram we get

$$\frac{P_1}{P_2} = \frac{V_2}{V_1} = \text{Isothermal expansion ratio}$$

Similarly $$\frac{P_3}{P_4} = \frac{V_4}{V_3} = \text{Isothermal compression ratio}$$

For isentropic expansion process 2–3

$$\frac{P_2}{P_3} = \left(\frac{V_3}{V_3}\right)^{\gamma}; \qquad \frac{T_2}{T_3} = \left(\frac{V_4}{V_1}\right)^{\gamma-1}$$

For isentropic compression process, 4–1

$$\frac{P_1}{P_4} = \left(\frac{V_4}{V_1}\right)^{\gamma}; \qquad \frac{T_1}{T_4} = \left(\frac{V_4}{V_1}\right)^{\gamma-1}$$

$$\therefore \quad \left(\frac{V_3}{V_2}\right)^{\gamma-1} = \left(\frac{V_4}{V_1}\right)^{\gamma-1} \quad \text{as } T_1 = T_2 \text{ and } T_3 = T_4$$

$$\therefore \quad \frac{V_3}{V_2} = \frac{V_4}{V_1}$$

or $$\frac{V_3}{V_4} = \frac{V_2}{V_1}$$

$$r_{comp} = r_{exp} \quad \text{The required condition for the carnot cycle.}$$

3.11 CARNOT THEOREM

Two cyclic heat engines E_A and E_B operating between the same source and sink out of which E_B is reversible (Fig. 3.15).

Carnot's theorem states that "*No engine operating between two thermal reservoirs can have an efficiency more than that of a Carnot's engine operating between the same two reservoirs*".

That is *"Carnot's engine will be more efficient than any other engine operating between the same temperature reservoirs"*.

Let two heat engines E_A and E_B operate between the source at temperature T_1 and the sink at temperature T_2 as shown in Fig. 3.15.

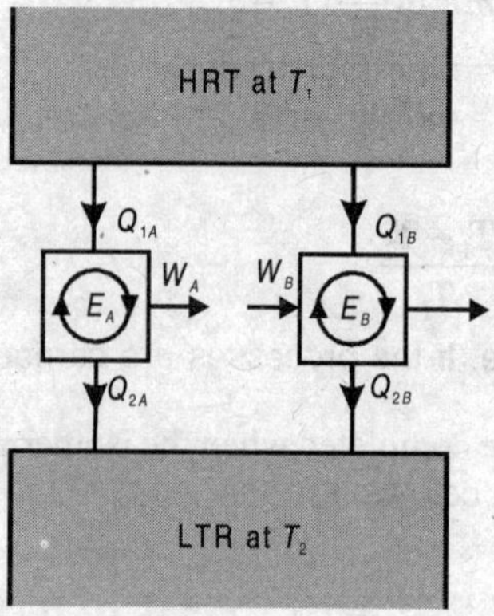

Fig. 3.15

Let E_A be any irreversible engine and E_B be any reversible heat engine. We have to prove that the efficiency of E_B is more than that of E_A. Now, let us assume that this is not true and η_A is greater than η_B. Let the rates of heat transfer are such that,

$$Q_{1A} = Q_{1B} = Q_1$$

and since

$$\eta_A > \eta_B$$

$$\frac{W_A}{Q_{1A}} > \frac{W_B}{Q_{1B}}$$

$\therefore$

$$W_A > W_B$$

Now, let E_B be reversed, since E_B is a reversible heat engine.

Note. Here the magnitudes of heat and work quantities will remain the same, but their directions will be reversed as shown in Fig. 3.16.

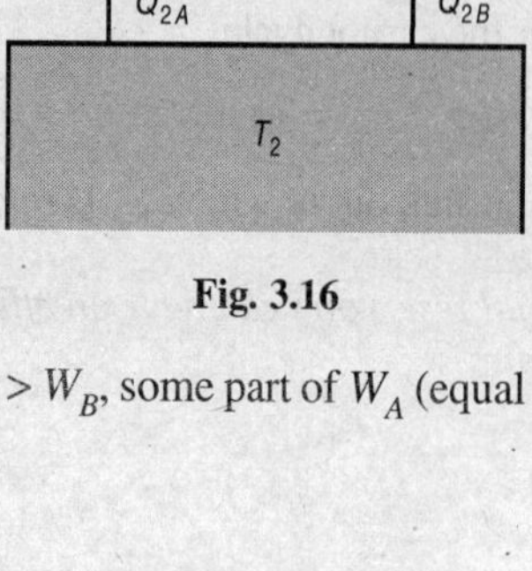

Fig. 3.16

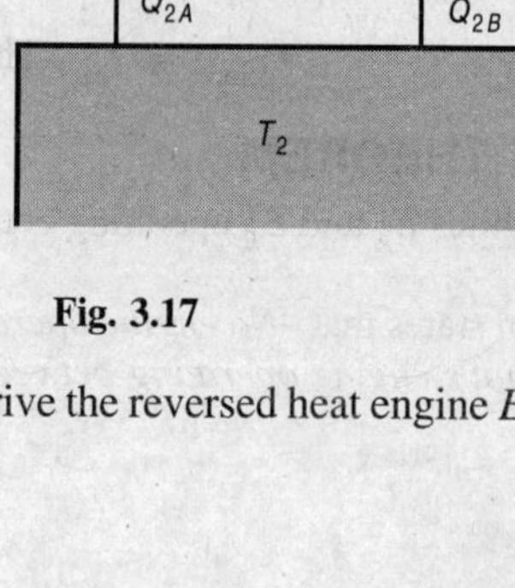

Fig. 3.17

Since $W_A > W_B$, some part of W_A (equal to W_B) may be fed to drive the reversed heat engine E_B.

Since $Q_{1A} = Q_{1B} = Q_1$ the heat discharged by E_B (reversed heat engine B) may be supplied to E_A. The source may therefore be eliminated as shown in the Fig. 3.17.

The net result is that E_A and E_B together will constitute a heat engine, which operating in a cycle, produces, network $(W_A - W_B)$ while exchanging heat with a single reservoir at T_2.

Thus violates the Kelvin Planck's statement of the second law. Hence the assumption than $h_A > h_B$ is wrong.

$$\therefore \qquad \eta_B \geq \eta_A$$

3.12 REVERSIBLE CYCLE

If a process can occur in a reverse order, and if the initial state all energies transferred or transformed during the process can be completely restored in both system and environment, the process is called *reversible.* If a process really reversible, then no after effects or changes are evident in the system or in the environment when the process occurs in the forward and then in the reverse direction. An irreversible process that is not reversible. A quasi-static process implies an infinitely slow process since all potential differences acting on the system are infinitesimally small. Such a process may be thought of as an infinite succession of equilibrium states. It may be stopped at any time and made to proceed in the opposite direction, thereby reversing the original process in every detail and restoring the system and environment to its original state.

We know that the thermodynamic cycle is defined as a cyclic process end states of which are same. If all the processes in a cycle are reversible process, then the cycle is called a *reversible cycle.*

Reversible process and consequently reversible cycle is an ideal one. There are no losses in the form of friction, temperature differences for heat transfer to take place. Presence of fraction makes the process mechanically irreversible and the temperature difference for heat transfer during the process makes the process thermally irreversible.

Therefore reversible cycle is an ideal cycle.

Engines working on reversible cycle are called reversible engines and we have seen that:

(a) No heat engine operating in cycles between two reservoirs at different temperatures can have a greater efficiency than a reversible engine operating between the same two reservoirs and

(b) All reversible engines operating between two reservoirs at given temperatures have the same efficiency.

Carnot cycle is a reversible cycle and its efficiency depends only on the temperature.

In reversible engines, heat is received while the working substance is at the same constant temperature as the source and heat is rejected while the working substance is at the same constant temperature as the sink.

3.13 POINTS TO REMEMBER

$$\eta_{\text{Carnot}} = \frac{T_1 - T_2}{T_1}$$

From this equation we can see that, efficiency can be increased either by increasing T_1 or by decreasing T_2. But in case of IC engines since T_2 is the temperature of exhaust gases, hence it cannot be decreased less than a certain value when high temperatures are used, then the metals used in the construction of engine cylinder, cylinder head etc. will not withstand high temperature stresses and will lose some of the properties. So if we have to increase T_1, then the metals which can withstand high temperature stresses are to be used in the construction of the engine even though they are costly.

(a) It is very very difficult to construct, operate and to measure the efficiency of carrot engine. It is not possible to construct such an engine. Since during the same stroke of the piston, isothermal and adiabatic processes are to be achieved.

For isothermal process the engine has to run very slowly so that the temperature during the process is maintained constant. Such a slow speed is not desirable as the size of the engine operating at very low speed is very large. The adiabatic process needs complete insulation and the engine has to run very fast.

As in this cycle isothermal and adiabatic processes take place during the same stroke, the piston has to move very slowly for the first half stroke and very fast for the second half stroke, but this is not possible to design.

(b) The cylinder head is perfect conductor for half part of the stroke and is a perfect insulator during the remaining stroke. This is also very difficult to obtain in actual practice.

(c) It is not possible to perform frictionless reversible process.

LIST OF FORMULAE

1. Thermal efficiency of the heat engine (any engine)

$$\eta_{Th} = \frac{\text{Energy output}}{\text{Energy input}}$$

$$= \frac{\text{Work done}}{\text{Heat supplied}} = \frac{W}{Q_1} = \frac{Q_1 - Q_2}{Q_1}$$

2. $$[\text{COP}]_{ref} = \frac{Q_2}{W} = \frac{Q_2}{Q_1 - Q_2} = \frac{T_2}{T_1 - T_2}$$

3. $$[\text{COP}]_{HP} = \frac{Q_1}{W} = \frac{Q_1}{Q_1 - Q_2} = \frac{T_1}{T_2 T_2}$$

Also, $$[\text{COP}]_{HP} = [\text{COP}]_{ref} + 1$$

4. $$\eta_{Carnot} = \frac{T_1 - T_2}{T_1} = 1 - \frac{T_2}{T_1} \quad \text{...(a)}$$

Also $$\eta_{any\ engine} = \frac{Q_1 - Q_2}{Q_1} = 1 - \frac{Q_2}{Q_1} \quad \text{...(b)}$$

For maximum efficiency

$$\eta_{any\ engine} = \eta_{Carnot}$$

$$\therefore \quad \frac{Q_2}{Q_1} = \frac{T_2}{T_1}$$

5. $$\eta_{Carnot} > \eta_{any\ engine}$$

SOLVED EXAMPLES

Example 3.1 *What will be the maximum efficiency of a heat engine operating between 227° C and 27° C?*

Data:

$$T_1 = 227 + 273 = 500\ K$$

$$T_2 = 27 + 273 = 300\ K$$

$$\eta_{Carnot} = ?$$

Solution

Maximum efficiency or Carnot engine is given by,

$$\eta_{Carnot} = \frac{T_1 - T_2}{T_1} = \frac{500 - 300}{500} = 0.4$$

$$\therefore \quad \eta_{Carnot} = 40\%$$

Example 3.2 *What would be the COP of the carnot refrigerator and heat pump working between temperature limits of 27° C and 600 K?*

Find: COP of the carnot refrigerator and heat pump.

Data: $T_1 = 600$ K, $T_2 = 273 + 27 = 300$ K

Solution

We know that,

$$[\text{COP}]_{\text{ref}} = \frac{Q_2}{Q_1 - Q_2} = \frac{T_2}{T_1 - T_2} = \frac{300}{600-300}$$

$$\therefore \quad \mathbf{[COP]_{ref} = 1}$$

$$[\text{COP}]_{\text{HP}} = \frac{Q_2}{Q_1 - Q_1} = \frac{T_2}{T_1 - T_2} = \frac{600}{600-300}$$

$$\therefore \quad \mathbf{[COP]_{HP} = 2}$$

Alternatively

$$[\text{COP}]_{\text{HP}} = [\text{COP}]_{\text{ref}} + 1 = 1 + 1 = 2$$

Example 3.3 *A heat engine operates on Carnot cycle between source and sink temperatures 227°C and 27°C respectively. If the heat engine receives 400 kJ from the source, find the net work done, heat rejected to the sink and efficiency of the heat engine.*

Find: $W_{net} = ?$, $Q_2 = ?$, $\eta_{Carnot} = ?$

Data: $T_1 = 227°C + 273 = 500$ K, $T_2 = 27°C + 273 = 300$ K, $Q_1 = 400$ kJ

Solution

(i)
$$\eta_{\text{Carnot}} = \frac{T_1 - T_2}{T_1}$$

$$= \frac{500-300}{500}$$

$$\eta_{\text{Carnot}} = \mathbf{0.4 = 40\%}$$

(ii)
$$\eta = \frac{\text{Work done}}{\text{Heat supplied}} = \frac{W_{\text{net}}}{Q_1}$$

$$\therefore \quad W_{\text{net}} = \eta_{\text{Carnot}} \times Q_1$$

$$= 0.4 \times 400 \text{ kJ}$$

$$\mathbf{W_{net} = 160\ kJ}$$

(iii)
$$W_{\text{net}} = Q_1 - Q_2$$

$$160 \text{ kJ} = 400 - Q_2$$

$$\therefore \quad \mathbf{Q_2 = 240\ kJ}$$

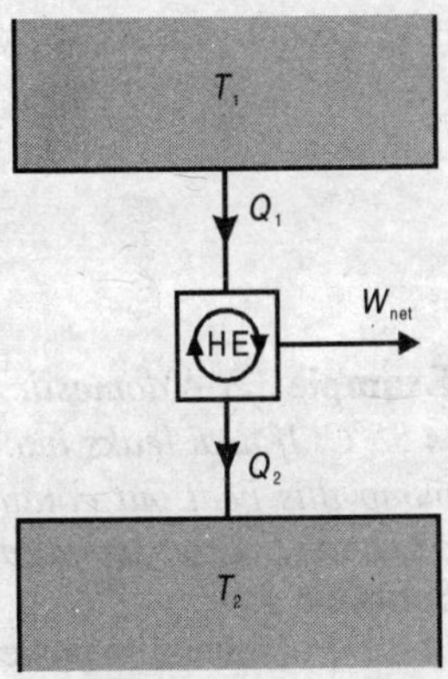

Fig. Ex. 3.3

Example 3.4 *A Carnot cycle heat engine operates between a source and sink temperatures of 227°C and 27°C. If the heat engine receives 25 kJ of heat from the source, calculate:*

1. Work done. 2. Heat rejected. 3. η_{Carnot}

What will be the COP if it operates as a heat pump and when it works as a refrigerator?

Find: $W_{net} = ?$, $Q_2 = ?$, $\eta_{Carnot} = ?$, $[COP]_{HP} = ?$, $[COP]_{ref} = ?$

Data: $T = 227 + 273 = 500$ K, $T_2 = 27 + 273 = 300$ K, $Q_1 = 425$ kJ

Solution

(i)
$$\eta_{\text{Carnot}} = \frac{T_1 - T_2}{T_1} = \frac{500-300}{500}$$
$$\boldsymbol{\eta_{\text{Carnot}} = 0.4 = 40\%}$$

(ii)
$$\eta = \frac{W_{\text{net}}}{Q}$$
$$0.4 = \frac{W_{\text{net}}}{425}$$
$$\therefore \quad \mathbf{W_{net} = 170\,kJ}$$

(iii)
$$W_{\text{net}} = Q_1 - Q_2$$
$$170 = 425 - Q_2$$
$$\therefore \quad \mathbf{Q_2 = 255\,kJ}$$

(iv) When it operates as a heat pump:
$$[\text{COP}]_{\text{HP}} = \frac{Q_1}{W} = \frac{Q_1}{Q_1 - Q_2}$$
$$= \frac{T_1}{T_1 - T_2} = \frac{500}{500-300}$$
$$\mathbf{[COP]_{HP} = 2.5}$$

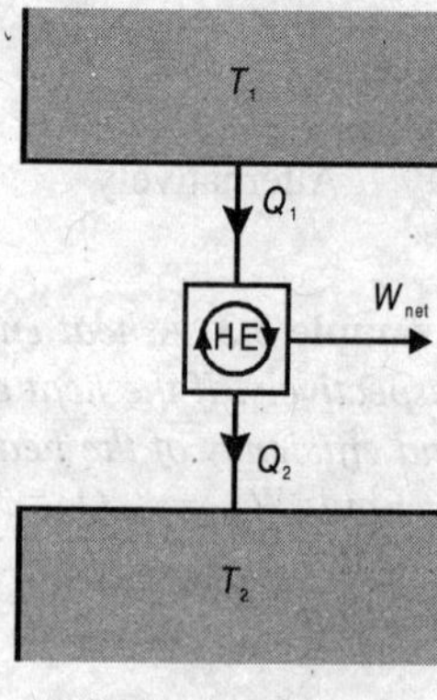

Fig. Ex. 3.4

(v) When its operates as a refrigerator :
$$[\text{COP}]_{\text{ref}} = \frac{Q_2}{W} = \frac{Q_2}{Q_1 - Q_2} = \frac{T_1}{T_1 - T_2}$$
$$= \frac{300}{500-300}$$
$$\mathbf{[COP]_{ref} = 1.5}$$

Example 3.5 *A domestic refrigerator maintains a temperature of – 10°C. The ambient (atmospheric) temperature is 35°C. If heat leaks into the freezer at the continuous rate of 1.75 kJ/sec. What is the least power necessary to pump this heat out continuously?*

Data: Freezer temperature $T_2 = -10 + 273 = 263$ K, $T_1 = 35 + 273 = 308$ K.

Solution

For minimum power requirements or for maximum COP, we have,
$$\frac{Q_2}{Q_1} = \frac{T_2}{T_1}$$
$$\therefore \quad Q_1 = \frac{T_1}{T_2} \times Q_2 = \frac{308}{263} \times 1.75 \text{ kJ/sec}$$
$$Q_1 = 2.049 \text{ kJ/sec}$$
$$\therefore \quad W = Q_1 - Q_2 = 2.049 - 1.75$$
$$\mathbf{W = 0.2994\,kJ/sec}$$

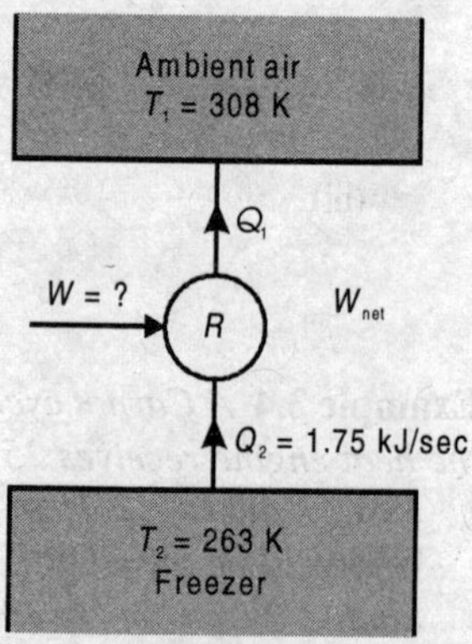

Fig. Ex. 3.5

Example 3.6 *A heat pump is used to maintain an auditorium hall at 25°C, when the atmospheric temperature is 10°C. The heat leakage from the hall is 1500 kJ/min. Calculate the power required to run the actual heat pump, if the COP of the actual heat pump is 30% of the COP of the Carnot heat pump working between the same temperature limits.*

Data: $T_1 = 25 + 273 = 298$ K, $T_2 = 10 + 273 = 283$ K, heat leakage $Q_1 = 1500$ kJ/min = 25 kJ/sec.

Solution

It is given that,

$$[\text{COP}]_{\text{actual}} = 0.3\times[\text{COP}]_{\text{Carnot HP}}$$

We know that,

$$[\text{COP}]_{\text{Carnot HP}} = \frac{T_1}{T_1 - T_2} = \frac{298}{298-283}$$

$$= 19.867$$

$$\therefore \quad [\text{COP}]_{\text{actual HP}} = 0.3\times 19.867 = 5.96$$

Also, $$[\text{COP}]_{\text{HP}} = \frac{Q_1}{W}$$

$$5.96 = \frac{25 \text{ kJ/sec}}{\text{W.D.}}$$

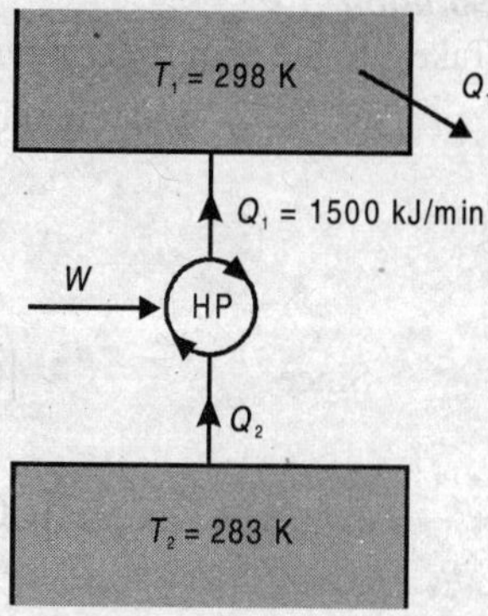

Fig. Ex. 3.6

Work done or power required,

$$W = \frac{25}{5.96}$$

$$\mathbf{W = 4.195 \text{ kJ/sec}}$$

Example 3.7. *The C.O.P. of a Carnot refrigerator is 6, when it maintains the temperature of –3°C in the evaporator. Determine the condenser temperature and refrigerating effect if the power required to run the refrigerator is 7.5 kW.*

Data: $T_2 = 273 - 3 = 270$ K, $[COP]_{ref.} = 6$, $T_1 = ?$

$W = 7.5$ kW

Solution

Now $$[\text{COP}]_{\text{ref}} = \frac{T_2}{T_1 - T_2} = \frac{\text{Evaporator temperature}}{\text{Condenser temperature} - \text{Evaporator temperature}}$$

$$6 = \frac{270}{T_1 - 270}$$

$$6(T_1 - 270) = 270$$

$$6T_1 = 270 + 6\times 270$$

$$6T_1 = 270 + 1620$$

$$\therefore \quad T_1 = 315 \text{K}$$

It is also given that the power required is,
= 7.5 kW = 7500 watts = 7500 J/sec.
We also know that,

$$[\text{COP}]_{\text{ref}} = \frac{Q_2}{W} = \frac{\text{Desired effect}}{\text{Work input}}$$

$$= \frac{\text{Refrigerating effect/sec}}{\text{Work supplied/sec}}$$

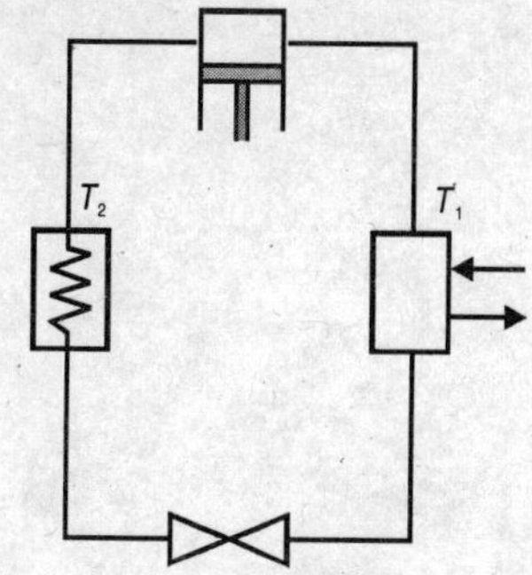

Fig. Ex. 3.7

$$6 = \frac{\text{Refrigerating effect}}{7500}$$

$\therefore$ Refrigerating effect = 45000 J/sec.

Example 3.8 *A fish freezing plant of 100 tonnes capacity is to be maintained at – 50°C when the outside atmospheric temperature is 40°C. The actual C.O.P. of the refrigeration system used is 1/5 of the theoretical Carnot refrigerator working between the same temperature limits. Calculate the power required to run the plant.*

Solution

Take 1 tonne of refrigeration in SI units = 3.516 kJ/sec of heat to be removed (Q_2).

$\therefore$ 100 ton of capacity $= 100 \times 3.516 = 351.6 \text{ kJ/sec} = Q_2$

$$T_1 = 40 + 273 = 313 \text{ K}$$

$$T_2 = -50 + 273 = 223 \text{ K}$$

Since $[COP]_{ref} = \frac{1}{5}[COP]_{\text{Carnot ref.}}$

$$\therefore \quad [COP]_{Carnot} = \frac{T_2}{T_1 - T_2} = \frac{223}{313 - 223} = 2.478$$

But $[COP]_{\text{actual ref.}} = \frac{1}{5} \times 2.478 = 0.4956$

Also we know that, $[COP]_{ref} = \frac{Q_2}{W}$

$$W = \frac{Q_2}{[COP]_{ref}}$$

$$\mathbf{W} = \frac{\mathbf{351.6}}{\mathbf{0.4956}} = \mathbf{709.44 \text{ kJ/sec.}}$$

Example 3.9 *The efficiency of a Carnot engine rejecting heat to a cooling pond at 28°C is 30%. If the cooling pond receives 1050 kJ/min. What is the power developed by the cycle in kW. Also find the temperature of the source.*

Find: $W = ?$, $T_1 = ?$

Data: $T_2 = 28 + 273 = 301$ K, $Q_0 = 1050$ kJ/min $= 17.5$ kJ/sec., $\eta_{Carnot} = 0.3$

Solution

We have, $\eta_{Carnot} = \frac{T_1 - T_2}{T_1}$

$$0.3 = \frac{T_1 - 301}{T_1} = 1 - \frac{301}{T_1}$$

$$\frac{301}{T_1} = 1 - 0.3$$

$$\frac{301}{0.7} = T_1$$

$$\therefore \quad \mathbf{T_1 = 430 \text{ K}}$$

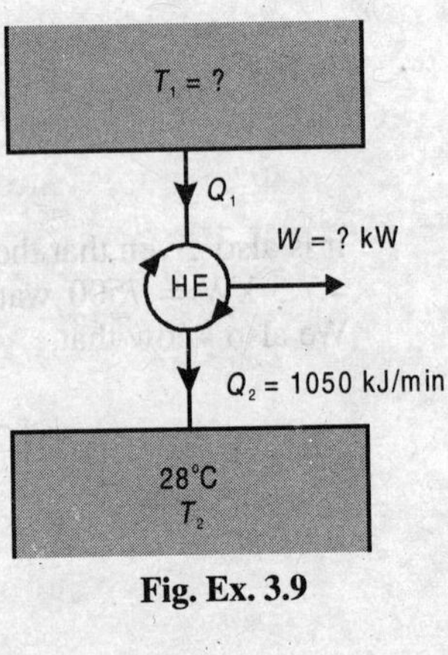

Fig. Ex. 3.9

We also know that, $\eta = \frac{Q_1 - Q_2}{Q_1}$

$$= 1-\frac{Q_2}{Q_1}$$

$$0.3 = 1-\frac{17.5}{Q_1}$$

$$\frac{17.5}{Q_1} = 1-0.3$$

$$\frac{17.5}{0.7} = Q_1 = 25 \text{ kJ/sec} = 25 \text{ kW}$$

$\therefore$ Power = Rate of doing work,

$$W = Q_1 - Q_2$$

$$\mathbf{W = 25 - 17.5 = 7.5 \text{ kW}}$$

Example 3.10 *300 kg of fish at 5° C is to be frozen at – 2°C. Specific heat above freezing point is 3.182 kJ/kg-K and latent heat of fusion is 234.5 (kJ/kg Freezing paint is – 2°C. A refrigerator is to reject heat to surrounding at 40°C. A refrigerator has actual COP 60% of Carnot refrigerator operating in the same temperature limits. How much power must be supplied to remove the heat in 10 hours.*

Solution

We know that,

$$[COP]_{ref.} = \frac{\text{Desired effect}}{\text{Work input}} = \frac{Q_2}{W}$$

$$= \frac{Q_2}{Q_1 - Q_2} = \frac{T_2}{T_1 - T_2} \quad ...(1)$$

Also

$$Q_2 = m\left[C_p \cdot \Delta t + \text{Latent heat of fusion}\right]$$

$$= m\left[C_P \ \Delta t + L_f\right]$$

$$= 300\left[3.187 \times [5-(-2)] + 234.5\right]$$

$$= 300[3.187 \times 7 + 234.5]$$

$$Q_2 = 77032.2 \text{ kJ}$$

$$\eta_{\text{Carnot ref}} = \frac{T_2}{T_1 - T_2} = \frac{271}{313-271}$$

$$= 6.452$$

$$\therefore \quad [COP]_{\text{actual ref.}} = 0.6 \times 6.452 = 3.871$$

Also

$$[COP]_{\text{actual}} = \frac{Q_2}{W} = \frac{77032.2}{W} = 3.871$$

$$\therefore \quad W = 19897.616 \text{ kJ}$$

$$\therefore \quad \text{Rate of work output} = \frac{19897.616}{10 \times 3600}$$

$$\textbf{Rate of work input} = \mathbf{0.552 \text{ kW}}$$

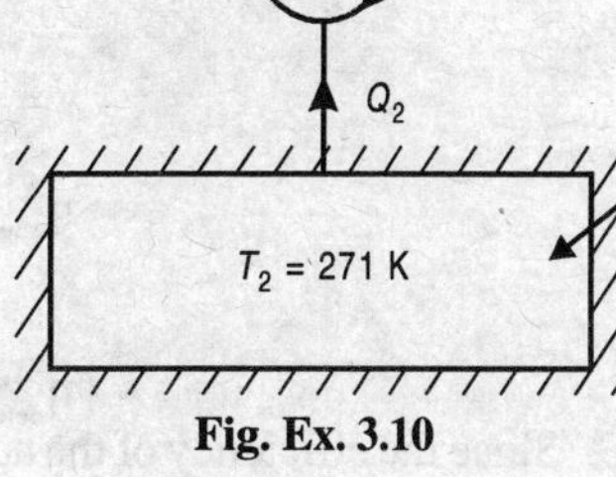

Fig. Ex. 3.10

Example 3.11 *An engineer claims his engine to develop 3.68 kW, on testing the engine consumes 0.44 kg of fuel in one hour. It has the calorific value (C.V.) of 41, 870 kJ/kg. The maximum temperature recorded in the cycle is 1400°C and minimum is 350° C. Is the engineer justified in his claim ?*

Solution

$$W = 3.68\text{ kW} = 3.68\text{ kJ/sec}$$

$$T_1 = 1400 + 273 = 1673\text{ K}$$

$$T_2 = 350 + 273 = 623\text{ K}$$

We know that,
$$\eta_{Carnot} = \frac{T_1 - T_2}{T_1}$$

$$= \frac{1673 - 623}{1673} = 0.6276 = 62.76\%$$

For actual engine, heat supplied/hr = C.V. $\times$ Mass of fuel consumption

= 41870 kJ/kg $\times$ 0.44 kg/hr = 18422.8 kJ/hr.

Work development = 3.68 kW = 3.68 kJ/sec

Work development = 13248 kJ/hr

$$\therefore \text{ Efficiency of actual engine} = \frac{\text{W.D.}}{\text{Heat supplied}} = \frac{13248}{18422.8}$$

Efficiency of actual engine = 71.91%

The actual efficiency of the engine is greater than the efficiency of carnot engine operating between the same temperature limits. So the claim of the engineer is not justified.

Example 3.12 *An inventor claims that his newly developed heat engine will produce 0.184 kW power, when the heat is supplied to engine at the rate of 70 kJ/min. The cycle operates between the maximum temperature of 1092°C and minimum temperature of 149°C. Make complete analysis of the inventors claim and clearly state whether it is worth investing any amount of money on such a project. Give reasons for your suggestions.*

Solution

$$W = 0.184\text{ kW power}$$

$$Q = 70\text{ kJ/min} = \frac{70}{60}\text{ kJ/sec} = 1.17\text{ kW}$$

$$T_1 = 1092°\text{C} + 273 = 1365\text{ K}$$

$$T_2 = 149° + 273 = 422\text{ K}$$

$$\therefore \quad \eta_{Carnot} = \frac{T_1 - T_2}{T_1}$$

$$= \frac{1365 - 422}{1365} = 0.6908 = 69.08\%$$

$$\therefore \quad \eta_{actual} = \frac{\text{W.D.}}{\text{Heat supplied}} \quad \frac{0.184}{1.17}$$

$$= 0.1577$$

$$\boldsymbol{\eta_{actual} = 15.77\%}$$

Since the efficiency of the actual engine is too low there is no point in investing any amount in such a project.

Example 3.13 *An inventor claims to have invented refrigeration machine operating between – 23°C and 27°C. It consumes 1 kW electrical power and gives 20000 kJ of refrigerating effect in one hour. Comment on his claim.*

Data: $T_1 = 300\ K \qquad T_2 = 250\ K$

$$W_{input} = 1\ kW \qquad Q_2 = \text{Refrigerating} = \frac{20000}{3600}\ kJ/sec$$

$$Q_2 = 5.55\ kJ/sec$$

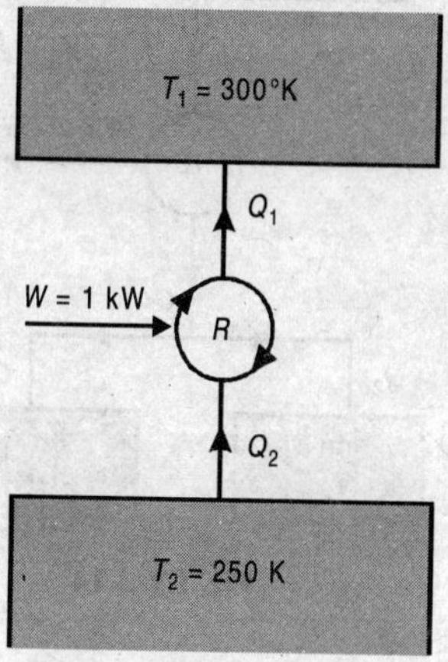

Fig. Ex. 3.13

Solution

We know that,
$$[\text{COP}]_{\text{Carnot ref}} = \frac{T_2}{T_1 - T_2} = \frac{250}{300 - 250}$$

$$[\text{COP}]_{\text{Carnot ref}} = 5 \qquad \text{...(1)}$$

$$\therefore \quad [\text{COP}]_{\text{actual ref}} = \frac{\text{Refrigerating effect}}{\text{Work input}} = \frac{5.55}{1}$$

$$= 5.55 \qquad \text{...(2)}$$

$$[\text{COP}]_{\text{actual.ref.}} > [\text{COP}]_{\text{Carnot ref.}}$$

∴ Claim of the inventor is not valid.

Example 3.14 *A reversible engine working in a cycle takes 4800 kJ of heat from a source at 800 K per minute and develops 20 kW power. The engine rejects heat to two reservoirs at 300 K and 360 K. Determine the heat rejected to each sink in kJ/min.*

Solution

It is given that,

W.D. by the reversible engine $= 20\,\text{kW} = 20\,\text{kJ/sec} = 20 \times 60\,\text{kJ/min} = 1200\,\text{kJ/min}.$

(Since we have to find heat rejected in kJ/min)

$$Q_1 = 4800\,\text{kJ/min} \qquad \therefore\ W = Q_1 - Q_2$$

$$Q_2 = Q_1 - W$$

$$Q_2 = 4800 - 1200 = 3600\,\text{kJ/min}$$

Since this heat is rejected to two sinks (a) and (b)

$$\left.\begin{aligned} Q_2 &= Q_{2a} + Q_{2b} \\ \therefore\quad Q_{2b} &= Q_2 - Q_{2a} \\ Q_{2b} &= 3600 - Q_{2a} \end{aligned}\right\} \qquad \text{...(1)}$$

We also know that,
$$\frac{Q_1}{T_1} = \frac{Q_2}{T_2}$$

$$\therefore \quad \frac{Q_1}{T_1} - \frac{Q_2}{T_2} = 0$$

$$\therefore \quad \frac{Q_1}{T_1} - \left(\frac{Q_{2a}}{T_{2a}} + \frac{Q_{2b}}{T_{2b}}\right) = 0$$

Let Q_{2a} be the heat rejected to sink a

Q_{2b} be the heat rejected to sink b

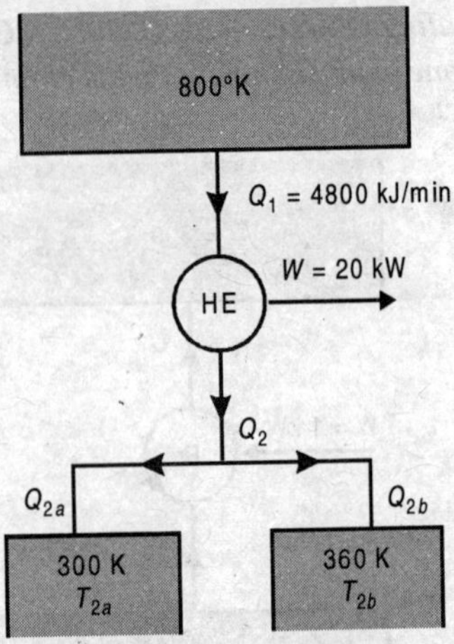

Fig. Ex. 3.14

$$\frac{4800}{800} - \left(\frac{Q_{2a}}{300} + \frac{Q_{2b}}{360}\right) = 0$$

$$\frac{4800}{800} - \left(\frac{Q_{2a}}{300} + \frac{3600 - Q_{2a}}{360}\right) = 0 \text{ ... (From 1)}$$

$$6 - \left[\frac{Q_{2a}}{300} + \frac{3600}{360} - \frac{Q_{2a}}{360}\right] = 0$$

i.e. $$6 - \frac{Q_{2a}}{300} - 10 + \frac{Q_{2a}}{360} = 0$$

i.e. $$-4 + \left[\frac{Q_{2a}}{360} - \frac{Q_{2a}}{300}\right] = 0$$

i.e. $$Q_{2a}\left[\frac{1}{360} - \frac{1}{300}\right] = 4$$

$$Q_{2a} \times \left[\frac{300 - 360}{360 \times 300}\right] = 4$$

$$Q_{2a} \times \left[\frac{-60}{108000}\right] = 4$$

$$Q_{2a}[-0.000555] = 4$$

$$\mathbf{Q_2 = -7200.00\ kJ/min}$$

$\therefore$ Since $$Q_2 = Q_{2a} + Q_{2b}$$

$$3600 = -7200 + Q_{2b}$$

$\therefore$ $$Q_{2b} = 3600 + 7200$$

$$\mathbf{Q_{2b} = 10800\ kJ/min}$$

Example 3.15 *A reversible engine working on reversible cycle is supplied with 5000 kJ/min of heat from a source at 527°C while it develops 25 kW of power. This engine rejects heat to two reservoirs at two different temperatures equal to 27°C and 97°C respectively. Determine the efficiency of the engine and heat rejected to each reservoir.*

Solution

$$T_1 = 527 + 273$$

$$Q_1 = 5000 \text{ kJ/min}$$

$$W = 25 \text{ kW} = 25 \text{ kJ/sec}$$

$$= 25 \times 60 = 1500 \text{ kJ/min}$$

and $$Q_2 = \text{Heat rejected/min}$$

We know that, $$W = Q_1 - Q_2$$

$\therefore$ $$Q_2 = Q_1 - W$$

$$Q_2 = 5000 - 1500 = 3500 \text{ kJ/min}$$

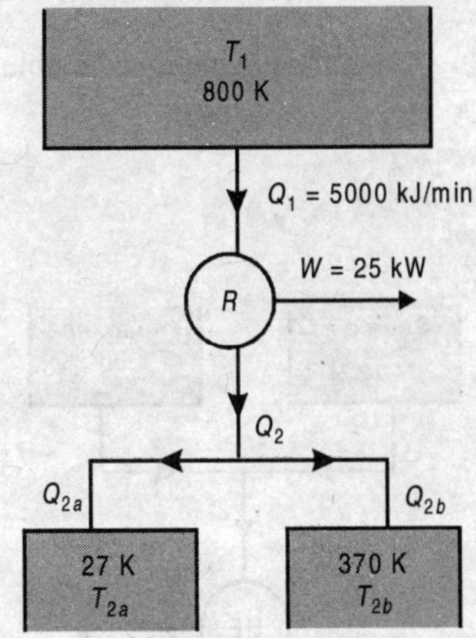

Fig. Ex. 3.15

$$Q_2 = Q_{2a} + Q_{2b}$$

$$\therefore \quad Q_{2a} = Q_2 - Q_{2b}$$

$$Q_{2a} = 3500 - Q_{2b}$$

$$\text{Efficiency of the engine} = \frac{\text{W.D.}}{\text{Heat supplied}}$$

$$= \frac{W}{Q} = \frac{1500}{5000} \times 100 = 30\%$$

For reversible engines, we have,

$$\frac{Q_1}{T_1} = \frac{Q_2}{T_2}$$

i.e.

$$\frac{Q_1}{T_1} - \frac{Q_2}{T_2} = 0$$

i.e.

$$\frac{Q_1}{T_1} - \left[\frac{Q_{2a}}{T_{2a}} + \frac{Q_{2b}}{T_{2b}}\right] = 0$$

$$\frac{5000}{800} - \left[\frac{3500 - Q_{2b}}{370} + \frac{Q_{2b}}{300}\right] = 0$$

$\therefore$ We get, $\mathbf{Q_{2b} = -5090.14}$ **kJ/min**

$$\therefore \quad Q_{2a} = 3500 - Q_{2b} = 3500 + 5090.14$$

$\mathbf{Q_{2a} = 8590.14}$ **kJ/min**

Example 3.16 *A reversible engine is supplied with heat from two constant temperature sources at 900 K and 600 K and rejects heat to a constant temperature sink at 300 K. If the engine executes a number of complete cycles while developing 100 kW, and rejecting 3600 kJ of heat per min. Determine the heat supplied by each source per minute and efficiency of the engine.*

Solution

Now the work developed by the engine

$$= 100 \text{ kW}$$

$$= 100 \text{ kJ/sec} = 100 \times 60 \text{ kJ/min}$$

$$= 6000 \text{ kJ/min}$$

We know that, $W = Q_1 - Q_2$

Given $Q_2 = 3600 \text{ kJ/min}$

$$\therefore \quad W + Q_2 = Q_1$$

$$600 + 3600 = Q_1$$

$$\therefore \quad Q_1 = 9600 \text{ kJ/min}$$

Assume Q_{1a} is the heat supplied by the source (*a*)

and Q_{1b} is the heat supplied by the source (*b*)

And therefore $Q_1 = Q_{1a} + Q_{1b}$

$$\therefore \quad Q_{1b} = Q_1 - Q_{1a}$$

i.e., $Q_{1b} = 9600 - Q_{1a}$ ...(1)

As the engine is reversible, the required condition for the engine to work is, $\frac{Q_1}{T_1} = \frac{Q_2}{T_2}$

$$\frac{Q_{1a}}{T_{1a}} + \frac{Q_{1b}}{T_{1b}} = \frac{Q_2}{T_2} \quad ...(2)$$

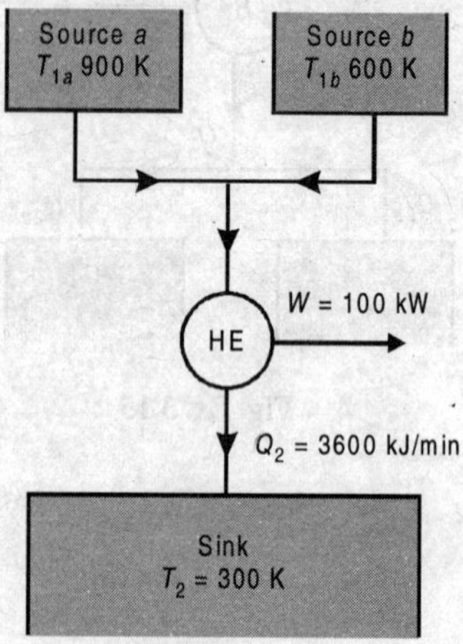

Fig. Ex. 3.16

$$\frac{Q_{1a}}{T_{1a}} + \frac{9600 - Q_{1a}}{T_{1b}} = \frac{Q_2}{T_2}$$

$$\frac{Q_{1a}}{900} + \frac{9600 - Q_{1a}}{600} = \frac{3600}{300}$$

$$\frac{Q_{1a}}{900} + \frac{9600}{600} - \frac{Q_{1a}}{600} = 12$$

$$\frac{Q_{1a}}{900} - \frac{Q_{1a}}{600} + 16 = 12$$

$$Q_{1a}\left[\frac{1}{900} - \frac{1}{600}\right] = -4$$

$$Q_{1a}\left[\frac{600 - 900}{900 \times 600}\right] = -4$$

$$Q_{1a}\left[\frac{-300}{900 \times 600}\right] = -4$$

$$Q_{1a}[-0.000555] = -4$$

$$Q_{1a} = 7200 \text{ kJ/min}$$

$\therefore$

$$Q_{1b} = 9600 - 7200$$

$$Q_{1b} = 2400 \text{ kJ/min}$$

i.e., Heat supplied by the source *b*

$$\text{Now, the Engine efficiency} = \frac{\text{W.D.}}{\text{Heat supplied}}$$

$$= \frac{6000}{Q_1}$$

$$= \frac{6000}{9600} = 0.625$$

i.e., η = **62.5%**

Example 3.17 *Two reversible engines R_1 and R_2 are connected in series between a heat source S and cold body C. R_1 receives 900 kJ from S, at 540 K and R_2 rejects heat to C at 210 K. Assuming equal thermal efficiencies from R_1 and R_2, determine,*

(i) The temperature at which heat is rejected by R_1 and received by R_2.
(ii) The work output W_1 and W_2 of the two engines.
(iii) The heat rejected by R_2 to the cold body C.

Data: Q_1 = 900 kJ , T_1 = 540 K, T_2 = 210 K

Solution

Thermal efficiencies of R_1 and R_2 are equal to

∴ (i) $$\eta_{R_1} = \frac{T_1 - T_3}{T_1}$$

and $$\eta_{R_2} = \frac{T_3 - T_2}{T_3}$$

Since $$\eta_{R_1} = \eta_{R_2}$$

∴ $$\frac{T_1 - T_3}{T_1} = \frac{T_3 - T_2}{T_3}$$

$$T_3 T_1 - T_3^2 = T_1 T_3 - T_1 T_2$$

$$T_1 T_3 - T_1 T_3 + T_1 T_2 = T_3^2$$

∴ $$T_3^2 = T_1 T_2 = 540 \times 210$$

∴ $$T_3^2 = 113400$$

∴ $$T_3 = 336.75 \text{ K}$$

(ii) We know that, $$W_1 = Q_1 - Q_3 \quad \text{...(1)}$$

We also know that, $$\frac{Q_1}{T_1} = \frac{Q_2}{T_2} = \frac{Q_3}{T_3}$$

∴ $$Q_3 = \frac{T_3}{T_1} \times Q_1$$

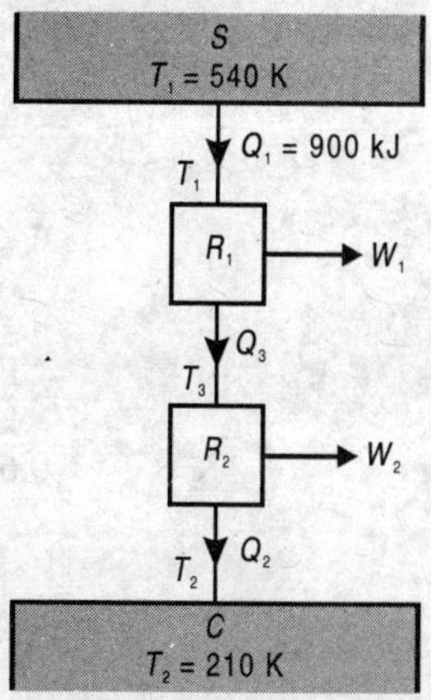

Fig. Ex. 3.17

Substituting this in Eq. (1)
We get,

∴ $$W_1 = Q_1 - Q_1 \frac{T_3}{T_1}$$

$$= 900 - 900 \times \frac{336.75}{540}$$

$$\mathbf{W_1 = 338.75 \text{ kJ}}$$

Similarly $$W_2 = Q_3 - Q_2 = Q_1 \times \frac{T_3}{T_1} - Q_2$$

From $$\frac{Q_1}{T_1} = \frac{Q_2}{T_2}$$

$$Q_2 = \frac{T_2}{T_1} \times Q_1 \text{ (i.e., } Q_2 \text{ in terms of } Q_1)$$

$$W_2 = Q_1 \times \frac{T_3}{T_1} - Q_1 \times \frac{T_2}{T_1} = 900 \times \frac{336.75}{540} - 900 \times \frac{210}{540}$$

∴ $$\mathbf{W_2 = 211.25 \text{ kJ}}$$

(iii) Heat rejected by R_2 to cold body (Q_2)

Since, $$\frac{Q_1}{T_1} = \frac{Q_2}{T_2}$$

$$Q_2 = \frac{T_2}{T_1} \times Q_1$$

$$= \frac{900 \times 210}{540}$$

$$\mathbf{Q_2 = 350\ kJ}$$

Example 3.18 *A reversible heat engine operates between two reservoirs at temperatures of 600°C and 40°C. The engine drives a reversible refrigerator, which operates between reservoirs at temperature 40°C and – 20°C. The heat transfer to the heat engine is 2000 kJ and the net work output of the combined engine refrigerator plant is 360 kJ,*

(a) Evaluate the heat transfer to the refrigerator and the net heat transfer to the reservoir at 40°C.

(b) Reconsider (a) given that the efficiency of the heat engine and the C.O.P. of the refrigerator are each 40% of their maximum possible values.

Solution

(a) Maximum efficiency of the heat engine

$$\eta_{max} = \frac{\text{W.D.}}{\text{Heat supplied}}$$

$$= \frac{W_1}{Q_1} = \frac{T_1 - T_2}{T_1}$$

i.e. $$\eta_{Carnot} \text{ or } \eta_{max} = 1 - \frac{T_2}{T_1} = 1 - \frac{313}{873} = 0.642$$

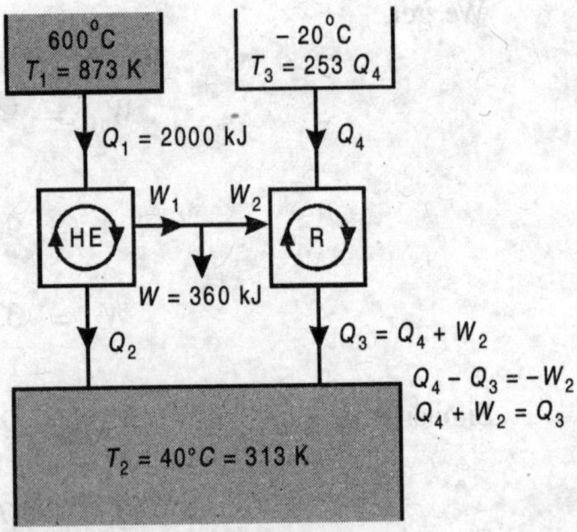

Fig. Ex. 3.18

∴ From $$\eta_{max} = \frac{W_1}{Q_1} = 0.642$$

∴ $$0.642 \times 2000 = W_1$$

∴ $$1284\ \text{kJ} = W_1$$

Since $$Q_1 - Q_2 = W_1$$

$$Q_2 = Q_1 - W_1$$

$$Q_2 = 2000 - 1284 = 716\ \text{kJ}$$

Since $$W_1 - W_2 = W = 360\ \text{kJ}$$

∴ $$W_2 = W_1 - W = 1284 - 360$$

$$W_2 = 924\ \text{kJ}$$

Now we know that,

$$[\text{COP}]_{ref} = \frac{\text{Desired effect}}{\text{Work supplied}} = \frac{Q_2}{W} = \frac{Q_2}{Q_1 - Q_2} = \frac{T_2}{T_1 - T_2}$$

But from Fig. Ex. 3.18 $$[\text{COP}]_{ref} = \frac{T_3}{T_2 - T_3} = \frac{253}{313 - 253} = 4.22$$

Also $[\text{COP}]_{ref} = \frac{Q_4}{W_2}$ (Ref. Fig. Ex. 3.18) $= 4.22$

$\therefore$ $Q_4 = W_2 \times [\text{COP}]_{ref}$

$Q_4 = 924 \times 4.22 = 3899$ kJ

$\therefore$ $Q_3 = Q_4 + W_2 = 924 + 3899$

$Q_3 = 4823$ kJ

$\therefore$ Heat rejected to the 40°C reservoir

$= Q_2 + Q_3$

$= 716 + 4823 = 5539$ kJ

(b) Efficiency of the actual heat engine cycle,

$\eta_{actual} = 0.4\,\eta_{max}$

$= 0.4 \times 0.642 = 0.2568$

Also since $\eta_{actual} = \frac{W_1}{Q_1}$

$\therefore$ $W_1 = 0.2568 \times 2000 = 513.6$ kJ

Since $W_1 - W_2 = W$

$\therefore$ $W_2 = W_1 - W$

$W_2 = 513.6 - 360 = 153.6$ kJ

and since $Q_1 - Q_2 = W_1$

$\therefore$ $Q_2 = Q_1 - W_1 = 2000 - 513.6$

$Q_2 = 1486.4$ kJ

[COP] of actual refrigerator cycle,

$$[\text{COP}]_{\text{actual ref.}} = \frac{Q_4}{W_2} = 0.4 \times 4.22 = 1.69$$

$\therefore$ $Q_4 = 153.6 \times 1.69$

$Q_4 = 259.6$ kJ

$\therefore$ $Q_2 = Q_4 + W_2$

$Q_2 = 159.6 + 153.6$

$Q_2 = 413.2$ kJ

$\therefore$ Heat rejected to 40°C reservoir

$= Q_2 + Q_3$

$= 413.2 + 1486.4 = 1899.6$ kJ.

Example 3.19 *A heat pump working on the Carnot cycle takes in heat from a reservoir at 5°C and delivers heat to a reservoir at 60° C. The heat pump is driven by a reversible heat engine which takes heat from a reservoir at 840° C and rejects heat at 60°C. The reversible heat engine also drives a machine that absorbs 36 kW. If the heat pum extracts 17 kJ/sec from 5°C reservoir, determine,* (*i*) *The rate of heat supply from 840°C source.*

(*ii*) *The rate of heat rejection to 60°C sink.*

Data: $T_1 = 60 + 273 = 333$ K, $T_2 = 5 + 273 = 278$ K, $T_3 = 840 + 273 = 1113$ K.

Solution

We know that, $[COP]_{HP} = \dfrac{Q_1}{Q_1 - Q_2} = \dfrac{T_1}{T_1 - T_2}$

∴ $[COP]_{HP} = \dfrac{T_1}{T_1 - T_2} = \dfrac{333}{333 - 278} = 6.05$

Also, $6.05 = \dfrac{Q_1}{Q_1 - Q_2}$

$6.05 = \dfrac{Q_1}{Q_1 - 17}$

⇒ $Q_1 = 20.37$kW or kJ/sec.

For the engine $\eta_{Carnot} = \dfrac{T_3 - T_1}{T_3} = \dfrac{1113 - 333}{1113} = 0.7$

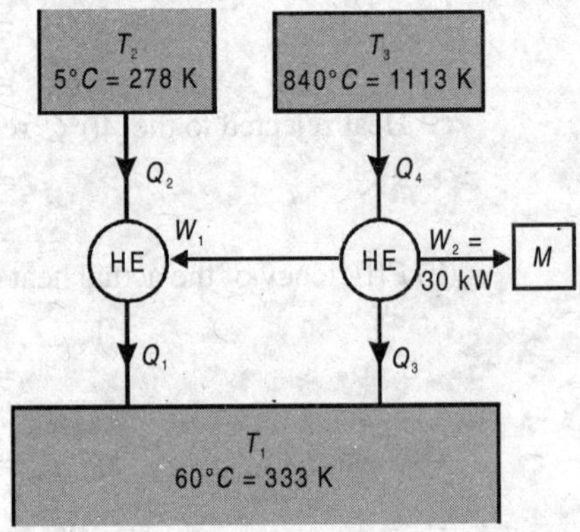

Fig. Ex. 3.19

Also, $0.7 = \dfrac{W_1 + 30}{Q_4}$ ...(a)

We know that, $W_1 = Q_1 - Q_2 = 20.37 - 17$

$W_1 = 3.37$ ... (b)

From (a) and (b) we have,

$Q_4 = 47.67$ kW

and $W_E = W_1 + W_2 = 3.37 + 30 = 33.37$

Also $W_E = Q_4 - Q_3$

⇒ $Q_3 = Q_4 - W_E = 47.67 - 33.37$

$\mathbf{Q_3 = 14.3}$ **kW or kJ/sec**

∴ Heat rejected to 333 K sink $= Q_1 + Q_3 = 20.37 + 14.3 = 34.67$ kJ

Example 3.20 *A heat pump is used to heat a bunglow in the winter and then reversed to cool the bunglow in the summer. The interior temperature is to be maintained at 25°C. Heat transfer through the walls and roof is estimated to be 2800 kJ/hr°C temperature difference between the inside and outside. Estimate:*

(*i*) *If the outside temperature in winter is 5°C, what is the minimum power required to drive the heat pump ?*

(*s*) *If the power input is the same as in part* (*i*), *what is the maximum outside temperature for which the inside temperature can be maintained at 25°C ?*

Solution

(i) In winter: $T_1 = 25 + 273 = 298$ K

$T_2 = 5 + 273 = 278$ K

Now heat loss through walls = 2800 kJ / hr°C

∴ Net heat loss $= 2800 \times (25 - 5)$

$Q_1 = 56000$ kJ/hr

For minimum power requirements of for maximum efficiency [COP] of heat pump should be maximum,

∴ Heat Pump work on reverse Carnot cycle

$$[\text{COP}]_{HP} = \frac{T_1}{T_1 - T_2} = \frac{298}{298 - 278} = 14.9$$

∴ Also,

$$14.9 = \frac{Q_1}{W}$$

$$W = \frac{Q_1}{14.9} = \frac{56000}{14.9} = 3758.45 \text{ kJ/hr}$$

$$\therefore \quad \text{Minimum Power} = \frac{3758.4}{3600}$$

$$\textbf{Minimum Power} = \mathbf{1.045\ kW}$$

(ii) For summer

$$T_2 = 25 + 273 = 298 \text{ K}$$

$$T_1 = \text{outside temperature} > T_2$$

$$Q_2 = \text{Heat to be removed/hr} = \text{Heat loss/hr}$$

$$Q_2 = 2800\,(T_1 - 298) \text{ kJ/hr}$$

As the heat pump is reversed to work as refrigerator in the summer,

$$[\text{COP}] = \frac{T_2}{T_1 - T_2} = \frac{Q_2}{W}$$

$$\frac{298}{T_1 - 298} = \frac{2800(T_1 - 298)}{3758.45}$$

$$(T_1 - 298)^2 = 460$$

$$\mathbf{T_1 = 318\ K = 45°C}$$

Example 3.21 *Three Carnot engines R_1, R_2, R_3 operate in series between two heat reservoirs which are at temperatures of 1000 K and 300 K as shown.*

Calculate intermediate temperatures if amount of work produced by these engines is in the proportions of 5 : 4 : 3.

Solution

From Fig. Ex. 3.21 by applying Carnot's Theorem we can write,

$$\frac{Q_1 - Q_2}{Q_1} = \frac{W_1}{Q_1} = \frac{T_1 - T_2}{T_1} \qquad \text{...(1)}$$

and

$$\frac{Q_2 - Q_3}{Q_2} = \frac{W_2}{Q_2} = \frac{T_2 - T_3}{T_2} \qquad \text{...(2)}$$

From (1) and (2) we get

$$\frac{W_1}{W_2} = T_1 - T_2 \times \frac{Q_1}{T_1} \times \frac{T_2}{Q_2} \times \frac{1}{T_2 - T_3}$$

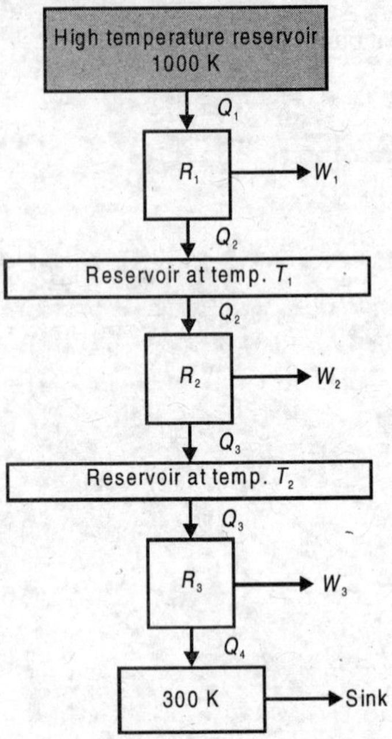

Fig. Ex. 3.21

$$= \frac{T_1 - T_2}{T_2 - T_3}\left(\text{as } \frac{Q_1}{T_1} = \frac{Q_2}{T_2}\right)$$

Therefore for Carnot engines operating in series, the temperature difference between the reservoirs, is proportional to the work ratios.

$$\therefore 1000 - T_1 : T_1 - T_2 : T_2 - 300 = 5:4:3$$

$$\therefore \quad \frac{1000 - T_1}{T_1 - T_2} = \frac{5}{4}$$

and
$$\frac{T_1 - T_2}{T_2 - 300} = \frac{4}{3}$$

From Eq. (1), we get,

$$4000 - 4\,T_1 = 5\,T_1 - 5\,T_2$$

or
$$4000 - 9\,T_1 = -5\,T_2 \qquad \text{...(3)}$$

Similarly from Eq. (2), we get,

$$3\,T_1 - 3\,T_2 = 4\,T_2 - 1200$$

or
$$3\,T_1 = 7\,T_2 - 1200$$

Multiplying by 3 throughout we get,

or
$$9\,T_1 = 21\,T_2 - 3600 \qquad \text{...(4)}$$

Adding Eqs (3) and (4), we get,

$$4000 - 9\,T_1 + 9\,T = -5\,T_1 + 21\,T_2 - 3600$$

$$\therefore \quad 16\,T_2 = 7600$$

$$\therefore \quad \mathbf{T_2 = 475\ K}$$

Substitute $T_2 = 475$ K in Eq. (3) we get,

$$\therefore \quad 4000 - 9\,T_1 = (-5)(475)$$

$$\therefore \quad 9\,T = (4000) + (2375)$$

$$\therefore \quad \mathbf{T_1 = 708.4\ K}$$

Example 3.22 *Which is more effective way to increase the efficiency of Carnot engine, either to increase T_1 keeping T_2 constant; or to decrease T_2 keeping T_1 constant.*

Solution

We know that, the efficiency of Carnot cycle working between the temperature limits T_1 and T_2 is given by,

$$\eta = \frac{T_1 - T_2}{T_1}$$

Consider that the source temperature is increased by ΔT maintaining the sink temperature T_2 constant, then,

$$\eta_1 = \frac{(T_1 + \Delta T) - T_2}{T_1 + \Delta T}$$

$$= \frac{T_1 + \Delta T - T_2}{T_1 + \Delta T} \qquad \text{...(a)}$$

Now consider that the sink temperature is decreased by ΔT maintaining source temperature T_1 constant, then,

$$\eta_2 = \frac{T_1 - (T_2 - \Delta T)}{T_1}$$

$$= \frac{T_1 + \Delta T - T_2}{T_1} \quad \text{...(b)}$$

Comparing Eqs. (a) and (b) it is obvious that the numerators of both the equations are same and denominator of Eq. (b) is less than the denominator of Eq. (a) as $T_1 < (T_1 + \Delta T)$

$$\therefore \quad \eta_2 > \eta_1$$

In order to increase the efficiency of Carnot cycle, decreasing to T_2 will be more effective.

Example 3.23 *A reversible engine operates between T_1 and T (T_1 > T). The energy rejected by this engine is received by a second reversible engine at the same temperature T. The second engine rejects the energy at temperature T_2 (T_2 < T). Show that*

(i) temperature T is the arithmetic mean of temperature T_1 and T_2 if the engine produces same amount of work output and

(ii) the temperature T is the geometric mean of T_1 and T_2 if the engines have the same cycle efficiency.

Solution

(i)
$$\frac{Q_1}{T_1} = \frac{Q_1 - W}{T} = \frac{Q_1 - 2W}{T_2}$$

$$= C \text{ some constant}$$

$$Q_1 = CT_1, Q_1 - W = CT,$$

$$Q_{1-2} W = CT_2$$

$$\therefore \quad T = \frac{T_1 + T_2}{2} = \text{Arithmetic mean}$$

(ii)
$$\eta = \frac{T_1 - T}{T_1} = \frac{T - T_2}{T}$$

$$\therefore \quad T = \sqrt{T T_2} = \text{Geometric mean}$$

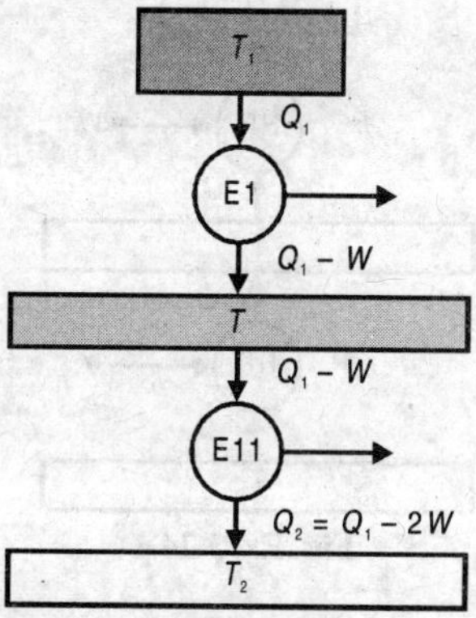

Fig. Ex. 3.23

Example 3.24 *Two heat pumps are connected in series between two heat reservoirs at T_1 and T_2. Heat pump A pumps heat from a reservoir at T_2 and rejects heat to a reservoir at T, while the heat pump B pumps heat from reservoir at T to n reservoir at T_1. If T_1 > T_2, show that (a) intermediate temperature T is the arithmetic mean of temperatures T_1 and T_2 if the work input to both the pumps is equal and (b) the temperature is the geometric mean of temperature T_1 and T_2 if the heat pumps have equal C.O.P.*

Solution

(a) For equal work inputs, we can write,

$$\frac{Q_2}{T_2} = \frac{Q}{T} = \frac{Q_1}{T_1} \quad \text{...(a)}$$

But
$$Q = Q_2 + W$$

and
$$Q_1 = Q + W = Q_2 + 2W$$

$$\therefore \quad \frac{Q_2}{T_2} = \frac{Q_2 + W}{T} = \frac{Q_2 + 2W}{T_1}$$

$$= C \text{ some constant}$$

$$\therefore \quad Q_2 = CT_2 \qquad \text{...(b)}$$

$$Q_2 + W = CT \qquad \text{...(c)}$$

$$Q_2 + 2W = CT_1 \qquad \text{...(d)}$$

Now from Eq. (c),

$$W = CT - Q_2 \text{ But from Eq. (b) } Q_2 = CT_2$$

$$\therefore \quad W = CT - CT_2 = C(T - T_2) \qquad \text{...(e)}$$

From Eq. (d)

$$2W = CT_1 - CT_2 \qquad \text{...(f)}$$

$$W = C\frac{(T_1 - T_2)}{2}$$

From Eqs (e) and (f)

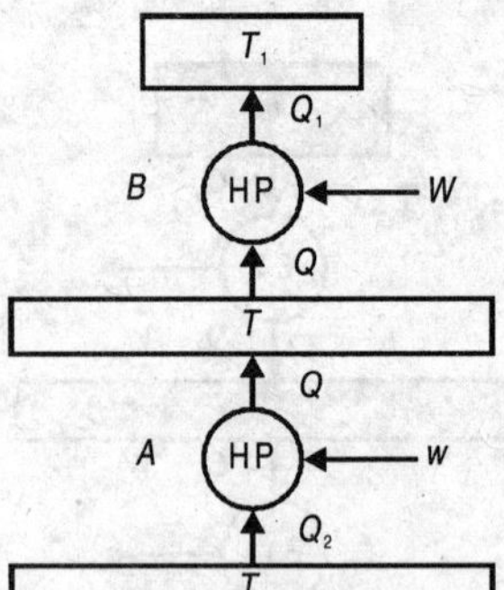

Fig. Ex. 3.24

$$C(T - T_2) = C\frac{(T_1 - T_2)}{2}$$

or
$$(T - T_2) = \frac{T_1 - T_2}{2}$$

$$\therefore \quad 2T - 2T_2 = T_1 - T_2$$

or
$$T = \frac{T_1 + T_2}{2} \text{ which is arithmetic mean}$$

(b) For equal C.O.P.

$$[\text{C.O.P.}]_{\text{HP(A)}} = \frac{T_1}{T_1 - T}$$

and
$$[\text{C.O.P.}]_{\text{HP(B)}} = \frac{T}{T - T_2} \text{ and for equal C.O.P.}$$

$$\therefore \quad \frac{T_1}{T_1 - T} = \frac{T}{T - T_2}$$

or
$$T = \sqrt{T_1 T_2} \text{ which is the geometric mean.}$$

Example 3.25 *A reversible engine works between three thermal reservoirs A, B, C. The engine absorbs an equal amount of heat from the thermal reservoirs. A and B kept at temperatures T_A and T_B respectively, and rejects heat to the thermal reservoir C kept at temperature T_C. The efficiency of the engine is μ times the efficiency of the reversible engine, which works between the two reservoirs A and C. Prove that*

$$\frac{T_A}{T_B} = (2\mu - 1) + 2(1 - 2\mu)\frac{T_A}{T_C}.$$

Solution

Given that, efficiency of the Engine = $\mu \times \eta$ of engine working between T_A and T_C

$\therefore$ Efficiency of engine $A = \dfrac{T_A - T_C}{T_A}$

and Output of Engine $A = Q \times \dfrac{T_A - T_C}{T_A}$

Similarly,

Efficiency of Engine $B = \dfrac{T_B - T_C}{T_B}$

and Output of Engine $B = Q \times \dfrac{T_B - T_C}{T_B}$

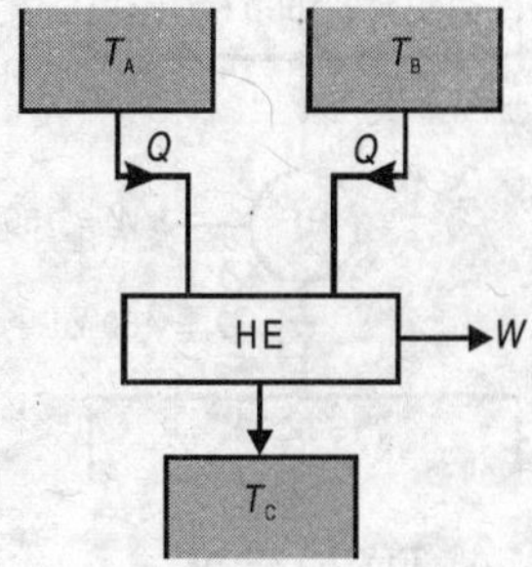

Fig. Ex. 3.25

$\therefore$ Net work output $= Q \times \left(\dfrac{T_A - T_C}{T_A}\right) + Q \times \left(\dfrac{T_B - T_C}{T_B}\right)$

and Net input $= Q + Q = 2Q$

$\therefore$ Efficiency $= \dfrac{\text{Net work output}}{\text{Net input}}$

$$= \frac{Q\left(\dfrac{T_A - T_C}{T_A}\right) + Q\left(\dfrac{T_B - T_C}{T_B}\right)}{2Q}$$

$$= \frac{1}{2}\left(1 - \frac{T_C}{T_A} + 1 - \frac{T_C}{T_B}\right)$$

But since $\eta = \mu \times \eta$ of Engine working between T_A and T_C

$$\therefore \quad \frac{1}{2}\left(1 - \frac{T_C}{T_A} + 1 - \frac{T_C}{T_B}\right) = \mu\left(1 - \frac{T_C}{T_A}\right)$$

Multiplying both sides by $\dfrac{T_A}{T_C}$ and rearranging terms we get the required result

$$\frac{T_A}{T_B} = (2\mu - 1) + 2(1 - 2\mu)\frac{T_A}{T_C}$$

Example 3.26 *A heat engine receives 600 kJ of heat from a high temperature source at 1000 K during a cycle. It converts 150 kJ of this heat to net work and rejects the remaining 450 kJ to a low temperature sink at 300 K. Determine if this heat engine violates the second law of thermodynamics on the basis of*

(a) the Clausius inequality and

(b) the Carnot Principle

Solution

Clausius inequality :

$$\oint \frac{\delta Q}{T} = \frac{Q_1}{T_1} - \frac{Q_2}{T_2}$$

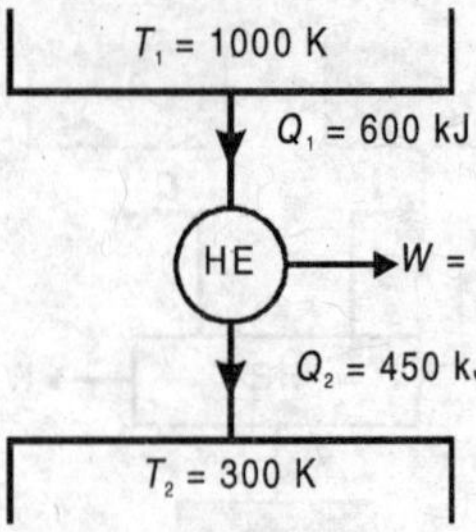

Fig. Ex. 3.26

$$= \frac{600}{1000} - \frac{450}{300}$$
$$= -0.9 \text{ kJ/K}$$

Since $\oint \frac{\delta Q}{T}$ is negative, the heat engine does not violate second law of thermodynamics.

Carnot Principle :

$$\eta_{\text{Th (HE)}} = \frac{Q_1 - Q_2}{Q_1} = \frac{150}{600} = 25\%$$

$$\eta_{\text{Th (Carnot HE)}} = \frac{T_1 - T_2}{2T_1} = \frac{1000 - 300}{1000} = 70\%$$

Since η_{Th} of actual engine working between the same temperature limits is less than that of Carnot engine, the heat engine does not violate second law of thermodynamics.

Example 3.27 *A heat pump is used to maintain a house at a constant temperature of 23°C. The house is losing heat to the outside air through the walls and the windows at a rate of 60,000 kJ/h while the energy generated within the house from people, lights and appliances amounts to 4000 kJ/h. For a COP of 2.5, determine the required power input to the heat pump.*

Solution

$$[\text{COP}]_{\text{HP}} = \frac{Q_1}{W}$$

Now, $$Q_1 = \begin{pmatrix}\text{Heat energy lost through} \\ \text{walls and windows}\end{pmatrix} - \begin{pmatrix}\text{Energy generated} \\ \text{from people, lights etc.}\end{pmatrix}$$

$\therefore$ $$Q_1 = 60{,}000 - 4{,}000$$
$$= 56{,}000 \text{ kJ/h}$$
$$= 15.556 \text{ kJ/s}$$

$\therefore$ $$\mathbf{W} = \frac{\mathbf{15.556}}{\mathbf{2.5}} = \mathbf{6.22 \text{ kW}}$$

Example 3.28 *A heat pump is used to maintain an auditorium hall at 24°C when the atmospheric temperature is 10°C. The heat lost from the hall is 1500 kJ/min. Calculate the power required to run the heat pump if its COP is 30 percent of Carnot machine, working between the same temperature limits.*

Solution

$$\text{COP}_{\text{HP(Carnot)}} = \frac{T_2}{T_2 - T_1} = \frac{297}{297 - 283}$$
$$= 21.21$$
$$\text{COP}_{\text{HP (Actual)}} = 0.3 \times \text{COP}_{\text{HP(Carnot)}}$$
$$= 0.3 \times 21.21 = 6.36$$
$$\text{COP} = \frac{Q}{W}$$

$\therefore$ $$\mathbf{W} = \frac{\mathbf{1500/60}}{\mathbf{6.36}} = \mathbf{3.928 \text{ kW}}$$

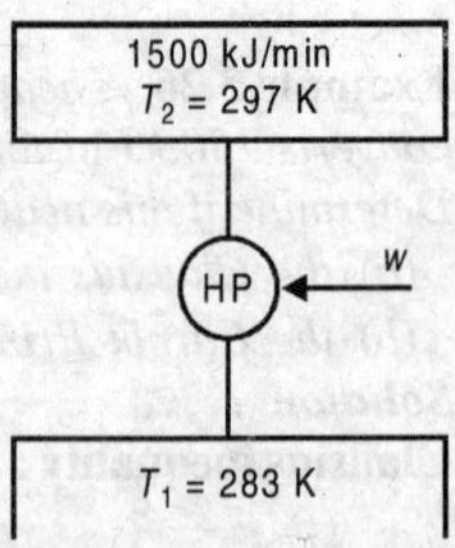

Fig. Ex. 3.28

THEORY QUESTIONS

3.1. Explain Limitation of first law of thermodynamics.

3.2. Explain Heat engine and Heat engine cycle.

3.3. What is a heat reservoir ? Explain its types.

3.4. Explain a refrigerator with a neat sketch.

3.5. Explain a heat pump with a suitable sketch.

3.6. Why we shall study Second Law of Thermodynamics? Give an example and explain the important statements of Second Law of Thermodynamics.

3.7. Explain perpetual motion machine of first and second kind.

3.8. Is it possible for a heat engine to operate without rejecting any waste heat to a low temperature reservoir? Explain with neat diagram.

3.9. In a refrigerator heat is transferred from a lower temperature medium to higher temperature one. Is this a violation of the second law of thermodynamics ?

3.10. Prove that Kelvin Planck and Clausius statements are equivalent in all respects.

3.11. State and explain Kelvin–Planck constant and Clausius statements of Second Law of Thermodynamics. Prove that violation of Kelvin–Planck statement leads to violation of Clausius statement.

3.12. What is a Quasi-static process ? Explain reversibility and irreversibility. Enumerate the factors making the process irreversible.

3.13. Define thermodynamic reversibility. Under what conditions a process is said to be reversible?

3.14. Explain Carnot Engine and Carnot Cycle. Prove, that $\eta_{\text{Carnot}} = \dfrac{T_1 - T_2}{1}$.

3.15. State and explain Carnot's theorem.

3.16. Explain why Carnot cycle cannot be used in practical engine?

3.17. How the efficiency of IC engines is increased ?

PROBLEMS FOR PRACTICE

3.1. Two reversible heat engines *A* and *B* are operating between the same temperature limits. Engine *A* develops 37.5 kW with 30% efficiency. The engine *B* receives 7535 kJ/min of heat from source. Determine the power developed, by engine *B* and heat rejected by each engine.
(*Ans.* $W_B = 37.675$ kW, $Q_{2A} = 5250$ kJ/min $Q_{2B} = 5274.5$ kJ/min)

3.2. An inventor of a heat engine claims that his engine is 75% efficient which receives heat at 100ºC and rejects at 20ºC. How do you rate his claim ? What is the maximum efficiency you would thihk to be realistic ? (*Ans.* Claim is wrong, 21.4%)

3.3. A heat engine is used to drive a heat pump. The heat transfers from the heat engine and from the heat pump are used to heat the water circulating through the radiators of a building. The efficiency of the heat engine is 27% and the C.O.P. of the heat pump is 4. Evaluate the Ratio of heat transfer to the circulating water to the heat transfer to the engine. (*Ans.* 1.81)

3.4. The C.O.P. of a refrigerator is 6, when it maintains the temperature of 3°C in the evaporator. Determine the condenser temperature and refrigerating effect if the power required to run the refrigerator is 7.5 kW.

3.5. An engineer claims his engine to develop 3.68 kW on testing, the engine consumes 0.44 kg of fuel in one hour. It has calorific value of 41870 kJ/kg. The maximum temperature recorded in the cycle is 1400°C and minimum is 350ºC. Is the engineer justified in his claim ?

3.6. An inventor claims that his newly developed heat engine will produce 0.184 kW power, when the heat is supplied to engine at the rate of 70 kJ/min. The cycle operates between maximum temperature of 1092°C and minimum temperature of 149°C. Make complete analysis of the inventors claim and clearly state whether it is worth investing any amount of money on such a project. Give reasons for your suggestions.

3.7. A heat engine is used to drive a heat pump. The heat transfer from the heat engine and from the heat pump are used to heat the water circulating through the radiators of a building. The efficiency of the heat engine is 27% and C.O.P. of the heat pump is 4. Evaluate the ratio of heat transfer to the circulating water to the heat transfer to the engine.

3.8. A reversible heat engine operates between two reservoirs at temperature of 650°C and 35°C. The engine drives a reversible refrigerator. Which operates between temperature of 35°C and – 20°C. The heat transfer to the engine is 1000 kJ/min. The work output (net) of the combined device is 180 kJ/min.

3.9. Estimate the refrigerating effect in kW and net heat transfer to the reservoir at 35°C. What is the power developed by the engine in kW.

3.10. A reversible heat engine used for a satellite, operates between a hot reservoir at T_1 and radiating panel at T_2. The radiation from the panel is proportional to its area and T_2^4. For a given work output and fixed T_1, show that the area of the panel will be minimum when

$$\frac{T_1}{T_2} = \frac{4}{3}$$

3.11. An inventor claims to have developed a refrigerator machine, which operates between – 20°C and 30°C and consumes and kW of power. The machine gives refrigerating effect of 21.6 MJ/hr. Comment on the claim of the inventor.

3.12. A heat pump is used to heat at room in winter and to cool in summer. The interior temperature is to be maintained at 20°C. The heat transfer through the wall and roof is 2400 kJ/hr per degree temperature difference between inside and outside.

(i) If the outside temperature in winter is 0°C, what is the maximum power required to drive the heat pump ?

(ii) If the power input is the same as that in part (i) what is the maximum outside temperature for which inside can be maintained at 20°C.

A Carnot engine between 537°C and 37°C develops 5 kW of power. Determine,

(i) the thermal efficiency,

(ii) heat supplied per sec.

(iii) the change of entropy per second during heat rejection.

3.13. A heat pump working on Carnot cycle takes in, heat from a reservoir at 5°C and delivers heat to a reservoir at 60°C. The heat pump is driven by a reversible heat engine which takes in heat from a reservoir at 840°C and rejects heat at 60°C. The reversible heat engine also drives a machine that absorbs 30 kW. If pump extracts 17 kJ/sec from 5°C reservoir determine.

(i) The rate of heat supply from 840°C source.

(ii) The rate of heat rejection to 60°C sink.

3.14. A Carnot cycle receives heat at 527°C causing increase in entropy equal to 5 kJ/kg K. The engine delivers 2000 kJ/kg of work. Determine the efficiency of the cycle and lowest temperature in the cycle.

3.15. A small metallic object 4 kg in mass of sp. heat C_p = 0.5 kJ/kg K at a temperature of 227°C is thrown into a lake at temperature of 27°C. Calculate the change in entropy of the universe.

3.16. An inventor claims that he has designed an engine which will produce a power output of 170 kW while consuming 0.5 kg/min of fuel having C.V. of 42000 kJ/kg. He states that the upper and lower temperature of the engine are 670 K and 230 K respectively. Do you believe his claim? Make your comments.

3.17. A Carnot engine operates between the temperature limits of 180°C and 60°C and produces 200 kN m/sec of work. Determine :

(i) Heat supplied in kW

(ii) Change of entropy during heat rejection

(iii) Thermal efficiency of the engine.

3.18. Consider two Carnot heat engines operating in series. The first engine receives heat from a reservoir at 1000 K and rejects the waste heat to another reservoir at temperature *T*. The second engine receives this energy rejected by the first one, converts some of it to work and rejects the rest to a reservoir at 300 K. If the thermal efficiencies of both engines are the same, determine the temperature *T*.

3.19. A house is to be maintained at a temperature of 15°C by means of a heat pump by pumping heat from the atmosphere. Heat losses through the walls are estimated to be 0.75 kW per unit temperature difference between the inside of the house and the outside atmosphere. If the atmosphere temperature is –10°C, calculate the power required to drive the heat pump if C.O.P. of the actual heat pump is 25% of C.O.P. of Carnot's heat pump working between the same temperature limits.

3.20. A heat engine receives half of its heat at a temperature of 1000 K and the rest at 5000 K while rejecting heat to a sink at 300 K. What is the maximum possible thermal efficiency of this heat engine ?

3.21. Source *A* can supply energy at the rate of 12000 kJ /min at 320°C. A second source *B* can supply energy at the rate of 12000 kJ/min at 70°C. Which source *A* or *B* would you choose, to supply energy to an ideal reversible heat engine that is to produce large amount of flow. If the temperature of surroundings is 35°C.

3.22. A domestic refrigerator maintains a temperature of –12°C. The ambient air temperature is 35°C. If heat leaks into the freezer at the continuous rate of 2 kJ/sec, determine the least power necessary to maintain the refrigerator.

3.23. A reversible engine takes 1190 kJ/min from a reservoir at 700 K and develops 190 kJ of work/min when executing complete cycle. The engine rejects heat to two reservoirs at 600 K and 500 K. Find the heat rejected to each sink.

3.24. A system undergoes a reversible cycle during which it exchanges heat with three thermal reservoirs *A*, *B*, *C* at T_1, T_2 and T_3. The amount of heat exchanged is Q_1, Q_2 and Q_3. Show that

$$\frac{Q_1}{\frac{1}{T_2}-\frac{1}{T_3}}=\frac{Q_1}{\frac{1}{T_3}-\frac{1}{T_1}}=\frac{Q_1}{\frac{1}{T_1}-\frac{1}{T_2}}$$

3.25. Two Carnot heat pumps are arranged in series. Pump 1 takes 150 kJ/cycle from heat sink at 300 K. The heat output from this heat pump serves as heat input to pump 2, which delivers its output to a heat reservoir at 1200 K. If the heat pumps have same C.O.P., how much energy (kJ/cycle) must be delivered to pump 1 and to pump 2.

IMPORTANT POINTS TO BE REMEMBERED

1. *Heat engine converts* heat energy into mechanical energy.
   (a) External combustion engines — steam turbine, steam engines.
   (b) Internal cmbustion engines — Petrol, Diesel, gas engines.
2. *Heat reservoir* is a body which is capable of absorbing (sink) or rejecting (source) quantity of heat. It may be *infinite* heat reservoir in which case there is no change in temperature or *finite* heat reservoire in which case temperature changes.
3. *Refrigerator* is a device which produces and maintains the temperature below atmospheric temperature.
4. *Heat pump* is a device operating in a cycle, maintains a body at a temperature higher than the atmosphere or surroundings.
5. *Kelvin–Plank* statement of second law of thermodynamics states that it is impossible for a heat engine to produce net work in a cycle, when it exchanges heat with a single heat reservoire.
6. *Clausium statement* of the second law of thermodynamics states that it is impossible to construct a device, which operating in a cycle will produce no effect other than transfer of heat from a low temperature body to a high temperature body without any external work.
7. Both these statements are equivalent.
8. *Conditions for reversibility are*
   (a) A process should be quasistatic
   (b) There should be no friction
   (c) Both the system and the surrounding should restore their initial state after the process is reversed
9. *Carnot cycle* is a reversible cycle.
10. *Carnot cycle* theorm states that no engine operating between two thermal reservoires can have an efficiency more than that of a Carnot engine operating between the same two reservoires i.e.
    Carnot engine will be more efficient than any other engine operating between the same temperature reservoires

$$\eta_{carnot} = \frac{T_1 - T_2}{T_1}.$$

CHAPTER 4

Ideal Gases and Ideal Gas Processes

4.1 INTRODUCTION

An ideal gas is one, which obeys all the gas laws and the characteristic gas equation $PV = mRT$, at all temperatures and pressures.

But actually there is no perfect or Ideal gas in reality. But all the real gases like Air, O_2, H_2, N_2, He, behave as ideal gases, at very low pressures and high temperatures.

4.2 IDEAL GAS LAWS

The behaviour of ideal gas is governed by following gas laws:

1. Boyle's law; 2. Charle's law ; 3. Avagadro's law.

1. Boyle's Law. Boyle experimentally established that, the volume of a given mass of a gas is inversely proportional to the absolute pressure, when the temperature is constant.

i.e., $$V \propto \frac{1}{P}, \text{ for particular temperature } T$$

$\therefore$ $$V = \frac{\text{Constant of proportionality}}{P} = \frac{C}{P}$$

or $$PV = C$$

where C is a constant.

Also $$P_1V_1 = P_2V_2 = C$$

Consider a gas in a cylinder fitted with a frictionless piston. Let the initial properties be P_1 and V_1. Now when the gas is heated at constant temperature, then the gas expands and the pressure falls to P_2 and volume increases to V_2. This is represented on PV-diagram as shown in Fig. 4.1.

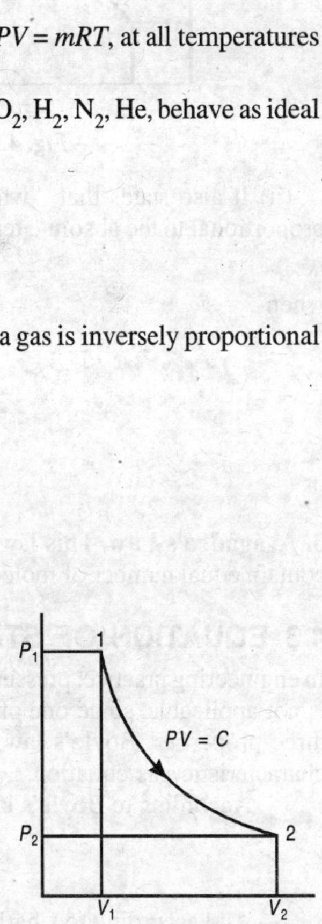

Fig. 4.1

2. Charle's Laws. (i) It states that, "when the pressure of a given mass of a gas is kept constant, then volume is directly to the absolute temperature". (Ref. Fig. 4.2)

i.e., When P = Constant

Then $V \propto T$

or V = Constant of proportionality T

or $\frac{V}{T}$ = Constant of proportionality

or $\frac{V_1}{T_1} = \frac{V_2}{T_2}$ = Constant of proportionality

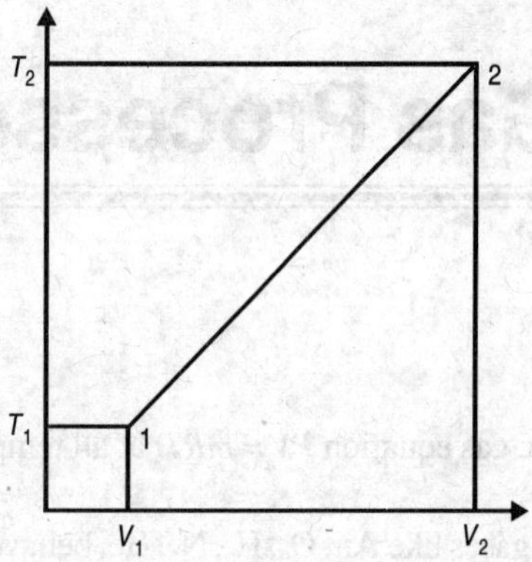

Fig. 4.2

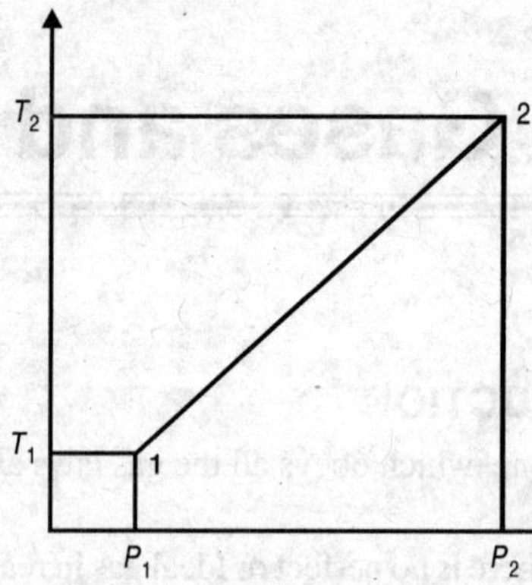

Fig. 4.3

(ii) It also states that, "when the volume of a given mass of a gas is kept constant, the pressure is directly proportional to the absolute temperature". (Ref. Fig. 4.3)

i.e. $P \propto T$

when V = Constant

or P = Constant of proportionality T

$\frac{P}{T}$ = Constant of proportionality

or $\frac{P_1}{T_1} = \frac{P_2}{T_2}$ = Constant of proportionality

3. Avagadro's Law. This law states that at the same pressure and temperature, equal volumes of different gases contain equal number of molecules.

4.3 EQUATION OF STATE OR CHARACTERISTIC GAS EQUATION

In engineering practice, pressure, volume, temperature all vary simultaneously. So, Boyle's law or Charle's law alone is not applicable, since one of three properties is kept constant. In order to establish a relationship between these three properties, Boyle's law and Charle's laws are combined together, which gives a general equation called characteristic gas equation.

According to Boyle's law,

$$V \propto \frac{1}{P} \text{ when } T = \text{Constant} \quad \ldots(1)$$

And according to Charle's law,

$$V \propto T \text{ when } P = \text{Constant} \quad \ldots(2)$$

It is obvious that,

$$V \text{ is } \propto \text{ to } T \text{ and } \frac{1}{P}$$

i.e. $$V \propto \frac{T}{P} \quad \text{...(3)}$$

or $$PV \propto T$$

or $$PV = CT \text{ where } C \text{ is a constant}$$

or $$\frac{PV}{T} = C \quad \text{...(4)}$$

or In general $$\frac{P_1V_1}{T_1} = \frac{P_2V_2}{T_2} = \frac{P_3V_3}{T_3} = \text{........} = C$$

or Since $$\upsilon = \text{specific volume in } \frac{\text{m}^3}{\text{kg}}$$

Then equation will be, $$\frac{P\upsilon}{T} = C \quad \text{...(5)}$$

When 1 kg of gas is considered, then the constant for the Eq. (4) is written as R and is called *Characteristic gas constant.*

$$\therefore \quad \frac{Pv}{T} = R \quad \text{...(6)}$$

Now consider m kg of gas. Multiply both sides of Eq. (6) by m

$$\frac{P\cdot(m\cdot v)}{T} = m.R$$

$$m.v = V = \text{Total volume.}$$

$$\therefore \quad \mathbf{PV} = \mathbf{mRT}$$

This is known as characteristic equation of an ideal gas.

Note that the value of R for air $= 0.287 \dfrac{\text{kJ}}{\text{kg-K}}$

4.4 UNIVERSAL GAS CONSTANT

Frequently it is required to express the equation of state, $PV = mRT$. (1) on mole basis. The mass of a substance m is equal to the product of number of moles n and the molecular weight M.

i.e., $$m = n.M \quad \text{...(1)}$$

Hence susbtituting the value of m in Eq. (1) we get,

$$PV = n.M.RT \quad \text{...(2)}$$

Now the molar specific volume is defined as the volume/unit mole and is denoted by $\bar{v}$,

$$\therefore \quad \bar{v} = \frac{V}{n}$$

Hence Eq. (2) becomes,

$$P\bar{v} = MRT \quad \text{...(3)}$$

Now if we consider two gases, *a* and *b* occupying equal molar volumes at the same temperature and pressure, Eq. (3) becomes,

$$P_a \cdot \bar{v}_a = M_a R_a T_a$$

and

$$P_b . \bar{v}_b = M_b \cdot R_b \cdot T_b$$

According to Avagardo's law, perfect gases at the same temperature and pressure, occupy the same molar volumes.

i.e.

$$\bar{v}_a = \bar{v}_b \text{ when } P_a \text{ and } T_a = T_b$$

We have,

$$\frac{P_a . \bar{v}_a}{T_a} = \frac{P_b . \bar{v}_b}{T_b}$$

$\therefore$

$$M_a . R_a = M_b . R_b = M.R$$

The product *M.R* is called the *Universal Gas Constant* and it is denoted by $\bar{R}$

$\therefore$ Equation (3) can be written as,

$$P\bar{v} = MRT$$

$$P\bar{V} = \bar{R}T$$

and Eq. (2) can be written as,

$$P\bar{V} = nMRT$$

$$P\bar{V} = n\bar{R}T$$

and the value of $\bar{R}$ for air $= 8.3143 \dfrac{\text{kJ}}{\text{k mol K}}$

4.5 JOULE'S EXPERIMENT TO PROVE $U = f(T)$

Internal Energy-Joule's Law

Joule's Law states that the internal energy of an ideal gas depends only on the temperature of gas and is independent of changes in pressure and volume i.e., $U = f(T)$.

Joule's Experiment

The apparatus for Joule's experiment is shown in Fig. 4.4. It consists of two copper vessels *A* and *B* connected by a valve *V*. Initially *A* contained high pressure air and *B* was evacuated. Both were placed in an insulated water bath with a thermometer to measure the temperature.

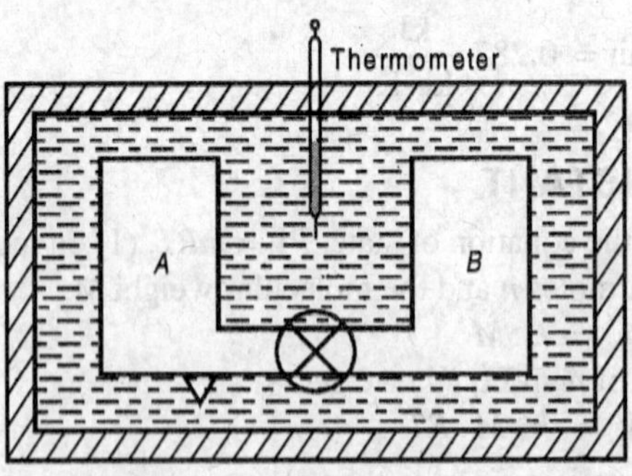

Fig. 4.4. *Joule's experiment*

The initial temperature of water was recorded. The valve *V* was then opened and air is allowed to flow from vessel *A* to vessel *B*.

The operation is slow enough for air to attain equilibrium states. Eventually pressure in both the vessels will become equal and flow will be stopped. Joule observed that there was no change in the temperature of water during or after the process.

Since there was no change in temperature, there was no heat transfer between air and water. Therefore $\delta Q = 0$. Also no work is done during the process. Therefore $\delta W = 0$. By first law of thermodynamics for closed system we know that $\delta Q = dU + \delta W$.

Therefore $dU = 0$ or change in internal energy is zero. The internal energy therefore remains constant.

Again we observe that pressure and volume both changed during the process and only temperature remained constant. Joule therefore concluded that internal energy is the function of temperature only and independent of pressure and volume changes.

Thus, $$U = f(T)$$

4.6 RELATIONS BETWEEN C_P AND C_V

We have studied specific heat at constant pressure and specific heat at constant volume in the I-law of thermodynamics.

Since C_p and C_v are the properties of a system, there is a relationship between these.

We know that, specific enthalpy h is given by,

$$h = u + PV$$

$$dh = du + d(P \cdot v) \qquad \ldots(1)$$

from the definition of,

$$\mathbf{C_p} = \left(\frac{\delta \mathbf{q}}{\mathbf{dT}}\right)_p = \left(\frac{\mathbf{dh}}{\mathbf{dT}}\right)_p$$

$$\therefore \quad dh = C_p \cdot dT \qquad \ldots(2)$$

and $$\mathbf{C_v} = \left(\frac{\delta \mathbf{q}}{\mathbf{dT}}\right)_v = \left(\frac{\mathbf{du}}{\mathbf{dT}}\right)_v$$

$$\therefore \quad du = C_v \cdot dT \qquad \ldots(3)$$

Since, $$PV = RT$$

$$d(PV) = d(RT)$$

$$d(PV) = R \cdot dT, \text{ since } R \text{ is constant} \qquad \ldots(4)$$

Substituting Eqs (2), (3) and (4) in (1) we get,

$$C_p \cdot dT = C_v\, dT + R \cdot dT$$

$$\therefore \quad C_p = C_v + R \quad \text{(Since } dT \text{ is common in both sides)}$$

or $$\mathbf{C_p - C_v = R} \qquad \ldots(5)$$

We also know that the ratio $\dfrac{C_p}{C_v} = \gamma$

$$\therefore \quad C_p = \gamma \cdot C_v$$

Substituting in, $$C_p - C_v = R$$

we get, $$\gamma \cdot C_v - C_v = R$$

$$\mathbf{C_v} = \frac{\mathbf{R}}{\boldsymbol{\gamma} - \mathbf{1}} \qquad \ldots(6)$$

and since $$C_p = \gamma \cdot C_v$$

$$\mathbf{C_p} = \frac{\boldsymbol{\gamma} \cdot \mathbf{R}}{\boldsymbol{\gamma} - \mathbf{1}} \qquad \ldots(7)$$

4.7 IDEAL GAS PROCESSES

4.7.1 Constant Volume Process

During this process, volume remains constant ($V = C$) and is represented on a PV-diagram by means of vertical line as shown.

When unit mass of a gas is heated in a closed vessel (i.e., $V = C$), then, since volume remains constant, no external work is done. But as temperature of the gas increases I.E. increases.

(a) Work Done. Work done in a non-flow system = $\int PdV$.

Here, $V = C$ $\qquad \therefore\ dV = 0$

$$\therefore\ W_{1-2} = 0$$

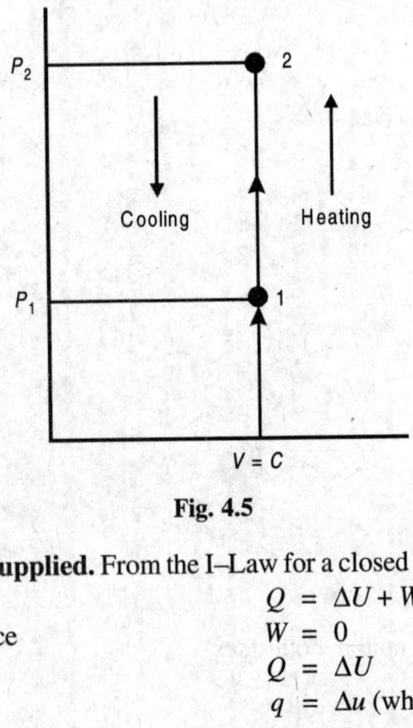

Fig. 4.5

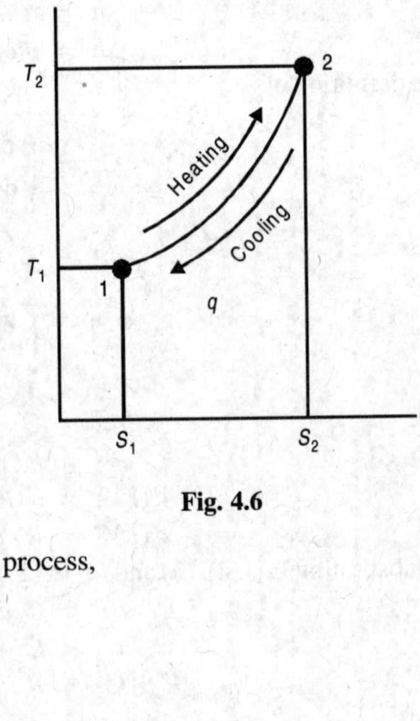

Fig. 4.6

(b) Heat Supplied. From the I–Law for a closed system undergoing a process,

$$Q = \Delta U + W \cdot \text{ In this case}$$

Since $\qquad W = 0$

$\therefore \qquad Q = \Delta U$

or $\qquad q = \Delta u$ (where q is kg)

$\therefore \qquad \delta q = du = C_v \cdot dT$

From the definition of C_v. For total mass,

$$\mathbf{\delta Q = dU = m \times C_v \times dT}$$

4.7.2 Constant Pressure Process

During this process pressure remains constant ($P = C$) and this process is represented by means of a horizontal line on the PV-diagram as shown.

When a unit mass of a gas is taken in a cylinder fitted with a frictionless piston and is heated. Then the piston moves up maintaining same pressure (i.e., $P = C$). As the volume of the gas increases, the work is done by the gas on the piston. As the temperature of the gas increases, Internal Energy increases. So, heat supplied in a constant pressure heating process is utilised for two purposes:

(i) For doing some external work.

(ii) For increasing the I.E. of the gas.

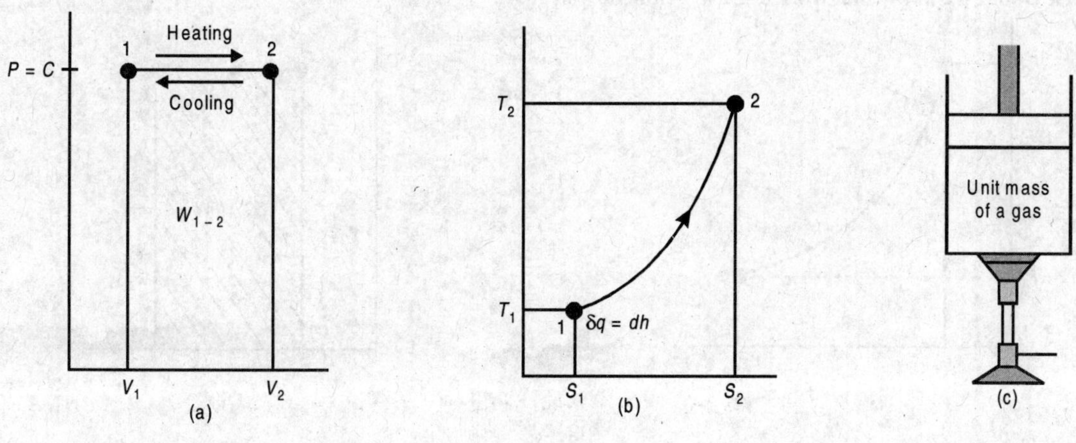

Fig. 4.7

(a) Work Done

Work done in non-flow system.

$$W_{1-2} = \int_1^2 PdV$$

Since P is constant.

$$\therefore \quad W_{1-2} = P\int_1^2 dV$$

$$W_{1-2} = P(V_2 - V_1)$$

If the pressure P is in $\frac{\text{N}}{\text{m}^2}$ and volume is in m^3 then the resultant unit of work will be N-m or J.

(b) Change in Internal Energy

$$(\Delta U) = U_2 - U_1$$

(c) Heat Transferred. From the I-law,

$$Q - W = \Delta U$$

$$Q = \Delta U + W$$

or

$$\delta Q = dU + \delta W$$

$$\delta Q = dU + P\,dV$$

$$\delta Q = dU + d(PV) \text{ since } P = C$$

$$\delta Q = d(U + PV)$$

$$U + PV = H \text{ which is enthalpy}$$

$$\delta Q = dH$$

or

$$\delta q = dh$$

where h is enthalpy/kg

i.e. heat supplied = change in enthalpy

4.7.3 Isothermal Process

During this process, temperature remains constant ($T = C$). The law for the process is $PV = C$ and is represented by means of a curve as shown on the PV-diagram. It is represented by means of a horizontal line on T–S diagram.

(a) Work Done. We know that work done in a non-flow system.

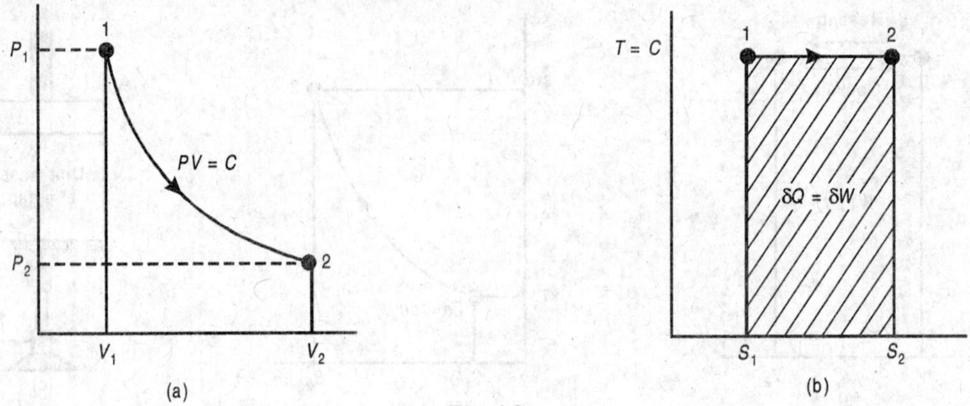

Fig. 4.8

$$W_{1-2} = \int_1^2 P.dV \quad \text{...(1)}$$

The law for the process is,

$$PV = C$$

i.e.,

$$PV = P_1 V_1 = P_2 V_2 = C$$

$$\therefore \quad P = \frac{P_1 V_1}{V} \text{ Substitute this in (1) we get,}$$

$$\mathbf{W_{1-2} = P_1 V_1 \ln \frac{V_2}{V_1}}$$

$$\mathbf{W_{1-2} = P_1 V_1 \ln \frac{P_1}{P_2} \text{ since } \frac{V_1}{V_2} = \frac{P_1}{P_2}}$$

(b) Change in IE $\quad dU = mC_v\, dT = 0 \text{ as } T = C,\, dT = 0$

(c) Heat Transferred. From the first law for a closed system undergoing a process,

$$Q = \Delta U + W$$

$$\delta Q = dU + \delta W$$

$$\delta Q = 0 + \delta W$$

$$\mathbf{\delta Q = \delta W = P_1 V_1 \ln \frac{V_2}{V_1} = P_1 V_1 \ln \frac{P_1}{P_2}}$$

4.7.4 Reversible Adiabatic Process

A process is said to be reversible adiabatic, when the heat added or rejected during the process is zero. In this case the process should be very fast so that there is very very little time for the exchange of heat to take place.

The system for this process may be considered as piston and cylinder mechanism.

We know that first law for the process is,

$$Q - W = \Delta U$$

$$Q = \Delta U + W$$

$$\delta Q = dU + \delta W$$

$$\delta Q = dU + PdV$$

$$0 = C_v \cdot dT + PdV \quad \text{...(1)}$$

Since heat transfer in zero.

We also know that, for an ideal gas,

$$PV = RT$$

Differentiating this we get,

$$P \cdot dV + V\,dP = R \cdot dT$$

$$\frac{P \cdot dV + VdP}{R} = dT \qquad \text{...(2)}$$

Substituting the value of dT in Eq. (1)

We get,

$$O = C_v \frac{(P \cdot dV + V \cdot dP)}{R} + P \cdot dV$$

$$= P \cdot dV\left(1 + \frac{C_v}{R}\right) + \frac{C_v}{R} \cdot VdP$$

$$= C_p \cdot \frac{P}{R} \cdot dV + \frac{C_v}{R} \cdot V \cdot dP$$

Dividing this equation by PV we get,

$$O = \frac{C_p}{R} \cdot \frac{dV}{V} + \frac{C_v}{R} \cdot \frac{dP}{P}$$

Now dividing by C_v throughout, we get,

$$O = \frac{C_p}{C_v R} \cdot \frac{dV}{V} + \frac{1}{R} \cdot \frac{dP}{P}$$

$$= \frac{C_p}{C_V} \cdot \frac{dV}{V} + \frac{dP}{P}$$

$$= \gamma \cdot \frac{dV}{V} + \frac{dP}{P}$$

$$\therefore \quad \frac{dP}{P} = -\gamma \cdot \frac{dV}{V} \qquad \text{...(3)}$$

Integration of this equation gives

$$\int_1^2 \frac{dP}{P} = -\gamma \int_1^2 \frac{dV}{V}$$

$$\mathbf{\log_e P = -\gamma \cdot \log_e V}$$

$$\therefore \quad [\log_e P]_1^2 = -\gamma[\log_e V]_1^2$$

$$\therefore \quad \log_e P_2 - \log_e P_1 = -\gamma[\log_e V_2 - \log_e V_1]$$

$$\therefore \quad \log_e \frac{P_2}{P_1} = -\gamma \log_e \frac{V_2}{V_1}$$

$$\therefore \quad \log_e \frac{P_2}{P_1} = \log_e \left(\frac{V_2}{V_1}\right)^{-\gamma}$$

$$\therefore \quad \left(\frac{V_2}{V_1}\right)^{-\gamma} = \frac{P_2}{P_1}$$

$$\therefore \quad \left(\frac{V_1}{V_2}\right)^{\gamma} = \frac{P_2}{P_1}$$

$$\therefore \quad P_1V_1^{\gamma} = P_2V_2^{\gamma}$$

or $$\mathbf{PV^{\gamma} = Constant} \qquad \ldots(4)$$

Combining this equation with ideal gas equation, we get,

$$\frac{P_1V_1}{T_1} = \frac{P_2V_2}{T_2}$$

$$\therefore \quad \frac{T_1}{T_2} = \frac{P_2V_2}{P_1V_1}$$

But $$\frac{P_2}{P_1} = \left(\frac{V_1}{V_2}\right)^{\gamma}$$

$$\therefore \quad \frac{T_2}{T_1} = \left(\frac{V_1}{V_2}\right)^{\gamma} \cdot \frac{V_2}{V_1} = \left(\frac{V_1}{V_2}\right)^{\gamma-1}$$

$$\mathbf{\frac{T_2}{T_1} = \left(\frac{V_1}{V_2}\right)^{\gamma-1}} \qquad \ldots(5)$$

Similarly, $$\frac{V_2}{V_1} = \left(\frac{P_1}{P_2}\right)^{\frac{1}{\gamma}}$$

$$\therefore \quad \frac{T_2}{T_1} = \frac{P_2}{P_1} \times \left(\frac{P_1}{P_2}\right)^{\frac{1}{\gamma}}$$

$$= \left(\frac{P_2}{P_1}\right)^{\frac{\gamma-1}{\gamma}}$$

$$\therefore \quad \mathbf{\frac{T_2}{T_1} = \left(\frac{P_2}{P_1}\right)^{\frac{\gamma-1}{\gamma}}} \qquad \ldots(6)$$

$$\therefore \quad \mathbf{\frac{T_2}{T_1} = \left(\frac{V_1}{V_2}\right)^{\gamma-1} = \left(\frac{P_2}{P_1}\right)^{\frac{\gamma-1}{\gamma}}} \qquad \ldots(7)$$

Now for Adiabatic Process ($PV^{\gamma} = C$) and during this process neither heat enters the system nor heat leaves the system.

(a) Work Done

Note. Derivation is just similar to work done in polytropic process, put γ in place of n.

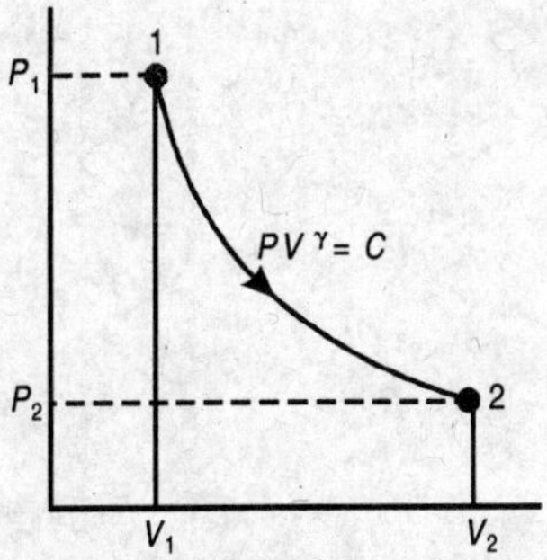

Fig. 4.9

$\therefore$ Work done, $$W_{1-2} = \int_1^2 PdV$$

$$W_{1-2} = \frac{P_2V_2 - P_1V_1}{1-\gamma}$$

$$W_{1-2} = \frac{P_1V_1 - P_2V_2}{\gamma - 1}$$

(b) Heat supplied. Here Heat supplied = 0 (Since neither heat enters nor leaves the system during adiabatic the process).

4.7.5 Polytropic Process

Polytropic Process is given by the law ($PV^n = C$) and is represented on the PV-diagrams as shown.

(a) Work done. Work done in a non-flow process

$$W_{1-2} = \int PdV \qquad \text{...(1)}$$

For a polytropic process,

$$PV^n = P_1V_1^n = P_2V^n = C$$

$$\therefore \quad P = \frac{C}{V^n}$$

Substituting in Eq. (1), we get,

$$\int Pdv = \int \frac{CdV}{V^n} = C\int V^{-n} dv$$

$$= \left[V^{1-n}\right]_1^2 = \left(V_2^{1-n} - V_1^{1-n}\right)$$

$$W_{1-2} = \frac{P_2V_2 - P_1V_1}{1-n} \text{ or } \frac{P_1V_1 - P_2V_2}{n-1} \qquad \text{...(2)}$$

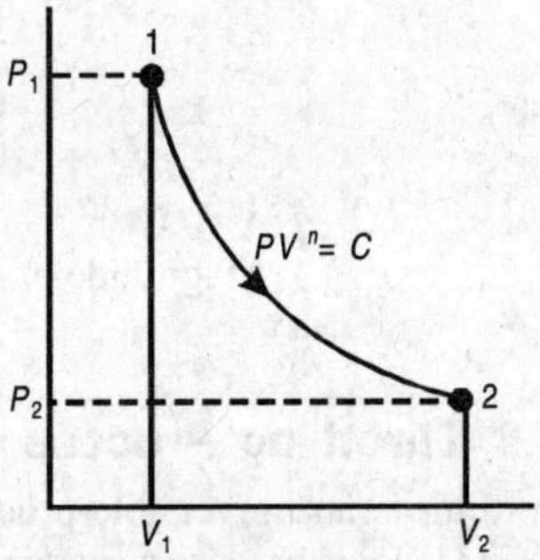

Fig. 4.10

Since, $$P_1V_1 = mRT_1$$

and $$P_2V_2 = mRT_2$$

$$W_{1-2} = \frac{mR(T_2 - T_1)}{1-n}$$

And for unit mass, $$W_{1-2} = \frac{R(T_2 - T_1)}{1-n}$$

(b) Heat supplied. From the I-law for closed system undergoing a process,

$$Q = \Delta U + W$$

For unit mass, $q = \Delta u + w$

In the differential form, $$\delta q = du + \delta w$$

$$\delta q = du + Pdv$$

$$q_{1-2} = C_v(T_2 - T_1) + \frac{R(T_2 - T_1)}{1-n} \text{ [From Eq. (3)]}$$

$$= \left[C_v + \frac{R}{1-n}\right](T_2 - T_1)$$

$$= \left[\frac{C_v - nC_v + C_p - C_v}{1-n}\right](T_2 - T_1) \text{ since } C_p - C_v = R$$

$$= \left[\frac{C_p - nC_v}{1-n}\right](T_2 - T_1)$$

Taking C_v common, we can write,

$$q_{1-2} = C_v\left[\frac{\gamma - n}{1-n}\right](T_2 - T_1), \text{ as } \frac{C_p}{C_v} = \gamma$$

$$\mathbf{q_{1-2} = C_n.(T_2 - T_1)}$$

where, $$C_n = C_v.\left[\frac{\gamma - n}{1-n}\right] \text{ or we can write as, } C_n = -C_v\left[\frac{\gamma - n}{n-1}\right]$$

i.e., $$\text{Magnitude of } C_n = C_v\left[\frac{\gamma - n}{n-1}\right] \text{ is known as Polytropic specific heat} \quad \ldots(5)$$

4.7.6 Throttling Process

Throttling is an irreversible process in which a fluid, flowing across a restriction, undergoes a drop in a total pressure, such a process occurs in the flow through a porous plug, a partially closed valve, or a small orifice. Joule and Thomson performed the basic throttling experiments in the period 1852–62, and their experiments clarified the process and let to use of throttling as a method for determining certain properties of gaseous substances.

The apparatus used in Joule Thomson experiment is shown in Fig. 4.11.

A steady stream of gas flows through a porous plug contained in a horizontal tube. This system is open, is thermally insulated ($Q = 0$) and does not exchange work with its environment ($W = 0$). At sections 1 and 2, both the

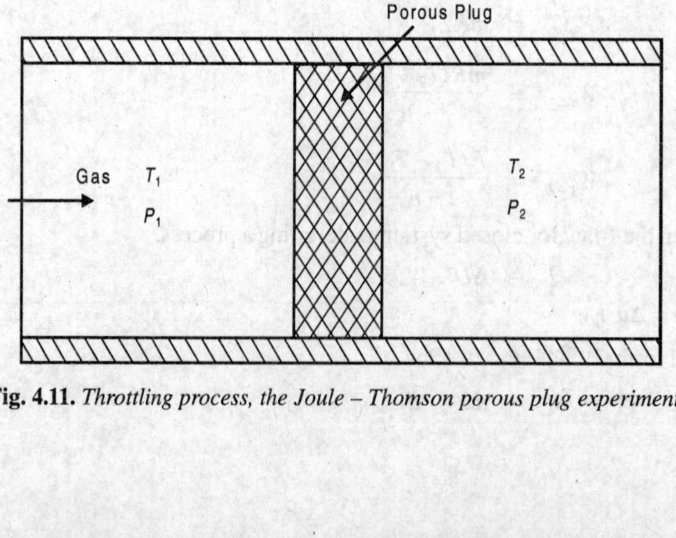

Fig. 4.11. *Throttling process, the Joule – Thomson porous plug experiment*

Table 4.1 shows the various quantities involved in various processes

Table 4.1

Process	Relation between P, V, T	$\int_1^2 P.dv$	$-\int_1^2 V.dP$	ΔU	ΔH	Q	Value of n	C	Work	ΔH
Constant volume	$V_1 = V_2$ $\frac{P_1}{P_2} = \frac{T_1}{T_2}$	0	$V_1(P_2 - P_1)$	$mC_v(T_2 - T_1)$	$mC_p(T_2 - T_1)$	$mC_v(T_2 - T_1)$	∞	C_v	0	$mC_v \log_e \frac{T_2}{T_1}$
Constant pressure	$P_1 = P_2$ $\frac{T_2}{T_1} = \frac{V_2}{V_1}$	$P_1(V_2 - V_1)$	0	$mC_v(T_2 - T_1)$	$mC_p(T_2 - T_1)$	$mC_p(T_2 - T_1)$	0	C_p	$mP_1(V_2 - V_1)$	$mC_p \log_e \frac{T_2}{T_1}$
Isothermal	$T_1 = T_2$ $P_1V_1 = P_2V_2$	$P_1V_1 \log_e \frac{V_2}{V_1}$ or $P_1V_1 \log_e \frac{P_1}{P_2}$	$P_1V_1 \log_e \frac{V_2}{V_1}$ or $P_1V_1 \log_e \frac{P_1}{P_2}$	0	0	$P_1V_1 \log_e \frac{V_2}{V_1}$ or $P_1V_1 \log_e \frac{P_1}{P_2}$	1	∞	$P_1V_1 \log_e \frac{V_2}{V_1}$	$R \log_e \frac{V_2}{V_1}$
Reversible adiabatic	$P_1V_1^r = P_2V_2^r$ $\frac{T_2}{T_1} = \left(\frac{V_1}{V_2}\right)^{r-1}$ $= \left(\frac{P_2}{P_1}\right)^{\frac{r-1}{r}}$	$\frac{P_1V_1 - P_2V_2}{r-1}$	$\frac{\gamma(P_1V_1 - P_2V_2)}{\gamma - 1}$	$mC_v(T_2 - T_1)$	$mC_p(T_2 - T_1)$	0	γ	0	$\frac{P_1V_1 - P_2V_2}{\gamma - 1}$ for NF $\frac{\gamma(P_1V_1 - P_2V_2)}{\gamma - 1}$ for SF	0
Polytropic	$P_1V_1^n = P_2V_2^n$ $\frac{T_2}{T_1} = \left(\frac{V_1}{V_2}\right)^{\frac{n-1}{1}}$ $= \left(\frac{P_2}{P_1}\right)^{\frac{n-1}{n}}$	$\frac{P_1V_1 - P_2V_2}{n-1}$	$\frac{n}{n-1}(P_1V_1 - P_2V_2)$	$mC_v(T_2 - T_1)$	$mC_p(T_2 - T_1)$	$mC_v.\frac{\gamma - 1}{n-1}(T_2 - T_1)$ $= mC_n(T_2 - T_1)$	n	$C_v\left(\frac{\gamma - 1}{n-1}\right)$ $= C_n$	$\frac{P_1V_1 - P_2V_2}{n-1}$ for NF $\frac{n(P_1V_1 - P_2V_2)}{n-1}$ for SF	$mC_n \log_e \frac{T_2}{T_1}$

temperature and the pressure are measured. If the kinetic energy does not change significantly as the fluid passes through the porous plug, the steady flow energy equation reduces to

$$Q = \Delta P + \Delta K + \Delta H + W_{sf}$$

$$\therefore \quad O = 0 + 0 + \Delta H + 0$$

or $$\Delta H = 0$$

or $$H_2 - H_1 = 0$$

or $$H_2 = H_1 \text{ or } h_1 = h_2 \text{ or } t_1 = t_2$$

Hence, in an adiabatic throttling process the enthalpy remains constant. In actual experiments with different p_2, t_2 does not remain constant but is different than t_1.

4.8 EFFECT OF VARYING '*n*'

Since it is helpful to be able to sketch curves of various thermodynamic process on *P–V* and *T–S* planes about as they should appear Fig. 4.12 will be of assistance in learning to make these sketches. It shows also the effect of varying *n*. Note that the isentropic curve on the *PV* plane is steeper than the isothermal curve and that, on *TS*-plane, the constant volume curve is steeper than the constant pressure curve when both the curves are drawn for the same temperature limits.

Expansions or compressions are imagined to take place from some common point '1'.

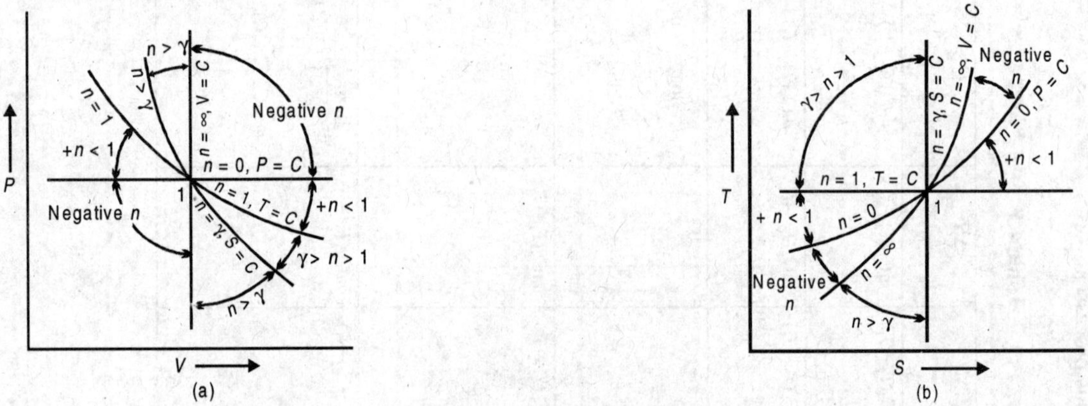

Fig. 4.12. *Effect of varying n*

SOLVED EXAMPLES

Example 4.1 *A gas whose original pressure, volume and temperature were 150 kN/m²; 0.1 m³; and 30°C is compressed such that its new pressure is 750 kN/m² and its new temperature is 70°C. Determine the new volume of the gas.*

Solution

By the characteristic equation i.e., $PV = mRT$

or
$$\frac{P_1V_1}{T_1} = \frac{P_2V_2}{T_2}$$

$$\therefore \quad V_2 = \frac{P_1}{P_2}\cdot\frac{T_1}{T_2}V_1 = \frac{150}{750}\times\frac{343}{303}\times 0.1$$

$$\mathbf{V_2 = 0.0226\ m^3}$$

Note. The temperatures must always be in degrees Kelvin.

Example 4.2 *A quantity of gas has a pressure of 300 kN/m² when its volume is 0.025 m² and its temperature is 37°C. If the value of R is 0.287 kJ/kg K, determine the mass of the gas present. If the pressure of gas is now increased to 0.9 MN/m² while the volume remains constant what will be the new temperature of the gas.*

Solution

(1) By characteristic gas equation,

$$PV = mRT$$

$$\therefore \quad m = \frac{PV}{RT} = \frac{300\times10^3\times0.025}{0.287\times10^3\times310} = 0.084 \text{ kg}$$

(2) We have,
$$\frac{P_1V_1}{T_1} = \frac{P_2V_2}{T_2} \text{ and in this case } V_1 = V_2$$

$$\therefore \quad T_2 = \frac{P_2T_1}{P_1} = 310\times\frac{0.9\times10^6}{300\times10^3}$$

$$\mathbf{T_2 = 930\ K}$$

$$\therefore \quad T_2 = 930 - 273$$

$$\mathbf{T_2 = 675^\circ\ C}$$

Example 4.3 *10 kg of air is heated at constant pressure, from a temperature of 100°C to 200°C. Calculate the heat added during the process and also the change in I.E. Take gas constant* $R = 0.287\frac{kJ}{kg\text{-}K}$ *and g = 1.4 for air.*

Data: $m = 10$ kg, $T_1 = 100°C$, $T_2 = 200°C$

$$R = 0.287\frac{kJ}{kg\ K};\ \gamma = 1.4$$

Solution

We know that, heat added during constant pressure process $= m\,C_P(T_2 - T_1)$ ...(1)

Also,
$$C_V = \frac{R}{\gamma - 1} = \frac{0.287}{1.4-1}$$

$$C_V = 0.7175$$

and
$$C_P - C_V = R \qquad \text{...(2)}$$

$$\Rightarrow \quad C_P = R + C_V$$
$$C_P = 0.287 + 0.7175$$
$$C_P = 1.0045 \frac{\text{kJ}}{\text{kg K}}$$

∴ Substituting (2) and (3) in (1) we get,

$$\text{Heat added} = m\,C_P (T_2 - T_1)$$
$$= 10 \times 1.0045\,(200 - 100)$$
$$\textbf{Head added} = \textbf{1004.5 kJ}$$
$$\text{Charge in I.E.} = \Delta U = m\,C_V (T_2 - T_1) = 10 \times 0.7175\,(200 - 100)$$
$$\Delta U = 717.5 \text{ kJ}$$

Example 4.4 *A mass of 0.8 kg of air at 1 bar and 25°C is contained in a gas tight frictionless piston cylinder device. The air is now compressed to a final pressure of 5 bar. During the process heat is transferred from the air such that the temperature inside the cylinder remains constant. Calculate the heat transfer and work done during the process and direction of each in the process.*

Data :

$m = 0.8$ kg $P_1 = 1$ bar $= 10^5$ N/m²

$T_1 = 298$ K $P_2 = 5$ bar $= 5 \times 10^5$ N/m²

$T = C$, Isothermal process

Solution

∴ We know that, for an isothermal process,

$$Q = W = P_1 V_1 \ln \frac{P_1}{P_2} = mRT_1 \ln \frac{P_1}{P_2}$$
$$\mathbf{Q = W = -110.11\ kJ}$$

as Q is – ve ∴ W is also – ve

Example 4.5 *1 kg of air at an initial condition of 3 bar and 175°C expands adiabatically without friction to a final condition of 1 bar. Calculate,*

(a) External work done during expansion.

(b) Change in I.E.

Take $\gamma = 1.4$ and $R = 0.287 \frac{kJ}{kg\text{-}K}$

Data :

$m = 1$ kg $P_1 = 3 \times 10^5 \frac{N}{m^2} = 3 \times 10^2$ kN/m²

$T_1 = 175 + 273 = 448$ K $P_2 = 1 \times 10^2$ kN/m²

$\gamma = 1.4, R = 0.287 \frac{kJ}{kg\text{-}K}$

Solution

(a) We know that the work done in an adiabatic expansion process is given by,

$$W_{1-2} = \frac{P_1 V_1 - P_2 V_2}{\gamma - 1} \quad ...(1)$$

Here V_1 and V_2 are not given.

Since it is an adiabatic process,

$$P_1 V_1^\gamma = P_2 V_2^\gamma$$
$$\frac{P_1}{P_2} = \left(\frac{V_2}{V_1}\right)^\gamma$$

$$\left(\frac{P_1}{P_2}\right)^{\frac{1}{\gamma}} = \frac{V_2}{V_1}$$

$$\therefore \quad V_2 = V_1 \times \left(\frac{P_1}{P_2}\right)^{\frac{1}{\gamma}} \qquad ...(2)$$

To find V_1, $\quad P_1V_1 = mRT_1$

$$V_1 = \frac{mRT_1}{P_1} = \frac{1\times 0.287\times 448}{3\times 10^2}$$

Note units : $\left[\dfrac{\text{kg}\dfrac{\text{kJ}}{\text{kg}-\text{K}}\times\text{K}}{\text{kN/m}^2} = \dfrac{\text{kJ}}{\text{kN/m}^2} = \dfrac{\text{kNm}}{\text{kN}}\times\text{m}^2 = \text{m}^3\right]$

$$V_1 = 0.429 \text{ m}^3$$

$\therefore$ From Eq. (2), $\quad V_2 = 0.429\times\left(\dfrac{300}{100}\right)^{\frac{1}{1.4}} = 0.429\times(3)^{\frac{1}{1.4}}$

$$\mathbf{V_2 = 0.940\ m^3}$$

$\therefore$ Now from Eq. (1), work done,

$$W_{1\text{-}2} = \frac{P_1V_1 - P_2V_2}{\gamma - 1}$$

$$= \frac{300\times 0.429 - 100\times 0.940}{1.4-1} = \frac{34.7}{0.4}$$

$$\mathbf{W_{1\text{-}2} = 86.75\ kJ}$$

(b) We also know from the I–law, that for a process, $Q = \Delta U + W$. Since it is adiabatic process $Q = 0$

$\therefore$ $\quad$ **W = – ΔU = 86.75 kJ ⇒ ΔU= – 86.75 kJ**

$\therefore$ I.E. decreases by 86.75 kJ.

Example 4.6 *Find the change of Internal energy and enthalpy of 3 kg of air, when the temperature changes from 25°C to 100°C.*

Take $\quad C_p = 1\dfrac{kJ}{kg\text{-}K}$ *and* $C_v = 0.71\dfrac{kJ}{kg\text{-}K}$

Solution

In the problem, the type of process is not given, but since temperature changes, it is not an isothermal process.

$$\therefore \quad \Delta U = m\cdot C_v\cdot(T_2 - T_1)$$

$$= 3\text{ kg}\times 0.71\frac{\text{kJ}}{\text{kg-K}}\times(373-298)\text{ K}$$

$$= 3\times 0.71\times 75$$

$$\mathbf{\Delta U = 159.75\ kJ}$$

and change of enthalpy $\Delta H = m\cdot C_P\cdot(T_2 - T_1) = 3\times 1\times(75)$

$$\mathbf{\Delta H = 225\ kJ}$$

Example 4.7 *0.44 kg of a gas having volume 0.28 m³ and a pressure of 1.4 bar is compressed to a pressure of 14 bar according to $PV^{1.25} = C$. Find the amount of heat transferred during the process by calculating the work done and the change of internal energy.*

Take $C_p = 1.041 \frac{kJ}{kg\text{-}K}$ and $C_v = 0.743 \frac{kJ}{kg\text{-}K}$

Data:

$m = 0.44$ kg, $V_1 = 0.28\ m^3$

$P_1 = 1.4$ bar $= 140$ kPa, $P_2 = 14$ bar $= 1400$ kPa

$PV^{1.25} = C$, $Q = ?$

$W = ?$, $\Delta U = ?$

Solution

$$C_P = 1.041 \text{ and } C_V = 0.743 \frac{\text{kJ}}{\text{kg-K}}$$

$$\text{W.D.} = \frac{P_2 V_2 - P_1 V_1}{1-n} \qquad ...(1)$$

Here only V_2 is not given and since it is a polytropic process,

$$P_1 V_1^n = P_2 V_2^n$$

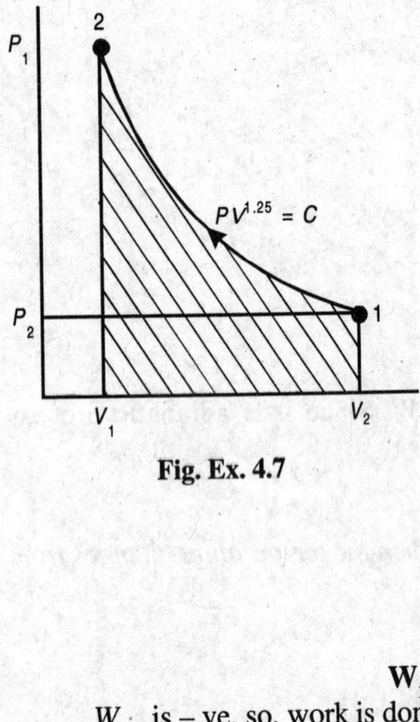

Fig. Ex. 4.7

$$\frac{P_1}{P_2} = \left(\frac{V_2}{V_1}\right)^n$$

i.e.

$$\frac{V_2}{V_1} = \left(\frac{P_1}{P_2}\right)^{\frac{1}{n}}$$

$$V_2 = V_1 \times \left(\frac{P_1}{P_2}\right)^{\frac{1}{n}} = 0.28 \times \left(\frac{140}{1400}\right)^{\frac{1}{1.25}}$$

$$= 0.28 \times (0.1)^{0.8}$$

$$V_2 = 0.044377\ \text{m}^3 \approx 0.044\ \text{m}^3$$

Now work done from Eq. (1),

$$W_{1-2} = \frac{1400 \times 0.0444 - 140 \times 0.28}{1 - 1.25}$$

$$= \frac{62.16 - 39.2}{-0.25}$$

$$\mathbf{W_{1-2} = -91.84\ kJ}$$

W_{1-2} is – ve, so, work is done on the gas.

We know that heat transferred in a polytropic process is given by,

We know that, $$q = C_n \cdot (T_2 - T_1)$$

$$= C_V \cdot \left(\frac{\gamma - n}{1-n}\right)(T_2 - T_1)$$

$$= \frac{R}{\gamma - 1}\left(\frac{\gamma - n}{1-n}\right)(T_2 - T_1)$$

$$q = \frac{R}{(1-n)}\frac{(\gamma-n)}{(\gamma-1)}\times(T_2-T_1)$$

$$q = \frac{(\gamma-n)}{(\gamma-1)}\times\frac{R}{(1-n)}\times(T_2-T_1)$$

And total heat transfer, $$Q = \frac{\gamma-n}{\gamma-1}\times\frac{mR\,(T_2-T_1)}{1-n}$$

i.e. $$Q = \frac{\gamma-n}{\gamma-1}\times \text{W.D. in a Polytropic process.}$$

Note: Remember this equation and derivation.

To find Q,

Let us find, $$\gamma = \frac{C_P}{C_V} = \frac{1.041}{0.743} = 1.401$$

$$Q = \frac{1.401-1.25}{1.104-1}\times -91.84 = \frac{0.15}{0.4}\times -91.84$$

$$\mathbf{Q = -34.583\ kJ}$$

∴ From the first law for a process,

$$Q - W = \Delta U$$

$$-34.583+91.84 = \Delta U$$

∴ $$\mathbf{\Delta U = 57.257\ kJ\ (Increases)}$$

Example 4.8 *A closed system containing 2 kg of working substance undergoes a reversible constant pressure process, during which the I.E. of the system decreases by 80 kJ. If*

$C_p = 1120\dfrac{kJ}{kg\text{-}K}$ *and* $C_v = 800\dfrac{kJ}{kg\text{-}K}$. *Find the work done during the process.*

Data :

$$m = 2\ kg$$
$$P = C$$
$$\Delta U = -80\ kJ$$
$$C_P = 1.120\frac{kJ}{kg\text{-}K}$$
$$C_V = 0.8\frac{kJ}{kg\text{-}K}$$
$$W_{1-2} = ?$$

Solution

We know that work done in a constant pressure process,

$$W_{1-2} = P(V_2 - V_1) \quad \text{...(1)}$$

Also from the I-law, $Q - W = \Delta U$ ...(2)

We know that heat supplied is utilised for,

(i) increasing I.E.,

(ii) doing external work and heat transfer,

$$\delta q = dh$$

or $$\delta Q = dH = m\,C_P\,\Delta T \quad \text{...(3)}$$

Also, $$\Delta U = m.\,C_V.\,\Delta T \quad \text{...(4)}$$

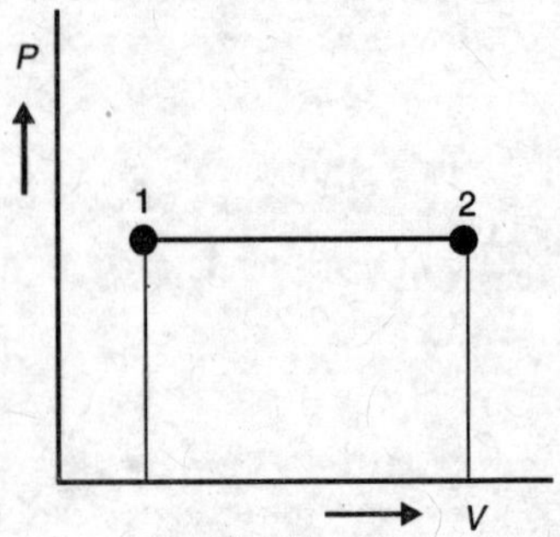

Fig. Ex. 4.8

$$\frac{\Delta U}{m \cdot C_V} = \Delta T$$

$$\frac{-80 \text{ kJ}}{2 \text{ kg} \times 0.8 \frac{\text{kJ}}{\text{kg-K}}} = \Delta T$$

$$\frac{-0.8}{1.6} = \Delta T$$

$$-50K = \Delta T \quad ...(5)$$

$\therefore$ From (3), $\delta Q = m.\, C_P.\, \Delta T = 2 \times 1.12 \times (-50)$

δQ or $Q = -112 \text{ kJ}$

$\therefore$ From Eq. (2), $Q - W = \Delta U$

$$Q - \Delta U = W$$

$$-112 + 80 = W$$

$\therefore$ $\mathbf{W = -32 \text{ kJ}}$

Example 4.9 *A cylinder contains 0.12 m³ of air at 1 bar and 100°C. The air is compressed to 0.03 m³. The final pressure being 6 bar.*

Determine:

(i) *The value of index n.*

(ii) *Mass of air in the cylinder.*

(iii) *Increase in I.E.*

Take $\gamma = 1.4$, $R = 0.287$ kJ/kg-K *and* $C_v = 0.72 \frac{kJ}{kg\text{-}K}$

Data:

$V_1 = 0.12 \text{ m}^3$

$P_1 = 100 \text{ kPa}$

$T_1 = 100 + 273 = 373K \qquad V_2 = 0.03 \text{ m}^3$

$P_2 = 600 \text{ kPa}$

Solution

(i) To find n.

So,
$$P_1 V_1^n = P_2 V_2^n$$

$$\frac{P_2}{P_1} = \left(\frac{V_1}{V_2}\right)^n$$

$$\left(\frac{600}{100}\right) = \left(\frac{0.12}{0.03}\right)^n$$

$$6 = (4)^n$$

$$\log_e 6 = n \log_e (4)$$

$$\frac{\log_e 6}{\log_e 4} = n$$

$$\frac{1.79175}{1.38629} = n$$

$$\mathbf{n = 1.29248 \approx 1.3}$$

(ii) To find mass of air in the cylinder.

We know that, $P_1V_1 = mRT_1$

$100 \times 0.12 = m \times 0.287 \times 373$

$\therefore$ **m = 0.112096 kg**

(iii) To find the increase in I. E.,

We have, $\Delta U = m.C_V(T_2 - T_1)$

Here T_2 is not known, so to find T_2, since it is a polytropic process,

$$\therefore \quad \frac{T_2}{T_1} = \left(\frac{V_1}{V_2}\right)^{n-1} = \left(\frac{P_2}{P_1}\right)^{n-1/n}$$

$$T_2 = T_1 \times \left(\frac{V_1}{V_2}\right)^{n-1}$$

$$= 373 \times \left(\frac{0.12}{0.03}\right)^{1.3-1}$$

$$= 373 \times (4)^{0.3} = 373 \times 1.516$$

$$T_2 = 565.36 \text{ K}$$

$$\therefore \quad \Delta U = m.C_V.(T_2 - T_1)$$

$$= 0.112 \text{ kg} \times 0.72\,(565.36 - 373)$$

$$\mathbf{\Delta U = 15.51 \text{ kJ (Increases)}}$$

Example 4.10 *Calculate the change of,*

(i) IE and

(ii) enthalpy of 2 kg of air when the temperature changes from 20°C to 90°C. Take

$$C_p = 1 \text{kJ/kg-K and } C_v = 0.71 \frac{\text{kJ}}{\text{kg-K}}$$

Solution

(i) $\Delta U = m\,C_V(T_2 - T_1) = 2 \times 0.71(90 - 20)$

$\mathbf{\Delta U = -99.4 \text{ kJ}}$

(ii) $\Delta H = m \cdot C_V \cdot (T_2 - T_1) = 2 \times 1 \times (90 - 20)$

$\mathbf{\Delta H = 140 \text{ kJ}}$

Example 4.11 *10 kg of air are heated in a rigid vessel from 20°C to 100°C. If the ratio of specific heats is 1.4, estimate the value's of C_p and C_v, the change in internal energy and enthalpy.*

Data: T_1 = 20°C, T_2 = 100°C

Solution

$$C_V = \frac{R}{\gamma - 1}$$

and $\gamma = 1.4$ and $R = 0.287$

$$\therefore \quad \mathbf{C_V = \frac{0.287}{0.4} = 0.7175 \frac{kJ}{kg\text{-}K}}$$

$$\mathbf{C_P = R + C_V = 0.287 + 0.7175 = 1.0045 \text{ kJ/kg-K}}$$

We know that,

Change in internal energy,

$$dU = m\,C_V\,(T_2 - T_1)$$

$$= (10)\,0.7175\,(100\text{–}20)$$

$$\mathbf{dU = 574 \text{ kJ} = 574 \text{ kJ}}$$

and change in enthalpy,

$$dH = m\,C_p(T_2 - T_1)$$
$$= (10)(1.0045)(100-80)$$
$$\mathbf{dH = 803.6\ kJ}$$

Example 4.12 *10 kg of Carbon monoxide at 40°C occupies 3 m³. Determine the gas pressure in bar. An additional mass of CO is then very slowly added to raise the tank pressure to 10 bar. Assuming that the gas temperature remains constant, how much extra mass must have been added. Assume R = 0.297 KJ/kg-K.*

Solution

We know that,

$$PV = mRT$$
$$P = \frac{m\,RT}{V} = \frac{10\times0.297\times313}{3}$$
$$P = 309.87\text{ kPa} = 3.0987\text{ bar}$$

Since additional CO is added to the tank, $V = C$ and as given $T = C$.

$\therefore$ $m \propto$ Pressure.

$$\therefore \quad \frac{m_2}{m_1} = \frac{P_2}{P_1} = \frac{10}{3.0987}$$

$$\therefore \quad m_2 = m_1\times\frac{10}{3.0987} = 10\times\frac{10}{3.0987}$$

$$\mathbf{m_2 = 32.2768\ kg}$$

$\therefore$ Additional mass, $\mathbf{\Delta m = 32.2768 - 10 = 22.2768\ kg}$

Example 4.13 *A piston and cylinder machine contains 1 kg of air, initially v = 0.8 m³/kg and T = 290 K. The air is then compressed in a slow frictionless process to a specific volume 0.2 m³/kg and a temperature 580 K. The law for compression is $PV^{1.5} = 0.75$ with P in bar and V in m³/kg. Determine work done and heat transferred during the process. Assume for air $C_p = 1.0\frac{kJ}{kg\text{-}K'}$ and $C_v = 0.743\frac{kJ}{kg\text{-}K}$ and $R = 0.287\frac{kJ}{kg\text{-}K}$.*

Data: $m = 1$ kg, $v_1 = 0.8$ m³/kg, $T_1 = 290$ K, $v_2 = 0.2$ m³/kg $T_2 = 580$ K

W = ? Q = ?, law for the Process = $PV^{1.5} = C = 0.75$

Solution

We know that, for a Polytropic process,

$$W = \frac{P_2V_2 - P_1V_1}{1-n} = \frac{m\,R(T_2 - T_1)}{1-n}$$
$$= \frac{1\times0.287\,(580-290)}{1-1.5}$$
$$= \frac{83.23}{-0.5}\text{ kJ}$$
$$\mathbf{W = -166.46\ kJ}$$

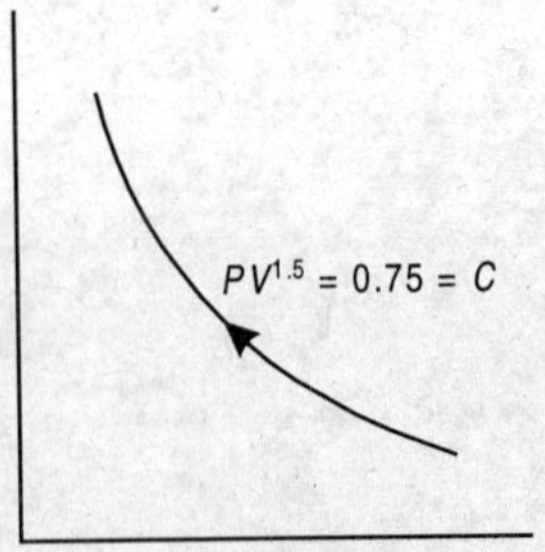

Fig. Ex. 4.13

The sign implies work is done on the air.

Also,

$$\Delta U = m\cdot C_V\cdot dT$$
$$= 1\times0.743\times(580-290)$$
$$\mathbf{\Delta U = 206.77\ kJ}$$

From I-law, for a process,

$$Q - W = \Delta U$$
$$Q = \Delta U + W$$
$$= 206.77 + (-166.46)$$
$$\mathbf{Q = 40.31\ kJ}$$

Example 4.14 *2 kg of air at 150°C and 3 bar expands according to law $PV^{1.2} = C$ to final pressure of 1 bar. Calculate: heat supplied, work done and change in internal energy.*

Take $R = 0.287$ kJ/kg-K and = 1.41.

Data: $P_1 = 3$ bar, $P_2 = 1$ bar, $T_1 = 150°C = 423$ K

Solution

(1) Non flow work done in polytropic process

$$= \frac{mR(T_2 - T_1)}{n-1}$$

We have,
$$\frac{T_2}{T_1} = \left(\frac{P_2}{P_1}\right)^{\frac{n-1}{n}} = \left(\frac{1}{3}\right)^{\frac{0.2}{1.2}} = 0.832$$

$$T_2 = 423 \times 0.832 = 351.936 \text{ K}$$

$$\therefore \quad \text{W.D.} = \frac{0.2 \times 0.287 \times (423 - 351.936)}{1.2 - 1}$$

$$\textbf{W.D.} = \textbf{203.954 kJ}$$

(2) Heat supplied in polytropic process,

$$Q = \left(\frac{\gamma - n}{\gamma - 1}\right) \times \text{W.D. in non flow process.}$$

$$= \frac{1.41 - 1.2}{1.41 - 1} \times 203.954$$

$$\therefore \quad \mathbf{Q = 104.464 \text{ kJ}}$$

(3) Change in internal energy.

$$\Delta U = Q - W$$

$$= 104.464 - 203.954$$

$$\mathbf{\Delta U = 99.49 \text{ kJ decrease}}$$

Alternatively :

(2) Change in internal energy,

$$\Delta U = m\,C_V\,(T_2 - T_1)$$

$$= 2 \times \frac{0.287}{1.41 - 1} \times (351.936 - 423)$$

$$\therefore \quad \mathbf{\Delta U = -99.49 \text{ kJ}}$$

(3) Heat supplied
$$Q = W + \Delta U = 203.954 - 99.49$$

$$\mathbf{Q = 104.464 \text{ kJ}}$$

Example 4.15 *One kg of certain gas undergoes a thermodynamic constant pressure process whereby the volume changes from 1 m³ to 1.8 m³ while the temperature changes from 50°C to 450°C. The specific heat at constant pressure is given by,*

$$C_p = \left(2.5 + \frac{40}{T + 20}\right) kJ/kg\ °C \text{ where } T \text{ is in } °C.$$

Find out (1) heat supplied, (2) work done, (3) change in internal energy and (4) change in enthalpy: Take $\gamma = 1.4$.

Solution

(1) Heat supplied in constant pressure process,

$$= \int m\,.\,C_P\,dT, \quad \text{As given } m = 1$$

$$= \int_{50}^{450} \left(2.5 + \frac{40}{T+20} \right) dT$$

$$= \int_{50}^{450} 2.5\, dT + \int_{50}^{450} \frac{0}{T+20} dT$$

$$= 2.5[T]_{50}^{450} + 40[\ln(T+20)]_{50}^{450}$$

$$= 2.5(450-50) + 40 \ln\left(\frac{470}{70}\right)$$

$$= 1000 + 76.169$$

Heat supplied in constant pressure process = 1076.169 kJ

(2) Change in internal energy is given by,

$$U_2 - U_1 = \int m\, C_V\, dT \text{ where } m = 1\; C_P = \gamma\, C_V$$

$$= \frac{1}{\gamma} \int C_P\, dT = \frac{1076.169}{1.4}$$

$$\mathbf{U_2 - U_1 = 768.692\ kJ}$$

(3) Work done = Heat supplied – Change in internal energy

$$= 1076.169 - 768.692$$

Work done = 307.477 kJ

(4) Change in enthalpy = Heat supplied in constant pressure process

Change in enthalpy = 1076.169 kJ

Example 4.16 *One kg of certain working substance undergoes a reversible constant pressure process at 1.2 bar during which its volume changes from 1 m³ to 2.8 m³ and temperature changes from 50°C to 370°C. The specific heat for the substance at constant pressure is given by,*

$$C_p = \left(1.1 + \frac{40}{t+30}\right) kJ/kg\ °C \text{ where, } t \text{ is in } °C. \text{ Find :}$$

(i) Heat supplied (ii) Work done
(iii) Change in I. E. (iv) Change in Enthalpy.

Data: $P = C.$ Mass = 1 kg

$$P = 1.2\ bar = 1.2 \times 10^5\ \frac{N}{m^2}$$

$V_1 = 1\ m^3$ $V_2 = 1.8\ m^3$

$T_1 = 50°C$ $T_2 = 370°C$

Solution

Given, $$C_P = \left[1.1 + \frac{40}{t+30}\right] \frac{kJ}{kg\ °C}$$

For $P = C$ process $\delta Q = dH$

$$\therefore \quad \delta Q = \text{Change in enthalpy} = \int_{T_1}^{T_2} m\, C_P\, dT$$

$$= \int_{T_1}^{T_2} 1 \times \left(1.1 + \frac{40}{t+30}\right) dT$$

$$= \int_{50}^{370} 1.1\, dT + \int_{50}^{370} \frac{40}{t+30}\, dT$$

$$= [1.1\,t]_{50}^{370} + 40\{\ln(t+30)\}_{50}^{370}$$

Change in enthalpy = 416.38 kJ

Also, for a constant pressure process,

Heat supplied = Change in enthalpy

$\therefore \quad Q = 416.38\,\text{kJ}$

Also, $\quad$ W.D. $= P(V_2 - V_1) = 120(1.8 - 1)$

$\mathbf{W_{1-2} = 96\ kJ}$

According to I-law for a process,

$Q - W = \Delta U$

$\Rightarrow \quad \Delta U = 416.38 - 96$

$\mathbf{\Delta U = 320.38\ kJ}$

Example 4.17 *60 litres of hydrogen at 20ºC and 1 bar is compressed adiabatically to 9.8 bar. It is then cooled at constant volume to a pressure P_3 such as to have temperature of 20ºC an further expanded isothermally so as to reach initial state. Find:*

(i) Work done during each process.

(ii) Change in internal energy during each process. Take

$C_p = 14.4$ kJ/kg-K

$C_v = 10.28$ kJ/kg-K

for hydrogen and show the process on PV and T-S plane.

Data: $V_1 = 0.06\ m^3\ (H_2), T_1 = 20 + 273 = 293\ K, P_1 = 100\ kPa, PV^\gamma = c, P_2 = 980\ kPa,$

$P_3 = T_3 = 20 + 273 = 293\ K, Pv = C, \left.\begin{matrix} C_P = 14.4\ kJ/kg\text{-}K \\ C_V = 10.28\ kJ/kg\text{-}K \end{matrix}\right\} \frac{C_P}{C_V} = \gamma = 1$

Find : (i) W. D. during processes, (ii) during processes.

Solution

(i) Work done :

For process 1– 2, $\quad W_{1-2} = \dfrac{P_2V_2 - P_1V_1}{1-\gamma}$ ...(1)

To find V_2 and since process 1 – 2 is an adiabatic process,

$$P_1V_1^\gamma = P_2V_2^\gamma$$

$$\left(\frac{P_1}{P_2}\right)^{\frac{1}{\gamma}} \times V_1 = V_2$$

$\therefore \quad \mathbf{V_2 = 0.01175\ m^3}$

$\therefore \quad W_{1-2} = \dfrac{980 \times 0.01175 - 100 \times 0.06}{1 - 1.4}$

$\mathbf{W_{1-2} = -13.788\ kJ}$

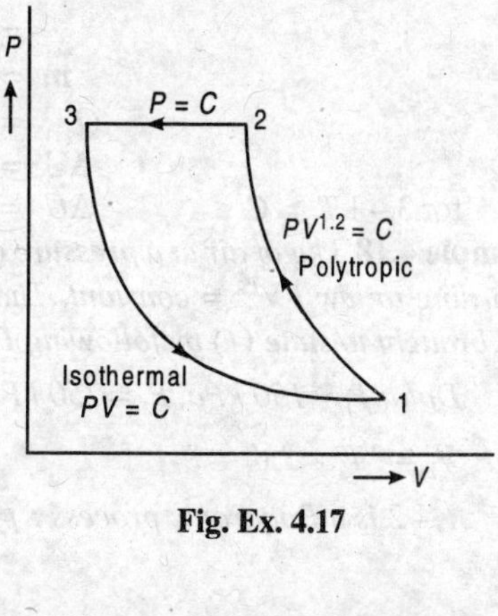

Fig. Ex. 4.17

Process 2 – 3 : Constant volume process $W_{2-3} = 0$

Process 3 – 1 : (Isothermal process) :

$$W_{3-1} = P_3 V_3 \ln\frac{V_1}{V_3} \qquad \text{...(2)}$$

Note, $V_2 = V_3$

and for process 2 – 3,

$$\frac{P_2V_2}{T_2} = \frac{P_3V_3}{T_3}$$

$$\frac{P_2}{T_2} = \frac{P_3}{T_3}$$

$$\frac{980}{T_2} = \frac{P_3}{298(\text{i.e.}, T_1)} \qquad \text{...(3)}$$

For isothermal, $P_1V_1 = P_3V_3$

$$100\times0.06 = P_3\times0.01175 \text{ (as } V_2 = V_3)$$

$\therefore$ $\mathbf{P_3 = 510.64\ kPa}$ ...(4)

From (3) and (4), $\mathbf{T_2 = 562.31\ K}$

$$\therefore \quad W_{3-1} = P_3 V_3 \ln\frac{V_1}{V_3}$$

$$= 510.64\times0.01175\times\ln\frac{0.06}{0.01175}$$

$$\mathbf{W_{3-1} = 9.7829\ kJ}$$

(ii) Change in I. E.:

For process 1 – 2, $Q = 0$

$\therefore$ $Q - W = \Delta U$

$0 - W = \Delta U$

$\therefore$ $\mathbf{\Delta U = 13.788\ kJ\ (i.e., U_2 - U_1)}$

For process 2 – 3, $W = 0$

$Q - W = \Delta U$

$Q = \Delta U = m \cdot C_V (T_3 - T_2)$

$P_1V_1 = mRT_1$

$100\times0.06 = m\,(C_P = C_V)\,T_1$

$= m\,(14.4 - 10.28)\times293$

$\mathbf{m = 0.00488\ kg}$

$\therefore$ $Q_{2-3} = \Delta U = 0.0048\times10.28\,(298 - 562.3)$

$\mathbf{\Delta U = Q_{2-3} = -13.27\ kJ}$

for 3 – 1 $T = C$ $\quad \Delta U = m \cdot C_V \cdot dT = 0$ as $T = C$.

Example 4.18 *3 kg of air at a pressure of 150 kPa and temperature 360 K is compressed polytropically to 750 kPa according to law $PV^{1.2}$ = constant. The air is then cooled to initial temperature at constant pressure. The air is then brought to state (1) by following PV = C. Draw the cycle on PV-diagram and determine net work and heat.*

Data: $P_1 = 150$ kPa, $P_2 = 750$ kPa, $P_2 = P_3$, $T_1 = 360$ K, $V_2 = ?$, $T_3 = T_1 = 360$ K

$V_1 = ?\ T_2 = ?\ V_3 = ?$

1 – 2 is a Polytropic process : $PV^{1.2} = C$

2 – 3 is a Constant pressure process
3 – 1 is a Isothermal process

Solution

We know that

$$PV = mRT$$
$$P_1 V_1 = mRT_1$$
$$\Rightarrow \quad V_1 = \frac{mRT_1}{P_1} = \frac{3 \times 0.287 \times 360}{150}$$
$$\mathbf{V_1 = 2.0664\ m^3}$$

1 – 2 adiabatic process

$$\frac{T_1}{T_2} = \left(\frac{P_1}{P_2}\right)^{\frac{n-1}{n}} \Rightarrow \frac{360}{T_2}$$
$$= \left(\frac{150}{750}\right)^{\frac{0.2}{1.2}} \Rightarrow T_2 = 470.76\ \text{K}$$
$$V_2 = \frac{mRT_2}{P_2} = \frac{3 \times 0.287 \times 470.76}{750}$$
$$= 0.540\ \text{m}^3$$

3 – 1 Isothermal process $PV = C$

$$P_3 V_3 = P_1 V_1 \Rightarrow V_3 = \frac{P_1 V_1}{P_3}$$
$$= \frac{150 \times 2.0664}{750}$$
$$\Rightarrow \quad V_3 = 0.413\ \text{m}^3$$
$$W_{1-2} = \frac{P_1 V_1 - P_2 V_2}{n-1} = \frac{mR(T_2 - T_1)}{1-n}$$
$$\mathbf{W_{1-2} = -476.82\ kJ}$$
$$W_{2-3} = P_2 (V_3 - V_2) = 750\,(0.413 - 0.54)$$
$$= -95.25\ \text{kJ}$$
$$W_{3-1} = P_1 V_1 \ln\left(\frac{P_3}{P_1}\right)$$
$$= 150 \times 2.0664 \ln \frac{750}{150}$$
$$\mathbf{W_{3-1} = 498.66\ kJ}$$

Net work done $= W_{1-2} + W_{2-3} + W_{3-1}$

$$= -73.21\ \text{kJ}$$
$$Q_{1-2} = \left(\frac{\gamma - n}{\gamma - 1}\right) \times \text{Work done}$$
$$= \frac{1.4 - 1.2}{1.4 - 1.0} \times W_{1-2}$$
$$\mathbf{Q_{1-2} = -238.41\ kJ}$$

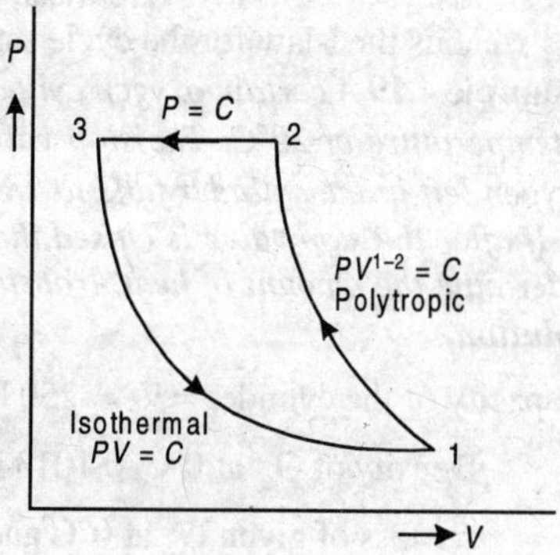

Fig. Ex. 4.18

$$Q_{2-3} = \text{change in enthalpy} = m\,C_P\,\Delta t = m\,C_P\,(T_3 - T_2)$$

$$= \frac{m\gamma R}{\gamma - 1}(T_3 - T_2)$$

$$= \frac{3 \times 1.4 \times 0.287}{1.4 - 1}(360 - 470.76)$$

$$\mathbf{Q_{2-3} = -\ 333.775\ kJ}$$

$$Q_{3-1} = \text{work done} = 498.86\ \text{kJ}$$

$$\therefore \quad \text{Net heat} = Q_{1-2} + Q_{2-3} + Q_{3-1}$$

$$\textbf{Net heat} = \mathbf{-\ 73.21\ kJ}$$

$$\therefore \quad \text{Net heat} = \text{Net work done}$$

Thus the I-law for the cycle satisfies.

Example 4.19 *A certain oxygen cylinder has a capacity 250 litres and contains oxygen at a pressure of 3 MPa and a temperature of 20°C. The stop valve is opened and some oxygen is used. If the pressure and temperature of oxygen left in the cylinder falls to 1.8 MPa and 16°C respectively, determine the mass of oxygen used.*

If after the stop valve is closed the oxygen remaining in the cylinder attains its original temperature of 20°C, determine the amount of heat, transferred through the cylinder walls from the atmosphere.

Solution

Capacity of the cylinder = V = 250 litre = 0.250 m^3 as 1000 litre = 1 m^3

Density of O_2 at 0°C, 0.1013 MPa = 1.43 kg/m^3

∴ Mass of given O_2 at 0°C and 0.1013 MPa = 1.43×0.25

$$\therefore \quad m_1 = 0.3575\ \text{kg at STA}$$

$$P_2 = 3\ \text{MPA}$$

$$P_3 = 1.8\ \text{MPa}$$

$$T_2 = 20°\text{C} = 293\ \text{K}$$

$$T_3 = 16°\text{C} = 289\ \text{K}$$

$$PV = mRT$$

Since V is constant, as volume of the cylinder does not change, we have,

$$\frac{P}{T.m} = \frac{R}{V} = \text{Constant}$$

$$\therefore \quad \frac{P_1}{T_1 m_1} = \frac{P_2}{T_2 m_2}$$

$$\therefore \quad \frac{0.1013}{273 \times 0.3575} = \frac{3}{293 \times m_2}$$

$$\therefore \quad m_2 = 9.865\ \text{kg}$$

Similarly,

$$\frac{P_1}{T_1 m_1} = \frac{P_3}{m_3 T_3}$$

$$= \frac{0.1013}{273 \times 0.3575} \times \frac{1.8}{289 \times m_3}$$

$$\therefore \quad m_3 = 6.001\ \text{kg}$$

$$\Delta m = m_2 - m_3$$

$$= 9.865 - 6.001$$

$$\mathbf{\Delta m = 3.864\ kg}$$

Also we have, $\frac{PV}{mT} = R$

$\therefore \quad R = \frac{3000 \times 0.25}{9.865 \times 293} = 0.26 \text{ kJ/kg-K}$

Heat transfer, $Q = m\,C_V (T_4 - T_3)$

{where, $T_4 = T_2$, the original temp.}

$$C_V = \frac{R}{\gamma - 1}$$

$$= \frac{R}{\left(\frac{C_P}{C_V} - 1\right)}$$

$$= \frac{0.26}{1.4-1} = 0.65 \text{ kJ/kg-K}$$

$\therefore \quad Q = 6.001 \times 0.65 (293 - 289)$

$\mathbf{Q = 15.603 \text{ kJ}}$

Example 4.20 *A gas initially at pressure 510 kPa and volume 142 litres undergoes: a process and final pressure of 170 kPa and volume of 275 litres during which enthalpy decreased by 65 kJ. Take $C_V = 0.718$ kJ/kg-K. Determine:*

(i) Change in internal energy.

(ii) Specific heat at constant pressure.

(iii) Specific gas constant.

Data: Intial pressure $P_1 = 510 \text{ kPa} = 510 \text{ kN/m}^2$

Intial volume $V_1 = 142\ l = 142 \times 10^{-3} \text{ m}^3$

Final pressure $P_2 = 170 \text{ kPa} = 170 \text{ kN/m}^2$

Final volume $V_2 = 275\ l = 275 \times 10^{-3} \text{ m}^3$

Change in enthalpy $= -65$ kJ [$\because$ Enthalpy is being decreased]

$C_V = 0.718$ kJ/kg-K

Solution

(i) We have, from first law of thermodynamics,

$$\Delta H = \Delta U + \Delta(PV)$$

$$\Delta U = \Delta H - \Delta(PV)$$

$$\Delta U = -65 \times 10^3 - (170 \times 275 - 510 \times 142)$$

$$= -65 \times 10^3 - (46750 - 72420)$$

$$= -65000 + 25670 = -39330$$

$\therefore \quad \mathbf{\Delta U = -39.33 \text{ kJ}}$

(ii) Also we know that,

$$\frac{\Delta H}{\Delta U} = \frac{C_P}{C_V} = \gamma$$

$\therefore \quad C_P = C_V \times \frac{\Delta H}{\Delta U} = 0.718 \times \frac{-65}{-39.33}$

$$\mathbf{C_P = 1.186\ kJ/kg\text{-}K}$$

(iii) Also,

$$C_P - C_V = R$$

$$R = 1.1866 - 0.718$$

$$\mathbf{R = 0.4686\ kJ/kg\text{-}K}$$

Example 4.21 *An engine has a piston area of 0.1 m² contains gas at a pressure of 1.5 MPa. The gas expands according to a process which is represented by a straight line on P-V plane. The final pressure is 0.15 MPa. Calculate work done by gas on piston if the stroke is 0.3 m.*

Data: Area of the piston $= 0.1\ m^2$

$P_1 = 1.5\ MPa$ $\qquad P_2 = 0.15\ MPa$

Stroke $= 0.3\ m$ $\qquad W = ?$

Solution

According to the formula,

Change in volume = Area of the piston × Stroke

$= 0.1 \times 0.3 = 0.03\ \text{m}$

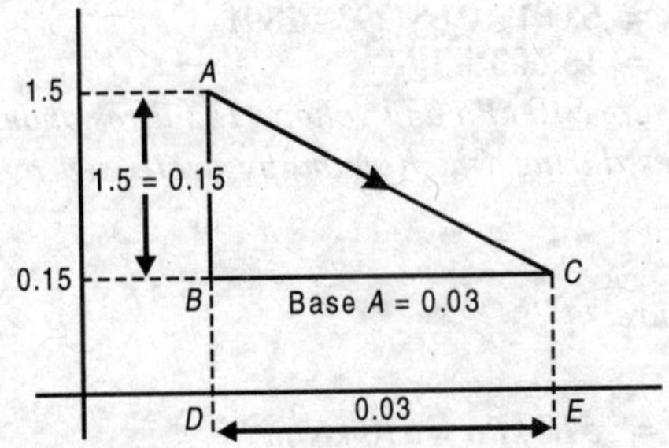

Fig. Ex. 4.21

Also we know that,

Work done by gas = Area under the curve (in this case a straight line) on the *P-V* diagram

∴ Net work done = Area of the rectangle *BCED* + Area of triangle *ABC*

$$W = \left[\left\{0.15 \times 10^6 \times 0.3\right\} + \left\{\frac{1}{2}(1.5 - 0.15) \times 10^6 \times 0.03\right\}\right]$$

$$\mathbf{W = 24.75\ kJ}$$

Example 4.22 *In a piston cylinder arrangement the pressure varies inversely proportional to the square of volume during a process. The initial pressure in the cylinder is 20 bar and the initial volume is 0.1 m³. The final pressure after the expansion is 200 kPa. Estimate the work done.*

Data:

$$P \propto \frac{1}{V^2} \Rightarrow P = \frac{C}{V^2} \text{ and } PV^2 = C$$

$$P_1 = 20\ bar = 20 \times 10^5\ \frac{N}{m^2}$$

$$V_1 = 0.1\ m^3$$

$$P_2 = 200\ kPa = 200 \times 10^3\ \frac{N}{m^2}$$

$$W = ?,\ V_2 = ?$$

Solution

We know that for a Polytropic process,

$$\text{Work done} = \frac{P_1 V_1 - P_2 V_2}{n-1} \quad ...(1)$$

Since V_2 is not given,

$$P_1V_1^2 = P_2V_2^2 = C$$

$$V_1^2 \times \frac{P_1}{P_2} = \left(V_2^2\right)$$

$$0.1^2 \times \frac{20\times10^5}{200\times10^3} = V_2^2$$

$$0.1 = V_2^2$$

$$0.3162 \text{ m}^3 = V_2 \quad ...(2)$$

Now from Eq. (1),

$$W = \frac{20\times10^5 \times 0.1 - 200\times10^3 \times 0.3162}{2-1}$$

$$W = \frac{200000 - 63245.55}{1}$$

$$\mathbf{W = 136754.45 \text{ kJ}}$$

or

$$\mathbf{W = 136.75445 \text{ kJ}}$$

Second Method :

We know that work done in a closed system,

$$= \int P \cdot dV \quad ...(3)$$

But Law for the process is,

$$PV^2 = C$$

or

$$P = \frac{C}{V^2}$$

$\therefore$ From Eq. (3),

$$W = \int P \cdot dV - \int_{V_1}^{V_2} \frac{C}{V^2} \cdot dV$$

$$= C\int_{V_1}^{V_2} V^{-2} \cdot dV = C\left[\frac{V^{-2+1}}{-2+1}\right]_{V_1}^{V_2} = C\left[-\frac{1}{V}\right]_{V_1}^{V_2}$$

$$W = C\left[-\frac{1}{V_2} + \frac{1}{V_1}\right] \quad ...(4)$$

Now,

$$P_1V_1^2 = C$$

$$\therefore \quad 20\times10^5 \times 0.1^2 = C = 20{,}000$$

Now, from Eqs (2), (4), (5),

$$W = 20000\left[-\frac{1}{0.316} + \frac{1}{0.1}\right] = 20000[-3.162 + 10]$$

$$= 20000\,[6.838]$$

$$\mathbf{W = 136754.45}$$

$$\mathbf{W = 136.754 \text{ kJ}}$$

Example 4.23 *A cylinder contains 0.12 cubic m of air at 1 bar and 90°C. It is compressed to 0.03 cubic m, the final pressure being 6 bar. Find the index of compression, increase in internal energy and heat transferred. Take:*
Characteristic gas constant = 0.287 kJ/kg-K; and
Specific heat at constant volume = 0.7176 kJ/kg-K.

Data:

$$V_1 = 0.12\ m^3 \qquad P_1 = 1\ bar = 1\times 10^5\ \frac{N}{m^2}$$

$$T_1 = 90 + 273 = 363\ K \qquad V_2 = 0.03\ m^3$$

$$P_2 = 6\times 10^5\ \frac{N}{m^2} \qquad n = ?$$

$$\Delta U = ? \qquad Q = ?$$

$$R = 0.287\ \frac{kJ}{kg\text{-}K} \qquad C_V = 0.7176\ \frac{kJ}{kg\text{-}K}$$

Solution

(i) We know that for the Polytropic process,

$$P_1V_1^n = P_2V_2^n = C$$

$$\frac{P_2}{P_1} = \left(\frac{V_1}{V_2}\right)^n$$

$$\left(\frac{6\times10^5}{1\times10^5}\right) = \left(\frac{0.12}{0.03}\right)^n$$

$$6 = (4)^n$$

$$\ln 6 = n \ln 4$$

$$\frac{\ln 6}{\ln 4} = n$$

$$\frac{1.79175}{1.38629} = n$$

$$\mathbf{n = 1.29248 = 1.3}$$

(ii) Now change in IE $= \Delta u = C_V \cdot (T_2 - T_1)$

To find T_2, since it is a polytropic process,

$$\frac{T_2}{T_1} = \left(\frac{V_1}{V_2}\right)^{n-1}$$

$$T_2 = T_1\left(\frac{V_1}{V_2}\right)^{n-1}$$

$$T_2 = 363\left(\frac{0.12}{0.03}\right)^{1.3-1}$$

$$= 363\times4^{0.3} = 363\times1.5157$$

$$\mathbf{T_2 = 550.199\ K}$$

$$\therefore \quad \Delta u = C_V\,(T_2 - T_1)\,\frac{\text{kJ}}{\text{kg}}$$

$$= 0.7176(550.199 - 363)$$

$$\mathbf{\Delta u = 134.334\ kJ/kg}$$

(iii) We know from the I-law,

$$q - w = \Delta u$$

$$q = \Delta u + w$$

And now to find W, $$W_{1-2} = \frac{P_2V_2 - P_1V_1}{1-n} = \frac{mR\,(T_2 - T_1)}{1-n}$$

For unit mass, $$W_{1-2} = \frac{R\,(T_2 - T_1)}{1-n}$$

$$= \frac{0.287(550.199 - 363)}{1 - 1.3}$$

$$= \frac{53.70}{-0.3}$$

$$\mathbf{W_{1-2} = -\ 179.01\ kJ/kg}$$

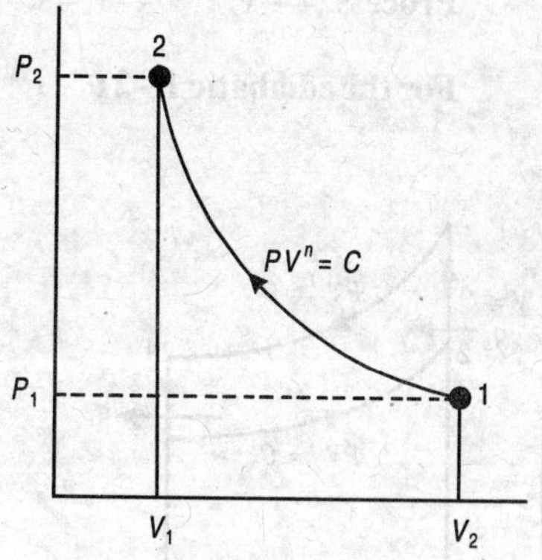

Fig. Ex. 4.23

Now heat transfer, from Eq. (1) will be,

$$q = \Delta u + W$$

$$= 134.334 + (-179.01)$$

$$\mathbf{q = -44.6765\ \frac{kJ}{kg}}$$

Example 4.24 ***Process 1:*** *Air initially at 100 kPa and 50° C undergoes reversible adiabatic compression such that its volume is reduced to 1/5th of initial volume.*

Process 2: *Then 940 kJ/kg of heat is added to this air at constant volume.*

Process 3: *Process 2 is followed by reversible adiabatic expansion up to initial volume.*

Process 4: *Finally heat is rejected at constant volume so as to reach the initial condition.*

Draw the four processes on one PV-diagram. Determine the maximum temperature, and heat rejected per kg of air. Assume adiabatic index of compression and expansion of 1.4 and constant volume specific heat as 0.717 kJ/kg-K.

Determine : (i) T_{max} = ?, (ii) Heat rejected/kg of air ?

Solution

Process (1 – 2) : $$P_1 = 100\text{ kPa}$$

$$PV^{\gamma} = C$$

$$V_2 = \frac{1}{5}V_1 = 0.2\,V_1$$

$$T_1 = 50 + 273 = 523\text{ K}$$

Process (2 –3) : $$q_s = 940\,\frac{\text{kJ}}{\text{kg}}$$

$$= C_V \cdot dT$$

at $\quad V = C$

Process (3 –4) : $\quad PV^{\gamma} = C$

$V_1 = V_4$

Process (4 – 1) : $\quad V = C: \quad \gamma = 1.4, \quad C_V = 0.717 \dfrac{\text{kJ}}{\text{kg-K}}$

For the adiabatic 1 – 2 :

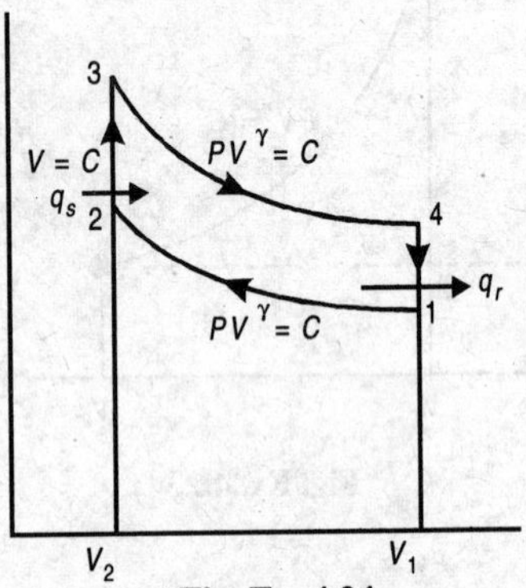

Fig. Ex. 4.24

$$\frac{T_2}{T_1} = \left(\frac{V_1}{V_2}\right)^{\gamma-1}$$

$$T_2 = \left(\frac{V_1}{V_2}\right)^{\gamma-1} \times T_1$$

$$= \left(\frac{1}{0.2}\right)^{1.4-1} \times 323 \text{ as } \frac{V_2}{V_1} = 0.2$$

$$T_2 = \left(\frac{1}{0.2}\right)^{0.4} \times 323 = (5)^{0.4} \times 323$$

$$= 0.90 \times 323$$

$$\mathbf{T_2 = 614.88\ K}$$

For Process (2 – 3) :

$$q_s = C_V (T_3 - T_2)$$

$$940 = 0.717(T_3 - 614.88)$$

$$\frac{940}{0.717} = T_3 - 614.88$$

$$1311.0181 = T_3 - 614.88$$

$$\mathbf{1925.898\ K = T_3}$$

This is the maximum temperature.

Now for the adiabatic 3 – 4:

$$\frac{T_4}{T_3} = \left(\frac{V_3}{V_4}\right)^{\gamma-1}$$

$$T_4 = \left(\frac{V_2}{V_1}\right)^{\gamma-1} \times T_3$$

$$= (0.2)^{1.4-1} \times 1925.9$$

$$= 0.2^{0.4} \times 1925.9$$

$$\mathbf{T_4 = 1011.686\ K}$$

Now heat rejected, $\quad q_r = C_V (T_1 - T_4) = 0.717(323 - 1011.69)$

$$\mathbf{q_r = -493.7878 \frac{kJ}{kg}}$$

– ve sign implies heat rejection.

Example 4.25 *Two vessels A and B, both containing Nitrogen are connected by a valve which is opened to allow the contents to mix and achieve an equilibrium temperature of 27°C. Before mixing, the following information is known about the gases in the two vessels.*

Vessel A: *Pressure = 1.5 MPa,*
Temperature = 50°C
and *Contents = 0.5 kg mol.*
Vessel B: *Pressure = 0.6 MPa,*
Temperature = 20°C
and *Contents = 2.5 kg*

Calculate the final equilibrium pressure, and the heat transferred to the surroundings. If the vessels had been perfectly insulated, calculate the final temperature and pressure which would have been reached. Take $\dfrac{C_p}{C_v}$ = *1.4, molecular weight = 28 kg/kg mol and R = 0.297 kJ/kg for Nitrogen.*

Solution

For the gas in vessel *A*

$$m_A = n_A \cdot M = 0.5 \times 28 = 14 \text{ kg}$$

$$P_A V_A = m_A RT_A$$

$$\therefore \quad V_A = \frac{14 \times 0.297 \times 323}{1.5 \times 10^3}$$

$$V_A = 0.895 \text{ m}^3$$

For vessel *B*,

$$V_B = \frac{m_B R T_B}{P_B} = \frac{2.5 \times 0.297 \times 293}{0.6 \times 10^3} = 0.362 \text{ m}^3$$

$\therefore$ Total volume of *A* and *B* $= V_A + V_B = 1.258 \text{ m}^3$

Also, Total mass = 16.5 kg = m

Final temperature after mixing = 300 K

For the final condition after mixing,

$$P.V. = mRT$$

$$P = \frac{16.5 \times 0.297 \times 300}{1.258} = 1168.64 \text{ kPa}$$

Since there is no work transfer,

Heat transfer = Change in internal energy

$= U_2 - U_1$

Taking $U = 0 \text{ kJ} \quad \text{at } T = 0 \text{ K}$

Initial internal energy, $U_1 = m_A C_V T_A + m_B C_V T_B$

where $C_V = \dfrac{R}{\gamma - 1} = \dfrac{0.297}{0.4}$

$\mathbf{C_v = 0743 \text{ kJ/kg-K}}$

$\therefore \quad U_1 = \left[(14 \times 323) + (2.5 \times 293)\right] \times 0.743$

$\mathbf{U_1 = 3904.09 \text{ kJ}}$

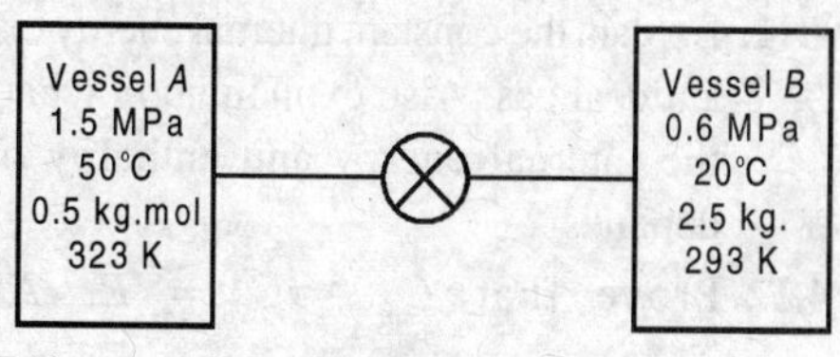

Fig. Ex. 4.25

Final Internal Energy, $U_2 = m \cdot C_V \cdot T = 16.5 \times 0.743 \times 300$

$\mathbf{U_2 = 3677.85 \text{ kJ}}$

$\therefore$ Heat transfer = 3677.85 – 3904.05

Heat transfer = – 226.24 kJ

If the vessels had been perfectly insulated,

$$U_1 = U_2$$

$$m_A C_V T_A + m_B C_V T_B = m C_V T$$

$$\therefore \quad T = \frac{m_A T_A + m_B T_B}{m}$$

$$= \frac{(14 \times 323) + (2.5 \times 293)}{16.5}$$

$$\mathbf{T = 318.45\ K = 45.45°C}$$

$$\text{The final pressure } P = \frac{mRT}{V} = \frac{16.5 \times 0.297 \times 318.45}{1.258}$$

$$\therefore \quad \mathbf{P = 1240.51\ kPa}$$

THEORY QUESTIONS

4.1. Define an Ideal Gas.

4.2. Explain the concept of an Ideal Gas.

4.3. Define an Ideal gas, give the approximate conditions for the gaseouss phase of a pure substance to behave as an ideal gas.

4.4. What are the ideal gas laws ?

4.5. Write a note on Equation of State or Characteristic Gas Equation.

4.6. Write a short note on Universal Gas Constant.

4.7. Define universal gas constant. What are its units?

4.8. What is the difference between specific gas constant and universal gas constant? How are these two related ?

4.9. With the help of Joule's experiment prove that I.E. is a function of temperature.

4.10. With the help of Joule's experiment, prove that I.E. is a function of temperature.

4.11. Explain the constant internal enemy experiment of an ideal gas. Also explain upon what property is the internal energy and enthalpy of ideal gas depends ?

4.12. Prove that $C_p - C_v = R$. Also prove

$$C_v = \frac{R}{\gamma - 1}; C_p = \frac{\gamma \cdot R}{\gamma - 1}.$$

4.13. Prove that,

(a) $C_p - C_v = R$ (b) $C_v = \dfrac{R}{\gamma - 1}$

(c) $C_p = \dfrac{\gamma \cdot R}{\gamma - 1}$

4.14. Derive the relation between C_p, C_v and R.

4.15. Derive the relation between C_v, γ, and R and also C_p, γ and R.

4.16. Prove that $\gamma = \dfrac{C_p}{C_v}$.

4.17. Explain Constant Volume Process with P–V and T–S diagrams. Also write the equation for work done and derive the equation for heat transferred and change of I.E.

4.18. Explain constant pressure process with P–V and T–S diagrams. Write the equation for work done and derive the equations for heat transferred.

4.19. What is an Isothermal process ? Represent this on P–V and T–S diagrams. Write the equation for work done, change in I.E. and heat transferred.

4.20. What is a polytropic process, show it on P–V diagram. Write the equation for work done and derive the equation for heat supplied.

4.21. Define polytropic specifc heat. What is the significance of negative sign?

4.22. Show that the polytropic sp. heat is given by

$$C_n = C_v \left(\frac{k - n}{n - 1} \right)$$

where k is specific heat ratio

4.23. What is an adiabatic process? Write the equations for work done, heat supplied.

PROBLEMS FOR PRACTICE

4.1. 2 kg of air at 247°C expands isothermally from a volume of 1 m^3 to a volume of 3 m^3. Determine the work done during the process. Take: $R = 287$ Nm/kg-K (*Ans.* 328 kJ)

4.2. A gas at 10 bar, 327°C occupies a volume of 0.02 m^3. This gas expands according to $PV^{14} = C$ upto a pressure of 1 bar. Find the temperature and work done during the process.

(*Ans.* $T_2 = 321.16$ K; 24.1 kJ)

4.3. 3 litre of air at 1200 kN/m^2 and 87°C expand polytropically according to $PV^{1.3} = C$ to a pressure of 120 kN/m^2. If for air $\gamma = 1.4$, $C_v = 0.718$ find,

(i) Polytropic specific heat.

(ii) Heat transfer per kg of air during the process and state its direction.

(iii) Change of specific IE of air.

(*Ans.* (1) – 0.239 kJ/kg-K; 35.51 kJ/kg; –106.27 kJ/kg)

4.4. 0.22 kg of a gas having a volume of 0.14 m^3 and a pressure of 1.4 bar is compressed to a pressure of 14 bar according to the, law $PV^{1.25} = C$. Estimate the amount of heat transferred during the process by calculating the work done and change in I.E. Take $C_p = 1.041$ kJ/kg-K and $C_v = 0.743$ kJ/kg-K.

4.5. A cylinder has a capacity of 300 litre and contains oxygen at a pressure of 3.1 MN/m^2 and temperature 18°C. The stop valve is opened and some gas is used. If the pressure and temperature of the gas left in the cylinder falls to 1.7 MN/m^2 and 15°C respectively, determine the mass of the oxygen used. If after the stop valve is closed, the oxygen remaining in the cylinder gradually attains its initial temperature of 18°C, determine the amount of heat transferred through the cylinder walls from atmosphere.

The density of oxygen at 0°C and 0.1013 MN/m^2 be taken as 1.43 kg/m^3 and $\frac{C_p}{C_v} = 1.4$.

4.6. 0.2 m^3 of mixture of fuel and air at 1.2 bar and 60°C is compressed till its pressure becomes 12 bar and temperature 270°C. Then it is ignited suddenly at constant volume and its pressure becomes twice the pressure at the end of compression. Find the maximum temperature reached and change in I.E. Also find the heat transfer during the compression process consider the mixture as perfect gas. Take $C_p = 1.005$ kJ/kg-K, and $\gamma = 1.4$

4.7. A gas initially at a pressure of 510 kPa and volume 142 litres undetgoes a process and final pressure of 170 kPa and volume of 275 litres during which enthalpy decreased by 65 kJ. Take $C_v = 0.718$ kJ/kg-K, determine,

(a) Change in I.E.

(b) C_p

(c) Specific gas constant

4.8. 10 kg of air is heated at constant pressure from a temperature of 100°C to 200°C. Calculate the heat added during the process and also change in I.E. Take $R = 0.287$ kJ/kg-K, $\gamma = 1.4$.

4.9. 50 litres of air at 1.013 bar and 100°C temperature is compressed to 28 bar. Volume of air at the end of polytropic compression is found to be 4 litres. Air is now heated at constant volume till pressure rises to 56 bar.

Assuming $C_p = 1.00$ kJ/kg-K and $C_v = 0.71$ kJ/kg-K, determine

(a) Polytropic index of compression

(b) Entropy change in each process.

Sketch the process on P–V and T–S diagram.

4.10. 3 kg of an ideal gas expanded from a pressure of 7 bar and volume 1.5 m^3 to a pressure of 1.4 bar and volume 4.5 m^3. If the change in I.E. during the process is 520 kJ. Determine,

(a) Specific gas constant

(b) Initial and final temperature

(c) Change in entropy in heat exchange. Take $C_p = 1.05$ kJ/kg-K

4.11. 0.01 kg of certain gas occupies a volume of 0.003 m^3 at a pressure of 7 bar and a temperature of 131°C. Calculate the molar mass of the gas. When the gas is allowed to expand until the pressure is 1 bar and final volume is 0.02 m^3, calculate final temperature.

4.12. 1 kg of perfect gas is compressed from 1.1 bar and 27°C, according. to the law, $PV^{1.3} = C$, until the pressure is 6.6 bar. Calculate the heat flow to or from the cylinder watts when the gas is argon having molecular mass of 40 kg/mol and $C_p = 0.52$ kJ/kg.

4.13. A certain quantity of gas has a volume of 0.11 m^3 at 0.15 MN/m^2 and 300 K. It is compressed to a pressure of 1.5 MN/m^2 according to the law $PV^{1.2}$ = constant. If the molecular mass of the gas is 27.9 and $C_p = 29.04$ kJ/k mol K determine heat transfer and its direction.

4.14. An ideal gas having $r = 1.4$ and volume 1.5 m^3 at 1 bar and 27°C is compressed according to the law $PV^{1.25} = C$, to a pressure of 6 bar. Calculate volume, temperature at the end of the process. Evaluate heat transfer, work done and the change of entropy during the process.

4.15. Following pressure – volume data is available for a process:

Estimate the work done during the process.

P (bar)	20	17.5	15	12.5	10	7.5	5	2.5
V (m^3)	1.00	1.1	1.2	1.3	1.4	1.5	1.6	1.7

4.16. Air with initial condition as 27°C, 1 bar and 1.5 m^3 is compressed according to the law $PV^{1.3} = C$, to a pressure of 7.5 bar. Calculate final volume, temperature at the end of the process. Evaluate heat transfer, W.D. and change of entropy during the process. Assume $r = 1.4$, $R = 0.287$ kJ/kg-K.

IMPORTANT POINTS TO BE REMEMBERED

1. *Ideal gas* obey's all gas law namely Charle's law, Boyles law and Avagadro's law.
2. Boyle's law states that volume of the gas is inversely proportion to its pressure when temperature is kept constant.
3. *Charles's law* states that (i) volumes are directly proportional to their absolute temperatures and when presure is constant and (ii) pressures are directly proportional to the absolute temperatures when volume remains constant.
4. *Avagadro's law* states that at the same pressure and temperature, equal volumes of different gases contain equal number of molecules.
5. *Equation of state or characteristic gas equation* is an equation relating pressure, volume and absolute temperature of gas and is given by $PV = RT$ for unit mass. R is called characteristic gas constant $R = 287$ Nm/kg/K for air.
6. *Universal gas constant* R is same for all gases and is 83143 N-m/K mole/K.
7. *Internal energy* of an ideal gas is a function of temperature. $U = f(T)$
8. *Relation between* C_p, C_v and R is

$$C_p - C_v = R$$

$$C_v = \frac{R}{\gamma-1}, C_p = \frac{\gamma}{\gamma-1}R.$$

9. Work done

Non-flow process:

$= O$ for constant volume process

$= P(V_2 - V_1)$ for constant pressure process

$= P_1V_1 \log e \dfrac{V_2}{V_1}$ for constant volume isothermal process

$= \dfrac{P_1V_1 - P_2V_2}{\gamma-1}$ for constant adiabatic process

$= \dfrac{P_1V_1 - P_2V_2}{n-1}$ for constant general law process

Work (For steady-flow process):

$= (P_1 - P_2)V$ for constant volume process

$= O$ for constant pressure process

$= P_1V_1 \log e \dfrac{V_2}{V_1}$ for constant temperature (isothermal) process

$= \dfrac{\gamma}{\gamma-1}[P_1V_1 - P_2V_2]$ for constant isentropic process

$= \dfrac{n}{n-1}[P_1V_1 - P_2V_2]$ for constant general law process

Heat $Q = C_v(T_2 - T_1)$ for constant volume process

$= C_p (T_2 - T_1)$ for constant pressure process

$= P_1 V_1 \log e \dfrac{V_2}{V_1}$ for constant temperature process

$= O$ for constant isentropic process

$= C_n (T_2 - T_1)$ for constant general law process

$= C_v - \dfrac{\gamma - n}{1 - n}(T_2 - T_1)$ for general law process

Internal energy $u = C_v (T_2 - T_1)$ for all processes

CHAPTER 5

Entropy

5.1 INTRODUCTION

We have studied the I–law of thermodynamics. Analysis of I–law leads to the definition of a derived property known as Internal energy. Analysis of second law will lead to the definition of another derived property known as Entropy.

Clausius discovered that, when a small amount of heat δQ is supplied to a system, which is at a absolute temperature T, then it will undergo a process, and the ratio $\frac{\delta Q}{T}$ is same for all reversible processes. He assigned the value $\frac{\delta Q}{T} = dS$ and called S as entropy.

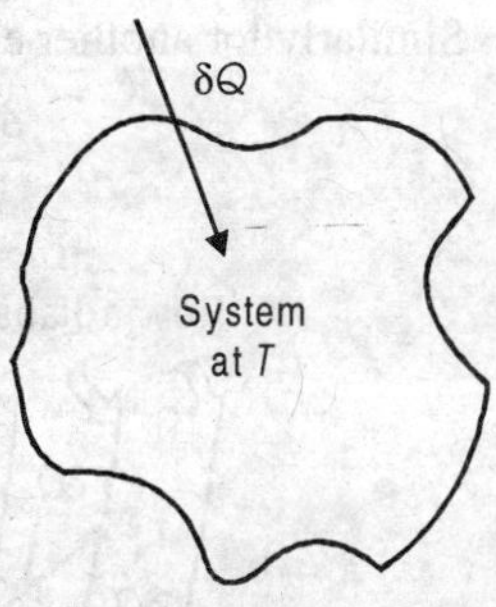

Fig. 5.1

The term entropy is taken, from Greek word 'tropee' meaning transformation. Thus, when a small amount of heat δQ is transformed to a system, the entropy changes by an amount dS. This change in entropy is regarded as the transformation content of the system.

Note that entropy is a thermodynamics property, it increases with the addition of heat and it decreases with the removal of heat.

5.2 CLAUSIUS THEOREM FOR A PROCESS AND FOR A CYCLE

Let a system be taken from an equilibrium state (1) to another equilibrium state (2) following the reversible path 1–2. Let (*a*) and (*b*) be two reversible adiabatic, which pass through the points (1) and (2) respectively. A reversible isotherm (*c*) is drawn, such that the area under 1–3–4–2 is equal to the area under the curve 1–2.

From the first law,

$$Q = \Delta U + W$$

For process 1–2, $\quad Q_{1-2} = U_2 - U_1 + W_{1-2}$ and for the process 1 – 3 – 4 – 2

$$Q_{1-3-4-2} = U_2 - U_1 + W_{1-3-4-2}$$

As we have assumed that the area under 1– 3 – 4 – 2 is equal to the area under 1–2.

i.e. $\quad W_{1-2} = W_{1-3-4-2}$

$\therefore \quad Q_{1-2} = Q_{1-3-4-2}$

$$= Q_{1-3} + Q_{3-4} + Q_{4-2}$$

But $\quad Q_{1-3} = 0$

and $\quad Q_{4-2} = 0 \ (\because \text{adiabatic})$

$\therefore \quad Q_{1-2} = Q_{3-4}$

Thus any reversible path can be replaced by a zig-zag path between the same end states, consisting of a reversible adiabatic process followed by a reversible isothermal process, which is again followed by a reversible adiabatic process, such that the heat transferred during the isothermal process is the same as that transferred during the original process.

Consider a reversible cycle as shown in Fig. 5·2. It is divided into a large number of strips by means of reversible adiabatic. Each strip may be closed at the top and bottom by means of reversible isotherms. The original closed cycle is thus replaced by a zig-zag closed path, consisting of alternate adiabatic and isothermal processes, such that the heat transferred during all the isothermal processes is equal to the heat transferred in the original cycle.

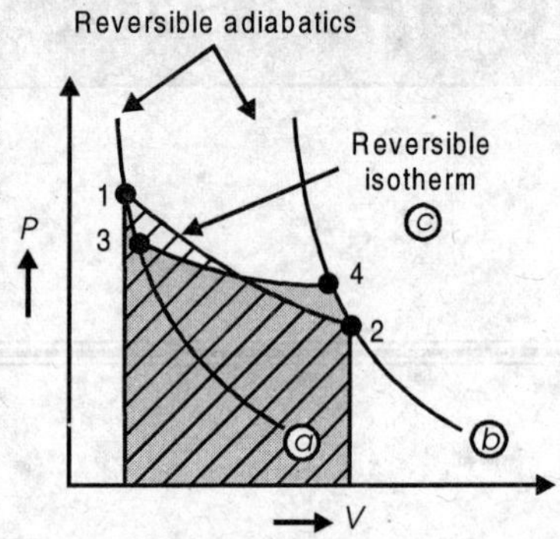

Fig. 5.2

Now, for cycle 1 – 2 – 3 – 4, δQ_1 is heat absorbed reversibly at T_1 and δQ_2 is heat rejected reversibly at T_2.

Then, $$\frac{\delta Q_1}{T_1} = \frac{\delta Q_2}{T_2}$$

If heat supplied is taken as positive and heat rejected as negative, then

$$\frac{\delta Q_1}{T_1} + \frac{\delta Q_2}{T_2} = 0$$

Similarly for another elemental cycle 5 – 6 – 7 –8

$$\frac{\delta Q_3}{T_3} + \frac{\delta Q_4}{T_4} = 0$$

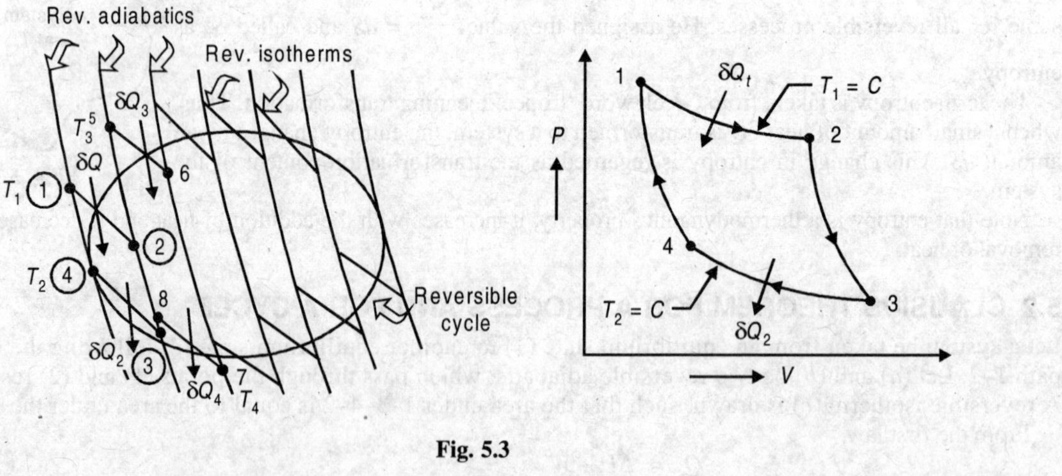

Fig. 5.3

If similar equation are written for all the elemental Carnot cycles, then for the whole (complete) original cycle.

$$\left(\frac{\delta Q_1}{T_1} + \frac{\delta Q_2}{T_2}\right) + \left(\frac{\delta Q_3}{T_3} + \frac{\delta Q_4}{T_4}\right) + \dots = 0$$

or $$\oint_R \frac{dQ}{T} = 0$$

Thus the cyclic integral of $\delta Q/T$ for a reversible cycle is equal to zero. This is known as **Clausious Theorem.**

5.3 ENTROPY—A PROPERTY

To prove this, we have to prove that the change of entropy does not depend upon path but it depends upon end states. Then, we will able to say that, entropy is a property of the system.

Consider a system which changes its state from state point (1) to state point (2) by following the reversible path a and returns from state point (2) to state point (1) by following the reversible path b.

Then the two paths $1-a-2$ and $2-b-1$ together will form a cycle.

Now from Clausius theorem,

$$\oint_{(1-a-2-b-1)} \frac{\delta Q}{T} = 0$$

Note. May be read as integral for the reversible path $1-a-2-b-1$.

The above integral may be replaced as the sum of two integrals one for the path a and other for the path b.

$$a\int_{R1}^{2} \frac{\delta Q}{T} + b\int_{R2}^{1} \frac{\delta Q}{T} = 0$$

Note. May be read as integral $\delta Q/T$ for the reversible path $1-a-2$.

$$\therefore \quad a\int_{R1}^{2} \frac{\delta Q}{T} = -b\int_{R1}^{2} \frac{\delta Q}{T}$$

Since path b is reversible

$$a\int_{R1}^{2} \frac{\delta Q}{T} = b\int_{R2}^{1} \frac{\delta Q}{T}$$

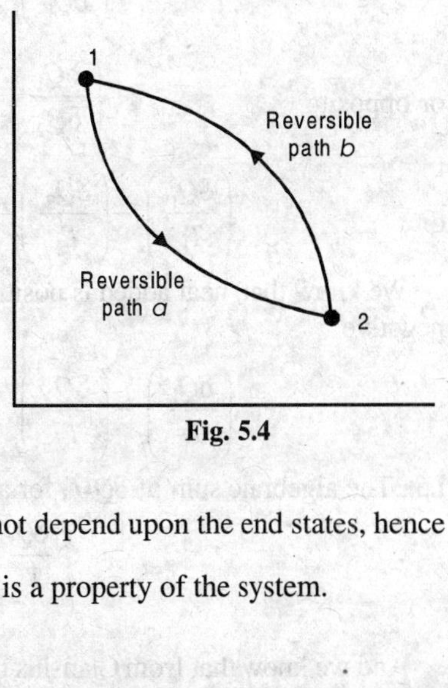

Fig. 5.4

The magnitude of $\frac{\delta Q}{T}$ (i.e., dS) is same for the paths a and b and it does not depend upon the end states, hence it is point function and we know that properties are point-functions, hence it is a property of the system.

5.4 CLAUSIUS INEQUALITY

According to Clausius Theorem,

$$\oint_R \frac{\delta Q}{T} = 0$$

In the II-Law, during Carnot theorem, we have proved that the efficiency of the Reversible engine is more than that of an Irreversible engine.

$$\eta_R > \eta_I$$

or $$\eta_I < \eta_R \quad \text{...(1)}$$

We also know that,

$$\eta \text{ of any engine} = \frac{\text{Heat supplied} - \text{Heat rejected}}{\text{Heat supplied}} = \frac{Q_1 - Q_2}{Q_1}$$

or For small amount of heat δQ,

$$\eta = \frac{\delta Q_1 - \delta Q_2}{\delta Q_1} = 1 - \frac{\delta Q_2}{\delta Q_1}$$

Hence the Eq. (1) will therefore be,

$$\left[1-\frac{\delta Q_2}{\delta Q_1}\right]_I < \left[1-\frac{\delta Q_2}{\delta Q_1}\right]_R$$

But for reversible engine,

$$\frac{Q_1}{Q_2} = \frac{T_1}{T_2}$$

$$\left[1-\frac{\delta Q_2}{\delta Q_1}\right] = 1-\frac{T_2}{T_1}$$

$$\therefore \quad 1-\left(\frac{\delta Q_2}{\delta Q_1}\right)_I < 1-\frac{T_2}{T_1}$$

$$\therefore \quad \left(\frac{\delta Q_2}{\delta Q_1}\right)_I > \frac{T_2}{T_1}$$

or opposite

$$\left(\frac{\delta Q_1}{\delta Q_2}\right) < \frac{T_1}{T_2}$$

or

$$\left(\frac{\delta Q_1}{T_1}\right)_I - \left(\frac{\delta Q_2}{T_2}\right)_I < 0$$

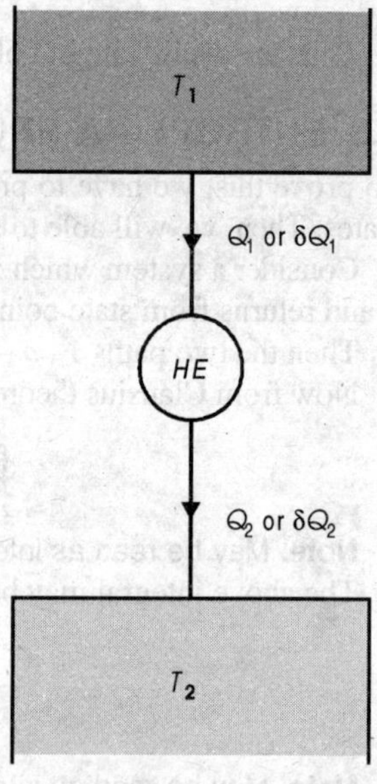

Fig. 5.5

We know that, heat added is positive. Therefore δQ_1 is positive and heat rejected is negative. Therefore δQ_2 is negative

$$\therefore \quad \left(\frac{\delta Q_1}{T_1}\right)_I + \left(\frac{\delta Q_2}{T_2}\right)_I < 0$$

i.e., The algebraic sum of $\delta Q/T$, for an irreversible cycle is always less than zero.

$$\therefore \quad \oint_R \frac{\delta Q}{T} < 0 \text{ for an Irreversible cycle.}$$

And we know that from Clausius theorem $\oint \frac{\delta Q}{T} = 0$ for a reversible cycle. Combining results for reversible and irreversible cycle, we may write,

$$\oint \frac{\delta \mathbf{Q}}{\mathbf{T}} \leq \mathbf{0}$$

Thus expression is known as Clausius Inequality. It implies whether any cyclic process is reversible or irreversible or impossible.

Note. According to Clausius Inequality

(i) $\oint \frac{dQ}{T} = 0$ for a reversible cycle

(ii) $\oint \frac{dQ}{T} < 0$ for a irreversible cycle

(iii) $\oint \frac{dQ}{T} > 0$ for a impossible cycle

(iv) All temperatures must be in K

(v) Heat received by the engine is positive

(vi) Heat rejected by the engine is negative

5.5 CHANGE OF ENTROPY IN AN IRREVERSIBLE PROCESS

We know that, change in entropy for a reversible process is given by,

$$\left(\frac{\delta Q}{T}\right)_R = (dS)_R \quad \text{...(1)}$$

Now to find the value of change in entropy in an Irreversible process.

Consider a system, which change its state from state point (1) to state point (2) by following the reversible path *a* and returns from state point (2) to state point (1) by following the irreversible path *b* as shown in Fig. 5.6.

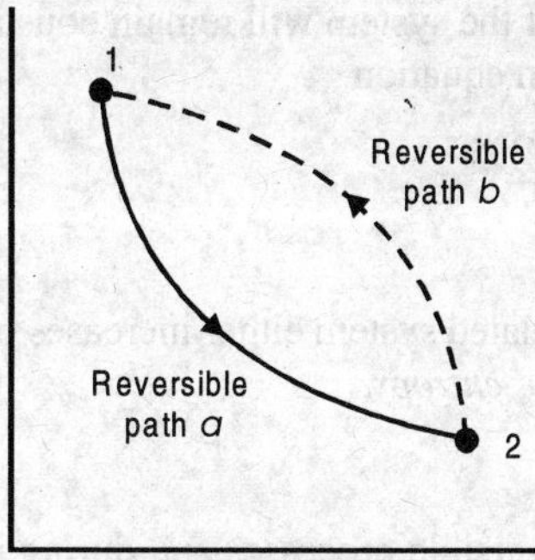

Fig. 5.6

Since cyclic integral of any property is zero and entropy is a property we can write,

$$\oint dS = a\int_1^2 (dS)_R + b\int_2^1 (dS)_I = 0 \quad \text{...(2)}$$

Now, from Eq. (1), for a reversible process we have,

$$a\int_1^2 (dS)_R = a\int_1^2 \left(\frac{\delta Q}{T}\right)_R \quad \text{...(3)}$$

Substituting this in Eq. (2) we get,

$$a\int_1^2 \left(\frac{\delta Q}{T}\right)_R + b\int_2^1 (dS)_I = 0 \quad \text{...(4)}$$

Since the processes 1 – *a* – 2 and 2 – *b* – 1 together will form Irreversible cycle. Applying Clausius inequality,

$$\oint_I \frac{\delta Q}{T} < 0$$

We get from Eq. (4),

$$a\int_1^2 \left(\frac{\delta Q}{T}\right)_R + b\int_2^1 \left(\frac{\delta Q}{T}\right)_I < 0 \quad \text{...(5)}$$

Subtracting Eq. (5) from Eq. (4) we get,

$$b\int_1^2 (dS)_I > a\int_2^1 \left(\frac{\delta Q}{T}\right)_I$$

In general, $\quad (dS)_I > \left(\frac{\delta Q}{T}\right)_I \quad$ ...(6)

Combining Eqs (1) and (6), we can write,

$$dS \geq \frac{\delta Q}{T}$$

where equality sign is for reversible process and inequality sign is for an Irreversible process (from Eq. 6).

Note. The effect of irreversibility is always to increase the entropy of the system.

5.6 CHANGE OF ENTROPY FOR AN ISOLATED SYSTEM

We know that, in an isolated system, matter, work or heat cannot cross the boundary of the system. Hence according to the first law of thermodynamics, the IE of the system will remain constant.

Since for an isolated system $\delta Q = 0$, from equation,

$$dS = \frac{\delta Q}{T} \quad \ldots(1)$$

We get, (dS) isolated ≥ 0 ...(2)

Equation (2) states, that entropy of an isolated system either increases or remains constant and never decreases. This is known as the *Principle of increase of entropy.*

5.7 ENTROPY OF UNIVERSE

We know that all natural processes are irreversible processes and during irreversible processes entropy increases and hence entropy of the universe always increases.

5.8 ENTROPY AND DISCORDER

Consider an isolated system comprising of two gases O_2 and H_2 in a separated box as shown in Fig. 5.7. When the partition or barrier is removed, the molecules of both gages get more space to move randomly and therefore collision take place between the mole molecules of the same gas as well as of both the gases, and after some time, equilibrium will be established.

If we assume that initially both gases have the same volume and temperature then after removing the partition each gas occupies double the previous volume while the temperature and the I.E. of the system remains the same, but we find that the entropy increase.

The reason is obvious, as the mixing process is irreversible and irreversible process is always associated with the increase in entropy.

Thus an irreversible process always tends to take the system (isolated) to state of greater disorder. It is a tendency on the part of nature to proceed to a state of greatest disorder. And an isolate system always tends to a state of greater entropy. So there is a close link between entropy and disorder.

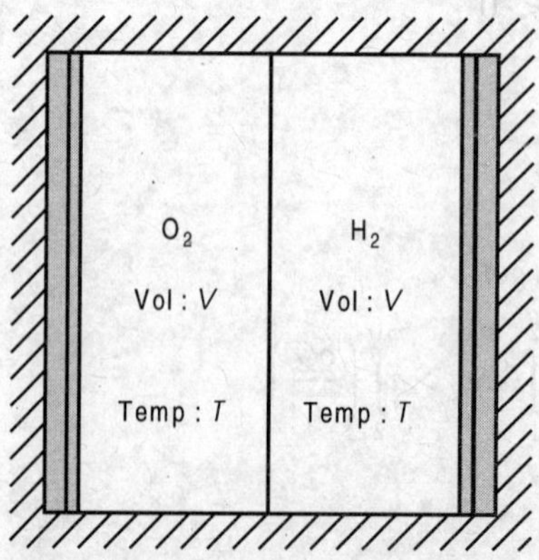

Fig. 5.7

It may be stated roughly that "**The entropy of a system is a measure of the degree of molecular disorder existing in the system**". When heat is supplied to the system, the disorderly motion of molecules increases and so the entropy of the system increases. The reverse occurs when heat is removed from the system.

5.9 PROPERTY RELATIONS FROM ENERGY EQUATIONS

For deriving the equation for change of entropy it is essential to know equation based on I and II law of thermodynamics.

We know that, change in entropy for a reversible process is given by,

$$(dS)_R = \left(\frac{\delta Q}{T}\right)_R$$

i.e.,
$$(\delta Q)_R = T \cdot dS \quad \ldots(1)$$

Also we know that, the first–law applied to a closed system undergoing a process gives,

$$Q - W = \Delta U$$

or
$$Q = \Delta U + W$$

In the differential form,
$$\delta Q = dU + \delta W$$

i.e.,
$$\delta Q = dU + PdV \quad \ldots(2)$$

∴ From Eqs. (1) and (2) we can write,

$$T \cdot dS = dU + PdV \quad \ldots(3)$$

This is very important equation because it gives relationship between all thermodynamic properties and does not involve any path functions.

We also know from the definition of enthalpy,

$$H = U + PV$$

or
$$dH = dU + PdV + VdP$$

$$dH = T.dS + V.dP \text{ from Eq. (3)}$$

or
$$T.dS = dH - V{\cdot}dP \quad \ldots(4)$$

From Eq. (3) i.e.,

$$T.dS = dU + P.dV$$

∴
$$dS = \frac{dU}{T} + \frac{PdV}{T}$$

∴
$$\mathbf{S_2 - S_1 = \int_1^2 \frac{dU}{T} - \int_1^2 \frac{PdV}{T}} \quad \ldots(5)$$

From Eq. (4) i.e.
$$T.dS = dH - V.dP$$

$$dS = \frac{dH}{T} + \frac{V.dP}{T}$$

∴
$$\mathbf{S_2 - S_1 = \int_1^2 \frac{dH}{T} - \int_1^2 \frac{V \times dP}{T}} \quad \ldots(6)$$

5.10 GENERAL EQUATIONS FOR CHANGE IN ENTROPY

We know that, the firlst law applied to a closed system undergoing a process gives,

$$Q - W = \Delta U$$

Now consider a unit mass of a gas which changes its states from pressure P_1, specific volume v_1, and temperature T_1 to a new state P_2, V_2, T_2.

For unit mass $q = \Delta u + w$

In the differential form

$$\delta q = du - \delta w \quad \text{...(1)}$$

Now, from the definition of specific heat at constant volume i.e.,

$$C_v = \left(\frac{\delta q}{dT}\right)_v = \left(\frac{du}{dT}\right)_v$$

From this $\quad du = C_v \cdot dT$

$\therefore$ Eq. (1) becomes, $\quad \delta q = C_v \cdot dT + P \cdot dv \quad \text{...(2)}$

Dividing throughout by T, $\quad \frac{\delta q}{T} = C_v \frac{dT}{T} + \frac{P}{T} dv \quad \text{...(3)}$

But $\quad \frac{\delta q}{T} = ds$

$$\therefore \quad \frac{\delta q}{T} = ds = C_v \frac{dT}{T} + \frac{P}{T} dv \quad \text{...(4)}$$

From perfect gases, for unit mass, $Pv = RT$ or $\frac{P}{T} = \frac{R}{v}$

Substituting this in Eq. (4) $\quad ds = C_v \frac{dT}{T} + R \frac{dv}{v} \quad \text{...(5)}$

$$\therefore \quad \int_1^2 ds = \int_1^2 C_v \frac{dT}{T} + \int_1^2 R . \frac{dv}{v}$$

$$\mathbf{s_2 - s_1 = C_v \ln \frac{T_2}{T_1} + R \ln \frac{v_2}{v_1}} \quad \text{...(6)}$$

This equation gives change in entropy for unit mass, in terms of temperature ratio and volume ratio.

From perfect gases, $C_p - C_v = R$ and susbtituting this in Eq. (6) we get,

$$\begin{aligned} s_2 - s_1 &= C_v \ln \frac{T_2}{T_1} + C_p \ln \frac{v_2}{v_1} - C_v \ln \frac{v_2}{v_1} \\ &= C_p \ln \frac{v_2}{v_1} + C_v \left[\ln \frac{T_2}{T_1} - \ln \frac{v_2}{v_1} \right] \\ &= C_p \ln \frac{v_2}{v_1} + C_v \left[\ln \frac{T_2}{T_1} \cdot \frac{v_1}{v_2} \right] \end{aligned} \quad \text{...(7)}$$

From characteristic equation $P \cdot v = RT$

or $\quad \frac{P_1 v_1}{T_1} = \frac{P_2 v_2}{T_2}$

$$\therefore \quad \frac{P_2}{P_1} = \frac{T_2 v_1}{T_1 v_2} \quad \text{...(8)}$$

Substituting Eq. (8) in Eq. (7) we get,

$$\mathbf{s_2 - s_1 = C_p \ln \frac{v_2}{v_1} - C_p \ln \frac{P_2}{P_1}} \quad \text{...(9)}$$

Again substituting $C_p - C_v = R$ or $C_p - R = C_v$ in Eq. (6) we get,

$$s_2 - s_1 = C_p \cdot \ln\frac{T_2}{T_1} - R \cdot \ln\frac{T_2}{T_1} + R \cdot \ln\frac{T_2}{T_1}$$

$$= C_p . \ln\frac{T_2}{T_1} - R.\left(\ln\frac{T_2}{T_1} - \ln\frac{v_2}{v_1}\right)$$

$$= C_p \cdot \ln\frac{T_2}{T_1} - \left(\ln\frac{T_2}{T_1} \cdot \frac{v_1}{v_2}\right)$$

From Eq. (8) $\quad \mathbf{s_2 - s_1 = C_p . ln\frac{T_2}{T_1} - R\,ln\frac{P_2}{P_1}}$ ...(10)

It is to be noted that for a given change, Eqs (6), (9) and (10) will give the same result. The choice of equation is a matter of convenience.

5.11 CHANGE IN ENTROPY DURING VARIOUS PROCESSES

We have studied the equations for work done, heat transferred change in I.E. in the ideal gas chapter. Now to derive the equations for change of entropy during,

(i) Constant Volume Process

Change of entropy:

Here $\quad v_1 = v_2$ hence $\ln\frac{v_2}{v_1} = \ln 1 = 0$

Hence from Eq. (6) (of last article)

$$s_2 - s_1 = C_v \ln\frac{T_2}{T_1} + R\ln\frac{v_2}{v_1}$$

$$\mathbf{s_2 - s_1 = C_v ln\frac{T_2}{T_1}}$$

And from Eq. (9) i.e., $\quad s_2 - s_1 = C_p \ln\frac{v_2}{v_1} - C_v \ln\frac{P_2}{P_1}$

$$\mathbf{s_2 - s_1 = C_v ln\frac{P_2}{P_1}}$$

(ii) Constant Pressure Process

Change in Entropy:

Here $\quad P_1 = P_2$

$\therefore \quad \ln\frac{P_1}{P_2} = \ln 1 = 0$

Hence from equation which are in terms or pressure ratio (From equations derived in last article) i.e., from Eq. (9),

$$\mathbf{s_2 - s_1 = C_p\ ln\frac{v_2}{v_1}}$$

And from Eq. (10), $\quad \mathbf{s_2 - s_1 = C_p ln\frac{T_2}{T_1}}$

(iii) Isothermal Process

Change in Entropy,

Here $$T_1 = T_2$$

$$\therefore \quad \ln\frac{T_1}{T_2} = \ln 1 = 0$$

Hence from equations which are in terms of temperature ratio.

i.e., from Eq. (6)

$$\mathbf{s_2 - s_1 = R\ln\frac{v_2}{v_1}}$$

And from Eq. (10) $$\mathbf{s_2 - s_1 = -R\ln\frac{P_2}{P_1}}$$

$$\mathbf{s_2 - s_1 = R\ln\frac{P_2}{P_1}}$$

(iv) Polytropic Process

(a) We know that heat transfer in a polytropic process,

$$q = \left[\frac{\gamma - n}{1-n}\right] C_v (T_2 - T_1)$$

In the differential form, $$\delta q = \left[\frac{\gamma - n}{1-n}\right] C_v \cdot dT$$

Since $$\frac{\delta q}{T} = ds$$

$$\therefore \quad ds = \frac{\delta q}{T} = \left[\frac{\gamma - n}{1-n}\right] C_v \cdot \frac{dT}{T}$$

or $$\int_1^2 ds = \int_1^2 \left[\frac{\gamma - n}{1-n}\right] \cdot C_v \frac{dT}{T}$$

$$\therefore \quad \mathbf{s_2 - s_1 = \left[\frac{\gamma - n}{1-n}\right] . C_v \ln\frac{T_2}{T_1}} \qquad \ldots(1)$$

This is the change in entropy, for unit mass in a polytropic process in terms of temperature ratio.

(b) *Change of entropy:* In terms of volume ratio

We know that from (1),

$$s_2 - s_1 = \left[\frac{\gamma - n}{1-n}\right] C_v \ln\frac{T_2}{T_1}$$

But, for a polytropic process

$$\frac{T_2}{T_1} = \left(\frac{V_1}{V_2}\right)^{n-1-}$$

$\therefore$ Substituting this value of $\frac{T_2}{T_1}$ in Eq. (1), we get

$$s_2 - s_1 = \left[\frac{\gamma - n}{1-n}\right] C_v \ln\left(\frac{V_1}{V_2}\right)^{n-1}$$

$$= \frac{\gamma - n}{1-n} C_v (1-n) \ln \frac{V_2}{V_1}$$

$$= (\gamma - n) C_v \ln \frac{V_2}{V_1}$$

$$\mathbf{s_2 - s_1 = \frac{\gamma - n}{1-n} R \ln \frac{V_2}{V_1}} \qquad \left(\text{As } C_v = \frac{R}{\gamma - 1}\right)$$

(c) *Change of entropy polytropic process:* In terms of pressure ratio:
We know that from (1)

$$s_2 - s_1 = \frac{\gamma - n}{1-n} C_v \ln \frac{T_2}{T_1}$$

But, for a polytropic process

$$\frac{T_2}{T_1} = \left(\frac{P_2}{P_1}\right)^{\frac{n-1}{n}}$$

Substituting this value of $\frac{T_2}{T_1}$ in Eq. (1), we get

$$s_2 - s_1 = \frac{\gamma - n}{1-n} C_v \ln\left(\frac{P_2}{P_1}\right)^{\frac{n-1}{n}}$$

$$s_2 - s_1 = \frac{\gamma - n}{1-n} \cdot \frac{R}{\gamma - 1} \cdot \frac{1-n}{n} \ln \frac{P_2}{P_1}$$

$$= \frac{\gamma - n}{1-n} \cdot \frac{R}{n} \cdot \ln \frac{P_2}{P_1}$$

$$\mathbf{s_2 - s_1 = \frac{\gamma - n}{n} C_v \ln \frac{P_2}{P_1}} \qquad \ldots(3)$$

(d) *Adiabatic Process: change of entropy for a reversible adiabatic process.* As we have studied during a reversible adiabatic process, entropy reamin constant ($S = C$) and the process is called as Isentropic process. Since $S = C$, $dS = 0$ i.e., change in entropy is zero.

5.12 REPRESENTATION OF VARIOUS PROCESSES ON T–S CHART

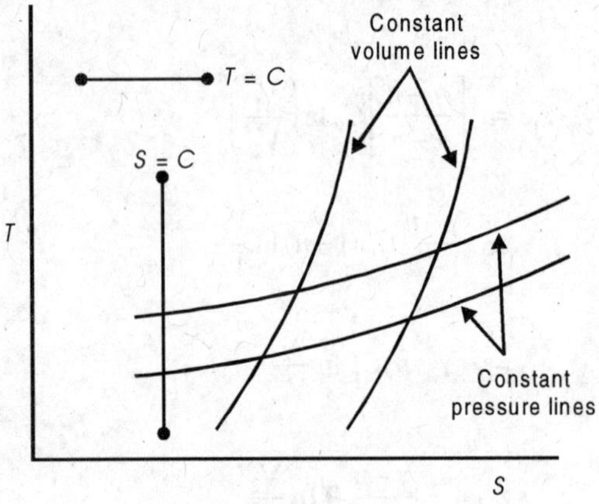

Fig. 5.8

Various processes are shown on *T–S* diagram in Fig. 5.8. Note that constant volume lines on *T–S* diagram have steep slope compared to constant pressure lines.

5.13 PURE SUBSTANCE AND CHANGE OF ENTROPY

We define a pure substance as a system which is homogeneous in composition and chemical aggregation that remains invariable with time.

In many cases we have to deal with system of fixed chemical composition with changing phases, as for example, in case of vapour power generation and referigeration cycles. In the former system water is commonly used whereas in the latter system ammonia, freon etc. are used. In these applications we find the working substance as liquid in one part of the cycle and as vapour or mixture of vapour and liquid in another part of the cycle and as vapour or mixture of vapour and liquid in another part of the cycle. The change in phase of the substance in the cycle is of physical nature only as its chemical composition remains unchanged. These fall in the category of pure substance.

We define a pure substance as a system which is homogeneous in composition and chemical aggregation than remain invariable with time.

A pure substance can exist in three different states of aggregation viz. solid, liquid and gas. It may exist in a single phase like solid, liquid, gas or more than one phase which are in equilibrium with each other. Thus a mixture of water and ice, water and steam, water and ice are all examples of pure substance.

Third law of thermodynamics states "*The entropy of a pure substance in complete thermodynamic equilibrium becomes zero at the absolute zero of temperature*".

This law enables the absolute entropies of pure substances to be calculated from the fundamental definition of entropy, with S_O set equal to zero.

$$\therefore \qquad S = \int_O^T \frac{\delta Q_{rev}}{T} \quad \text{pure substance in equilibrium}$$

Therefore, the change of entropy of a pure substance when it undergoes a reversible process, can be calculated as in case of ideal gases.

Thus, ΔS_w = Change in entropy of 1 kg water when heated at constant pressure from 0° C to saturation temperature.

$$= \int_{273}^{T_{sat}} \frac{\delta Q_{rev}}{T}$$

$$= C_{p_w} \int_{273}^{T_{sat}} \frac{dT}{T}$$

$$= C_{p_w} \log_e \frac{T_{sat}}{273} = C_{p_w} \ln \frac{T_{sat}}{273}$$

During evaporation at constant saturation temperature, change of entropy is given by

$$\Delta S_{fg} = \frac{\text{Latent heat}}{\text{Saturation temperature}} = \frac{h_{fg}}{T_{sat}}$$

Again for superheating process, the change of entropy (temperature increases from T_{sat} to T_{sup}).

$$\Delta S_{sup} = C_{p_v} \log_e \frac{T_{sup}}{T_{sat}} = C_{p_v} \ln \frac{T_{sup}}{T_{sat}}$$

5.14 THIRD LAW OF THERMODYNAMICS

Statistical analysis suggests that the entropy of a substance tends to zero as absolute zero temperature is reached.

This point has been so intensly investigated, that it is probably safe to say that at absolute zero temperature, **the entropy of a pure substance in some perfect crystalline form becomes zero, a generalisation known as the Third law.**

It is found that the specific heats approach zero as temperature tends to zero and the difference $C_p - C_v$ also approaches zero, becuase the co-efficient of thermal expansion approaches zero.

LIST OF FORMULAE

1. I–law for a closed system undergoing a cycle,

$$\oint \delta W = \oint \delta Q$$

2. I–law for a closed system undergoing a process,

$$Q - W = \Delta U$$

where Q = Heat transfer in J
W = Work transfer in J
ΔU = Change of Internal Energy in J.

3. Specific heat at constant pressure,

$$C_p = \left(\frac{\delta q}{dT}\right)_P = \left(\frac{dh}{dT}\right)_P$$

$\Rightarrow$ $\delta q = C_p \cdot dT$ for unit mass

And for total mass $\delta q = m \cdot C_p \cdot dT$

4. Specific heat at constant volume

$$C_v = \left(\frac{\delta q}{dT}\right)_V = \left(\frac{du}{dT}\right)_V$$

$\delta q = C_v \cdot dT$ for unit mass

And for total mass $\delta Q = m \cdot C_v \cdot dT$

5. Adiabatic index k or $\gamma = \dfrac{C_p}{C_V}$

Note for air $C_p = 1.005 \dfrac{\text{kJ}}{\text{kg-K}}$; $C_V = 0.718 \dfrac{\text{kJ}}{\text{kg-K}}$

and $\gamma = 1.4$

6. Continuity equation or law of conservation of mass

$$\rho_1 A_1 C_1 = \rho_2 A_2 C_2$$

or

$$\frac{A_1 C_1}{v_1} = \frac{A_2 C_2}{v_2} \text{ as } \rho = \frac{1}{v}$$

where ρ = Density (kg/m³)
A = Cross sectional area (m²)
C = Velocity of fluid in (m/sec)

7. Equation of state $PT = mRT$
and R for air = 0.287 kJ/kg-K

8. Relations between C_p, C_V and R, γ

(i) $C_p - C_V = R$

(ii) $C_V = \dfrac{R}{\gamma - 1}$

(iii) $C_P = \dfrac{\gamma \cdot R}{\gamma - 1}$

9. For constant volume process ($V = C$)

(i) Work done in a closed system,

$$W_{1-2} = \int P . dV = 0 \text{ as } V = C \text{ and } dV = 0$$

(ii) From the I–law $Q - W = \Delta U$

$$Q - 0 = \Delta U = m \cdot C_V \cdot dT$$

$$\Rightarrow \quad Q = \Delta U = m \cdot C_V \cdot dT$$

10. For constant pressure process ($P = C$)

(i) Work done in a closed system,

$$W_{1-2} = \int_1^2 P . dV$$

$$W_{1-2} = P(V_2 - V_1)$$

(ii) Change in Internal energy

$$\Delta U = U_2 - U_1 = m \cdot C_V \cdot dT$$

(iii) Heat transfer $\delta Q = dH = m \cdot C_P \cdot dT$

11. For Isothermal process ($PV = C$)

(i) Work done $W_{1-2} = \int P . dV = P_1 V_1 \ln \dfrac{V_2}{V_1}$

or

$$W_{1-2} = P_1 V_1 \ln \frac{V_2}{V_1}$$

(ii) Change in I.E. $= \Delta U = m \cdot C_V \cdot dT = 0$ as

$$T = C$$

$$\therefore \quad dT = 0$$

(iii) From I–law $Q - W = \Delta U$ as $\Delta U = 0$

$$Q = W = P_1V_1 \ln\frac{V_2}{V_1}$$

12. Polytropic process $(PV^{\gamma}) = C$

(i) Work done $W_{1-2} = \int P.dV$

$$W_{1-2} = \frac{P_2V_2 - P_1V_1}{1-n}$$

$$W_{1-2} = \frac{mR(T_2 - T_1)}{1-n}$$

As $PV = mRT$

(ii) Heat transfer during polytropic process,

$$q = C_V\left[\frac{\gamma - n}{1-n}\right](T_2 - T_1)$$

$$q = C_n(T_2 - T_1)$$

where C_n = polytropic specific heat $= C_V\left[\frac{\gamma - n}{1-n}\right]$

Also $Q = \frac{\gamma - n}{\gamma - 1} \times$ work done in a polytropic process

13. For adiabatic process $PV^{\gamma} = C$

(i) $W_{1-2} = \int P.dV = \frac{P_2V_2 - P_1V_1}{1-\gamma}$

(ii) Heat transfer $Q = 0$

14. For,

(i) Constant pressure process,

(ii) Constant volume process,

(iii) Constant temperature process.

To find the property at second state use :

$$\frac{P_1V_1}{T_1} = \frac{P_2V_2}{T_2}$$

15. For polytropic process to find the properties at the second state use :

$$\frac{T_2}{T_1} = \left(\frac{P_2}{P_1}\right)^{\frac{n-1}{n}} = \left(\frac{V_1}{V_2}\right)^{n-1}$$

and for adiabatic process, in place of n, use γ.

SOLVED EXAMPLES

Example 5.1 *A heat engine is supplied with 278 kJ/sec of heat at constant fixed temperature of 283° C and rejection takes place at 5°C. The following results were reported,*

(i) 208 kJ/sec are rejected; (ii) 139 kJ/sec are rejected; (iii) 70 kJ/sec are rejected.

Classify which of the results report a reversible cycle or irreversible cycle or impossible cycle.

Solution

(i)
$$\oint \frac{\delta Q}{T} = \left(\frac{Q_1}{T_1}\right) + \left(\frac{Q_2}{T_2}\right)$$
$$= \left(\frac{278}{283+273}\right) + \left(\frac{-208}{5+273}\right)$$
$$= 0.5 - 0.748 = -2.48 < 0$$

∴ The cycle is Irreversible.

(ii)
$$\oint \frac{\delta Q}{T} = \left(\frac{278}{556}\right) + \left(-\frac{139}{278}\right)$$
$$= \frac{278}{556} - \frac{139}{278} = 0.5 - 0.5 = 0$$

∴ Cycle is reversible.

(iii)
$$\oint \frac{\delta Q}{T} = \left(\frac{278}{556}\right) + \left(-\frac{70}{278}\right)$$
$$= 0.5 - 0.252 = 0.248 > 0$$

This cycle is impossible according to Clausius inequality.

Example 5.2 *2 kg of a gas is heat at constant volume in a non flow reversible process, till its pressure is doubled. The initial temperature of gas is 50°C. Determine change in entropy.*

Take $C_v = 0.652$ kJ/kg-K and $C_p = 0.91$ kJ/kg-K.

Data: $m = 2$ kg, $P_2 = 2P_1$, or $\frac{P_2}{P_1} = 2$ (given), $T_1 = 50°C + 273$, $S_2 - S_1 = ?$, $C_v = 0.652$ kJ/kg-K, $C_p = 0.91$.

Solution

We know that change of entropy for a constant volume process is,

$$S_2 - S_1 = m \cdot C_V \cdot \ln\frac{P_2}{P_1} \qquad \left[\text{Units} : \text{kg} \times \frac{\text{kJ}}{\text{kg}-\text{K}} \times \text{ratio} = \frac{\text{kJ}}{\text{K}}\right]$$
$$= 2 \times 0.652 \times \ln = 2 \times 0.652 \times 0.6931$$
$$\mathbf{S_2 - S_1 = 0.904 \ kJ/K}$$

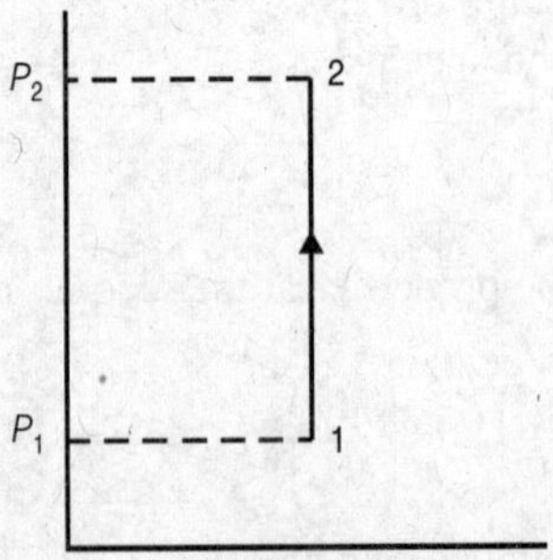

Fig. Ex. 5.2

Example 5.3 *A quantity of gas has a pressure of 8 bar and it occupies a volume of 0.015 m³ at 160°C. The gas is expand isothermally to a volume of 0.085 m³. Determine change of entropy.*

Data: $P_1 = 8$ bar $= 800$ kN/m² $\quad V_1 = 0.015$ m³

$T_1 = 160 + 273 = T_2 \quad V_2 = 0.085$ m³, $S_2 - S_1 = ?$

Solution

We know that change in entropy during isothermal process is given by,

$$S_2 - S_1 = mR \,.\ln \frac{v_2}{v_1}$$

$$S_2 - S_1 = mR \,.\ln \frac{V_2}{V_1}$$

Note. $\frac{v_2}{v_1}$ is a ratio, we can write ratio of specific volume is equal to the ratio of total volume.

i.e.,
$$\frac{v_2}{v_1} = \frac{V_2}{V_1}$$

We know that
$$P_1V_1 = mRT_1$$

or
$$\frac{P_1V_1}{T_1} = mR$$

Substituting this in above equation we can write,

$$S_2 - S_1 = \frac{P_1V_1}{T_1} \ln \frac{V_2}{V_1}$$

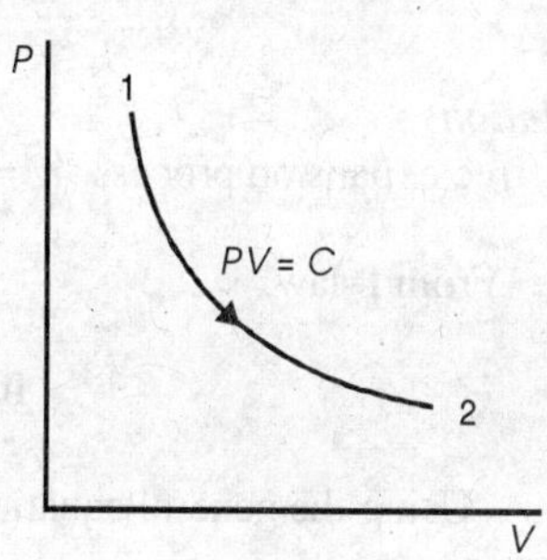

Fig. Ex. 5.3

Note.
$$\frac{\text{kN/m}^2 \times \text{m}^3}{\text{K}} = \frac{\text{kNm}}{\text{K}}$$

$$= \frac{800 \times 0.015}{433} \times \ln \frac{0.0585}{0.015} = \frac{\text{kJ}}{\text{K}}$$

$$\mathbf{S_2 - S_1 = 0.048\ kJ/K}$$

Example 5.4 *3 kg of air at a pressure of 5 bar and 250°C expands reversibly in a polytropic process. The final pressure is 1.5 bar and index of expansion, is 1.25. Find the change in entropy during the process. Take $C_v = 0.652$ kJ/kg-K and $\gamma = 1.4$.*

Data: $m = 3$ kg $\quad P_1 = 15$ bar $= 1500$ kN/m²

$T_1 = 250 + 273 \quad PV^n = C$

$P_2 = 1.5$ bar $= 150$ kN/m²

$n = 125 \quad S_2 - S_1 = ?$

Solution

Change in entropy in a polytropic process is given by (equation in terms of pressure ratio).

$$S_2 - S_1 = m\left[\frac{\gamma - n}{n} \cdot C_V \cdot \ln \frac{P_1}{P_2}\right]$$

$$= 3 \times \left[\frac{1.4 - 1.25}{1.25} \times 0.652 \times \ln \frac{15}{1.5}\right]$$

$$= 3\,[0.12 \times 0.652 \times 2.303]$$

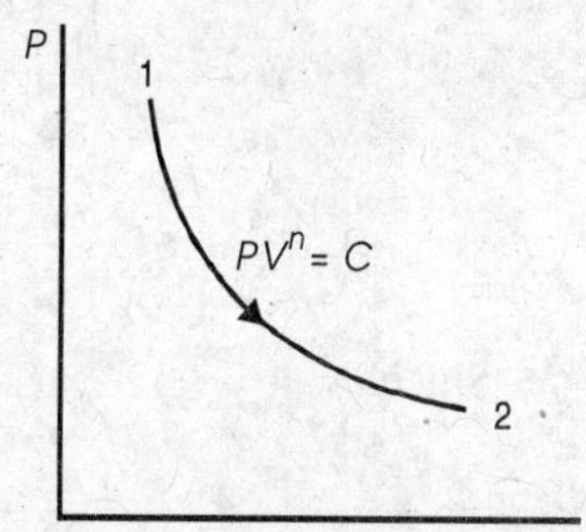

Fig. Ex. 5.4

$$S_2 - S_1 = 3\,[0.18015]$$
$$S_2 - S_1 = 0.5405 \text{ kJ/K}$$

Note. Unit of m is kg and

$$C_V = \frac{\text{kJ}}{\text{kg-K}}$$

$$\therefore \quad \text{kg} \times \frac{\text{kJ}}{\text{kg K}} = \left.\frac{\text{kJ}}{\text{K}}\right]$$

Example 5.5 *2 kg of certain gas with $C_p = 0.85$ kJ/kg-K and $C_v = 0.70$ kJ/kg-K is allowed to expand adiabatically through a partly opened valve, where by the its volume changes from $V_1 = 0.0113$ m³ to some higher value and during the process entropy increases by 0.8 kJ/K. Find the volume after expansion of gas.*

Data: $m = 0.85$ $\quad C_p = 0.85$
$C_v = 0.70$ $\quad PV = C$
$V_1 = 0.113$ m³ $\quad V_2 = ?$
$S_2 - S_1 = 0.8$ kJ/K

Solution

For free expansion process, $W = 0$ and given the process is adiabatic

$$Q = 0$$

$\therefore$ From I–law,

$$Q = \Delta U + W$$
$$0 = \Delta U + 0 = \Delta U = 0$$

$\therefore$ $\quad \text{Initial} = T_{\text{final}}$

i.e. $\quad T_1 = T_2$

Using the general equation for change in entropy,

$$S_2 - S_1 = m\left[C_V \ln\frac{T_2}{T_1} + R\ln\frac{v_2}{v_1}\right]$$

Since $\quad T_1 = T_2$

$$S_2 - S_1 = mR\ln\frac{v_2}{v_1}$$

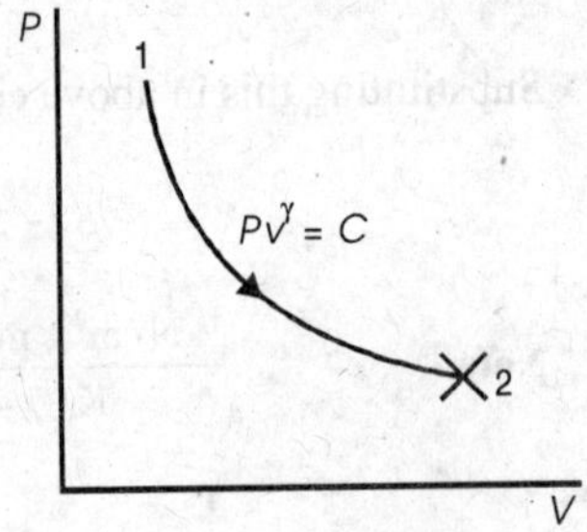

Fig. Ex. 5.5

$$0.8 = 2\times[0.85 - 0.70]\ln\frac{v_2}{v_1}$$

$$\frac{0.8}{2\times 0.15} = \ln\frac{v_2}{v_1}$$

$$\ln\frac{v_2}{v_1} = 2.667$$

$$\therefore \quad \frac{v_2}{v_1} = 14.397 = \frac{V_2}{V_1}$$

Since $\quad \dfrac{v_2}{v_1} = \dfrac{V_2}{V_1}$ ratio

$$\therefore \quad V_2 = 13.397 \times 0.0113 \text{ m}^3$$
$$\mathbf{V_2 = 0.1627\ m^3}$$

Example 5.6 *A quantity of gas has a pressure of 300 kN/m² when its volume is 0.025 m² and its temperature is 37°C. If the value of is 0.287 kJ/kg-K determine the mass of the gas present.*

If the pressure of gas is now increased to 0.9 MN/m² while the volume remains constant what will be the new temperature of the gas.

Solution

(1) By characteristic gas equation

$$PV = mRT$$

$$\therefore \quad m = \frac{PV}{RT} = \frac{300\times10^3\times0.025}{0.287\times10^3\times310} = 0.084 \text{ kg}$$

(2) We have

$$\frac{P_1V_1}{T_1} = \frac{P_2V_2}{T_2} \text{ and in this case } V_1 = V_2$$

$$\therefore \quad T_2 = \frac{P_2T_1}{P_1} = 310\times\frac{0.9\times10^6}{300\times10^3} = 930 \text{ K}$$

$$\therefore \quad T_2 = 930 - 273 = 657°\text{C}$$

Example 5.7 *10 kg of air is heated at constant pressure, from a temperature of 100°C to 200°C. Calculate the heat added during the process and also the change in I.E. Take gas constant R = 0.287 kJ/kg-K and γ = 1.4 for air.*

Data: $m = 10$ kg, $t_1 = 100$ °C, $t_2 = 200$°C, $T_1 = 373$ K, $T_2 = 473$ K, $R = 0.287$ kJ/kg-K, $\gamma = 1.4$

Solution

We know that, heat added during constant pressure process.

$$= mC_P(T_2 - T_1)$$

Also, $$C_V = \frac{R}{\gamma-1} = \frac{0.287}{1.4-1} = 0.7175$$

and $$C_P - C_V = R$$

$$\Rightarrow \quad C_P = R + C_V$$

$$C_P = 0.287 + 0.7175$$

$$\mathbf{C_p = 1.0045 \text{ kJ/kg-K}}$$

∴ Substituting (2) and (3) in (1) we get,

$$\text{Heat added} = m\,C_P(T_2 - T_1)$$

$$= 10 \times 1.0045\,(200 - 100) = 1004.5 \text{ kJ}$$

$$\text{Charge in I.E.} = \Delta U = mC_V(T_2 - T_1) = 10 \times 0.7175\,(200 - 100)$$

$$\mathbf{\Delta U = 717.5 \text{ kJ}}$$

Example 5.8 *A mass of 0.8 kg of air at 1 bar and 25°C is contained in a gas tight frictionless piston cylinder device. The air is now compressed to a final pressure of 5 bar. During the process heat is transferred from the air such that the temperature inside the cylinder remains constant. Calculate the heat transfer and work done during the process and direction of each in the process.*

Data: $m = 0.8$ kg, $P_1 = 1 \text{ bar} = 10^5 \text{ N/m}^2$, $T_1 = 298$ K, $P_2 = 5 \text{ bar} = 5\times10^5 \text{ N/m}^2$, $T = C$, *Isothermal process.*

Solution

∴ We know that, for an isothermal process,

$$Q = W = P_1V_1 \ln\frac{P_1}{P_2} = mRT_1 \ln\frac{P_1}{P_2} = -110.11 \text{ kJ}$$

As Q is –ve ∴ W is also –ve.

Example 5.9. *1 kg of air at an initial condition of 3 bar and 175°C expands adiabatically without friction to a final condition of 1 bar. Calculate:*

(a) *External work done during expansion.*

(b) *Change in Internal Energy.*

Take $\gamma = 1.4$ and $R = 0.287$ kJ/kg-K

Data:

$m = 1$ kg; $P_1 = 3 \times 105$ N/m² $= 3 \times 10^2$ kN/m²

$T_1 = 175 + 273 = 448$ K; $P_2 = 1 \times 10^2$ kN/m²

$\gamma = 1.4$; $R = 0.287$ kJ/kg-K

Solution

(a) We know that the work done in an adiabatic expansion process is given by,

$$W_{1-2} = \frac{P_1V_1 - P_2V_2}{\gamma - 1}$$

Here V_1 and V_2 are not given,

Since it is an adiabatic process,

$$P_1V_1^\gamma = P_2V_2^\gamma$$

$$\frac{P_1}{P_2} = \left(\frac{V_2}{V_1}\right)^\gamma$$

$$\left(\frac{P_1}{P_2}\right)^{1/\gamma} = \frac{V_2}{V_1}$$

∴ $$V_2 = V_1 \times \left(\frac{P_1}{P_2}\right)^{1/\gamma}$$

To find V_1 $$P_1V_1 = mRT_1$$

$$V_1 = \frac{mRT_1}{P_1} = \frac{1 \times 0.287 \times 448}{3 \times 10^2}$$

$$\text{Note units : } \left[\frac{\text{kg}\dfrac{\text{kJ}}{\text{kg-K}} \times \text{K}}{\text{kN/m}^2} = \frac{\text{kJ}}{\text{kN/m}^2} = \frac{\text{kNm}}{\text{kN}} \times \text{m}^2 = \text{m}^3\right]$$

∴ From Eq. (2), $$V_2 = 0.429 \times \left(\frac{300}{100}\right)^{1/1.4} = 0.429 \times (3)^{1/1.4}$$

$$\mathbf{V_2 = 0.940\ m^3}$$

∴ Now from Eq. (1), work done,

$$W_{1-2} = \frac{P_1V_1 - P_2V_2}{\gamma - 1}$$

$$= \frac{300\times0.429 - 100\times0.940}{1.4-1} = \frac{34.7}{0.4}$$

$$\mathbf{W_{1-2} = 86.75\ kJ}$$

(b) We also know from the I–law, that for a process, $Q = \Delta U + W$. Since it is adiabatic process $Q = 0$

∴ $W = -\Delta U = 86.75\text{ kJ} \Rightarrow \mathbf{\Delta U = -86.75\ kJ}$

Therefore internal energy decreases.

Example 5.10 *Find the change of Internal energy and enthalpy of 3 kg of air, when temperature changes from 25°C to 100°C.*

Take $C_p = 1$ kJ/kg-K and $C_v = 0.71$ kJ/kg- K.

Solution

In the problem, the type of process is not given, but since temperature change, it is not an isothermal process.

∴

$$\Delta U = m \cdot C_V \cdot (T_2 - T_1)$$

$$= 3\text{ kg} \times 0.71\text{ kJ/kg-K} \times (373 - 298)\text{ K}$$

$$= 3 \times 0.71 \times 75$$

$$\mathbf{\Delta U = 159.75\ kJ}$$

As change of enthalpy $= \Delta H = m\, C_P \cdot (T_2 - T_1)$

$$= 3 \times 1 \times (75)$$

$$\mathbf{\Delta H = 225\ kJ}$$

Example 5.11 *0.44 kg of a gas having volume 0.28 m³ and a pressure of 1.4 bar is compressed to a pressure of 14 bar according to $PV^{1.25} = C$. Find the amount of heat transferred during the process by calculating the work done and the change of internal energy.*

Take $C_p = 1.041$ kJ/kg-K and $C_v = 0.743$ kJ/kg-K

Data:

$m = 0.44$ kg $\qquad V_1 = 0.28\ m^3$

$P_1 = 1.4$ bar $= 140$ kPa

$P_2 = 14$ bar $= 1400$ kPa

$PV^{1.25} = C \qquad Q = ?\ W = ?\ \Delta U = ?$

$C_p = 1.041$ and $C_v = 0.743$ kJ / kg-K

Solution

$$\text{W.D.} = \frac{P_2V_2 - P_1V_1}{1-n} \qquad ...(1)$$

Here only V_2 is not given and since it is a polytropic process,

$$P_1V_1^n = P_2V_2^n$$

$$\frac{P_1}{P_2} = \left(\frac{V_2}{V_1}\right)^n$$

i.e.,

$$\frac{V_2}{V_1} = \left(\frac{P_1}{P_2}\right)^{1/n}$$

$$V_2 = V_1 \times \left(\frac{P_1}{P_2}\right)^{1/n}$$

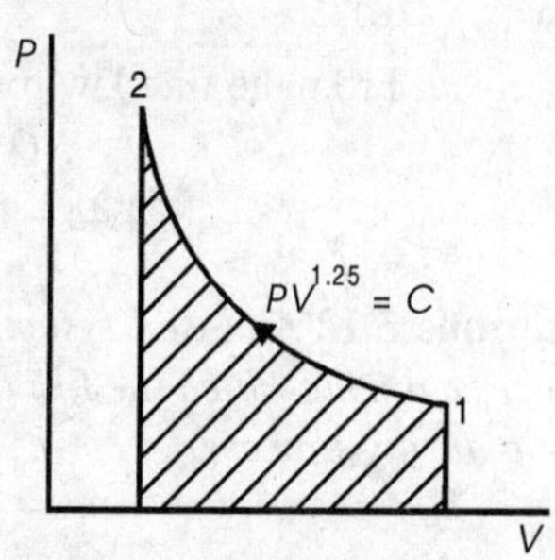

Fig. Ex. 5.11

$$= 0.28\times\left(\frac{140}{1400}\right)^{\frac{1}{1.25}}$$

$$= 0.28\times(0.1)^{0.8}$$

$$V_2 = 0.044377\ m^3 = 0.0444\ m^3$$

Now work done from Eq. (1),

$$W_{1-2} = \frac{1400\times0.0444-140\times0.28}{1-1.25}$$

$$= \frac{62.16-39.2}{-0.25}$$

$$\mathbf{W_{1-2} = -91.84\ kJ}$$

W_{1-2} is –ve, so work is done on the gas.

(ii) Heat transferred in a polytropic process

We know that, $$q = C_n(T_2-T_1) = C_V\cdot\left(\frac{\gamma-n}{1-n}\right)(T_2-T_1)$$

$$= \frac{R}{\gamma-1}\left(\frac{\gamma-n}{1-n}\right)(T_2-T_1)$$

$$= \frac{R}{\gamma-1}\left(\frac{\gamma-n}{1-n}\right)\times(T_2-T_1)$$

$$q = \frac{(\gamma-n)}{\gamma-1}\times\frac{R}{1-n}\times(T_2-T_1)$$

And total heat transfer,

$$Q = \frac{\gamma-n}{\gamma-1}\times\frac{mR(T_2-T_1)}{1-n}$$

i.e., $$Q = \frac{\gamma-n}{\gamma-1}\times W,\quad \text{in a polytropic process}$$

To find Q, let us find $$\gamma = \frac{C_P}{C_V} = \frac{1.041}{0.743} = 1.401 = 1.4$$

$$Q = \frac{1.4-1.25}{1.4-1}\times-91.84 = \frac{0.15}{0.4}\times-91.84$$

$$\mathbf{Q = -34.44\ kJ}$$

∴ From the first law for a process,

$$Q - W = \Delta U$$

$$-34.44+91.84 = \Delta U$$

∴ $$\mathbf{\Delta U = 57.4\ kJ\ (Increase)}$$

Example 5.12 *A closed system containing 2 kg of working substance undergoes a reversible constant pressure process, during which the I.E. of the system decreases kJ. If C_p = 1120 J/kg-K and C_v = 800 J/kg-K. Find the work done during process.*

Data: m = 2 kg, P = C

ΔC = –80 kJ, C_P = 1.120 kJ/kg-K

C_v = 0.8 kJ/kg-K, W_{1-2} = ?

Solution

We know that work done in a constant pressure process,

$$W_{1-2} = P(V_2 - V_1) \quad ...(1)$$

Also from the first law, $Q - W = \Delta U$

We know that heat supplied is utilised for,

(i) Increasing I.E.

(ii) For doing external work

And heat transfer,

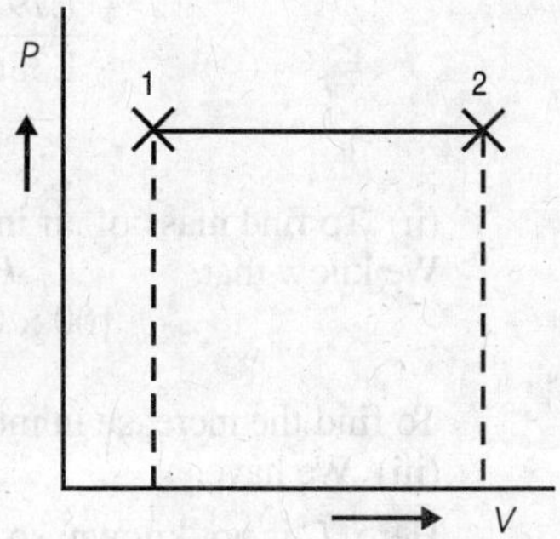

Fig. Ex. 5.12

$$\delta q = dh$$

or $$\delta Q = dH = mC_P\Delta T \quad ... (2)$$

Also, $$\Delta U = m\,C_P\,\Delta T$$

$$\frac{\Delta U}{m \cdot C_V} = \Delta T$$

$$\frac{-80 \text{ kJ}}{2 \text{ kg} \times 0.8 \text{ kJ/kg-K}} = \Delta T \quad ...(3)$$

$$= \frac{-80}{1.6} = \Delta T$$

$$-50 \text{ K} = \Delta T$$

$\therefore$ From (3) $$\delta Q = m \cdot C_P \cdot \Delta T = 1 \times 1.12 \times (-50)$$

$$\delta Q \text{ or } Q = -112 \text{ kJ}$$

$\therefore$ From Eq. (2), $$Q - W = \Delta U$$

$$Q - \Delta U = W$$

$$-112 + 80 = W$$

$\therefore$ $$\mathbf{W = -32 \text{ kJ}}$$

Example 5.13 *A cylinder contains 0.12 m³ of air at 1 bar and 100°C. The air is compressed to 0.03 m³. The final being 6 bar.*

Determine,

(i) The value of index n

(ii) Mass of air in the cylinder.

(iii) Increase in internal energy.

Take $\gamma = 1.4$ $R = 0.287$ kJ/kg-K and $C_v = 0.72$ kJ/kg-K

Data: $V_1 = 0.12 \text{ m}^3$ $\quad P_1 = 100 \text{ kPa}$

$T_1 - 100 + 273 = 373 \text{ K}$ $\quad V_2 = 0.03 \text{ m}^3$

$P_2 = 600 \text{ kPa}$

Solution

So $$P_1V_1^n = P_2V_2^n$$

$$\frac{P_2}{P_1} = \left(\frac{V_1}{V_2}\right)^n$$

$$\left(\frac{600}{100}\right) = \left(\frac{0.12}{0.03}\right)^n$$

$$6 = (4)^n$$

$$\log_e 6 = n \log_e (4)$$

$$\frac{\log_e 6}{\log_e 4} = n$$

$$\frac{1.79175}{1.38629} = n$$

$$n = 1.29248$$

$$n = 1.3$$

(ii) To find mass of air in the cylinder

We know that, $P_1 V_1 = mRT_1$

$$100 \times 0.12 = m \times 0.287 \times 383$$

$\therefore$ $\mathbf{m = 0.112096\ kg}$

To find the increase in internal energy

(iii) We have, $\Delta U = m \cdot C_V (T_2 - T_1)$

Here T_2 is not known, so to find T_2, since it is a polytropic process,

$$\therefore \quad \frac{T_2}{T_1} = \left(\frac{V_1}{V_2}\right)^{n-1} = \left(\frac{P_2}{P_1}\right)^{\frac{n-1}{n}}$$

$$T_2 = T_1 \times \left(\frac{V_1}{V_2}\right)^{n-1} = 373 \times \left(\frac{0.12}{0.03}\right)^{1.3-1}$$

$$= 373 \times (4)^{0.3}$$

$$= 373 \times 1.516$$

$$\mathbf{T_2 = 565.36\ K}$$

$$\therefore \quad \Delta U = m \cdot C_V (T_2 - T_1)$$

$$= 0.112 \text{ kg} \times 0.72 (565 - 373)$$

$$\mathbf{\Delta U = 15.51\ kJ\ (Increase)}$$

Example 5.14 *Calculate the change of,*

(i) Internal energy and

(ii) Enthalpy of 2 kg of air when the temperature changes from 20°C to 90°C.

Take C_p = 1 kJ/kg-K and C_v = 0.71 kJ/kg-K

Solution

(i) $\Delta U = m \cdot C_V \cdot (T_2 - T_1) = 2 \times 0.71 (90 - 20)$

$\Delta U = 99.4$ kJ

(ii) $\Delta H = m \cdot C_P \cdot (T_2 - T_1) = 2 \times 1 (90 - 20)$

$\mathbf{\Delta H = 140\ kJ}$

Example 5.15 *10 kg of air are heated in a right vessel from 20°C to 100°C. If the ratio of specific heats is 1.4, estimate the values of C_p and C_v the changes in internal energy and enthalpy.*

Solution

We have, $T_1 = 20°$, $T_2 = 100°C$

$$C_V = \frac{R}{\gamma - 1}$$

and $\gamma = 1.4$ and $R = 0.287$

$$\therefore \quad C_V = \frac{0.287}{0.4} = 0.7175 \text{ kJ/kg-K}$$

$$C_P = R + C_V = 0.287 + 0.7175 = 1.0045 \text{ kJ/kg°K}$$

We know that,
Change in internal energy

$$dU = m\,C_V\,(T_2 - T_1)$$
$$= (10)\,0.7175\,(100 - 20)$$
$$\mathbf{dU = 574\ kJ}$$

And change in enthalpy

$$dH = mC_P(T_2 - T_1) = (10)(1.0045)(100 - 80) = \mathbf{803.6kJ}$$

Example 5.16 *A piston and cylinder machine contains 1 kg of air, initially v = 0.8 m³ and T = 290 K. The air is then compressed in a slow frictionless process specific volume 0.2 m³/kg and a temperature 580 K. The law for compression is $PV^{1.5} = 0.75$ with P in bar and V in m³/kg. Determine work done and heat transferred during the process. Assume for air $C_p = 1.0$ kJ/kg-K, $C_v = 0.743$ kJ/kg-K and R = 0.287 kJ/kg-K.*

Data:

$m = 1$ kg	$V_1 = 0.8$ m³/kg
$T_1 = 290$ K	$V_1 = 0.2$ m³/kg
$T_2 = 580$ K	W.D. = ? Q = ?

Solution

Law for the process = $PV^{1.5} = C = 0.75$

We know that, for a polytropic process,

$$\text{W.D.} = \frac{P_2V_2 - P_1V_1}{1-n} = \frac{mR\,(T_2 - T_1)}{1-n}$$
$$= \frac{1\times 0.287(580-290)}{1-1.5} = \frac{83.23}{-0.5}\text{ kJ}$$
$$= -166.46\text{ kJ}$$

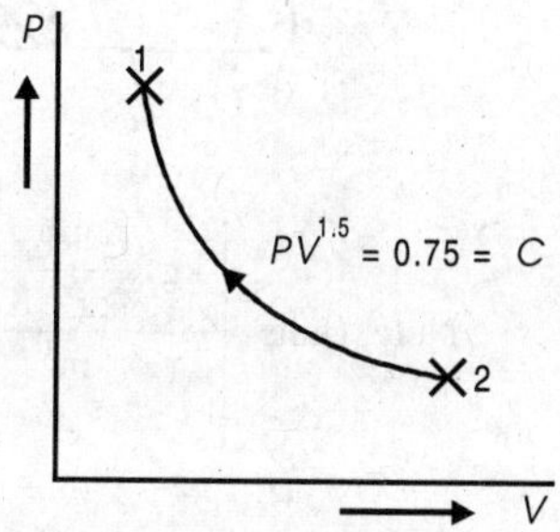

Fig. Ex. 5.16

The sign implies work is done on the air.

Also,
$$\Delta U = m \cdot C_V \cdot dT = 1 \times 0.743 \times (580 - 290)$$
$$= 206.77\text{ kJ}$$

From I-law, for a process,

$$Q - W = \Delta U$$
$$Q = \Delta U + W$$
$$= 206.77 + (-166.46)$$
$$\mathbf{Q = 40.31\ kJ}$$

Example 5.17 *One kg of N_2 at a temperature of 250° C occupies a volume of 0.2 m³. The gas undergoes a constant pressure expansion without friction to a final volume of 0.36 m³. Find final pressure, final temperature, work done, change in internal energy, heat transfer and change of entropy.*

Take $C_v = 0.743$ kJ/kg-K and R = A297 kJ/kg-K and also represent the process on P–V and T–S diagrams.

Data:

$m = 1$ kg	$T_1 = 250 + 273 = 523$ K	
$V_2 = 0.36$ m³	$V_1 = 0.2$ m³.	
$P_2 = ?$	$P = C$	$T_2 = ?$
$W = ?$	$\Delta U = ?$	$Q = ?$ $\Delta S = ?$

Solution

We know that,
$$\frac{P_1V_1}{T_1} = \frac{P_2V_2}{T_2}$$

Since
$$P = C,\ P_1 = P_2 = C$$

∴
$$\frac{V_1}{T_1} = \frac{V_2}{T_2}$$

$$\therefore \quad T_2 = T_1 \times \frac{V_2}{V_1} = 523 \times \frac{0.36}{0.20}$$

$$T_2 = 941.4 \text{ K}$$

Since

$$P_1 V_1 = mRT_1$$

$$P_1 = \frac{mRT_1}{V_1} = \frac{1 \times 0.297 \times 523}{0.2}$$

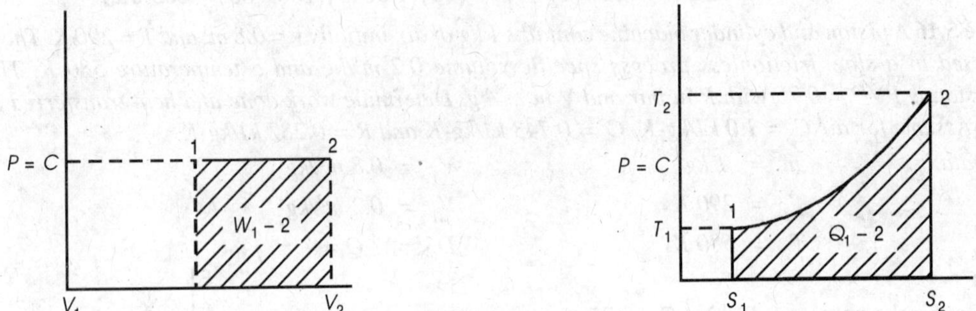

Fig. Ex. 5.17

Note. Units : $\left[\dfrac{\text{kg} \times \dfrac{\text{kJ}}{\text{kg}-\text{K}} \times \text{K}}{\text{m}^3} = \dfrac{\text{kJ}}{\text{m}^3} = \dfrac{\text{kNm}}{\text{m}^2} = \text{kPa}\right]$

$$P_1 = 776.655 \text{ kPa} = P_2$$

(iii) Work done in a non-flow process $= \int PdV$

(since it is a constant pressure process).

$$= P(V_2 - V_1)$$

$$= 776.655\,(0.36 - 0.20)$$

$$W_{1\text{-}2} = 124.26 \text{ kJ}$$

(iv)

$$\Delta U = m \cdot C_V \cdot (T_2 - T_1) = 1 \times 0.743 \times (941.4 - 523)$$

$$= 310.87 \text{ kJ (+ve so I.E. increases)}$$

Note. This $\left[\text{kg} \times \dfrac{\text{kJ}}{\text{kg}-\text{K}} \times \text{K} = \text{kJ}\right]$

(v) We also know from the I-law, for a closed system undergoing a process,

$$Q = \Delta U + W = 310.87 + 124.26$$

$$Q = 435.13 \text{ kJ}$$

(vi) Change in entropy

$$= m \cdot C_P \cdot \ln \frac{T_2}{T_1}$$

$$= m(C_V + R) \ln \frac{T_2}{T_1} \quad (\text{since } C_P - C_V = R)$$

$$= 1 \times (0.743 + 0.297) \ln \frac{941.4}{523} \; 1 \times (1.04) \times \ln \frac{941.4}{523}$$

Note. This $\left[\text{kg} \times \dfrac{\text{kJ}}{\text{kg}-\text{K}} \times \text{ratio} = \dfrac{\text{kJ}}{\text{K}}\right]$

$$S_2 - S_1 = 0.66112 \text{ kJ/K}$$

Example 5.18 *1 m³ of gas is filled in a closed tank. The initial condition of the gas is 3 bar and 50°C. The gas is heated until the pressure becomes 5 bar. Find the change in internal energy, work done, heat supplied, change in entropy. Take R = 0.287 kJ/kg-K and C_v = 0.711 kJ/kg-K.*

Solution

Since it is a constant volume heating process,

$$V_1 = V_2 = C = 1\text{m}^3$$
$$P_1 = 3 \text{ bar} = 3 \times 10^5 \text{ N/m}^2$$
$$T_1 = 50 + 273 = 323 \text{ K}$$
$$P_2 = 5 \text{ bar} = 5 \times 10^5 \text{ N/m}^2$$
$$R = 0.287 \text{ kJ/kg-K} = 287 \text{ J/kg-K}$$
$$C_V = 0.711 \text{ kJ/kg-K} = 711 \text{ J/kg-K}$$

(i) We know that change in IE in a constant volume process is given by,

$$\Delta U = m \cdot C_V (T_2 - T_1) \quad \ldots(1)$$

Here m and T_2 are not given

∴ To find T_2

Since $$\frac{P_1V_1}{T_1} = \frac{P_2V_2}{T_2}$$

as $$V_1 = V_2 = C$$

$$T_2 = \frac{P_2}{P_1} \times T_1$$

$$T_2 = \frac{5\times10^5}{3\times10^5} \times 323$$

$$T_2 = 538.33 \text{ K}$$

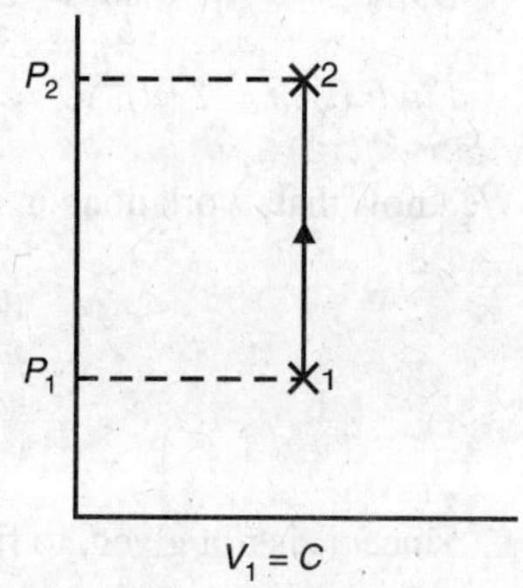

Fig. Ex. 5.18

And since $$P_1V_1 = mRT_1$$

$$m = \frac{R_1V_1}{RT_1} = \frac{3\times10^5\times1}{287\times323}$$

Note units: $$\left[\frac{\frac{\text{N}}{\text{m}^2}\times\text{m}^3}{\frac{\text{J}}{\text{kg-K}}\times\text{K}} = \frac{\text{Nm}}{\text{J/kg}} = \frac{\text{J}}{\text{J/kg}} = \text{kg}\right]$$

$$m = 32362 \text{ kg}$$

∴ From Eq. (1), $$\Delta U = m \cdot C_V (T_2 - T_1)$$
$$= 3.236 \times 711 \times (538.33 - 323)$$

Note this $$\left[\text{kg} \times \frac{\text{J}}{\text{kg-K}} \times \text{K} = \text{J}\right]$$

$$\Delta U = 595462.7 \text{ J} = 495.4627 \text{ kJ}$$

(ii) Work done. In a constant volume heating process, heat supplied is completely utilised for increasing the I.E. of the gas and no external work is done since volume remains constant.

$$W_{1-2} = \int Pdv = 0$$

As $$V = C, \; dV = 0$$

(iii) From first law

$$Q - W = \Delta U$$
$$Q = \Delta U + W$$

$$Q = \Delta U \text{ as } W = 0$$
$$Q = \Delta U = 495.4627 \text{ kJ}$$

(iv)
$$S_2 - S_1 = mC_V \ln\frac{P_2}{P_1}$$
$$= 3.23 \times 0.711 \times \ln\frac{5\times10^5}{3\times10^5}$$

Note this $\left[\text{kg} \times \frac{\text{J}}{\text{kg-K}} \times \text{ratio} = \frac{\text{kJ}}{\text{K}}\right]$

$$S_2 - S_1 = 1.173 \text{ kJ/K}$$

Example 5.19 *2 kg of air at 150°C and 3 bar expands according to $PV^{1.2} = C$ to a final pressure of 1 bar. Find W, Q, and ΔU.*

Take $R = 0.287$ kJ/kg-K and $\gamma = 1.4$

Data: $m = 2$ kg $\quad T_1 = 150 + 273 = 423$ K

$P_1 = 3 \times 10^3$ N/m² $\quad n = 1.2 \quad P_2 = 1 \times 10^5$ N/m²

Find: (i) m = ? (ii) ΔU = ? Q = ? R = 0.287 kJ/kg-K, γ = 1.4

Solution

(i) We know that, work done in a polytropic process is given by,

$$W_{1-2} = \int PdV = \frac{P_1V_1 - P_2V_2}{n-1}$$
$$= \frac{mR(T_1 - T_2)}{n-1}$$

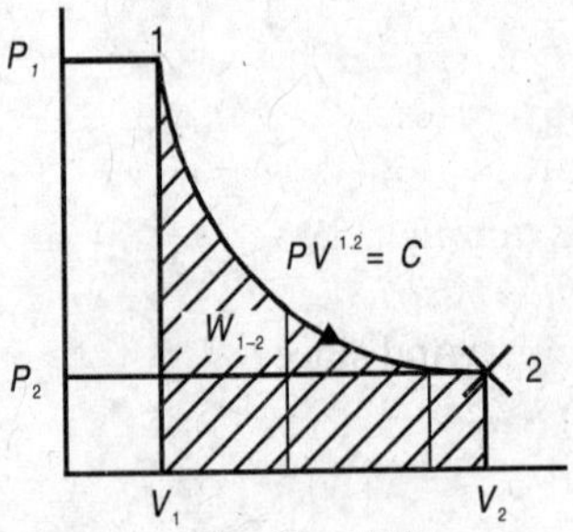

Fig. Ex. 5.19

Since T_2 is not given, to find T_2, use the relation

$$\frac{T_2}{T_1} = \left(\frac{P_2}{P_1}\right)^{\frac{n-1}{n}}\left[\frac{V_1}{V_2}\right]^{n-1}$$

Since it is a polytropic process.

[Note that if it is $P = C$; $V = C$; $T = C$ process then use $\frac{P_1V_1}{T_1} = \frac{P_2V_2}{T_2}$ to find the unknown]

$$\therefore \quad \frac{T_2}{T_1} = \left(\frac{P_2}{P_1}\right)^{\frac{n-1}{n}}$$

$$T_2 = T_1 \times \left(\frac{P_2}{P_1}\right)^{\frac{n-1}{n}} = 423 \times \left[\frac{1\times10^5}{3\times10^5}\right]^{\frac{1.2-1}{1.2}}$$

$$= 423 \times \left[\frac{1}{2}\right]^{\frac{0.1}{1.2}}$$

$$\mathbf{P_2 = 325.25\ K}$$

W_{1-2} from Eq. (1)

$$= \frac{mR(T_1 - T_2)}{n-1} = \frac{2\times0.287(423 - 352.25)}{1.2 - 0.2}$$

$$W_{1-2} = 203 \text{ kJ}$$

(ii) Change in IE,

$$\Delta U = m \cdot C_V (T_2 - T_1)$$

Now, $$\frac{C_P}{C_V} = \gamma = 1.4$$

$$C_p - C_V = T = 0.287$$

And $$C_V = \frac{R}{\gamma - 1} = \frac{0.287}{1.4 - 1} = 0.7175 \text{ kJ/kg-K}$$

∴ From Eq. (2)

∴ $$\Delta U = 2 \times 0.7175 \times (352.25 - 423)$$

$$\mathbf{\Delta U = -101.52 \text{ kJ (Decreases)}}$$

(iii) ∴ Heat transfer $$Q = \Delta U + W$$

$$= -102.52 + 203 \text{ kJ}$$

$$Q = 101.48 \text{ kJ}$$

(iv) $$S_2 - S_1 = m\left[\frac{\gamma - n}{n} \times C_V \times \ln \frac{P_2}{P_1}\right]$$

$$= 2\left[\frac{1.4 - 1.2}{1.2} \times 0.7175 \times \ln \frac{3}{1}\right]$$

$$\mathbf{S_2 - S_1 = 0.26916 \text{ kJ/K}}$$

For selecting the equation to find the change of entropy, note that it is a polytropic process and pressure ratio is given, also we can select the equation in which temperature ratio is given for this problem, since temperature are given. Both the equations will give the same answer.

Example 5.20 *One kg of Nitrogen at a temperature of 150°C occupies a volume of 0.2 m³. The gas undergoes a fully restricted constant pressure expansion without friction to a final volume of – 0.36 m³ . Evaluate the final pressure, final Take C_p = 743 J/kg-K, R = 0.297 kJ/kg-K.*

Solution

(a) We know that,

$$P_1 V_1 = mRT_1, \; m = 1$$

∴ $$P_1 = \frac{RT_1}{V_1} = \frac{0.297 \times 10^3 \times 423}{0.2}$$

$$= 628.155 \text{ kPa}$$

$$= P_2 = \text{final pressure}$$

(b) We know that for a constant pressure process,

$$\frac{P_1 V_1}{T_1} = \frac{P_2 V_2}{T_2} \; (P_1 = P_2)$$

∴ $$\frac{V_1}{T_1} = \frac{V_2}{T_2}$$

∴ $$T_2 = \frac{V_2}{V_1} \cdot T_1 = \frac{0.36}{0.2} \times 423$$

$$= 761.4 \text{ K or } 488.4°\text{C}$$

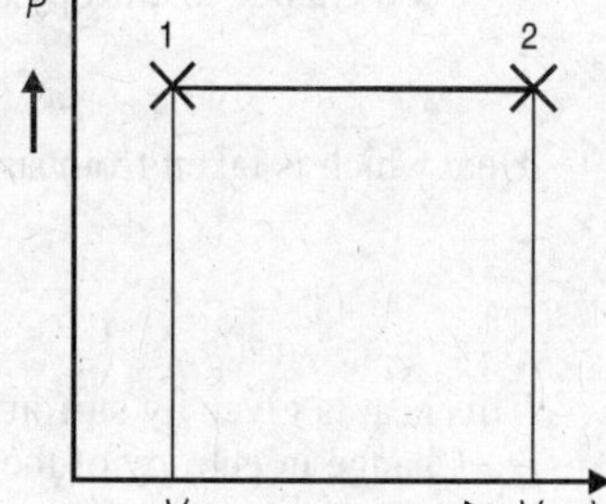

Fig. Ex. 5.20

(c) Work done in constant pressure process is given by

$$\delta Q = dH$$

∴ For unit mass, $$\delta q = dh = C_P \, dT = C_P (T_2 - T_1)$$

$$= 0.743 \times (761 - 423) = 251.43 \text{ kJ/kg}$$

(d) Change in entropy

$$\therefore \quad s_2 - s_1 = C_P \ln \frac{T_2}{T_1}$$

$$= 0.743 \times \ln \frac{761.4}{423}$$

$$s_2 - s_1 = \mathbf{0.437\ kJ/kg.K}$$

Example 5.21 *1 kg of ice at –5°C is exposed to the surrounding atmosphere which is at 20°C. The ice melts and comes into thermal equilibrium with the atmosphere which is at 20°C. The ice melts and comes into thermal equilibrium with the atmosphere. Determine the entropy increase of the universe. Show the melting process on T–S diagram. Take, C_p of ice = 2.093 kJ/kg-K, Latent heat of fusion of ice = 333.3 kJ/kg.*

Solution

The melting is shown on *T–S* diagram $T_1 = 26$ K, $T_2 = T_3 = 273$ K, $T_4 = 293$ K

With reference to *T–S* diagram we have

$$S_2 - S_1 = C_P \log_e \frac{T_2}{T_1}$$

$$= 2.093 \log_e \frac{273}{268}$$

$$= 2.093 \times \log_e 1.0186$$

$$= 0.0387 \text{ kJ/kg-K}$$

$$S_3 - S_2 = h_{fg}/T_3$$

$$= \frac{33.3}{273} = 1.221 \text{ kJ/kg-K}$$

$$S_4 - S_3 = C_{Pw} \log_e \frac{T_4}{T_3}$$

$$= 4.187 \log_e \frac{293}{273}$$

$$= 0.296 \text{ kJ/kg-K}$$

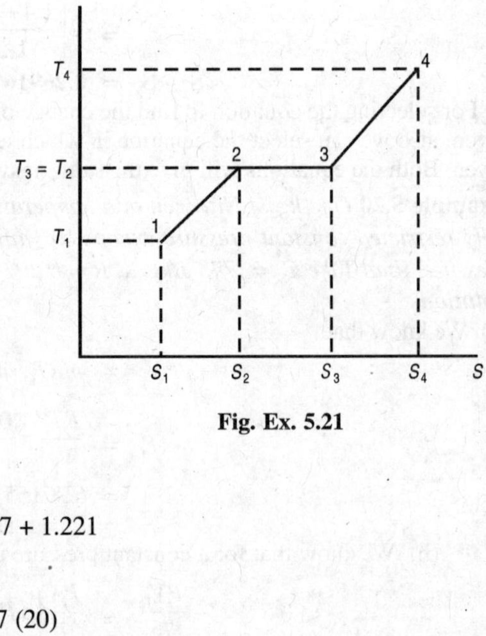

Fig. Ex. 5.21

∴ Total change in entropy of ice/water system

$$= S_4 - S_1 = 0.296 + 0.0387 + 1.221$$

$$= 1.556 \text{ kJ/kg-K}$$

Heat which is taken from surroundings

$$= 2.093\ (5) + 33.3 + 4.187\ (20)$$

$$= 10.465 + 333.3 + 83.74$$

$$= 427.505 \text{ kJ}$$

This heat is given by surrounding at constant temperature 293 K

∴ Change in entropy of the surrounding

$$= \frac{427.505}{293} = 1.4591 \text{ kJ/kg-K}$$

∴ Increase of entropy of universe

$$= \Delta S_P = 1.5560 - 1.4591$$

Increase of entropy of universe = 0.0969 kJ/kg-K

THEORY QUESTIONS

5.1. State and prove the principle of increase of entropy of the universe. Give some practical examples.

5.2. Prove that entropy is a property.

5.3. State and explain Clausius Theorem.

5.4. Explain Clausius inequality.

5.5. Write a short note on: Entropy and disorder.

5.6. Prove for polytropic process:

$$S_2 - S_1 = \frac{\gamma - n}{1 - n} C_v \ln\left(\frac{T_2}{T_1}\right)$$

5.7. Explain the principle of increase of entropy. What is its significance ?

5.8. Give the criteria of reversibility, irreversibility and impossibility of a thermodynamic cycle.

5.9. State and prove the principle of increase of entropy of the universe. Give some practical examples.

5.10. Derive and express the Clausius Inequality.

5.11. Straight change of entropy during a polytropic process for an ideal gas is given by,

$$\Delta S = C_v [k - n] \ln \frac{V_2}{V_1}$$

where k = superheat ratio, n = Polytropic index.

5.12. What is the principle of increase of entropy? Explain with two practical examples.

5.13. P.T. for a poly thermo process:

$$\Delta S = \frac{\zeta - n}{\zeta - 1} . R . \ln \frac{V_2}{V_1}$$

5.14. What do you understand by the term reversibility, when need in the thermodynamic source?

List to the best of your knowledge and explain all the possible irreverbilities that may be present in a system.

PROBLEMS FOR PRACTICE

5.1. 0.5 kg of air is compressed reversibly and adiabatically from 80 kPa 60°C to 0.4 MPa and is then expanded at constant pressure to the original volume. Sketch the process on *P–V* and *T–S* diagram, compute work transfer, heat transfer and change in entropy for whole path.

Take $R = 0.287$ kJ/kg-K and $\gamma = 1.4$.

(*Ans.* 93.5 kJ, 0.578 kJ /K)

5.2. A certain mass of air initially at a pressure of 480 kPa and temperature 190°C is expanded adiabatically to a pressure of 94 kPa. It is then heated at constant volume until it attains its initial temperature when the pressure is found. to be 150 kPa. State the type of compressor necessary to bring the system back to its original pressure and volume. Determine (1) the index of adiabatic expansion, (2) The work done per kg of air and (3) the change in specific entropy of air.

R for air = 0.29 kJ/kg K.

(*Ans.* 1.402, –31.59 kJ/kg)

5.3. An ideal gas having γ or $K = 1.35$ and volume is 1.5 m^3 at 100 kPa and 77°C. The gas is compressed according tot he law $PV^{1.25} = C$ to a pressure 3 MPa. Calculate the volume and temperature at the end of compression. Evaluate the heat transfer, work done, ΔU and change of entropy during the process.

Take $C_p = 0.98$ kJ/kg-K.

(*Ans.* –167 kJ, –584.5 kJ, 417.5 kJ, –0.322 kJ/K)

5.4. Show that when 1 kg of perfect gas is expanded through a volume ratio *r* the law of expansion being PV^n = constant, the change of entropy is given $(C_p - nC_v)$ Inr.

5.5. 50 litres of air at 1.013 bar and 100ºC temperature is compressed to 28 bar. Volume of air at the end of polytropic compression is found to be 4 litres. Air is now heated at constant volume till pressure rises to 56 bar. Assuming $C_p = 1.00$ kJ/kg K and $C_v = 0.71$ kJ/kg K over range of working determine.

(*i*) Polytropic index of compression.

(*ii*) Entropy change in each process. Sketch the process on *P–V* and *T–S* diagram.

(*Ans.* (i) 1.31, (ii) –2.987, 10.5 kJ/K)

5.6. One kg of Nitrogen at a temperature of 150°C

occupies a volume of 0.2 m³. The gas undergoes a fully restricted constant pressure expansion without friction to a volume of 0.36 m³. The gas is then expanded isothermally to a volume of 0.5 m³. Represent the process on *P–V* and *T – S* diagrams and determine the change in entropy for each process and overall change in entropy.

Take C_v = 743 J/kg-K and *R* = 0.297 kJ/kg K.

(*Ans.* 0.6089, 0.0962, 0.7051)

5.7. 4 kg of air are compressed from 40ºC and 125º kPa to 250ºC and 875 kPa. It is then throttled to 57 kPa. Finally it is cooled to a pressure of 125 kPa and 180ºC. Calculate the overall change in entropy and also for each process.

Take C_p = 1.055 kJ/kg K, C_v = 0.717 kJ/kg K.

(*Ans.* 1.4859)

5.8. A solar thermal power plant flat plate collects or parabolic concentrators of sufficient area on which solar energy is incident are used. Flat plate collectors can give highest temperature in cycle as 90ºC while parabolic concentrators give a temperature of 250ºC. If the ambient temperature is 27ºC. Find the minimum theoretical, area required of the solar system developing 10 kW power.

(i) By using flat plate collector having 60% collection efficient.

(ii) By using parabolic concentrator having 50% collection efficiency.

Assume incident plate solar energy collector, we can write. (*Ans.* (i) 96 m; (ii) 46.7 m²)

5.9. A novel reversible heat engine plot in (*T–S*) diagram as a circle. The maximum and minimum temperatures are 1100 K and 200 K respectively and the maximum entropy change in the cycle is 2 kJ/K. Calculate the heat added to the cycle, heat rejected, the net work output and thermal efficiency of the cycle.

(*Ans.* 2006.86 kJ, 593.119, 1413.7Z kJ, 70.45)

5.10. Certain gas at a pressure of 1.4 MN/m² and 360º C is expanded adiabatically to a pressure of 100 kJ/m². The gas is then heated at constant volume until it attains 300ºC when the pressure is found to be 220 kN/m² and finally it is compressed isothermally to the original pressure of 1.4 MN/m². Sketch the process on *P–V* and *T–S* diagrams. For 0.23 of gas and evaluate.

(i) ΔU during adiabatic expansion.

(ii) ΔS for each process.

Assume: C_v = 0.705 kJ/kg K.

(*Ans.* (i) 55.987 kJ; (ii) 0, 0.1278 kJ/K)

5.11. 60 litres of hydrogen at 20ºC and 1 bar is compressed adiabatically to 9.8 bar. It is then cooled at constant volume to a pressure P_3 such as to have temperature of 20°C and further expanded isothermally so as to reach initial state: Find,

(i) Work done during each process.

(ii) Change in internal energy during each process.

Take C_p = 14.4 kJ/kg-K

C_v = 10.28 kJ/kg-K

for hydrogen and show the process on *P–V* and *T–S* plane.

(*Ans.* (i) –13.788 kJ; (ii) 13.788 kJ)

5.12. 3 kg of air at a pressure of 150 kPa and temperature 360 K is compressed polytropically to 750 kPa according to law $PV^{1.2}$ = constant. The air is then cooled to initial temperature at constant pressure. The air is then heated at constant temperature till it reaches original pressure of 150 kPa. Draw the cycle on *PV* diagram and determine net work and heat. (*Ans.* –73.21 kJ, –73.21 kJ)

5.13. Process 1: Air initially at 100 kPa and 50ºC undergoes reversible adiabatic compression such that its volume is reduced to 1/5th of initial volume.

Process 2: Then 940 kJ/kg of heat is added to this air at constant volume.

Process 3: Process 2 is followed by reversible adiabatic expansion up to initial volume.

Process 4: Finally heat is rejected at constant volume so as to reach the initial condition.

Draw the four processes on one *PV* diagram. Determine the maximum temperature, and heat rejected per kg of air. Assume adiabatic index of compression and expansion of 1.4 and constant volume specific heat as 0:717 kJ.kg K.

(*Ans.* 1926 K, – 493.8 kJ/kg)

IMPORTANT POINTS TO BE REMEMBERED

1. *Clausium theorem* states that the cyclic integral of $\frac{\delta Q}{T}$ for a reversible cycle is equal to zero $\oint_{Rev} \frac{\delta T}{T} = 0$

2. For irreversible cycle $\oint \frac{\delta Q}{T} < 0$ is known as Clausius inequality.

3. $\oint \frac{\delta Q}{T} = 0$ for reversible cycle

   < 0 for irreversible cycle

   > 0 for an impossible cycle.

4. $ds \geq \frac{\delta Q}{T}$ equality sign for reversible process inequality sign for irreversible process

5. Entropy of the universe always increases.

6. $$\Delta S = S_2 - S_1 = C_v \log_e \frac{T_2}{T_1} + R \log_e \frac{V_2}{V_1}$$

   $$= C_p \log_e \frac{V_2}{V_1} - C_p \log_e \frac{P_2}{P_1}$$

   $$= C_p \log_e \frac{T_2}{T_1} - R \log_e \frac{P_2}{P_1}$$

   $= C_v \log_e \frac{T_2}{T_1}$ for constant vol. process

   $= C_p \log_e \frac{T_2}{T_1}$ for constant pre. process

   $= R \log_e \frac{V_2}{V_1}$ for constant temperature process

   $= O$ for constant isentropic porcess

   $= C_v \cdot \frac{\gamma - n}{1 - n} \log_e \frac{T_2}{T_1}$ for constant general law (Polytropic)

7. *Third law of thermodynamics*, states that the entropy of a pure substance in some perfect crystalline form becomes zero.

CHAPTER 6

Real Gases

6.1 INTRODUCTION

As we have studied in ideal gases that, an Ideal gas is one which obeys all the gas laws and the characteristic gas equation $PV = RT$ at all temperatures and pressures.

But actually there is no ideal or perfect gas in reality. But all the real gases like Air, O_2, N_2, H_2, He behave as ideal gases at very low pressures and high temperatures.

So a real gas does not obey gas laws $PV = RT$ for the entire range of pressure and temperature except when the pressure tends to zero and at a very high temperature. However real gas behaviour deviates from perfect gas equation at high pressures and low temperatures.

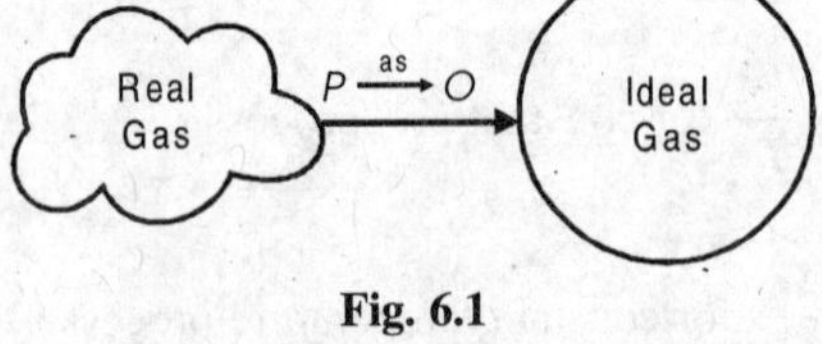

Fig. 6.1

In order to study the behaviour of real gases under all the conditions of pressures and temperatures Van der Waals' equation is most commonly used.

6.2 VAN der WAAL'S EQUATION OF STATE

The ideal gas equation $PV = RT$ is developed from kinetic theory of gases and is based on two important assumptions viz.; (i) There is very little or no attraction between the molecules of the gas (ii) The volume of the gas molecules themselves is very very small compared with the total volume of gas.

At low pressures and at very high temperatures, the intermolecular attractions and the volume of molecules compared to total volume of gas are not of much significance and the real gases obey the ideal gas equation. But when the pressure increases, the intermolecular attraction and the volume of the molecules becomes considerable when compared to total volume of gas. Then the real gases deviate from the ideal gas equation.

So, Van der Waal suggested the corrections for the ideal gas equations, so that it can be used for real gases at all pressures and temperatures.

6.2.1 Correction for Intermolecular Attractive Forces

To find a correction term for intermolecular attractive forces of the molecules, let us consider the molecule *A*, which is surrounded by the other molecules as shown and is attracted equally in all the directions and is in equilibrium. If we consider the molecule *B*, which is about to strike two walls of the containing vessel. Then this molecule will strike with less velocity or less momentum because of attractive force of the molecules on the inner side. When the momentum or velocity is less, then the pressure exerted by the molecules is less.

Van der Waal experimentally found that this decrease in pressure because of molecular attractive force is proportional to the square of the molal density i.e., Decrease in pressure α (Density of gas)^2.

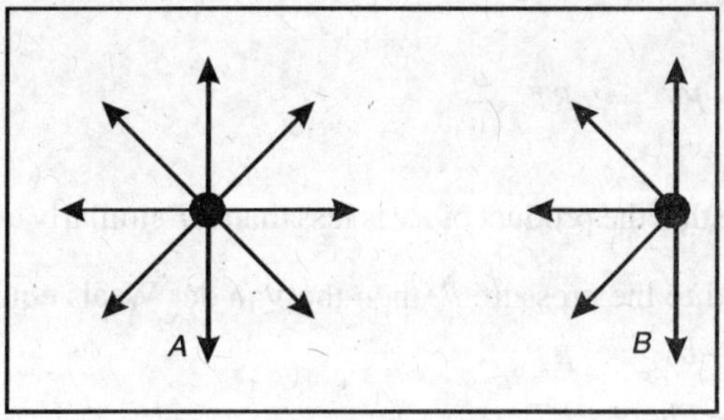

Fig. 6.2

$$p \propto (\rho_{gas})^2$$

$$p \propto \frac{1}{v^2} \quad \left(\because \rho = \frac{1}{v} \right)$$

$$p = \frac{a}{v^2}$$

where v is molal volume and a is constant of proportionality.

Now if P, is the measured pressure, the corrected, pressure will be $(P + p) = \left(P + \frac{a}{v^2} \right)$

6.2.2 Correction for Volume

This correction is to consider the self volume of the gas molecules, which is generally neglected in comparison with total volume. To correct this self volume, Van der Waal experimentally found and suggested that specific volume in ideal gas equation must be replaced by $(v - b)$. Where constant b takes care of reduction in volume due to finite size of molecules.

When we consider both these corrections, the equation of state for the Real gases can be written as,

$$\left(\mathbf{P} + \frac{\mathbf{a}}{\mathbf{v}^2} \right)(\mathbf{v} - \mathbf{b}) = \mathbf{RT} \qquad \ldots(1)$$

or
$$\left(P + \frac{a}{v^2} \right) = \frac{RT}{(v - b)}$$

or
$$P = \frac{RT}{(v - b)} - \frac{a}{v^2}$$

where P is the pressure in Pa or N/m^2
R is the Gas Const. = 8314 J/kg mole K
T temperature in K

v molar volume in m³/kg mole
b is constant in m³/kg mole
a is constant N-m⁴/(kg mole)².

At low pressures, b may be neglected,

$$\left(p + \frac{a}{v^2}\right)v = RT$$

$$PV + \frac{a}{v} = RT$$

$$PV = RT - \frac{a}{v} \qquad \ldots(2)$$

From this equation we can see that the product of Pv is less than RT similarly at very high pressures, the term $\frac{a}{v^2}$ can be neglected when compared to the pressure P, then the Van der Waals equation can be written as,

$$P(v-b) = RT$$

$$PV - Pb = RT$$

$$PV = RT + Pb \qquad \ldots(3)$$

Equation (3) is known *Clausius Equation of State* and it can be seen that the product of pressure and volume PV is more than RT.

6.3 DETERMINATION OF CONSTANTS OF VAN der WAAL'S EQUATION

Eventhough the Van der Waal's equation is generally used for real gases, it has some limitations, that is the constants a and b vary with temperature and pressure. One of the methods generally used to find the value of the constants a and b is based on the fact that in case of real gases the critical isotherm on P-v diagram has zero slope at critical point. (refer Fig. 6.3).

It is to be noted that the properties at the critical state are known as critical properties such as critical pressure P_C, critical temperature T_C, critical volume v_C.

Since the critical isotherm has zero slope at critical point, we can write,

$$\left(\frac{dP}{dv}\right)_c = 0 \text{ and } \left(\frac{d^2P}{dv^2}\right)_c = 0.$$

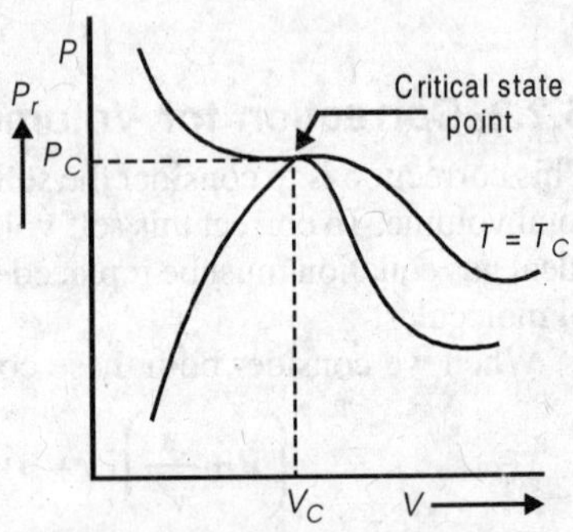

Fig. 6.3. *Critical properties on Pv diagram*

Then from the Van der Waal's equation,

$$\left(P + \frac{a}{v^2}\right)(v-b) = RT$$

i.e.
$$P + \frac{a}{v^2} = \frac{RT}{v-b}$$

i.e.
$$P = \frac{RT}{v-b} - \frac{a}{v^2}$$

At critical point we can write,
$$P_c = \frac{RT_c}{v_c - b} - \frac{a}{v^2} \qquad \ldots(1)$$

So, differentiating this w.r.t. v_c and assuming T_c = constant we get,

$$\left(\frac{dP}{dv}\right)_c = -\frac{RT_c}{(v_c-b)^2} + \frac{2a}{v_c^3} = 0$$

or
$$\frac{-2a}{v_c^3} = \frac{RT_c}{(v_c-b)^2}$$

$$2\frac{a}{v_c^3}(v_c-b)^2 = RT_c \quad \text{...(2)}$$

Then
$$\left(\frac{d^2P}{dv^2}\right)_c = \frac{2RT_c}{(v_c-b)^3} - \frac{6a}{v_c^4} = 0$$

or
$$\frac{3_a}{v_c^4}(v_c-b)^3 = RT_c \quad \text{...(3)}$$

From Eqs (2) and (3) we get, $\mathbf{v_c = 3b}$ ...(4)

Substituting v_c in Eq. (3) we get,

$$\mathbf{T_c = \frac{8a}{27Rb}} \quad \text{...(5)}$$

Substituting v_c and T_c in Eq. (1) we get,

$$\mathbf{P_c = \frac{a}{27b^2}} \quad \text{...(6)}$$

Equations (4), (5) and (6) give the values of critical properties.

Because of practical limitations it is very very difficult to measure the critical volume. So, the constants a and b are generally expressed in terms of T_c and P_c.

From Eqs (5) and (6) we get,

$$\mathbf{a = \frac{27}{64}\frac{R^2T_c^2}{P_c}} \quad \text{...(7)}$$

and
$$\mathbf{b = \frac{RT_c}{8P_c}} \quad \text{...(8)}$$

Note: When the Van der Waals equation is compared with ideal gas equation at critical state we get,

$$\frac{P_cv_c}{RT_c} = \frac{3}{8} \quad \text{...(9)}$$

So, the ratio of $\frac{PV}{RT}$ at critical point is a constant 3/8 is same for all real gases and is unity for ideal gases.

Table below gives the constants of Van der Wall's equation.

S. No.	*Substance*	*a Nm⁴/(kmol)²*	*b m³/kmol*
1	Hydrogen (H_2)	25105	0.0262
2	Oxygen (O_2)	139250	0.0314
3	Carbon dioxide (CO_2)	362850	0.0423
4	Helium (He)	3417620	0.0228
5	Air	135522	0.0362
6	Water (H_2O) Vapour	551130	0.0300
7	Mercury (Hg) Vapour	2031940	0.0657

6.4 REDUCED CO-ORDINATES (VAN der WAAL'S EQUATION) IN REDUCED CO-ORDINATES

As we have studied in the previous section that the properties at the critical state are known as critical properties such as critical pressure P_c, critical temperature T_c, critical volume v_c.

Reduced co-ordinates or properties are defined as those properties which are obtained by dividing co-ordinates/properties by corresponding co-ordinates/properties at the critical state for example,

$$\text{Critical Pressure } P_r = \frac{P}{P_c} \quad \text{...(1)}$$

$$\text{Critical Temperature } T_r = \frac{T}{T_c} \quad \text{...(2)}$$

and

$$\text{Critical Volume } v_r = \frac{v}{v_c} \quad \text{...(3)}$$

The Relationship Between Reduced Co-ordinates is known as the Law of Corresponding States:

From previous section we know from Eqs (4), (5) and (6)

$$v_c = 3b \quad \text{...(4)}$$

$$T_c = \frac{8a}{27Rb} \quad \text{...(5)}$$

$$P_c = \frac{a}{27 \cdot b^2} \quad \text{...(6)}$$

From Eqs (1), (2) and (3) we can write,

$$P = P_r \cdot P_c = P_r \times \frac{a}{27 \cdot b^2} \quad \text{...(7)}$$

$$T = T_r \cdot T_c = T_r \times \frac{8.a}{27 \cdot R \cdot b} \quad \text{...(8)}$$

$$v = v_r \cdot v_c = v_r \times 3b \quad \text{...(9)}$$

Substituting these values in Van der Waals equation we get,

$$\left(\mathbf{P_r + \frac{3}{v_r^2}}\right)\mathbf{(v_r - \tfrac{1}{3}) = \frac{8}{3}T_r}$$

or

$$\left(\mathbf{P_r + \frac{3}{v_r^2}}\right)\mathbf{(3v_r - 1) = 8T_r}$$

This is the Van der Waal's equation in terms of reduced co-ordinates and its is independent of R, a, and b.

6.5 COMPRESSIBILITY FACTOR; COMPRESSIBILITY CHART, GENERALISED COMPRESSIBILITY CHART

Compressibility Factor. The deviation of real gas from ideal gas is measured by a factor known as compressibility factor Z, and defined as the ratio of the actual volume to the volume predicted by the ideal gas law at the same temperature and pressure is given by $Z = \dfrac{\text{Actual volume}}{\text{Volume predicted by the ideal gas}} = \dfrac{v}{\frac{RT}{P}}$

But $\frac{RT}{P}$ is the ideal gas volume V_{ideal}. So, compressibility factor may also be defined as, the ratio of specific volume of real gas to specific volume of ideal gas i.e.,

$$\text{Compressibility Factor} = Z = \frac{v_{\text{real gas}}}{v_{\text{ideal gas}}}$$

For ideal gas $$Z = \frac{v_{\text{ideal}}}{v_{\text{ideal}}} = 1$$

As we know all the gases behave as ideal gases at very low pressures and high temperatures. So when the pressures are lowered, then the value of Z tends to unity as the gas behaves as ideal.

It is to be noted that the value of Z may be less than unity or greater than unity depending upon the pressure and temperature. Figure 6.4 shows the compressibility factor chart which shows the values of Z corresponding to pressure.

Note that a graph of compressibility factor Z versus P_r for different values of T_r for different gases is known as *Generalised Compressibility Chart* which is shown in Fig. 6.5.

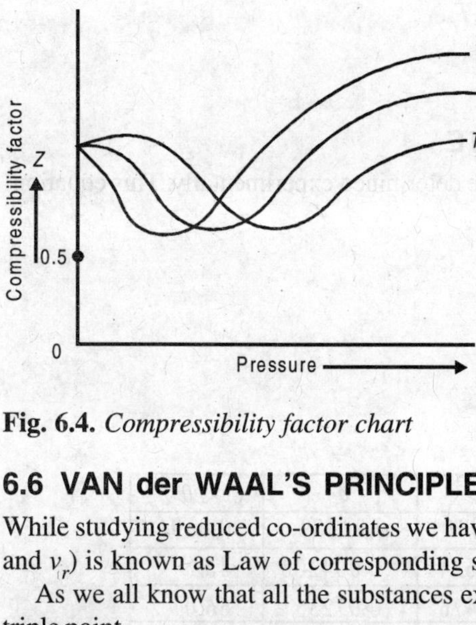

Fig. 6.4. *Compressibility factor chart*

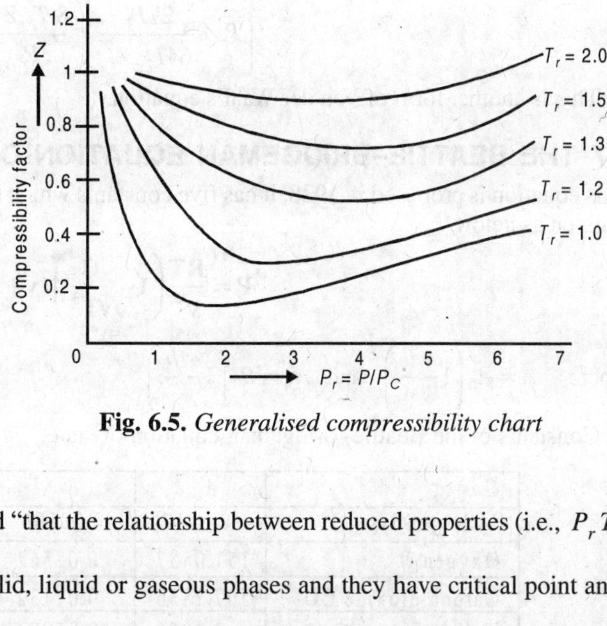

Fig. 6.5. *Generalised compressibility chart*

6.6 VAN der WAAL'S PRINCIPLE

While studying reduced co-ordinates we have studied "that the relationship between reduced properties (i.e., $P_r T_r$ and v_r) is known as Law of corresponding states.

As we all know that all the substances exist in solid, liquid or gaseous phases and they have critical point and triple point.

And Van der Waal suggested that $P\ v\ T$ relationship can be shown by a single model and he has used critical point as reference point and critical properties as reducing properties and the principle of corresponding states may be expressed as,

$$f(P_r, v_r, T_r) = 0 \quad \text{.... (1)}$$

From the definition of compressibility factor, we know that,

$$Z = \frac{P_v}{RT} \quad \text{...(2)}$$

Since $P_r = \frac{P}{P_c}; v_r = \frac{v}{v_c}; T_r = \frac{T}{T_c}$

We can write, $$Z = \frac{P_r \cdot P_c}{R} \cdot \frac{v_r \cdot v_c}{T_r \cdot T_c}$$

or $$Z = \frac{P_r \cdot v_r}{T_r} \cdot \frac{P_c \cdot v_c}{R \cdot T_c}$$

$$Z = \frac{P_r \cdot v_r}{T_r} \cdot Z_c \qquad \ldots(3)$$

where $Z_c = \frac{P_c \cdot v_c}{R \cdot T_c} = \frac{3}{8}$ according to Van der Waal.

$$\therefore \qquad \mathbf{Z = \frac{P_r \cdot v_r}{T_r} \times \frac{3}{8}} \qquad \ldots(4)$$

Now from Eq. (10) of reduced co-ordinates article and above Eq. (4) we get

$$\left[P_r + \frac{27 P_2^r}{64 T_2^r . Z^2}\right]\left(\frac{8.T_r.Z}{3P_r} - \frac{1}{3}\right) = \frac{8}{3} T_r.$$

This is another form of Van der Waal's equation.

6.7 THE BEATTIE–BRIDGEMAN EQUATION OF STATE

This equation is proposed in 1928, it has five constants which have to be determined experimentally. This equation is given as below,

$$\mathbf{P = \frac{RT}{v^2}\left(1 - \frac{C}{vT^3}\right)(v + B) - \frac{A}{v^2}}$$

where $A = A_0\left(1 - \frac{a}{v}\right)$ and $B = B_0\left(1 - \frac{b}{v}\right)$

Constants of the Beattie–Bridgeman equation of state.

Gas	A_o	a	B_o	b	$C \times 10^{-4}$
Hydrogen H_2	20.0117	−0.00506	0.02096	−0.04359	0.0504
Oxygen O_2	151.0857	0.02562	0.04624	0.004208	4.8
Carbon dioxide CO_2	507.2836	0.07132	0.10476	0.07235	66.0
Helium He	2.1886	0.05984	0.01400	0.0	0.0040
Air	131.8441	0.01931	0.04611	−0.001101	4.34
Nitrogen N_2	136.2315	0.02617	0.05046	−0.00691	4.20

6.8 VIRIAL EQUATION OF STATE

We know that the equation of state of a substance is $PV = RT$ and may be expressed as,

$$P = \frac{RT}{v} + \frac{a(T)}{v^2} + \frac{b(T)}{v^3} + \frac{c(T)}{v^4} + \ldots$$

Then this equation is known as Virial Equation of state. And the co-efficients $a\,(T)$, $b\,(T)$, $c\,(T)$ which are the functions of temperature are called as Virial Co-efficients.

6.9 P-v-T SURFACE OF AN IDEAL GAS

We have studied the equation of state for ideal gases and real gases. These equations contain three variables *P*, *v*, *T*. If we plot these three variables along the three mutually perpendicular axes, then a surface is developed and is known as *P-v-T* surface.

Figure 6.6 shows the *P-v-T* surface of an ideal gas. On this figure constant pressure, constant volume, and isothermal lines are shown. It is to be noted that on this surface each point is an equilibrium state and the processes proceed from one equilibrium state to another equilibrium state till the end of the process and the processes are quasi-static.

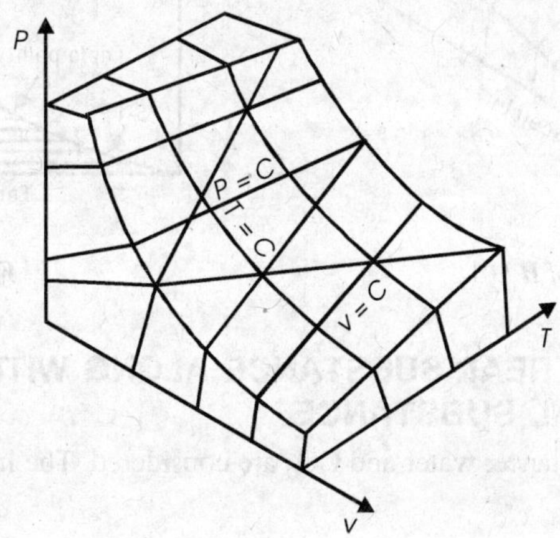

Fig. 6.6

From Fig. 6.6 we can draw *P.v* and *PT* plans for an ideal gas as shown in Figs. 6.7 and 6.8.

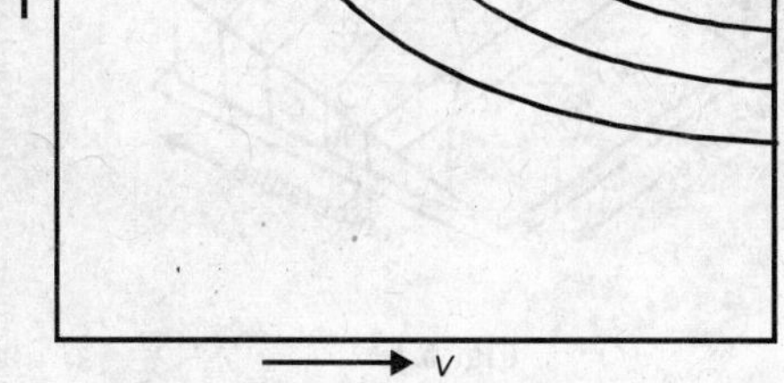

Fig. 6.7

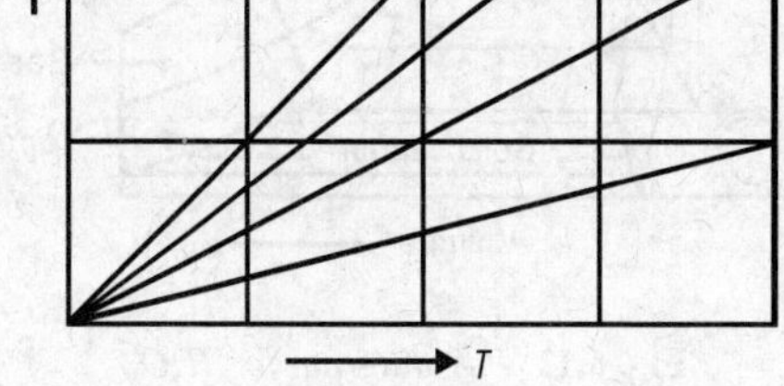

Fig. 6.8

Figure 6.9 shows the isothermal processes i.e., it graphically shows Boyle's law whereas Fig. 6.10 shows graphically Charle's law.

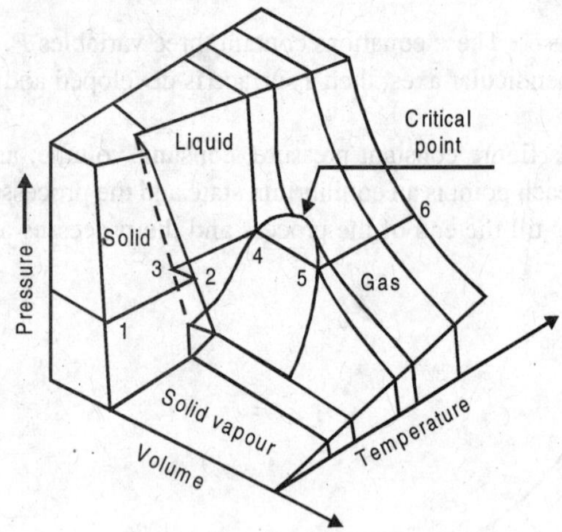

Fig. 6.9. *P-v-T surface of H_2O*

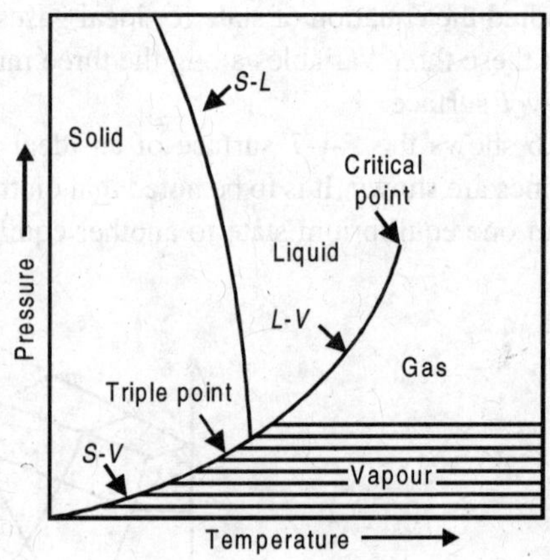

Fig. 6.10

6.10 P-v-T SURFACE FOR REAL SUBSTANCE ALONG WITH Pv AND PT DIAGRAMS FOR REAL SUBSTANCE

To study *P-v-T* surfaces of real substance, water and CO_2 are considered. The important features of the two are:

(i) Water expands on freezing

(ii) CO_2 contracts on freezing.

Figure 6.10 shows *P. v. T* surface of H_2O. Also Figs 6.10 and 6.11 show *PT* and *Pv* diagrams.

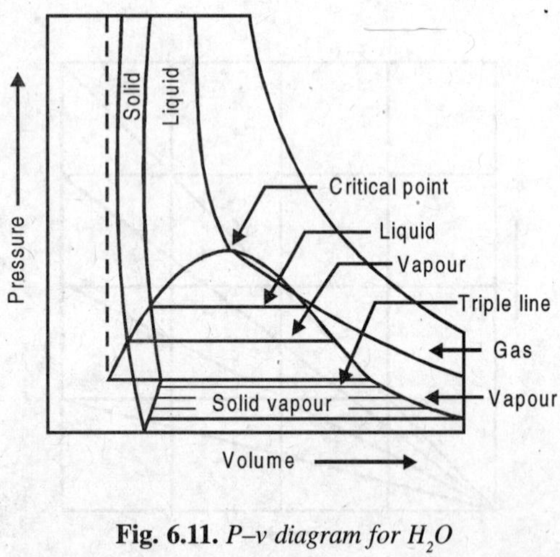

Fig. 6.11. *P–v diagram for H_2O*

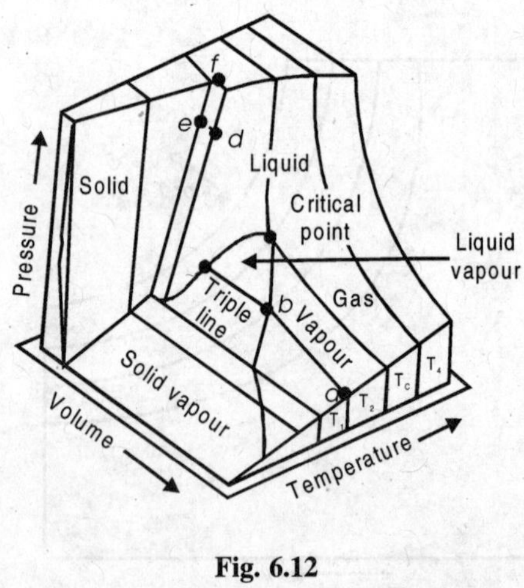

Fig. 6.12

As we have discussed in previous section that every point on the *P-v-T* surface is a equilibrium point and the processes are quasi-static.

In certain portions the substance exists in single phase i.e., it may be solid, liquid or a gas and accordingly they are marked. And if the substance has two phases, then the region is marked as solid-liquid, liquid-vapour, solid-vapour.

Note that triple point line appears as a point on PT-diagram.

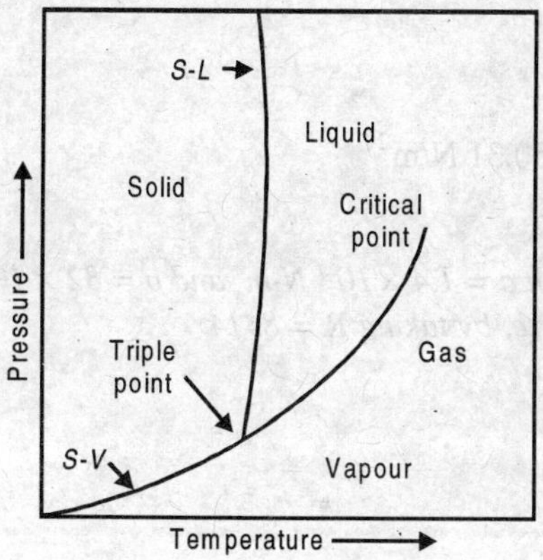

Fig. 6.13. *p-T diagram for* CO_2

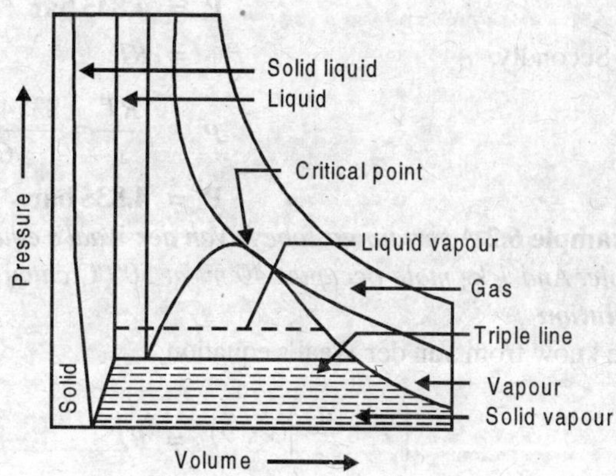

Fig. 6.14. *p-v diagram for* CO_2.

The line 1 – 2 – 3 – 4 – 5 – 6 is a constant pressure line on the P-v-T surface. To understand this, consider water in a closed vessel and let heat be supplied. The temperature of water will increase upto point 4 is reached. Then the steam generation starts. And the heat supplied is utilised for converting water into steam. Once the last particle of water is converted into steam the steam is known as dry saturated steam. After this point 5, if the steam is further heated, then it will be superheated along 5 – 6.

Figures 6.13 and 6.14 show *P.v.T.* surface for CO_2 and *PT* and *Pv*.

SOLVED EXAMPLES

Example 6.1 *Determine pressure of 1 kg of oxygen at 100°C if the specific volume is 0.2. m³/kg using. (i) Van der Waal's equation and (ii) Ideal gas laws.*

Take for O_2,

$a = 13.93 \times 10^4$ N. m⁴ (kg mole)²; $b = 0.0314$ m³/kg mole. and $R = 8314$ J/kg mole K.

Data:

mass of $O_2 = 1$ kg

$T = 100 + 273 = 373$ K

$v = 0.2 \times 32 = 6.4$ m³/kg mole.

Solution

Since M of Oxygen = 32

w.r.t., from Van der Waal's equation.

$$\left(P + \frac{a}{v^2}\right)(v - b) = RT$$

$$P = \frac{RT}{(v-b)} - \frac{v}{v^2}$$

$$= \frac{8314 \times 373}{6.4 - 0.0314} - \frac{13.93 \times 10^4}{(6.4)^2}$$

$$= 486939.36 - 3400.8789$$
$$= 483538.48 \text{ N/m}^2$$
$$\mathbf{P = 4.845 \text{ bar}}$$

Secondly, $Pv = RT$

$$P = \frac{RT}{v} = \frac{8314 \times 373}{6.4} = 484550.31 \text{ N/m}^2$$
$$\mathbf{P = 4.835 \text{ bar.}}$$

Example 6.2 *A certain gas obeys Van der Waal's equation for which a = 1.4 × 10⁻¹ N-m⁴ and b = 32 × 10⁻³ m³/kg mole. And 1 kg mole occupies 40 m³ at 500°C calculate the pressure, by taking R = 8314.*

Solution

We know from Van der Waal's equation,

$$\left(P + \frac{a}{v^2}\right)(v - b) = RT$$

$$P = \frac{RT}{v-b} - \frac{a}{v^2} = \frac{8314 \times (500+273)}{(40 - 32 \times 10^{-3})} - \frac{1.4 \times 10^{-1}}{(40)^2}$$
$$= 160796.69 - 8.75 \times 10^{-5}$$
$$= 160796.69 \text{ N/m}^2$$
$$P = 1.607 \text{ bar.}$$

IMPORTANT POINTS TO BE REMEMBERED

1. Van-der-Waal's equation of state is given by

$$P = \frac{\bar{R}T}{v-b} - \frac{a}{v_2}$$

$$\bar{R} = \text{Universal gas constant} = 8314 \cdot \frac{\text{N-M}}{\text{Kmol-K}}$$

$$a = \text{constant } \frac{\text{N}}{\text{m}^4} \Big/ (\text{kg mol})^2$$

b = constant m³/kmol

v = molar volume

2. $V_e = 3b,\quad T_c = \frac{8a}{27Rb},\quad P_c = \frac{a}{27b^2},\quad \frac{P_c V_c}{RTc} = \frac{3}{8}$

3. Reduced co-ordinates

$$P_r = \frac{P}{P_c},\ T_r = \frac{T}{T_c},\ v_c = \frac{v}{v_c}$$

∴ Van-der-Waal's equation becomes (in reduced co-ordinates)

$$\left(P_r + \frac{3}{Vr^2}\right)(3V_r - 1) = 8T_r$$

4. The Beattie–Bridgeman equation of state is

$$P = \frac{RT}{v^2}\left(1 - \frac{c}{vT^3}\right)(v+b) = \frac{A}{v^2}$$

$$A = A_0\left(1 - \frac{G}{v}\right) \text{ and } B = B_0\left(1 - \frac{b}{v}\right)$$

5. Virial equation of state

$$P = \frac{RT}{v} + \frac{a(T)}{v^2} + \frac{b(T)}{v^3} + \frac{c(T)}{v^4} + \ldots..$$

CHAPTER 7

Non-Reactive Ideal Gas Mixtures

7.1 INTRODUCTION

The thermodynamic behaviour of a mixture of gases depends upon the individual properties of its constituent gases. Because substances in the gaseous phase are so miscible with each other, there is no limit to the number of different gaseous mixtures. Furthermore, individual constituents of gaseous mixtures often react chemically with each other, and these reactions introduce another factor that exerts a strong influence on the properties of a gas. This chapter deals only with mixtures of inert gases, in which changes in phase of constituents may take place.

A gas mixture is frequently treated as if it consisted of a single component rather than many properties of the individual constituents of an inert gas tend to be submerged, so that the gas behaves, in certain ways, as though it were a single, pure substance. The main constituents of substance are, for example, oxygen, nitrogen, argon and water vapour. But air is commonly said to have a molecular weight of 28.97 kg/K-mol, even though this figure represents a composite value based on the molecular weights and the proportions of the constituent species.

In the first case of mixtures of ideal or perfect gases, equations are derived which express the properties of mixtures in terms of properties of the constituents. Mixtures are treated in which no chemical reactions, condensation or evaporation take place. The derived expressions apply, in general to gas mixtures over a wide range of temperatures and pressures. Although the study centers about perfect gases, mixtures containing non-perfect gases often show only negligible deviation from perfect gas behaviour.

7.2 DALTON'S LAW OF PARTIAL PRESSURES

It was John Dalton (1766 – 1844) who first stated that the total pressure p_m exerted by a mixture of gases is the sum of the pressures which each gas (or vapour), would exert were it to occupy the vessel alone at the volume V_m and temperature T_m of the mixture true only of ideal gases. Thus, if p_x, p_y, p_z ...... represent respectively the individual pressures of the mixed gases x, y, z.

∴ According to Dalton's Law of partial pressures, we have

$$p_m = p_x + p_y + p_z + ...$$

at

$$t_m = t_x + t_y + t_z = ...$$

and

$$V_m = V_x + V_y = V_z .....$$

This is also a part of Gibbs – Dalton principle.

Similarly, the number of moles of the mixture is the sum of the moles of the individual components.

Thus,

$$n_m = n_x + n_y + n_z + ...$$

Applying the ideal or perfect gas equation to a component x and to the mixture, we get,

$$p_x V_x = p_x V_m = 8315\, n_x t_m$$

and $$p_x V_m = 8315\, n_x t_m$$

$$\therefore \quad \frac{P_x V_m}{P_m V_m} = \frac{8315\, n_x T_m}{8315\, n_m T_m}$$

or $$\frac{P_x}{P_m} = \frac{n_x}{n_m}$$

We write this also as, $$\frac{p_x}{p_m} = \frac{n_x}{n_m} = B_x \text{ (Volume fraction)}$$

Thus volume fraction is the ratio of the partial pressure to the total pressure the mixture.

Figure 7.1 shows the mixtures and constituents.

Constituents
$x : m_x, p_x, V_x, T_x$
$y : m_y p_x v_y T_y$
$z : M_x p_y V_z T_z$
Mixture $\rightarrow$ m_m, p_m, V_m, T_m

Fig. 7.1. *Mixture and constituents*

Note: Here that the individual pressure exerted by a **real gas** is not necessarily given by these equations.

7.3 AMAGAT'S LAW AND VOLUMETRIC ANALYSIS

Amagat's Law states that *the total volume of a mixture of gases is equal to the sum of the volumes that would be occupied by each component at the pressure p_m and temperature T_m of the mixture.*

This law is found to be true for ideal gases only, though it is widely used for all kinds of gaseous components. Figure 7.2 shows the mixture of gases following this Law.

Gas a	Gas b	Gas c	
m_a	m_b	m_c	$V_m = v_a + v_b + v_c$
p_m	p_a	p_m	$p_m = p_a = p_b = p_c$
V_a	V_b	V_c	$T_m = T_a = T_b = T_c$
T_m	T_m	T_m	$m_m = m_a + m_b + m_c$ (Mixture)
Mixture: p_m, V_m, T_m, m_m			

Fig. 7.2. *System for Amagat's experiment*

It may be noted here that Dalton's Law and Amagat's Law are coincident with the kinetic theory of gases. These laws assume that no intermolecular forces exist in a mixture of gases and that each constituent acts as if no other constituents were present.

When this mixture is analyzed, the perfect gas law leads to :

$$p_m V_a = n_a \bar{R} T_m$$

$$p_m V_b = n_b \bar{R} T_m$$

$$p_m V_c = n_c \bar{R} T_m$$

where n_a, n_b and n_c are the number of moles of individual gas in the mixture.

∴ Addition of these equations, we get

$$p_m (V_a + V_b + V_c) = \bar{R} T_m (n_a + n_b + n_b)$$

or

$$p_m V_m = n\bar{R} T_m$$

where n is the sum of moles of the individual gases.

By dividing the perfect gas equation for each constituent by the equation for the mixture, one obtains the following :

$$\frac{V_a}{V_m} = \frac{n_a}{n}; \frac{V_b}{V_m} = \frac{n_b}{n} \text{ and } = \frac{V_c}{V_m} = \frac{n_c}{n}$$

The ratio of the partial volume of any constituent to the total volume, then, is equal to the mole fraction of that constituent.

This $\left[\frac{V_a}{V_m} = \frac{n_a}{n}\right]$ gives the volumetric analysis of the constituent gas a. Similarly, the volumetric analysis of the other gas can be determined by $\frac{n_b}{n}$ for gas b and $\frac{n_c}{n}$ for gas c. Volumetric analysis is generally given by B like B_a, B_b, B_c etc.

7.4 GRAVIMETRIC ANALYSIS

Constituents or *components* of the mixture of gases are the individual gases and vapours in a mixture. Generally, the description of a mixture of any substances like solids, liquids or gases—can be given by a *gravimetric analysis* which is the percentage or fraction by weight or mass of each constituent.

Thus if a, b and c are the constituents in a mixture and they are having weights or masses as m_a, m_b and m_c respectively , then the total mass of the mixture is

$$m_m = m_a + m_b + m_c$$

Therefore, the gravimetric analysis of the mixture is given by

% of gas a in a mixture is $G_a = \frac{m_a}{m_m} \times 100$

Similarly the percentage composition of gas b and c are given as

$$G_b = \frac{m_b}{m_m} \times 100$$

$$G_c = \frac{m_c}{m_m} \times 100$$

Here G stands for the percentage of the gas mass or for the fractional part of the said constituent.

7.5 CONVERSION BETWEEN GRAVIMETRIC AND VOLUMETRIC ANALYSIS

For converting the gravimetric analysis of the mixture of gases into volumetric analysis and vice-versa, two tables are given below.

(a) Gravimetric to Volumetric Analysis

Sr. No.	Name of the gas	Molecular Weight (a)	Composition of gas by weight (b)	Relative volume $c=\frac{b}{a}$	% composition by volume $=\frac{c}{\Sigma c}\times 100$
1	x	M_x	m_x	$\frac{m_x}{M_x}$	$\frac{m_x}{M_x}\times\frac{100}{\Sigma c}$
2	y	M_y	m_y	$\frac{m_y}{M_y}$	$\frac{m_y}{M_y}\times\frac{100}{\Sigma c}$
3	z	M_y	m_y	$\frac{m_z}{M_z}$	$\frac{m_z}{M_z}\times\frac{100}{\Sigma c}$
			$\Sigma b=100$	Σc	100

(b) Volumetric to Gravimetric Analysis

Sr. No.	Name of the gas	Molecular weight (a)	% Composition of gas by weight (b)	Relative volume $c=(a)\times(b)$	% composition by volume $=\frac{c}{\Sigma c}\times 100$
1	x	M_x	V_x	$V_x.M_x$	$\frac{V_x.M_x}{\Sigma c}\times 100$
2	y	M_y	V_y	$V_y.M_y$	$\frac{V_y.M_y}{\Sigma c}\times 100$
3	z	M_y	V_z	$V_z.M_z$	$\frac{V_z.M_z}{\Sigma c}\times 100$
			$\Sigma b=100$	Σc	100

The use of these tables will be best understood when the numerical problems are solved.

7.6 MOLECULAR WEIGHT AND GAS CONSTANT FOR MIXTURE

Consider the mixture of three gases x, y, and z having the molecular weights as M_x, M_y and M_z. Let the moles of each of these gases be $-n_x$, n_y, and n_z.

$\therefore$ Total weight or mass of the mixture can be found as

$$\text{Weight of } x \text{ gas} = m_x = n_x M_x$$
$$\text{Weight of } y \text{ gas} = m_y = n_y M_y$$
$$\text{Weight of } z \text{ gas} = m_z = n_z M_z$$

$\therefore$ Total mass of the mixture is given by

$$m_m = m_x + m_y + m_z$$
$$= n_x M_x + n_y M_y + n_z M_z$$

and the total number of moles in the mixture is

$$n_m = n_x + n_y + n_z$$

If M_m is the molecular weight of the mixture, then

$$\begin{bmatrix}\text{Total mass of}\\ \text{mixture}\end{bmatrix} = \begin{bmatrix}\text{Number of moles}\\ \text{in mixture}\end{bmatrix} = \begin{bmatrix}\text{Molecular weight}\\ \text{of mixture}\end{bmatrix}$$

$$M_m = n_m \times M_m$$

$$M_m = \text{Molecular weight of the mixture}$$

$$= \frac{m_m}{n_m}$$

$$\mathbf{M_m = \frac{m_m}{n_m}}$$

Similarly, specific gas constant for the mixture is given by

$$\bar{R} = R \times M_m$$

$$\therefore \quad \bar{R} = \text{Specific gas constant}$$

$$= \frac{\bar{R}}{M_m} \text{ kJ/kg-K}$$

$$\mathbf{R = \frac{\bar{R}}{M_m}} \quad \text{where } \bar{R} = 8314.3 \frac{\text{N-m}}{\text{Kmol-K}} \text{ Universal gas constant}$$

7.7 INTERNAL ENERGY, ENTHALPY, SPECIFIC HEATS AND ENTROPY OF MIXTURE OF IDEAL GASES

According to the Gibbs–Dalton Law, an extensive property of a mixture of perfect gases is the sum of the contributions of the individual constituents. Properties of a mixture can be expressed on either a mass or a mole basis and are referred to as per kg or K mol, respectively. Expressions for some of the extensive properties are presented here on a mass basis. The internal energy per unit mass of a mixture is given by

$$u = \frac{m_a U_a + m_b U_b + m_c U_c}{(m_m = 1)}$$

where m_a, m_b and m_c are the masses of gases a, b and c in a mixture of 1 kg and U_a, U_b and U_c are the internal energies of gases a, b and c in a mixture of 1 kg.

We know that internal energy is a function of temperature and for constant volume specific heats of individual gas in mixture, you can calculate the specific heat at constant volume for the mixture is given by

$$\mathbf{(C_v)_m = m_a C_{v_a} + m_b C_{v_b} + m_c C_{v_c} \text{ kJ/kg-K}}$$

on the same lines we can get the specific heat at constant pressure of a mixture containing gases a, b, and c.

$$\mathbf{(C_p)_m = m_a C_{p_a} + m_b C_{p_b} + m_c C_{p_c}}$$

and

$$\mathbf{h_m = m_a h_a + m_b h_b + m_c h_c \text{ kJ/kg}}$$

Similarly, the entropy of the mixture is

$$\mathbf{S_m = m_a S_a + m_b S_b + m_c S_c \text{ kJ/kg-K}}$$

7.8 PROCESSES OF GASEOUS MIXTURE

There are two basic mixing processes for ideal gases:

1. Mixing process when temperature of each gas is same. Consider a system of gases as shown in Fig. 7.3 (a) and (b).

Let the system be ideally insulated so that the mixing of two gases is *adiabatic mixing*.

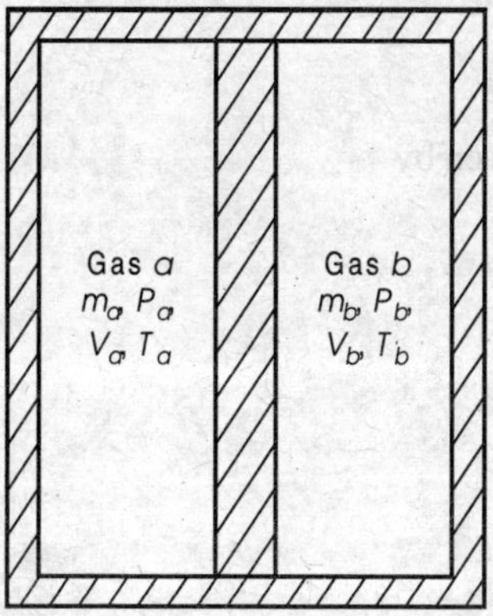

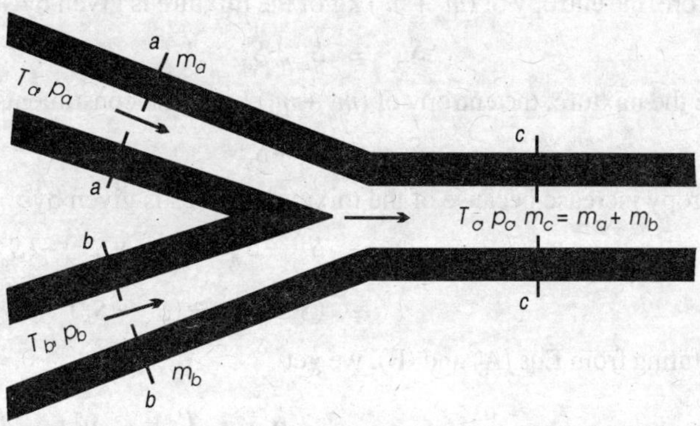

Fig. 7.3. *Adiabatic mixing*

Let there are two gases a and b having the states as p_a, T_a, m_a and p_b, T_b, m_b as shown in Fig.7.3, Here, $T_a = T_b$. These states are shown on T–s diagram in Fig 7.4.

Let $p_a > p_b$ so that the pressure of the mixture is p_m and $p_a > p_m$ and $p_m > p_b$ and $t_m = t_a = t_b$.

After mixing, entropy of gas (a) increases and entropy of gas (b) decreases. Let the state of the mixture be given by the point m.

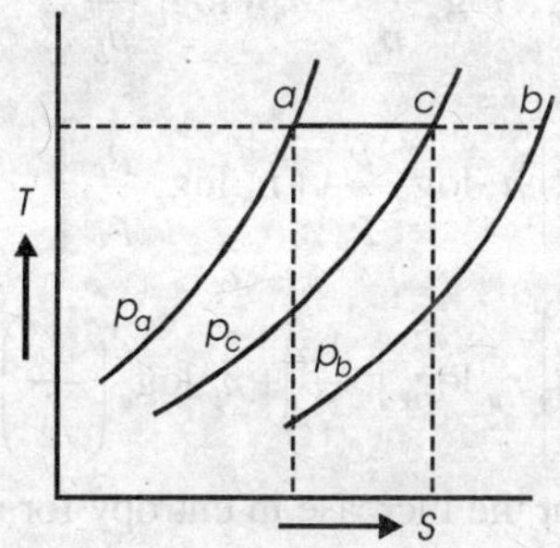

Fig. 7.4

$\therefore$ For gas a

$$S_a - S_{c_a} = m_a R_a \log \frac{p_a}{p_c}$$

or

$$S_{c_a} - S_a = m_a R_a \log \frac{p_c}{p_a} \quad \ldots(A)$$

and for gas b

$$S_b - S_{c_b} = m_b R_b \log_e \frac{p_b}{p_c}$$

or $$S_{c_b} - S_b = m_b R_b \log_e \frac{p_c}{p_b} \qquad \text{...(B)}$$

Then by Daltons–Gibbs law, we have the entropy of the mixture of gases is the sum of the partial entropies of the constituents and the partial entropy is calculated as if each constituent alone the occupied the total volume of the mixture at the temperature of the mixture.

Therefore, the entropy of $(m_a + m_b)$ kg of the mixture is given by

$$S_{m_2} = S_{c_a} + S_{c_b}$$

Before the mixture, the entropy of $(m_a + m_b)$ kg of the constituents is given by

$$S_{m_1} = S_a + S_b$$

∴ Entropy increase because of the mixing of gases is given by

$$\Delta S = S_{m_2} - S_{m_1} = (S_{c_a} + S_{c_b}) - (S_a + S_b)$$

$$= (S_{c_a} - S_a) + (S_{c_b} - S_b)$$

Substituting from Eqs (A) and (B), we get

$$\Delta S = m_a R_a \log_e \frac{p_c}{p_a} + m_b R_b \log_e \frac{p_c}{p_b}$$

$$= m_a \times \frac{\bar{R}}{M_a} \log_e \frac{p_c}{p_a} + m_b \frac{\bar{R}}{M_b} \log_e \frac{p_c}{p_b}$$

$$= n_a m_a . \frac{\bar{R}}{M_a} \log_e \frac{p_c}{p_a} + n_b M_b \frac{\bar{R}}{M_b} \log_e \frac{p_c}{p_b}$$

$$= n_a \bar{R} \log_e \frac{p_c}{p_a} + n_b \bar{R} \log_e \frac{p_c}{p_b}$$

$$= \bar{R} \left[n_a \log_e \frac{p_c}{p_a} + n_b \log_e \frac{p_c}{p_b} \right]$$

$$= \bar{R} \left[n_a \log_e \left(\frac{1}{f_1} \right) + n_b \log_e \left(\frac{1}{f_2} \right) \right]$$

Thus we observe from this equation that the increase in entropy for a mixture depends only on the number of moles of the gases in the mixture and does not depend on the gases themselves. Mixing of gases is an irreversible process and therefore the entropy increases according to second law of thermodynamics.

Again, we observe from these equations that even if the gases approach to similarity (same gases) but the pressure are different, the change of entropy does not become zero.

2. Different gases at different initial pressures and temperatures. The system considered for this problem may be represented by the two vessels, each containing an ideal gas, at different pressures and temperatures. Let these vessels be connected to each other by means of pipe containing a valve as shown in Fig. 7.5.

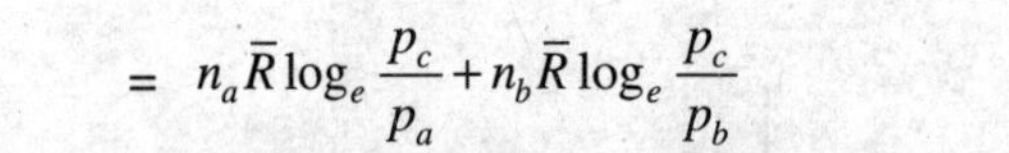

Fig. 7.5

When the valve is open, then the gases will be allowed to diffuse and the pressure and temperatures of the mixture will be in between p_1 and p_2 and T_1 and T_2.

Let
$$m = \text{Mass of the mixture}$$
$$= (m_1 + m_2)$$

Internal energy is a function of the temperature and the total internal energy is the sum of the total internal energies of the individual gases.

$$\therefore \quad mu = m_1 u_1 + m_2 u_2$$

$$\therefore \quad u = \frac{m_1}{m} \cdot u_1 + \frac{m_2}{m} \cdot u_2$$

$$= m_{f_1} u_1 + m_{f_2} u_2$$

where
$$m_{f_1} = \text{mass fraction of gas 1}$$

and
$$m_{f_2} = \text{mass fraction of gas 2}$$

$$\therefore \quad u = \Sigma m_{f_1} u_1$$

$$= \Sigma \left(m_{f_1} C_{v_i} T_i \right)$$

Now we apply the specific gas equation to the individual gases and the mixture.

$$pV = mRT \text{ (for mixture)}$$

and
$$p_1 V_1 = mR_1 T_1$$
$$p_1 V_2 = m_2 R_2 T_2$$

or
$$p_1 V_1 + p_2 V_2 = m_1 R_1 T_1 + m_2 R_2 T_2$$

or
$$(p_1 + p_2)\, V = m_1 R_1 T_1 + m_2 R_2 T_2$$

and
$$p = p_1 + p_2$$

$$\therefore \quad pV = (p_1 + p_2)\, V = mRT$$
$$= m_1 R_1 T_1 + m_2 R_2 T_2$$

$$\therefore \quad T = \text{temperature of the mixture}$$

$$= \frac{m_1 R_1 T_1 + m_2 R_2 T_2}{mR}$$

$$= \frac{\Sigma m_1 R_i T_1}{mR}$$

But
$$R = \Sigma m_1 R_1 T_1$$

$$T = \frac{\Sigma m_1 R_1 T_1}{m \Sigma m_{f_1} R_i}$$

$$= \frac{\Sigma m_1 R_1 T_1}{\Sigma m_i R_i}$$

and
$$mRT = (n_m) \frac{\bar{R}}{M} T = n\bar{R}T$$

$$\therefore \quad nRT = n_1 \bar{R} T_1 + n_2 \bar{R} T_2$$

$$\therefore \quad T = \frac{n_1}{n} T_1 + \frac{n_2 \bar{R}}{n \bar{R}} T_2$$

$$= f_1T_1 + f_2T_2$$
$$= \Sigma f_iT_i$$

Again $$pV = mRT = m_1R_1T_1 + m_2R_2T_2$$
$$= p_1V_1 + p_2V_2$$
$$\therefore \quad p = \frac{p_1V_1 + p_2V_2}{V}$$

Note here that the final volume is the sum of the individual volumes (of the vessels)

$$\therefore \quad V = V_1 + V_2$$

The procedure followed so far for two gases can be applied to any number of gases.

Again, the change of entropy of each constituent is given by

$$\Delta S = m\left[\int_{T_1}^{T_2} C_p \frac{dT}{T} - R\log_e \frac{p_2}{p_1}\right]$$

In this equation, T_1 and p_1 and T_2 and p_2 are the temperature and pressure before mixing and after mixing. T_2 represents the mixture temperature and p_2 represents the partial pressure of the constituent in the mixture after mixing.

After calculating the changes of entropy of each constituent, addition of all changes gives the change of entropy due to mixing.

SOLVED EXAMPLES

Example 7.1 *If the analysis by volume of a product of combustion is 20% CO_2, 3.8% CO, 70% N_2 and 6.2% O_2, find the analysis by weight.*

Solution

The example can be solved by tabulating the various quantities.

Constituent	*Volume m^3* (*a*)	*Mol. Weight* (*b*)	*Proportional weight* (*a*) × (*b*)	*% by weight*
CO_2	0.2	44	$0.2 \times 44 = 8.8$	27.984
CO	0.038	28	$0.038 \times 28 = 1.064$	3.383
N_2	0.7	28	$0.7 \times 28 = 19.6$	62.32
O_2	0.062	32	$0.062 \times 32 = 1.984$	6.31
			$\Sigma c = 31.448$	100

Thus the percentage composition of the products by weight is

$$CO_2 = 27.984$$
$$CO = 3.383$$
$$N_2 = 62.32$$
$$O_2 = 6.31$$
$$= 100$$

Example 7.2 *A sample of water-gas contains H_2 = 38%, CO = 23%, CH_4 = 17%, C_2H_6 = 9%, H_2O = 6%, CO_2 = 4% and N_2 = 3%. Convert this to analysis by weight.*

Solution

As in Example 7.1, we will make a table for solving the problem.

S. No.	*Constituents*	*Volume m^3 of gas (a)*	*Mol. weight (b)*	*Proportional weight (a) × (b) = (c)*	*% by weight $\frac{c}{\Sigma c} \times 100$*
1	H_2	0.38	2	0.76	4.6626
2	CO	0.23	28	6.44	39.5100
3	CH_4	0.17	16	2.72	16.6871
4	C_2H_6	0.09	30	2.07	16.5644
5	H_2O	0.06	18	1.08	6.6258
6	CO_2	0.04	44	1.76	10.7975
7	N_2	0.03	28	0.84	5.1534
				$\Sigma c = 16.3$	100.00

∴ Percentage analysis of water gas, by weight, is given as

$$\begin{bmatrix} H_2 = 4.6626\%; \; CO = 39.51\%; \; CH_4 = 16.6871\% \\ C_2H_6 = 16.5644\%; \; H_2O = 6.6258\%; CO_2 = 10.7975\% \\ N_2 = 5.1534\% \end{bmatrix}$$

Example 7.3 *The composition of a dry flue gas, as obtained by using Orsat apparatus was CO_2 = 10%, CO = 1.8%, O_2 = 7% and N_2 = 81.2% by volume. What will be the percentage composition of the flue gas by weight?*

Solution Table Calculations are as given in Table below.

S. No.	*Constituents*	*Volume m^3 of gas (a)*	*Mol. weight (b)*	*Proportional weight (a) × (b) = (c)*	*% by weight $\frac{c}{\Sigma c} \times 100$*
1	CO_2	0.1	44	4.400	14.77
2	CO	0.018	28	0.504	1.687
3	O_2	0.07	32	2.240	7.497
4	N	0.812	28	22.736	76.091
				$\Sigma c = 29.88$	100.00

Example 7.4 *The diesel oil used in an engine consisted of Carbon 82%, Oxygen 7%, Hydrogen 8%, and H_2O 3% by weight. Convert this to analysis by volume.*

Solution

First we can convert weight into volume by dividing weight by the molecular weight of the constituent. We get

$$82 \text{ kg carbon} = \frac{82}{12} = 6.8333 \text{ moles}$$

$$7 \text{ kg oxygen} = \frac{7}{32} = 0.2188 \text{ moles}$$

$$8 \text{ kg hydrogen} = \frac{8}{02} = 4.0000 \text{ moles}$$

$$3 \text{ kg } H_2O = \frac{3}{18} = 0.1667 \text{ moles}$$

∴ **Total volume = 11.2188 moles**

∴ % analysis by volume is

$$\% \text{ carbon} = \frac{6.8333}{11.2188} \times 100$$
$$= \mathbf{60.9\ \%}$$

$$\% \text{ oxygen} = \frac{0.2188}{11.2188} \times 100$$
$$= \mathbf{1.95\ \%}$$

$$\% \text{ hydrogen} = \frac{4.0}{11.2188} \times 100$$
$$= \mathbf{35.65\ \%}$$

$$\% H_2O = \frac{0.1667}{11.2188} \times 100$$
$$= \mathbf{1.486\ \%}$$

Example 7.5 *In a boiler trial, the products of combustion were found to contain*

Carbon dioxide = 3.227 kg
Sulphur dioxide = 0.010 kg
Oxygen = 1.296 kg and
Nitrogen = 13.005 kg

Estimate the percentage analysis of the dry flue: (a) by weight and (b) by volume.

Solution

(a) Percentage analysis by weight. From the given data, the total weight of the dry products of combustion is

$$\Sigma p = 3.277 + 0.010 + 1.296 + 13.005$$
$$= 17.538$$

∴ Percentage analysis by weight is given as

$$\text{Carbon dioxide } \frac{3.277}{17.538} \times 100 = \mathbf{18.68\ \%}$$

$$\text{Sulphur dioxide } \frac{0.01}{17.538} \times 100 = \mathbf{0.057\ \%}$$

$$\text{Oxygen } \frac{1.296}{17.538} \times 100 = \mathbf{7.389\ \%}$$

$$\text{Nitrogen } \frac{13.005}{17.538} \times 100 = \mathbf{74.15\ \%}$$

$$\text{Total} = \mathbf{100.076}$$

Now we can convert these weight into moles, by dividing by molecular weights of the particular gases and we, then, get the relative volumes.

∴ Relative volumes of the gases are

$$\text{Carbon dioxide } \frac{18.68}{44} = 0.4245$$

$$\text{Sulphur dioxide } \frac{0.057}{64} = 0.0009$$

$$\text{Oxygen} \quad \frac{7.389}{32} = 0.2309$$

$$\text{Nitrogen} \quad \frac{74.15}{28} = 2.6482$$

$$\Sigma = 3.3045$$

$\therefore$ Percentages of volumes of the products of combustion are

$$\text{Carbon dioxide} = \frac{0.4245}{3.3045} \times 100 = 12.85$$

$$\text{Sulphur dioxide} = \frac{0.0009}{3.3045} \times 100 = 0.027$$

$$\text{Oxygen} = \frac{0.2309}{3.3045} \times 100 = 6.099$$

$$\text{Nitrogen} = \frac{2.6482}{3.3045} \times 100 = 80.14$$

$$\text{Total} = 99.116$$

Example 7.6 *Air contains 21% of O_2 and 79 % of N_2 by volume. Determine the molecular weight of air and specific gas constant.*

Solution

Molecular weight of O_2 = 32

Molecular weight of N_2 = 28

$\therefore$ Relative weights of oxygen and nitrogen in the mixture are

Oxygen $\rightarrow 21 \times 32 = 672$ kg

Nitrogen $\rightarrow 79 \times 28 = 2212$ kg

$\therefore$ Total weight of mixture = 672 + 2212 = **2884 kg**

Total number of moles are given by

$$n = n_1 + n_2 = 21 + 79 = 100$$

$\therefore$ Molecular weight of air is given by

$$nM = n_1M_1 + n_2M_2$$

$$100 \times M = 2884$$

or M = Molecule weight of air

$$= \frac{2884}{100}$$

$$= \mathbf{28.84\ kg/Kmol}$$

We know that

$$mR = \bar{R}$$

or R = specific gas constant

$$= \frac{\bar{R}}{M} = \frac{8314.3}{28.84}$$

$$= \mathbf{288.29\ N\text{-}m/kg\text{-}K}$$

Example 7.7 *10 kg of oxygen, 5 kg of N_2 and 15 kg of CO_2 are contained in a vessel where the total pressure is 300 kPa and temperature of 100°C. Calculate (a) partial pressure of each constituent in the vessel (b) the volume of the vessel and (c) if the temperature is increased to 150° C, find the total pressure.*

Solution

(a) We know that mole fraction is equal to the ratio of partial pressure to total pressure, therefore first find the mole fractions of the individual constituent of the mixture

$$\therefore \quad \text{Moles of } O_2 \text{ present} = \frac{10}{32} = 0.3125$$

$$\text{Moles of } N_2 \text{ present} = \frac{5}{28} = 0.1786$$

$$\text{Moles of } CO_2 \text{ present} = \frac{15}{44} = 0.3409$$

$$\therefore \quad \text{Total numbers of moles} = 0.3125 + 0.1786 + 0.3409 = \mathbf{0.8320}$$

$$\therefore \quad \text{Partial pressure of } O_2 = \frac{0.3125}{0.8320} \times 300 = \mathbf{112.68\ kPa}$$

Similarly,

$$\text{Partial pressure of } N_2 = \frac{0.1786}{0.8320} \times 300 = \mathbf{64.399\ kPa}$$

and

$$\text{Partial pressure of } CO_2 = \frac{0.3409}{0.8320} \times 300 = \mathbf{122.92\ kPa}$$

(b) Volume of the vessel. For the mixture, we can write

$$pV = mR_m T$$

where

$$R_m = \frac{\bar{R}}{M_m}$$

where

$$M_m = \text{molecular weight of the mixture}$$

$$= \frac{\text{Total mass of mixture}}{\text{Total number of moles}}$$

$$= \frac{10+5+15}{0.8320} = \frac{30}{0.8320}$$

$$= \mathbf{36.06}$$

$\therefore$ Specific gas constant for the mixture

$$R_m = \frac{0.8320}{36.06} = \mathbf{230.568}$$

$$V = \frac{mR_m T}{p}$$

$$= \frac{30 \times 230.568 \times 373}{300 \times 10^3}$$

$$= \mathbf{7.609\ m^3}$$

(c) Pressure after heating. Heating is at constant volume, and therefore, we have

$$\frac{p_1}{T_1} = \frac{p_2}{T_2}$$

where

$$p_1 = 300\,\text{kPa}$$
$$T_1 = 373\,\text{K}$$
$$T_2 = 150 + 273 = 423\,\text{K}$$

$$\therefore \quad p_2 = \frac{T_2}{T_1} \times p_1 = \frac{423}{373} \times 300$$
$$= \mathbf{340.21\ kPa}$$

Example 7.8 *A perfect gas mixture consists of 3 kg of nitrogen and 5 kg of carbon dioxide at a pressure of 3 bar and a temperature of 20°C. Find: (a) the mole fraction of each constituent, (b) the equivalent molecular weight of the mixture, (c) the equivalent gas constant of the mixture, (d) the partial pressures and the partial volumes, and (e) the volume and density of the mixture.*

Solution

(a) Mole fraction

Moles of nitrogen present in the mixture

$$n_{N_2} = \frac{\text{Mass}}{\text{Molecular weight}}$$
$$= \frac{3}{28}$$
$$= \mathbf{0.1071}$$

Similarly $\quad n_{CO_2}$ = Moles of carbon dioxide

$$= \frac{5}{44}$$
$$= \mathbf{0.1136}$$

$\therefore$ Total moles in the mixture

$$= 0.1071 - 0.1136$$
$$= \mathbf{0.2207}$$

$$\text{Mole fraction of } N_2 = \frac{0.1071}{0.2207}$$
$$= \mathbf{0.4853}$$

$$\text{Mole fraction of } CO_2 = \frac{0.1136}{0.2207}$$
$$= \mathbf{0.5147}$$

(b) Equivalent molecular weight of mixture

$$\text{Total moles} = 0.2207$$
$$\text{Total mass} = 3 + 5$$
$$= 8\,\text{kg}$$

$$\text{Molecular weight} = \frac{\text{Total mass}}{\text{Total number of moles}}$$
$$= \frac{8}{0.2207}$$
$$= \mathbf{36.2483\ kg\ m/K\ mol.}$$

(c) Equivalent gas constant of mixture

We know that $\overline{R} = MR$

where $\overline{R}$ is the universal gas constant

$$R = \frac{\overline{R}}{M}$$

$$= \frac{8314.3}{36.2483}$$

$$= \mathbf{229.37\ N\text{-}m/kg\text{-}K}$$

(d) Partial pressures and partial volumes

$$\text{Partial pressure of } N_2 = 3 \times 0.4853$$

$$= \mathbf{1.4559\ bar}$$

$$\text{Partial pressure of } CO_2 = 3 \times 0.5147$$

$$= \mathbf{1.5441\ bar}.$$

For partial volume we write specific gas equation for the gases.

$$\therefore \quad V_N = \frac{(mRT)_{N_2}}{p}$$

$$= \frac{3 \times \frac{8314.3}{28} \times 293}{3 \times 10^5}$$

$$= \mathbf{0.87\ m^3\ cu.m}$$

$$\text{Similarly} \quad V_{CO_2} = \frac{5 \times \frac{8314.3}{28} \times 293}{3 \times 10^5}$$

$$= \mathbf{0.9228\ m^3}$$

(e) Volume and density of mixture

Volume of the mixture is given by

$$v = \left(\frac{mRT}{p}\right)_{mix}$$

$$= \frac{8 \times 229.37 \times 293}{3 \times 10^5}$$

$$= \mathbf{1.7921\ m^3}$$

$$\text{Density of mixture} \quad \delta_m = \frac{m_m}{V_m}$$

$$= \frac{8}{1.7921}$$

$$= \mathbf{4.467\ kg/m^3}$$

Example 7.9 *For the mixture of Example 7.8, calculate the constant volume and constant pressure specific heats. If the mixture is heated at constant volume to 35° C, find the change in internal energy, enthalpy and entropy of the mixture. Assume the specific heats remain constant and the perfect gas laws apply.*

Also, find the foregoing changes, if the heating process is preformed at constant pressure.

Solution

Constant volume process :

Taking C_v for N_2 = 0.741 kJ/kg-K

and C_v for CO_2 = 0.6615 kJ/kg-K

$\therefore$ C_v for the mixture is given by

$$C_{v_m} = \frac{3}{8}\times 0.7411 + \frac{5}{8}\times 0.6615$$
$$= 0.2779 + 0.4134$$
$$= \mathbf{0.6913\ kJ/kg\text{-}K}$$

Again taking C_p for N_2 = 1.0384 kg-K

and for CO_2 = 0.85 kJ/kg-K

$$C_{p_m} = \text{specific heat of the mixture}$$
$$= \frac{3}{8}\times 1.0384 + \frac{5}{8}\times 0.85$$
$$= 0.3894 + 0.5312$$
$$= \mathbf{0.9206\ kJ/kg\text{-}K}$$

For the mixture

Change in internal energy

$$(\Delta u)_m = C_V \Delta T$$
$$= 0.6913\times(35-20)$$
$$= \mathbf{10.3695\ kJ/kg\text{-}K}$$

Total change in internal energy

$$\Delta U = 8\times 10.3695$$
$$= \mathbf{82.956\ kJ/kg\text{-}K}$$

Change in enthalpy of the mixture

$$\Delta H = m_m C_{p_m} \Delta T$$
$$= 8\times 0.9206\times(35-20)$$
$$= 110.472\ \text{kJ}$$

The change in entropy of the mixture

$$\Delta S = m_m C_{pm} \log_e \frac{T_2}{T_1}$$
$$= 8\times 0.6913\times \log_e\left(\frac{308}{293} = 1.0512\right)$$
$$= \mathbf{0.2761\ kJ/K}$$

Constant Pressure Process. As internal energy and enthalpy are the functions of temperatures only, the change in internal energy and change in enthalpy are the same as in constant volume process.

$$\mathbf{\Delta W = 19.812\ kJ}$$
$$\mathbf{\Delta H = 26.4\ kJ}$$

$$\Delta S = m_m C_{p_m} \log_e \frac{T_2}{T_1}$$
$$= 8\times 0.9206 \log_e \frac{308}{293}$$
$$= \mathbf{0.3677\ kJ/K}$$

Example 7.10 *Find the increase in entropy when 2 kg of oxygen at 63°C are mixed with 6 kg of nitrogen at the same temperature. The initial pressure of each constituent is 1 bar and is the same as that of the mixture. Assume perfect gases.*

Solution

Since the mixing process is isothermal the change of entropy of the mixture is

$$\Delta S = -m_{O_2} R_{O_2} \log_e \frac{P_{O_2}}{P} - m_{N_2} R_{N_2} \log_e \frac{p_{N_2}}{P}$$

where

$$\frac{p_{O_2}}{p} = \frac{2/32}{\frac{2}{32}+\frac{6}{28}} = \frac{0.0625}{0.0625+0.2143}$$

$$= \frac{0.0625}{0.2768}$$

$$= \mathbf{0.2258}$$

Similarly

$$\frac{p_{N_2}}{p} = 1-0.2258$$

$$= \mathbf{0.7742}$$

$\therefore$

$$\Delta S = -2\times\frac{8314.3}{32}\log_e 0.2258$$

$$= -6\times\frac{8314.3}{28}\log_e 0.7742$$

$$= 773.285 + 455.965$$

$$= \mathbf{1229.250\ N\text{-}m/K}$$

$$= \mathbf{1.22925\ kJ/K}$$

Example 7.11 *The number of moles, the pressures and the temperatures of gases a, b, c of figure given below are given in the following table:*

Gas	*Number of moles*	*Pressure bar*	*Temperature °C*
N_2	1	3	90
CO	3	4	200
O_2	2	6	317

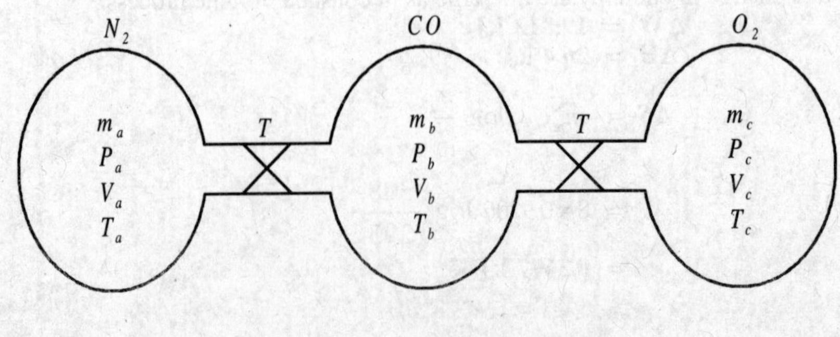

If the containers are connected, allowing the gases to mix freely, find (a) the pressure and temperature of the resulting mixture at equilibrium, (b) the change of entropy of each constituent and that of the mixture.

Solution

(a) Temperature and pressure of the mixture

Temperature of the mixture is given by

$$T_m = \frac{n_a}{n}T_a + \frac{n_b}{n}T_b + \frac{n_c}{n}T_c$$

$$= \frac{1}{6}\times 363 + \frac{3}{6}\times 473 + \frac{2}{6}\times 590$$

$$= 60.5236.5 + 196.7$$

$$= \mathbf{493.7\ K}$$

Also,

$$V_{N_2} = \frac{n_{n_2}\bar{R}T_{N_2}}{p_{N_2}}$$

$$= \frac{1\times 8314.3\times 363}{3\times 10^5}$$

$$= \mathbf{10.06\ m^3}$$

$$V_{CO} = \frac{n_{CO}\bar{R}\,T_{CO}}{P_{CO}}$$

$$= \frac{3\times 8314.3\times 473}{4\times 10^5}$$

$$= \mathbf{29.495\ m^3}$$

$$V_{O_2} = \frac{n_{O_2}\bar{R}\,T_{O_2}}{p_{O_2}}$$

$$= \frac{2\times 8314.3\times 590}{6\times 10^5}$$

$$= \mathbf{16.35\ m^3}$$

∴ Total volume

$$V_m = 10.06 + 29.495 + 16.350$$

$$= \mathbf{55.905\ m^3}$$

∴ Pressure of the mixture from the perfect gas law is

$$p_m = \frac{n\bar{R}T_m}{V_m}$$

$$= \frac{6\times 8314.3\times 493.7}{55.905\times 10^5}$$

$$= \mathbf{4.405\ bar}$$

(*b*) *Change of Entropy.* The partial pressures of the constituents are

$$p_{N_2} = \frac{1}{6}\times 4.405 = \mathbf{0.7341\ bar}$$

$$p_{CO} = \frac{3}{6}\times 4.405 = \mathbf{2.2025\ bar}$$

and $$p_{O_2} = \frac{2}{6}\times 4.405 = \mathbf{1.4683\ bar}$$

$$\Delta s_{N_2} = \left(\int_{t_1}^{t_2} C_p \frac{d_T}{T} - R\log_e \frac{p_2}{p_1}\right)_{N_2}$$

$$= 1.0384\log_e \frac{493.7}{363} - \frac{8314.3}{28}\log_e \frac{0.7341}{3}$$

$$= 1.0384\log_e 1.36 - 296.94\log_e 0.2447$$

$$= 1.0384\times 0.3075 - \frac{296.94}{1000}\times -1.4077$$

$$= \mathbf{0.3193 + 0.418}$$
$$= \mathbf{0.7373\ kJ/kg\text{-}K}$$
$$= \mathbf{0.7373 \times 28}$$
$$= \mathbf{20.644\ kJ/kg}$$

$$\Delta S_{CO} = 1.0384\log_e \frac{493.7}{473} - \frac{8314.3}{28\times 1000}\log_e \frac{2.2025}{4}$$

$$= 1.0384\times 0.0429 + \frac{296.94\log_e \times 0.309}{1000}$$

$$= 0.04455 + \frac{91.7545}{1000}$$

$$= 0.04455 + 0.0917545$$
$$= \mathbf{0.1353045\ kJ/kg\text{-}K}$$
$$= \mathbf{0.135045 \times 84}$$
$$= \mathbf{11.3652\ kJ/K}$$

$$\Delta S_{O_2} = 0.9165\log_e \frac{493.7}{590} - \frac{8.314}{32}\log_e \frac{1.4683}{6}$$

$$= 0.9169\times\log_e 0.8368 - 0.2598\log_e 0.2477$$
$$= 0.9169 + 0.1782 + 0.2598\times 1.3955$$
$$= -0.1634 + 0.3625$$
$$= \mathbf{0.1991\ kJ/kg\text{-}K}$$
$$= \mathbf{0.1991 \times 64}$$
$$= \mathbf{12.7466\ kJ/K}$$

$\therefore$ $$\Delta S_{mix} = 20.644 + 11.3652 + 12.7466$$
$$= \mathbf{44.7558\ kJ/K}$$

■■■

THEORY QUESTIONS

7.1. What is partial pressure of a gas in a mixture of gases ?

7.2. Define the terms — mass fraction and mole fraction of the gas in a mixture.

7.3. Give the procedure to find the gravimetric analysis of the mixture of gases and to convert this gravimetric analysis into volumetric analysis and vice-versa.

7.4. Explain Dalton's Law of partial pressures.

7.5. What is Amaga's Law and its application.

7.6. How will you find the molecular weight and gas constant for mixture of number of ideal non-reactive gases?

7.7. What is Gibbs–Dalton Law and its use?

PROBLEMS FOR PRACTICE

7.1. A mixture of perfect gases at 20°C has the following composition by volume :

$N_2 = 55\%, O_2 = 20\%, CH_4 = 25\%$

If the partial pressure of CH_4 = 0.5172 bar, determine:

(a) The partial pressures of N_2 and O_2

(b) The mass proportions of the mixture

(c) R_{mix} and volume per mole of mixture.

7.2. Five moles of perfect gas mixture has the following mass analysis.

$N_2 = 45\%$, He = 27%, and $C_6H_6 = 28\%$

Find

(a) the analysis by volume and the number of moles of each constituents

(b) The molecular weight of the mixture

(c) The volume of the mixture at 3 bar and 20° C

7.3. The analysis by weight of a perfect gas mixture at 20°C and 1.5 bar is 10% O_2, 70% N_2, 15% CO_2, and 5% CO. For a reference state of –15°C and atmospheric pressure, determine

(a) Partial pressures of the constituents

(b) Gas constant of the mixture

(c) Enthalpy and internal energy of the mixture

(d) Constant pressure specific heat of the mixture

(e) Entropy of the mixture.

7.4. A rigid vessel has a volume of 0.5 cu-m and contains air at bar and 20°C, Helium at 20°C is added to the vessel with the pressure is raised to 1.5 bar. Compute,

(a) the partial pressures of the final components of the mixture composed of O_2, N_2 and He.

(b) the masses of the components of the mixture.

(c) the gas constant of the mixture.

7.5. The dry products of combustion of a coal have the following volumetric composition: CO_2 11%, O_2 3.5%, N_2 85.5% . Calculate (a) the gravimetric analysis (b) the equivalent molecular weight (c) the mixture's gas constant (d) the partial pressures of each constituents for p_m = 1 bar. (e) the specific heat at constant pressure for the mixture (f) the molar specific heat using the result in (e).

IMPORTANT POINTS TO BE REMEMBERED

1. Dalton's Law of partial pressure

$$P_m = P_x + P_y + P_z + ...$$
$$n_m = \text{Total number of moles}$$
$$= n_x + n_y + n_z + ...$$

2. $$\frac{P_x}{P_m} = \frac{n_x}{n_m} = B_x \text{ (volume fraction);}$$

$$\frac{P_y}{P_m} = \frac{n_y}{n_m} = B_y$$

$$\frac{P_z}{P_m} = \frac{n_z}{n_m} = B_z$$

3. $$m_x = n_x M_x, \quad m_y = n_y M_y, \quad m_z = n_z M_z$$

$$\therefore \quad m_m = n_m M_m, \quad M_m = \frac{m_m}{n_m}$$

$$R = \frac{\bar{R}}{M_m}$$

4. $$(C_v)_m = m_a Cv_a + m_b + Cv_b + m_c Cv_c + ...$$
$$m_a + m_b + m_c = 1$$
$$(C_p)_m = m_a Cp_a + m_b Cp_b + m_c Cp_c + ...$$
$$m_a + m_b + m_c = 1$$

5. When $t_m = t_a = t_b = t_c = ...$

then $\Delta S = m_a R_a \log e \frac{P_m}{P_a} + m_b R_b \log e \frac{P_m}{P_b} + ...$

6. When temperature are different

$$\Delta S = \left[\int_{T_1}^{T_2} C_p \frac{dT}{T} - R \log_e \frac{P_2}{P_1} \right]$$

CHAPTER 8

General Thermodynamic Relations

8.1 INTRODUCTION

We have studied the six thermodynamic properties of a system so far. These are, pressure (P), volume (V), temperature (T), internal energy (U), enthalpy (H) and entropy (S). In addition to the above, two more important properties, viz. Helmholtz function (A) and Gibbs function (G) will be discussed in this chapter.

Of the above eight properties, only three i.e., P, V and T can be directly measured and hence it becomes necessary to develop thermodynamic relations to express the other properties in terms of these. These combinations of properties are called 'thermodynamic gradients' and are defined as the rate of change of one property with another while a third is kept constant.

8.2 HELMHOLTZ FUNCTION OR WORK FUNCTION

Helmholtz function (A) is a combination of properties and is mathematically expressed as

$$A = U - TS \qquad \ldots(1)$$

For unit mass, specific Helmholtz function is given by,

$$a = u - TS$$

where, u = Specific internal energy, and

s = Specific entropy

From the fist law, for a closed system, the work dW is given by

$$\delta W = \delta Q - du \qquad \ldots(2)$$

Thus, the energy converted to work is provided in part by the system, whose internal energy decreases by $-du$, and in part by the heat reservoirs with which the system is in contact and which give up a quantity of heat dQ.

The expressions for maximum amount of work can be derived when a system undergoes a process between two equilibrium states. It is assumed that system exchanges heat only with single heat reservoirs at temperature T_o.

From the principle of increase of entropy, we know that

$$dS + dS_o \geq 0 \qquad \ldots(3)$$

where, dS = Increase in entropy of system.

dS_o = Increase in entropy of reservoir.

Now, heat dQ is absorbed by the system from the reservoir.

Hence, $$dS_0 = \frac{-dQ}{T_o} \quad \text{(Reservoir has given out heat. } \therefore dQ \text{ is –ve)}$$

Substituting above value of dS_o in Eq. (3)

$$dS_o - \frac{dQ}{T_o} \geq 0$$

$$\therefore \quad T_o dS \geq dQ \quad \ldots(4)$$

From Eqs (2) and (4), we get

$$dW \leq T_o\, dS - du \quad \ldots(5)$$

For a finite process between two equilibrium states, this equation can be integrated to give,

$$\int_1^2 dW \leq T_o \int_1^2 dS - \int_1^2 du$$

$$\therefore \quad W_{1-2} \leq T_o (S_2 - S_1) - (u_2 - u_1)$$

$$\therefore \quad W_{1-2} \leq (u_2 - u_1) - T_o (S_2 - S_1) \quad \ldots(6)$$

For a special case, where initial and final temperatures are equal and same as that of the reservoir, Eq. (6) can be written as,

$$(W_T)_{1-2} \leq (u_1 - u_2)_T - T (S_1 - S_2)_T \quad \ldots(7)$$

For a small change, the Helmholtz function can be written as,

$$dA = du - d\,(TS) \quad \ldots(8)$$

Integrating the above equation between two equilibrium states considering temperature (T) as constant, we get,

$$\int_1^2 dA = \int_1^2 du - T \int_1^2 dS$$

$$\therefore \quad (A_2 - A_1)_T = (u_2 - u_1) - T (S_2 - S_1)_T$$

$$\therefore \quad (A_1 - A_2)_T = (u_1 - u_2)_T - T (S_1 - S_2)_T \quad \ldots(9)$$

Therefore, from Eqs (7) and (9), we have,

$$(W_T)_{1-2} \leq (A_1 - A_2)_T \quad \ldots(10)$$

Thus, the above equation implies that in any process between two equilibrium states at the same temperature during which the sytem exchanges heat only with the reservoir, the work done is equal to or less than the decrease in Helmholtz function of the system during the process.

The maximum work is done when the process is reversible and it is less than maximum for irreversible process.

$$(W_T)_{max} = (W_T)_{rev} = (A_1 - A_2)_T$$

8.3 GIBBS FUNCTION (THERMODYNAMIC POTENTIAL)

Gibbs function (G) is a property of a system, which is given as

$$G = U - TS + PV$$

$$= (U + Pv) - TS$$

$$\therefore \quad G = H - TS$$

where, H = Enthalpy of system

T = Temperature of system

S = Entropy of system

For unit mass, we have

$$g = h - TS$$

Consider a system that can do work in other form e.g., electrical, magnetic works etc., in addition to PdV

Let

$$\delta W = \text{Work done by the system}$$
$$\delta W_o = \text{Form of work other than } PdV$$

We get $\delta W = Pdv + \delta W_o$

Let P = external pressure (P_o) = Constant

We now integrate the above equation from state 1 to state 2.

$$\therefore \quad \int_1^2 \delta W = P_o \int_1^2 dv + \int_1^2 \delta W_o$$

$$\therefore \quad W_{1-2} = P_o (v_2 - v_1) + W_o$$

Here, we may be electrical work (EdI) for a voltaic cell, or may be magnetic work (mdH) in case of a magnetic material.

The work done by a system exchanging heat with a reservoir at temperature T_o satisfies the following condition:

$$W_{1-2} \le (U_1 - U_2) - T_o (S_1 - S_2) \quad \ldots(2)$$

From Eqs (1) and (2), we get

$$P_o (v_2 - v_1) + W_o \le (U_1 - U_2) - T_o (S_1 - S_2)$$

$$\therefore \quad W_o \le (U_1 - U_2) - T_o (S_1 - S_2) - P_o (v_2 - v_1) \quad \ldots(3)$$

When the initial and final equilibrium states are at same temperature and pressure, then we get

$$T_o = T \text{ and } P_o = P$$

From Eq. (3) we have $\quad (W_o)_{P,T} \le (U_1 - U_2)_{P,T} - T(S_1 - S_2) + P(v_1 - v_2)_{P,T} \quad \ldots(4)$

We have,

Gibbs function, $G = U - Ts + Pv$

Writing in differential form, we get

$$dG = dU - d(TS) + d(Pv)$$

Keeping P and T constant and integrating above equation from state 1 to state 2, we get:

$$\int_1^2 dG = \int_1^2 dU - T\int_1^2 dS + p\int_1^2 dV$$

$$\therefore \quad (G_2 - G_1) = (U_2 - U_1) - T(S_2 - S_1) - P(v_2 - v_1)$$

i.e., $\quad (G_1 - G_2)_{p,T} = (U_2 - U_1)_{p,T} - T(S_1 - S_2)_{p,T} + P(v_2 - v_1)_{p,T} \quad \ldots(5)$

From Eqs (4) and (5) we get,

$$(W_o)_{p,T} \le (G_1 - G_2)_{p,T} \quad \ldots(6)$$

It is observed that the difference between Gibbs function of a system between two equilibrium states sets the maximum limit to the work that can be performed by the system in addition to the *Pdv* work. This is observed when the two equilibrium states are at the same temperature and pressure and when the system exchanges heat with a single heat reservoir.

The work done is maximum when the process is reversible and is less for an irreversible process.

∴ For maximum work, we have

$$\left[(W_o)_{P,T}\right]_{\max} = \left[(W_o)_{P,T}\right]_{\text{rev}} = (G_1 - G_2)_{P,T}$$

Note. Gibbs function is also known as Free Energy Function.

8.4 MATHEMATICAL FUNDAMENTALS

We will now discuss briefly certain mathematical manipulations which are repeatedly useful and important from the point of view of the subject.

(*a*) *Exact Differential*: Let,

$$Z = f(x, y) \quad \ldots(1)$$

where *x*, *y*, *z* represent any three properties of a system.

Then,
$$dz = \left(\frac{\partial z}{\partial x}\right)_y dx + \left(\frac{\partial z}{\partial y}\right)_x dy \quad \ldots(2)$$

where,

$\left(\frac{\partial z}{\partial x}\right)_y$ = Partial derivative of *z* with respect to *x*, keeping *y* as constant.

$\left(\frac{\partial z}{\partial y}\right)_x$ = Partial derivative of *z* with respect to *y*, keeping *x* constant.

The differential *dz* in Eq. (2) must be an exact differential (i.e., a point function), and an equation relating point functions must meet the test for exactness. This test is that the magnitude of second mixed partial derivatives of a function are independent of order of differentiation and hence equal.

i.e.,
$$\frac{\partial^2 z}{\partial x \partial y} = \frac{\partial^2 z}{\partial y \partial x} \quad \ldots(3)$$

Now, Eq. (2) can also be written as,

$$dz = Mdx + Ndy$$

where $M = \left(\frac{\partial z}{\partial x}\right)_y$ and $N = \left(\frac{\partial z}{\partial y}\right)_x$

$$\therefore \quad \left(\frac{\partial M}{\partial y}\right)_x = \frac{\partial^2 z}{\partial x \partial y} \text{ and } \left(\frac{\partial N}{\partial x}\right)_y = \frac{\partial^2 z}{\partial x \partial y}$$

Hence, from Eq. (3), we have

$$\left(\frac{\partial M}{\partial y}\right)_x = \left(\frac{\partial N}{\partial x}\right)_y$$

(*b*) *Cyclic Relations*. We know that,

$$Z = f(x, y) \quad \ldots(1)$$

Since *Z* is a function of two independent variables *x* and *y*, *x* can similarly be expressed in terms of the other two variables i.e., *y* and *z*

$$\therefore \quad x = f(y, z) \quad \ldots(2)$$

Similarly, we can write

$$y = f(x, z) \quad \ldots(3)$$

According to the condition of exact differentials, we can write,

$$dx = \left(\frac{\partial x}{\partial y}\right)_z dy + \left(\frac{\partial x}{\partial z}\right)_z dz \quad \ldots(4)$$

and
$$dy = \left(\frac{\partial y}{\partial x}\right)_z dx + \left(\frac{\partial y}{\partial z}\right)_z dz \quad \ldots(5)$$

Substituting the value of *dy* from Eq. (5) in Eq. (4), we get,

$$dx = \left(\frac{\partial y}{\partial z}\right)_y dz + \left(\frac{\partial x}{\partial y}\right)_z \left[\left(\frac{\partial y}{\partial x}\right)_z dx + \left(\frac{\partial y}{\partial z}\right)_x dz\right]$$

$$= \left(\frac{\partial x}{\partial z}\right)_y dz + \left(\frac{\partial x}{\partial y}\right)_z \left(\frac{\partial y}{\partial x}\right)_z dx + \left(\frac{\partial x}{\partial y}\right)_z \left(\frac{\partial y}{\partial z}\right)_x dz$$

$$= \left[\left(\frac{\partial x}{\partial z}\right)_y + \left(\frac{\partial x}{\partial y}\right)_z \left(\frac{\partial y}{\partial z}\right)_x\right] dz + dx$$

$$\therefore \quad = \left(\frac{\partial x}{\partial y}\right)_z \left(\frac{\partial y}{\partial z}\right)_x + \left(\frac{\partial x}{\partial z}\right)_y = 0$$

$$\therefore \quad \left(\frac{\partial x}{\partial y}\right)_z \left(\frac{\partial y}{\partial z}\right)_x = -\left(\frac{\partial x}{\partial z}\right)_y$$

$$\therefore \quad \left(\frac{\partial x}{\partial y}\right)_z \left(\frac{\partial y}{\partial z}\right)_x \left(\frac{\partial z}{\partial x}\right)_y = -1$$

The above relation is called 'cyclic relation'. We can write the cyclic relation for thermodynamic properties P, V and T as

$$\left(\frac{\partial P}{\partial V}\right)_T \left(\frac{\partial V}{\partial T}\right)_P \left(\frac{\partial T}{\partial P}\right)_V = -1$$

8.5 MAXWELL'S EQUATIONS

The Maxwell relations are derived by applying the exactness test to the functions for the various properties:

(*a*) *Internal Energy*. General gas equation gives us

$$du = dQ - dW$$
$$= TdS - PdV$$

The above equation is in the form

$$dz = Mdx + Ndy$$

where $M = T, N = -P, x = S, y = V$

Hence, we can write,

$$\left(\frac{\partial T}{\partial V}\right)_S = \left(\frac{\partial P}{\partial S}\right)_V \qquad \ldots(1)$$

(*b*) *Enthalpy*. We know that,

$$dH = dU + d(pV)$$
$$\therefore \quad dH = du + PdV + VdP$$
$$= (TdS - PdV) + PdV + VdP$$
$$\therefore \quad dH = TdS + VdP$$

In the above expression, $M = T, N = V, x = S, y = P$

Hence, we can write

$$\left(\frac{\partial T}{\partial P}\right)_S = \left(\frac{\partial V}{\partial S}\right)_T \qquad \ldots(2)$$

(*c*) *Helmholtz function* (*A*). As discussed earlier,

$$dA = dU - d(TS)$$
$$= dU - Tds - SdT$$
$$= TdS - PdV - TdS - SdT$$
$$= -PdV - SdT$$

Thus, $M = -P, N = -S, x = V, y = T$

$$\therefore \quad -\left(\frac{\partial P}{\partial T}\right)_V = -\left(\frac{\partial S}{\partial V}\right)_T$$

$$\text{i.e.} \quad \left(\frac{\partial P}{\partial T}\right)_V = \left(\frac{\partial S}{\partial V}\right)_T \quad \ldots(3)$$

(d) Gibbs Function

Now,

$$\begin{aligned} dG &= dH - d(TS) \\ &= (TdS + Vd) - TdS - SdT \\ &= vdp - SdT \end{aligned}$$

$\therefore$ $M = V, N = -S, x = P, y = T$

$$\therefore \quad \left(\frac{\partial V}{\partial T}\right)_P = \left(\frac{\partial S}{\partial P}\right)_T \quad \ldots(4)$$

The Eqs. (1), (2), (3) and (4) are called Maxwell's equations. They apply to pure substances in process involving heat, work and internal energy.

8.6 CO-EFFICIENT OF VOLUME EXPANSION OR VOLUME EXPANSIVITY (β)

The co-efficient of volume expansion is defined as the change in volume per unit volume for a unit change of temperature at constant pressure.

It is mathematically expressed as,

$$\beta = \frac{1}{V}\left(\frac{\partial V}{\partial T}\right)_p$$

When β is positive, an isothermal increase in pressure results in rejected heat. When β is negative, an isothermal increase in pressure results in heat added.

8.7 ISOTHERMAL COMPRESSIBILITY

The isothermal compressibility or the isothermal co-efficient of compressibility is defined as the change in volume per unit volume for unit change in pressure, at constant temperature.

It is denoted by K and given by,

$$K = -\frac{1}{V}\left(\frac{\partial V}{\partial P}\right)_T$$

Note. The negative sign indicates that volume decreases as pressure increases.

8.8 ADIABATIC COMPRESSIBILITY

Adiabatic compressibility is defined as the change in volume per unit volume for unit change in pressure at constant entropy. It is denoted by K_S and given by the expression.

$$K_S = -\frac{1}{V}\left(\frac{\partial V}{\partial P}\right)_S$$

Note. Negative sign indicates that the volume decreases as pressure increases.

The ratio of isothermal and adiabatic compressibilities is equal to the ratio of specific heats.

$$\text{i.e.} \quad \gamma = \frac{C_P}{C_V} = \frac{K}{K_S} = \frac{-(1/V)(\partial V/\partial P)_T}{-(1/V)(\partial V/\partial P)_S} = \frac{(\partial V/\partial P)_T}{(\partial V/\partial P)_S}$$

Thus, $$\gamma = \frac{(\partial P/\partial V)_S}{(\partial P/\partial V)_T}$$

As γ is always greater than unity, K is always greater than K_S.

8.9 RATIO OF COEFFICIENT OF VOLUME EXPANSION AND ISOTHERMAL COMPRESSIBILITY

As studied earlier, coefficient of volume expansion is given by,

$$\beta = \frac{1}{V}\left(\frac{\partial V}{\partial T}\right)_P$$

$$\therefore \quad \left(\frac{\partial V}{\partial T}\right)_P = \beta V \qquad \ldots(1)$$

Also, isothermal compressibility,

$$K = -\frac{1}{V}\left(\frac{\partial V}{\partial P}\right)_T$$

i.e. $$\left(\frac{\partial V}{\partial P}\right)_T = -KV \qquad \ldots(2)$$

By cyclic property of differentials, we get

$$\left(\frac{\partial P}{\partial T}\right)_V\left(\frac{\partial T}{\partial V}\right)_P\left(\frac{\partial V}{\partial P}\right)_T = -1$$

Substituting Eqs (1) and (2) in above equation, we get

$$\left(\frac{\partial P}{\partial T}\right)_V\left(\frac{1}{\beta V}\right)(-KV) = -1$$

$$\therefore \quad \left(\frac{\partial P}{\partial T}\right)_V = \frac{\beta V}{KV} = \frac{\beta}{K}$$

Thus, the ratio of coefficient of volume expansion and isothermal compressibility is equal to the change in pressure with temperature at constant volume.

8.10 GENERAL EQUATION FOR CHANGE IN INTERNAL ENERGY

We will now derive certain equations for change in internal energy, in terms of pressure, volume and temperature, by considering the following conditions:

(a) Pressure (P) and volume (V) are independent

The change in internal energy in this case is given by,

$$dU = \left(\frac{\partial U}{\partial P}\right)_V dP + \left(\frac{\partial U}{\partial V}\right) dV \qquad \ldots(1)$$

Now, we can write

$$\left(\frac{\partial U}{\partial P}\right)_V = \left(\frac{\partial U}{\partial T}\right)_V + \left(\frac{\partial T}{\partial P}\right)_V = mC_V \times \frac{K}{\beta} \quad \left\{\left(\frac{\partial U}{\partial T}\right)_V = mC_V \text{ and } \left(\frac{\partial T}{\partial P}\right)_V = \frac{K}{\beta}\right\} \qquad \ldots(2)$$

Also, we know that, $$H = U + PV \qquad \ldots(3)$$

Taking partial derivative of above equation with respect to V keeping P constant, we get

$$\left(\frac{\partial H}{\partial v}\right)_P = \left(\frac{\partial u}{\partial v}\right)_P + \frac{\partial}{\partial v}(PV)$$

$$= \left(\frac{\partial u}{\partial v}\right)_P + P$$

$$\therefore \quad \left(\frac{\partial u}{\partial v}\right)_P = \left(\frac{\partial H}{\partial v}\right)_P - P \qquad \text{...(4)}$$

We can write,

$$\left(\frac{\partial H}{\partial V}\right)_P = \left(\frac{\partial H}{\partial T}\right)_P \left(\frac{\partial T}{\partial V}\right)_P = \frac{mC_V}{\beta V} \qquad \text{...(5)}$$

$$\left\{\left(\frac{\partial H}{\partial T}\right)_P = mC_P \text{ and } \left(\frac{\partial T}{\partial v}\right)_P = \frac{1}{\beta V}\right\}$$

Substituting Eq. (5) in Eq. (4), we get

$$\left(\frac{\partial U}{\partial V}\right)_P = \frac{mC_V}{\beta V} - P \qquad \text{...(6)}$$

Now, substituting Eqs (2) and (6) in Eq. (1), we get,

$$dU = \left[mC_V \times \frac{K}{\beta}\right] dP + \left[\frac{mC_P}{\beta V} - P\right] dV$$

(b) Volume (V) and temperature (T) are independent

Change in internal energy is given by,

$$dU = \left(\frac{\partial U}{\partial V}\right)_T dV + \left(\frac{\partial u}{\partial T}\right)_T dT \qquad \text{...(1)}$$

Substituting the value of $\left(\frac{\partial U}{\partial T}\right)_v$, as mC_V, we get

$$dU = \left(\frac{\partial U}{\partial V}\right)_T dV + mC_V dT \qquad \text{...(2)}$$

From first law, we have

$$dQ = dU + dW = dU + PdV \qquad \text{...(3)}$$

and from second law, we have

$$dQ = TdS \qquad \text{...(4)}$$

Thus from Eqs (3) and (4),

$$TdS = dU + PdV \qquad \text{...(5)}$$

Now,

$$H = U + pV$$

Differentiating above equation, we get

$$dH = dU + PdV + VdP$$

From Eq. (5), we get

$$dH = TdS + VdP$$

$$\therefore \quad TdS = dH - VdP \qquad \text{...(6)}$$

The change in enthalpy, if P and T are taken as independent variables, is

$$dH = \left(\frac{\partial H}{\partial T}\right)_p dT + \left(\frac{\partial H}{\partial P}\right)_T dP$$

$$= mC_p dT + \left(\frac{\partial H}{\partial P}\right)_T dP \qquad \left\{\because \left(\frac{\partial H}{\partial T}\right)_p - mC_p\right\} \qquad \ldots(7)$$

From Eqs (5) and (6), we get

$$dU + PdV = dH - VdP$$

Substituting Eqs. (2) and (7) in above equation, we get

$$\left(\frac{\partial U}{\partial V}\right)_T dV + mC_V dT + PdV = mC_P dT + \left(\frac{\partial H}{\partial p}\right)_T dP - VdP$$

i.e.

$$mC_V dT + \left[P + \left(\frac{\partial u}{\partial v}\right)_T\right] dV = mC_P dT - \left[V + \left(\frac{\partial H}{\partial P}\right)_T\right] dP$$

If pressure is assumed constant i.e., $dP = 0$, we get

$$m(C_P - C_V)(dT)_P = \left[P + \left(\frac{\partial U}{\partial V}\right)_T\right](dV)_P$$

$$\therefore \quad m(C_P - C_V) = \left[P + \left(\frac{\partial U}{\partial V}\right)_T\right]\left(\frac{\partial V}{\partial T}\right)_P$$

$$\therefore \quad m(C_P - C_V) = \left[P + \left(\frac{\partial U}{\partial V}\right)_T\right]\beta V \qquad \left\{\because \left(\frac{\partial V}{\partial T}\right)_P = \beta V\right\}$$

$$\therefore \quad \frac{m(C_P - C_V)}{\beta V} = p + \left(\frac{\partial U}{\partial V}\right)_T$$

$$\therefore \quad \left(\frac{\partial U}{\partial V}\right)_T = \frac{m(C_P - C_V)}{\beta V} - P \qquad \ldots(8)$$

Substituting Eq. (8) in Eq. (2), we get

$$dU = \left[\frac{m(C_P - C_V)}{\beta V} - P\right] dV + mC_V dT$$

(c) When temperature (T) and pressure (P) are independent

The change in internal energy is given by,

$$dU = \left(\frac{\partial U}{\partial T}\right)_P dT + \left(\frac{\partial U}{\partial P}\right)_T dP \qquad \ldots(1)$$

We know that,

$$mC_p = \left(\frac{\partial H}{\partial T}\right)_P = \left[\frac{\partial (U + PV)}{\partial T}\right]_P$$

$$= \left(\frac{\partial U}{\partial T}\right)_P + V\left(\frac{\partial P}{\partial T}\right)_P + P\left(\frac{\partial V}{\partial T}\right)_p$$

Since pressure is constant, we have

$$mC_P = \left(\frac{\partial U}{\partial T}\right)_P + P\left(\frac{\partial V}{\partial T}\right)_P$$

$$\therefore \quad \left(\frac{\partial U}{\partial T}\right)_P = mC_P - P\left(\frac{\partial V}{\partial T}\right)_P$$

$$= mC_P - P\beta V \qquad \left\{\because \left(\frac{\partial V}{\partial T}\right)_P = \beta V\right\} \qquad \ldots(2)$$

From the first law,

$$dQ = dU + dW = dU + PdV$$

Since P and T are independent variables,

$$dQ = \left(\frac{\partial U}{\partial T}\right)_P dT + \left(\frac{\partial U}{\partial P}\right)_T dp + p\left[\left(\frac{\partial V}{\partial T}\right)_P dT + \left(\frac{\partial V}{\partial P}\right)_T dP\right]$$

$$= \left[\left(\frac{\partial U}{\partial T}\right)_P + P\left(\frac{\partial V}{\partial T}\right)_P\right] dT + \left[\left(\frac{\partial U}{\partial P}\right)_T + P\left(\frac{\partial V}{\partial P}\right)_T\right] dP$$

Now, we know that $\quad mC_P = \left(\frac{\partial U}{\partial T}\right)_P + P\left(\frac{\partial V}{\partial T}\right)_P$

$$\therefore \quad dQ = mC_p dT + \left[\left(\frac{\partial u}{\partial P}\right)_T + P\left(\frac{\partial V}{\partial P}\right)_T\right] dP$$

For constant volume, we have

$$(dQ)_V = mC_V (dT)_V$$

$$= mC_P (dT)_V + \left[\left(\frac{\partial U}{\partial P}\right)_T + P\left(\frac{\partial V}{\partial P}\right)_T\right] (dP)_V$$

$$\therefore \quad mC_V = mC_P + \left[\left(\frac{\partial U}{\partial P}\right)_T + P\left(\frac{\partial V}{\partial P}\right)_T\right]\left(\frac{\partial P}{\partial T}\right)_V$$

But $\quad \left(\frac{\partial V}{\partial P}\right)_T = -KV$ and $\left(\frac{\partial P}{\partial T}\right)_V = \frac{\beta}{K}$

$$\therefore \quad mC_V = mC_P + \left[\left(\frac{\partial U}{\partial P}\right)_T - pKV\right]\frac{\beta}{K}$$

$$\therefore \quad \left(\frac{\partial U}{\partial P}\right)_T = pKV - \frac{m(C_P - C_V)K}{\beta} \qquad \ldots(3)$$

Thus, from Eqs (2) and (3) we get

$$dU = (mC_P - P\beta V)dT + \left[pKV - \frac{m(C_P - C_V)K}{\beta}\right] dP$$

8.11 TdS EQUATIONS

From the laws of thermodynamics, we have

$$dQ = dU + PdV$$

and $$dS = \frac{dQ}{T}$$

$\therefore$ $$TdS = dU + PdV$$

We will now derive some useful relations for the above equation which is sometimes known as 'general *TdS* equation'.

(*a*) *When temperature* (*T*) *and volume* (*V*) *are independent*

Change in internal energy, in this case, is given by,

$$dU = \left(\frac{\partial u}{\partial T}\right)_v dT + \left(\frac{\partial u}{\partial v}\right)_T dV \qquad \ldots(1)$$

We have $$TdS = dU + PdV$$

Substituting Eq. (1) in above equation, we get,

$$TdS = \left[\left(\frac{\partial u}{\partial T}\right)_v dT + \left(\frac{\partial u}{\partial v}\right)_T dV\right] + PdV$$

$\therefore$ $$dS = \frac{1}{T}\left[\left(\frac{\partial u}{\partial T}\right)_v dT + \left(\frac{\partial u}{\partial v}\right)_T dV\right] + \frac{P}{T} dV \qquad \ldots(2)$$

We can write

$$dS = \left(\frac{\partial S}{\partial T}\right)_V dT + \left(\frac{\partial S}{\partial V}\right)_T dV \qquad \ldots(3)$$

From Eqs (2) and (3), we get

$$\left(\frac{\partial S}{\partial T}\right)_V = \frac{1}{T}\left(\frac{\partial u}{\partial T}\right)_V \qquad \ldots(4)$$

and $$\left(\frac{\partial S}{\partial V}\right)_T = \frac{1}{T}\left[P + \left(\frac{\partial u}{\partial V}\right)_T\right] \qquad \ldots(5)$$

Taking the partial derivative of Eq. (4) with respect to *V*, we get

$$\frac{\partial^2 S}{\partial V\,\partial T} = \frac{1}{T}\left[\frac{\partial^2 u}{\partial V \partial T}\right] \qquad \ldots(6)$$

Taking the partial derivative of Eq. (5) with respect to *T*, we get

$$\frac{\partial^2 S}{\partial T.\partial V} = -\frac{1}{T_2}\left[P + \left(\frac{\partial u}{\partial T}\right)_T\right] + \frac{1}{T}\left[\left(\frac{\partial P}{\partial T}\right)_V + \frac{\partial^2 u}{\partial T \partial V}\right] \qquad \ldots(7)$$

By the test of exactness, we have

$$\frac{\partial^2 S}{\partial V.\partial T} = \frac{\partial^2 S}{\partial T.\partial V}$$

i.e. $$\frac{1}{T}\left[\frac{\partial^2 u}{\partial V\,\partial T}\right] = -\frac{1}{T_2}\left[P + \left(\frac{\partial U}{\partial V}\right)_T\right] + \frac{1}{T}\left[\left(\frac{\partial P}{\partial T}\right)_V + \frac{\partial^2 U}{\partial T \partial V}\right]$$

$\therefore$ $$\frac{1}{T_2}\left[P + \left(\frac{\partial u}{\partial V}\right)_T\right] = \frac{1}{T}\left(\frac{\partial P}{\partial T}\right)_V$$

$\therefore$ $$\frac{1}{T}\left[P + \left(\frac{\partial U}{\partial V}\right)_T\right] = \left(\frac{\partial P}{\partial T}\right)_V = \frac{\beta}{K} \cdots \left\{\because \frac{\beta}{K} = \left(\frac{\partial P}{\partial T}\right)_V\right\} \qquad \ldots(8)$$

We know that

$$mC_V = T\left(\frac{\partial S}{\partial T}\right)_V$$

i.e.

$$\left(\frac{\partial S}{\partial T}\right)_V = \frac{mC_V}{T}$$

From Eq. (4), we have

$$\left(\frac{\partial S}{\partial T}\right) = \frac{1}{T}\left(\frac{\partial u}{\partial T}\right)_V$$

i.e.

$$\frac{1}{T}\left(\frac{\partial u}{\partial T}\right) = \frac{mC_V}{T} \qquad \ldots(9)$$

Substituting Eqs (8) and (9) in Eq. (2), we get,

$$dS = \frac{mC_V}{T}dT + \frac{\beta}{K}dV$$

$$\therefore \quad TdS = mC_V dT + \frac{T\beta}{K}dV \qquad \ldots(10)$$

This is known as the 'first *TdS* equation.'

(*b*) *When temperature* (*T*) *and pressure* (*P*) *are independent*

Change in internal energy,

$$dU = \left(\frac{\partial U}{\partial T}\right)_P dT + \left(\frac{\partial U}{\partial T}\right)_T dP \qquad \ldots(1)$$

Change in volume is given by,

$$dV = \left(\frac{\partial V}{\partial T}\right)_P dT + \left(\frac{\partial V}{\partial P}\right)_T dP \qquad \ldots(2)$$

For independent variables *T* and *V*, *dU* can be expressed as,

$$dU = (mC_P - P\beta V)dT + \left[pKv - \frac{m(C_P - C_V)K}{\beta}\right]dP \qquad \ldots(3)$$

Also,

$$m(C_P - C_V) = \left[P + \left(\frac{\partial u}{\partial V}\right)_T\right]\left(\frac{\partial V}{\partial T}\right)_P$$

$$= \frac{T\beta}{K}\left(\frac{\partial V}{\partial T}\right)_P$$

$$= \frac{VT\beta^2}{K} \qquad \left\{\because P + \left(\frac{\partial U}{\partial V}\right)_T = \frac{T\beta}{K}\left(\frac{\partial V}{\partial T}\right)_p = \beta V\right\} \qquad \ldots(4)$$

Substituting Eq. (4) in Eq. (3),

$$dU = (mC_P - P\beta V)dT + (PK_V - VT\beta)\,dP \qquad \ldots(5)$$

We know that,

$$\left(\frac{\partial V}{\partial T}\right)_P = \beta V \text{ and } \left(\frac{\partial V}{\partial P}\right)_T = -KV$$

Substituting above values in Eq. (2), we get

$$dV = \beta V dT - KV dP \quad \text{...(6)}$$

Now, $TdS = dU + PdV$

Substituting Eqs (5) and (6) in above equation, we get

$$TdS = (mC_P - P\beta V)dT + (PKV - VT\beta)dP + p(\beta VdT - KVdP)$$

$$\therefore \quad TdS = mC_P dT - VT\beta dP \quad \text{...(7)}$$

This is known as Second TdS equation.

(c) *When pressure (P) and volume (V) are independent*

The change in internal energy,

$$dU = \left(\frac{\partial U}{\partial P}\right)_V dP + \left(\frac{\partial U}{\partial V}\right)_P dV \quad \text{...(1)}$$

But as discussed earlier, for independent variables p and v, we have

$$dU = \frac{KmC_V}{\beta} dP + \left(\frac{mC_P}{\beta C_{VP}} - p\right) dV \quad \text{...(2)}$$

Now, $TdS = dU + PdV$

$$= \frac{KmC_V}{\beta} dp + \left(\frac{mC_P}{\beta V} - p\right) dV + PdV$$

$$\therefore \quad TdS = \frac{KmC_V}{\beta} dP + \frac{mC_P}{\beta V} dV \quad \text{...(3)}$$

This is known as third TdS equation.

8.12 SPECIFIC HEAT

Specific heat of a substance is defined as amount of heat required to raise a unit mass of a substance through unit rise in temperature. There are two processes whose specific heats are important viz. constant volume C_V and constant pressure C_P.

Specific heat at constant volume (C_V)

At constant volume, the specific heat is given by,

$$C_V = \left(\frac{\partial U}{\partial T}\right)_V = \left(\frac{\partial U}{\partial T}\right)_V \left(\frac{\partial S}{\partial T}\right)_V$$

$$= T\left(\frac{\partial S}{\partial T}\right)_V \qquad \left\{\because T = \left(\frac{\partial u}{\partial S}\right)_V\right\}$$

Specific heat at constant pressure (C_P)

At constant pressure, specific heat is given by,

$$C_P = \left(\frac{\partial h}{\partial T}\right)_p = \left(\frac{\partial h}{\partial S}\right)_p \left(\frac{\partial S}{\partial T}\right)_p$$

$$= T\left(\frac{\partial S}{\partial T}\right)_p \qquad \left\{\because T = \left(\frac{\partial h}{\partial S}\right)_p\right\}$$

8.13 SPECIFIC HEAT RELATIONS

We know that,

The first TdS equation is given by

$$TdS = mC_V dT + \frac{T\beta}{K} dV$$

$$= mC_V dT + T\left(\frac{\partial P}{\partial T}\right)_V dV \qquad \left[\because \frac{\beta}{K} = \left(\frac{\partial P}{\partial T}\right)_V\right] \qquad \ldots(1)$$

The second TdS equation is given by

$$TdS = mC_P dT - VT\beta dP$$

$$= mC_P dT - T\left(\frac{\partial V}{\partial T}\right)_P dP \qquad \left[\because \beta V = \left(\frac{\partial P}{\partial T}\right)_P\right] \qquad \ldots(2)$$

From Eqs (1) and (2) we get

$$mC_V dT + T\left(\frac{\partial P}{\partial T}\right)_V dV = mC_P dT - T\left(\frac{\partial V}{\partial T}\right)_P dp$$

$$\therefore \quad m(C_P - C_V)dT = T\left(\frac{\partial P}{\partial T}\right)_V dV + T\left(\frac{\partial V}{\partial T}\right)_P dP$$

$$\therefore \quad dT = \frac{T\left(\frac{\partial P}{\partial T}\right)_V}{m(C_P - C_V)} dV + \frac{T\left(\frac{\partial P}{\partial T}\right)_P}{m(C_P - C_V)} dP \qquad \ldots(3)$$

The variables P and V are treated as independent variables. Thus, we get

$$dT = \left(\frac{\partial T}{\partial V}\right)_P dV + \left(\frac{\partial T}{\partial P}\right)_V dP \qquad \ldots(4)$$

From Eqs (3) and (4) we get

$$\left(\frac{\partial T}{\partial V}\right)_P = \frac{T\left(\frac{\partial P}{\partial T}\right)_V}{m(C_P - C_V)} \qquad \ldots[5(a)]$$

and

$$\left(\frac{\partial T}{\partial P}\right)_V = \frac{T\left(\frac{\partial V}{\partial T}\right)_P}{m(C_P - C_V)} \qquad \ldots[6(a)]$$

or

$$m(C_P - C_V) = T\left(\frac{\partial P}{\partial T}\right)_V \left(\frac{\partial V}{\partial T}\right)_P \qquad \ldots[5(b)]$$

$$m(C_P - C_V) = T\left(\frac{\partial V}{\partial T}\right)_P \left(\frac{\partial P}{\partial T}\right)_V \qquad \ldots[6(b)]$$

From Eqs 5 (a), 5 (b) and 6 (a), 6(b), it is observed that

$$m(C_P - C_V) = T\left(\frac{\partial P}{\partial T}\right)_V \left(\frac{\partial V}{\partial T}\right)_P \qquad \ldots(7)$$

From the cyclic property of differentials, we have

$$\left(\frac{\partial P}{\partial T}\right)_V \left(\frac{\partial T}{\partial V}\right)_P \left(\frac{\partial V}{\partial P}\right)_T = -1$$

i.e. $$\left(\frac{\partial P}{\partial T}\right)_V = \frac{1}{\left(\frac{\partial T}{\partial V}\right)_P\left(\frac{\partial V}{\partial P}\right)_T}$$

or $$\left(\frac{\partial P}{\partial T}\right)_V = -\left(\frac{\partial V}{\partial T}\right)_P\left(\frac{\partial P}{\partial V}\right)_T$$

If we put the above value in Eq. (7), we get,

$$m(C_P - C_V) = T\left[-\left(\frac{\partial v}{\partial T}\right)_p\left(\frac{\partial p}{\partial v}\right)_T\right]\left(\frac{\partial V}{\partial T}\right)_P$$

$$= -T\left(\frac{\partial V}{\partial T}\right)_P^2\left(\frac{\partial P}{\partial V}\right)_T \quad \text{...(8)}$$

But we know that $\left(\frac{\partial V}{\partial T}\right)_P = \beta V$ and $\left(\frac{\partial V}{\partial P}\right)_T = -K_V$

Substituting the above value in Eq. (8), we get,

$$m(C_P - C_V) = -T(\beta V)^2\left(-\frac{1}{KV}\right)$$

$$= \frac{T\beta^2 V}{K}$$

$$\therefore \quad (C_P - C_V) = \frac{T\beta^2 V}{Km}$$

But $\frac{V}{m} = V_S =$ Specific volume of gas (m^3 /kg)

$$\therefore \quad C_P - C_V = \frac{T\beta^2 V_S}{Km}$$

The values of T, V_S and K are always positive. Also, even if β is negative, β^2 is always positive

$$\therefore \quad C_P - C_V > 0 \text{ i.e. } C_P > C_V$$

8.14 JOULE THOMSON COEFFICIENT

Consider a throttling process as shown in Fig. 8.1.

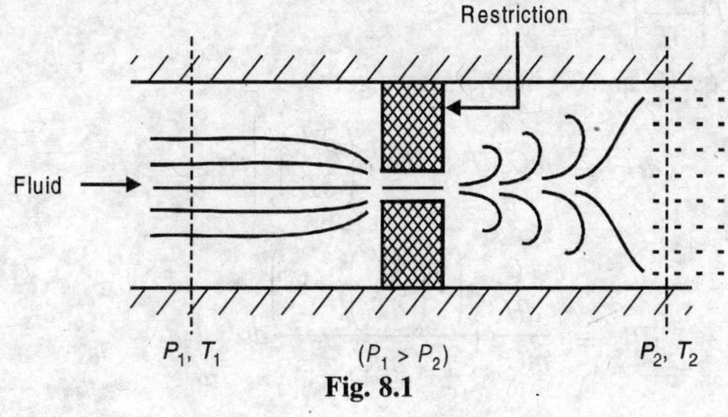

Fig. 8.1

The velocity of fluid, whose pressure is reduced by a throttling process, increases at the restriction with a consequent decrease in enthalpy. But this increase in kinetic energy is dissipated by friction, as the eddies die down after restriction.

The steady flow energy equation implies that if the flow is adiabatic and if the fluid velocity before restriction is equal to that downstream of it, the enthalpy of fluid is restored. Throttling is therefore an Isenthalpic process.

The change in temperature with drop in pressure, at constant enthalpy, is defined as the Joule–Thomson Coefficient (μ).

$$\therefore \quad \mu = \left(\frac{\partial T}{\partial P}\right)_h \quad \text{...(1)}$$

μ varies with both temperature and pressure of the gas. The magnitude of μ is a measure of deviation of a gas from perfect gas behaviour.

For real gases, μ may be either positive or negative, depending on the thermodynamic state of the gas. If μ is greater than zero, it implies that the gas temperature decreases with decrease in pressure. If μ is less than zero, the gas temperature increases with decreases in pressure.

When $\mu = 0$, the temperature of gas remains constant with throttling. The temperature at which $\mu = 0$ is called as Inversion temperature for a given pressure.

Using usual notation, we know that enthalpy is given by

$$H = U + PV \quad \text{...(2)}$$

Writing Eq. (2) in differential form, we get

$$dH = dU + d(PV)$$

$$= dU + PdV + VdP \quad \text{...(3)}$$

We know that

$$TdS = dU + PdV$$

or

$$dU = TdS - PdV \quad \text{...(4)}$$

Substitute Eq. (4) in Eq. (3)

$$\therefore \quad dH = TdS - PdV + PdV + VdP$$

$$\therefore \quad dH = TdS + VdP$$

Taking P and T as independent variables, we get

$$dH = T\left[\left(\frac{\partial S}{\partial T}\right)_p dT + \left(\frac{\partial S}{\partial p}\right)_T dp\right] + Vdp$$

$$= T\left(\frac{\partial S}{\partial T}\right)_p dT + \left[T\left(\frac{\partial S}{\partial p}\right)_T + V\right]dp \quad \text{...(5)}$$

From Maxwell relations,

$$\left(\frac{\partial S}{\partial T}\right)_T = -\left(\frac{\partial v}{\partial T}\right)_P \quad \text{...(6)}$$

Also

$$T\left(\frac{\partial S}{\partial T}\right)_P = mC_P \quad \text{...(7)}$$

Substituting value from Eqs. (6) and (7) in Eq. (5), we get

$$dH = mC_p dT + \left[V - T\left(\frac{\partial V}{\partial T}\right)_p\right]dp \quad \text{...(8)}$$

$$\therefore \quad dT = \frac{dH}{mC_p} + \frac{\left(T\left(\frac{\partial V}{\partial T}\right)_P - V\right)}{mC_p}dp$$

$$= \frac{dh}{C_p} + \frac{\left(T\left(\frac{\partial Vs}{\partial T}\right)_p - Vs\right)}{C_p} dp \qquad \text{...(9)}$$

$$\left[\because dh = \frac{dH}{m} = \text{Change in sp. enthalpy (kJ/kg)}\right]$$

$$\left[\text{Also, } V_s = \frac{V}{m} = \text{Specific volume of gas } (\text{m}^3/\text{kg})\right]$$

Assuming h and p as independent variables, we get

$$dT = \left(\frac{\partial T}{\partial h}\right)_p dh + \left(\frac{\partial T}{\partial p}\right)_h dp \qquad \text{...(10)}$$

From Eqs (9) and (10) we get

$$\left(\frac{\partial T}{\partial h}\right)_P = \frac{1}{C_P}$$

Multiplying the numerator and denominator by T^2, we get

$$\left(\frac{\partial T}{\partial p}\right)_h = \frac{T^2}{C_P}\left[\frac{T\left(\frac{\partial V_S}{\partial T}\right)_P - V_S}{T^2}\right] = \mu \qquad \left[\because \mu = \left(\frac{\partial T}{\partial p}\right)_h\right]$$

$$\therefore \qquad \mu = \frac{T^2}{C_P}\left[\frac{\partial}{\partial T}\left(\frac{V_S}{T}\right)\right]_P$$

Note. For an ideal gas, $PV_S = RT$

$$\therefore \qquad \frac{V_S}{T} = \frac{R}{P} \quad \text{i.e.} \quad \frac{\partial}{\partial T}\left(\frac{V_S}{T}\right)_p = 0 \quad (\because R \text{ and } P \text{ are constants})$$

$$\therefore \qquad \mu = \frac{T^2}{C_p}\left[\frac{\partial}{\partial T}\left(\frac{V_S}{T}\right)\right]_p = 0 \text{ i.e. } \mu = 0 \text{ for ideal gas}$$

8.15 EQUATION OF STATE

The ideal gas equation of state $PV_S = RT$ can be established from the postulates of kinetic theory of gases developed by Clerk Maxwell. The two main assumptions made in deriving the perfect gas laws are:

(1) The gas moleculas are mere mass points occupying no space or negligibly small space compared to the volume of the gas.

(2) There is very little or no attraction or repulsion between the molecules.

In practice, all real gases deviate from the ideal gas laws. The deviation is small at ordinary temperatures and pressures but high at high pressure and low temperature. This is because all real gas molecules do occupy some space and to attract or repulse each other. At high pressure and low temperature, the intermolecular attraction and repulsion forces increase and also the volume of molecules becomes appreciable compared to total gas volume. Hence the ideal gas equation is not suitable for real gases at high pressure and low temperature. Some equations used to correct the ideal gas equation are given below.

(a) Van der Waal's Equation. J. D. Van der Waal, a Dutch physicist, was the first to correct the ideal gas equation $PV_S = RT$. He applied the laws of mechanics to individual molecules and introduced two correction terms in the ideal gas equation.

Van der Waal's equation is given by,

$$\left(P+\frac{a}{V_S^2}\right)(V_S-b) = RT$$

Coefficient a was introduced to account for the existence of mutual attraction between the molecules.

Coefficient b was introduced to account for the volume of molecules, and is known as co-volume.

The term $\left(a/V_S^2\right)$ is called the force of cohesion.

Real gases conform more closely with Van der Waal's equation of state than the ideal gas equation of state, particularly at high pressures. But is not obeyed by real gases in all ranges of pressures and temperatures.

(b) Clausius Equation. The Clausius equation is a modification of Van der Waal's equation and it neglects the mutual attraction between molecules. Thus, the Clausius equation is of the form:

$$P(V_S-b) = RT$$

where b = Volume occupied by all molecules.

(c) Beattie Bridgeman Equation. In 1928, Beattie–Bridgeman developed an equation of state which gives accurate results. The equation is of the form:

$$P = \frac{RT(1-\varepsilon)}{V_S}(V_S+B)-\frac{A}{V_S}$$

where,

$$A = A_o\left(1-\frac{a}{V_S}\right)$$

$$B = B_o\left(1-\frac{b}{V_S}\right)$$

$$\varepsilon = \left(\frac{C}{V_S T^3}\right)$$

The constants *Ao*, *Bo*, *a*, *b* and *c* have to be determined experimentally for each gas.

The Beattie–Bridgeman equation does not give satisfactory results in the critical point region, but is fairly accurate (within 2 percent) in regions where density is less than 0.8 times the critical density.

(d) Bertholet Equation. Bertholet equation is similar to the Van der Waal's equation and uses the same constants a as in Van der Waal's equation. The equation is of the form

$$\left(P-\frac{a}{TV_S^2}\right)(V_S-b) = RT$$

The Bertholet equation is of limited accuracy and gives properties accurate within 1 percent.

(e) Dieterici Equation. Dieterici equation of state gives very accurate properties in the neighbourhood of the critical point and on the critical isotherm. The equation is of limited accuracy and may produce large errors when used away from critical region. The equation is of the form

$$P = \frac{RT}{(V_S-b)}\cdot e^{-a/RTV_S}$$

where a and b are constants.

(f) Redlich-Kwong Equation. The equation developed by Redlich–Kwong has the following form

$$\left[P - \frac{a}{\sqrt{T}\,V_S(v_S + b)}\right](V_S - b) = RT$$

where a and b are constants.

The Redlich–Kwong equation gives good results at high pressures and is fairly accurate for temperature above the critical value.

(g) Benedict–Webb–Rubin Equation. The Benedict–Webb–Rubin equation is used for regions of high density i.e., where the density is approximately twice the critical density.

The equation has the form:

$$P = \frac{RT}{V_s} + \frac{B_o RT - A_o - C_o/T^2}{V_S^2} + \frac{bRT - a}{V_S^3} + \frac{a\alpha}{V_S^6} + \frac{C}{V_S^3 T^2}\left(1 + \frac{\gamma}{V_S^2}\right)e^{-\gamma/v_S^2}$$

The eight constants A_o, B_o, C_o, a, b, c, x and γ are to be determined experimentally.

(h) Virial Equation. The relation between PV_s and P may be expressed in the form of power series, as shown below.

$$PV_S = A(1 + B'P + C'P^2 + D'P^3 + \ldots)$$

It can be proved that for any gas,

$$\lim_{p\to 0} PV_S = RT = A$$

$\therefore$ We get,

$$PV_S = RT(1 + B'P + C'P^2 + D'P^3 + \ldots)$$

or

$$\frac{PV_S}{RT} = 1 + B'P + C'P^2 + D'P^3 + \ldots \quad \ldots(i)$$

An alternate expression for the above is

$$\frac{PV_S}{RT} = 1 + \frac{B}{V_S} + \frac{C}{V_S^2} + \frac{D}{V_S^3} + \ldots \quad \ldots(3)$$

Both the expressions in Eqs. (i) and (ii) are known as virial expansions or virial equations of state. These equations were first introduced by Dutch physicist Kammerlingh Onnes.

In the above equations, B, C, D, B, C, etc. are called virial coefficients. B and B are known as second viral coefficients, C and C are known as third virial coefficients and so on.

There exists a relation between the virial coefficients in Eqs (i) and (ii) as follows:

$$B' = \frac{B}{RT}, \quad C' = \frac{C - B^2}{(RT)^2},$$

$$D' = \frac{D - 3BC + 2B^3}{(RT)^3} \ldots \text{ and so on.}$$

8.16 COMPRESSIBILITY FACTOR

We know that the virial equation of state is of the form

$$\frac{PV_S}{RT} = 1 + \frac{B}{V_S} + \frac{C}{V_S^2} + \frac{D}{V_S^3} + \ldots$$

Here, the ratio (PVs/RT) is known as the compressibility factor, denoted by Z.

For ideal gas, $Z = 1$.

For real gas, Z is determined experimentally. The magnitude of Z for a particular gas at a particular pressure and temperature is a measure of deviation of the gas from ideal gas behaviours.

IMPORTANT POINTS TO BE REMEMBERED

1. *Helmhottz function* (*A*)

$$A = U - TS \text{ or } a = u - TS$$

$$(W_T)_{max} = (W_T)_{rev} = (A_1 - A_2)_T$$

2. *Gibbs function* (*G*)

$$G = U - TS + PV$$

or

$$G = H - TS$$

$$[(W_o)_{P,T}]_{Max} = [(W_o)_{P,T}]_{rev} = (G_1 - G_2)_{P,T}$$

3. *Maxwell's equations*

(a) *Internal energy* $\left(\frac{\partial T}{\partial V}\right)_S = \left(\frac{\partial P}{\partial S}\right)_V$

(b) *Enthalpy* $\left(\frac{\partial T}{\partial P}\right)_S = \left(\frac{\partial V}{\partial S}\right)_T$

(c) *Helmholtz function* $\left(\frac{\partial P}{\partial T}\right)_V = \left(\frac{\partial S}{\partial V}\right)_T$

(d) *Gibb's function* $\left(\frac{\partial V}{\partial T}\right)_P = \left(\frac{\partial S}{\partial P}\right)_T$

Maxwell's equations

4. *Coefficient of volume expansion* (β)

$$\beta = \frac{1}{V}\left(\frac{\partial V}{\partial T}\right)_P$$

5. *Isothermal compressibility*

$$K = \frac{1}{V}\left(\frac{\partial V}{\partial P}\right)_T$$

6. *Adiabatic compressibility*

$$K_s = -\frac{1}{V}\left(\frac{\partial V}{\partial P}\right)_S$$

$$\gamma = \frac{(\partial P/\partial V)_S}{(\partial P/\partial V)_T}$$

7. $$\frac{\beta}{K} = \left(\frac{\partial P}{\partial T}\right)_V$$

8. *Joule-Thomson coefficient* (μ)

$$\mu = \frac{T_2}{Cp}\left[\frac{\partial}{\partial T}\left(\frac{V_s}{T}\right)\right]_P \quad \mu = 0 \text{ for ideal gas.}$$

CHAPTER 9

Availability

9.1 INTRODUCTION

Energy exists in many forms like kinetic energy, potential energy, electrical energy, heat energy etc. While it is possible to convert most of the energies like potential energy, kinetic energy, electricity etc. into work, only some part of heat energy can be converted into work. The part of energy that can be converted into work is *Available energy* and the remainder is *Unavailable portion.*

Availability is about work potential or quality of the energy. When thermal energy is used to do work entire heat energy cannot be converted to useful work as per the second law of thermodynamics. The part that can be converted to useful work is referred to as the *Available energy* or *Availability* or *Energy*. The quality of energy is measured by Availability. As energy is used in a process it loses quality, its availability or Energy decreases.

9.2 HIGH/LOW GRADE OF ENERGY

Energies like electrical energy or potential energy which can be converted almost entirely to do work are high-grade energy. Heat on the other hand is low-grade energy since only a part of it can be converted into work. Heat is of the lowest grade in which energy exists in a very disordered state. The lower the temperature at which the heat energy exists the lower would be the grade of energy.

However low grade energy is used to produce high-grade energy. For example electricity a high-grade energy is produced in thermal or nuclear power plants using heat a low-grade energy source.

High Grade Energy Sources	**Low Grade Energy Sources**
Mechanical work	Heat
Electricity	Heat derived from combustion
Kinetic energy	Kinetic energy
Tidal power	Heat from nuclear fission

9.3 AVAILABILITY

Available energy is that of energy which is available for doing work. It is either work or that portion of energy that can be converted wholly into work. Concept of availability comes from the Second Law of Thermodynamics where some portion of heat has to be rejected to a sink to produce work. The work has to be with reference to some datum. Generally earth's atmosphere is taken as reference and is called the dead state designated by '0'. Dead state implies the temperature and pressure existing in the surroundings and are designated by T_0 and P_0.

Availability or Exergy is the maximum portion of energy that can be converted into work by ideal process that reduces the system to dead state.

9.3.1 Dead State

Any system that has temperature T and pressure P can do useful work till the temperature and pressure are reduced to T_0 and P_0. When the temperature and pressure are equal to that of the earth or dead state all transfer of energy, stops although the system contains internal energy which would be unavailable. When a system is in equilibrium with the surroundings its potential to do any work ceases. Thus when the pressure P of the system reduces to the atmospheric pressure P_0, temperature T becomes equal to T_0 and likewise the KE and PE become equal to that of the surroundings no more work can be obtained.

The properties of the system in dead state are denoted by subscript 0 i.e., as P_0, T_0, H_0, S_0, U_0, C_0 etc. Generally dead state temperature is taken at 25°C (298 K) and P_0 as 1.01325 bar unless otherwise indicated.

9.4 AVAILABILITY OF VARIOUS SYSTEMS

We can study the availability of

(a) Work Reservoir
(b) Heat Reservoir
(c) Closed System
(d) Steady Flow System

9.4.1 Work Reservoir

A work reservoir is a source of infinite work and it would not come to equilibrium with the surroundings. However the amount of W withdrawn from the work reservoir can be converted fully into useful work in the absence of any dissipative effects. Therefore Availability of work reservoir is $A = \Delta W$.

9.4.2 Heat Reservoir

(i) Infinite Heat Reservoir (at constant temperature).
(ii) Finite Heat Reservoir (where temperature changes).

9.5 IRREVERSIBILITY

Most of the real processes in nature are irreversible due to mixing, friction and heat transfers with finite temperature difference. In a reversible process there is no net increase in the entropy of the universe. In a reversible process the change in the entropy of a system equals the change in the entropy of the universe. However in the case of an irreversible process the reduction in the entropy of a system is less than the increase of entropy of the heat receiving system. This leads to loss of availability or an equivalent increase in unavailability— which however are equal. In any process, aim is always to get maximum work. In an expansion process we try to get the maximum work output. In compression process aim is to have minimum work input. Thus both in expansion as well as in compression processes aim is to maximize the work as per sign convention of work.

Maximum work (W_{rev}) during a process between two states can be obtained by reversible process. However in actual practice the actual work (W_{act}) between these two states is always less than the W_{rev}. The difference between the two is called Irreversibility.

$$I = W_{rev} - W_{act}$$

In irreversibility (I)/Exergy destruction = wasted work potential in a process. It is lost opportunity to do work. The greater the irreversibility the greater is the loss of work that could have been performed. Thus irreversibility gives the quality of the process.

Let there be two processes, one reversible and another irreversible between two states and let heat δQ_{rev} and δQ be added giving work output of W_{rev} and W_u respectively.

Thus $\qquad \delta Q_{rev} = dE + \delta W_{rev} \qquad \ldots(1)$

and $\qquad \delta Q = dE + \delta W \qquad \ldots(2)$

$$\delta I = (1) - (2)$$

$$\delta I = \delta Q_{rev} - \delta Q = TdS - \delta Qs$$

$$= \delta W_{rev} - \delta W$$

Integrating $\qquad I = W_{rev} - W_{act} = \text{Lost work}$

9.6 EFFECTIVENESS

The effectiveness of a system is the ratio of the useful or actual work done to the maximum or reversible work.

Thus, $\qquad \text{Effectiveness} = \delta W/\delta W_{rev}$

9.7 SECOND LAW EFFICIENCY

So far we have been using the efficiencies and COP based on the First Law of Thermodynamics. Based on the increasing use of availability which gives indication about possibility of maximum work output or minimum work input if the process is carried out reversibly — concept of Second Law efficiency is being used these days. Second law efficiency is a measure of reversible operation.

Let us assume that an engine operating between say 1200 K and 300 K has – η_{th} of 30% (by measuring brake power and dividing by the heat input—based on First Law of Thermodynamics).

However, had the work been done in a reversible manner i.e., by Carnot Cycle.

$$\eta_{rev} = \left(1 - \frac{To}{T}\right)$$

$$= (1 - 300/200) = 0.75 = 75\%$$

The ηth is far less than the η_{rev}. We can infer that η_I is not realistic measure of the performance. To overcome we define η_{II} for various heat-work machines

$$\eta_{II} = \eta_{th}/\eta_{rev} \text{ (For heat engines)}$$

$$\eta_{II} = W/W_{rev} \text{ (For power or work producing devices)}$$

$$\eta_{II} = W_{rev}/W \text{ (For power absorbing devices)}$$

where $\qquad W = \text{useful work and}$

$$W_{rev} = \text{maximum or reversible work}$$

$$\eta_{II} = COP/COP_{rev} \text{ (For refrigerators and heat pumps)}$$

9.8 IMPORTANCE OF AVAILABILITY/IRREVERSIBILITY/EFFECTIVENESS/η_{II}

These concepts are based on the second law of thermodynamics and are applicable to both cycle and process. These indicate the departure of an actual process from an ideal or reversible process.

9.9 HELMONTZ AND GIBBS FUNCTIONS

Both these functions are properties (based on combination of properties just like $h = u + pv$.

Helmontz Function $\qquad F = U - TS$

or $\qquad f = u + Ts$ (per unit mass)

This function is useful for closed systems that undergo reversible, isothermal process. Decrease in the value of F equals the max work that can be done by system undergoing an isothermal process at temperature equal to that of the surroundings.

9.10 GIBBS FUNCTION

$$G = H - Ts$$

or $\qquad g = h - Ts \qquad$ (For unit mass)

9.11 SUMMARY

Availability of Energy is the maximum useful work that can be obtained from a system under ideal conditions by reducing it to the dead state. Availability A of various systems are

(a) Heat Reservoir $\quad A = \Delta Q - T_0\ \Delta S$

(b) Closed System $\quad A = A - A_0 = (U - T_0S + P_0V) - (U_0 - T_0S_0 - P_0 + P_0V_0)$

(c) Steady Flow System $\quad A = A - A_0 = m\left\{\left(h - T_0S + \frac{C^2}{2} + gZ\right) - \left(h_0 - T_0S_0 + \frac{C_0^2}{2} + gZ_0\right)\right\}$

$$\text{Irreversibility} = W_{rev} - W = T_0 - T_0\,\Delta S$$

$$\text{Effectivness} = W/W_{rev}$$

9.12 ENTROPY CHANGE ΔS

The unavailable energy $UA = T_0\,\Delta S$. Thus change in entropy during a process is required to find out avilability and unavailability.

From Ist Law

$$\delta Q = \delta u + \delta w$$

$$\delta Q = Au + Pdv$$

Divide by T

$$\frac{\delta Q}{T} = \frac{du}{T} + Pdv$$

$$= C_v\frac{dt}{T} + R\frac{dv}{V} \quad \left(\text{Since}\ \frac{P}{T} = \frac{R}{V}\right)$$

But

$$\frac{\delta Q_{rev}}{T} = ds$$

$$\therefore \quad ds = C_V\frac{dT}{T} + R\frac{dv}{V}$$

Integrating between 1 and 2

$$s_2 - s_1 = C_V \ln\frac{T_2}{T_1} + R\frac{V_2}{V_1}$$

$$\Delta S = C_V \ln\frac{P_2V_2}{P_1V_1} + R\ln\frac{V_2}{V_1} \quad \ldots(1)$$

$$= C_V\left(\ln\frac{P_2}{P_1} + \ln\frac{V_2}{V_1}\right) + R\ln\frac{V_2}{V_1}$$

$$= Cv\left(\ln\frac{P_2}{P_1}\right) + \ln\frac{v_2}{v_1} + (R + Cv)$$

But

$$C_v + R = C_p$$

$$\therefore \quad \Delta S = C_V \ln \frac{P_2}{P_1} + C_p \ln \frac{V_2}{V_1} \quad \text{...(2)}$$

We can also put $\frac{v_2}{v_1} = \frac{P_1 T_2}{P_2 T_1}$ in Eq. (1)

$$\therefore \quad \Delta S = C_V \ln \frac{T_2}{T_1} + R \ln \frac{P_1 T_2}{P_2 T_1}$$

$$= C_V \ln \frac{T_2}{T_1} + R\left(\ln \frac{P_1}{P_2} + \ln \frac{T_2}{T_1} \right)$$

$$= (C_V + R) \ln \frac{T_2}{T_1} + R \ln \frac{P_1}{P_2}$$

$$\Delta S = C_p \ln \frac{T_2}{T_1} + R \ln \frac{P_1}{P_2} \quad \text{...(3)}$$

Entropy Changes for Various Processes are

(1) Constant volume $\quad s_2 - s_1 = CV \ln \frac{T_2}{T_1}$

(2) Constant pressure $\quad s_2 - s_1 = C_P \ln \frac{T_2}{T_1}$

(3) Iso thermal $\quad s_2 - s_1 = R \ln \frac{V_2}{V_1}$

(4) Adiabatic $\quad s_2 - s_1 = 0$

(5) Polytropic $\quad s_2 - s_1 = C_V \left(\frac{n-r}{n-1} \right) \ln \frac{T_2}{T_1}$

9.13 AVAILABILITY OF HEAT RESERVOIR

(a) Infinite Heat Reservoir. In such a reservoir temperature remains constant when small amount of heat δQ is withdrawn from it.

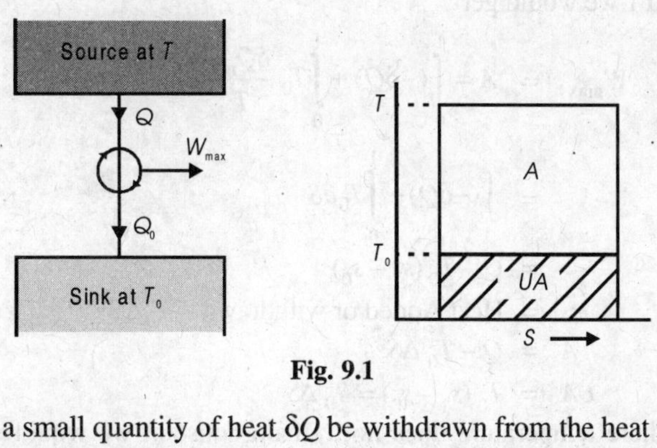

Fig. 9.1

As shown in Fig. 9.1, let a small quantity of heat δQ be withdrawn from the heat reservoir which is at average temperature T. For getting maximum work, let there be a Carnot engine operating between temperatures T and T_0 (sink temperature being of the surroundings).

$$\frac{\delta W}{\delta Q} = \left(1-\frac{T_0}{T}\right)$$

or
$$\delta W = \left(1-\frac{T_0}{T}\right)(-\delta Q) = T_0\frac{\delta Q_{rev}}{T}$$

Integrating,
$$W_{max} = \int_{T_0}^{0} = A = -\delta Q - \int_{0}^{1} T_0 dS$$
$$= Q + T_o(s_0 - s_1)$$
$$= Q - T_o(S_1 - S_0)$$

Thus
$$W_{max} = Q - T_0 \Delta S$$
$$= \text{Heat Added} - T_0 \text{ (change in entropy)}$$

(b) Finite Heat Reservoir. In this case when heat is added or withdrawn from the reservoir which has finite capacity the temperature does not remain constant.

Consider a small quantity δ heat being withdrawn from the reservoir (1) at temperature *T* and producing entropy change *ds* as shown. Using a Carnot engine to derive maximum work between *T* and T_0; we have

$$\delta W'_{carnot} = \left(1-\frac{T_0}{T}\right)(-\delta Q)$$

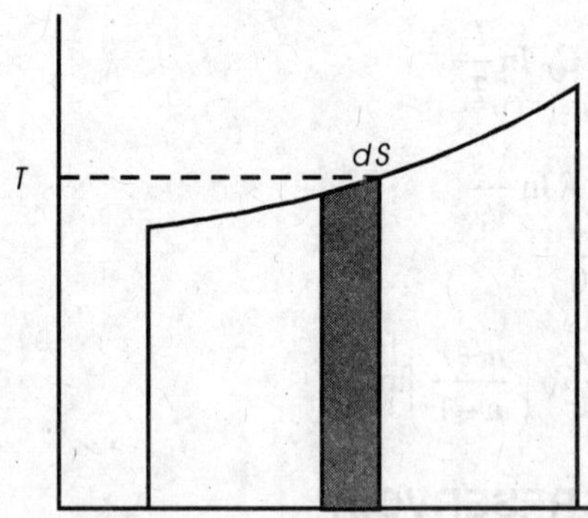

Fig. 9.2

Integrating between 0 and 1 we would get

$$W_{max} = A = \int_0^1 (-\delta Q) + \int_0^1 T_0 \frac{\delta Q_{rev}}{T}$$
$$= \int_0^1 (-\delta Q) + \int_0^1 T_0 dS$$
$$= Q - T_0(s_1 - s_0)$$
$$= \text{Heat Added or withdrawn} - T_0 \Delta S$$
$$A = Q - T_0 \Delta S$$
$$UA = T_0(s_1 - s_0) = T_0 \Delta S$$

For the heat reservoir where temperature does not remain constant we would use appropriate equation for entropy change as applicable.

SOLVED EXAMPLES

Example 9.1 *0.5 kg of air at 500 K is heated reversibly at constant pressure to 2000 K. Find the available and unavailable energies and the heat added. Take T_0 = 300 K and C_P = 1.005 kJ/kg-K.*

Solution

$$\Delta Q = m\,C_P\,(T_2 - T_1)$$
$$= 0.5 \times 1.005(2000 - 500)$$
$$= 753.75 \text{ kJ}$$

Entropy changes at constant pressure

$$\Delta S = mC_P \ln \frac{T_2}{T_1}$$
$$= 0.5 \times 1.005 \ln \frac{2000}{500}$$
$$= 0.6966 \text{ kJ/kg-K}$$
$$UA = T_0 \times \Delta S = 300 \times 0.6966$$
$$= 208.98 \text{ kJ}$$
$$\mathbf{A = 753.75 - 208.98 = 544.77 \text{ kJ}}$$

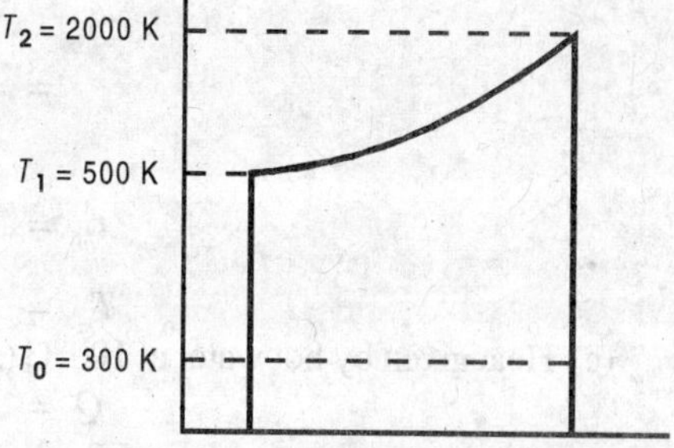

Fig. Ex. 9.1

Example 9.2 *One kg of air is heated from 70°C to 270°C at constant pressure. Determine (a) Heat supplied (b) Entropy change (c) Available energy for sink temperature of 20°C and – 20°C. Assume C_P = 1.005 kJ/kg-K for air.*

Solution

$$T_1 = 273 + 70 = 343 \text{ K}$$
$$T_2 = 543 \text{ K}$$

(a) Heat supplied $= mC_P\Delta T$
$$= 1 \times 1.005 \times (270.70) = 201 \text{ kJ}$$

(b) ΔS at const. pressure $= m\,Cp \ln \frac{T_2}{T}$
$$= 1 \times 1.005 \ln \frac{543}{343}$$
$$= 0.46167 \text{ kJ/kg.K}$$

(c) (i) Available energy for sink temperature and $20 + 273 = 293$ K
$$UA = T_0 \Delta S$$
$$= 293 \times 0.808 = 135.27 \text{ kJ}$$
$$A = \Delta Q - UA = 201 - 135.27 = 65.73 \text{ kJ}$$

(ii) Available and unavailable energy at sink temperature of
$$273 - 20 = 253 \text{ K}$$
$$T_0 = 253 \text{ K}$$
$$UA = T_0 \Delta S = 116.80$$
$$\mathbf{A = 201 - 116.80 = 84.2 \text{ kJ}}$$

Example 9.3 *100 kg of water at 100°C is mixed with 50 kg of water at 50°C, while the surrounding temperature is 27°C. Determine the decrease in availability due to mixing.*

Data:

$$T_0 = 273 + 27 = 300\ K; \qquad m_1 = 100\ kg$$
$$T_1 = 273 + 100 = 373\ K; \qquad m_2 = 50\ kg$$
$$T_2 = 273 + 50 = 323\ K$$

Solution

Temperature after mixing will be

$$(100+50)\,t_m = 100\,t_1 + 50\,t_2$$
$$= \frac{100\times100+50\times50}{150} = \frac{2\times100+50}{3}$$
$$t_m = \frac{250}{3} = 83.33°C$$
$$T_m = 273+83.33 = 356.33\ K$$

(c) Heat given by hot water at 100°C (373 K)

$$Q = m_1 \times C_p \times (100-83.33)$$
$$= 100\times4.187\times16.67 = 6979.73\ kJ$$

$$\text{Entropy decrease} = m_1 C_P \ln\frac{T_1}{T_m}$$
$$= 100\times4.187\times\ln\frac{373}{356.33}$$
$$\Delta S = 19.1434\ kJ/K$$
$$UA_{100} = T_o\Delta S = 300\times19.1434 = 5743.0\ kJ$$

$\therefore$ Availability

$$A_{100} = \Delta Q_{100} - UA_{100}$$
$$= 6979.73 - 5753.6 = 1226.13$$

(d) Heat received by cold water at 50°C

$$Q_{50} = m_2\,C_P\,\Delta T$$
$$= 50\times4.187\times(83.33-50) = 6977.64\ kJ$$

$$\text{Entropy gain by cold water} = m_2 C_P \ln\frac{T_m}{T_2}$$
$$= 50\times4.187\ \ln\frac{356.33}{323} = 20.56\ kJ/K$$
$$UA_{50} = T_o\Delta S = 300\times20.56$$
$$= 6168\ kg$$
$$A_{50} = \Delta Q_{50} - UA_{50}$$
$$= 6977.64 - 6168 = 809.64\ kJ$$

(e) Decrease in Availability = (1) – (2)

$$= 1226.13 - 809.64$$

Decrease in Availability = 416.49 kJ

Example 9.4 *Exhaust gas leave an IC engine at 827° and 1.0 atmosphere after doing 1000 kJ of work per kg of gas in the engine (C_{pg} = 1.1). The surrounding temperature is 25°C.*

(a) How much available energy is cost by throwing away exhaust gases.

(b) What is the ratio δ the lost available energy to the engine work.

Solution

$$T_1 = 827+273=1100$$
$$T_0 = 25+273=298$$

Entropy change $\Delta S = ?$

$$= m \cdot C_{Pg} \cdot \ln\frac{T_1}{T_0}$$

$$= 1\times1.1\times\ln\frac{1100}{298}$$

$$= 1.4365 \text{ kJ/kg.K}$$

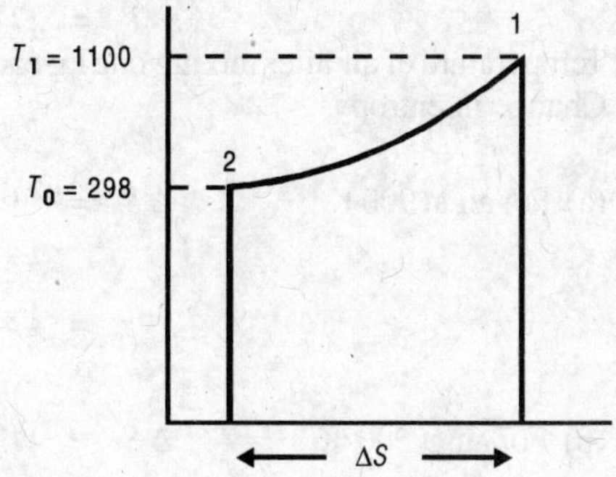

Fig. Ex. 9.4

(a) $UA = T_0\times\Delta S$

$$= 298\times1.4365$$
$$= 428.1 \text{ kJ/kg-K}$$

Thus lost available energy = 428 kJ/kg

(b) Ratio of lost available energy/useful work

$$= \frac{428.1}{1000} = 0.428$$

Example 9.5 *A hot spring produce water at a temperature δ 57°C, at the rate of 0.1 m³/minute. The water into a large lake with minimum temperature of 15°C. What is the rate of working of an ideal heat engine which use all the available energy.*

Solution

Now

$$T_1 = 273+57=330 \text{ K}$$
$$T_0 = 273+15=288 \text{ K}$$
$$\dot{m} = 0.1 \text{ m}^3/\text{min}$$
$$= \frac{0.1\times1000}{60} \text{kg/sec} = 1.667 \text{ kg/s}$$

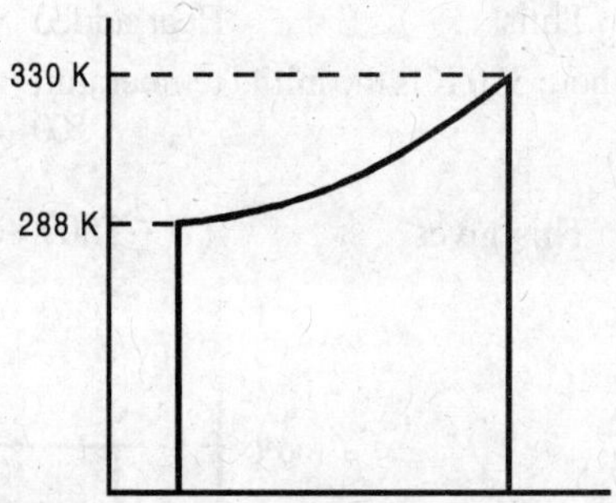

Fig. Ex. 9.5

(a) Heat given by the cooling water/s

$$\Delta Q = m\, C_P\,(T_1 - T_0)$$
$$= 1.6667\times4.187\times(57-15)$$
$$= 293.09 \text{ kJ.s}$$

(b) Change of entropy of spring water

$$= \dot{m}\, C_P \ln\frac{T_1}{T_0} = 1.667\times4.187\times\ln\frac{330}{288}$$

$$\Delta S = 0.94999 \text{ kJ/K}$$
$$UA = T_0\Delta S = 288\times\Delta S = 273.6 \text{ kJ}$$

(c) Available energy $= Q - UA = 293.09 - 273$

$$\cong 19.5 \text{ kJ/s} = 19.5 \text{ kW}$$

Thus rate of working of an ideal heat engine which are all energy is 19.5 kW.

Example 9.6 *1 kg of air of pressure p_1 and temperature 900 K is mixed with 1 kg of air at the same pressure and 500 K. Determine the loss in availability if surrounding temperature is 27°C. Take C_P for air = 1.0 J/kg-K.*

Solution

$$T_0 = 27+273=300\text{ K}$$

Temperature of air after mixing one kg each will be (900 + 500) / 2 = 700 K

Change in entropy.

(a) For air at 900 K $\quad \Delta S = \dot{m}\, C_P \ln\frac{T_2}{T_1}$

$$= 1\times1\times\ln\frac{700}{900} = -\,0.2513 \qquad ...(1)$$

(b) For air at 500 K $\quad \Delta S = \dot{m}C_P\ln\frac{700}{500} = 1\times1\times\ln\frac{700}{500}$

$$= 0.3365$$

(c) Net change in entropy after mixing is

$$(1)+(2) = 0.0852$$

Loss in availability $= m\times T_0$ (Net change in entropy)

$$= (1+1)\,300\times0.0852$$

$$= 2\times300\times0.0852$$

Loss in Availability = 51.12 K

Example 9.7 *500 kJ of heat from a finite temperature source at 1000 K is supplied to 2 kg of gas initially at 2 bar and 350 K in a closed tank. Find the loss in available energy due to the above heat transfer. Take C_V of gas on 0.8 kJ/kg-K and temperature of surrounding as 27°C.*

Solution

With the addition of 500 kJ heat, the temperature of the gas will increase.

Thus $\quad$ Heat added $= \dot{m}_{gas}\times C_V\,(T_2-350)$

where 350 K is the initial temperature

$$500 = 2\times0.8\times(T_2-350)$$

This gives $\quad (T_2-350) = \frac{500}{2\times0.8} = 312.5$

or $\quad T_2 = 350+312.5 = 662.5\text{ K}$

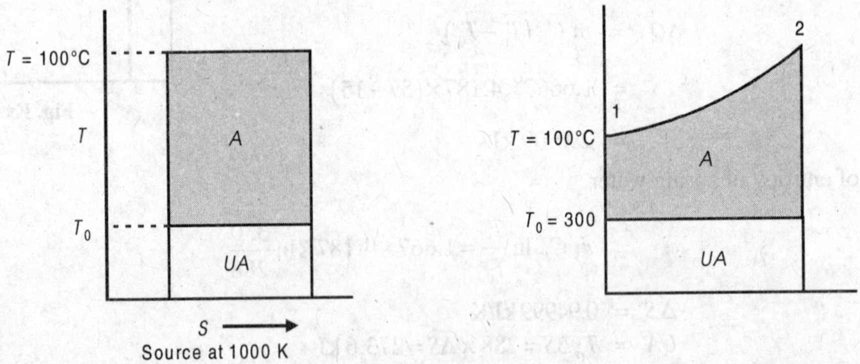

Fig. Ex. 9.7

Available UA Energy with Source

Now $\quad UA = T_0\,\Delta S$

$$T_0 = 27+273 = 300\text{ K}$$

Change in entropy of the source at temperature of 1000 while 500 kJ is taken out is

$$\Delta S = \frac{\Delta Q}{T} = \frac{500}{1000} = 0.5 \text{ kJ/kg-K}$$

$$UA = T_0 \Delta S = 300 \times 0.5$$

$$= 150 \text{ kJ}$$

$$A = Q - UA = 500 - 150$$

$$= 350 \text{ kJ}$$

An UA of the gas

Change in entropy of the gas at constant volume is given by

$$\Delta S = m_g \; C_V \ln \frac{T_2}{T_1}$$

$$= 2 \times 0.8 \times \ln \frac{662.5}{350}$$

$$= 1.0209398$$

$$UA = T_0 \times \Delta S = 300 \times \Delta S = 306.28 \text{ K}$$

$$A = Q - UA$$

$$= 500 - 306.28 = 193.7 \text{ kJ}$$

Loss in Availability due to heat transfer is

$$(1) + (2) = 350 - 193.7$$

$$\mathbf{A = 156.3 \text{ kJ}}$$

Example 9.8 *Calculate the availability of a system which absorbs 1000 kJ of heat from a source at following constant temperature, while the environment temperature all three cases remain 27°C. (a) 600 K ; (b)1000 K ; (c) 1500 K.*

Solution

Now $\quad A = Q - T_0 \Delta S$

where $\quad Q = 1 \text{ H added} = 1000 \text{ kJ}$

ΔS = Change in entropy, it will be different in all three cases.

Entropy change at constant temperature is

$$\Delta S = \int \frac{TQ}{T} = \frac{\Delta Q}{T} = \frac{1000}{T}$$

Thus $\quad A = Q - T_0 \Delta S$

$$= 1000 - 300 \times \frac{1000}{T} = 1000\left(1 - \frac{300}{T}\right)$$

At 600 K $\quad A_{600} = 1000\left(1 - \frac{300}{600}\right) = 1000(1 - 0.5) = 500 \text{ kJ}$

At 1000 K $\quad A_{1000} = 1000\left(1 - \frac{300}{1000}\right) = 700 \text{ kJ}$

At 1500 K $\quad A_{1500} = 1000\left(1 - \frac{300}{1500}\right) = 800 \text{ kJ}$

Thus A increaseas temperature increases.

Example 9.9 *800 kJ of heat is added to a system during vaporisation process occurring at 150°C while pressure remain constant. Determine the available and unavailable energy of the surrounding temperature is 25°C.*

Solution

$$UA = \text{Unavailable Energy} = T_0\,\Delta S$$
$$A = \text{Available Energy} = Q - UA$$

Now

$$Q = 800\,\text{kJ}$$
$$T_0 = 273 + 25 = 298\text{ K}$$
$$T = 273 + 150 = 423$$
$$\Delta S = \int \frac{\delta Q_{rev}}{T}$$
$$= \frac{Q}{T} \text{(since temperature remain constant)}$$
$$UA = T_0 \times \Delta S$$
$$= 298 \times \Delta S = 563.59\text{ kg}$$
$$A = Q - UA = 800 - 563.59$$
$$\mathbf{A = 236.40\ kg}$$

Example 9.10 *20 kg of water at 90°C is mixed with 32 kg of water at 30°C or constant pressure. Calculate the decrease in availability of surrounding air at 27°C.*

Solution

Decrease in Availability $= A)_{20} + A)_{32} - A)_{52}$

i.e., Sum of the original availability of the unmixed 20 kg and 32 kg water less the availbility of the mixture.

Temperature the mixture t_3

$$m_1 C_p t_1)_{20} + m_2 C_p t_2 = m_3 C_p t_3)_{52}$$

or

$$20 \times 90 + 32 \times 30 = 52 \times t_3$$

$\Rightarrow$

$$t_3 = 53°\text{C (Approximately)}$$

Now $T_0 = 300$ K; $T_1 = 273 + 90 = 363$ K, $T_2 = 300$ and $T_3 = 273 + 53 = 326$ K

We have to find net change in entropy. Resulting due to 20 kg water at 363 K after mixing getting lower temperature 326

$$\Delta S_{20} = m_{20} C_P \ln\frac{363}{326} = 20 \times 4.187 \times \ln\frac{363}{326}$$
$$= 9.0025 \quad \text{...(1)}$$

Similarly for 32 kg work at 303

$$\Delta S_{32} = m_{32} C_P \ln\frac{303}{326} = -9.8028 \quad \text{...(2)}$$

$$\text{Net change in entropy} = (1) + (2)$$
$$= -0.80037$$
$$\text{Loss in Availability} = T_0\,\Delta S = 300 \times 0.80037$$
$$\textbf{Loss in Availability} = \mathbf{240.112\ kJ}$$

On mixing the water at 90°C (363 K)

Example 9.11 *From a fire box of a boiler 1000 kg of heat leave hot gases at 1400°C and is given to steam at 250°C. Atmospheric temperature is 20°C. Divide the energy into available and unavailable portions (a) As it leave the hot gas (b) As it reaches the steam (c) Also show that the equation that*

Increase in unavailability = $T_0\ \Delta S$ holds good.

Solution

Now
$$T_0 = 273+20=293\text{ K}$$
$$T_2 = 273+250=523\text{ K}$$
$$T_1 = 273+1400=1673\text{ K}$$

Case (*a*) Hot Gas at 1673 K

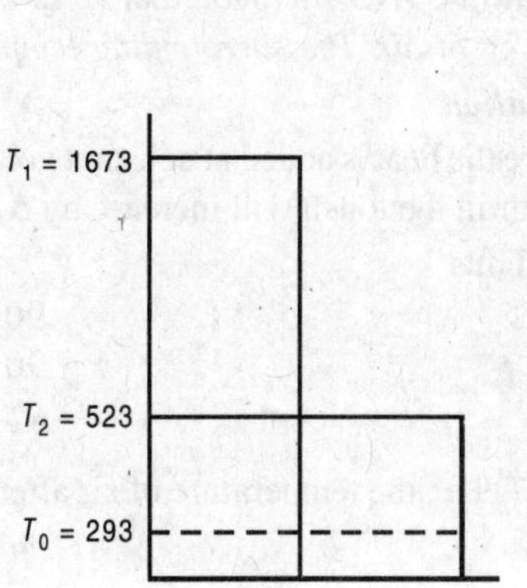

Fig. Ex. 9.11

$$UA = T_0\,\Delta S$$

Since heat is taken out at constant temperature

$$\Delta S_{\text{Gas}} = \frac{Q}{T}=\frac{1000}{1673}\text{ kJ/kg-K}$$

$$UA = 293\times\frac{1000}{1673}=175.13\text{ kJ}$$

$$A = -1000-(-UA)=-175.13$$
$$= -824.87\text{ kg}$$

(Heat is leaving the system)

Case (*b*) Heat reaching steam at 523 K

$$\Delta S_{\text{steam}} = \frac{Q}{T}=\frac{1000}{523}$$

$$UA = T_0\,\Delta S = 293\times\frac{1000}{523}=560.23\text{ kJ}$$

$$A = Q-T_0\Delta S=1000-560.23=439.77\text{ kJ}$$

(c) To prove that increase in $UA = T_0\,\Delta S$, we have to find

$$\Delta S_{\text{total}} = \Delta S_{\text{gas}}+\Delta S_{\text{steam}}$$

$$= -\frac{1000}{1673}+\frac{1000}{523}=1.3143$$

$$T_0\,\Delta S = 293\times1.3143=385.1$$

Increase in unavailability $= (1)+(2)$

$$= 175.13+560.23$$
$$= 385.1$$

Since (3) = (4), hence proved.

Example 9.12 *900 kJ of heat is added to a system during vaporisation process occurring at 200°C and at a constant pressure. Determine the available energy if the surrounding are at 17°C.*

Solution

$$\text{Available } A = Q-T_0\,\Delta S$$

where
$$Q = \text{Heat added}=900\text{ kJ}$$
$$T_0 = 17+273=290\text{ K}$$
$$\Delta S = \text{Entropy changes}$$
$$= \frac{\Delta Q}{T}\text{ since temperature is constant}$$
$$= \frac{900}{273+200}=\frac{900}{473}\text{ kJ / kg-K}$$

$$\therefore \quad A = 900-290\times\frac{900}{473}$$
$$= 900-551.797$$
$$\mathbf{A = 348.203\text{ kJ}}$$

Example 9.13 *An amount of 90 kJ of heat is added to a touch of air that is originally at 26°C. The touch contains 2.8 kg of air. The surrounding temperature is 15°C. Find available and unavailable energy.*

Solution

Here the heat is added at constant volume. C_V for air = 0.718 kJ/kg-K with the addition of 90 kJ heat, the temperature of air in the touch will increase by ΔT

Thus

$$90 = mC_V \Delta T$$

or

$$90 = 2.8 \times 0.718 \, \Delta T$$

$$\Delta T = 44.76$$

T_2 i.e., the temperature of air after heat addition

$$T_2 = 26 + 44.76 = 70.76$$
$$= 70.76 + 273 = 343.76$$
$$\cong 444 \text{ K}$$

Now change of entropy ΔS for constant volume increase

$$\Delta S = C_V \ln \frac{T_2}{T_0}$$

$$= 0.718 \ln \frac{444}{273+15}$$

$$\mathbf{\Delta S = -0.31079}$$

Example 9.14 *A temperature of 2000°C is obtained in a furnace by burning fuel in air at atmosphere pressure and ambient tempeature of 27°C. The gases can be assumed to be perfect gas with C_p = 1.0 kJ/kg-K. Determine the availability of heat in the products of combustion.*

Solution

$$T_0 = 27 + 273 = 300 \text{ K}$$

For unit mass of gas –

(a) Heat addition $\quad Q = mC_p \Delta T$

$$= 1 \times 1.00 \times (2000 - 27)$$
$$= 1 \times 1983 = 1983 \text{ kJ/kg}$$

(b) We have now to find change is entropy during the process

$$\Delta S = mC_P \ln \frac{T_1}{T_2}$$

$$= 1 \times 1.0 \ln \frac{2000+273}{27+273}$$

$$= \ln \frac{2773}{300} = 2.025 \text{ kJ/kg-K}$$

(c) $\quad UA = T_0 \Delta S = 300$

$$= 300 \times 2.025 = 607.522 \text{ kJ/kg}$$

(d) $\quad A = Q - UA = 1983 - 607.522$

$$= 1375.478 \text{ kJ/kg}$$

THEORY QUESTIONS

9.1. Explain
(a) Dead state (b) Availability
(c) Unavailable energy

9.2. Write short notes on:
(a) Irreversibility (b) Effectiveness
(c) Second law efficiency

9.3. Comment on 'There is no energy crisis but available energy crisis'.

9.4. Derive expressions' for available energies for heat source
(a) At constant temperature
(b) At variable temperature

PROBLEMS FOR PRACTICE

9.1. Determine available and unavailable apart of energy of a system which absorbs 15 MJ of heat from a source at 227°C. The surroundings are at 17°C.

(*Ans.* $UA = 8.7$ MJ, $A = 6.3$ MJ)

9.2. A source at 227°C receives 7200 kJ/min from a source at a temperature of 727°C. The atmospheric temperature is 27°C. Assuming that the temperature of the system and that of the source remain constant, find the decrease in availability after the heat transfer. (*Ans.* 2160 kJ)

9.3. A steel ball of 800 kg at 1250 K is cooled to 500 K. Using steel ball as source of energy, compute available and unavailable energies. Assume specific heat of steel to be 0.5 kJ/kg-K and ambient temperature to be 27°C.

(*Ans.* $UA = 110$ MJ, $A = 190$ MJ)

9.4. Air 0.2 kg initially at a temperature of 575 K receives 300 kJ heat reversibly at constant pressure. Find available and unavailable portions of the heat added. Take C_p for air to be 1.005 kJ/kg K and ambient temperature to be 27°C.

(*Ans.* $UA = 77.16$ kJ, $A = 222.54$ kJ)

9.5. During a vaporization process at constant pressure taking place at 180 C, 850 kJ of heat is added while ambient temperature is 299 K. Find available and unavailable energies.

(*Ans.* $UA = 55.53$, $A = 296.46$ kJ)

9.6. A steam stable condenses at 1000° C rejecting 3000 kJ of heat while another sample of heat condenses at 500°C rejecting the same quantity of heat. Determine the difference in the available energy between the two processes. Take sink temperature to be 27°C. (*Ans.* $A = 2840$ kJ, $UA = 1160$ kJ)

9.7. A constant temperature source is maintained at 1000 K and the surroundings are at 290 K. If 4000 kJ heat is transferred from the source reversibly, determine available and unavailable portions of heat.

(*Ans.* $A = 2840$ kJ, $UA = 1160$ kJ)

9.8. From a heat reservoir 1000 kJ of heat is withdrawn at 600 K. Find the availability and unavailability of heat if surroundings are at 290 K.

(*Ans.* $A = 516.7$ kJ, $UA = 483.3$ kJ)

9.9. 500 kJ of heat was removed at constant temperature at 527°C and supplied a system at constant temperature at 427°C. Find the loss in available energy as a result of this irreversible heat transfer. Take ambient temperature as 27°C.

(*Ans.* 26.79 kJ)

IMPORTANT POINTS TO BE REMEMBERED

1. *Irreversibility (I)*

$$I = W_{rev} - W_{actual} = \text{Lost work}$$

2. *Effectiveness*

$$\varepsilon = \frac{\delta w}{\delta w_{rev}}$$

3. *Second law efficiency*

$$\eta_{II} = \eta_{th}/\eta_{rev} \text{ for heat engines (Power producing)}$$
$$= W/W_{rev}$$
$$= W_{rev}/W \text{ for power absorbing devices}$$
$$= \frac{COP}{COP_{rev}} \text{ for refrigerators and heat pump}$$

4. *Infinite heat reservoir—Availability*

$$W_{max} = Q - T_o\,\Delta S$$

5. *Finite heat reservoir —Availability*

$$W_{max} = Q - T_o\,\Delta S$$

PART–2

CHAPTER 10

Steam Generation and Properties, Processes of Steam

10.1 INTRODUCTION

Upto now we have considered a fluid, which under normal conditions remains in a gaseous state. Such substances are oxygen, nitrogen, air etc. Now we consider fluids, which under normal working conditions may be either in a liquid state or in a gaseous state. Such substances are known as vapours. Examples of vapours, employed in engineering practices are steam, ammonia, freon, mercury etc. The behaviour of vapours is typically represented by steam and water.

With vapours, the simple laws of perfect gas does not even approximately apply and therefore sometimes the definition of vapour is given as: "A vapour is any substance in the gaseous state which does not even approximately follow the general gas laws". The reason for this departure of the performance of vapours from the general gas law is that the vapours from the general gas law is that the vapours when heated or cooled experience some degeneration work while the general gas laws are based on a perfect gas, which would experience no degeneration work when heated or cooled. It should be remembered that all known actual gases do experience a small amount of degeneration work. So we can use the formula derived for perfect gas, for vapour calculations. For this purpose we use values, which have been determined by experiments and listed in tables.

Highly superheated vapours approximately follow the perfect gas laws. Vapour obeys its own Laws.

10.2 FORMS OF MATTER

There are three forms of matter, viz. (i) Solid, (ii) Liquid and (iii) Gaseous.

In order to explain the changes in form of matter, a hypothetical experiment may be described with the help of Fig. 10.1.

As substance in the solid state is placed in a cylinder, which is fitted with a frictionless right fitting piston being loaded by a weight as shown in Fig. 10.1. Heat is supplied to the cylinder. The substance may change in volume during the heating process but the pressure is kept constant by the piston and weights on it. The change in temperature of the substance will be recorded by a thermometer (Not shown).

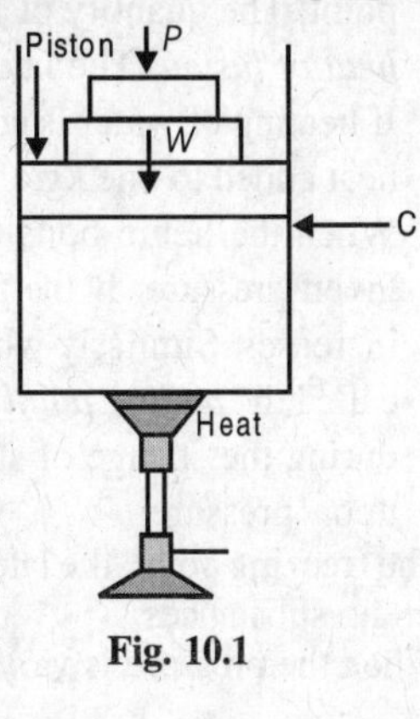

Fig. 10.1

At first, when the heat is added, the temperature of the solid rises and a certain value of temperature will be reached after which a considerable amount of heat is added, without a change in temperature. Through a sight window provided in the cylinder wall it can be seen that the solid is melting. When the solid has been completely transformed into liquid, further addition of heat again causes a rise in temperature. This rise in temperature continues till a point is reached where further addition of heat does not change the temperature of the substance but on observation we see that the liquid is boiling. When the liquid has been completely transformed into a vapour, the temperature again rises when the heat is added. The relation between the heat added, entropy change and the temperature of the substance is shown in Fig. 10.2.

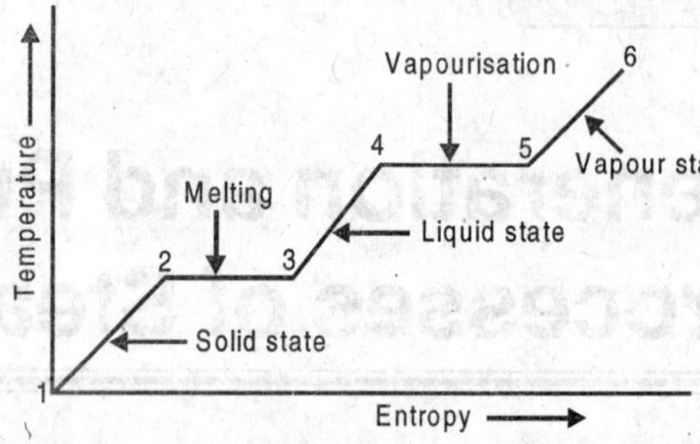

Fig. 10.2

1–2 Shows the stage during which the rise in temperature of a substance in solid state takes place.

2–3 Shows the stage during which the substance is being transformed from a solid state to a liquid state without rise in temperature.

3–4 Shows the stage during which rise in temperature of the substance in liquid state takes place.

4–5 Shows the stage during which the substance is being transformed from a liquid state to a vapour state without rise in temperature.

5–6 Shows the stage, during which rise in temperature of the substance may take place in a vapour state.

If the hot vapour is cooled the curve will be exactly retraced with all changes taking place in reverse manner. If the weight *W* is increased and the experiment is repeated the same general behaviour is obtained but the temperature will be higher than that in the first case. If the weight is reduced still the behaviour will be the same but temperature will be lower than that in first case.

The above hypothetical experiment brings out certain important facts.

- When a solid melts the temperature remains constant and is always the same for a given substance at a given pressure. When a liquid freezes the temperature remains constant. This temperature is called the freezing point. The quantity of heat removed or added during the change of state is constant and is known as *latent heat of fusion*. The latent heat of fusion depends upon pressure.
- If heating of water is continued after total melting, temperature increases till it starts boiling. The amount of heat added to one kg of water from freezing point to boiling point is called *sensible heat*.
- When the liquid boils the temperature remains constant and is always the same for a given substance at a given pressure. If the pressure is increased, the temperature at which the change of state takes place also increases. Similarly when the vapour condenses the temperature also remains constant. This temperature is called the *boiling point* or saturated temperature of the liquid. The definite quantity of heat removed or added during the change of state is called the *latent heat of vaporisation*. The latent heat of vaporisation depends upon pressure.

The freezing point, the latent heat of fusion, the boiling point and the latent heat of vaporisation vary widely with different substances.

When the pressure is vary low, an interesting phenomenon is noticed. A solid when heated directly transforms into a vapour without passing through an intermediate liquid state. Such a phenomenon is known as *sublimation.* A familiar example is the conversion of solid CO_2 (dry ice) directly into vapour. Figure 10.3 shows a temperature pressure curve.

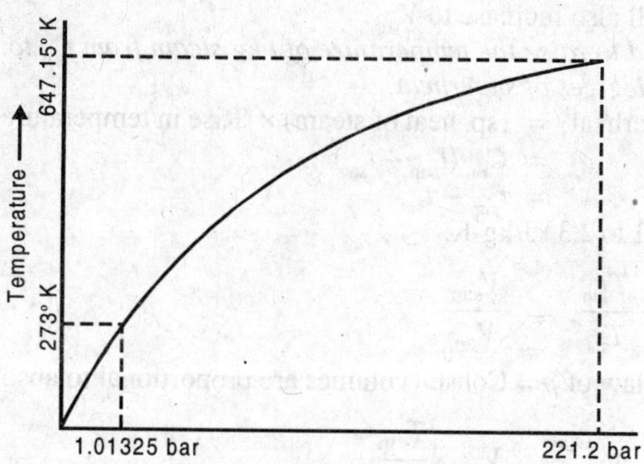

Fig. 10.3

In practice, steam is generated in a device called steam boiler.

10.3 GENERATION OF STEAM AT A GIVEN PRESSURE

Consider one kg of water at 0°C. Also let us decide upon the pressure under which one kg of water is to be heated. Consider a cylinder with a frictionless piston loaded such that the pressure P under consideration is acting on water. See Fig. 10.1.

Heat is added to the water so that its temperature increases from 0°C to t_{sat} (saturation temperature) at which the water boils at pressure P. Volume of water increases to V_f. This is shown in Fig. 10.4(a).

Heat supplied or added to 1 kg water to raise its temperature from 0°C to saturation temperature t_{sat} *is called sensible heat* because it can be detected by the sense of touch and produces a rise in temperature to be seen on a thermometer. Generally this sensible heat is denoted by h_f.

Note. V_f means volume of fluid which is taken as liquid.

Heating is continued after the water reaches saturation temperature. Evaporation of water takes place and steam is generated at pressure P. This heating is continued till all the water is evaporated and the temperature of the steam is saturation temperature. V_g or V_{sat} denotes the volume of steam at this condition. The external work done during evaporation is given by $W = P(V_{sat} - V_f)$ or $p(V_g - V_f)$ with proper units.

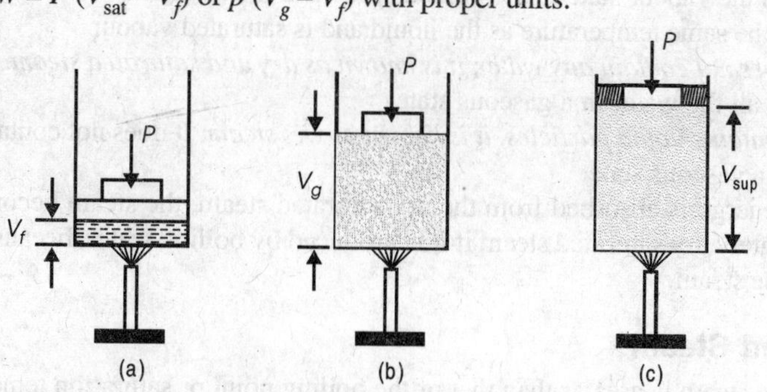

Fig. 10.4

Because the heat during this stage of evaporation cannot be recorded by a rise in temperature on the thermometer, it is called Latent heat (hidden heat).

The amount of heat required to evaporate 1 kg of water at saturation temperature to steam or vapour at saturation temperature is called Latent heat or enthalpy of evaporation and is generally denoted by h_{fg}.

This is shown in Fig. 10.4(b).

Once all the water is evaporated to saturated steam, if heat is added further, the steam behaves approximately as an ideal gas and the heating of saturated steam increases its temperature. *The steam whose temperature is greater than saturation temperature at the given pressure, is called a superheated steam*; and let its temperature be denoted by t_{sup}. Its volume will also increase to V_{sup}.

The amount of heat required to raise the temperature of l kg steam from t_{sat} to t_{sup} is called superheat and the temperature difference called *degrees of superheat.*

$$\therefore \quad \text{Superheat} = (\text{sp. heat of steam}) \times (\text{Rise in temperature})$$

$$= C_{pv} (t_{sup} - t_{sat})$$

$$\text{Degree of superheat} = t_{sup} - t_{sat}$$

The value of C_{pv} varies from 2.1 to 2.3 kJ/kg-K.

Similarly,

$$\frac{T_{sup}}{T_{sat}} = \frac{V_{sup}}{V_{sat}}$$

This is according to Charles's law of p = Const. (volumes are proportional to absolute temperatures)

$$\therefore \quad V_{sup} = V_{sat} \times \frac{T_{sup}}{T_{sat}}$$

This is shown in Fig. 10.4(c).

Note: V_{sat} and V_g also means volume of gas which is in this case, steam

where

$$h_f = \text{sensible heat} = C_{pw} (t_{sat} - 0) \text{ kJ/kg}$$

$$C_{pw} = \text{specific heat of water which is taken as 4.187 kJ/kg-K if not given.}$$

$$h_g = h_f + h_{fg} \text{ for 1 kg of dry and saturated steam.}$$

Total heat or enthalpy of superheated steam

$$h = h_g + C_{pv} (t_{sup} - t_{sat})$$

10.4 CONDITIONS OF STEAM

Steam may occur in any one of the following three conditions:

(a) Saturated steam, which may be either dry or wet, (b) Superheated steam and

(c) Supersaturated steam

10.4.1 Saturated Steam

Saturated steam is a vapour at the temperature corresponding to the boiling point of the liquid at the given pressure. A substance, which is in the vapour state in a confined space and is in contact with some of the same substance in liquid state is always at the same temperature as the liquid and is saturated vapour.

If saturated steam does not contain any water, it is known as dry and saturated steam. It contains just sufficient heat energy to maintain all the water in a gaseous state.

If saturated steam contains liquid particles, it is known as wet steam. It does not contains sufficient heat energy to maintain all water in a gaseous state.

If some of the heat energy is absorbed from the dry saturated steam, the steam becomes wet. In practice it is difficult to get an absolutely dry saturated steam if it is produced by boiling water, because some of water particles are carried away with the steam.

10.4.2 Superheated Steam

If the temperature of the steam is greater than that of the boiling point or saturation temperature corresponding to the pressure of steam generation, the steam is known as superheated steam.

The amount of superheat in the steam is given in terms of the difference between its temperature and that of the saturated steam at the pressure of steam generation. If the temperature of the steam is 100 degrees higher than the saturation temperature corresponding to the pressure of steam generation, we say that the steam has 100 degrees of superheat.

10.4.3 Supersaturated Steam

Supersaturated steam at a particular saturation pressure has temperature less and density greater than the corresponding values given in the steam tables. This condition of steam is obtained when it is cooled by its own expansion until it contains less heat energy than the saturated steam under the same conditions. This state of steam is varying, unstable and the steam soon returns to the saturation condition. Such a state of steam occurs in expansion in a nozzle.

10.5 PROPERTIES OF STEAM

The properties of steam are interrelated. If we know certain properties, the other properties may be found out or derived. For example, if the pressure of saturated steam is observed by the pressure gauge, its temperature can be found from the steam tables in which the results of the various experiments have been tabulated.

(a) Dryness Fraction of Saturated Steam. We have seen that steam in contact with water contains liquid particles in suspension. Thus the steam consists of dry saturated and water particles in suspension.

The dryness fraction of steam is defined as the ratio of the weight of dry steam in a certain quantity of steam to the weight of total wet steam. It is generally denoted by the letter x. If W is the weight of dry steam in the total steam W_t, then $x = \dfrac{W_s}{W_t}$.

The dryness fraction of a wet steam may also be defined as the amount of dry steam in unit amount of the steam. When $x = 1$, the steam is said to be dry and saturated.

(b) Sensible heat (h_f): Refer Sec.10.3

(c) Latent heat or enthalpy of evaporation h_{fg} Refer Sec. 10.3

(d) Superheat and degrees of superheat: Refer Sec. 10.3

(e) Volume of wet steam: $V = xV_g$

(f) Volume of dry and saturated steam: $V = V_g$

(g) Volume of superheated steam: $V_{sup} = V_{sat} \times \dfrac{T_{sup}}{T_{sat}} = V_g \times \dfrac{T_{sup}}{T_{sat}}$

10.6 STEAM TABLES

Experimentally, accurately determined, values are tabulated in tables that are called as Steam tables. These values form the basis for many calculations concerned with steam engineering. These tables are to be used because vapours do not obey general gas laws. The values given in the tables are for one kg of dry and saturated steam but these values can also be employed for wet steam calculations. Properties of dry and saturated steam and superheated steam at various pressures and temperatures for SI units are given at the end of this book.

In order to determine the properties of steam at some pressure between those given in tables, we interpolate assuming the linear relation between these values. The method is very simple and at the same time accurate.

Steam Table: *Properties of Dry and Saturated Steam.*

Pressure bar	*Sat. temp. °C*	*Volume m³/kg*		*Enthalpy kJ/kg K*			*Entropy kJ/kg K*		
		V_f	V_g	H_f	H_{fg}	H_g	S_f	S_{fg}	S_g
10.0	179.88	0.0011274	0.19429	762.61	2013.6	2776.2	2.1382	4.4443	6.5828
10.5	182.02	0.0011303	0.18545	772.03	2005.9	2778.0	2.1588	4.4071	6.5659

Method of linear interpolation to find, say, enthalpy at 10.2 bar pressure.

Difference between two values at 10 and 10.5 is 772.03 – 762.61= 009.42 kJ / kg.

This is the difference for 0.5 bar pressure.

$\therefore$ For 0.2 bar difference, value is $\left(\frac{0.2}{0.5} \times 9.42\right) = 3.768$ kJ/kg.

Now, this is to be added to the value at 10 bar.

(This addition should be done if the value is increasing with pressure. If the value is decreasing, it should be deducted. This is seen in case of volumes.)

$\therefore$ Enthalpy at 10.2 bar pressure

$$= 762.61 + 3.768$$
$$= 766.378 \text{ kJ/kg.}$$

10.7 INTERNAL ENERGY AND ENTROPY OF STEAM

10.7.1 Internal Energy of Steams

As in case of ideal gases, internal energy of steam is given by ($u = h - pV$).

$\therefore$ For wet steam $\quad u = (h_f + xh_{fg}) - p.xVg$

For dry and saturated steam $u = h_g - p.V_g$

and for superheated steam $u = [h_g + C_{pv}(t_{sup} - t_{sat})] - \left[p.V_g . \frac{T_{sup}}{T_{sat}}\right]$

10.7.2 Entropy of Steam

The change of entropy during any reversible process is defined as the ratio of heat added or rejected to the absolute temperature at which the heat is added or rejected.

We can apply the definition for getting entropy of steam above 0°C.

Consider one kg of water at 0°C and pressure p. Let heat be added so that the temperature increases to saturation temperature corresponding to pressure p. Here the temperature changes. Let during the heating, process, T be the temperature and an infinitesimal heat δQ is added and for the process 1–2; where the temperature changes from 0°C (273K) to T_{sat} (saturation temperature at p).

$$\therefore \quad \delta Q = Cp_w dT \text{ and } dS = \frac{\delta Q}{T}$$

$$\therefore \quad dS = \frac{Cp_w \cdot dT}{T}$$

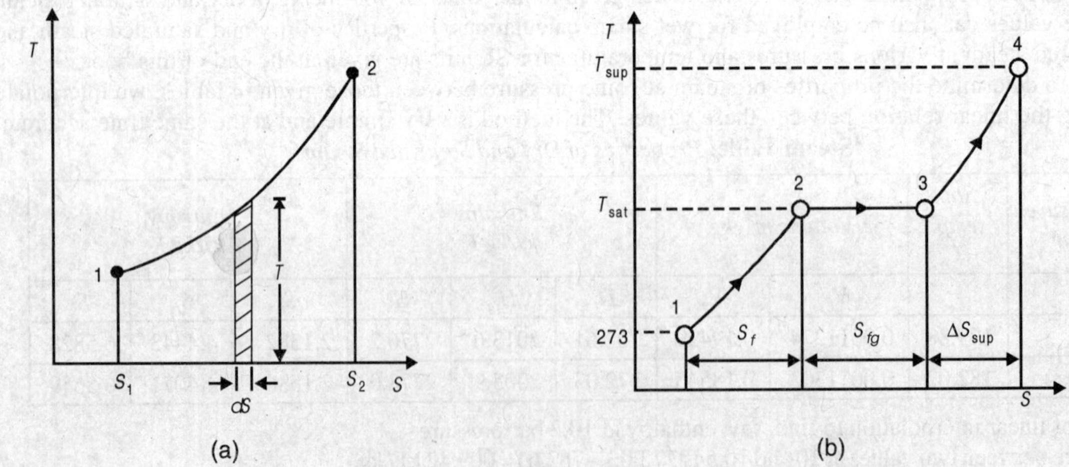

Fig. 10.5

∴ Change of entropy during sensible heating of water from 273 K to T_{sat},

$$\int_1^2 dS = S_2 - S_1 = \Delta S_{1-2} = C_{p_w} \int_{273}^{T_{sat}} \frac{dT}{T}$$

This in denoted by s_f and is given in steam tables.

$$\therefore \quad s_f = C_{p_w} \ln . \frac{T_{sat}}{273}$$

Water, at saturation temperature, if heated further, evaporation takes place. Here the heat added is latent heat or enthalpy of evaporation and temperature is saturation temperature; heat added is h_{fg}, and the temperature is constant

∴ Change of entropy during evaporation 2–3 is given by

$$\Delta S_{2-3} = \frac{h_{fg}}{T_{sat}} = s_{fg} \text{ and is given in steam tables.}$$

and the entropy of dry and saturated steam is given by

$$\Delta S_{1-3} = \left[C_{p_w} \ln \frac{T_{sat}}{273} + \frac{h_{fg}}{T_{sat}} \right] = s_g.$$

It is given in steam tables.

Similarly, if the steam is further heated so that its temperature is T_{sup} (temperature of superheated steam), then the entropy change during superheating is given by

$$\Delta S_{3-4} = Cp_v \ln . \frac{T_{sup}}{T_{sat}}$$

∴ Total change of entropy of superheated steam or simply entropy of superheated steam is given by

$$\Delta S_{1-4} = S = C_{p_w} \ln . \frac{T_{sat}}{273} + \frac{h_{fg}}{T_{sat}} + Cp_v \ln . \frac{T_{sup}}{T_{sat}}.$$

$$= S_f + S_{fg} + C_p \ln . \frac{T_{sup}}{T_{sat}}.$$

$$= S_g + C_{pv} \ln \frac{T_{sup}}{T_{sat}}$$

If the steam is wet having a dryness fraction x, then entropy of steam is given by

$$S = S_f + x . S_{fg}.$$

Temperature – Entropy diagram (T–S diagram) is shown in Fig. 10.5.

10.8 PHASE DIAGRAM

Pure substance is a substance consisting only of a single molecular species such as He, H_2, H_2O or NH_3, phases of these substances being different namely solid, liquid and gaseous.

We have considered the phase change for the matter in Secs 10.2 and 10.3.

We will consider a pressure temperature co-ordinates to study the constant pressure process discussed in Secs 10.2 and 10.3.

Figure 10.3 shows such a diagram where the phase changes are shown. Such a diagram is called a phase diagram for pure substance.

For a particular pressure, the horizontal line *a, b,c,d,e* is drawn as shown in Fig.10.6.

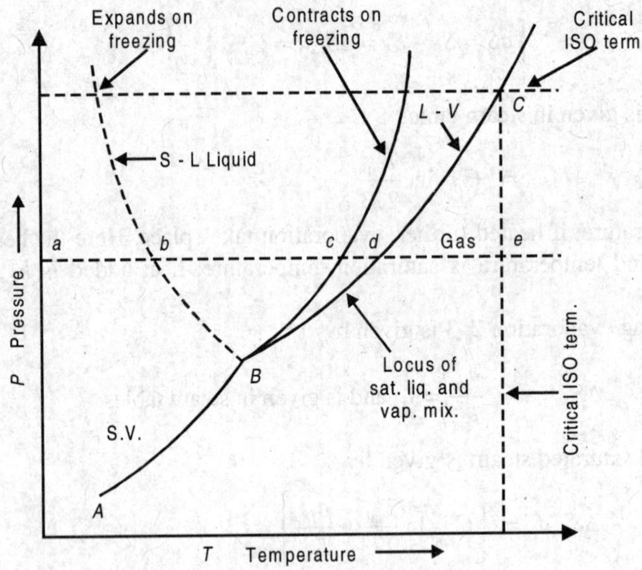

Fig. 10.6

a = Solid state
b = Melting point of ice
c = Boiling or saturation temperature point, *C* does not apply to substance that expand on freezing
d = dry and saturation point
de = Superheated vapour
bd = Liquid vapour region

This process is repeated at numerous pressure and the point analogous to *b* and *d* duly noted, say down to point *B*. A smooth curve through all the *d* points is the phase boundary between liquid and vapour; it is the locus of all points representing saturated mixtures. The smooth curve through all points *b* is the phase boundary for solid and liquid; points on this line represent all equilibrium mixtures of solid and liquid.

The states *B* and *C* in Fig. 10.6 are particularly significant. Point *B*, the junction of the three phase boundaries, is called the triple point because it represents all equilibrium mixtures of solid, liquid and vapour.

The critical point *C* where the curve *BC* ends is a state where liquid and vapour are indistinguishable. As this point is approached the latent heat of evaporation decreases. At *C* it is zero. The temperature T_C here is called the critical temperature and the corresponding pressure is the critical pressure (P_C). The specific volume at P_C and T_C is the critical volume.

Density of water and vapour at the point *C* is same.

For water,

$$P_C = 221.2 \text{ bar}$$
$$T_C = 374.15°\text{C}$$
$$V_C = 0.00317 \text{ m}^3/\text{kg}$$

10.8.1 Triple Point

We know that to determine the temperature of any system, the thermometer must be brought to thermal equilibrium with an arbitrarily chosen standard system in an easily reducible state. The temperature of the standard system in the chosen state is called the fixed point. Before 1954 there were two fixed points : (i) the ice point and (ii) steam point.

The use of two fixed points was found unsatisfactory, partly because of the difficulty of achieving equilibrium between pure ice and air-saturated water. When ice melts, it sounds itself with pure water and thus prevents intimate contact of ice and air saturated water. Also the steam point is extremely sensitive to a change in pressure.

The temperature scale in use since 1954 is based on one fixed point only. This is the temperature at which ice, liquid water, and water vapour coexist in equilibrium, a state known as the triple point of water.

The temperature of the triple point of water is the standard fixed point of thermometry.

The Celsius temperature of the triple point of water is 0.01 °C and the pressure at triple point for water is 0.006114 bar.

10.9 THE *P–V* AND *T–S* DIAGRAMS FOR H_2O (VAPOURS)

10.9.1 P–V Diagram for Saturated Liquid and Vapour

The saturated water curve (liquid line) is obtained by plotting: *P* and V_f values form the steam tables on *P–V* co-ordinates and $T–S_f$ values on *T–S* coordinates. Similarly saturated vapour (steam) line or curves are obtained from steam tables.

(i) 221.2 bar, (ii) 374.15°C, (iii) $V_C = 0.00317\ m^3/kg$, (iv) $V_b = 0{,}0014526\ m^3/kg$, (v) $V_a = 0.0014526\ m^3/kg$.

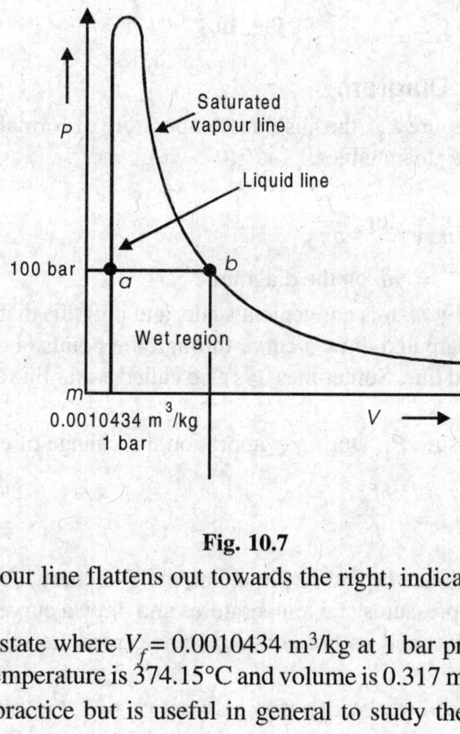

Fig. 10.7

Notice here that saturated vapour line flattens out towards the right, indicating that the volume increases at an increasing rate at low pressure.

In Fig. 10.7, *m* represents the state where $V_f = 0.0010434\ m^3/kg$ at 1 bar pressure. *C* is the critical point where pressure is 221.2 bar, saturation temperature is 374.15°C and volume is 0.317 m^3/kg.

This diagram is not used in practice but is useful in general to study the property relations and the vapour behaviour.

10.9.2 The Temperature–Entropy Diagram for Steam

It is possible to plot on a Temperature Entropy diagram, a chart which gives the entropy of water and steam at all temperatures and pressures. Such a chart is very convenient for solving problems containing an adiabatic (reversible) expansion of steam. A Temperature Entropy Chart (*T–S* Chart) is plotted in Fig. 10.8 with co-ordinates representing the absolute temperature and entropy. The units of entropy are reckoned from the freezing point of water.

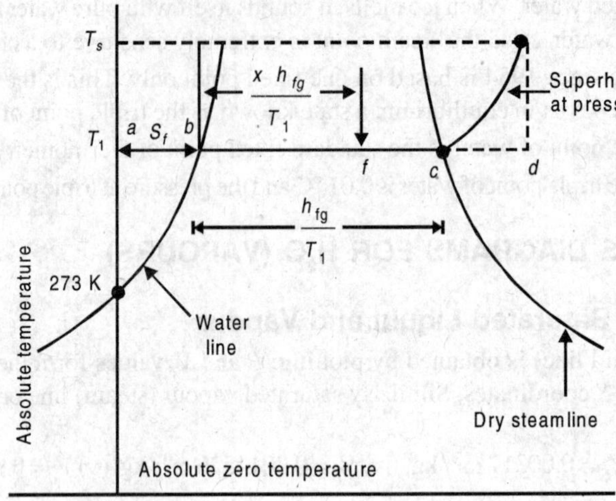

Fig. 10.8

10.9.3 Construction of T–S Diagram

Consider 1 kg steam at absolute pressure P_1, the absolute temperature of formation at this pressure will be T_1; the value of T_1 can be obtained from the steam tables.

Then
$$S_{f1} = \log_e \frac{T_1}{273}$$
$$= ab \text{ on the diagram.}$$

Let this value of S_{f1} be represent by *ab* to a convenient scale, and plot this distance on *T–S* diagram of Fig. 10.8. Repeat this for other pressure of steam and draw a curve through the points obtained. This curve is known as the saturation liquid line or simply liquid line. Sometimes it is also called water line and will give the values of S_f at all temperatures and pressure.

Again consider the steam at pressure P_1. During evaporation the change of entropy of this steam will be

$$S_{fg} = \frac{hf_g}{T_1}$$

Let this be represented by the distance *bc* and *be* plotted on the *T–S* chart. The point *C* will represent dry steam at pressure P_1. Repeat this for other pressures and temperatures and draw a curve through the points obtained. This curve is known as the dry steam line, and its horizontal distance from the vertical ordinate represents the total entropy of steam at any pressure.

If the steam at pressure P_1 had been wet, with a dryness fraction of *x*, the entropy of evaporation would be

$$\frac{x.hf_g}{T_1}$$

Let this be represented by *bf* and *be* plotted to the same scale. Then *af* is the total entropy of the wet steam, and the dryness fraction is the ratio of *bf* to *bc*.

$$\therefore \quad x = \frac{bf}{bc}$$

Again, consider 1 kg of dry steam at condition *c*. Let steam be now superheated by increasing the absolute temperature from T_1 to T_S. Then, increase in entropy due to superheating is

$$= C_{p_w} \log_e \frac{T_s}{T_1}$$

$$= C_{p_w} \log_e \frac{T_{sup}}{T_{sat}}$$

Let this increase of entropy be represented by the horizontal distance *cd*. Draw a horizontal through T_S and a vertical through point *d* let these lines intersect at *e*. Then the point *e* represents the condition of the superheated steam at pressure P_1. Repeat this for other valves of T'_s the temperature T_1 remaining the same, as the pressure is constant.

Draw the curve *ce* through the points so obtained. Then this curve represents the entropy of superheated steam at pressure P_1, for all temperatures for superheat. Actually, the curve *ce* is a constant pressure line. Similar constant pressure lines can be drawn for superheated steam at other pressures.

It should be noticed that the area of the chart between the water line and the dry steam line represents wet steam. The area of the chart to the right of the dry steam represents superheated steam, and is known as the area of superheat, or superheat region.

Thus a complete *T–S* diagram for pressure P_1 is shown in Fig. 10.9.

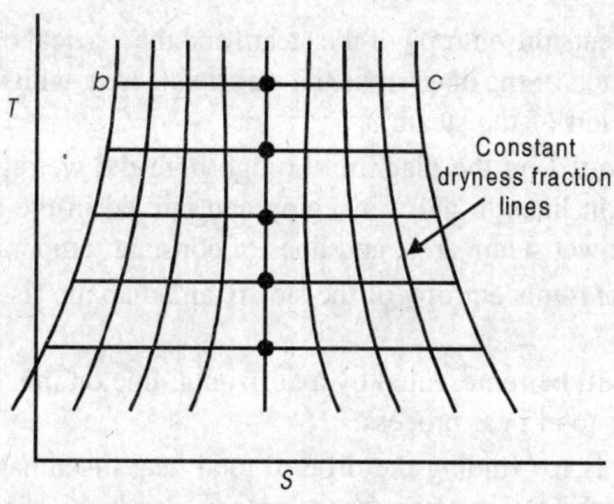

Fig. 10.9

In this diagram, the evaporation lines *bc* at all pressure can be divided into ten equal parts. Curves can then be drawn through each corresponding tenth part of these lines. These curves represent the dryness fraction of the steam at any temperature and pressure. These curves or lines are known as constant dryness fraction lines. Thus, any condition of steam can be represented by a corresponding point on the *T–S* diagram. Such a point is known as state point.

Dryness fraction lines are shown on *T–S* diagram of Fig. 10.9.

10.10 MOLLIER DIAGRAM OR TOTAL–HEAT–ENTROPY OR ENTHALPY–ENTROPY DIAGRAM (h–s DIAGRAM)

Another type of entropy diagram which is used by engineers is the Mollier diagram or Enthalpy – Entropy (*h–s*) diagram. This diagram represents the entropy and total heat or enthalpy of steam. It is shown in Fig. 10.10.

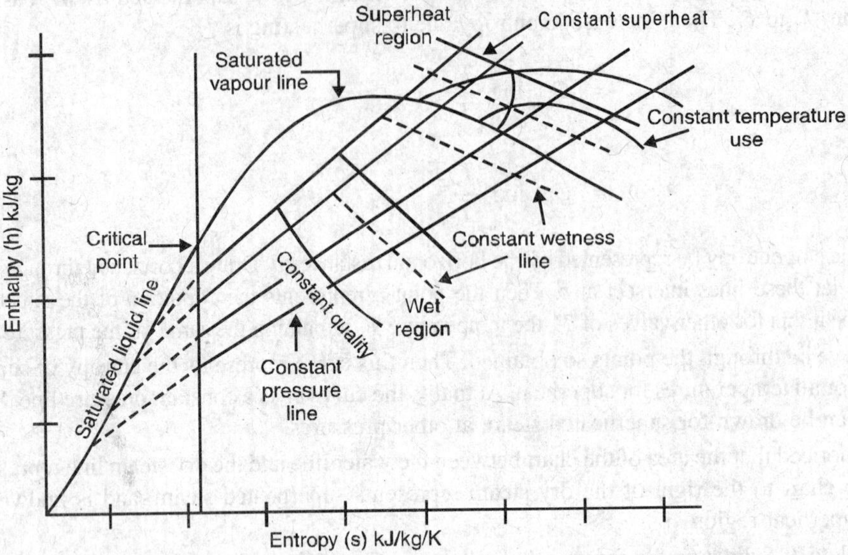

Fig. 10.10

The base of the diagram represents the entropy of the steam and the vertical ordinate represents the total heat or enthalpy of steam. The lower portion of the diagram is the wet steam area, whilst the upper portion of the diagram represents the superheated condition of the steam.

Constant pressure lines are plotted on the diagram through both the wet steam area and the superheat area. Constant pressure lines are straight lines in the wet region and curved in the superheat area. Lines of constant dryness fraction are plotted in the wet steam area, and lines of constant temperature in the superheat area.

As the base of the diagram represents entropy of the steam, an adiabatic (isentropic) expansion will be represented by a vertical line.

Again a throttling expansion will be represented by a horizontal line on this diagram, as throttling process is a constant enthalpy (isenthalpic) or total heat process.

The main use of this diagram is for finding the drop in total heat or enthalpy of steam during an isentropic expansion. The final condition of the steam after expansion can also be read off from the diagram. The drop in enthalpy during an adiabatic or isentropic expansion. The final condition of the steam after expansion can also be read off from the diagram. The drop in enthalpy during an adiabatic or isentropic expansion of steam is known as isentropic enthalpy drop.

10.11 NON-FLOW AND STEADY FLOW PROCESSES

In the previous course of thermodynamics we have studied about the different thermodynamic systems. Here we will first revise to certain extent about the system and the process.

Thermodynamic system is classified as:

1. Closed or Non-Flow system
2. Open onflow system and
3. Isolated system

In closed or non-flow system the boundary allows the energy to flow in or flow out i.e., the energy crosses the boundary of the system. Secondly, the boundary does not allow the mass to cross the boundary of the system.

When the energy crosses the boundary of the system, the boundary may or may not move. The process that will take place in this closed or Non-flow system is called *Non-Flow Process.*

The energy equation for this Non-Flow Process is, with usual relations, as

$$Q = \Delta U + W_{nf}$$

where Q = Heat exchange during the process

ΔU = Change of internal energy during the process

W_{nf} = Work of non-flow process

$$= \int_1^2 p \cdot dv$$

In open or Flow system, both energy and mass will cross the boundary of the system. This open or flow system can further be classified as *Steady-Flow System* or *Unsteady-Flow-System.*

At this stage, we are required to consider steady flow system. In this *steady flow* system the rate of mass flow and energy flow into the system is equal to the rate of flow of mass and energy out of the system i.e., $m_1 = m_2$, and $E_1 = E_2$. This means that there is no accumulation of mass as well as energy, within the system.

The process that takes place in a Steady-Flow-System is called *Steady Flow Process.*

Steady flow energy equation can be written as

$$Q = \Delta P + \Delta K + \Delta H + W_{sf}$$

where Q = Heat exchange during the process

ΔP = change in potential energy

ΔK = change in kinetic energy

ΔH = change in enthalpy

W_{sf} = work of steady flow process

$= h_1 - h_2$ for isentropic process when $\Delta K = 0$

and $h_1 - h_2 = K_1 - K_2$ for isentropic process and $W_{sf} = 0$

10.12 PROCESSES OF VAPOURS

The basic energy relations for the processes as defined for perfect gases also hold for vapours all previous equations in terms of the general symbols *W, Q, H, h, U, u, K, P* apply to any substance under the circumstances specified. The equations derived from the assumption of an ideal gas do not hold.

Remember that the areas on the *P–V* diagram under the curve at an internally reversible process represent $\int p.dv$, and that this area is the work of a non-flow process. The area behind the same curve is the $-\int v.dp$.

The vapour processes that are to be studied here are:

(i) Constant pressure process (Isobaric process)
(ii) Constant volume process (Isochoric process)
(iii) Reversible adiabatic process (Isentropic process)
(iv) Irreversible adiabatic process (Throttling process)
(v) Isothermal process
(vi) Polytropic process (Reversible)
(vii) Hyperbolic process

10.12.1 Constant Pressure Process

A constant pressure, also called an isobaric process, is a change of state during which the pressure remains constant. On the *P–V* plane, Fig.10.11, the process is represented by a horizontal line, and on the *T–S* plane the process is represented by a horizontal line in the wet region and by a curve obtained from the indefinite integral of

$$ds = \frac{\delta q}{T}$$

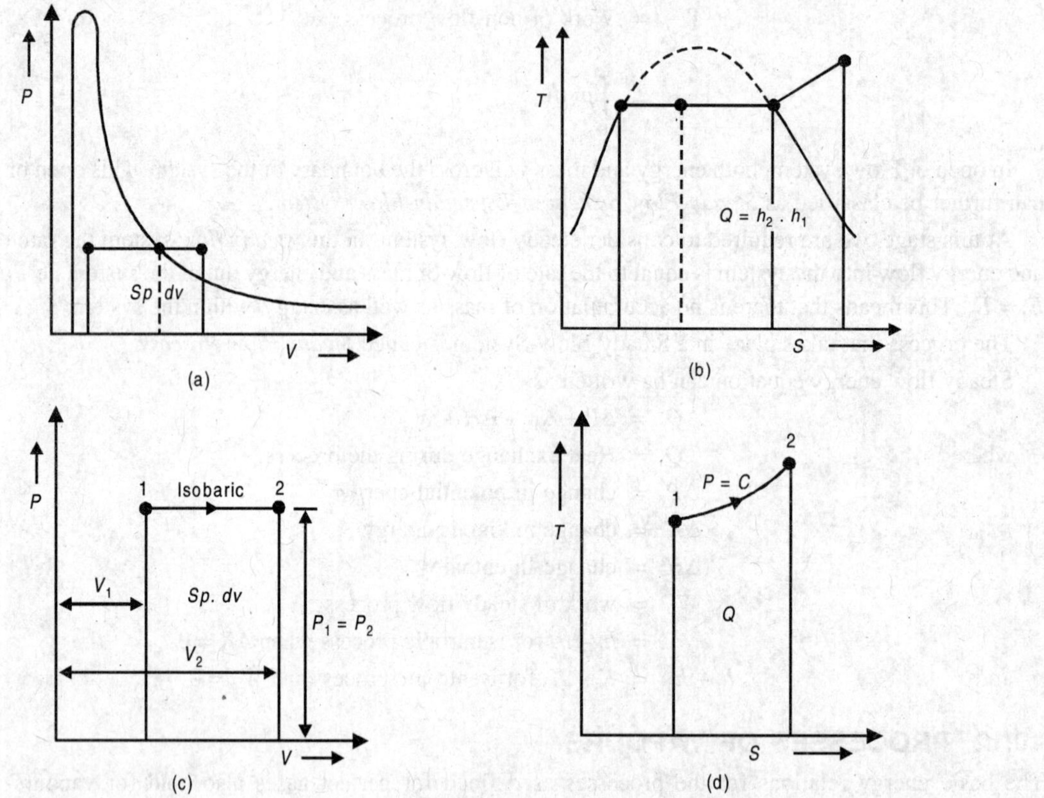

Fig. 10.11

In addition to defining relation, such as $p = c$, other quantities defining the limits of the process are needed. Suppose, in this case that point 1 is in the wet region and point 2 in the superheat region. With the pressure known, a point in the wet region is generally defined by giving its quality *x* or percentage moisture *y*. Similarly, a point 2 in the superheat region is generally, but not necessarily, defined by giving its temperature.

The work of reversible non-flow process at $p = c$ is

$$W_{nf} = \text{work of a non-flow system}$$
$$= \int p \cdot dV$$
$$= p(V_2 - V_1)$$

Also

$$Q = \text{Heat} = \Delta u + \int (p.dV)$$
$$= u_2 - u_1 + P_1V_1$$
$$= (P_2V_2 + u_2) - (u_1 + P_1V_1)$$
$$= (P_2V_2 + U_2) - (P_1V_1 + U_1)$$
$$= (h_2 - h_1)$$

10.12.2 Constant Volume Process

A constant volume process, also called an *isometric process* is a change of state during which the volume remains constant. This process is represented by a vertical line 1–2 in Fig. 10.12.

For non-flow, constant volume process we have

$$Q = \Delta U + \int p \cdot dV$$

$$Q = U_2 - U_1 \text{ Non-flow process } \int p.dv = 0$$

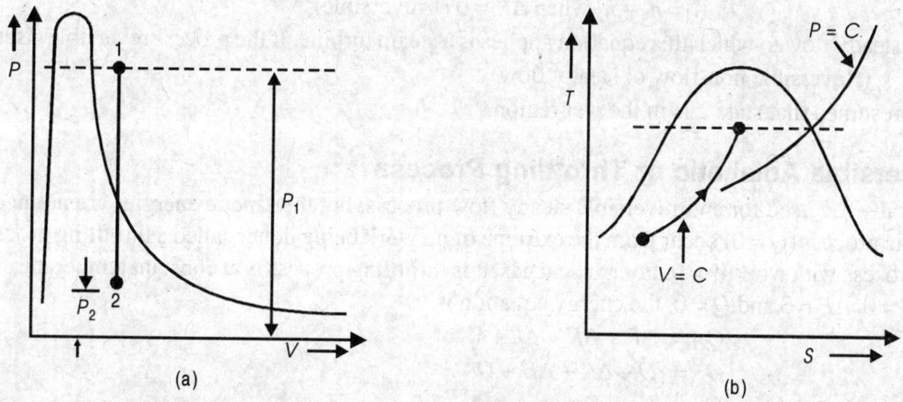

Fig. 10.12

For a steady flow isometric process, we have,

$$\begin{aligned} Q &= \Delta U + \Delta W_f + \Delta K + \Delta P + W_{sf} \\ &= \Delta P + \Delta K + \Delta H + W_{sf} \end{aligned}$$

For a simple flow process, (like nozzle), $W_{sf} = 0$ and assuming $\Delta P = 0$, we have

$$Q = \Delta K + \Delta H$$

10.12.3 Reversible Adiabatic Process (Isentropic Process)

Reversible adiabatic process is that process during which heat exchange is not taking place and $Q = 0$. This process is also called Isentropic process during which heat is zero and entropy change is also zero i.e., $\Delta S = 0$.

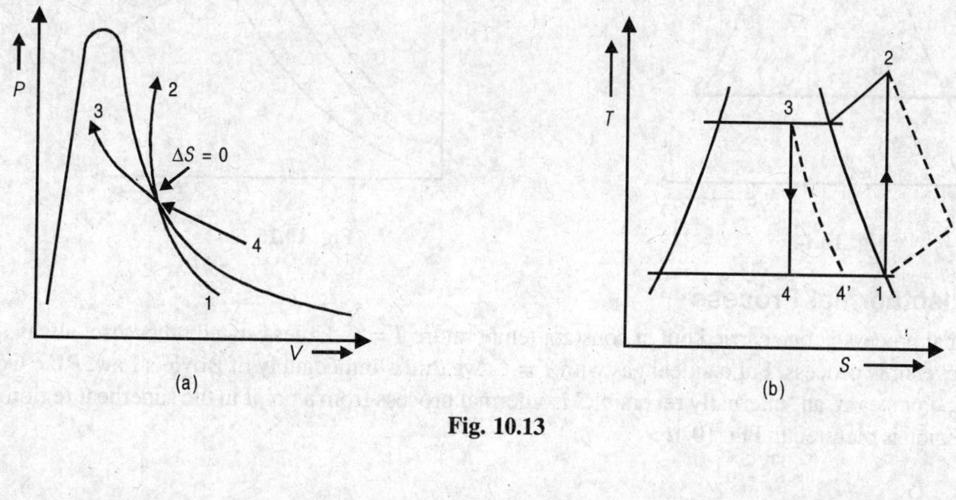

Fig. 10.13

This process is represented on *PV* plane , 1–2 Isentropic Compression, 3–4 Isentropic expansion and $T-S$ plane of Fig. 10.13.

The isentropic and irreversible adiabatic can either be non-flow or steady flow. Let the second state along isentropic be designated along an irreversible adiabatic 2'.

Then, from the non-flow energy equations with $Q = 0$ for isentropic process, we write $Q = \Delta U + W_{nf}$

$\therefore\ 0 = \Delta U + W_{nf}$ or $W_{nf} = U_1 - U_2$ or $W = U_1 - U_2$ (Reversible) and $W = U_1 - U_2'$ (Irreversible)

For steady flow process we get, in general,

$$W = (h_3 - h_4) \text{ when } \Delta K = 0 \text{ (Reversible)}$$

or

$$W = h_3 - h_4 \text{ when } \Delta K = 0 \text{ (Irreversible)}$$

An example of steady flow to which this equation applies is a steam turbine. If the process reversible (Isentropic), $S_1 = S_2 = (S_f + xS_{fg})_2$ (Reversible non-flow or steady flow)

In which it is presumed that state 2 is in the wet region.

10.12.4 Irreversible Adiabatic or Throttling Process

This term is generally reversed for an irreversible steady flow process, but the kinetic energies are not necessarily zero. This adiabatic process ($Q = 0$) occur from the extreme of no work being done, called a throttling process. This is a simple flow process with work $W = 0$. For an ideal gas, this throttling process is at constant temperature process, therefore when $W = 0$, $\Delta P = 0$ and $Q = 0$, the energy equation is

$$Q = \Delta P + \Delta K + \Delta h + W$$

$$O = O + \Delta K + \Delta H + O$$

or

$$h_1 + K_1 = h_2 + K_2$$

If velocity changes are negligible, then $K_1 = K_2$ and $h_1 = h_2$. This is a constant enthalpy process or *Isenthalpic Process*.

Thus the throttling process is defined by $h_1 - h_2$ and it is commonly used in steam power practice to determine the quantity of steam (dryness fraction of steam). Generally this process is shown by a dotted line on *T–S* plane as shown in Fig.10.14 and by a vertical line on *h–s* diagram of Fig. 10.15.

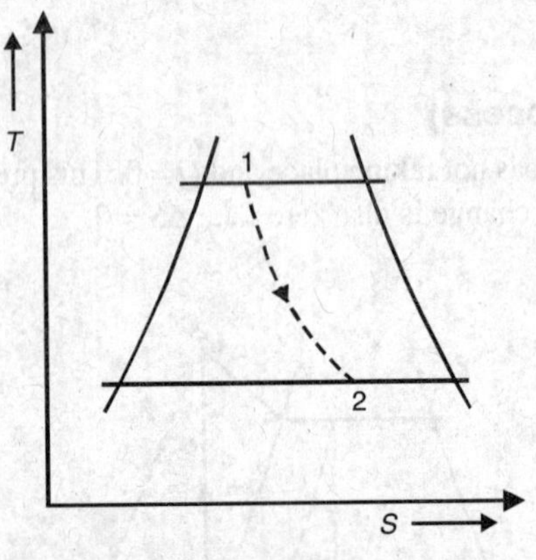

Fig. 10.14

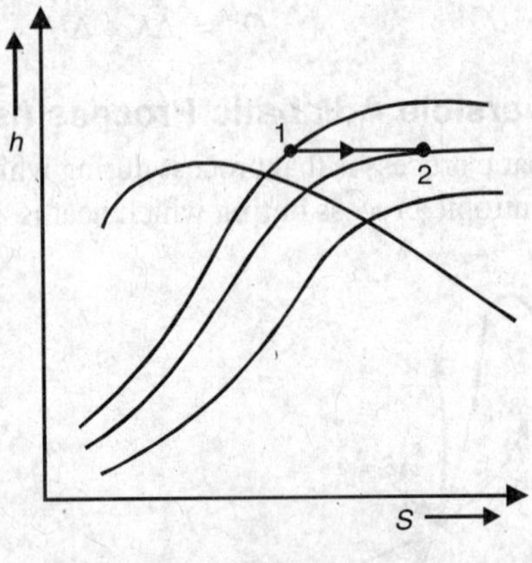

Fig. 10.15

10.12.5 Isothermal Process

An Isothermal process is one carried out at constant temperature $T = C$. Unless stated otherwise, always it will be taken as a reversible process. For an ideal gas with $T = C$, we think immediately of Boyle's Law, PV = Constant or $P_1V_1 = P_2V_2$. For steam, an 'internally reversible' isothermal process from a point in the superheat region to a point in the wet region is pictured in Fig. 10.16.

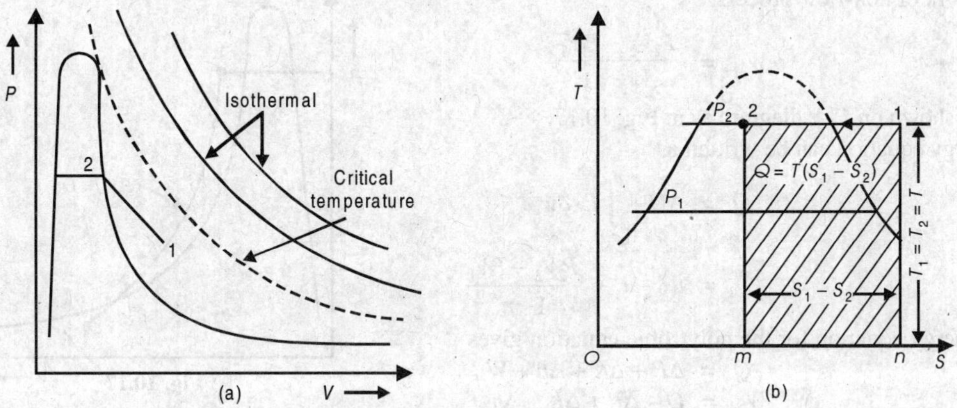

Fig. 10.16

The area m –2-1 -n, $\int T.ds = T\int ds$, represents heat

or $$Q = T(S_2 - S_1) \text{ kJ/kg}$$

where $(S_2 - S_1)$, being a negative term or number indicates that heat is passing from the substance. Then non-flow energy equation will be

$$Q = \Delta U + \int p.dv = \Delta U + W$$

or $$W = Q - \Delta U$$
$$W = Q - U_1 - U_2 \text{ kJ/kg}$$

For a steady flow isothermal process, the energy equation can be given as

$$Q = \Delta P + \Delta K + \Delta h + W$$

with $$\Delta P = 0$$
$$W = Q - \Delta K - \Delta h \text{ and with } \Delta K = 0$$
$$= Q - \Delta h$$
$$W = Q + h_1 - h_2$$

10.12.6 Polytropic Process

A polytropic process is an internally reversible process, which conforms to the relation $pv^n = 0$ or $P_1V_1{}^n = P_2V_2{}^n$ which n is any constant.

If a process is known or assumed to be polytropic and if two state points are defined (say, pressures and volumes are known), the value of n can be found by use of logarithms.

$$\therefore \quad P_1V_1{}^n = P_2V_2{}^n \text{ or } \frac{P_1}{P_2} = \left(\frac{V_2}{V_1}\right)^2$$

or $$\log(P_1 / P_2) = n.\log\left(\frac{V_2}{V_1}\right)$$

or $$n = \frac{\log_e(P_1/P_2)}{\log(V_2/V_1)}$$

Using $pv^n = C$, we get,

$$\int_1^2 p.dv = \frac{P_2V_2 - P_1V_1}{(1-n)} \frac{\text{N-m}}{\text{kg}}$$

which is the work of non-flow process.

$$\therefore \quad W_{nf} = \frac{P_2V_2 - P_1V_1}{(1-n)}$$

The process is shown on P–V diagram as in Fig. 10.17.
Non-flow energy equation can be written as

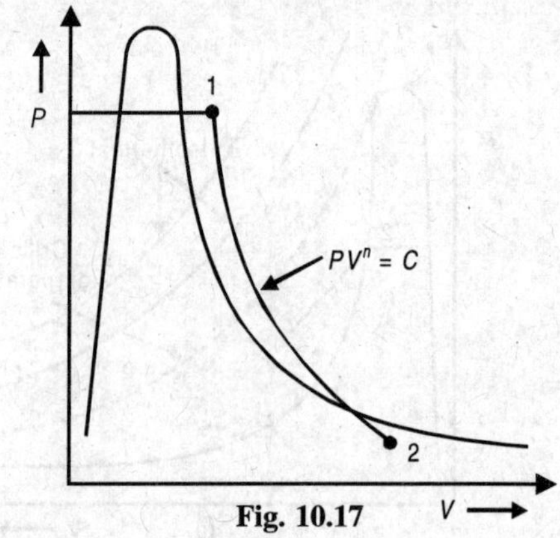

Fig. 10.17

$$Q = \Delta U + \int P \cdot dv$$

$$= u_2 - u_1 + \frac{P_2V_2 - P_1V_1}{1-n}$$

steady flow energy equation for the polytropic equation gives

$$Q = \Delta P + \Delta K + \Delta h + W_{sf}$$

or

$$W_{sf} = Q - \Delta P + \Delta K + \Delta h$$

when $\Delta P = 0$, $W = Q - \Delta K - \Delta h$ and again $\Delta K = 0$ gives

$$W = Q - \Delta K - \Delta h.$$

Also for flow process

$$W = -\int VdP$$

and for polytropic process $PV^n = C$

$$W = -\int_1^2 VdP = \frac{n}{n-1}(P_2V_2 - P_1V_1)$$

Besides constant volume, constant pressure, constant temperature, adiabatic (reversible and irreversible) processes the following are the other important operations:

(1) Hyperbolic and (2) Free expansion.

(1) Hyperbolic Process. When a gas or steam expands or is compressed in such a manner that the product of pressure and volume remains constant during the process of expansion or compression, the process is known as hyperbolic process because such a process on the P–V diagram will give a rectangular hyperbola. Hence a hyperbolic expansion or compression process follows the law,

$$\text{Pressure} \times \text{Volume} = \text{Constant}$$

$$P\,V = \text{Constant}$$

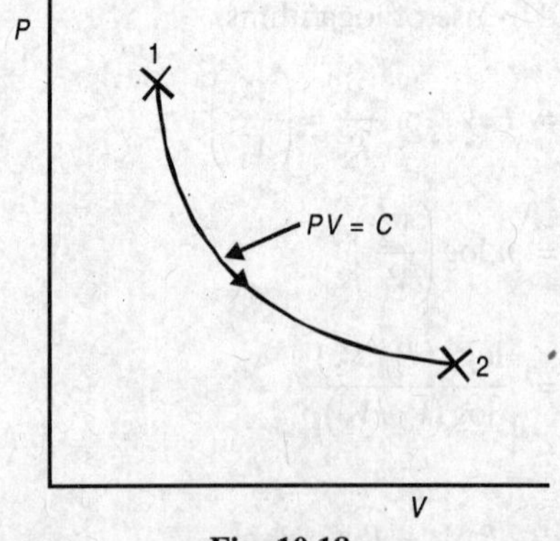

Fig. 10.18

or

$$P_1V_1 = P_2V_2$$

For ideal gas, this process is similar to isothermal process
For such a process,

$$\int_1^2 P\cdot dV = \text{work done of non-flow process}$$

$$= P_1V_1 \log_e\left(\frac{P_1}{P_2} = \frac{V_1}{V_2}\right)$$

$$-V.dp = P_1V_1 \log_e\left(\frac{P_1}{P_2} = \frac{V_1}{V_2}\right)$$

For steam, we can find the volume of steam after the process. Thus

$$P_1V_1 = P_2V_2$$

$$\therefore \quad V_2 = \frac{P_1}{P_2} - V_1$$

If $V_2 > V_{g2}$, the steam is superheated.
$V_2 < V_{g2}$, the steam is wet.
$V_2 = V_{g2}$,
the steam is dry and saturated.

when $V_2 > V_{g_2}$,

$$V_2 = V_{sat_2} \times \frac{T_{sup\,2}}{T_{sat\,2}}$$

$\therefore$ T_{sup} is determined.

$$\therefore \quad h_2 = \text{enthalpy} = h_{g2}\,{}^{T}Cpv(T_{sup} - T_{sat})$$

and $U_2 = h_2 - P_2V_2$

and hence Q heat can be calculated for non-flow process. Similarly,

when $V_2 < V_{g2}$

Then $V_2 = x_2V_{g_2}$,

$$\therefore \quad x_2 = \frac{V_2}{V_{g_2}}$$

$$\therefore \quad h_2 = h_{f2} + x_2hf_{g2}$$

$$\therefore \quad U_2 = h_2 - P_2V_2$$

and hence Q heat can be calculated. Again,

when

$$V_2 = V_{g2}$$

$$x_2 = 1$$

$$h_2 = h_{g2}$$

$$U_2 = h_{g2} - P_2V_2$$

and thus Q can be calculated.

Note: Remember we will have to calculate h_1, U_1 etc. at the condition of steam before the process commences.

(2) Free Expansion. When a fluid is allowed to expand suddenly into a vacuum chamber through an orifice of large dimensions, the free expansion of the fluid occurs. During this operation no heat has been supplied or rejected and no external work has been done; hence it follows that the total enthalpy of the fluid remains constant. This type of expansion is also called the constant total heat expansion.

After studying the different vapour process, we can conclude that for any process, it will be required to determine or calculate

(i) Heat during the process (Q)
(ii) Change in Internal energy (ΔU)
(iii) Change in Enthalpy (ΔH)
(iv) Work done (W)
(v) $\int p.dV$
(vi) $-\int V.dP$
(vii) Change in entropy (ΔS)

10.13 STEAM DRYNESS FRACTION AND ITS DETERMINATION WITH STEAM CALORIMETERS

We have defined the dryness fraction of steam as the ratio of (dry steam) to (total steam containing it) and is represented by x.

$$\therefore \qquad x = \frac{\text{Dry steam (kg)}}{\text{Total steam (kg)}}$$

Dryness fraction of steam is determined with the help of steam calorimeters.

These calorimeters are as follows.

(i) Barrel Calorimeter (ii) Separating Calorimeter (iii) Throttling Calorimeter and
(iv) Combined separating and throttling Calorimeter

Any one combination these Calorimeters can be used for finding the dryness fraction of steam. Each of these methods will be dealt within term.

10.13.1 Barrel Calorimeter

To get the approximately dryness fraction of steam, a very simple appe called Barrel Calorimeter is used. This consists of a wooden barrel containing cold water.

Steam is taken from the pipe through which it is flowing and is condensed in the cold water.

Temperatures of water before and after condensation are recorded. Knowing the quantities of steam and water, we can very easily calculate the dryness fraction steam. The calorimeter is shown in Fig. 10.19.

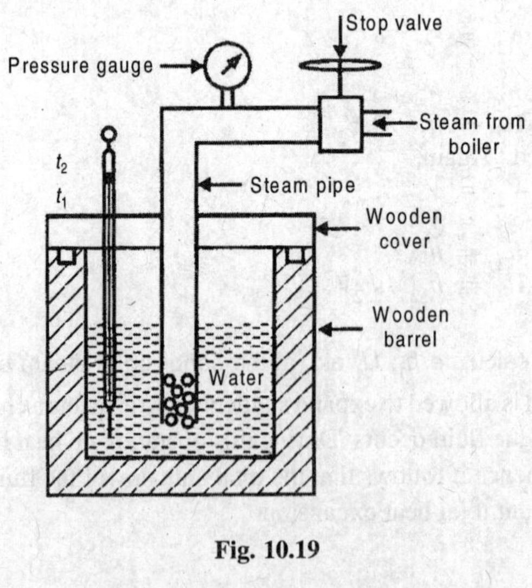

Fig. 10.19

Let, ms = weight of steam condensed
m_w = weight of cold water in barrel
C_{pw} = sp. heat of water
x = dryness fraction of steam
h_{fg} = latent heat of the steam at pressure of steam P
t_1 = initial temperature of water before mixing steam
t_2 = final temperature of steam after mixing of steam
t = saturation the temperature of steam at P.

Then we can find the dryness fraction of steam by using the equations given below.

[Heat lost by steam] = [Heat gained by coal water]

$$Ms\left[x.h_{fg} + C_{pw}(t - t_2)\right] = M_w C_{pw}(t_2 - t_1)$$

From this equation, we can find the dryness fraction of steam (x)

10.13.2 Separating Calorimeter

The separating calorimeter is quite simple in its action and it may be used for testing steam of practically any degree of wetness. Figure 10.20 shows a design of separating calorimeter by **Prof Carpenter.**

The sample of steam to be tested enters the calorimeter at D and passes down through the central passage DF into the perforated metal cup H. The current of steam then reverses and flows as shown by the arrows in the jacket J surrounding the inner chamber C. From the jacket the steam escapes into the atmosphere through the small orifice at the bottom and exhaust pipe E.

The water in the steam is thrown out in the cup H and collected in the chamber C. A gouge G has two scale on its dial, the inner one showing the pressure of the steam in the jacket while the outer one shows the weight of steam passing through the jacket for each pressure in ten minutes. The graduations on the outer scale are determined by experiment with steam of different pressures, the escaping steam being condensed and weighted, but in using the instrument it is only necessary to take during a ten-minute test the average reading on the outer scale to obtain the weight of steam passing through the jacket.

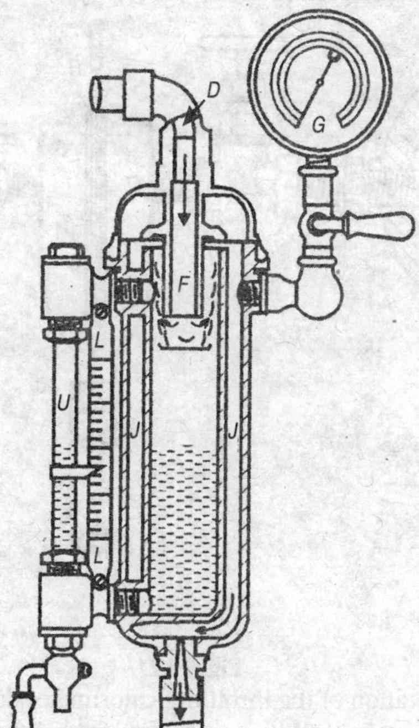

Fig. 10.20

The water separated from the steam is shown by means of the glass gauge U. A pointer P having a frictional grip of the glass tube is set to the water level at the beginning of the taste. The change in level of the separated water during a test is read off on the scale L. Which is graduated in fractions of mols.

Let W be the weight of steam passing through the jacket during a test and W is the weight of water separated in the same time, then x, the dryness fraction of the steam tested is $x = \frac{W}{W + W}$.

The weight of steam passing through the jacket may also be readily determined by passing the escaping steam into a bucket of water where it is condensed, the increase in the weight of the bucket and its contents giving the value of W. In that case gauge G is not required, but it used its reading may be taken as a check.

10.13.3 Throttling Calorimeter

The throttling calorimeter was first introduced by Professor Peabody. Figure 10.21 shows Professor Carpenter's modification of the Peabody throttling calorimeter.

P is the pipe carrying the steam to be tested. AB is the sampling tube through which the sample of steam flows to the vessels S which is the principle part of the calometer. V is the stop valve. The sample of steam enters the vessel S through a small orifice O and expands to a pressure a little above that of atmosphere into which it escapes freely through the exhaust pipe E.

The pressure of the steam in S is measured by means of the manometer or siphon gauge G which is a glass U-tube containing mercury. G is connected to S by 2a rubber tube T. Steam to the gauge may be cut-off by means of the stop cock C. A deep pocket K, containing cylinder oil, takes the thermometer R which indicates the temperature of the steam in S. The vessel S, the stop valve V, and the exposed part of the sampling tube AB must be carefully lagged to prevent undue loss of heat by radiation.

In using this instrument four observations have to be made

(1) The pressure P_1 or temperature t_1 of the steam in the pipe P.
(2) The pressure of the steam in the vessel S as shown by the gauge G in inches/centimeters of mercury.
(3) Reading of thermometer R
(4) The height of the barometer in inches/centimeters of mercury.

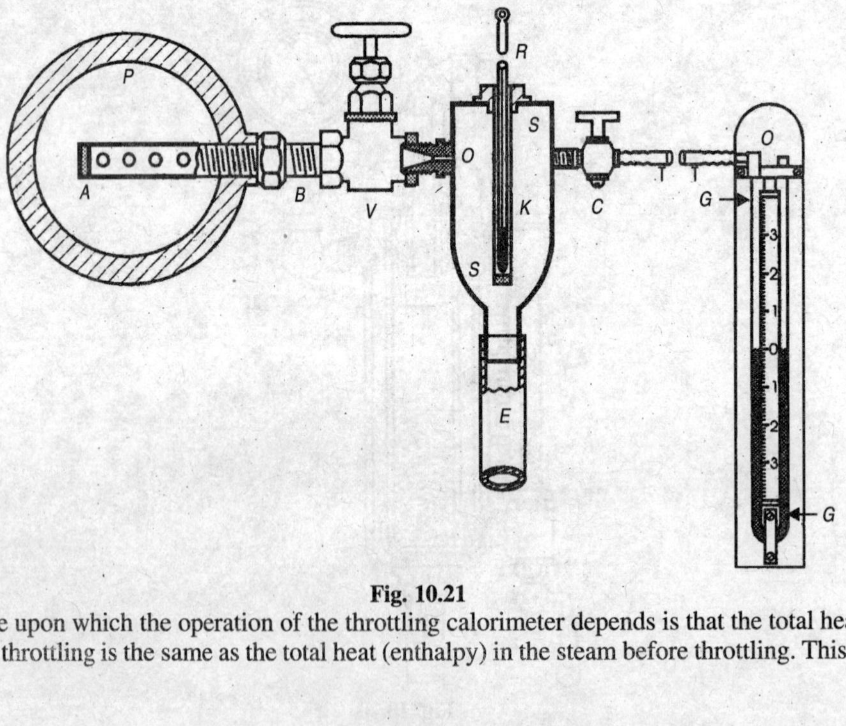

Fig. 10.21

The principle upon which the operation of the throttling calorimeter depends is that the total heat (enthalpy) in the steam after throttling is the same as the total heat (enthalpy) in the steam before throttling. This is given by the equation

$$h_{f1} + xh_{fg1} = h_{g2} + Cp_v(t_3 - t_2)$$

$$\therefore \quad x = \frac{h_{g_2} - h_{f1} + C_{pv}(t_3 - t_2)}{h_{fg1}}$$

where

x = dryness fraction of steam before throttling
h_{f1} = sensible heat of steam before throttling
h_{fg1} = latent heat of steam before throttling
h_{g2} = total heat or enthalpy of saturated steam at pressure after throttling
t_2 = saturated temperature of steam at pressure after throttling
t_3 = temperature of steam after throttling as indicated by thermometer R.
C_{pv} = mean specific heat of steam between t_3 and t_2 at constant.

Since the pressure after throttling is in the neighbourhood of that of the atmosphere, and since the specific heat of steam at that pressure is little affected by the temperature, C_{PV} may be taken as 2 or 2.01 kJ/kg-K.

The operation of the throttling calorimeter depends on the steam being superheated after throttling and it will fail in its purpose when the steam is so wet before throttling that it is wet after throttling. This means that t_3 must be greater than t_2. Putting $t_3 = t_2$ the value of x which may be determined by the throttling calorimeter must be greater than

$$= \frac{h_{g_2} - h_{f1}}{h_{fg1}}$$

10.14 LIMITATIONS OF CALORIMETERS

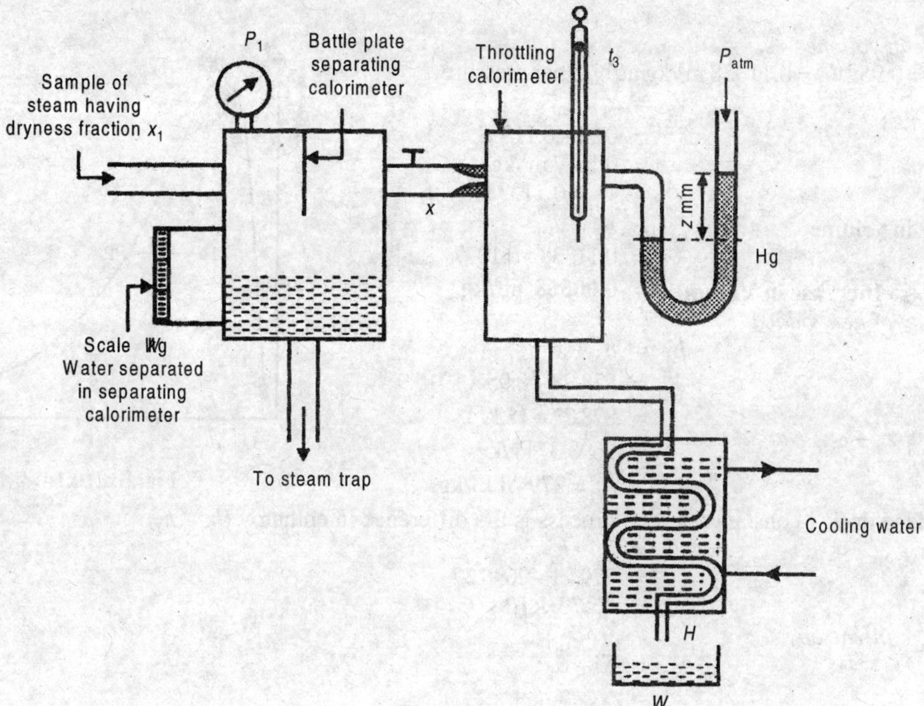

Fig. 10.22

10.14.1 Barrel Calorimeter

Basically, this calorimeter is used when we want to know the approximate dryness fraction of steam, or to have a rough idea about the dryness fraction. This calorimeter gives better results if the dryness fraction of steam is greater than 0.95.

10.14.2 Separating Calorimeter

This calorimeter will give better results when the dryness fraction of steam to be determined is above 0.95.

10.14.3 Throttling Calorimeter

Basically the condition of steam after throttling must be known so that the unknown *x*-dryness fraction before throttling can be calculated. For this purpose the steam after throttling must be separated and its temperature should be more than the saturated temperature of steam corresponding to the pressure after throttling . To get such a result dryness fraction before throttling should be more than 0.9.

If the dryness fraction of steam is less than 0.9, then the steam will not be superheated after throttling and the dryness before throttling cannot be found.

In such a case, a combined separating and throttling calorimeter is used. In this, steam is first admitted to separating calorimeter where some portion of the water particles are removed and the quality of steam is improved to the required level for throttling calorimeter.

A Schematic diagram of a combined separating and throttling calorimeter is shown in Fig. 10.22 and now is self explanatory.

SOLVED EXAMPLES

I. Constant Pressure Process

Example 10.1 *One kg of steam at a pressure of 17.5 bar and dryness 0.95 is heated at constant pressure, until it is completely dry. Determine*

(i) Increase in volume. (ii) Quantity of heat added. (iii) Change in entropy.

Solution

(i) Increase in volume

We have $x_1 = 0.95$, $V_{g1} = 0.11338$ m³/kg at 17.5 bar pressure

$$\therefore \quad V_1 = x_1 V_{g1} = 0.95 \times 0.11338 = 0.1077 \text{ m}^3/\text{kg}$$

$$V_2 = V_{g2} = 0.11338 \text{ m}^3/\text{kg}$$

$$\therefore \text{ Increase in volume } = V_2 - V_1 = 0.11338 - 0.1077$$

$$\textbf{Increase in Volume} = \mathbf{0.00568\ m^3/kg}$$

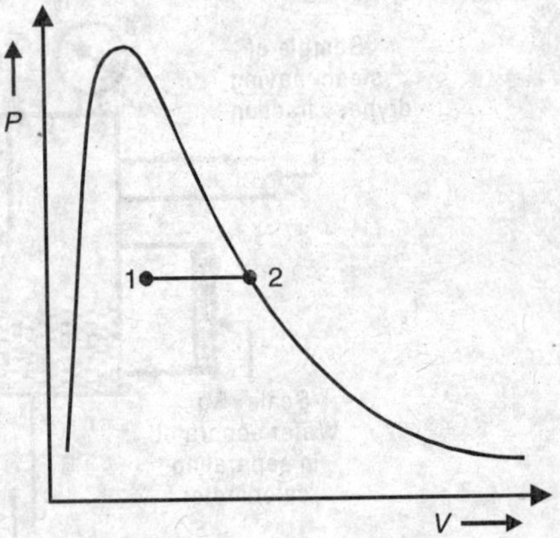

Fig. Ex. 10.1

(ii) Quantity of heat added

$$h_1 = h_{f1} + x_1 h_{fg1} = 878.27 + 0.95 \times 1915.9 = 878.27 + 1820.1 = 2698.37 \text{ kg/kg}$$

$$h_2 = h_{g2} = 2794.1 \text{ kJ/kg}$$

$\therefore$ Heat added during constant pressure process is the difference in enthalpy $(h_2 - h_1)$

$$\therefore \quad Q = h_2 - h_1 = 2794.1 - 2698.37$$

$$\mathbf{Q = 95.73\ kJ/kg}$$

(iii) Change in entropy

$$\Delta S = S_2 - S_1 = S_{g2} - (sf_1 + x_1 S_{fg1}) = (S_{f2} + S_{fg2}) - (Sf_1 + x_1 S_{fg1})$$

But $\quad S_{f1} = S_{f2}$ and $S_{fg1} = S_{fg2}$ for $P = C$

$$\therefore \quad \Delta S = S_{fg1}(1 - x_1) = 0.05 \times 4.0007$$

$$\mathbf{\Delta S = 0.200035\ kJ/kg\text{-}K.}$$

Example 10.2 *A piston cylinder arrangement is filled with a wet steam of quality 0.8 at a pressure of 0.1 MPa. Energy is added at constant pressure till the temperature of steam rises to 300°C. Calculate the heat added and the work done.*

Solution

$$0.1 \text{ MPa} = 1 \text{ bar}$$

Volume of one kg before heat is added

$$= X V_g$$
$$= 0.8 \times 1.6938$$
$$= 1.35504 \text{ m}^3/\text{kg}$$

Heat is added at constant pressure. The temperature of steam after heat added is while saturated temperature is 170.17°C

∴ Steam is superheated.

$$V_2 = \text{Volume of 1 kg steam} = 2.639 \text{ m}^3/\text{kg}$$

∴ $$\text{Work done/kg} = P(V_2 - V_1)$$
$$= 100 \times 1\,(2.639 - 1.35504)$$
$$= 100 \times 1.28396$$
$$\textbf{Work done/kg} = \textbf{128.396 kJ/kg}$$

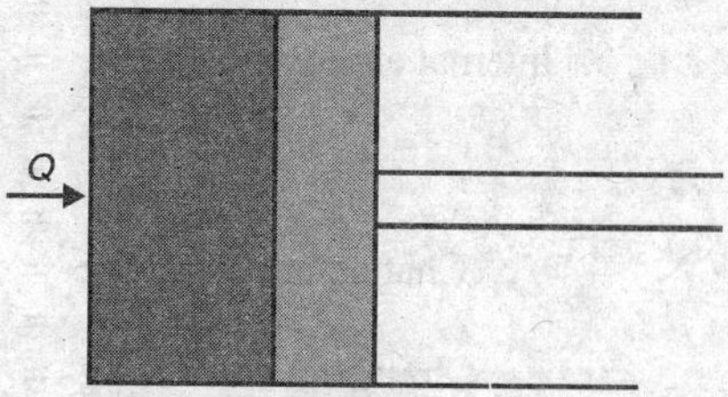

Fig. Ex. 10.2

For constant pressure process,

$$Q = \text{Heat} = h_2 - h_1 \text{ (for non-flow process)}$$
$$= 3074.5 - (h_{f1} + x_1 h_{fg1})$$
$$= 3074.5 - (417.5 + 0.8 \times 2257.9)$$
$$= 3074.5 - 2223.82$$
$$\mathbf{Q = 850.68 \text{ kJ/kg}}$$

Example 10.3 *Steam having a dryness fraction 0.8 at a pressure of 10 bar changes its state so that the temperature of steam become 275°C. The pressure of steam remains the same. Calculate per kg of steam, the change in total heat, change in volume and change in internal energy.*

Solution

Initial Condition:

$$\text{Enthalpy of steam/kg} = h_1 = h_{f1} + x_1 h_{fg1}$$
$$= 762.61 + 0.8 \times 2013.6$$
$$= 762.61 + 1610.88$$
$$= 2373.59 \text{ kJ/kg}$$

Volume $$V_1 = x_1 v_{g1}$$
$$= 0.8 \times 0.19429$$
$$\mathbf{V_1 = 0.1554 \text{ cu.m/kg}}$$

Internal energy $$u_1 = h_1 - P_1 V_1$$
$$= 2373.59 - \frac{10^5}{1000} \times 10 \times 0.1554$$
$$= 2373.59 - 155.4$$
$$\mathbf{u_1 = 2218.19 \text{ kJ/kg}}$$

Final condition

$$P = 10 \text{ bar}$$
$$\text{Temperature} = 275°\text{C}$$

$$\text{Enthalpy of 1 kg steam} = hg + Cp_v\,(T_{sup} - T_{sat})$$

$$h_2 = \frac{3052.1 + 2943}{2} \qquad \text{(From table Interpolation)}$$

$$= \frac{5995}{2}$$

$$= 2997.5\ \text{kJ/kg}$$

$$V_2 = \frac{0.2327 + 0.2580}{2} \qquad \text{(Volume from tables)}$$

$$= \frac{0.4907}{2}$$

$$= 0.24535\ \text{m}^3$$

$$\therefore \text{Internal energy} \quad u_2 = h_2 - P_2V_2$$

$$= 2997.5 - 100 \times 0.24535 \times 10$$

$$= 2997.5 - 245.35$$

$$= 2752.15\ \text{kJ/kg}$$

$$\therefore \text{Change in volume/kg} = V_2 - V_1$$

$$= 0.24535 - 0.1554$$

Change in volume/kg = 0.18995 m³/kg

Change in internal energy/kg

$$= 2752.15 - 2218.19$$

Change in internal energy/kg = 2533.96 kJ/kg

Example 10.4 (*a*) *The internal energy of 4.5 kg of wet steam at 7 bar is 11388 kJ. What is the dryness fraction of steam?*

(*b*) *By how many heat units, the internal energy of the above steam shall increase if it is superheated at constant pressure to a temperature of 250°C ?*

Assume specific heat of steam as 2.1 kJ/kg K.

Solution

(*a*) *Dryness Fraction*

Let x be the dryness fraction of steam

$$\text{Internal energy/kg} = \frac{11388}{4.5}$$

$$= 2530.67\ \text{kJ}$$

$$= h - PV$$

$$= (h_f + xh_{fg}) - p \times xvg$$

$$\therefore \quad 2530.67 = 697.06 + x \times 2064.9 - 100 \times 7 \times x \times 0.27268$$

$$\therefore \quad 2530.67 - 697.06 = 2064.9\,x - 190.88\,x$$

$$\therefore \quad 1833.61 = 1874.02\,x$$

$$x = \frac{1833.61}{1874.02}$$

$$\mathbf{x = 0.9784}$$

Dryness fraction of steam

(*b*) *Increase in Internal energy*

After superheating, enthalpy of 1 kg steam is given by

$$h = h_g + C_{pv}\,(t_{sup} - t_{sat})$$

This equation used C_p is given

$$h = 2762.0 + 2.1\,(250 - 164.96)$$

$$= 2762.0 + 2.1 \times 85.04$$
$$= 2762.0 + 178.584$$
$$= 2940.584 \text{ kJ/kg}$$

$$V = v_{sat} \times \frac{T_{sup}}{T_{sat}}$$
$$= 0.27268 \times \frac{250 + 273}{273 + 16496}$$
$$= \frac{0.27268 \times 523}{437.96}$$
$$= 0.32563 \text{ m}^3/\text{kg}$$
$$u = \text{Internal energy}$$
$$= h - PV$$
$$= 2940.584 - 100 \times 7 \times 0.32563$$
$$= 2940.584 - 227.94 = 2712.644 \text{ kJ/kg}$$

∴ Rise in heat units of Internal energy

$$= u_2 - u_2$$
$$= 2712.644 - 2530.67 = 181.974 \text{ kJ/kg}$$

Rise in heat units of Internal energy = 818.883 kJ

Example 10.5 *1.5 cu.m of steam at a pressure 9 bar and having dryness fraction 0.95 loses 9420 kJ at constant pressure. Calculate dryness fraction of the steam at the end of heat loss.*

Solution

We will have calculations on kg basis and therefore first we will find mass of steam in 1.5 cu.m at 9 bar and 0.95 dry

$$= x \cdot Vg \qquad \text{(Volume of 1 kg steam at 9 bar and 0.95 dry)}$$
$$= 0.95 \times 0.21481$$
$$= 0.2041 \text{ cu.m}$$

∴ Mass of steam in 1.5 cu.m

$$= \frac{\text{Total volume}}{\text{Sp. volume}}$$
$$= \frac{1.5}{0.2041} = 7.35 \text{ kg}$$

$$\therefore \quad \text{1 kg steam loses} = \frac{9420}{7.35} \text{ kJ} = 1281.63 \text{ kJ at constant pressure}$$

$$\text{Initial enthalpy/kg steam} = h_1 = h_{f1} + x_1 h_{fg1}$$
$$= 742.64 + 0.95 \times 2029.5$$
$$= 742.64 + 1924 = 2670.64 \text{ kJ}$$

$$\therefore \quad \text{Final enthalpy/kg} = 2670.64 - 1281.63$$
$$= 1389.01 \text{ kJ}$$

$$\therefore \quad 1389.01 = h_{f2} + x_2 h_{fg2} = 742.64 + x_2 h_{fg2}$$
$$= 742.64 + x_2 \times 2029.5$$

$$\therefore \quad x_2 = \frac{1389.01 - 742.64}{2029.5}$$

$$\mathbf{x_2 = 0.3185}$$

Dryness fraction of steam at the end of heat loss.

Example 10.6 *A vessel contains 4 kg of steam at 10 bar and 220°C. Find the volume of the vessel. If the vessel is cooled till the steam pressure drops to 3 bar. Find the final condition of steam and the heat transfer during cooling.*

Solution

Volume of 1 kg steam at 10 bar and 220°C (from superheated steam tables and by interpolation)

$$V_1 = 0.2059 + 0.4\,(0.2328 - 0.2059)$$

$$= 0.21666\ \text{m}^3/\text{kg}$$

$$\text{Total volume of vessel} = \text{Mass of steam} \times \text{Specific volume}$$

$$= 4 \times 0.21666$$

$$\textbf{Total volume of vessel} = \mathbf{0.86664\ m^3}$$

Volume of 1 kg steam at 3 bar after cooling

$$V_{g2} = 0.60533\ \text{m}^3/\text{kg}$$

This is greater than that of V_1

∴ Steam is wet

∴

$$0.60533\,x_2 = V_1 = 0.21666$$

$$x_2 = \frac{0.21666}{0.60533}$$

$$\mathbf{x_2 = 0.3579}$$

For constant volume process, we have

$$Q = \Delta u + \int P \cdot dv$$

$$= \Delta u \text{ as } \int P.dV = 0 \text{ (Constant volume)}$$

$$= u_2 - u_1$$

$$u_2 = h_2 - P_2V_2$$

$$= h_{f2} + x_2 h_{fg2} - P_2V_2$$

$$= (561.5 + 0.3579 \times 2163.2) - \frac{3 \times 10^5}{1000} \times 0.21666$$

$$= 1335.7 - 65$$

$$= 1270.7\ \text{kJ/kg}$$

$$u_1 = h_1 - p_1 v_1$$

$$= 2873.28 - \frac{10 \times 10^5}{1000} \times 0.21666$$

$$= 2873.28 - 216.66$$

$$= 2656.62\ \text{kJ/kg}$$

∴

$$Q = u_2 - u_1$$

$$= 1270.7 - 2656.62$$

$$= -1385.92\ \text{kJ/kg}$$

∴ Total heat transfer during cooling

$$= -4 \times 1385.92$$

Total heat transfer during cooling = –5543.68 kJ

– ve sign indicates that the heat is rejected by the system.

Example 10.7 *1 kg of steam at initial condition of 6 bar and 0.2 dry is heated at constant volume until the pressure is 20 bar. Determine the final state of steam, heat added and change of energy.*

Solution

Initial volume of 1 kg steam

$$V_1 = x_1 V_{g1} = 0.2 \times 0.31546$$
$$= 0.063092\ \text{m}^3/\text{kg} = V_2$$

After cooling pressure is 20 bar

$$V_{g2} = 0.09955 \text{ and is greater than } V_1$$

∴ Steam after cooling is wet.

$$\therefore \quad 0.063092 = x_2 \times 0.09955$$

$$x_2 = \frac{0.063092}{0.09955}$$

$$\mathbf{x_2 = 0.634}$$

Initial entropy per kg is given by

$$S_1 = S_{f1} + x_1 S_{fg1}$$
$$= 1.931 + 0.2 \times 4.827$$
$$= 1.931 + 0.9654$$
$$= 2.8964\ \text{kJ/kg°K}$$

Similarly

$$S_2 = S_{f2} + x_2 S_{fg2}$$
$$= 2.447 + 0.634 \times 3.890$$
$$= 2.447 + 2.4663$$
$$= 4.9133\ \text{kJ/kg/K}$$
$$\Delta S = S_2 - S_1$$
$$= \text{Change of entropy}$$
$$= 4.9133 - 2.8964$$
$$\mathbf{\Delta S = 2.0169\ kJ/kg\text{-}K}$$

Now

$$u_1 = h_1 - P_1 V_1 = h_{f1} + x_1 h_{fg1} - P_1 V_1$$
$$= 670.4 + 0.2 \times 2085.1 - 100 \times 6 \times 0.063092$$
$$= 1087.42 - 37.86$$
$$= 1049.56\ \text{kJ/kg} \qquad \text{...(I)}$$
$$u_2 = h_2 - P_2 V_2$$
$$= h_{f2} + x_2 h_{fg2} - P_2 V_2$$
$$= 908.5 + 0.634 \times 1888.7 - 100 \times 20 \times 0.063092$$
$$= 2105.94 - 126.184$$
$$= 1979.756\ \text{kJ/kg} \qquad \text{...(II)}$$

For constant volume process

$$Q = \Delta u + \int P.dV = \Delta u \text{ as } \int P.dV = 0$$
$$= u_2 - u_1 + 0$$
$$= 1979.756 - 1049.56$$
$$\mathbf{Q = 930.196\ kJ/kg}$$

Example 10.8 *A closed vessel of 0.24 cu. m capacity, constant steam at a pressure of 11 bar and at a temperature of 200°C.*

Find the quantity of steam in the vessel.

At what pressure in the vessel will the steam be dry and saturated if the vessel is cooled? The cooling is further continued till the temperature of steam in the vessel becomes 116.93°C. Find out the pressure and condition of the steam at the end of operation.

Solution

The capacity of the vesel is 0.24 m^3. Saturation temperature of steam corresponding to pressure of 11 atm. bar is 184.07ºC. Therefore steam is superheated steam can be found out by Charle's Law steam behaves as a perfect gas and as the pressure of steam is constant, the volume of superheated steam can be found out by the formula

$$v_{sup} = v_{sat} \times \frac{T_{sup}}{T_{sat}}$$

$$= \frac{0.17738 \times 473}{184.07 + 273} = \frac{0.17738 \times 473}{457.07}$$

$$= 0.1836 \text{ m}^3/\text{kg}$$

As the vessel is closed, the volume of steam/kg will always remain constant whatever the pressure will be

$$\therefore \text{ Quantity of steam in vessel} = \frac{\text{Total volume}}{\text{Sp. volume}}$$

$$= \frac{0.24}{0.1836} = 1.307 \text{ kg}$$

When the steam is cooled, its quality will not be improved and the saturated steam having volume of steam = 0.1836 will have a pressure ≈10.5 bar.

Now the steam is further cooled so that its temprature is 116.93°C. The saturated pressure is 1.8 bar.

At 1.8 bar, volume of dry and saturated steam

$$= 0.97723 \text{ m}^3/\text{kg}$$

$$\therefore \quad 0.97723x = 0.1836$$

$$\therefore \quad \text{Condition of steam} = x = \frac{0.1836}{0.97723}$$

Condition of steam = 0.188

∴ Quantity steam in vessel = 1.307 kg

Pressure after cooling = 10.5 bar

Pressure after further cooling = 1.8 bar

Condition after further cooling D.F. = 0.188

Example 10.9 *Calculate the total energy and internal energy of 1 kg of steam at a pressure of bar and dryness 0.95. One cu.m of this steam is cooled at constant volume until its pressure has fallen to 5.4 bar. Determine the heat abstracted from steam.*

Solution

Total energy is enthalpy of steam

$$\therefore \quad h_1 = h_{f1} + x_1 h_{fg1}$$

$$= 742.64 + 0.95 \times 2029.5$$

$$= 742.64 + 1928.02$$

$$\mathbf{h_1 = 2670.66 \text{ kcal/kg}}$$

$$\text{Internal energy kg} \quad u_1 = h_1 - P_1 V_1$$

$$= 2670.66 - 100 \times 9 \times 0.95 \times 0.21481$$

$$= 2670.66 - 183.33$$

$$\mathbf{u_1 = 2497.00 \text{ kJ/kg}}$$

$$\text{Mass of one cu.m steam} = \frac{\text{Total volume}}{\text{Sp. volume}}$$

$$= \frac{1}{0.95 \times 0.21481} = 4.99 \text{ kg}$$

After cooling, the pressure is 5.4 bar at 5.4 bar, $V_{g2} = 0.34846 \text{ m}^3/\text{kg}$

$$\text{Actual volume/kg} = xv_{g2} = 0.95 \times 021481$$

$$\therefore \quad x_2 = \text{Dryness fraction of steam at 5.4 bar}$$

$$= \frac{0.95 \times 0.21481}{0.34846} = 0.586$$

$$\therefore \quad h_2 = h_{f2} + x_2 h_{fg2}$$

$$= 652.76 + 0.586 \times 2098.1$$

$$= 652.76 + 1229.5 = 1882.26 \text{ kJ/kg}$$

$$\therefore \quad u_2 = h_2 - p_2 v_2$$

$$= 1882.26 - 100 \times 5.4 \times 0.95 \times 0.21481$$

$$= 1882.26 - 110.2$$

$$= 1672.06$$

$$\therefore \quad u_2 - u_1 = \text{Change in internal energy} = \text{Heat}$$

$$\therefore \quad Q = 1672.03 - 2497.00 = -824.94 \text{ kJ/kg}$$

Total heat $Q = -499 \times 824.94$

$$\mathbf{Q = -4116.45 \text{ kJ}}$$

Example 10.10 *A closed vessel of 0.2 cu.m capacity contains steam at a pressure of 8 bar and a temperature of 200°C.*

(a) *What is the weight of steam contained in the vessel ?*

(b) *The vessel is cooled and the steam becomes just dry and saturated. What will be the pressure of steam at this state? Estimate the specific entropy.*

(c) *The vessel is further cooled till the temperature drops to 160.8°C. Find the pressure and condition of steam at this state.*

Solution

(a) *Weight of Steam*

Specific volume of superheated steam

Here pressure = 8 bar

$$V_{sup} = V_{sat} \times \frac{T_{sup}}{T_{sat}}$$

$$V_{sat} = 0.24026 \text{ m}^3$$

$$T_{sat} = 273 + 170.41$$

$$= 443.41°\text{K}$$

$$\therefore \quad V_{sup} = 0.24026 \times \frac{200 + 273}{170.41 + 273}$$

$$= \frac{0.24026 \times 473}{443.41} = 0.2563 \text{ m}^3/\text{kg}$$

$\therefore$ Weight of steam in the vessel

$$= \frac{\text{Actual volume}}{\text{Sp. volume}} = \frac{0.2}{0.2563}$$

Weight of steam in the vessel = 0.78 kg

(*b*) *Pressure and Entropy after cooling*

After cooling, the steam becomes dry and saturated.

∴ Volume of 1 kg dry and saturated steam = 0.2563 m^3

∴ **Saturation pressure of steam ≈ 7.5 bar**

(*c*) *Pressure and Entropy*

Saturation temperature of steam = 160.8°C

∴ Saturation pressure corresponding to this temperature ≈ 6.3 bar

$$= 6.3 \text{ bar}$$

$$V_g = \frac{0.30585 + 0.29681}{2}$$

$$= 0.30133 \text{ m}^3$$

$$\therefore \quad 0.2563 = x \times 0.30133$$

$$x = \frac{0.2563}{0.30133}$$

$$\mathbf{x = 0.85}$$

= Condition of steam (final)

II. Isothermal Process

Example 10.11 *One kilogram of 0.3 dry steam at 10 bar is heated isothermally so as to make it dry and saturated. Calculate the mechanical work, change in internal energy, enthalpy and entropy during the process.*

Solution

Note: Here the heating process is taking place in the wet region of *T–S* diagram. During this heating process, both temperature and pressure are constant

$$\therefore \quad P_1 = P_2 = 10 \text{ bar};$$

$$x_1 = 0.3 \text{ and } x_2 = 1$$

$$\therefore \quad V_1 = \text{Sp. volume of 1 kg steam before heating}$$

$$x_1 V_{g1} = 0.3 \times 0.1946$$

$$= 0.05838 \text{ m}^3/\text{kg}$$

∴ Mechanical work available

$$W = P(V_2 - V_1)$$

$$= 100(0.1946 - 0.05838) \times 10$$

$$= 1000 \times 0.13622$$

$$\mathbf{w = 136.22 \text{ kJ/kg}}$$

$$h_1 = \text{Enthalpy of 1 kg}$$

$$= h_{f1} + x_1 h_{fg1}$$

$$= 762.2 + 0.3 \times 2013.8$$

$$= 1366.34 \text{ kJ/kg}$$

$$h_2 = h_{g2} = 2776.1 \text{ kJ/kg}$$

$$h_2 - h_1 = \text{Change in enthalpy}$$

$$= 2776.1 - 1366.4$$

$$\mathbf{h_2 - h_1 = 1409.7 \text{ kJ/kg}}$$

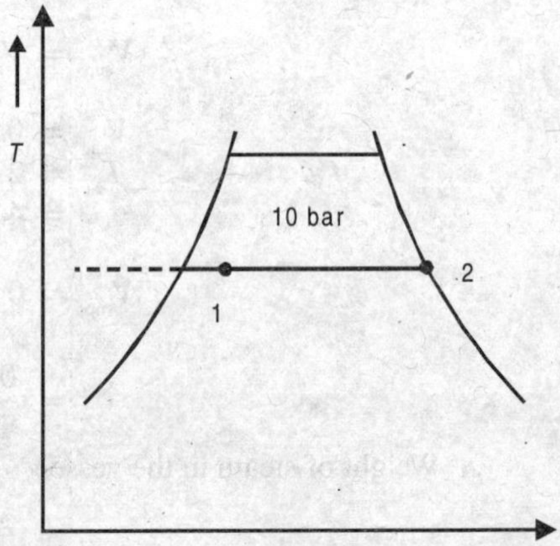

Fig. Ex. 10.11

$$u_1 = h_1 - P_1V_1$$
$$= 1366.34 - 100 \times 10 \times 0.05838$$
$$= 1366.34 - 58.38$$
$$= 1307.96 \text{ kJ/kg}$$
$$u_2 = h_2 - P_2V_2$$
$$= 2776.1 - 100 \times 10 \times 0.1946$$
$$= 2776.1 - 194.6$$
$$= 258.5 \text{ kJ/kg}$$

$\therefore$ $u_2 - u_1$ = Change in internal energy

$$= 2581.5 - 1307.96$$
$$\mathbf{u_2 - u_1 = 1273.54 \text{ kJ/kg}}$$

Again
$$S_1 = S_{f1} + x_1 S_{fg1}$$
$$= 2.138 + 0.3 \times (6.582 - 2.138)$$
$$= 2.138 + 1.3332$$
$$= 3.4712 \text{ kJ/kg-K}$$
$$S_2 = S_{g2} = 6.582 \text{ kJ/kg-K}$$

$\therefore$ $S_2 - S_1$ = Change in entropy

$$= 6.582 - 3.4712$$
$$\mathbf{S_2 - S_1 = 3.1108 \text{ kJ/kg-K}}$$

Example 10.12 *2 kg steam at 7 bar and 190°C is compressed isothermally to 11 bar. Find the change in entropy, heat added or removed and final condition of steam. Assume C_p = 2.3 kJ/kg-K.*

Solution

Here saturation temperature at 7 bar is 164.92°C and that at 11 bar 184.06°C. Actual temperature is 190°C. Therefore, steam is superheated before and after isothermal compression. The process is shown on *T–S* diagram.

$\therefore$ Final condition of steam is superheated steam with 5.94°C of superheat

$$S_1 = S_{g1} + C_p \log_e \frac{T_{sup}}{T_{sat}}$$
$$= 6.705 + 2.3 \log_e \frac{273 + 190}{273 + 164.92}$$
$$= 6.705 + 2.3 \log_e \frac{463}{437.92}$$
$$= 6.705 + 0.1281$$
$$= 6.8330 \text{ kJ/kg-K}$$
$$S_2 = S_{g2} + C_p \log_e \frac{T_{sup}}{T_{sat}}$$
$$= 6.55 + 2.3 \log_e \frac{463}{457.06}$$
$$= 6.55 + 0.0297 = 6.5797 \text{ kJ/kg-K}$$

$\therefore$ Change in entropy

$$= S_2 - S_1 = 6.5797 - 6.8331$$
$$= -0.2534 \text{ kJ/kg-K}$$
$$= -0.5068 \text{ kJ/K.}$$

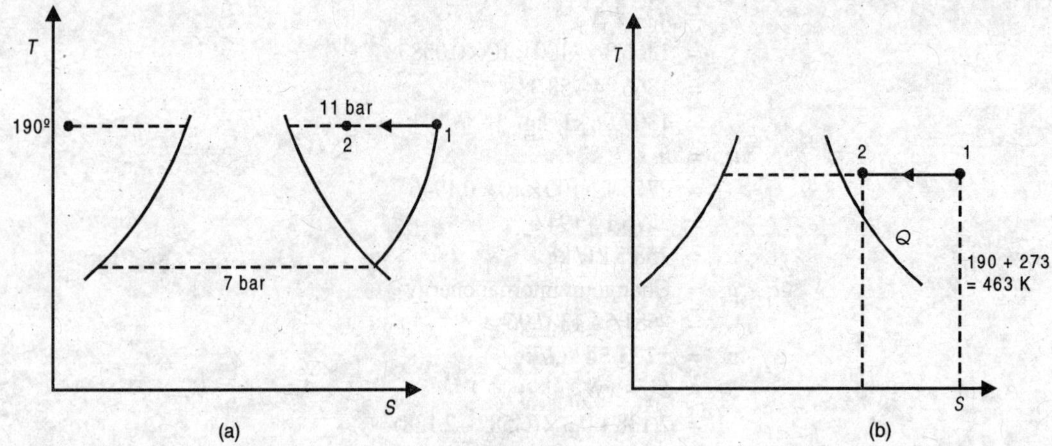

Fig. Ex. 10.12

Heat during the process

$$Q = T \times (S_2 - S_1)$$
$$= (273 + 190) \times (-0.50680)$$
$$= -463 \times 0.5068$$
$$\mathbf{Q = -234.65\ kJ}$$

– ve sign indicates that heat is rejected or removed.

Example 10.13 *Dry and saturated steam at 6.5 bar is expanded isothermally to 0.5 bar in a special apparatus. Calculate the external work available from 2 m³ of steam initially taken, as well as the change in entropy during the process.*

Solution

The process is shown in Fig. Ex. 10.13

$$T_1 = T_2 = 161.96 + 273 = 434.96\ \text{K}$$
$$S_1 = \text{Entropy before expansion}$$
$$= S_{g1} = 6.730\ \text{kJ/kg.K}$$
$$S_2 = S_{g2} + C_p \ln \frac{T_{sup}}{T_{sat}}$$
$$= 7.597 + 2.3 \log_e \frac{434.96}{273 + 81.31}$$
$$= 7.597 + 2.3 \ln \frac{434.96}{354.31}$$
$$= 7.597 + 0.4717$$
$$= 8.0687\ \text{kJ / Kg-K}$$

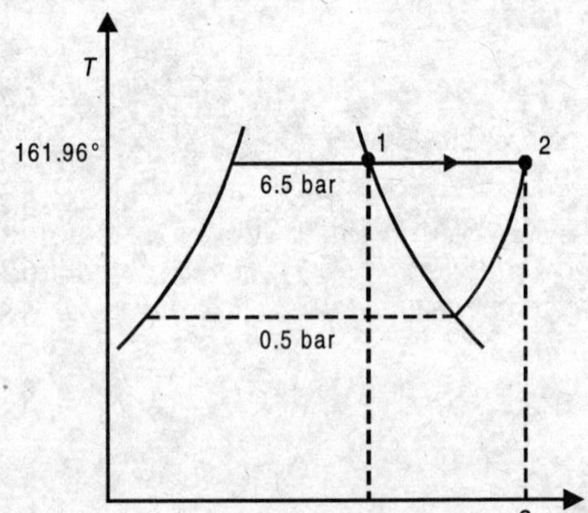

Fig. Ex. 10.13

$\therefore$ Q = Heat during the expansion process

$$= T_1 \times (S_2 - S_1)$$
$$= 434.96 \times (8.0687 - 6.730)$$
$$= 434.96 \times 1.3387$$
$$= 582.28\ \text{kJ / kg}$$

It is positive. i.e., heat is added.

Initial volume of steam $= V_1 = 0.292847 \text{ m}^3/\text{kg}$

$$\therefore \text{Mass of steam processed} = \frac{2}{0.292847} = 6.8295 \text{ kg}$$

$$\therefore \text{Total heat} = 6.8295 \times 582.28$$

Total Heat = 3976.68 kJ

Also change in entropy $= (S_2 - S_1) = 6.8295 \times 582.28$

$= 9.1426 \text{ kJ/K}$

$$h_1 = h_{g1} = 2758.8 \text{ kJ/kg}$$

$$\therefore u_1 = h_1 - P_1V_1 = 2758.8 - 100 \times 6.5 \times 0.29284$$
$$= 2758.8 - 190.346$$
$$= 2568.454 \text{ kJ/kg}$$

$$h_2 = h_{g2} + C_p(t_{sup} - t_{sat})$$
$$= 2645.3 + 2.3(161.96 - 81.31)$$
$$= 2645.3 + 185.495$$
$$= 2830.795 \text{ kJ / kg}$$

$$V_2 = V_{g2} \times \frac{T_{sup}}{T_{sat}} = 3.24827 \times \frac{434.96}{354.31}$$
$$= 3.9877 \text{ m}^3/\text{kg}$$

$$\therefore u_2 = h_2 - P_2V_2$$
$$= 2830.795 - 100 \times 0.5 \times 3.9877$$
$$= 2830.795 - 199.383$$
$$= 2631.412 \text{ kJ / kg}$$

$$\therefore u_2 - u_1 = 2631.412 - 2568.454$$
$$= 62.958 \text{ kJ/kg}$$

$$\therefore Q = \Delta u + \text{Work}$$

$$3976.68 = 6.8295 \times 62.958 + \text{Work}$$

$$\therefore \text{Work} = 3976.68 - 429.97$$

Work = 3546.71 kJ

Example 10.14 *A quantity of dry and saturated steam occupies 0.2876 cu.m at 14 bar. Determine the final condition of steam if it is compressed until its volume is halved, if the compression is carried out isothermally. Determine the heat rejected during compression.*

Solution

$$\text{Mass of steam} = \frac{\text{Total steam volume}}{\text{Sp. volume}}$$

$$= \frac{0.2876}{0.14072}$$

Consider 1 kg steam. Then we have

$P_1 = 14$ bar; $V_1 = 0.14072 \text{ m}^3$;

$t_1 = 195.04°\text{C}$; $V_2 = 0.07036 \text{ m}^3$;

$t_2 = 195.04°\text{C}$

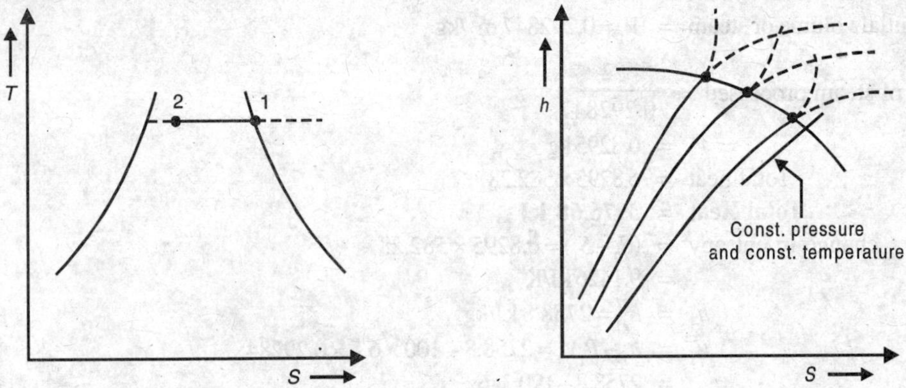

Fig. Ex. 10.14

As the temperature remains same and as the steam is initially dry and saturated, the compression of steam is at the same pressure because 195.04°C is going to be the saturation temperature at p_2 or $p_1 = p_2$.

Now as the volume is halved, the dryness fraction is given by

$$V_2 = x_2 V_{g2}$$

$$\therefore \quad 0.07036 = x_2 \times 0.14072$$

$$\mathbf{x_2 = 0.5}$$

Compression takes place in the wet region and therefore that is also a constant pressure (in addition to constant temperature process) process.

$$\therefore \quad \text{Heat rejected} = (1 - x_2)\, h_{fg2}$$

$$= (1 - 0.5) \times 1957.7$$

$$= 0.5 \times 1957.7$$

$$= 978.85 \text{ kJ/kg}$$

$$\therefore \quad \text{Total heat rejected} = 2.044 \times 978.85$$

$$\textbf{Total heat rejected} = \textbf{2000 kJ}$$

III. Isentrpoic Process

Example 10.15 *Steam at 12 bar and 100°C superheat is expanded adiabatically and reversibly till becomes dry and saturated. Calculate the lower pressure: Calculate also the work done if the process is non-flow type.*

Solution

For reversible adiabatically process, entropy before and after process is unchanged.

$$\therefore \quad S_1 = S_2$$

Assuming specific heat of steam as 2.2 kJ / kg.K

We have,

$$S_1 = S_{g1} + C_{pv} \log_e \frac{T_{\text{sup}}}{T_{\text{sat}}}$$

$$= 6.52 + 2.2 \log_e \frac{273 + 187.92 + 100}{273 + 187.92}$$

$$= 6.52 + 2.2 \log_e \frac{560.92}{460.92}$$

$$= 6.52 + 0.432$$

$$= 6.952 \text{ kJ/kg-K}$$

This can also be found from superheat steam tables

$\therefore \quad S_2 = 6.952 = S_{g2}$

as steam is dry and saturated after expansion.

$\therefore$ Pressure corresponding to $S_{g2} = 6.952$ is approximately 3.4 bar.

$\therefore \quad$ **P = 3.4 bar**

For non-flow process, $Q = \Delta u + \int P.dV$

$\therefore \quad O = \Delta u + \int P.dV$

$\therefore \quad$ Work done/kg $= -\Delta u = u_1 - u_2$

h_1 = Enthalpy of 1 kg steam before expansion

$= h_{g1} + C_{pv}$ (Degree of superheated)

$= 2782.6 + 2.2 \times 100$

$= 2782.6 + 220$

$= 3002.6 \text{ kJ/kg}$

V_1 = Volume of 1 kg steam before expansion

$$= V_{g1} \times \frac{T_{\text{sup }1}}{T_{\text{sat }1}}$$

$$= 0.163446 \times \frac{5609.92}{460.92}$$

$= 0.1989 \text{ m}^3\text{/kg}$

$\therefore \quad u_1$ = Internal energy of 1 kg steam before expansion

$= h_1 - P_1V_1$

$= 3002.6 - 100 \times 12 \times 0.1989$

$= 3002.6 - 238.68$

$= 2763 \text{ kJ/kg}$

$\therefore \quad h_2 = h_{g2}$ = Enthalpy of 1 kg steam after expansion

$= 2730.4 \text{ kJ/kg}$

$\therefore \quad u_2 = h_2 - P_2V_2$

$= 2730.4 - 100 \times 3.4 \times 0.539$

$= 2730.4 - 183.26$

$= 2547.14 \text{ kJ/kg}$

$\therefore \quad$ Work done/kg $= u_1 - u_2$

$= 2763.92 - 2547.14$

Work done/kg = 216.78 kJ/kg

Example 10.16 *Steam flows into a steam nozzle with a pressure of 14 bar when it is dry and saturated. If the expansion is isentropic to 2 bar, find the final condition of steam and change in internal energy, taking C_P = 2.2 kJ/kg.K.*

Find also the velocity of steam leaving the nozzle.

Solution

This is an example of steady flow system. Flow is isentropic.

$P_1 = 14$ bar, $P_2 = 2$ bar

$S_1 = S_{g1} = 6.465 \text{ kJ/kg-K}$

$= S_2 = S_{f2} + x_2 S_{fg2}$

$$= 1.529 + x_2(7.127 - 1.529)$$
$$= 1.529 + 5.598\,x_2$$
$\therefore$ x_2 = Find dryness fraction of steam

$$= \frac{6.465 - 1.529}{5.598} = \frac{4.936}{5.598}$$
$$\mathbf{x_2 = 0.8817}$$

Now $h_1 = h_{g1} = 2787.9 \text{ kJ/kg}$

$$h_2 = h_{f2} + x_2 h_{fg2}$$
$$= 504.5 + 0.8817 \times 2201.8$$
$$= 504.5 + 1941.33$$
$$= 244.83 \text{ kJ/kg}$$
$$V_1 = V_{g1} = 0.1409 \text{ m}^3\text{/kg}$$
$$V_2 = x_2 V_{g2} = 0.8817 \times 0.8872$$
$$= 0.7822 \text{ m}^3\text{/kg}$$
$$\therefore \quad u_1 = h_1 - P_1 V_1$$
$$= 2787.9 - 100 \times 14 \times 0.1409$$
$$= 2787.9 - 197.26$$
$$= 2590.64 \text{ kJ/kg}$$
$$u_2 = h_2 - P_2 V_2$$
$$= 2445.83 - 100 \times 2 \times 0.7822$$
$$= 2445.83 - 156.44$$
$$= 2289.39 \text{ kJ/kg}$$
$$\therefore \quad \Delta u = u_2 - u_1 = \text{change in internal energy}$$
$$= 2289.39 - 2590.64$$
$$\mathbf{\Delta u = -301.25 \text{ kJ/kg}}$$

Assuming change in potential energy and initial velocity, we have

$$Q = \Delta P + \Delta K + \Delta H + W_f$$
$$O = O + \Delta K + \Delta H + O$$
$$\therefore \quad \Delta K = -\Delta H = H_1 - H_2$$
$$\frac{C_2^2 - C_1^2}{2000} = H_1 - H_2$$
$$= 2787.9 - 2445.83$$
$$= 342.07$$
$$C_2 = \sqrt{2000 \times 342.07}$$
$$\mathbf{C_2 = 827.13 \text{ m/s}}$$
= Velocity at exit

Example 10.17 *Steam at 15 bar and 250°C expands reversibly and adiabatically to 3 bar in a steam turbine, find the condition of steam after expansion and the work done per kg steam admitted to the turbine. Assume ΔP and ΔK to be zero.*

Solution

This is an other example of steady flow system. Here $P_1 = 3$ bar, $T_1 = 523$ K

$$h_1 = 2923.8 \text{ kJ/kg}$$
$$S_1 = 6.711 \text{ kJ/kg.K} = S_2$$
$$\therefore \quad 6.711 = s_{f2} + x_2 s_{fg2}$$

$$\therefore \quad 6.711 = 1.671 + x_2(6.991 - 1.671)$$

$$x_2 = \frac{6.711 - 1.671}{5.320}$$

$$= \frac{5.040}{5.320}$$

$$\mathbf{x_2 = 0.9474}$$

= Final condition of steam

$\therefore \quad h_2$ = Enthalpy of steam after expansion

$$= h_{f2} + x_2 f_{fg2}$$

$$= 561.3 + 0.9474 \times 2163.3$$

$$= 561.3 + 2049.5$$

$$= 2610.8 \text{ kJ/kg}$$

For steady flow system, we have

$$Q = \Delta P + \Delta K + \Delta H + W_{sf}$$

$$O = O + O + \Delta H + W_f$$

$\therefore \quad W_{sf}$ = work of steady flow system

$$= h_1 - h_2$$

$$= 2923.8 - 2610.8$$

$$\mathbf{W_{sf} = 313.0 \text{ kJ/kg}}$$

Example 10.18 *Find the internal energy of steam 0.9 dry at 16 bar reckoned from water at 0° C. If the steam is expanded isentropically in a cylinder to 0.9 bar, find the final internal energy of the steam.*

Solution

$$h_1 = h_{f1} + x_1 f_{fg1}$$

$$= 858.3 + 0.9 \times 1933.4$$

$$= 858.3 + 1740.06$$

$$= 2598.36 \text{ kJ/kg}$$

v_1 = Initial volume of steam

$$= X_1 V_{g1} = 0.9 \times 0.123843$$

$$= 0.1115 \text{ m}^3\text{/kg}$$

Neglecting volume of water at 0°C,

$$u_1 = h_1 - P_1 V_1 = 2598.36 - 100 \times 16 \times 0.1115$$

$$= 2598.36 - 178.33$$

$$\mathbf{u_1 = 2420.03 \text{ kJ/kg-K}}$$

= Internal energy reckoned from water at 0°C

Now $\quad S_1 = s_{f1} + x_1 s_{fg1} = 2.343 + 0.9(6.417 - 2.343)$

$$= 2.343 + 3.6666$$

$$= 6.0096 \text{ kJ/kg-K}$$

$$S_2 = S_{f2} + x_2 S_{fg2} = 1.269 + x_2(7.396 - 1.269)$$

$$= 1.269 + x_2 \times 6.127$$

$\therefore \quad S_1 = S_2$ for isentropic process

$$= 1.269 + 6.127$$

$$x_2 = \frac{6.0096 - 1.2690}{6.127} = \frac{4.7406}{6.127}$$

$$= 0.7737$$

$$\therefore \quad h_2 = h_{f2} + x_2 h_{fg2} = 405.1 + 0.7737 \times 2265.8$$
$$= 405.1 + 1753.1 = 2158.2\ \text{kJ/kg}$$
$$V_2 = x_2 v_{g2} = 0.7737 \times 1.87135$$
$$= 1.4479\ \text{m}^3/\text{kg}$$
$$\therefore \quad u_2 = \text{Final internal energy above } 8°\text{C}$$
$$= h_2 - P_2 V_2$$
$$= 2158.2 - 100 \times 0.9 \times 1.4479$$
$$= 2158.2 - 130.31$$
$$\mathbf{u_2 = 2027.89\ kJ/kg}$$

Example 10.19 *Calculate the index of adiabatic expansion if steam expanded adiabatically from 6 bar and 200°C to 1 bar has dryness fraction 0.93 at the end of expansion.*

Take $C_p = 2.3\ kJ/kg.K$; $p_1 = 6$ bar; $p_2 = 1$ bar; $x_2 = 0.93$

Solution

$$V_1 = V_{g1} \times \frac{T_{sup}}{T_{sat}}$$

$$= 0.3159 \times \frac{200 \times 273}{158.81 + 273} = \frac{0.3159 \times 473}{431.81}$$

$$= 0.346\ \text{m}^3/\text{kg}$$
$$V_2 = x_2 V_{g2}$$
$$= 0.93 \times 1.6949$$
$$= 1.5762\ \text{m}^3/\text{kg}$$

Let n be the index of adiabatic expansion.

$$\therefore \quad P_1 V_1^n = P_2 V_2^n$$

or
$$\left(\frac{V_2}{V_1}\right)^n = \frac{P_2}{V_2}$$

or
$$n \ln \frac{V_2}{V_1} = \ln \frac{p_1}{p_2}$$

$$N = \frac{\ln P_1/P_2}{\ln V_2/V_1} = \frac{\ln \frac{6}{1}}{\ln . \frac{1.5762}{0.346}}$$

$$= \frac{1.7918}{1.5163}$$

$$\mathbf{n = 1.1817}$$

Example 10.20 *1 kg of steam at 100 bar and 375° C expands isentropically upto 38 bar. At the end of expansion steam is dry and saturated. For both non flow and steady flow process with $\Delta K = 0$, find the work done.*

Data: $p_1 = 100$ bar; $t_1 = 375°$ C; $p_2 = 38$ bar; $x_2 = 1$.

Solution

From superheated steam tables –

$$V_1 = \frac{0.02242 + 0.2641}{2} = \frac{0.04883}{2} = 0.024415$$

$$h_1 = \frac{2925.8+3099.9}{2} = \frac{6025.7}{2} = 3012.85 \text{ m}^3/\text{kg}$$

$$\mathbf{h_1 = 3012.85}$$

$$V_2 = V_{g2} = 0.052438 \text{ cu.m/kg}$$

$$h_2 = 2801.1 \text{ kJ/kg}$$

Non-flow process :

$$u_1 = h_1 - P_1V_1 = 3012.85 - 100 \times 0.024415$$
$$= 3012.85 - 244.15$$
$$= 2768.70 \text{ kJ/kg}$$
$$u_2 = h_2 - P_2V_2 = 2801.1 - 100 \times 0.052438$$
$$= 2801.1 - 199.26$$
$$= 2601.84 \text{ kJ/kg}$$

$\therefore$ For non-flow process $= Q = \Delta u + \text{work}$

$$0 = (u_2 - u_1) + w_{nf}$$

$\therefore$ W_{nf} = Work done $= u_1 - u_2$

$$= 2768.70 + 2601.84$$
$$W_{nf} = 166.86 \text{ kJ/kg}$$

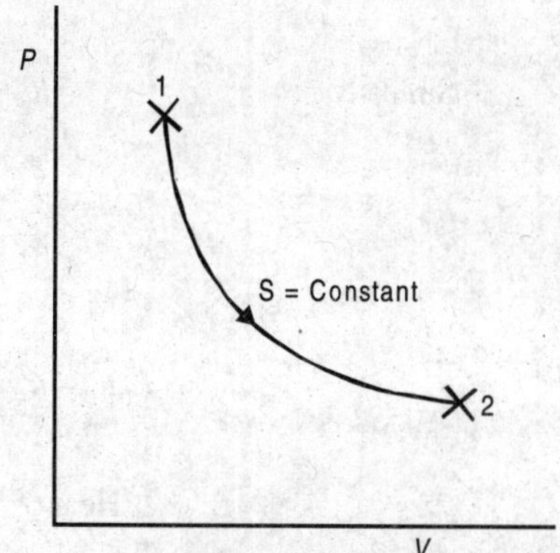

Fig. Ex. 10.20

For steady flow process

with $\Delta K = 0$

We have Work $= -\Delta H = h_1 - h_2$

$$= 3012.85 - 2801.1$$

Work = 277.75 kJ/kg

IV. Hyperbolic Process

Example 10.21 *Steam at 3.6 bar and 0.66 dry is compressed hyperbolically to 10 bar. Determine the heat removed from 2 m³ of steam.*

Solution

$$P_1 = 3.6 \text{ bar}; x_1 = 0.66$$

$\therefore$
$$V_1 = x_1V_{g1} = 0.66 \times 0.5108$$
$$= 0.3371 \text{ m}^3/\text{kg}$$
$$P_2 = 10 \text{ bar}$$

Law of expansion is $PV = C$

$\therefore$
$$P_1V_1 = P_2V_2$$
$$V_2 = V_1 \times \frac{P_1}{P_2} = 0.337 \times \frac{3.6}{10}$$
$$= 0.1214 \text{ m}^3/\text{kg}$$

This is less than V_{g2}. Therefore the steam is wet after compression.

$\therefore$
$$0.1214 = x_2 \times 0.1946$$
$$x_2 = \frac{0.1214}{0.1946}$$
$$= 0.6237$$

$\therefore$ Work done/kg $= P_1V_1 \ln\left(\frac{V_2}{V_1} = \frac{P_1}{P_1}\right)$

$$= -100 \times 3.6 \times 0.337 \times \ln \frac{10.0}{3.6}$$

$$= -367.8 \text{ kJ/kg}$$

Now $h_1 = h_{f1} + x_1 h_{fg1}$

$$= 588.4 + 0.66 \times 2144.5$$

$$= 2003.77 \text{ kJ/kg}$$

$\therefore$ $u_1 = h_1 - P_1V_1$

$$= 2003.77 - 100 \times 3.6 \times 0.3371$$

$$= 2125.126 \text{ kJ/kg}$$

Similarly $h_2 = h_{f2} + x_2 h_{fg2}$

$$= 762.2 + 0.6237 \times 2013.8$$

$$= 2018.2 \text{ kJ/kg}$$

$\therefore$ $u_2 = h_2 - P_2V_2$

$$= 2018.2 - 121.356$$

$$= 1896.844 \text{ kJ/kg}$$

$\therefore$ $u_2 - u_1 = 1896.844 - 2125.126$

$$= -228.282 \text{ kJ/kg}$$

$\therefore$ $\text{Heat } Q = \Delta u + \int P \cdot dV$

$$= -228.282 - 367.8$$

Heat Q = –596.082 kJ/kg

= Heat removed.

$$= -596.082 \times \frac{2}{0.3371} = -3536.53 \text{ kJ}$$

Example 10.22 *Steam at 8 bar and 175°C is throttled down to 4 bar and then expanded hyperbolically to 1.2 bar. Determine*

(i) *change in entropy during throttling*

(ii) *change in enthalpy during expansion and*

(iii) *work done*

Solution

(i) Change in entropy during throttling : (1 – 2)

h_1 = Enthalpy before throttling

$$= h_{g1} + c_p (t_{\text{sup}} - t_{\text{sat}})$$

$$= 2676.3 + 2.3 (175 - 170.38)$$

$$= 2767.3 + 10.626$$

$$= 2777.923 \text{ kJ/kg}$$

$$S_1 = S_{g1} + C_p \ln \frac{T_{\text{sup}}}{T_{\text{sat}}} = 6.659 + 2.3 \ln \frac{175 + 273}{170.38 + 273}$$

$$= 6.659 + 0.0238$$

$$= 6.6828 \text{ kJ/kg.K}$$

For throttling process,

$$h_1 = h_2 > h_{g2}$$

$\therefore$ $2777.926 = 2737 + 2.3 (t_{\text{sup2}}) - 143.6$

$$\therefore \quad t_{sup2} = \frac{2777.962 - 2737.7}{2.3} + 143.6$$

$$= \frac{40.226}{2.3} + 143.6$$

$$= 17.49 + 143.6 = 161.09°C$$

$$\therefore \quad S_2 = S_{g2} + C_p \ln \frac{T_{sup}}{T_{sat}}$$

$$= 6.894 + 2.3 \ln \frac{273 + 161.09}{273 + 143.6}$$

$$= 6.894 + 2.3 \ln \frac{434.09}{416.6}$$

$$= 6.894 + 0.0946$$

$$= 6.9886 \text{ kJ/kg}$$

$\therefore$ Change in entropy during throttling

$$= S_2 - S_1 = 6.9886 - 6.6828$$

$$= 0.3058 \text{ kJ/kg-K}$$

(i) Change in enthalpy during expansion (2 – 3)

$$h_2 = 2777.926 \text{ kJ/kg}$$

$$V_2 = V_{g2} \times \frac{T_{sup}}{T_{sat}}$$

$$= 0.462626 \times \frac{434.09}{416.6}$$

$$\mathbf{V_2 = 0.482 \text{ m}^3\text{/kg}}$$

$P_2 = 4$ bar; $P_3 = 1.2$ bar

$$\therefore \quad P_2V_2 = P_3V_3$$

$$\therefore \quad V_3 = V_2 \times \frac{P_2}{P_3} = \frac{0.482 \times 4}{1.2}$$

$$= 1.607 \text{ m}^3\text{/kg}$$

$$> v_{g3}$$

$\therefore$ Steam is superheated after expansion

$$\therefore \quad 1.607 = 1.43016 \times \frac{T_{sup_3}}{104.77 + 273}$$

$$\therefore \quad T_{sup3} = \frac{1.607}{1.43016} \times \frac{377.77}{1}$$

$$= 424.436 \text{ K}$$

$$\therefore \quad h_{g3} = h_{g3} + C_p (T_{sup3} - T_{sat3})$$

$$= 2684.9 + 2.3 (424.436 - 377.77)$$

$$= 2684.9 + 2.3 \times 46.666$$
$$= 2684.9 + 107.33$$
$$= 2792.23 \text{ kJ/kg}$$

∴ Change in enthalpy during expansion

$$h_3 - h_2 = 2792.23 - 2777.926$$
$$\mathbf{h_3 - h_2 = 14.304 \text{ kJ/kg}}$$

(iii) Work done

$$\text{Work done} = P_2 V_2 \ln\left(\frac{V_3}{V_2} = \frac{P_2}{P_3}\right)$$
$$= 100 \times 4 \times 0.482 \ln \frac{4}{1.2}$$
$$\textbf{Work done} = \mathbf{232.126 \text{ kJ/kg}}$$

Example 10.23 *The steam in the steam engine cylinder expands according to the hyperbolic law from 8.5 bar and 0.85 dry to 0.7 bar.*

Find the final condition of steam and work done per kilogram assuming C_p = 2.01 kJ/kg-K.

Solution

We have $P_1 = 8.5$ bar; $P_2 = 0.7$ bar; $x_1 = 0.85$ dry

$$\therefore \quad V_1 = x_1 V_{g1} = 0.85 \times 0.227037$$
$$= 0.193 \text{ m}^3/\text{kg}$$
$$\therefore \quad P_1 V_1 = P_2 V_2$$
$$V_2 = \frac{P_1}{P_2} \cdot V_1 = \frac{8.5}{0.7} \times 0.193$$
$$= 2.3433 \text{ m}^3/\text{kg}$$
$$< V_{g2}$$

∴ The steam is wet

$$\therefore \quad 2.3433 = x_2 \times 2.36795$$
$$\therefore \quad x_2 = \frac{2.3433}{2.36795}$$
$$\mathbf{x_2 = 0.9896}$$
$$\text{Work done/kg} = P_1 V_1 \log_e\left(\frac{V_2}{V_1} = \frac{P_1}{P_2}\right)$$
$$= 100 \times 8.5 \times 0.193 \ln \frac{8.5}{0.7}$$
$$\textbf{Work done/kg} = \mathbf{409.59 \text{ kJ/kg}}$$

Example 10.24 *10 kg of steam at 8 bar and 0.9 dry expanded in a steam engine cylinder hyperbolically to 0.4 bar. Find the final condition of steam and the external work done during the process.*

Solution

$$P_1 = 8 \text{ bar};$$
$$P_2 = 0.4 \text{ bar};$$
$$V_1 = xV_{g1} = 0.9 \times 0.2405 = 0.21645 \text{ m}^3/\text{kg}$$
$$\therefore \quad P_1 V_1 = P_2 V_2$$

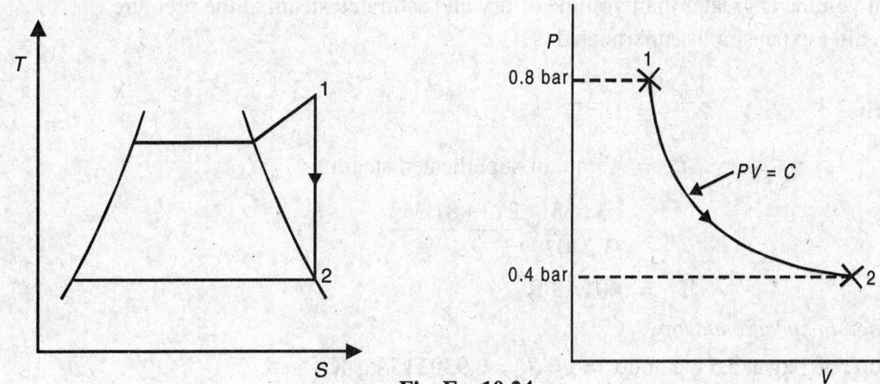

Fig. Ex. 10.24

$$\therefore \quad V_2 = \frac{P_1}{P_2} \times V_1 = \frac{8}{0.4} \times 0.21645$$

$$= 4.329\ \text{m}^3/\text{kg}$$

$$> V_{g2}\,(3.993)$$

$\therefore$ Steam is superheated.

$$\therefore \quad 4.329 = V_{g2} \times \frac{T_{\sup 2}}{T_{sat\,2}} = \frac{3.993 \times T_{\sup 2}}{75.9 + 273}$$

$$\therefore \quad T_{\sup 2} = \frac{4.329}{3.993} \times 348.9$$

$$= 378.26°\text{K}$$

$$\mathbf{T_{sup2} = 105.26°\ C}$$

$$\therefore \quad \text{Work done/kg} = P_1 V_1 \ln \frac{P_1}{P_2}$$

$$= 100 \times 8 \times 0.21645 \ln \frac{8}{0.4}$$

$$= 518.74\ \text{kJ/kg}$$

$$\textbf{Work done/kg} = \textbf{5187.4 kJ}$$

Example 10.25 *Dry and saturated steam of 3.5 bar pressure is expanded hyperbolically to 0.5 bar. Using steam tables, determine*

(i) *Final condition of steam*

(ii) *Change of specific entropy during the process*

Solution

(i) Hyperbolic process is defined by PV = Const. or $P_1V_1 = P_2V_2$

Here $P_1 = 3.5$ bar and $P_2 = 0.5$ bar; $x_1 = 1$

$$\therefore \quad V_1 = V_{g1} = 0.524\ \text{m}^3/\text{kg}$$

$$\therefore \quad V_2 = \text{volume after axpansion}$$

$$= \frac{P_1 V_1}{P_2} = \frac{3.5 \times 0.524}{0.5}$$

$$= 3.668\ \text{cu.m/kg}$$

$$V_{g2} = 3.2402\ \text{m}^3/\text{kg}$$

∴ Actual volume is greater than volume of dry and saturated steam at the pressure.

∴ Steam after expansion is superheated.

∴ We write, $\frac{V_2}{V_{g_2}} = \frac{T_2}{T_{sat_2}}$

or T_2 = Temp. of superheated steam

$$= \frac{3.668}{3.2402} \times \frac{273+81.343}{1}$$

$$\mathbf{T_2 = 401.13\ K}$$

(ii) *Change in specific entropy*:

S_2 = Entropy of 1 kg at 3.5 bar and $x_1 = 1 = S_{g1} = 6.9392$ kJ/kg-K

Entropy of 1 kg steam at 0.5 bar

$$S_2 = S_{g2} + C_{pv} \ln \frac{T_{sup}}{T_{sat}}$$

$$= 7.5947 + 2 \ln \frac{401.13}{354.343}$$

$$= 7.5947 + 2 \times 0.1240$$

$$= 7.5947 + 248$$

$$= 7.8427 \text{ kJ/kg-K}$$

∴ Change in entropy $= S_2 - S_1$

$= 7.8427 - 6.9392$

Change in entropy = 0.9035 kJ/kg-K

Example 10.26 *A quantity of dry and saturated steam occupies 0.2876 cu.m at 14 bar. Determine the final condition of the steam if it is compressed until its volume is halved, and if the compression is carried out hyperbolically.*

Determine the heat rejected during compression.

Solution

Volume of 1 kg dry and saturated steam at 14 bar = 0.14072 cu.m.

∴ Mass of steam $= \frac{0.2876}{0.14072} = \frac{\text{Total volume}}{\text{Sp. volume}}$

$= 2.044$ kg

Hyperbolic process follows the law $PV = C$

∴ $P_1V_1 = P_2V_2$ Consider 1 kg of steam

∴ $\frac{V_1}{V_2} = \frac{P_2}{P_1} = \frac{0.14072}{0.14072/2} = 2$

∴ $P_2 = P_1 \times 2 = 14 \times 2 = 28$ bar

$V_2 = 0.14072/2 = 0.07036$ cu.m/kg

$< V_{g2}\ (0.071389)$

∴ Steam after compression is wet.

∴ $V_2 = X_2V_{g2} = x_2 \times 0.071389$

∴ x_2 = Dryness fraction after compression

$$= \frac{0.07036}{0.071389}$$

$$x_2 = 0.9856$$

Calculation for heat $h_1 = 2787.8 \text{ kJ/kg}$

$$\therefore \quad u_1 = h_1 - P_1V_1 = 2787.8 - \frac{10^5 \times 14 \times 0.14072}{1000}$$

$$= 2787.8 - 197$$

$$= 2590.8 \text{ kJ/kg}$$

Similarly, $h_2 = h_{f2} + x_2 h_{fg2}$

$$= 990.48 + 0.9856 \times 1811.5$$

$$= 990.48 + 1785.41$$

$$= 2775.89 \text{ kJ/kg}$$

$$\therefore \quad u_2 = h_2 - P_2V_2 (= P_1V_1)$$

$$= 2775.89 - 197$$

$$= 2578.89 \text{ kJ/kg}$$

$$\text{Work done kg} = P_1 V_1 \ln \frac{V_2}{V_1}$$

$$= \frac{10^5 \times 14 \times 0.14072}{1000} \times \ln \frac{1}{2}$$

$$= 197 \ln \frac{1}{2}$$

$$= -136.55 \text{ kJ/kg}$$

$$\therefore \quad Q = \Delta u + \int P \cdot dV$$

$$= (u_2 - u_1) + (-136.55)$$

$$= (2578.89 - 2590.8) - 136.55$$

$$= 11.91 - 136.55$$

$$= -148.46 \text{ kJ/kg}$$

$\therefore$ Total heat rejected $= -148.46 \times 2.044$

$\therefore$ **Total heat rejected $= -303.45$ kJ**

V. Polytropic Process

Example 10.27 *Steam originally at 16 bar and 0.9 dry expands in a cylinder according to law $pV^{1.1}$ = const. Until the pressure is 1.6 bar. Determine the amount of heat put into or extracted from the steam per kg during process stating the direction of heat transfer.*

Given: P_1 = 16 bar; x_1 = 0.9 dry; P_2 = 1.6 bar; $PV^{1.1}$ = constant.

Solution

From steam table V_1 = Initial volume of steam/kg

$$= x_1 V_{g1}$$

$$= 0.9 \times 0.12369$$

$$= \frac{0.1113}{n} \text{ m}^3\text{/kg}$$

We have, $P_1 V^n_1 = P_2 V_2$

$$\therefore \quad \left(\frac{V_2}{V_1}\right)^n = \frac{P_1}{P_2}$$

or

$$V_2 = V_1 \cdot \left(\frac{P_1}{P_2}\right)^{1/n}$$

$$= 0.1113\left[\left(\frac{16}{1.6}^{1/1.17}\right)\right]$$

$$= 0.1113 \times (10)^{0.9091}$$

$$= 0.1113 \times 8.1115$$

$$= 0.9028 \text{ m}^3\text{/kg}$$

This is less than $V_{g2} = 1.0911$

Therefore, steam after expansion is wet and

$$x_2 = \frac{0.9028}{1.0911}$$

$$= 0.8274$$

$\therefore \quad h_1$ = Enthalpy of 1 kg steam before expansion

$$= h_{f1} + x_1 h_{fg2}$$

$$= 858.56 + 0.9 \times 1933.2$$

$$= 858.56 + 1739.88$$

$$= 2598.44 \text{ kJ/kg}$$

$$\therefore \quad u_1 = h_1 - P_1 V_1$$

$$= 2598.44 - 100 \times 16 \times 0.1113$$

$$= 2598.44 - 178.08$$

$$= 2420.36 \text{ kJ/kg}$$

Similarly, h_2 = Enthalpy of 1 kg steam after exp.

$$= hf_2 + x_2 h_{fg2}$$

$$= 475.38 + 0.8274 \times 2220.9$$

$$= 2302.95 \text{ kJ/kg}$$

$$\therefore \quad u_2 = h_2 - P_2 V_2$$

$$= 2302.95 - 100 \times 1.6 \times 0.9028$$

$$= 2158.50 \text{ kJ/kg}$$

$$\text{Work done/kg} = \int P.dV$$

$$= \frac{P_1 V_1 - P_2 V_2}{n-1}$$

$$= \frac{178.08 - 144.45}{1.1 - 1}$$

$$= \frac{33.63}{0.1}$$

$$= 336.3 \text{ kJ/kg}$$

Energy equation gives

$$Q = \Delta u + \int P.dV$$
$$= (u_2 - u_1) + 336.3$$
$$= (2158.5 - 2420.36) + 336.3$$
$$\mathbf{Q = 74.44\ kJ/kg}$$

Here Q is positive. This means that heat is added to the system i.e., steam.

Example 10.28 *0.9 kg steam initially at a pressure if 15 bar and temperature of 250°C expands reversibly and polytropically to 1.5 bar. Find (i) the final condition (ii) work done (iii) heat transferred and (iv) change in entropy if the index of expansion is 1.25.*

Show the process on P–V and T–S diagrams.

Given: $P_1 = 15$ bar; $T_1 = 250°C$; $P_2 = 1.5$ bar; $n = 1.25$

Solution

The process is shown on *P–V* and *T–S* diagram in Fig. Ex. 10.28.

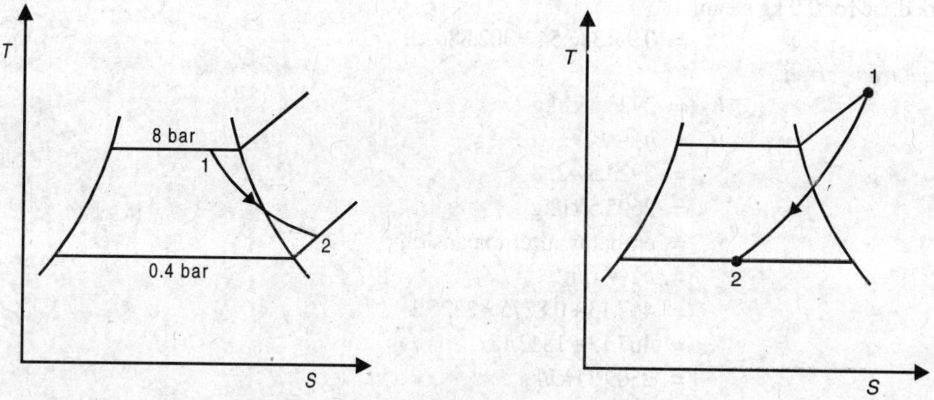

Fig. Ex. 10.28

First we will calculate on kg basis.

(i) Final condition : $V_1 = 0.1520\ m^3/kg$; $h_1 = 2923.5$ kJ/kg; $S_1 = 6.7099$ kJ/kg-K

$$\therefore \quad P_1 V_1^n = P_2 V_2^n$$

or

$$V_2 = V_1 \times \left(\frac{P_1}{P_2}\right)^{1/n}$$
$$= 0.1520 \times \left(\frac{15}{1.5}\right)^{1/1.25}$$
$$= 0.152 \times 6.3096$$
$$= 0.9591\ m^3/kg$$
$$V_{g2} = 1.1590\ m^3/kg$$

$\therefore$ Steam is wet

$$\therefore \quad x_2 = \text{Dryness fraction of steam after expansion}$$
$$= \frac{0.9591}{1.159}$$
$$\mathbf{x_2 = 0.8275}$$

(*ii*) *Work done* :

Work done for non-flow process = $\int P \cdot dV$

$$= \frac{P_1 V_1 + P_2 V_2}{n-1}$$

$$= \frac{100}{0.25}\left[\frac{15\times0.152-1.5\times0.9591}{1}\right]$$

$$= \frac{228-143.865}{0.25}$$

$$= \frac{84.135}{0.25}$$

$$= 336.54 \text{ kJ/kg}$$

∴ Work done for 0.9 kg steam

$$= 0.9\times336.54 = 302.886 \text{ kJ.}$$

(*iii*) *Heat transferred*

$$h_1 = 2923.5 \text{ kJ/kg}$$

∴ $u_1 = h_1 - P_1V_1$

$$= 2923.5 - 228$$

$$= 2695.5 \text{ kJ/kg}$$

h_2 = enthalpy after expansion

$$= h_{f2} + x_2 h_{fg2}$$

$$= 467.13 + 0.8275\times2226.2$$

$$= 467.13 + 1842.18$$

$$= 2309.31 \text{ kJ/kg}$$

∴ $u_2 = h_2 - P_2V_2 = 2309.31 - 143.865$

$$= 2165.445 \text{ kJ/kg}$$

∴ For non-flow process,

$$Q = \Delta u + \int P.dV = u_2 - u_1 + \int P.dV$$

$$= (2165.445 - 2695.5) + 336.54$$

$$= -530.055 + 336.54$$

$$= -193.575 \text{ kJ/kg}$$

∴ For 0.9 kg $Q = -0.9\times193.515$

Q = –174.1635 kJ/kg

– ve sign indicates that heat is rejected by the steam.

(*iv*) *Change in entropy*

S_1 = Entropy before expansion

$$= 6.7099 \text{ kJ/kg-K}$$

S_2 = Entropy after expansion

$$= S_{f2} + x_2 S_{fg2}$$

$$= 1.4336 + 0.8275\times5.7898$$

$$= 6.2247 \text{ kJ/kg-K}$$

∴ Change in entropy

$$\Delta S = S_2 - S_1 = 6.2247 - 6.7099$$
$$= -0.4852 \text{ kJ/kg-K}$$

∴ Total change in entropy

$$= -(0.9 \times 0.4852)$$
$$= -0.43668 \text{ kJ-K}$$

– ve sign indicates that entropy is decreasing.

Example 10.29 *1 kg of steam, initially saturated at 11 bar, expands non flow according to* $pV^{1-13} = C$ *to 1 bar. Determine: (a) final volume* v_2 *(b)* x_2 *(c) work (d)* Δu *and (e) Q. (f) If the process is steady flow which* $\Delta K = +4$ *kJ/kg , what is the work?*

Data: $p_1 = 11$ *bar;* $x_1 = 1$*;* $P_2 = 1$ *bar;* $PV^{1.3} = C$

Solution

(a) *Final volume* V_2

From tables,

$$V_1 = 0.17738 \text{ m}^3/\text{kg}$$
$$P_1 V_1^n = P_1 V_2^n$$
$$\therefore \quad V_2 = \left(\frac{P_1}{P_2}\right)^{1/n} \cdot V_1$$

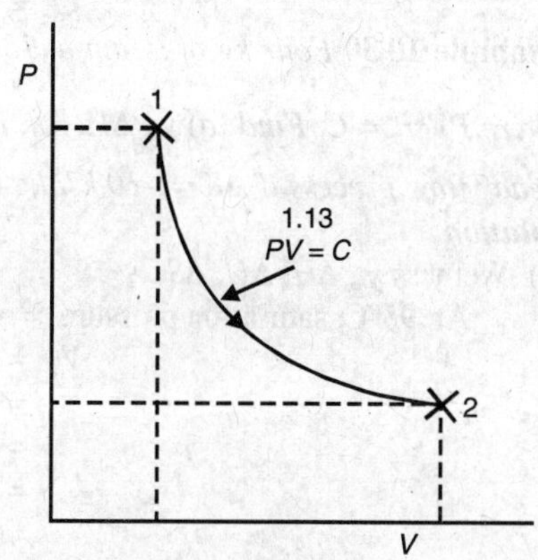

Fig. Ex. 10.29

(b) *Final dryness fraction* x_2

$$V_{g2} = 1.6937 \text{ m}^3/\text{kg}$$

∴ Steam is wet

$$\therefore \quad 1.481 = x_2 \times 1.6937$$
$$\therefore \quad x_2 = \frac{1.481}{1.6937}$$
$$\mathbf{x_2 = 0.8743}$$

(c) *Work* :

$$\text{Work of non-flow process} = \int P \cdot dV$$
$$= \frac{P_1 V_1 - P_2 V_2}{n-1}$$
$$= \frac{100}{1.13-1}[11 \times 0.17738 - 1 \times 1.481]$$
$$= \frac{100}{0.13}[1.9512 - 1.481]$$
$$= 769.231 \times 0.4702$$
$$= 361.69 \text{ kJ/kg}$$

(d) *Change in internal energy*

$$h_1 = h_{f1} + x_2 h_{fg1} = h_{g1}$$
$$= 2779.7 \text{ kJ/kg}$$
$$\therefore \quad u_1 = h_1 - P_1 V_1 = 2779.7 - 195.12$$
$$= 2574.58 \text{ kJ/kg}$$
$$h_2 = h_{f2} + x_2 h_{fg2} = 417.51 + 0.8743 \times 2257.9$$
$$= 417.51 + 1974.08$$
$$= 2391.59 \text{ kJ/kg}$$

$$\therefore \quad u_2 = h_2 - P_2V$$
$$= 2391.59 - 100(1.481) = 2391.59 - 148.1$$
$$= 2243.49 \text{ kJ/kg}$$
$$\therefore \quad \Delta u = u_2 - u_1 = 2243.49 - 2574.58$$
$$\mathbf{\Delta u = -331.09 \text{ kJ/kg}}$$

(e) *Work for steady flow process*

$$\text{Work done} = -\int P.dV + \Delta K = 1.13\int P.dV - 4$$
$$= 1.13 \times 361.69 - 4 = 408.71 - 4$$
$$\text{Work done} = 404.71 \text{ kJ/kg}$$

(f) *Heat Q*

For non-flow process,
$$Q = \Delta u + \int P.dV$$
$$Q = -331.09 + 361.69$$
$$\mathbf{Q = 30.6 \text{ kJ/kg}}$$

Example 10.30 *Four kg of steam at 13 bar and 260°C expand into wet region to 95°C in a polytropic process where $PV^{1.21} = C$. Find (a) y_2, ΔH, Δu, ΔS. (b) $\int P.dV$ (c) $-\int V.dP$ (d) work for nonflow process and (e) work for steady flow process if $\Delta K = -10$ kJ/ kg (f). Find Q for both non-flow and steady flow processes.*

Solution

(a) Wetness y_2, ΔH, ΔU, ΔS :

At 95°C, saturation pressure $= P_2 = 0.8453$ bar; $P_1 = 13$ bar;

$$V_1 = 0.1769 + 0.2\,(0.1969 - 0.1769)$$
$$= 0.1769 + 0.2 \times 0.0200$$
$$= 0.1769 + 0.004$$
$$= 0.1809 \text{ cu.m/kg}$$

$$\therefore \quad V_2 = \left(\frac{P_1}{P_2}\right)^{1/n} \times V_1 = 0.1809 \times \left(\frac{13}{0.845}\right)^{1/1.20}$$
$$= 0.1809 \times (15.38)^{0.8264}$$
$$= 1.7312 \text{ m}^3\text{/kg}$$

But
$$V_{g2} = 1.982 \text{ m}^3\text{/kg}$$

Actual volume is less than v_{g2}. Therefore steam is wet

$$\therefore \quad 1.7312 = x_2.V_{g2}$$
$$x_2 = \frac{1.7312}{1.982}$$
$$= 0.8734$$
$$\therefore \quad y_2 = \text{Wetness} = 0.1216$$
$$\therefore \quad h_1 = 2931.5 + 0.2\,(3044.3 - 2931.5)$$
$$= 2931.2 + 0.2 \times 112.8$$
$$= 2931.5 + 22.56 = 2954.06 \text{ kJ/kg}$$
$$h_2 = h_{f2} + x_2 h_{fg2} = 397.99 + 0.8734 \times 2270.2$$
$$= 397.99 + 1982.79$$
$$= 2380.78 \text{ kJ/kg}$$
$$\therefore \quad \Delta H = h_2 - h_2 = 2380.78 - 2954.06$$
$$\mathbf{\Delta H = -573.28 \text{ kJ/kg}}$$

$$S_2 = S_{f2} + x_2 S_{fg2} = 1.2501 + 0.8734 \times 6.1665$$
$$= 1.2581 + 5.3858$$
$$= 6.6359 \text{ kJ/kg-K}$$
$$S_1 = 6.7878 + 0.2\,(6.9938 - 6.7878)$$
$$= 6.7878 + 0.2 \times 0.1060 = 6.7878 + 0.02012$$
$$= 6.80792 \text{ kJ/kg-K}$$
$$\therefore \quad \Delta S = S_2 - S_1 = 6.63590 - 6.80792$$
$$\mathbf{\Delta S = 0.17202 \text{ kJ/kg-K}}$$
$$u_1 = h_1 - P_1V_1 = 2954.06 - 100 \times 13 \times 0.1809$$
$$= 2954.06 - 235.17$$
$$= 2718.89 \text{ kJ/kg}$$
$$u_2 = h_2 - P_2V_2 = 2380.78 - 100 \times 0.8453 \times 1.1712$$
$$= 2380.78 - 99.0$$
$$= 2281.78 \text{ kJ/kg}$$
$$\therefore \quad \Delta u = u_2 -{}_1 = 2281.78 - 2718.89$$
$$\mathbf{\Delta u = -437.11 \text{ kJ/kg}}$$

(b) $\int P.dV$:
$$\int P.dV = \frac{P_1V_1 - P_2V_2}{n-1} = \frac{100[13 \times 0.1809 - 0.8453 \times 1.7312]}{0.21}$$
$$= \frac{1}{0.21}[295.17 - 99] = \frac{136.17}{0.21}$$
$$\mathbf{\int P \cdot dV = 648.43 \text{ kJ/kg}}$$

(c) $-\int V.dP$:
$$-\int V.dP = n\int P.dV = 1.21 \times 648.43$$
$$\mathbf{\int P \cdot dV = 784.6 \text{ kJ/kg}}$$

(d) *Work for non-flow process*
$$W_{nf} = \int P.dV = 648.43 \text{ kJ/kg}$$

(e) *Work for steady flow process*
$$W_{sf} = -\int V.dP + \Delta k = 784.6 + 10 = 794.6 \text{ kJ/kg}$$

(f) *Q – Heat*

For non-flow process
$$Q = (u_2 - u_1) + \int P.dV$$
$$= -437.11 + 648.43 = 211.32 \text{ kJ/kg}$$

For steady flow process
$$Q = \Delta K + W_{sf} + \Delta H.\ 10 + 794.6 - 573.28$$
$$\mathbf{Q = 784.6 \text{ kJ/kg}}$$

Example 10.31 *One kg per second of steam expands in a polytropic proces from P_1 = 10 bar and V_1 = 0.2059 cu.m to 1 bar and 1.58 cu.m. Find (a) n in $PV^n = C$ (b) $\int P.dV$ (c) $-\int V.dP$ (d) W and Q for non-flow process (e) w and Q for steady flow process with ΔK = + 9 kJ/kg.*

Data: P_1 =10 bar; V_1= 0.2059 m³/ kg, P_2 =1 bar , V_2 = 1.58 m³/ kg

Solution

(a) Index n in $PV^n = C$

We have $P_1 V^n_1 = PV^n_2$

$$\therefore \quad \left(\frac{V_2}{V_1}\right)^n = \frac{P_1}{P_2} = \frac{10}{1} = 10$$

$$\therefore \quad \left(\frac{1.58}{0.2059}\right)^n = (7.6736)^n = 10$$

$$\therefore \quad n \log_{10} 7.6736 = \log_{10} 10$$

$$\therefore \quad n = \frac{1}{0.885}$$

$$\mathbf{n = 1.13}$$

(b) $\int P \cdot dV$:

$$\int P.dV = \frac{P_1 V_1 - P_2 V_2}{n-1}$$

$$= \frac{100}{0.13}[10 \times 0.2059 - 1 \times 1.58]$$

$$= \frac{100}{0.13}[2.059 - 1.580]$$

$$= \frac{100}{0.13} \times 0.479$$

$$\int \mathbf{P \cdot dV} = \mathbf{368.46\ kJ/kg/sec}$$

(c) $-\int V \cdot dP$:

$$-\int V.dP = n \int P \cdot dV$$

$$= 1.13 \times 368.46$$

$$-\int \mathbf{V.dP} = \mathbf{416.36\ kJ/kg/sec}$$

(d) Work and Heat for non-flow process :
For non-flow process

$$\text{Work} = \int P \cdot dV$$

$$\textbf{Work} = \mathbf{368.46\ kJ/kg/sec}$$

$$V_1 = 0.2059\ \text{m}^3/\text{kg}$$

$$V_{g1} = 0.19429$$

$\therefore$ Steam is superheated

$$\therefore \quad 0.2059 = 0.19429 \times \frac{T_{sup}}{T_{sat}}$$

$$= \frac{0.19429 \times T_{sup}}{179.88 + 273}$$

$$\therefore \quad T_{sup1} = \frac{0.2059}{0.19429} \times 452.88$$

$$= 480°\text{K} = 207.°\text{C} \approx 200°\text{C}$$

$$\therefore \quad h_1 = 2826.8\ \text{kJ/kg}$$

$$\therefore \quad u_1 = h_1 - P_1 V_1 = 2826.8 - 100 \times 10 \times 0.2059$$

$$= 2826.8 - 205.9$$
$$= 2620.9 \text{ kJ/kg/sec}$$
$$V_2 = 1.58 \text{ m}^3\text{/kg}$$
$$V_{g2} = 1.6937$$
$$\therefore \quad x_2 = \frac{1.58}{1.6937} = 0.9329$$
$$\therefore \quad h_2 = h_{f2} + x_2 h_{fg2} = 417.51 + 0.9329 \times 2257.9$$
$$= 417.51 + 2106.32 = 2523.83 \text{ kJ/kg/sec}$$
$$\therefore \quad u_2 = h_2 - P_2V_2 = 2523.83 - 158$$
$$= 2365.83 \text{ kJ/kg/sec}$$
$$\therefore \quad Q = \Delta u + \int P \cdot dV$$
$$= (u_2 - u_1) + 368.46$$
$$= (2365.83 - 2620.9) + 368.46$$
$$\mathbf{Q = 113.39 \text{ kJ/kg/sec}}$$

(e) W and Q for steady flow process :

$$\text{Work} = -\int V \cdot dP + \Delta K$$
$$= 416.36 + 9$$
$$\textbf{Work} = \mathbf{425.36 \text{ kJ/kg/sec.}}$$
$$Q = \Delta k + \Delta H + W_{sf}$$
$$= 9 + (2523.83 - 2826.8) + 425.36$$
$$= 9 + 425.36 - 302.97$$
$$\mathbf{Q = 131.39 \text{ kJ/kg/sec}}$$

Example 10.32 *Steam at 20 bar and 300°C is expanded to four times its initial volume according to the law $PV^{1.3}$ = constant. Find the heat during the process.*

Data: P_1 = 20 bar; t_1 = 300°C ; $V_2 = 4V_1$; $PV^{1.3}$ = constant.

Solution

$$\therefore \quad P_1V_1^{1.3} = P_2V_2^{1.3}$$
$$\therefore \quad \left(\frac{V_2}{V_1}\right)^{1.3} = \frac{P_1}{P_2} = \left(\frac{4V_1}{V_1}\right)^{1.3} = 4^{1.3} = 6.063$$
$$\therefore \quad P_2 = \frac{20}{6.063} = 3.3 \text{ bar}$$

From tables, $V_1 = 0.126 \text{ m}^3\text{/kg}$

$\therefore \quad V_2 = 4 \times 0.126 = 0.504 \text{ m}^3\text{/kg}$

$$\therefore \quad \text{Work done/kg} = \frac{P_1V_1 - P_2V_2}{n-1} \quad \text{(For non-flow)}$$
$$= \frac{100}{1.3-1}[20 \times 0.126 - 3.3 \times 0.504]$$
$$= \frac{100}{0.3}[2.52 - 1.6632]$$
$$= \frac{100 \times 0.8568}{0.3}$$

$$= 285.6 \text{ kJ/kg}$$
$$h_1 = 3025.2 \text{ kJ/kg}$$
$$u_1 = h_1 - P_1V_1 = 3025.2 - 100 \times 20 \times 0.126$$
$$= 3025.2 - 252$$
$$= 2773.2 \text{ kJ/kg}$$
$$V_2 = 0.504 \text{ m}^3\text{/kg}$$
$$V_{g2} = \frac{0.570627 + 0.539025}{2}$$
$$= \frac{1.1099652}{2}$$
$$= 0.554826$$
$$< V_{g2}$$
$$= \frac{1.109652}{2} = 0.554826$$

$\therefore$ Steam is wet after expansion

$$\therefore \quad x_2 = \frac{0.504}{0.554826} = 0.910$$
$$\therefore \quad h_2 = h_{f2} + x_2 h_{fg2}$$

where

$$h_{f2} = \frac{570.8 + 579.8}{2}$$
$$= \frac{1150.6}{2} = 575.3 \text{ kJ/kg}$$
$$h_{fg2} = \frac{2156.9 + 2150.5}{2} = \frac{4307.4}{2}$$
$$= 2153.7 \text{ kJ/kg}$$
$$\therefore \quad h_2 = 575.3 + 0.910 \times 2153.7$$
$$= 575.3 + 1959.87$$
$$= 2535.17 \text{ kJ/kg}$$
$$\therefore \quad u_2 = h_2 - P_2V_2 = 2535.17 - 100 \times 3.3 \times 0.504$$
$$= 2535.17 - 166.32$$
$$= 2368.85 \text{ kJ/kg}$$

Applying the non-flow energy equation,

$$Q = \Delta u + \int P.dV$$
$$= \text{Heat during the process}$$
$$= (u_2 - u_1) + W_n$$
$$= (2368.85 - 2773.2) + 285.6$$
$$= -404.35 + 285.6$$
$$\mathbf{Q = -118.75 \text{ kJ/kg}}$$

– ve sign indicates that heat is rejected.

Example 10.33 *Steam expands from the condition 16 bar and 250°C to 0.9 bar according to the law $pV^{1.1}$ = constant. Find for 1 kg of steam, the expansive work and change in internal energy of steam.*

Data: P_1 = 16 bar; P_2 = 0.9 bar; T_1 = 523 K and $pV^{1.1}$ = C.

Solution

$$V_1 = \text{volume of 1 kg steam before expansion}$$

$$= V_{g1}\frac{T_{\text{sup}_1}}{T_{\text{sat}_1}} = 0.123843 \times \frac{523}{474.33}$$

$$= 0.1366\ \text{m}^3/\text{kg}$$

Similarly,

$$h_1 = \text{Enthalpy before expansion}$$

$$= h_{g1} + C_p\,(T_{\text{sup}} - T_{\text{sat}})$$

$$= 2791.7 + 2.3\,(250 - 201.33)$$

$$= 2791.7 + 111.941 = 2903.641\ \text{kJ/kg}$$

$\therefore$

$$u_1 = \text{Internal energy}$$

$$= h_1 - P_1V_1$$

$$= 2903.641 - 100 \times 16 \times 0.1366$$

$$= 2903.641 - 218.56$$

$$= \frac{2685.081}{1.1}\ \text{kJ/kg}$$

Now we have

$$P_1V_1 = P_1V_2$$

or

$$V_2 = \left(\frac{P_1}{P_2}\right)^{1/1.1} \times V_1$$

$$= \left(\frac{16}{0.9}\right)^{1/1.1} \times 0.1366$$

$$= (17.78)^{0.9091} \times 0.1366$$

$$= 13.69 \times 0.1366$$

$$= 1.87\ \text{m}^3/\text{kg}$$

$$\approx V_{g2}\ \text{approx.}$$

$\therefore$ The steam is dry and saturated after expansion.

$\therefore$

$$h_2 = h_{g2} = 2670.9\ \text{kJ/kg}$$

$$v_2 = 1.87\ \text{m}^3/\text{kg}$$

$\therefore$

$$u_2 = h_2 - P_2V_2 = 2670.9 - 100 \times 0.9 \times 1.87$$

$$= 2670.9 - 168.3 = 2502.6\ \text{kJ/kg}$$

$\therefore$ Change in internal energy/kg

$$= u_2 - u_1 = 250.2.6 - 2685.081$$

$$= -182.481\ \text{kJ/kg}$$

Work done/kg (non-flow)

$$W = \frac{P_1V_1 - P_2V_2}{n-1} = \frac{218.56 - 168.3}{0.1}$$

$$= \frac{50.26}{0.1}$$

$$\mathbf{W = 502.6\ kJ/kg}$$

Example 10.34 *At a particular instant during expansion stroke of the steam engine, the volume of the steam was 0.05 m³, the pressure and condition of steam being 8 bar and 0.9 dry respectively. Determine (i) the index of expansion (ii) the work done per kg and (iii) the change in entropy during the process.*

Data: $P_1 = 8$ *bar;* $P_2 = 2$ *bar;* $V_1 = 0.05\ m^3$*;* $V_2 = 0.15\ m^3$*;* $x_1 = 0.9$

Solution

(*i*) *Index of expansion*

Let *n* be the index of expansion.

$$\therefore \quad P_1 V_1^n = P_2 V_2^n$$

$$\therefore \quad \left(\frac{V_2}{V_1}\right)^n = \frac{P_1}{P_2} = \frac{8}{2} = 4$$

$$\therefore \quad \left(\frac{0.15}{0.05}\right)^n = 4 = 3^n$$

$$\therefore \quad n \log 3 = \log 4$$

$$n = \frac{\log 4}{\log 3} = \frac{0.6021}{0.4771}$$

$$\mathbf{n = 1.262}$$

(*ii*) *The work done* :

$$\text{Work done} = \frac{P_1 V_1 - P_2 V_2}{n-1} = \frac{100}{0.262}[8\times0.05 - 2\times0.15]$$

$$= \frac{40-30}{0.262} = \frac{10}{0.262} = 38.168 \text{ kJ}$$

Specific volume $V_1 = x_1 V_{g1} = 0.9 \times 0.240506$

$= 0.2165$ kJ/kg

$$\therefore \quad \text{Mass of steam} = \frac{0.05}{0.2165} = 0.231 \text{ kg}$$

$$\therefore \quad \text{Work done/kg} = \frac{38.168}{0.231}$$

Work done/kg = 165.23 kJ/kg

(*iii*) *Change in entropy*:

Initial entropy $S_1 = S_{f1} + x_1 S_{fg1}$

$= 2.045 + 0.9\,(6.659 - 2.045) = 6.1976$

$$\text{Final volume/kg} = \frac{0.15}{0.231}$$

$= 0.6493$ m³/kg

$< V_{g2}\,(0.8872)$

$$\therefore \quad x_2 = \frac{0.6493}{0.8872}$$

$= 0.732$

Final entropy $S_2 = S_{f2} + x_2 S_{fg2}$

$\therefore \quad S_2 = 1.529 + 0.732\,(7.127 - 1.529)$

$= 1.529 + 4.098$

$= 5.627$ kJ/kg-K

$\therefore$ Change in entropy/kg $\Delta S = S_2 - S_1$
$= 5.627 - 6.1976$
$= -5706$ kJ/kg-K

$\therefore$ Change in entropy $\Delta S = -0231 \times 0.5706$ kJ/K

$$\mathbf{\Delta S = -0.1318\ kJ/K}$$

Example 10.35 *Dry and saturated steam at 20 bar passes through an expander where it is expanded to a pressure of 500 mm Hg (vacuum) (barometer reads 725 mm Hg) according to the law $pV^{1.07}$ = constant. Calculate*

(*i*) *Condition of steam after expansion*

(*ii*) $\int P.dV$

(*iii*) $-\int V.dP$

(*iv*) *Work of non-flow process*

(*v*) *Work of steady flow system with $\Delta P = \Delta K = 0$*

(*vi*) *Change in internal energy*

(*vii*) *Change in internal entropy.*

(*viii*) *Heat Q.*

Solution

(*i*) *Condition of steam expansion*:

Abs. pressure of steam after expansion

$$P_2 = \text{barometer reading vacuum} = 725 - 500 = 225 \text{ mm Hg}$$

In bar
$$P_2 = \frac{225 \times 1.01325}{760} = 0.3 \text{ bar}$$

$$P_1 = 20 \text{ bar and } V_1 = 0.099623 \text{ m}^3/\text{kg}$$

$$\therefore \quad P_1 V^n{}_1 = P_2 V_2^n \text{ or } V_2 = \left(\frac{P_1}{P_2}\right)^{1/n} \cdot V_1$$

$$\therefore \quad V_2 = \left(\frac{20}{0.3}\right)^{1/1-0.7} \times 0.099623$$

$$= \left(\frac{20}{0.3}\right)^{0.9346} \times 0.099623$$

$$= 5.0464 \text{ m}^3/\text{kg}$$

$$< V_{g2}(5.229)$$

$$\therefore \quad x_2 = \frac{5.0464}{5.229}$$

$$\mathbf{x_2 = 0.9651}$$

(ii) $\int P.dV$:
$$\int P.dV = \frac{P_1V_1 - P_2V_2}{n-1}$$

$$= \frac{100}{0.07}[20 \times 0.099623 - 0.3 \times 5.0464]$$

$$= \frac{199.246 - 151.392}{0.07}$$

$$= \frac{47.854}{0.07}$$

$$\int \mathbf{P.dV} = \mathbf{683.63\ kJ/kg}$$

(iii) $-\int V.dP$:

$$-\int V.dP = n\int P.dV$$

$$= 1.07 \times 683.63$$

$$-\int \mathbf{V.dP} = \mathbf{731.48\ kJ/kg}$$

(iv) Work for non-flow process

$$W_{nf} = \text{Work of non-flow process}$$

$$= \int P.dV$$

$$\mathbf{W = 683.63\ kJ/kg}$$

(v) Work of steady flow process

We have $W_{sf} = -\int V.dP$

$$\mathbf{W_{sf} = 731.48\ kJ/kg}$$

(vi) Change in internal energy

$$h_1 = h_{g1} = 2797.3 \text{ kJ/kg}$$

$$\therefore \quad u_1 = h_1 - P_1V_1 = 2797.3 - 199.246$$

$$= 2598.054 \text{ kJ/kg}$$

$$h_2 = h_{f2} + x_2 h_{fg2}$$

$$= 289.3 + 0.9651 \times 2336.1$$

$$= 289.3 + 2254.57 = 2543.87 \text{ kJ/kg}$$

$$\therefore \quad u_2 = h_2 - P_2V_2$$

$$= 2543.87 - 151.392$$

$$= 2392.478 \text{ kJ/kg}$$

$$\therefore \quad \Delta u = u_2 - u_1 \text{ change in internal energy}$$

$$= 2392.478 - 2598.054$$

$$\Delta u = -205.576 \text{ kJ/kg}$$

(vii) Change in entropy

$$S_1 = S_{g1} = 6.337 \text{ kJ/kg-K}$$

$$S_2 = S_{f2} + x_2 S_{fg2}$$

$$= 0.944 + 0.9651\,(7.767 - 0.944)$$

$$= 0.944 + 0.9651 \times 6.823$$

$$= 0.944 + 6.5849$$

$$= 7.5289 \text{ kJ/kg-K}$$

$$\therefore \quad S_2 - S_2 = \text{Change in entropy}$$

$$= 7.5289 - 6.3370$$

$$\mathbf{S_2 - S_2 = 1.1919\ kJ/kg\text{-}K}$$

(viii) Heat Q

Heat during the non-flow process is given by

$$Q = \Delta u + \int p.dv$$
$$= -205.576 + 683.63$$
$$\mathbf{Q = 478.054 \text{ kJ/kg-K}}$$

For Steady Flow Process

$$Q = \Delta P + \Delta K + \Delta H + W_{sf}$$
$$= 0 + 0 + (2543.87 - 2797.3) + 731.48$$
$$\mathbf{Q = 478.05}$$

VI. Combined Processes

Example 10.36 *Steam at 2 MN/m² and 250°C expands isentropically to 3.5 bar. It is then further expanded hyperbolically to 0.5 bar. Using steam tables, find out*

(i) *condition of steam after each process.*

(ii) *work done kg/of steam in each process.*

Data: $P_1 = 2\ MN/m^2 = 20\ bar;\ T_1 = 250°C;\ P_2 = 3.5\ bar;\ P_3 = 05\ bar.$

Solution

The process are shown in Fig. Ex. 10.36

Isentropic Process (1 – 2) :

From steam tables : $h_1 = 2902.4 \text{ kJ/kg}.$

$S_1 = 6.5454 \text{ kJ/kg-K}$

For isentropic process, $S_1 = S_2 = S_{f2} + x_2 S_{fg2}.$

$\therefore \quad 6.5454 = 1.7273 + x_2 \times 5.2119$

$$\therefore \quad x_2 = \frac{6.5454 - 1.7273}{5.2119}$$

$$x_2 = 0.9244$$

$$\therefore \quad h_2 = h_{f2} + x_2 h_{fg2}$$
$$= 584.27 + 0.9244 \times 2147.4$$
$$= 2569.32 \text{ kJ/kg}.$$

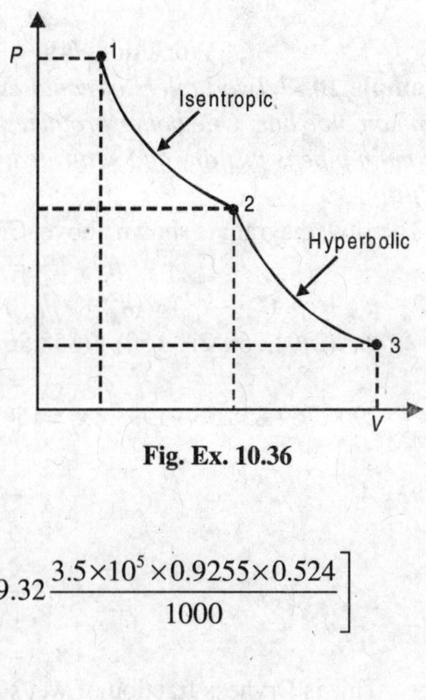

Fig. Ex. 10.36

Assuming non-flow process,

$$\text{Work done/kg} = u_1 - u_2$$
$$= (h_1 - P_1V_1) - (h_2 - P_2V_2)$$
$$= \left[290234 \frac{-10^5 \times 0.1114}{1000}\right] - \left[2569.32 \frac{3.5 \times 10^5 \times 0.9255 \times 0.524}{1000}\right]$$

$$\textbf{Work done/kg} = \mathbf{279.81 \text{ kJ/kg}}$$

For steady flow process with $\Delta P = \Delta K = 0$

$$\text{Work done / kg} = h_1 - h_2$$
$$\textbf{Work done/kg} = \mathbf{2902.4 - 2569.42 = 333.08 \text{ kJ/kg}}$$

Hyperbolic Expansion (2 – 3): The law of expansion is $pV = \text{Const.}$

$$\therefore \quad P_2V_2 = P_3V_3$$

where $P_2 = 3.5$ bar, $P_3 = 0.5$ bar.

$$V_2 = x_2 V_{g2} = 0.9244 \times 0.524 = 0.4844 \text{ m}^3/\text{kg}$$

$$\therefore \quad V_3 = \frac{P_2 V_2}{P_3} = \frac{3.5 \times 0.4844}{0.5} = 3.391 \text{ m}^3/\text{kg}$$

From tables, $V_3 = 3.391 > (V_{g3} = 3.2402)$

$$\therefore \quad \frac{V_3}{V_{g3}} = \frac{T_3}{T_{sat_3}}$$

$$\therefore \quad \frac{3.391}{3.2402} = \frac{T_3}{273+81.345}$$

$$\therefore \quad T_3 = \frac{3.391 \times 354.345}{3.2402}$$

$$\mathbf{T_3 = 370.8\ K = 97.8°C}$$

Superheated condition

For pV = Const., the work done of non-flow process and steady flow process with

$$\Delta P = \Delta K = 0, \text{ is same}$$

$$\text{Work done/kg} = P_2 V_2 \ln \frac{P_2}{P_3}$$

$$= \frac{3.5 \times 105 \times 0.4844}{1000} \ln \frac{3.5}{0.5}$$

$$\textbf{Work done/kg} = \mathbf{329.91\ kJ/kg}$$

Example 10.37 *Two boilers of equal evaporative capacities generate steam at the same pressure of 16 bar to a common pipe line. One boiler produces steam with 50°C superheat and other produces wet steam. If the steam in the main pipe is just dry and saturated, find the dryness fraction of wet steam.*

Solution

The flow diagram is shown above. Consider 1 kg steam from each boiler

$$h_A + h_B = 2hC$$

$$h_{gA} + cp_v (t_{sup} - t_{sat}) + (h_{fB} + x_B h_{fgB}) = 2h_{gc}$$

$$\therefore \quad (2791.76 + 2.3 \times 50) + (858.56 + x_B \times 1933.2)$$

$$= 2 \times 2791.76$$

$$2906.76 + 858.56 + 1933.2 x_B = 5583.52$$

$$\therefore \quad x_B = \frac{5583.56 - 2906.76 - 858.56}{1933.2}$$

$$= \frac{1818.2}{1933.2}$$

$$\mathbf{x_B = 0.9405}$$

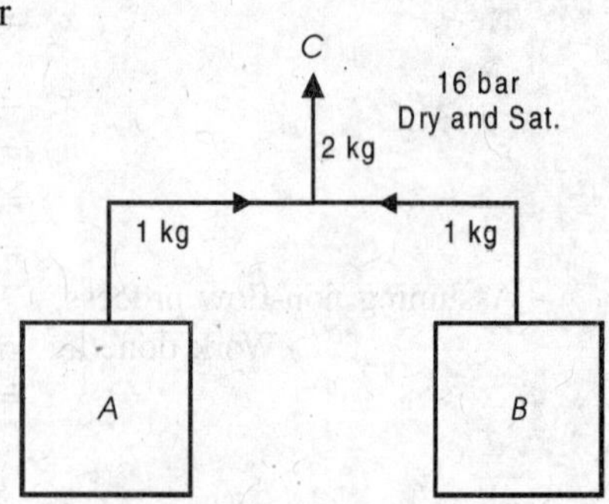

Fig. Ex. 10.37

This is Dryness fraction of wet steam.

Example 10.38 *A closed vessel of 0.3 cu.m capacity contains steam at a pressure of 1.2 bar and 0.8 dry. It is connected to a steam pipe supplying steam of 0.9 dryness fraction at a pressure of 4.5 bar. Neglecting the volume of moisture in suspension and thermal capacity of the vessel, determine, (a) the weight of steam which will pass through the steam pipe to the vessel, when the valve in the pipe is opened and (b) the final dryness of the steam in the vessel when the flow ceases.*

Solution

When the valve is opened, the steam will flow into the vessel till the pressure in the vessel pressure increases to 4.5 bar and then the flow will cease.

Initially,

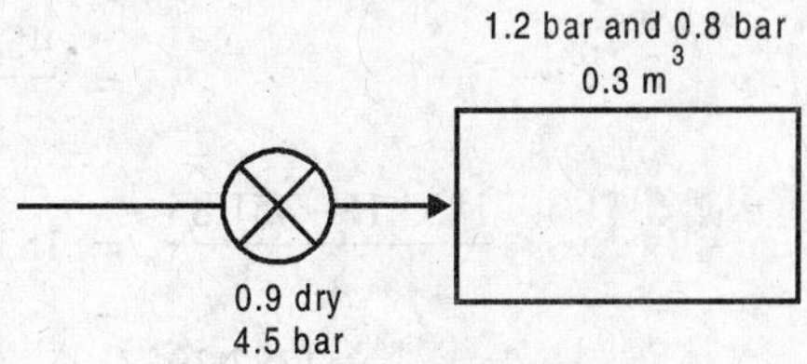

Fig. Ex. 10.38

Sp. vol. of steam in vessel $= V_1 = x_1 V_{g1}$

$= 0.8 \times 1.4281$

$= 1.14248 \text{ m}^3/\text{kg}$

$\therefore$ Initial wt. of steam in the vessel

$$W_1 = \frac{0.3}{0.14248} = 0.2626 \text{ kg}$$

After the steam is admitted, the pressure is 4.5 bar

Let W_2 = wt. of steam dryness

x_2 = Dryness fraction of steam admitted

x_3 = Dryness fraction of after mixing

Total heat of steam in vessel before mixing.

$$h_1 = 0.2626[h_{f1} + x_1 h_{fg1}]$$
$$= 0.2626[439.36 + 0.8 \times 2244.1]$$
$$= 586.82 \text{ kJ} \quad ...(1)$$

Total heat of steam admitted

$$h_2 = W_2[h_{f2} + x_2 h_{fg2}]$$
$$= W_2[623.16 + 0.9 \times 2119.7]$$
$$= 2530.89\, w_2 \text{ kJ} \quad ...(2)$$

Total heat of steam after mixing

$$h_3 = (W_1 + W_2)\,[h_{f3} + x_3 h_{fg3}]$$
$$= (0.2626 + W_2)\,[623.16 + x_3 \times 2119.7] \quad ...(3)$$

$\therefore$ 1 + 2 + 3

$$586.82 + 2530.89\, W_2 = (0.2626 + W_2)\,(623.16 + x_3 \times 2110.7) \quad ...(4)$$

Again volume of mixture is 0.3 m^3

$$\therefore \quad = (W_1 + W_2)\, x_3 V_{g3} = 0.3$$
$$= (0.2626 + W_2)\, x_3 \times 0.41375 = 0.3$$

$$(2626 + W_2)\, x_3 = \frac{0.3}{0.41375} = 0.7251 \quad ...(5)$$

$$\therefore \quad (2626 + W_2) = \frac{0.7251}{x_3} \quad ...(6)$$

$$\therefore \quad W_2 = \frac{0.7251}{x_3} - 0.2626 \quad ...(7)$$

$\therefore$ Equation (4) gives

$$= 586.82 + 2530.89\left(\frac{0.7251}{x_3} - 0.2626\right)$$

$$= \frac{0.7251}{x_3}(623.16 + 2119.7\, x_3)$$

$$= 586.82 + \frac{1835.15}{x_3} - 664.61$$

$$= \frac{451.85}{x_3} + 1537$$

$$\frac{1835.15 - 451.85}{x_3} = 1537 + 664.61 - 586.82$$

$$\frac{1383.30}{x_3} = 2201.61 - 586.82 = 1614.79$$

$$\therefore \quad x_3 = 0.8566$$

$$\therefore \quad W_2 = \frac{0.7251}{0.8566} - 0.2626$$

$$= 0.5839 \text{ kg}$$

(a) The weight of steam admitted = 0.5839 kg

(b) Final dryness fraction = 0.8566

Example 10.39 *Steam initially at 14 bar and having specific volume of 0.12 m³/ kg is throlled to 7 bar and then expanded hyperbolically to 1.1 bar. For each process determine change in entropy.*

Solution

Throttling Process

Volume of steam at 14 bar and dry and saturated is 0.14073 m³/kg. Actual volume is 0.12 m³/kg. Therefore the steam is wet before throttling.

$$\therefore \quad x_1 = \frac{0.12}{0.14073} = 0.8527$$

$$\therefore \quad h_1 = \text{Enthalpy of 1 kg steam before throttling}$$

$$= h_{f1} + x_1 h_{fg1}$$

$$= 830 + 0.8527 \times 1957.7$$

$$= 2799.32 \text{ kJ/kg}$$

This enthalpy is less than h_{g2}. Therefore steam is wet after throttling.

$$\therefore \quad h_2 = 2499.32 = h_{f2} + x_2 h_{fg2} \text{ at 7 bar}$$

$$= 697.1 + x_2 \times 2064.9$$

$$\therefore \quad x_2 = \frac{2499.32 - 697.1}{2064.9} = \frac{1802.22}{2064.9}$$

$$= 0.8728$$

Entropy of wet steam before throttling is

$$S_1 = S_{f1} + x_1 S_{fg1}$$

$$= 2.2837 + 0.8527 \times 4.1814$$

$$= 5.8492 \text{ kJ/kg-K}$$

Similarly $\quad S_2 = \text{Entropy after throttling}$

$$= S_{f2} + x_2 S_{fg2}$$

$$= 1.9918 + 0.8728 \times 4.7134$$

$$= 6.1056 \text{ kJ/kg-K}$$

$\therefore$ Change of entropy during throttling

$$\Delta S = S_2 - S_1 = 6.1056 - 5.8492$$

$$= 0.2564 \text{ kJ/kg-K}$$

Hyperbolic Process

$P_2 = 7$ bar; V_2 = volume of 1 kg steam before hyperbolic expansion

$$V_2 = x_2 V_{g2}$$
$$= 0.8728 \times 0.27268 = 0.2379\, m^3/kg$$

$P_3 = 1.1$ bar = Pressure after hyperbolic expansion

$$\therefore \quad P_2V_2 = P_3V_3$$
$$\therefore \quad V_3 = \frac{P_2 V_2}{P_3}$$
$$= \frac{7}{1.1} \times 0.2379 = 1.5147\, m^3/kg$$

This is less than $V_{g3} = 1.5492$

∴ Steam is wet after hyperbolic expansion.

$$\therefore \quad 1.5147 = x_3 \times 1.5492$$
$$x_3 = \frac{1.5147}{1.5492}$$
$$= 0.9774\, m^3/kg$$
$$\therefore \quad S_3 = S_{f3} + x_3 S_{fg3}$$
$$= 1.3330 + 0.9774 \times 5.9947$$
$$\mathbf{S_3 = 7.1922\ kJ/kg\text{-}K}$$

Example 10.40 *Two boilers discharge equal amounts of steam into the same main. The steam from one boiler is at 17 bar and 270°C and other at 17 bar and 0.95 quality. What is (a) the equilibrium condition after mixing. ($\Delta K = 0$) (b) the change of entropy by the higher temperature steam, (c) the change of entropy of the lower temperature steam? (d) is there a net increase or decrease of entropy?*

Solution

(a) Energy diagram is shown in Fig. Ex. 10.40 $\Delta K = 0$;

$$h_A + h_B = 2\, h_c$$

From table

$$h_A = \frac{2915.3 + 3033.5}{2}$$
$$= \frac{5948.8}{2}$$
$$= 2974.4\, kJ/kg$$
$$S_{A1} = \frac{6.6398 + 6.8557}{2} = \frac{13.4955}{2}$$
$$= 6.74775\, kJ/kg\text{-}K$$
$$h_B = h_f + xh_{fg}$$
$$= 871.84 + 0.95 \times 1921.5$$
$$= 2697.26\, kJ/kg$$
$$S_{B1} = S_f + xS_{fg}$$
$$= 2.3713 + 0.95 \times 4.0244$$
$$= 6.1945\, kJ/kg.K$$

After mixing in the main pipes

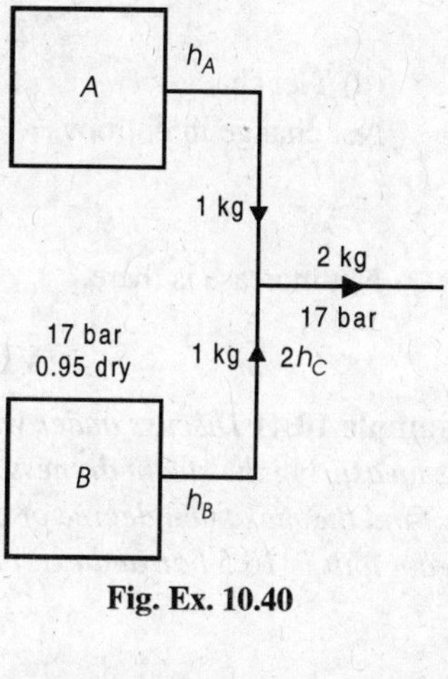

Fig. Ex. 10.40

$$2h_c = h_A + h_B = 2974.4 + 2697.26$$
$$= 5671.66$$

$$\therefore \quad h_C = \frac{5671.66}{2} = 2835.83 \text{ kJ/kg}$$

The steam is superheated in the main pipe.

Assuming $C_{pv} = 2.3 \text{ kJ/kg-K}$

$$2835.83 = h_g + C_{pv}(t_{sup} - t_{sat})$$
$$= 2793.4 + 2.3(t_{sup} - 204.31)$$

$$\therefore \quad t_{sup} = \text{Temp of steam in main pipe}$$
$$= \frac{2835.83 - 2793.4}{2.3} + 204.31$$
$$= 222.76°C$$

$\therefore$ Equilibrium condition is

$$P = 17 \text{ bar.}$$
$$t = 222.76°C$$

Entropy in main pipe is

$$S_c = S_{gc} C_{pv} \log_e \frac{T_{sup}}{T_{sat}}$$

$$= 6.3957 + 2.3 \log_e \frac{222.76 + 273}{204.31 + 273}$$

$$\mathbf{S_c = 6.48293 \text{ kJ/kg-K}}$$

(b) Change in entropy of A stream :

$$\Delta S_A = S_C - S_A$$
$$= 6.48293 - 6.74775$$
$$\mathbf{\Delta S_A = -0.26482 \text{ kJ/kg-K}}$$

(c) Change in entropy of B stream

$$\Delta S_B = S_C - S_B$$
$$= 6.48293 - 6.19450$$
$$\mathbf{\Delta S_B = 0.28843 \text{ kJ/kg-K}}$$

(d) Net change

Net change in entropy

$$= 0.28843 + (-0.26482)$$
$$= 0.02361 \text{ kJ/kg-K}$$

Net increase is there.

VII. Separating and Throttling Calorimeter

Example 10.41 *Discuss under what circumstances, a throttling calorimeter is capable of being used as a device for measuring the steam dryness.*

Find the maximum degree of wetness that can be determined by a throttling calorimeter if the inlet pressure of steam to it is 10.5 bar and exit pressure is 1.2 bar.

Solution

Condition of the use of throttling calorimeter for finding the dryness fraction of steam is that the condition of steam after throttling must be known. For this purpose the steam should be slightly superheated so that temperature can be measured by the thermometer. In the limit, the steam should be dry and saturated after throttling.

Let x, be the dryness fraction of steam to be found.

For throttling process, enthalpies are same before and after throttling.

$$\therefore \quad h_1 = h_2$$

or

$$h_{f1} + x_1 h_{fg1} = h_{g2}$$

$$\therefore \quad 772.03 + x_1 \times 2005.9 = 2683.4$$

$$\therefore \quad x_1 = \frac{2683.4 - 772.03}{2005.9}$$

$$= \frac{1911.37}{2005.9}$$

$$= 0.9529$$

$$\therefore \quad \text{Maximum wetness} = 1 - 0.9529 = 0.0471 \text{ or } 4.71\ \%$$

Example 10.42 *The following, data were obtained with a separating and throttling calorimeter:*

Pressure in the pipe line 1.5 MPa

Condition after throttling 0.1 MPa and 110°C

Moisture collected in separating

Calorimeter/5 min 0.15 lit at 70°C

Steam condensed after throttling 3.24 kg

Find the quality of steam in the pipe.

Solution

If w is the wt. of steam separated in separating calorimeter and w the wt. of condensed steam after throttling, then dryness fraction of steam is given by

$$x = \frac{Wx_1}{W + w}$$

where x_1 is the dryness fraction of steam entering the throttling calorimeter

$$\therefore \quad x = \frac{3.24\, x_1}{3.24 + 0.15}$$

$$= \frac{3.24\, x_1}{3.39} \qquad ...(1)$$

For throttling calorimeter,

$$[\text{Enthalpy before}] = [\text{Enthalpy after throttling}]$$

$$h_{f1} + x_1 h_{fg1} = h_{g2} + C_V (t_{sup} - t_{sat})$$

$$844.66 + x_1 \times 1945.2 = 2676.2 + 2.0[110 - 99.632]$$

$$= 2676.2 + 2 \times 10.368$$

$$= 2676.2 + 20.736$$

$$= 2696.936 \text{ kJ/kg}$$

$$\therefore \quad x_1 = \frac{2696.936 - 844.66}{1945.2}$$

$$= \frac{4852.276}{1945.2} \quad ...(2)$$

$$= 0.9522$$

∴ Dryness fraction of steam in the pipe

$$x = \frac{3.24}{3.39} \times 0.9522$$

$$\mathbf{x = 0.91}$$

Example 10.43 (*a*) *Explain the influence of pressure of the following properties of steam with the help of T–S diagram.*

(*i*) *Enthalpy of evaporation*

(*ii*) *Temperature of saturation*

(*iii*) *Enthalpy of dry saturated steam*

(*b*) *With a separating and throttling calorimeter arranged in series, the following observations, were taken. Estimate the quality of steam sampled.*

Water separated = 2 kg

Steam discharged from throttling calorimeter = 20.5 kg

Temperature of steam after throttling = 110º C

Initial pressure = 11 bar (gauge)

Barometer = 760 mm of Hg

Final pressure = 28 mm of Hg

Solution

(a) The *T–S* diagram for steam is shown in Fig. Ex. 10.43. Effect of pressure on

(i) Enthalpy of esvaporation: Horizontal distance on *T–S* diagram represents to scale, the enthalpy of evaporation.

If the pressure increases, i.e., saturation temperature increases, enthalpy of evaporation decreases and vice-versa.

(ii) Temperature of saturation: As the pressure increases, saturation temperature increases and vice-versa.

(iii) Enthalpy of dry saturated steam: Enthalpy of dry saturated steam decreases with the increase in pressure and vice-versa. This is beyond certain pressure only otherwise it increases.

(b) The dryness fraction of steam sampled is given by

$$x = \frac{W_1}{W_1 + W_2} .x$$

here W_1 = 20.5 kg, W_2 = 2 kg, x_1 = Dryness fraction of steam leaving the separating calorimeter and entering the throttling calorimeter Atm. pressure = 760 mm Hg = 1.01325 bar

∴ Abs. pressure of steam sampled = 11 + 1.01325 ≈ 12.01325 bar

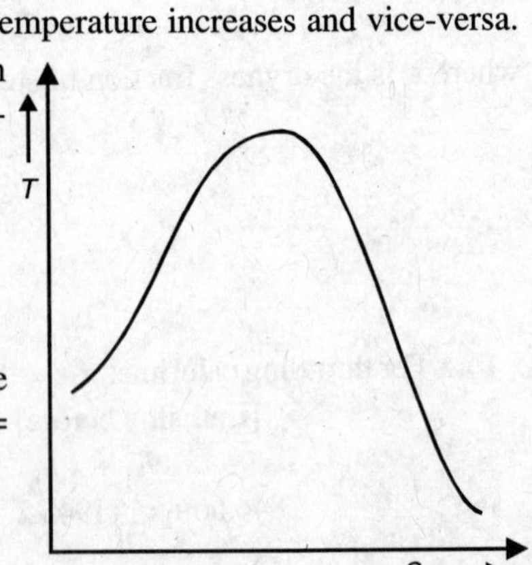

Fig. Ex. 10.43

Similarly, $$28 \text{ mm Hg} = \frac{28}{700} \times 1.01325 = 0.03733 \text{ bar}$$

Abs. pressure = 0.03733 + 1.01325 = 1.05058 bar, considering 1.1 bar

For throttling calorimeter, we have

$$h_1 = h_2$$

$$\therefore \quad 798.43 + x_1 \times 1984.3 = \frac{2679.6 + 2.1(110 - 102.32)}{1984.3}$$

$$= \frac{1897.298}{1984.3} = 0.9561$$

$$x = \frac{09561 \times 20.5}{22.5} = 0.871$$

PROBLEMS FOR PRACTICE

10.1. The internal energy of 2 kg of steam at a pressure of 9 bar is 4605 kJ. Calculate the dryness fraction of the steam. By how many heat units the internal energy of the above steam will increase if it is superheated, at constant pressure, to a temperature of 225°C.
Take $C_p = 2.26\ \text{kJ/kg-K}$

10.2. 1 kg steam at a pressure of 10 bar and having a dryness fraction of 0.2 receives 1046 kJ heat at a constant pressure. Calculate:
(a) The dryness fraction of steam
(b) The total heat
(c) Specific volume
(d) Internal energy of steam at the end of constant pressure heat supply :
Constant Pressure Process

10.3. Two boilers one with a superheater and the other without a superheater are delivering equal quantities of steam into a common main. The pressure in the boilers and the main is 14 bar absolutely. The temperature of steam from the boiler with superheater is 325°C and that of the steam in the main is 250°C. Estimate the quality of steam supplied by other boiler.

[*Ans.* 0.9785 dry]

10.4. In a Babcock Wilcox boiler, the wet steam at 20 bar and 0.93 dry is taken, from the steam drum to the superheater, where it is superheated to 150°C superheat at constant pressure. Find the heat during the process, work done and the change of entropy during the process. Take C_P for superheated steam as 2.3 kJ / kg-K.

[**Ans.** 477.2 kJ / kg; 75.5 kJ / kg; 0.8922 kJ / kg-K]

Constant Volume Process

10.5. A pressure vessel contains 1.25 kg of steam at 5 bar and 0.7 dry. The vessel is cooled until the steam attains the final dryness fraction of 0.5. Find (a) The final pressure of steam and (b) Heat transfer during the process.

[*Ans.* (a) 2.676 bar (b) –1034.25 kJ]

10.6. A closed vessel of 0.2 m³ capacity contains steam at a pressure of 8 bar and a temperature of 200°C.
(i) What is the weight of steam contained in the vessel ?
(ii) The vessel is cooled till the steam becomes just dry and saturated. What will be the pressure of steam at this state ? Estimate the specific entropy.
(iii) The vessel is further cooled till the temperature drops; to 160.8°C. Find the pressure and condition of steam at this stage.

[*Ans.* (i) 0.78 kg (ii) 7.5 bar; 6.686 kJ/ kg-K (iii) 6.32 bar, (iv) 0.849 dry]

Isentropic Process

10.7. Steam initially at 1.5 mPa and 300°C expands reversibly and adiabatically in a steam turbine to 40°C. Determine:
(i) condition of steam after expansion
(ii) work done by the turbine per kg of steam

[*Ans.* (i) 0.818 (ii) 870 kJ / kg-K]

10.8. Steam at 2 MN/m² and 250°C is expanded isentropically to 0.35 MN/m². Using the steam tables, determine
(a) condition of steam after expansion

(b) work done for non-flow process

[*Ans.* (a) 0.915 (b) 297.38 kJ/ kg]

10.9. In a steam engine cylinder steam expands hyperbolically from 8 bar pressure and dryness fraction 0.9 to 1 bar. Find the final condition of steam and change in specific volume. Find also the non-flow work during the process.

[*Ans.* (i) 107.61C; (ii) 1.51515 m^3/kg; (iii) 360 kJ / kg]

10.10. Five kilogram steam having a pressure of 16 bar and 0.9 dry is expanded in a cylinder to a pressure of 0.8 bar. Determine the final condition of steam mechanical work done and the heat transfer during the process if the process is hyperbolic.

[*Ans.* (1) 117.82C; 534.055 kJ / kg (2670.276 kJ) ; (2) 657.445 kJ]

10.11. 1 kg of wet steam at 250°C and with enthalpy of 2500 kJ is confined in a rigid vessel. Heat is supplied until the steam becomes saturated. Determine

(a) initial pressure of steam

(b) final pressure of steam

(c) heat supplied

[*Ans.* (a) 39.7894 bar (b) 48 bar (c) 262.15 kg]

10.12. A sample of steam at 15 bar has an internal energy, of 2500 kg. Find the condition of steam.

[*Ans.* $x = 0.946$]

10.13. A rigid vessel of volume 0.01 m^3 contains 90% (by volume) water and 10% steam at a pressure of 2 bar. Deterrnine

(i) temperature of water

(ii) total mass of water

(iii) steam in the vessel

[*Ans.* (i) 120.2°C (ii) 9 kg (iii) .0:00113 kg]

10.14. Determine the specific volume, enthalpy of steam under the following conditions

(i) 4 MPa pressure 360°C

(ii) 150 kPa, 10% quality

(iii) 350 kPa, 10% moisture

[*Ans.* (i) 0.062 m^3, 3053.31 kJ/kg (ii) 0.1159 m^3, 689.6 kJ/kg (iii) 0.4717 m^3;1517.2 kJ kg]

10.15. A closed vessel of 0.75 cu.m capacity contains dry and saturated steam at 5 bar. The vessel is cold until the pressure is reduced to 2 bar. Determine,

(i) mass of steam present in the vessel

(ii) final dryness fraction

(iii) amount of heat transferred during the process.

[*Ans.* (i) 2.001 kg, (ii) 0.4232, (iii) –23.78.52 kg]

10.16. 0.125 cu.m of wet steam at 10 bar has an enthalpy of 1800 kJ/kg. Calculate mass and dryness fraction of steam. [*Ans.* m = 1.249 kg, x = 0.5152]

10.17. One kg of steam at an initial condition of 6 bar and 0.2 dry is heated at constant volume until the pressure is 20 bar. Determine the final state of steam, heat added.

[*Ans.* $x_2 = 0.635$, $Q = 932$ kJ/kg]

10.18. Three boilers *A*, *B*, *C* working at 12 bar absolute supply steam in equal quantities to a plant through a common steam pipe. The boiler *A* supplies, dry saturated steam while the steam supplied by boiler *B* is superheated to 350°C. The steam from boiler *C* is of dryness fraction 0.96. Calculate the condition of steam in common pipe. Assume that there is no loss of heat in pipes and specfic heat of superheated steam is 2.1 kJ/kg°K. [*Ans.* 229.2 °C]

10.19. Steam is stored in a container 0.45 m^3 capacity. The pressure of steam is 8 bar and temperature is 200°C. The container is cooled and pressure drops to 25 bar. Find

(i) initial internal energy in kJ

(ii) quality of steam after cooling

(iii) heat lost to atmosphere due to cooling in kJ

Neglect volume of water.

[*Ans.* (i) 4533.77 kJ, (ii) 0.3625, (iii) – 2360.42 kJ]

10.20. 1 kg of steam at 20 bar and 300°C is cooled at constant pressure till the dryness fraction of steam becomes 0.5. Find the change in enthalpy work done and the heat transfer during the process.

[*Ans.* $\Delta H = -1172$ kJ, $W = -151.5$ kJ, $Q = -1172$ kJ]

IMPORTANT POINTS TO BE REMEMBERED

1. *Critical pressure* $P_c = 221.2$ bar
2. *Critical temperature* $T_c = 374.15 + 273 = 647.15°K$
3. *Critical volume* $V_c = 0.00317$ m³/kg
4. *Enthalpy*

$$h = h_f + x\,h_{fg} \text{ wet steam}$$
$$= h_f + h_{fg} = h_g \text{ Dry and saturated steam}$$
$$= h_g + C_{PV}(t_{sup} - t_{sat}) \text{ superheated steam}$$

5. *Volume*

$$V = x\,V_g \text{ wet steam}$$
$$= V_g \text{ Dry and sat. steam}$$
$$= V_g \times \frac{T_{sup}}{T_{sat}} \text{ Superheated steam}$$

6. *Dryness fraction* $x = \dfrac{\text{Dry steam}}{\text{Total steam}} = \dfrac{W_S}{W_T}$
7. *Internal energy of steam*

$$u = (h_f + x\,h_{fg}) - P \cdot x\,V_g \text{ Wet steam}$$
$$= h_g - PV_g \text{ Dry and saturated steam}$$
$$= [h_g + C_{PV}(t_{sup} - t_{sat})] - P \cdot V_g \frac{I_{sup}}{I_{sat}} \text{ Superheated steam}$$

8. *Entropy of steam*

$$S_p = C_{pw} \cdot \log e \frac{T_{sat}}{273} \text{ Sensible entropy}$$
$$= S_f + \frac{h_{fg}}{T_{sat}} \times x \text{ Wet steam}$$
$$= S_f + \frac{h_{fg}}{T_{sat}} = S_g \text{ Dry and sat. steam}$$
$$= S_f + S_{fg}$$
$$= S_g + C_{PV} \log e \frac{T_{sup}}{T_{sat}} \text{ Superheated steam}$$

9. *Triple point:* Pressure = 0.00641 bar

   Temperature = 0.01°C
10. *Constant pressure process*

$$\text{Work} = P(V_2 - V_1)$$
$$Q = h_2 - h_1 = (u_2 - u_1) + P(V_2 - V_1)$$
$$\Delta S = (S_2 - S_1)$$

11. *Constant volume process*

$$\text{Weak} = 0$$
$$Q = u_2 - u_1$$
$$\Delta S = S_2 - S_1$$

12. *Constant temperature process*

$$Q = T\Delta S = T(S_2 - S_1)$$
$$= (u_2 - u_1) + \text{Work} \quad \text{Non-flow process}$$
$$= \Delta P + \Delta K + \Delta h + W \quad \text{Flow process}$$

13. *Isentropic process*

$$Q = 0, \quad W = u_1 - u_2 \quad \text{Now-flow process}$$
$$= h_1 - h_2 \quad \text{Flow process when } \Delta K = 0$$
$$S_2 = S_1$$

14. *Polytropic process*

$$PV^n = C$$
$$W_{nf} = \frac{P_1V_1 - P_2V_2}{n-1}$$
$$W_{sf} = \frac{n}{n-1}[P_1V_1 - P_2V_2] = -\int_t^2 Vdp \quad \text{Flow process}$$
$$Q = (u_2 - u_1) + W_{nf} \quad \text{Now-flow process}$$
$$= \Delta P + \Delta K + \Delta h + W_{sf} \quad \text{Flow process}$$

15. *Hyperobolic process* ($PV = C$)

$$W_{nf} = W_{sf} = P_1V_1 \log e \frac{V_2}{V_1}$$

16. *Throttling process* (*Free expansion process*)

$$h_1 = h_2$$

STEAM TABLES

Introduction

1. The following steam tables are based on the Mollier Tables and give all the necessary data with regard to saturated and superheated steam required by Engineering students.
2. The various tables are arranged as follows:
   (*a*) *Saturated Steam*
   Table 1 : It provides data with reference to varying *temperatures.*
   Table 2 : It provides data with reference to varying *pressures.*
   (*b*) *Superheated Steam*
   Table 3: It provides *Enthalpy* of steam superheated to various temperatures at different pressures.
   Table 4: It provides *Entropy* of steam superheated to various temperatures at different pressures.
   Table 5: It provides *Volume* of steam superheated to various temperatures at different pressures.
3. All heat energy quantities expressed in these tables are in terms of kJ/kg of steam.
4. All pressures are in terms of N/m^2.

Important Relationships

(i) 1 J is equivalent of 1 Nm or 0.10197 m kg_f of work.

(ii) $T = t + 273$, where T = Temperature expressed in centrigrade absolute scale, (K).

(iii) Absolute pressure = Gauge pressure + Atmospheric pressure.

(iv) Volume of wet steam = $x \times V$,

where x = dryness fraction.

V_s = Specific volume of saturated Steam.

(v) Enthalpy of wet steam = Enthalpy of liquid + (Latent heat × dryness fraction).

or $H = h_L \times x L$

(This also applies to *Entropy* and *Internal Energy*)

(vi) 1 N/m^2 = 10^{-5} bar = 0.10197 kg_f/m^2 of pressure.

Table 1. Saturated Steam (Temperature Table)

Temperature		Pressure N/m² $p \times 10^{-5}$	Volume m²/kg		Density of Vapour kg/m²	Enthalpy kJ/kg		Latent heat L kJ/kg	Entropy kJ\kg–K		L/T kJ/kg-K
°C	K		Liquid v_i	Vapour v_i		Liquid h_i	Vapour h_s		Liquid s_i	Vapour s_v	
1	2	3	4	5	6	7	8	9	10	11	12
0	273	0.00610	0.001000	206.3	0.004847	0.00	2501.6	2501.6	0.00	9.1577	9.1577
1	274	0.00656	0.001000	192.6	0 005192	4.17	2503.4	2499.2	0.0152	9.1311	9.1159
2	275	0.00705	0.001000	179.9	0.005558	8.39	2505.2	2496.8	0.0306	9.1047	9.0741
3	276	0.00757	0.001000	168.2	0.005946	12.60	2507.1	2494.5	0.0459	9.0785	9.0326
4	277	0.00813	0.001000	157.3	0.006358	16.80	2508.9	2492.1	0.0611	9.0526	8.9915
5	278	0.00872	0.001000	147.2	0.006795	21.01	2510.7	2489.7	0.0762	9.0269	8.9507
6	279	0.00934	0.001000	137.8	0.007258	25.21	2512.6	2487.4	0.0913	9.0015	8.9102
7	280	0.01001	0.001000	129.1	0.007748	29.41	2514.4	2485.0	0.1063	8.9762	8.8699
8	281	0.01072	0.001000	121.0	0.008267	33.60	2516.2	2482.6	0.1213	8.9513	8.8300
9	282	0.01147	0.001000	113.4	0.008816	37.80	2518.1	2480.3	0.1361	8.9265	8.7904
10	283	0.01227	0.001000	106.4	0.009396	41.99	2519.9	2477.9	0.1510	8.9020	8.7510
11	284	0.01311	0.001000	99.91	0.01001	46.19	2521.7	2475.5	0.1658	8.8776	8.7118
12	285	0.01401	0.001000	93.84	0.01066	50.38	2523.6	2473.2	0.1805	8.8536	8.6731
13	286	0.01496	0.001000	88.18	0.01134	54.57	2525.4	2470.8	0.1952	8.8297	8.6345
14	287	0.01597	0.001000	82.90	0.01206	58.75	2527.2	2468.5	0.2098	8.8060	8.5962
15	288	0.01703	0.001000	77.98	0.01282	62.94	2529.1	2466.1	0.2243	8.7826	8.5583
16	289	0.01817	0.001001	73.38	0.01363	67.13	2530.9	2463.8	0.2388	8 7593	8.5205
17	290	0.01936	0.001001	69.09	0.01447	71.31	2532.7	2461.4	0.2533	8.7363	8.4830
18	291	0.02062	0.001001	65.09	0.01447	75.50	2534.5	2459.0	0.2677	8.7135	8.4453
19	292	0.02196	0.001001	61.34	0.01630	79.68	2536.4	2456.7	0.2820	8.6908	8.4088
20	293	0.02337	0.001002	57.84	0.01729	83.86	2538.2	2454.3	0.2963	8.6684	8.3721
21	294	0.02485	0.001002	54.56	0.01833	88.04	2540.0	2452.0	0.3105	8.6462	8.3357
22	295	0.02642	0.001002	51.49	0.01942	92.23	2541.8	2449.6	0.3247	8.6241	8.2994
23	296	0.02808	0.001002	48.62	0.02057	96.41	2543.6	2447.2	0.3389	8.6023	8.2634
24	297	0.02982	0.001002	45.93	0.02177	100.59	2545.5	2444.8	0.3530	8.5806	8.2276
25	298	0.03166	0.001003	43.40	0.02304	104.77	2547.3	2442.5	0.3670	8.5592	8.1922
26	299	0.03360	0.001003	41.03	0.02437	108.95	2549.1	2440.2	0.3810	8.5379	8.1569
27	300	0.03564	0.001003	38.81	0.02576	113.13	2550.9	2437.8	0.3949	8.5168	8.1219
28	301	0.03778	0.001004	36.73	0.02723	117.31	2552.7	2435.4	0.4088	8.4959	8.0871
29	302	0.04004	0.001004	34.77	0.02876	121 48	2554.5	2433.1	0.4227	8.4751	8.0524
30	303	0.04241	0.001004	32.93	0.03037	125.66	2556.4	2430.7	0.4363	8.4546	8.0181
31	304	0.04491	0.001005	31.20	0.03205	129.84	2558.2	2428.3	0.4503	8.4342	7.9339
32	305	0.04753	0.001005	29.57	0.03382	134.02	2560.0	2425.9	0.4640	8.4140	7.9500

(Contd.)

Table 1 *Contd.*: Saturated Steam (Temperature Table)

1	2	3	4	5	6	7	8	9	10	11	12
33	306	0.05029	0.001005	28.04	0.03566	138.20	2561.8	2423.6	0.4777	8.3939	7.9162
34	307	0.05318	0.001006	26.60	0.03759	142.38	2563.6	2421.2	0.4913	8.3740	7.8827
35	308	0.05622	0.001006	25.24	0.03961	146.56	2565.4	2418.8	0.5049	8.3543	7.8494
36	309	0.05940	0.001006	23.97	0.04172	150.74	2567.2	2416.4	0.5184	8.3348	7.8164
37	310	0.06274	0.001007	22.76	0.04393	154.91	2569.0	2414.1	0.5319	8.3155	7.7835
38	311	0.06624	0.001007	21.63	0.04624	159.04	2570.8	2411.7	0.5453	8.2962	7.7509
39	312	0.06991	0.001007	20.56	0.04865	163.27	2572.6	2409.3	0.5588	8.2772	7.7184
40	313	0.07375	0.001008	19.55	0.05116	167.45	2574.4	2406.9	0.5721	8.2583	7 6862
41	314	0.07777	0.001008	18.59	0.05379	171.63	2576.2	2404.5	0.5854	8.2395	7.6541
42	315	0.08198	0.001009	17.69	0.05652	175.81	2577.9	2402.1	0.5987	8.2209	7.6222
43	316	0.08639	0.001009	16.84	0.05938	179.99	2579.7	2399.7	0.6120	8.2025	7.5305
44	317	0.09100	0.001009	16.04	0.06236	184.17	2581.5	2397.3	0.6252	8.1342	7.5590
45	318	0.09582	0.001010	15.28	0.06546	188.35	2583.3	2394.9	0.6383	8.1661	7.5278
46	319	0.10086	0.001010	14.56	0.06869	192.53	2585.1	2392.5	0.6514	3.1481	7.4967
47	320	0.10612	0.001011	13.88	0.07206	196.71	2586.9	2390.1	0.6645	3.1302	7.4737
48	321	0.11162	0.001012	13.23	0.07557	200.89	2588.6	2387.7	0.6776	8.1125	7.4349
49	322	0.11736	0.001012	12.62	0.07922	205.07	2590.4	2385.3	0.6906	8.0950	7.4044
50	323	0.12335	0.001012	12.05	0.08302	209.26	2592.2	2382.9	0.7035	8.0776	7.5741
55	328	0.15741	0.001014	9.579	0.1044	230.17	2601.0	2370.8	0.7677	7.9926	7.2249
60	333	0.1992	0.001017	7.679	0.1302	251.09	2609.7	2358.6	0.8310	7.9108	7.0798
65	338	0.2501	0.001020	6.202	0.1612	272.02	2618.4	2346.3	0.8933	7.8322	6.9389
70	343	0.3116	0.001023	5.046	0.1982	292.97	2626.9	2334.0	0.9548	7.7565	6.8017
75	348	0.3855	0.001026	4.134	0.2419	313.94	2635.4	2321.5	1.0154	7.6835	6.6681
80	353	0.4736	0.001030	3.409	0.2933	334.92	2643.8	2308.8	1.0753	7.6132	6.5379
85	358	0.5780	0.001032	2.829	0.3535	355.92	2652.0	2296.5	1.1343	7.5454	6.4111
90	363	0.7011	0.001036	2.361	0.4235	376.94	2660.1	2283.2	1.1925	7.4799	6.3074
95	368	0.8453	0.001040	1.982	0.5045	397.99	2668.1	2270.2	1.2501	7.4166	6.1665
100	373	1.0133	0.001043	1.673	0.5977	419.06	2676.0	2256.9	1.3069	7.3554	6.0485
105	378	1.2080	0.001050	1.419	0.7096	440.17	2683.7	2243.6	1.3630	7.2962	5.9332
110	383	1.4327	0.001052	1.210	0.8265	461.32	2696.3	2230.0	1.4185	7.2388	5.8203
115	388	1.6906	0.001056	1.036	0.9650	482.50	2698.7	2216.2	1.4733	7.1832	5.7099
120	393	1.9854	0.001060	0.8915	1.122	503.72	2706.0	2202.2	1.5276	7.1293	5.6017
125	398	2.3210	0.001065	0.7702	1.298	524.99	2713.0	2188.0	1.5813	7.0769	5.4956
130	403	2.7013	0.001070	0.6681	1.497	546.31	2719.9	2173.6	1.6344	7.0261	5.3917
135	408	3.131	0.001075	0.5818	1.719	567.68	2726.6	2158.9	1.6869	6.9766	5.2897
140	413	3.131	0.001080	0.5085	1.967	589.10	2733.1	2144.0	1.7390	6.9284	5.1894

(Contd.)

Table 1*Contd.* : Saturated Steam (Temperature Table)

1	2	3	4	5	6	7	8	9	10	11	12
145	418	4.155	0.001085	0.4460	2.242	610.60	2739.3	2128.7	1.7906	6.8815	5.0919
150	423	4.760	0.001090	0.3924	2.548	632.15	2745.4	2113.2	1.8416	6.8358	4.9942
155	428	5.433	0.001096	0.3464	2.886	653.78	2751.2	2097.4	1.8923	6.7911	4.8988
160	433	6.181	0.001100	0.3068	3.260	675.47	2756.7	2081.3	1.9425	6.7475	4.8050
165	438	7.008	0.001108	0.2724	3.671	697.25	2762.0	2064.8	1.9923	6.7048	4.7125
170	443	7.920	0.001110	0.2426	4.123	719.12	2767.1	2047.9	2.0416	6.6630	4.6214
175	448	8.924	0.001120	0.2165	4.618	741.07	2771.8	2030.7	2.0906	6.6221	4.4715
180	453	10.027	0.001127	0.1938	5.160	763.12	2776.3	2013.1	2.1393	6.5813	4.4426
185	458	11.233	0.001134	0.1739	5.752	785.26	2780.4	1995.2	2.1876	6.5424	4.3548
190	463	12.551	0.001141	0.1563	6.397	807.52	2784.3	1976.7	2.2356	6.5036	4.2680
195	468	13.987	0.001149	0.1408	7.100	829.88	2787.8	1957.9	2.2833	6.4654	4.1821
200	473	15.549	0.001565	0.1272	7.864	852.37	2790.9	1938.6	2.3307	6.4278	4.0971
205	478	17.243	0.001164	0.1150	8.694	874.99	2793.8	1918.8	2.3778	6.3906	4.0128
210	483	19.077	0.001173	0.1042	9.593	897.74	2796.2	1898.5	2.4247	6.3539	3.9292
215	488	21.060	0.001181	0.0946	10.57	920.63	2798.3	1877.6	2.4713	6.3176	3.8463
220	493	23.198	0.001190	0.0860	11.62	943.67	2799.9	1856.2	2.5178	6.2817	3.7639
225	498	25.501	0.001199	0.0783	12.76	966.89	2801.2	1834.3	2.5641	6.2461	3.6820
230	503	27.976	0.001209	0.0714	14.00	990.26	2802.0	1811.7	2.6102	6.2107	3.5995
235	508	30.632	0.001218	0.0652	15.33	1013.8	2802.3	1788.5	2.6562	6.1756	3.5194
240	513	33.478	0.001229	0.0596	16.76	1037.6	2802.3	1764.6	2.7020	6.1406	3.4386
245	518	36.523	0.001240	0.0546	18.31	1061.6	2801.6	1740.0	2.7478	6.1057	3.3579
250	523	39.776	0.001251	0.0500	19.99	1085.5	2800.4	1714.6	2.7935	6.0708	3.2773
255	528	43.246	0.001263	0.0459	21.79	1110.2	2798.7	1688.5	2.8392	6.0359	3.1967
260	533	46.943	0.001276	0.04213	23.73	1134.9	2796.4	1661.5	2.8848	6.0010	3.1162
265	538	50.877	0.001288	0.03871	25.83	1159.9	2793.5	1633.6	2.9306	5.9658	3.0352
270	543	55.058	0.001302	0.03559	28.10	1185.2	2789.9	1604.6	2.9763	5.9304	2.9541
275	548	59.496	0.001317	0.03274	30.55	1210.9	2785.5	1574.7	3.0223	5.8947	2.8724
280	553	64.202	0.001332	0.03013	33.19	1236.8	2780.4	1543.6	3.0683	5.8536	2.7903
285	558	69.186	0.001348	0.02773	36.06	1263.2	2774.5	1511.3	3.1146	5.8220	2 7074
290	563	74.461	0.001366	0.02554	39.16	1290.0	2767.6	1477.6	3.1611	5.7848	2.6237
295	568	80.037	0.001384	0.02351	42.53	1317.3	2759.8	1442.6	3.2079	5.7469	2.5390
300	573	85.927	0.001404	0.02165	46.19	1345.0	2751.0	1406.0	3.2552	5.7081	2.4529
305	578	92.144	0.001425	0.01993	50.18	1373.4	2741.1	1367.7	3.3029	5.6685	2.3656
310	583	98.700	0.001448	0.01833	54.54	1402.4	2730.0	1327.6	3.3512	5.6278	2.2766
315	588	105.67	0.001472	0.01686	59.33	1432.1	2717.6	1285.6	3.4002	5.5858	2.1856
320	593	112.89	0.001499	0.01548	64.60	1492.6	2703.7	1241.1	3.4500	5.5423	2.0923

(Contd.)

Table 1*Contd.* : Saturated Steam (Temperature Table)

1	2	3	4	5	6	7	8	9	10	11	12
325	598	120.56	0.001529	0.01419	70.45	1490.0	2688.0	1194.0	3.5008	5.4969	1.9961
330	603	128.63	0.001561	0.01299	76.99	1526.5	2670.2	1143.6	3.5528	4.4490	1.8962
335	608	137.12	0.001597	0.01185	84.36	1560.3	2649.7	1089.5	3.6063	5.3979	1.7916
340	613	146.05	0.001638	0.01078	92.76	1595.5	2626.2	1030.7	3.6616	5.3427	1.6811
345	618	155.45	0.001685	0.00976	102.4	1632.5	2598.9	966.4	3.7193	5.2828	1.5635
350	623	165.35	0.001741	0.00880	113.6	1671.9	2567.7	895.7	3.7800	5.2177	1.4377
355	628	175.77	0.001808	0.00786	127.2	1716.6	2530.4	813.8	3.8489	5.1442	1.2953
360	633	186.75	0.001896	0.00694	144.1	1764.2	2485.4	712.3	3.9210	5.0600	1.1390
365	638	198.33	0.002016	0.00601	166.3	1818.0	2428.0	610.0	4.0021	4.9579	0.9558
370	643	210.54	0.002136	0.00497	201.1	1890.2	2342.8	452.6	4.1108	4.8144	0.7036
374.15	647.15	221.20	0.00317	0.00317	315.5	2107.4	2107.4	0.0	4.4429	4.4429	0.0

Table 2: Saturated Steam (Pressure Table)

Pressure N/m² $p \times 10^{-5}$	Temperat-ure °C	Volume, v m²/kg		Density of Vapour, ρ kg/m²	Enthalpy, h, kJ/kg		Latent heat L kJ/kg	Entropy, kJ/kg–K		L/T kJ/kg-K
		Liquid v_i	Vapour v_s		Liquid h_i	Vapour h_s		Liquid s_i	Vapour s_v	
1	2	3	4	5	6	7	8	9	10	11
0.010	6.983	0.001000	129.2	0.00774	29.34	2514.4	2485.0	0.1060	8.9767	8.8707
0.015	13.036	0.001000	87.98	0.01137	54.71	2525.5	2470.7	0.1957	8.8288	8.6331
0.020	17.513	0.001001	67.01	0.01492	73.46	2533.6	2460.2	0.2607	8.7246	8.4639
0.025	21.096	0.001002	54.26	0.01843	88.45	2540.2	2451.7	0.3119	8.6440	8.3321
0.030	24.100	0.001003	45.67	0.02190	101.00	2545.6	2444.6	0.3544	8.5785	8.2241
0.035	26.694	0.001003	39.48	0.02533	111.85	2550.4	2438.5	0.3907	8.5232	8.1325
0.040	28.983	0.001004	34.80	0.02873	121.41	2554.5	2433.1	0.4225	8.4755	8.0530
0.045	31.035	0.001005	31.14	0.03211	129.99	2558.2	2428.2	0.4507	8.4335	7.9828
0.050	32.898	0.001005	28.19	0.03547	137.77	2561.6	2423.8	0.4763	8.3960	7.9197
0.055	34.605	0.001006	25.77	0.03880	144.91	2564.7	2419.8	0.4995	8.3621	7.8626
0.060	36.183	0.001006	23.74	0.04212	151.50	2567.5	2416.0	0.5209	8.3312	7.8103
0.065	37.651	0.001007	22.02	0.04542	157.64	2570.2	2412.5	0.5407	8.3029	7.7622
0.070	39.025	0.001007	20.53	0.04871	163.38	2572.6	2409.2	0.5591	8.2767	7.7176
0.075	40.316	0.001008	19.24	0.05198	168.77	2574.9	2406.2	0.5763	8.2523	7.6760
0.080	41.534	0.001008	18.10	0.05523	173.86	2577.1	2403.2	0.5925	8.2296	7.6371
0.085	42.689	0.001009	17.10	0.05848	178.69	2579.2	2400.5	0.6079	8.2082	7.6003
0.090	43.787	0.001009	16.20	0.06171	183.28	2581.1	2397.9	0.6224	8.1881	7.5657
0.095	44.833	0.001010	15.40	0.06493	187.65	2583.0	2395.3	0.6361	8.1691	7.5330
0.10	45.833	0.001010	14.67	0.06814	191.83	2584.8	2392.9	0.6493	8.1511	7.5018
0.15	53.997	0.001014	10.02	0.09977	225.97	2599.2	2373.2	0.7549	8.0093	7.2544
0.20	60.086	0.001017	7.650	0.1307	251.45	2609.0	2358.4	0.8321	7.9094	7.0773
0.25	64.992	0.001019	6.204	0.1612	271.99	2618.3	2346.4	0.8932	7.8323	6.9391
0.30	69.124	0.001023	5.229	0.1912	289.30	2625.4	2336.1	0.9441	7.7695	6.8254
0.35	72.651	0.001024	4.529	0.2210	304.20	2631.5	2327.3	0.9880	7.7168	6.7288
0.40	75.886	0.001026	3.993	0.2504	317.65	2636.9	2319.2	1.0261	7.6709	6.6448
0.45	78.743	0.001028	3.576	0.2796	329.64	2614.7	2312.0	1.0603	7.6307	6.5704
0.50	81.345	0.001030	3.240	0.3086	340.56	2646.0	2305.4	1.0912	7.5947	6.5035
0.55	83.737	0.001031	2.964	0.3374	350.61	2649.9	2299.3	1.1194	7.5623	6.4429
0.60	85.954	0.001033	2.732	0.3661	359.93	2653.6	2293.6	1.1454	7.5327	6.3873
0.65	88.021	0.001035	2.535	0.3945	368.62	2656.9	2288.3	1.1696	7.5055	6.3359
0.70	89.959	0.001036	2.365	0.4229	376.77	2660.1	2283.3	1.1921	7.4804	6.2883
0.75	91.785	0.001037	2.217	0.4511	384.45	2663.0	2278.6	1.2131	7.4570	6.2439

(Contd.)

Table 2 *Contd.* : Saturated Steam (Pressure Table)

1	2	3	4	5	6	7	8	9	10	11
0.80	93.512	0.001038	2.087	0.4792	391.72	2665.8	2274.0	1.2330	7.4352	6.2022
0.85	95.152	0.001040	1.972	0.5071	398.63	2668.4	2269.8	1.2518	7.4147	6.1629
0.90	96.713	0.001041	1.869	0.5350	405.21	2670.9	2265.6	1.2696	7.3954	6.1258
0.95	98.204	0.001042	1.777	0.5627	411.49	2673.2	2261.7	1.2865	7.3771	6.0906
1.00	99.632	0.001043	1.694	0.5904	417.51	2675.4	2257.9	1.3027	7.3598	6.0571
1.50	111.37	0.001053	1.159	0.8628	467.13	2693.4	2226.2	1.4336	7.2234	5.7898
2.00	120.23	0.001060	0.8854	1.129	504.17	2706.3	2201.6	1.5301	7.1268	5.5967
2.50	127.43	0.001067	0.7184	1.392	535.34	2716.4	2181.0	1.6071	7.0520	5.4449
3.00	133.54	0.001073	0.6056	1.651	561.43	2724.7	2163.2	1.6716	6.9906	5.3190
3.50	138.87	0.001079	0.5240	1.908	584.27	2731.6	2147.4	1.7273	6.9392	5.2119
4.00	143.62	0.001084	0.4622	2.163	604.67	2737.6	2133.0	1.7764	6.8943	5.1179
4.50	147.92	0.001088	0.4132	2.417	623.16	2742.9	2119.7	1.8204	6.8647	5.0343
5.00	151.84	0.001092	0.3747	2.669	640.12	2747.5	2107.4	1.8604	6.8192	4.9588
5.50	155.46	0.001097	0.3425	2.920	655.66	2751.7	2096.1	1.8969	6.7869	4.8900
6.00	158.84	0.001101	0.3155	3.170	670.42	2755.5	2085.0	1.9308	6.7575	4.8267
6.50	162.00	0.001104	0.2923	3.419	684.16	2758.8	2074.7	1.9623	6.7304	4.7681
7.00	164.96	0.001108	0.2727	3.667	697.06	2762.0	2064.9	1.9918	6.7302	4.7134
7.50	167.76	0.011116	0.2554	3.916	709.30	2764.8	2055.5	2.0191	6.6812	4.8621
8.00	170.41	0.001115	0.2403	4.162	720.94	2767.5	2046.5	2.0457	6.6596	4.6139
8.50	172.95	0.001118	0.2268	4.410	732.05	2769.9	2037.9	2.0670	6.6388	4.5718
9.00	175.36	0.001121	0.2148	4.655	742.64	2772.1	2029.5	2.0941	6.6192	4.5251
9.50	177.66	0.001125	0.2040	4.890	752.82	2774.2	2021.4	2.1164	6.6078	4.4914
10.00	179.88	0.001127	0.1943	5.147	762.61	2776.2	2013.6	2.1382	6.5828	4.4446
11.00	184.07	0.001133	0.1774	5.637	781.13	2779.7	1998.5	2.1786	6.5497	4.3711
12.00	187.96	0.001138	0.1632	6.127	798.43	2782.7	1984.3	2.2161	6.5194	4.3033
13.00	191.61	0.001144	0.1511	6.617	814.70	2785.7	1970.7	2.2510	6.4913	4.2403
14.00	195.04	0.001149	0.1407	7.106	830.08	2787.8	1957.7	2.2837	6.4651	4.1814
15.00	198.29	0.001154	0.1317	7.596	844.67	2789.9	1945.7	2.3145	6.4406	4.1261
16.00	201.37	0.001159	0.1237	8.085	858.56	2791.7	1933.2	2.3436	6.4175	4.0739
17.00	204.31	0.001163	0.1166	8.575	871.84	2793.4	1921.5	2.3713	6.3957	4.0244
18.00	207.11	0.001168	0.1103	9.065	884.58	2794.8	1910.3	2.3976	6.3751	3.9775
19.00	209.80	0.001172	0.1047	9.555	896.81	2796.1	1899.3	2.4228	6.3554	3.9326
20.00	212.37	0.001176	0.0995	10.05	908.59	2797.2	1886.6	2.4469	6.3367	3.8898
21.00	214.85	0.001180	0.09489	10.54	919.96	2798.2	1878.2	2.4700	6.3187	3.8487
22.00	217.24	0.001185	0.09065	11.03	930.95	2799.1	1868.1	2.4922	6.3015	3.8093
23.00	219.55	0.001189	0.08677	11.52	941.60	2799.8	1858.2	2.5136	6.2849	3.7713
24.00	221.78	0.001193	0.0832	12.02	951.93	2800.4	1848.5	2.5343	6.2690	3.7347

(*Contd.*)

Table 2 *Contd.*: Saturated Steam (Pressure Table)

1	2	3	4	5	6	7	8	9	10	11
25.00	223.94	0.001197	0.07991	12.51	961.96	2800.9	1839.0	2.5543	6.2536	3.6993
26.00	226.04	0.001201	0.07686	13.01	971.72	2801.4	1829.6	2.5736	6.2387	3.6651
27.00	228.07	0.001205	0.07402	13.51	981.22	2801.7	1820.5	2.5924	6.2244	3.6320
28.00	230.05	0.001209	0.07139	14.01	990.48	2802.0	1811.5	2.6106	6.2104	3.5998
29.00	231.97	0.001212	0.06893	14.51	999.53	2802.2	1802.6	2.6283	6.1969	3.5686
30.00	233.84	0.001216	0.06663	15.01	1000.84	2802.3	1793.9	2.6455	6.1837	3.5382
31.00	235.67	0.001220	0.06447	15.51	1017.0	2802.3	1785.4	2.6623	6.1709	3.5086
32.00	237.45	0.001223	0.06244	16.02	1025.4	2802.3	1776.9	2.6786	6.1585	3.4799
33.00	239.18	0.001227	0.06053	16.52	1033.7	2802.3	1768.6	2.6945	6.1463	3.4518
34.00	240.88	0.001231	0.05873	17.03	1041.8	2802.1	1760.3	2.7101	6.1344	3.4243
35.00	242.54	0.001234	0.05703	17.54	1049.8	2802.0	1752.2	2.7253	6.1228	3.3975
36.00	244.16	0.001238	0.05541	18.05	1057.6	2801.7	1744.2	2.7401	6.1115	3.3714
37.00	245.75	0.001241	0.05389	18.56	1065.2	2801.4	1736.2	2.7545	6.1004	3.3457
38.00	247.31	0.001245	0.05244	19.07	1072.7	2800.1	1728.4	2.7689	6.0896	3.3207
39.00	248.84	0.001248	0.05106	19.58	1080.1	2800.8	1720.6	2.7829	6.0789	3.2960
40.00	250.33	0.001252	0.04975	20.10	1087.4	2800.3	1712.9	2.7965	6.0885	3.2720
41.00	251.80	0.001255	0.04850	20.62	1094.6	2799.9	1705.3	2.8099	6.0583	3.2484
42.00	253.24	0.001258	0.04731	21.14	1101.6	2799.4	1697.8	2.8231	6.0482	3.2251
43.00	254.66	0.001262	0.04617	21.66	1108.5	2798.9	1690.3	2.8360	6.0863	3.2023
44.00	256.05	0.001265	0.04508	22.18	1115.4	2798.3	1682.9	2.8487	6.0286	3.1799
45.00	257.41	0.001269	0.04404	22.71	1122.1	2797.7	1675.6	2.8612	6.0191	3.1579
46.00	258.75	0.001272	0.04304	23.24	1128.8	2797.0	1668.3	2.8735	6.0097	3.1362
47.00	260.07	0.001276	0.04208	23.76	1135.3	2796.4	1661.1	2.8855	6.0004	3.1149
48.00	261.37	0.001279	0.04116	24.29	1141.8	2795.7	1653.9	2.8974	5.9913	3.0939
49.00	262.65	0.041282	0.04028	24.83	1148.2	2794.9	1646.8	2.9091	5.9824	3.0733
50.00	263.91	0.001286	0.03943	25.36	1154.5	2794.2	1639.7	2.9206	5.9735	3.0529
52.00	266.37	0.001292	0.03782	26.44	1166.8	2792.6	1625.7	2.9431	5.9561	3.0130
54.00	268.76	0.001299	0.03633	27.52	1178.9	2790.8	1611.9	2.9650	5.9392	2.9742
56.00	271.09	0.001305	0.03495	28.62	1190.8	2789.0	1598.2	2.9863	5.9227	2.9364
58.00	273.35	0.001312	0.03365	29.72	1202.3	2787.0	1584.7	3.0071	5.9066	2.8995
60.00	275.55	0.001319	0.03244	30.83	1213.7	2785.0	1571.3	3.0273	5.8908	2.8635
62.00	277.70	0.001325	0.03130	31.95	1224.8	2782.9	1558.0	3.0471	5.8753	2.8282
64.00	279.79	0.001331	0.03023	33.08	1235.7	2780.6	1544.9	3.0664	5.8601	2.8057
66.00	281.84	0.001338	0.02922	34.22	1246.5	2778.3	1531.9	3.0853	5.8452	2.7601
68.00	283.84	0.001345	0.02827	35.37	1257.0	2775.9	1518.9	3.1038	5.8306	2.7268
70.00	285.79	0.001351	0.02737	36.53	1267.4	2773.5	1506.0	3.1219	5.8162	2.6943

(*Contd.*)

Table 2 *Contd.* : Saturated Steam (Pressure Table)

72.00	287.70	0.001358	0.02652	37.70	1277.6	2770.9	1493.3	3.1397	5.8020	2.6637
74.00	289.57	0.001364	0.02572	38.89	1287.7	2768.3	1480.5	3.1571	5.7880	2.6311
76.00	291.41	0.001371	0.02495	40.08	1297.6	2765.5	1467.9	3.1742	5.7742	2.6000
78.00	293.21	0.001377	0.02422	41.29	1307.4	2762.8	1455.3	3.1911	5.7606	2.5695
80.00	294.97	0.001384	0.02353	42.51	1317.1	2759.9	1442.8	3.2076	5.7471	2.5395
82.00	296.70	0.001390	0.02286	43.74	1326.6	2757.0	1430.3	3.2239	5.7338	2.5099
84.00	298.39	0.001397	0.02223	44.98	1336.1	2754.0	1417.9	3.2399	5.7207	2.4808
86.00	300.06	0.001404	0.02163	46.24	1345.4	2750.9	1405.5	3.2558	5.7076	2.4518
88.00	301.70	0.001411	0.02105	47.51	1354.6	2747.8	1393.2	3.2713	5.6948	2.4235
90.00	303.31	0.001418	0.02050	48.79	1363.7	2744.6	1380.9	3.2867	5.6820	2.3953
92.00	304.89	0.001425	0.01996	50.09	1372.8	2741.4	1368.6	3.3018	5.6694	2.3676
94.00	306.44	0.001432	0.01945	51.40	1381.7	2738.0	1356.3	3.3168	5.6568	2.3400
96.00	307.97	0.001438	0.01897	52.73	1390.6	2734.7	1344.1	3.3315	5.6444	2.3129
98.00	309.48	0.001445	0.01849	54.07	1399.3	2731.2	1331.9	3.3461	5.6321	2.2860
100.00	310.96	0.001453	0.01804	55.43	1408.0	2727.7	1319.7	3.3605	5.6198	2.2593
102	312.42	0.001460	0.01760	56.80	1416.7	2724.2	1307.5	3.3748	5.6076	2.2328
104	313.86	0.001467	0.01718	58.19	1425.2	2720.6	1295.3	3.3889	5.5955	2.2134
106	315.27	0.001474	0.01678	59.60	1433.7	2716.9	1283.1	3.4029	5.5835	2.1806
108	316.67	0.001481	0.01639	61.03	1442.2	2713.1	1270.9	3.4167	5.5715	2.1548
110	318.05	0.001488	0.01601	62.48	1450.6	2709.3	1258.7	3.4304	5.5595	2.1291
112	319.40	0.001496	0.01564	63.94	1458.9	2705.4	1246.5	3.4440	5.5476	2.1036
114	320.74	0.001503	0,01528	65.43	1467.2	2701.4	1234.3	3.4575	5.5358	2.0783
116	322.06	0.001511	0.01494	66.93	1475.4	2697.4	1222.0	3.4708	5.5239	2.0531
118	323.36	0.001519	0.01461	68.46	1483.6	2693.3	1209.7	3.4840	5.5121	2.0281
120	324.65	0.001526	0.01428	70.01	1491.8	2689.2	1197.4	3.4972	5.5002	2.0030
122	325.91	0.001534	0.01397	71.59	1499.9	2684.9	1185.0	3.5102	5.4484	1.9382
124	327.17	0.001542	0.01366	73.19	1508.0	2680.6	1172.6	3.5232	5.4765	1.9533
126	328.40	0.001550	0.01337	74.81	1516.0	2676.1	1160.1	3.5361	5.4646	1.9285
128	329.62	0.001559	0.01308	76.46	1524.0	2671.6	1147.6	3.5488	5.4527	1.9039
130	330.83	0.001567	0.01280	78.14	1532.0	2667.0	1135.0	3.5616	5.4408	1.8792
132	332.02	0.001576	0.01252	79.85	1540.0	2662.3	1122.3	3.5742	5.4288	1.8546
134	333.19	0.001584	0.01226	81.59	1547.9	2657.4	1109.5	3.5868	5.4168	1.8300
136	334.36	0.001593	0.01200	83.36	1555.8	2652.5	1096.7	3.5993	5.4047	1.8054
138	335.51	0.001602	0.01174	85.16	1563.7	2647.5	1083.8	3.6118	5.3925	1.7807
140	336.64	0.001610	0.01150	86.09	1571.6	2642.4	1070.7	3.6242	5.3803	1.7561
142	337.76	0.001620	0.01125	88.86	1579.5	2637.1	1057.6	3.6366	5.3679	1.7313
144	338.87	0.001630	0.01102	90.77	1587.4	2631.8	1044.4	3.6490	5.3555	1.7065

(*Contd.*)

Table 2 *Contd.* : Saturated Steam (Pressure Table)

1	2	3	4	5	6	7	8	9	10	11
144	338.87	0.001630	0.01102	90.77	1587.4	2631.8	1044.4	3.6490	5.3555	1.7065
146	339.97	0.001638	0.01079	92.71	1595.3	2626.3	1031.0	3.6613	5.3431	1.6818
148	341.06	0.001648	0.01056	94.69	1603.1	2620.7	1017.6	3.6736	5.3305	1.6569
150	342.13	0.001658	0.01034	96.71	1611.0	2615.0	1004.0	3.6859	5.3178	1.6319
152	343.19	0.001668	0.01012	98.77	1618.9	2609.2	990.3	3.6981	5.3051	1.6070
154	344.24	0.001678	0.00991	100.9	1626.8	2603.3	976.5	3.7104	5.2922	1.5818
156	345.28	0.001688	0.00970	103.0	1634.7	2597.3	962.6	3.7226	5.2793	1.5567
158	346.31	0.001699	0.00950	105.2	1642.6	2591.1	948.5	3.7348	5.2663	1.5315
160	347.33	0.001710	0.00931	107.4	1650.5	2584.9	934.3	3.7471	5.2531	1.5060
162	348.34	0.001722	0.00912	109.7	1658.5	2578.5	920.0	3.7594	5.2399	1.4805
164	349.34	0.001733	0.00892	112.0	1666.5	2572.1	905.6	3.7717	5.2267	1.4550
166	350.32	0.001748	0.00874	114.4	1674.5	2565.5	891.0	3.7842	5.2132	1.4290
168	351.30	0.001757	0.00855	116.9	1683.0	2558.6	875.6	3.7974	5.1994	1.4020
170	352.26	0.001770	0.00837	119.5	1691.7	2551.6	859.9	3.8107	5.1855	1.3748
172	353.22	0.001782	0.00819	122.1	1700.4	2544.4	844.1	3.8240	5.1713	1.3473
174	354.17	0.001796	0.00801	124.8	1709.0	2537.1	828.1	3.8372	5.1570	1.3198
176	355.11	0.001810	0.00784	127.6	1717.6	2529.5	811.9	3.8504	5.1425	1.2921
178	356.04	0.001825	0.00767	130.4	1726.2	2521.8	795.6	3.8635	5.1278	1.2643
180	356.96	0.001840	0.00749	133.4	1734.8	2513.9	779.1	3.8765	5.1128	1.2363
182	357.87	0.001856	0.00733	136.4	1743.4	2505.8	762.3	3.8896	5.0975	1.2079
184	358.77	0.001872	0.00716	139.6	1752.1	2497.4	745.3	3.9028	5.0820	1.1792
186	359.67	0.001890	0.00700	142.8	1760.9	2488.8	727.9	3.9160	5.0661	1.1501
188	360.55	0.001907	0.00684	146.2	1769.7	2479.8	710.1	3.9294	5.0498	1.1204
190	361.43	0.001926	0.00668	149.8	1778.7	2470.6	692.0	3.9429	5.0332	1.0903
192	362.30	0.001945	0.006517	153.4	1787.8	2461.1	673.3	3.9566	5.0160	1.0594
194	363.16	0.001966	0.006358	157.3	1797.0	2451.1	654.1	3.9706	4.9983	1.0277
196	364.02	0.001988	0.006198	161.3	1806.5	2440.7	634.2	3.9849	4.9801	0.9952
198	364.86	0.002012	0.006038	165.6	1816.3	2429.8	613.5	3.9996	4.9610	0.9624
200	365.70	0.002037	0.005877	170.2	1826.5	2418.4	591.9	4.0149	4.9412	0.9263
202	366.53	0.002064	0.005714	175.0	1837.0	2406.2	569.2	4.0308	4.9204	0.8896
206	368.17	0.002125	0.005379	185.9	1859.9	2379.3	519.5	4.0651	4.8750	0.8099
210	369.78	0.002201	0.005023	199.1	1886.3	2347.6	461.3	4.1048	4.8223	0.7175
214	317.37	0.002306	0.004624	216.3	1919.0	2307.4	388.4	4.1543	4.7569	0.6026
218	379.92	0.002483	0.004115	243.0	1967.2	2248.0	280.8	4.2276	4.6622	0.4346
221.20	374.15	0.00317	0.00317	315.5	2107.4	2107.4	0.0	4.4429	4.4429	0.0

Table 3: Enthalpy of Superheated Steam

Pressure $p \times 10^{-5}$ N/m^2	Saturated Temp. °C	Saturated Enthalpy kJ/kg	Enthalpy of Superheated System at t°C										
			T = 100	150	200	250	300	350	400	450	500	550	600
1	2	3	4	5	6	7	8	9	10	11	12	13	14
0.01	69.83	2514.4	2688.6	2783.7	2808.1	2977.7	3076.8	3177.5	3279.7	3383.6	3489.2	3596.5	3705.6
0.05	32.90	2561.6	2688.1	2783.4	2879.9	2977.6	3076.7	3177.4	3279.7	3383.6	3489.2	3596.5	3705.6
0.10	45.83	2584.8	2687.5	2783.1	2879.6	2977.4	3076.6	3177.3	3279.6	3383.5	3489.1	3596.5	3705.6
0.20	60.09	2609.9	2686.3	2782.3	2879.2	2977.4	3076.4	3177.1	3279.4	3383.4	3489.0	3596.4	3705.4
0.30	69.12	2625.4	2685.1	2781.6	2878 7	2976.8	3076.1	3176.9	3279.3	3383.3	3488.9	3596.3	3705.4
0.40	75.89	2636.9	2683.8	2780.9	2878.2	2976.5	3075.9	3176.8	3279.1	3383.1	3488.8	3596.2	3705.1
0.50	81.35	2646.0	2682.6	2780.1	2877.7	2976.1	3075.7	3176.6	3279.0	3383.0	3488.7	3596.1	3705.2
0.60	85.95	2653.6	2681.3	2779.4	2877.3	2975.8	3075.4	3176.4	3278.8	3382.9	3488.6	3596.0	3705.2
0.70	89.96	2660.1	2680.0	2778.6	2876.8	2975.5	3075.2	3176.2	3278.7	3382.7	3488.5	3596.9	3705.0
0.80	93.51	2665.8	2678.8	2778.8	2876.3	2975.2	3075.0	3176.0	3278.5	3382 6	3488.4	3596.9	3705.0
0.90	96.71	2670.9	2677.5	2771.1	2875.8	2974.8	3074.7	3175.8	3278.4	3382.5	3488.3	3595.7	3704.9
1.0	99.63	2675.4	2676.2	2776.3	2875.4	2974.5	3074.5	3175.6	3278.2	3382.4	3488.1	3595.6	3704.8
1.2	104.81	2683.4	–	2774.8	2874.4	2973.9	3074.0	3175.3	3277.9	3382.1	3487.9	3595.4	3704.6
1.4	109.32	2690.3	–	2773.2	2873.4	2973.2	3073.5	3174.9	3277.6	3381.8	3487.7	3595.2	3704.5
1.6	113.32	2696.2	–	2771.7	2872.5	2972.5	3073.0	3174.5	3277.3	3381.6	3487.5	3595.0	3704.5
1.8	116.93	2701.5	–	2770.1	2871.5	2971.9	3072.6	3174.1	3777.0	3381.3	3487.3	3594.9	3704.1
2.0	120.23	2706.3	–	2768.5	2870.5	2971.2	3072.1	3173.8	3276.7	3381.1	3487.0	3594.7	3704.0
2.5	127.43	2716.4	–	2764.5	2868.0	2969.6	3070.9	3172.8	3275.9	3380.4	3486.5	3594.2	3703.6
3.0	133.54	2724.7	–	2760.4	2865.5	2967.9	3069.7	3171.9	3275.2	3379.8	3486.0	3593.7	3703.2
3.5	138.87	2731.6	–	2756.2	2862.9	2966.2	3068.5	3170.9	3274.4	3379.1	3485.4	3593.2	3702.7
4.0	143.62	2737.6	–	2752.0	2860.4	2964.5	3067.2	3170.0	3273.6	3378.5	3484.9	3592.8	3702.3
4.5	147.92	2742.9	–	2747.6	2857.8	2962.7	3065.9	3169.0	3272.8	3377.8	3484.3	3592.3	3701.9
5.0	151.84	2747.5	–	–	2855.1	2961.1	3064.8	3168.1	3272.1	3377.2	3483.8	3591.8	3701.5
5.5	155.47	2751.7	–	–	2852.5	2959.3	3063.5	3167.2	3271.3	3376.6	3483.2	3591.4	3700.1
6.0	158.84	2755.5	–	–	2849.7	2957.6	3062.3	3166.2	3270.6	3376.0	3482.7	3590.9	3700.7
6.5	161.99	2758.9	–	–	2847.0	2955.8	3061.0	3165.3	3269.8	3375.3	3482.1	3590.4	3700.3
7.0	164.96	2762.0	–	–	2844.2	2954.0	3059.8	3164.3	3269.0	3374.7	3481.6	3589.9	3699.9
7.5	167.76	2764.8	–	–	2841.4	2952.2	3058.5	3163.4	3268.3	3374.0	3481.0	3589.5	3699.5
8.0	170.41	2767.5	–	–	2838.6	2950.4	3057.3	3162.4	3267.5	3373.4	3480.5	3589.0	3699.1
8 5	172.94	2769.9	–	–	2835.7	2948.6	3056.0	3161.4	3266.7	3372.7	3479.9	3538.5	3698.6
9.0	175.36	2772.1	–	–	2832.7	2946.8	3054.7	3160.5	3266.0	3372.1	3479.4	3588.1	3698.2
9.5	177.67	2774.2	–	–	2829.8	2944.9	3053.4	3159.5	3265.2	3371.5	3478.8	3587.6	3697.8
10.0	179.88	2776.2	–	–	2826.8	2943.0	3052.1	3158.5	3264.4	3370.8	3478.3	3587.1	3697.4

(*Contd.*)

Table 3 *Contd.*: Enthalpy of superheated steam

1	2	3	4	5	6	7	8	9	10	11	12	13	14
11.0	184.07	2779.7	–	–	2820.7	2939.3	3049.6	3156.6	3262.9	3369.5	3477.2	3586.2	3696.6
12.0	187.96	2782.7	–	–	2814.4	2935.4	3046.9	3154.6	3261.3	3368.2	3476.1	3585.2	3695.8
13.0	191.61	2785.4	–	–	2808.0	2931.5	3044.3	3152.7	3259.7	3366.9	3475.0	3584.3	3695.5
14.0	195.04	2787.8	–	–	2801.4	2927.6	3041.6	3150.7	3258.2	3365.6	3473.9	3383.3	3694.0
16.0	201.37	2791.7	–	–	–	2919.4	3036.2	3146.7	3255.0	3363.0	3471.7	3581.4	3692.1
18.0	207.11	2794.8	–	–	–	2911.0	3030.7	3142.7	3251.4	3360.4	3469.5	3579.5	3690.6
20.0	212.37	2797.2	–	–	–	2902.4	3025.0	3138.6	3248.7	3357.8	3467.3	3577.6	3689.2
22.0	217.24	2799.1	–	–	–	2893.4	3019.3	3134.5	3245.5	3355.2	3465.1	3575.7	3687.6
24.0	221.78	2800.4	–	–	–	2884.2	3013.4	3130.4	3242.3	3452.6	3562.9	3573.8	3685.9
26.0	226.04	2801.4	–	–	–	2874.7	3007.4	3126.1	3239.0	3349.0	3460.6	3571.9	3684.3
28.0	230.05	2802.0	–	–	–	2864.9	3001.3	3121.9	3235.8	3347.3	3458.4	3570.0	3682.5
30.0	233.84	2802.3	–	–	–	2854.8	2995.1	3117.5	3232.5	3344.6	3456.2	3568.1	3681.0
35.0	242.54	2802.0	–	–	–	2828.1	2979.0	3106.5	3224.2	3338.0	3450.6	3563.4	3676.9
40.0	250.33	2800.3	–	–	–	–	2962.0	3095.1	3215.7	3331.2	3445.0	3558.6	3672.8
45.0	257.41	2797.7	–	–	–	–	2944.2	3083.3	3324.4	3324.4	3439.3	3553.8	3668.6
50.0	263.91	2794.2	–	–	–	–	2925.5	3071.2	3198.3	3317.5	3433.7	3549.0	3664.5
55.0	269.93	2789.9	–	–	–	–	2905.7	3058.7	3189.3	3310.5	3430.3	3544.1	3660.3
60.0	275.55	2785.0	–	–	–	–	2885.0	3045.8	3180.1	3303.5	3422.2	3539.3	3656.2
65.0	280.82	2779.5	–	–	–	–	2862.9	3032.4	3170.8	3296.3	3416.4	3534.4	3652.0
70.0	285.79	2773.5	–	–	–	–	2839.4	3018.7	3161.2	3289.1	3410.6	3529.6	3647.9
75.0	290.50	2766.9	–	–	–	–	2814.0	3004.5	3151.6	3281.8	3404.7	3524.7	3643.7
80.0	294.97	2759.9	–	–	–	–	2786.8	2989.9	3141.6	3274.3	3398.8	3519.7	3639.5
85.0	299.23	2752.5	–	–	–	–	2758.2	2974.7	3131.4	3266.8	3392.8	3514.8	3635.3
90.0	303 31	2744.6	–	–	–	–	–	2959.0	8121.2	3259.2	3386.8	3509.8	3631.1
95.0	307.21	2736.4	–	–	–	–	–	2942.7	3110.7	3251.8	3380.7	3504.9	3626.9
100.0	310.96	2727.7	–	–	–	–	–	2925.8	3099.9	3243.6	3374.6	3499.8	3622.7
110.0	318.05	2709.3	–	–	–	–	–	2889.6	3077.8	3227.7	3362.2	3489.7	3614.2
			300	350	400	450	500	550	600	650	700	750	800
120.0	324.65	2689.2	–	2849.7	3054.8	3211.4	3349.6	3479.6	3605.2	3730.2	3854.3	3978.3	4102.7
130.0	330.83	2667.0	–	2805.0	3030.7	3194.6	3336.8	3469.3	3597.1	3722.9	3618.0	3972.9	4098.0
140.0	336.64	2642.4	–	2754.2	3005.6	3177.4	3323.8	3458.8	3588.5	3715.6	3841.7	3967.5	4093.3
150.0	342.13	2615.0	–	2694.8	2979.1	3159.7	3310.6	3448.3	3579.8	3708.3	3835.4	3962.1	4088.6
160.0	347.33	2584.9	–	2620.8	2951.3	3141.6	3297.1	3437.7	3571.0	3700.0	3829.1	3956.7	4084.0
170.6	352.26	2551.6	–	–	2921.7	3123.1	3283.5	3427.0	3557.8	3689.8	3819.7	3998.5	4077.0
180.0	356.96	2513.9	–	–	2890.3	3104.0	3269.6	3416.1	3553.4	3686.1	3816.5	3945.8	4074.6
190.0	361.43	2470.6	–	–	2856.7	3084.4	3255.4	3405.2	3544.5	3678.6	3810.2	3940.4	4070.0

(*Contd.*)

Table 3 *Contd.*: Enthalpy of superheated steam

1	2	3	4	5	6	7	8	9	10	11	12	13	14
200.0	365.70	2418.4	–	–	2820.5	3064.3	3241.1	3394.1	3535.5	3671.1	3803.8	3935.0	4065.3
210.0	369.78	2347.6	–	–	2781.3	3043.6	3226.5	3382.9	3526.5	3663.6	3797.5	3925.5	4060.6
220.0	373.69	2195.6	–	–	2738.8	3022.3	3211.7	3371.6	3517.6	3656.1	3791.1	3924.1	4055.9
221.20	374.15	2107.4	–	–	2732.5	3012.3	3202.7	3363.7	3510.4	3650.2	3786.1	3920.1	4052.9
250.0	–	–	–	–	2582.0	2954.3	3165.9	3337.0	3489.9	3633.4	3771.9	3907.7	4041.9
300.0	–	–	–	–	2161.8	2825.6	3085.0	3277.4	3443.0	3595.0	3739.7	3880.3	4018.5
350.0	–	–	–	–	–	2676.4	2998.3	3215.4	3395.1	3556.1	3707.3	3852.9	3995.1
400.0	–	–	–	–	–	2515.6	2906.8	3151.6	3346.4	3517.0	3674.8	3825.5	3971.7
450.0	–	–	–	–	–	2384.2	2813.5	3086.5	3297.4	3477.8	3642.4	3798.1	3948.4
500.0	–	–	–	–	–	2293.2	2723.0	3021.1	3248.3	3438.9	3610.2	3770.9	3925.3
550.0	–	–	–	–	–	2231.4	2639.8	2956.5	3199.3	3400.2	3578.2	3744.1	3904.1
600.0	–	–	–	–	–	2187.1	2570.6	2896.2	3151.6	3362.4	3547.0	3717.4	3879.6
700.0	–	–	–	–	–	–	2476.1	2790.7	3060.4	3289.0	3480.3	3665.8	3833.3
800.0	–	–	–	–	–	–	2397.4	2708.0	2980.3	3220.3	3428.7	3616.7	3798.4
900.0	–	–	–	–	–	–	2349.9	2643.0	2913.5	3158.2	3374.6	3570.1	3752.3
1000.0	–	–	–	–	–	–	2316.1	2593.8	2857.5	3105.3	3324.4	3526.0	3714.8

Table 4: Entropy of Superheated Steam

Pressure $p \times 10^{-5}$ N/m^2	Saturated Temp. °C	Saturated Entropy kJ/kg	Enthalpy of Superheated System at t°C										
			T = 100	150	200	250	300	350	400	450	500	550	600
1	2	3	4	5	6	7	8	9	10	11	12	13	14
0.01	6.98	8.9767	9.5136	9.7527	9.9679	10.1641	10.3450	10.5133	10.6711	10.8200	10.9612	11.0957	11.2243
0.05	32.90	8.3960	8.7098	9.0094	9.2248	9.4211	9.6021	9.7704	9.9283	10.0772	10.2184	10.3529	10.4815
0.10	45.83	8.1511	8.4486	8.6888	8.9045	9.1011	9.2820	9.4504	9.6083	9.7572	9.8984	10.0329	10.1616
0.20	60.09	7.9094	8.1261	8.3676	8.5839	8.7806	8.9618	9.1303	9.2882	9.4372	9.5784	9.7130	9.8416
0.30	69.12	7.7695	7.9363	8.1791	8.3960	8.5930	8.7744	8.9430	9.1010	9.2499	9.3912	9.5257	9.6544
0.40	75.89	7.6709	7.8009	8.0450	8.2625	8.4598	8.6413	8.8100	8.9680	9.1170	9.2583	9.3929	9.5216
0.50	81.35	7.1947	7.6953	7.9406	8.1587	8.3564	8.5380	8.7068	8.8649	9.0139	9.1552	9.2898	9.4185
0.60	85.95	7.5327	7.6085	7.8551	8.0738	8.2718	8.4536	8.6224	8.7806	8.9296	9.0710	9.2056	9.3343
0.70	89.96	7.4804	7.5346	7.7825	8.0019	8.2002	8.3822	8.5511	8.7093	8.8594	8.9997	9.1344	9.2631
0.80	93.51	7.4352	7.4703	7.7195	7.9395	8.1381	8.3202	8.4892	8.6475	8.7966	8.9380	9.0727	9.2041
0.9	96.71	7.3954	7.4132	7.6638	7.8843	8.0832	8.2656	8.4347	8.5929	8.7421	8.8835	9.0182	9.1469
1.0	99.63	7.3598	7.3618	7.6137	7.8349	8.0342	8.2166	8.3858	8.5442	8.6934	8.8348	8.9695	9.1982
1.2	104.81	7.2984	–	7.5267	7.7492	7.9491	8.1319	8.3012	8.4597	8.6090	8.7505	8.8852	9.0141
1.4	109.32	7.2465	–	7.4527	7.6765	7.8770	8.0601	8.2297	8.3882	8.5376	8.6791	8.8139	8.9429
1.6	113.32	7.2017	–	7.3882	7.6133	7.8144	7.9979	8.1676	8.3263	8.4757	8.6173	8.7521	8.8809
1.8	116.93	7.1622	–	7.3309	7.5574	7.7591	7.9429	8.1128	8.2716	8.4211	8.5627	8.6975	8.8264
2.0	120.23	7.1268	–	7.2794	7.5072	7.7096	7.8937	8.0638	8.2226	8.3722	8.5139	8.6487	8.7776
2.5	127.43	7.0520	–	7.1689	7.4001	7.6042	7.7891	7.9597	8.1188	8.2686	8.4104	8.5453	8.6743
3.0	133.54	6.9909	–	7.0771	7.3119	7.5176	7.7034	7.8744	8.0388	8.1838	8.3257	8.4608	8.5898
3.5	138.87	6.9392	–	6.9993	7.2364	7.4442	7.6309	7.8024	7.9622	8.1122	8.2539	8.3890	8.5182
4.0	143.62	6.8943	–	6.9285	7.1708	7.3800	7.5675	7.7395	7.8994	8.0497	8.1919	8.3271	8.4563
4.5	147.92	6.8547	–	6.8661	7.1121	7.3232	7.5116	7.6841	7.8445	7.9945	8.1369	8.2722	8.4012
5.0	151.84	6.8192	–	–	7.5092	7.2721	7.4614	7.6343	7.7948	8.9454	8.0879	8.2233	8.3526
5.5	155.47	6.7870	–	–	7.0108	7.2256	7.4158	7.5892	7.7500	7.9008	8.0433	8.1789	8.3083
6.0	158.84	6.7575	–	–	6.9662	7.1829	7.3740	7.5479	7.7090	7.8600	8.0027	8.1383	8.2678
6.5	161.99	6.7304	–	–	6.9247	7.1434	7.3355	7.5099	7.6712	7.8224	7.9652	8.1009	8.2305
7.0	164.96	6.7052	–	–	6.8849	7.1066	7.2997	7.4745	7.6362	7.7875	7.9305	8.0663	8.1959
7.5	167.76	6.6817	–	–	6.8491	7.0721	7.2662	7.4416	7.6035	7.7550	7.8981	8.0340	8.1637
8.0	170.41	6.6596	–	–	6.8148	7.0391	7.2348	7.4107	7.5729	7.7246	7.8678	8.0038	8.1333
8.5	172.94	6.6388	–	–	6.7820	7.0091	7.2051	7.3851	7.5441	7.6960	7.8395	7.9754	8.1056
9.0	175.36	6.6192	–	–	6.7508	6.9800	7.1771	7.3540	7.5109	7.6689	7.8124	7.9486	8.0785
9.5	177.67	6.6005	–	–	6.7209	6.9523	7.1505	7.3279	7.4911	7.6433	7.7869	7.9232	8.0532

(*Contd.*)

Table 4 *Contd.*: Entropy of superheated Steam

1	2	3	4	5	6	7	8	9	10	11	12	13	14
10.0	179.88	6.5828	–	–	6.6922	6.9259	7.1251	7.3031	7.4665	7.6190	7.7627	7.8991	8.0292
11.0	184.07	6.5497	–	–	6.6397	6.8764	7.0778	7.2569	7.4209	7.5737	7.7177	7.8543	7.9845
12.0	187.96	6.5193	–	–	6.5872	6.8305	7.0342	7.2144	7.3790	7.5323	7 6765	7.8132	7.9436
13.0	191.61	6.4913	–	–	6.5394	6.7878	6.9938	7.1751	7.3404	7.4940	7.6385	7.7754	7.9060
14.0	195.04	6.4651	–	–	6.4941	6.7477	6.9561	7.1386	7.3045	7.4585	7.6032	7.7404	7.7710
16.0	201.37	6.4175	–	–	–	6.6740	6.8873	7.0723	7.2394	7.3942	7.5395	7.6770	7.8080
18.0	207.11	6.3751		–	–	6.6071	6.8257	7.0131	7.1816	7.3372	7.4830	7.6290	7.7522
20.0	212.37	6.3306	–	–	–	6.5454	6.7696	6.9596	7.1296	7.2859	7.4323	7.5706	7.7022
22.0	217:24	6.3015	–	–	–	6.4879	6.7179	6.9107	7.0821	7.2394	7.3862	7.5249	7.6568
24.0	221.78	6.2690	–	–	–	6.4338	6.6699	6.8656	7.0384	7.1964	7.3439	7.4830	7.6152
26.0	226.04	6.2387	–	–	–	6.3823	6.6249	6.8236	6.9979	7.1568	7.3048	7.4443	7.5768
28.0	230.05	6.2104	–	–	–	6.3331	6.5824	6.7824	6.9601	7.1199	7.2685	7.4083	7.5412
30.0	233.84	6.1837	–	–	–	6.2857	6.5422	6.7471	6.9246	7.0854	7.2345	7.3748	7.5079
35.0	242.54	6.1228	–	–	–	6.1732	6.4491	6.6626	6.8443	7.0074	7.1580	7.2993	7.4332
40.0	250.33	6.0685	–	–	–	–	6.3642	6.5870	6.7733	6.9388	7.0909	7.2333	7.3680
45.0	257.41	6.0191	–	–	–	–	6.2852	6.5182	6.7093	6.8774	7.0311	7.1746	7.3100
50.0	263.91	5.9735	–	–	–	–	6.2105	6.4545	6.6508	6.8217	6.9770	7.1215	7.2578
55.0	269.93	5.9309	–	–	–	–	6.1388	6.3949	6.5968	6.7070	6.9276	7.0732	7.2101
60.0	275.55	5.8908	–	–	–	–	6.0692	6.3386	6.5492	6.7230	6.8818	7.0285	7.1664
65.0	280.82	5.8527	–	–	–	–	6.0008	6.2849	6.4987	6.6786	6.8390	6.9872	7.1256
70.0	285.79	5.8162	–	–	–	–	5.9327	6.2333	6.4536	6.6368	6.7993	6.9485	7.0880
75.0	290,50	5.7811	–	–	–	–	5.8641	6.1836	6.4107	6.5960	6 7615	6.9121	7.0527
80.0	294.97	5.7471	–	–	–	–	5.7942	6.1349	6.3694	6.5597	6.7262	6.8778	7.0191
85.0	299.23	5.7141	–	–	–	–	5.7211	6.0875	6.3306	6.5238	6.6924	6.8455	6.9873
90.0	303.31	5.6820	–	–	–	–	–	6.0408	6.2915	6.4894	6.6600	6.8143	6.9573
95.0	307.21	5.6506	–	–	–	–	–	5.9947	6.2537	6.4563	6.6292	6.7848	6.9287
100.0.	310.96	5.6198	–	–	–	–	–	5.9489	6.2182	6 4243	6.5994	6.7564	6.9013
110.0	318.05	5.5595						5.8571	6.1483	6.3633	6.5432	6.7031	6.8499
		$t =$	300	350	400	450	500	550	600	650	700	750	800
120.0	324.65	5.5002	–	5.7636	6.0810	6.3056	6.4906	6.6535	6.8022	6.9409	7.0718	7.1961	7.3147
130.0	330.83	5.4408	–	5.6661	6.0155	6.2506	6.4409	6.6069	6.7577	6 8979	7.0298	7.1550	7.2743
140.0	336.64	5.3803	–	5.5618	5.9513	6.1978	6.3937	6.5630	6.7159	6.8575	6.9906	7.1166	7.2367
150.0	342.13	5.2531	–	5.4467	5.8876	6.1468	6.3487	6.5213	6.6764	6.8195	6.9536	7.0806	7.2013
160.0	347.33	5.3178	–	5.3109	5.8240	6.0972	6.3054	6.4816	6.6389	6.7835	6.9188	7.0466	7.1681
170.0	352.26	5.1855	–	–	5.7599	6.0489	6.2636	6.4435	6.5858	6.7328	6.8698	6.9989	7.1215
180.0	356.96	5.1128	—	–	5.6947	6.0015	6.2232	6.4069	6.5688	6.7166	6.8542	6.9838	7.1067

(*Contd.*)

Table 4 *Contd.*: Entropy of superheated Steam

1	2	3	4	5	6	7	8	9	10	11	12	13	14
190.0	361.43	5.0332	–	–	5.6278	5.9549	6.1839	6.3716	6.5360	6.6854	6.8241	6.9546	7.0783
200.0	365.70	4.9412			5.5585	5.9089	6.1456	6.3374	6.5043	6.6554	6.7953	6.9267	7.0511
210.0	369.78	4.8223	–		5.4863	5.8633	6.1082	6.3043	6.4737	6.6265	6.7677	6.9000	7.0251
220.0	373.69	4.5799	–	–	5.4102	5.8179	6.0716	6.2721	6.4441	6.5986	6.7410	6.8743	7.0001
221.20	374.15	4.4429	–	–	5.2955	5.7932	5.9839	6.2643	6.4042	6.5286	6.6910	6.7859	6.9823
250.0	–	–	–	–	5.1455	5.6821	5.9655	6.1801	6.3604	6.5203	6.6664	6.8025	6.9306
300.0	–	–	–	–	4.4896	5.4495	5.7972	6.0386	6.2340	6.4033	6.5560	6.6970	6.8288
350.0	–	–	–	–	4.2214	5.2031	5.6349	5.9074	6.1194	6.2988	6.4584	6.6043	6.7400
400.0	–	–	–	–	4.1190	4.9511	5.4762	5.7835	6.0135	6.2035	6.3701	6.5210	6.6606
459.0					4.0554	4.7469	5.3226	5.6654	5.9143	6.1154	6.2890	6.4451	6.5885
500.0	–	–	–	–	4.0083	4.6026	5.1782	5.5525	5.8207	6.0331	6.2138	6.3749	6.5222
550.0	–	–	–	–	3.9705	4.5000	5.0470	5.4421	5.7321	5.9558	6.1436	6.3095	6.4600
600.0	–	–	–	–	–	4.4246	4.9374	5.3463	5.6477	5.8827	6.0775	6.2483	6.4031
700.0	–	–	–	–	–	4.3182	4.7688	5.1748	5.4931	5.7480	5.9562	6.1361	6.2979
800.0	–	–	–	–	–	4.2434	4.6488	5.0382	5.3595	5.6270	5.8470	6.0354	6.2034
900.0	–	–	–	–	–	4.1852	4.5602	4.9276	5.2468	5.5195	5.7479	5.9439	6.1179
1000.0						4.1373	4.4913	4.8393	5.1505	5.4267	5.6579	5.8600	6.0397

Table 5: Volume of Superheated Steam

Pressure $p \times 10^{-5}$ N/m^2	Saturated Temp. °C	Saturated Volume m^3/kg	Volume of Superheated Steam at t°C										
			t = 100	150	200	250	300	350	400	450	500	550	600
1	2	3	4	5	6	7	8	9	10	11	12	13	14
0.01	6 983	129.20	172.20	195.30	218.40	241.40	264.50	287.60	310.70	333.70	356.80	379.90	403.00
0.05	32.90	28.19	34.420	39.04	43.66	48.28	52.90	57.51	62.13	66.74	71.36	75.98	80.59
0.10	45.83	14.67	17.200	19.51	21.83	24.14	26.45	28.75	31.06	33.37	35.68	37.99	40.29
0.20	60.09	7.650	8.585	9.748	10.907	12.064	13.219	14.374	15.529	16 684	17.838	18.992	20.146
0.30	69.12	5.229	5.714	6 493	7.268	8.040	8.811	9.581	10.351	11.121	11.891	12.661	13.430
0.40	75.89	3.993	4.279	4.865	5.447	6.027	6.606	7.184	7.7625	8.340	8.917	9.494	10.072
0.50	81.35	3.240	3.418	3.889	4.356	4.821	5.284	5.747	6.209	6.671	7.133	7.595	8.057
0.60	85.95	2.732	2.844	3.238	3.628	4.016	4.402	4.788	5.174	5.559	5.944	6.329	6.714
0.70	89.96	2.365	2.434	2.773	3.108	3.441	3.772	4.103	4.434	4.764	5.095	5 425	5.425
0.80	93.51	2.087	2.126	2.425	2.718	3.010	3.300	3.590	3.879	4.168	4.457	4.746	5.035
0.90	96.71	1.869	1.887	2.153	2.415	2.674	2.933	3.190	3.448	3.705	3.962	4.219	4.475
1.0	99.63	1.694	1.696	1.936	2.172	2.406	2.639	2.871	3.102	3.334	3.565	3.797	4.028
1.2	104.81	1.428	—	1.611	1.808	2.004	2.198	2.391	2.585	2.778	2.971	3.163	3.356
1.4	109.32	1.236	—	1.378	1.548	1.716	1.883	2.049	2.215	2.380	2.546	2.711	2.876
1.6	113.32	1.091	—	1.204	1.353	1.501	1.647	1.792	1.937	2.082	2.227	2.372	2.517
1.8	116.93	0.972	—	1.068	1.202	1.333	1.463	1.593	1.722	1.851	1.979	2.108	2.237
2.0	120.23	0 8854	—	0.9595	1.0804	1.1989	1.3162	1.4328	1.5492	1.6653	1.7812	1.8971	2.0129
2.5	127.43	0.7185	—	0.7641	0.8620	0.9574	1.0516	1.1452	1.2385	1.3315	1.4244	1.5172	1.6099
3.0	133.54	0.6056	—	0.6337	0.7164	0.7964	0.8753	0.9535	1.0314	1.1090	1.1865	1.2639	1.3412
3.5	138.87	0.5240	—	0.5410	0.6120	0.6810	0.7495	0.8172	0.8848	0.9506	1.0181	1.0830	1.1504
4.0	143.62	0.4622	—	0.4707	0.5343	0.5952	0.6549	0.7139	0.7725	0:8309	0.8892	0.9474	1.0054
4.5	147.92	0.4138	—	0.4167	0.4738	0.5284	0.5816	0.6341	0.6862	0.7386	0.7908	0.8425	0 8939
5.0	151.84	0.3747	—	—	0.4250	0.4744	0.5226	0.5701	0.6172	0.6640	0.7108	0.7574	0.8039
5.5	155.47	0.3425	—	—	0.3852	0.4305	0.4745	0.5178	0.5607	0.6034	0.6459	0.6883	0.7307
6.0	158.84	0.3155	—	—	0.3520	0.3939	0.4344	0.4742	0.5136	0.5528	0.5918	0.6308	0.6696
6.5	161.99	0.2925	—	—	0.3240	0.3629	0.4005	0 4373	0.4738	0.5100	0.5461	0 5821	0.6180
7.0	164:96.	0.2727	—	—	0.2999	0.3364	0.3744	0.4057	0.4396	0.4733	0.5069	0.5403	0.5737
7:5	167.76	0.2554	—	—	0.2791	0.3134	0.3462	0.3783	0 4100	0.4416	0.4729	0.5041	0.5353
8.0	170.41	0.2403	—	—	0.2608	0.2932	0.3241	0.3543	0.3842	0 4137	0.4432	0 4725	0.5017
8 5	172.94	0.2268	—	—	0.2447	0.2754	0.3047	0.3332	0.3613	0.3892	0.4169	0.4446	0.4721
9.0	175.36	0.2148	—	—	0.2303	0.2596	0.2874	0.3144	0.3410	0.3674	0.3936	0.4197	0.4458
9.5	177.67	0.2040	—	—	0.2175	0.2455	0.2719	0.2976	0 3228	0.3479	0.3727	0.3975	0.4222
10.0	179.94	0.1943	—	—	0.2059	0.2327	0.2580	0.2824	0.3065	0.3303	0.3540	0.3775	0.4010
11.0	184.07	0.1774	—	—	0.1859	0.2107	0.2339	0.2563	0.2782	0.3000	0.3215	0.3430	0.3643
12.0	—	0.1632	—	—	0.1692	0.1924	0.2139	0.2345	0.2547	0.2747	0.?945	0.3142	0.3338
13.0	191.61	0.1511	—	—	0.1551	0.1769	0.1969	0.2160	0.2348	0.2533	0.2716	0.2898	0.3080
14 0	195.06	0.1407	—	—	0.1429	0.1635	0.1823	0.2002	0.2177	0.2349	0.2520	0.2690	0.2859
16.0	201.37	0.1237	—	—	—	0.1419	0.1587	0.1745	0.1900	0.2051	0.2202	0.2351	0.2499
18.0	207.11	0.1103	—	—	—	0.1250	0.1402	0.1546	0.1684	0.1820	0.1954	0.2087	0.2219
20.0	212.37	0.0995	—	—	—	0.1114	0.1255	0.1386	0.1511	0.1634	0.1756	0.1876	0.1995
22 0	217.24	0.0906	—	—	—	0.10035	0.11343	0.12547	0.13700	0.14825	0.15934	0.17030	0.18119
24.0	221.78	0.08320	—	—	—	0.09107	0.10336	0.11455	0.12522	013561	0.14582	0.15591	0.18991
26.0	226.04	0.07680	—	—	—	0.08320	0.09483	0.10532	0.11526	0.12491	0.13438	0.14374	0.15301

(*Contd.*)

Table 5 *Contd.*: Volume of superheated Steam

1	2	3	4	5	6	7	8	9	10	11	12	13	14
28.0	230.05	0.07139	—	—	—	0.07644	0.08751	0.09740	0.10671	0.11574	0.12458	0.13330	0.14194
30.0	233.84	0.06663	—	—	—	0.07055	0.08116	0.09053	0.09931	0.10779	0.11608	0.12426	0.13234
35.0	242.54	0.05703	—	—	—	0.05869	0.06842	0.07678	0.08449	0.09189	0.09909	0.10617	0.11315
40.0	250.33	0.04975	—	—	—	—	0.05883	0.06645	0.07338	0.07996	0.08634	0.09260	0.09876
45.0	257.41	0.04404	—	—	—	—	0.05134	0.05840	0.06472	0.07068	0.07643	0.08204	0.08757
50.0	263.91	0.03943	—	—	—	—	0.04530	0.05194	0.05779	0.06325	0.06342	0.07360	0.07862
55.0	269.93	0.03563	—	—	—	—	0.04033	0.04665	0.05212	0.05719	0.06202	0:06666	0.07131
60.0	275.55	0.03244	—	—	—	—	0.03614	0.04222	0.04738	0.05210	0.05659	0.06094	0.06518
65.0	280.82	0.02972	—	—	—	—	0.03225	0.03850	0.04340	0.04783	0.05203	0.05598	0.06012
70.0	285.79	0.02737	—	—	—	—	0.02946	0.03523	0.03992	0.04413	0.04809	0.05189	0.05559
75.0	290.50	0.02533	—	—	—	—	0.02672	0.03242	0.03694	0.04095	0.04469	0.04825	0.05323
80.0	294.97	0.02353	—	—	—	—	0.02426	0.02995	0.03431	0.03431	0.03814	0.04170	0.04510
85.0	299.23	0.02193	—	—	—	—	0.02200	0:02774	0.03199	0.03567	0.03909	0.04232	0.04544
90.0	303.31	0.02050	—	—	—	—	—	0.02579	0.02993	0.03348	0.03674	0.03982	0.04280
95.0	307.21	0.01921	—	—	—	—	—	0.02402	0.02805	0.03151	0.03465	0.03759	0.04045
100.0	310.96	0.01804	—	—	—	—	—	0.02244	0.02641	0.02974	0.03276	0.03560	0.03832
110.0	318.05	0.01601	—	—	—	—	—	0.01961	0.02351	0.02668	0.02950	0.03214	0.03466
				$t = 350$	**400**	**450**	**500**	**550**	**600**	**650**	**700**	**750**	**800**
120.0	324.65	0.01428	—	0.01721	0.02108	0.02412	0.02679	0.02926	0.03160	0.03387	0.03607	0:03822	0.04033
130.0	330.83	0.01280	—	0.01510	0.01902	0.02194	0.02440	0.02682	0.02902	0.03114	0 03319	0.03519	0.03716
140.0	336.64	0.01150	—	0.01321	0.01723	0.02008	0.02251	0.02472	0.02680	0.02880	0.03072	0.03260	0.03444
150.0	342.13	0.01034	—	0.01146	0.01566	0.01845	0.02080	0.02291	0.02488	0.02677	0.02859	0.03036	0.03209
160.0	347.33	0.009308	—	0.009764	0.01427	0.01703	0.01929	0.02132	0.02320	0.02499	0.02672	0.02839	0.03002
170.0	352.26	0.008371	—	—	0.01303	0.01576	0.01797	0.01992	0.02104	0.02271	0.02431	0.02587	0.02738
180.0	356.96	0.007498	—	—	0.01191	0.01464	0.01678	0.01867	0.02040	0.02204	0.02360	0.02512	0.02659
190.0	361.43	0.006677	—	—	0.01089	0.01362	0.01573	0.01755	0.01922	0.02079	0.02229	0.02374	0:02515
200.0	365.70	0.005877	—	—	0.009947	0.01271	0.01477	0.01655	0.01816	0.01967	0.02111	0.02250	0.02385
210.0	369.78	0.005023	—	—	0.009071	0.01187	0.01391	0.01564	0.01720	0.01866	0.02004	0.02138	0.02267
220.0	373 69	0.003728	—	—	0:008251	0.0111	0.01312	0.01481	0.01633	0.01774	0.01907	0.02036	0.02160
221.20	374.15	0.00317	—	—	0.008041	0.009878	0.01212	0.01395	0.01553	0.01708	0.01847	0.0200	0.01987
250.0	—	—	—	—	0.006014	0.009171	0.01113	0.01272	0.01413	0.01542	0.01663	0.01779	0.01891
300.0	—	—	—	—	0.002831	0.006735	0.008681	0.01017	0:01144	0.01258	0.01365	0.01465	0.01562
350.0	—	—	—	—	0.002179	0.004950	0.006925	0.008364	0.009519	0.01056	0.01152	0.01242	0.01327
400.0	—	—	—	—	0.001909	0.003675	0.005616	0.006982	0.008088	0.009053	0.009935	0.01075	0.01152
450.0	—	—	—	—	0.001801	0.002913	0.004625	0.005934	0.006984	0.007886	0.008699	0.009452	0.01016
500.0	—	—	—	—	0 001729	0 002492	0 003882	0.005113	0.006111	0.006960	0.007720	0.008420	0.00907
550.0	—	—	—	—	0.001672	0.002246	0.003320	0.004484	0.005454	0 006200	0.006925	0.007581	0.008195
600.0	—	—	—	—	0.001632	0.002084	0.002952	0.003947	0.004835	0.005596	0.006269	0:006885	0.007460
700.0	—	—	—	—	—	0.001890	0.002467	0.003922	0.003972	0.004652	0.005257	0.005808	0 006321
800.0	—	—	—	—	—	0.001772	0.002188	0.002764	0.003379	0.003974	0.004519	0.005017	0.005481
900.0	—	—	—	—		0.001691	0.002013	0.002458	0.002967	0.003476	0.003964	0.004419	0.004841
1000.0	—	—	—	—		0.001629	0.001893	0.002246	0.002668	0.003106	0.003536	0.003952	0.004341

CHAPTER 11

Steam Generators

11.1 INTRODUCTION

Amongst a number of power plants fulfilling the power requirements of the world, major share is contributed by thermal power plants. As this type of power plants use steam as the working fluid, the generator of the steam can be said to be heart of it.

The steam generator is also known as 'Boiler'. The steam produced in the boiler is used in the industries for different purposes viz:

1. For generation of electrical energy as in power plants.
2. For generating mechanical energy in steam turbines.
3. For process heating purpose as heating of ground nuts in Oil Mills.
4. For sizing and bleaching in textile industries.

Normally a boiler produces steam at high pressure. This steam is first used to drive turbines for generating power. Then the steam exhausted from turbines, which is at lower pressure is used for process heating purpose.

A typical boiler has two different parts known as pressure part and combustion chamber. Pressure part is closed vessel, strongly fabricated out of steel, this contains water in it. The combustion chamber is constructed around pressure part with the help of fire bricks. The fuel undergoes combustion in this combustion chamber producing high temperature gases. These gases are also known as flue gases, hot furnace gases, or products of combustion. When flue gases come in contact with the water container, they give their heat to water, causing generation of steam. Afterwards flue gases are exhausted high into the atmosphere through the chimney.

11.2 HISTORICAL DEVELOPMENT OF THE BOILER

As we know boiler is steam generator, or it is a closed vessel in which water is heated and converted into steam. The boiler was initially developed in the very first century by Mr. Hero of Alexandria. However it was too small and it was used as toy only.

Consider a tea kettle on stove or burner, which has water or tea boiling in it, we can see that steam spreads as it leaves the spout. If the spout is closed by means of a cork, after some time the cork will be blown off.

This shows that the steam generated in a closed vessel keeps on the developing pressure. Such steam at high pressure is utilised to do work in expanding in a steam engine.

At the beginning a small closed vessel was used to generate steam by the application of heat. At the beginning the quantity of steam generated and subsequently used in an engine was not sufficient. Therefore larger vessel called **Boilers** were developed for giving a sufficient quantity of steam at the required pressure. The pressure was maintained constant in the boiler by opening the outlet valve such that the steam generated is the same as the steam supplied to the engine.

At the initial stages boilers used were of the fire tube type or tank boilers.

In our everyday life we used to have hot water for bathing purpose a device similar to a tank boiler but on a very smaller scale as shown in Fig. 11.1.

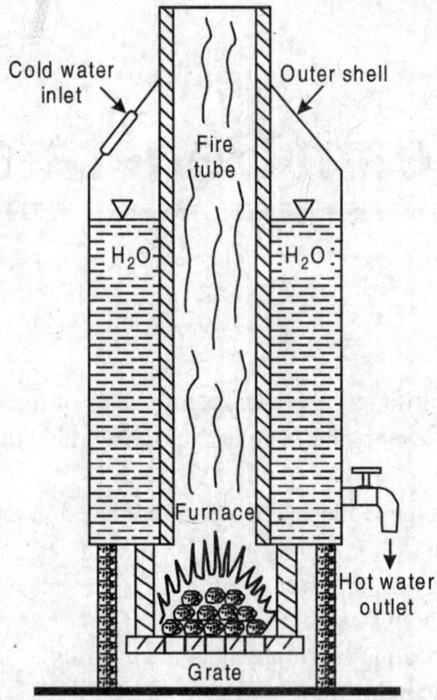

Fig. 11.1. *Small tank boiler used for heating water*

Taking into consideration the safety, the development of simple boiler took place over the years. Till the 17th century, there was no much development in the boiler design and manufacturing. Mr. Denis Papin of France designed and developed the first boiler with safety valve in 1679 A.D.

The major developments took place after 1769 A.D., when James Watt of Scotland developed the practical steam engine.

Afterwards taking into consideration the quantity of steam generated with respect to time, water tube boilers were developed. Subsequently high pressure boilers and supercritical boilers for power generation were developed.

The steam generated in the boilers is utilised for 2-purposes viz:

(i) Power Generation. The boilers generating steam for power generation purpose are bigger in size, generate steam at high pressure and this high pressure steam is first expanded in the turbines for generating power. Then the exhausted steam is utlised for process heating purpose.

(ii) Process Heating Purpose. For example, heating of the oil seeds in the oil mills before they are being crushed.

11.3 CLASSIFICATION

Boilers are mainly classified according to the following factors.

11.3.1 By the Relative Position of Flue Gases and Water (or Tube Contents)

(*a*) *Fire tube.* In this case flue gasses flow through the tubes, which are surrounded by the water which is to be evaporated. Example: Lancashire, Cornish, Cochran, Locomotive, Simple vertical, Scotch marine boiler, etc.

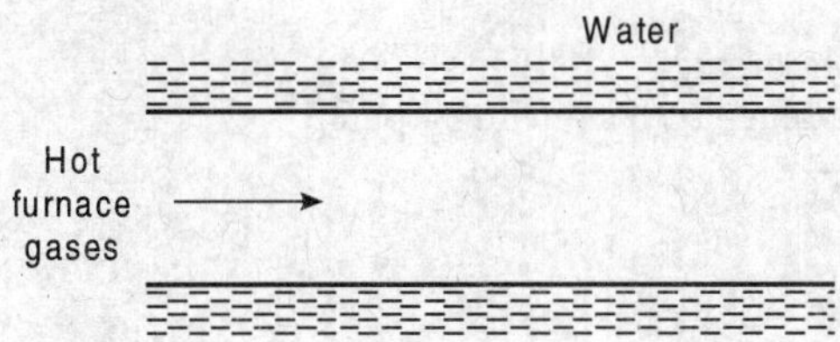

Fig. 11.2

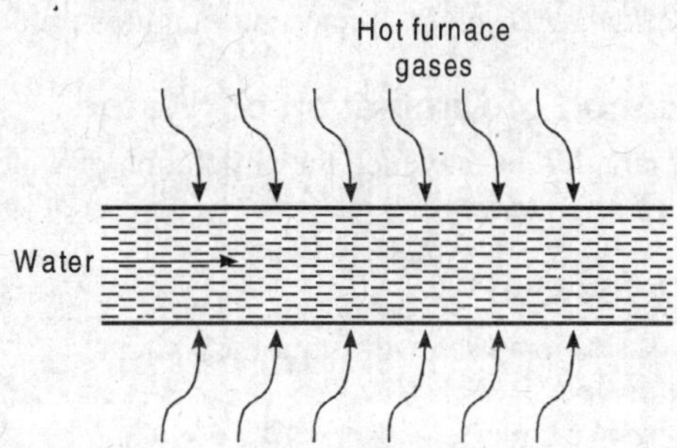

Fig. 11.3

(*b*) *Water tube.* In this case water flows through the tubes and hot gases heat the tubes from out side e.g., Babcock Wilcox, Stirling Boiler, etc.

11.3.2 By the Method of Firing

(*a*) *Internally fired.* In this case the furnace is located inside the boiler shell as in case of Lancashire boiler.

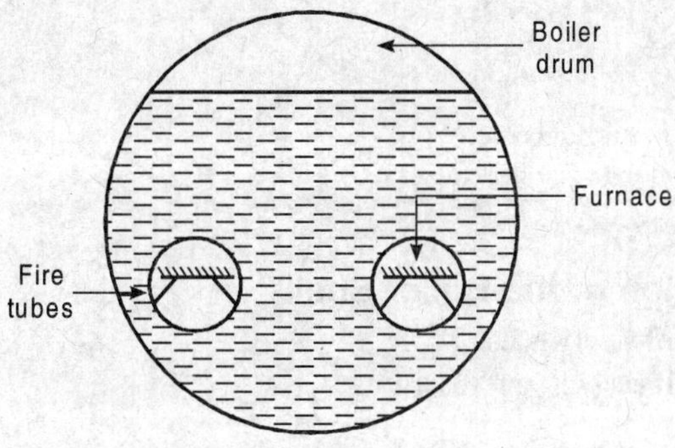

Fig. 11.4. *Internal fired*

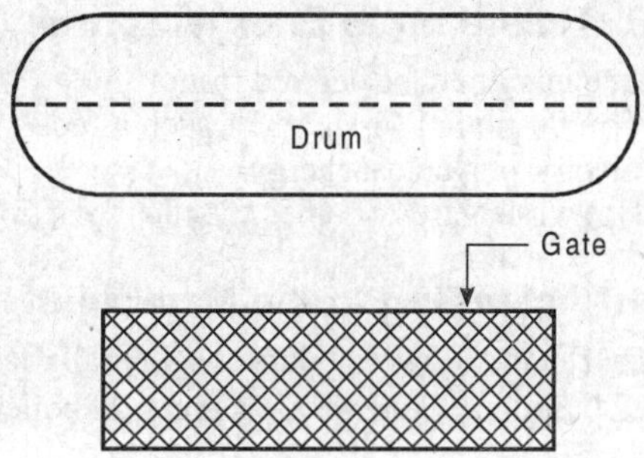

Fig. 11.5. *External fired*

(*b*) *Externally fired.* The furnace is located outside the boiler shell as in case of Babcock Wilcox boiler. (Fig. 11.5).

11.3.3 According to Pressure of Steam

(*a*) *Low pressure boiler.* Boilers which can develop pressures below 80 bar are known as low pressure boilers e.g., Lancashire, Cochran, Cornish etc.

(*b*) *High pressure boilers.* Boilers which can develop pressures above 80 bar are known as high pressure boilers e.g., Babcock, Lamount, Velox, Benson etc.

11.3.4 Nature of Draught

Draught is the pressure difference, which is necessary to draw the sufficient quantity of it for combustion. It is of following types:

(*a*) *Natural draught.* In this case the draught is produced by mean of chimney only. The amount of draught directly depends upon the height of chimney.

(*b*) *Artificial draught.* Artificial draught is produced by means of fans.

(i) *Induced Draught.* Induced Draught is produced by sucking the flue gases from chamber. It is obtained by providing a fan between the boiler and the chimney. This fan is known as I.D. fan.

(ii) *Forced Draught.* Forced draught is produced by forcing the air into the combustion chamber with the help of a fan, known as F.D. fan.

(iii) *Balanced Draught.* Balanced draught is produced by means of I.D. fan and F.D. fan.

11.3.5 Method of Circulation of Water

(*a*) *Natural circulation.* In which the circulation of water is due to gravity.

(*b*) *Forced circulation.* In which the circulation of water is obtained by a centrifugal pump.

11.3.6 By the Use

(*a*) *Land type.* Boilers used with stationary plants.

(*b*) Marine type

(*c*) Locomotive boilers.

11.3.7 By the Design of Flue Gas Passages

The flue gas may follow a single pass, return pass or multipass.

11.3.8 By the Number of Drums

There are 2 types (a) Single drum (b) Multidrum boilers.

11.3.9 According to Energy Source (Fuel) Used

The heat energy may be derived from,

(a) Combustion of fossile fuels such as coal, wood, oil or natural gas etc.

(b) Electric or Nuclear energy.

(c) Hot waste gases of other chemical reactions.

11.3.10 According to the Material of Construction of the Boiler Shell

(*a*) *Steel Boilers.* Power boilers are generally fabricated out of steel plates.

(*b*) *C.I. Boilers.* Generally low pressure boilers are, sometimes built out of castings.

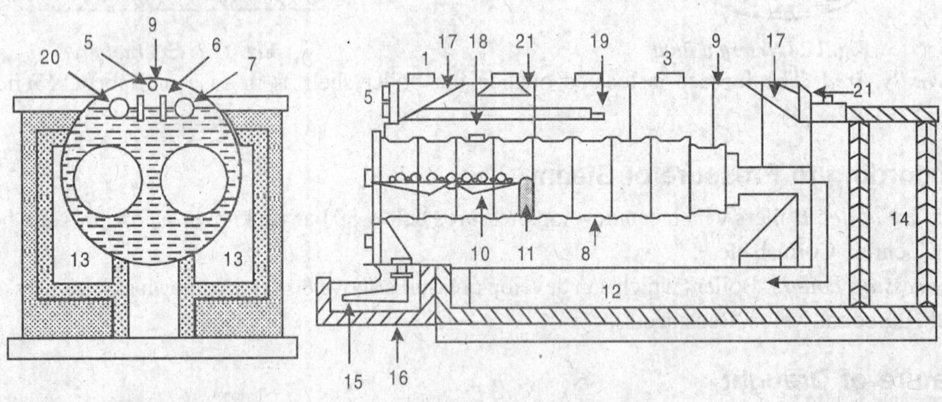

1. Dead weight safety valve.
2. High steam low water alarm.
3. Manhole.
4. Steam stop valve and Antipriming Device.
5. Pressure gauge.
6. Water level indicator.
7. Feed check valve and perforated feed pipe.
8. Fire tubes.
9. Boiler shell.
10. Fire bars.
11. Brick work Bridge.
12. Bottom flue.
13. Side flues.
14. Main flue.
15. Blow-off cock.
16. Blow-off pit.
17. Gusset stays.
18. Fusible plug.
19. Scum tray.
20. Scum valve.
21. End plate.

11.4 LANCASHIRE BOILER

It is fire tube, stationary, horizontal straight tube, internally fired, natural circulation boiler. Normal working pressure of this boiler is about 15 bar and steam generation capacity is upto 8*T* /hr.

11.4.1 Constructional Details

The main parts of the boiler are as follows.

(1) Boiler Drum. This boiler consists of a very large boiler drum. The size of the boiler drum is approximately 7–9 metres in length and 2–3 metres in diameter. This is a very old type of boiler, in this boiler, the meeting edges of the plate are connected by means of rivetted joints. Steel plates are rolled to form shell and the meetings edges of the shell are connected by means of rivetted joints. The alternate shells are made smaller in a diameter, so that the longer shells of required length can be obtained by inserting one shell into the other and rivetting over the circumference.

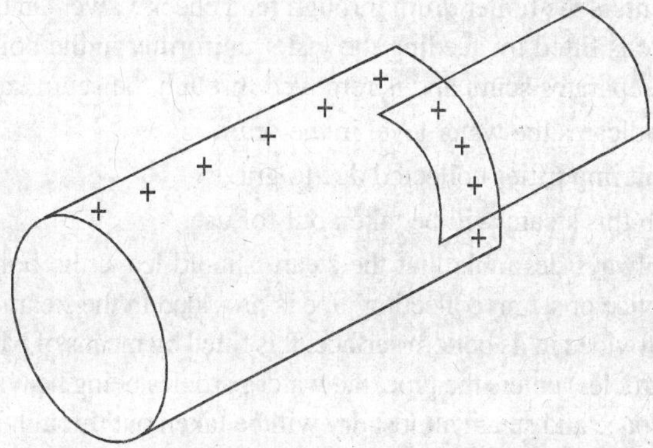

Fig. 11.6 *Lancashire boiler*

The end plates of the boiler are flat. These flat end plates, when they are subjected to steam pressure from inside, they will have a tendency to bulge out. In order to prevent them from bulging, they are strengthened by means of stays. There are 2 types of stays.

(i) *Longitudinal stay*. It consist of round bar which extends from one end plate to the other and are connected to the end plates by means of lock nuts.

(ii) *Diagonal gusset stay*. Consists of a triangular plate, which is rivetted to the shell plate and end plate as shown in Fig. 8.6(a).

(2) Fire Tubes or Flue Tubes. Through the boiler drum passes 2 fire tubes.

Part of each of fire tubes is corrugated, so as to take up expansion and contraction when the boiler gets hot and cold. Also by providing corrugations surface area increases. These tubes are also made of number of shells. The ends of each of the shell are flanged outwards and the flanges of the two adjacent shells are connected by means of rivets, so as to get shells of required length.

It is to be noted that, the centre line of fire tubes is slightly below that of the centre line of boiler shell. Because fire tubes are provided approximately in the centre of water, so that heat can be transferred uniformly in all directions.

(3) Grate, Brickwork Bridge etc. Combustion of fuel takes place over the Grate, Top surface of the grate is made up of fire-bars. Below the grate Ash-pit is provided.

When the combustion of fuel takes place, hot gases are generated and the Brickwork bridge deflects the gases to move upwards.

(4) Manhole. It gives access to the inside of the boiler drum. It will be sufficiently large enough for a man to pass through it, for cleaning and inspection purpose.

(5) Mountings. The boiler is fitted with various mountings. Mountings are necessary for the safe and efficient operation of the boiler.

The term mountings refers to the items such as,

(*i*) *Safety valves*. (a) Dead weight safety valve. (b) Spring loaded safety valve. (c) High steam low water alarm. When the pressure of steam in the boiler drum increases beyond safe value, then these valves will open and allow the surplus steam to escape to the atmosphere, till the pressure of steam in the boiler drum comes to normal working pressure.

(*ii*) *Pressure gauge*. It indicates pressure inside the boiler drum.

(*iii*) *Feed check valve*. Water enters the boiler drum through feed check valve. On the inner side of the feed check valve, perforated feed pipe is fitted for feeding the water uniformly in the boiler drum. Below the feed pipe scum tray is fitted which separates scum and is removed through the scum valve.

(*iv*) *Water level indicator*. It indicates the water level in the drum.

(*v*) *Blow-off valve*. It blows out impurities collected due to gravity.

(*vi*) *Steam stop valve*. Through this steam will be taken out for use.

(*vii*) *Antipriming device*. It is always desirable that the steam should leave the boiler as dry as possible, for that purpose an antipriming device or steam collecting pipe is provided in the steam space. It consists of a pipe and rectangular openings are provided in its bottom surface. It is fitted by means of baffles from inside. When the wet steam (steam and water particles) enters the pipe, the water particles being heavier than steam, after striking the baffles fall back to water space and steam almost dry will be taken out through the steam stop valve.

(6) Accessories. Accessories are used for improving the efficiency of boiler plant. Generally Super heater and Economiser are used as accessories.

(7) Path of Flue Gases. When the combustion of fuel takes place, hot gases are generated and the brick work bridge deflects these gases to move upwards.

During their first pass, these gases flow from the front end of the boiler to the rear end through central tubes. During their I-pass, they pass through the superheater, then they enter the common bottom flue at the rear end. Now their II-pass starts, during this gases flow from rear end to front end through the common bottom flue. At the end of II-pass, the gases are bifurcated into two side flues, when the flue gases enter side flues, their III-pass starts end during their III-pass, they flow from the front end to rear end through side flues and during this pass they heat the water from sides.

After their third pass they meet in main common flue, from where they pass through the economiser and then they are exhausted through the chimney.

11.5 CORNISH BOILER

The only difference between Cornish boiler and Lancashire boiler is that, in case of Cornish boiler there is only one fire tube, whereas in case of Lanchashire boiler there are 2 fire tubes, rest all the arrangement is same.

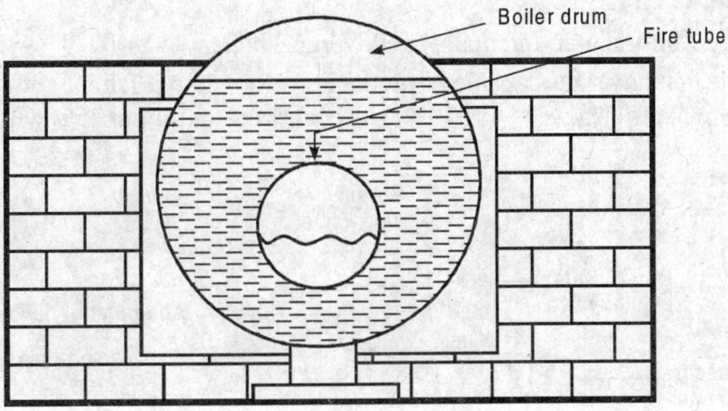

Fig. 11.7. *Cornish boiler*

11.6 LOCOMOTIVE BOILER

It is a Fire tube, Horizontal, Multitubular, Internally fired, Natural circulation boiler.

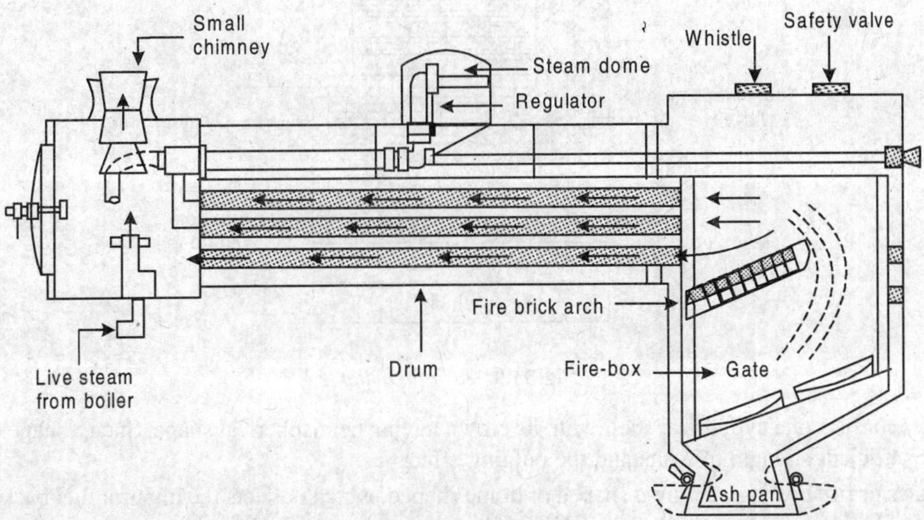

Fig. 11.8. *Locomotive boiler*

As shown in Fig. 11.8 it consist of a large cylindrical shell. This shell is fitted to a rectangular fire box at one end and smoke box at other end. In between the smoke box and fire box a number of tubes are fitted.

When the combustion of fuel (coal) takes place over the grate, hot gases are generated. These hot gases rising from the grate are deflected by the Fire brick arch, so that they come in contact with the complete heating area of the fire box. The dampers as shown, control the flow of air for combustion. These hot gases then flow through the tubes, heat will be transferred to water and water will be converted into steam and accumulates in the steam dome and gases afterward are exhausted to the atmosphere through the small chimney.

The regulator can be operated from the cabin by turning the lever. In order to have superheated steam, the steam is passed through the superheater tubes. These superheater tubes are of smaller diameter and are laid in the large diameter fire tubes as shown. Then this superheated steam is supplied to the engine. This boiler is fitted with various mountings as shown in Fig. 11.8.

11.7 COCHRAN BOILER

It is a Fire tube, Stationary, Multitubular, Internally fired, Vertical boiler.

Cochran boiler is made in many sizes capable of producing steam upto 3 T/hour and working pressures upto 17 bar. This boiler gives Thermal efficiency (η_{th}) of 70% with coal firing and about 75% with oil firing.

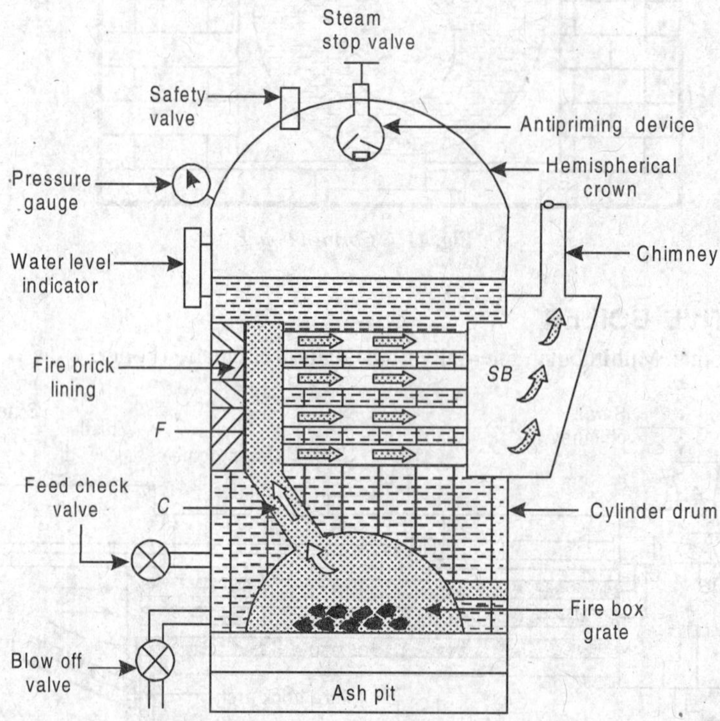

Fig. 11.9. *Cochran boiler*

The boiler consists of a cylindrical shell, with its crown having hemispherical shape. Such a shape of the crown plate, gives sufficient strength to withstand the bulging effect.

The furnace or fire box is also crown shaped or dome shaped, which deflects the unburnt fuel back to the grate. Combustion of fuel takes place over the grate. Hot gases generated in the fire box, enter into the chamber *C*, though a short flue pipe *F* and strikes the boiler shell plate or Back plate which is lined by fire bricks. The back plate directs the gases into the smoke or fire tubes. The gases after passing through the horizontal tubes enter the smoke box *SB* and from where they are exhausted through the chimney.

The hot gases passing through the horizontal tubes, give their heat to the water and in doing so, convert water into steam, which gets accumulated in the steam space.

11.8 BABCOCK WILCOX BOILER

It is a water tube boiler. It mainly consists of 3 parts.

(i) A Horizontal Steam and Water Drum. This is the main part of the Babcock Wilcox boiler. The size of the boiler drum is very much smaller, when compared with boiler drums of fire tube boiler of same capacity. Hence these are simpler from transportation point of view.

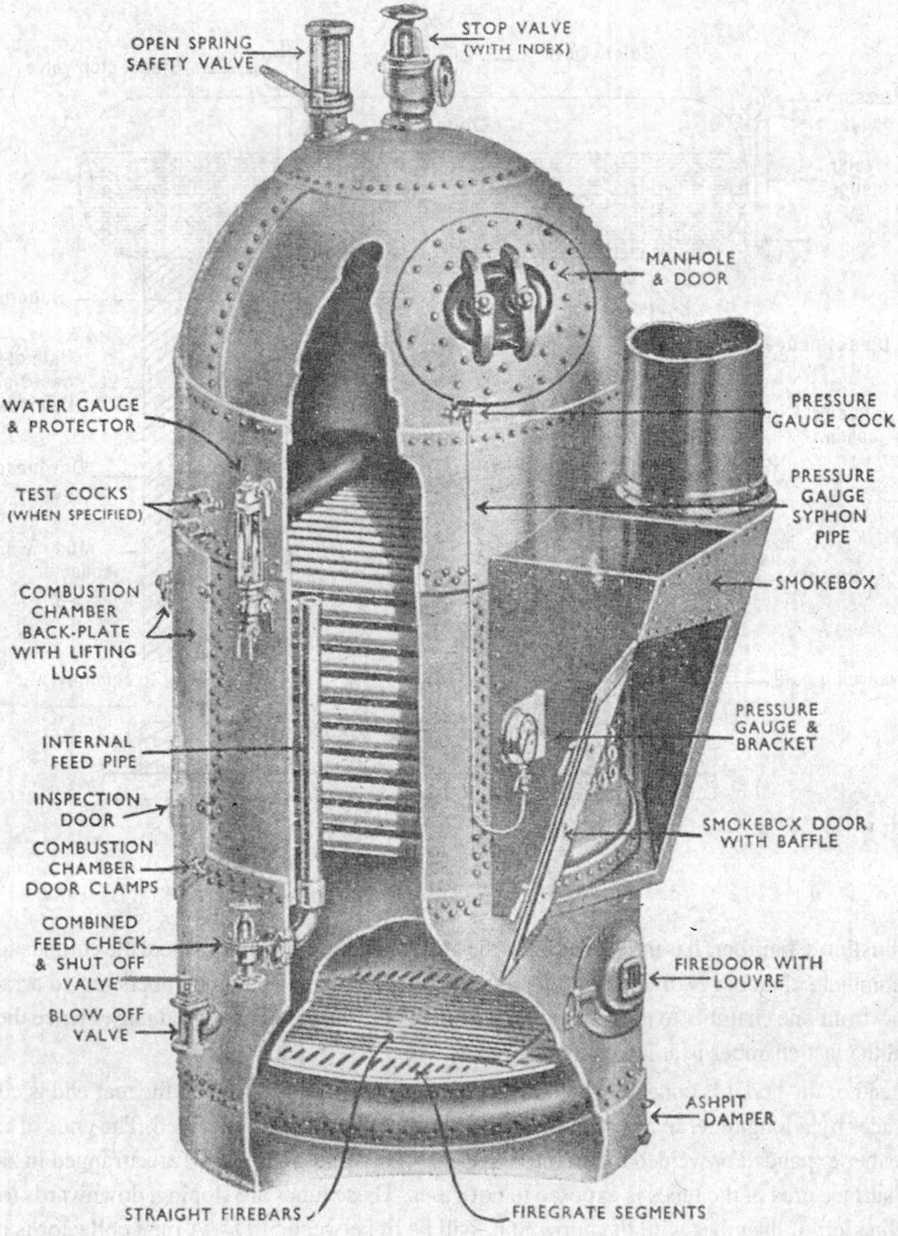

Fig : Cochran Boiler

(ii) A Bundle of Steel Tubes. It is provided in the combustion chamber.

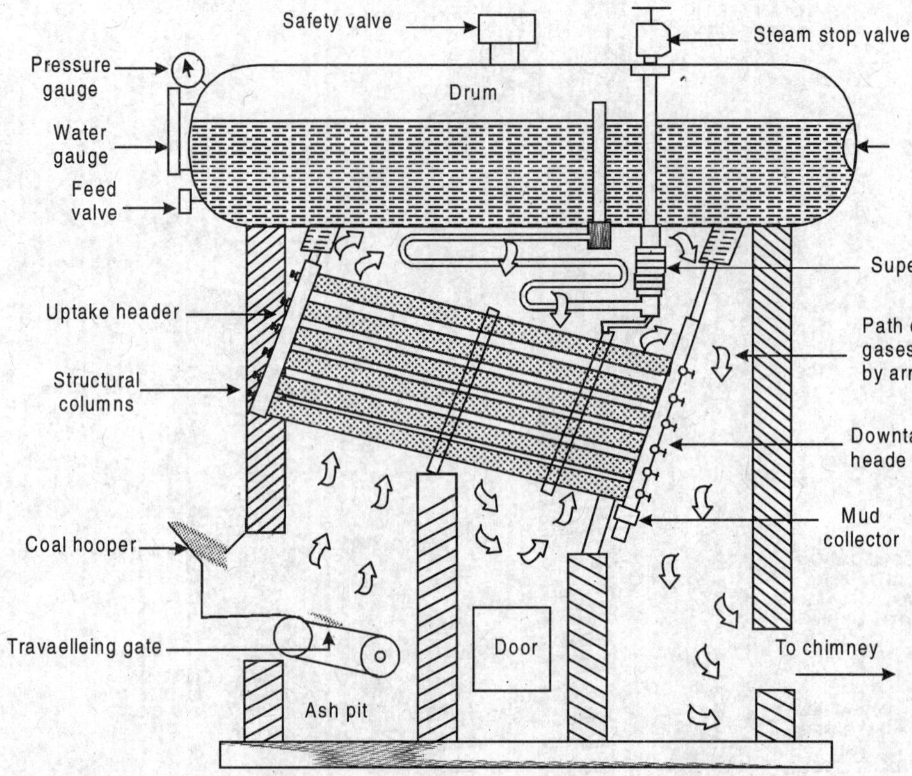

Fig. 11.10. *Babcock Wilcox boiler*

(iii) Combustion Chamber. It is the space above the grate and below the drum where the combustion of fuel takes place. This complete chamber is divided into three separate chambers. These chambers are so arranged, that, the gases can flow from one chamber to the other and give out heat in each chamber gradually. Hence the first chamber is hottest and the last chamber is at lowest temperature.

The front end of the boiler is connected to the uptake header by a short tube and the rear end is connected to the downtake header by a long tube. In between the headers a number of tubes are fitted. The ends of the tubes in the headers are either expanded or welded in order to have gas tight joint. These tubes are arranged in zig-zag manner, so that more surface area of the tubes is exposed to hot gases. These tubes are sloping, downwards towards the rear end. The inclination of the tubes with the horizontal, will be in between 5–15°. A mud collector is provided at the bottom most position of down take header for collecting and disposing the mud or any other suspended impurities. The entire boiler except the furnace in suspended on steel structure, so that the boiler can expand or contract freely without producing the cracks into the brick work, which encloses the boiler and furnace. This brickwork is lined from inside by means of fire bricks. Doors provided give access for cleaning inspection and repair purpose. Dampers provided are used for regulating draught.

This boiler is fitted with various mountings viz., safety valve, steam stop valve, pressure gauge, water level indicator, fusible plug, feed check valve, blow-off valve etc. This boiler is also fitted with superheater as shown in Fig. 11.10.

11.9 STERLING BENT TUBE BOILER

It is a,

(a) water tube,

(b) multi drum and

(c) externally fired boiler.

The boiler mainly consists of,

(i) Steam drum *A*.

(ii) Water drum/Mud drum *B*.

(iii) Bottom beaders *C* and *D*.

(iv) A bundle of tubes and

(iv) Burners.

The steam drum *A* and water drum *B* are connected of by means of a number of tubes. The ends of the tubes are bent so that they can be connected to the drums radially as shown in Fig. 11.11. The ends of the tubes in the drums are expanded in order to have gas tight joint. The water drum is connected to the water headers *C* and *D* as shown in Fig. 11.11 by means of 2 pipes on either end (front and rear end). The upper steam drum is connected to the bottom headers by means of a number of tubes on either side.

Liquid fuel such as L.D.O. (Light Diesel Oil) can be used as a fuel and burners are provided at the front end.

The boiler is suspended over the steel structure. The steel structure is enclosed in red brickwork and which is lined from inside by means of fire bricks.

E—are the doors which give access to the inside of the combustion chamber for inspection and repair purpose.

F—are the baffles which divide the combustion chamber into 3 separate chambers (I, II and III). These 3-chambers are so connected that the hot gases can flow from the I-chamber to III-chamber and they are made to pass through economiser and finally they are exhausted through chimney. The water in the tubes will absorb heat in each chamber, so the temperature of flue gases will go on decreasing towards the rear end.

The boiler is fitted with various mountings viz. :

(a) Spring loaded safety valves

(b) Pressure gauges

(c) Thermometers

(d) Water level indicators

(e) Feed check valve.

(f) Steam stop valve.

(g) Blow off valve.

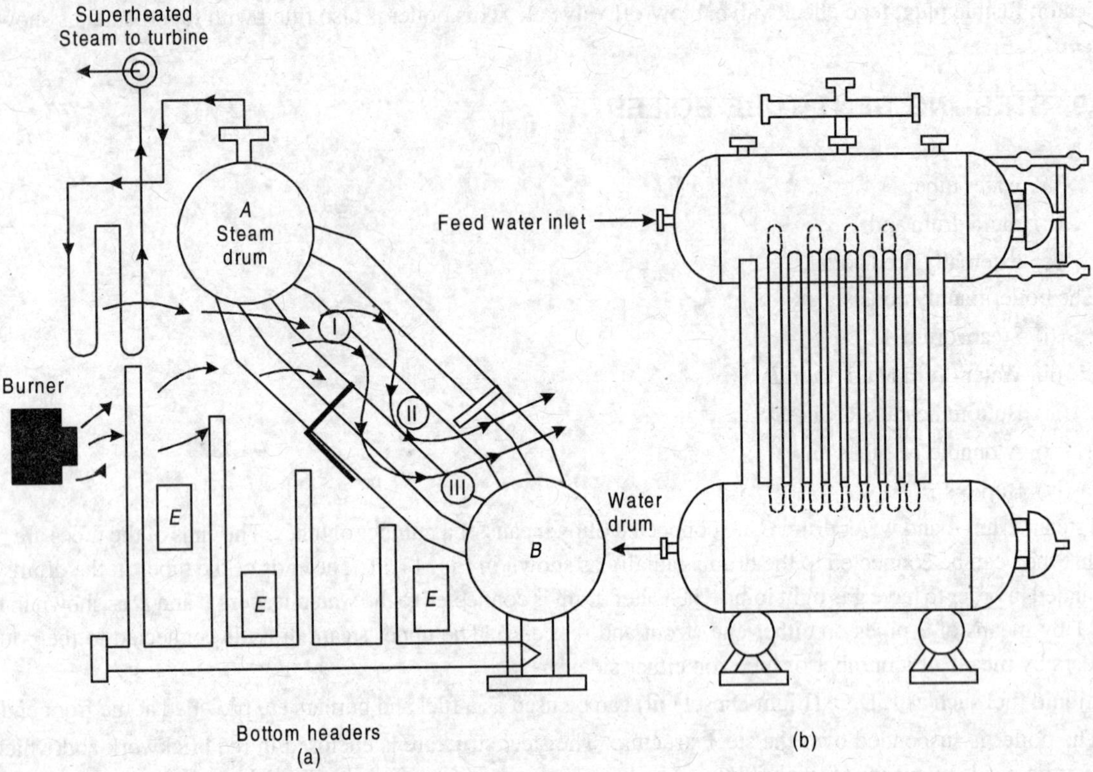

Fig. 11.11

Also superheater and economiser are connected in the boiler circuit as accessories.

Water will be pumped by means of a centrifugal multi-stage feed pump. Water first enter the bottom header of the economiser and when it rises to through the tubes, it recovers some of the heat of the flue gases and becomes hot. The hot water will then be collected into the upper header of the economiser and from where it goes to the upper drum.

Water is filled in the drum upto a pre-specified level, so that tubes, water drum, headers and vertical tubes are flooded by means of water.

We know that the I-chamber is at the hottest temperature, II-chamber is at the medium hot temperature and III-chamber is at the lowest temperature. So, due to this difference, in temperature, difference in densities of water is created and water will rise from water drum to steam drum. As the water rises, some of the water gets converted into steam. Water and steam then enters the steam drum *A*. In this drum steam gets accumulated in the steam space and the water which has not converted into steam; flows down to bottom headers through vertical tubes, and from bottom headers water goes to the water drum and again the cycle repeats.

The steam so generated will be wet, in order to have superheated steam, it is made to pass through the superheater and then it is utilised for power generation purpose.

11.10 COMPARISON BETWEEN WATER TUBE AND FIRE TUBE BOILERS

Water Tube	Fire Tube
1. 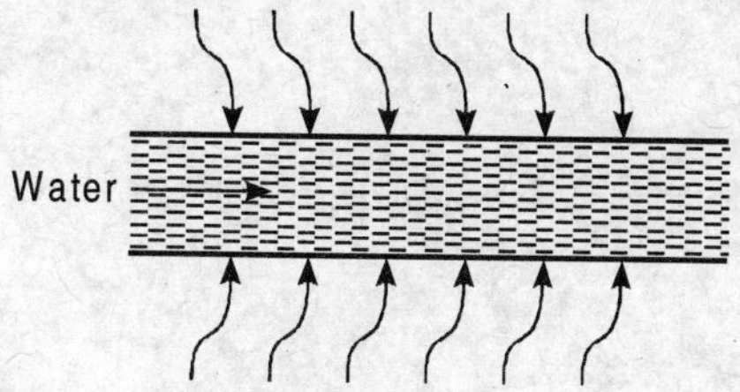 Fig. 8.12(a)	1. 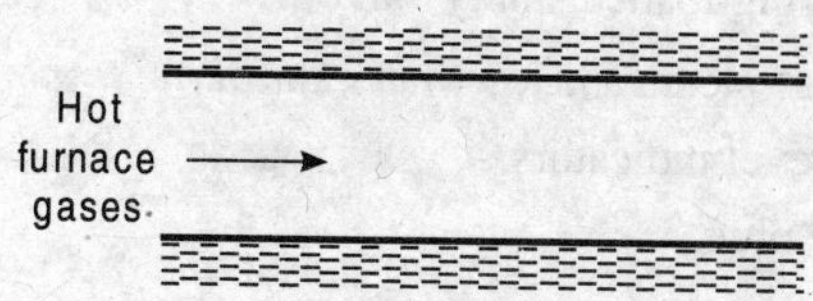 Fig. 8.12 (b)
2. Water tube boilers can generate steam upto a pressure of 225 bar	2. These generate steam only upto a pressure of 20–25 bar
3. The rate of steam generation is more (upto 45000 kg/hr)	3. The rate of steam generation is less (upto 9000 kg/hr)
4. It can produce steam upto a maximum temperature of 560°C	4. It can produce steam upto a maximum temperature of 340°C
5. It has small shell that is area required by this type of boiler is less	5. It has large shell and the area required by this boiler is more
6. Since the size of the boiler drum is small it is simpler from transportation point of view	6. Its transportation and erection is difficult
7. Overall efficiency with economiser and air-preheater is upto 90%	7. Overall efficiency is only upto 75%
8. Construction is very much complex, hence more skill is required for operation	8. Construction is relatively simpler and hence less skill is required
9. Operating costs are more	9. Operating costs are less
10. Steam generation capacity is more, hence large power plants use these boilers	10. It is not suitable for power plants, but is used in Dairy, Laundry and some process industries
11. Boiler won't get blown off, even if some of the tubes fail	11. Danger of explosion of complete drum

11.11 ESSENTIALS OF A GOOD BOILER

1. With minimum fuel consumption, the boiler should be capable of generating steam at the required quality.
2. It should be light and occupy less space.
3. It should be capable of quick starting.
4. The initial cost, installation cost and maintenance cost of the boiler should be minimum.
5. The different parts of the boiler should be easily approachable for repairs.
6. The boiler should have minimum number of joints and those too should be away from direct flames.
7. There should be no deposition of mud and other deposits on the inner surface of the boiler drum.
8. Boiler should conform to safety regulations, laid down in Boiler act [Indian Boiler Regulations (IBR) Act].

11.12 BOILER MOUNTINGS AND ACCESSORIES

Mountings. Mountings are the fittings mounted on the pressure part of the boiler. Mountings are nècessary for the safe and efficient operation of the boiler. Without mountings boiler cannot work safely. Special provisions are made on the pressure part to mount the mountings.

The term mountings refers to the items such as,

(1) Safety valves

(a) Dead weight safety valve

(b) Spring loaded safety valve

(c) High steam and low water alarm..

(2) Water level indicators

(3) Fusible plug

(4) Pressure gauges

(5) Steam stop valve

(6) Feed check valve

(7) Blow-off valve etc.

The above mountings are usually installed in accordance with IBR.

Accessories. Accessories are used to improve the efficiency of the boiler plant. Without accessories boiler can work safely. These are not mounted on the boiler, but these are connected in the boiler circuit.

Usually following accessories are provided with the boiler.

1. Superheater
2. Economiser
3. Air-preheater
4. Steam injector
5. Feed pump
6. Steam trap
7. Induced draught fan
8. Forced draught fan etc.

11.12.1 Dead Weight Safety Valve

As the name implies, this valve is used to assure safe working of the boiler. When the pressure of steam inside the boiler drum increases beyond the safety valve, it may cause damage to the pressure part, or even a severe accident. It can be prevented by allowing the surplus steam to escape to the atmosphere, by means of the safety valve.

This valve was first introduced by M/s. J.J. Hopkinson and Co. Ltd. It consists of a vertical pipe *A*, bottom end of which is flanged and through this flanged end it is mounted on the mounting blocks of the boiler. *B* is the Gun metal valve, resting over the valve seat *C*. Valve seat *C* is fixed on the top of the pipe *A* by means of a securing ring and screw. *F* is the feather which guides the valve *B*. Valve *B* is also guided by a ring in steam discharge casing. The weight carrier is suspended from the top of the valve *B*. Dead weights are placed on the weight carrier and these are covered by the Cast Iron cover.

Weight of the valve *B*, spindle, carrier, dead weights and cover will be acting downwards. During normal working conditions, this downward acting load will be balanced by the steam pressure from below. When the steam pressure exceeds pre specified or safe valve, then the valve will be lifted up and the surplus steam will be exhausted to the atmosphere.

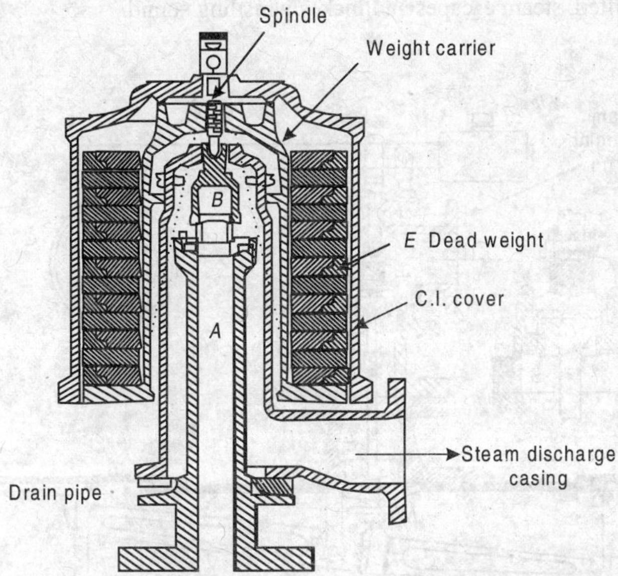

Fig. 11.13. *Dead-weight safety valve*

Example, Dead weight safety valve used on pressure cookers.

11.12.2 High Steam Low Water Alarm

This valve was also introduced by M/s Hopkinson and Co. Ltd. This is a two-in-one valve, i.e., it consists of 2 valves.

(i) *High steam valve.* During high steam conditions this valve opens and allows surplus steam to escape to the atmosphere.

(ii) *Low water valve.* When the water level in the boiler drum falls below the safe level, then this valve opens and allows the steam to escape. When the steam is allowed to escape, it makes whistling sound i.e., it gives warning to the boiler attendant. The whistling sound for high steam conditions will be louder than low water conditions, so that one can distinguish between the two. Since it makes whistling sound it is known as alarm.

As shown in Fig. 11.13, *A* is the high steam valve, resting over the valve seat *B*. Valve seat *B* in secured in position by means of a ring and feather *S* is cost on it. This feather guides the valve *A*. *C* is the low water valve, which is resting over the inner edge of the valve *A* as shown. *R* is the rib which guides the valve *C*. Low water valve. *C* will be held in position by means of the Dead weight, *D* which is connected to the rod *E*, which position by means of the weight *D* and partly by means of weight and counter weight outside the steam discharge casing.

In the steam space lever *FK* is provided. A large fire brick float *L* is suspended from the end *K*.

During normal working conditions, the float will be submerged in water and the weight of this float in partly balanced by means of counter weight *M* and partly by means of buoyant force due to water, which is acting in the upward direction. But when the water level goes below the safe level, the float gets uncovered by means of water, then there won't be buoyant force and counter, weight. *M* alone will not be sufficient to balance the float. So, the lever *FK* tilts in the clockwise direction. Then the knife edge projection of the lever *FK* comes in contact with the collar *N* and in turn lifts the hemispherical valve *C*. When the valve *C* is lifted, steam escapes and makes whistling sound.

During high steam conditions, both the valves are lifted since the low water valve is resting over inner edge of the high steam valve *C* is lifted, steam escapes and makes whistling sound.

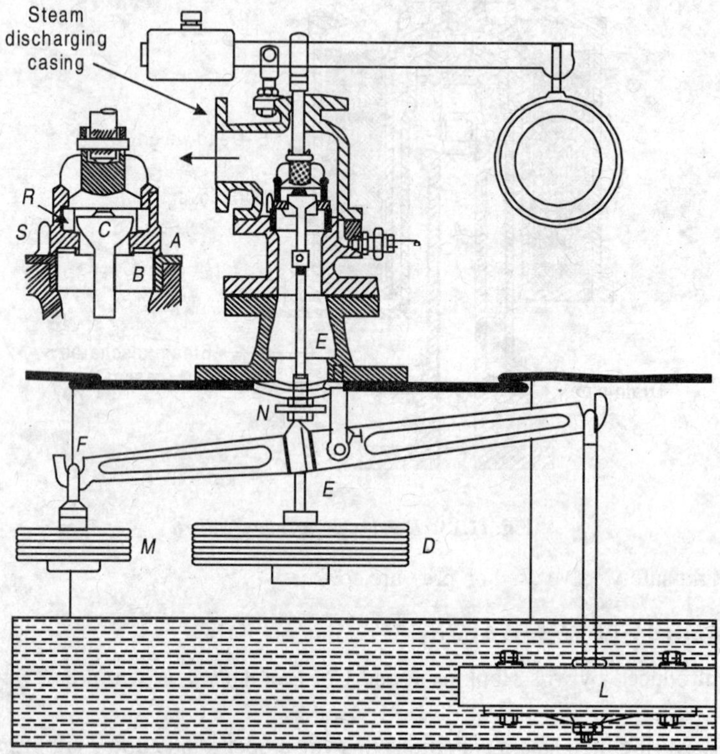

Fig. 11.14. *High steam low water alarm*

During high steam conditions, both the valves are lifted since the low water valve is resting over inner edge of the high steam valve, and surplus steam will be exhausted.

11.12.3 Water Level Indicator

It indicates the water level inside the boiler drum. This water level can be made visible from the boiler room floor by using mirrors. This unit was also introduced by M/s Hopkinson and Co. Ltd.

The unit consist of two hollow gun-metal stuffing boxes. One end of each of the stuffing box will be flanged and through this flanged end it is connected to the front end plate of the boiler. Upper stuffing box is connected to the steam space and lower stuffing box is connected to the water space. In between the two stuffing boxes a strong glass tube is fitted. (In between the glass and metal, rubber packing is provided in order to have gas tight joint).

Thus two valves *A* and *B* which control the passage of steam and water between the boiler and the glass tube. When these valves are open, the handles are vertical and the water level in the glass tube will be the water level in the boiler drum. A third valve *C* called as Blow-off valve is generally closed. When this valve is closed, its handle is vertical and to blow out sediments and impurities handle is to be made horizontal.

An arrangement is also provided, in order to shut off automatically the supply of steam and water to the glass tube, if the glass tube breaks due to any reason. This arrangement consists of a hollow gun-metal column *D* and two steel balls as shown.

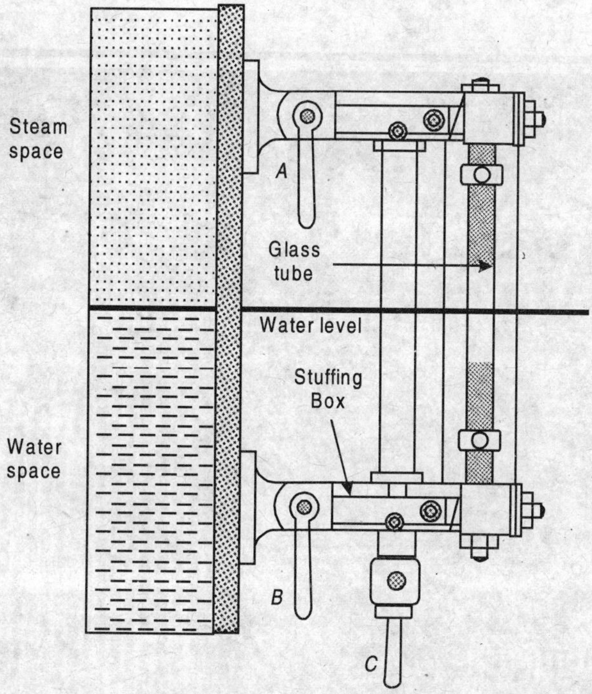

Fig. 11.15. *Water level indicator*

During normal working conditions these balls will be in position shown by full circles if the glass tube breaks the rush of water in the bottom passage carries the ball into the position shown by dotted circle and shuts off the supply of water. At the same time the steam rushing through the upper passage aided by the water rushing upwards through the column *D* carries the ball into the position shown by dotted circle and shuts off the steam supply. Then the boiler attendant can safely close the valves *A* and *B* and can replace the tube.

Screw plugs provided at the end of stuffing boxes, give access for cleaning of passages. The glass tube is generally covered from the front and from sides by Guard glass in order to protect the boiler attendant from the flying fragments of glass tube, if it breaks.

11.12.4 FUSIBLE PLUG

It actuates itself as the last resort, when the water level in the boiler drum falls below the minimum permissible water level. As a result of which, fusible metal melts by the heat of steam, the plug drops out and the steam rushes into the combustion chamber and extinguishes the fire. Thus it safeguards the combustion chamber crown from burning and also warns the boiler attendant about the fall of water level in the drum below minimum permissible level.

Fig : Sterling Bent tube boiler.

Courtesy : LIPI BOILERS LTD. PUNE-1

Sterling Bent tube boiler.

Fire Tube, 6 TPH, LIPI Make Boiler

Water Tube, 90TPH, LIPI Make Boiler

Function :

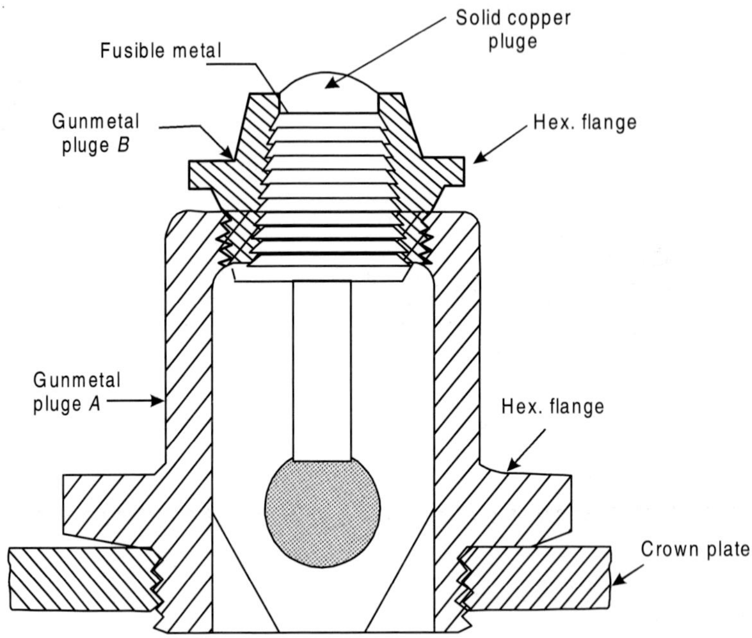

Fig. 11.16. *Fusible plug*

Fusible plug is fitted on the combustion chamber cr own, but in the drum. As shown in Fig. 11.16, *A* is the hollow gun metal plug screwed into the crown plate. *B* is the second hollow gun metal screwed into the pl ug *A* and *C* is the third solid copper conical plug. The inner surface of *B* and *C* for removal and refitting purpose by using spanner s.

Under normal working conditions, fusible plug will be covered by means of water in the drum. This wate r keeps the temperature of the fusible metal below its melt ing point. But when due to any reason, water level in the drum falls below the minimum permissible water level, the plug gets uncovered from water and is exposed to steam. Due to heat of steam, fusible metal melts and copper plug drops and the steam rushes through the access into the combustion chamber. Before referring the boiler, the attendant removes and refits the plug after interposing the fusible metal between the plugs *B* and *C*.

11.12.5 Bourdon Pressure Gauge

This is the most common type of pressure gauge. This is used to measure pressure inside the containers and pipe lines etc.

The basic element of this gauge is the tube *A*. It is elliptical in cross section and is bent int o and arc of circle as shown in Fig. 11.17.

End*B* of this tube is sealed, while open end *C* is connected to the connecting union *D*. Through the connecting union the gauge will be mounted over the containers or pipes of whose pressures are to be measured.

When it is mounted on the systems and if the pressu re of the system is more than atmosphere (i.e., + v e pressure) then the tube will tend to curl out. Conversely if the pressure of the system is less than the atmosphere, (i.e., –ve pressure) the tube will tend to curl in. Thus the c hange of pressure will therefore appear as a moveme nt of the end *B*. This end *B* is connected by means of a link *E* to a quadrant gear *F*. The quadrant gear engages with a small gear *G* on to which a point *H* is attached.

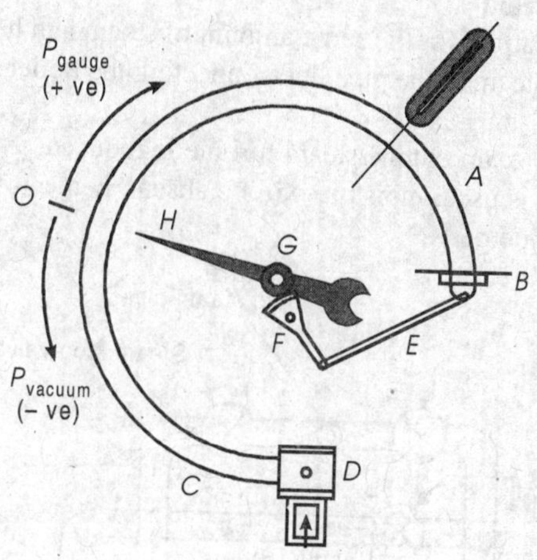

Fig. 11.17. *Bourdon pressure guage*

Thus then change of pressure will appear as the movement of the end *B*, which will be transmitted to the quadrant gear, which in turn rotate the pinion and hence the pointer. The pointer rotates over the calibrated scale, which directly gives the pressure readings.

11.12.6 Steam Stop Valve

Through the steam stop valve, steam will be taken out for power generation purpose.

As shown in Fig. 11.18 valve body is made up of C.I. Through its bottom flange *A* it is mounted in the boiler drum, and its side flange *F* is connected to the pipe through which steam goes to the steam engine or steam turbine.

It is clear from full figure that when the hand wheel is rotated in clockwise or anticlockwise direction, the valve will move up and down. And the required quantity of steam can be taken out for power generation.

11.12.7 Feed Check Valve

Through the feed check valve, water enters the boiler drum. It allows the water to flow into the drum and it will not allow the water to flow out of the drum, so it is called as check valve.

11.12.8 Blow Off Valve

Mud or any other settled impurities are blown out through this valve.

11.12.9 Superheater

The steam generated from a simple low pressure boiler is generally wet. So, for superheating the steam, superheaters are used. In the superheater, wet steam is first dried at the same temperature and pressure and then it is superheated at constant pressure.

Generally, same heat of the flue gases is used for superheating purpose and hence superheaters are placed in the path of flue gases. However in bigger installations superheaters are provided with independent furnaces. And such superheaters with independent furnaces are known as Independently Fired Superheaters.

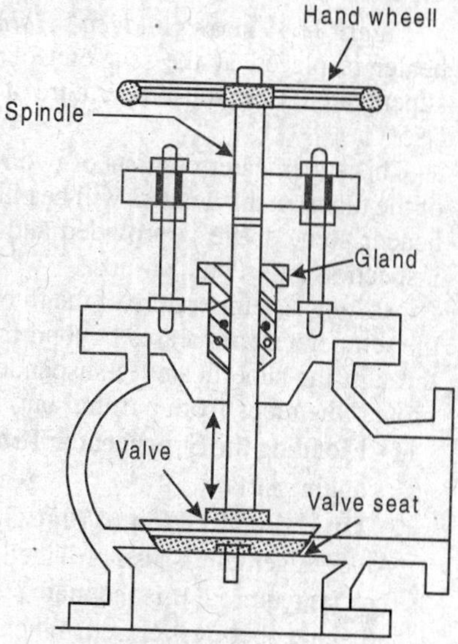

Fig. 11.18. *Steam stop valve*

Advantages of Superheating the Steam

(1) Increase in amount of work output, for the same amount of steam and hence increase in cycle efficiency.

(2) Loss due to condensation of steam in the pipe line connecting the boiler to the steam engine or steam turbine is reduced.

(3) Loss due to condensation of steam engine/steam turbine is reduced.

(4) When the superheated steam is used, moisture will be absent, hence it will reduce corrosion and erosion of steam engine and steam turbine parts.

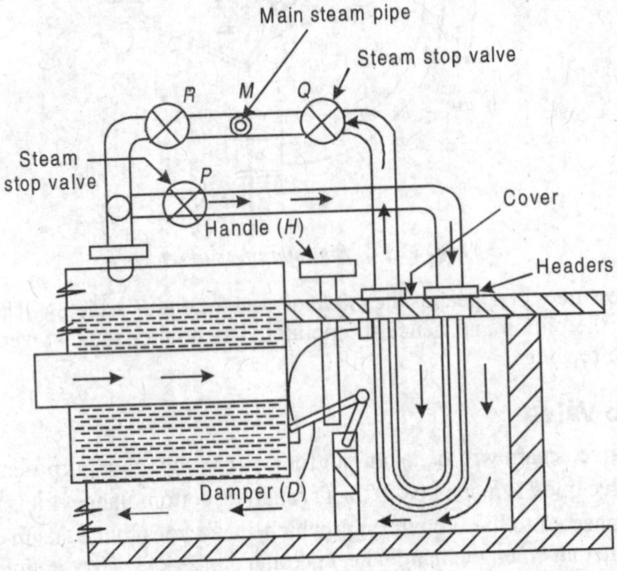

Fig. 11.19. *Superheater*

Figure 11.19 shows *Sudgen's Hair pin type of superheater* arranged with Lancashire/Cornish boiler. The superheater is placed at the rear end of the boiler and the flue gases after their first pass, they pass through the superheater. Here the temperature of the flue gases is not less than 550°C. This superheater gives a superheat of 40 –95°C.

This superheater consists of two Mild Steel boxes or headers and a number of tubes bent to U-shape. The ends of the tubes, in the headers will be either expanded or welded in order to have gas tight joint. Between front end rear header some space is provided and this space is covered by means of a cover. This gives access for cleaning, inspection and repair purpose.

As long as temperature of the flue gases is less than 550°C then there is no danger of burning of tubes. But when the temperature increases beyond this and also when the delivery of steam from the boiler to the steam turbine through the superheater is suspended for along time, then the following arrangements are to be made so as to protect the tubes from burning out.

(1) **Flooding the Superheater Tubes.** This water is to be drained before the supply of steam through the superheater again starts.

(2) **Diverting the Flow of Flue Gases.** This is done by Damper D which is operated by means of handle H.

An arrangement is also provided so as to pass the steam through superheater or directly to the main steam pipe as may be required M is the main steam pipe, P, Q and R are stop valves. When the superheater is in action valves P and Q are open R is closed. When steam is to be taken out directly from the boiler to the steam pipe, then valves P and Q are closed and R is opened.

11.12.10 Economiser

One of the major heat losses in the boiler plant is the heat carried away by the exhaust gases. An economiser is a device or appliance, which recovers some of the heat of the exhaust gases. This recovered heat in utilised for preheating the water. This preheated water when it is supplied to the boiler drum, it requires less heat for its conversion into steam. This reduces the amount of fuel required and thus the efficiency of boiler plant increases.

Figure 11.20 shows the *Green's economiser*. It is used with stationary low pressure boilers.

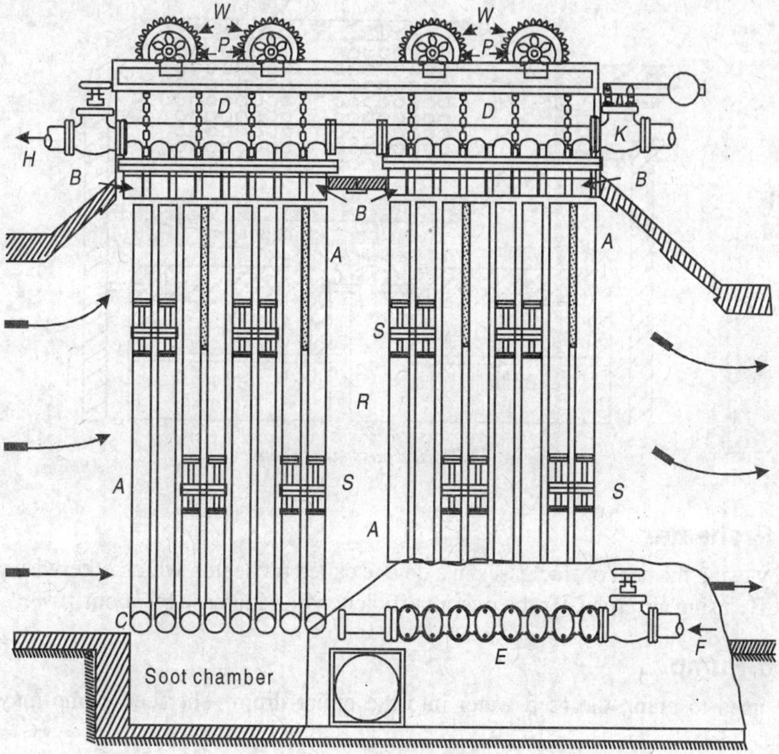

Fig. 11.20. *Green's economiser*

It consists of a number of vertical pipes *A*. These pipes are arranged in groups of 4, 8, 12,16 etc. Upper ends of the pipes are connected to the upper boxes *B*. These boxes *B* are in turn connected to the common header *D*. Lower ends of the pipes are connected to lower boxes. These lower boxes are connected to the horizontal pipe *C* through the cross pipes. All the tubes are enclosed in the brick work of the economiser except header *D* and pipe *C*.

Water will be pumped by means of a feed pump, water enters the economiser through the stop valve at the inlet *F* and passes into the pipe *C*, from where it flows through the cross pipe and then rises through the vertical tubes *A*. Since the economiser is provided in the path of flue gases the water rising through the tubes recovers heat and becomes hot. Hot water will be then collected into the top common header.

This hot water from the top header is taken out through the stop valve *H*, from where it goes to the boiler drum. *K* is the safety valve for the economiser.

Since the economiser is provided in the path of flue gases, soot (or ash) is likely to be collected over the tubes. It retards the rate of heat transfer from the flue gases to the water inside the tubes. In order to avoid this, soot is scrapped by means of scrapers. Scrapers of the two-adjacent tubes are coupled to form one pair and two such pairs of scrapers are connected by means of a chain, which passes over pulley *P*. Pulley *P* is connected to worm shaft, in such a way that when one pair of scrappers comes to top most position, another pair will be at the bottom most position, then it automatically reverses the direction of motion. The scrapers are kept in motion as long as economiser is in action.

By Pass Arrangement. By Pass arrangements are provided for the flue gases and for feed water, so that the economiser may be put out of action when not required. Figure 11.21 shows such an arrangement for two Lancashire boilers fitted with an economiser. When the economiser is in service, then dampers *L* and *M* are open and the damper *N* is closed. For isolating the economiser damper *N* is opened and dampers *L* and *M* are closed.

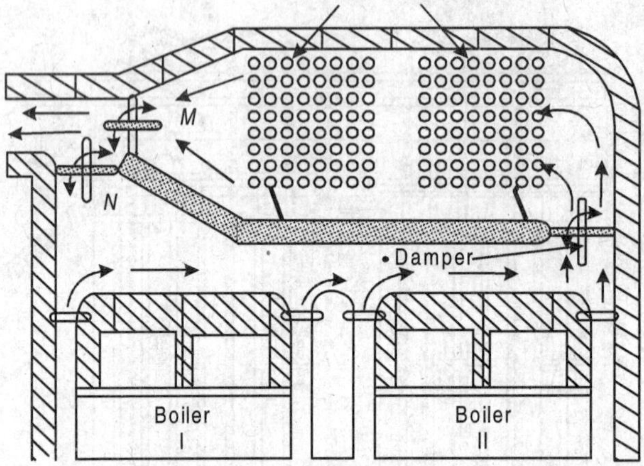

Fig. 11.21. *By pass arrangement*

11.12.11 Air Preheater

The air is heated, by using the heat of flue gases in a device called preheater, which otherwise would have been lost to the atmosphere. By using air preheater the overall efficiency will increase by about 10%.

11.12.12 Feed Pump

It is an appliance used to pump the feed water into the boiler drum. The feed pump may be either rotary or reciprocating type.

11.12.13 Steam Injector

When the feed pump fails due to any reasons, then steam injector, injects or pumps or delivers feed water to the boiler drum, by the use of steam.

11.12.14 Steam Trap

Steam trap automatically collects and returns to the boiler, the water resulting from partial condensation of steam, without allowing any steam to escape.

11.13 BOILER DRAUGHT

Draught is the pressure difference which is necessary to draw the required quantity of air for combustion and to remove the flue gases out of the system.

Thus the object of producing draught in a boiler is,

(i) To provide sufficient quantity of air for combustion

(ii) To make the resulting hot gases, to flow through the system

(iii) To discharge these gases to the atmosphere through the chimney.

Usually this drought (pressure difference) in boiler is of small magnitude and is measured in mm of water column by means of draught gauge/manometer.

Note. The amount of draught depends upon,

(i) Nature and depth of fuel on the grate.
(ii) Design of combustion chamber/firebox.
(iii) Rate of combustion required.
(iv) Resistance offered in the system due to baffles, tubes, superheater, economiser, air pre-heater, etc.

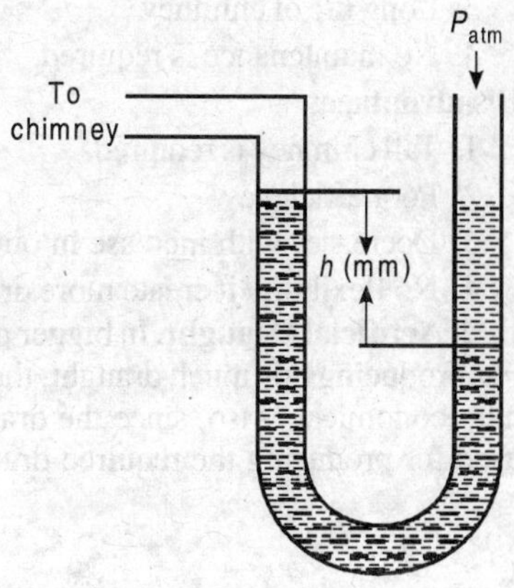

Fig. 11.22

11.14 CLASSIFICATION

Draught is broadly classified into 2-types,

(1) Natural or chimney draught (2)Artificial draught
(a) Fan Drought (Produced by Mechanical Fans)
(i) Forced draught (ii) Induced draught (iii) Balanced draught
(b) Steam jet draught (Produced by steam jet)
(i) Induced draught (ii)Forced draught

1. Natural or Chimney Draught. In this case the amount of draught directly depends upon the height of chimney. It is produced due to the difference in densities between the column of hot gases in the chimney and a similar column of cold air out side the chimney.

Let us first consider the case when fires are not lighted.

Let, the atmospheric pressure at grate level be P_1 and P_2 be the atmospheric pressure at an altitude H. The pressure P_2 is lower than the pressure P_1 because with the altitude pressure goes on decreasing.

Fig. 11.23

Now let us consider the case when fires are lighted and the chimney is full of hot gases. Under these circumstances, the pressure at the base of the chimney is the sum of pressure P_2 at the top and the pressure due to hot gas column H. But pressure P_1 at grate is the sum of pressure P_2 and the pressure due to similar cold column of air H.

Since, ρ cold air > ρ hot gases

i.e., $P_A > P_B$

$\therefore$ P_2 + Pressure due to cold column $H > P_2$ + Pressure due to hot column H.

$\therefore$ Pressure at grate due to cold column > Pressure at the chimney base due to hot column H.

This difference is called **static draught** and because of the pressure difference, (draught) air will rush to the combustion chamber, where combustion of air and fuel takes place and hot gases are generated. Then these hot gases because of draught, flow through the system and finally they are exhausted to the atmosphere through the chimney.

Advantages of Natural Draught

1. Easy to construct.
2. No power is required for producing the draught.
3. Long life of chimney.
4. No maintenance is required.

Disadvantages

1. Tall chimney is required.
2. Poor efficiency.
3. Decreases with increase in outside temperature.
4. No flexibility it create more draught to take peak loads.

2. Artificial Draught. In bigger power plants, the draught of the order of 25–350 mm of H_2O column is required. Far producing this much draught, the chimney height has to be increased considerably, which is neither convenient nor economical. Also, since the draught depends upon the climatic conditions, some mechanical equipments are used for producing the required draught and the draught so produced is called as the Artificial draught.

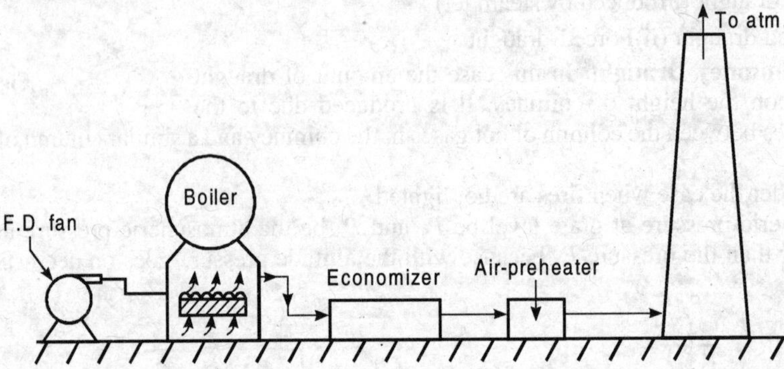

Fig. 11.24

3. Forced Draught. In a Forced draught system, a Fan or Blower is provided is shown in figure which forces the air in the combustion chamber. In the combustion chamber combustion of air and fuel takes place and hot gases generated. These gases are forced to pass through the flues, economiser, air pre-heater and then they are exhausted after recovering heat of flue gases. This draught system is known positive draught system, since the pressure of gases throughout the system is above atmospheric pressure.

It is to be noted that, the function of chimney use is to discharge the gases high in the atmosphere to reduce air pollution and it is not much significant for producing draught.

4. Induced Draught. In this system, the Blower or Induced Draught fan is located near the base of chimney. The air is sucked in the system, by reducing the pressure through the system below atmosphere. The flue gases, generated after combustion are drawn through the system and after recovering heat in the economiser, air-preheater, they are exhausted through the chimney to the atmosphere.

Here it is to be noted that the draught produced is independent of the temperature of hot gases, so the gases may be discharged as cold as possible after recovering as much heat as possible.

Advantages of Forced Draught (F.D.) over Induced Draught (I.D.)

1. The size and power required by I.D. fan is more because this fan handles more gases.
2. Since the I.D. fan handles hot gases, water cooled or air cooled bearings are to be used.
3. F.D. fan consumes less power and normal bearing can be used.

5. Balanced Draught. It is always preferable to use combinations of I.D. and F.D. instead of Forced or Induced draught alone.

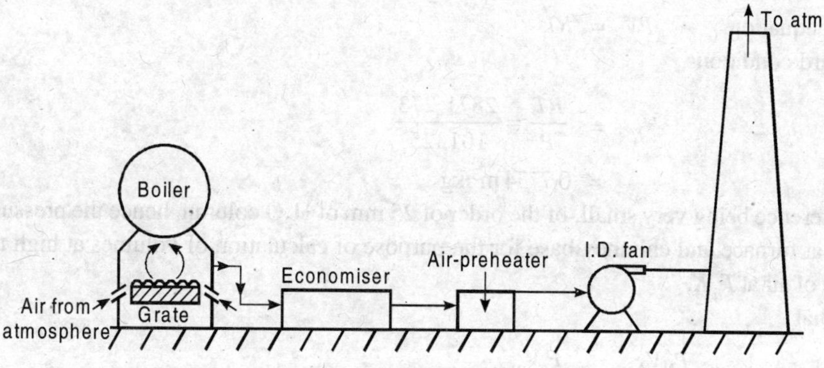

Fig. 11.25

If Forced Draught alone is used then the furnace cannot be opened for firing or for inspection. Because the high pressure air/gases inside the furnace will try to blow out, and there is every chance of blowing out of the fire completely and the furnace may stop.

If Induced Draught fan alone is used, then also furnace cannot be opened either for firing or for inspection. Because the cold air will try to rush into the furnace, which reduces the effective draught.

To overcome both these difficulties Balanced Draught is used. In this case I.D. fan and F.D. fan are provided as shown.

6. Steam Jet Draught. It may be Forced or Induced type, (Refer Locomotive boiler Fig. 11.8). If the steam jet is directed into the smoke box, near, the chimney, the air is sucked through the system, into the smoke box. If the jet is Located before grate, air is forced through the system. Induced type is favoured as forced draught increases heat losses.

Induced draught is produced by steam jet in case of Locomotive boiler. When the Locomotive in stationary, steam from the boiler may be supplied to the smoke box through the nozzle to create draught. While locomotive is running, due to motion, the air flows to the furnace.

11.15 CALCULATION OF CHIMNEY HEIGHT

We know that,

$$C + O_2 = CO_2 \qquad \ldots(1)$$

$$1\,\text{vol} + 1\,\text{vol} = 1\,\text{vol}$$

and

$$2H_2 + O_2 = 2H_2O$$

$$2\,\text{vol} + 2\,\text{vol} = 2\,\text{vol} \qquad \ldots(2)$$

Therefore, for Eqs (1) and (2), one *volume* of O_2 from air is used for the combustion of carbon in the fuel, produces one volume of gaseous products of combustion CO_2. But one vol. of O_2 from air produces 2 volumes of steam (H_2O) in the gaseous products and since the hydrogen content in solid or liquid fuel is quite small, so steam formed is negligible compared to other constituents. N_2 and other constituents does not flake part in combustion.

So, the assumption that the volume of air required for combustion is approximately the same as the volume of the products of combustion is fairly accurate.

Let

m = Mass of air used in kg/kg of fuel.

T_f = Mean absolute temperature of chimney on flue gases in K.

T_a = Absolute temperature of outside air K

$\therefore$ Mass of chimney gases = $(m + 1)$ kg/kg of fuel

Volume per kg of flue gases = 273 K or 0°C

$$= \text{Volume per kg of air at 273 K}$$

From Perfect gas equation, $PV = RT.$

Volume at standard conditions,

$$V_N = \frac{RT}{P} = \frac{287 \times 273}{101325}$$

$$= 0.7734 \text{ m}^3/\text{kg}$$

The pressure difference being very small, of the order of 25 mm of H_2O column, hence the pressure can be considered as constant at furnace and chimney base for the purpose of calculation of volumes at high temperatures.

∴ Volume per kg of air at $T_a K$,

Since we know that

$$\frac{P_N V_N}{273} = \frac{P_a V_a}{T_a}$$

$$V_a = V_N \times \frac{T_a}{273} \times \frac{P_N}{P_a}$$

$$V_a = 0.7734 \times \frac{T_a}{273} \times \frac{P_N}{P_a}$$

∴ Volume of m kg of air at

$$T_a K = 0.7734 \times m \times \frac{T_a}{273} \times \frac{P_N}{P_a} \text{ and}$$

Volume of $(m + 1)$ kg of chimney gases at $T_f K$,

$$= \frac{0.7734 \times m \times T_f}{273} \times \frac{P_N}{P_a} \text{ and}$$

∴ Density of air at

$$T_a K = \frac{\text{Mass}}{\text{Volume}} = \frac{m \times 273}{0.7734 \times m \times T_a} \times \frac{P_a}{P_N}$$

$$= \frac{273}{0.7734 \times T_a} \times \frac{P_a}{P_s}$$

$$\rho_a = 1.293 \left[\frac{273}{T_a}\right] N \frac{\text{kg}}{\text{m}^3} \times \frac{P_a}{P_N}$$

And density of flue gases at T_f at K,

$$= \frac{273}{0.7734 \times T_a} \times \frac{P_a}{P_s}$$

$$\rho_a = 1.293 \frac{273}{T_a} \frac{\text{kg}}{\text{m}^3} \times \frac{P_a}{P_N}$$

And density of flue gases at $T_f K$

$$= \frac{(m+1) \times 273}{0.7734 \times m \times T_f} \times \frac{P_a}{P_N}$$

$$\rho_g = 1.293 \left[\frac{m+1}{m}\right] \times \frac{273}{T_f} \times \frac{P_a}{P_N}$$

Let H = chimney height in metre from grate level.

Then pressure exerted by the column of hot gases.

$$H = \rho_g \cdot g \cdot H$$

$$P_f = 1.293\left[\frac{m+1}{m}\right]\times\frac{273}{T_1}\times g\times H \text{ N/m}^2 \quad \ldots(1)$$

Pressure exerted due to cold air column,

$$\rho_a \times g\times H = 1.293\times\frac{273}{T_a}\times g\times H \text{ N/m}^2 \quad \ldots(2)$$

Draught pressure, in terms of static pressure difference between hot gas column in chimney and a similar cold column.

$$P_d = \text{Eq. (2) – Eq. (1)}$$

$$= g\times H[P_a - P_f]$$

$$= 1.293\times 273\times g\times H\left[\frac{1}{T_a}-\left[\frac{m+1}{m}\right]\times\frac{1}{T_f}\right]\text{N/m}^2$$

$$P_d = 353\times g\times H\left[\frac{1}{T_a}-\left[\frac{m+1}{m}\right]\times\frac{1}{T_f}\right]\text{N/m}^2$$

If P_a and P_N are different, then

$$P_d = 353\times g\times H\times\frac{P_a}{P_N}\left[\frac{1}{T_a}-\left[\frac{m+1}{m}\right]\times\frac{1}{T_f}\right]\text{N/m}^2$$

To express in mm of H_2O column

$$P_d = \rho_w . g . h_w \frac{\text{N}}{\text{m}^2}$$

where ρ_w = Density of water in kg/m^3

g = Acceleration due to gravity in m/sec^2

h_w = Water column in mm

Then, P_d = Pressure difference wil be in N/m^2

i.e.,

$$P_d = \frac{\text{kg}}{\text{m}^3}\times\frac{\text{m}}{\text{sec}^2}\times m$$

$$= \left[\text{kg}\times\frac{\text{m}}{\text{sec}^2}\right]\times\frac{1}{m^2} = \text{Force}\times\frac{1}{m^2}$$

$$\therefore \quad = \frac{\text{N}}{\text{m}^2} \text{ resultant unit.}$$

$$= h_a = \frac{P_d}{P_w . g}$$

$$= \frac{353\times g\times H\times\frac{P_a}{P_N}\left[\frac{1}{T_a}-\left[\frac{m+1}{m}\right]\times\frac{1}{T_f}\right]}{1000\times g} \text{ metres of } H_2O$$

$$\therefore \quad h_w = 353 \times H \times \frac{P_a}{P_w}\left[\frac{1}{T_a} - \left[\frac{m+1}{m}\right] \times \frac{1}{T_f}\right] \text{mm of water}$$

If $\quad P_a = P_N$ then,

$$h_w = 353 \times H \left[\frac{1}{T_a} - \left[\frac{m+1}{m}\right] \times \frac{1}{T_f}\right] \text{mm of } H_2O$$

To express, P_d in metres of hot flue gas

$$\therefore \quad P_d = \rho_f \cdot g \cdot h_f$$

where ρ_f = Density of hot flue gas

h_f = height of flue gas column

$$\therefore \quad P_d = \rho_f \cdot g \cdot H$$

$$H = \frac{P_d}{\rho_f \cdot g}$$

$$= \frac{353 \times g \times H \times \frac{P_a}{P_N}\left[\frac{1}{T_a}\left[\frac{m+1}{m}\right] \times \frac{1}{T_f}\right]}{g \times 353 \times \left[\frac{m+1}{m}\right] \times \frac{1}{T_f} \times \frac{P_d}{P_N}}$$

On simplification we get,

$$H' = H\left[\frac{m}{m+1} \times \frac{T_f}{T_a} - 1\right] \text{metre of hot gas column}$$

Condition for maximum discharge through the chimney.

Mass of the flue gases discharged through the chimney is proportional to the density, cross sectional areas and velocity.

i.e., $M \propto$ density $\times$ cross section area $\times$ velocity

Note. Cross section areas is constant because of construction,

And we know that,

$$\rho_f = \frac{1}{T_f}$$

$$M = \text{Constant} \times \frac{1}{T_f} \times \text{Velocity}$$

$$= k \times \frac{1}{T_f} \times \sqrt{2gH}$$

$$= \text{Constant} \times \frac{1}{T_f} \times \sqrt{H}$$

$$\text{Mass of gas} = \frac{1}{T_f} \times \left[H \times \left[\frac{m}{m+1} \times \frac{T_f}{T_a} - 1\right]\right]^{1/2} \times \text{Constant}$$

$$= \frac{1}{T_f} \times \left[\frac{m}{m+1} \times \frac{T_f}{T_a} - 1\right]^{1/2} \times \text{Constant (Since } H = \text{Constant)}$$

$$M = \left[\frac{m}{m+1} \times \frac{1}{T_f . T_a} - \frac{1}{T^2{}_f}\right]^{1/2} \times \text{Constant}$$

For maximum discharge,

$$\frac{dM}{dT_f} = 0$$

$$\frac{dM}{dT_f} = \frac{1}{2}\left[\frac{m}{m+1} \times \frac{1}{T_f . T_a} - \frac{1}{T_f^2}\right]^{1/2} \times \left[\frac{m}{m+1} \times \frac{1}{T_a} \times \frac{1}{T_f^2} - \frac{1}{T_f^3}\right] = 0$$

$$\therefore \quad \left[-\frac{m}{m+1} \times \frac{1}{T_a . T_f{}^3}\right] = 0$$

$$\frac{m}{m+1} \times \frac{1}{T_a \cdot T_f^2} = \frac{2}{T_f^3}$$

$$T_f = 2\left[\frac{m}{m+1}\right] \times T_a$$

We know that,

$$H = \left[\frac{m}{m+1} \times \frac{T_f}{T_a} - 1\right] - H$$

Substituting T_f in above equation we get,

$$H' = H[2-1] = H$$

$$\therefore \quad H' = H$$

11.16 DRAUGHT LOSSES

The actual or available draught will be less than the theoretical Draught obtained by calculations, because of the draught losses viz.

(1) Head loss due to fuelled resistance h_{fb}: Fuel bed resistance depends upon the size of the fuel, depth of fuel on grate and combustion rate.

(2) Head loss due to frictional resistance to the flow of flue gases because of the equipments like superheater, economiser, air preheater, etc. (h_e)

(3) Head loss due to friction in bends and curves in the flue gas passages (h_b).

Thus, total draught loss = $h_{\text{Total}} = h_{fb} + h_e + h_b$

11.17 INTRODUCTION TO IBR OR REGULATIONS FOR CHIMNEY (WITH REFERENCE TO IBR/MSNC BOMBAY)

To keep control on boiler accidents, Indian Government after independence made following rules and regulations, which the boiler should conform. They are as under:

1. A plan of the boiler house and chimney is to be approved by the chief inspector of Indian Boiler Regulations (IBR)/Maharashtra Chief Inspector of Nuisance Commission (MSNC)—Bombay.
2. Chimney fabrication drawings are to be approved before fabrication.
3. A plan of 1m =1cm in blue print showing the exact position of furnace, flues, chimney, duly certified by the chartered engineer/owner should be submitted.

4. Substantial structure certificate for the chimney stating that the chimney in self supported/supported by wire ropes and with stands a wind load of 207 kg/m^2 is to be submitted (Note this should be certified by the chartered engineer).
5. Chimney height minimum required as per IBR/MSNC is 100 feet that i.e., 30.48 m from the firing floor of the boiler.
6. An undertaking for the height of chimney is to be submitted by the owner to MSNC.
7. An undertaking for the use of specific fuel is to be sumitted by the owner.
8. No objection certificate (NOC) is to be obtained from Air Force Deptt.
9. Smoke generated should conform to Bombay Smoke Nuisance Act 1912.
10. Chimney should be dry painted. (i.e., Red and white strips of equal width in total 7 strips are to be painted)
11. Anti collision lights are to be fitted, for sighting in the night and poor visibility.
12. Following details are to be submitted,
    (a) Height of own factory building/boiler house.
    (b) Height of tallest building within a radius of 300 m around ground level.
    (c) Capacity of boiler.
    (d) Working pressure
    (e) Type of fuel
    (f) Quantity of fuel/hour
    (g) Type of chimney self-supported/supported by wire ropes
    (h) Material of construction.

11.18 BOILER CALCULATIONS

In case of boilers it is very important to find,

(*a*) *Heat transfer required to form steam.* We have studied steam generation process,

Let h_2 = Specific enthalpy of steam formed in kJ/kg

h_1 = Specific enthalpy of feed water in kJ/kg

Then, since the steam is generated ($P = C$) at constant pressure,

And for a constant pressure process $\delta Q = dH$

i.e., Heat supplied = Change of enthalpy,

∴ Heat transfer required to form 1 kg of steam in the boiler,

$$= (h_2 - h_1) \text{ kJ/kg}$$

If m_s is the mass of steam formed in kg, then heat transfer required to form total steam.

$$= m_s.(h_2 - h_1)\left[\text{kg}\times\frac{\text{kJ}}{\text{kg}}=\text{kJ}\right] \quad ...(1)$$

(*b*) *Energy received from fuel.* This is obtained from the mass of fuel burnt and its calorific value of fuel,

If mass of fuel used = m_f kg

And calorific value of fuel = C.V. kJ/kg

Then, Energy received $= m_f \times \text{C.V.}\left[\text{kg}\times\frac{\text{kJ}}{\text{kg}}=\text{kJ}\right] \quad ...(2)$

(*c*) *Boiler thermal efficiency.* It is the ratio of the energy received by the steam to the energy supplied by the fuel to produce steam.

Thus, Boiler Thermal Efficiency;

$$\eta_{th} = \frac{\text{Energy to steam}}{\text{Energy from fuel}}$$

$$\eta_{th} = \frac{\text{Energy received by the steam}}{\text{Energy supplied by the fuel}}$$

From Eqs (1) and (2)

$$\textbf{Boiler Thermal Efficiency } h_{th} = \frac{m_s(h_2 - h_1)}{m_f \times C.V.} \times 100\%$$

(*d*) *Boiler performance.* As we have studied, the main function of the boiler is to produce steam from feed water, with the use of heat liberated by the combustion of fuel. It is to be noted the number of kg of water evaporated per hour is not an exact measure of the performance of the boiler. Since different boilers may be operating at different at different conditions, so their comparison may be made on the basis of some reference condition known as Equivalent of Evaporation.

11.18.1 Equivalent of Evaporation from and at 100° C

"The evaporation which would be obtained if the feed water were supplied at 100°C and converted into dry and saturated steam at 100ºC, at standard atmospheric pressure, is known as equivalent of evaporation from and at 100ºC".

We know that from Eq. (1) energy received by the steam

$$= m_s (h_2 - h_1) \text{ kJ}$$

It is to be noted that, at 100°C, it is the enthalpy of evaporation which is supplied and the specific enthalpy of evaporation at 100°C, = 2256.9 kJ/kg. (From steam tables h_{fg} at 100° C and 1 atm.)

The equivalent of evaporation of a boiler from and at 100°C,

$$= \frac{m_s(h_2 - h_1)}{2256.9} \text{ kg/kg of coal.}$$

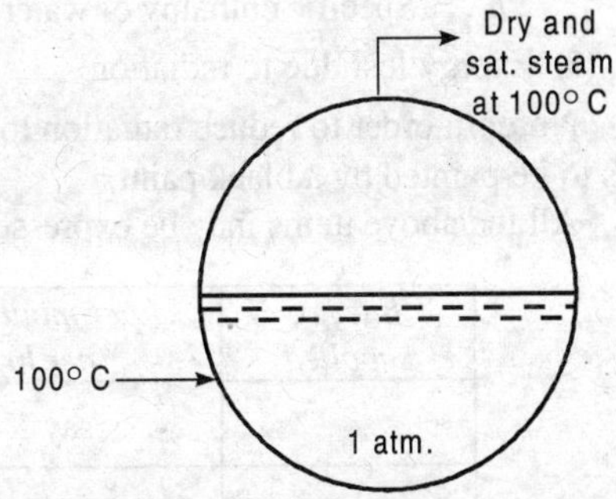

Fig. 11.27

11.18.2 Evaporative Capacity

"It is defined as the amount of water and water evaporated into steam per hour". The evaporative capacity may be expressed in kg/hr or kg/kg, of fuel or kg/hr-m^2 of heating surface area.

Actual Evaporation

It is defined as the amount of water evaporated into steam, at actual working conditions, per kg of fuel burnt.

8.19 ENERGY BALANCE

Energy released due to the combustion of fuel will be utilised for various purposes and is tabulated in heat balance sheet.

Heat balance sheet gives the account of heat energy released and heat energy utilised

(1) Energy released due to the combustion of fuel

$$= m_f \times \text{C.V.} = \text{kg} \times \frac{\text{kJ}}{\text{kg}} = \text{kJ}$$

(2) This energy released will be utilised for various purposes, viz.

(i) Heat energy utlised to form the steam

$$= m_s (h_2 - h_1)$$

$$\text{Units} = \text{kg} \times \frac{\text{kJ}}{\text{kg}} = \text{kJ}$$

(ii) Heat energy lost to the flue gases

$$= \text{Mass of flue gases} \times C_p \text{ of flue gases} \times (T_f - T_a)$$

$$= m_{f_g} \times C_{pg} \times (T_f - T_a)$$

where T_f = temperature of flue gases
and T_a = atmospheric temperature

$$\left[\text{units} = \frac{\text{kg}}{\text{kg of coal}} \times \frac{\text{kJ}}{\text{kg.K}} \times \text{K} = \frac{\text{kJ}}{\text{kg of coal}}\right]$$

(3) Energy lost due to the unburnt fuel falling through the grate bars. When solid fuels are used, some of the fuel falls through the grate bars and is lost with ash and this is

$$= m_{uf}\ C.V_{uf}$$

(4) Energy lost due to the water vapour formed due to the combustion of H_2 in the flue gases

$$= m_1 (h'_2 - h'_1)$$

where m_1 = mass of steam formed due to the combustion of H_2 in kg/kg of coal burnt.

h'_2 = Specific enthalpy of steam formed due to the combustion of H_2 in the flue gases at the temperature of flue gases and at 1 atmospheric pressure or, at 1 bar pressure or at the given partial pressure.

h'_1 = Specific enthalpy of water at the boiler room temperature

(5) Energy lost due to radiation

Note. In order to reduce radiation losses, Asbestos lagging is to be provided on the exposed boiler surface and is to be painted by a black paint.

All the above items may be expressed in the balance sheet as follows.

Energy supplied	*kJ per minutes or per hr*	*%*	*Energy utilised*	*kJ*	
	say X	100	(i) To form steam	x_1	.....
			(ii) Energy lost to, flue gases	x_2	.....
			(iii) Energy lost unburnt fuel	x_3	.....
			(iv) Energy lost to water vapour	x_4	.....
			(v) Every unaccounted by difference	x_5	.....
Total	X	100	Total	X	100

LIST OF FORMULAE

1. Draught produced in mm of H_2O

$$h_w = 353 \times H \left\{ \frac{1}{T_a} - \left[\frac{m+1}{m}\right] \times \frac{1}{T_f} \right\} \frac{P_a}{P_N}$$

where H = Height of chimney in metre

P_a = Actual pressure at boiler location.

Contd.

P_N = Standard atmospheric pressure

T_a = Actual atmospheric temperature K

T_f = Temperature of flue gases in K

m = Mass of air in kg to burn 1 kg of fuel

$\therefore$ $m+1$ = Mass of flue gases in kg/kg of fuel.

(i.e., when 1 kg of fuel is burnt with m kg of air, then total flue gases produced = $m + 1$)

2. Density of air (ρ_a)

$$\rho_a = 1.293\left[\frac{273}{T_a}\right]\frac{P_a}{P_N} \text{ in } \frac{\text{kg}}{\text{m}^3}$$

3. Density of flue gases

$$\rho_f = 1.293\left[\frac{m+1}{m}\right].\frac{273}{T_f}.\frac{P_a}{P_N}$$

or

$$\rho_f = \frac{m+1}{m}\times\frac{353}{T_f}\times\frac{P_a}{P_N} \text{ kg/m}^3$$

4. Draught pressure in terms of static pressure difference between hot gas column and a parallel cold air column.

$$P_d = g\times H\times[P_\alpha - P_f]\frac{\text{N}}{\text{m}^2}$$

and in terms of mm of water column

$$h_w = H[P_\alpha - P_f] \text{ mm of } H_2O$$

5. Velocity of flue gases = $C = \sqrt{2gH'}$

6. Equivalent height of hot gas column H_{gases} or H'

$$H' = H\left[\frac{m}{m+1}\times\frac{T_f}{T_a}-1\right] \text{ m of hot gas column.}$$

7. Condition for maximum discharge through chimney $H' = H$ and during maximum discharge conditions, Maximum temperature of flue gases,

$$T_f = 2\left[\frac{m+1}{m}\right]\times T_a$$

■■■

SOLVED EXAMPLES

(Problems on Boiler Draught)

Example 11.1 *Find the minimum height of chimney required to produce a draught of 16 mm of H_2O, if 19 kg of air is required per kg of fuel burnt on the grate. The mean temperature of flue gases inside the chimney is 330°C.*

Data: $H = ?$ $\quad h_w = 16$ mm of H_2O

$m = 19$ kg/kg of fuel $\quad T_f = 330 + 273 = 603$ K

$T_a = 30 + 273 = 303$ K

Solution

We know that, the drought produced in mm of H_2O column is given by,

$$h_w = 353 \times H \left\{ \frac{1}{T_a} - \frac{m+1}{m} \times \frac{1}{T_f} \right\} \frac{P_a}{P_N} \qquad \text{...(1)}$$

Since in the problem they have not mentioned anything about P_a and P_N

∴ Assuming $P_a = P_N$

Now substituting the values in Eq. (1) we get,

$$h_w = 353 \times H \left\{ \frac{1}{303} - \frac{20}{19} \times \frac{1}{603} \right\} \text{mm of } H_2O$$

$$16 = 353 \times H \{0.00330033 - 0.0017456577\}$$

$$\mathbf{H = 29.155 \text{ m}}$$

Example 11.2 *Readings from a boiler trial,*

(i) Height of chimney : 100 m

(ii) Ambient temperature : 32 °C

(iii) Mass of air : 18 kg/kg of fuel

(iv) Flue gas temperature : 300°C

Find natural draught in mm of H_2O

Data: $H = 100$ m $\quad T_a = 32 + 273 = 305$ K

$m = 18$ kg/kg of fuel $\quad T_f = 300 + 273 = 573$ K

$h_w = ?$

Solution

We know that, $$h_w = 353 \times H \left\{ \frac{1}{T_a} - \left[\frac{m+1}{m} \right] \times \frac{1}{T_a} \right\} \text{ mm of } H_2O$$

Assuming $P_a = P_N$

Now $$h_w = 353 \times 100 \left\{ \frac{1}{305} - \frac{19}{18} \times \frac{1}{573} \right\}$$

$$h_w = 35300 \times \{0.0014365322\}$$

$$\mathbf{h_w = 50.7096 \text{ m}}$$

Example 11.3 *A boiler is equipped with a chimney of 30.48 m height. Atmospheric conditions are 755 mm of Hg and 35°C. Average temperature of flue gases in the Chimney is 300°C. If the boiler is supplied with 20 kg of air kg of fuel.*

Calculate theoretical draught created in mm of H_2O. Take standard atmospheric pressure as 760 mm of Hg.

Solution

We know that $$h_w = 353 \times H \times \frac{P_a}{P_N} \left\{ \frac{1}{T_a} - \left[\frac{m-1}{m} \right] \times \frac{1}{T_f} \right\}$$

$$= 353 \times 30.48 \times \frac{755}{760}\left\{\frac{1}{308} - \frac{21}{20} \times \frac{1}{573}\right\}$$

$$\mathbf{h_w = 15.1168 \text{ mm of } H_2O}$$

Example 11.4 *Find the amount of air used in a boiler/kg of fuel, when chimney height is 50 m. Draught produced is 25 mm of H_2O. The temperature of flue gases is 50 m. Draught produced is 25 mm of H_2O. The temperature of flue gases is 353°C and that of room is 35°C.*

Also determine the draught produced in metres of hot gas column and flue gas temperature for maximum discharge conditions.

Data: $m = ?$ $\quad H = 50$ m

$h_w = 25$ mm of H_2O $\quad T_f = 353 + 273 = 626$ K

$T_a = 35 + 273 = 308$ K $\quad H = ?$

$T_f = ?$ for max. discharge conditions

Solution

We know that,
$$h_w = 353 \times H\left\{\frac{1}{T_a} - \left[\frac{m+1}{m}\right] \times \frac{1}{T_f}\right\} \text{mm of } H_2O$$

$$25 = 353 \times 50\left\{\frac{1}{308} - \left[\frac{m+1}{m}\right] \times \frac{1}{626}\right\} \text{mm of } H_2O$$

$$\therefore \quad \mathbf{m = 6.8595 \text{ kg} = 6.86 \text{ kg}}$$

(ii) And we know that equivalent height of hot gas column,

$$H' = H\left\{\frac{m}{m+1} \times \frac{T_f}{T_a} - 1\right\} m \text{ of hot gas column}$$

$$= 50\left\{\frac{6.86}{7.86} \times \frac{626}{308} - 1\right\}$$

$$\mathbf{H' = 38.6942 \text{ m of hot gas column}}$$

(iii) We know that for max. discharge conditions

T_f is given by,
$$T_f = 2 \times \frac{(m+1)}{m} \times T_a = 2 \times \frac{7.86}{6.86} \times 308$$

$$\mathbf{T_f = 705.7959 \text{ K}}$$

Example 11.5 *Find the minimum temperature of the flue gases required to produce a draught of 1.5 cm of water by a chimney of 30 m height, when air-fuel ratio required in the combustion is 20. Atmospheric air temperature is 27°C.*

Data: $T_f = ?$ $\quad h_w = 15$ mm

$H = 30$ m $\quad \frac{Air}{Fuel} = \frac{20}{1}$

$\therefore$ *Mass of air,* $m = 20$ kg

and mass of flue gases $m + 1 = 21$ kg/ kg of fuel

$T_a = 27 + 273 = 300$ K

Solution

We know that,
$$h_w = 353 \times H \times \left[\frac{1}{T_a} - \frac{(m+1)}{m} \times \frac{1}{T_f}\right] \text{mm of } H_2O$$

$$15 = 353\times30\times\left[\frac{1}{300}-\frac{21}{20}\times\frac{1}{T_f}\right]$$

$\therefore$ $$\mathbf{T_f = 547.75\ K}$$

Example 11.6 *How much air is used/kg of coal burnt in a boiler having a chimney of 35 m height to create a draught of 20 mm of water, when the temperature of flue gas in the chimney is 370° C and the boiler house temperature is 34°C. Does this chimney satisfy the condition of max. discharge ?*

Data: $m = ?$ $\quad H = 35\ m$

$h_w = 20\ mm\ of\ H_2O$ $\quad T_f = 643\ K$

$T_a = 307\ K$

Solution

We know that,
$$h_w = 353\times H\times\left\{\frac{1}{T_a}-\frac{m+1}{m}\times\frac{1}{T_f}\right\}\ \text{mm of } H_2O$$

$$20 = 353\times35\left\{\frac{1}{307}-\frac{m+1}{m}\times\frac{1}{643}\right\}$$

$\therefore$ $$\mathbf{m = 18.66\ kg/kg\ of\ coal}$$

For max. discharge,
$$T_f = 2\times\left[\frac{m+1}{m}\right]\times T_a$$

$$= 2\times\frac{19.66}{18.66}\times307$$

$$\mathbf{T_f = 646.905\ K}$$

Whereas actual T_f given is 643 K which is less than T_f max.

$\therefore$ Condition is satisfied.

Example 11.7 *The draught produced in a certain boiler is 18.5 mm of water when the temperature of the flue gases in the chimney is 370°C and the temperature of the boiler house is 30°C. Estimate the air fuel ratio. Also state whether chimney operates under the maximum discharge condition or not. The height of the chimney is 32 m.*

Data: $h_w = 18.5\ mm\ of\ H_2O$ $\quad T_f = 370 + 273 = 643\ K$

$T_a = 30 + 273 = 303\ K$ $\quad m = ?\ in\ \frac{kg\ of\ air}{kg\ of\ fuel}$

$H = 32\ m$

Solution

Check T_f for maximum discharge conditions.

We know that, the draught produced in mm of H_2O column is given by,

$$h_w = 353\times H\left\{\frac{1}{T_a}-\left[\frac{m+1}{m}\right]\times\frac{1}{T_f}\right\}$$

$$18.5 = 353\times32\left\{\frac{1}{303}-\left[\frac{m+1}{m}\right]\times\frac{1}{643}\right\}$$

$$\mathbf{m = 14.52\frac{kg\ of\ air}{kg\ of\ fuel}}$$

$\therefore$ Air fuel ratio is 14.52:1

Now to check for T_f maximum discharge conditions,

$$T_f = 2\left[\frac{m+1}{m}\right]T_a = 2\times\left[\frac{14.52+1}{14.52}\right]\times 303$$

$$\mathbf{T_f = 647.76\ K}$$

Since T_f actual given is 643 which is less than T_f for maximum discharge condtitions. So the chimney operates under maximum discharge conditions.

Example 11.8 *In a certain boiler installation a chimney of 42.5 m height produces a natural drought equivalent of 33.25 m of hot gas column. Boiler house temperature is 33°C. The mass of air supplied per kg of fuel is 19 kg, determine,*

(i) Temperature of gas leaving the chimney.

(ii) Temperature of flue gases and corresponding draught for maximum discharge.

Data: $H = 42.5$ m $H' = 33.25$ m of hot gas column

$T_a = 33 + 273 = 306$ K $m = 19$ kg/kg of fuel

$m + 1 = 20$ kg of gases/kg of fuel $T_f = ?$

and T_f for max discharge consitions = ?

Solution

We know that, equivalent height of hot gas column,

$$H' = H\left\{\frac{m}{m+1}\times\frac{T_f}{T_a}-1\right\} \text{ metre of hot gas column}$$

$$33.25 = 42.5\left\{\frac{19}{20}\times\frac{T_f}{306}-1\right\} \text{ metre of hot gas column}$$

$$33.25 = 42.5\,\{3.1\times 10^{-3}\,T_f - 1\}$$

$$T_f = 547.105\ \text{K}$$

We know that T_f for maximum discharge conditions,

$$T_f = 2\left[\frac{m+1}{m}\right]\times T_a$$

$$T_f = \left[\frac{20}{19}\right]\times 306$$

$$\mathbf{T_f = 644.211\ K}$$

Example 11.9 *Determine the height of chimney required for a boiler having an average coal consumption of 1500 kg/hr and producing 15 kg of dry flue gases per kg of coal fired. The pressure drop in mm of water at various stages in a boiler flue passes are (i) fuel bed 4.9 mm, (ii) Boiler flue 7.8 mm, (iii) Boiler tube passage 2.7 mm. (iv) Through bends, dampers etc. 0.6 mm.*

Actual draught may be taken as 25% greater than the theoretical draught. The flue gases have a mean temperature of 277°C and ambient temperature is. 27°C.

Does this chimney fulfill the condition of maximum discharge?

Data: $H = ?$ $m_f = 1500$ kg/hr

$$m_{fg} = (m+1) = \frac{15\ \text{kg of gases}}{\text{kg of coal}}$$

Solution

Total pressure drop or Draught = 4.9 + 7.8 + 2.7 + 0.6 = 16 mm of H_2O

Since, Actual draught in 25% more than the theoretical draught,

$$\therefore \quad h_w = 16\times 1.25 = 20\ \text{mm of } H_2O$$

$$T_f = 277°C + 273 = 500\text{ K} \quad \text{(Actual)}$$
$$T_a = 27 + 273 = 300\text{ K} \quad \text{(Mean)}$$

Now we know that the draught,

$$h_w = 353H\left\{\frac{1}{T_a} - \left[\frac{m+1}{m}\right]\times\frac{1}{T_f}\right\}$$

$$20 = 353\times H\left\{\frac{1}{300} - \frac{15}{14}\times\frac{1}{550}\right\}$$

$$H = 40.91\text{ m}$$

And the condition for max. discharge is,

$$T_f = 2\left\{\frac{m+1}{m}\right\}\times T_a$$

$$= 2\left\{\frac{15}{14}\right\}\times 300$$

$$\mathbf{T_f = 642.8571\ K}$$

As T_f given in 550 K which is within T_f maximum. So the chimney satisfies the condition for maximum discharge.

Example 11.10 *Find the draught produced in mm of water by a chimney of 40 m height. Let the mass of flue gas be 20 kg/kg of fuel burnt in the combustion chamber. The temperature of flue gases and the ambient air are 270° C and 23° C. Assuming the diameter of the chimney is 150 cm and 30% of the theoretical draught is lost in friction, find the mass of the flue gases passing through the chimney/min.*

Data: $h_w = ?$ $h = 40\text{ m}$

$m + 1 = 20$ *kg of flue gases/kg of fuel*

$T_f = 543\text{ K}$ $T_a = 296\text{ K}$

Also mass of flue gases passing through the chimney/min = ?

Solution

We know that,

$$h_w = 353H\left\{\frac{1}{T_a} - \left[\frac{m+1}{m}\right]\times\frac{1}{T_f}\right\}$$

$$= 353\times 40\left\{\frac{1}{296} - \frac{20}{19}\times\frac{1}{543}\right\}$$

$$h_w = 20.33\text{ mm}$$

For the 2nd part from the continuity equation,
Mass of flue gases discharged

$$= \text{c/s area} \times P_f \times \text{velocity of flue gases} \quad \text{...(1)}$$
$$= A\times P_f\times C$$

$$\text{Units} = \text{m}^2\times\frac{\text{kg}}{\text{m}^3}\times\frac{\text{m}}{\text{sec}} = \frac{\text{kg}}{\text{sec}} \quad \text{...(2)}$$

We also know that,

$$A = \frac{\pi}{4}D^2 = \frac{\pi}{4}\times 1.5^2 = 1.766\text{ m}^2$$

$$P_f = \frac{m+1}{m}\times\frac{353}{T_f} = \frac{20}{19}\times\frac{353}{543}$$

$$P_f = 0.6843 \frac{\text{kg}}{\text{m}^3} \quad \text{...(3)}$$

and $$c = \sqrt{2gH'}$$

To find H', we know that,

$$H' = H\left\{\frac{m}{m+1} \times \frac{T_f}{T_a} - 1\right\} \text{ metres of hot gas column.} \quad \text{...(4)}$$

$$= 40\left\{\frac{19}{20} \times \frac{543}{296} - 1\right\} = 40 \times 0.7427364$$

$$H' = 29.7095 \text{ m}$$

Since 30% of draught is lost frcition only 70% of the draught will be available

$\therefore$ Available draught $= 0.7 \times 29.7095$

H' available $= 20.7966 \text{ m}$

$\therefore$ Substitute this in Eq. (4) we get,

$$C = \sqrt{2 \times g \times H'} = \sqrt{2 \times 9.81 \times 20.7966}$$

$$C = 20.199745 \text{ m/sec}$$

Now mass of the flue gases discharged from Eq. (1)

$$= 1.766 \times 0.6843 \times 20.199745$$

$$= 24.4109 \text{ kg/sec}$$

Mass of flue gases = 1464.65 kg/min

Example 11.11 *In a boiler trial following data were noted,*

Height of chimney : 25 m

Temperature of flue gases : 320°C

Ambient temperature : 22°C

Mass of air : 17 kg/kg of coal

Atmospheric pressure : 720 mm of Hg

Calculate $h_w = ?$

Solution

We know that draught produced in mm of H_2O column is given by,

$$h_w = 353 \times H\left\{\frac{1}{T_a} - \frac{m+1}{m} \times \frac{1}{T_f}\right\}\frac{P_a}{P_N} \text{ mm of } H_2O$$

$$= 353 \times 25\left\{\frac{1}{295} - \frac{18}{17} \times \frac{1}{593}\right\}\frac{720}{760}$$

$$h_w = 13.40 \text{ mm of } H_2O$$

This problem can also be solved by using fundamental equations.

We know that, $$h_w = H[P_\alpha - P_f]$$

Volume of 17 kg of air at NTP

$$P_0V_0 = mRT_0$$

$$V_0 = \frac{17\times287\times273}{101325}$$

$$V_0 = 13.145\,\text{m}^3$$

Note. P_0 = 760 mm of Hg = 101325 N/m² and T_0 = 273

Volume of air at T_a = 22° C = 295 K

and pressure of 720 mm of Hg is 95992.105 N/m²

$$\therefore \quad \frac{P_0V_0}{T_0} = \frac{P_aV_a}{T_a}$$

$$V_a = \frac{101325\,\text{N/m}^2\times13.145\,\text{m}^3\times295\,\text{K}}{273\,\text{K}\times95992.105\,\text{N/m}^2}$$

$$V_a = 15\,\text{m}^3$$

$$\therefore \quad \rho_a = \frac{\text{Mass of air}}{\text{Volume of air at 295 K}}$$

$$\rho_a = \frac{17}{15} = 1.133\,\text{kg/m}^3$$

Note. Volume of flue gas at NTP = 13.145 m³

and to find volume of flue gases at 320ºC.

$$\frac{P_0V_0}{T_0} = \frac{P_fV_f}{T_f}$$

$$V_f = \frac{101325\times13.145\times593}{273\times95992.105}$$

$$V_f = 30.137\,\text{m}^3$$

$$\rho_f = \frac{\text{Mass of flue gases}}{\text{Volume of flue gases}}$$

$$\rho_f = \frac{18}{30.137} = 0.5972\,\text{kg/m}^3$$

$$h_w = H[\rho_\alpha - \rho_\tau]\,25\,[1.133 - 0.5972]$$

$$h_w = 13.40\,\text{mm of H}_2\text{O}$$

LIST OF FORMULAE

1. Heat transfer required to form the steam

$$= m_s (h_2 - h_1) \text{ kJ}$$

m_s = Mass of steam generated in kg

h_2 = Specific enthalpy of steam formed in kJ/kg

h_1 = Specific enthalpy of feed water in kJ/kg

2. Energy received from fuel

$$= (m_f \times \text{C.V.}) \text{ kJ}$$

where m_f = mass of fuel in kg

C.V. = Calorific value in kJ/kg

3. Boiler thermal efficiency,

$$\eta_{th} = \frac{m_s (h_2 - h_1)}{m_f \times \text{C.V.}}$$

4. Equivalent of Evaporation from and at 100°C

$$= \frac{\dot{m}_s (h_2 - h_1)}{2256.9} \quad \frac{\text{kg}}{\text{kg of coal}}$$

Note. Here $\dot{m}_s$ should be in kg/kg of fuel burnt and it is written as $\dot{m}_s$ or m_a.

PROBLEMS ON BOILER PERFORMANCE

Example 11.12 *A boiler working at a pressure of 14 bar evaporates 8 kg of water/kg of coal burnt from the feed water entering at 39°C. The steam at the steam stop valve is 0.95 dry. Determine the equivalent of evaporation from and at 2009°C.*

Data: $P = 14$ bar $\qquad \frac{m_s}{m_f} = 8$ kg of water/kg of coal $= m_s$

Feed water temperature = 39°C $\qquad x = 0.95$

Equivalent of evaporation?

Solution

We know that the equivalent of evaporation form and at 100°C

$$= \frac{\dot{m}_s (h_2 - h_1)}{2256.9} \qquad \ldots(1)$$

where h_2 is specific enthalpy if steam formed at 14 bar pressure and 0.95 dry and the steam formed is wet,

$$\therefore \quad h_2 = h_f + x \cdot h_{fg}$$

From steam tables at 14 bar pressure

$$h_f = 830.07 \text{ kJ/kg}$$

$$h_{fg} = 1957.7 \text{ kJ/kg}$$

$$\therefore \quad h_2 = 830.07 + 0.95 \times 1957.7$$

$$h_2 = 2689.885 \text{ kJ/kg} \quad ...(2)$$

and h_1= specific enthalpy of feed water at 39°C from steam tables,

$$h_1 = h_f = 163.27 \text{ kJ/kg} \quad ...(3)$$

Also $h_1 = h_f = m.C_p.\Delta t = 1 \times 4.187 \times 39°C$

$$h_1 = 163.29 \text{ kJ/kg}$$

∴ From Eq. (1),

Now, equivalent of evaporation = $\dfrac{\dot{m}_s \times (h_2 - h_1)}{2256.9}$

m_s should be in kg/kg of fuel burnt

$$= \frac{8 \times [2689.885 - 163.27]}{2256.9}$$

Equivalent of Evaporation = 8.9560548 kg of steam/kg of coal

Example 11.13 *A boiler plant supplies 5400 kg of steam /hr at 7502 and 0.98 dry from feed water at 41.5°C, when using 670 kg of coal/hr having a calorific value of 31000 kJ/kg.*

Determine,

(i) The efficiency of the boiler;

(ii) The equivalent of evaporation from and at 100°C.

Solution

We know that the thermal efficiency of the boiler is given by

$$\eta_{th} = \frac{m_s(h_2 - h_1)}{(m_f \times \text{C.V.})} \times 100\%$$

Now Actual Evaporation

or Steam raised/kg of coal = m_a or m_s= $\dfrac{m_s}{m_f} = \dfrac{5400}{670}$

$$m_a \text{ or } \dot{m}_s = 8.06 \text{ kg/kg of coal}$$

and h_2 = specific enthalpy of steam formed in kJ/kg at 7.5 bar and 0.98 dry,

$$\therefore \quad h_2 = h_f + x \,.\, h_g$$
$$= 709.3 + 0.98 \times 2055.55$$
$$= 2723.74 \text{ kJ/kg}$$

Note h_f and h_{fg} are to be found at a pressure of 7.5 bar by interpolation, and h_1 = Specific enthalpy of feed water 41.5°C = 173.7 kJ/kg

$$\therefore \quad \eta_{th} = \frac{8.06 \times (2723.74 - 173.7)}{31000} \times 100$$

$$\boldsymbol{\eta_{th} = 66.30\ \%}$$

(ii) Equivalent of evaporation from and at 100°C

$$= \frac{\dot{m}_s \times (h_2 \cdot h_1)}{2256.9}$$

$$= \frac{8.06 \times (2723.74 - 173.7)}{2256.9}$$

$$= \mathbf{9.11\ kg/kg\ of\ coal.}$$

Example 11.14 *Duration of a boiler trial the following data was obtained,*

During of trial = 8 hrs
Pressure of steam = 14 bar
Dryness fraction = 0.973
Feed water evaporated = 26700 kg
Temperature of water at inlet = 50°C
Coal used = 4260 kg
Calorific value of coal = 28900 kJ/kg
Air used/kg of coal = 17 kg
Temperature of flue gases = 344°C
Boiler room temperature = 21 °C
C_p of flue gases = 1.1 kJ/kg-K
Determine :

1. *Boiler efficiency*
2. *Equivalent of evaporation*
3. *Heat lost to flue gases in kJ/kg of coal and in percentage*

Solution

(i) Boiler efficiency

We know that,
$$\eta_{th} = \frac{m_s[h_2 - h_1]}{m_f \times \text{C.V.}} \times 100\%$$

And steam raised/kg of coal = Actual evaporation

$$\dot{m}_s = \frac{m_s}{m_f} = \frac{26700}{4260}$$

$$= 6.2676 \frac{\text{kg}}{\text{kg of coal}}$$

Specific enthaply of steam formed at a pressure of 14 bar and 0.973 dry,

$$h_2 = h_f + \times h_{fg}$$

$$= 830.3 + 0.973 \times 1957.7$$

$$h_2 = 2735.14\,\text{kJ/kg}$$

and
$$h_1 = \text{specific enthaply of feed water at 50°C}$$

$$h_1 = 209.26\,\text{kJ/kg}$$

$$\therefore \quad \eta_{th} = \frac{6.2676 \times [2735.14 - 209.26]}{28900} \times 100$$

$$\boldsymbol{\eta_{th} = 54.77\%}$$

(ii) Equivalent of evaporation

$$= \frac{\dot{m}_s[h_2 - h_1]}{2256.9} \quad \left[\text{here } m_s \text{ should be in } \frac{\text{kg}}{\text{kg of coal}} \text{ i.e., } \dot{m}_s \text{ is to be used}\right]$$

$$= \frac{6.2676 \times [2735.14 - 209.26]}{2256.9}$$

$= 7.01$ kg/kg of coal

(iii) Heat to flue gases = Mass of flue gases × C_p of flue gases × temperature rise

$$= m_g \times C_{pg} \times (T_f - T_a)$$

$$\left[\textbf{Note.}\ \text{Units} = \frac{\text{kg}}{\text{kg of coal}} \times \frac{\text{kJ}}{\text{kg K}} = \frac{\text{kJ}}{\text{kg of coal}}\right]$$

$$= 18 \times 1.1 \times (344 - 21)$$

$$= 6395.4 \frac{\text{kJ}}{\text{kg of coal}}$$

Now to get heat to flue gases in % divide by C.V.

i.e., $$= \frac{m_g \times C_{pg}(T_f - T_a)}{\text{C.V.}} \times 100$$

$$= \frac{6395.4}{28900} \times 100 = 22.129\ \%$$

Example 11.15 *During a boiler trial following readings were noted,*

Chimney height = 90 m
Feed water = 1400 kg/hr
Temperature of feed water = 30°C
Boiler working pressure of dry and saturated steam = 20 bar
Coal used = 200 kg/hr
Calorific value of coal = 28000 kJ/kg
Flue gas temperature in chimney = 300°C
Flue gas formed = 18 kg/kg of coal burnt
Ambient temperature = 32° C
C_p of products of combustion = 1.028 kJ/kg K
Find

(1) η_{th}
(2) Equivalent of evaporation
(3) Heat carried away by flue gases
(4) Natural draught produced in mm of H_2O

Solution

(i) η_{th}

We know that, $$\eta_{th} = \frac{m_s(h_2 - h_1)}{m_f \times \text{C.V.}} \times 100$$

From steam tables,
Specific enthalpy of dry and saturated steam at 20 bar
i.e., $h_2 = h_g = 2797.2$ kJ/kg
And specific enthalpy of fluid at 30° C.

$$= h_1 = h_f \text{ at } 30°C = 125.7 \text{ kJ/kg}$$

And steam generator per kg of fuel

$$\frac{m_s}{m_f} = \frac{1400}{200} = 7\frac{\text{kg}}{\text{kg of fuel}} = m_a = \dot{m}_s$$

$$\therefore \quad \eta_{th} = \frac{7\times(2797.2-125.7)}{28000}\times 100$$

$$\boldsymbol{\eta_{th} = 66.781\%}$$

(ii) Equivalent of evaporation

$$= \frac{\dot{m}_s(h_2-h_1)}{2256.9}$$

$$\text{Equivalent of evaporation} = \frac{(2797.2-125.7)}{2256.9}$$

$$= 8.2859\frac{\text{kg}}{\text{kg of coal}}$$

$$= 8.2859\times 200 \text{ kg/hr}$$

Equivalent of evaporation = 1657.18 kg/hr

(iii) Heat to flue gases = $m_g \times C_{pg} \times (T_f - T_a)$

$$= 18\frac{\text{kg}}{\text{kg of coal}}\times 1.028\frac{\text{kJ}}{\text{kg K}}\times(300-32)\text{ K}$$

$$= 4959.072\frac{\text{kJ}}{\text{kg of coal}}$$

or

$$\text{Heat of the flue gases} = 4959.072\frac{\text{kJ}}{\text{kg of coal}}\times 200\frac{\text{kg of coal}}{\text{hr}}$$

$$= 991814.4\frac{\text{kJ}}{\text{hr}}$$

or

$$= 17.71\%$$

(iv) Chimney draught

$$h_w = 353\times H\left\{\frac{1}{T_a}-\frac{m+1}{m}\times\frac{1}{T_f}\right\}$$

$$= 353\times 90\left\{\frac{1}{305}-\frac{18}{17}\times\frac{1}{573}\right\}$$

$$h_w = 45.46 \text{ mm of } H_2O$$

Note.

m = Mass of air supplied/kg of fuel

$m+1$ = Mass of flue gases in kg/kg of fuel

Example 11.16 *In a boiler trial, following readings were noted,*

Mean temperature of feed water : 15°C

Mean boiler working pressure : 1.2 MPa

Mean dryness fraction : 0.95

Mass of coal burnt/hr : 250 kg

C. V. of coal burnt : 32400 kJ/kg

Mass of water supplied to the boiler in 7 hrs and 14 minutes = 16500 kg.

Mass of water in the boiler at the end of the test was less than at the commencement by 1000 kg.

Calculate,

(i) *Actual evaporation/kg of coal*

(ii) *Equivalent of evaporation from and at 100°C*

(iii) η_{th}

Solution

(i) Water will be filled into the boiler, upto the prespecified level before firing.

Then the water is supplied continuously during the test period of 7 hrs and 14 minutes = 16500 kg and at the end of the trial water in the drum was less by 1000 kg.

∴ Total mass of steam raised in 7 hrs and 14 minutes

$$= 16500 + 1000 = 17500 \text{ kg}$$

Mass of steam raised/hr $= \dfrac{17500}{7\frac{14}{60}} = 2419.35 \text{ kg/hr (Note)}$

∴ Actual evaporation or steam raised/kg of coal

$$m_a = \frac{m_s}{m_f} = \dot{m}_s = \frac{2419.35}{250} = 9.677 \text{ kg/kg of coal}$$

(ii) Equivalent of evaporation

$$= \frac{m_s\,(h_2 - h_1)}{2256.9}$$

h_2 at a pressure of 12 bar and 0.92 dry from steam tables

$$h_2 = h_f + x h_{fg} = 798.43 + 0.95 \times 1984.3$$

$$h_2 = 2683.515 \text{ kJ/kg}$$

And h_1 at 15°C from steam tables

$$h_1 = h_f$$

$$= 62.94 \text{ kJ/kg}$$

∴ Equivalent of evaporation

$$= \frac{9.667[2683.515 - 62.94]}{2256.9}$$

$$= 11.2363 \text{ kg/kg of coal}$$

(iii) Boiler thermal efficiency,

$$\eta_{th} = \frac{m_s(h_2 - h_1)}{m_f \times C.V.} \times 100$$

$$= \frac{9.677 \times [2683.515 - 62.94]}{32400} \times 100$$

$$\eta_{th} = \mathbf{78.2695\%}$$

Example 11.17 *A boiler is supplied with 15,900 kg of water at 15°C during a trial for a period of 8 hours 20 minutes. The mass of water at the end of trial was 680 kg less than that of the commencement of trial. The steam is produced at a pressure of 125 N/cm² and is 0.9 dry. The coal having calorific value of 30,000 kJ/kg was fired at the rate of 275.8 kg per hour. Calculate*

(i) The actual evaporation
(ii) The equivalent evaporation
(iii) The thermal efficiency of boiler.

Data: $P = 125 \frac{N}{cm^2} = 12.5 \text{ bar}$

Note.
$$\left[\begin{array}{l} 100\,cm \times 100\,cm = 1000\,cm^2 = 1\,m^2 \\ P = 125\frac{N}{cm^2} = 125 \times 10000 \\ P = 1250000\frac{N}{m^2} - 12.5\,bar \end{array}\right]$$

$$m_s = 15900 \text{ kg of water in 8 hours 20 min.}$$

Solution

Mass of water at the end of test was less by 680 kg. So, this much water is also converted into steam . So total m_s = 15900 + 680 = 16500 kg of water in 8 hr and 20 min.

$$C.V. = 30000\frac{kJ}{kg}$$

$$m_f = 275.8\frac{kg}{h_r}$$

Calculate :

(i) Actual evaporation $= m_a = \frac{m_s}{m_f} = m_s$

(ii) Equivalent of evaporation

(iii) η_{th}

Now, mass of steam produced/hr

$$= \frac{16500}{8\frac{20}{60}}$$

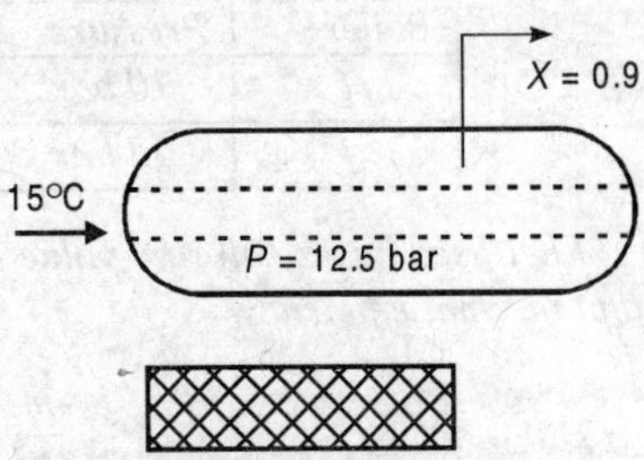

Fig. Ex. 11.17

$$m_s \doteq 1980\,\text{kg/hr}$$

(i) Now actual evaporation

$$m_a = \frac{m_s}{m_f} = \dot{m}_s$$

$$m_a = \frac{1980}{275.8}$$

$$m_a = 7.1791 \frac{\text{kg of steam}}{\text{kg of fuel}}$$

(ii) Equivalent of evaporation

$$= \frac{\dot{m}_s (h_2 - h_1)}{2256.9}$$

and $$h_2 = h_f + \times h_{fg} \text{ at } 12.5 \text{ bar and } 0.9 \text{ dry}$$

$$= 806.7 + 0.9 \times 1977.5$$

$$h_2 = 2586.45\,\text{kJ/kg}$$

and $$h_1 = h_f \text{ at } 15°\text{C} = 62.9\,\text{kJ/kg}$$

Now from Eq. (1)
Equivalent of evaporation

$$= \frac{1980}{275.8} \times \frac{2586.45 - 62.9}{2256.9}$$

(iii) Thermal efficiency η_{th}

$$= \frac{m_s (h_2 - f_1)}{m_f \times \text{C.V.}} \times 100\,\%$$

$$= \frac{1980 \times (2586.45 - 62.9)}{275.8 \times 30000} \times 100$$

$$\boldsymbol{\eta_{th} = 60.389\,\%}$$

Example 11.18 *Following readings were obtained during a trial on two boilers.*

Boiler	*Pressure*	*Quality and steam*	*Evaporation rate*	*Feed water temp.*
I	*10 bar*	*0.9 dry*	*8.5 kg/kg of fuel*	*40°C*
I	*15 bar*	*300°C*	*8.0 kg/kg of fuel*	*60°C*

Fuel used has a calorific value of 30,000 kJ/kg. Compare these boilers in respect of equivalent evaporation and thermal efficiency.

Data. $$\frac{m_s}{m_f} = 8.5 \frac{\text{kg of steam}}{\text{kg of fuel}} = m_a = \dot{m}_s$$

$$C.V. = 30{,}000\,\frac{kJ}{kg}$$

Equivalent of evaporation = ?

$$\eta_{th\,I} = ?$$

Solution

I –Boiler We know that equivalent of evaporation from and at 100°C

$$= \frac{\dot{m}_s(h_2 - h_1)}{2256.9} \qquad ...(1)$$

$$h_2 = h_f + xh_{fg} \text{ at 10 bar and 0.98 dry}$$

$$= 762.6 + 0.98 \times 2013.6$$

$$h_2 = 2735.928 \text{ kJ/kg}$$

and $h_1 = h_f \text{ at } 40°\text{C} = 167.5 \text{ kJ/kg}$

Now from Eq. (1)

$$= \frac{8.5\dfrac{\text{kg of steam}}{\text{kg of fuel}} \times (2735.928 - 167.5)}{2256.9}$$

$$= \frac{21831.638}{2256.9}$$

$$\textbf{Equivalent of evaporation} = \mathbf{9.673}\,\frac{\textbf{kg of steam}}{\textbf{kg of fuel}}$$

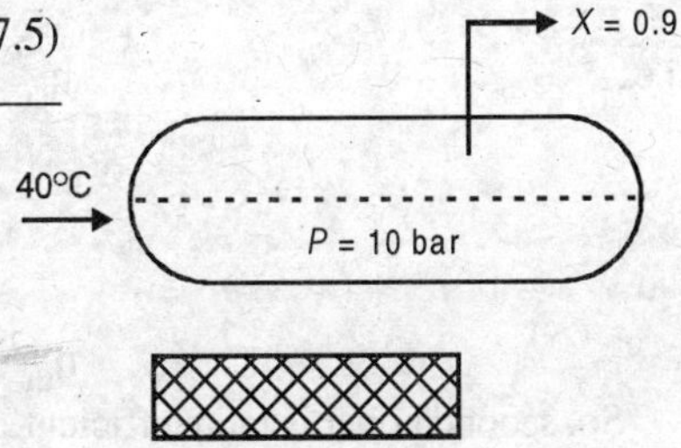

Fig. Ex. 11.18 (a)

And Boiler thermal efficiency,

$$\eta_{th} = \frac{m_s(h_2 - h_1)}{m_f \times \text{C.V.}} \times 100$$

$$= \frac{m_s}{m_f} \times \frac{[h_2 - h_1]}{\text{C.V.}} \times 100\%$$

$$= \frac{8.5 \times (2735.928 - 167.5)}{30000} \times 100$$

$$\boldsymbol{\eta_{th} = 72.772\%}$$

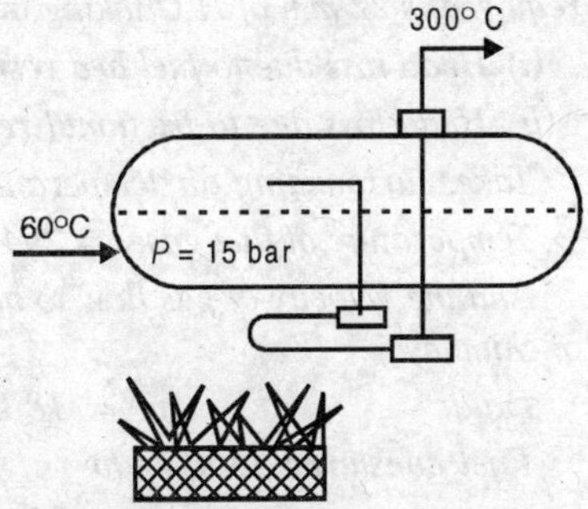

Fig. Ex. 11.18 (b)

II–Boiler $\dfrac{m_s}{m_f} = \dfrac{8\,\text{kg of steam}}{\text{kg of fuel}} = m_a = \dot{m}_s$

At 15 bar pressure T_{sat} is 198.3, and the given temperature of steam is 300° C, so the steam is superheated.

Now $h_2 = h_{sub}$ at 15 bar and 300°C

$$= 3038.9 \text{ kJ/kg}$$

and $h_1 = h_f \text{ at } 60°\text{C} = 251.1 \text{ kJ/kg}$

Now equivalent of evaporation from Eq. (1),

$$= \frac{\dot{m}_s(h_2 - h_1)}{2256.9}$$

$$= \frac{8\dfrac{\text{kg of steam}}{\text{kg of fuel}}\times(3038.9-251.1)}{2256.9}$$

$$= \frac{22302.4}{2256.9}$$

Equivalent of evaporation $= \mathbf{9.8818\dfrac{kg\ of\ steam}{kg\ of\ fuel}}$

and Thermal efficiency from Eq. (2),

$$\eta_{th} = \frac{\dot{m}_s(h_2-h_1)}{m_f\times \text{C.V.}}\times 100\%$$

i.e.,

$$\eta_{th} = \frac{m_s}{m_f}\times\frac{(h_2-h_1)}{\text{C.V.}}\times 100\%$$

$$= \frac{8\times(3038.9-251.1)}{30000}\times 100$$

$$\eta_{th} = \mathbf{74.34\%}$$

So, second boiler is more efficient.

Example 11.19 *Design for principal dimensions i.e., determine the height and diameter of a chimney of a boiler which uses 18 kg of air/kg of coal burnt, when the coal is consumed at the rate of 1800 kg/hr. Actual draught required is 22 mm of H_2O taking into account all losses viz.,*

(i) Head loss due to fuel bed resistance

(ii) Head loss due to frictional resistance because of equipments and various bends/curves

Take, surrounding air temperature = 37°C

Temperature of flue gases = 294°C

And the velocity of gas flow to be 35% of the theoretical velocity because of roughness at the interior surface of chimney.

Data: $m = 18$ *kg/kg of coal*

Fuel consumed 1800 kg/hr

$h_w = 22$ mm of H_2O $\qquad T_a = 37 + 273 = 310$ K

$T_f = 294 + 273 = 567$ K $\qquad C_{actual} = 0.35 \times C_{theoretical}$

$H = ?$ and $D = ?$

Solution

We know that

$$h_w = 353\times H\left\{\frac{1}{T_a}-\frac{m+1}{m}\times\frac{1}{T_f}\right\}$$

$$22 = 353\times\left\{\frac{1}{310}-\frac{19}{18}\times\frac{1}{567}\right\}$$

$$\mathbf{H = 45.68607\ metres}$$

We also know that, $\rho_f = \frac{m+1}{m} \times \frac{353}{T_f}$

$$\rho_f = \frac{19}{18} \times \frac{353}{567} = 0.6572 \text{ kg/m}^3$$

And note mass flow rate of flue gases = $(m + 1) \times$ Mass of coal burnt/hr

$$\left[\textbf{Note.} \text{ Units} = \frac{\text{kg}}{\text{kg of coal}} \times \frac{\text{kg of coal}}{\text{hr}} = \frac{\text{kg}}{\text{hr}} = \frac{\text{kg}}{\text{hr}} \times \frac{1}{3600} = \text{kg/sec}\right]$$

∴ Mass flow rate of the flue gases

$$= 19 \times \frac{1800}{3600} = 9.5 \text{ kg/sec}$$

We also know that, $C = \sqrt{2 g H'}$

To find H' $= H\left\{\frac{m}{m+1} \times \frac{T_f}{T_a} - 1\right\}$

$$= 45:687\left\{\frac{18}{19} \times \frac{567}{310} - 1\right\}$$

$$\mathbf{H' = 33.479 \text{ m of hot gas column}}$$

Since $C_{actual} = 0.35$

$$C_a = 0.35\sqrt{2 \times 9.81 \times 33.4799}$$

$$C_a = 8.97008 \text{ m/sec}$$

From continuity equation, we know that,

$$\text{Mass flow rate} = \rho_f \times A \times C_f$$

$$9.5 \frac{\text{kg}}{\text{sec}} = 0.6572 \frac{\text{kg}}{\text{m}^3} \times A \times 8.97 \frac{\text{m}}{\text{sec}}$$

$$A = 1.6115122 \text{ m}^2 = \frac{\pi}{4} D^2$$

$$\mathbf{D = 1.4324 \text{ m}}$$

Example 11.20 *Following data was available in a boiler trial.*

Feed water = 2400 kg/hr

Ambient temperature = 33° C

Temperature of feed water = 42°C

Fuel used = 205 kg/hr

Constituents of fuel,

Carbon 84%, Hydrogen 9.27% and Oxygen = 6.73%

Calorific value of fuel = 39500 kJ/kg

Average chimney gas temperature = 307°C

Height of chimney = 32 m
Steam pressure = 11.054 bar (Gauge)
Barometric pressure = 710 mm of Hg
Air supplied = 50% in excess of stoichiometric (theoretical) air,
Steam condition = 0.96 dry.
Determine
(i) *Boiler efficiency*
(ii) *Equivalent of evaporation from and at 100° and*
(iii) *Draught produced in min of H_2O.*

Solution

Since in the problem Gauge pressure is given, we know that

$$P_{abs} = P_{atm} + P_g$$

Since $760 \text{ mm of Hg} = 1.01325 \text{ bar}$

$710 \text{ mm of Hg} = ?$

$$\therefore \quad \frac{710}{760} \times 1.01325 = P_{atm}$$

$$\therefore \quad P_{atm} = 0.9465888 \text{ bar}$$

$$\therefore \quad P_{abs} = 12.00 \text{ bar i.e., } (P_{atm} + P_g)$$

(i) We know that $\eta_{th} = \dfrac{m_s(h_2 - h_1)}{m_f \times \text{C.V.}} \times 100\%$

Now specific enthalpy of steam formed, h_2 at 12 bar and 0.96 dry

$$\therefore \quad h_2 = h_f + xh_{fg}$$

$$= 798.43 + 0.96 \times 1984.3$$

$$h_2 = 2703.3 \text{ kJ/kg}$$

And specific enthalpy of feed water at 42°C

$$h_1 = 175.81 \text{ kJ/kg}$$

Actual evaporation $ma = \dot{m}_s = \dfrac{m_s}{m_f} = \dfrac{2400}{205}$

$$= 11.7073 \frac{\text{kg of steam}}{\text{kg of fuel}}$$

$$\therefore \quad \eta_{th} = \frac{11.7073(2703.3 - 175.81)}{39500} \times 100$$

$$\boldsymbol{\eta_{th} = 74.912\%}$$

(ii) Equivalent of evaporation from and at 100°C

$$= \frac{\dot{m}_s(h_2 - h_1)}{2256.9} = \frac{11.7073(2703.3 - 175.81)}{2256.9}$$

$$= 13.1 \text{ kg/kg of coal}$$

(iii) Mass of theoretical air required to burn 1 kg of fuel. From fuels and combustion chapter,

$$= \frac{100}{23}\left[\frac{8}{3}C + 8H - O_2\right]$$

$$= \frac{100}{23}\left[\frac{8}{3}\times 0.84 + 8\times 0.0927 - 0.0673\right]$$

$$= 12.6709 \text{ kg/kg of coal}$$

∴ As given in the example

Actual air supplied $= 1.5 \times$ theoretical air

$$m = 12.6709 \times 1.5$$

$$m = 19.006 \text{ kg/kg of fuel}$$

$$\therefore \quad h_w = 353 \times H\left\{\frac{1}{T_a} - \frac{m+1}{m}\times\frac{1}{T_f}\right\}$$

$$= 353 \times H\left\{\frac{1}{306} - \frac{20.006}{19.006}\times\frac{1}{580}\right\}$$

$$\mathbf{h_w = 16.415 \text{ mm of } H_2O}$$

Example 11.21 *Equivalent of evaporation = 10.4 kg/kg of fuel. Boiler products 15 T/hr of steam at 2.00 MN/m². Feed water temperature is 40°C. Fuel consumption is 1650 kg/hr find boiler efficiency and condition of steam produced. Take calorific value = 29800 kJ/kg.*

Data: Equivalent of evaporation = 10.4 kg/kg of coal

$$m_s = 1500 \text{ kg/hr}$$

Pressure = 20 bar

Temperature of feed water at the inlet = 40°C

$$\eta_{th} = ? \qquad x = ?$$

$$C.V. = 29800 \text{ kJ/kg}$$

Solution

We know that,

$$\text{Equivalent of evaporation} = \frac{\dot{m}_s(h_2 - h_1)}{2256.9}\ \frac{\text{kg}}{\text{kg of coal}}$$

$$10.4 = \frac{1500}{1650}\times\frac{(h_2 - h_1)}{2256.9}$$

$$(h_2 - h_1) = 2581.894\frac{\text{kJ}}{\text{kg}} \qquad ...(1)$$

We also know that,

$$\eta_{th} = \frac{m_s(h_2 - h_1)}{m_f \times \text{C.V.}}\times 100$$

$$\eta_{th} = 78.76\%$$

To determine the condition of steam.

We have already found that

$$(h_2 - h_1) = 2581.894$$

and h_1 is the specific enthalpy of feed water at 40°C = 167.45 kJ/kg from steam tables

$$\therefore \quad (h_2 - h_1) = 2581.894$$

$$h_2 = 2581.894 + 167.45$$

$$h_2 = 2749.33 \text{ kJ/kg}$$

At a pressure of 20 bar

$$h_2 = 908.6 \text{ kJ/kg}$$

$$h_{fg} = 1.888.7 \text{ kJ/kg}$$

$$\therefore \quad h_2 = h_f + xh_{fg}$$

$$2749.334 = 908.6 + \times 1887.7 \text{ kJ/kg}$$

$$\mathbf{x = 0.9746}$$

Example 11.22 *The observation recorded during a boiler trial are given below,*

Duration : 60 min
Steam generated : 5250 kg
Coal burnt : 695 kg
Calorific value of coal burnt : 30200 kJ/kg
Boiler pressure : 12 bar
Dryness fraction : 0.94
Temperature of steam leaving the superheater : 250 °C
Temperature of hot well : 45°C
Calculate without superheater
(i) Equivalent of evaporation (ii) η_{th} of boiler
Also calculate with superheater
(i) Equivalent of evaporation in kg/kg of coal (ii) η_{th} of boiler
(iii) Heat supplied by the superheater

Solution

(i) Without superheater

We know that equivalent of evaporation, $= \dfrac{\dot{m}_s (h_2 - h_1)}{2256.9} \dfrac{\text{kg of steam}}{\text{kg of coal}}$

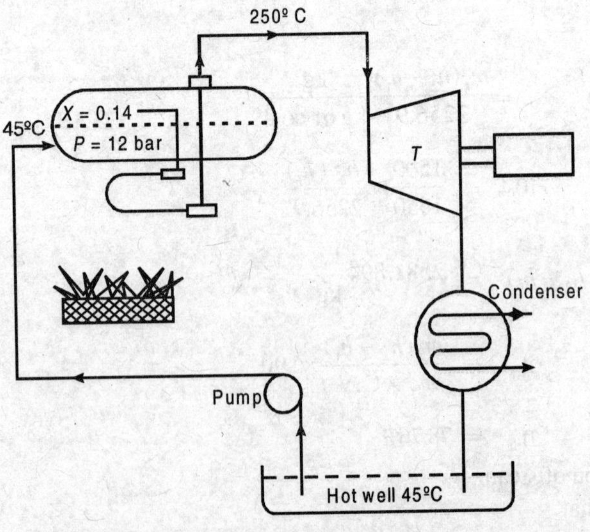

Fig. Ex. 11.22

Also steam raised/kg of coal $= \dfrac{m_s}{m_f} = m_a = \dot{m}_s$

$$= \frac{5250}{695} = 7.554 \text{ kg/kg of coal}$$

h_2, corresponding to a pressure of 12 bar and 0.94 dry

$$h_2 = h_f + x \cdot h_{fg}$$

$$= 798.43 + 0.94 \times 1984.3$$

$$h_2 = 2663.672 \text{ kJ/kg}$$

And h_1 corresponding to a temperature of hot well 45°C = 188.35 kJ/kg
(Since from the hot well water goes to the boiler drum)
∴ Equivalent of evaporation

$$= \frac{7.554 \times [2663.672 - 188.35]}{2256.9}$$

$$= 8.2851 \text{ kg/kg of coal}$$

(ii) Boiler Thermal Efficiency,

$$\eta_{th} = \frac{m_s (h_2 - h_1)}{m_f \times \text{C.V.}} \times 100$$

$$= \frac{7.554 \times [2663.67 - 188.35]}{30200} \times 100\%$$

$$\eta_{th} = 61.91\ \%$$

(iii) With superheater
Here h_2 should be corresponding to a pressure of 12 bar and temperature of 250°C from superheated steam tables,

$$h_2 = 2935.4 \text{ kJ/kg}$$

and Note h_1 will be the same.
∴ Now,
(i) Equivalent of evaporation,

$$= \frac{7.554 \times [2935.4 - 188.35]}{2256.9}$$

$$= 9.1945 \text{ kg/kg of coal}$$

(ii)

$$\eta_{th} = \frac{m_s (h_2 - h_1)}{m_f \times \text{C.V.}} \times 100\%$$

$$= \frac{7.554 \times [2935.4 - 188.35]}{30200} \times 100$$

$$\eta_{th} = 68.71\ \%$$

(iii) Heat added in the superheater

$= \text{Enthalpy of superheated steam} - \text{Enthalpy of wet steam}$

$= h_{sub} - h_{wet}$

$= (2935.4 - 2663.672)\ \text{kJ/kg}$

$$= 271.728 \frac{\text{kJ}}{\text{kg of steam}}$$

∴ Total heat added in the superheated/hr

$$= 271.728 \frac{\text{kJ}}{\text{kg of steam}} \times 5250 \frac{\text{kg of steam}}{\text{hr}}$$

$= 1426572\ \text{kJ/hr}$

Example 11.23 *The observation recorded during a boiler trial are as under:*

Duration : 60 min
Steam generated : 35,500 kg
Steam pressure : 12 bar
Steam temperature : 250°C
Temperature of water entering economiser = 17°C
Temperature of water leaving economiser = 77°C
Fuel burnt = 3460 kg
Calorific value of fuel used = 39500 kJ/kg
Calculate:

(*1*) *Equivalent of evaporation from and at 100°C per in kg per kg of fuel.*
(*2*) *Thermal efficiency of the plant.*
(*3*) *The % of heat energy of the fuel utilised by the economiser.*

Solution

We know that,

$$\text{(i) Equivalent of evaporation from and at 100°C} = \frac{\dot{m}_s (h_2 - h_1)}{2256.9} \frac{\text{kg of steam}}{\text{kg of fuel}}$$

Here h_2 is the specific enthalpy of steam formed at 12 bar and 250°C. The steam is superheated so h_2 from superheated steam tables = 2935.4 kJ/kg.

And h_1 at the feed water inlet temperature of economiser i.e.,

$$17°\text{C} = 71.31\ \text{kJ/kg}$$

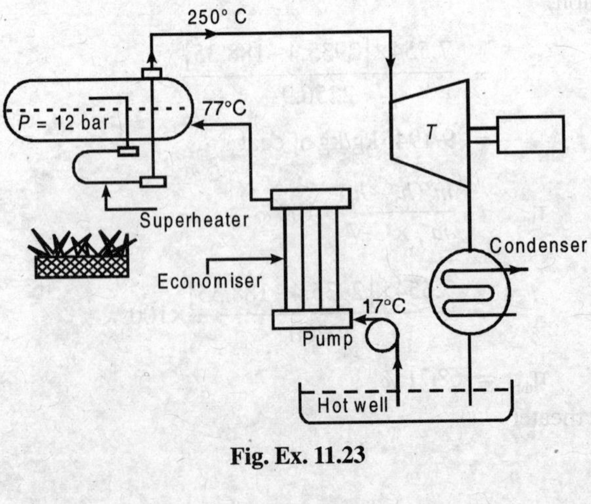

Fig. Ex. 11.23

And h_1 at the economiser outlet temperature i.e., inlet temperature of the water to the boiler

i.e., $77°C = 322.33$ kJ/kg

$$\text{Now steam raised/kg of coal} = \frac{m_s}{m_f} = \frac{35500}{3460} = 10.26 = \dot{m}_s = m_a$$

$$\text{Equivalent of evaporation} = \frac{\dot{m}_s(h_2 - h_1)}{2256.9}$$

$$= \frac{10.26 \times (2935.4 - 322.33)}{2256.9}$$

$$= 11.879 \text{ kg /kg of coal}$$

Note. Here h_1 should be the enthaply of water entering the drum, from the definition of equivalent of evaporation.

(ii) Thermal efficiency of the plant

$$\eta_{th} = \frac{m_s(h_2 - h_1)}{m_f \times \text{C.V.}} \times 100$$

i.e.,

$$\eta_{th} = \frac{m_s(h_2 - h_1)}{m_f \times \text{C.V.}} \times 100\%$$

$$= \frac{10.26 \times (2935.4 - 71.31)}{39500}$$

$$= 74.395\%$$

Note. Since η_{th} of the plant is to be found, h_1 should correspond to the temperature of water entering the economiser

(iii) The % of used in the economiser

= Mass of water passing through Economiser/kg of coal

$$\times \frac{C_p \text{ of water} \times \text{temperature rise}}{\text{C.V.}} \times 100\%$$

$$\left[\textbf{Note.} \text{ Unit} = \frac{\frac{\text{kg}}{\text{kg of coal}} \times \frac{\text{kJ}}{\text{kg K}} \times \text{K}}{\frac{\text{kJ}}{\text{kg of coal}}} = \frac{\text{kJ/kg of coal}}{\text{kJ / kg of coal}} \right]$$

$$= \frac{10.26 \times 4.187 \times (77.17)}{39500} = 6.52\%$$

Example 11.24. *During a boiler trial, following readings were noted*

Steam pressure = 15 bar

Amount of water evaporated = 5400 kg/hr

Condition of steam = 0.92

Feed water is supplied at 50°C

Determine the rate of coal consumption in kg/hr if η_{th} = 70%
Assume C.V. of coal = 40000 kJ/kg.
Also determine the equivalent of evaporation from and at 100°C in kJ/hr
If the superheater is used then the temperature of steam leaving the super heater is 250° C. Find η_{th} = ? Assume C_p for superheated steam as 2.27 kJ/kg-K.

Solution

We know that,

$$\eta_{th} = \frac{m_s (h_2 - h_1)}{m_f \times \text{C.V.}} \times 100 \quad \text{...(1)}$$

Now, h_2 = specific enthaply of steam formed at 15 bar and 0.92 dry.

$$= h_f + x \cdot h_{fg}$$
$$= 844.66 + 0.92 \times 1945.2$$
$$h_2 = 2634.244 \text{ kJ/kg}$$

And h_1 at 50°C $= 209.26 \text{ kJ/kg} = h_f$

∴ From Eq. (1),
$$\eta_{th} = \frac{\dot{m}_s (h_2 - h_1)}{m_f \times \text{C.V.}} \times 100$$

∴
$$m_f = \frac{5400 \text{ kg/hr } (2634.24 - 209.26)}{0.7 \times 40000}$$

$$m_f = 467.675 \text{ kg/hr}$$

(ii) Equivalent of evaporation,

$$= \frac{\dot{m}_s (h_2 - h_1)}{2256.9}$$

Note. Here m_s should be in kg/hr so as to get Equivalent of Evaporation in $\frac{\text{kg of steam}}{\text{hr}}$

$$= \frac{5400(2634.24 - 209.26)}{2256.9} = 5802.159 \text{ kg/hr}$$

(iii) With superheater, since C_p of superheated steam is given,

∴
$$h_2 = h_g + c_p (T_{sup} - T_{sat}) \text{ at 15 bar}$$
$$h_2 = 2789.9 + 2.27(250 - 198.29)$$
$$h_2 = 2907.282 \text{ kJ/kg}$$

$$\eta_{th} = \frac{m_s (h_2 - h_1)}{m_f \times \text{C.V.}} \times 100 \quad \frac{5400(2907.28 - 209.26)}{467.675 \times 40000}$$

$$= 77.88\ \%$$

Example 11.25 *During a boiler trial on fire tube boiler, following readings were noted,*
Duration of trial = 24 hrs
Amount of water evaporated = 20 T
Feed water temperature = 50°C
Condition of steam = 0.92

Absolute pressure = 15 bar
Coal burnt = 6T
Grate area = 4 m²
Heating surface = 3200 m²
C.V. of coal = 20000 kJ/kg
Calculate:

(1) *Mass of coal burnt in kg per m² of grate area per hr* $\left[i.e., \frac{kg}{m^2\text{-}hr} \right]$

(2) *Equivalent of evaporation from and at 100° C in kg/kg of coal.*

(3) η_{th}

(4) *Equivalent of evaporation in kg/m² of heating surface per hour.*

Solution

$$\text{Now, mass of coal burnt /hr} = \frac{6000\,\text{kg}}{24\,\text{hrs}} = 250\,\text{kg/hr}$$

$$\text{And steam reduced/hr} = \frac{20000}{24}$$

$$= 833.33\ \text{kg/hr}$$

Therefore,

(1) Mass of coal burnt in kg/m² of grate area/hr

$$= \frac{250\ \text{kg/hr}}{4\text{m}^2} = 62.5\,\text{kg/m}^2\text{-hr.}$$

(2) Mass of steam produced/kg of coal

$$= \frac{20000}{6000} = 3.333\,\text{kg/kg of coal}$$

∴ Equivalent of evaporation from and at 100°C in kg/kg of coal

$$= \frac{\dot{m}_s\,(h_2 - h_1)}{2256.9} \qquad \text{...(1)}$$

Now from Eq. (1),

And h_2 = specific enthalpy of steam formed at 15 bar and 0.92 dry

$$= h_f + x \cdot h_{fg}$$

$$= 844.66 + 0.92 \times 1945.2 = 2634.244\ \text{kJ/kg}$$

$$h_1 \text{ at } 50°\text{C} = 209.26\,\text{kJ/kg}$$

Now from Eq. (1), Equivalent of Evaporation,

$$= \frac{3.333(2634.24 - 209.26)}{2256.9}$$

$$= 3.5815795\ \text{kg/kg of coal}$$

(3)
$$\eta_{th} = \frac{m_s\,(h_2 - h_1)}{m_f \times \text{C.V.}} \times 100$$

$$= \frac{3.33(2634.24 - 209.26)}{20000} \times 100$$

$= 40.41\,\%$

(4) Mass of steam produced per m² of heating surface/hr

$$= \frac{833.33 \text{ kg/hr}}{3200\,\text{m}^2 \text{ of heating surface area}}$$

$$= 0.2604156 \frac{\text{kg}}{\text{hr - m}^2}$$

∴ Equivalent, of evaporation in $\frac{\text{kg}}{\text{hr - m}^2}$

$$= \frac{m_s\,(h_2 - h_1)}{2256.9}$$

∴ Equivalent, of evaporation in $\frac{\text{kg}}{\text{hr - m}^2}$

$$= \frac{0.2604156\,(2634.24 - 209.26)}{2256.9}$$

Equivalent of evaporation = 0.2798097 kg/hr-m²

PROBLEMS ON ENERGY BALANCE

Example 11.26 *The following results were obtained in a boiler trial,*

Feed water/hr : 700 kg
Feed water inlet temperature : 27°C
Steam produced at a pressure : 8 bar
Dryness fraction : 0.97
Coal used : 100 kg/hr
C.V. of coal : 25000 kJ/kg
Ash and unburnt coal collected from beneath the grate bars = 7.25 kg/hr
C. V. of unburnt fuel = 2000 kJ/kg
Flue gases formed / kg of fuel = 17.3 kg
Flue gas temperature = 325° C
Temperature of air in the room = 16° C
C_p of flue gases = 1. 025 kJ/kg-K
Draw up energy balance on minute basis and find the boiler efficiency.

Solution

(1) Energy supplied/released due to the combustion of fuel

$= m_f \times \text{C.V. of fuel}$

$= 100 \text{ kg/hr} \times 25000 \text{ kJ/kg}$

$= 25{,}00{,}000 \text{ kJ/hr}$

$= 41666.667 \text{ kJ/min} = 100\% \text{ (Say)}$

(2) Energy utilised.

(i) Energy utilised to form the steam

$$= m_s (h_2 - h_1)$$

where,

h_2 = specific enthalpy of steam formed in kJ/kg corresponding to a pressure of 8 bar and since the steam is wet.

$$h_2 = h_f + x \cdot h_{fg} = 720.9 + 0.97 \times 2046.5$$

$$\therefore \quad h_2 = 2706.005$$

And h_1 specific enthaply of feed water at 27°C

$$= 113.13 \text{ kJ/kg}$$

$\therefore$ Energy or Heat to form steam

$$= m_s (h_2 - h_1)$$
$$= 700 \text{ kg/hr} (2706.005 - 113.13) \text{ kJ/kg}$$
$$= 1815012.5 \text{ kJ/kg}$$
$$= 30250.21 \text{ kJ/min}$$

Total energy released $= 41666.67 \text{ kJ/min} = 100\%$

for 30250.21 kJ/min $= ?$

$$= \frac{30250.21}{41666.67} \times 100 = 72.60\% = \eta_{th}$$

(ii) Heat to flue gases

$$= \text{Mass of flue gases} \times C_p \text{ of flue gases} \times (T_f - T_a)$$
$$= 17.3 \times 1.025 \times (325 - 16)$$
$$= 5479.3425 \text{ kJ/kg of coal}$$

Since 100 kg of coal is burnt/hr

$$\therefore \quad = 5479.3425 \frac{\text{kJ}}{\text{kg of coal}} \times 100 \frac{\text{kg of coal}}{\text{hr}}$$

$$= 547934.25 \text{ kJ/hr}$$
$$= 9132.2375 \text{ kJ/min}$$

$$\therefore \quad \text{In } \% = \frac{9132.24}{41666.67} \times 100$$

(iii) Heat lost to unburnt fuel

$$= m_{uf} \times \text{C.V. of unburnt fuel}$$
$$= 7.25 \text{ kg/hr} \times 2000 \text{ kJ/kg}$$
$$= 14500 \text{ kJ/hr} = 241.667 \text{ kJ/min}$$
$$= 0.58\%$$

(iv) Heat unaccounted by difference

$$= \text{Total heat generated} - \text{Total heat utilised}$$
$$= 41666.67 - (30250.21 + 9132.24 + 241.67)$$
$$= 2024.55 \text{ kJ/min}$$
$$= 4.1\%$$

Energy Supplied	*kJ/min*	*%*	*Energy utilised*	*kJ/min*	*%*
	41666.67	100%	(i) To steam (ii) To flue gases (iii) To unburnt fuel (iv) To unaccounted	30250.21 9132.24 241.67 2042.55	72.60 21.92 0.58 4.10
	41666.67	100%		41666.67	100%

Example 11.27 *The following data were obtained in a boiler trial,*

Feed water supplied : 680 kg/hr
Temperature of feed water 20° C
Steam pressure : 15 bar
Steam temperature : 300° C
Coal used : 95 kg/hr
C.V. of coal used : 34000 kJ/kg
Ash and unburnt coal collected in the ash pit = 4 kg/hr
C.V. of unburnt coal = 2200 kJ/kg
Flue gas formed = 18 kg/kg of coal burnt
Flue gas temperature = 343° C
Ambient temperature = 28° C
Mean C_p of flue gases = 1.025 kJ/kg-K
Determine:
(1) η_{th}
(2) Equivalent of evaporation from and at 100°C
(3) % of heat unaccounted
(4) Drought produced in mm of H_2O, if the chimney height is 50 m.

Solution

(1) We know that,

$$\eta_{th} = \frac{m_s(h_2 - h_1)}{m_f \times \text{C.V.}} \times 100\%$$

At 15 bar and 300°C, from superheated steam tables,

$$h_2 = 3038.9 \text{ kJ/kg}$$

and

$$h_1 \text{ at } 20°\text{C} = 83.86 \text{ kJ/kg} = h_f$$

$$\eta_{th} = \frac{680 \times (3038.9 - 83.86)}{95 \times 34000} \times 100$$

$$\mathbf{\eta_{th} = 62.21\%}$$

(2) Equivalent of evaporation from and at 100°C

$$= \frac{\dot{m}_s (h_2 - h_1)}{2256.9}$$

$$= \frac{680}{95} \times \frac{(3038.9 - 83.86)}{2256.9}$$

$$\textbf{Equivalent of Evaporation} = 9.3721 \frac{\textbf{kg}}{\textbf{kg of coal}}$$

(3) (A) Heat supplied from coal

$$= m_f \times \text{C.V.}$$
$$= 95 \text{ kg/hr} \times 34000 \text{ kJ/kg}$$
$$= 32{,}30{,}000 \text{ kJ/hr}$$
$$= 53833.333 \text{ kJ/min}$$
$$= 100\% \text{ (Say)}$$

(B) (i) Heat to steam $= m_s (h_2 - h_1)$

$$= 680 (3038.9 - 83.89) \text{ kJ/hr}$$
$$= 2009406.8 \text{ kJ/hr}$$
$$= 33490.113 \text{ kJ/min}$$
$$= 62.210\%$$

(ii) Heat carried away by the flue gases,

$$= m_{\text{flue gas}} \times C_{p \text{ of flue gases}} \times (T_f - T_a)$$
$$= 18 \times 1.025 (343 - 28)$$
$$= 5811.77 \frac{\text{kJ}}{\text{kg of coal}}$$
$$= 5811.75 \text{ kJ/kg of coal} \times 95 \text{ kJ of coal/hr}$$
$$= 552116.25 \text{ kJ/hr}$$
$$= 9201.9375 \text{ kJ/min}$$
$$= 17.09\%$$

(iii) Heat lost to unburnt fuel

$$= m_{\text{uf}} \times \text{C.V.}_{\text{uf}}$$
$$= 4 \frac{\text{kg}}{\text{hr}} \times 2200 \frac{\text{kJ}}{\text{kg}}$$
$$= 8800 \text{ kJ/hr}$$
$$= 146.667 \text{ kJ/min}$$
$$= 0.2724\,\%$$

(iv) Heat unaccounted by difference

$$= 53833.333 - [33490.113 + 9201.938 + 146.667]$$
$$= 10994.615 \text{ kJ/min}$$
$$= 20.423\%$$

(4) Chimney height $= 50$ m

Mass of flue gases $= 18$ kg/kg of coal

$\therefore$ Mass of flue air $= 18 - 1 = 17$ kg/kg of coal

$$T_a = 28+273=301\text{ K}$$
$$T_f = 343+273=616\text{ K}$$
$$\therefore \quad h_w = 353\times H\times\left\{\frac{1}{T_a}-\frac{m+1}{m}\times\frac{1}{T_f}\right\}\text{ mm of H}_2\text{O}$$
$$h_w = 353\times 50\left\{\frac{1}{301}-\frac{18}{17}\times\frac{1}{616}\right\}$$
$$\mathbf{h_w = 28.3\ mm\ of\ H_2O}$$

Example 11.28 *Following observations were made during a boiler trial,*

Feed water = 1500 kg/hr
Inlet temp. of feed water = 20°C
Steam generated = 15 bar and 300°C
Coal consumed = 200 kg/hr
C.V. = 34000 kJ/kg
Ash in the ash pit contain unburnt products of 10 kg/hr of C.V. 2000 kJ/kg.
Flue gases formed = 18 kg/kg of coal burnt.
Flue gas temperature = 343° C
Atmospheric temp. = 28° C
Mean specific heat of fuel gases = 1.025 kJ/kg-K
Determine
(i) Boiler thermal efficiency
(ii) Equivalent of evaporation in
(iii) Heat balance in kJ/min for the boiler and
(iv) Drought produced in mm of H_2O if chimney height H = 100 m.

Solution

$$m_s = 1500\text{ kg/hr}$$
$$m_f = 200\text{ kg/hr}$$
$$\text{C.V.} = 34000\text{ kJ/kg}$$
$$m_{uf} = 10\text{ kJ/kg}$$
$$\text{C.V.}_{uf} = 2000\text{ kJ/kg}$$
$$m_{fg} = (m+1) = 18\frac{\text{kg}}{\text{kg of coal}}$$

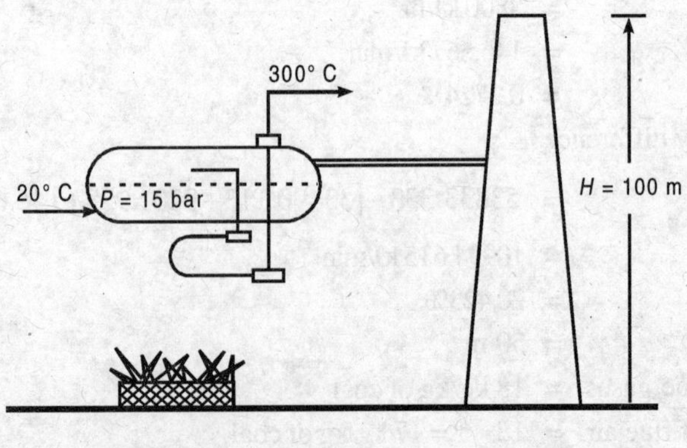

Fig. Ex. 11.28

$$T_f = 343 + 273 = 616\text{ K}$$
$$T_a = 28 + 273 = 301\text{ K}$$

Mean $C_{pg} = 1.025$ kJ/kg-K

At 15 bar pressure $T_{sat} = 198.3$°C and the steam temperature given is 300°C so the steam is superheated.

Now $h_2 = h_{sup}$ at 15 bar and 300°C from superheated steam tables

$= 3038.9$ kJ/kg

and $h_1 = h_f$ at 20°C = 83.9 kJ/kg

(i) Now we know that Boiler thermal efficiency,

$$\eta_{th} = \frac{m_s(h_2 - h_1)}{m_f \times \text{C.V.}} \times 100\%$$

$$= \frac{1500\frac{\text{kg}}{\text{hr}}(3038.9 - 83.9)}{200\frac{\text{kg}}{\text{hr}} \times 34000}$$

$$= \frac{4432500}{200 \times 34000}$$

$$\boldsymbol{\eta_{th} = 65.1838\%}$$

(ii) Equivalent of evaporation

$$= \frac{\dot{m}_s(h_2 - h_1)}{2256.9}$$

$$= \frac{1500}{200} \times \frac{(3038.9 - 83.9)}{2256.9}$$

Equivalent of evaporation = 9.8198 kg of st/kg of coal

(iii) Heat Balance sheet

(A) Heat energy released due to the combustion of fuel

$= m_f \times$ C.V.

$= 200$ kg/hr × 34000 kJ/kg

$= 6800000$ kJ/hr

$= 113333.33$ kJ/min

$= 100\%$ (Say)

(B) Heat Energy Utilised

(a) To form steam $= m_s(h_2 - h_1)$

$= 1500$ kg/hr (3038.9 − 83.9) kJ/kg

$= 443250$ kJ/hr

$= 73875.0$ kJ/min

(b) Heat to flue gases $= m_{fg} \times C_{Pg} \times (T_f - T_a)$

$$= 18\frac{\text{kg of gases}}{\text{kg of coal}} \times 1.025\frac{\text{kJ}}{\text{kg-K}}(616 - 301)\text{ K}$$

$$= 5811.75 \text{ kJ/kg of coal}$$

$$= 5811.75 \frac{\text{kJ}}{\text{kg of coal}} \times 200 \frac{\text{kg of coal}}{\text{hr}}$$

$$= 1162350 \text{ kJ/hr}$$

Heat to flue gases $= 19372.5$ kJ/min

(b) Heat lost due to unburnt fuel

$$= m_{uf} \times \text{C.V.}_{uf}$$

$$= 10 \text{ kg/hr} \times 2000 \text{ kJ/kg}$$

$$= 2000 \text{ kJ/hr}$$

$$= 333.333 \text{ kJ/min}$$

(d) Heat unaccounted by difference

$$= \{113333.33\} - \{73875 + 5811.75 + 19375.5 + 333.33\}$$

$$= (113333.33 - 99392.58) \text{ kJ/min}$$

$$= 13940.75 \text{ kJ/min}$$

(iv) Since $H = 100$ m

We know that,
$$h_w = 353 \times H \left\{ \frac{1}{T_a} - \frac{m+1}{m} \times \frac{1}{T_f} \right\}$$

$$= 353 \times 100 \{353 \times 100\} \left\{ \frac{1}{301} - \frac{18}{17} \times \frac{1}{616} \right\}$$

$$\mathbf{h_w = 56.61 \text{ mm of } H_2O}$$

Example 11.29 *A boiler generates 8 kg of steam/kg of coal at 30 bar and 350° C. Water is fed at 40°C. The mass of dry flue gas and water vapour/kg of coal burnt are 18.75 kg and 0.7 kg respectively. Boiler house temperature is 27°C and flue gas temperature leaving the boiler is 127°C.*

Draw-up heat balance sheet on kJ/kg of coal burnt basis and % basis.

Assume C_p of gases = 1.1 and C_p for the steam formed due to the combustion of H_2 = 2.1 kJ/kg K. Partial pressure of steam as 0.07 bar and C.V. of coal = 32000 kJ/kg.

Solution

(I) Heat energy liberated due to the combustion of 1 kg of fuel

$$= m_f \times \text{C.V.}$$

$$= 1 \times 32000$$

$$= 32000 \text{ kJ/kg of coal} = 100\% \text{ (say)}$$

(II) Energy utilised

(i) Heat to steam
$$= m_s (h_2 - h_1)$$

$$= 8 \text{ kg/kg of coal} \times (3117.5 - 167.45) \text{ kJ/kg}$$

$$= 23600.4 \text{ kJ/kg of coal} = 73.75\%$$

(ii) Heat lost to the flue gases

$$= m_g \times C_{pg} \times (T_f - T_a)$$

$$= 18.75 \times 1.1 (127 - 27)$$

$$= 2062.5 \text{ kJ/kg of coal}$$

$$= 6.45\,\%$$

(iii) Heat lost due to the water vapour formed due to the combustion of H_2

$$= m_1(h_2' - h_1')$$

Here h_2' – is the specific enthalpy of steam formed due to the combustion of H_2 at flue gas temperature and at the given partial pressure.

Since C_p of steam formed is given, from steam tables corresponding to 0.07 bar pressure,

$$\therefore \quad h_2' = h_g + C_p(T_{sup} - t_{sat})$$

$$= 2572.6 + 2.1\,(127 - 39)$$

$$h_2' = 2557.4 \text{ kJ/kg}$$

and $\quad h_1'$ = specific enthalpy of water at room temperature

i.e., at $\quad 27°C$ = 113.13 kJ/hr

Heat lost to water vapour

$$= m_1(h_2' - h_1')$$

$$= 0.7\,\{2757.4 - 113.13\}$$

$$= 1850.99 \text{ kJ/kg}$$

$$= 5.79\%$$

(iv) Heat unaccounted by difference

$$= 32000 - \{23600.4 + 2062.5 + 1850.99\}$$

$$= 4486.11 \text{ kJ/kg}$$

$$= 14.02\,\%$$

■■■

THEORY QUESTIONS

11.1. What is a boiler? Explain how they are classified.

11.2. Explain the following boilers with the help of neat sketches.
(a) Landcashire boiler
(b) Cornish boiler
(c) Locomotive boiler
(d) Cochran boiler
(e) Bacock Wilcox boiler
(f) Stirling boiler

11.3. Compare water tube and fire tube boilers.

11.4. What are the essentials of good boiler.

11.5. What is the difference between the Mountings and Accessories ? Mention the various Mountings and Accessories.

11.6. Explain with neat sketches the working of the following Mountings and Accessories
(a) Dead weight safety valve
(b) High steam low water alarm
(c) Water level indicator
(d) Fusible Plug
(e) Bourdon pressure gauge
(f) Steam top valve
(g) Super heater
(h) Economiser

11.7. What is boiler draught ?

11.8. Explain various types of draught used in a boiler.

11.9. Compare Artificial draught versus Natural (Chimney) draught.

11.10. Derive, $h_w = 353H\left[\frac{1}{T_a} - \frac{m+1}{m} \times \frac{1}{T_f}\right]$ mm of H_2O.

11.11. Derive the condition for maximum discharge of flue gases through the chimney.

11.12. What is a Steam Jet Draught ?

11.13. Explain,
(i) Equivalent evaporation from and at 100°C
(ii) Boiler Efficiency

11.14. What are the draught losses ?

11.15. Write a note on IBR regulation.

11.16. State the essentials of a good boiler.

11.17. Draw a neat sketch of any one type of boiler you have studied. Label the parts and describe its working.

11.18. State the function and location of the following :
(a) Fusible plug
(b) Feed check valve
(c) Blow-offcock
(d) Pressure gauge

11.19. Staff the factors considered for selection of a boiler, for a specific application.

11.20. Describe with a neat indicator
(a) Water level indicator
(b) Superheater
(c) Fusible plug
(d) Economiser
(e) Spring loaded safety valve
(f) Feed check valve

11.21. Describe briefly the advantages of incorporating an economiser, air-preheater and superheater in a steam boiler. Indicate the position of these in the passage of flue gases by a line diagram.

11.22. Define :
(i) Equivalent evaporation
(ii) Boiler accessories

11.23. What is a boiler draught ? Compare natural draught versus artificial draught.

11.24. Define :
(i) Boiler efficiency,
(ii) Equivalent evaporation and explain the working parameters which influence them.

11.25. Define the term 'equivalent evaporation' and explain its significance.

11.26. Write a short note on 'classification of boiler'.

11.27. Write a short note on 'regulations for chimney height'.

11.28. State the working principle of
(a) Fusible plug,
(b) Water level indicator,
(c) Air preheater.

11.29. Prove that under maximum discharge condition of chimney, the draught produced in terms of the hot gas column is same as the chimney height.

11.30. State the advantages of high pressure boilers.

11.31. Explain Evaporative capacity of a boiler.

11.32. Explain Function and working of the following:
(a) Superheater
(b) Spring loaded safety valve
(c) Air preheater

11.33. What is IBR? What are the provisions made in IBR?

11.34. Compare water tube boilers with five tube boilers.

11.35. Explain:

(i) Grate	(ii) Shell
(iii) Mounting	(iv) Heating surface
(v) Manhole	(vi) Damper

11.36. Make a list of any five boiler mountings and write down their functions in brief.

11.37. Make 'a list of any five boiler accessories and write down their functions in brief.

11.38. With the help of neat sketch explain working of Benson boiler.

11.39. What is a boiler? How they are classified?

11.40. Explain the following brifers with the help of neat sketches.
(a) Lancashire Boiler
(b) Cornish Boiler
(c) Locomotive Boiler
(d) Cochram Boiler
(e) Babcock Wilcox Boiler
(f) Stirling Boiler

11.41. Compare water tube and fire tube brifers.

11.42. Whar are the essentials of a good brifer.

11.43. What is the difference between mountings and accessories mountings and accessories.

11.44. Explain with neat sketches working of the following mountings and accessories.
(a) Dead water safety valve
(b) High-steam low water alarm
(c) Water level indicator
(d) Fusible plug

(e) Bourden pressure gauge
(f) Steam stop valve
(g) Superheater
(h) Economiser

11.45. What is boiler draught?

11.46. Explain various types of draught used in boiler?

11.47. Compare Artificial draught verses Natural draught.

11.48. Derive $h_w = 353 \times H\left[\frac{1}{T_a} - \frac{me1}{m} \times \frac{1}{T_f}\right]$ mm of H_2O.

11.49. Derive the condition for maximum discharge of the gases through the chimney?

11.50. What is a steam jet draught?

11.51. Explain
(i) Equivalent of evaporation from and @ 100°C
(ii) Boiler efficiency

11.52. What are the draught losses?

11.53. Write a note on IBR Regulations.

11.54. Explain the function of
(i) Scum valve
(ii) Steam stop valve
(iii) Steam trap

11.55. Define the Boiler accessories.

11.56. Explain
(i) Feed check valve
(ii) Blow off cock

11.57. State the difference between
(i) Externally fired and internally fired
(ii) Forced circulation and Natural circulation
(iii) High pressure and Lowe pressure
(iv) Stationary Portable
(v) Single tube and multi tube.
Give at least one example of each type.

11.58. What do you understand by steam jet draught? Where is it generally employed. State advantages of it.

11.59. Draw neat pencil sketch of any one type of boiler you have studied. Label the parts and describe its working.

11.60. Explain the function and working of
(i) Superheater
(ii) Spring loaded safety valve
(iii) Air preheater

11.61. Bring out a comprehensive comparison between FT and WT Boilers.

11.62. Explain the following terms an applied, to a St. generator
(i) Grate
(ii) Shell
(iii) Mounting
(iv) Heating surface
(v) Marble
(vi) Demper.

11.63. What is IBR? Explain some of its provisions.

11.64. Explain in detail classification of boilers.

11.65. Draw-neat sketches of the following and explain their working
(i) Feed check valve
(ii) Water level indicator
(iii) Economiser

11.66. Enumerate the factors to be considered in selections of a boiler.

11.67. Write the location and function of the following components used in a boiler
(i) Fusible plug
(ii) Blow off cock
(iii) Superheater
(iv) St. stop valve

11.68. What is equivalent of evaporation of a boiler? What is its importance?

11.69. What ave the factors involved in the selection of a boiler?
In what circumstances W.T. boilers are used in preferance to F.T. boilers?
Give reasons.
Suggest a boiler for a small chemical industry which is in need of 500 kg/hrof steam at 12 atm pressure.

11.70. Write short note on :
(i) Classification of boilers
(ii) IBR

PROBLEMS FOR PRACTICE

11.1. A boiler generates steam at 18 bar and 325° C feed water is supplied at 41.45° C. The thermal efficiency of the boiler is 80%. It uses furnace oil of C.V. 45500 kJ/kg. The steam is supplied to a turbine producing 100 kW power, having a specific steam consumption of 10 kg/hr-kW. Calculate the furnace oil consumption in kg/hr and equivalent of evaporation.

(*Ans.* 395.71 kg/hr, 6382.55 kg/hr)

11.2. A boiler has a chimney of 35 m height. It produces a draught of 20 mm of water column. The mean temperature of flue gases is 370°C and the boiler house temperature is 34° C. Estimate the air-fuel ratio. Does this boiler satisfy the condition for maximum discharge? Also find the height of hot gas column under maximum discharge condition.

[*Ans.* (i) The chimney nearly fulfils the condition for maximum discharge, (ii) $H' = H = 35$ m of hot gas columb]

11.3. A boiler generates 5000 kg/hr of steam at 16 bar and 300°C from feed water at 30°C. Coal used in 600 kg/hr of C.V. 30,000 kJ/kg. Find

(a) Equivalent of evaporation

(b) Boiler efficiency.

[*Ans.* (a) 10.75 kg/kg of coal (b) 80.9%]

11.4. The equivalent of evaporation of a boiler from a 100°C is 10.4 kJ/kg of fuel. The C.V. of fuel is 29800 kJ/kg.

Determine efficiency of boiler. If the boiler produces 15000 kg of steam per hour at 20 bar from a feed water at 40° C and the fuel used 1650 kg/hr. Determine the condition of steam produced.

[*Ans.* (a) 78.76 % (b) $x = 0.9734$]

11.5. The following readings were noted during an hour trial on a boiler.

Coal burnt : 750 kg

Feed water : 7000 kg

C.V. of coal : 33500 kJ/kg

Feed water temperature entering the economiser : 30°C

Leaving the economiser : 90°C

Boiler pressure : 10 bar

Steam dryness : 0.96

Temperature of steam leaving superheater : 250°C

Determine the thermal efficiency of the plant and also calculate the enthalpy received by feed water in various components as a % of total enthalpy increase.

[**Ans.** (i) $\eta_{th} = 77.92\%$ (ii) Enthalpy received in economizer = 230.0 kJ/kg i.e., 8.23%

(iii) Enthalpy received in boiler = 2138.8 kJ/kg i.e., 82.91%

(iv) Enthalpy received in superheater = 247.4 kJ/kg i.e., 8.86%]

11.6. The following observations were obtained in a boiler trial:

Steam condition = 15 bar and 300°C

Coal used = 95 kg/hr having C.V. = 34,000 kJ/kg

Feed water supplied = 680 kg/hr at 20°C.

Ash and unburnt coal collected in ash pit = 4 kg hr.

C.V. of unburnt coal = 2200 kJ/kg

Flue gas formed = 18 kg/kg of coal burn

Flue gas formed = 18 kg/kg of coal burnt

Flue gas temperature = 343°C

Boiler room temperature = 28°C

Mean specific heat of flue gases = 1.025 kJ/kg-K

Determine:

(i) Boiler efficiency (*Ans.* 62.21%)

(ii) Heat carried away by the flue gases in kJ/min (*Ans.* 9201.9375 kJ/min)

(iii) Heat unaccounted in kJ/min and in percentage.

(*Ans.* 10994.615 kJ/min, 20.4 %)

IMPORTANT POINTS TO BE REMEMBERED

1. *Height of chimney (H)*

$$h_w = \text{mm of water column}$$

$$= 353H\left[\frac{P}{T_a} - \frac{m+1}{m}\cdot\frac{1}{T_f}\right] \quad \text{when } P_a = P_N$$

or

$$= 353H\cdot\frac{P_a}{P_N}\left[\frac{1}{T_a} - \frac{m+1}{m}\cdot\frac{1}{T_P}\right]$$

$$= H' \text{ metres of hot flue gas}$$

$$= H\left[\frac{m}{m+1}\cdot\frac{T_f}{T_a} - 1\right]$$

2. *Condition for maximum discharge through the chimney*

$$H' = H \quad \text{and } T_f = 2\left(\frac{mH}{m}\right)T_a$$

3. *Boiler thermal efficiency*

$$\eta_{th} = \frac{m_s(h_2 - h_1)}{m_f \times \text{C.V.}} \times 100$$

4. Equivalent evaporation from and at 100°C

$$= \frac{m_s(h_2 - h_1)}{m_f\, 2256.9} \text{ kg of steam per kg of fuel.}$$

5. *Items in heat balance sheet*
   (i) Heat supplied/min = $m_f \times$ C.V.
   (ii) Heat utilized to form steam/min = $m_s\,[(h_2 - h_1) - C_{pw}\,(t_w)]$
   (iii) Heat lost to flue gases/min = $(m_a + m_f)\, C_{Pg}\,(t_g - t_k)$
   (iv) Heat lost of water vapour/min = $[m_w = 9H](h_{2'} - h_{1'})$
   (v) Heat unaccounted for
   (vi) Heat lost to unburnt fuel.

CHAPTER 12

High Pressure Boilers

12.1 INTRODUCTION

12.1.1 Steam Generator

As we have studied, **boiler is a steam generator** or it is a closed vessel, in which water is heated and converted into steam.

Actually steam generator is a complex combination of boiler, economiser, air preheater, superheater, re-heater, etc. In addition it has various auxiliary equipments like stokers, pulverisers, burners, fans, emission control devices, water treatment plant, ash handling equipments etc.

12.1.2 High Pressure Boilers

These are used to increase the efficiency of the plant. These are developed because of rising cost of fuel and restrictions on air pollution. Modern power generation plants generally use high pressure boilers. These high pressure boilers have pressures above 140 bar and the temperature may be up to 540 – 610°C. High pressure boilers use both natural and forced circulations. Forced circulation is achieved with the help of centrifugal pumps. By using this forced circulation, because of high velocity of water scale formation in the tubes and boiler drums is reduced.

12.2 ADVANTAGES OF HIGH PRESSURE BOILERS

1. Because of high velocities, tendency of scale formation in the tubes is reduced.
2. Small light weight tubes of better heating surface arrangements can be used. This makes the unit compact and reduces erection time and cost of boiler.
3. Due to forced circulation, for heating of all the parts is uniform, which avoids overheating and development of thermal stresses.
4. There is more flexibility in the arrangement of furnace, tubes and other boiler components.
5. The steam can be raised quickly to meet the variable load requirements.
6. The efficiency of the plant is increased.
7. A very rapid start from cold is possible if an external power supply is available. Hence the boiler can be used for carrying peak loads or standby purposes with hydraulic power stations.

12.3 HIGH PRESSURE TYPE BOILERS

Following are the types of high pressure boilers:

(i) La Mont (ii) Benson

(iii) Schmidt–Hartmann boiler (iv) Loeffler boiler

(v) Velox boiler.

12.3.1 La Mont Boiler

This is a modern high pressure, water tube boiler employs forced circulation. This was first introduced by La Mont in 1925. This is generally used in Europe and USA (Ref. Fig. 12.1).

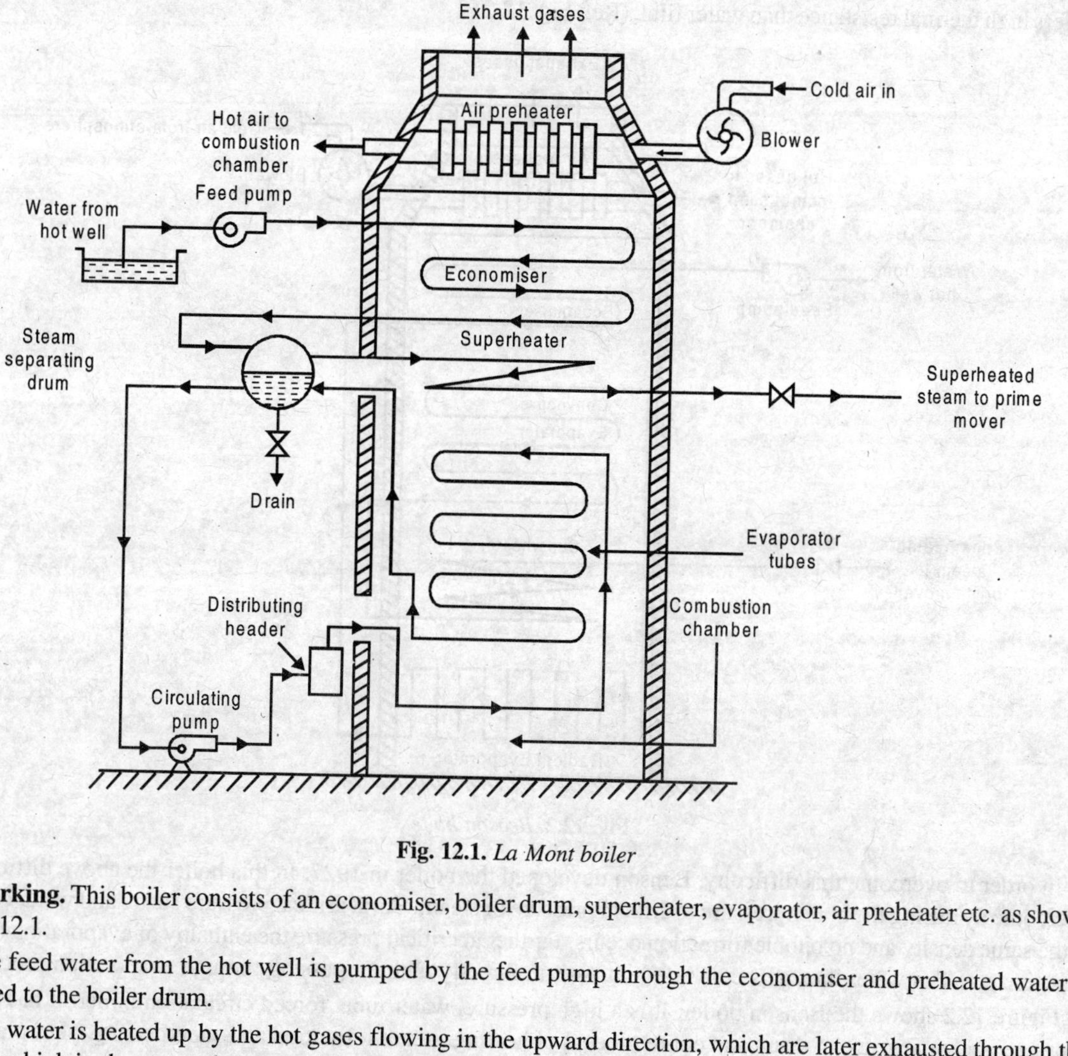

Fig. 12.1. *La Mont boiler*

Working. This boiler consists of an economiser, boiler drum, superheater, evaporator, air preheater etc. as shown in Fig. 12.1.

The feed water from the hot well is pumped by the feed pump through the economiser and preheated water is supplied to the boiler drum.

The water is heated up by the hot gases flowing in the upward direction, which are later exhausted through the chimney high in the atmosphere to reduce air pollution.

This boiler drum also acts as steam separating drum. The circulating pump, which is operating at a pressure of about 2.5 bar above the boiler pressure, pumps the water to the distributor header.

This distributor header has nozzles, which distribute the feed water into the evaporator.

The water in the evaporator tubes gets evaporated. And the mixture of water and steam is forced into the boiler drum, where the steam is separated. This steam is passed through the superheater, so as to have superheated steam and the superheated steam goes to the steam turbine for power generation.

The air preheater is also provided in the flue gas passage for the economic combustion of fuel in the furnace.

These boilers can generate steam up to critical pressure.

12.3.2 Benson Boiler

The main problem experienced with the La Mont boiler is the formation and attachment of bubbles on the inner surfaces of the tubes. These attached bubbles to the tube surfaces reduce the heat flow and steam generation, as it offers high thermal resistance than water film. (Ref. Fig. 12.2)

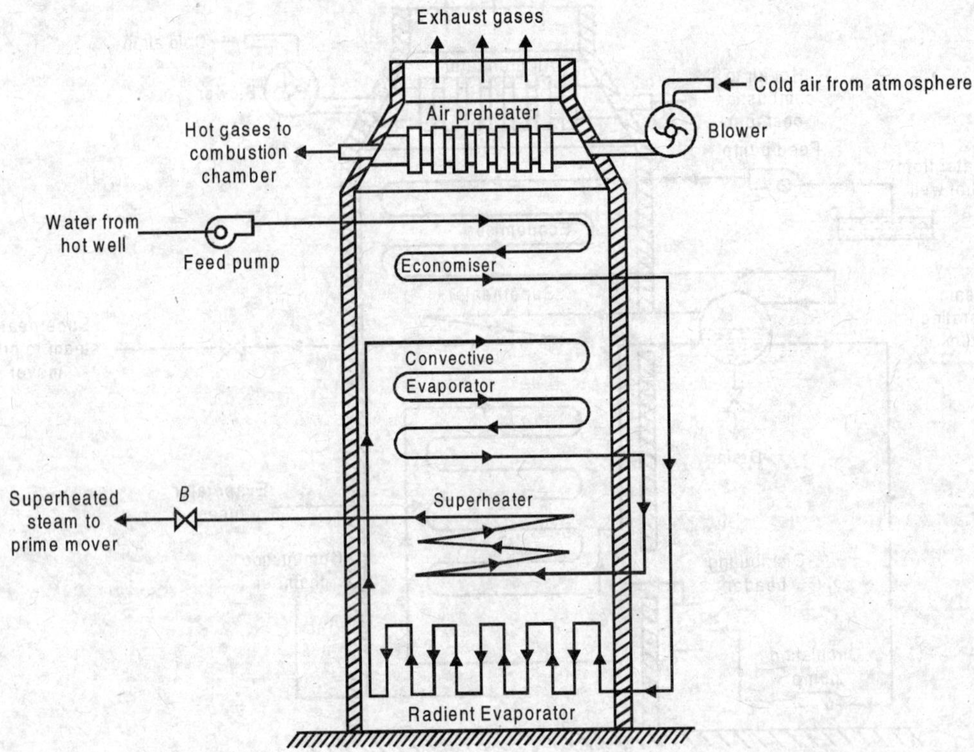

Fig. 12.2. *Benson boiler*

In order to overcome this difficulty, Benson developed the boiler in 1927. In this boiler the above difficulty is overcome by raising the boiler pressure to critical pressure of steam. As at critical pressure steam and water co-exist at the same density and no bubble formation occurs. Further at critical pressure the enthalpy of evaporation (Latent heat) is zero. The water thus gets converted into steam directly without passing through evaporation process.

Figure 12.2 shows the Benson boiler. It is a high pressure, water tube, forced circulation boiler. It has unique characteristic of absence of drum and entire process of heating, steam formation and superheating takes place in a single continuous tube.

Working. Feed water pump feeds water to the economiser, from where it goes to the radient heating section. Here most of the heat is transmitted and water gets heated almost to the critical temperature. In the evaporative section i.e., in radient evaporator and connective evaporator, evaporation is completed and superheating begins. In the superheater the final desired temperature is obtained.

Advantages

1. Due to the absence of drum, initial cost is low.
2. Circulation pump is absent.
3. The boiler is light in weight because of absence of drum.
4. The boiler can be started within 15 minutes and can be put in service.
5. The boiler can be operated at supercritical pressure also.

Disadvantages

1. It requires specific means of washing the tubes to avoid solid deposition.
2. The tubes are likely to get overheated when the flow is not sufficient.
3. It requires skilled operators to co-ordinate the operations.

12.3.3 Schmidt–Hartmann Boiler

This boiler uses distilled water for its operation for generating high pressure steam. This steam is recirculated without any wastage in the circuit. Figure 12.3 shows the schematic diagram of the boiler.

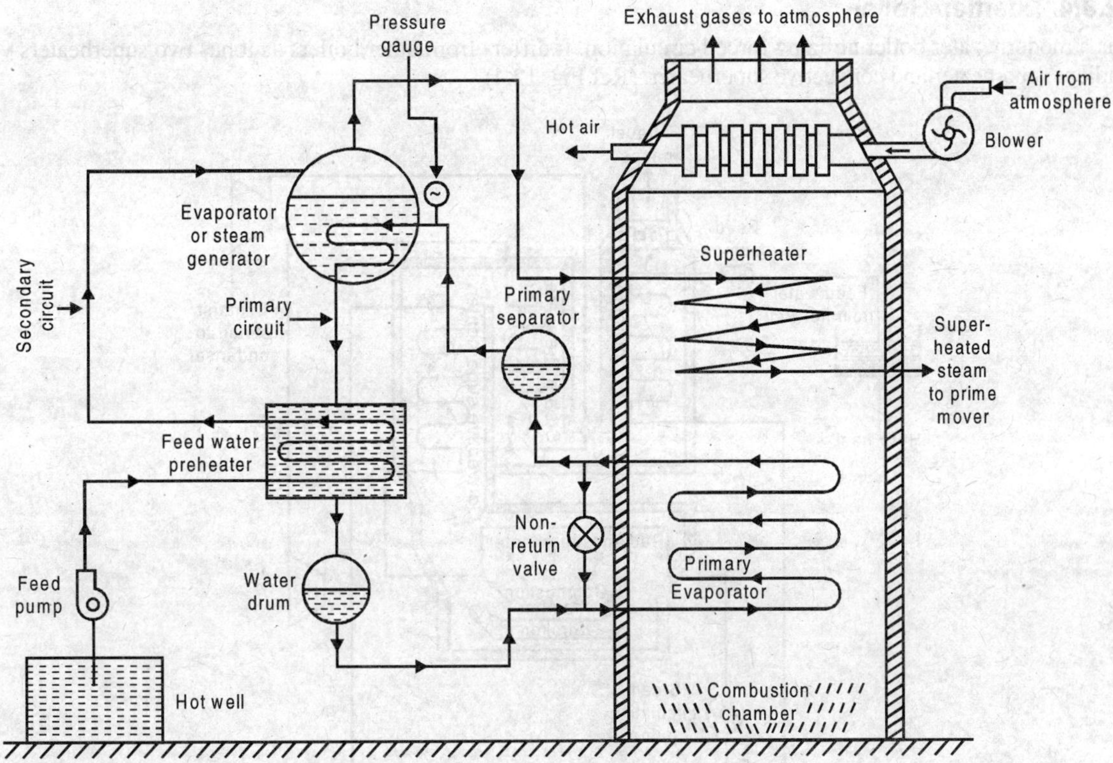

Fig. 12.3. *Schmidt-Hartmann boiler*

In this case high pressure steam generated from pure water is used for generating low pressure steam from impure water or normal water and is used for power generation.

As shown, distilled water from the water drum flows into the primary evaporating tubes by natural circulation. The hot gases from the cumbustion chamber passes over the evaporator tubes and the steam upto a pressure of 100 bar is generated from distilled water. This steam then goes to the primary separator. And then this steam passes through the tubes provided in the steam generator, which are submerged in impure water in the steam generator. Here heat will be transferred from steam to impure water and the impure water is converted into steam at about 60 bar pressure.

This low pressure steam generated from the impure water is then passed through the superheater for superheating. Finally the superheated steam is supplied to the steam turbines for power generation.

Note that the condensate of high pressure steam from below the boiler drum passes into the feed water heater and then to the water drum.

It is also to be noted that the impure water from the hot well will be pumped by the feed pump and is made to pass through the tubes submerged. Where primary heating of this impure water takes place and preheated water is supplied to the steam generator.

As shown air preheater is also provided, which supplies preheated air for combustion.

Advantages

1. It can use impure water for steam generation.
2. Water softening plant is not required.
3. There is rare chance of overheating of components of primary circuit (distilled water circuit) as there is no danger of salt deposition. Hence long life of these parts.

12.3.4 Loeffler Boiler

It is a modern water boiler utilizing forced circulation. It differs from other boilers as it has two superheaters viz., Radient superheater and convective superheater. (Ref. Fig. 12.4)

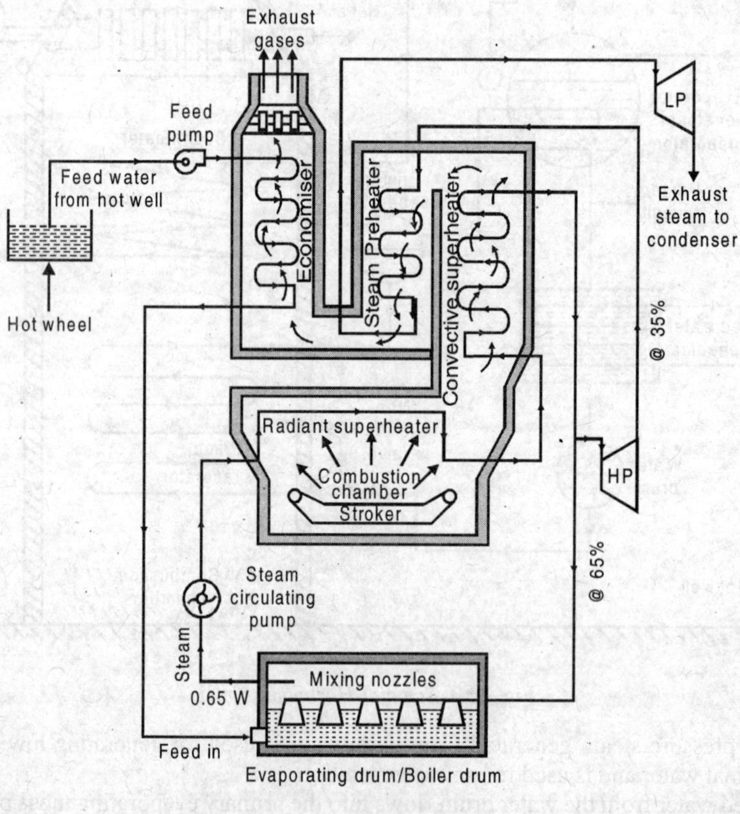

Fig. 12.4. *Loeffler boiler*

Radient superheater is located above the combustion chamber, so that it receives intense radiations heat.

The feed pump supplies feed water to the economiser, where the water is heated by the hot gases and the hot water will be supplied to evaporating/boiler drum. Note that a part of superheated steam (@ 65%) generated is passed to the evaporating drum through the nozzles where it mixes with feed water and steam is generated.

The saturated steam from the boiler or evaporating drum is drawn by steam circulating pump and is passed through the radient superheater and then convective superheater. About 35% of steam coming out from the superheater is supplied to high pressure turbine. The steam coming out from the turbine is then passed through steam reheater before supplying to low pressure turbine as shown in Fig. 12.4.

As already mentioned about 35% of total superheated steam is used for power generation and remaining 65% is used in the evaporating drum for generating steam.

It has a steam-generating capacity of 90–95 T/hr and steam pressure upto 150–155 bar.

12.3.5 Velox Boiler

It is known fact that, the heat transfer rate from the flue gasses can be increased by increasing the velocity of gases.

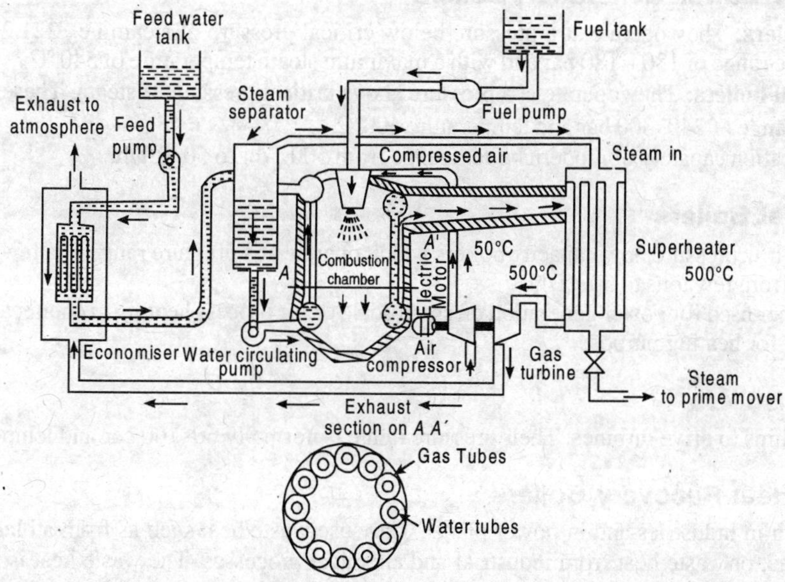

Fig. 12.5. *Velox boiler*

Thus to have high transfer rate, with reduced surface area, is by using high velocity (approximately or above sonic velocity) of flue gases. This concept is used in Vilox boiler. And in order to get high velocity of flue gases, pressurised combustion is used. (Ref. Fig. 12.5)

Air is compressed at about of 3 bar pressure with the help of a compressor driven by the gas turbine before supplying to the combustion chamber. In the combustion chamber, combustion of high pressure air and fuel takes place to produce supersonic velocity of hot gases. These gases are then passed through the annulus of tubes as shown. The heat is transferred from gases to water while passing through the annulus to generate the steam. The mixture of water and steam thus formed then enters the steam separator in a tangential manner. It forms a vortex in the separator and due to centrifugal action, the moisture content of the steam is separated. This moisture separated is sent back to feed water line.

The dry steam so obtained in the steam separator is sent to the superheater, where it is further heated by means of hot flue gases coming out from combustion chamber.

The gases coming out of superheater are used to run the gas turbine, which drives the compressor.

The exhaust gases coming out from the gas turbine are passed through the economiser to preheat the water and finally they are exhausted through the chimney.

Advantages

(1) If can start quickly.

(2) The boiler design is compact and have flexibility towards varying load.

Disadvantages

Solid fuel can not be used. It operates only with liquid or gaseous fuel.

12.4 TYPES OF BOILERS

Boilers are broadly of the following types:

(1) Utility boilers or Power Generation boilers. (2) Industrial boilers. (3) Marine boilers.

(4) Waste heat recovery boilers.

12.4.1 Utility or Power Generation Boilers

(*a*) **Subcritical boilers:** They operate at a pressure below critical pressure of steam i.e., 221.2 bar. They usually operate in a pressure range of 130 – 180 bar and with a maximum steam temperature of 540 °C.

(*b*) **Supercritical boilers:** They operate at a pressure above critical pressure of steam. These usually operate in the pressure range of 240–300 bar and temperature 600°C.

The steam generation capacity of modern boilers range from 500 T/hr to 5000 T/hr.

12.4.2 Industrial Boilers

They are relatively low pressure, low capacity boilers which operate at a pressure range of a few bar to 100 bar and capacities ranging from few tons to 500 T/hr.

The steam is either used for power generation on small scale or for process heating in chemical, textile industries and in refineries or for heating purpose.

12.4.3 Marine Boilers

They are used on ships to drive turbines. Their pressure range is normally 60–100 bar and temperature 550 °C.

12.4.4 Waste Heat Recovery Boilers

These are used both in industries and in power plants. They use waste heat such as from a blast furnace, exhaust from the gas turbine, or waste heat from industrial and chemical processes. The waste heat is passed over a heat exchanger surface to produce steam or hot water for the required use.

12.5 CHARACTERISTICS OF MODERN HIGH PRESSURE BOILERS

Even though the modern boilers are termed as high pressure boilers, it does not mean them fully. Because high pressure operation is only one of the characteristics of modern boilers used for power generation.

Other aspects/characteristics are as follows,

(1) Modern power boilers are water tube.

(2) They use forced type of water circulation.

(3) They use feed water heaters.

(4) They use modern methods of firing viz. pulverised firing; cyclone furnace; fluidised bed combustion.

(5) They use F.D. and I.D. fans for creating the necessary draught.

(6) They use water treatment plant, demineralisation plant to purify boiler feed water to minimise scale formation and salt deposits in the boiler tubes.

(7) They use various controls for pressure temperature water level etc.

(8) They use complex assembly of evaporater superheaters, reheaters, economisers and air heaters etc.

12.6 FLUIDISED BED COMBUSTION SYSTEM (FBCS)

12.6.1 Introduction

(a) Oil and gas prices increased in recent decades.

(b) Coal is available in large quantities.

(c) Pulverized coal burning system is costly and the combustion takes place at 1400–1600°C at which NO_x is produced.

(d) At high temperaure, ash forms a slag on superheater tubes and causes fouling, corrosion and reduces heat transfer rates.

To overcome these difficulties a new coal combustion system is developed and is called as Fluidized Bed Combustion System, (FBCS).

In this system any type of coal can be burnt efficiently in the temperature range of 900–1000°C. At these temperatures production of pollutant like NO_x is less, and slag formation is almost nil.

It is to be noted that FBCS is introduced in 1950s and aims at removing sulphur from coal during combustion process itself and reducing to formation of NO_x.

12.6.2 Principle of Working of Fluidised Bed Combustion

Figure 12.6 shows the simple fluidised bed combustion which has a distributor or perforated support plate on which finely divided particles of fuel (6 mm–20 mm) are provided and the particles are in collapsed state. Air

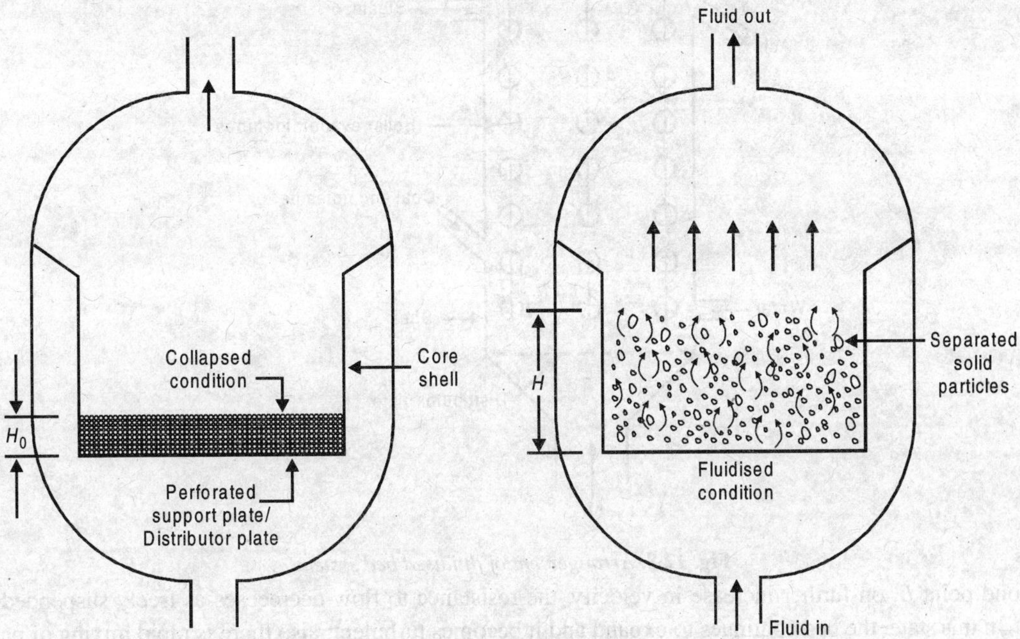

Fig. 12.6. *Schematic diagram of a fluidized bed*

is passed in the upwards direction with low velocities. The pressure drop across the combustion chamber can be measured with the help of a manometer.

The velocities of air is gradually increased and the logarithmic plot of ΔP and velocities C is shown in Fig. 12.7.

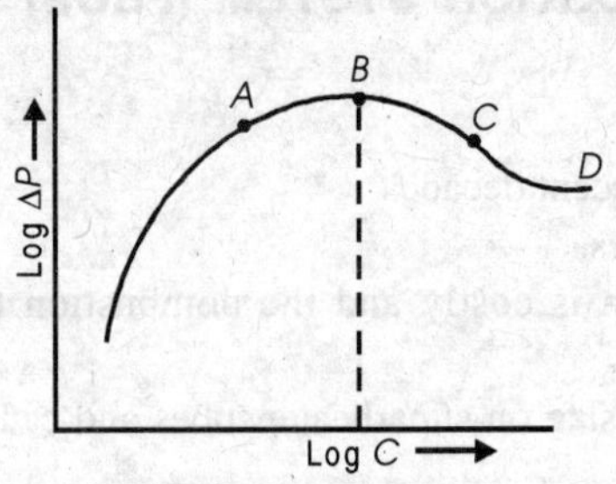

Fig. 12.7

Up to A — The bed is stable.

Between A and B — The bed is unstable particles adjust their position to present as little résistance to flow as possible.

At B — The particles are just in contact with each other.

Beyond B – Beyond this point – on further increase in velocity, the particles start moving freely in the bed with frequent collisions with each other. *Point B is called the point of fluidisations and the corresponding velocity is called the minimum fluidisation velocity.*

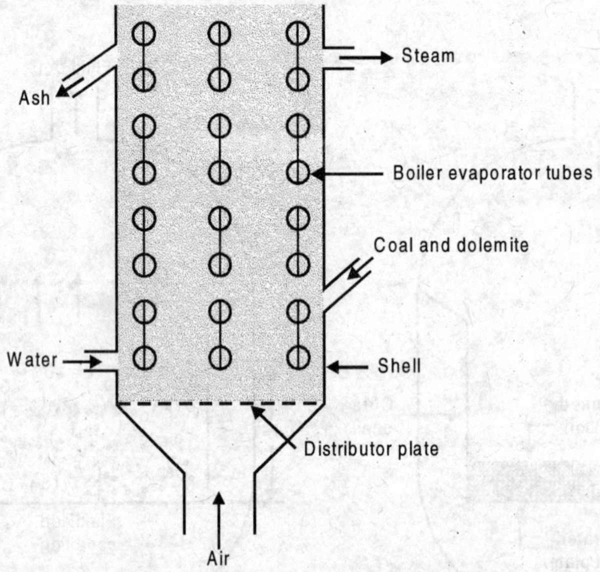

Fig. 12.8. *Arrangement of fluidised bed system*

Beyond point B, on further increase in velocity, the resistance to flow decreceses as freely suspended in gas stream. In this state, the bed continues to expand and it becomes turbulent; also there is rapid mixing of particles. Under these conditions the mixuture of bed and gas behaves like a fluid and the burning of fuel in such a state is called **fluidized bed combustion.**

The simple schematic diagram of fluidised bed combustion system is shown in Fig. 12.8.

It consists of a shell with a distributor plate at its bottom with holes in it. The shell carries the boiler evaporator tubes which remain immersed in the fluidized bed as shown. Coal of 6 to 20 mm size with *dolemite* or lime stone are fed on the distributor plate. Air is supplied from the bottom at a velocity more than the minimum velocity of fluidisation. The heat energy released due to combustion is rapidly transferred at high rates to water in the evaporator tubes and the generated steam is taken out. The sulphur content of the coal is converted into SO_2 on burning and SO_2 is mostly absorbed by dolemite. Hence the flue gases leaving the bed are almost free from SO_2 gas. The molten ash formed is taken out from the top as shown. Such a system reduces the size of the boiler considerably due to high rates of heat transfer.

12.7 ADVANTAGES OF FBC SYSTEM

Following are the advantages of the FBC system:

1. Size of the plant and its cost is considerably reduced.
2. Coals with high ash contents can be used.
3. The low temperatures of combustion in FBC system allows ash to remain soft and less abrasive.
4. SO_2 emissions are much less due to addition of dolemite.
5. Combustion efficiencies obtained are as high as 99.5 %.
6. Its initial and operational costs are low.
7. The rates of heat transfer are much higher since the system behaves like a violently burning liquid.
8. Volatilisation of alkali compounds does not occur, therefore there deposits on boiler tubes is almost absent.
9. Coal used is 2 to 3 % of bed volume. Remained space is occupied by the inert material like sintered ash, fused alumina, sand etc. These materials help in controlling the bed temperatures in the range of 800°C to 1000°C.
10. NO_x emissions are low since the amount of excess air used is less and temperatures maintained are low in FBC system in the range of 800 – 1100°C compared to pulverized coal combustion system. This is so as the fixation of free nitrozen at low temperatures does not occur.

12.8 TYPES OF FBC SYSTEMS

The FBC systems are classified as:

(1) Atmospheric FBC system. (2) Pressurised FBC system.

Atmospheric fluidized bed combustion systems are presently in fully developed state and these are being used in many power plants. In these systems, the pressures inside the combustion chamber is kept atmospheric. These are for further classified as *underfeed* and *overfeed* systems based on whether the loading of the fuel and supply of air is from the bottom of the bed or on the top of the bed respectively. Underfeed atmospheric FBC systems are compact in design, but, they have high operational costs. While the overfeed atmospheric FBC systems are simple in operation with less operational cost but these give lower heat transfer rates per m^2 of area of the bed.

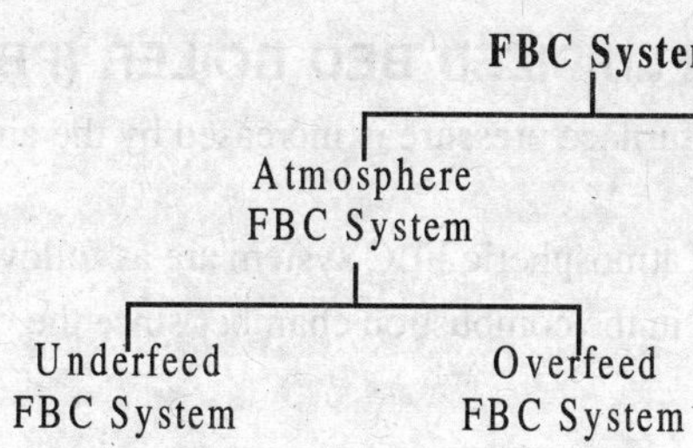

Pressurised FBC system is the latest development used for generation of power in which air is supplied under pressure of 10 bar approximately. With the help of an air compressor driven by a gas turbine as shown in Fig. 12.11. The supply of air under pressure results into better rates of heat transfer compared to atmospheric FBC system.

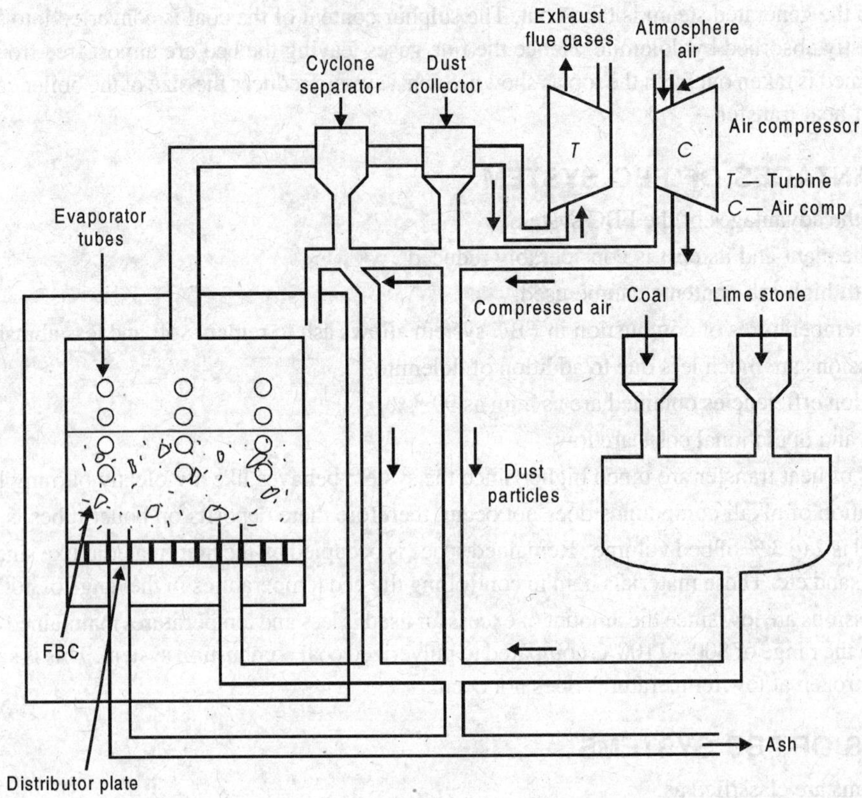

Fig. 12.9

The sketch is self explanatory. Note that the cleaned gas is supplied to the gas turbine producing power required by the air compressor which supplied pressurised air to the combustion chamber. FBC systems have various designs e.g.,

(a) FBC system with gas turbine (Ref. Fig. 12.9)

(b) FBC system with steam turbine as below.

10.9 PRESSURIZED BUBBLING FLUIDIZED BED BOILER (PBFB)

Figure 12.10 shows the PBFB boiler. In the furnace pressure is increased by the air compressor driven by the gas turbine. (Fig. 12.10 is self-explanatory)

Advantages of pressurized FBC system over atmospheric FBC system are as following :

(a) It allows increased coal loading per m^2 in the combustion chamber since the burning rate of fuel is higher.

(b) It required reduced air velocity.

(c) No_x, emissions are reduced.

(d) It has improved overall thermal efficiency of the plant.

(e) The overall size of the plant is considerably reduced, hence the overall cost of the plant.

(f) Better desulphurisation is accomplished.

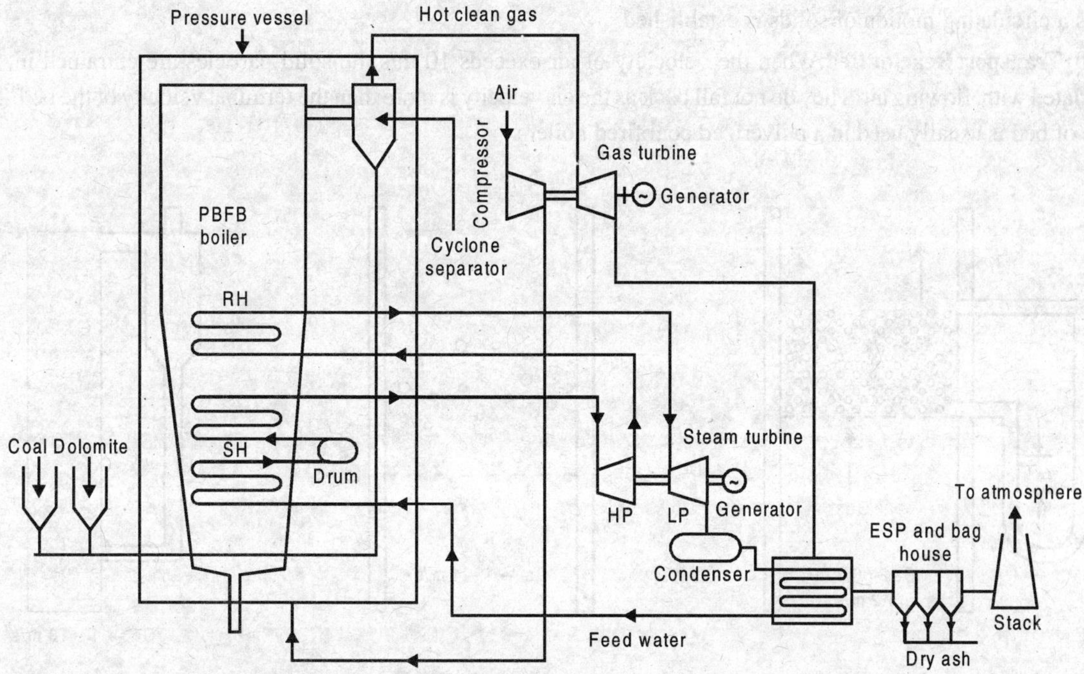

Fig. 12.10. *Schematic diagram of a pressurized bubbling fluidized bed boiler with steam turbine*

It allows to use reduced flue gas temperature before supplied to the fluidized bed may also be classified or described as—

(a) Packed bed or Fixed Bed.

(b) Bubbling Fluidised Bed.

(c) Fast or Circulating Fluidized Bed.

(d) Transport Reactor Bed.

(a) Packed or Fixed Bed. It consists of a bed of stationary particles on a perforated grid or distributor plate, through which gas is flowing. In a moving packed bed which is used in stoker fired boilers, the solid particles move with respect to walls, but not with respect to each other. The velocity of air is 0.3 to 1.2 m/sec.

(b) Bubbling Fluidized Bed. When air velocity is more than the minimum fluidisation velocity, the bed behaves like a liquid fluid. The pressure drop across the bed equals the weight of the bed per unit area (m^2). The air flows through the bed in the form of bubbles, which pass through—what is called—the emulsion phase bed. The size of the bubbles increases as it rises up in the emulsion phase. They may carry some solid particles which they throw up into the space above the bed, called *Free Board.* Some particles fall back to the bed. Beyond a certain height above the bed called transport disengaging height (TDH), the particles do not full back and are carried away by the gas. This is called *Elutriation* of the solids. Air velocity in this zone is 1-1–4 m/s.

(c) Flast Fluidized Bed (Circulating Fluidized Bed). The air velocity in circulating bed combustion is between 4–10 m/s. In circulating bed the solid particles elutriated with are return back to the bed at sufficiently high rate. The bed column is divided into two regions. The core of column consists of ditute gas which carries the particles upwards. The walls or annulus of the bed consists of denser medium in which the solid particles form clusters and fell down to the bed. The cluster fall down a certain distance and disintegrate to be carried away upwards again. Thus a circulating motion of solids is established.

(d) Transport Reactor Bed. When the velociety of air exceeds 10 m/s the solid particles are entrained in, and elutriated with, flowing air. They do not fall back as the air velocity is more than the terminal velocity of the bed. This type of bed is usually used in a pulverized coal fired boilers.

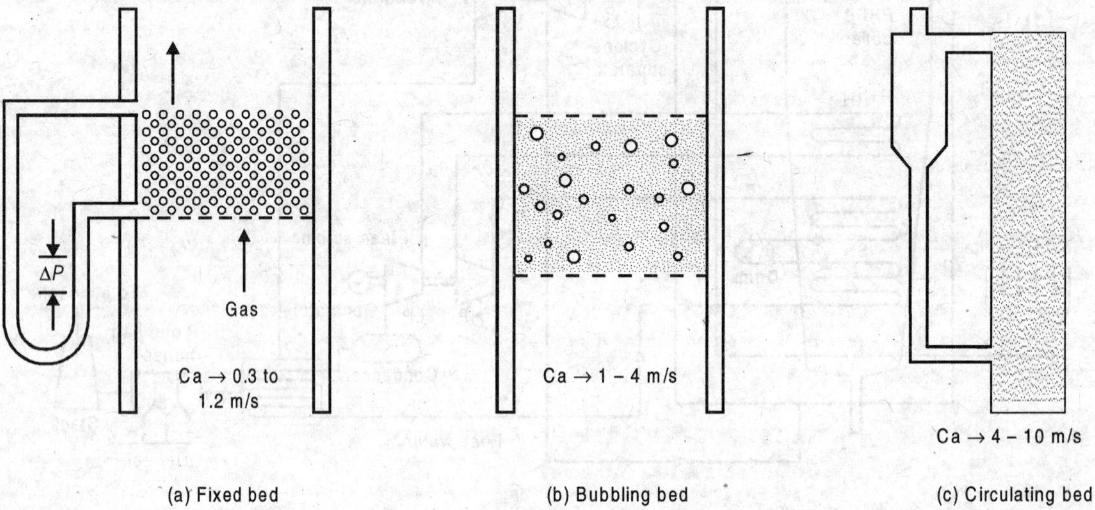

Fig. 12.11

Disadvantages

(a) Feeding of lime stone and coal into the bed in the required proportion is difficult especially under variable load conditions.
(b) Control of carbon carry over with the gases is also difficult.
(c) Erosion of furnace walls and immersed water tubes takes place.
(d) The cold water tube may quench the combustion process.

12.10 SPECIFICATIONS OF SOME BOILERS USED IN THERMAL POWER PLANTS IN INDIA

The specifications of some of the boilers used in India in thermal power plants are as follows.

Table 12.1

Name of the power plant	*Working pressure, bar*	*Maximum temperature, °C*	*Steaming capacity tons/hr*	*Electrical power output MW*
Bokaro (2 units)	60	500	150	50
Ramagundam	100	500	300	66
Chandrapura	135	540	500	140
Trombay old unit	175	670	520	150
Trombay (new unit)	178	540	1674	500

PROBLEMS FOR PRACTICE

12.1. What do you mean by high pressure boiler? Give the advantages of high pressure boilers.

12.2. Explain following high pressure boilers with net schesmatic diagram.

(a) La-Mont boiler
(b) Benson boiler
(c) Schsmidt-Hartmann boiler
(d) Loeffler boiler
(e) Velox boiler.

12.3. What do you mean by supercritical boilers and give advantages of these boilers.

12.4. Explain the principle of fluidized bed combustion with sketches.

12.5. Explain advantages of FBC system.

12.6. Explain types of FBC system.

12.7. Explain bubbling (Pressurised) fluidised bed ambition boiler with a neat sketch.

CHAPTER 13

Fuels and Combustion

13.1 INTRODUCTION

The fuel is material which when once raised to its ignition temperature continues to burn if sufficient oxygen or air is available. The main constituents of any fuel are carbon and hydrogen. These constituents are called combustibles. Small traces of sulphur are also present in certain fuels.

13.1.1 CLASSIFICATION OF FUELS

Fuels may be classified as given below:

(i) According to origin (mineral and organic)

(ii) According to mode of occurrence (natural, synthetic)

(iii) According to state (solid, liquid and gaseous)

(iv) According to condition (pulversied, briquetted, cooked)

(v) According to calorific value (high and low)

(vi) According to combination of C and H (paraffin based, aromatic, benzoic)

Another way to classifying the fuels is as given on next page.

13.2 ALTERNATIVE FUELS

13.2.1 Biogas

Biogas is produced by anaerobic fermentation of wet organic waste like animal dung, municipal garbage etc. Biogas is mostly (CH_4(50 – 70%) and CO_2 (remaining). Biogas can be conveniently used as alternate fuel. India has much scope to tap the large animal waste available in rural areas. India has very large animal population and its dung can be used not only as manure that is a byproduct in Gobar gas plant but also to produce biogas.

Another way of classifying the fuels is as follows.

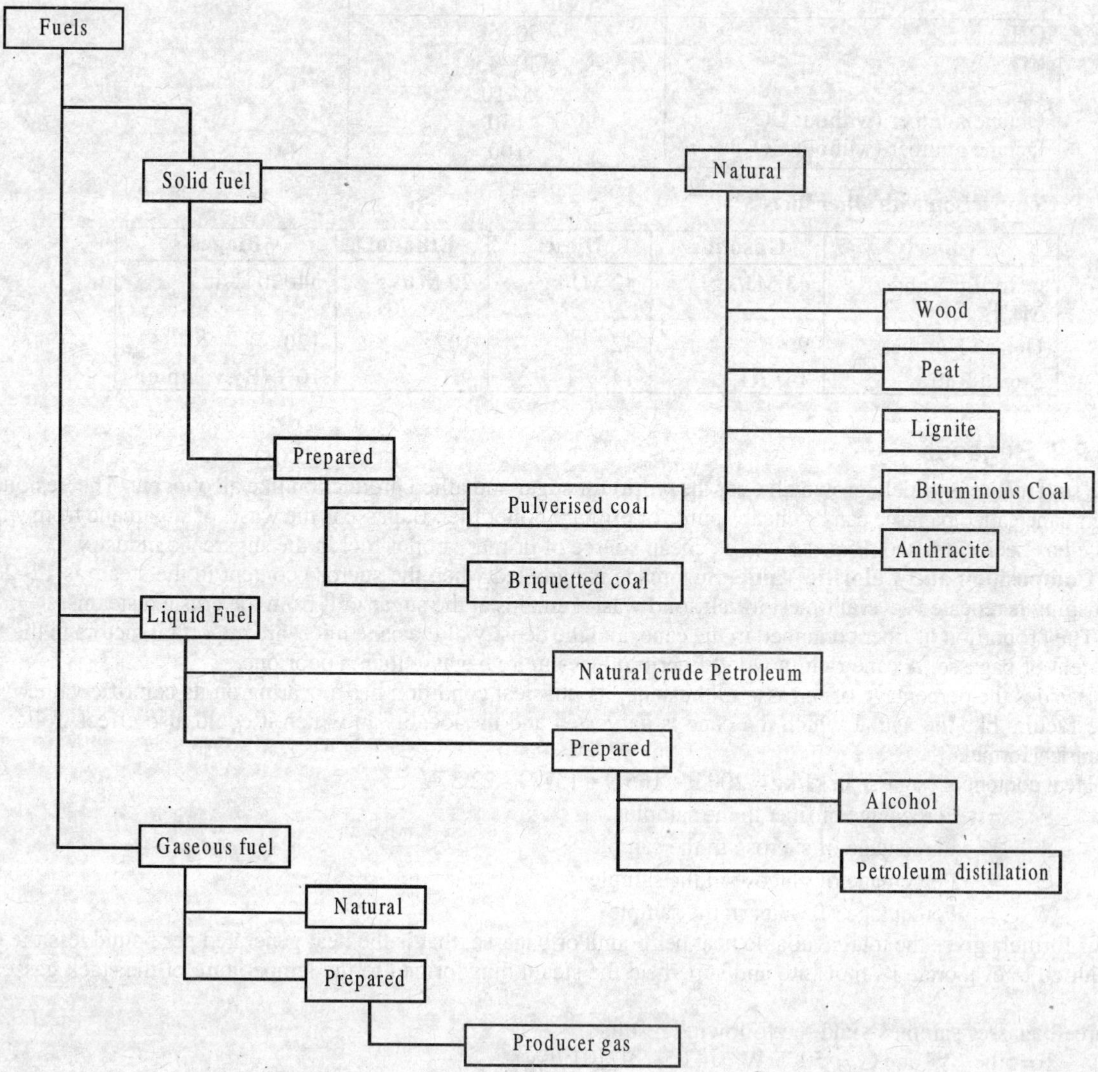

The biogas being lighter than air will rise up in case of any leak without any harm. Its ignition temperature is also higher than many fuels that make it quite safe. Recent advance in storage and transportation also make it quite useful as attractive fuel for IC engines apart from domestic use.

The biogas is produced usually in a gobar gas plant. Factors affecting Biogas production are:

(a) Animal dung and organic wastes should contain adequate nitrogen. Best carbon to nitrogen (C/N) ratio is about 30.

(b) Best temperature is about 35°C.

(c) pH ratio of the slurry should be between 7.2 to 8.2.

(d) The percentage of solids in the slurry should not exceed 10%.

The calorific value of Biogas is low but it has excellent knock resistance. In a typical Biogas composition and Octane No. is as follows.

Component	% by volume
CH_4	50–60
CO_2	30–45
H_2	5–10
Octane number (without CO_2)	130
Octane number (without CO_2)	100

Comparison with other fuels

Property	Gasoline	Diesel	Ethanol	Biogas
Calorific Value MJ/kg	43 MJ/kg	42 MJ/kg	20 MJ/kg	34000 kJ/m^3
Octane Number	90	42	107	120
Stoich Ratio	14, 2:1	14, 7:1	9:1	10:1 (By volume)

13.2.2 Bagasse

The sugarcanse is widely grown all over the world for sugar and allied production like alcohol etc. The residue of sugarcane called bagasee can be cheap source of primary boiler fuel. Bagasse is the waste of sugarcane from which juice has been extracted. Bagasse can be cheap source of primary boiler fuel in the sugarcane industry.

Composition and Calorific Value. Sugarcane is harbeted when the sucrose content in the steam is 12–20%. Crushing is repeated several times. Much solid waste remains at the sugar mill from the crushed steams.

The proportion of fiber contained in the cane, and the density of the cane juice, are important factors in the heat content of bagasse. A cane rich in wood fiber produces more bagasse than a poor one.

Besides the percentage of bagasse in the cane, its physical condition has a bearing on its calorific value. The factors like the age at which the cane is harvested and the locality in which it could also effect following empirical formula

Heat content of bagasse in kJ/kg = 200 F + 165 S + 157 G – 22.5 W

F = Percentage of fiber in the sample

S = Percentage of sucrose in the sample

G = Percentage of glucose in the sample

W = Percentage o£ water in the sample

This formula gives the total available heat per pound of bagasse, that is the heat generated per pound less the heat required to evaporate its moisture and superheat the steam thus formed to the temperature of the stack gasses.

Three bagasses samples yield the following results

F = 50	S and G = 50	W = 42.5	9710 kJ/kg
F = 33.3	S and G = 6.0	W = 51.0	7780 kJ/kg
F = 33.3	S and G = 7.0	W = 56.7	6500 kJ/kg

The bagasses with a high moisture contents that will not burn properly but which smolders and produces a large quantity of products of destructive distillation mainly heavy hydrocarbons, which escape unburned.

The Energy Aspect. The steam is raised in bagasse fired boilers which usually have a secondary fuel to accommodate imbalances in bagasse supply and steam or power demand. The factory designer attempts to balance the site such that bagasse is neither left over not insufficient any secondary fuel costs money and a large surplus of bagasse may cost money to dispose. Today, more and more factories are considering power export as another by product of sugar production. To do this they are improving the efficiency of their thermodynamic cycles and converting equipment drives to optimise power output.

Factories are establishing frequently in very undevelop d places and have no connection to an external power supply. This requires special techniques to start the factory and means that any breakdown in the powerhouse impacts on the entire neighbourhoods.

13.2.3 Natural Gas

It is the most important fuel belonging to gaseous fuels and is found in the vicinity of coal mines or oil reservoirs. Natural gas is a mixture of methane, ethane, propane, butane, pentane, nitrogen and carbondioxide.

Natural gas is obtained from oil wells as underground reservoirs, with or without petroleum oil, hence, it is either dry or wet. When there is no oil but only gas in the petroleum well, the natural gas is said to be dry whereas if natural gas occurs along with petroleum in oil wells, it is called wet gas. The wet contains gaseous hydrocarbons like methane and its mixture with higher hydrocarbons like propane, butane, isopentane pentane whereas the dry gas consists of methane, ethane with small amount of CO, CO_2, H_2S, N_2, H_2 and inert gases. The calorific value of wet natural gas is higher than that of dry one because of higher percentage of heavier unsaturated molecules. Dry gas does not form a liquid phase during production conditions whereas a gas is said to be wet if a liquid phase is produced at the surface without retrograde condensation in the reservoir. The wet gas is suitably treated to remove propane, propene, butane and butene and is called as LPG. For using the natural gas commercially it is necessary to remove, H_2S (toxic and corrosive), CO_2 (corrosive and has no heating value), Mercury (toxic), water (forms hydrates), heavy hydrocarbons (condenses in the transport system), Nitrogen (with no heating value).

Composition of Natural Gas. The approximate composition of natural gas is $CH_4 = 70 - 90\%$, $C_2H_6 = 5 - 10\%$, $C_3H_8 = 1 - 2\%$, $C_4H_{10} = 0.55 - 1\%$, Pentane $= 5 - 1\%$, $H_2 = 3\%$, $CO + CO_2$ = Rest and the calorific value varies from 1200 – 1400 kcal/m^3. This composition of natural gas varies widely as it depends on the underground conditions methods used for winning the gas etc. Figure 13.1 shows the temperature scale of Natural gas.

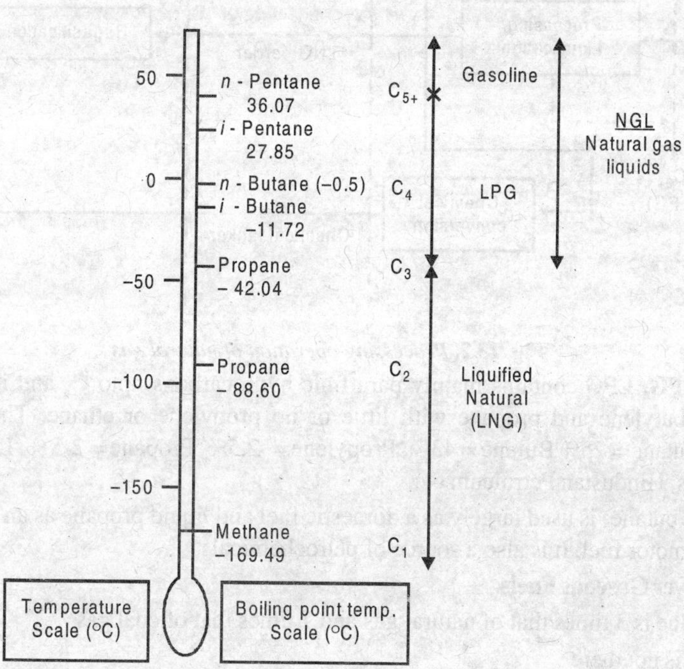

Fig. 13.1. *Temperature scale of natural gas*

Application of Natural Gas. Natural gas has got wide application as it is used not only for domestic and industrial purpose but also as a chemical raw material for various synthesis.

(i) It is an excellent domestic fuel, for example LPG.

(ii) Various byproducts are obtained from raw natural gas like methane, ethane, propane, butane, natural gasoline etc.

(iii) It is used in manufacturing of carbon black (used as a filler) and hydrogen (used in ammonia synthesis)

(iv) Microbiological fermentation of methane yields synthetic proteins which is used as animal feed.

(v) Natural gas and its fractions can be transported in various forms like CNG, LNG, LPG and chemicals like methanol, amonia, area.

(i) Liquified Petroleum Gas (LPG). It is generally manufactured from natural gas or the gas from the cracking units of petroleum refineries. LPG in general contains hydrocarbons of such volatility that they can exist as gases under atmospheric pressure, but can be readily liquified under pressure. The mixture of lower hydrocarbons is condensed under high pressure in refrigeration units and passed through de-ethaniser and debutaniser. The gas is finally passed through washing column and mixed with mercaptans (which acts as a warning gas on leaking). It is then supplied under pressure in containers. The pressure of the gas at room temperature within the cooking cylinder is about 3 bar. Figure 13.2 gives a schematic diagram showing processing operation of natural gas.

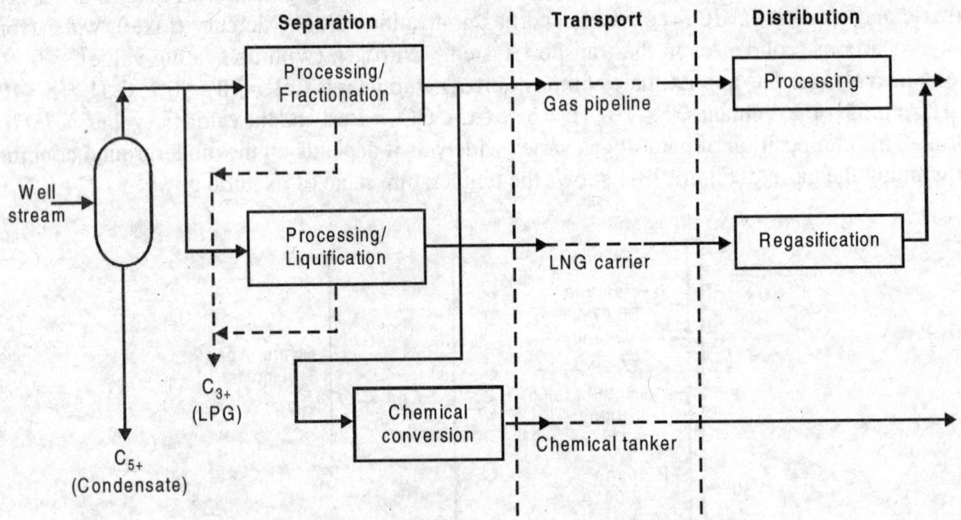

Fig. 13.2. *Processing operation of natural gas*

Composition of LPG. LPG contains mainly paraffinic hydrocarbons upto C_4 and its main constituents are *n*-butane, isobutane, butylene and propane with little or no propylene or ethane. The exact composition in *n*- butane = 27%, 2- Butane = 25% Butane = 43%, Propylene = 2.5%, Propane = 2.5 %. LPG is available under the trade name Bharat Gas, Hindustan Petroleum etc.

Uses. LPG (Liquid butane) is used largely as a domestic fuel and liquid propane as an industrial fuel. Research is on to use LPG as a motor fuel. It is also a source of petrochemicals.

Advantages of LPG over Gaseous Fuels

1. The calorific value is 3 times that of natural gas and 7 times that of coal gas
2. Smoke problem is not there
3. Low maintenance
4. Easy transportation in cylinders or pipelines
5. High efficiency and heating rate

Advantages of LPG over Gasoline as a Motor Fuel

1. LPG is cheaper than gasoline
2. It mixes easily with air as compared to gasoline
3. It burns cleanly without giving any residue

Disadvantages of LPG over Gasoline as a Motor Fuel

1. Its octane number is quite low
2. It has to be handled under pressure
3. It cannot be blended with other fuels
4. Due to its faint colour, one cannot easily detect its leakage.

(ii) Liquid Natural Gas (LNG). It is mostly used as an industrial fuel and in power generation projects. Liquified natural gas is predominantly methane and it has been reduced to liquid state by cooling to –161°C. It is colourless, odourless, non-toxic and has a specific gravity of about 0.5. On liquifaction the reduction in the volume is in the ratio 1 : 600. At –161°C and atmospheric pressure the volume of the gas decreases significantly so that it can be liquified.

(i) One cubic meter of LNG contains 584 cubic metre equivalent of natural gas.

(ii) It is a clean burning fuel and more abundantly available than oil.

(iii) It emits low levels of CO_2, NO_2, SO_2 etc. The gas is produced from the gas well. The raw gas is transported through pipelines for liquifaction. Before liquifaction the impurities like CO_2 sulphur compounds and water are removed along with heavier hydrocarbons leaving the pure natural gas i.e. mainly consisting of CH_4. Then it is cooled to –161°C at atmospheric pressure till the gas becomes liquid and then it is stored in insulated tanks. For using, it has to be heated up so as to regasify it and then transport it through pipelines. Figure 13.2 gives its processing and transportation scheme.

Applications

1. LNG can be used as a fuel where pollution control, and efficient heat control are major requirements.
2. Used in Industries like fertilizer units, glass industry, ceramic industry, and general transport.
3. Used in Boiler fuel for steam generation.

(iii) Compressed Natural Gas (CNG). This is another alternative fuel obtained from natural gas which is much more fuel efficient and requires less maintenance.

Compared to Diesel and Petrol, CNG cause lower pollution without affecting the performance of the vehicle. The exhaust level of CO_2 in CNG vehicle is about 40% lower than that of gasoline or diesel engine. Since CNG does not operate on fuel containing high lead and sulphur hence these are not emitted in CNG emission. Even though CNG has a higher self ignition temperature as compared to petrol but because of its low inflammability it is much safer than petrol. CNG is regarded as one of the cleanest fuel as it emits less CO and has lower potential of ozone formation as compared to petroleum gasoline. Costwise also it is less expensive than Diesel and petrol. The carcinogenic potential of even the best quality diesel in vehicles is 5 times that of CNG. CNG operated vehicles emit less particulate matter than Diesel operated vehicles, Catalytic converters which are used to reduce the emissions from automobiles cannot function effectively because of high sulfur content in both petrol and diesel.

Composition. CNG is compressed natural gas i.e., natural gas is compressed at high pressure (100 – 200 bar). Composition is same as natural gas.

Applications

1. As a substitute for petrol/diesel.
2. On blending with diesel (15% diesel and rest CNG) the fuel can be more cost effective.
3. Less polluting than gasoline.
4. Least contribution towards smog as compared to gasoline.

13.3 COAL ANALYSIS

Coal as found in nature is neither a pure substance nor of uniform composition. A definite chemical formula cannot be therefore written for a coal found in coal mines. Consequently two different methods of analysis are employed to know the composition of coal. They are known as *Ultimate analysis* and *Proximate analysis*.

In ultimate analysis a complete chemical breakdown of coal into its chemical constituents is carried out by chemical process. This analysis is required when important large scale trials are being performed. The analysis serves the basis of calculations of the amount of air required for complete combustion of the kg of fuel. The analysis gives the percent content on mass basis of carbon, hydrogen, nitrogen, oxygen, sulphur and ash, and their sum is taken as equal to 100%. Moisture is expressed as a separate item. The analysis also enables us to determine calorific value or heating value of coal.

Proximate analysis is the separation of coal into its physical components and can be made without the knowledge of analytical chemistry. The analysis is made by means of a chemical balance and a temperature controlled furnace. The sample of fuel is heated in the furnace or oven. The component in the analysis we fixed carbon, volatile matter, moisture and ash. These components are expressed in percent on mass basis and their sum if taken as 100%. Sulphure is determined separately. This analysis also enables us to determine the C.V. of coal. Compacted vegetation, in the absence of air and under the influence of pressure and temperature, is converted into coal. The various stages of conversion are as follows:

Wood → Peat → brown → coal → lignite → sub-bituminous coal → bituminous coal → anthracite.

This conversion takes 10 – 300 million years — the peat being the youngest and the anthracite the oldest one.

13.3.1 Proximate Analysis

(1) Powdered coal –m gm (weighted) → heated at 110°C for 20 min. Sample is again weighted, (m_1)

$$(m - m_1) = \text{moisture evaporated}$$

$$\% \text{ of moisture } (M) = \frac{m - m_1}{m} \times 100$$

(2) m_1 is then heated in a close container at 954°C for seven (7) minutes. Sample is again weighed. Volatile matter evolves at about 600°C and leaves the sample of m_2.

$$\% \text{ of volatile volume } (MV) = \frac{m_1 - m_2}{m} \times 100$$

(3) m_2 is heated at 732°C in an open crucible. It is completely burnt. The residue is the ash fraction → m_3

$$\% \text{ Ash (A)} = \frac{m_3}{m} \times 100$$

13.3.2 Ultimate Analysis

The ultimate analysis consists of determining the percentages of the ultimate constituents

(a) Carbon
(b) Hydrogen
(c) Oxygen
(d) Sulphur
(e) Nitrogen
(f) Ash

Ultimate analysis is used for calculative and scientific work.

13.4 CALORIFIC VALUE OF FUELS : (HCV AND LCV)

The calorific value of a fuel is amount of heat liberated by its complete combustion. For solid and liquid fuels, calorific value is expressed in kJ/kg, whereas for gaseous fuels it is expressed as kJ/m^3 where m^3 is normal cubic metre measured at NTP conditions i.e., at 0°C temperature and 760 mm Hg barometric pressure (1.01325 bar). Sometimes, calorific value of gaseous fuels can also be expressed as kJ per cubic metre expressed at STP conditions. STP (standard temperature and pressure) conditions are taken as 15°C and 760 mm Hg barometric pressure (1.01325 bar).

A fuel consists of one or more combustile components like carbon, hydrogen, carbon monoxide, hydrocarbons, sulphur etc. Of the above mentioned combustibles, sulphur is not a desirable ingredient due to corrosive properties of SO_2 formed by its combustion. The above mentioned combustibles liberate heat by their combustion, but to evaporate the water formed by combustion of hydrogen or hydrocarbons as well as to evaporate the moisture content of the fuel, its sensible heat from temperature to the saturation temperature at combustion pressure and its latent heat and superheat, if any, have to be supplied to bring it to the temperature of the products of combustion. This heat is naturally taken from the heat of combustion of the fuel. Thus the total heat of combustion of the fuel is not available to do external work. The heat liberated by the combustion of the fuel, neglecting the heat required to evaporate the water is called the *Higher Calorific Value* of the fuel (HCV). Higher calorific value is the maximum heat energy liberated by the complete combustion of the fuel. This is also called as the gross calorific value of the fuel. If we subtract from the higher calorific value, an amount of heat required to evaporate the water formed, we get *Lower Calorific Value* (LCV) or *Net Calorific Value* (LCV) or *Net Calorific Value* of the fuel.

No definite agreement is to be found in the literature on fuel as to whether the lower calorific value shall be found simply by subtracting latent heat of steam or both the latent heat and sensible heat in cooling from 100°C, from the gross calorific value of the fuel; in the latter case it would be necessary to fix the temperature to which the products are finally reduced. In literature the net calorific value of the fuel is obtained by subtracting from higher calorific value the amount 2466 kJ/kg (latent heat of dry and saturated steam at STP (15°C) conditions) (amount of water formed because of the combustion of 1 kg fuel.)

Calorific values of solid and liquid fuels are found experimentally with the help of **Bomb Calorimeter**.

13.4.1 Theoretical Determination of Calorific Value (Provided Composition by Weight is known)

The calorific value of a fuel is the heat liberated from the combustion of 1 kg of the fuel. This value can be calculated from the analysis of the fuel if the calorific value of its constituents are known. Any oxygen contained in the fuel is assumed to be already combined with the hydrogen; the heat of this hydrogen is not, therefore, available for further combustion.

The calorific values of the individual combustible in the fuel are given below:

Carbon C burns to CO_2 $\rightarrow$ 33915 kJ/kg
Carbon C burns to CO_2 $\rightarrow$ 10200 kJ/kg
Hydrogen H_2 burns to H_2O $\rightarrow$ 1444515 kJ/kg
Sulphure S burns to SO_2 $\rightarrow$ 9630 kJ/kg

Now let C, H, O and S% be the carbon, hydrogen, oxygen and sulphur contents of the fuel respectively.

∴ Assume or consider 100 kg of fuel.

∴ The quantity of heat evolved or liberated due to combustion of carbon is $C \times 33915$ kJ when burnt to CO_2.

The quantity of heat evolved due to combustion of hydrogen is $\left[H - \frac{0}{8}\right] \times 144451.5$ kJ. Here the assumption is made that any oxygen present in the fuel is already wholly in combination with hydrogen. Since it is known that 8 parts of oxygen by weight are combined with 1 part of hydrogen, it is customary to deduct from the total hydrogen an amount equal to 1/8 of the oxygen present, calling the remainder available as hydrogen.

The quantity of heat evolved due to combustion of sulphur is $S \times 9630$ kJ

Therefore, total heat evolved due to combustion of 100 kg of fuel

$$= 33915C + 144452.5\left(H - \frac{0}{8}\right) + 9630\,S \text{ kJ}$$

∴ Higher Calorific Value of the fuel

$$HCV = \frac{33915C + 144452.5\left(H - \frac{0}{8}\right) + 9630\ S}{100}\ \text{kJ/kg}$$

The amount of steam formed = $\frac{9H}{100}$ kg/kg fuel because from 1 kg of hydrogen 9 kg of water vapour is formed.

∴ Lower Calorific Value of the fuel is

$$LCV = \frac{33815C + 144451.5\left(H - \frac{0}{8}\right) + 9635S - 9H \times 2466}{100}\ \text{kJ/kg}$$

$$\left[HCV = \frac{33815C + 144451.5\left(H - \frac{0}{8}\right) + 9635S}{100}\right] \text{ is } \textit{Dulong formula}$$

This higher value is based on ultimate analysis. Similarly, for gaseous combustibles we have

H_2 — 12800 kJ/m^3

CO_2 — 12620 kJ/m^3

CH_2 — 40070 kJ/m^3

Goutel suggested the following formula from calculating the higher calorific value when the percentage proximate analysis of fuel is known. The formula is, cal. value = 343.3 × fixed carbon % + α × % volatile matter kJ/kg.

The value of the factor α depends on the percent amount of volatile matter on the dry ash free basis. The value of α decreases with the rise in volatile matter as can be seen from the table below.

Volatile matter % (dry ash free)	5	10	15	20	25	30	35	40
α	607	544	490	456	431	410	394	335

Goutel formula is unreliable for fuels having high percentage in oxygen.

13.4.2 Experimental Determination of Calorific Value of Fuel

An apparatus which is used for determining the calorific value of a fuel is known as a fuel calorimeter.

The principle of all the calorimeters is the transference of heat of combustion of the given weight of fuel to water and the vessel. From the observed rise of temperature of the water and the container the calorific value of the fuel can be determined by equating the heat given out by the fuel to the heat taken by the water and the container. In order to know the heat taken by the container, water equivalent of the container should be known.

In this method of determining the calorific value of the fuel the following conditions should be satisfied:

(i) the combustion of the fuel must be complete

(ii) the heat must be transferred completely to the water

(iii) cooling losses from the calorimeter must be corrected

(iv) the rise of temperature of water must be correctly determined because the mass of the fuel is very small in comparison with the quantity of the water heated.

(a) Boy's Gas Calorimeter. This is the standard instrument used for measuring the calorific value of gas. The sectional view of the calorimeter is shown in Fig. 13.3. It consists of two burners in which a known volume of gas is burnt.

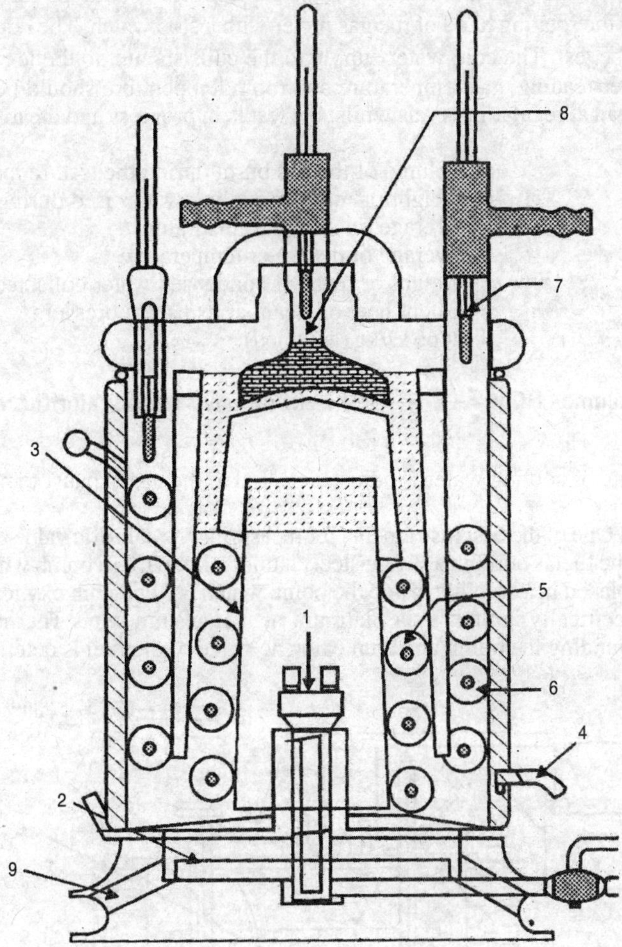

Fig. 13.3. *Boy's gas calorimeter*

1. Gas burners, 2. Metal vessel, 3. Copper chimney, 4. Condensed water outlet, 5. Inner coil, 6. Outer coil, 7. Inlet water box, 8. outlet water box, 9. Holes in base

The hot gases produced by combustion pass up the copper chimney. This is surrounded by a double coil of metal tubing through which a known weight of cooling water is circulating. After passing upwards the hot gas are deflected downwards through the space containing the inner coil. By the time the gases reach the top of this passage, practically the whole of their heat has been absorbed by the circulating water. The gases are then passed away into the atmosphere.

The cooling water enters the outer coil. After circulating through both coils it leaves at the water box. Its temperature is measured at inlet and outlet by thermometers at inlet and outlet by thermometers shown. These can be read to a fraction of a degree by the reading leases. For this purpose special thermometers giving least count of 0.01 can also be used. The temperature of the exhaust gases is measured by another thermometer. Any water formed by the combustion of hydrogen is condensed on the coils and is collected at the base of the instrument. From here it is drained through the pipe and is collected in a measuring glass.

In order to increase the cooling surface of the coils, the outsides of the tubes are ribbed or finned as shown in the cross-sectional view. The outer cylinder of the instrument is lined with a felt jacket to prevent any leakage of heat. For measuring the volume of the gas consumed, a meter is used reading 1/100 cu. meter for one revolution of the hand.

The test is carried out after the gas has been burning under uniform conditions for 45 minutes, the rate of gas flow is obtained by timing the rotating hand of the gas meter with a stop watch. The cooling water is collected in a measuring vessel during the test. The cold water supply to the coils should not be less than 5°C, below the room temperature. The barometer reading, gas temperature and room temperature should be taken during the test. The thermometers should be read at regular intervals whilst the test is in progress and the average values obtained from these readings.

Let V = Volume of the gas burnt during the test, reduced to NTP

m_w = Weight or mass of cooling water used during the test

t_1 = Average inlet water temperature

t_2 = Average outlet water temperature

w = Weight or mass of condensed water collected

h_{fg} = Latent heat of steam at its partial pressure = 2466 kJ/kg (assumed)

∴ Higher calorific volume $HCV = \dfrac{m_w C_{pw} (t_2 - t_1)}{V}$ kJ/m^3 and lower calorific value $LCV = HCV - w.h_{fg}$ kJ/m^3.

Generally mass flow rate of cooling water is adjusted such that the gas exhaust temperature is same as that of atmosphere.

(b) Bomb Calorimeter. One of the best instruments for measuring the calorific value of powdered and liquid fuels is the bomb calorimeter. The fuel is burnt in a strong steel chamber, known as a bomb which is immersed in a known mass of water. The fuel is placed in a crucible inside the bomb which is filled with oxygen under a pressure of 25–30 atmospheres. It is then electrically ignited by a platinum or magnesium wire. The heat liberated by the rise in temperature of water surrounding the bomb and then calorific value of the fuel is determined.

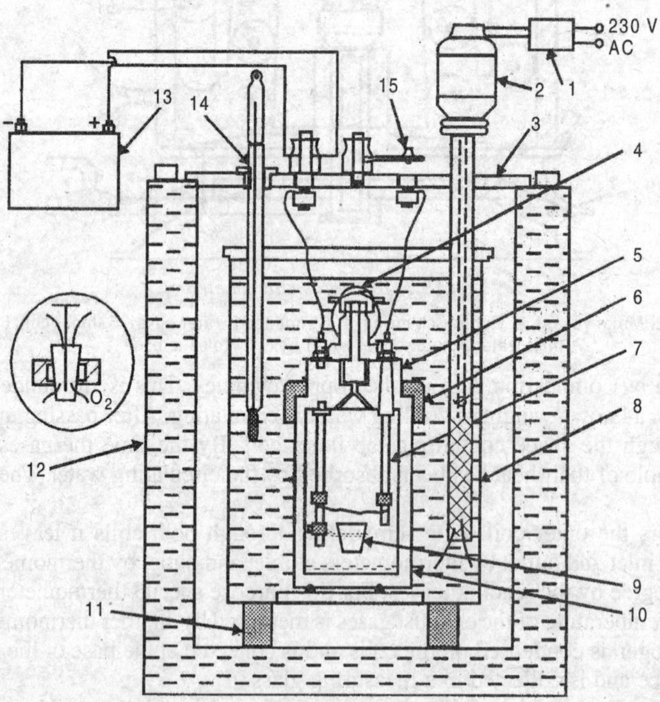

Fig. 13.4. *Bomb calorimeter*

1. Rheostat, 2. Motor, 3. Cover, 4. Cap, 5. Lid, 6. Union Nut, 7. Electrode, 8. Calorimeter vessel, 9. Crucible, 10. Bomb, 11. Support, 12. Outer vessel, 13 Battery, 14. Thermometer, 15. Key

The advantages of using such a procedure for the determination of calorific value are as given below:

(a) Due to high pressure of oxygen supplied, large excess oxygen is provided and therefore the combustion is complete.

(b) All heat liberated will be given to the surrounding water as the combustion is at constant volume.

(c) The addition of oxygen and very small amount of water (generally 10 cc) in the bomb do not affect the combustion.

(d) The test procedure and the arrangement employed favour the accurate computation of temperature loss correction (cooling correction).

The general arrangement of the Bomb calorimeter is shown in Fig. 13.2. The calorimeter consists of a stainless steel vessel called the bomb which is placed in a calorimeter vessel. The bomb is also made of monel metal. The calorimeter is placed in a double walled chromium plated jacket vessel containing water. The calorimeter is closed on top with ebonite cover. This is because of the reduction of radiation losses to the surroundings. An electrically driven stirrer is provided to agitate water in the vessel. A mercury in glass thermometer having accuracy not less than 0.01°C is used to measure the temperature of water surrounding the bomb. A thermometer called Beckman's thermometer is used for this purpose. Generally silica–crucible is used to hold the fuel. For filling the bomb with oxygen, necessary arrangement is provided such as copper tubing, a pressure gauge, control valve on the oxygen cylinder are provided. For test 4 or 12 volt battery is required.

About 1 gm of the fuel is weighed in the crucible. Then 10 cc of water is taken in the bomb. When solid fuel as coal is used, then it is ground to a fine powder and then in a hand press this powder is made into a pallet and while preparing the pallet, a cotton thread will be embedded and connected to the two electrodes. The crucible is kept in a ring connected to the two electrodes. The lid or a screwed cover is then tightened. Then bomb is connected to an oxygen cylinder through a pressure gauge and copper tubing. Oxygen at 25–30 bar pressure is filled in the bomb. Then measured quantity of water is taken such that water will not come to the electric wire leads, (generally this quantity may be 1500 cc or 2500 cc). Motor of the electrodes are connected to the main power supply and the electrodes are connected to the battery through a tap-key.

After this arrangement is done, the set up is ready for the experiment or test.

(1) Preliminary Period. Start the stirrer so that temperature of water in calorimeter will be same and uniform throughout. Take the readings after every one minute till the temperature becomes uniform for 2-3 consecutive readings. This generally requires 5 minutes.

(2) Main Period. Once the temperature in the calorimeter becomes uniform, then the fuel is ignited by pressing the tap-key so that current will flow through the fuse wire and it will glow. A cotton thread is connected to the wire so that fuel pallet will be ignited giving out heat. Continue to take readings of temperature at the interval of 1 minute till the maximum temperature is reached.

(3) After Period. After the maximum temperature is reached, the temperature will start decreasing. Continue to take temperature readings at the interval of one minute. Take 10–15 readings, and then stop the test.

Take the bomb out of the calorimeter and release the pressure by opening the valve. After removing the lid of the bomb, measure the quantity of water in the bomb. The difference in this water will be the weight of H_2O formed because of the combustion of hydrogen in the fuel. This weight is very much essential to measure or determine the lower calorific value of the fuel. Ash content of the fuel is also obtained after weighing the crucible.

Observation Table :

1. Weight of empty crucible — W_1 gm
2. Weight of crucible + fuel — W_2 gm
3. Weight of the fuel taken — $(W_2 - W_1)$ gm
4. Weight of water in the bomb — 10 gm (generally)
5. Weight of water after test — $10 + x$ gm
6. Weight of moisture condensed — x gm
7. Weight of water in the calorimeter — m_w gm
8. Weight of water equivalent of calorimeter — m_e gm

9. Weight of fuse wire — mf_w gm
10. Calorific value of fuse-wire — (C.V.) FW
11. Weight of crucible + ash
12. Weight of ash — $W_4 = W_3 - W_1$ gm
13. Weight of ash.
14. Observed temperature rise during main period — $\Delta t = (T_n - T_0)$

Temperature Record

Initial Period		Main Period		After or Find	
Time	Temperature	Time	Temperature	Time	Temperature
0 time (min)		6 (min)		11	
⋮		⋮		$\}t_{0_1}$	
⋮		⋮		12	
⋮		⋮		$\}t_{0_2}$	
⋮		⋮		13	
5	To	10 say		⋮	
				20	

Cooling correction by Whitaker method

Average rate of change of temperate in after period

$$DT_2 = \frac{t_{0_1} + t_{0_2} + \ldots}{\text{No. of reading taken}}$$

$$\text{Then cooling correction } T_c = (DT_2) \times (\text{No. of reading in main period})$$

$$\therefore \text{ Corrected rise in temperature} = (T_n - T_0) + T_c + T_t$$

where T_t = Minimum temperature that can be read on thermometer
= 0.01 with Beckman thermometer.

Calculations for Calorific Value of Fuel

Total heat liberated = Total heat absorbed

Heat released by fuel + Heat released by fuse wire + Heat released by cotton thread = Total heat absorbed by (Water + Bomb + Calorimeter)

$(m_f + \text{C.V})_{\text{fuel}} + (m \times \text{C.V})_{\text{fuese wire}} + (m \times \text{CV})_{\text{Thread}}$ = (Total weight of water and water equivalent of calorimeter) × C_{Pw} × (Total rise of temperature of water)

$$(g_m \times \text{J/gm})_{\text{fuel}} + (g_m \times \text{J/gm})_{\text{fuse wire}} + (g_m \times \text{J/gm})_{\text{thread}} = g_m \times \text{J/kg°C} \times \text{°C}$$

$$\text{Joules/gm} = \text{Joules/gm}$$

or

$$\text{kJ/kg} = \text{kJ/kg}$$

Calorific value obtained with observation taken on Bomb Calorimeter Test, is the higher calorific value of the fuel (HCV).

Lower calorific value (LCV) of the fuel is then obtain as

$$\text{LCV} = \text{HCV} - (\text{weight of moisture formed per kg fuel} \times 2466)$$

For liquid fuels, the same procedure is followed except of making a pallet of a coal powder. One end of the cotton thread will be connected to the fuse wire and the other end is dipped in the liquid fuel.

13.5 FUELS AND COMBUSTION

The fuel is a material which when once raised to its ignition temperature continues to burn if sufficient oxygen or air is available. The principal constituents of any fuel are carbon and hydrogen. The materials which evolve heat after burning, are called combustibles. Carbon and hydrogen are combustibles. Sulphur is also a combustible material.

When anything slowly combines chemically with oxygen, the process is called *oxidation*. When the same process occurs with a considerable swiftness and exotherm chemical reaction, it is called *combustion*; whereas such a process with almost instantaneous action is called *detonation*.

The mechanical engineers are interested in combustion. Chemists are interested in oxygenation while designers of arms, ammunition, missiles etc. are interested in the process of detonation.

Main priorities of fuels that in general interest to an engineer are

1. Heat liberated by complete combustion i.e., the calorific value of the fuel
2. Combustion temperature
3. Air required for complete combustion and
4. Volume and composition of the products of combustion.

Combustion Temperature

The combustion temperature indicates the degree of calorific intensity of the fuel. Quicker the rate of burning of a fuel, larger is the calorific intensity. The combustion temperature is the temperature of the burning fuel as measured on the products of combustion. If C_{Pm} is the mean specific heat in kJ/kg-K of the products, W_P kg is the mass of the products of combustion per kg fuel burnt, Q kJ the amount of heat released by combustion, t_a the temperature of atmosphere, then the combustion temperature t_c is given by the equation

$$Q = m_p C_{p_m} (t_c - t_a)$$

The amount of air supplied per kg fuel, as well as the completeness of combustion influence the combustion temperature of fuels.

13.5.1 Chemical Equations (Combustion Equations) Involved in Combustion

The process of combustion of elements and compounds can be represented by the chemical equations. These equations are also called as combustion equations. The equations can be treated as those dealing with molar weights or molar volumes (kmol). These dealings can further be simplified by the fact, that equal volumes of different gases at same pressure and at any temperature sufficiently above the liquifying temperature of the respective gas, always contain the same number of molecules. Thus the relative weights of equal volumes of gases will be the same as their molar weights. Thus, at constant pressure, in case of complete combustion on carbon, is represented by

$$C + O_2 = CO_2$$

So that, 1 kmol of carbon + 1 kmol of oxygen = 1 kmol of carbon dioxide

This equation represents molecules.

Similarly, (negligible volume of C) + (1 volume of O_2) = (1 vol of CO_2)

This equation represents volumes.

Again, (1 molar weight of carbon) + (1 molar weight of oxygen) = (1 molar weight of carbndioxides)

This equation represents weights or masses.

Thus, (12 kg of carbon) + (32 kg of oxygen) = (44 kg of carbon dioxide)

or (1 kg of carbon) + (8/3 kg of oxygen) = (11 /3 kg of carbon dioxide)

From this, we see that 1 kg of carbon requires 8/3 = 2.667 kg of oxygen to procedure 11 /3 3.667 kg of carbon dioxide or negligible volume of carbon combines with 1 cu.m (m^3) oygen and produces 1 m^3 of carbon dioxide. In these two cases, both total weight and total volume of the gases before combustion have remained the same even after combustion.

Let us now consider the case of hydrogen chemical or combustion equation is

$$2H_2O + O_2 = 2H_2O$$

i.e., $2\,\text{kmol} + 1\,\text{kmol} = 2\,\text{kmol}$

or $2\,m^3 + 1\,m^3 = 2\,m^3$ by volume

or $4\,kg + 32\,kg = 36\,kg$ by weight

or $1\,kg + 8\,kg = 9\,kg$ by weight

Thus, the total volume before combustion was ($2 + 1 = 3\,m^3$) and became $2\,m^3$ after combustion. Thus contraction took place during combustion. According to the principle of conservation of mass, the total weight before and after combustion is unaltered. Similarly, in case of some other combustion process expansion takes place during the process.

The above type of equations can also be made to show the amount of heat liberated or absorbed during the process, depending upon whether the process is exothermic or endothermic Thus, for example,

$$C + O_2 = CO_2 + 406976\,kJ$$

This means that complete combustion of 1 kmol (12 kg) of carbon produces 1 kmol (44 kg) of carbondioxide, thereby liberating 406976 kJ of heat. Similarly,

$$2H_2 + O_2 = 2H_2O + 573619\,kJ$$

means that two kmols (4 kg) of hydrogen liberate 573619 kJ heat during its complete combustion.

We will now, consider some more examples of combustions.

If carbon burns in insufficient supply of oxygen, the product of combustion will be carbon monoxide instead of carbon dioxide, and the process will be represented by the chemical equation

$$2C + O_2 = 2CO$$

$$\text{Carbon} + \text{Oxygen} = \text{Carbon monoxide}$$

i.e. $2 \times 12 + 32 = 56$

$$1\,kg(C) + \frac{4}{3}kg(O_2) = \frac{7}{3}kg(CO)$$

The carbon monoxide produced during the above combustion is a fuel and therefore it should not be allowed to escape otherwise it would be a loss. It is highly poisonous and has no smell, therefore it is dangerous if it escapes.

Also, combustion of carbon monoxide

$$2CO + O_2 = 2CO_2$$

$$2 \times 28 + 32 = 2 \times 44 \text{ by weight}$$

$$1 + \frac{4}{7} = \frac{11}{7} \text{ kg by weight}$$

Sulphur is also one of the combustibles. It is of very minor importance in contributing to the calorific value of a fuel because only small quantities are present and its calorific value is very low being 9260 kJ. It has got injurious effects because it forms, on burning, sulphuric acid which corrodes the metal and therefore fuel having higher percentage of sulphar is undesirable particularly for steam raising purposes.

When sulphur burns with oxygen, it produces sulphur dioxide. The combustion of sulphur can be represented by a chemical equation

$$S + O_2 = SO_2$$

i.e.,

$$\text{Sulphur} + \text{Oxygen} = \text{Sulphur dioxide}$$
$$32\,\text{kg} + 32\,\text{kg} = 64\,\text{kg}$$
$$1\,\text{kg} + 1\,\text{kg} = 2\,\text{kg}$$
$$\therefore \quad \text{Sulphur} + \text{Oxygen} = \text{Sulphur dioxide}$$

13.6 MINIMUM QUANTITY OF AIR REQUIRED FOR THE COMPLETE COMBUSTION OF ONE KG OF SOLID OR LIQUID FUEL

It will be seen from the above discussion that an adequate supply of oxygen is essential for complete combustion. If the combustion is complete, then and then only maximum heat is available from the given fuel. The theoretically exact amount of oxygen required can be calculated with the help of various chemical or combustion equations discussed above.

These calculations will give us the minimum or theoretical quantity of oxygen if we know the ultimate analysis of the fuel.

The oxygen required for combustion is taken from the atmospheric air although in some cases a certain amount of oxygen is a constituent of the fuel. Air is a mixture of oxygen, nitrogen, a very amount of carbon dioxide and very small traces of rare gases such as neon, argon, krypton etc. For all practical purposes we assume that air is made up of 23% by weight of oxygen and the remaining 77% by weight of nitrogen. If considered by volume, air contains 21% of oxygen and 79% of nitrogen. Once we know the amount of oxygen necessary for complete combustion of 1 kg of fuel, we can determine the weight of air necessary for complete combustion of 1 kg fuel. This minimum or theoretical amount of air required for complete combustion of 1 kg, 1 kg fuel is also known as *stoichiometric air.*

Let C, H and S be the weights of carbon, hydrogen and sulpher contained in 1 kg of fuel

$$\therefore \quad O_2 \text{ required for C} = \frac{8}{3}\text{C kg} = 2.667\,\text{kg}$$

$$O_2 \text{ required for H} = 8\,\text{H kg}$$

and $$O_2 \text{ required for S} = \text{S kg}$$

Therefore, total oxygen required for 1 kg fuel,

$$= \left(\frac{8}{3}\text{C} + 8\text{H} + \text{S}\right)\text{kg}$$

$\therefore$ Minimum quantity of air required for complete combustion of 1 kg fuel

$$= \frac{100}{23}\left(\frac{8}{3}\text{C} + 8\text{H} + \text{S}\right) \text{kg/kg fuel}$$

$$\therefore \quad \text{Minimum air} = \frac{100}{23}\left(\frac{8}{3}\text{C} + 8\text{H} + \text{S}\right) \text{kg/kg fuel}$$

If fuel contains 0 kg of oxygen then total minimum oxygen required/kg fuel

$$= \left(\frac{8}{3}\text{C} + 8\text{H} + \text{S} - 0\right)$$

$$= \left[\frac{8}{3}\text{C} + 8\left(\text{H} - \frac{0}{8}\right) + \text{S}\right]$$

Here $\left(\text{H} - \frac{0}{8}\right)$ is called the available hydrogen for combustion.

∴ Minimum amount of air required for complete combustion of 1 kg fuel.

$$(\text{Air})_{\min} = \frac{100}{23}\left[\frac{8}{3}+\left(\text{H}-\frac{0}{8}\right)+\text{S}\right]$$

13.7 EXCESS AIR

In the previous section we have calculated the minimum quantity of air required for complete combustion of one kg fuel or 1 cu.m of gaseous fuel. In actual practice we supply air more than the theoretical minimum amount in order to ensure the complete combustion of the fuel because all of the air supplied does not come into intimate contact with the particles of the fuel. Weight of air supplied over and above the theoretical air supplied, is called excess air. Generally this is given as a percentage of the theoretical air.

A large amount of excess air has a cooling effect on the process of combustion and represents a loss, and in order to avoid this cooling effect air is preheated before it enters the furnace of a boiler. With natural draught system the excess air is more in comparison with the artificial draught system. The excess air may approach 100% but the modern practice is to use 20–25% or even 50%. When the Lancashire type boiler uses hand firing with natural draught system 10 to 12% CO_2 in the flue gases would be considered good practice. With mechanical stokers and artificial draught it would be quite reasonable to expect 12 to 15% CO_2.

In case of I.C. engines all the air taken in during the induction, or suction stroke will not come in contact with the fuel particles. Therefore excess air is supplied and it should be reduced to minimum to get more specific output.

If the supply of air is less than the minimum required then the mixture is rich and if the supply of air is more than 30% in excess of the theoretical minimum, the mixture is known as lean or weak mixture.

13.8 MASS FRACTION AND MOLE FRACTION

Let *A*, *B*, and *C* be three gases in a mixture. Let each of these gases having a mass of m_a, m_b and m_c.

Then,

$$m = \text{Total mass of the mixture considered}$$

Then, mass fraction of the mass of the constituent is defined as the ratio of the mass of the constituent to that of the mixture and is denoted by m_f.

Thus, $$m = m_a + m_b + m_c$$

and mass fraction of gas *A* is given by

$$m_{fa} = \frac{m_a}{m}$$

Similarly $$m_{fb} = \frac{m_b}{m}$$

$$m_{fc} = \frac{m_c}{m}$$

∴ $$\sum m_{fa} = \frac{m_a}{m}+\frac{m_b}{m}+\frac{m_c}{m} = \frac{m_a+m_b+m_c}{m}$$

Mole is an unit used to express a quantity of gas whose mass is equal to its molecular weight in magnitude. Thus mass is given in kg, the one mole is represented by K-mol. Similarly, when the molecular weight is expressed in gm or pound, mole will be represented by gram-mole or g mole and pound-mole or p.mol. Thus one g mol of O_2 is 32 gm and p.mol, if H_2O is 18.0156 pounds and so on. The number of moles *n*, is obtained by dividing the mass *m*, in grams or pound–mass or kilograms, by the molecular weight *M*.

$$\therefore \qquad n = \frac{m}{m}$$

Thus in a mixture of gases *A*, *B* and *C* if n_a, n_b and n_c are the moles, then the total number of moles is $n_a + n_b + n_c = n$ sag.

Then, mole fraction of a component in a gas mixture is defined as the ratio of number of moles of the component to the total number of moles of the mixture.

$$n_{fa} = \frac{n_a}{n}, n_{fb} = \frac{n_b}{n} \text{ and so on.}$$

13.9 FLUE GAS (EXHAUST GAS) ANALYSIS

When the combustion of the fuel takes place, the products of combustion will be carbon dioxide, (some times carbon monoxide may also be present) sulphur dioxide, vapour and nitrogen all the while, oxygen also will be present in the flue gases.

Generally, the composition of solid or liquid fuels is expressed by weight, whereas that of the gaseous fuels is expressed by volume. Products of combustion are in the gaseous state and the analysis of the products of combustion will be in volumes or volumetric analysis will be obtained.

In order to get the analysis of the exhaust gases by volume, first the analysis of the exhaust gases is found by weight. i.e., the weights of the individual products are first determined and then the percentage analysis of the gas is found. When the weight analysis is known, it can be converted into volumetric analysis by the help of Avogadro's hypothesis.

13.9.1 Analysis by Weight or Gravimetric Analysis

(a) Theoretical Air Supplied. As before, let C, H, S and O be the weight of carbon, Hydrogen sulphur and oxygen per kg of fuel. Then the chemical equations can be written as

$$C + O_2 = CO_2$$

$$1\,\text{kg} + \frac{8}{3}\,\text{kg} = \frac{11}{3} CO_2$$

i.e., 1 kg carbon produces $\frac{11}{3}$ kg of CO_2

Similarly
$$2H + O_2 = 2H_2O$$

$$1\,\text{kg} + 8\,\text{kg} = 9\,\text{kg}$$

or 1 kg hydrogen produces 9 kg of H_2O

$$S + O_2 = O_2$$

$$1\,\text{kg} + 1\,\text{kg} = 2\,\text{kg}$$

or 1 kg of sulphur produces 2 kg of SO_2

$$\text{Theoretical air required} = \frac{100}{23}\left[\frac{8}{3}C + 8H + S - O\right]$$

∴ Nitrogen present in the gas will be kg of fuel

$$N = \frac{77}{23}\left[\frac{8}{3}C + 8H + S - O\right] \text{kg/kg fuel}$$

∴ The total weight of the products per kg of fuel

$$= \frac{11}{3}C + 9H + 2S + \frac{77}{23}\left[\frac{8}{3}C + 8H + S - O\right]$$

$$= P \text{ says}$$

∴ Gravimetric analysis of the products of combustion of 1 kg fuel is

$$CO_2 \rightarrow \frac{11}{3}C \times \frac{100}{P}$$

$$H_2O \rightarrow \frac{9H \times 100}{P}$$

$$SO_2 \rightarrow \frac{2S \times 100}{P}$$

$$N_2 \rightarrow \frac{77 \times 100}{23P}\left[\frac{8}{3}C + 8H + S - O\right]$$

All these quantities can be tabulated as given below.

Constituent of fuel	*Wt. per kg/fuel*	*O_2 required*	*Product*	*% by weight of product*
Carbon	C	$\frac{8}{3}C$	$CO_2 = \frac{11}{3}C$	$\frac{11}{3}C \times \frac{100}{P}$
Hydrogen	H	8 H	$H_2O = 9H$	$\frac{9H \times 100}{P}$
Sulphur	S	S	$SO_2 = 2H$	$2S \times \frac{100}{P}$
Oxygen	O	O	—	
Nitrogen	–	–	$N_2 = \frac{77}{23}\sum O_2$	$\frac{77}{23} \times \frac{100}{P}\sum O_2$

When H_2O vapours are included in the exhaust or flue gases, then the products of combustion are wet. When H_2O vapours are excluded from the analysis, then the products of combustion are known as Dry Products of Combustion and in practice we consider the dry products of combustion. With the exclusion of the H_2O vapours, the table will be slightly modified accordingly. Note here that hydrogen has to be considered for finding the total oxygen required. N_2 will depend on this total oxygen required.

(b) Excess Air Supplied. Let *x* be the excess air supplied per kg of fuel. Then as before, consider 1 kg fuel to contain carbon (C), hydrogen (H), sulphur (S) and oxygen (O) and the products of combustion will be

$$CO_2 \rightarrow \frac{11}{3}C$$

$$H_2O \rightarrow 9H$$

$$SO_2 \rightarrow 2S$$

$$O_2 \rightarrow \frac{23}{100}x$$

$$N_2 \rightarrow \frac{77}{23}\left[\frac{8}{3}C+8H+S-O+\frac{23}{100}x\right]$$

Total weight of wet products = P

∴ Percentage analysis by weight of the wet products of combustion will be

$$CO_2 = \frac{11}{3}C\times\frac{100}{P}$$

$$H_2O = 9H\times\frac{100}{P}$$

$$SO_2 = \frac{2S\times100}{P}$$

$$O_2 = \frac{23}{100}x\times\frac{100}{P}$$

$$N_2 = \frac{77}{23}\left[\frac{8}{3}C+8H+S-O+\frac{23}{100}x\right]$$

Similarly, the percentage analysis by weight, of the dry products of combustion will be

$$CO_2 = \frac{11}{3}C$$

$$SO_2 = 2S$$

$$O_2 = \frac{23}{100}x$$

$$N_2 = \frac{77}{23}\left[\frac{8}{3}C+8H+S-O+\frac{23}{100}x\right]$$

Total weight of dry products = P_1

∴ Percentage analysis is given as

$$CO_2 = \frac{11}{3}C\times\frac{100}{P_1}$$

$$SO_2 = 2S\times\frac{100}{P_1}$$

$$O_2 = \frac{23}{100} \times \frac{100}{P_1}$$

$$N_2 = \frac{77}{23}\left[\frac{8}{3}C + 8H + S - O + \frac{23}{100}x\right] \times \frac{100}{P_1}$$

13.9.2 Analysis by Volume or Volumetric Analysis

Once we have the analysis of the products of combustion by weight, we can convert this into volumetric analysis of the products of combustion. Remember, for this purpose we make use of Avagadro's law or hypothesis (molecular weights of all gases at NTP occupy equal volumes). Therefore, weight or gravimetric analysis can be converted into volumetric analysis by dividing the weight of the gas by its molecule weight to get the volume of the gas.

Now consider the percentage analysis of the wet products.

Product Weight	*Molecular Weight*	*Relative Volumes* V
$CO_2 = \frac{11}{3}C \times \frac{100}{P}$	44	$\frac{11}{3}C \times \frac{1}{44}$
$H_2O = 9H \times \frac{100}{P}$	18	$9H \times \frac{1}{18}$
$SO_2 = 2S \times \frac{100}{P}$	64	$2S \times \frac{1}{64}$
$O_2 = \frac{23}{100}x \times \frac{100}{P}$	32	$\frac{23}{100}x \times \frac{1}{32}$
$N_2 = \frac{77}{23}\left[\frac{8}{3}C + 8H + S - O + \frac{23}{100}x\right] \times \frac{100}{P_1}$	28	$\frac{N_2}{28}$
		Total $= \Sigma V$

$\therefore$ % analysis by volume of the wet products of combustion is given as

$$CO_2 \rightarrow \frac{C}{12} \times \frac{100}{\Sigma V'} H_2O \rightarrow \frac{H}{2} \times \frac{100}{\Sigma V'} SO_2 \rightarrow \frac{S}{32} \times \frac{100}{\Sigma V}$$

$$O_2 \rightarrow \frac{23}{100} \times \frac{100}{32\Sigma V}$$

$$N_2 \rightarrow \frac{77}{23}\left[\frac{8}{3}C + 8HS - O + \frac{23}{100}x\right]\frac{1}{28} \times \frac{100}{\Sigma V}$$

Similarly, volumetric analysis of the dry products of combustion will be as

% analysis

$$CO_2 \rightarrow \frac{11}{3}C \times \frac{1}{44} \qquad \frac{C}{12} \times \frac{100}{\Sigma V_1}$$

$$SO_2 \rightarrow \frac{2S}{64} \qquad \frac{S}{32} \times \frac{100}{\Sigma V_1}$$

$$O_2 \rightarrow \frac{23}{100} x \times \frac{1}{32} \qquad \frac{23}{100} \times \frac{x}{32} \times \frac{100}{\Sigma V_1}$$

$$N_2 \rightarrow \frac{77}{23}\left[\frac{8}{3}C + 8HS - O + \frac{23}{100}x\right] \times \frac{1}{28} \qquad \frac{77}{23}\left[\frac{8}{3}C + 8HS - O + \frac{23x}{100}\right] \frac{100}{28x} \div \Sigma V_1$$

Now, when the volumetric analysis of the products of combustion is known, we can convert the volumetric analysis into gravimetric analysis in converting volume into weight of the gas by multiplying the volume with the molecular weight of the gas. Thereby, weights of all constituents of the flue or exhaust gases are determined and therefore total weight of the gas and then the percentage composition of the gas can be determined. Table given below will show this procedure of converting volumetric analysis into gravimetric analysis.

The illustrative examples will give the reader idea about the conversion of gravimetric into volumetric analysis and vice-versa.

13.9.3 Conversion between Gravimetric and Volumetric Analysis

For converting the gravimetric analysis of the products of combustion into volumetric analysis and vice-versa, two tables are given below.

(a) Gravimetric to Volumetric Analysis

Sr. No.	*Name of the gas*	*Molecular weight (a)*	*% composition by weight (b)*	*Relative volume* $c = \frac{b}{a}$	*% composition volume* $= \frac{c}{\Sigma c} \times 100$
1 2					
			$\Sigma b = 100$	Σc	100

(b) Volumetric to Gravimetric Analysis

Sr. No.	*Name of the gas*	*Molecular weight (a)*	*% composition by weight (b)*	*Relative volume* $c = a \times b$	*% composition volume* $= \frac{c}{\Sigma c} \times 100$
1 2					
				Σc	100

13.9.4 Flue Gas Analysis (Using Orsat Apparatus)

A simple and convenient apparatus used for analysis of dry flur gases by volume is called an Orsat Appartus (Fig. 13.5). It consists of a graduated measuring bottle also called as eudiometer, an aspirating bottle, three double reagant pipettes to absorb CO_2, O_2 and CO. Eudiometer is connected to aspirating bottle by means of rubber tube.

The first reagent pipette next to eudiometer contains the KOH solution (33% KOH + 67% water by weight) to absorb CO_2. The second pipette contains alkaline solution of pyrogallic acid (5 gm of pyrogallic acid in 15 cc of water + 33 gm KOH in 67 gm H_2O) to absorb O_2 and the third pipette contains acidic solution of cuprous chloride (5 cc of cuo + 1000 cc of HCl) to absorb CO.

Each pipette is made of two pipettes joined together by a glass tube at the bottom as shown. Front row of pipette is connected to the front flue gas tube and rear row of pipette are connected to another tube which runs parallel to flue gas sample tube. Each front pipette is provided with number of glass tubes inside to increase the wetted surface area, hence it accelerates action of the gas absorption.

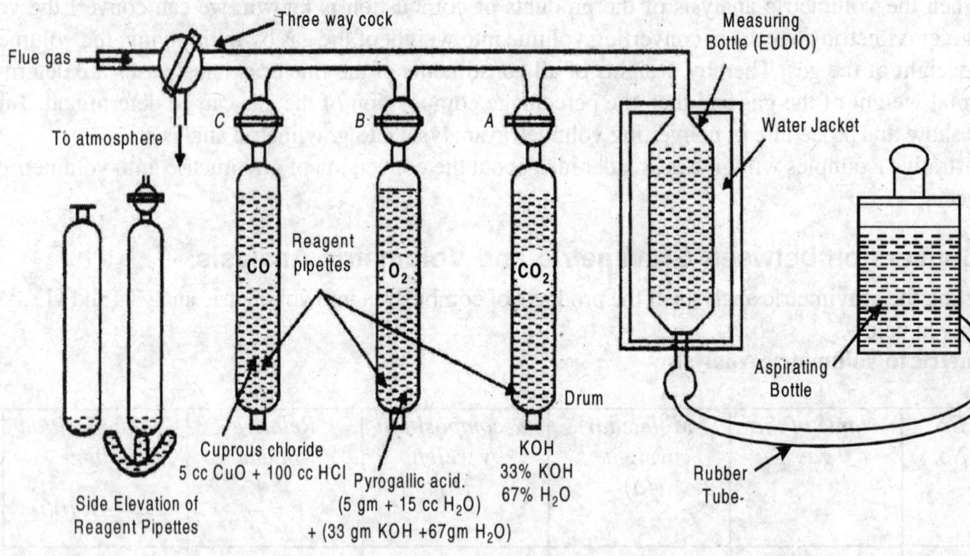

Fig. 13.5. *Orsat apparatus*

Firstly, the existing air or the gas is expelled from the eudiometer by raising the aspirating bottle and keeping three way valve open to atmosphere. The three cocks *a, b, c* and a three way cock are closed. The aspirating bottle is lowered so that the eudiometer reads zero on graduated scale. The three way cock is open and the sample of flue gas is drawn in the measuring bottle. The flue gas is expelled by raising the aspirating bottle. Now the three way cock is kept in the closed position.

Now the cock *a* is opened and the sample of the gas is forced into the first reagent pipette containing KOH by raising the aspirating bottle. Here the CO_2 component of the flue gas is absorbed. Aspirating bottle is moved up and down several times to ensure the complete absorption of CO_2 by KOH solution in reagent pipette. The sample of gas is returned to eudiometer and cock *a* is closed by keeping the reagent to its original level of solution. The eudiometer reading is taken and the difference of two readings gives the percentage of CO_2 by volume in the gas sample.

The same procedure is repeated with the pipettes *b* and *c* to find the percentage analysis of O_2 and CO respectively in the flue gas. The remainder is N_2 gas.

Thus, Orsat apparatus gives analysis of the dry gas.

13.9.5 Determination of the Quantity of Air Supplied per kg of Fuel from the Analysis of Flue Gas given in Weight (Gravimetric Analysis given)

Let C be the percentage of carbon in one kg of fuel and the percentage analysis of the dry flue gas by weight is as given below

$$\text{Carbon dioxide} = C_1\%$$
$$\text{Carbon monoxide} = C_2\%$$
$$\text{Oxygen} = 0\%$$
$$\text{Nitrogen} = N_2$$

Weight of carbon present in carbon dioxide

$$= \frac{12C_1}{44} = \frac{3C_1}{11}$$

Similarly weight of carbon in carbon monoxide

$$= \frac{12C_2}{28} = \frac{3C_2}{7}$$

∴ For flue gas–ratio of carbon to nitrogen

$$\frac{\text{Carbon}}{\text{Nitrogen}} = \frac{\frac{3C_1}{11} + \frac{3C_2}{7}}{N}$$

If we neglect the nitrogen present in the fuel, the above ratio is also the ratio of the weight of Carbon in the fuel to the nitrogen in the air supplied. Let W be the weight of air supplied per kg of fuel.

∴ Weight of carbon in fuel = $C/100$

Weight of nitrogen in air = $0.77W$

$$\therefore \quad \text{Ratio} = \frac{C/100}{0.77W}$$

$$\therefore \quad \frac{C}{77W} = \frac{\frac{3C_1}{11} + \frac{3C_2}{7}}{N}$$

$$\therefore \quad W = \frac{NC}{77\left[\frac{21C_1 + 33C_2}{77}\right]}$$

$$= \frac{NC}{21\,C_1 + 33\,C_2}$$

13.9.6 Determination of Actual air Supplied peı kg Fuel when Volumetric Analysis of the Dry Flue Gases is given

In the power plants the amount of fuel used is very large and consequently the amount of air supplied to a boiler furnace is also great and so it cannot be measured directly. Its indirect measurement can be done if we know the volumetric analysis of the dry flue gas. The main constituents of the flue gases are carbon dioxide, carbon monoxide, oxygen and nitrogen. Quantity of sulphur dioxide is negligibily small and the water vapours are condensed and do not enter the flue gas.

Let C be the percentage by weight of carbon in the fuel. And the percentage volumetric analysis of the dry products of combustion is given as

$$\begin{aligned} \text{Carbon Dioxide} &= C_1 \\ \text{Carbnon Monoxide} &= C_2 \\ \text{Oxygen} &= 0 \\ \text{Nitrogen} &= N \end{aligned}$$

We will convert this volumetric analysis into gravimetric analysis and then the procedure is same as the above article.

∴ To convert volumetric analysis into gravimetric analysis, multiply volume by its molecular weights to get relative weights.

∴ Relative weights of these constituents are

$$\begin{aligned} \text{Carbon dioxide} &= 44\,C_1 \\ \text{Carbon monoxide} &= 28\,C_2 \end{aligned}$$

∴ Carbon content of carbon dioxide

$$= \frac{12C_1}{44}\times\frac{44}{1} = 12\,C_1$$

Carbon contents of carbon monoxide

$$= \frac{12\times 28C_2}{28} = 12\,C_2$$

∴ Total carbon contents $=12(C_1+C_2)$

∴ In exaust or flue gases,

$$\text{Ratio}\frac{\text{Carbon}}{\text{Nitrogen}} = \frac{12(C_1+C_2)}{28N} \qquad \text{...(I)}$$

Let W be the weight of air supplied / kg fuel

$$\therefore \quad \text{Ratio}\frac{\text{Carbon}}{\text{Nitrogen}} = \frac{C/100}{0.77W} \qquad \text{...(II)}$$

$$\therefore \quad \text{I} = \text{II}$$

$$\therefore \quad \frac{12(C_1+C_2)}{28N} = \frac{C}{77W}$$

$$\therefore \quad W = \frac{C}{77}\times\frac{28N}{12(C_1+C_2)}$$

$$= \frac{NC}{33(C_1+C_2)}$$

$$\therefore \quad W = \text{Wt. of air supplied/kg fuel} = \frac{NC}{33(C_1+C_2)}$$

Note: All values are in percentage

To find excess air supplied, first we have to consider the oxygen required for combustion of carbon monoxide to carbon dioxide and after deducting this oxygen from the oxygen present in the flue gases, the excess oxygen is obtained. Once this excess oxygen is obtained, then we can find the weight of excess air supplied.

The illustrative example given below will give you an procedure for accounting oxygen for the combustion of carbon monoxide to carbon dioxide.

Approximate equation for excess air supplied is

$$\text{Excess air supplied/kg fuel} = \frac{79°C}{21 \times 33(C_1 + C_2)}$$

This equation does take into consideration the complete combustion of carbon to carbon dioxide and hence approximation.

Note: All values are in percentages.

13.10 DETERMINATION OF AIR FUEL RATIO WITH THE HELP OF DRY FLUE GAS ANALYSIS

When the volumetric analysis of dry gas is given, then the amount of air supplied per kg of fuel can be obtained by any one of the following methods (Carbon content of the fuel is not given directly)

1. Carbon balance
2. Hydrogen balance
3. Carbon-hydrogen balance
   (i) with known composition of fuel
   (ii) with unknown composition of fuel
4. Oxidized product method

Of these method (2) and (3) required that three should be sufficient proportion of hydrogen in the fuel and hence they are not applicable for boiler fuel as coal.

We can write the chemical equation by considering a certain moles of fuel and corresponding moles of air such that the analysis of dry products given will be obtained. For balancing purpose, we have the consider the moles of H_2O produced.

(i) Carbon Balance. In this method, carbons in the reactants (fuel) and the products are balanced and thereby getting the moles of fuel taken. The condition required for this method is that the formation of free carbon does not take place.

(ii) Hydrogen Balance. After writing the combustion equation, we can proceed to balance the hydrogen also and the unknown.

(iii) Carbon-Hydrogen Balance. This method is very much useful when the composition of the fuel is now known.

These methods will be better understood with the help of illustrative examples.

13.11 INTRODUCTION OF CHEMISTRY OF COMBUSTION

It is seen before that thermodynamic relationships determine the equilibrium composition of product mixtures resulting from chemical reaction. The nature of the product species, their proportions, their temperature, and their pressure depend upon three major controlling factors. For *one thin*, the composition of the products of reaction is affected by material balance considerations. This requires that the total quantity of each chemical element in the product mixture be the same at that in reactants. We shall therefore consider the calculation of elemental composition of various mixtures and examine the laws of stoichiometry which apply in chemical reactions. *Second*, the resultant temperature and pressire of product mixtures depend upon the first law of thermodynamics. Thus dictates that the total energy be conserved, just as the total ass is. Hence, this gives attention to calculation of *flame temperatures* of combustible mixtures. *Third*, equilibrium composition is determined by relationships that arise indirectly from the second law of thermodynamics. The second law dictates that the composition of the products depends on temperature and pressure, whereas the first law dictates that the temperature and pressure of the product depend on their composition. Clearly, then, complex interactions are involved, and problems of calculation of product composition can be handled only by considering both laws simultaneously.

Although, the principles outlined here apply to any chemical reaction, particular attention will be given to an important class of chemical reactions, the combustion processes. Compustion may be defined as a rapid exo-thermic reaction between a fuel and an oxidizer in which chemical energy is liberated.

13.12 ENTHALPY OF FORMATION

To establish an enthalpy scale of elements and compounds, it is first necessary to choose an arbitrary datum for zero enthalpy. It was conventionally agreed to assign a zero value to the enthalpy of elements in their most stable forms at a standard state of 1 standard atmospheric pressure and 25°C (77°F). If an element has two or more stable forms, that form which is stable at 25°C is chosen as reference. With this standard reference state established, enthalpies of all species have a common base. The *enthalpy of formation* of a compound is defined as the enthalpy change that accompanies the formation of 1 mole of the compound at the standard reference state from elements at the same state. The enthalpy of formation is given by the symbol $\bar{h}_f^0$, where the subscript *f* refers to the formation of the compound from its elements, and superscript (0) indicates that all reactants and products are at the standard state. Thus, if elements A_a, B_b, and C_c at the same reference state, then the chemical equation can be written

$$aA(g) + bB\,(g) + cC\,(g) \rightarrow A_a\,B_a\,C_c\,(g)$$

An energy balance gives

$$a\bar{h}_A^0 + b\bar{h}_B^0 + c\bar{h}_C^0 + \Delta\bar{h}_{P0T_0} \rightarrow \bar{h}_{A_aB_bC_c}^0$$

where $\Delta\bar{h}_{P0,T_0}$ is the enthalpy change at the standard reference state per mole of products. By definition, $\bar{h}_A^0$, $\bar{h}_B^0$, and $\bar{h}_C^0$ are zero, and therefore, the enthalpy of formation is equal to the enthalpy of the compound at the standard reference state,

or
$$\bar{h}_f^0 \;=\; \bar{h}_{A,B,C}^0 = \Delta\bar{h}_{P_0T_0}$$

A positive value of $\bar{h}_f^0$ means that the enthalpy of formation of the compound is more than the enthalpy of its elements at the reference state. Such compounds are often unstable.

A compound at a temperature *T* above 25°C (77°F) will then have an enthalpy equal to the sum of its enthalpy of formation plus the sensible enthalpy above the standard state

or
$$\bar{h} \;=\; \bar{h}_f^0 + (\bar{h}_T - \bar{h}_{298})$$

13.13 ENTHALPY AND INTERNAL ENERGY OF REACTION

Chemical energy is part of the internal energy of a system and is attributed to the binding energy between the atoms of the substance. So far we have not considered chemical reactions and consequently chemical energy was excluded. Now, however, chemical energy constitutes the major part of the internal energy term in the first law.

In chemical reactions, one important concern is the amount of heat transfer due to the reaction. Two cases will be considered : (1) when the reaction takes place at constant pressure, (2) when it takes place at constant volume. Consider first, the constant-pressure process.

If no work other than reversible work is done, the first law of thermodynamics applied to a closed system is

$$dQ \;=\; dH - VdP$$

The differential of the enthalpy $H = H(p, T, \lambda)$ is

$$dH = \left(\frac{\partial H}{\partial P}\right)_{T,\lambda} \cdot dP + \left(\frac{\partial H}{\partial T}\right)_{p,X} \cdot dT + \left(\frac{\partial H}{\partial \lambda}\right)_{p,T} \cdot d\lambda$$

Substituting in the first law and letting $dp = 0$ for a constant pressure process, yields

$$dQ_p = \left(\frac{\partial H}{\partial P}\right)_{p,\lambda} \cdot dT + \left(\frac{\partial H}{\partial \lambda}\right)_{p,T} \cdot d\lambda$$

The coefficient of dT,

$$\left(\frac{\partial H}{\partial P}\right)_{p,\lambda} = (n\bar{C}_P)_\lambda$$

where $\bar{C}_P$ is the specific heat per mole at constant pressure. The coefficient of $d\lambda$, $\left(\frac{\partial H}{\partial \lambda}\right)_{p,T}$ is called *enthalpy of reaction* and is given the symbol $\Delta H_{p,T}$. It is equal to the amount of heat interaction when to reaction is carried out at constant pressure with no change in temperature.

Therefore,

$$dQ_p = (n\bar{C}_P)_\lambda \cdot dT + \Delta H_{p,T} d\lambda$$

Integrating the foregoing equation between a reference state (0) and any other state, gives

$$Q_P = \int_{T_0}^{T} (n\bar{C}_p)_\lambda \cdot dT + \int_{\lambda_0}^{\lambda} \Delta H_{p,T} d\lambda$$

If the reaction is stoichiometric where λ changes from 0 to 1, and if, further, temperature at the end of the reaction is equal to the temperature at the beginning of the reaction, then,

$$Q_P = \int_0^1 \Delta H_{p,T} \cdot d\lambda = \Delta H_{p,T}$$

which means that the heat interaction is equal to the enthalpy of reaction subject to the conditions previously state. Accordingly,

$$\Delta H_{p,T} = \left(\frac{\partial H}{\partial \lambda}\right)_{p,T} = \sum_t v_i \bar{h}_1 = \left(\sum_r v_i h_1\right)_p - \left(\sum_i v_i \bar{h}_1\right)_r$$

where subscripts r and p refer to reactants and products respectively. In the case of exothermic reaction, $\Delta H_{p,T}$ is negative and

$$\left(\sum_i v_i \bar{h}_i\right)_p > \left(\sum_i v_i \bar{h}_i\right)_T$$

Similarly, if H_{pT} is positive, the reaction is *endothermic* and

$$\left(\sum_i v_i \bar{h}_i\right)_p > \left(\sum_i v_i \bar{h}_i\right)_T$$

The second case to be considered is constant volume process. The precedure is analogous to the constant pressure case. The first law applied to a closed system when only reversible work is involved is

$$dQ = dU + p \cdot dV$$

The differential of $U = U(V, T, \lambda)$ is

$$dU = \left(\frac{\partial U}{\partial V}\right)_{T,\lambda} dV + \left(\frac{\partial U}{\partial T}\right)_{v,\lambda} dT + \left(\frac{\partial U}{\partial \lambda}\right)_{v,T} d\lambda$$

Substituting for dU in the first law and considering constant volume, then

$$dQ_r = \left(\frac{\partial U}{\partial T}\right)_{v,\lambda} dT + \left(\frac{\partial U}{\partial \lambda}\right)_{u,T} d\lambda$$

where $\left(\frac{\partial U}{\partial T}\right)_{v,\lambda} = \left(n\bar{C}_v\right)_\lambda$ $\bar{C}_v$ being the specific heat per mole at constant volume, and $\left(\frac{\partial U}{\partial \lambda}\right)_{v,T} = \Delta U_{v,T}$ is called the *internal energy of the reaction.*

Therefore,

$$dQ_v = \left(n\bar{C}_v\right)_\lambda dT + \Delta U_{v,T} d\lambda$$

where $\Delta U_{v,T}$, according to the previous equations, is given by

$$\Delta U_{V,T} = \left(\frac{\partial U}{\partial T}\right)_{V,T} = \sum_i v_i \bar{u}_i = \left(\sum_i v_i \bar{u}_i\right)_p - \left(\sum_i v_i \bar{u}_i\right)_r$$

The relation between $\Delta H_{p,T}$ and $\Delta U_{v,T}$ is obtained as follows :

$$\Delta H_{p,T} = \left[\sum_i v_i(\bar{u}_i + p_i v_i)\right]_p - \left[\sum_i v_i(\bar{u}_i + p_i \bar{v}_i\right]_r$$

$$= \left[\left(\sum_i v_i \bar{u}_i\right)_p - \left(\sum_i v_i \bar{u}_i\right)_r\right] + \left[\sum_i v_i p_i \bar{v}_i\right]_p - \left[\sum_i v_i p_i \bar{v}_i\right]_r$$

$$= \Delta U_{V,T} + \left[\sum_i v_i p_i \bar{v}_i\right]_p - \sum v_i p_i \bar{v}_i\Big]$$

In the case of reactions involving liquids and solids, the changes in volume are small and can be neglected in comparison to gaseous substances in the reaction. If, further, the gaseous reactant and products are considered perfect gases, then $p_i\bar{v}_i = \bar{R}T$. Therefore,

$$\Delta H_{p,T} = \Delta U_{V,T} + \bar{R}T\sum_i (v_{i_p} - v_i)$$

Note that the term in parantheses is the change in the number of moles due to reaction. Therefore, $\Delta H_{p,T} = \Delta U_{V,T}$ if there is no change in the number of moles of the gaseous substances due to the reaction.

13.14 ADIABATIC FLAME TEMPERATURE

Under adiabatic conditions, the temperature attained in a reaction is called the *adiabatic reaction or flame temperature*. In actual processes the combustion temperature is considerably less, owing to the inevitable heat interaction with the environment. Also, some of the reactants may fail to react, or may decompose only partially. Dissociation of the products of combustion also results in a further reduction of the combustion temperature. The adiabatic reaction temperature is an upper limit of temperature and therefore is a useful parameter in the design of combustion chambers.

The adiabatic reaction (flame) temperature may be determined from the first law which reduces to

$$\Sigma H_T = \Sigma H_p$$

The adiabatic reaction temperature can be computed, making use of (1) gas tables – % of excess air + Stoichometric air – 100% excess air means gas tables to be used as 200%. (2) enthaply of formation (3) enthalpy of compustion (d) specific heat equations.

13.15 FREE ENERGY AND CHEMICAL EQUILIBRIUM

During the course of chemical reaction, many intermediate compositions are formed before equilibrium is reached and the final composition is attained. In a chemical reaction, the products can react in the reverse direction, forming reactants, just as the reactants form products in the forward direction. At equilibrium, the rates of the forward reaction and the reverse reaction are equal. Now the factors governing the final composition of a chemical system will be considered. The Gibbs function is the main criterian for equilibrium.

The reference state for the Gibbs face energy of formation is 25°C (77°F) and 1 standard atm. The stable form of each element at standard reference state is arbitrarily assigned a value of zero free energy of formation. The free energy change of any reaction is given by

$$\Delta G^\circ = \Sigma(G^\circ_f) - \Sigma(G^\circ_f)_r$$

When a compound is formed from its elements, the resultabt free-energy change at the standard state represents the standard free energy of formation G°_f of the compound. A negative value of ΔG°_f indicates that reactants at the standard state proceeed spontaneously to products at the standard state. Conversely a positive value of ΔG° indicates the reaction does not take place spontaneously. Stand tables are available for the free energy of formation of several substances.

The free energy change of a chemcial reaction under isothermal standard state conditions may be determined according to the equation

$$\Delta G^\circ = \Delta H^\circ - T_o \Delta S^\circ$$

The change of entraply ΔS° is determined from the absolute entropies of the reactants and products at the standard state according to the equation

$$\Delta S^\circ = \Sigma S^\circ_p - S^\circ_i$$

Absolute values of entropy of several pure substances at the standard state (1 atm. and 25°C) are also listed in the same tables. According to the third law of thermodynamics, the entropy of a pure substance in equilibrium approaches zero as the temperature reaches zero. Thus a datum point for entropy is established at zero absolute temperature. Note that elements as well as compounds have entropies greated than zero at 25°C and 1 atm pressure. At other temperatures and pressures the absolute entropy of a perfect gas can be calculated from the Sackur-Tetrode equation which developed from statistical mechanics. When the entropy of a perfect gas is known at one pressure and temperature, its entropy at another set of temperature and pressure conditions can be calculated from

$$S = \bar{S}^\circ + \bar{C}_p \ln \frac{T}{T_o} - \bar{R} \ln \frac{p}{p_o}$$

As in the case of enthalpy of formation, the difference between the entropy of a compound at the standard state and the entropy of its elements at the standard state is the entropy of formation :

$$\Delta S^\circ_f = S^\circ_{\text{compound}} - \Sigma S^\circ_{\text{elements}}$$

The standard state is 25°C and 1 atm pressure.

The Gibbs function can be expressed as

$$dG = dH \; TdS - SdT$$

But is was shown that entropy could be expressed in terms of reversible heat interaction and affinity :

$$T \cdot dS = dQ + Zd\lambda$$

From the first law of thermodynamics, this becomes

$$T \cdot dS = dH - V \cdot dp + Zd\lambda$$

Therefore, the Gibbs function becomes

$$dG = -SdT + V \cdot dp - Z\lambda$$

Since the Gibbs function is a property determined by temperature, pressure and extent of reaction,

$$\left(\frac{dG}{d\lambda}\right)_{p,T} = -Z$$

But the Gibbs function is given by

$$\left(\frac{dG}{d\lambda}\right)_{p,T} = \sum_{i=1}^{i=r} v_i\bar{u} = G_p - \Delta G_r$$

These two equations, when combined, lead to

$$Z = -\Delta G_{p\cdot T} = (\Delta G_r - \Delta G_p)_{p,\,T}$$

A relationship can be shown to exit between affinity and enthalpy of reaction.

$$\therefore \quad Z = -\Delta H_{p,\,T} + T\Delta\, S_{p,\,T}$$

But $\quad Z = -\Delta G_{p,\,T}$

Also since

$$S = -\left(\frac{\partial G}{\partial T}\right)_{p,\lambda}$$

Therefore,

$$\Delta S_{p,T} = \left(\frac{-\partial \Delta G}{\partial T}\right)_{p,\lambda}$$

The substitutions, therefore, lead to

$$\Delta G_{p,T} = \Delta H_{p,T} + T\left(\frac{\partial \Delta G}{\partial T}\right)_{p,T}$$

which is called the *Gibbs–Helmhltz* relation. It shows the effect of changes in temperatures on the Gibbs function.

The last equation may be transformed into can integrable form. Thus

$$\frac{Z}{T} = \frac{\Delta H_{p,T}}{T} + \Delta S_{p,T}$$

Differentiation with respect to temperature alone leads to

$$\frac{\partial}{\partial T}\left(\frac{Z}{T}\right)_{p,\lambda} = \left(\frac{\Delta H_{p,T}}{T^2}\right)_{p,\lambda} = \frac{1}{T}\left(\frac{\partial \Delta H_{p,T}}{\partial T}\right)_{p,\lambda} + \left(\frac{\partial \Delta S_{p,T}}{\partial T}\right)_{p,\lambda}$$

But form the first and second laws of thermodynamics, it can be shown that

$$\left(\frac{\partial \Delta S_{p,T}}{\partial T}\right)_{p,\lambda} = \frac{1}{T}\left(\frac{\partial \Delta H_{p,T}}{\partial T}\right)_{p,\lambda}$$

Therefore,

$$\frac{\partial}{\partial T}\left(\frac{Z}{T}\right)_{p,\lambda} = \left(\frac{\Delta H_{p,T}}{T^2}\right)$$

which when integrated, gives

$$\frac{Z}{T} = \int \frac{\Delta H_{p,T}}{T^2}\, dT + C(dp = 0 \text{ and } d\lambda = 0)$$

where C is a constant of integration.

13.16 EQUILIBRIUM CONSTANT

This topic investigates the condition of equilibrium for a reaction given by the equation

$$v_a A + v_b B + ... + v_c C + v_d D + ... \quad ...(i)$$

The differential of the Gibbs functions is

$$dG = V \cdot dp - SdT.$$

Two conditions are now imposed on the preceding expression. The first is that either the reaction is isothermal; that is, the temperature during the reaction remains unchanged, or, alternatively, the temperature is the same at only the beginning and end of reaction. The second conditions is that only perfect gases are involved. The differential of the Gibbs function then becomes

$$dG = Vdp = \frac{n\bar{R}T}{p} dp$$

Integrating the foregoing equation from the standard state to any other state at the same temperature gives

$$G - G^\circ = n\bar{R}T \ln \frac{p}{p_o}$$

where G° is the free energy at the standard state and p_o is the reference pressure of 1 atm. The above equation on a mole basis may be written

$$\bar{g} = \bar{g}^\circ + \bar{R}T \ln p$$

where p is the pressure in atmospheres.

The change of the Gibbs function, when reactants A, B, ... are transformed to products C, D, ..., is

$$\Delta G = \Sigma G_p - \Sigma G_r$$

$$= (v_c \bar{g}_c + v_a \bar{g}_a + ...)\ (v_a \bar{g}_a + v_b \bar{g}_b + ...)$$

Substituting for $\bar{g}$, this equation gives

$$\Delta G = (v_c \bar{g}_c^0 + v_d \bar{g}_d^0 + ...) - (v_a \bar{g}_a^0 + v_b \bar{g}_b^0 + ...) + (v_c \bar{R}T \ln P_e + v_d \bar{R}T \ln p_a + ...) - (v_a \bar{R}T \ln p_a + v_b \bar{R}T \ln p_b + ...)$$

But $$\Delta G^\circ = (v_c \bar{g}_c^0 + v_d \bar{g}_d^0 + ...) - (v_a \bar{g}_a^0 + v_b \bar{g}_b^0 + ...)$$

then, $$\Delta G = \Delta G^\circ + \bar{R}T \ln \frac{p_c^{v_a}\, p_d^{v_d} \ .....}{p_A^{v_a}\, p_B^{v_b} \ ....}$$

At equilibrium $dG = 0$ and the Gibbs function at the reference state may than be expressed as

$$-\Delta G^\circ = Z^\circ = \bar{R}T \frac{p_c^{v_a}\, p_d^{v_d} \ .....}{p_A^{v_a}\, p_B^{v_b} \ ....}$$

$$= RT \ln K_p$$

K_p is called the equilibrium constant defined by the equation $K_p = e^{-\Delta G/\bar{R}T} = e^{Z^\circ/\bar{R}T} = \dfrac{p_c^{v_c}\, p_d^{v_d} \ .....}{p_A^{v_a}\, p_B^{v_b} \ .....}$

The subscript p is to emphasize that K_p is defined in terms of the partial pressure of reactants and products. Note that ΔG° and K_p are independent of the total pressure. They only depend on temperature and the nature of the reactants and products of the reaction. Recall also that in choosing the reference state, the will of pressure for the components was expressed in atmospheres. Therefore, the value of K_p is based on this unit of pressure. In the preceding definition of K_p the partial pressures of the products are placed in the numerator and the partial pressures of the reactants are placed in the denominator. When the reaction is reversed, the value of the equilibrium constant becomes the reciprocal of that of the forward reaction. Values of K_p as a function of temperature are usually given in the form of an equation, table or graph.

The equilibrium constant may also be expressed in terms of mole fractions. Noting that the mole fraction of a component in a perfect gas mixture is equal to the ratio of the partial pressure of the component to the total pressure, equilibrium constant becomes

$$K_p = \frac{n_C^{\nu_c} n_D^{\nu_a} \ldots}{n_A^{\nu_a} n_B^{\nu_b} \ldots}$$

$$= \left(\frac{p}{\sum_i n_i}\right)^{\nu_C + \nu_a + \ldots \nu_a - \nu_b \ldots}$$

or in terms of the mole fraction,

$$K_p = \frac{x_C^{\nu_c} x_D^{\nu_d} \ldots}{x_A^{\nu_a} x_B^{\nu_b} \ldots} p^{\nu_c + \nu_d + \ldots - \nu_a - \nu_b \ldots}$$

$$= \frac{x_C x_D}{x_A^{\nu_a} x_B^{\nu_b}} \text{ if } \nu_a + \nu_b = \nu_c + \nu_d$$

that is, if the number of moles remain unchanged during the reaction. This means further that at constant temperature and pressure the volume of the gaseous components remains constant. Thus the equilibrium composition of a reacting system at a given temperature and pressure can be determined if the equilibrium constant is known.

The affinity Z can be expressed in terms of K_p as

$$Z = \bar{R}T \ln K_p - \bar{R}T \sum_i \nu_i \ln p_i$$

But

$$Z^\circ = \bar{R}T \ln K_p \text{ then } Z = Z^\circ - \bar{R}T \sum_i \nu_i \ln p_i$$

At equilibrium, $Z = 0$ so that

$$\ln K_p = \sum_i \ln p_i^{\nu_i}$$

or

$$K_p = p_1^{\nu_1} + p_2^{\nu_2} + \ldots + p_r^{\nu_r}$$

$$= \sum_{i=1}^{i=r} p_i^{\nu_i}$$

A relation between the equilibrium constant and the enthalpy of reaction may be derived as follows :

The equilibrium constant is given by $\ln K_p = \dfrac{-\Delta G^\circ}{\bar{R}T}$. But the free energy change in an isothermal process when both reactants and products are at the standard state is $\Delta G° = \Delta H° - T\Delta S°$

Therefore,
$$\ln K_p = \frac{-(\Delta H° - T\Delta S°)}{\bar{R}T} = \frac{-\Delta H°}{\bar{R}T} + \frac{\Delta S°}{\bar{R}}$$

Differentiation of the preceding equation with respect to T and noting that $\Delta H°$ and $\Delta S°$ are constant gives

$$\frac{d \ln K_p}{dT} = \frac{\Delta H°}{\bar{R}T^2}$$

$$\frac{d(\ln K_p)}{d(1/T)} = \frac{-\Delta H°}{\bar{R}}$$

If $\Delta H°$ is positive (endothermic reaction), K_p increases with increasing temperature. Above equation is called *Van't Hoff* equation. This equation may be used to determine the value of $\Delta H°$ as a function of temperature, provided that the variation of K_p with temperature is known.

SOLVED EXAMPLES

(Theoretical Method)

Example 13.1. *A boiler uses anthracite coal containing 8% moisture by weight. The higher calorific value of the coal is observed to be 26000 kJ/kg. Calculate the heat available from complete combustion of the fuel to the boiler flue gases, if the rate of firing in the boiler is 1500 kg per hour. Neglect the hydrogen content of the coal.*

Solution

In case of complete combustion of a fuel, its lower calorific value is available as heat to the products of combustion. In the problem here, the moisture content of the coal is 8%.

∴ Weight of moisture/kg coal = 0.08 kg

∴ Lower calorific value of the coal

$$\begin{aligned} \text{LCV} &= \text{HCV} - 2466 \times 0.08 \\ &= 26000 - 197.28 \\ &= 25802.72 \text{ kJ/kg} \end{aligned}$$

∴ Total heat available to the flue gases per hour

$$\begin{aligned} &= \frac{\text{Coal}}{\text{Hour}} \times \text{LCV} \\ &= 1500 \times 25802.72 \\ &= 38.7041 \times 10^6 \text{ kJ/kg} \end{aligned}$$

Example 13.2. *The composition of motor car petrol was found to be 86%C, 12% H_2, 0.1 % O_2 and 0.5% moisture by weight. Calculate the lower and higher calorific values of petrol.*

Solution

Higher calorific value (HCV) given by,

$$\begin{aligned} \text{HCV} &= 33915 \times \text{C} + 144452.5\left(\text{H} - \frac{0}{8}\right) + 9630\,\text{S} \\ &= 33915 \times 0.86 + 144451.5\left[0.12 - \frac{0.001}{8}\right] + 0.0001 \times 9630 \\ &= 29166.9 + 17332.374 + 0.963 \end{aligned}$$

$$\textbf{HCV} = \textbf{46500.237 kJ/kg}$$

$$\simeq 46500 \text{ kJ/kg Approx.}$$

$$\text{Moisture formed} = 0.005+9\times 0.12$$
$$= 0.005+1.08=1.085 \text{ kg}$$

∴ Lower calorific value is given by

$$LCV = HCV-2466\times 1.085$$
$$= 46500-2675.61$$
$$\textbf{LCV} = \textbf{43924.39 kJ/kg}$$

Example 13.3. *An analysis of a simple Diesel oil fuel is given by H_2 = 12.4%, carbon = 84.6%, sulphur =1.3% and incombustible substances 1.7%. Calculate the lower and higher calorific values of this fuel.*

Solution

Higher calorific value of fuel is given by

$$HCV = 33915\,C+144451.5\left(H-\frac{0}{8}\right)+9630\,S$$
$$= 333915\times 0.846+144451.5\times 0.124+9630\times 0.013$$
$$= 28692.09+17911.986+125.19$$
$$\textbf{HCV} = \textbf{46729.266 kJ/kg}$$

$$\text{Moisture Formed} = 9H=9\times 0.124=1.116 \text{ kg}$$

Lower calorific value is given by

$$LCV = HCV-2466\,(H_2O)$$
$$= 46729.266-2466\times 1.116$$
$$= 46729.266-2752.056$$
$$\textbf{LCV} = \textbf{43977.210 kJ/kg}$$

Example 13.4. *If the higher calorific value of carbon and hydrogen are 36000 kJ/kg and 142300 kJ/kg calculate the lower calorific value of methane, CH_4 kJ/kg.*

Solution

One 1 kmol of CH_4 contains $12\times 1 = 12$ kg of carbon and $1\times 4 = 4$ kg of hydrogen.

$$\therefore \text{Wt. of 1 kmol of } CH_4 = 12+4=16 \text{ kg}$$

$$\text{Carbon in 1 kg } CH_4 = \frac{12}{16}=0.75 \text{ kg}$$

$$\text{Hydrogen in 1 kg } CH_4 = \frac{04}{16}=0.25 \text{ kg}$$

Moisture or H_2O formed per kg of CH_4 is

$$9H = 9\times 0.25=2.25 \text{ kg}$$

Now, heat produced by 0.75 kg carbon

$$= 0.75\times 36000$$
$$= 27000 \text{ kJ}$$

Heat produced by 0.25 kg hydrogen

$$= 0.25\times 142300$$
$$= 35575 \text{ kJ}$$

∴ Total heat produced by combustion of 1 kg

$$= 2700+35575$$
$$= 62575 \text{ kJ/kg}$$

$$\therefore \quad LCV = HCV - 9H \times 2466$$
$$= 62575 - 2.25 \times 2466$$
$$\textbf{LCV} = \textbf{57026.5 kJ/kg}$$

Example 13.5. *Calculate the higher calorific value of the fuel whose proximate analysis is given below as moisture 5% volatile matter 29%, fixed carbon 59%, ash 7%.*

Solution

Goutel formula is to be used when the proximate analysis of the fuel is known.

$$\text{Percentage of fixed carbon} = \frac{59 \times 100}{59 + 29} = \frac{59}{88} \times 100 = 67\%$$

∴ Percentage of volatile matter on dry ash free basis = 100 – 67 = 33%

For dry as free basis volatile matter % of 33 will be

$$\alpha = 410 - (410 - 394) \times 0.6 \text{........ (Refer data Given)}$$
$$= 410 - 0.96$$
$$= 400.04$$
$$\text{Calorific value of fuel} = 343.3 \times 59 + 400.4 \times 29 = 20254.7 + 11611.6$$
$$\text{Calorific value of fuel} = \textbf{31866.3 kJ/kg}$$

Example 13.6. *Estimate the higher and lower calorific values of a gaseous fuel whose volumetric composition is CH_4 = 77%, C_2H_6 – 22.5%, nitrogen = 0.5%. The higher calorific values of the gases are given below – CH_4 = 40,000 kJ/m³, C_2H_6 = 695000 kJ/m³.*

Solution

Let us consider one cu-metre of gas. The volume of methane is 0.77 cubic metre and heat evolved = 0.77 × 40000 = 30800 kJ. The volume of ethane is 0.225 cubic metre and the heat evolved = 0.225 × 69500 = 15637.5 kJ. Nitrogen does not evolve any heat. Therefore the higher calorific value of gaseous fuel is 30800 + 15637.5 = **46437.5 kJ/Nm³**.

Volume of water vapour formed due to combustion of 0.77 cu-m of CH_4 is 2 × 0.77 = 1.54 cu-m.

Volume of water vapour formed due to combustion of 0.225 cu-m of C_2H_6 = 0.225 × 3 = 0.675 cu-m

∴ Total volume of water vapour formed due to combustion of 1 cu-m gaseous fuel

$$= 1.54 + 0.675$$
$$= 2.215 \text{ cu-m.}$$

$$\text{Wt. of one cu-m of water vapour} = \frac{18}{22.41} \quad \text{(1 molar volume = 22.41cu-m. (This ways 18 kg } (H_2O)\text{))}$$

$$\therefore \quad \text{Wt. of 1 cu-m} = \frac{18}{22.41} = 0.803$$

$$\therefore \quad \text{Total wt. of water vapour} = 2.215 \times 0.803 = 1.7786 \text{ kg}$$

∴ Heat of water vapour formed/cu-m of gas fuel

$$= 1.7786 \times 2466$$
$$= 4386.14 \text{ kJ}$$

∴ Lower calorific value of the gaseous fuel is given by

$$LCV = HCV - 4386.14$$
$$= 46437.5 - 4386.14$$
$$\textbf{LCV} = \textbf{42051.36 kJ/Nm}^3$$

Example 13.7. *The analysis of a sample producer gas shows 60 % CO, 20% H_2, 10% N_2, 1 % O_2, 2% CO_2 and rest moisture, by weight. If the higher calorific value of the gas was 1465 kJ/m³ and if the density of the gas was 1.4 kg/m³, calculate the lower calorific value of the producer gas. Neglect the steam formed by the combustion of hydrogen.*

Solution

1 m³ gas weighs 1.3 kg/m³ and its moisture content is 7% by difference i.e.

$$1.3 \times 0.07 = 0.091 \text{ kg/m}^3$$

∴ Heat required to evaporate this moisture is

$$= 2466 \times 0.091$$

$$= 224.4 \text{ kJ}$$

∴ Lower calorific value of 1 Nm³ of producer gas is given by LCV = 1465 – 224.4

$$\textbf{LCV} = \textbf{1240.6 kJ/m}^3$$

Example 13.9. *Estimate the higher and lower calorific value of a gas whose volumetric analysis is 75% CH_4, 24.5 % C_2H_2, and 0.5% N_2. Corresponding higher calorific values of CH_4, C_2H_2 and N_2 are 43000, 48600 and 0 kJ/ m³ respectively. Heat required to evaporate 1 kg of water is 2466 kJ/kg for the condition after combustion.*

Solution

Let us consider 1m³. The volume of CH_4 = 0.75 m³ and the heat evolved by its combustion is 0.75 × 43000 = 32250 kJ.

The volume of C_2H_2 is 0.245 m³ and heat evolved by its combustion is 0.245 × 48600 = 11907 kJ. Nitrogen evolves no heat. Hence the higher calorific value of the gas is given by

$$\text{HCV} = 32250 + 11907$$

$$= 44157 \text{ kJ/m}^3$$

Similarly, volume of water vapour by burning of 0.75 m³ of CH_4 is given by

$$CH_4 + 2O_2 = 2H_2O + CO_2$$

$$1 \text{ vol} = 2 \text{ vol}$$

$$(\text{Vol})_{CH_4} = 2 \times 0.75 = 1.5 \text{ Nm}^3$$

Also, volume of water vapour formed by burning 0.245 m³ of C_2H_2 is given by

$$C_2H_2 + 3O_2 = CO_2 + H_2O$$

$$1 \text{ Vol} = 1 \text{ Vol}$$

$$(\text{Vol})_{C_2H_2} = 0.245 \text{ m}^3$$

∴ Total water vapour in the product of combustion

$$= 1.5 + 0.245 = 1.745 \text{ Nm}^3$$

∴ Weight of this vapour is

$$= 1.745 \times \frac{18}{22.41}$$

$$= 1.4 \text{ kg}$$

∴ Heat required to evaporate this water

$$= 1.4 \times 2466$$

$$= 3456.36 \text{ kJ}$$

∴ Lower calorific value is given by,

$$\text{LCV} = 44157 - 3456.36$$

$$\textbf{LCV} = \textbf{40700.64 kJ/Nm}^3$$

(Practical Methods)

Example 13.9. *In a bomb calorimeter test following observations were recorded:*

Weight of coal tested 2 gm

Weight of water in the calorimeter 1.2 kg

Water equivalent of the calorimeter 0.8 kg

Rise in temperature of jacket water 8.125°C

If the coal contains 2% moisture by weight, the room temperature is 20°C and 1 kg moisture at 0°C requires 2466 kJ to evaporate to form dry and saturated steam, calculate calorific value of the coal.

Solution

Total water equivalent of the calorimeter is

$$= 1.2 + 0.8$$

$$= 2 \text{ kg}$$

∴ Heat received by water

$$= m_w C_p \Delta T = 2 \times 4.187 \times 8.125$$

$$= 68.04 \text{ kJ} \quad ...(1)$$

Again heat given by coal

$$= \text{HCV} \times \text{wt. of fuel} \quad ...(2)$$

$$= \text{HCV} \times 0.002$$

∴ Therefore by balancing these two equations we get

$$\text{HCV} \times 0.002 = 68.04$$

or
$$\text{HCV} = \frac{68.04}{0.002}$$

$$\textbf{HCV} = \textbf{34020 kJ/kg}$$

Now the heat that would be consumed to evaporate the moisture per kg coal

$$= 2466 \times 0.02$$

$$= 49.32 \text{ kJ}$$

∴ The lower calorific value of the coal is given by

$$\text{LCV} = \text{HCV} - 49.32$$

$$\textbf{LCV} = \textbf{34020 – 49.32 = 33970.68 kJ/kg}$$

Example 13.10. *During bomb calorimeter test on diesel oil according to the BSS specifications, following data were recorded:*

Room temperature 25°C

Weight of the crucible 8.116 gm

Weight of the crucible and oil 8.702 gm

Weight of can 1.051 kg

Water equivalent of the can 0.559 kg

Weight of can and water 3.492 kg

Rise in temperature of the can and water 2.305°C

Fuse correction for the heat supplied by the fuse to the bomb 0.12 kJ

Find the higher calorific value of diesel oil.

Data:

$$\text{Weight of oil taken} = 8.702 - 8.116 = 0.586 \text{ gm} = 0.000586 \text{ kg}$$

Solution

∴ Heat released by the combustion of oil

$$= \text{HCV} \times 0.000586 \text{ kJ}$$

Heat received by the bomb from fuse = 0.12 kJ

weight of water in the can

$$= 3.492 - 1.051 = 2.441 \text{ kg}$$

∴ Total equivalent of can and water

$$= 2.441 + 0.559 = 3.000 \text{ kg}$$

∴ Heat received by can and water

$$= m_w C_{P_w} \times \text{Rise in temp.}$$

$$= 3 \times 4.187 \times 2.305$$

$$= 28.9531 \text{ kJ}$$

∴ Equating heat evolved and received by water we get

$$\text{HCV} \times 0.000586 + 0.12 = 28.9531$$

$$\therefore \quad \text{HCV} = \frac{28.9531 - 0.12}{0.000586} = \frac{28.8331}{0.000586}$$

$$\therefore \quad \textbf{HCV} = \textbf{49203.242 kJ/kg}$$

Example 13.11. *The following particulars refer to an experimental determination of the calorific value of a sample of dry coal containing 88% carbon and 4.2 % hydrogen by weight.*

Weight of coal taken 0.78 gm
Weight of fuse wire 0.02 gm
Calorific value of the fuse 6.7 kJ/gm
Weight of water in the calorimeter 1.9 kg
Water equivalent of calorimeter 0.35 kg
Observed temperature rise 2.98°C
Cooling correction 0.016°C
Room temperature 20°C

Find the lower calorific value of the sample. Assume the hydrogen content of the fuel requires, 2600 kJ heat to convert 1 kg water at 0°C to dry and saturated steam.

Solution

$$\text{Observed temp. rise} = 2.98°\text{C}$$

$$\therefore \quad \text{Corrected temp. rise} = 2.98 + 0.16 = 2.996°\text{C}$$

Similarly corrected weight of water

$$= 1.9 + 0.35 = 2.25 \text{ kg}$$

∴ Heat received by water and calorimeter

$$= \text{weight} \times \text{sp. ht} \times \text{Rise in temp.}$$
$$= 2.25 \times 4.187 \times 2.996$$
$$= 28.255\,\text{kJ}$$

$$\text{Heat supplied by fuse wire} = 0.02 \times 6.7 = 0.134\,\text{kJ}$$
$$\text{Heat evolved by coal} = 0.78 \times \text{HCV}$$

Then equating the heat evolved and heat gained by water, we get

$$0.78 \times \text{HCV} + 0.134 = 28.225$$

or
$$\frac{0.78\text{HCV}}{1000} = 28.225 - 0.134$$
$$= 28.091$$

$$\therefore \quad \text{HCV} = \frac{28.091}{0.78} \times 1000$$

$$\therefore \quad \text{HCV} = 36014\,\text{kJ/kg}$$

Heat required to evaporate water above 20ºC

$$= 2466 \times \text{Mass of water}$$
$$= 9 \times 0.042 \times [2466 - (20 \times 4.18)]$$
$$= 9 \times 0.042 \times [2466 - 83.74]$$
$$= 900.49\,\text{kJ}$$

$\therefore$ Lower calorific value of coal is given by

$$\text{LCV} = \text{HCV} - 900.49$$
$$= 36014 - 900.49$$
$$\textbf{LCV} = \textbf{3513.51 kJ}$$

Example 13.12. *In a test to determine the calorific value of a produce gas, the following observations were made.*

Volume of gas burned per hour at 18°C and 100 mm H_2O gas pressure above atmosphere pressure of 750 mm Hg = 500 litres.

Volume of water flowing per minute =2360 cm^3

Temperature of incoming water = 15°C

Temperature of outgoing water = 16.5°C

Rate of condensation water = 4.2 cm^3/min

Calculate the lower calorific value of the gas at NTP. Allow 2466 kJ/kg steam evaporated at 0°C.

Solution

Heat taken or received by water/min

$$= \frac{2360}{1000} \times 4.18 \times (16.5 - 15)$$
$$= 2.36 \times 4.187 \times 1.5$$
$$= 14.822\,\text{kJ/min}$$

Now we will convert actual conditions of gas to NTP conditions

$$\text{Absolute pressure of gas} = 750 + \frac{100}{13.6}$$
$$= 750 + 5.353$$
$$= 757.353\,\text{mm Hg}$$

∴ We can write
$$\frac{P_0V_0}{T_0} = \frac{PV}{T}$$

∴ Volume of gas
$$\text{NTP} = V_0 = \frac{PV}{T}\times\frac{T_0}{P_0}$$
$$= \frac{757.353\times500}{273+18}\times\frac{273}{760}$$
$$V_0 = 467.44 \text{ litres/hour}$$

∴ Heat supplied by the fuel/hour
$$= \frac{\text{HCV}\times497.44}{1000} = \text{HCV}\times0.46744 \text{ kJ/hr}$$

∴
$$\text{Heat given by gas} = \text{Heat received by water}$$
$$\text{HCV}\times0.46744 = 14.822\times60$$
$$\text{HCV} = \frac{14.822\times60}{0.46744}$$

Again heat supplied by fuel to steam/min
$$= \left(\frac{4.2}{106}\times1000\right)\times(2466-18\times4.187)$$
$$= 0.0042\times(2466\,75.366)$$
$$= 0.0042\times2390.634 = 10.041 \text{ kJ/min}$$
$$\text{LCV} = \frac{(14.822-10.041)\times60}{0.46744}$$
$$= \frac{4.781\times60}{0.46744}$$
$$\textbf{LCV} = \textbf{613.68 kJ/Nm}^3$$

Example 13.13. *In a Boy's Gas calorimeter 500 litres of water gas at 27°C, 75 mm of Hg about the atmosphere pressure of 750 mm Hg, was burned during a test conducted to determine the lower calorific value of water gas. During the test 0.3 kg of water was collected from the products of combustion. The circulating was admitted at 30°C and left at 47°C, the rate of its flow being 250 kg during the test.*

Calculate the higher and lower calorific values of water gas. Assume that 2466 kJ of heat of required to convert 1 kg water at 0°C to dry and saturated steam.

Solution

$$\text{Pressure of gas} = 750+75 = 825 \text{ mm Hg}$$

From the relation
$$\frac{P_0V_0}{T_0} = \frac{PV}{T}$$
$$V_0 = \text{Volume of gas at NTP}$$
$$= \frac{PV}{T}\times\frac{T_0}{P_0}$$

$$= \frac{825}{760} \times \frac{273}{303} \times \frac{500}{1000} \text{ Nm}^3$$

$$= 0.489 \text{ Nm}^3$$

Heat given by gas is given by

$$= \text{HCV} \, 0.489$$

$$\therefore \quad \text{HCV} = \text{Higher calorific value of gas}$$

$$= \frac{177794.95}{0.489}$$

$$\textbf{HCV} = \textbf{36390 kJ}$$

Condensed water from the products of combustion leaving the calorimeter is at 27°C

$\therefore$ Heat rejected by 1 kg steam in condensing to water at 27°C is

$$= 2466 - 4.187 \times 27$$

$$= 2466 - 113.049$$

$$= 2352.951 \text{ kJ}$$

$\therefore$ Total heat recovered from condensate during the test

$$= 2352.951 \times 0.3$$

$$= 705.885 \text{ kJ}$$

Therefore heat supplied by gas to water

$$= 177794.75 - 705.885$$

$$= 17088.865 \text{ kJ}$$

$$\text{LCV} = \text{Lower calorific value}$$

$$= \frac{17088.865}{0.489}$$

$$\textbf{LCV} = \textbf{34946 kJ/Nm}^3$$

Example 13.14. *Write the equations of combustion for CH_3, C_6H_{14} and CH_3OH. Find the amount of oxygen required in terms of kg/kg fuel for complete combustion in each case.*

Assume constant pressure. Find also air to fuel ratio in each case.

Solution

(a) CH_3: We will require 4 K moles of O_2

$$4CH_3 + 7O_2 = 4CO_2 + 6H_2O$$

$$4(12+3) + 7 \times 32 = 4(44) + 6 \times 18$$

or $$60 \text{ kg} + 224 \text{ kg} = 176 + 108$$

$\therefore$ Required ratio of oxygen to fuel

$$= \frac{224}{60} = 3.733$$

$\therefore$ Air to fuel ratio

$$\text{A/F} = \text{Mass of Air/kg fuel}$$

$$= \frac{100}{23} \times 3.733$$

$$\textbf{A/F} = \textbf{16.232 kg/kg fuel}$$

(b) C_6H_{14}: Chemical equation will be as given

$$2C_6H_{14} + 19O_2 = 12CO_2 + 14H_2O$$

$$\therefore \quad 2(12 \times 6 + 1 \times 14) + 19 \times 32 = 12 \times 44 + 14 + 18$$

$$\therefore \quad 2(72 \times 14) + 608 = 528 + 252$$

$\therefore \quad 2\times 84+608\,(O_2) = 780$

$\therefore$ Oxygen to Fuel ratio

$$= \frac{608}{172} = 3.535$$

$$\therefore \quad \text{A/F ratio} = \frac{100}{23}\times 3.535$$

A/F ratio = 15.37 kg/kg fuel

(c) CH_3OH : Chemical equation is as given

$$2CH_3OH + 3O_2 = 2CO_2 + 4H_2O$$

$$2(12+4+16)+3\times 32 = 2+44+4\times 18$$

$$(64)+(96) = 88+72$$

$$\text{(fuel)}\ (O_2)$$

$$\text{Required}\ \frac{\text{Oxygen}}{\text{Fuel}} = \frac{96}{64} = 1.5$$

$$\therefore \quad \text{A/F} = 1.5\times\frac{100}{23}$$

A/F = 6.522

Example 13.15. *Calculate the volume of oxygen required for complete combustion of 1m³ of C_2H_2, C_2H_4 and CO at constant pressure.*

Solution

(a) C_2H_2

The chemical equation will be

$$2C_2H_2 + 5O_2 = 4CO_2 + 2H_2O$$

i.e. $\quad 2\text{ K mol} + 5\text{ K mol} = 4\text{ K mol} + 2\text{ K mol}$

or $\quad 1\text{ m}^3 + 2.5\text{ m}^3 = 4\text{ m}^3 + 2\text{ m}^3$

$\therefore$ Volume of O_2 required is 2.5 m³/m³ of C_2H_2

(b) CO

Again the chemical equation will be

$$2CO + O_2 = 2CO_2$$

i.e. $\quad 2\text{ K mol} + 1\text{ K mol} = 2\text{ K mol}$

$$1\text{m}^3 + \frac{1}{2}\text{m}^3 = 2\text{ m}^3$$

$\therefore$ Volume of O_2 required per 1 m³ of CO is 0.5 m³.

Example 13.16. *Calculate the amount of air required for combustion of producer gas containing 25% CO, 10% CO_2, 17% H_2, 5% O_2 and 43%, N_2 by volume. Assume that the combustion takes place at constant pressure and air contains 21% oxygen by volume.*

Solution

We will write chemical or combustion equations for individual constituent of the produced gas.

CO_2: $\quad 2CO + O_2 = 2CO_2$

$\quad 1\text{ m}^3 + 0.5\text{ m}^3$

H_2: $\quad 2H + O_2 = 2H_2O$

$\quad 1\text{ m}^3 + 0.5\text{ m}^3 = 2\text{ m}^3$

∴ Total volume of O_2 required for 1 m^3 gas

$$= 0.25\times0.5+0.17\times0.5$$
$$= 0.125+0.085$$
$$= 0.210\,m^3 \text{ of } O_2$$

Oxygen present in the gas is 0.05 m^3.

∴ Net volume of O_2 required

$$= 0.210-0.05$$
$$= 0.160\,m^3$$

∴ Air required for complete combustion

$$= \frac{0.16}{0.21} = 0.762\ m^3 \text{ per cu-m of producer gas}$$

Air required for complete combustion = 0.762 cu-m of air/m^3 gas

Example 13.17. *A diesel fuel contains 70% C, 10% H_2, 5% O_2, 1% S and rest incombustible by weight. If air contains 23% oxygen by weight, find the amount of air required for complete combustion of 1 kg fuel.*

Solution

Total oxygen required for complete combustion of 1 kg fuel is given by the relation

$$= \left(\frac{8}{3}C+8H+S-O\right)$$

$$= \frac{8}{3}\times0.7+8\times0.1+0.01-0.05$$

$$= 1.867+0.8+0.01+0.05 = 2.627\text{ kg}$$

∴ Weight of air required for complete combustion of 1 kg fuel

$$W_a = \frac{100}{23}\times2.627$$

$$\mathbf{W_a = 11.422\ kg/kg\ fuel}$$

Example 13.18. *Determine the theoretical weight of air required for complete combustion of 1 kg of coal whose analysis by weight is given as under:*

Carbon – 83%, Hydrogen – 5%, Oxygen – 2%, Sulphur – 0.2% remainder being incombustible.

Solution

In one kg of coal, the constituents by weight are

Carbon = 0.83 kg
Hydrogen = 0.05 kg
Oxygen = 0.02 kg
Sulphur = 0.002 kg

∴ Minimum quantity of air required for complete combustion of 1 kg fuel

$$= \frac{100}{23}\left[\frac{8}{3}C+8H+S-O\right]$$

$$= \frac{100}{23}\left[\frac{8}{3}\times0.83+8\times0.05+0.002-0.02\right]$$

$$= \frac{100}{23}[2.2133+0.4+0.002-0.02]$$

$$= \frac{100}{23} \times 2.5953 = 11.284 \text{ kg/kg fuel}$$

Example 13.19. *Determine the theoretical weight of air required for the complete combustion of one litre of petrol; given that the composition of petrol by weight is 86% carbon, and 14% hydrogen specific gravity of petrol being 0.8. If one kg or air at atmospheric temperature and pressure occupies 0.77 cu-m. Find also the volume of air supplied.*

Solution

Constituent of 1 kg petrol are 0.86 kg carbon, 0.14 kg of hydrogen.

∴ Minimum quantity of air required /kg fuel

$$= \frac{100}{23}\left[\frac{8}{3}C + 8H\right]$$

$$= \frac{100}{23}\left[\frac{8}{3} \times 0.86 + 8 \times 0.14\right]$$

$$= \frac{100}{23}[2.2933 + 1.12] = \frac{100}{23} \times 3.4133$$

$$= 14.84 \text{ kg/kg fuel}$$

$$1 \text{ litre of petrol} = 0.8 \text{ kg}$$

∴ Minimum quantity of air required/litre

$$= 0.8 \times 14.84$$

$$\text{Minimum quantity of air} = 11.87 \text{ kg}$$

∴ Minimum volume of air required/litre

$$= 11.87 \times 0.77$$

$$\text{Minimum quantity of air} = 9.142 \text{ m}^3\text{/litre fuel}$$

Example 13.20. *If the analysis by volume of a product of combustion is 20% CO_2, 3.8% CO 70% N_2 and 6.2% O_2, find the analysis by weight.*

Solution

This example can be solved by tabulating the various quantities.

Substance	*Volume/m³ (b)*	*Mol. Weight (b)*	*Proportional wt =(a) × (b)*	*Wt/kg of gas × 100*
CO_2	0.2	44	$0.2 \times 44 = 88$	27.983
CO	0.038	28	$0.038 \times 28 = 1.064$	3.383
N_2	0.7	28	$0.7 \times 28 = 19.6$	62.32
O_2	0.062	32	$0.062 \times 32 = 1.984$	6.31
			$\sum P = 31.448$	99.996

Thus the percentage composition of the products by weight is

$$CO_2 = 27.983\%$$

$$CO = 3.383\%$$

$$N_2 = 62.32\%$$

$$O_2 = 6.31\%$$

Example 13.21. *A sample of water-gas contains 38% H_2, 23% CO, 17% CH_4, 9% C_2H_6, 6% H_2O, 4% CO_2 and 3% N_2. Convert this to analysis by weight.*

Solution

We will make a table as shown below

S. No.	Substance	Vol/m³ of gas (a)	Mol. Weight (b)	Proportional weight (a)×(b)	% by weight = $\frac{(a)\times(b)}{\sum(a\times b)}\times 100$
1	H_2	0.38	2	0.76	4.6626
2	CO	0.23	28	6.44	39.5100
3	CH_4	0.17	16	2.72	16.6781
4	C_2H_6	0.09	30	2.70	16.5644
5	H_2O	0.06	18	1.08	6.6258
6	CO_2	0.04	44	1.76	10.7975
7	N_2	0.03	28	0.84	5.1534
				$\sum(a\times b)=163$	100.00

∴ Percentage analysis of water gas by weight is

$H_2=4.6626\%, CO=39.5100\%, CH_4=16.6871\,\%, C_2H_6=165644\%, H_2O=6.6258\%, CO_2=10.7975\,\%, N_2=5.1534\%.$

Example 13.22. *The composition of a dry flue gas, as obtained by using Orsat apparatus was 10% CO_2, 1.8 % CO, 7% O_2, and 81.2 % N_2 by volume. What will be the percentage composition of the flue gas by weight ?*

Solution

$$\begin{aligned}
\text{Relative weight of } CO_2 &= 0.1\times 44=4.4\,\text{kg}\\
\text{Relative weight of CO} &= 0.018\times 28=0.504\,\text{kg}\\
\text{Relative weight of } O_2 &= 0.07\times 32=2.24\,\text{kg}\\
\text{Relative weight of } N_2 &= 0.812\times 28=22.736\,\text{kg}\\
\therefore\quad \text{Total weight of the gas} &= 29.88\,\text{kg}
\end{aligned}$$

$$\therefore \text{Percentage of } CO_2 \text{ in gas} = \frac{4.4\times 100}{29.88} = 1477$$

$$\therefore \quad \text{Percentage of CO in gas} = \frac{0.504\times 100}{29.88} = 1.687$$

$$\therefore \quad \text{Percentage of } O_2 \text{ in gas} = \frac{2.24\times 100}{29.88} = 7.497$$

$$\therefore \quad \text{Percentage of } N_2 \text{ in gas} = \frac{22.736\times 100}{29.88} = 76.091$$

$$\text{Total} = 100$$

Example 13.23. *The diesel oil used in an engine consisted of 82% C, 7% O_2, 8% H_2 and 3% H_2O by weight. Convert this to analysis by volume.*

Solution

First convert weight into volume by dividing weight of the substance by its molecular weight. Thereby relative volumes will be obtained.

$$\therefore \quad 82\% \text{ of carbon} = \frac{82}{44} = 1.8636 \text{ volume}$$

$$7 \text{ kg of } O_2 = \frac{7}{32} = 0.2188 \text{ volume}$$

$$8 \text{ kg } H_2 = \frac{8}{2} = 4.0000 \text{ volume}$$

$$3 \text{ kg } H_2O = \frac{3}{18} = 0.1667 \text{ volume}$$

$$\text{Total volume} = 6.2491$$

∴ Volumetric analysis will be

$$\text{Carbon} = \frac{1.8636 \times 100}{6.2491} = 29.822\%$$

$$\text{Oxygen} = \frac{0.2188 \times 100}{6.2491} = 3.501\%$$

$$\text{Hydrogen} = \frac{4 \times 100}{6.2491} = 64.009\%$$

$$H_2O = \frac{0.1667 \times 100}{6.2491} = 2.668\%$$

Example 13.24. *In a boiler trial the fuel analysis was carbon 88%, hydrogen 3%, sulphur 0.5% by weight and the remainder being ash. Determine the weight of air required per kg of fuel for complete combustion. If the actual supply of air is 50% in excess of this, estimate the percentage analysis of the dry fuel gas by*

(a) by weight

(b) by volume

Solution

We can write the chemical equations for the minimum, theoretical air required for complete combustion of 1 kg fuel.

(i) $C + O_2 = CO_2$

1 kg of carbon requires 8/3 kg of carbon dioxide.

$$\therefore \quad O_2 \text{ required} = \frac{8}{3} \times 0.88 = 2.347 \text{ kg}$$

and produces $\frac{11}{3} \times 0.88 = 3.227$ kg

(ii) $H_2 + O = H_2O$

1 kg of hydrogen requires 8 kg of xygen and produces 9 kg H_2O.

$$\therefore \quad O_2 \text{ required} = 8 \times 0.03 = 0.24 \text{ kg}$$

and H_2O produced $= 9 \times 0.03 = 0.27$ kg

(iii) $S + O_2 = SO_2$

∴ 1 kg of sulphur requires 1 kg of oxygen

∴ Oxygen required for 0.005 kg of sulphur

$0.005 \times 1 = 0.005$ kg and produces

$2S = 2 \times 0.005 = 0.01$ kg of sulphurdioxide.

∴ Total minimum oxygen required for complete combustion of 1 kg coal

$$= 2.347 + 0.24 + 0.005$$
$$= 2.592 \text{ kg}$$

Now 50% of the minimum air is supplied as excess air

∴ Actual weight of air supplied/kg fuel

$$= 1.5 \times 11.261$$
$$= 16.89 \text{ kg}$$

With this actual air supplied, the weights of the dry products of combustion are

$$\text{Carbon dioxide} = 3.227 \text{ kg}$$
$$\text{Sulphur dioxide} = 0.010 \text{ kg}$$
$$\text{Oxygen} = 0.5 \times 2.592$$
$$= 1.296 \text{ kg}$$
$$\text{Nitrogen} = 0.77 \times 16.89$$
$$= 13.005 \text{ kg}$$

Total weights of the dry products

$$= 3.227 + 0.01 + 1.296 + 13.005$$
$$= 17.5383 \text{ kg}$$

∴ Percent analysis by weight of the dry products is

$$\text{Carbon Dioxide} = \frac{3.227}{17.5383} \times 100 = 18.402$$

$$\text{Sulphur Dioxide} = \frac{1}{17.5383} = 0.057$$

$$\text{Oxygen} = \frac{129.6}{17.5383} = 7.389$$

$$\text{Nitrogen} = \frac{1300.5}{17.5383} = \frac{74.152}{100.00}$$

We can get relative volumes by dividing weights by molecular weights.

∴ Relative volumes are

$$\text{Carbon dioxide} = \frac{18.402}{44} = 0.4182$$

$$\text{Sulphur dioxide} = \frac{1}{64} = 0.0156$$

$$\text{Oxygen} = \frac{7.389}{32} = 0.2309$$

$$\text{Nitrogen} = \frac{74.152}{28} = \frac{2.6483}{\Sigma = 3.313}$$

Percent analysis by volume of the products

$$\text{Carbon dioxide} = \frac{0.4182}{3.13} \times 100 = 12.6230$$

$$\text{Sulphur dioxide} = \frac{0.0156}{313}\times 100 = 0.4708$$

$$\text{Oxygen} = \frac{0.2309}{313}\times 100 = 6.9695$$

$$\text{Nitrogen} = \frac{2.6483}{313}\times 100 = \frac{79.9367}{1.00}$$

Example 13.25. *A sample of coal has the following analysis by weight : Carbon 84%, Hydrogen 5%, incombustible material 11%. This coal is supplied to a furnace and burnt with a volume of air 50% in excess of that theoretically required for complete combustion. Estimate the volumetric composition of the dry flue gas.*

Solution

We now can write directly the equation for theoretical air required for complete combustion of 1 kg fuel.

$$\text{Theoretical air/kg fuel} = \frac{100}{23}\left[\frac{8}{3}C + 8H\right]$$

$$= \frac{100}{23}\left[\frac{8}{3}\times 0.84 + 8\times 0.05\right]$$

$$= 11.4783 \text{ kg}$$

Here excess air supplied is 50 % by volume. Volume of theoretical air can be determined and 50 % of that can be obtained. This excess volume again will be converted into weight which will be 50 %, by weight, of the theoretical air.

∴ We will consider 50 % by weight, excess air.

∴ Actual weight of air supplied/kg fuel

$$= 1.5\times 11.4783$$

$$= 17.2174 \text{ kg}$$

Now first find the flue gas analysis by weight and then this is to be converted into volumetric analysis.

$$\text{Weight of Carbon Dioxide} = \frac{11}{3}C = \frac{11}{3}\times 0.84$$

$$= 3.08 \text{ kg}$$

$$\text{Weight of oxygen} = 0.5\times 2.64 = 1.32 \text{ kg}$$

$$\text{Weight of nitrogen} = 0.77\times 17.2174$$

$$= 13.26 \text{ kg}$$

∴ Total weight of flue gases produced per kg of fuel

$$= 3.08 + 1.32 + 13.26$$

$$= 17.66 \text{ kg}$$

$$\text{Volume of carbon dioxide} = \frac{3.08}{44} = 0.07$$

$$\text{Weight of oxygen} = \frac{1.32}{32} = 0.04125$$

$$= 1.32 \text{ kg}$$

$$\text{Weight of nitrogen} = \frac{13.26}{28} = 0.4736$$

$$= 13.26\,\text{kg}$$

$$\text{Total} = \Sigma = 0.58485$$

$\therefore$ Percentage analysis of the dry products of combustion is given as

$$\text{Carbon dioxide } CO_2 = \frac{0.07 \times 100}{0.58485} = 11.9$$

$$\text{Oxygen } O_2 = \frac{0.04125 \times 100}{0.58485} = 7.053$$

$$\text{Nitrogen } N_2 = \frac{0.4738 \times 100}{0.58485} = 81.0122$$

Example 13.26. *A petrol has a composition by weight of 86% carbon and 14% hydrogen. When used in an engine the air supply is 90% of that theoretically required for complete combustion. Assuming that all the hydrogen is burnt and that the carbon burns to carbon monoxide and carbon dioxide so that there is no free carbon left. Calculate the percentage analysis of dry exhaust gases by volume.*

Solution

Theoretically air required for complete combustion of 1 kg fuel

$$= \frac{100}{23}\left[\frac{8}{3}C + 8H\right] = \frac{100}{23}\left[\frac{8}{3} \times 0.86 + 8 \times 0.14\right]$$

$$= \frac{100}{23} \times [2.2933 + 1.12]$$

$$= \frac{100}{23} \times 3.4133$$

$$= 14.84\,\text{kg}$$

$\therefore$ Actual air supplied/kg fuel

$$= 0.9 \times 14.84$$

$$= 13.3564\,\text{kg}$$

$\therefore$ Actual oxygen supplied/kg fuel

$$= 0.9 \times 3.4133$$

$$= 3.072\,\text{kg}$$

$\therefore$ Actual nitrogen supplied/kg fuel

$$= 0.77 \times 13.3564$$

$$= 10.2844\,\text{kg}$$

Oxygen used for burning of hydrogen

$$8H = 8 \times 0.14$$

$$= 1.12\,\text{kg}$$

$\therefore$ Oxygen left for combustion of carbon

$$= 3.072 - 1.12$$

$$= 1.952\,\text{kg}$$

Let C be weight of carbon burnt to CO_2

∴ (0.86 – C) will be weight of carbon burnt carbon monoxide.

∴ Total oxygen required for this combustion

$$= \frac{8}{3}C + \frac{4}{3}(0.86 - C)$$

$$= \frac{8}{3}C + \frac{3.44 - 4C}{3} = \frac{4C + 3.44}{3}$$

$$\therefore \quad 1.952 = \frac{4C + 3.44}{3}$$

$$\therefore \quad 4C = 3 \times 1.952 - 3.44$$

$$\therefore \quad = 5.86 - 3.44 = 2.416$$

$$\therefore \quad C = \frac{2.416}{4} = 0.604 \text{ kg}$$

∴ Weight of carbon burnt carbon monoxide.

$$= 0.86 - 0.604 = 0.256 \text{ kg}$$

Weight of carbon dioxide produced/kg fuel.

$$= \frac{11}{3} \times 0.604 = 2.215 \text{ kg}$$

Weight of carbon monoxide produced/kg fuel..

$$= \frac{7}{3} \times 0.256 = 0.597 \text{ kg}$$

$$\text{Weight of nitrogen} = 10.2844 \text{ kg}$$

∴ Relative volumes produced are

$$CO_2 = \frac{2.215}{44} = 0.0503$$

$$CO = \frac{0.597}{28} = 0.0213$$

$$N_2 = \frac{10.2844}{28} = 0.3673$$

$$\text{Total relative volumes} = 0.4389$$

∴ Percentage analysis of dry products of combustion is as

$$CO_2 = \frac{0.0503}{0.4389} \times 100 = 11.46\%$$

$$CO = \frac{0.0213}{0.4389} \times 100 = 4.85\%$$

$$N_2 = \frac{0.3673}{0.4389} \times 100 = 86.69\%$$

$$\text{Total} = 100.00$$

Example 13.27. *A sample of coal burnt in a Lancashire boiler has the following composition by weight: carbon 80%, hydrogen 5%, oxygen 2%, sulphur 1%, and the remainder ash. The coal burns with 60% of excess air over the minimum quantity of air required for the complete combustion of the fuel. Estimate the volumetric composition of the dry flue gas. If the mean temperature of the gas leaving the boiler plant is 315°C and that of the boiler house be 27°C, estimate the heat carried away by flue gases leaving the boiler plant. Assume C_p of gas to be 1.01 kg/kg-K and total heat of water vapour in the gases to be 2658 kJ/kg.*

Solution

Theoretical or minimum air required for complete combustion of 1 kg fuel is similarly carbon content of CO_2 / kg gas

$$= \frac{22}{44} \times 0.1618 = 0.04413 \text{ kg}$$

$\therefore$ Total carbon/kg gas $= 0.441 + 0.00802$

$= 0.05215$ kg

$\therefore$ Weight of gas produced per kg fuel

$$= \frac{0.81}{0.05215} = 15572 \text{ kg}$$

$\therefore$ Excess air/kg fuel $= 15.532 \times 0.193$

Example 13.28. *In a boiler trial, dry fuel as burnt contained 84% carbon and 3% free hydrogen. The flue gas analysis by volume gave CO_2 – 11.5% , oxygen – 8.4% and nitrogen – 80.1%. Calculate per kg of dry fuel the weight of air supplied and.weight of excess air.*

Solution

We will solve this problem by tabulating the data as

Constituent	*% Volume (a)*	*Mol. Wt (b)*	*Rel. wt (a x b)*	*Wt. per kg gas*	*Carbon Content kg. gas*
Carbon Dioxide	$C_1 = 11.5$	44	506	0.1677	0.04574
Oxygen	O = 8.4	32	268.8	0.0891	
Nitrogen	N = 80.1	28	2242.8	0.7432	
			Σ3017.6	1.0000	

$\therefore$ Weight of gas/kg fuel $= \dfrac{0.84}{0.04574} = 18.36$ kg

Theoretical air per kg $= \dfrac{100}{23} \times \left[\dfrac{8}{3}C + 8H + S - O\right]$

$$= \frac{100}{23} \times \left[\frac{8}{3} \times 0.8 + 8 \times 0.05 + 0.01 - 0.02\right]$$

$= 10.97$ kg

∴ Actual weight of air supplied/kg fuel

$$= 1.6 \times 10.97 = 17.55 \text{ kg}$$

Dry products of combustion by weight are

$$\text{Carbon dioxide} = \frac{11}{3}C = \frac{11}{3} \times 0.08 = 2.933 \text{ kg}$$

$$\text{Sulphur dioxide} = 2S = 2 \times 0.01 = 0.02 \text{ kg}$$

$$\text{Oxygen} = 0.6 \times 2.523 = 1.5138 \text{ kg}$$

$$\text{Nitrogen} = 0.77 \times 17.55 = 13.5135 \text{ kg}$$

∴ Relative volumes of the dry products are

$$\text{Carbon dioxide} = \frac{2.933}{44} = 0.0667$$

$$\text{Sulphur dioxide} = \frac{0.02}{64} = 0.0003125$$

$$\text{Oxygen} = \frac{1.5138}{32} = 0.04731$$

$$\text{Nitrogen} = \frac{13.5135}{28} = 0.4826$$

$$\text{Total} = 0.59691$$

Percentage composition of dry products of combustion by volume is

$$\text{Carbon dioxide} = \frac{0.0667}{0.59691} \times 100 = 11.174\%$$

$$\text{Sulphur dioxide} = \frac{0.0003}{0.59691} \times 100 = 0.050\%$$

$$\text{Oxygen} = \frac{0.04731}{0.59691} \times 100 = 7.926\%$$

$$\text{Nitrogen} = \frac{0.4826}{0.59691} \times 100 = 80.850\%$$

$$\text{Total} = 100.00$$

Heat carried away by the flue gases leaving the boiler plant consists of

(i) Heat with dry gases.

(ii) Heat with vapours.

Heat carried by dry gas = $mg\ Cpg\ (Tg - T_R)$

where

$$mg = \text{Weight of the gases/kg fuel}$$

$$= \text{Air} + \text{Fuel} - H_2O \text{ formed} - \text{Ash.}$$

$$= 17.55 + 1 - 9 \times 0.05 - 12$$

$$= 18.43 - 0.45 = 17.98 \text{ kg}$$

∴ Heat carried by gases $= 17.98 \times 1.01(315 - 27)$

$$= 5230 \text{ kJ/kg fuel}$$

Heat carried by vapours

$$= 9H \times 2658$$

$$= 0.45 \times 2658 = 1196.1 \text{ kJ}$$

$\therefore$ Total heat carried by exhaust or flue gases per kg of fuel

$$= 5230 + 1196.16 = 6426.1 \text{ kJ/kg fuel}$$

Example 13.29. *The analysis of fuel used in an engine is carbon 50%, hydrogen 16.6% and oxygen 33.4% by weight. It is burnt with twice the theoretical minimum quantity of dry air required for complete combustion. Give the analysis of the exhaust gases (a) by weight and (b) by volume.*

Solution

Minimum weight of air required for complete combustion of 1 kg fuel is

$$= \frac{100}{23}\left[\frac{8}{3}C + 8H - O\right]$$

$$= \frac{100}{23}\left[\frac{8}{3} \times 0.5 + 8 \times 0.166 - 0.334\right]$$

$$= 10.12 \text{ kg/kg fuel}$$

$\therefore$ Weight of air supplied / kg

$$= 10.12 \times 2 = 20.24 \text{ kg}$$

$\therefore$ The weight of the products/kg fuel are

$$\text{Carbon dioxide} = \frac{11}{3} \times 0.5 = 1.8333 \text{ kg}$$

$$\text{Water vapour} = 9H = 9 \times 0.166 = 1.494 \text{ kg}$$

$$\text{Oxygen} = 2.3273 \text{ kg}$$

$$\text{Nitrogen} = 0.77 \times 20.24 = 15.5848$$

$$\text{Total Weight} = 21.2394$$

$\therefore$ Percentage analysis by weight of the products of combustion is as

$$\text{Carbon dioxide} = \frac{1.8333}{21.2394} \times 100 = 0.6317$$

$$\text{Water vapour} = \frac{1.494}{21.2394} \times 100 = 7.0341$$

$$\text{Oxygen} = \frac{2.3273}{21.2394} \times 100 = 10.9575$$

$$\text{Nitrogen} = \frac{15.5848}{21.2394} \times 100 = 73.3768$$

Relative volumes of the products are

$$CO_2 = \frac{1.8333}{44} = 0.04167$$

$$H_2O = \frac{1.494}{18} = 0.08300$$

$$O_2 = \frac{2.3273}{32} = 0.07273$$

$$N_2 = \frac{15.5848}{28} = \underline{0.55660}$$

$$\text{Total} = 0.754$$

∴ Percentage analysis of the products of combustion by volume is

$$\text{Carbon dioxide} = \frac{0.04167}{0.754} \times 100 = 5.5265\%$$

$$\text{Water vapour} = \frac{0.083}{0.754} \times 100 = 9.6459\%$$

$$\text{Oxygen} = \frac{0.07273}{0.754} \times 100 = 9.6459\%$$

$$\text{Nitrogen} = \frac{0.5566}{0.754} \times 100 = 73.8196\%$$

$$\text{Total} = 100.0000$$

Example 13.30. *In a boiler trail the composition of dry flue gases by volume was as under : CO_2 = 11%, CO = 2%, O_2 = 4%, N_2 = 83 %. Determine the quantity of air passing to the furnace per kg of fuel burnt if the carbon in the fuel was 81% by weight. Also determine the weight of excess air per kg of gas.*

Solution

$$\text{Actual air supplied per kg fuel} = \frac{NC}{33(C_1 + C_2)} \text{ with usual notations}$$

$$\therefore \quad \text{Air / kg fuel} = \frac{83 \times 81}{33(11+2)} = \frac{83 \times 81}{33 \times 13}$$

$$\text{Air / kg fuel} = \mathbf{15.67\ kg}$$

Oxgen present in 1 kg flue gas can be found as

Gas	*Quantity m^3*	*Mol. Wt.*	*Rel Wt.*	*Wt/kg gas*
CO_2	11	44	44 × 11 = 484	$\frac{484}{\Sigma} = 0.16176$
CO	2	28	28 × 2 = 056	$\frac{56}{\Sigma} = 0.01872$
O_2	4	32	32 × 4 = 128	$\frac{128}{\Sigma} = 0.04278$
N_2	83	28	28 × 83 = 2324	$\frac{2324}{\Sigma} = 0.77674$
			Σ = 2992	Σ = 1.0000

We have $\quad CO + O = CO_2$

∴ Oxygen required to burn 0.01872 kg CO to carbondioxide

$$= \frac{16}{28} \times 0.01872$$

$$= 0.01067 \text{ kg}$$

∴ Excess oxygen, after accounting for CO,

$$= 0.04278 - 0.0107$$

$$= 0.03208 \text{ kg/kg gas}$$

∴ $$\text{Excess air} = \frac{100}{23} \times 0.03208 = \mathbf{0.1395\ kg}$$

Example 13.31. *During a boiler trial, the average volumetric analysis of the flue gases gave the following results. CO_2 = 10.7%, CO = 1.1 %, O_2 = 6.85 % and N_2 = 81.35 %. The chemical analysis of coal as fired gave, C = 87.6 %, H_2 = 3.4 %, O_2 = 4,5 %, ash etc. = 4.5 %. Determine per kg of coal*

(i) The weight of flue gas.

(ii) The weight of air supplied.

Solution

(i) The weight of flue gas

CO_2	10.7×44 = 470.8	0.1570	C = 0.04282
CO	1.1×28 = 30.8	0.0103	C = 0.00421
O_2	6.85×32 = 219.2	0.0731	0.04723
N_2	81.35×28 = 2277.8	0.7596	
	2998.6	0.9997 = 1 kg gas	

∴ $$\text{Weight of gases/kg fuel} = \frac{0.876}{0.04723}$$

$$\text{Weight of gases/kg fuel} = \mathbf{18.59\ kg}$$

$$\text{Wt. of wet gases} = 18.55 + 9 \times 0.034$$

$$= 18.55 + 0.306 = 19.906$$

∴ $$\text{Weight of flue gas/kg fuel} = \mathbf{19.906\ kg}$$

(ii) Weight of air supplied

Weight of air supplied /kg fuel

$$= \frac{NC}{33(C_1 + C_2)} = \frac{8.135 \times 87.6}{33(10.7 + 1.1)}$$

$$= \mathbf{18.3\ kg/kg\ fuel}$$

Example 13.32. *A SI engine uses octane (C_8H_{18}) as the fuel and the exhaust gas analysis gave the following composition. CO_2 = 9.9%, CO = 7.2%, H_2 = 3.3%, CH_4 = 0.3%, N_2 =79.3%. Calculate the air fuel ratio.*

Solution

For volumetric analysis of air 1 mol of O_2 will be accompanied by 3.76. mol. of N_2 $\left(\frac{79}{21} = 3.76\right)$

Now we can write the chemical equation for combustion of this fuel. Thus,

$$a(C_8 + H_8) + b(O_2 + 3.76N_2) = 9.9\,CO_2 + 7.2\,CO + 3.3\,H_2 + 0.3\,CH_4 + 79.3\,N_2 + DH_2O$$

Last term in this equation is because of the combustion of hydrogen in the fuel.

Values of a and b in the above equation should be such that we will get the analysis as given on the right hand side.

First we will balance carbon.

$$8a = 9.9+7.2+03=17.4$$

$$\therefore \quad a = \frac{17.4}{8} = 2.175$$

Again, we can balance nitrogen. Then

$$3.76b = 79.3$$

$$b = \frac{79.3}{3.76} = 21.09$$

Then we will balance hydrogen

$$18a = 6.6+1.2+2D = 7.8+2D$$

$$\therefore \quad 18\times 2.175 = 7.8+2D = 39.15$$

$$\therefore \quad D = \frac{39.15-7.8}{2} = \frac{31.35}{2}$$

$$= 15.675$$

Thus all unknowns a, b and D in the equation are known. Therefore, the combustion equation is

$$2.175(C_8H_{18})+21.09(O_2+3.76N_2)$$

$$= 9.9\,CO_2+7.2\,CO+3.3\,H_2+0.3\,CH4-79.3\,N_2+15.675\,H_2O$$

$$\therefore \quad \frac{\text{Air}}{\text{Fuel}} = \frac{21.09(32+3.76\times 28)}{2.175(96+18)}$$

$$= \frac{2.109\times(32+105.78)}{2.175\times 114}$$

$$\frac{\text{Air}}{\text{Fuel}} = \mathbf{11.68\ kg\ air/kg\ fuel}$$

Example 13.33. *Volumetric analysis of products of combustion of a hydrocarbon fuel is as under : CO_2 = 12.5 %, CO = 0.3%, O_2 = 5% N_2 is found by difference. Determine (a) air : fuel ratio (b) carbon and hydrogen percentage by mass (c) percentage of excess air.*

Solution

Let the fuel be denote by C_xH_y. Nitrogen present in the exhaust gases = 100 – 17.8 = 82.2

The combustion equation is

$$C_xH_y + b(O_2+3.76N_2) = 12.5CO_2+0.3CO+5O_2+82.2N_2+DH_2O$$

∴ Nitrogen balance gives

$$3.76b = 82.2$$

$$\therefore \quad b = \frac{82.2}{3.76} = 21.862$$

Carbon balance gives

$$x = 12.5+0.3 = 12.8$$

Oxygen balance gives

$$b = 12.5+\frac{0.3}{2}+5+\frac{D}{2}$$

$$21.862 = 17.65+\frac{D}{2}$$

or
$$D = \left(\frac{21.862-17.654}{1}\right)2 = 4.208\times 2$$
$$= \mathbf{8.416}$$

Hydrogen balance gives

$$y = D$$
$$= 8.416$$

$\therefore$ Actual combustion equation is given by

$$(C_{12.8}H_{8.416})+21.862O_2+21.862\times 3.76N_2$$
$$= 12.5CO_2+0.3CO+5O_2+82.2N_2+8.416H_2O$$

(i) Air fuel ratio:

$$\frac{\text{Air}}{\text{Fuel}} = \frac{(21.862\times 32)+(21.862\times 3.76\times 28)}{12.8\times 12+8.416\times 1}$$
$$= \frac{699.584+2301.631}{153.6+8.416} = \frac{3001.215}{162.016}$$
$$\frac{\textbf{Air}}{\textbf{Fuel}} = \mathbf{18.524\ kg/kg\ of\ fuel}$$

(ii) Composition of fuel:

Carbon weight = 153.6 kg
Hydrogen weight = 8.416
Total weight = 162.016 kg

$\therefore$ % of carbon in fuel $= \frac{153.6}{162.016}\times 100 = 94.8$

$\therefore$ **% of hydrogen in fuel = 5.2**

(iii) Percentage of excess air

$$\text{Theoretical air/kg fuel} = \frac{100}{23}\left[\frac{8}{3}\times 0.9012+8\times 0.0988\right]$$
$$= 13.885 \text{ kg/kg fuel}$$

Actual air supplied/kg fuel = 18.524 kg/kg fuel

$\therefore$ Excess air = 18.524 – 13.885
= 4.639 kg

$\therefore$ % of excess air $= \frac{4.639}{13.885}\times 100 = \mathbf{33.41\ \%}$

Example 13.34. *(a) Derive an equation for minimum quantity of air required for complete combustion of 1 kg of solid or liquid fuel, assuming that the fuel contains carbon, hydrogen, sulphur and oxygen as constituents.*

(b) The analysis by mass of the coal fired in a boiler is 82 % carbon, 8% hydrogen, 3% oxygen, 7% ash. The boiler uses 0.19 kg/sec of air and is supplied to the furnace by a fan. The air supplied to the furnace by a fan. The air supplied is 30% in excess of that require for theoretical correct, combustion. Calculate.

(i) Volume of air taken in by the fan per sec. when the pressure and temperature of the air at the fan intake are 100 kN/m² and 18°C respectively. R for air 0.287 kJ/kg-K.

(ii) Percentage composition by mass of the dry flue gases. Air contains 23% oxygen by mass.

Solution

(b) (i) Volume of air sec

$$\text{Volume of air supplied/sec} = \frac{mRT}{p} = \frac{0.19\times287\times291}{100\times1000}$$

$$= 0.1587 \text{ cu-m/sec}$$

(ii) Gravimetric analysis of dry products:

Minimum air required/kg fuel

$$= \frac{100}{23}\left[\frac{8}{3}C+8H+S-O\right]$$

$$= \frac{100}{23}[2.187+0.64\times0.03]$$

$$= 12.16 \text{ kg/kg fuel}$$

∴ Actual air supplied /kg fuel

$$= 1.3\times 12.16 = 15.81 \text{ kg/kg fuel}$$

Dry products of combustion

$$CO_2 = \frac{11C}{3} = \frac{11}{3}\times0.82 = 3.001 \text{ kg}$$

$$N_2 = 0.77\times15.81$$

$$= 12.1737 \text{ kg}$$

$$O_2 = (0.3\times12.16)\times0.23$$

$$= 0.839 \text{ kg}$$

Total mass of products (assuming ash remain in the furnace only)

$$= 3.001+12.1737+.839$$

$$= 16.014 \text{ kg}$$

$$\therefore \text{Percentage of } CO_2 \text{ in products} = \frac{3.001\times100}{16.014} = 18.74\%$$

$$\text{Percentage of } O_2 \text{ in products} = \frac{0.839}{16.014}\times100 = 5.24\%$$

$$\text{Percentage of } N_2 \text{ in products} = \frac{12.1737}{16.014}\times100 = \underline{76.02\%}$$

$$= 100.00$$

Example 13.35. *(a) Describe briefly a method of obtaining the higher calorific value of a sample of fuel oil, showing how the result is calculated from the data obtained. State the possible sources of inaccuracy in the result.*

(b) A fuel consists of the following percentage analysis by mass

$C = 84$ %, $H_2 = 10\%$, $O_2 = 2\%$, $S = 1$ % and $N_2 = 3\%$.

Find out the amount of air required to completely burn 1 kg of this fuel and determine the mass of products of combustion.

Solution

Minimum amount of air required for complete combustion of 1 kg fuel is given by

$$m_a = \frac{100}{23}\left[\frac{8}{3}C + 8H + S - O\right]$$

$$= \frac{100}{23}\left[\frac{8}{3}\times 0.84 + 8\times 0.1 + 0.1 + 0.01 - 0.02\right]$$

$$= \frac{100}{23}[2.24 + 0.84 - 0.01]$$

$$= \frac{100}{23}\times 3.03$$

$$\mathbf{m_a = 13.174\ kg/kg\ fuel}$$

Products of combustion are

$$CO_2 = \frac{11}{3}C = \frac{11}{3}\times 8.84 = 3.08\text{ kg}$$

$$H_2O = 9H = 9\times 0.1 = 0.90\text{ kg}$$

$$O_2 = 2S = 2\times 0.01 = 0.02\text{ kg}$$

$$N_2 = 0.03 + \frac{77}{23}\times 0.03 = 10.144$$

$$= 10.174\text{ kg}$$

$\therefore$ Total mass $= 14.174$ kg

$\therefore$

$$CO_2 = \frac{3.08\times 100}{14.174} = 21.73\%$$

$$H_2O = \frac{0.9\times 100}{14.174} = 6.35\%$$

$$SO_2 = \frac{0.02\times 100}{14.174} = 0.14\%$$

$$\mathbf{N_2 = \frac{10.174\times 100}{14.174} = 71.78\%}$$

Example 13.36. (*a*) *Derive an equation for minimum quantity of air required for complete combustion of 1 kg of solid or liquid fuel, assuming that the fuel contains carbon, hydrogen, sulphur and oxygen as constituents.*

(b) The analysis by mass of the coal fired in a boiler is 82% carbon, 8% hydrogen, 3% oxygen, 7% ash. The boiler uses 0.19 kg/s of air and is supplied 30% in excess of that required for theoretically correct combustion. Calculate,

(i) Volume of air taken in by the fan per sec. when the pressure and temperature of the air at the fan intake are 100 kN/m² and 18°C respectively. R for air 0.287 kJ/kg-K.

(ii) Percentage composition by mass of the dry flue gases. Air contains 23% oxygen by mass.

Solution

(i) Volume of air/sec:

$$\text{Volume of air supplied/sec} = \frac{mRT}{P} = \frac{0.19\times 287\times 291}{100\times 1000}$$

Volume of air supplied/sec = 0.1587 cu.m/sec

(ii) Gravimetric analysis of dry products:

$$\text{Minimum air required/kg fuel} = \frac{100}{23}\left[\frac{8}{3}C+8H+S-O\right]$$

$$= \frac{100}{23}\left[\frac{8}{3}\times 0.82+8\times 0.08-0.03\right]$$

$$= 12.16 \text{ kg/kg fuel}$$

$\therefore$ Actual air supplied/kg fuel = 1.3×12.16

$$= 15.81 \text{ kg/kg fuel}$$

Dry products of combustion

$$CO_2 = \frac{11C}{3} = \frac{11}{3}\times 0.82 = 3.001 \text{ kg}$$

$$N_2 = 0.77\times 15.81 = 12.1737 \text{ kg}$$

$$O_2 = (0.3\times 12.16)\times 0.23$$

$$= 0.839 \text{ kg}$$

Total mass of products (assuming ash remain in the furnace only)

$$= 3.001+12.1737+0.839$$

$$= 16.014 \text{ kg}$$

Total mass of products assuming ash remain in the furnace only

$$\textbf{\% of CO}_2 \textbf{ in exhaust by mass} = \frac{3.001}{16.014}\times 100 = 18.73\%$$

$$\textbf{\% of O}_2 \textbf{ in exhaust by mass} = \frac{0.839}{16.014}\times 100 = 5.24\ \%$$

$$\textbf{\% of N}_2 \textbf{ in exhaust by mass} = \frac{12.1737}{16.014}\times 100 = 76.02\ \%$$

Example 13.37. (*a*) *Define excess air, stoichiometric air fuel ratio.*

(*b*) *Methods of coal analysis*

(*c*) *The following data was obtained during experimental determination of calorific value of fuel by bomb calorimeter:*

Mass of fuel = 0.8 gm

Mass of fuse wire = 0.03 gm

Calorific value of fuse wire = 7 kJ/gm

Mass of water in calorimeter = 2 kg

Water equivalent of calorimeter = 0.4 kg

Rise in temperature of calorimeter = 3.2°C

Cooling correction = 0.02°C

Determine HCV and LCV of coal. Given the coal contains 90% of carbon and 5% of hydrogen.

Solution

We can use the energy balance principle.

Heat given by combustion of fuel + Heat of fuse wire = Heat taken by water

$$\therefore \quad \frac{0.8}{1000}\times CV+\frac{0.03}{1000}\times\frac{7\times1000}{1} = \left[(2+0.4)(3.2+0.02)\times4.187\right]$$

$$\therefore \quad \frac{0.8\times CV}{1000}+\frac{210}{1000} = 2.4\times4.187\times3.22=32.36$$

$$\therefore \quad 0.8\times CV+210 = 32.36\times1000$$

$$\therefore \quad CV = \frac{32.36\times1000-210}{0.8}$$

$$= 40187.5 \text{ kJ/kg}$$

Bomb calorimeter readings or observations give us high calorific value HCV.

$$\therefore \quad \textbf{HCV} = \textbf{40187.5 kJ/kg}$$

In one kg coal, 5% is hydrogen

$$\therefore \quad H_2O \text{ formed/kg coal} = 0.05\times 9=0.45 \text{ kg}$$

$$\therefore \quad LCV = \text{Lower calorific value}$$

$$= HCV-\frac{H_2O}{\text{kg coal}}\times2466$$

$$= 40187.5-0.45\times 2466$$

$$\textbf{LCV} = \textbf{39077.8 kJ/kg}$$

Example 13.38. (*a*) *Define the following terms:*

(i) Higher calorific value of a fuel

(ii) Lower calorific value of a fuel

(b) A fuel has the following percentage composition by mass : C = 85 %, H_2 = 15%

Determine:

(i) The stoichiometric mass of air required for complete combustion of 1 kg of this fuel.

(ii) The percentage composition by mass of the dry products of combustion air contain 23.3% oxygen by mass.

(c) The dry exhaust gas from an oil engine had the following composition by volume CO_2 8.85%, CO = 1.2%, O_2 = 6.8 %, N_2 = 83.15.

Convert the volumetric analysis of the exhaust gas into a mass analysis.

Solution

(i) Stoichiometric air:

Mass of air required for complete combustion of 1 kg of fuel

$$= \left[\frac{8}{3}C + 8H - O + S\right]\frac{100}{23.2}$$

$$= \left[\frac{8}{3}\times 0.85 + 8\times 0.15\right]\frac{100}{23.2}$$

$$= \textbf{14.94 kg/kg of fuel}$$

(ii) Gravimetric analysis: (Dry products)

$$CO_2 \rightarrow \frac{11}{3}C = \frac{11}{3}\times 0.85 = 3.117\,\text{kg}$$

$$N_2 \rightarrow \frac{76.7}{23.3}\times 3.4667 = 11.476\,\text{kg}$$

$$\text{Total mass of products} = 3.117 + 11.476$$

$$= \textbf{14.593 kg/kg fuel}$$

$$\%\,CO_2 \rightarrow \frac{3.117}{14.593}\times 100 = 21.3658\,\%$$

$$\%\,N_2 \rightarrow \frac{12.8}{14.593}\times 100 = 78.64\,\%$$

(c) Volumetric to gravimetric analysis

			Rel. wts	%
CO_2	8.85×44	=	089.4	13.116
CO	1.2×28	=	33.6	1.132
O_2	6.8×32	=	217.6	7.330
N_2	83.15×28	=	2328.2	78.442
			2968.8	100.00

Example 13.39. *A sample of dry anthracite has the following composition by mass:*

C = 90%, H = 3%, O =2.5%, N = 1%, S = 0.5%, Ash = 3%.

Calculate

(i) The stoichoimetric A/F ratio

(ii) The A/F ratio and the dry and wet analysis of the products of combustion by mass and by volume, when 20% excess air is supplied.

Solution

(i) Stoichoimetric A/F ratio:

Theoretical or stoichiometric air supplied per kg of fuel

$$= \frac{100}{23}\left[\frac{8}{3}C + 8H + S - O\right]$$

$$= \frac{100}{23}\left[\frac{8}{3}\times 0.9 + 8\times 0.03 + 0.005 - 0.025\right]$$

$$= \textbf{11.3913 kg}$$

(ii) Product analysis:

Products are

$$CO_2 = \frac{11}{3}\times 0.9 = 3.3\text{ kg} \div 44 = 0.075\text{ mol}$$

$$H_2O = 9H = 9\times 0.03 = 0.27\text{ kg} \div 18 = 0.015\text{ mol}$$

$$SO_2 = 2S = 2\times 0.005 = 0.01\text{ kg} = 0.001563\text{ mol}$$

$$O_2 = 0.2\times 2.62 = 0.524\text{ kg} = 0.01638\text{ mol}$$

$$N_2 = 1.2\times 0.77\times 11.3913 = 10.5256 + 0.01$$

$$= 10.5356\text{ kg} = 0.376\text{ mol}$$

Gravimetric Analysis (Dry Products)

$$\text{Total} = 3.3 + 0.01 + 0.524 + 10.5356 = 14.3696\text{ kg}$$

$$CO_2\,\% \to \frac{3.3}{14.3696}\times 100 = 22.965\%$$

$$O_2\,\% \to \frac{0.524\times 100}{14.3696} = 3.6466\%$$

$$SO_2\,\% \to \frac{0.01\times 100}{14.3696} = 0.0696\%$$

$$N_2\,\% \to \frac{10.5356\times 100}{14.3696} = 73.32\%$$

Wet Products:

Total products are $14.3696 + 0.27 = 14.63$ kg

$$\%\,CO_2 \to \frac{330}{14.6396} = 22.5416\%$$

$$O_2\,\% \to \frac{52.4}{14.6396} = 3.5793\%$$

$$SO_2\,\% \to \frac{1}{14.6396} = 0.06830\%$$

$$N_2 \to \frac{1053.56}{14.6396} = 71.964\%$$

$$H_2 \to \frac{27}{14.6396} = 1.8443\%$$

Volumetric Analysis

Dry products

$$\text{Total mol} = 0.0752 + 0.3763 + 0.0001563 + 0.01638$$

$$\cong 0.46784$$

$$\therefore \quad CO_2 \rightarrow \frac{0.075 \times 100}{0.46784} = 16.031\%$$

$$SO_2 \rightarrow \frac{0.1563}{0.46784} = 0.0334\%$$

$$O_2 \rightarrow \frac{1.638}{0.46784} = 3.501\%$$

$$N_2 \rightarrow \frac{37.63}{0.46784} = 80.4335\%$$

Wet Products

$$\text{Total} = 0.46784 + 0.015 = 0.48284$$

$$\therefore \quad CO_2 \rightarrow \frac{7.5}{0.48284} = 15.5331\%$$

$$SO_2 \rightarrow \frac{0.01563}{0.46284} = 0.0324\%$$

$$O_2 \rightarrow \frac{1.638}{0.48284} = 3.3924\%$$

$$N_2 \rightarrow \frac{37.63}{0.48284} = 77.9342\%$$

$$H_2O \rightarrow \frac{1.5}{0.48284} = 3.1066\%$$

PROBLEMS FOR PRACTICE

13.1. A sample of coal when analyzed, gave following composition by weight—Carbon 85%, hydrogen 12%, ash 1% and moisture 2%. Find the higher and lower calorific values of the coal.

(*Ans.* 46141 kJ/kg, 43498 kJ/kg)

13.2. Find the higher and lower calorific value of a gaseous fuel whose volumetric composition is 7% CO_2, 25% CO, 15% H_2, 4% CH_4 and 49% N_2. The latent heat of water may be assumed to be 2466 kJ/kg at the condition after combustion. The higher calorific value of CO_2, CO, H_2 CH_4 and N_2 is zero, 12600, 12800, 43000 and zero kJ/Nm³ respectively.

13.3. A sample of lignite coal contains 6% hydrogen and 3% moisture by weight. If the higher calorific value of the sample was found to be 29300 kJ/kg, calculate its lower calorific value. Assume 2466 kJ of heat is required to evaporate water at room temperature to dry and saturated steam.

13.4. A producer gas contains 16% H_2, 20% CO, 6% CO_2 and 58% N_2 by volume. Determine the HCV and LCV of the gas in kJ/Nm³.

Assume calorific value of H_2 = 144000 kJ/Nm²

Assume calorific value of CO = 10000 kJ/Nm²

13.5. In a test carried on diesel oil in a bomb calorimeter following observations were recorded

Wt. of crucible = 2.5 gm

Wt. of crucible + oil = 3.079 gm

Wt. of can = 1 kg

Wt. of can + water = 2.4 kg

Wt. of cotton used in igniting oil = 0.005 gm

Calorific value of cotton = 16750 kJ/kg

Room temperature = 20°C

Temp. correction due to radiation.= 0.012C

If the oil contains 14% H_2 and 3% moisture, calculate HCV and LCV of the oil.

13.6. During an engine trial, the gas as used was tested in a Boy's calorimeter with the following results

Gas supply 18°C and 1.04 bar

Gas burnt 0.5 m^3

Temp. rise of water 16°C to 24.5°C

Wt. of cooling water 24.5 kg

Condensate from the products 0.004 kg and 18°C

Calculate the higher and lower calorific value of the gas.

13.7. The dry exhaust gases from an Internal combustion engine burning a pure hydrocarbon fuel were found to contain CO_2 =125%, CO = 2.7 %, N_2 = 83.4 % with small quantities of other gases. The fuel consumption was observed to be 11 kg/hour.

Estimate the ratio of carbon to hydrogen by weight in fuel. Find the air to fuel ratio and thus the air consumption of the engine.

(*Ans.* A/F = 14.12, $\frac{C}{H}$ = 5.4672, 155.32 kg/hr)

IMPORTANT POINTS TO BE REMEMBERED

1. *D-ulong formula for HCV*

$$\text{HCV} = \frac{33915C + 1444451.5\left(H - \frac{0}{8}\right) + 9630S}{100}$$

2. *For gaseous fuel – Boy's Calorimeter*

$$\text{HCV} = \frac{m_a C_{Pw}(t_2 - t_1)}{V}, \ \text{LCV} = \text{HCV} - wh_{fg}$$

3. *For Bomb calorimeter*

$$DT_2 = \text{Average rate of change of temperature in after period}$$

$$= \frac{t_{01} + t_{02} + t_{03} + \ldots\ldots}{\text{No. of readings taken}}$$

Cooling correction $t_c = (DT_2) \times$ No. of readings in main period

$\therefore$ Rise in temperature $= (T_n - T_0) + T_e + T_i$

T_i = Minimum temperature on thermometer

= 0.01 with Beckman thermometer

4. *Minimum weight of air for complete combustion of 1 kg fuel*

$$(\text{Air})_{\min} = \frac{100}{23}\left[\frac{8}{3}C + 8\left(H - \frac{0}{8}\right) + S\right]$$

5. 1 mol of fuel = Molecular weight

$$\frac{\text{Total weight}}{\text{Molecular wt}} = \text{No. of moles}$$

6. Actual air supplied/kg fuel

$$= \frac{NC}{33\,(CO_2 + CO)} \quad \text{All in \%}$$

CHAPTER 14

Simple Vapour Power Cycles

14.1 INTRODUCTION OR IDEAL STEAM PLANT CYCLES

Figure 14.1 shows the simple steam power plant. The heat energy released due to the combustion of fuel will be utilised in the boilers for converting water into steam (i.e., vapour) and this steam is then expanded into the steam engine/steam turbines to obtain useful work. The steam after producing work output, is generally condensed in the condensers. If the condensate is pure, then it is pumped as feed water to the boiler by means of a feed pump.

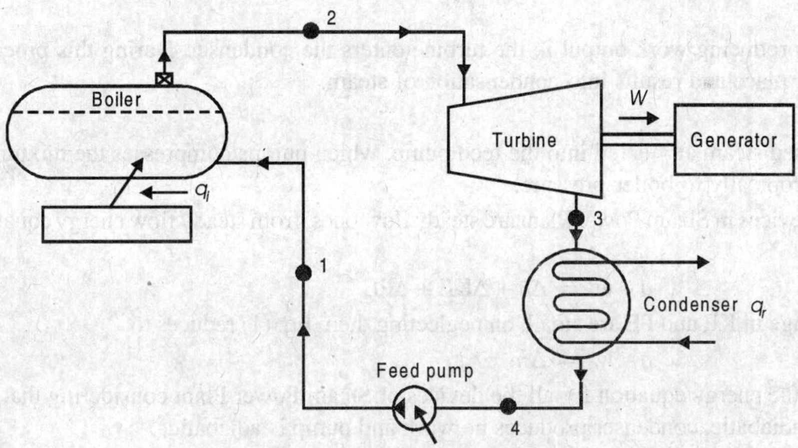

Fig. 14.1. *General arrangement of steam power plant*

14.2 CARNOT CYCLE AND STEAM PLANT

As we know, the Carnot cycle has the greatest possible efficiency between any two given temperature limits T_1 and T_2 and is given by $\eta_{th} = \dfrac{T_1 - T_2}{T_1}$. It is important therefore to see whether Carnot cycle can be successfully applied to the steam plant.

Referring the T–S diagram as shown in Fig. 14.2.

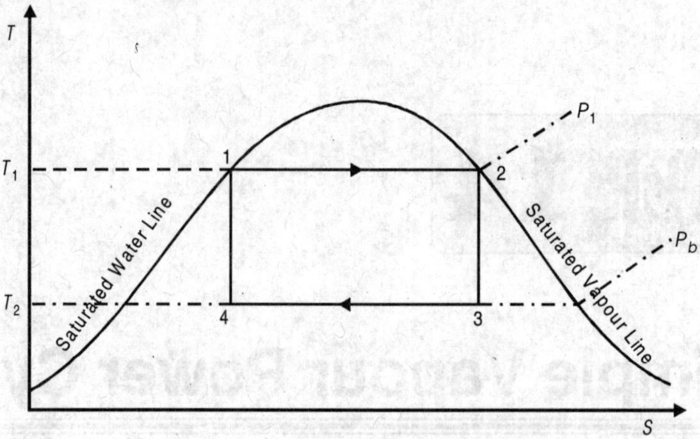

Fig. 14.2. *Carnot cycle on T–S diagram*

Process 1–2:

This is an isothermal heat addition process and during this, water in the boiler gets converted into steam at constant temperature and pressure.

At point 2, steam is dry saturated.

Process 2–3:

The steam is expanded adiabatically isentropically in the steam turbine or steam engine from pressure P_1 to P_b.

Process 3–4:

The steam after producing work output in the turbine enters the condenser. During this process isothermal heat rejection q_r takes place and results into condensation of steam.

Process 4–1:

Partially condensed steam at 4 is fed into the feed pump, which pumps/compresses the mixture of wet steam and condensate isentropically to boiler pressure.

Since all the devices in Steam Power Plant are steady flow ones, from steady flow energy equation (SFEE) for unit mass,

$$q - w = \Delta h + \Delta \text{KE} + \Delta \text{PE} \qquad \text{...(1)}$$

Since the change in KE and PE are small, on neglecting them Eq. (1) reduces to

$$q - w = \Delta h \qquad \text{...(2)}$$

We can write the energy equation for all the devices of Steam Power Plant considering that boiler produces no work, turbine is adiabatic, condenser produces no work and pump is adiabatic.

$\therefore$ Heat added in the boiler q_1 during 1 – 2

$$q_i = h_2 - h_1 \qquad (\because W = 0 \text{ for boiler})$$

i.e.

$$q_i = h_2 - h_1 \text{ kJ/kg}$$

$$= \text{Area under } 1 - 2$$

$$= T_1 \times (S_2 - S_1) \qquad \text{...(3)}$$

Work produced by the turbine W_T during the expansion process 2–3 and since expansion adiabatic.

$$q = 0$$

$$\therefore \quad 0 - W_T = h_3 - h_2$$

$$W_T = h_2 - h_3 \text{ kJ/kg} \quad \text{...(4)}$$

During the process 3–4 isothermal heat rejection takes place and results into condensation of steam (since $W = 0$)

$$q_r = h_4 - h_3$$

$$q_r = -(h_3 - h_4) \text{ kJ/kg} \quad \text{...(5)}$$

$$= \text{Area under } 3-4$$

$$= T_2 \times (S_3 - S_4)$$

$$= T_2 \times (S_2 - S_1)$$

Negative sign implies that the heat is transferred to the cooling water from steam in the condenser by an amount equal to $(h_3 - h_4)$.

Partially condensed steam is then pumped during 4–1, then from Eq. (2), since the pump is adiabatic $q = 0$

$$\therefore \quad -W_p = h_1 - h_4$$

or

$$W_P = -(h_1 - h_4)$$

Negative sign implies that work is to be supplied to the pump for pumping by amount equal to $(h_1 - h_4)$.

$$\eta_{th} = \frac{\text{Net work output}}{\text{Heat supplied}} = \frac{\text{Shaft work } (W_s)}{\text{Heat supplied } q_i}$$

$$\eta_{th} = \frac{\text{Difference between turbine and pump work}}{q_i}$$

$$= \frac{W_T - W_p}{q_i}$$

$$\eta_{th} = \frac{[h_2 - h_3] - [h_1 - h_4]}{[h_2 - h_1]} \quad \text{...(7)}$$

$$= \frac{(h_2 - h_1) - (h_3 - h_4)}{h_2 - h_1}$$

$$= \frac{T_1(S_1 - S_2) - T_2(S_1 - S_1)}{T_1(S_2 - S_1)}$$

$$= \frac{T_1 - T_2}{T_1} \text{ as for ideal gas.}$$

Even though, the Carnot efficiency is maximum for a given value T_1 and T_2. But because of some inherent practical difficulties, as discussed below, Carnot cycle cannot be applied for the steam plant.

Carnot cycle when applied to S.P.P. is practical upto a point that, Isothermal heat addition and the adiabatic expansion of steam in the turbine, which is quite reasonable.

The impractical part is in handling of steam in the condenser and pump. In the condenser, the steam is only partially condensed and the condensation must be stopped at 4. Also the feed pump must be capable of handling both wet steam and condensate.

A slight modification is made to this Carnot cycle which however produces a more practical cycle, called Rankine cycle. This cycle is accepted as, the practical but ideal cycle for the steam plant even though it has reduced thermal efficiency.

14.3 RANKINE CYCLE

To obtain Rankine cycle, modification made to Carnot cycle, is that, instead of stopping the condensation at some intermediate condition, the condensation is completed as shown in Fig. 11.3.

On T–S diagram Carnot cycle, would be 1'–2–3–4'. For Rankine cycle, however, condensation is continued until it is complete at 4. At this point there is all water. This can be successfully dealt with by the feed pump.

Process 4–1:

Pumping process, where water of condenser pressure P_b is raised to boiler pressure P_1. The pumping compression process is isentropic. At point (1) temperature is lower than T_{sat}. Subcooled state:

Process 1–2: (1 – 1' and 1' – 2)

During 1 – 1' feed water is heated to T_{sat} (may occur in economiser/boiler itself.) During 1' – 2 heat added q_t at constant temperature and pressure P_1 convert water into steam in the boiler.

Process 2–3:

Isentropic expansion of steam in the turbine from boiler pressure P_1 to back pressure P_b, which results into turbine work W_T

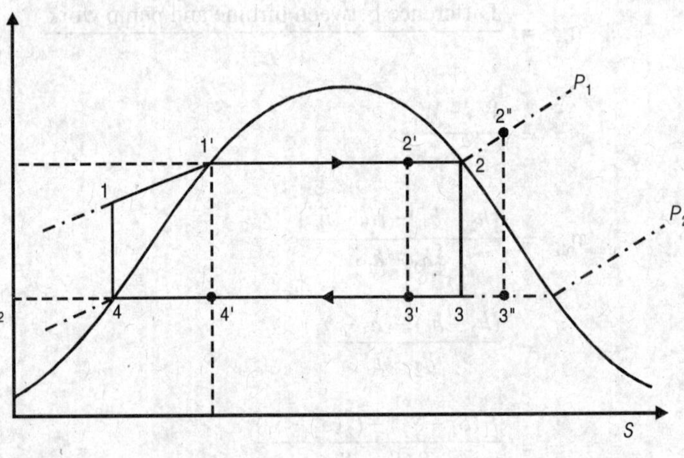

Fig. 14.3. *Rankine cycle*

Process 3–4:

The steam is condensed at constant pressure P_b in the condenser. The steam rejects heat q_r to the cooling water.

It should be noted that the steam generated in the boiler or steam entering the turbine may be wet, dry-saturated and superheated. Correspondingly the expansion process in the turbine will be 2'–3', 2–3, 2"–3" respectively.

Analysis of Rankine Cycle (Assuming 1 kg of steam)

From Eq. (5) we get $q - w = \Delta h$

Process 1–2:

In the boiler heat q_i converts water into steam

$$q_i = h_2 - h_1 \text{ (since } q = 0)$$

Process 2–3:

In the turbine, when the steam expands, work will be produced.

$$W_T = h_2 - h_3 \text{ kJ/kg (since } q = 0)$$

Process 3–4:

In the condenser condensation will be completed and an amount of heat q_r is rejected to the cooling water.

$$q_r = h_4 - h_3$$

i.e., $q_r = -[h_3 - h_4]$ Negative sign implies heat rejection.

Process 4–1:

Feed pump pumps the water and increases the pressure of water from P_b to P_1.

$$0 - W_p = h_1 - h_4 \text{ (Since } q = 0)$$

or $W_P = -(h_1 - h_4)$

Negative sign implies work to be supplied to pump.

Also since pump is a steady flow equipment, we know that work done in a steady flow system.

$$= \int vdP$$

$$\therefore \quad W_P = -\int_{P_b}^{P_1} vdP$$

The specific volume of water remains constant since the effect of pressure is negligible on liquids.

$$\therefore \quad W_P = -V\int_{P_b}^{P_1} dP$$

$$\therefore \quad \text{Magnitude of } W_P = V[P_1 - P_b]$$

We also know that

$$\text{Specific volume} = \frac{1}{\rho}$$

and for water $\rho_w = 1000 \text{ kg/m}^3$

$\therefore$ Specific volume $= 0.001 \text{ m}^3\text{/kg}$. If the pressure are measured in bar then,

$$W_P = 0.001 \text{ m}^3\text{/kg } [P_1 - P_b]\, 10^5 \text{ N/m}^2$$

$$= 0.001\,(P_1 - P_b)\, 10^5 \text{ Nm/kg or J/kg}$$

$$= \frac{0.001(P_1 - P_b)10^5 \text{ kJ}}{1000} \text{ kJ/kg}$$

$$W_P = \left[\frac{P_1 - P_b}{10} \text{ kJ/kg}\right] \qquad \text{...(8)}$$

And the efficiency of Rankine cycle,

$$\eta_R = \frac{\text{Shaft work } W_s}{\text{Heat supplied } q_i} = \frac{W_T - W_p}{q_i}$$

$$\therefore \quad \eta_R = \frac{[h_2 - h_3] - [h_1 - h_4]}{[h_2 - h_1]}$$

Also we can write this as,

$$\therefore \qquad \eta_R = \frac{[h_2 - h_3] - [h_1 - h_4]}{[h_2 - h_4] - [h_1 - h_4]}$$

Since the pump work $W_P = h_1 - h_4$ is very small quantity, it can be neglected,

$$\therefore \qquad \eta_{th} = \frac{[h_2 - h_3]}{[h_2 - h_4]}$$

Note: For solving the problems we can read the value of $h_4 = h_f$ from steam tables at back pressure P_b, h_2 and h_3 can be determined with the help of Mollier diagram or by calculations from steam tables. Since $W_P = h_1 - h_4$ the volume of h_l can be determined by using Eq. (8) i.e., by finding W_p.

It should be noted that $\eta_R < \eta_{carnot}$ because the mean temperature of heat addition process in Rankine cycle is lower than the mean temperature of heat addition process in Carnot cycle. Whereas the mean temperature of heat rejection process in both the cycles is same.

In order to improve the efficiency of Rankine cycle, superheated steam has to be used, as represented by the process 2–2". It improves the cycle efficiency, as it increases the mean temperature of heat addition process in the boiler. Also when the superheated steam is used, it reduces the corrosion of turbines.

The Rankine efficiency has the same form as before, from Fig. 14.3 on similar lines

$$\eta_R = \frac{h_2'' - h_3''}{h_2'' - h_4''}$$

and when the steam is wet, then the cycle would be 1 – 1' – 2' – 3' – 4 –1

and $$\eta_R = \frac{h_2' - h_3'}{h_2' - h_4'}$$

14.4 EFFECT OF INLET PRESSURE (BOILER PRESSURE) AND BACK PRESSURE (EXHAUST PRESSURE) AND SUPERHEAT ON RANKINE EFFICIENCY

The effect of inlet pressure, back pressure and the superheat on the performance of Rankine cycle can be studied with the help of T–S diagram as shown in Fig. 14.4.

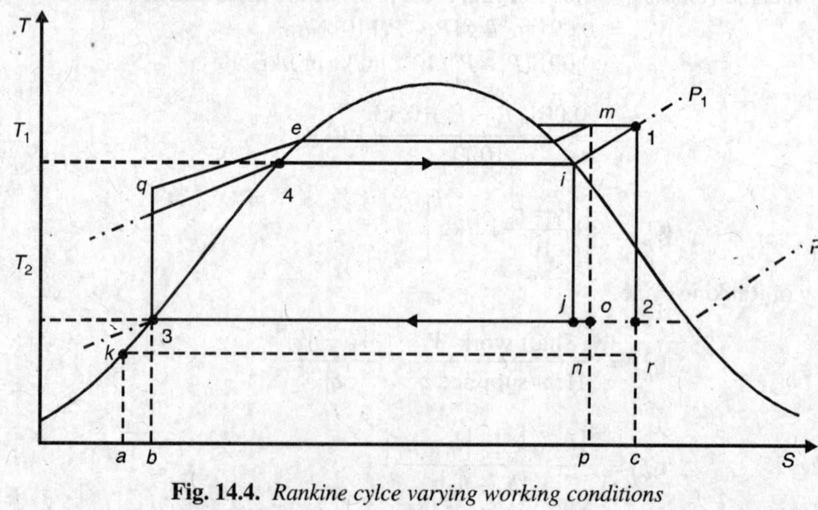

Fig. 14.4. *Rankine cylce varying working conditions*

14.4.1 Boiler or Inlet Pressure

If the boiler pressure is raised, the evaporation line is raised as the saturation temperature is increased from 4–1 to *em*. The maximum temperature t_1 is kept constant i.e., $t_m; = t_1$. $1-2-3-4-1$ is the basic cycle taken for comparison.

With increased boiler pressure, keeping the condenser pressure and temperature constant, the new cycle will be m 03 *qem*. The amount of heat rejected is decreased given by C20 pc and difference in the heat added –*b*341*cb* and *bqempb* is small and the thermal efficiency increases. It is experimentally found that at $p_1 = 170$ bar and $p_2 = 0.14$ bar, the efficiency is maximum and decrease with increase in p_1.

14.4.2 Condenser Pressure and Temperature is Lowered

The new cycle will be 1*dk* 41. Work done is increased and heat rejected will be less so that the efficiency will be increased. Nature, however, precisely defines the limit of improvement that may be obtained by this means. Atmospheric temperature and therefore temperature of the cooling water puts the limits of the steam temperature and pressure.

14.4.3 Superheat

If the steam is dry and saturated at the inlet of turbine, and if the steam is superheated then the work is increased and the proportional rise of work is more than the heat supplied so that the efficiency of the Rankine cycle increases.

It is interesting to note that condensing plants commonly operate more efficiently in the winter than in summer because of the lower cooling water temperature in the winter.

SOLVED EXAMPLES

Example 14.1 *A steam turbine operates on the Rankine cycle. It receives steam from the boiler at 30 bar, 250°C and it is exhausted into a condenser at 0.5 bar. Condensate is returned to the boiler by a feed pump. Calculate,*

(*1*) *Heat supplied in the boiler drum.*

(*2*) *Dryness fraction of steam entering the condenser.*

(*3*) *Work done.*

(*4*) *Steam rate/kWh and*

(*5*) *Efficiency of the cycle.*

Solution

Since it operates on Rankine Cycle,

Refer to Rankine cycle 1–2–3–4, (Fig. Ex. 14.1)

h_4 from tables at 0.5 bar condenser pressure $h_4 = h_f = 340.56$ kJ/kg.

We know that,

$$W_P = \frac{[P_1 - P_b]}{10}\frac{\text{kJ}}{\text{kg}} = \left[\frac{30-0.5}{10}\right]\frac{\text{kJ}}{\text{kg}}$$

$$W_P = 2.95 \text{ kJ/kg}$$

Also,
$$W_P = [h_1 - h_4] \text{ kJ/kg}$$

or
$$h_1 = W_p + h_4 = 2.95 + 340.56 \text{ kJ/kg}$$

$$h_1 = 343.51 \text{ kJ/kg}$$

From steam tables, h_2 at 30 bar and 250ºC

$$h_2 = 2855 \text{ kJ/kg (2854.8)}$$

and $$s_2 = 6.28 \text{ kJ/kg-K}$$

For isentropic process (2–3)

Entropy before expansion s_2 = Entropy after expansion s_3

$$6.28 = s_{f3} + x_3.s_{fg3} = 1.0912 + x_3.(6.5035)$$

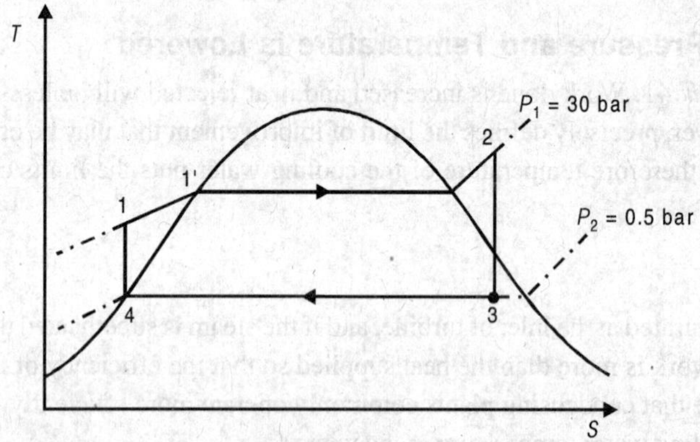

Fig. Ex. 14.1

∴ Dryness fraction of steam entering the condenser

$$x_3 = 0.797 \quad \text{...(A)}$$

∴ $$h_3 = h_{f3} + x_3.h_{fg3} \text{ corresponding to condenser pressure}$$

$$= 340.56 + 0.797 \times 2305.4$$

$$h_3 = 2177.9 \text{ kJ/kg}$$

Now,

(1) Heat supplied in the boiler drum

$$q_i = h_2 - h_1 = 2855 - 343.51$$

$$q_i = 2511.49 \text{ kJ / kg}$$

(2) Dryness fraction of steam entering condenser

$$x_3 = 0.797 \text{ as above}$$

(3) W.D. or shaft work

$$W_s = W_T - W_P = [h_2 - h_3] - [h_1 - h_4]$$

$$= (2855 - 2177.9) - 2.95 = 674.15 \text{ kJ/kg}$$

(4) Since 1 kWh = 3600 kJ

$$\text{Steam rate} = \frac{3600}{W_s\,(\text{kJ/kg})} = \frac{3600}{674.15}$$

$$= 5.34 \frac{\text{kg}}{\text{kWh}}$$

(5) $$\eta_R = \frac{W_s}{q_1} = \frac{674.15}{2511.49}$$

$$\mathbf{\eta_R = 26.84\ \%}$$

Example 14.2 *Steam at 20 bar and 360°C expands in a steam turbine to 0.08 bar. It is then condensed in a condenser to saturated water. The pump feeds back the water to the boiler. Assume ideal Rankine cycle and determine.*

(*i*) *The net work done/kg of steam*

(*ii*) η_R

Solution

Refer Fig. Ex. 11.2. From steam tables corresponding to condenser pressure 0.08 bar $h_4 = h_f = 173.86$ kJ/kg.

We know that,
$$W_P = \left[\frac{P_1 - P_b}{10}\right]\frac{\text{kJ}}{\text{kg}} = h_1 - h_4$$

$$W_P = \left[\frac{20-0.08}{10}\right]\frac{\text{kJ}}{\text{kg}}$$

$$W_P = 1.992\frac{\text{kJ}}{\text{kg}} = h_1 - h_4$$

or
$$h_1 = W_P + h_4 = 1.992 + 173.86$$

$$\mathbf{h_1 = 175.852\ kJ/kg}$$

or corresponding to 20 bar and 360° C, $h_2 = 3160.62$

At 350° C, $h = 3138.6$ kJ/kg and 400° C = h 3248.7 kJ/kg

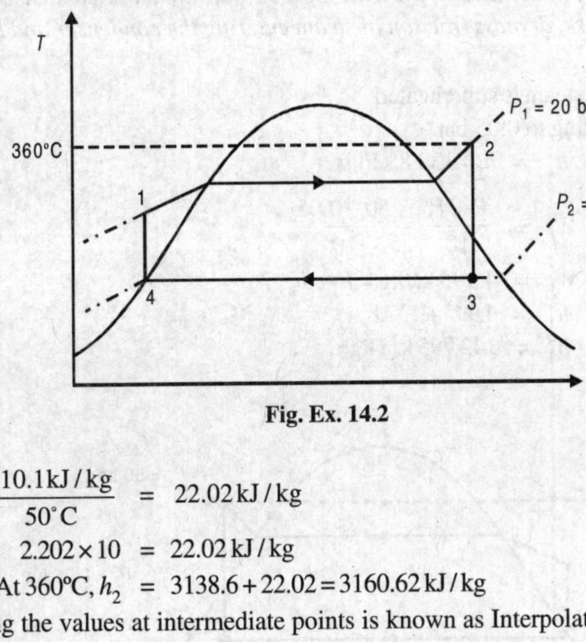

Fig. Ex. 14.2

$$\text{Difference is } \frac{110.1\,\text{kJ/kg}}{50^\circ\text{C}} = 22.02\,\text{kJ/kg}$$

$$\therefore \quad 2.202 \times 10 = 22.02\,\text{kJ/kg}$$

$$\therefore \quad \text{At } 360^\circ\text{C}, h_2 = 3138.6 + 22.02 = 3160.62\,\text{kJ/kg}$$

This method of finding the values at intermediate points is known as Interpolation.

To find h_s:

For isentropic process 2–3

Entropy before expansion s_2 = Entropy after expansion s_3

To find $s_2 = 7.1296$ at 400° C

6.9596 at 350° C

$$0.17/50 = 0.0034 \text{ kJ/kg-K}$$

$$0.0034 \times 10 = 0.034 \text{ kJ/kg-K}$$

$$\therefore \quad 6.9596 + 0.034 = 6.9936 \text{ kJ/kg-K}$$

$$6.9936 = f_{f3} + x_3.s_{fg3} \text{ at condenser pressure}$$

$$= 0.5925 + x_3.7.6371$$

$$\therefore \quad x_3 = 0.8382$$

$$h_3 = h_{f3} + x_3\, h_{fg3} \text{ at condenser pressure}$$

$$= 173.86 + 0.8382 \times 2403.2$$

$$h_3 = 2188.15 \text{ kJ/kg}$$

Now,

$$\text{Work done} = W_s = W_T - W_P = (h_2 - h_3) - W_P$$

$$= (3160.62 - 2188.15) - 1.992$$

$$= 970.478 \text{ kJ/kg}$$

and

$$\eta_R = \frac{W_S}{q_i} = \frac{970.478}{h_2 - h_1} = \frac{970.478}{3160.62 - 175.852}$$

$$\boldsymbol{\eta_R = 32.51\%}$$

Example 14.3 *A steam turbine operates on Rankine cycle. Steam entering the turbine has a pressure and temperature of 50 bar and 500°C, expands to a condenser pressure of 0.05 bar. Steam is condensed completely. Find heat supplied, turbine work, pump work, dryness fraction of steam entering the condenser and Rankine cycle efficiency.*

Solution

Refer Fig. Ex. 11.3. Since the steam is superheated,
From steam tables corresponding to 0.05 bar.

$$h_4 = h_f = 137.8 \text{ kJ/kg}$$

Also pump work

$$W_P = \frac{P_1 - P_b}{10} = \frac{50 - 0.05}{10}$$

$$W_P = 4.995 \text{ kJ/kg} = h_1 - h_4$$

$$\therefore \quad h_1 = 4.995 + 137.8$$

$$h_1 = 142.795 \text{ kJ / kg}$$

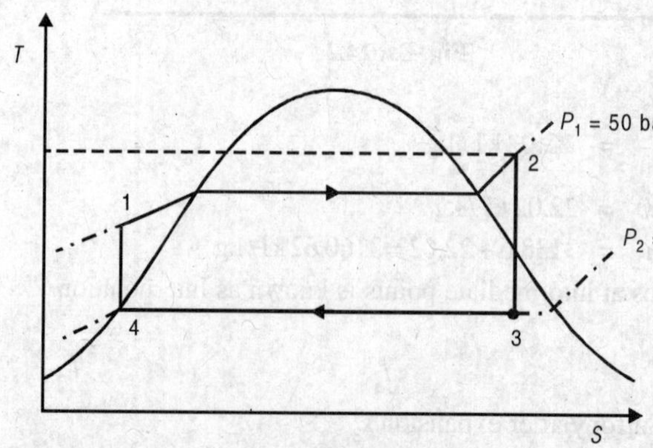

Fig. Ex. 14.3

Now, h_2 corresponding to 50 bar and 500° C from superheated steam tables,

$$h_2 = 3433.7 \text{ kJ/kg}$$

To find h_3; from the isentropic process 2–3

Entropy before expansion = Entropy after expansion.

$$s_2 = s_3$$

$$s_2 = s_{f3} + x_3 \cdot s_{fg3} \quad \text{...(A)}$$

Now s_2 at 50 bar and 500° C = 6.977 kJ/kg-K

$\therefore$ From (A), $\quad 6.977 = 0.4763 + x_3 \times (7.92)$

(At condenser pressure 0.05 bar)

$$\Rightarrow \quad x_3 = 0.8208 = 0.82$$

$$h_3 = h_{f3} + x_3 . h_{fg3} = 137.77 + 0.82 \times 2423.8$$

$$\mathbf{h_3 = 2125.286 \ kJ/kg}$$

Now,

(i) Heat supplied, $q_i = h_2 - h_1 = 3433.7 - 142.795$

$$\mathbf{q_i = 3290.905 \ kJ/kg}$$

(ii) Turbine work $W_T = h_2 - h_3 = 3433.7 - 2125.286$

$$\mathbf{W_T = 1308.414 \ kJ/kg}$$

(iii) Pump work W_P = 4.995 kJ/kg, already found in the beginning.

(iv) Dryness fraction of steam entering the condenser.

$$x_3 = 0.8208 \text{ already found}$$

(v) Rankine cycle efficiency

$$\eta_R = \frac{W_s}{q_i}$$

$$\eta_R = \frac{W_T - W_p}{q_i} = \frac{1308.414 - 4.995}{3290.905}$$

$$\boldsymbol{\eta_R = 39.606 \ \%}$$

Example 14.4 *Steam at 20 bar and 360°C is expanded in a steam turbine to 0.08 bar. It then enters a condenser where it is condensed to saturated liquid. It is then fed back to the boiler. Determine,*

(*i*) *Net work (shaft work) per kg of steam.*

(*ii*) *Dryness fraction of steam entering the condenser.*

Solution

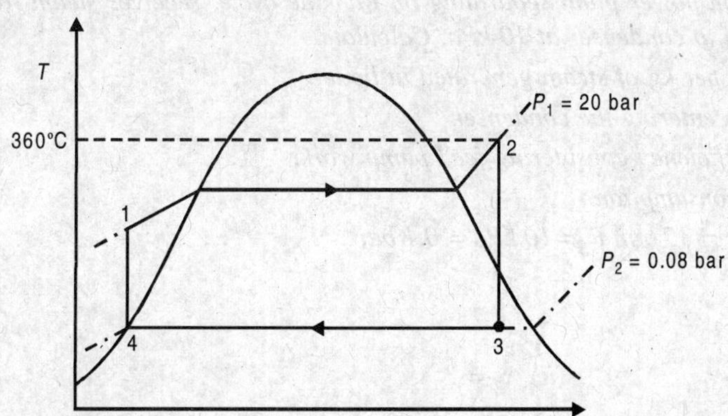

Fig. Ex. 14.4

From steam tables at 0.08 bar

$$h_4 = h_f = 173.9 \text{ kJ/kg}$$

And pump work $\quad W_P = \left[\frac{P_1 - P_b}{10}\right] = \left[\frac{20 - 0.08}{10}\right]$

$$W_P = 1.992 \text{ kJ/kg} = h_1 - h_4$$

$$h_1 = 1.992 + h_4$$

$$h_1 = 1.992 + 173.9$$

$$h_1 = 175.892 \text{ kJ/kg}$$

h_2 corresponding to 20 bar and 360° C from superheated steam tables = 3160.62 kJ/kg (By interpolation)

To find h_3 from the isentropic process 2–3

Entropy before expansion = Entropy after expansion

$$s_2 = s_3$$

$$s_2 = s_{f3} + x_3 \cdot s_{fg3} \quad \ldots(A)$$

Now s_2 at 20 bar and 360° C = $6.994 \dfrac{\text{kJ}}{\text{kg K}}$ (By interpolation)

Now from (A) $\quad 6.994 = 0.593 + x_3 \times 7.637$ at 0.08 bar

$$6.994 - 0.593 = 7.637\, x_3$$

$$0.8381 = x_3$$

$$\therefore \quad h_3 = h_{f3} + x_3 \cdot h_{fg3}$$

at $\quad$ 0.08 bar $= 173.9 + 08381 \times 2403.2$

$$\mathbf{h_3 = 2188.157 \text{ kJ/kg}}$$

(i) ∴ Net work or shaft work = $W_s = W_T - W_P$

$$W_s = [h_2 - h_3] - [h_1 - h_4]$$

$$= (3160.62 - 2188.157) - (175.892 - 173.9)$$

$$= (972.4626) - (1.992)$$

$$\mathbf{W_s = 970.4706 \text{ kJ/kg}}$$

(ii) Dryness fraction of steam entering the condenser $x_3 = 0.8381$ as found earlier.

Example 14.5 *A steam power plant separating on Rankine cycle, receives steam from boiler at 3.5 MPa and 350°C. It is exhausted to condenser at 10 kPa. Calculate,*

(i) Energy supplied per kg of steam generated in boiler.

(ii) Quality of steam entering the condenser.

(iii) Rankine cycle efficiency considering feed pump work.

(iv) Specific steam consumption.

Data: P = 3.5 MPs = 35 bar, P_b =10 kPa = 0.1 bar

As $100 \dfrac{kN}{m^2} = 1 bar$

(i) $q_i = ?$ *(ii)* $x_3 = ?$

(iii) $\eta_R = ?$ *(iv) SR = ?*

Solution

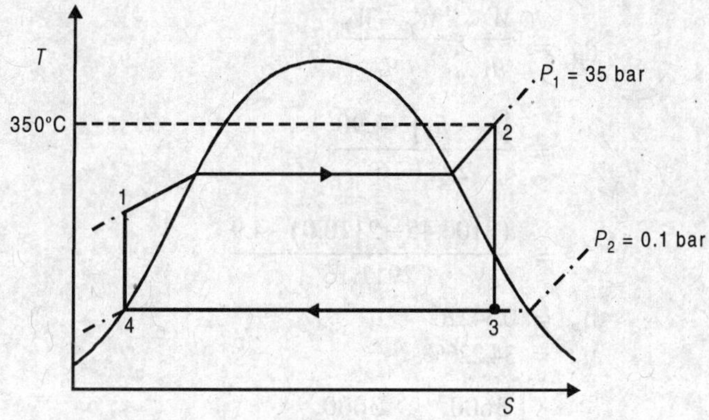

Fig. Ex. 14.5

From steam tables corresponding to 0.1 bar,

$$h_4 = h_f = 191.8 \text{ kJ/kg}$$

Also pump work $\quad W_p = \dfrac{P_1 - P_b}{10}$

$$= \frac{35-0.1}{10} = 3.490 \text{ kJ/kg} = h_1 - h_4$$

i.e.

$$3.499 + h_4 = h_1$$
$$3.490 + 191.8 = h_1$$
$$195.29 = h_1$$

And h_2 corresponding to 35 bar and 350° C from superheated steam tables = 3106.45 kJ/kg by interpolation.

To find h_3 from the isentropic process 2–3

$$s_2 = s_3$$
$$s_2 = s_{f3} + x_3 \, . \, s_{fg3} \qquad \ldots(A)$$

Now s_2 at 35 bar and 350° C = $6.663 \dfrac{\text{kJ}}{\text{kg-K}}$ by interpolation.

∴ From (A) $6.663 = 0.649 + x_3 \times 7.502$ at 0.1 bar

$$\frac{6.663 - 0.649}{7.502} = x_3$$
$$0.8016 = x_3$$

∴ $\quad h_3 = h_{f3} + x_3 . h_{fg3}$ at 0.1 bar

$$= 191.8 + 0.8016 + 2392.9$$
$$h_3 = 2110.075 \text{ kJ/kg}$$

Now,

(i) Heat supplied $\quad q_i = h_2 - h_1 = 3106.45 - 195.29$

$$q_i = 2911.16 \text{ kJ/kg}$$

(ii) $x_3 = 0.8016$ dryness fraction of steam which is already found.

(iii) Rankine cycle efficiency η_R considering feed pump work,

$$\eta_R = \frac{W_s}{q_i} = \frac{W_T - W_P}{q_i}$$

$$= \frac{[h_2 - h_3] - 4.90}{2911.16}$$

$$= \frac{(3106.45 - 2110.0) - 4.9}{2911.16}$$

$$\eta_R = 0.34226$$

$$\eta_R = 34.226\%$$

(iv)

$$\text{Steam Rate} = \frac{3600}{W_S} = \frac{3600}{[W_T - W_P]}$$

$$= \frac{3600}{[h_2 - h_3] - 4.9} = \frac{3600}{996.375 - 4.9}$$

$$= \frac{3600}{991.475}$$

S.R. = 3.630 kg/kWh

Example 14.6 *Find heat supplied, turbine work, pump work, shaft work, efficiencies, steam rate of Carnot and Rankine cycle using steam between 30 bar and 0.04 bar.*

Steam is dry and saturated at 30 bar. Compare the results.

Data: $q_i = ?$ $W_T = ?$

$W_P = ?$ $W_S = ?$

$\eta_c = ?$ $\eta_R = ?$

$S.R. = ?$

$P_1 = 30$ bar $P_b = 0.04$ bar.

Solution

Refer Fig. Ex. 14.6

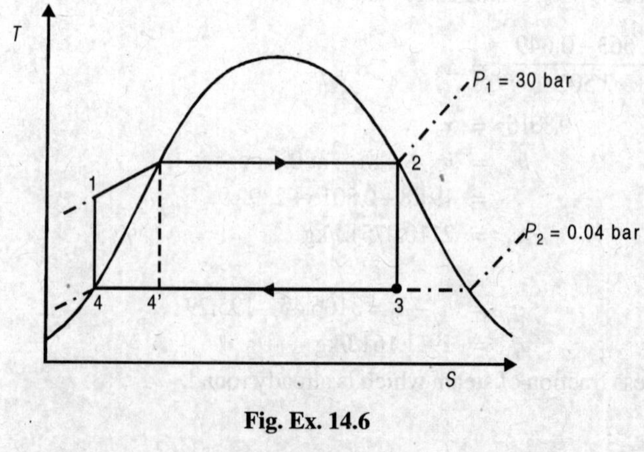

Fig. Ex. 14.6

First consider the Carnot cycle 1' – 2 – 3 – 4'

From steam tables at 30 bar,

$$h_{1'} = h_f = 1008.35 \text{ kJ/kg}$$
$$s_{1'} = s_f = 2.6455 \text{ kJ/kg-K}$$

and

$$h_2 = h_g \text{ at } 30 \text{ bar} = 2802.3 \text{ kJ/kg}$$
$$s_2 = s_g = 6.1837 \text{ kJ/kg-K}$$

For the isentropic process 2–3

$$s_2 = s_{f3} + x_3.s_{fg3} \text{ at } 0.04 \text{ bar}$$
$$6.1837 = 0.423 + x_3 \times 8.053$$
$$x_3 = 0.7154$$
$$\therefore \quad h_3 = h_{f3} + x_3 h_{fg3}$$
$$= 121.41 + 0.7154 \times 2433.1$$
$$h_3 = 1862.04 \text{ kJ/kg}$$

Now for Isentropic Compression process 4'–1'

$$s_1' = s_4' = S'_{f_4} + x_4' s'_{fg_4}$$
$$2.6455 = 0.4225 + x_4' \times 8.0530$$
$$x_4' = 0.276 \qquad \text{Nearer to liquid line}$$
$$\therefore \quad h_4' = h'_{f_4} + x'_4 h'_{fg_4}$$
$$= 121.41 + 0.276 \times 2433.1$$
$$h_4' = 792.94 \text{ kJ/kg}$$

Now,

(i) Heat supplied $\quad q_i = h_2 - h_{1'}$

$$= 2802.3 - 1008.35$$
$$q_i = 1793.95 \text{ kJ/kg}$$

(ii) Turbine work $\quad = W_T = h_2 - h_3 = 2802.3 - 1862.04$

$$W_T = 940.26 \text{ kJ/kg}$$

(iii) Pump work $\quad W_P = h_{1'} - h_{4'} = 1008.35 - 792.94 = 215.41 \text{ kJ/kg}$

(iv) Shaft work $\quad W_s = W_T - W_P = (940.26) - (215.41) = 724.85 \text{ kJ/kg}$

(v) Carnot cycle efficiency

$$\eta_{\text{carnot}} = \frac{W_s}{q_i} = \frac{724.85}{1793.95} = 0.4040$$

$$\boldsymbol{\eta_{\text{carnot}} = 40.40\%}$$

(iii) Steam Rate $= \dfrac{3600}{W_s} = \dfrac{3600}{724.85}$

$$\mathbf{SR = 4.96654 \frac{kg}{kWh}}$$

II. Rankine Cycle 1 – 2 – 3 – 4

From Fig. Ex. 14.6 it is to be noted that

h_2 and h_3 will remain same.

$$\therefore \quad h_2 = 2802.3\,\text{kJ/kg}$$
$$h_3 = 1862.04\,\text{kJ/kg}$$

and $$h_4 = h_f \text{ at } 0.04\,\text{bar} = 121.41\frac{\text{kJ}}{\text{kg}}$$

Now to find h_1,

We also know that, pump work

$$W_P = \frac{P_1 - P_b}{10} = \frac{30 - 0.04}{10} = 2.996$$
$$\cong 3.0\,\text{kJ/kg}$$

Also, $$W_p = h_1 - h_4$$
$$W_p + h_4 = h_1$$
$$3 + 121.41 = h_1$$
$$\Rightarrow \quad h_1 = 124.41\,\text{kJ/kg}\,124.41\,\text{kJ/kg}$$

Now,

(a) Heat supplied $$q_i = h_2 - h_1 = 2802.3 - 124.41$$
$$q_i = 2677.89\,\text{kJ/kg}$$

(b) Turbine work $W_T = h_2 - h_3 = 940.26\frac{\text{kJ}}{\text{kg}}$ · It will remain same.

(c) Pump work $$W_P = h_1 - h_4 = 124.41 — 121.41$$
$$\mathbf{W_p = 3\ kJ/kg}$$

(d) Shaft Work $W_S = W_T = h_2 - h_3 = 940.26 - 3 = 937.26$ kJ/kg

(e) Rankine Cycle Efficiency

$$\eta_R = \frac{W_S}{q_i} = \frac{937.26}{2677.89}$$
$$\eta_R = 0.34999$$
$$\eta_R = 34.999\%$$

(f) Steam Rate,

$$SR = \frac{3600}{W_s} = \frac{3600}{937.26}$$

SR = 3.8409 kg/kWh

Comparison:

(1) Pump work for Carnot Cycle is more, since it has to handle large amount of mixture of wet steam and condensate.

(2) Turbine work in both the cases is same as the state points are same.

(3) Carnot cycle efficiency is more than Rankine cycle efficiency.

Example 14.7 *Calculate the cycle efficiency and steam consumption in kg/kWh for Carnot cycle and Rankine cycle using steam between two pressure of 20 bar and 0.1 bar. The steam is dry and saturated at 20 bar.*

Find: $\eta_{carnot} = ?\ \ \eta_R = ?\ \ SR = ?$

Solution

Refer Fig. Ex. 14.7

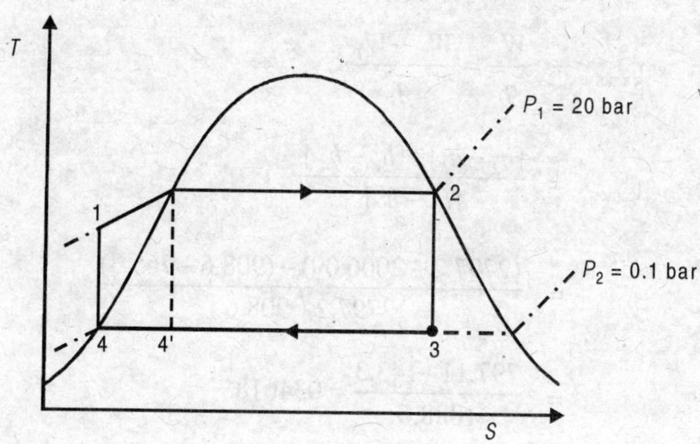

Fig. Ex. 14.7

(1) Consider carnot Cycle 1' – 2 – 3 – 4'

$$h_{1'} = h_f \text{ at 20 bar} = 908.6 \frac{\text{kJ}}{\text{kg}}$$

$$s_{1'} = s_f = 2.447 \frac{\text{kJ}}{\text{kg-K}}$$

and

$$h_2 = h_g \text{ at 20 bar} = 2797.2 \frac{\text{kJ}}{\text{kg}}$$

$$s_2 = s_g \text{ at 20 bar} = 6.337 \frac{\text{kJ}}{\text{kg-K}}$$

For Isentropic Process 2–3

$$s_2 = s_3$$
$$s_2 = s_{f3} + x_3 \cdot s_{fg3}$$
$$6.337 = 0.649 + x_3 \times 7.502$$
$$x_3 = 0.75819 \cong 0.76$$
$$\therefore \quad h_3 = h_{f3} + x_3 . s_{fg3} = 191.8 + 0.76 \times 2392.9$$
$$h_3 = 2006.09 \text{ kJ/kg}$$

Now for Isentropic Compression process 4' – 1'

$$s_{1'} = s_{4'} = s_{f4'} + x_{4'} \times s_{fg_{4'}}$$

$$2.447 = 0.649 + x_{4'}\, 7.502 \text{ at } 0.1 \text{ bar}$$

$$0.2397 = x_{4'}$$

$$h_{4'} = h_{f_{4'}} + x_{4'} \times h_{fg_{4'}}$$

$$h_{4'} = 191.8 + 0.2397 \times 2392.9$$

$$h_{4'} = 765.305 \text{ kJ/kg}$$

Now $$\eta_{carnot} = \frac{W_s}{q_i} = \frac{W_T - W_P}{q_i}$$

$$\eta_{carnot} = \frac{[h_2 - h_3] - [h_{1'} - h_{4'}]}{[h_2 - h_{1'}]}$$

$$= \frac{(2797.2 - 2006.09) - (908.6 - 765.3)}{2797.2 - 908.6}$$

$$= \frac{797.11 - 143.3}{1888.6} = 034618$$

$$\eta_{carnot} = \mathbf{34.618\ \%}$$

And Steam Rate $$= \frac{3600}{W_S} = \frac{3600}{W_T - W_P}$$

$$= \frac{3600}{[h_2 - h_3] - [h_{1'} - h'_{4'}]}$$

$$S.R = \frac{3600}{653.81} = 5.506 \text{ kg/kWh}$$

(II) Now Consider Rankine Cycle = 1 – 2 – 3 – 4

$$h_4 = h_f \text{ at } 0.1 \text{ bar} = 191.8 \frac{\text{kJ}}{\text{kg}}$$

From Fig. Ex. 14.7 h_2 and h_3 will remain same.

i.e.
$$h_2 = 2797.2 \text{ kJ/kg}$$
$$h_3 = 2006.09 \text{ kJ/kg}$$

Now to find $h_{1'}$ we also know that,

$$W_P = \frac{P_1 - P_b}{10} = \frac{20 - 0.1}{10}$$

$$W_P = 1.99 \text{ kJ/kg} = h_1 - h_4$$

$$\therefore \quad h_1 = 1.99 + 191.8$$

$$h_1 = 193.79 \text{ kJ/kg}$$

$$\text{Now, } \eta_g = \frac{W_s}{q_i} = \frac{W_T - W_P}{q_i}$$

$$\eta_R = \frac{[h_2 - h_3] - W_P}{[h_2 - h_1]}$$

$$= \frac{(2797.2 - 2006.09) - 1.99}{(2797.2 - 193.79)}$$

$$= \frac{791.11 - 1.99}{2603.41} = \frac{789.12 (\text{i.e. } W_s)}{2603.41}$$

$$= 0.30311$$

$$\eta_R = 30.311\ \%$$

and $$\text{Steam Rate} = \frac{3600}{W_s} = \frac{3600}{[h_2 - h_3] - W_p} = \frac{3600}{789.12}$$

S.R = 4.562 kg/kWh

Example 14.8 *A steam plant operating on Rankine cycle generates superheated steam at 10 bar and 380°C. Condensation occurs at 0.06 bar. Calculate the thermal efficiency of the plant and compare this with the thermal efficiency of Carnot cycle working between the same temperature limits.*

Find: $\eta_R = ?$ $\eta_{carnot} = ?$

Solution

Refer Fig. Ex. 14.8.

(I) Rankine Cycle (1 – 2 – 3 – 4)

From steam tables, at 0.06 bar,

$$h_4 = h_f = 151.5 \text{ kJ/kg}$$

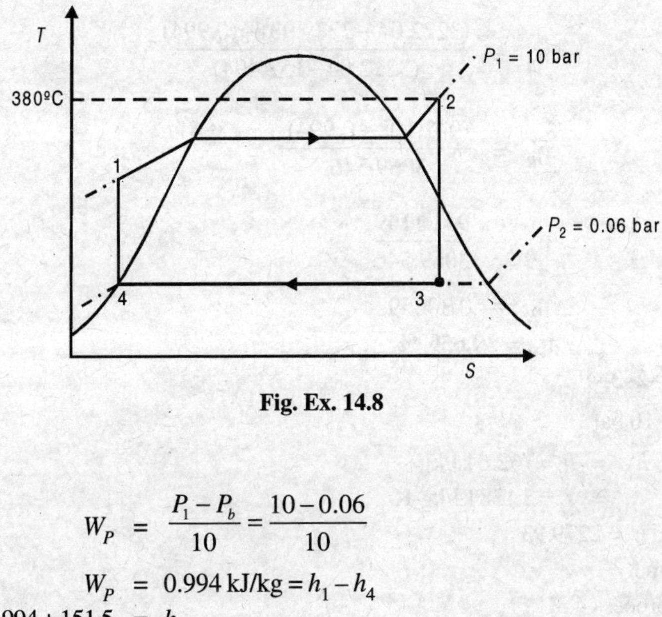

Fig. Ex. 14.8

Also Pump work

$$W_P = \frac{P_1 - P_b}{10} = \frac{10 - 0.06}{10}$$

$$W_P = 0.994 \text{ kJ/kg} = h_1 - h_4$$

$$0.994 + 151.5 = h_1$$

$$h_1 = 152.494 \text{ kJ/kg}$$

And h_2 at 10 bar and 380ºC from superheated steam tables = 3222.04 kJ/kg by interpolation.

To find h_3; from the Isentropic process 2–3

$$s_2 = s_3$$

$$s_2 = s_{f3} + x_3 s_{fg3} \quad \text{...(A)}$$

Now s_2 at 10 bar and 380ºC = 7.4.014 kJ/kg-K

Now from Eq. (A),

$$7.4014 = 0.521 + x_3 \times 7.81 \text{ at } 0.06 \text{ bar}$$

$$\frac{7.4014 - 0.521}{7.81} = x_3$$

$$0.8809 = x_3$$

Now
$$h_3 = h_{f3} + x_3 \,.\, h_{fg3} \text{ at } 0.06 \text{ bar}$$

$$= 151.5 + 0.8809 \times 2416$$

$$h_3 = 2279.93 \text{ kJ/kg}$$

Now, (i)
$$\eta_R = \frac{W_S}{q_i}$$

$$\eta_R = \frac{W_T - W_p}{q_i}$$

$$\eta_R = \frac{[h_2 - h_3] - W_p}{[h_2 - h_1]}$$

$$\eta_R = \frac{(3222.04 - 2279.93) - (0.994)}{(3222.04 - 152.494)}$$

$$\eta_R = \frac{(942.108 - 0.994)}{3069.546}$$

$$\eta_R = \frac{941.1149}{3069.546}$$

$$\eta_R = 0.30659.$$

$$\boldsymbol{\eta_R = 30.659\ \%}$$

(II) Carnot Cycle 1'–2–3–4'

From steam tables at 10 bar

$$h_{1'} = h_f = 762.6 \text{ kJ/kg}$$

$$s_1 = s_f = 2.138 \text{ kJ/kg-K}$$

And $h_2 = 3222.04$ and $h_3 = 2279.93$

These will remain same.

Now for Isentropic process 4'–1'

$$s_{1'} = s_{4'} = s_{f4'} + x_{4'} \times s_{fg_{4'}}$$

$$2.138 = 0.521 + x \times 7.81 \text{ at } 0.06 \text{ bar}$$

$$\frac{2.138 - 0.521}{7.81} = x_{4'}$$

$$0.2070 = x_{4'}$$

$$h_{4'} = h_{f_{4'}} + x_{4'} \times h_{fg_{4'}} \text{ at } 0.06 \text{ bar}$$

$$= 151.5 + 0.2070 \times 2416$$

$$h_{4'} = 651.714 \text{ kJ/kg}$$

Now,

$$\eta_{\text{carnot}} = \frac{W_s}{q_i} = \frac{W_T - W_P}{q_i}$$

$$= \frac{[h_2 - h_3] - [h_{1'} - h_{4'}]}{h_2 - h_{1'}}$$

$$= \frac{(3222.04 - 2279.93) - (762.6 - 651.714)}{3222.04 - 762.6}$$

$$= \frac{(942.11) - (110.886)}{(2459.44)}$$

$$= 0.33797$$

$$\eta_{\text{carnot}} = 33.797\%$$

Hence Carnot Cycle efficiency is more than Rankine cycle for an engine working between same temperature limits.

THEORY QUESTIONS

14.1. Derive the expression for efficiency of Carnot cycle.

14.2. Draw Carrot cycle on *P–V* and *T–S* diagram.

14.3. Explain Rankine cycle with *P–V* and *T–S* diagram.

14.4. Derive an expression for efficiency of Rankine cycle.

14.5. Compare Carnot cycle with Rankine cycle.

14.6. Define the terms : (i) Work ratio (ii) SSC.

14.7. Write a short note on factors affecting performance of Rankine cycle.

14.8. Explain why Rankine cycle and not Carnot cycle is considered as the thermodynamic cycle for the steam power plant.

14.9. Discuss the effects of the following parameters on the performance of Rankine cycle:

(i) Steam pressure at turbine inlet.

(ii) Steam temperature at turbine inlet.

(iii) Steam pressure at turbine exhaust.

14.10. Draw a neat sketch of Rankine cycle of steam plant, label the parts and describe the working of each component, giving the formula for energy exchange for the same.

14.11. Enumerate the advantages of Rankine cycle over the Cannot cycle for steam power plant.

14.12. Explain the drawbacks of Carnot cycle for a steam power plant. How are they overcome in a Rankine cycle ?

14.13. In what respect does the Rankine cycle differ from the Carnot cycle. Draw Carnot and Rankine cycle on *T–S* plane.

14.14. Enumerate the advantages of Rankine cycle over the Carnot cycle for steam power plant.

14.15. Why is Carnot cycle not practicable for a steam power plant.

14.16. Draw a neat sketch of Rankine cycle steam plant, label the part and describe the working of each component, giving formula for energy exchange for the same.

14.17. Explain clearly how the following affects the performance of a Rankine cycle.

(i) Lowering the condensation pressure.

(ii) Superheating the steam to high temperature.

(iii) Increasing the boiler pressure.

14.18. Parameters affecting the performance of Rankine cycle.

14.19. Explain the factors affecting the efficiency of Rankine cycle.

14.20. Explain with *T–S* diagram, factors affecting performance of Rankine cycle.

14.21. Explain the effect of following on the efficiency of Rankine cycle.

(i) Increasing the average temperature at which heat is supplied.

(ii) Decreasing the temperature at which heat is rejected.

14.22. Show the Rankine cycle on

(i) Wet (ii) Dry saturated

(iii) Superheated

PROBLEMS FOR PRACTICE

14.1. A steam power plant operating on Rankine cycle recieves steam from boiler at 35 bar and 350°C. It is exhausted to condenser at 0.1 bar, calculate

(i) Energy supplied/kg of steam generated in boiler

(ii) Quality of steam entering the condenser

(iii) Rankine cycle efficiency considering feed pump work.

[*Ans.* (i) 2911.16 kJ/kg, (ii) 0.8016, (iii) 34.226%]

14.2. Consider a steam power plant operating on the ideal Rankine cycle. The steam enters the turbine at 3 MPa and 350ºC and is condensed in the condenser at a pressure of 10 kPa. Determine

(i) The thermal efficiency of this power plant.

(ii) The thermal efficiency if the steam is superheated to 600°C instead of 350°C.

(iii) The thermal efficiency if the boiler pressure is raised to 15 MPa while the turbine inlet temperature is maintained at 600° C.

IMPORTANT POINTS TO BE REMEMBERED

1. Carnot cycle for steam remains in the wet region of T–S diagram and its efficiency is given by

$$\eta_{carnot} = \frac{T_1 - T_2}{T_1} \text{ as for ideal gas}$$

2. *Rankine cycle*

$$\eta = \text{Rankine cycle efficiency}$$

$$= \frac{h_1 - h_2 - w_p}{h_1 - h_2 - w_p}$$

$$= \frac{h_1 - h_2}{h_1 - h_f} \text{ when pump-work is neglected}$$

PART–3

CHAPTER 15

Reciprocating Air Compressors

15.1 INTRODUCTION

Air compressor is a power absorbing machine which provides high pressure air. It takes the air from the atmosphere, compresses it to a high pressure and high pressure air will be stored in a storage vessel (reservoir) from where it can be taken out for use.

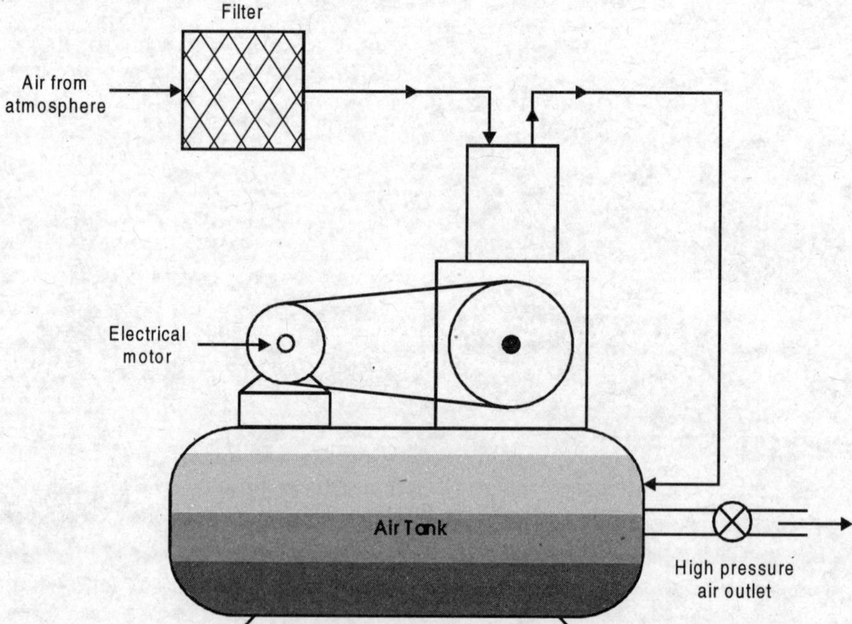

Fig. 15.1

There are many uses of high pressure air in industry. The main uses of compressed air are,

1. For inflating automobile tyres.

2. To clean workshop machines, generators etc.
3. To operate air operated drills, hammers (Pneumatic tools).
4. To inject fuel in the diesel engine cylinder.
5. In spray painting.
6. To operate air brakes in automobiles.
7. To operate compressed air engines/air motors in mines.

(In mines IC engines and electricity are not used because of fire risks. So high pressure air operated machines are used).

15.2 CLASSIFICATION

Air compressors may be broadly classified as given below: (i) Reciprocating and (ii) Rotary.

The principle parts of a reciprocating air compressor are same as that of a reciprocating I.C. engine. The air compressor may also be as follows,

(i) Single Acting or Double Acting. The air is admitted to one side of the piston or air is admitted to either side of the piston alternately.

(ii) Single Stage/Multistage. In case of single stage air is compressors, air is compressed in a single cylinder completely, whereas in case of multistage air compressor, air where air is compressed in several stages. And each stage in pressure of air increases.

15.3 SINGLE STAGE/SINGLE ACTING RECIPROCATING AIR COMPRESSOR

It can be seen that the construction is just similar to reciprocating I.C. engine. In this case crank of the compressor is driven by means of electrical motor or engine output shaft.

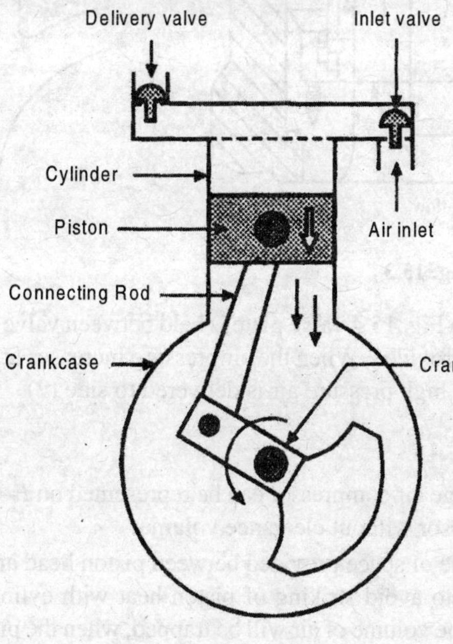

Fig. 15.2

During suction stroke, piston will be moving downwards and the pressure in the cylinder falls below the atmospheric pressure. As a result of which inlet valve opens and the air is drawn from the atmosphere. Then during compression stroke air will be compressed by means of upward moving piston and delivered through the delivery valve, when the pressure in the cylinder increases beyond the prespecified value of delivery valve.

15.3.1 Compressor Valves and Types

Compressor valves are automatic i.e., there is no valve operating mechanism provided, but they are operated because of the pressure difference.

Types of compressor valves:

(i) Reed types
(ii) Plate type
(iii) Poppet type,
(iv) Thimble valves

(i) Reed Type Valves. Figure 15.3 shows the reed type valve during compression the pressure of air increases and the pressure developed at (1) increases beyond the delivery chamber pressure, (2) then the valve plate (which has 6–8 reeds) lifts and the air is delivered.

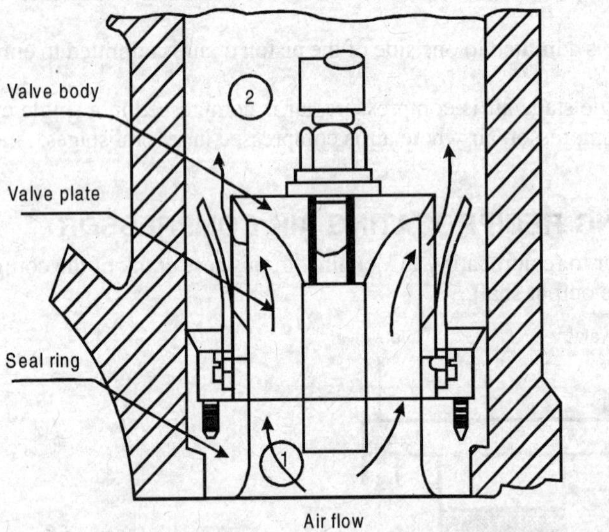

Fig. 15.3

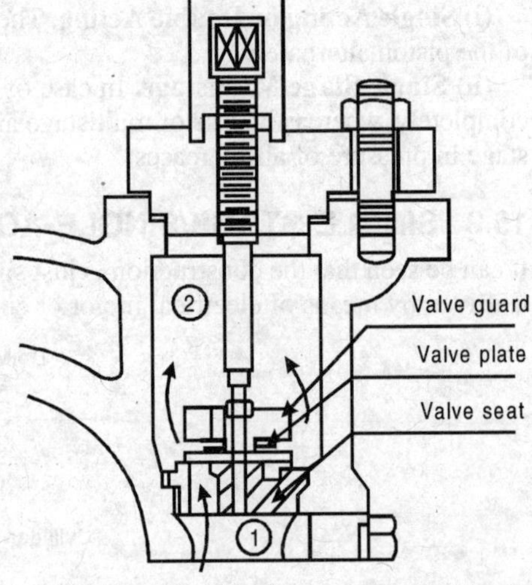

Fig. 15.4

(ii) Plate Type Valves. As shown in Fig. 15.4 valve plate is held between valve seat and valve guard. In between valve plate and valve guard spring is provided. When the air pressure increases at (1), then the valve plate is lifted against the compression of spring and high pressure air is delivered to side (2).

15.4 INDICATOR DIAGRAM

The events described for the single stage air compressor can be represented on *P–V* diagram as shown in Fig. 15.5. The diagram is drawn for the compressor without clearance volume.

Clearance Volume. It is the volume of space provided between piston head and cylinder head when the piston is at TDC. This is provided in order to avoid striking of piston heat with cylinder head. Also in this clearance volume, during the delivery stroke some volume of air will be trapped, when the piston comes to TDC. This volume of air during the downward stroke expands and produces the necessary vacuum.

During suction stroke, the charge of air is drawn along 4–1 at constant pressure P_1, which is slightly below P_{atm}. At point 1 piston completes suction stroke and starts compression stroke. During starting both the valves being closed air will be compressed along 1–2. At point 2 pressure P_2 is reached which is slightly higher than the pressure in the receiver.

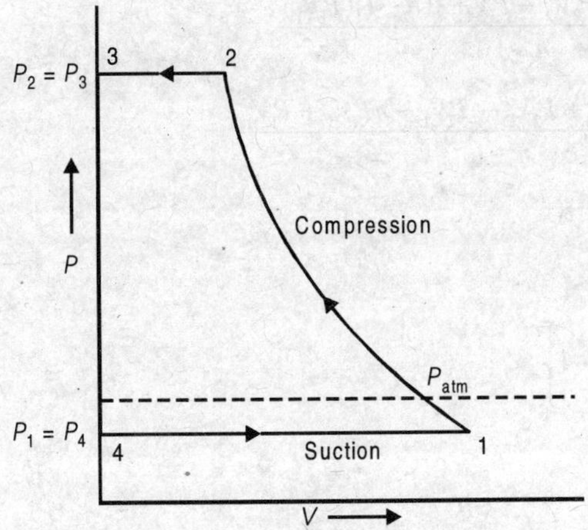

Fig. 15.5

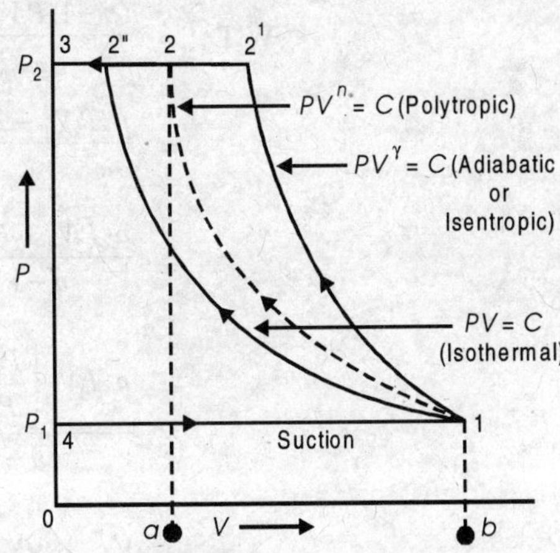

Fig. 15.6

The delivery valve at this point opens and the air is delivered along 2–3 into the receiver.

The net work input required for compression and delivery of air per cycle is given by the area 1–2–3–4 on the *P–V* diagram.

The amount of work done on air depends upon the nature of compression curve. If the compression occurs very rapidly in a non-conducting cylinder so that heat transfer is zero, then the compression will be practically isentropic.

If it is carried out very slowly and heat of compression is extracted from air by jacket cooling water, then the compression will approach isothermal.

However in actual practice neither of these conditions can be fulfilled and the actual compression will be in between these two and is called as the polytropic compression as shown in Fig. 15.6.

From Fig. 15.6 we can see that the isentropic work done = Area 4–1–2'–3 and Isothermal work done = Area 4–1–2"–3. So, if we achieve isothermal compression curve, the work input required will be minimum.

Note: For isothermal compression process $PV = C$, and $n = 1$

For isentropic compression process $PV^\gamma = C$ and $\gamma = 1.4$ for air.

In actual practice we achieve the process in between these two and for water cooled cylinders the value of n is 1.3.

15.5 WORK INPUT FOR COMPRESSOR

(i) If the Compression Process Follows $PV^n = C$

As shown in Fig. 19.4 work done during polytropic compression and delivery is given by the area 1–2–3–4 on the *P–V* diagram.

Now

$$\text{Area } 1-2-3-4 = [\text{Area } 2-3-0-a] + [\text{Area } 1-2-a-b] - [\text{Area } 4-1-0-b]$$

$$= P_2(V_2V_3) + \frac{P_2V_2 - P_1V_1}{n-1} - P_1(V_1 - V_4)$$

As

$$V_C = V_3 = V_4 = 0$$

$$= P_2V_2 + \frac{P_2V_2 - P_1V_1}{n-1} - P_1V_1$$

$$\text{W. D.} = \frac{(n-1)P_2V_2 + P_2V_2 - P_1V_1 - (n-1)(P_1V_1)}{n-1}$$

$$= \frac{n.P_2V_2 - P_2V_2 + P_2V_2 - P_1V_1 - nP_1V_1 + P_1V_1}{n-1}$$

$$= \frac{n(P_2V_2 - P_1V_1)}{n-1}$$

$$= \frac{n.P_1V_1\left\{\dfrac{P_2V_2}{P_1V_1} - 1\right\}}{n-1}$$

But as $PV^n = C; = \dfrac{V_2}{V_1} = \left(\dfrac{P_1}{P_2}\right)^{1/n} = \left(\dfrac{P_2}{P_1}\right)^{-1/n}$

$$\therefore \quad \text{W.D.} = \frac{n}{n-1}P_1V_1\left\{\frac{P_2}{P_1}\left(\frac{P_2}{P_1}\right)^{-1/n} - 1\right\}$$

Work Done or Indicated Power Required to Drive Compressors

$$\textbf{W.D or I.P.} = \frac{\mathbf{n}}{\mathbf{n-1}}\mathbf{P_1 . V_1}\left\{\left(\frac{\mathbf{P_2}}{\mathbf{P_1}}\right)^{\frac{\mathbf{n-1}}{\mathbf{n}}} - \mathbf{1}\right\}$$

Note generally,

W = in kW $\quad$ n = index of compression no units

P_1 = kN/m² $\quad$ V_1 = m³/sec

P_2 = kN/m² (delivery pressure)

Also $\quad \textbf{W.D.} = \dfrac{\mathbf{n}}{\mathbf{n-1}}\mathbf{mRT_1}\left\{\left(\dfrac{\mathbf{P_2}}{\mathbf{P_1}}\right)^{\frac{\mathbf{n-1}}{\mathbf{n}}} - \mathbf{1}\right\}$

As $\quad P_1V_1 = mRT_1$ and as air is the ideal gas.

(ii) If the Compression Process follows PV$^\gamma$ = C (i.e., Adiabatic Process). The replace n by γ and the derivation is same as above and

$$\text{W.D.} = \frac{\gamma}{\gamma-1}P_1V_1\left[\left(\frac{P_2}{P_1}\right)^{\frac{\gamma-1}{\gamma}} - 1\right]\text{kW}$$

(iii) If the Compression is Isothermal i.e., it follows *PV* = const. Then W.D = Area 1 – 2 – 3 – 4 = Area under 2 – 3 + Area under 1 –2 – Area under 4 – 1

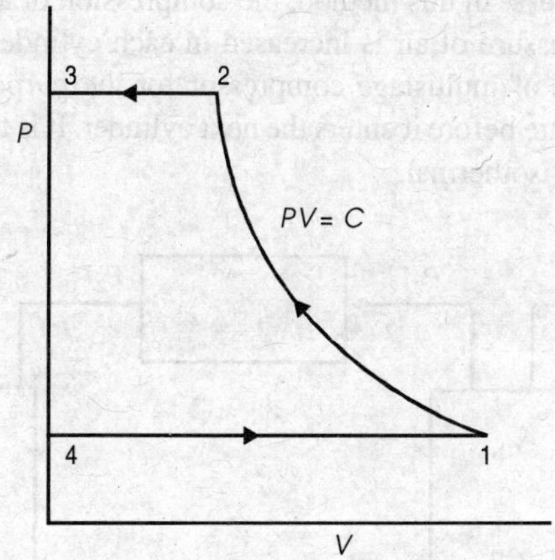

Fig. 15.7

$$= P_2V_2 + P_1V_1 \ln \frac{V_1}{V_2} - P_1V_1$$

$$\textbf{W.D.} = \mathbf{P_1V_1 \ln \frac{V_1}{V_2}}$$

or $$\textbf{W.D.} = \mathbf{P_1V_1 \ln \frac{P_2}{P_1}}$$

As $$P_1V_1 = P_2V_2 = C$$

Note. All the expressions are arranged to give positive value, since we are interested in the magnitude of work input.

15.6 ISOTHERMAL EFFICIENCY

It is defined as the ratio of,

$$\eta_{iso} = \frac{\text{Isothermal work of compressor}}{\text{Actual work input}}$$

And Isothermal work input $= P_1V_1 \ln \frac{P_2}{P_1}$ or $= mRT_1 \ln \frac{P_2}{P_1}$ as we have studied in earlier article.

Note. η_{iso} gives an indication of closeliness of the compression process to isothermal compression i.e., lower the value of n higher is η_{iso}.

15.7 MEASURES OR METHODS TO IMPROVE ISOTHERMAL EFFICIENCY

(i) Effective Cooling System. This can be provided by providing fins on the cylinder and cylinder head and air blown on the fins. When the air is blown, it removes heat from the compressed air and dissipates it to the atmosphere.

Secondly cooling water jackets may also be provided around the cylinder and cylinder head for large sized compressors.

(ii) Multistaging with Intercoolers. In this method, the compression of air is carried out in two or more stages i.e., two or more cylinders. The pressure of air is increased in each cylinder. It is a common practice to provide inter-coolers between the cylinders of multistage compressor for the purpose of cooling the compressed air to intake (or atmosphere) air temperature before it enters the next cylinder. It is this cooling between the cylinders that keeps the compression very near to isothermal.

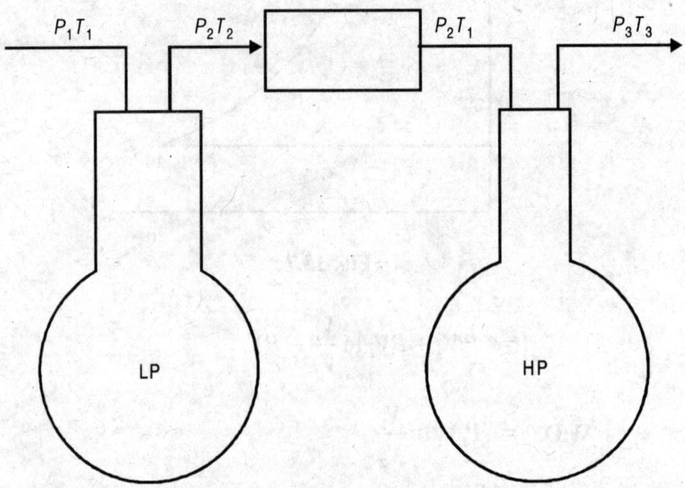

Fig. 15.8

(iii) Spray Injection. In this method a small amount of water is injected or sprayed into the compressed air. It immediately vapourises by absorbing heat from air and thereby keeps the temperature of air low.

(iv) Suitable Cylinder Dimensions. To have effective heat transfer, surface to volume ratio should be higher i.e., a cylinder with large dia. and short stroke should be selected. But this has the limitation since increased dia increases clearance volume and hence reduces volumetric efficiency.

15.8 EFFECT OF CLEARANCE ON VOLUMETRIC EFFICIENCY

The volume of space provided between piston heat and cylinder head when the piston is at TDC is known as clearance volume. In actual compressors it is provided to safeguard the piston from striking the cylinder head.

The events taking place in a compressor with clearance are same as those taking place in a compressor without clearance.

Figure 15.9 shows the *P–V* diagram for a single stage/single acting reciprocating air compressor with clearances volume. In this case both compression and expansion curves are assumed to be polytropic following the law $PV^n = C$.

As shown in Fig. 15.9, 1– 2 is polytropic compression, 2 – 3 is delivery stroke. At point 3, piston reaches TDC and delivery of high pressure air stops and some air is trapped in the clearance volume at the delivery pressure P_2.

As the piston moves downwards this air which is trapped in the clearance volume expands according to $PV^n = C$ and produces the necessary vacuum. At point 4 the inlet valve opens and actual volume of air taken is along 4–1.

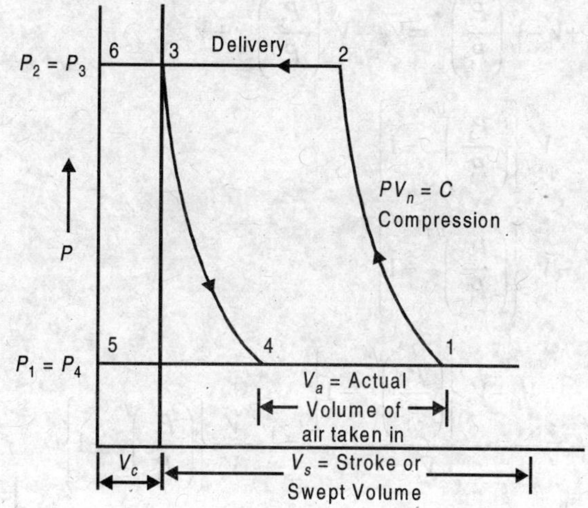

Fig 15.9

Hence $V_1 - V_4 = V_a$ = Actual volume of air taken in and $V_1 - V_3$ = stroke volume = Swept volume. And the ratio of actual volume of air taken into the stroke volume is known as volumetric efficiency i.e.,

$$\eta_{vol.} = \frac{\text{Actual volume of air taken in}}{\text{Stroke volume}}$$

$$= \frac{V_1 - V_4}{V_1 - V_3} = \frac{V_a}{V_s}$$

i.e.,

$$\boldsymbol{\eta_{vol.} = \frac{V_a}{V_s}} \qquad \text{...(1)}$$

For the process 3–4

$$P_3 V_3^n = P_4 V_4^n$$

$$\left(\frac{P_3}{P_4}\right)^{1/n} = \frac{V_4}{V_3}$$

$$V_3 \times \left(\frac{P_3}{P_4}\right)^{1/n} = V_4$$

$$V_c \left(\frac{P_2}{P_1}\right)^{1/n} = V_4 \; [\because V_3 = V_c \text{ and } P_3 = P_2 \text{ and } P_4 = P_1] \qquad \text{...(2)}$$

As we know actual volume of air taken in,

$$V_a = V_1 - V_4$$

$$= V_1 - V_c \left(\frac{P_3}{P_4}\right)^{1/n} = (V_c + V_s) - V_c \left(\frac{P_3}{P_4}\right)^{1/n} \text{ as } V_1 = V_c + V_s$$

or $$= V_s + V_c - V_c\left(\frac{P_3}{P_4}\right)^{1/n} = V_s - V_c\left(\frac{P_3}{P_4}\right)^{1/n} + V_c$$

$$V_a = V_s - V_c\left\{\left(\frac{P_2}{P_1}\right)^{1/n} - 1\right\}$$

i.e., $$V_a = V_s - V_c\left\{\left(\frac{P_2}{P_1}\right)^{1/n} - 1\right\} \quad \ldots(3)$$

$\therefore$ From (1), $$\eta_{vol} = \frac{V_a}{V_s} = \frac{V_s - V_c\left\{\left(\frac{P_2}{P_1}\right)^{1/n} - 1\right\}}{V_s} = 1 - \frac{V_c}{V_s}\left\{\left(\frac{P_2}{P_1}\right)^{1/n} - 1\right\}$$

$$\mathbf{\eta_{vol} = 1 - C\left\{\left(\frac{P_2}{P_1}\right)^{1/n} - 1\right\}}$$

or $$\mathbf{\eta_{vol} = 1 - C\times\left(\frac{P_2}{P_1}\right)^{1/n} + C}$$

or $$\mathbf{\eta_{vol} = 1 + C - C\left(\frac{P_2}{P_1}\right)^{1/n}}$$

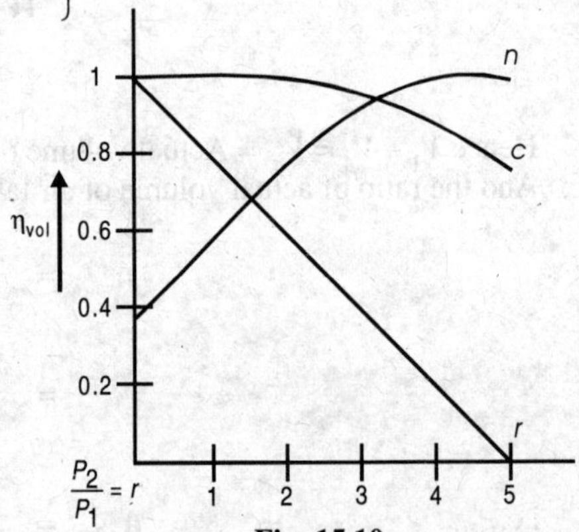

Fig. 15.10

Let the ratio $\frac{V_c}{V_s} = C$ = clearance ratio and $\frac{P_2}{P_1}$ = pressure ratio r.

The variation of volumetric efficiency with clearance volume and polytropic index is shown in Fig. 15.10. We note from this figure that

(i) As r increases η_{vol} decreases.
(ii) As n increases η_{vol} increases.
(iii) As c increases η_{vol} decreases.

Note.

(i) η_{vol} ranges from 60 to 85% and
(ii) Clearance ratio ranges from 4 to 10%
(iii) n ranges from 1.25 to 1.35 for air.

15.9 EXPRESSION FOR WORK INPUT WHEN CLEARANCE IS CONSIDERED

The work input required to compress the air with clearance is given by the area 1 – 2 – 3 – 4. (Ref. Fig. 15.9)

i.e., $$W_{input} = \text{Area } 1-2-3-4$$
$$= [\text{Area } 1-2-6-5] - [\text{Area } 3-4-5-6]$$

Assuming index of compression to expansion to be same as n.

$$W_{input} = \frac{n}{n-1}P_1V_1\left[\left(\frac{P_2}{P_1}\right)^{\frac{n-1}{n}} - 1\right] - \frac{n}{n-1}P_4V_4\left\{\left(\frac{P_3}{P_4}\right)^{\frac{n-1}{n}} - 1\right\}$$

But $\qquad P_4 = P_1,\ P_3 = P_2$

$$\therefore \qquad W_{input} = \frac{n}{n-1} P_1 V_1 \left\{ \left(\frac{P_2}{P_1} \right)^{\frac{n-1}{n}} - 1 \right\} - \frac{n}{n-1} P_1 V_4 \left\{ \left(\frac{P_2}{P_1} \right)^{\frac{n-1}{n}} - 1 \right\}$$

$$\mathbf{W_{input} = \frac{n}{n-1} P_1 (V_1 - V_4) \left\{ \left(\frac{P_2}{P_1} \right)^{\frac{n-1}{n}} - 1 \right\}}$$

Therefore, it can be seen that from above equation work will depend upon volume of air taken/cycle and will not depend upon clearance. If clearance is neglected then $V_1 - V_4 = V_1$ and we get the expression for work without clearance as,

$$\mathbf{W_{input} = \frac{n}{n-1} P_1 V_1 \left[\left(\frac{P_2}{P_1} \right)^{\frac{n-1}{n}} - 1 \right]}$$

$$\therefore \qquad IP = \text{Indicated work/cycle} \times \frac{N}{60}$$

If V_1 is in m^3, then resultant unit will be kJ. Then multiply by $N/60$ to get power in kW.

15.10 ACTUAL INDICATOR DIAGRAM

In actual air compressor, some air is present in the clearance volume. This air expands during suction stroke.

At point 4, the inlet valve will not open in actual practice because of valve inertia and pressure differential required to open the valve. Thus the pressure drops away until the valve is forced to open. Some valve bounce will taken place and slowly intake will become nearly steady. This negative pressure, is known as *Intake depression.*

A similar situation occurs at point 2 i.e., at point 2 pressure has to increase otherwise valve will not open.

$\therefore$ Actual work required to compress and deliver the air will be greater than theoretical work.

And, $$\eta_{mech} = \frac{\text{Indicator work}}{\text{Actual work}} = \frac{\text{Indicator work}}{\text{Brake work}} = \frac{IP}{BP}$$

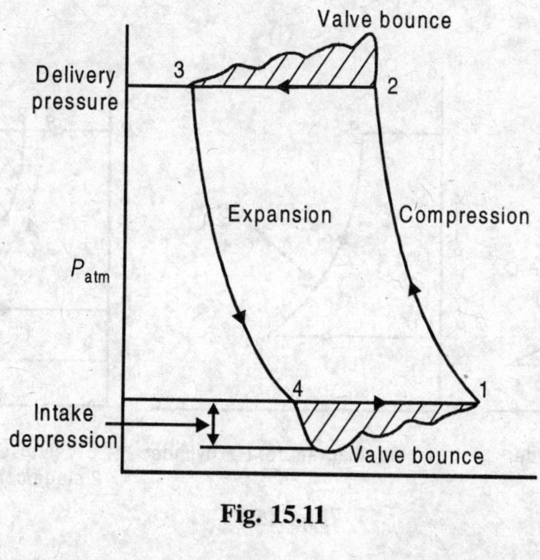

Fig. 15.11

15.11 FREE AIR DELIVERY (FAD)

It is the actual volume of air delivered by an air compressor reduced to either NTP or STP conditions or intake conditions. We know that from continuity equation.

Mass flow rate at the inlet of compressor = Mass flow rate at outlet of compressor.

Since $$PV = \dot{m}RT$$

$$\frac{PV}{RT} = \dot{m}$$

$$\therefore \quad \frac{P_1V_1}{RT_1} = \frac{P_2V_2}{RT_2} = \frac{P_fV_f}{T_f} \text{ neglecting clearance}$$

where P_fV_f and T_f are free air conditions and V will be FAD. For convenience take $P_f = 101.325$ kPa, $T_f = 288$ K if not given and V_f can be calculated.

Note. $\frac{P_1(V_1 - V_4)}{T_1} = \frac{P_2(V_2 - V_3)}{T_2}$ if clearance is considered.

15.12 MULTISTAGE AIR COMPRESSOR

A 2 stage air compressor with water jacketed cylinder and inter cooler is shown in Fig. 15.10 (i).

The suction in LP cylinder ends at 1 and the air is compressed polytropically to 2'. The LP cylinder then delivers

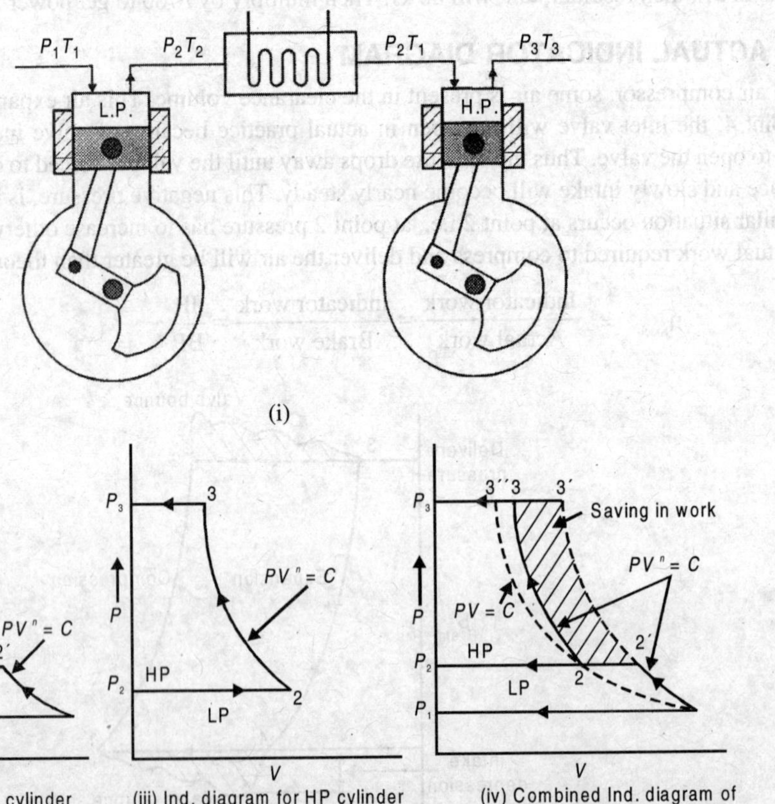

(ii) Ind. diagram for LP cylinder (iii) Ind. diagram for HP cylinder (iv) Combined Ind. diagram of 2 stage/comp. with perfect intercooling

Fig. 15.12

the air along 2' – P_2 to the intercooler. Where air is cooled to initial intake temperature T_1 at constant pressure P_2 by circulating cold water in the inter cooler. When the air is cooled in the intercooler to initial temperature T_1, the cooling is perfect. The air when it is cooled at constant pressure, suffers reduction in volume from 2' – P_2 to P_2 – 2. The cooled air is drawn into the HP cylinder along P_2 –2 for II-stage compression, where it is compressed polytropically to final pressure P_3 along 2 – 3 and then delivered at constant pressure P_3 along 3 – P_3. (Ref. Fig. 15.10 (iii)).

Figure 15.10 (iv) shows the combined indicator diagram for LP and HP cylinder of a single acting 2-stage air compressor with perfect intercooling.

LP diagram is shown as P_1 – 1 – 2' – P_2 and HP diagram is shown as P_2 – 2 – 3 – P_3. The reduction of work required due to intercooling is shown by shaded are 2 – 3 – 3' – 2', when the intercooling is perfect, the point 2 lies on the isothermal line 1 – 3".

Work required in LP cylinder/cycle.

$$= \frac{n}{n-1} P_1V_1 \left\{ \left(\frac{P_2}{P_1}\right)^{\frac{n-1}{n}} - 1 \right\}$$

And work required in HP cylinder/cycle

$$= \frac{n}{n-1} P_2V_2 \left\{ \left(\frac{P_3}{P_2}\right)^{\frac{n-1}{n}} - 1 \right\}$$

∴ Total work required /cycle,

$$W = \frac{n}{n-1} P_1V_1 \left\{ \left(\frac{P_2}{P_1}\right)^{\frac{n-1}{n}} - 1 \right\} + \frac{n}{n-1} P_2V_2 \left\{ \left(\frac{P_3}{P_2}\right)^{\frac{n-1}{n}} - 1 \right\} \quad \ldots(1)$$

and

$$\text{IP of compressor} = \frac{WN}{60} \text{ J/sec or W.}$$

If the intercooling is perfect, then point 2 will lie on isothermal line, then $P_1V_1 = P_2V_2$ and Eq. (1) will be,

$$W = \frac{n}{n-1} P_1V_1 \left\{ \left(\frac{P_2}{P_1}\right)^{\frac{n-1}{n}} + \left(\frac{P_3}{P_2}\right)^{\frac{n-1}{n}} - 2 \right\} \text{J} \quad \ldots(2)$$

and

$$\text{IP} = \frac{WN}{60} \text{ J/sec or W} \quad \ldots(3)$$

Intercoolers and Aftercoolers. In the intercoolers air is cooled in-between the cylinders of a multistage compressor. And in the afterwards air is cooled after the last cylinder and before it enters the reservoir.

By aftercooling there is no saving in work as such, but by aftercooling volume of air decreases. So more amount of air can be stored in the reservoir. Hence capacity (mass of air to be stored) of air increases.

Ideal Intercooler Pressure. It may be noted that, saving in work increases as intercooling is increased. When the intercooling is perfect, saving of work is maximum and work input to compressor is given by Eq. (2).

It may further be noted that saving in work also varies with the chosen intercooler pressure P_2. When P_1 and P_3 are fixed, the best value, of intercooler pressure P_2 shall be fixed to give minimum work.

This value of P_2 can be found by differentiating the Eq. (2) w.r.t. P_2 and equating it to zero.

$$\therefore \quad W = \frac{n}{n-1} P_1V_1 \left[\left(\frac{P_2}{P_1}\right)^{\frac{n-1}{n}} + \left(\frac{P_3}{P_2}\right)^{\frac{n-1}{n}} - 2 \right] \quad \ldots(3)$$

$$W = \frac{n}{n-1} P_1V_1 \left[\left(\frac{P_2}{P_1}\right)^{a} + \left(\frac{P_3}{P_2}\right)^{a} - 2 \right] \text{ where } a = \frac{n-1}{n}$$

$$W = \text{Constant} \times \left\{ \left(\frac{P_2}{P_1}\right)^{a} + \left(\frac{P_3}{P_2}\right)^{a} - 2 \right\}$$

Differentiating w.r.t. P_2 and equating it to zero for minimum work,

$$\frac{dW}{dP_2} = \frac{aP_2^{a-1}}{P_1^a} - \frac{aP_3^a}{P_2^{a+1}} = 0$$

Dividing by a throughout,

$$\frac{P_2^{a-1}}{P_1^a} = \frac{P_3^a}{P_2^{a+1}}$$

$$\therefore \quad P_2^{2a} = (P_1P_3)^a$$

Taking a^{th} root throughout,

$$P_2^2 = P_1P_3 \Rightarrow P_2\sqrt{P_1P_3} \text{ or } \frac{P_2}{P_1} = \frac{P_3}{P_2}$$

This shows that for minimum work required or maximum efficiency, the intercooler pressure is the geometric mean of initial and final pressure, i.e., pressure ratio in each stage is same.

$\therefore$ Equation (3) will be,

$$W = \frac{n}{n-1} P_1V_1 \left\{ \left(\frac{P_2}{P_1}\right)^{\frac{n-1}{n}} + \left(\frac{P_3}{P_2}\right)^{\frac{n-1}{n}} - 2 \right\}$$

$$W = \frac{2n}{n-1} P_1V_1 \left\{ \left(\frac{P_2}{P_1}\right)^{\frac{n-1}{n}} - 1 \right\} \text{since} \frac{P_2}{P_1} = \frac{P_3}{P_2}$$

$$= \frac{2n}{n-1} P_1V_1 \left\{ \left(\frac{(P_1P_3)^{1/2}}{P_1}\right)^{\frac{n-1}{n}} - 1 \right\}$$

$$= \frac{2n}{n-1} P_1V_1 \left\{ \left[\left(\frac{P_3}{P_1} \right)^{1/2} \right]^{\frac{n-1}{n}} - 1 \right\}$$

$$\mathbf{W} = \frac{\mathbf{2n}}{\mathbf{n-1}} \mathbf{P_1V_1} \left\{ \left(\frac{\mathbf{P_3}}{\mathbf{P_1}} \right)^{\frac{n-1}{2n}} \mathbf{-1} \right\}$$

and

$$\text{min IP} = \frac{WN}{60} \text{ J/sec}$$

Similarly for 3 stage compressor

$$\mathbf{W} = \mathbf{3}\frac{\mathbf{n}}{\mathbf{n-1}} \left\{ \left(\frac{\mathbf{P_4}}{\mathbf{P_1}} \right)^{\frac{n-1}{3n}} \mathbf{-1} \right\}$$

And similarly for x-stage compress

$$W = \frac{xn}{n-1} P_1V_1 \left\{ \left(\frac{P_{x+1}}{P_1} \right)^{\frac{n-1}{xn}} - 1 \right\}$$

Now in general,

$$\frac{P_2}{P_1} = \frac{P_3}{P_2} = \frac{P_4}{P_3} = \frac{P_{x+1}}{P_1} = k$$

where k is the pressure ratio is each stage.

From this,

$$P_2 = k P_1$$

$$P_3 = kP_2 = k\ k P_1 = k^2 P_1$$

$$P_4 = kP_3 = k.k^2 P_1 = k^3 P_1$$

$$P_{x+1} = kP_x = k^x P_1$$

$$\therefore \quad k^x = \left(\frac{P_{x+1}}{P_1} \right) \text{ or } k = \left(\frac{P_{x+1}}{P_1} \right)^{\frac{1}{x}}$$

i.e.,

$$k = (\text{Pressure Range})^{1/x}$$

15.13 CAPACITY CONTROL OF COMPRESSORS

Generally compressors are not operated at full capacity. Depending upon the demand for high pressure air, they are operated.

Various types of controls for compressors are:

(i) Throttle control

(ii) Blowing air to atmosphere

(i) Throttle Control. When the demand for high pressure air is more and when less amount of air is present is the reservoir tank, then less pressure from the reservoir will act on the piston (1). Hence because of less pressure piston moves upwards and more amount of air is taken during suction stroke opposite operation will take place when the demand for high per air is less.

(ii) Blowing Air to Atmosphere. When the pressure in the receiver increases beyond the required limit, then because of high pressure air piston (1) of relay cylinder (2) in lifted and high pressure air enters cylinder (3) and presses the piston (4) downwards and high pressure air from the compressor is directly blown off to atmosphere.

And when the pressure comes below normal in the receiver, then piston (1) of relay cylinder moves down because of dead weights (5). Because of compressor air pressure check valve (6) opens and air enters receiver (7).

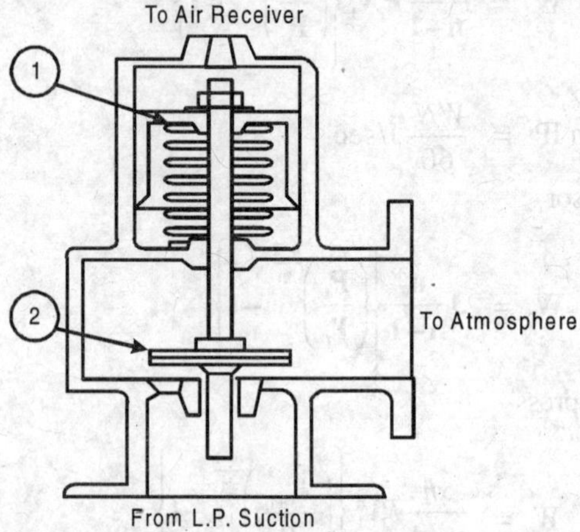

Fig. 15.13. *Throttle control*

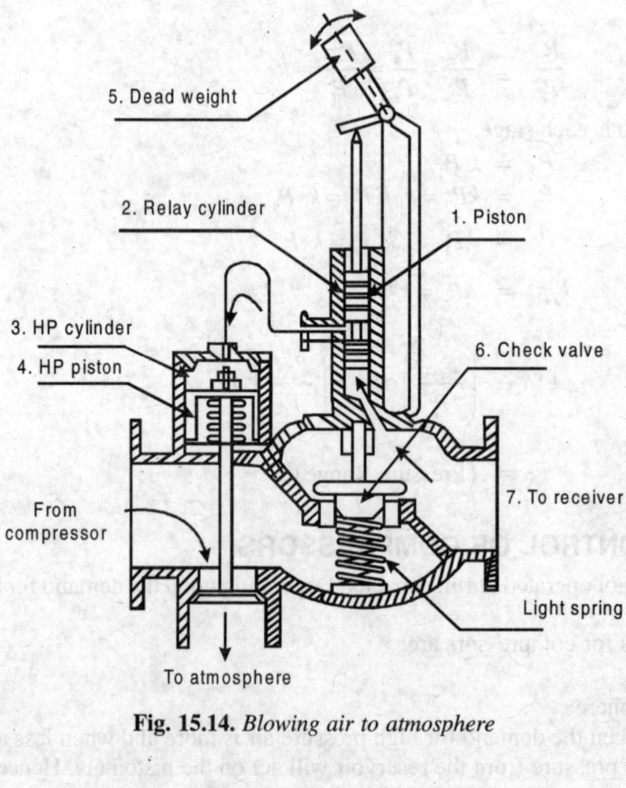

Fig. 15.14. *Blowing air to atmosphere*

■■■

SOLVED EXAMPLES

(Problems on Single Stage Air Compressor)

Example 15.1. *An air compressor is required to compress air from a pressure of 1 bar to 10 bar. Calculate the amount of work required per kg of air, when n = 1.2 for compression. Assume initial temperature as 15°C and R for air is 0.287 kJ/kg-K.*

Data: P_1 = 1 bar 100 kPa, P_2 = 10 bar = 1000 kPa

W.D/kg = ?, n = 1.2

T_1 = 15 + 273 = 288 K, R = 0.287 kJ/kg-K.

Solution

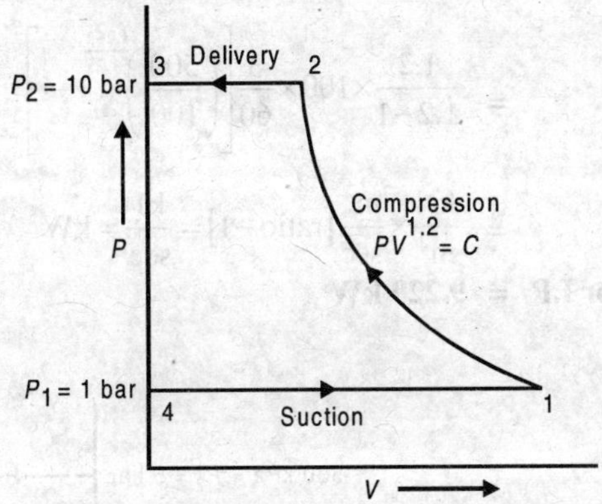

Fig. Ex. 15.1

We know that amount of work input required for compression and delivery of air polytropically/cycle,

$$W = \frac{n}{n-1} P_1 V \left\{ \left(\frac{P_2}{P_1} \right)^{\frac{n-1}{n}} - 1 \right\}$$

$$W = \frac{n}{n-1} RT_1 \left\{ \left(\frac{P_2}{P_1} \right)^{\frac{n-1}{n}} - 1 \right\} \quad \text{as } PV_1 = mRT_1$$

Note. m = 1 kg as work input/kg is required

$$W = \frac{1.2}{1.2-1} \times 0.287 \times 288 \left\{ \left(\frac{1000}{100} \right)^{\frac{1.2-1}{1.2}} - 1 \right\}$$

$$= \frac{1.2}{0.2} \times 0.287 \times 288 \{ (10)^{0.2/1.2} - 1 \}$$

$$= 495.936 \{1.46779 - 1\}$$

$$= 495.936 \{0.46779\}$$

$$\mathbf{W = 231.9985 \text{ kJ/kg}}$$

Example 15.2. *A single stage, single acting air compressor delivers air at 5 bar. The suction temperature is 20°C and suction pressure is 1 bar volume of air entering the compressor is 3 m³/min. Index of compression is 1.2. Calculate isothermal efficiency and power required to drive the compressor. Neglect clearance volume.*

Data: $T_1 = 20 + 273 = 293\ K$ $V_1 = 3\ m^3/min$

$n = 1.2$ $\eta_{iso} = ?$

$W.D. = ?$

Solution

We know that,

$$\text{W.D. or I.P} = \frac{n}{n-1} P_1 V_1 \left\{ \left(\frac{P_2}{P_1} \right)^{\frac{n-1}{n}} - 1 \right\}$$

$$= \frac{1.2}{1.2-1} \times 100 \times \frac{3}{60} \left[\left(\frac{500}{100} \right)^{\frac{1.2-1}{1.2}} - 1 \right]$$

$$= \frac{\text{kN}}{\text{m}^2} \times \frac{\text{m}^3}{\text{sec}} [\text{ratio} - 1] = \frac{\text{kJ}}{\text{sec}} = \text{kW}$$

W.D. or I.P = 9.229 kW

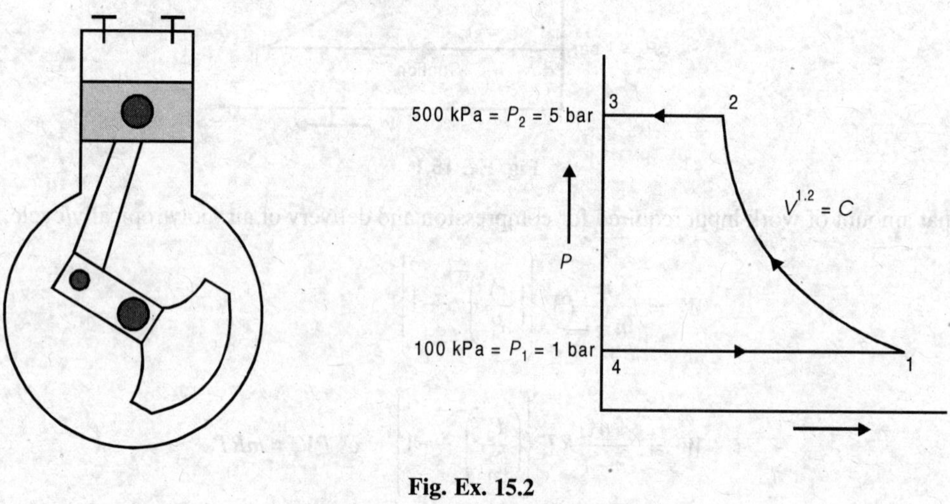

Fig. Ex. 15.2

Also

$$\eta_{iso} = \frac{\text{Isothermal work of compression}}{\text{IP}}$$

$$= \frac{P_1 V_1 \ln \frac{P_2}{P_1}}{9.229}$$

$$= \frac{100 \times \frac{3}{60} \times \ln \frac{500}{100}}{9.229} = \frac{8.05}{9.229}$$

$$\eta_{iso} = \mathbf{87.2\ \%}$$

Example 15.3. *A single stage single acting/compressor with bore 15 cm and stroke 20 cm run at 200 rpm. Air is drawn in at 1 bar and 25°C. Air is delivered at 6 bar, law of compression is $PV^{1.3} = C$. Determine FAD and indicated power required to drive the compressor in kW.*

Data: $D = 0.5\ m$ $\quad L = 0.2\ m$

$N = 200\ rpm$ $\quad P_1 = 100\ kPa$

$T_1 = 25 + 273 = 298\ K$

$FAD = ?$ $\quad IP = ?$

Solution

We know that,

$$\frac{P_1V_1}{T_1} = \frac{P_2V_2}{T_2} = \frac{P_fV_f}{T_f}$$

Now V_1 = volume of air handled/min

$$= \frac{\pi}{4}D^2LN$$

$$= \frac{\pi}{4}\times 0.15^2\times 0.2\times 200$$

$$= 0.707\ m^3/min$$

To find FAD i.e., V_f

$$\frac{P_1V_1}{T_1} = \frac{P_fV_f}{T_f}$$

Since nothing is mentioned take,

$$P_f = 101325\ kPa$$

$$T_f = 15 + 273 = 288\ K$$

$$\therefore \frac{100\times 0.7087}{283} = \frac{101.325\times V_f}{288}$$

$$\mathbf{0.674\ m^3/min = V_f}$$

$$IP = \frac{n}{n-1}P_1V_1\left[\left(\frac{P_2}{P_1}\right)^{\frac{n-1}{n}} - 1\right]$$

$$= \frac{1.3}{1.3-1}\times 100\times\frac{0.707}{60}\left[\left(\frac{600}{100}\right)^{\frac{0.3}{1.3}} - 1\right]$$

$$\mathbf{IP = 2.615\ kW}$$

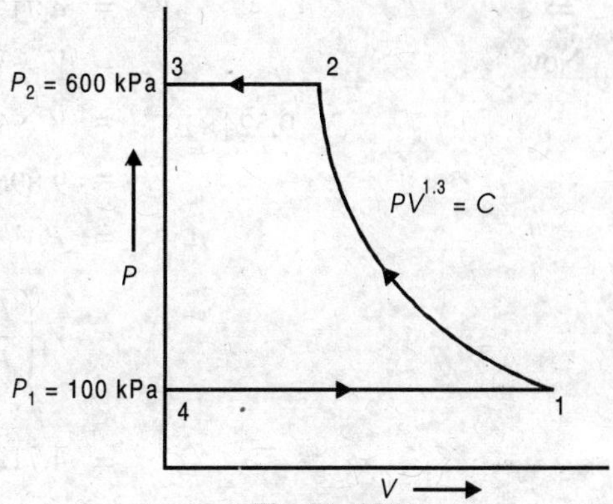

Fig. Ex.15.3

Example 15.4. *A single stage, single acting reciprocating air compressor has a bore of 200 mm and a stroke of 300 mm. It runs at a speed of 500 rpm. The clearance volume is 5% of swept volume and polytropic index is 1.3 throughout. Intake pressure and temperature are 97 kPa, 20°C and the compression pressure is 550 kPa. Detertermine:*

(i) FAD in m³/min

(ii) Air delivery temperature

(iii) Cycle power

(iv) η_{iso} neglecting clearance volume.

Solution

$$\frac{V_c}{V_s} = 0.05 = \frac{V_3}{V_1 - V_3} \quad \ldots(1)$$

$$V_s = \text{Stroke volume}$$

$$= \frac{\pi}{4} 0.2^2 \times 0.3$$

$$V_s = V_1 - V_3 = 9.425 \times 10^{-3}\,\text{m}^3$$

$$\therefore \quad 0.05 = \frac{V_3}{V_1 - V_3} = \frac{V_3}{9.425 \times 10^{-3}}$$

$$\Rightarrow \quad V_3 = 4.712 \times 10^{-4}\,\text{m}^3$$

Now

$$V_s = V_1 - V_3$$

$$9.524 \times 10^{-3} = V_1 - 4.712 \times 10^{-4}$$

$$\therefore \quad V_1 = 9.896 \times 10^{-3}\,\text{m}^3$$

$$p_3 v^n{}_3 = P_4 V_4^n$$

$$\therefore \quad V_4 = V_3 \left(\frac{P_3}{P_4}\right)^{1/n} = V_3 \left(\frac{P_2}{P_1}\right)^{1/n}$$

$$= 4.712 \times 10^{-4} \left(\frac{550}{97}\right)^{1/1.3}$$

$$V_4 = 1.79 \times 10^{-3}\,\text{m}^3$$

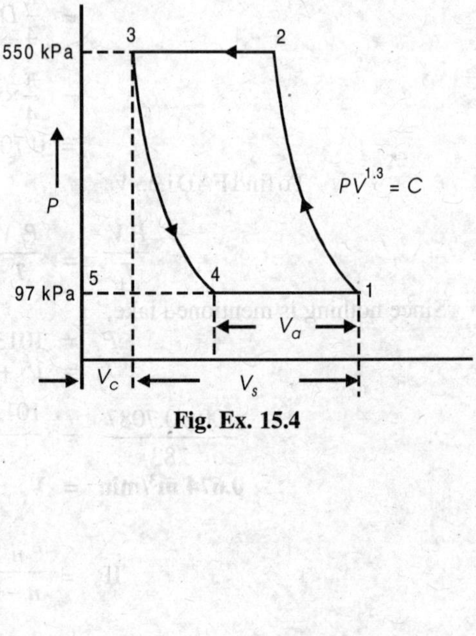

Fig. Ex. 15.4

Then

$$\frac{P_1(V_1 - V_4)}{T_1} = \frac{P_f V_f}{T_f}$$

$$\mathbf{V_f = 7.625 \times 10^{-3}\,m^3}$$

(i) $$\text{FAD} = V_f \times N = 3.814\,\text{m}^3/\text{min}$$

(ii) $$\eta_{\text{vol}} = \frac{V_1 - V_4}{V_1 - V_3} = 0.8656 \Rightarrow 86\%$$

(iii) $$\frac{T_1}{T_2} = \left(\frac{P_1}{T_2}\right)^{\frac{n-1}{n}} = \left(\frac{V_2}{V_1}\right)^{n-1}$$

$$\mathbf{T_2 = 437.29\,K}$$

(iv) $$\text{Cycle power} = \frac{n}{n-1} P_1(V_1 - V_4)\left[\left(\frac{P_2}{P_1}\right)^{\frac{n-1}{n}} - 1\right] \times \frac{N}{60}$$

$$= \frac{1.3}{0.3} \times 97\,(8.106 \times 10^{-3})\left[\left(\frac{550}{97}\right)^{\frac{0.3}{1.3}} - 1\right] \times \frac{500}{60}$$

$$\mathbf{I.P. = 13.983\,kW}$$

Example 15.5. *A single stage, single acting air compressor works between 1 bar and 16 bar compression follows $PV^{1.3} = C$. The piston speed is 200 m/min. It runs at 350 rpm. It has an IP consumption of 30 kW and η_{vol} is 85%. Estimate the cylinder bore and stroke.*

Data: $P_1V_1^{1.3} = P_2V_2^{1.3}$ Piston speed = 2 LN = 200 m/min

N = 350 rpm IP = 30 kW

η_{vol} = 85% D = ?

L = ?

Solution

Since Piston speed = 2 LN = 200 m/min

$$= 2 \times L \times 350 = 200 \text{ m/min}$$

$$L = 0.286 \text{ m}$$

∴ Stroke L = 286 mm

$$\eta_{vol} = \frac{V_a}{V_s} = \frac{V_1 - V_4}{V_1 - V_3}$$

$$\text{IP} = \frac{n}{n-1} P_1(V_1 - V_4)\left[\left(\frac{P_2}{P_1}\right)^{\frac{n-1}{n}} - 1\right] \times \frac{N}{60} \frac{\text{kJ}}{\text{sec}}$$

$$30\,\text{kW} = \frac{1.3}{0.3} \times 100\,(V_1 - V_4)\left[\left(\frac{1600}{100}\right)^{\frac{0.3}{1.3}} - 1\right] \times \frac{350}{60}$$

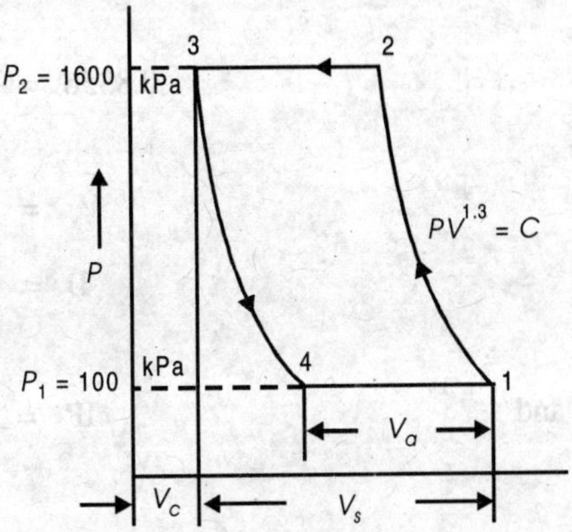

Fig. Ex. 15.5

$$(V_1 - V_4) = 0.01324 \text{ m}^3$$

Now from (1)

$$0.85 = \frac{0.01324}{V_1 - V_3}$$

$$(V_1 - V_3) = 0.01558 \text{ m}^3 = V_s$$

$$\therefore \quad V_s = 0.01558\,\text{m}^3 = \frac{\pi}{4} D^2 L$$

$$\therefore \quad D = 0.263 \text{ m}$$

D = 263 mm

Example 15.6. *Actual volume of air taken in at 0.98 bar and 27°C by a single stage, acting reciprocating air compressor is 8.5 m³/min. The delivery conditions are 6.5 bar. The clearances volume is 5% of stroke volume compression and expansion follows $PV^{1.3} = C$. If the speed is 300 rpm. L/D ratio is 1:1. Find the cylinder dimensions by finding the volumetric efficiency required to run the compressor. Also find I.P.*

Data: V_a = 8.5 m³/min $P_2 = 6.5 \times 10^2$ kPa

T_1 = 27 + 273K $P_1 = 0.98 \times 10^2$ kPa

$\frac{V_c}{V_s}$ = 0.05 n = 1.3, N = 300 rpm

Solution We know that,

$$\eta_{vol} = 1 - C\left\{\left(\frac{P_2}{P_1}\right)^{1/n} - 1\right\}$$

$$= 1 - 0.05\left\{\left(\frac{6.5\times10^2}{0.98\times10^2}\right)^{1/1.3} - 1\right\}$$

$$\eta_{vol} = \mathbf{0.8352 = 83.57\%}$$

Also volumetric efficiency,

$$\eta_{vol} = \frac{V_a}{V_s}$$

$$\therefore \quad 0.8356 = \frac{8.5}{V_s}$$

$$V_s = 1.7712\ m^3/min = \frac{\pi}{4}D^2LN$$

$$\mathbf{D = L = 0.3508\ m}$$

and

$$IP = \frac{n}{n-1}P_1(V_1 - V_4)\left[\left(\frac{P_2}{P_1}\right)^{\frac{n-1}{n}} - 1\right]$$

$$= \frac{n}{n-1}P_1V_a = \left\{\left(\frac{P_2}{P_1}\right)^{\frac{n-1}{n}} - 1\right\}$$

$$= \frac{1.3}{1.3-1}\times\frac{0.98\times10^2\times8.5}{60}\left[\left(\frac{6.5}{0.98}\right)^{\frac{0.3}{1.3}} - 1\right]$$

$$\mathbf{IP = 33.6\ kW}$$

Example 15.7. *Atmospheric air at 1 bar and 27°C taken into a single stage single acting air compressor. It is compressed according to $PV^{1.3}$ = constant to the discharge pressure of 7 bar. The compressor takes 1 m^3 of air per min. Calculate:*

(1) IP

(2) Power of motor required to drive compressor.

(3) η_{iso}, and

(4) The cylinder dimensions, assuming speed of motor = 300 rpm, stroke to bore ratio = 1.5: 1; η_{mech} of compressor = 85%, Motor transmission efficiency 90%.

Data: $\eta = 90\%$ $\quad P_1 = 100\ kPa$

$T_1 = 300\ K \quad PV^{1.3} = C$

$V_1 = 1\ m^3/min$ $\qquad N = 300\ rpm$

$\frac{L}{D} = 1.5 \quad \Rightarrow \quad L = 1.5\,D$

$P_2 = 700\ kPa$

Solution Neglecting clearance

$$\text{IP} = \frac{n}{n-1} P_1 V_1 \left[\left(\frac{P_2}{P_1}\right)^{\frac{n-1}{n}} - 1\right]$$

$$= \frac{1.3}{0.3} \times 100 \times \frac{1}{60} \times \frac{1}{60}\left\{\left(\frac{7}{1}\right)^{\frac{0.3}{1.3}} - 1\right\}$$

$$\textbf{IP} = \textbf{4.09 kW}$$

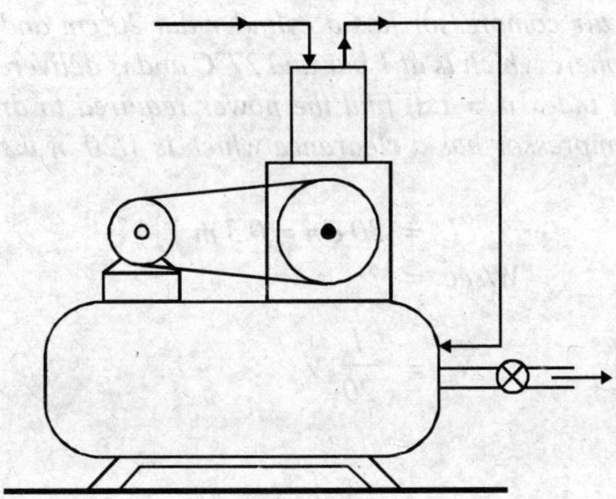

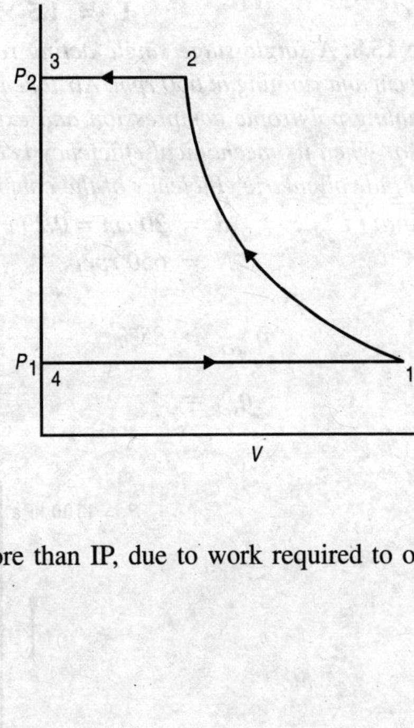

Fig. Ex. 15.7

Note. Actual work IP required to drive the compressor is more than IP, due to work required to overcome frictional lossess in compressor.

$\therefore$ Shaft power = IP + FP for compressor

And

$$\text{Compressor } \eta_{mech} = \frac{\text{IP}}{\text{shaft power}}$$

$$\therefore \quad \text{Shaft Power} = \frac{\text{IP}}{\eta_{mech} \text{ of compressor}} = \frac{4.09}{0.85}$$

$$= 4.81 \text{ kW}$$

And $\quad$ Power input to motor $= \dfrac{\text{Shaft power}}{\text{Transmission } \eta \text{ (inclusive of motor and drive } \eta)}$

$$= \frac{4.81}{0.9} = 5.35 \text{ kW}$$

Also Iso η, $\eta_{iso} = \frac{\text{Iso. work}}{\text{Indi. work}}$

$$\text{Iso. Work} = P_1V_1 \ln\frac{P_2}{P_1} = 150 \times \frac{1}{60} \times \ln\frac{7}{1} = 3.24 \text{ kW}$$

$$\therefore \quad \eta_{iso} = \frac{3.24}{4.09} = 0.792 \text{ i.e., } 79.2\%$$

Volume of air taken in/min

$$= \frac{\pi}{4} \cdot D^2 LN = \frac{\pi}{4} D^2 (1.5D) \times 300$$

$$\mathbf{D = 0.1414 \ m}$$

$$\mathbf{L = 1.5 \times 0.1414 = 0.2121 \ m}$$

Example 15.8. *A single stage single acting reciprocating air compressor has a cylinder dia 20 cm and piston stroke 30 cm and running at 650 rpm. Air taken from atmosphere which is at 1 bar and 27°C and is delivered at 11 bar. Assuming polytropic compression and expansion with index n = 1.3, find the power required to drive the compressor when its mechanical efficiency is 85%. The compressor has a clearance which is 1/20 of the stroke volume. Find volumetric efficiency of the compressor.*

Data: $d = 20 \text{ cm} = 0.2 \text{ m}$ $\quad L = 30 \text{ cm} = 0.3 \text{ m}$

$N = 650 \text{ rpm}$ $\quad W/sec = ?$

$\eta_{mech} = 85\%$ $\quad V_c = \frac{1}{20} V_c$

$\eta_{vol} = ?$

Solution

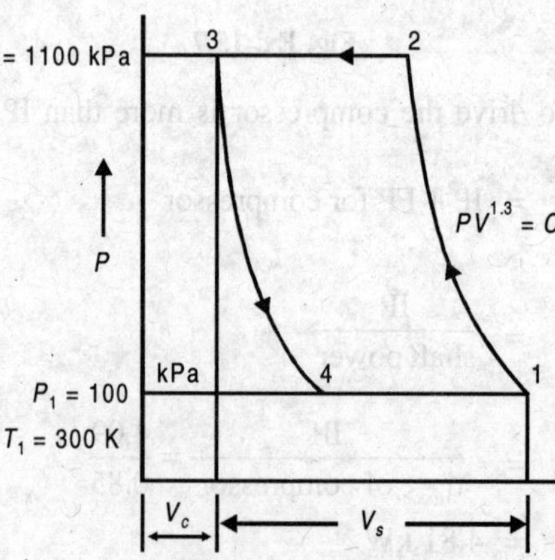

Fig. Ex. 15.8

We know that, work input to a compressor or I.P. of a compressor with clearance,

$$= \frac{n}{n-1} P_1 (V_1 - V_4) \left\{ \left(\frac{P_2}{P_1} \right)^{\frac{n-1}{n}} - 1 \right\} \quad ...(1)$$

Now find V_1 and V_4. We also know that,

$$\text{Stroke volume} = V_s = V_1 - V_3 = \frac{\pi}{4} d^2 L$$

$$V_s = \frac{\pi}{4} \times (0.2)^2 \times 0.3$$

$$V_s = 0.094278 \text{ m}^3$$

As given $$V_c = \frac{1}{20} V_s$$

$\therefore$ $$V_c = \frac{1}{20} \times 0.0094278$$

$\therefore$ $$V_c = \frac{1}{20} \times 0.0094278$$

$\therefore$ $$V_c = 0.0004712389 \text{ m}^3$$

From Fig. Ex. 15.8 $$V_1 = V_c + V_s = 0.0004712389 + 0.009427$$

$$\mathbf{V_1 = 0.0098990 \text{ m}^3}$$

Now to find V_4 consider polytropic 3 – 4 for which

$$P_3 V_3^n = P_4 V_4^n$$

$$V_4 \text{ ss} = V_3 \times \left(\frac{P_3}{P_4} \right)^{1/n}$$

$$V_4 = V_c \left\{ \frac{P_2}{P_4} \right\}^{1/n}$$

As $$P_3 = P_2 \text{ and } P_4 = P_1$$

$\therefore$ $$V_4 = 0.0004712389 \left\{ \frac{1100}{100} \right\}^{1/1.3}$$

$$\mathbf{V_4 = 0.0029894139 \text{ m}^3}$$

Now from Eq. (1),

$$\text{IP} = \frac{n}{n-1} ((V_1 - V_4) \left\{ \left(\frac{P_2}{P_1} \right)^{\frac{n-1}{n}} - 1 \right\}$$

$$= \frac{1.3}{1.3-1} \times 100 \times (0.009890 - 0.0029804139) \times \left\{ \left(\frac{1100}{100} \right)^{\frac{1.3-1}{1.3}} - 1 \right\}$$

$$= 4.33 \times 100 \times (0.069186) \times \{1.739 - 1\}$$

$$\textbf{IP} = \textbf{2.2141 kJ/cycle}$$

Since $\quad N = 650 \text{ rpm}$

$$\text{IP} = 2.2141 \times \frac{650}{60}$$

$$\textbf{IP} = \textbf{23.9835 kJ/sec or kW.}$$

Since Mechanical efficiency is given

$$\text{Actual power input} = \frac{\text{IP}}{\eta_{mech}} = \frac{23.98365}{0.85} = 28.2158 \text{ kW}$$

Also we know that, volumetric efficiency.

$$\eta_{vol} = \frac{V_a}{V_s} = \frac{V_1 - V_4}{V_1 - V_3}$$

$$= \frac{0.0098990 - 0.0029804139}{0.0098990 - 0.000471239} = \frac{0.0069186}{0.0094278}$$

$$\eta_{vol} = \textbf{0.73385} \text{ or } \textbf{73.385 \%}$$

Example 15.9. *Calculate the volumetric efficiency of the compressor having a cylinder diameter 410 mm and stroke 610 mm. Compressor makes 420 rpm and delivers 30 kg/min of air at 1.01325 bar and 15° C.*

Data: $\eta_{vol} = ?$ $\qquad d = 410 \text{ mm} = 0.41 \text{ m}$

$L = 610 \text{ mm} = 0.61 \text{ m}$

$N = 420 \text{ rpm}$ $\qquad m = 30 \text{ kg/min}$

$P = 1.01325 \text{ bar} = 101.325 \text{ kPa}$

$T = 15 + 273 = 288 \text{ K}$

Solution

We know that volumetric efficiency,

$$\eta_{vol} = \frac{V_a}{V_s}$$

and $\quad V_s = \text{Stroke volume} = \frac{\pi}{4} d^2 \times L \times N$

Note. $m^2 \times m \times rmp = \text{m}^3/\text{min}$

$$= \frac{\pi}{4} \times 0.41^2 \times 0.61 \times 420$$

$$V_s = 33.824 \text{ m}^3/\text{min}$$

As 30 kg of air is delivered at 101.325 kPa of 288 K,

As $PV = mRT$

Actual volume of air handled

$$V_a = \frac{mRT}{P} = \frac{30\,\text{kg/min} \times 0.287 \times 288}{101.325}$$

Note. $V = \dfrac{\dfrac{\text{kg}}{\text{min}} \times \dfrac{\text{kJ}}{\text{kg-k}} \times \text{K}}{\text{k}\dfrac{\text{N}}{\text{m}^2}} = \dfrac{\text{k-Nm}}{\text{min}} \times \dfrac{\text{m}^2}{\text{kN}}$

$$V = \text{m}^3/\text{min}$$

$$V_a = 24.472539 \text{ m}^3/\text{min}$$

Now $\eta_{vol} = \dfrac{V_a}{V_s} = \dfrac{24.472539}{38.8249} \doteq 0.7235062$

$$\eta_{vol} = 72.35\ \%$$

Example 15.10. *A single stage single acting reciprocating air compressor delivers 10 m³ of free air per minute and it is running at 200 rpm. It takes air at a pressure of 1 bar and 303 K and delivers at 6 bar. Take n = 1.25 and clearance volume is 6% of stroke volume.*

Determine

(i) *Delivery temperature*

(ii) *Volumetric efficiency*

(iii) *Stroke volume*

(iv) *Power required and*

(v) *Dimensions of cylinder if L = 1.2 D*

Data: $FAD = 10 \text{ m}^3/\text{min} = V_f$

$N = 200$ rpm $\qquad V_c = 0.06\ V_s$

$T_2 = ?$ $\qquad \eta_{vol} = ?$

$V_s = ?$ $\qquad IP = ?$

$L = ?$ $\qquad$ *and* $D = ?$

Solution

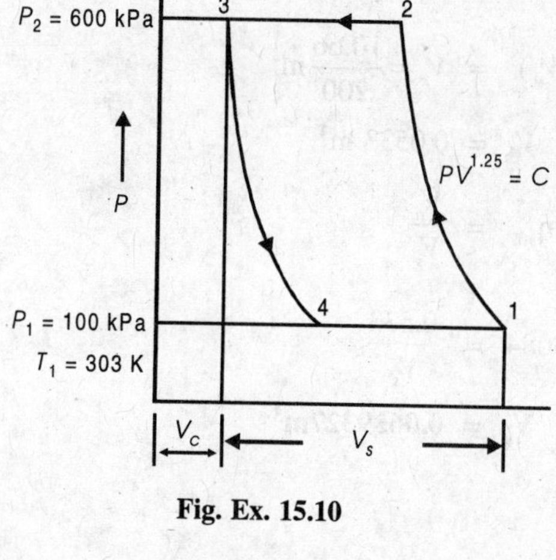

Fig. Ex. 15.10

(i) We know that,

$$\frac{T_2}{T_1} = \left(\frac{P_2}{T_1}\right)^{\frac{n-1}{n}}$$

$$T_2 = T_1\left(\frac{P_2}{P_1}\right)^{\frac{n-1}{n}} = 303\times\left(\frac{600}{100}\right)^{\frac{1.25-1}{1.25}}$$

$$T_2 = 303\times\left(\frac{6}{1}\right)^{\frac{0.25}{1.25}}$$

$$\mathbf{T_2 = 433.584\ K}$$

(ii) We also know that, volumetric efficiency,

$$\eta_{vol} = 1+C-C\times\left(\frac{P_2}{P_1}\right)^{1/n} = 1+0.06-0.06\times\left(\frac{6}{1}\right)^{\frac{1}{1.25}}$$

$$= 1 + 0.06 - 0.06\ \times 4.19 = 0.8084$$

$$\boldsymbol{\eta_{vol} = 80.84\ \%}$$

(iii) To find stroke volume, we know that

$$\frac{P_f\,V_f}{T_f} = \frac{P_1(V_1 - V_4)}{T_1}$$

$$\frac{P_f V_f}{T_f}\times\frac{T_1}{P_1} = (V_1 - V_4)$$

$$\frac{101.325\times 10\,\text{m}^3/\text{min}}{288}\times\frac{303}{100} = (V_1 - V_4)$$

$$\mathbf{(V_1 - V_4) = 10.86\ m^3/min = V_a}$$

Since the compressor is running at 200 rpm.

$$(V_1 - V_4) = V_a = \frac{10.66}{200}\text{m}^3$$

$$V_a = 0.0533\ \text{m}^3$$

Also since $\eta_{vol} = \dfrac{V_a}{V_s}$

$$0.8084 = \frac{0.533}{V_s}$$

$$\mathbf{V_s = 0.0659327 m^3}$$

(iv) I.P. required $= \frac{n}{n-1} P_1(V_1 - V_4)\left[\left(\frac{P_2}{P_1}\right)^{\frac{n-1}{n}} - 1\right]\frac{\text{rpm}}{60}$

$$= \frac{1.25}{1.25-1} \times 100 \times 0.533\left\{\left(\frac{6}{1}\right)^{\frac{0.25}{1.25}} - 1\right\} \times \frac{200}{60}$$

$$= 5 \times 100 \times 0.0533(1.43 - 1) \times \frac{200}{60}$$

I.P. = 38.284 kW

(v) To find cylinder dimensions :

We know that, stroke volume,

$$V_s = 0.0659327 = \frac{\pi}{4} D^2 \times L = \frac{\pi}{4} D^2 \times 1.2\,D$$

$$0.0659327 = D^3 \times 0.9425$$

$$\frac{0.0659327}{0.9425} = D^3$$

$$0.0699567 = D^3 \qquad \therefore \mathbf{0.4120\ m = D}$$

As $L = 1.2\text{ D}$ $\therefore$ $L = 1.2 \times 0.4120$

L = 0.49445 m

Example 15.11. *The following data were recorded for a single acting single stage reciprocating air compressor; speed : 200 rpm.*

Free air delivered = 6 m³/min

Volume of intake air = 6 m³/min

n = 1.25

Clearance volume = 6% of stroke volume

Suction pressure = 1 bar

Delivery pressure = 7 bar

Temperature of air at inlet = 288 K

Determine:

(i) Temperature of air at delivery

(ii) Volumetric efficiency

(iii) Volume of air delivered at the outlet stroke

(iv) Compressor indicated power in kW neglecting clearance.

(v) Isothermal efficiency of the compressor.

Data: $N = 200$ rpm, $FAD = V_f = 6\ m^3/min$

$V_1 = 6\ m^3/min$, $PV^{1.25} = C$

$V_c = 0.06\ V_s$, $P_1 = 100$ kPa

P_2 = 700 kPa $\qquad$ T_1 = 288 K

Find: T_2 = ? $\qquad$ η_{vol} = ?

V_s = ? $\qquad$ η_{iso} = ?

Solution

(1) We know that, for a polytropic process,

$$\frac{T_1}{T_2} = \left(\frac{P_2}{P_1}\right)^{\frac{n-1}{n}}$$

$$\therefore \quad T_2 = T_1 \times \left(\frac{P_2}{P_1}\right)^{\frac{n-1}{n}} = 288 \times \left(\frac{7}{1}\right)^{\frac{1.25-1}{1.25}}$$

$$= 288 \times (7)^{\frac{0.25}{1.25}} = 288 \times 1.475$$

$$\mathbf{T_2 = 425.022\ K}$$

(2) For volumetric efficiency, we know that,

$$\eta_{vol} = 1 + C - C\left(\frac{P_2}{P_1}\right)^{\frac{1}{n}} = 1 + 0.06 - 0.06(7)^{\frac{1}{1.25}}$$

$$= 1 + 0.06 - 0.06 \times 4.73 = 1 + 0.06 - 0.2859$$

$$= 0.7754034$$

$$\therefore \quad \boldsymbol{\eta_{vol} = 77.54\%}$$

(3) For volume of air delivered we know that,

$$\frac{T_2}{T_1} = \left(\frac{P_2}{P_1}\right)^{\frac{n-1}{n}} = \left(\frac{V_2}{V_1}\right)^{n-1}$$

$$\left(\frac{V_2}{V_1}\right)^{n-1} = \left(\frac{P_1}{P_2}\right)^{\frac{n-1}{n}}$$

or

$$\frac{V_2}{V_1} = \left(\frac{P_1}{P_2}\right)^{\frac{1}{n}}$$

$$V_2 = V_1 \times \left(\frac{P_1}{P_2}\right)^{\frac{1}{n}} = 6\,\text{m}^3/\text{min} \times \left(\frac{1}{7}\right)^{1/1.25}$$

$$= 6 \times 0.2108 = 1.2649\ \text{m}^3/\text{min}$$

$$= 1.2649\,\text{m}^3/\text{min} = \frac{1.2649}{200}\,\text{m}^3/\text{stroke}$$

$$\mathbf{V_2 = 0.006324742\ m^3/stroke}$$

(4) We know that, indicated power,

$$\text{I.P.} = \frac{n}{n-1} P_1 V_1 \left\{ \left(\frac{P_2}{P_1} \right)^{\frac{n-1}{n}} - 1 \right\}$$

$$= \frac{1.25}{1.25-1} \times 100 \times \frac{6}{10} \text{m}^3/\text{sec}\left\{(7)^{0.25/1.25} - 1\right\}$$

$$= 5 \times 100 \times 0.1\{1.4757 - 1\}$$

Note. $$\frac{\text{kN}}{\text{m}^2} \times \frac{\text{m}^3}{\text{sec}} = \frac{\text{kNm}}{\text{sec}} = \frac{\text{kJ}}{\text{sec}} = \text{kW}$$

$$\textbf{IP} = \textbf{23.785 kW}$$

(5) For isothermal efficiency, we know that,

$$\eta_{iso} = \frac{\text{Isothermal work}}{\text{Actual work (or IP)}}$$

$$\text{Isothermal work} = P_1 V_1 \ln \frac{P_2}{P_1}$$

$$= 100 \times \frac{6}{60} \times \ln \frac{7}{1}$$

$$\text{Isothermal work} = 19.459 \text{ kW}$$

Then Isothermal efficiency from Eqs (1) and (2)

$$\eta_{isothermal} = \frac{19.459}{23.785} = 0.8181294$$

$$\eta_{iso} = \textbf{81.8125 \%}$$

Example 15.12. *A single stage single acting air compressor is working between 1 bar and 7 bar, initial intake temperature is 27°C. Take FAD as 7.5 m³/min and clearance ratio as 0.04, L : D = 1.3 and N = 300 rpm and n = 1.3,*

Determine:

(a) Volumetric efficiency

(b) Bore and stroke dimensions

(c) Indicated power

(d) Isothermal efficiency

Data: $V_f = 7.5$ *m min* $\quad C = 0.04$

$LD = 1.3 \quad N = 300$ *rpm*

$\eta_{vol} = ? \quad D = ?$

$L = ? \quad IP = ?$

$\eta_{iso} = ?$

Solution

(a) We know that, volumetric efficiency of an air compressor is given by,

$$\eta_{vol} = 1+C-C\left\{\frac{P_2}{P_1}\right\}^{1/n} = 1+0.04-0.04\left\{\frac{700}{100}\right\}^{1/1.3}$$

$$= 1+0.04-0.04\times 4.467$$

$$\eta_{vol} = 0..8612955$$

$$\therefore \quad \boldsymbol{\eta_{vol} = 86.13\ \%}$$

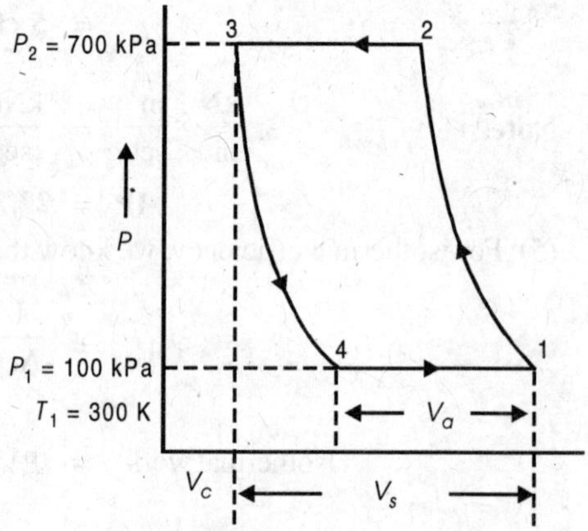

Fig. Ex. 15.12

(b) We also know that,

$$\frac{P_f V_f}{T_f} = \frac{P_1(V_1 - V_4)}{T_1}$$

$$\frac{P_f V_f}{T_f}\times\frac{T_1}{p_1} = V_1 - V_4 = V_a$$

$$\frac{101.325\times 7.5\ \text{m}^3/\text{min}}{288}\times\frac{300}{100} = (V_1 - V_4) = V_a$$

$$\therefore \quad \mathbf{V_a = 7.9160\ m^3/min}$$

Since the compressor is running at 300 rpm

$$V_a = \frac{7.9160}{300}\text{m}^3$$

$$\mathbf{V_a = 0.0263866\ m^3}$$

Also

$$\eta_{vol} = \frac{V_a}{V_s} = 0.8613$$

$$\therefore \quad 0.8613 = \frac{0.263867}{V_s}$$

$$\therefore \quad \mathbf{V_s = 0.0306358\ m^3}$$

and

$$= 0.0306358 = \frac{\pi}{8}D^2 L = \frac{\pi}{4}D^2\times 1.3D$$

$$0.0306358 = 1.0210\ D^3$$

$$0.0300051 = D^3$$

$$\mathbf{0.310741\ m = D}$$

and

$$\mathbf{L = 1.3D = 0.4039634\ m}$$

(c) Now to find induced power,

$$\text{W or IP} = \frac{n}{n-1}P_1 V_a\left\{\left(\frac{P_2}{P_1}\right)^{\frac{n-1}{n}} - 1\right\}$$

$$= \frac{1.3}{1.3-1} \times 100 \times 0.26386 \left\{ \left(\frac{7}{1} \right)^{\frac{0.3}{1.3}} - 1 \right\}$$

Note. Unit kN/m² × m³ = kJ

$$= 4.33 \times 100 \times 0.26386 \{1.5668 - 1\}$$

$$\text{IP} = 6.476289 \text{ kJ}$$

As the compressor is running at 300 rpm.

$$\text{Power} = \frac{WN}{60} = 6.476289 \times \frac{300}{60}$$

$$\textbf{Power = 32.38145 kW}$$

(d) Finally to find Isothermal efficiency

$$\eta_{vol} = \frac{\text{Isothermal work}}{\text{Actual work}} = \frac{P_1 V_1 \ln \frac{P_2}{P_1}}{\frac{n}{n-1} P_1 V_1 \left\{ \left(\frac{P_2}{P_1} \right)^{\frac{n-1}{n}} - 1 \right\}}$$

$$= \frac{\ln \frac{P_2}{P_1}}{\frac{n}{n-1} \left\{ \left(\frac{P_2}{P_1} \right)^{\frac{n-1}{n}} - 1 \right\}}$$

$$= \frac{\ln \left(\frac{7}{1} \right)}{\frac{1.3}{1.3-1} \left\{ \left(\frac{7}{1} \right)^{0.3-1.3} - 1 \right\}} = \frac{1.9459}{2.456275} = 0.7922$$

$$\eta_{vol} = \mathbf{79.22\%}$$

Example 15.13. *For a single stage single acting air compressor, actual volume of air taken in is 10 m³/min, initial intake pressure 1.013 bar, initial temperature 27°C. Final pressure 900 kN/m². Clearance is 6% of stroke volume, L : D = 1.25 and compressor is running at 400 rpm, take n =1.3.*

Determine:

(i) Volumetric efficiency

(ii) Cylinder dimensions

(iii) Indicated power

Data: V_a = *10 m³/m* P_1 = *1.013 bar = 101.3 kN/m²*

T_1 = *27 + 273 = 300 K* P_2 = *900 kN/m²*

$V_c = 0.06\ V_s$ $N = 400\ rpm$

(i) η_{vol} = ? (ii) L = ? D = ? (iii) IP = ?

Solution

(i) We know that the volumetric efficiency of a compressor is given by,

$$\eta_{vol} = 1+C-C\times\left(\frac{P_2}{P_1}\right)^{1/n}$$

$$= 1+0.06-0.06\times\left(\frac{900}{101.3}\right)^{1/1.3}$$

$$= 1+0.06-0.06\times5.366$$

$$\eta_{vol} = 0.73799 \cong 0.7380$$

$$\eta_{vol} = \mathbf{73.80\%}$$

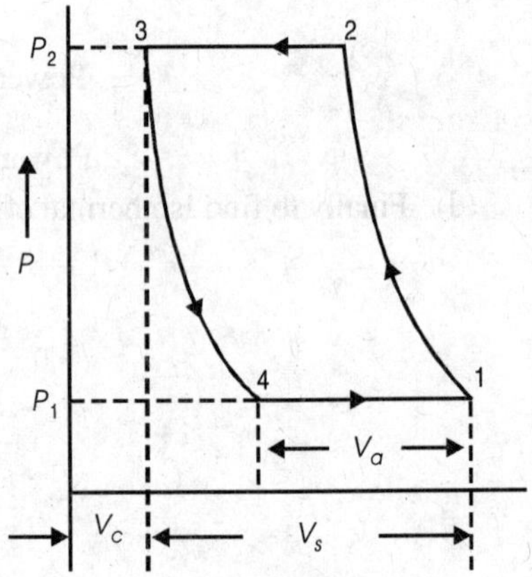

Fig. Ex. 15.13

(ii) We also know that,

$$\eta_{vol} = \frac{\text{Actual volume of air taken in}}{\text{Stroke volume}}$$

$$\therefore \quad 0.7380 = \frac{10\ m^3/min}{V_s}$$

$$V_s = \frac{10\ m^3/min}{0.7380}$$

$$V_s = 13.550\ m^3\ /min$$

Since N = 400 rpm

$$V_s = \frac{13.550}{400}$$

$$V_s = 0.03387 = \frac{\pi}{4}D^2\times L = \frac{\pi}{4}D^2\times 1.25\,D$$

$$0.03387 = 0.9817\times D^3$$

$$D^3 = 0.0344996$$

$$\frac{0.3387}{0.9817} = D^3$$

$$(0.0344996)^{1/3} = D$$

$$\mathbf{D = 0.3255\ m}$$

and $$\mathbf{L = 0.4069\ m}$$

(iii) We also know that, IP of compressor is given by

$$= \frac{n}{n-1}P_1V_a\left\{\left(\frac{P_2}{P_1}\right)^{\frac{n-1}{n}}-1\right\}$$

$$= \frac{1.3}{1.3-1} \times 101.3 \times \frac{10}{60} \left\{ \left(\frac{900}{101.3} \right)^{\frac{1.3-1}{1.3}} - 1 \right\}$$

Note. $\left[\frac{\text{kN}}{\text{m}^2} \times \frac{\text{m}^3}{\text{sec}} = \frac{\text{kNm}}{\text{sec}} = \frac{\text{kJ}}{\text{sec}} = \text{kW} \right]$

$$= 4.33 \times 101.3 \times 0.166 \times \left\{ (8.88)^{0.3/1.3} - 1 \right\}$$

$$= 4.33 \times 101.3 \times 0.166 \times \{1.655 - 1\}$$

$$\mathbf{IP = 47.71\ kW}$$

Example 15.14. *A single stage double acting compressor running at 120 rpm and power input = 75 kW, piston speed = 200 m/min. Suction pressure 1 bar and delivery pressure 10 bar,* η_{vol} *= 85%. Assuming* $PV^{1.25} = C$ *for expansion and compression find cylinder bore and* V_c *as a percent of* V_s.

Data: Double acting; $N = 120$ *rpm* $\qquad P = 75$ *kW*

Piston speed $= 2\,LN$ *m/min = 200 m/min* $\qquad \eta_{vol} = 0.85$

$D = ?$ $\qquad \frac{V_c}{V_s} = ?$

Solution

Note. N = Number of complete cycles/min
= RPM for single acting
= 2 × rpm for double acting

We know that,

$$\text{Piston speed} = 2\,LN = 200$$

$$L = \frac{200}{2 \times N} = \frac{200}{2 \times 120}$$

$$L = 0.8333 \text{ m}$$

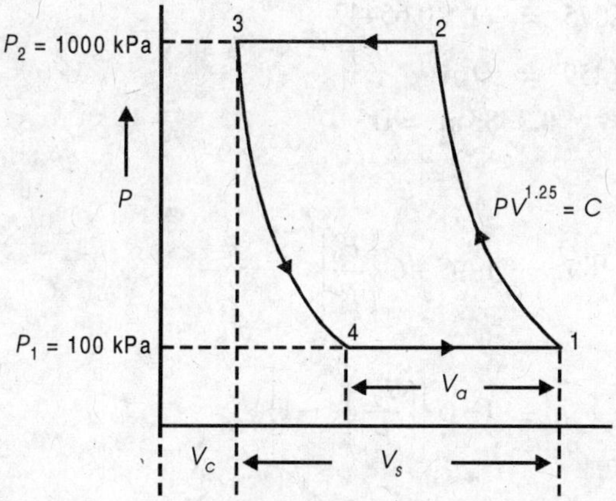

Fig. Ex. 15.14

Also Power $= \dfrac{WN}{60}$

$$75 = \frac{W\times(2\times120)}{60}$$

$$\mathbf{W = 18.75\ kJ/cycle}$$

Also work done per cycle is given by

$$W = \frac{n}{n-1}P_1V_a\left\{\left(\frac{P_2}{P_1}\right)^{\frac{n-1}{n}}-1\right\}$$

$$18.75 = \frac{1.25}{1.25-1}\times100\times V_a\left\{\left(\frac{10}{1}\right)^{\frac{0.25}{1.25}}-1\right\}$$

$$\Rightarrow \quad \mathbf{V_a = 0.064m^3 = Actual\ vol.}$$

Then, we know that volumetric efficiency is given by,

$$\eta_{vol} = \frac{V_a}{V_s}$$

$$0.85 = \frac{0.064}{V_s}$$

$$V_s = 0.075\,\text{m}^3 = \frac{\pi}{4}D^2\times L$$

$$0.075 = \frac{\pi}{4}\times D^2\times0.833$$

$$0.075 = D^2\times0.65447$$

$$0.11459 = D^2$$

$$\mathbf{0.3385\,m = D}$$

Now to find C, we know that,

$$\eta_{vol} = 1+C-C\left\{\frac{P_2}{P_1}\right\}^{1/n}$$

or

$$= 1-C\left\{\left(\frac{P_2}{P_1}\right)^{1/n}-1\right\}$$

∴

$$0.85 = 1-C\left\{(10)^{1/0.25}-1\right\}$$

$$0.85 = 1 - C\{5.31\}$$
$$0.85 = 1 - 5.31\ C$$
$$5.31\ C = 1 - 0.85$$
$$C = \frac{0.15}{5.31} = 0.0282$$

or
$$C = \frac{V_c}{V_s} = 0.0282$$

$$\mathbf{V_c = 2.82\ \%\ V_s}$$

Example 15.15. *A single stage, double acting air compressor has a cylinder diameter of 400 mm and a stroke of 300 mm. The clearance volume is 6% of V_s. The intake pressure is 0.95 bar and temperature is 25°C. If the delivery pressure is, 5 bar and speed is 250 rev/min, calculate the mass rate of air compressed. The index of compression and expansion is 1.3 and assume R = 0.287 kJ/kg-K.*

Data: Double Acting, so use N = 2 × rpm

$D = 0.4$ m $\qquad L = 0.3$ m

$V_c = 0.06\ V_s \qquad P_1 = 95$ kPa

$T_1 = 25 + 273 = 298$ K

$P_2 = 500$ kPa $\qquad N = 250$ rpm

$m = ?$ kg/sec $\qquad n = 1.3$

and $R = 0.287$ kJ/kg-K for air.

Solution

We know that, $P_1V_1 = mRT_1$

$$\dot{m} = \frac{P_1V}{RT_1} \qquad \text{...(1)}$$

Also, we know that, stroke volume

$$V_s = \frac{\pi}{4}D^2 \times L = \frac{\pi}{4} \times 0.4^2 \times 0.3$$

$$\mathbf{V_s = 0.037699\ m^3}$$

As $V_c = 0.06\ V_s$

$\therefore$ $V_c = 0.06 \times 0.0377$

$\mathbf{V_c = 0.002262\ m^3}$

$\therefore$ $V_1 = V_c + V_s$

$V_1 = 0.002262 + 0.0377$

$V_1 = 0.0399619\ m^3$

$V_1 = 0.002262 + 0.0377$

$\mathbf{V_1 = 0.0399619\ m^3}$

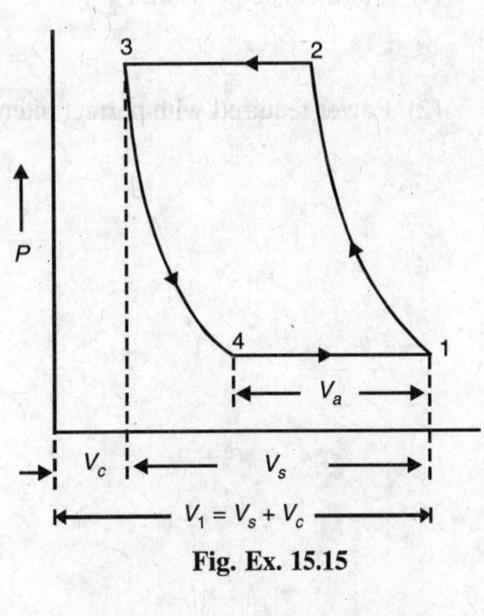

Fig. Ex. 15.15

Now from Eq. (1)

$$\dot{m} = \frac{P_1V_1}{RT_1} = \frac{95\times0.0399619}{0.287\times298}$$

$$\mathbf{\dot{m} = 0.0443887\,kg}$$

Now to find mass of air handled in kg/sec

$$\dot{m} = 0.0443887\times\frac{2N}{60}$$

(Double acting)

$$= 0.0443887\times\frac{2\times250}{60}$$

$$\mathbf{\dot{m} = 0.3699\,kg/sec}$$

Problems on Multistage Air Compressors

Example 15.16. *A 2-stage reciprocating air compressor handles 2 kg/min of air corresponding to condition of 1 bar and 27°C. It is delivered of 25 bar. The polytropic index of compression is 1.25 and the intercooling is perfect. Neglect in clearance find,*

(1) *Intermediate pressure for maximum* η.

(2) *Power required in kW.*

(3) $\eta_{isothermal}$

(4) *Saving in power required compared with that for single stage compression.*

Data: $m = 2$ kg/min, $P_1 = 1$ bar $= 100$ kPa

$T_1 = 300$ K, $P_3 = 25$ bar $= 2500$ kPa

$n = 1.25$

Solution

Perfect intercooling

(I) Intermediate pressure P_2

$$= \sqrt{P_1 P_3} = \sqrt{1\times25} = 5\,\text{bar} = 500\text{ kPa}$$

(2) Power required with perfect intercooling

$$= \frac{2n}{n-1}\dot{m}RT_1\left\{\left(\frac{P_3}{P_1}\right)^{\frac{n-1}{2n}} - 1\right\}$$

$$= \frac{2\times1.25}{1.25-1}\times\frac{2}{60}\times\times0.287\times300\left\{\left(\frac{2500}{100}\right)^{\frac{0.25}{2.5}-1}\right\}$$

$$= 10.89\text{ kW}$$

(3) Isothermal Power input $= P_1V_1 \ln\frac{P_3}{P_1} = \dot{m}RT_1 \ln\frac{P_3}{P_1}$

$$= \frac{2}{60}\times 0.287\times 300\ln\frac{2500}{100} = 9.24\,\text{kW}$$

$\therefore$ $$\eta_{iso} = \frac{\text{Isothermal power input}}{\text{Actual work input}} = \frac{9.24}{10.89}$$

$$= 0.848$$

$$\eta_{iso} = \mathbf{84.8\%}$$

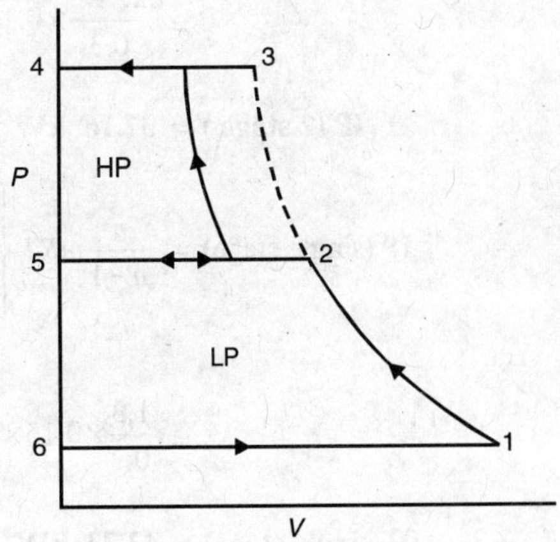

Fig. Ex. 15.16

(4) Power input (single stage)

$$= \frac{n}{n-1}\times \dot{m}RT_1\left\{\left(\frac{P_3}{P_1}\right)^{\frac{n-1}{n}} - 1\right\}$$

$$= \frac{1.25}{0.25}\times\frac{2}{60}\times 0.287\times 300\left\{\left(\frac{2500}{100}\right)^{\frac{0.25}{1.25}} - 1\right\}$$

$$= 12.96\ \text{kW}$$

Saving in power = 12.96 – 10.89

Saving in Power = 2.07 kW

Example 15.17. *A 2 stage single acting reciprocating air compressor draws in air at 1 bar and 300 K. The delivery pressure is 12 bar and intermediate pressure is ideal for minimum work and the intercooling is perfect. The compression follows the law $PV^{1.3} = C$. Flow rate of air through the compressor is 0.15 kg/sec.*

Determine:

(1) Power required to drive the compressor.

(2) Saving in power required to drive the compressor compared with single stage working.

(3) Isothermal η.

Data: P_1 = 1 bar = 100 kPa T_1 = 300 K

P_3 = 12 bar = 1200 kPa $PV^{1.3} = C$

m = 0.15 kg/sec

Solution

Since ideal intermediate pressure is there,

$$P_2 = \sqrt{P_1P_3} = 3.46\,\text{bar} = 346\ \text{kPa}$$

With ideal intermediate pressure and perfect intercooling,

$$\text{IP (2 stage)} = \frac{2n}{n-1} mRT_1\left\{\left(\frac{P_3}{P_1}\right)^{\frac{n-1}{2n}} - 1\right\}$$

$$= \frac{2\times1.3}{0.3}\times0.15\times0.287\times300\left\{\left(\frac{1200}{100}\right)^{\frac{0.3}{2.6}}-1\right\}$$

IP (2 stage) = 37.16 kW

$$\text{IP (single stage)} = \frac{n}{n-1}mRT\left\{\left(\frac{P_3}{P_1}\right)^{\frac{n-1}{n}}-1\right\}$$

$$= \frac{1.3}{0.3}\times0.15\times0.287\times300\left\{\left(\frac{1200}{100}\right)^{\frac{0.3}{1.3}}-1\right\}$$

IP (single stage) = 43.33 kW

Saving in power = 43.33 – 37.16

Saving in Power = 6.17 kW

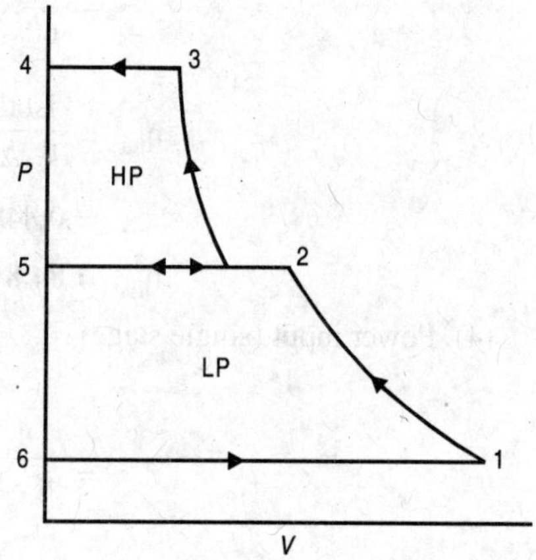

Fig. Ex. 15.17

And Isothermal work $= \dot{m}RT_1 \ln\frac{P_3}{P_1}$

$$= 0.15\times0.287\times300\ln\frac{1200}{100} = 32.09\,\text{kW}$$

$$\therefore \quad \eta_{iso} = \frac{\text{Isothermal work}}{\text{IP}} = \frac{32.09}{37.16}$$

$$\boldsymbol{\eta_{iso} = 0.863 = 86.3\%\ \text{(for multistage)}}$$

Example 15.18. *A 2 stage single acting air compressor takes in air at 1 bar and 300 K. Air is discharged at 10 bar. The intermediate pressure is ideal and intercooling is perfect. The law of compression is $PV^{1.3} = C$. The rate of discharge is 0.1 kg/sec. Find.*

1. *Power required to drive compressor.*
2. *Saving in work compared with single stage.*
3. *$\eta_{isothermal}$ for multistage and single stage.*
4. *Heat rejected in intercooler.*

Take: R = 0.287 and C_p = 1 kJ/kg-K

Solution

For ideal conditions,

$$P_2 = \sqrt{P_1P_3} = \sqrt{1\times10} = 3.162\,\text{bar}$$

and

$$\text{IP (2 stage)} = \frac{2n}{n-1}\dot{m}RT_1\left\{\left(\frac{P_3}{P_1}\right)^{\frac{n-1}{2n}}-1\right\}$$

$$= \frac{2\times1.3}{1.3-1}\times0.1\times0.287\times300\left\{\left(\frac{10}{1}\right)^{\frac{1.3\times1}{2\times1.3}}-1\right\} = 22.62\,\text{kW}$$

$$\text{IP single in work} = \frac{n}{n-1}\dot{m}RT_1\left\{\left(\frac{P_3}{P_1}\right)^{\frac{n-1}{n}}-1\right\}$$

$$= \frac{1.3}{0.3}\times0.1\times0.28\times300\left\{\left(\frac{10}{1}\right)^{\frac{0.3}{1.3}}-1\right\} = 26.05\,\text{kW}$$

Saving in work = 26.05 – 22.62

Saving in work = 3.43 kW

(iii) Isothermal work $= m.RT_1 \ln\frac{P_3}{P_1} = 0.1\times0.287\times300\times\ln\frac{10}{1}$

= 19.82 kW

$$\eta_{iso} = \frac{19.82}{22.62} = 0.8765$$

η_{iso} = **87.65% with multistage**

$$\eta_{iso} = \frac{19.82}{26.05} = 0.7617$$

η_{iso} = **76.17 % with single stage.**

(4) Heat rejected in intercooler.

Temperature rise after I stage

$$T_2 = T_1\left(\frac{P_2}{P_1}\right)^{\frac{n-1}{n}} = 300(3.162)^{0.3/1.3} = 390.95\,\text{K}$$

The heat lost in intercooler.

$$= \dot{m}C_p(T_2 - T_1) = 0.1\times1\times(390.95-300)$$

= 9.095 kJ/sec

Example 15.19. *A single acting two stage reciprocating air compressor with complete intercoooling delivers 10 kg/min of air at 16 bar pressure. The suction occurs at 1 bar and 15°C. The compression and expansion processes are reversible polytropic with polytropic index n = 1.25. Calculate*

(i) The power required

(ii) The isothermal efficiency

(iii) The free air delivered

(iv) Heat transferred in intercooler

Data: m = 10 kg/min, P_3 = 16 bar,

P_1 = l6 bar, T_1 = 15 + 273

n = 1.25.

Solution

For ideal condition

$$P_2 = \sqrt{P_1P_3} = \sqrt{1\times16} = 4\,\text{bar}$$

(i) $$\text{Power required} = \frac{2n}{n-1}\dot{m}RT_1\left\{\left(\frac{P_2}{P_1}\right)^{\frac{n-1}{2n}}-1\right\}$$

$$= \frac{2\times1.25}{1.25-1}\times10\times0.287(15+273)\left\{(4)^{\frac{1.25-1}{2\times1.25}}-1\right\}$$

Power = 44 kW

$$\text{Isothermal work} = mRT_1\ln\frac{P_3}{P_1}$$

$$= \frac{10}{60}\times0.287\times(15+273)\ln\left(\frac{16}{1}\right)$$

= 38.2 kW

(ii) $$\text{Isothermal efficiency} = \frac{\text{Isothermal work}}{\text{Actual work}}$$

$$\eta_{iso} = \frac{38.2}{44} = 0.868 = 86.8\%$$

(iii) $$\text{FAD} = \frac{mRT_1}{P_1} = \frac{10\times0.287\times288\times1000}{1\times10^5}$$

= 8.266 m³/min

(iv) Temperature after end of compression upto intermediate pressure

$$\Rightarrow \quad T_2 = T_1\times\left(\frac{P_2}{P_1}\right)^{\frac{n-1}{n}}$$

$$T_2 = 288\times(4)^{\frac{0.25}{1.25}}$$

$$T_2 = 380\text{ K}$$

Heat transferred in the intercooler $Q = m\,C_p(T_2 - T_1)$

$$= \frac{10}{60}\times1.005\times1.005(380-288)$$

Q = 15.14kW

Example 15.20. *In a 3 stage compressor air is compressed from 98 kPa to 500 kPa. Calculate for 1m³ of air per sec.*

(1) Work under ideal condition for n = 1.3,

(2) *Isothermal work*
(3) *Saving in work due to multistaging*
(4) *Isothermal efficiency.*

Data: $P_1 = 98$ *kPa* $\quad$ $P_4 = 500$ *kPa*
$V = 1\ m^3/sec$

Solution

Work input or IP for 3-stage

$$= \frac{3n}{n-1} P_1 V_1 \left\{ \left(\frac{P_4}{P_1} \right)^{\frac{n-1}{n}} - 1 \right\}$$

$$= \frac{3 \times 1.3}{1.3-1} \times 98 \times 1 \left\{ \left(\frac{500}{98} \right)^{\frac{0.3}{3 \times 1.3}} - 1 \right\} = \mathbf{170\,kW}$$

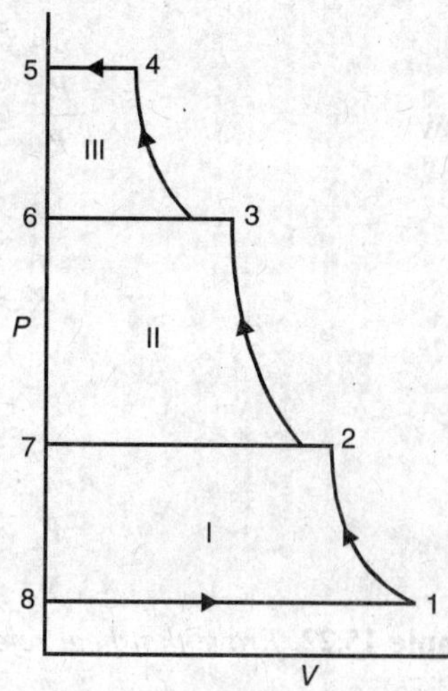

Fig. Ex. 15.20

For single stage

$$\text{IP} = \frac{n}{n-1} P_1 V_1 \left\{ \left(\frac{P_4}{P_1} \right)^{\frac{n-1}{n}} - 1 \right\}$$

$$= \frac{1.3}{1.3-1} \times 98 \times 1 \left\{ \left(\frac{500}{98} \right)^{\frac{0.3}{1.3}} - 1 \right\} = \mathbf{193.1}$$

$\therefore$ **Saving in work 193.1 – 170 = 23.1 kW**

$$\text{Isothermal work} = P_1 V_1 \ln \frac{P_{\text{final}}}{P_1}$$

$$= 90 \times 1 \times \ln \frac{500}{98} = 159.7\,\text{kW}$$

$\therefore$ Isothermal efficiency for multistage,

$$\eta_{\text{iso}} = \frac{\mathbf{159.7}}{\mathbf{170}} = \mathbf{93.89\%}$$

$$\eta_{\text{iso}} \textbf{ for single stage} = \frac{\mathbf{159.7}}{\mathbf{193}} = \mathbf{82.7\%}$$

Example 15.21. *A 3 stage reciprocating air compressor compresses air from 1 bar and 17°C to 35 bar. The law for compression is $PV^{1.25} = C$ for all stages. Assuming perfect intercooling and neglecting clearance find the minimum work required to compress 15 m³/min of free air. Also find intermediate pressures.*

Data: 3-stage

$P_1 = 100\ kN/m^2$ $\quad$ $T_1 = 290\ K$
$P_4 = 3500\ kN/m^2$ $\quad$ $n = 1.25$
$V_1 = 15\ m^3/min$

Solution

$$\text{IP (3 stage)} = \frac{3n}{n-1} P_1V_1 \left\{ \left(\frac{P_4}{P_1} \right)^{\frac{n-1}{3n}} - 1 \right\}$$

$$= \frac{3 \times 1.25}{1.25 - 1} \times 100 \times \frac{15}{60} \left\{ \left(\frac{3500}{100} \right)^{\frac{0.25}{3 \times 1.25}} - 1 \right\}$$

$$= 100.29 \text{ kJ/sec or kW}$$

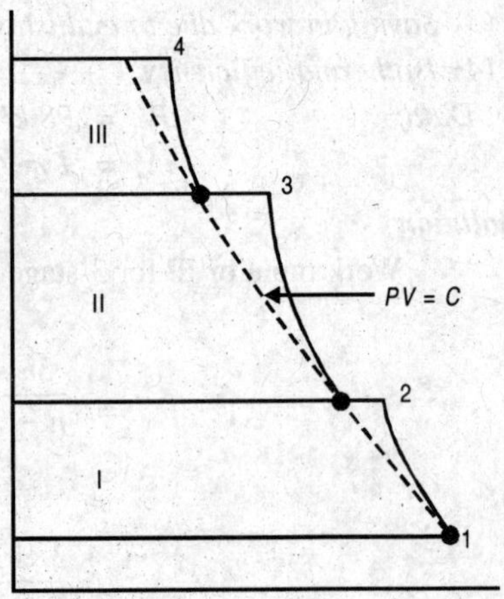

Fig. Ex. 15.21

Now
$$\frac{P_2}{P_1} = \frac{P_3}{P_2} = \frac{P_n}{P_3} = k$$

and
$$K = \left(\frac{P_{x+1}}{P_1} \right)^{1/x}$$

$$= (\text{Pressure Range})^{1/x} = \left(\frac{3500}{100} \right)^{\frac{1}{3}} = 3.27$$

∴
$$\mathbf{P_2 = kP_1 = 3.27\ bar}$$

$$\mathbf{P_3 = P_1k^2 = 10.674\ bar}$$

Example 15.22. *Prove that heat rejected per stage per kg of air in a reciprocating air compressor with perfect intercooling is given by* $\left[C_p + C_v \left(\frac{\gamma - n}{n - 1} \right) \right]$ $(T_2 - T_1)$ *where* $(T_2 - T_2)$ *is the temperature rise during each stage.*

Solution

Total heat rejected in a reciprocating compressor

= Heat rejected during compression + Heat rejected during intercooling ...(1)

Now heat rejected during compression per kg,

$$Q\text{ (poly).} = \frac{\gamma - n}{\gamma - 1} \times W_{poly}\text{/kg (W.D. in a poly compression process)}$$

$$= \frac{\gamma - n}{\gamma - 1} \times \frac{P_2V_2 - P_1V_1}{n - 1}$$

$$= \frac{\gamma - n}{\gamma - 1} \times \frac{R(T_2 - T_1)}{n - 1} \qquad ...(2)$$

And heat rejected during intercooling/kg of air

$$= C_p(T_2 - T_1)$$

$$\text{Total heat rejected} = (2) + (3)$$

$$= \frac{\gamma - n}{\gamma - 1} \times \frac{R(T_2 - T_1)}{n-1} + C_p(T_2 - T_1)$$

But $$R = C_p - C_v \text{ and } \gamma = \frac{C_p}{C_v}$$

$\therefore$ $$R = \gamma C_v - C_v$$

$$R = C_v(\gamma - 1)$$

$\therefore$ $$C_v = \frac{R}{\gamma - 1}$$

$\therefore$ Total heat rejected/kg/stage

$$= \frac{\gamma - n}{n-1} \times C_v(T_2 - T_1) + C_p(T_2 - T_1)$$

$$= \left\{C_p + C_v\left(\frac{\gamma - n}{n-1}\right)\right\}(T_2 - T_1)$$

Example 15.23. *A 4 stage compressor works between limits of 1 bar and 115 bar. The index of compression in each stage is 1.29. The temperature at the start of compression in each stage is 35°C and the intermediate pressure are so chosen that the work is divided equal among the stages (∴ pressure ratio same). Neglecting clearance find,*

(1) Pressures P_2, P_3 and P_4
(2) Delivery temperature in each stage.
(3) Individual work/kg and
(4) Isothermal work and, $h_{isothermal}$

Data: $P_1 = 1$ bar, $n = 1.29$, $P_5 = 115$ bar, $T_1 = 35°C = 308$ K

Solution

For ideal intermediate pressure

$$\frac{P_2}{P_1} = \frac{P_3}{P_2} = \frac{P_4}{P_3} = \frac{P_5}{P_4} = k$$

$\therefore$ $$P_2 = kP_1$$

$$P_3 = kP_2 = k^2 P_1$$

$$P_{x+1} = k^x P_1$$

i.e., $$P_5 = k^4 P_1$$

$$\frac{P_5}{P_1} = k^4$$

$$\left(\frac{P_5}{P_1}\right)^{1/4} = k; \left(\frac{115}{1}\right)0.25 \quad \Rightarrow k = 3.27$$

$$P_2 = 327.47 \text{ kPa}$$

$$P_3 = 1072.38 \text{ kPa}$$

$$P_4 = 3511.74 \text{ kPa}$$

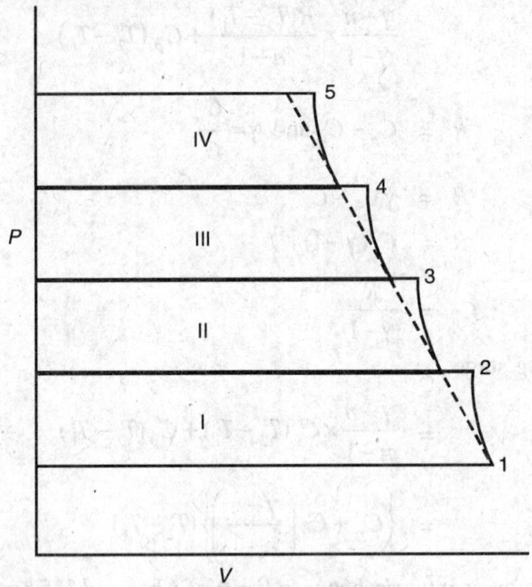

Fig. Ex. 15.23

Since compression is polytropic,

$$\frac{T_2}{T_1} = \left(\frac{P_2}{P_1}\right)^{\frac{n-1}{n}}$$

$$T_2 = 308(3.27)^{0.29/1.29}$$

$$= 401.39 = 402\,\text{K}$$

Delivery temperature from each stage.

$$\text{IP (4 stage)} = \frac{4n}{n-1} 1.RT_1\left[\left(\frac{P_5}{P_1}\right)^{\frac{n-1}{4n}} - 1\right]$$

$$= \frac{4\times1.29}{0.29}\times0.287\times308\{(115)^{0.29/5.26} - 1\}$$

$$= 480.7\ \text{kJ/kg}$$

$$\text{Isothermal work} = mRT_1 1\text{n}\frac{P_5}{P_1} \text{ and } m = 1\,\text{kg (for kJ/kg)}$$

$$= 0.287\times3081\text{n}\left(\frac{115}{1}\right)$$

$$= 419.43\ \text{kJ/kg}$$

$$\therefore \text{ Isothermal efficiency} = \eta_{iso} = \frac{419.45}{480.7}$$

$$\eta_{iso} = \mathbf{0.872 = 87.2\%}$$

Example 15.24. *A single acting, two stage RAC running at 280 rpm delivers air at a pressure of 18 bar, and the conditions at the commencement of compression being 0.98 bar and 305 K.*

The intermediate pressure is 4 bar and the clearance volume of LP cylinder is 5% of stroke volume (swept volume). Equation for compression and expansion for each cylinder is $PV^{1.25} = C$. The compressor has a free air capacity of 2.25 m³/min measured under free conditions of 1 bar and 290 K. Take mass of air handled as 2.7 kg/min.

Determine

(i) η_{vol} of L.P. cylinder.

(ii) Theoretical work spent in driving the compressor.

(iii) Dimension of L.P. cylinder if L = D

The given compressor has a perfect intercooling.

Data: $V_c = 0.05\ V_s$ *for L.P. cylinder*

$V_f = 2.25\ m^3/min$ $\qquad P_f = 100\ kPa$

$T_f = 290\ K$ $\qquad \dot{m} = 2.7\ kg/min$

Solution

(i) Now volumetric efficiency of L.P. cylinder

$$\eta_{vol}\,(\text{L.P.}) = 1 - C\left\{\left(\frac{P_2}{P_1}\right)^{\frac{1}{1.25}} - 1\right\}$$

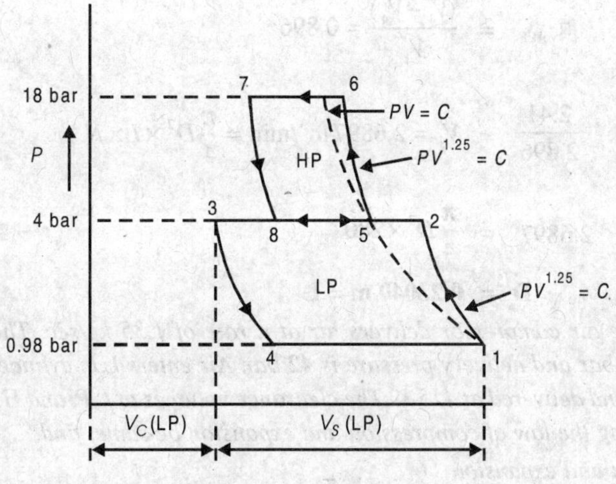

Fig. Ex.15.24

$$= 1 - 0.05\left\{\left(\frac{4}{0.98}\right)^{\frac{1}{1.25}} - 1\right\} = 0.896 \text{ and } 89.6\,\%$$

(ii) Theoretical work input, for 2-stage air compressor, we know that,

$$\text{IP (2 stage)} = \frac{2n}{n-1}\dot{m}RT_1\left\{\left(\frac{P_3}{P_1}\right)^{\frac{n-1}{2n}} - 1\right\}$$

$$= \frac{2\times1.25}{1.25-1}\times\frac{2.7}{60}\times0.287\times305\left\{\left(\frac{18}{0.98}\right)^{\frac{1.25-1}{2\times1.25}} - 1\right\} = 13.307 \text{ kW}$$

(iii) We also know that,

$$\frac{P_f V_f}{T_f} = \frac{P_1(V_1 - V_4)}{T_1}$$

$$\frac{P_f V_f}{T_f}\times\frac{T_1}{P_1} = (V_1 - V_4)$$

$$\frac{100\times2.25\,\text{m}^3/\text{min}}{210}\times\frac{305}{0.98\times100} = (V_1 - V_4)$$

$$2.41\,\text{m}^3/\text{min} = (V_1 - V_4)$$

$$\eta_{\text{vol LP}} = \frac{(V_1 - V_4)}{V_s} = 0.896$$

$$\frac{2.41}{2.896} = V_s = 2.6897\,\text{m}^3/\text{min} = \frac{\pi}{4}D^2\times L\times N$$

$$2.6897 = \frac{\pi}{4}D^3\times280$$

D = 0.23040 m = L

Example 15.25. *A 2 stage air compressor delivers air at a rate of 1.35 kg/sec. The suction pressure is 1 bar, intermediate pressure is 7 bar and delivery pressure is 42 bar. Air enters L.P. cylinder at 290 K is intercooled to 305 K before inlet to H.P. and delivered at 415 K. The clearance volumes of L.P. and H.P. cylinders are 6% and 8% of stroke volumes. Assuming the law of compression and expansion be same, find,*

(i) Index of compression and expansion

(ii) Volumetric efficiency of L.P. and H.P. cylinders

(iii) Total power

Data:

$m = 1.35$ kg/sec

$V_c = 0.06\, V_s$ (L.P)

$V_c = 0.08\, V_s$ (H.P)

$V_{s\,LP} = V_1 - V_3$

$V_{s\,HP} = V_5 - V_7$

Solution

For the process 1–2,

$$\frac{T_2}{T_1} = \left(\frac{P_2}{P_1}\right)^{\frac{n-1}{n}}$$

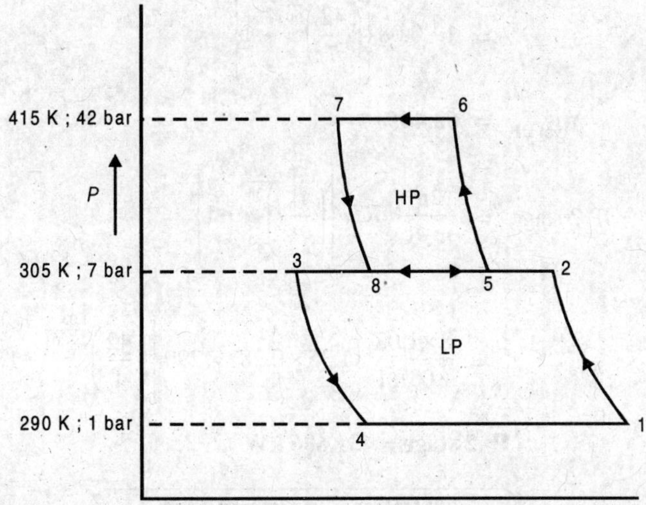

Fig. Ex.15.25

$$\frac{305}{290} = (7)^{\frac{n-1}{n}}$$

$$1.051 = (7)^{n-1/n}$$

$$\ln 1.051 = \frac{n-1}{n}\ln 7$$

$$0.255 = \frac{n-1}{n}$$

$$\Rightarrow \qquad n = 1.026$$

$$\mathbf{n \simeq 1.03}$$

$$\eta_{vol}(\text{L.P.}) = 1 - C\left\{\left(\frac{P_2}{P_1}\right)^{1/n} - 1\right\}$$

$$= 1 - 0.06\left\{\left(\frac{7}{1}\right)^{1/n} - 1\right\}$$

$$= 1 - 0.06\left\{\left(\frac{7}{1}\right)^{1/n} - 1\right\} \Rightarrow 66.31\%$$

$$\eta_{\text{vol HP}} = 1 - C\left\{\left(\frac{P_3}{P_2}\right)^{1/n} - 1\right\}$$

$$= 1 - 0.08\left\{\left(\frac{42}{7}\right)^{\frac{1}{1.03}} - 1\right\}$$

$$\eta_{\text{vol HP}} = \mathbf{62.44\%}$$

$$\text{IP 2 stage} = \frac{2n}{n-1}\dot{m}RT_1\left\{\left(\frac{P_3}{P_2}\right)^{\frac{n-1}{2n}} - 1\right\}$$

$$= \frac{2 \times 1.03}{1.03 - 1} \times 1.35 \times 0.287 \times 290\left\{\left(\frac{42}{1}\right)^{\frac{1.03-1}{2\times 1.03}} - 1\right\}$$

IP 2 stage = 431.606 kW

THEORY QUESTIONS

15.1. State the uses of compressed air.

15.2. Explain classification of air compressor.

15.3. Explain:
(i) Volumetric efficiency,
(ii) Isothermal efficiency,
(iii) FAD.

15.4. Obtain an expression for volumetric efficiency of an air compressor in terms of clearance ratio index of compression and pressure ratio.

15.5. Explain the reasons for low volumetric efficiency of a reciprocating air compressor.

15.6. Discuss the reasons for multistage compression. Also explain the additional benefits of multistage compression, if any.

15.7. Bring out the factors that influence the volumetric efficiency of a reciprocating. air compressor.

15.8. What are the advantages of multistage compression ?

15.9. Explain the working of a two stage reciprocating air compressor with intercooler.

15.10. Prove that the volumetric efficiency of air compressor is given by,

$$\eta_{\text{vol}} = 1 + C - C\left[\frac{P_d}{P_s}\right]^{1/n}$$

where C = Clearance ratio, P_d = Discharge pressure, P_s = Suction pressure.

15.11. Explain the effect of clearance on volumetric efficiency.

15.12. Explain the methods to improve the isothermal efficiency of air compressor.

15.13. Obtain an expression for IP of single stage air compressor.

15.14. Derive the expression for the volumetric efficiency of a reciprocating air comp. in terms of clearance ratio; pressure ratio; and index of compression.

15.15. Explain why staging is done in reciprocating compression.

15.16. Explain clearly how the actual indicator design for a single stage reciprocating compression is different than theoretical indicator design.

15.17. What are the advantages of multistage compressor ?

15.18. Explain the reasons for low volumetric efficiency of reciprocating air compressor.

15.19. Obtain an express for how volumetric efficiency of reciprocating air compressor.

15.20. State difference uses of compress air.

15.21. Why are multistage lamp are preferred in single stage ? Explain.

15.22. When is multistage compressor used for air ? What are its advantage ?

15.23. With the help of *PV* design, explain how work is seemed by a 2 stage compressor in the same pressure limit. Write expansion for amount of work seemed because of 2-state opm. State ending for major maximum work saving.

15.24. Explain clearly how actual *P–V* diagram for a single stage air compressor deviates from the theoretical diagram.

PROBLEMS FOR PRACTICE

15.1. A 2 stage single acting air compressor takes in air at 1 bar and 300 K. Air is discharged at 10 bar. The intermediate pressure is ideal and intercooling is perfect. The law of compression is $PV^{1.3} = C$. The rate of discharge is 0.1 kg/sec. Find

(i) Power required to drive the compressor. (*Ans.* 22.62 kW)

(ii) Saving in work compared with single stage. (*Ans.* Saving in work = 3.43 kW)

(iii) Isothermal efficiency for single and multi-stage. (*Ans.* Iso η_{single} = 76.17°%, Iso η_{mult} = 87.65%)

15.2. A single acting single stage compressor is belt driven from an electric motor at 400 rpm. The cylinder diameter is 15 cm and stroke is 17.5 cm. The air is compressed from 1 bar to 7 bar and law of compression is $PV^{1.3} = C$. Find power of motor, if transmission efficiency is 97%, and η_{mech} = 90%. Neglect clearance effects.

IMPORTANT POINTS TO BE REMEMBERED

(A) Single cylinder compressor

I. *No clearance*

(a) *Polytropic compression*

Work input or work done or Indicated work

$$W = \frac{n}{n-1}\bar{p}_1 v_1\left[\left(\frac{P_2}{P_1}\right)^{\frac{n-1}{n}} - 1\right]$$

$$= \frac{n}{n-1}\cdot mRT_1\left[\left(\frac{P_2}{P_1}\right)^{\frac{n-1}{n}} - 1\right]$$

(b) *Isentropic compression*

$$\text{Work done} = W\frac{\gamma-1}{\gamma}P_1V_1\left[\left(\frac{P_2}{P_1}\right)^{\frac{\gamma-1}{\gamma}} - 1\right]$$

$$= \frac{\gamma-1}{\gamma}\cdot mRT_1\left[\left(\frac{P_2}{P_1}\right)^{\frac{\gamma-1}{\gamma}} - 1\right]$$

$$\text{Isentropic efficiency} = \frac{\text{Isentropic work}}{\text{Actual work done}} = \eta_{isen}$$

(c) *Isothermal compression*

$$\text{Work done} = P_1V_1 \log e \frac{V_1}{V_2} \quad \text{and is – ve}$$

$$\text{Isothermal efficiency} = \eta_{iso} = \frac{\text{Isothermal work}}{\text{Actual work input}}$$

II. *With clearance*

$$\text{Volumetric efficiency} = \eta_{vol} = \frac{\text{Actual volume of air taken in}}{\text{Stroke volume}}$$

$$= 1 - C\left\{\left(\frac{P_2}{P_1}\right)^{1/n} - 1\right\} \quad \text{Ploytropic compressions expansion}$$

$$= 1 - C\left\{\left(\frac{P_2}{P_1}\right)^{1/\gamma} - 1\right\} \quad \text{Isentropic compression and expansion}$$

$$C = \text{Clearence ratio} = \frac{\text{Clearance volume}}{\text{Stroke volume}}$$

$$\frac{P_2}{P_1} = \text{Pressure tatio} = r_p$$

As r_p increases η_{vol} decrease
As n increasaes η_{vol} increases
As C increases η_{vol} decreases

$$\text{Work input } W_{\text{input}} = \frac{n}{n-1} P_1(V_1 - V_1)\left[\left(\frac{P_2}{P_1}\right)^{\frac{n-1}{n}} - 1\right]$$

$$= \text{Indicated work/cycle}$$

$$\text{IP} = \text{Indicated work/cycle} \times \frac{\text{No. of cycles}}{\text{sec.}}$$

(B) *Multistage compressor*
I. *No Clearance*

$$\text{Work} = \frac{n}{n-1}\left\{P_1V_1\left[\left(\frac{P_2}{P_1}\right)^{\frac{n-1}{n}} - 1\right] + P_2V_2\left[\left(\frac{P_3}{P_2}\right)^{\frac{n-1}{n}} - 1\right]\right\}$$

If point 2 lies on isothermal through 1, then $P_1V_1 = P_2\,V_2$

$$\text{and} \quad \text{Work done} = \frac{n-1}{n} P_1V_1\left[\left(\frac{P_2}{P_1}\right)^{\frac{n-1}{n}} - 1\left(\frac{P_3}{P_2}\right)^{\frac{n-1}{n}} - 2\right]$$

If compression is isentropic, replace n by γ

Condition for minimum work is $P_2 = \sqrt{P_1P_3}$ or $\dfrac{P_2}{P_1} = \dfrac{P_3}{P_2}$

$$\text{Minimum work} = W_{\min} = \frac{2n}{n-1} P_1V_1\left\{\left(\frac{P_3}{P_1}\right)^{\frac{n-1}{2n}} - 1\right\}$$

For x stages –

$$W_{\min} = \frac{xn}{n-1} P_1V_1\left\{\left(\frac{P_{xH}}{P_1}\right)^{\frac{n-1}{xn}} - 1\right\}$$

CHAPTER 16

Rotary Air Compressors

16.1 INTRODUCTION

Compressor is a device which compresses air/gases or vapours from low pressure to high pressure. It needs external energy input in the form of work. Out of the total work input to the compressor, some work is utilised to compress the fluid while the remaining is lost in overcoming friction, some work is lost to cooling medium, etc.

Compressors are mainly grouped into two categories:

(i) Positive displacement compressor and

(ii) Dynamic compressors.

In a positive displacement compressor, the pressure of the gas is increased by reduction of its volume i.e., by positive displacement of the gas to the delivery side.

While in a dynamic compressor, the kinetic energy imparted to the gas by the rotation of the rotor (impeller) is converted into pressure energy partly in the rotor and the remaining in the diffuser. Thus, the pressure rise is developed due to the dynamic action of gas.

In the subsequent article, various types of compressors are listed. Compressors are extensively used for a variety of applications (Refer Sec. 16.15).

16.2 CLASSIFICATION OF COMPRESSORS

Compressors are classified on the basis of several criteria as follows.

(a) On the basis of design and principle of operation. Based on the design and principle of operation, compressors are classified into two main groups and further sub-divided into sub-groups as shown in flow chart given on next page.

(b) According to delivery pressure. Low pressure (upto 10 bar), medium pressure (10 – 80 bar), high pressure (80 – 1000 bar).

Hyper-compressors are multistage reciprocating compressors with delivery pressure upto 1000 bar.

(c) According to the pressure ratio (as per ASME code). Fans — Pressure ratio upto 1.1

Blowers — Pressure ratio between 1.1 to 2.3

Compressor — Pressure ratio above 2.3

(d) According to free air delivered (capacity). Small capacity — Upto 9 m^3/min

Medium capacity — Between 9 – 3000 m^3/min

Large capacity — Above 3000 m^3/min

(e) According to number of stages adopted. Single stage, Multi-stage

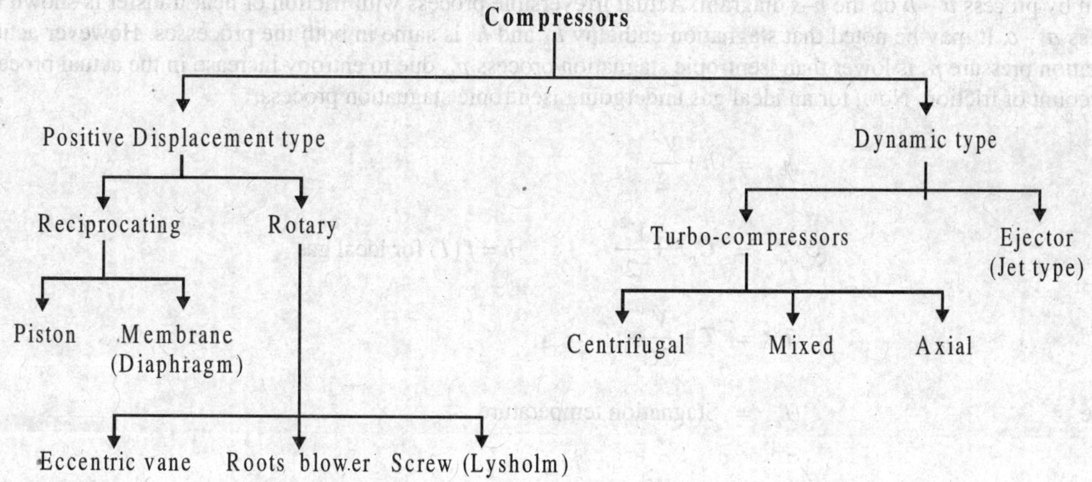

(f) According to the drive (prime mover). Electric motor driven, I.C. engine driven.

The present chapter deals with Rotary positive displacement compressors and dynamic compressors.

16.2.1 Stagnation Properties of a Flowing Fluid

When a flowing fluid having same velocity is brought to rest, it is said to have attained a stagnation state. The final stagnation state is governed by the manner in which it is attained. A reversible adiabatic or isentropic process has a great significance.

For an isentropic stagnation process, the steady flow energy equation simplifies to

$$h + \frac{V^2}{2} = h_o, \text{ kJ/kg}$$

where h_o is stagnation enthalpy and h is its initial enthalpy when fluid is flowing with a velocity of V m/s.

During the stagnation process, the kinetic energy of the fluid is converted to enthalpy, thereby pressure and temperature of the fluid increases as shown below.

The properties of the fluid at the stagnation state are referred as stagnation properties e.g., stagnation pressure, stagnation temperature, stagnation density, etc. The stagnation state and stagnation properties are denoted by lower suffix o.

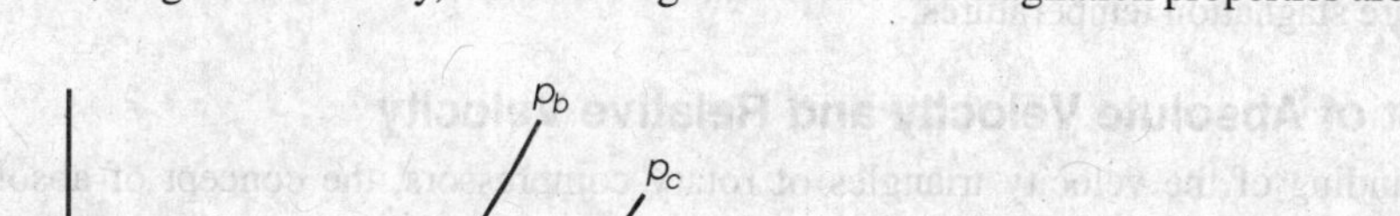

Fig. 16.1

An isentropic stagnation state is the state, fluid attains when the stagnation is reversible adiabatic or isentropic, shown by process $a - b$ on the h–s diagram. Actual irreversible process with friction or heat transfer is shown by process $a - c$. It may be noted that stagnation enthalpy h_b and h_c is same in both the processes. However actual stagnation pressure p_c is lower than isentropic stagnation process p_b, due to entropy increase in the actual process on account of friction. Now, for an ideal gas undergoing isentropic stagnation process,

$$h_o = h + \frac{V^2}{2}$$

$$\therefore \quad C_p T_o = C_p T + \frac{V^2}{2} \qquad h = f(T) \text{ for ideal gas}$$

$$\therefore \quad T_o = T + \frac{V^2}{2C_p}$$

where T_o = Stagnation temperature

$\frac{V^2}{2C_p}$ = Dynamic temperature

Now, stagnation pressure p_o can be found as

$$p_o = p_b = p_a \left(\frac{T_o}{T_1}\right)^{\frac{\gamma-1}{\gamma}}$$

Also, we have,

$$\frac{p_o}{p_a} = \frac{p_b}{p_a} = \left(\frac{T_o}{T_1}\right)^{\frac{1}{\gamma-1}}$$

Similarly, steady flow energy equation can be written as

$$q - w_s = \left(h_{o_b} - h_{o_a}\right) + g\left(z_b - z_a\right)$$

$$= C_p\left(T_{o_b} - T_{o_a}\right) + g\left(z_b - z_a\right)$$

where T_{o_b} and T_{o_a} are stagnation temperatures.

16.2.2 Concept of Absolute Velocity and Relative Velocity

For better understanding of the velocity triangles of rotary compressors, the concept of absolute velocity and relative velocity is necessary.

Absolute velocity and relative velocity

Absolute velocity. It is defined as the velocity of a moving object as sensed by a stationary observer. In a true sense, no observer can be stationary on the earth as the earth is constantly moving slowly. However, its slow motion can be ignored.

Relative velocity. It is defined as the velocity of a moving object as sensed by an observer who is moving with its own velocity.

To illustrate this concept further, consider an object moving with a velocity V m/s. When a stationary observer looks at this object, the observer can get a correct feel about the magnitude and direction of the moving object. However, if the observer also has its own velocity, V m/s, then the observer gets only a apparant feel about the magnitude and direction of the moving object. This apparant feel of velocity is the relative velocity. Let us consider the following examples.

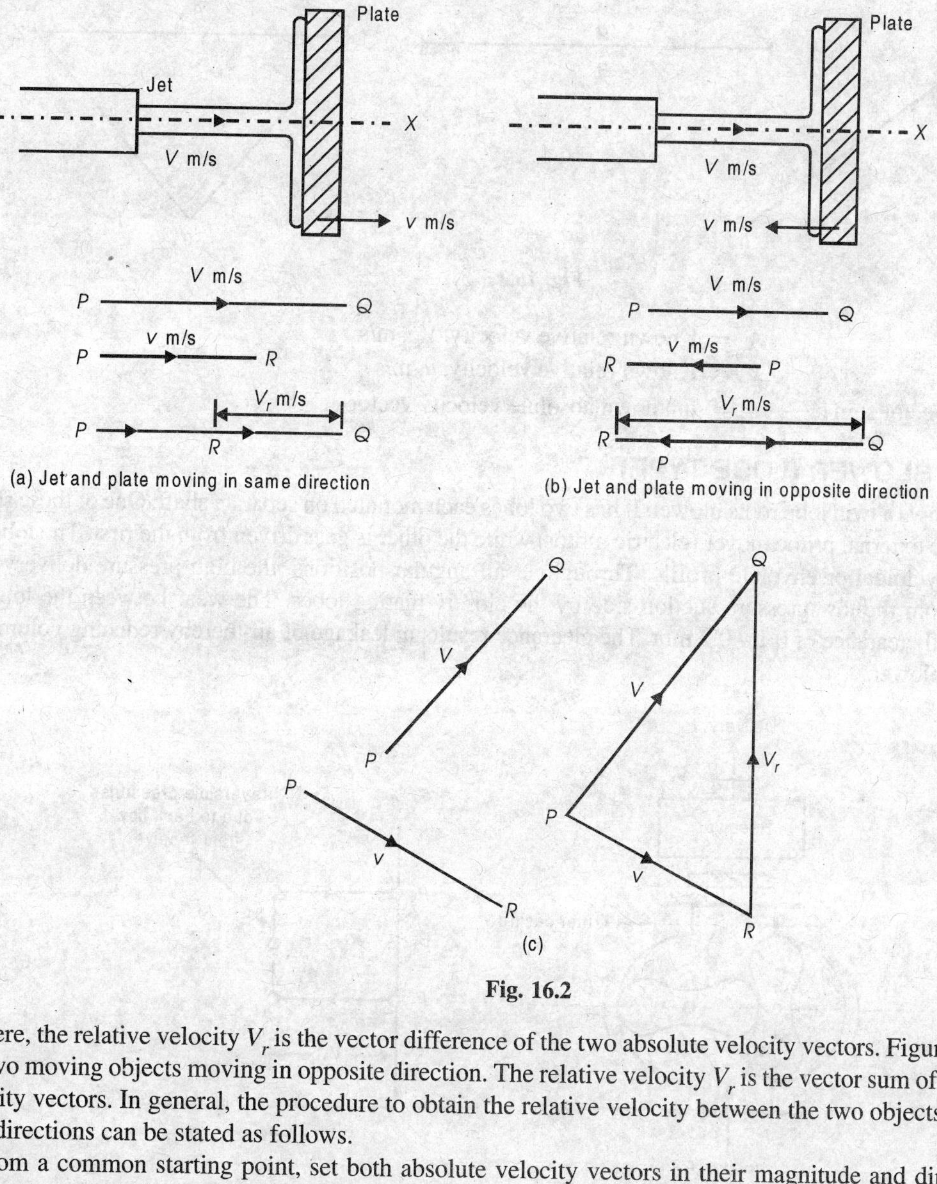

Fig. 16.2

Here, the relative velocity V_r is the vector difference of the two absolute velocity vectors. Figure 16.2 (b) shows the two moving objects moving in opposite direction. The relative velocity V_r is the vector sum of the two absolute velocity vectors. In general, the procedure to obtain the relative velocity between the two objects moving in their own directions can be stated as follows.

From a common starting point, set both absolute velocity vectors in their magnitude and direction. The line joining the ends of the two absolute velocity vectors represents the relative velocity which is the vector difference of the known absolute velocity vectors, see Fig. 16.2 (c).

Similarly, when a relative velocity and one of the absolute velocities are known, the unknown absolute velocity is given by the vector sum of the known relative velocity and known absolute velocity. The graphical procedure is as follows.

Draw the known relative velocity vector to some scale, in its magnitude and direction. In succession i.e., starting from the end of the relative velocity vector draw to same scale, the known absolute velocity vector in its magnitude and direction. Then the line joining the starting point to end point, representing vector sum, gives the unknown absolute velocity, as shown in Fig. 16.3.

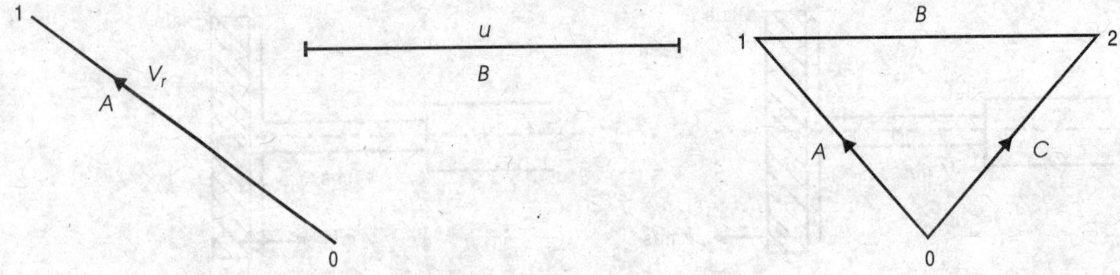

Fig. 16.3

$$A = \text{Known relative velocity, } V_r \text{, m/s}$$
$$B = \text{Known relative velocity, } u \text{, m/s}$$
$$\text{Then } C = \text{Vector sum } (V_r + u) = \text{unknown absolute velocity vector}$$

16.3 ROOTS BLOWER (LOBE TYPE)

Figure 16.4 (a) shows a twin lobe roots blower. It has two lobes each mounted on separate shaft. One of these shafts is connected to the external prime mover (electric motor) while the other is gear driven from the first. The lobes of the rotor are of cycloidal or envolute profile. Throughout all angular positions, the high pressure delivery side remains sealed from the low pressure suction side by the closely mating lobes. The wear between the lobes is avoided by a small clearance of 0.1 – 0.2 mm. The clearance results in leakage of air thereby reducing volumetric efficiency of the blower.

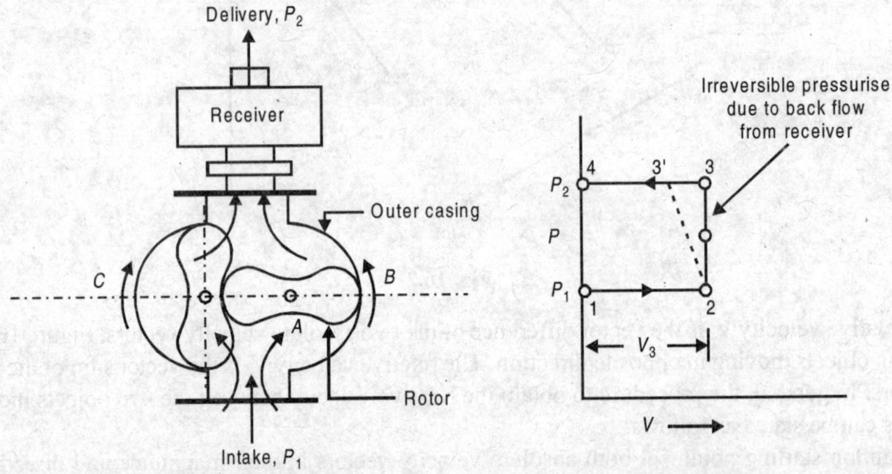

(a) Twin lobe roots blower

(b) p-V diagram

Fig. 16.4

Suction takes place through the intake port. The entrapped air between the lobes and casing is carried forward during the rotation and is finally discharged to the delivery port. There is no change in the flow area and no reduction of volume of air. However, when the delivery port is open, blower discharges air into the high pressure reservoir causing irreversible pressurerise as shown in the *p-V* diagram [Fig. 16.4 (b)]. The dotted line in *p-V* diagram shows compression process of a reciprocating compressor. The area represents excess work required due to irreversible pressure rise.

Let,

V = Volume of entrapped air in space A i.e., lobe face and casing, m^3

p_1 = Suction pressure, kPa

p_2 = Delivery pressure, kPa

N = Speed, r.p.m.

Volume of Free Air delivered/revolution = $4V$

$$\therefore \quad \text{Work required/rev.} = (p_2 - p_1) \times 4V \text{ kJ/rev.} \qquad \ldots(1)$$

$$\therefore \quad \text{Power required} = (p_2 - p_1) \times 4V \times \frac{N}{60} \text{ kW}$$

$$\text{Further isentropic work reqd./s} = \frac{\gamma}{\gamma - 1} p_1 \times 4V \left[\left(\frac{p_2}{p_1}\right)^{\frac{\gamma-1}{\gamma}} - 1\right] \times \frac{N}{60} \text{ kW} \qquad \ldots(2)$$

Roots efficiency η_{root} is defined as the ratio of isentropic work required to the actual work required.

$$\therefore \quad \eta_{roots} = \frac{\frac{\gamma}{\gamma-1} \times p_1 4V \left[\left(\frac{p_2}{p_1}\right)^{\frac{\gamma-1}{\gamma}} - 1\right]}{(p_2 - p_1) \times 4V}$$

$$= \frac{\frac{\gamma}{\gamma-1} \times p_1 \times 4V \left[\left(\frac{p_2}{p_1}\right)^{\frac{\gamma-1}{\gamma}} - 1\right]}{p_1 (r_p - 1) \times 4V} = \frac{\frac{r}{r-1}\left[r_p^{\frac{\gamma-1}{\gamma}} - 1\right]}{r_p - 1}$$

where $\quad r_p$ = pressure ratio = $\frac{p_2}{p_1}$

$$\therefore \quad \eta_{roots} = \frac{C_p}{R} \frac{\left[r_p^{\frac{\gamma-1}{\gamma}} - 1\right]}{r_p - 1} \qquad \ldots(3)$$

From Eq. (3), it is seen that with increase of pressure ratio r_p, roots efficiency η_{roots} decreases.

Roots blower have capacity ranging between 0.14 m^3/min to 1400 m^3/min and can develop pressure ratio of 2 in a single stage and of 3 in two stage.

The roots (lobes) are made of steel and are finished with close precision. Extremely fine clearances are required to be maintained during their manufacture and assembly.

16.4 ROTARY VANE COMPRESSOR

This compressor is also called as 'sliding vane compressor'. Figure 16.5 shows a multi-vane rotary compressor. It consists of a rotor eccentrically housed in the casing. Rotor has several radial slots in it, each housing a spring loaded vane. These vanes are made of steel or synthetic fibrous material. Larger the number of vanes, internal leakage of air decreases due to small pressure difference prevailing between the adjacent spaces around the rotor.

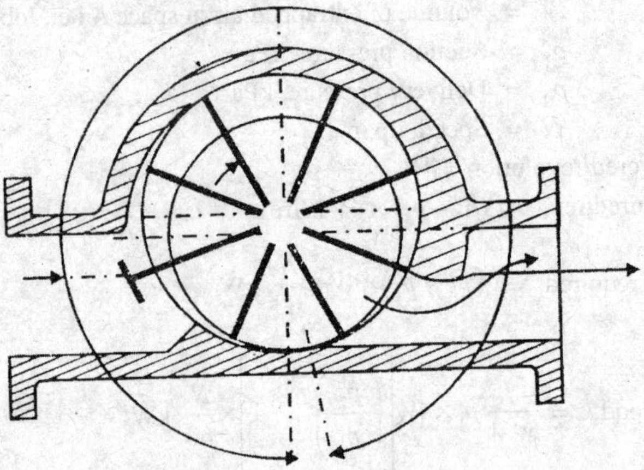

Fig. 16.5. *Vanes compressors*

High pressure ratio requires large number of vanes (20 – 30). The casing has intake and delivery openings. These compressors are often used for capacities upto 150 m³/min and for pressure ratios upto 8.5. For a given pressure ratio and FAD, vane compressor requires less work input than that for roots blower.

When the rotor rotates, vanes are driven out of the rotor towards the casing due to centrifugal force. The space between the two adjacent vanes, rotor and the casing increases creating vacuum. Thus, the gas is drawn in, from the suction opening. When the rotor crosses the point just opposite to its eccentricity, suction starts. As the rotor continues to rotate, the entrapped gas is compressed due to reduction in volume. The high pressure gas is then discharged through delivery opening. The *p-V* diagram of vane compressor is shown in the Fig. 16.6. Usually, half of the total pressure rise is developed during the internal reversible compression and the remaining pressure rise occurs irreversibly when the entrapped gas is released to the delivery side, due to back flow of high pressure air from the receiver.

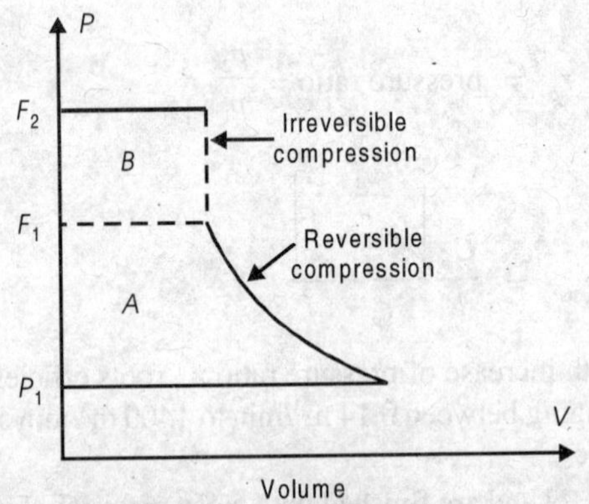

Fig. 16.6. *p-V diagram*

Due to uniform torque, the compressor does not require any flywheel and is light weight.

Let,

ε = eccentricity of rotor, m

L = axial length of rotor, m

t = thickness of each vane, m

D = diameter of rotor, m

n = number of vanes

$\therefore$ Swept volume/revolution = $2L \times \varepsilon\,(\pi D - nt)$ m³/rev. ...(4)

16.5 SCREW (LYSHOLM) COMPRESSOR

Figure 16.7 shows a screw (Lysholm) compressor. It consists of two rotors meshing with each other closely. Each rotor is helical with large pitch. The male rotor has three or four lobes while the female rotor has four, five or six recesses. The entrapped air is continuously compressed as it flow axially, due to narrowing of passages formed by the rotors. The inlet and discharge openings are somewhat oblique as shown in the 16.7.

The cycle of working of screw compressor consists of four processes namely:

(i) Suction. When the meshing rotors begins to disengage near the suction opening, gas is drawn in due to low pressure created. This is continuous so long as rotors unmesh.

(ii) Transportation. As the rotation of the rotor continues, the space filled with gas isolates from suction opening. This space is then gradually transported around the outer periphery at constant pressure until the male lobe starts meshing with the space being transported.

(iii) Compression. When rotors remesh from suction end gradually, the entrapped gas is compressed.

(iv) Delivery. Soon after the space filled with entrapped gas is exposed to discharge opening, the gas is driven out into the delivery line due to continuous penetration of male rotor into the space.

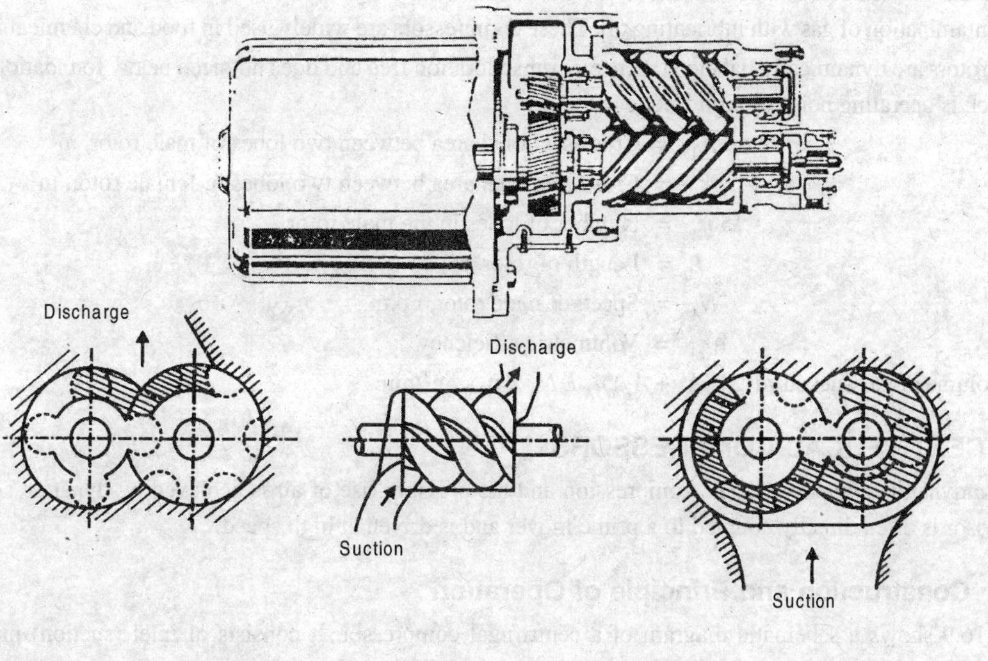

Fig. 16.7

The compressor is run at very high speeds, 3000 r.p.m. for large compressors and upto 32000 r.p.m. for small units. Rotors have peripheral velocity of 50 m/s to 100 m/s. The rotors made of Carbon steel need precision machining. The variation of volumetric efficiency and adiabatic efficiency versus pressure ratio is shown in Fig. 16.8. High pressure ratio increases internal leakage of gas back to suction opening and reduces volumetric efficiency.

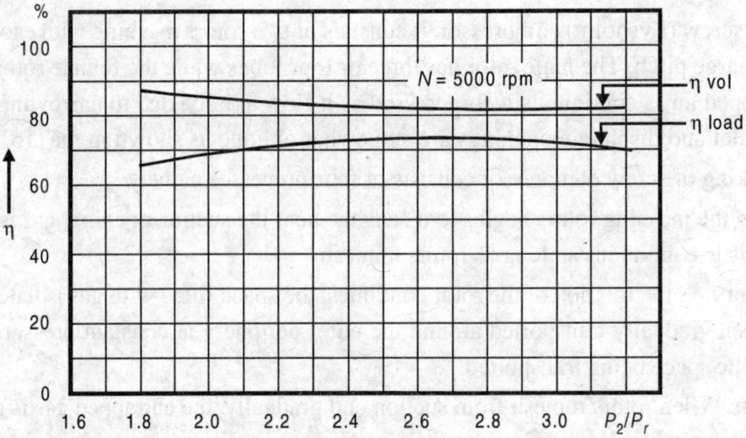

Fig. 16.8. *Variation of efficiencies*

These compressors have capacity ranging between 3 m³/min to 1000 m³/min and pressure ratios of 4 (in single stage) and 8 – 11 (in double stage). The compressor does not need any lubrication, except for bearings. Thus, there is no contamination of gas with lubricating oil. These compressors are widely used in food and chemical industries.

The rotors are dynamically balanced, hence, it runs vibration free and does not need heavy foundation. Its main drawback is operating noise of high frequency.

Let, A_A = Cross-sectional area between two lobes of male rotor, m²

A_B = Cross-sectional area between two lobes of female rotor, m²

n_A = Number of lobes in the male rotor

L = Length of rotor, m

N_A = Speed of male rotor, r.p.m.

η_{vol} = Volumetric efficiency

$\therefore$ Volume of gas sucked/rev. $= (A_A + A_B)\, n_A\, L\, N_A \times \eta_{vol}$, m³/min. ...(5)

16.6 CENTRIFUGAL COMPRESSORS

This is a dynamic compressor. The compression and the pressure rise of air is achieved by dynamic action. The compressor is often directly coupled to a prime mover and is driven at high speed.

16.6.1 Construction and Principle of Operation

Figure 16.9 shows a schematic diagram of a centrifugal compressor. It consists of inlet (suction) pipe (1), an impeller (rotor) with blades or vanes (2), a vaned or vanless diffuser (3), a volute or scroll casing (4), an outlet or delivery pipe (5), a shaft (6) and shaft seal (7). The impeller is keyed to the shaft and the rotating assembly is dynamically balanced.

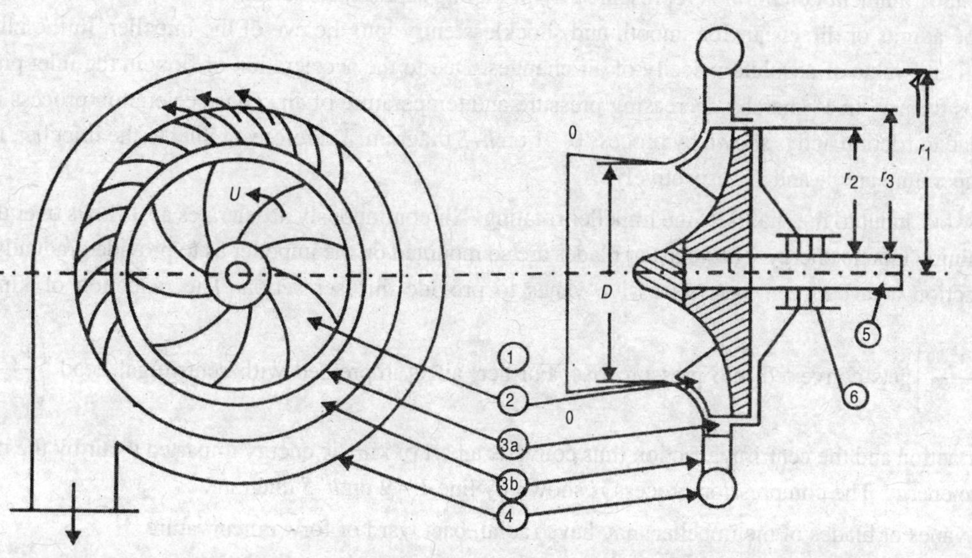

1. Inlet pipe, 2. Impeller, 3a. Vaneless diffuser, 3b. Vaned diffuser, 4. Volute casing, 5. Shaft, 6. Sealing.

Fig. 16.9. *Schematic diagram of a centrifugal compressor*

The thermodynamic process of compression and pressure-rise over the rotor and in the diffuser is shown on *h–S* diagram in Fig. 16.10.

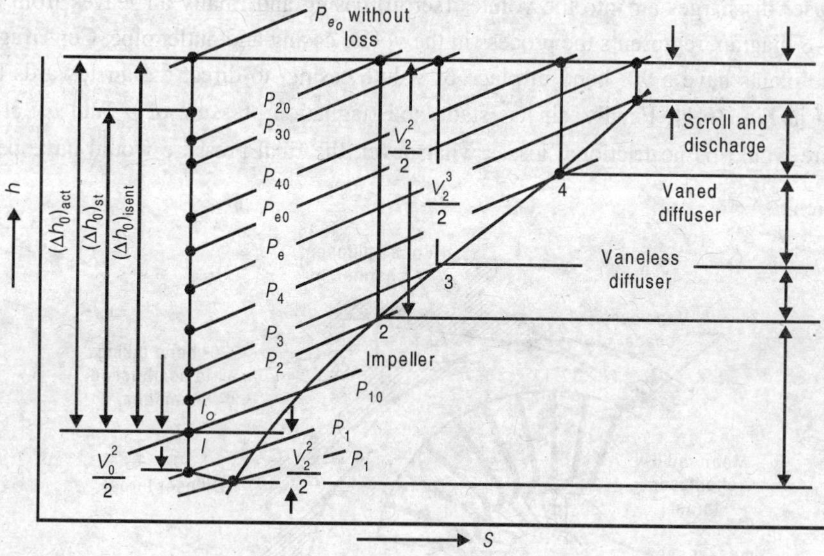

Fig. 16.10. *Centrifugal compressor process on h–S chart*

Due to the rotation of impeller, vacuum is created at the eye of the impeller. Thus, air is sucked in from the surrounding ambient, at initial stagnation pressure p_{I_o} and stagnation temperature T_{I_o}, represented by point *a* on *h–S* diagram.

The static ambient condition is represented by point o. The air is somewhat accelerated in the inlet pipe. The inlet pipe admits or directs air for smooth and shockless entry into the eye of the impeller. In the inlet pipe, the direction and value of absolute velocity of air changes. Due to the acceleration of flow in the inlet pipe, velocity increases from V_o to V_1 thereby decreasing pressure and temperature of air. The acceleration process is non-isentropic due to friction and is shown by process 0 – 1 on h–S diagram. Therefore, at inlet to the impeller, the pressure and temperature are p_1 and T_1 respectively.

The work input to the shaft sets the impeller rotating. Air continuously accelerates as it flows over the impeller, thus, gaining kinetic energy. The vanes or blades are so mounted on the impeller as to provide gradually increasing cross-section area between two successive vanes to provide diffuser action. The reduction of kinetic energy $\left(\frac{V_{r_2}^2 - V_{r_1}^2}{2}\right)$ gets converted into pressure rise. Further, air is impressed with centrifugal head $\left(\frac{u_2^2 - u_1^2}{2}\right)$. The diffuser action and the centrifugal action thus converts a part of kinetic energy imparted to air by the impeller into pressure energy. The compression process is shown by line 1 – 2 on h–S diagram.

The vanes or blades of the impeller may have radial, backward or forward curvature.

Air leaves the impeller outlet at very high absolute velocity V_2 which is further converted into pressure in a vaneless or vaned diffuser and volute (scroll) casing. In the vaneless diffuser the absolute velocity of air decreases from V_2 to V_3, process 2–3 on h–S diagram. It also stabilises the flow leaving the vaned impeller tips for shockless entry into the vaned diffuser. The velocity of air further decreases from V_3 to V_4 in the vaned diffuser to increase pressure further. This process is represented by line 3–4 on h–S diagram.

The vaned diffuser discharges air into the volute (scroll) casing and finally air leaves from the outlet pipe. Process 4 – e on h–S diagram represents the process in the volute casing and outlet pipe. Centrifugal compressors used in gas turbine plants have a 90° bend in place of volute casing, to direct the air towards the combustion chamber as shown in Fig. 16.11. Finally, air has static and stagnation pressure of p_e and p_{e_o} at the exit of the compressor. If there would be no frictional losses whatsoever, the final pressure would have been together, as represented by $p_{e_{o(wl)}}$.

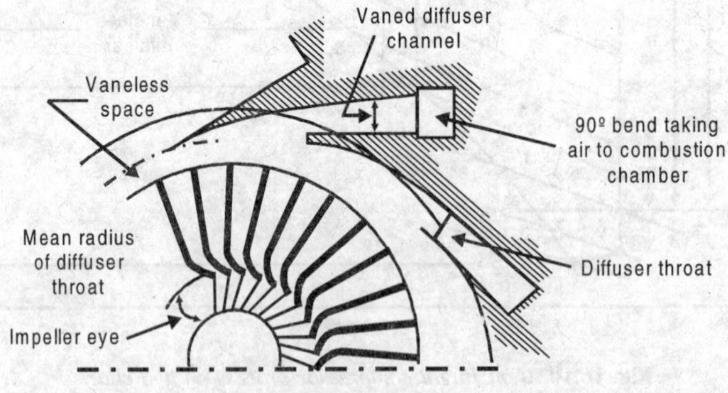

Fig. 16.11. *Radial centrifugal compressor coupled with gas turbines*

A typical multistage centrifugal compressor is shown in Fig. 16.12.

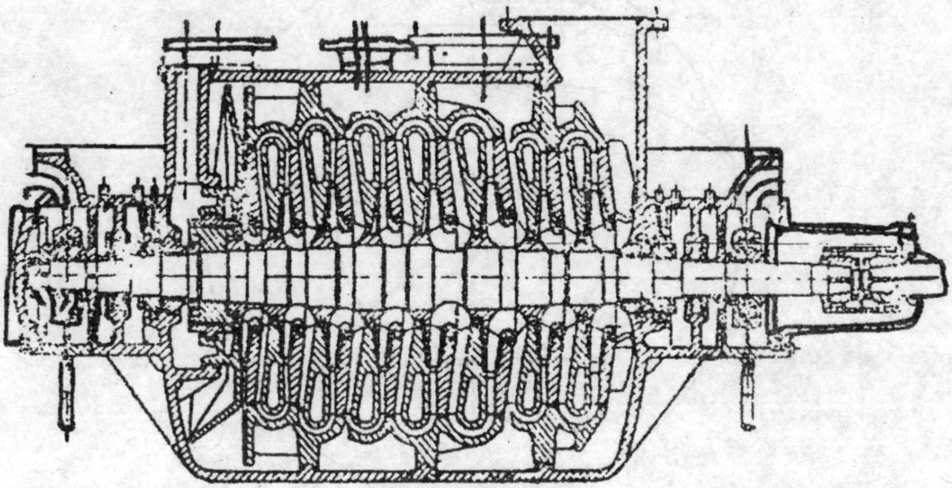

Fig. 16.12. *Multistage centrifugal compressor*

16.6.2 Velocity Diagrams of a Centrifugal Compressor

Figure 16.13 shows velocity diagrams at the inlet and outlet of impeller and at inlet and outlet of diffusers.

Nomenclature. The following notations shall be adopted hereafter for energy analysis of the centrifugal compressor. Suffix '1' denotes parameters at inlet and '2' at outlet.

β_1 = Angle of the rotor blade at inlet
β_2 = Angle of the rotor blade at outlet
α_1 = Angle made by entering air or exit angle of guide blade
α_2 = Angle made by the outgoing air from rotor blade
V_1 and V_2 = Absolute velocity of air at inlet and outlet of rotor, m/s
V_{r_1} and V_{r_2} = Relative velocity of air at inlet and outlet of rotor, m/s
V_{f_1} and V_{f_2} = Velocity of flow at inlet and outlet of rotor, m/s
V_{w_1} and V_{w_2} = Velocity of whirl at inlet and outlet of rotor, m/s
u_1 and u_2 = Mean peripheral velocity of blade tip at inlet and outlet, m/s
r_1 and r_2 = Inner and outer radii of rotor, m
m = Mass flow rate of air, kg/s
α_3 = Vaned diffuser inlet angle or vaneless diffuser outlet angle

At the inlet to rotor, air enters with absolute velocity V_1 making an angle α_1 to the direction of motion of blade (usually $\alpha_1 = 90°$), without any shock and its whirl component $V_{w_1} = 0$.

Inlet triangle of velocities is now drawn to scale setting V_1 in the radial direction, u_1 in the tangential direction to inlet periphery. The vector joining end points of V_1 and u_1 represents relative velocity V_{r_1} at inlet. The blade tip at inlet has angle β_1 i.e., its curvature at inlet lies in the direction of V_{r_1}.

At the outlet from rotor, air leaves with a relative velocity V_{r_2} at an angle β_2 with the direction of motion. Now, as u_2 is known, V_2 can be found by vectorial addition.

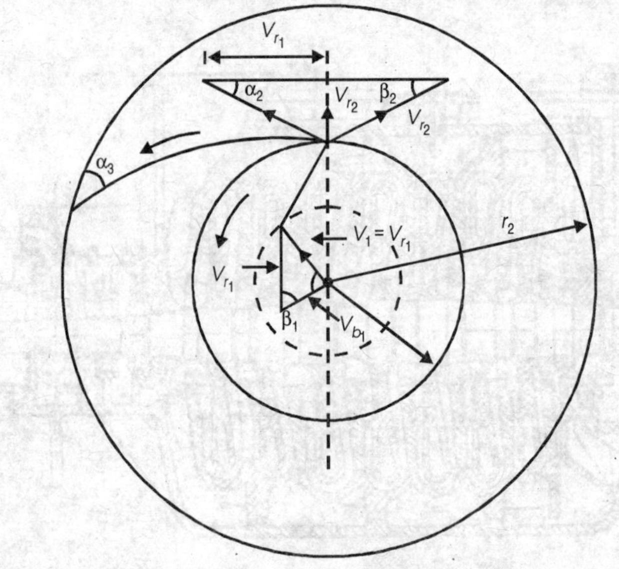

Fig. 16.13. *Velocity at inlet and outlet*

In a vaneless diffuser, the flow is assumed to be logarithmic spiral and free-vortex. Air enters the vaned diffuser with velocity V_3 at an angle α_3 and leaves the diffuser with a velocity.

16.6.3 Work Requirement (Euler's work) for a Centrifugal Compressor

The work required/kg of air in a stage of a centrifugal compressor can be found by applying the moment of momentum theorem.

As per the Newtonian equation, force is given by rate of change of momentum.

Similarly, rate of change of moment of momentum about the centre of rotation gives torque.

Consider 1 kg of air flowing through the impeller (rotor). Theoretical torque supplied by impeller,

$$T = \left(V_{w_2} r_2 - V_{w_1} r_1\right) \text{ N-m/kg}$$

Now, for an angular velocity of impeller w rad/s

Theoretical work required/kg of air = $T \times w$ w/kg

$$= W\left(V_{w_2} r_2 - V_1 r_1\right) \text{J/kg} \quad \ldots(6)$$

Equation (6) is known as Euler's equation and gives Euler's work.

Further, if air enters radially i.e., when $\alpha_1 = 90°$ and $V_{w_1} = 0$,

Theoretical work reqd./kg = $V_{w2}\, u_2$ J/kg ...(7)

The theoretical work reqd./kg also represents theoretical or virtual head developed in the absence of all losses. Thus, virtual or theoretical head; $H_{\text{virt.}}$ is given by

$$H_{\text{virt.}} = \frac{V_{w_2} u_2}{g} \quad \ldots(8)$$

From inlet and outlet velocity triangles, we have

$$V_{r_1}^2 = V_1^{\,2} + u_1^{\,2} - 2\, V_{w_1}, u_1 \text{ vectorially}$$

and

$$V_{r_1}^2 = V_2^{\,2} + u_2^{\,2} - 2\, V_{w_2}, u_2 \text{ vectorially}$$

Substituting the value of $V_{w1} u_1$ and $V_{w2} u_2$ from the above equations in Eq. (6), we get theoretical work required/kg.

$$= \frac{V_{r_2}^2 - V_1^2}{2} + \frac{V_{r_1}^2 - V_{r_2}^2}{2} + \frac{u_2^2 - u_1^2}{2} \text{ w/kg} \quad \ldots(9)$$

In the Eq. (9), first term $\left(\frac{V_2^2 - V_1^2}{2}\right)$ represents the increase in kinetic energy per kg of fluid in the impeller which has to be converted into pressure energy in the diffuser.

The last two terms signify the pressure rise in the impeller due to diffusion action and centrifugal action respectively. The second term is known as diffusion action as the relative velocity of fluid decreases from inlet to outlet of impeller. The last term represents centrifugal action due to increase of peripheral velocity of the vane tips from inlet to outlet.

Therefore, out of kinetic energy imparted by the impeller to the fluid, the part of the kinetic energy converted into pressure energy in impeller is

$$\int_1^2 \frac{dp}{p} = \left(\frac{V_{r_1}^2 - V_{r_2}^2}{2}\right) + \left(\frac{u_2^2 - u_1^2}{2}\right) \quad \ldots(10)$$

Similarly, if velocity at outlet of the diffuser is V_4 then,

Decrease of kinetic energy in the diffuser

$$\frac{V_2^2 - V_4^2}{2} = \int_2^4 \frac{dp}{p} + \Delta h_{\text{diff-loss}}$$

Thus, decrease of the kinetic energy of fluid in the diffuser is partly converted into pressure and the remaining is lost irreversibly as heat losses in the diffuser.

16.6.4 Dimensionless Parameters of Centrifugal Compressors

For the design and performance analysis of centrifugal compressor, several dimensionless parameters are very useful. These are defined and explained in what follows.

(a) Flow coefficient (ϕ). It is defined as the actual mass flow rate to the mass flow rate referred to the tip of the impeller, corresponding to tip peripheral velocity.

$$\therefore \quad \text{Flow coefficient, } \phi = \frac{\dot{m}}{\rho_2 A_2 u_2} \quad \ldots(11)$$

where, $\dot{m}$ = actual mass flow rate, kg/s

ρ_2 = density of fluid at the outlet tip of impeller, kg/m^3

A_2 = flow area at the exit tip of impeller, m^2

Applying continuity equation,

$$\text{Flow cofficient, } \phi = \frac{\rho_2 \pi D_2 b_2 V_{f_2}}{\rho_2 \pi D_2 b_2 u_2} = \frac{V_{f_2}}{u_2} \quad \ldots(12)$$

Value of flow coefficient often lies between 0.28 and 0.32, its optimum value being 0.3.

(b) Head coefficient 'λ'. It is defined as ratio of enthalpy increase in the stage to the kinetic energy corresponding to the tip peripheral velocity.

Thus,

$$\text{Head coefficient, } \lambda = \frac{\Delta h}{u_2^2/2} \quad \ldots(13)$$

where Δh = Enthalpy increase in the stage, per kg of fluid. The enthalpy increase includes the work supplied /kg of fluid compressed, as given by the Euler's Eq. (6) plus half of the heat corresponding to rotor friction losses. So we get,

$$\text{Head coefficient, } \lambda = \frac{2\left[\left(V_{w_2}u_2 - V_{w_1}u_1\right) + \frac{1}{2}\alpha u_2^2\right]}{u_2^{\,2}} \quad \text{...(14)}$$

$$\lambda = 2\left[\mu \bar{V}_{w_{2\infty}} - \left(\frac{D_1}{D_2}\right)^2 \bar{V}_{w_1}\right] + \alpha$$

where, μ = slip factor

$$\bar{V}w_{2\infty} = \frac{V_{w_{2\infty}}}{u_2} \text{ and } \bar{V}w_1 = \frac{V_{w_1}}{u_1}$$

For inlet velocity in axial direction,

$$V_1 = V_{f_1} \text{ and } V_{w_1} = 0$$

$$\therefore \quad \lambda = 2\,\mu\bar{V}_{w_{2\infty}} + \alpha \quad \text{...(15)}$$

Now, if $\alpha = 0$, head coefficient for infinite number of blades is given by

$$\lambda = 2\mu\bar{V}_{w_{2\infty}} \quad \text{...(16)}$$

and head coefficient with finite number of blades is

$$\lambda = 2\mu\bar{V}_{w_{2\infty}} \quad \text{...(17)}$$

(c) Pressure coefficient (Ψ). It is defined as the ratio of isentropic enthalpy rise to the kinetic energy corresponding to tip peripheral velocity. Thus,

$$\text{Pressure coefficient, } \psi = \frac{\Delta h_{\text{isen}}}{u_2^2/2}$$

Considedring isentropic efficiency, η_{isent}, we get

$$\psi = \frac{\eta_{\text{isent}} \times \Delta h}{u_2^2/2} = \lambda \times \eta_{\text{isent}} \quad \text{...(18)}$$

$$\text{where} \quad \eta_{\text{isent}} = \frac{\Delta h_{\text{isent}}}{\Delta h}$$

(d) Reaction (Ω). It is defined as the ratio of temperature rise in the rotor to the temperature rise in the stage. Thus,

$$\text{Reaction } \Omega = \frac{\Delta T_R}{\Delta T_{\text{stage}}} = 1 - \frac{C_p \Delta T_{\text{stator}}}{C_p \Delta T_{\text{stage}}}$$

$$= 1 - \frac{V_2^2 - V_1^2}{2C_p \Delta T_{\text{stage}}}$$

where it is assumed that $V_1 = V_4$ i.e., initial and final velocity is same. Now, for axial entry at inlet, $V_{w1} = 0$.

$$\text{Reaction } \Omega = 1 - \frac{V_{w_2}^{\,2} + V_{f_2}^{\,2} - V_{f_1}^{\,2}}{2u_2 V_{w_2}} \quad \text{...(19)}$$

16.6.5 Slip Factor of a Centrifugal Compressors

Under ideal conditions, fluid particles follow exactly the same path of blade profile such that relative velocity at impeller outlet tip is inclined with the tangential direction at blade tip angle β_2 only, irrespective of mass flow rate, speed etc.

Such an ideal flow is possible when impeller has infinite number of blades of no thickness. Figure 16.14 shows velocity triangle for radial blades. Dotted lined diagram represents ideal conditions while full lined diagram represents actual conditions.

In actual practice, when impeller has finite number of blades, fluid is trapped between the impeller vanes due to its inertia and the fluid is reluctant to flow over the impeller.

This causes a pressure difference across the blades. There is a high pressure at the leading face and low pressure at the trailing face. This pressure difference generates a relative velocity gradient and formation of eddies.

The fluid is thus discharged at a certain average angle β'_2 which is less than β_2.

Therefore fluid is said to have slipped with respect to impeller during its flow across it.

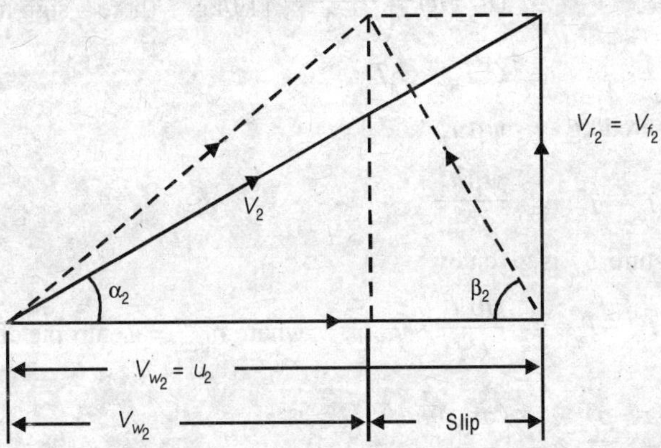

Fig. 16.14. *Slip over backward curved bladed impeller*

The slip factor μ is defined as the *ratio of actual velocity of whirl to the ideal velocity of whirl.*

$$\text{Slip factor, } \mu = \frac{\text{Actual velocity of whirl}}{\text{Ideal velocity of whirl}}$$

$$\mu = \frac{V_{wa}}{V_{wi}} = \frac{V_{w_2a}}{u_2} \text{ (for radial blades)} \qquad \ldots(20)$$

Slip factor depends upon number of blades and is usually 0.9. It does not reduces the efficiency but only reduces the head developed.

16.6.6 Pressure Ratio of Compression in a Centrifugal Compressor

Slip factor is also defined as the ratio of work required under actual conditions to the work required under ideal conditions.

$$\text{Thus, } \quad \mu = \frac{V_{w_2} u_2 - V_{w_1} u_1}{V_{w_{2i}} u_2 - V_{w_{1i}} u_1}$$

But with radial entry, $V_{w_1} = V_{w_{1i}} = 0$

$$\mu = \frac{V_{w_2}}{V_{w_{2i}}}$$

$$\text{Work} = \mu \times V_{w_2} \times u_2$$

$$= \mu u_2^2 \quad \ldots(21)$$

Actual work required is greater than theoretical work given by Eq. (21) due to friction, windage etc.

Power input factor Ψ is introduced now such that

$$\text{Actual Work req./kg} = \psi \mu u_2^2 \quad \ldots(22)$$

The value of power input factor Ψ is about 1.04.

As per steady flow energy equation applied at inlet and outlet of impeller and assuming no heat transfer,

$$Q = \Delta P + \Delta K + \Delta h + W_s \text{ kJ/kg}$$

With $Q = 0$, $\Delta P = 0$, we get

$$w_s = -(\Delta h + \Delta K) \text{ kJ/kg}$$

$$= (h_2 - h) + \left(\frac{V_2^2 - V_1^2}{2}\right) \text{ kJ/kg} \quad \text{(here – sign indicator work required)}$$

$$= C_p T_{o_2} - C_p T_{o_1} \quad \ldots(23)$$

But, $\quad$ work required/kg $= \psi \mu u_2^2$

$$\therefore \quad T_{o_2} - T_{o_1} = \frac{\psi \mu u_2^2}{C_p} \quad \ldots(24)$$

Actual stagnature temperature T_{o_1} is given by

$$T'_{o_2} - T_{o_1} = \frac{\psi \mu u_2^2}{C_p} \times \eta_{\text{isent}} \quad \text{where } \eta_{\text{isent}} = \text{isentropic efficiency}$$

Therefore, pressure ratio

$$\frac{p_{o_2}}{p_{o_1}} = \left(\frac{T'_{o_2}}{T_{o_1}}\right)^{\frac{\gamma}{\gamma-1}}$$

$$= \left[1 + \frac{T'_{o_2} - T_{o_1}}{T_{o_1}}\right]^{\frac{\gamma}{\gamma-1}}$$

$$= \left[1 + \frac{\eta_{\text{isent}} \; \psi \mu u_2^2}{C_p T_{o_1}}\right]^{\frac{\gamma}{\gamma-1}} \quad \ldots(25)$$

If all factors are unity as in case of isentropic compression, then,

$$\frac{p_{o_2}}{p_{o_1}} = \left[1 + \frac{u_2^2}{C_p T_{o_1}}\right]^{\frac{\gamma}{\gamma-1}} \quad \ldots(26)$$

The slip factor μ and the power input factor Ψ are different and independent from each other. Whereas the power input factor Ψ signifies the increase in the work input required which is lost in overcoming friction. This energy lost gets converted into heat and causes temperature rise. The temperature rise of compressed air is often not a loss as in gas turbine plants, the hot compressed air elevates the temperature in the combustion chamber without increase of fuel supply. Yet the power input factor should be low and compression should ideally be isentropic.

But the slip factor limits the work capacity of the compressor even with isentropic working. The slip factor should be as high as possible. High slip factor increases V_{w_2}. When $V_{w_2} = u_2$, work input is maximum which can be gainfully utilised.

Slip factor increases with greater number of vanes which in turn decreases area of flow passage for air. A compromise is often necessary. The present practice is to adopt about 20 vanes to obtain a slip factor of 0.9.

An impirical formula proposed by Stanitz is

$$\eta = 1 - \frac{0.63\pi}{n}$$

where, n = number of blades

Higher tip speed increases the pressure ratio.

Maximum tip speed is restricted to 460 m/s for the common materials.

Also reducing the inlet temperature T_{o_1} raises the pressure ratio for a given work input.

16.6.7 Influence of Impeller Blade Geometry

The blades of the impeller have various shapes which can be described based on the blade angle at outlet β_2. Depending on the value of β_2, the blades have different curvatures compared with the forward direction of rotation of the impeller. The various shapes of blades are:

(i) Backward curved blades, $\beta_2 < 90°$

(ii) Radial blades; $\beta_2 = 90°$

(iii) Forward curved blades; $\beta_2 > 90°$

Figure 16.15 shows the velocity triangle for each case of the above.

In all cases, blade tip velocity at exit u_2 is kept same.

It is seen from Fig. 16.15 that for the backward curved blades, the tangential component V_{w_2} is least compared with those in other cases. Therefore, for a given speed work input required for the impeller is low, as given by Euler's work (Eq. 6).

In the case of forward curved blades, V_{w_2} is maximum, thus the impeller needs maximum work input. Too, the value of V_2 is highest.

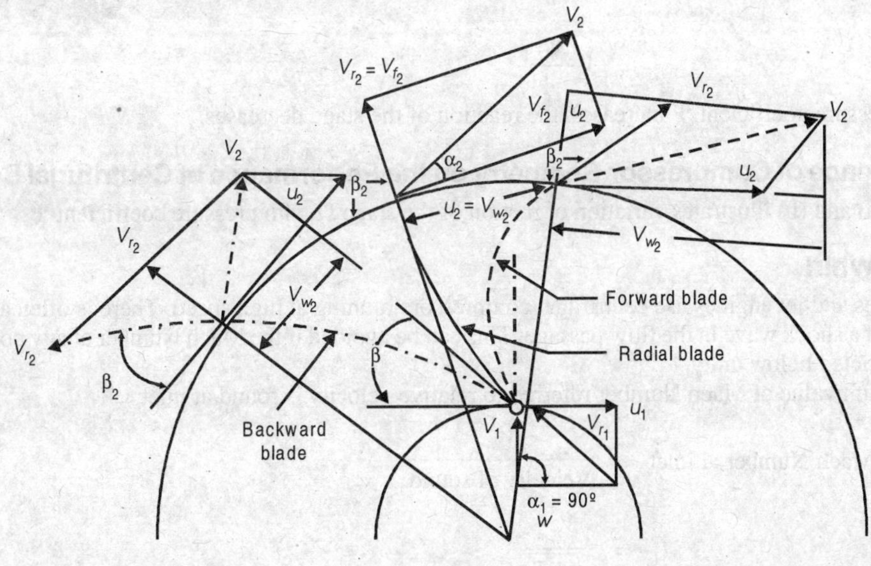

Fig. 16.15. *Velocity diagram for each blade shape*

And in case of radial blades, V_{w_2} lies in between, that is, backward and forward curved blades.

Therefore, backward curved blades are best giving high efficiency.

The influence of blade geometry can also be explained by dimensionless parameters as follows.

The head coefficient is one dimensionless parameter which is used to express the work required per stage of centrifugal compressor.

For a working fluid treated as ideal gas, isentropic efficiency $\eta_{isent} = 1$. As such head coefficient λ and pressure coefficient Ψ are same, the work input to the fluid is then given by,

$$\psi_i = 2\overline{V_w} = 2\frac{V_{w_{2i}}}{u_2} \qquad \ldots(27)$$

But,

$$\beta_2 = \frac{V_{f_2}}{u_2 - V_{w_{2i}}} = \frac{V_{f_2}/u_2}{1-\dfrac{V_{w_{2i}}}{u_2}} \qquad \ldots(28)$$

where,

$$\frac{V_{f_2}}{u_2} = \phi = \text{flow coefficient}$$

Therefore,

$$\psi_i = 2\left(1-\frac{\phi}{\tan\beta_2}\right) \qquad \ldots(29)$$

It is seen from Eq. (29) that for a given value of flow coefficient ϕ, the work input required per stage increases as blade angle at outlet β_2 increases.

For backward curved blades, $\beta_2 < 90°$ and with $\tan\beta_2 = \phi$, the $\Psi_i = 0$.

For radial blades, $\beta_2 = 90°$, $\Psi_i = 2$ and is independent of flow coefficient ϕ.

For forward curved blades, $\beta_2 > 90°$, $\tan\beta_2$ is negative and when $\phi = -\tan\beta_2$, $\Psi_i = 4$.

Further blade angle at outlet β_2 influences the reaction Ω of the stage. For an impeller with infinite number of blades,

$$\text{Reaction } \Omega = 1-\frac{V_{w_2}}{u_2}$$

$$= 1-\frac{\psi_i}{4}$$

Thus, as pressure coefficient Ψ increases, the reaction of the stage decreases.

16.6.8 Influence of Compressor Geometry on the Performance of Centrifugal Compressor

Figure 20.20 (a) and (b) illustrates variation of rection of the stage Ω with pressure coefficient Ψ.

16.6.9 Pre-Whirl

The gas turbines and jet engines use centrifugal compressors running at high speed. There is often a possibility of development of a shock wave in the flow passage. This can be avoided if the Mach Number at any point in the flow passage is restricted below unity.

The maximum value of Mach Number referred to relative velocity is found at inlet.

$$\therefore \quad \text{Mach Number at inlet} = \frac{V_{r_1}}{\text{Velocity of sound}}$$

$$M_1 = \frac{V_{r_1}}{\sqrt{\gamma R T_1}} \qquad \ldots(30)$$

where, T_1 = Static temperature at inlet, K

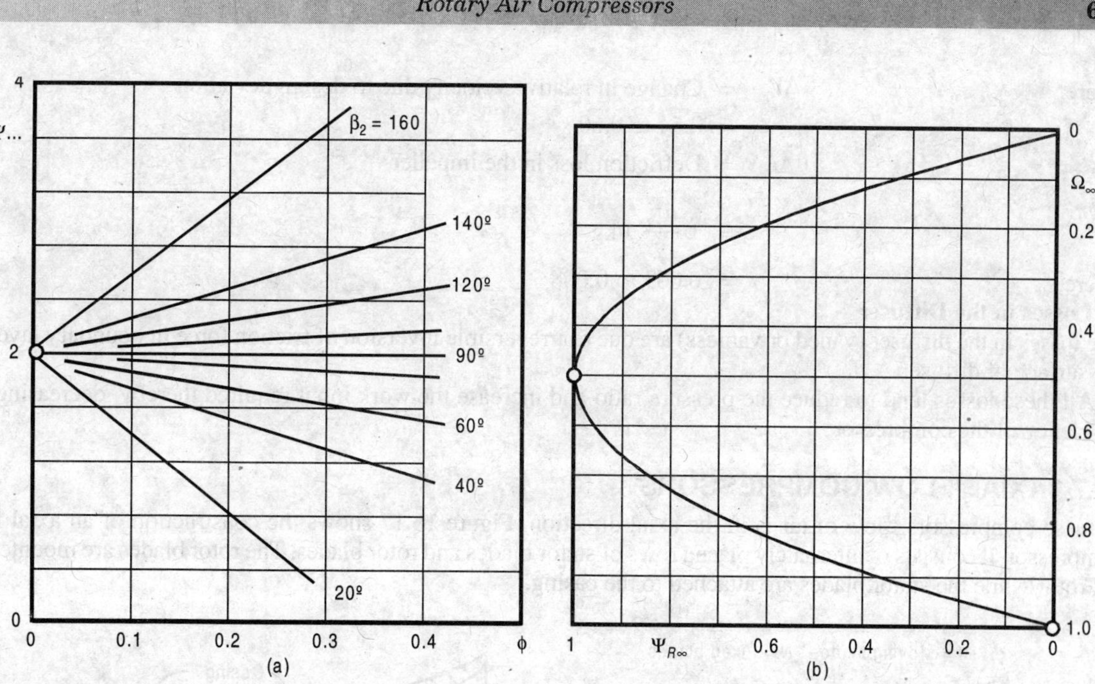

Fig. 16.20. *Influence of blade shape*

For a given flow rate, if V_{r_1} and u_1 is high for shockless operation of the compressor, the fluid is given an initial pre-rotation or pre-whirl with the help of fixed guide blades attached to the casing at inlet.

The entering air passing through the fixed guide blades is given pre-whirl. Thus V_{r_1} is reduced without affecting V_{f_1} or mass flow rate.

16.6.10 Losses in a Centrifugal Compressor

The frictional losses in a centrifugal compressor occur mainly at two places viz. over the impeller and in the diffuser.

I. Losses Over the Impeller

The fluid friction losses over the impeller are given by

$$\Delta h_{imp} = \Delta h_1 + \Delta h_2 + \Delta h_3 + \Delta h_4 + \Delta h_d$$

where Δh_1 = Loss in the inducer $= \xi_1 \dfrac{V_{r1}^2}{2}$ J/kg where, $= \xi_1 = 0.2$ to 0.3

Δh_2 = Loss due to change of direction from axial to radial at the impeller inlet

$$= \xi_2 \frac{V_1^2}{2} \text{ J/kg where, } \xi_2 = \text{loss coefficient} = 0.1 \text{ to } 0.2$$

Δh_3 = Loss due to friction in the blade passages

$$= \xi_3 \frac{Vr_1^2}{2} \text{ J/kg where, } \xi_3 = \text{loss coefficient } 0.2 \text{ to } 0.4$$

Δh_4 = Loss due to design deviations

$$= \xi_4 \frac{(\Delta Vr_1)^2}{2} \text{ J/kg}$$

where, ΔV_{r_1} = Change in relative velocity due to design deviation

= 0.54 to 0.60

Δh_d = Defriction loss in the impeller

$$= \sigma \frac{u_2^2}{2} \text{ J/kg}$$

where, σ = 0.003 to 0.008

II. Losses in the Diffuser

The losses in the diffuser (vaned or vanless) are due to irreversible inversion of friction force in boundary layer on the surface of diffuser.

All these losses tend to reduce the pressure ratio and increase the work input required thereby decreasing the efficiency of the compressor.

15.7 AXIAL FLOW COMPRESSORS

In these compressors, flow of air is in the axial direction. Figure 16.17 shows the construction of an axial flow compressor. It consists of alternately placed rows of stator blades and rotor blades. The rotor blades are mounted on the rotor while the stator blades are attached to the casing.

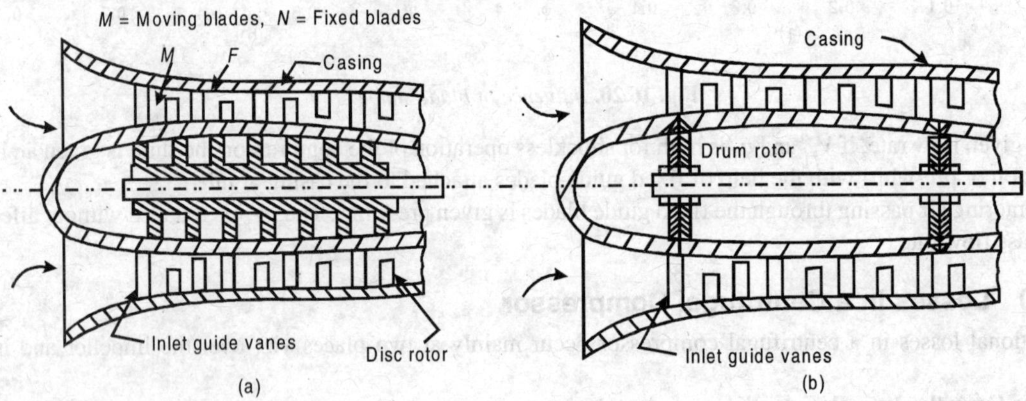

Fig. 20.17. *Axial flow compressor*

The conversion of kinetic energy into pressure head i.e., the diffusion action takes place partly over the rotor blades and partly over the stator blades. The stator blades also direct air over the successive rotor blades for a shockless entry. The flow passages between rotor blades as well as stator blades provide diverging area. One row of rotor blades preceded by a row of stator blades is known as a stage of axial flow compressor. A single stage compressor can achieve a pressure ratio of 1.12 to 1.2 only. Therefore, for higher pressure ratio, often several stages of axial flow compressor are necessary.

16.7.1 Construction and Principle of Operation

Rotors are of two types viz. drum type and disc type (shown in Fig. 16.17). The disc type rotors being light weight are preferred for aircraft applications. The drum type rotors are used for static industrial applications. (Volume of casing converges from low pressure suction end to high pressure discharge end). It helps to maintain more or less a constant velocity throughout the length of compressor, though the density of fluid increases gradually. Constant axial velocity throughout the stage is an important design consideration of axial flow compressor. Contraction of the flow annulus is achieved by contraction of casing and diverging the rotor.

The materials used for various components of the compressor are listed below:

(i) Rotor blades. Fibrous composites, Aluminium, Titanium, Steel, Nickel alloy.

(ii) Rotor. Steel for shaft, disc. In aircrafts, first stage uses Titanium while the later stages use Nickel steel.

(iii) Stator blades. These are made of same materials as those for rotor blades.

(iv) Casing. It is either a casting of magnesium, aluminium, steel, iron or fabricated from Titanium or steel.

The blades of compressor are of aerofoil section. The axial flow compressors can be of impulse type or reaction type. The blade profile and variation of pressure and velocity in the stage are shown in the Fig. 16.17 (a). While in Fig. 16.17 (b), the difference in the shape of Axial flow turbines and Axial flow compressor is illustrated.

16.7.2 Velocity Diagrams of Axial Flow Compressors

As explained earlier, in the axial flow compressor, fluid is imparted with kinetic energy from the rotor. This kinetic energy of the fluid is converted into pressure energy by diffuser action. The process occurs over several stages.

The energy analysis is considered for one stage of the compressor. The flow is assumed to occur tangential to the mean blade speed of *u*.

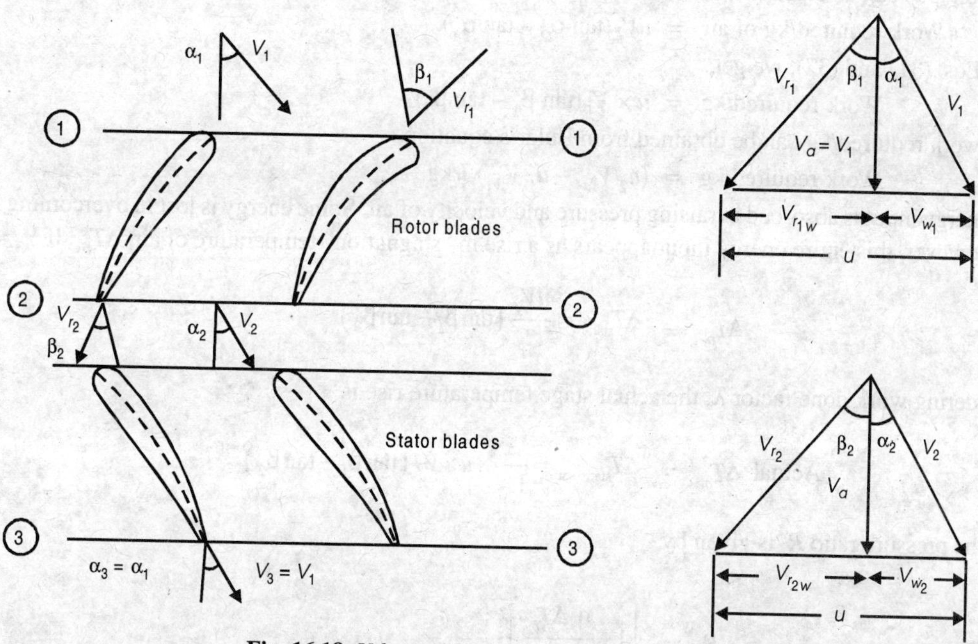

Fig. 16.18. *Velocity traingle for compressor stages*

Figure 16.18 shows velocity triangles for one stage of the compressor. Air approaches rotor blade with absolute velocity V_1 at an angle α_1 to the axial direction. Combining V_1 with blade velocity u_1 gives relative velocity V_{r_1} at inlet, inclined at angle β_1 with the direction of motion.

While air flows through the diverging passage between two successive rotor blades, its absolute velocity increases and air leaves the rotor with relative humidity V_{r_2} at an angle β_2. The reduction in the relative velocity from V_{r_1} at inlet to V_{r_2} at exit causes some pressure rise over the rotor. The absolute velocity at exit V_2 can be found by vector difference between V_{r_2} and u_2. The absolute velocity at exit V_2 makes an angle α_2 with the direction of motion. The air then enters the next stage via stator blades. Over the stator blades, the pressure of air further increases due to diffusion action. Air leaves the stator blades with absolute velocity V_3 (less than V_2) making an angle α_3 with the axial direction. The stator blades are so designed that $V_3 = V_1$ and $\alpha_3 = \alpha_1$ for a shockless entry over the next stage.

It may particularly be noted that unlike in centrifugal compressors, in this case all angles are measured with axial direction and not from tangential direction.

Referring to the velocity triangles shown in Fig. 16.18 we have,

$$\frac{u}{V_f} = \tan\alpha_1 + \tan\beta_1 \quad \text{...(31)}$$

and

$$\frac{u}{V_f} = \tan\beta_2 + \tan\alpha_2 \quad \text{...(32)}$$

where, $V_{f1} = V_{f2} = V_f$ is axial velocity of flow assumed to be constant throughout the stage. Now, for 1 kg of air flowing through the stage

$$\text{Work required/kg of air} = u\left(V_{w_2} - V_{w_1}\right) = C_p\left(T_{o_2} - T_{o_1}\right) \text{ J/kg} \quad \text{...(33)}$$

Equation (33) can also be written as

$$\text{Work required/kg of air} = uV_f(\tan\alpha_2 - \tan\alpha_1) \quad \text{...(34)}$$

From Eqs. (31) and (32), we get,

$$\text{Work required/kg} = u \times V_f(\tan\beta_1 - \tan\beta_2) \quad \text{...(35)}$$

Also, work required/kg can be obtained from Euler's equation,

$$\text{Work required/kg} = (u_2 V_{w2} - u_1 V_{w1}) \text{ J/kg} \quad \text{...(36)}$$

This energy input is absorbed in raising pressure and velocity of air. Some energy is lost in overcoming frictional losses. However, this entire energy input appears as a rise in "stagnation" temperature of air, ΔT_{os}. If $V_3 = V_1$, then

$$\Delta T_{o_s} = \Delta T_{\text{static}} = \frac{uV_f}{C_p}(\tan\beta_1 - \tan\beta_2) \quad \text{...(37)}$$

Considering work done factor λ, the actual stage temperature rise is

$$\text{Actual } \Delta T_{o_s} = \Delta T_{\text{static act}} = \frac{\lambda}{C_p} \times u \times V_f(\tan\beta_1 - \tan\beta_2)$$

And the pressure ratio R_s is given by,

$$R_s = \left[1 + \frac{\eta_s \Delta T_{o_s}}{T_{o_i}}\right]^{\frac{\gamma}{\gamma-1}} \quad \text{...(38)}$$

where, T_{oi} = inlet stagnation temperature and η_s is isentropic efficiency of the stage

$$\eta_s = \frac{\text{Isentropic temperature rise}}{\text{Actual temperature rise}}$$

Using velocity triangles and cosine rule, Eq. (38) can be written as

$$\text{Work required/kg} = \frac{V_{r_1}^2 - V_{r_2}^2}{2} + \frac{V_2^2 - V_3^2}{2} \quad \text{...(39)}$$

where, $V_3 = V_1$

Equation (39) is represented graphically on h–S diagram in Fig. 16.19.

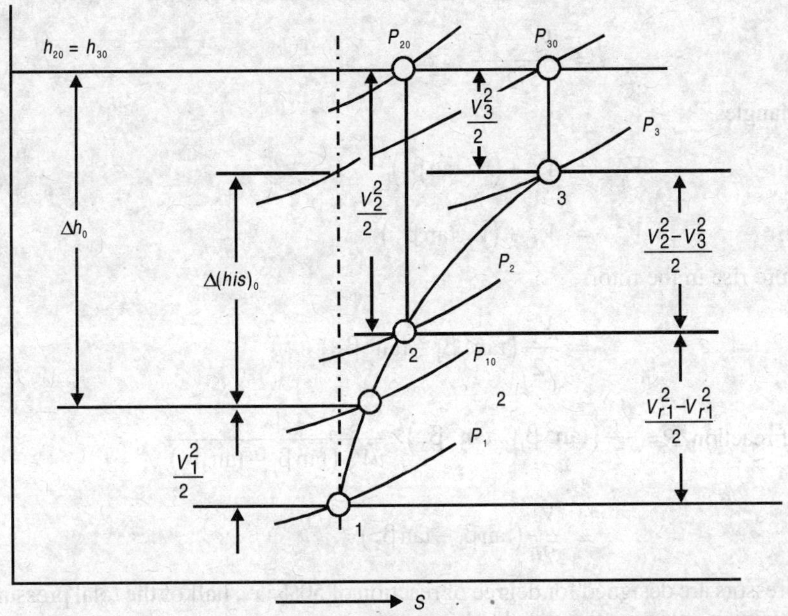

Fig. 16.19. *Process in axial flow compressor without GV on h-S diagram*

The first term on the right hand side of Eq. (39) represents the increase of kinetic energy in the rotor which is converted into pressure due to diffusion action in the rotor itself while the second term represents increase of kinetic energy over the rotor that has to be converted into pressure energy in stator blades.

Comparing this Eq. (39) of work required/kg with Eq. (39) applicable for centrifugal compressor, it is seen that the term centrifugal head $\left(\frac{u_2^2 - u_2^2}{2}\right)$ is missing for axial flow compressor. Therefore, pressure ratio per stage in axial flow compressor is much less than that of centrifugal compressor.

16.7.3 Degree of Reaction of Axial Flow Compressor

The degree of reaction is defined as the ratio of pressure rise in the rotor to the pressure rise in the stage.

$$\text{Degree of reaction, } R = \frac{\text{Pressure rise in the door}}{\text{Pressure rise in the stage}} \quad \text{...(40)}$$

Total pressure rise in the stage of a compressor is given by work input per stage

$$= u\,(V_{w_2} - V_{w_1}) \quad \text{...(41)}$$

From velocity triangles,

$$V_{w_2} = u - V_f \tan\beta_2$$

and

$$V_{w_1} = u - V_f \tan\beta_1$$

Hence, pressure rise in the stage

$$= uV_f\,(\tan\beta_1 - \tan\beta_2) \quad \text{...(42)}$$

Further, pressure rise over the rotor is due to change in relative kinetic energy.

$$= \frac{V_{r_1}^2 - V_{r_2}^2}{2} \quad \text{...(43)}$$

From velocity triangles,

$$V_{r_1}^2 = V_f^2 + (V_f \tan\beta_1)^2$$

and

$$V_{r_2}^2 = V_f^2 + (V_f \tan\beta_2)^2$$

Therefore, pressure rise in the rotor

$$= \frac{V_f^2}{2}(\tan^2\beta_1 - \tan^2\beta_2) \quad \text{...(44)}$$

Hence, degree of reaction, $R = \frac{V_f^2}{2}(\tan^2\beta_1 - \tan^2\beta_2) \times \frac{1}{uV_f\,(\tan\beta_1 - \tan\beta_2)}$

$$= \frac{V_f}{2u}(\tan\beta_1 - \tan\beta_2) \quad \text{...(45)}$$

Axial flow compressors are designed for degree of reaction of 50% i.e., half of the total pressure rise occurs over the rotor and the remaining half over the state blades.

$$\therefore \quad R = 0.5 = \frac{V_f}{2u}(\tan\beta_1 - \tan\beta_2) \quad \text{... (46)}$$

$$\therefore \quad \frac{u}{V_f} = (\tan\beta_1 - \tan\beta_2) \quad \text{...(47)}$$

Also, from velocity traingles,

$$\frac{u}{V_f} = (\tan\alpha_1 - \tan\beta_1) \quad \text{...(48)}$$

and

$$\frac{u}{V_f} = (\tan\alpha_2 - \tan\beta_2) \quad \text{...(49)}$$

From Eqs (47), (48) and (49), it follows that

$$\alpha_1 = \beta_2 \text{ and } \alpha_2 = \beta_1$$

Also constant velocity of flow through the stage,

$$V_f = V_1 \cos\alpha_1 = V_3 \cos\alpha_3$$

But,

$$V_3 = V_1 \text{ (assumed)}$$

$$\therefore \quad \alpha_1 = \alpha_3$$

As $\alpha_1 = \beta_2 = \alpha_3$ and $\alpha_2 = \beta_1$, rotor blade and stator blade are symmetrical. In the above deductions, work done factor is assumed to be unity.

For work done factor lower than unity, the degree of reaction differs slightly.

16.7.4 Cascade or Diffuser Efficiency

It is defined as the ratio of actual pressure rise to the isentropic pressure rise. Thus,

$$\eta_c = \frac{P_2 - P_1}{P_{2_{is}} - P_1} \quad \ldots(50)$$

Friction in rotor and stator blade passages, shock losses, etc. reduce the efficiency of the cascade (stage).

16.7.5 Polytropic Efficiency

Consider the compression of air in a multistage axial flow rotary compressor shown on T–S diagram in Fig. 16.20.

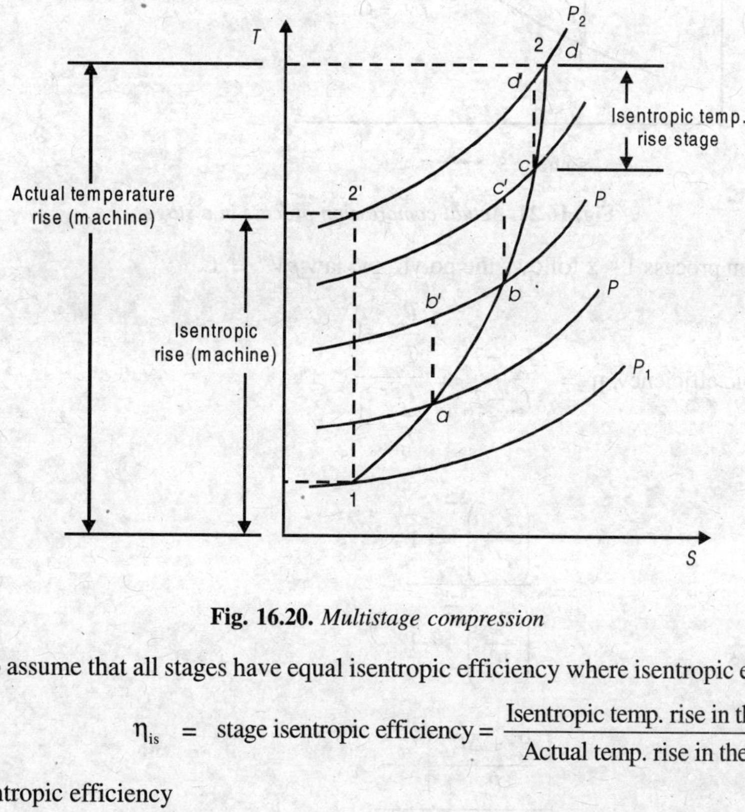

Fig. 16.20. *Multistage compression*

It is customary to assume that all stages have equal isentropic efficiency where isentropic efficiency is given by

$$\eta_{is} = \text{stage isentropic efficiency} = \frac{\text{Isentropic temp. rise in the stage}}{\text{Actual temp. rise in the stage}}$$

Now, overall isentropic efficiency

$$\text{Overall} \quad \eta_{is} = \frac{T_{o_{2'}} - T_{o_1}}{T_{o_2} - T_{o_1}}$$

$$\text{and,} \quad \text{Polytropic efficiency} = \frac{\Sigma(\text{Isentropic temp. rise in the stage})}{T_{o_2} - T_{o_1}}$$

$$= \frac{(T_{o_{a'}} - T_{o_1}) + (T_{o_{b'}} - T_{o_a}) + (T_{o_{c'}} - T_{o_c}) + (T_{o_{d'}} - T_{o_d})}{T_{o_2} - T_{o_1}}$$

16.7.6 Polytropic Efficiency η Terms of *n*

Consider an actual compression process in a stage of an axial flow compressor, as shown in Fig. 16.21.

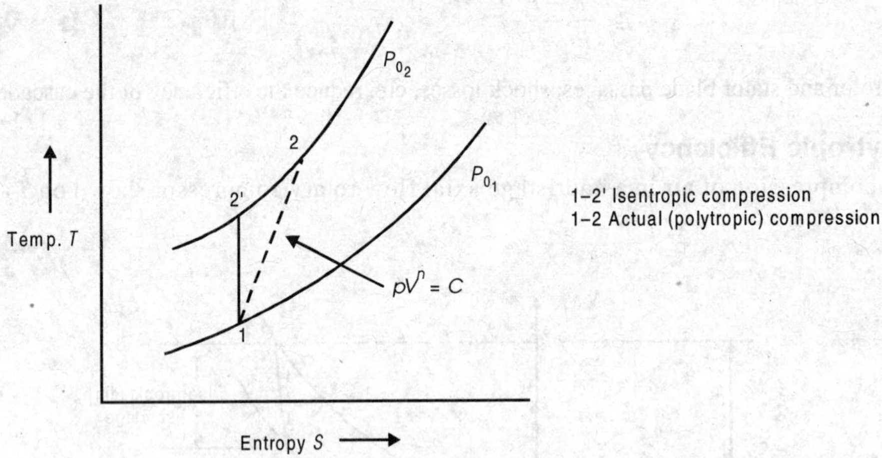

Fig. 16.21. *Actual compression process in a stage*

Actual compression process 1 – 2 follows the polytropic law $pV^n \cong C$.

$$\therefore \quad \text{Polytropic efficiency, } \eta = \frac{T_{o_{2'}} - T_{o_1}}{T_{o_2} - T_{o_1}} = \frac{\dfrac{T_{o_{2'}}}{T_{o_1}} - 1}{\dfrac{T_{o_2}}{T_{o_1}} - 1}$$

$$= \frac{\left(\dfrac{P_{o_2}}{P_{o_1}}\right)^{\frac{\gamma-1}{\gamma}} - 1}{\left(\dfrac{P_{o_2}}{P_{o_1}}\right)^{\frac{n-1}{n}} - 1}$$

$$= \frac{\left(\dfrac{P_{o_1} + \Delta P_o}{P_{o_1}}\right)^{\frac{\gamma-1}{\gamma}} - 1}{\left(\dfrac{P_{o_1} + \Delta P_o}{P_{o_1}}\right)^{\frac{n-1}{n}} - 1}$$

$$= \frac{\left(1 + \dfrac{\Delta P_o}{P_{o_1}}\right)^{\frac{\gamma-1}{\gamma}} - 1}{\left(1 + \dfrac{\Delta P_o}{P_{o_1}}\right)^{\frac{n-1}{n}} - 1}$$

Expanding the above equation binomially and ignoring higher order terms, we may write,

$$\eta_{\text{poly}} = \frac{\left(\frac{\gamma-1}{\gamma}\right)\left(\frac{\Delta P_o}{P_{o_1}}\right)}{\left(\frac{\gamma-1}{\gamma}\right)\left(\frac{\Delta P_o}{P_{o_1}}\right)} = \left(\frac{\gamma-1}{\gamma}\right)\left(\frac{n-1}{n}\right) \quad \text{...(51)}$$

The concept of isentropic efficiency cannot be used to compare the performance of two compressors having different pressure ratio or different number of stages. For such a comparison, polytropic efficiency is used.

Further, the work input required to a multistage compressor is given by

$$\text{Work input required/kg of air} = C_p\left(T_{o_2} - T_{o_1}\right) = C_p \times \left(T_{o_2} - T_{o_1}\right) \times \eta_{o_{\text{isent}}}$$

$$= C_p \times \eta_{o_{\text{isent}}} \times T_{o_1}\left[\left(\frac{P_{o_2}}{P_{o_1}}\right)^{\frac{\gamma-1}{\gamma}} - 1\right] \quad \text{...(52)}$$

16.7.7 Work Done Factor

It is rather difficult to draw the actual velocity triangles at the root, mean and tip of the blades as accurate estimation of contraction of main stream cannot be done.

The axial velocity at the mean section is somewhat greater than the average axial velocity.

The actual temperature rise is somewhat lower than that estimated from the velocity triangles.

An empirical relation suggested by Hotwell, for stagnation enthalpy rise across a stage is

$$C_p \Delta T_o = \lambda u\left(V_{f_2} \tan\alpha_2 - V_{f_1} \tan\alpha_1\right)$$

where λ is work done factor which is approximately 0.85. Hotwell further suggested that the work done factor depends upon the number of stages. Greater the number of stages, the work done factor reduces.

16.7.8 Losses in Axial Flow Compressor Stage

During the flow of fluid through a stage of a compressor, various losses occur. The pressure loss is due to three factors, namely

(a) profile losses on the blade surface; (b) skin friction on the annulus walls and (c) secondary flow losses.

Figure 16.22 illustrates various losses influencing the stage efficiency.

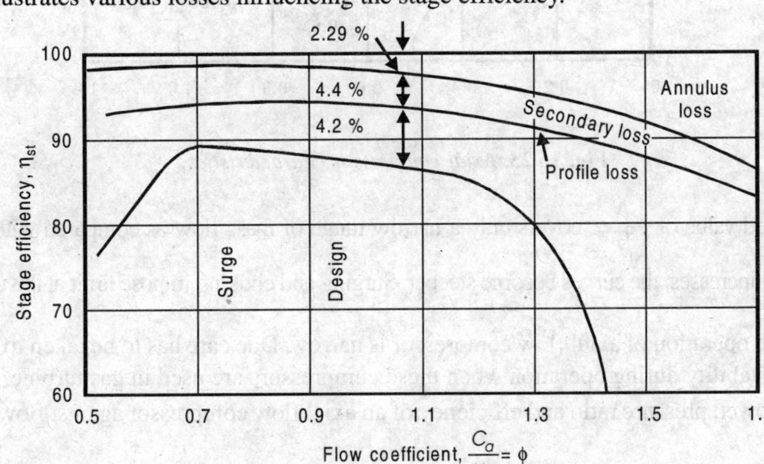

Fig. 16.22. *Losses in compressor stage*

(a) Profile losses. This is a pressure loss of two dimensional cascade arising from the skin friction on the blade surface and due to mixing of fluid particles after the blade.

(b) Skin friction losses on the annulus walls. The total pressure loss arise from the skin friction on the annulus walls and some secondary loss.

(c) Secondary flow losses. These losses occur due to combined effects of curvature and boundary layer.

16.7.9 Performance of Axial Flow Compressors

Figure 20.23 shows the performance characteristics of an axial flow compressor using dimensionless parameters namely stagnation pressure ratio $\frac{P_{o_2}}{P_{o_1}}$, dimensionless mass flow $\frac{m\sqrt{T_{o_1}}}{P_{o_1}}$ for dimensionless constant speeds $\frac{N}{\sqrt{T_{o_1}}}$ relative to design value.

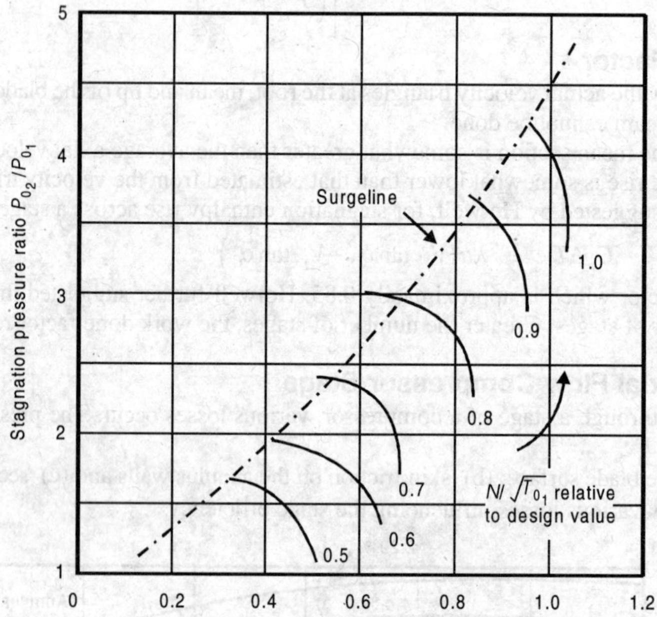

Fig. 16.23. *Axial compressor characteristics*

It is seen that a fixed value of $\frac{N}{\sqrt{T_{o_1}}}$ covers only a narrow range of mass flow as compared with that in centrifugal compressors. As speed increases, the curves become steeper. Surging and choking impose limit at the upper end of $\frac{N}{\sqrt{T_{o_1}}}$.

The range of stable operation of axial flow compressor is narrow. Due care has to be taken to match the components for avoiding instability during operation when these compressors are used in gas turbine.

The variation of power, pressure ratio and efficiency of an axial flow compressor against flow rate is depicted in the Fig. 16.24.

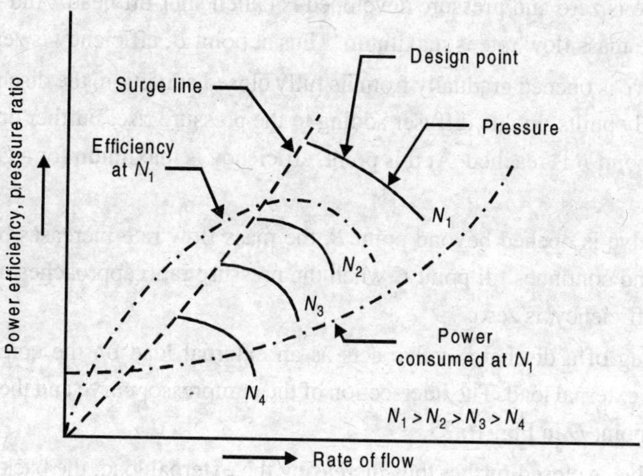

Fig. 16.24. *Performance of axial flow compressor*

Figure 16.24 shows variation of pressure ratio versus volume flow rate for axial flow compressor.

16.8 SURGING

In the operation of centrifugal and axial flow compressor, there is an instability known as surging. It is caused due to unsteady, periodic and reversal of flow through the compressor when it is operated at less mass flow rate than that corresponding to maximum pressure.

Surging may lead to mechanical failure. The rotor is subjected to alternating stresses during this irregular working and may result in damage to the bearings, rotor blades and seals. In the extreme case, the rotor shaft may bend.

Consider the pressure ratio (head) versus mass flow rate curve of a centrifugal or axial flow compressor as shown in Fig 16.25. The curve consists of part *AB* having a positive slope and part *BC* having a negative slope. Point *A* represents fully closed discharge valve while *B* represents full open discharge valve.

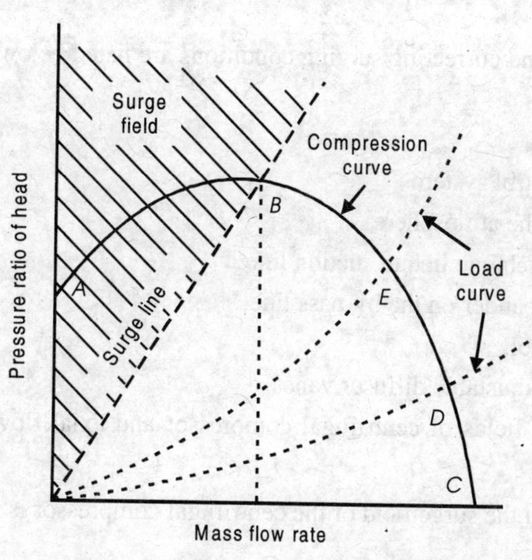

Fig. 16.25. *Surging*

At point A, the flow rate is zero and pressure developed is called shut-off head. And at point B, pressure head developed is zero when the mass flow rate is maximum. Thus at point B, efficiency is zero.

When the discharge valve is opened gradually from its fully closed condition, the discharge of fluid commences and static pressure gradually builts-up due diffuser adding to the pressure rise. Further opening of discharge valve increases the pressure till point B is reached. At this point, efficiency is maximum for a given speed, inlet pressure and temperature.

Now, when delivery valve is opened beyond point B, the mass flow rate increases but pressure rise and efficiency decreases. This trend continues till point C when the pressure ratio approaches unity while the mass flow rate is maximum but the efficiency, is zero.

The opening and closing of a discharge valve acts as an external load on the compressor. The compressor operates in response to the external load. The intersection of the compressor curve and the load curve represents the operating point, say as at point D in Fig. 16.25.

When the discharge valve is closed further thus increasing the external load, the back pressure in the discharge line increases. As a result, new operating point shifts to E. The new operating point is possible and stable as the compressor develops greater pressure head to offset the increased back pressure in the discharge line.

With further closing of the discharge valve, external load increases, the operating point shifts in the region A–B. During this operation, mass flow rate is less than the estimated value, say corresponding to point B. The compressor develops less head than that existing in the discharge line. As such, no discharge of fluid is possible. This leads to momentary flow reversal. A short while after, the fluid from the discharge line leaves and the back pressure would reduce. The delivery of fluid from the compressor would restart and the cycle would repeat with instability of operation again.

Therefore, when the flow rate from the compressor is less than the estimated value, a surge or pulsation commences. Air surges back and forth through the whole compressor rather than giving a unidirection steady stream of flow. The unstable operation of the compressor prevails in the region of positive slope of the curve shown in Fig. 16.25.

Various methods of avoiding and correcting surging conditions are listed below. It can be controlled,

(a) by speed control.

(b) by throttling inlet.

(c) by provision of flow control system.

(c–1) by venting discharge to the atmosphere.

(c–2) by passing fluid from discharge line to suction line.

(c–3) by providing integral expander on the by-pass line.

(d) by cut-off capacity control.

(e) by inlet guide vanes or adjustable diffuser vanes.

Figure 16.26 shows the surge fields of centrifugal compressor and axial flow compressor on pressure ratio $\frac{P_{o_2}}{P_{o_1}} - \frac{m\sqrt{T_{o_1}}}{P_{o_1}}$ plane. It is seen that the surge field of the centrifugal compressor is narrow as compressed with that of the axial flow compressor.

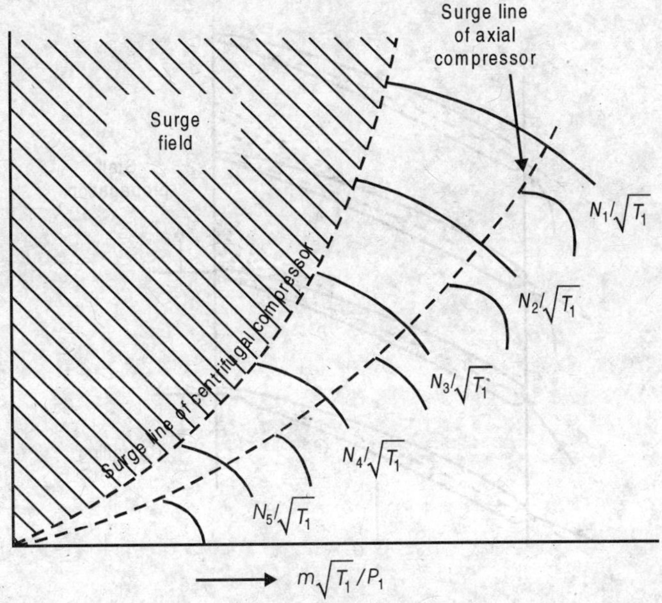

Fig. 20.26. *Surge line*

16.9 CHOKING

Referring to Fig. 16.25, when the mass flow rate is increased beyond *B*, the pressure ratio decreases and efficiency falls, as the air angle differs significantly from the vane angle causing the break-away of the air-stream.

This continues till point *C* where the pressure ratio becomes unity and efficiency zero. Entire power input is lost in overcoming internal friction. The point *D* on the curve *BC*, represents the maximum mass flow rate, known as 'choking mass flow'. The phenomena of choking limits the maximum mass flow rate.

As pressure ratio falls to unity, the theoretical mass flow rate is maximum. This occurs when the Mach number corresponding to inlet relative velocity becomes sonic.

16.10 STALLING

Stalling of a stage of axial flow compressor is defined as the *aerodynamic stall or breakaway of the flow stream from the suction side of the blade aerofoil*. It may be due to less mass flow rate than the designed value or due to non-uniformity in the blade profile.

The phenomena of stall preceds surging. A multistage compressor may operate in the stable unsurged region with some of its stages stalling. Thus, stalling is a local phenomena while surging is a complete system phenomena. The phenomena of stalling has been extensively researched by Smith and Fletcher.

A rotating stall is illustrated in Fig. 16.27.

The non-uniformity of the flow or geometry of blade profile causes blade *B* to stall. The air now flows on to the blade *A* at an increased incidence due to blockage of channel *AB* while blade *C* receives air at reduced incidence. As a result blade *A* stalls while blade *C* may unstall. In this way, stall 'cell' shifts along the stage in the direction of lift of the blades. Thus, stall can be stated as a reduction in the lift force at higher angles of incidence.

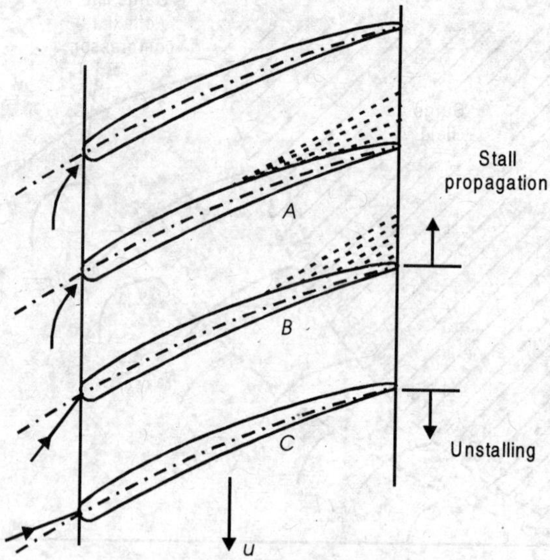

Fig. 16.27. *Mechanism of stall propagations*

A rotating stall may rotate in the opposite direction to that of the rotor at about half the rotational speed. It may lead to aerodynamically induced vibrations resulting in fatigue failure of the attached components. Axial flow compressors are more prone to stalling.

16.11 PERFORMANCE CHARACTERISTICS OF CENTRIFUGAL FLOW COMPRESSOR

The performance of centrifugal compressors is specified by illustrating the variation of delivery pressure and temperature with mass flow rate for various speeds. However, these characteristics are further influenced by other variables like inlet pressure and temperature.

To understand the interdependence of these variables, dimensionless parameters are often useful, such as

$$\frac{P_{o_2}}{P_{o_1}}, \frac{T_{o_2}}{T_{o_1}}, \frac{m\sqrt{T_{o_1}}}{P_{o_1}}, \frac{ND}{\sqrt{RT_{o_1}}}$$

Even isentropic efficiency is also expressed in a similar way as

$$\eta_{isent} = \frac{T_{o_2'} - T_{o_1}}{T_{o_2} - T_{o_1}} = \frac{\left(\frac{P_{o_2}}{P_{o_1}}\right)^{\frac{\gamma-1}{\gamma}} - 1}{\left(\frac{T_{o_2}}{T_{o_1}} - 1\right)}$$

Figure 16.28 shows the plot of such parameters.

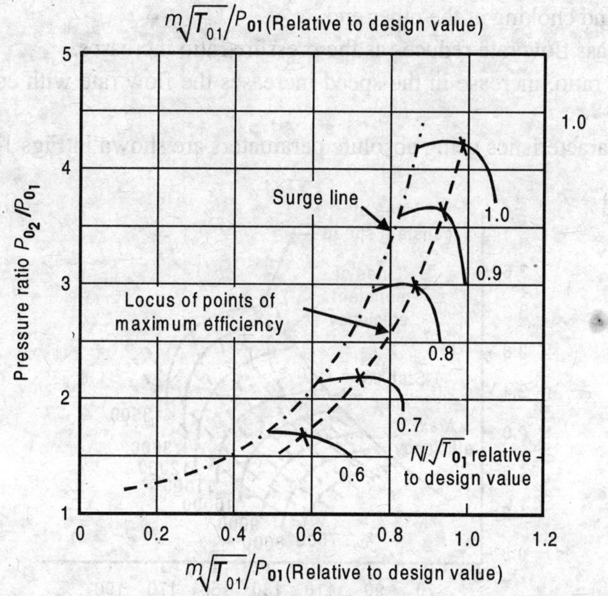

Fig. 16.28. *Centrifugal compressor charcateristics*

Points of low mass flow at the extremities of constant speed curves can be joined to obtain surge line. While the curve obtained by joining extremities on the right hand side of constant speed curves represents choking limit occurring at low pressure ratio. The compressor is operated only between these extreme limits.

Figure 16.29 shows the variation of isentropic efficiency with $\dfrac{m\sqrt{T_{o_1}}}{P_{o_1}}$ for various values of $\dfrac{N}{\sqrt{T_{o_1}}}$ relative to the design value. The maximum efficiency is almost same at all speeds.

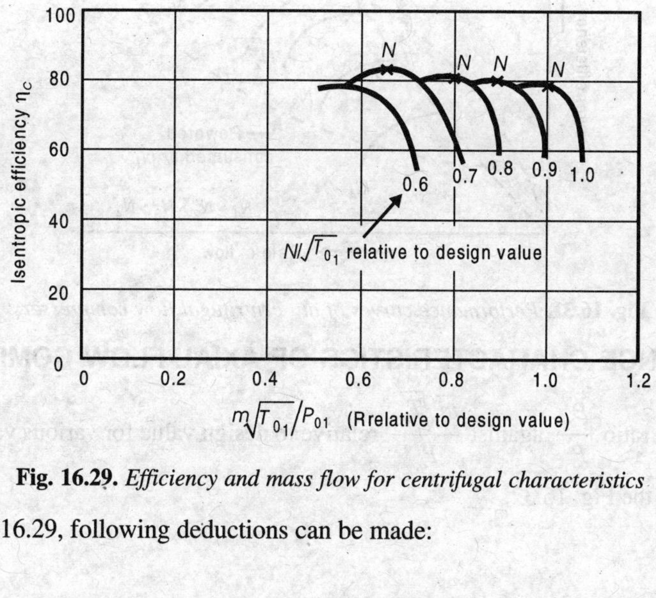

Fig. 16.29. *Efficiency and mass flow for centrifugal characteristics*

From Figs 16.28 and 16.29, following deductions can be made:

1. The curves are fairly flat at lower speeds and are limited by surging. At high speed, the range is limited by surging at one end and choking at the other end.
2. At a given speed, mass flow rate reduces as the pressure ratio increases.
3. At a given pressure ratio, increase in the speed increases the flow rate with considerable reduction in the efficiency.

Similar performance characteristics using absolute parameters are shown in Figs 16.30 and 16.31.

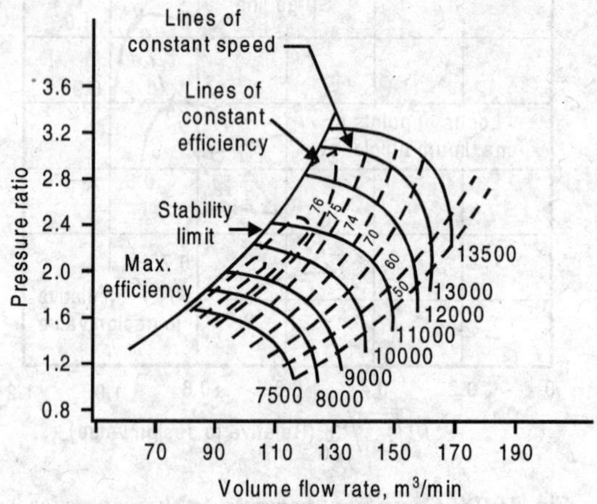

Fig. 16.30. *Performance charcateristics of centrifugal flow compressors*

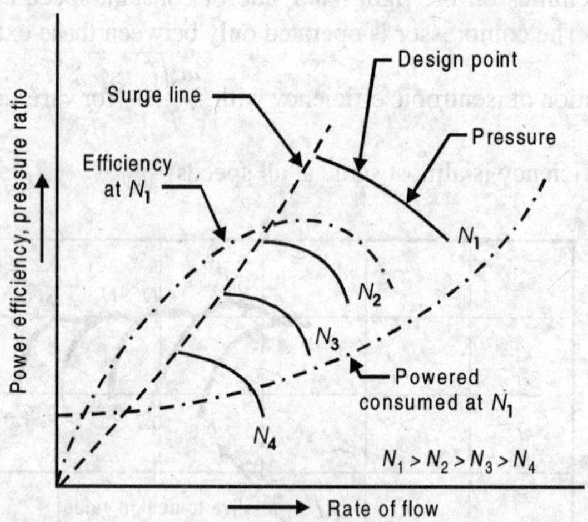

Fig. 16.31. *Performance curves of an centrifugal flow compressors*

16.12 PERFORMANCE CHARACTERISTICS OF AXIAL FLOW COMPRESSORS

The variation of pressure ratio $\frac{P_{o_2}}{P_{o_1}}$ aganist $\frac{m\sqrt{T_{o_1}}}{P_{o_1}}$ relative to design value for various values of $\frac{N}{\sqrt{T_{o_1}}}$ relative to design value is shown in the Fig. 16.32.

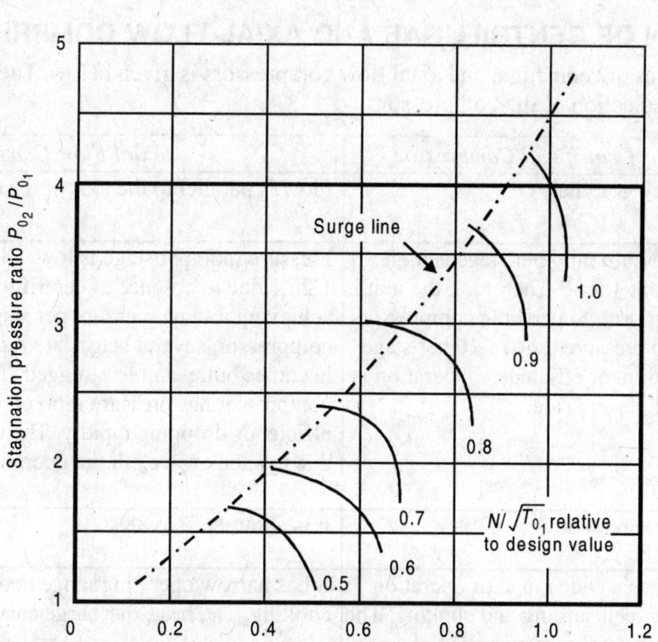

Fig. 16.32. *Axial compressor characteristics*

It is seen that, for a given value of $\dfrac{N}{\sqrt{T_{0_1}}}$, these curves cover a much narrow range of mass flow rate as compared with those for centrifugal compressors. Also at higher rotational speeds, the curves become steep, nearly vertical. Therefore, the range of stable operation of axial flow compressor is much narrower. Hence, great care is necessary to match the components of a gas turbine plant for avoiding instability of operation. The phenomena of surging and stalling can hardly be distinguished as occurrence of one may lead to occurrence of the other. Stalling of these compressors results in severe vibrations of blade.

The performance characteristics using absolute parameters are also shown in Fig. 16.33.

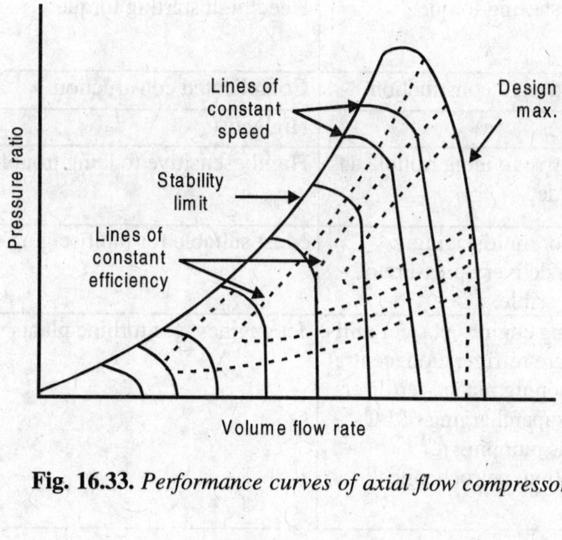

Fig. 16.33. *Performance curves of axial flow compressors*

16.13 COMPARISON OF CENTRIFUGAL AND AXIAL FLOW COMPRESSORS

Comparison of main features of centrifugal and axial flow compressors is given below. The comparison enables to arrive at some criteria for selection of the compressor.

The	*Item*	*Centrifugal Compressor*	*Axial Flow Compressor*
1.	Direction of flow	Flow is radial	Flow is parallel to the axis of the compressor.
2.	Pressure Ratio	Pressure ratio per stage is high about 4.5 : 1. Therefore, the unit is compact. Supersonic compressors use pressure ratio of 10 per stage with poor efficiency. Operation is simple and safe.	Pressure ratio per stage is low, about 1.2 : 1 due to absence of centrifugal action. Hence, for achieving a same pressure per stage of a centrifugal compressor, several stages are required. Thus, the unit becomes bulky and less rugged. The supersonic compressor has pressure ratio of 10 per stage with efficiency dropping rapidly. The compression is achieved due to shock wave but the operation is risky.
3.	Isentropic efficiency	It is about 80% to 82%.	It is about 86% to 88%.
4.	Operating range	It has a wide range of operation between surging and choking. The head-discharge curve is flat.	It has narrow operating range between surging and choking. The head-discharge curve is steep.
5.	Part load performance	It is better.	It is poor.
6.	Frontal area	It has a large frontal area.	It has smaller frontal area for a given mass floor rate and pressure than that of centrifugal compressor. Therefore, it is more suitable for jet engines.
7.	Effect of deposits on the blade surfaces	Accumulated deposits on the blade surfaces due to the flow of contaminated fluid has no adverse effect on the performance.	Performance is adversely affected due to accumulated deposits on the blade surfaces when contaminated fluid flows.
8.	Starting torque required	Need low starting torque.	Need high starting torque.
9.	Construction	Simple and rigid construction.	Complicated construction.
10.	Cost	Low cost.	High cost.
11.	High altitude operation	Less sensitive to icing trouble at high altitude.	Highly sensitive to icing trouble at high altitude.
12.	Multistaging	Difficult for multistaging. Maximum delivery pressure of 400 bar possible.	Most suitable for multistaging.
13.	Field of application	For blowing engines in steel mills, low pressure refrigeration, central air conditioning plants, fertiliser industry, supercharging of I.C. engine, gas pumping for transportation, petrochemical industries.	Jet engines, gas turbine plants, steel mills, etc.

16.14 COMPARISON OF CENTRIFUGAL AND RECIPROCATING COMPRESSORS

The distinguishing features of these compressors are tabulated below.

Item	*Reciprocating Compressor*	*Centrifugal Compressor*
1. Mechanical balance	Poor mechanical balance and greater vibrations due to presence of reciprocating masses.	Absence of reciprocating masses provides better mechanical balance and vibration free operation.
2. Pressure ratio per stage	High upto 5–8.	Comparatively low upto 4.5.
3. Multistaging capability	Suitable for high delivery pressures with multistaging, delivery pressure upto 5000 bar possible.	Suitable for medium delivery pressure. Delivery pressure of 400 bar possible with multistaging.
4. Volume flow rate	Capable of delivering small volume flow rates.	Capable of delivering large volume of flow rates.
5. Operational flexibility	Greater flexibility in pressure range and capacity.	No flexibility in pressure range and capacity.
6. Maintenance cost	High.	Low.
7. Effect of pressure ratio	For pressure ratios above 2, efficiency is high.	High efficiency for pressure ratio below 2.
8. Drive	Suitable for slow speed drive only.	Suitable for direct coupling to high speed drives like turbines.
9. Attention required	Greater attention required.	Less attention is adequate.
10. Contamination with lubricating oil	Greater chance of mixing working fluid with lubricating oil.	No chance of contamination of working fluid with lubricating oil.
11. Suitability	Suitable for low and medium and high delivery pressures with low/medium volume flow rates.	Suitable for low and medium pressures with large volume flow rates.

16.15 APPLICATIONS OF COMPRESSED AIR/GASES

The various applications of compressed air/gases are

1. For industrial furnaces for supporting combustion.
2. For infloating the vehicle tubes.
3. For actuating pneumatic tools such as hammers, chisels, drills, screw drivers, etc.
4. For industrial and household cleaning of floors, equipments, machine tools, components etc.
5. For pneumatic conveyor belts.
6. For pneumatic brakes of automobiles.
7. For air conditioning plants.
8. For ventilation of mines.
9. For production of aerated soft drinks.
10. For the production of synthetic ammonia, hydrogenation of brown coal, polymerisation of ethylene.
11. For air lift pumps, glass blowing.
12. For supercharging of IC engines.
13. For long distance transportation of industrial gases.
14. In gas turbine plants and jet engines.
15. For cabin pressurisation of air crafts.

■■■

SOLVED EXAMPLES

Example 16.1. *A centrifugal compressor has a total pressure head ratio 4:1 and discharges 18.2 kg of air per second. It runs at 15000 rpm. This inlet stagnation temperature is 15° C. The slip factor is 0.9 and power input factor is 1.4. It has isentropic efficiency of 80%.*

Determine the outer diameter of the impeller.

Solution

Let suffix *o* denote the stagnation temperature.

$$\frac{T_{O_2}}{T_{O_1}} = \left(\frac{P_{O_2}}{P_{O_1}}\right)^{\frac{\gamma-1}{\gamma}} = 4^{\frac{0.1}{1.4}}$$

$$\therefore \quad T_{O_2} = 288\times 4^{\frac{0.1}{1.4}} = 427.97 \text{ K}$$

$$\therefore \text{ Now, isentropic efficiency } \eta_{is} = \frac{T_{O_2}-T_{O_1}}{T_{O_{2'}}-T_{O_1}}$$

$$\therefore \text{ Actual temperature rise } \left(T_{O_2}-T_{O_1}\right) = \frac{427.97-288}{0.8} = 174.96 \text{ K}$$

$$\text{Work required/kg of air} = \frac{\sigma\mu u_2^2}{1000} = C_p\left(T_{O_2}-T_{O_1}\right)$$

$$\therefore \text{ Blade speed at outlet } u_2 = \sqrt{\frac{1.005\times1000\times174.96}{1.04\times0.9}} = 433.42 \text{ m/s}$$

But

$$u_2 = \frac{\pi D_2 N}{60}$$

$$D_2 = \frac{60\times433.42}{\pi\times15000} = 0.5518 = 55.18 \text{ cm}$$

Example 16.2. *A rotary vane compressor has pressure ratio of 1.4 and the intake pressure of 1.0 bar. The compressor discharges 00.6 m³/rev. of FDA.*

Calculate the work input required per revolution to drive the compressor when

(1) there is no internal compression (Roots blower).

(2) there is 50% pressure rise due to internal isentropic compression before back flow occurs (vane compressor).

Also find the efficiency of compressor.

Data: $V_1 = 0.6$ m³/rev $\quad p_1 = 1.0$ bar $\quad p_2 = 1.4$ bar

Solution (i) When there is no internal compression (Roots blower):

The p–V diagram is shown in Fig. Ex. 16.1 (a) and (b).

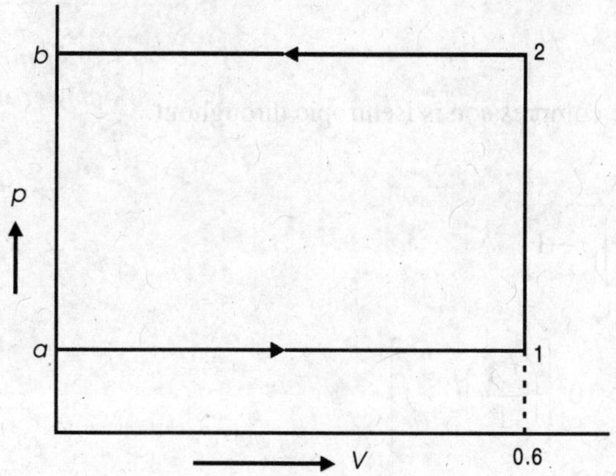

Fig. Ex. 16.1 (a)

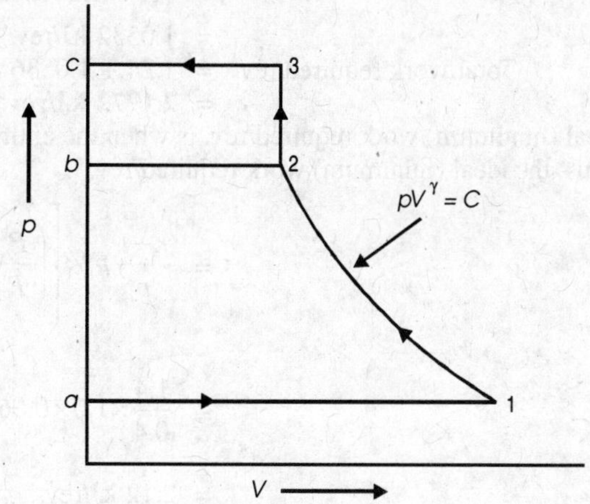

Fig. Ex. 16.1 (b)

$$\text{Work req./rev.} = (P_2 - P_1) \times V_1$$
$$= (1.4 - 1.0) \times 10^2 \times 0.06$$
$$= \mathbf{2.4\ kJ/rev.}$$

(ii) When half the pressure rise occurs due to internal compression (vane compressor). The *p–V* diagram is shown below.

Data: $p_1 = 1$ bar
$p_3 = ?$

$$\text{The intermediate pressure } P_2 = P_1 + \frac{(P_2 - P_1)}{2}$$
$$= 1 + \frac{1.4 - 1.0}{2}$$
$$= 1.2 \text{ bar}$$

Now, work required/rev. for isentropic compression 1 – 2

$$\text{Area } a-1-2-b = \frac{\gamma}{\gamma - 1} p_1 V_1 \left[\left(\frac{P_2}{P_1}\right)^{\frac{\gamma-1}{\gamma}} - 1\right]$$
$$= \frac{1.4}{0.4} \times 1.0 \times 10^2 \times 0.06 \left[\left(\frac{1.2}{1.0}\right)^{\frac{r-1}{r}} - 1\right]$$
$$= 1.24 \text{ kJ/rev.}$$

and
$$V_2 = V_1 \left(\frac{P_2}{P_1}\right)^{\frac{\gamma-1}{\gamma}} = \left(\frac{1.2}{1}\right)^{\frac{0.4}{1.4}} \times 0.06 = 0.05266 \text{ m}^3$$

Work required/rev. for irreversible pressure rise

$$= \text{area } 2\text{–}3\text{–}c\text{–}b\text{–}2$$
$$= (P_3 - P_2) \times 10^2 \times V_2$$

$$= (1.4-1.2)10^2 \times 0.05266$$

$$= 1.0532 \text{ kJ/rev.}$$

∴ Total work required/rev. $= 1.24 + 1.0536$

$$= \mathbf{2.1772 \ kJ/rev.}$$

Ideal (minimum) work required/rev. is when the entire compression is isentropic throughout.
Thus, the ideal (minimum) work required/rev.

$$= \frac{r-1}{r} p_1 V_1 \left[\left(\frac{P_2}{P_1} \right)^{\frac{\gamma-1}{\gamma}} - 1 \right]$$

$$= \frac{1.4}{0.4} \times 1.0 \times 0.06 \times 10^2 \left[\left(\frac{1.4}{1} \right)^{\frac{0.4}{1}} - 1 \right]$$

$$= 2.12 \text{ kJ/rev.}$$

Therefore, efficiency of Roots blower $= \dfrac{2.12}{2.4} \times 100$

$$= \mathbf{88.3\%}$$

Similarly, efficiency of vane compressor $= \dfrac{2.12}{2.1772} \times 100$

$$= \mathbf{97.3\%}$$

Example 16.3. *A centrifugal compressor delivers 10 m/s free air while running at 9000 r.p.m. The air is compressed from 1 bar and 20°C to 4 bar with an isentropic efficiency of 0.82.*

Blades are radial at outlet of the impeller and the flow velocity is 62 m/s which is constant throughout. The ratio of outer to inner r 'ii of the impeller is 2 and the slip factor is 0.9. The blade area coefficient at inlet is 0.9.

Determine

(1) temperature of air discharged
(2) theoretical power required
(3) diameters of the impeller at inlet and outlet
(4) the breadth of impeller at inlet
(5) impeller blade angle at inlet
(6) diffuser blade angle at inlet

Data:

$N = 9000$ *r. p. m.*   *F. A. D.* $= 10 \ m^3/s$

$P_1 = 1.0$ *bar* $T_{o_2} = 293$ *K*

$P_2 = 4.0$ *bar* $\eta_{isent} = 0.82$

Solution

Slip factor, $\mu = 0.9$

$$V_{f1} = V_{f2} = 62 \text{ m/s} \qquad \frac{R_2}{R_1} = 2.0$$

$$k_1 = 0.9$$

$$T_{o_2} = T_{o_1} \left(\frac{P_{o_2}}{P_{o_1}} \right)^{\frac{\gamma-1}{\gamma}} = 293 \, (4)^{\frac{0.4}{1.4}}$$

$$= \mathbf{433.56 \ K}$$

$$\text{Actual temp. rise} = \left(T_{O_2} - T_{O_1}\right) = \frac{T_{O_2} - T_{O_1}}{\eta_{\text{isent}}}$$

$$= \frac{433.56 - 293}{0.82}$$

$$= 173.85$$

$$\therefore \quad T_{o_2} = 293 + 173.85 = 466.85 \text{ K}$$

$$\text{Mass flow rate } \dot{m} = \frac{p_1 V}{RT_{o_1}} = \frac{1 \times 10^2 \times 10 \times 60}{0.287 \times 293}$$

$$= 713.5 \text{ kg/min.}$$

$$\text{Now, work required/s} = \text{Theoretical power input required.}$$

$$= \dot{m}\, C_P \left(T_{o_2} - T_{o_1}\right)$$

$$= \frac{713.5}{60} \times 1.005 \times 173.85$$

$$= 2077.7 \text{ kW}$$

For radial blades, the work input required is given by

$$\text{Work input required/kg} = \frac{\sigma \mu u_2^2}{1000} = C_p\left(T'_{O_2} - T_{O_1}\right)$$

$$\therefore \quad u_2 = \sqrt{\frac{1000 \times 1.005 \times 173.85}{0.9}}$$

$$= 440.6 \text{ m/s} = \frac{\pi D_2 N}{60}$$

$$\therefore \quad D_2 = \frac{60 \times 440.6}{\pi \times 9000} = 0.9349 \text{ cm}$$

$$= \mathbf{93.49 \text{ cm}}$$

$$\text{and} \quad D_1 = \frac{D_2}{2} = \mathbf{46.745 \text{ cm}}$$

where k = blade area coefficient at inlet

$$b_1 = \frac{600}{60 \times \pi \times 0.46745 \times 62 \times 0.9}$$

$$= 0.122 \text{ m} = 12.2 \text{ cm}$$

Referring to inlet velocity triangle (Fig. 16.18)

$$\tan \beta_1 = \frac{V_{f_1}}{u_1} = \frac{62 \times 2}{440.0} \left[\because u_1 = \frac{1}{2} u_2\right]$$

$$= 0.2814$$

$$\therefore \quad \beta_1 = 15.7°$$

Also, $$\tan\alpha_2 = \frac{V_{f_2}}{\mu u_2} = \frac{62}{0.9\times 440.6}$$

$$= 0.1563$$

$$\alpha_2 = 8.9^\circ.$$

Example 16.4. *A centrifugal air compressor delivers air with a stagnation pressure ratio of 4 :1. Its speed is 15000 r.p.m. The stagnation temperature at inlet is 17°C and the power input factor is 1.04. It has overall diameter 0.5538 m and isentropic efficiency of 0.8.*

Determine the slip factor.

Data: $N = 15000$ *r.p.m.* $\eta_{isent} = 0.8$

$\psi = 1.04$ $T_{o_1} = 280$ K

$\frac{P_{o_2}}{P_{o_1}} = 4$ $D_2 = 0.5538$ *m*

Solution

$$\therefore \quad T_{o_2} = T_{o_1}\left(\frac{P_{o_2}}{P_{o_1}}\right)^{\frac{\gamma-1}{\gamma}} = 290\times 4^{\frac{0.4}{1.4}}$$

$$= 431 \text{ K}$$

Actual stagnation temperature after compression T'_{o_2} can be calculated as

$$T'_{o_2} - T_{o_1} = \frac{T_{o_2} - T_{o_1}}{\eta_{isent}} = \frac{431-290}{0.8}$$

$$= 176.25 \text{ K}$$

Now, Work required/kg $$\frac{\Psi\mu u_2^2}{1000} = C_p\left(T'_{o_2} - T_{o_1}\right)$$

But, $$u_2 = \frac{\pi D_2 N}{60} = \frac{\pi\times 0.5538\times 15000}{60}$$

$$= 435 \text{ m/s}$$

$$\text{Ship factor } \mu = \frac{1.005(176.25)\times 1000}{1.04\times 435^2}$$

$$= \mathbf{0.9}$$

Example 16.5. *An aeroengine uses centrifugal compressor for supercharging. The compressor delivers 3 kg/s of air. Air enters the compressor at 1 bar, 280 K and 90 m/s after compression in the impeller, air has pressure 1.5 bar temperature 335 K and 230 m/s.*

Determine:

(1) isentropic efficiency

(2) power required to drive the compressor

(3) overall efficiency.

Assume that all the kinetic energy of air gained in the impeller is entirely converted into pressure in the diffuser.

Solution

$$\dot{m} = 3 \text{ kg/s}$$

$p_1 = 1$ bar $\quad T_1 = 280$ K, $\quad V_1 = 90$ m/s

$p_2 = 1.5$ bar $\quad T_2 = 335$ K, $\quad V_2 = 230$ m/s

Referring to the *T–S* diagram,

Actual work required/kg of air in the impeller

$$= C_p(T_2 - T_1) + \left(\frac{V_2^2 - V_1^2}{2\times1000}\right)$$

$$= 1.005(335 - 280) + \frac{230^2 - 90^2}{2\times1000}$$

$$= 81.71 \text{ kJ/kg}$$

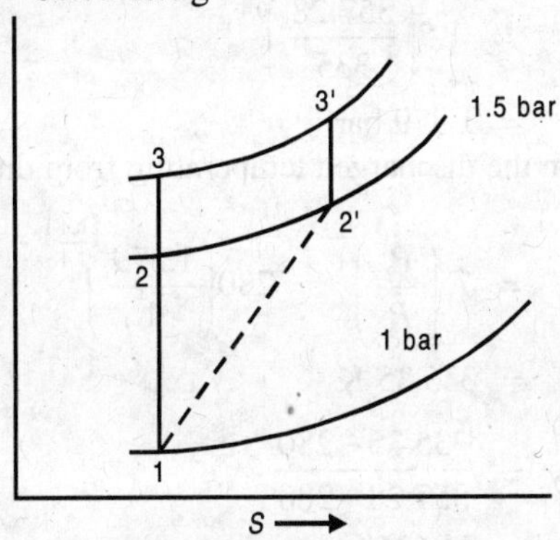

Fig. Ex. 16.5

$$\frac{T_2}{T_1} = \left(\frac{P_2}{P_1}\right)^{\frac{\gamma-1}{\gamma}} \qquad \therefore \quad T_2 = 280\times1.5^{\frac{0.4}{1.4}} = 314.4 \text{ K}$$

Isentropic work required/kg of air

$$= C_p(T_2 - T_1) + \frac{V_2^2 - V_1^2}{2\times1000}$$

$$= 1.005(314.4 - 280) + \frac{230^2 - 90^2}{2\times1000}$$

$$= 56.97 \text{ kJ/kg}$$

$$\therefore \quad \text{Isentropic efficiency} = \frac{\text{Isentropic work reqd./kg}}{\text{Actual work reqd./kg}}$$

$$= \frac{56.97}{87.71}$$

$$= 69.72\%$$

$$\text{Power required} = \text{Actual work required/kg} \times \text{m}$$

$$= 81.71 \times 3 = \mathbf{245.13 \text{ kW}}$$

As kinetic energy gained in the impeller is converted into pressure, therefore,

$$C_p\,(T_3'-T_2') = \frac{V_2^2-V_1^2}{2\times1000}$$

$$1.005\,(T_3'-335) = \frac{230^2-90^2}{2000}$$

$$\mathbf{T_3' = 357.278\ K}$$

Now, the pressure of air after leaving the diffuser

$$= P_3 = P_2\left(\frac{T_3'}{T_2'}\right)^{\frac{\gamma-1}{\gamma}}$$

$$= 1.5\left(\frac{357.28}{335}\right)^{\frac{1.4}{0.4}}$$

$$= 1.879 \text{ bar}$$

In case of isentropic compression the discharged temperature from diffuser is

$$T_2' = T_1\left(\frac{P_2}{P_1}\right)^{\frac{r-1}{r}} = 280\left(\frac{1.879}{1}\right)^{\frac{0.4}{1.4}}$$

$$= 335.35 \text{ K}$$

$$\text{Overall isentropic efficiency} = \frac{335.35-280}{357.28-280}$$

$$= \mathbf{71.62\%}$$

Example 16.6. *A single suction centrifugal compressor delivers 6 kg/s of air. The ambient air conditions are 1 bar and 20°C. It runs at 22000 r.p.m. with an isentropic efficiency 82%. Air is compressed from 1 bar to 4.2 bar total (stagnation) pressure. Air enters the eye of the impeller with a velocity of 150 m/s without any prewhirl. Assuming the ratio of whirl speed to tip speed as 0/9, calculate*

(a) rise in stagnation temperature during compression if the change in kinetic energy is negligible
(b) tip diameter of the impeller
(c) power required
(d) diameter of the eye of the impeller if the hub diameter is 10 cm.

Data: $T_1 = 293$ K, $V_1 = 150$ m/s

$N = 22000$ r. p.m., $\eta_{isent} = 82\%$

$\frac{V_{w_2}}{u_2} = 0.9$, $P_{o_2} = 4.2$ bar

$p_1 = 1.0$ bar, $\dot{m} = 6$ kg/s

$V_{w_1} = 0$

Solution

$$T_{o_1} = T_1 + \frac{V_1^2}{2C_P} = 293 + \frac{150^2}{2\times1.005\times1000}$$

$$= 304.19 \text{ K}$$

$$P_{o_1} = P_1\left(\frac{T_{o_1}}{T_1}\right)^{\frac{\gamma-1}{\gamma}}$$

and $$T_{o_2} = T_{o_1}\left(\frac{P_{o_2}}{P_{o_1}}\right)^{\frac{\gamma-1}{\gamma}} = T_{o_1}\left[\frac{P_{o_2}}{P_1\left(\frac{T_{o_1}}{T_1}\right)^{\frac{\gamma-1}{\gamma}}}\right]^{\frac{\gamma-1}{\gamma}}$$

$$= T_{o_1}\left(\frac{p_{o_2}}{p_{o_1}}\right)^{\frac{\gamma-1}{\gamma}}$$

$$= 293\left(\frac{4.2}{1}\right)^{\frac{0.4}{1.4}} = 441.68 \text{ K}$$

Referring to the Fig. Ex. 16.6

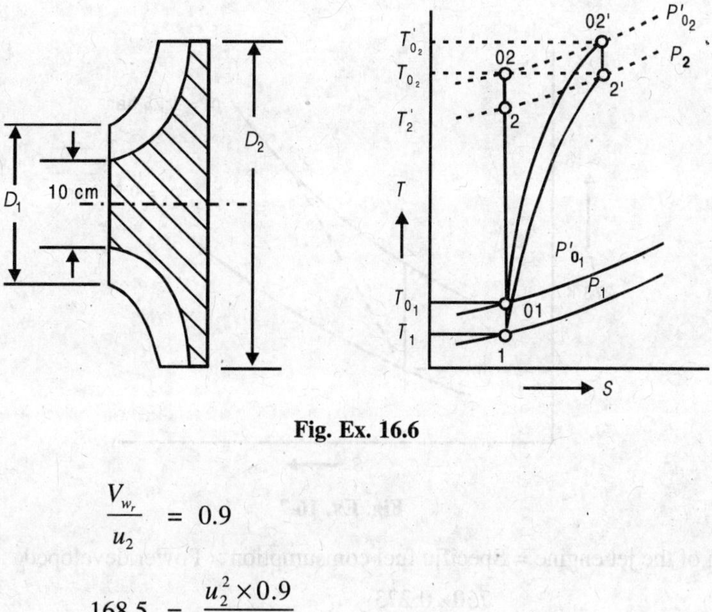

Fig. Ex. 16.6

But, $$\frac{V_{w_r}}{u_2} = 0.9$$

$\therefore$ $$168.5 = \frac{u_2^2 \times 0.9}{1000}$$

$\therefore$ $$u_2 = 432.69 \text{ m/s}$$

and $$u_2 = 432.69 = \frac{\pi D_2 N}{60}$$

$\therefore$ $$D_2 = 0.3756 \text{ m}$$
$$= \mathbf{37.56\ cm}$$

Similarly, power required = Work required/kg $\times \dot{m}$

$$= 168.5 \times 6$$
$$= \mathbf{1011.00\ kW}$$

Now, using continuity equation,

$$\dot{m} = \frac{\pi}{4}\left(D^2{}_1 - D^2{}_h\right)V_1 \times \rho_1$$

where,
$$\rho_1 = \frac{P_1}{RT_1} = \frac{1\times10^5}{0.287\times293}$$
$$= 1.189 \text{ kg/m}^3$$

$$\therefore \quad \rho = \frac{\pi}{4}\left(D_1^2 - 0.1^2\right)\times90\times1.189$$

$$\therefore \quad D_1 = 0.2298 \text{ m}$$
$$= \mathbf{22.98\ cm}$$

Example 16.7. *A centrifugal compressor is used as a supercharger for an aircraft engine developing 750 kW power and having specific fuel consumption of 0.273 kg/kW-h. The supercharger supplies fuel : air mixture at 1.25 bar. The air : fuel ratio is 17. Air enters the supercharger at 0.55 bar and 0°C if the adiabetic efficiency of the supercharger is 85%, find the volume of mixture to be supplied and the power required to drive the supercharger.*

Assume C_p = 1.0 kJ/kg-K and C_v = 0.277 kJ/kg-K.

Solution

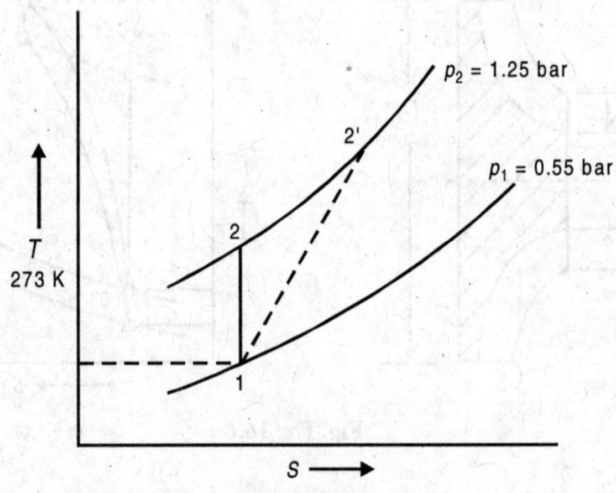

Fig. Ex. 16.7

Fuel consumption of the jet engine = Specific fuel comsumption × Power developed

$$= \frac{750\times0.273}{3600} \text{ kg/s}$$

$$\dot{m}_f = 0.05688 \text{ kg/s}$$

$$\text{Air supplied/s} = \dot{m}_f \times \text{air : fuel ratio}$$
$$= 0.5688 \times 17$$
$$\dot{m}_a = 0.967 \text{ kg/s}$$

Hence, mass of mixture = $\dot{m}_f + \dot{m}_a$

$$m_m = 1.0239 \text{ kg/s}$$

Now,
$$r = \frac{C_p}{C_v} = \frac{C_p}{C_p + R} = \frac{1.0}{1.0 - 0.277} = 1.383$$

$$T_{o_2} = T_{o_1}\left\{\frac{P_{o_2}}{P_{o_1}}\right\}^{\frac{\gamma-1}{\gamma}} = 273\left\{\frac{1.25}{0.55}\right\}^{\frac{0.4}{1.4}}$$

$$= 342.7 \text{ K}$$

Isentropic efficiency, $\eta_{\text{isent}} = \dfrac{T'_{o_2} - T_{o_1}}{T_{o_2} - T_{o_1}}$

$$0.85 = \frac{342.7 - 273}{T'_{o_2} - 273}$$

$\therefore$ $$T_{o_2} = 355 \text{ K}$$

$\therefore$ Volume of mixture to be supplied, $V_m = \dfrac{\dot{m}_m R T''_{o_{2'}}}{P_o}$

$$= \frac{1.0239 \times 0.277 \times 355}{1.25 \times 10^2}$$

$$= \mathbf{0.8055\ m^3/s}$$

And, Power required $= \dot{m}_m C_p \left(T'_{o_2} - T_{o_1}\right)$

$$= 1.0239 \times 1 \times (355 - 273)$$

$$= \mathbf{84.0\ kW}$$

Example 16.8. *The following data refers to a centrifugal compressor:*

Ratio of total pressure	*= 3.6 : 1*
Diameter of the eye of impeller	*= 35 cm*
Axial velocity at inlet	*= 140 m/s*
Mass flow rate of air	*= 12 m kg/s*
Velocity in the delivery duct	*= 120 m/s*
Tip speed of impeller	*= 460 m/s*
Speed	*= 20000 r.p.m.*
Total head isentropic efficiency	*= 80%*
Pressure coefficient	*= 0.73*
Ambient conditions	*= 1.013 bar and 15°C*

Calculate:

(i) Static pressure ratio

(ii) Static pressure and temperature at inlet and outlet of compressor

(iii) Work reqd./kg of air

(iv) Theoretical power required.

Data: $r_p = 3.6$

$T_{o_1} = 288$ K, $\quad \eta_{isent} = 0.80$

$\dot{m} = 12$ kg/s $\quad \psi = 0.73$

Solution

$$T_{o_2} = T_{o_1}(r_p)h^{\frac{\gamma-1}{\gamma}} = 288(3.66)^{\frac{0.4}{1.4}}\text{ K}$$

$$\therefore \quad T'_{o_2} - T_{o_1} = \frac{T_{o_2} - T_{o_1}}{\eta_{\text{isent}}} = \frac{288\left[3.6^{\frac{0.4}{1.4}} - 1\right]}{0.80} = 159.7°\text{C}$$

$$T_{o_2} = 288 + 159.7 = \mathbf{447.1\ K}$$

Now static temperature at exit $= T'_{o_2} - \dfrac{120°}{2C_p}$.

$$= 447.1 - \frac{120°}{2 \times 1.005 \times 10^3}$$

$$T_2 = 440\text{ K}$$

Also static pressure at exit

$$P_2 = P_{o_2} - \frac{\rho_2 \times 120^2}{2}$$

where

$$= 3.65 - \frac{P_2 \times 120^2}{0.287 \times 440 \times 2 \times 10^3} \qquad \rho_2 = \frac{P_2}{RT_2}$$

$$= 3.45\text{ bar} \quad \text{and} \quad P_{o_2} = 1.013 \times 3.6 = 3.65\text{ bar}$$

Now static condition at inlet

$$T_{o_1} = T_1 + \frac{140^2}{2 \times C_p} = T_1 + \frac{140^2}{2 \times 1.005 \times 10^3} = T_1 + 9.75$$

$$T_1 = 288 - 9.5$$
$$= 278.25\text{ K}$$

Also,

$$P_1 = P_{o_1} - \frac{140^2 \times P_1}{2 \times 0.287 \times 278.25 \times 10^3}$$
$$= 1.013 - 0.123\,P_1$$

$$\therefore \quad \mathbf{P_1 = 0.9\ bar}$$

$$\therefore \quad \textbf{Static pressure ratio} = \frac{\mathbf{P_2}}{\mathbf{P_1}} = \frac{\mathbf{3.45}}{\mathbf{0.9}} = \mathbf{3.83}$$

Furthur, work required/kg of air $= C_p\left(T'_{o_2}, T_{o_1}\right)$

$$= 1.005\,(159 - 1)$$
$$= \mathbf{159.89\ kJ/kg}$$
$$= \dot{m}C_p\left(T'_{o_2}, T_{o_1}\right)$$
$$= 10 \times 159.89\text{ kW}$$
$$= \mathbf{1598.9\ kW}$$

Example 16.9. *A centrifugal compressor running at 18000 r.p.m. admits at 300 K. The other data is as follows:*

Isentropic total head efficiency = 0.76

Outer diameter of blade tip = 550 mm

Slip factor = 0.82

Calculate:

(i) Temperature rise of air

(ii) Static pressure ratio.

Assume that absolute velocity of air at inlet and exit to be same.

Take C_p *= 1.005 kJ/kg-K*

Solution

$T_1 = 300\ K,\qquad N = 18000\ r.p.m.$

$D_2 = 0.55\ m\qquad \eta_{isent} = 0.76$

$\mu = 0.82$

$$u_2 = \frac{\pi D_2 N}{60} = \frac{\pi\times 0.55\times 18000}{60}$$

$$= 518.36 \text{ m/s}$$

Work required/kg of air = $\left(V_{w_2}u_2 - V_{w_1}u_1\right)$ N-m/kg

But, $\quad V_{w_1} = 0$ and $\mu = \dfrac{V_{w_2}}{u_2}$

∴ $\quad$ Work reqd./kg $= \mu u_2^2 = 0.82\times 518.36^2$ N-m/kg

$= \mathbf{220.331\ kN/kg}$

Also, work reqd./kg of air = $C_p\left(T_{o_2} - T_{o_1}\right)$

$$220.331 = 1.005\left(T'_{o_2} - T_{o_1}\right)$$

∴ $$\left(T'_{o_2} - T_{o_1}\right) = \frac{220.331}{1.005} = 219.23°C$$

And Isentropic efficiency = $\dfrac{T_{o_2} - T_{o_1}}{T'_{o_2} - T_{o_1}}$

∴ $$\mathbf{T_{o_2} = T_{o_1} + 0.76(219.36) = 466.6\ K}$$

And, Static pressure ratio $\left(\dfrac{P_2}{P_1}\right) = \left(\dfrac{T_{o_2}}{T_{o_1}}\right)^{\frac{\gamma-1}{\gamma}} = \mathbf{4.69}$

Example 16.10. *An axial flow compressor has 8 stages and with 50 % degree of reaction compresses air with a pressure ratio of 4 : 1. Air enters the compressor at 20°C and flows through it with a constant speed of 90 m/s. The mean blade speed is 180 m/s. The isentropic efficiency is 82 %.*

Calculate:

(i) Work required/kg of air

(ii) Blade angles.

Assume $\gamma = 1.4$ *and* $C_p = 1.005$ *kJ/kg-K*

Solution

No. of stages = 8

D. R. = 0.50, $\qquad \therefore \alpha_1 = \beta_2$ and $\beta_1 = \alpha_2$

$r_p = 4.0, \qquad T_1 = 293$ K

$u = 180$ m/s

$$\eta_{isent} = \frac{T'_2 - T_1}{T'_2 - T_1} \qquad \text{But} \qquad T_2 = T_1 (r_p)^{\frac{\gamma-1}{\gamma}}$$

$$= 293 \times 4^{\frac{0.4}{1.4}}$$

$$0.82 = \frac{435.4 - 293}{T'_2 - 293} \qquad = 435.4 \text{ K}$$

$$T'_2 = 466.6 \text{ K}$$

$$\text{Total work reqd./kg} = C_p (T'_2 - T_1) = 1.005\,(466.6 - 293) \text{ kJ/kg}$$

$$= \mathbf{174.47\ kJ/kg}$$

$$\text{Total work reqd./kg} = \text{No. of stages} \times u\left(V_{w_2} - V_{w_1}\right) = 8 \times u \times V_f\,(\tan\alpha_2 - \tan\alpha_1)$$

$$\therefore \quad 174.47 = 8 \times 180 \times 90 (\tan\alpha_2 - \tan\alpha_1) \times 10^3$$

$$\therefore \quad \tan\alpha_2 - \tan\alpha_1 = 1.346 \qquad \text{...(i)}$$

With 50% degree of reaction,

$$\alpha_1 = \beta_2 \text{ and } \beta_2 = \alpha_1$$

Also,

$$\tan\alpha_1 + \tan\beta_1 = \frac{u}{V_f} = \frac{180}{90} = 2 \qquad \text{...(ii)}$$

Solving Eqs (i) and (ii)

$$\beta_1 = \alpha_2 = \mathbf{59.1^\circ}$$

$$\alpha_1 = \beta_2 = \mathbf{18.1^\circ}$$

Example 16.11. *Determine the pressure rise and work done by a rotary axial flow compressor for which the following data is available.*

$$u = 200 \text{ m/s}, \; V_w = V_f = 186 \text{ m/s},$$

$$\alpha_1 = 45^\circ, \; \alpha_2 = 14^\circ,$$

and $\rho = 1.0 \text{ kg/m}^3$.

Solution

Pressure rise through a ring of axial flow compressor

$$\Delta p = \rho V_w^2 \left(\tan^2\alpha_1 - \tan^2\alpha_2\right)$$

$$= \frac{1 \times 186^2}{2 \times 10^5}\left(\tan^2 45 - \tan^2 14\right)$$

$$= \mathbf{0.2022\ bar}$$

Work required/kg of air on the blades

$$= uV_w (\tan\alpha_1 - \tan\alpha_2)$$

$$= \frac{200 \times 186}{10^3} (\tan 45 - \tan 14)$$

$$= \mathbf{27.92\ kW}$$

Example 16.12. *An eight stage axial flow compressor admits air at 1.0 bar and 20°C. It has an overall pressure ratio of 5:1 and overall isentropic efficiency of 90%. The work is equally divided between the stages. The mean blade speed is 175 m/s and 50% reaction design is used. The axial velocity of flow through the compressor is contant at 100 m/s.*

Calculate the power required and blade angles.

Data: Degree of reaction = 50%

$P_1 = 1.0$ bar $\qquad T_1 = 293$ K

$\eta_{isent} = 0.90, \qquad r_p = 5.0$

$u = 175$ m/s $\qquad V_f = 100$ m/s

$\alpha_1 = \beta_2$ and $\alpha_2 = \beta_1$

Solution

Let sufix z denote the number of stages.

$$T_{o_2} = T_{o_1} \times r_p^{\frac{\gamma-1}{\gamma}} = 293 \times 5^{\frac{1.4}{1.4}}$$

$$= 464 \text{ K}$$

But
$$\eta_{isent} = \frac{T_{o_2} - T_{o_1}}{T'_{o_2} - T_{o_1}}$$

$\therefore$
$$0.90 = \frac{464 - 293}{T'_{o_2} - 293} \qquad \therefore \quad T'_{o_2} = 483 \text{ K}$$

$$\text{Work reqd./kg} = C_p\left(T'_{o_2} - T_{o_1}\right) = \left(V_{w_2} - V_{w_1}\right) u \times z$$

$$= V_f\left(\tan\alpha_2 - \tan\alpha_1\right) u \times z$$

$\therefore$
$$\left(\tan\alpha_2 - \tan\alpha_1\right) = \frac{1.004(483-293)\times 10^3}{100 \times 175 \times 8}$$

$$= 1.3639 \qquad \ldots(i)$$

Referring to velocity triangles,

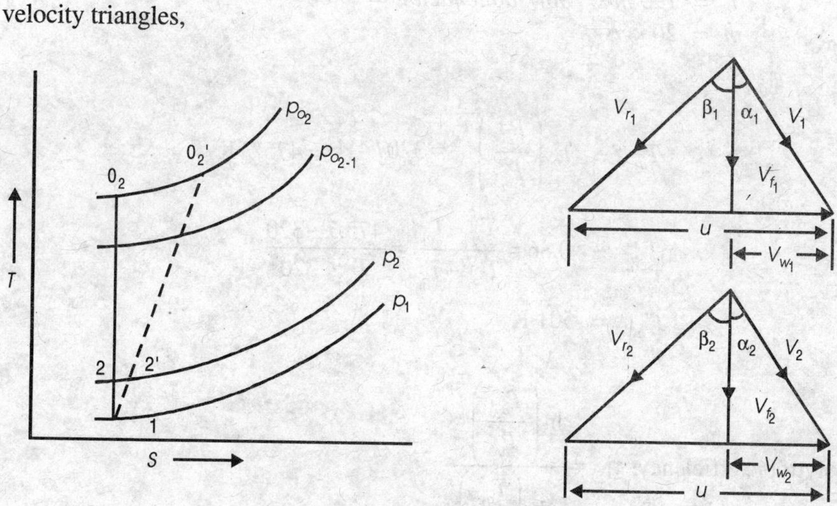

Fig. Ex. 16.12

$$\frac{u}{V_f} = \tan\alpha_1 - \tan\beta_1 = \frac{175}{100}$$
$$= 1.75$$

From Eqs (i) and (ii), we get

$$\tan\beta_1 = \frac{1.365+1.75}{2}$$
$$= 1.5569$$
$$\therefore \quad \beta_1 = \alpha_2 = 57.4^\circ$$

and substituting in Eq. (i),

$$\alpha_1 = \beta_2 = 10.95^\circ$$

And power required to drive the compressor

$$= \dot{m}C_p\left(T'_{o_2} - T_{o_1}\right)$$
$$= 1.005\ (483 - 293)/1$$
$$= \mathbf{190.93\ kW}$$

Example 16.13. *An axial flow compressor operates with an overall stagnation pressure ratio of 4 and overall stagnation isentropic efficiency 86 %. The inlet stagnation pressure and temperature are 1.0 bar and 320 K. The mean blade speed is 190 m/s. The degree of reaction is 0.5 at the mean radius with relative air angles of 30° and 10° at inlet and outlet of rotor respectively. The work done factor is 0.88.*

Determine the stagnation temperature. If the hub : tib ratio is 0.4 and mass flow rate is 20 kg/s, calculate the blade height in the first stage and stagnation polytropic efficiency.

Data: $\frac{P_{o_2}}{P_{o_1}} = 4$, $\quad \eta_{isent} = 0.86$

$D.R. = 0.50$, $\quad \alpha_2 = \beta_2 = 10^\circ$

$\alpha_2 = \beta_1 = 30^\circ$

$p^1 = 1.0$ bar, $\quad T_1 = 320$ K

$u = 190$ m/s $\quad$ Work done factor $= 0.88$

$\dot{m} = 20$ kg/s

Solution

$$T_{o_2} = T_{o_1}\left(\frac{P_{o_2}}{P_{o_1}}\right)^{\frac{\gamma-1}{\gamma}} = 320/4^{\frac{0.4}{1.4}} = 475.7\ \text{K}$$

$$\eta_{isent} = 0.86 = \frac{T_{o_2} - T_{o_1}}{T'_{o_2} - T_{o_1}} = \frac{475.7 - 320}{T_{o_2} - 320}$$

$$T'_{o_2} = 501\ \text{K}$$

$$\text{Stagnation polytropic efficiency, } \eta_p = \frac{\ln\left(\frac{P_{o_2}}{P_{o_1}}\right)^{\frac{\gamma-1}{\gamma}}}{\ln\left(\frac{T_{o_2}}{T_{o_1}}\right)}$$

$$= \frac{\ln(4)^{\frac{0.4}{1.4}}}{\ln\left(\frac{501}{302}\right)}$$

$$= \mathbf{88.4\%}$$

Referring to velocity triangles shown in Fig. Ex. 16.12.

$$\frac{u}{V_f} = \tan\alpha_1 + \tan\beta_1 = \tan 10° + \tan 30°$$

$$= 0.7536$$

$$\therefore \quad V_f = \frac{190}{0.7536} = 252.12 \text{ m/s}$$

Also

$$V_{w_2} = V_f \tan\alpha_2 = 252.12 \tan 30° = 145.56 \text{ m/s}$$

$$V_{w_1} = V_f \tan\alpha_1 = 252.12 \tan 10° = 44.45 \text{ m/s}$$

$\therefore$ Work required/kg of air per stage

$$= u\left(V_{w_2} - V_{w_1}\right)\times\lambda\times 10^{-3} \text{ kJ/kg}$$

$$= \frac{190\times 0.88(145.56\text{-}44.45)}{1000}$$

$$= 16.906 \text{ kJ/kg per stage}$$

Total work required/kg for all stages

$$= C_p\left(T'_{o_2} - T_{o_1}\right)$$

$$= 1.005\,(501 - 320)$$

$$= 181.9 \text{ kJ/kg}$$

$$\therefore \quad \text{No. of stages } z = \frac{181.9}{16.906} = 10.76$$

$$= 11 \text{ stages}$$

Now,

$$V_1 = \frac{V_f}{\cos\alpha_1} = \frac{252.12}{\cos 10°} = 256 \text{ m/s}$$

and

$$T_1 = T_{o_1} - \frac{V^2}{2C_p} = 320 - \frac{190^2}{2\times 1.005\times 1000} = 302.03 \text{ K}$$

$$\frac{P_1}{P_{o_1}} = \left(\frac{T_1}{T_{o_1}}\right)^{\frac{\gamma}{\gamma-1}}$$

$$\therefore \quad \frac{P_1}{P_{o_1}} = \left(\frac{302.03}{320}\right)^{\frac{1.4}{0.4}}$$

$$\therefore \quad P_1 = 0.816 \text{ bar}$$

and

$$\rho_1 = \frac{P_1}{RT_1} = \frac{0.816\times 10^2}{0.287\times 302.03}$$

$$= 0.941 \text{ kg/m}^3$$

From the continuity equation,

$$\rho_1 A_1 V = \dot{m} = 20 \text{ kg/s}$$

$$\therefore\ 0.941 \times \pi r_1^2 \left[1 - 0.4^2\right] \times 252.12 = 20$$

$$r_1 = 0.1787 \quad m = 17.87 \text{ cm}$$

But,

$$\frac{r_2}{r_1} = 0.4$$

$$\therefore \quad r_1 = 17.87 \times 0.4 = 7.1 \text{ cm}$$

∴ Hence height of blade in the first stage

$$z = r_2 - r_1$$
$$= 17.87 - 7.148$$
$$= 10.722 \text{ cm}$$

PROBLEMS FOR PRACTICE

16.1. Air at 101.3 kPa and 288 K is to be compressed at a rate of 5.6 m^3/min to 11.75 bar. Two compressors are considered (i) Roots blower and (ii) Sliding vane rotary compressor. Compare the power required assuming that for vane type compressor, 75% of the pressure rise occurs due to internal compression before the delivery takes place and that compressor is uncooled.

[*Ans.* (i) 6.88 kW, (ii) 5.75 kW]

16.2. Free air of 30 m^3/min. is compressed from 101.3 kPa to 2.23 bar. Determine the power required (i) if the roots blower is used, (ii) if vane blower is used. Assume that there is 25% reduction in volume before the back flow occurs. Also find the isentropic efficiency in each case.

[*Ans.* (i) 60.85 kW, (ii) 48.46 kW, (iii) 73.69%, (iv) 92.53%]

16.3. The design pressure ratio of a centrifugal compressor is 3.5 : 1. The diameter of the impeller eye is 30 cm. The axial velocity of flow at inlet is 130 m/s and the mass flow rate of air is 5 kg/s. The velocity in the delivery duct is 115 m/s. The tip speed of the impeller is 450 m/s and runs at 20000 r.p.m. with a total head isentropic efficiency of 78%. The pressure coefficient is 0.72.

Determine:

(i) Static pressure ratio
(ii) Static pressure and temperature at the inlet and outlet of the compressor
(iii) Work required/kg of air
(iv) Theoretical power required.

Assume ambient conditions as 1.013 bar and 15°C.

[*Ans.* (i) 4.21, (ii) 0.917 bar, 279.6 K, 3.86 bar, 461.67 K, (iii) 180.29 kJ/kg, (iv) 901.45 kW]

16.4. A centrifugal compressor runs at 20000 r.p.m. draws in air at 290 K. It has slip factor 0.80, total head isentropic efficiency 0.75 and outer diameter of blade tip 50 cm.

Calculate,

(i) temperature rise of air flowing through the compression
(ii) static pressure ratio.

Assume that velocities of air at inlet and outlet to remain constant.

[*Ans.* (i) 218.62°C, (ii) 4.8]

16.5. Air at a temperature of 17°C flows axially into the eye of the impeller of an uncooled centrifugal compressor running at 18000 r.p.m. and leaves the impeller with a velocity of whirl 0.88 times the velocity of the blade tip. The outer diameter of the blade tip is 46 cm and the kinetic energy of air at the outlet from the impeller is equal to that at the inlet to the impeller. Assuming isentropic efficiency of 0.75, determine

(i) Rise in temperature of the air compressed
(ii) Static head pressure ratio.

[*Ans.* (i) Actual Δt = 205°C, (ii) Pressure ratio = 3.48]

16.6. A single sided centrifugal compressor of a gas turbine plant has to deliver 8.2 kg/s of air with a pressure ratio of 4.4 : 1 at 18000 r.p.m. The intake conditions are static pressure 1.013 bar and 15°C. Assuming isentropic efficiency of 78%, ratio of whirl speed to tip speed 0.94. Calculate rise in stagnation temperature, tip speed, external diameter of impeller, external diameter of the eye for which the internal diameter is 12.5 cm and the axial velocity is to be 140 m/s with no pre-whirl.

[*Ans.* Outer dia. of impeller D_2 = 48.5 cm, Eye external dia. = 28.6 cm, ΔT_O, = 195 ºC]

16.7. An eight stage axial flow compressor takes in air at a temperature of 20°C at a rate of 3 kg/s. The pressure ratio is 6 and the isentropic efficiency is 0.89. The compression process is adiabetic. All stages of the compressor are similar and operate with 50 reaction. In each stage, the mean blade speed is 180 m/s. The constant axial velocity of flow of air is 105 m/s.

Determine the power required and the direction of air at the entry to and exit from the rotor and stator. Assume air as an ideal gas.

[*Ans.* Power = 660.3 kW, $\alpha_1 = \beta_2 = 7.4°$. $\alpha_2 = \beta_1 = 57.8°$]

16.8. The mean blade speed in the second stage of an uncooled axial flow compressor is 230 m/s and the axial velocity of flow is 150 m/s. The shape of the fixed and moving blades is same and the difference between the tangents of the angles at inlet and outlet is 0.56. If the adiabetic efficiency of the stage is 85% and the air inlet temperature to the stage is 290 K, find the number of required for a pressure rise from 1 bar to 3.8 bar. Also find the power required to compress a mass flow of 1000 kg/min. Assume C_p = 1.005, mechanical efficiency 99%. Neglect the difference in the kinetic energy entering successive stages is negligible.

[*Ans.* W = 19.32 kJ/kg, ΔT = 19.22ºC, r_p = 1.211, No. of stages = 7, power = 2276 kW]

16.9. An axial flow compressor compresses air from an inlet condition of 1 bar and 290 K so a delivery pressure of 5 bar with an overall isentropic efficiency 87%. The degree of reaction is 0.5 and the blade angles at inlet and outlet are 44° and 13° respectively. The mean blade speed and axial velocity of flow are constant throughout the compressor. Assuming a blade velocity of 180 m/s and work done factor of 0.85,

Determine:

(i) Number of stages,

(ii) Change in entropy.

[*Ans.* (i) 12, (ii) 0.4898 kJ/kg-K]

16.10. The first stage of an axial flow compressor is to be designed for an axial velocity of flow 133 m/s, mass flow rate of air 22.7 kg/s and for ambient conditions of 1 bar an 10°C.

At the mean radius, both moving and fixed blade sections are to be suitable for relative air angles of 50° at inlet and 30º at outlet. A row of guide vanes giving an outlet air angle of 30° is to be fitted before each stage. Draw the triangles of velocity assuming the ratio of the hub to tip diameters as 0.68, free vortex conditions.

Calculate:

(i) Speed

(ii) Theoretical temperature rise across the stage

(iii) Number of similar stages theoretically required to give a total head delivery, pressure of 8.4 bar if the stage isentropic efficiency is 85%.

[*Ans.* (i) 9000 r.p.m., (ii) 20ºC, (iii) 13]

16.11. A six stage axial flow compressor take air at a temperature of 25ºC at a rate of 5 kg/s. The pressure ratio is 5 and isentropic efficiency 0.85. The compressor stages are similar and has 50% degree of reaction. In each stage, the mean blade speed is 190 m/s and axial velocity of flow is 100 m/s. Calculate the power required and blade angles.

[*Ans.* 1074.675 kW, $\beta_1 = 72.23°$, $\beta_2 = 62.72°$]

16.12. In an axial flow air compressor has a pressure ratio 6 : 1. Air enters at 20º, mean velocity of rotor blades is 200 m/s and the inlet and exit angles of both moving and fixed blades are 45° and 15° respectively. The degree of reaction is 50%, the work done factor is 0.86 throughout. There are 12 stages and axial velocity remains constant throughout.

Calculate the isentropic efficiency of the compressor. Explain why the overall pressure ratio is not equal to the stage pressure ratio raised to the power of n, is the number of stages.

[*Ans.* η_{isent} = 0.83]

16.13. Sketch the velocity diagram for an axial flow compressor and derive the relation,

$$\frac{p_{o_2}}{p_{o_1}} = \left[1 + \eta_c \frac{V_b(V_{w_2} - V_{w_1})}{C_p T_{o_1}}\right]^{\frac{\gamma}{\gamma-1}}$$

IMPORTANT POINTS TO BE REMEMBERED

1. For 2-lobes (Roots blower)

$$\text{Isentropic work} = \frac{\gamma-1}{\gamma}\cdot P_1(\text{No. lobes}\times 2V)\left[\left(\frac{P_2}{P_1}\right)^{\frac{\gamma-1}{\gamma}}-1\right]\times\frac{N}{60}\text{ kW}$$

$$= \frac{\gamma-1}{\gamma}P_1\times 4V\left[\left(\frac{P_2}{P_1}\right)^{\frac{\gamma-1}{\gamma}}-1\right]\times\frac{N}{60}$$

$$= C_P\cdot P_1 4V\left[(r_p)^{\frac{\gamma-1}{\gamma}}-1\right]\times\frac{N}{60}\text{ kW}$$

$$\eta_{root} = \frac{\frac{\gamma-1}{\gamma}P_1\times 4V\left[(r_p)^{\frac{\gamma-1}{\gamma}}-1\right]}{(P_2-P_1)\times 4V} = \frac{C_P\left[(r_p)^{\frac{\gamma-1}{\gamma}}-1\right]}{r_p-1}$$

Centrifugal compressor

2. *Euler's work for centrifugal compressor*

$$\text{Torque} = T = V_{w_2}\cdot r_2 - V_{w_1}\cdot r_1\ \cdot\text{N-m/kg}$$

$$\text{Work/kg} = \frac{2\pi N}{60}\times T\ \ \frac{\text{N-m}}{\text{sec}}\Big/\text{kg}$$

$$= \frac{2\pi NT}{60\times 1000}\text{ kW/kg}$$

For axial inlet, $V_{w_1} = 0$

$$\therefore\ \text{Work/kg} = \frac{2\pi N}{60{,}000}\times V_{w_2} r_2\ \text{kW}$$

3. *Flow coefficient* (ϕ)

$$\phi = V_{f2}/V_{f2}$$

4. *Head coefficient* (λ)

$$\lambda = \frac{\Delta h}{V_{b2}^2/2}$$

5. *POressure coefficient* (ψ)

$$\psi = \Delta h_{isen}\Big/\frac{V_{b2}^2}{2}$$

6. *Degree of reaction* (*R*)

$$R = \Delta T_{rotor}/\Delta T_{stage}$$

7. *Slip factor* (*μ*)

$$\mu = \frac{Vw_{2'}}{Vw_2}$$

Axial-flow compressor

1. *Isentropic pressure ratio* $\left(r_{p_s}\right)$

$$r_{p_s} = \left[1+\frac{\eta_s \Delta T_{os}}{T_{o_1}}\right]^{\gamma/\gamma-1}$$

2. *Degree of reaction* (*R*)

$$R = \frac{V_f}{2u}(\tan\beta_1 - \tan\beta_2)$$

3. *Polytropic efficiency* (η_{Poly})

$$\eta_{Poly} = \frac{\gamma-1}{\gamma}\times\frac{n-1}{n}$$

CHAPTER 17

Introduction to Heat Transfer

17.1 DEFINITION

We have repeatedly used the term heat (Q) and heat exchanges between bodies without discussing the details of the phenomena by which heat is transferred. The purpose of this chapter is to give the student a general idea of the nature of the problems and to provide an introductory background for further study.

Heat transfer is a name of general process applied to any device that effects a transfer of heat from one substance to another.

The various devices that employ the principles of heat transfer are: (i) steam boilers (ii) surface condensers, evaporators, closed feed water heaters and the automobile radiator etc.

Whenever a temperature gradient exists within a system, or when two systems at different temperatures are brought into contact, energy is transferred. The process by which the heat energy transport takes place is known as heat transfer.

17.2 MODES OF HEAT TRANSFER

Heat transfer can be defined as the transmission of energy from one region to another as a result of a temperature difference between them. Since temperature differences exist all over the universe, the phenomena of heat flow over the universe, the phenomena of heat flow are as universal as those associated with gravitational attractions. Unlike gravity, however, heat flow is governed not by a unique relationship, but rather by a combination of various independent laws of physics.

The literature of heat transfer generally recognises three distinct modes of heat transmission:

(i) Conduction
(ii) Radiation and
(iii) Convection

Strictly speaking, only conduction and radiation should be classified as heat transfer processes, because only these two mechanisms depend for their operation on the mere existence a temperature difference. The last of the three, convection, does not strictly comply with the definition of heat transfer because it depends for its operation on mechanical mass transport also. But since convection also accomplishes transmission of energy from regions of higher temperatures to regions of lower temperatures, the term "heat transfer by convection", has been generally accepted.

17.3 FOURIER'S LAW OF CONDUCTION

The basic relation for heat transfer by conduction was proposed by French scientist JBJ Fourier in 1882. It states that q_k the rate of heat flow by conduction in a material, is equal to the product of the following three quantities:

(i) K – the thermal conductivity of the material

(ii) A – the area of the section through which heat flows by conduction and

(iii) $\frac{dT}{dx}$ – the temperature gradient at the section i.e., the rate of change of temperature T with respect to distance in the direction of heat flow or

$$q_k = -K.A\frac{dT}{dx}$$

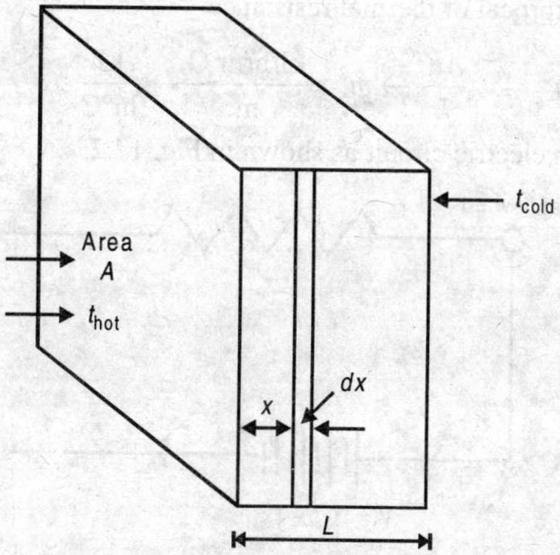

Fig. 17.1 *One dimensional heat flow*

$$q_k \rightarrow \text{kJ/hr}; A \rightarrow m^2; \frac{dT}{dx} \rightarrow \text{°C/m}$$

$$K \rightarrow \frac{q_k}{A \cdot dT/dx} \rightarrow \frac{\text{kJ}}{\text{hr}} \times \frac{1}{\text{m}^2} \times \frac{1}{\text{°C/m}} \rightarrow \frac{\text{kJ}}{\text{hr.m°C}} \text{ units of K}$$

Refer Fig. 17.1.

Consider a wall or slab of thickness L, let A be the area of heat flow across the wall. Let the temperatures on both sides of the wall be T_{hot} and T_{cold}. Consider a section at a distance x from hot surface and let dx and dT be the thickness of the strip and temperature drop across the strip as shown. Therefore, applying Fouriers Law, we get

$$q_k = \frac{-KA.dT}{dx}$$

We have

$$\frac{q_k}{A} = -\int K.\frac{dT}{dx}$$

$$\therefore \quad \frac{q_k}{A}\int_o^l dx = -K\int_{Hot}^{Cold} dT \quad K.\int_{Cold}^{Hot} dT$$

$$\therefore \quad \frac{q_k}{A}\cdot L = K(T_{hot} - T_{cold})$$

or $$\mathbf{q_k = \frac{AK}{L}\left(T_{hot} - T_{cold}\right)}$$

We can write this also as,

$$q_k = \frac{AK}{L}\cdot\Delta T = \frac{\Delta T}{L/AK}$$

Here we can compare with simple electric circuit as shown in Fig.17.2.

Here ΔT = Driving force for heat transfer similar to (Voltage)

$\frac{L}{AK}$ = Thermal resistance R_k similar to (Electric Resistance)

q_k = Heat flow similar to (Electric current)

Thermal conductance is the reciprocal of thermal resistance

$$K_k = \frac{AK}{L} \to m^2 \times \frac{\text{kJ/hrm°C}}{\text{m}} \to \frac{\text{kJ}}{\text{hr°C}}$$

∴ This can be looked upon as an electric circuit as shown in Fig. 17.2.

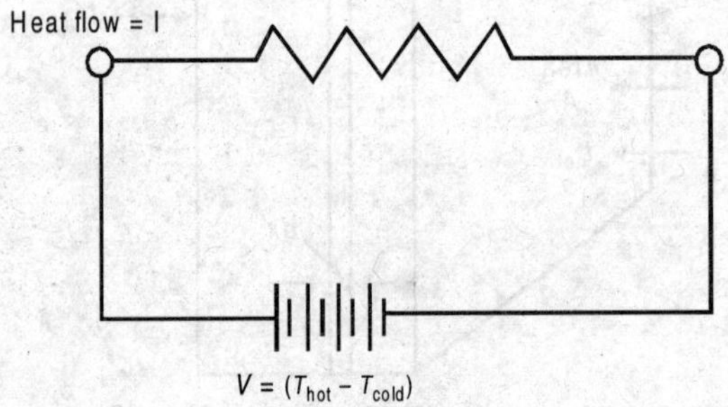

Fig. 17.2. *Electric analog*

17.3.1 Composite Walls

A composite wall, typical of the type used in a large furnace, is shown in Fig.17.3. The inner layer, which is exposed to the high temperature gases, is made of firebrick. The intermediate layer consists of an insulating brick and is followed by an outer layer of ordinary red bricks.

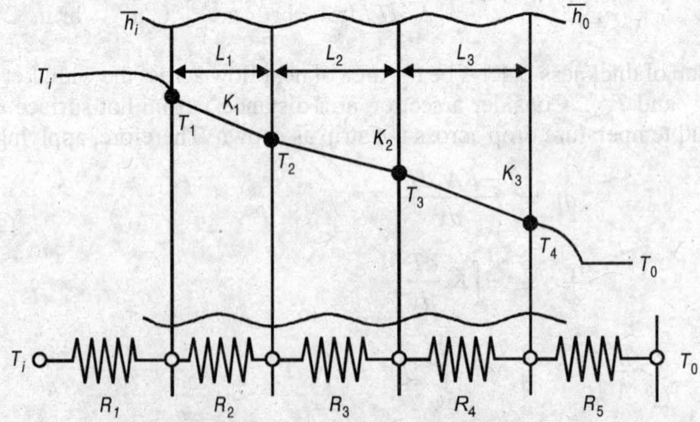

Fig. 17.3

The temperature of the hot gases is T_i and the unit surface conductance over the interior surface is $\bar{h}_i$. The atmosphere surrounding the furnace is at a temperature t_0 and the unit surface conductance over the exterior surface is $\bar{h}_0$.

Under these conditions there will be a continuous heat flow from the hot gases through the walls to the surroundings. Since the heat flow through a given area A is the same for any section, we obtain

$$q = \bar{h}_i A(T_i - T_1) = \frac{K_1 A}{L_1}(T_1 - T_2) = \frac{K_2 A}{L_2}(T_2 - T_3)$$

This equation can be written in terms of the thermal resistance of the various section as

$$q = \frac{T_i - T_1}{R_1} = \frac{T_1 - T_2}{R_2} = \frac{T_2 - T_3}{R_3} = \frac{T_3 - T_4}{R_4} = \frac{T_4 - T_0}{R_5}$$

where the resistances may be determined. Solving for the various temperature differences, we have

$$T_i - T_1 = qR_1$$
$$T_1 - T_2 = qR_2$$
$$T_2 - T_3 = qR_3$$
$$T_3 - T_0 = qR_4$$
$$T_4 - T_0 = qR_5$$

Adding, we get

$$T_i - T_0 = q(R_1 + R_2 + R_3 + R_4 + R_5)$$

$$\therefore \quad q = \frac{T_i - T_0}{R_1 + R_2 + R_3 + R_4 + R_5}$$

$$\therefore \quad \text{Heat flow} = \frac{\text{Overlap temp. potential}}{\text{Sum of resistance in the path}}$$

17.3.2 Conduction of Heat through Hollow Cylinder

The heat flow through a cylinder is considered along radial direction. Therefore the Fourier equation of conduction for cylinder can be written as

$$Q = -KA\frac{dT}{dr}$$

Figure 17.4 (a) shows hollow pipe having inside radius R_1 and outside radius R_2. Let the inside temperature be T_1 and outside temperature T_2. The conductivity of the pipe material is say K.

Consider an elementary ring at a radius of r and having a thickness dr as shown in Fig. 17.4(a). The heat flow equation through an element dr under steady state condition can be written as

$$Q = -K \cdot (2\pi r)\frac{dT}{dr}$$

where L is the length of pipe and $A = 2\pi rL$. Integrating this equation between the limits of R_1 and R_2 and T_1 to T_2 as the temperature changes from T_1 and T_2 through the thickness of the cylinder $(R_2 - R_1)$,

$$\int_{R_1}^{R_2} Q.\frac{dr}{r} = \int_{T_1}^{T_2} -K2\pi L.dT$$

$$\therefore \quad Q.\log_e \frac{R_2}{R_1} = -2\pi LK(T_2 - T_1)$$

$$= 2\pi LK(T_1 - T_2)$$

$$\therefore \quad Q = \frac{2\pi LK(T_2 - T_1)}{\log_e \frac{R_2}{R_1}}$$

$$= \frac{T_1 - T_2}{\frac{1}{2\pi LK}\log_e \frac{R_2}{R_1}}$$

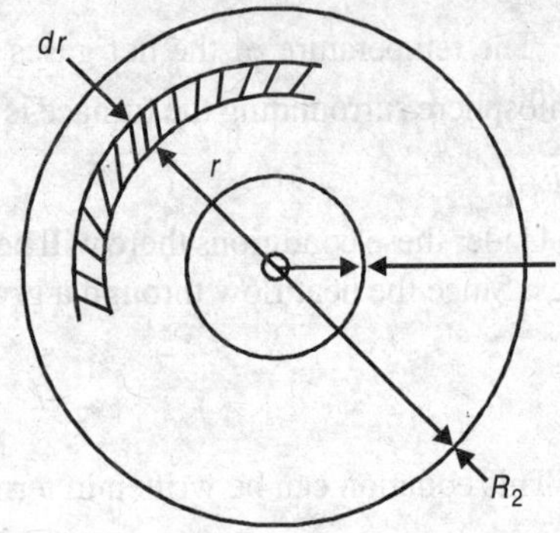

Fig. 17.4

As before $\left[\frac{1}{2\pi LK}\right]\log_e \frac{R_2}{R_1}$ is known as thermal resistance of the pipe of length L.

17.4 CONVECTION HEAT TRANSFER

Convection heat transfer is defined as the transfer of heat through the agency of particles of fluid which receive heat from a high temperature source and move to the locality of a lower temperature sink to reject heat. The sink may be some body exposed to the fluid, or it may be cooler particles of fluid. It should be noted that convection heat transfer occurs by means of fluid motion, in contrast to conduction heat transfer, which takes place entirely by means of intermolecular energy transfers.

Convection heat transfer may be further classified according to the mode of motivating the flow. If the convection currents are caused by density differences brought about by temperature gradients within the fluid, the convection is said to be *natural*. If the flow is aided by some mechanical device such as a pump (fan) the resulting convection currents are forced.

17.5 NEWTON'S LAW OF COOLING

The rate of heat transfer by convection between a surface and a fluid may be computed by the relation

$$q_c = \bar{h}CA.\Delta T$$

where q_c = rate of heat transfer by convection (kJ/hr)

A = heat transfer area (m^2)

ΔT = difference between the surface temperature T_S and the temperature of the fluid $T\infty$ at some specified location (°C).

$\bar{h}_c$ = average unit thermal convective conductance often called the surface coefficient of heat transfer or the convective heat transfer coefficient kJ/hr.m^2.°C.

Thermal conductance K_c for convective heat transfer $K_c = \bar{h}_c \,.\, A$ and thermal resistance

$$R_c = \frac{1}{\bar{h}_c A}$$

The relation expressed by this equation was originally proposed by the British scientist, Issac Newton in 1701. This is also known as Newton's law of cooling. The evaluation of the convective heat transfer coefficient is difficult because convection is a very complex phenomenon.

17.6 FILM COEFFICIENT OF HEAT TRANSFER

In heat exchangers, we are mostly concerned with walls separating liquids or gases from each other. In these cases we do not know the temperatures of both surfaces of the separating walls, but only the temperatures of the liquids or fluids on both sides of the walls. These temperatures are indicated as t_1 and t_2 as shown in Fig. 17.5. By measuring the temperature field in the liquids one obtains the curves shown.

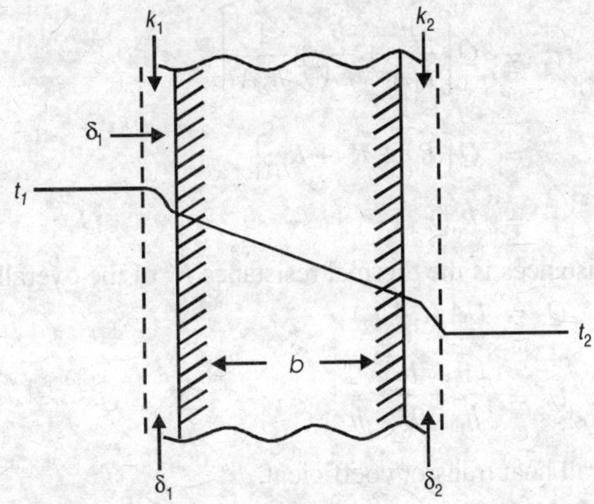

Fig. 17.5

The temperature gradient is confined to a relatively narrow layer of thickness δ quite close to the wall, whereas at a grater distance from the wall in most cases only small temperature differences exist. Simplified, the temperature curve can be replaced by the dashed broken line. This can be explained by assuming that a thin film of liquid (of the thickness δ') adheres to (i) the wall whereas outside thin film, temperature differences vanish as a result of mixing motions of the liquid. This picture over simplifies the actual process considerably as we shall see later, but it brings out the salient features. Within the film, the heat transfer takes place by conduction, as in a solid wall. The temperature in the film, therefore, is again linear, and the flow of heat follows the equation

$$Q = \frac{K}{b}.A.(t_1 - t_2)$$

into which the thermal conductivity K of the liquid or of the gas and the thickness of the film δ' are to be inserted.

Thus, for the heat flow to the wall, the equation $Q = \frac{K}{\delta'} A(t - t_w)$ is obtained.

From this equation the rate of heat flow Q can be calculated if the film thickness δ' is known. The latter, however, depends to a great extent upon the external flow conditions, for instance upon the velocity with which the liquid flows along the wall, upon the shape of the wall, upon the structure of the wall surface and similar factors. In engineering practice it has become customary to calculate with the expression K/δ, rather (2) than directly with the film thickness δ'. This value is called film heat transfer coefficient and is designated by the letter h. Thus the expression

$$Q = hA\,(t - t_w)$$

as formulated by Issac Newton is obtained. The value $(1/hA)$ is called thermal resistance R_t, of the film heat transfer process

$$R_t = 1/hA$$

From Fig. 17.3 we have,

$$t_1 - t_{w1} = \frac{1}{h_1 A} \cdot Q$$

$$t_{w1} - t_{w2} = \frac{b}{kA}.Q$$

$$t_{w2} - t_2 = \frac{1}{h_2 A}.Q$$

∴ Adding, we get

$$t_1 - t_2 = Q\left[\frac{1}{h_1 A} + \frac{b}{kA} + \frac{1}{h_2 A}\right]$$

$$= Q\left[R_{t_1} + R_c + R_{t_2}\right]$$

$$= R_0 Q$$

The sum of the individual resistances is the thermal resistance R_0 of the overall heat transfer process. We have

$$Q = UA\,(t_1 - t_2)$$

where

$$U = \frac{1}{h_1} + \frac{b}{k} + \frac{1}{h_2}$$

The value U is called the overall heat transfer coefficient.

17.7 THERMAL RADIATION

It will be recalled that conduction and convection heat transfer depend on the existance of a medium to convey heat. The third mode of heat transfer, *radiation*, differs from the other two in that a medium is not required. Radiation is the term applied to the process by which energy is continually emitted from the surfaces of all bodies. This energy is called *radiant energy* and is in the form of electromagnetic waves. These waves travel with the velocity of light and are transmitted through a vacuum as well as through substances which are transparent to them. In fact, transmission through a vacuum is better, since intervening media absorb some, if not all, of the radiant energy.

When the electromagnetic waves fall on the body which is not completely transparent to them, they are absorbed and there energy is converted to heat. The process by which radiant energy is converted to heat is not completely understood. However, it is known that all substances emit and absorb radiant energy at a rate which depends on the absolute temperature and physical properties of the substance.

The waves incident upon the surface of the substance may be partly absorbed, partly reflected and partly transmitted through the substance. The fraction of the radiant energy which is absorbed is called *absorptivity* (α). The fraction reflected is termed the *reflectivity* ρ and the fraction transmitted through the substance is known as the *transmissivity* τ. The relationship between the absorptivity, reflectivity and transmissivity is written as

$$\alpha + \rho + \tau = 1$$

When the transmissivity is zero, as it is for most solids *opaque* to light, the substance is said to be *opaque* to radiation. Conversely, if transmissivity is unit, the substance is transparent to radiation. No substance is perfectly transparent but the less dense fluids, such as gases, have a high transmissivity.

An ideal *reflector* is a body whose surface reflects all radiant energy incident upon it. Highly polished surfaces are good approximations to an ideal reflector. An *ideal absorber* absorbs all radiant energy incident upon its surface and hence its absorptivity is unity.

Only the radiant energy absorbed can contribute to increasing the internal energy of the substance, and since the rate at which radiant energy is emitted depends on the temperature of the emitter, a good absorbed is also a good emitter. Therefore, we define an *ideal radiator* as a body which absorbs all radiant energy incident upon its surface. We see that the characteristics of an ideal radiator are an absorptivity of unity, and reflectivity and transmissivity of zero.

The rate at which energy is radiated at a wavelength λ, per unit area of surface is called *monochromatic emissive power*, E_λ. The *total emissive power*, *E*, is the energy radiated per unit time and per unit area of surface at all wavelengths.

The total emissive power of an ideal radiator may be found by integrating the monochromatic emissive power over all wavelengths.

or

$$E = \int_{\lambda=0}^{\lambda=\infty} E_\lambda \, d\lambda.$$

17.8 KIRCHHOFF'S RADIATION LAW

No true ideal radiator exist in nature, although some substances, such as lampblack and platinum black, are good approximations. Kirchhoff conceived of a method of producing a close approximations to an ideal radiator through the use of a hollow enclosure, as shown in Fig. 17.6.

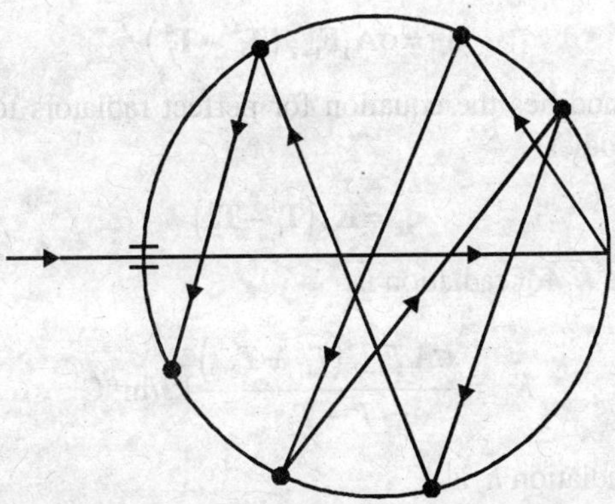

Fig.17.6. *Hollow sphere for producing black-body radiation*

A small hole is cut in a large hollow enclosure, such as a sphere, and the radiant energy incident upon this hole is partially absorbed by the inside surface of the enclosure. The rest of the radiant energy is diffused within the enclosure. The radiant energy enetring through the hole continues to be absorbed reflected and reradiated, but only a very small amount finds its way out through the hole. Hence, the hole behaves very much like an ideal radiator.

17.9 STEFAN BOLTZMANN'S LAW OF RADIATION

The quantity of energy leaving a surface as radiant heat depends upon the absolute temperature and the nature of the surface. A perfect radiator or black-body emits radiant energy from its surface at a rate q_r given by

$$q_r = \sigma A_1 T_1^4 \text{ kJ/hr}$$

where A_1 is the surface area in sq.m. T_1 is the surface temperature in degrees Kelvin and σ is a dimensional constant and is known as Stefan–Boltzmann constant after two Austrian scientists, J. Stefan who in 1879 found the above equation experimentally and L. Boltzmann, who in 1884 derived it theoretically. Here $6 = 20.42 \times 10^{-8}$ kJ/hr. m^2 T^4 = 5.671 W/m^2.T^4.

Stefan–Boltzmann Law of radiation states that any black-body surface above a temperature of absolute zero radiates heat at a rate proportional to the fourth power of the absolute temperature.

While the rate of emission is independent of the conditions of the surroundings, a net transfer of radiant heat requires a difference in the surface temperature of any two bodies between which the exchange is taking place. If the black-body radiates to an enclosure which completely surrounds it and whose surface is also black i.e. absorbs all the radiant energy incident upon it, the net rate of radiant heat transfer is given by,

$$q_r = \sigma A_1 \left(T_1^4 - T_2^4\right)$$

where T_2 is the surface temperature of the enclosure in K.

Real bodies do not meet the specifications of an ideal radiator but emit radiation at a lower rate than the black-bodies. If they emit, at a temperature equal to that of a black-body, a constant fraction of the black-body emission at each wavelength, they are called *Gray Bodies*. The net rate of heat transfer from a gray body at T_1 to a black surrounding body at T_2 is

$$q_r = \sigma A_1 \varepsilon_1 \left(T_1^4 - T_2^4\right)$$

where ε is the emissivity of the gray surface and is equal to the ratio of emission from the gray surface to the emission from a perfect radiator at the same temperature.

If neither of the two bodies is a perfect radiator and if two bodies possess a given geometrical relationships to each other, the net heat transfer by radiation between them is given by

$$q_r = \sigma A_1 F_{1-2} \left(T_1^4 - T_2^4\right)$$

where F_{1-2} is a modulus which modifies the equation for perfect radiators to account for the emissivities and relative geometries of the actual bodies.

$$q_r = K_r \left(T_1^4 - T_2^4\right)$$

By comparison the conductance K_r for radiation is

$$K_r = \frac{\sigma A_1 F_{1-2}\left(T_1^4 - T_2^4\right)}{T_1 - T_2} \text{ kJ/hr°C}$$

Unit thermal conductance for radiation $\bar{h}_r$ is

$$\bar{h}_r = \frac{K_r}{A_1} = \frac{\sigma F_{1-2}\left(T_1^4 - T_2^4\right)}{T_1 - T_2} \frac{\text{kJ}}{\text{hr}} \text{ m}^2\text{/°C or } \frac{\text{kJ}}{\text{hr.m}^2\text{°C}}$$

Similarly, the thermal resistance for radiation is given by

$$R_r = \frac{T_1 - T_2}{\sigma A_1 F_{1-2}\left(T_1^4 - T_2^4\right)}$$

17.10 HEAT EXCHANGER

A heat exchanger is an apparatus in which the processes of heating or cooling occur i.e., heat is transferred from one substance to another. There is a large number of different heat exchangers or heat exchange apparatuses varying both in application and design. In respect to the principle of operation, heat exchangers may be divided in three categories as follows.

17.10.1 Direct Contact Heat Exchanger

The simplest type of heat exchanger is a container in which a hot and a cold fluids are mixed directly. In such a system both fluids will reach the same final temperature and the amount of heat transferred can be estimated by equating the energy lost by the hotter fluid to the energy gained by the hotter fluid to the energy gained by the cooler one. Open feed water heaters, desuperheaters, water cooling towers, scrubbers and jet condensers are the examples of heat transfer equipment employing direct mixing of fluids.

17.10.2 Recuperative Heat Exchangers

In heat exchangers of the recuperative variety, hot and cold fluids flow simultaneously through the heat exchanger and the heat is transferred through a wall separating the fluids. This group unites such heat transfer equipments such as steam boilers, water heaters, condensers etc. There are many forms of such equipment ranging from a simple pipe—within a pipe with a few sq.m of heat transfer surface upto complex surface condensers and evaporators with many hundreds of sq.m of heat transfers surface. In between these extremes is a broad field of common shell and tube exchangers. These units are widely used because they can be constructed with large heat transfer surfaces in a relatively small volume, can be farbricated from alloys to resist corrosion, and are suitable for heating, cooling, evaporating or condensing all kinds of fluids.

17.10.3 Regenerative Heat Exchanger

A regenerative heat exchanger is an apparatus in which one and the same heating surface is alternately exposed to the hot and cold fluids. The heat carried by the hot fluid is taken away by and accumulated in the walls of the apparatus and is then transferred to the cold fluid flowing through the heat exchanger. Regenerators of open hearth and glass-melting furnaces, air preheaters of blast furnaces are some of the specimens of regenerative heat-exchange equipment.

The process of heat transfer in recuperative and regenerative heat exchangers is invariably bound with the surface of solid. That is why they are known as *surface exchanger*.

The special names given to heat exchanger are usually determined by their application and designation for example, steam boilers, furnaces, water heaters, evaporators, super-heaters, condensers etc. In spite of the great variety of shapes, layouts, principles of operation and working media, heat exchange apparatuses—ultimately serve one and the same purpose—transfer of heat from one, hot fluid to another cold fluid. The design fundamentals are, therefore common to all.

The complete design of a heat exchanger can be broken down into three major phases:

1. The thermal analysis
2. The preliminary mechanical design
3. Design for manufacture.

Here the emphasis will be on the thermal design. This phase of the design is primarily concerned with the determination of the heat-transfer surface area required to transfer heat at a specified rate for given flow rates and temperatures of the fluids.

The mechanical design involves considerations of the operating temperatures and pressures, the corrosive characteristics of one or both fluid, the relative thermal expansions and accompanying thermal stresses, and the relation of the heat exchanger to other equipment concerned.

The design for manufacture requires the translation of the physical characteristics and dimensions into a unit, which can be built at a low cost. Selection of materials, seals, enclosure, and the optimum mechanical arrangement have to be made and the manufacturing procedures must be specified.

17.11 BASIC TYPES OF HEAT-EXCHANGERS

The simple type of shell and tube heat exchanger is shown in Fig. 17.7. It consists of a tube or a pipe located concentricity inside another tube which forms the shell for this arrangement. One of the fluids flows through the inner tube and the other through the annulus formed between the inner and the outer tube. Since both fluid streams traverse the exchanger only once, this arrangement is called *single pass* heat exchanger. If both fluids flow in the same direction, the heat exchanger is a *parallel flow type*; if the fluids move in the opposite direction, the exchanger is of the counterflow type.

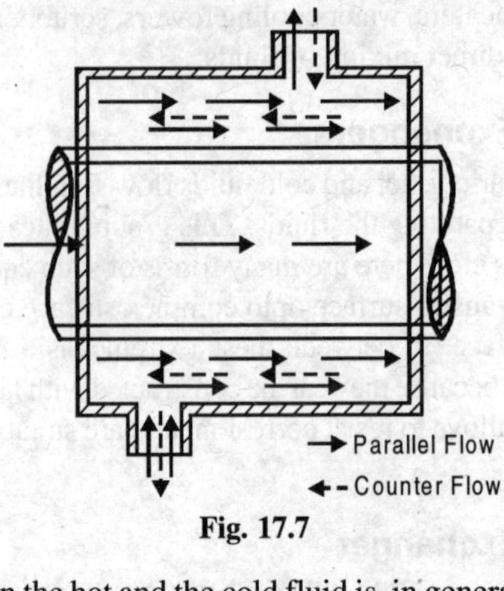

Fig. 17.7

The temperature difference between the hot and the cold fluid is, in general, not constant along tube, and the rate of heat flow vary from section to section. To determine the rate of heat flow one must therefore use an appropriate mean temperature difference.

When the two fluids flowing along the heat transfer surface move at right angles to each other, the heat exchanger is of the *cross flow type*. Three separate arrangements of this type of exchanger are possible. In the first case, each of the fluids is unmixed as it passes through the exchanger and therefore, the temperatures of the fluids leaving the heater section are not uniform, being hotter one side than on the other. A flat-type heater, a design used for turbine regenerators to claim the energy of the exhaust gases, or an automobile radiator approximates this type of exchanger.

In the second case, one of the fluids is unmixed and the other is perfectly mixed as it flows through the exchanger. The temperature of the mixed fluids will be uniform across any section and will vary in the direction of flow. An example of this type is the cross flow air heaters. The air flowing over the bank of tubes is mixed, while the hot gases inside the tubes are confined and therefore do not mix.

In the third case, both of the fluids are mixed as they flow through exchanger, that is the temperature of both fluids will be uniform across the section, and will vary only in the direction of flow. Thus type of arrangement is less important than the other two.

In order to increase the effective heat transfer, optimum heat transfer surface area per unit volume, is provided. Most commercial heat exchangers provide for more than a single pass through the tubes, and the fluid flowing outside the tubes in the shell is routed back and forth by means of baffles. Such an exchanger is known as "*multi-pass*" or *multiflow heat exchanger*. Figure 17.7(a) shows a three pass surface condenser.

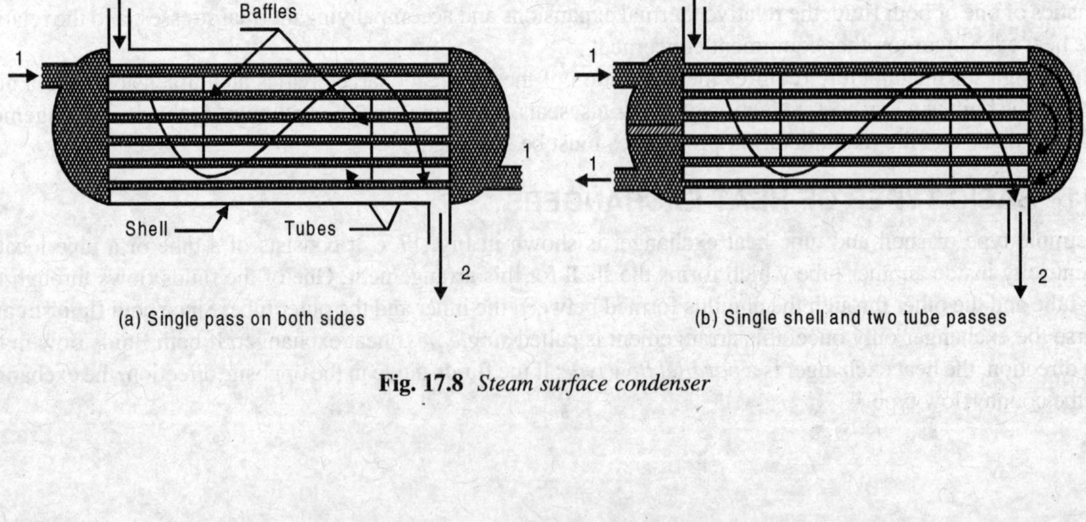

Fig. 17.8 *Steam surface condenser*

17.12 INSULATION

17.12.1 Purpose of Insulation

The insulation is defined as a material which retards the heat flow with reasonable effectiveness. It is not always necessary to have low thermal conductivity for insulating material because aluminum foil has five thousand times greater conductivity than low temperature insulation. It also works as insulator when used with air spaces.

Heat is transferred through insulation by conduction, convection and radiation or by combination of these three modes. There is no insulation which is 100% effective to prevent the flow of heat transfer under temperature gradient.

The purpose of insulation is two fold

(a) to prevent the flow of heat from the system to the surroundings as in the case of steam and hot water pipes which are used for air-conditioning in winter.

(b) to prevent the flow of heat from the surroundings to the system as in the case of brine pipes which are used for air conditioning in summer or refrigerator.

The insulations are commonly used for the following industrial purposes.

(i) Air-conditioning systems
(ii) Refrigerators and food preserving halls
(iii) Preservation of liquid gases
(iv) Boilers and steam pipes
(v) Insulating bricks in all types of furnaces.

17.12.2 Required Properties of Insulating Material

The desired properties of insulators are:

1. Thermal conductivity. Some insulators have higher thermal conductivity than other. It is best to choose insulator having the lowest conductivity. The choice of the insulating material also depends upon outer practical requirements.

2. Structural strength. Some insulating materials tend themselves readily to various types of construction and easily handled under all circumstances. Structural strength generally is obtained by the use of wood or steel frame work. For curved surfaces, material such as ground cork or mineral wool has been used extensively.

3. Weight. The weight of insulation for use in household cabinets is factor of small importance but when insulation under consideration is for use in truck, marine work, rail road, refrigerated car, aeroplane and similar application, weight often becomes a deciding factor.

4. Origin of insulation. Insulators derived from vegetable or animal sources are always subject to decomposition over a period of time. Such materials may develop odours unless special consideration is given to the method of applications.

5. Resistance to vermin. Insulators of the vegetable origin are regarded as food by certain forms of vermin. In tropical and semitropical countries, this difficulty may be troublesome on account of the presence of roaches, ants, mites and other forms of vermin, so that the vegetable type insulators need special protection.

6. Freedom from odour. Many foods absorb odours when they are exposed to odours. For this reason, the insulation used in refrigerator cabinets should be odourless when it is dry. In the presence of moisture, certain forms of vegetable insulation favour the growth of mold which produces odours and soon destroys the insulation. Insulation of this type when used in cabinets must be kept dry to avoid such difficulties.

7. Resistance to moisture absorption. An insulator which absorbs moisture deteriorates rapidly, especially if used on low temperature equipments. The rate of deterioration depends on the following factors:

(a) the thickness of insulation
(b) the temperature and humidity of atmosphere
(c) method of application of the insulation
(d) the temperature difference between interior and exterior faces of the insulation.

8. Resistance to fire. A few insulators are extensively resistant to combustion and some burn very readily. Resistance to fire is not deciding factor, but in some application, such as, fire proof warehouse, is important.

17.13 TYPES OF INSULATIONS

Insulating materials are frequently classified in accordance with the temperatures at which they are used, such as low temperature (<100°C), moderate or steam temperature (100–600°C) and high temperature (> 600°C). This classification is not entirely satisfactory because many insulating materials can be properly employed in two or more of these temperature ranges. It is better to classify these insulators according to physical structure.

1. Loose Fill Insulators. This type of material is generally made in powered, granulated, cellular or fibrous form: (a) Mineral wool (b) Diatomaceous Earth (c) Venniculate (d) Silica Aerogal (e) Crushed insulating brick, asbestos fibres, granulated cork (f) Shredded redwood bark (g) Batts or Flexible Insulation.

2. Slab Insulation. Slab insulation consists usually of wood or sugar cane fibres, processed with an adhesive: (a) Cork board slabs (b) Rock cork (c) Balsa wood (d) Rubatex (e) Foam glass (f) Celotex (g) Hair felts (h) Kapok (i) Palco Bark (j) Insulating papers.

3. Insulating Fire-Brick. The insulating fire bricks are generally made by mixing an organic material with the refractory before moulding and the organic material is subsequently burned during firing leaving voids. The moulded brick is dried, fired. Standard size of brick is 6 × 12 × 25 cm.

4. Reflective Insulation. The effectiveness of metallic surface for retarding heat transfer was well-known to physicists early in the nineteenth century. Conduction and convection are practically eliminated by maintaining a high vacuum in the space.

(a) Aluminum Foil (Alfol)
(b) Low temperature reflective insulation
(c) Lead Tin alloy coated steel sheets.

17.14 EXTENDED SURFACES (FINS)

The equations of conduction heat transfer presented so far have not included the influence of heat losses or gains at the surfaces of the conducting body. When such concurrent heat flow must be considered, the surface coefficient of heat transfer h enters the basic equations. Also it is seen that the overall heat transfer resistance of a plane wall is determined principally by the greatest single resistance. If this resistance is one of the convection resistances, the heat flow through the wall can be increased by putting fins (extended surfaces) on the surface where this large resistance occurs. Such extended surfaces are widely used, for example, in economizers in steam power plants, convectors for steam and hot water heating systems, or electrical transformers, for the cylinders of air-craft engines, two wheeler engines etc.

There are many forms and shapes of these extended surfaces. Some of these are as follows :

1. A thin rod. This is transferring heat at its surface to a surrounding fluid and which is connected at its base to a heated wall. This system is shown in Fig. 17.9.

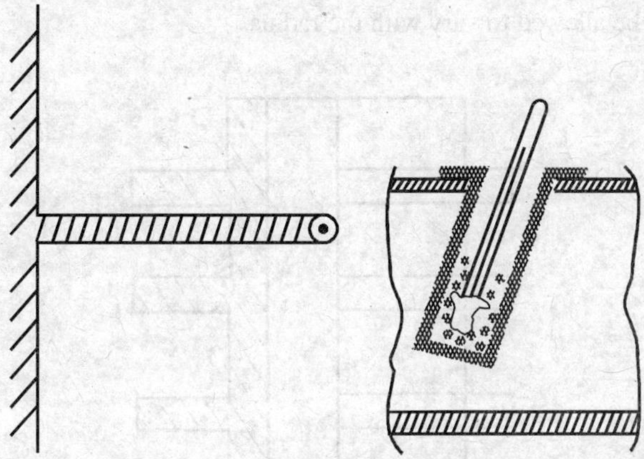

Fig. 17.9. *Thermometer packet or well*

2. The Rectangular Fin. This is the simplest case for the plane finned surface. As long as the height of the fins on a tube is comparatively small with respect to the tube diameter, the formula derived from the plane wall can also be used for the tube. This system is shown in Fig. 17.10.

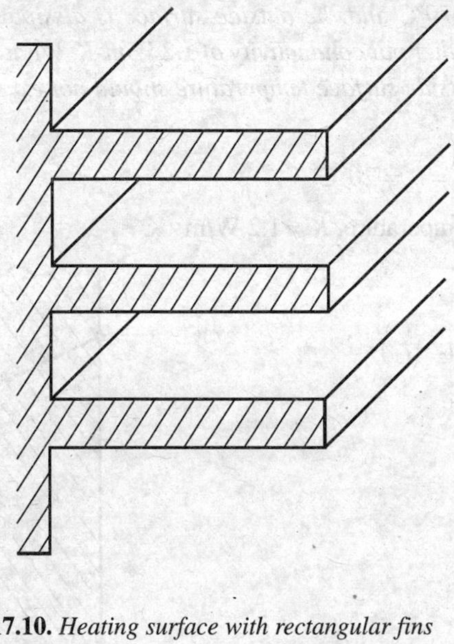

Fig. 17.10. *Heating surface with rectangular fins*

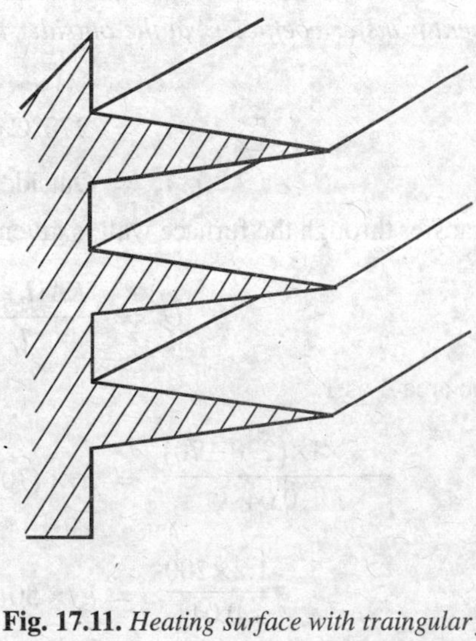

Fig. 17.11. *Heating surface with traingular fins*

3. Straight Fins of Triangular Profile. In determining the optimum fin, the question arises as to whether or not weight advantage can be gained by using a shape other that rectangular for the fin cross-section. Generally, a straight fin of triangular cross-section is considered. Such a fin is shown in Fig. 17.11.

The mathematical treatment in this case is similar to the case of the fin of rectangular cross-section except that the area nornal to the heat flow is a function of the distance along the fin, decreasing as the fin length increases.

4. Cylindrical Fins. Fins which are arranged around tubes are called cylindrical fins and are quite important from an engineering point of view.

Such a fin system is shown in Fig. 17.12. Here again the treatment is substantially the same as for the rectangular fin except that the area must be allowed to vary with the radius.

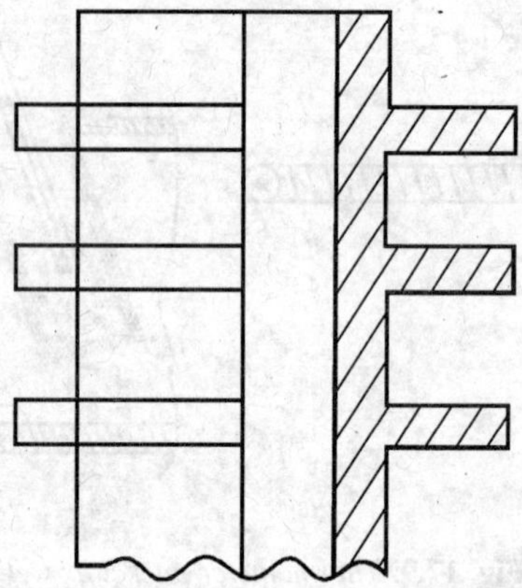

Fig. 17.12. *Circular fins.*

SOLVED EXAMPLES

Example 17.1 *The inside surface of a furnace wall is at 270°C and the outside surface is dissipating heat by convection into air at 20°C. The wall is 40 mm thick and has thermal conductivity of 1.2 W/m-K. What is minimum value of heat transfer coefficient at the outside surface if outside surface temperature should not exceed 70°C.*

Solution

Given:

$$t_1 = 270°\text{C} = t_2 = 70°\text{C} = t_3 = 20°\text{C}$$

$$t_2 = \text{Outside Surface Temperature, } K = 1.2 \text{ W/m} \circ \text{K}$$

Heat transfer through the furnace wall is given by

$$Q = \frac{KA(t_1 - t_2)}{L} = h_c A(t_2 - t_3)$$

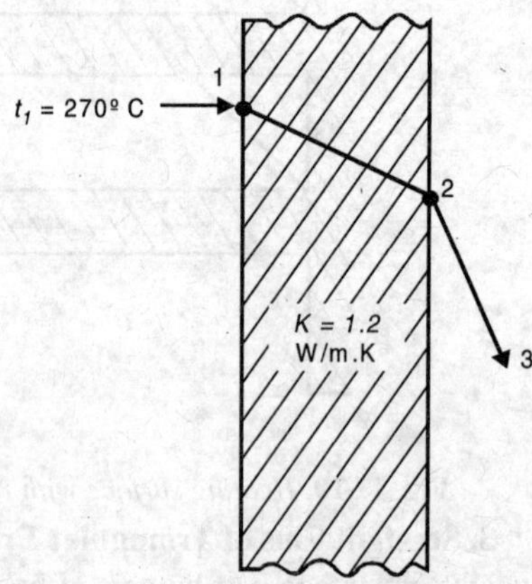

Fig. Ex. 17.1

Assume area A as 1

$$\therefore \quad \frac{1.2 \times 1 \times (270 - 70)}{0.04} = h_c \times (70 - 20)$$

$$\frac{1.2 \times 200}{0.04} = h_c \times 50$$

$$h_c = \text{heat transfer coefficient}$$

$$= \frac{1.2 \times 200}{50 \times 0.04}$$

$$= \mathbf{120\ W/m^2K}$$

Example 17.2 *A black body at 30°C is heated to 100°C. Calculate the change in its emissive power.*

Solution

(c) Emissive power of block body $= \sigma A T^4$

where $\sigma = 20.52 \times 10^{-8}$ kJ/hr K^4.

$\therefore$ For unit area, $A = 1$ m^2

Increase of emmissive power

$$= \sigma\left(T_2^4 - T_1^4\right)$$

$$= 20.52 \times 10^{-8} \,(100 + 273)^4 - (30 + 273)^4$$

$$= 20.52 \times 10^{-8}\left[373^4 - 303^4\right]$$

$$= 20.52 \times 109.28$$

$$= \mathbf{2242.4256 \text{ kJ/hr.}}$$

Example 17.3 (*a*) *A mild steel tank of wall thickness 12 mm contains water at 100°C. Calculate the rate of heat loss per square metre area of the surface of the tank when the atmospheric temperature is 20°C. Also calculate temperature of the outside surface of the tank and overall heat transfer coefficient.*

Assume the following :

Conductivity of mild steel 50 W/m-K

Convection heat transfer coefficient

(*i*) *on water side 2850 W/m² L*

(*ii*) *on air side 10 W/m² K*

(*b*) *A 60 W incandescent lamp has coil surface temperature 2500 K and room temperature 300 K. Estimate the surface area of the coil in mm².*

(*c*) *Derive an equation of conduction heat transfer rate in terms of temperatures dimensions and thermal conductivity for a thick, hollow cylinder.*

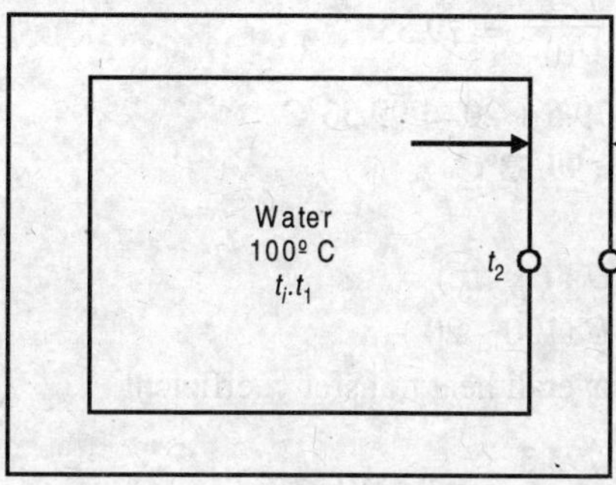

Fig. Ex. 17.3

Solution

Consider 1 sq. m of tank surface. The system is given in Fig. Ex. 17.3. Let Q be the heat rate

$$Q = h_i A\,(t_1 - t_2) \text{ (Water to inner surface)}$$

$$= \frac{KA(t_2 - t_3)}{x} \text{ (Inner to outer surface)}$$

$$= h_0 A\,(t_3 - t_4) \text{ (Outer surface to air)}$$

$$\therefore \quad t_1 - t_2 = \frac{Q}{h_i} \quad \text{for unit area}$$

$$t_2 - t_3 = \frac{Q.x}{K} \quad \text{for unit area}$$

$$t_3 - t_4 = \frac{Q}{h_0} \quad \text{for unit area}$$

$$t_1 - t_4 = Q\left[\frac{1}{h_i} + \frac{x}{K} + \frac{1}{h_0}\right]$$

$$\therefore \quad 100 - 20 = Q\left[\frac{1}{2850} + \frac{12/1000}{50} + \frac{1}{10}\right]$$

$$80 = Q[0.0003509 + 0.00024 + 0.1] = Q \times 0.1006$$

$$Q = \text{Rate of heat transfer/m}^2$$

$$= \mathbf{\frac{80}{0.1005989} = 795.3\ W}$$

Again
$$795.3 = h_0 A(t_3 - t_4) = 10[t_3 - 20]$$

$$t_3 - 20 = \frac{795.3}{10}$$

$$t_3 = 79.5 + 20$$

$$t_3 - 20 = \frac{795.3}{10} = 79.53$$

$$t_3 = 79.5 + 20 = 99.53°\text{C}$$

Temperature of the outside surface = 99.53°C

For overall heat transfer coefficient

We have
$$Q = UA\,(t_i - t_0)$$

$$795.3 = U\,(100 - 20)$$

$$U = \text{overall heat transfer coefficient}$$

$$= \mathbf{\frac{795.3}{80} = 9.94\ W/m^2/K}$$

(b) Heat transfer from the incandescent lamp is 60 W.

$$\therefore \quad 60 = \sigma\left(T_2^4 - T_1^4\right)$$

where
$$\sigma = 5.671 \times 10^{-8}\ \text{W/m}^2/\text{K}^4$$

$$A = \text{Area sq. m.}$$

$$T_1 = \text{Lamp surface temperature}$$

$$T_2 = \text{Room temperature}$$

$$\therefore \quad 60 = 5.671 \times 10^{-8} \times A \times (2500^4 - 300^4)$$

$$A = \frac{60\times10^8}{5.671\times(2500^4-300^4)}$$

$$= \frac{60\times10^8}{5.671\times10^8(25^4-3^4)}\text{m}^2 = \frac{10.58\times10^6}{(625-9)(625+9)}\text{mm}^2$$

$$= \mathbf{27.091\ mm^2}$$

$$= \text{Coil/surface area}$$

(c) For derivation of the required equation, **Refer Ex. 17.5.**

Example 17.4 (*a*) *Determine the rate of heat loss per hour through a 3.5 m² boiler plate if it is 20 mm thick with thermal conductivity 50 W/m-K the inner surface of the wall is covered with a layer of scale and deposits 2 mm thickness having a conductivity of 1 W/m-K. The hot gas temperature of the plate is 250°C while that on the cold air (outside) is 200°C. Also determine interface temperature.*

(*b*) *determine the loss by radiation from a steel tube 70 mm diameter and 3 m long at a temperature of 227°C the tube is located in a huge room having wall temperature 27°C.*

Solution

The plate with scale and deposit is shown in Fig. Ex. 17.4

With reference to this figure,

We write
$$Q = \frac{K_1A_1(t_1-t_2)}{x_1} = \frac{K_2A_2(t_2-t_3)}{x_2}$$

$$\therefore \quad t_1 - t_2 = \frac{Q.x_1}{K_1A_1}$$

$$t_2 - t_3 = \frac{Q.x_2}{K_2A_2}$$

$$t_1 - t_3 = \frac{Q}{A_1}\left[\frac{X_1}{K_1}+\frac{X_2}{K_2}\right]$$

$$250 - 200 = \frac{Q}{3.5}\left[\frac{2}{1000\times1}+\frac{20}{1000\times50}\right]$$

$$= \frac{Q}{3.5\times1000}[2+0.4]$$

$$Q = \frac{50\times3.5\times1000}{2.4} = 72916.7\text{ W}$$

$$= \text{Rate of heat loss}$$

$$= \mathbf{72.92\ kJ/s}$$

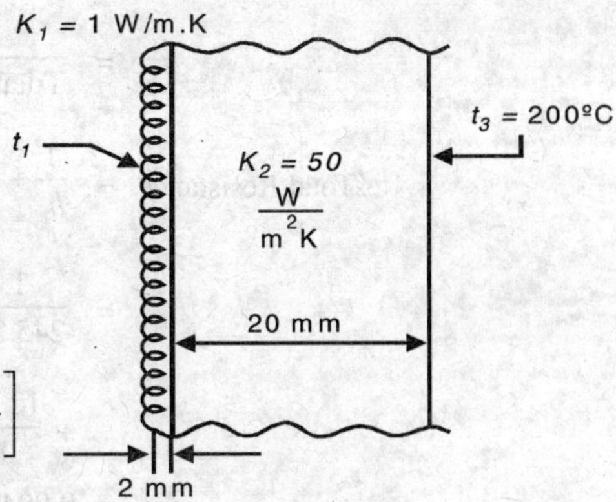

Fig. Ex. 17.4

Rate of heat per hour

$$= 72.92 \times 3600 \text{ kJ/hr} = 26512 \text{ kJ/hr}$$

$$= \mathbf{26.25 \times 10^4 \text{ kJ/hr}}$$

Example 17.5 *A furnace wall consists of two lawyer 22.5 cm of brick (K = 4.984 kJ /hr.m °C/m) and 12.5 cm of insulating brick (K= 0.623 kJ/hr.m °C/m) the temperature inside the furnace is 1650 °C and the unit surface conductance at the inside wall is 245.28 kJ/hr.m °C the temperature of the surrounding atmosphere is 27 °C and the unit surface conductance at the outer wall is 40.88 kJ /hr. m² °C. Neglecting the thermal resistance of the morter joints. Estimate (a) the rate of heat loss per sq.m of wall and the temperatures at the (b) intersurface and (c) outersurface.*

Solution

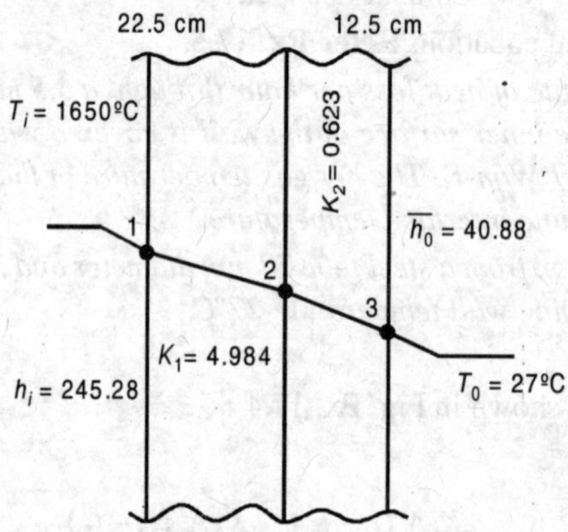

Fig. Ex. 17.5

(a) The rate of heat flow is given by

$$\frac{q}{A} = \frac{T_I - T_O}{\text{Total Resistance}}$$

$$\text{Total Resistance} = \frac{1}{h_I} + \frac{L_1}{K_1} + \frac{L_2}{K_2} + \frac{1}{h_O}$$

$$= \frac{1}{245.28} + \frac{22.5/100}{4.984}$$

$$+ \frac{12.5/100}{0.623} + \frac{1}{40.88}$$

$$= 0.00408 + 0.04514 + 0.20064 + 0.02446$$

$$= 0.27432$$

$$\therefore \quad q = \frac{1650 - 27}{0.27432} = \frac{1623}{0.27432}$$

$$= \mathbf{5916.9 \text{ kJ/hr.m}^2}$$

(b) We have $\frac{q}{A} = 5916.9 \frac{T_i - T_1}{R_1} = \frac{1650 - T_1}{0.00408}$

$$\therefore \quad T_1 = \text{Inner wall temperature}$$
$$= 1650 - 5916.9 \times 0.00408 = 1650 - 24.14$$
$$= \mathbf{1625.86 \,^\circ C}$$

(c) Again for outer surface temperatures

We have
$$\frac{q}{A} = \frac{T_i - T_3}{R_1 + R_2 + R_3} = \frac{1650 - T_3}{0.00408 + 0.04514 + 0.20064}$$

$$5916.9 = \frac{1650 - T_3}{0.24986}$$

$$\therefore \quad T_3 = 1650 - 5916.9 \times 0.24986$$
$$= 165\ 0 - 1478.38$$
$$= \mathbf{171.62\ ^\circ C}$$

Example 17.6 *A brick wall of a building is 30 cm thick and has and inside surface temperature of 24°C and an outside surface temperature of –6°C the wall is 2.75 m high and 6.1 m long. Estimate (a) the heat transfer by conduction through the wall per hour and (b) the conductance and resistance of the wall. The coefficient of conductivity of the brick material is 2.6 kJ/hr.m °C.*

Solution

(a) Heat transfer by conduction

We have
$$q = \frac{KA\Delta T}{x}$$

Where
$$K = 2.6 \text{ kJ/hrm °C}$$
$$\Delta T = t_2 - t_1$$
$$= -(t_1 - t_2)$$
$$= -(24 + 6) = -30\ ^\circ\text{C}$$

$$A = \text{Area of the wall}$$
$$= 2.75 \times 6.1 = 16.8 \text{ m}^2$$
$$x = \text{Thickness of wall in m}$$
$$= \frac{30}{100} = 0.3 \text{ m}$$

Substituting these values, we get

$$q = \frac{2.6 \times 16.8 \times 30}{0.3}$$
$$= \mathbf{4368\ kJ/hr}$$

(b) Conductance and resistance of the wall

$$\text{Resistance} = \frac{t_1 - t_2}{q} = \frac{30}{4368}$$
$$= \mathbf{0.00687\ C.hr/kJ}$$

$$\text{Conductance} = \frac{1}{\text{Resistance}}$$
$$= \frac{4368}{30}$$
$$= \mathbf{145.6\ kJ/m\ ^\circ C}$$

Example 17.7 *A temperature of the inner surface of a furnace wall is 450ºC and that of the outer surface 200ºC. The ambient temperature is 50ºC. It is required to reduce the heat loss from the outer surface by doubling the thickness of the brickwork. Assuming constant values convective heat transfer coefficient of air and thermal conductivity of wall material calculate percent decrease in heat loss.*

Solution

Let K be the conductivity of the wall material and $\bar{x}$ be the thickness.

Let T_0 be outside temperature and h_0 be the unit coefficient of heat transfer for outer surface.

∴ Heat transfer

$$q = \frac{450 - T_0}{\frac{x}{K} + \frac{1}{h_0}} \text{ per unit area} \quad \ldots\text{(i)}$$

When the wall is doubled in thickness, heat transfer.

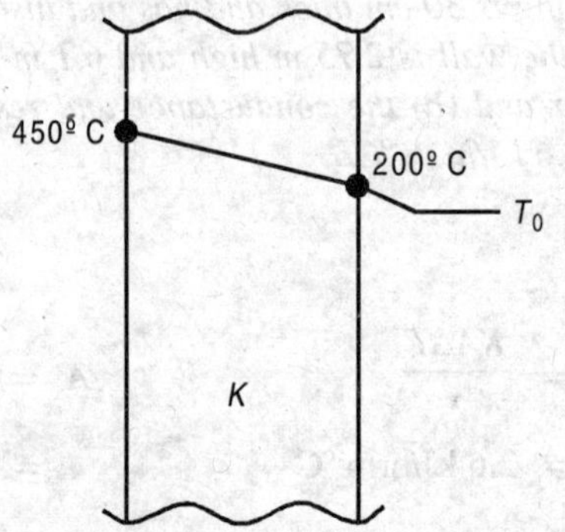

Fig. Ex. 17.7

$$q = \frac{450 - T_0}{\frac{2x}{K} + \frac{1}{h_0}} \quad \ldots\text{(ii)}$$

$$\text{Decrease in heat loss} = (\text{i} - \text{ii})$$

$$= (450 - T_0)\left\{\frac{450 - T_0}{\frac{x}{K} + \frac{1}{h_0}} - \frac{450 - T_0}{\frac{2x}{K} + \frac{1}{h_0}}\right\}$$

$$\text{Percentage decrease in heat loses} = \frac{(450 - T_0)\left\{\frac{1}{\frac{x}{K} + \frac{1}{h_0}} - \frac{1}{\frac{2x}{K} + \frac{1}{h_0}}\right\}}{(450 - T_0)\left\{\frac{1}{\frac{x}{K} + \frac{1}{h_0}}\right\}}$$

$$= 1 - \frac{\frac{1}{2x/K} + \frac{1}{h_0}}{\frac{1}{x/K} + \frac{1}{h_0}} = 1 - \frac{\frac{x}{K} + \frac{1}{h_0}}{\frac{2x}{K} + \frac{1}{h_0}}$$

$$= \frac{\frac{2x}{K} + \frac{1}{h_0} - \frac{x}{K} - \frac{1}{h_0}}{\frac{2x}{K} + \frac{1}{h_0}}$$

$$= \frac{\frac{x}{K}}{\frac{2x}{K} + \frac{1}{h_0}}$$

$$= \frac{\mathbf{x}}{\mathbf{K}\left[\frac{\mathbf{2x}}{\mathbf{K}} + \frac{\mathbf{1}}{\mathbf{h_0}}\right]} \times \mathbf{100}$$

Example 17.8 *Assuming earth as a block body at a temperature of 300K. Estimate the energy received by unit area of the earth's surface perpendicular to the solar rays.*

Solution

According to Stefan–Boltzamn's law heat absorbed = Heat radiated

For a black-body

∴ Heat received by unit area of the earth's surface perpendicular to the solar ray

$$Q = \sigma T^4 = 20.52 \times 10^{-8} \times (300)^4$$

$$= 20.52 \times 10^{-8} \times 10^8 \times 3^4$$

$$= 20.52 \times 81$$

$$= \mathbf{1662.12\ kJ/hr}$$

Example 17.9 *Determine the loss of heat by radiation from a steel tube 70 mm diameter and 3 m long at a temperature of 227 °C the tube is located in a large room having wall temperature 27°C.*

Solution

$$Q = A.\sigma\left(T_H^4 - T_C^4\right)$$

$$A = 2\pi DL = 2\pi \times \frac{70}{1000} \times 3\,\text{m}^3$$

$$= \frac{42\pi}{100} = 0.42\,\pi\text{m}^2$$

$$= \frac{22}{7} \times \frac{42}{100} = \frac{132}{100} = 1.32$$

$$= 1.32 \times 20.52 \times 10^{-8} [500^4 - 300^4]$$
$$= 1.32 \times 20.52 [5^4 - 3^4]$$
$$= 1.32 \times 20.52 [625 - 81]$$
$$= 1.32 \times 20.25 \times 544$$
$$= 14730 \text{ kJ/hr}$$
$$= \textbf{4094 W or 4.094 kW}$$

Example 17.10 *The inside temperature of refrigerator is required to be 7 °C the walls are constructed with two mild steel sheets 3 mm thick with 5 cm of glass wool insulation between them. The heat transfer coefficient on inner and outer surface of refrigerator are 10 W/m°C and 12.5 W/m²C. Find the rate of heat removed from the refrigerator when it is kept in a kitchen room. The temperature in kitchen is 28°C and K for mild steel is 40W/°C and glass wool is 0.04 W/m°C.*

Solution

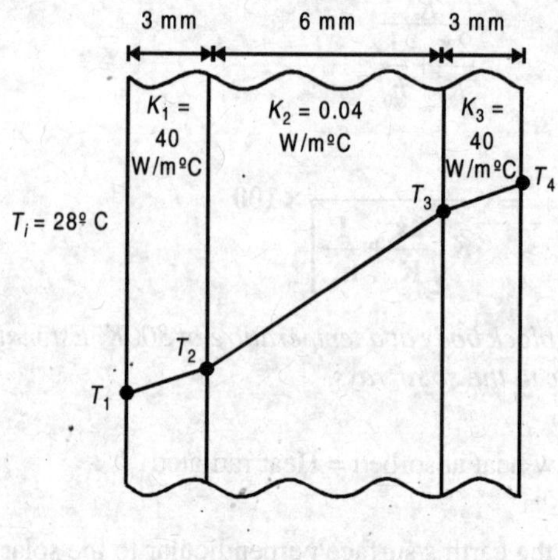

Fig. Ex. 17.10

This is a problem of composit walls and the wall in problem is shown Fig. Ex. 17.10.

T_1 = 7°C, T_0 = 28°C, K_1 = 40 W.m°C K_3; x_1 = 3 mm = x_3 and x_2 = 50 mm, h_i = 10 W/m²°C and h_0 = 12.5 W/m²/°C. Consider unit area and we write

$$Q = h_i A (T_i - T_1) \quad \therefore \quad T_i - T_1 = \frac{Q}{h_i A}$$

$$= \frac{K_1 A_1 (T_1 - T_2)}{x_1} \quad \quad T_1 - T_2 = \frac{Q \times x_1}{K_1 A_1}$$

$$= \frac{K_2 A_2 (T_2 - T_3)}{x_2} \quad \therefore \quad T_2 - T_3 = \frac{Q \times x_2}{K_2 A_2}$$

$$= \frac{K_3 A_3 (T_3 - T_4)}{x_3} \quad \therefore \quad T_3 - T_4 = \frac{Q \times x_3}{K_3 A_3}$$

$$= h_0 A_0 (T_4 - T_0) \quad \therefore \quad T_4 - T_0 = \frac{Q}{h_0 A_0}$$

Adding these we get $T_i - T_0 = Q\left[\frac{1}{h_iA_i}+\frac{1}{\frac{K_1A_1}{x_1}}+\frac{1}{\frac{K_2A_2}{x_2}}+\frac{1}{\frac{K_3A_3}{x_3}}+\frac{1}{h_0A_0}\right]$

Assuming unit area of heat transfer

$$Q = \frac{T_i - T_0}{\frac{1}{h_i}+\frac{x_1}{K_1}+\frac{x_2}{K_2}+\frac{x_3}{K_3}+\frac{1}{h_0}}$$

$$= \frac{7-28}{\frac{1}{10}+\frac{0.003}{40}+\frac{0.05}{0.04}+\frac{0.003}{40}+\frac{1}{12.5}}$$

$$= \frac{-21}{1.43015}$$

$$= \mathbf{-14.684\ W/m^2}$$

The –ve sign indicates that the heat flows from the room to the refrigerator.

Example 17.11 *A furnace wall is made of 20 cm of brick and 20 cm of common brick. The fire brick is exposed to hot gases at 1400°C and common brick outer surface is exposed to 50°C. The convection and radiation heat transfer coefficient towards gas side are 16.5 and 17.5 W/m°C respectively. The convection end radiation heat transfer coefficient towards outside are 12.5 and 6. W/m°K respectively. Thermal conductivity's of fire brick and common brick are 4 and 0.065 W/m°C respectively. Determine (i) heat loss par m² area of the furnace wall, (ii) maximum temperature to which common brick is subjected.*

Solution

1. The heat flow through the furnace composite wall per unit area is given by

$$Q = h_i(T_i - T_1)$$

$$= \frac{K_1.(T_1 - T_2)}{x_1}$$

$$= \frac{K_2.(T_2 - T_3)}{x_3}$$

$$= h_0\ (T_3 - T_0)$$

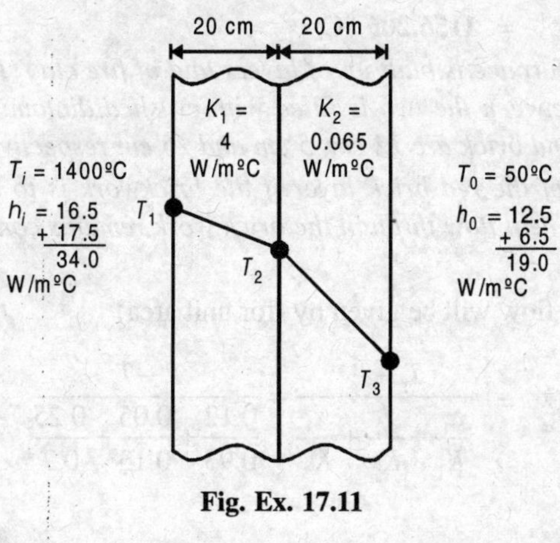

Fig. Ex. 17.11

$$\therefore \qquad Q = \frac{T_i - T_0}{\frac{1}{h_i}+\frac{x_1}{K_1}+\frac{x_2}{K_2}+\frac{1}{h_0}}$$

We take here $h_i = 34 = \overline{h_c} + \overline{h_r}$

$$= 16.5 + 17.5 = 34$$

$$= \frac{1400-50}{\frac{1}{34}+\frac{0.2}{4}+\frac{0.2}{0.65}+\frac{1}{19}} = 1350.$$

$$= \frac{1350}{0.0294+0.05+0.3077+0.0526}$$

$$= \frac{1350}{0.4397} = 3070 \text{ W/m}^2$$

$$= \mathbf{3.07\ kW/m^2}$$

2. Maximum temperature to which common brick is subjected is T_2

$$\therefore \qquad Q = h_i\,(T_i - T_1)$$

$$\therefore \qquad 3070 = 34\,(1400 - T_1)$$

$$\therefore \qquad \frac{3070}{34} = 90.294 = 1400 - T_1$$

$$\therefore \qquad T_1 = 1400 - 90.294$$

$$= 1309.706\ °C$$

Similarly $Q = \dfrac{K_1(T_1 - T_2)}{x_1}$

$$\therefore \qquad T_1 - T_2 = \frac{3070\times 0.2}{4} = 153.5$$

$$T_2 = 1309.706 - 153.5$$

$$= \mathbf{1156.206\ °C}$$

Example17.12 *A brick work of a furnace is built up of layers laid of fire clay (K_1= 0.93 W/m °C) and red brick (K_3= 0.7W/m °C and the space between the two is filled with crushed diatomite brick (K= 0.13W/m °C). The thickness of fireday, diatomobile and brick are 12 cm, 5 cm and 25 cm respectively.*

What should be the thickness of the red brick layer if the brickwork is to be laid without diatomite filling between the two layers so that the heat flow through the brick work remains constant.

Solution

With the diatomite filling, the heat flow will be given by (for unit area)

$$Q = \frac{T_1 - T_4}{\frac{x_1}{K_1}+\frac{x_2}{K_2}+\frac{x_3}{K_3}} = \frac{T_1 - T_4}{\frac{0.12}{0.93}+\frac{0.05}{0.13}+\frac{0.25}{0.7}}$$

$$= \frac{T_1 - T_4}{0.1290 + 0.3846 + 0.3571}$$

$$= \frac{T_1 - T_4}{0.8708} \text{ W/m}^2 \quad \ldots(i)$$

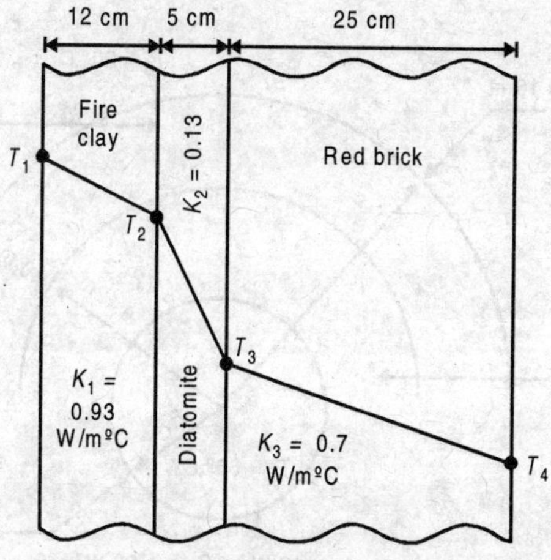

Fig. Ex.17.12

For the same Q-heat flow and without diatomite fillings, we have

$$Q = \frac{T_1 - T_4}{\frac{x_1}{K_1} + \frac{x_3}{K_3}} = \frac{T_1 - T_4}{\frac{0.12}{0.93} + \frac{x_3}{0.7}}$$

$$\therefore \quad \frac{T_1 - T_4}{0.8708} = \frac{T_1 - T_4}{0.1290 + \frac{x_3}{0.7}}$$

$$\therefore \quad \frac{1}{0.8708} = \frac{1}{0.1290 + \frac{x_3}{0.7}}$$

$$\therefore \quad 0.1290 + \frac{x_3}{0.7} = 0.8707$$

$$\therefore \quad \frac{x_3}{0.7} = 0.8707 - 0.1290$$

$$= 0.7417$$

$\therefore \quad x_3$ = Thickness of the red brick layer

$= 0.7417 \times 0.7$

$= 0.51919$ m

$=$ **51.919 cm**

Example 17.13 *Hot air at a temperature of 656°C is flowing through a steel pipe of 120 mm diameter. The pipe is covered with two layers of different insulating materials of thickness 60 mm and 40 mm and their corresponding thermal conductivities are 0.24 and 0.4 W/m°C the inside and outside heat transfer coefficients are 60 and*

12 W/m² °C. The atmosphere is at 20 °C. Find the rate of heat loss from a 60 m length of pipe. Neglect the resistance of the steel pipe.

Solution

The pipe with insulations is shown in Fig. Ex. 17.13 we have

$$R_1 = 60 \text{ mm} = 0.06 \text{ m}$$
$$R_2 = 60 + 60 = 120 \text{ mm} = 0.12 \text{ m}$$

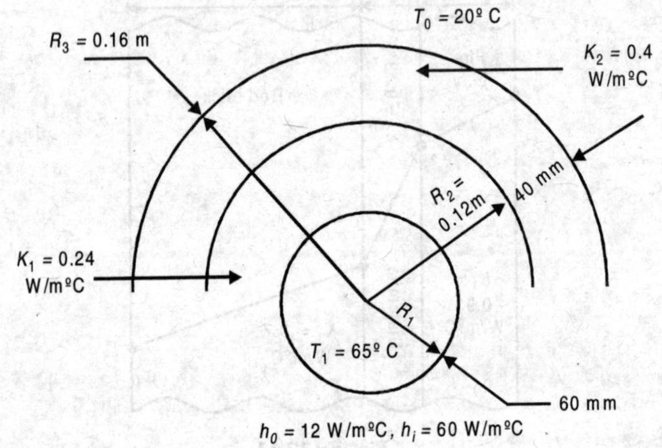

Fig. Ex.17.13

$$R_3 = 60 + 60 + 40 = 160 \text{ mm} = 0.16 \text{ m}$$
$$K_1 = 0.24 \text{ W/m °C} \qquad K_2 = 0.4 \text{ W/m °C}$$
$$h_i = 60 \text{ W/m °C} \qquad h_0 = 12 \text{ W/m °C}$$

Length of the pipe to be considered is 60 m

Heat transfer through a circular, hollow pipe with raddi r_1 and r_2 is given by

$$Q = \frac{K.2\pi(t_1 - t_2)}{\log_e \frac{r_2}{r_1}} = \frac{t_1 - t_2}{\frac{1}{2\pi L}.\frac{1}{K}\log_e \frac{r_2}{r_1}}$$

Applying this equation to the problem, we have

$$Q = \frac{t_i - t_o}{\frac{1}{2\pi L}\left[\frac{1}{h_i R_1} + \frac{1}{K_1}\log_e \frac{R_2}{R_1} + \frac{1}{K_2}\log_e \frac{R_3}{R_2} + \frac{1}{h_0 R_3}\right]}$$

$$= \frac{65 - 20}{\frac{1}{2\pi L}\left[\frac{1}{60\times 0.06} + \frac{1}{0.24}\log_e \frac{0.12}{0.06} + \frac{1}{0.4}\log_e \frac{0.16}{0.12} + \frac{1}{12\times 0.16}\right]}$$

$$= \frac{45}{0.002653[0.2778 + 2.888 + 0.7192 + 0.5208]}$$

$$= \frac{45}{0.002653 \times 4.4058}$$

$$= \frac{45}{0.01169}$$

$$= \mathbf{3.8494\ kW}$$

Example 17.14 *A steam pipe of 16 cm inside diameter and 17 cm of outside diameter is covered with two layers of insulation. The thickness of first and second layers are 3 cm and 5 cm respectively and their corresponding conductivities are 0.15 and 0.08 W/m °C respectively. The conductivity of the pipe material is 50 W/m °C. The temperature of inner surface of the steam pipe is 300 °C and that of the outer surface of insulation is 50 °C. Find*

(i) the quantity of heat lost per metre length of the steam pipe.

(ii) if the steam is fully saturated at the entrance of the pipe find the quality of the steam coming out of one metre length of pipe. The quantity of steam flow is 40 gm/hr. Neglect the pressure loss.

Solution

The insulated pipe is shown in Fig. Ex. 17.14.

(i) Referring to this Fig. Ex. 17.14.

R_1 = 8 cm, R_2 = 8.5 cm, R_3 = 11.5 cm, R_4 = 16.5 cm, T_1 = 300 °C, T_4 = 50 °C

Length of pipe = L = 1m. K_1 = 50 W/m °C, K_2 = 0.15 W/m °C, K_3 = 0.08 W/m °C.

The heat loss through a composite, cylinder is given by

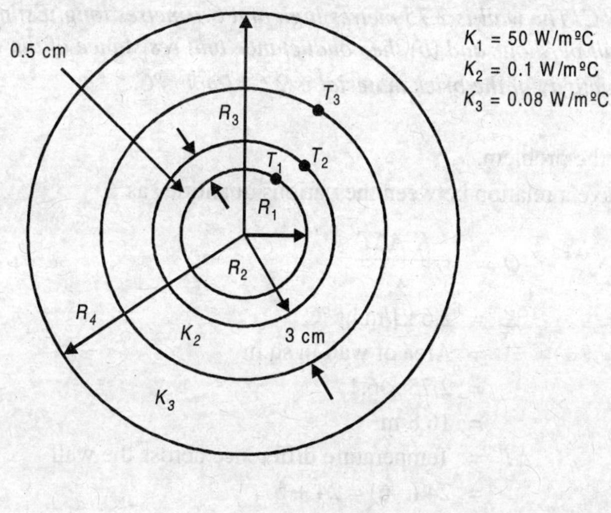

Fig. Ex. 17.14

$$Q = \frac{2\pi L(T_1 - T_4)}{\frac{1}{K_1}\log_e \frac{R_2}{R_1} + \frac{1}{K_2}\log_e \frac{R_3}{R_2} + \frac{1}{K_3}\log_e \frac{R_4}{R_3}}$$

$$= \frac{2\pi \times 1(300-50)}{\frac{1}{50}\log_e \frac{8.5}{8} + \frac{1}{0.15}\log_e \frac{11.5}{8.5} + \frac{1}{0.08}\log_e \frac{16.5}{11.5}}$$

$$= \frac{1570.8}{0.001212+2.0152+4.513}$$

$$= \frac{1570.8}{6.5291} = 240.58 \text{ W/m}$$

$$= 0.24058 \text{ kW} = \mathbf{866 \text{ kJ/hr}}$$

(ii) In enthalpy or total heat of dry and saturated steam at 300 ºC is obtained from steam tables and is 2751 kJ/kg total heat of steam at 1 m ahead of the entrance is given by

$$2751 - \frac{866}{40} = 2751 - 21.65$$

$$= 2729.35 \text{ kJ/kg}$$

$$= h_f + xhfg$$

$$= 1345 + x \times 1406$$

$\therefore$ x = Dryness fraction of steam

$$= \frac{2729.35 - 1345}{1406} = \frac{1384.35}{1406}$$

$$= \mathbf{0.9846}$$

Example 17.15 *A brickwall of a building is 30 cm thick and has an inside surface temperature of 24 ºC, an outside surface temperature of –6 ºC. The wall is 2.75 metres high and 6.1 metres long. Estimate* (a) *the heat transfer by conduction through the wall per hour and* (b) *the conductance and resistance of the wall.*

The coefficient of conductivity of the brick material is 2.6 kJ/m.hr ºC.

Solution

Refer Fig. Ex. 17.15 from the problem.

(a) Heat transfer: We have a relation between the various quantities as

$$Q = \frac{-K.A\Delta T}{x}$$

where K = 2.6 kJ/m.hr ºC

A = Area of wall in sq.m

$= 2.75 \times 6.1$

$= 16.8 \text{ m}^2$

ΔT = Temperature difference across the wall

$= 24 (-6) = 24 + 6$

$= 30$ ºC

x = Thickness if wall in metres

$$= \frac{30}{100} \text{ m}$$

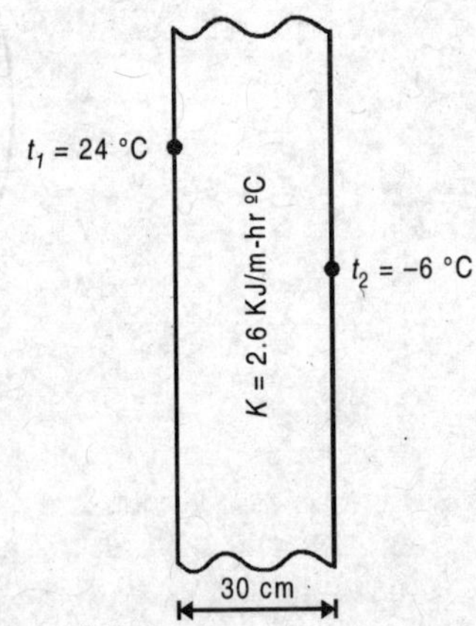

Fig. Ex. 17.15

Substituting these values in the above equation, we get

$$Q = \frac{2.6 \times 16.8 \times 30}{0.3} = 62 \times 16.8$$

$$= \mathbf{4368 \text{ kJ/hr}}$$

(b) Conductance and resistance of the wall:

$$\text{Resistance} = \frac{t_1 - t_2}{Q} = \frac{30}{1068}$$

$$= 0.00687 \text{ hr.C/kJ}$$

$$\text{Conductance} = \frac{1}{\text{Resistance}} = \frac{4368}{30}$$

$$= \mathbf{145.6\ kJ/hr\ ^\circ C}$$

Example 17.16 *The inside and outside surface of a furnace wall 30 cm thick of refractory bricks (K = 5.66 kJ/hr.m°C) are at 1650°C and 320°C respectively. Find the reduction in heat loss through the wall to be obtained by adding 30 cm thickness of insulating bricks for which K = 1.26 kh/hr m°C. Assuming the inside surface temperature of refractory bricks to remain fixed at 1650°C. The temperature of outside surface of the brick may be taken as 27°C.*

Solution

Arrangement are shown in Fig. Ex. 17.16 (a) and (b)

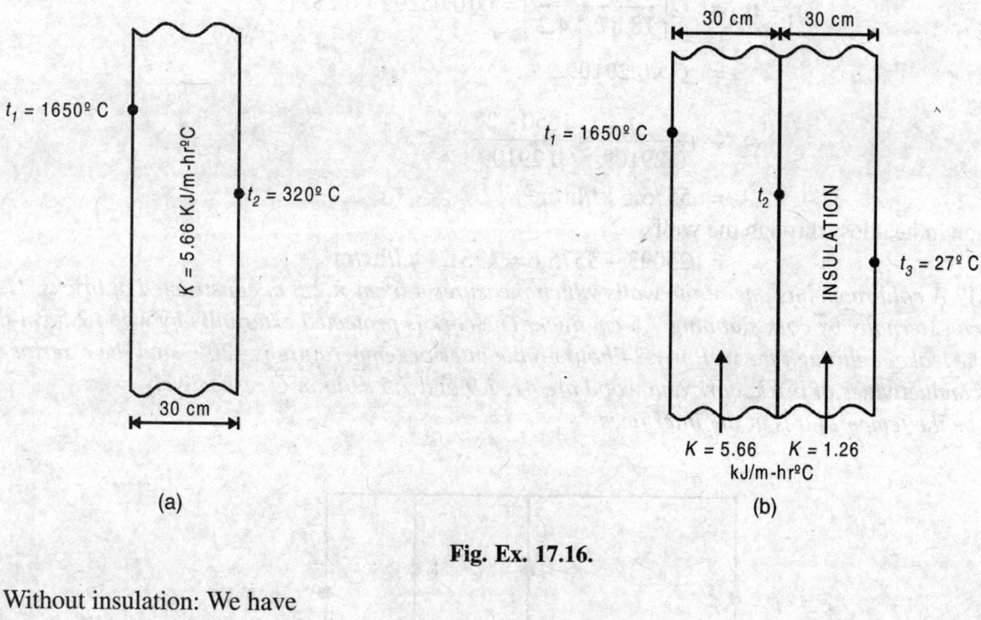

Fig. Ex. 17.16.

(a) Without insulation: We have

$$Q = \frac{-K\,A\Delta T}{x}$$

Assuming area A as unity we get

$$Q = \frac{566 \times 1 \times (320 - 1650)}{0.3}$$

$$= 25093 \text{ kJ/hr/m}^2$$

(b) With insulation: For the refractory brick wall, we have

$$Q = -\frac{K_1 A\Delta T}{x_1} = \frac{5.66 \times (t_1 - t_2)}{0.3}$$

$$= \frac{5.66\times(t_1 - t_2)}{0.3} = 18.87\ (t_1 - t_2)$$

or $$t_1 - t_2 = \frac{Q}{18.87} \quad \text{...(i)}$$

Similarly for insulating brick wall,

$$Q = \frac{-K_2 A(t_3 - t_2)}{x_2} = \frac{1.26\times 1\times t_2 - t_3}{0.3}$$

$$= 4.2\ (t_2 - t_3)$$

$\therefore$ $$t_2 - t_3 = \frac{Q}{4.2} \quad \text{...(ii)}$$

Adding Eqs (i) and (ii) we get,

$$t_1 - t_3 = Q\left[\frac{1}{18.87} + \frac{1}{4.2}\right] = Q[0.05299 + 0.2381]$$

$$= Q\times 0.29109$$

$\therefore$ $$Q = \frac{t_1 - t_3}{0.29109} = \frac{1650 - 27}{0.29109}$$

$$= 5575.6 \text{ kJ/hr/m}^2$$

$\therefore$ Reduction in heat loss through the wall

$$= 25093 - 5575.6 = \mathbf{1951.4\ kJ/hr/m^2}$$

Example 17.17 *A cold room has one of the walls which measures 4.6 cm × 2.3 m constructed of bricks, 11.5 cm thick, insulated externally by cork slabbing 7.5 cm thick. The cork is protected externally by wood 2.5 cm-thick.*

Estimate the leakage through the wall per 24 hours if the interior temperature is –20°C and the exterior 18°C. The thermal conductivities of brick, cork and wood are 41, 1.9 and 7.5 kJ/hr.m°C respectively.

What will be the temperatures at the interfaces ?

Solution

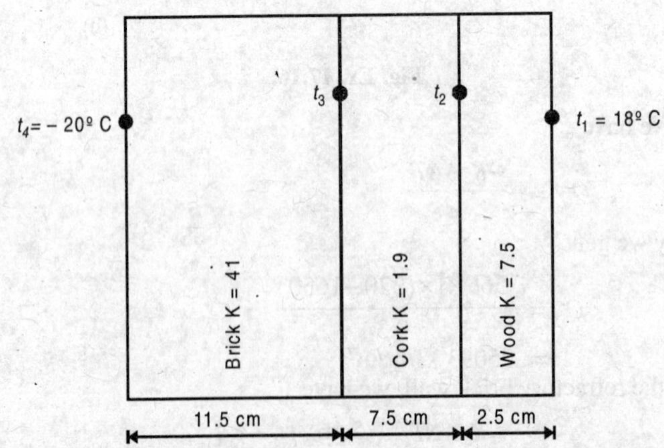

Fig. Ex. 17.17

Figure Ex. 17.17 shows the arrangement of the various materials in the example.

For wood we have
$$Q = \frac{KA(t_2 - t_1)}{x}$$

$$= 7.5 \times \frac{(4.6 \times 2.3)}{2.5/100}(t_1 - t_2)$$

$$= 3174\,(t_2 - t_1)$$

$$\therefore \quad t_2 - t_1 = \frac{Q}{3174}$$

$$= Q \times 0.000315$$

$$t_2 - t_3 = Q \times 0.003731$$

Similarly for cork we have

$$Q = \frac{1.9 \times 4.6 \times 2.3(t_2 - t_3)}{7.5/100}$$

$$= 268.03\,(t_3 - t_4)$$

$$\therefore \quad t_3 - t_4 = Q \times 0.00373 \qquad \text{...(ii)}$$

And for bricks, we have

$$Q = \frac{41 \times 4.6 \times 2.3(t_3 - t_4)}{11.5/100}$$

$$= 3772\,(t_3 - t_4)$$

$$\therefore \quad t_3 - t_4 = \frac{Q}{3772}$$

$$= Q \times 0.000265 \qquad \text{...(iii)}$$

Adding these three equations, we get

$$t_2 - t_4 = Q[0.000315 + 0.003731 + 0.000265]$$

$$= Q \times 0.00431$$

$$\therefore \quad Q = \frac{t_1 - t_4}{0.00431}$$

$$= \frac{18 + 20}{0.00431} = \frac{38}{0.00431}$$

$$= 8816.7 \text{ kJ/hr}$$

$$= \mathbf{211600.93 \text{ kJ/24 hr.}}$$

For interface temperatures

$$t_1 - t_2 = Q \times 0.000315$$

$$= 8816.7 \times 0.000315$$

$$= 2.7773$$

$$\therefore \quad t_2 = 18 - 2.7773$$
$$= \mathbf{15.2277\ °C}$$

Similarly $$t_2 - t_3 = \mathbf{8816.7 \times 0.00373}$$
$$= 32.89$$
$$\therefore \quad t_3 = 15.2227 - 32.89$$
$$= \mathbf{-17.667\ °C}$$

Example 17.18 *Prove that temperature distribution. Along the thickness of slab is uniform that is linear, if the thermal conductivity of the material of the slab is constant.*

Compute the heat loss per sq. metre of surface area for a furnace wall, 30 cm thick. The inner outer surface temperatures are 320°C and 38°C respectively. The variation of the thermal conductivity with the temperature is given by the following relation

$$K = \left(0.01256T - 4.2\times10^{-6}T^2\right) kJ/hr.m°C.$$

Solution

Consider a wall having a thickness as L and consider a plane surface at a distance x from the left face of the wall. Let the temperature at this surface be t and let dx be the elementary thickness of the slab at x. This is shown in Fig. Ex. 17.18.

$$Q = -\frac{KA.dt}{dx}$$

$$Q.dx = -KA.dt$$

Integrating this equation,

$$Q\int_o^L dx = -KA\int_{t_1}^{t} dt$$

$$\therefore \quad QL = -KA(t - t_1)$$

$$\therefore \quad Q = -\frac{KA(t-t_1)}{L}$$

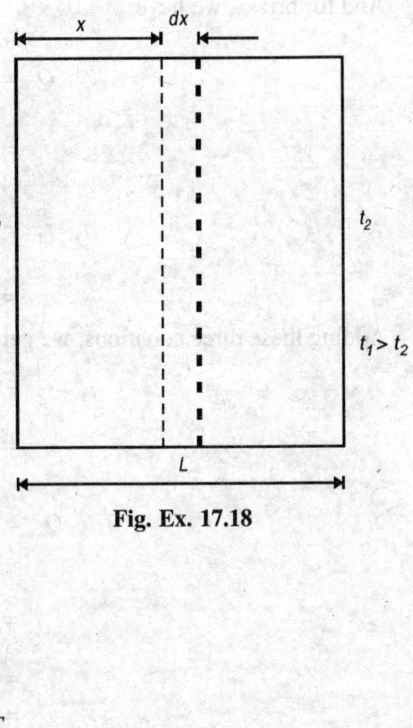

Fig. Ex. 17.18

This shows that the temperature variation is linear one or uniform.

$$\frac{QL}{AK} = t_1 - t$$

or $$\mathbf{t = t_1 - \frac{QL}{AK}}$$

For the problem given, thermal conductivity of the material is varying.

$$\therefore \quad \text{We have } Q = \frac{AK.dt}{dx}$$

$$= -A\left[0.01256T - 4.2T_2 \times 10^{-6}\right]\frac{dT}{dx}$$

Integrating, we get $$Q'L = A\left[\frac{0.01256T^2}{2}.\frac{4.2T^3}{3}\times10^{-6}\right]_{38}^{320}$$

$$QL = A\left[0.00628\left(320^2-38^2\right)-\frac{4.2}{3}\times10^{-6}\left(320^3-38^3\right)\right]$$

$$= 1\left[0.00628(358\times282)-1.4\times10^{-6}\times10^{6}(32.768-0.055)\right]$$

$$= 634 - 1.4 \times 32.713$$

$$= 634 - 45.78$$

$$= 588.2 \text{ kJ/hr/m}^2$$

$$\therefore \quad Q = \frac{588.2}{0.3}$$

$$= \mathbf{1960\ kJ/hr/m^2}$$

Example 17.19 *A steam pipe line 11.5 cm outside diameter, is covered with two layers of different materials. The first layer is 5 cm thick and a thermal conductivity of 0.222 kJ/hr.m °C. The second layer is 3 cm thick and as a conductivity of 3.14 kJ/hr.m °C.*

Outside surface temperature of steam pipe is 235 °C and that of the outer surface of lagging is nm 38 °C calculate the heat loss per metre length of pipe per hour, and the temperature between the two layers of insulation.

Solution

Figure Ex. 17.19 shows the arrangement of the pipe with layers of different materials.

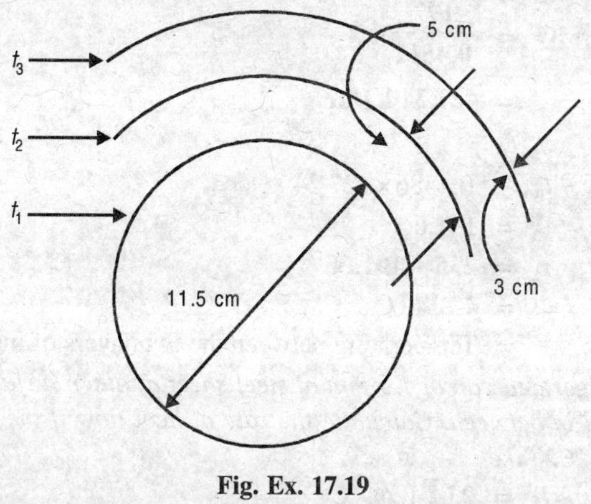

Fig. Ex. 17.19

We have for cylindrical pipes,

$$Q = \frac{2\pi l \times K\left(t_1 - t_2\right)}{\log_e \dfrac{D}{d}}$$

For the inner layer. $\log_e \dfrac{D}{d} = \log_e \dfrac{11.5+10}{11.5}$

$$= \log_e 1.87$$

$$= 0.6257$$

Similarly for outer layer

$$\log_e \frac{D}{d} = \log_e \frac{21.5+6}{21.5}$$
$$= \log_e 1.28$$
$$= 0.2409$$

Taking the length of the pipes as 1 metre,

$$t_1 - t_2 = \frac{Q \times 0.6257}{2\pi \times 0.22}$$
$$= 0.4486\ Q \qquad \ldots(i)$$

Similarly,
$$t_2 - t_3 = \frac{Q \times 0.2569}{2\pi \times 3.14}$$
$$= 0.0125\ Q \qquad \ldots(ii)$$

Adding Eqs (i) and (ii), we get,

$$t_1 - t_3 = Q \times [0.4486 + 0.0125]$$
$$= Q \times 0.4611$$
$$\therefore \quad 235 - 38 = 0.4611\ Q$$
$$\therefore \quad \mathbf{Q} = \frac{\mathbf{197}}{\mathbf{0.4611}}$$
$$= \mathbf{427.24\ kJ/hr}$$

From Eq. (i) we have,

$$235 - t_2 = 0.4486 \times 427.24$$
$$= 191.66$$
$$t_2 = 235 - 191.66$$
$$= \mathbf{43.34\ °C}$$

= Temperature between the two layers of insulation.

Example 17.20. *A wall of refrigerated van of 1.5 mm of steel sheet at outer surface, 10 mm plywood at a inner surface and 2 cm of glass wool in between. Calculate the rate of heat flow if the temperature at the inside and outside surface are –15 and 24 °C. Take*

$$K\ (steel) = 23.2\ W/m\ °C$$
$$K\ (glasswood) = 0.014\ W/m\ °C$$
$$K\ (plywood) = 0.052\ W/m\ °C$$

Solution

This is shown in Fig. Ex. 17.20 consider unit area of flow of heat (1 m^2)

With refrence to Fig. Ex. 17.20, we have

$$Q = \frac{KA(t_1 - t_2)}{x_1}$$

$$= \frac{23.2 \times 1(24 - t_2)}{1.5/1000}$$

$$\therefore \quad (24 - t_2) = \frac{Q \times 1.5}{1000 \times 23.2} = \frac{0.0646Q}{1000}$$

$$t_2 - t_3 = \frac{Q \times 20}{1000} \times \frac{1}{0.014}$$

$$t_3 - t_4 = \frac{Q \times 10}{1000} \times \frac{1}{0.052}$$

$$= \frac{Q \times 192.31}{1000}$$

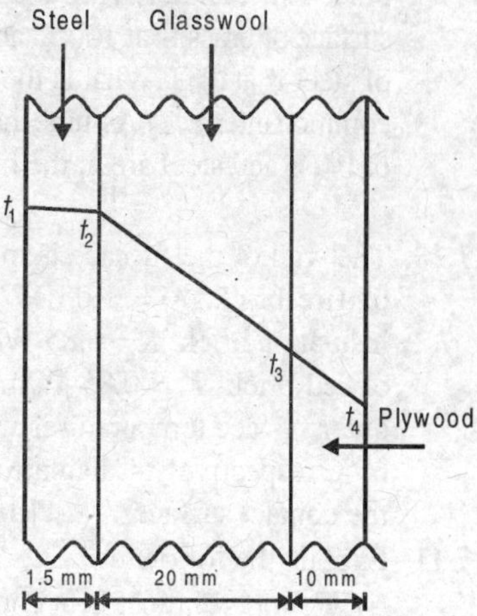

Fig. Ex. 17.20

Adding we get

$$24 + 15 = \frac{Q}{1000}[0.0646 + 1428.57 + 192.31]$$

$$39 = \frac{Q}{1000} 1620.9446$$

$$Q = \frac{39000}{1620.9446} = \mathbf{24.06\ W/m^2}$$

PROBLEMS FOR PRACTICE

17.1. What are the modes of heat transfer. Explain very briefly each one mentioned.

17.2. State the following laws
(i) Fouriers' law
(ii) Newton' law of cooling and
(iii) Stefan–Boltzmann law of radiation.

17.3. Explain the terms
(i) convective heat transfer coefficient and
(ii) overall heat transfer coefficient.

17.4. Derive an expression for the heat loss in W/m^2 through a composite wall of *n* layers.
(a) without considering convective heat transfer coefficients and
(b) with considering the convective heat transfer coefficient.

17.5. Prove that the heat flow through the hollow cylinder is given by

$$Q = \frac{2\pi LK(T_1 - T_2)}{\log_e\left(\frac{R_2}{R_1}\right)} \quad \text{where } T_1 > T_2$$

17.6. If the inner surface and outer surface temperatures of simple brickwall are 40ºC and 20ºC, calculate the rate of heat transfer per (metre)^2 of surface area of the wall having a thickness of 250 mm. Assume *K* for the brick is 0.52 W/m °C. [*Ans.* 41.6 W/m^2]

17.7. Determine the rate of heat transfer/flow through the boiler wall of 20 nun thickness steel plate which is covered with an insulating material of 5 mm thick. The temperature at the inner and outer surfaces of the all are 300ºC and 50°C respectively. Assume *K* for steel = 58 W/m°C and for insulation as 0.116 W/m.K. [*Ans.* 5.8 kW/m^2]

17.8. A steel tank of wall thickness 10 mm contains water at 90ºC. Calculate the rate of heat loss per unit area of tank surface when the atmospheric temperature is 15ºC. *L* for steel is 50 W/m°C and the heat transfer coefficients for inside and outside the tank are 2800 and 11W/m^{2}K respectively. Calculate also the temperature of the outside surface of the tank. [*Ans.* 820 W/m^2, 89.6ºC]

17.9. Sheets of brass and steel, each 10 mm thick, are

placed in contact with each other. The outer surface of brass is at 100°C and the outer surface of steel is at 0°C. What is the temperature of the common interface? Assume that the conductivities of brass and steel are in the ratio of 2 : 1.

[*Ans.*. 66.7°C]

17.10. The wall of the furnace is made up of 250 mm of fire brick, K = 1.05 W/m°C; 120 mm of insulation brick, K = 0.85 W/m°C, and 200 mm of red brick, K = 0.85 W/m°C. The inner and outer surface temperatures walls are 850°C and 65°C respectively. Calculate the temperatures at the contact surfaces. [*Ans.* 703°C, 210°C]

17.11. Explain the following
(i) Fouriers' law of heat conduction
(ii) Newton's law of cooling
(iii) Stephan–Boltzmann law of radiation heat transfer.

17.12. State different types of heat exchangers.

17.13. Explain the phenomenon of heat transfer by convection. State the Newton's Law of cooling.

17.14. State any three exchangers normally observed in thermal power plant, mentioning clearly the type of heat exchange.

[*Ans.* The heat exchangers that are normally used in thermal power plant are 1. Air preheater—Multipass, shell and tube type. 2. Economiser—Multipass, tube assembly kept in the flue gas passage. 3. Superheater—Steam carrying tubes kept in the flow gas passages]

17.15. Steam Generator:
(i) Fire tube boiler furnace tubes through which gases flow and tubes are surrounded by water
(ii) Water tube boiler-water tubes carrying water are placed in gas passage. Gases and water both are flowing.

17.16. Explain the following:
(a) Conduction heat transfer and its associated property
(b) Convection heat transfer and its associated property.

IMPORTANT POINTS TO BE REMEMBERED

1. *Rate of heat flow by conduction* (q_k)

$$q_k = -KA \cdot \frac{dT}{dx} = \frac{KA}{L}(T_{hot} - T_{cold})$$

$$q_k = \text{Hear flow similar to electric current}$$

$$\Delta T = T_{hot} - T_{cold} = \text{Driving force for heat transfer similar to voltage}$$

$$\frac{L}{AK} = \text{Thermal resistance similar to electric resistant}$$

2. *Heat flow through composite wall*

$$q = \frac{T_1 - T_0}{\sum R} \quad \text{and } R = \sum \frac{L}{AK}$$

3. *Hollow cylinder*

$$Q = \frac{2\pi LK(T_1 - T_2)}{\log e \dfrac{R_2}{R_1}}$$

4. *Heat transfer by convection*

$$q_c = \bar{h}_c \cdot A \cdot \Delta T \quad \bar{h}_c - \text{Overall heat transfer coeff.}$$

5. *Stefan Boltzmann's Law of radiation*

$$q_r = \sigma A_1 T_1^4 \qquad \sigma = 2042 \times 10^{-8} \text{ kJ}/\eta r \cdot \text{m}^2 \cdot T^4$$

$$= 5.671 \text{ W/m}^2 \cdot T^4$$

$$= \sigma A_1 \, (T_1^4 - T_2^4) \text{ between two black bodies}$$

$$= \sigma A_1 \, \varepsilon_1 \, (T_1^4 - T_2^4) \text{ Gray body at } T_1 \text{ and black surrounding at } T_2$$

$$= \sigma A_1 \, F_{1-2} \, (T_1^4 - T_2^4) \text{ No one perfect radiator and given geometric relationship}$$

and

$$R_r = \frac{T_1 - T_2}{\sigma A_1 F_{1-2}(T_1^4 - T_2^4)}$$

Table 17.1. Properties of Metals at 20°C

Metal	Density	Thermal	Specific Heat	Thermal
	ρ kg/m^3	α m^2/s × 10^6	c J/kg-K	k W/mK
Aluminium, Pure	2707	94.44	896	204.2
Steel 0. 5% carbon	7833	14.72	465	53.6
Nickel Steel 20% Ni	7983	5.28	461	19.1
Chrome Steel 20% Cr	7683	6.67	461	22.5
Constantan 60% Cu	8922	6.11	410	22.7
Magnesium, pure	1746	96.94	1013	171.3
Nickel, pure	8906	22.78	444	90.0
Nickel Chrome	8666	4.44	44	17.2
Tungsten Steel 10% W	8313	13.61	419	48.5
Copper, pure	8954	12.22	381	386.0
Bronze	8666	8.61	343	25.9
Brass	8522	33.89	385	110.7
Silver, pure	10524	165.56	235	406.8
Tungsten	19350	62.78	134	162.7
Zinc, pure	7144	41.11	385	112.1
Tin, pure	7304	38.61	226	4.1
Cr Ni steel 15% Cr 10% Ni	7865	5.28	461	19.1
1 W/mK = 0.86 kcal/m hr°C 1 J/kg-K = 238.9 × 10^{-6} kcal/kg°C				

Table 17.2. Properties of Metals at 20°C

Metal	Density	Thermal Diffusivity	Specific Heat	Thermal Conductivity
	ρ kg/m^3	α m^2/s × 10^6	c J/kgK	k W/mK
Berylium	1840	50.97	1675	157
B	2500	10.90	1047	28.6
Cadmium	8660	46.67	230	93
Carbon (graphite)	1700–2300	122.22	670	116.3–174.5
Chromium	7150	218.06	448	69.8
Cobalt	8800	17.64	448	69.8
Gold	19300	127.00	129	317.0
Lithium	5340	40.28	3308	68.6
Molybdenum	10200	54.44	253	140.7
Platinum	21460	24.58	132	69.8
Potassium	870	155.56	737	100.0
Rhodium	12450	48.6	248	150.0
Sodium	975	94.44	1197	109.3
Silicon	2330	93.4	703	153
Thorium	11700	38.17	118	54.0
Uranium	19100	12.70	113	27.4
Vanadium	5900	11.94	496	34.9
Titanium	4540	6.22	532	15.12
Zirconium	6570	12.50	278	22.7
1 W/mK=0.86 kcal/m hr°C 1 J/kg-K = 238.9 × 10^{-6} kcal/kg°C				

Table 17.3. Properties of Insulating, Building, and Other Materials

Substance	*Temperature* °C	*k,* W/m°C	*ρ,* kg/m³	*C* kJ/kg°C	*α,* $m^2/s \times 10^7$
	Structural and heat resistant materials				
Asphalt	20–55	0.74–0.76			
Brick:					
Building brick, common	20	0.69	1600	0.84	5.2
Face		1.32	2000		
Carborundum brick	600	18.5			
	1400	11.1			
Chrome brick	200	2.32	3000	0.84	9.2
	550	2.47			9.8
	900	1.99			7.9
Diatomaceous earth,					
moulded and fired	200	0.24			
	870	0.31			
Fireclay brick, burnt	500	1.04	2000	0.96	5.4
1330°C	800	1.07			
	1100	1.09			
Burnt 1450°C	500	1.28	2300	0.96	5.8
	800	1.37			
	1100	1.40			
Missouri	200	1.00	2600	0.96	4.0
	600	1.47			
	1400	1.77			
Magnesite	200	3.81		1.13	
	650	2.77			
	1200	1.90			
Cement, portland		0.29	1500		
Mortar	23	1.16			
Concrete, cinder	23	0.76			
Stone 1–2–4 mix	20	1.37	1900–2300	0.88	8.2–6.8
Glass, window	20	0.78 (avg)	2700	0.84	3.4
Corosilicate	30–75	1.09	2200		
Plaster, gypsum	20	0.48	1440	0.84	4.0
Metal lath	20	0.47			
Wood lath	20	0.28			
Stone					
Granite		1.73–3.98	2640	0.82	8–18
Limestone	100–300	1.26–1.33	2500	0.90	5.6–5.9
Marble		2.07–2.94	2500–2700	0.80	10–13.6
Sandstone	40	1.83	2160–2300	0.71	11.2–11.9
Wood: (across the grain)					
Balsa,	30	0.055	140		
Cypress	30	0.097	460		
Fir	23	0.11	420	2.72	0.96
Maple or oak	30	0.166	540	2.4	1.28
Yellow pine	23	0.147	640	2.8	0.82
White pine	30	0.112	430		

Table 17.3. *Contd.*

Substance	*Temperature* °C	*k,* W/m°C	*ρ,* kg/m³	*C* kJ/kg°C	*α,* m²/s × 10⁷
		Insulating material			
Asbestos:					
Loosely packed	–45	0.149			
	0	0.154	470–570	0.816	3.3–4
	100	0.161			
Asbestos–cement boards	20	0.74			
Sheets	51	0.166			
Felt, 16 laminations/cm	38	0.057			
	150	0.069			
	260	0.083			
8 laminations/cm	38	0.078			
	150	0.095			
	260	0.112			
Corrugated, 150 plies/m	38	0.087			
	93	0.100			
	150	0.119			
Asbestos cement	–	2.08			
Balsam wool,	32	0.04	35		
Cardboard, corrugated	–	0.064			
Celotex	32	0.048			
Corkboard,	30	0.043	160		
Cork, regranulated	32	0.045	45–120	1.88	2–5.3
Ground	32	0.043	150		
Diatomaceous earth	0	0.061	320		
(Sil–o–cel)					
Felt, hair	30	0.036	130–200		
Wool	30	0.052	330		
Fiber, insulating board	20	0.048	240		
Glass wool,	23	0.038	24	0.7	22.6
Insulex, dry	32	0.064			
		0.144			
Kapok .	30				
Magnesia, 85%	38	0.067	270		
	93	0.071			
	150	0.074			
	204	0.080			
Rock wool,	32	0.040	160		
Loosely packed	150	0.067	64		
	260	0.087			
Sawdust	23	0.059			

PART–4

CHAPTER 18

Steam Engine

18.1 INTRODUCTION

Steam engine is a prime mover. It utilises steam as a working medium. The heat energy in steam engine is converted into the mechanical work and this conversion is done in the cylinder of the steam engine.

Steam engine was first developed by James Watt (1736–1819) and were in continuous use upto 1930. Still, also we can see some of the steam engines in railway locomotives. But nowadays they are getting replaced by Diesel or Electric Locomotives.

Generally it has been stated that the steam engine owes little to science, than science owes to the steam engine. Nowadays steam turbines and IC engines have largely replaced to steam engines because of their higher power outputs, higher efficiencies and smaller bulk.

18.2 STEAM ENGINE PARTS AND THEIR FUNCTION

Frame. It is made of cast iron. It support the moving parts; and gives rigidity to various members, It rests upon the foundation.

Cylinder. It is made of cast iron and is bolted to the frame. In small engines it is cast integral with the frame. It forms a chamber in which a piston moves to and fro due to the action of steam. One end of the cylinder is closed by means of a cover. This end is known as cover end. The other end is known as crank end. It carries a cover in which a stuffing box is provided. The piston rod passes through the stuffing box.

Steam Chest. It is cast integral with the cylinder. The high pressure steam is first admitted to the steam chest. The steam chest accommodates the D slide valve. On one end it carries a stuffing box through which valve rod comes out. It is closed from top by a cover. The steam chest is connected to the cylinder by means of passage made in the casting ports, through which the steam passes from the steam chest to the cylinder.

Piston. It is made of cast iron. The steam under pressure acts on the face of the piston and creates force. Piston transmits this force to the main shaft via piston rod, and connecting rod.

Piston Rings. They are made of cast iron. The piston is provided with grooves on its circumference. The piston rings are fitted into these grooves. The piston rings prevent leakage of steam from one side of piston to the other sides. This is known as leakage past piston.

Piston Rod. It is made of mild steel. It connects the piston to the cross head.

Stuffing Box and Gland. This is fitted on the crank end cover of the cylinder at a point where piston rod passes through the cylinder cover. It prevents leakage of steam from the cylinder to the atmosphere. A similar stuffing box is also fitted on the steam chest where valve rod passes through the steam chest side.

Cross Head. This moves between the guides. It guides the motion of piston and prevents the bending of piston rod. The cross head is essential for double acting steam engines. For single acting steam engine cross head is not needed.

Connecting Rod. It is made of forged steel. One end of it is connected to the piston rod by a gudgeon pin and the other end is connected to the crank by crank pin. It converts reciprocating motion of the piston (and of cross head) into rotary motion of the crank.

Crank Shaft. It is made of steel. It is supported on the main bearings, but is free to rotate in the bearings. The crank webs are keyed or pressed on to it. The crank webs may be made integral with the shaft also. In that case the entire shaft is made as a forging. The crank shaft carries the eccentric and flywheel.

Eccentric. It converts rotary motion of the crank shaft into the reciprocating motion of the slide valve.

Valve Rod and Eccentric Rod . Both of them are made of mild steel. The valve rod is connected at one end to the valve and at the other to the valve guide. The valve guide, guides the valve rod and prevents it from bending. The eccentric rod connects the valve rod and the eccentric. It converts rotary motion of eccentric into reciprocating motion of valve guide and therefore of valve itself.

Slide Valve. It is placed in the steam chest. The steam chest is divided from the cylinder by a wall. The valve has ports through which steam is admitted to the engine cylinder. The slide valve controls the amount of steam admitted by closing and opening the steam ports in predetermined sequence.

In double action steam engine there are two ports which are used alternately as admission and exhaust port.

Fly Wheel. It is made of cast iron and has heavy mass and therefore large inertia. The pressure on the piston face during a stroke is not uniform. The torque produced also is not uniform. To reduce the fluctuations in the torque a fly wheel is used. It absorbs energy from the engine when it is in excess. Stores it in the form of inertia and gives it back when it is deficit.

18.3 TERMS CONNECTED WITH STEAM ENGINE

1. **Bore.** This is cylinder diameter.

2. **Stroke.** It is distance travelled by a piston from one dead centre to the other.

3. **Dead Centres.** These are end position of piston travel. At dead centre the piston rod, connecting rod and crank are in straight line. There are two positions of dead centres. For vertical engine they are Top Dead Centre (TDC) and Bottom Dead Centre (BDC) and for horizontal engine they are—Outer Dead Centre (ODC) and Inner Dead Centre (IDC).

4. **Swept Volume.** It is volume swept by a piston during one stroke.

5. **Clearance Volume.** It is volume included between the piston and the cylinder cover when the piston is at dead centre. This is provided to avoid the direct hitting of cylinder cover by the piston, and to provide initial space for incoming steam.

6. **Mechanical Clearance.** This is a distance between the piston face and nearest cylinder cover when the piston is at dead centre.

7. **Engine Speed.** This is measured in terms of revolutions of crank shaft per minute and is denoted by rpm.

8. **Piston Speed.** This is the linear speed of the piston and is measured in metre/minute, or metre/second.

$$\text{Piston speed} = 2 \times L \times N \text{ m/min}$$

where L is stroke length and N is rpm.

18.4 WORKING OF A STEAM ENGINE

We describe here the working of a simple, vertical, double acting, non condensing, *D* slide valve type, steam engine.

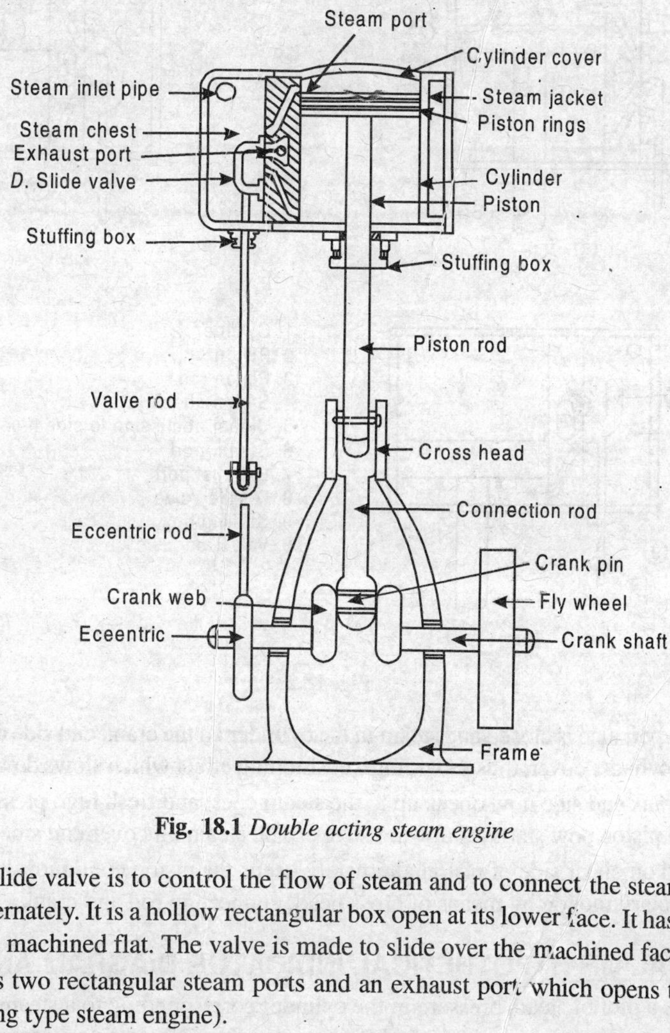

Fig. 18.1 *Double acting steam engine*

The function of a *D* slide valve is to control the flow of steam and to connect the steam ports to the admission side and exhaust side alternately. It is a hollow rectangular box open at its lower face. It has projections on its lower face and the surfaces are machined flat. The valve is made to slide over the machined face of the cylinder.

The cylinder wall has two rectangular steam ports and an exhaust port which opens to atmosphere (or to the condenser in a condensing type steam engine).

Figure 18.2 (a) shows the beginning of a stroke (on cover side) The steam chest is full of high pressure steam. The valve has just opened the steam port and the high pressure steam is admitted to the top side of the piston. The steam pressure acting on the piston pushes the piston down.

At the same time the cranker port on crank end side is just uncovered and the steam on the crank end side of piston goes out of the cylinder via exhaust port.

As the piston moves downwards the *D* slide valve completes its stroke and comes back to close the admission port (Fig.18.2 (b)). This is known as cut off point. The supply of steam is cut off and rest of the stroke is completed by the expansion of steam. The pressure drops, volume increases and expanding steam does work. When expansion continuous till the end of the stroke, just before the end of the stroke the valve uncovers the cover end steam port and connects it to the exhaust port. The pressure of steam inside the cylinder suddenly drops. This is known as point of release.

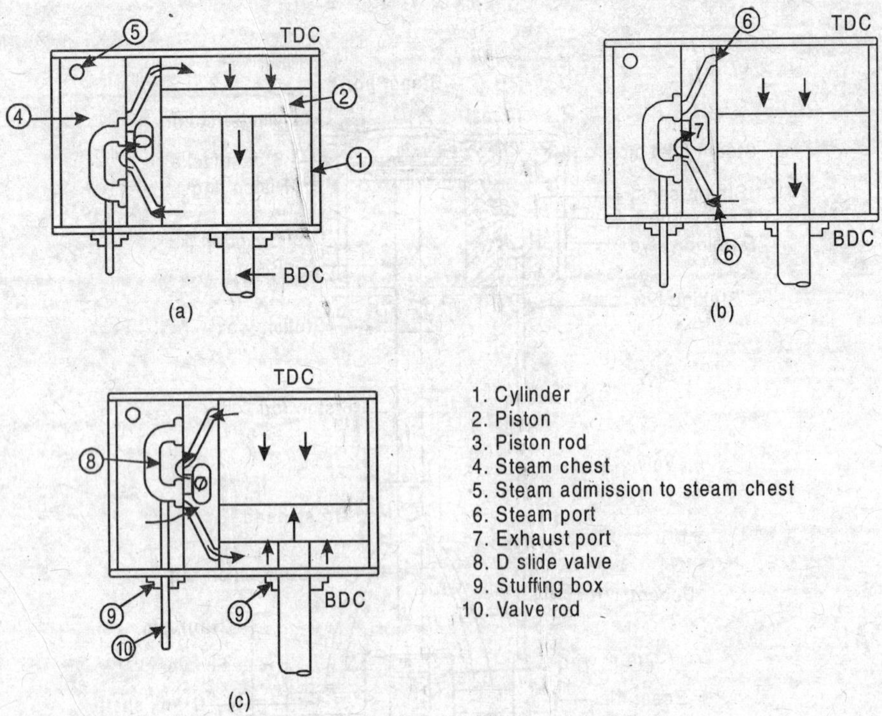

Fig. 18.2

The crank end steam port also is closed and steam in the cylinder on the crank end side of piston gets compressed between the piston and cylinder cover, thus providing a cushioning effect which slows down piston speed gradually.

The steam port on crank end side now opens up to the steam chest and fresh high pressure steam is admitted on the crank end side. The piston now starts its upward stroke. The steam on cover end side now is exhausted.

Thus steam admitted on either side of piston alternately keeps the piston continuously in reciprocating motion which is converted to rotary motion by means of cross head, connecting rod and crank.

18.5 THEORETICAL OR HYPOTHETICAL INDICATOR DIAGRAM AND MEP

An indicator diagram is a plot of steam pressure in the cylinder corresponding to a steam volume during the cycle of operation. The indicator diagram for a single cycle of steam engine (Neglecting clearance) is shown in Fig. 18.3.

Assumptions made for the construction of theoretical indicator diagram :

1. Clearance volume is neglected.
2. Steam is admitted at constant pressure
3. Closing of port at cut off point 2 is instantaneous. So we get a well defined point 2.
4. Expansion of steam is hyperbolic
5. We get a well defined point 3 as opening of exhaust port is instantaneous.
6. Steam will be exhausted at constant pressure through the return stroke.
7. 5-1 is constant volume of steam addition

During starting in case of stationary steam engines, the flywheel will be rotated manually, so that the marking on the flywheel coincides with the pointer. Then the position of *D*-slide valve with respect to the ports will be adjusted (as shown in Fig. 18.3) so that it uncovers the head end port.

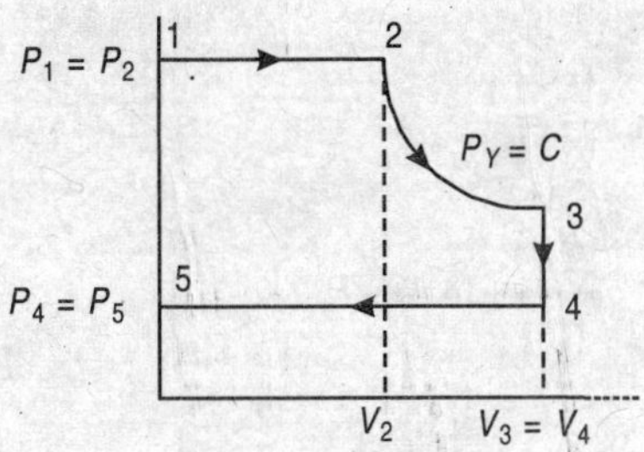

Fig. 18.3

As shown in Fig. 18.3 the cycles of operations are as follows.

Process 1–2. At point 1 the steam is admitted and it continuous to do so until the point 2 is reached, where the steam is cut off. Point 1 is called the point of admission Point 2 is called the point of cut off.

Process 2–3. After cut off, steam locked in the cylinder, expands hyperbolically from 2–3.

Process 3–4. Just before the piston reaches to Bottom dead center, the *D*-slide valve will uncover the head end port. Now the pressure of steam, from above the piston will be released at constant volume along 3–4.

At the same time steam will be entering from below the piston through crank end port. Note that till the end of downward stroke, this steam gives cushioning effect and after the piston reaches Bottom dead center its upward stroke starts and the steam entering from below will force the piston to move upwards.

Process 4–5. The steam from above the piston will be exhausted along this. Work done/cycle will be given by the area 1–2–3–4–5–1 on the *pV*-diagram.

i.e.

$$\text{Work done/cycle} = W = (\text{Area under 1–2}) + (\text{Area under 2–3}) — (\text{Area under 4–5})$$

$$= P_2V_2 + P_2V_2 \ln\cdot\frac{V_2}{V_2} - P_4V_4$$

$$= P_2V_2\left[1 + \ln\frac{V_2}{V_2}\right] - P_4V_4$$

During the process 2–3 the volume V_2 increases to V_3 and ratio $\frac{V_3}{V_2}$ is generally called as expansion ratio and is denoted by *r*.

$$\therefore \quad \text{Work done /cycle} = P_2\,V_2\,(1 + \text{In}.r) - P_4\,V_4$$

Secondly, the mean effective pressure (MEP) or (Average effective pressure on the piston or mean vertical intercept of *pV*-diagram) is defined as the ratio of area of indicator diagram to length of indicator diagram.

$$\therefore \quad \text{MEP} = \frac{\text{Area of indicator diagram}}{\text{Length of indicator diagram}}$$

$$\therefore \quad = \frac{P_2V_2(1 + \ln r) - P_4V_4}{V_4}$$

$$= \frac{P_2 V_2}{V_4}(1+\ln r) - P_4$$

and as $V_4 = V_3$

$$\therefore \quad \frac{V_2}{V_4} = \frac{V_2}{V_3} = \frac{1}{r}$$

$$\text{MEP} = \frac{P_2}{r}(1+\ln r) - P_4$$

In general this is written as

$$P_m = \frac{P}{r}(1+\ln r) - P_b \quad \frac{N}{m^2} \qquad \text{... (1)}$$

Where

P_m = Mean effective pressure in N/m²
P = Intake pressure in N/m²
P_b = Back or exhaust pressure N/m²
r = Expansion ratio.

If P and P_b are given in bar, then P_m also will be in bar.

If the clearance volume is taken into account, theoretical Indicator diagram will be represented as shown in Fig. 18.4.

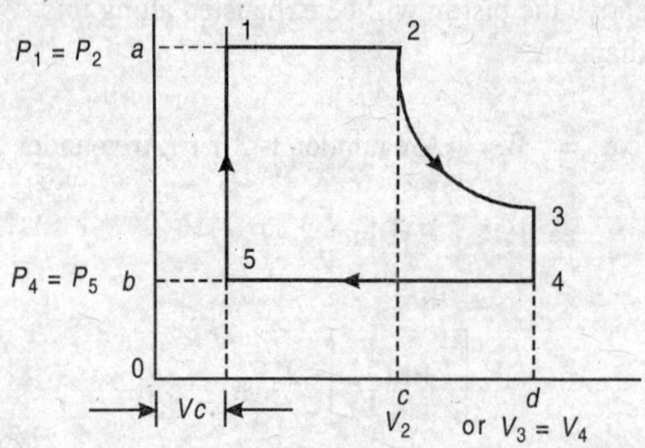

Fig. 18.4

$$\begin{aligned}
\text{Work done/cycle} &= \text{Area } 1-2-3-4-5-1 \\
&= (\text{Area } a-2-c-0) + (\text{Area } 2-3-d-c) - (\text{Area } b-4-d-0) \\
&\quad - (\text{Area } a-1-5-b) \\
&= P_2 V_2 + P_2 V_2 \ln \cdot \frac{V_3}{V_2}\; P_4 V_A - (P_1 - P_4) V_c \\
&= P_2 V_2 (1 + \ln \cdot r) - P_4 V_4 - (P_1 - P_4) V_c
\end{aligned}$$

where

$$V_c' = \text{clearance volume and } r = \frac{V_3}{V_2}$$

and

$$\text{MEP} = \frac{P_2V_2(1+\ln \cdot r) - P_4V_4 - (P_1 - P_4)V_c}{(V_4 - V_c)} \quad \text{... (2)}$$

If we substitute $V_c = 0$, Eq. (2) reduces to Eq. (1).

18.6 ACTUAL INDICATOR DIAGRAM AND DIAGRAM FACTOR

Figure 18.5 shows the actual indicator diagram (showing continuous line) and the dotted line diagram shows the theoretical one as discussed earlier.

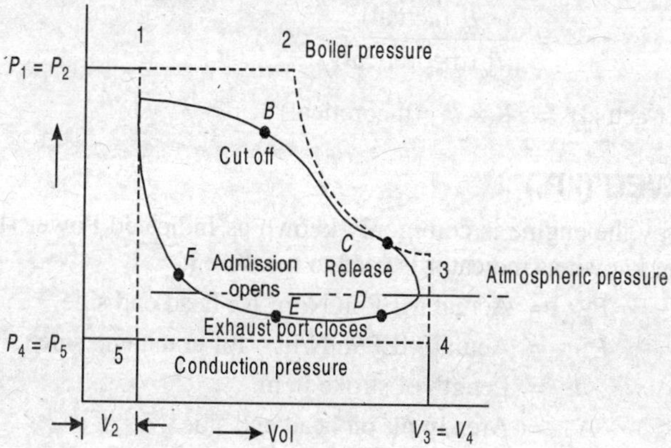

Fig. 18.5

The difference between the diagrams is on account of certain practical factors given below, which were not considered in arriving at the theoretical diagram.

1. **Point A .** From the Fig. 18.5 it can be seen that point *A* is below '1' that is the steam pressure drops considerably between the boiler and engine cylinder. This is due to the condensation of steam in the pipeline connecting between the boiler and engine (or wire drawing through valve).

2. **Points.** *A–B*. There is a gradual pressure drops that is $P \neq C$, before the cut off point. This is mainly because the steam entering the cylinder, comes in contact with the cold cylinder surfaces and condenses. The cylinder walls can be kept hot to avoid excessive condensation by passing live steam through the cylinder jackets.

3. **Points.** *B*–Corner at *B* is rounded off because (when the piston comes to some intermediate position the *D*-slide valve will return retrains its path and in the reverse direction and closes the head end port) closing of admission port is not instantaneous.

4. **Points.** *B–C*–It is not hyperbolic due to condensation of steam in the cylinder.

5. **Points.** *C–D*–Rounded off because exhaust does not open instantaneously.

6. **Points.** *D–E*–The exhaust pressure is slightly above the condenser pressure or back pressure as the steam has to be forced out of the cylinder.

7. **Points.** *EF*–The exhaust port closes at *E* and the extrapped steam compresses to point, *F*. At *F* admission port opens and the extrapped steam or cushion steam mixes with the admission steam and reaches the initial starting point, *A*, and the cycle continues.

It will be seen from Fig. 18.5 that the area of actual Indicator diagram is less than the area of theoretical Indicator diagram. The ratio between the areas of these two diagrams is known as Diagram factor and is denoted by *K* i.e.,

$$K = \frac{\text{Area of Actual Indicator, Diagram}}{\text{Area of Theoretical Indicator, Diagram}}$$

or

$$\text{Actual Work done} = K \times \text{Theoretical work done}$$

Also Diagram factor

$$K = \frac{\text{Mean height of Actual Indicator Diagram}}{\text{Mean height of Theoretical Indicator Diagram}}$$

$$= \frac{P_m\text{(actual)}}{P_m\text{(theoretical)}}$$

$$\therefore \quad P_m\text{(actual)} = K \times P_m\text{ (theoretical)}$$

18.7 INDICATED POWER (I.P.)

The total power developed by the engine is commonly known as Indicated Power (I.P). It is called as indicated power because an instrument known as indicator is used to measure it.

If, P_{m_1} = Actual MEP in N/m^2 for head end side
P_{m_1} = Actual MEP in N/m^2 for crank end side
L = Length of stroke in m.
A_1 = Area in m^2 on head end side
A_2 = Area in rn^2 on crank end side
N = Number of power or working stroke/min.

Then force on the piston = $P_{m_1} \times A_1$ Newtons for head end side and work done/working stroke for head end side

$$= \text{Force} \times \text{Displacement}$$

$$= (P_{m_1} \times A_1 \times L) \text{ Nm or J}$$

and work done/min for head end side = work done per working stroke × Number of strokes/minute.

$$= P_{m_1} \times A_1 \times L \times N \frac{\text{J}}{\text{min}}$$

$$\text{Work done/sec} = \frac{P_m A_1 \times L \times N}{60} \frac{\text{J}}{\text{sec}} \text{ or Watts}$$

For single acting in general

$$\text{IP} = \frac{P_m L\,A\,N}{60} \text{ Watts}$$

Since steam engines are double acting,

$$\text{Total IP} = \text{IP}_{\text{head end}} + \text{IP}_{\text{crarck end}}$$

and we can write

$$\text{IP}_{\text{crank end}} = \frac{P_{m_2} \cdot L \cdot A_2 N}{60} \text{ Watts}$$

∴ Total I.P. for Double acting steam engine .

$$= \frac{(P_{m_1} A_1 + P_{m_2} A_2)LN}{60} \text{ Watts}$$

Also $A_2 = A_1 -$ Area of piston rod.

Assuming mean effective pressure to be same for crank head end and neglecting area of piston end

$$\text{IP} = \frac{2P_m L\,A\,N}{60} \text{ W}$$

The total power developed by the engine will not be available for doing useful work, but some of the power will be lost in over coming the frictional resistance and is termed as frictional power (FP.) The power available for doing useful work is termed as Brake Power (BP) or shaft power. The BP of the engine can be determined by a brake of some kind applied to the brake pulley of the engine. The arrangement for the determination 1 of BP is termed as dynamometer. The various types of dynamometers are, Rope brake, prony brake, hydraulic dynamometers, etc.

Rope brake, prony brake, hydraulic dynamometers etc.

18.8 ROPE BRAKE DYNAMOMETER

It is generally used for the determination of brake power in the laboratories, since it is easy to fabricate and cheap.

It consist of a belt or rope wrapped round the flywheel of an engine of whose Brake power is to be measured. The upper end is connected to the spring balance suspended from overhead and the lower end carries the load *W*.

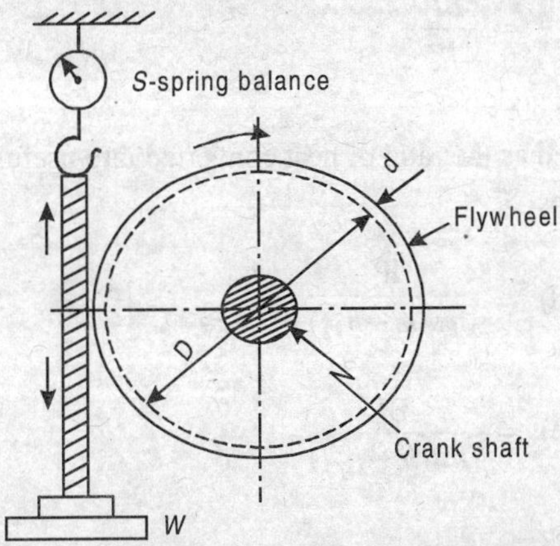

Fig. 18.6

Let,

W = Dead weight or load suspended in N

S = Spring balance reading in N

R = Mean radius of brake drum

$$= \frac{D+d}{2}$$

and N = speed in RPM

The net frictional force or the net load

$= (W - S)$ Newtons

and if R is the mean radius of the drum or wheel, then

$$\text{Frictional torque} = \text{Net frictional force} \times R = (W - S) \times R \text{ Nm and}$$

if N = speed of the engine in RPM, then the number of radians/second $= \dfrac{2\pi N}{60}$

$$\therefore \quad \text{The work absorbed} = \frac{(W-S)R \cdot 2\pi N}{60} \frac{\text{Nm}}{\text{sec.}}$$

or

$$\text{B.P.} = \frac{(W-S)R \cdot 2\pi N}{60} \frac{\text{J}}{\text{sec.}} \text{ or Watts}$$

Note that the power lost in overcoming the friction of rotating and sliding parts of the engine, is the difference between the IP and BP

i.e.

$$\text{FP} = \text{IP} - \text{BP}$$

18.9 EFFICIENCIES

1. Mechanical efficiency. It is defined as the ratio of BP to IP

$$\therefore \quad \eta_{\text{mech}} = \frac{BP}{IP} \qquad \text{... (1)}$$

Generally η_{mech} will be between 80–90 %

2. Thermal efficiency. Is defined as the ratio of heat converted into useful work to heat supplied.

(i) Indicated thermal of efficiency

$$\eta_{\text{th}} \text{ (indicated)} = \frac{\text{IP}}{ms(h_1 - h_{f2})} \qquad \text{... (2)}$$

$$\eta_{\text{th}} \text{ (brake)} = \frac{\text{BP}}{ms(h_1 - h_{f2})} \qquad \text{... (3)}$$

where,

m_s = mass of steam supplied is kg/sec

h_1 = specifc enthalpy of steam entering the cylinder in kJ/kg.

h_f = specific liquid enthalpy of water returned to hot well in kJ/kg.

from Eqs. (1) (2) and (3)

$$\eta_{\text{mech}} = \frac{\text{Brake thermal efficiency}}{\text{Indicalcd thermal efficiency}} \qquad \text{... (4)}$$

3. Overall efficiency. It is the ratio of brake power to the energy supplied by fuel for the steam generation.

$$\eta_{\text{overall}} = \frac{\text{BP}}{m_f \times \text{C.V.}}$$

where m_f = mass of fuel in kg/sec and C.V. = calorific value in J/kg or $\frac{kJ}{kg}$

Accordingly whether BP is in $\frac{J}{sec}$ or $\frac{kJ}{sec}$

4. Relative efficiency or efficiency ratio. It is defined as the ratio of Indicated thermal efficiency to the corresponding efficiency.

Indicated thermal efficiency

$$\text{Relative efficiency (IP bass)} = \frac{\text{Indicated thermal efficiency}}{\text{Rankine efficiency}}$$

or $$\text{Relative efficiency on BP basis} = \frac{\text{Brake thermal } \eta}{\text{Rankine efficiency}}$$

18.10 STEAM CONSUMPTION

Steam Consumption /hr. Amount of steam consumed by the steam engine in kg/hr.

Specific steam consumption : It is the steam consumption per unit of power produced.

$$\text{Specific Steam Consumption (SSC)} = \frac{\text{kg/hr}}{\text{kW}} = \frac{\text{kg}}{\text{kWh}}$$

When SSC is based on kW (Indicated) then it is known as SSC/indicated kWh and when SSC is based on kW (brake) then it is known as SSC/brake kWh.

18.11 CYLINDER CONDENSATION AND THE METHOD FOR REDUCING IT

When the steam is admitted and allowed to expand in the engine cylinder some of the steam condenses, when it comes in contact with the colder cylinder surface. This reduces the power to be developed. So, the methods adopted to reduce condensation are,

1. Using superheated steam.
2. The pipe line connecting between boiler and steam engine should be Lagged by asbestis in order to reduce, steam condensation in it.
3. Jacketing the cylinder with hot steam from the boiler, which reduces condensation of steam in the cylinder.
4. Using of compound steam engines, which reduces temperature range/cylinder.
5. Using increased engine speeds.

18.12 GOVERNING OF STEAM ENGINES

The function of a governor is to maintain the speed of the engine within the prespecified limits, depending upon the load on the engine. In other words, the engine should be able to adjust its power developed according to the load on the engine.

There are two methods used for maintaining the constant speed of the engine by the action of the governor.

(i) Throttle governing,

(ii) Cut-off governing.

(i) Throttle governing : Refer Fig. 18.7.

In this method of governing the point of cut-off is kept constant and the steam is throttled from boiler pressure. P_1 to a lower pressure according to the reduction in load by partial closure of throttle valve under the control of a governer. The decreasing pressure of steam during admission process reduces the work developed by the engine as shown in Fig. 18.7

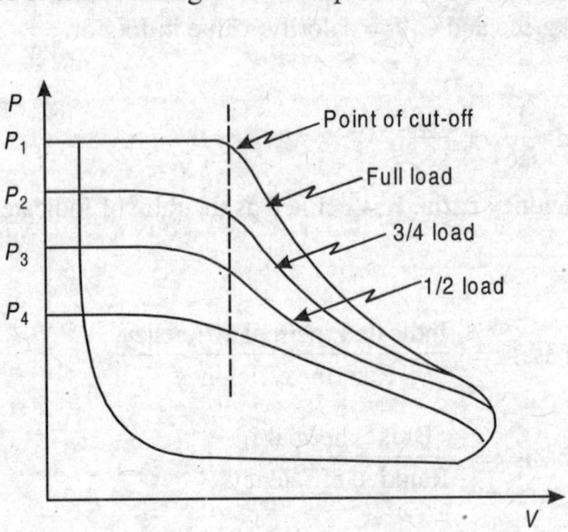

Fig. 18.7 *Throttle governing*

(ii) **Cut-off governing.** In this method of governing, the steam inlet pressure P_1 is kept constant and the work developed by the engine is varied by alternating the point of cut-off with the help of a slide valve operating under the control of a centrifugal governor.

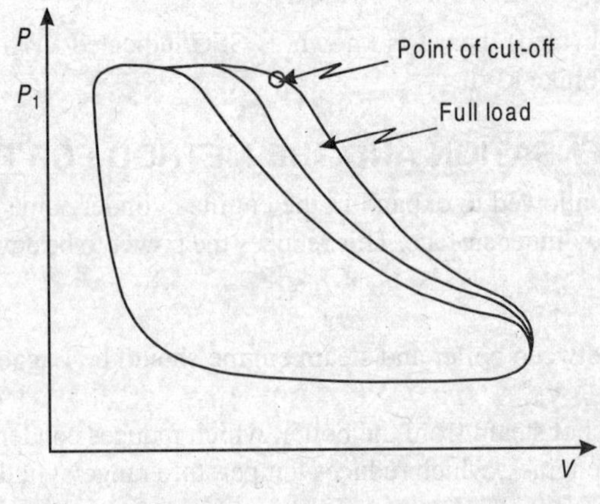

Fig. 18.8 *Cut-off governing*

18.12.1 Comparison between Throttle and Cut-off Governing

In case of cut-off governing since the admission pressure is corresponding to the boiler pressure and only the volume/mass of steam supplied is varied, it efficiency is not affected. But, in case of throttle governing since the admission pressure is lowered, it lowers the thermal efficiency. Also, the steam consumption is lower in case of cut-off governing compared to throttle governing.

18.13 WILLAN'S LINE

In case of throttle governing, if we plot a graph between the steam consumption in kg/hr taken on y-axis against the indicated power (I.P.) taken on X-axis, we get a straight line variation as shown in Fig. 18.9. This particular relationship was first observed by Mr. Willan and after his name, this straight line relationship between steam consumption (kg/hr) and indicator power is called the Willan's line.

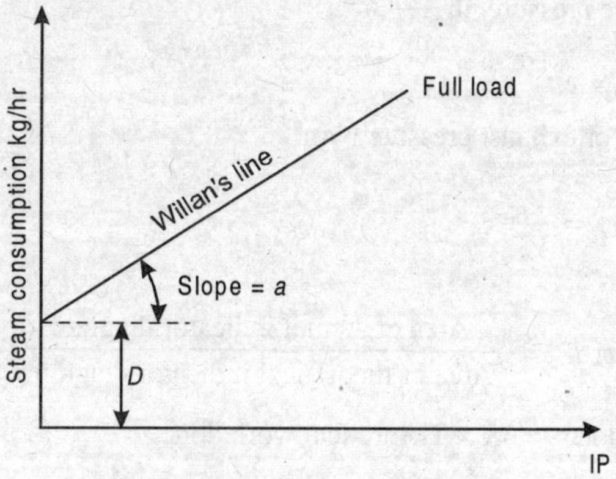

Fig. 18.9 *Willan's line*

Let,

$$\text{Steam consumption in kg/hr} = m$$
$$\text{Indicator power} = I$$
$$m = aI + b$$

18.14 ENERGY BALANCE (HEAT BALANCE SHEET)

This gives an account of total heat energy supplied in, per minute or per second and its utilization as under,

Heat supplied for min or per sec	*kJ*	*%*	*Heat Expenditure/min a per sec*	*kJ*	*%*
1. Energy supplied by steam = ms ($h_1 - h_f$)	×	100	(i) Energy to useful work BP	x_1	—
			(ii) Energy rejected to the cooling water in the end	x_2	–
			(iii) Energy rejected to hot well	x_3	–
			(iv) Energy lost by friction, leakage etc. by deference or energy unaccounted	x_4	–
	×	100		×	100

It should be noted that, there will be one energy balance sheet for each load condition of the engine.

18.15 LIST OF FORMULAE

1. MEP or Average Effective Pressure or mean vertical intercept of

$$\text{PV diagram} = \frac{\text{Area of indicator diagram}}{\text{Length of indicator diagram}}$$

$$P_m = \frac{P}{r}(1 + \ln \cdot r) - P_b \frac{\text{N}}{\text{m}^2}$$

where

$$P_m = \text{Mean effective pressure in } \frac{\text{N}}{\text{m}^2}$$

$$P = \text{Intake pressure}$$

$$P_b = \text{Back pressure or exhaust pressure N/m}^2$$

$$r = \text{Expansion ratio } = \frac{V_3}{V_2}$$

2. $$\text{Diagram factor } K = \frac{\text{Area of Actual indicator diagram}}{\text{Area or theoretical indicator diagram}}$$

or $$\text{Actual work done} = K \times \text{Theoretical work done}$$

Also,

$$K = \frac{P_m\text{(actual)}}{P_m\text{(theo.)}}$$

i.e. $$P_m\text{(actual)} = K \times P_m\text{(theo.)}$$

3. $$\text{IP for single acting} = \frac{P_m L A N}{60}\text{ W}$$

and

IP for double acting Total IP = $\text{IP}_{\text{head end}} + \text{IP}_{\text{crank end}}$

$$\text{IP} = \frac{(P_{m_1} A_1 + P_{m_2} A_2) L N}{60}\text{ W}$$

Also

$$A_2 = A_1 - \text{Area of piston rod}$$

If P_m is assumed to be same and neglecting the area of piston rod,

$$\text{Total IP} = \frac{2 P_m L A N}{60}\text{ W}$$

4. $$\text{Work absorbed /sec or BP} = \frac{(W - S) \cdot R \times 2\pi N}{60}\text{ W}$$

$$\text{BP} = \frac{2\pi N T}{60}\text{ W}$$

5. $$\eta_{\text{mech}} = \frac{\text{BP}}{\text{IP}}$$

6. $$\text{Indicated } \eta_{\text{the.}} = \frac{\text{IP}}{m_s (h_1 - h_{f_2})}$$

$$\text{Brake } \eta_{the.} = \frac{BP}{m_s(h_1 - h_{f_2})}$$

where

m_s = mass of steam in kg/sec.

h_1 = specific enthalpy of steam entering the cylinder

h_f = specific enthalpy of fluid or water returned to hot well

and

$$\eta_{mech} = \frac{\text{Brake } \eta_{th}}{\text{Indicated } \eta_{th}}$$

7. $$\eta_{over\ all} = \frac{BP}{m_r \times C.V.}$$

8. $$\eta_{relative} \text{ or efficiency ratio } = \frac{\text{Ind } \eta_{thermal}}{\text{Rankine efficiency}}$$

9. Steam consumption : kg/hr

$$\text{Specific stem consumption} = \frac{\text{kg/hr}}{\text{kW}} = \frac{\text{kg}}{\text{kWh}}$$

18.16 COMPOUND STEAM ENGINES

Nowadays high pressure steam from the high pressure boilers is generally used for power generation purpose because,

1. It gives higher cycle efficiency
2. Plant requires less space/kW of power generation.

If the high pressure steam is used with a single cylinder engine, then it will have the following limitations.

1. A huge and robust cylinder is required to with stand and accommodate high pressure steam.
2. Hence reciprocating parts are very large.
3. So, large variation, in torque.
4. To evenout this variation, a larger flywheel is required.
5. More balancing problems.
6. More condensation losses—Due to large pressure difference between the inlet and exhaust conditions, by the properties of steam, there will be large temperature difference and which results in more condensation losses.

To overcome these difficulties a compound steam engine is used whenever pressure ranges are high.

A compound steam engine consists of more than one cylinder. If two cylinders are used, then it is called as Double expansion compound steam engine.

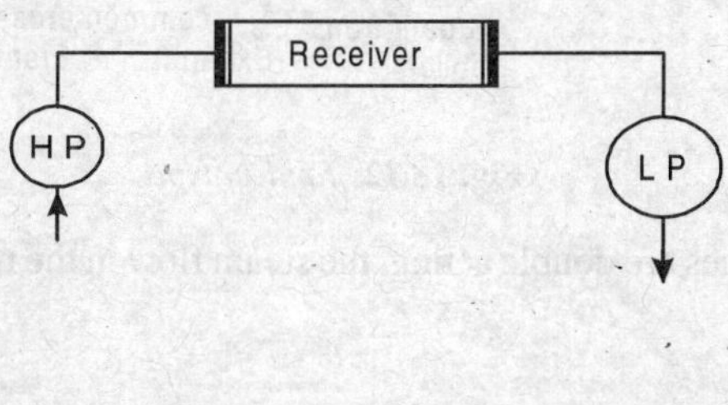

Fig. 18.10

Here high pressure steam from high pressure boiler will be first admitted in the HP cylinder, where partial expansion of steam takes place and the remaining expansion will be completed in LP cylinder.

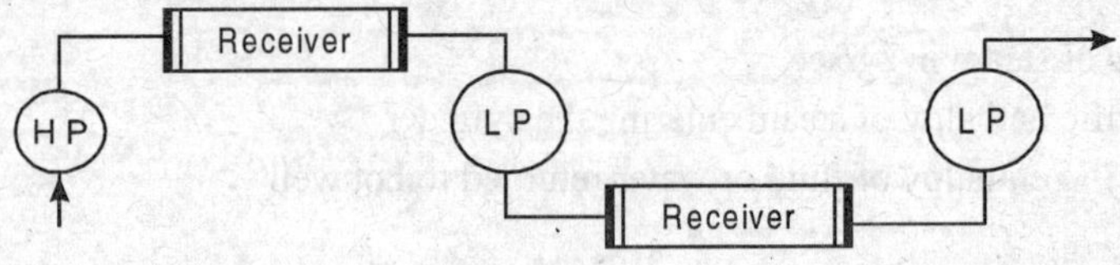

Fig. 18.12

If three cylinders are used, then it is called as Tripplé expansion compound steam engine. In this case an Intermediate Pressure (IP) cylinder is provided in between the HP and LP cylinders as shown in Fig. 18.11.

18.17 CLASSIFICATION OF COMPOUND STEAM ENGINES

I. Tandem Compounding

II. Cross compounding

(a) Woolfe type

(b) Receiver type.

I.Tandem Compound Steam Engine:

In this case there is only one piston rod, one cross head, one connecting rod and one crank shaft. In this case piston for the HP cylinder and LP cylinder are mounted on the same piston rod as shown in Fig. 18.12.

Let the high pressure steam, from the high pressure boiler be admitted to one side of the piston of HP cylinders as shown by solid arrow heads. Steam from the other side of the HP piston, which is shown by solid arrow heads will be simultaneously exhausted/admitted into the LP cylinder. The valve setting should be such that admission of steam continues simultaneously in both the cylinders. The steam from the LP cylinder will be exhausted either to the condenser or to the atmosphere.

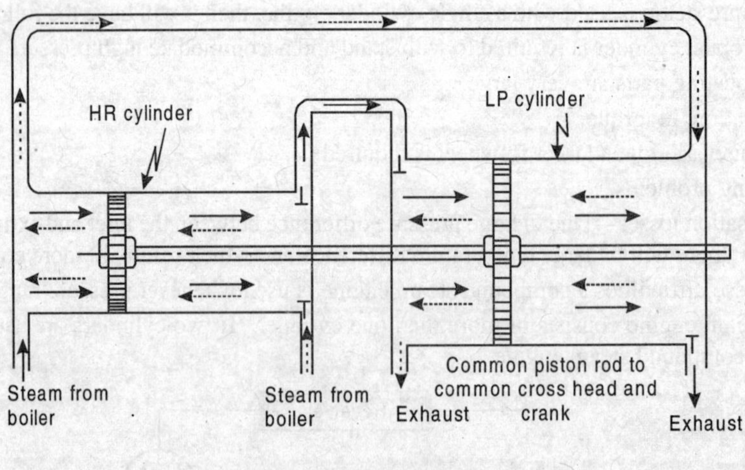

Fig. 18.12 *Tandem type*

Since normally the steam engines are double acting, the steam flow in the return stroke is shown by dotted arrow heads.

Since the pistons are mounted on the common piston rod, so that when the piston of HP cylinder is at IDC (Inner Dead Centre), the piston of LF cylinder will be at IDC, (Inner Dead Centre) and similarly when the piston of HP cylinder is at ODC (Outer Dead Centre) then the piston of LP cylinder will be at ODC, so the operation in both the cylinders are in phase. So, the maximum and minimum torques will occur approximately at the same crank angle.

As shown in Fig. 18.13 there is large variation in turning moment, so a large fly wheel is required to even out this variation.

(II) (a) Woolfe type : In Woolfe type two-cylinder CSE, the cranks of the two cylinders are 180° out of phase, i.e.,

(i) When the crank of HP cylinder is at TDC, the crank of LP cylinder will be at BDC and

(ii) When the piston of HP cylinder is at TDC, the piston of LP cylinder will be at BDC and

(iii) when the piston of HP cylinder is moving downwards, the piston of LP cylinder will be moving upwards.

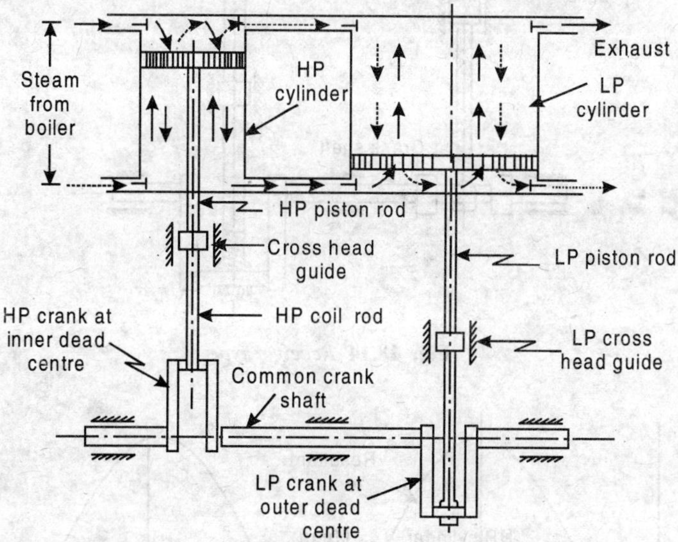

Fig. 18.13 *Woolfe Type*

As shown in Fig.. 18.13 .cylinders are arranged side by side. The pistons are connected to separate piston rods, which are in turn connected to crank through separate connecting rods.

In this case the HP cylinder directly exhausts into LP cylinder so no receiver is required. The turning moment variation is exactly similar to that of Tandem type. So this arrangement also requires a heavy Flywheel to evenout the variation.

(II) (b) Receiver Type : In this type of engine the cranks of the cylinders are placed at 90° to each other, the strokes of the two cylinders are thus out of phase. The HP cylinder cannot exhaust directly to the LP cylinder. In order to overcome this, a receiver is provided in between HP and LP cylinder as shown in Fig. 18.14. The HP cylinder exhausts in to the receiver while the LP cylinder draws from the receiver. Therefore to minimise fluctuations in pressure, the volume capacity of the receiver must be sufficiently large.

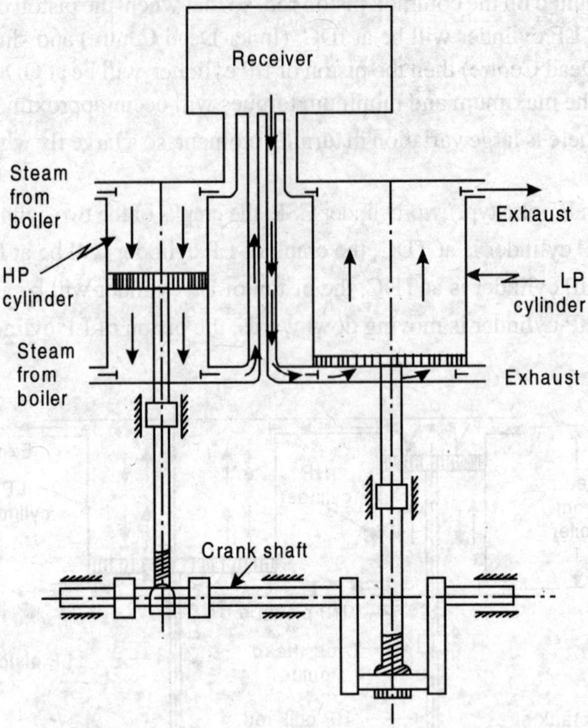

Fig. 18.14 *Receiver type*

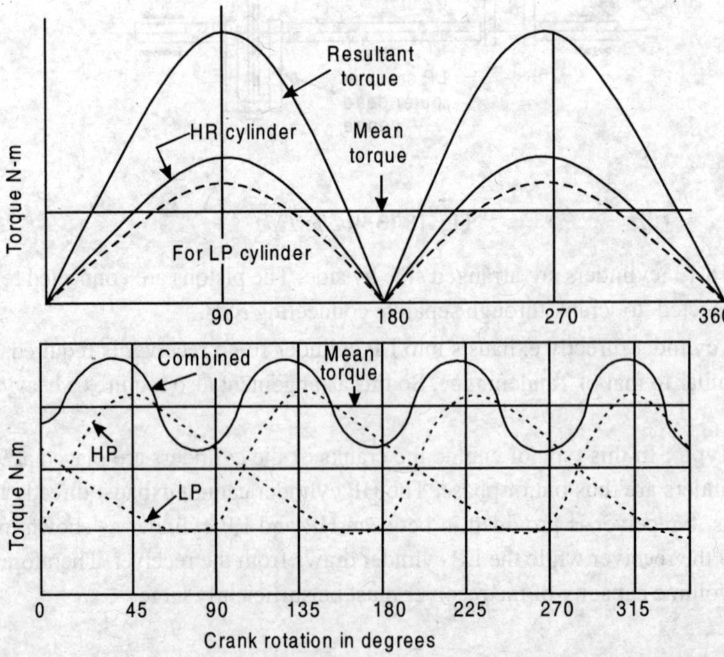

Fig. 18.15 *For Receiver Type*

The turning moment diagram is more uniform as shown in Fig. 18.15 and hence these engines need a much smaller flywheel.

18.17 ADVANTAGES OF COMPOUND STEAM ENGINES

1. By compound pressure range/cylinder is reduced and in turn the temperature range/cylinder is reduced—so less stress and less condensation losses.

2. Only HP cylinder is subjected to high pressure steam and therefore can be made of stronger material. But cheaper materials can be used for LPG cylinder because of low pressure range.

3. Since pressure range/cylinder is reduced, leakage of steam will be less.

4. Steam can be reheated in a separate heater after expansion in each cylinder, which increases the efficiency of the engine.

5. More uniform, torque can be obtained in receiver type engines and a smaller flywheel can be used.

6. Engine vibrations are reduced because of lighter reciprocating parts. Therefore Balancing will be simpler.

7. Engines can start in any position.

8. Cost of CSE/kW of power produced is less than single cylinder engines.

9. During breakdown, the engine may be modified to continue to operate on reduced loads.

18.18 DISADVANTAGES OF COMPOUND STEAM ENGINES

1. The arrangement is complex, needs greater care and maintenance.

2. Increased wear and tear.

3. More lubrication problems.

18.19 CHOOSING THE INTERMEDIATE PRESSURE FOR THE COMPOUND STEAM ENGINE

There are two possibilities

1. Equal initial piston loads 2. Equal work per cylinder

1. *Equal initial piston loads*

In a double-acting steam engine, the net pressure on the piston will be the difference between the pressures on either side of the piston.

For the high-pressure cylinder.

$$\text{Net pressure} = P_1 - P_7$$

For the low-pressure cylinder,

$$\text{Net pressure} = P_7 - P_5$$

Now

Force on piston = Net pressure on piston × piston area

Thus, for equal initial piston loads.

$$(P_1 - P_7)\,A_{HP} - (P_7 - P_5)\,A_{LP} \quad ...(1)$$

where A_{HP} = area of high-pressure piston.

A_{LP} = area of low-pressure piston.

Now, as already stated, the strokes of the pistons of a compound engine are usually made the same.

Thus, if the stroke = L, then from Eq. (1).

$$(P_1 - P_7)\,A_{HP}\,L = (P_7 - P_5)\,A_{LP}\,L \quad ...(2)$$

But neglecting clearance.

$A_{HP}L = V_{HP}$ = volume of high-pressure cylinder also

$A_{LP}L = V_{LP}$ = volume of low-pressure cylinder hence, Eq. (2) becomes,

$$(P_1 - P_7)\,V_{HP} - P = (P_7 - P_5)\,V_{LP} \quad ...(3)$$

2. *Equal work per cylinder*

For this to be the case, the area of the high-pressure and low-pressure diagrams must be the same.

For the high-pressure cylinder.

Work done = Area 1 2 6 7

(Area under 1 – 2) + area (under area 2–6) – (area under 6–7)

$$P_2 V_2 + P_2 V_2 \ln \frac{V_6}{V_2} - P_6 V_6$$

But $$P_2 V_2 = P_2 V_2 \text{ since } PV = C$$

$\therefore$ Work done $$= P_2 V_2 + P_2 V_2 \ln \frac{V_6}{V_1} - P_2 V_2$$

$$= P_2 V_2 \text{ in } r \quad ...(4)$$

where $r = V_6/V_2$ = number of expansions in high-pressure cylinder.

The low-pressure cylinder diagram is of the conventional shape, hence

Work done in low-pressure cylinder

$$= P_6 V_6 (1 + \ln R) - P_4 V_4 \quad ...(5)$$

where $R = \dfrac{V_4}{V_6}$ = number of expansions in the low-pressure cylinder.

Hence, for equal work per cylinder

$$P_2 V_2 \ln r = P_6 V_6 (1 + \ln R) - P_4 V_4 \quad ...(6)$$

Note that the high-pressure cylinder volume is the same as the cut-off volume, 7–6, of the low-pressure cylinder.

Note also that the total work done is given by the total area 1 2 3 4 5. This would also be the work done if the steam were admitted and expanded in the low-pressure cylinder only. This is a point which is made use of in the solution of some steam engine problems. ■■■

SOLVED EXAMPLES

Example 18.1. *A single cylinder double acting steam engine having the cylinder diameter 20 cm and stroke 30 cm is working with 40 % cut off and running at 200 rpm. The supply of steam is at a pressure of 8.5 bar and back pressure is 0.25 bar. Estimate the indicated power assuming a diagram factor of 0.75 for the engine. If the mechanical efficiency of the engine is 80 %, what is its brake power ?*

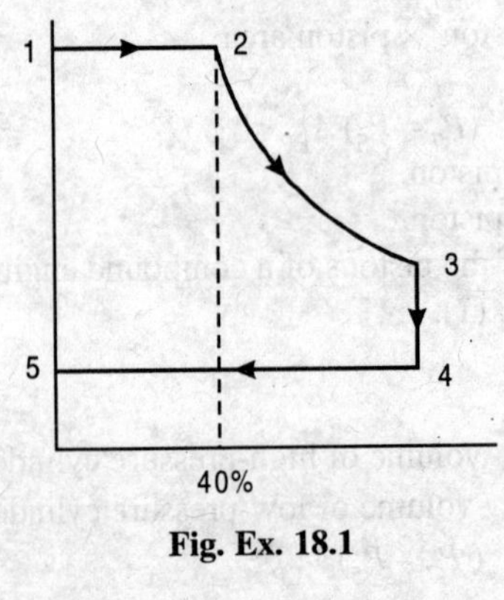

Fig. Ex. 18.1

Data :

$$D = 20\ cm = 0.2\ m$$

$$L = 30\ cm = 0.3\ m$$

$$r = \frac{1}{0.4} = 2.5 \text{ i.e., } \frac{1}{\text{Cutt-off } \%}$$

$$N = 200\ rpm$$

$$P = 8.5\ bar = 850\ kPa$$

$$P_b = 0.25\ bar = 25\ kPa$$

$$IP = ? \text{ Diagram factor } k = 0.75,\ \eta_{mech} = 80\ \%$$

$$BP = ?$$

Solution We know that

$$IP = \frac{2P_m LAN}{60}$$

To find

$$P_m = \frac{P}{r}(1 + \ln \cdot r) - P_b$$

$$= \frac{8.5}{2.5}(1 + \ln 2.5) - 0.25$$

$$= 6.515 - 0.25$$

$$= 6.2653885 \text{ bar}$$

$$= 626.53885 \frac{\text{kW}}{\text{m}^2}$$

$$= 626538.85 \text{ N/m}^2$$

$\therefore$ $$P_{m\,(\text{actual})} = 0.75 \times 626.539 = 469.90 \text{ kPa}$$

Then $$IP = \frac{2P_m L\,A\,N}{60}$$

$$= \frac{2 \times 469.90 \times 0.3 \times 200}{60} \text{ kW}$$

$$IP_{(\text{actual})} = 939.80 \text{ kW}$$

We known that

$$\eta_{mech} = \frac{BP}{IP}$$

or $$BP = \eta_{mech} \times IP$$

$$= 0.8 \times 29.52$$

$$BP = 23.6197 \text{ kW}$$

Example 18.2 *Find L and D ?*

Data

$$\textit{Boiler pressure} = 10\ bar = 1000\ \frac{kN}{m^2}$$

$$\text{Back pressure} = 0.15 \text{ bar} = 15 \text{ kW/m}^2$$

$$\text{Cut off at 40 \% of stroke, } N = 120 \text{ rpm.}$$

$$\text{D.F.} = 0.85$$

and $D = 0.75\,L$

Solution

$$r = \frac{1}{\text{cut off}} = \frac{1}{0.4} = 2.5$$

$$P_m = \frac{P}{r}(1+\ln r) - P_b$$

$$= \frac{1000}{2.5}(1+\ln 2.5) - 15$$

$$P_m = 751.516\ \frac{\text{kN}}{\text{m}^2}$$

$\therefore$

$$P_m\ \text{actual} = 751.516 \times 0.85$$

$$= 638.789 \text{ kN/m}^2$$

$$\text{IP} = \frac{2 \times P_m \times L \times A \times N}{60}\ \text{W}$$

$$A = \frac{\pi}{4} D^2$$

$$= \frac{\pi}{4}(0.75L)^2$$

$$\because \quad \frac{D}{L} = 0.75$$

$$200 \text{ kW} = \frac{2 \times 638.789 \times L \times \frac{\pi}{4}(0.75L)^2 \times 120}{60}\ \text{kW}$$

$$200 \text{ kW} = \frac{2 \times 638.789 \times \frac{\pi}{4} \times (0.75L)^2 \times L \times 120}{60}$$

$$= 1128.83\ L^3$$

$$0.1772 = L^3$$

$$L = 0.56165 \text{ m or } 56.17 \text{ cm}$$

$\therefore$ Since

$$\frac{D}{L} = 0.75$$

$$D = 42.13$$

Example 18.3 *During a test on double acting steam engine, following observations were made.*

Duration of test = *20 min*
Cylinder diameter = *20 cm*
Stroke = *28 cm*
Speed = *120 rpm*
Spring balance reading = *100 N*
Dead load = *650 N*
Effective brake radius = *56 cm*

Average indicated mean effective pressure for both the diagrams = 1.4 bar

Steam consumption = *18 kg*

Calculate IP, BP, η_{mech} and steam consumption in kg kWh.

Solution 1.

$$\text{IP} = \frac{2P_m(\text{actual})L\,A\,N}{60}$$

Average P_m or actual = 1.4 bar = 140 $\frac{\text{kN}}{\text{m}^2}$

$$\text{IP} = \frac{2\times140\times0.28\times\frac{\pi}{4}(0.2)^2\times120}{50}\ \text{kW}$$

$$\text{IP} = 4.9269\ \text{kW}$$

2.

$$\text{BP} = \frac{(w-s)R-2\,\text{kN})}{60}\text{W}$$

$$= \frac{(650-100)\times0.56\times2\pi\times120}{60}\text{W}$$

$$= 3870.44\ \text{W}$$

$$BP = 3.870\ \text{kW}$$

3.

$$\eta_{\text{mech}} = \frac{\text{BP}}{\text{IP}} = \frac{3.870}{4.926}$$

$$= 0.7856$$

$$= 78.56\%$$

4. Steam used in 20 min = 18 kg

$\therefore$ Steam used/hr $= \frac{60}{20}\times18$

$= 54$ kg/hr

$\therefore$ Steam consumption/Brake kWh

$$= \frac{54\ \text{kg/hr}}{3.87\ \text{kW}}$$

$$= 13.95 \frac{\text{kg}}{\text{kWh (BP)}}$$

Example 18.4 *During a trial on vertical double acting steam engine following observations were taken.*

Indicator card area (cover end stroke) = 10.5 cm²

Indicator card area crank = 10.2 cm²

Length of indicator card = 7.6 cm

Spring constant for indicator used = 0.8 bar/cm

Dia. of piston/cylinder diameter = 21.5 cm

Dia. of pistion rod = 3.7 cm

Stroke = 30.5 cm

Speed = 120 rpm

Calculate the IP, if the η_{mech} *of the engine is 80% at the given load find the brake torque required.*

Solution We know that

$$\text{Total IP} = \text{IP}_{\text{cover/head end}} + \text{IP}_{\text{crank end}}$$

$$= \frac{P_{m_1} A_1 LN}{60} + \frac{P_{m_2} A_2 LN}{60}$$

We know that

$$\text{Total } P_m = \frac{\text{Area of indicator Diagram}}{\text{Length of indicator Diagram}}$$

or

$$= \frac{\text{Ind. car area}}{\text{Ind. card length}} \times \text{Spring scale constant}$$

$$P_{m_j} = \frac{10.5 \text{ cm}^2}{7.6 \text{ cm}} \times 0.8 \text{ bar/cm}$$

(For cover/head end side of piston)

$$P_{m_1} = 1.1052 \text{ bar} = 110.52 \text{ kN/m}^2$$

$$P_{m_2} = \frac{10.2}{7.6} \times 0.8$$

(For crank end side)

$$= 1.0737 \text{ bar} = 107.37 \frac{\text{kN}}{\text{m}^2}$$

We know that A_1 = Area on cover/head end side

$$A_2 = \text{Area crank end side}$$

and

$$A_2 = A_1 - \text{Area of piston rod.}$$

$\therefore$

$$A_1 = \frac{\pi}{4} D^2$$

$$= \frac{\pi}{4}(0.215)^2$$

$$= 0.036305 \text{ m}^2 \qquad \dots (1)$$

and $$A_2 = A_1 - \text{Area of piston rod}$$

$$= 0.036305 - \frac{\pi}{4}(0.037)^2$$

$$= 0.001075$$

$$= 0.0352297 \text{ m}^2$$

$$\therefore \quad \text{IP total} = \frac{[P_{m_1} A_1 + P_{m_2} A_2]LN}{60}$$

$$= \left[110.52 \frac{\text{kN}}{m^2} \times 0.036305 + 107.37 \times 0.0352297\right] \frac{0.305 \times 120}{60}$$

$$= [4.012486 + 3.7826129] \times \frac{0.305 \times 120}{60}$$

$$= 4.754975 \text{ kW}$$

We know that

$$\eta_{mech} = \frac{\text{BP}}{\text{IP}}$$

$$0.8 = \frac{\text{BP}}{4.755}$$

$$\therefore \quad \text{BP} = 3.804 \text{ kW} = 3804 \text{ W}$$

and $$\text{BP} = \frac{2\pi N}{60} T (\text{Brake torque}) \quad \frac{(W - s)R \cdot 2\pi N}{60}$$

$$\therefore \quad 3.804 \times 1000 = \frac{2 \times \pi \times 120}{60} \times T$$

$$\therefore \quad \text{Brake torque } T = 302.71 \text{ Nm}$$

Example 18.5 *The following data was obtained during a trial on a double acting steam engine.*

Avergae Mean effective pressure on cover and crank and crank end = 105 N/cm²

Speed = 240 rpm

Stroke = 1.5 bore = 30 cm

Net brake load = 1.5 kN

Effective brake diameter = 160 cm

Supply steam condition = Dry -saturated at 9 bar

Condensate collected (Pure) = 800 kg/hr

Condensate pressure = 0.07 bar

Exhaust steam quality = 0.7 dry

Define (i) IP, (ii) BP, (iii) η_{mech}, (iv) Indicated $\eta_{thermal}$ and (v) brake h_{th}.

Solution Given data :

$$P_m \text{ (actual)} = 105 \text{ N/cm}^2 = 105 \times 10^4 \frac{N}{m^2}$$

$$N = 240 \text{ rpm}$$

$$L = 1.5\,D = 30 \text{ cm}$$

$$\therefore \quad D = 20 \text{ cm}$$

$$(W - S) = 1.5 \text{ kN} = 1500 \text{ N}$$

$$\text{Effective brake radius} = 80 \text{ cm} = 0.8 \text{ m}$$

(i)

$$\text{IP} = \frac{2P_m LAN}{60} \text{ W}$$

$$= \frac{2 \times 105 \times 10^4 \times 0.3 \times \frac{\pi}{4}(0.2)^2 \times 240}{60} \text{ W}$$

$$= 79168.135 \text{ W}$$

$$= 79.168135 \text{ kW}$$

(ii)

$$\text{BP} = \frac{(W - S) \cdot R \cdot 2\pi N}{60} \text{ W}$$

$$= \frac{150 \times 0.8 \times 2 \times \pi \times 240}{60} \text{ W}$$

$$= 30159.289 \text{ W}$$

$$= 30.159289 \text{ kW}$$

(iii)

$$\therefore \quad \eta_{mech} = \frac{\text{BP}}{\text{IP}} = \frac{30.159}{79.168}$$

$$= 38.095\%$$

(iv)

$$\text{Ind. } \eta_{th} = \frac{I_p}{m_s(h_1 - h_{f_2})}$$

where h_1 = specific enthalpy of steam entry the engine cylinder corresponding to a pressure of 9 bar (dry sat.)

$$h_1 = h_2 = 2772.1 \frac{\text{kJ}}{\text{kg}}$$

and h_f = sp. fluid enthalpy of water returned to hot well corresponding pressure is 0.07 bar

$$h_{f_2} = 163.38 \text{ kJ/kg}$$

and m_s = mass of steam entry to engine cylinder in kg/sec

$$= \frac{\text{cond. collected kg/hr}}{3600} = \frac{800 \text{ kg/hr}}{3600} = \text{in kg/sec}$$

$$\therefore \quad \text{Ind. } \eta_{th} = \frac{\text{IP}}{m_s(h_1 - h_{f_2})}$$

$$= 79.168 \text{ kW}$$

$$= \frac{800}{3600}(2772.1 - 163.4)$$

$$= 13.66\%$$

and (v)

$$\text{Break } \eta_{th} = \frac{\text{BP}}{m_s(h_1 - h_{f_2})}$$

$$= 30.159 \text{ kW}$$

$$= \frac{800}{3600}(2772.1 - 163.4)$$

$$= 0.052 = 5.2\ \%$$

Example 18.6 *A throttled governed steam engine uses 2820 kg of steam per hour when developing an output power of 150 kW and 5000 kg/hr when developing 375 kW. Estimate the thermal efficiency of the engine when developing an output of 225 kW assuming the steam supply is dry and saturated at 10 bar and exhaust is at 0.25 bar.*

Solution Refer Fig. of Willan's line (Fig. 18.9)

By Willan's law,

$$m = a\,I + B$$

$$\therefore \quad 2820 = a \times 150 + b \qquad \text{...(i)}$$

$$5000 = a \times 375 + b \qquad \text{...(ii)}$$

On solving Eqs. (i) and (ii), we get

$$a = 9.6889 \text{ kg/kWh}$$

$$b = 13.66.667 \text{ kg/hr}$$

∴ The law becomes,

$$m = 9.6889\ \text{I} = 1366.667 \qquad \text{...(iii)}$$

at I = 225 kW, the steam consumption is,

$$m = 9.6889 \times 225 + 1366.667$$

$$= 3546.67 \text{ kg/hr}$$

Thermal efficiency,

$$\eta = \frac{\text{Power}}{m(h_1 - h_2)}$$

From steam tables,

$$h_1 = 2776.2 \text{ kJ/kg} \qquad \text{(at 10 bar, dry-saturated)}$$

$$h_2 = 272.0 \text{kJ/kg} \qquad \text{(liquid enthalpy at 0.25 bar)}$$

∴ Thermal efficiency,

$$\eta = \frac{225}{\frac{3546.67}{3600}(2776.2 - 272.0)}$$

$$= \mathbf{0.0912 \text{ or } 9.12\ \%}$$

Example 18.7 *A double-acting, compound, reciprocating steam engine has two cylinders and is supplied with steam at a pressure of 1.4 MN/m² and 0.9 dry. Exhaust from the engine is into a condenser working at a pressure of 35 kN/m². Each cylinder has equal initial piston loads. The diagram factor referred to the low-pressure cylinder is 0.8. The stroke in both cylinders is 350 mm. The diameter of the high and low pressure cylinders are 200 mm and 300 mm respectively. Expansion in the high-pressure cylinder is complete. The engine runs at 300 rev/min. Expansion in the engine is hyperbolic.*

Neglecting clearance, determine

(i) the intermediate pressure;

(ii) the indicated power output;

(iii) the steam consumption of the engine in kg/h.

Solution

(*i*) For equal initial piston loads,

$$(P_1 - P_7)\,A_{HP} = (P_7 - P_5)\,A_{LP}$$

$$(1400 - P_7)\frac{\pi\times(0.2)^2}{4} = (P_7 - 35)\frac{\pi\times(0.3)^2}{4}$$

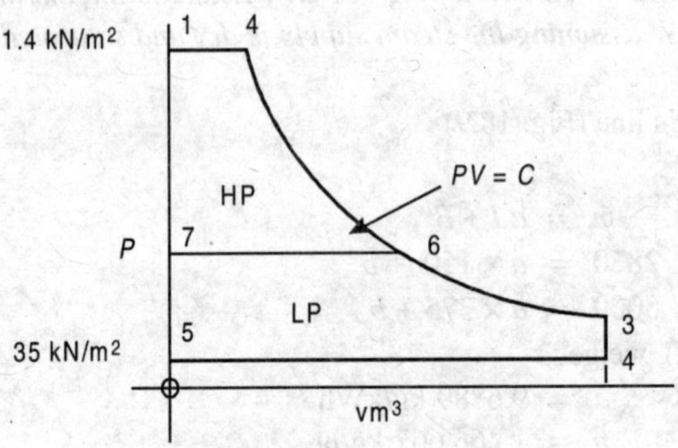

Fig. Ex. 18.7

$$\therefore \quad 1400 - P_7 = (P_7 - 35)$$
$$= (P_7 - 35)\,2.25$$
$$= 2.25\,P_7 - 78.75$$
$$\therefore \quad 3.25\,P_7 = 1478.75$$
$$P_7 = 1478.75/3.25$$
$$= 455 \text{ kN/m}^2$$

This is the intermediate pressure,

(*ii*) Now

$$P_2V_2 = P_6 - V_6$$
$$= \text{cut-off volume in the high-pressure cylinder.}$$

Volume of low-pressure cylinders

$$= \left(\frac{\pi\times(0.3)^2}{4}\times 0.35\right)\text{m}^3$$

$\therefore$ Number of expansions through the engines

$$R = \frac{\text{Volume of low-pressure cylinder}}{\text{Cut – off volume in the high-pressure cylinder}}$$

$$= \frac{\pi\times(0.3)^2\times0.34\ 4}{455\times\pi\times(0.2)^2\times0.35/(1400\times4)}$$

$$= \frac{(0.3)^2\times100}{(0.2)^2\times455} = 2.25\times\frac{1400}{455} = 6.92$$

P = referred the low-pressure cylinder

$$= \frac{P}{R}(1+\text{In } R) - P_6 = \frac{1400}{6.92}(1+\text{In } 6.92) - 35$$

$= 594 - 35 = 559$ kN/m^2

Actual $P_m = 0.8 \times 559 = 447.2$ kN/m^2

Indicated power $= P_m \text{LAN} = 447.2 \times 0.35 \times \frac{\pi\times(0.3)^2}{4}\times2\times\frac{300}{60}$

$= 110.6$ kW

(iii) Volume of the low-pressure cylinder

$$= \frac{\pi\times(0.3)^2}{4}\times0.35 = 0.0247\text{m}^3$$

$\therefore$ cut-off volume in high pressure cylinder

$$= \frac{0.0247}{6.92} = 0.00357\text{ m}^3$$

This is the volume of steam admitted/stroke

Hence, volume of steam admitted/h

$= (0.00357 \times 300 \times 2 \times 60)$ m^2

Specific volume of admission steam

$= xv = 0.9 \times 0.1407 = 0.1266$ m^2/kg

$\therefore$ Steam consumption $= \frac{0.00357\times300\times2\times60}{0.1266}$

$= 1015$ kg/h

Example 18.8 *A double-acting, two cylinder, compound reciprocating steam engine has a low-pressure cylinder 600 mm diameter with a stroke of 600 mm. Steam is supplied at a pressure of 1.1 MN/m^2 and it exhausts from the engine at a pressure of 28 kN/m^2. The engine runs at 4 rev/s and has a diagram factor referred to the low pressure cylinder of 0.82. Expansion is hyperbolic throughout, clearance is neglected and expansion in the high-pressure cylinder is complete. The total expansion ratio through the engine is 8 and equal work is developed in each cylinder. Determine*

(*a*) *the indicated power output of the engine;*

(*b*) *the diameter of the high-pressure cylinder if the stroke is the same as that of the low pressure cylinder;*

(c) *the intermediate pressure.*

Solution

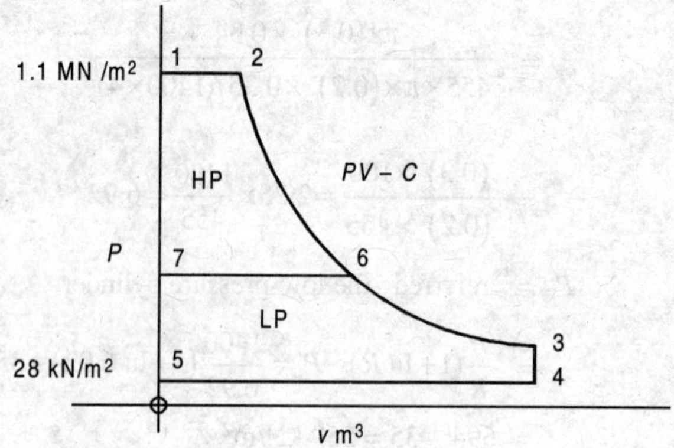

Fig. Ex. 18.8

(a) P_m referred to low-pressure cylinder

$$= \frac{P}{R} - (1 + \ln R) - P_6 = \frac{1100}{8}(1 + \ln 8) - 28$$

$$= \frac{1100}{8}(1 + 9.08) - 28 - \frac{1100 \times 3.08}{8} - 28$$

$$= 424 - 28 = 396 \text{ kN/m}^2$$

Actual $\quad P_m = 0.82 \times 396 = 325 \text{ kN/m}^2$

$\therefore$ indicated power

$$= P_m \text{LAN} = 325 \times 0.6 \times \frac{\pi \times (0.6)^2}{4} \times 2 \times 4$$

$$= 441 \text{ kW}$$

(b) Work done by two cylinders/stroke

$$= \text{Total area of diagram} = \left(\frac{325 \times 0.6 \times \pi \times (0.6)^2}{2}\right) \text{kJ}$$

Since equal work is performed/cylinder, then,
Work done in high-pressure cylinder

$$= \frac{\text{Total area of diagram}}{2} = \frac{325 \times 0.6 \times \pi \times (0.6)^2}{2 \times 4}$$

$$= 27.6 \text{ kJ}$$

Work done in high-pressure cylinder

$$= \text{area } 1267 = P_2 V_2 \ln \frac{V_6}{V_2} = 27.6 \text{ kJ}$$

$$V_2 = \frac{V_3}{8} = \frac{p \times (0.6)^2}{4} \times \frac{0.6}{8} = 0.0212 \text{ m}^3$$

$$\therefore \quad 1100 \times 0.0212 \ln \frac{V_2}{V_2} = 27.6$$

In

$$\frac{V_6}{V_2} = \frac{27.6}{1100 \times 0.0212} = 1.18$$

$$\therefore \quad \frac{V_6}{V_2} = 3.25$$

From this,

$$V_6 = 3.25\, V_2 = 3.25 \times 0.0212 = 0.069 \text{ m}^2$$

= volume of high – pressure cylinder

Let d = diameter of high-pressure cylinder in m.

Stroke = 0.6 m

Then

$$\frac{\pi d^2}{0.6} = 0.069$$

$$d^2 = \frac{0.069 \times 4}{\pi \times 0.6} = 0.1464$$

$$\therefore \quad d = (0.1464) = 0.383 \text{ m} = 383 \text{ mm}$$

(c) For expansion 2-6

$P_2 V_2 = P_6 V_6$ since expansion is hyperbolic

$$\therefore \quad P_6 = \frac{P_2 V_2}{V_6} = \frac{1100 \times 0.0211}{0.069} = 338 \text{ kN/m}^2$$

= intermediate pressure

Example 18.9 *A compound steam engine consists of one high-pressure cylinder and one low-pressure cylinder, both of which are double-acting. The engine develops 230 kW. The admission pressure is 1.4 MN/m² and the exhaust pressure is 35 kN/m . The total expansion ratio referred to the low pressure cylinder is 12.5:1. The diameter of the low-pressure cylinder is 400 mm and the stroke in both cylinders is 500 mm. The diagram factor is 0.78, expansion is assumed to be hyperbolic and clearance is neglected. The expansion ratio in the high pressure cylinder is 2.5:1. Determine,*

(*a*) *the speed of the engine in rev/s*

(*b*) *the diameter of the high-pressure cylinder*

(*a*) P_m *referred to the low-pressure cylinder*

$$= \frac{P}{R}(1 + \ln R) - P_b$$

$$= \frac{1400}{12.5}(1 + \ln 12.5) - 35 = \frac{1400}{12.5}(1 + 2.53) - 35$$

$$= \frac{1400 \times 3.53}{12.5} - 35 = 395 - 35 = 360 \text{ kN/m}^2$$

$$\text{Theoretical indicated power} = \frac{230}{0.78} = 295 \text{ kW}$$

$$\text{Indicated power} = P_m \text{LAN}$$

$$\therefore \quad 295 = 360 \times 0.5 \times \frac{\pi \times (0.4)^2}{4} \times 2 \times \text{rev/s}$$

$$\therefore \quad \text{rev/s} = \frac{295 \times 4}{360 \times 0.5 \times \pi \times (0.4)^2 \times 2} = 6.52 = \text{engine speed.}$$

(b) Volume of low-pressure cylinder $= \frac{\pi \times (0.4)^2}{4} \times 0.5 = 00.0628 \text{ m}^2$

$\therefore$ cut-off volume in high-pressure cylinder

$$= \frac{0.0628}{12.5} = 0.00502 \text{ m}^2$$

Total volume of high-pressure cylinder

$$= 0.00502 \times 2.5 = 0.01255 \text{ m}^3$$

If d = diameter of high-pressure cylinder in m.

$$\frac{\pi d^2}{4} \times 0.5 \times 0.01255$$

$$d^2 = \frac{4 \times 0.01255}{\pi \times 0.5} = 0.032$$

$$d = (0.032) = 0.179 \text{ m} = 179 \text{ mm}$$

Example 18.10 *A double-acting, compound steam engine has cylinders of 300 mm and 600 mm diameter and a common stroke of 400 mm. Steam is supplied at a pressure of 1.1 MN/m² and cut-off in the high- pressure cylinder is at one-third stroke. The back pressure in the loss pressure cylinder is 32 kN/m². Average areas of indicator diagrams from the engine are : H.P., 12.5 cm²; spring 270 kN/m²/cm; L.P., 11.4 cm²; spring 80 kN/m²/cm. The length of both diagrams is 75 cm. The engine speed is 2.7 rev/s. Determine,*

(a) the actual and hypothetical mean effective referred to the low-pressure cylinder,

(b) the overall diagram factor,

(c) the indicated power.

Solution

(a) For high-pressure cylinder,

$$P_{mH} = \frac{12.5}{7.5} \times 270 = 450 \text{ kN/m}^2$$

Force on high-pressure piston $= P_{mH} 4 = \left[450 \times \frac{\pi \times (0.3)^2}{4}\right]$ kN

Since the stroke and rev/s are the same for both cylinders, then this force can be referred to the low-pressure cylinder by calculating from this force the pressure required over the low-pressure piston to produce this force.

$\therefore$ P_{mH} referred to the low pressure cylinder

$$= 450 \times \frac{\pi \times (0.3)^2 / 4}{\pi \times (0.6)^2 / 4} = 450 \times \frac{(0.3)^2}{(0.6)^2} = 450 \times \frac{0.09}{0.36}$$

$$= \frac{450}{4} = 112.5 \text{ kN/m}^2$$

For the low-pressure cylinder,

$$P_{mL} = \frac{11.4}{7.5} \times 80 = 121.6 \text{ kN/m}^2$$

Actual P_{mA} referred to the low-pressure cylinder

$$= 112.5 + 121.6 = 234.1 \text{ kN/m}^2$$

Volume of high-pressure cylinder $= \left[\frac{\pi \times (0.3)^2}{4} \times 0.4\right]$ m^3

Cut-off volume of high-pressure cylinder $= \left[\frac{\pi \times (0.3)^2}{4} \times \frac{0.4}{3}\right]$ m^3

Volume of low-pressure cylinder $= \left[\frac{\pi \times (0.6)^2}{4} \times 0.4\right]$ m^3

Number of expansion through the engine

$$= \frac{\text{Volume of low-pressure cylinder}}{\text{Cut-off volume of high-pressure cylinder}} = R$$

$$= \frac{\left[\pi \times \frac{(0.6)^2}{4}\right] \times 0.4}{\left[\pi \times \frac{(0.3)^2}{4}\right] \times \frac{0.4}{3}} = \frac{(0.6)^2 \times 3}{(0.3)^2} = \frac{0.36}{0.09} \times 3 = 4 \times 3 = 12$$

Hypothetical $P_m = \frac{P}{R}(1 + \ln R) - P_6$

Hypothetical $P_m = \frac{1100}{12}(1 + \ln 12) - 32 = \frac{1100}{12}(1 + 2.48) - 32$

$$= \left(\frac{1100}{12}\times 3.48\right) - 32 = 319 - 32 = 287 \text{ kN/m}^2$$

(b) Overall diagram factor $= \dfrac{P_{mA}}{P_m} = \dfrac{234.1}{287} = 0.816$

(c) Indicated power $= P_{mA}\,\text{LAN}$

$$= 234.1 \times 0.4 \times \frac{\pi\times(0.6)^2}{4} \times 2 \times 27$$

$$= 143 \text{ kW}$$

Example 18.11 *During a trial on a triple-expansion engine the following results were recorded:*

Cylinder	*Diameter*	*Actual* P_m
HP	*300 mm*	*590 kN/m²*
IP	*1500 mm*	*214 kN/m²*
LP	*900 mm*	*88 kN/m²*

Steam supply pressure, 1.4 MN/m².
Cut-off in H.P. cylinder, 0.6 stroke.
Back pressure in LP cylinder, 20 kN/m².
All cylinders have the same stroke. Determine,
(a) the actual and hypothetical mean effective pressures referred to the low-pressure cylinder,
(b) the overall diagram factor,
(c) the percentage of the total indicated power developed in each cylinder.

Solution

(*a*) For the high-pressure cylinder,

$$P_{mH} \text{ referred to LP cylinder} = 590\times\frac{(0.3)^2}{(0.9)^2} = 500\times\frac{0.09}{0.81}$$

$$= \frac{590}{9} = 65.55 \text{ kN/m}^2$$

For the intermediate-pressure cylinder,

$$P_m \text{ referred to LP cylinder} = 214\times\frac{0.25}{0.81}$$

$$= 66 \text{ kN/m}^2$$

∴ Actual P_{mA} referred to the low-pressure cylinder

$$= 65.55 + 66 + 88 = 219.55 \text{ kN/m}^2$$

Total number of expansion through the engine

$$= \frac{900^2}{0.60\times 300^2} = \frac{9}{0.6} = 15$$

Hypothetical P_m referred to the low-pressure cylinder

$$= \frac{P}{R}(1+\ln R) - P_b = \frac{1400}{15}(1+\ln 15) - 20$$

$$= \frac{1400}{15}(1+2.7) - 20\left(\frac{1400}{15}\times 3.7\right) - 20 = 345 - 20$$

(b) Overall diagram factor $= \frac{219.55}{325} = 0.675$

(c) The percentage of the total indicated power developed in each cylinder is the same as the percentage analysis of the actual P_{mA} referred to the low-pressure cylinder.

Example 18.12 *A double acting compound steam engine has cylinder bore, LP = 40 cm, H.P. = 30 cm, stroke = 45 cm. Dry saturated steam at 1.6 MPa is admitted. The back pressure is 0.08 bar. The diagram factor = 0.8. Engine RPM = 240. Expansion is hyperbolic throughout, clearance is neglected and expansion is complete in HP cylinder. Assume equal initial piston load. Calculate the intermediate pressure, the output in kW and specific steam consumption.*

Data given,

$$D_{LP} = 40 \text{ cm}$$
$$D_{HP} = 30 \text{ cm}$$
$$L = 45 \text{ cm}$$
$$P_1 = 1600 \text{ kPa}$$
$$P_4 = P_5 = 8 \text{ kPa}$$
$$\text{Diagram Factor} = 0.8$$
$$\text{RPM} = 240$$

Solution Refer Fig. Ex. 18.12

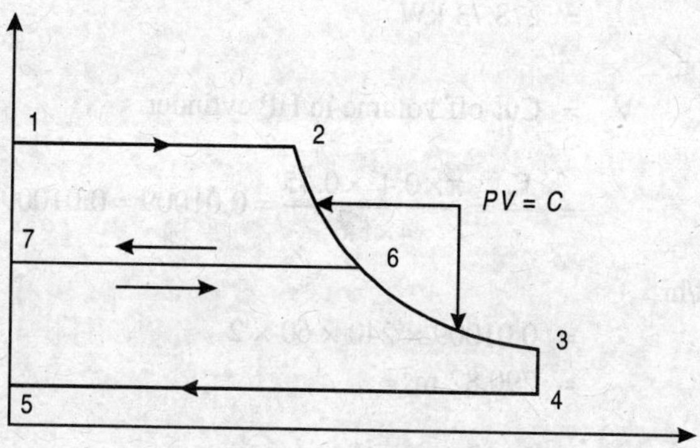

Fig. Ex. 18.12

For equal initial Piston Load

$$A_{HP}(P_1 - P_7) = A_{LP}(P_7 - P_5)$$

$$\frac{\pi}{4}D^{2HP}(P_1 - P_7) = \frac{\pi}{4}D^{2LP}(P_7 - P_5)$$

$$\therefore \quad 0.3^2(1600 - P_1) = 0.4^2(P_7 - 8)$$

$$\therefore \quad P_7 = 470.88 \text{ kPa}$$

Mean effective pressure referred to LP cylinder

We have, $V_3 = V_4 = V_{LP} = \frac{\pi}{4} \times 0.4^2 \times 0.45$

$= 0.05654 \text{ m}^3$

and V_2 = Volume in HP cylinder at cut off

$$= \frac{P_6 V_6}{P_2} = \frac{570.88}{1600} \times \frac{\pi}{4} \times 0.3^2 \times 0.4$$

$= 0.010088 \text{ m}^3$

∴ Expansion ratio referred to LP cylinder

$$r = \frac{V_3}{V_2} = \frac{0.05654}{0.010088} = 5.6$$

∴ P_m referred to LP cylinder

$$= \frac{1600}{5.6}[1 + \ln 5.6] - 8$$

$= 770 \text{ kPa}$

and indicated power developed

$$\text{IP} = P_m \times L \times A_{LP} \times N \times 2 \times \text{Dia. factor}$$

$$= 770 \times 0.45 \times \frac{\pi}{4} \times \frac{0.4^2}{1} \times \frac{240}{60} \times 2 \times 0.8$$

$= 278.73 \text{ kW}$

Specific steam consumption

V_2 = Cut-off volume in HP cylinder

$$= \frac{V_3}{r} = \frac{\pi \times 0.4^2 \times 0.45}{4 \times 5.6} = 0.01009 = 0.01009$$

Volume of steam admitted/hr

$= 0.01009 \times 240 \times 60 \times 2$

$= 290.82 \text{ m}^3$

Specific volume of dry steam at 1600 kPa = $Vg = 0.1237 \frac{\text{m}^3}{\text{kg}}$

∴ Steam consumption

$$= \frac{\text{Total volume of steam used}}{\text{Special volume}} = \frac{290.82}{0.1237} = 2351.02$$

$$= 2351.02 \frac{\text{kg}}{\text{hr}}$$

Example 18.13 *A Steam locomotive has two HP and LP cylinders. Steam enters at 15 bar and 300° C. The power developed is 5000 kW IP The overall diagram factor is 0.65. RPM = 175. Overall expansion ratio 6.8, stroke length 0.8 metre, HP cylinder cut off at 3/8 of stroke. Exhaust at 1.25 bar from LP cylinder. Neglecting clearance and cushioning calculate the diameter of cylinders.*

Solution

Actual P_m referred to LP cylinder

$$= \text{DF} \times \left(\frac{P_1}{R}(1 + \ln R) - P_b \right)$$

$$= 0.65 \times \left(\frac{15}{6.8}(1 + \ln 6.8) - 1.25 \right)$$

$$= 3.3695 \text{ bar} = 336.95 \text{ kPa}$$

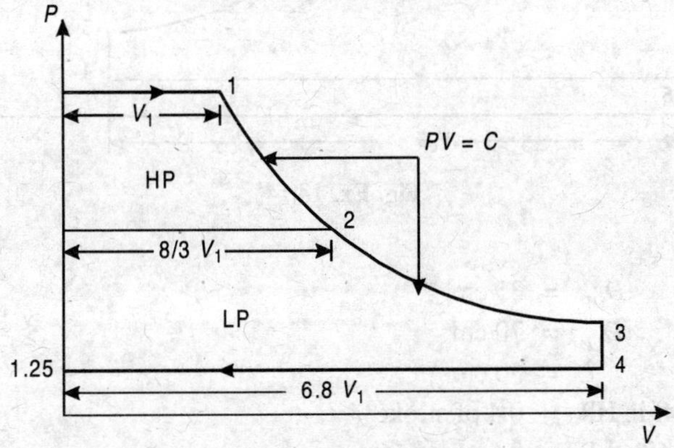

Fig. Ex. 18.13

$$\text{IP} = 2\, P_m \times L \times A \times N \times \text{No. of cylinders.}$$

$$5000 = 2 \times 336.95 \times 0.8 \times \frac{\pi}{4} \times D^{2\text{LP}} \times \frac{175}{60} \times 2$$

$$\therefore \quad D_{\text{LP}} = 1.423 \text{ m or } 142.3 \text{ cm}$$

$$\text{Cylinder Volume Ratio} = \frac{V_{\text{LP}}}{V_{\text{HP}}} = 6.8k_{\bar{8}} = 255$$

$$\therefore \quad D_{\text{HP}} = \frac{D_{\text{LP}}}{\sqrt{2.55}} = 89.11 \text{ cm}$$

Example 18.14 *A double acting compound steam engine has 35 cm HP cylinder diameter and 70 cm LP cylinder diameter with common stroke 50 cm. Cut-off in HP cylinder is at 40 % of stroke and that in LP cylinder is at 50 % of stroke. The dry saturated steam is supplied at 10 bar and is exhausted at 0.1 bar. Assume a diagram factor of 0.8 for both the cylinders and find power developed when engine runs at 120 r.p.m. assuming mechanical efficiency of 90 %. Find overall expansion ratio. Neglect clearance. Assume hyperbolic expansion in both cylinders.*

Solution

As cut-off in each cylinder is given, the example can be solved by determining P_m for each cylinder separately and IP also separately.

Refer to Fig. Ex. 18.14

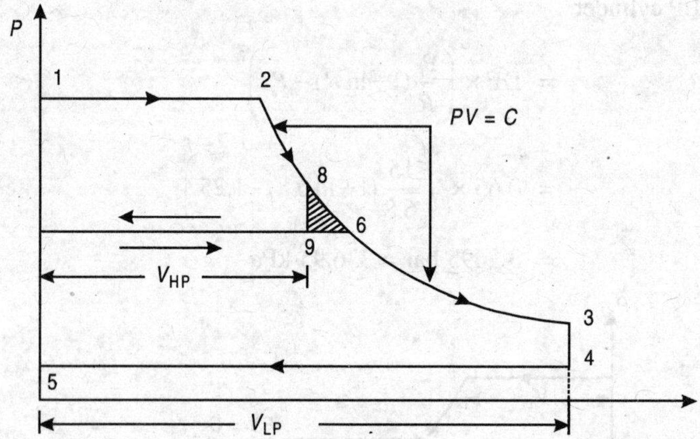

Fig. Ex. 18.14

Given that

$$D_{HP} = 35 \text{ cm}$$
$$D_{LP} = 70 \text{ cm}$$
$$L = 50 \text{ cm}$$
$$\text{Cut-off in HP} = 0.4 \text{ of stroke}$$
$$\text{Cut-off in LP} = 0.5 \text{ of stroke}$$
$$P_1 = 10 \text{ bar}$$
$$P_4 = P_s = 0.1 \text{ bar}$$
$$\text{Dia. factor} = 0.8$$
$$\text{RPM} = 120$$
$$\text{RPM} = 120$$
$$\text{Mech. } \eta = 0.9$$

$$\text{Ratio of expansion in HP cylinder} = \frac{1}{0.4} = 2.5 = \frac{V_8}{V_2} = r_{HP}$$

$$\text{Ratio of expansion in LP cylinder} = \frac{1}{0.5} = 2.5 = \frac{V_3}{V_6} = r_{LP}$$

$$\text{Ratio of cylinder volume} = \frac{V_{LP}}{V_{HP}} = \frac{V_4}{V_8} = \frac{V_3}{V_8}$$

$$= \frac{\frac{\pi}{4} \times 0.7^2 \times 0.5}{\frac{\pi}{4} \times 0.35^2 \times 0.5} = 4$$

Since expansion process is hyperbolic, intermediate pressure P_3

$$P_6 = \frac{P_1 V_2}{V_6} = \frac{P_1}{1} \times \frac{V_2}{V_8} \times \frac{V_8}{V_3} \times \frac{V_3}{V_6}$$

$$= 10 \times \frac{10}{2.5} \times \frac{1}{4} \times \frac{2}{1}$$

$$= 2 \text{ bar}$$

Now mean effective pressure of HP cylinder

$$P_m \text{(HP)} = \frac{P_1}{r\text{(HP)}} (1 + \ln r\text{(HP)}) P_3$$

$$= \frac{10}{2.5} \,(1 + \ln (2.5)) - 2$$

$$= 4.7725 \text{ bar}$$

Actual mean effective pressure in HP cylinder

$$P_m\text{(HP)}\ a = 4.7725 \times 0.8 = 3.818 \text{ bar}$$

Power developed in HP cylinder

$$= 2\, P_m\text{(HP)}\ a \times L \times A_{\text{HP}} \times N$$

$$= 3.818 \times 100 \times 0.5 \times \frac{\pi}{4} \times 0.3^2 \times \frac{120}{60}$$

$$= 26.9878 \text{ kW}$$

Now on LP cylinder

Mean effective pressure of LP cylinder

$$= \frac{P_3}{r_{(\text{LP})}} [1 + \ln (r_{\text{LP}})] - P_4$$

$$= \frac{2}{2} [1 + \ln 2] - 0.1$$

$$= 1.5931 \text{ bar.}$$

Actual mean effective pressure

$$= 1.593 \times 0.8 = 1.2745 \text{ bar}$$

$$\text{Power developed} = 2\, P_m (LP)_a \times L \times A_{\text{LP}} \times N$$

$$= 2 \times 1.2745 \times 100 \times 0.5 \times \frac{\pi}{4} \times 0.7^2 \times \frac{120}{60}$$

$$= 98.097 \text{ kW}$$

$$\text{Total power developed} = 26.9878 + 98.097$$

$$= 125.085 \text{ kW}$$

THEORY QUESTIONS

18.1. With a neat sketch explain the working of a steam engine.

18.2. Describe the hypothetical and actual indicator diagram for a steam engine and explain what do you understand by diagram factor.

18.3. How the Indicted Power is calculated ?

18.4. Explain Rope brake dynamometer.

18.5. Explain η_{mech}, ηth, $\eta_{overall}$, and $\eta_{relative}$.

18.6. What is specific steam consumption ?

18.7. Explain with neat diagrams the methods of governing of simple steam engine.

18.8. Explain 'Willan's line' for steam engine.

18.9. Differentiate between Carnot cycle of Rankine cycle as applied to steam power plants. Why is Rankine cycle modified.

18.10. What is the criteria for comparison of Carnot cycle and Rankine cycle when used for steam power plants ? Which cycle is most ideal for steam power plants.

18.11. Discuss the necessity of modifying the Rankine cycle for working with steam engine.

18.12. What are the merits and demerits of compound steam engines over simple steam engine ?

18.13. Explain how the compound steam engines are classified.

18.14. Describe the working of a receiver type compound engine. State its advantages over the other compound engine.

PROBLEMS FOR PRACTICE

18 1. The following observations were made during a trial on a single cylinder double acting steam engine.

Cylinder dimensions : 250 mm bore and 200 mm stroke.

Effective broke load : 800 N.

Effective radius of brake drum : 50 cm.

Steam supply at : 7 bar and 200°C

MEP : 250 kPa

Speed : 200 rpm.

Condensate collected : 200 kg/hr

Temperature of condensate: 50°C

Determine,

(i) Brake Power

(ii) η_{mech}

(iii) Indicated specific consumption of steams

(iv) Brakes thermal efficiency.

[*Ans.* (i) 8.377 kw ; (ii) 51.2% ; (iii) 12.22 $\frac{\text{kg}}{\text{kWh}}$; (iv) 5.72%]

18.2. Steam is supplied to a Rankine cycle employing reheating at 30 bar and 400°C. It is exhausted into a condenser at 0.04 bar. It is expanded in the first stage to such a pressure till it is dry saturated. Steam is then reheated at this pressure till its original temperature of 400° C

Find,

(i) Thermal efficiency of the reheat cycle,

(ii) Specific steam consumption,

(iii) Thermal efficiency without reheating,

(iv) Steam flow rate necessary for a power output of 400 kW.

Neglect feed pump work.

[*Ans.* (i) 44.5 %; (ii) 2.214 kg/kWh (iii) 36.97 %; (iv) 885.6 kg/hr]

18.3. A Rankine cycle operates between a boiler pressure of 25 bar and a condenser pressure of 0.5 bar. Steam is dry saturated at entry to the turbine. Neglecting the feed pump work and find the thermal efficiency and the specific steam consumption.

If the boiler pressure is reduced to 15.5 bar with same temperature of steam as in the earlier working. Find the change in the thermal efficiency and specific steam consumption.

[*Ans.* A-(i) 25.86 %; A-(ii) 5.6604 kg/kWh ; B-(i) 23.02%; B-(ii) 6.2069 kg/kWh]

18.4. A double acting steam engine is supplied with steam at 8.25 bar while the condenser is 0.27 bar. Cut off in HP cylinder is at 0.5 of the stroke. Ratio of cylinder volumes is 4. Both pistons have a same stroke length. Explain in the HP cylinder is incomplete. Assuming hyperbolic expansion and neglecting dearance, find

(i) Cut off ratio in L.P. for equal initial loads on both printing,

(ii) MEP in both cylinders,

(iii) % loss of work done to incomplete expansion in H.P. cylinder.

[*Ans.* (i) 0.552; (ii) P_{mHP} = 5.118 bar and P_{mLP} = 1.37 bar; (iii) 8.82 %]

18.5. A double acting compound steam engine develop brake power of 295 kW under following conditions.

Steam admission pressure = 9 bar

Condenser pressure = 0.2 bar

Mean piston speed = 180 m/minute

Engine speed = 120 r.p.m.

Ratio of cylinder volumes = 4

Overall expansion ratio = 14

D.F. for each cylinder = 0.85

η_{mech} = 0.8

Determine,

(i) Cylinder diameters

(ii) Cut-off in L.P. cylinder

(iii) Ratio of work done in LP to HP cylinder.

Assume equal initial loads on printers and hyperbolic expansion throught.

[*Ans.* (i) D_{LP} = 0.9324 m ; D_{HP} = 0.466 m (ii) 32.79 %; (iii) 1.21]

CHAPTER 19

Steam Nozzles

19.1 INTRODUCTION

Turbomachines like steam turbines, water turbines and gas turbines produce power by utilising the kinetic energy of the jets produced by passing high pressure steam, water and gas through the devices called nozzles. Corresponding to the fluids used, the nozzles are called steam nozzles, water nozzles and gas nozzles.

These nozzles serves two purposes:

(1) To convert pressure energy and thermal energy into kinetic energy and

(2) To direct the fluid jet at the specific angle known as nozzle angle.

19.2 DEFINITION

A nozzle is a device, a duct of varying cross-section area in which a steadily flowing fluid can be made to accelerate by a pressure drop along the duct.

So when a fluid flows through a nozzle, its velocity increases continuously and pressure decreases continuously.

19.3 SOME APPLICATIONS OF A NOZZLE

There are following applications of a nozzle are:

1. Nozzles are used in steam turbines, gas turbines, water turbines and in jet engines, Jet propulsion.
2. Nozzles are used for flow measurement e.g. in venturimeter.
3. Nozzles are used to remove air from a condenser.
4. Injectors for pumping feed water to boilers.
5. Artificial fountains.

There are three types of nozzles:

1. Convergent nozzle,
2. Divergent nozzle, and
3. Convergent-Divergent nozzle.

If the cross-section of the nozzle decreases continuously from entrance to exit, it is called a convergent nozzle.

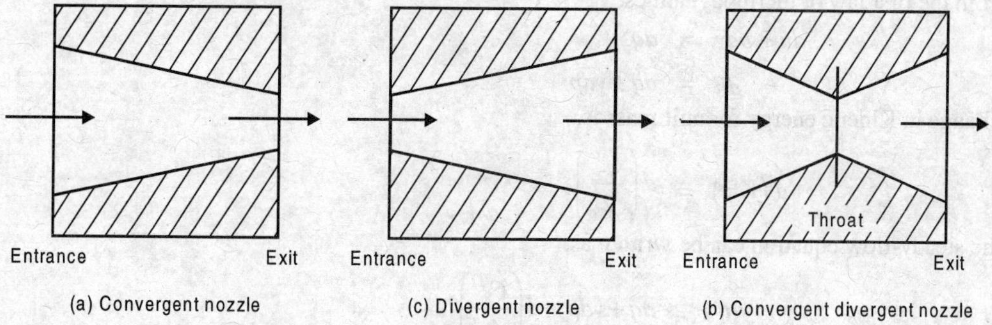

Fig. 19.1

If the cross-section of the nozzle increases continuously from entrance to exit, it is called a divergent nozzle.

If the cross-section of the nozzle first decreases and then increases, it is called a convergent-divergent nozzle. The section where cross-sectional area is minimum is called 'throat' of the nozzle.

Diffuser. When a steadily flowing fluid is decelerated in a duct causing rise in pressure along the stream, then the duct is called a diffuser.

So in a diffuser, velocity of the fluid decreases continuously and pressure increases continuously.

Similar to nozzle, there are three types of diffusers.

1. Convergent diffuser
2. Divergent diffuser
3. Convergent-divergent diffuser.

The given duct will work as a diffuser or a nozzle depending upon the fluid velocity at the inlet of a duct.

Some Applications

1. Diffusers are used in centrifugal compressor.
2. Diffusers are used in ram-jet engines to increase the pressure of incoming fresh-air.

19.4 GENERAL-FLOW ANALYSIS

Assumptions

1. The flow of the fluid is assumed to be one dimensional.
2. The change in area and curvature along the axis of the duct are gradual.
3. Thermodynamic and mechanical properties are uniform across planes normal to the axis of a duct.
4. The flow of the fluid is assumed to be steady flow.

The steady-flow equation is,

$$Q - W = \Delta \text{H} + \Delta \text{PE} + \Delta \text{KE}$$

or differential form of the equation is,

$$\partial q - \partial w = dh + d(\text{PE}) + d(\text{KE})$$

We will assume heat the nozzle is horizontal,

$$\therefore \quad d(\text{PE}) = 0$$

The fluid is just flowing through a duct. So obviously work-done is zero

$$\therefore \quad \partial w = 0$$

The definition of enthalpy is

$$h = u + PV$$

$$\therefore \quad dh = du + d(PV)$$

$$= du + pdv + Vdp$$

But from the first law of thermodynamics,

$$du + pdv = dq$$

$$\therefore \quad dh = dq + vdp$$

The change in Kinetic energy for unit mass is,

$$d(\text{KE}) = d\left(\frac{C^2}{2}\right)$$

∴ The steady-flow equation can be written as,

$$\partial q = \partial q + vdp + d\left(\frac{C^2}{2}\right)$$

$$\therefore \quad d\left(\frac{C^2}{2}\right) = -vdp$$

$$\therefore \quad \frac{2C}{2} \cdot dC = -vdp$$

$$\therefore \quad CdC = -vdp \qquad \text{...(1)}$$

The fluid flows through the duct under steady-flow conditions. Therefore the mass-flow rate is constant.

$$m = \rho A \cdot C = \text{Constant}$$

or

$$m = \frac{A \cdot C}{v} = \text{Constant}$$

where

A : cross-sectional area,
C : velocity of fluid
v : specific volume of the fluid.

$$\therefore \quad \frac{A \cdot C}{v} = \text{Constant}$$

$$\log A + \log C - \log v = \text{Constant}$$

Differentiating the above equation, we get

$$\frac{dA}{A} + \frac{dC}{C} - \frac{dv}{v} = 0$$

From Eq. (1)

$$dC = \frac{-vdp}{C}$$

$$\therefore \quad \frac{dA}{A} - \frac{vdp}{C^2} - \frac{dv}{v} = 0$$

$$\therefore \quad \frac{dA}{A} = \frac{vdp}{C^2} + \frac{dv}{v}$$

$$= vdp\left(\frac{1}{C^2} + \frac{dv}{v^2 dp}\right)$$

From the knowledge of fluid-dynamics, the velocity of sound wave is given by,

$$a^2 = -v^2\left(\frac{dp}{dv}\right)_s$$

where a : velocity of sound (or sonic velocity)

$$\therefore \quad \frac{dA}{A} = vdp\left(\frac{1}{C^2}-\frac{1}{a^2}\right)$$

$$\therefore \quad \frac{dA}{A} = \frac{vdp}{C^2}\left(1-M^2\right)$$

where M : mach number $= c/a$

$$\therefore \quad \frac{dA}{A} = \frac{-vdp}{C^2}\left(M^2-1\right).$$

But from Eq. (1) $CdC = vdp$

$$\therefore \quad \frac{dA}{A} = \frac{dC}{C}\left(M^2-1\right)$$

This equation gives information whether the given duct will act as a nozzle or a diffuser if the inlet fluid velocity is known. It also gives information which type of duct should be used for a particular application.

For a nozzle, velocity of the fluid should increase continuously from entrance to exit. So if the inlet fluid condition is known, we can select the nozzle as below.

1. When the velocity of fluid is less than sonic velocity (i.e. when the flow is sub-sonic), the mach no. (M) is less than 1.

$$\therefore \quad M < 1$$

$$\therefore \quad \left(M^2-1\right) \text{ is negative.}$$

For a nozzle change in velocity (dC) has to be positive.

$$\therefore \quad \frac{dC}{C}\left(M^2-1\right) \text{ is negative.}$$

$$\therefore \quad \frac{dA}{A} \text{ is negative.}$$

i.e. change in across sectional area along the duct should be negative.

This tells that, for *sub-sonic flow*, the duct must be convergent. By using this convergent nozzle, the flow of the fluid can be increased to sonic velocity.

But by using convergent nozzle we can not obtain super-sonic flow.

2. When the flow is super-sonic, $M > 1$

i.e. for super-sonic flow, the duct must be divergent.

3. The convergent-divergent nozzle is used for convert sub-sonic flow into super-sonic flow.

In the convergent part the velocity of fluid is increased from sub-sonic to sonic condition. At throat, the velocity is sonic. In the divergent part, the velocity is increased from sonic to super-sonic.

So only in convergent-divergent nozzle, the sub-sonic flow is converted into super-sonic flow.

For diffuser the velocity should decrease continuously so a diffuser is selected as below,

1. For sub-sonic flow

$$M < 1 \;\therefore\; \left(M^2-1\right) < 0$$

$$\therefore \quad \frac{dC}{C}\left(M^2-1\right) > 0$$

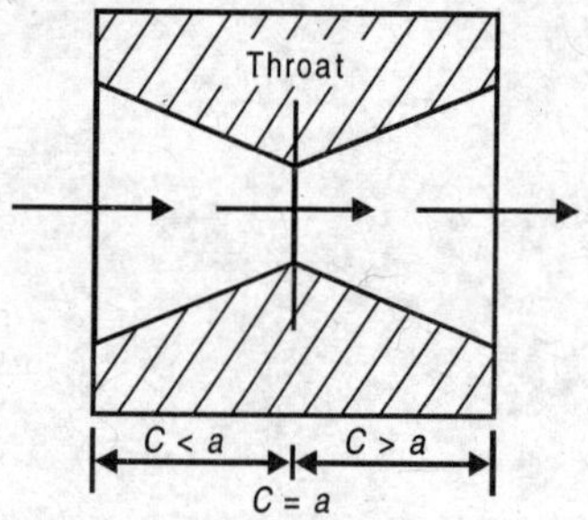

Fig. 19.2

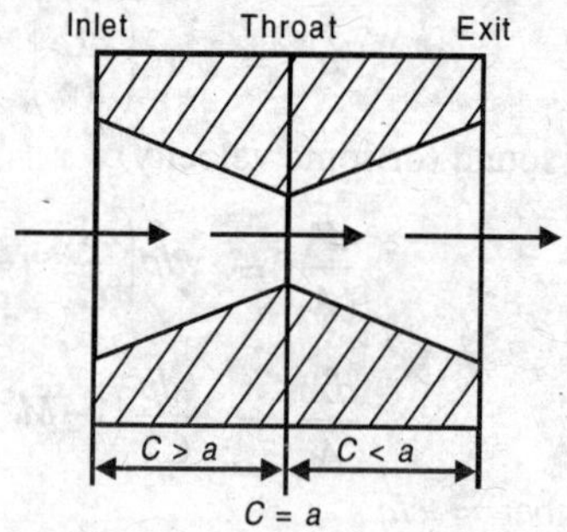

Fig. 19.3

$\therefore \quad \frac{dA}{A}$ is positive i.e., duct must be divergent.

2. For super-sonic flow: $M > 1$

$$\therefore \quad \frac{dC}{C}\left(M^2-1\right)<0$$

$\therefore \quad \frac{dA}{A}$ is positive i.e., duct must be divergent.

The convergent-diffuser will decrease the velocity of fluid to sonic velocity.

3. The convergent-divergent diffuser is used to convert super-sonic flow into sub-sonic flow.

19.5 VELOCITY IN A NOZZLE

For unit mass,

The steady flow equation is,

$$q - w = \Delta h + \Delta \text{PE} + \Delta \text{KE}$$

For a horizontal nozzle, $\Delta \text{PE} = 0$

There is no work-done in nozzle therefore $W = 0$. In the nozzle, the velocity of the fluid is so high that there is hardly any time available for fluid to exchange heat with the surroundings. Therefore for nozzle, it is assumed that heat transfer is zero i.e., flow in Isentropic

$$\therefore \quad q = 0$$

$$\therefore \quad \Delta h + \Delta \text{KE} = 0$$

$$\therefore \quad h_2 - h_1 + \frac{1}{2}\frac{\left(C_2^2 - C_1^2\right)}{1000} = 0$$

$$C_2 = \sqrt{2000(h_1 - h_2) + C_1^2}$$

where C_1 : velocity of fluid at inlet

h_1 : specific enthalpy of fluid at inlet

h_2 : specific enthalpy of fluid at exit condition.

For convergent-divergent nozzle, the velocity of fluid at throat is obtained by the equation,

$$C_t = \sqrt{2000(h_1 - h_2) + C_1^2}$$

where, h_t = specific enthalpy at the throat conditions.

For steam nozzles the values of enthalpy $(h_1, h_2, h_t$ etc.$)$ are normally obtained by using Mollier Chart.

If C_1, the initial or approach velocity is neglected, then

$$C_2 = \sqrt{2000(h_1 - h_t)} = 44.72\sqrt{h_1 - h_t}$$

$$= 44.72\sqrt{(h_1 - h_2) + \frac{C_1^2}{2000}} \text{ when } C_1 \text{ is considered.}$$

19.6 MASS-FLOW RATE IN A NOZZLE

The steady flow energy equation is

$$q - w = \Delta h + \Delta PE + \Delta KE$$

In the general analysis, we have obtained the equation,

$$d\left(\frac{C^2}{2}\right) = -vdp$$

Integrating the equation

$$\frac{C_2^2 - C_1^2}{2} = -\int_1^2 vdp$$

The simple relation, $PV^n =$ const. holds good for any process. So we assume that the expansion in a nozzle is polytropic.

$$\therefore -\int_1^2 vdp \text{ will become } \frac{n}{n-1}(P_1 V_1 - P_2 V_2)$$

$$\therefore \quad \frac{C_2^2 - C_1^2}{2} = \frac{n}{n-1}(P_1 V_1 - P_2 V_2)$$

For nozzle, generally C_1 is too small.

$\therefore$ Neglecting C_1, we get

$$C_2 = \sqrt{\frac{2n}{n-1}(P_1 V_1 - P_2 V_2)}$$

$$C_2 = \sqrt{\frac{2n}{n-1} P_1 V_1 \left(1 - \frac{P_2 V_2}{P_1 V_1}\right)}$$

The mass flow rate is given by the continuity equation

$$m = \rho AC = \frac{AC}{v}$$

$$\therefore \quad \frac{m}{A_2} = \frac{C_2}{v_2} = \frac{\sqrt{\frac{2n}{n-1} P_1 V_1 \left(1 - \frac{P_2 V_2}{P_1 V_1}\right)}}{V_2}$$

For polytropic process, $P_1 v_1^n = P_2 v_2^n$

$$\therefore \quad \frac{m}{A_2} = \frac{\sqrt{\frac{2n}{n-1} P_1 v_1 \left(1 - \left(\frac{P_2}{P_1}\right)^{1-\frac{1}{n}}\right)}}{v_1 \cdot \left(\frac{P_1}{P_2}\right)^{1/n}}$$

$$\therefore \quad \frac{m}{A_2} = \sqrt{\frac{2n}{n-1} \cdot \frac{P_1}{v_1}\left[\left(\frac{P_2}{P_1}\right)^{\frac{2}{n}} - \left(\frac{P_2}{P_1}\right)^{\frac{n+1}{n}}\right]}$$

This is the equation for the mass-flow rate of fluid per unit area of nozzle.

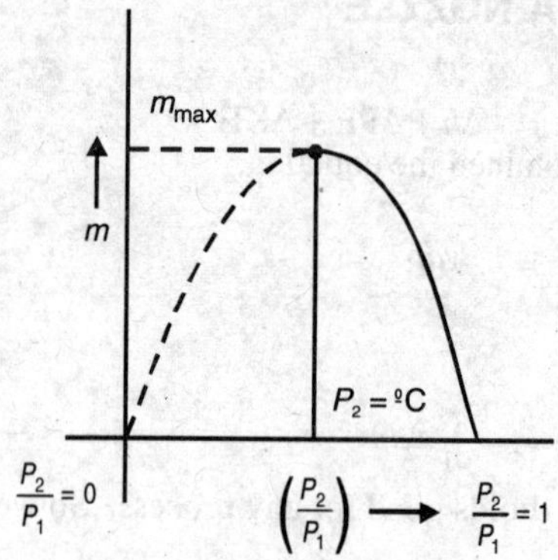

Fig. 19.4

The equation states that the mass flow-rate depends upon pressure-ratio $\left(\frac{P_2}{P_1}\right)$.

If $P_1 = P_2$ the mass-flow rate is zero.
Also if $P_2 = 0$, the mass flow rate is zero.
If the graph is plotted for mass flow rate v_s pressure ratio, it will be as shown in the figure. It is observed that at some value of (P_2/P_1) the velocity and mass-flow rate reaches to its maximum value.

19.7 CRITICAL PRESSURE RATIO

The nozzle is designed for maximum mass-flow rate per unit area. It is observed from the above discussion that the mass-flow rate depends on pressure-ratio. The equation for maximum mass-flow rate can be found out by differentiating the above equation w.r.t. $\left(\frac{P_2}{P_1}\right)$ and equating it to zero. The other parameters n, p, v are constant.

The equation is

$$\frac{m}{A_2} = \sqrt{\frac{2n}{n-1} \frac{P_1}{v_1}\left[\left(\frac{P_2}{P_1}\right)^{2/n} - \left(\frac{P_2}{P_1}\right)^{\frac{n+1}{n}}\right]}$$

$$\therefore \quad \frac{d\left(\frac{m}{A_2}\right)}{d\left(\frac{P_2}{P_1}\right)} = \frac{d}{d\left(\frac{P_2}{P_1}\right)}\left[\frac{2n}{n-1} \frac{P_1}{v_1}\left[\left(\frac{P_2}{P_1}\right)^{\frac{2}{n}} - \left(\frac{P_2}{P_1}\right)^{\frac{n+1}{n}}\right]\right]^{1/2}$$

$$= \frac{1}{\left[\frac{2n}{n-1}\frac{P_1}{v_1}\left[\left(\frac{P_2}{P_1}\right)^{\frac{2}{n}} - \left(\frac{P_2}{P_1}\right)^{\frac{n+1}{n}}\right]\right]^{1/2}} \times \left[\frac{2n}{n-1} \times \frac{P_1}{v_1}\left[\frac{2}{n}\left(\frac{P_2}{P_1}\right)^{\frac{2}{n}-1} - \frac{n+1}{n} \times \left(\frac{P_2}{P_1}\right)^{1/n}\right]\right]$$

$$= 0$$

$$\therefore \quad \frac{2}{n}\left(\frac{P_2}{P_1}\right)^{\frac{2-n}{n}} - \left(\frac{n+1}{n}\right) \times \left(\frac{P_2}{P_1}\right)^{1/n} = 0$$

$$\therefore \quad \frac{2}{n}\left(\frac{P_2}{P_1}\right)^{\frac{2-n}{n}} = \left(\frac{n+1}{n}\right)\left(\frac{P_2}{P_1}\right)^{1/n}$$

$$\therefore \quad \frac{2}{n} \times \frac{n}{n+1} = \left(\frac{P_2}{P_1}\right)^{\frac{1}{n}-\frac{2}{n}+1}$$

$$\therefore \quad \left(\frac{2}{n+1}\right) = \left(\frac{P_2}{P_1}\right)^{n-\frac{1}{n}}$$

$$\therefore \quad \left(\frac{P_2}{P_1}\right) = \left(\frac{2}{n+1}\right)^{\frac{n}{n-1}}$$

This is the equation for critical pressure ratio for maximum discharge. This pressure ratio will accelerate the flow to a velocity equal to sonic velocity.

For convergent-divergent nozzle, at throat the critical pressure is maintained to get maximum discharge. For steam nozzles value of n depends on condition of steam at inlet.

Initial condition	**value of n**
1. Superheated steam	1:3 $\quad \therefore \frac{P_2}{P_1} = 0.546$
2. Wet-steam	$1.035 + 0.1 \times x$ where, x is the dryness fraction of steam.
3. Dry-saturated steam	1.135 $\quad \therefore \frac{P_2}{P_1} = 0.58$
4. For air	$\gamma = 1.4 \; \therefore \frac{P_2}{P_1} = 0.5283$

19.8 EFFECT OF FRICTION

The flow of steam through nozzle is assumed to be isentropic. The Mollier Chart shows the isentropic flow $(1 - t - 2)$ of steam through a convergent-divergent nozzle.

But in actual case, the friction losses occur. The frictional losses are due to,

(i) Friction between sides of nozzle (wall of nozzle) and fluid.

(ii) Internal fluid friction and

(iii) Due to eddies in the flow.

Most of the friction in convergent-divergent nozzle is assumed to occur between the throat and exit. The expansion upto throat is taken to be isentropic. This is due to low initial velocity. In Mollier Chart, 1 – t – 2' is the actual expansion of steam through nozzle. Figure 19.5 shows the actual expansion of steam through nozzle.

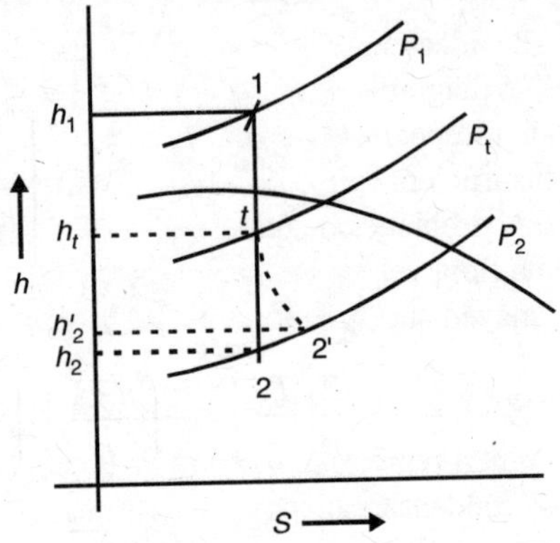

Fig. 19.5. *h-S diagram with friction*

Effects

1. Reduction in actual enthalpy drop.
2. Reduction in exit velocity of fluid.
3. Increase in the dryness-fraction of steam.
4. Increase in the specific volume and
5. Decrease in the mass-flow rate.

19.9 NOZZLE EFFICIENCY

The nozzle efficiency is given as,

$$\text{Nozzle efficiency} = \frac{\text{Actual enthalpy drop}}{\text{Isentropic enthalpy drop}}$$

$$\eta_{\text{nozzle}} = \frac{h_1 - h_2'}{h_1 - h_2}$$

19.10 VELOCITY COEFFICIENT

Velocity coefficient is defined as the ratio of actual velocity of steam to an isentropic or theoretical velocity.

$$\text{Velocity coefficient } (C_V) = \frac{\text{Actual velocity}}{\text{Isentropic velocity}}$$

Velocity of steam is given as,

$$C_2 = \sqrt{2(h_1 - h_2)} \text{ when inlet velocity is neglected}$$

$$\therefore \quad C_V = \frac{\sqrt{2(h_1 - h_2')}}{\sqrt{2(h_1 - h_2)}} = \sqrt{\frac{h_1 - h_2'}{h_1 - h_2}}$$

$$\therefore \quad \text{Velocity coefficient} = \sqrt{\text{nozzle efficiency}}$$

19.11 SUPER SATURATED OR METASTABLE FLOW THROUGH NOZZLE

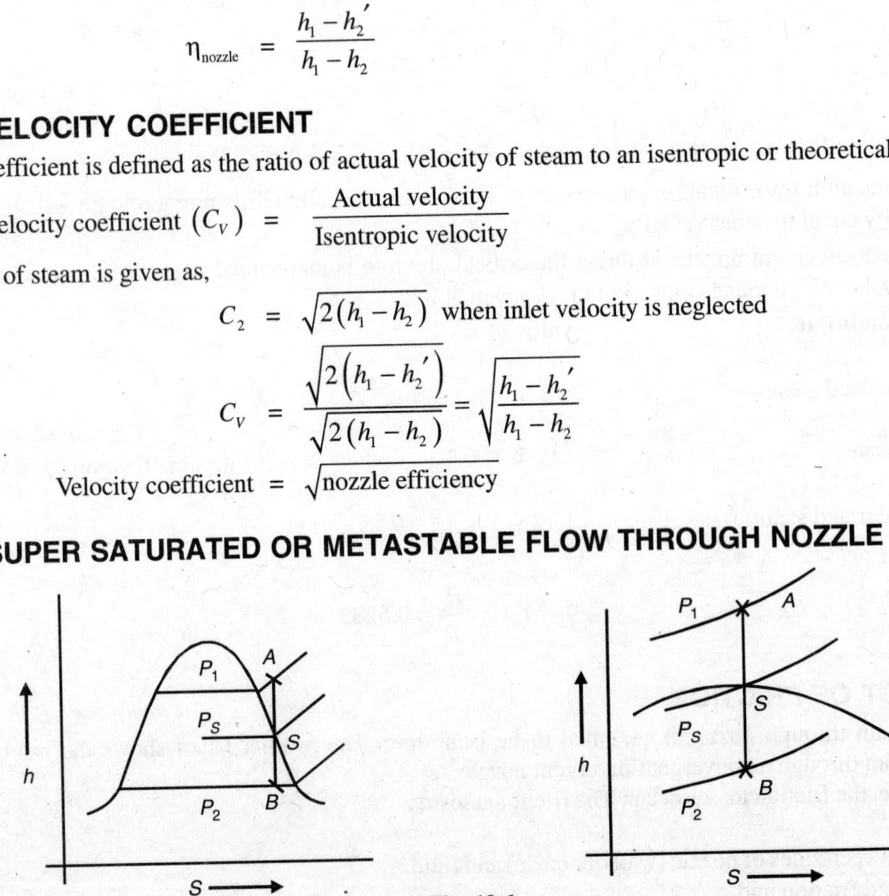

Fig. 19.6

19.11.1 Normal or Equilibrium Expansion

The ideal case of isentropic expansion of a superheated vapour to a state in the wet region is shown in T–S diagram and h–s diagram of Fig. 19.6. At the point in the expansion where the pressure is P_s, a change of phase should begin to occur. At this point, the random kinetic energy of the molecules has fallen to a level which is insufficient to overcome the attractive forces of the molecules and some of the slow moving molecules start to form tiny droplets to condensate. Now although this process is rapid, it does not have time to occur in the nozzle where the flow velocity is very great. The achievement of equilibrium between the liquid and vapour phase is therefore delayed and vapour continues to expand in a superheated or dry state.

19.11.2 Super-saturated Expansion

Point A represents a steam in superheated region at pressure P_i. When at point s, due to the rapid expansion, instead of condensation commencing, the steam continues to behave as a super-heated vapour down to point B, at same intermediate pressure P_2. This non-equilibrium behaviour as a superheated vapour does not continue indefinitely and at point B, restoration of equilibrium quickly occurs and is after the throat in divergent portion of the nozzle. It is accompanied by a small increase in pressure.

In equilibrium flow, the energy released by condensing the molecules is provided for increasing the kinetic energy of the steam as it passes through the nozzle. If the steam does not condense, then the energy for this increase in kinetic energy come by reduction in the temperature and therefore the steam is called super-cooled.

Thus we see that condensation does not start immediately after S is passed, no drops of liquid are formed until some state B is reached, where condensation suddenly occurs, a phenomenon sometimes called *Condensation Shock*. The steam in states between S and B is *supersaturated* or a *metastable state*.

Since the collaspe of the metastable state has been observed not in a converging nozzle, but always, in the diverging part of the De-Laval nozzle, one is probably safe assuming that the supersaturation, if it occurs at all, will persist to some point beyond throat. Of course the flow should be computed for the throat section since this is where it is limited.

In general, the velocity of supersaturated steam is less than the value computed for the equilibrium flow. The density of supersaturated steam is greater than the ewuilibrium density at the same pressure. However, the density increases by a large percentage that the velocity decreases, with the result that the computed mass flow of supersaturated steam is greater than the computed flow of equilibrium steam, given the initial state and throat area. When discharge from a nozzle is actually measured it is found to be from 2 to 5% greater than the calculated discharge.

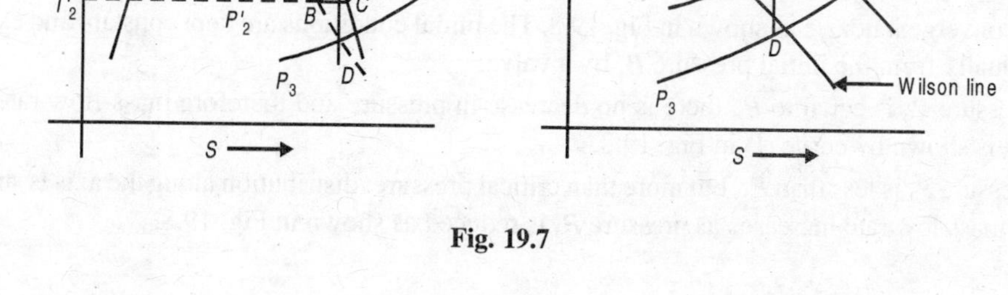

Fig. 19.7

Wilson line: Calender found that the Wilson line approximately follows the 97% dryness line.

Degree of under-cooling: Difference in the temperature at point C and temperature at point B is known as degree of under-cooling or difference in saturation temperatures at pressure P_2 end P_B is degree of under-cooling.

The temperature of the supersaturated steam at B will be below the normal temperature of steam for that pressure; this state is known as *undercooled state*, the amount of undercooling being the difference between the normal temperature and the actual temperature.

$$\therefore \quad \text{Degree of under-cooling} = T_2 - T_2'$$

$$= \begin{bmatrix} \text{Saturation temperature} \\ \text{corresponding to point } B \end{bmatrix} - \begin{bmatrix} \text{Actual temperature} \\ \text{at } B \end{bmatrix}$$

For this purpose we have

$$\frac{T_B}{T_A} = \left(\frac{P_B}{P_A}\right)^{\frac{0.3}{1.3}}$$

Degree of Supersaturation: Note that the actual pressure at the supersaturated state B is P_2, but the saturation pressure corresponding to the temperature t_B is P_2. The ratio of these pressures P_d/P_b is called the *Degree of Supersaturation* or *Supersaturation ratio.*

$$\text{Degree of supersaturation} = \frac{\text{Actual pressure}}{\text{Saturation pressure corresponding to actual temperature } (T_2')}$$

$$= \frac{P_2}{P_2'}$$

19.11.3 Wilson's Line

There is generally a limit to super-saturation. It is upto 96% dryness and beyond it, steam condensation occurs suddenly and irreversibly at constant enthalpy and remains in stable condition thereafter. This limit line is known as Wilson's line.

Effects of Super-saturated Flow

1. Increase in discharge by 2 to 5% due to increase in density due to super cooling.
2. Decrease in exit velocity.
3. Increase in final dryness-fraction and increase in enthalpy.

19.12 PHENOMENON IN NOZZLES OPERATING OFF THE DESIGN PRESSURE RATIO

(a) For Convergent Nozzle

Consider a convergent nozzle as shown in Fig. 19.8. The initial conditions are kept constant and exit pressure P_2 is reduced gradually from the initial pressure P_1 by a valve.

(i) When pressure P_2 is equal to P_t, there is no decrease in pressure and therefore mass-flow rate is zero. This condition is shown by curve (I) in Fig. 19.8.

(ii) When pressure P_2 is less than P_1, but more than critical pressure; distribution along the axis is shown by curve (II). The mass low rate increases as pressure P_2 is reduced as shown in Fig. 19.8.

(iii) When exit pressure P_2 is equal to critical pressure, the nozzle operates with maximum mass flow rate and the pressure distribution is shown by curve (III).

This state is called as 'chocked flow' or the nozzle is said to be 'chocked'.

(iv) When pressure P_2 is less than critical pressure, there is no change in mass-flow through nozzle and also pressure distribution along the nozzle is same. The pressure-drop from critical pressure to P_2 takes place after the nozzle. This expansion is irreversible and gives rise to pressure oscillations as shown by curve (IV).

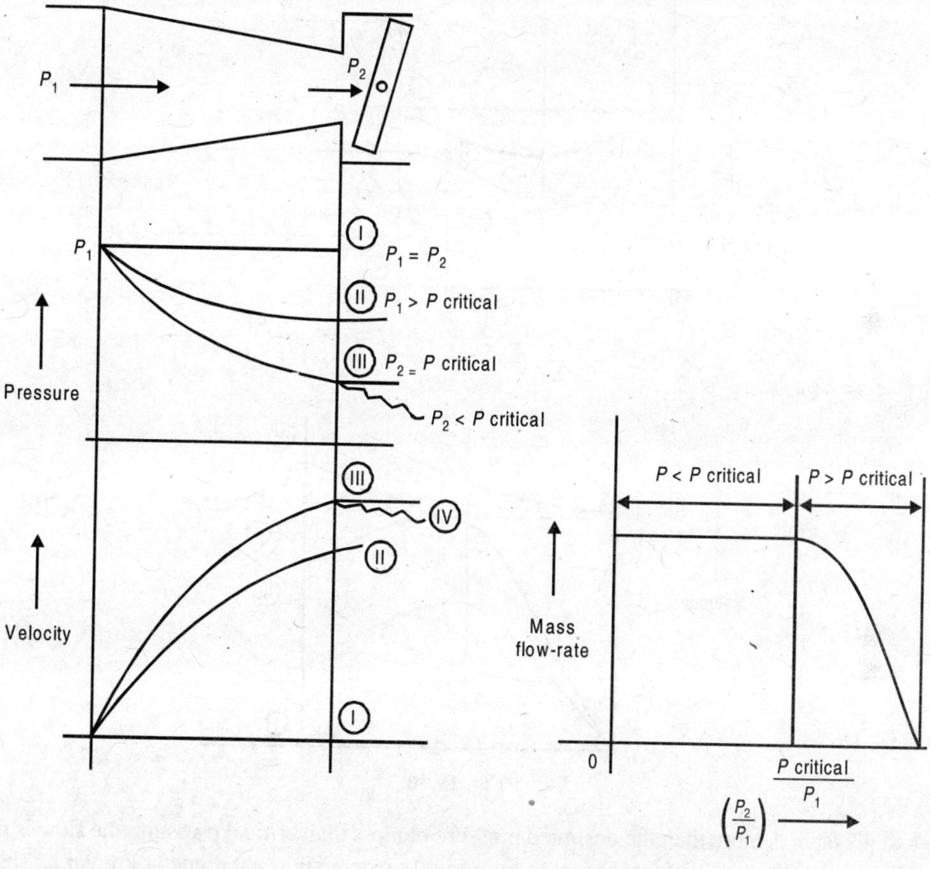

Fig. 19.8

(b) For Convergent-Divergent Nozzle

Let us consider a convergent-divergent nozzle as shown in Fig. 19.9

1. When pressure P_e is equal to inlet pressure P_t, there is no flow. Refer Fig. 19.9 and Fig. 19.10.
2. When P_e is reduced to the pressure denoted by curve (II). So that P_e/P_t is less than 1 but greater than critical pressure ratio, the velocity increases in the convergent region of the nozzle, but mach number (m) is less than 1 at throat. So the divergent section acts as a sub-sonic diffuser in which the pressure increases and velocity decreases.
3. In case (IV), pressure is critical at throat and exit pressure P_e is design pressure. Therefore the flow is isentropic throughout the nozzle and velocity continuously increases along the nozzle. At throat, velocity is equal to sonic velocity. The divergent portion acts as a super-sonic nozzle with a continuous decrease in pressure and continuous increase in velocity.

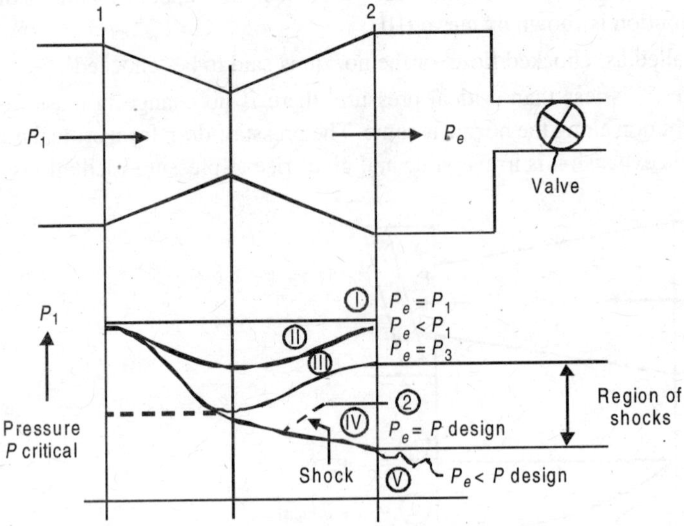

Fig. 19.9

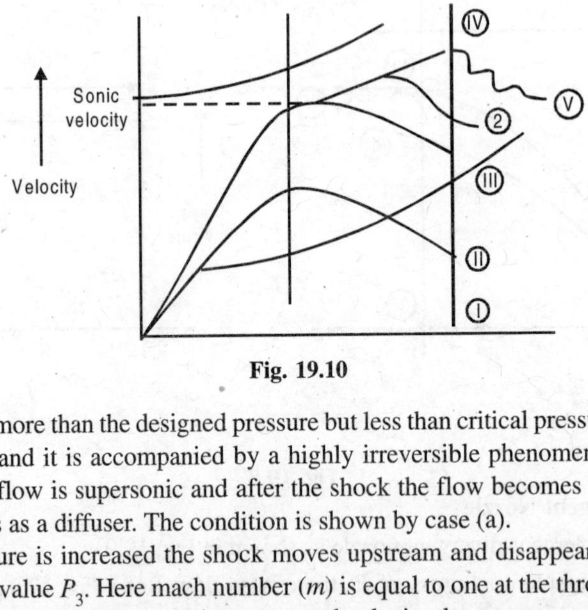

Fig. 19.10

4. If the exit pressure is more than the designed pressure but less than critical pressure, the flow is not isentropic in the divergent part and it is accompanied by a highly irreversible phenomena known as shocks. Shocks occur only when the flow is supersonic and after the shock the flow becomes sub-sonic and the rest of the diverging portion acts as a diffuser. The condition is shown by case (a).
5. When the back pressure is increased the shock moves upstream and disappears at the nozzle throat where pressure P_e has some value P_3. Here mach number (m) is equal to one at the throat but divergent portion acts as a sub-sonic diffuser in which pressure increases and velocity decreases.
6. If the pressure P_e is less than the design pressure, no further decrease in exit pressure occurs and drop of pressure from design pressure to P_e occurs outside the nozzle giving pressure fluctuations as shown by case (V).

Overexpansion and Underexpansion

If the area of the exit section of a nozzle is such that the fluid expands to a pressure at this section less than that of the discharge region. *Overexpansion* has occurred. As this lower pressure stream emerges into the higher pressure discharge region, there is a sudden increase in pressure, an act hat sets up compression pressure waves, much stronger than sound waves. This jump in pressure outside the nozzle occurs when the back-pressure is above the exit pressure.

If on the other hand the area of the exit section is such that the fluid expands to a pressure at this section greater than that in the discharge region, *underexpansion* has occurred. Since there is now a sudden decrease in pressure on the jet, expansion waves are initiated.

Both situations involve an increase in irreversibilities and loss of efficiency. It is best for the expansion in the nozzle to occur to just the right (designed) discharge pressure.

SOLVED EXAMPLES

Example 19.1. *Steam at 30 bar, 300° C expands to 5 bar through a nozzle isentropically. Calculate exit velocity. Assume steam is entering at velocity of 200 m/s.*

Also calculate area at exit of the nozzle for flow of 2 kg/sec.

Solution

From Mollier Chart,

$$h_1 = 2995 \text{ kJ/kg}$$

$$h_2 = 2629.8 \text{ kJ/kg}$$

$$x_2 = 0.94$$

The equation for exit velocity is,

$$C_2 = \sqrt{2(h_1 - h_2) + C_1^2} = 44.72\sqrt{(h_1 - h_2) + \frac{C_1^2}{2000}}$$

$$C_1 = 200 \text{ m/sec}$$

$$\therefore \quad C_2 = \sqrt{2(2995 - 2629.8)\times 10^3 + 200^2}$$

$$= \mathbf{877.75 \text{ m/s}}$$

Applying continuity equation at exit conditions of nozzles, we get,

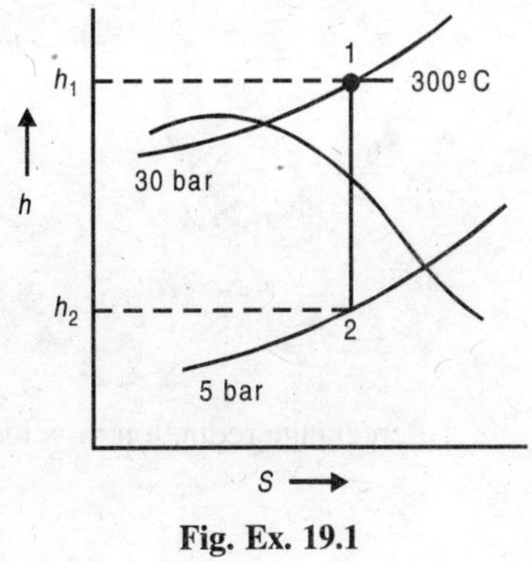

Fig. Ex. 19.1

$$m = \frac{A_2 - C_2}{V_2}$$

$$V_2 = \text{specific volume at exit}$$

$$= x_2 \times v_{g2}$$

From Steam Table, at 5 bar, $V_g = 0.3747 \text{ m}^3\text{/kg}$

$$\therefore \quad v_2 = 0.94 \times 0.3747$$

$$= 0.3522$$

$$\text{Area at exit}(A_2) = \frac{m \times v_2}{c_2}$$

$$= \frac{2 \times 0.3522}{877.75}$$

$$= 8.025 \times 10^{-4} \text{ m}^2$$

$$= \mathbf{8.025 \text{ cm}^2}$$

Example 19.2. *Steam expands isentropically through a nozzle from 15 bar to 490 kPa with initial dry-saturated state.*

Calculate the exit velocity, neglecting initial velocity. If 10% heat is lost in friction, calculate the percentage reduction in velocity and improvement in the final quality of the steam.

Solution

$$P_1 = 15 \text{ bar}$$

$$P_2 = 4.9 \text{ bar}$$

From Mollier Chart,

$$h_1 = 2790 \text{ kJ / kg}$$

$$h_2 = 2590 \text{ kJ /kg}$$

$$C_2 = \sqrt{2(h_1 - h_2) + C_1^2}$$

$$C_1 = 0 \text{ (given)}$$

$$\therefore \quad C_2 = \sqrt{2(h_1 - h_2)} = \sqrt{2 \times (2790 - 2590) \times 10^3}$$

$$= 632.45 \text{ m/s}$$

$$= \text{Velocity without friction.}$$

When the friction is considered, the expansion will be similar to process $1-2'$ on *h–S* diagram of Fig. Ex. 19.2.

First find h_2'.

$$\text{heat lost in friction} = 10 \%$$

$$= 0.1(h_1 - h_2)$$

So the enthalpy available $= 0.9(h_1 - h_2)$

$$\therefore \quad (h_1 - h_2') = 0.9(h_1 - h_2)$$

$$\therefore \quad h_1 - h_2' = 0.9 \times (2790 - 2590)$$

$$= 180 \text{ kJ / kg}$$

$$C_2' = \sqrt{2(h_1 - h_2')}$$

$$= \sqrt{2(180) \times 10^3} = 600 \text{ m / s.}$$

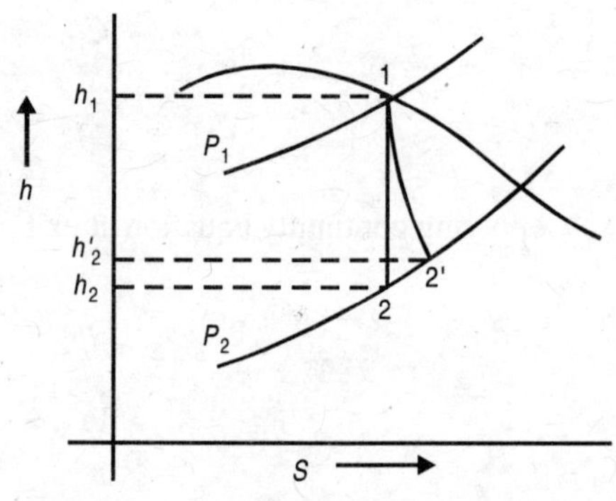

Fig. Ex. 19.2

$$\therefore \text{ Percentage reduction in velocity} = \left(\frac{C_2 - C_2'}{C_2}\right) \times 100$$

$$= \left(\frac{632.45 - 600}{632.45}\right) \times 100$$

$$= 5.13 \ \%$$

From Mollier Chart

$$x_2 = 0.92$$

and

$$x_2' = 0.94$$

∴ Improvement in the final quality of steam, when friction considered

$$= x_2' - x_2 = 0.02$$

Example 19.3. *Determine the throat area, exit velocity and exit area for convergent divergent steam nozzle to pass a mass of 0.2 kg/s, when the inlet conditions are 10 bar and 250° C and final pressure is 2 bar. Assume expansion is isentropic without friction and neglect initial velocity.*

Solution

$$P_1 = 10 \text{ bar}$$
$$P_2 = 2 \text{ bar}$$
$$P_t = \text{throat pressure}$$
$$m = 0.2 \text{ kg/sec}$$

First find throat pressure.

$$\frac{P_t}{P_1} = \left(\frac{2}{n+1}\right)^{n/n-1}$$

$$n = 1.3 \text{ (since steam is super-heated at inlet)}$$

$$\therefore \quad P_t = P_1 \times \left(\frac{2}{1.3+1}\right)^{\frac{1.3}{0.3}}$$

$$= 10 \times 0.5457 = 5.457 \text{ bar}$$

From Mollier Chart,

$h_1 = 2944$ kJ/kg, $h_2 = 2628$ kJ / kg, $v_2 = 0.8537 \text{ m}^3/\text{kg}$

$h_t = 2813$.kJ/kg $V_t = 0.3735 \text{ m}^3/\text{kg}$

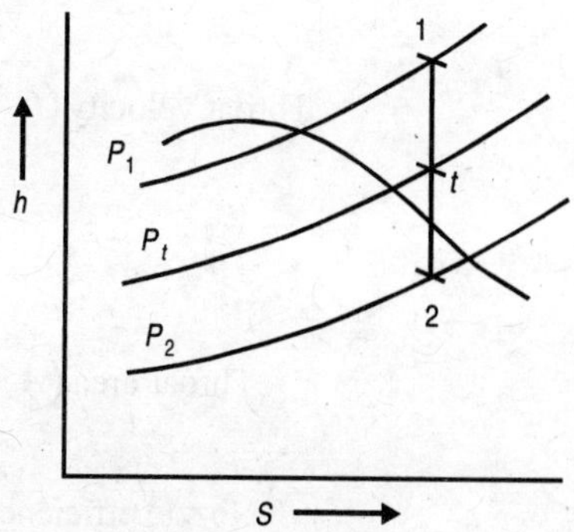

Fig. Ex. 19.3

$$\text{Velocity at throat } (C_1) = \sqrt{2(h_1 - h_t)}$$

$$= \sqrt{2 \times (2944 - 2813) \times 10^3}$$

$$= 511.86 \text{ m / s}$$

$$\text{Area at throat } (A_t) = \frac{m \cdot v_t}{C_t}$$

$$= \frac{0.2 \times 0.3735}{511.86} = 1.46 \times 10^{-4} \text{ m}^2$$

$$= \mathbf{1.46 \text{ cm}^2}$$

$$\text{Exit velocity } (C_2) = \sqrt{2(h_1 - h_2)}$$

$$= \sqrt{2 \times (2944 - 2628) \times 10^3}$$

$$= 795 \text{ m/s}$$

$$\text{Exit area } (A_2) = \frac{m \cdot v_2}{C_2} = \frac{0.2 \times 0.8537}{795}$$

$$= 2.15 \times 10^{-4} \text{ m}^2 = \mathbf{2.15 \text{ cm}^2}$$

Example 19.4. *An adiabatic steam nozzle is to be designed for a discharge rate of 10 kg/sec at steam from 10 bar and 400° C to a back pressure of 1 bar. The nozzle efficiency is 92% and frictional loss is assumed to take place in the diverging portion of the nozzle only. Assume a critical pressure ratio of 0.545. Determine throat and exit areas.*

Solution

$$P_1 = 10 \text{ bar}$$

$$T_1 = 400°\text{C}$$

$$P_2 = 1 \text{ bar}$$

$$P_t = \text{throat pressure}$$

$$\eta_{nozzle} = 0.92$$

$$m = 10 \text{ kg/s}$$

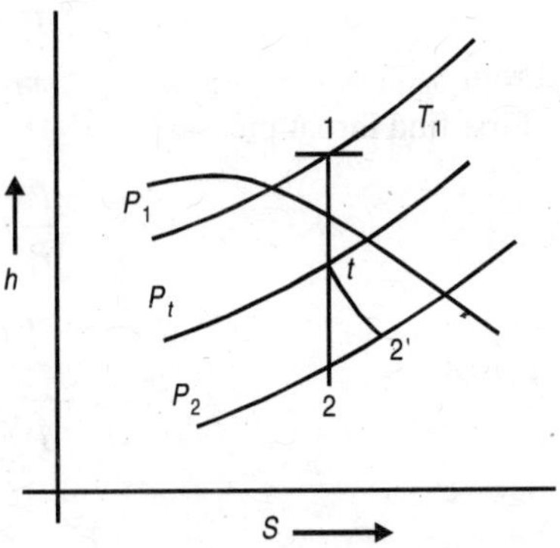

Fig. Ex. 19.4

$$\text{Critical pressure ratio} = \frac{P_t}{P_1}$$

$$\therefore \quad P_t = P_1 \times 0.545 = 10 \times 0.545$$

$$= 5.45 \text{ bar}$$

From Mollier Chart,

$$h_1 = 3263.9 \text{ kJ / kg}$$

$$h_t = 3110 \text{ kJ / kg},$$

$$V_t = 0.5 \text{ m}^3\text{/kg}$$

$$h_2 = 2700 \text{ kJ/kg},$$

$$V_2 = 1.8 \text{ m}^3\text{/kg}$$

$$\text{Throat velocity}\,(C_t) = \sqrt{2(h_1 - h_t)}$$

$$= \sqrt{2(3263.9 - 3110)\times 10^3}$$

$$= \mathbf{554.8 \ m/s}$$

$$\text{Throat area}\,(A_t) = \frac{(m \cdot v_t)}{C_t} = \frac{10\times0.5}{554.8} = 9.012\times10^3 \text{ m}^2$$

$$\text{Nozzle efficiency} = \frac{h_1 - h_2'}{h_1 - h_2}$$

$$h_1 - h_2' = 0.92\,(\,3263.9 - 2700\,)$$

$$= 518.79 \text{ kJ/kg}$$

$$\text{Exit velocity}\,(C_2) = \sqrt{2(h_1 - h_2')}$$

$$= \sqrt{2(518.79)\times10^3} = \mathbf{1018.62 \ m/s}$$

$$\text{Exit area}\,(A_2) = \frac{m \cdot v_2}{C_2} = \frac{10\times1.8}{1018.62}$$

$$= \mathbf{0.0176 \ m^2}$$

Example 19.5. *A nozzle is supplied with steam at 7 bar and 180°C. The discharge takes place against back pressure of 1 bar. Approach velocity is 75 m/s. The pressure at throat is 4 bar. No frictional losses take place between inlet and throat. Determine velocity at throat.*

If exit velocity is 742 m/s, what is overall nozzle efficiency?

Solution

$$P_1 = 7 \text{ bar}$$

$$P_2 = 1 \text{ bar}$$

$$P_t = \text{throat pressure} = 4 \text{ bar}$$

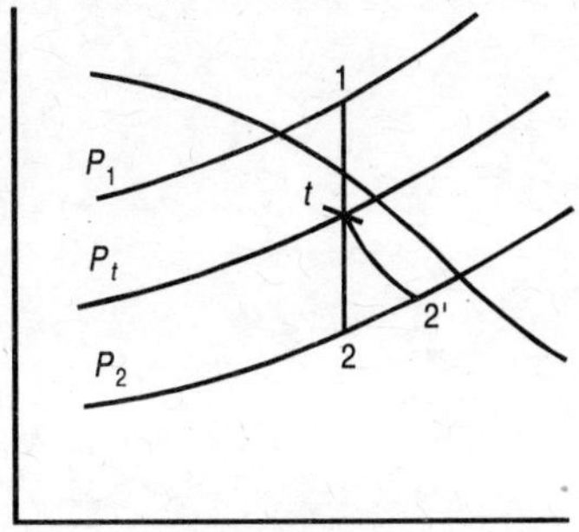

Fig. Ex. 19.5

From Mollier Chart,

$$h_1 = 2800 \text{ kJ/kg}$$

$$h_t = 2680 \text{ kJ/kg}$$

$$h_2 = 2457 \text{ kJ/kg}$$

Throat velocity $(C_t) = \sqrt{2(h_1 - h_t) + C_1^2}$

$$= \sqrt{2(2800-2680)\times 10^3 + 75^2}$$

$$= 464 \text{ m/s}$$

Exit velocity $(C_2') = \sqrt{2(h_1 - h_2')\times 10^3 + C_1^2}$

$$\therefore \quad 742 = \sqrt{2(2800 - h_2')\times 10^3 + 75^2}$$

$$\therefore \quad h_2' = 2530 \text{ kJ/kg}$$

$$\text{Overall nozzle efficiency} = \left[\frac{h_1 - h_2' + \left(\dfrac{C_1^2}{2}\right)}{h_1 - h_2 + \left(\dfrac{C_1^2}{2}\right)}\right]\times 100$$

$$= \left[\frac{(2800-2530) + \dfrac{75^2}{2\times 10^3}}{(2800-2457) + \dfrac{75^2}{2\times 10^3}}\right]\times 100$$

$$= 78.89 \%$$

Example 19.6. *Steam enters a convergent-divergent nozzle at 10 bar and 240°C, with a velocity of 50 m/s. It is discharged at 0.5 bar with a velocity of 978 m/s. The expansion is isentropic upto throat and with friction in the diverging part.*

Determine:

1. *Final quality of steam.*
2. *Exit diameter required if the diameter at throat is 10 mm.*
3. *Number of nozzles required for a steam flow rate of 60 kg/min to a turbine.*
4. *Percentage of overall isentropic drop lost in the friction in the diverging part.*

Solution

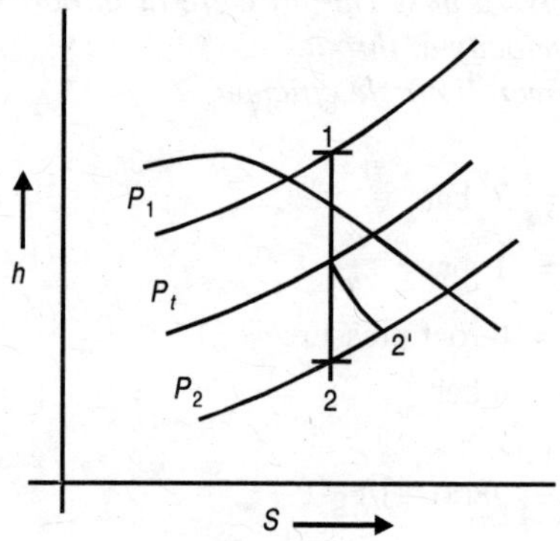

Fig. Ex. 19.6

$$P_1 = 10 \text{ bar}$$
$$T_1 = 240°C$$
$$P_2 = 0.5 \text{ bar}$$
$$C_1 = 50 \text{ m/s}$$
$$C_{2'} = 978 \text{ m/s}$$

First found throat pressure (P_t)

$$\frac{P_t}{P_1} = \left(\frac{2}{n+1}\right)^{n/n-1}$$

$$n = 1.3 \text{ (since steam is superheated at inlet)}$$

$$\frac{P_t}{10} = \left(\frac{2}{1.3+1}\right)^{\frac{1.3}{0.3}}$$

$$\therefore \quad P_t = 5.45 \text{ bar}$$

From Mollier Chart,

$$h_1 = 2910 \text{ kJ/kg}$$

$$h_t = 2775 \text{ kJ/kg}, \qquad V_t = 0.4 \text{ m}^3\text{/kg}$$

$$h_2 = 2390 \text{ kJ/kg}, \qquad V_2 = 0.35 \text{ m}^3\text{/kg}$$

$$\text{Exit velocity } C_2' = \sqrt{2(h_1 - h_2') + C_1^2}$$

$$\therefore \quad 978 = \sqrt{2(2910 - h_2') + 50^2}$$

$$h_2' = 2433 \text{ kJ/kg}$$

Now by locating point 2′ Mollier Chart for $h_2' = 2433$ kJ / kg and $P_2' = 0.5$ bar, dryness fraction can be found.

$$\therefore \quad \mathbf{x_2' = 0.905}$$

$$\text{Throat velocity } (C_t) = \sqrt{2(h_1 - h_t) + C_1^2}$$

$$= \sqrt{2(2910-2775)\times10^3+50^2}$$

$$= 522 \text{ m/sec}$$

Mass flow rate

$$m = \frac{A_t \cdot C_t}{V_t} = \frac{\frac{\pi}{4}\times(10^{-2})^2\times522}{0.4} = 0.1025 \text{ kg/s}$$

$$\text{Total steam flow rate} = 1 \text{ kg/s}$$

$$\therefore \quad \text{No. of nozzles required} = \frac{\text{Total flow rate}}{\text{Flow through a single nozzle}}$$

$$= \frac{1}{1.1025} = 9.756$$

$$\approx \textbf{10 Nozzles}$$

Now find exit diameter.

$$m = \frac{A'_2 - C'_2}{V'_2}$$

$$\therefore \quad A'_2 = \frac{m \cdot v_2'}{C'_2}$$

$$\therefore \quad \frac{\pi}{4}(d_2^2) = \frac{0.1025\times0.35}{978} \quad \therefore\ d_2 = \mathbf{21.6\,mm}$$

$$\text{Percentage enthalpy drop} = \frac{h_2' - h_2}{h_1 - h_2}\times100$$

$$= \frac{2433-2390}{2910-2390}\times100$$

$$= 8.27\ \%$$

Example 19.7. *A convergent-divergent nozzle is supplied with steam at a pressure of 1 MN/m² and 225°C. Supersaturated expansion occurs according to $pV^{1.3} = C$ in the nozzle down to an exit pressure of 0.32 MN/m². The exit diameter of the nozzle is 25mm. If the flow through the nozzle is chocked, determine (i) exit velocity, (ii) mass flow, (iii) throat diameter.*

Solution

$$1 \text{ MN/m}^2 = \frac{1\times10^6}{10^5} = 10 \text{ bar}$$

$$0.32 \text{ MN/m}^2 = \frac{0.32\times10^6}{10^5} = 3.2 \text{ bar}$$

(i) Exit velocity

$$\text{From steam table, } h_1 = \frac{2826.8+2943}{2}$$

$$= \frac{5769.8}{2} = \mathbf{2884.9}$$

$$v_1 = \frac{0.2059+0.2327}{2} = \frac{0.4386}{2} = \mathbf{0.2193\ m^3/kg}$$

$$\text{Work done/kg} = \text{Change in KE}$$
$$= \text{Area of } pV \text{ diagram}$$
$$= \frac{n}{n-1}[p_1v_1 - p_2v_2]$$

$$\therefore \left(\frac{v_2}{v_1}\right)^{1.3} = \frac{p_1}{p_2} = \frac{10}{3.2}$$

$$\frac{v_2}{v_1} = \left(\frac{10}{3.2}\right)^{\frac{1}{1.3}} = (3.125)^{0.7692} = 2.4024$$

$$\therefore v_2 = 0.2193\times2.4024 = \mathbf{0.5268\ m^3/kg}$$

$$\therefore \text{Work done/kg} = \frac{1.3}{0.3}\times100[10\times0.2193-3.2\times0.5258]$$
$$= 433.33[2.1930-1.6858]$$
$$= 433.33\times0.5072$$
$$= \mathbf{219.78\ kJ/kg}$$

$$\therefore \text{Exit velocity} = 44.72\sqrt{219.78}$$
$$= \mathbf{662.93\ m/sec}$$

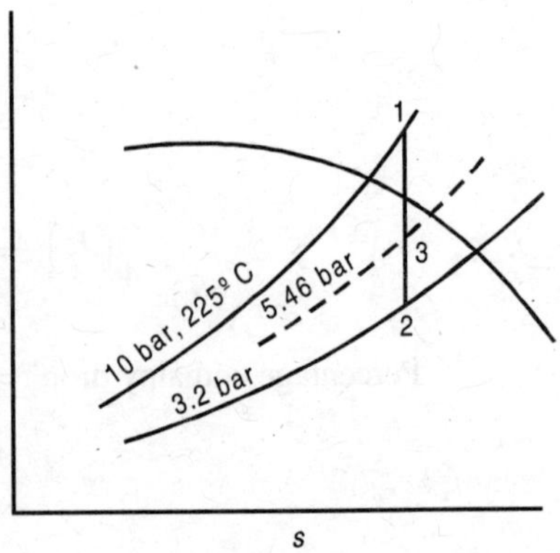

Fig. Ex. 19.7

(ii) Mass flow

$$\text{Mass flow at exit} = \frac{A_3V_3}{v_3} = \frac{A_2V_2}{v_2}$$

$$A_2 = \frac{\pi}{4}\times\frac{25\times25}{1000\times1000} = \frac{\pi}{4\times1600}\ \text{m}^2$$

$$V_2 = 662.93\ \text{m/s},\ v_2 = 0.5268\ \text{m}^3/\text{kg}$$

$$\therefore \dot{m} = \frac{\pi}{4\times1600}\times\frac{662.93}{0.5268}$$
$$= \mathbf{0.6177\ kg/sec}$$
$$= \mathbf{2223.72\ kg/hour}$$

(iii) Throat diameter :

$$\text{Throat pressure} = 0.546\ p_1 = 5.46\ \text{bar}$$

$$\therefore v_3 = v_1\left(\frac{P_1}{P_3}\right)^{1/n} = 0.2193\left(\frac{10}{5.46}\right)^{0.7692}$$

$$v_3 = 0.2193\times1.5923$$
$$= \mathbf{0.3493\ m^3/kg}$$

Change in K.E. upto throat $= \frac{n}{n-1}[p_1v_1 - p_3v_3]$

$$= \frac{1.3}{0.3} \times \frac{10^5}{1000}[10 \times 0.2193 - 5.46 \times 0.3493]$$

$$= 433.33[2.193 - 1.907]$$

$$= \mathbf{123.93\ kJ/kg}$$

$$\therefore \quad V_3 = 44.72\sqrt{123.93}$$

$$= \mathbf{497.845\ m/s}$$

$$\therefore \quad \dot{m} = \frac{A_3V_3}{v_3} = \frac{A_3 \times 497.845}{0.3493}$$

$$\therefore \quad A_3 = \frac{0.6177 \times 0.3493}{497.845}$$

$$= 4.334 \times 10^{-4}$$

$$= 0.0004334\ m^2$$

$$= \frac{\pi d_3^2}{4}$$

$$\therefore \quad d_3^2 = 0.0004334 \times \frac{4}{\pi}$$

$$= 0.0005518$$

$$d_3 = 0.0235\ m$$

$$= \mathbf{2.35\ cm}$$

Example 19.8. *1.2 kg/s of steam enters a nozzle at 12 bar and 250°C and expands upto 2 bar. Condensation does not occur while the steam is in the nozzle. Find the throat area, degree of undercooling, degree of supersaturation, exit area, actual heat drop and isentropic heat drop.*

Solution

p_1 = Inlet pressure = 12 bar

and temperature = 250°C

$\therefore \quad p_2$ = Throat pressure = 0.546×12

$= 6.552$ bar

$p_3 = 2$ bar

From diagram $(h - s)$,

$$h_1 = 2935\ kJ/kg$$

$$h_2 = 2800\ kJ/kg$$

The flow throughout is supersaturated and $n = 1.3$.

$$\therefore \quad \frac{T_1}{T_3} = \left(\frac{p_1}{p_3}\right)^{\frac{n-1}{n}} = \left(\frac{12}{2}\right)^{\frac{0.3}{1.3}} = 6^{0.2308}$$

$$= 1.531$$

$$\therefore \quad T_3 = \frac{250+273-523}{1.512} = 345.90 \text{ K} = \mathbf{72.90°C}$$

Saturation temprature at 2 bar = **120.23° C**

$\therefore$ Degree of undercooling =120.23 – 72.9

= **47.33° C**

Pressure corresponding to 72.9° C = 0.3543 bar.

$$\therefore \quad \text{Degree of supersaturation} = \frac{2}{0.3543} = \mathbf{5.645}$$

Upto throat the steam is superheated.

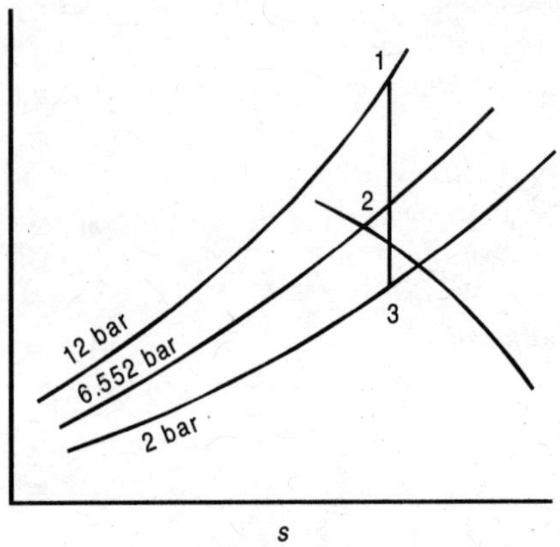

Fig. Ex. 19.8

$$\therefore \quad h_1 - h_2 = 2935 - 2800$$

$$= \mathbf{135 \ kJ/kg}$$

$$\therefore \quad V_2 = 44.72\sqrt{135}$$

$$= \mathbf{519.6 \ m/s}$$

$$V_2 = 0.35 \text{ m}^3/\text{kg}$$

$$\therefore \quad m_2 = \frac{A_2 V_2}{V_2}$$

$$\therefore \quad 1.2 = \frac{A_2 \times 519.6}{0.35}$$

$$\therefore \quad A_2 = \frac{1.2 \times 0.35}{519.6} \text{ m}^3 = 0.000808$$

$$= \mathbf{8.08 \ cm^2}$$

$$h_1 - h_3 = \frac{n}{n-1} p_1 v_1 \left[1 - \left(\frac{p_3}{p_1}\right)^{\frac{n-1}{n}}\right]$$

$$= \frac{1.3}{0.3} \times \frac{12 \times 10^5 \times 0.1924}{1000} \left[1 - \left(\frac{2}{12}\right)^{\frac{0.3}{1.3}}\right]$$

$$= 1000.48 \left[1 - \left(\frac{1}{6}\right)^{\frac{0.3}{1.3}}\right]$$

$$= 1000.48(1 - 0.6616)$$

$$= 1000.48 \times 0.3384$$

Actual heat drop = **338.56 kJ/kg**

Isentropic enthalpy drop =2935 – 2590

= **345 kJ/kg**

$$\therefore \quad V_3 = 44.72\sqrt{338.56} = \mathbf{822.85 \ m/s}$$

$$\therefore \quad 1.2 = \frac{A_3 V_3}{v_3}$$

or $$A_3 = \frac{1.2 \times v_3}{V_3}$$

$$= \frac{1.2 \times 0.779}{822.85}$$

$$= \mathbf{11.36\ cm^2}$$

Example 19.9. (*a*) *What do you understand by superstaurated flow and metastable state?*

(*b*) *Derive the following property, ratios for isentropic flow in a duct in terms of specific heat ratio* (γ) *and Match number* (*M*)

(*i*) *Stagnation pressure/static pressure.*

(*ii*) *Stagnation density/static density.*

(*c*) *Air at 7.8 bar, 180°C expands through a convergent-divergent nozzle into a space at 1.03 bar. The flow rate of air is 3.2 kg/s. Assuming isentropic flow throughout, calculate the throat and exit areas of the nozzle. Neglect inlet velocity.*

Solution

$$p_1 = 7.8 \text{ bar}$$

$$p_3 = 1.03 \text{ bar}$$

$$T_1 = 180 + 273 = 453 \text{ K}$$

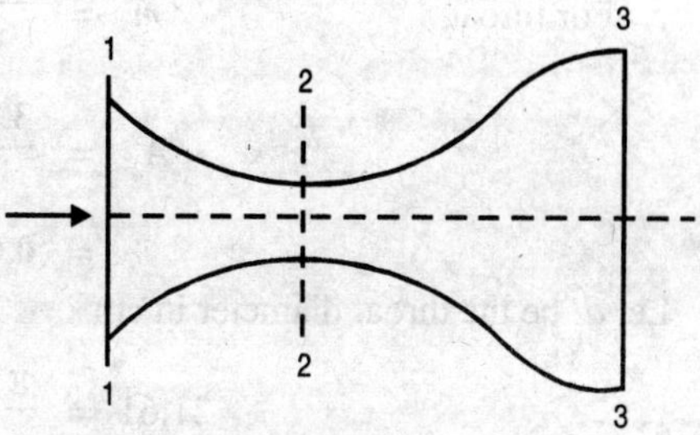

Fig. Ex. 19.9

Critical pressure ratio,

$$\frac{p_2}{p_1} = 0.5283 \text{ for air}$$

$\therefore$ $$p_2 = 0.5283$$
$$= 4.12 \text{ bar}$$

$\therefore$ $$\frac{T_2}{T_1} = \left(\frac{p_2}{p_1}\right)^{\frac{\gamma-1}{\gamma}} = (0.5283)^{0.2857} = 0.8333$$

$\therefore$ $$T_2 = 0.8333 \times 453 = 377.51 \text{ K}$$

Similarly, $$\left(\frac{p_1}{p_3}\right)^{\frac{\gamma-1}{\gamma}} = \frac{T_1}{T_3} = \left(\frac{7.8}{1.03}\right)^{0.2857} = 1.7832$$

$\therefore$ $$T_3 = \frac{453}{1.7832} = \mathbf{254.04\ K}$$

Enthalpy drop between 1–1 and 2–2.

$$-\Delta H_{1-2} = h_1 - h_2 = C_{pa}(T_1 - T_2)$$

$$= 1.005(453 - 377.51) = 1.005 \times 75.49$$

$$= 75.87 \text{ kJ / kg}$$

$\therefore$ $$V_2 = 44.72\sqrt{75.87} = 44.72 \times 88.1$$
$$= 389.52 \text{ m / s}$$

Similarly, $$-\Delta H_{1-3} = h_1 - h_3 = C_{pa}(T_1 - T_3)$$

$$= 1.005(453 - 254.04)$$

$$= 1.005 \times 198.96$$
$$= \mathbf{199.955\ kJ/kg}$$

$$\therefore \quad V_3 = 44.72\sqrt{199.955}$$
$$= \mathbf{632.36\ m/s}$$

Specific volume of air at throat conditions

$$v_2 = \frac{RT_2}{p_2} = \frac{287 \times 377.51}{4.12 \times 10^5}$$

$$v_2 = 0.2631\ \text{m}^3/\text{kg}$$

Similarly, specific volume at exit conditions,

$$v_3 = \frac{RT_3}{p_3} = \frac{287 \times 254.04}{1.03 \times 10^5}$$

$$= \mathbf{0.7079\ m^3/kg}$$

$\therefore$ For throat, $$\dot{m} = \frac{A_2 V_2}{v_2}$$

$$\therefore \quad A_2 = \frac{3.2 \times 0.2631}{389.52}$$

$$= \mathbf{0.002161\ m^2}$$

Let d_2 be the throat diameter in cm.

$$\therefore \quad 21.61 = \frac{\pi}{4} \times d_2^2$$

$$d_2^2 = \frac{21.6}{\pi} \times 4 = 27.5$$

$$\therefore \quad d_2 = \text{throat diameter} = \mathbf{5.244\ cm}$$

Similarly, exit area, $$A_3 = \frac{\dot{m}\, v_3}{V_3} = \frac{3.2 \times 0.7079}{632.36}$$

$$= 0.003582\ \text{m}^2$$
$$= \mathbf{35.82\ cm^2}$$

$$\therefore \quad d_3 = \left(\frac{35.82 \times 4}{\pi}\right)^{1/2} = (45.61)^{1/2}$$

$$= \mathbf{6.75\ cm}$$
$$= \text{Exit diameter}$$

Example 19.10. (*a*) *Prove that the maximum discharge in a steam nozzle per unit area at the throat is given by*

$$m_{\max} = \left\{1000n.\frac{p_1}{v_1}\left(\frac{2}{n+1}\right)^{\frac{n+1}{n-1}}\right\}^{1/2}$$

(*b*) *A convergent-divergent nozzle is supplied with dry saturated steam at 10 bar. The divergent portion of the nozzle is 5 cm along and the throat diameter is 6 cm. Calculate the semi cone angle of the divergent part of the nozzle so that the steam leaves the nozzle at 0.5 bar. Neglect friction.*

Solution

(a) Discharge through any section is given by,

$$\dot{m} = \frac{\text{Area} \times \text{Velocity}}{\text{Specific volume}}$$

$$= \frac{A\,V}{v}$$

Consider the flow follows the law $pv^n = V$

$$\therefore \quad p_1 v_1^n = p_2 v_2^n$$

or

$$v_2 = \left(\frac{p_1}{p_2}\right)^{1/n} \cdot v_1 \text{ and } v_1 = v_2 \times \left(\frac{p_2}{p_1}\right)^{1/n}$$

For the flow through throat, we have

$$\dot{m} = \frac{A_2 \cdot V_2}{v_2}$$

and

$$\frac{V_2^2 - V_1^2}{2 \times 1000} = h_1 - h_2 = \frac{n}{n-1} \cdot p_1 v_1 \left[\left(\frac{p_2}{p_1}\right)^{\frac{n-1}{n}} - 1\right]$$

$$\left(\frac{p_2}{p_1}\right) = \left(\frac{2}{n+1}\right)^{\frac{n}{n-1}} \text{ for maximum discharge}$$

$$\dot{m} = A_2 \left\{\frac{2000n}{n-1} \cdot \frac{p_1}{v_1}\left[\left(\frac{p_2}{p_1}\right)^{\frac{2}{n}} \cdot \left(\frac{p_2}{p_1}\right)^{\frac{n+1}{n}}\right]\right\}^{1/2}$$

$$\dot{m}_{\max} = \left\{\frac{2000 \cdot n}{n-1} \cdot \frac{p_1}{v_1}\left[\left(\frac{2}{n+1}\right)^{\frac{n}{n+1} \times \frac{2}{n}} - \left(\frac{2}{n+1}\right)^{\frac{n}{n-1} \times \frac{n+1}{n}}\right]\right\}^{1/2}$$

$$= \left\{\frac{2000\,n}{n-1}\,\frac{p_1}{v_1}\right\}\left[\left(\frac{2}{n+1}\right)^{\frac{2}{n-1}} - \left(\frac{2}{n+1}\right)^{\frac{n+1}{n-1}}\right]^{1/2}$$

$$= \left\{\frac{2000\,n}{n-1} \cdot \frac{p_1}{v_1} \cdot \left(\frac{2}{n+1}\right)^{\frac{n+1}{n-1}}\left[\left(\frac{2}{n+1}\right)^{\frac{2}{n-1} - \frac{n+1}{n-1}} - 1\right]\right\}^{1/2}$$

$$= \left\{\frac{2000\,n}{n-1} \cdot \frac{p_1}{v_1} \cdot \left(\frac{2}{n+1}\right)^{\frac{n+1}{n-1}}\left[\left(\frac{2}{n+1}\right)^{\frac{2-n-1}{n-1}} - 1\right]\right\}^{1/2}$$

$$= \left\{\frac{2000\,n}{n-1}\cdot\frac{p_1}{v_1}\cdot\left(\frac{2}{n+1}\right)^{\frac{n+1}{n-1}}\left[\left(\frac{2}{n+1}\right)^{\frac{1-n}{n-1}}-1\right]\right\}^{1/2}$$

$$= \left\{\frac{2000\,n}{n-1}\cdot\frac{p_1}{v_1}\cdot\left(\frac{2}{n+1}\right)^{\frac{n+1}{n-1}}\left[\left(\frac{2}{n+1}\right)^{\frac{1-n}{n-1}}-1\right]\right\}^{1/2}$$

$$= \left\{\frac{2000\,n}{n-1}\cdot\frac{p_1}{v_1}\cdot\left(\frac{2}{n+1}\right)^{\frac{n+1}{n-1}}\left[\frac{n+1}{2}-1\right]\right\}^{1/2}$$

$$= \left\{\frac{2000\,n}{n-1}\cdot\frac{p_1}{v_1}\cdot\left(\frac{2}{n+1}\right)^{\frac{n+1}{n-1}}\left[\frac{n+1-2}{2}\right]\right\}^{1/2}$$

$$= \left\{\frac{2000\,n}{n-1}\cdot\frac{p_1}{v_1}\cdot\left(\frac{2}{n+1}\right)^{\frac{n+1}{n-1}}\times\frac{n-1}{2}\right\}^{1/2}$$

$$= \left\{1000\,n\cdot\frac{p_1}{v_1}\cdot\left(\frac{2}{n+1}\right)^{\frac{n+1}{n-1}}\right\}^{1/2}$$

Neglecting approach velocity V_1, we have

$$V_2 = \left\{\frac{n}{n-1}\cdot p_1 v_1\cdot\left[\left(\frac{p_2}{p_1}\right)^{\frac{n-1}{n}}+1\right]\right\}^{1/2}$$

$$= \sqrt{\frac{n}{n-1}\,p_1 v_1\left[1-\left(\frac{p_2}{p_1}\right)^{\frac{n-1}{n}}\right]\times 2000}$$

$$\therefore \qquad \dot{m} = \frac{A_2\times\sqrt{2\times 1000\cdot\frac{n}{n-1}\cdot p_1 v_1\left[1-\left(\frac{p_2}{p_1}\right)^{\frac{n-1}{n}}\right]}}{\left(\frac{p_1}{p_2}\right)^{1/n} v_1}$$

$$= A_2\times\sqrt{\frac{2000\times n}{n-1}\cdot\frac{p_1}{v_1}\left[\left(\frac{p_2}{p_1}\right)^{2/n}-\left(\frac{p_2}{p_1}\right)^{\frac{n-1}{n'}\frac{2}{n}}\right]}$$

$$= A_2 \cdot \left\{ \frac{2000\,n}{n-1} \cdot \frac{p_1}{v_1} \cdot \left[\left(\frac{p_2}{p_1} \right)^{2/n} - \left(\frac{p_2}{p_1} \right)^{\frac{n+1}{n}} \right] \right\}^{1/2}$$

For maximum discharge,

$$\frac{d\dot{m}}{d\left(\frac{p_2}{p_1}\right)} = 0 = \frac{2}{n} \cdot \left(\frac{p_2}{p_1}\right)^{\frac{2-n}{n}} - \frac{n+1}{n}\left(\frac{p_2}{p_1}\right)^{\frac{n+1-n}{n}}$$

$$\therefore \quad \frac{2}{n} \cdot \left(\frac{p_2}{p_1}\right)^{\frac{2-n}{n}} = \frac{n+1}{n}\left(\frac{p_2}{p_1}\right)^{1/n}$$

$$\frac{2}{n+1} = \left(\frac{p_2}{p_1}\right)^{\frac{1}{n}-\frac{2-n}{n}=\frac{1-2+n}{n}=\frac{n-1}{n}}$$

$$\therefore \quad \frac{p_2}{p_1} = \left(\frac{2}{n+1}\right)^{\frac{n}{n-1}}$$

Condition for maximum discharge.

Substituting this condition in the above equation for $\dot{m}$, we get for unit area

$$\dot{m}_{\max} = \left\{ \frac{2000 \cdot n}{n-1} \cdot \frac{p_1}{v_1} \left[\left(\frac{2}{n+1} \right)^{\frac{2}{n-1}} - \left(\frac{2}{n+1} \right)^{n+1} \right] \right\}$$

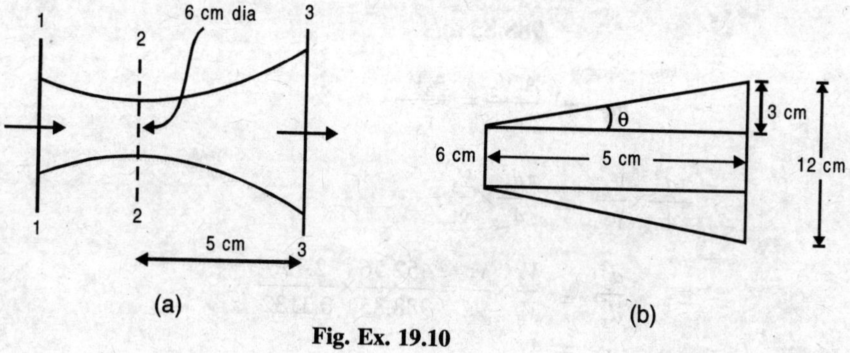

Fig. Ex. 19.10

(b)

$$p_1 = 10 \text{ bar}$$

$$x_1 = 1$$

$$p_3 = 0.5 \text{ bar}$$

$$p_2 = 0.58\ p_1 = 0.58 \times 10 = 5.8 \text{ bar}$$

From table, $h_1 = 2776.2$ kJ/kg

$$s_1 = 6.5828 \text{ kJ/kg-K}, \quad s_2 = 1.9176 + x_2 \times 4.8514$$

$$\therefore \quad x_2 = \frac{6.5828 - 1.9176}{4.8514} = \frac{4.6652}{4.8514} = \mathbf{0.9616}$$

$$\therefore \quad h_2 = 664.69 + 0.9616 \times 2089.3$$

$$= 664.69 + 2009.1 = \mathbf{2673.79 \text{ kJ/kg}}$$

and $$v_2 = x_2 v_{g_2} = 0.9616 \times 0.32574 = \mathbf{0.3132 \text{ m}^3\text{/kg}}$$

$$\therefore \quad V_2 = 44.72\sqrt{2776.2 - 2673.79} = 44.72\sqrt{102.41}$$

$$= \mathbf{452.56 \text{ m/s}}$$

$s_1 = s_3$ $$\therefore 6.5828 = 1.0912 + x_3 \times 6.5035$$

$$\therefore \quad x_3 = \frac{6.5825 - 1.0912}{6.5035} = \frac{5.4916}{6.5035}$$

$$= 0.8444$$

$$\therefore \quad h_3 = 340.56 + 0.8444 \times 2305.4$$

$$= 340.56 + 1946.7$$

$$= \mathbf{2287.26 \text{ kJ/kg}}$$

$$v_3 = x_3 v_{g_3} = 0.8444 \times 3.2402$$

$$= \mathbf{2.736 \text{ m}^3\text{/kg}}$$

$$V_3 = 44.72\sqrt{(h_1 - h_3)} = 2776.2 - 2287.26$$

$$= 44.72\sqrt{488.94}$$

$$= \mathbf{988.85 \text{ m/s}}$$

$$\therefore \quad \dot{m}_S = \frac{A_2 V_2}{v_2} = \frac{A_3 V_3}{v_3}$$

$$\therefore \quad \frac{\pi d^2}{4} \times \frac{V_2}{v_2} = \frac{\pi d_3^2}{4} \times \frac{V_3}{v_3}$$

$$\frac{d_3^2}{d_2^2} = \frac{V_2}{V_3} \times \frac{v_3}{v_2} = \frac{452.56}{988.85} \times \frac{2.736}{0.3132}$$

$$= 4$$

$$\therefore \quad \frac{d_3}{d_2} = 2 \quad \therefore d_3 = 2 \times 6 = \mathbf{12 \text{ cm}}$$

$$\therefore \quad \tan\theta = \frac{3}{5} = 0.6$$

$$\therefore \quad \theta = \text{semi-cone angle} = \tan^{-1} 0.6$$

$$= \mathbf{30.964^o}$$

$$= \mathbf{30^o.58'}$$

Example 19.11. (*a*) *What do you understand by (i) critical pressure ratio and (ii) choked flow for a nozzle?*

(*b*) *Dry saturated steam at 5 bar enters a converging-diverging nozzle at a velocity of 100 m/s. The exit pressure is 1.5 bar. The throat and exit area are 1200 m² and 1600 mm² respectively. Assuming isentropic flow upto the throat and taking critical pressure ratio as 0.58, estimate the mass flow rate of steam and nozzle efficiency.*

Solution

$$V_1 = 100 \text{ m/s}$$

$$p_1 = 5 \text{ bar}$$

$$x_1 = 1$$

$$p_3 = 1.5 \text{ bar}$$

$$A_2 = 1200 \text{ mm}^2$$

$$A_3 = 1600 \text{ mm}^2$$

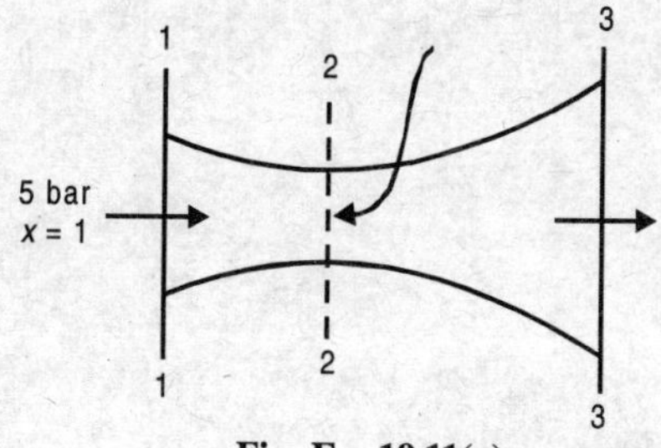

Fig. Ex. 19.11(a)

$$\frac{p_2}{p_1} = 0.58 \quad \therefore p_2 = 5 \times 0.58 = \mathbf{2.90 \text{ bar}}$$

$$h_1 = 2747.5 \text{ kJ/kg}, \; s_1 = 6.8192 = s_2 = 1.6595$$

$$\therefore \quad x_2 = \frac{6.8192 - 1.6595}{5.3428} = \frac{5.1597}{5.3428} = \mathbf{0.9657}$$

$$\therefore \quad h_2 = h_{f_2} + x_2 \times h_{fg_2} = 556.5 + 0.9657 \times 2166.6$$

$$556.5 + 0.209235 = \mathbf{2648.85 \text{ kJ/kg}}$$

$$\therefore \quad h_1 - h_2 = 2747.5 - 2648.85 = 98.65 \text{ kJ/kg}$$

$$\therefore \quad V_2 = 44.72\sqrt{98.65}$$

$$= \mathbf{455.3 \text{ m/s}}$$

$$v_2 = x_2 \, v_{g_2} = 0.9657 \times 0.62513$$

$$= 0.604 \text{ m}^3/\text{kg}$$

$$\therefore \quad \dot{m} = \text{mass flow rate} = \frac{A_2 V_2}{v_2}$$

$$= \frac{1200}{10^6} \times \frac{455.4}{0.604} = \mathbf{0.905 \text{ kg/s}}$$

Now,

$$p_3 = 1.5 \text{ bar} \quad \therefore s_3 = 1.4336 + x_3 \times 5.7898$$

$$\therefore \quad x_3 = \frac{6.8192 - 1.4336}{5.7898} = \frac{5.3856}{5.7898}$$

$$= 0.9302$$

$$\therefore \quad h_3 = 467.13 + 0.9302 \times 2226.2$$

$$= 467.13 + 2070.78$$

$$= 2537.91 \text{ kJ/kg}$$

Now, m is constant.

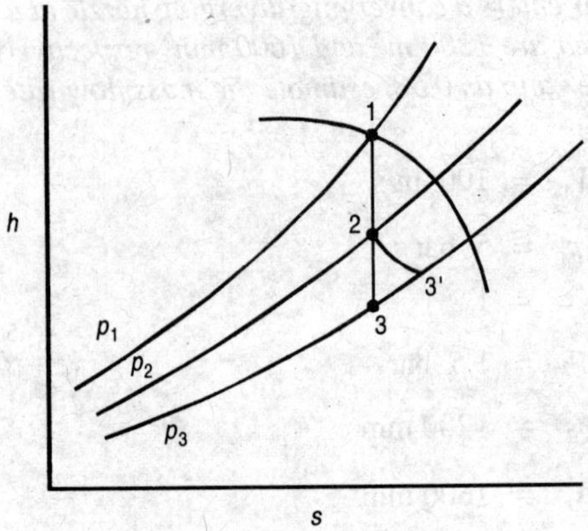

Fig. Ex. 19.11(b)

$$\therefore \quad 0.905 = \frac{A_3 V_{3'}}{v_{3'}}$$

$$= \frac{1600}{10^6} \times 44.72 \frac{\sqrt{2747.5 + 5 - 467.13}}{x_3' \times 1.59}$$

$$\therefore \quad \frac{0.905 \times 100 \times 1.159\, x_3'}{0.16 \times 44.72} = \sqrt{2285.37 - 2226.2\, x_3'}$$

$$(14.66\, x_3')^2 = 2285.37 - 2226.2\, x_3'$$

$$214.9\, x_{3'}^2 = 2285.37 - 2226.2\, x_3'$$

$$x_{3'}^2 = 10.63 - 10.36\, x_3'$$

$$x_{3'}^2 = 10.36\, x_3' - 10.63 = 0$$

$$\therefore \quad x_3' = \frac{-10.36 \pm \sqrt{(10.36)^2 + 4 \times 10.63}}{2}$$

$$= \frac{-10.36 \pm \sqrt{107.32 + 42.52}}{2}$$

$$= \frac{-10.36 \pm \sqrt{149.84}}{2}$$

$$= \frac{-10.36 \pm 12.241}{2}$$

$$= \frac{12.241 - 10.360}{2}$$

$$= \frac{1.881}{2}$$

$$= \mathbf{0.9405}$$

$$\therefore \quad h_{3'} = 467.13 + 0.9405 \times 2226.2$$
$$= 467.13 + 2093.74$$
$$= \mathbf{2560.87\ kJ/kg}$$

$\therefore$ Nozzle efficiency,

$$\eta_{nozzle} = \frac{h_1 - h_{3'}}{h_1 - h_3}$$
$$= \frac{2747.5 - 2560.87}{2747.5 - 2537.91}$$
$$= \frac{186.63}{209.59}$$
$$= 0.89$$

or $\quad = \mathbf{89\%}$

Example 19.12. *A steam turbine is developing 185 kW power with a specific steam consumption of 7 kg/kWh. The steam supplied to the turbine is dry and saturated and at a pressure of 13 bar and exausted at a pressure of 0.14 bar. The diameter at the throat of each nozzle is 7 mm. Assume that 12% of the total heat drop in lost in friction. Expansion of steam is ideal upto throat and all the friction loses occur in diverging part of nozzles. Determine the total number of nozzles required and find the diameter of nozzles at their exit end. Assume the ratio of throat pressure to inlet pressure as 0.578.*

Solution

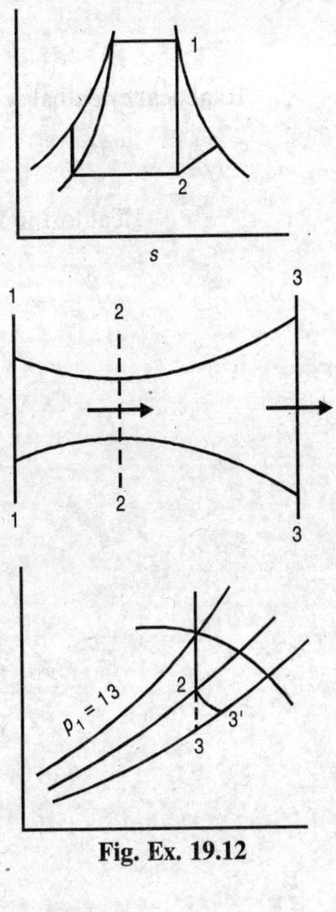

Fig. Ex. 19.12

$$\text{Total steam required} = \text{SSC} \times \text{Power developed}$$
$$= 7 \times 185$$
$$= 1295 \text{ kg / hour}$$
$$= \mathbf{0.3597\ kg/sec}$$
$$h_1 = 2785.4 \text{ kJ / kg}$$
$$p_2 = 7.514 \text{ bar} = 13 \times 0.578 \text{ kJ / bar}$$
$$s_1 = 6.4913 = s_2 = \frac{4.0390}{2} + x_2 \cdot \frac{9.3243}{2}$$
$$= 2.0195 + 4.6621 \times x_2$$
$$x_2 = \frac{6.4913 - 2.0195}{4.6621} = \frac{4.4718}{4.6621} = \mathbf{0.9592}$$
$$\therefore \quad h_2 = \frac{706.9 + 711.67}{2} + x_2 \left(\frac{2057.4 + 2053.7}{2}\right)$$
$$= 709.83 + 0.9592 \times 2055.55 = \mathbf{2681.5}$$
$$\therefore \quad v_2 = \left(\frac{0.25870 + 0.25224}{2}\right) \times 0.9592$$
$$= 0.9592 \times \left(\frac{0.51094}{2} = 0.25547\right) = \mathbf{0.245}$$
$$V_2 = 44.72\sqrt{h_1 - h_2} = 44.72\sqrt{2785.4 - 26875}$$

$$= 44.72\sqrt{103.9} \approx 455.8 \text{ m/s}$$

$$\text{Throat area } A_2 = \frac{\pi}{4}\times(0.007)^2 \text{ m}^2 = 0.0000385 \text{ m}^2$$

$$\therefore \text{ Mass of steam flowing/sec} = \frac{A_2 V_2}{v_2}$$

$$= \frac{0.0000385\times455.8}{0.245} = \mathbf{0.0716 \text{ kg/s}}$$

$\therefore$ No. of nozzles required

$$= \frac{\text{Total steam/sec}}{\text{Steam/nozzle}} = \frac{0.3597}{0.0716} = \mathbf{5.02 \text{ say } 5}$$

For divergent portion:

We have,

$$s_1 = s_3 = 6.4913$$

$$= 0.7367 + x_3\times7.2967$$

$$\therefore \quad x_3 = \frac{6.4913-0.7367}{7.2967} = \mathbf{0.7886}$$

$$\therefore \quad h_3 = 220.02+0.7886\times2376.7$$

$$= \mathbf{2094.32 \text{ kJ.kg}}$$

$$\therefore \text{ Total heat or enthalpy drop} = h_1 - h_3 = 2785.4 - 2094.32$$

$$= \mathbf{691.08 \text{ kJ/kg}}$$

$$\therefore \quad \text{Heat lost in friction} = \frac{12}{100}\times691.08 = \mathbf{73.33 \text{ kJ/kg}}$$

$$\therefore \quad h_{3'} = 2094.32+73.33 = \mathbf{2167.65 \text{ kJ/kg}}$$

$$= h_{f_3} + x_{3'} h_{fg_3} = 220.02 + x_{3'}\times2376.7$$

$$\therefore \quad x_{3'} = \frac{2167.65-220.02}{2376.7} = \mathbf{0.824}$$

$$\therefore \quad v_{3'} = 0.824\times10.694 = \mathbf{8.812 \text{ m}^3\text{/kg}}$$

$$\therefore \quad V_{3'} = 44.72\sqrt{h_1 - h_{3'}} = 44.72\sqrt{2785.4-2167.65}$$

$$= \mathbf{1111.5 \text{ m/s}}$$

$$\therefore \quad m = \frac{A_3 V_{3'}}{v_{3'}}$$

$$\therefore \quad A_3 = \frac{0.0716\times8.812}{1111.5} = \frac{\pi\, d_3^2}{4}$$

$$\therefore \quad d_3 = \mathbf{2.73 \text{ cm} = 27.3 \text{ mm}}$$

Example 19.13. (*a*) *Explain why a subsonic nozzle is convergent while a supersonic nozzle is divergent.*

(b) A nozzle is supplied with dry steam at 12 bar and back pressure 1 bar. Whether the nozzle is convergent or divergent? Calculate the throat and exit diameters if it discharges 15 kg steam per minute. Solve by using steam tables only.

Solution

(a) Combining continuity and energy equations, we get

$$\frac{dA}{A} = \frac{dV}{V}\left(-1+M^2\right) \quad \text{with usual notations.}$$

In the subsonic region, $M < 1$

$\therefore$ $\left(-1+M^2\right)$ is negative .

$\therefore$ dV is to be positive i.e. increase in velocity, dA has to be negative i.e. area is decreasing i.e. the nozzle is convergent.

$\therefore$ Subsonic nozzle is a convergent nozzle.

Similarly, a supersonic nozzle is a diverging nozzle.

(b) $\quad p_1 = 12$ bar and $p_3 = 1$ bar.

$\therefore$ Critical pressure p_2 at the throat is given by

$$p_2 = 0.578, p_1 = 0.578 \times 12 = \mathbf{6.936\ bar}$$

which is greater than outlet pressure.

$\therefore$ The nozzle is **convergent-divergent.**

$$h_1 = 2782.7 \text{ kJ/kg}$$

$$s_1 = 6.5194 \text{ kJ/kg-°K}$$

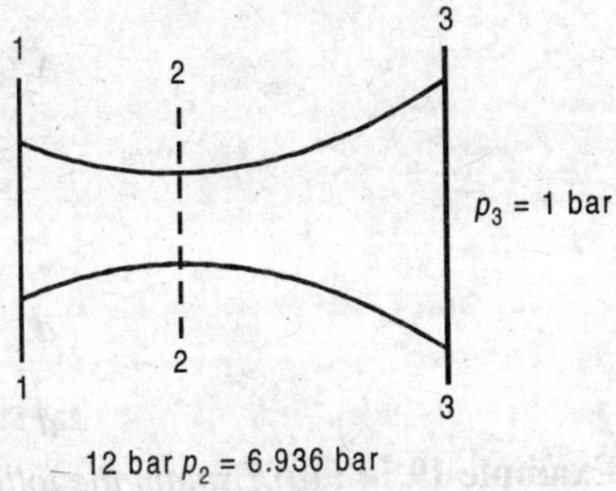

Fig. Ex. 19.13

For finding properties from steam tables, for p_2, p_2 is taken as 7 bar.

$$\therefore \quad s_2 = 1.9918 + x_2 \times 4.7134 \text{ and } s_1 = s_2$$

$$\therefore \quad x_2 = \frac{6.5194 - 1.9918}{4.7134} = \frac{4.5276}{4.7134}$$

$$\therefore \quad h_2 = 679.06 + 0.9606 \times 2064.9$$

$$= \mathbf{2681\ kJ/kg}$$

$$v_2 = \text{specific volume} = 0.9606 \times 0.27268$$

$$= \mathbf{0.26194\ m^3/kg}$$

$$h_1 - h_2 = \text{Enthalpy drop upto throat}$$

$$= 2782.7 - 2681 = 101.7 \text{ kJ/kg}$$

$$\therefore \quad V_2 = \text{Velocity of steam at throat}$$

$$= 44.72\sqrt{h_1 - h_2} = 44.72\sqrt{101.7}$$

$$= 450.99 \text{ m/s} \approx \mathbf{451\ m/s}$$

$$\therefore \quad m_2 = \frac{15}{60} = \frac{\text{AreaxVelocity}}{\text{Specific volume}} = \frac{A_2 \times 451}{0.26194}$$

$$= 1721.77\ A_2$$

$$\therefore \quad A_2 = \frac{151.26194}{60 \times 451} \times 14 = \mathbf{1.452\ cm^2} = \frac{\pi}{4} d^2$$

$$\therefore \quad d_2 = \text{Diameter at throat}$$

$$= \mathbf{1.36\ cm}$$

$$d_2 = 1.36 \text{ cm}$$

$$p_3 = 1 \text{ bar}$$

$$\therefore \quad s_3 = 1.3027 + x_3 \times 6.0571$$

$$\therefore \; s_1 = s_3 \qquad \therefore 6.5194 = 1.3027 + 6.0571\, x_3$$

$$x_3 = \frac{6.5194 - 1.3027}{6.0571} = \frac{5.2167}{6.0571}$$

$$= \mathbf{0.8613}$$

$$\therefore \quad h_3 = 417.51 + 0.8613 \times 2257.9$$

$$= \mathbf{2362.13\ kJ/kg}$$

$$\therefore \quad v_3 = 0.8613 \times 1.6937 = 1.4588 \text{ m}^3$$

$$\therefore \quad V_3 = 44.72\sqrt{2782.7 - 2362.13}$$

$$= 44.72\sqrt{420.57} = \mathbf{917.11\ m/s}$$

$$\therefore \quad \dot{m}_3 = \frac{15}{60} = \frac{A_3 V_3}{v_3} = \frac{A_3 \times 917.11}{1.4588}$$

$$\therefore \quad A_3 = \frac{1}{4} \times \frac{1.4588}{917.11} \text{ m}^2$$

$$= \frac{14588}{4 \times 917.11} \text{ cm}^2 = \mathbf{3.98\ cm^2}$$

$$\therefore \quad d_3 = \sqrt{3.98 \times \frac{4}{\pi}} = \mathbf{2.25\ cm}$$

$$d_3 = 2.25 \text{ cm}$$

Example 19.14 : (*a*) *Explain the following terms as referred to a system nozzle*

(*i*) *under expanding nozzle*
(*ii*) *chocked nozzle*
(*iii*) *over expanding nozzle.*

(*b*) *Steam enters a convergent-divergent nozzle at 10 bar and 240°C with the approach velocity of 50 m/s. It is discharged at 0.5 bar with a velocity of 978 m/s. The expansion is isentropic upto throat and with friction in the divergent part.*

Determine:

(*i*) *Final quality of steam.*
(*ii*) *Diameter at exit if the throat diameter is 10 mm.*
(*iii*) *Number of nozzles required for a steam flow rate of 60 kg/min.*
(*iv*) *% of overall isentropic enthalpy drop lost in the friction the diverging part.*

Solution

(b) **(i) Final quality of steam:**

Initially steam is superheated.

$$\therefore \quad n = 1.3$$

$$\therefore \quad \text{Critical pressure ratio} = \frac{p_2}{p_1} = 0.546$$

$p_c = p_2 = 10 \times 0.546 = $ **5.46 bar ≈ 5.5 for calculation** 1–2 expansion upto throat.

$$h_1 = 2925 \text{ kJ/kg}.$$

For the divergent part :

$$V_3 = 44.72\sqrt{h_1 - h_3}$$

$$\therefore \quad 978 = 4.72\sqrt{(h_1 - h_3)}$$

$$= 44.72\sqrt{(h_1 - h_3) + 0.25}$$

$$\therefore \quad \sqrt{h_1 - h_3} = \frac{978}{44.72} = 21.87$$

$$h_1 - h_3 + 1.25 = 478.27$$

$$\therefore \quad h_3 = h_1 - 478.27 = 2446.73$$

$$\therefore \quad x_3 = \text{Final quality of steam}$$

$$= \frac{2446.73 - 340.56}{2305.4} = \frac{2106.17}{2305.4} = \mathbf{0.914}$$

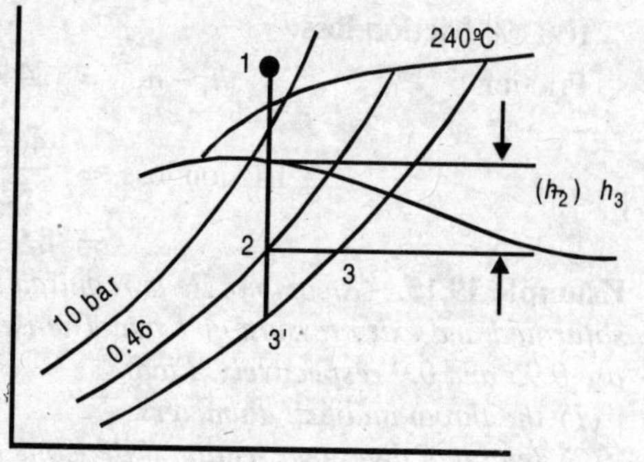

Fig. Ex. 19.14(a)

(ii) Exit diameter:

$$\therefore \quad h_1 + \frac{V_1^2}{2} = h_2 + \frac{V_2^2}{2}$$

$$\therefore \quad h_1 - h_2 = 2925 - 2800 = 125$$

$$= \frac{V_2^2 - V_1^2}{2 \times 1000}$$

$$\therefore \quad (V_2^2 - V_1^2) = 2000 \times 125$$

$$\therefore \quad V_2^2 = 2000 \times 125 + V_1^2$$

$$V_2 = \sqrt{2000 \times 125 + 2500} = 100\sqrt{0.2 \times 125 + 0.25}$$

$$= 100 \times \sqrt{25.25} = \mathbf{502.5\ m/s}$$

Fig. Ex. 19.14(b)

$$v_2 = \text{specific volume from chart} = 0.4046$$

$$v_3 = 0914 \times 3.2402 = 2.96$$

$$\therefore \quad \frac{A_2 V_2}{v_2} = \frac{A_3 V_3}{v_3} \quad \therefore A_3 = \frac{A_2 V_2}{v_2} \cdot \frac{v_3}{V_3}$$

$$\therefore \quad \frac{d_3^2}{d_2^2} = \frac{d_2^2 \times 502.5}{0.4046} \times \frac{2.96}{978} = 3.76$$

$$\therefore \quad \frac{d_3}{d_2} = 1.94 \ \therefore d_3 = 1.94 \times 10 = \mathbf{19.4\ mm}$$

(iii) Number of nozzles: Mass of steam flowing through one nozzle

$$= \frac{V_3 A_3}{v_3} = \frac{978 \times \frac{\pi}{4} \cdot \left(\frac{19.4}{1000}\right)^2}{2.96} = \mathbf{0.391\ kg/sec}$$

Actual flow is 60 kg/min i.e. **1 kg/sec.**

$$\therefore \quad \text{Number of nozzles required} = \frac{1}{0.391}$$

$$2.56 \approx \mathbf{3\ nozzles}$$

(iv) % friction loss:

Friction : $h_3 - h_{3'} = 2446.73 - 2400 = \mathbf{46.73 kJ/kg}$

$$\therefore \quad \% \text{ friction loss} = \frac{46.73}{525} \times 100$$

$$= \mathbf{8.9\ \%}$$

Example 19.15. *A steam nozzle is required to supply steam at rate 1 kg/sec. from inlet condition of bar abs, dry saturated and exit pressure of 1 bar. The isentropic efficiencies for the convergent section and divergent section are 0.95 and 0.9 respectively. Find:*

(1) the throat and exit diameters,

(2) length of divergent section if the cone angle of the divergent section is 14,

(3) Power in kW corresponding to the exit velocity of steam.

Solution

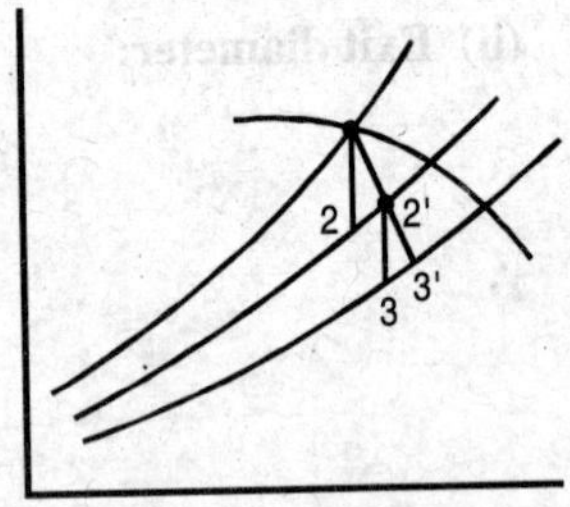

Fig. Ex. 19.15(a)

$$h_1 = 2776.2 \text{ kJ / kg}, \quad s_1 = 6.583 \text{ kJ / kg-K}$$

$$p_2 = 0.58\, p_1 = 10 \times 0.58 = 5.8 \text{ bar}$$

$$\therefore \quad s_2 = 1.918 + x_2 \times 4.851$$

$$\therefore \quad x_2 = \frac{6.583 - 1.918}{4.851} = \frac{4.665}{4.851} = \mathbf{0.9617}$$

$$h_2 = 664.7 + 0.9617 \times 2089.3$$

$$= \mathbf{2009.28\ kJ/kg}$$

$$\therefore \quad \text{Isentropic enthalpy drop} = h_1 - h_2 = 2776.2 - 2009.28 = 766.92$$

$$\therefore \quad \text{Actual enthalpy drop} = 776.92 \times 0.95 = \mathbf{738.1\ kJ/kg}$$

$$h_{2'} = 2776.2 - 738.1 = 2038.1$$

$$= h_{f_2} + x_{2'}\, h_{fg_2}$$

$$= 664.7 + x_{2'} \times 2089.3$$

$$\therefore \quad x_{2'} = \frac{2038.1 - 664.7}{2089.3}$$

$$= 0.6573$$

$$\therefore \quad v_{2'} = 0.6573 \times 0.32572$$

$$\approx \mathbf{0.72157\ m^3/kg}$$

$$V_2 = 44.72\sqrt{h_1 - h_{2'}} = 44.72\sqrt{738.1}$$

$$= \mathbf{1214.95\ \ m/s}$$

$$\therefore \quad \dot{m} = \frac{A_2 V_2}{v_2} = \frac{A_2 \times 1214.95}{0.2157} = 1$$

$$\therefore \quad A_2 = \text{Throat area} = \frac{0.2157}{1214.95} \text{ m}^2$$

$$= \frac{2157}{1214.95} \text{ cm}^2 = \mathbf{1.78\ cm^2}$$

$$\therefore \quad d_2 = \sqrt{\frac{1.78}{\frac{\pi}{4}}} = \mathbf{2.27\ cm}$$

$$s_{2'} = 1.918 + 0.6622 \times 4.851$$

$$= \mathbf{5.130}$$

$$s_3 = 1.303 + x_3 \times 6.057$$

$$\therefore \quad x_3 = \frac{5.130 - 1.303}{6.057} = \mathbf{0.6318}$$

$$\therefore \quad h_3 = 417.5 + 0.6318 \times 2257.9 = 1844.04 \text{ kJ/kg}$$

∴ Actual enthalpy drop in divergent portion

$$= 0.9(h_{2'} - h_3) = 0.9(2048.13 - 1844.04)$$

$$= \mathbf{183.68\ kJ/kg}$$

$$\therefore \quad h_{3'} = 1844.04 + 20.41 = \mathbf{1864.45\ kJ/kg}$$

$$\therefore \quad h_1 - h_{3'} = 2776.2 - 1864.45$$

$$= \mathbf{911.75\ kJ/kg}$$

$$\therefore \quad V_3 = 44.72\sqrt{911.75} = \mathbf{1350.33\ m/s}$$

$$x_{3'} = \frac{1864.45 - 417.5}{2257.9}$$

$$= \frac{1446.96}{2257.9} = \mathbf{0.641}$$

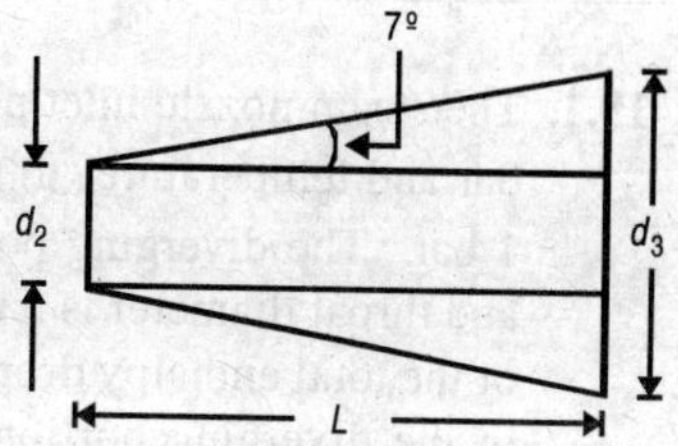

Fig. Ex. 19.15(b)

$$\therefore \quad v_{3'} = x_{3'} v_{s_3} = 0.641 \times 1.6938 = 1.085 \text{ m}^3\text{/kg}$$

$$\therefore \quad A_3 = \text{Exit area} = \frac{1 \times 1.085}{1350.33} \text{ m}^2 = \frac{1.085 \times 10^4}{1350.33} \text{ cm}^2$$

$$= \mathbf{8.04\ cm^2}$$

$$\therefore \quad d_3 = \text{Exit diameter} = \sqrt{\frac{8.04}{0.785}} = \mathbf{3.2\ cm}$$

$$\frac{d_3 - d_2}{2} = \frac{3.2 - 2.27}{2} = \frac{0.93}{2} = 0.465 \text{ cm}$$

$$\sin 7^\circ = \frac{0.465}{L} = 0.1219$$

$$\therefore \quad L = \text{Length of the divergent portion}$$

$$= \frac{0.465}{0.1219} = \mathbf{3.8\ cm}$$

(3) Power in kW corresponding to the exit velocity of steam

$$= \frac{\dot{m} V_3^2}{2} = \frac{1 \times 1350.33 \times 1350.33}{2 \times 1000}$$

$$= \mathbf{911.7\ kW}$$

THEORY QUESTIONS

19.1. What is a nozzle and what is its function? [4.1]

19.2. Derive the expression for velocity and mass flow rate in a nozzle. [*Ans.* 4.3 and 4.4]

19.3. Derive the expression of critical pressure ratio for the maximum discharge through a nozzle. [*Ans.* 4.5]

19.4. What do you understand by supersaturated flow in case of nozzles. [*Ans.* 4.9]

19.5. Explain the various terms. [*Ans.* 4.9]
(a) Degree of supersaturation
(b) Degree of undercooling.

19.6. An nozzle is supplied with steam at 7 bar and 180° C. The discharge take place against back pressure of 1 bar. The initial velocity is 75 m/s. The pressure at throat is 4 bar and there is no friction between inlet and throat. Determine the velocity at the throat. If exit velocity is 742 m/sec. What is overall nozzle efficiency. (*Ans.* 78.62%)

19.7. Ten nozzles are required to discharge steam to an impulse turbine from a pressure of 8 bar and 350° C. The back pressure of the turbine is maintained at 0.14 bar. The nozzle efficiency is 90%. If the throat diameter for each nozzle is 0.8 cm find the total mass flow rate and exit diameter of each nozzle.

PROBLEMS FOR PRACTICE

19.1. The steam nozzle inlet pressure of steam is 7 bar and temperature 300°C and discharge at 1 bar. The diverging portion is 50 mm long and throat diameter is 5 mm. Assuming 10 % of the total enthalpy drop to be lost in friction in the diverging portion of nozzle, find the cone angle of the divergent portion. (*Ans.* 2° – 12')

19.2. The inlet condition of steam to a convergent-divergent nozzle is 15 bar pressure and 250°C temperature. The exit pressure is 0.4 bar. Assuming 85% nozzle efficiency and throat area of 30 cm^2, determine the exit area for the nozzle. (*Ans.* 221.5 sq.cm)

19.3. The nozzles in the stage of an impulse turbine receive steam at 12 bar, superheated by 62°C. The pressure in the wheel chamber is 4 bar. State whether the nozzle is convergent, divergent or convergent-divergent. Assuming negligible approach velocity and 10% frictional loss, suggest suitable number of nozzles to be used for a flow rate of 450 kg/min. Approximately, with the exit area of 2.4 cm^2. Also state the exact rate of discharge through the nozzle. (*Ans.* 24, 0.3125 kg/sec/nozzle)

19.4. Steam enters the nozzles of an impulse turbine at 10 bar, 250°C and leaves at 1 bar. The turbine develops 225 kW with the specific steam consumption of 16. If the throat diameter is 8 mm, find the number of nozzles and the exit diameter, assuming that 10% of the total enthalpy drop is lost in friction in the divergent portion only, velocity of approach may be neglected. (*Ans.* 15, 12.3 mm)

19.5. A dry saturated steam is expanded in a nozzle from pressure 10 bar to 4 bar. If the expansion is supersaturated, find :
(i) the degree of undercooling
(ii) the degree of supersaturation.
(*Ans.* (i) 10.4°C ; (ii) 1.354)

IMPORTANT POINTS TO BE REMEMBERED

1. *Equation deciding the shape of the nozzle*

$$\frac{dA}{A} = \frac{dC}{C}(M^2 - 1) \text{ with usual notations}$$

2. *Velocity of steam at any section of the nozzle*

C_2 = Velocity at any secting of the nozzle

$$= \sqrt{2000(h_1 - t_2) + C_1^2}$$

$$= 44.72\sqrt{(h_1 - h_2) + \frac{C_1^2}{2000}}$$

$$= 44.72\sqrt{(h_1 - h_2)} \text{ if } C_1 \text{ is neglected}$$

$$\text{Rate of Mass flow of steam} = m_s = \frac{\text{Section area} \times \text{Velocity at the sec.}}{\text{Sp. volume}}$$

$$= \frac{A_2 C_2}{v_2}$$

3. *Condition for maximum discharge*

$$\text{Critical pressure ratio} = \frac{Pt}{P} = \frac{P_2}{P_1} = \left(\frac{2}{n+1}\right)^{\frac{n}{n-1}}$$

where n = 1.3 if initial steam is spuerheated.

and $\frac{P_c}{P_1} = \frac{P_t}{P_1} = 0.546$

n = 1.135 for dry and saturated steam at inlet

$\frac{P_c}{P_1} = \frac{P_t}{P_1} = 0.58$

$n = 1.035 + 0.1x$ for wet steam and corresponding pressure can be found.

4. *Nozzle efficiency*

$$\eta_{noz} = \frac{\text{Actual enthalpy drop}}{\text{Isentropic enthalpy drop}} = \frac{h_1 - h_{2'}}{h_1 - h_2}$$

∴ Actual velocity of steam

$$C_2 = 44.72\sqrt{h_1 - h_{2'} + \frac{C_1^2}{2000}}$$

$$= 44.72\sqrt{h_1 - h_{2'}} \text{ if } C_1 \text{ is neglected}$$

and $$\dot{m} = \frac{A_2 (or A_+)}{v_2} \cdot \sqrt{h_1 - h_{2'}} \times 44.72$$

5. *Supersaturated/metastable flow*

$$\frac{T_A}{T_B} = \left(\frac{P_A}{P_B}\right)^{\frac{n-1}{n}} \quad n = 1.3$$

Degree of undercooling = Saturation temperature at point P] – [Actual temperature at B T_B]

$$\text{Degree of supersaturation} = \frac{\text{Actual pressure at } B}{\text{Saturated pressure corresponding to } B}$$

CHAPTER 20

Steam Turbines

20.1 INTRODUCTION

Turbine is a device having rotating wheel to which is fastened a series of blades uniformly placed on its periphery. The expansion of the steam in nozzle or bucket converts its enthalpy into kinetic energy. The steam from nozzles or guide passage is directed continuously against the moving blades. The high velocity steam leaving the nozzle looses its velocity as it moves over moving blades causing rotation of moving blades.

The steam turbine has various applications. It is essentially a high speed device. It is used to run the electric generators, fans, blowers centrifugal pumps and with gear connections to run marine propeller for propulsion of ships. Turbines are also used in rolling mills.

20.1.1 Advantages

There are some advantages as given below:

1. Low initial cost smaller floor space required. Low expenses for maintenance. Light foundation is required.
2. The turbines have large over-load capacity.
3. The exhaust steam is free from oil.
4. The turbines have uniform velocity of rotation. So there is absence of the vibration.
5. High efficiency over a wide range.
6. Due to absence of reciprocating parts, perfect balancing is possible.
7. It can be built in much greater capacity than practical steam-engine.
8. Due to uniform speeds, turbines are suitable for power generation.

20.2 COMPARISON BETWEEN STEAM-TURBINE AND STEAM-ENGINE

Steam turbines are compared with steam engines as given below.

1. In steam engine, reciprocating motion has to be converted into rotary motion by connecting rod and crank but in steam turbine direct rotational motion is available. Steam engines are inherently low speed engines becuase of the balancing problems. As against this, steam turbines being rotary ones there are no balancing problems.

2. It is difficult to couple steam-engine to electric generator. Steam turbines are most suitable to be coupled to electric generators so turbines are used for power generation.

3. In steam-engine, steam gets mixed with oil, so condensate is not suitable as feed-water. For steam turbine, internal lubrication is not required so steam is free from oil and condensate can be used as boiler feed.

4. The wear and tear and maintenance cost of steam engine is more than turbine.

5. It is difficult to design large units for steam-engine. But for steam-turbines large units (250 MW to 500 MW) are feasible.

6. The thermal efficiency of steam-turbine is more than that of steam engine.

20.3 CLASSIFICATION OF STEAM-TURBINES

Steam turbines are classified as follows:

1. By principal of operation

(a) Impulse turbine

(b) Reaction-turbine

(c) Impulse-Reaction Turbine

2. By the method of compounding

(a) Pressure compounded turbine

(b) Velocity compounded turbine

(c) Pressure-velocity compounded turbine

3. By the pressure of inlet-steam

(a) High pressure turbine

(b) Low pressure turbine

4. By the exhaust pressure

(a) Condensing

(b) Non-condensing

(c) High back pressure

(d) Bleeder

5. By shaft position

(a) Vertical-shaft

(b) Horizontal-shaft

6. By field service

(a) Stationary, (b) variable speed, (c) locomotive, (d) ship, (e) industrial etc.

20.4 SIMPLE IMPULSE TURBINE

The impulse turbine depends completely upon the dynamic action of the steam. The direction of motion of the steam is changed as it passes across the blade. As a result of change of direction of steam across the blade, it will impart a force on the blade. Now if the blade is free, it will move in the direction of force.

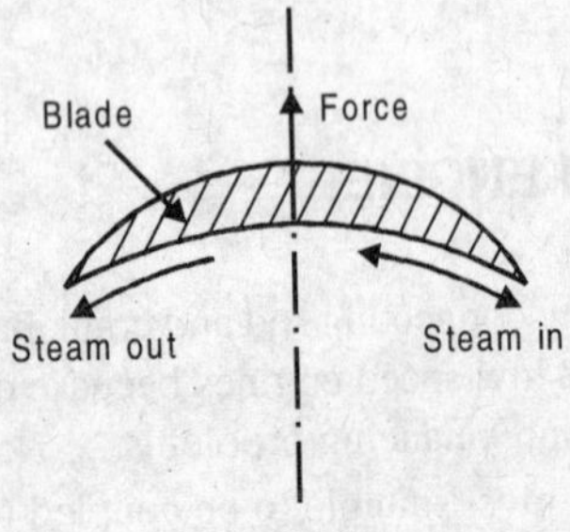

Fig. 20.1

In impulse turbine, steam first passes through the nozzle. In nozzle, the enthalpy is reduced and is converted into kinetic energy. The high velocity steam passes over moving blades and force is developed as shown in Fig. 20.1. The number of blades are fixed around the circumference of the shaft. So the steam will cause the turbine shaft to rotate. Thus mechanical work is obtained. The velocity of steam is reduced as it moves over the moving blades. But the pressure remain unchanged.

20.4.1 Working

The steam turbine shown in Fig. 20.2 is simple impulse turbine or De-Laval turbine. There is only one set of nozzles. The complete expansion of steam from the steam chest pressure to the exhaust or condenser pressure takes place in the one set of nozzles. Thus the pressure in the moving blades chamber is approximately equal to condenser pressure.

The steam enters the moving blades chamber with a high velocity. The pressure in the set of moving blades remains constant. The velocity of steam is reduced as it passes over moving blades as some of the kinetic energy of the steam is used in producing work on the turbine shaft.

For simple, impulse turbine, rotational speeds of the magnitudes of 20000 rpm may be obtained. This high speed of rotation will produce high centrifugal force which will restrict the size of the turbine. Thus the De-Laval type turbine is of relatively small size and hence has a small power-output. Also due to the high speed of rotation, a direct drive between the turbine disc and external equipment such as generator is not possible. For this reason, a reduction gear box is installed between the turbine-shaft and the external equipment. Also the velocity of steam leaving the turbine is quite appreciable resulting in a energy loss. It is as high as 10 to 12 % of the initial kinetic energy of steam.

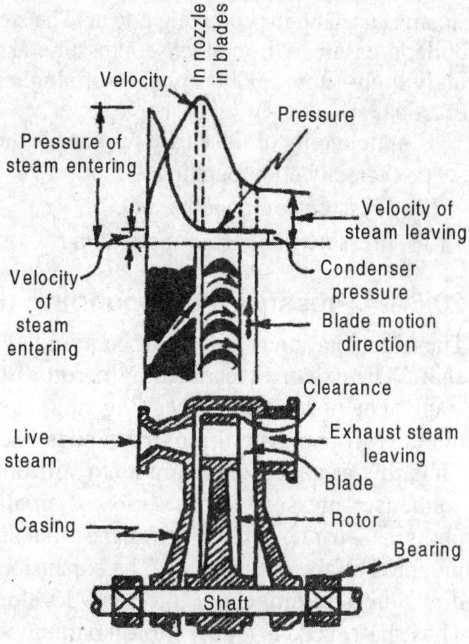

Fig. 20.2

20.5 IMPULSE REACTION TURBINE

Principal of operation: In reaction turbines, the pressure of the steam is reduced as it moves over the moving blades that gives rise to an increase in kinetic energy. This pressure drop also gives rise to a reaction in the direction opposite to that of increased velocity. Thus driving force is obtained. In impulse reaction turbine, impulse principle plus reaction principle is used.

20.5.1 Working

There are number of rows of moving blades fixed to the shaft and equal number of fixed blades attached to the casing. Refer Fig. 20.3. The fixed blades in this arrangement correspond to the nozzles referred to in the impulse turbine. Instead of a set of nozzles, steam is admitted for whole of the circumference and therefore there is all round admission. In passing through the first row of fixed blades, the steam undergoes a small drop in pressure and, its velocity is increased. It then enters the first row of moving blades and just as in impulse turbine, it suffers change in direction and momentum. This gives an impulse to the blades. But here the passage of the blades is so designed that there is a small drop in pressure in the moving blades that gives rise to the reaction force. Thus the driving force is a vector sum of impulse and reaction forces. The steam velocities in this type of turbine are comparatively low.

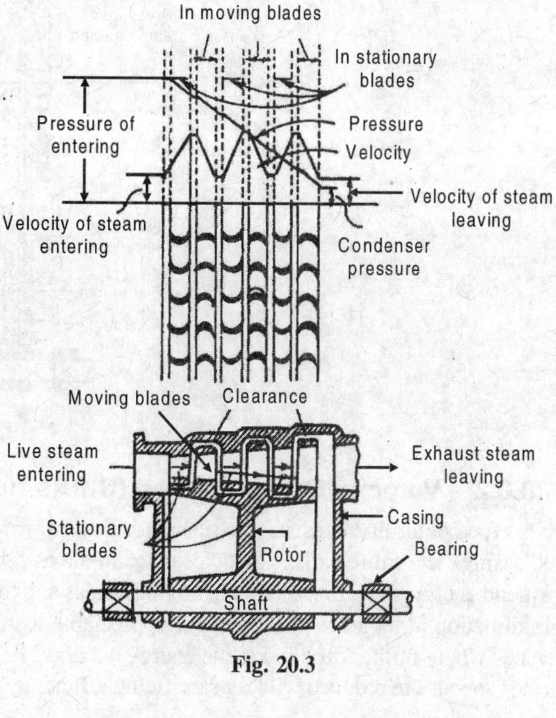

Fig. 20.3

20.6 COMPOUNDING OF STEAM TURBINES

Impulse turbines require nozzles and pressure drop of steam takes place in nozzles. The steam enters the turbine with a high velocity. There is no pressure drop in the turbine. The velocity reduces as some of the kinetic energy of steam is used up in producing power. The steam velocity entering the turbine is very high if the pressure drop from boiler pressure to the condenser pressure takes place in a single nozzle row. The speed of rotation becomes exceedingly high—upto 30000 r.p.m. — if single row of turbine blading is used. Hence, this speed needs reduction to practical speed.

Compounding of the steam turbines is the utilisation of the energy in stages. The methods used are:

(i) Velocity compounding.

(ii) Pressure compunding and

(iii) Pressure-velocity compounding

20.6.1 Pressure Compounding (Reateu Stage)

The high rotational speed can be avoided by combining the impulse stages in series and on the same output shaft. The exhaust steam from the rotor blades of the first, stage enters the nozzle of the succeeding impulse stage. The nozzles in succeeding stage are built in and locked to fixed diaphragm. Each set of nozzles combined with one disc of moving blades is called a stage of a turbine. The arrangement is shown in Fig. 20.4. Thus this arrangement amounts to splitting up the whole pressure drop from the steam chest pressure to the condenser pressure into a series of smaller pressure drops across the successive impulse stages. The diaphragms separate one stage from the other and all the moving blades are mounted on wheels which are individually mounted on the output shaft. The expansion of steam takes place only in the nozzles. The blades of rotor change direction of steam, apply force and develop torque. With drop in pressure per stage, steam velocity is reduced. Thus shaft speed is reduced for maximum work done.

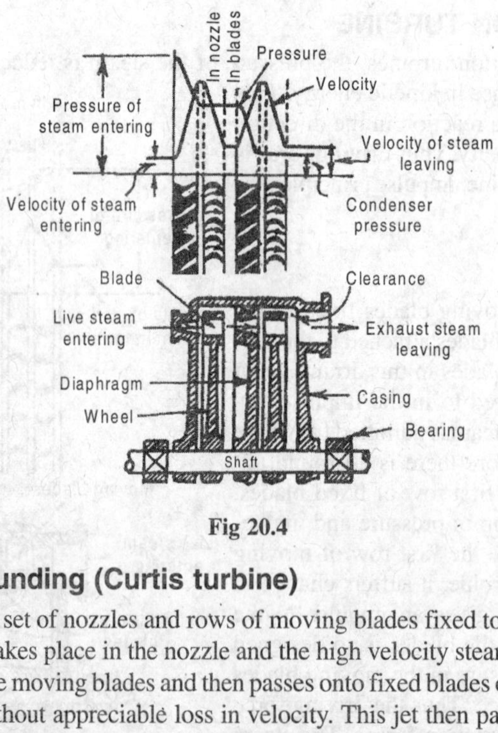

Fig 20.4

20.6.2 Velocity Compounding (Curtis turbine)

This type of turbine consists of a set of nozzles and rows of moving blades fixed to the shaft and fixed blades fixed to casing. The entire expansion takes place in the nozzle and the high velocity steam parts with only a portion of the kinetic energy in the first set of the moving blades and then passes onto fixed blades or guide vanes where only change in direction of jet takes place without appreciable loss in velocity. This jet then passes on to another set of moving vanes where further drop in kinetic energy occurs. The arrangement is shown in Fig. 20.5. By this method also high rotor speeds are reduced without sacrificing efficiency or work output.

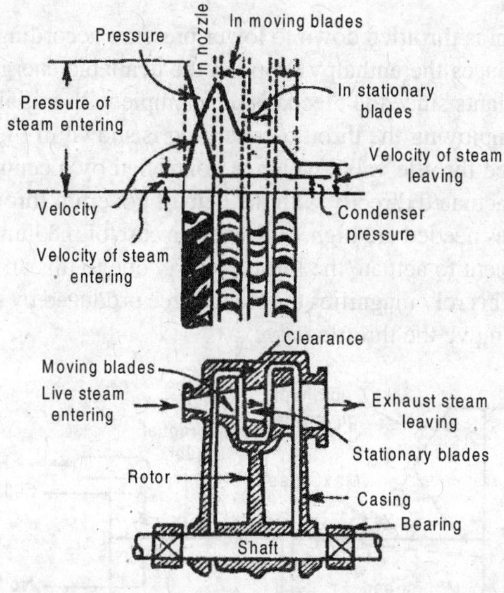

Fig. 20.5

The effect of velocity staging may be obtained with a single row of buckets by especially contrived means. In one such turbine, the blading is of the conventional type, but as the steam leaves the wheel the first time trough, it enters a reversing chamber that directs it back into the blades. Emerging for the second time from this row of blades, the steam may again be reversed and directed through the blades, thus providing three velocity stages. Turbines in which the steam enters a single wheel two or three times are called *re-entry turbines*.

20.6.3 Pressure Velocity Compounding

This type of arrangement combines the pressure compounding and velocity compounding. The arrangement for two stage is shown in Fig. 20.6. First stage and second stage taken separately are identical to a velocity compounding explained before. But both stages taken together will mean that the pressure drop from the chest pressure to the condenser pressure occurs at two stages. This type of arrangement is very popular due to simple construction but its efficiency is very low.

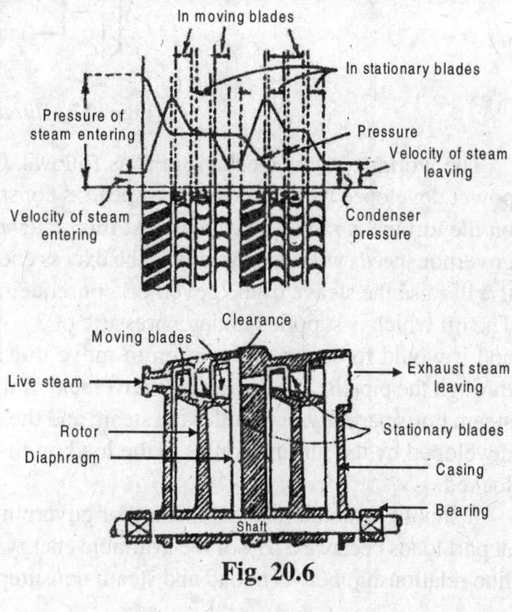

Fig. 20.6

20.7 GOVERNING OF STEAM TURBINES

The function of a governor is to control the fluctuation of speed of a prime mover within the prescribed limits with the variation of loads on it. In case of steam turbines when it is connected to drive an alternator for converting its mechanical energy into electrical energy, device is used to vary the turbine output according to the load on the alternator with very small fluctuations in speed, called governor.

The methods used for governing the steam turbines are:

(a) Throttle governing.
(b) Nozzle governing.
(c) By-pass governing.
(d) Combined throttle and nozzle governing.
(e) Combined throttle and by-pass governing.

20.7.1 Throttle Governing

In this type of governing the steam is throttled down to lower pressure according to the load on the turbine before it is supplied to the turbine. It reduces the enthalpy drop i.e. the available energy of the steam. Such a method is useful for small capacity power plants since the mechanism is simple with initial low cost.

The schematic arrangement employing the throttle governing is shown in Fig. 20.7. The throttling of steam is achieved with the help of balanced throttle valve which is controlled by a centrifugal governor. In case of small turbines the throttle valve can be actuated directly with the help of governor through linkages since the steam flow rates would be small and the valves needed are high. However, in case of medium and large power plants the effort of the governor may not be sufficient to actuate the throttle valve, due to this an oil operated relay is incorporated in the circuit as shown in figure.This rely magnifies the small force produced by the governor for a small change of speed to produce a large force to move the throttle valve.

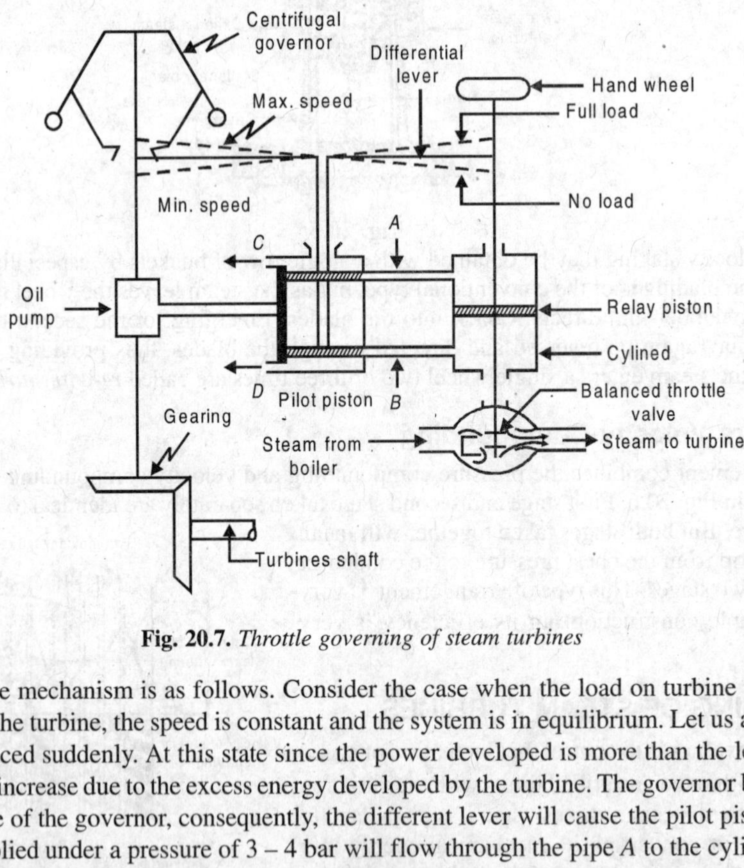

Fig. 20.7. *Throttle governing of steam turbines*

The working of the mechanism is as follows. Consider the case when the load on turbine shaft is equal to the power developed by the turbine, the speed is constant and the system is in equilibrium. Let us assume that the load on the turbine is reduced suddenly. At this state since the power developed is more than the load, the turbine and governor speeds will increase due to the excess energy developed by the turbine. The governor balls will fly out and it will raise the sleave of the governor, consequently, the different lever will cause the pilot piston to be raised. The oil which is supplied under a pressure of 3 – 4 bar will flow through the pipe *A* to the cylinder of relay piston and it would force the relay piston to move downwards, while the oil from relay piston cylinder is drained out through the pipe *B*. The downward movement of the relay piston operates the throttle valve which in turn closes the steam ports partially. It throttles the steam and the steam pressure at inlet to the turbine is reduced. When the power developed by the turbine equals to the load on the turbine, the oil ports *C* – *D* are covered and the relay piston is locked.

It should be noted that this method of governing is through simple but it reduces the efficiency of power plants at part loads because a part of the available energy is lost in the irreversible throttling process. The Willan's straight line relationship between load and steam consumption is also followed as discussed in case of steam engines.

20.7.2 Nozzle-Control Governing

In this method, the first stage nozzles are split-up groups which are controlled by individual throttle valves. Various arrangements of valves and group of nozzles are employed. Three such arangements are shown diagramatically in the Fig. 20.8. An arrangement often adopted with large steam turbines using high-pressure steam is shown in Fig. 20.8(a). The number of nozzle groups may vary from three to five or more.

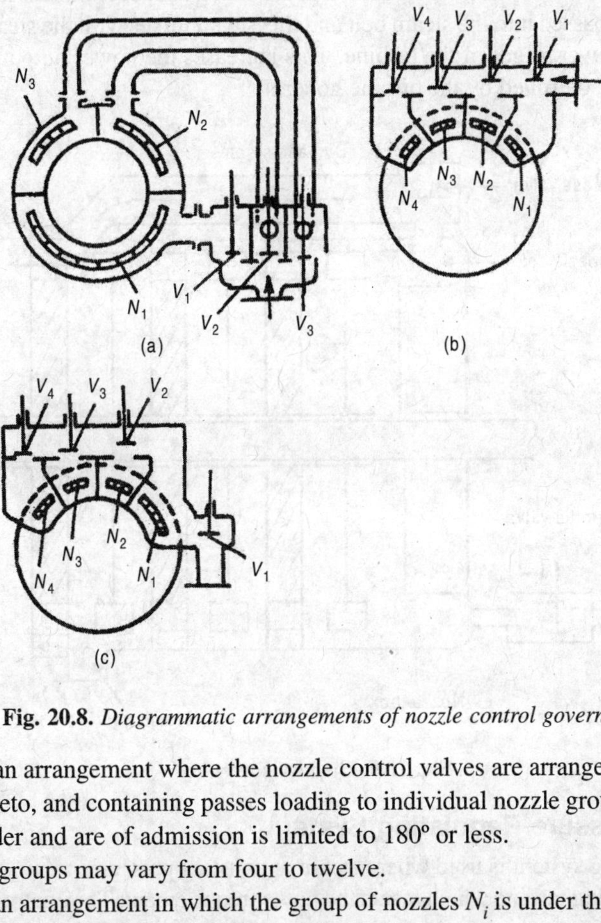

Fig. 20.8. *Diagrammatic arrangements of nozzle control governing*

Figure 20.8(b) shown an arrangement where the nozzle control valves are arranged in a casting forming part of the cylinder of bolted thereto, and containing passes loading to individual nozzle groups. The nozzles are confined to upper half of the cylinder and are of admission is limited to 180° or less.

The number of nozzle groups may vary from four to twelve.

Figure 20.8(c) shows an arrangement in which the group of nozzles N_i is under the control of valve V_1 through which all the steam entering the turbine passes. Further admission of stem is through the valves V_2 V_3 etc. in turn.

The nozzle-control is necessarily restricted to first stage and the absolute pressure of steam in front of the second stage nozzles will be directly proportional to the rate of steam flow through the turbine.

There is greater heat drop available when nozzle control is employed but the efficiency decreases when load is reduced. When there is a comparatively large heat drop in the first stage, nozzle control leads to reduction in steam consumption.

20.7.3 By-pass Governings

Modern steam power plants employing impulse turbines at very high pressure of steam at admission are usually designed to operate usually at an economical load which is bout 75–80% of the maximum load. In such cases it is desirable to have full admission of steam in first few stages of the turbine operating at very high pressure since the enthalpy drop is very small in the initial stages and the nozzle governing cannot be used effectively. Therefore the load regulation is achieved by throttle governing upto the stage of economical loads. However at the maximum loads the additional amount of steam required cannot be passed through the first-stages since the required additional number of nozzles are not available. This difficulty of steam regulation is overcome by employing by-pass governing as shown in Fig. 20.9.

Steam required upto economical loads is passed through the inlet valve and it is collected in the nozzle-chest. The governing is affected by throttle valve. However at loads more than economical load, the by-pass valve lifts and a part of the steam is by-passed into the steam belt and this steam mixes with the steam of high pressure turbine stages and it is supplied to lower stages of the turbine. This increases the power output of the turbine. The movement of the by-pass valve is controlled by the turbine governor.

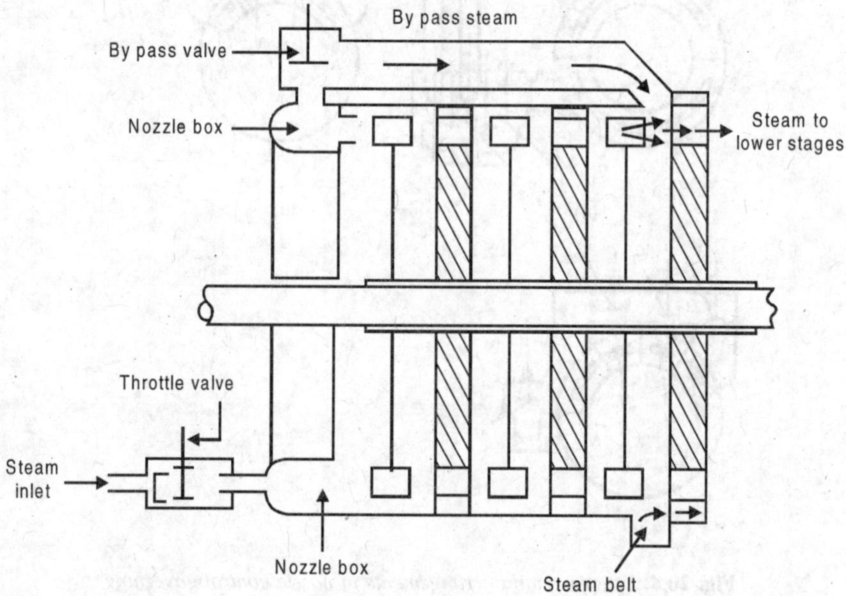

Fig. 20.9. *By-pass governing*

20.7.4 Extraction Pressure Regulating System

Extraction pressure regulating system is used where steam at constant pressure is extracted for heating or process requirements from a condensing turbine which meets the power requirement. The control gear must be designed to maintain speed and pressure of extracted steam regardless of the variation in power and heating loads. These requirements are met by using a speed governor which controls the admission of h.p. steam, to the turbine and a pressure regulator controlling the admission of steam to the L.P. stages. A change of either power load or heating load will cause both governors to operate.

Figure 20.10 shows the diagrammatic arrangement of the system. When the generator load increases the throttle valve opens causing a pressure rise in the extracted steam. The pressure regulator acts to open the regulating valve to maintain the steam pressure. If the heating load increases, the regulating valve closes and consequently the speed falls due to less flow of steam through L.P. stages. The speed governor acts to open the throttle valve. When the speed and pressure regulators operate independently, there is a tendency to hunting and this is overcome by inter-connecting the governor gear and pressure regulating system. A drop in speed causes both the valves to lift whereas an increase in heat load and fall in heating steam pressure causes pressure, regulating valve to close particularly and the throttle valve to open further. In order to simplify the diagram oil relays have been omitted.

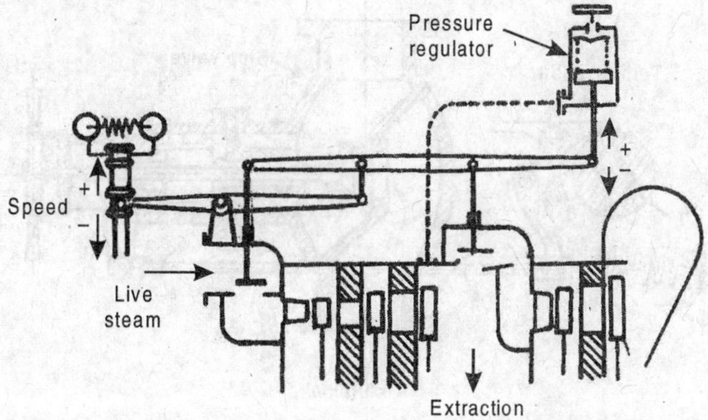

Fig. 20.10. *Diagrammatic arrangement of speed governor and pressure regulator on pass-out turbine*

20.7.5 Overspeed Tripping System/Emergency Governor

If a turbine has a not too satisfactory speed regulating system the sudden increase in the shaft speed at times of load tripping may reach dangerous figures. The usual overspeed limits is taken as about 10 to 12% of the normal operating speed. Hence every turbine is provided with one or two overspeed trips which shut off the supply of steam to the turbine if the r.p.m. exceeds a certain limit. The overspeed tripping device consists of an unstable centrifugal governor (static regularity). Figure 20.11 shows the main details of construction of such a ring-type static regulator. The eccentric ring 1 is directly mounted on the turbine shaft. The eccentric right is held in the position shown in Fig. 20.11 spring 2. The eccentricity (e) of the regulator is given by the distance between the axis of the turbine shaft and the center of gravity of the regulator ring 1. Distance a shows the regulator travel. The regulator ring 1 is displaced through a when the shaft speed exceeds the limiting speed. Overspeed regulators are made of various different constructions as well. One of the most widely used overspeed tripping devices as shown in Fig. 20.12. Regulator pin 1, when overspeeding, trips lever 7 disconnecting the interlock 8. The tensile force of the helical spring rotates lever 2 and segment 3 in the clockwise direction. Segment 3 now leaves its notch in sleeve 4 and allows the valve to close under the pressure of its own spring 5, thus shutting off steam supply to the turbine.

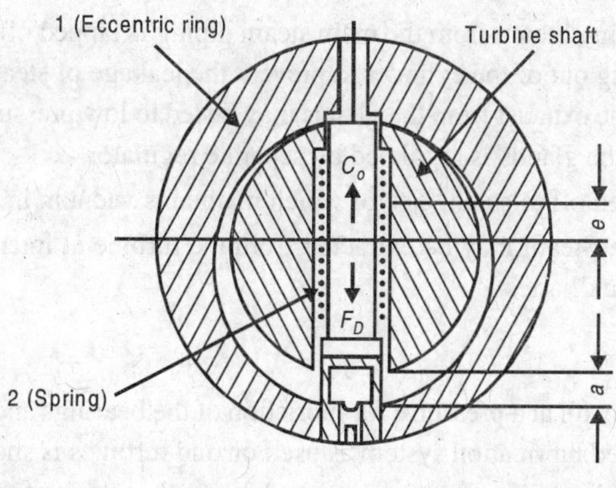

Fig. 20.11

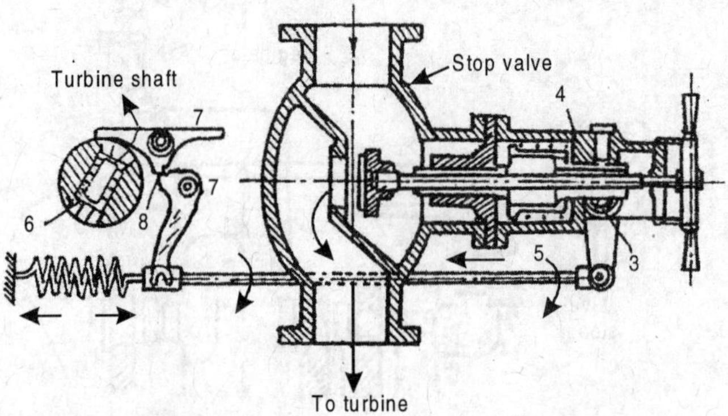

Fig. 20.12. *Overspeed trip with lever control*

If the stop valve is to be re-opened the following operations have to be carried out. The hand wheel of the stop valve is rotated in the direction of closing. During this operation sleeve 4 is displaced upward. Levers 7, 2 and segments 3 are now brought back to their original position. The stop valve is opened by means of the handwheel.

20.8 STUDY OF STEAM TURBINE SYSTEMS

20.8.1 Steam Circuit

The steam piping to or from the turbine consists of:

1. Main steam supply to turbine. Valves in this steam piping are isolating valve, the stop valve, and the governor or throttle valve.
2. Drain piping which is provided to drain out the water condensed during the start-up of the turbine. This is normally provided at the lowest point at,
   (a) before the isolating valve,
   (b) between the stop valve and the throttle valve, and
   (c) the turbine casing inlet. Valves are provided in the drain piping which will be opened during warming-up and closed after the turbine has been loaded to certain output.
3. Gland steam supply piping. Steam from the main steam piping is tapped off and supplied to the glands at the turbine shaft ends coming out of the cylinder to prevent the leakage of steam to the atmosphere and leakage of air into the turbine. The exhaust from the glands may be led to low pressure stages of turbine or condenser. The supply of steam to the glands is regulated by separate regulator.
   Steam may also be tapped off to steam ejector which maintains vacuum in the condenser.
4. Extraction steam Piping. Steam may be extracted from the turbine at intermediate stage for use as heating medium or for driving auxiliaries.

20.8.2 Oil Circuit

Oil circuit is provided to supply oil at a pressurre for lubriction of the bearings and also to operate the servo-motors through relays. A typical forced lubrication system as used on and turbines is shown in the Fig. 20.13. The pump which is normally of the positive action gear type usually delivers the whole of the oil required for lubrication and governing at the pressure required for the governor gear oil relays.

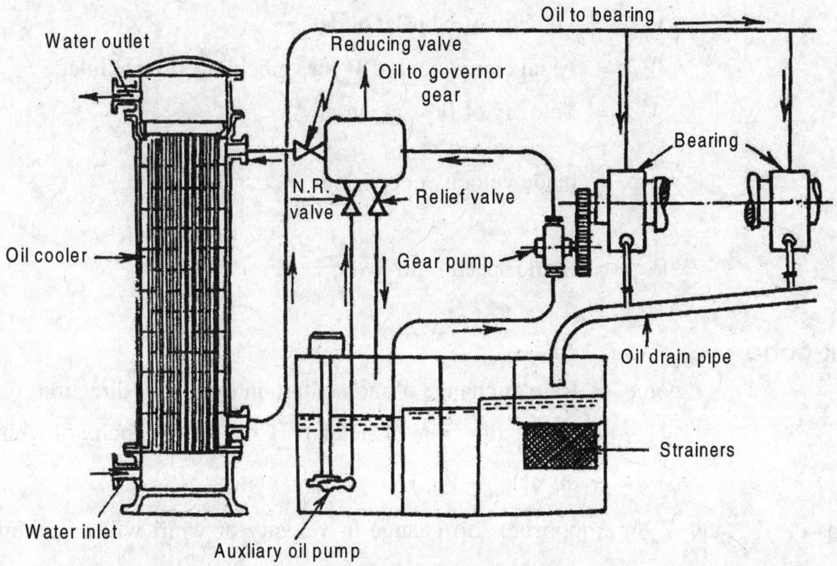

Fig. 20.13. *Diagrammatic arrangement of forced lubrication system*

As the oil for operating the relays is only required intermittently the surplus is allowed to passes through relief valve set to lift at the relay pressure. The oil used in the governor relays may or may not be cooled. Oil for lubrication is passed through a reducing valve which is so adjusted that the oil supply pressure at the bearings is about 0.5 to 1 bar. It is passed through all the bearings. Similarly a large bus pipe collects the oil from all the bearings and leads it to the oil drain tank. This pipe must be large enough to carry the oil with the pipe only partly filled so that air bubbles have a chance to escape. Oil strainers are placed on the oil inlet side of the tank.

The auxiliary oil pump comes into operation at starting. Stopping when the oil pressure falls below a certain valve to supply oil to the bearings when the shaft is being rotated by turning gear, and also as emergency stand-by. It is capable of supplying same quantity as the main pump. The type-most commonly used is electric/turbo driven centrifugal pump.

Water is used to cool the lubricating oil. Oil tanks should be large, enough to allow considerable time to elapse between the oil draining back into the tank and re-entering the pump.

20.9 VELOCITY DIAGRAM FOR SIMPLE IMPULSE TURBINE

Nomenclature used:

α = Angle of absolute velocity measured at inlet with the plane of rotation or Nozzles angle.

β = Angle of absolute velocity at outlet

θ = Blade inlet angle or angle of relative velocity at inlet

Φ = Blade outlet angle or angle of relative velocity at outlet

V_b = Linear blade speed or blade velocity

V_{ai} = Absolute velocity of steam at inlet

V_{ae} = Absolute velocity of steam at outlet

V_{ri} = Relative velocity of steam at inlet

V_{re} = Relative velocity of steam at outlet

V_{wi} = Tangential component of V_{ai} or velocity of whirl at inlet

V_{we} = Velocity of whirl at outlet

V_{fi} = Axial component of V_{ai} or velocity of flow at inlet

V_{fe} = Velocity of flow at outlet

$$K = \text{blade velocity coefficient} = \frac{V_{re}}{V_{ri}}$$

$$\rho = \text{Blade speed ratio} = \frac{V_b}{V_{ai}}$$

20.9.1 Work-done

$$\text{Force} = \text{Rate of change of momentum in tangential direction}$$
$$= (\text{mass flow rate of steam}) \times (\text{Change in velocity of whirl})$$
$$= m \times (V_{wi} - V_{we})$$

But directions of V_{wi} and V_{we} are opposite. So change in velocity of whirl will be arithmetic addition of V_{wi} and V_{we}

$$\therefore \quad F = m \times (V_{wi} + V_{we})$$

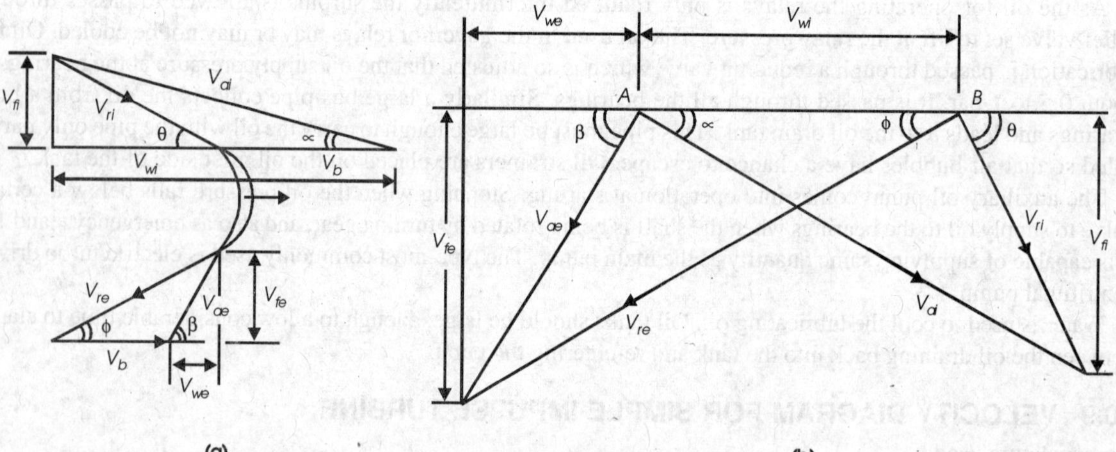

Fig. 20.14

$$\text{Work done} = \text{Force} \times \text{Velocity}$$

$$\therefore \quad \text{Work done on blade} = m(V_{wi} + V_{we}) \times V_b$$

20.9.2 Diagram Efficiency (or Blade Efficiency)

$$\text{Diagram efficiency} = \frac{\text{Work done on the blade}}{\text{Kinetic energy supplied to the blade}}$$

$$\text{Kinetic energy supplied to the blade} = \frac{V_{ai^2}}{2} \times m$$

$$\therefore \text{ Diagram efficiency } (\eta_{\text{dia}}) = \frac{m \times V_b (V_{wi} + V_{we})}{m \times \frac{V_{ai}^2}{2}}$$

$$= \frac{2 \times V_b \cdot (V_{wi} + V_{we})}{V_{ai}^2}$$

20.9.3 Stage Efficiency

$$\text{Stage efficiency} = \frac{\text{Work done per stage}}{\text{Energy supplied per stage}}$$

The stage consists of set of nozzles and set of moving blades and stage efficiency will also include the performance of nozzle.

$$\text{Stage efficiency} = \frac{m \times V_b \times \Delta V_w}{m \times \Delta h_s}$$

where Δh_s = change in enthalpy of steam per stage per unit mass.
= Isentropic enthalpy drop per stage.

20.9.4 Nozzle Efficiency

We know that,

$$\text{Nozzle efficiency} = \frac{\text{Kinetic energy of steam leaving the nozzle}}{\text{Change in enthalpy of steam in the nozzle}}$$
Isentropic enthalpy drop

$$= \frac{\frac{1}{2} m V_{ai}^2}{m \times \Delta h_s} = \frac{V_{ai}^2}{2 \times \Delta h_s}$$

$$\therefore \quad \text{Stage efficiency} = \text{Diagram efficiency} \times \text{Nozzle efficiency}$$

20.9.5 End Thrust or (Axial Thrust)

= Change of momentum in axial direction
Axial thrust = Mass flow rate × Change of velocity in axial direction
$= m \times (V_{fi} - V_{fe})$

20.10 OPTIMUM OPERATING CONDITIONS

$$\text{Work done by blade} = m \times V_b (V_{wi} + V_{we})$$

But

$$V_{wi} = V_{ai} \cos\alpha = V_b + V_{ri} \cos\theta$$

$$V_{we} = V_{ae} \cos\beta = V_{ro} \cos\Phi - V_b$$

$$\therefore \quad \text{work done by blade} = m \times V_b \times (V_{ai} \cos\alpha + V_{ae} \cos\beta)$$

$$= m \times V_b \times (V_{re} \cos\Phi + V_{ri} \cos\theta)$$

$$= m \times V_b \times V_{ri} \cos\theta \left(\frac{V_{re}}{V_{ri}} \times \frac{\cos\Phi}{\cos\theta} + 1 \right)$$

$$\text{But } k = \frac{V_{re}}{V_{ri}}$$

and $$\text{let } c = \frac{\cos\Phi}{\cos\theta}$$

$$\therefore \quad \text{work done} = m \times V_b \times V_{ri}\cos\theta(1+kc)$$

$$= m \times V_b (V_{ai}\cos\alpha - V_b)(1+kc)$$

$$\text{So the diagram efficiency} = \frac{m \times V_b \times (V_{ai}\cos\alpha - V_b)(1+kc)}{m \times \frac{V_{ai}^2}{2}}$$

$$\rho = \text{blade speed ratio} = \frac{V_b}{V_{ai}}$$

$$\therefore \quad \eta_{dia} = \frac{2m}{m} \times \frac{V_b}{V_{ai}} \times \left(\cos\alpha - \frac{V_b}{V_{ai}}\right) \times (1+kc)$$

$$= 2 \times \rho(\cos\alpha - \rho) \times (1+kc)$$

Now if the impulse turbine has symmetric blades then $\theta = \Phi$ and $k = 1$ for frictionless flow

$$\therefore \quad \dot{\eta}_{dia} = 4(\rho\cos\alpha - \rho^2)$$

∴ For fixed nozzle angle (α) the optimum value of diagram efficiency can be obtained by differentiating efficiency w.r.t. ρ and then equating it to zero.

$$\therefore \quad \frac{d}{d\rho}(\eta_{dia}) = 0$$

$$\therefore \quad \frac{d}{d\rho}\left[4(\rho\cos\alpha - \rho^2)\right] = 4(\cos\alpha - 2\rho) = 0$$

$$\therefore \quad \rho = \frac{\cos\alpha}{2},$$

Which is the condition for maximum diagram efficiency

$$(\eta_{dia})_{max} = 4 \times \left(\frac{\cos\alpha \times \cos\alpha}{2} - \frac{\cos^2\alpha}{4}\right)$$

$$= \cos^2\alpha$$

$$\text{Maximum work} = 2 \times m \times V_b^2$$

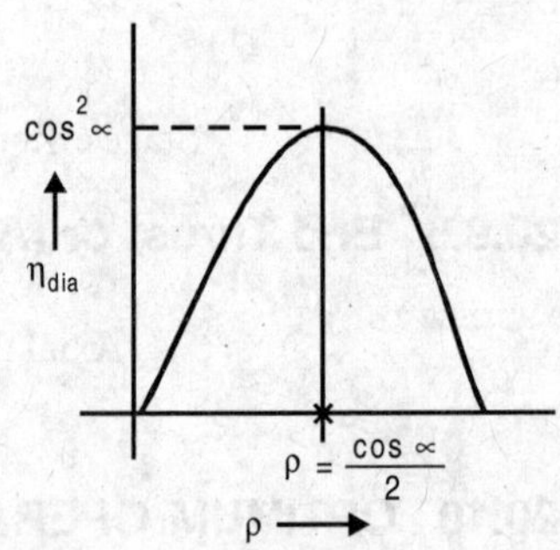

Fig. 20.15

20.11 BLADE HEIGHT

For impulse turbine the blade height at a particular section of a turbine will be function of mass flow rate (m), specific volume (v), area (A), and velocity of steam

Let,

P = pitch = distance between two blades.
N = number of blades covered by nozzle.
H = blade height (in metres)

We know the equation,

$$\text{Mass flow rate } (m) = \frac{A \times C}{v}$$

Applying the same equation for a turbine, we get

$$m = \frac{N \times P \times H \times V_{fi}}{v}$$

$$\therefore \quad \text{Blade height } (H) = \frac{m \times v}{N \times P \times V_{fi}}, \text{ for impulse turbine}$$

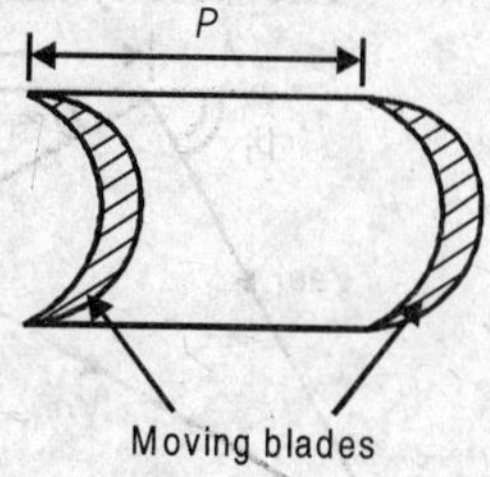

Fig. 20.16

For reaction turbine, there is full admission of steam through all nozzles

$$\therefore \quad N \times P = \text{Circumference at the blade mean diameter}$$

$$= \pi \times D$$

$$\therefore \quad \text{Blade height } (H) = \frac{m \times v}{\pi \times D \times V_{fi}}, \text{ for reaction turbine}$$

20.12 VELOCITY DIAGRAM FOR VELOCITY COMPOUNDED STEAM TURBINE CURTIS STAGE

The velocity compounded steam turbine has been already discussed. In this type of turbine, steam is first expanded in nozzles and is passed through set of moving blades fixed blades and then through second set of moving blades. The function of fixed blades is to change the direction of motion of steam received from earlier set of moving blades and to re-direct it to the next set of moving blades. The complete expansion of steam takes place in nozzle.

Nomenclature:

α_1 = Nozzle angle
θ = Inlet angle for first set of moving blades
ϕ_1 = Outlet angle for first set of moving blades
β_1 = Inlet angle for first set of fixed blades
α_2 = Exit angle for fixed blades
θ_2 = Inlet angle for second set of moving blades
ϕ_2 = Outlet angle for second set of moving blades
β_2 = Inlet angle for second set of fixed blades
k = Friction coefficient

The velocity diagram for the first set of moving blades can be drawn easily if the required data is known. It will be similar to velocity diagram of simple impulse turbine.

Following points should be noted for drawing a velocity diagram for second set of moving blades.

(1) Angle α_2 is exit angle of fixed-blades. The steam leaving fixed blades will enter smoothly in second set of moving blades. So α_2 is similar to nozzle angle, if compared with velocity diagram of simple impulse turbine.

(2) If the friction is considered in blades, the relative velocity will decrease. If friction coefficient is k, we know that,

$$V_{re_1} = k.V_{ri_1}$$

The steam leaving the first set of moving blades will enter into fixed blades. Due to friction relative velocity will decreased.

$$\therefore \quad V_{ri_2} = k.V_{re_2}$$

$$\text{and} \quad V_{re_2} = k.V_{ri_2}$$

(3) Angle β_1 is exit angle of steam leaving first moving blade. It is same as blade angle at inlet for the first set of fixed blades.

The velocity diagrams, for first set of moving blades is given by Fig. 20.17(a)

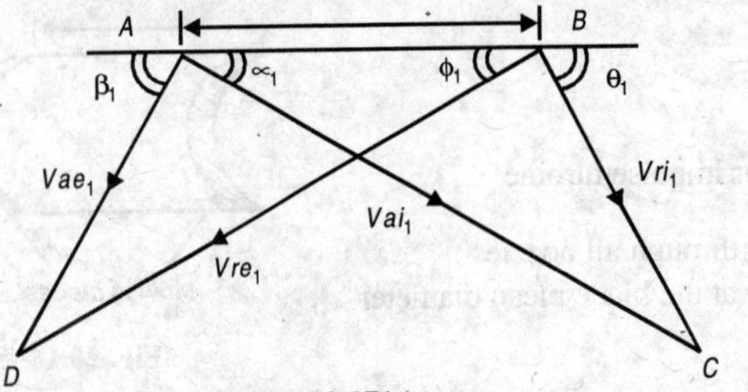

Fig. 20.17(a)

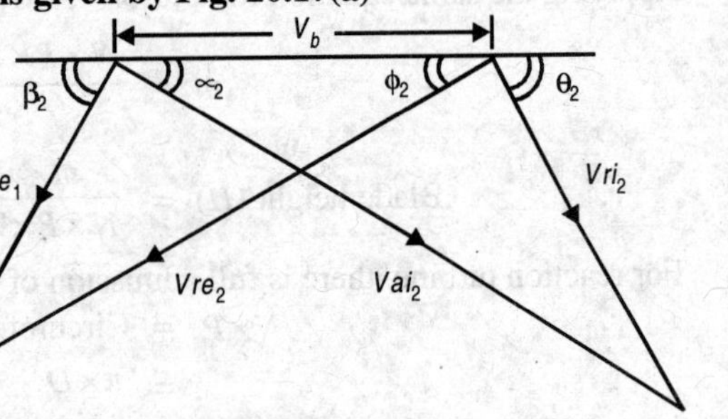

Fig. 20.17(b)

Blade velocity V_b is same in all sets of moving blades.

Total work done is work done in the first row of moving blades plus in the second set of moving blades.

The velocity diagram for second stage is shown in Fig. 20.17(b).

$$\therefore \quad \text{Total work } (W) = W_1 + W_2$$

$$W_1 = \text{Work done in the first set of moving blades.}$$

$$= V_b\left(V_{wi_1} + V_{we_1}\right) \times m = m \,.\, V_b \,\Delta V_{w_2}$$

and

$$W_2 = V_b\left(V_{wi_2} + V_{w_2}\right) \times m = m \,.\, V_b \,\Delta V_{w_2}$$

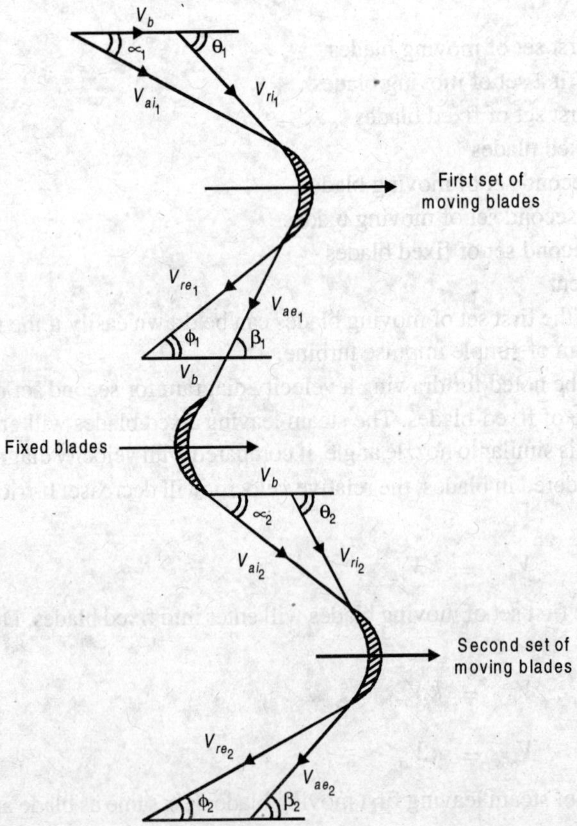

Fig. 20.18

$$\therefore \quad W = m \,.\, V_b \left(\Delta V_{w_1} + \Delta V_{w_2}\right)$$

$$\therefore \quad \text{Blade Efficiency} = \frac{m \,.\, V_b \left(\Delta V_{w_1} + \Delta V_{w_2}\right)}{\frac{1}{2} m \,.\, V_{ai_1}^2}$$

$$= \frac{2 \,.\, V_b \left(\Delta V_{w_1} + \Delta V_{w_2}\right)}{V_{ai_1}^2}$$

$$\text{Stage Efficiency} = \frac{m \,.\, V_b \left(\Delta V_{w_1} + \Delta V_{w_2}\right)}{\Delta h \times m}$$

$$= \frac{V_b \left(\Delta V_{w_1} + \Delta V_{w_2}\right)}{\Delta h}$$

where, Δh = specific enthalpy drop in nozzle.

20.13 CALCULATIONS OF MAXIMUM EFFICIENCY

$$W = W_1 + W_2$$

For unit mass,

$$W_1 = V_b \left(V_{wi_1} + V_{we_1}\right)$$

$$= V_b \left(V_{ai_1} \cos \alpha_1 + V_{ae1} \cos \beta_1\right)$$

$$= V_b \left(V_{ri_1} \cos \theta_1 + V_{re_1} \cos \phi_1\right)$$

Similarly,

$$W_2 = V_b (V_{ai_2} \cos \alpha_2 + V_{ai_2} \cos \beta_2)$$

$$= V_b (V_{ri_2} \cos \theta_2 + V_{re_2} \cos \phi_2)$$

If the moving blades and fixed blades are symmetric, $\theta_1 = \phi_1, \theta_2 = \phi_2$ and $P_1 = \alpha_2$.

If the friction in moving blades and fixed blades is neglected, $V_{ri1} = V_{re1}$ and $V_{ri_2} = V_{re_2}, V_{ae_1} = V_{ai_2}$.

For maximum efficiency, energy loss to exit steam in the last stage has to be minimum. So angle has to be equal to 90°. Therefore, consider $\beta_2 = 90°$.

$$\therefore \quad W_1 = V_b \left(V_{ri_1} \cos \theta_1 + V_{re_1} \cos \phi_1\right)$$

$$= V_b \,.\, V_{ri_1} \cos \theta_1 \left(1 + \frac{V_{re_1}}{V_{ri_1}} \cdot \frac{\cos \phi_1}{\cos \theta_1}\right)$$

But $V_{rei} = V_{ri_1}$ and $\theta_1 = \phi_1$

$$\therefore \quad W_1 = V_b \,.\left(V_{ri_1} \cos \theta_1\right) \times 2$$

$$= 2 \cdot V_b \cdot \left(V_{ai_1} \cos \alpha_1 - V_b\right)$$

Similarly,

$$W_2 = 2 \,.\, V_b \,.\left(V_{ri_2} \cos \theta_2\right)$$

$$= 2 \cdot V_b \cdot \left(V_{ai_2} \cos \alpha_2 - V_b\right)$$

But,

$$\alpha_2 = \beta_2 \text{ and } V_{ai_2} = V_{ae_1}$$

$$\therefore \quad W_2 = 2 - V_b\left(V_{ae_1} \cos\beta_1 - V_b\right)$$

But $V_{ae_1} \cos\beta_1 = \left(V_{re_1} \cos\phi_1 - 2\,V_b\right)$... see velocity diagram for first set of moving blades.

$$\therefore \quad W_2 = 2\,.\,V_b\left(V_{re_1} \cos\phi_1 - 2\,V_b\right)$$

But

$$V_{ri_1} = V_{re_1} \text{ and } \phi_1 = \theta_1$$

$$\therefore \quad W_2 = 2\,.\,V_b\left(V_{ri_1} \cos\theta_1 - 2V_b\right)$$

$$V_{ri_1} \cos\theta_1 = V_{ai_1} \cos\alpha_1 - V_b$$

$$\therefore \quad W_2 = 2 - V_b\left(V_{ai_1} \cos\alpha_1 - 3V_b\right)$$

$$\text{Total work done } (W) = W_1 + W_2$$

$$= 2\,.\,V_b\left(V_{ai1} \cos\alpha_1 - V_b\right) + 2\,.\,V_b\left(V_{ai_1} \cos\alpha_1 - 3V_b\right)$$

$$= 2\,.\,V_b\left(2V_{ai_1} \cos\alpha_1 - 4V_b\right)$$

$$\text{Diagram efficiency} = \frac{\text{Total work done}}{\text{Kinetic energy of steam}}$$

$$= \frac{2\,.\,V_b\left(2V_{ai_1} \cos\alpha_1 - 4V_b\right)}{\frac{1}{2}\cdot V_{ai}^2}$$

$$= 4\cdot\frac{V_b}{V_{ai}}\cdot\left(2\cos\alpha_1 - \frac{4V_b}{V_{ai_1}}\right)$$

But,

$$\frac{V_b}{V_{ai}} = \rho = \text{Blade speed ratio.}$$

$$\therefore \quad \text{Diagram efficiency} = 4\cdot\rho\cdot(2\cos\alpha_1 - 4\,\rho) = 8\rho\cos\alpha_1 - 16\rho^2.$$

For fixed nozzle angle (α_1), the maximum value of diagram efficiency can be obtained by differentiating efficiency w.r.t. ρ and equating to zero.

$$\therefore \quad \frac{d(\eta_{\text{dia}})}{d\rho} = (8\cos\alpha_1 - 32\rho) = 0$$

$$\therefore \quad \rho = \frac{\cos\alpha_1}{4} \text{ – Condition for maximum efficiency.}$$

$$\text{Maximum diagram efficiency} = 4\times\frac{\cos\alpha_1}{4}\times\left(2\cos\alpha_1 - \frac{4\cos\alpha_1}{4}\right)$$

$$(\eta_{\text{dia}})_{\max} = \cos^2\alpha_1$$

The work done for maximum diagram efficiency

$$= 2\times V_b\left(2\,V_{ai_1}\cos\alpha_1 - 4\,V_b\right)$$

$$\rho = \frac{V_b}{V_{ai_1}} = \frac{\cos\alpha_1}{4}$$

$$\therefore \quad V_{ai_1} = \frac{4\cdot V_b}{\cos\alpha_1}$$

$$\therefore \quad W = 2\times V_b\left(\frac{2\times 4\times V_b}{\cos\alpha_1}\cdot\cos\alpha_1 - 4\,V_b\right)$$

$$= 2\cdot V_b\left(8\,V_b - 4\,V_b\right) = 8\cdot V_{b2}$$

The work done in first stage $= 6V_b^2$.

The work done in the second stage $= 2\,V_b^2$.

The above analysis is for two rows of moving blades. Similarly analysis can be done for more number of rows. If there are n number of rows of moving blades, then optimum blade speeds ratio,

$$\rho = \frac{\cos\alpha_1}{2n}$$

and work done in the last row is $\frac{1}{2^n}$ of total work

i.e. $$\rho_{\text{optimum}} = \frac{\cos\alpha_1}{2n}$$

and work done by last row $= \frac{1}{2^n}\times(\text{Total work done})$

It means that in two row turbine, the work obtained in second row is one-fourth of total work and if there are three rows of moving blades, work obtained in the row is one eighth of total work. Thus we see that the increase in number of rows the work obtained in the last decreases. So the number of rows are limited.

20.14 STEAM TURBINES PERFORMANCE

In steam turbines when steam flows over the blades, the friction takes place. It results into reduction in the useful enthalpy drop. So the useful or effective enthalpy-drop is less than isentropic enthalpy drop. The energy lost in friction reheats the steam and the quality of steam at exit from the stage gets improved.

Figure 20.19 shows $h_A - h_{\beta_1}$ isentropic enthalpy drop per stage for a pressure range form P_1 to P_2. The useful of enthalpy drop per stage is $\left(h_{A_1} - h_{c_1}\right)$ or $\left(h_{A_1} - h_{A_1}\right)C_1$.

So the stage efficiency is $\eta_{\text{stage}} = \dfrac{h_{A_1} - h_{C_1}}{h_{A_1} - h_{B_1}}$

Thus the quality of steam at exit from the first stage of a pressure compounded or reaction turbine is obtained as at point A_2 at pressure P_2 and enthalpy h_{c_1} or h_{A_2}. All such state points can be obtained for successive stages. The line joining such state points (i.e. A_1, A_2, A_3, A_4 and A_5) is called the condition-line.

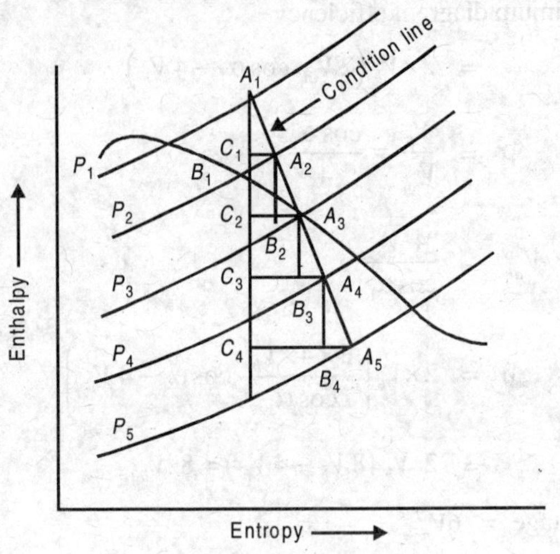

Fig. 20.19

Now,

$$\Sigma(h_A - h\beta) = \left[\left(h_{A_1} - h_{B_1}\right) + \left(h_{A_2} - h_{B_2}\right) + \left(h_{A_3} - h_{B_3}\right) + \left(h_{A_4} - h_{B_4}\right)\right]$$

The reheat-factor is defined as,

$$\text{Reheat factor (R.F.)} = \frac{\text{Cumulative enthalpy drop}}{\text{Isentropic enthalpy drop}}$$

$$= \frac{\Sigma(h_A - h_B)}{h_{A_1} - h_D}$$

The reheat factor is always greater than one. It depends on stage efficiency, initial temperature and pressure and exit pressure. The value normally lies from1.02 to 1.06

Now,

$$\eta_{\text{normal}} = \frac{\Sigma(h_A - h_C)}{\left(h_{A_1} - h_D\right)}$$

But

$$\left(h_{A_1} - h_D\right) = \frac{\Sigma(h_A - h_B)}{\text{R.F.}}$$

$$\therefore \quad \eta_{\text{internal}} = \frac{\Sigma(h_A - h_C)}{\Sigma(h_A - h_B)} \times \text{R.F.}$$

$$\eta_{\text{internal}} = \eta_{\text{stage}} \times \text{R.F.}$$

20.15 VELOCITY DIAGRAM FOR REACTION TURBINE

In case of a reaction turbine, the expansion of steam takes place in fixed blades as well as in moving blades. The figure shows the expansion of steam in a reaction turbine. The steam enters the turbine at pressure P_o. In fixed blades it expands to pressure P_1. So $(h_o - h_1)$ is enthalpy drop in fixed blades. Then steam flows over the moving blades and expands from pressure P_1 to pressure P_2. The enthalpy drop in moving blades is $(h_1 - h_2)$. Due to expansion in moving blades, the relative velocity of steam continuously increases (i.e. $V_{r_e} > V_{r_i}$).

The velocity diagram for a reaction turbine is as shown in figure.

(i) Work-done $= m \times V_b \left(V_{wi} + V_{we}\right)$

(ii)
$$\text{Stage efficiency} = \frac{\text{Work done in a stage}}{\text{Enthalpy drop in a stage}}$$

$$= \frac{m \times V_b \left(V_{wi} + V_{we}\right)}{m \times \Delta h}$$

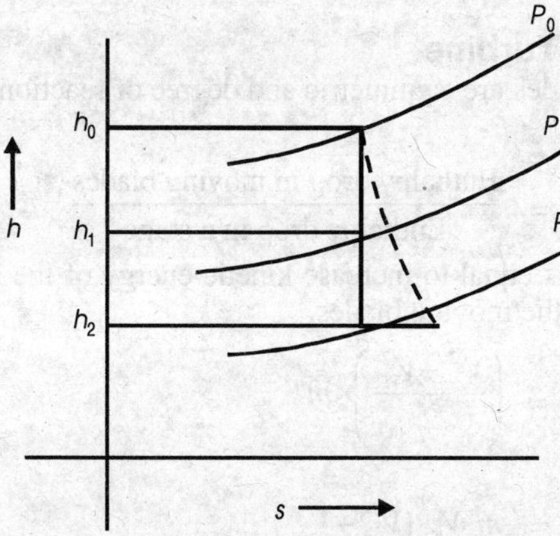

Fig. 20.20

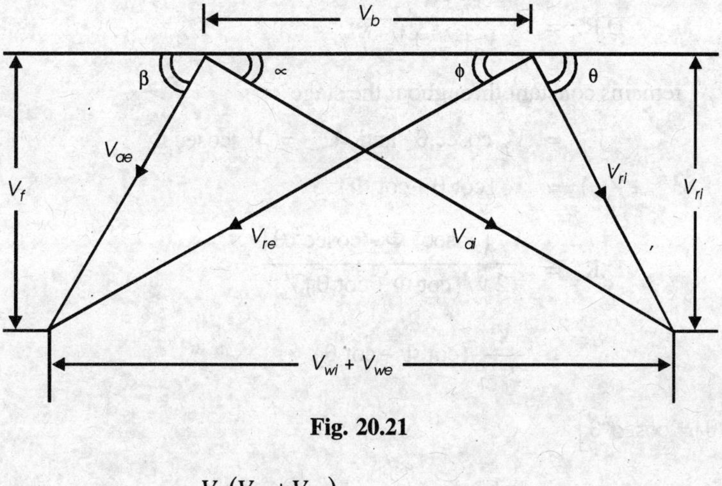

Fig. 20.21

$$= \frac{V_b \left(V_{wi} + V_{we}\right)}{\Delta h}$$

Where Δh – Specific enthalpy drop in a stage.

(iii)
$$\text{Blade efficiency} = \frac{\text{Work done}}{\text{Energy supplied}}$$

$$= \frac{m\, V_b \left(V_{wi} + V_{we}\right)}{\dfrac{V_{ai}^2}{2} + \left(\dfrac{V_{re}^2 - V_{ri}^2}{2}\right)}$$

20.15.1 Degree of Reaction

The term 'degree of reaction' is used for reaction turbine. It measures the proportion of the work-done by the effect (due to pressure drop in moving blades)

The degree of reaction is defined as,

$$\text{D.R.} = \frac{\text{Enthalpy drop in moving blades}}{\text{Total enthalpy drop in a stage}}$$

When D.R. = 0, it is an impulse turbine, since no enthalpy drop is taking place in moving blades.
When D.R. = 1, it is 100% reaction turbine.

20.15.2 Parson's Reaction Turbine

For Parson's reaction turbine, the blades are symmetric and degree of reaction is 50%.

We know that,

$$\text{D.R.} = \frac{\text{Enthalpy drop in moving blades}}{\text{Enthalpy drop in a stage}}$$

Enthalpy drop in moving blades is equal to increase kinetic energy of the steam to corresponding to relative velocity while, the steam flows over the moving blades.

$$\therefore \quad (\Delta h)_{\text{moving}} = \left(\frac{V_{re}^2 - V_{ri}^2}{2}\right) \times m$$

The enthalpy in a stage

$$(\Delta h)_{\text{fixed}} = m \cdot V_b \cdot (V_{wi} + V_{we})$$

$$\therefore \quad \text{D.R.} = \frac{V_{re}^2 - V_{ri}^2}{2 \cdot V_b (V_{wi} + V_{we})}$$

Velocity of flow (V_f) remains constant throughout the stage

$$V_{ri} = V_f \text{ cosec } \theta \quad \text{and} \quad V_{re} = V_f \text{ cosec } \Phi$$

$$(V_{wi} + V_{we}) = V_f (\cot \theta + \cot \Phi)$$

$$\therefore \quad \text{D.R.} = \frac{V_f^2 (\text{cosec}^2 \Phi - \text{cosec}^2 \theta)}{2 v_f (\cot \Phi + \cot \theta) V_b}$$

$$= \frac{V_f}{2 \cdot V_b} (\cot \Phi - \cot \theta)$$

$$\left[\text{since} \left(1 + \cot^2 \theta\right) = \text{cosec}^2 \theta\right]$$

For 50 % reaction turbine, $\text{D.R.} = \frac{1}{2}$

$$\therefore \quad \frac{1}{2} = \frac{V_f}{2 \cdot V_b} (\cot \Phi - \cot \theta)$$

$$V_b = V_f (\cot \Phi - \cot \theta) \qquad \text{....(I)}$$

From velocity triangles of velocity diagram,

$$V_b = V_f (\cot \Phi - \cot \beta) \qquad \text{....(II)}$$

and

$$V_b = V_f(\cot\alpha - \cot\theta) \qquad ....(III)$$

Comparing Eqs. (I), (II) and (III), we get,

$$\alpha = \Phi \text{ and } \theta = \beta$$

Thus for 50 % reaction turbine i.e. for Parson's turbine, the fixed and moving blades have the same shape. This gives rise to symmetric velocity diagram.

20.15.3 Maximum Efficiency of Reaction Turbine

For 50 % reaction turbine, the velocity diagram is symmetric i.e. inlet velocity triangle is same as outlet velocity triangle.

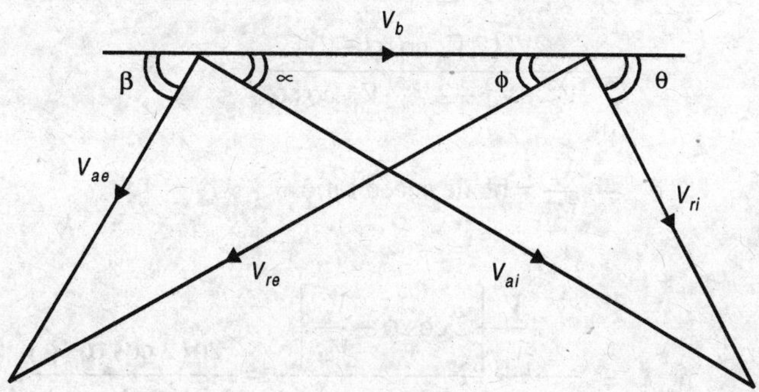

Fig. 20.22

$$\text{Kinetic energy obtained in fixed blades} = \frac{V_{ai}^2}{2}$$

$$\text{Kinetic energy obtained moving blades} = \frac{V_{re}^2 - V_{ri}^2}{2}$$

$$\text{Total energy supplied} = \frac{V_{ai}^2}{2} + \frac{V_{re}^2 - V_{ri}^2}{2}$$

From symmetric velocity triangles, we get $V_{re} = V_{ai}$

$$\therefore \quad \text{Total energy supplied} = \frac{V_{ai}^2}{2} + \frac{V_{ai}^2}{2} - \frac{V_{ri}^2}{2} = V_{ai}^2 - \frac{V_{ri}^2}{2}$$

From inlet velocity triangle,

$$V_{ri}^2 = V_{ai}^2 + V_b^2 - 2\,V_{ai} - V_b \cdot \cos\alpha$$

$$\therefore \quad \text{Total energy supplied} = V_{ai}^2 - \left(\frac{V_{ai}^2 + V_b^2 - 2\,V_{ai}\,V_b\cos\alpha}{2}\right)$$

$$= \left(\frac{V_{ai}^2 - V_b^2 + 2\,V_{ai} \cdot V_b\cos\alpha}{2}\right)$$

Work done per kg of steam $= V_b\left(V_{wi}+V_{we}\right)$

$$= V_b\left(V_{ai}\cos\alpha+V_{re}\cos\Phi-V_b\right)$$

But $\alpha=\Phi$

and $V_{ai}=V_{re}$

$\therefore$ Work done per kg of steam $= V_b\left(2V_{ai}\cos\alpha-V_b\right)$

Stage efficiency is given by,

$$\eta_{stage} = \frac{\text{Work done by the stage}}{\text{Energy supplied the stage}}$$

$$= \frac{2V_b\left(2V_{ai}\cos\alpha-V_b\right)}{V_{ai}^2-V_b^2+2V_{ai}\cdot V_b\cdot\cos\alpha}$$

Let us define, $\rho = \dfrac{V_b}{V_{ai}}$ = blade speed ratio

$$\therefore \quad \eta_{stage} = \frac{2\dfrac{V_b}{V_{ai}}\left[2\cos\alpha-\dfrac{V_b}{V_{ai}}\right]}{1-\left(\dfrac{V_b}{V_{ai}}\right)^2+2\left(\dfrac{V_b}{V_{ai}}\right)\cos\alpha} = \frac{2\rho\left(2\cos\alpha-\rho\right)}{1-\rho^2+2\rho\cos\alpha}$$

$$= \frac{2\left[\left(\rho\cos\alpha-\rho^2+1\right)-1\right]}{\left(1-\rho^2+2\rho\cos\alpha\right)}$$

$$= 2-\frac{2}{1-\rho^2+2\rho\cos\alpha}$$

The stage efficiency is maximum, when $\left(1-\rho^2+2\rho\cos\alpha\right)$ is maximum. So the required condition is,

$$\frac{d}{d\rho}\left(1-\rho^2+2\rho\cos\alpha\right) = 0$$

$$\therefore \quad -2\rho+2\cos\alpha = 0$$

$$\therefore \quad \rho = \cos\alpha$$

This is the condition for maximum stage efficiency

$$\therefore \quad \left(\eta_{stage}\right)_{maximum} = \frac{2\cos\alpha\left(2\cos\alpha-\cos\alpha\right)}{1-\cos^2\alpha+2\cos\alpha\cdot\cos\alpha}$$

$$= \frac{2\cos^2\alpha}{1+\cos^2\alpha}$$

SOLVED EXAMPLES

Example 20.1. *A De-Laval turbine has a nozzle angle of 20° and blade inlet angle 30° and blade outlet angle 25°. Enthalpy drop available is 350 kJ/kg. Nozzle efficiency is 82 % and blade friction coefficient is 0.8.*

Determine,

1. Absolute velocity of steam at inlet (V_{ai})

2. Blade velocity (V_b)

3. Absolute velocity at outlet (V_{ae})

Solution

Given

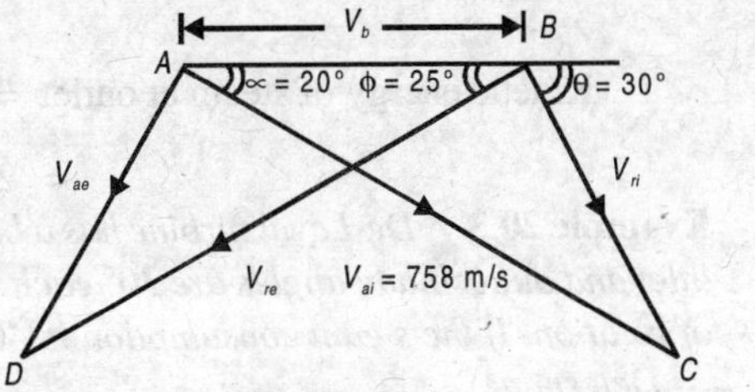

Fig. Ex. 20.1

$$\alpha = 20°$$

$$\theta = 30°, \Phi = 25°$$

$$\Delta h = 350 \text{ kJ/kg}, \frac{V_{re}}{V_{ri}} = 0.8$$

$$\eta_{nozzle} = 0.82$$

$$\eta_{nozzle} = \frac{V_{ai}^2}{2 \times \Delta h}$$

$$V_{ai} = \sqrt{2 \times \eta_{nozzle} \times \Delta h} = \sqrt{2 \times 0.82 \times 350 \times 10^3} = 783 \text{ m/s}$$

Draw the velocity diagram to the scale

From velocity diagram,

$$V_{ri} = 520 \text{ m/s}$$

$$V_{re} = 0.8 \times V_{ri} = 416 \text{ m/s}$$

Draw $BD = V_{re}$ to the scale so the velocity diagram is completed.

1. $V_{ai} = 578$ m/s

2. $V_b = AB = 265$ m/s

3. $V_{ae} = AD = 210$ m/s

Example 20.2. *A simple impulse turbine has a blade speed of 350 m/s and blade speed to steam velocity (inlet) ratio is 0.45. Nozzles are inclined at 20° to plane of rotation and steam leaves the stage at an angle of 70° to the plane of rotation.*

Determine,

1. Blade inlet angle (θ)

2. Kinetic energy of steam at outlet.

Solution

Given

$$V_b = 350 \text{ m/s}$$

$$\frac{V_b}{V_{ai}} = 0.45$$

$$\alpha = 20°, \beta = 70°$$

$$V_{ai} = \frac{V_b}{0.45} = 777 \text{ m/s}$$

Frictional losses are not given. Therefore assume it nil

$$\therefore \quad V_{re} = V_{ri}$$

Draw the velocity diagram.
From velocity diagram,

$$\theta = 35.5°$$

$$V_{ae} = 207 \text{ m/s}$$

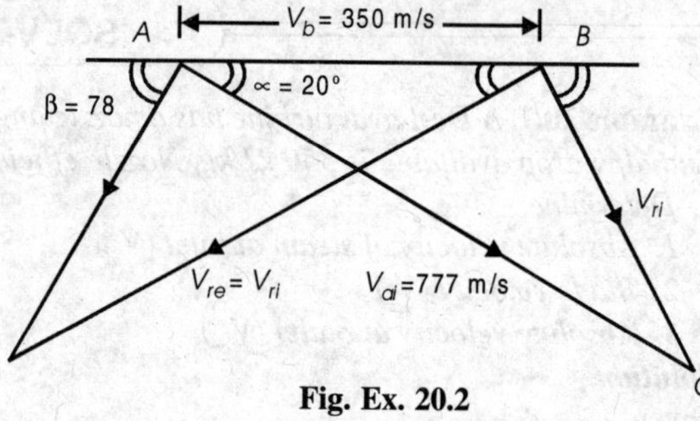

Fig. Ex. 20.2

$$\text{Kinetic energy of steam at outlet} = \frac{V_{ae}^2}{2} = \frac{(207)^2}{2 \times 1000}$$

$$= 21.4 \text{ kJ/kg}$$

Example 20.3. *A De-Laval turbine has a blade speed of 200 m/s and relative velocity of steam at inlet is 300 m/s. The inlet and outlet blade angles are 30° each, while the steam leaves the wheel inclined at an angle of 85° to the plane of rotation. If the steam consumption is 4000 kJ/hr,*

Determine,

1. *Absolute velocity of steam at inlet,*
2. *Blade friction coefficient,*
3. *Power developed.*

Solution

Given,

$$V_b = 200 \text{ m/s}$$

$$V_{ri} = 300 \text{ m/s}$$

$$\theta = \Phi = 30°$$

$$\beta = 85°$$

$$m = 4000 \text{ kg/hr}$$

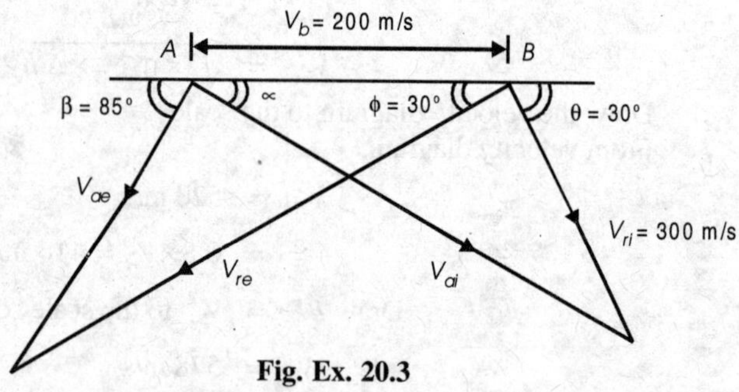

Fig. Ex. 20.3

Draw the velocity diagram
From velocity diagram,

$$V_{ai} = 487 \text{ m/s}$$

$$V_{re} = 247 \text{ m/s}$$

$$\Delta V_w = (V_{wi} + V_{we}) = 473 \text{ m/s}$$

$$\text{Blade velocity coefficient} = \frac{V_{re}}{V_{ri}} = \frac{247}{300} = 0.823$$

$$\text{Power development} = m \times (\Delta V_w) \times V_b$$

$$= \frac{4000}{3600} \times \frac{473 \times 200}{10^3} = 105 \text{ kW}$$

Example 20.4. *A De-Laval turbine is supplied with steam at pressure of bar and 250°C and back pressure is 0.12 bar.*

The nozzle efficiency is 0.9, blade velocity coefficient is 0.8. Mechanic efficiency of turbine is 0.9. The nozzle angle is 20°. The blades are symmetric and blade angle is 30°.

Draw the velocity diagram and calculate

1. *Speed of the rotor if mean diameter of wheel is 75 cm*
2. *Steam consumption per kW-hr.*
3. *Blade efficiency*
4. *Stage efficiency*

5. *Axial thrust per kg of steam*
6. *Percentage loss in exit*

Solution

Given,

$P_1 = 15$ bar and $T_1 = 250°C$, $P_2 = 0.12$ bar

$$\eta_{nozzle} = 0.9, \frac{V_{re}}{V_{ri}} = 0.8, \alpha = 20°$$

$$\eta_{mech} = 0.9$$

$$\theta = \Phi = 30°$$

From Mollier-chart,

For $P_1 = 15$ bar and $T_1 = 250°C$. Mark point 1. Draw a vertical-downward line from point 1(since expansion in nozzle is isentropic). At exit pressure is 0.12 bar. So get point 2.

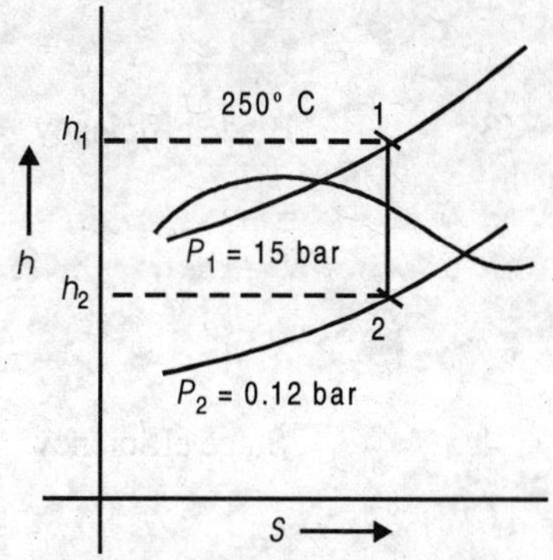

Fig. Ex. 20.4(a)

$$h = \Delta(h_1 - h_2) = 764.94 \text{ kJ/kg}$$

$$\eta_{nozzle} = \frac{V_{ai}^2}{2 \times \Delta h}$$

$$\therefore \quad V_{ai} = \sqrt{2 \times \eta_{nozzle} \times \Delta h}$$

$$V_{ai} = \sqrt{2 \times 0.9 \times 764.94 \times 1000}$$

$$= 1173.4 \text{ m/s}$$

Draw a velocity diagram,

$$k = \frac{V_{re}}{V_{ri}} = 0.8$$

$$\therefore \quad V_{re} = 0.8\, V_{ri}$$

$$\therefore \quad BD = 0.8 \times BC$$

Use this while drawing the velocity diagram.

From velocity diagram,

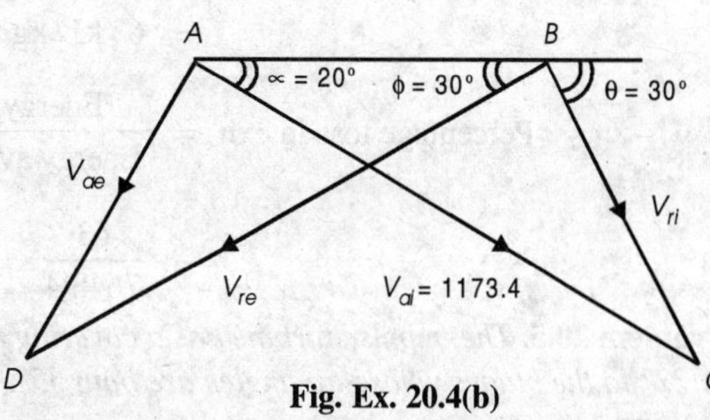

Fig. Ex. 20.4(b)

$$V_b = 407 \text{ m/s}, (V_{wi} + V_{we}) = 1252 \text{ m/s}$$

$$V_{fi} = 402 \text{ m/s}, V_{ae} = 355 \text{ m/s}$$

$$V_{fe} = 322 \text{ m/s}$$

1. Rotor speed : $V_b = \frac{\pi D N}{60}$, where D = mean rotor diameter.

$$\therefore \quad N = \frac{60 \times V_b}{\pi \times D} = \frac{60 \times 407}{\pi \times 0.75} = 10364 \text{ rpm}$$

2. Work done $= m \times V_b \times \Delta V_w$

$$= \frac{407 \times 1252}{1000} = 509.564 \text{ kJ/kg}$$

$$\text{Shaft output} = (\text{work done}) \times (\text{mechanical efficiency})$$
$$= 509.564 \times 0.9$$
$$= 458.6 \text{ kJ/kg}$$

$$\text{Steam consumption per kW-hr} = \frac{1}{458.6} \times 3600$$
$$= 7.85 \text{ kg/kW-hr}$$

3. $$\text{Blade efficiency} = \frac{\text{Work done by the blade}}{\text{Energy supplied to the blade}}$$
$$= \frac{509.564}{\frac{(1173.4)^2}{2 \times 1000}} \times 100 = 74\ \%$$

4. $$\text{Stage efficiency} = \eta_{\text{blade}} \times \eta_{\text{nozzle}} = (0.74 \times 0.9) \times 100$$
$$= 66.6\ \%$$

5. $$\text{Axial thrust per kg} = (V_{fe} - V_{fi}) = -80 \text{ N/kg}$$

6. $$\text{Energy lost in exit} = \frac{V_{a^2e}}{2} = \frac{(355)^2}{2 \times 1000}$$
$$= 63 \text{ kJ/kg}$$

$$\therefore \quad \text{Percentage loss in exit} = \frac{\text{Energy lost in exit}}{\text{Energy available i.e. } \Delta h}$$
$$= \frac{63}{764.94} \times 100 = 8.2\ \%$$

Problem 20.5. *The impulse turbine has a covering nozzle and one ring of moving blades. The nozzles are inclined at 22° to the blades whose tip angles are both 35°. The velocity of steam leaving the nozzle is 660 m/s.*

Find,

1. *Blade speed (V_b) and diagram efficiency assuming no friction along the turbine blades.*
2. *If the relative velocity of steam to the blade is reduced by 15 % due to friction, find blade efficiency and end thrust.*

Assume power developed = 170 kW

Solution

Given,

$$\alpha = 22°$$
$$\therefore \quad \theta = \Phi = 35°$$
$$V_{ai} = 660 \text{ m/s}$$

1. There is no friction

$$\therefore \quad V_{re} = V_{ri}$$

From velocity diagram, $V_b = 260$ m/s

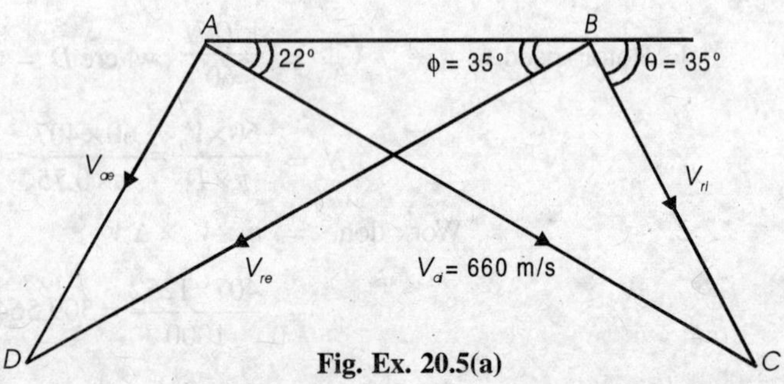

Fig. Ex. 20.5(a)

$$\Delta V_w = (V_{wi} + V_{we}) = 715 \text{ m/s}$$

$$\text{Blade efficiency} = \frac{2 \times V_b \times \Delta V_w}{V_{ai}^2}$$

$$= \frac{2 \times 260 \times 715}{(660)^2} \times 100 = 85\ \%$$

2. When the blade friction is considered

$$k = \frac{V_{re}}{V_{ri}} = 0.85$$

Draw the velocity diagram,
From the velocity diagram,

$$\Delta V_w = 655 \text{ m/s}$$

$$V_b = 260 \text{ m/s}$$

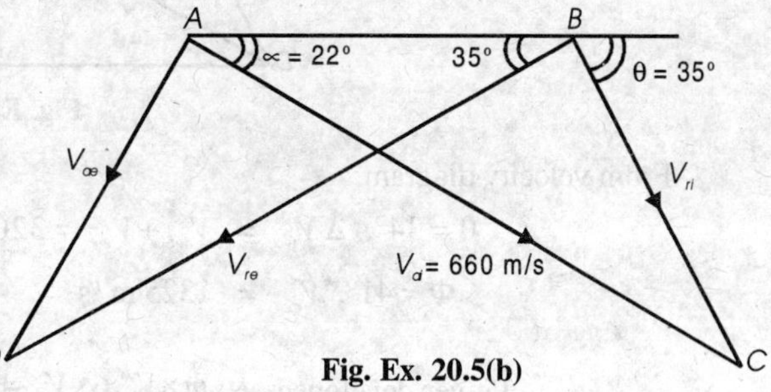

Fig. Ex. 20.5(b)

$$\text{Blade efficiency} = \frac{2 \times \Delta V_w \times V_b}{V_{ai}^2} = \frac{2 \times 655 \times 260}{(660)^2} \times 100$$

$$= 78.25\ \%$$

$$\text{Power developed} = m \times \Delta V_w \times V_b = 170 \text{ kW}$$

$$\therefore \quad m = \frac{170 \times 1000}{655 \times 260} = 10 \text{ m/s}$$

$$\text{End thrust} = m(V_{fi} - V_{fe})$$

$$= 10 \times (250 - 205)$$

$$= 45 \text{ N}$$

Example 20.6. *In a simple impulse turbine, nozzles are inclined at an angle of 20º to the plane of rotation. Steam leaves the nozzle with velocity of 375 m/s. The blade speed is 165 m/s. Find the suitable inlet and outlet blade angles in order that there should be no axial thrust and allowing velocity reduction of 15 % , while steam flows over the blade. Determine also the power developed for now of 5 kg/s and kinetic energy of steam finally leaving the wheel.*

Solution

Given,

$$\alpha = 20°,\ V_b = 165 \text{ m/s}$$

$$V_{ai} = 375 \text{ m/s},\ V_{fe} = V_{fi}$$

$$\frac{V_{re}}{V_{ri}} = 0.85,\ \text{m} = 5 \text{ kg/s}$$

$$BD = BC \times 0.85$$

For no-axial thrust $V_{fe} = V_{fi}$ i.e. point C and D should be on the same horizontal line parallel to AB so first drum a horizontal line from point C. Then mark an arc on that horizontal line, taking distance ($BC \times 0.85$) from point B. So point D is obtained and velocity diagram is complete.

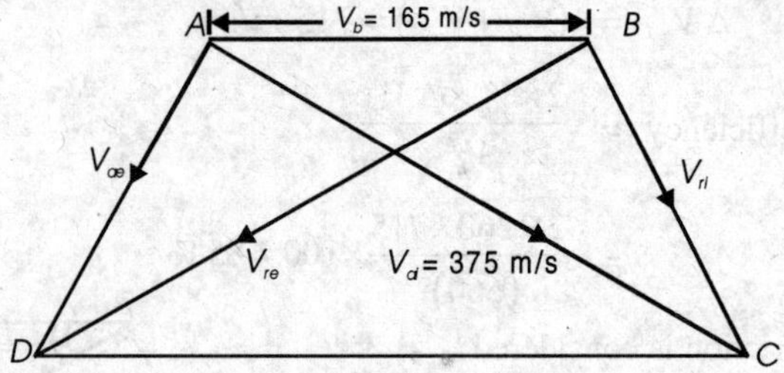

Fig. Ex. 20.6

From velocity diagram,

$$\theta = 34°,\ \Delta V_w = V_{wi} + V_{we} = 320 \text{ m/s}$$

$$\Phi = 41°,\ V_{ae} = 1325 \text{ m/s}$$

$$\text{Power developed} = m \times V_b \times \Delta V_w = \frac{5 \times 165 \times 320}{1000} = 264 \text{ kW}$$

$$\text{Kinetic energy of steam at exit} = \frac{m \times V_{ae}^2}{2}$$

$$= \frac{5 \times (1325)^2}{2 \times 1000} = 43.5 \text{ kW}$$

Example 20.7. *The following particulars refer to a velocity compounded impulse turbine with two rows of moving blades and a fixed row of guide blade between them. The steam velocity leaving the nozzle is 1000 m/s and nozzle angle is 20°. The blade speed is 200 m/s. The moving blades for the first row are symmetric. The exit angle for fixed blade is 23°. The exit angle of second row of moving blades is 30°. The blade friction coefficient 0.9 for each row.*

Determine power developed in kW.

Solution

Given,

$$V_{ai_1} = 1000 \text{ m/s},\ V_b = 200 \text{ m/s}\ \theta_1 = \Phi_1$$

$$\alpha_1 = 20°\ \alpha_2 = 23°\ \Phi_2 = 30°$$

$$K = 0.9$$

The velocity diagram for first row of moving blades.

For drawing the velocity diagram for first row of moving blades, use

$$\theta_1 = \Phi_1$$

and

$$\frac{V_{re_1}}{V_{ri}} = 0.9$$

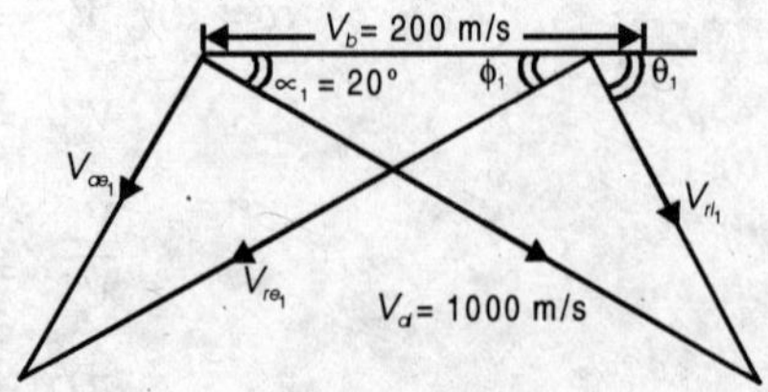

Fig. Ex. 20.7(a)

From the velocity diagram, we get,

$$V_{ri_1} = 813.33 \text{ m/s}$$

$$\therefore\ V_{re_1} = 0.9 \times 813.33 = 732 \text{ m/s}$$

$$V_{ae_1} = 556.66 \text{ m/s},\ \Delta_{w_1} = V_{wi_1} + V_{we_2} = 1433.33 \text{ m/sec}$$

Also for fixed blades, $K = 0.9$

$$\therefore \quad \frac{V_{ai_2}}{V_{ae_1}} = 0.9$$

$$\therefore \quad V_{ai_2} = 0.9 \times 556.66 = 501 \text{ m/sec}$$

Now draw the velocity diagram for second row of moving blades.
Use.

$$\frac{V_{re_2}}{V_{ri_2}} = K$$

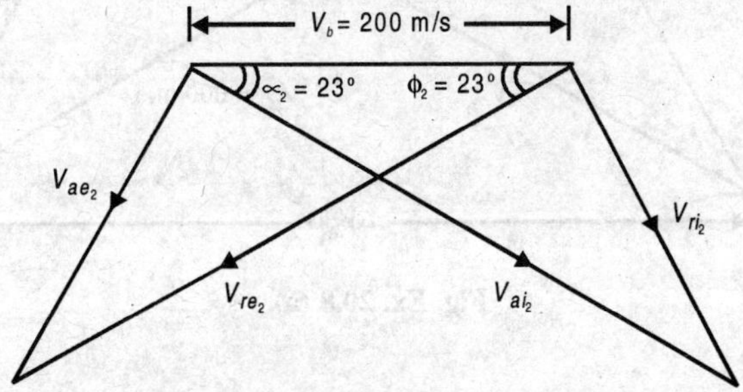

Fig. Ex. 20.7(b)

$$V_{ri_2} = 330 \text{ m/sec}$$

$$V_{re_2} = K \times V_{ri_2} = 0.9 \times 330 = 297 \text{ m/s}$$

$$\Delta V_{w_2} = \left(V_{wi_2} + V_{we_2}\right) = 520 \text{ m/sec}$$

$$\text{Assume mass flow rate } (m) = 1 \text{ kg/sec}$$

$$\text{Power developed} = m \cdot V_b \left(\Delta V_{w1} + \Delta V_{w_2}\right)$$

$$= 1 \times 200 \times (1433.33 + 520)$$

$$= 390660 \text{ W}$$

$$= 390.66 \text{ kW}$$

Example 20.8. *The steam in a two row velocity compounded impulse turbine leaves the nozzle at 600 m/s. The blade speed is 120 m/s. The stage consists of two rows of moving and one fixed blades between the two. The nozzle angle is 16°, while the discharge angles are 18° for the first row of moving blades, 21° for fixed blades and 35° for the second row of moving blades. Assume 10% loss of velocity during the passage of steam through each ring of blades.*

Determine

(a) Blade inlet angle for each row

(b) Driving force and axial thrust for stage in Newton

(c) Power developed in kW

(d) Diagram efficiency.

What would be maximum diagram efficiency for given steam inlet velocity and nozzle angle?

Solution

Given,

$$V_{ai_1} = 600 \text{ m/s}, \quad V_b = 120 \text{ m/s}, \ \alpha_1 = 16°$$

$\Phi_1 = 18°,\ \alpha_2 = 21°,\ \Phi_2 = 35°$

$k = 0.9$

Draw the velocity diagram for first row of moving blades

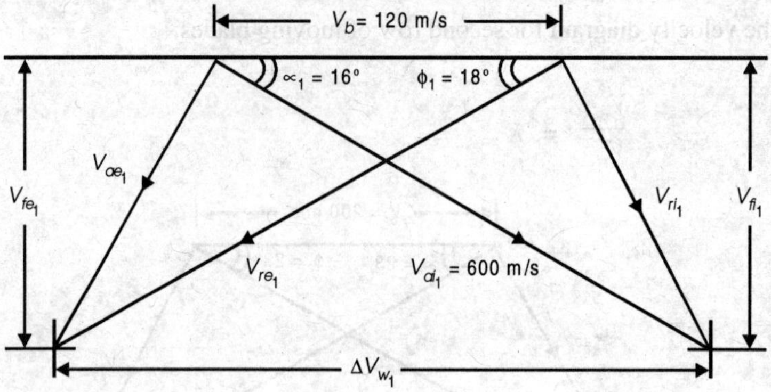

Fig. Ex. 20.8 (a)

Use

$$\frac{V_{re1}}{V_{ri1}} = 0.9$$

From velocity diagram,

$$\Delta V_{w_1} = (V_{wi_1} + V_{we_1}) = 872 \text{ m/sec}$$

$$\theta_1 = 21°$$

$$\Phi_1 = 35°$$

$$V_{fi_1} = 40 \times \frac{872}{21.8} = 1600 \text{ m/s}$$

$$V_{ae_1} = 7.9 \times \frac{872}{21.8} = 316 \text{ m/s}$$

$$V_{fe_1} = 35 \times \frac{872}{21.8} = 1400 \text{ m/s}$$

$$\frac{V_{ai_2}}{V_{ae_1}} = 0.9$$

$$V_{ai_2} = 0.9 \times 316 = 284.4 \text{ m/s}$$

Draw the velocity diagram for second row of moving blades.
From the velocity diagram,

$$\Delta V_{w_2} = (V_{wi_2} + V_{we_2}) = 272 \text{ m/s}$$

(i) Power developed $= m \times V_b (\Delta V_{w_1} + \Delta_{w_2})$

$$= 1 \times 120 \times (872 + 272) = 137.28 \text{ kW}$$

(ii) Axial thrust :
For first row of moving blades

$$F_1 = m\times(V_{fi_1} - V_{fe_1}) = 1(20) = 20 \text{ N}$$

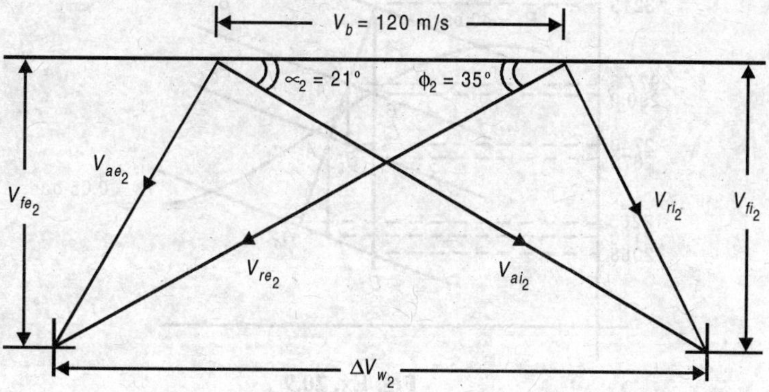

Fig. Ex. 20.8 (b)

For second row of moving blades,

$$F_2 = m\times(V_{fi_2} - V_{fe_2}) = 16 \text{ N}$$

$$\text{Total axial thrust} = 36 \text{ N}$$

(iii) Diagram efficiency

$$\eta_{dia} = \frac{2\times V_b \Delta V_w}{V_{ai_1}^2}$$

$$= \frac{2\times120\times(872+272)}{(600)^2} = 76.27\ \%$$

(iv) Driving force : $\left[(V_{wi_1} + V_{we_1}) + (V_{wi_2} + V_{we_2})\right]\times m$

$$= (872+272)\times1 = 1144 \text{ N}$$

(v) Maximum diagram efficiency :

$$\eta_{dia} = (1+KC)\times\frac{\cos^2\alpha}{2}$$

For maximum value, assume $K = 1$ and $C = 1$

$$\therefore \quad \eta_{dia} = \cos^2\alpha = 92.3\ \%$$

Example 20.9. *Steam is supplied to a 3-stage turbine at 40 bar and 400° and the exhaust to the condenser takes place at 50 milli-bar, with a wetness fraction of 12%. The work developed in three stages is 1:1:2. The condition line may be assumed as straight line.*

Determine,

1. *Condition at entry of each stage*
2. *Stage efficiencies*
3. *Reheat factor*
4. *Internal efficiency of turbine*

Solution

$$(h_{A_1} - h_{A_4}) = (3215 - 2265) = 950 \text{ kJ/kg.}$$

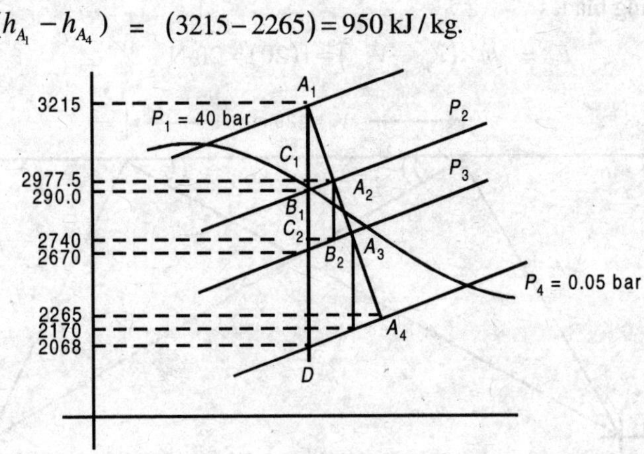

Fig. Ex. 20.9

$\therefore$ Enthalpy drop for first stage $= \dfrac{950}{4} = 237.5$ kJ/kg

Enthalpy drop for second stage = 237.5 kJ/kg

Enthalpy drop for third stage = 475 kJ/kg

1. Condition of steam at entry to each stage

From Mollier Chart

At first stage the steam is superheated with pressure = 40 bar and temp = 400°C.

Second stage, the steam is superheated

Pressure = 11.4 bar, Temp. = 265°C

For third stage, the steam is superheated and pressure = 2.5 bar

Temperature = 135°C

2. Stage efficiencies

$$\eta_{\text{stage}} = \frac{(h_A - h_C)}{(h_A - h_B)}$$

$$\eta_{1^{\text{st}}\text{stage}} = \frac{h_{A_1} - h_{C_1}}{h_{A_1} - h_{B_1}} = \frac{237.5}{315} = 0.754$$

$$= 75.4\ \%$$

$$\eta_{2^{\text{nd}}\text{stage}} = \frac{h_{A_2} - h_{C_2}}{h_{A_2} - h_{B_2}} = \frac{237.5}{307.5} = 0.7724$$

$$= 77.24\ \%$$

$$\eta_{3^{\text{rd}}\text{stage}} = \frac{h_{A_3} - h_{C_3}}{h_{A_3} - h_{B_3}} = \frac{475}{570} = 0.8333$$

$$= 83.33\ \%$$

3. Reheat factor $= \dfrac{\Sigma(h_A - h_B)}{h_{A_1} - h_D}$

$$= \frac{315+307.5+570}{3215-2068} = 1.04$$

$$\text{Internal efficiency} = \frac{\Sigma(h_A - h_C)}{h_{A_1} - h_D}$$

$$= \frac{950}{1147} = 0.8282$$

$$= 82.82\ \%$$

Example 20.10. *The three-stage steam turbine receives steam at 35 bar and 400°C. The steam is exhausted to condenser at 0.05 bar and dryness fraction = 0.9. The work developed by each stage is same.*

Determine,

(i) Condition of steam at entry to each stage

(ii) Stage efficiency

(iii) Reheat factor

(iv) Internal efficiency of turbine,

Solution

Given,

$P_1 = 35$ bar

$T_1 = 400°C$

$P_b = 0.05$ bar

$X = 0.9$

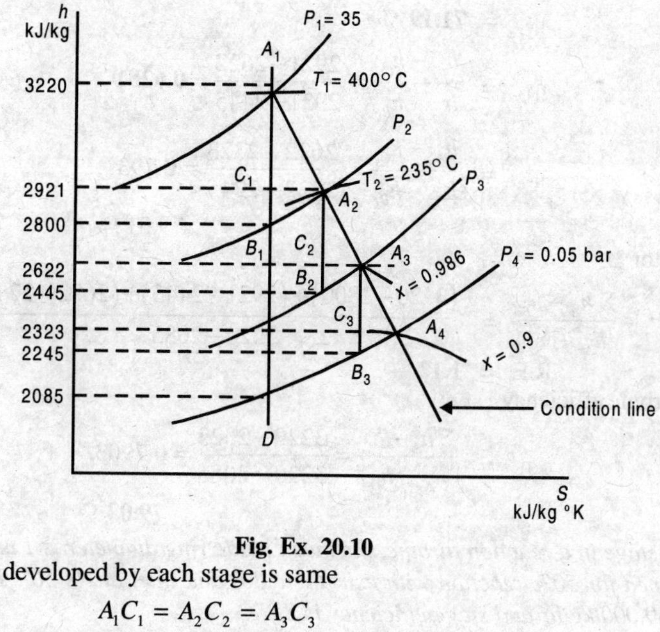

Fig. Ex. 20.10

It is given that work developed by each stage is same

$$\therefore \quad A_1C_1 = A_2C_2 = A_3C_3$$

$$\therefore \quad A_1C_1 + A_2C_2 + A_3C_3 = 8220 = 2323$$

$$\therefore \quad 3A_1C_1 = 897$$

$$\therefore \quad A_1C_1 = A_2C_2 = A_3C_3 = 299 \text{ kJ/kg}$$

$$\therefore \quad h_{A_2} = h_{C_1} = 3220 - 299 = 2921 \text{ kJ/kg}$$

$$\therefore \quad \text{From chart } h_{B_1} = 2800 \text{ kJ/kg}$$

$$h_{C_2} = h_{A_3} = 2921 - 299 = 2622 \text{ kJ/kg}$$

$$\therefore \quad \text{From chart } h_{B_2} = 2445 \text{ kJ/kg}$$

$$h_{A_4} = h_{C_3} = 2622 - 299 = 2323 \text{ kJ/kg}$$

$$\text{From chart } h_{B_3} = 2245 \text{ kJ/kg}$$

(a) Condition at entry to each stage

1st stage superheated $P_1 = 35$ bar, $T_1 = 400°C$

2nd stage superheated $T_2 = 235°C$

3rd stage wet steam $x = 0.986$

(b) The stage efficiency

$$\text{Stage efficiency } \eta_s = \frac{\Sigma(h_A - h_C)}{\Sigma(h_A - h_B)}$$

$$= \frac{h_{A_1} - h_{C_3}}{\Sigma(h_A - h_B)} = \frac{3220 - 2323}{(3220-2800)+(2921-2445)+(2622-2245)}$$

$$\therefore \quad \eta_s = 0.7046 \text{ or } 70.46\ \%$$

For 1st stage
$$\eta s_1 = \frac{h_{A_1} - h_{C_1}}{h_{A_1} - h_{B_1}} = \frac{3220 - 2921}{3220 - 2800}$$

$$\therefore \quad \eta s_1 = \mathbf{71.19\ \%}$$

For 2nd stage
$$\eta s_2 = \frac{h_{A_2} - h_{C_2}}{h_{A_2} - h_{B_2}} = \frac{2921 - 2622}{2921 - 2445} = \mathbf{0.6281}$$

For 3rd stage
$$\eta s_3 = \frac{h_{A_3} - h_{C_3}}{h_{A_3} - h_{B_3}} = \frac{2622 - 2323}{2622 - 2245} = 0.7931$$

or 79.31 %

(c) Reheat factor

$$\text{R.F.} = \frac{\Sigma\, h_A - h_B}{h_{A_1} - h_D} = \frac{(3220-2800)+(2921-2445)+(2622-2245)}{3220-2085}$$

$$\therefore \quad \text{R.F.} = \mathbf{1.12}$$

(d) Internal turbine efficiency

$$\eta_i = \frac{\Sigma\, h_A - h_C}{h_{A_1} - h_D} = \frac{3220 - 2323}{3220 - 2085} = 0.7903$$

or **79.03 %**

Example 20.11. *At a stage in a reaction turbine, the mean blade ring diameter is 1 m and turbine runs at 50 rps. The blades are designed for 50% reaction with exit angle 30º and inlet angle 50º. The turbine is supplied with steam at the rate of 60,000 kg/hr and stage efficiency is 85%.*

Determine

(a) Power output

(b) Specific enthalpy drop in the stage in (kJ/kg)

(c) % increase in relative velocity.

Solution

$$V_b = \text{mean blade speed} = \pi D \cdot n = \pi \times 1 \times 50$$

$$= 157 \text{ m/s}$$

For 50 % reaction, $\alpha = \Phi$ and $\theta = \beta$

So draw the velocity diagram

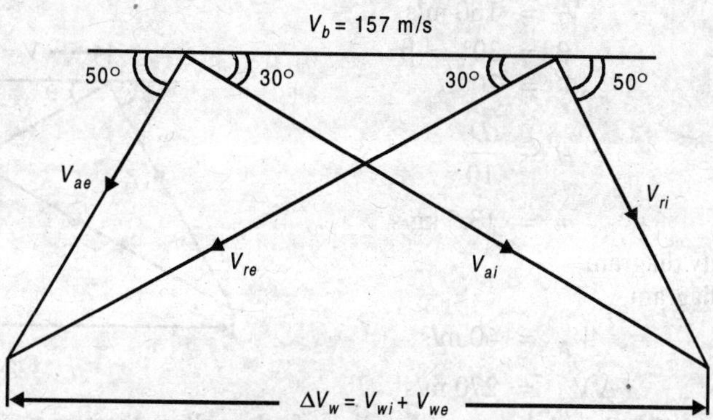

Fig. Ex. 20.11

For velocity diagram, $V_b = 157 \text{ m/s}, \Delta V_w = 444 \text{ m/s}$

$$V_{ri} = 224 \text{ m/s}, V_{re} = 341 \text{ m/s}$$

$$\text{Power} = m \times V_b \times (V_{wi} + V_{we})$$

$$= \frac{60000}{3600} \times \frac{157 \times 444}{10^3}$$

$$= 11.6 \times 10^3 \text{ kW}$$

$$\eta_{stage} = \frac{V_b \times \Delta V_w}{\Delta h}$$

$$\therefore \quad \Delta h = \frac{V_b \times \Delta V_w}{\eta_{stage}}$$

$$= \frac{157 \times 444}{0.85} = 82 \times 10^3 \text{ J/kg} = 82 \text{ kJ/kg}$$

$$\% \text{ increase in relative velocity} = \frac{V_{re} - V_{ri}}{V_{ri}} \times 100$$

$$= \frac{341 - 224}{224} \times 100 = 52.2 \%$$

Example 20.12. *At the particular stage of reaction turbine, the mean blade speed is 160 m/s and the steam is at the pressure of 350 kN/m² with the temperature of 175° C fixed and moving blades at the stage have inlet angle 30° and exit angle 20°. The blade height =* $\frac{D}{10}$.

The steam flow rate is 13.5 kg/s.
Determine:
1. *Blade height,*
2. *Power developed by a pair of fixed and moving blades at this stage*
3. *Specific enthalpy drop if stage efficiency is 85%.*

Solution

Given,

$$V_b = 160 \text{ m/s}$$
$$\theta = 30^\circ = \beta$$
$$\Phi = 20^\circ\ \alpha$$
$$H = \frac{D}{10}$$
$$m = 13.5 \text{ kg/s}$$

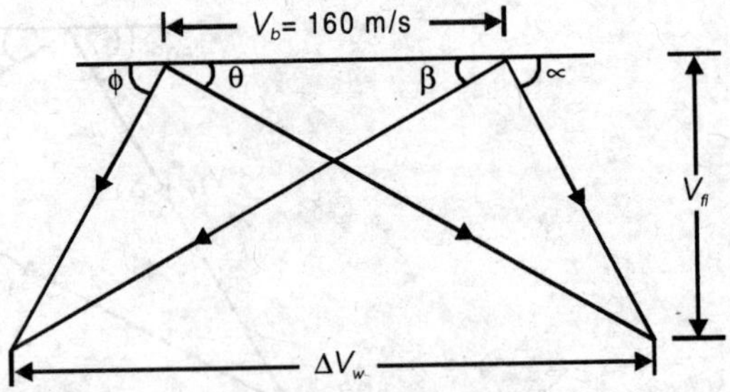

Fig. Ex. 20.12

Draw the velocity diagram.
From velocity diagram,

$$V_{fi} = 60 \text{ m/s}$$
$$\Delta V_w = 270 \text{ m/s}$$

At 350 kN/m² and 175°C find the specific volume from mollier-chart or steam-table.

$$V_s = 0.572 \text{ m}^3\text{/kg}$$

For reaction turbine, blade height is given as,

$$H = \frac{m \times V_s}{\pi D V f_i}$$

$$H = \frac{D}{10} \text{ (Given)}$$

$$\therefore \quad D = 10\text{ H}$$

$$\therefore \quad H = \frac{13.5 \times 0.572}{\pi \times 10 \times H \times 60}$$

$$\therefore \quad H = 0.064 \text{ m} = 64 \text{ mm}$$

$$\text{Power developed} = m \times V_b \times \Delta V_w$$
$$= \frac{13.5 \times 60 \times 270}{1000}$$
$$= 218 \text{ kW}$$

$$\eta_{stage} = \frac{V_b \times \Delta V_w}{\Delta h}$$

$$\therefore \quad \Delta h = \frac{V_b \times \Delta V_w}{\eta_{stage}}$$

$$\therefore \quad \Delta h = \frac{60 \times 270}{0.85 \times 10^3} = 19.1 \text{ kJ/kg}$$

Example 20.13. *Steam leaves the nozzle of a De-Laval turbine with a velocity of 1160 m/s. The nozzle is 20°. The blades are symmetrical with an angle of 30°. If the blade velocity coefficient is 0.8 and mechanical efficiency of 0.9. Calculate*

(i) *Rotational speed if the rotor diameter is 75 cm,*
(ii) *Specific steam consumption in kg/kW-hr,*
(iii) *Blade efficiency, and*
(iv) *Axial thrust.*

Solution

Drawing the velocity diagram to scale.

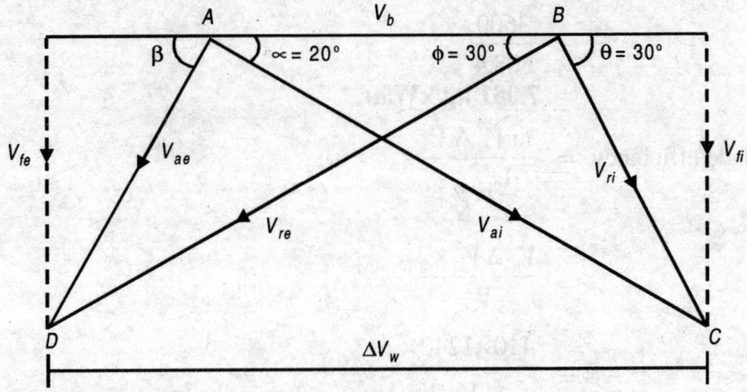

Fig. Ex. 20.13

From the velocity triangle,

$$V_{ai} = 1160 \text{ m/s}$$
$$V_b = 410 \text{ m/s}$$
$$V_{ri} = 800 \text{ m/s}$$
$$K = 0.8$$
$$\therefore \quad V_{re} = K\,V_{ri}$$
$$= 0.8 \times 800$$
$$= 640 \text{ m/s}$$
$$V_{fi} = 397.7 \text{ m/s}$$
$$V_{fe} = 320 \text{ m/s}$$

We have,

$$V_b = \frac{\pi DN}{60}$$
$$\therefore \quad N = \frac{60 \times V_b}{\pi \times D}$$
$$= \frac{60 \times 410}{\pi \times 7.5 \times 10^{-1}}$$
$$N = 10440 \text{ rpm}$$
$$\text{Power consumption} = \dot{m}\,V_b\,\Delta V_w$$
$$\Delta V_w = 1240 \text{ m/s}$$

$$\therefore \quad \text{Power} = \frac{1\times410\times1240}{1000}$$
$$= 508.4 \text{ kW}$$

Specific steam consumption

$$= \frac{3600}{\text{Work done}}$$
$$= \frac{3600}{508.4}$$
$$= \mathbf{7.081 \text{ kg/kW.hr.}}$$

$$\text{Blade efficiency} = \frac{\dot{m}\, V_b\, \Delta V_w}{\frac{1}{2}\dot{m}\, V_{ai}^2}$$
$$= \frac{V_b\, \Delta V_w \times 2}{V_{ai}^2}$$
$$= \frac{410\times1240\times2}{(1160)^2}\times100$$
$$\eta_{\text{blade}} = 75.56\ \%$$
$$\text{Axial thrust} = \dot{m}\left(V_{fi} - V_{fe}\right)$$
$$= 1\times(397.7-320)$$
$$= 77.7 \text{ N}$$

Example 20.14. *In a stage of an impulse turbine, the mean diameter of blade is 1.5 m. It runs at 3000 rpm. The velocity of steam at inlet to blade is 500 m/s and nozzle angle is 20° moving blade angle at inlet and exit are same. If blade friction factor is 0.9, find power in kW. Axial thrust on the bearing is 80 N.*

Solution

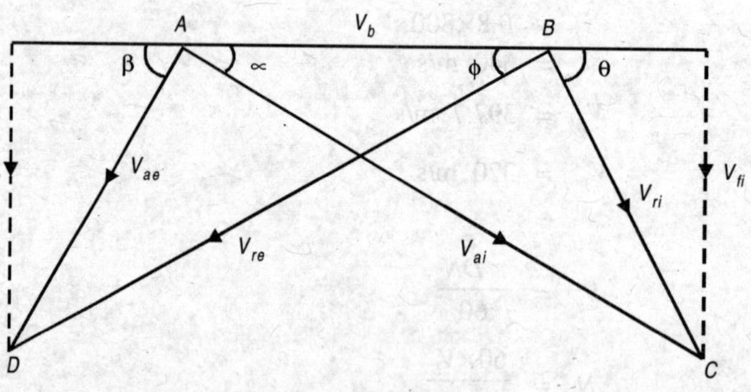

Fig. Ex. 20.14

Given,

$$N_b = 3000 \text{ rpm}$$
$$V_{ai} = 500 \text{ m/s}$$
$$\alpha = 20°$$
$$\phi = \theta$$
$$k = 0.9$$

$$\text{Axial thrust} = 180 \text{ N}$$

$$\text{Blade speed } V_b = \frac{\pi DN}{60}$$

$$= \frac{\pi \times 1.5 \times 3000}{60} = 235.62 \text{ m/s}$$

$$V_{ai} \cos\alpha = V_{ri} \cos\theta + C_b$$

$$V_{ai} \sin\alpha = C_{ri} \sin\theta$$

$$\therefore \quad 500\cos 20° - 235.6 = V_{ri}\cos\theta$$

$$\therefore \quad 500\sin 20° - 235.6 = V_{ri}\sin\theta$$

$$\therefore \quad \tan\theta = \frac{500\sin 20°}{500\cos 20° - 235.6}$$

$$= 0.7299$$

$$\therefore \quad \theta = 36.12° = \phi = \text{blade angles}$$

$$V_{ri} = \frac{500\sin 20°}{\sin 36.12°} = 290.1 \text{ m/s}$$

We have

$$\frac{V_{ri}}{V_{re}} = \frac{1}{0.9} \qquad K = 0.9$$

$$\therefore \quad V_{re} = 0.9\,V_{ri}$$

$$= 0.9 \times 290.1$$

$$= 261.1 \text{ m/s}$$

$$V_{wi} = V_{ai}\cos 20°$$

$$= 500 \times \cos 20°$$

$$= 469.84 \text{ m/s}$$

$$V_{we} = V_{re}\cos\phi - V_b$$

$$= 261.1\cos 36.12 - 235.9$$

$$= -23.8 \text{ m/s}$$

This shows that is greater than 90º, so the line *AD* will be inclined towards right side.

$$\therefore \quad V_{fi} = V_{ai}\sin\alpha$$

$$= 500\sin 20$$

$$= 171 \text{ m/s}$$

$$V_{fe} = V_{re}\sin\phi$$

$$= 261\sin 36.12$$

$$= 153.85 \text{ m/s}$$

Given axial thrust on the bearing is 180 N

$$\text{Axial thrust} = \dot{m}\left(V_{fi} - V_{fe}\right)$$

$$\therefore \quad 180 = \dot{m}(171-153.85)$$

$$\therefore \quad \dot{m} = \mathbf{10.497\ kg/s}$$

$$\text{Power} = \dot{m}\, V_b\,(V_{wi} + V_{we})$$

$$= 10.497 \times 235.62\,(469.84 + (-23.8))$$

$$\therefore \quad \text{Power} = 1103320.4 \text{ J/s or Nm/s}$$

$$= \mathbf{1103.32\ kW}$$

Example 20.15. *In a stage of an impulse turbine, steam leaves the nozzles at 1000 m/s. Nozzle angle is 20°. Mean blade speed is 400 m/s. Efficiency of nozzle is 0.95. Friction factor for blades is 0.8. Mass flow rate of steam is 0.5 kgs. Find: (i) Blade angle, (ii) Axial thrust, (iii) Output in kW, (iv) Diagram efficiency (v) Stage efficiency.*

Solution

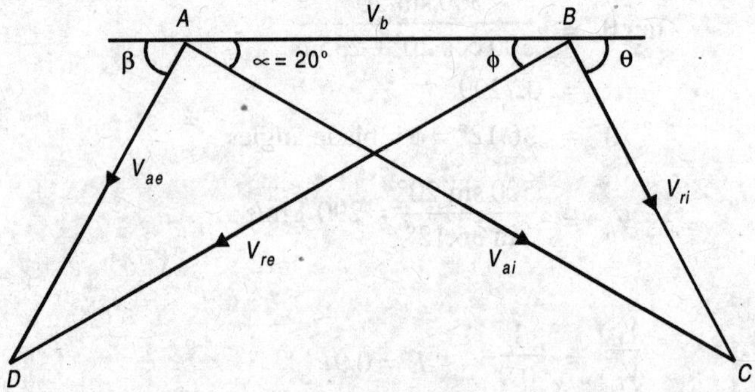

Fig. Ex. 20.15

Given,

$$\alpha = 20°$$

$$V_{ai} = 1000 \text{ m/s}$$

$$V_b = 400 \text{ m/s}$$

$$k = 0.8$$

$$\dot{m} = 0.5 \text{ kg/s}$$

Assuming symmetrical blades and drawing velocity diagram to scale

$$\textbf{Blades angles } \boldsymbol{\theta} = \boldsymbol{\phi = 32°}$$

$$V_{ri} = 640 \text{ m/s}$$

$$V_{re} = 0.8\,V_{ri}$$

$$= 512 \text{ m/s}$$

$$V_{fi} = 340 \text{ m/s}$$

$$V_{fe} = 270 \text{ m/s}$$

$$V_{wi} + V_{we} = 980 \text{ m/s}$$

$$\text{Axial thrust} = \dot{m}(V_{fi} - V_{fe})$$

$$= 0.5(340-270)$$

$$= \mathbf{35\ N}$$

$$\text{Output in kW} = \frac{\dot{m}\, V_b\, \Delta V_w}{1000}\ \text{kW}$$

$$= \frac{0.5 \times 400 \times 980}{1000}$$

$$= \mathbf{196\ kW}$$

$$\text{Diagram efficiency} = \frac{\dot{m}\, V_b \left(V_{wi} + V_{we}\right)}{\frac{\dot{m}}{2} V_{ai}^2}$$

$$= \frac{2\, V_b\, \Delta V_w}{V_{ai}^2}$$

$$= \frac{2 \times 400 \times 980}{(1000)^2} \times 100$$

$$\therefore \quad \eta_{dia} = \mathbf{78.4\%}$$

$$\text{Stage efficiency} = \text{Diagram efficiency} \times \text{Nozzle efficiency}$$

$$= 0.784 \times 0.95$$

$$\eta_{stage} = \mathbf{74.48\%}$$

Example 20.16. *The data refers to single stage impulse turbine. Isentropic nozzle drop is 190 kJ/kg friction in nozzle is 10 % of isentropic heat drop. Nozzle angle is 20°. Ratio of blade speed to which component of steam speed is 0.4. Velocity component for blade is 0.9. The velocity of steam at entry to nozzle is 35 m/s. Find the blade angles so that steam may enter then without shock and leave them in axial direction.*

Solution

Given,

$$\alpha = 20°,$$

$$\beta = 90°$$

$$k = 0.9$$

$$\frac{V_b}{V_{wi}} = 0.4$$

Velocity at entry to nozzle = 35 m/s

Isentropic enthalpy drop in nozzle = 190 kJ/kg

Because of friction,

$$\text{actual enthalpy drop} = 190 \times 0.9$$

$$\Delta h = 171\ \text{kJ/kg}$$

Steam velocity at inlet

$$V_{ai} = \sqrt{2 \times \Delta h \times 1000 + (35)^2}$$

$$= 585.85\ \text{m/s}$$

Whirl velocity at inlet

$$V_{wi} = V_{ai} \cos\alpha$$

$$= 585.85 \times \cos 20$$

$$= 550.52\ \text{m/s}$$

$$V_b = 0.4\,V_{wi}$$

$$\therefore \quad V_b = 0.4\times 550.52$$

$$= 220.21 \text{ m/s}$$

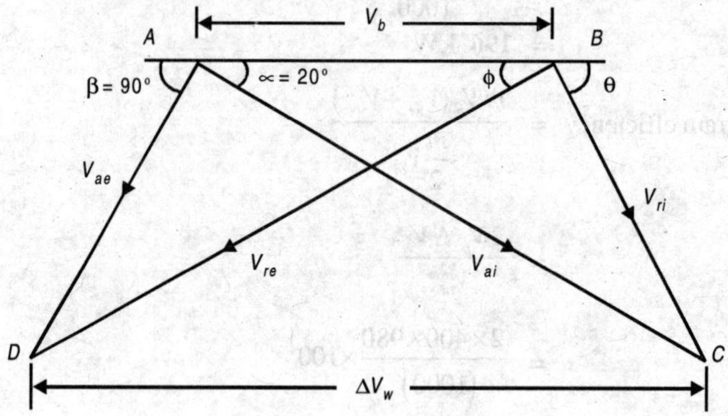

Fig. Ex. 20.16

From inlet triangle,

$$V_{ai}\sin 20 = V_{ri}\sin\theta$$

$$\therefore \quad V_{ri}\sin\theta = 585.52\sin 20$$

$$= 200.37 \text{ m/s}$$

$$V_{ri}\cos\theta = V_{wi} - V_b$$

$$= 550.52 - 220.21$$

$$= 330.31 \text{ m/s}$$

$$\tan\theta = \frac{\sin\theta}{\cos\theta} = \frac{V_{ri}\sin\theta}{V_{ri}\cos\theta}$$

$$= \frac{200.37}{330.31}$$

$$= 0.6066$$

$$\therefore \quad \theta = 31.24^\circ$$

$$V_{ri} = \frac{200.37}{\sin 31.24} = 386.33 \text{ m/s}$$

$$V_{re} = 0.9\,V_{ri}$$

$$= 0.9\times 386.33$$

$$= 347.7 \text{ m/s}$$

$$\cos\phi = \frac{V_b}{V_{re}} \text{ as } \beta = 90^\circ$$

$$= \frac{220}{347.25}$$

$$= 0.5752$$

$$\therefore \quad \boldsymbol{\phi = 54.88^\circ}$$

Example 20.17. *The following particulars refer to a two row velocity compounded impulse steam turbine : nozzle angle = 17º, blade speed = 125 m/s, exit angle of the first row moving blade = 22º, exit angle of the fixed blade = 30º, blade velocity coefficient = 0.9 for all the blades. Assume that the absolute velocity of steam leaving the stage is in the axial direction find (i) absolute velocity of steam leaving the stage, (ii) Diagram efficiency, (iii) Axial thrust per kg of steam, (iv) Power output per kg of steam.*

Solution

Drawing the velocity diagram to a suitable scale,

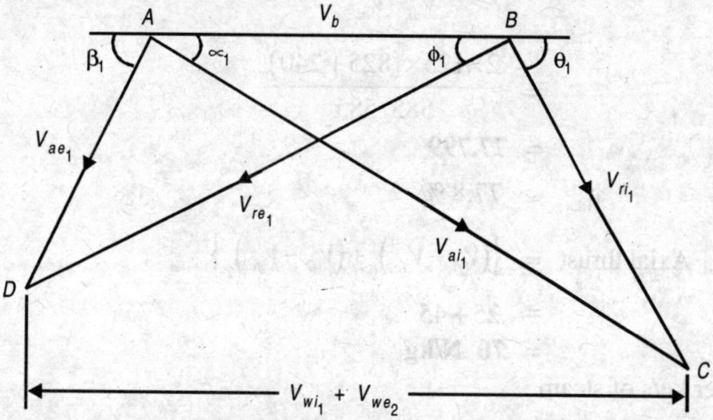

Fig. Ex. 20.17 (a)

Given :

$$\alpha_1 = 17^\circ \qquad \alpha_2 = 26^\circ$$
$$V_b = 125 \text{ m/s} \qquad K = 0.9$$
$$\phi_1 = 22^\circ \qquad \phi_2 = 30^\circ$$
$$\beta_2 = 90^\circ$$

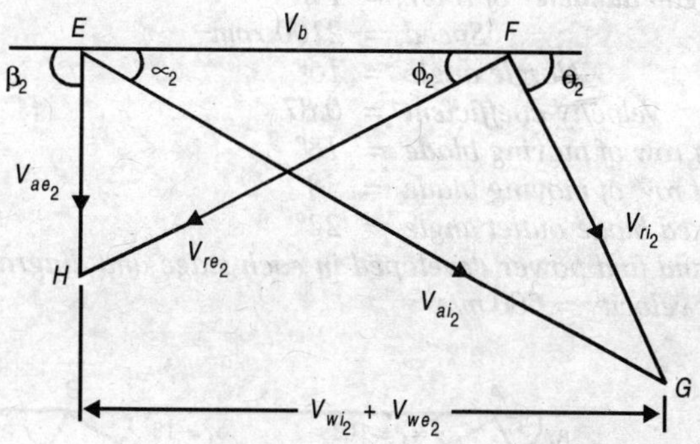

Fig. Ex. 20.17 (b)

Second row

We get $\quad V_b = 125$ m/s

From outlet triangle

$$V_{ae_2} = V_{fe_2} = 72 \text{ m/s}$$

i.e. exit velocity of steam at second stage.

Also we have

$$V_{ai_2} = V_{ae_1} \times 0.9 \quad (\text{as } K = 0.9)$$

From the velocity triangles,

$$\text{Diagram efficiency} = \frac{V_b\left(\Delta V_{w_1} + \Delta V_{w_2}\right)}{\frac{1}{2}V_{ai_1}^2}$$

$$= \frac{2V_b\left(\Delta V_{w_1} + \Delta V_{w_2}\right)}{V_{ai_1} \cdot V_{ai_1}}$$

$$= \frac{2\times125\times(825+240)}{585\cdot585}$$

$$= \mathbf{77.799}$$

$$\eta_{dia} \approx 77.8\,\%$$

$$\text{Axial thrust} = \left\{(V_{fi} - V_{fe})_1 + (V_{fi} - V_{fe})_2\right\}$$

$$= 25 + 45$$

$$= \mathbf{70\ N/kg}$$

Power output per kg/s of steam

$$= \frac{V_b\left\{(V_{wi} + V_{we})_1 + (V_{wi})_2\right\}}{1000}$$

$$= \frac{125\{825+240\}}{1000}$$

$$= \mathbf{133.125\ kW}$$

Example 20.18. *A two row velocity compounded impulse turbine has the following specifications*

Mean diameter of rotor = *1 m*
Speed = *2100 rpm*
Nozzle angle = *16°*
Velocity coefficient = *0.87*
Outlet angle of first row of moving blade = *18°*
Outlet angle of second row of moving blade = *38°*
Fixed blade outlet angle = *22°*

Draw velocity diagram and find power developed in each stage and diagram efficiency. Assume steam flow rate of 5 kg/s and steam jet velocity = 600 m/s.

Solution

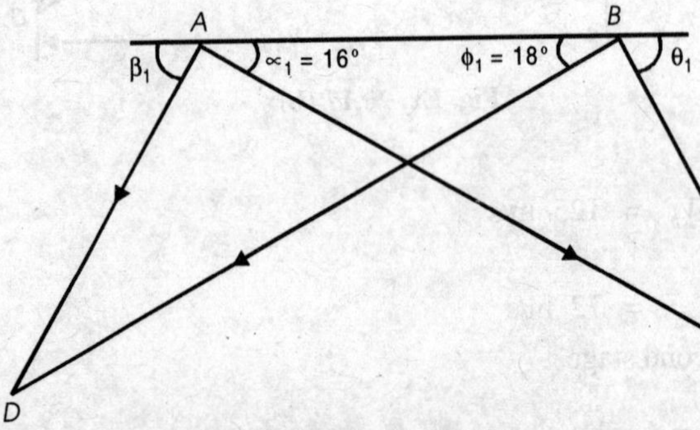

Fig. Ex. 20.18 (a)

Given,

$$V_{ai_1} = 600 \text{ m/s}$$

$$\alpha_1 = 16°$$

$$\phi_1 = 18°$$

$$\phi_2 = 38°$$

$$\alpha_2 = 22°$$

$$V_b = \frac{\pi DN}{60}$$

$$= \frac{\pi \times 1 \times 2100}{60}$$

$$= 109.95 \text{ m/s}$$

$$V_b \approx 110 \text{ m/s}$$

Velocity triangle for the first stage is drawn as shown.
We have

$$\Delta V_{w_1} = 875 \text{ m/s}$$

$$V_{re_1} = 0.87\, V_{ri_1}$$

Second row velocity triangle:

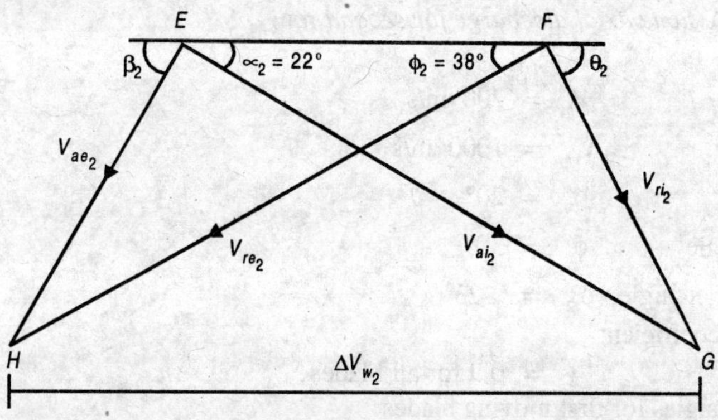

Fig. Ex. 20.18 (b)

$$\alpha_2 = 22° \qquad \phi_2 = 38°$$

$$V_{ai_2} = 0.87\, V_{ae_1}$$

From velocity diagram

$$\Delta V_{w_2} = 245 \text{ m/s}$$

$$V_{re_2} = 0.87\, V_{ai_2}$$

Power developed in first stage

$$P_1 = \dot{m} \cdot V_b \cdot \Delta V_{w_1}$$

$$= \frac{(5 \times 110 \times 875)}{1000} = \mathbf{481.25 \text{ kW}}$$

Power developed in second stage

$$P_2 = \dot{m} \cdot V_b \cdot \Delta V_{w_2}$$

$$= \frac{(5 \times 110 \times 245)}{1000}$$

$$= \mathbf{134.7\ kW}$$

$\therefore$ Total power output

$$= 481.25 + 134.7$$

$$= \mathbf{615.95\ kW}$$

Diagram efficiency

$$= \frac{V_b \cdot (\Delta V_{w_1} + \Delta V_{w_2})}{\frac{V_{ai_1}^2}{2}}$$

$$= \frac{110 \times [875 + 245] \times 2}{(600)^2}$$

$$\therefore \quad \eta_{dia} = \mathbf{68.4\ \%}$$

Example 20.19. *The following particulars refer to a velocity compounded impulse turbine with 2 rows of moving blades with a fixed row of guide blade between them. The velocity of steam leaving nozzle is 1000 m/s. Nozzle angle is 20°. Blade speed is 200 m/s. Blade angles of first moving row are symmetrical and blade outlet angle of fixed blade is 23°, blade outlet angle of second moving blade is 30°. Friction factors for all rows is 0.9. Determine power output in kW. Assume axial discharge for second row.*

Solution

Given, $V_b = 200$ m/s

$V_{ai} = 1000$ m/s

Nozzle angle $\alpha_1 = 20°$

Blade outlet angle $\phi_2 = 30°$

Fixed blade outlet angle $\alpha_2 = 23°$

Blade velocity coefficient

$k = 0.9$ for all blades

(i) Velocity triangles for first moving blades

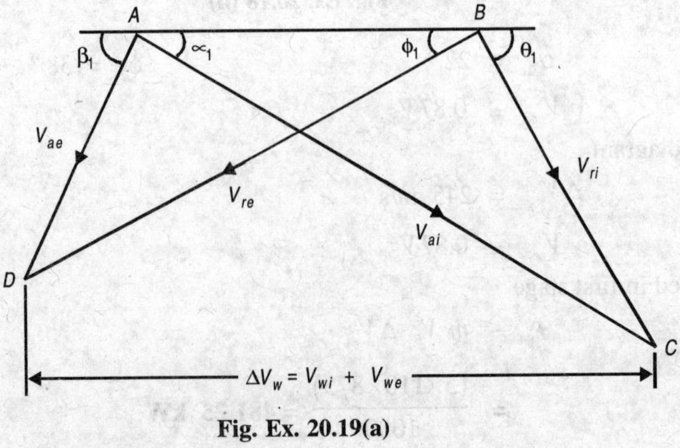

Fig. Ex. 20.19(a)

From velocity triangle of first row,

$$V_{ai_1} \cos \alpha_1 = V_{ri_1} \cos \theta_1 + C_b$$

and $$V_{ai_1} \sin \alpha_1 = V_{ri_1} \sin \theta_1$$

$$\therefore \quad \tan \theta_1 = \frac{V_{ai_1} \sin \alpha_1}{V_{ai_1} \cos \alpha_1 - C_b}$$

$$= \frac{1000 \sin 20}{1000 \cos 20 - 200}$$

$$= 0.462$$

$$\therefore \quad \theta_1 = 24.81° = \phi_1 \qquad (\because \text{ blades are symmetrical})$$

$$\therefore \quad V_{ri_1} = \frac{V_{ai_1} \sin 20}{\sin \theta_1}$$

$$= \frac{1000 \times \sin 20}{\sin 24.81}$$

$$= 815.1 \text{ m/s}$$

$$V_{re_1} = k \cdot V_{ri_1}$$

$$= 0.9 \times 815.1$$

$$= 733.5 \text{ m/s}$$

$$V_{wi_1} = V_{ai_1} \cos \alpha_1$$

$$= 1000 \cos 20$$

$$= 939.7 \text{ m/s}$$

$$V_{we_1} = V_{re_1} \cos \phi - V_b$$

$$= 489.26 \text{ m/s}$$

and $$\Delta V_{w_1} = 939.7 + 489.26$$

$$= 1428.96 \text{ m/s}$$

For second row:

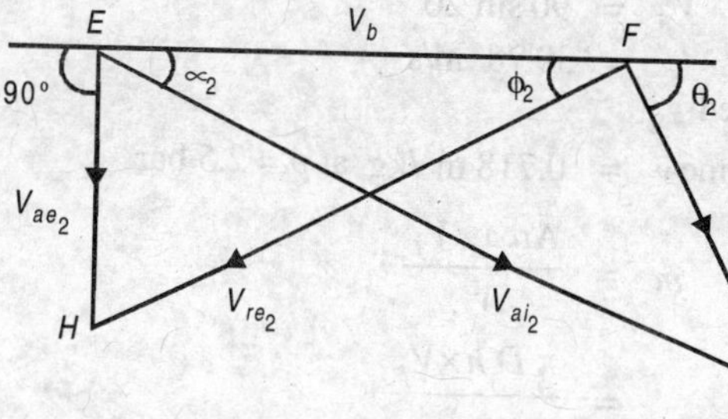

Fig. Ex. 20.19(b)

$$V_{ai_2} = V_{ae_1} \times 0.9$$

But $$V_{ae_1}^2 = V_{re_1}^2 + V_b^2 - 2\, V_{re_1}\, V_b \cos \phi_1$$

$$= 733.5^2 + 200^2 - 2 \times 733.5 \times 200 \times \cos 24.81°$$

$$V_{ae_1} = 558.3 \text{ m/s}$$

$$\therefore \quad V_{ai_2} = 0.9\,V_{ae_1}$$
$$= 502.47 \text{ m/s}$$
$$V_{wi_2} = V_{ai_2} \cos\alpha_2$$
$$= 502.47 \times \cos 23$$
$$= 462.53 \text{ m/s}$$
$$V_{we_2} = 0 \text{ as discharge is radial}$$

$$\therefore \quad \text{Power output} = \frac{V_b \times \left[\Delta V_{w_1} + \Delta V_{w_2}\right]}{1000}$$
$$= \frac{200 \times [1428.96 + 462.53]}{1000}$$
$$= \mathbf{378.3 \text{ kW}}$$

Example 20.20. *A reaction turbine has identical blading. It delivers dry, saturated steam at 2.5 bar. Velocity of steam is 90 m/s. Mean blade height is 3 cm and exit angle of the moving blade is 20°. At the mean radius, the axial flow velocity equals 3/4 blade speed. For a steam flow rate of 9000 kg/hr, calculate (i) power output, (ii) rotor speed in rpm, (iii) diagram efficiency, (iv) enthalpy drop in a stage, (v) percentage increase in relative velocity in moving blade.*

Solution

Given,

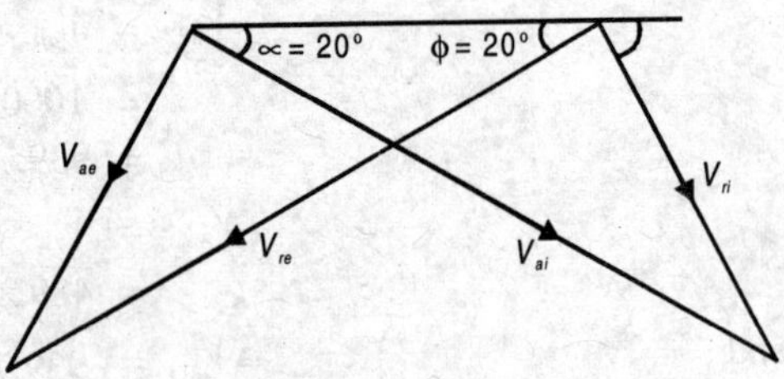

Fig. Ex. 20.20

$$V_{ai} = 90 \text{ m/s}$$
$$h = 3 \text{ cm}$$
$$= 0.03 \text{ m}$$
$$\text{Stage pressure} = 2.5 \text{ bar}$$
$$\dot{m} = 9000 \text{ kg/hr}$$
$$= \frac{9000}{3600} \text{ kg/s}$$
$$= 2.5 \text{ kg/s}$$
$$V_{fi} = 90 \sin 20$$
$$= 30.78 \text{ m/s}$$

Steam is dry and saturated

$$\text{Specific volume } v = 0.718 \text{ m}^3\text{/kg at } p = 2.5 \text{ bar}$$

$$\dot{m} = \frac{\text{Area} \times V_f}{v}$$
$$= \frac{\pi D h \times V_f}{v}$$

$$\therefore \quad 2.5 = \frac{\pi \times D \times 0.03 \times 30.78}{0.718}$$

$$\therefore \quad D = 0.6187 \text{ m}$$

$$V_b = \frac{4}{3} \cdot V_{fi}$$

$$= \frac{4}{3} \times 30.78$$

$$= 41.04 \text{ m/s}$$

But $$V_b = \frac{\pi DN}{60}$$

$$41.04 = \frac{\pi \times 0.6187 \times N}{60}$$

$$\therefore \quad N = 1266.86 \text{ rpm}$$

Solving the velocity triangles analytically,

$$\mathbf{V_{ai} \cos \alpha = V_{ri} \cos \theta + V_b}$$

$$V_{ai} \sin \alpha = V_{ri} \sin \theta$$

$$\therefore \quad \tan \theta = \frac{V_{ai} \sin \alpha}{V_{ai} \cos \alpha - V_b}$$

$$= \frac{90 \times \sin 20}{90 \times \cos 20 - 41.04}$$

$$= 0.707$$

$$\therefore \quad \theta = 35.26° = \beta \text{ (as blades are identical)}$$

$$V_{ri} = \frac{V_{ai} \sin \alpha}{\sin \theta}$$

$$= 53.31 \text{ m/s}$$

$$V_{wi} = V_{ai} \cos \alpha = 90 \cos 20°$$

$$= 84.57 \text{ m/s}$$

$$V_{we} = V_{ae} \cos \beta = 90 \cos 35.26$$

$$= 73.48 \text{ m/s}$$

$$\text{Blade speed ratio} = \rho = \frac{V_b}{V_{ai}} = \frac{41.04}{90}$$

$$= 0.456$$

$$\text{Diagram efficiency} = \frac{2\rho(2\cos\alpha - \rho)}{1 - \rho^2 + 2\rho\cos\alpha}$$

$$= \frac{2 \times 0.456(2 \times 0.9396 - 0.456)}{1 - (0.456)^2 + 2 \times 0.456 \times 0.9396}$$

$$= 0.787$$

$$\boldsymbol{\eta_{dia} = 78.7\ \%}$$

Enthalpy drop per stage

$$= \frac{2\left(V_{ai}^2 - V_{ri}^2\right)}{2 \times 1000}$$

$$= \frac{2 \times (90)^2 - (53.31)^2}{2 \times 1000}$$

$$= \mathbf{5.258 \ kJ/kg.}$$

Percentage increase in relative velocity

$$= \left(\frac{V_{re} - V_{ri}}{V_{ri}}\right) \times 100$$

$$= \left(\frac{90 - 53.31}{53.31}\right) \times 100 \quad (\text{as } V_{re} = V_{ai})$$

$$= 0.6888$$

$$= \mathbf{68.88\ \%}$$

$$\text{Power developed} = \frac{V_b (V_{wi} + V_{we})}{1000} \times \frac{\dot{m}}{1}$$

$$= \frac{41.04(84.57 + 73.48)}{1000} \times 2.5$$

$$= \mathbf{16.216\ kW}$$

Example 20.21. *In a stage of reaction turbine, the mean diameter of rotor is 1.4 m. Velocity ratio is 0.7. Turbine rotates at 3000 rpm. Blade outlet angle is 20º. Find diagram efficiency and also find value of maximum diagram efficiency.*

Solution

$$V_b = \frac{\pi DN}{60}$$

$$= \frac{\pi \times 1.4 \times 3000}{60}$$

$$= 219.9 \text{ m/s}$$

$$V_{ai} = \frac{V_b}{0.7}$$

$$= 314.15 \text{ m/s}$$

$$\text{Blade efficiency} = \frac{2\rho(2\cos\alpha - \rho)}{1 - \rho^2 + 2\rho\cos\alpha}$$

$$= \alpha = 20°, \rho = 0.7$$

$$= \frac{2 \times 0.7(2 \times \cos 20 - 0.7)}{1 - (0.7)^2 + 2 \times 0.7 \cos 20}$$

$$\therefore \quad \eta_{blade} = 90.4\ \%$$

Maximum diagram efficiency

$$= \frac{2\cos^2\alpha}{1 + \cos^2\alpha}$$

$$= \frac{2 \times \cos^2 20}{1 + \cos^2 20}$$

$$= 93.78\ \%$$

Example 20.22. *At a stage of a reaction turbine, the mean ring diameter of 140 cm and speed ratio is 0.7. Speed is 3000 rpm. Determine required entrance angle to blading if exit angle is 20º. Also determine work done per kg of steam per second. Calculate percentage increase in diagram efficiency if the blades are desingned for and run at the best theoretical speed, the exit angle being kept 20º.*

Solution

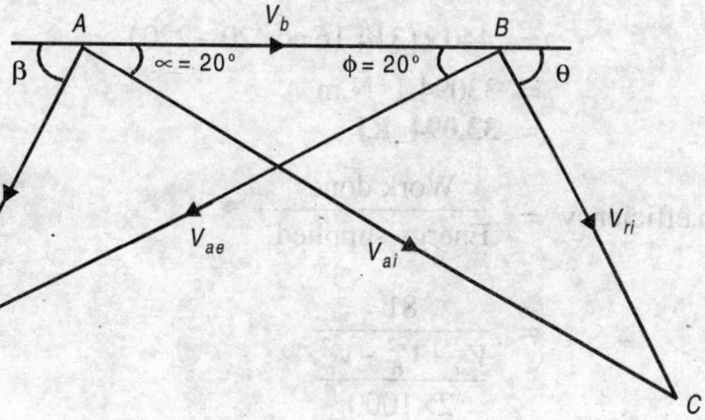

Fig. Ex. 20.22

Drawing velocity diagram, for 50% reaction, inlet and exit triangles are symmetrical

$$\alpha = \phi,\ \theta = \beta,\quad V_{ai} = V_{re}$$

$$V_{fi} = V_{fe}$$

$$\text{Blade speed } V_b = \frac{\pi D N}{60}$$

$$= \frac{\pi \times 1.4 \times 3000}{60}$$

$$= 220 \text{ m/s}$$

$$V_{ai} = \frac{V_b}{\text{Blade speed ratio}}$$

$$= \frac{220}{0.7}$$

$$= 314.16 \text{ m/s}$$

$$V_{ri}^2 = V_{ai}^2 + V_b^2 - 2V_b \cdot V_{ai} \cdot \cos\alpha$$

$$\therefore \quad V_{ri}^2 = 314.16^2 + 220^2 - 2 \times 314.16 \times 220 \cos 20$$

$$\therefore \quad V_{ri} = 131.15 \text{ m/s}$$

Using sin rule,

$$\frac{V_{ri}}{\sin\alpha} = \frac{V_{ai}}{\sin(\pi - \theta)}$$

$$\therefore \quad \sin\theta = \frac{V_{ai}}{V_{ri}} \sin\alpha$$

$$= \frac{314.16}{131.15} \sin 20$$

$$\therefore \quad \boldsymbol{\theta = 55^\circ}$$

$$\text{Work done per kg} = V_b\left(V_{wi} + V_{we}\right)$$

$$= V_b\left(V_{ai}\cos\alpha - V_b + V_{re}\cos\phi - V_b\right)$$

$$= 220 \times 2(V_{ai} \cos \alpha - V_b)$$

$$= 440 \times (314.16 \cos 20 - 220)$$

$$= 33094.1 \text{ N.m}$$

$$= \mathbf{33.094 \text{ kJ}}$$

$$\text{Diagram efficiency} = \frac{\text{Work done}}{\text{Energy supplied}}$$

$$= \frac{81}{\frac{V_{ai}^2 + V_{re}^2 - V_{ri}^2}{2 \times 1000}}$$

$$= \frac{81 \times 2 \times 1000}{(314.16)^2 + (314.16)^2 - (131.15)^2}$$

$$= 0.899 \text{ i.e. } 89.9\ \%$$

For best theoretical condition

$$V_b = V_{ai} \cos \alpha$$

$$= 314.16 \times \cos 20$$

$$= 295.2 \text{ m/s}$$

$$V_{ri} = \sqrt{(V_{ai})^2 + (V_b)^2 - 2 \cdot V_{ai} \cdot V_b \cos 20}$$

$$\therefore \quad V_{ri} = 107.45 \text{ m/s}$$

$$= V_{ae}$$

Maximum diagram efficiency

$$= \frac{2 \cos^2 \alpha}{1 + \cos^2 \alpha}$$

$$= \frac{2 \cos^2 20°}{1 + \cos^2 20°}$$

$$= 0.9378$$

$$= 93.78\ \%$$

∴ Percentage increase in efficiency

$$= \frac{93.78 - 89.9}{89.9} \times 100$$

$$= \mathbf{4.324\ \%}$$

Example 20.23. *A 3 stage turbine is fed at 26 bar and 370°C. The exhaust takes place at 0.05 bar. Inter stage pressure are 5 bar and 1 bar. The stage efficiency for all the stages is 80%. Assuming a condition line to be straight.*

Determine:

(*i*) *Rankine efficiency,*
(*ii*) *Quality of steam leaving each stage,*
(*iii*) *Reheat factor,*
(*iv*) *Work dine per kg of steam in each stage,*
(*v*) *Overall efficiency.*

Solution

From the Mollier Chart

Pressure P_{bar}	*Temperature or Quality*	*Iso inthalpy drop kJ/kg*	*Actual enthalpy drop kJ/kg*
$P_1 = 26$	370ºC	3170 – 2676	3170 – 2854.6
$P_2 = 5$	198ºC at 2′		
$P_3 = 1$	$x_3 = 0.981$	2854.8 – 2568.8	2854.4 – 2626
$P_4 = 0.05$	$x_4 = 0.9$	2626 – 2200	2626 – 2285.2
		$\sum \Delta h_{\text{ise}} = 1106 \dfrac{\text{kJ}}{\text{kg}}$	$\sum \Delta h_{\text{act}} = 884.8 \dfrac{\text{kJ}}{\text{kg}}$

Also $\quad h_4'' = 2100$ kJ/kg

$h_{f_4} = 137.7$ kJ/kg

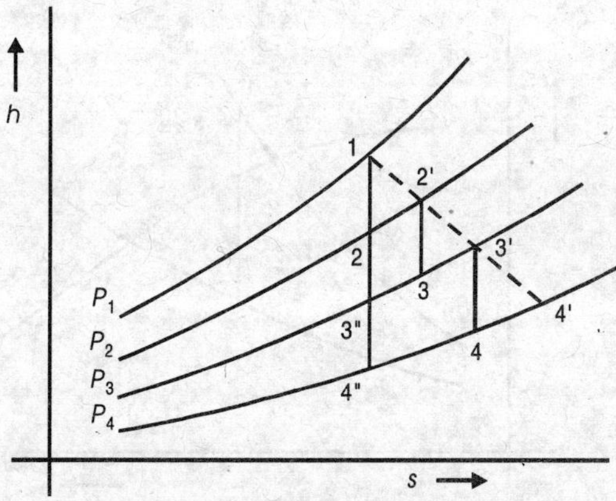

Fig. Ex. 20.23

We haves

$$\text{Rankine efficiency} = \frac{h_1 - h_4''}{h_1 - h_{f_4}}$$

$$= \frac{3170 - 2100}{3170 - 137.7} \times 100\%$$

$$= \mathbf{35.28\ \%}$$

$$\text{Reheat factor} = \frac{\Sigma\,\Delta\,h_{isen}}{h_1 - h_4''}$$

$$= \frac{1106}{3170 - 2100}$$

$$= \mathbf{1.0336}$$

$$\text{Overall efficiency} = \text{Stage efficiency} \times \text{Reheat factor}$$

$$= \frac{\Sigma\,\Delta\,h_{\text{act}}}{h_1 - h_4''} \times 100$$

$$= \mathbf{82.62\ \%}$$

Work done per stage :

$$\text{W.D. in stage I} = h_1 - h_2' = 315.12 \text{ kJ/kg}$$

$$\text{W.D. in stage II} = h_2' - h_3' = 228.5 \text{ kJ/kg}$$

$$\text{W.D. in stage III} = h_3' - h_4' = 340 \text{ kJ/kg}$$

Example 20.24. *Steam at 15 bar and 300°C is supplied to a stage Parson's reaction turbine, the steam is exhausted from the turbine at 0.1 bar. The stage efficiency is 0.75 and reheat factor is 1.5. Total power developed by turbine is 11025 kW. Find steam flow rate through turbine assuming all stages develop equal power. At a certain stage of turbine, the pressure of steam is 1.0 bar and steam is dry saturated. Blade exit angle is 20° and blade speed ratio is 0.7. Find mean diameter of this stage and rotor speed. Take blade height as 1/10th of mean diameter of the rotor. Neglect thickness of blade. Velocity of flow is 60 m/s.*

Solution

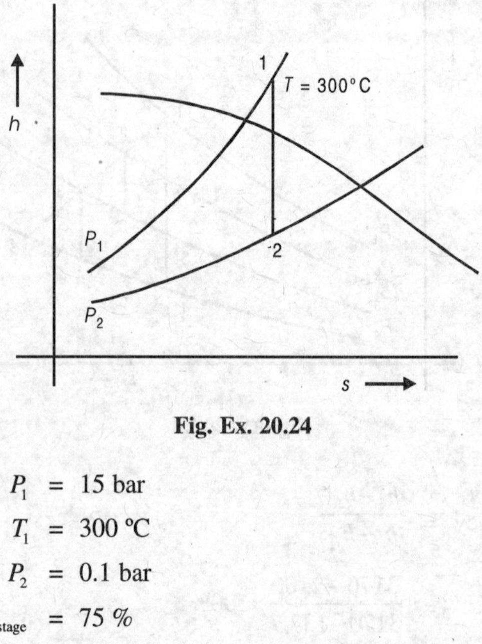

Fig. Ex. 20.24

$$P_1 = 15 \text{ bar}$$

$$T_1 = 300 \text{ °C}$$

$$P_2 = 0.1 \text{ bar}$$

$$\eta_{stage} = 75 \text{ \%}$$

From Mollier Chart,

We get isentropic enthalpy drop as

$$\Delta h = h_1 - h_2 = 3042 - 2185 = 857 \text{ kJ/kg}$$

We have

$$\text{Overall efficiency} = \text{Stage efficiency} \times \text{Reheat factor} = 0.75 \times 1.05 = 0.7875 \text{ or } 78.75 \text{ \%}$$

$$\text{Actual enthalpy drop} = \text{Work done/kg} = \Delta h_{ise} \times \text{Overall } \eta$$

$$= 857 \times 0.7875$$
$$= 647.88 \text{ kJ/kg}$$

Work done per kg of steam per stage

$$= \frac{647.88}{20}$$
$$= 33.74 \text{ kJ/kg}$$

$$\text{Power output/stage} = \text{Work done per stage} \times \dot{m}$$

$$\therefore \quad \frac{11025}{20} = 33.74 \times \dot{m}$$

$$\therefore \quad \dot{m} = 16.336 \text{ kg/s}$$

Work done/kg per stage

$$= (V_{wi} + V_{we}) \cdot V_b$$

We have $\quad \dfrac{V_b}{V_{ai}} = 0.7$

$$\therefore \quad V_{ai} = 1.43\, V_b$$

$$33.74 \times 10^3 = V_b \{2 \times (1.43\, V_b \cos 20 - V_b)\}$$
$$= 2\, V_b \times 0.3424\, V_b$$

$$\therefore \quad V_b = 183.68 \text{ m/s}$$

Specific volume of steam at 1 bar of dry steam $= 1.694 \text{ m}^3 \text{ kg} = V_g$

$$\dot{m} = \frac{\pi \times D \times h \times V_{fi}}{V_g}$$

$$\therefore \quad 16.336 = \frac{\pi \times D \times \frac{D}{10} \times 60}{1.694}$$

$$\therefore \quad \boldsymbol{D = 1.211 \text{ m}}$$

$$V_b = \frac{\pi D N}{60}$$

$$183.68 = \frac{\pi \times 1.211 \times N}{60}$$

$$\therefore \quad N = \frac{183.68 \times 60}{\pi \times 1.211}$$

$$\therefore \quad \mathbf{N = 2896.8 \text{ rpm}}$$

THEORY QUESTIONS

20.1. Explain the principal of the operation of steam turbine?

20.2. Discuss the advantages of steam turbines?

20.3. Compare steam turbine and steam engine?

20.4. How the steam turbines are classified?

20.5. Differentiate between pressure and velocity compounding.

20.6. Write short notes on:

1. Governing
2. Types of steam turbines

20.7. State various methods of governing and describe any two methods of governing of steam turbines.

20.8. Derive the relation for condition for maximum efficiency of

(a) Impulse turbine

(b) Reaction turbine

(c) Two stage velocity compounded steam turbine.

PROBLEMS FOR PRACTICE

20.1. A simple impulse turbine has one ring of moving black running at 150 m/s the absolute velocity of steam at exit from stage is 85 m/s at an angle of 80° with tangent of wheel. The black velocity coefficient is 0.82. The steam rate is 2 kg/sec. Assume moving blade to be equiangular, find

(a) The blade angle

(b) Absolute velocity of steam at inlet

(c) Axial Thrust

(d) Power developed

[*Ans.* (a) 26.5° (b) 365.59 m/s (c) 37 N (d) 108.9 kW]

20.2. At a stage in a reaction turbine the mean blade ring diameter is 1m and turbine runs at a speed 50 rpm. The blade are designed for 50% reaction with exit angle of 30° and inlet angle of 50°. The turbine is supplied with the rate of 6,00,000 kg/hr and η_{isent} = 85%

Determine

(a) Power developed

(b) Sp. enthalpy drop per stage per kg

(c) % increase in relative velocity

[*Ans.* (a) 11.58 MW (b) 82.5 kJ/kg (c) 52.2%]

IMPORTANT POINTS TO BE REMEMBERED

A. *Impulse turbine*

(a) Force in the tangential direction

$$F = \text{(mass flow rate of steam)} (V_{w1} + V_{w2})$$
$$= m_s (V_{w1} - V_{w2}) \times V_b \text{ N-m/s}$$

Energy input to blade= $\dfrac{m_s V_1^2}{2}$

∴ Blade efficiency = Diagram efficiency

$$\eta_b = \frac{2V_b (V_{w1} + V_{w2})}{V_1^2}$$

(b) Axial thrust $= m_s (V_{f1} - V_{f2})$

(c) Stage efficiency $= \dfrac{\text{Actual enthalpy drop}}{\text{Isentropic enthalpy drop}}$

$$\eta_s = \frac{h_1 - h_{2'}}{h_1 - h_2}$$

(d)Nozzle efficiency = $\eta_{mz} = \dfrac{\text{Actual velocity at outer}}{\text{Theoretical velocity at outer}}$

$$= V_{2'} / V_2$$

$$= \frac{44.72\sqrt{h_1 - h_2}}{44.72\sqrt{h_1 - h_2}}$$

$$= \sqrt{(h_1 - h_{2'}) / (h_1 - h_2)}$$

(e) Condition for maximum blade or diagram efficiency

$$\frac{V_b}{V_1} = \frac{\cos x}{2}$$

∴ Maximum diagram or blade efficiency

$$(\eta_b)_{max} = \cos^2 \alpha$$

B. *Reaction turbine*

(a) Work done $= m_s (V_b) (V_{w1} + V_{w2})$

(b) Eneregy supplied = $m_s \left[\dfrac{V_1^2}{2} + \dfrac{Vr_1^2 - Vr_1^2}{2} \right]$

(c) Blade or diagram efficiency

$$\eta_p = \frac{V_b (V_{w1} + V_{w2})}{\dfrac{V_1^2}{2} + \dfrac{Vr_2^2 - Vr_1^2}{2}}$$

(d) Condition for maximum efficiency

$$\frac{V_b}{V_1} = \cos\alpha \quad \text{for 50\% reaction turbine}$$

(e) Max. efficiency of 50% reaction turbine

$$(\eta_b)_{max} = \frac{2\cos^2\alpha}{1+\cos^2\alpha}$$

(f) Degree of reaction R

$$R = \frac{\text{Enthalpy drop in moving blades}}{\text{Total enthalpy drop in stage}}$$

For 50% reaction turbine, velocity diagrams at inlet and outlet are same and $\alpha = \phi$, $\theta = \beta$.

(g) Internal efficiency = Stage efficiency × Reheat factor

$$\eta_i = \eta_s \times RF$$

CHAPTER 21

Steam Condensers

21.1 INTRODUCTION: NECESSITY OF CONDENSERS

As we know steam will be generated in the boilers, after it is being superheated, it is expanded in the turbines for generating power. After generating the power, it is either exhausted into the atmosphere or it is exhausted into the condenser. In the condensers this steam will be condensed i.e., its phase will be changed from vapour to liquid by means of cooling water (Ref. Fig. 21.1).

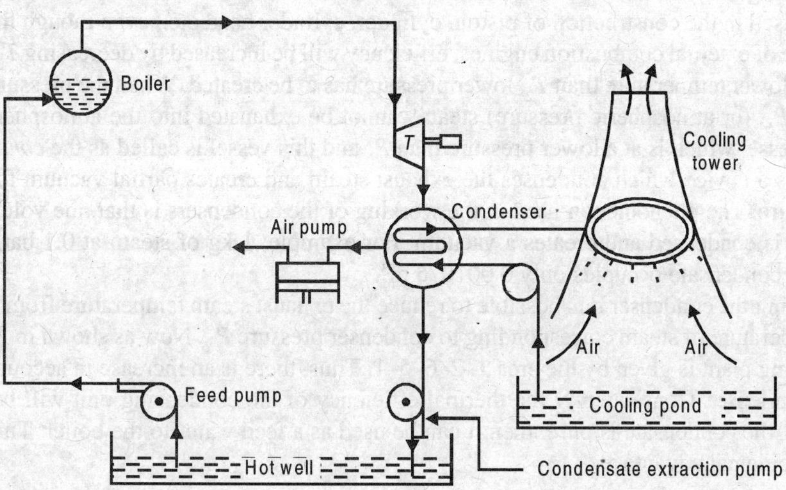

Fig. 21.1. *Elements of steam condensing plant*

Figure 21.2 shows the theoretical indicator diagram for the steam turbine. Let the steam available in the turbine have the initial condition of pressure P_1 and temperature T_1. This steam is expanded in the turbines along 2–3 for generating work output and then it is exhausted into the atmosphere along 3–4 (slightly above atmospheric pressure). Let the final condition of steam be P_b and temperature T_2. Then the work output is given by the area 1–2–3–4–1 on the indicator diagram.

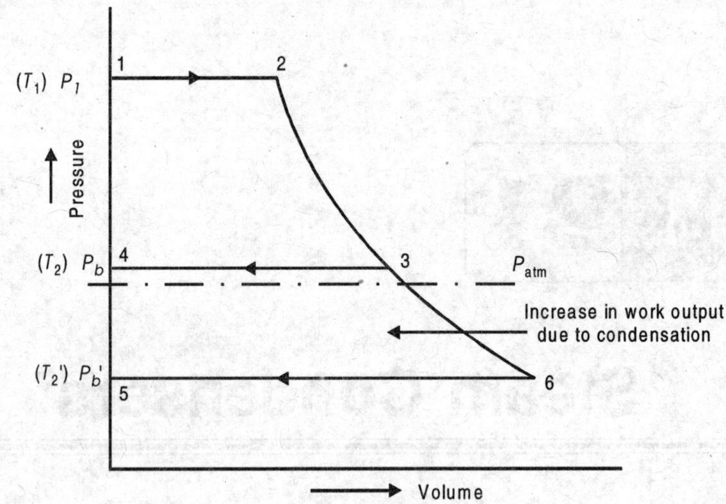

Fig. 21.2. *Theoretical/Ideal indicator diagram*

According to Carnot's maximum efficiency of any heat engine cycle is given by $\eta_{carnot} = \dfrac{T_1 - T_2}{T_1}$.

From this equation it is clear that the efficiency can be increased either by increasing T_1 or by decreasing T_2.

In the case of IC engines, efficiency is increased by increasing T_1. (In this case T_1 is temperature of products of combustion in the engine cylinder). When T_1 is increased, then the metals which can withstand high temperature stresses are to be used in the construction of piston, cylinder, cylinder head etc., even though they are costly.

Whereas in case of external combustion engines, efficiency will be increased by decreasing T_2. By the properties of steam to obtain lower temperature than T_2, lower pressure has to be created. When the pressure of exhaust steam is lowered below P_b, (or atmospheric pressure) steam cannot be exhausted into the atmosphere. So, it has to be exhausted into a vessel which is at a lower pressure than P_b and this vessel is called as the *condenser.*

So, condenser is a device which condenses the exhaust steam and creates partial vacuum for the steam below atmospheric pressure. The phenomenon used in the working of the condensers is that, the volume of steam drops down heavily if it is condensed and creates a vacuum. For example, 1 kg of steam at 0.1 bar pressure occupies 14.93 m^3 while its condensate occupies only 0.001076 m^3.

Thus, by installing the condenser it is possible to reduce the exhaust steam temperature from T_2 to T_2' where T_2' is the saturation temperature of steam corresponding to condenser pressure P'_b. Now as shown in P–V diagram, work done in a condensing plant is given by the area 1–2–6–5–1. Thus, there is an increase in account of work by 3–4–5–6 by using a condenser. Consequently the thermal efficiency of the condensing unit will be higher than non-condensing unit..If the condensate is pure, then it can be used as a feed water to the boiler. This recovery of feed water is also an incidental advantage.

21.2 CLASSIFICATION

Steam condensers are broadly classified as,

I. Jet Condensers

(a) Low Level Jet Condensers

(i) Counter-Flow Type

(ii) Parallel-Flow Type

(b) High Level Jet Condensers
(c) Ejector Condensers

II. Surface Condensers

(a) Steam flowing through the tubes and water from outside
(b) Water flowing through the tubes and steam from outside
(c) Evaporative condensers

The main difference between the Jet condensers and Surface condensers is as given below.

In case of Jet condensers, the steam to be condensed directly mixes with the cooling water and the condensation will be effected with less amount of water.

Whereas in the case of surface condensers, the steam to be condensed and the cooling water are separated by means of a tube well and through which heat transfer takes place and the condensation will be effected. Here the amount of cooling water required will be more.

In case of Jet condensers, the temperature of condensate and cooling water are same, whereas, in case of surface condensers temperature of condensate will be more than the temperature of cooling water at the outlet.

In case of Jet condensers, the condensate collected will be impure. (Since the steam and cooling water mix with each other). So it has to be wasted. Whereas in case of surface condensers, it is pure and can be used as a feed water to the boiler.

The size, space required and cost of surface condenser is more, because it has to handle large quantity of water for the same capacity.

The most commonly used condensers are counterflow type Jet and Water tube type surface condensers.

(a) (i) Low level Counter-Flow Type Jet Condensers. In this type, the direction of flow of steam and cooling water are opposite.

In this condenser, the cold water is drawn up in the condenser from the cooling pond, due to the vacuum head created in the condenser shell. This condenser consists of a cylindrical shell arranged with 2–3 water trays with perforations to break up water into small jets. The exhaust steam and any mixed air enters the condenser at lower portion as shown in Fig. 21.3 and tries to ascend up through the falling sprays. Thus, when the steam comes in contact with the cold water particles gets condensed and the air gets cooled.

This cooled air will be removed through the air pump at the top. The mixture of condensate and cooling water descends down through the vertical pipe to the centrifugal extraction pump and is pumped to the hot well. If the condensate in the hot well is pure then it can be pumped to the boiler as a feed water otherwise it has to be wasted.

The capacity of air pump required is small, since it has to handle cold air and water vapour. These condensers can be directly installed below the turbines. Such condensers have a disadvantage of flooding the steam turbine, if the condensate extraction pump fails.

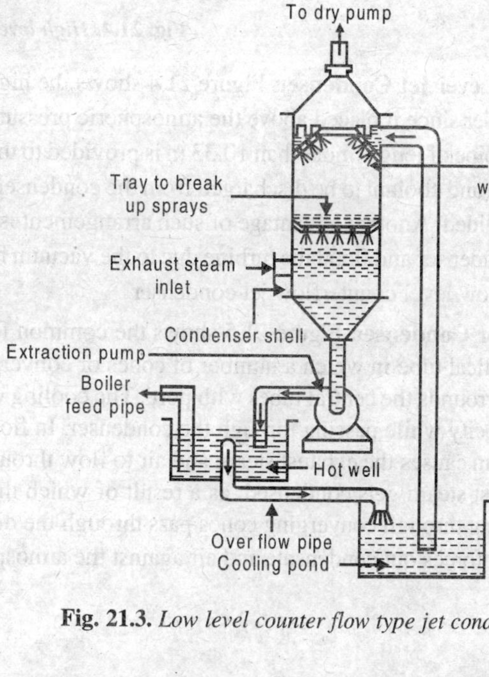

Fig. 21.3. *Low level counter flow type jet condenser*

(ii) Low Level Parallel Flow Jet Condenser. In this case, the direction of flow of exhaust steam and cooling water are parallel. In this case, the steam enters from top of the condenser and the cooling water just below it from sides as shown in Fig. 21.4 and rest all the arrangements are similar to counter flow Jet condenser. A wet air pump is provided at the bottom which removes air, condensate and cooling water.

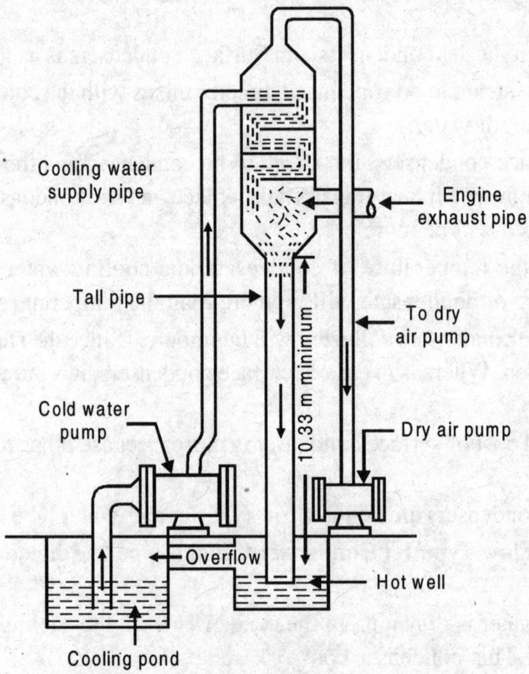

Fig. 21.4. *High level jet condenser*

(b) High Level Jet Condenser. Figure 21.4 shows the high level Jet condenser. This is also called as barometric Jet condenser since it placed above the atmospheric pressure equivalent to 10.33 m of water pressure. In this case, a long tail pipe of length more than 10.33 m is provided to the bottom of the condenser as shown. This facilitates the condensate and coolant to be discharged from the condenser under gravity, hence condensate extraction pump need not be provided. Another advantage of such arrangement is that the water from the hot well will not be able to rise into the condenser and flood the turbine due to the vacuum head in the condenser. The other details and working are similar to low level counterflow jet condenser.

(c) Ejector Condenser. Figure 21.5 shows the common form of ejector condenser. This condenser consists of a central vertical tube in which a number of cones or converging nozzles are fitted. The exhaust steam enters at the left and surrounds the central tubes with ports.The cooling water enters at the top under a head of 5–6 m and attains a high velocity while passing through the condenser. In flowing port the steam ports, the water produces vacuum. The vacuum causes the exhaust steam and air to flow through the ports in the tube and mix with the cooling water. The exhaust steam gets condensed, as a result of which the vacuum is further increased. The condensate and air after passing through converging cones pass through the diverging cones and in doing so water gains the momentum which forces out condensate and air against the atmospheric pressure. So in this case, no air pump is required.

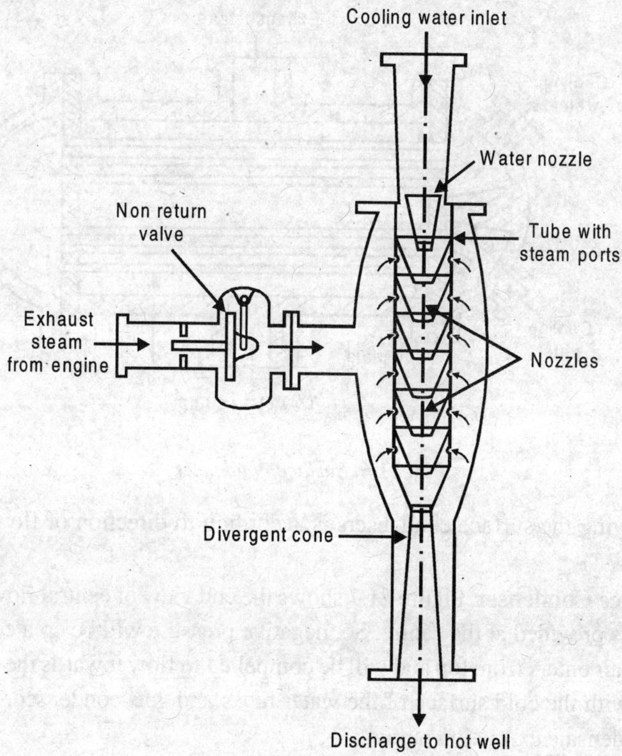

Fig. 21.5. *Ejector condenser*

This type of condenser is usually fitted with a non-return value to prevent a sudden backward flow of water into the engine exhaust pipe, in case of sudden failure of water supply to the condenser.

(c) Surface Condensers. (Cooling water flowing through tubes and steam from outside)

A surface condenser essentially consists of a Mild Steel rolled shell, closed at both the ends by dished ends (also called as water boxes) as shown. Ends of the shell and dishes are fitted with flanges. A tube sheet is provided in between the shell flanges and dish end flange on either side. A number of water tubes are fitted into these tube sheets. The ends of the tubes are either expanded or welded in order to have leak-proof joint.

The exhaust steam from the engine enters at the top of the condenser and gets condensed when it comes in contact with the cold surfaces of water tubes, through which cooling water is being circulated.

The cooling water enters at one end of the tubes (as shown in Fig. 21.6) situated at the lower half of the condenser and after flowing to the other end it returns in the opposite direction through the tubes situated in the upper half of the condenser. This is obtained by providing pass partition plates in the dished ends. In this type since the water traverses 2 times, it is called as 2-flow surface condenser. If the water traverses more than two times then it is called Multi-flow surface condenser.

As shown in Fig. 21.6 this condenser requires two pumps viz., wet air pump to remove air and condensate and secondly water circulating pump.

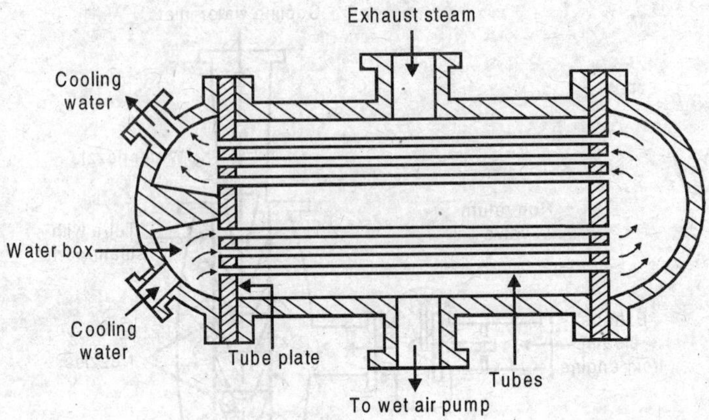

Fig. 21.6. *Surface condenser*

Another way of classifying the surface condensers is according to direction of flow of exhaust steam viz., (i) central flow (ii) down flow.

(i) Central Flow Surface Condenser. Figure 21.7 shows the end view of central flow surface condenser. In this case, air extraction pump is provided at the centre. So, negative pressure will be created at the centre. In this case, the exhaust steam and any air enters from top and will be compelled to flow towards the centre as shown. During its flow, it comes in contact with the cold surface of the water tubes and gets condensed, and the condensate will be removed by means of condensate extraction pump.

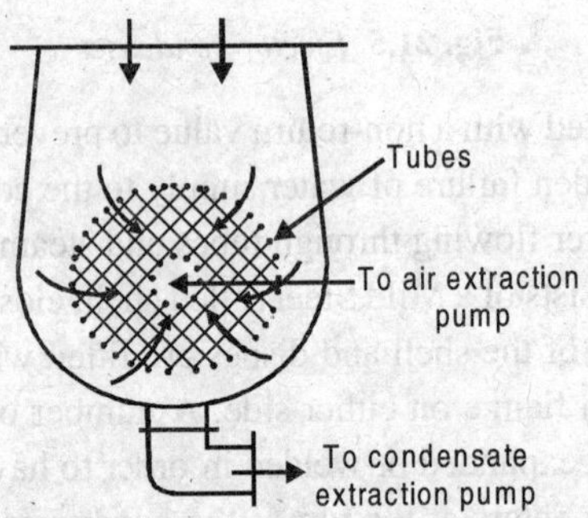

Fig. 21.7. *Central flow surface condenser*

(ii) Down Flow Surface Condenser. Figure 21.8 shows the end view of the down flow surface condenser. In this case a baffles is provided throughout the length of the condenser. It consists of two pumps, one is for removing the condensate and other is the air pump. In this case exhaust steam mixed with any air enters from top and decends down as shown by arrows. The steam when comes in contact with the cold surfaces of tubes it gets condensed and the air gets cooled. Air is further cooled when it flows from below the baffle. So the air is cooled to the minimum temperature before it is extracted. Due to this cooling of air, its specific volume will be reduced, so it reduces the pump capacity. So, it reduces the energy consumption for running the air pump.

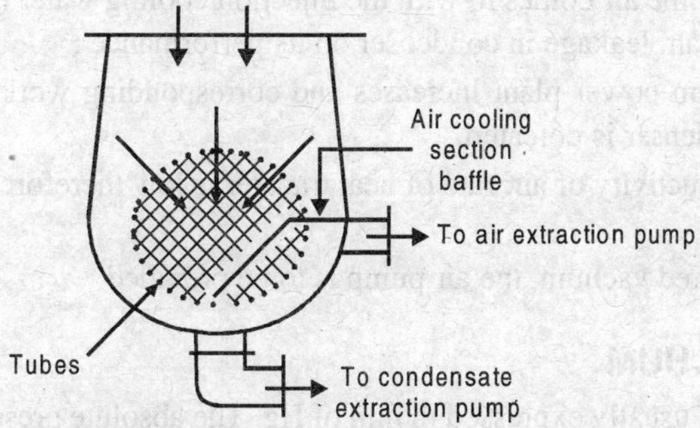

Fig. 21.8. *Down flow surface condenser*

(iii) Evaporative Condenser. When the supply of cold water is extremely limited, then the evaporative condenser is used since it uses very small amount of circulating and make up water (since the coolant is used again and again). However these condensers are used with small power plants.

In this case the exhaust steam from the steam engine enters into a coil of gilled pipes or series of tubes, the outlet of which is connected to the wet air pump.

The water which is pumped up by means of a pump is sprayed from the top and decends down. The water falling downwards forms a thin film over the pipes as it falls from one level to another and remainder water will be collected in the water tank.

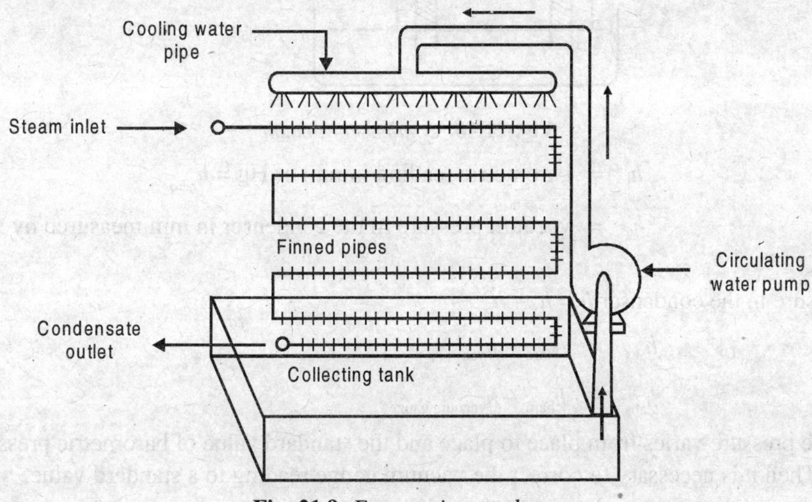

Fig. 21.9. *Evaporative condenser*

A natural or forced air current causes rapid evaporation of this film of water. The rapid evaporation of film of water results into condensation of steam. The evaporated vapour will be removed by means of air.

21.3 SOURCES OF AIR IN THE CONDENSER

The sources of air in the condenser are as follows :

1. Leakage through packing glands and microscopic holes in shell joints since the pressure inside the condenser is less than the atmospheric pressure.
2. Leakage through relief valves and other accessories.
3. Feed water contains dissolved air, which is liberated in the boiler during steam formation, the exhaust steam from the steam engine carries air.

4. In case of jet condenser some air comes in with the injection/cooling water (in which it is dissolved).

Following are the effects of air, leakage in condenser on its performance :

1. Back pressure in the steam power plant increases and corresponding work output decreases and the very purpose of using the condenser is defeated.
2. Due to poor thermal conductivity of air, rate of heat transfer is low therefore more cooling water is required to be supplied.
3. For maintaining the required vacuum, the air pump is to be provided.

21.4 CONDENSER VACUUM

The vacuum in the condenser is usually expressed in mm of Hg. The absolute pressure in the condenser is equal to the difference of barometric pressure and the vacuum pressure as shown in Fig. 21.10.

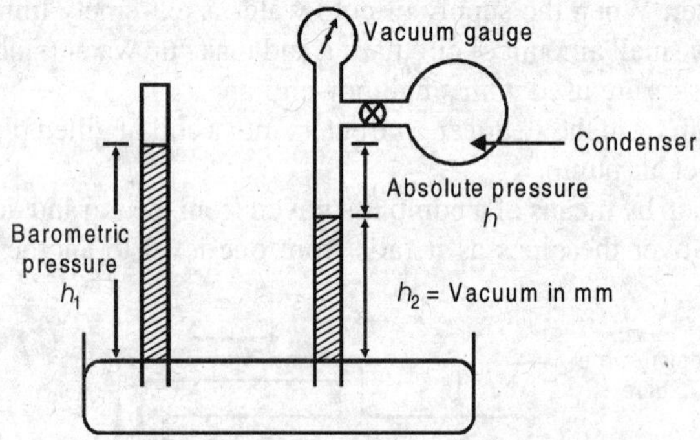

Fig. 21.10. *Condenser vacuum*

Let h_1 = barometric reading in mm of Hg = h_{baro}

h_2 = vacuum pressure in the condenser in mm measured by vacuum gauge = h_{vac}

∴ Absolute pressure in the condenser $h = h_1 - h_2$

or $h_{abs} = h_{baro} - h_{vac}$

$h_{abs} = h_{baro} - h_{vacuum\ gauge}$

The barometric pressure varies from place to place and the standard value of barometric pressure is 760 mm of Hg, at sea level. Then it is necessary to correct the vacuum gauge reading to a standard value.

The standard or corrected vacuum in mm of Hg.

$$h_s = 760 - (h_1 - h_2) = 760 - (h_{baro} - h_{vac})$$

Note 760 mm of Hg = 1.01325 bar

21.5 DALTON'S LAW OF PARTIAL PRESSURE

This law states that "Total pressure exerted by mixture of gases or mixture of gas (air) and vapour (steam) is equal to the sum of partial pressure of constituents, if they would occupy the same volume and are at the same temperature".

∴ Pressure of mixture in condenser = Partial pressure of steam + Partial pressure of air.

$$P_c = P_s + P_a$$

If the condenser temperature is known, P_s is found from steam tables and P_c can be read from the vacuum gauge.

∴ We can write, $$P_a = P_c - P_s$$

Knowing the partial pressure of air P_a its mass m_a can now be calculated with the help of ideal gas equation $P_a V_c = m_a R T_a$ where V_c = volume of condenser; T_a = temperature; R = characteristic gas constant of air.

Also note that : Each constituent is considered to occupy condenser volume V_c.

and $$V_c = \text{specific volume} \times \text{mass}$$

∴ $$V_c = m_s \times V_s = m_a \times V_a$$

$$\therefore \text{Mass of air/m}^3 \text{ of condenser} = \frac{m_a}{V_c} = \frac{1}{V_a}$$

$$\therefore \text{Mass of steam/m}^3 \text{ of condenser} = \frac{m_s}{V_c} = \frac{1}{V_s}$$

By the conservation of mass

$$m_c = m_s + m_a$$

21.6 VACUUM EFFICIENCY

It may be defined as the ratio of actual vacuum as recorded by the vacuum gauge to the ideal vacuum.

∴ Vacuum Efficiency

$$\eta_{\text{vacuum}} = \frac{\text{Actual vacuum recorded by the gauge}}{\text{Ideal vacuum}}$$

The ideal vacuum, means the vacuum due to steam alone when air is absent. In that case total pressure in the condenser will approach the pressure of the steam corresponding to saturated temperature of steam.

∴ $$\text{Ideal vacuum} = \text{barometric pressure} - \text{absolute pressure of steam}$$

$$\eta_{\text{vac}} = \frac{\text{Actual vacuum recorded by the gauge}}{\text{Barometric Pressure} - \text{Absolute Pressure of Steam}}$$

Let P_c = Actual absolute pressure in the condenser.

$$P_c = P_s + P_a$$

and P_{baro} = Barometric pressure

Then $$\text{Actual vacuum} = P_{\text{baro}} - P_c$$
$$= P_{\text{baro}} - (P_s + P_a)$$

and $$\text{Ideal vacuum} = P_{\text{baro}} - P_s$$

∴ $$\eta_{\text{vac}} = \frac{P_{\text{baro}} - P_c}{P_{\text{baro}} - P_s}$$

∴ $$\boldsymbol{\eta_{\text{vac}} = \frac{P_{\text{baro}} - (P_s + P_a)}{P_{\text{baro}} - P_s}}$$

21.7 CONDENSER EFFICIENCY

There is no standard method of determining the condenser efficiency but a method adopted by the well-known makers of steam turbine M/s Parson & Co. has been widely used in engineering practice.

Let t_i = Inlet temperature of cooling water.

t_o = Exit temperature of cooling water.

t_s = Saturated temperature of steam corresponding to actual absolute pressure in the condenser.

Then condenser efficiency

$$\eta_{cond} = \frac{\text{Temperature rise of cooling water}}{\text{Temperature of steam} - \text{Inlet temperature of condenser}}$$

$$\therefore \quad \eta_{condenser} = \frac{t_o - t_i}{t_s - t_i}$$

21.8 COOLING WATER REQUIREMENTS

In a condenser, cooling water absorbs heat from steam to be condensed and the temperature of cooling water increases. In jet condensers steam to be condensed directly mixes with cooling water and hence temperature of condensate and cooling water are same.

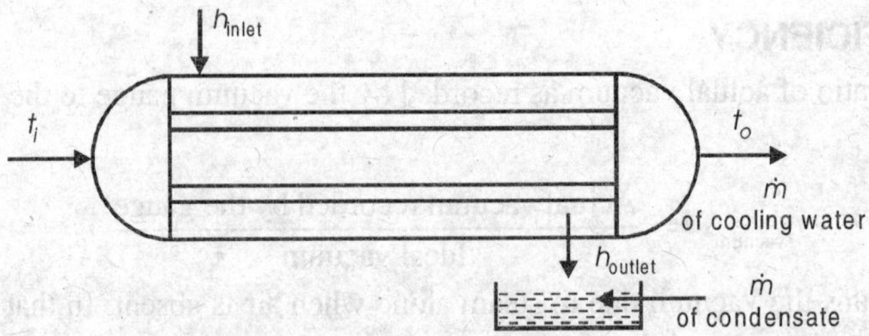

Fig. 21.11

Let t_i = Inlet temperature of cooling water.

t_o = Outlet temperature of cooling water.

$\dot{m}_{cw}$ = Mass flow rate of cooling water.

h_{inlet} = Enthalpy of steam entering condenser.

h_{outlet} = Enthalpy of condensate leaving condenser.

$\dot{m}_{condensate}$ = Mass flow rate of condensate.

Then heat lost by steam = Heat gained by cooling water

$$\dot{m}_{condensate}\left(h_{inlet} - h_{outlet}\right) = \dot{m}_{cooling\ water} \times C_{Pw} \times (t_o - t_i)$$

$$\dot{m}_{condensate}\left\{\left(h_f + x\,h_{f_g}\right) - h_{f_{outlet}}\right\} = \dot{m}_{cooling\ water} \times C_{Pw} \times (t_o - t_i)$$

From this equation quantity of cooling water or condition of steam entering the condenser can be calculated if other quantities are given.

The quantity of water used varies from 20–60 kg/kg of steam. And temperature rise of circulating water may vary from 10 – 20 °C.

21.9 CAPACITY OF AIR EXTRACTION PUMPS

We know that, absolute pressure in the condenser

$$P_c = P_s + P_a \qquad \text{(from Dalton's law)} \qquad \ldots(1)$$

or

$$P_a = P_c - P_s \qquad \ldots(2)$$

Also absolute pressure in the condenser P_c or $h_{abs} = h_{baro} - h_{vac}$ ...(3)

At the condenser temperature t°C, we can find P_s and volume of steam v_g from steam tables.

$v_g = v_a$ = volume of condenser if air and steam occupy condenser volume and are at the condenser temperature.

Then mass of air $$m_a = \frac{p_a \cdot v_a}{RT_a}$$

$$m_a = \frac{p_a \times x \cdot v_g}{RT_a} \quad ...(4)$$

Note: Generally steam entering the condenser is wet.

Units $$m_a = \frac{\frac{\text{kN}}{\text{m}^2} \times \frac{\text{m}^3}{\text{kg of steam}}}{\frac{\text{kJ}}{\text{kg of air K}} \times \text{K}}$$

$$m_a = \frac{\text{kg of air}}{\text{kg of steam}}$$

Note that along with condensate air is also removed by the wet air pump. If Z is the quantity of steam consumption of turbine in kg/hr then,

Mass of air to be handled by air pump

$$m_a = \frac{p_a \cdot x \cdot v_g}{RT_a} \times Z$$

$$= \frac{\text{kg of air}}{\text{kg of steam}} \times \frac{\text{kg of steam}}{\text{hr}}$$

Mass of air to be removed by air pump $m_a = \frac{\text{kg of air}}{\text{hr.}}$

21.10 AIR PUMPS

Air pump is a device which is used to remove air from the condenser and to maintain necessary vacuum in the condenser. (The sources of air leakage and their effects on condenser have already been discussed in earlier articles).

Air pumps are of two types:

(i) Dry air pump —Which removes only cold air

(ii) Wet air pump — It is used to remove cold air and condensate.

These pumps may be reciprocating or rotary type. We shall discuss now a very common form of reciprocating air pump called Edward's air pump.

As shown in Fig. 21.12, it consists of head or delivery valves, one of them is shown at 1. These valves are provided in the cover 2, which on the top of the pump barrel 3. The piston 4 has the conical bottom 5 and the bottom of the pump casing also follows the conical shape as that of piston. Ports 6 are provided in the pump barrel near its bottom end. Connection with the condenser is made at 7..

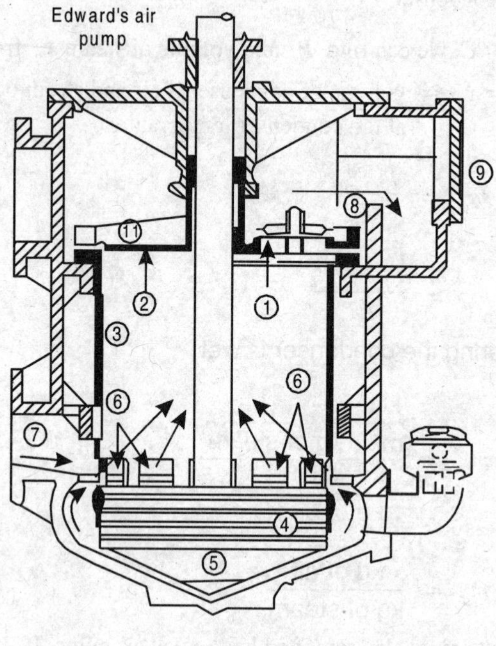

Fig. 21.12

During the downward stroke of the piston, a vacuum is produced in the barrel above the piston, the head valves being closed and as soon as the piston begins to uncover the ports 6, the air and water vapour from the condenser rush into the space above the piston. The condensate flows down in the pump casing. The conical end of downward moving piston displaces the condensate to the top of piston till the end of downward stroke.

Now during the upward stroke of the piston, some of the water flows down in the pump casing through the ports till the piston closes the ports. When the piston closes the ports the pressure of the charge above the piston will go on increasing and the charge of air, water vapour and condensate will be expelled through the head valves and then over the weir 8 on its way to the hot well. The weir ensures a head of water over the valves, so that if there is any leakage through these valves into the barrel, then it will be a leakage of water and not air.

The door 9 gives access to the head valves and 10 is the relief valve which opens to atmosphere, to release the pressure in case the pressure inside increase above the atmospheric pressure.

21.11 COOLING TOWERS

The function of the cooling tower is to cool the cooling water of the condenser, by the current of air flowing in the opposite direction.

A large amount of cooling water is required for condensation in large capacity power plants. If the water is freely available either from the river or lake, then the water can be directly pumped from the river to the condenser as shown in Fig. 21.13 (a).

If the water is not freely available, then a cooling tower has to be used for cooling the hot water of condenser. The cooled water can be used again as cooling water for the condenser.

Figure 21.13 (b) shows the Hyperbolic cooling tower. It is usually made of steel Reinforced cement concrete. The hot water from the condenser is supplied to the ring troughs which are placed at 8–10 m above the ground level. The nozzles are provided on the bottom side of troughs to break up water into sprays. The air rises up from the pond in the opposite direction of water flow and absorbs heat from the falling water spray. The cooled water is collected into a pond built below the tower. This type of cooling tower is generally used since it is very efficient, however, it needs about 3–5 % of make up water for compensating the evaporation losses.

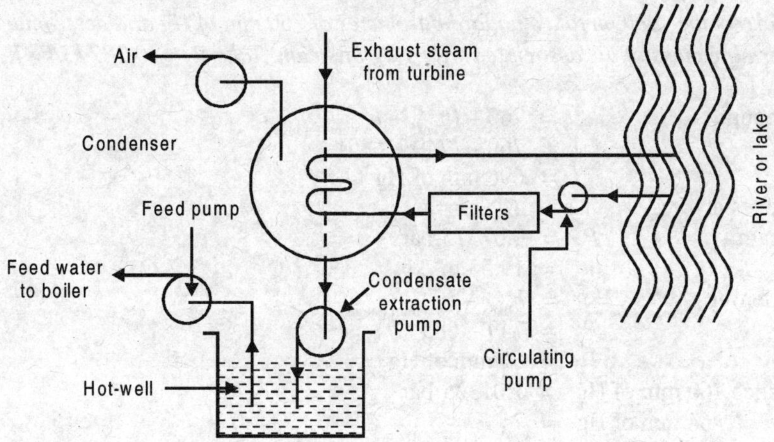

(a)

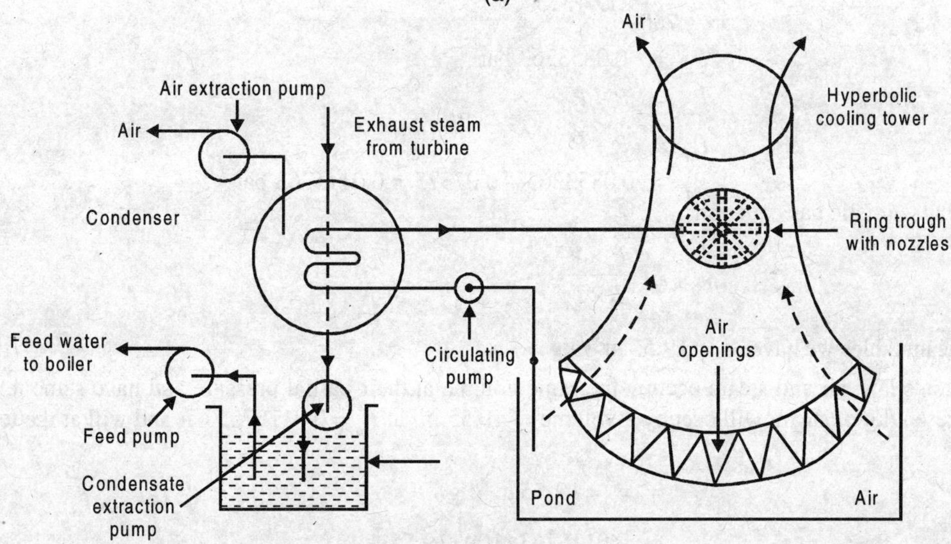

(b)

Fig. 21.13

The other types of cooling towers are:

(i) Natural draught towers. In this type the circulation of air is obtained by virtue of pressure difference of the air inside and outside the tower. Here no fan is required.

(ii) Forced draught tower. In this case, the circulation of air is obtained by means of fans provided at the bottom of the tower.

(iii) Induced draught tower. In this case, the circulation of air is obtained by providing a fan at the top of the tower.

■■■

SOLVED EXAMPLES

Example 21.1 *The temperature in a surface condenser is 40 °C and the vacuum is 685 mm of Hg. Barometer reads 749 mm of Hg. Correct the vacuum to a standard barometer of 760 mm of Hg and determine the pressure of steam and air. Also determine mass of air associated with 1 kg of steam. Take R = 0.287 kJ/kg K.*

Solution

Corrected vacuum $= 760 - (h_{\text{baro}} - h_{\text{vacuum gauge}})$
$= 760 - (749 - 685)$
$= 696$ mm of Hg

From steam tables at 40°C,

Pressure of steam $P_S = 0.07375$ bar

and $v_g = 19.55$ m³/kg

Absolute Condenser $= h = h_{\text{baro}} - h_{\text{vacuum gauge}}$

or $P_c = 749 - 685$

$P_c = 64$ mm of Hg

Also since 760 mm of Hg = 1.01325 bar

64 mm of Hg = ?

$$\frac{64}{760}\times 1.01325$$

$$P_c = 0.0853263 \text{ bar}$$

Since, $P_c = P_s + P_a$

$$P_a = P_c - P_s$$

$$P_a = 0.0853263 - 0.07375 = 0.0115763 \text{ bar}$$

From characteristic gas equation,

$$m_a = \frac{P_a \times V_a}{R \times T_a}$$

From steam tables we have $v_g = 19.55\,\text{m}^3/\text{kg}$

By Dalton's law, air and steam occupy the same volume at their partial pressure and have same temperature, have air present/kg of steam will occupy a volume of 19.55 m³ at $P_a = 0.0115763$ bar and will at the temperature 40 °C

i.e. $v_g = v_s = 19.55\,\text{m}^3/\text{kg}$

$$m_a = \frac{0.0115763\times 100\times 19.55}{0.287\times 273+40} \qquad \text{(Since 1 bar = 100 kN/m}^2\text{)}$$

∴ Resultant unit will be,

$$= \frac{\dfrac{\text{kN}}{\text{m}^2}\times\dfrac{\text{m}^3}{\text{kg}}}{\dfrac{\text{kJ}}{\text{kg K}}\times\text{K}} = \frac{\dfrac{\text{kJ}}{\text{kg of steam}}}{\dfrac{\text{kJ}}{\text{kg}}} = \frac{\text{kg}}{\text{kg of steam}}$$

$$m_c = \frac{0.0115763\times 100\times 19.55}{0.287\times 313}$$

$$\mathbf{m_a = 0.251936\ kg/kg\ of\ steam.}$$

Example 21.2 *In a surface condensating plant following data were obtained,*

Temperature of exhaust steam entering the condenser	= *42.32 °C*
Temperature of condensate leaving the condenser	= *35 °C*
Inlet temperature of cooling water	= *16.5 °C*
Outlet temperature of cooling water	= *30 °C*
Quantity of water/hour	= *46250 kg*
Quantity of condensate/hour	= *1190 kg*

Calculate the dryness fraction of steam entering the condenser.

Solution

$$\text{Heat lost by steam} = \text{Heat gained by cooling water}$$

$$\dot{m}_{\text{condensate}}\left[h_{\text{inlet}} - h_{\text{outlet}}\right] = \dot{m}_{cw} \times C_{p\ \text{water}} \times \Delta T$$

Now,

$$h_{\text{inlet}}\ \text{(at inlet it will be wet steam)} = h_f + x \times h_{fg}\ \text{at}\ 42.32^\circ\text{C}$$

$$= 177.14 + x \times 2401.33\ \text{by interpolation}$$

and

$$h_{\text{outlet}}\ \text{(at outlet it will be water)} = h_f\ \text{at}\ 35^\circ\text{C}$$

$$= 146.6\ \text{kJ/kg}$$

$$\therefore \quad = \frac{46250}{3600} \times 4.187 \times (30 - 16.5)$$

$$\therefore\ [(177.14 + 2401.33\,x) - 146.6 = 2193.18$$

$$177.14 + 2401.3\,x = 2339.78$$

$$x = 0.9006$$

$$\mathbf{x \cong 0.9}$$

Example 21.3 *The surface condenser has a vacuum of 685 mm of Hg and mean temperature 35°C. The barometer reads 760 mm of Hg. Dryness fraction of steam entering the condenser is 0.9. The condensate collected is 15.5 kg/min. Calculate the vacuum efficiency and mass of air to be removed by the dry air pump/min.*

Solution

From steam tables at 35 °C,

$$P_s = 0.5622\ \text{bar}$$

and

$$v_s = v_g = 25.24\ \text{m}^3/\text{kg}$$

Absolute pressure in the condenser $= h_{\text{baro}} - h_{\text{vac}} = P_c$

$$= 760 - 685$$

$$= 75\ \text{mm of Hg}$$

$$760\ \text{mm of Hg} = 1.01325\ \text{bar}$$

$$75\ \text{mm of Hg} = ?$$

$$= \frac{75}{760} \times 1.01325$$

$$P_c = 0.0999917\ \text{bar}$$

We also know that,

$$\eta_{\text{vac}} = \frac{P_{\text{baro}} - P_c}{P_{\text{baro}} - P_s} = \frac{P_{\text{baro}} - (P_s + P_a)}{P_{\text{baro}} - P_s}$$

$$= \frac{760 - 75}{760 - P_s}$$

$$P_s = 0.05622\ \text{bar}$$

$$1.01325\ \text{bar} = 760\ \text{mm}$$

$$0.05622 \text{ bar} = ?$$

$$= \frac{0.05622}{1.01325} \times 760$$

$$P_s = 42.168468 \text{ mm of Hg}$$

$$= 42.17 \text{ mm of Hg}$$

$$\eta_{vac} = \frac{760-75}{760-42.17}$$

$$= \frac{685}{760-42.17}$$

$$= 0.9543 = 95.43\,\%$$

Also,

$$P_c = P_s + P_a$$

$$P_a = P_c - P_s$$

$$= 0.0999917 - 0.05622$$

$$P_a = 0.0437717 \text{ bar}$$

$\therefore$ Mass of air present/kg of steam $= \dfrac{P_a v_a}{RT_a}$

$$v_a = x \times v_g$$

$$= 0.9 \times 25.24$$

$$\therefore \quad m_a = \frac{0.0437717 \times 100 25.24 \times 0.9}{0.287 \times (35+273)}$$

$$= 1.1248 \text{ kg of air/kg of steam.}$$

Given rate condensate $= 15.5$ kg/min

Air to be removed by air pump

$$= 15.5 \text{ kg/min} \times 1.1243 \text{ kg/kg of steam}$$

$$= 17.4351 \text{ kg/min}$$

Example 21.4 *A vacuum of 720 mm was obtained in a condenser when the barometer reads 765 mm. The temperature of condensate was 25°C. Determine the pressure of air and steam in the condenser and mass of air/kg of steam. Also determine the vacuum efficiency.*

Solution

Absolute pressure in the condenser $= h = h_{baro} - h_{vac} = P_c$

$$P_c = 765 - 720$$

$$= 45 \text{ mm of Hg}$$

Since 760 mm of Hg $= 1.01325$

45 mm of Hg $= ?$

$$\frac{45}{760} \times 1.01325 = P_c = 0.05995 \text{ bar}$$

Partial pressure of steam at saturation temperature of 25°C from steam tables,

$$P_s = 0.03166 \text{ bar}$$

Since 1.01325 bar $= 760$ mm of Hg

0.03166 bar $= ?$

$$= 23.25 \text{ mm of Hg}$$

Since,

$$P_c = P_s + P_a$$

$$P_a = P_c - P_s$$

$$P_a = 0.059995 - 0.03166$$

$$P_a = 0.028335 \text{ bar}$$

From steam tables, $v_g = v_s = v_a = 43.40\ m^3/kg$

$$\therefore \quad m_a = \frac{P_a v_a}{RT_a}$$

$$= \frac{0.028335 \times 100 \times 43.40}{0.287 \times (25+273)}$$

$$m_a = 1.4379 \text{ kg/kg cf steam}$$

$$\eta_{vac} = \frac{P_{baro} - P_c}{P_{baro} - P_s}$$

$$= \frac{765-45}{765-23.75} = \frac{720}{741.25}$$

$$= 97.13\%$$

Example 21.5 *The following observations were made during a test on surface condenser*

Barometric reading	= *760 mm of Hg*
Condenser vacuum	= *705 mm of Hg*
Mean temperature of condensate	= *35 ºC*
Condensate collected	= *2000 kg/hr*
Quantity of cooling water circulated	= *60,000 kg/hr*
Rise in temperature of cooling water	= *16 ºC*
Hot well temperature	= *28 ºC*

Determine :

1. η_{vac}

2. $\eta_{condenser}$

3. Quantity of steam entering the condenser

4. Mass of air present/m³ of condenser volume

Assume inlet temperature of water as 20 ºC

Solution

Absolute pressure in the condenser = $h_{baro} - h_{vac} = P_c$

$$P_c = 760 - 705$$

$$P_c = 55 \text{ mm of Hg}$$

$$P_c = 0.0733273 \text{ bar}$$

as 760 mm of Hg = 1.01325 bar

From steam tables at 35ºC

$$P_s = 0.05622 \text{ bar}$$

$$P_s = 42.168 \text{ mm of Hg}$$

as 1.01325 bar = 760 mm of Hg

0.05622 bar = ?

$$\therefore \quad P_a = P_c - P_s$$

$$= 0.0733273 - 0.05622 = 0.0171073 \text{ bar}$$

(i) $$\eta_{vac} = \frac{P_{baro} - P_c}{P_{baro} - P_s}$$

$$= \frac{760-55}{760-42.168}$$

$$= \frac{705}{717.832}$$

$$= 98.21\%$$

(ii) Saturated temperature of steam corresponding to condenser pressure.

$$P_c = 0.0733273 \text{ bar } t_s = 39.88\ ^{o}\text{C, by interpolation}$$

$$\therefore \quad \text{Condenser efficiency} = \eta_{cond} = \frac{t_o - t_i}{t_s - t_i}$$

$$= \frac{16}{39.88-20}$$

$$= 80.48\ \%$$

(iii) Heat lost by steam = Heat gained by cooling water

$$\dot{m}_{condensate}\left[h_{inlet} - h_{outlet}\right] = \dot{m}_{cooling\ water} \times C_p \times \Delta T$$

$$\dot{m}_{condensate}\left[(h_f + x \times h_{fg}) - h_f\right] = \dot{m}_{cooling\ water} \times C_p \times \Delta T$$

$$\frac{2000}{3600}\left[(146.56 + x \times 2418.8) - 4.187 \times 28\right] = \frac{60000}{3600} \times 4.187 \times 16$$

Note: If hot well temperature is not given the corresponding to Mean temperature of condensate.

$$2000[(146.56 + x \times 2418.8) - 117.24] = 4019520$$

$$(146.56 + x \times 2418.8) - 117.24 = 2009.76$$

$$146.56 + 2418.8 \times x - 117.24 = 2009.76$$

$$2418.8x = 2009.76 - 146.56 + 117.24$$

$$= 1980.44$$

$$\therefore \quad \mathbf{x = 0.8187696}$$

(iv)

$$m_a = \frac{P_a\ v_a}{RT_a}$$

$$= \frac{0.0171073 \times 100 \times 1}{0.287 \times (35+273)}$$

$$= 0.019\ \text{kg/m}^3 \text{ of condensate volume}$$

Note Units

$$= \frac{\frac{kN}{m^2} \times \frac{m^3}{m^3}\left(\frac{m^3 \text{ of air}}{m^3 \text{ of cond. volume}}\right)}{\frac{kJ}{kg\ K} \times K}$$

$$= \frac{kg}{m^3 \text{ cond. volume}}$$

Example 21.6 *Following readings were noted during a trial on surface condenser*

Mean temperature of condenser	= *30 ºC*
Condensate temperature	= *23 ºC*
Barometer reading	= *72 cm = 720 mm of Hg*
Condenser vacuum	= *66.375 cm = 663.75 mm of Hg*
Condensate collected	= *500 kg/hr*
Cooling water circulated	= *18000 kg/hr*
Rise in temperature of cooling water	= *14 ºC*

Calculate:ss

(i) Dryness fraction of steam entering the condenser

(ii) Capacity of air pump in m³/min.

(iii) Mass of air discharged in kg/min

Solution

Absolute Pressure in the condenser $= h_{\text{baro}} - h_{\text{vac}} = P_c$

$$P_c = 720 - 663.75$$
$$= 56.25 \text{ mm of Hg.}$$
$$= 0.0749938 \text{ bar}$$

Since 760 mm of Hg = 1.01325 bar

(i) Heat lost by steam = Heat gained by cooling water

$$\dot{m}_{\text{condensate}}\left[h_{\text{inlet}} - h_{\text{outlet}}\right] = \dot{m}_{\text{cooling water}} \times C_p \times \Delta T$$

$$\frac{500}{3600}\times\left[(h_f + x\times h_{fg}) - h_f\right] = \dot{m}_{\text{cooling water}} \times C_p \times \Delta T$$

$$= \frac{500}{3600}\times[(125.66 + x\times 2430.7) - 4.187\times 23]$$

$$\simeq \frac{1800}{3600}\times 4.187\times 14\,°\text{C}$$

∴ $x = 0.856$

(ii) Rate of steam flow in condenser

$$\dot{V} = \frac{500}{60}\times 32.929\times 0.856$$

$$= \frac{\text{kg}}{\text{min}}\times\frac{\text{m}^3}{\text{kg}} = \frac{\text{m}^3}{\text{min}}$$

Or note specific volume of steam/kg $v = x \times v_g$

$$= 0.856 \times 32.929$$
$$= 28.187 \text{ m}^3/\text{kg}$$

∴ $$\dot{v} = \frac{500}{60}\times 28.187 \text{ m}^3/\text{min}$$

∴ Capacity of air pump = 234.89 m³/min.

(iii) Mass of air $m_a = \dfrac{P_a v_a}{R T_a}$

From steam tables at 30ºC,

$$P_s = 0.04242 \text{ bar}$$

and P_c as we have found = 0.0749938 bar

Since $$P_a = P_c - P_s$$
$$= 0.0325738 \text{ bar}$$

$$\therefore \quad m_a = \frac{0.0325738 \times 100 \times 234.89}{0.287 \times 303}$$
$$= 8.7934 \text{ kg/min}$$

Note:

$$\frac{\frac{\text{kN}}{\text{m}^2} \times \frac{\text{m}^3}{\text{min}}}{\frac{\text{kJ}}{\text{kg K}} \times \text{K}} = \frac{\text{kg}}{\text{min}}$$

Example 21.7 *The absolute pressure in the condenser is 11.56 kPa. When the barometer reads 1 bar, the condenser temperature is 40 °C. Determine the partial pressure of air, η_{vac}, and the mass of air present in the condenser/kg of steam.*

Solution

Given Absolute pressure in the condenser = 11.56 kPa

$$P_c = 0.1156 \text{ bar}$$

∴ Vacuum in the condenser = Barometric pressure – Absolute pressure in the condenser

$$= 1 - 0.1156$$
$$= 0.8844 \text{ bar}$$

Partial pressure of steam at saturated temperature of 40°C,

From steam tables, $P_s = 0.07375$ bar

$$\therefore \quad P_a = P_c - P_s = 0.04185 \text{ bar}$$

$$\eta_{vac} = \frac{P_{baro} - P_c}{P_{baro} - P_s}$$
$$= \frac{1 - 0.1156}{1 - 0.07375}$$
$$= 95.59\ \%$$

Specific volume of steam at 40 °C = 19.546 m³/kg

= Specific volume of air = v_a

$$m_a = \frac{P_a v_a}{R T_a} = \frac{0.04185 \times 100 \times 19.546}{0.287 \times (40 + 273)}$$

$$\mathbf{m_a = 0.9106 \text{ kg/kg of steam}}$$

Example 21.8 *The following observations were made on a steam condensing plant,*

Barometer reading = 760 mm of Hg
Recorded vacuum = 700 mm of Hg
Mean temperature of condensation = 34 °C
Hot well temperature = 27 °C
Mass of condensate = 2120 kg/hr
Mass of cooling water = 66000 kg/hr
Rise in temperature of cooling water = 16 °C

Find :

(i) The state of steam entering the condenser
(ii) Mass of air present in kg/m³ of condenser
(iii) Vacuum efficiency

Solution

(i) Heat lost by steam = Heat gained by cooling water.

$$\dot{m}_{\text{condensate}}\left[h_{\text{inlet}} - h_{\text{outlet}}\right] = \dot{m}_{\text{cooling water}} \times C_p \times \Delta T$$

$$\dot{m}_{\text{condensate}}\left[(h_f + x \times h_{fg}) - h_{f\,(\text{hotwell})}\right] = \dot{m}_{\text{cooling water}} \times C_p \times \Delta T$$

$$\frac{2120}{3600}\left[(142.38 + x \times 2421.2) - 4.187 \times 27\right] = \frac{66000}{3600} \times 4.187 \times 16$$

$\therefore$ x = 0.8492 i.e., state of steam entering the condenser.

(ii) Absolute pressure in the condenser = $P_c = h_{\text{baro}} - h_{\text{vac}}$

$$= 760 - 700$$

$$= 60 \text{ mm of Hg}$$

$$P_c = 0.0799934 \text{ bar}$$

And P_s corresponding to condenser at temperature of 34 °C = 0.05318 bar

$\therefore$ $P_a = P_c - P_s$ = 0.0268134 barometer

$\therefore$
$$m_a = \frac{P_a v_a}{R T_a}$$

$$= \frac{0.0268134 \times 100 \times 1}{0.287 \times (34 + 273)}$$

$$\mathbf{m_a = 0.0304\ kg/m^3\ of\ condenser\ volume}$$

(iii)
$$\eta_{\text{vac}} = \frac{P_{\text{baro}} - P_c}{P_{\text{baro}} - P_s}$$

$$= \frac{760 - 60}{760 - P_s}$$

and $P_s = 0.05318$ bar = 39.89 mm of Hg

$\therefore$
$$= \frac{760 - 60}{760 - 39.89}$$

$$\eta_{\text{vac}} = \mathbf{97.21\%}$$

Example 21.9 *The following observations were recorded during a test on a steam condenser*

Recorded vacuum = *71 cm of Hg* = *710 mm of Hg*
Barometer reading = *76.5 cm* = *765 mm of Hg*
Mean condenser temperature = *34 °C*
Condensate collected = *1800 kg/hr*
Mass of cooling water = *57500 kg/hr*
Cooling water temperature rise = *17.5 °C*

Calculate:

(*a*) *Corrected vacuum to standard barometer of 76 cm of Hg*
(*b*) *Vacuum efficiency*
(*c*) *Condenser efficiency*
(*d*) *State of steam entering the condenser*

Assume inlet temperature of cooling water = *8.5 °C.*

Solution

(a) Corrected vacuum $= 760 - [h_{\text{baro}} - h_{\text{vac}}]$

$= 760 - [765 - 710]$

$= 705$ mm of Hg

From steam tables at saturated temperature of 34 °C, we get

$$P_s = 0.05318 \text{ bar} = 39.888 \text{ mm of Hg}$$

$$h_f = 142.4 \text{ and } h_{fg} = 2421.2 \text{ kJ/kg}$$

(b)
$$\eta_{\text{vac}} = \frac{P_{\text{baro}} - P_c}{P_{\text{baro}} - P_s}$$

$$P_c = h_{\text{baro}} - h_{\text{vac}}$$

$$= 765 - 710 = 55 \text{ mm of Hg}$$

$$P_c = 0.0733 \text{ bar}$$

$$\therefore \quad \eta_{\text{vac}} = \frac{765 - 55}{765 - 39.89}$$

ss
$$\eta_{\text{vac}} = \mathbf{97.92\%}$$

(c) Condenser Efficiency $= \eta_{\text{cond}} = \dfrac{t_o - t_i}{t_s - t_i}$

Now, saturated temperature corresponding to condenser pressure

$$P_c = 0.0733 \text{ bar}$$

From steam tables by interpolation $t_s = 39.88$°C

$$\eta_{\text{cond}} = \frac{17.5}{39.88 - 8.5}$$

$$= 55.76\%$$

(d) Heat lost by steam = Heat gained by cooling water

$$\dot{m}_{\text{condensate}}\left[h_{\text{inlet}} - h_{\text{outlet}}\right] = \dot{m}_{\text{cooling water}} \times C_p \times \Delta T$$

$$\dot{m}_{\text{condensate}}\left[(h_f + x \times h_{fg}) - h_{f\,(\text{hotwell})}\right] = \dot{m}_{\text{cooling water}} \times C_p \times \Delta T$$

$$= \frac{1800}{3600}[(142.4 + x \times 2421.2) - 4.187 \times 34]$$

$$= \frac{57500}{3600} \times 4.187 \times 17.5$$

$$\mathbf{x = 0.9667}$$

Note. Corresponding to condenser temperature since hot well temperature is not given.

Example 21.10 *Following observations recorded during a condenser test,*

Mean temperature of condenser = *35 °C*

Hot well temperature = *30 °C*

Condenser vacuum = *70 cm = 700 mm of Hg*

Barometer reading = *76 cm = 760 mm of Hg*
Condensate = *16.75 kg/min*
Cooling water = *39600 kg/min*
Rise in temperature of cooling water : 14 ºC
Inlet temperature of cooling water : 25 ºC
Find :
(*a*) *Vacuum efficiency*
(*b*) *Condenser efficiency*
(*c*) *Dryness fraction of steam entering*
(*d*) *Mass of air/m³ of condenser volume*

Solution

From steam tables at 35 ºC,

$$P_s = 0.05622 \text{ bar} = 42.17 \text{ mm of Hg}$$

$$v_g = 25.24 \text{ m}^3/\text{kg}, \quad h_f = 146.6, \quad h_{fg} = 2418.8$$

Absolute pressure in the condenser $P_c = h_{\text{baro}} - h_{\text{vac}}$

$$= 760 - 700 = 60 \text{ mm of Hg}$$

i.e., $P_c = 0.0799934$ bar

$\therefore$ $P_a = P_c - P_s = 0.0799934 - 0.05622 = 0.0237734$ bar

(a) Vacuum efficiency

$$\eta_{\text{vac}} = \frac{P_{\text{baro}} - P_c}{P_{\text{baro}} - P_s} = \frac{760-60}{760-42.17} = 97.52\%$$

(b) Condenser efficiency

$$\eta_{\text{cond}} = \frac{t_o - t_i}{t_s - t_i} = \frac{14}{41.54-25} = 84.64\ \%$$

Note. Saturated temperature corresponding to pressure $P_c = 0.0799934$ bar, $t_s = 41.54$, by interpolation.

(c)

$$m_a = \frac{P_a \times v_a}{R \times T_a} = \frac{0.0237734 \times 100 \times 1}{0.287 \times (35+273)} = 0.02715 \text{ kg/m}^3 \text{ of condenser volume}$$

(d)

$$\dot{m}\left[h_{\text{inlet}} - h_{\text{outlet}}\right] = \dot{m} \times C_p \times \Delta T$$

$$\frac{16.75}{60}\left[(h_f + x \times h_{fg}) - h_{f\,(\text{houtlet})}\right] = \dot{m} \times C_p \times \Delta T$$

$$\frac{16.75}{60}[(146.6 + x \times 2418.8) - 4.187 \times 30] = \frac{39600}{3600} \times 4.187 \times 14$$

$\therefore$ $x = 0.9469$

THEORY QUESTIONS

21.1. Enumerate the advantage of using steam condenser in a steam power plant. Explain the significance of Vacuum efficiency and Condenser efficiency.

21.2. How the condenser are classified?

21.3. Give merits and demerits of a surface condenser over the Jet condenser.

21.4. How does air come into the condenser? What are its effects on the performance of a condenser? List the merits of air extraction.

21.5. Explain Dalton's law of Partial Pressures.

21.6. Write a note on cooling towers.

PROBLEMS FOR PRACTICE

21.1. The following observations were recorded during a test on a surface condenser.

Condenser vacuum	: 70 cm of Hg
Barometer	: 76.5 cm of Hg
Mean condenser temperature	: 35 °C
Hot well temperature	: 28 °C
Condensate collected	: 1800 kg/hr
Air leakage	: 1 kg/1000 kg of steam
Cooling water circulated	: 80,000 kg/hr
Inlet temperature of cooling water	: 120 °C
Rise in temperature of cooling water	: 15 °C

Calculate

(i) η_{vacuum}

(ii) $\eta_{condenser}$

(iii) Quality of steam entering the condenser.

(iv) Capacity of wet extraction pump required is m³/hr

[*Ans.* 96.84 %; 60 %; 0.8886; 53.985 m³/hr]

21.2. The following data were obtained during a test on a steam condenser.

Condenser vacuum recorded	: 705 mm of Hg
Barometer reading	: 750 mm of Hg
Mean condenser temperature	: 35 °C
Hot well temperature	: 30 °C
Mass of cooling water circulated	: 54000 kg/hr
Inlet temperature of cooling water	: 9 °C
Outlet temperature of cooling water	: 28.5°C
Mass of condensate	: 1850 kg/hr

Find:

(i) η_{vacuum} corrected to standard barometric pressure

(ii) $\eta_{condenser}$

(iii) Quality of steam

(iv) Mass of air present is per m³ of condenser volume.

[*Ans.* 99.6 %; 75 %; 0.9769; 0.004315 kg/m³]

21.3. Following data refers to a surface condenser.

Vacuum in condenser	: 710 mm
Barometric reading	: 765 mm of Hg
Temperature in condenser	: 32 °C
Hot well temperature	: 30 °C
Cooling water circulated	: 48000 kg/hr
Cooling water inlet temperature	: 20 °C
Cooling water outlet temperature	: 34 °C
Condensate	: 1500 kg/hr

(a) Mass of air in kg/m³ of condenser volume.

(b) Dryness fraction.

(c) η_{vacuum}

[*Ans.* 0.029 kg/m³; 0.7617; 97.35 %]

21.4. The data below refers to a surface condenser

Condensate temperature	: 30 °C
Mean temperature of condenser	: 35 °C

Barometer : 76 cm of Hg
Condenser vacuum : 70 cm of Hg
Condensate collected : 930 kg/hr
Cooling water circulated : 36800 kg/hr
Rise in temperature of cooling water : 12.5 ºC

Calculate:

(i) Dryness fraction.

(ii) Capacity of air pump in m^3/min.

(iii) Mass of air handled in kg/hr.

[*Ans.* 0.845; 391.23 m^3/min; 259 kg/hr]

21.5. The following observations were made during the test on surface condenser.

Barometer reading : 760 mm of Hg
Condenser vacuum : 705 mm of Hg
Measure condenser temperature : 35 ºC
Condensate collected : 2000 kg/hr
Cooling water circulated : 60000 kg/hr
Rise in temperature of cooling water : 16 ºC
Hot well temperature : 28 ºC
Inlet water temperature : 15 ºC

Determine:

(i) η_{vacuum}

(ii) $\eta_{condenser}$

(iii) Mass of air present per m^3 of condenser volume.

[*Ans.* 98.2 %; 80 %; 0.1933 kg/m^3]

IMPORTANT POINTS TO BE REMEMBERED

1. *Atmospheric pressure at sea level*

$$P_{atm} = 760 \text{ mm of Hg}$$
$$= \mathbf{1.01325 \text{ bar}}$$

2. Pressure in the condenser

$$P_c = \text{Partial pressure of steam + Partial at pressure of air}$$

3. Volume of steam = Volume of air in condenser
4. *Vacuum efficiency* (η_{vac})

$$\eta_{vac} = \frac{\text{Actual vacuum recorded by the gauge}}{\text{Idea vacuum (with no air present)}}$$

$$= \frac{P_{bar} - (P_s + P_a)}{P_{bar} - P_s}$$

5. *Condenser efficiency* (η_{cond})

$$\eta_{cond} = \frac{\text{Temperature rise of cooling water}}{\text{Temperature of steam} - \text{Inlet temp. of water}}$$

$$= \frac{t_o - t_i}{t_s - t_i}$$

6. *Capacity of air extraction pump*

$$m_a = \frac{p_a \times xv_g}{RT_a} \times m_s \text{ (kg/hour)}$$

$$= \frac{\text{kg of air}}{\text{hour}}$$

7. Dryness fraction of steam inlet to condenser

$$m_{condensate} \begin{pmatrix} h_f + xh_{fg} \\ h_{fout} \end{pmatrix} = \begin{matrix} m_w C_{p_a} (t_0 - t_i) \\ \text{[Cooling water]} \end{matrix}$$

CHAPTER 22

Improved Rankine Cycle

22.1 INTRODUCTION

In chapter 14 we have observed that Carnot cycle is the most efficient thermodynamic cycle for all the heat engines utilising the heat energy and producing mechanical energy. We have also observed that there are limitations for the use of this cycle in practice.

The ideal practical cycles for IC Engines are Otto cycle (for Petrol and gas engines) and Diesel cycle (for Diesel or oil engines). Similarly the ideal practical cycle for steam—is the Rankine Cycle.

22.2 DIFFERENCE IN CARNOT AND RANKINE CYCLE

Figure 22.1 shows saturated cycle for both Carnot and Rankine cycle. Carnot cycle is 1'–2–3–4' and Rankine cycle 1–2–3–4.

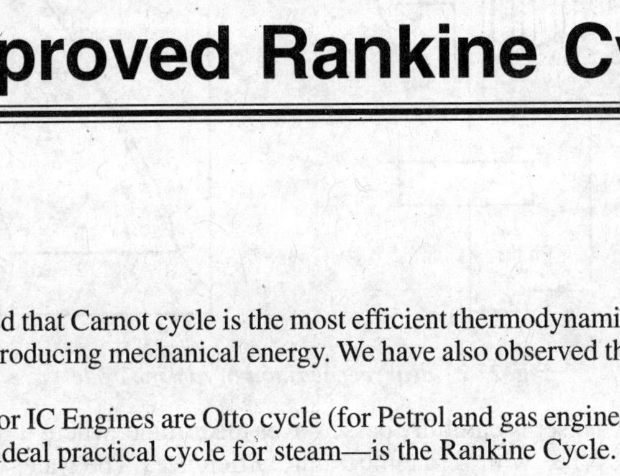

Fig. 22.1

Comparing these two cycles we observe that all the processes except water heating 1–1' are reversible processes. Rankine cycle efficiency is less than Carnot efficiency because of only one irreversible process 1–1'. If we could eliminate this irreversibility, we can get the Rankine cycle efficiency same as that of Carnot efficiency.

To eliminate this irreversible process, we have to heat water from 1–1' by some internal sources only. This type of the heating of water within the cycle is called Regenerative heating of water.

22.3 REGENERATIVE FEED HEATING CYCLE

Regenerative feed water heating can be achieved with the help of steam extracted or bled from the turbine, in the device called feed water heaters.

This irreversibility can be eliminated by the process of regeneration i.e., by internal and reversible heat exchange between the steam expanding in the turbine and water in the economizer. Figure 22.2 shows a flow diagram and a *T-s* diagram of the process for a saturated Rankine cycle.

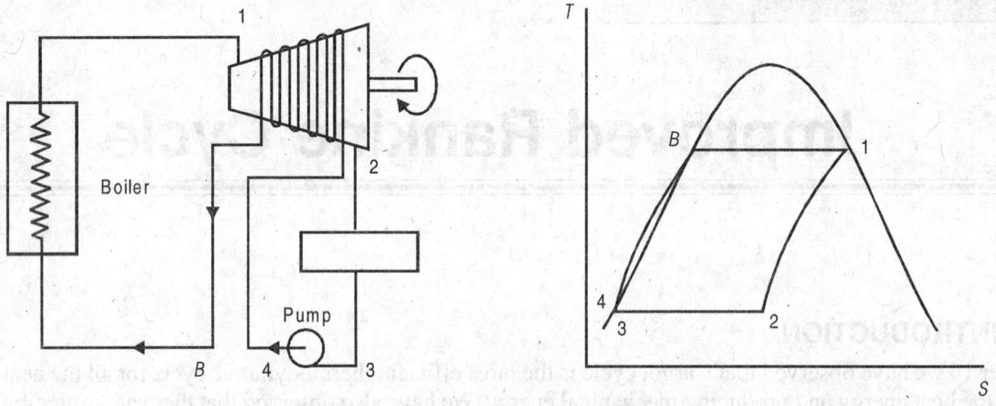

Fig. 22.2. *Ideal regeneration of rankine cycle*

The feed water after the condenser is carefully passed over the turbine, where it receives heat from the expanding steam reversibly at all times (i.e., with zero temperature difference). The water is progressively heated until it enters the steam generator at point *B*. The economizer and the irreversibilities introduced along with it, are thus eliminated. The cycle would receive and reject heat at constant temperature and would have same efficiency as that of Carnot cycle, working between same temperature limits.

This heating of feed water is called *ideal regenerative feed water heating*. It is however not employed as it has some serious limitations. They are as follows:

Limitations of Ideal Regeneration

1. Water is flowing outside the turbine in tubes while the steam is expanding over turbine blades placed inside. In this situation, no adequate surface is available for effective heat transfer.
2. The mass flow rates of water and steam are very large. Hence effectiveness of heat exchange is very low.
3. The final dryness of steam leaving the turbine is very low and amount of moisture content is unacceptable for proper turbine operation and efficiency.

22.4 PRACTICAL–REGENERATIVE FEED WATER HEATING

In practical method of feed water heating, economizer is not eliminated totally but its size is reduced. The steam expands adiabatically through the turbine as usual. The feed water at 4 (Refer Fig. 22.2) is heated in steps and not continuously, by the steam bled from the turbine at selected stages. Heating of liquid takes place in heat exchangers, called *feed water heaters.*

Feedwater heating was introduced in early 1920's and since then has become an important and essential component of large thermal power plants, which use between five to eight such feed water heaters.

Feedwater heating is done in steps and not continuously. The feedwater therefore enters the boiler below point *B*. Also, feedwater heaters themselves are heat exchangers (They are practically very much like condensers) and have their own irreversibilities and losses. Thus we cannot attain an ideal situation of Fig. 22.2 and Rankine cycle cannot attain Carnot efficiency. A well designed Rankine cycle is however a closest practical cycle to Carnot and hence is widely accepted for most power plants.

22.5 TYPES OF FEEDWATER HEATERS

There are three types of feedwater heaters in use. They are,

1. Open or direct contact type
2. Closed type with drains cascaded backward
3. Closed type with drains pumped forward.

These types are now discussed and analysed in detail.

22.5.1 Open or Direct Contact Feedwater Heaters

In open or direct contact feedwater heater, the extraction steam is mixed directly with incoming subcooled feedwater; to produce saturated water at extraction steam pressure Fig. 22.3 (a) and (b) shows a 'schematic flow diagram and corresponding *T-s* diagram for a Rankine cycle using two such feedwater heaters.

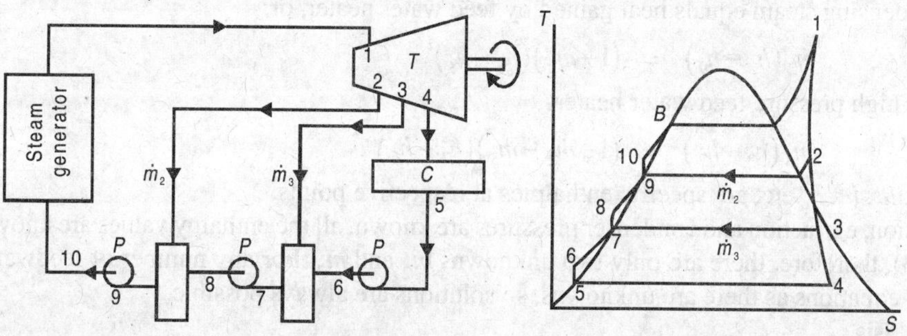

Fig. 22.3. *Schematic flow and T-s diagrams of a non-ideal superheat rankine cycle with two open-type feedwater heaters*

Condensate water leaves the condenser at 5 and is pumped to 6 to a pressure equal to that of the extraction steam at 3. The now-subcooled water at 6 and wet steam at 3 mix in the low pressure feed water heater to produce saturated water at 7.

Amount of bled steam $\dot{m}_3$ should essentially be such that subcooled water at 6, should be raised close to its saturation point. If it were much less, the water will be heated to below its saturation temperature and advantage of feedwater heating would be partially lost. If it were more there will be surplus steam at point 7. This would mean loss of turbine work and presence of steam-water mixture at 7 which is difficult to pump.

The pressure at 6-7 should be less than extraction steam pressure at 3 or else reverse flow of condensate water would enter the turbine at 3. The second pump pressurises saturated water at 7 to a subcooled condition at 8, which is at the pressure of extraction steam at 2. In the high pressure feed water heater superheated steam at 2, mixes with subcooled water at 8 to produce saturated water at 9. This is pressurised to 10 by a pump to enter the steam generator at its working pressure.

Steam at 2 or 3 loses its latent heat of vaporisation which is much larger than the gain of sensible heat by water at 8 or 6. Therefore the quantity of extracted steam $\dot{m}_2$ and $\dot{m}_3$ is only a small fraction of the steam passing through the turbine.

In addition to a condensate pump, one pump for each open feedwater heater is required. This is a disadvantage.

Open feed water heaters because of vigorous action of mixing remove non-condensible gases like Air, O_2, H_2, CO_2 from the feed water. They are therefore known as *deareating heaters* or *DA*.

Mass Flow Rates and Energy Balance

This is done on the basis of unit mass flow rate at the turbine inlet (point 1). The mass balance therefore is as follows:

Mass flow between points 1 and 2 = 1 kg/s
Mass flow between points 2 and 9 = $\dot{m}_2$
Mass flow between points 2 and 3 = $1-\dot{m}_2$
Mass flow between points 3 and 7 = $\dot{m}_3$
Mass flow between points 3 and 4 = $1-\dot{m}_2-\dot{m}_3$ ...(22.1)
Mass flow between points 4 and 7 = $1-\dot{m}_2-\dot{m}_3$
Mass flow between points 6 and 7 = $1-\dot{m}_2-\dot{m}_3$
Mass flow between points 7 and 9 = $1-\dot{m}_2$
Mass flow between points 9 and 1 = 1

where $\dot{m}_2$ and $\dot{m}_3$ are fraction of 1.

For energy balance, consider the low and high pressure feed water heaters. At low pressure feed water heater, heat lost by condensing steam equals heat gained by feed water heater, or,

$$\dot{m}_2(h_2-h_9) = (1-\dot{m}_2)(h_9-h_8) \quad ...(22.2)$$

Similarly for high pressure feed water heaters,

$$\dot{m}_3(h_3-h_7) = (1-\dot{m}_2-\dot{m}_3)(h_7-h_6) \quad ...(22.3)$$

where h_2, h_3, h_6, h_7, h_3, h_9 etc. are specific enthalpies at respective points.

Since admission, extraction and condenser pressures are known, all the enthalpy values are known. For two Eqs. (22.2) and (22.3), therefore, there are only two unknowns $\dot{m}_2$ and $\dot{m}_3$. For any number of feedwater heaters there will be as many equations as there are unknowns, so solutions are always possible.

The Cycle Analysis

Based on the mass flow rates, cycle analysis can now proceed as follows:

1. Heat added $\quad q_s = v(h_1-h_{10})$ kJ/kg

2. Turbine work $\quad W_T = 1(h_1-h_2)+(1-\dot{m}_2)(h_2-h_3)$
$+(1-\dot{m}_2-\dot{m}_3)(h_3-h_4)$ kJ/kg

3. Pump work $\quad |\sum W_P| = (1-\dot{m}_2-\dot{m}_3)(h_6-h_5)$
$+(1-\dot{m}_2)(h_8-h_7)+1(h_{10}-h_9)$ kJ/kg ...(22.4)
$= (1-\dot{m}_2-\dot{m}_3)v_5(p_6-p_5)+(1-\dot{m}_2)v_7(p_8-p_7)$
$+1\times v_9(p_{10}-p_9)$ kJ/kg

4. Heat rejected $\quad |q_r| = (1-m_2-m_3)(h_4-h_5)$ kJ/kg

5. Net cycle work $\quad = \Delta W_{|net|} = |W_T - W_P|$ kJ/kg

6. Cycle thermal efficiency $\quad \eta_{th} = \dfrac{\Delta W_{net}}{q_s}$

7. Work ratio $\quad$ W. R. $= \dfrac{\Delta W_{net}}{W_T}$ s

8. Specific steam consumption, SSC $= \dfrac{3600}{\Delta W_{net}}$ kg/k Wh

If turbine and pump efficiencies are considered then,

$$W_T\,.\,\text{actual} = \eta_{th} \times W_T$$

where η_t = turbine efficiency and η_p = pump efficiency

and
$$W_P\,.\,\text{actual} = \frac{W_P}{\eta_P}$$

22.5.2 Closed Type Feed Water Heaters with Drains Cascaded Backward

This is the simplest and most commonly used type in power plants. It is a shell and tube type heat exchanger. The feedwater passes through the tubes, and the bled steam, on the shell side transfers its heat energy to it and condenses. Thus, they are like surface condensers, operating at higher pressures than those in the main condenser. Since feedwater does not mix with the steam and passes through the tubes of successive close feed water heaters, only one condensate pump is sufficient to pressurise the water from condenser to the boiler. However, to reduce the pressure rise in one pump, one condensate pump before the feed water heaters and one feed water pump after the feed water heaters are used. If deareater heater is used, one boiler feed pump is automatically required after the same.

Figure 22.4 shows a simplified flow diagram and corresponding *T-s* diagram of a non-ideal superheat Rankine cycle. There are two feedwater heaters. One pump 5–6 pressurises the condensate, sufficient to pass through the two feedwater heaters and enters the steam generator at 8.

The bled steam condensed at each FWH is fed back to the next lower pressure FWH. The condensate from the last fwh is led back to the main condenser. One can imagine, then, a cascade from high pressure to lower-pressure heaters. Hence, the name.

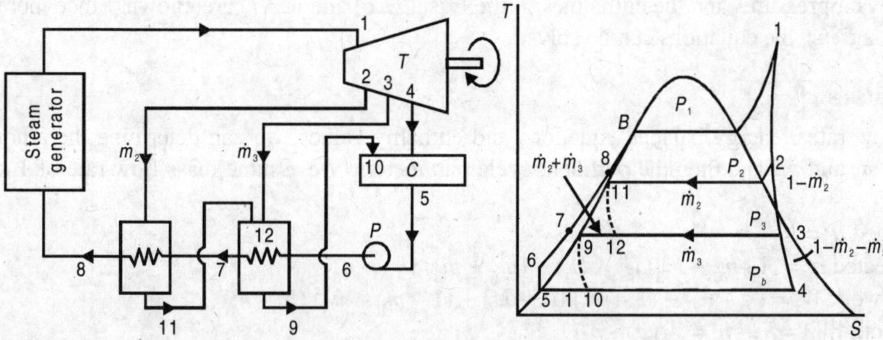

Fig. 22.4. *Schematic flow and T-s diagrams of a non-ideal superheat rankine cycle with two closed-type feed water heaters with drains cascaded backward*

Operation

Wet steam bled at 3 at the rate of $\dot{m}_3$ enters the low pressure FWH and heats the subcooled water at 6. The temperature of water at 7 is less than saturation temperature at 3. This difference is called the *terminal temperature difference* (TTD) and is defined for all closed FWHS.

TTD or TD = saturation temperature of bled steam – exit water temperature

The value of TTD for low pressure heaters is positive and is of the order of 3°C. For high pressure heaters it may be negative and lies between 0 and – 3°C. In case no data is available, temperature at 7 is taken equal to saturation temperature at 3.

The drain from low pressure FWH is led to the main condenser and enters it as two phase mixture at 10. The process 9 –10 is throttling process from the pressure corresponding to 9 and the main condenser pressure. Due to throttling there is some loss of availability. Also enthalpy at 9 and 10 are equal, as the process is constant enthalpy one.

The high pressure FWH receives superheated steam bled from the turbine at 2 that flows on the shell side at the rate of $\dot{m}_2$ and heats the water entering the tubes at 7. The inlet steam at 2 is superheated, hence temperature at 8 may be higher than saturation temperature of that steam and TTD may be negative. The drain at 11 may be slightly subcooled; and is throttled to the low pressure fwh which it enters at 12 as a two phase mixture. There it joins with the steam bled at 3 and aids the heating of water in low pressure FWH. The drain from low pressure fwh is thus $\dot{m}_2 + \dot{m}_3$, which is throttled to the main condenser at 10. The higher pressure exit water at 8 is led to the steam generator.

Mass Flow Rate and Energy Balance

A mass balance based on unit flow rate at turbine inlet point 1, is given clockwise by

$$\begin{aligned}
\text{Mass flow between 1 and 2} &= 1 \\
\text{2 and 3} &= 1 - \dot{m}_2 \\
\text{3 and 10} &= 1 - \dot{m}_2 - \dot{m}_3 \\
\text{10 and 1} &= 1 \\
\text{2 and 12} &= \dot{m}_2 \\
\text{3 and 12} &= \dot{m}_3 \\
\text{12 and 10} &= \dot{m}_2 + \dot{m}_3
\end{aligned} \quad \text{...(22.5)}$$

The energy balances on the high and low pressure heaters are given respectively by,

$$\dot{m}_2 (h_2 - h_{11}) = h_8 - h_7 \quad \text{...(22.6)}$$

$$\dot{m}_3 (h_3 - h_9) + \dot{m}_2 (h_{12} - h_9) = h_7 - h_6 \quad \text{...(22.7)}$$

For throttling processes, we have, $h_{12} = h_{12}$ and $h_{10} = h_9$.

All the relevant pressures and the enthalpies of the Eqs. (22.6) and (22.7) are known. Since there are only two unknowns $\dot{m}_2$ and $\dot{m}_3$, the equations can be solved.

Cycle Analysis

From mass flow rates, energy balance equations and enthalpy values, we can determine the amounts of steam extracted i.e., $\dot{m}_2$ and $\dot{m}_3$ and then the pertinent cycle parameters. We assume mass flow rate of 1 kg/s at turbine inlet (point 1).

1. Heat added $q_s = h_1 - h_8$
2. Heat rejected $q_r = (1 - \dot{m}_2 - \dot{m}_3)(h_4 - h_5) + (\dot{m}_2 + \dot{m}_3)(h_{10} - h_5)$
3. Turbine work $W_T = (h_1 - h_2) + (1 - \dot{m}_2)(h_2 - h_3) + (1 - \dot{m}_2 - \dot{m}_3)(h_3 - h_4)$
4. Pump work $|W_p| = h_6 - h_6 = v_5 (p_6 - p_5)$
5. Net cycle work $\Delta W_{net} = W_T - |W_p|$ ...(22.8)
6. Cycle thermal efficiency $= \eta_{th} = \dfrac{\Delta W_{net}}{q_s}$
7. Work ratio = W. R. = $\dfrac{\Delta W_{net}}{W_T}$
8. Specific steam consumption, SSC = $\dfrac{3600}{\Delta W_{net}}$

22.5.3 Closed Type Feed Water Heaters with Drains Pumped Forward

This is a second closed type FWH. It avoids throttling and the irreversibilities therein but has to use a small additional pump. Like previous closed type FWH, it is also a shell and tube type heat exchanger. The feedwater passes through the tubes, the steam passes over them; transfers heat and condenses. They do not mix and only one pump is sufficient to carry water from the condenser to the boiler. If, however, *DA* is used one more pump would be necessary.

The drain from these FWHS is pumped forward into the main feedwater line. Figure 22.5 shows a simplified flow diagram and corresponding *T-s* diagram for a non-ideal superheat Rankine cycle with two such FWHS. Like open FWH these heaters also require an additional pump per heater. But the pump is small and carry only fractional flows corresponding to $\dot{m}_2$ and $\dot{m}_3$ while pumps in open FWH carry full steam and water flow.

Operation

The drain of low pressure FWH at 13 is pumped forward to the main feedwater line and enters it at 14. It mixes with exit water from l.p. heater at 7 and forms mixture at 8. Point 8 is closer to 7 than 1.4 on the *T-S* diagram because flow at 7 is much greater than the drain flow $\dot{m}_3$.

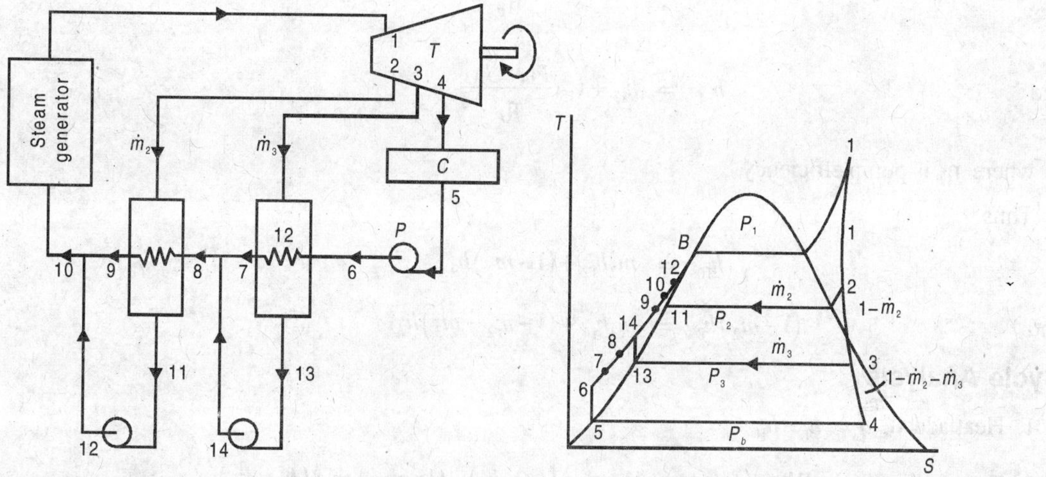

Fig. 22.5. *Schematic flow and T-s diagrams of non-ideal superheat Rankine cycle with two closed-type feedwater heaters with drains pumped forward*

The water at 8 enters h.p. heater and is heated to 9. The drain leaves the heater at 11, is pumped to 12. It mixes with the feedwater at 9 to form full feedwater flow at 10, which crow goes to boiler.

Mass Balance and Energy Balance

A mass balance based on a mass flow rate of 1 kg/s at turbine inlet (point 1) is given clockwise on *T-S* diagram by,

$$\begin{aligned}
\text{Mass flow between 1 and 2} &= 1 \\
\text{2 and 12} &= \dot{m}_2 \\
\text{2 and 3} &= 1 - \dot{m}_2 \\
\text{3 and 14} &= \dot{m}_3 \\
\text{3 and 7} &= 1 - \dot{m}_2 - \dot{m}_3 \\
\text{at 14} &= \dot{m}_3
\end{aligned} \qquad \ldots(22.9)$$

$$8 \text{ and } 9 = 1 - \dot{m}_2$$
$$\text{at } 12 = \dot{m}_2$$
$$10 \text{ and } 1 = 1$$

The energy balance equations are,

h.p. heater $\dot{m}_2 (h_2 - h_{11}) = (1 - \dot{m}_2)(h_9 - h_8)$ ...(22.10)

and l.p. heater $\dot{m}_3 (h_3 - h_{13}) = (1 - \dot{m}_2 - \dot{m}_3)(h_7 - h_6)$ ...(22.11)

The values of h_9 and h_7 are obtained from the temperatures t_9 and t_7 which are equal to the saturation temperature of the steam at each heater minus its terminal temperature difference, or

$$t_9 = t_{11} - \text{TTD h. p. heater} \quad \text{...(22.12a)}$$

and

$$t_7 = t_{13} - \text{TTD I. P. heater} \quad \text{...(22.12b)}$$

h_{10} needed for q_A and h_8 to be used in Eq. (22.10) are obtained from h_{11} and h_{14} respectively, or

$$h_{12} = h_{11} + v_{11}\frac{p_{12} - p_{11}}{\eta_p} \quad \text{...(22.13a)}$$

and

$$h_{14} = h_{13} + v_{13}\frac{p_{14} - p_{13}}{\eta_P} \quad \text{...(22.13b)}$$

where η_P is pump efficiency.

Thus

$$h_{10} = m_2 h_{12} + (1 - m_2) h_9 \quad \text{...(22.14a)}$$

and

$$(1 - m_2) h_8 = m_3 h_{14} + (1 - m_2 - m_3) h_7 \quad \text{...(22.14b)}$$

Cycle Analysis

1. Heat added, $q_s = h_1 - h_{10}$
2. The turbine work, $W_T = (h_1 - h_2) + (1 - \dot{m}_2)(h_2 - h_3) + (1 - \dot{m}_2 - \dot{m}_3)(h_3 - h_4)$
3. Pump work, $W_P = (1 - \dot{m}_2 - \dot{m}_3)(h_6 - h_5) + \dot{m}_3 (h_{14} - h_{13}) + \dot{m}_2 (h_{12} - h_{11})$
4. Net work $\Delta W_{\text{net}} = |W_T - W_P|$ ...(22.15)
5. Thermal Efiiciency, $\eta_{\text{th}} = \dfrac{\Delta W_{\text{net}}}{q_s}$
6. Work Ratio W. R. $= \dfrac{\Delta W_{\text{net}}}{W_T}$
7. Specific steam consumption, SSC $= \dfrac{3600}{\Delta W_{\text{net}}}$

22.6 COMPARISON OF CLOSED FEED WATER HEATERS

Table 22.1 comparison of closed feed water heaters.

Drains cascaded backward	*Drains pumped forward*
1. There is a loss of availability due to throttling.	1. No throttling, no loss of availability.
2. Number of pumps are less-one or two.	2. Number of pumps are more; one for each 3. feedwater heater.
3. Simple and less expensive system.	4. Complex and expensive system.
4. Drain is slightly subcooled. This imparts more energy to the water and reduces loss of availability.	5. Is used as the lowest pressure feed water heater in an otherwise all cascaded system. This prevents throttling of the combined cascaded flows to the condenser.

22.7 THE PLACEMENT OF FEEDWATER HEATERS

When number of FWHS are used, a natural question arises, as to where to place these FWHS in the cycle or what are the pressures at which steam is to be bled from the turbine, that will result in the maximum increase in the efficiency ?

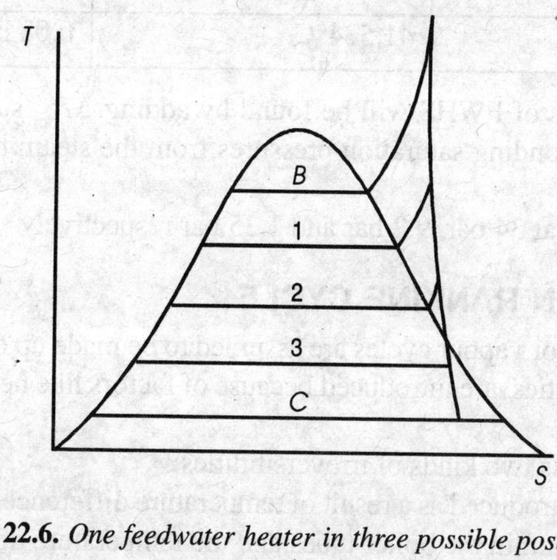

Fig. 22.6. *One feedwater heater in three possible positions*

To answer this let us consider for the sake of simplicity that only one FWH is used. The aim of using any feedwater heater is to heat the feedwater as close to the boiler temperature as possible. We may now place the feedwater heater at position 1, 2 or 3. (Fig. 22.6) T_B and T_C are saturation temperatures at boiler and condenser pressures respectively. Now at position 1 the heat transfers are because of temperature differences $T_B - T_1$ and $T_1 - T_C$. At position 3 the heat transfers are because of $T_B - T_3$ and $T_3 - T_C$. In either case one of the temperature differences is very large. To minimise the temperature differences, we have to select position 2 which is in the middle i.e., $T_B - T_2 = T_2 - T_C$. Thus we find the temperature which is *halfway between* T_B and T_C and then obtain the saturation pressure at that temperature. Feed water placed at this pressure gives us optimum position of maximum efficiency. Note that the temperature at which steam is actually bled from the turbine may be superheated.

If two FWHS are used, the temperature range $T_B - T_C$ is divided into three equal parts and for three FWHS it is divided into 4 equal parts.

Thus, in general, for n FWHS the optimum temperature rise per heater would be given by

$$\Delta T_{\text{opt}} = \frac{T_B - T_C}{n+1} \qquad \text{...(22.14)}$$

Consider for example a cycle where boiler pressure is 100 bar and condenser pressure is 0.08 bar and three FWHS are to be used. We find the pressures at which they are placed. From steam tables we get saturation temperatures at 100 bar and 0.08 bar. They are $T_B = 310.96°C$ and $T_C = 41.534°C$ respectively. From above Eq. (22.14) we have,

$$\Delta T_{\text{opt}} = \frac{310.96 - 41.534}{4} = 67.36$$

Table 22.2

Component	*Saturation temperature*	*Saturation pressures at sat. temperature*
The boiler	310.96°C	100 bar
The high pressure FWH	240.60°C	34 bar
The intermediate, pressure FWH	176.24°C	9.2 bar
The low pressure FWH	108.88°C	1.35 bar
The condenser	41.534°C	0.08 bar

Thus the saturation temperatures of FWHS will be found by adding ΔT_{opt} successively from $T_C = 41.534$ onwards and then finding the corresponding saturation pressures from the steam tables. Table 22.1 summarises the results.

Thus, the FWHS will be placed at 34 bar, 9.2 bar and 1.35 bar respectively.

22.8 IRREVERSIBILITIES IN RANKINE CYCLE

All ideal cycles whether gas cycles or vapour cycles are assumed to be made up of reversible processes only. In the actual cycles, however, irreversibilities are introduced because of factors like heat transfer through finite temperature difference. friction, mixing etc.

In Rankine cycle, we will discuss two kinds of irreversibilities:

1. **External irreversibilities.** Introduced as a result of temperature differences between combustion gases in the furnace and feed water or steam in the boiler tubes and the temperature differences between the condensing steam and condenser cooling water.
2. **Internal irreversibilities.** Introduced because of fluid friction and throttling in the turbine and the pump and mixing.

22.8.1 External Irreversibilities

In Fig. 22.4 (a), line ab represents combustion gases in a counter-flow furnace and $4 - B - 1$ represents water in a saturated Rankine cycle. Line *cd* represents condenser cooling water and $2 - 3$ represents condensing steam.

The temperature differences between line *ab* and $4 - B - 1 - 1'$ and between $2 - 3$ and line *cd* are not constant.

Figure 22.7 also shows temperature *vs* heat exchanger path length diagrams for (b) parallel flow and (c) counter flow steam generators and effect of flow directions in the steam generators. $L - 1$ and $e - B$ represent minimum approach point called *pinch point*; and is finite. For very small pinch point temperature difference, the overall temperature differences are low and hence reversibilities are lower but the size and cost of steam generators are

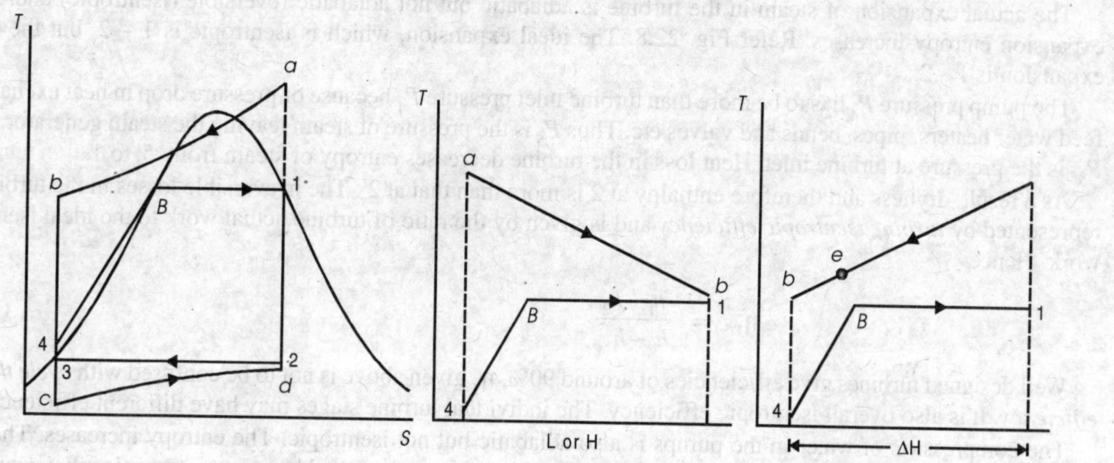

(a) External irreversibility with Rankine cycle

(b) Effect of flow direction on external irreversibility: (i) parallel flow, (ii) counter flow

Fig. 22.7

high. For very large pinch point temperature differences, size and cost of steam generators are low but overall temperature differences and irreversibilities are higher and plant efficiencies are lower. The most economical pinch point temperature difference is obtained by optimization of fixed cost (based on capital costs), and operating costs (based on fuel costs).

Figure 22.7 also shows that overall temperature differences between the combustion gases and water and steam are greater for parallel flow than counter flow heat exchange in steam generators. Heat transfer considerations also favour counter flow as it gives higher overall heat transfer coefficients. The counter flow is thus favoured over parallel flow.

22.8.2 Internal Irreversibilities

They are result of fluid friction, throttling and mixing in turbines, pumps, and pressure losses in pipes, bends, valves etc.

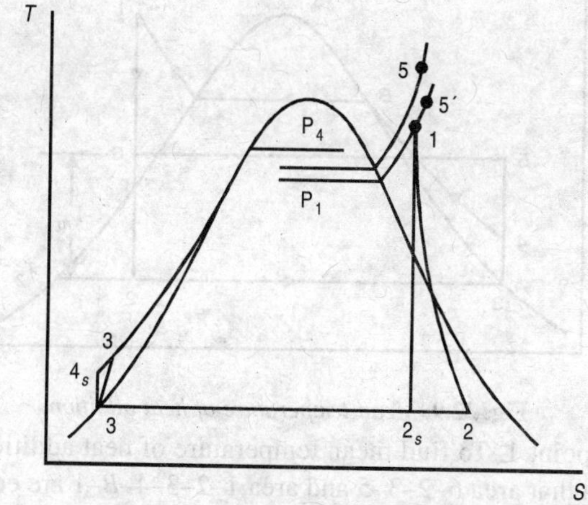

Fig. 22.8. *A T-s diagram of an internally irreversible superheat rankine cycle*

The actual expansion of steam in the turbine is adiabatic but not adiabatic reversible (isentropic) and during expansion entropy increases. Refer Fig. 22.8. The ideal expansion, which is isentropic is $1 - 2_s$ but the actual expansion is $1 - 2$.

The pump pressure P_4 has to be more than turbine inlet pressure P_1 because of pressure drop in heat exchangers, feed water heaters, pipes, bends and valves etc. Thus P_5 is the pressure of steam leaving the steam generator, while $P_{5'}$ is the pressure at turbine inlet. Heat loss in the turbine decreases entropy of steam from 5' to 1.

As a result, dryness and therefore enthalpy at 2 is more than that at 2_s. The irreversible losses in the turbine are represented by *turbine isentropic efficiency* and is given by the ratio of turbine actual work to the ideal isentropic work. Hence,

$$\eta_T = \frac{h_1 - h_2}{h_1 - h_{2s}} \quad \text{...(22.15)}$$

Well designed turbines give efficiencies of around 90%. η_T given above is not to be confused with *cycle thermal efficiency*. It is also overall isentropic efficiency. The individual turbine stages may have different efficiencies.

The compression of water in the pumps is also adiabatic but not isentropic. The entropy increases. The ideal compression is $3 - 4_s$ while the actual one is $3 - 4$. The losses are represented by *pump isentropic efficiency* which is the ratio of ideal isentropic work to the actual work taken by the pump. Hence

$$\eta_P = \frac{h_3 - h_{4s}}{h_3 - h_4} \quad \text{...(22.16)}$$

In both Eqs. (22.14) and (22.16), the smaller quantity is in the numerator. The actual pump work can be obtained by

$$|W_p| = \frac{h_{4s} - h_3}{\eta_P} = \frac{v_3(p_4 - p_3)}{\eta_P}$$

Thus we pay penalty for the turbine and pump irreversibilities. The turbine produces less work and the pump absorbs more work.

22.9 MEAN TEMPERATURE OF HEAT ADDITION

In Rankine cycle, heat is added at constant pressure but not at constant temperature. Refer Fig. 22.9. There is a large difference between temperature of water at the entry of the economiser at point 4 and that of steam at

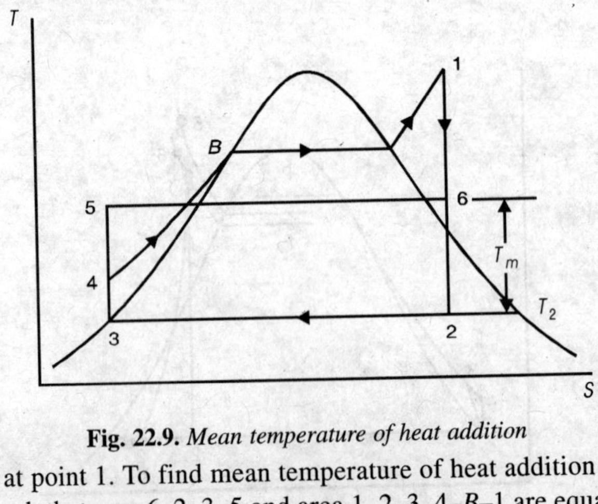

Fig. 22.9. *Mean temperature of heat addition*

the exit of the superheater at point 1. To find mean temperature of heat addition T_m, construct a rectangle 6–2–3–2–5 on the base 2–3 such that area 6–2–3–5 and area 1–2–3–4–*B*–1 are equal. Recall a similar construction of a rectangle on indicator diagram of an IC engine to find the mean effective pressure P_m.

Now heat supplied is,

$$q_s = h_1 - h_4 = T_m(s_1 - s_4)$$

$$\therefore \quad T_m = \frac{q_s}{s_1 - s_4}\ ^\circ\text{C} = \frac{h_1 - h_4}{s_1 - s_4}$$

Now heat rejected is

$$q_r = T_2(s_1 - s_4)$$

$$\therefore \quad \eta_{\text{Rankine}} = 1 - \frac{q_r}{q_s} = 1 - \frac{T_2}{T_m}$$

Since T_2 is fixed by condenser pressure for which very little improvement is possible over present practices, the efficiency largely depends on T_m. Larger the value of T_m greater is the cycle efficiency.

It will be realised now that superheating of steam will raise the value of T_m over the saturated cycle and hence will increase the cycle efficiency.

22.10 REHEAT CYCLE

Raising the turbine inlet pressure i.e., boiler outlet pressure or boiler working pressure as well as raising the temperature of superheat improves the thermal efficiency of the Rankine cycle. Higher working pressures and temperatures increase mean temperature T_m of heat addition; and takes T_m closer to gas line *ab* reducing the irreversibilities. Refer Fig. 22.7 (b).

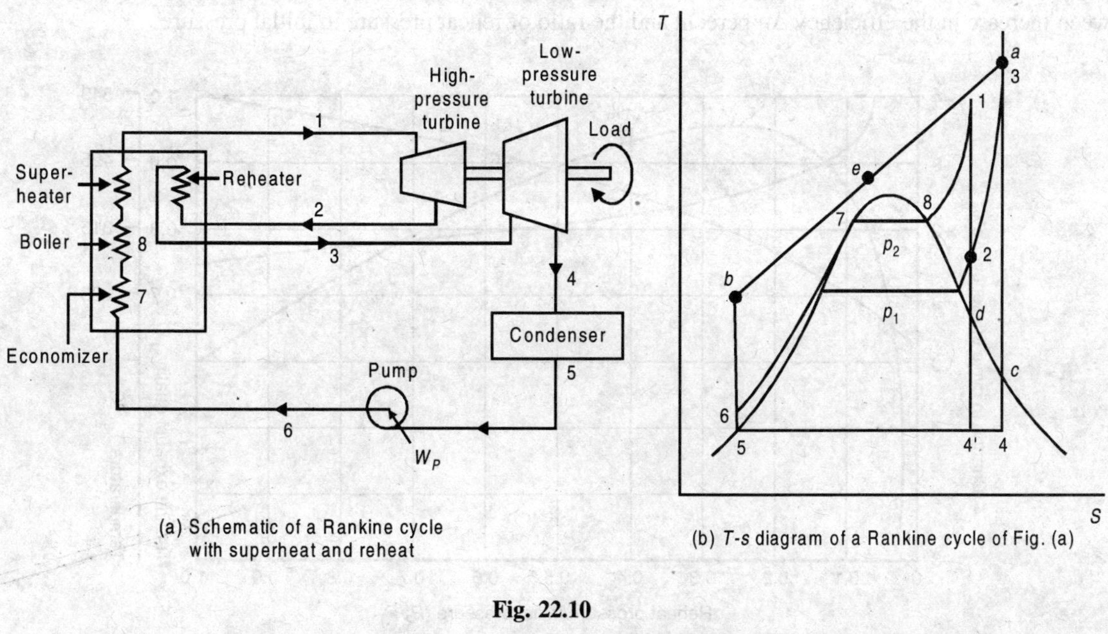

(a) Schematic of a Rankine cycle with superheat and reheat

(b) *T-s* diagram of a Rankine cycle of Fig. (a)

Fig. 22.10

Increasing the inlet pressure from P_1 to P_2 however results in the lowering of the final dryness fraction from 4 to 4' and all the stages after point *d* on the expansion line get progressively wet steam. The moisture present in the steam, in the form of suspended water particles hit the high speed turbine blades with a large relative velocities and erode them.

To improve the thermal efficiency of the cycle over and above that of superheat cycle as well as to improve the final dryness of steam, reheating of steam is used.

Figure 22.10 (a) and (b) shows a schematic flow diagram and *T-s* diagram of Rankine cycle that uses superheat and resuperheat or reheat.

Reheating is resuperheating of steam at point 2 after it expands in high pressure section of turbine. The steam after expansion is sent to the steam generator where it is reheated (resuperheated) to the original temperature with the help of hot flue gases. The reheated steam now expands in the low pressure section of the turbine to the condenser pressure.

22.10.1 Advantages of Reheat Cycle

There are following advantages of reheat cycle is,

1. In reheat cycle, heat is added twice; from 6 to 1 and from 2 to 3. This increases the value of T_m—the temperature of heat addition and keeps the evaporator, superheat reheat portion from 7 to 3 closer to the flue gas line *ae*. This results in higher cycle efficiency.
2. It improves the final dryness of steam (from 4' to 4) and reduces turbine blade erosion.
3. It increases cycle work by increasing the cycle area. This increases the turbine power.

Modern fossil fuel power plants employ superheat and at least one stage of reheat. Some employ two. More than two reheats however, result in cycle complications and increased capital costs, that are not justified by improvements in efficiency.

22.10.2 Effect of Reheat Pressure Ratio on Cycle Efficiency

The pressure p_2 at which the steam is reheated affects the cycle efficiency. Figure 22.11 shows the relation between increase in the efficiency $\Delta\eta$ percent and the ratio of reheat pressure to initial pressure.

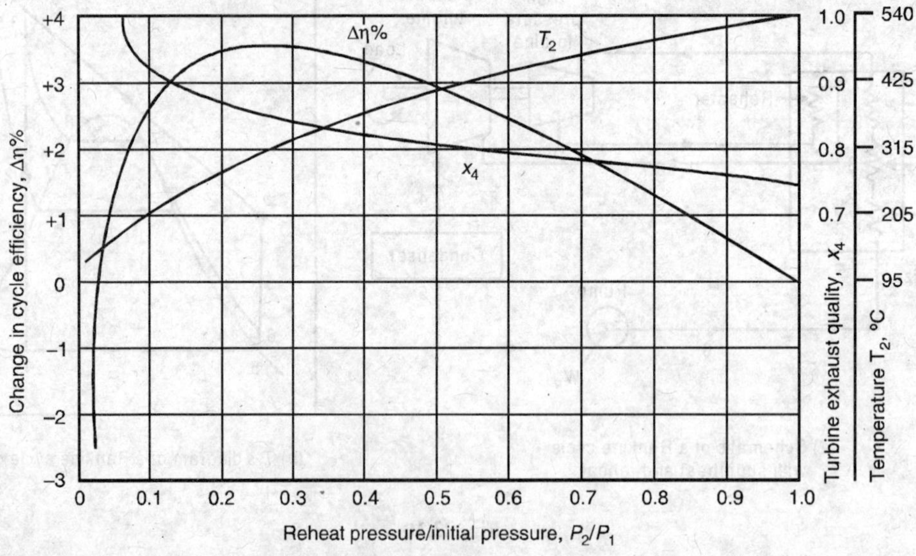

Fig. 22.11 *Effect of reheat-to-initial pressure ratio on efficiency, high-pressure turbine exit temperature, and low-pressure turbine exit quality*

At $p_2/p_1 = 1$ there is no reheat and $\Delta\eta = 0$. As p_2 is lowered the efficiency improves and reaches the maximum at p_2 equal to about 20 percent of p_1. Lowering p_2 further decreases the efficiency and at p_2/p_1 less than say 0.025 the efficiency decreases and $\Delta\eta$ becomes negative. The optimum pressure ratio is between 0.2 and 0.25.

Figure 22.8 is plotted for initial pressure of 170 bar and initial and reheat temperature of 500° C, but the optimum value of p_2/p_1 holds good for most modern power plants. The graph shows variation of T_2 and X_4 also. Note that reheating improves quality of steam.

The superheat reheat plant is often designated by $p_1/T_1/T_3$ in bar and degrees centigrade. The above case for example is 170/500/500 while a double reheat may be designated as 150/490/500/510 etc.

22.11 TYPICAL LAYOUT OF A STEAM POWER PLANT

Figure 22.12 shows a typical layout of a steam power plant. It shows steam and water circuits of the plant which includes reheaters and feedwater heaters. Note the positions of four closed feedwater heaters and one open feedwater heater (deaerator). Note also that there are two *hp*, two *ip* and one *lp* feedwater heater.

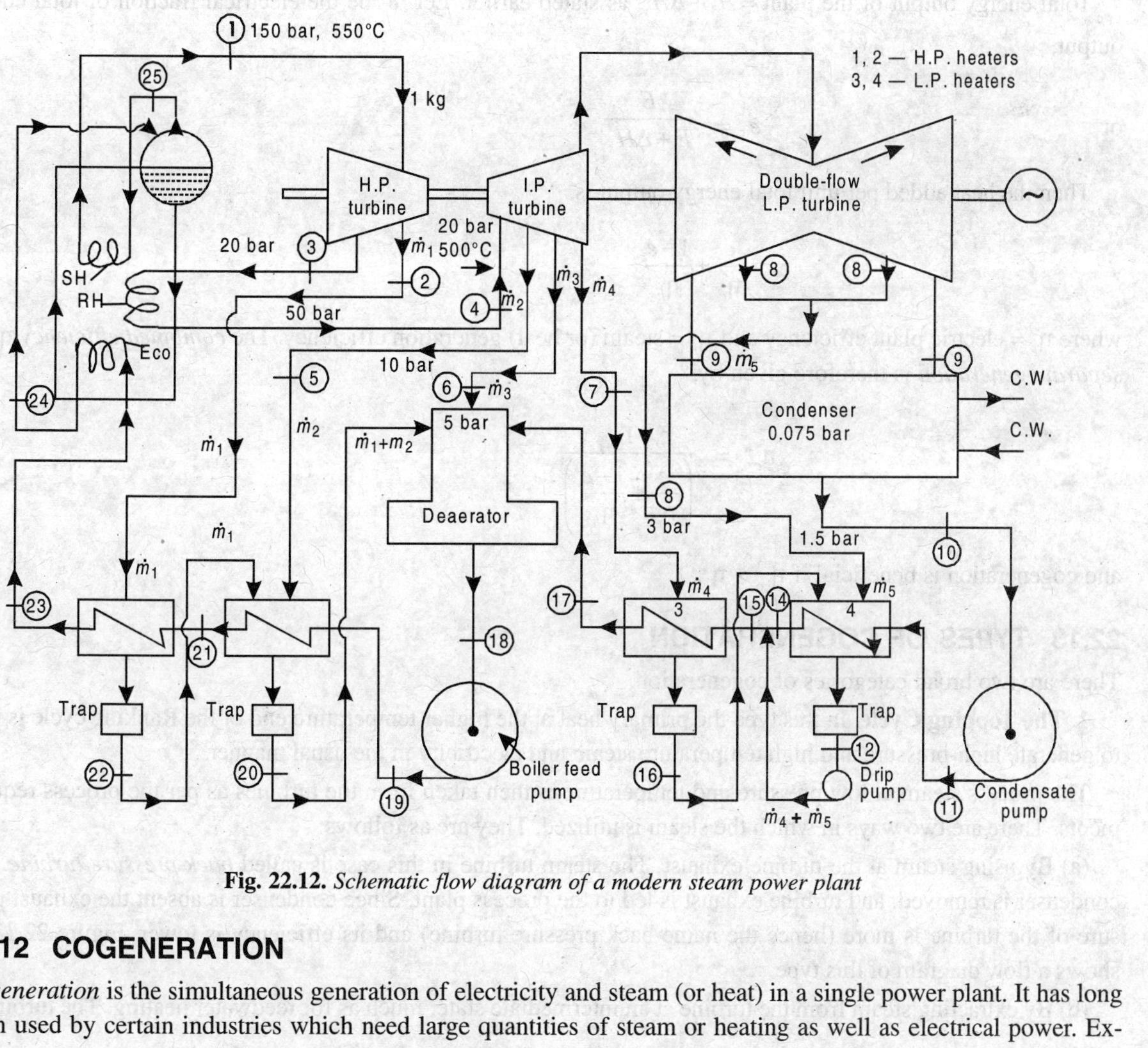

Fig. 22.12. *Schematic flow diagram of a modern steam power plant*

22.12 COGENERATION

Cogeneration is the simultaneous generation of electricity and steam (or heat) in a single power plant. It has long been used by certain industries which need large quantities of steam or heating as well as electrical power. Examples are cement factories, paper mills, chemical industries, textile plants, fertilizer plants, extraction plants, edible oil industries, refineries etc. Cogeneration is used if the electricity thus produced is cheaper, than that purchased from the state electricity boards.

In a cogeneration plant if,

$$E = \text{electrical energy generated}$$

$$\Delta H_s = \text{heat energy generated or heat in the process steam}$$

and

$$Q_s = \text{total heat added to the plant}$$

then the cogeneration plant efficiency is given by

$$\eta_{co} = \frac{E + \Delta H_s}{Q_A} \quad \text{...(22.17)}$$

Now consider the generation of electricity and steam separately.

Total energy output of the plant = $E + \Delta H_s$ as stated earlier. Let 'e' be the electrical fraction of total energy output,

or

$$e = \frac{E}{E + \Delta H_s}$$

Then the heat added per unit total energy output is,

$$\frac{e}{\eta_e} + \frac{1-e}{\eta_h}$$

where η_e = electric plant efficiency and η_h = steam (or heat) generation efficiency. The *combined efficiency* η_c *for separate generation* is therefore given by,

$$\eta_c = \frac{1}{\left(\frac{e}{\eta_e} + \frac{1-e}{\eta_h}\right)}$$

and cogeneration is beneficial if $\eta_{co} > \eta_c$.

22.13 TYPES OF COGENERATION

There are two broad categories of cogeneration.

1. The Topping Cycle. In this type the primary heat at the higher temperature end of the Rankine cycle is used to generate high-pressure and high temperature steam and electricity in the usual manner.

The process steam at low pressure and temperature is then taken from the turbines as per the process requirements. There are two ways in which the steam is utilized. They are as follows:

(a) By using steam at the turbine exhaust. The steam turbine in this case is called *back pressure turbine*. The condenser is removed, and turbine exhaust is led to the process plant. Since condenser is absent the exhaust pressure of the turbine is more (hence the name back pressure turbine) and its efficiency is lower. Figure 22.13 (a) shows a flow diagram of this type.

(b) By extracting steam from the turbine at an intermediate state; much as for feedwater heating. The turbine of this type is known as *extraction turbine* or *passout turbine*. Since condenser is used the efficiency of the cycle is higher; and steam extracted has much higher pressure and temperature. Figure 22.13 (b) shows a flow diagram of this type.

The main advantage of this cycle is its ability to use wide range of fuels including wood, coal, petroleum, coke, refuse etc. and hence is used where low quality refuse and byproduct fuels are readily available.

This type however is seldom used because of its high initial cost and low power-to-heat ratio; especially at high process-steam pressures.

2. Bottoming Cycle. Here the first primary heat at high temperature is directly used for process requirements. The low grade (low temperature and availability) waste heat from process plant is used to generate electricity. Figure 22.14. The efficiency of the cycle is obviously low, –10 percent for low and upto 20 percent at high temperature applications. Given that the waste heat is free, the cycle economics may be favourable.

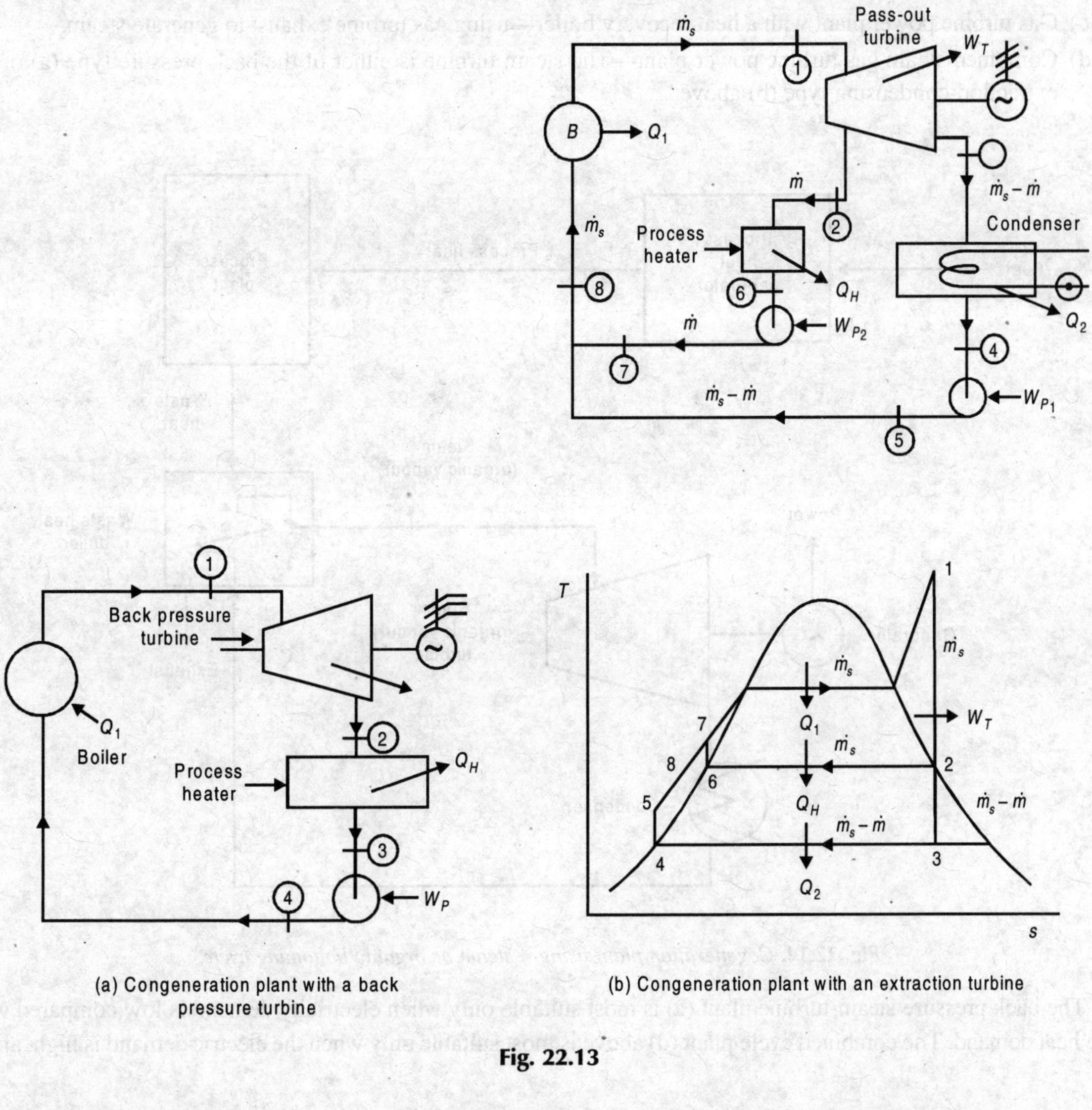

(a) Congeneration plant with a back pressure turbine

(b) Congeneration plant with an extraction turbine

Fig. 22.13

The main application of bottoming cycles are in the processes of the chemical and material processing industries such as cement, limestone and steel.

Since the temperature of waste heat is low, hydrocarbon fluids are often used as working medium.

22.13.1 Types of Topping Cycle

Since the bottoming cycle has very low combined efficiency, it is of little thermodynamic or economic interest and only the topping cycle therefore can provide true savings in primary energy. In addition, most process applications require a low grade (low temperature and availability) steam which can be conveniently produced in a topping cycle.

Some of the arrangements for cogeneration in topping cycle are as under.

(a) Steam turbine power plant with a back pressure turbine—refer section 22.13.
(b) Steam turbine power plant with steam extraction from condensing turbine—refer section 22.13.
(c) Gas turbine power plant with a heat recovery boiler—using gas turbine exhaust to generate steam.
(d) Combined steam gas turbine power plant—The steam turbine is either of the back pressure type (a) or the extraction-condensing type (b) above.

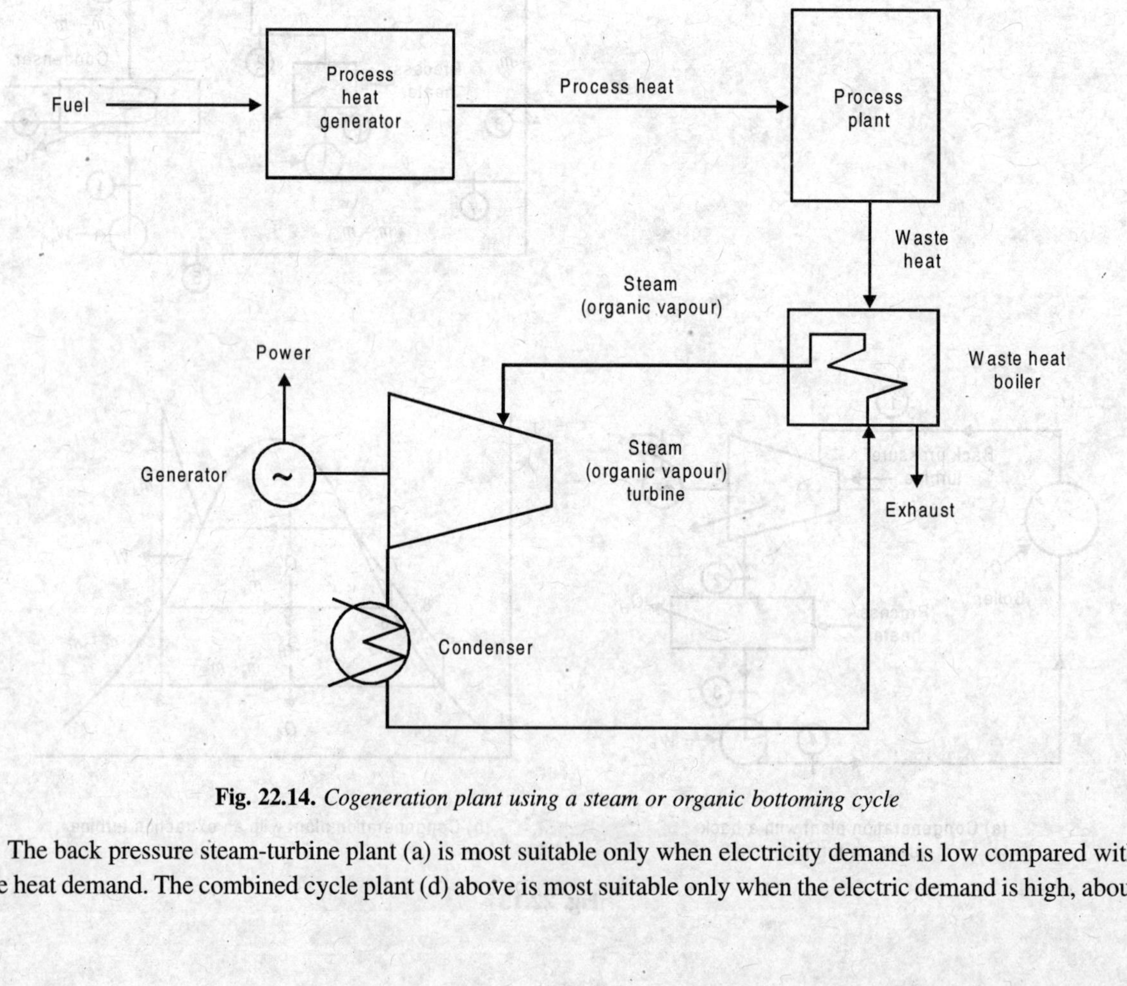

Fig. 22.14. *Cogeneration plant using a steam or organic bottoming cycle*

The back pressure steam-turbine plant (a) is most suitable only when electricity demand is low compared with the heat demand. The combined cycle plant (d) above is most suitable only when the electric demand is high, about comparable to heat demand or higher. The gas turbine cycle (c) lies in between. Only the extraction condensing plant (b) above is suitable over a wide range of ratios.

22.14 COGENERATION WITH A GAS TURBINE PLANT

Gas turbines are used by themselves in wide range of services. Some of them are:

1. As a powering unit for all types of air-crafts.
2. For driving mechanical equipment such as pumps, compressors in industrial plants.
3. As a powering unit for railways and racing cars.
4. To drive electric generators in electric power plants — To produce power for peak loads.

There is also growing interest in using gas turbines for cogeneration and in combined cycle plants. The working temperatures of Gas Turbine are quite high (Max. temperature about 1200 – 1300°C) and hot exhaust gases leaving the turbine at about 500°C, still have a great deal of heat energy stored in them. This heat energy can be used for various purposes such as,

1. Drying and direct heating—In food industry.
2. Heating of fluids in process industries—to heat air, oil, water which is a working medium in a process plant.
3. Producing steam by using a waste recovery boiler—The steam produced may be used for process application or for generation of electricity by using steam turbines. This is known as "cogeneration" for the former and "combined cycle" for the latter.

Figure 22.15 shows a schematic flow diagram of a gas turbine plant used for cogeneration, with a heat recovery boiler.

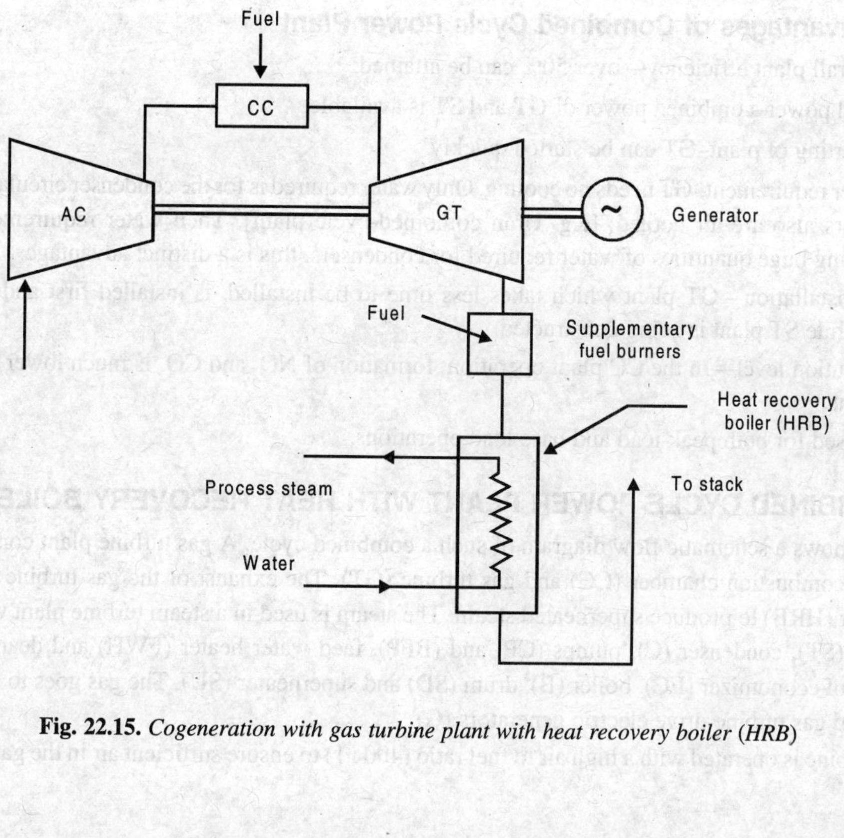

Fig. 22.15. *Cogeneration with gas turbine plant with heat recovery boiler (HRB)*

A simple gas turbine cycle consisting of an air compressor (AC), combustion chamber (CC) and gas turbine (GT) is used. The turbine exhaust gases are passed through a heat recovery boiler (HRB) to generate process steam which is then taken out for process plant. Supplementary fuel burners are provided to burn additional fuel to increase the mass flow rate of process steam for a short period, if required.

22.15 COMBINED CYCLE POWER PLANTS

In *combined cycle* power plants both gas and steam turbines supply electrical power to the grid. The idea of combined cycle has grown out of the need to improve the simple Brayton-cycle efficiency by utilising the waste heat in the turbine exhaust gases. This can be done by regeneration also. The regeneration however hassome serious drawbacks as given below.

1. Regeneration does not increase output. In fact regeneration reduces turbine pressure ratio and hence the net plant output by a few percent.
2. Because of large heat transfer surfaces and large gas and air piping; regeneration makes the plant costly.
3. With regeneration the optimum pressure ratio for maximum efficiency reduces sharply (from 20 – 30 to 7 – 10). This reduces the power output.

Thus raising the efficiency of a gas turbine plant by regeneration is costly. A means was saught whereby both efficiency and power are increased. A solution was found in using the large quantity of energy in the turbine exhaust, to generate steam for a steam turbine plant. This is a natural solution. Gas turbine is a high temperature machine (1100°C – 1625°C) while steam turbine is a low temperature machine (550 – 650ºC). The joint operation with gas turbine at "hot end" and a steam turbine at "cold end" is called a *combined-cycle power plant.*

22.15.1 Advantages of Combined Cycle Power Plant

1. High overall plant efficiency—over 50% can be attained.
2. High total power–combined power of GT and ST is available.
3. Quick starting of plant–GT can be started quickly.
4. Low water requirement–GT needs no cooling. Only water required is for the condenser circulation. Sometimes condensers also are air cooled. [e.g. Uran combined-cycle plant]. Then water requirement is very low. Considering huge quantities of water required for condensers, this is a distinct advantage.
5. Phased installation—GT plant which takes less time to be installed, is installed first and stars producing power, while ST plant is being constructed.
6. Low pollution level—In the CC plant operation, formation of NO_x and CO_2 is much lower than that in coal fired plant.
7. Can be used for both peak load and base load operations.

22.16 COMBINED CYCLE POWER PLANT WITH HEAT RECOVERY BOILER

Figure 22.16 shows a schematic flow diagram of such a combined cycle. A gas turbine plant consists of air compressor (AC), combustion chamber (CC) and gas turbine (GT). The exhaust of the gas turbine goes to the heat recovery boiler (HRB) to produce superheated steam. The steam is used in a steam turbine plant which consists of steam turbine (ST), condenser (C), pumps (CP) and (BFP), feed water heater (FWH) and deareator (DA). The HRB consists of economizer (EC), boiler (B), drum (SD) and superheater (SU). The gas goes to stack after HRB. Both steam and gas turbine drive electric generators (G).

The gas turbine is operated with a high air to fuel ratio (400 : 1) to ensure sufficient air in the gas turbine exhaust for further combustion. Plant output can be increased for short periods during load peaks by burning, additional full in supplementary fuel burners. HRB has no refractory lining nor water walls. Hence, its temperature is limited to 750°C.

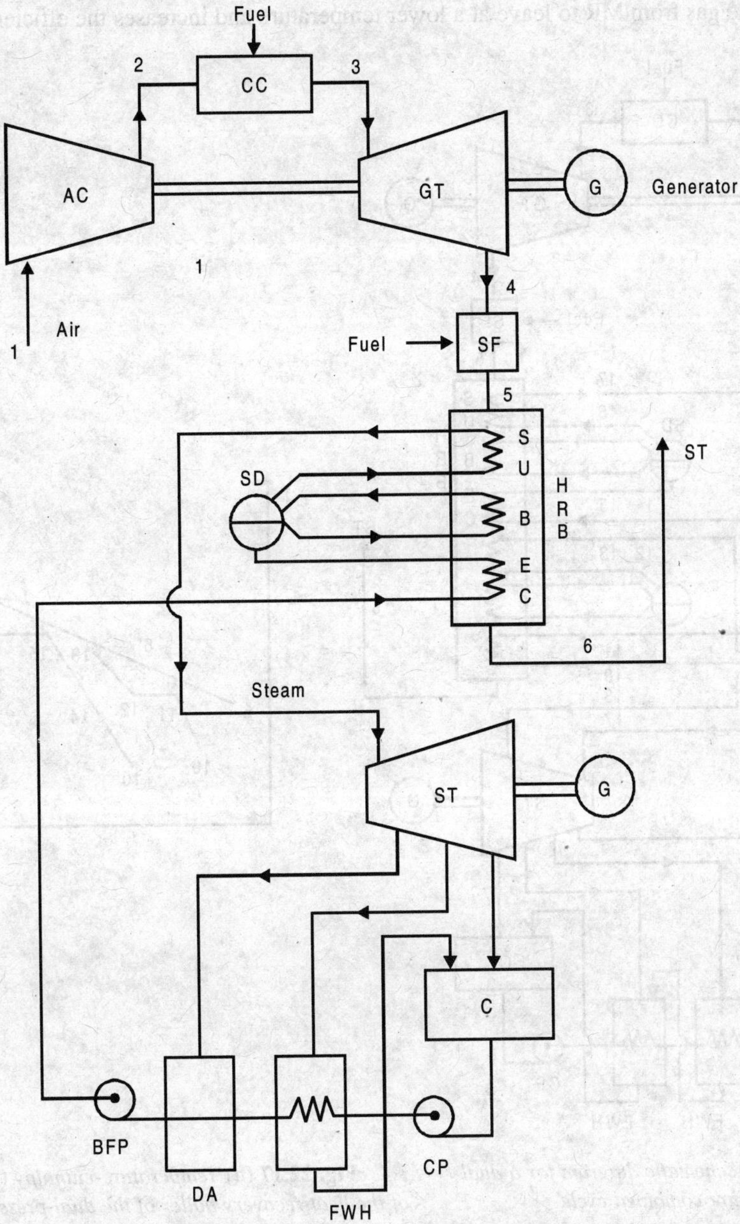

Fig. 22.16. *Schematic flow diagram of a combined cycle with heat-recovery boiler (HRB)*

In small plants steam turbine output is less than the gas turbine output by as much as 500 percent. In large plants, steam turbine output is greater than the gas turbine output by upto 8 : 1.

The supplementary firing equipment (SF) is interposed between the GT and HRB. The steam cycle is designed for high efficiency with reheat and full set of feed water heaters.

22.17 COMBINED CYCLE PLANT WITH MULTIPRESSURE STEAM

This plant allows the gas from MR to leave at a lower temperature and increases the efficiency of the plant.

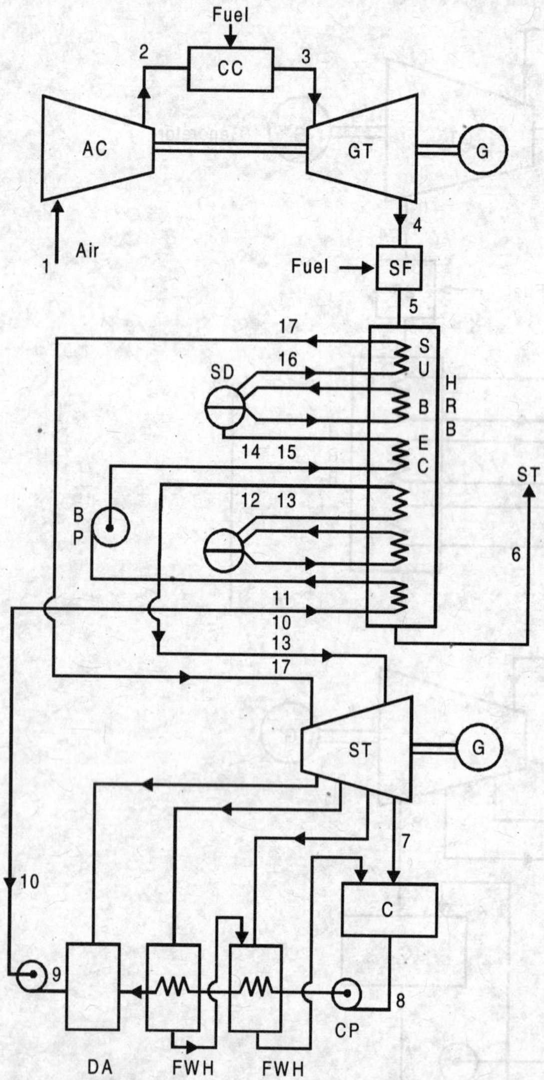

Fig. 22.17 (a) *A schematic diagram for a dual-pressure combined cycle*

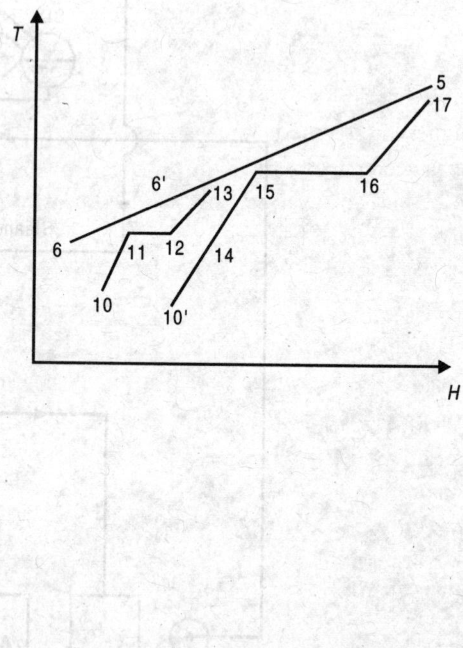

Fig. 22.17 (b) *Temperature-enthalpy (T-H) diagram of the heat recovery boiler of the dual-pressure combined cycle*

Figure 22.17(a) shows a schematic flow diagram of the plant. The "dual-pressure cycle" has a HRB which has two steam circuits. The high pressure circuit supplies steam to the high pressure stages of the turbine and low pressure circuit supplies steam to the low pressure stages of the same turbine.

Figure 22.17(b) shows the temperature-enthalpy diagram of both gas and steam circuits in the HRB.

Exhaust gas leaving the gas turbine enters the supplementary firing (SF) at 4 and the HRB at 5, leaving it to the stack at 6. Condensate leaving the condenser at 8 enters the low pressure economizer at 10 via condensate pump, two feedwater heaters, one DA and boiler feed pump. Process 10–11 is feed heating of low pressure circuit; followed by evaporation to 12 and superheat to 13. Superheated steam at 13 enters the steam turbine at low pressure stage.

Water from 11 is pumped by booster pump (BP) to 14 and goes to the high pressure economizer. Evaporation occurs from 15 to 16 and superheat to 17. High pressure superheated steam at 17 enters the steam turbine first stage.

In *T-H* diagram we see that low pressure steam boils at a lower temperature (12) below that of high pressure steam (15). Single high pressure circuit is shown by 10'–15–16–17 with gas leaving at 6'. Adding low pressure circuit allows the gas to leave at a lower temperature (6), thus extracting more energy from it and increasing overall cycle efficiency.

22.18 BINARY VAPOUR CYCLE

The efficiency of a reversible cycle is given by $\eta_{rev.} = 1 - \dfrac{T_1}{T_2}$, where T_1 is the average temperature of heat addition and T_2 is the average temperature of heat rejection. Higher the value of T_1 or lower the value of T_2, higher the efficiency.

The irreversibilities of Rankine cycle are least during evaporation than during feed heating or superheating. Thus, higher the saturation temperature of steam, higher the efficiency. However, for steam higher saturation temperature is accompanied by high pressure. Recall that at critical temperature of 374°C, the critical pressure of steam is 221 bar—a very high value. This high pressure poses some design problems and results in thick and heavy boiler and turbine components.

Fluids other than water having better thermodynamic properties were tried. Mercury was most successful of them. The metallurgical limit of steam turbine materials is 560°C. Mercury at 560°C has saturation pressure of only 12 bar. Thus mercury is a better fluid at high pressure end.

But at low pressure end, mercury is a bad fluid. At 30°C which is normal lower limit of a steam cycle, its saturation pressure is only 3.6×10^{-6} bar; and its specific volume is enormously large.

Therefore a plant is designed to use mercury vapour in high pressure cycle which transfers its heat to water vapour in a low pressure cycle. Such a cycle is called *Binary vapour cycle*.

Figure 22.18 shows a schematic flow diagram and a *T-s* diagram of a binary vapour cycle.

Description

The "Mercury cycle" abcde is a simple Rankine cycle using saturated mercury vapour. The heat rejected by mercury during condensation (process *b–c*) is transferred to convert saturated water of "steam cycle" into saturated steam (process 5–6). The saturated steam is superheated in the superheater, using furnace heat (process 6–1) and then admitted to steam turbine. The condensate is heated in the economizer (process 4–5) to convert it into saturated liquid which is then admitted to steam boiler—mercury condenser, where it absorbs its latent heat.

Let $\dot{m}$ represent the flow rate of mercury in the mercury cycle per kg of steam circulating in the steam cycle. Then for 1 kg of steam,

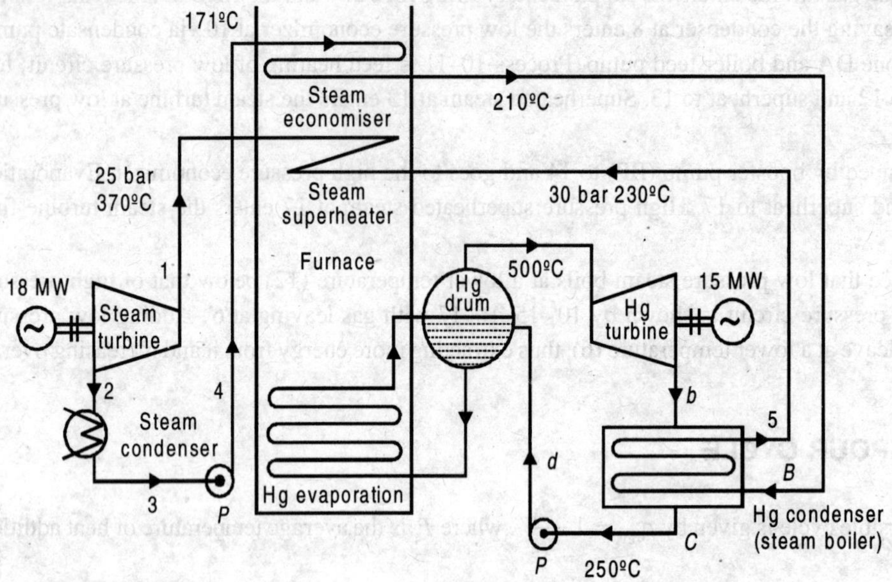

Fig. 22.18(a) *Binary vapour cycle*

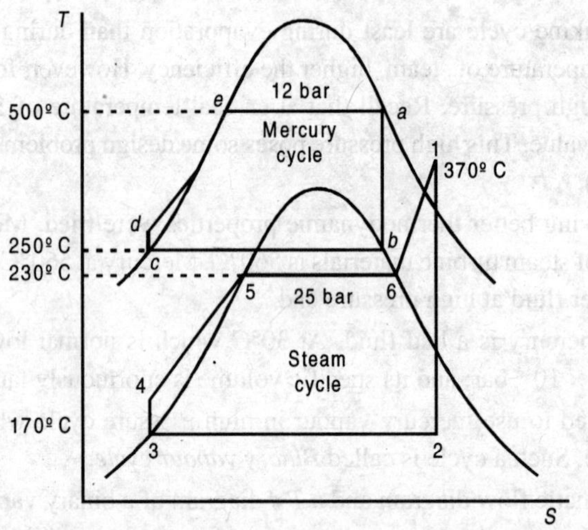

Fig. 22.18(b) *T-s diagram of binary vapour cycle*

1. Heat supplied, $q_5 = \dot{m}(h_a - h_d) + (h_1 - h_6) + (h_5 - h_4)$ kJ/kg
2. Heat rejected, $q_r = (h_2 - h_3)$ kJ/kg
3. Turbine work, $W_T = \dot{m}(h_a - h_b) + (h_1 - h_2)$ kJ/kg
4. Pump work, $W_P = \dot{m}(h_d - h_c) + (h_4 - h_3)$
5. Cycle efficiency,

$$\eta_{\text{cycle}} = \frac{q_s - q_r}{q_s} = \frac{W_T - W_P}{q_s} = \frac{W_{\text{net}}}{q_s} \quad \ldots(22.14)$$

6. Specific steam consumption,

$$\text{S.S.C.} = \frac{3600}{W_{net}} = \text{kg/kWh}$$

7. Quantity of mercury per kg of steam.

The energy balance equation for mercury condenser steam boiler is,

$$\dot{m}(h_b - h_c) = (h_6 - h_5)$$

$$\dot{m} = \frac{h_6 - h_5}{h_b - h_c} \text{ kg/kg of steam}$$

To vapourise 1 kg of water from its saturation temperature, about 7 to 8 kg of mercury need to be condensed. Other fluids which are used in place of mercury are:

1. Diphenyl ether $(C_6H_5)_2O$
2. Aluminium bromide $AlBr_3$.
3. Zinc ammoniun chloride $Zn(NH_3)_2Cl$.
4. Liquid metals like sodium, potassium.

However, mercury vapour cycles are not very widely used because:

1. Mercury is toxic, highly expensive and difficult to procure.
2. It has low latent heat and hence about 8–10 kg of mercury is required to evaporate 1 kg of water.
3. It does not wet the tube surface, so addition of magnesium or titanium is necessary.

22.19 BINARY VAPOUR ENGINE (CYCLE)

We know that maximum possible efficiency of any engine working between the temperature limits of T_2 and T_1 is given by the equation

$$\eta = \frac{T_1 - T_2}{T_1}$$

where T_1 = Higher temperature

T_2 = Lower temperature

(Atmospheric temperature)

Since T_2 is fixed by atmospheric conditions, the thermal efficiency of a plant can only be improved by increasing T_1. In a steam plant an increase of T_1 may entail the use of a corresponding higher pressure, which is one of the limiting factors in its design. By using mercury vapour in place of steam in the high temperature range of the cycle, an increase in T_1 is obtained without any increase in maximum pressure. The heat rejected by mercury in condensing can be utilized in raising superheated steam for the low temperature range of the cycle. A power plant using two vapours in this way is known as binary vapour plant.

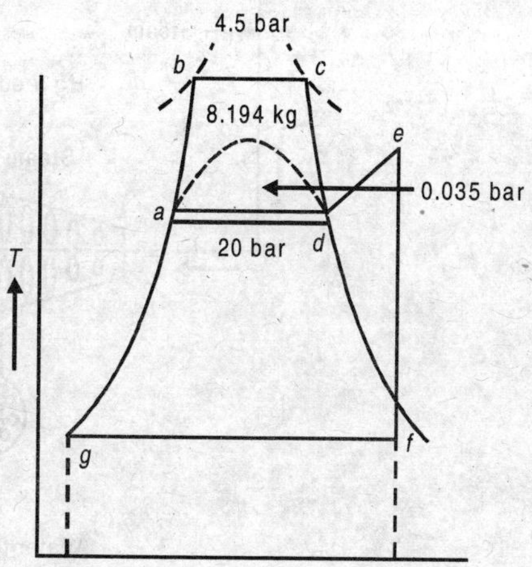

Fig. 22.19

On the *T-s* diagram; are shown the liquid and saturated steam lines for water and steam, meeting at the critical temperature which occurs at a steam pressure of 221 bar. The lines *ab* and *bc* represent the heating and evaporation of liquid mercury plotted to the same temperature scale as the steam, the scale of corresponding pressures is lower for mercury. The mercury vapour at *c* has a much higher temperature than steam at the same pressure. At *c* the mercury expanded adiabatically through a separate mercury vapour turbine to *d*; from *d* it is condensed to *a*, its latent heat being utilized for evaporating a corresponding amount of steam. The mercury has thus described the cycle *abcd*.

The steam cycle is represented by *gadef*, *ga* represents the heating of the feed water, *ad* its evaporation by the condensing mercury, and *de* the superheating of the steam by the flue gases At *e* the steam is expanded adiabatically through a steam turbine to *f*; *fg* represents the condensing of the exhaust steam in the condenser. This completes the steam cycle.

To obtain the correct amount of heat from the condensing mercury for evaporating the steam, it is found that 8.196 (8.2 Approx.) kg of mercury are required per kg of steam. It will be seen from the areas of *T-s* diagram that the use of mercury for the high temperature range of the cycle gives a higher efficiency than that obtainable from steam with the same addition of heat. Using pressure limits of 24 bar and. 0.07 bar, a steam plant gives an efficiency of 24–25% whilst that of the binary vapour plant is 50%.

A diagrametic view of a binary vapour plant is shown below. The hot gases from the furnace pass through the mercury boiler, through the liquid mercury heater, then through the steam superheater and finally through the feed water economiser, after which they are exhausted to the chimney.

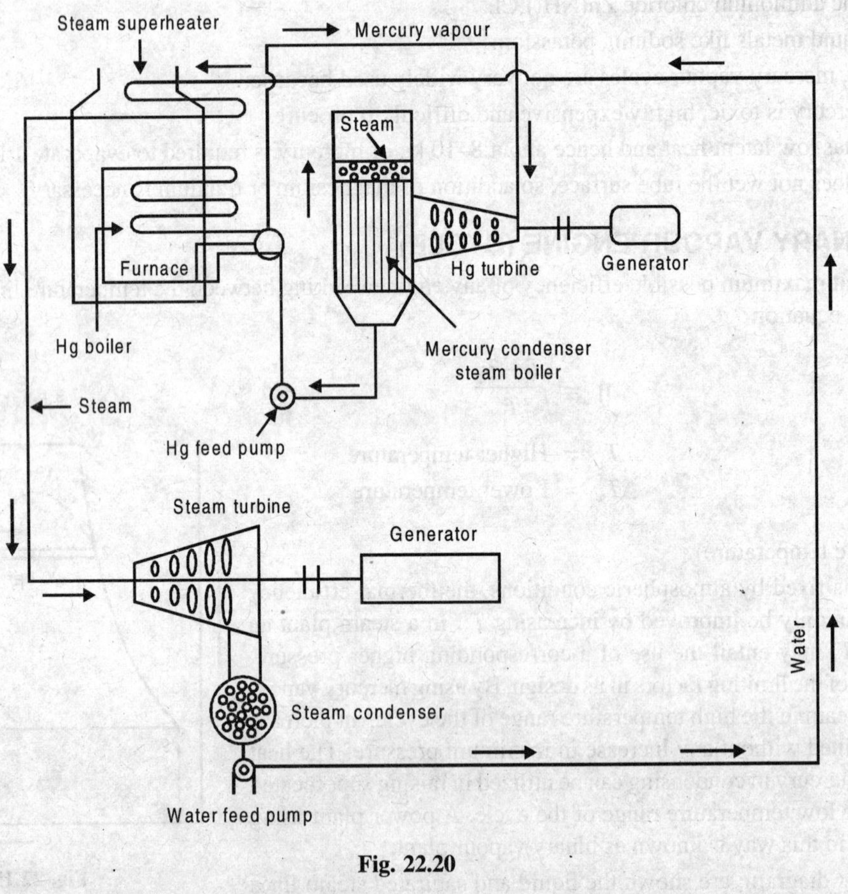

Fig. 22.20

The liquid mercury passes from the mercury liquid heater to the mercury vapour boiler where it is evaporated. It then flows to the mercury turbine through which it is expanded to its low pressure limit. From here it exhausts to the mercury condenser steam boiler where its latent heat is given upto the hot feed water; this operation condenses the mercury whilst at the same time the feed water is evaporated to steam. The mercury is then returned to the mercury liquid boiler, thus completing its cycle.

Referring to figure, the feed water from the economiser (not shown) is evaporated to steam in the mercury condenser steam boiler and then passes to the superheater where it is superheated by the hot flue gases. From the superheater the superheated steam passes to the steam turbine in which it is expanded down to the steam condenser pressure, it is then condensed in the steam condenser and passes back to the economiser, thus completing the cycle of the water and steam.

At the Hartford-USA–binary vapour plant, erected by the General Electric Co., the weight of mercury in use in the plant is 60 tonnes. A higher efficiency is obtained than is possible from a steam plant of the same output. Daily saving of 150 T of coal compared with the consumption of an equivalent steam plant. Five binary vapour plants have been erected by the General Electric Co.

In the early stages of the development of the mercury boiler, it was found that the liquid mercury did not wet the heating surface of the tubes. There was a tendency for a film of mercury vapour to form at the tube surface which prevented a good heat transfer to the liquid. The difficulty was overcome by a 0.002 percent solution of magensium and titanium being used in the boiler as a surface wetting agent.

The chief objection to the mercury vapour plant is the danger due to poisonous fumes if any leakage of mercury vapour occurs, and special pecautions are taken to prevent this.

SOLVED EXAMPLES

Example 22.1. *(a) A superheat Rankine cycle works with a boiler pressure of 50 bar and condenser pressure of 0.08 bar. The steam is superheated to 500°C.*

Analyse the cycle.

(b) For the above cycle, reheat is added at 1.0 bar and 500°C.

Analyse the cycle and compare the results with those of the cycle given in (a) above.

Solution

Case (a) Superheat Rankine Cycle:

Figure Ex. 22.1 (a) shows the superheat Rankine cycle.

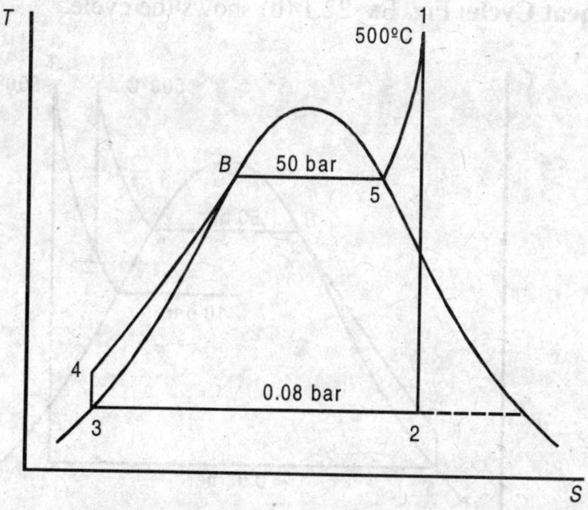

Fig. Ex. 22.1(a) *Superheat rankine cycle*

Our first aim is to determine enthalpies at various salient points of the cycle. From the steam tables we have,

(i) at 50 bar 500°C

$$h_1 = 3433.7 \text{ kJ/kg}$$
$$s_1 = 6.977 \text{ kJ/kg-K}$$

(ii) at 0.08 bar

$$s_{f3} = 0.5925 \text{ kJ/kg-K}$$
$$s_{fg2} = 7.6371 \text{ kJ/kg-K}$$
$$h_{f3} = 173.86 \text{ kJ/kg}$$
$$h_{fg2} = 2403.2 \text{ kJ/kg}$$

1. To find X_2 equate entropies at point 1 and point 2.

$$6.977 = 0.5925 + x_2 (7.6371)$$
$$\therefore \quad x_2 = 0.386$$
$$\therefore \quad h_2 = 173.86 + 0.836 \times 2403.2 = 2182.93 \text{ kJ/kg}$$

2. Turbine work = $W_T = h_1 - h_2 = 3433.7 - 2182.93 = 1250.76$ kJ/kg
3. Pump work = $W_P = 0.001\ [50 - 0.08] \times 100 = 5$ kJ/kg.
4. Net work = $W_{net} = W_T - W_P = 1250.76 - 5.0 = 1245.76$ kJ/kg
5. Heat supplied = $q_s = h_1 - h_4 = h_1 - (h_3 - W_P)$

$$= 3433.7 - 178.76 = 3254.84 \text{ kJ/kg}$$

6. Rankine efficiency = $h_r = \dfrac{W_{net}}{q_s} = \dfrac{1250.76}{3254.84} = 0.3843$ i.e. 38.43%

7. Work ratio = $\dfrac{W_{net}}{W_T} = \dfrac{1245.76}{1250.76} = 0.996$

8. Specific steam consumption = $\dfrac{3600}{1250.76} = 2.878$ kg/kWh

Case (b) Superheat Reheat Cycle: Fig. Ex. 22.1 (b) shows the cycle.

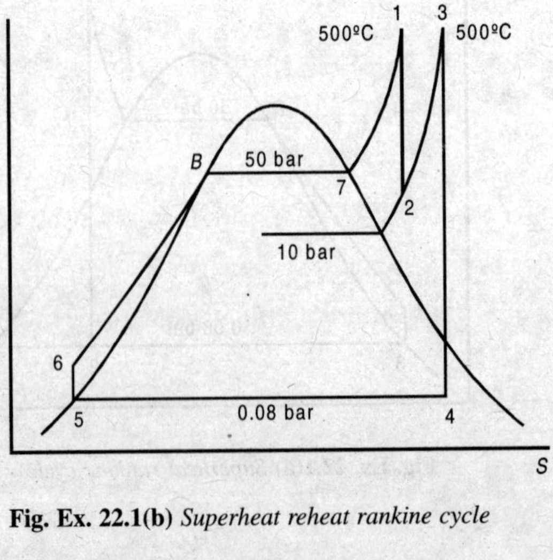

Fig. Ex. 22.1(b) *Superheat reheat rankine cycle*

The steam expands from 50 bar to 10 bar before reheating. s_{g2} at 10 bar is 6.5828. Since $s_1 > s_{g2}$ the condition of steam at point 2 is superheated. To find T_{sup2} we equate entropies at point 1 and point 2.

or
$$s_1 = s_{g2} + C_p \ln \frac{T_{sup2}}{T_{sat2}}$$

$$\therefore \quad 6.997 = 6.5825 + 2.1 \ln \frac{T_{sup2}}{273 + 179.88}$$

$$\therefore \quad T_{sup2} = 551.62 \text{ K} = 278.62 \text{ °C}$$

$\therefore$ At 10 bar and 278.62°C, we have from steam tables (by interpolation)

$$h_2 = 3005.45 \text{ kJ/kg}$$

At 10 bar and 500 °C, we have from steam tables,

$$h_3 = 3478.3 \text{ kJ/kg and } s_3 = 7.7627$$

To find x_4 equate entropies at 3 and 4.

$$\therefore \quad s_3 = s_4$$

$$\therefore \quad 7.7627 = 0.5925 + x_4\, 7.6371$$

$$x_4 = 0.939$$

$$\therefore \quad h_4 = 173.86 + 0.939 \times 2403.2 = 2430.14 \text{ kJ/kg}$$

Let us analyse the cycle.

1. Heat supplied = $q_{s1} = h_1 - h_6 = 3433.7 - (h_5 + W_P)$ $[W_p = 5 \text{ kJ/kg}]$

$$= 3433.7 - 178.86$$
$$= 3254.84 \text{ kJ/kg}$$

also heat supplied under reheat

$$= q_{s2} = h_3 - h_2$$
$$= (3478.3 - 3005.45)$$
$$= 472.85 \text{ kJ/kg}$$

$$\therefore \quad \text{Total heat supplied} = q_s = q_{s1} + q_{s2}$$
$$= 3254.84 + 472.85$$
$$= 3727.7 \text{ kJ/kg}$$

2. Turbine work
$$W_T = (h_1 - h_2) + (h_3 - h_4)$$
$$= (3433.7 - 3005.45) + (3478.3 - 2430.14)$$
$$= 428.25 + 1048.16$$
$$= 1476.41 \text{ kJ/kg}$$

3. Net work
$$W_{net} = W_T - W_P = 1476.41 - 50$$
$$= 1471.41 \text{ kJ/kg}$$

4. Cycle efficiency
$$h_r = \frac{W_{net}}{q_s} = \frac{1471.41}{3727.7} = 0.3497 \text{ i.e. } 39.47\%$$

5. Work ratio = $\dfrac{W_{net}}{W_T} = \dfrac{1471.41}{1476.41} = 0.9966$

6. Specific steam consumption = $\dfrac{3600}{1471.41} = 2.447 \text{ kg/kWh}$

Comments: Superheat reheat cycle shows following improvements:

1. Final dryness is improved from 0.836 to 0.939.
2. Rankine efficiency improves from 38.43% to 39.47% increased by 2.7%.
3. Specific steam consumption improves from 2.878 kg/kWh to 2.447 kg/kWh; reduction by 15%.

Example 22.2. (*a*) *What is a combined cycle power plant? State the advantages.*

(*b*) *In a Rankine cycle, steam is supplied to a turbine at 50 bar. 400°C and is expanded to a pressure of 0.2 bar. Find the thermal efficiency of the plant if the reheat pressure is 7 bar and reheated steam temperature 400°C. Assume isentropic efficiency of the turbine as 80% throughout.*

Solution

The cycle is shown on *T–S* diagram.

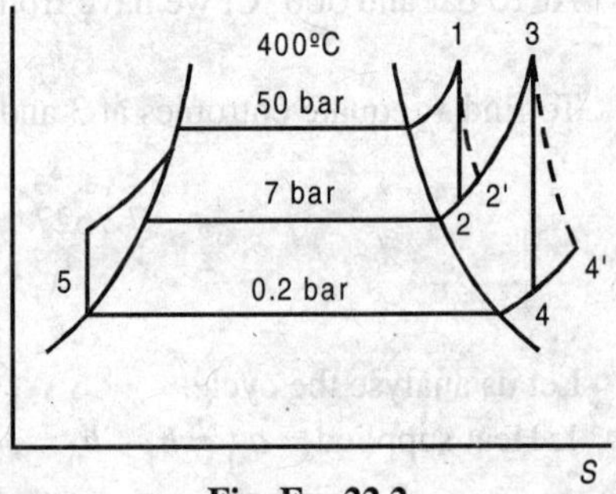

Fig. Ex. 22.2

Given:

$$P_1 = 50 \text{ bar}$$

$$T_1 = 673 \text{ K} = T_3$$

$$P_3 = 7 \text{ bar}$$

$$P_4 = 0.2 \text{ bar}$$

$$\eta_t = 80\ \%$$

From Steam tables,

$$h_1 = 3198.3 \text{ kJ/kg};\ h_3 = 3269 \text{ kJ/kg}$$

$$h_{f_5} = 251.45 \text{ kJ/kg}$$

$$s_1 = 6.6508;\ s_3 = 7.6362$$

$$s_2 = 1.9918 + x_2 \times 4.7134 \qquad \because s_1 = s_2$$

$$x_2 = \frac{6.6508 - 1.9918}{4.7134} = \frac{4.6590}{4.7134}$$

$$= 0.9885$$

$$h_2 = 697.06 + 0.9885 \times 2064.9$$

$$= 697.06 + 2041.07$$

$$= 2738.13 \quad h_2 = 2738.13$$

From *T–s* diagram,

$$\eta_t = \frac{h_1 - h_{2'}}{h_1 - h_2} \qquad \therefore 0.8 = \frac{3198.3 - h_{2'}}{3198.3 - 2738.13}$$

$$\therefore \quad 0.8 \times 460.17 = \mathbf{368.14\ kJ/kg}$$

= work of turbine I

$$\therefore \quad h_{2'} = 3198.3 - 368.14 = 2830.16 \text{ kJ/kg}$$

Similarly, for 2nd stage turbine (after reheating)

$$s_3 = s_4 \qquad \therefore 7.6362 = 0.8321 + x_4 \times 7.0773$$

$$\therefore \quad x_4 = \frac{7.6362 - 0.8321}{7.0773} = \frac{6.8041}{7.0773}$$

$$= \mathbf{0.9614}$$

$$\therefore \quad h_4 = 251.45 + 0.9614 \times 2358.4$$
$$= 251.45 + 2267.36$$
$$= \mathbf{2518.81\ kJ/kg}$$

$$\therefore \quad 0.8 = \frac{h_3 - h_{4'}}{h_3 - h_4} = \frac{3269 - h_{4'}}{3269 - 2518.81}$$

$$\therefore \quad h_{4'} = 3269 - 0.8 \times 750.19$$
$$= 3269 - 600.15$$
$$= \mathbf{2668.85\ kJ/kg}$$

$$W_{t_2} = 600.15 \text{ kJ/kg}$$

$$\therefore \quad \text{Total work done/kg} = 368.14 + 600.15 - W_p$$
$$= 968.15 - W_p$$

$$\text{Heat supplied/kg} = (h_1 - h_{f_5}) + (h_3 - h_{2'}) - W_p$$
$$= (3198.3 - 251.45) + (3269 - 2830.16) - W_p$$
$$= 2946.85 + 439.84 - W_p$$
$$= 3389.69 - W_p$$

$$\therefore \quad W_p = \frac{p_1 - p_4}{10} = \frac{50 - 0.2}{10} = \frac{49.8}{10} = \mathbf{4.98\ kJ/kg}$$

$$W = (968.29 - 4.98) = \mathbf{963.31\ kJ/kg}$$
$$Q = 3386.69 - 4.98 = \mathbf{3381.71\ kJ/kg}$$

$$\therefore \quad \text{Thermal efficiency} = \frac{W}{Q} = \frac{963.31}{3381.71}$$
$$= \mathbf{0.285} \text{ or } \mathbf{28.5\%}$$

Example 22.3. *In a two stage steam turbine working on ideal Rankine cycle, steam enters the first stage at 90 bar and 550°C temperature. Second stage entry temperature after reheating is 550°C. Steam is never wet during expansion in both sides. Find the cycle efficiency and specific steam consumption. Also find the steam flow rate in the turbine in kg/sec if the turbine is directly connected to an alternator of 150 MW output. Assume 0.9 as alternate efficiency. Draw T-s diagram for the cycle. Neglect pump work.*

Solution

Assume dry and saturated steam after expansion.

$$h_1 = 3518.95$$
$$s_1 = 6.8087$$
$$P_2 = 5.5 \text{ bar}$$
$$h_2 = 2750$$
$$h_3 = 3600$$
$$h_4 = 2590$$

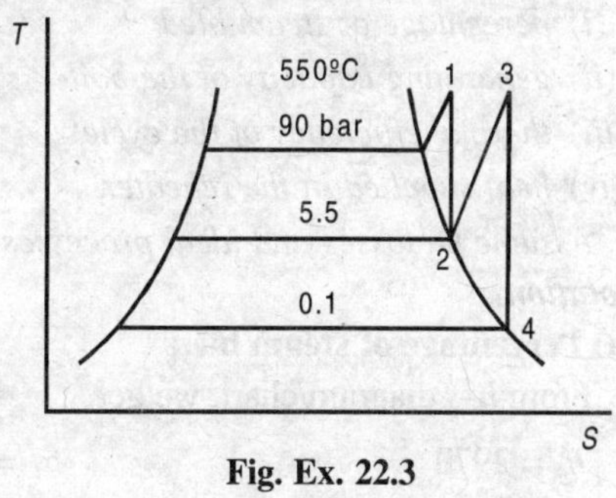

Fig. Ex. 22.3

$p_1 = 90$ bar, $p_2 = 5.5$ bar $= p_3$; $p_4 = 0.1$ bar.

$$\text{Work done/kg} = (h_1 - h_2) + (h_3 - h_4)$$
$$= (3518.95 - 2750) + (3600 - 2590)$$
$$= 768.95 + 1010$$
$$= \mathbf{1778.95\ kJ/kg}$$

$$\text{Heat supplied/kg} = (h_1 - h_{f4}) + (h_3 - h_2)$$
$$= (3518.95 - 191.83) + (3600 - 2750)$$
$$= 3397.12 + 850$$
$$= \mathbf{4247.12\ kJ/kg}$$

$$\therefore \quad \text{Cycle efficiency} = \frac{W}{Q_A} = \frac{1778.95}{4247.12}$$
$$= 0.4189$$
$$= \mathbf{41.89\%}$$

$$\text{Specific steam consumption} = \frac{3600}{\text{work/kg}}$$
$$= \frac{3600}{1778.95}$$
$$= \mathbf{2.027\ kg/kWh}$$

$$\text{Alternator output} = 150\ \text{MW}$$

$$\therefore \quad \text{Alternator input} = \frac{150}{0.9} = 166.67\ \text{MW} = \mathbf{166670\ kW(kJ/s)}$$

$$\text{Steam flow rate} = \frac{166670}{1778.95} = \mathbf{93.69\ kg/s}$$

Example 22.4. (*a*) *How can a combined cycle plant be used for co-generation ? What is its thermodynamic advantage?*

(*b*) *Why is the efficiency of a condensing steam power plant is less in a warm region than in a cold region?*

(*c*) *A steam power plant of 120 MW capacity is equipped with regenerative as well as reheat arrangement. The steam is supplied to the turbine at 80 bar and 50°C superheat. Steam is extracted at 7 bar for feed heating and the remaining steam is reheated to its initial temperature before it is expanded to 0.36 bar in the L.P. stage. Assuming indirect contact type of feed water heater, determine*

(*i*) *percentage of steam bled*

(*ii*) *generating capacity of the boiler*

(*iii*) *thermal efficiency of the cycle*

(*iv*) *heat supplied in the reheater.*

Assume no losses and ideal processes of expansion.

Solution

(a) Percentage of steam bled

From *h–s* diagram/chart, we get

$h_1 = 2970$ $\qquad h_2 = 2490$

$h_3 = 2880$ $h_4 = 2380$

$h_{f_2} = 697.06$ kJ/kg $h_{f_4} = 307.12$ kJ/kg

Let m_2 be the weight of steam bled for feed heating

$$\therefore \quad m_2\left(h_2 - h_{f_4}\right) = 1\left(h_2 - h_{f_4}\right)$$

$$\therefore \quad m_2\,(2490 - 307.12) = (697.06 - 307.12)$$

$$= 389.94$$

$$m_2 = \frac{389.94}{2182.88}$$

$$= 0.1786 \text{ kg/kg supplied}$$

$$\therefore \quad \% \text{ of steam bled} = \mathbf{17.86\ \%}$$

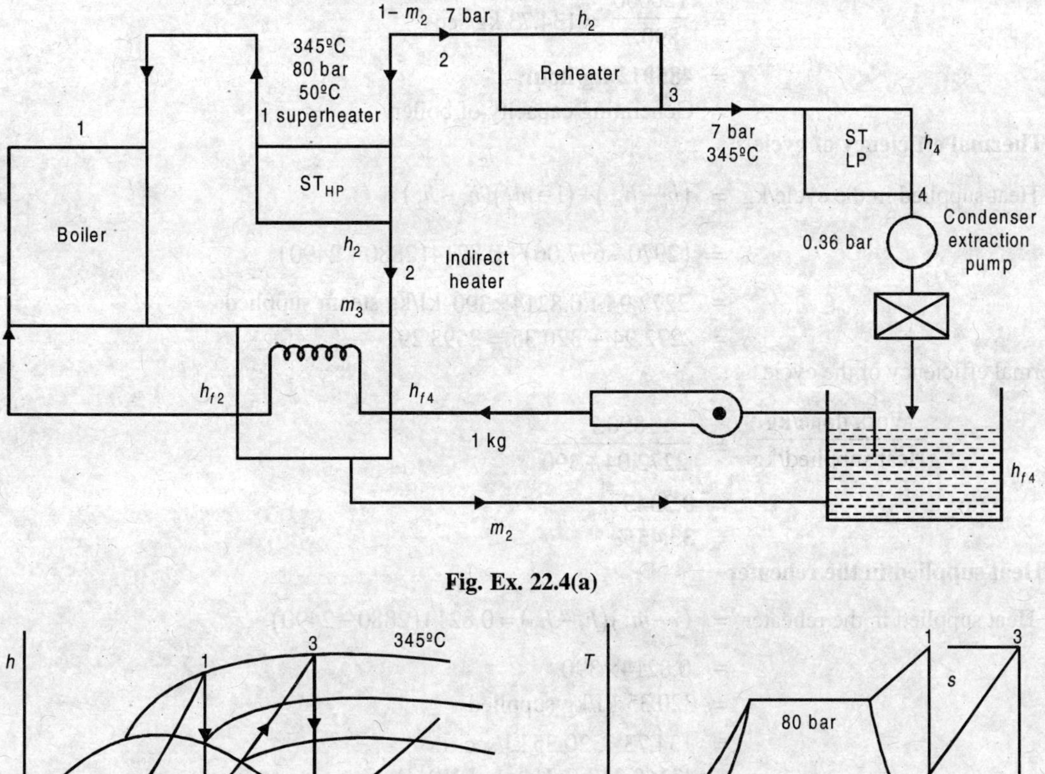

Fig. Ex. 22.4(a)

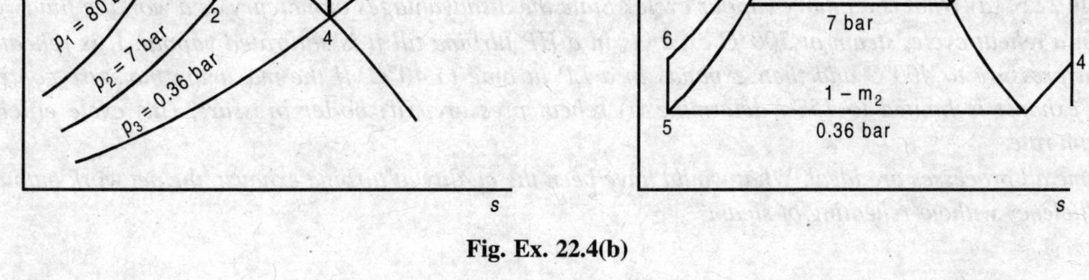

Fig. Ex. 22.4(b)

(b) Generating capacity of boiler

Weight of steam expanded in HP turbine = 1 kg

$\therefore$ Work done in HP turbine $= 1(h_1 - h_2) = 2970 - 2490 =$ **480 kJ/kg**

Weight of steam expanded in LP turbine = 1 – 0.1786 = **0.8214 kg**

$$\therefore \text{Work done in LP turbine} = 0.8214(h_3 - h_4)$$
$$= 0.8214(2880 - 2380) = 0.8214 \times 500$$
$$= 410.70 \text{ kJ/kg supplied}$$

$\therefore$ Total work done/kg steam supplied

$= 480 + 410.7 =$ **890.7 kJ/kg**

Capacity of the plant = 120 MW = **120,000 kW**

$\therefore$ Steam consumption/sec

$$= \frac{120000}{890.7} = 134.73 \text{ kg/sec}$$

= **485012 kg/hour**

= Generating capacity of boiler

(c) Thermal efficiency of cycle

$$\text{Heat supplied in the cycle/kg} = (h_1 - h_{f_2}) + (1 - m_2)(h_3 - h_2)$$
$$= (2970 - 697.06) + 0.8214(2880 - 2490)$$
$$= 2272.94 + 0.8214 \times 390 \text{ kJ/kg steam supplied}$$
$$= 2272.94 + 320.35 = 2593.29$$

Thermal efficiency of the cycle

$$\frac{\text{Work done/kg}}{\text{Heat supplied/kg}} = \frac{890.7}{2272.04 + 390}$$

= **0.3345**

= **33.45%**

(d) Heat supplied in the reheater

$$\text{Heat supplied in the reheater} = (1 - m_2)(h_3 - h_2) = 0.8214(2880 - 2490)$$
$$= 0.8214 \times 390$$
$$= 320.35 \text{ kJ/kg supplied}$$
$$= 134.73 \times 320.35 \text{ kJ/sec}$$

= **43160.217 (kJ/sec = kW)**

Example 22.5. *(a) What is a binary vapour cycle? State the disadvantages of mercury as a working fluid.*

(b) In a reheat cycle, steam at 500°C expands in a HP turbine till it is saturated vapour. It is reheated at constant pressure to 400°C and then expands in a LP turbine to 40°C. If the maximum moisture content at turbine exhaust is limited to 15%, determine (i) reheat pressure, (ii) boiler pressure, (iii) cycle efficiency, (iv) steam rate.

Assume all processes are ideal. What would have been the quality at turbine exhaust, the net work output and cycle efficiency without reheating of steam?

Solution

Locate point 4. Draw vertical to point 3 on 400ºC temperature line. Point 3 will give the pressure of reheating. Follow this pressure line upto saturation curve. We will get point 2. From point 2, draw a vertical line to meet a temperature line of 500ºC and we will get point 1 which will give the boiler pressure.

From diagram:

$h_4 = 2220$ kJ/kg, $p_3 = 20$ bar, $h_3 = 3240$

$h_2 = 2800$ kJ/kg; 0 $h_1 = 3300$ kJ/kg

$p_1 = 150$ bar $h_{f_4} = 167.45$ kJ/kg

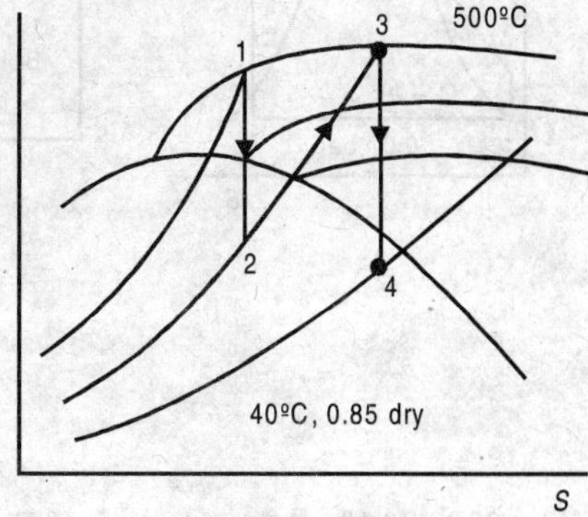

Fig. Ex. 22.5

(i) Reheat pressure

Reheat pressure = p_2 = **20 bar**

(ii) Boiler pressure

Boiler pressure = p_1 = **150 bar**

(iii) Cycle efficiency

$$\text{Cyclic work/kg} = (h_1 - h_2) + (h_3 - h_4)$$

$$= (3300 - 2800) + (3240 - 2220)$$

$$= 500 + 1020$$

$$= \mathbf{1520\ kJ/kg}$$

$$\text{Heat supplied/kg} = (h_1 - h_{f_4}) + (h_3 - h_2)$$

$$= (3300 - 167.45) + (3240 - 2800)$$

$$= 3132.55 + 440$$

$$= \mathbf{3572.55\ kJ/kg}$$

$$\text{Cycle efficiency} = \frac{\text{Work}}{Q_A} = \frac{1520}{3572.55} = 0.4255$$

$$= \mathbf{42.55\ \%}$$

(iv) Steam rate

$$\text{Steam rate} = \frac{3600}{\text{Work done/kg}} = \frac{3600}{1520}$$

$$= \mathbf{2.368\ kg/kWh}$$

Example 22.6. *The net power output of an ideal reheat regenerative steam cycle is 80 MW. Steam enters the h.p. turbine at 80 bar, 500°C and expands till it becomes saturated vapour. Some of the steam then goes to an open feed water heated and the balance is reheated to 400°C, after which it expands in an LP turbine to 0.07 bar.*

Compute: (i) the reheat pressure, (ii) the steam flow rate to h.p. turbine, (iii) the cycle efficiency, (iv) the rate of flow of cooling water in the condenser if the temperature rise of water is 8°C, (v) if the velocity of steam flowing from the turbine to the condenser is limited to 130 m/s, find the diameter of the connecting pipe.

Solution

The schematic diagram for the arrangements of the equipments (plant) is shown in Fig. Ex. 22.6.

From *h-s* diagram, we get the following ethalpies.

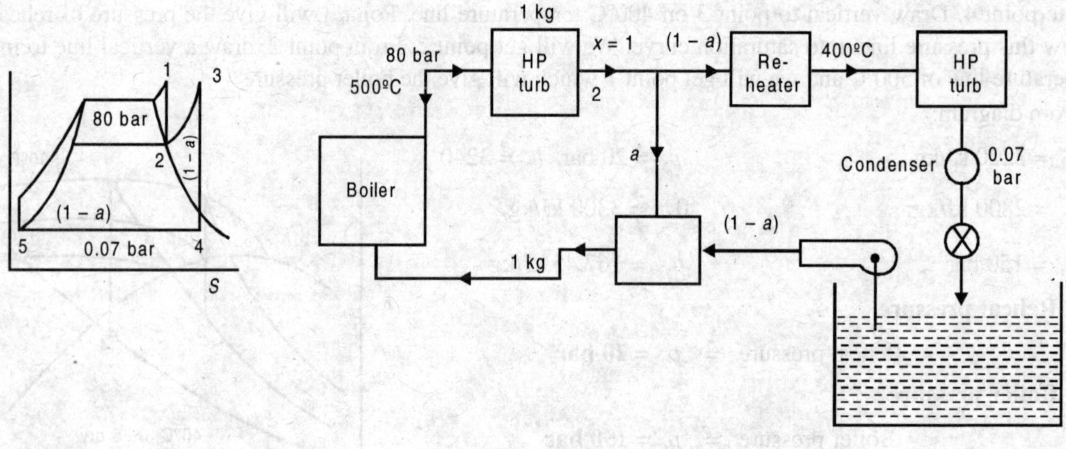

Fig. Ex. 22.6

$h_1 = 3398.8$ kJ/kg ∴ $p_2 = 6.6$ bar

$h_2 = 2759.5$ kJ/kg $s_1 = 6.7262 = s_{g_2}$

$h_3 = 3269.1$ kJ/kg $h_{f_2} = 686.78$ kJ/kg

$h_4 = 2380$ kJ/kg $h_{f_4} = 163.38$ kJ/kg

$x_4 = 0.914$

Specific volume at LP turbine outlet = $0.914 \times 20.531 = 18.77 \text{ m}^3$

(i) Reheat Pressure = 6.6 bar

(ii) Work done/kg in HP turbine

$$= W_{HP} = h_1 - h_2 = 3398.8 - 2759.5$$

$$= \mathbf{639.3 \text{ kJ/kg}}$$

∴ Steam flow rate to H.P. turbine

$$= \frac{3600}{639.3} = \mathbf{5.631 \text{ kg/kWh}}$$

(iii) Cycle efficiency : Consider the open heater

$$m_a \times (h_2 - h_{f_2}) = (1 - m_a)(h_{f_2} - h_{f_4})$$

$$m_a (2759.5 - 686.78) = (1 - m_a)(686.78 - 163.38)$$

$$m_a \times 2072.72 = (1 - m_a) 523.4$$

$$1 - m_a = m_a \times \frac{2072.72}{523.4}$$

$$= 3.96\, m_a$$

$$1 = 4.96\, m_a$$

or $$m_a = \frac{1}{4.96} = \mathbf{0.2016 \text{ kg}}$$

∴ Work done in L.P. stage/kg steam admitted

$$= (1-m_a)(h_3-h_4) = (1-0.2016)(3269.1-2380)$$

$$0.7916 \times 889.1 = 703.81 \text{ kJ/kg admitted}$$

$$\therefore \quad \text{Total work/done/kg} = W_{HP} + W_{LP}$$

$$= 639.3 + 703.81$$

$$= \mathbf{1343.11 \text{ kJ/kg admitted}}$$

$$\text{Heat supplied/kg} = 1 \times (h_r - h_{f_5} - W_p) + (1-9)(h_3 - h_2)$$

$$= \left(3398.8 - 163.68 - \frac{80-0.07}{10}\right) + 0.7916(3269.1 - 2759.5)$$

$$= 3235.12 - 7.993 + 0.7916 \times 509.6$$

$$= 3227.127 + 403.4$$

$$= 3630.527 \text{ kJ/kg}$$

$$\therefore \quad \text{Cycle efficiency} = \eta_{cyc} = \frac{\text{Work done/kg}}{\text{Heat supplied/kg}}$$

$$= \frac{1343.11}{3630.527}$$

$$= \mathbf{0.37 \text{ or } 37\%}$$

(iv) Flow rate of cooling water at condenser

0.7916 kg if steam leaves the LP turbine and enters the consdenser per kg of steam admitted to the HP turbine or the system.

$$\text{But, steam flow/sec to the system} = \frac{80 \times 100}{1343.11}$$

$$= \mathbf{59.56 \text{ kg/sec}}$$

$$\therefore \text{Steam to condenser/sec} = (1-9) \times 59.56$$

$$= 0.7916 \times 59.56$$

$$= \mathbf{47.15 \text{ kg/s}}$$

∴ Heat gained by cooling water/sec = Heat given by steam/sec

$$\therefore \quad M_W C_{PW} \Delta T = 47.15 \times (h_4 - h_{f_4})$$

$$\therefore \quad M_W = \text{Mass of cooling water/sec}$$

$$= \frac{47.15 \times (2380 - 163.38)}{4.187 \times 8}$$

$$= \mathbf{3120 \text{ kg/s}}$$

(v) Total volume of steam coming to condenser/sec

$$= \text{Mass of steam/s} \times \text{Specific volume of steam}$$

$$= 47.15 \times 18.77 \text{ m}^3\text{/s}$$

$$= 885 \text{ m}^3\text{/s}$$

$$= \text{Velocity} \times \text{Area of pipe}$$

$$= 130 \times \frac{\pi}{4} d^2$$
$$d^2 = 8.666$$
$$d = \mathbf{2.944\ m}$$

Example 22.7. (*a*) *Write a short note on combined gas steam power cycle.*

(b) A steam power plant is operating on the ideal reheat Rankine cycle. Steam enters the high pressure turbine at 15 MPa and 600ºC and is condensed in the condenser at F pressure of 1.0 kPa. If the moisture content of the steam at the exit of the low pressure turbine is not to exceed 10.4%, determine:

(*i*) *the pressure at which the steam should be reheated before supplying the same to the low pressure turbine, and*

(*ii*) *the thermal efficiency of the cycle.*

Assume the steam is reheated at constant pressure to a temperature of 600°C. Assume both steam turbines are mounted on same shaft.

Solution

Neglecting pump work,

$$15 \text{ MPa} = 150 \text{ bar}$$
$$10 \text{ kPa} = \frac{10 \times 1000}{10^5}$$
$$= \frac{1}{10} = \mathbf{0.1\ bar}$$

10.4% wetness means 89.6% dryness

$$x_4 = 0.896$$
$$= \mathbf{0.9\ Approx.}$$
$$h_1 = 3580,\ h_2 = 3160,\ h_3 = 3675,\ h_4 = 2340 \text{ kJ/kg}$$

(i) Reheat pressure = 40 bar

(ii) Thermal Efficiency

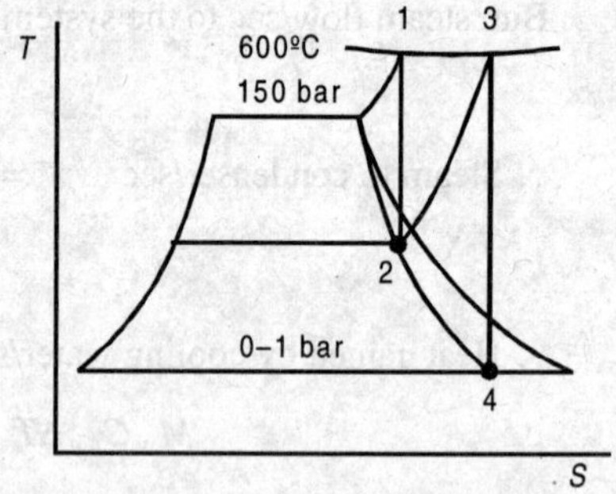

Fig. Ex. 22.7

$$\text{Work done/kg} = (h_1 - h_2) + (h_3 - h_4)$$
$$= (3580 - 3160) + (3675 - 2340)$$
$$= 420 + 1335$$
$$= \mathbf{1755\ kJ/kg}$$

$$\text{Heat added/kg} = (h_1 - h_{f_4}) + (h_3 - h_2)$$
$$= (3580 - 191.83) + (3675 - 3160)$$
$$= 3388.17 + 515$$
$$= \mathbf{3903.17\ kJ/kg}$$

$$\therefore \quad \text{Thermal efficiency} = \frac{\text{Work done/kg}}{\text{Heat added/kg}}$$
$$= \frac{1755}{3903.17}$$
$$= 0.4496$$
$$= \mathbf{or\ 0.45\ or\ 45\%}$$

Example 22.8. (*a*) *Write a short note on cogeneration.*

(b) In a single feed water, regenerative cycle, the steam enters the turbine at a pressure of 30 bar and 400°C. The exhaust pressure of steam is 0.1 bar. The exhaust pressure of steam is 0.1 bar. The feed water heater is direct-

contact type which operates at a pressure of 5 bar. Find the thermal efficiency of the cycle and steam rate. Show the flow diagram, this regenerative cycle on h-s and T-s diagram. (Neglect pump work).

Solution

The flow diagram and *T-s*, *h-s*, diagrams are shown. Given data:

$p_1 = 30$ bar, $p_2 = 5$ bar $\quad p_3 = 0.1$ bar, $T_1 = 400$°C.

For steam table, $\quad h_{f_2} = 640.12$ kJ, $h_{f_4} = 191.83$ kJ

From chart, $\quad h_1 = 3225$ kJ, $p_2 = 2780$ kJ, $h_3 = 2200$ kJ

Consider the heater:

$$\text{Heat given by system} = \text{Heat gained by water}$$

$$\dot{m}_2\left(h_2 - h_{f_2}\right) = \left(1-\dot{m}_2\right)\left(h_{f_2} - h_{f_3}\right)$$

$$\dot{m}_2(2780 - 640.12) = (1-\dot{m}_2)(640.12 - 181.83)$$

$$\dot{m}_2(2139.88) = (1-\dot{m}_2)(458.29)$$

$$\frac{2139.88}{458.29} = 4.6693 = \frac{1-\dot{m}_2}{\dot{m}_2} = \frac{1}{\dot{m}_2} - 1$$

$$\frac{1}{\dot{m}_2} = 5.6693 \text{ or } \dot{m}_2 = 0.1764$$

$$\dot{m}_2 = \mathbf{0.1764}$$

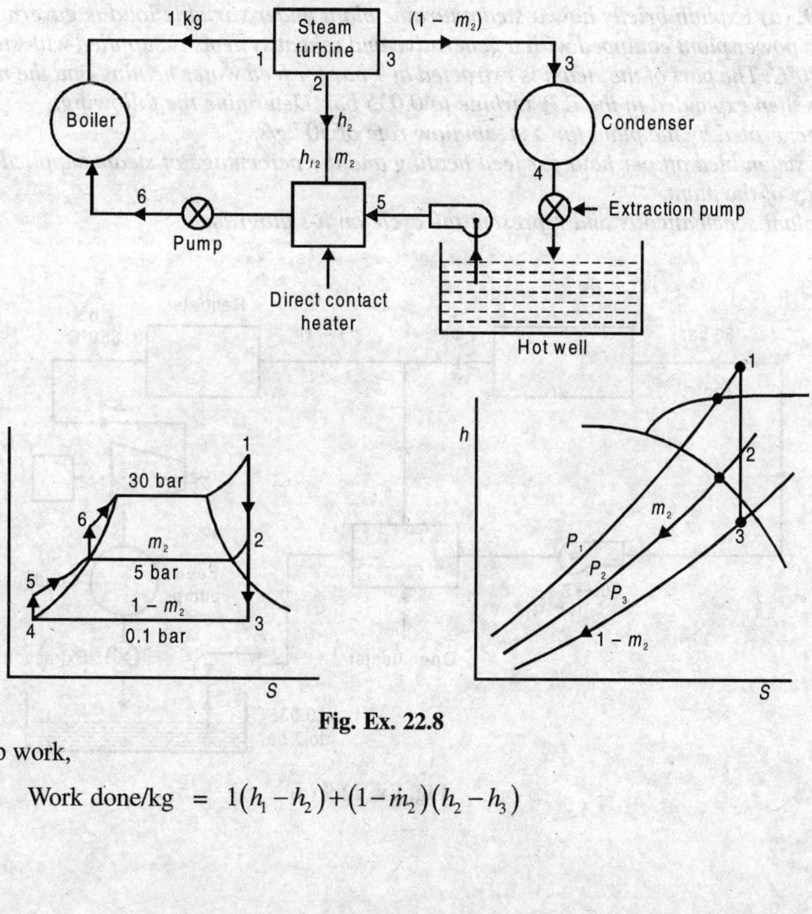

Fig. Ex. 22.8

Neglecting pump work,

$$\text{Work done/kg} = 1(h_1 - h_2) + (1-\dot{m}_2)(h_2 - h_3)$$

$$= (3232-2780)+(1-0.1764)\times(2780-2200)$$
$$= 452+0.8236\times580$$
$$= 452 + 477.69$$
$$= \mathbf{929.69\ kJ/kg}$$

$$\text{Heat supplied/kg} = h_1 - h_{f_2}$$
$$= 3232 - 640.12$$
$$= \mathbf{2591.88\ kJ/kg}$$

∴ Thermal Efficiency of the cycle

$$= \frac{\text{Work done/kg}}{\text{Heat added/kg}}$$
$$= \frac{929.69}{2591.88}$$
$$= \mathbf{0.3587}$$
$$= \mathbf{35.87\%}$$

$$\text{Steam rate} = \frac{3600}{\text{Work done/kg}}$$
$$= \frac{3600}{929.69} = \mathbf{3.8722\ kg/kWh}$$

Example 22.9. (*a*) *Explain briefly how a steam turbine plant under variable load is governed.*

(*b*) *A steam power plant equipped with a generative and reheat systems is supplied with steam to HP turbine at 80 bar and 470°C. The part of the steam is extracted at 7 bar for feed water heating and the remaining is reheated to 350°C. It is then expanded in the L.P. turbine to 0.035 bar. Determine the following:*

(*i*) *Power generated by the plant for a steam flow rate of 50 kg/s.*

(*ii*) *Mass of steam bled off per hour for feed heating and the percentage of steam supplied, bled.*

(*iii*) *Efficiency of the plant.*

Draw the plant schematically and represent the cycle on h-s diagram.

Solution

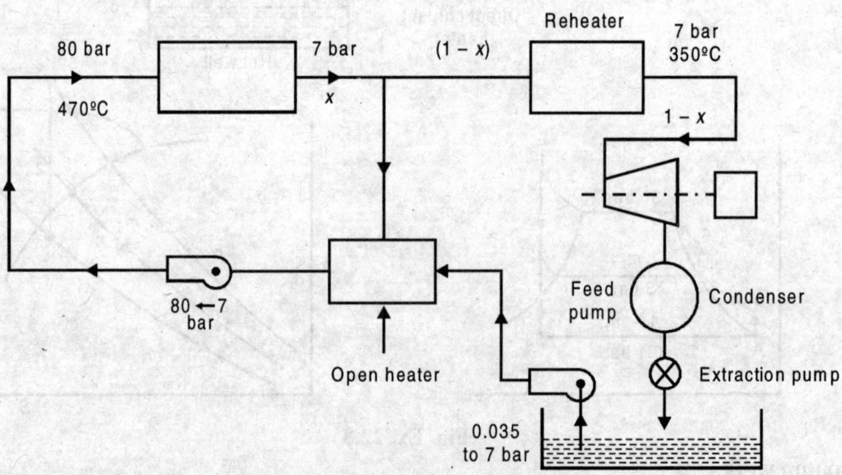

Fig. Ex. 22.9 (a)

Plant in equation is shown on a schematic diagram.

$$p_1 = 80 \text{ bar}$$
$$T_1 = 470^\circ \text{ C}$$
$$p_2 = 7 \text{ bar}$$
$$T_3 = 350^\circ \text{ C}$$
$$p_4 = 0.035 \text{ bar}$$
$$\text{Steam flow rate} = 50 \text{ kg/s}$$

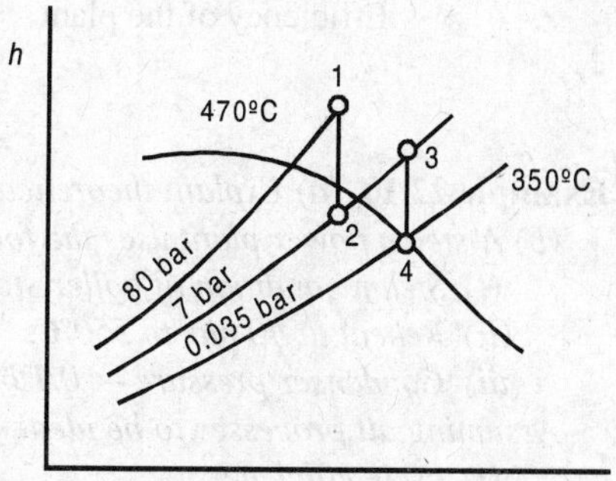

Fig. Ex. 22.9 (b)

(i) Power generated : (ii) Steam bled

$h_1 = 3325$ kJ/kg $\quad h_2 = 2725$ kJ/kg

$h_3 = 3170$ kJ/kg $\quad h_4 = 2250$ kJ/kg

$h_{f_4} = 111.85$ kJ/kg

$h_{f_2} = 697.06$ kJ/kg

Assume 1 kg steam supplied to H.P. turbine and x kg is bled for feed water heating. Again assume a open (direct) feed heater (steam and water are mixed together).

For the feed heater

$$\text{Heat lost by steam} = \text{Heat gained by feed water}$$

$$x(h_2 - h_{f_2}) = (1-x)(h_{f_2} - h_{f_4})$$

$$\therefore \quad (2725 - 697.06)x = (1-x)(697.06 - 111.85)$$

$$2027.94x = (1-x)585.21$$
$$= 585.21 - 585.21\,x$$

$$\therefore \quad x\,(2027.94 + 585.21) = 585.21$$

$$x = \frac{585.21}{2613.15}$$

$$= \mathbf{0.224 \ kg/kg \ steam = 22.4\%}$$

$$\text{Power generated} = (h_1 - h_2) + (1-x)(h_3 - h_4)$$
$$= (3325 - 2725) + (1 - 0.224)(3170 - 2250)$$
$$= 600 + 0.776 \times 920$$
$$= 600 + 713.92$$
$$= \mathbf{1313.92 \ kJ/kg}$$
$$= \mathbf{65696 \ kJ/s = kW}$$
$$= \mathbf{66 \ MW}$$

(iii) Efficiency of the plant

$$\text{Heat added/kg} = (h_1 - h_{f_2}) + (1-x)(h_3 - h_2)$$
$$= (3325 - 697.06) + (1 - 0.224)\,(3170 - 2725)$$
$$= 2627.94 + 0.776 \times 445$$
$$= 2627.94 + 345.32$$
$$= \mathbf{2973.26 \ kJ/kg}$$

$\therefore$ Efficiency of the plant $= \dfrac{W}{Q_A} = \dfrac{1313.92}{2973.26}$

$= \mathbf{0.4419}$

$= \mathbf{44.19\%}$

Example 22.10. (*a*) *Explain theoretical and actual regenerative feed heating as applied to Rankine cycle plant.*

(*b*) *A steam power plant uses the following cycle:*

(*i*) *Steam condition at boiler stop valve — 150 bar, 550°C.*

(*ii*) *Reheat at 40 bar to 550°C.*

(*iii*) *Condenser pressure — 0.1 bar.*

Assuming all processes to be ideal, find

(*i*) *cycle efficiency*

(*ii*) *specific steam rate (kg/kWh of the plant).*

Solution

From Mollier chart,

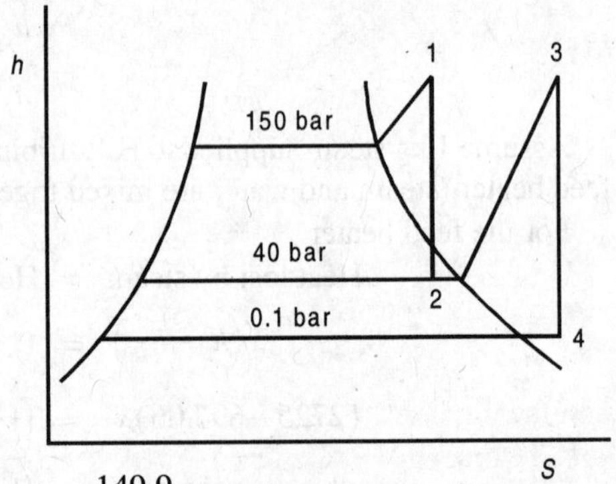

Fig. Ex. 22.10

$h_1 = 3445.2$

$h_2 = 3050$

$h_3 = 3558.9$

$h_4 = 2280$

$h_{f_4} = 191.83$

Expansion work/kg $= (h_1 - h_2) + (h_3 - h_4) - W_p$

$= (3445.2 - 3050) + (3558.9 - 2280) - \dfrac{149.9}{10}$

$= 395.2 + 1278.9 - 14.99$

$= \mathbf{1659.11\ kJ/kg}$

Heat supplied $= (h_1 - h_{f_4}) + (h_3 - h_2) - W_p$

$= (3445.2 - 191.83) + (3558.9 - 3050) - 14.99$

$= 3253.37 + 508.9 - 14.99$

$= \mathbf{3747.28\ kJ/kg}$

$\therefore$ Cycle efficiency $= \dfrac{\text{Work}}{\text{Heat supplied}} = \dfrac{1659.11}{3747.28}$

$= 0.4428$

$= \mathbf{44.28\%}$

Specific steam rate $= \dfrac{3600}{1659.11} = \mathbf{2.17\ kg/kWh}$

Example 22.11. (*a*) *Explain the working of a modern reheat regenerative cycle used for steam power plants with the help of T-s and h-s diagrams and plant schematic.*

(*b*) *A steam power plant working on reheat regenerative cycle has the following data:*

(*i*) *Steam inlet to HP turbine 80 bar 470°C.*

(*ii*) *Steam extraction for feed heating at 7 bar.*

(*iii*) *Remainder steam reheated to 350°C in reheater.*

(iv) Exhaust pressure from LP turbine = 0.035 bar.

Determine: (i) Fraction of steam bled off for reheating. (ii) Heat supplied/kg of steam in boiler and in reheater. (iii) Power output of the turbine. (iv) Thermal efficiency.

Solution

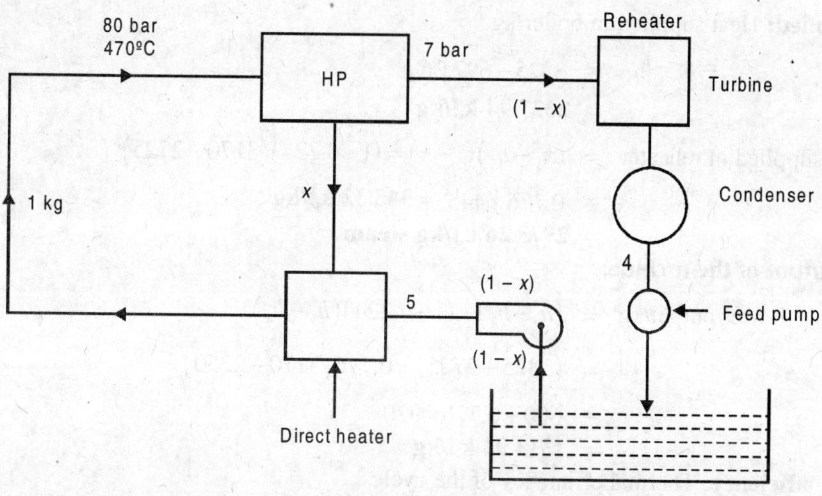

Fig. Ex. 22.11(a)

Assume direct heater.

Referring to Mollier chart,

$h_1 = 3325$ kJ/kg, $h_2 = 2725$, $h_{f_2} = 697.06$, $h_{f_3} = 3170$, $h_4 = 2250$, $h_{f_4} = 111.85$

(i) Fraction of steam bled off or for feed heating: Let x be the steam bled per kg of steam supplied to turbine.

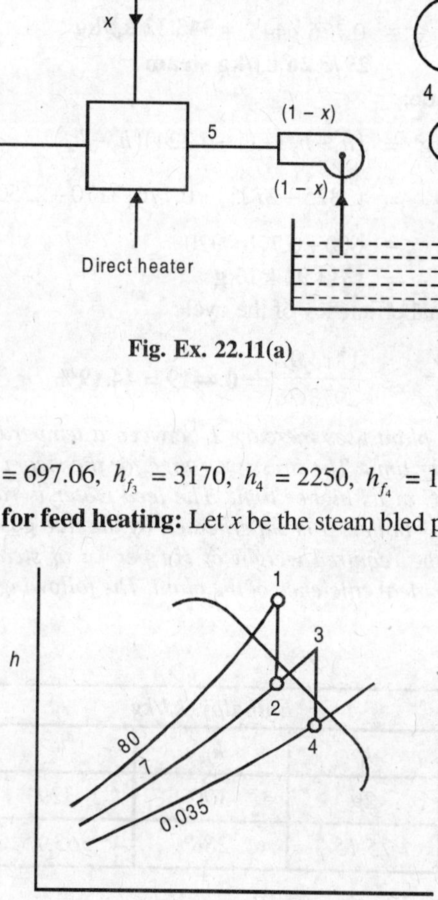

Fig. Ex. 22.11(b)

$$\therefore \quad \text{Heat given by steam} = \text{Heat gained by water}$$

$$x\left[h_1 - h_{f_2}\right] = (1-x)\left(h_{f_2} - h_{f_4}\right)$$

$$\therefore \quad x(2725 - 697.06) = (1-x)(697.06 - 111.85)$$

$$2027.94x = 585.21 - 585.21x$$

$$x = \frac{585.21}{2613.15} = \mathbf{0.224}$$

x = 0.224 kJ/kg steam supplied to turbine

i.e. **22.4%**

(ii) Heat supplied: Heat supplied in boiler/kg

$$= h_1 - h_{f_2} = 3325 - 697.06$$
$$= \mathbf{2627.94\ kJ/kg}$$

$$\text{Heat supplied in reheater} = (h_3 - h_2)(1 - x) = (1 - 0.224)(3170 - 2725)$$
$$= 0.776 \times 445 = \mathbf{345.32\ kJ/kg}$$
$$= \mathbf{2973.26\ kJ/kg\ steam}$$

(iii) Power putput of the turbine:

$$\text{Power output/kg} = (h_1 - h_2) + (1 - 0.224)(h_3 - h_4)$$
$$= (3325 - 2725) + 0.776(3170 - 2250)$$
$$= 600 + 0.776 \times 920$$
$$= \mathbf{1313.92\ kJ/kg}$$

(iv) Thermal efficiency: Thermal efficiency of the cycle

$$= \frac{W}{Q_A} = \frac{1313.92}{2973.26} = \mathbf{0.4419 = 44.19\%}$$

Example 22.12. *A binary vapour plant uses mercury 1 between a temperature range of 204°C and 537°C, the mercury being just dry at the higher limit. The pressure range for the steam is between 73.66 cm vacuum and 17 bar and it is superheated to 371°C at its higher limit. The feed water is raised to 204°C in economiser and is evaporated to dry steam in Hg-H_2O boiler; it is superheated by the flue gases. Assuming adiabatic expansion in Hg and steam turbine, find. (1) the required weight of Hg per kg of steam used, (2) work done by mercury, (3) work done by steam and (4) the ideal efficiency of the plant. The following are the required values of total heats and entropies of Hg and steam.*

Mercury:

Temperature, °C	*Enthalpy, kJ/kg*			*Entropy, kJ/kg.K*	
	h_f	h_{fg}	h_g	s_f	s_g
204°C	*29*	*300*	*329*	*0.044*	*0.3883*
537°C	*75.15*	*288*	*363.15*	*0.08413*	*0.2820*

Steam:

Pressure bar	*Enthalpy*			*Entropy*	
	h_f	h_{fg}	h_g	s_f	s_g
0.03 bar 73.66 cm Hg	*101.0*	*2444*	*2545*	*0.354*	*8.576*
16.9 bar	*8.72*	*1923*	*2792*	*2.372*	*6.4*

Solution

Steam at 371ºC and 17 bar, $h = 3200$ kJ/kg, $s = 7.1259$.

Refer the T–s diagram drawn.

$$s_c = 0.2820 = s_d = 0.044 + x_d \times (0.3883 - 0.044)$$

$$= 0.044 + x_d \times 0.3443$$

$$\therefore \quad x_d = \frac{0.2820 - 0.044}{0.3443}$$

$$= \frac{0.2380}{0.3443}$$

$$= \mathbf{0.6913}$$

(i) Weight of mercury

Let, W_m = Weight of mercury required per kg of steam

Then, Loss of heat of Hg between d and a = Gain of latent heat of steam from a to d

$$W_m \times x.h_{f_g} = L_s \text{ at 17 bar}$$

$$W_m \times 0.6913 \times 300 = 1923$$

$$\therefore \quad W_m = \frac{1923}{0.6913 \times 300}$$

$$= \mathbf{9.27\ kg}$$

(ii) Work done by mercury

$$\text{Work} = W_m[H_c - H_d]$$

$$= 9.27[363.15 - 237.29]$$

$$= 9.27 \times 125.86$$

$$= \mathbf{1166.72\ kJ/kg\text{-}steam}$$

where, $H_c = 363.15$

$$H_d = 29.9 + 0.6913 \times 300$$

$$= 29.9 + 207.39$$

$$= 237.29$$

(iii) Work done by steam

$$H_e = 3200 \text{ kJ/kg}, \qquad s_e = 7.1259$$

$$s_f = 0.354 + x_f\,(8.576 - 0.354)$$

$$= 0.354 + 8.222\,x_f$$

$$\therefore \quad x_f = \frac{7.1259 - 0.354}{8.222} = \frac{6.7719}{8.222} = \mathbf{0.8236}$$

$$\therefore \quad H_f = 101 + 0.8236 \times 2444$$

$$= 101 + 2012.88 = 2113.88$$

$$\therefore \quad \text{Work done/kg} = 3200 - 2113.88$$

$$= \mathbf{1086.12}$$

(iv) Total work done/kg of steam = 1166.72 + 1086.12 = **2252.84**

$$[\text{Heat supplied by fuel}] = \begin{bmatrix}\text{Heat supplied}\\ \text{to mercury}\end{bmatrix} + \begin{bmatrix}\text{Heat to}\\ \text{superheated steam}\end{bmatrix} + \begin{bmatrix}\text{Heat of}\\ \text{feed water}\end{bmatrix}$$

$$= W_m[H_c - H_a] + [H_e - H_d] + [H_a - H_g]$$

$$= 9.27(363.15 - 29) + (3200 - 2795) + (872 - 101)$$

$$= 3097.57 + 405 + 771$$

$$= \mathbf{4273.57\ kJ/kg\ steam}$$

$$\therefore \text{Ideal efficiency of the plant} = \frac{\text{Total work done}}{\text{Total heat supplied}}$$

$$= \frac{2252.84}{4273.57} = \mathbf{0.5272 \text{ or } 52.72\%}$$

Example 22.13. *A cyclic steam power plant is to be for a steam temperature at the turbine inlet of 360°C and an exhaust pressure of 0.08 bar isentropic expansion of steam in the turbine, the moisture content at the turbine exhaust is not to exceed 15%. Determine the greatest allowable steam pressure at the turbine inlet and calculate the Rankine cycle efficiency for these steam conditions. Also estimate the mean temperature of heat addition.*

Solution

$$T_1 = 360^\circ \text{ C}$$

$$P_2 = 0.08 \text{ bar} \quad x_2 = 0.85$$

$$\text{From steam tables} - t_{sat} = 42.689$$

$$\therefore \text{Mean temperature of heat addition} = \frac{360 + 41.5}{2} = \frac{401.5}{2}$$

$$= 200.75^\circ\text{C}$$

Corresponding to this temperature as saturation temperature, pressure is given as

$$p = \frac{15.6 \text{ bar}}{16}$$

$$= \mathbf{0.975\ bar}$$

$$\text{Rankine efficiency} = \frac{h_1 - h_2}{h_1 - h_2} \text{ neglecting } W_p$$

$$h_1 = 2795.0 - 2.1\ (360 - 200.75)$$

$$= 2795.0 - 334.425$$

$$= 3129.425 \text{ kJ/kg}$$

$$h_2 = 174 + 0.85 \times 2402$$

$$= 174 + 2041.7$$

$$= 2215.7 \text{ kJ/kg}$$

$$h_{f_2} = 174 \text{ kJ/kg}$$

$$\therefore \quad \eta_f = \frac{3129.425}{3129.425 - 174}$$

$$= \frac{913.725}{2955.425}$$

$$= \mathbf{0.3092 \text{ or } 30.92\%}$$

Example 22.14. *Find the ideal cycle efficiency and steam consumption of a reheat cycle operating between the pressures of 30 bar and 0.04 bar with initial superheated temperature of 400°C, expanded to the point where the steam is dry and then it is reheated at that pressure upto initial temperature of 400°C.*

The feed pump work may be neglected.

Solution

The cycle represented on *T-s* diagram as shown.

Reffering to *h-s* diagram,

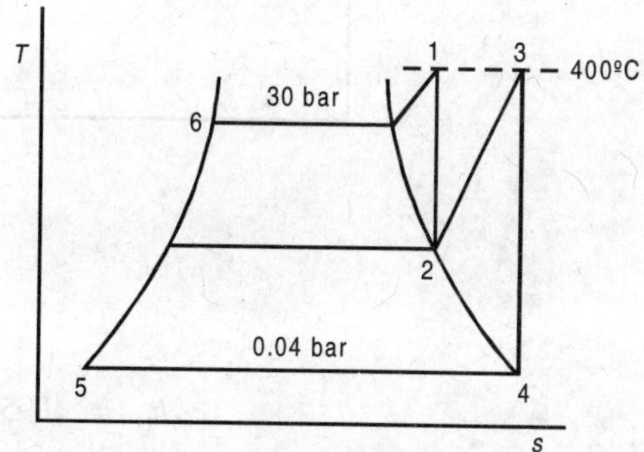

Fig. Ex. 22.14

$$h_1 = 3230 \text{ kJ/kg}$$

$$h_2 = 2730 \text{ kJ/kg}$$

$$h_3 = 3270 \text{ kJ/kg}$$

$$h_4 = 2395 \text{ kJ/kg}$$

$$\text{Ideal work done/kg steam} = (h_1 - h_2) + (h_3 - h_4)$$

$$= (3230 - 2730) + (3270 - 2395)$$

$$= 500 + 875$$

$$= \mathbf{1375 \text{ kJ/kg}}$$

$$\text{Heat supplied/kg of steam} = (h_1 - h_{f_5}) + (h_3 - h_2)$$

$$= (3230 - 121.4) + (2370 - 2730)$$

$$= 3108.6 + 540$$

$$= \mathbf{3648.6 \text{ kJ/kg}}$$

$$\therefore \quad \text{Ideal cycle efficiency} = \frac{\text{Work done}}{\text{Heat supplied}}$$

$$= \frac{1375}{3648.6} = 0.3769$$

$$= \mathbf{37.69\%}$$

$$\text{Specific steam consumption} = \frac{3600}{1375} = \mathbf{2.618 \text{ kg/kWh}}$$

Example 22.15. *In a thermal power plant of 60 MW capacity, boiler supplies steam at 100 bar and 400°C. The steam expands to 20 bar in the HP turbine. Steam is then reheated to 380°C in a reheater. Thereafter, steam expands to a condenser pressure of 0.05 bar. The pressure loss in the reheater is 1 bar. Assuming isentropic efficiency in both stages of expansion is 80%, find the steam generating capacity required in ton/hour. Take* *transmission efficiency from turbine to generator = 98% and generator efficiency of 95%. Neglect feed pump work. Also find the overall efficiency of the plant.*

Draw a neat line diagram of the plant and represent the cycle on T-s diagram.

Solution

From Mollier diagram, we have

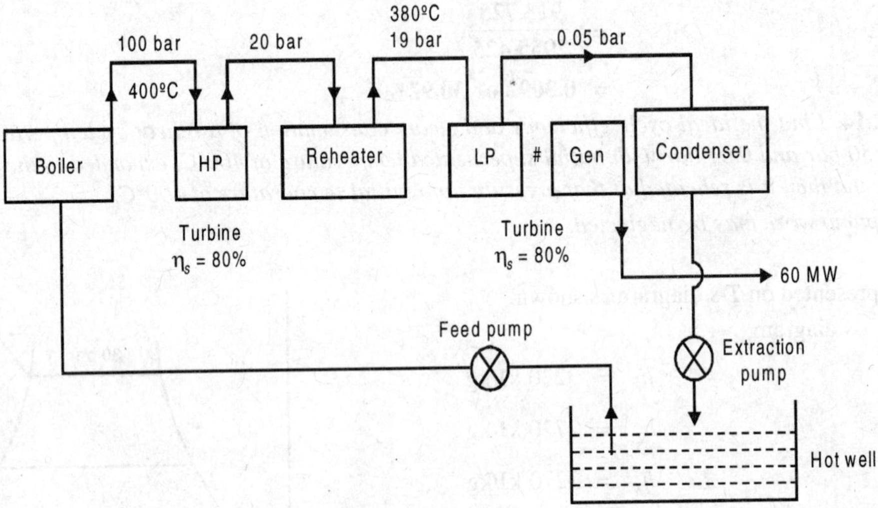

Fig. Ex. 22.15(a)

$$h_1 = 3095 \text{ kJ/kg}$$
$$h_2 = 2730 \text{ kJ/kg}$$
$$h_3 = 3200 \text{ kJ/kg}$$
$$h_4 = 2140 \text{ kJ/kg}$$
$$h_{f_5} = 137.77 \text{ kJ/kg}$$

We have, $\eta_s = 0.8 = \dfrac{h_1 - h_{2'}}{h_1 - h_2} = \dfrac{3095 - h_{2'}}{3095 - 2730} = \dfrac{3095 - h_{2'}}{365}$

$\therefore \quad 3095 - h_{2'} = 0.8 \times 365 = 292.0$

$\therefore \quad h_{2'} = 3095 - 292 = \mathbf{2803\ kJ/kg}$

Again $0.8 = \dfrac{h_3 - h_{4'}}{h_3 - h_4} = \dfrac{3200 - h_{4'}}{3200 - 2140} = \dfrac{3200 - h_4}{1060}$

$\therefore \quad 3200 - h_{4'} = 0.8 \times 1060 = \mathbf{848\ kJ/kg}$

$\therefore$ Work done in HP turbine $= h_1 - h_2 = 292 \text{ kJ/kg}$

$\therefore$ Work done in LP turbine $= h_3 - h_4 = 848 \text{ kJ/kg}$

$\therefore$ Net work/kg $= 292 + 848 = \mathbf{1140\ kJ/kg}$

Work transmitted to generator $= 1140 \times 0.98 = \mathbf{1117.2 kJ/kg}$

$\therefore$ Work delivered by generator $= 1117.2 \times 0.95 = 1061.34$ kJ/kg

Capacity of the plant is 60 MW $= 60 \times 1000$ kW

$$\therefore \quad \text{Mass flow rate} = \frac{60 \times 1000}{1061.34}$$

$$= 56.53 \text{ kg/sec}$$

$$= \frac{56.53 \times 3600}{1000} \text{ tons/hour}$$

$$= \mathbf{203.5 \text{ tons/hour}}$$

$$= \text{Capacity of steam generation}$$

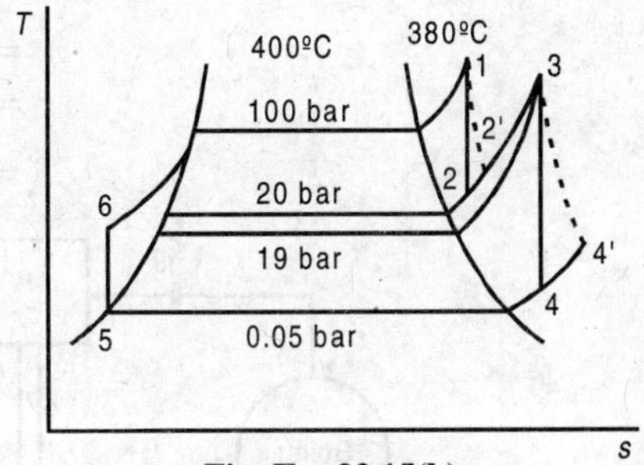

Fig. Ex. 22.15(b)

$$\text{Heat supplied/kg} = (h_1 - h_{f_5}) + (h_3 - h_{2'})$$

$$= (3095 - 137.77) + (3200 - 2803)$$

$$= 2957.23 + 397 = \mathbf{3354.23 \text{ kJ/kg}}$$

$$\therefore \text{Overall efficiency of the plant} = \frac{\text{Work delivered/kg}}{\text{Heat supplied/kg}} = \frac{1061.34}{3354.23}$$

$$= \mathbf{0.3164 = 31.64 \%}$$

Example 22.16. *The steam at 392 N/cm² and 300°C is supplied to steam turbine. The steam is bled for regenerative heating at 137.2 N/cm² and 33.3 N/cm². The condenser pressure is 0.686 N/cm². Calculate*

(i) At each stage the optimum weight of bled beam per kg of steam supplied to the turbine and

(ii) the cycle efficiency.

Assume turbine efficiency of 0.8 for each portion of the expansion and that the feed water leaving each heater is raised to the steam temperature entering the heater.

The heaters used are indirect heaters and the condensate coming out from the first heater is passed through the second heater is passed to the hot well. Neglect pump work.

Solution

Indirect heater

$\therefore$ Steam does not mix with feed water

$$p_1 = 40 \text{ bar}(39.2)$$
$$p_2 = 14 \text{ bar}(13.72)$$
$$p_3 = 3.50 \text{ bar}(3.33)$$
$$p_4 = 0.07 \text{ bar}(0.0686)$$
$$h_1 = 2965 \text{ kJ/kg}$$
$$h_a = 1740 \text{ kJ/kg}$$
$$h_{2'} = 2740 + (2965 - 2740) \times 0.2$$
$$= 2740 + 45 = \mathbf{2785 \text{ kJ/kg}}$$

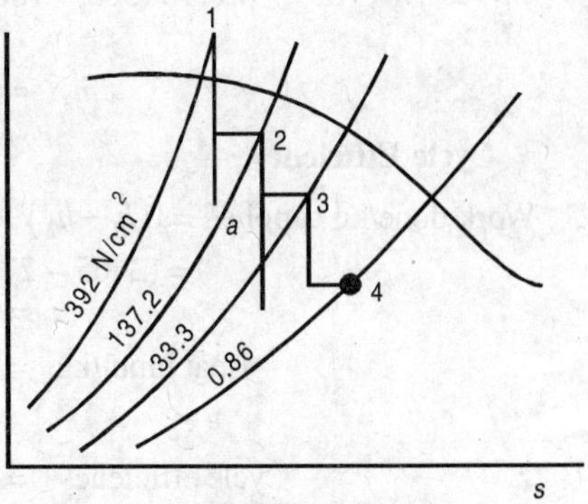

Fig. Ex. 22.16 (a)

$$h_b = 2540 \text{ kJ/kg}$$
$$h_3 = (2785 - 2540) \times 0.2 + 2540$$
$$= 2540 + 0.2 \times 245 = 2540 + 49 = \mathbf{2589}$$

$$h_{f_3} = 584 \text{ kJ/kg}$$
$$h_c = 2330 \text{ kJ/kg}$$
$$h_4 = 2330+0.2(2589-2330)$$
$$= 2330+51.8$$
$$= \mathbf{2381 \text{ kJ/kg}}$$
$$h_{f_4} = 160 \text{ kJ/kg}$$

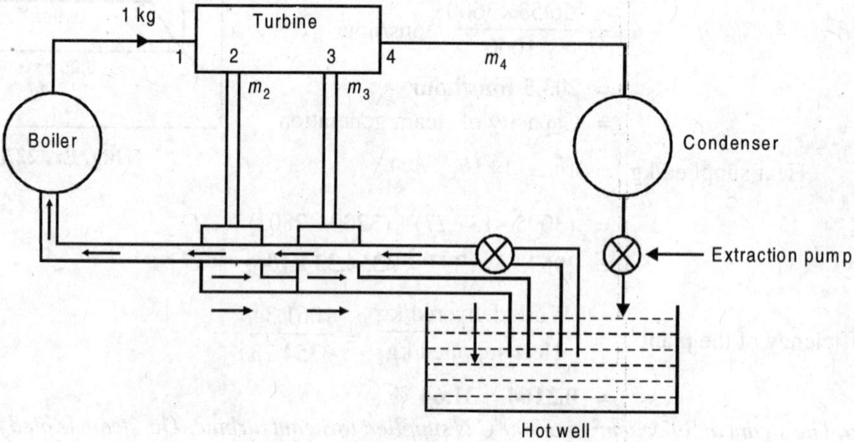

Fig. Ex. 22.16 (b)

(i) Heater at 2

$$\dot{m}_2(h_2 - h_{f_3}) = 1(h_{f_2} - h_{f_3}) = 830 - 584 = 246$$

$$\therefore \quad \dot{m}_2 = \frac{246}{2785-584} = \frac{246}{2201} = \mathbf{0.1118 kg}$$

(ii) Heater at 3

$$\dot{m}_3(h_3 - h_{f_4}) + \dot{m}_2(h_{f_3} - h_{f_4}) = 1(h_{f_3} - h_{f_4})$$

$\dot{m}_3$ (2589 – 160) + 0.1118 (584 – 160) = 584 – 160 = **424**

$$\therefore \quad \dot{m}_3 = \frac{424-47.4}{2429} = \mathbf{0.155 \text{ kg}}$$

Cycle Efficiency

$$\text{Work done/kg supplied} = 1(h_1 - h_2) + (1-\dot{m}_2)(h_2 - h_3) + (1 - \dot{m}_2 - \dot{m}_3)(h_3 - h_4)$$

= (2965 – 2785) + (1 – 0.1118) (2785 – 2589)+ (1 – 0.1118 – 0.155) (2589 – 2381.8)

= 506 kJ/kgs

$$\text{Heat input/kg} = 1(h_1 - h_{f_2}) = 2965 - 830 = 2135 \text{ kJ/kg}$$

$$\therefore \quad \text{Cycle efficiency} = \frac{\text{Work output}}{\text{Heat input}} = \frac{506}{2135} = 0.237$$

$$= \mathbf{23.7\%}$$

Example 22.17. *The steam turbine of a power station using regenerative cycle receives dry and saturated steam at 32 bar. The steam is bled at 2.8 bar and the condenser vacuum is 723.1 mm of Hg. The heating of feed water is done in the direct contact heater. Find (i) the quantity of steam bled per kg of steam supplied to the turbine, (ii)* *cycle efficiency and (iii) specific steam consumption.*

Solution

$$\text{Condenser pressure} = 760 - 723.1 = 36.9 \text{ mm Hg}$$
$$= 0.04916 \text{ bar} = \mathbf{0.05 \text{ bar}}$$

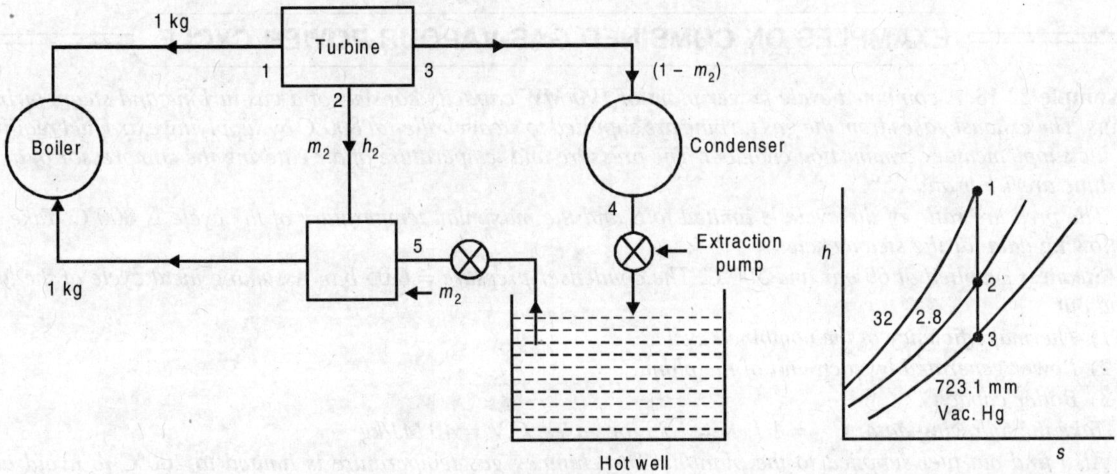

Fig. Ex. 22.17

From Mollier chart,

$$h_1 = 2803 \text{ kJ/kg}, h_2 = 2490 \text{ kJ/kg} = 1880 \text{ kJ/kg}, h_{f_2} = 551, h_{f_3} = 138 \text{ kJ/kg}$$

(i) For heater

$$\text{Heat given by steam} = \text{Heat gained by feed water}$$

$$\dot{m}_2(h_2 - h_{f_2}) = (1 - \dot{m}_2)(h_{f_2} - h_{f_3})$$

$$\therefore \quad \dot{m}_2\ (2490 - 551) = (1 - \dot{m}_2)(551 - 138)$$

$$1939\ \dot{m}_2 = 413\ (1 - \dot{m}_2)$$

$$\therefore \quad \dot{m}_2 = \frac{413}{1939 + 413} = \frac{413}{2352}$$

$$= \mathbf{0.1756 \text{ kg}}$$

(ii)

$$\text{Work done/kg} = (h_1 - h_2) + (1 - m_2)(h_{f_2} - h_{f_3})$$
$$= (2803 - 2490) + (1 - 0.1756)\ (2490 - 1880)$$
$$= 313 + 0.8244 \times 610 = 313 + 502.88$$
$$= \mathbf{815.88 \text{ kJ/kg}}$$

$$\text{Heat supplied/kg} = h_1 - h_2 = 2803 - 551$$
$$= \mathbf{2232 \text{ kJ}}$$

$$\therefore \quad \text{Cycle thermal efficiency} = \frac{\text{Work output}}{\text{Heat input}} = \frac{815.88}{2252}$$

$$= \mathbf{0.3623 \text{ or } 36.23\%}$$

(iii) Specific steam consumption $= \dfrac{3600}{815.88}$

$$= \mathbf{4.412 \text{ kg/k Wh}}$$

EXAMPLES ON COMBINED GAS-VAPOUR POWER CYCLE

Example 22.18. *A combined cycle power plant of 190 MW capacity consists of a gas turbine and steam turbine units. The exhaust gases from the gas turbine are supplied to steam boiler at 800°C by supplying extra fuel required in the supplementary combustion chamber. The pressure and temperature of air entering the compressor of a gas turbine are 1 bar and 15°C.*

The pressure ratio of the cycle is limited to 8 and the maximum temperature of the cycle is 800°C. Take the following data for the steam cycle.

Steam is supplied at 60 bar and 500°C. The condenser pressure = 0.05 bar. Assuming ideal cycle of the unit, find out

(1) Thermal efficiency of the combined cycle.

(2) Power generated by each unit of the plant.

(3) Boiler capacity.

Take the following data: $C_{pg} = 1.1$, $K = 1.33$ *for gases, C.V. = 40 MJ/kg*

Also find out fuel supplied to the plant/hr. The chimney gas temperature is limited to 200°C to avoid acid condensation.

Sketch the schematic of the entire plant and represent the processes on T-s diagram for both the units.

Solution

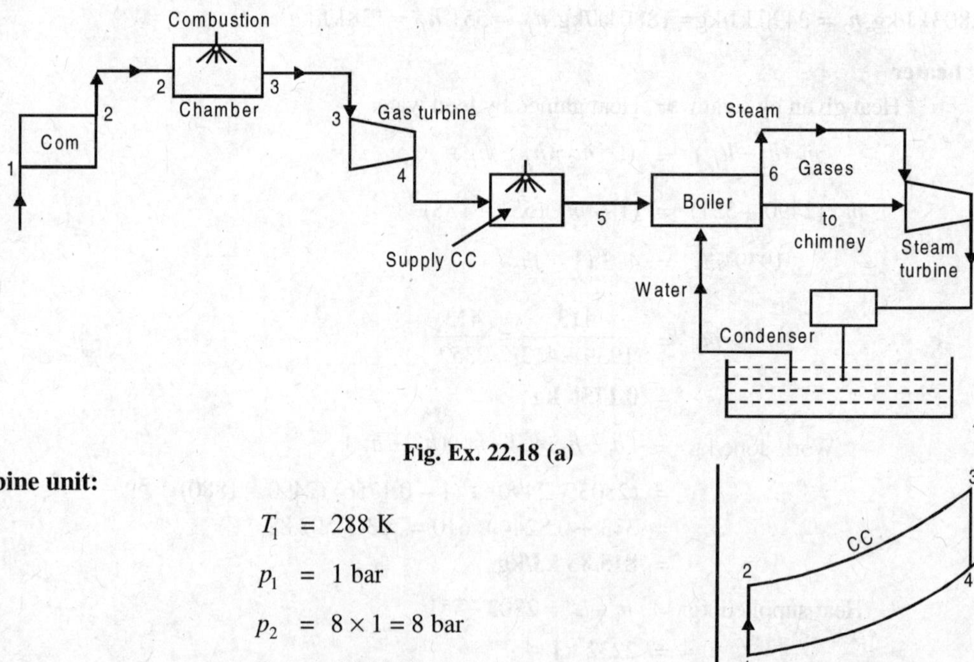

Fig. Ex. 22.18 (a)

Gas turbine unit:

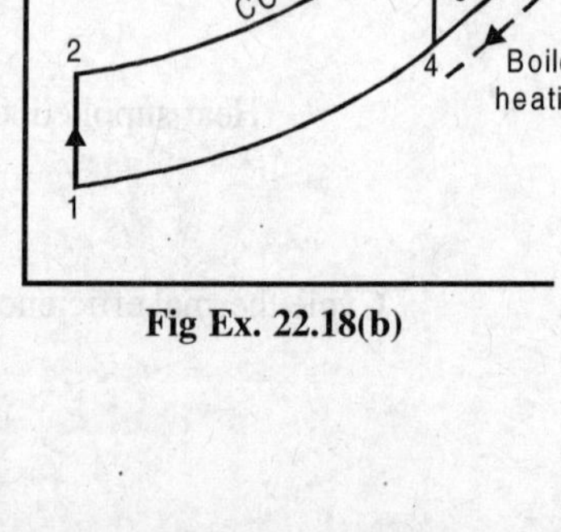

Fig Ex. 22.18(b)

$$T_1 = 288 \text{ K}$$

$$p_1 = 1 \text{ bar}$$

$$p_2 = 8 \times 1 = 8 \text{ bar}$$

$$T_3 = 1073° \text{ K}$$

Assume 1 kg air, $\quad C_{Pa} = 1.005 \text{ kJ/kg}$

$$\gamma \text{ for air} = 1.4$$

$$\frac{T_2}{T_1} = \left(\frac{p_2}{p_1}\right)^{\frac{\gamma-1}{\gamma}} = 8^{\frac{1.4-1}{1.4}} = 8^{0.286} = 1.8125$$

$$\therefore \quad T_2 = 288 \times 1.8125 = \mathbf{522\ ^\circ K}$$

$$\therefore \quad T_4 = \frac{T_3}{8^{\frac{133-1}{1.33}}} = \frac{1073}{(8)^{0.24812}}$$

$$= \frac{1073}{1.67523}$$

$$= \mathbf{640.51^\circ\ K}$$

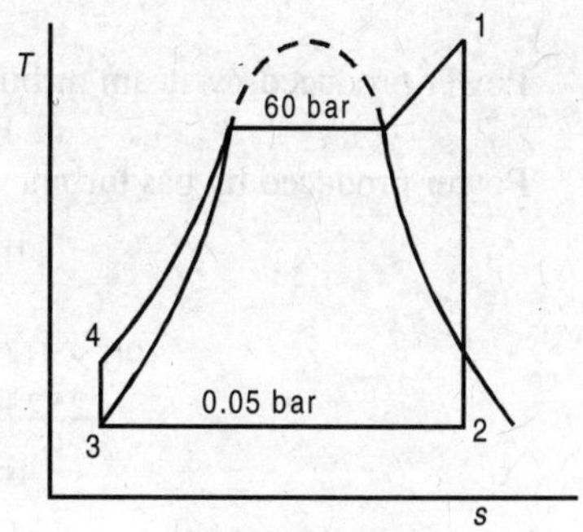

Fig Ex. 22.18(c)

$$\therefore \quad \text{Compressor work/kg} = C_{p_a}(T_2 - T_1) = 1.005\ (522 - 288)$$

$$= \mathbf{235.17\ kJ/kg}$$

$$\text{Turbine work/kg} = C_{p_g} = (T_3 - T_4)$$

$$= 1.1\ (1073 - 640.51)$$

$$= 1.1\ (432.49)$$

$$= \mathbf{475.74\ kJ/kg}$$

$$\therefore \text{Net work of gas/turbine/kg} = W_t - W_c = 475.74 - 235.17$$

$$= \mathbf{240.57\ kJ/kg}$$

$$\text{Heat supplied air} = C_{p_g} T_3 - C_{p_a} T_2 = 1.1 \times 1073 - 1.005 \times 522$$

$$= 1180.3 - 524.61$$

$= \mathbf{655.69\ kJ/kg}$ i.e. **0.01164 kgf/kg air**

Steam turbine h_1 = 3422.2 kJ/kg from table and 3410 from chart,

$$h_{f_3} = 137.77\ \text{kJ/kg}$$

$$\text{Heat required/kg steam} = 3410.2 - 137.77$$

$$= \mathbf{3272.43\ kJ/kg}$$

$$h_2 = 2080\ \text{kJ/kg from chart}$$

$$\therefore \quad \text{Work done/kg steam} = h_1 - h_2$$

$$= 3410 - 2080$$

$$= \mathbf{1330\ kJ/kg}$$

Gases required to produce 1 kg steam

$$= \frac{3274.43}{1.1 \times (800 - 200)} = \frac{3274.43}{660}$$

$$= \mathbf{4.961\ kJ/kg\ steam}$$

Heat in supplimentary combustion chamber/kg steam

$$= m_g c_{pg}(T_5 - T_4) = 4.961 \times 1.1(1073 - 640.51)$$

$$= 4.961 \times 1.1 \times 432.49$$

$$= \mathbf{2360.14}$$

i.e. **0.05918 kgf/kg steam**

$$\text{Total power generated} = 190\times10^6$$
$$= 190 \times 1000 \text{ kW} \quad \ldots\text{(I)}$$

Let m_s is mass of steam/sec

$$\dot{m}_a = \dot{m}_g = \text{mass of air required/sec}$$
$$= \mathbf{4.961 \times m_s}$$

Power produced by steam turbine

$$W_s = \dot{m}_s \times 1330 \text{ kW}$$

Power produced by gas turbine

$$W_g = \dot{m}_a \times 240.57 = 4.961 \times \dot{m}_s \times 240.57$$
$$= \mathbf{1193.46\ \dot{m}_s\ kW} \quad \ldots\text{(II)}$$
$$\therefore \quad 190 \times 1000 = 1330\ \dot{m}_s + 1193.46\ \dot{m}_s$$
$$2523.5 = 2527.58\ \dot{m}_s$$
$$\therefore \quad \dot{\mathbf{m}}_\mathbf{s} = \mathbf{75.29\ kg/sec}$$
$$\dot{\mathbf{m}}_\mathbf{a} = \mathbf{373.52\ kg/sec}$$
$$\text{Heat supplied/sec} = 2360.14 \times 75.29 + 373.52 \times 655.69$$
$$= 178235.5 + 244913.32$$
$$= 423148.82 \text{ kJ/s}$$
$$\therefore \quad \text{Fuel supplied/hour} = \frac{\text{Heat supplied/sec}\times3600}{\text{C. V. of fuel}} = \frac{423148.82\times3600}{40000}$$
$$= 38083.4 \text{ kg/hour} = \mathbf{38.083\ T/hour}$$

$\therefore$ Thermal efficiency of the plant

$$= \frac{190\times100}{423148.82} = 0.449 = \mathbf{44.9\%}$$

$$\text{Power produced by GT} = \frac{1197.08\times75.29}{1000}$$
$$= 90.13 = \mathbf{90\ MW}$$
$$\text{Steam turbine} = 75.29 \times 1330$$
$$= 99.87 = \mathbf{100\ MW}$$
$$\text{Heat supplied in GT plant} = \dot{m}_a\times655.69 = 373.52\times655.69 = 244913.32 \text{ kJ/sec}$$

Example 22.19. *In a combined cycle power plant, the air is supplied at the rate of 2000×10^3 kg/hr at 20°C. The pressure ratio used in gas turbine plant is 7. The suction pressure of the compressor and the exhaust pressure of turbine is 1 bar. The maximum temperature in the gas turbine is limited to 1000°C. The isentropic efficiencies of the gas turbine and compressor are 85% and 80% respectively. The following data is given for the steam power plant.*

t_g of exhaust from G.T. is increased to 1200°C by burning fuel before passing into the boiler. The steam is generated at 50 bar and 500°C. The condenser pressure is maintained at 0.1 bar. Temperature of gases entering the chimney is 200°C.

Find the following:

(i) Total power generating capacity of the combined plant.
(ii) Overall efficiency of the plant.
(iii) Fuel consumption of two combined plant/hr.
(iv) Air to fuel ratio used for gas turbine plant. Use the following data:
C.V. of fuel = 45 MJ/kg, C_{pa} = 1 kJ/kg °K, γ_a = 1.4, C_{pg} = 1.1 γg = 1.33

Do not neglect the mass of the fuel used.

Draw neat line diagram for the components.
Represent the process on T-s diagram.

Solution

Gas turbine

$$\frac{T_2}{T_1} = \left(\frac{p_2}{p_1}\right)^{\frac{\gamma-1}{\gamma}} = 7^{0.286} = 1.745$$

$$\therefore \quad T_2 = 293\times1.745 = \mathbf{511.17^\circ\ K}$$

$$\frac{T_3}{T_4} = (7)^{\frac{1.33-1}{1.33}} = 7^{0.2481} = 1.621$$

$$\therefore \quad T_4 = \frac{1273}{1.621} = 785.32^\circ K$$

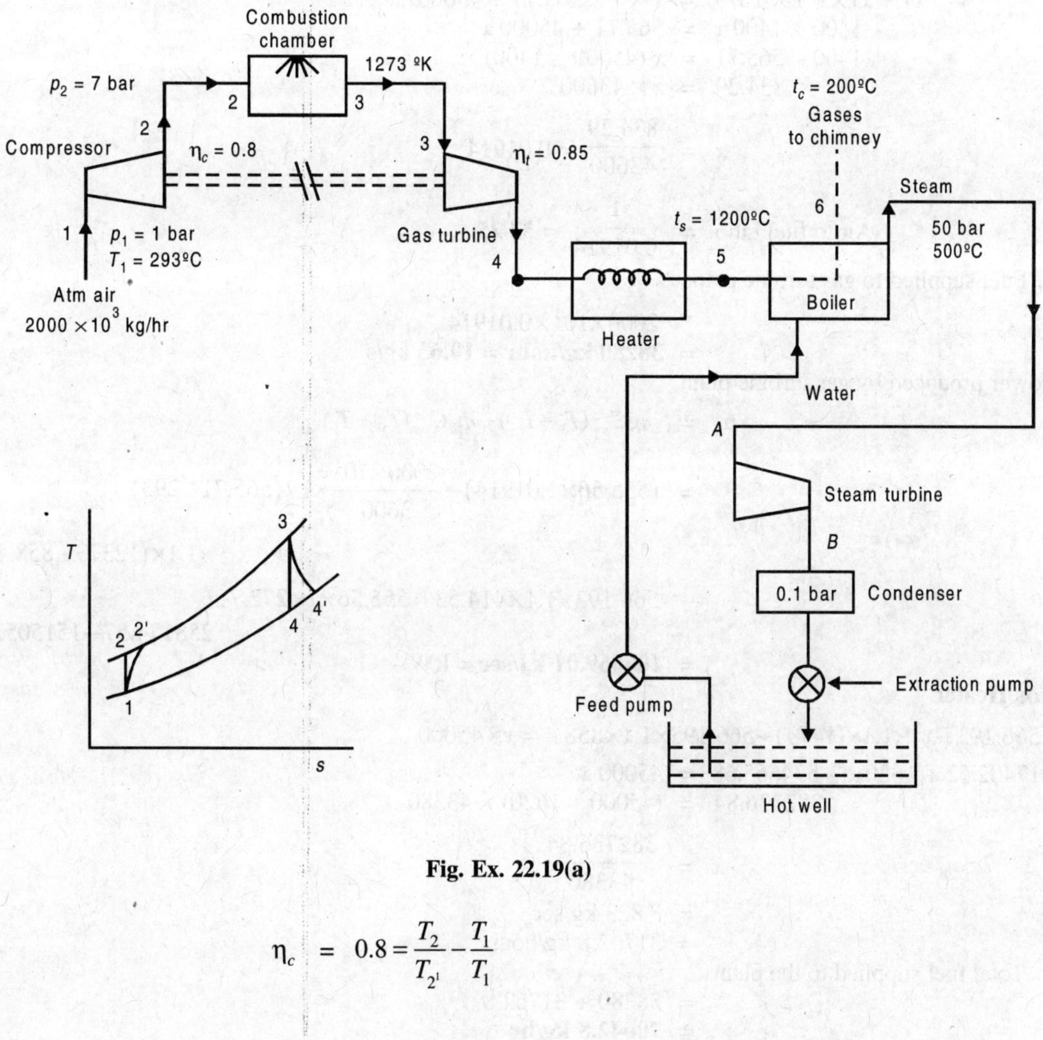

Fig. Ex. 22.19(a)

$$\eta_c = 0.8 = \frac{T_2}{T_{2'}} - \frac{T_1}{T_1}$$

$$0.8 = \frac{511.17-293}{T_{2_1}-293} = \frac{218.17}{T_{2_1}-293}$$

$$\therefore \quad T_{2'} = 293+\frac{218.17}{0.8} = 293 + 272.71$$

$$= \mathbf{565.71^{\circ}\ K}$$

Similarly $$\eta_t = 0.85 = \frac{T_3-T_{4'}}{T_3-T_4} = \frac{1273-T_{4'}}{1273-785.32}$$

$$\therefore \quad T_{4'} = 1273 - 0.85 \times 487.68 = 1273 - 414.53$$

$$= \mathbf{858.47\ ^{\circ}K}$$

Consider combustion chamber. Let x be the mass of fuel per kg of air.

$$C_{pg}(1-x)T_3 = C_{pa}T_{2^1} + x\times \text{C.V.}$$

$$(1 + x) \times 1.1 \times 1273 = 1 \times 1 \times 565.71 + 45000x$$

$$1400 + 1400\,x = 565.71 + 45000\,x$$

$$1400 - 565.71 = x\,(45000 - 1400)$$

$$834.29 = x \times 43600$$

$$x = \frac{834.29}{43600} = 0.01914$$

$$\text{Air to fuel ratio} = \frac{1}{0.01914} = \mathbf{52.25}$$

$\therefore$ Fuel supplied to gas turbine plant

$$= 2000\times 10^3 \times 0.01914$$

$$= \mathbf{38280\ kg/hour = 10.63\ kg/s}$$

Power produced by gas turbine plant

$$p_1 = \dot{m}gC_{pg}(T_3-T_{4^1}) - \dot{m}_aC_{pa}(T_{2^1}-T_1)$$

$$= (555.56\times 1.01914) - \frac{2000\times 10^3}{3600}\times 1\times(565.71-293)$$

$$\times 1.1\times(12373-858.47)$$

$$= 566.193\times 1.1\times 414.53 = 555.56\times 1\times 272.71$$

$$258174.57-151505.56$$

$$= \mathbf{106669.01\ kJ/sec = kW} \quad \text{...(I)}$$

For Heater

$$(566.193+x)\times 1.1\times(1473)-566.193\times 1.1\times 858.4 = x\times 45000$$

$$917402.52 + 1620\,x - 534665.68 = 45000\,x$$

$$\therefore \quad 382736.84 = (45000 - 1620) \times 43380$$

$$x = \frac{382736.84}{43380}$$

$$= 8.823 \text{ kg/sec}$$

$$= 31762.8 \text{ kg/hour}$$

$\therefore$ Total fuel supplied to the plant

$$= 38280 + 31762.8$$

$$= \mathbf{70042.8\ kg/hr}$$

Steam turbine:

$p_1 = 50\,\text{bar}, t_1 = 500°\text{C}, p_2 = 0.1\,\text{bar}$

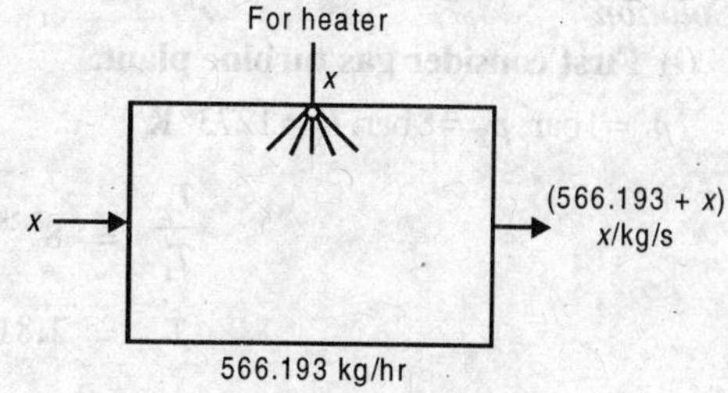

Fig. Ex. 22.19(b)

∴ From *H-s* chart, $h_1 = 3420$ kJ/kg, $h_2 = 2190$

$h_{f_2} = 191.83$ kJ/kg

∴ Heat required for 1 kg steam

$$h_1 - h_{f_2} = 3420.0 - 191.83$$
$$= \mathbf{3228.17\ kJ/kg}$$

Available heat for steam generation

$$= \dot{m}_g C_{pg}(t_5 - t_6)$$
$$= (555.56 + 10.63 + 4.87) \times 1.1 \times (1200 - 200)$$
$$= 571.056 \times 1100 \text{ kJ/sec}$$
$$= 628161.6 \text{ kJ/sec}$$

∴ Amount of steam generated/sec

$$= \frac{628161.6}{3228.17} = \mathbf{194.59\ kg/sec}$$

∴ Work done in steam turbine/plant

$$= \dot{m}_s \times (h_1 - h_2) = 194.59 \times (3420 - 2190)$$
$$= 194.59 \times 1230$$
$$= \mathbf{239342.65\ kJ/sec = kW}$$

Total power produced in the combined plant

$$= 106669.01 + 239342.65$$
$$= \mathbf{346011.66\ kW}$$

$$\text{Overall efficiency of the plant} = \frac{\text{Total work produced}}{\text{Fuel used} \times \text{Calculated value}}$$

$$= \frac{36011.66}{\frac{70042.8}{3600} \times 45000} = \frac{7.689}{19.46}$$

$$= \mathbf{0.3951\ or\ 39.51\%}$$

Example 22.20. *A combined cycle power plant is to be designed to develop 100 MW. Out of this 60 % power is to be developed in a gas turbine plant and the remaining in a steam power plant. The pressure and temperature of the air entering the compressor are 1 bar and 27°C.*

The maximum pressure and temperature in the gas turbine cycle are limited to 8 bar and 1000°C. Steam is to be generated at 50 bar and 600°C by using supplementary fuel. The condenser pressure is maintained at 0.05 bar.

Take the given data and find:

(i) Overall efficiency of the plant.

(ii) Ratio of air flow rate to steam flow rate.

(iii) Quantity of cooling water required in ton/hour if the rise in temperature of the cooling water is limited to 10 °C.

$\eta_c = 0.85$, $\eta_t = 0.90$, $\eta_{comb} = 0.95$.

$C_{pa} = 1$ kJ/kg-°K, $\gamma_a = 1.4$, $C_{pg} = 1.1$ kJ/kg-°K, $\gamma_g = 1.33$

C.V. of fuel used = 40 MJ/kg

Gas temperature going to chimney = 200°C.

Neglect W_p. *Consider mass of fuel in gas turbine cycle.*

Solution

(i) First consider gas turbine plant.

$p_1 = 1\,\text{bar}, p_2 = 8\,\text{bar}, T_3 = 1273°\,\text{K}$

$$\frac{T_2}{T_1} = 8^{0.286} = 1.813$$

$$\therefore \quad T_2 = 1.813 \times 300 = 543.76°\,\text{K}$$

$$\therefore \quad 0.85 = \eta_c = \frac{T_2 - T_1}{T_{2'} - T_1}$$

$$= \frac{543.76 - 300}{T_{2'} - 300}$$

$$\therefore \quad T_{2'} = 300 + \frac{243.76}{0.85}$$

$$= 300 + 286.8 = \mathbf{586.8\ °K}$$

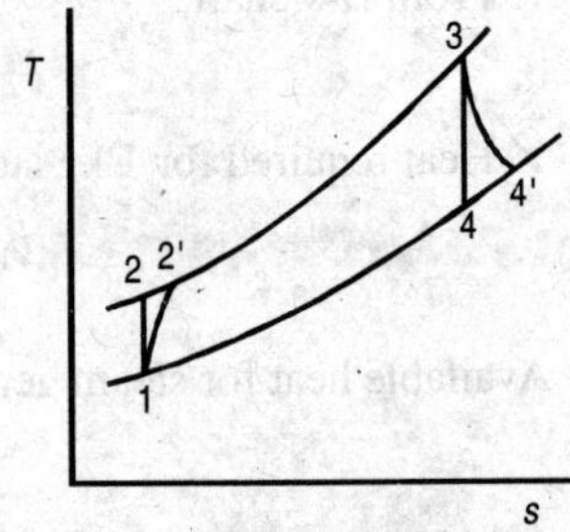

Fig. Ex. 22.20(a)

For combustion chamber ,

$$1 \times C_{pa} \times t_{2'} + x_f \times \text{C.V.} = (1 + x_f) C_{pg} \times t_3$$

$$1 \times 1 \times (586.8 - 273) + (x_f \times 40000 \times \eta_{\text{comb}}) = (1 + x_f) \times 1.1 \times 1000$$

$$313.8 + x_{\frac{1}{2}} \times 0.95 \times 40000 = (1 + x_{\frac{1}{2}})1100 = 1100 + 1100 x_{\frac{1}{2}}$$

$$38000\, x_{\frac{1}{2}} - 1100\, x_{\frac{1}{2}} = 1100 - 313.8 = 786.2$$

$$\therefore \quad x_{\frac{1}{2}} = \frac{786.2}{36900} = \mathbf{0.0213\ kg/kg\ air}$$

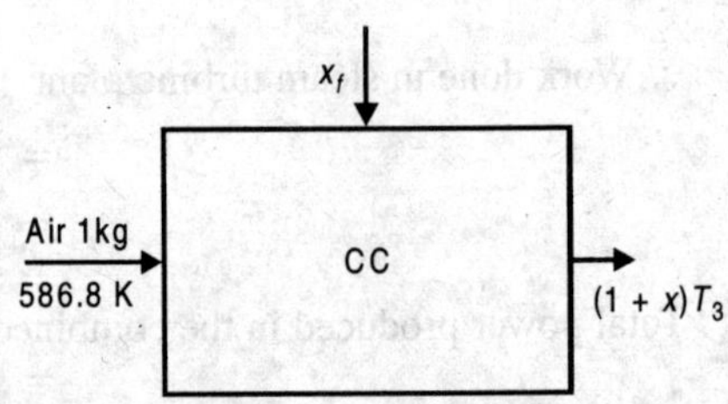

Fig. Ex. 22.20(b)

$$\text{Compressor work/kg air} = C_{pa}(T_{2'} - T_1) = 1\ (586.6 - 300) = \mathbf{286.8\ kJ/kg}$$

Now consider gas turbine

$$\left(\frac{T_3}{T_4}\right) = 8^{0.2481} = 1.675$$

$$\therefore \quad T_4 = \frac{1273}{1.675} = \mathbf{760\ °K}$$

$$\therefore \quad \eta_t = 0.9 = \frac{T_3 - T_{4'}}{T_3 - T_4} = \frac{1273 - T_{4'}}{1273 - 760} = \frac{1273 - T_{4'}}{513}$$

$$\therefore \quad T_{4'} = 1273 - 461.7 = \mathbf{811.3\ °K}$$

$$\therefore \quad \text{Turbine Work} = W_t = (1 + x_f) \times C_{pg}(T_3 - T_{4'})$$

$$= 1.1\ (1273 - 811.3)$$

$$= 1.1 \times 461.7 \times 1.0213 = \frac{507.87}{1.0213}\ \text{kJ/kg}$$

$$= \mathbf{518.7\ kJ/kg}$$

$\therefore$ Net gas turbine work/kg air admitted

$$= 518.7 - 286.8 = \mathbf{231.9\ kJ/kg}$$

To produce 60 MW, quantity of air required

$$\dot{m}_a = \frac{60 \times 1000}{231.9} = 258.73 \text{ kg/sec}$$

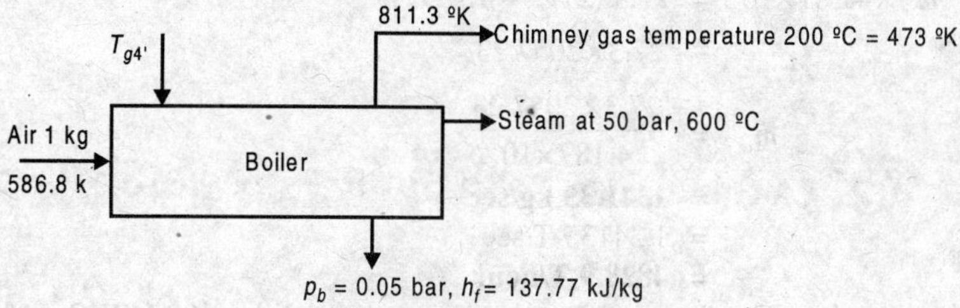

Fig. Ex. 22.20(c)

Now, consider steam power plant

$$h_1 = 3660 \text{ kJ/kg}$$

$$h_2 = 2195 \text{ kJ/kg}$$

$$h_{f_2} = 137.77 \text{ kJ/kg}$$

$$\text{Work done/kg steam} = h_1 - h_2 = 3660 - 2195$$

$$= \mathbf{1465 \text{ kJ/kg}}$$

$$\therefore \quad \text{Mass of steam/sec} = \frac{40 \times 1000}{1465}$$

$$= \mathbf{27.3 \text{ kg/sec}}$$

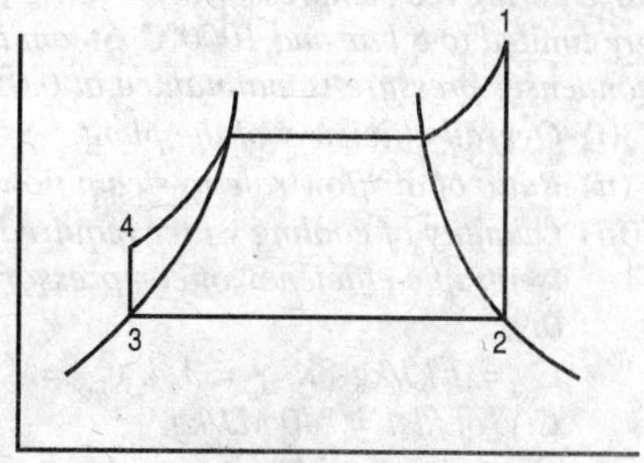

Fig. Ex. 22.20(d)

Let t_g be the temperature of the gases coming out of the boiler combustion chamber.

$$\therefore \quad \text{Heat given by gases/sec} = \text{Heat given by steam/sec}$$

$$m_g C_{pg}(T_{g_{4'}} - T_{\text{chim}}) = \dot{m}_s = (h_1 - h_{f_3})$$

$$(258.73 \times 1.0213) \times 1.1 \times (811.3 - 473) = 27.3(3660 - 137.77)$$

$$98331.98 = 96156.9$$

Temperature of gases coming from gas turbine is sufficient to produce steam in the boiler and therefore, no extra fuel is required to be burned in boiler combustion chamber.

$$\therefore \quad \text{Heat supplied/sec} = \dot{m}f \times \text{C.V.}$$

$$= 258.73 \times 0.0213 \times 40000$$

$$= 220438 \text{ kJ/sec}$$

$$\text{Net work/sec} = 100{,}000 \text{ kW}$$

$$\therefore \quad \text{Overall efficiency} = \frac{100{,}000}{220438}$$

$$= 0.4536$$

$$= \mathbf{45.36\%}$$

(ii) Ratio of air flow to steam flow rate

$$= \frac{\dot{m}_a}{\dot{m}_s} = \frac{258.73}{27.3}$$

$$= \mathbf{9.48}$$

(iii) Quantity of cooling water: We have

$$\dot{m}_w \times C_{pw} \text{ (Rise in temperature)} = \dot{m}_s (h_2 - h_{2f})$$

$$\dot{m}_w \times 4.187 \times 10 = 27.3\,(2195 - 137.77)$$

$$= 27.3 \times 2057.23$$

$$\therefore \quad \dot{m}_w = \frac{27.3 \times 2057.23}{4.187 \times 10}$$

$$= \mathbf{1341.35\ kg/sec}$$

$$= 1.34135 \text{ T/sec}$$

$$= \mathbf{4828.9\ T/hour}$$

Example 22.21. *A combined cycle power plant is to be designed to develop 100 MW. Out of this, 60% power is to be developed in a gas turbine plant and the remaining in a steam power plant. The pressure and temperature of the air entering the compressor are 1 bar and 27°C. The maximum pressure and temperature in the gas turbine cycle are limited to 8 bar and 1000°C. Steam is to be generated at 50 bar and 600°C by using supplementary fuel. The condenser pressure is maintained at 0.05 bar. Taking the following data, find*

(i) Overall efficiency of the plant.

(ii) Ratio of air flow rate to steam flow rate.

(iii) Quantity of cooling water required in tons/hr. If the rise in temperature of cooling water is limited to 10°C, isentropic efficiency of compressor = 85%, isentropic efficiency of turbine = 90%, combustion efficiency = 0.95.

C_{pa} = 1 kJ/kg-°K, γ_a = 1.4, C_{pg} = 1.1 kJ/kg-°K, γ_g = 1.33.

C.V. of fuel = °40 MJ/kg.

Gas temperature going to chimney = 200°C.

Neglect feed pump work.

Consider mass fuel in gas turbine.

Solution

$$p_1 = 1 \text{ bar} = p_4$$

$$T_1 = 300^\circ \text{ K}$$

$$C_{pa} = 1 \text{ kJ/kg}^\circ\text{K}$$

$$\gamma_a = 1.4$$

$$p_2 = p_3 = 8 \text{ bar}$$

$$\eta_o = 0.85$$

$$C_{pg} = 1.1 \text{kJ/kg-}^\circ\text{K}, \gamma_g = 1.33, \eta_t = 90\%$$

$$\text{C.V. of fuel} = 40 \text{ mJ/kg} = 40000 \text{ kJ/kg}$$

$$\eta_{\text{comp.}} = 95\%, \; T_3 = 1273^\circ\text{K}$$

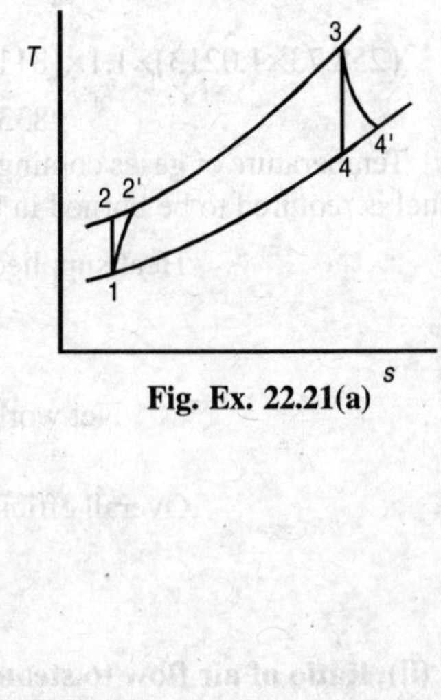

Fig. Ex. 22.21(a)

$$\frac{T_2}{T_1} = \left(\frac{p_2}{p_1}\right)^{\frac{\gamma_a - 1}{\gamma_a}} = (8)^{0.287} = 1.8163$$

$$T_2 = 300 \times 1.8163 = 544.9\ ^\circ\text{K}$$

$$\eta_c = \frac{T_2 - T_1}{T_{2'} - T_1} = 0.85 = \frac{544.9 - 300}{T_{2'} - 300}$$

$$\therefore \quad T_{2'} = 300 + \frac{244.9}{0.85} = 300 + 288.12 = \mathbf{588.12\ °K}$$

Similarly,
$$\frac{T_3}{T_4} = \left(\frac{p_3}{p_4}\right)^{\frac{\gamma-1}{\gamma}} = (8)^{\frac{0.33}{1.33}} = 8^{0.24812} = \mathbf{58.12\ °K}$$

$$\therefore \quad T_4 = \frac{1273}{1.6752} = \mathbf{759.91°K}$$

$$\therefore \quad \eta_t = 0.9 = \frac{T_3 - T_{4'}}{T_3 - T_4} = \frac{1273 - T_{4'}}{1273 - 759.91} = \frac{1273 - T_{4'}}{513.09}$$

$$\therefore \quad T_{4'} = 1273 - 0.9 \times 513.09$$
$$= 1273 - 461.78$$
$$= \mathbf{811.22°K}$$

Now, we will consider combustion chamber of gas turbine. Let x be fuel supplied per kg of air.

$$\therefore \quad C_{pg} \times \dot{m}_g \times t_3 - \dot{m}_a C_{pa} T_{2'} = x \times \text{C.V.} \times \eta_{comb}$$

$$(1+x) \times 1.1 \times 1273 - 1 \times 1 \times 588.12 = x \times 40000 \times 0.95$$

$$\therefore 1400.3 + 1400.3\,x - 588.12 = 40000x \times 0.95$$

$$812.18 = x(40000 \times 0.95 - 1400.3) = 36599.7$$

$$\therefore \quad x = \frac{812.18}{36599.7} = 0.022191 \frac{\text{kg of fuel}}{\text{kg air}}$$

$$\therefore \quad \text{Air fuel ratio} = \frac{1}{0.022191}$$

$$\frac{A}{F} = 45.06 \text{ kg air/kg fuel}$$

$$\therefore \quad \frac{\text{Net work done}}{\text{kg air supplied}} = W_t - W_c = m_g C_{pg}(T_3 - T_4) - m_a C_{pa}(T_{2'} - T_1)$$

$$= 1.021041 \times 1.1(1273 - 811.22) - 1 \times 1 \times (588.12 - 300)$$

$$= 1.123145 \times 461.78 - 288.12$$

$$= 518.65 - 288.12 = \mathbf{230.53\ kJ/kg}$$

Power produced by gas turbine

$$= 0.6 \times 100 \times 1000 \text{ kW} = 60{,}000 \text{ kW}$$

$\therefore$ Mass of air required/sec

$$\frac{60000}{230.53} = \mathbf{260.27\ kg/sec}$$

In gas-turbine plant, fuel used per second

$$= 260.27 \times 0.022191$$
$$= \mathbf{5.7756\ kg/s}$$

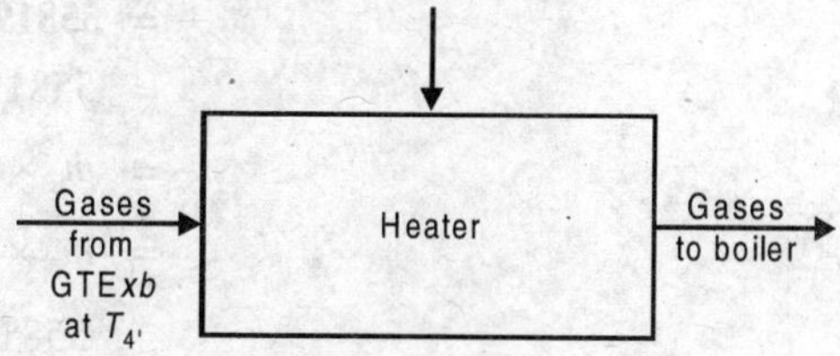

Fig. Ex. 22.21(b)

From chart,
$$h_1 = 3680 \text{ kJ/kg}$$
$$h_2 = 2200 \text{ kJ/kg}$$

$\therefore$ Work done/kg steam $= h_1 - h_2$

$= 3680 - 2200$

$=$ **1480 kJ/kg**

Power produced by steam turbine

$= 0.4 \times 100 \times 1000 = 40,000$ kW

$\therefore$ Amount of steam required per sec

$$= \frac{40,000}{1480} = \mathbf{27.03\ kg\ steam/sec}$$

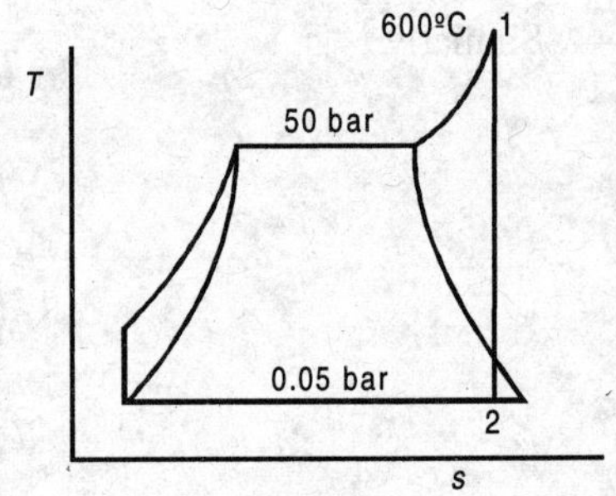

Fig. Ex. 22.21(c)

$\therefore$ Amount of heat required/sec $= 27.03\left[h_1 - h_{f_2})\right]$

$= 27.03\ [3680 - 137.77]$

$= 27.03 \times 3542.23 =$ **95746.5 kJ/sec**

$$\therefore\ 266.0456 \times 1.1 \times \left[\begin{pmatrix}\text{Temperature at}\\ \text{boiler net}\end{pmatrix} - \begin{pmatrix}\text{Temperature of gases}\\ \text{at chimney inlet}\end{pmatrix}\right] = 95746.5$$

$$\therefore\ \begin{bmatrix}\text{Temperature of}\\ \text{gases at boiler inlet}\end{bmatrix} - 200 = \frac{95746.5}{265.75 \times 1.1} = 327.53°\text{C}$$

$\therefore$ Temperature of gases at boiler inlet or heater exhaust $= 327.53 + 200$

$= 537.53$ °C

$= 810.53$ ° K

This is more or less equal to the temperature of gases leaving the turbine and hence no heat is added in the heater.

$\therefore$ Ratio of air flow rate to steam flow rate

$$= \frac{260.27}{27.03} = 9.63$$

$$\frac{m_a}{m_s} = 9.63$$

$$\text{Overall efficiency of the plant} = \frac{100 \times 1000}{5.48 \times 40000} = \frac{\text{Work}}{\text{Heat supplied}}$$

$$= \frac{10}{5.7756} = 0.4328$$

$\eta_o =$ **43.48%**

Heat to be rejected to cooling water in condenser/sec

$= 27.03 \times DF \times h_{fg}$

$= 27.03 \times 0.852 \times 2423.8$

$=$ **55819 kJ/sec**

$= 55819 \times 3600$ kJ/hr

$= \dot{m}_w \times C_{pw} \times$ Rise in temperature

$= \dot{m}_w \times 4.187 \times 10$

$$\dot{m}_w = \frac{55819 \times 3600}{4.187 \times 10}\ \text{kg/hr}$$

$$= \frac{55819 \times 3600}{4.187 \times 10 \times 1000} \text{ tons/hr}$$

$$= 4799.34 \text{ tons/hr} = 4800 \text{ tons/hr}$$

$$\dot{m}_w = \mathbf{4800\ tons/hr}$$

Example 22.22. *In a power plant, the efficiencies of the electric generator, turbine (mechanically) boiler. cycle and the overall plant are 0.97, 0.95, 0.92, 0.42 and 0.33 respectively. What percentage of the total electricity generated is consumed in running the auxiliaries?*

Solution

The overall plant efficiency is given by the multiplication of all the other efficiencies mentioned in the text of the example. Thus,

$$\eta_{plant} = \eta_{boiler} \times \eta_{cycle} \times \eta_{turbine(mech)} \times \eta_{generator} \times \eta_{auxiliaries}$$

$$\therefore \quad \eta_{auxiliaries} = \frac{0.33}{0.92 \times 0.42 \times 0.95 \times 0.97} = 0.9268$$

$\therefore$ The losses in the auxiliaries are = 1 – 0.9268 = 0.0732 i.e. **7.32%**

or 7.32% of total electrical power generated is consumed in running the accessories.

Example 22.23. *A cyclic steam power plant is to be designed for a steam temperature at turbine inlet of 360°C and an exhaust pressure of 0.08 bar. After isentropic expansion of steam in the turbine, the moisture content at the turbine exhaust is not to exceed 15%. Determine the greatest allowable steam pressure at the turbine inlet, and calculate the Rankine cycle efficiency for these steam conditions. Estimate also the mean temperature of heat addition.*

Solution

Refer Fig. Ex. 22.23 (a)

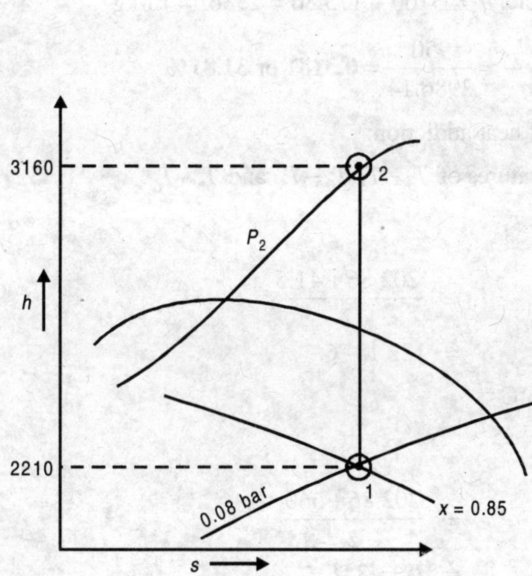

Fig. Ex. 22.23(a)

The example is best solved with the help of the Mollier diagram. Locate point 1 on 0.08 bar and $x = 0.85$. Go straight up on constant entropy line till you get temperature equal to 360ºC and locate point 2. Read off pressure on pressure line passing through point 2. It is 16.5 bar which is required pressure.

$\therefore$ Inlet pressure = **16.5 bar**

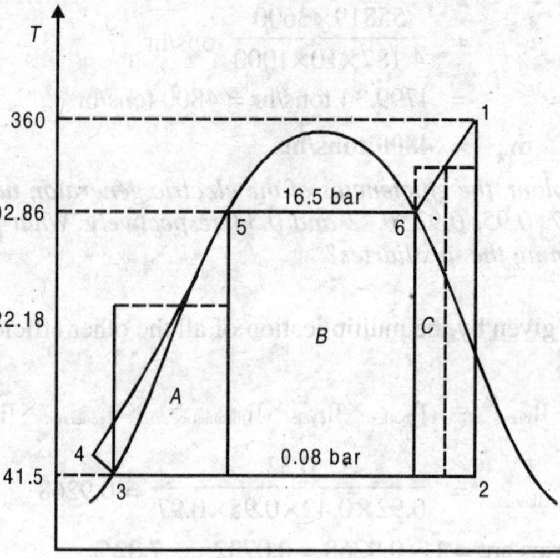

Fig. Ex. 22.23 (b)

Now to find cycle efficiency read off h_2 at point 2 = 3160 kJ/kg and h_1 at point 1 = 2210 kJ/kg then work done or turbine work,

$$W_T = 3160 - 2210 = 950 \text{ kJ/kg}$$

To find heat supplied, find h_{f1} = 173.86 at 0.08 bar from the steam tables then, neglecting the pump work,

Heat suupplied during the cycle, q_s = 3160 – 173.86 = 2986.14 kJ/kg

The cycle efficiency, $\eta_R = \dfrac{W_T}{q_s} = \dfrac{950}{2986.14}$ = **0.3181** or **31.81%**

To find mean temperature of heat addiction:

We first find average temperatures of $T_5 - T_3, T_6 - T_5$ and $T_1 - T_6$.

Average temperature of $T_5 - T_3$

$$= \frac{202.86 + 41.5}{2}$$

$$= 122.18\ ^\circ\text{C}$$

Average of $T_5 - T_6$ is 202.86°C

Average of $T_6 - T_1$

$$= \frac{202.86 + 360}{2}$$

$$= 281.43\ ^\circ\text{C}$$

We now find total area of *A, B* and *C*. Fig. Ex. 22.23(b)

$$\text{Total Area} = \text{Area } A + \text{Area } B + \text{Area } C$$

$$= 122.18(s_5 - s_3) + 202.86(s_6 - s_5) + 281.43(s_1 - s_6)$$

$$= 122.18(2.3576 - 0.5925) + 202.86(4.0489) + 281.43(7.084 - 6.4065)$$

$$= 215.66 + 821.34 + 190.67$$
$$= 1227.69 \text{ kJ/kg}$$

$\therefore$ To find mean temperature of heat addition, we divide the area by the base $s_2 - s_3$

$$\therefore \quad s_2 - s_3 = x_2\left(s_{f_{g2}}\right) = 0.85 \times 7.6371 = 6.4915 \text{ kJ/kg°C}$$

$$T_m = \frac{1227.69}{6.4915} = \mathbf{189.12\ °C}$$

Example 22.24. *A textile factory requires 10 T/hr of steam for process heating at 3 bar saturated and 1000 kW of power, for which a back pressure turbine of 70% internal efficiency is to be used. Find the steam condition required at inlet of the turbine.*

Solution

If $\dot{m}$ = mass flow rate of steam,

$$\dot{m}\left(h_1 - h_{2^1}\right) = 1000 \text{ kW}$$

$$\therefore \quad h_1 - h_{2^1} = \frac{1000 \times 3600}{10{,}000} = 360 \text{ kJ/kg}$$

With reference to Fig. Ex. 22.24

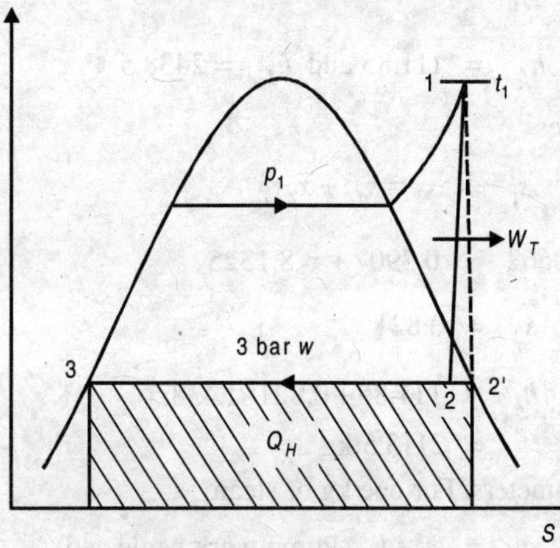

Fig. Ex. 22.24

$$\therefore \quad h_{2^1} = 2725.3 = h_{g_2} \text{ at 3 bar}$$

$$\therefore \quad h_1 = 3085.3 \text{ kJ/kg}$$

Now,
$$h_1 - h_2 = \frac{360}{0.7} = 514.286 \text{ kJ/kg}$$

$$h_2 = 3085.3 - 514.286$$
$$= 2571.014 \text{ kJ/kg}$$

$$\therefore \quad h_2 = h_f + x_2 h_{f_g} = 561.47 + x_2 \times 2163.8 = 2571.014$$

$$x_2 = 0.9287$$

$$s_2 = 1.6718 + 0.9287\,(6.9919 - 1.6718) = 6.6125 \text{ kJ/kg °K}$$

$$= s_1$$

Corresponding to $h_1 = 3085.3$ kJ/kg

and $s_1 = 6.6125$ kJ/kg °K, from the mollier diagram,

$p_1 =$ **37.3** bar, $t_1 =$ **344°C**, which is the state of steam at turbine inlet.

Example 22.25. *A steam power plant operates between a boiler pressure of 42 bar and condenser pressure of 0.035 bar. The steam is superheated to 500°C. Neglecting feed pump work, compare the performance of the plant if reheat is included. The steam condition at inlet to the turbine is same in reheat cycle and also condenser pressure. Assume that steam is just dry saturated on leaving the first turbine and is reheated to its initial temperature.*

Solution

Here performance means cycle efficiency and specific steam consumption with and without reheat.

Case (a) without reheat:

At 42 bar and 500°C, we have

$h_1 = 3442.7$ kJ/kg and $s_1 = 7.0662$ kJ/kg°K

At 0.035 bar we have

$$s_{f_2} = 0.3907 \text{ and } s_{f_{g2}} = 8.1325$$

Also $h_{f_2} = 111.85$ and $h_{f_{g2}} = 2438.5$

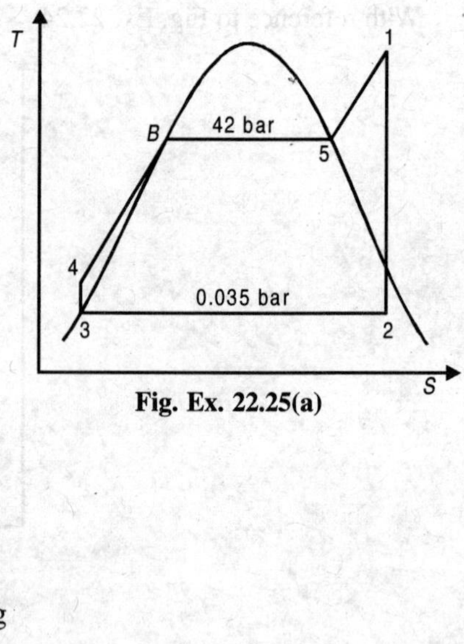

Fig. Ex. 22.25(a)

To find x_2, we equate entropies

$$\therefore \quad s_1 = s_2 = s_{f_2} + x_2 s_{f_{g2}}$$

$$\therefore \quad 7.0662 = 0.3907 + x_2 8.1325$$

$$x_2 = 0.821$$

$$\therefore \quad h_2 = 111.85 + 0.821 \times 2438.5$$

$$= 2113.5 \text{ kg}$$

We now determine cycle parameters. For one kg of steam

1. Heat supplied, $q_s = h_1 - h_3$ (Pump work neglected)

$$= 3442.7 - 111.85 = 3330.85 \text{ kJ/kg}$$

2. Turbine work, $W_T = h_1 - h_2 = 3442.7 - 2113.5 = 1329.2$ kJ/kg

3. Cycle efficency, $\eta_R = \dfrac{W_T}{q_s} = \dfrac{1329.2}{3330.85} =$ **0.3991 = 39.91%**

4. Specific steam consumption, S. S. C.

$$= \frac{3600}{W_T} = \frac{3600}{1329.2} = \mathbf{2.708\ kg/kWh}$$

Case (b) with reheat:

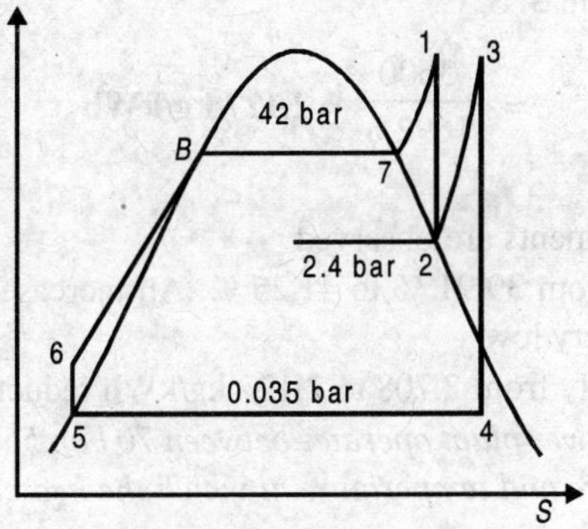

Fig. Ex. 22.25(b)

Steam condition at 2 is dry saturated.

$$\therefore \quad s_1 = s_2 = s_{g_2} = 7.0662$$

$$\therefore \quad \text{From steam tables, } P_2 = 2.4 \text{ bar}$$

Also from steam tables,

$$h_2 = h_{g_2} = 2714.5 \text{ kJ/kg}$$

At 2.4 bar and 500°C

$$h_3 = 3486.6 \text{ kJ/kg}$$

and

$$s_3 = 8.4293 \text{ kJ/kg-K}$$

To find x_4, we equate entropies at 3 and 4.

$$\therefore \quad s_3 = s_4 = s_{f_4} + x_4 s_{fg_4}$$

$$\therefore \quad 8.4293 = 0.3907 + x_4 8.1325$$

$$x_4 = 0.988$$

$$\therefore \quad h_4 = 111.85 + 0.988 \times 2438.5 = 2522.2$$

The cycle parameters are,

1. Heat supplied during the cycle,

$$q_s = (h_1 - h_5) + (h_3 - h_2)$$

$$= (3442.7 - 111.85) + (3486.6 - 2714.5)$$

$$= 3330.85 + 772.1 = 4102.95 \text{ kJ/kg}$$

2. Turbine work,

$$W_T = (h_1 - h_2) + (h_3 - h_4)$$

$$= (3442.7 - 2714.5) + (3486.6 - 2522.2)$$

$$= 728.2 + 964.4 = 1692.6 \text{ kJ/kg}$$

3. Cycle efficiency

$$= \frac{W_T}{q_s} = \frac{1692.95}{4102.95}$$

$= 0.4125 = \mathbf{41.25\%}$

4. Specific steam consumption. S. S. C.

$$= \frac{3600}{1692.6} = \mathbf{2.127\ kg/kWh}$$

Comments:

With reheat following improvements are observed.

1. Cycle efficiency increases from 39.91 % to 41.25 %. An increase of 3.36 %. The increase in efficiency is small as reheat pressure is very low.
2. S.S.C. is reduced considerably from 2.708 to 2.127 kg/kWh reduction of 21.45 %.

Example 22.26. *An ideal steam power plant operates between 70 bar, 550°C and 0.075 bar. It has seven feedwater heaters. Find the optimum pressure and temperature at which the heaters operate.*

Solution

With reference to Fig. Ex. 22.26

T_B = Saturation temperature at 70 bar
= 285.9 °C

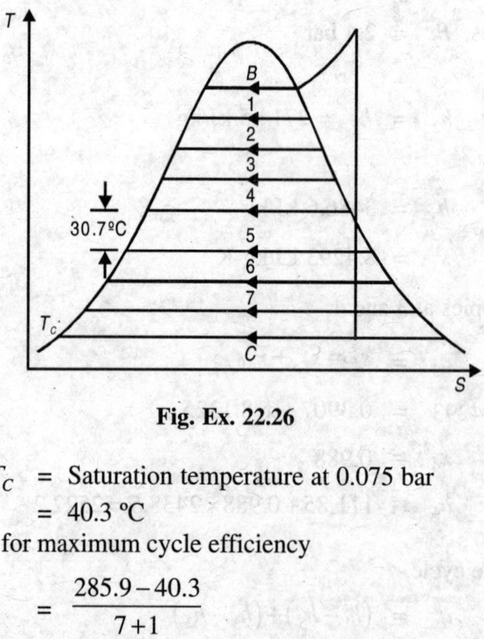

Fig. Ex. 22.26

T_C = Saturation temperature at 0.075 bar
= 40.3 °C

The temperature rise per heater for maximum cycle efficiency

$$= \frac{285.9 - 40.3}{7+1}$$

= 30.7°C

Heater 1 : $t_1 = 285.9 - 30.7 = 255.2°\text{C}$

: $p_1 = 43.3$ bar (from steam tables)

Heater 2 : $t_2 = 255.2 - 30.7 = 224.5°\text{C}$; $p_2 = 25.32$ bar

Heater 3 : $t_3 = 193.8°\text{C}$; $p_3 = 13.67$ bar

Heater 4 : $t_4 = 163.1°\text{C}$; $p_4 = \mathbf{6.71\ bar}$

Heater 5 : $t_5 = 132.4°\text{C}$; $p_5 = 2.91$ bar

Heater 6 : $t_6 = 101.7°C; p_6 = 1.1$ bar

Heater 7 : $t_7 = 71.0°C; p_7 = 0.327$ bar

Example 22.27. *A steam power plant operating on the ideal regenerative Rankine cycle with one open feed water heater. Steam enters the turbine at 150 bar and 600°C and is condensed at a pressure of 10 kPa. Steam is bled at 12 bar and enters the open feed water heater. Determine the fraction of steam extracted from the turbine and the thermal efficiency of the cycle and specific steam consumption. Consider pump work.*

Solution

First we determine enthalpies at various points. See Fig. Ex. 22.27

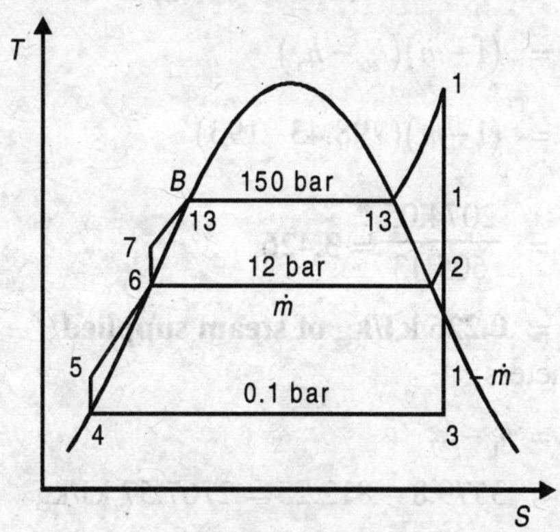

Fig. Ex. 22.27

State 1: At 150 bar 600°C

$$s_1 = 6.6764 \text{ kJ/kg-°K and } h_1 = 3579.8 \text{ kJ/kg}$$

State 2: At 12 bar, $s_{g_2} = 6.5194$ kJ/kg-°K

Since $s_1 > s_{g_2}$, steam at 2 is superheated.

To find T_{sup} 2, equate s_1 and s_2.

$$6.6764 = 6.5194 + 2.1 \text{ In} \frac{T_{sup} 2}{273+188}$$

$$T_{sup} 2 = 496.78°K = 223.78°C \sim 224°C$$

At 12 bar 224°C, we find h_2 by interpolation.

$$h_2 = 2814.4 + \frac{2935.4 - 2814.4}{50} \times 24$$

$$= 2872.48 \text{ kJ/kg}$$

State 3: To find s_3 equate $s_1 = s_3$ (at 0.1 bar).

$$\therefore \quad 6.6764 = 0.6493 + x_3 7.5018$$

$$\therefore \quad x_3 = 0.803$$

$$\therefore \quad h_3 = 191.83 + 0.803 \times 2392.9 = 2114.3 \text{ kJ/kg}$$

State 4: $h_4 = 191.83$ kJ/kg

State 5: W_P at 4 = $0.001(12-0.1)\times100 \sim 1.2$ kJ/kg

$$\therefore \quad h_5 = h_4 + W_P = 193.83 + 1.2 = 193 \text{ kJ/kg}$$

State 6: $h_6 = 798.43$

State 7: W_P at 6 = $0.001(150-12)\times100 = 13.8$ kJ/kg

$$\therefore \quad h_7 = h_6 + W_P = 798.43 + 13.8 = 812.23 \text{ kJ/kg}$$

Let $\dot{m}$ be the quantity of extracted. To find $\dot{m}$, we write energy balance for the open fwh.

$$\therefore \quad \dot{m}(h_2 - h_6) = (1-\dot{m})(h_6 - h_5)$$

and
$$\dot{m}(2872.48 - 798.43) = (1-m)(798.43 - 193)$$

$$\therefore \quad \frac{1-\dot{m}}{\dot{m}} = \frac{2074.05}{605.43} = 3.426$$

$$\therefore \quad \dot{m} = \textbf{0.226 kJ/kg of steam supplied}$$

We now determine the cycle parameters.

1. Heat supplied in the cycle, $q_s = h_1 - h_7$

$$= 3579.8 - 812.23 = 2767.57 \text{ kJ/kg}$$

2. Turbine work, $W_T = (h_1 - h_2) + (1-\dot{m})(h_2 - h_3)$

$$= (3579.8 - 2872.48) + (0.774)(2872.48 - 2114.3)$$

$$= 707.32 + 586.83 = 1294.15 \text{ kJ/kg}$$

$$\text{Net work} = \Delta W_{net} = W_T - (W_{P_1} + W_{P_2}) = 1294.15 - (1.2 + 13.8)$$

$$= 1279.15 \text{ kJ/kg}$$

3. Cycle efficiency = $\dfrac{\Delta W_{net}}{q_s} = \dfrac{1279.15}{2767.57} = 0.4621 \sim$ **46.21%**

4. Specific steam consumption, S. S. C.

$$= \frac{3600}{1279.15}$$

$$= \textbf{2.814 kg/kWh}$$

Example 22.28. *In a cogeneration plant, the power load is 5.6 MW and the heating load is 1.163 MW. Steam is generated at 40 bar and 500°C and is expanded isentropically through a turbine to a condenser at 0.06 bar. The heating load is supplied by extracting steam from the turbine at 2 bar, which condensed in the process heater to saturated liquid at 2 bar and then pumped back to the boiler. Compute (a) the steam generation capacity of the boiler in t/h, (b) the heat input to the boiler in kW, (c) the fuel burning rate of the boiler in t/h if a coal of calorific value 25 MJ/kg is burned and the boiler efficiency is 88%, (d) the heat rejected to the condenser, (e) the rate of flow of cooling water in the condenser if the temperature rise of water is 6°C. Neglect pump work.*

Solution

With reference to Fig. Ex. 22.28

$$h_1 = 3445.3 \text{ kJ/kg}, \; s_1 = 7.0901 = s_2 - s_3$$

$$7.0901 = 1.5301 + x_2 \times 5.5970 \text{ or } x_2 = \frac{5.56}{5.597} \approx 1.0$$

$$\therefore \quad h_2 = h_g = 2706.7 \text{ kJ/kg}$$

$$h_2 - h_6 = h_{f_g} = 2201.9 \text{ kJ/kg}$$

If $\dot{m}$ is the rate of steam extraction for process heating,

$$\dot{m}(h_2 - h_6) = 1.163 \times 10^3$$

$$\dot{m} = \frac{1.163 \times 10^3}{2201.9} = 0.528 \text{ kg/s} = 1901.4 \text{ kg/h}$$

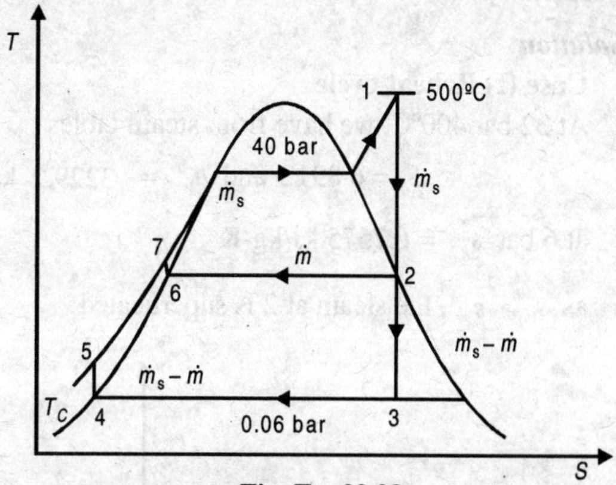

Fig. Ex. 22.28

$$s_1 = 7.0901 = s_f + x_3 s_{f_g}$$

$$= 0.52 + x_3 \times 7.815$$

$$\therefore \quad x_3 = 0.84$$

$$\therefore \quad h_3 = 149.79 + 0.84 \times 2416 = 2180.59 \text{ kJ/kg}$$

Total work output, $W_T = \dot{m}_s(h_1 - h_2) + (\dot{m}_s - \dot{m})(h_2 - h_3)$

$$\therefore \quad 5.6 \times 10^3 = \dot{m}_s \times 738.6 + \dot{m}_s \times 526.11 - 277.8$$

$$\therefore \quad \dot{m}_s = \frac{5877.8}{1264.7} = 4.648 \text{ kg/s} = 16731 \text{ kg/h} = \mathbf{16.73 \text{ t/h}}$$

$$h_7 = 5.407 + 1.061 \times 10^{-3}(40 - 2) \times 100 = 508.73 \text{ kJ/kg}$$

$$h_5 = 149.79 + 1.006 \times 100 \times 40 \times 10^{-3} = 153.8 \text{ kJ/kg}$$

$$Q_1 = (\dot{m}_s - \dot{m})(h_1 - h_5) + \dot{m}(h_1 - h_7)$$

$$= (4.648 - 0.528)(3445.3 - 153.8) + 0.528(3445.3 - 508.73)$$

$$= 4.120 \times 3291.5 + 0.528 \times 2936.57 = 15111.5 \text{ kJ/s}$$

$$= \mathbf{15.111 \text{ MW}}$$

$$\eta_{boiler} = \frac{Q_1}{\dot{m}f \times \text{C.F.}} = \frac{15.111}{\dot{m}f \times 25} = 0.88$$

$$\dot{m}f = 0.687 \text{ kg/s} = 2473.2 \text{ kg/h} = \mathbf{2.473 \text{ t/h}}$$

$$Q_2 = (\dot{m}_s - \dot{m})(h_3 - h_4) = 4.12 \times 2030.8 = 8367 \text{ kW}$$

$$= \mathbf{8.367 \text{ mW}}$$

If m_w = water flow rate in condensor,

then, $Q_2 = \dot{m}_w C_p (t_2 - t_1)$

$$\dot{m}_w = \frac{8367}{4.187 \times 6} = 333.05 \text{ kg/s} = \mathbf{0.333 \text{ m}^3/\text{s}}$$

Example 22.29. *In a Rankine cycle plant steam is supplied to turbine at 32 bar, 400°C and is expanded to a pressure of 0.16 bar. Calculate percent moisture and thermal efficiency if the steam is reheated at 6 bar to a temperature of 350°C. Assume isentropic efficiency of turbine 80% throughout. If the turbine is put up with one extraction point (for optimum results) and an open. feed heater, recalculate thermal efficiency.*

Solution

Case (I) Reheat cycle

At 32 bar 400°C, we have from steam tables

$$S_1 = 6.8912 \text{ and } h_1 = 3229.2 \text{ kJ/kg}$$

at 6 bar $s_{g_2} = 6.7575$ kJ/kg-K

as $s_1 > s_{g_2}$, the steam at 2 is superheated

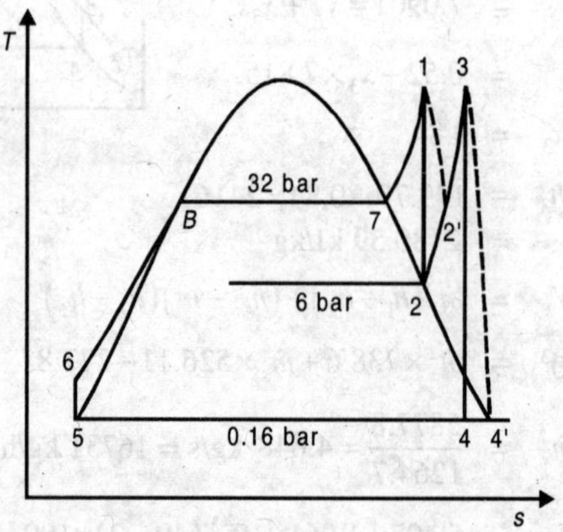

Fig. Ex. 22.29 (a)

To find T_{sup} at 2, we have

$$6.8192 = 6.7545 + 2.1 ln \frac{T_{sup^2}}{273 + 158.84}$$

$$\therefore \quad T_{sup^2} = 460.23 - 273 = 187.23\text{°C}$$

$$\therefore \quad h_{sup^2} = 2755.5 + 2.1\,(187.23 - 158.84)$$

$$\therefore \quad h_2 = 2815.1 \text{ kJ/kg}$$

At 6 bar 350°C $h_3 = 3166.2$ kJ/kg and $s_3 = 7.5479$ kJ/kg-°K

To find x_4 equate $s_3 = s_4$

$$7.5479 = 0.7721 + x_4 7.2148$$

$$\therefore \quad x_4 = 0.939$$

∴ Moisture = 1 – 0.939 = 0.061 or 61%

$$\therefore \quad h_4 = 231.5 + 0.939 \times 2370 = 2457.38 \text{ kJ/kg}$$

Considering turbine efficiencies, we find h_{2^1} and h_{4^1} as follows

$$0.8 = \frac{h_1 - h_{2^1}}{h_1 - h_2} \qquad \therefore\ h_{2'} = h_1 - 0.8(h_1 - h_2)$$

$$= 3229.2 - 0.8\ (3229.2 - 2815.1)$$

$$\therefore\ h_{2'} = 2898 \text{ kJ/kg}$$

also $$0.8 = \frac{h_3 - h_{4^1}}{h_3 - h_4} \qquad \therefore\ h_{4^1} = h_3 - 0.8(h_3 - h_4)$$

$$= 3166.2 - 0.8\ (3166.2 - 2457.38)$$

$$\therefore\ h_{4^1} = 2599.14 \text{ kJ/kg}$$

Let us determine cycle parameters.

1. Pump work $= |W_p| = (32 - 0.16)100 \times 0.001 = 3.184$ kJ/kg.
2. Heat supplied $= q_s = h_1 - h_6 = 3229.2 - 234.77\ = 2994.43$ kJ/kg.
3. Turbine work $= W_T = (h_1 - h_{2^1}) + (h_3 - h_{4^1})$

$$= (3229.2 - 2898) + (3166.2 - 2599.14)$$

$$= 331.2 + 567.06 = 898.26 \text{ kJ/kg}$$

4. Thermal efficiency, η_{th}

$$= \frac{W_{Tnet}}{q_s} = \frac{W_T - W_P}{q_s}$$

$$= \frac{898.26 - 3.18}{2994.43} = 0.2989 \text{ or } \mathbf{29.89\%}$$

Case (II) Regenerative cycle:

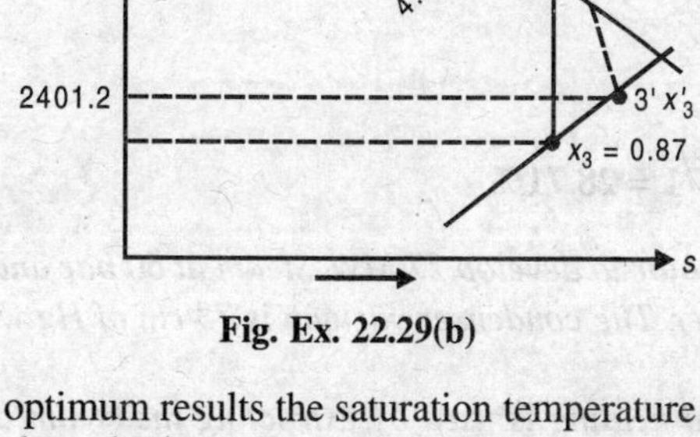

Fig. Ex. 22.29(b)

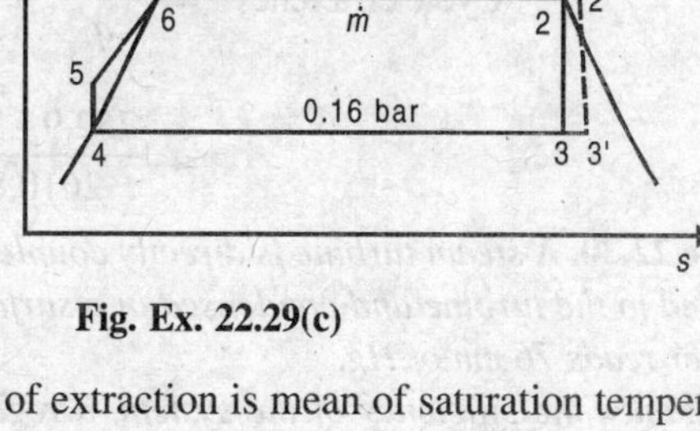

Fig. Ex. 22.29(c)

For optimum results the saturation temperature at the point of extraction is mean of saturation temperatures at inlet and condenser pressures.

$$\therefore \qquad T_{opt} = \frac{T_{sat} \text{ at 32 bar} + T_{sat} \text{ at 0.16 bar}}{2}$$

$$= \frac{237.45 + 55.341}{2} = 146.4°C$$

Saturation pressure at 146.4 is ~ 4.3 bar. (from steam tables)
Thus extraction pressure is 4.3 bar.
The example is best solved by the Mollier chart Fig. Ex. 22.29 (b).

To find $h_{2'}$ with turbine efficiency = 0.8

$$h_{2'} = 3230 - 0.8\,(3230 - 2750)$$
$$= 2846 \text{ kJ/kg}$$
$$h_3 = 2290 \text{ kJ/kg}$$
$$h_{3'} = 2846 - 0.8\,(2846 - 2290)$$
$$= 2401.2 \text{ kJ/kg}$$

From Mollier chart,

$$x_3' = \mathbf{0.918}$$

To find $\dot{m}$ we write energy balance equation,

$$\dot{m}(h_{2'} - h_6) = (h_6 - h_5)$$
$$h_5 = h_3 + W_P = 231.59 + 3.18$$
$$= 234.77 \text{ kJ/kg}$$
$$\therefore \quad \dot{m}(2846 - 615.97) = (615.97 - 234.77)$$
$$\therefore \quad \dot{m} = 0.171 \text{ kg/kg of steam}$$
$$1 - \dot{m} = 0.829$$

We now find cycle parameters,

1. Heat supplied $q_s = h_1 - h_7 = 3230 - 619.15 = 2610.85$ kJ/kg

2. Turbine work, $W_T = (h_1 - h_{2'}) - (1 - m)(h_2 - h_{3'})$
$$= (3230 - 2846) + 0.829\,(2846 - 2401)$$
$$= 384 + 368.9 \text{ kJ/kg}$$

3. Cycle efficiency $= \dfrac{W_{Tnet}}{q_s} = \dfrac{W_T - W_P}{q_s}$
$$= \frac{752.9 - 3.18}{2610.85} = 0.2871 = \mathbf{28.71\%}$$

Example 22.30. *A steam turbine is directly coupled to a generator to develop 27 MW. Steam at 60 bar and 450°C is supplied to the turbine and condensed in a surface condenser. The condenser vacuum is 73 cm of Hg when the barometer reads 76 cm of Hg.*

To enhance the efficiency of the system, a regenerator feed heating is used by extracting the steam at 3 bar pressure, and directly mixing with the condensate. Assuming turbine efficiency for each portion of expansion as 87%, find:

(i) mass of steam bled, per kg of steam supplied to the turbine.
(ii) steam flow rate supplied to the turbine in ton per hour.

(iii) *assuming 10 % generator power is used to run the pumps and accessories, find overall efficiency of the plant.*

Take the following:

Boiler efficiency = 0.9, Alternator efficiency = 0.95, Mechanical efficiency = 0.98

Draw a neat line diagram of all components rents of plant and represent the cycle on h-s diagram.

Solution

A schematic flow diagram and the cycle on *h-s* diagram are drawn. refer Fig. Ex. 22.30 (a) and (b) . The example is solved by using Mollier chart. Refer Fig. Ex. 22.30 (c).

1. The condenser vacuum is 73 cm of Hg and barometer is 76 cm of Hg.

∴ Absolute pressure in the condenser

$$= \frac{76-73}{75} = 0.04 \text{ bar.}$$

2. The inlet conditions are 60 bar and 450°C. Mark point 1 and read h_1 = 3300 kJ/kg. Refer Fig. Ex. 22.30 (c).

3. Draw vertical line to 3 bar. The pressure at which steam is bled for open fwh. Read $h_{2'}$ = 2610 kJ/kg. The turbine efficiency for this expansion is 0.87. We find h_2 by $h_{2'}$ = 3300 – 0.87 (3300 – 2610) = 2699.7 – 2700 kJ/kg.

4. Further expansion is upto 0.04 bar. Draw line 2' – 3 from 3 bar to 0.04 bar and read h_3 = 2085 kJ/kg. Find h_3' by h_3' = 2700 – 0.87 (2700 – 2085) = 2164.95 – 2615 kJ/kg. Other entalpies can be found out from the steam tables.

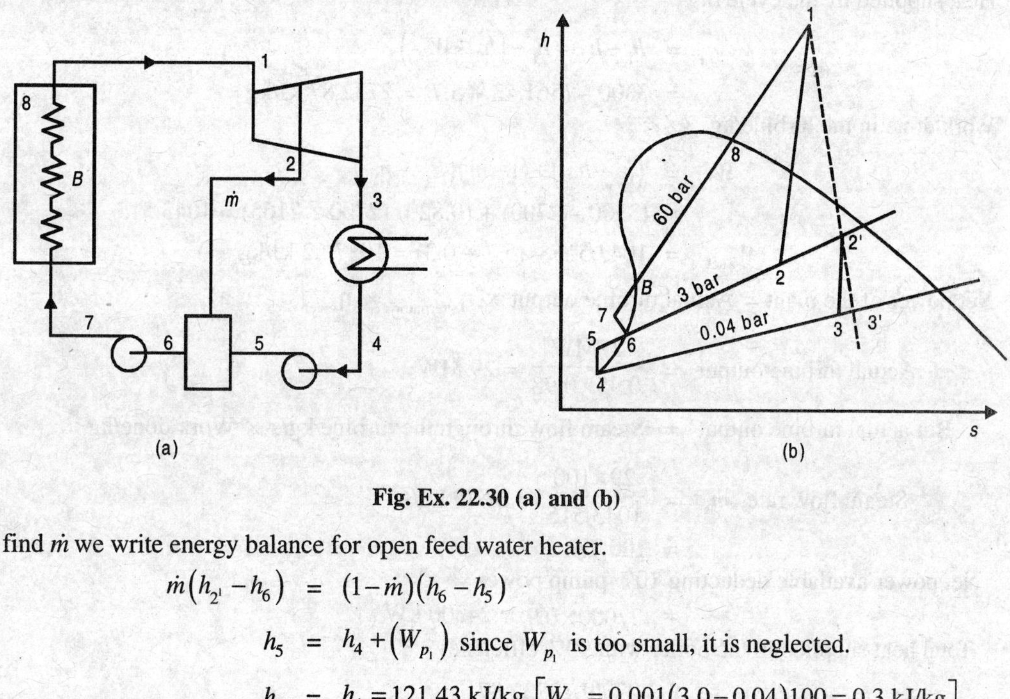

Fig. Ex. 22.30 (a) and (b)

5. To find $\dot{m}$ we write energy balance for open feed water heater.

$$\dot{m}\left(h_{2'} - h_6\right) = (1-\dot{m})(h_6 - h_5)$$

Now $\quad h_5 = h_4 + \left(W_{p_1}\right)$ since W_{p_1} is too small, it is neglected.

and $\quad h_5 = h_4 = 121.43 \text{ kJ/kg} \left[W_{P_1} = 0.001(3.0-0.04)100 = 0.3 \text{ kJ/kg}\right]$

$$\therefore \quad \dot{m}(2700-561.43) = (1-\dot{m})(561.43-121.43)$$

Solving $\quad \dot{m}$ = **0.171 kg/kg of steam**

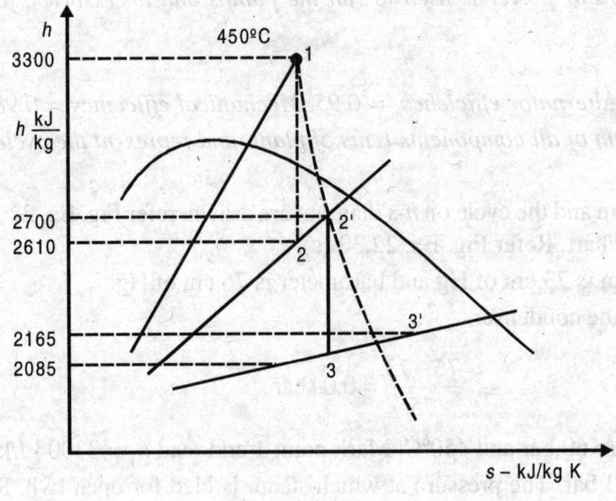

Fig. Ex. 22.30 (c)

6. $W_{p_2} = 0.001(60-3)\times 100 = 5.7\text{ kJ/kg and } W_{p_1} = 0.3\text{ kJ/kg}$

We now find the cycle parameters.

7. Heat supplied by the cycle /kg

$$q_s = h_1 - h_7 = h_1 - \left(h_6 + W_{p_2}\right)$$

$$= 3300 - (561.43 + 5.7) = 2732.87 \text{ kJ/kg}$$

8. Work done in the turbine/kg

$$W_T = \left(h_1 - h_{2'}\right) + (1-m)\left(h_{2'} - h_{3'}\right)$$

$$= (3300 - 2700) + (0.829)(2700 - 2165) = 1043.515$$

$$W_{T\text{net}} = 1043.515 - (5.7 + 0.3) = 1037.52 \text{ kJ/kg}$$

9. Net power of the plant = Actual turbine output × $\eta_{\text{alternator}}$ × η_{mech}

$$\therefore \quad \text{Actual turbine output} = \frac{27\text{MW}}{0.95\times 0.98} = 29 \text{ MW}$$

But actual turbine output = Steam flow through the turbine kg/s × Work done/kg

$$\therefore \quad \text{Steam flow rate, } \dot{m}_s = \frac{29\times 100}{1043.515} = \mathbf{27.79\ kg/s}$$

$$= \mathbf{100\ T/h}$$

10. Net power available deducting 10% pump power

$$= 27000\times 0.9 = 24300 \text{ kW}$$

Total heat supplied in the boiler with 90% efficiency

$$= \frac{27.79\times 2732.87}{0.9} = 84347.84 \text{ kW}$$

11. $$\text{Overall efficiency} = \frac{\text{Net power}}{\text{Heat supplied}} = \frac{24300}{84347.84} = 0.2881 = \mathbf{28.81\%}$$

Example 22.31. *A steam power plant with inlet steam to the h.p. turbine at 90 bar and 500°C, and condensation at 40°C produces 500 MW. It has one stage of reheat optimally placed which raises the steam temperature back to 500°C. One closed feedwater heater with drains cascaded back to the condenser receives bled steam at the reheat pressure, and the remaining steam is reheated and then expanded in the l.p. turbine. The h.p. and l.p. turbines have isentropic efficiencies of 92% and 90% respectively. The isentropic efficiency of the pump is 75%. Calculate (a) the mass flow rate of steam at turbine inlet in kg/s, (b) the cycle efficiency, and (c) the cycle work ratio. Use TTD = – 1.6°C.*

Solution

The flow and *T-S* diagrams are shown in Fig. Ex. 22.31. The optimum reheat pressure is taken to be 20% of teh boiler pressure, which becomes 0.2 × 90 = 18 bar. Now, h_1 = 3396.1 kJ/kg, ε_1 = 6.6576 kJ/kg-K = s_2, h_2 = 2915 kJ/hg, h_3 = 3469.8 kJ/kg.

$$s_3 = 7.4825 = s_4 = s_{f_4} + x_4 s_{f_4}$$

$$\therefore \quad x_4 = \frac{7.4825 - 0.5725}{7.6845} = 0.8992$$

$$\therefore \quad h_4 = 167.57 + 0.8992 \times 2406.7 = 2331.7 \text{ kJ/kg}$$

$$h_5 = 167.57 \text{ kJ/kg}, \; h_7 = 883.42 \text{ kJ/kg}$$

$$W_{p_5} = \int vdp = 0.001008 \times 90 \times 10 = 9.072 \text{ kJ/kg}$$

$$h_6 = 176.64 \text{ kJ/kg}$$

$$h_1 - h_2 = 0.92\,(3386 - 2915) = 433.3 \text{ kJ/kg, or } h_2 = 3386.1 - 433.3$$

$$= 2952.8 \text{ kJ/kg}$$

$$h_3 - h_{4'} = 0.9\,(3469.8 - 2331.7) = 1024.29 \text{ kJ/kg}$$

$$h_{4'} = 3169.8 - 1024 = 2445.5 \text{ kJ/kg}$$

$$W_P = h_6' - h_5 = \frac{9.072}{0.75} = 12.1 \text{ kJ/kg, or } h_6' = 167.15 + 12.1$$

$$= 179.67 \text{ kJ/kg}$$

$$t_{sat} \text{ at 18 bar} = 207.15°\text{C}$$

$$t_9 = 207.15 + 1.6 = 207.75°\text{C} \quad h_9 = 875 \text{ kJ/kg}$$

$$1(h_9 - h_{6'}) = \dot{m}(h_2 - h_7) \quad \dot{m} = \frac{875 - 179.67}{2952.8 - 883.4} = 0.336 \text{ kg}$$

$$W_T = (h_1 - h_2') + (1 - \dot{m})(h_3 - h_4') = 433.3 + 0.664 \times 1024.3$$

$$= 1113.435 \text{ kJ/kg}$$

$$W_{net} = W_T - W_P = 1113.435 - 12.1 = 1101.335 \text{ kJ/kg}$$

$$\dot{m}_s = \frac{500 \times 10^3}{1101.335} = \mathbf{454 \text{ kg/s}}$$

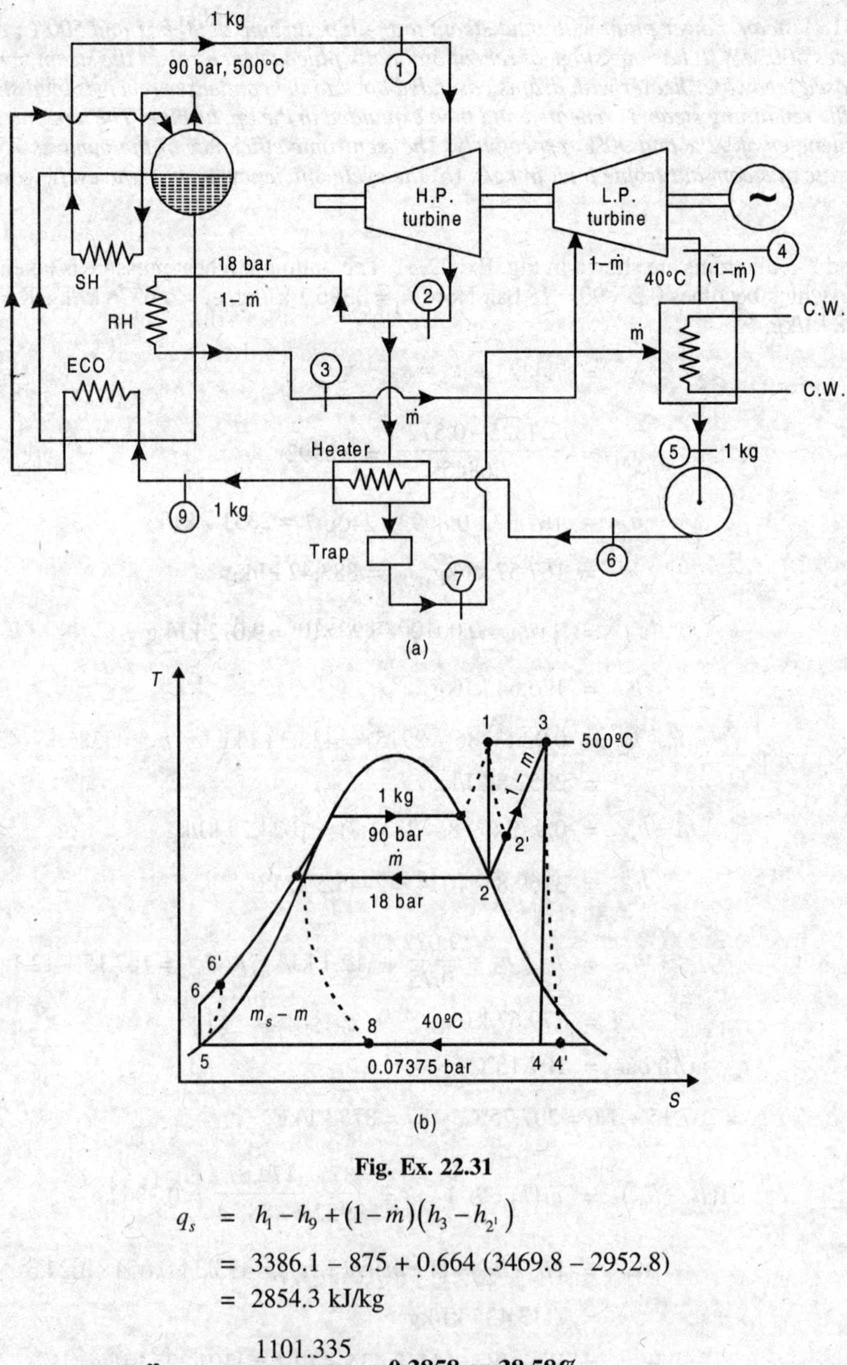

Fig. Ex. 22.31

$$q_s = h_1 - h_9 + (1-\dot{m})(h_3 - h_{2'})$$

$$= 3386.1 - 875 + 0.664\,(3469.8 - 2952.8)$$

$$= 2854.3 \text{ kJ/kg}$$

$$\eta_{cycle} = \frac{1101.335}{2854.3} = \mathbf{0.3858 \text{ or } 38.58\%}$$

$$\text{Work ratio} = \frac{W_{net}}{W_T} = \frac{1101.335}{1113.435} = \mathbf{0.989}$$

Example 22.32. *A steam power plant of 120 MW capacity is equipped with regenerative as well as reheat arrangements. The steam is supplied to the turbine at 86 bar and 50° superheat. The steam is extracted at 7 bar for feed heating and remaining steam is reheated to its initial temperature, before it is expanded to 0.35 bar in the LP stage.*

Assume indirect contact type feed heater. Determine :

1. *Percentage of steam bled from the turbine*
2. *Generating capacity of the boiler*
3. *Thermal efficiency of the cycle*
4. *Heat supplied in the reheater in kJ/min.*

Assume no losses and ideal processes of expansion.

Solution

The flow diagram and the *T-s* diagram of the plant are given below.

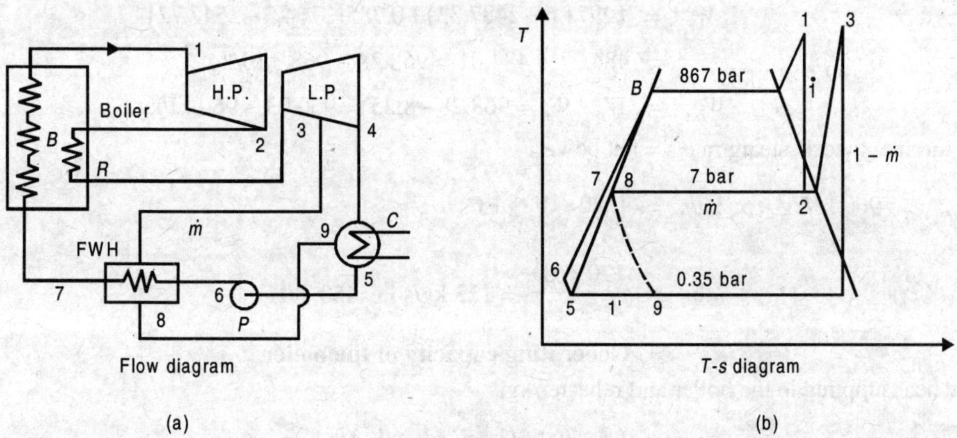

Fig. Ex. 22.32

At point 1, pressure is 86 bar and temperature = 350°C

1. From steam table, we have h_1 = 2971.6 kJ/kg and s_1 = 6.0781 kJ/kg°K
2. Since s_{g_2} at 7 bar = 6.7052 > s_1 condition at 2 is wet.

   We have at 7 bar, s_{f_2} = 1.9918 and sf_{g_2} = 4.7134

$$\therefore \qquad x_2 = \frac{6.0781 - 1.9918}{4.7134} = 0.867$$

$\therefore$ h_2 at 7 bar = 697.06 + 0.867 × 2064.9 = 2487.33 kJ/kg

3. Steam at 7 bar is reheated to 350°C at 7 bar 350°C from steam tables,

$$h_3 = 3164.3 \text{ kJ/kg and } s_3 = 7.4745 \text{ kJ/kg°K}$$

4. To find dryness at point 4 equate s_3 and s_4

$$\therefore \qquad 7.4745 = 0.98765 + x_4 \times 6.72915$$

$$\therefore \qquad x_4 = 0.964$$

$\therefore$ At 0.35 bar $h_4 = 304.3+0.964\times2327.2 = 2547.72$ kJ/kg

5. To find quantity of steam $\dot{m}$ extracted at 7 bar, we write the energy balance for feedwater heater. Thus,

$$\dot{m}[h_2 - h_8] = h_8 - h_6$$

Now, $$W_P = 0.001\times(86-0.35)\times100 = 8.25 \text{ kJ/kg}$$

$$\therefore \dot{m}[2487.33-697.06] = 697.06-[304.03+8.25]$$

$$\therefore \quad \dot{m} = 0.215 \text{ i.e. } \mathbf{21.5\%}$$

6. Turbine work is given by,

$$W_T = (h_1 - h_2)+(1-\dot{m})(h_3 - h_4)$$

$$\therefore \quad W_T = (2971.6-2487.33)+0.785(3164.3-2547.72)$$

$$= 484.27 + 484.01 = 968.28 \sim 968.3 \text{ kJ/kg}$$

7. $$W_{net} = W_T - W_P = 968.28-8.25 = 960.03 \sim 960 \text{ kJ/kg}$$

8. Net turbine × work steam rate/s = net power

$$960 \text{ kJ/kg} \times \dot{m}_s \text{ kg/s} = (120\times10^3) \text{ kJ/s}$$

$$\therefore \quad \dot{m}_s = \frac{120\times10^3}{960} = \mathbf{125 \text{ kg/s i.e. } 450 \text{ T/H}}$$

$$= \textbf{Generating capacity of the boiler}$$

9. Now heat supplied in the boiler and reheater/kg,

$$q_s = (h_1 - h_8)+(1-\dot{m})(h_3 - h_2)$$

$$= (2971.6-697)+0.785(3164.3-2487.3)$$

$$= 2274.6 + 531.45 = 2806 \text{ kJ/kg}$$

10. $$\text{Cycle efficiency} = \frac{W_{net}}{\text{heat supplied}} = \frac{960}{2806} = \mathbf{0.3421 \text{ or } 34.21\%}$$

11. Total heat supplied in reheater = $531.45\times125 = 66.43$ MW

$\therefore$ Heat supplied in reheater/min. = 66.43×60 = **3985.875 MJ/min**

Example 22.33. *In a thermal power plant of 60 MW capacity, boiler supplies steam at 100 bar and 400°C. The steam expands to 20 bar in the H.P. turbine. Steam is then reheated to 380°C in a reheater. Thereafter, steam expands to a condenser pressure of 0.05 bar. The pressure loss in the reheater is 1 bar. Assuming isentropic efficiency in both stages of expansion is 80%, find the steam generating capacity required in ton/h. Take transmission efficiency from turbine to generator = 98%, and generator efficiency of 95%. Neglect feed pump work. Also find the overall efficiency of the plant.*

Draw a neat line diagram of the plant and represent the cycle on T-S diagram.

Solution

The schematic flow diagram, the *T-s* diagram and *h-s* diagram are given below.

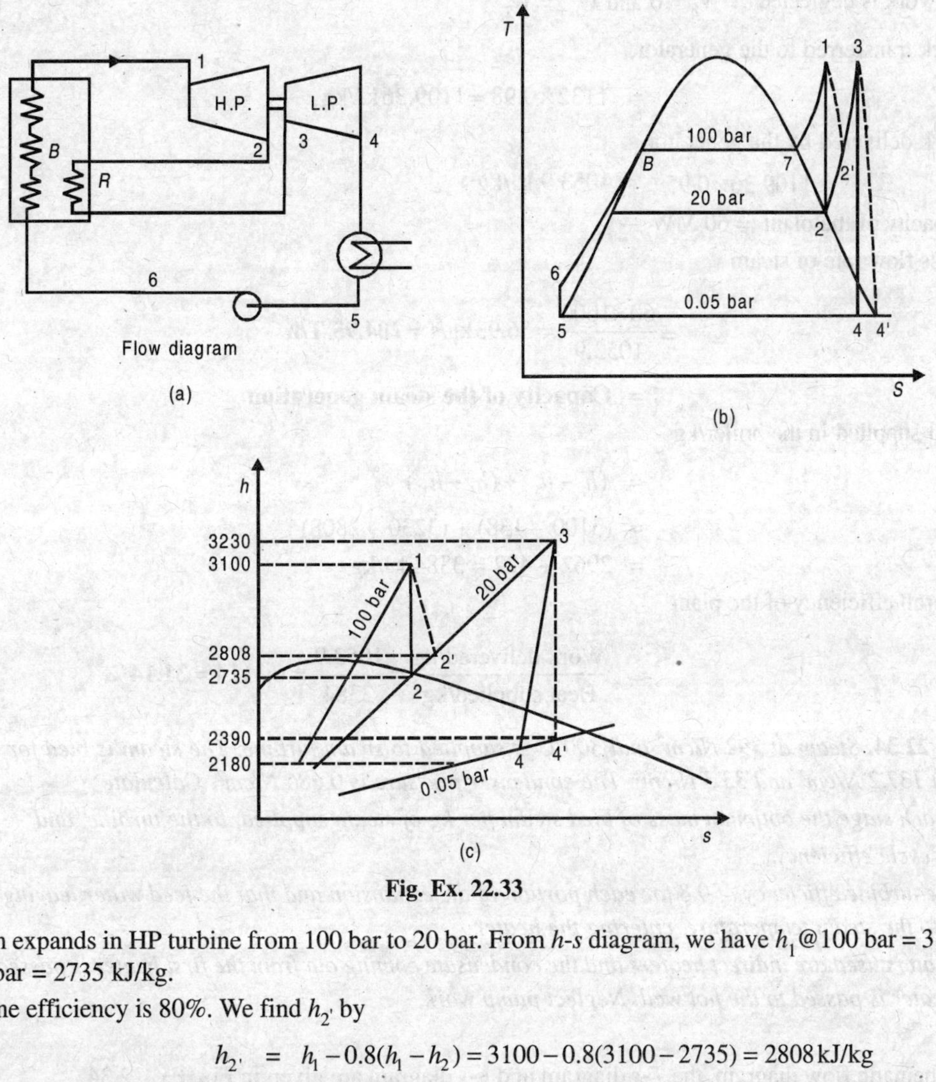

Fig. Ex. 22.33

The steam expands in HP turbine from 100 bar to 20 bar. From *h-s* diagram, we have h_1@100 bar = 3100 kJ/kg and h_2@20 bar = 2735 kJ/kg.

The turbine efficiency is 80%. We find $h_{2'}$ by

$$h_{2'} = h_1 - 0.8(h_1 - h_2) = 3100 - 0.8(3100 - 2735) = 2808 \text{ kJ/kg}$$

The steam is then reheated at 20 bar to 350°C

$$h_3 \text{ at 20 bar } 350°\text{C} = 3230 \text{ kJ/kg}$$

The steam then expands to 0.05 bar.

$$h_4 = 2180 \text{ kJ/kg. We find } h_{4'} \text{ by.}$$

$$h_{4'} = h_3 - 0.8(h_3 - h_4) = 3230 - 0.8(3230 - 2180) = 2390 \text{ kJ/kg}$$

We now find cycle parameters:

1. Work done in turbine,

$$W_T = (h_1 - h_2) + (h_3 - h_4')$$

$$= (3100 - 2808) + (3230 - 2390) = 1132 \text{ kg}$$

Pump work is neglected $\therefore$ $W_p = 0$ and $W_{net} = W_T$

2. Work transferred to the generator,

$$= 1132 \times 0.98 = 1109.36 \text{ kJ/kg}$$

3. Work delivered by the generator,

$$1109.36 \times 0.95 = 1053.9 \text{ kJ/kg}$$

4. Capacity of the plant = 60 MW

$\therefore$ Mass flow rate of steam

$$= \frac{60 \times 1000}{1053.9} = 56.93 \text{ kg/s} = \mathbf{204.95\ T/h}$$

= **Capacity of the steam generation**

5. Heat supplied in the boiler/kg

$$= (h_1 - h_5) + (h_3 - h_{2'})$$

$$= (3100 - 138) + (3230 - 2808)$$

$$= 2962 + 422 = 3384 \text{ kJ/kg}$$

6. Overall efficiency of the plant

$$= \frac{\text{Work delivered/kg}}{\text{Heat supplied/kg}} = \frac{1053.9}{3384} = 0.3114 = \mathbf{31.14\%}$$

Example 22.34. *Steam at 392 N/cm² and 300°C is supplied to steam turbine. The steam is bled for regenerative heating at 137.2 N/cm² and 33.3 N/cm². The condenser pressure is 0.686 N/cm². Calculate*

1. At each stage the optimum mass of bled steam per kg of steam supplied, to the turbine, and

2. The cycle efficiency.

Assume turbine efficiency of 0.8 for each portion of the expansion and that the feed water leaving each heater, is raised to the steam temperature, entering the heater.

The heaters used are indirect heaters and the condensate coming out from the first heater is passed through the second heater is passed to the hot well. Neglect pump work.

Solution

The schematic flow diagram, the *T-s* diagram and *h-s* diagram are given in Fix. Ex. 22.34.

1. The turbine η is 0.8

$\therefore$ $\quad h_2' = 2740 + (2907 - 2740) \times 0.2 = 2773 \text{ kJ/kg}$

also $\quad h_3' = 2540 + (2773 - 2540) \times 0.2 = 2587 \text{ kJ/kg}$

also $\quad h_4' = 2120 + (2587 - 2120) \times 0.2 = 2218 \text{ kJ/kg}$

also $\quad h_8 = h_9 = h_{10} = 826.28 \text{ kJ/kg}$ from steam tables by interpolation

and $\quad h_{11} = h_7 = 577.69 \text{ kJ/kg}$ from steam tables by interpolation

and $\quad h_6 = h_5 = W_p = 160.51 \text{ kJ/kg}$

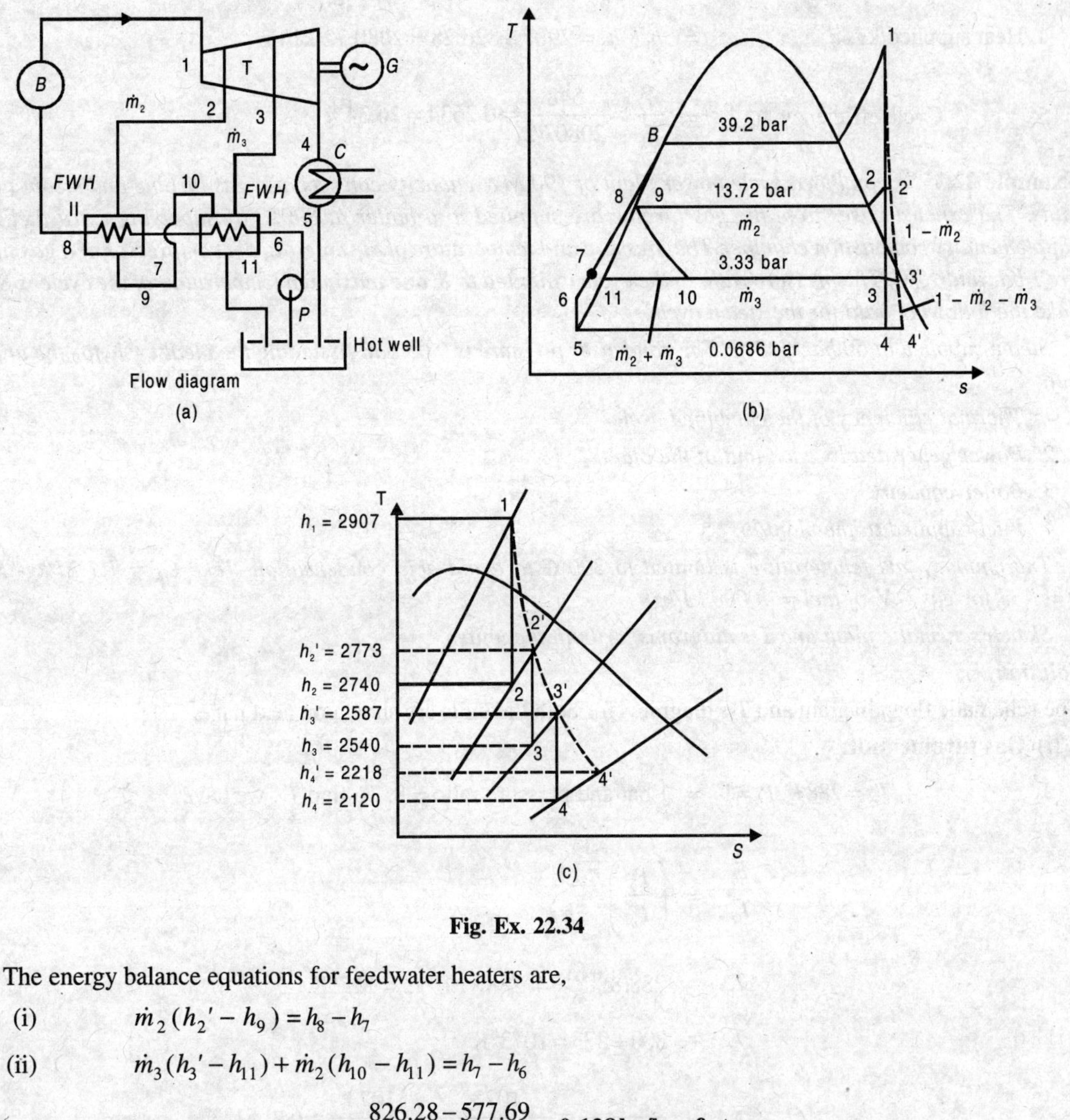

Fig. Ex. 22.34

2. The energy balance equations for feedwater heaters are,

(i) $\dot{m}_2 (h_2' - h_9) = h_8 - h_7$

(ii) $\dot{m}_3 (h_3' - h_{11}) + \dot{m}_2 (h_{10} - h_{11}) = h_7 - h_6$

$$\therefore \quad \dot{m}_2 = \frac{826.28 - 577.69}{2773.4 - 826.28} = \mathbf{0.128\,kg/kg\ of\ steam}$$

and $\dot{m}_3(2587 - 577.79) + 0128(826.28 - 577.69)$

$$= 577.69 - 160.51 - 31.82$$

$$\therefore \quad \dot{m}_3 = \frac{385.36}{2009.2} = \mathbf{0.19\ kJ/kg\ of\ steam}$$

3. Turbine Work, W_T

$$= (h_1 - h_{2'}) + (1 - m_2)(h_{2'} - h_{3'})(1 - \dot{m}_2 - \dot{m}_3)(h_{3'} - h_{4'})$$

$$= (2907 - 2773) + (1 - 0.128)(2773 - 2587)$$

$$+ (1 - 0.128 - 0.19)(2587 - 2218)$$

$$= 134 + 162.2 + 251.8 = 548 \text{ kJ/kg}$$

4. Heat supplied/kg, q_s $= h_1 - h_8 = 2907 - 826.28 = 2080.72$ kJ/kg

5. Cycle efficiency $\eta_{cycle} = \frac{W_t}{q_s} = \frac{548}{2080.72} = 0.2634 = \mathbf{26.34\%}$

Example 22.35. *A combined cycle power plant of 190 MW capacity consists of a gas turbine and steam turbine units. The exhaust gases from the gas turbine are supplied to a boiler at 800°C by supplying extra fuel in the supplementary combustion chamber. The pressure and temperature of air entering the compressor of a gas turbine are 1 bar and 15°C. The pressure ratio of the cycle is limited to 8, and maximum temperature of the cycle is 800°C. Take the following data for the steam cycle.*

Steam supplied at 60 bar, 500°C. The condenser pressure is 0.05 bar. Assuming the ideal cycle for the unit find out,

1. *Thermal efficiency of the combined cycle.*
2. *Power generated by each unit of the plant.*
3. *Boiler capacity.*
4. *Fuel supplied to the plant/hr.*

The chimney gas temperature is limited to 200°C to avoid acid condensation. Take C_p = 1.1 kJ/kg-°K and γ = 1.33 for gas C.V. of fuel = 40000 kJ/kg.

Sketch schematic plant and T-s diagrams for both the units.

Solution

The schematic flow diagram and *T-s* diagrams for both the units are shown on next page.

(i) Gas turbine unit:

1. $T_a = 288\,K, P_1 = 1$ = 1 bar and pressure ratio = 8. To find T_b, we have

$$\frac{T_b}{T_a} = \left(\frac{P_b}{P_a}\right)^{\frac{r-l}{r}}$$

$$\therefore \quad T_b = 288(8)^{0.4/1.4} = 288 \times 1.8125 = 522°K$$

also

$$T_c = 800 + 273 = 1073°K$$

$$\therefore \quad T_d = \frac{T_c}{(r_p)^{\gamma-1/\gamma}} = \frac{1073}{(8)^{1.33-1/1.33}} = \frac{1073}{1.67523} = 640.51°K$$

2. Compressor work per kg,

$$W_c = C_{pa}(T_b - T_a) = 1.005(522 - 288) = 235.17 \text{ kJ/kg}$$

and turbine work per kg

$$W_{GT} = C_{pg}(T_c - T_d) = 1.1(1073 - 640.51) = 475.74 \text{ kJ/kg}$$

$\therefore$ Net work produced by the gas turbine.

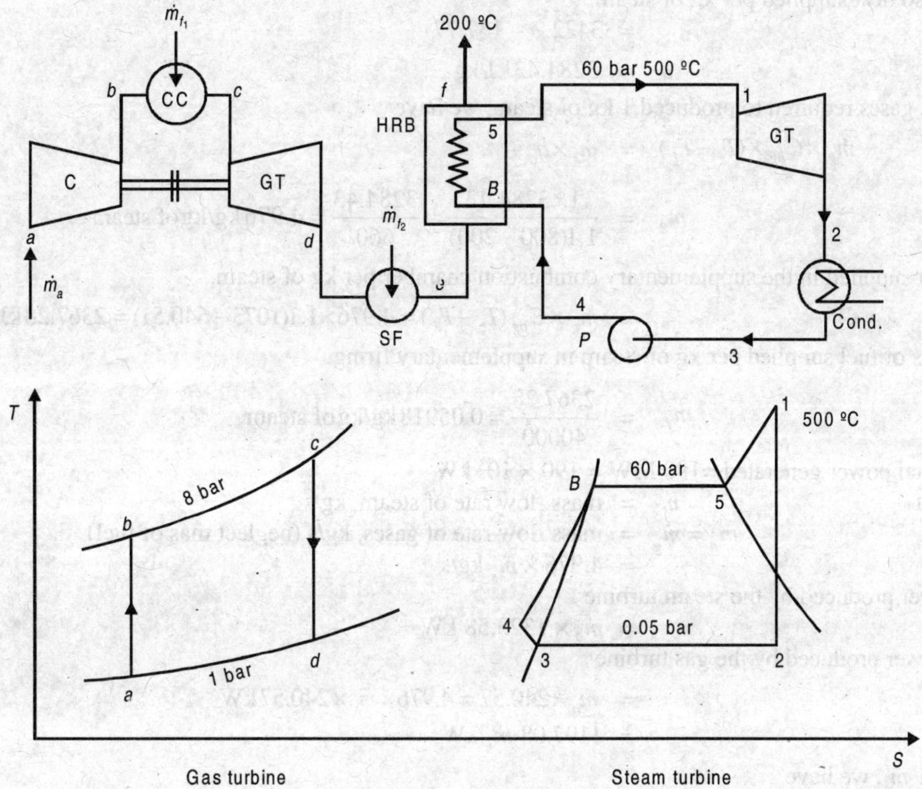

Fig. Ex. 22.35

$$W_{GT\,\text{net}} = 475.74 - 235.17 = 240.57 \text{ kJ/kg}$$

3. Heat supplied in the combustion chamber/kg of air

$$= C_{pg} \cdot T_c - C_{pa} T_b = 1.1 \times 1073 - 1.005 \times 522$$

$$= 655.69 \text{ kJ/kg}$$

$\therefore$ Fuel supplied/kg of air $\dot{m}_{f_1} = \dfrac{655.69}{40000} = 0.0164$ kg/kg of air

Steam Turbine Unit:

4. At 60 bar and 500ºC,

$$h_1 = 3422.2 \text{ kJ/kg and}$$

$$s_1 = 6.8818 \text{ kJ/kg}$$

$$\therefore \quad x_2 = \frac{6.8818 - 0.4763}{7.9197} = 0.809$$

$$\therefore \quad h_2 = 137.77 + 0.809 \times 2423.8 = 2098.62 \text{ kJ/kg}$$

$\therefore$ Work done by steam turbine/kg

$$= 3422.2\ 0 - 2098.62 = 1323.58 \text{ kJ/kg}$$

5. Also heat supplied per kg of steam

$$= 3422.2 - 137.77$$

$$\therefore \quad q_s = 3284.43 \text{ kJ/kg}$$

To find gases required to produced 1 kg of steam, we have

$$\dot{m}_g \times C_{pg} \times (T_e - T_f) = \dot{m}_s \times q_s$$

$$\therefore \quad \dot{m}_g = \frac{1 \times 3284.43}{1.1(800-200)} = \frac{3284.43}{660} = 4.976 \text{ kg/kg of steam}$$

∴ Heat supplied in the supplementary combustion chamber per kg of steam,

$$= \dot{m}_g \times C_{pg}(T_c - T_d) = 4.976 \times 1.1(1073 - 640.51) = 2367.28 \text{ kJ}$$

∴ Mass of fuel supplied per kg of steam in supplementary firing

$$\dot{m}_{f_2} = \frac{2367.28}{40000} = 0.05918 \text{ kg/kg of steam}$$

6. Total power generated =190 MW = 190×10^3 kW

Let $\dot{m}_s$ = mass flow rate of steam, kg/s

and $\dot{m}_a = \dot{m}_g$ = mass flow rate of gases, kg/s (neglect mas of fuel)

$= 4.976 \times \dot{m}_s$ kg/s

∴ Power produced by the steam turbine

$$= \dot{m}_2 \times 1323.58 \text{ kW} \qquad \text{...(I)}$$

and Power produced by the gas turbine

$$= \dot{m}_a \times 240.57 = 4.976 \times \dot{m}_s \times 240.57 \text{ kW} \qquad \text{...(II)}$$

$$= 1197.08\ \dot{m}_s \text{ kW}$$

To find $\dot{m}_s$, we have

$$190 \times 1000 = 1323.58\ \dot{m}_s + 1197.08\ \dot{m}_s$$

$$\therefore \quad \dot{m}_s = 75.38 \text{ kg/s}$$

$$\therefore \quad \dot{m}_a = 4.976 \times 75.38 = 375 \text{ kg/s}$$

7. Heat supplied/sec for the total plant

$$= \dot{m}_s \times \text{heat supplied/kg of steam} + \dot{m}_a \times \text{heat supplied/kg of air}$$

$$= 75.38 \times 23.67.28 + 375 \times 655.69$$

$$= 178445.57 + 245883.75 \text{ kJ/s}$$

8. Thermal efficiency of the combined plant

$$= \frac{190 \times 1000}{424329.32} = 0.4478. \text{ i.e. } \mathbf{44.78\%}$$

9. Power produced by gas turbine

$$= 1197.08 \times 75.38 = 90.236 \text{ MW} - \mathbf{90\ MW}$$

and Power produced by steam turbine

$$= 75.38 \times 1323.58 = 99.764 \text{ MW} - \mathbf{100\ MW}$$

10. Fuel supplied for total plant/hr

$$= \frac{\text{total heat supplied}}{W} = \frac{424329.32 \times 3600}{40000}$$

$$= 38189.64 \text{ kg/hr.} = \mathbf{3819\ T/hr}$$

Example 22.36. *In a combined cycle power plant the air is supplied at the rate of 2000×10^3 kg/hr. at 20°C. The pressure ratio used in the gas turbine plant is 7. The suction pressure of the compressor and the exhaust pressure of the turbine is 1 bar. The maximum temperature in the gas turbine is limited to 1000°C. The isentropic efficiencies of the gas turbine and the compressor are 85% and 80% respectively. The following data is given for the steam power plant.*

The temperature of exhaust gas from the gas turbine is increased to 1200°C by burning fuel before passing into the boiler. The steam is generated at 50 bar and 500°C. The condenser pressure is maintained at 0.1 bar. Temperature of the gases entering the chimney is 200°C. Find the following :

1. Total power generating capacity of the combined plant
2. Overall efficiency of the plant
3. Fuel consumption of the combined plant/hour
4. Air/fuel ratio for the gas turbine plant. Use the following data,

Calorific value of the fuel = 45 MJ/kg

C_{pa} = 1 kJ/kg-°K γ for air = 1.4

C_{pg} = 1.1 kJ/kg-°K γ for gas = 1.33

Do not neglect the mass of the fuel used.
Draw neat line diagram for the components.
Represent the process on T-s diagram.

Solution

The schematic flow diagram and *T-S* diagrams of the combined plant are given below.

1. The GT plant: We find the temperature $T_{2'}$ and $T_{4'}$. We have $T_1 = 20 + 273 = 293°K$

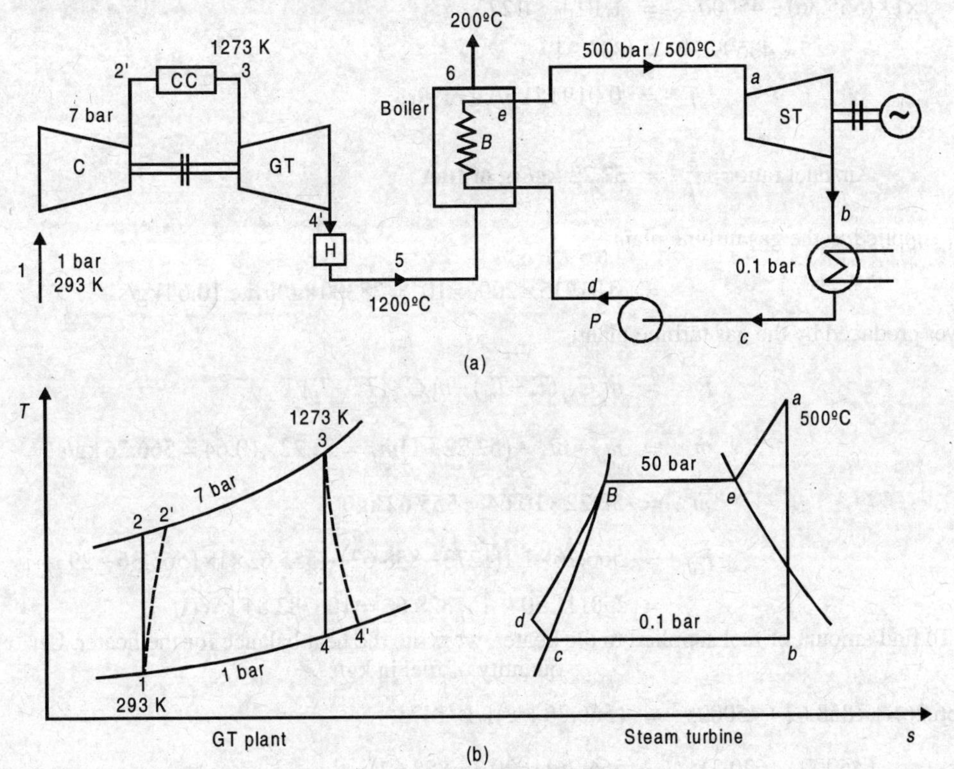

Fig. Ex. 22.36

$$\therefore \quad T_2 = T_1\left(\frac{P_2}{p_1}\right)^{\frac{y-1}{y}} = 293(7)^{0.4/1.4} = 510.886°K$$

$$\eta_c = 0.8 = \frac{T_2 - T_1}{T_{2'} - T_1} = \frac{510.886 - 293}{T_{2'} - 293} \quad \therefore T_{2'} = 565.36°K$$

For GT, $T_3 = 1000 + 273 = 1273°K$

We have, $\frac{T_4}{T_3} = \left(\frac{P_4}{P_3}\right)^{\frac{\gamma-1}{\gamma}} \quad T_4 = 1273\left(\frac{1}{7}\right)^{\frac{0.33}{1.3}} = 785.49°K$

As GT has efficiency of 0.85

$$\therefore \quad T_{4'}T_3 - 0.85(T_3 - T_4) = 1273 - 0.85(1273 - 785.49)$$

$$\therefore \quad T_{4'} = 858.62°K$$

2. To find A/F ratio, we write energy balance for the combustion chamber. Let *t* be the mass of fuel per kg of air

$$\therefore \quad C_{ga} \times T_{2'} \times C.V. \times f = C_{pg}(1+f)T_3$$

Substituting the values,

$$\therefore \quad 1 \times 1 \times (565.36) + 45000f = 1.1(1+f)1273$$

$$43599.7f = 834.94$$

$$\therefore \quad f = 0.01915 \text{ kg/kg of air}$$

$$\therefore \quad \text{Air/Fuel ratio} = \frac{1}{f} = \mathbf{52.22\ kg/kg\ of\ fuel}$$

3. Fuel supplied to the gas turbine plant,

$$= 0.01915 \times 2000 \times 10^3 = 38300 \text{ kg/hr.i.e.} 10.64 \text{ kg/s}$$

4. Power produced by the gas turbine plant,

$$P_{GT} = \dot{m}_g C_{pg}(T_3 - T_{4'}) - m_a C_{pa}(T_{2'} - T_1)$$

Now $\dot{m}_g = \dot{m}_a + \dot{m}_f - (52.22+1)\dot{m}_f = 53.22 \times 10.64 = 566.26 \text{ kg/s}$

and $\dot{m}_a = 52.22 \times 10.64 = 555.62 \text{ kg/s}$

$$\therefore \quad P_{GT} = 566.26 \times 1.1(1273 - 858.62) - 555.62 \times 1 \times (565.36 - 293)$$

$$= 258111.50 - 151328.66 = 106782.84 \text{ kW(I)}$$

5. To find amount of fuel supplied to the heater, we write the heat balance for the heater. Let '*x*' be the quantity of fuel in kg/s.

$$\therefore \quad 566.26 \times 1 \times 858.62 + 45000x = (566.26 + x)1.1 \times 1473$$

$$\therefore \quad (45000 - 1620.3)x = 566.26(1620.3 - 858.62)$$

$$\therefore \quad x = \frac{431308.92}{43379.7}$$

$$= 9.9426 \text{ kg/s}$$

$$= 35793.52 \text{ kg/hr. (II)}$$

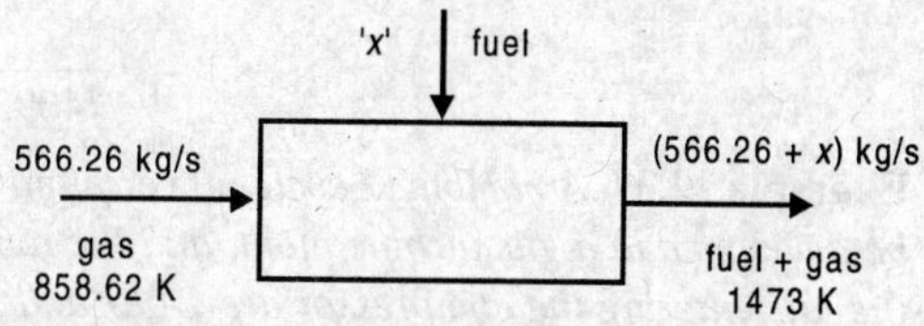

Fig. Ex. 22.36 (c)

6. Total fuel supplied to the plant,

$$= 38300 + 35793.52 = \mathbf{74093.52 kg/hr.}$$

$$= \mathbf{74.1 \text{ T/hr}}$$

7. For steam turbine

At 50 bar 500°C $\quad h_1 = 3433.7 \text{ kJ/kg. and } s_1 = 6.977 \text{ kJ/kg°K}$

At 0.1 bar $\quad s_{f_2} = 0.6493 \text{ kJ/kg °K. and } s_{f_{g^2}} = 7.5018 \text{ kJ/kg °K}$

also $\quad h_{f_2} = 191.83 \text{ and } h_{f_{g2}} = 2392.9 \text{ kJ/kg}$

$$\therefore \quad x_b = \frac{6.977 - 0.6493}{7.5018} = 0.846$$

$$\therefore \quad h_b = 191.83 + 0.846 \times 2392.9 = 2216.6 \text{ kJ/kg}$$

Pump work, $\quad W_p = 0.001(50 - 0.1) \times 100 = 5 \text{ kJ/kg}$

$$\therefore \quad h_d = h_e + w_p = 191.83 + 5 = 196.83 \text{ kJ}$$

Heat required per kg of steam $= h_a - h_d = 3433.7 - 196.83 = 3236.87 \text{ kJ/kg}$

8. Heat supplied in the boiler $= \dot{m}_g \times C_{pg}\,(1200 - 200)$

$$= (566.26 + 9.94) 1.1 \times 1000$$

$$= 633820 \text{ kJ/s}$$

9. Amount of steam generated/sec.

$$= \frac{633820}{3236.87} = 195.8 \text{ kg/s. or. } 704.9.\text{T/h}$$

10. Power produced by the steam turbine plant

$$= 195.8(h_a - h_b)$$

$$= 195.8(3433.7 - 2216.6)$$

$$= 238308.18 \text{ kW}$$

11. Total power produced by combined plant

$$= 06782.84 + 238308.18 = \mathbf{345091 \text{ kW}}$$

$$= \mathbf{345.1 \text{ MW}}$$

12. Overall efficiency of the combined plant

$$= \frac{\text{Total work produced}}{\text{Fuel used} \times \text{Cal. value}}$$

$$= \frac{345091}{74093.52/3600 \times 45000} = 0.3726 \text{ or } \mathbf{37.26\%}$$

Example 22.37. *A combined cycle power plant is to be designed to develop 100 MW. Out of this 60% power is to be developed in a gas turbine plant, and the remaining in a steam turbine plant. The pressure and temperature of the air entering the compressor are 1 bar and 27°C.*

The maximum pressure and temperature in the gas turbine cycle are limited to 3 bar and 1000°C. Steam is to be generated at. 50 bar and 600°C, by using supplementary fuel. The condenser pressure is maintained at 0.05 bar.

Take the given data and find:

1. Overall efficiency of the plant

2. Ratio of air flow rate to steam flow rate

3. Quantity of cooling water required in tons/hr, if the rise in temperature of cooling, water is limited to 10°C.

Solution

Data:

Isentropic epic efficiency of the compressor = 85%

Isentropic efficiency of the turbine = 90%

Combustion efficiency = 95%

$C_{pa} = 1\,\text{kJ/kg°K}, \qquad \gamma_a = 1.4$

$C_{pg} = 1.1\,\text{kJ/kg°K}, \qquad \gamma_g = 1.33$

C. V. of fuel = 40 mJ/kg

Gas temperature going to chimney = 200°C. Neglect the pump work. Consider the mass of fuel in the gas turbine.

The schematic diagram and the *T-s* diagrams are given on next page.

(I) Gas turbine plant:

1. We find $T_{b'}$ and $T_{d'}$ by usual method.

$P_a = 1\,\text{bar}, t_a = 300°\text{K} \; P_d = 1\,\text{bar}, P_b = P_c = 8\,\text{bar}.$

$\eta_c = 0.85, C_{pa} = 1\,\text{kJ/kg °K}, \gamma_a = 1.4, T_c = 1273°\text{K}$

$\eta_t = 0.9, C_{pg} = 1.1\,\text{kJ/kg °K}, \gamma_g = 1.33$

$$\frac{T_b}{T_a} = \left(\frac{p_b}{p_a}\right)^{\frac{\gamma-1}{\gamma}}$$

$\therefore\ T_b = 300 \times (8)^{0.4/1.4} = 543.43\,\text{K}$

$$T_{b'} = 300 + \frac{543.43 - 300}{0.85} = 586.4°\text{K}$$

Similarly, $$T_d = \frac{T_c}{(8)^{0.33/1.33}} = \frac{1273}{1.675} = 759.89°\text{K}$$

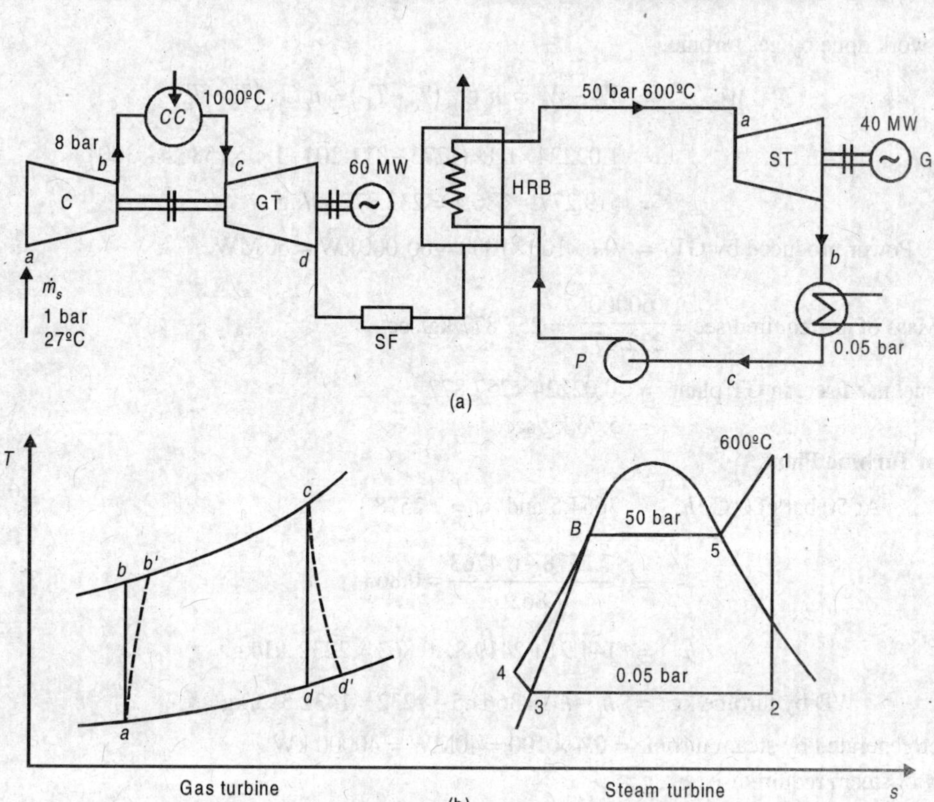

Fig. Ex. 22.37

Now, $T_{d'} = 1273 - 0.9(1273 - 759.89) = 811.20°\text{K}$

2. Now consider combustion chamber of Gas turbine.

Let x be fuel supplied per kg of air.

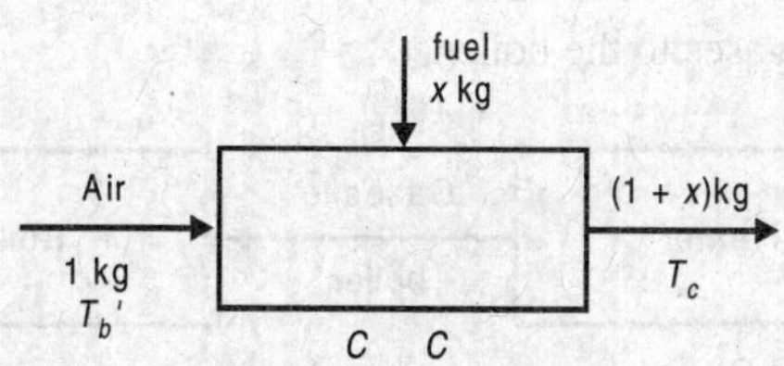

Fig. Ex. 22.37 (c)

Writing energy balance for combustion chamber, we get

$$C_{pg} \times \dot{m}_g \times (T_3) - \dot{m}_a C_{pa} T_{2'} \;=\; C.V.\,x.\eta_{\text{comb}}$$

$1.1\ (1 + x)\ 1273 - 1 \times 1 \times 586.4 = 40000x \times 0.95$

$$x \;=\; \frac{813.9}{36600} = 0.02224\,\text{kg of fuel/kg of air}$$

$$\text{Air fuel ratio} = \frac{1}{x} \;=\; \frac{1}{0.2224} = 44.969 - 45 : 1\,\text{kg of air/kg of fuel}$$

3. Net work done by gas turbine,

$$W_{GT\,net} = W_T - W_C = \dot{m}_g C_{pg}(T_c - T_{d'}) - \dot{m}_a C_{pa}(T_{b'} - T_a)$$

$$= 1.02224 \times 1.1 \times (1273 - 811.20) - 1 \times 1 \times (586.4 - 300)$$

$$= 519.277 - 286.4 = 232.877 \text{ kJ/kg}$$

$$\text{Power produced by GT} = 0.6 \times 100 \times 1000 = 60,000 \text{ kW} = 60 \text{ MW}$$

$$\therefore \quad \text{Mass of air required/sec} = \frac{60000}{232.877} = 257.877 \text{ kg/sec}$$

$$\therefore \text{Fuel used/sec. in GT plant} = 0.02224 \times 257.877$$

$$= 5.73 \text{ kg/sec.}$$

(II) Steam Turbine Plant

At 50 bar 600°C, $h_1 = 3664.5$ and $s_1 = 7.2578$

$$\therefore \quad x^2 = \frac{7.2578 - 0.4763}{7.8626} = 0.863$$

$$\therefore \quad h_2 = 144.91 + 2419.8 \times 0.863 = 2232 \text{ kJ/kg}$$

$$\text{WD by turbine/kg} = h_1 - h_2 = 3664.5 - 2232 = 1432.5 \text{ kJ/kg}$$

1. Power generated by steam turbine = 0.4 × 100 = 40MW = 40000 kW
Quantity of steam required/sec.

$$= \frac{40000}{1432.5} = 27.923 \text{ kg of steam/sec.}$$

2. Amount of heat required to generate steam/sec.

$$27.923(h_1 - h_4) = 27.923\,[3664.5 - (137.7 + 5)]$$

$$= 98337.267 \text{ kW.}$$

3. We find inlet temperature of gases to the boiler.

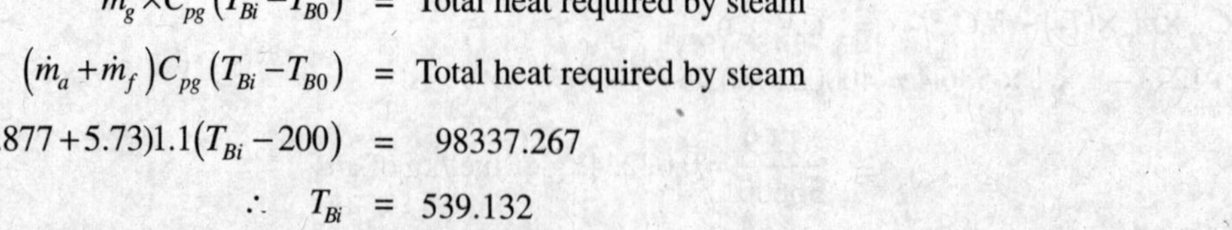

Fig. Ex. 22.37(d)

$$\therefore \quad \dot{m}_g \times C_{pg}(T_{Bi} - T_{BO}) = \text{Total heat required by steam}$$

$$(\dot{m}_a + \dot{m}_f)C_{pg}(T_{Bi} - T_{BO}) = \text{Total heat required by steam}$$

$$(257.877 + 5.73)1.1(T_{Bi} - 200) = 98337.267$$

$$\therefore \quad T_{Bi} = 539.132$$

$$= 812.132 \text{ °K}$$

This is very nearly equal to the inlet temperature of gases to the heater and hence no heat added in the heater.

4. Ratio of air flow to steam flow rate,

$$\frac{\dot{m}_a}{\dot{m}_s} = \frac{257.877}{27.923} = \mathbf{9.235}$$

5. Overall efficiency of the plant

$$= \frac{\text{Power developed}}{\text{Total heat supplied}} = \frac{\text{Power}}{m_f \times \text{C.V.}}$$

$$= \frac{100 \times 1000}{5.73 \times 40000} = \mathbf{0.436} \text{ i.e. } \mathbf{43.60\%}$$

6. Heat rejected by steam in the condenser to its cooling water

$$= \dot{m}_s \times \text{dryness fraction} \times h_{f_g}$$

$$= 27.923 \times 0.863 \times 2423.8$$

$$= 58407.639 \text{ kW}$$

This is equal to heat gained by water

$$\therefore \quad \dot{m}_w C_{pw} \Delta T = 58407.639$$

$$\dot{m}_w = \frac{58407.639}{4.1867 \times 10} = 1395.076 \text{ kg/s}$$

$$= \mathbf{5022.273 \text{ ton/hr.}}$$

THEORY QUESTIONS

22.1. Explain. Why Rankine cycle rather than Carnot cycle is used as a standard cycle for steam power plants.

22.2. What is work ratio? What is its importance in power cycles?

22.3. Explain the effect of following parameters on the performance of Rankine cycle:
(a) admission pressure
(b) inlet temperature
(c) exhaust pressure.

22.4. Explain the "external" and "internal" irreversibilities of Rankine cycle. What are their effects on the performance of the cycle?

22.5. What is 'reheating' employed in Rankine cycle? How does reheating improve the cycle efficiency?

22.6. What are the advantages of superheat-reheat cycle?

22.7. How Rankine cycle efficiency is affected by reheat pressure ratio?

22.8. Explain ideal regenerative feed heating. Why it is not used in practice?

22.9. What is practical method of regenerative feed heating?

22.10. What are the advantages and limitations of regenerative feed heating?

22.11. What are different types of feed water heaters?

22.12. What is DA? Where and why it is used in power plants?

22.13. Compare following closed feed water heaters.
(i) With drains cascaded backwards
(ii) With drains pumped forward.

22.14. How feed water heaters are placed ideally in a power plant?

22.15. Sketch a typical layout of modern power steam power plant with superheating, repeating and regenerative feed heating.

22.16. What is cogeneration? What are the categorise of cogeneration?

22.17. What are the basic types of topping cycles in cogeneration?

22.18. Explain with a sketch the gas turbine plant with cogeneration.

22.19. Explain the concept of a combined cycle operation. What are the advantages of combined cycle power plant?

22.20. Explain with a sketch the combined cycle power plant with a heat recovery boiler.

22.21. What is a multipressure combined cycle plant? What are its advantages?

22.22. What is binary vapour cycle ? Explain with a sketch the working of binary vapour power plant. What are its limitations?

PROBLEMS FOR PRACTICE

22.1. Find the ideal cycle efficiency and steam consumption of reheat cycle operating between pressure of 30 and 0.04 bar with a superheat temperature of 450°C. Assume that the first expansion is carried out to the point where the steam is dry saturated and that the steam is reheated to the original temperature. The feed pump work may be neglected.

[*Ans.* $\eta = 0.387$, s.s.c. = 2.39 kg/kWh]

22.2. Steam is supplied to a two stage turbine at 40 bar and 350°C. It expands in the first turbine until it is just dry saturated, then it is reheated to 350°C and expanded through the second stage turbine. The condenser pressure is 0.035 bar. Calculate the work output and the heat supplied per kg of steam for the plant. Calculate also the specific steam consumption and the cycle efficiency.

[*Ans.* (i) 1290 kJ, (ii) 3362 kJ, (iii) 2.79 kg/kWh, (iv) 38.4 %]

22.3. In a reheat cycle, the initial steam pressure and the maximum temperature are 150 bar and 550°C respectively. If the condenser pressure is 0.1 bar and the moisture at the condenser inlet is 15%, and assuming ideal processes, determine (a) the reheat pressure, (b) the cycle efficiency and (c) the steam rate.

[*Ans.* (a) 13.5 bar, (b) 43.6 %, (c) 2.05 kg/kWh]

22.4. In a reheat steam cycle, the maximum steam temperature is limited to 500°C. The condenser pressure is 0.1 bar and the quality at turbine exhaust is 0.8778. Had there been no reheat the exhaust quality would have been 0.7592. Assuming ideal processes, determine (a) the reheat pressure, (b) the boiler pressure, (c) the cycle efficiency, and (d) the steam rate.

[*Ans.* (a) 30 bar, (b) 150 bar, (c) 50.51 %, (d) 1.9412 kg/kWh]

22.5. Find the cycle efficiency and steam consumption of regenerative cycle with one open feed heater. The steam leaves the boiler at 30 bar and superheated to 450°C. The bleed pressure is so that the temperature difference is divided into approximately equal steps.

[*Ans.* $\eta = 0.373$, s.s.c. = 4.29 kg/kWh]

22.6. Calculate the efficiency of regenerative cycle using a single feed water heater, neglecting the feed pump work. The steam leaves the boiler at 30 bar and superheated to 450ºC and the condenser pressure is 0.035 bar. The heater is of closed type and bled steam is flashed into the condenser.

[*Ans.* Bleeding pressure = 2.7 bar, Steam extracted = 0.165 kg, Efficiency = 0.401]

22.7. A steam turbine is supplied with steam at 40 bar and 450°C and exhausted at 0.035 bar. Neglecting the feed pump work, find the ideal Rankine cycle efficiency of the plant and compare it with the ideal efficiency, which could be obtained if regenerative feed heating were employed. Assume the use of open feed water heater at one bleeding point at 3 bar. Also calculate the amount of steam extracted per kg of steam supplied.

[*Ans.* Ideal efficiency = 0.391, Ideal efficiency with feed water heater = 0.415, Steam extracted = 0.173 kg of steam supplied]

22.8. In a single-heater regenerative cycle the steam enters the turbine at 30 bar, 400°C and the exhaust pressure is 0.10 bar. The feedwater heater is a direct contact type which operates at 5 bar. Find (a) the efficiency and the steam rate of the cycle, and (b) the increase in mean temperature of heat addition, efficiency and steam rate, as compared to the Rankine cycle (without regeneration). Neglect pump work.

[*Ans.* (a) 35.36 %, 3.93 kg/kWh, (b) 27.4°C, 1.18%, 0.47 kg/kWh]

22.9. A regenerative steam cycle employing three closed heaters the steam is supplied to the turbine at 42 bar and 500°C and is exhausted to the condenser at 0.035 bar. The bled steam for heating is taken at pressure of 15.4 and 0.5 bar. Assuming the ideal processes and neglecting pump work, calculate:

(i) the fraction of steam bled at each stage
(ii) power output per kg of steam flow rate
(iii) the cycle efficiency.

[*Ans.* (1) 0.0969, 0.0902, (ii) 1133.6 kW/kg, (iii) 43.6%]

22.10. A simple steam power cycle uses solar energy for the heat input. Water in the cycle enters the pump as a saturated liquid at 40°C, and is pumped to 2 bar. It then evaporates in the boiler at this pressure, and enters the turbine as saturated vapour. At the turbine exhaust the conditions are 40°C and 10% moisture. The flow rate is 150 kg/h. Determine (a) the turbine isentropic efficiency, (b) the net work output, (c) the cycle efficiency, and (d) the area of the solar collector needed if the collectors pick up 0.58 kW/m^2.

[*Ans.* (a) 76.7 %, (b) 15.5 kW, (c) 12.78 %, (d) 182 m^2]

22.11. The net power output of an ideal reheat regenerative steam cycle is 80 MW. Steam enters the h.p. turbine at 80 bar, 50W and expands till it becomes saturated vapour. Some of the steam then goes to an open feedwater heater and the balance is reheated to 400°C, after which it expands in an L.P. turbine to 0.07 bar. Compute: (a) the reheat pressure, (b) the steam flow rate to the h.p. turbine, (c) the cycle efficiency, and (d) the rate of flow of cooling water in the condenser if the temperature rise of water is 8°C. (e) If the velocity of steam flowing from the turbine to the condenser is limited to 130 in/s, find the diameter of the connecting pipe

[*Ans.* (a) 6.5 bar, (b) 58.4 kg/s, (c) 43.7%, (d) 3146.5 kg/s, (e) 2.97 m]

22.12. A binary vapour cycle operates on mercury and steam. Saturated mercury vapour at 4.5 bar is supplied to the mercury turbine, from which it exhausts at 0.04 bar. The mercury condenser generates saturated steam at 15 bar which is expanded in a steam turbine to 0.04 bar. (a) Find the overall efficiency of the cycle. (b) If 50,000 kg/h of steam flows through the steam turbine, what is the flow through the mercury turbine? (c) Assuming that all processes are reversible, what is the useful work done in the binary vapour cycle for the specified steam flow? (d) If the steam leaving the mercury condenser is superheated to a temperature of 300°C in a superheater located in the mercury boiler, and if the internal efficiencies of the mercury and steam turbines are 0.85 and 0.87 respectively, calculate the overall efficiency of the cycle. The properties of saturated mercury are given below:

p (bar)	t (°C)	h_f (kJ/kg)	h_g (kJ/kg)	s_f (kJ/kg-°K)	s_g (kJ/kg-°K)	v_f (m^3/kg)	v_g (m^3/kg)
4.5	450	62.93	355.98	0.1352	0.5397	79.9×10^{-6}	0068
0.04	216.9	29.98	329.85	0.0808	0.6925	76.5×10^{-6}	5.178

[**Ans.** (a) 53%, (b) 59.35 × 10^4 kg/h, (c) 28.5 MW, (d) 46.2%]

22.13. In a combined cycle, the gas turbine portion has a pressure ratio of 16. Air enters the compressor at 300 °K at a rate of 14 kg/s and heated to 1500 °K in the combustion chamber. The combustion gases leaving the gas turbine are used to heat steam to 460°C at 100 bar in a heat exchanger. The combustion gases leave the heat exchanger at 420°C. The steam leaving the turbine is condensed at 15 kPa. Assuming all the compression and expansion processes to be isentropic, determine (1) the mass flow rate of the steam, (2) the net power output and (3) the thermal efficiency of the combined cycle. Assume constant specific heat for air at room temperature.

[*Ans.* (1) 1.275 kg/s, (2) 7881 kW, (3) 66.3%]

22.14. A pass-out two-stage turbine receives steam at 50 bar, 350°C. At 1.5 bar, the h.p. steam exhausts and 12,000 kg of steam per hour are taken at this stage for process purposes. The remainder is reheated at 1.5 bar to 250°C and then expanded through the i.p. turbine to a condenser pressure of 0.05 bar. The power output from the turbine unit is 3750 kW. Take isentropic efficiency for h.p. and i.p. stages as 0.84 and 0.81 respectively.

Calculate the boiler capacity required.

[*Ans.* 18.5 t/h]

22.15. Steam is generated at 70 bar, 500°C and expands in a turbine to 30 bar with an isentropic efficiency of 77%. At this condition, it is mixed with steam twice its mass at 30 bar, 400°C. The mixture then expands with an isentropic efficiency of 80% to 0.06 bar. At a point in the expansion where the pressure is 5 bar, steam is bled for feedwater heating in a direct contact heater, which raises the feedwater to the saturation temperature of the bled steam. Calculate the mass of steam bled per kg of high pressure steam and the cycle efficiency. Assume that the L.P. expansion condition line is straight.

[*Ans.* 0.53 kg, 31.9 %]

IMPORTANT POINTS TO BE REMEMBERED

1. *Types of feed-water heaters*
   (a) Open or direct contact type (Deareating heater DA)
   (b) Closed type with drains cascated backward
   (c) Closed type with drains pumped forward.
2. *In open water heaters*
   (i) amount of steam extracted or bled such that the water coming to the heater is heated to the saturation temperature of steam entering the heater.
   (ii) steam entering the heater is condensed to the saturation temperature of steam.
3. *Closed water heater-cascading backwards*
   (i) amount of steam extracted or bled is such that the water or condensate is heated to the saturation temperature of steam entering that heater.
   (ii) Steam entering the heater is condensed and cooled to the temperature of water entering that heater
   (iii) The difference between the saturation temperature of steam entering the heater and the temperature of water leaving the heater is known as terminal temperature difference (TTD or TD).
4. *Closed type water heaters with drains pumped forward*
   (i) amount of steam bled is such that the water entering the heater is heated to the saturation temperature of steam entering the heater
   (ii) Steam entering the heater is condensed to the saturation temperature of the steam.
5. *Placement of Feed-water heaters*
   For n number of FWH are used, the optimum temperature per heater is given by

$$\Delta T_{opt} = \frac{T_B - T_C}{n+1}$$

T_B is boiler saturation temperature

T_C is condenser saturation temperature

6. *Reheat cycle*
   Generally, the reheater pressure of steam is 20% of the pressure of steam entering the stage before reheater.

PART–5

CHAPTER 23

Air Standard Cycles

23.1 INTRODUCTION

In all heat engines — particularly Internal Combustion Engines — the common working substance is air. In addition to air, these different engines use different fuels and in different quantities. To avoid the effect of different properties of fuels used, and because the quantity of fuel is very very small as compared to air, (generally air to fuel ratio is 12 to 16) we generally neglect the fuel quantity and hence the effect of fuel, and the study is theoretical only. Therefore, the working substance in the study is assumed to be air only. The cycle of working of the engine is called as theoretical or standard cycle and as the working substance is air, the cycle is called *Air Standard Cycle*.

We will study in this chapter four different air standard cycles viz.

(i) Otto Cycle
(ii) Diesel Cycle
(iii) Combination of Otto and Diesel Cycles
(iv) Stirling Cycle

23.2 ASSUMPTIONS MADE FOR AIR STANDARD CYCLES

The analysis of all air standard cycles is based upon the following assumptions:

(1) The working substance used in the engine is air.
(2) Air is assumed to behave as an ideal gas and therefore obeys all gas laws.
(3) The physical properties of air in the cylinder remain unchanged for the pressures and temperatures encountered. in practice i.e specific heats (C_p and C_V) do not vary with the temperature. Also specific heat ratio $\gamma = \dfrac{C_P}{C_V}$ is constant.
(4) Compression and expansion process are isentropic (reversible and adiabatic).
(5) No chemical reaction takes place in the cylinder. Heat is supplied or rejected by bringing a hot body or a cold body in contact with the cylinder cover at the appropriate points during the cycle.
(6) The cycle is considered or assumed to be closed cycle, with the same air always remaining in the cylinder to repeat the cycle.

Because of many simplifying assumptions, it is clear that the air standard cycle approximations do not closely represent the conditions within the cylinder of an actual engine. Although the quantities calculated from this approximations are considerably in error, the trends shown are usually correct, so that the general effect on the efficiency of such variables as inlet pressure or compression ratio may be calculated. Because of the simplicity of air standard cycle calculations, it is often used to obtain rough answers to complex engine problems.

23.3 OTTO – CYCLE: (CONSTANT VOLUME CYCLE)

To avoid high values of pressures and compression ratios, a practical cycle was introduced by a German Scientist Dr. A .N. Otto in 1876 and it was successfully applied for the working of petrol and gas engines. This cycle consists of two isentropic processes and two constant volume processes as shown in Fig. 23.1.

The sequences of operations is as follows: air as the working substance in the cylinder is at a pressure P_1, volume V_1 and temperature T_1 and is shown by a point 1 on both *P–V* and *T–S* diagram in Fig. 23.1 (a) and (b).

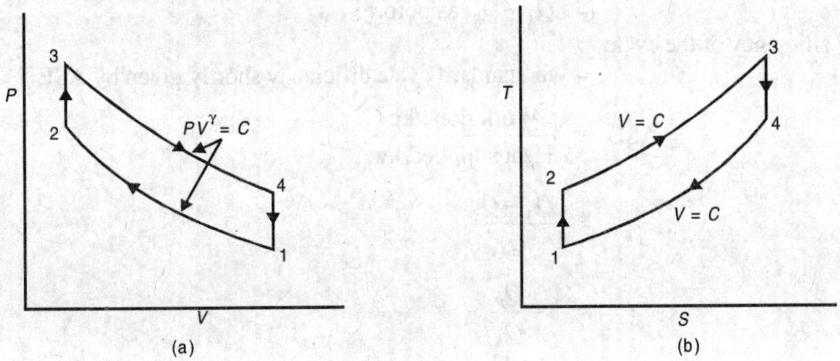

Fig. 23.1

Process 1–2. Air is compressed isentropically to 2. Ratio V_1 to V_2 is called compression ratio donated by r_k. Cover is adiabatic. At the end of compression, diabatic cover replaces adiabatic cover.

Process 2–3. At the end of compression, heat is added at constant volume by bringing a hot body near the cylinder diabatic cover. Heat is added at constant volume and hence pressure and temperature increase. Because the heat is added at constant volume, the cycle is called as *Constant Volume Cycle*.

Process 3–4. After heat is added, air expands isentropically to state 4. Here the expansion ratio is given by

$r_e = \frac{V_4}{V_3}$. For Otto cycle $r_k = r_e$. Adiabatic cover is kept during expansion.

Process 4–1. At the end of isentropic expansion, adiabatic cover is replaced by diabatic cover and the cold body is brought near the cover. Heat is rejected at constant volume thus decreasing the pressure and temperature to initial values.

Thus the cycle is completed.

(a) Thermal Analysis of Cycle:

$$\text{Heat added/kg} = \text{heat added during 2–3}$$

$$Q_A = C_V(T_3 - T_2) \text{ kJ/kg}$$

$$\text{Heat rejected/kg} = \text{heat rejected during 4–1}$$

$$Q_R = C_V(T_4 - T_1) \text{ kJ/kg}$$

Because $Q = 0$ during 1–2 and 3–4

$$\text{Work done/kg} = Q_A - Q_R$$

$$= C_V(T_3 - T_2) - C_V(T_4 - T_1)$$

or Work done/kg = Area of $P–V$ diagram

$$= (\text{Area under } 3–4) - (\text{Area under } 2–1)$$

$$= \frac{p_3v_3 - p_4v_4}{\gamma-1} - \frac{p_2v_2 - p_1v_1}{\gamma-1}$$

We can replace pv product by RT for unit mass.

$$\therefore \quad \text{Work done/kg} = \frac{R}{\gamma-1}[T_3 - T_4 - T_2 + T_1]$$

$$= C_V\left[(T_3 - T_2) - (T_4 + T_1)\right]$$

$$= C_V(T_3 - T_2) - (T_4 + T_1)$$

$$= Q_A - Q_R \text{ as before.}$$

∴ Thermal efficiency of the cycle

= air standard cycle efficiency shortly given by ASE

$$\text{ASE} = \frac{\text{Work done/kg}}{\text{Heat supplied/kg}}$$

$$= \frac{Q_A - Q_R}{Q_A}$$

$$= 1 - \frac{Q_R}{Q_A}$$

$$= 1 - \frac{C_V(T_4 - T_1)}{C_V(T_3 - T_2)}$$

$$\mathbf{ASE} = \mathbf{1 - \frac{T_4 - T_1}{T_3 - T_2}} \qquad ...(1)$$

This is the efficiency given in terms of temperatures.

For compression process, we have

$$\frac{\text{Higher temperature}}{\text{Lower temperature}} = (\text{Compression ratio})^{\gamma-1}$$

$$\therefore \quad \frac{T_2}{T_1} = r_k^{\gamma-1}$$

$$\therefore \quad T_2 = T_1 r_k^{\gamma-1}$$

Similarly, for expansion process 3–4

We have

$$\frac{T_3}{T_4} = r_e^{\gamma-1} = r_k^{\gamma-1}$$

$$T_3 = T_4 \cdot r_k^{\gamma-1}$$

Substituting these values of T_2 and T_3 in Eq. (1) we get

$$\text{ASE} = 1 - \frac{T_4 - T_1}{T_4 \cdot r_k^{\gamma-1} - T_1 \cdot r_k^{\gamma-1}}$$

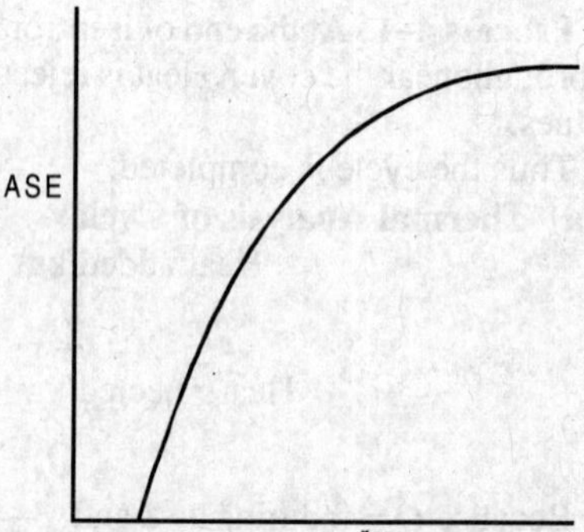

Fig. 23.2

$$= 1-\frac{T_4-T_1}{r_k^{\gamma-1}(T_4-T_1)}$$

$$\mathbf{ASE} = \mathbf{1-\frac{1}{r_k^{\gamma-1}}} \qquad ...(2)$$

We note from these equations that the air-standard efficiency of Otto–Cycle depends upon the compression ratio only and it increases with the increase in compression ratio. The variation of ASE with compression ratio r_k is shown in Fig. 23.2. For $r_k = 1$, efficiency is zero.

The compression ratio used in actual engine working on this cycle lies between 5 and 9 and depends upon the type of the fuel used.

Example. *Show that the compression ratio for the maximum work to be done per kg of air in an Otto cycle is given by*

$$r_k = \sqrt{\left(\frac{T_3}{T_1}\right)^{1/\gamma-1}}$$

where T_3 and T_1 are the upper and lower limits of absolute temperatures in the cycle.

Solution

$$\text{We have, Work done/kg} = Q_A - Q_R$$

$$= C_V\left[(T_3-T_2)-C_V(T_4-T_1)\right]$$

$$= C_V(T_3-T_2-T_4+T_1)$$

We get $T_2 = T_1 r_k^{\gamma-1}$ and $T_3 = T_4 r_k^{\gamma-1}$

$$\therefore \quad T_4 = \frac{T_3}{r_k^{\gamma-1}}$$

$$\therefore \quad \text{Work done/kg} = C_V\left(T_3 - T_1 r_k^{\gamma-1} - \frac{T_3}{r_k^{\gamma-1}} + T_1\right)$$

$\therefore$ Work done is a function of r_k only as T_3 and T_1 are constants

$\therefore$ Work is maximum when $\frac{dw}{dr_k} = 0$

$$\therefore \quad 0 = 0-(\gamma-1)T_1 \cdot r_k^{\gamma-1} - 1 + T_3 \frac{\gamma-1}{T_k^{(\gamma-1)^2}} \cdot r_k^{\gamma-2}$$

$$= -T_1(\gamma-1)+T_3 \cdot \frac{\gamma-1}{T_k^{(\gamma-1)^2}}$$

$$\therefore \quad T_1 = T_3 . \frac{1}{r_k^{(\gamma-1)^2}}$$

$$\text{or} \quad r_k^{(\gamma-1)^2} = \frac{T_3}{T_1}$$

$$\mathbf{r_k} = \sqrt{\left(\frac{\mathbf{T_3}}{\mathbf{T_1}}\right)^{1/\gamma-1}}$$

$$= \left(\frac{T_3}{T_1}\right)^{\frac{1}{2(\gamma-1)}}$$

$$= \left(\frac{T_3}{T_1}\right)^{\frac{1}{2\times0.4}}$$

$$= \left(\frac{T_3}{T_1}\right)^{1.25}$$

$$\therefore \quad \text{Max. Eff.} = 1-\frac{1}{r_k^{\gamma-1}}$$

$$= 1-\frac{1}{\sqrt{\frac{T_3}{T_1}}}$$

$$\boldsymbol{\eta_{\max}} = \mathbf{1}-\frac{\mathbf{1}}{\sqrt{\frac{\mathbf{T_1}}{\mathbf{T_3}}}} = \mathbf{1}-\sqrt{\frac{\mathbf{T_3}}{\mathbf{T_1}}}$$

Example. *An engine working on Otto–cycle in which the salient points 1, 2, 3, 4 has upper and lower temperature limits T_3 and T_1. If the maximum work per kg air is to be done, then show that the intermediate temperature is given by*

$$T_2 = T_4 = \sqrt{T_1 T_3}$$

Solution

$$\text{Work done/kg} = Q_A - Q_R$$

$$= C_V(T_3-T_2)-C_V(T_4-T_1)$$

$$= C_V(T_3-T_2-T_4+T_1)$$

As $\quad r_k = r_e$ then $\dfrac{T_2}{T_1} = \dfrac{T_3}{T_4}$

or $\quad T_4 = \dfrac{T_1 T_3}{T_2}$

Substituting this value of T_4 in the equation for work done, we get

$$W = C_V\left[T_3 - T_2 - \frac{T_1T_3}{T_2} + T_1\right]$$

Here T_3 and T_1 are the upper and lower limits and are constants.

∴ Work done is a function of T_2 only.

∴ Work is maximum when $\frac{dw}{dT_2} = 0$

$$\therefore \quad \frac{dw}{dt_2} = 0 = 0 - 1 + \frac{T_1T_3}{T_2^2} + 0$$

$$= -1 + \frac{T_1T_3}{T_2^2}$$

$$\therefore \quad T_1T_3 = T_2^2$$

$$\mathbf{T_2} = \sqrt{\mathbf{T_1T_3}}$$

Similarly $$T_4 = \frac{T_1T_3}{T_2} = \frac{T_1T_3}{\sqrt{T_1T_3}}$$

$$\mathbf{T_4} = \sqrt{\mathbf{T_1T_3}}$$

$$\therefore \quad \mathbf{T_2} = \mathbf{T_4} = \sqrt{\mathbf{T_1T_3}}$$

(b) Mean Effective Pressure–MEP

Mean effective pressure is defined as that constant pressure, acting on the piston throughout the stroke, which will produce the work equal to that of the cycle.

If d and L are the diameter and stroke of the piston, then the work produced by the mean effective pressure is given by

$$= P_m \times A \times L = P_m \times \text{stroke volume}$$

$$= P_m \times V_s$$

This is equal to the work of the cycle.

$$\therefore \quad P_m = \text{Mean effective pressure}$$

$$= \frac{\text{Work done during cycle}}{\text{Stroke volume}}$$

Now stroke volume $= V_s$ from P–V diagram

$$= V_1 - V_2$$

$$= V_1\left[1 - \frac{V_2}{V_1}\right]$$

$$= V_1\left[1 - \frac{1}{r_k}\right]$$

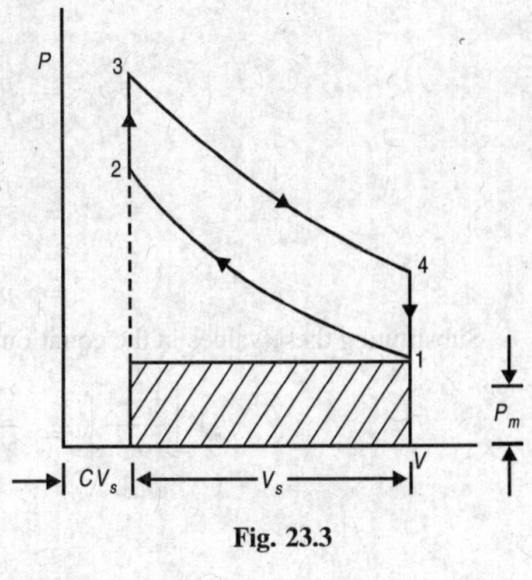

Fig. 23.3

$$= V_1\left[\frac{r_k - 1}{r_k}\right]$$

Also work done during cycle

$$= \text{Area of } P\text{–}V \text{ diagram 1–2–3–4–1}$$
$$= \text{Area under } (3-4) - \text{Area under } (2-1)$$
$$= \frac{P_3V_3 - P_4V_4}{\gamma - 1} - \frac{P_2V_2 - P_1V_1}{\gamma - 1}$$

We will write all pressures in terms of p_1 and volumes in terms of V_1
First consider volumes,

$$\therefore \quad V_1 = V_4 \text{ and } V_2 = V_3$$

$$r_k = \frac{V_1}{V_2} \quad \therefore \; V_2 = V_3 = \frac{V_1}{r_k}$$

Now consider pressures

For process 1 – 2 $P_1V_1^{\gamma} = P_2V_2^{\gamma}$

$$\therefore \quad P_2 = P_1 \times \frac{V_1^{\gamma}}{V_2^{\gamma}} = P_1 \times r_k^{\gamma}$$

Then for process 3 – 4

$\frac{P_3}{P_2}$ is the explosion ratio say, denoted by $\propto$,

$$\therefore \quad \frac{P_3}{P_2} = \alpha \text{ and } P_3 = P_2 \propto$$
$$= P_1 \cdot r_k^{\gamma} \cdot \propto$$

For process 3 – 4 $p_3v_3^{\gamma} = p_4v_4^{\gamma}$

$$\text{or} \quad p_4 = p_3 \times \left(\frac{V_3}{V_4}\right)^{\gamma} = p_3 \times \left(\frac{V_2}{V_1}\right)^{\gamma} = p_3 \times \frac{1}{r_k^{\gamma}}$$

$$\therefore \quad = \frac{p_1 \times r_k^{\gamma}.\alpha}{r_k^{\gamma}}$$

$$= p_1\alpha$$

Substituting these values in the equation for $p_{m'}$ we get

$$p_mV_1\left[\frac{r_k - 1}{r_k}\right] = \frac{1}{\gamma - 1}\left[p_1\alpha r_k^{\gamma}\frac{V_1}{r_k} p_1\alpha V_1 - p_1\frac{V_1}{r_k}r_k^{\gamma} + p_1V_1\right]$$

$$p_m = \frac{r_k \cdot p_1}{(r_k-1)(\gamma-1)}\left[\alpha.r_k^{\gamma-1} - \alpha - r_k^{\gamma-1} + 1\right]$$

$$= p_1 \cdot \frac{r_k}{(r_k-1)(\gamma-1)} \cdot \left[r_k^{\gamma-1}(\alpha-1) - (\alpha-1)\right]$$

$$= p_1 \cdot \frac{r_k}{(r_k-1)(\gamma-1)} \cdot \left[(\alpha-1)\left(r_k^{\gamma-1}-1\right)\right]$$

Alternative Method

This method considers temperatures instead of pressures and volumes

$$\text{Work done/kg} = C_v(T_3 - T_2) - C_v(T_4 - T_1)$$

We have, $T_2 = T_1 - r_k^{\gamma-1}, T_3 = \alpha T_2 = \alpha.T_1 r_k^{\gamma-1}$

$$T_4 = \frac{T_3}{r_k^{\gamma-1}} = \frac{\alpha.T_1.r_k^{\gamma-1}}{r_k^{\gamma-1}} = \alpha T_1$$

$$\therefore \quad p_m \times v_1 . \frac{r_k - 1}{r_k} = \frac{R}{\gamma-1}\left[\alpha.T_1 r_k^{\gamma-1} - T_1 r_k^{\gamma-1} - \alpha T_1 + T_1\right]$$

$$V_1 = \text{Volume of kg} = \frac{RT_1}{p_1}$$

$$\therefore \quad p_m \times \frac{RT_1}{p_1} . \frac{r_k - 1}{r_k} = \frac{RT_1}{\gamma-1}\left[r_k^{\gamma-1}(\alpha-1) - (\alpha-1)\right]$$

$$\mathbf{p_m} = \frac{\mathbf{p_1 \cdot r_k}}{(\gamma-1)(\mathbf{r_k}-1)}\left[(\gamma-1)\left(\mathbf{r_k}^{\gamma-1}-1\right)\right]$$

Same as before

$$\therefore \quad \mathbf{p_m} = \mathbf{p_1} \cdot \frac{\mathbf{r_k}\left(\mathbf{r_k}^{\gamma-1}-1\right)}{(\mathbf{r_k}-1)} \times \frac{(\alpha-1)}{(\gamma-1)}$$

If area of p–v diagram and length of diagram are given, then mean effective pressure can be found by

$$p_m = \frac{\text{Area of diagram (mm)}^2}{\text{Length of diagram (mm)}} \times \text{Pressure scale}$$

The unit for pressure scale is bar/mm

So that unit of p_m is bar.

If the scale is N/m²/mm, then p_m-N/m² and so on.

23.4 DIESEL CYCLE: (CONSTANT PRESSURE CYCLE)

This cycle was introduced by Rudolph Diesel in 1893 and it is successfully used in all Diesel engines or oil engines.

This cycle consists of two isentropics, one constant pressure and one constant volume processes.

The cycle is shown on *P–V* and *T–S* diagrams as in Fig. 23.4.

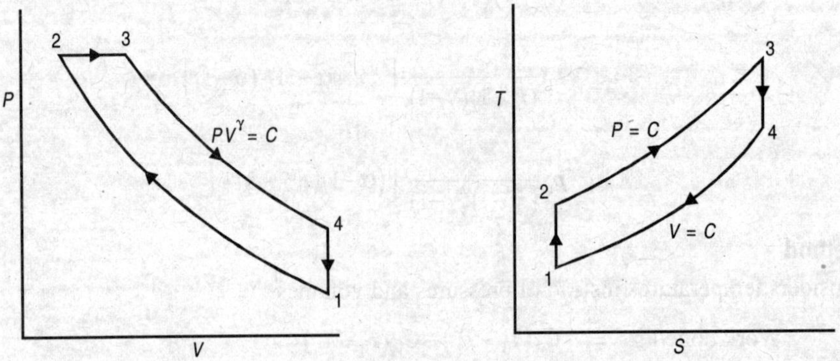

Fig. 23.4

The sequence of operations is as follows:

As before, air as the working substance in the cylinder is at pressure P_1, volume V_1 and temperature T_1 and is shown by state point 1 on both *P–V* and *T–S* diagram of Fig. 23.4 (a) and 23.4 (b).

1–2 : Air is compressed isentropically to 2. Ratio V_1 to V_2 is called compression ratio denoted by r_k. Cylinder cover is adiabatic.

2–3: At the end of compression, diabetic cover is placed and heat is added at constant pressure by bringing a hot body near the cylinder cover. Heat is added for a part of the stroke. At the point 3, the heat addition is stopped or cut-off. The ratio of the volume of cut-off point to the initial volume before heat addition is called the cut–off ratio and is donated by r_c. Therefore, $r_c = \frac{V_3}{V_2}$. Because the heat is added at constant pressure, the cycle is also called as constant pressure cycle.

3–4: After the heat is added, air expands isentropically to state 4. Here the expansion ratio is given by $r_e = \frac{V_4}{V_3} = \frac{V_1}{V_3}$.

Here the cover is adiabatic.

$$\therefore \quad r_e = \frac{V_1}{V_3} = \frac{V_1}{V_2} \times \frac{V_2}{V_3} = r_k / r_e$$

$$\text{or} \quad \mathbf{r_k = r_e \cdot r_c}$$

4 – 1: Adiabatic cover is replaced by diabatic cover at the end of isentropic expansion, the cold body is brought near the cover. Heat is rejected at constant volume thus decreasing the pressure and temperature to initial values.

Thus the cycle is completed.

(a) Thermal analysis of cycle:

$$\text{Heat added kg} = Q_A = C_p (T_3 - T_2) \text{ kJ/kg}$$

$$\text{Heat rejected kg} = Q_R = C_v (T_4 - T_1) \text{ kJ/kg}$$

$$\therefore \quad \text{Work done/kg} = W = Q_A - Q_R$$

$$= C_p (T_3 - T_2) - C_V (T_4 - T_1)$$

∴ Air standard efficiency

$$\text{ASE} = \frac{\text{Work done/kg}}{\text{Heat added/kg}}$$

$$= \frac{Q_A - Q_R}{Q_A} = 1 - \frac{Q_R}{Q_A}$$

$$= 1 - \frac{C_p(T_3 - T_2) - C_V(T_4 - T_1)}{C_P(T_3 - T_2)}$$

$$= 1 - \frac{C_V(T_4 - T_1)}{C_P(T_3 - T_2)}$$

$$= 1 - \frac{(T_4 - T_1)}{\gamma(T_3 - T_2)}$$

Now we have $\quad \frac{T_2}{T_1} = r_k^{\gamma-1} \quad \therefore T_2 = T_1.r_k^{\gamma-1}$

$$\frac{T_3}{T_2} = \frac{V_3}{V_2} = r_c \text{ for constant pressure process.}$$

∴ $\quad T_3 = T_2 \cdot r_c = T_1 \cdot r_k^{\gamma-1} \cdot r_c$

Again $\quad \frac{T_3}{T_4} = r_e^{\gamma-1} = \left(\frac{r_k}{r_c}\right)^{\gamma-1}$

∴ $\quad T_4 = T_3 \times \left(\frac{r_k}{r_c}\right)^{\gamma-1}$

$$T_4 = T_1 \cdot r_k^{\gamma-1} \cdot r_c \times \frac{r_c^{\gamma-1}}{r_k^{\gamma-1}}$$

$$= T_1 \cdot r_c^{\gamma-1}$$

Substituting these values of temperatures in the equation for ASE we get

$$\text{ASE} = 1 - \frac{T_1 r_c^{\gamma-1} - T_1}{\gamma\left[T_1 \cdot r_k^{\gamma-1} \cdot r_c - T_1 \cdot r_k^{\gamma-1}\right]}$$

$$= 1 - \frac{r_c^{\gamma} - 1}{\gamma \cdot r_k^{\gamma-1}(r_c - 1)}$$

∴ $\quad \mathbf{ASE} = \mathbf{1 - \frac{r_c^{\gamma} - 1}{\gamma \times r_k^{\gamma-1}(r_c - 1)}}$

From this equation we see that the efficiency decreases as cut-off ratio increases.

Again as the compression ratio is equal to that for Otto-cycle, ASE of Otto-cycle is more than ASE of Diesel cycle.

In pratice, the compression ratio of Diesel engines ranges from 12 to 22.

(b) Mean Effective Pressure of Diesel Cycle:

$$\begin{aligned}\text{Work done} &= \text{Area of } P\text{–}V \text{ diagram}\\ &= \text{Area 1–2–2–3–4–1}\\ &= \text{Area under (2–3–4) Area under (2–1)}\\ &= p_2(V_3 - V_2) + \frac{p_3V_3 - p_4V_4}{\gamma - 1} - \frac{p_2V_2 - p_1V_1}{\gamma - 1}\end{aligned}$$

As before, we will write all pressure in terms of P_1 and all volumes in terms of V_1.

First consider volumes

$$V_2 = \frac{V_1}{r_k}, \frac{V_3}{V_2} - r_c$$

$$\therefore \quad V_3 = V_2 r_c = \frac{V_1 \cdot r_c}{r_k}$$

$$V_4 = V_1$$

Again for pressures

$$p_2 = p_1 \cdot r_k^{\gamma} = p_3$$

$$\frac{p_3}{p_4} = r_e^{\gamma} = \left(\frac{r_k}{r_c}\right)^{\gamma}$$

$$\therefore \quad p_4 = \frac{p_3 \cdot r_c^{\gamma}}{r_k^{\gamma}} = \frac{p_1 r_k^{\gamma} \cdot r_c^{\gamma}}{r_k^{\gamma}} = p_1 r_c^{\gamma}$$

Substituting these values we get,

$$p_m \times V_1 \cdot \frac{r_k - 1}{r_k} = \left(\frac{V_1 r_c}{r_k} - \frac{V_1}{r_k}\right) p_1 r_k^{\gamma} + \left(\frac{p_1 r_k^{\gamma} \cdot \frac{V_1 r_c}{r_k} - p_1 r_c^{\gamma} \cdot V_1}{\gamma_k}\right) - \left(\frac{p_1 r_k^{\gamma} \cdot \frac{V_1}{r_k} - p_1 V_1}{\gamma - 1}\right)$$

$$p_m \times \frac{r_k - 1}{r_k} = \left[r_k^{\gamma-1}(r_c - 1) + \frac{r_k^{\gamma-1} . r_c - r_c^{\gamma}}{\gamma - 1} - \frac{r_k^{\gamma-1} - 1}{\gamma - 1}\right] p_1$$

$$= \frac{p_1}{\gamma - 1}\left[r_k^{\gamma-1}(r_c - 1)(\gamma - 1) + \left(r_k^{\gamma-1} - r_c^{\gamma-1}\right) - r_c\left(r_k^{\gamma-1} - 1\right)\right]$$

$$= \frac{p_1 \cdot r_k^{\gamma-1}}{\gamma-1}\left[(r_c-1)(\gamma-1)+(r_c-1)-\left(\frac{r_c^{\gamma}-1}{r_k^{\gamma-1}}\right)\right]$$

$$p_m = \frac{p_1 \cdot r_k^{\gamma-1} \cdot r_k}{(\gamma-1)(r_k-1)}\left[(r_c-1)(\gamma-1+1)-\frac{r_c^{\gamma}-1}{r_k^{\gamma-1}}\right]$$

$$= \frac{p_1 r_k^{\gamma-1}}{(\gamma-1)(r_k-1)}\left[\gamma(r_c-1)-\frac{r_c^{\gamma-1}-1}{r_k^{\gamma-1}}\right]$$

$$\mathbf{p_m = \frac{p_1}{(\gamma-1)(r_k-1)}\left[r_k^{\gamma-1}\times\gamma(r_c-1)-r_k\left(r_c^{\gamma}-1\right)\right]}$$

Alternative Method:

$$\text{Work done/kg} = Q_A - Q_R$$

$$= C_P(T_3-T_2)-C_V(T_4-T_1)$$

Here

$$C_p = \frac{R\gamma}{\gamma-1}, \qquad C_v = \frac{R}{\gamma-1}$$

$$\frac{T_2}{T_1} = r_k^{\gamma-1}$$

$$\therefore T_2 = T_1 \cdot r_k^{\gamma-1}$$

$$\frac{T_3}{T_4} = r_e^{\gamma-1} = \left(\frac{r_k}{r_c}\right)^{\gamma-1}$$

$$\therefore \quad T_4 = \frac{T_3 \times r_c^{\gamma-1}}{r_k^{\gamma-1}} = T_1 \cdot r_c^{\gamma-1}$$

Substituting these values in the equation for work done, we get

$$p_m \cdot V_1 \cdot \frac{r_k-1}{r_k} = p_m \cdot \frac{RT_1}{p_1} \cdot \frac{r_k-1}{r_k}$$

$$= \frac{\gamma R}{\gamma-1}\left[T_1 r_c \cdot r_k^{\gamma-1} - T_1 r_k^{\gamma-1}\right]-\frac{R}{\gamma-1}\left[T_1 r_c^{\gamma}-T_1\right]$$

$$\therefore \quad P_m \cdot \frac{r_k-1}{r_k} \times \frac{1}{P_1} = \frac{\gamma}{\gamma-1}\left[r_k^{\gamma-1}(r_c-1)\right]-\frac{1}{\gamma-1}\left[r_c^{\gamma}-1\right]$$

$$\therefore \quad p_m = \frac{p_1 . r_k}{(r_k-1)(\gamma-1)}\left[\gamma . r_k^{\gamma-1}(r_c-1)-(r_c^{\gamma}-1)\right] \text{ as before}$$

23.5 DUAL COMBUSTION OR SEMI-DIESEL CYCLE

From the expression for the air standard efficiency of the Diesel cycle, it is seen that the efficiency decreases as the value of cut-off ratio (r_c) increase and therefore, it is desirable to keep the value of r_c small as possible to increase the efficiency.

This can be done by supplying the heat partly at constant pressure. This type of the cycle is called Dual Combustion cycle or semi-Diesel cycle. The cycle is represented on *P-V* and *T-S* diagram as in Fig. 23.5.

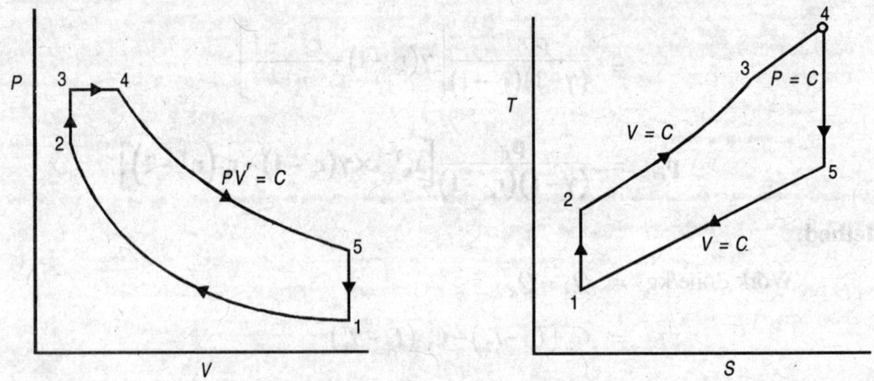

Fig. 23.5

As before the sequential operations for this cycle are

1–2 Isentropic compression
2–3 Constant volume heat addition
3–4 Constant pressure heat addition
4–5 Isentropic expansion
5–1 Constant volume heat rejection

$$Q_A = \text{Heat added/kg}$$

$$= C_V(T_3 - T_2) + C_P(T_4 - T_3)$$

$$Q_R = \text{Heat rejected/kg}$$

$$= C_V(T_5 - T_1)$$

$\therefore$
$$W = \text{Work done/kg} = Q_A - Q_R$$

$$= C_V(T_3 - T_2) + C_P(T_4 - T_3) - C_V(T_5 - T_1)$$

$\therefore$ Air standard efficiency is given by

$$\text{ASE} = \frac{W}{Q_A} = \frac{C_V(T_3 - T_2) + C_P(T_4 - T_3) - C_V(T_5 - T_1)}{C_V(T_3 - T_2) + C_P(T_4 - T_3)}$$

$$= 1 - \frac{C_V(T_5 - T_1)}{(T_3 - T_2) + C_P(T_4 - T_3)}$$

As before, all the temperatures in the above equation are to be expressed in terms of T_1.

Students are to derive the expression of Air Standard Efficiency ASE terms of compression ratio r_k cut-off ratio r_c explosion ratio α and the specific heat ratio γ and get ,

$$\text{ASE} = 1 - \frac{1}{r_k^{\gamma-1}}\left[\frac{\alpha \times r_c^{\gamma} - 1}{(\alpha-1)+\gamma-\alpha(r_c-1)}\right]$$

It is obvious from this equation that the efficiency of the dual-combustion cycle can be increased by increasing the value of explosion ratio (i.e. more heat is added at constant volume) and decreasing the value of r_C (less heat is added during constant pressure process) keeping the total heat added constant .

The Otto and Diesel cycles are special cases of Dual cycle. If $r_c = 1$, then $V_3 = V_4$ and the cycle becomes Otto-cycle. Again if $\alpha = 1$, then $P_2 = P_3$ and the cycle becomes Diesel cycle.

Again applying the same procedures used in Otto-cycle and Diesel cycles, the mean effective pressure of the dual combustion can be calculated (students have to derive this relation as an exercise).

$$p_m = \frac{p_1}{(r_k-1)(\gamma-1)}\left[r_k^{\gamma} \cdot \alpha\gamma(r_c-1) + r_k^{\gamma} - 1\right]$$

23.6 THE REGENERATOR

The regenrator is a contrivance which alternately stores and rejects heat in a manner which, theoretically, is thermodynamically reversible. It was made use of in the working of the Stirling and Ericson hot air engines. By the incorporation of a regenerator the cycles used by both of these engines were made to conform to thermodynamic reversibility; their ideal efficiencies were, therefore, the same as the Carnot cycle efficiency when working between the same temperature limits.

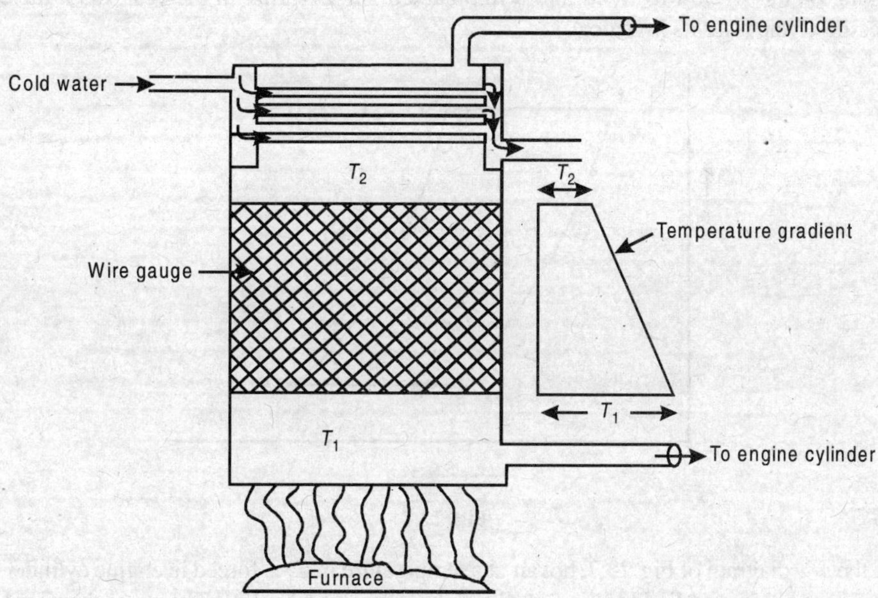

Fig. 23.6

A view of one form of a regenerator is shown in Fig. 23.6. The air in the space at the lower end is kept at a higher temperature T_1 by the furnace underneath it. The low temperature air the upper space is kept cool at a temperature of T_2 by a supply of cold water passing through tube in contact with it. The space between the two end is filled with wire gauge for storing the heat. The temperature of the gauge varies uniformly from T_1 at the base to T_2 at the top, as shown by the temperature diagram to the right of the figure.

The hot air in the base in contact with the engine cylinder during a portion of the cycle. The air is forced upwards through the gauze, by means of a pump, and is thus cooled at constant volume to a temperature T_2. In passing through the gauze, it is cooled at a uniform rate, the temperature gradient being approximately a straight line. This process can be regarded as thermodynamically reversible as the heat is assumed to flow from the air to the gauze when at the same temperature. The heat is thus stored in the layers of wire gauze; the increase in temperature of the latter due to this storage is extermely small on account of its large bulk.

During another portion of the cycle the cool air in the upper space is in contact with engine cylinder. The cool air is then forced downwards through the wire gauze; it thus receives heat from the gauze at constant volume by a graduated method of heating which makes the process thermodynamically reversible. The air is now in the lower space at a temperature T_1 and has absorbed from the gauze the whole of the heat stored during its previous upward flow.

It will be seen from this that the regenerator is able to store and reject heat alternately, in a reversible manner, owing to its uniformly varying temperature gradient. In order to approximate to the assumed processes the regenerator must be very large compared with the volume of air used in the engine cylinder. The pump forcing the air through the regenerator is driven off the engine crank-shaft.

23.6 THE ERICSSON CYCLE

This cycle consists of two isothermal processes and two constant pressure processes, as shown in Fig. 23.7. It is made to be thermodynamically reversible by the action of a regenerator, during the two constant pressurers. A hot-air engine working on this cycle was fitted into a ship named *The Ericsson* in the year 1853; the cycle has been termed the Ericsson Cycle after its inventor.

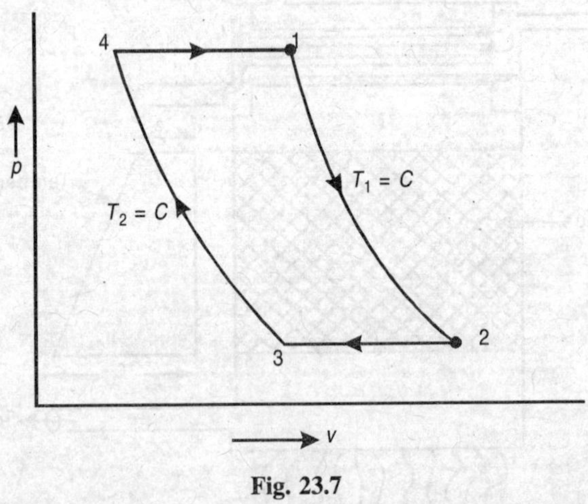

Fig. 23.7

Referring to the p–v diagram of Fig.23.7, hot air at a temperature of T_1 is forced in engine cylinder and expanded isothermally, the heat being supplied by the furnace; this represented by 1–2. The air is then cooled at constant pressure to T_2 by passing it through a regenerator. The process is thus made thermodynamically reversible by this graduated method of cooling. By this method the air is brought to the condition 3. It is now compressed isothermally by a seperate pump, the heat being rejected to a cold water supply; this operation is represented by 3–4. The air is next heated at constant pressure to T_1 by being passed through the regenerator, in the reversed direction, whilst the pressure is maintained constant; this is represented by 4–1. The cycle is thus completed.

It should be noticed that the whole of the heat given to regenerator during the constant pressure operation 2–3 is abstracted from it during the constant pressure operation 4–1. Hence there is no interchange of heat from an external source during these two operations.

Heat supplied from an external source = Heat supplied during isothermal 1–2

$$= RT_1 \log_e r$$

Heat rejected to an external body = Heat rejected during isothermal 3–4

$$= RT_2 \log_e r$$

the volume ratio r being the same for both processes

$$\therefore \quad \text{work done} = Q_A - Q_R$$

$$= \text{Heat added} - \text{Heat rejected}$$

$$= CT_1 \log_e r - CT_2 \log_e r$$

$$= C \cdot \log_e r\,[T_1 - T_2]$$

$\therefore$ Efficiency of the cycle

$$= \frac{\text{Work done}}{\text{Heat supplied}}$$

$$= \frac{C \cdot \log_e r[T_1 - T_2]}{T_1 C \log_e r}$$

$$= \frac{T_1 - T_2}{T_1}$$

which is the same as the Carnot efficiency.

The hot air engine of the Ericsson had four cylinders each of 4.2 m diameter (14 ft); the storke was 1.8 m (6 ft). This engine developed 300 IP (indicated power) and its speed was 9 rpm, the temperature range being from 212°C (414°F) to 50°C (122°F). Its mean effective pressure MEP was 0.1406 × 10^5 N/m^2 (2 psi) and each cylinder was fitted with compression pump. The heating surfaced consisted of the bottom of the cylinders, and although the total heating surface occupied an area of 65 m^2 (700 sq. ft.) it was found to be insufficient.

23.7 THE ATKINSON CYCLE

This cycle was used in the Atkinson gas engine and is show, in its ideal form, in Fig. 23.8. It consists of two adiabatics, a constant volume process and a constant pressure process. The heat is supplied during the constant volume process 4–1 and rejected during constant pressure process 2–3. There is no interchange of heat during the adiabatics.

The air standard efficiency for this cycle is obtained by imagining the cylinder to contain 1 kg of air at point 3. The air is then compressed isentropically to 4 at which condition the hot body is placed in contact with the cylinder cover. Heat is now taken in at constant volume and the pressure rises; this operation is represented by 4–1. The hot body is then removed and the air allowed to expand adiabatically to point 2. During this process work is done on the piston. At point 2 the cold body is placed at the cylinder cover; the air now contracts at constant pressure to point 3 rejecting heating to the cold body during the process. The cold body is then removed and the cycle is thus completed.

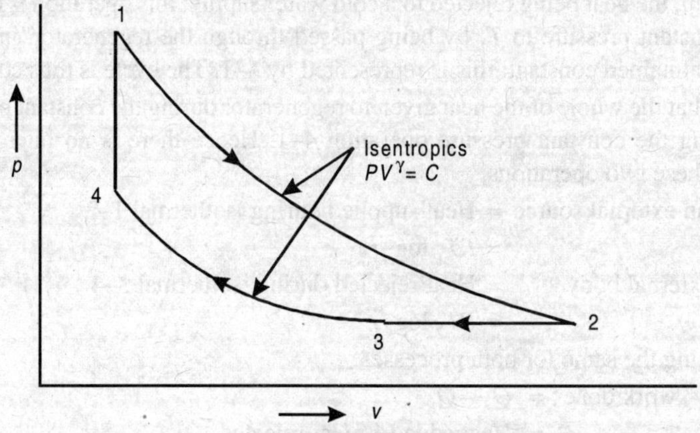

Fig. 23.8

Concept. The cyliner cover is replacable. When heat addition and rejection process take place, the cover is *diabatic* and during compression and expansion the cover will be *adiabatic*.

There are two methods for calculating efficieny of Atkinson Cycle. One method is to find efficiency in terms of pressure ratios and the other method is to find efficiency in terms of volume ratio.

23.7.1 Effiviency in Terms of Pressure Ratios

Assuming clearance volume as unity, then

$$v_1 = v_4 = 1$$

Let r = compression ratio during compresssion process 3–4

$$= \frac{v_3}{v_4} \text{ but } v_4 = v_1 = 1$$

$$\therefore \quad v_3 = r$$

Again α = Pressure ratio during constant volume process 4–1

$$= \frac{p_1}{p_4}$$

$$Q_A = \text{Heat supplied}$$
$$= \text{Heat absorbed during constant volume process}$$
$$= C_v (T_1 - T_4)$$
$$Q_R = \text{Heat rejected during constant pressure process 2–3}$$
$$= C_p (T_2 - T_3)$$
$$\therefore \quad W = \text{Work done}$$
$$= Q_A - Q_R$$
$$= C_v (T_1 - T_4) - C_p (T_2 - T_3)$$

$$\therefore \quad \text{Efficiency} = \frac{\text{Work done}}{\text{Heat supplied}}$$

$$= \frac{C_v(T_1 - T_4) - C_p(T_2 - T_3)}{C_v(T_1 - T_4)}$$

$$= 1 - \frac{C_p}{C_v}\left(\frac{T_2 - T_3}{T_1 - T_4}\right)$$

$$= 1 - \gamma\left(\frac{T_2 - T_3}{T_1 - T_4}\right)$$

As usual, find all temperatures in terms of T_1.

Constant pressure process 2–3

$$\frac{v_2}{v_3} = \frac{T_2}{T_3} \qquad \therefore\ v_3 = v_2 \times \frac{T_3}{T_2} = r \qquad \therefore\ v_2 = r \cdot \frac{T_2}{T_3}$$

From adiabatic or isentropic process 1–2

$$\frac{T_1}{T_2} = \left(\frac{v_2}{v_1}\right)^{\gamma-1} = v_2^{\gamma-1} \quad \text{as } v_1 = 1$$

Substituting this value of v_2, we get

$$\frac{T_1}{T_2} = \left(\frac{v_2}{1}\right)^{\gamma-1}$$

$$= \left(\frac{r \cdot T_2}{T_3}\right)^{\gamma-1}$$

$$\therefore \qquad T_1 = T_2 \cdot \left[\frac{r \cdot T_2}{T_3}\right]^{\gamma-1}$$

$$= T_2^{\gamma}\left[\frac{r}{T_3}\right]^{\gamma-1}.$$

From constant volume process 4–1

$$\frac{T_1}{T_4} = \frac{p_1}{p_4}$$

$$= \alpha$$

$$\therefore \qquad T_4 = \frac{T_1}{\alpha}$$

From 3–4 process

$$\frac{T_4}{T_3} = \left(\frac{v_3}{v_4}\right)^{\gamma-1}$$

$$\therefore \qquad T_4 = T_3 \cdot r^{\gamma-1}$$

$$\therefore \qquad \frac{T_1}{\alpha} = T_3 \cdot r^{\gamma-1}$$

$$\therefore \qquad T_3 = \frac{T_1}{\alpha} \cdot \frac{1}{r^{\gamma-1}}$$

We know that
$$T_1 = T_2^{\gamma}\left[\frac{r}{T_3}\right]^{\gamma-1}$$

$$= T_2^{\gamma}\cdot\left[\frac{r}{\frac{T_1}{\alpha}\cdot\frac{1}{r^{\gamma-1}}}\right]^{\gamma-1}$$

$$= T_2^{\gamma}\left[\frac{r\cdot r^{\gamma}}{T_1}\right]^{\gamma-1}$$

$$\therefore \quad T_2^{\gamma} = T_2^{\gamma}\left(\frac{T_1}{\alpha r^{\gamma}}\right)^{\gamma-1}\cdot T_1$$

$$= \frac{T_1^{\gamma}}{(\alpha\cdot r^{\gamma})^{\gamma-1}}$$

$$\therefore \quad T_2 = \frac{T_1^{\gamma/\gamma}}{(\alpha\cdot r^{\gamma})^{\gamma-1/\gamma}} - \frac{T_1}{\alpha^{\gamma-1/\gamma}\cdot r^{\gamma-1}}$$

The values of T_2, T_3 and T_4 are all found in terms of T_1

$$\therefore \quad \text{Ideal efficiency} = 1 - \frac{\gamma(T_2 - T_3)}{T_1 - T_4}$$

$$= 1 - \frac{\gamma\left(\frac{T_1}{\alpha^{\gamma-1/\gamma}\cdot r^{\gamma-1}} - \frac{T_1}{\alpha r^{\gamma-1}}\right)}{\left(T_1 - \frac{T_1}{\alpha}\right)}$$

$$= 1 - \frac{\gamma\left(\frac{1}{\alpha^{\gamma-1/\gamma}\cdot r^{\gamma-1}} - \frac{1}{\alpha\cdot r^{\gamma-1}}\right)}{1 - \frac{T_1}{\alpha}}$$

$$= 1 - r^{\gamma-1}\,\frac{\gamma\alpha}{(\alpha-1)}\left[\frac{1}{\alpha^{\gamma-1/\gamma}} - \frac{1}{\alpha}\right]$$

$$= 1 - \frac{r}{r^{\gamma-1}(\alpha-1)}\left[\frac{1}{\alpha^{\gamma}} - 1\right]$$

$$= 1 - \frac{r\,(1-\gamma^{\gamma})}{r^{\gamma-1}\,(\alpha-1)\alpha^{\gamma}}$$

$$= 1 - \frac{\gamma}{r^{\gamma-1}\,(\alpha-1)}\left[\frac{\alpha}{\alpha^{\gamma-1/\gamma}} - 1\right]$$

$$\boldsymbol{\eta = 1 - \frac{\gamma}{r^{\gamma-1}\cdot(\alpha-1)}\left[\frac{\alpha}{\alpha^{\gamma/\gamma}} - 1\right]}$$

Now $\left[\dfrac{\alpha}{\alpha^{\gamma-1/\gamma}} - 1\right] = \alpha\cdot\alpha^{-\gamma+1/\gamma} - 1$

$$= \alpha^{-\gamma+1/\gamma} - 1$$

$$= \alpha^{\gamma-\gamma+1/\gamma} - 1$$

$$= \alpha^{1/\gamma} - 1$$

$\therefore$ $$\eta = 1 - \frac{\gamma}{r^{\gamma-1}\cdot(\alpha-1)} \times \frac{\alpha^{1/\gamma} - 1}{1}$$

$$\boldsymbol{\eta = 1 - \frac{\gamma}{r^{\gamma-1}}\cdot\frac{\alpha^{1/\gamma} - 1}{\alpha - 1}}$$

23.7.2 Efficiency of Atkinson cycle in Terms of Volume Ratios

Efficiency is given by

$$\eta = 1 - \frac{\gamma\,(T_2 - T_3)}{T_1 - T_4}$$

Consider the P–V diagram as shown in Fig. 23.9.

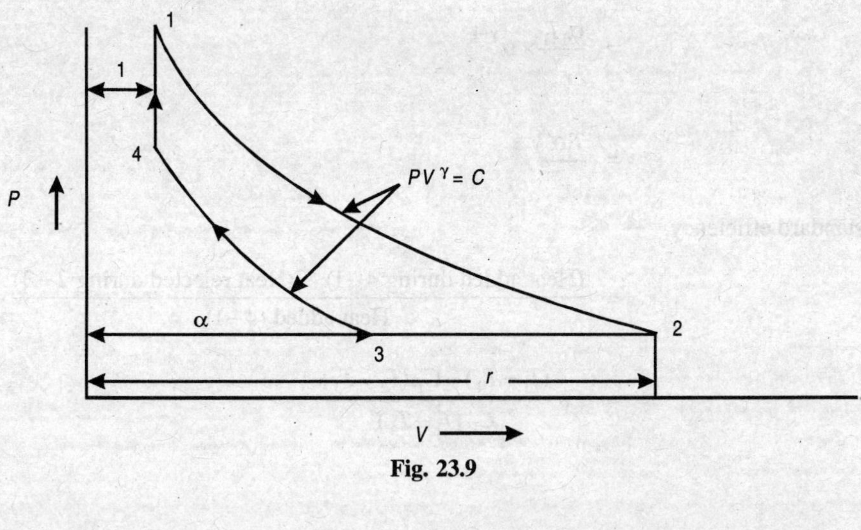

Fig. 23.9

Let α = Volume at the beginning of adiabatic compression (3)

$= V_3$

r = Volume at the end of adiabatic expansion (2)

$= v_2$

$v_1 = v_4 = 1$

Obtain all temperatures in terms of T_1.

Adiabatic expansion 1–2

$$\frac{T_1}{T_2} = \left(\frac{v_2}{v_1}\right) = v_2^{\gamma-1} \text{ as } v = 1$$

$$\therefore \quad T_2 = \frac{T_1}{r^{\gamma-1}} \quad \text{... (1)}$$

Constant pressure process 2–3

$$\frac{T_2}{T_3} = \frac{v_2}{v_3} = \frac{r}{\alpha}$$

$$\therefore \quad T_3 = \frac{\alpha \cdot T_2}{r} \quad \text{... (2)}$$

$$= \frac{\alpha}{r} \cdot \frac{T_1}{r^{\gamma-1}}$$

$$= \frac{\alpha T_1}{r^{\gamma}} \quad \text{... (3)}$$

Adiabatic compression process 3–4

$$\frac{T_4}{T_3} = \alpha^{\gamma-1}$$

$$\therefore \quad T_4 = T_3 \times \alpha^{\gamma-1}$$

$$= \frac{\alpha_1 T_1}{r^{\gamma}} \times \alpha^{\gamma-1}$$

$$= \frac{T_1 \alpha^{\gamma}}{r^{\gamma}} \quad \text{... (4)}$$

$\therefore$ Air standard efficiency

$$= \frac{(\text{Heat added during } 4-1) - (\text{Heat rejected during } 2-3)}{\text{Heat added } (4-1)}$$

$$= C_V \frac{(T_1 - T_4) - C_p (T_2 - T_3)}{C_V (T_1 - T_4)}$$

$$= 1 - \frac{\gamma(T_2 - T_3)}{T_1 - T_4} \quad \text{.... (5)}$$

Substituting the values of T_2, T_3 and T_4 in terms of T_1, we get

Air standard efficiency

$$= 1 - \frac{\gamma(T_2 - T_3)}{(T_1 - T_4)}$$

$$= 1 - \frac{\gamma\left[\frac{T_2}{r^{\gamma-1}} - \frac{\alpha T_1}{r^{\gamma}}\right]}{T_1 - \frac{T_1 \cdot \alpha^{\gamma}}{r^{\gamma}}}$$

$$= 1 - \frac{\gamma\left[\frac{1}{r^{\gamma-1}} - \frac{\alpha}{r^{\gamma}}\right]}{1 - \frac{\alpha^{\gamma}}{r^{\gamma}}}$$

$$= 1 - \frac{\gamma \cdot r^{\gamma}\left[\frac{1}{r^{\gamma-1}} - \frac{\alpha}{r^{\gamma}}\right]}{r^{\gamma} - \alpha^{\gamma}}$$

$$= 1 - \frac{\gamma}{r^{\gamma} - \alpha^{\gamma}}\left[r^{\gamma-\gamma+1} - \alpha\right]$$

$$= 1 - \gamma\left[\frac{r - \alpha}{r^{\gamma} - \alpha^{\gamma}}\right]$$

The Atkinson cycle was used in the Atkinson gas engine, but it was slightly modified in practice by the cutting off the toe of the p–v diagram; this was due to the opening of the exhaust valve before the end of the expansion or working stroke.

The Atkinson gas engine proved more efficient than the ordinary gas engine working on the Otto = cycle, but it was not a commercial success on account of the excessive wear of the mechanism due to the toggle joint drive required for a constant pressure process.

The Atkinson cycle is now used by Gas-turbines of the explosive type.

23.8 STIRLING CYCLE (REGENERATIVE CYCLE)

Stirling constructed a hot air engine in 1845 which works on stirling cycle. Theoretically, the efficiency of the Stirling cycle is equal to Carnot cycle. The constructuion of the generator and its function is explained below:

Regenerator: The construction of the regenerator is shown in Fig. 23.10 which alternately stores and rejects heat in a manner which theoretically, is thermodynamically reversible.

The air near the lower end of the filling (important part of the generator) is maintained at a higher temperature

T_2 by supplying the heat from external heat source. The air near the upper end of the wire mesh is maintained at a lower temperature T_1 with the help of external heat sink (circulation of water as shown). The wire mesh filling is used to store the heat. The temperature of the wire gauge varies uniformly from T_1 to T_2 as shown. The heat storing capacity of the wire mesh should be large.

Stirling cycle: The stirling cycle is represented on *P–V* and *T–S* diagrams as shown.

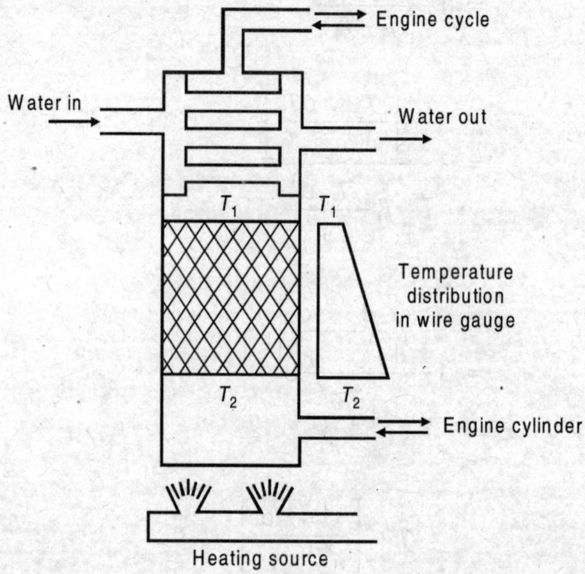

Fig. 23.10

The sequence of operations is given below:

(1) The air is compressed isothermally from 1–2.

(2) The air state 2 is passed into the regenerators from the top at a temperature T_1. The air passing through the matrix gets heated from T_1 to T_2 at constant volume. This heat transfer from the matrix to air is considered thermodynamically reversible because in each element of the matrix heat is transfer due to infinitesimal difference in temperature.

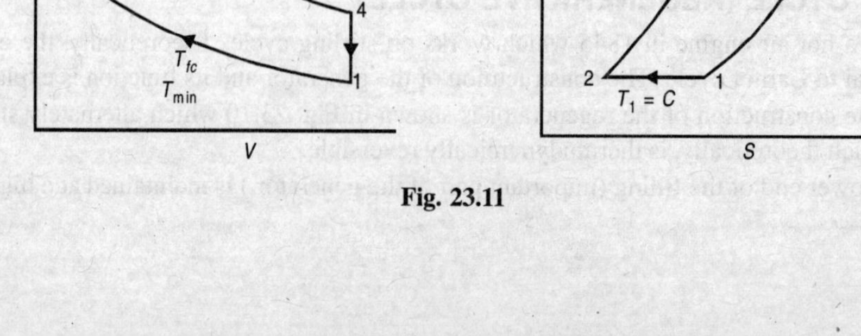

Fig. 23.11

(3) Then the air at '2' expands isothermally into the cylinder until it reaches to the state '4'.

(4) The air coming out of the engine at temperature T_2 (at 4) enters the generator from the bottom and gets cooled while passing through the generator at constant volume and it comes out at a temperatures T_1 and the cycle is repeated. This heat transfer during 4–1 processes is also considered thermodynamically reversible.

The work done by the cycle per kg air and air standard efficiency of the cycle can be calculated as follows:

$$Q_A = \text{Heat supplied during isothermal expansion process } 3-4$$

$$= p_3 v_3 \log_e \frac{V_4}{V_3} = RT_2 \log_e \left(\frac{V_4}{V_3} = \frac{V_1}{V_2}\right)$$

$$= RT_1 \log_e r_k$$

$$Q_A = \text{Heat rejected during isothermal expansion compression } 1-2$$

$$= RT_1 \log_e r_k$$

$$\therefore \quad \text{Work done} = Q_A - Q_R = r \log_e r_k [T_2 - T_1]$$

$$\text{ASE} = \frac{W}{Q_A} = \frac{R(T_2 - T_1)\log_e r_k}{RT_1 \log_e r_k}$$

$$= \frac{T_2 - T_1}{T_1}$$

Thus the efficiency of the Stiriling cycle is equal to that of Carnot cycle when both are working within same temperature limits.

Heat transferred during 2–3 and 4–1 are equal.

In practice, 10–20% heat losses occur in the generator and the efficiency of the generator lies between 90% and 80%.

If the generator efficiency (η_r) is considered then

$$Q_A = RT_2 \log_e r_k + (1-\eta_r) C_V (T_2 - T_1)$$

$$Q_R = RT_1 \log_e r_k + (1-\eta_r) C_V (T_2 - T_1)$$

$$\text{Efficiency} = \frac{R(T_2 - T_1)\log_e r_k}{RT_1 \log_e r_k + (1-\eta_r) C_v (T_2 - T_1)} \quad \text{...(b)}$$

$$= \frac{T_2 - T_1}{T_2} \quad \text{when } \eta_r = 100\%$$

The cycle has not become very successful even though its ideal efficiency is sufficiently high because the regenerator volume has to be very large compared with the engine size.

$$p_1 V_2 = p_2 V_2$$

$$p_2 = p_1 \cdot \frac{V_1}{V_2} = p_1 \cdot r_k$$

$$\frac{p_3}{p_2} = \alpha = \text{pressure ratio}$$

$$= \text{expansion ratio}$$

$$\therefore \quad p_3 = \alpha \cdot p_2 = \alpha \cdot p_1 r_k$$

$$p_m \times V_s = p_m (V_1 - V_2) = p_m V_1 \left[\frac{r_k - 1}{r_k}\right]$$

$$= p_3 V_3 \log_e \frac{V_4}{V_3} - p_2 V_2 \log_e \frac{V_1}{V_2}$$

$$= \log_e r_k \cdot [p_3 - p_2] V_2$$

$$= \log_e r_k \cdot [p_3 - p_2] \frac{V_1}{r_k} = \frac{V_1}{r_k} \log_e r_k [\alpha p_1 r_k - p_1 r_k]$$

$$= V_1 \log_e r_k \cdot p_1 (\alpha - 1)$$

$$p_m = p_1 \cdot \frac{r_k}{r_k - 1} (\alpha - 1) \log_e r_k$$

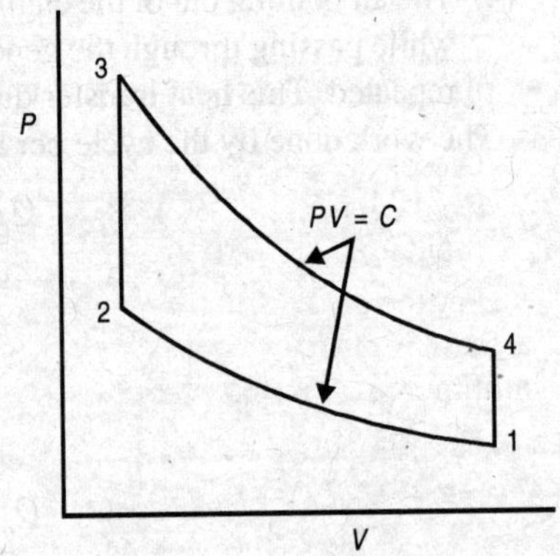

Fig. 23.12

23.9 COMPARISON OF OTTO, DIESEL AND DUAL COMBUSTION CYCLE

The important parameters in cycle analysis are compression ratio, maximum pressure, maximum temperature, heat input, output etc. The cycle can be analysed for the different parameters.

(a) For same compression ratio and same heat input. Figure 23.13 shows the three cycles on *P–V* and *T–S* diagrams. All the cycles start from the same initial state 1 and air is compressed isentropically to the same point 2 as compression ratio is same.

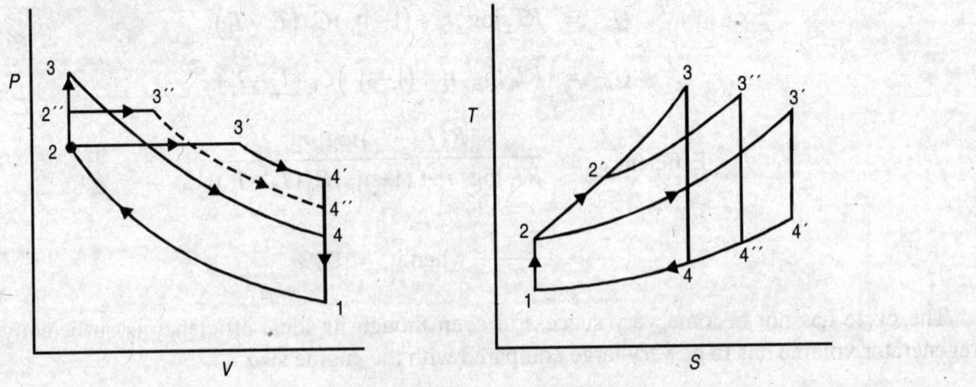

Fig. 23.13

It is seen that adding the heat at constant volume resulting the highest maximum temperature and pressures (Otto–cycle). Adding the heat at constant pressure results in the lowest maximum temperatures (Diesel cycle) while the values for the limited pressure cycle (Dual cycle) lie between Otto and Diesel cycles.

For the same heat input and compression ratio, minimum heat is rejected by the Otto-cycle, maximum heat is rejected by Diesel cycle and the heat rejected is in between Otto and Diesel in case of dual cycle. Consequently, work output and efficiency is maximum in Otto-cycle and the least work done and efficiency in case of Diesel cycle. Dual cycle work and efficiency lie in between the two–Otto and Diesel cycles.

(b) For constant maximum pressure and same heat input :

Otto-cycle 1–2–3–4–1

Diesel cycle 1–2'–3'–4'–1

Dual cycle 1–2"–3"–5"–1

Figure 23.14 compares the three cycles on the basis of same maximum pressure and same heat input.

For the same maximum pressure the points 2, 3', and 3", must be on the same pressure line and the heat input, area 2356 (Otto) = area 2'3'5'6 = area 2". 3" and 5"6. From the diagram, heat rejected its maximum in Otto-cycle, minimum in Diesel cycle and intermediate in dual cycle.

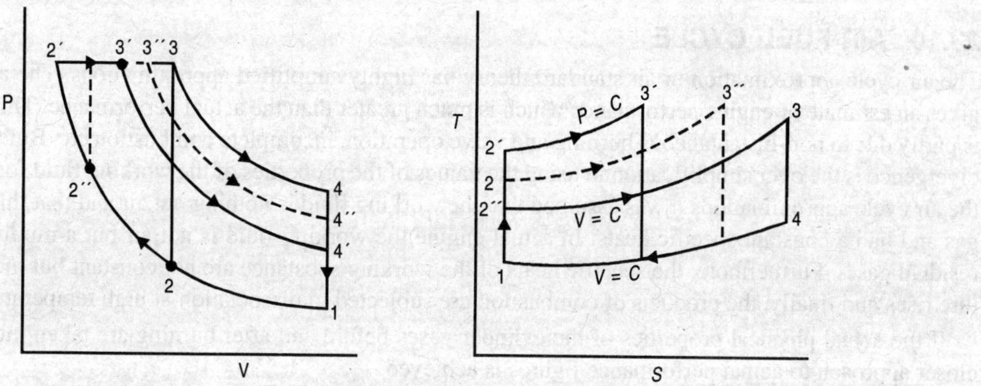

Fig. 23.14

Therefore, for the same heat input, Diesel cycle produces more work, Otto-cycle less work and dual cycle intermediate work. Therefore, for maximum pressure and same heat input. Diesel cycle is the most efficient, dual cycle come the next and otto-cycle has the least efficiency.

(c) For the same maximum pressure and temperature. Figure 23.15 compares Otto -Diesel cycles for the same maximum pressure and maximum temperature. It is clear from the Figures diagram that the heat rejected by both Otto and Diesel cycles is same. But the heat supplied to Diesel cycle is more than that for Otto-cycle. Hence in this case, Diesel cycle is more efficient than Otto cycle. The dual cycle efficiency again lies / falls between the two to others.

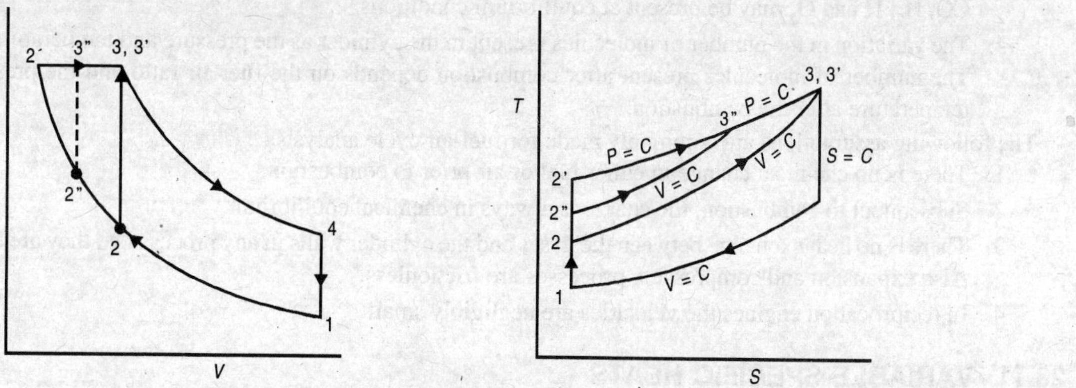

Fig. 23.15

(d) For the same maximum pressure and output. For same work output, point 3' will be on left side of 3. Therefore heat rejected in Diesel is less.

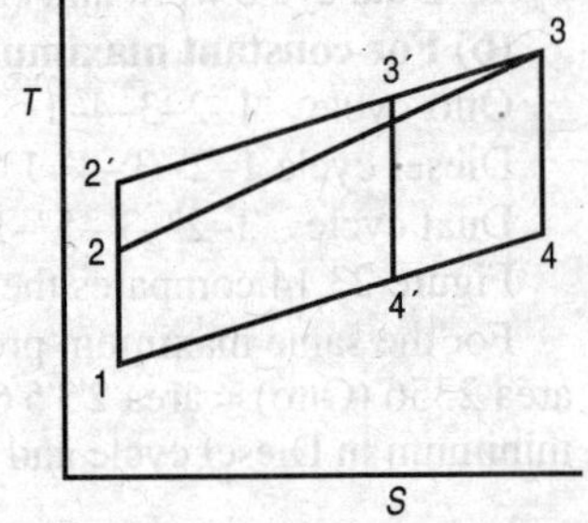

Fig. 23.16

$$\therefore \quad \eta = \frac{\text{Work output}}{\text{Work output + Heat rejected}}$$

= 12342 – Otto

= More in case of Diesel cycle than Otto -cycle 12'3'4'1–Diesel

23.10 AIR-FUEL CYCLE

The air cycle approximation or air standard theory has highly simplified approximations. The air standard theory gives an estimate of engine performance which is much greater than the actual performance. This large divergence is partly due to non-instantaneous burning and valve operation, incomplete combustion etc. But the main reason of divergence is the over simplification in using the values of the properties of the working fluid for cycle analysis. In the air cycle approximations it was assumed that the working fluid is nothing but air and that this air was a perfect gas and had a constant specific heats. In actual engine the working fluid is not air but a mixture of air, fuel and residual gases. Furthermore, the specific heats of the working substance are not constant but increase as temperature rises and finally, the products of combustion are subjected to dissociation at high temperatures.

If the actual physical properties of the cylinder gases before and after burning are taken into account a much closer approach to actual performance figures is achieved.

The theoretical cycle based on the actual properties of the cylinder gases on the actual properties of the cylinder gases is called the fuel-air cycle and represents a nearly attainable ideal for comparison with actual performance.

The fuel-air cycle calculations take into consideration the following –

1. The actual composition of the cylinder gases i.e. fuel + air + combustion gases. The fuel air ratio is changed during the operation of the engine which changes the relative amounts of CO_2, CO, water vapour etc.
2. The variation in the specific heat of these gases with temperature. Specific heat increases with temperature except the monoatomic gases.
3. The fact that the fuel and air do not completely combine chemically at high temperatures (above 1600 K), CO, H_2, H and O_2 may be present at equilibrium conditions.
4. The variation in the number of molecules present in the cylinder as the pressure and temperature change. The number of molecules present after combustion depends on the fuel-air ratio and the pressure and temperature after the combustion.

The following assumptions are commonly made for fuel-air cycle analysis :

1. There is no chemical change in either fuel or air prior to combustion.
2. Subsequent to combustion, the charge is always in chemical equilibrium.
3. There is no heat exchange between the gases and the cylinder walls in any process i.e., they are adiabatic. Also expansion and compression processes are frictionless.
4. In reciprocation engines the velocities are negligibly small.

23.11 VARIABLE SPECIFIC HEATS

All gases, except monoatomic gases, show an increase in specific heat at temperatures. The increase in specific heat does not follow any particular law. However, over the temperature range in general use for gases in heat engines (300 K to 1300 K) the specific heat curve is nearly a straight line which may be approximately expressed in the form,

$$C_p = a + KT; C_v = b + KT$$

where a, b and K are constants.

$$\therefore \quad R = C_p - C_v = (a+KT) - (b+KT)$$

$$= (a-b)$$

Above 1500 K, the specific heat increases much more rapidly and may be approximately expressed in the form

$$C_p = a + K_1T + K_2T^2$$

$$C_v = b + K_1T + K_2T^2$$

The physical explanation of increases in specific heat is that as the temperature is raised, larger and larger fractions of the heat input go to produce motion of the atoms within the molecules. Since temperature is the result of motion of the molecules as a whole the energy which goes into moving the atoms does not contribute to the temperature rise. Hence more heat is required to raise the temperature of unit mass through one degree. This heat—by definition—is specific heat.

For air C_p = 1.005 at 0°C and 1.2645 at 2000°C.

SOLVED EXAMPLES

(Constant Volume Cycle—Otto-Cycle)

Example 23.1. *An engine is operating on ideal Otto cycle for which the following information is available.*

Maximum temperature = 1277 °C

Exhaust temperature = 447 °C

Ambient condition = 1 bar and 37 °C

Air consumption = 2 kg/min

Estimate (i) Compresion ratio (ii) Air Standard efficiency (iii) Power output.

Assume : C_p = 1.005 kJ/kg-K, C_v = 0.718 kJ/kg.K

Solution

Ideal Otto-cycle is shown on P–V diagram given Fig. Ex. 23.1

Referring to Fig. Ex. 23.1

$$T_1 = 37 + 273 = 310 \text{ K}$$

$$p_1 = 1 \text{ bar}$$

$$T_3 = \text{Max. temp.}$$

$$= 1277 + 273$$

$$= 1550 \text{ K}$$

$$T_4 = \text{Exhaust temperature}$$

$$= 447 + 273 = 720 \text{ K}$$

$$C_p = 1.005 \text{ kJ/kg.K}$$

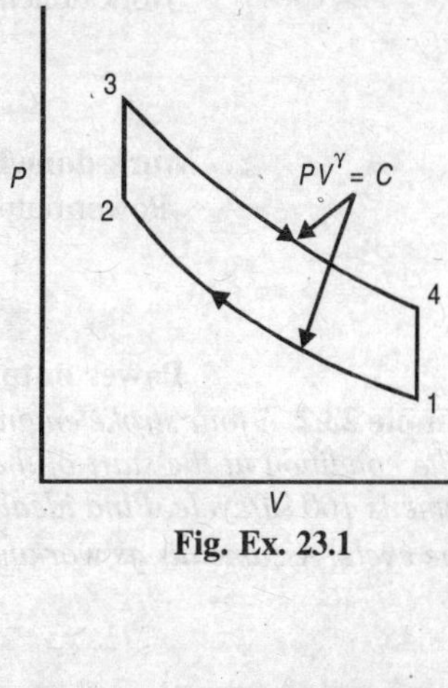

Fig. Ex. 23.1

$$C_V = 0.718 \text{ kJ/kg.K}$$

$$\therefore \quad \gamma = \frac{C_p}{C_V} = \frac{1.005}{0.718} = 1.4$$

(i) Compression ratio :

For Otto-cycle $\quad r_k$ = Compression ratio

$$= r_e$$

$\therefore$ For expansion process

$$\frac{T_3}{T_4} = r_e^{\gamma-1} = r_k^{\gamma-1}$$

$$= \frac{1550}{720} = 2.153$$

$$\therefore \quad r_k^{0.4} = 2.153$$

$$r_k = (2.153)^{1/0.4} = (2.153)^{2.5}$$

$$\mathbf{r_k = 6.8}$$

(ii) Air standard efficiency

$$\text{ASE} = 1 - \frac{1}{r_k^{\gamma-1}}$$

$$= 1 - \frac{1}{2.153} = 1 - 0.4645$$

$$\textbf{ASE} = \mathbf{0.5355 = 53.55\%}$$

(iii) Power output

$$\text{Work done/kg} = Q_A - Q_R$$

$$= C_V(T_3 - T_2) - C_V(T_4 - T_1)$$

We have

$$\frac{T_3}{T_2} = \frac{T_4}{T_1} \text{ or } \frac{T_3}{T_2} = \frac{T_2}{T_1}$$

$$\therefore \quad T_2 = T_1 \times \frac{T_3}{T_4} = \frac{310 \times 1550}{720}$$

$$\mathbf{T_2 = 667.36 \text{ K}}$$

$$\therefore \quad \text{Work done/kg} = 0.718(1550 - 667.36) - 0.718(720 - 310)$$

$$= 0.718 \times 882.64 - 0.718 \times 410$$

$$= 633.74 - 294.38$$

$$\textbf{Work done/kg} = \mathbf{339.36 \text{ kJ/kg}}$$

$$\therefore \quad \text{Power output} = \text{mass of air/sec} \times \text{work per kg}$$

$$= \frac{2}{60} \times 339.36 = 11.312 \text{ kJ/sec}$$

$$\textbf{Power output} = \mathbf{11.312 \text{ kW}}$$

Example 23.2. *A four stroke engine working on Otto-cycle has a swept volume of 0.1 m³. The compression ratio is 7. The condition at the start of the cycle is pressure 1 bar and temperature 90°C. The heat addition at constant volume is 100 kJ/cycle. Find ideal efficiency, mean effective pressure and pressure and temperature at key points in the cycle. Assume air as working substance with* $C_V = 0.718$ *kJ/kg.K and* $\gamma = 1.4$.

Solution

The cycle is shown on *P–V* diagram of Fig. Ex. 23.13 given

We have
$$P_1 = 1 \text{ bar}$$
$$T_1 = 273 + 90 = 363 \text{ K}$$
$$Q_A = 100 \text{ kJ/cycle}$$
$$r_k = 7$$

(i) Ideal efficiency :

$$\text{Ideal efficiency} = \text{Air standard efficiency}$$
$$= 1-\frac{1}{r_k^{\gamma-1}}$$
$$= 1-\frac{1}{7^{0.4}}$$
$$= 1-\frac{1}{2.1780}$$
$$= 1-0.459 = 0.541$$
$$\textbf{Ideal efficiency} = \textbf{54.1\%}$$

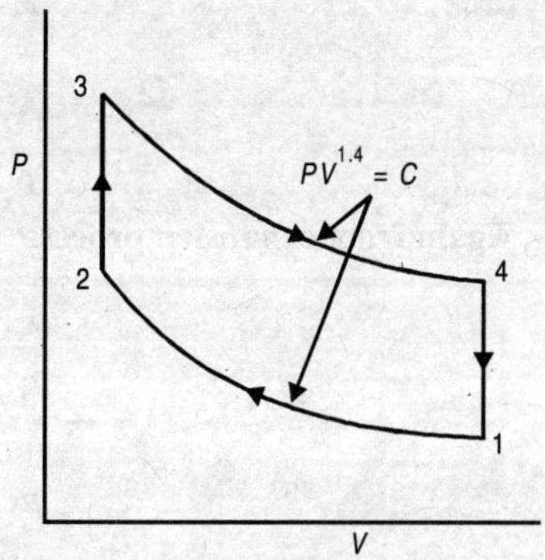

Fig. Ex. 23.2

(ii) Pressure and temperature at key points :

$$\frac{T_2}{T_1} = r_k^{\gamma-1} = 7^{0.4} = 2.178$$
$$T_2 = 363\times 2.178$$
$$\mathbf{T_2 = 790.61\ \ K}$$
$$\frac{p_2}{p_1} = \left(\frac{V_1}{V_2}\right) = 7^{1.4} = 15.245$$
$$\mathbf{p_2 = 1\times 15.245 = 15.245 \text{ bar}}$$
$$Q_A = mC_V\left(T_3-T_2\right) = \frac{p_2v_2}{RT_2}C_V\left(T_3-T_2\right)$$
$$\text{Now compression ratio} = \frac{V_1}{V_2} = \frac{V_2+V_S}{V_2}$$
$$= 1+\frac{V_S}{V_2}$$
$$\therefore \quad \frac{V_S}{V_2} = 7-1 = 6\, V_S = 0.1 \text{ cu.m}$$
$$\therefore \quad V_2 = \frac{V_S}{6} = \frac{0.1}{6} = 0.0167 \text{ m}^3$$
$$\therefore \quad 100 = \frac{10^5\times 15.245\times 0.0167}{287\times 790.61}\times 0.718\left(T_3-T_2\right)$$
$$100 = 0.1122\times 0.718\left(T_3-T_2\right)$$
$$T_3-T_2 = \frac{100}{0.1122\times 0.718}$$

$$= 1241.32$$

$$T_3 = 1241.32 + 790.61$$

$$\mathbf{T_3 = 2031.93\ K}$$

$$\therefore \quad \frac{T_3}{T_2} = \frac{p_3}{p_2} \text{ for constant volume heating}$$

or P_3 = pressure after heat addition

$$= \frac{2031.92}{790.61} \times 15.245$$

$$\mathbf{P_3 = 39.18\ bar}$$

Again from expansion process

$$\frac{p_3}{p_4} = \left(\frac{V_4}{V_3}\right)^{\gamma} = 7^{1.4} = 15.243$$

$$\therefore \quad p_4 = \frac{p_3}{15.243} = \frac{29.18}{15.243}$$

$$\mathbf{p_4 = 2.57}$$

$$\therefore \quad T_4 = T_3 \times \left(\frac{1}{7}\right)^{0.4} = \frac{2031.93}{2.178} \doteq \mathbf{932.93\ K}$$

(iii) Mean effective pressure

$$\text{Heat rejected/cycle} = mC_v(T_4 - T_1)$$

$$= 0.11220 \times 0.178(932.92 - 363)$$

$$= 45.91 \text{ kJ}$$

$$\therefore \quad \text{Work done} = Q_A - Q_R$$

$$= 100 - 45.91$$

$$= 54.09 \text{ kJ}$$

$$\therefore \quad 54.09 = p_m \times V_1 \cdot \left(\frac{r_k - 1}{r_k}\right)$$

$$= p_m \times 0.1167 \times \frac{7-1}{7}$$

$$= p_m \times 0.7/7$$

If p_m is bar, them

$$54.09 = 100 p_m \times 0.1167 \times \frac{6}{7}$$

$$p_m = \frac{54.09 \times 7}{100 \times 6 \times 0.1167}$$

$$\mathbf{p_m = 5.41\ bar}$$

II. CONSTANT PRESSURE CYCLE (DIESEL CYCLE)

Example 23.3. *A Diesel cycle operates at a pressure of 1 bar at the beginning of compression and the volume is compressed to $\frac{1}{16}$ of the initial volume. Heat is supplied until volume is twice that of the clearance volume. Calculate mean effective pressure of the cycle. Take K= 1.4.*

Solution

With reference to *P–V* diagram shown

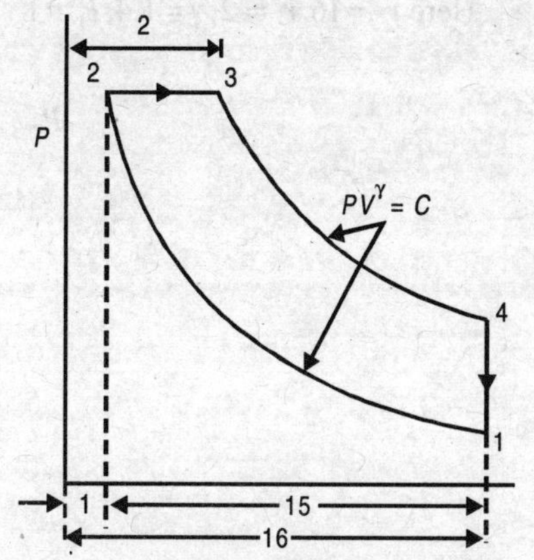

Fig. Ex. 23.3

$$\text{Compression ratio} = r_k = 16$$

$$\text{Cut-off ratio} = \frac{V_3}{V_2} = 2$$

Work done/Cycle/kg air

$$= p_2(V_3 - V_2)$$

$$= \frac{p_3V_3 - p_4V_4}{\gamma - 1} - \frac{p_2V_2 - p_1V_1}{\gamma - 1}$$

$$\text{We have } \frac{p_2}{p_1} = r_k^{\gamma} \therefore p_2 p_1 r_k^{\gamma} = p_3$$

$$\frac{p_3}{p_4} = \left(\frac{V_4}{V_3}\right)^{\gamma} = r_e^{\gamma} = \left(\frac{r_k}{r_e}\right)^{\gamma}$$

$$p_4 = \frac{p_3 - r_c^{\gamma}}{r_k^{\gamma}} = \frac{p_1 \times r_k^{\gamma} r_c^{\gamma}}{r_k^{\gamma}} = p_1 r_e^{\gamma}$$

$$\text{Similarly} \quad V_2 = \frac{V_1}{r_k}, \frac{V_3}{V_2} = r_c$$

$$\therefore \quad V_3 = V_2 r_c = \frac{V_1 r_c}{r_k}; V_4 = V_1$$

Substituting these values in the equation of work we get

$$\text{Work done} = P_1 r^k K\left(\frac{V_1 r_c}{r_k} - \frac{V_1}{r_k}\right) + \left(\frac{p_1 r^k K \frac{V_1 r_c}{r_k} - \frac{P_1 r^k{}_c V_1}{1}}{\gamma - 1}\right) - \left(\frac{p_1 r^k K \frac{V_1}{r_k} - p_1 V_1}{\gamma - 1}\right)$$

$$= p_1\left[\frac{r^{\gamma}{}_K . V_1}{r_k}(r_c - 1). + \frac{V_1}{\gamma - 1}\left(\frac{r^{\gamma}{}_K r_c}{r_k} - r^{\gamma}{}_e\right) - \frac{V_1}{K - 1}\left(\frac{r^{\gamma}{}_K}{r_k} - 1\right)\right]$$

$$= p_m \times (V_1 - V_2) = p_m \times \left(V_1 - \frac{V_1}{r_k}\right) = p_m V_1\left(\frac{r^{\gamma}{}_K}{r_k} - 1\right)$$

$$p_m \times \frac{r_K^\gamma - 1}{r_k} = \frac{p_1}{(K-1)}\left[\frac{(\gamma-1)(r_c-1)r_K^\gamma}{r_k} + \frac{r_c^\gamma r_c - r_k r_c^\gamma}{\gamma-1} - \frac{r_K^\gamma - r_k}{r_k}\right]$$

$$p_m = \frac{P_1}{(r_k-1)(\gamma-1)}\left[(\gamma-1)(r_c-1)r_j^\gamma + r_k^\gamma(r_c-1) - r_k\left(r_c^\gamma-1\right)\right]$$

Here $r_k = 16, r_c = 2, \gamma = 1.4, P_1 = 1$

$$p_m = \frac{1}{(16-1)(1.4-1)}\left[0.4\times1\times16^{1.4} + 16^{1.4}\times1 - 16\left(2^{1.4}-1\right)\right]$$

$$= \frac{1}{15\times0.4}\left[0.4\times16^{1.4} + 16^{1.4} - 16\left(2^{1.4}-1\right)\right]$$

$$= \frac{1}{6}\left[0.4\times4.851 + 4.57 - 16(2.636-1)\right]$$

$$= \frac{1}{6}[19.404 + 48.51 - 16.1636] = \frac{1}{6}p[67.914 - 26.176]$$

$$= \frac{1}{6}\times41.738$$

$$\mathbf{p_m = 6.96\ bar}$$

Example 23.4. *In an air standard Diesel cycle, compression begins at 103 kPa and 300°K. After compression heat addition is of 545 kJ/kg of air, the peak pressure reached in the cycle is 4.7 MPa. Calculate*

(i) fuel cut-off ratio
(ii) compression ratio
(iii) maximum temperature in the cycle
(iv) air standard efficiency

Assume $\gamma = 1.4$ and $C_p = 1.004$ kJ/kg.

Solution

$$P_1 = 1.03 \text{ bar}$$

$$Q_A = 545 \text{ kJ/kg}$$

$$P_2 = p_3 = 47 \text{ bar}$$

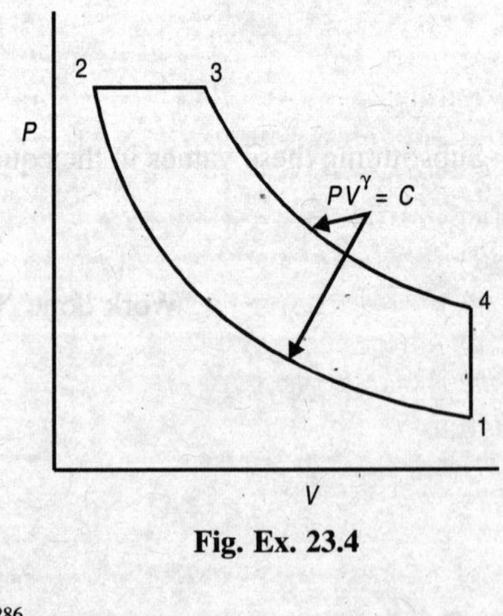

Fig. Ex. 23.4

(i) With reference to P–V diagram

$$\frac{P_2}{P_1} = \left(\frac{V_1}{V_2}\right)^\gamma = r_k^{1.4} = \frac{47}{1.03} = 45.63$$

$$r_k = (45.63)^{1/1.4} = 45.63^{0.7143}$$

$$= 15.32$$

$$\mathbf{r_k = Compression\ ratio = 15.32}$$

(i) Fuel cut off ratio

$$\frac{T_2}{T_1} = \left(\frac{P_2}{P_1}\right)^{\frac{\gamma-1}{\gamma}} = (45.63)0^{04/1.4}(45.63)^{0.286} = 2.983$$

$$\therefore \quad T_2 = 300\times2.983$$

$$= 894.9 \text{ K}$$

$$\therefore \quad Q_A = 545 = C_P(T_3 - T_2) = 1.004(T_3 - 894.9)$$

$$T_3 = \frac{545}{1.004} + 894.9 = 542.83 + 894.9$$

$$= 1437.73 \text{ K}$$

$$\therefore \quad \frac{T_3}{T_2} = \frac{V_3}{V_2} = r_c \text{ (cut-off ratio)} = \frac{1437}{894.9} = 1.61$$

$$\mathbf{r_c = \text{cut-off ratio} = 1.61}$$

(iii) Maximum temperature in the cycle

$$T_3 = \text{Maximum. temp. in the cycle}$$

$$= \mathbf{1437.73 \text{ K} = 1164.73°C}$$

(iv) Air standard efficiency

$$r_e = r_k / r_c = \frac{15.32}{1.61} = 9.52$$

$$\therefore \quad (T_3/T_4) = (r_e)^{\gamma-1} = (9.52)^{0.4} = 2.462$$

$$\therefore \quad T_4 = \frac{1437.73}{2.462} = 584 \text{ K}$$

$$\therefore \quad Q_R = \text{Heat rejected} = 1.004(584 - 300)$$

$$= 1.004 \times 284 = 285 \text{ kJ/kg}$$

$$\therefore \quad \text{Work done/kg} = Q_A - Q_R = 545 - 285$$

$$= 260 \text{ kJ/kg}$$

$$\therefore \quad \text{Thermal efficiency} = \text{ASE} = \frac{W}{Q_A} = \frac{260}{545} = 0.477$$

$$= 47.7 \%$$

$$\textbf{ASE} = \mathbf{47.7 \%}$$

Example 23.5. *In an air standard Diesel Cycle, compression begins at 1 bar and 300ºK. The compression ratio is 16. Heat added per kg of air of 1000 kJ/kg. Determine:*

(i) fuel cut-off ratio
(ii) pressure and temperature at the end of expansion
(iii) thermal efficiency
(iv) mean effective pressure
(v) power output for an air flow rate of 0.2 kg/sec

Take R = 0.286 kJ/kg K and C_p = 1.0 kJ/kg K.

Solution

(i) Fuel cut-off ratio :

$$\frac{T_2}{T_1} = r_k^{\gamma-1} = 16^{0.4}$$

$$R = C_p - C_V = 3.031$$

$$\therefore \quad C_V = C_P - R = 1.0.286$$

$$\therefore \quad T_2 = 300 \times 3.031 \quad \therefore \quad \gamma = \frac{1}{0.714} = 1.4$$

$$= 909.3 \text{ K}$$

Now $\quad Q_A = C_p(T_3 - T_2)$

$$\therefore \quad 1000 = 1(T_3 - 909.3)$$

$$\therefore \quad T_3 = 1000 + 909.3 = 1909.3 \text{ K}$$

$$\therefore \quad \frac{T_3}{T_2} = \frac{V_3}{V_2}$$

$\therefore \quad r_c$ = Cut-off ratio = 1909.3/909.3 = 2.1

$\therefore \quad$ **$\mathbf{r_c}$ = Cut-off ratio = 2.1**

(ii) Pressure and temperatures at the end of expansion:

$$\frac{P_3}{P_4} = \left(\frac{V_4}{V_3}\right)^{\gamma} = (r_e)^{\gamma} = \left(\frac{r_k}{r_c}\right)^{\gamma} = \left(\frac{16}{2.1}\right)^{1.4} = (7.619)^{1.4} = 17.17$$

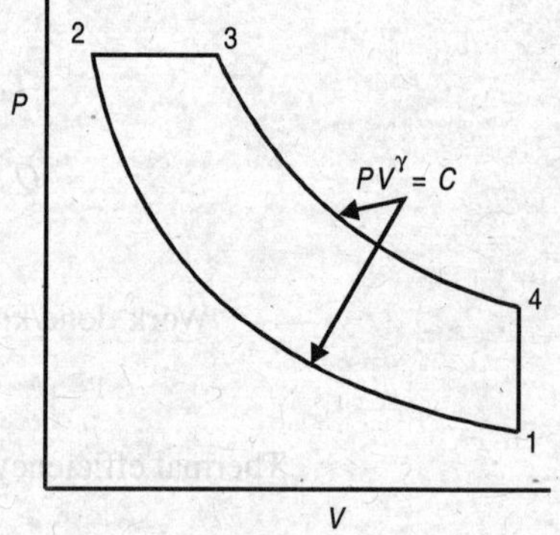

Fig. Ex. 23.5

Also $\quad \dfrac{P_2}{P_1} = r_k^{1.4} = 16^{1.4} = 48.51$

$$\therefore \quad P_2 = 1 \times 4851 = 48.51 \text{ bar} = P_3$$

$$\therefore \quad P_4 = \frac{P_3}{17.17} = \frac{48.51}{17.17} = 2.825 \text{ bar}$$

$$\frac{T_3}{T_4} = (r_e)^{\gamma-1} = (7.619)^{1.4} = 2.253$$

$$\therefore \quad T_4 = \frac{T_3}{2.253} = \frac{1909.3}{2.253} = 847.4 \text{ K}$$

$\therefore \quad$ **$\mathbf{P_4}$ = 2.285 bar**

$\mathbf{T_4}$ = 847.4 K

(iii) Thermal efficiency :

$$Q_R = \text{Heat rejected} = C_V(T_4 - T_1) = 0.714(847.4 - 300)$$

$$= 0.714 \times 574.4 = 390.84 \text{ kJ/kg}$$

$$\therefore \quad \text{Thermal efficiency:} = \frac{Q_A - Q_R}{Q_A} = \frac{1000 - 390.84}{1000} = \frac{609}{1000} = 0.60916$$

$$= 60.916 \text{ \%}$$

ASE = 60.916 %

(iv) MEP

$$\text{Work done/kg} = 608.16 \text{ kJ} = P_m \times V_s = P_m(V_1 - V_2)$$

But $$V_1 - V_2 = V_1\frac{V_1}{r_k} \text{ as } \frac{V_1}{V_2} = r_k$$

$$= V_1\frac{(r_k - 1)}{r_k} = \frac{RT_1}{P_1}.\frac{15}{16} = \frac{286\times300}{10^5}\times\frac{15}{16}$$

$$\therefore P_m \times \frac{286\times300}{10^5\times1000}\times\frac{15}{16} = 609.16$$

$$\therefore \quad P_m = \frac{609.16\times16}{15}\times\frac{1000\times10^5}{286\times300} = 10^5\times7.573 \text{ N/m}^2$$

$$= 7.573 \text{ bar}$$

$$\therefore \quad \mathbf{p_m = 7.573 \text{ bar}}$$

(v) Power output

$$\text{Work done/kg} = Q_A - Q_R = 1000 - 390.84 = 609.16 \text{ kJ}$$

$$\text{Flow rate of air} = 0.2 \text{ kg/sec}$$

$$\text{power developed} = 0.2\times609.16$$

$$= \mathbf{121.832 \text{ (kJ/sec = kW)}}$$

III. DUAL COMBUSTION CYCLE

Example 23.6. *An oil engine works on the ideal dual combustion cycle with a volume compression ratio of 10. The temperature at the beginning of compression is 80°C and the maximum pressure of the cycle is 40 bar. At what percentage of stroke must heat reception of constant pressure cease in order that the temperature of the beginning of expansion is 1700°C? What is the ratio of the work done during expansion to that done during compression?*

Solution

$$r_k = 10$$

$$T_1 = 80 + 273 = 353 \text{ K}$$

$$P_3 = P_4 = 40 \text{ bar}$$

$$T_4 = 1700 + 273 = 1973 \text{ K}$$

Let $$P_1 = 1 \text{ bar}$$

We have, $$\frac{T_2}{T_1} = r_k^{\gamma-1} = 10^{0.4}$$

$$= 25.12 \text{ bar}$$

$$P_3 = 40 \text{ bar}$$

$$\therefore \quad \frac{P_3}{P_2} = \frac{T_3}{T_2} = \frac{40}{25.12}$$

$$\therefore \quad T_3 = \frac{40}{25.12}\times886.7$$

$$= 1411.94 \text{ K}$$

$$\therefore \quad \text{Cut-off ratio } r_c = \frac{T_4}{T_3} = \frac{1973}{1411.94}$$

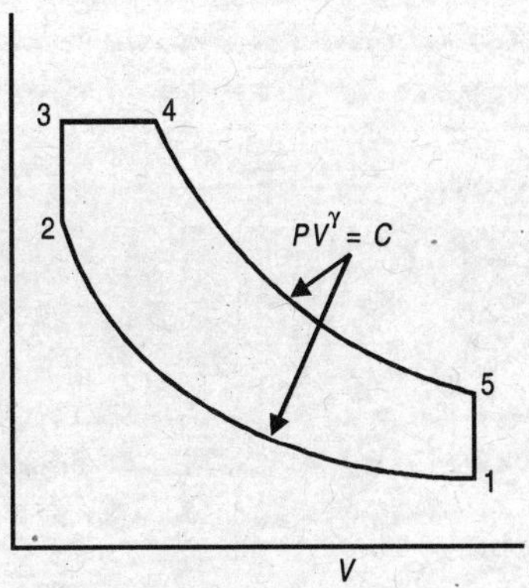

Fig. Ex. 23.6

$$= 1.3974$$

Now $$r_k = 10 = \frac{V_1}{V_2}$$

$$\therefore \quad V_1 = 10V_2$$

$$\therefore \quad V_s = V_1 - V_2 = 10V_2 - V_2 = 9V_2$$

Again $$r_c = \frac{V_4}{V_3} = 1.3974V_2$$

$$\therefore \quad V_4 - V_3 = V_4 - V_2 = 1.3974V_2 - V_2$$

$$= 0.3874V_2$$

% of stroke at which heat addition is stopped

$$= \frac{V_4 - V_3}{V_1 - V_2} \times 100$$

$$= \frac{0.3974V_2}{9V_2} \times 100$$

$$= 0.04416 \times 100 = 4.416\%$$

Work done during expansion

$$W_e = \frac{P_4V_4 - P_SV_S}{\gamma - 1} = \frac{R}{\gamma - 1}[T_4 - T_5]$$

For expansion $$r_e = \frac{r_k}{r_c} = \frac{10}{1.3974}$$

$$= 7.156$$

$$\therefore \quad \frac{T_4}{T_5} = (7.156)^{0.4} = 2.1972$$

$$\therefore \quad T_s = \frac{1973}{2.1972} = 897.96 \text{ K}$$

$$\therefore \quad W_e = \frac{R}{\gamma - 1}[T_4 - T_5] = C_v(T_4 - T_5)$$

$$= 0.718[1973 - 897.96]$$

$$= 0.718 \times 1076.04 = 772.6 \text{ kJ/kg}$$

Work during compression

$$W_e = \frac{P_2V_2 - P_1V_1}{\gamma - 1} = \frac{R}{\gamma - 1}(T_2 - T_1)$$

$$= 0.718(886.7 - 353)$$
$$= 0.718 \times 533.7$$
$$= 383.197 \text{ kJ/kg}$$

$$\therefore \quad \textbf{Ratio} \frac{W_e}{W_c} = \frac{772.6}{383.197} = 2.019$$

Example 23.7. *The following data refer to an oil engine operating on dual cycle i.e. with heat reception partly at constant volume and partly at constant pressure and with heat rejection at constant volume. Compression ratio 11.6, pressure and temeparture at the beginning of compression 1 bar and 320°K, percentage increase of pressure during constant volume burning 53% and percentage volume increase during constant pressure during 38%. If it is assumed that C_p = 1.089 and C_v = 0.795 and that the compression and expansion curved are isentropic, find mep of the cycle.*

Solution

Here $$\gamma = \frac{C_p}{C_v} = \frac{1.089}{0.795}$$
$$= 1.37$$

Given: $$r_k = \frac{V_1}{V_2} = 11.6$$
$$P_1 = 1 \text{ bar}$$
$$T_1 = 320 \text{ K}$$
$$\frac{P_3 - P_2}{P_2} \times 100 = 53 = \frac{P_3}{P_2} - 1 = 0.53$$
$$\therefore \quad \frac{P_3}{P_2} = 1.53$$
$$\frac{V_4 - V_3}{V_3} \times 100 = 38$$
$$\frac{V_4}{V_3} - 1.38 = r_c$$

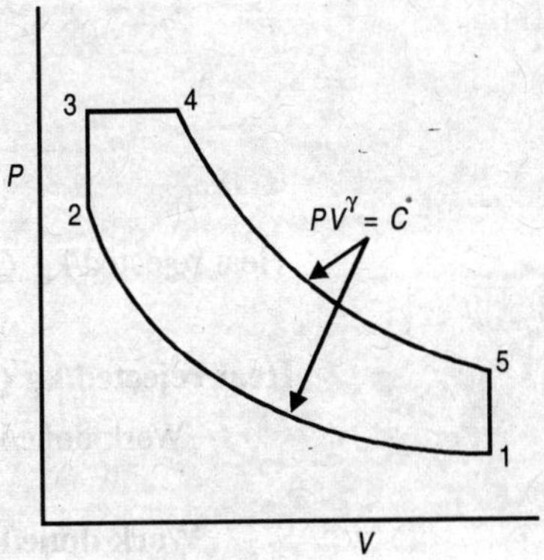

Fig. Ex. 23.7

$$\therefore \quad P_2 = P_1 \times r_k^{1.37} = 1 \times 28.73 \text{ bar} = 2873 \text{ bar}$$
$$\therefore \quad P_3 = 28.73 \times 1.53 = 43.954 \text{ bar} = P_4$$
$$\therefore \quad r_e = \text{expansion ratio} = \frac{r_k}{r_c} = \frac{11.6}{1.38}$$
$$= 8.406$$
$$\therefore \quad \frac{P_4}{P_5} = r_e^\gamma = (8.406)^{1.37}$$
$$= 18.48 \text{ bar}$$
$$\therefore \quad P_5 = \frac{43.954}{18.48} = 2.378 \text{ bar}$$

Also $\frac{T_4}{T_5} = r_e^{\gamma-1} = (8.406)^{0.37}$

$= 2.1984$

and $\frac{T_2}{T_1} = r_e^{\gamma-1} = 11.6^{0.37} = 2.4766$

$\therefore$ $T_2 = 320\times 2.4766 = 792.5 \text{ K}$

$T_3 = 792.5\times 1.53$

$= 1212.5255 \text{ K}$

$T_4 = 1212.5255 \text{ K}$

$T_5 = \frac{1673.28}{2.1984}$

$= 761.1377 \text{ K}$

Heat added/kg $Q_A = C_V(T_3 - T_2) + C_P(T_4 - T_3)$

$= 0.795(1212.5255 - 792.5) + 1.089(1673.28 - 792.5)$

$= 0.795\times 420.0255 + 1.089\times 460.7545$

$= 333.92 + 501.76$

$= 835.68 \text{ kJ/kg}$

Heat rejected/kg $Q_R = C_V(T_5 - T_1)$

$= 0.795\times 441.1377$

Heat rejected/kg Q_R = 350.70 kJ/kg

$\therefore$ Work done/kg $= Q_A - Q_R$

$= 835.68 - 350.7$

Work done/kg = 484.98 kJ/kg

$\therefore$ $P_m V_1 \frac{r_k - 1}{r_k} = 484.98$

$P_m \times \frac{287\times 320}{1000} \times \frac{11.6-1}{11.6} = 484.98$

P_m = 5.78 bar

IV. STIRLING CYCLE

Example 23.8. *A striling regenrative air engine works between temperatures of 400°C and 15°C, the ratio of isothermal expansion is 3. Calculate the ideal efficiency when (a) the engine is fitted with a regenerator (b) when the efficiency of the regenerator is 0.8. Take C_p = 1 kJ/kg-K and C_v = 0.718 kJ/kg-K.*

Solution

(a) When regenerator is 100% efficient.

In this case, the efficiency is the same as that for Carnot cycle

$\therefore$ $\eta = \frac{T_1 - T_2}{T_1} = 1 - \frac{T_2}{T_1}$

where $T_1 = 400+273 = 673 \text{ K}$

$\therefore$ $T_2 = 15+273 = 288 \text{ K}$

$\therefore$
$$\eta = \frac{673\text{-}288}{673}$$
$$= \frac{386}{673}$$
$$= 0.5721$$
$$\eta = \mathbf{57.21\%}$$

(b) When regenerator efficiency is 80%

When the regenerator is not 100% efficient, more heat is taken by the working fluid and more heat is rejected by the working fluid. The net output remains the same.

Heat added during the cycle

$$Q_A = RT_2 \log_e r_e + (1-\eta_r)C_V(T_2 - T_1)$$

where r_e = Isothermal expansion ratio = 3

η_r = Regenerator efficiency = 80% or 0.8

$$\text{Work done/cycle} = (T_1 - T_2)R\log_e r_e$$
$$= (673-288)(1-0.718)\log_e 3$$
$$= 385\times0.282\times1.0986$$
$$= 119.276 \text{ kJ/kg}$$

Heat supplied during cycle/kg

$$= T_1 R\log_e r_e + (1-\eta_r)C_v(T_1 - T_2)$$
$$= 673\times0.282\times1.0986 + (1-0.8)\times0.718\times(673-288)$$
$$= 208.5 + 0.2\times0.718\times385$$
$$= 208.5 + 55.286$$

$\therefore$ Efficiency of the engine

$$= \frac{\text{Work done}}{\text{Heat supplied}}$$
$$= \frac{119.276}{263.786} = 0.4522$$

Efficiency of the engine = 45.22%

Example 23.9. *A stirling cycle has a temperature ratio of 3 and volumetric compression ratio of 80% and has only 75% effectiveness in heat exchnages. Calculate the efficiency of the cycle.*

Solution

$$\frac{T_1}{T_2} = 3$$
$$r_k = 8$$
$$\eta_r = 0.75$$

Work done/kg = Area under (3 – 4) – Area under (2 – 1)

$$= P_3V_3 \log_e \frac{V_4}{V_3} - P_2V_2 \log_e \frac{V_1}{V_2}$$

$$= RT_1 \log_e 8 - RT_2 \log_e 8$$

$$= R(T_1 - T_2)\log_e 8$$

$$= RT_2\left(\frac{T_1}{T_2} - 1\right)\log_e 8$$

$$= 0.287T_1(3-1)\times 2.0794$$

$$= 2\times 0.287\times 0.0794T_2$$

Work done/kg $=$ $\mathbf{1.1936\ T_2\ kJ/kg}$

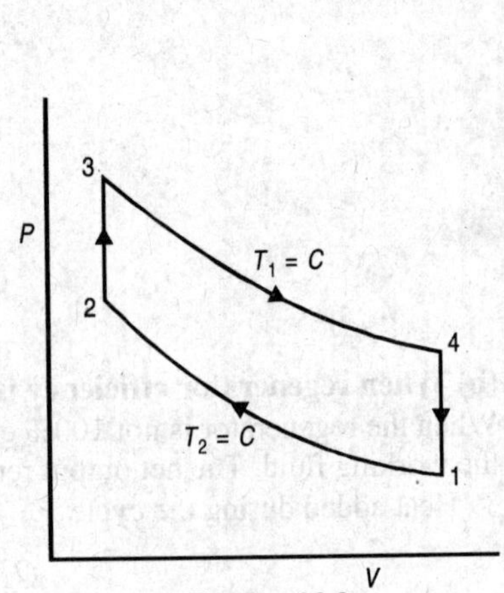

Fig. Ex. 23.9

Heat supplied/kg when $\eta_r = 100\%$

$$= P_3V_3 \log \frac{V_4}{V_3} RT_1 \log_e 8$$

When regenerator effectiveness is 0.75, additional heat required to be supplied

$$= (1-\eta_r)C_V(T_1 - T_2)$$

$$= 0.25\times 0.718T_2(3-1)$$

$$= 2\times 0.25\times 0.718T_2$$

$$= 0.359\,T_2$$

∴ Total heat supplied/kg

$$= 0.287\times T_2 \log_e 8 + 0.359T_2$$

$$= 0.287\times 3T_2\times 2.0794 + 0.359T_2$$

$$= 1.7904T_2 + 0.359T_2$$

Total heat supplied/kg $=$ $\mathbf{1.8263\ T_2\ kJ/kg}$

∴ Efficiency of the cycle

$$\text{ASE} = \frac{W}{Q_A} = \frac{1.1936T_2}{1.8263T_2}$$

$$= 0.6536$$

ASE = 65.36%

ADDITIONAL EXAMPLES

Example 23.10. (*a*) *Show that for maximum work, the intermediate temperature in Otto-cycle is given by* $T_2 = T_4 = \sqrt{T_1T_3}$ *where* T_1 *and* T_3 *are initial and maximum temperature respectively. Also find out maximum efficiency of Otto cycle having maximum temperature 1000°C and minimum temperature 15°C. Take* $\gamma = 1.4$.

(b) A four stroke engine working on Otto-cycle has a swept volume of 0.1 m^3. The compression ratio is 7. The condition at the start of the cycle is 1 bar and temperature 90°C. The heat addition at constant volume is 100 kJ/cycle. Find ideal efficiency, mean effective pressure and temperature at the key points in the cycle. Assume air at working substance with C_1 = 0.718 kJ/kg.K and γ = 1.4.

Solution

(a) For derivation of $T_2 = T_4 = \sqrt{T_1 T_3}$ refer theory

$$T_1 = 15 + 273 = 288 \text{ K},$$

$$T_3 = 1000 + 273 = 1272$$

$$\therefore \quad T_2 = \sqrt{1273 \times 288} = 605.5 \text{ K}$$

$$\therefore \quad \text{Max. eff.} = 1 - \frac{T_4 - T_1}{T_3 - T_2} = 1 - \frac{605 - 288}{1273 - 605}$$

$$= 1 - \frac{317}{668} = 1 - 0.4746$$

$$\textbf{Max. Eff.} = \textbf{52.54\%}$$

$$\therefore \quad \frac{T_2}{T_1} = (r_k)^{\gamma-1} = \frac{605}{288} = 2.1$$

$$\therefore \quad \text{ASE} = 1 - \frac{1}{2.1} = 1 - 0.4762$$

$$= 0.5238 = 52.38\%$$

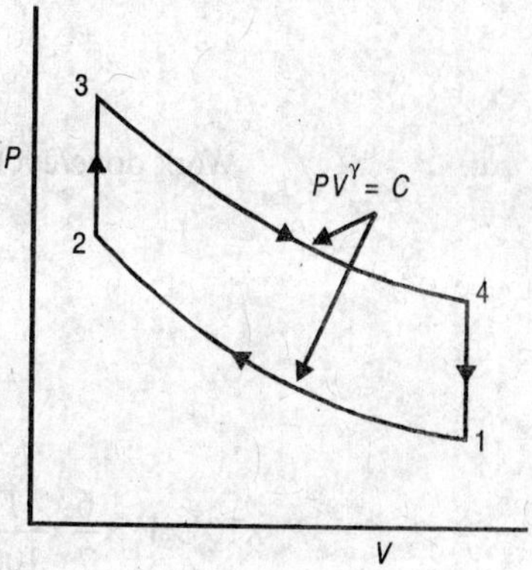

Fig. Ex. 23.10

$$T_1 = 273 + 90 = 363 \text{ K}$$

$$P_1 = 1 \text{ bar}$$

$$r_k = 7$$

$$\therefore \quad T_2 = T_1 r_k^{\gamma-1} = 363 \times 7^{0.4}$$

$$= 363 \times 2.1779$$

$$= 7906 \text{ K}$$

$$P_2 = P_1 \cdot r_k^{\tau} = 1 \times 7^{1.4} = 15.24 \text{ bar}$$

Mass of air in the cycle

$$m = \frac{P_1 V_2}{RT_1} = \frac{1 \times 10^5 \times 0.1}{287 \times 363} = 0.096 \text{kg}$$

$$\therefore \quad \text{Heat added/cycle} = 100 \text{ kg} = m(T_3 - T_2)$$

$$= 0.096(T_3 - 790.6)$$

$$\therefore \quad T_3 = \frac{100}{0.096} + 790.6 = 1041.67 + 790.36$$

$$= 1832.27 \text{ K}$$

$$\therefore \quad P_3 = P_2 \times \frac{T_3}{T_2} = 15.245 \times \frac{1832.27}{790.6}$$

$$\mathbf{P_3 = 35.33 \text{ bar}}$$

Again $$P_4 = \frac{P_3}{15.245} = \frac{35.33}{15.245}$$

$$\mathbf{P_4 = 2.1376 \text{ bar}}$$

$$T_4 = \frac{T_3}{2.1779} = \frac{1832.27}{2.1779}$$

$$\mathbf{T_4 = 841.3 \text{ K}}$$

(c) Mean Effective Pressure (MEP)

$$\text{Heat rejected/cycle} = mC_4(T_4 - T_1)$$

$$= 0.096 \times 0.718(841.3 - 363)$$

$$= 32.97 \text{ kJ/cycle}$$

$$\therefore \quad \text{Work done/cycle} = Q_A - Q_R$$

$$= 100 - 32.97$$

$$= 67.07 \text{ kJ}$$

$$= P_m \times V_1 . \frac{r_k - 1}{r_k}$$

$$\therefore \quad P_m \times 10^5 \times V_1 \times \frac{6}{7} \times \frac{1}{100} = 67.03$$

$$P_m = \frac{67.03 \times 7}{0.6 \times 100}$$

$$\mathbf{P_m = 7.82 \text{ bar}}$$

Example 23.11. (*a*) *With the help of P–V and T–S diagrams compare Otto, Diesel and Dual combustion cycles on the basis of same compression ratio and same heat added to each cycle.*

(*b*) *In an engine working on Diesel cycle inlet pressure and temperature are 1 bar and 17°C respectively. Pressure at the end of adiabatic compression is 35 bar. The ratio of expansion i.e., after constant pressure heat addition is 5. Calculate:*

(*i*) *Heat addition*

(*ii*) *Heat rejection*

(*iii*) *Efficiency of the cycle*

Take $\gamma = 1.4$, $C_p = 1.004$, $C_v = 0.717$ *kJ/kg K*

Data

$$P_1 = 1 \text{ bar},$$

$$T_2 = 17 + 273 = 290 \text{ K}$$

$$P_2 = P_3 = 35 \text{ bar}$$
$$r_e = 5$$

Solution

(*a*) The cycle is shown on *P–V* diagram.

First we will find the temperatures at the salient points

We have $$P_1V_1^{\gamma} = P_2V_2^{\gamma}$$

$$\therefore \quad \left(\frac{V_1}{V_2}\right)^{\gamma} = \left(\frac{P_2}{P_1}\right) = 35$$

$$\therefore \quad \frac{V_2}{V_1} = 35^{1/1.4} = 35^{0.7143}$$

$$= 12.674 \ r_k$$

$$= \text{compression ratio } r_k$$

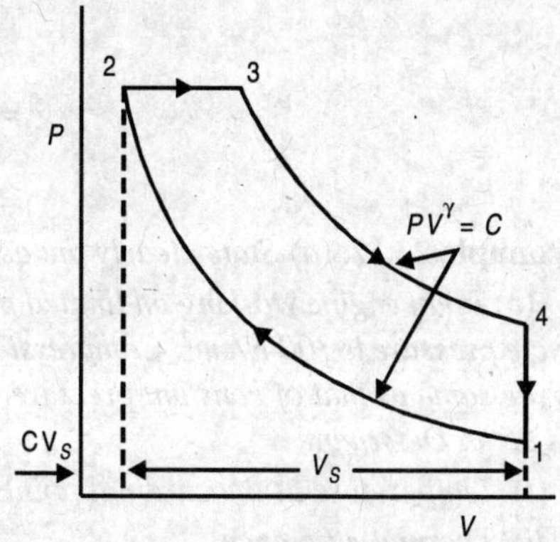

Fig. Ex. 23.11

$$\therefore \quad T_2 = T_1 \times (35)^{0.286} = 2.7644$$

$$= 290 \times 2.7644 = 301.68 \text{ K}$$

How, $$r_k = r_c \times r_e$$

$$\therefore \quad r_c = \text{cut-off ratio}$$

$$= \frac{12.674}{5} = 2.5348$$

$$\therefore \quad r_c = \frac{V_3}{V_2} = \frac{T_3}{T_4}$$

$$T_3 = 2.5348 \times T_2 = 2.5348 \times 801.68$$

$$= 2032.1 \text{ K}$$

Again, $$\frac{T_3}{T_4} = r_e^{\gamma-1} = 5^{0.4}$$

$$= 1.9037$$

$$T_4 = \frac{2032.1}{1.9037}$$

$$= 1067.45 \text{ K}$$

(i) Heat addition:

$$\text{Heat added/kg} = Q_A = C_p\left(T_4 - T_1\right)$$

$$= 1.004\left(2032.1 - 801.68\right)$$

$$= 1.004 \times 1230.42$$

$$\textbf{Heat added/kg} = \textbf{1235.34 kJ/kg}$$

(ii) Heat rejected:

$$\text{Heat rejected/kg} = C_v\left(T_4 - T_1\right)$$

$$= 0.717\left(106.45 - 290\right)$$

$$= 0.717 \times 777.45$$

Heat rejected/kg = 557.43 kJ/kg

(iii) Efficiency of cycle:

Thermal efficiency of the cycle

$$= \eta_{th} = 1 - \frac{Q_R}{Q_A}$$

$$= 1 - \frac{557.43}{1235.34} = 1 - 0.4512$$

$$= 0.5488 \text{ or } 54.88\ \%$$

Example 23.12. (*a*) *State clearly the assumptions of air standard cycle analysis.*

(*b*) *In an engine working on limited pressure at the end of adiabatic compression is 164.5 N/cm². The maximum cycle pressure is 300 N/cm². Compression begins at 1 bar and 27°C. The amount of heat added at constant volume is the same as that of constant pressure. Assume heating volume of the fuel as 41840 kJ/kg, and relative efficiency as 55%. Determine*

(*i*) *Temperature at the cardinal points of the cycle.*

(*ii*) *Thermal efficiency*

(*iii*) *Fuel consumption in kg/IP/hour*

(*iv*) *Power developed per cu-m, displacement at engine speed of 1800 rpm.*

The engine works on four stroke cycle.

Solution

Given:

$$P_1 = 1 \text{ bar}$$

$$T_1 = 27 + 273 = 300°\text{K}$$

$$P_2 = 164.5 \text{ N/cm} = 164.5 \times 10^4 \text{ N/m}^2 = 16.45 \text{ bar}$$

$$P_3 = 300 \text{ N/cm}^2 = 30 \text{ bar}$$

Cal. value of fuel used = 41840 kJ/kg and Rel. eff = 55%

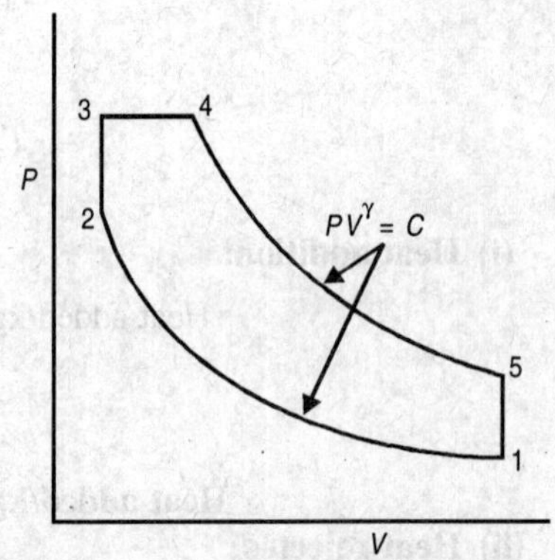

Fig. Ex. 23.12

$$\therefore \quad T_2 = T_1 \times \left(\frac{16.45}{1}\right)^{0.4/1.4} = 300 \times 2.2275$$

$$T_2 = 668.26°\text{K}$$

$$\therefore \quad r_k = (2.2275)^{3.5} = 16.5$$

Again,

$$\frac{P_3}{P_2} = \frac{T_3}{T_2} = 16.5$$

$$T_3 = T_2 \times \frac{30}{16.45} = 668.26 \times \frac{30}{16.45}$$

$$\mathbf{T_3 = 1278.71\ K}$$

$$C_v(T_3 - T_2) = 0.718(1278.71 - 668.26)$$

$$= 0.718 \times 610.45$$

$$= 438.3 \text{ kJ/kg}$$

$$= Q_{A_P} = C_p(T_4 - T_3) = 1.005[T_4 - 1278.71]$$

$$T_4 = \frac{438.3}{1.005} + 1278.71 = 436.12 + 1278.71$$

$$\mathbf{T_4 = 1714.83\ K}$$

$$\frac{T_4}{T_3} = \frac{V_4}{V_3} = r_c = \frac{1714.83}{1278.71}$$

$$= 1.341$$

$$r_e = r_k / r_c = \frac{16.5}{1.341} = 12.3$$

$$\frac{T_4}{T_5} = (r_e)^{0.4} = (12.3)^{0.4} = 2.729$$

$$T_5 = (r_e)^{0.4} = (12.3)^{0.4} = 2.729$$

$$\mathbf{T_5 = \frac{1714.83}{2.279} = 628.37\ K}$$

(ii) Thermal efficiency

Heat added/kg $= 2 \times 438.3 = 876.6$ kJ/kg

Heat rejected/kg $= 1.005\ (628.37 - 300)$

$= 330$ kJ/kg

Work done/kg $= 876.6 - 330$

$= 546.6$ kJ/kg

$$\text{ASE} = \frac{546.6}{876.6} = 0.6235$$

$$\textbf{ASE} = \mathbf{62.35\%}$$

Actual thermal efficiency $= 0.55 \times 0.6235 = 0.343$ or 34.3%

(iii) Fuel consumption

Indicated thermal efficiency

$$= 0.343$$

$$= \frac{\text{IP} \times 3600}{\text{Fuel/hr} \times \text{cal.value}}$$

$$= \frac{\text{IP} \times 3600}{\text{Fuel/hr} \times 41840}$$

$$\therefore \quad \text{Fuel consumption} = \frac{3600}{41840 \times 0.343}$$

Fuel consumption = 0.2508 kg/IP/hour

(iv) Power developed per m³ displacement

$$\text{Actual work/done kg} = 546.6 \times 0.55$$

$$= \frac{300.60}{P_m \times V_s} \text{ kJ/kg}$$

$$\therefore \quad 300.6 = P_m \times V_1 \frac{r_k - 1}{r_k}$$

$$= P_m \times \frac{10^5}{1000} \times \frac{1 \times 287 \times 300}{1 \times 10^5} \times \frac{1.2275}{2.2275}$$

$$= P_m \times 47.45$$

$$\therefore \quad P_m = \frac{300.6}{47.45} = 6.335 \text{ bar}$$

∴ Work done/m³ displacement

$$= P_m \times \frac{\pi}{4} \frac{d^2 \times L \times 1800/2 \times 60}{1000}$$

$$= \frac{6.335 \times 10^5 \times 1 \times 1800/2 \times 60}{1000}$$

$$= \frac{6.335 \times \frac{\pi}{4} d^2 \times L \times 1800/2 \times 60}{1000}$$

$$= 6.335 \times 100 \times 15 \text{ J/S}$$

Work done/m³ displacement = 9.502 kW

Example 23.13. (*a*) *Compare air standard Otto and Diesel cycle for:*

(*i*) *Same compression ratio and same heat input.*

(*ii*) *Same maximum pressure and same heat input*

(*b*) *Explain in brief why actual cycle deviates from theoritical cycle*

(*c*) *What will be loss in the ideal efficiency of a Diesel engine with compression ratio 16 if the fuel cut-off is delayed from 6 to 8%.*

Solution

First consider the case when fuel cut-off takes place at 6% stroke

$$V_3 - V_2 = 0.06\, V_s$$

$$\therefore \quad r_c = \frac{CV_s + 0.06V_s}{CV_s}$$

$$= \frac{C+0.06}{C}$$

and $$\frac{V_1}{V_2} = r_k = 16 = \frac{CV_s + V_s}{CV_s}$$

$$= \frac{C+1}{C} = 1 + \frac{1}{C}$$

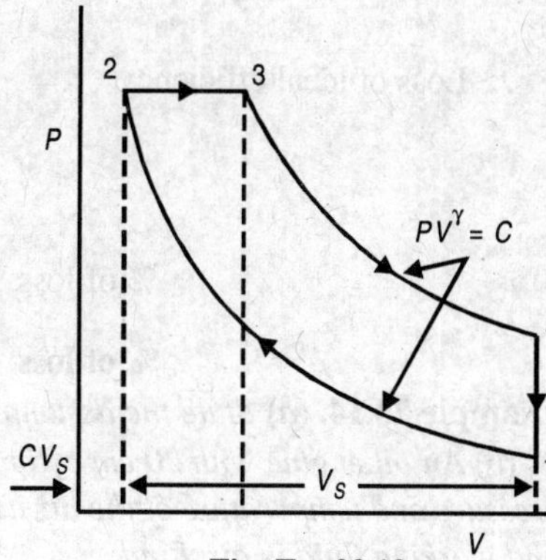

Fig. Ex. 23.13

$$\therefore \quad \frac{1}{C} = 15$$

or $$C = \frac{1}{15}$$

$$r_c = \frac{1/15 + 0.06}{1/15}$$

$$= 1 + 15 \times 0.06$$

$$= 1 + 0.90 = 1.9$$

$$\frac{V_3}{V_2} = \text{cut-off ratio}$$

$$(\text{ASE})_1 = 1 - \frac{1}{\gamma \cdot r_k^{\gamma-1}}\left[\frac{r_c^{\gamma} - 1}{r_c - 1}\right]$$

$$= 1 - \frac{1}{\gamma \cdot r_k^{\gamma-1}}\left[\frac{1.9^{1.4} - 1}{19 - 1}\right]$$

$$= 1 - \frac{1}{1.4 \times 16^{0.4}}\left[\frac{2.456 - 1}{0.9}\right]$$

$$= 1 - \frac{1}{1.4 \times 3.031}\left[\frac{1.456}{0.9}\right]$$

$$= 1 - 0.3812$$

$$= 0.6188 \text{ or } 61.88\%$$

When cut off is delayed to 8%

$$r_c = 1 + (r_k - 1)0.08$$

$$= 1 + 1.20 = 2.2$$

$$\text{ASE}_2 = 1 - \frac{1}{1.4 \times 3.031}\left[\frac{2.2^{1.4}}{2.2 - 1}\right]$$

$$= 1 - \frac{1}{4.2434} \times \frac{3.016}{1.2}$$

$$= 1 - 0.5922$$

$$= 0.4078 \text{ or } 40.78\%$$

∴ Loss of ideal efficiency

$$= 61.88 - 40.78$$

$$= 21.1\ \%$$

$$\% \text{ of loss} = \frac{21.1}{61.88} \times 100$$

$$\textbf{\% of loss} = \textbf{34.1\%}$$

Example 23.14. *(a) State the assumptions made for Air Standard Cycles.*

(b) An oil engine with 20 cm cylinder diameter and 30 cm stroke works on theoretical Diesel cycle. The initial pressure and temperature of the air used are 1 bar and 30°C. Cut-off ratio is 2.6. Assume compression ratio as 16 and working fluid is air. Find:

(i) Pressure and temperature at all salient points

(ii) Theoretical air standard efficiency

(iii) Mean Effective pressure

(iv) Power developed by the engine if the working cycle per minute are 400. Take ideal conditions.

Solution

$$P_1 = 1 \text{ bar}$$

$$T_1 = 303°K$$

$$r_k = 16$$

$$r_c = 2.6$$

$$r_e = \frac{16}{2.6} = 6.1538$$

$$P_2 = P_1 \times r_k^{\gamma} = 1 \times 16^{1.4}$$

$$= 48.503 \text{ bar}$$

$$T_2 = T_1 \times r_k^{\gamma-1} = 303 \times 16^{0.4}$$

$$= 303 \times 3.031 = 918.52$$

$$P_2 = P_3 = 48.503 \text{ bar}$$

$$\frac{V_3}{V_2} = r_c = \frac{T_3}{T_2}$$

$$\mathbf{T_3 = 2.6 \times 918.52 = 2388.152\ K}$$

$$\frac{P_3}{P_4} = r_e^{\gamma} = (6.01538)^{1.4} = 12.7292$$

$$P_4 = \frac{48.503}{12.7292} = 3.8104 \text{ bar}$$

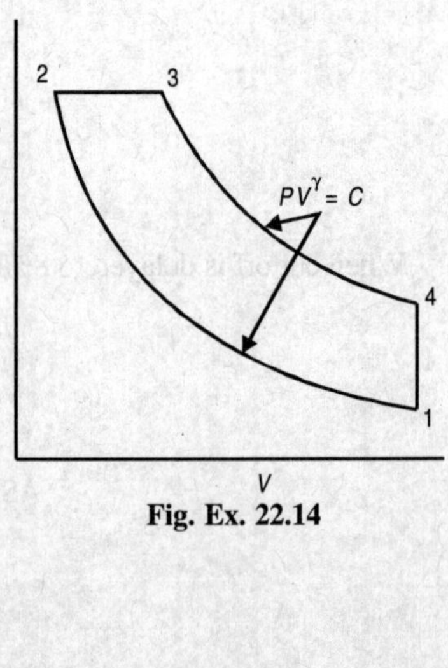

Fig. Ex. 22.14

Also $$\frac{T_4}{T_1} = \frac{P_4}{P_1} = 3.8104$$

$$T_4 = 303 \times 3.8104 = 1145.55 \text{ K}$$

(iii) Theoretical ASE

$$Q_A = \text{Heat rejected/kg} = C_p (T_3 - T_2)$$

$$= 1.005(2388.152 - 918.52)$$

$$= 1.005 \times 1469.532 = 1476.88 \text{ kJ}$$

$$\mathbf{Q_R = Heat\ rejected/kg = C_v (T_4 - T_1)}$$

$$= 0.718\ (1154.55 - 303)$$

$$\mathbf{Q_R = 609.26\ kJ}$$

$$\text{ASE} = 1 - \frac{Q_R}{Q_A}$$

$$= 1 - \frac{609.26}{1426.88}$$

$$= 1 - 0.4125$$

$$= 0.5875$$

or $= \mathbf{58.75\%}$

(iii) Mean effective pressure

$$\text{Work done/kg} = 1476.88 - 609.26$$

$$= 867.62$$

$$= P_m \times V_s$$

$$= P_m \cdot V_1 \cdot \frac{r_k - 1}{r_k}$$

$$= P_m \times \frac{RT_1}{P_1} \times \frac{r_k - 1}{r_k}$$

$$= P_m \times \frac{287 \times 10^5 \times 303}{1000 \times 10^5} \times \frac{15}{16}$$

$$P_m = \frac{867.62 \times 16}{15} \times \frac{1}{303 \times 0.287}$$

$$\mathbf{P_m = 10.64\ bar}$$

(iv) Power developed

$$\text{Mass of air/second} = \frac{P_1 V_1}{RT_1}$$

$$= \frac{1 \times 10^5}{287 \times 303} \times \frac{\pi d^2 L}{4} \times \frac{400}{60}$$

$$= 1.15\times\frac{\pi}{4}\times(0.2\times0.2)\times0.3\times\frac{400}{60}$$

$$= 1.15\times\frac{\pi}{4}\times0.04\times0.3\times\frac{400}{60}$$

$$= 0.07226 \text{ kg/sec}$$

$$\therefore \quad \text{Work/sec} = 867.62\times0.07226$$

$$\textbf{Work/sec} = \mathbf{62.7\ kJ/sec = kW}$$

Example 23.15. *(a) Derive the expression for thermal efficiency for air standard Otto cycle and prove that, for maximum work, compression rate should be*

$$r_k = \left(\frac{T_3}{T_1}\right)^{1.25}$$

where T_3 and T_1 are upper and lower and temperature limits of cycle respectively.

Take adiabatic index = 1.4

(b) Find air standard efficiency of 4-stroke engine working on otto-cycle for given data:

Bore diameter = 13.7 cm

Stroke length = 13.0 cm

Clearance vol. = 280 cm³

(c) Compare Otto, Diesel and Dual air standard cycle for some comp. ratio, same heat input. Explain with neat P–V and T–S diagram.

Solution

$$\text{Work done/kg} = Q_A - Q_R$$

$$= C_V(T_3 - T_2) - C_V(T_4 - T_1)$$

$$\text{Heat supplied/kg} = Q_A = C_V(T_3 - T_2)$$

$$\therefore \quad \text{Thermal efficiency} = \frac{Q_A - Q_R}{Q_A}$$

$$= 1 - \frac{Q_R}{Q_A} = 1 - \frac{T_4 - T_1}{T_3 - T_2}$$

We have

$$\frac{T_2}{T_1} = r_k^{\gamma-1}$$

$$\frac{T_3}{T_4} = (r_e)^{\gamma-1} = r_k^{\gamma-1}$$

$$T_3 = T_4 r_c^{\gamma-1} \text{ and } T_2 = T_1 r_k^{\gamma-1}$$

$$\text{Thermal efficiency} = 1 - \frac{T_4 - T_1}{T_4 r_k^{\gamma-1} - T_1 r_k^{\gamma-1}}$$

$$= \frac{1}{r_k^{\gamma-1}}$$

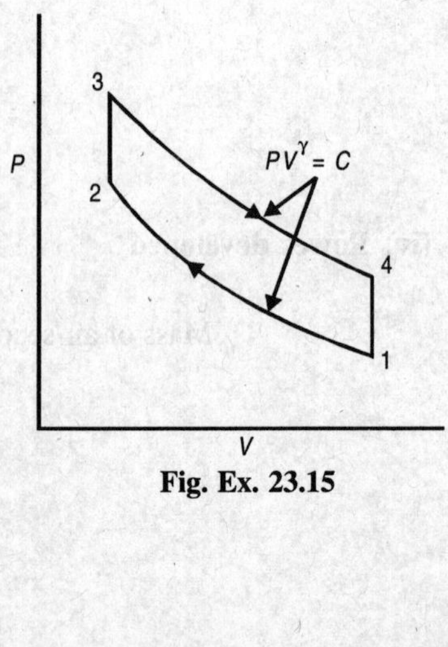

Fig. Ex. 23.15

$$\text{Work done} = C_V\left[T_3 - T_2 - T_4 + T_1\right]$$

From above $$T_2 = T_1.r_k^{\gamma-1}$$

and $$T_4 = \frac{T_3}{r_k^{\gamma-1}}$$

$$C_V\left[T_3 - T_1 r_k^{\gamma-1} - \frac{T_3}{r_k^{\gamma-1}} + T_1\right] = W$$

Work done is maximum when $\frac{dW}{dr_k} = 0$

$$-(\gamma-1)T_1 r_c^{\gamma-1-1} + T_3(\gamma-1)^{(-\gamma+1-1)r_c} = 0$$

$$T_1 r_k^{\gamma-1} = T_3 r_k^{-\gamma}$$

$$\therefore \quad \frac{T_3}{T_1} = \frac{r_k^{\gamma-2}}{r_k^{-\gamma}} = r_k^{\gamma-2+\gamma} = r_k^{2(\gamma-1)}$$

$$\therefore \quad r_c = \left(\frac{T_3}{T_1}\right)^{1/2(\gamma-1)} = \left(\frac{T_3}{T_1}\right)^{1/2(1.4-1)}$$

$$= \left(\frac{T_3}{T_1}\right)^{1/0.8} = \left(\frac{T_3}{T_1}\right)^{1.25}$$

$$\mathbf{r_k} = \left(\frac{\mathbf{T_3}}{\mathbf{T_1}}\right)^{\mathbf{1.25}}$$

(b) Air standard efficiency $= 1 - \frac{1}{r_k^{\gamma-1}}$

$$r_k = \frac{V_s + \text{cl. vol}}{\text{cl. vol}}$$

where $$V_s = \text{Stroke volume}$$

$$= \frac{\pi}{4} d_2 \times l$$

$$= \frac{\pi}{4} \times (13.7)^2 \times 3$$

$$= 1916.348 \text{ cm}^3$$

$$\therefore \quad r_k = \frac{1916.348 + 280}{280}$$

$$= \frac{2196.348}{280} = 7.844$$

$$\eta_{th} = 1-\frac{1}{(7.844)^{0.4}} = 1-\frac{1}{2.2794}$$

$$= 1 - 0.4387$$

$$\boldsymbol{\eta_{th} = 0.5613 \text{ or } 56.13\%}$$

(ii) Heat supplied/kg air

$$\text{Heat supplied/kg} = C_p (T_3 - T_2)$$

$$= 1.005[1753 - 873.05]$$

$$= 1.005 \times 879.95$$

$$\textbf{Heat supplied/kg} = \mathbf{884.35\ kJ/kg}$$

(iii)
$$\frac{T_3}{T_4} = r_e^{\gamma-1} = \left(\frac{r_k}{r_c}\right)^{\gamma-1}$$

$$= \left(\frac{16}{2.0079}\right) = (7.9685)^{0.4} = 2.294$$

$$\therefore \quad T_4 = \frac{1753}{2.294} = 764.17 \text{ K}$$

$$\therefore \quad Q_R = \text{Heat rejected} = C_V (T_4 - T_1)$$

$$= 0.718(764.17 - 288) = 0.718 \times 476.17$$

$$= 341.89 \text{ kJ/kg}$$

$$\therefore \quad \text{Work done/kg} = Q_A - Q_R = (884.35 - 341.89)$$

$$= 542.46 \text{ kJ/kg}$$

The cycle efficiency

$$= 1-\frac{Q_R}{Q_A} = 1-\frac{341.89}{884.35} = 1 - 0.3866$$

The cycle efficiency = 0.6134 i.e. 61.344%

(iv) MEP

$$\text{Work done} = P_m \times \text{Stroke volume}$$

$$\text{Stroke volume} = V_s = V_1 - V_2 = V_1\left[1-\frac{V_2}{V_1}\right]$$

$$= V_1\left[1-\frac{1}{r_k}\right] = V_1 \times \frac{r_k - 1}{r_k}$$

$$\therefore \quad 542.46 \times 1000 = P_m \times 10^5 \times \frac{16-1}{16}\left[\frac{RT_1}{P_1}\right]$$

$$= P_m \times 10^5 \times \frac{15}{16}\left[\frac{287\times 288}{1\times 10^5}\right]$$

$$= P_m \times 10^5 \times 0.7749$$

$$P_m = \frac{542.46\times 1000}{10^5 \times 0.7749} = \frac{5.4246}{0.7749}$$

$$\mathbf{P_m = 7}$$

Example 23.16. (*a*) *Compare the limitations of air standard cycle. What are the assumptions of fuel air cycle?*

(*b*) *The parameters of the initial stage of 1 kg of air in the cycle of an IC Engine are 0.95 bar, 65°C. The compression ratio is 11. Compare the values of ideal thermal efficiency for*

(*i*) *Constant volume cycle*

(*ii*) *Constant pressure cycle*

(*iii*) *Dual combustion cycle with equal heat supplied in constant pressure and constant volume heat supply process.*

Assume total heat supplied in the cycle 800 kJ.

Solution

(i) Constant volume cycle:

$$\text{Thermal efficiency} = 1 - \frac{1}{r_k^{\gamma-1}}$$

$$= 1 - \frac{1}{11^{0.4}} = 1 - \frac{1}{2.6095}$$

$$= 1 - 0.3832$$

$$\textbf{Thermal efficiency} = \mathbf{0.6168 \text{ or } 61.68\%}$$

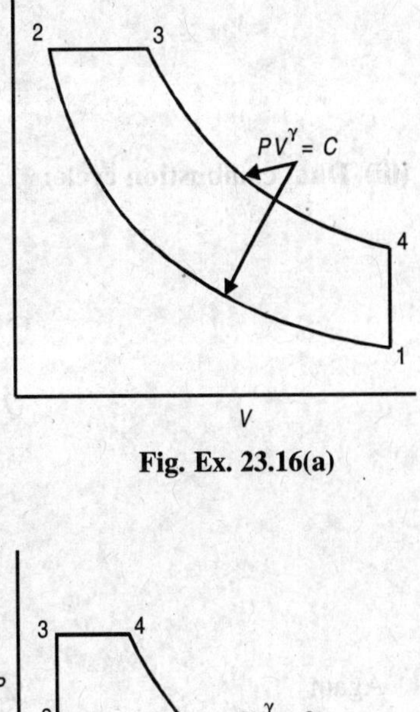

Fig. Ex. 23.16(a)

(ii) Constant pressure cycle

$$P_1 = 0.95 \text{ bar}$$

$$T_1 = 273 + 65 = 338 \text{ K}$$

$$\frac{T_2}{T_1} = r_k^{\gamma-1} = 11^{0.4} = 2.6095$$

$$T_2 = 338 \times 2.6095$$

$$= 882.01 \text{ K}$$

$$\eta = \frac{W}{Q_A} = \frac{Q_A - Q_R}{Q_A}$$

$$= \frac{800 - 314.81}{800} = \frac{485.19}{800}$$

$$= \mathbf{0.6065 \text{ or } 60.65\%}$$

$$mC_p(T_3 - T_2) = 1.005(T_3 - 882.011) = 800$$

$$T_3 - 882.011 = \frac{800}{1.005} = 796.02$$

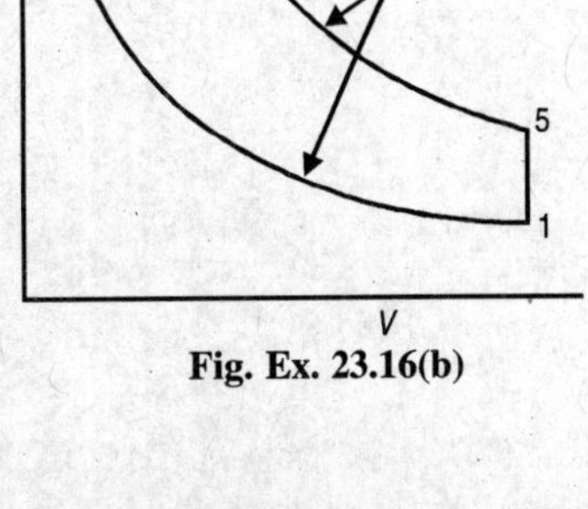

Fig. Ex. 23.16(b)

$$T_3 = 796.02 + 882.01$$
$$= 1678.031 \text{ K}$$

$$\frac{V_3}{V_2} = \frac{T_3}{T_4} = r_e = \frac{1678.031}{882.011} = 1.9025$$

Expansion ratio
$$r_e = \frac{r_k}{r_c} = \frac{11}{1.9025}$$
$$= 5.782$$

$$\therefore \quad \frac{T_3}{T_4} = r_e^{\gamma-1} = 5.782^{1.4-1} = 2.0176$$

$$\therefore \quad T_4 = \frac{T_3}{2.0176} = \frac{1678.031}{2.0176} = 831.7 \text{ K}$$

$$Q_R = C_V(T_4 - T_1) = 0.718[831.7 - 338]$$
$$= 0.781 \times 493.7$$
$$= 354.48 \text{ kJ}$$

$$\therefore \quad \eta = 1 - \frac{Q_R}{Q_A} = 1 - \frac{354.48}{800} = 1 - 0.4431$$

$$\boldsymbol{\eta = 0.5569 \text{ or } 55.69\%}$$

(iii) Dual combustion cycle:

$$P_2 = 0.95 \times 11^{1.4} = 0.95 \times 28.7$$
$$= 27.27 \text{ bar}$$
$$T_2 = 882.11 \text{ K}$$
$$Q_{AV} = 400\text{kJ} = (T_3 - T_2) \times C_V$$
$$= 0.718(T_3 - 882.011)$$
$$T_3 = \frac{400}{0.718} + 882.011$$
$$= 557.1 + 882.01 = 1439.11 \text{ K}$$

Again
$$Q_{A_p} = 400 = 1.005[T_4 - T_3]$$
$$= 1.005[T_4 - 1439.11]$$
$$T_4 = \frac{400}{1.005} + 1439.11 = 398 + 1439.11$$
$$= 1837.11 \text{ K}$$

$$R_c = \frac{11}{1.277} = 9.614$$

$$\frac{T_4}{T_5} = 8.164^{0.4} = 2.366$$

$$T_5 = \frac{1837.11}{2.366} = 776.46°C$$

$$Q_R = 0.718[776.46-338] = 0.718 \times 438.46 = 314.81 \text{ kJ}$$

$$\therefore \quad \eta = \frac{W}{Q_A} = \frac{Q_A - Q_R}{Q_A}$$

$$= \frac{800-314.81}{800} = \frac{485.19}{800}$$

$$= \mathbf{0.6065 \text{ or } 60.65\%}$$

EFFECT OF VARIABLE SPECIFIC HEAT ON OTTO-CYCLE EFFICIENCY

Example 23.17. *What is the effect of percentage change in the efficiency of Otto-cycle having a compression ratio of 7, if the specific heat at constant volume increases by 1%?*

Solution

The Otto-cycle efficiency is given by

$$\eta = 1 - \frac{1}{r_K^{\gamma-1}}$$

We have $C_p - C_v = R$ and $\gamma - 1 = \frac{R}{C_v}$ as $C_v = \frac{R}{\gamma-1}$

$$\therefore \quad \eta = 1 - \frac{1}{r_K^{R/C_v}}$$

$$\therefore \quad 1-\eta = r_K^{R/C_v} \quad \text{or} \quad \log_e(1-\eta) = -\frac{R}{C_v}\cdot \log_e r_K$$

Differentiating,

$$-\frac{1}{1-\eta}\cdot d\eta = \frac{R}{C_v^2}\cdot \log_e r_K \cdot dC_v$$

or

$$d\eta = -(1-\eta)\frac{R}{C_v^2}\cdot \log_e r_K \cdot dC_v$$

or

$$\frac{d\eta}{\eta} = -\frac{(1-\eta)}{\eta}\cdot(\gamma-1)\cdot \log_e r_K \cdot \frac{dC_v}{C_v} \quad \text{....(1)}$$

$$\eta = 1 - \frac{1}{r_K^{\gamma-1}} = 1 - \frac{1}{T^{1.4-1}} = 0.54 \text{ or } 54\ \%$$

$$\therefore \quad \frac{d\eta}{\eta} = -\frac{1-0.54}{0.54}\times(1.4-1)\log_e 7\times\frac{1}{100}$$

$$= -\frac{0.46\times0.4}{54}\times[\log_e 7 = 1.946] = -0.00663$$

$$= \mathbf{-0.663\ \%\ (approx.)}$$

OTTO-CYCLE: EFFECT OF VARIABLE SPECIFIC HEAT ON MAXIMUM PRESSURE

Example 23.18. *A petrol engine of compression ratio 6 uses a fuel of calorific value* 44000 kJ/kg. *The air to fuel ratio is 15 : 1. The pressure and temperature of the charge at the end of suction stroke are 1 bar and 60°C respectively. Determine the maximum pressure in the cylinder if the index of compression is 1.32 and the specific heat at constant volume is expressed by the relation* $C_v = \frac{0.712}{0.712} + \frac{20.94T^{-5}}{20.94}$ *where T is the temperature in K. Compare this value with that of constant specific heat* $C_v = (0.712\ \text{kJ/kg°K})$.

Solution

We have

$$p_2 = p_1,\ r_K^{n-1} = 1\times6^{1.32}$$

$$= 10.64 \text{ bar}$$

$$T_2 = T_1\times r_K^{n-1} = 333\times6^{0.32}$$

$$= 333\times1.772$$

$$= \mathbf{590\ K}$$

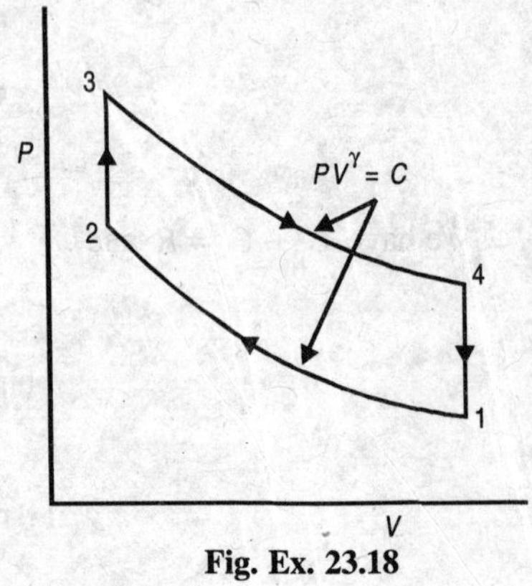

Fig. Ex. 23.18

Assume 1 kg air : Fuel $= \frac{1}{15}$ kg

$$\therefore \quad \text{heat supplied} = \frac{44000}{15} = \mathbf{2933.33\ kJ}$$

$$\text{Mass of charge} = 1+\frac{1}{15} = \frac{16}{15} \text{ kg}$$

$$\therefore \quad Q = \text{Heat added} = \text{mass}\times\text{specific heat}\times(T_3 - T_2)$$

But we have

$$dQ = m\cdot C_v\cdot dT = m\cdot\left(0.712+20.94\,T^{-5}\right)\cdot dT$$

$$\therefore \quad Q = m\int\left(0.712+20.94\,T^{-5}\right)\cdot dT$$

$$\therefore \quad 2933.33 = \frac{16}{15}\left[0.712\,T - \frac{20.94\times T^{-4}}{4}\right]_{T_2}^{T_3}$$

$$\therefore \quad \frac{2933.33\times15}{16} = 0.712(T_3 - 590) - 5.235\left(T_3^{-4} - T_2^{-4} - 590\right)$$

We can solve this by another method:

If T_3 is the maximum temperature after heat addition – then mean temperature during heating is $\frac{T_3+T_2}{2}$.

$$\therefore \quad \frac{2933.33\times15}{16} = \left[0.712+20.94\times10^{-5}\left(\frac{T_3+590}{2}\right)\right]\times(T_3-590)$$

$$2750 = \left(1.424+20.94\,T_3\times10^{-5}+12354.6\times10^{-5}\right)(T_3-590)$$

$$2750 = 10^{-5}\left(\frac{1.424}{10^{-5}}+20.94\,T_3+12354.6\right)(T_3-590)$$

$$10^5\times2750 = (142400+20.94\,T_3+12354.6)(T_3-590)$$

$$= (20.94\,T_3+154754.6)(T_3-590)$$

$$= 20.94\,T_3^2+154754.6\,T_3-91305214-12354.6\,T_3$$

$$= 20.94\,T_3^2+142400\,T_3-91305014$$

$$10^5\times131.3 = T_3^2+6800\,T_3-4.36\times10^6$$

$$0 = T_3^2+6800\,T_3-10^6(4.36+13.13)$$

$$= T_3^2+6800\,T_3-17.49\times10^6$$

$$T_3 = 3095 \text{ K by trial and error}$$

$$\therefore \quad p_3 = 10.64\times\frac{3095}{590}$$

For normal cycle, (without variation of specific heat)

$$2750 = 0.712(T_3-590)$$

$$\therefore \quad T_3 = \frac{2750}{0.712}+590$$

$$= 3862.36+590$$

$$= 4452.36\ ^\circ\text{K}$$

$$\therefore \quad \frac{p_3}{p_2} = \frac{T_3}{T_2}$$

or

$$p_3 = p_2\times\frac{T_3}{T_2} = 10.64\times\frac{4452.36}{590}$$

$$= \mathbf{80.3\ bar}$$

$$\frac{2933.33 \times 15}{16} = \left[0.712 + 20.94 \times 10^{-5}\left(\frac{T_3 + 590}{2}\right)\right](T_3 - 590)$$

$$2750 \times 10^5 = \left[71200 + 20.46\left(\frac{T_3 + 590}{2}\right)\right](T_3 - 590)$$

$$= [71200 + 10.46\,T_3 + 6171.4](T_3 - 590) + 6171.4$$

$$= [77371.4 + 10.46\,T_3](T_3 - 590)$$

$$= 77371.4\,T_3 + 10.46\,T_3^2 - 45.65 \times 10^6 - 6171.4\,T_3$$

$$= 71200\,T_3 + 10.46\,T_3^2 - 45.65 \times 10^6$$

$$= 71200\,T_3 + 10.46\,T_3^2 - (45.65 + 275)10^6$$

$$= 71200\,T_3 + 10.46\,T_3^2 - 320.65 \times 10^6$$

$$= 6807\,T_3 + T_3^2 - 30.65 \times 10^6$$

$$= T_3^2 + 6807\,T_3 - 30.65 \times 10^6$$

By trial and error method

$$T_3 = 3095 \text{ K}$$

$$\therefore \quad p_2 = p_1 \times \frac{3095}{590}$$

$$= \frac{10.64 \times 3095}{590}$$

$$= \mathbf{55.8 \text{ bar}}$$

DIESEL ENGINE – % STROKE IN CONE

Example 23.19. *In a Diesel engine combustion is assumed to begin at IDC and to be at constant pressure. The air to fuel ratio is 28 to 1, the calorific value of the fuel is 42000 kJ/kg and the specific heat of the products of combustion is given by –*

$$C_v = 0.712 + 18.85 \times 10^{-5} T$$

$$R \text{ for the products} = 287 \text{ N-m/kg/K}$$

If the compression ratio is 14 : 1, and the temperature at the end of compression is 800 K. What percentage of the stroke during combustion is completed.

Solution

For one kg fuel the charge is (28 + 1 = 29 kg) and the heating value is 42000 kJ/kg.

$$\therefore \quad dQ = \int_2^3 C_p \cdot dT$$

But $\quad C_p = R + C_v$

$$= 0.287 + 0.712 + 18.82 \times 10^{-5} T$$

$$= 0.999 + 18.85 \times 10^{-5} T$$

$$\therefore \quad \frac{42000}{29} = \int_2^3 \left(0.999 + 18.85 \times 10^{-5} T\right) dT$$

$$1448.3 = \left[0.999\,T + \frac{18.85 \times 10^{-5}\,T^2}{2}\right]_{800}^{T_3}$$

$$= 0.999(T_3 - 800) + 9.425 \times 10^{-5}\left(T_3^2 - 800^2\right)$$

$$= 0.999\,T_3 - 800 + 9.425 \times 10^{-5}\,T_3^2 - 60.32$$

$$0 = 9.425 \times 10^{-5}\,T_3^2 + 0.999\,T_3 - (860.32 + 1448.3)$$

$$= 9.425 \times 10^{-5}\,T_3^2 + T_3 - 2308.62$$

$$= T_3^2 + 10610\,T_3 - 244.95 \times 10^5$$

$$= T_3^2 + 10610\,T_3 - 24.495 \times 10^6$$

For $T_3 = 2000$, $\quad 0 = 4 \times 10^6 + 21.22 \times 10^6 - 24.495 \times 10^6$

$$= 25.22 \times 10^6 - \mathbf{24.495 \times 10^6}$$

For $T_3 = 1990 \rightarrow \quad 0 = \left(3.96 \times 10^6 + 21.11 \times 10^6\right)$

$$= 25.07 \times 10^6$$

$1980 \rightarrow \quad = (3.92 + 21.01 = 24.93) \times 10^6$

$1975 \rightarrow \quad = 3.90 + 20.95 = 24.85 \times 10^6$

$1960 \rightarrow \quad = (3.84 + 20.8 = 24.64) \times 10^6$

$1950 \rightarrow \quad = 3.80 + 20.69 = \mathbf{24.49 \times 10^6}$

$\therefore \quad \mathbf{T_3 = 1950^o}$

$\therefore$ Cut off ratio $= \dfrac{V_3}{V_2} = \dfrac{T_3}{T_2} = \dfrac{1950}{800}$

$$= \mathbf{2.4375}$$

$\therefore$ Combustion occupies

$$= \frac{1.4375}{13} \times 100$$

$$= \textbf{11.06 \% stroke}$$

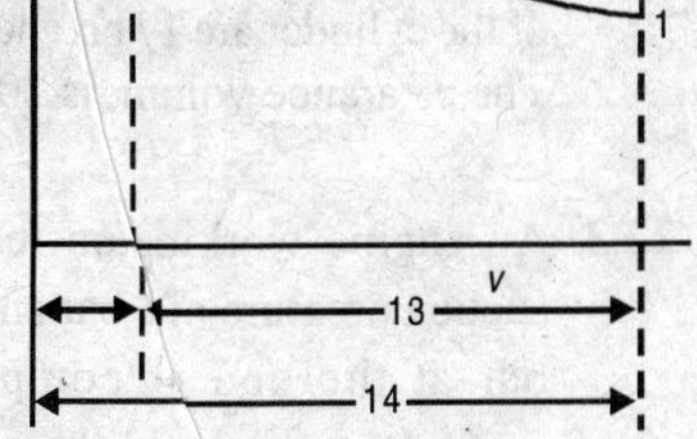

Fig. Ex. 23.19

ACTUAL CYCLE

Actual cycle efficiency is much lower than the air standard efficiency due to various losses occurring in the actual engine operation. These losses are —

(1) Losses due to variation of specific heats with temperature.
(2) Chemical equilibrium losses or dissociation losses.
(3) Time losses.
(4) Losses due to incomplete combustion.
(5) Direct heat losses.
(6) Exhaust blow-down losses.
(7) Pumping losses.

ASE > Fuel-Air Cycle efficiency > Actual cycle efficiency

Air Cycle → Correction for the characteristics of fuel-air medium becomes → Fuel-Air Cycle

i.e., Air Cycle → Variable specific heat losses + dissociation losses → Fuel-Air cycle

Fuel-air cycle → losses (listed above) → Actual Cycle

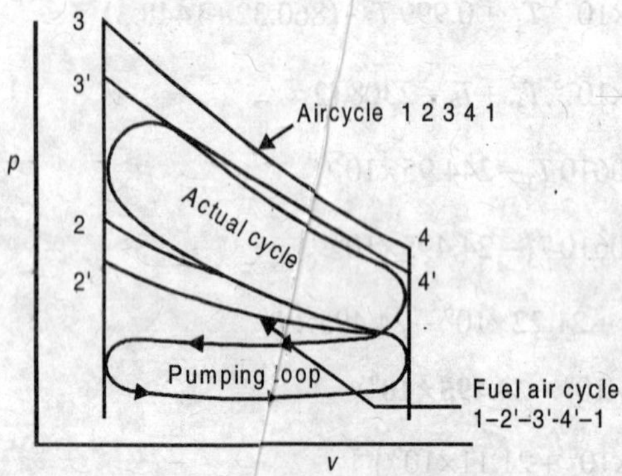

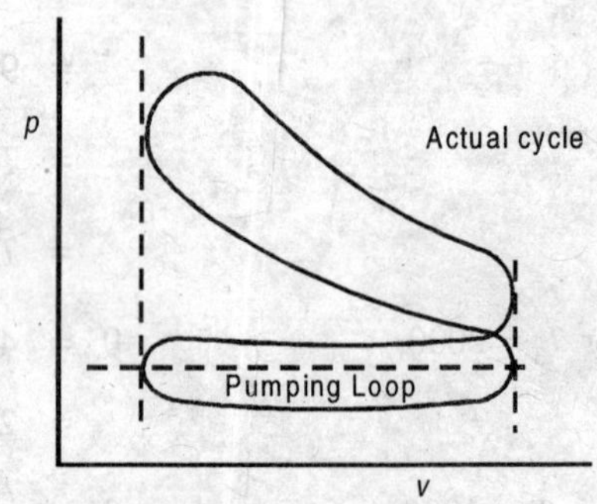

PROBLEMS FOR PRACTICE

23.1. An air engine working on Otto cycle the compression ratio raised from 5 to 6. Compare the change in efficiency due to the rise in compresion ratio. [*Ans.* 8.12% rise]

23.2. Calculate the thermal eficiency of an engine working on the Otto cycle. The bore and stroke of the cylinder are 17 cm and 30 cm respectively. The clearance volume is 2025 cm^3.

[*Ans.* 44.45%]

23.3. An engine working on the Otto cycle has a suction pressure of 1 bar and a pressure of 13.5 bar at the end of compression. Find (a) compression ratio (b) the percentage clearance (c) thermal efficiency (d) if the initial temperature is 27°C, find the temperature at the end of compression (e) if the pressure at the end of combustion is 23 bar, what is the mean effective pressure?

[*Ans.* (a) 6.65% (c) 53.18% (d) 639 K (e) 2.385 bar]

23.4. The maximum and minimum pressure in an Otto cycle are 21 bar and 1 bar. With thermal efficiency of 47.5 %, determine (a) compression ratio (b) mean effective pressure. The working substance has $C_p = 1.055$, $C_V = 0.754$.

[*Ans.* (a) 5 (b) 3.4 bar]

23.5. Prove that for Otto cycle the optimum compression ratio for maximum work done is given by

$$r_k = \left(\frac{T_3}{T_1}\right)^{1.25}$$

where T_3 = Max. temperature

T_1 = Min. temperature

23.6. In an air standard Diesel cycle, compression ratio is 16 and compression begins at 1 bar and 300. Heat added pre kg is 1500 kJ/kg. Find:

(a) Cut-off ratio

(b) Presure and temperature at the end of expansion

(c) Thermal efficiency

(d) Mean effective pressure

(e) Power output if flow rate of air is 0.1 kg/sec.

[*Ans.* (a) 0.1832 (b) 1905.86 K, 6.36 bar (c) 35.77% (d) 11.11 bar (e) 89.42 kW]

23.7. A diesel engine has a cylinder of 25 cm diameter and 40 cm of stroke. Cut-off takes place at 5% of the stroke. Estimate (a) Air standard efficiency (b) Mean effective pressure. Assume clearance volume 1.2 litres and pressure at the start of suction 1 bar. For 4 stroke operation at the start of suction 1 bar. For 4 stroke operation, determine power developed by the engine at 5 rev/sec.

[*Ans.* (a) 63.7% (b) 6.05 bar (c) 29.7 kW]

23.8. In an engine working on air standard Diesel cycle, the compression ratio is 16. At the beginning of compression (isentropic), the conditions are pressure 1 bar and temperature 15°C. Heat is added at constant pressure till temperature is 1480°C. Calculate:

(i) Cut-off ratio (ii) Heat supplied (iii) Cycle efficiency (iv) Mean effective pressure.

[*Ans.* (i) 2.008 (ii) 884.35 kJ/kg (iii) 61.33% (iv) 7 bar]

23.9. A gas engine working on Otto cycle has 200 mm bore and stroke 300 mm. Clearance volume of 1.78 litres. Engine when running at 300 rpm the number of firing cycles per min. are 135. Indicated mep is 518 kPa. Gas consumption is 8.8 m^3/hour and C_v of gas = 16350 kJ/m^3. Determine:

(i) Air standard efficiency

(ii) Indicated power and indicated thermal efficiency.

[*Ans.* (i) 52 % (ii) 12.21 kW, 30.56 5]

23.10. An air standard Otto-cycle with a compression ratio of 3.55. At the begining of the compression the air is at 98 kPa and 32°C. During the heat addition process the pressure is trippled. The cylinder volume is 2800 cm^3.

Determine:

(i) The thermal efficiency of the cycle by computing heat added and heat rejected.

(ii) Mean efective pressure of the cycle.

Take for air C_p = 1.005 kJ/kg–°K

C_V = 0.718 kJ/kg–°K

[*Ans.* (i) 57.63% (ii) 7.551 bar]

23.11. An air standard dual cycle has a compression ratio of 15:1. The intake conditions of air are 0.95 bar and 17°C. The pressure ratio during constant volume heating is 15:1 and the volume ratio during the constant pressure part is 1.8:1. Calculate:

(i) Maximum pressure and maximum temperature attained in the cycle.

(ii) The heat input and heat rejection

(iii) The thermal efficiency of the cycle per kg per cycle.

Assume of air γ = 1.4, R = 0.287 kJ/kg–°K

[*Ans.* (i) 63.146 bar, 2313.2 K (ii) 1337.4 kJ/kg (iii) 62.81%]

23.12. A dual combustion cycle has compression ratio of 15 to 1 with air suction at 1 bar, 25°C. The maximum pressure of the cycle is limited to 55 bar. If the heat supplied at constant volume is twice that supplied at constant pressure,

Determine:

(i) Explosion ratio

(ii) Cut-off ratio and

(iii) Thermal efficiency

[*Ans.* (i) 1.24 (ii) 65.95%]

23.13. The following data refers to an air standard Diesel cycle –

Maximum temperature 1600 K

Temperature after expansion 700 °K

Condition of air before compression 90 kPa, 300°K

Determine:

(i) Ratio of compression

(ii) Air standard efficiency

[*Ans.* (i) 14.44 (ii) 61.9%]

23.14. An engine develops 50 kW shaft output at full load with mechanical eff. 75%. If mechanical loss remain constant, determine its mechanical efficiency at half load. [*Ans.* 60%]

23.15. A stirling cycle works between temperature of 500°C and 50°C, the ratio of isothermal expansion being 401. Calculate the cycle efficiency if regenerator returns only 85% of heat supplied to it. [*Ans.* 58.21%]

23.16. In the analysis of Diesel cycle the temperatures recorded at the salient points are as follows. T_1 = 300 °K, T_2 = 909 °K, T_3 = 1500 °K, T_4 = 604.6

(i) With cut-off ratio held fixed, to what value must the compression ratio be changed if the thermodynamic efficiency of the cycle is to be increased by 5%?

(ii) By how much [kJ/kg] is heat added per cycle changed?

Assume for air C_p = 1.005 kJ/kg.°K, γ = 1.4

[*Ans.* (i) 20.13 (ii) 53.34 kJ/kg]

23.17. The mean effective pressure of an ideal Diesel cycle is 8 barand the ratio of compression is 12. If suction pressure is 100 kPa, determine the cut-off ratio.

Derive the formula you may use. [*Ans.* 2.42]

23.18. The initial conditions for an Otto cycle operating with a compression ratio of 8 are 1 bar, 27°C. At the beginning of compression stroke, the volume of air is 3.8 litres and 5.7 kJ of heat energy is supplied to the air during constant volume heating process.

Calculate the pressure and temperature at the end of each process and determine mean effective pressure of the cycle.

[*Ans.* 18.38 bar, 690 K, 66.33 bar, 2490 K, 3.61 bar, 1082 K]

23.19. In an ideal Diesel cycle the pressure and temperature at the beginning of the adiabatic compression are 98.5 KN/m² and 60°C respectively. The maximum pressure attained during the cycle is 4.5 MN/m² and the heat received during the cycle is 580 kJ/kg gas.

Determine, for the cycle,

(i) Volume ratio of compression.

(ii) The temperature at the end of compression.

(iii) The temperature at the end of heat reception.

(iv) The temperature at the end of adiabatic expansion.

(v) The net work done/kg of gas.

(vi) The thermal efficiency.

Take g = 1.4, C_P = 1.004

[*Ans.* (i) 15.33 (ii) 992.4 (iii) 1577 (iv) 635.2 k (v) 363.3 kJ/kg (vi) 62.698]

23.20. (a) Derive an expression for thermal efficiency of an ideal dual combustion cycle in terms of compression ratio, explosion ratio, fuel cut-off ratio and specific heat ratio. Hence obtain an expression for ASE of Otto-Cycle and that of diesel cycle.

(b) An engine works on the constant volume cycle. It has a bore, of 80 mm and a stroke of 85 mm. Clearance volume of the engine is 0.06 litter. The actual thermal efficiency of the engine is 22 %.

Determine the relative efficiency of the engine. Take γ = 1.4 [*Ans.* 38.8 %]

23.21. An air standard dual cycle has a compression ratio 16 and compression begins at 1 bar, 50°C. The maximum pressure is 70 bar. The heat transferred to air at constant pressure is equal to that at constant volume.

Estimate:

(i) The pressure and temperature at the points of the cycle.

(ii) The cycle efficiency.

(iii) The mean effective pressure of the cycle.

Take C_V = 0.718, C_P = 1.005kJ/kg °K.

[**Ans.** (i) 48.5 bar, 979 K, 70 bar, 1413 k, 70, 1723 K, 1.91 bar, 615 K (ii) 66.32 % (iii) 4.458 bar]

IMPORTANT POINTS TO BE REMEMBERED

1. *Otto-cycle* (*Constant volume cycle*)
   Air standard cycle efficiency

$$\text{ASE} = 1 - \frac{1}{r_k^{\gamma-1}}$$

For maximum work, $T_2 = T_4 = \sqrt{T_1 T_3}$

Mean effective pressure = MEP = P_M

$$= \frac{\text{Work done during cycle}}{\text{Stroke volume}}$$

2. *Diesel cycle* : (*Constant Pressure Cycle*)

$$\text{ASE} = 1 - \frac{r_k^{\gamma-1}}{\gamma, r_k^{\gamma-1}(r_c - 1)}$$

$$r_c = \text{Cut off ratio}$$

$$= \frac{\text{Volume after heat addition}}{\text{Volume before heat addition}}$$

3. *Stirling cycle* (*Regenerative Cycle*)

$$\text{ASE} = \frac{T_2 - T_1}{T_1} = \frac{\text{Higher temp.} - \text{Lower temp.}}{\text{Higher temp}}$$

CHAPTER 24

Introduction to IC Engines

24.1 INTRODUCTION

Internal combustion engines or IC engines are extensively used in automobiles, locomotives, marine applicaiton, power generation etc. Here the working media is hot and high pressure products of combustion of air and gasoline/ diesel fuel. The combustion occurs internally within a cylinder and hence the name. The hot high pressure gases push down a piston in the cylinder producing power.

24.2 CLASSIFICATION

IC engines can be classified based on

(a) Type of fuel used—petrol, diesel, gas
(b) Type of mechanical cycle—2 stroke, 4 stroke
(c) Type of thermodynamic cycle used. Otto cycle (for gasoline engine and gas engine), diesel, dual cycle (for diesel engines)
(d) Type of ignition used. Spark ignition or SI engines/Compression Ignition or CI engines
(e) According to inlet conditions. Naturally aspirated or supercharged
(f) Type of cooling—air cooled, water cooled
(g) Number of cylinders—Single cylinder or multi cylinder
(h) Accoriding to axis of cylinders—Horizontal, Vertical or Radial
(i) According to fuel injection in Diesel engine—Air injection or solid injection.
(j) According to use—Automobile engines, Marine engines, Locomotive engines, Land engines.

24.3 IC ENGINE AND COMPONENTS

4-stroke SI engine is shown in Fig. 24.1. Main parts of the internal combustion engine are—

(a) **Cylinder.** It is the main body of the engine. It has finely machined bore in which piston with rings closely fits and is able to reciprocate. In the cylinder combustion takes place. It should be able to withstand high temperature and pressure. It is necessary to cool the engine to avoid damage.

(b) Cylinder Head. The cylinder head closes the top side of the cylinder. It has inlet and exhaust valves and the spark plug to initiate combustion, in case of petrol engine. In case of diesel engines it has fuel injector instead of the spark plug.

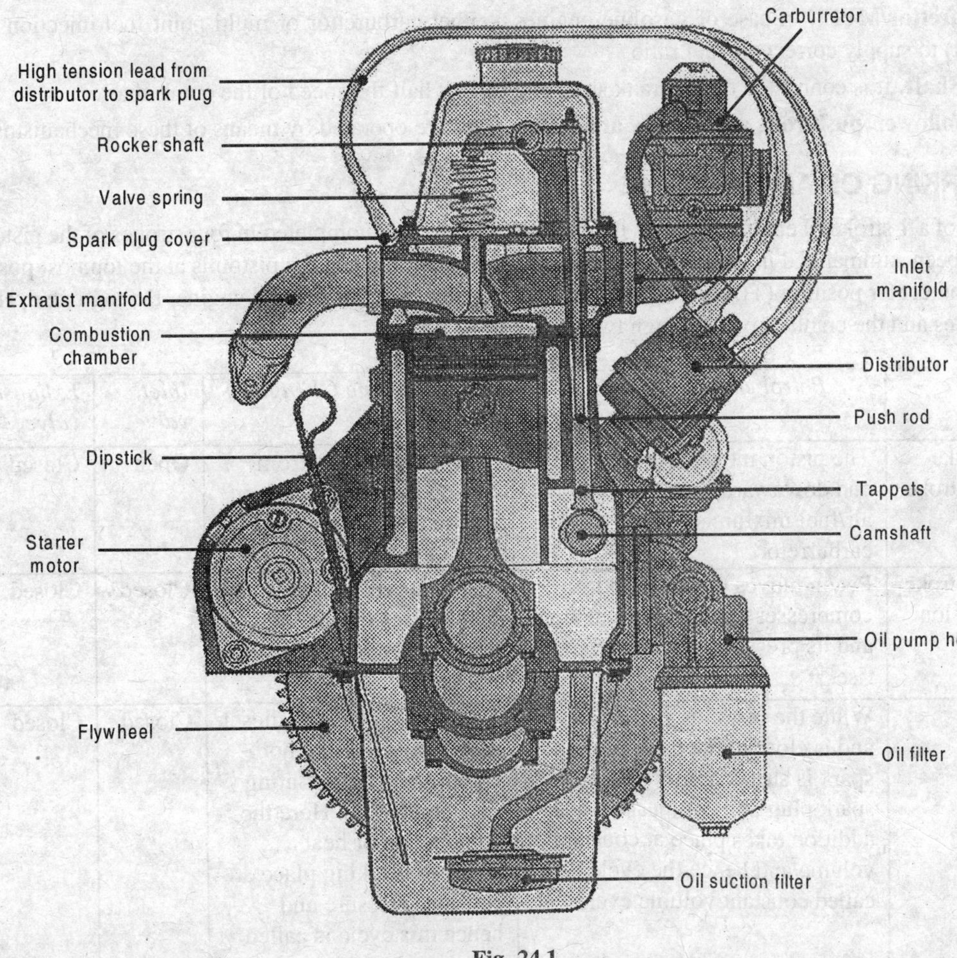

Fig. 24.1

(c) Piston. The piston is able to reciprocate i.e., move up and down in the cylinder. On the upward stroke i.e., while raising towards the cylinder head it compresses the charge of air/fuel. Also during and after combustion it pushes down the piston. For sealing purpose there are piston rings on the piston.

(d) Connecting Rod. The movement/force on the piston is passed to the crankshaft through the connecting rod that is attached to the piston by piston pin or gudgeon pin. The end of the connecting rod connecting the piston side is called the small end of the connecting rod.

(e) Crank Shaft. The big end of the connecting rod is attached to the crankshaft through crank that converts the reciprocating motion of the piston to rotary motion.

(f) Crank. The big end of the connecting rod is connected to the crank by means of a crank pin while the other end is connected to the crank shaft. Cranks may be centre crank in between two bearings or over hung crank on one side of the bearing.

(g) **Inlet and exhaust valve.** These allow fresh charge and let out spent gases respectively at proper times in the cycle.

(h) **Flywheel.** It stores energy during power stroke and when required distributes energy during other storkes i.e. suction, compression and exhaust.

(i) **Carburettor/MPFI.** In case of gasoline engines—either carburettor or multi point fuel injection system (MPFI) to supply correct air fuel ratio

(j) **Cam Shaft.** It is connected to the crank shaft and runs at half the speed of the crank shaft.

(k) **Cam follower, push rods and rocker arm.** The valves are operated by means of these mechanisms.

24.4 WORKING OF AN IC ENGINE

The working of a 4-stroke SI engine is shown in Fig. 24.2. The cycle is completed in four strokes of the piston. The working has been summarized in the table given below. The position when the piston is at the topmost position is called top dead center position (TDC)/IDC inner dead centre in case of horizontal engine. During one cycle there are four strokes and the crank moves through two revolutions.

Stroke	*Petrol engine SI engines*	*Diesel engine CI engines*	*Inlet valve*	*Exhaust valve*
First stroke suction stroke	The piston moves from the top end downwards and sucks in air/fuel mixture supplied by the carburretor.	The piston moves from the top end downwards and sucks only air in the case of Cl engine.	Open	Closed
Second stroke compression stroke	Piston moves upwards and compresses the charge of air/fuel and its pressure and temperature rise.	Piston moves upwards and compresses the air. Its pressure and temperature rises.	Closed	Closed
Ignition	While the piston is still rising and is close to the top end the spark is struck across is the spark plug and combustion heat addition takes place at constant volume and hence the cycle is called constant volume cycle.	Near the top end the diesel fuel is injected into hot compressed air initiating the combustion. Here the combustion or heat addition is taking place at constant pressure and hence this cycle is called constant pressure cycle.	Closed	Closed
Third stroke working or expansion stroke or power stroke	Due to high pressure and high temperature the piston is pushed and the gas expands and work on the piston.	Same as in petrol engine.	Closed	Closed
Fourth stroke exhaust stroke	The spent gases are exhausted out through the exhaust valve as the piston moves upwards.	– Do –	Closed	Open

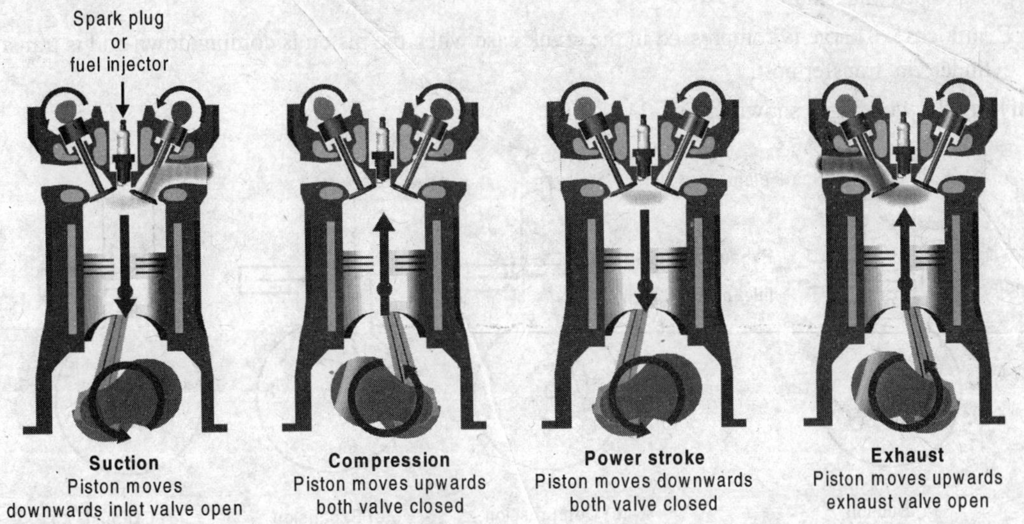

Fig. 24.2. *Working of 4-stroke IC engine*

24.5 TWO STROKE SI ENGINE

In this engine the cycle is completed in two strokes of the piston or one revolution of the crankshaft (in four stroke engine crank rotates through two revolution in one cycle).

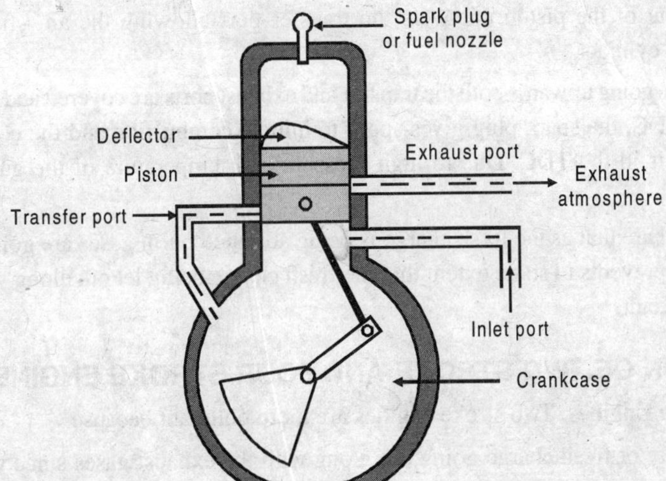

Fig. 24.3. *Two stroke engine*

(a) **Ports.** This engine has no valves but there are three ports inlet, exhaust and transfer ports as shown. These ports get opened and closed due to the position of the piston.

(b) **Piston.** The piston has a deflector on its crown which prevents loss of fresh charge while allowing the burnt gases to be exhausted.

(c) **Crank case.** Charge is compressed in the crank case when the piston is coming down and is transferred to cylinder via. transfer port.

Working. Its working is shown in Fig. 24.4.

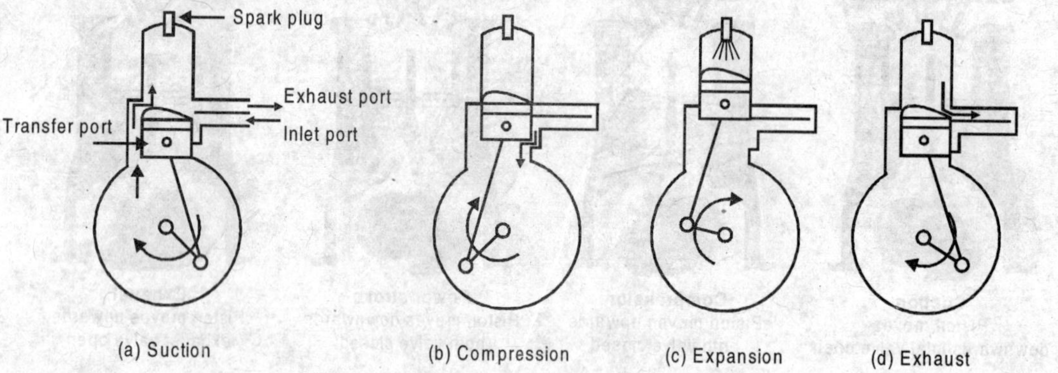

Fig. 24.4. *Two stroke petrol engine*

(a) Starting from top dead centre position (TDC), as the piston goes downwards first the exhaust port gets uncovered. The burnt gases are going out through this port.

(b) With further downward movement of the piston the inlet port is opened. Fresh air fuel mixture is admitted into the crank case. The downward movement of the piston acts as pump and slightly compresses the fresh charge in the crank case.

(c) Further movement of the piston uncovers the transfer port allowing the air—fuel mixture charge to be admitted into the cylinder.

(d) When the piston is going upwards both the transfer and exhaust ports are covered and the charge is compressed. Just before the TDC, the spark plug gives spark to initiate combustion and the combustion develops fully when the piston is in the TDC. Due to high pressure and temperature of the gases the piston is pushed downwards.

(e) Points to be noted are that as the fresh charge is being admitted burnt gases are going out. Specific shape of the piston crown prevents to some extent that the fresh charge being let out along with the exhaust but total prevention is difficult.

24.6 COMPARISON OF TWO STROKE AND FOUR STROKE ENGINES

Pollution in 2-stroke Engines. Two stroke engines are more pollutant because

(a) There is possibility of fresh charge going out along with the exhaust gases since while the exhaust port is open and fresh charge is being admitted.

(b) Usually the 2-stroke engines use mist lubrication. The lubricant mixed with the petrol does not get fully burnt and is pollutant.

Two stroke engines	*Four stroke engines*
1. Cycle completed in 2 strokes and one revolution of the crank shaft.	1. Cycle is completed in 4 strokes and two revolution of the crank shaft.
2. No valves, it has inlet, exhaust and transfer ports instead.	2. It has inlet and exhaust valves.
3. It has special raised deflector on the piston head to prevent fresh charge going out with the exhaust gases.	3. No special piston head required.
4. Have higher fuel consumption i.e., give less mileage per litre compared to similar 4-stroke engine.	4. Have better mileage.
5. Are more pollutant as some unbrunt charge can go out along with the exhaust.	5. Cause less pollution as compared to the 2-stroke engines.
6. Mostly use mist lubrication. Simple construction and relatively cheaper.	6. Costlier than 2-stroke engines
7. Theoretically, produces double power for the same speed of the engine.	7. Theoretically produces half the power.

CHAPTER 25

Testing and Performance of IC Engines

25.1 INTRODUCTION

Any machine which derives heat energy from the combustion of fuel and converts part of this energy into mechanical work is known as heat engine. Heat engines are mainly divided into two groups viz., external combustion engine and internal combustion engine.

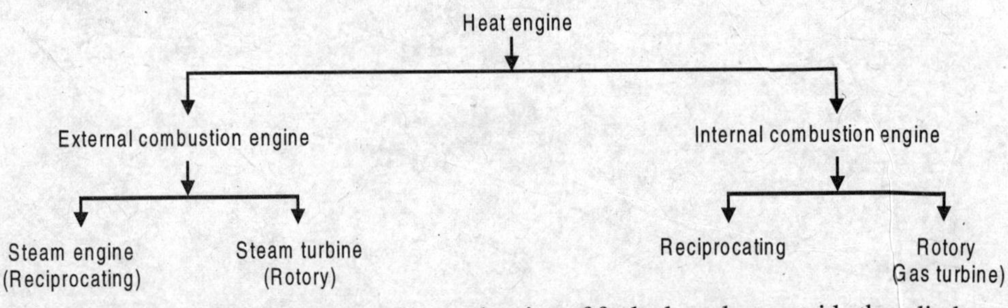

In the case of external combustion engines, the combustion of fuel takes place outside the cylinder as in case of steam engines (also hot air engine, steam turbines and closed cycle gas turbines). In external combustion engines, the combustion of fuels takes place in boilers in case of steam engines and steam turbines and in open combustion chambers in case of open-cycle gas turbines.

In internal combustion engines, the combustion of fuel, in the presence of air, takes place inside the cylinder and products of combustion directly act on the piston to develop the power.

In heat engines, there are three essentials as:

1. The source of heat
2. The working substance and
3. The engine

In internal combustion engine:

1. The source of heat is the fuel that is used in the engine. Combustion of fuel takes place in the cylinder and the heat is liberated. This heat is given to the air taken inside the engine, for the combustion and either the

temperature and pressure or temperature only increases. The products of combustion are the gases which act as the working substance.

2. The working substance in the engine is the gases. In IC engines this working substance does not change its phase during the working cycle. The working substance is made up of air plus a small amount of fuel resulting in the products of combustion.
3. The sink is always the atmosphere. All IC engines exhausts into the atmosphere i.e., the heat is rejected to the atmosphere.

The internal combustion engines offer some special advantages over external combustion engines in smaller power sector. They are:

1. Starting of engine is easy and quick.
2. Thermal efficiency is high.
3. Specific power developed is high i.e., power developed per unit weight of engine is high (kw/kg wt.).
4. It offers greater mechanical simplicity.
5. It requires less space.
6. The capital cost is low.

25.2 ENGINE TESTING AND PERFORMANCE

On the basis of the actual data available for the engine working, we can find the ideal cycle work and its efficiency. The actual work produced or developed in the engine cylinder will be less than the ideal work calculated from the ideal cycle.

The amount of actual work obtained depends upon where it is measured. In Fig. 25.1, for example, the work that is done in the cylinder can be measured with the use of a device or instrument called an *indicator*. This work, called the *indicated work*, is equal to the work of the fluid, which is the work obtained by the use of thermodynamic properties of the fluid. This indicated work, represented by W_I, is relevant only for a reciprocating engine.

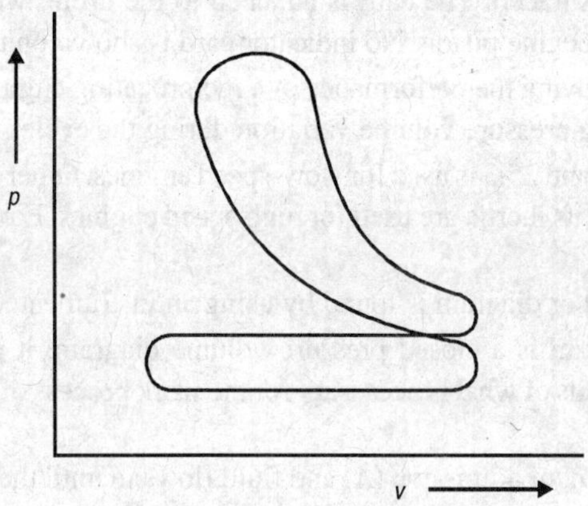

Fig. 25.1. *Indicator diagram*

We also have means of measuring the work on the shaft. This work is called *shaft work* or *brake work* and is represented by W_B. Since the shaft drives further machines, the work can similarly be measured further along as desired. In electric power generation, the shaft drives a generator, the output of the generator—being electricity, is called overall work or combined work represented by W_K. This work is easily measured with electrical instruments.

25.2.1 Indicator and Indicator Card

Indicators used in steam engine or internal combustion engines shown in Figs. 25.2 and 25.3. The different parts are shown on the diagram. Indicator is mounted on the cylinder.

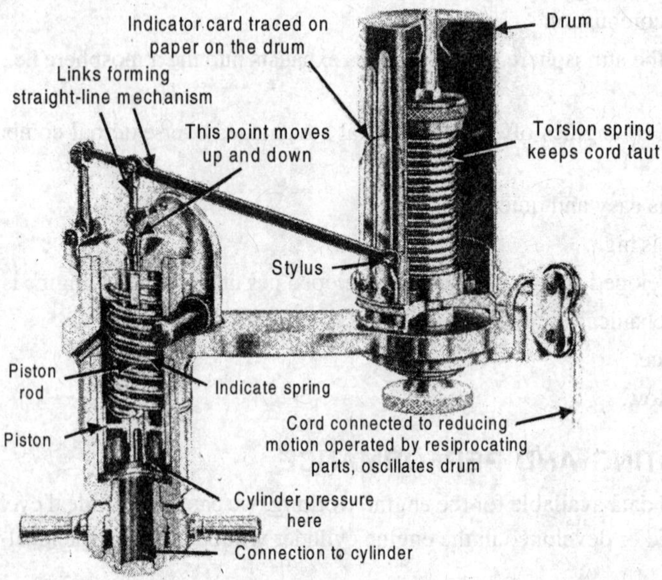

Fig. 25.2

Steam or fluid (air and gases) enters through the connection to the cylinder and acts on a small piston, whose motion is opposed by the indicator spring. The greater the pressure from the cylinder, the more the piston moves. The stylus is connected to the piston of the indicator. The stylus is actuated through an appropriate straight line linkage and the indicator card is traced. The card is attached to the drum, which moves in phase (the motion is reduced in magnitude) with the engine piston. No indicator card is shown on the drum.

A widely useful means of studying the performance of a reciprocating engines is the *indicator card* or *indicator diagram*. It is the diagram giving pressure volume variation during the cycle.

Indicator shown in Figs. 25.2 and 25.3 is used for slow-speed engines. Other types of indicators that eliminate the piston in Figs. 25.2 and 25.3 and its inertia are used for high speed engines. For a reciprocating engine, an indicator card is shown in Fig. 25.4.

The area of the indicator card or diagram is found by using an instrument known as the *planimeter*.

Although the indicator diagram is a closed pressure volume diagram, it is not a thermodynamic cycle. The indicator diagram shows the details of what is necessary for the work process of a cycle. The events in a power cycle are:

(i) Inlet valve opens at the point of admission (*A*) and fluid flows in until the end of stroke, where the inlet valve closes.

(ii) Compression of the fluid takes place at *B* and pressure rises and volume decreases.

(iii) Heat is added at *C*. Pressure increases, temperature increases.

(iv) Expansion of the products takes place at *D*. *DE* is expansion.

(v) At *E*—the point of release where the exhaust valve opens and the fluid (gases) begins to flow out of the cylinder.

Fig. 25.3

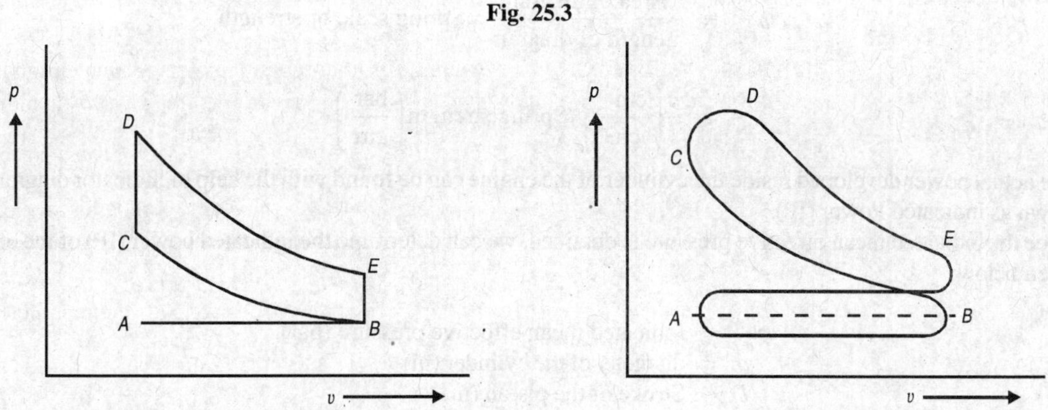

Fig. 25.4

Since the indicator card gives a record of the pressure on the piston and the volume behind the piston, areas enclosed within the indicator cards or diagrams represent work, known as indicated work W_I. This work includes the effects of getting the working substance into and out of the cylinder.

25.2.2 Planimeter

A simple planimeter used to determine the area of the indicator diagram is shown in Fig. 25.5.

Point F is fixed. Point P is moved around the boundaries of the area to be measured. Wheel D turns with the movement of P, and after the complete outline has been traced, the magnitude of the area may be read from the dial G and the vernier E.

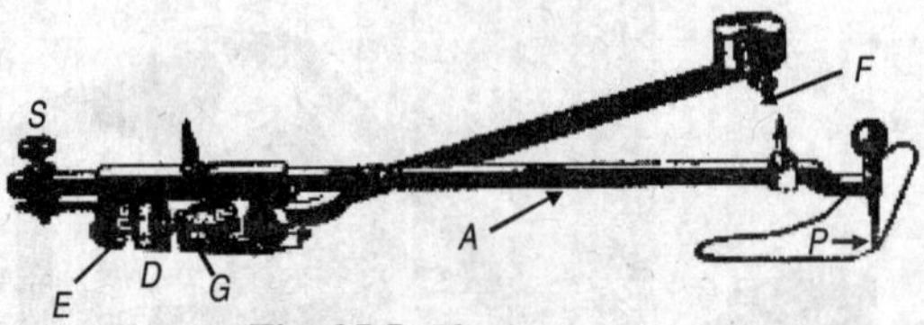

Fig. 25.5. *Planimeter*

25.2.3 Mean Effective Pressure (P_{mi}), Indicated Power (IP)

The indicator card is taken by an indicator. Figure 25.2 shows that it is fitted with a spring of a certain scale or strength. If the strength of the spring is given as 10 bar, in case of an indicator, we mean that the vertical movement of 1 cm of the stylus indicates a pressure of 10 bar. Therefore the actual heights of the indicator card multiplied by the scale of the indicator spring will be the actual pressure in bar or N/m^2.

We have studied the various air standard cycles and the corresponding mean effective pressure (p_m). By definition, mean effective pressure is that constant pressure, when acting on a piston for the complete storke, produces the work of the cycle. From the indicated diagram, we get the indicated work W_I. We, therefore can write

$$W_I = \text{Actual area of indicator diagram}$$
$$= \text{Stroke volume} \times \text{Mean effective pressure.}$$

To get indicated mean effective pressure MEP or p_{m_i}, find the area of the indicator card in square centimeter (cm^2), divide this area by the length of the card in cm and obtain the average height of card in (cm). This average height multiplied by the scale of the indicator spring is the average pressure called the indicated MEP $\left(p_{m_i}\right)$.

∴ Indicated Mean Effective Pressure p_{m_i}, is given by

$$p_{m_i} = \frac{\text{Area of diagram}}{\text{Length of diagram}} \times \text{Spring scale or strength}$$

$$= \frac{a}{1}\left(\frac{\text{cm}^2}{\text{cm}}\right) \times \text{Spring strength}\left(\frac{\text{bar}}{\text{cm}}\right).$$

The actual power developed inside the cylinder of the engine can be found with the help of indicator diagram and is known as Indicated Power (IP).

Once the indicated mean effective pressure is obtained, we can determine the indicated power (IP) of the engine, as given below:

Let,

$$p_{m_i} = \text{Indicated mean effective pressure (bar)}$$
$$d = \text{diameter of the cylinder (m)}$$
$$L = \text{Stroke of the piston (m)}$$
$$N = \text{Speed of the engine (r.p.m.)}$$
$$A = \text{Area of cylinder (m}^2)$$

∴

$$= \frac{\pi d^2}{4}$$

$$F = \text{Force acting on the piston (N)}$$

$$= p_{m_i} \times A \times 10^5 \text{ (N)}$$

$\therefore$ Work done/cycle $=$ Force $\times$ Distance travelled

$$= (p_{m_i} \times A \times 10^5) \times L(N - m)$$

Again, let n = Number of cycles/sec

$$= \frac{N}{2 \times 60} \text{ in case of 4-stroke cycle engine.}$$

$$= \frac{N}{60} \text{ in case of 2-stroke cycle engine}$$

$$\therefore \text{Work done/sec} = \frac{\text{Work done}}{\text{Cycle}} \times \text{No. of cycles/sec.}$$

$$= p_{m_i} \times A \times L \times n \times 10^5 \quad \frac{\text{N-m}}{\text{sec}}$$

$$= p_{m_i} LAn \times 10^5 \quad \frac{\text{N-m}}{\text{sec}}$$

$$= \text{J/sec}$$

$$= \text{Watt}$$

$$= \frac{p_{m_i} \times L \times An \times 10^5}{1000} \text{ kW}$$

$$\therefore \text{With proper units IP} = \frac{p_{m_i} \times L \times A \times n}{1000}$$

The engine size is given by two numbers, say 6×8 cm, where the first number is always the bore or diameter and the second one the stroke; 6 cm diameter, 8 cm stroke.

25.2.4 Brake Power (BP)

The portion of the power developed inside the cylinder is lost in friction (crankshaft and bearings, gears, valve mechanisms and other moving parts. Some of the power developed is used for running the fuel pumps, water circulating pumps, lubricating oil-pumps, governors etc. The net power available at the crankshaft for doing useful work is known as shaft power or generally called Brake Power (BP).

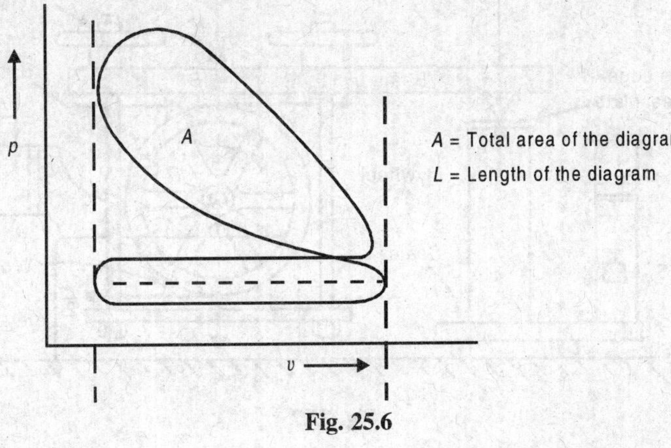

Fig. 25.6

The name brake—work or brake power came about because in the first determinations the power output was dissipated in the friction of a brake. Brakes are still used for this purpose over suitable ranges of power and speed.

25.2.5 Brake Power Measurement: Dynamometers

The device that is used for measuring the brake power is called a Dynamometer. A dynamometer is essentially a device for measuring the forces or couples which tend to change the state of rest or uniform motion of a body. There are many kinds of dynamometers, but reference can only be made to a few of the types used in measuring the power available from a uniformly revolving shaft.

Broadly, there are two types of dynamometers, namely, Absorption Dynamometers and Transmission Dynamometers. As the names imply, an absorption dynamometer absorbs the available power in doing work, usually against friction, whereas a transmission dynamometer transmits the available power unchanged.

(a) Absorption Dynamometers. These generally consist of some form of brake in which provision is made for measuring the friction torque on the drum.

The dynamometers that come under this catagory are:

(i) The Prony brake dynamometer.
(ii) The Rope brake dynamometer.
(iii) Eddy or current dynamometer.
(iv) The Heenan and Froude dynamometer.
(v) The Swinging field dynamometer.

(i) The Prony Brake Dynamometers. A simple type, known as the Prony brake is shown in Fig. 25.7. It consists of two blocks of wood, each of which embraces rather less than one half of the pulley rim. The two blocks can be drawn together by means of bolts, cushioned by springs, so as to increase the pressure on the pulley. One block carries an arm to the end of which pull can be applied by means of a dead weight or spring balance. The friction torque on the pulley may be increased until it balances the torque due to the available power.

When the brake is clamped to the fly wheel, the frictional force $F = \left(\frac{F}{2} + \frac{F}{2}\right)$ at the brake shoes tends to turn the brake with the flywheel. However, the knife edge on the beam rests on scales and prevents motion of the brake. The force P, which is weighed by the scales, consists of the reaction produced by friction and a portion of the weight of the brake, called the tare—unless there is a counterweight to balance the brake about the centre line of the flywheel. To find the tare, support the brake on the edge at B and weigh the tare on the scales. Thus the net force on the scales produced by the frictional moment is (P-tare).

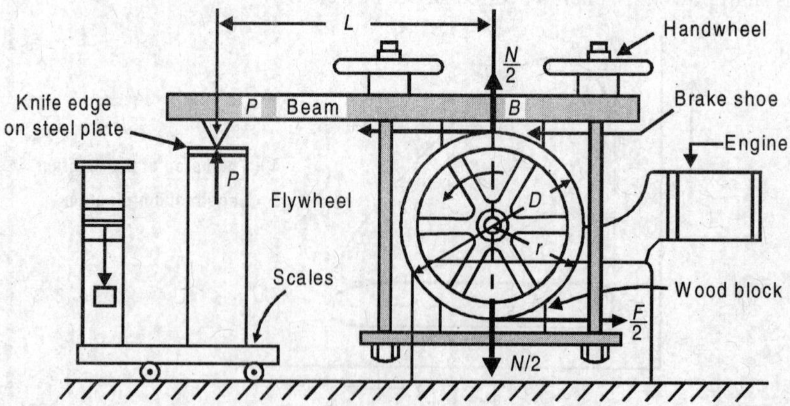

Fig. 25.7

The information obtained from a prony brake test is used to compute the brake-power. Thus, the frictional force F acting through one revolution of the flywheel does work $W = F \times \pi D$ where D is the flywheel diameter in metres.

This work multiplied by N rpm gives the work in N-m/min. $= \dfrac{F \times \pi D \times N}{60}$ N-m/sec i.e., Joules and $\dfrac{F \times \pi DN}{60 \times 1000}$ kW.

$$\therefore \qquad \text{BP} = \frac{F \times \pi DN}{60 \times 1000} \text{ kW.}$$

or

$$\text{Torque} = (P - \text{tare})\, L = T = \frac{FD}{2}$$

Where L = Brake arm in metres.

(ii) The Rope-Brake Dynamometer. Figure 25.8 shows a rope brake dynamometer. In this brake two or more ropes rest on the rim of a pulley. They are spaced evenly across the width of the rim by means of wooden blocks positioned at different points round the rim. The total pull on the slack ends of the ropes is registered on a spring balance, while the pull on the tight end is provided by dead weights. Friction torque on the pulley may be increased by increasing the dead load by addition of weights.

When the speed of the engine is constant, the friction Torque counterbalances the torque due to available power of the engine.

Let
W = Dead weight or load in Newtons
S = Spring balance reading in Newtons
R = Effective radius of brake drum in metre

$$= \frac{D + d}{2}$$

D = Diameter of brake pulley in metre
d = Diameter of the rope in metre.
N = Speed in RPM (revolutions per minute)

Then $W - S$ = Net load acting at a radius R

$\therefore$ T = Torque produced (friction)

$$= (W - S)\, R \text{ N-m}$$

$$\text{Work done/sec} = \frac{2\pi NT}{60} \text{ (N-m/sec = J/s = Watt)}$$

$$= \frac{2\pi N \times (W - S)R}{60 \times 1000} \text{ kW}$$

$$= \frac{\pi D_e N \times (W - S)}{60 \times 1000}$$

where D_e = Effective Diameter

$$= (D + d)$$

$$= \frac{\pi (D + d) N \times (W - S)}{60 \times 1000} \text{ kW}$$

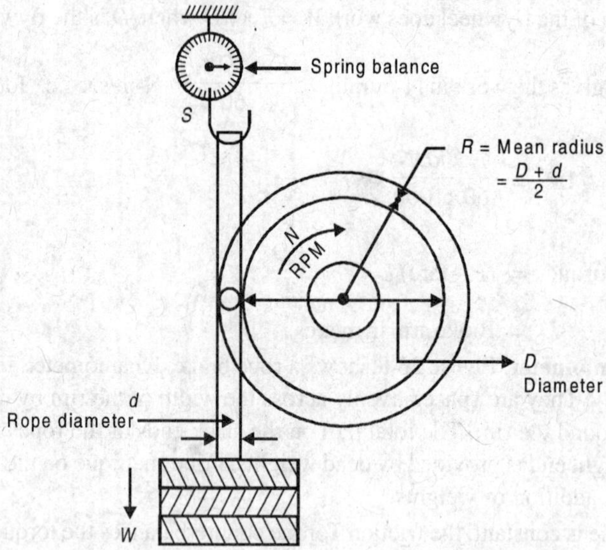

Fig. 25.8. *Rope brake*

In most of the cases the diameter of the rope d is negligible compared with the radius of the brake wheel and thus we get

$$\text{Work done/sec} = \frac{\pi DN \times (W - S)}{60 \times 1000} \text{kW}$$

The power is called Brake-Power written as BP.

$$\therefore \quad \text{BP} = \frac{\pi DN\ (W - S)}{60 \times 1000} \text{kW}$$

25.2.5 Mechanical Efficiency

We have seen above that the brake power BP is less than the indicated power IP because of the power lost in friction between the various parts of the engine. The power lost in friction is called Friction Power (FP).

The ratio of the power available at the engine crankshaft to the power developed in the cylinder is known as the Mechanical efficiency of the engine.

$$\therefore \quad \text{Mechanical Efficiency} = \frac{\text{Brake Power (BP)}}{\text{Indicated Power (IP)}} = \frac{\text{BP}}{\text{IP}}$$

But $\quad \text{IP} = \text{BP} + \text{FP}$

$$\therefore \quad \text{Mechanical Efficiency} = \frac{\text{BP}}{\text{IP}}$$

$$= \frac{\text{IP} - \text{FP}}{\text{IP}}$$

$$= 1 - \frac{\text{FP}}{\text{IP}}$$

$$= \frac{\text{BP}}{\text{BP} + \text{FP}}$$

Thus we see that the aim of the designer is to reduce the friction power of the engine as far as possible.

25.2.6 Methods of Finding Indicated Power (IP)

Generally, indicated power is calculated or determined if an indicator is connected to the engine and thereby an indicator diagram is available.

If an engine is not provided with an arrangement of connecting an indicator and only brake dynamometer is available, then we can determine an indicated power by carrying out a test on the engine.

There are three methods to find an indicated power of the engine. They are

(a) Willan's Line method
(b) Morse test method and
(c) Motoring test method.

(a) Willan's Line Method. The engine is tested at constant speed but at different loads and the corresponding fuel consumption is measured. From this data, the fuel consumption per hour at each load is calculated. A graph between fuel consumption per hour and Brake Power (BP) is now plotted as shown in Fig. 25.9.

The relation between BP and fuel per hour is represented approximately by a straight line indicating that BP is proportional to fuel consumption per hour.

It will be seen, with the line intersecting *Y* - axis at *A*, that some fuel is consumed even at no-load i.e., zero BP. The power developed at the piston head by the fuel consumed at zero BP is just sufficient to overcome the friction.

The friction power (FP) can thus be estimated from the constant of proportionality i.e., slope of the straight line. It has been pointed out earlier that at constant speed running, the friction power does not change with the load, but remains approximately constant.

Then the friction power FP is the negative intercept of the straight line on X-axis.

$\therefore$ Indicated Power developed at any particular load

$$= BP + FP.$$

$$IP = BP + FP$$

This test is applicable only to compression Ignition engines. (CI Engines/Diesel or oil engine)

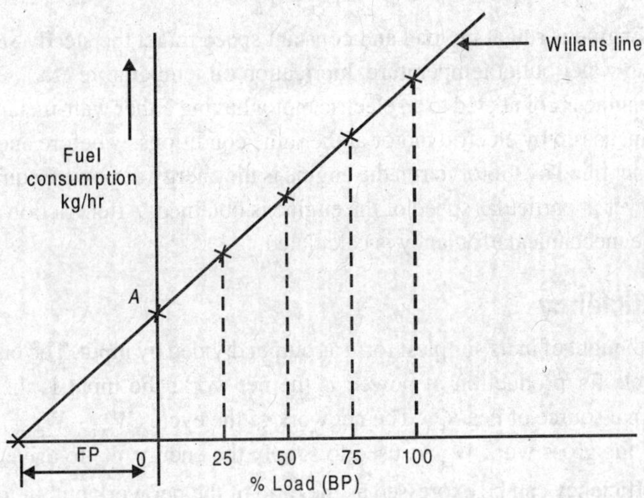

Fig. 25.9

(b) Morse Test for Multi-Cylinder Petrol Engine. This method is also based on the assumptions that the mechanical losses at constant speed are independent of load.

Requirements for the test are:

1. Multicylinder Engine
2. Constant speed at all loads
3. Throttle full – open.

Procedure

The engine is run at constant speed during the period of test. Initially, the brake power BP developed by the engine with all its cylinders firing or working is measured with the help of dynamometer. Now, No. 1 cylinder of the engine is cut out or the spart plug is short circuited so that no ignition takes place in that cylinder and consequently the engine speed falls off. The engine speed is brought back to its original value by adjusting the load on the engine. The (BP), now developed by the engine is less than that developed with all the cylinders firing and the difference ($BP - BP_1$) is equal to the IP of the cylinder No. 1. The same procedure is repeated with the remaining cylinders to get the IP of the engine.

Let there be 4-cylinders numbered 1, 2, 3 and 4. Knowing IP = BP + FP.

∴ When all cylinders are working, we have

$$(IP_1 + IP_2 + IP_3 + IP_4) - (FP_1 + FP_2 + FP_3 + FP_4) = BP_1 + BP_2 + BP_3 + BP_4 = BP \quad \text{...(i)}$$

With cylinder No. 1, cut out we have

$$(IP_2 + IP_3 + IP_4) - (FP_1 + FP_2 + FP_3 + FP_4) = BP_1 \quad \text{...(ii)}$$

∴ Equation (i) – (ii),

$$IP_1 = BP - BP_1 \quad \text{...(iii)}$$

In the same way, IP_2, IP_3 and IP_4 can be found.

and
$$\text{Mech. Eff.} = \frac{IP}{BP}$$

where
$$IP = IP_1 + IP_2 + IP_3 + IP_4$$

(c) Motoring Test. Engine is run at no load and constant speed to get the steady-state conditions—particularly all temperatures (cooling water outlet temperature, lubrication oil temperature etc.). Once this state is attained the engine is stopped. The engine is connected to an electric motor having either watt-meter (energy meter) or Voltmeter and Ammeter. The engine is run by electric motor at the same conditions as before and again the same steady state conditions. The energy supplied by motor to run the engine is the energy or power required to overcome the friction power. Thus friction power at particular speed of the engine is obtained. After friction power is obtained, IP can be calculated and hence the mechanical efficiency is calculated.

25.2.7 Thermal Efficiency

Efficiency may well be thought of in its simplest form as output divided by input. The output of a power cycle, which is a thermodynamic cycle for production of power, is the net work; the input is the heat added to the working substance from an external source of heat Q_A. The net work of the cycle ($W_{out} - W_{input}$) is taken as the output of the power cycle because of the gross work W_{out} is used to supply the energy W_{in} to drive the pump.

Therefore, thermal efficiency can be expressed as the ratio of the net work output to the heat energy required to produce that workoutput.

If the work output in question is the shaft work or shaft power or BP, then the thermal efficiency is called Brake Thermal Efficiency (B. Th. E.) η_{thb}

If the work output in question is the developed work called indicated work, the thermal efficiency is called Indicated Thermal Efficiency. (I. Th. E.)η_{thi}

∴ Indicated Thermal efficiency is given by

$$\text{I. Th. E} = \frac{\text{Indicated work/sec}}{\text{Heat added/sec}}$$

$$= \frac{\text{IP (kW = kJ/s)}}{\frac{\text{Mass of fuel}}{\text{sec}} \times \text{Cal. Val.}} = \frac{\text{IP}}{\frac{m_f}{\text{sec}_f} \times \text{Cal. Val.}}$$

$$= \frac{\text{IP} \times 3600}{\frac{\text{Mass of fuel}}{\text{hr.}} \times \text{Cal. Val.}} = \frac{3600}{\frac{\text{Mass of fuel}}{\text{hour} \times \text{IP}} \times \text{Cal. Val.}}$$

Mass of fuel used in kg/hour/IP is called Indicated specific fuel consumption (kg/kWH). Similarly, Brake thermal efficiency is given by

$$\text{B. Th. E} = \frac{\text{Brake Work/sec}}{\text{Heat added/sec}} = \frac{\text{BP (kW = kJs/sec)}}{\frac{\text{Mass of fuel}}{\text{sec}} \times \text{Cal. Value}}$$

$$= \frac{\text{BP}}{\frac{m_f}{\text{sec}} \times \text{Cal. Val.}} = \frac{\text{BP} \times 3600}{\frac{\text{Mass of fuel}}{\text{hour}} \times \text{Cal. Value}}$$

$$= \frac{3600}{\frac{\text{Mass of fuel}}{\text{hour} \times \text{BP}} \times \text{Cal. Value}}$$

Here, $\frac{\text{Mass of fuel}}{\text{hour} \times \text{BP}}$ is called **Brake Specific Fuel Consumption BSFC.**

Brake thermal efficiency = Indicated thermal efficiency × Mechanical efficiency.

(d) Eddy Current Dynamometer. One of the oldest forms of dynamometers is the eddy-current dynamometer. The simplest form consists of a disc which, driven by the engine under test, turns in magnetic field. The strength of the field is controlled by varying the current through a series of coils located on both sides of the disc. The revolving disc acts as a conductor cutting the magnetic field. Currents are induced in the disc and, since no external circuit exists, the induced currents heat the disc. For larger power absorptions, the heating of the disc becomes excessive and difficult to control.

Figure 25.10 shows the view of a simple eddy current dynamometer. In this device the eddy currents are induced in the stator for ease of cooling. A stator or a housing *a* is supported on trunnion bearings *b* so that any tendency of the stator to rotate is read on the fixed scale *c*. Inside the stator is rotor *d*, keyed to the shaft *e* and provided with the *f* not shown passing close to the smooth face of the stator. When the rotor is turned, the flux enters the rotor principally through the ends of the teeth. As these teeth move, the lines of magnetic flux are caused to sweep through the iron of the stator; the flux induces eddy currents in the stator and tends to rotate the stator in the same direction as the shaft. Coil *g* when energized with direct currect, magnetizes the stator and rotor, with flux lines encircling the coil. The temperature rise of the stator is controlled by the flow of water in channels h. The generator i supplies supplementary excitation for the field coils. If this unit is used, it can cause a steep change in excitation as the engine speed increases, thereby tending to hold the speed constant.

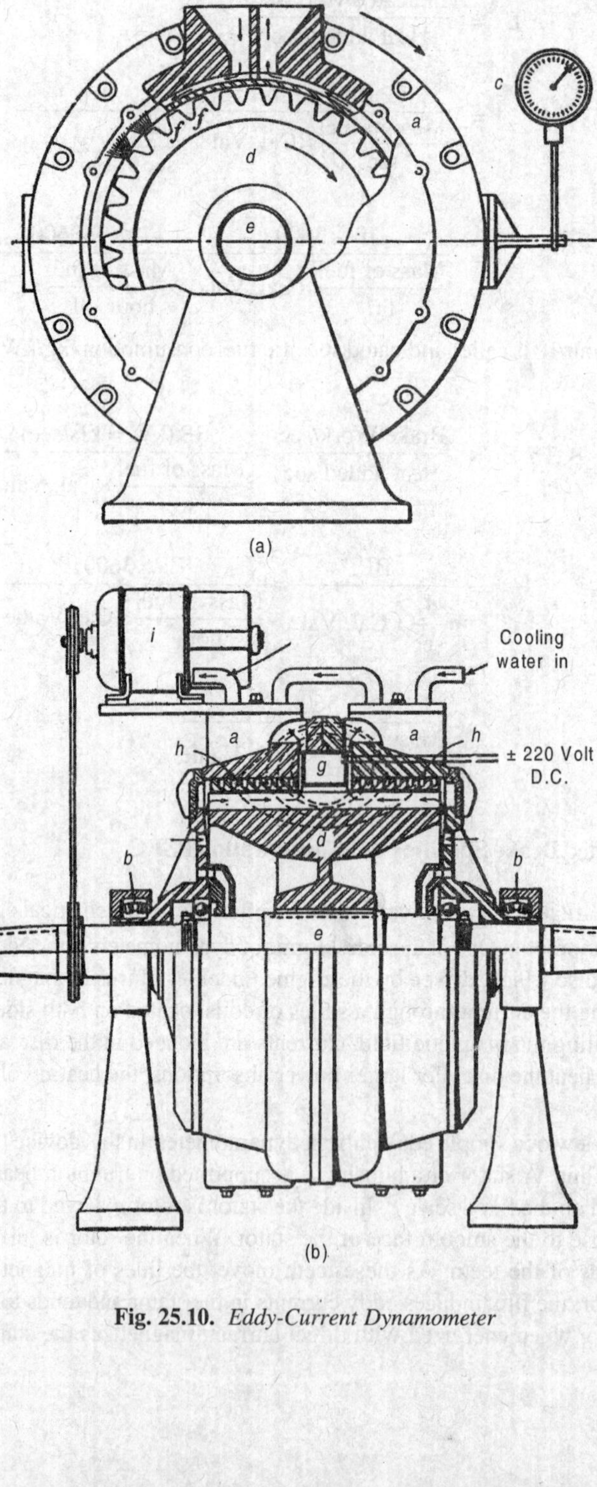

Fig. 25.10. *Eddy-Current Dynamometer*

(e) The Electric Dynamometer. The electric generator can be used for loading the engine, but the output of the generator must be measured by electrical instruments and corrected in magnitude for generator efficiency. The electric load connected to the generated may be Lamp-bank used in laboratory work.

(*f*) Swinging Field D.C. Dynamometer. DC shunt motors are used as swinging field dynamometer. DC motor is supported in trunnion bearing to measure the reacting torque that the outer stator and field coil tend to rotate due to magnetic drag. The torque is measured with an arm and swinging equipment and spring scale in the usual manner.

This shunt motor dynamometer can run as a generator and motor also. That is it is reversible and it works as a motoring and power generating unit. When it is used as a generator, the electrical energy produced is dissipated in resistor. When used as motoring device dc energy is supplied from outside and used as starting device for an engine. The load and operation is controlled by varying field current.

(g) Hydraulic Dynamometer. The hydraulic dynamomtere. Figure 25.10(a) works on the principle of dissipating the power in the fluid frcition rather than in dry friction. In principle its construction is similar to that of a fluid fly-wheel. It consists of an inner rotating member or impeller coupled to the output shaft of the engine. This impeller rotates in a casing filled with some hydraulic fluid. The outer casing, due to the centrifugal force developed, tends to revolve with the impeller, but is resisted by a torque arm supporting the balance weight.

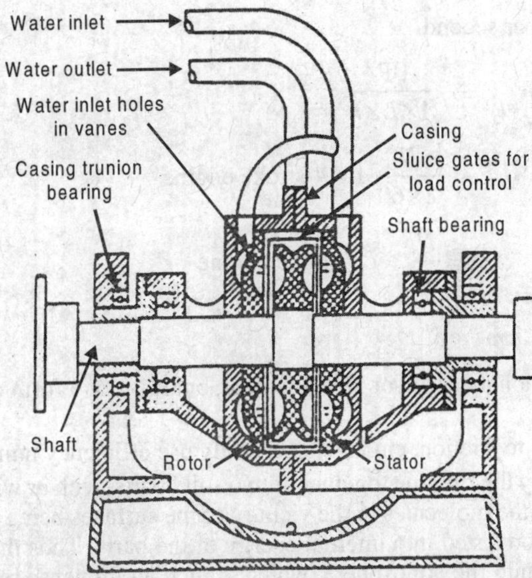

Fig. 25.10(a). *Hydraulic Dynamometer.*

The frictional forces between the impeller and the fluid is measured by the spring balance fitted on the casing. The heat developed due to the dissipation of power is carries away by a continuous supply of the working fluid, usually water. The outer can be controlled by regulating the sluice gates which can be moved in an out to partially or wholly obstract the flow of water between impeller and the casing.

25.3 BRAKE MEAN EFFECTIVE PRESSURE: BMEP

It is a fictitious but very useful concept. If we know or determine mean effective pressure from the indicator diagram, we can calculate or determine the indicated power IP from the engine cylinder dimensions and speed of the engine.

If we represent brake work per cycle on *P–V* diagram, Fig. 25.11, by a rectangle whose base is the same as that for indicated diagram, then the ordinate or vertical side of this rectangle will be called as brake mean effective pressure BMEP or p_{m_b}.

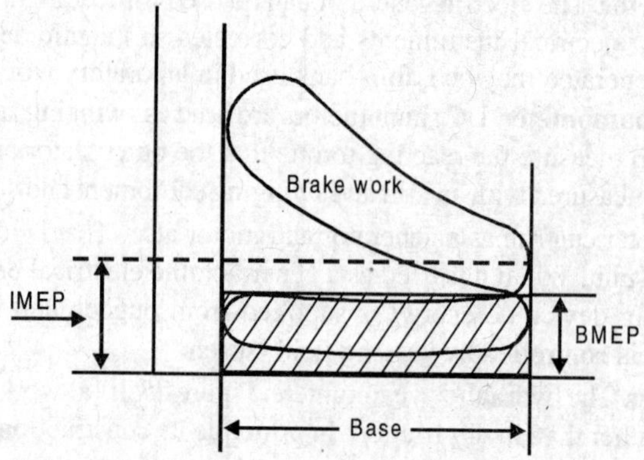

Fig. 25.11. *Energy diagram with friction*

$\therefore$ Brake work/cycle $= p_{m_b} \times A \times L$ with proper units.

$$BP = p_{m_b} \times A \times L \times n$$

where n is the number of cycles per second.

$$p_{m_b} = \frac{BP}{A \times L \times n}$$

Here, $n = \frac{N}{2 \times 60}$ for 4-stroke engine

$= \frac{N}{60}$ for 2-stroke engine

25.4 FRICTION POWER: FP

We frequently hear the term Frictional energy or its equivalent, but it is not a form of energy different from those forms already studied.

To allow for energy losses due to friction, engineers generally use efficiency numbers.

If the frictional loss is caused by the rubbing together of two solid parts, with or without lubricant, what happens is that work is expended to excite the molecules in the vicinity of the surfaces being rubbed, raising their temperature. This is to say that work is converted into internal energy of the parts. Thus the frictional energy eventually appears as heat—heat dissipated into the atmosphere, which is such an immense reservoir that it does not notice the difference except locally.

If the frictional loss is that caused by fluid friction, the substance has more internal energy at the end of the process that it would have had in the absence of friction. If the substance also gets hotter than its environment, at least a part of the frictional energy eventually leaves as heat to the surroundings. In all events, frictional energy, in effect, is eventually spilled into the surroundings, and always at the expense of work.

Let E_f represent the friction losses in system, then we get,

$$W_I = W_B + E_f$$

W_s = shaft work

E_f = frictional loss in the bearing etc.

If $Q = 0$, the energy equation is

$$h_1 + K_1 = h_2 + K_2 + W_s + e_f$$

The mechanical efficiency h_{mech} is a number that tells us the mechanical losses in the machine. For the reciprocating engine of any type that delivers work.

$$\eta_{mech} = \frac{W_B}{W_I}$$

Remember that engine efficiency is a dimensionless number or ratio. The difference [(IP) – (BP)] or $[W_I - W_B]$ represents the loss due to mechanical friction of the moving parts of the engine. This loss expressed in Power, is called the Friction Power written as FP.

$$\therefore \quad FP = (1 - \eta_{mech})\, IP$$

Mechanical efficiency is not a fixed number characteristic of machine. It depends on the operating conditions, especially output, speed and lubrication.

25.5 AIR STANDARD EFFICIENCY AND RELATIVE EFFICIENCY

(i) Air Standard Efficiency. Air standard efficiency is the efficiency of a engine working on standard, ideal cycle.

Air standard efficiency of an engine working on Otto cycle is given by

$$(ASE)_{Otto} = 1 - \frac{1}{r_k^{\gamma-1}}$$

where r_k = Compression ratio.

Air standard efficiency of an engine working on Diesel cycle is given by

$$(ASE)_{Diesel} = 1 - \frac{r_c^r - 1}{\gamma . e_k^{r-1} (r_c - 1)}$$

where r_c = Cut-off ratio

r_k = Compression ratio.

(ii) Relative Efficiency. Relative efficiency is defined as the ratio of thermal efficiency to the air standard efficiency. For this, thermal efficiency may be either brake thermal efficiency and indicated thermal efficiency. Accordingly, relative efficiency will be based on either BP or IP.

$$\therefore \quad \text{Relative efficiency based on BP} = \frac{\text{Brake Thermal Efficiency}}{\text{Air Standard Efficiency}}$$

$$\text{and} \quad \text{Relative efficiency based on BP} = \frac{\text{Indicated Thermal Efficiency}}{\text{Air Standard Efficiency}}$$

25.6 VOLUMETRIC EFFICIENCY

Other factors remaining the same, the power obtained from an engine that draws in air and fuel depends upon the mass of combustible mixture drawn into the cylinders—given a mixture with the correct air fuel ratio and anything that reduces the mass of fuel entering the engine reduces the power output. For example,

(1) The suction pressure is less than atmospheric; therefore the mass of gas $\left(m = \frac{PV}{RT}\right)$ is less than if atmospheric pressure were maintained.

(2) The internal surfaces and passages of the engine are relatively hot, so that the mixture is heated as it passes into the cylinder. According to Charle's law, the increase in temperature further reduces the mass of mixture that the given displacement can contain.

(3) The unpuged gases in the clearance space or volume of the real engine are at a pressure above atmospheric at the end of exhaust stroke and must expand during the suction stroke to the intake pressure before a new charge begins to enter.

(4) The pressure of the atmosphere decreases with altitude so that the mass of mixture drawn in at high altitudes is less than at sea level.

The mass of charge brought into the cylinder is sometimes defined in terms of volumetric efficiency. Thus, volumetric efficiency of an IC Engine is defined as the ratio of mass of air drawn into the engine to the mass of air in displacement or stroke volume at inlet pressure and temperature conditions.

$$\therefore \quad \eta_v = \frac{\text{Mass of air drawn into the engine}}{\text{Mass of air in swept volume at inlet conditions}}$$

Here standard conditions are the pressure and temperature at intake, say, in the test room. In a particular engine, volumetric efficiency is affected by the speed. As the speed increases, the volumetric efficiency decreases because of the greater throttling effect at higher speeds.

The value of the volumetric efficiency can be made greater than unity by the use of a supercharger.

25.7 HEAT BALANCE

It is clear or evident from the study of the second law of thermodynamics that all the heat released in the engine cylinders is not converted into useful work and an appreciable part of it is lost in different forms e.g. in exhaust gases, cooling water etc.

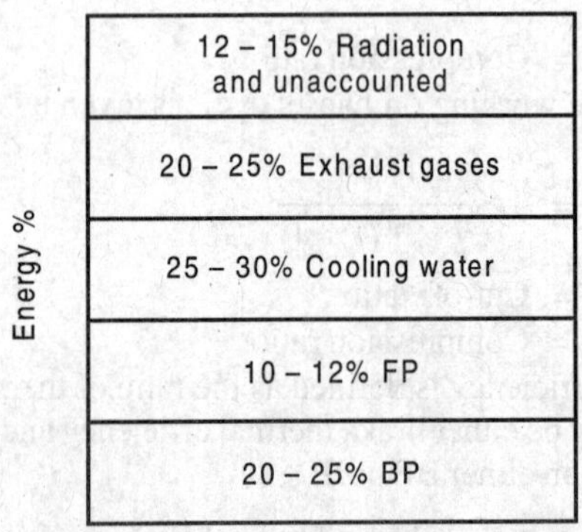

Fig. 25.12. *Heat energy distribution*

The power output and the efficiency of an engine can be estimated and improved upon only with a thorough understanding of the influence of various engine parameters on the losses. The losses are chiefly of three types, (i) Heat Losses, (ii) Friction Losses and (iii) Pumping losses. The sum of friction and pumping losses is commonly known as mechanical losses.

$$\therefore \quad \text{Engine power output} = \text{Heat input} \times \text{Thermal Indicated efficiency} \times \text{Mechanical efficiency}$$

A study of the heat distribution in a given engine will give sufficient indication as to how efficiently the engine is working and the general distribution of heat in an IC engine is shown below.

To draw up the heat balance sheet, the engine will be run at constant speed and load and the following quantities must be measured for the period of test:

1. Fuel Consumption
2. Indicated Power IP
3. Brake Power
4. Quantity of water circulated for cooling
5. The temperatures of the cooling water before entering and after leaving the engine cylinder jacket.
6. Quantity of exhaust gases and its temperature.

Generally the heat balance sheet will be on minute basis and percentage basis.

Calculations for the items that are to be included in the heat balance sheet are given below.

1. Heat supplied/min. = $\dfrac{\text{Mass of fuel}}{\text{min}} \times \text{Calorific Value} \ldots 100\%$

2. Heat equivalent of BP = BP × 60 kJ/min ... $\frac{(2)}{(1)} \times 100\%$

3. Heat equivalent of FP = FP × 60 kJ/min ... $\frac{(3)}{(1)} \times 100\%$

4. Heat to cooling water/min = Mass of water.min × sp. ht × $(t_0 - t_i)$ ... $\frac{(4)}{(1)} \times 100\%$

5. Heat carried way by exhaust gasses/min.

= Mass of gases per min. × sp. ht. × [engine exhaust temperature – Room temperature]

$$= m_g \times C_{pg} \times (t_e - t_R) \ldots = \frac{(5)}{(1)} \times 100 \ldots$$

6. Heat unaccounted for (by difference)

= 1 – [2 + 3 + 4 + 5]... %

Generally, for the good performance, the various percentages given above are found as:

(i) Heat equivalent to BP ...... 25 – 30%
(ii) Heat equivalent of FP ...... 5 – 8%
(iii) Heat to cooling water 30%
(iv) Heat in exhaust gases 30%
(v) Heat unaccounted for ...... 8 – 10%

Generally we are using exhaust gas calorimeter for finding the product $(m_g \times C_{p_g})$. Otherwise we should know the air to fuel ratio so that mass of the gases will be calculated. In that case, the specific heat will be assumed as 1.1 kJ/kg.K.

25.7.1 Use of Exhaust Gas Calorimeter

Exhaust gas calorimeter is a simple heat exchanger through which water is circulated and temperatures of water at inlet and outlet are recorded. Exhaust gases are passed through this calorimeter and again temperatures at inlet and outlet are recorded Fig. 25.13.

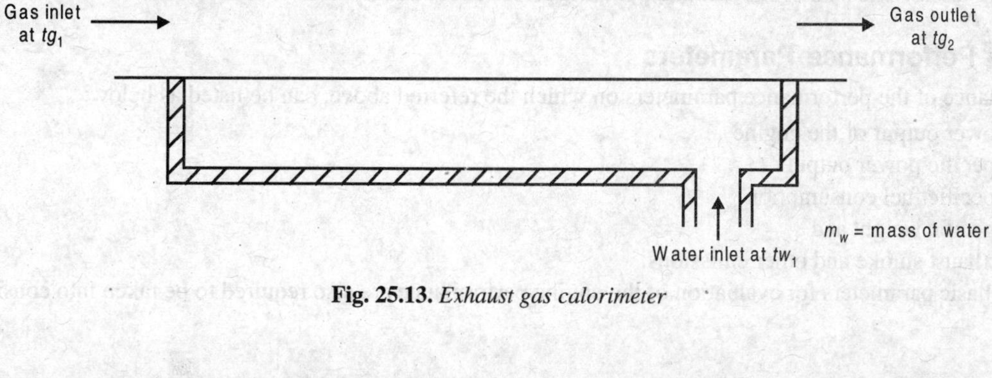

Fig. 25.13. *Exhaust gas calorimeter*

We can balance the heat gained by water and heat given by gases.

$\therefore$ Heat given by gases/min. = Heat gained by water/min.

$$m_g \times C_{p_g} \times (t_{g_1} - t_{g_2}) = m_w \ C_{p_w} (t_{w_2} - t_{w_1})$$

$$\therefore \quad m_g \ C_{p_g} = \frac{m_w \ C_{p_w} (t_{w_2} - t_{w_1})}{t_{g_1} - t_{g_2}}$$

This value of the product $m_g \ C_{p_g}$ is substituted in the proper equation and get the heat carried away by the exhaust gases.

Note: The readings or observations for exhaust gas Calorimeter are to be taken simultaneously.

$\therefore$ $m_g \ C_{p_g}$ will be per minute

All these quantities are tabulated as:

Heat balance sheet on minute basis

(i) Heat supplied/min $= \frac{\text{fuel}}{\text{min}} \times \text{Cal. Value}$

$= \frac{m_f}{\text{min}} \times \text{Cal. Value}$

(i) Heat equivalent of BP = BP × 60

(ii) Heat equivalent of FP = FP × 60

(iii) Heat to cooling water

$= m_w \ C_{p_w} (t_{w_2} - t_{w_1})$

(iv) Heat to exhaust gases = $m_g \ C_{p_g} (t_e - t_R)$

(v) Heat unaccounted for (by difference) = I − [(i) + (ii) + (iii) + (iv)]

25.8 ENGINE PERFORMANCE CHARACTERISTIC

25.8.1 Introduction

The word 'performance' is loosely used when referred to internal combustion engines and it is used here to study the dependence of engine power output, and efficiency on different parameters of the engine. It is clear that all the heat released in the engine cylinders is not converted into useful work and an appreciable part of it is lost in different forms, e.g. in exhaust of the engine, cooling water or cooling air etc. The power output and the efficiency of an engine can be estimated and improved upon only with a thorough understanding of the influence of the various engine parameters on the losses. The losses are chiefly of three types: (i) Heat losses, (ii) Friction losses and (iii) Pumping losses. The sum of friction and pumping losses is commonly known as mechanical losses.

25.8.2 Performance Parameters

Dependance of the performance parameters on which the referred above, can be listed as below:

(i) Power output of the engine
(ii) Specific power output
(iii) Specific fuel consumption
(iv) Specific weight and
(v) Exhaust smoke and other emissions.

The basic parameters for evaluation of the engine performance are also required to be taken into consideration.

These basic performance parameters are:

(a) Power and Mechanical efficiency
(b) Mean effective pressure and torque
(c) Volumetric efficiency
(d) Air : Fuel ratio
(e) Specific fuel consumption
(f) Thermal efficiency and Heat balance
(g) Specific power output, and
(h) Specific weight.

We will now study these basic parameters.

(a) Power and Mechanical Efficiency

The power obtained from an engine is most frequently called brake power (BP). Sometimes this is also called shaft power or merely delivered power. You have already studied the method of determining this brake power.

The total power actually developed on the pistons in the engine is called indicated power (IP).

A part of the indicated power developed by burning fuel and air does not appear as brake power but it is used in overcoming friction of the bearings, piston and other mechanical parts of the engine and also in indication of the fuel-air charge and delivering the exhaust gases. The power to perform these tasks is called the friction power (FP). These losses are sometimes called Mechanical Losses. The brake power is less than the indicated power by the amount of friction power consumed in the engine.

$$\therefore \quad \text{IP} = \text{BP} + \text{Mechanical losses}$$
$$= \text{BP} + \text{FP}$$

The friction power is difficult to determine experimentally because there is no direct method of measurement and there are variations under operating and test conditions for the engine.

The ratio of the power delivered by the engine (BP) to the total power developed within the engine (IP) is known the Mechanical Efficiency (η_m):

$$\therefore \quad \eta_m = \frac{\text{BP}}{\text{IP}}$$
$$= \frac{\text{IP} - \text{FP}}{\text{IP}}$$
$$= 1 - \frac{\text{FP}}{\text{IP}}$$

(b) Mean Effective Pressure and Torque

We have already defined mean effective pressure and in general it is defined as that constant pressure when acting on the piston throughout the stroke, produces the work equal to that of the cycle.

Depending on the work considered, mean effective pressure will be called either Indicated Mean Effective Pressure – IMEP – (p_{m_i}) or Brake Mean Effective Pressure – BMEP – (p_{m_b})

∴ If
A = Area of the piston in m²
L = Length of stroke in m
N = Speed of the engine rpm
BP = Brake Power kW
IP = Indicated Power kW.

Then P_{m_i} = Indicated mean effective pressure (bar).

$$= \frac{\text{IP} \times 1000}{A \times L \times n \times 10^5} \text{ bar}$$

Similarly
$$p_{m_b} = \frac{BP \times 1000}{A \times L \times n \times 10^5} \text{ bar}$$

where
$$n = \text{No. of cycles/sec.}$$
$$= \frac{N}{60} \text{ for 2-stroke engine}$$
$$= \frac{N}{60} \text{ for 4-stroke engine}$$

Also, we can write an equation for BP as
$$BP = \frac{2\pi NT}{1000 \times 60}$$

∴ From equations for mean effective pressure and brake power we write
$$\frac{2\pi NT}{1000 \times 60} = \frac{p_{m_b} \times 10^5 \times A \times L \times n}{1000}$$
$$= \frac{p_{m_b} \times 10^5 \times \frac{\pi}{4} \times D^2 \times L \times \frac{N}{60}}{1000} \text{ or } \frac{p_{m_b} \times 10^5 \times \frac{\pi}{4} D^2 \times L \times N}{60 \times 2 \times 1000}$$

$$\therefore \quad 2T = \frac{p_{m_b} \times 10^5 \times D^2 L}{4} = \frac{p_{m_b} \times 10^5 \times D^2 L}{8}$$

$$\therefore \quad p_{m_b} = \frac{8T}{10^5 \times D^2 L} \text{ or } \frac{16T}{10^5 \times D^2 L}$$

Here T = Brake torque

If we consider this torque corresponding to IP i.e., indicated torque, then the general equation giving relation between p_m and T will be
$$p_{m_b} = \frac{8T_b}{10^5 \times D^2 L} \text{ or } \frac{16T_b}{10^5 \times D^2 L}$$

and
$$p_{m_i} = \frac{8T_i}{10^5 \times D^2 L} \text{ or } \frac{8T_i}{10^5 \times D^2 L}$$

The parameter – mean effective pressure – shows how well the engine is using its size (displacement) to produce work and is generally used for comparative purpose.

On the contrary torque cannot be used for comparative purpose as it will be higher for high power engines. Therefore, the main objective in design of the engine is that the mean effective pressure is higher.

(c) Volumetric Efficiency

We have a definition of volumetric efficiency of an engine as ratio of the actual weight of air inducted by the engine on the intake stroke to the theoretical weight of air that should have been inducted by filling the piston displacement volume with air at intake pressure and temperature.

In supercharge engines this efficiency is greater than one.

$$\therefore \quad \eta_{vol} = \frac{m_a}{m_t}$$

where m_a = Actual weight of air

m_t = Theoretical weight of air

(d) Air : Fuel Ratio

This ratio shows the relative proportions of air and fuel in the combustion chamber.

$$\text{Air : Fuel ratio} = \frac{m_a}{m_f}$$

and

$$\text{Fuel : Air ratio} = \frac{m_f}{m_a}$$

(e) Specific Fuel Consumption : Thermal Efficiency

Let m_f = Weight of fuel used in time t min.

$$\therefore \quad \text{Wt. of } \frac{\text{fuel}}{\text{hour}} = \frac{m_f}{t} \times 60 \text{ kg/hr.}$$

Therefore, Specific fuel consumption is defined as the fuel consumption per hour per unit power. Its units will be kg/kW/Hour or kg/kWH. Depending on power under consideration this specific fuel consumption is Brake Specific Fuel Consumption – BSFC or Indicated Specific Fuel Consumption – ISFC.

This is the widely used (fuel consumption) parameter for comparative purposes.

As a revision, thermal efficiency is defined as the ratio of the output to the heat supplied by the burning of the fuel in cylinder. It gives us the fraction of the heat supplied which is converted into work. This is based on work developed i.e. indicated power (IP) or the shaft work available at the engine shaft (BP). Accordingly, the thermal efficiency may be either Indicated Thermal Efficiency or Brake Thermal Efficiency.

∴ Indicated Thermal Efficiency

$$= \frac{\text{Indicated work or Power IP (kW = kJ/s)}}{\text{Heat supplied power second}}$$

$$= \frac{\text{IP}}{\frac{m_f}{\text{sec}} \times \text{Cal. Value}}$$

$$\text{Also Brake Thermal Efficiency} = \frac{\text{Brake work or Power IP (kW = kJ/s)}}{\text{Heat supplied power second}}$$

$$= \frac{\text{BP}}{\frac{m_f}{\text{sec}} \times \text{Cal. Value}}$$

(f) Heat Balance

This gives us the distribution of heat supplied in the various ways and thus an indication of the working of the engine. General distribution of heat supplied in IC Engine.

(g) Specific Power Output and Specific Weight

Specific power output is defined by the ratio of the power produced by the engine to the displacement volume of the engine or it is the power produced per unit displacement of the piston.

$$\therefore \quad \text{Specific output} = \frac{\text{BP}}{A \times L} = \frac{p_{m_b} \times A \times L \times n}{A \times L}$$

$$= p_{m_b} \times \frac{N}{2\times 60} \text{ or } p_{m_b} \times \frac{N}{2}$$

$$= \text{Constant} \times p_{m_b} \times \text{Speed of engine.}$$

Thus specific output depends on p_{m_b} (bmep) and speed of the engine. By increasing the speed of the engine we can more work output but in that case the mechanical losses will be increased. In addition, increase in speed results in increase in mechanical stresses of the various engine parts. Similarly by increasing bmep, which requires better heat release and more load on engine, power output can be increased.

Specific weight is defined as the weight of the engine in kg to produce unit shaft power or BP developed. This gives the idea of the engine weight. This parameter plays an important role in the applications such as engine used in Automobiles and Aircrafts.

(h) Exhaust Smoke and Emissions

Smoke indicates that the combustion of the fuel is incomplete and naturally it limits the power output.

Exhaust emissions are oxides of nitrogen (NO_x), unburnt hydrocarbons etc. and are the nuisance to the public and environment in general.

Exhaust emissions are, nowadays, considered as the engine performance parameters.

25.8.3 Performance Curves

The actual performance of the engine is the most important factor to the user of the engine. Indicator cards may be taken that, together with the fuel consumption, reveal much more about the power and efficiency of the engine.

Tests on the engines are generally carried out either at constant speed or at variable speed. Tests at variable speed are generally carried on automotive and marine engines while that at constant speed are generally meant for the engines used for generators and pumps.

These tests are basically meant for determining the maximum power the engine can develop and the minimum specific fuel consumption and secondly to get the information in respect of the variation of all parameters with respect to each other.

(a) Test at Constant Speed : SI Engine. A constant speed test is run with variable throttle from no load to full load in suitable steps of load—generally zero load, 25 % of the full load, 50 % of the full load, 75 % of the full load and full

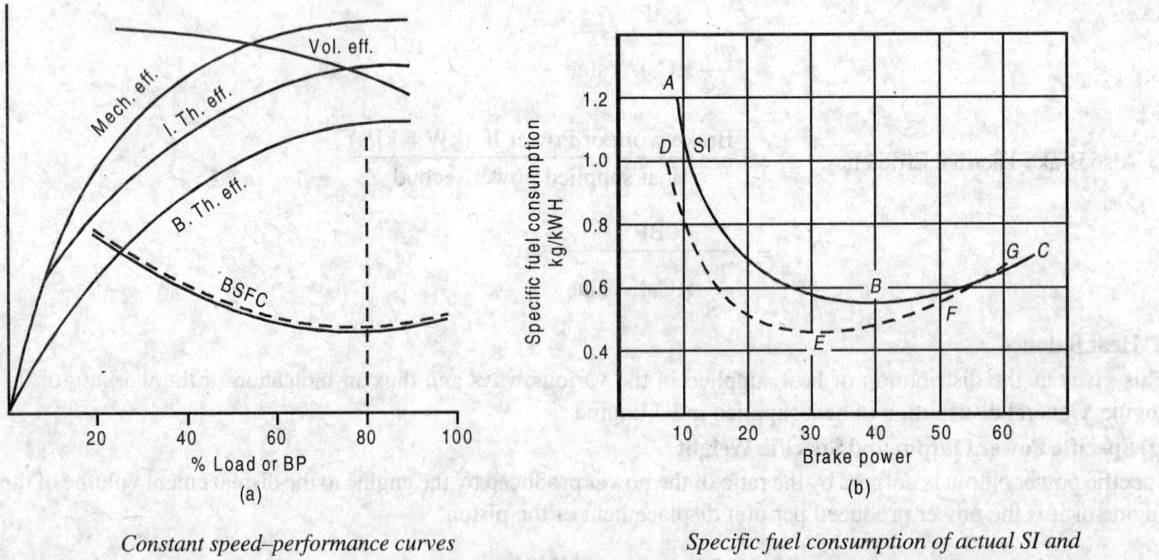

Constant speed–performance curves

Specific fuel consumption of actual SI and CI engine at constant speed

Fig. 25.14 (a) and (b)

load so that we will get the smooth curves. Starting at zero load, the throttle is opened to given the desired speed. The engine is run for a period of time until water, lubricating oil have been brought to definite operating temperature. Again when the temperature of the exhaust remains stable, the readings of different parameters are recorded. Then the same procedure is followed for other loads and the test is completed. Recorded items include all data necessary to calculate the required results.

The performance curves are plotted as shown in Fig. 25.14 on *X*-axis, different loads are plotted while on *Y*- axis, other quantities such as Mechanical efficiency, Brake thermal efficiency, fuel consumption, Brake power and Brake specific fuel consumption are plotted with different scales. Generally the last run will be at wide open throttle.

Note the following:

(1) Brake power, being directly proportional to load, variation is a straight line passing through origin.

(2) Fuel consumption is also a more or less straight line. At zero load, certain quantity of fuel is required to run the engine and account for the mechanical losses.

(3) Mechanical efficiency varies with load and is zero at zero load.

(4) Brake thermal efficiency again varies and is maximum at 75 % of maximum load approximately. This load is the economic load for the given engine.

(5) Brake specific fuel consumption decreases as the load increases and is minimum at the load where Brake thermal efficiency is maximum. This is the load at which the performance is best and is for general interest.

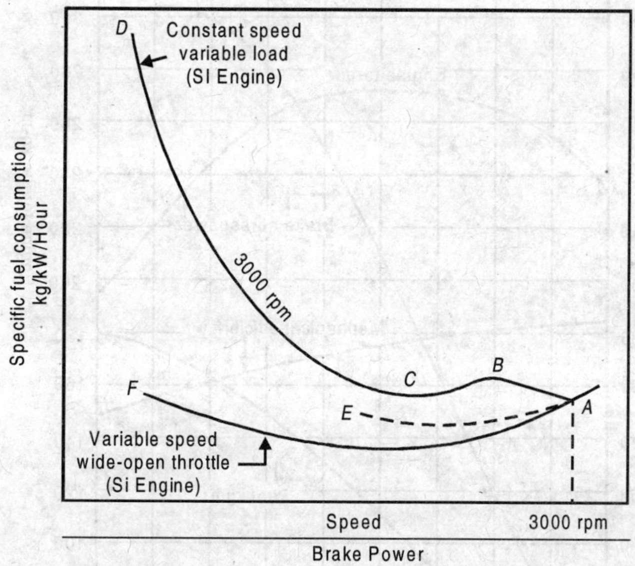

Fig. 25.14 (c) *Specific fuel consumption at constant and at variable speed*

(b) Test at Constant Speed – CI Engine. Constant speed test on CI Engine will be similar as that for SI Engine. In land engines, the speed is kept constant by the action of the governor which varies the quantity of fuel required for the load.

In this engine, the last run would show smoke in the exhaust gases.

The performance curves are plotted in the same manner as in SI Engine.

(c) Test at variable Speed – SI Engine. For a maximum power test on an SI Engine, the throttle is fully opened and the lowest desired speed is maintained by the brake or external load adjustment. The spart is adjusted (if manual) to give maximum power at this speed. The engine is run for a period of time until the water and lubricating oil have been brought to definite operating temperature.

When the engine is operating in approximate temperature equilibrium, the test is started and the average speed, brake load, temperatures, fuel weight etc. are recorded. Recorded items include all data necessary to calculate the required results.

After the completion of this run the brake load is adjusted until the speed has changed by the desired amount while the spark is adjusted for maximum torque. Equilibrium conditions of temperature are again obtained, and the procedure of the preceding paragraph is repeated.

Typical performance curves for automobile engine are shown in Fig. 25.15.

(i) Brake Power (BP). The power output increases to a peak after which the net output decreases. It is significant to note that although the friction power, which includes fluid frictional losses in pumping the gases into and out of the cylinder, is relatively small at low speeds, it increases rapidly at the high speeds.

(ii) Specific Fuel Consumption. For this type of the engine, the specific fuel consumption if a minimum and the thermal efficiency is a maximum at about 2200 to 2400 rpm. Over a fairly wide range, this fuel consumption and the efficiency vary little, a desirable characteristic, but the curve of fuel consumption increases rapidly at speeds, from about 3200 rpm for this engine. Thus the power necessary to drive the car at high speeds increases rapidly, about as the cube of the speed, and the efficiency of the engine falls off.

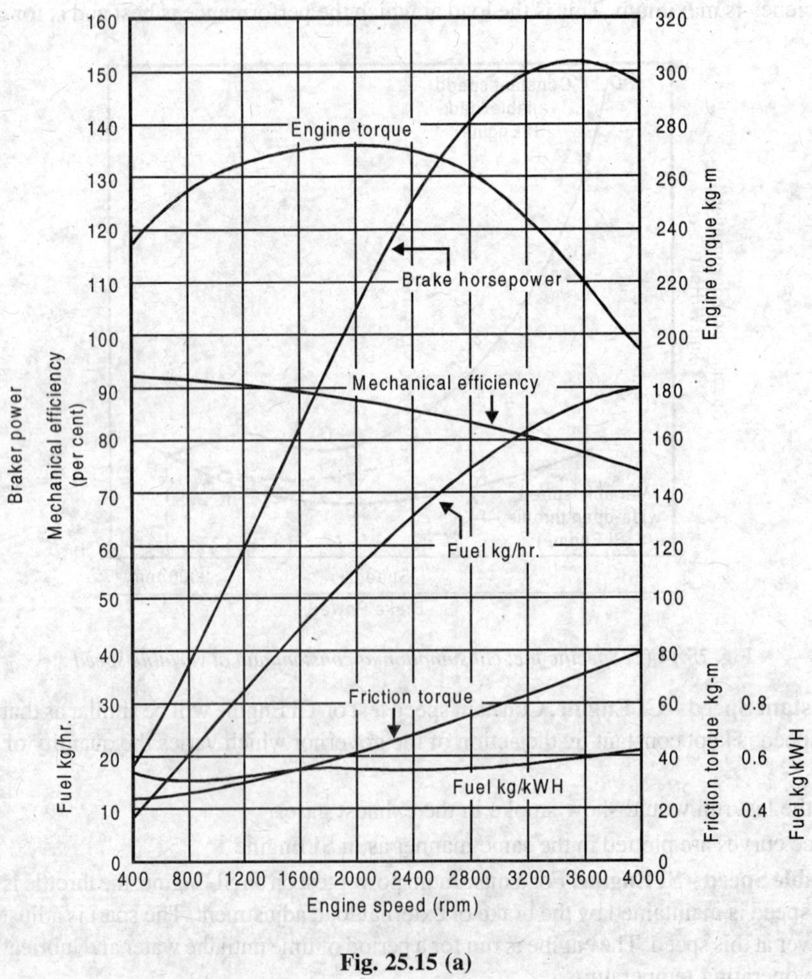

Fig. 25.15 (a)

(iii) Torque. The brake torque curve is significant for an automobile engine because the acceleration (pick - up) of the car depends upon the available torque, which is the maximum possible torque at a given speed minus the torque necessary to maintain the speed constant.

(iv) Mean Effective Pressure (MEP). Since the MEP indicates the relative power of an engine of a given size and speed, high values are desirable. In a particular type of engine, Otto or Diesel, an increase in the MEP is the result of burning more fuel for cycle, a change that may be brought about by improving the volumetric efficiency.

(v) Efficiency. The curve for the brake thermal efficiency characteristically rises to a peak and turns down, the curve for brake engine efficiency follows the same pattern.

(vi) Mainfold Vacuum. From the mainfold vacuum curve, we may check on the proper size of carburetor, the proper setting for the carburretor, and the proper timing of the values. Too high a vacuum (100 low pressure) suggests too much resistance to flow in the carburretor and other intake passages.

(d) Variable Speed Test With CI Engine. In a full power test of a CI engine at variable speed, the problem is more difficult than for the SI engine because there is no sharp limit of output at any speed. Following the procedure of the SI engine test, the brake is adjusted until the lowest operating speed is obtained with the fuel pump injecting a quantity of fuel sufficient to make the exhaust gas of the engine slightly coloured. This indicates that the engine is near full load, because some of the fuel is being wasted is smoke. Since the CI engine inducts a constant amount of air on the intake stroke, a small amount of fuel injected into the engine will not use all of the air in the cylinder. This occurs at part load. As the load is increased, greater amounts of fuel are injected and more and more of the air is required for combustion. At some stage, further injection of fuel leads to part of the fuel not being fully oxidized and to the production of smoke. Even at this condition, part of the air in the engine may not

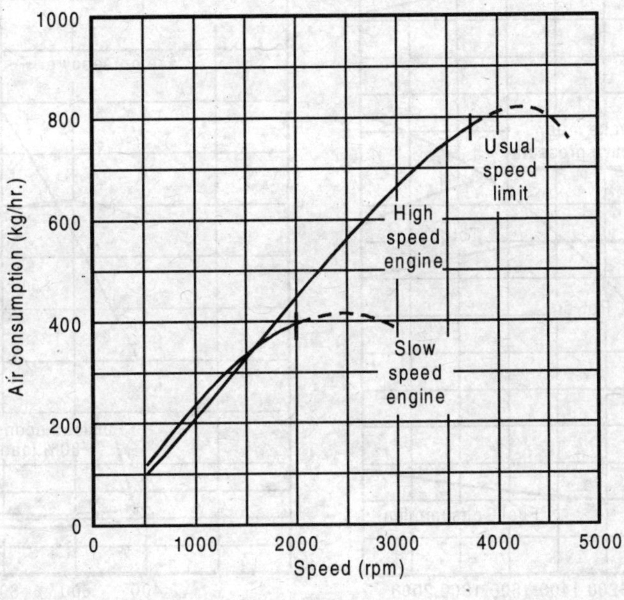

Fig. 25.15 (b) *Variation of the air consumption of high- and low-speed engines at wide-open throttle*

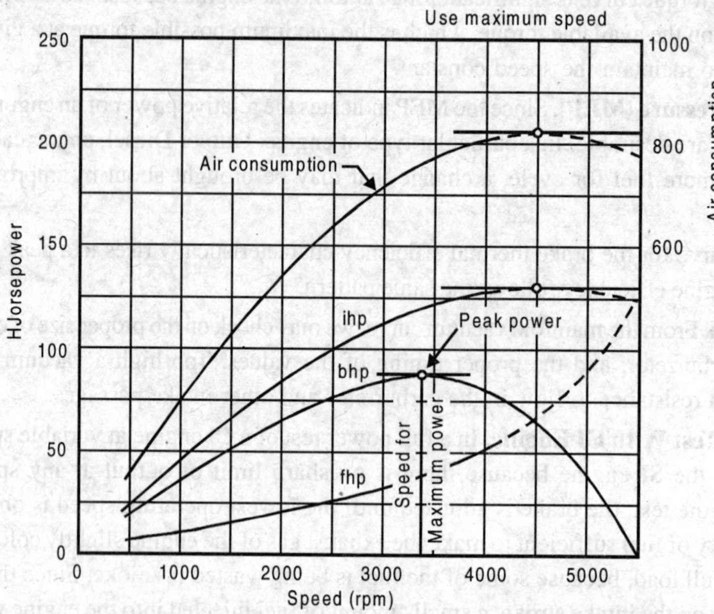

Fig. 25.15 (c) *variation of power and air consumption of high-speed engine at wide-open throttle*

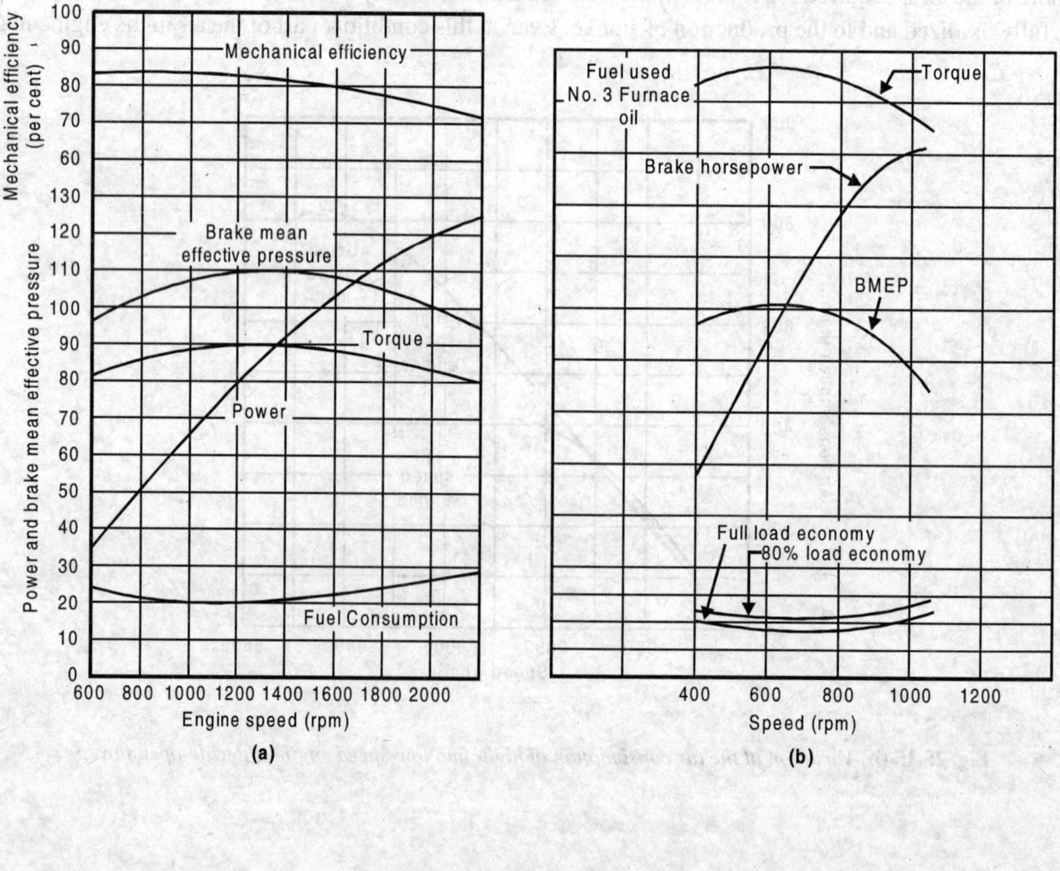

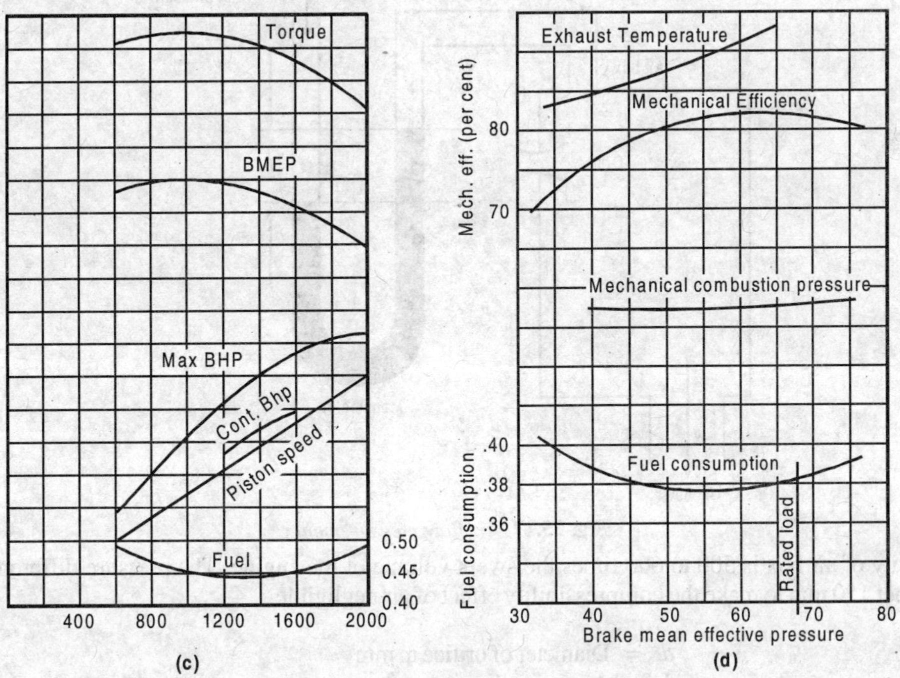

Fig. 25.16

be used because of failure of the injected fuel to find the air. In SI engine, the throttle was opened to the limit; in the CI engine, there is no sharp limit and the colour of the exhaust smoke is used to indicate that limit. It must be realised that smoke—colour observations are not an absolute index of degree of loading, because smoke may be the result of other conditions, such as poor atomization, very late injection, inadequate compression and unbalanced fuel feed to different cylinders. However, with an engine in good condition, the smoke indication may be considered a relatively satisfactory index of degree of loading.

Variable - speed tests of a CI engine at part load are run in the same manner as for SI engine.

Figure 25.16 gives the performance curves in different forms.

25.9 MEASUREMENTS

While performing a test on a engine, it is required to measure few quantitites in addition to quantities given or indicated by the indicating instruments. The quantitites that are to be measured are as follows:

1. Air consumption
2. Fuel consumption
3. Engine speed
4. Cooling water for both engine cooling and water flow through exhaust gas calorimeter.

1. Air Flow Measurement. The satisfactory measurement of the rate of flow of air is more difficult. It is essentially the measurement of the rate of flow of a compressible fluid, complicated by the fact that the flow is pulsating due to cyclic nature of the engine. The usual method of damping out the pulsation is to fit an air box of suitable volume to the engine with an orifice in the side of the box remote from the engine. The pressure difference across the orifice is measured by means of a suitable manometer as shown in Fig. 25.17.

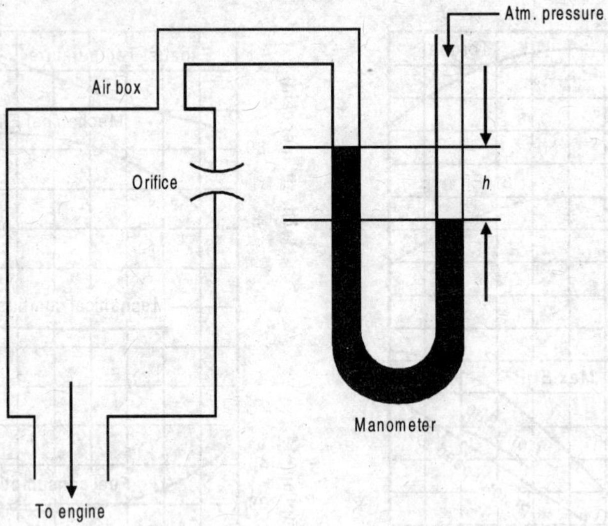

Fig. 25.17. *Air flow measurement*

The capacity of air box is 500 to 600 times the swept volume of the engine. The pressure difference should be limited to about 150 mm to make the compressibility effect of air negligible.

Let

d = Diameter of orifice in mm
A = Area of orifice in mm^2
C_d = Coefficient of discharge
h = Pressure difference in mm of water from U-tube manometer
H = Head causing flow of air in mm of air
C = Air velocity in m/sec
ρ_d = Density of air under atmospehric condition in kg/m^3
w = Density of water in kg/m^3

$\therefore$ Head causing the air flow $= H$

$$= h \times \frac{\delta w}{\rho_a} = \frac{h}{1000} \times \frac{1000}{\rho_a} \text{ m}$$

Velocity of air passing through orifice is given by

$$C = \sqrt{2gH} = \sqrt{2g \times \frac{h}{\rho_a}} \text{ m/s}$$

$\therefore$ Volume of air passing through the orifice is given by

$$V_a = A \times C \times C_d = \frac{A}{10^6} \times C_d \times \sqrt{\frac{2gh}{\rho_a}} \text{ m}^3/\text{s}$$

and mass flow of air $= V_a \times \delta_a = \frac{A}{10^6} C_d \sqrt{\frac{2gh}{\rho a}} \times \delta_a$

$$= \frac{A}{10^6} C_d \times \sqrt{2gh\rho_a}$$

2. Fuel Consumption. A knowledge of the fuel consumed by an engine and time it takes to consume this fuel is essential when assessing the qualities of the engine. For petrol and diesel engines, the fuel is run through a special measuring device. This can take the form of a reservoir of fuel of known quality and the time for the engine to consume this measured quantity of fuel. Several means are available for measurement of fuel consumption and one simple arrangement is available in Fig. 25.18.

The measuring vessel consists of two bulbs of known capacity in series, the capacities being measured between marks on the connecting capillary tubes. The fuel level falls quickly past the marks on the small box capillary tube. The fuel from the tank or measuring vessel passes through a three way valve to the engine. The three positions of the valve are shown in Fig. 25.18. Provision must be made to allow the fuel to fill the measuring bulbs upto the level in the tank. The time is taken for the consumption of a known volume of fuel and thus the rate of fuel consumption can be determined.

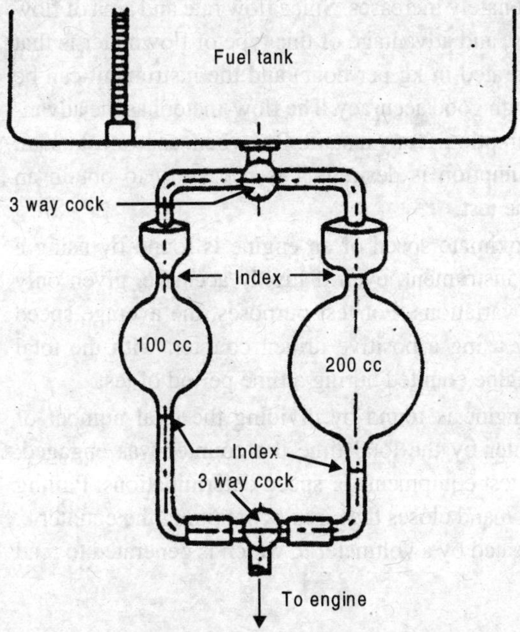

Fig. 25.18. *Fuel measuring system*

The other arrangement to measure the volume of fuel supplied to the engine is shown in Fig. 25.19.

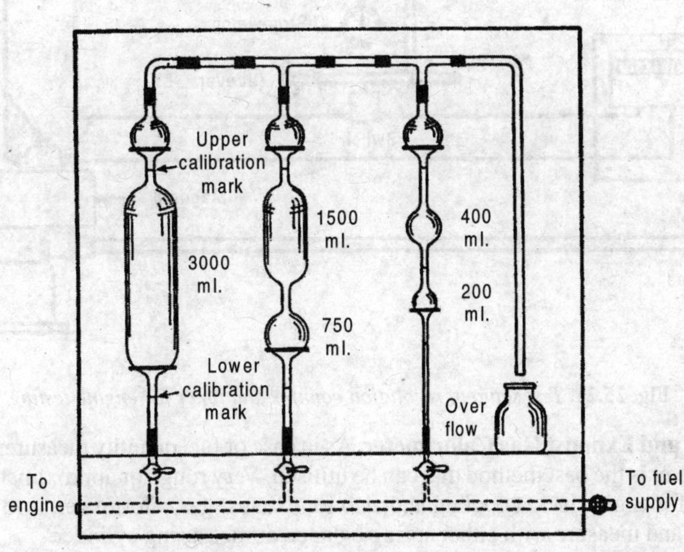

Fig. 25.19. *Test setup for volumetric determination of fuel consumption*

There are other types of instruments available called flow meters which give the rate of flow directly and the calibration can be made in litres per hour or kg/hour as required. The volume flow can be converted in mass flow by multiplying fuel density.

In many automotive laboratories and for air craft testing, flow meters are used. Figure 25.20 illustrates the principle of the Fisher-Porter *Rotameter*. As the flow increases through the meter, the float rises and the area between the float and the tapered graduated tube proportionately increases. Since flow rate and area of flow are directly related to each other, and advantage of this type of flowmeter is that the graduations are linear (calibrated in kg per hour) and the instrument can be used for a wide range of flows with good accuracy. The flow-meter has the advantage of indicating the fuel consumption at any instant. But when, as is most usual, the average or total fuel consumption is desired, it is necessary to obtain an average reading made during the test.

3. Engine Speed. The approximate speed of an engine is found by using a *tachometer*. However, such an instrument, even if highly accurate, given only instantaneous speed and speed variations. For test purposes, the average speed of the engine is determined by using a positive driven counter, with the total number of revolutions of the engine counted during a time period of test.

The average speed of the engine is found by dividing the total number of revolutions shown on the counter by the total time the counter was engaged. Figure 25.21 shows a common test equipment for speed determinations. Pulling the lever *a* starts the stop watch *b* and closes the circuit containing the counter *c*. The instantaneous speed is indicated by a voltmeter *d*, which is generated to read rpm.

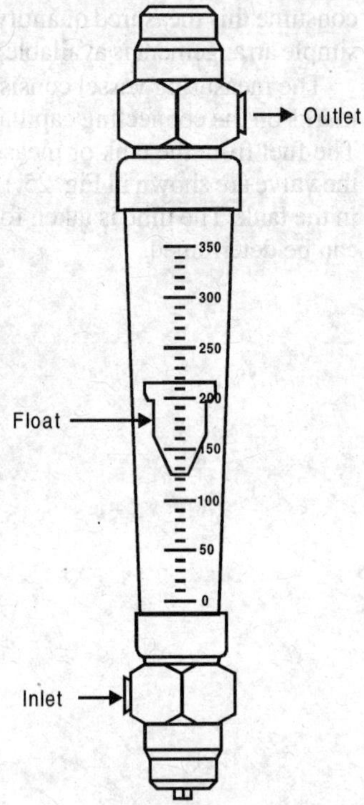

Fig. 24.20. *Fischer-Porter flowmeter. Taper of tube magnified*

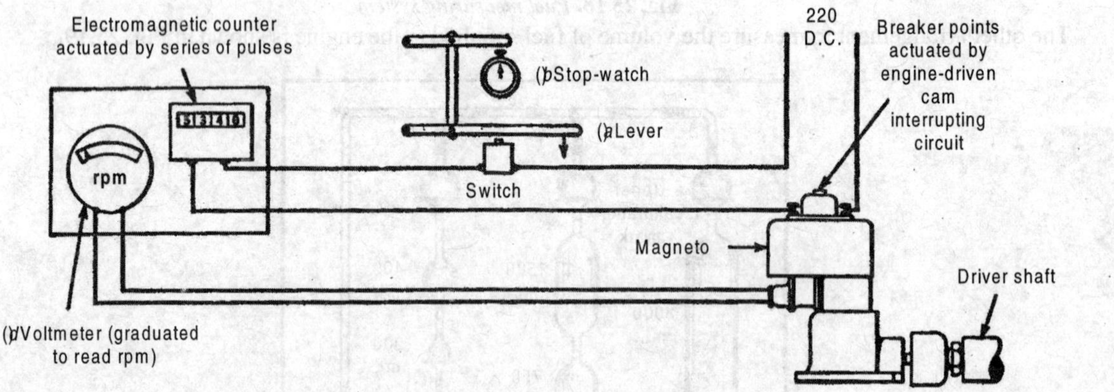

Fig. 25.21. *Test setup of revolution counter and timer for engine testing.*

4. Water for Jacket and Exhaust Gas Calorimeter. As in case of fuel quantity measurement, for water measurement nowadays, rotameter is the best method that can be utilised. Very rough or approximate method to measure the quantity of water used for engine jacket cooling and also for exhaust gas calorimeter, is to collect the water for the specific time in bucket and measure with either spring balance or measuring cylinder.

SOLVED EXAMPLES

Example 25.1. *A six cylinder, gasoline engines operates on the four stroke cycle. The bore of each cylinder is 80 mm and stroke 100 mm. The clearance volume per cylinder is 70 cc. At a speed of 4000 rpm the fuel consumption is 20 kg/hr and the torque developed is 150 N-m. Calculate*

(i) Brake power

(ii) The brake mean effective pressure

(iii) Brake thermal efficiency

If the calorific value is 43,000 kJ/kg and relative efficiency on a BP basis assuming the engine works on the constant volume cycle. γ for air = 1.4.

Solution

$$\text{Stroke Volume} = \frac{\pi d^2}{4} \times L$$

$$= \frac{\pi}{4} \times \left(\frac{80}{10}\right)^2 \times \frac{100}{10}$$

$$= \frac{\pi}{4} \times 64 \times 10 \text{ cc}$$

$$= 502.655$$

$$\therefore \quad \text{Compression ratio} = r_k = \frac{V_s + V_c}{V_c}$$

$$= \frac{502.655 + 70}{70}$$

$$= \frac{572.655}{70}$$

$$= 8.181$$

$\therefore$ Air Standard Efficiency

$$\text{ASE} = 1 - \frac{1}{r_k^{\gamma-1}} = 1 - \frac{1}{8.181^{0.4}}$$

$$= 1 - \frac{1}{2.318} = 1 - 0.4314$$

$$= 0.5686 = 56.68\ \%$$

(i) Brake Power :

$$\text{BP} = \frac{2\pi\, NT}{1000} = \frac{2\pi \times 4000}{60 \times 1000} \times \frac{150}{1}$$

$$= 20\,\pi = \mathbf{62.832\ kW}$$

(ii) Brake Mean Effective Pressure : BMEP

$$BP = p_{m_b} \times \frac{ALn}{1000} = \frac{p_{m_b} \times 10^5}{1000} \times \frac{\pi}{4} \times \left(\frac{80}{1000}\right)^2 \times \frac{100}{1000} \times \frac{4000}{2 \times 60}$$

$$62.832 = 100\ p_{m_b} \times \frac{\pi}{4} \times (0.08)^2 \times 0.1 \times \frac{200}{6_N}$$

$$= 100\ p_{m_b} \times 0.1676 = 16.76\ p_{m_b}$$

$$p_{m_b} = \frac{62.832}{16.76} = 3.75 \text{ bar}$$

$$BMEP = p_{m_b} = \mathbf{3.75\ bar}$$

(iii) Brake Thermal Efficiency :

$$B.Th.E = \frac{BP \times 3600}{\frac{m_f}{hr} \times \text{Cal. Val.}}$$

$$= \frac{62.832 \times 3600}{20 \times 43000}$$

$$= \mathbf{0.263}$$

$$= \mathbf{26.3\%}$$

$$\text{Relative Efficiency} = \frac{\text{B. Th. Eff.}}{\text{A. S. E.}}$$

$$= \frac{0.263}{0.5686}$$

$$= 0.4625$$

$$= \mathbf{46.25\ \%}$$

Example 25.2. (*a*) *Explain the laboratory method to determine the air-fuel ratio for an IC Engine.*

(*b*) *During the trial of a single cylinder, four stroke oil engine, the following results were obtained:*

Cylinder diameter	= *200 mm*
Stroke of piston	= *400 mm*
MEP	= *6 bar*
Torque	= *407 N-m*
Speed	= *250 RPM*
Oil Consumption	= *4 kg/hour*
C.V. of fuel	= *43 MJ/kg*
Cooling water flow rate	= *4.5 kg/min*
Air used/kg fuel	= *30 kg*
Rise in cooling water temperature	= *45° C*
Temperature of exhaust gases	= *420° C*

Room temperature = 20º C
Mean sp-heat of exhaust gases = 1 kJ/kg K
Sp-heat of water = 4.18 kJ/kg K
Barometric pressure = 1.01325 bar

Find IP, BP, and draw up a heat balance sheet for the test in kg/hour basis.

***Solution:* (*b*)**

$$\text{IP} = p_m \frac{LAn}{1000} \quad \text{with usual notations}$$

$$= \frac{6\times10^5}{1000}\times\frac{400}{1000}\times\frac{\pi}{4}\times\left(\frac{200}{1000}\right)^2\times\frac{250}{2\times60}$$

$$= 240\times\frac{\pi}{4}\times0.04\times\frac{25}{12}$$

$$= 5\pi$$

$$= \mathbf{15.71\ kW}$$

$$\text{BP} = \frac{2\pi\, NT}{1000}$$

$$= 2\pi\times\frac{250}{60}\times\frac{407}{1000}$$

$$= \mathbf{10.66\ kW}$$

Heat balance sheet on Hour basis:

(1) Heat supplied/hour $= \frac{m_f}{\text{hour}}\times\text{Cal. Val.}$
$= 4\times43000$
$= 172000$ kJ/hour…100 %

(2) Heat equivalent of IP (BP + FP) $= 10.66\times3600$
or $= 56556$ kJ/Hour…32.38 %

(2a) Heat equivalent of BP $= 10.66\times3600$
$= 0.2231$ or 22.31 %

(3) Heat equivalent of FP $= (15.71-1066)\times3600$
$= 18180$ kJ/hour
$= 10.57$ %

(4) Heat given to cooling water $= m_w\times C_p\times\text{Temperature rise}$
$= 4.5\times60\times4.18\times45$
$= 50787$ KJ/hour...29.53 %

(5) Heat carried away by exhaust gases $= m_g\times C_{pg}\times\text{Temperature drop}$
$= (4\times31)\times1\times(420-20)$

$= 49600$ kJ/hour...28.84 %

(6) Heat unaccounted for – by difference $= 172000-(38376+18180+50787+49600)$

$= 172000-156943$

$= 15057$ kJ/hr...8.75 %

Example 25.3. *Obtain the cylinder dimensions of a twin cylinder, two-stroke IC Engine from the following data:*

Engine speed	= *4000 rpm*
Volumetric efficiency	= *77%*
Mechanical Efficiency	= *75%*
Piston speed	= *600 m/min*
IMEP	= *5 bar*
Fuel consumption	$= \dfrac{10\, Lit.}{hr.}$
Sp. gr. of fuel	= *0.73*
Air to fuel ratio	= *18*

Also, determine the power output at 577 conditions.

Assume $R_{air} = 0.287$ *kJ/kg K.*

Solution

$$\text{Piston speed} = 600\frac{\text{m}}{\text{min}}$$

$$= 2L \times N$$

$$\therefore \quad 600 = 2L \times 4000$$

$$\therefore \quad L = \text{stroke of the piston}$$

$$= \frac{600}{2 \times 4000}$$

$$= \frac{3}{40}$$

$$= 0.075\,\text{m}$$

$$= 7.5\,\text{cm}$$

$$= \mathbf{75\,mm}$$

$$\text{Fuel Consumption} = 10 \text{ lit/hr.}$$

$$= 10 \times 0.73\frac{\text{kg}}{\text{hr}}$$

$$= 7.3 \text{ kg/hr}$$

$$\therefore \quad \text{Air Supplied} = 7.3 \times 18$$

$$= 131.4 \text{ kg/hr}$$

$$= 2.19 \text{ kg/min}$$

$$\therefore \quad = \frac{\text{Actual volume of air supplied}}{\text{min}}$$

$$V = \frac{m\,R\,T}{p}$$

$$= \frac{2.19\times 287\times 288}{1.01325\times 10^5}$$

$$= 1.7865\ \frac{\text{cu.m}}{\text{min}}$$

$$\therefore \qquad \frac{\text{Stroke volume}}{\text{min}} = \frac{1.7865}{\eta_{\text{vol.}}}$$

$$= \frac{2.32\times 4000}{260}\ \frac{\frac{\text{m}^3}{\text{cycle}}}{\text{cyl}}$$

$$= \frac{1.7865}{0.77\times 2}$$

$$= \frac{2.32}{2}\ \frac{\text{cu.m/min}}{\text{cyl.}}$$

$$\frac{\text{Stroke volume/cycle}}{\text{min}} = \frac{2.32}{2}$$

$$= 1.16\frac{\text{m}^3}{\text{min}} = 0.00029\ \frac{\text{m}^3}{\text{cyl.}}$$

$$= \frac{\pi d^2}{4}\times L$$

$$\therefore \qquad d^2 = \frac{0.00029\times 4}{\pi\times 0.075} = 0.004923$$

$$d = 0.07 \text{ m}$$

$$= 7 \text{ cm} = \mathbf{70\ mm}$$

$$\text{IP} = p_{m_i}\frac{\text{LAN}}{1000}\times 2$$

$$= \frac{5\times 10^5\times \frac{\pi}{4}\times (0.07)^2\times 0.075}{1000}\times 2\times \frac{4000}{3600}$$

$$= \frac{5\times 10^5\times \pi}{4}\times \frac{(0.075)\times 0.7^2\times 2}{1000}\times \frac{4000}{60}$$

$$= 5\times 8\times \frac{\pi}{4}\times 75\times 100\times 0.07\times \frac{0.070}{60}$$

$$= \frac{40\frac{\pi}{4}\times 75\times 0.70\times 0.70}{60}$$

$$= 19.24 \text{ kW}$$

$\therefore$ Output of the engine $= 19.24\times$ Mech. Efficiency

$$= 19.24\times 0.75 = \mathbf{14.43\,kW}$$

Example 25.4. *Calculate bore and stroke of a 4-stroke petrol engine for the following data:*

Compression ratio	*= 6*
Brake Power	*= 73.5 kW*
Speed	*= 400 rpm*
Brake mep	*= 8.5 bar*
Mechanical Efficiency	*= 80%*
Brake sp. fuel consumption	$= 0.346\ \frac{kg}{kWH}$
Calorific value of fuel	$= 44100\ \frac{kJ}{kg}$

Also calculate indicated and brake thermal efficiencies, air standard efficiency, relative efficiency and indicated mean effective pressure.

Assume bore = Stroke

Solution

Bore and Stroke:

We have, $$BP = \frac{p_{m_b}\ LAN}{1000}$$

$$73.5 = \frac{8.5\times 10^5}{1000}\times L\times \frac{\pi}{4}d^2\times \frac{400}{60\times 2}$$

$$= 850\times \frac{\pi}{4}\cdot d^3\times \frac{10}{3}$$

$$= 2225.3\,d^3$$

$$\therefore \quad d^3 = \frac{73.5}{2225.3}$$

$$= 0.033$$

$$\therefore \quad d = \text{Bore diameter}$$

$$= 3.2\times 10^{-1}\text{ m}$$

$$= \mathbf{32\ cm}$$

$$= L$$

Indicated mean effective pressure:

$$p_{m_i} = \text{IMEP} = \frac{p_{m_b}}{\eta_{\text{mech}}}$$

$$= \frac{8.5}{0.8}$$

$$= 10.625 \text{ bar}$$

$$\textbf{Brake thermal efficiency} = \frac{\text{BP}}{m_f \times \text{Cal. Value}}$$

$$= \frac{73.5 \times 3600}{\frac{m_f}{\text{hour}} \times 44100}$$

$$= \frac{3600}{\frac{m_f}{\text{bp}}\frac{}{\text{hr.}} \times \text{Cal. value}}$$

$$= \frac{3600}{0.346 \times 44100}$$

$$= 0.236$$

or $= \textbf{23.6 \%}$

$$\therefore \textbf{Indicated thermal efficiency} = \frac{\text{B. Th. E.}}{\text{Mechanical Efficiency}}$$

$$= \frac{0.236}{0.8}$$

$$= 0.285$$

or $= \textbf{29.5 \%}$

Air Standard Efficiency:

$$\text{ASE} = 1 - \frac{1}{r_k^{\gamma} - 1}$$

$$= 1 - \frac{1}{6^{0.4}}$$

$$= 1 - \frac{1}{2.048}$$

$$= \frac{1.048}{2.048}$$

$$= 0.5115$$

or $= \mathbf{51.15\ \%}$

$$\therefore \text{Relative efficiency (Based on BP)} = \frac{\text{B. Th. E.}}{\text{ASE}}$$

$$= \frac{0.236}{0.5115} = 0.4613$$

or $= \mathbf{46.13\ \%}$

$$\therefore \text{Relative efficiency based on IP} = \frac{\text{I. Th. Efficiency}}{\text{ASE}}$$

$$= \frac{0.295}{0.5115}$$

$$= 0.5767$$

$$= \mathbf{57.67\ \%}$$

Example 25.5. *A four cylinder petrol engine has bore of 5.7 cm and stroke of 9 cm. Its rated speed is 2800 rpm and it is tested at this speed against a broke which has a torque arm of 0.356 m. The net brake load is 155 N and fuel consumption is 6.74 lit./hr, with sp. gravity of 0.735 and it has low calorific value of 44200 kJ/kg. A Morse test is carried out and cylinders are count out in the order 1, 2, 3 and 4 with corresponding brake loads of 111, 106.5, 104.2 and 111 newtons respectively. Calculate for this speed—engine torque, brake mean effective pressure, brake thermal efficiency, specific fuel consumption, mechanical efficiency and indicated mean effective pressure.*

Solution

When all cylinders are working,

$$\text{BP} = \frac{2\pi TN}{1000\times 60}$$

$$= \frac{2\pi\times 2800\times 155\times 0.356}{60\times 1000}$$

$$= \mathbf{16.18\ kW}$$

$$= p_{m_b}\times\frac{\pi}{4}\times d^2\times L\times\frac{2800}{60\times 2}\times\frac{4}{1000}$$

$$= p_{m_b}\times 10^5\times\frac{\pi}{4}\times\left(\frac{5.7}{100}\right)^2\times\left(\frac{9}{100}\right)\times\frac{2.8}{120}\times 4$$

$$= p_{m_b}\times 2.14434$$

$$\therefore \quad p_{m_b} = \text{Brake MEP}$$

$$= \frac{16.18}{2.14434}$$

$$= \mathbf{7.545\ bar}$$

$$\frac{\text{Fuel consumption}}{\text{sec.}} = \frac{6.74}{3600}\times 0.735\frac{\text{kg}}{\text{sec.}}$$

$$\frac{\text{Heat Supplied}}{\text{sec.}} = \frac{6.74\times 0.735}{3600}\times 44200$$

$$= 60.82 \text{ kJ/sec}$$

$$\therefore \text{Brake thermal efficiency} = \eta_{bth} = \frac{16.18}{60.82} = 0.266$$

$$= 26.6\ \%$$

$$\text{Specific fuel consumption} = \frac{\text{Fuel /hour}}{\text{BP}} = \frac{6.74 \times 0.735}{16.18}$$

$$= \mathbf{0.3062\ \ kg/kWH}$$

Morse Test Calculations:

When all cylinders are working, BP = 16.18 kW

Cyl. No. 1 cut-out : }
Cyl. No. 4 cut-out : }

$$\text{BP} = \frac{2\pi \times 2800 \times 0.356 \times 111}{1000 \times 60}$$

$$= 11.59 \text{ kW}$$

$$\therefore \quad IP_1 = IP_4 = 16.18 - 11.59$$

$$= \mathbf{4.59\ \ kW}$$

When Cyl. No. 2 is cut-out :

$$B_2 = \frac{2\pi \times 2800 \times 0.356 \times 106.5}{1000 \times 60}$$

$$= 11.12 \text{ kW}$$

Similarly

$$B_3 = \frac{11.12}{106.5} \times \frac{104.2}{1}$$

$$= 10.88 \text{ kW}$$

$$IP_2 = 16.18 - 11.12$$

$$= 5.06 \text{ kW}$$

$$IP_3 = 16.18 - 10.88$$

$$= 5.30 \text{ kW}$$

$$\therefore \quad \text{Total IP of the engine} = 4.59 + 4.59 + 5.06 + 5.30$$

$$= 9.81 + 10.36$$

$$= \mathbf{19.54\ \ kW}$$

$$\therefore \quad \text{Mechanical efficiency} = \frac{\text{BP}}{\text{IP}}$$

$$= \frac{16.18}{19.54}$$

$$= 0.828$$

$$= \mathbf{82.8\ \ \%}$$

Let p_{m_b} be in bar

$$\therefore \quad 16.18 = p_{m_b} \times 10^5 \times \frac{\pi}{4} \times \left(\frac{5.7}{100}\right)^2 \times \frac{9}{100} \times \frac{2800}{120} \times \frac{4}{1000}$$

$$\therefore \quad p_{m_b} = 7.545 \text{ bar}$$

$$\text{Indicated MEP} = p_{m_i} = \frac{p_{m_b}}{\eta_{\text{mech}}}$$

$$= \frac{7.545}{0.828}$$

$$= \mathbf{9.112 \text{ bar}}$$

Example 25.6. *Test on a single cylinder oil engine with bore 30 cm and stroke 40 cm, working on 4-stroke cycle, gave the following observations*

(1) Duration of trial	=	*1 hour*
(2) Total fuel consumption	=	*8.5 litres*
(3) Sp-gravity of oil	=	*0.83*
(4) Calorific value	=	*44500 kJ/kg*
(5) Total revolutions	=	*12000*
(6) MEP	=	*6 bar*
(7) Brake load	=	*160 kg*
(8) Spring balance reading	=	*10 kg*
(9) Brake drum diameter	=	*180 cm*
(10) Rope diameter	=	*0.3 cm*
(11) Jacket cooling water	=	*550 kg*
(12) Water inlet and outlet temperatures	=	*15° C and 60° C respectively*
(13) Total air consumption	=	*360 kg*
(14) Specific heat of exhaust gas	=	*1.01 kJ/kg - OR*
(15) Exhaust gas outlet temperature	=	*300° C.*

Find:

(i) Mechanical efficiency,

(ii) Draw Heat balance sheet on min. basis,

(iii) Write the remarks.

Solution

(i) Mechanical efficiency:

$$\text{IP} = p_m \frac{LAn}{60}$$

$$= \frac{6 \times 10^5}{1000} \times \frac{40}{100} \times \frac{\pi}{4} \times (0.3)^2 \times \frac{12000}{2 \times 60 \times 60}$$

$$= 60 \times \pi \times 0.09 \times \frac{10}{6}$$

$$= 9\pi$$

$$= \mathbf{28.27 \text{ kW}}$$

$$\text{BP} = \frac{2\pi NT}{1000}$$

$$= \frac{2\pi \times 42000}{3600 \times 1000} \times (W - S)R$$

$$= \frac{\pi \times (160-10)9.81}{150} \times \frac{\left(\frac{180}{2}\right) + 0.15}{100}$$

$$= 9.81\pi \times \frac{90.15}{100}$$

$$= \mathbf{27.78\ kW.}$$

(ii) Heat balance on minute basis:

(1) $$\frac{\text{Heat supplied}}{\text{min}} = \frac{\text{fuel}}{\text{min}} \times \text{Cal. Value}$$

$$= \frac{8.5}{60} \times 0.83 \times 44500$$

$$= 5232.46 \text{ kJ/min......}100\ \%$$

(2) $$\text{Heat equivalent of IP} = 28.27 \times 60 \frac{\text{kJ}}{\text{min}}$$

$$= 1696.2 \text{ kJ/min......}32.42\ \%$$

(3) $$\frac{\text{Heat to cooling water}}{\text{min}} = m_w C_{p_w} (T_o - T_i)$$

$$= \frac{550}{60} \times 4.187 \times (60 - 15)$$

$$= 1727.14 \text{ kJ}$$

$$= 33\ \%$$

(4) $$\frac{\text{Heat in exhaust gases}}{\text{min}} = m_g C_{pg} (T_o - T_{\text{atm}})$$

$$= \left(\frac{360}{60} + \frac{8.5 \times 0.83}{60}\right) 1.01 \times (300 - 15)$$

$$= (6.1176) \times 1.01 \times 285$$

$$= 1760.95 \text{ kJ/min.}$$

$$= 33.65\ \%$$

(5) $$\text{Heat unaccounted for} = 5232.46 - [(1696.2\,[+1727.14 + 1760.952\,]$$

$$= 5232.46 - 5184.29$$

$$= 48.17 \text{ kJ/min.}$$

$$= \mathbf{0.9206\ \%}$$

Example 25.7. (a) *During a trial on a 4 cyl. petrol engine running at 50 rev/s, the brake load was 267 N when all the cylinders were working. When each cylinder was cut-out in turn and the speed returned to 50 rps, the brake readings were 176 N, 187 N, 182 N and 182 N. Using these readings, determine the brake power of the engine and* *estimate its indicated power and mechanical efficiency. For the brake,* $BP = \frac{FN}{455}$*. Where F = brake load in Newton and N is rps. The following results were also obtained during the trial.*

(1) *Fuel consumption* = 0.568 *litre/30 sec*

Calorific value C.V. = $43000 \frac{kJ}{kg}$

(2) *Sp-gr.* = 0.72

(3) *Air: Fuel ratio* = 14 : 1

(4) *Ex. temperature* = 760° *C*

(5) *Sp. ht. of gas* = $1.015 \frac{kJ}{kgK}$

(6) *Cooling water flow rate* = 0.28 *kg/sec.*

(7) *Inlet and outlet temperature of cooling water* = 18 *and* 56° *C*

(8) *Ambient temperature* = 21° *C*

From these results, draw up an energy balance in kJ/sec. and as a percentage of the energy supplied.

Solution

BP when all cylinder are working, is given by

$$BP = \frac{FN}{455} = \frac{267 \times 50}{455}$$

$$= \mathbf{29.34\ kW}$$

$$BP_1 = \frac{176 \times 50}{455}$$

$$= \mathbf{19.56\ kW}$$

$$BP_2 = \frac{187 \times 50}{455}$$

$$= \mathbf{20.55\ kW}$$

$$BP_3 = \frac{182 \times 50}{455}$$

$$= \mathbf{20\ kW}$$

$$BP_4 = \frac{182 \times 50}{455}$$

$$= \mathbf{20\ kW}$$

$$\therefore \quad IP_1 = 29.34 - 19.56 = 9.78$$

$$IP_2 = 29.34 - 20.55 = 8.79$$

$$IP_3 = 29.34 - 20 = 9.34$$

$$IP_4 = 29.34 - 20 = 9.34$$

Total IP of the engine is

$$IP = \mathbf{37.25\ kW}$$

$$\therefore \quad \text{Mechanical efficiency} = \frac{BP}{IP}$$

$$= \frac{29.34}{37.25}$$

$$= 0.7876$$

$$= \mathbf{78.76\ \%}$$

Heat balance on kJ/s basis and % basis

(1) $\frac{\text{Heat supplied}}{\text{sec.}} = \frac{0.568}{60} \times 0.72 \times 43000$

$= 586.176$ kJ/s...... 100 %

$= 293.88$ kJ/s

(2) Heat equivalent of BP $= 29.34$ kJ/sec...... 9.9 %

(3) Heat equivalent of FP $= 37.25 - 29.34$

$= 7.91$ kJ/s

$= 2.69$ %

(4) Heat cooling water $= 0.28 \times 4.187 \times (56 - 18)$

$= 44.55$ kJ/s

$= 15.16$ %

(5) Heat to exhaust gases $= m_g C_{pg} (t_e - t_R)$

$$= \frac{(14+1) \times 0.568 \times 0.72}{30} \times 1.015 \times (760 - 21)$$

$= \mathbf{153.38}$ **kJ/sec...... 52.19 %**

Total Heat expanded $= 235.18$ kJ/s

(6) Heat unaccounted for $= 293.88 - 235.18$

$= 58.70$ kJ/s...... 20 %

Example 25.8. *A four-cylinder, four-stroke engine of 100 × 120 mm cylinder size was tested at 1600 rpm and the following readings were obtained:*

Fuel consumption	= *0.27 lit./min*
Sp. gr. of fuel	= *0.74*
Net load on brake	= *400 N*
Constant of dynamometer K	= *20380* $\left[BP = \frac{WN}{K}\right]$
Calorific value of fuel	= *44 MJ/kg*

Assuming 80 % mechanical efficiency,

Calculate:

(*1*) *Brake thermal efficiency*

(*2*) *Indicated thermal efficiency*

(*3*) *BSFC*

(*4*) *IMEP*

Solution

(1) Brake thermal efficiency:

$$BP = \frac{WN}{K}$$

$$= \frac{400 \times 1600}{20380}$$

$$= \mathbf{31.4 \ kW}$$

$$Q_A = m_f \times \text{Cal. Value}$$

$$= \frac{0.27 \times 0.74}{60} \times \frac{44000}{1}$$

$$= \frac{8791.2}{60} \text{ kJ/min}$$

$$= 146.52 \text{ kJ/sec}$$

$$\text{Brake thermal efficiency} = \frac{BP}{Q_A}$$

$$= \frac{31.4 \text{ kJ/sec.}}{146.52 \text{ kJ/sec}}$$

$$= 0.2143$$

$$= \mathbf{21.43 \ \%}$$

(2) Indicated thermal efficiency

$$\text{Indicated thermal efficiency} = \frac{\text{Brake thermal efficiency}}{\text{Mechanical efficiency}}$$

$$= \frac{0.2143}{0.8}$$

$$= 0.2679 = \mathbf{26.79 \ \%}$$

(3) BSFC

$$\text{Fuel used per hour} = 0.27 \times 0.24 \times 60$$

$$= 11.988 \text{ kJ/hr.}$$

$$\therefore \quad BSFC = \frac{\text{Fuel/hr.}}{BP}$$

$$= \frac{11.988}{31.4}$$

$$= \mathbf{0.3818 \ kg/KWH}$$

(4) Indicalted Mean Effictive Pressure

$$IP = \frac{BP}{\eta_{mech}}$$

$$= \frac{31.4}{0.8}$$

$$= 39.25 \text{ kW}$$

$$= (\text{No. of cylinders}) \times \frac{p_{m_i} LAn}{1000}$$

$$= \frac{4 \times p_{m_i}}{1000} \times \frac{120}{1000} \times \frac{\pi}{4} \times \left(\frac{100}{1000}\right)^2 \times \frac{1600}{60 \times 2}$$

$$= \frac{p_{m_i}}{1000}\left[0.48 \times \frac{\pi}{4} \times \frac{1}{100} \times \frac{1600}{120}\right] \times 10^5$$

$$= 100\, p_{m_i}\left[\frac{0.04\pi \times 4}{10}\right] \text{ where } p_{m_i} \rightarrow \text{bar}$$

$$39.25 = \frac{50.265}{10} p_m$$

$$p_{m_i} = \frac{39.25 \times 10}{50.265}$$

$$= \mathbf{7.81\ bar}$$

Example 25.9. *During a trial on a four cylinder petrol engine running at 50. rps, the brake load was 267 N when all cylinders were working. When each cylinder was cut out in turn and the speed returned to 50 rps, the brake readings were 178 N, 187 N, 182 N and 182 N. Using these readings, determine the brake power of the engine and estimate its indicated power and mechanical efficiency. For the brake power =* $\frac{FN}{455}$ *where F = brake load in Newtons and N in rps. Following results were also obtained during the trial:*

Fuel consumption	*= 0.568 lit. in 138 seconds;*
sp.-gr. of fuel	*= 0.72*
Calculate value of fuel	*= 43000 KJ/kg;*
Air to fuel ratio	*= 14: 1*
Exhaust temperature	*= 760° C* C_{pg} *= 1.015*
Cooling water inlet and outlet temperature	*= 18° C and 56° C*
Cooling water flow rate	*= 0.28 kg/sec.*
Ambient temperature	*= 21° C*

From these results draw up an energy balance in kJ/sec. and as percentage.

Solution

$$\text{BP of the engine} = \frac{267 \times 50}{455}$$

$$= \mathbf{29.34\ kW}$$

$$\text{IP}_1 = 29.34 - \frac{178 \times 50}{455}$$

$$= 29.34 - 19.56$$
$$= \mathbf{9.78\ kW}$$

$$IP_2 = 29.34 - \frac{187 \times 50}{455}$$
$$= 29.34 - 20.55$$
$$= \mathbf{8.79\ kW}$$

$$IP_3 = 29.34 - \frac{182 \times 50}{455}$$
$$= 29.34 - 20$$
$$= \mathbf{9.34\ kW}$$

$$IP_4 = 29.34 - \frac{182 \times 50}{455}$$
$$= 29.34 - 20$$
$$= \mathbf{9.34\ kW}$$

$$\text{Total IP} = \mathbf{37.25\ kW}$$
$$\text{BP of the engine} = \mathbf{29.34\ kW}$$

$$\therefore \quad \text{Mechanical Efficiency} = \frac{29.34}{37.25}$$
$$= \mathbf{0.7876\ or\ 78.76\ \%}$$

Heat balance sheet on kJ/sec. basis

(I) Heat supplied $= \frac{0.568}{138} \times 0.72 \times 43000$

$= \mathbf{127.43\ kJ/sec}$

$= 100\ \%$ kJ/sec.

(II)(a) Heat equivalent of IP $= \mathbf{37.25}$ kW $\frac{\text{kJ}}{\text{sec.}} \rightarrow \mathbf{29.23\ \%}$

(b) Heat to cooling water $= 0.28 \times 4.187 \times (56 - 18)$

$= \mathbf{44.55\ kJ/sec. \rightarrow 34.56\ \%}$

(c) Heat to exhaust gases $= m_g C_{p_g} \Delta T$

$$= \left(\frac{15 \times 0.568}{138} \times \frac{720}{1000}\right) \times 1.015 \times (760 - 21)$$

$= 33.343$ kJ/s

$= \mathbf{26.16\ \%}$

(d) Heat unaccounted for $= 100 - (29.23 + 34.96 + 26.16)$

$= 100 - 90.35$

$= \mathbf{9.65\ \%}$

Example 25.10. *During a test on a four stroke cycle oil engine the following data and results were obtained*

Mean height of indicator diagram	=	*21 mm*
Indicator spring number	=	$27 \dfrac{kN/m^2}{mm}$
Swept volume of cylinder	=	*14 litres*
Speed of Engine	=	*6.6 rev. sec*
Effective brake load	=	*77 kg*
Effective brake radius	=	*0.7 m*
Fuel consumption	=	*0.0002 kg/sec*
Calorific value of fuel	=	*44000 kJ/kg*
Cooling water circulation	=	*0.15 kg/s*
Cooling water inlet temperature	=	*38° C*
Cooling water outlet temperature	=	*71° C*
Specific heat capacity of water	=	*4.18 kJ/kg-°K*
Energy to exhaust gases	=	*33.6 kJ/s*

Determine the indicated and brake outputs and the mechanical efficiency.
Draw up an overall energy balance in kJ/s and as a percentage.

Solution

$$\text{IMEP} = p_{mi} = 21 \times 27$$

$$= 567 \text{ kN/m}^2$$

$$= \frac{567 \times 1000}{10,000}$$

$$= \mathbf{5.67\ bar}$$

$$\text{IP} = p_{m_i} \frac{LAn}{1000}$$

$$= \frac{5.67 \times 10^5}{1000} \times \frac{14}{1000} \times \frac{6.6}{2}$$

$$= 2.62 \times 10$$

$$= \mathbf{26.2\ kW}$$

$$\text{BP} = \frac{2\pi\ NT}{60 \times 1000}$$

$$= \frac{2\pi \times 6.6 \times (W - S)R}{1000}$$

$$= \frac{2\pi \times 6.6 \times (77 \times 9.81) \times 0.7}{1000}$$

$$= \mathbf{21.93\ kW}$$

$$\therefore \quad \text{Mecahnical Efficiency} = \frac{\text{BP}}{\text{IP}}$$

$$= \frac{21.93}{26.2}$$

$$= \mathbf{0.837 \text{ or } 83.7\ \%}$$

Heat balance on Minute Basis:

(1) $\frac{\text{Heat supplied}}{\text{min}} = 0.002 \times 60 \times 44000$

$= 5280$ kJ/min... 100 %

(2) Heat equivalent of IP $= 26.2 \times 60$

$= 1572$ kJ/min

$= \mathbf{29.77\ \%}$

(3) $\frac{\text{Heat to circulating water}}{\text{min}} = m_w c_{p_w} (T_2 - T_1)$

$= 0.15 \times 60 \times 4.18(71 - 38)$

$= 1241.46$ kJ/min

$= \mathbf{23.51\ \%}$

(4) Heat to exhaust gases $= 33.6 \times 60$

$= 2016$ kJ/min

$= \mathbf{38.2\ \%}$

(5) $\% \frac{\text{Heat unaccounted}}{\text{min.}} = 100 - (29.77 + 23.51 + 38.2)$

$= 100 - 91.48$

$= \mathbf{8.52\ \%}$

Example 25.11. *The power of six cylinder, four stroke diesel engine is measured by a water brake dynamometer and air consumption by air box method. The following expression is applicable for the dynamometer* $BP = \frac{WN}{2000}$ *kW, where W = brake load in kg_f, and N = speed in rpm.*

The following readings were noted during the test

Bore	=	*100 mm*
Stroke	=	*120 mm*
Engine speed	=	*2500 rpm*
Fuel: $\frac{C}{H}$ ratio	=	$\frac{83}{17}$
Fuel oil density	=	*0.831 kg/lit.*
Air box orifice diameter	=	*30 mm*
C_d	=	*0.60*
Brake load	=	*55 kg_f*
Head across orifice	=	*14.5 cm of Hg*

Ambient conditions 750 mm Hg and 25° C Hg.

Time for 100 CC fuel consumption 18 seconds.

Calculate:

(i) *Brake mep.*

(ii) *Brake sfc.*

(iii) *Present excess air and volumetric efficiency.*

Solution

(i) Brake m.e.p.: Let p_{m_b} in bar

$$\text{BP} = \frac{WN}{2000}$$

$$= \frac{55 \times 2500}{2000}$$

$$= 68.75 \text{ kW}$$

$$\therefore \quad 68.75 = \text{No. of Cyl.} \times p_{m_b} LAN$$

$$= \frac{6 \times p_{m_b} \times 100 \times \frac{\pi}{4} \times \left(\frac{100}{1000}\right)^2 \times \frac{120}{1000} \times \frac{2500}{60 \times 2}}{1000}$$

$$\frac{68.75}{6} = \frac{p_{m_b} \times 10^5}{1000} \times \frac{120}{1000} \times \frac{\pi}{4}\left(\frac{100}{1000}\right)^2 \times \frac{2500}{120}$$

$$11.46 = p_{m_b} \times \frac{3}{25} \times \frac{\pi}{4} \times \frac{1}{100} \times \frac{125}{3 \times 2} \times \frac{100}{1}$$

$$p_{m_b} = \frac{11.46 \times 25}{3} \times \frac{400}{\pi} \times \frac{3 \times 2}{125 \times 100}$$

$$= \frac{2 \times 11.46 \times 80}{100 \times \pi}$$

$$= 2.92 \text{ bar} \times 2$$

$$= \mathbf{5.834 \text{ bar}}$$

(b) Brake SFC:

$$\frac{\text{Fuel consumption}}{\text{sec}} = \frac{100}{18} \frac{\text{CC}}{\text{sec}}$$

$$= \frac{100}{18} \times \frac{1}{1000} \text{ kg/sec}$$

$$= \frac{1}{180} \times 0.831 \text{ kg/sec}$$

$$= 0.00462 \text{ kg/sec}$$

$$= \mathbf{0.00462 \times 3600 \text{ kg/hour} = 16.632 \text{ kg/hour}}$$

$$\therefore \quad \text{BSFC} = \frac{16.632}{68.75}$$

$$= \mathbf{0.2419 \text{ kg/kWh}}$$

(c) Excess air supplied and volumetric efficiency:

$$\frac{\text{Theoretical air}}{\text{kg fuel}} = \frac{100}{23}\left[\frac{8}{3}C + 8H\right]$$

$$= \frac{100}{23}\left[\frac{8}{3}\times 0.83 + 8\times 0.17\right]$$

$$= \frac{100}{23}[2.2133 + 1.36] = \frac{100}{23}\times 3.5733$$

$$= 15.54 \text{ kg/kg fuel}$$

$$\rho_a h_a = \rho_m h_m$$

$$\frac{p_a}{R_a T_a} = \rho a$$

$$= \frac{75}{76}\times\frac{1.01325\times 180000}{287\times 298}$$

$$= 1.169$$

$$\therefore \quad h_a = 100\times\frac{13.6}{1.169}\times\frac{14.5}{100}$$

$$= 1686.9$$

$$\therefore \quad C_a = \sqrt{2\times 9.81\times 1686.9}$$

$$= 181.91 \text{ m/s}$$

$$Va = C_d A V_a$$

$$= 0.6\times\frac{\pi}{4}\times\left(\frac{30}{1000}\right)^2\times 181.91$$

$$= 0.07722 \text{ cu.m/sec}$$

$$\therefore \quad m_a = 0.07722\times 1.169$$

$$= 0.0027 \text{ kg/sec}$$

$$= 19.54 \text{ kg/kg fuel}$$

$$\therefore \quad \text{Excess air} = 19.54 - 15.54$$

$$= 4.00 \text{ kg/kg fuel}$$

$$= \mathbf{25.74\ \%}$$

$$\frac{\text{Stroke Volume}}{\text{sec.}} = \frac{\pi}{4}d^2\,\frac{L\times N}{60}$$

$$= \frac{\pi}{4}\times\left(\frac{100}{1000}\right)^2\times\frac{120}{1000}\times\frac{2500}{60\times 2}\times 6$$

$$= \frac{\pi}{4}\times\frac{1}{100}\times\frac{12}{100}\times\frac{125}{6}\times 6$$

$$= 0.1179 \text{ cu.m/sec.}$$

$$\therefore \quad \text{Vol. }\eta = \frac{V_a}{V_s}$$

$$= \frac{0.0835}{0.1179}$$

$$= \mathbf{0.7082 \rightarrow 70.82\ \%}$$

Example 25.12. *A CI engine consumes 21.2 kg of diesel per hour and produces 85 kW of brake power. Following observations were recorded on a test ring using exhaust gas calorimeter. The water leaving the engine jacket enters the calorimeter.*

(1)	*Water circulating rate*	*30 kg/min*
(2)	*Temperature of water entering engine*	*29° C*
(3)	*Temperature of leaving engine*	*70°*
(4)	*Temperature of leaving calorimeter*	*95° C*
(5)	*Temperature exhaust gas leaving engine*	*475°C*
(6)	*Temperature exhaust leaving calorimeter*	*181° C*
(7)	*Room temperature*	*27° C*

If air standard efficiency of this engine is 63 % using the data above find:

(a) relative efficiency

(b) draw up a heat balance sheet on minute basis and percent basis.

Take calorific value of fuel = 44100 kJ/kg.

Solution

(a) Relative efficiency:

$$\text{Brake thermal efficiency} = \frac{\text{BP}\times 3600}{\frac{\text{fuel}}{\text{hr}}\times\text{Cal. Value}}$$

$$= \frac{85\times 3600}{21.2\times 44100}$$

$$= \mathbf{0.3273 \text{ or } 32.73\ \%}$$

$$\therefore \quad \text{Relative efficiency} = \frac{\text{Brake thermal efficiency}}{\text{Air standard efficiency}}$$

$$= \frac{0.3273}{0.63}$$

$$= 0.5195$$
$$= \mathbf{51.95\ \%}$$

(b) Heat balance on minute basis:

(i) $$\frac{\text{Heat supplied}}{\text{min}} = \frac{21.2}{60} \times 44100$$
$$= 15582 \text{ kJ/min..... } 100\ \%$$

(ii) $$\text{Heat equivalent of BP} = \text{BP} \times 60$$
$$= 85 \times 60$$
$$= 5100 \text{ kJ..... } 32.73\ \%$$

(iii) $$\frac{\text{Heat in circulating water}}{\text{min}} = m_w C_{p_w} (t_o - t_a)$$
$$= 30 \times 4.187 \times (70 - 29)$$
$$= 5150 \text{ kJ/min...... } 33.05\ \%$$

(iv) Heat in exhaust gases :

$$\text{Heat to water in Calorimeter} = \text{Heat given by exhaust gases}$$
$$30 \times 4.187 \times (95 - 70) = m_g \times C_{p_g} (t_o - t_i)$$
$$= m_g C_{p_g} \times (475 - 181)$$
$$\therefore \quad m_g C_{p_g} = \frac{30 \times 4.187 \times 25}{294}$$
$$= 10.68$$

$$\therefore \text{Heat in exhaust gases to atmosphere} = m_g C_{p_g} [t_e - t_R]$$
$$= 10.68(475 - 27)$$
$$= 448 \times 10.68$$
$$= 4785.14 \text{ kJ/min...... } 30.71\ \%$$

$$\text{(v) Heat unaccounted for (by difference)} = 100 - (32.73 + 33.05 + 30.71)$$
$$= 100 - 96.49$$
$$= \mathbf{3.51\ \%}$$

Example 25.13. *A four cylinder petrol engine has an output of 52 kW at 2000 rpm. A Morse test is carried out and brake torque readings are 177, 170, 168 and 174 Nm. respectively. A fuel consumption of 500 CC was noted in 70 seconds. The fuel has relative density 0.74 and low calorific value 44000 kJ/kg.*

Calculate:

(1) Mechanical efficiency

(2) Brake thermal efficiency.

Solution

(1) BP of cylinder 1 (177 Nm)

$$= \frac{2\pi \times 2000}{60 \times 1000} \times \frac{177}{1000}$$

$$= 37 \text{ kW}$$

$$\therefore \quad IP_1 \text{ of cyl. No. } 1 = BP - BP_1$$

$$= 52 - 37$$

$$= 15$$

IP_2 of cylinder 2 (170 Nm)

$$BP_2 = \frac{2\pi \times 2000}{60 \times 1000} \times 170$$

$$= \frac{\pi}{15} \times 170$$

$$= \mathbf{35.60 \ kW}$$

$$\therefore \quad IP_2 = IP \text{ of cyl. No. } 2$$

$$= 52 - 35.6$$

$$= \mathbf{16.40}$$

IP_3 of cylinder 3 (168 Nm)

$$BP_3 = \frac{\pi}{15} \times 168$$

$$= 35.19 \text{ kW}$$

$$\therefore \quad IP_3 = 52 - 35.19$$

$$= \mathbf{16.81 \ kW}$$

IP_4 of cylinder 4 (174 Nm)

$$BP_4 = \frac{\pi}{5} \times 174$$

$$= 36.44 \text{ kW}$$

$$\therefore \quad IP_4 = 52 - 36.44$$

$$= \mathbf{15.56}$$

$$\text{Total IP} = (15 + 16.40 + 16.81 + 15.56)$$

$$= \mathbf{63.77 \ kW}$$

$$\therefore \quad \text{Mechanical Efficiency} = \frac{52}{63.77}$$

$$= \mathbf{0.8155 \ or \ 81.55 \ \%}$$

(2) Fuel Consumption per hour

$$= \frac{500}{70} \times \frac{3600 \times 740}{10^6}$$

$$= 19.03 \text{ kg/hr.}$$

$$\therefore \quad \text{Brake thermal efficiency} = \frac{BP\left(\frac{kJ}{sec} = kW\right)}{\frac{Fuel}{sec} \times \text{Cal. Value}}$$

$$= \frac{52 \times 3600}{19.03 \times 44000}$$

$$= 0.2236$$

$$= \mathbf{22.36\ \%}$$

Example 25.14. *A six cylinder four stroke diesel engine has the following specifications of rated full-load*

Engine displacement	*7392 cm³*
Speed	*1200 rpm*
Brake MEP	*7.3 bar*
Brake SFC	*216.0 gm/kwh*

Assuming 80 % volumetric efficiency, estimate brake power, engine torque and air-fuel ratio. You may assume suitable ambient conditions.

Solution

Assume ambient conditions as

$p = 1$ bar, $t = 27$°C

$$\therefore \quad \text{Actual volume} = 7392 \times 0.8$$

$$= 5913.6 \text{ m}^3$$

$$\therefore \quad \frac{\text{Actual air supplied}}{\text{Cycle}} = \frac{1 \times 10^5 \times 5913.6}{287 \times 300 \times 10^6}$$

$$= 0.006868$$

$$\therefore \quad \frac{\text{Air}}{\text{hour}} = \frac{1200}{2} \times 60 \times 0.006868$$

$$= 247.26 \text{ kg}$$

$$\text{BP} = \frac{p_{m_b} \times L \times A \times n}{1000 \times 60}$$

$$= \frac{7.3 \times 10^5 \times 7392}{60 \times 1000 \times 10^6} \times \frac{120}{2}$$

$$= \frac{3237.7}{60}$$

$$= \mathbf{54}$$

$$\frac{\text{Fuel supplied}}{\text{hour}} = 0.216 \times \frac{3237.7}{60}$$

$$= \frac{699.34}{60} \text{ kg/hour}$$

$$= 11.66$$

$$\therefore \quad \text{Air to fuel ratio} = \frac{\text{Air}}{\text{Fuel}}$$

$$= \frac{247.26\times 60}{699.34}$$

$$= 21.21$$

$\therefore$ **Air to fuel ratio** $= \mathbf{21.21 \frac{kg\ of\ air}{kg\ of\ fuel}}$

$$\text{We know BP} = \frac{2\pi NT}{60\times 1000}$$

$$54 = \frac{2\pi\times 1200T}{60\times 1000}$$

$$\therefore \quad T = \frac{54\times 60\times 1000}{2\pi\times 1200}$$

$$= \frac{54\times 25}{\pi}$$

$$= \mathbf{429.55\ N\text{-}m}$$

Example 25.15. *The following readings were noted during the trial on a single cylinder two stroke diesel engine. The engine is motored by an electric motor and the watt meter reading was 1.5 kW. The brake load was applied after starting and then the readings are noted*

Net load on the brake	*23.2 kg*
Diameter of brake-wheel	*100 cm*
Engine speed	*500 rpm*
Fuel consumption	*2.04 kg/hr*
Jacket cooling water	*1.365 kg/min*
Rise in temperature of Jacket cooling water	*40° C*
Ambient temperature	*25° C*
Exhaust gas temperature	*345° C*
A : F ratio	*32.1*
Sp. heat of exhaust gases	*1 kJ/kg °K*
C. V. of the fuel used	*42000 kJ/kg*

Find:

(*1*) *Mechanical efficiency*

(*2*) *Indicated thermal efficiency.*

(*3*) *Draw up heat balance sheet in kW.*

Solution

(1) Mechanical Efficiency

$$\text{BP} = \frac{2\pi NT}{60\times 1000}$$

$$= \frac{2\pi\times 500}{60}\times\frac{(23.2\times 9.81)}{10000}\times\frac{100}{2}$$

$$= \frac{\pi \times 23.2 \times 9.81}{60 \times 2}$$

$$= 5.96 \text{ KW}$$

$$\therefore \quad IP = BP + IP$$

$$= 5.96 + 1.5$$

$$= 7.46 \text{ KW}$$

$$\therefore \quad \text{Mechanical Efficiency} = \frac{BP}{IP}$$

$$= \frac{5.96}{7.46}$$

$= 0.7989$ or **79.89 %**

(2) Indicated Thermal Efficiency:

$$\text{Indicated Thermal Efficiency} = \frac{IP \times 3600}{\frac{\text{Fuel}}{\text{hr}} \times \text{Cal. Value}}$$

$$= \frac{7.46 \times 3600}{2.04 \times 42000}$$

$$= 0.3134$$

= **31.35 %**

(3) Heat balance sheet in kg/kW

(i) $\frac{\text{Heat supplied}}{\text{sec}} = \frac{2.04 \times 42000}{3600}$

= **23.8 kW**......100%

(ii) Heat equivalent of IP = IP

= **7.46 kW**...... 31.34 %

(iii) Heat to cooling water $= m_w C_{p_w} (t_o - t_i)$

$$= \frac{1.365}{60} \times 4.187 \times 40$$

= **3.81 kW**...... 16 %

(iv) Heat in exhaust gases $= m_g C_{p_g} (t_e - t_R)$

$$= \frac{3212.04}{3600} \times 1 \times (345 - 25 = 320)$$

= 3.026 kW...... 12.71 %

(v) Heat unaccounted for (by difference) $= 100 - (31.24 + 12.71 + 44.05 + 16)$

= 39.95 %

Example 25.16. *A six cylinder four stroke petrol engine running at 3960 rpm consumes 0.325 kg of petrol (calorific value 44000 kJ/kg) per minute. The cylinder size is 70 mm by 100 mm and dearance volume per cylinder is 67 cm³. Torque developed is 140 N-m. If the ambient temperature is 30°C and exhaust gas temperature 350° C with air to fuel ratio 16 to 1, calculate:*

1. *Brake power.*
2. *BMEP*
3. *Brake thermal efficiency*
4. *Percentage heat rejected throughout exhaust.*

Take C_{pg} *= 1.11 KJ/kg/°K.*

Solution

(1) Brake Power:

$$\text{BP} = \frac{2\pi NT}{60\times1000}$$

$$= \frac{2\pi\times3960\times140}{60\times1000} = \mathbf{58.03\ kW}$$

(2) Brake Mean Effective Pressure:

$$\text{BP} = \frac{(\text{BMEP})_{\text{bar}}\times10^5}{1000}\times\frac{\pi}{4}d^2L\times n\times\text{No. of cylinder}$$

$$\frac{58.08}{6} = \text{BMEP}\times100\times\frac{\pi}{4}\times(0.07)^2\times\frac{100}{10\times100}\times\frac{3960}{2\times60}$$

$$= \text{BMEP}\times100\times\frac{\pi}{4}\times\frac{0.0049}{10}\times33$$

$$= \frac{\pi}{4}\times33\times0.049\times\text{BMEP}$$

$$= p_{m_b}\times1.27$$

$$\therefore \quad \text{BMEP} = \frac{58.08}{6\times1.27}$$

$$= \mathbf{7.622\ bar}$$

(3) Brake Thermal Efficiency:

$$\text{Brake thermal }\eta = \frac{\text{BP}\times3600}{\frac{\text{Fuel}}{\text{Hour}}\times\text{Cal. Value}}$$

$$= \frac{58.08\times3600}{(0.325\times60)\times44000} = 0.2437$$

$$= \mathbf{24.37\ \%}$$

(4) Percentage of heat supplied that goes to exhaust :

$$\frac{\text{Heat supplied}}{\text{min.}} = 0.325 \times 44000$$

$$= 14300 \text{ kJ/min.}$$

$$\frac{\text{Mass of air supplied}}{\text{min.}} = 16 \times 0.325$$

$$= 5.2 \text{ kg}$$

$$\therefore \quad \frac{\text{Mass of exhaust gases}}{\text{min.}} = \text{mass of air + mass of fuel}$$

$$= 5.2 + 0.325$$

$$= 5.525 \text{ kg/min.}$$

$$\therefore \quad \frac{\text{Heat carried away by exhaust gases}}{\text{min.}} = m_g C_{p_g} \left[t_{gi} - t_{gR} \right]$$

$$= 5.525 \times 1.11 \times [350 - 30 = 320°\text{C}]$$

$$= 1962.48 \text{ kJ/min.}$$

$$= \frac{1962.48}{14300}$$

$$= 0.1372$$

$$= \mathbf{13.72\ \%}$$

(5) Relative efficiency :

$$V_s = \frac{\pi}{4} \times 49 \times 10 \text{ cm}^3$$

$$= 385 \text{ cm}^3$$

$$v = 67 \text{ cm}^3$$

$$\therefore \quad \text{Compression ratio } r_k = \frac{385 + 67}{67}$$

$$= 6.75$$

$$\therefore \quad r_k = 2.146$$

$$\therefore \quad \text{ASE} = 1 - \frac{1}{r_k^{\gamma} - 1}$$

$$= 1 - \frac{1}{21.46}$$

$$\text{ASE} = \frac{1.146}{2.146}$$

$$= \mathbf{0.534}$$

$$\therefore \quad \text{Relative efficiency} = \frac{\text{BP}}{\text{ASE}}$$

$$= \frac{0.2437}{0.534}$$

$$= 0.4564 \text{ or } \mathbf{45.64\ \%}$$

Example 25.17. *A trial was conducted on a two stroke cycle 4 cylinder gasoline engine having 10 cm bore and 12 cm stroke. The net dynamometer load was 310 N at a radius of 40 cm when the engine speed was 2050 rpm. At the same speed and throttle opening, the engine required 6 kW to motor it with the ignition switch off.*

During the 3 minute run at this speed and power the engine consumed 0.52 kg of fuel with Cal. value of 42100 kJ.kg and 24 kg of cooling water was circulated through the jacket. The temperature rise of cooling water was by 60° C. Determine

(1) Mechanical efficiency
(2) IMEP
(3) Brake thermal efficiency
(4) Relative efficiency if compression ratio used in the engines is 7.5 and
(5) The percentage heat lost to the cooling water.

Solution

(1) Mechanical Efficiency:

$$\text{BP} = \frac{2\pi \text{N}(W-S)R}{60\times1000}$$

$$= \frac{2\pi\times2050}{60\times1000}\times\left(310\times\frac{40}{100}\right)$$

$$= \frac{41\pi}{600}\times\frac{124}{1}$$

$$= \mathbf{26.62\ kW}$$

$$\text{FP} = \mathbf{6\ kW}$$

$$\therefore \quad \text{Mechanical Efficiency} = \frac{26.62}{26.62+6}$$

$$= \frac{26.62}{32.62}$$

$$= 0.8161$$

$$= \mathbf{81.61\ \%}$$

(2) Indicated M. E. P.:

$$\text{IP} = 32.62$$

$$= 4\ p_{m_i}\frac{LAn}{1000}$$

$$8.155 = \frac{p_{m_i} \times 10^5}{1000} \times \left(\frac{\pi}{4} \times 0.01 \times 0.12\right) \times \frac{2050}{2 \times 60}$$

$$8.155 = 100\ p_{m_i} \times \frac{\pi}{4} \times 0.0012 \times \frac{2050}{2 \times 60}$$

$$= p_{m_i} \times \frac{\pi}{4} \times \frac{0.12 \times 205}{12}$$

$$= p_{m_i} 1.61 \times 2$$

$$\therefore \quad p_{m_i} = \text{Indicated MEP} = \frac{8.155}{1.61 \times 2}$$

$$= \frac{5.065}{2} \text{ bar}$$

$$= \mathbf{2.5325}$$

(3) Brake thermal efficiency:

$$\text{Brake thermal efficiency} = \frac{\text{BP} \times 3600}{\frac{\text{fuel}}{\text{hr.}} \times \text{Cal. Value}}$$

$$= \frac{26.62 \times 3600}{0.52 \times 20 \times 42100}$$

$$= 0.2189$$

$$= \mathbf{21.89\ \%}$$

(4) Relative efficiency:

$$\text{Compression ratio} = r_k = 7.5$$

$$\therefore \quad \text{Air Standard Efficiency} = 1 - \frac{1}{7.4^{0.4}}$$

$$= 1 - \frac{1}{2.24}$$

$$= 0.5536$$

$$= \mathbf{55.36\ \%}$$

(5) Heat to cooling water:

$$\frac{\text{Heat supplied}}{3 \text{ min.}} = 0.52 \times 421000$$

$$= 21892 \text{ kJ}$$

$$\text{Heat lost to cooling water per 3 min.} = m_w C_{p_w} \Delta T$$

$$= 24 \times 4.187 \times 60$$

$$= 6029.28 \text{ kJ}$$

$$\% \text{ of heat supplied to cooling water} = \frac{6029.28}{21892}$$

$$= \mathbf{0.2752 \text{ or } 27.52 \%}$$

Example 25.18. *Calculate the bore and stroke of a four stroke single cylinder diesel engine designed to the following particulars:*

Brake power	*25 kW*
Speed	*300 rpm*
Fuel composition	*C = 85 % H_Z =15 %*
Lower Cal. Value	*41000 kJ/kg*
Air supplied	*25 % excess*
Volumetric efficiency	*80 % at 1.01325*
Mechanical efficiency	*0.9*
Indicated thermal efficiency	*0.35*
R for air	*0.287 kJ/kg - °K*
Stroke to bore ratio	*1.2*

Solution

$$\frac{\text{Theoretical air supplied}}{\text{kg fuel}} = \frac{100}{23}\left[\frac{8}{3}\times 0.85 + 8\times 0.15\right]$$

$$= \frac{100}{23}[2.267 + 1.20]$$

$$= \frac{100}{23}\times 3.467$$

$$= \mathbf{15.074 \text{ kg}}$$

$$\therefore \quad \frac{\text{Actual air supplied}}{\text{kg fuel}} = 1.25\times 15.074$$

$$= \mathbf{18.85 \text{ kg}}$$

$$\text{Mecahnical Efficiency} = \frac{BP}{IP}$$

$$= 0.9$$

$$= \frac{25}{IP}$$

$$IP = \frac{25}{0.9}$$

$$= \mathbf{27.78 \text{ KW}}$$

Indicated thermal efficiency = 0.35

$$= \frac{IP\times 3600}{\frac{\text{Fuel}}{\text{Hour}}\times \text{Cal. Value}}$$

$$= \frac{27.78 \times 3600}{\frac{\text{Fuel}}{\text{Hour}} \times 41000}$$

$$\therefore \quad \frac{\text{Fuel}}{\text{Hour}} = \frac{27.98 \times 3600}{0.35 \times 41000}$$

$$= 6.97 \text{ kg/hour}$$

$$\therefore \quad \frac{\text{Air supplied}}{\text{min.}} = \frac{6.97}{60} \times 18.84$$

$$= 2.19 \text{ kg/min.}$$

$$\frac{\text{Volume of this air}}{\text{min.}} = \frac{mRT}{p}$$

$$= \frac{2.19 \times 287 \times 283}{1.01325 \times 10^5}$$

$$= 1.7555 \text{ cu.m/min.}$$

$$\therefore \quad \frac{\text{Stroke volume}}{\text{min.}} = \frac{1.7555}{0.8}$$

$$= 2.194 \text{ m}^3\text{/min}$$

$$\therefore \quad 2.194 = \left(\frac{\pi}{4} d^2 \times L\right) \times \frac{N}{2}$$

$$\frac{L}{d} = \frac{1}{1.2}$$

$$= \left[\frac{\pi}{4} \times (1.2L)^2 \times L\right] \times \frac{300}{2}$$

$$d = 1.2\,L$$

$$\therefore \quad 2.194 = \frac{\pi}{4} \times \left(1.44\, L^2 \times L^2\right) \times 150$$

$$= L^3 \left[\frac{\pi}{4} \times 1.44 \times 150\right]$$

$$= L^3 \times 169.65$$

$$\therefore \quad L^3 = \frac{2.194}{169.65}$$

$$= 0.01293$$

$$\therefore \quad L = 0.235 \text{ m}$$

$$= \mathbf{23.5 \text{ cm}}$$

$$d_j = \text{Bore} = 1.2\times 23.5$$
$$= \mathbf{28.17\ \ cm}$$

Example 25.19. *A gasoline engine consisting of eight cylinders and working on four stroke cycle has bore of 10 cm and stroke of 9 cm and has compression ratio of 7. It is tested on a dynamometer which has 60 cm. arm. During a 10 minute test duration, the dynamometer scale beam reading was 420 N and the engine consumed 4.8 kg of fuel having a calorific value of 44100 kJ/kg. The air was supplied at 1 bar and 27° C. If the engine runs at 4500 rpm.*

Calculate:

(1) BP
(2) Brake specific air consumption
(3) BMEP
(4) Brake thermal efficiency
(5) BSFC
(6) Air fuel ratio.

Solution

(1) Brake Power BP:

$$\text{BP} = \frac{2\pi NT}{60\times 1000}$$
$$= \frac{2\pi\times 4500}{60\times 1000}\times 420\times\frac{60}{100}$$
$$= \frac{\pi\times 9\times 42}{10}$$
$$= \mathbf{118.75\ kW}$$

(2) Brake specific air consumption:

$$\frac{\text{Stroke volume}}{\text{hour}} = 8\times\frac{\pi d^2}{4}\times L\times\frac{N}{2}\times 60$$
$$= 8\times\frac{\pi}{4}\times(0.1)^2\times 0.09\times\left(\frac{4500}{2}\times 60\right)$$
$$= 2\pi\times 0.01\times 0.09\times(4500\times 30)$$
$$= 90\pi\times 01.109\times 30$$
$$= 9\pi\times 9\times 3$$
$$= 763.41\ \text{cu.m./hour}$$

$$\therefore \quad \frac{\text{Mass of air}}{\text{hour}} = \frac{1\times 10^5\times 763.41}{287\times 300}$$
$$= 886.66\ \text{kg/hour}$$

$$\therefore \text{Brake specific air consumption} = \frac{886.66}{118.75}$$
$$= \frac{\text{Air/hour}}{\text{BP(kW)}}$$
$$= \mathbf{7.47\ \ kg/kWh}$$

(3) BMEP:

$$\text{BP} = p_{m_b} \times \left(\frac{\pi}{4} d^2 \times L \right) \times \frac{D}{D1} \times 8 \times 10^5$$

$$118.75 = \frac{8 \times p_{m_b} \times 10^5}{1000} \times \left(\frac{\pi}{4} \times 0.01 \times 0.09 \right) \times \frac{4500}{2 \times 60}$$

$$\frac{118.75}{8} = 100 p_{m_b} \times \pi \times 0.0009 \times \frac{75}{2}$$

$$18.75 = 100 p_{m_b} \times \pi \times 0.09 \times 75$$

$$= 21.21\, p_{m_b}$$

$$p_{m_b} = \frac{118.75}{21.21}$$

$$= \mathbf{5.6\ bar}$$

(4) Brake thermal efficiency:

$$\text{Brake thermal efficiency} = \frac{\text{BP} \times 3600}{\frac{\text{fuel}}{\text{hr.}} \times \text{Cal. Value}}$$

$$= \frac{18.75 \times 3600}{4.8 \times 6 \times 44100}$$

$$= 0.3366$$

$$= \mathbf{33.66\ \%}$$

(5) BSFC:

$$\text{Brake specific fuel consumption} = \frac{\text{BP(kW)}}{\left(\frac{\text{fuel}}{\text{hr.}} \right) \frac{\text{kg}}{\text{hr.}}}$$

$$= \frac{4.8 \times 6}{118.75}$$

$$= \mathbf{0.2425\ kg/kWH}$$

(6) Air-fuel ratio:

$$\text{Air to fuel ratio} = \frac{\frac{\text{Mass of air}}{\text{hour}}}{\frac{\text{Mass of fuel}}{\text{hour}}}$$

$$= \frac{886.66}{4.8 \times 6}$$

$$= \mathbf{30.79}$$

Example 25.20. *A test conducted on a 4-stroke gas engine having diameter 18 cm and stroke 24 cm give following observations:*

Duration of trial	*30 min.*
Total revolutions	*9000*
No. of explosions	*4450*
Indicated mrp	*5 bar*
Net brake load	*440 N*
Effective diameter of brake wheel	*1 m*
Gas consumption	*2.4 m³ at NTP.*
Calorific Value of gas at NTP	*19 MJ/m³*
Air consumption	*36 m³*
Air intake pressure	*720 mm Hg*
Ambient temperature	*17° C*
Density of air at NTP	*1.29 kg/m³*
Exhaust gas temperature	*350° C*
Cooling water circulated	*80 kg*
Rise in temperature rise of cooling water	*30° C*
Barometer	*760 mm Hg*

Find:

(i) BSFC

(ii) Mechanical efficiency

(iii) Draw up a energy balance sheet on minute basis and percentage basis.

Solution

(i) Brake specific fuel consumption:

$$\text{BP} = \frac{2\pi NT}{60\times1000}$$

$$= \frac{2\pi\times9000\times400\times\frac{1}{2}}{30\times60\times1000}$$

$$= \frac{2\pi\times9}{9}$$

$$= 2\pi$$

$$= \mathbf{6.283}$$

$$\therefore \quad \text{BSFC} = \frac{2.4}{30}\times\frac{60}{6.283}$$

$$= \frac{4.8}{6.283}$$

$$= \mathbf{0.764 \ cu.m/kW\ hr}$$

(ii) Mechanical efficiency:

$$\text{IP} = p_{m_i}\frac{LAn}{1000}$$

$$= \frac{5\times10^5}{1000}\times\frac{24}{100}\times\frac{\pi}{4}\times(0.18)^2\times\frac{4450}{30\times60}$$

$$= 120\times\frac{\pi}{4}\times 0.0324\times\frac{445}{3}$$

$$= \pi\times 0.0324\times\frac{445}{60}$$

$$= \frac{452.955}{60}$$

$$= \mathbf{7.55}$$

$$\therefore \quad \text{Mechanical efficiency} = \frac{\text{BP}}{\text{IP}}$$

$$= \frac{6.283}{7.55}$$

$$= \mathbf{0.8322} \text{ or } \mathbf{83.22\ \%}$$

(iii) Heat balance sheet:

(a)
$$\frac{\text{Heat supplied}}{\text{min.}} = \frac{2.4}{30}\times\frac{19\times 10^6}{1000}$$

$$= \frac{1900\times 2.4}{3}$$

$$= \mathbf{1520\ kJ/min......\ 100\ \%}$$

(b) Heat equivalent of IP (BP + FP) $= 7.55\times 60$

$$\doteq \mathbf{453\ kJ......\ 29.8\ \%}$$

(c)
$$\text{Heat to cooling water} = m_g C_{p_g}\left(t_0 - t_i\right)$$

$$= \frac{80}{30}\times 4.187\times 30$$

$$= \mathbf{334.96\ KJ/min......\ 22.04\ \%}$$

(i) Wt. of gas neglected

(ii) C_{p_g} is assumed as 1.006

(d)
$$\text{Heat in exhaust gases} = m_g C_{p_g}\left(t_e - t_R\right)$$

$$= 1.381\times 1.006\times(350-17)$$

$$= 462.5\ \text{kJ/min......}\ 30.43\ \%$$

$$\frac{p_o v_o}{T_o} = \frac{p_a v_a}{T_a}$$

$$\frac{760\times 1}{1.29\times 273} = \frac{720\times V_a}{290}$$

$$V_a = \frac{760}{720} \times \frac{290}{273} \times \frac{1}{1.29}$$

$$= 0.8692 \text{ cu. m.}$$

$$m_a = \frac{36}{0.8692}$$

$$= 41.42 \text{ kg/30 min.}$$

$$\frac{m_a}{\text{min.}} = \frac{41.42}{30}$$

$$= \mathbf{1.381 \text{ kg/min.}}$$

(e) Heat unaccounted for (by difference)

$$= 1520 - [453 + 334.96 + 462.5]$$

$$= 1520 - 1250.46$$

$$= \mathbf{269.54 \text{ kJ/min......} \ 17.73 \ \%}$$

PROBLEMS FOR PRACTICE

25.1. The following observations were made during a test on a two-stroke cycle oil engine. Bore = 20 cm, stroke = 25 cm, speed = 6 rps, effective brake drum diameter = 1.2 m. Net brake load = 440 N, IMEP = 280 kPa fuel oil consumption = 3.6 kg/hour, calorific value of oil = 42500 kg/kg, Mass of jacket cooling water = 468 kg/hr. Rise in temperature of jacket cooling water = 28° C Air used = 34 kg/kg of fuel oil, Temp. Net brake load = 440 N, IMEP = 280 kPa fuel oil consumption = 3.6 kg/hour, calorific value of oil = 42500 kJ/kg, Mass of jacket cooling water = 468 kg/hr. Rise in temperature of jacket cooling water = 28° C. Air used = 34 kg/kg of fuel oil, Temperature of air in test house = 30°C, Temperature of exhaust gases = 400°C. Mean specific heat of exhaust gases = 1 kJ/kg.K.

Calculate: (i) BP (ii) IP (iii) Mechanical efficiency (iv) BMEP (v) BSFC.

Draw up a heat balance sheet in kJ/min and as percentage of the heat supplied to the engine.

[*Ans.* (i) 9.948 kW (ii) 13.188 kW (iii) 75.41% (iv) 211.15 kPa (v) 0.3616 kg/kWH. Heat unaccounted = 10.26 %%.]

25.2. In trial of a single cylinder oil engine working on dual cycle, the following observations were made Compression ratio = 15, Oil consumption = 10.2 kg/hr, Cal. value = 43890 kJ/kg, Air consumption = 3.8 kg/min speed = 1900 rpm, Torque on the brake drum = 186 N.m, Quantity of cooling water used = 15.5 kg/min Temperature rise = 36°C, Exhaust gas temperature = 410°C Room temperature = 20°C, C_p for exhaust gases = 1.17 kg/kg. K

Calculate : (i) BP (ii) BSFC (iii) Brake thermal efficiency. Draw up a heat balance on minute basis.

[*Ans.* (i) 37 kW (ii) 0.2756 kg/kWH (iii) 29.75 % unaccounted heat 14.7 %]

25.3. In a constant speed CI engine operating on four stroke cycle and fitted with band brake, the following observations were taken:

Brake wheel diameter = 60 cm, Band thickness = 5 mm, speed = 450 rpm, Load on band = 210 N, spring balance reading = 30,N, Area of indicator diagram = 4.15 cm², Length of indicator diagram = 6.25 cm, Spring No. 11 *i.e.* 11 bar/cm Bore 10 cm, stroke = 15 cm, BSFC = 0.3 kg/kWh, Cal. value of fuel = 41800 kJ/kg.

Determine: BP, TP, Mechanical efficiency, indicated and Brake thermal efficiency.

[*Ans.* 2.566, 3.225, 79.5%, 36%, 28.6 %.]

25.4. The following particulars were obtained on a single cylinder 4-stroke oil engine Cylinder diameter = 250 mm, stroke length = 400 mm, MEP = 6.5 bar, Engine speed = 250 rpm, Net load on the brake = 1080 N, Effective diameter of the brake wheel = 1.5 m, Fuel used = 10 kg/hr, C.V. of fuel = 44300 kg/kg.

Calculate : (i) IP (ii) BP (iii) Mech. efficiency (iv) Indicated thermal efficiency.

[*Ans.* (i) 26.6 kW (ii) 21.2 kW (iii) 79.7% (iv) 21.6 %]

25.5. A single cylinder 4-stroke diesel engine gave the following results while running on full load Area of indicator card = 300 mm, length of diagram = 40 mm spring constant = 1 bar/min, speed of the engine = 400 rpm, Load on the brake = 370 N, spring balance reading = 50 N, Diameter of brake drum = 1.2 m, Fuel consumption = 2.8 kg/hr, C. V. of fuel = 41800 kJ/kg, Diameter of cylinder = 160 mm, stroke of the piston = 200 mm.

Calculate: (i) IMEP (ii) BP (iii) BMEP (iv) BSFC (v) Brake thermal efficiency

[*Ans.* (i) 750 kPa (ii) 8.04 kw (iii) 599.81 kPa (iv) 0.348 kg/kWh (v) 24.73 %]

25.6. Following data relates to 4 cylinder 4-stroke petrol engine

A.F = 16 :1 (by weight), C.V. of fuel = 45200 kJ/kg, Mechanical efficiency 82 %, Air standard efficiency = 52%, Relative efficiency = 70%, volumetric efficiency = 78%, stroke to bore ratio = 1.25, Suction conditions = 1 bar, 25°C, RPM = 2400, Power at brake = 72 kW.

Calculate : (i) Compression ratio (ii) Indicated thermal efficiency (iii) Brake specific fuel consumption (iv) Bore and stroke.

[*Ans.* (i) 6.2 (ii) 36.4 (iii) 0.2665 kg/kwh (iv) 108 mm, 135 mm]

25.7. A 6-cylinder petrol engine has a volume compression ratio of 5 : 1. The clearance volume of each cylinder is 0.00115 m^3. The engine consumes 10.5 kg/hour fuel whose calorific value is 418000 kJ/kg. The engine runs at 2500 rpm the efficiency ratio is 0.65. Calculate the average indicated mean effective pressure eveloped.

[*Ans.* 6.52 bar]

25.8. In a test of a 4-cylinder, 4 stroke engine 75 mm bore and 100 mm stroke, the following results were obtained at full throttle at a particular constant speed with fixed setting of fuel supply of 6 kg/hr.

BP with all cylinders working = 15.6 kw BP with cylinder No. 1 cutout = 11.1 kw BP with cylinder No. 2 cutout = 11.03 kw BP with cylinder No. 3 cutout = 10.88 kw BP with cylinder no. 4 cutout = 10.66 kW

If the calorific value of fuel is 83600 kJ/kg and clearance volume is 0.0001 cu.m calculate

(i) Mechanical efficiency

(ii) Indicated thermal efficiency

(iii) Air standard efficiency

[*Ans.* (i) 83.3 % (ii) 13.44 % (iii) 49 %]

25.9. A four stroke petrol engine 7 cm bore and 9 cm stroke, is tested at full throttle at constant speed. The fuel supply is fixed at 0.065 kg/min and the plugs of the four cylinders are successively short circuited without change of speed, brake torque being correspondingly adjusted. The power measurements are the following

With all cylinders firing = 11.91 kW

With cylinder No. 1 short circuited = 8.46 kW

With cylinder No. 2 short circuited = 8.56 kW

With cylinder No. 3 short circuited = 8.6 kW

With cylinder No. 4 short circuited = 8.5 kW

Determine indicated power of the engine under these conditions. Also determine the indicated thermal efficiency if the cal. value of fuel is 43500 kJ/kg. Compare this efficiency with air standard efficiency, clearance volume of one cylinder being 69.5 cm^3.

[*Ans.* 13.52 kW, 28.7 %, ASE = 51.2 %, Rel. eff. = 56 %]

25.10. The following data were recorded during a trial on 4-stroke diesel engine

No. of cylinders = 4 diameter of cylinder = 10 cm, stroke = 15 cm, IMEP = 0.67 MPa, speed = 2000 rpm, No. of explosions = 980 min, Brake torque = 181.5 N-m, fuel consumption = 11.89 kg/hr. Cal. value = 41800 kJ/kg, Relative efficiency on BP basis = 50%, Quantity of cooling water = 1020 kg/hr. Rise in temperature of cooling water =

35°C, Specific heat of cooling water = 4.18 kJ/kg.K

Find : (i) Mechanical efficiency (ii) Brake thermal efficiency (iii) Air standard efficiency (iv) BSFC (v) % heat loss to cooling water.

[*Ans.* (i) 73.72 % (2) 27.52 % (3) 55.04 % (4) 0.3129 kg/kwh (5) 30.25%]

25.11. In a test on a single cylinder oil engine with 30 cm bore and 45 cm stroke and working on 4-stroke cycle, the following observations were made

Duration of trial = 1 hour, Total fuel consumption = 7.6 kg, Cal. value of fuel = 45000 kJ/kg, Total revolution made = 12000, Room temperature = 20°C, IMEP = 6 bar, Net brake load = 150 kg, Brake drum diameter = 180 cm, Rope diameter = 3 cm, Mass of jacket cooling water = 550 kg. Inlet temperature of jacket cooling water = 15°C, Outlet temperature of jacket cooling water = 60°C, Temperature of exhaust gases = 300°C, Total air consumption = 365 kg, sp. heat of exhaust gases = 1 kJ/kg.K.

Calculate : IP, BP, Mech Eff. indicated thermal efficiency, Draw up a heat balance sheet on minute.

[*Ans.* 31.8 kW, 28.1 kW, 88.3%, 35.5%, Heat in accounted = 9.62%]

25.12. The following observations were taken during a test on a single cylinder, lower speed, four stroke cycle oil engine having a bore of 30 cm and stroke of 45 cm Ambient temperature = 20°C, Duration of trial = 1 hour, speed = 300 rpm. Total fuel consumption = 11.4 kg/hr, Cal. val. of fuel = 42000 kJ/kg, BMEP = 6 bar. Brake drum diameter = 1.8 m, Brake rope diameter = 2 cm. Quantity of jacket cooling water = 660 kg/hr, Temperature rise of cooling water = 55°C, Quantity of air measured = 250 kg/hr. Sp. heat of gases = 1 kJ/kgK. Exhaust gas temperature = 420°C.

Determine : (i) IP (ii) BP (iii) Mech. efficiency (iv) Indicated thermal efficiency. Draw up a heat balance sheet on hour basis.

[*Ans.* (i) 47.7 kW (ii) 42.9 kW (iii) 89.9 % (iv) 35.86% Unaccounted = 14.16%]

IMPORTANT POINTS TO BE REMEMBERED

1. $$\text{MEP} = \frac{\text{Area of indicator diagram}}{\text{Length of diagram}} \times \text{Spring scale}$$

$$p_m = \frac{a}{l}\left(\frac{cm^2}{CM}\right) \times \text{ Spring strength } \left(\frac{\text{bar}}{CM}\right)[\cdot\textbf{bar}]$$

2. *Indicated Power (IP)*

$$IP = p_{m_i} \times A \times L \times n \times 10^5 \; \frac{N-M}{\text{Sec}}$$

$$= \frac{p_{m_i} \times A \times L \times n \times 10^5}{1000} \text{ kW}$$

$$n = \text{No. of cycles/sec.}$$

$$= \frac{N}{60 \times 2} \text{ for 4 stroke cycle}$$

$$= \frac{N}{60} \text{ for 2 stroke cycle}$$

A = Piston area m^2 = cyclinder

L = Piston stroke m

3. *Brake Power (BP)*

$$\text{BP} = \frac{\text{IIDN(W} - \text{S)}}{60 \times 1000} \text{kW}$$

D = Dia of brake wheel

W = Dead weight

S = Spring balance reading

4. *Mechanical efficiency* (η_{mech})

$$\eta_{\text{Mech}} = \frac{\text{BP}}{\text{IP}} = \frac{\text{BP}}{\text{BP} + \text{FP}}$$ FP = Friction power

5. *Thermal efficiency* (η_{th})

(a) Thermal efficiency based on break power

B.T.E. = Brake thermal efficiency

$$= \frac{\text{BP}}{\frac{m_f}{\text{sec.}} \times \text{Cal. Value}}$$

$$= \frac{\text{BP} \times 3600}{\frac{m_f}{\text{hour}} \times \text{Cal. Value}}$$

$$\frac{\text{Mass of fuel/hour}}{\text{BP}} = \text{Brake specific fuel consumption}$$

$$= \text{BSFC kg/hour/BP}$$

(b) Indicated thermal efficiency

$$\text{I. th. Eff} = \frac{\text{IP}}{\frac{m_f}{\text{sec}} \times \text{cal. value}}$$

$$= \frac{\text{IP} \times 3600}{\frac{m_f}{\text{hour}} \times \text{cal. value}}$$

$$\frac{m_f\ /\ \text{hour}}{\text{IP}} = \text{Indicated specific fuel consumption}$$

6. *Break MEP* (p_{m_b})

$$p_{m_b} = \frac{\text{Break work}}{ALn} \quad n = \frac{N}{120} \quad \text{for 4 stroke cycle}$$

$$= \frac{N}{60} \quad \text{for 2 stroke cycle}$$

7. *Relative efficiency*

Relative efficiency based on BP = Thermal efficiency

$$\text{Relative efficiency based on IP} = \frac{\text{Ind. Thermal efficiency}}{\text{ASE}}$$

8. Items included in heat balance sheet are (on minute basis)

(a) Heat supplied/min = $\frac{m_f}{\text{min}} \times$ cal. value

(b) Heat equivalent of BP = BP (kW) × 60

(c) Heat equivalent of FP = FP (kW) × 60

(d) Heat taken by cooling water/min = $m_w\ C_p\ (t_o - t_i)$

(e) Heat carried by exhaust gases = $mg\ C_{pg}\ (t_e - t_R)$

(f) Heat unaccounted for (by difference) = $a - (b + c + d + e)$

CHAPTER 26

Combustion in SI and CI Engines

26.1 ENGINES

26.1.1 Introduction

Combustion. Combustion is relatively fast exothermic chemical reaction of carbon and hydrogen in the fuel with the oxygen present in the air producing heat energy. When fuel and oxygen react chemically and the heat evolved is sufficient to sustain the reaction—combustion is said to take place. Combustion will sustain if heat released is sufficient and reaction rates are high. However combustion is a complex phenomenon requiring deep understanding.

Combustion in IC engines. The energy of combustion is used for operating the internal combustion engines.

Basic conditions necessary for combustion in IC engine are:

(a) The presence of a combustible mixture
(b) Means of initiating combustion
(c) Stabilisation and propagation of flame in the combustion chamber.

The combustible elements in fuels are generally carbon and hydrogen with small amount of sulphur and other additives. Both the SI and CI engines are designed to meet the above given requirements. In IC engine, combustion generally takes place when the fuel is in gaseous state. It is necessary to have equipment in IC engine to achieve the gaseous state e.g. in SI engine we have a carburettor or multi-point fuel injection (MPFI) system.

Gasoline that may contain more than 40 constituents can be approximated to Octane fuel (C_8H_{18}) and its combustion in the presence oxygen in the air is given as

$$\underset{\text{(Octane)}}{C_8H_{18}} + 12.5\,O_2 = 8\,CO_2 + 9H_2O$$

$$\underset{(114\text{ kg})}{12 \times 8 + 1 \times 18} + \underset{(400)}{12.5 \times 32} = 8(12+32) + 9(2+16)$$

Thus for 114 kg petrol we need 400 kg of oxygen for complete reaction.

Or for 1 kg of petrol O_2 required is 400/114 kg

And air having 23.2% O_2 by weight would be = 400/114 × 23.3 = 15.12

The ratio of Fuel to Air of 1:15.12 is called Stoichiometric ratio (stoichiometric ratio is chemically correct ratio of reactants required for complete reaction)

Thus for complete combustion of 1 kg of gasoline 15.12 kg air having 23.2% O_2 proportion by weight is required. This can be easily found by substituting the molecular weights in the above equation as shown above.

26.2 LIMITS OF IGNITION

In IC engines the ignition of hydrocarbon fuels is possible within certain range of air fuel mixtures as shown in Fig. 26.1. The flame propagates in IC engines if the temperature of burnt gases exceeds about 1200° C. Such temperatures are possible within Fuel to Air mixture ratios of 1 : 7 to 1 : 30. However practical limits of combustion are between 1 : 9 and 1 : 21 for SI engines using carburettor. However chemically correct mixture of fuel to air ratio is around 1 : 15 i.e., for complete combustion of 1 kg of fuel about 15 kg of air is required.

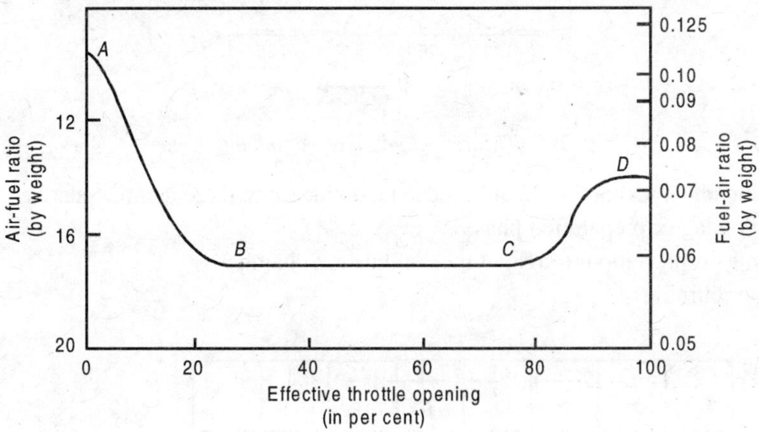

Fig. 26.1. *Limits of ignition in SI engines.*

26.3 COMBUSTION PHENOMENON IN SI ENGINES

In SI engines the combustion chambers have homogeneous air fuel mixture that is compressed after induction stroke. Towards the end of the compression stroke the spark plug gives out high temperature spark to initiate the ignition process. The spark may have temperature as high as 10,000°C. The combustion spreads from around the spark plug and spreads rapidly across the combustible mixture inside the chamber.

The flame speed would depend on the temperature of the flame front and to some extent on the temperature and density of the air fuel mixture. The molecules in and around the spark plug starts burning, when an intense and high temperature spark is generated between the electrodes of homogenous mixture of vaporised fuel, air and residual gases. As the charge near the spark burns, a flame is formed whose speed is extremely low because the reaction zone is yet to be established and heat loss is high since walls are cold. With the burning of adjacent layers, the flame travels the entire cylinder. We can broadly have two types of combustion in SI engines— Normal Combustion and Abnormal Combustion.

26.4 STAGES OF COMBUSTION IN SI ENGINES

(a) Theoretical pressure time or—crank angle curve would consist of adiabatic compression $(a - b)$, heat addition at constant volume—virtually instantaneously $(b - c)$ and expansion $(c - d)$ as given in Fig. 26.2. This is idealised situation and actual pressure time curve is quite different due to:

(i) Combustion taking finite time for completion

(ii) Presence of ignition lag.

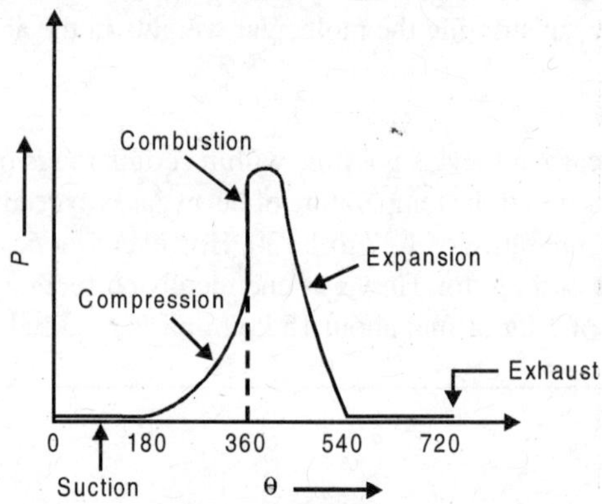

Fig. 26.2. *Theoretical pressure—crank angle curve*

(b) The combustion phenomenon in SI engines can be divided into three distinct phases:

(i) Phase I—Ignition lag or preparation phase

(ii) Phase II—Flame propagation throughout the combustion chamber

(iii) Phase III—After burning

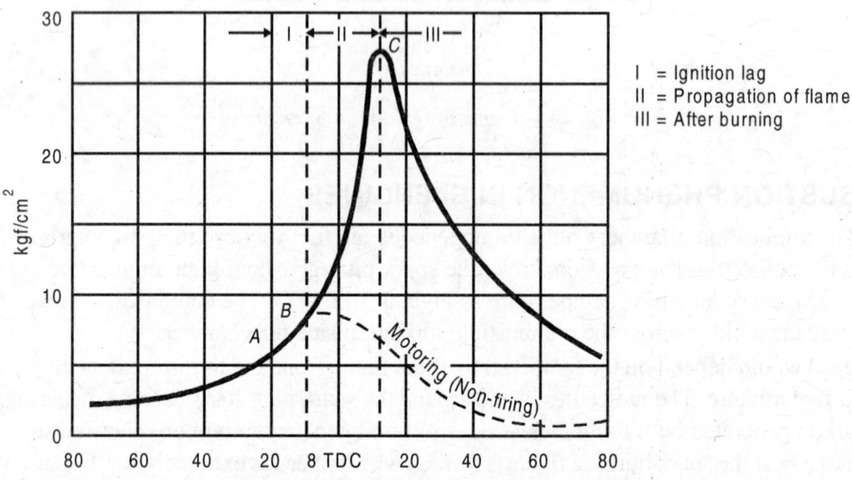

Fig. 26.3. *Stages of combustion in SI engine*

26.4.1 Phase I—Ignition Delay/Ignition Lag

Ignition lag is not a period of inactivity but it is preparation phase during which pre-flame chemical reactions take place in which growth and development of nucleus of flame occurs. It depends on various factors like nature of fuel; temperature, pressure and presence of residual exhaust gases. Ignition delay can be defined as the time between the initiation of the spark and distinct rise in pressure that can be seen from separation from the motoring curve. (Motoring curve is the pressure versus crank travel curve when the engine cylinder is not firing—shown by dotted lines in Fig. 26.3).

When the spark is occurs at point *A* (Fig. 26.3) the pre-flame reactions commence and the flame in the vicinity of the spark plug spreads very slowly and not much energy is released. The pressure rise is quite insignificant. But once the flame front gets established it accelerates at a very high speed. Point *B* where appreciable pressure rise can be seen marks the end of the ignition lag phase. Figure 26.3 shows *ab* as region of ignition lag or stage I.

Duration of the ignition lag is very small—0.15 to 2 millisecond and may correspond to 350 crank angle for 2 ms at 3000 rpm. It is some where about 10–200 crank angle rotation but depends on the RPM. In this region the pressure produced is around 1% of max combustion pressure and volume taken by burnt portion is approximately 5% of the total volume of combustion chamber.

26.4.2 Second Phase—Flame Propagation

In this phase the flame spreads and consumes the charge in the combustion chamber. This stage starts from point *B* as shown in Fig. 26.3 and ends at point *C*. In this stage the flame front travels with a maximum speed. This is the main stage of combustion accompanied by maximum release of energy. This is a primarily mechanical stage and is quite simple. The starting point of second stage is where ignition lag ends at *B* and appreciable rise of pressure can be seen on the indicator diagram. It ends at point *C* where the maximum pressure is attained. As in this stage the flame front moving at high velocity keeps the pressure and temperature rising till they reach maximum values point *C*. The burnt and un-burnt mixtures are compressed continuously with accompanying temperature rise. The flame front travels in a spherical shape. The main cause of high-speed travel of flame is turbulence in the engine. In this zone the reaction rate plays only a small role. Turbulence would depend on the velocity of the mixture entering the cylinder during intake, shape of the piston and cylinder head etc. The turbulence and hence the flame speed can be increased by adjusting these factors.

26.4.3 Third Phase—After Burning

Point *C* marks the completion of the second stage and represents the end of flame travel. However the energy continues to be released till point *d* although the burning has ceased. The reaction of gases continues with the release of additional energy. The heat of the fuel has been liberated because even after the passage of flame the reaction continues during this stage of 'After Burning'. Around 10% to 12% of heat may be evolved in this stage.

1st Stage-Ignition Lag	*2nd Stage—Flame Propagation*	*3rd Stage—After Burning*
Preparation stage, generally chemical reactions occur, growth of flame nuclei around spark plug.	Physical or mechanical stage, flame spreads virtually at constant velocity.	Energy release due to rearrangement of molecules, very low flame speed. At times in rich mixture combustion continues after point *C*.
Depends on pressure, temperature, presence of exhaust gases, type of fuel, spark plug gap.	Energy release depends on reaction rates, mixture strength, turbulence.	It continues during some part of the expansion stroke.

26.5 EFFECT OF ENGINE VARIABLES ON IGNITION LAG

Fuel. Since ignition lag is primarily chemical phenomenon the fuel quality—especially its self-ignition temperature has important bearing on ignition lag. The higher the self-ignition temperature of the fuel longer will be the ignition delay.

Air-fuel Ratio. The mixture strength giving highest temperature would result in lower ignition delay. Maximum temperature occurs in slightly richer mixtures. Figure 26.4 below shows relation between ignition lag and Fuel/Air mixtures.

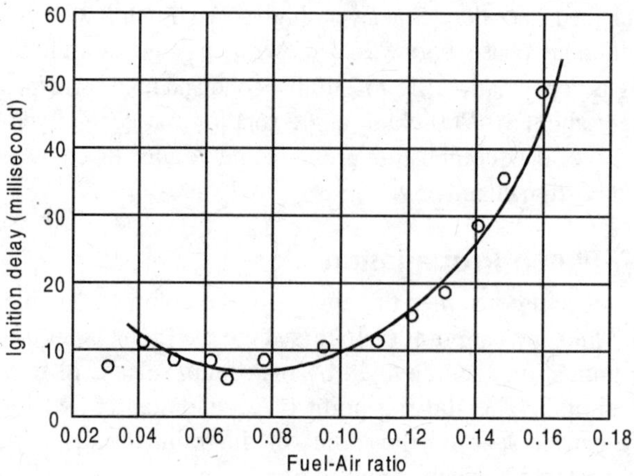

Fig. 26.4

The air fuel charge gets consumed as the flame travels across the combustion chamber. Velocity of flame propagation is of great importance in SI engine combustion because it—(i) Affects the rate of pressure rise in combustion chamber and (ii) it can cause abnormal combustion in engine. Factors on which flame velocity depends are:

(a) Turbulence. It plays a vital role in combustion; it is one of the major factors deciding flame speed. The flame velocity increases in proportion to the turbulence velocity. Flame velocity is very low for non-turbulent fuel mixtures. In such cases of low turbulence or no-turbulence, time for each explosion would be so high that IC engine would become impracticable.

(b) Fuel-Air Ratio. When the mixture is too lean or too rich the flame speed decreases because (a) too lean mixture produces less energy and (b) too rich mixture results in incomplete combustion. Therefore it has to be just correct for proper flame speed.

(c) Intake Temperature and Pressure. With increase in intake temperature and pressure the flame speed increases.

(d) Engine Load. There is increase in flame speed with increase in engine load. At higher loads with more throttling the working mixture is suited for combustion and hence smooth operation is obtained.

(e) Engine Size. In small size engines the flame has to travel less distance and hence it takes less time for the flame travel.

(f) Engine Speed. Flame speed increases almost linearly with engine speed. The engine speed higher would be the turbulence and greater would be the flame speed.

(g) Compression Ratio. At high compression ratios both pressure and temperature of the air fuel mixture would increase. These two conditions favour the flame speed. Thus flame speed increases with increase in compression ratio.

(h) Residual Gases. Increase of residual gases in the combustion chamber would reduce the flame speed. The presence of the burnt out gases would reduce the fresh charge in the combustion chamber and hence its capacity to produce energy.

26.6 COMBUSTION QUALITY

Combustion quality is a concept given by T H Ma. It is how closely the actual cycle matches the theoretical Otto cycle. Good quality combustion almost makes fast burn cycles whereas poor combustion quality consists of late and delayed burn cycles.

For best results the combustion must be completed close to the TDC for greater efficiency. Based on how combustion is completed following cycle grouping can be made.

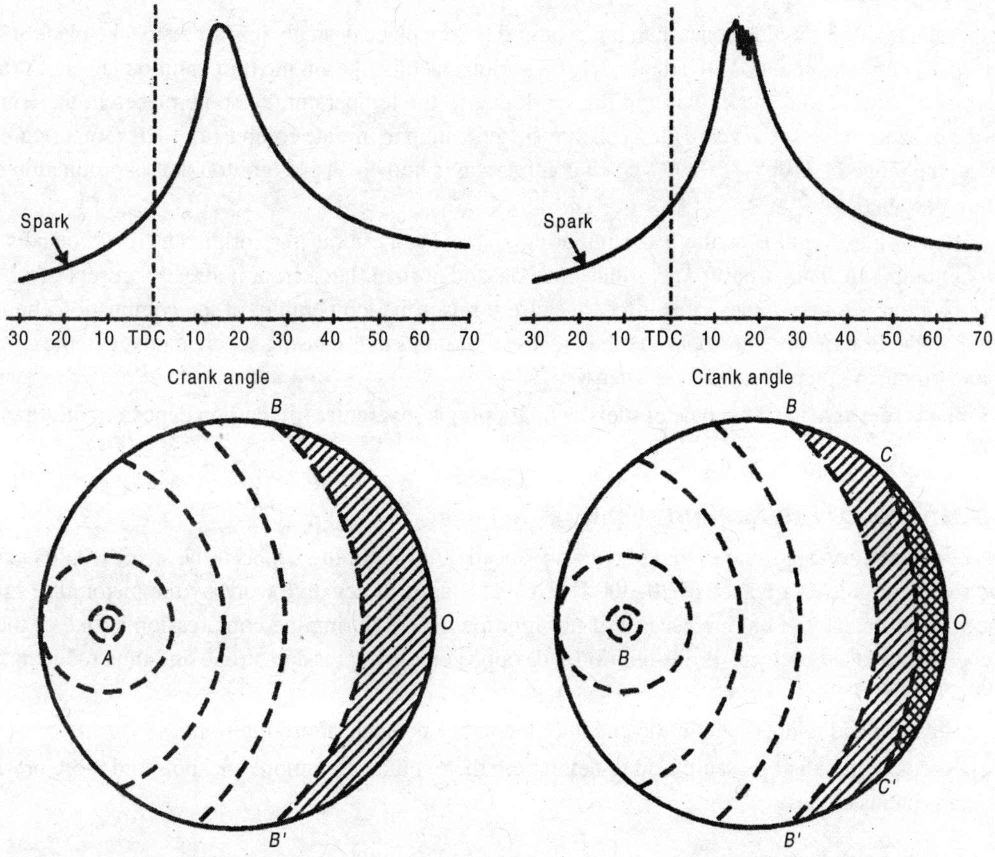

(a) Normal combustion (b) Detonation of knock

Fig. 26.5. *Combustion in the spark-ignition engine*

26.7 ABNORMAL COMBUSTION IN SI ENGINES

In normal combustion flame travels across the combustion chamber in a fairly even and uniform manner till entire charge gets consumed. However under certain operating conditions combustion in the cylinder can become uneven and sudden and may result in what is known as Abnormal Combustion. Abnormal combustion includes—preignition, knock, run-on, ping, rumbling etc. These are undesirable and harmful to the engine. Of these pre-ignition and knock or detonation are the most important. Detonation puts a limit on the compression ratio. For higher efficiency the compression ratio should be higher but in SI engines one cannot go beyond 10–11 due to onset of knocking.

Flame detection between	*Remarks*
-45° and TDC	Early burn cycle—low efficiency
TDC and 45°	Fast burn cycle—high efficiency
45° and 90°	Late burn cycle
90° and BDC	Delayed burn cycle

26.7.1 Pre-Ignition

In IC engines the ignition should occur when it is required to take place. Usually it is timed to take place at 20 – 300 before top dead centre position (BTDC) depending on various factors like engine rpm, ignition lag etc. At times the combustion can occur automatically without the spark due to the temperature at some places in the combustion chamber like exhaust valve or the spark plug etc. may be quite high to initiate combustion. The temperature for pre-ignition is in the range of 1100 – 12000°C. Such combustion is known as pre-ignition that is production of flame front without the spark.

Pre-ignition is auto ignition without the initiating spark. Pre-ignition may originate from hotspots. Due to formation of number of flame front from various hotspots and normal flame front if already present combustion is erratic. Pre-ignition is totally undesirable effect because it raises the temperature of the combustion chamber and this leads to production of more hotspots and the process multiplies thus causing severe damage to the combustion chamber and producing uneven amount of energy.

Pre-ignition is sensitive to (i) the type of fuel (ii) Spark plug temperature (iii) carbon deposit in the chamber (iv) engine speed.

26.7.1.1 Effects of Pre-ignition

(a) Effects of pre-ignition are same as that of over-advanced ignition timing i.e., as if the spark would occur much before the required ignition advance of 20–300 BTDC. This adds to negative work by the piston. Pre-ignition is also reduces the efficiency of engine because if pre-ignition occurs during the compression stroke of the engine then the energy produced because of pre-ignition will oppose the compression stroke and thus reducing the efficiency of the engine.

(b) Peak pressure and temperature in the chamber increases as result of pre-ignition.

(c) Higher temperature and pressure lead to detonation that would cause more hot spots and more pre-ignition. This becomes vicious circle.

26.7.2 Auto Ignition

Another abnormal combustion in SI engine is auto ignition. In auto ignition the pressure and temperatures especially during the end portion of the charge may reach such values where ignition could take place on its own. This phenomenon is known as auto ignition. Pre-ignition also plays important role in occurrence of auto ignition.

26.7.3 Detonation

Knock or detonation is phenomenon is signified by audible sharp metallic sound during abnormal combustion. Detonation is accompanied by very high pressures and uneven combustion resulting in mechanical damage and loss of power.

Theories of Detonation: There are possibly three theories that can explain the knock:

(i) Auto-Ignition of the end charge

(ii) Detonation or Shockwave theory

26.7.3.1 Auto-Ignition Approach

Consider a normal case in which the flame front is travelling across the chamber releasing the energy and increasing the pressure and temperature of the mixture ahead of it. The flame front travels say across the chamber as shown in Fig. 26.6.

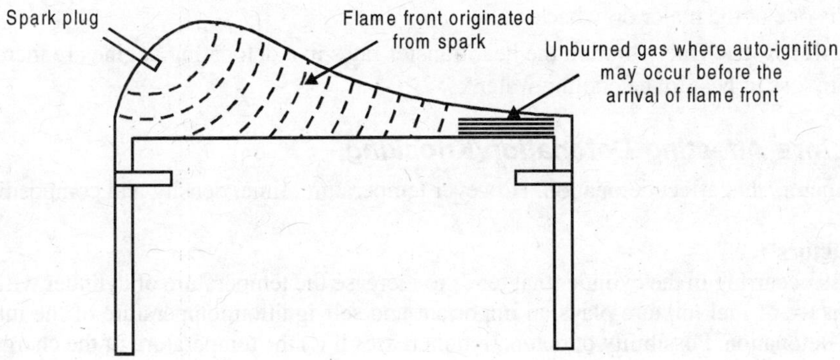

Fig. 26.6. *Detonation of unburned end gas in SI engines*

Suppose due to high pressure and temperatures the charge ahead of the flame front reaches self-ignition temperature further fuel mixture ahead it. Normally such conditions are obtained in the end charge i.e., that portion of the charge that is located towards the end portion of the combustion chamber. Following would occur:

(a) The ignition process will commence in the end portion due to self-ignition of the end charge if such conditions have been reached.

(b) In the mean time the flame front travelling from left to right would continue to consume the charge. Let us assume that the normal flame front is at *B–B* when autoignition conditions arise in the end charge.

(c) The combustion of the end portion at *B* would take place only after the ignition delay period.

(d) There are two possibilities—

  (i) If the time taken by the flame front *B–B* is less than the ID period the entire end charge would get consumed. No detonation would occur, or

  (ii) However if the end charge is not consumed before the combustion occurs in the end portion due to less ID there would be second flame front travelling from right to left. This would result in detonation or knocking.

During detonation very high local pressures of the order of 150–200 bar are reached. Detonation of even 5% of the charge can lead to very violent knocking.

26.7.3.2 Detonation Theory

In the Auto-ignition theory the combustion is supposed to be quite normal before the onset of ignition in the end charge. However in the Detonation theory it is assumed that there is shockwave travelling at about twice the speed of sound which would compress the charge to almost instantaneous reactions.

26.7.3.3 Chemical Changes during Detonation

When the combustion is normal products of combustion are CO_2, CO and H_2O. However if there is detonation—aldehydes and peroxides have been observed.

26.7.3.4 Effects of Detonation

The Detonation is harmful due to the following effects:

(a) Mechanical Damage. Severe knocking conditions may result in pressures of the order of 200 bar, vibrations of large amplitudes upto 5000 cycles per second and temperature of 2000–25000°C. Under such severe conditions there could be marked mechanical damage to the piston, valves and cylinder head. The damage may become cumulative and lead to eventual failure.

(b) Pre-Ignition. Due to high production of heat detonation becomes major factor, which causes pre-ignition.

(c) Noise and Vibration. Whereas mild knock is not audible and is not much harmful—the more violent detonation causes lot of noise and results in high vibration. Due to vibration the engine runs rough.

(d) Power Output and Efficiency. With the formation of more heat energy. Power output and efficiency also decreases which is one of the major drawbacks.

(e) Heat Transfer Rates. Due vibration the heat transfer rates in a detonating engine are increased. There is a marked increase in loss of heat to the cooling water.

26.7.3.5 Factors Affecting Detonation/Knocking

There are many factors that affect detonation. However temperature, time, density and composition are most important.

Temperature Factors

Any of the process occurring in the cylinder that tends to increase the temperature of cylinder will lead to Detonation. The temperature of fuel mixture plays an important and self-ignition temperature of the intake charge play important role in detonation. Possibility of detonation increases if (i) the temperature of the charge is high and (ii) the self-ignition temperature of the fuel is low.

(a) Raising Compression Ratio. Due to higher compression ratio the temperature and density increases. Higher temperature may lead to detonation.

(b) Supercharging. In super-charging or turbo-charging both the pressure and also the temperature of intake charge increases. This may lead to detonation.

(c) Increasing the Load. Due to this the cylinder temperature and combustion cylinder wall temperature increase. These leads to increment in end gas temperature that results in detonation.

Time Factors

The detonation can be avoided based on following time related factors—(i) increase the flame propagation speed, (ii) reducing the flame travel time by compact design and (iii) increase the ignition delay period.

Following time related factors that cause detonation can be controlled.

(a) Low turbulence, lead to less flame speed and more time is required to travel. This can be avoided by increasing flame speed.

(b) Distance travelled by the flame front should be reduced. Larger the distance, more the time and more the probability hence this can be reduced by reducing the path travelled.

(c) Flame speed is sometimes slow because of the fact that enough fuel is not present. Hence, having rich Air-fuel mixture can reduce this defect.

Density Factors

Due to increase in the density of the unburnt mixture (due to various causes) there is an increment in the probability of Detonation. Following are the causes.

(a) Increasing the Load. This lead to more opening of throttle valve thereby admitting more mixture enters and causes detonation.

(b) Supercharging the Engine. By doing so, when the compression ratio is increased the mixture becomes highly dense, which leads to detonation.

(c) Increasing the Inlet Pressure. As the inlet pressure increases overall pressure in the engine increases which leads to less ignition delay, more dense change and high probability of detonation.

(d) Advancing the Spark Timing. This leads to burning of fuel in the dense form and causing detonation.

Composition Factors

Quality of fuel and the fuel-air ratio are great importance for controlling knock. Once the compression ratio and engine size is decided, we can reduce knocking by having a change in composition that can be attained by:

(a) **Increasing octane rating or fuel.** Ring compounds like Benzene and addition of tetra ethyl lead reduces tendency to knock.

(b) **Having rich or lean mixture for combustion.** This leads to longer delay period and low temperature. Both these factors reduce possibility of knocking.

Methods to Reduce Knocking

(a) Good combustion chamber design to reduce the flame travel by central location of the spark plug or having more than one spark plug by proper placement, avoiding pockets of stagnant charge, having turbulence to increase flame speed.

(b) Cooling of spark plug and exhaust valve that are major hot spots in most of the combustion chambers. Increasing the number of spark plugs with proper placement.

(c) Reducing the temperature of the last portion of the charge by increasing the surface to volume ratio as in the wedge shaped chambers to increase the heat transfer to the walls.

(d) Using of lower compression ratio and high octane number fuel.

(e) In general rich mixtures are less prone to detonation due to cooling effect.

(f) Increasing engine rpm also leads reduction in detonation.

Increase in variable	*Effect on charge*	*Action required to reduce knocking*
CR	Increase in pressure and temperature	Reduce CR
A/F ratio	–do–	Making it rich
Spark advance	–do–	Retard the timing
Charge inducted	Increase in pressure	Reduce intake
Inlet temperature	Increase in temperature	Reduce temperature
Wall temperature	–do–	–do–
Turbulence/engine speed	Decreases time factor	Increase
Distance of flame travel	Increases time factor	Reduce

Other Abnormal Combustions

There are other abnormal combustion like Rumble, wild ping etc. but are not of much importance.

Rumble. Due to the presence of deposits on the combustion chamber ignition may start from hot spots before or after the normal ignition. This would lead heavy explosive burning of the charge. There is appreciable rise in pressure and a low pitch noise called Rumble—caused by multiple pre-ignitions is audible. Rumble is harmful and results in damage to the engine parts.

Ping. It is caused due to some burning part of the deposit—causing ignition now and then. It is somewhat similar to knocking. It can also be known as knock, which is sudden erratic, sharp and unpredictable.

The combustion progress in the above has been shown in table.

Run-on Ignition. Due to the presence of deposits on the surface at times there is combustion even when ignition is switched off.

Summary of Normal-ignition, Pre-ignition and Detonation

Progression	*Normal Combustion*	*Pre-ignition*	*Detonation*
1. Initiation	Ignition by spark plug	Ignition by hot spot-like hot exhaust valve /hot deposit	Ignition by spark plug
2. Continuation	Combustion continues with the development of flame front	Spark occurs after the ignition has ocurred at the hot spot	Normal combustion continues
3. Flame front	Single flame front	There would be two or more flame fronts which may collide	When the flame front reaches towards the end—the conditions may become suitable for auto ignition and another flame front may develop
4. Abnormality	None	Due to collision of the flame fronts there is loss of power and loss of power and uneven power generation	Detonation occurs which is harmful

26.8 RATING OF SI ENGINE FUELS/OCTANE NUMBER

Tendency to knocking depends quite significantly on the type and composition of the Hydrocarbon fuels. Resistance to knocking is of great significance in fuels. The fuels differ quite widely in their property of anti- knocking characteristic. Hydrocarbons have a tendency to knock when the engine operating conditions become severe. Detonation is undesirable and resistance to detonation or knocking is required for the fuels. This measure of fuel resistance to knock is known as Octane Rating.

The rating of fuels for their resistance to knocking is done by comparing them with standard reference fuel. The standard fuel is mixture of two primary fuels—(i) Iso Octane C_8H_{18} which has very high knock resistance properties and is assigned rating of 100 and (ii) normal heptane having very poor anti-knock characteristics and is assigned rating of 0. The resistance of a fuel to knock is indicated by Octane Number. Octane number is the percentage of iso-octane fuel in a mixture of iso-octane and *n*-heptane by volume which will give the same knocking as the fuel under standard test conditions. Octane number gives a fair indication of the anti-knock properties of the fuel. Gasoline with Octane No. 83 means that this fuel has the same knock characteristics as mixture of 83 % is—octane and with 17% *n*-heptane under standard test conditions.

Octane rating tests are carried out in standard Co-operative Fuel Research (CFR) engine of 82.6 mm bore and 114.3 mm stroke engine. It is robust 4-stroke engine where compression ratio can be varied from 3 to 30. The engine has multi-bowl carburettor in which fuel to be tested as well as iso-octane and *n*-heptane can be stored separately. The testing for Octane rating can be done under two different conditions.

(i) Firstly under normal conditions which is called Research Method and the corresponding Octane No. is called RON.

(ii) Secondly more severe conditions which are close to the actual operating conditions which is called Motoring Method. The Octane No. found by motoring method is called MON.

The details of testing under these two conditions are as tabulated below—

	Research method	*Motor method*
Intake air temperature	125° F	300° F
Engine speed	600 RPM	900 RPM
Spark advance	13° BTDC	14–260 BTDC (varies with compression ratio)
Air/Fuel ratio	Adjusted for maximum knock in both	
Coolant temperature	100° C in both methods	

Uses of RON and MON

Sensitivity. From RON and MON we can get the change that would occur under test condition and practical conditions. We define sensitivity as difference between RON and MON. Thus sensitivity can also be defined as the measure of the extent to which a gasoline is downgraded under severe conditions. Higher the sensitivity, poor will be its performance under severe conditions.

$$\text{Sensitivity} = \text{RON} - \text{MON}$$

Anti-knock Index. This is the newest rating system usually called Road Octane Number. It is the mean of the RON and MON. The Octane number displayed on petrol pumps is generally Road Octane Number. Our premium grade petrol is having Octane No. of 91.

$$\text{Road Octane Number} = (\text{RON} + \text{MON})/2$$

Performance Number. In many applications like aero engines high anti-knocking fuels are required. In such engines we require anti-knock property even higher than Iso-Octane. When Octane number increased beyond

100 with the additions of new additives like Tetra Ethyl Lead (TEL) and tetra methyl lead need was felt for other units to indicate knock resistance of such fuels. The discovery of TEL in 1923 lead to marked improvement of ON when blended with gasoline. The effect of TEL addition on Octane No. is given in Fig. 26.7.

The anti-knock rating of fuel of ON greater than 100 is given by Performance Number. Performance Number is found by carrying out test in a standard super charged engine under prescribed conditions without knocking. Performance Number is the maximum power that can be produced in such test by using the fuel under test and also by using iso-octane which is assigned Performance Number of 100. Thus if power using the fuel be 1.5 kW and with iso-octane it is 1.2 kW then

$$\text{Performance Number} = (1.5/1.2) \times 100 = 125$$

Where 100 represents Performance Number of iso-octane

Performance Number can also be defined as = KLMEP of the fuel/ KLMEP of iso-octane

Where KLMEP stands for knock limited mean effective pressure

Another method of Performance Number is to represent it in terms of ml of TEL per US gallon

Performance Number = 100 = millilitres of TEL per US gallon of gasoline

Performance Number can also be used for measuring ON less than 100 as shown in Fig. 26.7 but it is generally used for greater values only.

There are other methods for SI engine rating. Like Ricardois toluene no. in which toluene and heptanes are reference fuels and scale is drawn.

Summary. For anti-knock capability of fuels above ON 100 we make use of Performance Number

(a) Performance Number = 100 + ml of TEL per US gallons of fuel

(b) Performance Number = $(\text{KLMEP})_{\text{fuel}} / (\text{KLMEP})_{\text{iso-octane}}$

Combustion Chamber

The type of combustion chamber design has important bearing on the performance and anti-knock characteristics. Thus a combustion chamber must fulfil the following conditions:

(a) Maximum power output
(b) The operation should be as smooth as possible.
(c) Pressure rise should be uniform and smooth.
(d) Resistance to Knocking/Detonation.

To fulfil the above conditions, the chamber must be designed considering the following factors:

(a) It should have high compression ratio: The higher the compression ratio the higher would be the power output and the efficiency.
(b) There should be no excess air i.e., there should be maximum utilisation of inducted air and there should be no dead pockets.
(c) Adequate turbulence for high flame speed.
(d) Flame travel should be minimum hence engine should be compact; also compactness gives small surface volume ratio which leads to less heat loss during combustion.
(e) Absence of exhaust gases i.e., proper scavenging.
(f) Spark plugs should be properly located to have a good flame start and travel.
(g) Absence of hotspots i.e., no deposits and good cooling arrangement.
(h) Pressure rise in the combustion chamber should be uniform and moderate
(j) Valve head should have right rise and right cooling arrangement.
(k) Minimum time to complete combustion.
(l) The thermal and volumetric efficiencies should be high.
(m) Fuel efficient and non-polluting.

26.9 EVOLUTION OF COMBUSTION CHAMBERS FOR SI ENGINES

As early as 1908 Ford Co had used T-head type engines for its automobiles. The main disadvantage was presence of two camshafts and long distance of flame travel that resulted in detonation. Subsequently L-head and I-head or overhead valve type engines were developed.

Ricardo turbulent combustion chamber was designed to obtain fast flame speed while reducing knocking. This designed also reduced effective flame travel by reducing distance between piston crown and head. It was designed to dissipate heat at a much faster rate and thus was improvement to T-head and I-head chambers.

26.10 DIFFERENT TYPES OF COMBUSTION CHAMBERS

Different types of SI engine combustion chambers have been shown in Figs 26.7 and 26.8.

(a) T-head Combustion Chamber. In this design the valves are on the opposite side as shown in Fig. 26.7 (a). This combustion chamber is not in much use as compression ratio cannot exceed a specified limit due to fact that this chamber facilitates large flame travel. This type of combustion chamber was first time used by Ford Company.

(b) L-head Combustion Chamber. In this type of chamber both the inlet and outlet valves are on the same side as shown in Fig. 26.7 (b). We can remove cylinder head without disturbing valve assembly. This chamber has slow rate of combustion and has tendency to detonate because it has long air flow path and due to tendency to detonate, the compression ratio used is small. The major advantage of this type of chamber is that it has less maintenance that help using in those applications where regular checking is not feasible.

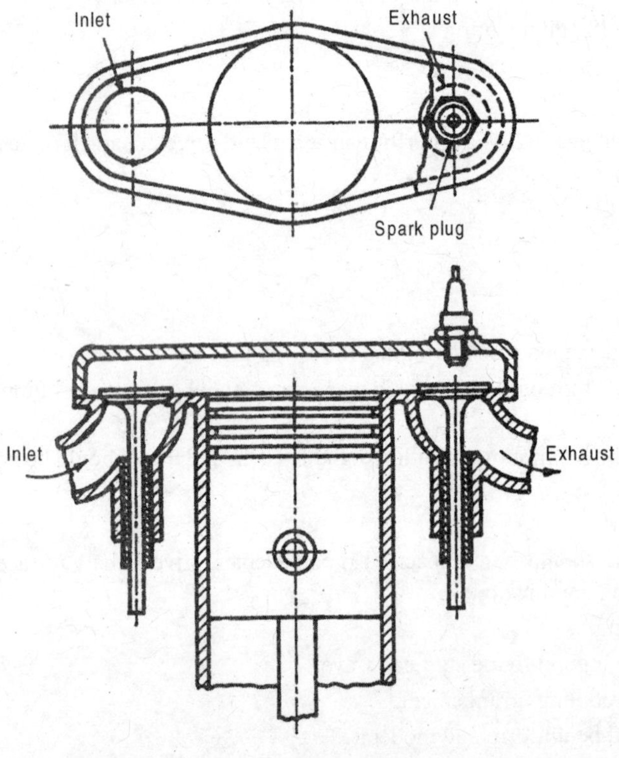

Fig. 26.7(a) *T-head combustion chamber*

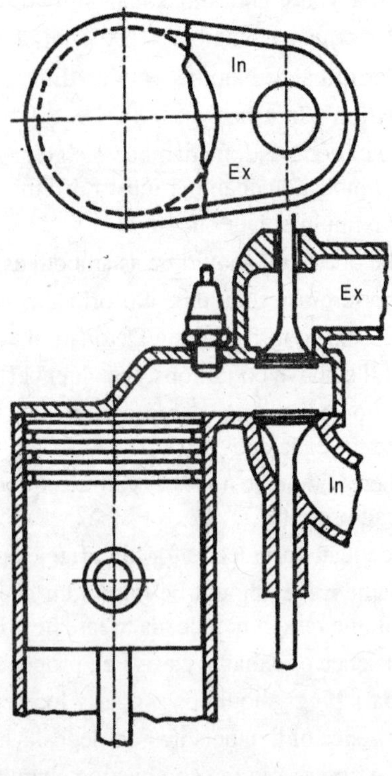

Fig. 26.7(b) *L-head combustion chamber*

(c) I-head Overhead Valve Combustion Chamber. In the I-head or overhead design, the valves are in the cylinder head. The design allows the use of large sized valves (resulting in improved volumetric efficiency. However as clearance volume increases, the compression ratio decreases. Different type of I-head combustion chambers are available as shown in Fig. 26.7(c), designs like bath tube design wedge designs etc. are available. These chambers have both valves on the cylinder head. The spark plug is on the top side of the chamber. These chambers have high volumetric efficiency and pressure rise is also smooth and uniform. Therefore these can be operated on high compression ratio. These chambers also have a very less tendency to detonate as the flame travel is not long due to compact design in which all available space is used. Most of the cars in India use I-head or its variants.

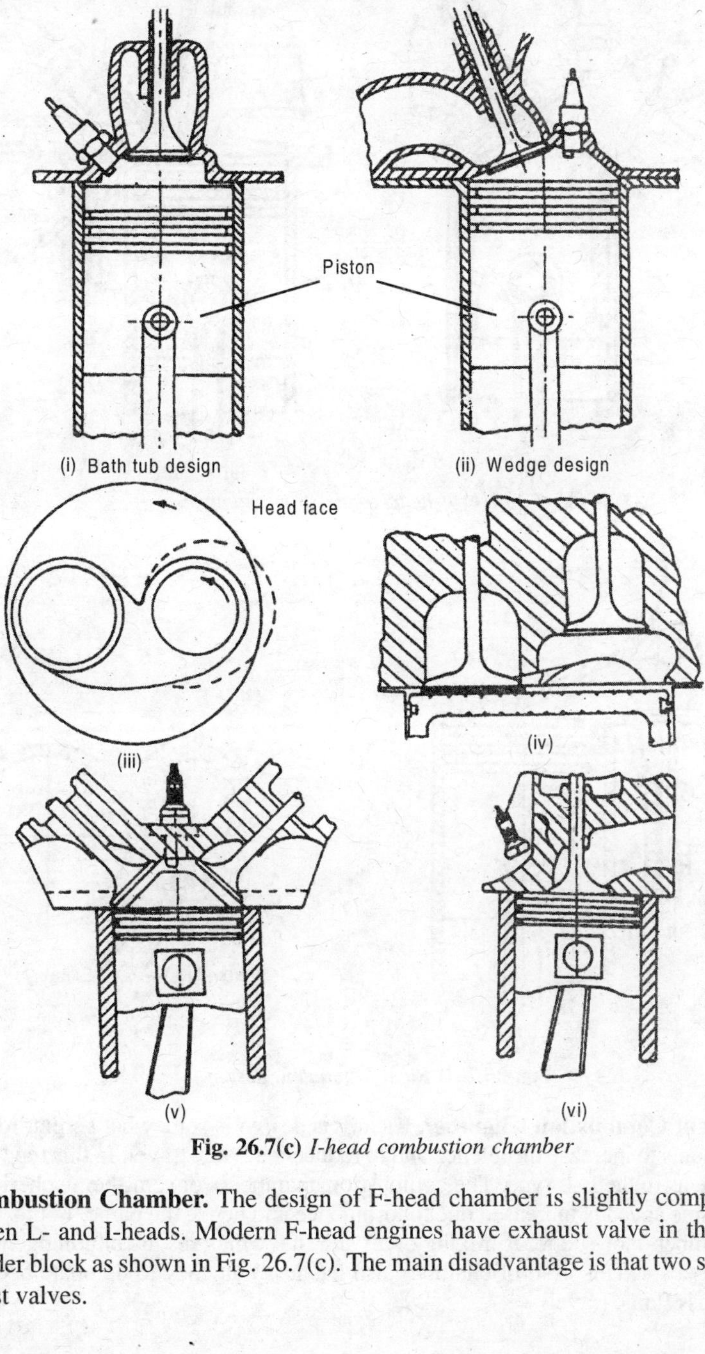

Fig. 26.7(c) *I-head combustion chamber*

(d) F-Head Combustion Chamber. The design of F-head chamber is slightly complex. Valve positioning is compromise between L- and I-heads. Modern F-head engines have exhaust valve in the cylinder head and inlet valve is in the cylinder block as shown in Fig. 26.7(c). The main disadvantage is that two separate camshafts actuate the inlet and exhaust valves.

The F-head chambers are quite compact. In these spark plugs are generally placed on the top of chamber. Due to location of the spark plug and compactness of the chamber such chambers have high efficiency, can operate on higher pressure and have almost negligible knocking tendency. However its complicated design makes it difficult to manufacturing.

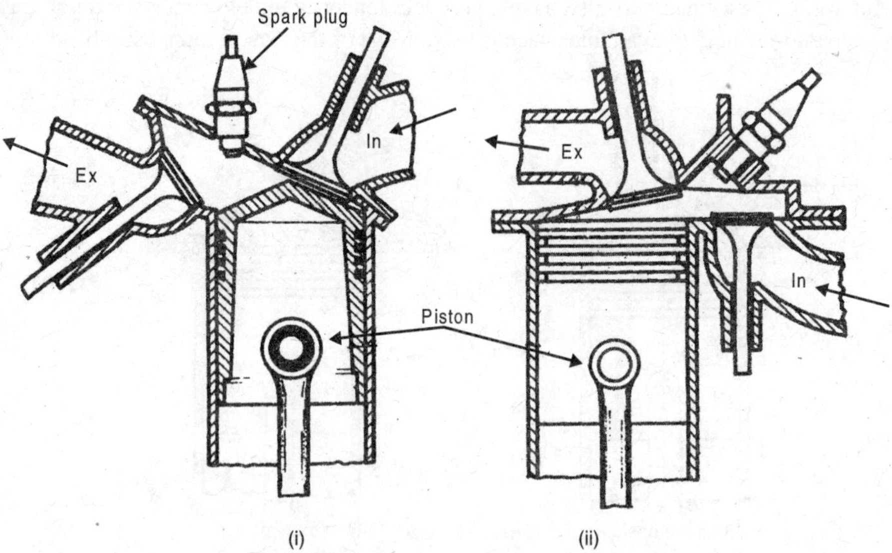

Fig. 26.7(c) *F-head combustion chamber*

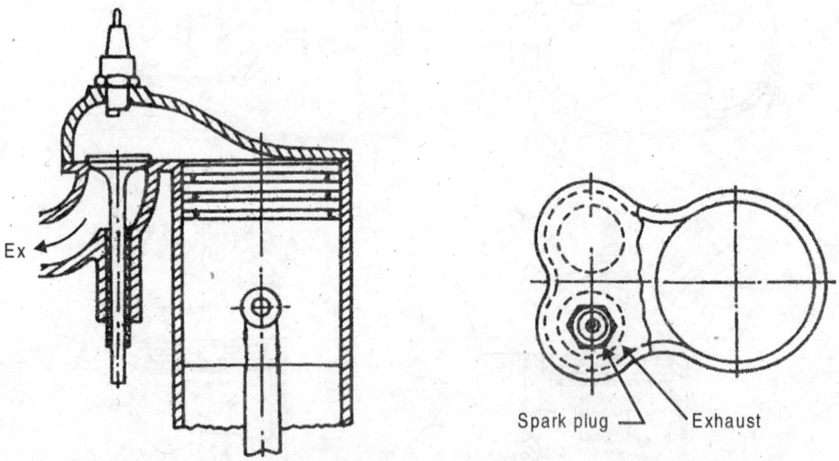

Fig. 26.7(d) *Ricardo tubulent design*

(e) Ricardois Turbulent Combustion Chamber. Ricardois design is somewhat similar to L-L type chambers but with major improvements to increase turbulence and to reduce the flame travel. In this type of chamber position of valve is on the one side as in the L-L type. The combustion chamber is of semi-hemispherical shape located on the cylinder head. The flame speed is increased due to turbulence and hence the name. In this type, due to peculiar shape of combustion chamber, more better is mixing of air and fuel which results in homogeneous and distributed fuel. This is used in many cars. This kind of chamber also has less tendency to detonate because of presence of higher flame speed and less flame travel.

	T-head	*L-head*	*Ricardo's*	*I-head/ Overhead*	*F-Head*
		Improvement over T-head	Improved version of L-head	Overhead valves	Valve position compromise between L- and I-heads
Valve position/ cams	Valves on either side of the cc/two cam-shafts needed	Valves on one side of the cc, one camshaft required	Valves on one side.	Both valves are in the cylinder head	One valve (usually EV) in the cylinder bead and IV in the cylinder block/two cam shafts are needed
General	Flame travel is more cc has high surface to volume ratio which increases cooling losses leading h the prone to knocking due to more flame travel	Has less surface to volume ratio which decreases thermal losses and has high h_{th}, more turbulence is advantageous. The end portion being close EV prone to detonation. Lower g_{vol} due to change of the direction of inlet air	Hemispherical combustion space located in cylinder head, turbulence increases flame speed, less tendency to knock	Large size valves can be used resulting in higher h_{vol}	Size of suction valve is increased for higher h_{vol}. Complex shape difficult in manufacturing

26.11 CI ENGINES

26.11.1 Introduction

Compression Ignition is generally known as Diesel Engine—named after its inventor Dr. Rudolf Diesel who invented it in 1899. The CI engines have high thermal efficiency and use relatively cheaper diesel fuel compared to gasoline. CI engines are extensively used for power generation, in commercial transportation, buses, marine engines, locomotives, tractors, pumping sets, stationary industrial application and machine applications. However due to its higher weight, smoke and odour its application in passenger cars is limited. The CI engines have been produced in wide power range.

26.12 COMBUSTION PHENOMENON IN CI ENGINE

Combustion in CI engines is entirely different than in SI engines. In CI engines are compressed to a much higher pressure than that in SI engine due to higher compression ratio.

Compression ratio is of the order of 12–22 due to which temperature and pressure of the air are quite high. The fuel is injected just before TDC in the form of high-pressure jet. The fuel enters the combustion chamber containing hot gases that vaporizes and after some time because of heat its combustion occurs. It is to be noted that time available for combustion is relatively quite small. Table 26.1 shows the time available for combustion in a typical CI engines.

Table 26.1. *Time availability for injection/combustion in a typical diesel engine operating at 3300 rpm.*

	Suction	*Compression*	*Injection*	*Combustion*	*Expansion*	*Exhaust*
Time in ms	11.4 ms	7.2	1.6	4	7.2	11.9

Unlike SI engines where we have homogeneous air fuel mixture of required quality delivered to the combustion chamber by carburretor or multi-point fuel injector (MPFI) system, in CI engines air and fuel are injected separately at different times. The mixture is at best heterogeneous and mixing takes place in the combustion chamber. After the fuel in injected at high pressure in the combustion chamber containing air at high pressure and high temperature the jet disintegrates. The fuel droplets vaporise while absorbing heat of vaporisation from the hot air resulting in decrease in the temperature of air in the vicinity of the fuel droplets. Fuel takes heat from the surrounding air in the combustion chamber. This air has to take heat from the nearby air. Until the temperature of air increases the combustion cannot occur till the full mixture reaches the ignition temperature. The delay period for combustion cannot occur till the full mixture reaches the ignition temperature. The delay period for combustion process on account of this is physical delay. After the physical delay there would be chemical delaying which reaction starts slowly and then accelerates until ignition takes place. Ignition delay would comprise of the physical and chemical delay periods. To understand the combustion process in CI engines it would be desirable to compare salient features of combustion in SI and CI engines.

Table 26.2

	SI engines	*CI engines*
Air fuel mixing	Homogeneous, mixed outside the engine	Heterogeneous, separate supply of air and fuel
Fuel : Air ratio	Correct ratio supplied by Carburretor/ MPFI Closed to Stoichiometric from part to full load	Fuel by injected by fuel injector/air only constant quantity of air Fuel : air ratio may vary widely from 1 : 20 at full load to 1 : 100 at no load
Turbulence	Disorderly turbulence	Well organised turbulence—swirl, squish, tumble etc. are essential
Vaporisation	Disorderly turbulence	Well organised turbulence— swirl, squish, tumble etc. are essential
Stages of Ignition delay	Physical delay as in CI engines is absent. Only chemical delay	Physical delay—in which fuel vaporises Chemical delay for onset of sustain combustion
Excess air	Not required	15 to 40% excess air supplied
Compression ratio	1 : 6–11	1 : 16–21
Spark ignition	Required	Ignition due to high temperature of compressed air
Flame spread	There is flame front which sweeps across the combustion chamber	There is no distinct flame front. Combustion can commence from many points in the chamber where fuel and air reach combustion stage after vaporisation

26.13 STAGES OF COMBUSTION CI ENGINE

The combustion in CI engines takes place in four stages as given in Fig. 26.8.

(a) Ignition delay

(b) Rapid or uncontrolled trolled combustion

(c) Controlled combustion
(d) After burning

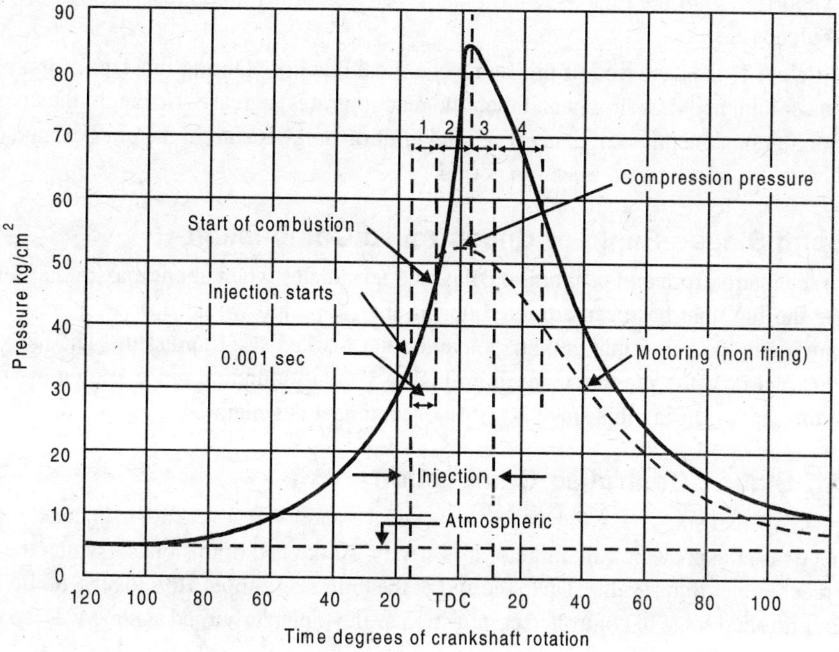

Fig. 26.8. *Stages of combustion in the CI engine*

26.13.1 First Stage—Ignition Delay

It is preparatory phase beginning from the time fuel injection commences and ends when the pressure-time curve separates from the motoring curve Figure (26.9). It is the stage in which fuel has been injected in the combustion chamber but has not ignited. The ignition delay stage should be as small as possible. However ignition delay cannot be reduced beyond a certain limit. The ignition delay can be divided into two distinct parts.

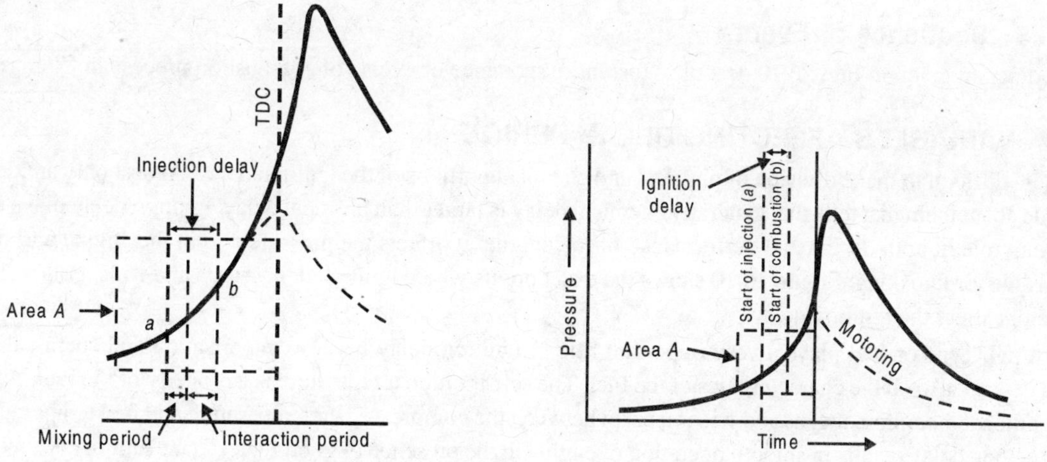

Fig. 26.9. *Pressure-time diagram showing delay period*

(a) Physical delay. It begins from the time of injection till the fuel in atomised, vaporised and get mixed with air till its temperature reaches the self ignition temperature. Physical delay will depend on the factors which affect the atomisation and vaporisation of the fuel—such as the type of fuel, injection pressure, type of nozzle, temperature and extent of turbulence etc.

(b) Chemical delay. This is a period of pre-flame chemical reaction. During this part of the ignition delay the chemical reactions commence first slowly and then the reaction rates increase. Normally the chemical delay portion of the ID is longer than the physical delay. The total ignition delay is sum of the physical and chemical delays. The ignition delay time is shorter than injection period.

26.13.2 Second Stage—Rapid or Uncontrolled Combustion

During this stage that starts from end of ignition delay and terminates when about maximum pressure is reached. During this phase the fuel that has been admitted during the ID period would have evaporated and is ready for combustion. There would be very rapid combustion and quiet marked rise in pressure. The rate of heat release is highest during this phase. If the pressures are quire high a violent pounding noise known as 'Diesel knock', is produced during this stage. During this stage 30–35% of total heat is generated.

26.13.3 Third Stage—Controlled Combustion

It starts at the end of second stage the combustion fairly steady state. Whatever fuel is injected virtually same gets burnt. During this phase the pressures are more or less steady. At the end of this phase, which the end of injection, the temperature reaches maximum value. Up to the end of the third stage about 70% to 80% of total heat is released. This is known as controlled stage of controlled combustion as the fuel gets burned as shown as it enters the chamber in this stage.

26.13.4 Fourth Stage—After Burning

Theoretically combustion should end after the third stage when fuel injection has ceased. However this does not happen because there would be still unbumt or partially burnt fuel particle which would continue to burn as soon as they come in contact with O_2 in the combustion chamber. Thus phase starting from the end of the third phase continues during part of the expansion stroke. The duration may last 70–80° from the TDC. During this phase 5% to 10% of total energy is released.

26.13.5 Sequence of Events

Block diagram given in Fig. 26.10 describes the entire sequence of events of combustion process in CI engine.

26.14 VARIABLES AFFECTING DELAY PERIOD

It is quite difficult to make division of physical and chemical portions of the ignition delay. At best only an idea can be made about them that tells that generally chemical delay is larger than physical delay. Ignition delay plays a very important role in both design and performance of the engine. It affects the pressure rise in the engine and thus is responsible for knocking. Figure 26.10 shows the exact points where ignition delay starts and ends. Following are the factors affect the ignition delay.

(a) Fuel. Type of fuel plays a very important role in ignition delay because vaporisation and chemical delay would depend also on the characteristics of the fuel. The self-ignition temperature is a property of the fuel. A lower self-ignition temperature means and wider margin between the compressed air temperature and fuel temperature. A short ignition delay results in smooth operation of engine in the presence of good fuel. Cetane number is a measure of the ignition delay.

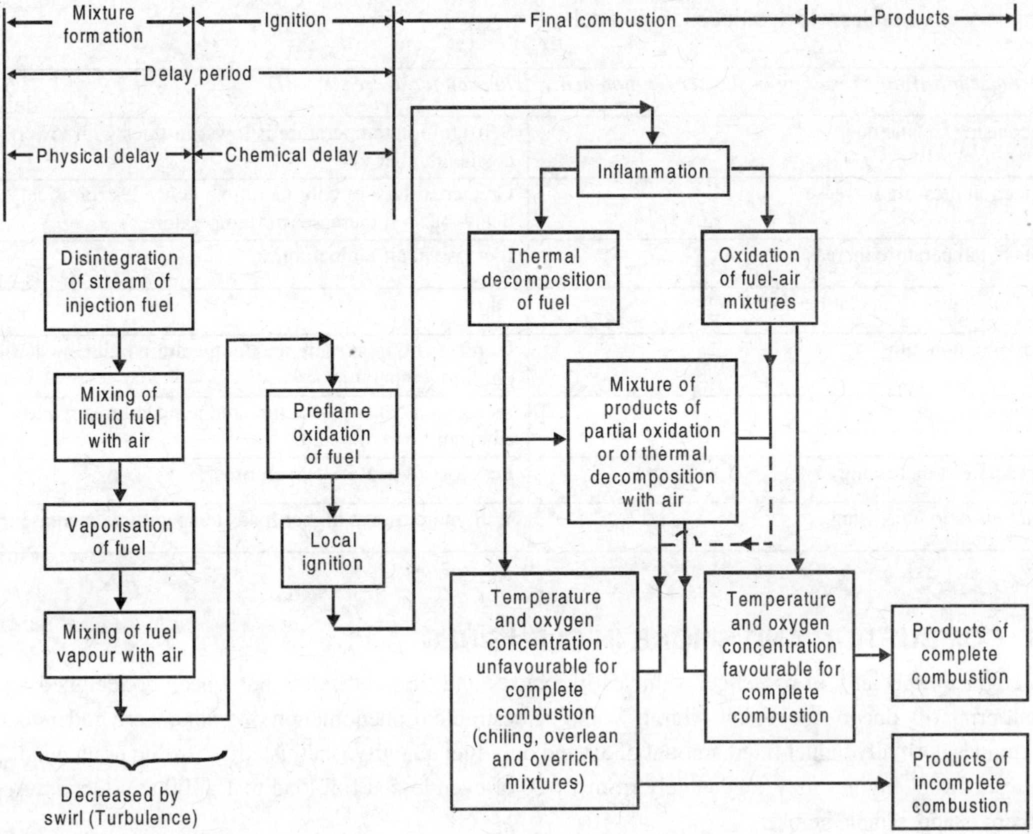

Fig. 26.10. *Combustion in the CI engine*

(b) Injection Advance. The delay period increases with increase in injection advance angle in high-speed diesel engine.

(c) Injection Pressure. Ignition delay increases with the increase in inlet pressure.

(d) Intake Temperature. Ignition delay increases with the increase in inlet pressure.

(e) Compression Ratio. Increase in CR would increase both pressure and temperature. The delay period is reduce with the increase in compression ratio.

(f) Cooling Water Temperature. Ignition delay reduces with increases in jacket cooling water temperature.

(g) Speed. With the increase in engine speed both temperature and pressure of air would increase thus reducing the delay period.

(h) Air Fuel Mixture. Fuel air to ratio increases with load and vice versa. Any reduction in the fuel air ratio will reduce the cylinder wall temperature and hence delay period increase.

(j) Engine Side. Large engines operate on slow speed because of size limitation. At lower speeds the ignition delay period increases.

(k) Combustion Chamber. The combustion chambers can have import bearing on ignition delay. The combustion chambers with pre-combustions chambers have shorter ignition delays as compared to the open type.

Summary of variables which reduce ignition delay.

Increase in variable resulting in shorter ignition delay	*Reason for decrease in ID*
Increasing Cetane no.	Self-ignition temperature is lower in flues with lower cetane number
Injection pressure increase	Greater surface to volume ratio—hence less physical delay—also it increases air temperature
Intake temperature increasing	Increase in air temperature
Jacket cooling temperature increasing	–do–
Compression ratio	Increases both pressure and temperature reducing auto-ignition temperature
Supercharging	Increase in density, pressure and temperatures reduce auto-ignition temperature
Speed/RPM increasing	Less loss of heat in a given time
Fuel-air ratio increasing	With more fuel at higher loads temperatures increase

26.15 COMBUSTION AND SMOKE IN CI ENGINES

Diesel engines especially in passenger vehicles like buses and three wheelers have been condemned for being pollutant primarily due to smoke emission. It would be desirable if phenomenon of smoke were understood.. The diesel engines normally induct fixed amount of air and only fuel quantity is adjusted according to the load. The air fuel ratio in diesel engines may vary widely from 1 : 20 or even less at full load to 1 : 100 at no loads. At higher loads there is appreciable smoke.

26.15.1 Diesel Knock

Fuel which beings at the beginning of the ignition delay period gets accumulated till combustion appears at the end of the ignition lag phase. This fuel would auto-ignite after end of delay period giving very high pressure rise. If the ignition delay has been long resulting in large amount of fuel accumulation rate of rise of pressure and maximum pressure during the second phase of burning could be quite high. Such pressure rates may cause diesel knock. Diesel knock primarily arises from long ignition delay for the following reasons.

(a) Accumulation of large amount of fuel and

(b) Homogenisation and preparation of fuel for combustion.

It is very difficult to distinguish between knocking in CI engines since normal working of these engines is also quite noisy. There are method to reduce or minimise it. In the CI engines knocking occur in the initial phase only, whereas in SI engine it occurs towards the last phase but though the phenomenon sounds same there is difference as follows.

	Detonation in SI	*Knocking in CI*
Commencement	Detonation occurs at the end of combustion	Detonation occurs at the starting of combustion
Pressures	Here the rate of pressure is very high as mixture is homogeneous also max pressure is high	Here the rate of pressure rise and max. pressure both are lower because of heterogeneous nature of mixture
Identifying combustion with knocking	The SI engine we can distinguish easily between knowing and non-knocking engine being audible	In CI engine it is difficult to distinguish between knocking and non knocking combustion process itself is quite noisy. The combustion have sufficiently rate of pressure-rise and accompanied by audible noise. Hence here a personal judgement and experience is required
Mitigating factors	The variables that tend to reduce knocking in SI engine increases possibility of knock in CI engines	

26.15.2 Factors Affecting Knock in CI Engines

The following factors affecting knock in CI engines are given below:

(a) Tendency to knock reduces if the air fuel mixture has

(i) High temperature

(ii) High density

(iii) Short delay

(v) Reactive mixture

(b) Also tendency of knock would increase if

(i) Compression ratio is lowered

(ii) Inlet air temperature is reduced

(iii) Coolant temperature is reduced

(iv) Load is reduced

(v) Decreasing the injection pressure

(vi) Decreasing the speed

(vii) Decreasing the turbulence

26.16 CETANE RATING OF CI ENGINE FUELS

(a) Cetane rating is quite similar Octane rating of gasoline which is measure of resistance to knock. The Cetane measure the ignition lag on which diesel knock depends. Cetane ($C_{16}H_{34}$) is straight can paraffin and has a good ignition quality. It is assigned arbitrarily a rating of 100 Cetane number (CN = 100). Alpha Methyl Napthalene ($C_{11}H_{10}$) which has poor ignition quality is assigned a Cetane value of size (CN = 0).

(b) Cetane Number is defined as the percentage by volume of Cetane in a mixture of Cetane and α-methyl naphthalene that produces the same ignition delay as the fuel being tested in the same standard engine under the same standard operating condition. Originally α-methyl naphthalene was used as reference fuel along with Cetane. However these days instead of α-methyl naphthalene heptamethyl none is used with CN = 15.

(c) Thus Cetane rating of 60 indicates that fuel would give the same ignition delay in a standard engine under standard test conditions as a mixture of 60 parts by volume of Cetane and 40 parts of α-methyl-naphthalene.

Thus Cetane No. calculated by

$$CN = (\% \text{ of Cetane}) + (\% \text{ α-methyl-naphthalene})$$

(d) Test Conditions. The test is carried out in a single cylinder standard engine at the following condition

(i) Speed — 900 RPM

(ii) Jacket water temperature —100° C

(iii) Injection advance — 23°C

(iv) Ignition delay — 13°C

The test fuel is first tested and the fuel pump delivery is adjusted to give particular air/fuel ratio. The injection timing is adjusted to give injection advance of 14°. Then compression ratio is adjusted so that combustion occurs at TDC. Thus the test fuel has ignition delay of 13° in this case. Now the charts showing relationship between compression ratio at 13° and the Cetane number are referred to find out the Cetane number. Thus the knock in the engine is directly related to the ignition delay of fuel.

26.17 ADDITIVES

Some additives serve to reduce the self-ignition temperature of fuel by acting as local ignition points. Such additives can be mixed in the diesel to reduce its tendency to knock. Thus they improve the ignition quality of fuel by reducing ignition delay. Some of these additives are given in Table 26.4.

Table 26.4

Additives	*Cetane Rating*	*Change*
Isopropyl nitrate	39	+18
Amyl nitrate	39	+15
Heptytyl Peroxide	77	+ 2
Butyl Peroxide	39	+ 2
Methyl Acetate	26	+16

26.18 IMPORTANCE OF TURBULENCE IN CI COMBUSTION CHAMBERS

(a) It is essential to have well organised movement of air within the combustion chamber for (i) speedy evaporation of fuel (ii) to enhance air fuel mixing (iii) to increase combustion speed and (iv) to increase efficiency. Due to high velocities involved the flow of air within the combustion chamber is turbulent. Due to turbulence the transfer rates like vaporisation of fuel, heat transfer rates, mixing and combustion rates are greatly increased.

(b) The main aim of the CI combustion chamber design is to ensure proper mixing of fuel with the air in a very short time. To achieve this well organised movement of air is essential. Turbulence in the CI engines is quite systematic and built into the design. It would be desirable to understand various types of movements of the air within the combustion chamber.

(i) Turbulence is quite high during suction and recedes towards BDC.

(ii) Again during compression the turbulence increases as swirl, squish and tumble.

26.18.1 Swirl

The rotational motion of air within the cylinder is called swirl. The swirl is achieved by a variety of methods like shaping the intake manifold, valve, valve port and at times by suitably contouring the piston. Swirl enhances mixing and makes the flue air mixtures homogeneous. Swirl is the main mechanism to spread the flame within the combustion chamber.

The swirl ratio used to indicate the extent of rotational motion within the combustion chamber. If the ratio of average angular speed to the engine RPM. Following are the main types of swirls.

(a) Induction Swirl. In this air flow is directed in a particular path while entering the cylinder. This method is generally employed for open combustion chamber.

(b) Compression Swirl. Here the swirl is produced during the compression stroke, air is forced through a passage into swirl chamber made separately during the compression stroke, as shown in Fig. 26.11.

(c) Combustion Induced Swirl. This swirl is produced by using the phenomenon of high initial pressure rise due to partial combustion.

26.18.2 Squish

Radial inward movement of air while the piston reaches the TDC is called Squish. "Squish" can be defined as radically inward flow of air towards the combustion recess by squeezing it out from between the piston and cylinder head during the end of compression stroke. This movement of air is due to the fact that most of the modern chambers have clearance volume in the central portional only. While the piston goes moves the TDC the air from the outeredges where there remains no space rushes towards the central portional giving rise to inwards radial motion called squish.

26.18.3 TUMBLE

As the piston reaches close to the TDC the squish generates secondary motion about circumferential axis near the outer edges. This motion is called 'tumble'.

Swirl, squish and tumble have been shown in Fig 26.11.

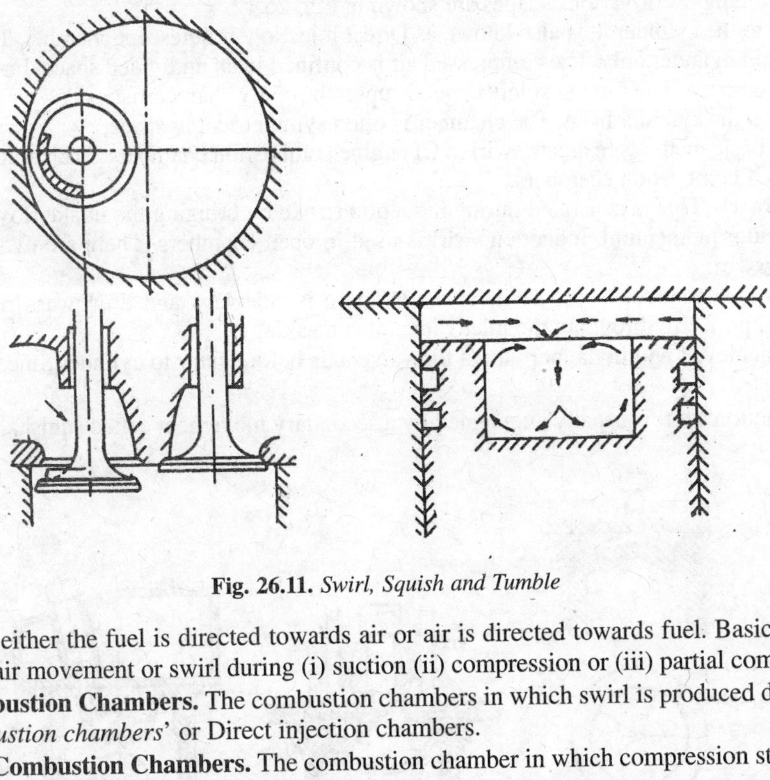

Fig. 26.11. *Swirl, Squish and Tumble*

To achieve this either the fuel is directed towards air or air is directed towards fuel. Basically there are three methods to create air movement or swirl during (i) suction (ii) compression or (iii) partial combustion.

(a) Open Combustion Chambers. The combustion chambers in which swirl is produced during induction are called '*open combustion chambers*' or Direct injection chambers.

(b) Turbulent Combustion Chambers. The combustion chamber in which compression stroke is used to produce swirl are called '*Turbulent or swirl combustion chambers*'.

(c) Pre-combustion Chambers. The combustion chambers in which partial combustion is used to create swirl are '*Turbulent or air-cell combustion chambers*'.

26.19 COMBUSTION CHAMBERS FOR CI ENGINES

In CI engine only air is admitted in the cylinder. The field is injected in the hot compressed air at abut 15° before TDC in the compression stroke. The mixing of air and fuel as well as combustion of the mixture is to be done in the cylinder with a short time of 15–25° of crank rotation after TDC in the working stroke for best efficiency. Good mixing requires entire mass of air in the combustion chamber must sweep past in the fuel jet in an ordinary fashion. Turbulence is not enough. Orderly movent is called **'Swirl'**. The main objective of CI combustion chamber is to create the swirl.

Basic functions of CI Engine combustion chamber are

(i) To assist and support the fuel injection system in preparation of cyl-charge.

(ii) To obtain efficient combustion for high thermal efficiency.

(iii) To assist in achieving smooth and noiseless combustion.

In general, combustion chambers are classified as follows:

Combustion Chamber
- Non-tubulent open combustion chamber
- Turbulent type combustion chamber
  - Precombustion chamber
  - Turbulent chamber
  - Lanova chamber

26.9.1 Open Combustion Chamber

Open combustion chambers of various shapes are shown in Fig. 26.12.

These are open to the cylinder. It is also known as Direct injection or quiescent chamber. This chamber is often an extension of main cylinder only. The compressed air is confined in an undivided space. Fuel is injected directly into the combustion space. The mixing solely depends upon the spray characteristics and air motion present. Fuel injector is placed on the cylinder head. The chamber is often symmetrical in space.

There are three basic methods to create swirl in CI engine combustion chambers. Each method is used in one of the three types of CI combustion chambers.

1. Induction Swirl. This is generated during induction stroke by bringing the intake flow of air into cylinder with an initial angular momentum. Induction swirl is used in open chambers. There are two general methods of creating induction swirl:

(i) The air flow is tangential to the cylinder wall. There is sideways and downward motion due to swirl: (a) Deflected port, (b) Shrouding the inlet valve (also masking),

(ii) The swirl is achieved within the port about the valve axis before entry to cylinder Since whole periphery of the valve is used.

(iii) *Squish.* Induction swirl is usually augmented by a secondary movement called squish.

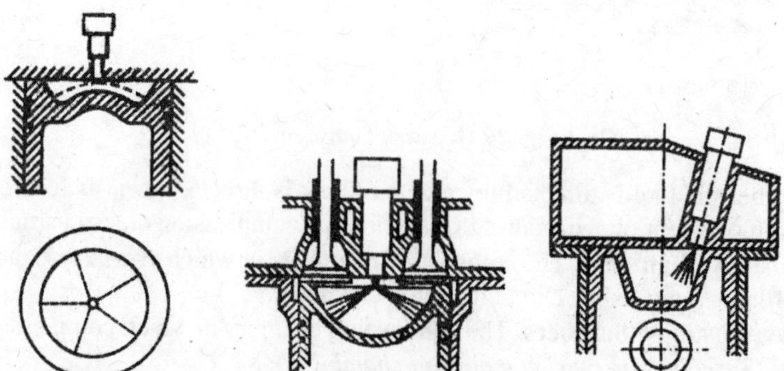

(a) Shallow depth chamber (b) Hemispherical chamber (c) Modified cylindrical chamber

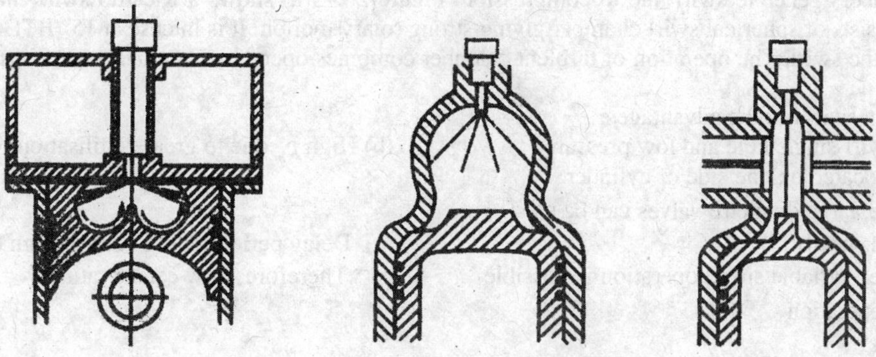

(d) Toroidal chamber (e) NACA design displacer on piston

Fig. 26.12. *Open combustion chambers*

2. Compression Swirl. This is generated in a separate chamber adjacent to main CC. It is called a divided chamber system and chamber is called swirl chamber. Compression creates a very strong swirl.

3. Combustion Induced Swirl. This also takes place in a separate chamber adjacent to the CC and connected by a small throat. It is called pre-combustion chamber. Fuel is injected in this chamber. Here part of combustion takes place. Pressure created forces the hot mixture along with burning fuel with high velocity. This creates a very strong swirl. This type of swirl is used in precombustion chamber and air-cell type turbulent CC.

Open combustion chamber using induction swirl have following advantages and disdavantages.

Advantages

(a) High thermal efficiency—low turbulence and low heat transfer.
(b) Cold starting easier.
(c) No work for producing swirl. Therefore, High Thermal efficiency and low BSFC.
(d) Suitable for large, slow speed engines ($d > 2000$ mm) cheap low cetane rating field can be used. Ignition delay unimp.

Disadvantages

(a) Weak swirl necessitate multihole nozzles and high injection pressure. More maintenance.
(b) Small nozzle openings make regulation of fuel supply.
(c) Shrouded valves—Low volumetric efficiency.
(d) Weak swirl requires excess air. Bulk engine.
(e) Induction swirl independent of speed. Difficult to maintain efficiency.

26.20 TURBULENT COMBUSTION CHAMBER USING COMPRESSION SWIRL

A chamber that depends upon primary turbulence of air to break up the fuel spray and form a homogeneous mixture is called a **turbulent chamber.**

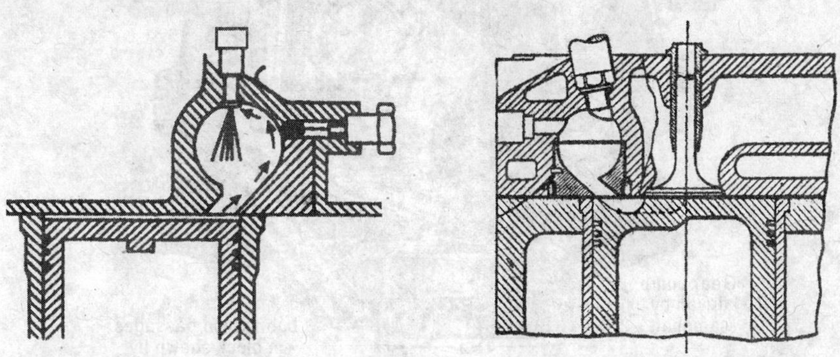

(a) Ricardo swirl chamber Comet, Mark II (b) Ricardo turbulent-chamber diesel

Fig. 26.13

These chambers generate swirl due to compression. Figure 26.13(a) shows a Ricardo swirl chamber Comet Mark-II. It consists of spherical swirl chamber giving strong rotary motion. It is intense at 15° BTDC. Higher the speed, greater the swirl. The operation of turblent chamber combines open chamber and precombustion chamber operation.

Compression swirl has the advantages:

(a) Strong swirl-single hole and low pressure

(b) High p_m due to greater utilisation of air

(c) Injector located on one side of cylinder
Therefore, large diametre valves can be used

(d) Swirl and speed
Therefore, Variable speed operation is possible

(e) Delay period reduces due to high temperature
Therefore, Low cetane rating

(f) Smooth operation

Disadvantages

(a) Compression work increases. Therefore, low η_m, η_{th} and high BSFC

(b) Complex construction of cylinder makes it costly

(c) Strong swirl increases heat losses starting difficult

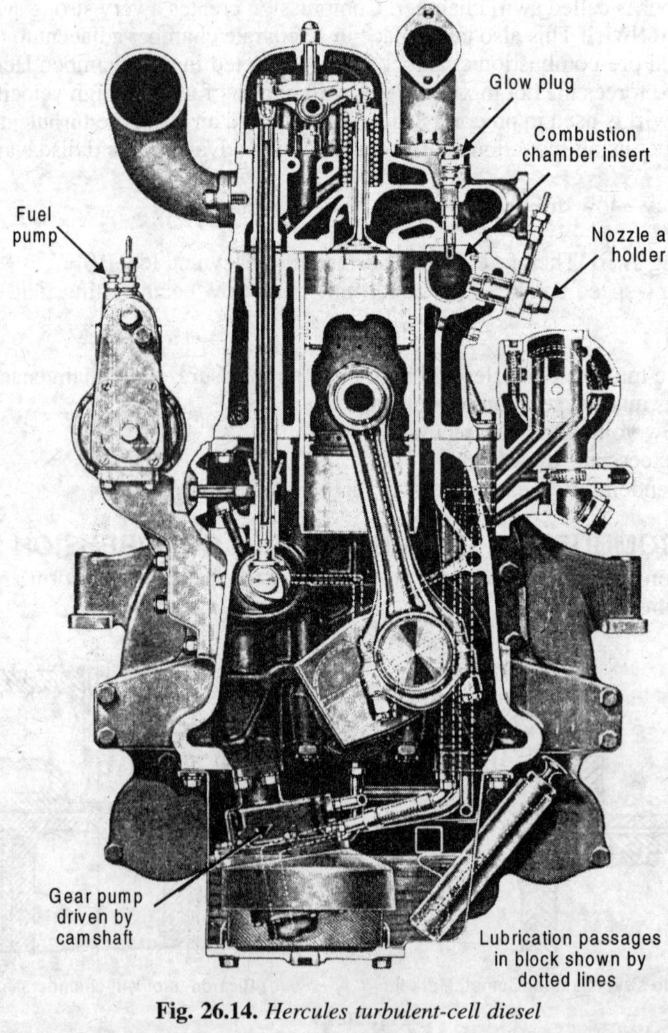

Fig. 26.14. *Hercules turbulent-cell diesel*

26.20.1 Divided (turbulent) Combustion Chamber using Combustion Swirl

These chambers are not widely used nowadays. They are

1. Precombustion chamber
2. Air cell combustion
3. MAN combustion chamber.

1. Pre-combustion chamber. The most popular design in the small, high speed engine range is the *precombustion chamber*. In this type of engine the clearance volume is divided into two parts by a restricting passageway

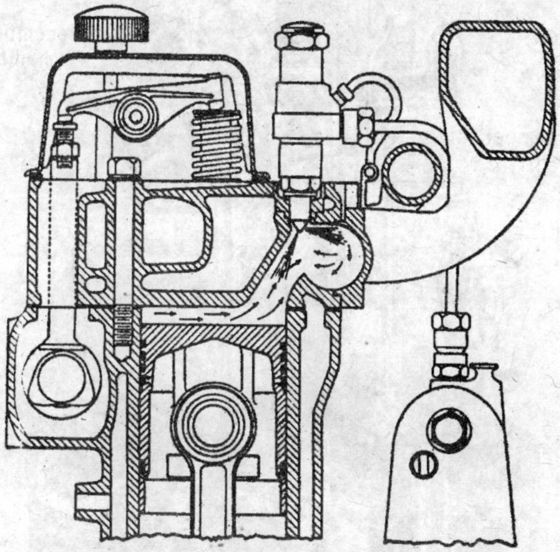

Fig. 26.15

as illustrated in Figs 26.14 and 26.15. The small ante chamber is called the pre-combustion chamber (25 to 40% of clearance volume).

Advantages

(1) Single orifice nozzle
(2) Takes initial stock of peak pressure
(3) Low cetane rating fuel can be used
(4) It has multifuel capability
(5) Tendency to knock is minimum—short or no delay period
(6) High injection pressure not required
(7) Suits for high speed engine

Disadvantages

(1) Cold starting difficult
(2) Heat loss is high

2. Air Cell Combustion Chamber. The air-cell diesel is essentially an open combustion chamber engine with a small ante-chamber (air cell) that is remote from the nozzle. The chamber requires high primary turbulence, if a single orifice nozzle is used or else requires a multiorifice nozzle.

An engine of the former type is illustrated in Fig. 26.14 and an American version of the latter type in Fig. 26.15. In operation of the engine, the sequence of events is similar to the open *chamber* until '*after combustion*' has been initiated. In Fig. 26.15 the air cell is called the cup wiper and contains about 5% of compressed volume.

Energy Cell

A combination of the pre-combustion chamber and air-cell principles in one design is illustrated in Fig. 26.16. *It is called the energy cell or lanova combustion chamber.*

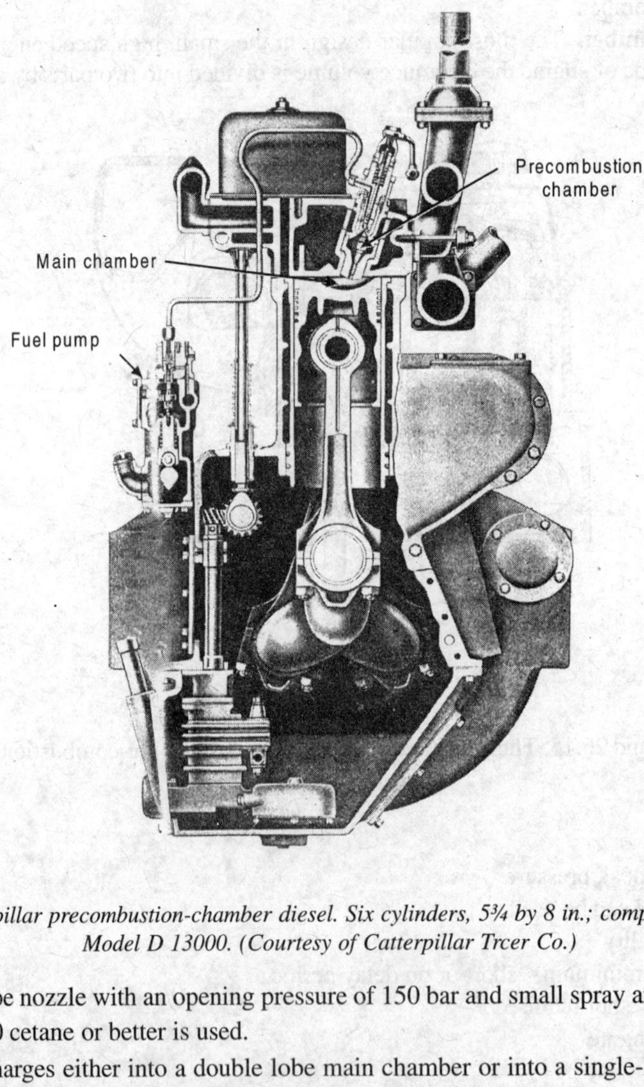

Fig. 26.16. *Caterpillar precombustion-chamber diesel. Six cylinders, 5¾ by 8 in.; compression ratio, 16:1; Model D 13000. (Courtesy of Catterpillar Trcer Co.)*

A throttling pintle type nozzle with an opening pressure of 150 bar and small spray angle is recommended, and a fuel of good quality 50 cetane or better is used.

The energy cell discharges either into a double lobe main chamber or into a single-lobe chamber. Either construction assists the formation of high velocity swirl in the main chamber.

Combustion probably begins in the main chamber, where the spray envelope resides, but spreads immediately to the fuel in the energy cell, and the initial high pressure is confined in that chamber. The blast from the minor cell is directed against the final portion of the fuel spraying from the nozzle, and therefore a mixture of fuel and burning gases is swept around the main chamber. As the piston descends, the pressure difference between cell and chamber will assist discharge from the energy cell. The discharge from the main major cell will pick up fuel in the minor cell and pass into the main chamber, to renew the turbulence and complete combustion.

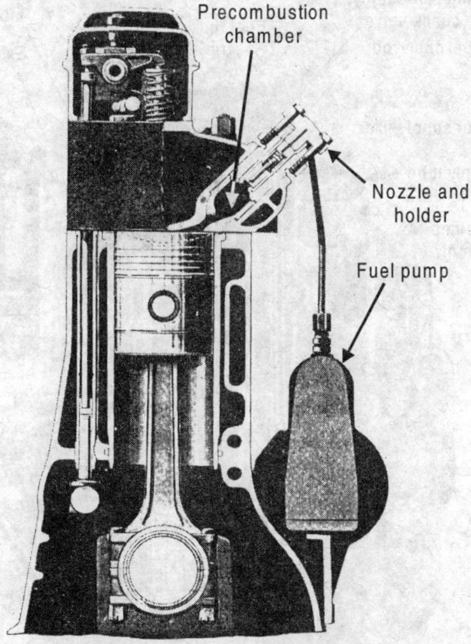

Fig. 26.17. *International Harvestor precombustion-chamber diesel. Compression ratio, 14:1*

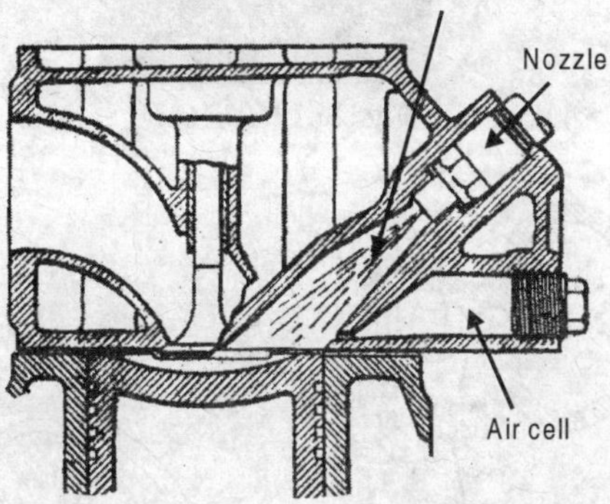

Fig. 26.18. *MAN air-cell diesel German army model, Six cylinder, 4.13 by 4.72 in*

3. M-combustion Chamber (MAN C.C.). This combustion chamber was developed by Dr. Meurer of MAN Germany. This is in the form of a spherical cavity in the piston. Contrary to the common belief that the fuel should not impinge on the combustion chamber walls, the fuel in this chamber is directed tangentially on to the combustion chamber walls. It is believed that a small but sufficient quality of fuel (particles) will burn before the fuel strikes the chamber walls so that the temperature and delay period is normal. The fuel that impinges must evaporate before it burns. This process being relatively slow there will be much of the fuel to take part in the second stage of combustion. This means the tendency to knock is greately reduced. This leads us to doubt that combustion during third

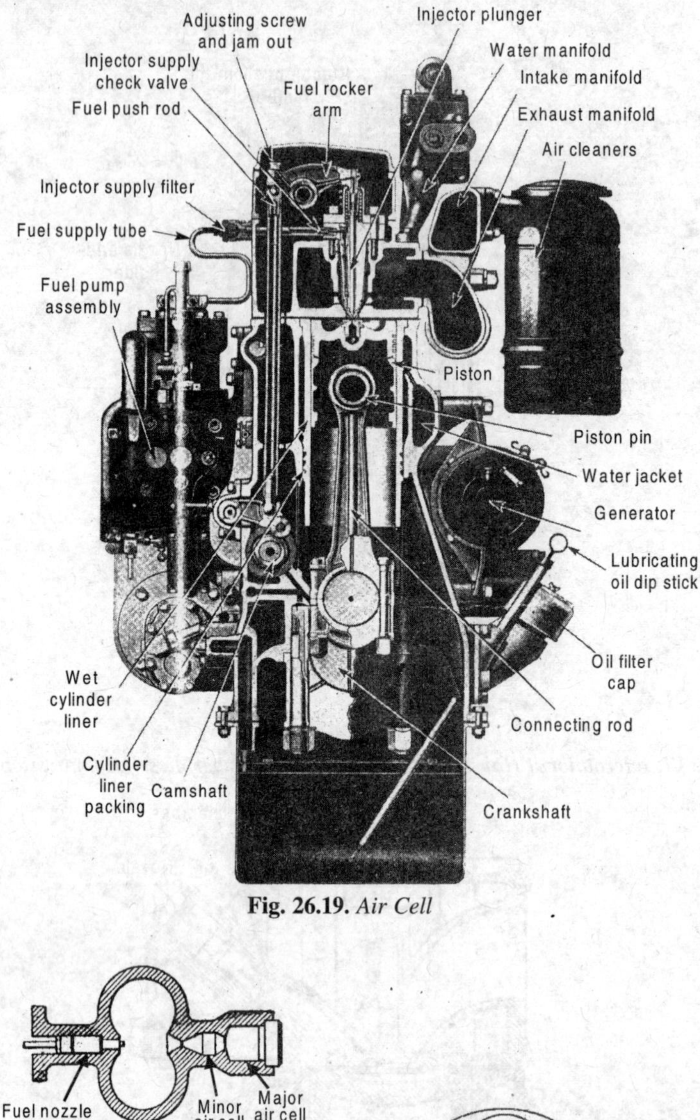

Fig. 26.19. *Air Cell*

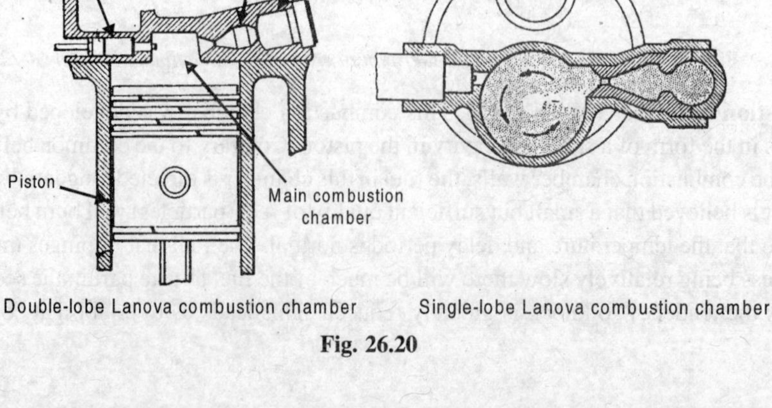

Fig. 26.20

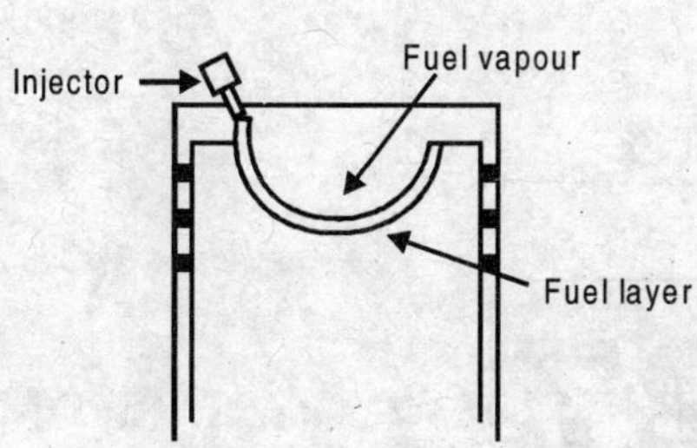

Fig. 26.21. *MAN combustion chamber.*

stage may occupy considerable time which lower the efficiency due to advancement of piston in the expansion stroke. But the temperatures are sufficiently high to cause rapid evaporation and combustion of fuel and the time losses will not increase significantly. This combustion chamber is not sensitive to quality of fuel and better performance can be expected from smaller engines.

CHAPTER 27

Emission and Pollution Control

27.1 INTRODUCTION

The atmosphere which is the largest fraction of biosphere continuously absorbs a wide range of solids, liquids and gases from both man-made and natural sources. These substances travel through air, disperse and react with each other both physically and chemically. The portion of these substances which interacts with the environment to cause toxicity, disease, aesthetic distress, physiological effects or environmental decay, has been labelled by man as a "**pollutant**".

Air pollution is basically the presence of foreign substances in air. Thus it is defined as the presence of one or more contaminants such as dust, fumes, gas, mist, odour, smoke or vapour which is injurious to human, plant or animal life or property. And the day man invented fire air pollution started.

27.2 CLASSIFICATION OF POLLUTANTS

Air pollutants can be classified as follows:

(1) Natural contaminants
  e.g., pollen grains, natural fog, bacteria etc.

(2) Aerosols
  e.g., dust, smoke, fog and fumes, mists, carbon particles etc.

(3) Gases and vapours
  (i) Oxides of Nitrogen (NO_x)
  (ii) Oxides of sulphur (SO_x)
  (iii) Oxygen compounds O_3, CO, CO_2
  (iv) Vapours from various chemical reactions e.g., paraffins, olefins, acetylenes, chlorinated hydrocarbons etc.
  (v) Radioactive compounds.

In this chapter we will study in detail the air-pollution, causes and remedies, as regards SI and CI engines. The main pollutants contributed by IC engines are CO, unburnt hydrocarbons (UBHC), oxides of nitrogen (NO_x), lead and other particulate emissions. Apart from IC engines, other sources of air pollution are electric power generating stations, industrial and domestic fuel consumption, refuse burning (incinerators), industrial processes etc. also contribute heavily to contamination of environment.

27.3 POLLUTANTS FROM SI ENGINES

The SI engine exhaust gases contain:

(i) Oxides of nitrogen e.g., NO, NO_2 etc. collectively called as NO_x

(ii) Carbon Monoxide (CO)

(iii) Organic compounds—unburned or partially burned hydrocarbons (UBHC).

The relative amounts depend on the engine design and operating conditions. Normally their presence is as follows:

(i) NO_x – 500 to 1000 ppm or 20 gm/kg of fuel.

(ii) CO – 1% to 2% or 200 gm/kg of fuel.

(iii) HC – 3000 ppm or 25 gm/kg of fuel.

(1 ml/m^3 = 1 ppm)

In gasolene engines other sources of unburned hydrocarbons (UBHC) are:

(a) Piston blow by gases. (Exhaust gases and vapours of fuel which leaks into the crank case are called blow by gases).

(b) Fuel evaporation from carburettor and fuel tank. These are called as non-exhaust sources.

27.3.1 Sources of Pollution from Gasolene Engine

The four possible sources of atmospheric pollution from a petrol engine are as follows.

(i) The fuel tank
(ii) The carburettor
] Together they are called as evaporative losses and their contribution is 15 to 20% of HC.

(iii) The crank case blow by—20 to 35% of HC.

(iv) Exhaust pipe—50 to 60% of HC and almost all CO and NO_x

27.3.2 Evaporative Losses

These losses are the direct losses of raw gasolene from the engine fuel system. These emissions amount to 15 to 20% of total HC emission and two main sources are (i) fuel tank, (ii) the carburettor.

(i) Fuel Tank Losses. While the petrol is being filled into the tank these losses occur. The vapours in the fuel tank are released to the atmosphere through breather vent. The tank temperature increases if the vehicle is parked in bright sun light or due to vehicle operation.

Parameters affecting the fuel tank losses are : (a) volatility of fuel (b) location, capacity and design of fuel tank (c) amount of fuel in the tank (d) ambient temperature (e) mode of vehicle operation etc.

(ii) Carburettor Losses. These losses are because of two things (a) float chamber being vented to atmosphere in order to relieve the internal pressure as the carburettor gets heated up and (b) "hot soak" losses which occur when the engine is stopped after the vehicle is run for some time. The loss from the carburettor takes place due to boiling of fuel in it. During hot soak the carburettor temperature rises from 16ºC to 48° C above the ambient.

(iii) Crank Case Blow By. The leakage of flue gases past the piston and piston rings from the cylinder to the crank case, is called as blow by. The blow by loss is about 18 to 20% of total HC loss from the engine. This loss may go upto 32% if the piston rings get worn out.

During compression and combustion, the increased cylinder pressure forces some of the gas, in the cylinder, in to crevices viz. volume between piston, rings and cylinder wall. Most of this gas is unburned fuel air mixture and much of the gas escapes into the atmosphere as unburned HC, because these crevices are too narrow for the flame to enter.

The top land clearance and position of the top ring greatly affects the blow by losses, because some of the quenched gas is re-cycled in the combustion chamber. The nearness of this un-burned charge to spark plug, flame speed, local temperature etc. decide the ability to burn, otherwise it goes as UBHC.

(iv) Exhaust Pipe. Mixture of various hydrocarbons is present in petrol fuel. Thus if the combustion is perfect, then exhaust emissions will have CO_2, water vapours and the air that did not take part in combustion process. However because of numerous reasons, the combustion is not complete and hence CO and UBHCs are also present in exhaust gases. The smog (the mixture of smoke and gas) is formed mainly because of presence of HCs in air.

Apart from CO and HCs, exhaust also contains the third main pollutant NO_x. The air in the combustion chamber used for combustion contains 78% of N_2. At low temperature N_2 is not so active but at about 1100 to 1150° C, the reactivity of N_2 with O_2 is very high and various oxides of nitrogen, collectively called as NO_x are formed.

In addition to the pollutants mentioned above, the exhaust also contains organic compounds namely ketones, aldehydes which form smog. To increase the antiknock quality of petrol, it is added with TEL (Tetra- Ethyl-lead). Thus because of TEL the exhaust emission also contains poisonous lead compounds.

Organic and inorganic compounds of higher molecular weights and lead compounds are exhausted in the form of very small size particles of the order of 0.02 to 0.06 μ.

Table 27.1 shows effect of engine operating conditions on exhaust (car):

Table 27.1

Description	*Engine Operation*			
	Idle	*Acceleration*	*Cruising*	*Deceleration*
A/F ratio	11–12.5	11–13	13–15	11–12.5
Engine speed (rpm)	400–500	3000–4000	2500–3500	3500–500
Exhaust emissions				
(i) CO%	4–5.5	0–5.8	0.9–3.8	2.1–4.8
(ii) NO_x ppm.	10–55	1000–3800	980–2800	10–53
(iii) HC ppm	450–1000	55–550	450–280	3850–1900
(iv) Unburned fuel as % of fuel supplied	3.5 to 5.5	1.5–4.5	1.5–4.5	19–59

27.4 DIESEL EMISSION

In diesel engine, emissions can be classified in the same categories as for the gasoline engines. In diesel exhaust concentrations of NO_x are comparable to those from SI engines. Diesel hydrocarbon emissions are significant though exhaust concentrations are lower by about a factor of 5 than typical SI engine levels. Smoke, odour and particulate emission are the other important factors of concern, which cause air pollution. Hydrocarbons in diesel exhaust may also condense to form white smoke during engine starting and warm-up. The particulate emissions are of the order of 0.2 to 0.5% of the fuel mass and diameter of the particles is as small as 0.1 μm. However diesel engines are not a significant source of CO.

Gasolene and diesel contain sulphur. In gasolene it is in small amounts ≤ 600 ppm by weight and in diesel fuel ≤ 0.5 P.C.

Sulphur is oxidised to produce SO_2 of which a fraction can be oxidised to SO_3 which combines with water to form a sulphuric acid-aerosol.

Table 27.2 shows approximately the possible variations in concentration of different constituents of diesel exhaust.

Table 27.2. *Range of Concentrations of Different Constituents of Diesel Exhaust*

Sr. No.	*Constituent*	*Range*
1.	HC	a few – 1000 ppm
2.	NO_x	100 – 2000 ppm
3.	CO	0 – 2 P.C.

(i) Engine type and mode of operation are two main factors which influence the diesel exhaust emissions. Emission levels of different engines at full load and rated speed are given in Table 27.3.

Table 27.3

Sr. No.	Emission	Engine			
		Two stroke air scavenged	Four stroke turbo charged	Four stroke normally aspirated	
				Medium speed	High speed
1.	CO%	0.34	0.03	0.26	0.14
2.	CO_2%	5.31	7.65	7.18	7.78
3.	UBHC (ppm)	540	250	370	1000
4.	NO_x (ppm)	1240	2550	800	790
5.	Smoke (Hartridge units)	10	8	60	90
6.	Odour (D.I. units)	3.5	3.4	3.3	3.5

Thus with reference to the above table the conclusions can be drawn as under:

1. The two stroke air scavenged engine produces high UBHC and intermediate NO_x emissions. The smoke level remains low.
2. The four stroke medium speed normally aspirated engine has lowest emissions in all categories except for a very high smoke intensity.
3. The four stroke high speed normally aspirated engine has high HC. The odour intensity is also high.
4. The turbocharged four stroke engine is notably low in UBHC and high in NO_x. Generally very low smoke levels are recorded.

(ii) Effect of mode of operation on engine exhaust. The various modes of operation like idle, full load at rated speed and acceleration at full rack and its effect on emission levels in diesel exhaust are shown in Table 27.4.

During idle mode concentration of HC, NO_x emissions are lower than other modes. Emissions at idle are less significant than during any other mode. Smoke and odour are highest during acceleration. Emissions at full load relative to emissions at other modes vary significantly with engine type. Two stroke and 4 stroke turbocharged engines smoke lightly at load, while 4 stroke normally aspirated engines smoke very much at rated full load.

Table 27.4

Sr. No.	Engine Exhaust Constituent	Con. Values of Exhaust Emission			
		Idle	Acceleration	Full load	Part load
	2-stroke cycle				
1.	HC (ppm)	250	500	550	350
2.	NO_x (ppm)	180	1200	1250	1. 00
3.	Smoke (Hartridge unit)	4	44	10	4
4.	Odour (D.I)	3.6	4.1	3.5	3.0
5.	CO %	0.01	0.24	0.34	0.01
6.	CO_2 %	0.83	5.42	5.29	3.79
	4-stroke cycle normally aspirated				
1.	HC (ppm)	180	330	150	210
2.	NO_x (ppm)	330	920	780	590
3.	Smoke	4	44	10	4
4.	Odour	3.6	4.1	3.5	3.4
5.	CO %	0.02	0.08	0.26	0.04
6.	CO_2 %	2.56	3.40	6.68	5.33

(iii) NO_x in diesel exhaust. The quantity of NO_x varies from few hundred to well over 1000 ppm. The highest local peak temperature and presence of sufficient O_2, causes highest NO_x concentration in diesel exhaust.

A precombustion engine gives rise to less NO_x level than a direct injection (DI) engine. For high fuel-air ratio the additional fuel tends to cool the charge, so the localised peak temperatures are lowered resulting in drop in NO_x level.

The NO_x concentration is also significantly affected by injection system and time. Also the fuel characteristics such as Cetane Number (CN), viscosity, rate of burning etc. all contribute to differences in NO_x levels obtained from different levels.

Diesel Smoke

The visible products of combustion is called as smoke. It originates early in the combustion cycle in a localised volume of rich fuel-air (F/A) mixture. If the fuel is burnt in a volume where F/A ratio is greater than 1.5, soot is produced. The amount of soot formed depends upon type of fuel, pressure and local F/A ratio. The soot formed if it finds sufficient O_2, it will burn completely and will become visible. The size of soot particles affects the appearance of smoke. The diesel engine smoke is basically of two types.

(a) Blue white smoke and (b) Black smoke.

(a) Blue white smoke. While engine is started from cold, this smoke is caused by liquid droplets of lubricating oil or fuel oil, due to lower surrounding temperature and the intermediate products of combustion do not burn. If the piston rings are worn out and when the lubricating oil leaks past the rings, it gives rise to this type of smoke.

(b) Black smoke. It is nothing but carbon particles suspended in the exhaust gas. It largely depends upon air-fuel ratio and increases rapidly as load is increased and available air is depleted.

27.5 DIESEL ODOUR

Till date no equipment is developed to measure the intensity of odour. But it is a general practice to have a human panel, who will decide the intensity of odour by comparison—which is an inaccurate method.

Note that aldehydes give a pungent odour for diesel engines, which also causes irritation to nose and eyes.

27.6 CONTROL OF EXHAUST EMISSIONS

The main approaches to minimise exhaust emissions are:

1. Modifications in the engine design and operating variables.
2. Treatment of exhaust gases.

1. Modifications in the Engine Design. The following modifications may help in cleaner exhaust—

(a) Use of lean mixtures and maximum spark retard compatible with good power output and drivability. It is a well-known fact that burning leaner mixtures greatly reduces HC and CO emissions refer Fig. 27.1. This reduction will result only if there is good mixture formation and distribution.

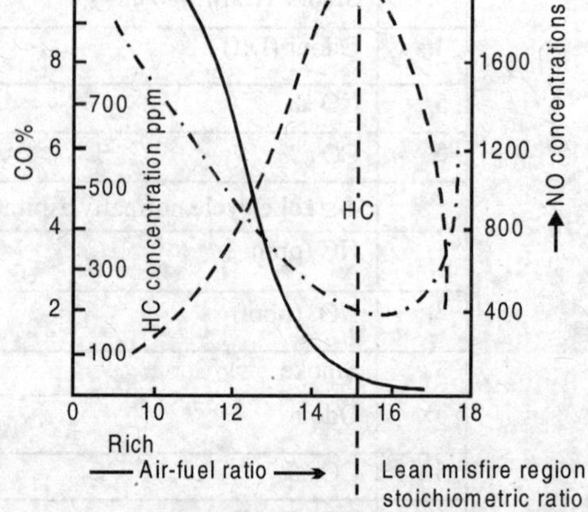

Fig. 27.1. *Effect of A/F ratio*

One of the methods of producing a uniform mixture and refined fuel metering is the adoption of triple venturi carburettor.

(b) Use of minimum valve over-lap.

(c) Pretreatment of the mixture to improve vaporisation and mixing of fuel with air. To achieve this narrow venturies are used, to produce higher air speeds and better fuel automisation. The exhaust heat can be used to pre-heat the mixture at part loads. Also to minimise the "Crevice volume" between topland and bore and to block the leakage path through the gas into the crank case, a system known as "Sealed ring orifice system" is adopted.

2. Exhaust Treatment Devices. Here the basic technique is to promote oxidation of HC and CO after emission from the engine. Exhaust oxidation devices fall into two categories:

(i) Promotion of after burning of the pollutants by exhaust heat conservation, introduction of supplimentary, air and providing sufficient volume to ensure adequate reaction time.

(ii) Use of catalytic converters.

Catalytic convertors depend on the action of a catalyst containing certain exotic chemicals to convert HC and CO emissions to their oxidised products.

CONTROL OF HC – EMISSIONS

(i) Blow by Control. Here the blow by vapours are re-circulated back to the intake manifold via air cleaner or inlet of the carburettor. This system is called as positive crank case ventilation (PCV). Thus the blow by gases are consequently re-introduced into the combustion chamber where they are burned along with fresh in coming air and fuel.

(ii) Evaporation Loss Control Device (ELCD). In this case the device collects the evaporative emissions, they are absorbed in an absorbent chamber and are re-circulated. The activated charcoal or foamed polyurethene is used as an adsorbent to hold the HC vapours before they escape to the atmosphere. The two main sources of HC emissions viz. fuel tank and the carburettor bowl, are directly connected to the adsorbent chamber when the engine is turned off. i.e., under hot soak. Apart from it the diurnal cycle loss from the tank and carburettor is also taken care of. Diurnal cycle is the daily cyclic variation in the temperature which causes tank breathing or forcing the gasolene out of the tank.

Adsorbent bed when saturated, is relieved of the vapours, by purging action i.e., by purging the air from air cleaner to the intake manifold via the adsorbent bed.

OTHER METHODS—TO CONTROL EXHAUST EMISSIONS

(a) Petrol Injection. This system provides the flexibility to meet the engine requirements and eliminates the problem of mixture distribution completely and hence the engine will have low levels and exhaust emissions and better fuel economy. Thus HC emissions are controlled by making use of optimum lean A/F ratios at all operating conditions. NO_x level is also controlled as maximum temperature reached in the combustion chamber is low.

(b) Stratified Charge Engine. It operates on very lean air-gasolene mixtures, depending on a localised rich mixture region near the spark plug to initiate combustion. The emission levels of all three pollutants CO, HC and NO_x are very much lower with this engine at part throttle due to the leanness of the mixture. Mixture as lean as 30 : 1 can be successfully used.

(c) Proper Maintenance of an Engine. The pollution from a poorly maintained engine is more. Because of mis-fire the entire A/F charge gets exhausted without combustion. If the air cleaner is not periodically cleaned, then its element gets choked and can reduce A/F ratio, generally giving rise to increased emissions of HC and CO. Similar problem is encountered if the choke gets automatically sticked.

(d) Fuel Variation. Methane and propane could burn at leaner ratios than gasoline at a given throttle. Also at leaner mixtures, lower throttle position can be used. Thus methane and propane will produce much lower CO as their A/F mix is lean i.e., it contains more O_2.

Hence if we change from the gasoline to propane as an engine fuel, CO emissions can be substantially reduced with reduced HC and NO_x and if we change propane by methane the CO as well as HC touch zero emission level and only NO_x remains as a significant factor. From point of view of emissions, these fuels are attractive but because of technological progress, we are unable to use these fuels.

The various methods like Catalytic converters, Exhaust gas Recirculation, Passive regeneration, Trap systems are used to control pollution. And Catalytic Converters, EGR systems are discussed in detail.

27.7 CATALYTIC CONVERTER

In order to control the air pollution from IC engines (or automobiles with diesel or gasolene fuel), the necessity arises in after treatment of exhaust gases. Thus it is required to reduce or oxidise HC, CO, NO_x to H_2O, CO_2 and N_2. The Figure 27.2 shows the anatomy of a catalytic converter, it starts with a ceramic or a metallic monolithic substrate. This substrate is coated with a wash coat which serves as a primer for the purpose of holding noble metal (Pt, Rh or Pd) and that noble metal molecules dispersed to allow maximum surface exposure to the exhaust gases.

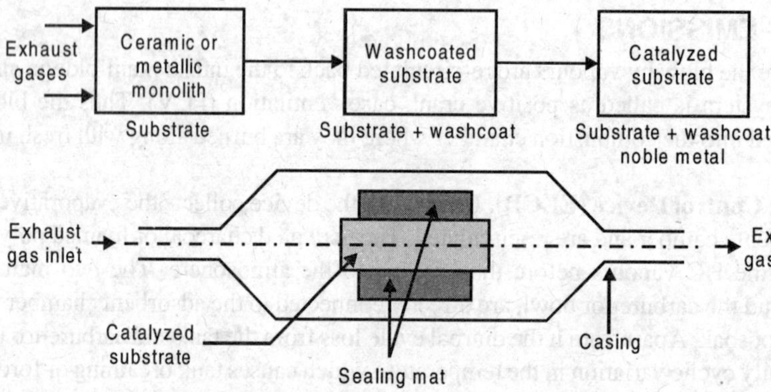

Fig. 27.2. *Diesel catalyst schematic*

After the washcoat substrate is catalysed, it is canned with a heat-expanding sealing mat between the catalysed substrate and its can. The substrate material is selected to provide adequate mechanical strength, surface area and low flow resistance. Careful consideration has to be given to the type and amount of wash coat used. Alumina (Al-oxide) is very popular because of its surface area and other favourable physical properties. However silica (silicon oxide) oxides are also used as wash coat. The container design, lub-oil additives, fuel sulphur level, application and particulate composition are other important design parameters. The performance of three catalyst viz. Pt, Pd and Rh to convert to SO_3 relative to exhaust temperature is also shown in Fig. 27.3 given on next page.

27.8 EXHAUST GAS RECIRCULATION FOR DIESELS (EGR SYSTEM)

The most accepted technology for NO_x reduction is exhaust gas recirculation. It is commonly being used in light duty applications but is not accepted in heavy duty diesel engines because of its negative impact on engine wear. EGR is defined by following formula.

$$\text{EGR\%} = \frac{\text{Recirculated exhaust}}{\text{Total air charge}} \times 100$$

$$= \frac{\text{Recirculated exhaust}}{\text{Fresh air + Recirculated exhaust}} \times 100$$

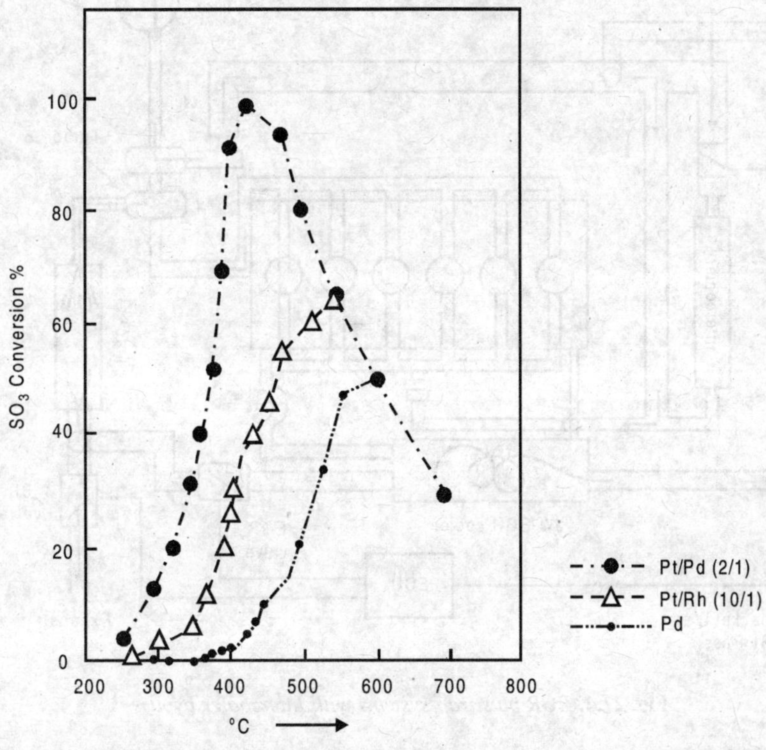

Fig. 27.3. *Performance of three catalyst conversion to SO_3 relative to exhaust temperature*

EGR is schematically as shown in Fig. 27.4. In this method the peak cylinder combustion temperature is reduced by circulating a portion of exhaust gas thereby reducing the quantity of O_2 required for combustion. Nevertheless the re-circulated exhaust gas serves as heat sink. The exhaust gas from the exhaust manifold goes through an EGR control valve, then a cooler and finally into intake manifold by way of air-to-air intercooler or a bypass system.

With 10% EGR, there is substantial reduction in NOx level. The EGR inter cooler is used in order to reduce the temperature of exhaust gases so that the temperature in the combustion chamber does not reach its peak value and there by reactivity of N_2 with O_2 is reduced.

However the disadvantage of the system is that there is increase in smoke as well as particulates.

The following Fig. 27.5 (a) and (b) shows the effect of EGR on NO_x and smoke level v/s load.

27.9 EFFECT OF AIR POLLUTION ON HUMAN AND PLANT LIFE

The greatest environmental evils is polluted air. The air we breathe has not only life saving but also life damaging properties. Under ideal conditions the air we inhale has a quantitative and qualitative balance that maintains the well being of a human life.

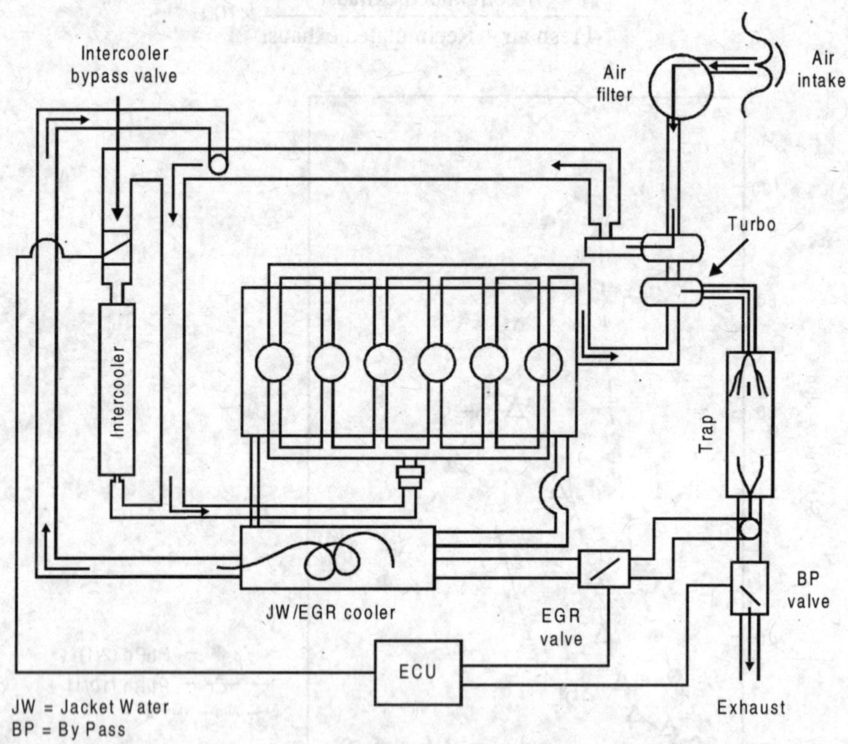

Fig. 27.4. *EGR plus trap systems with Intercooler bypass*

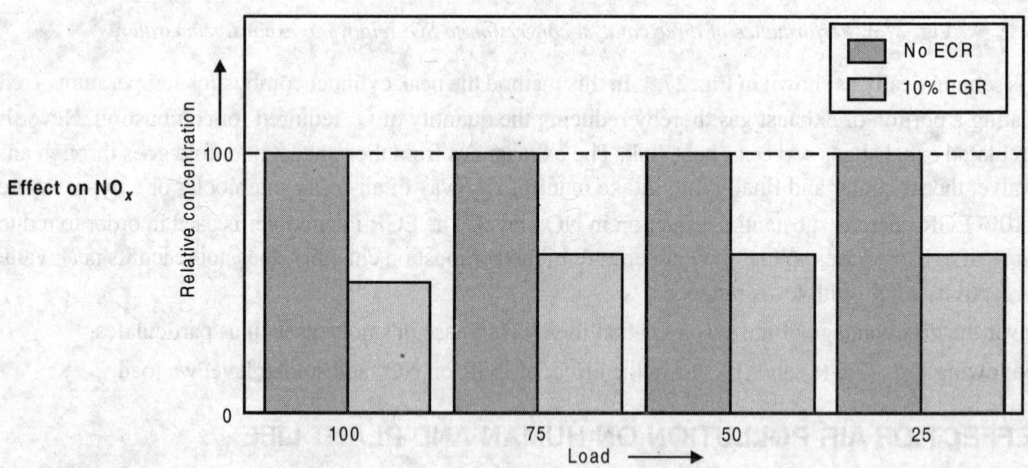

Fig. 27.5 (a)

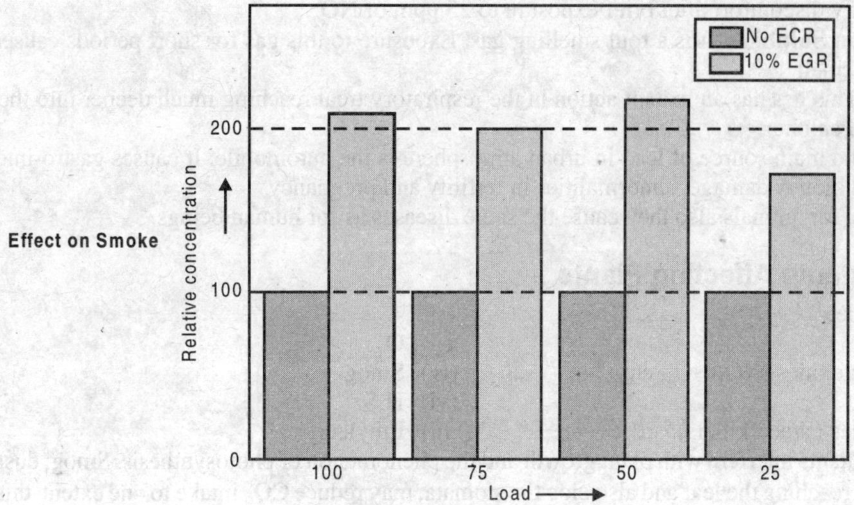

Fig. 27.5 (b)

An average man takes in 16 kg of air each day. The water and food far exceeds than air intake. It has been estimated that a man can live without food for 5 weeks, and five days without water but only for 5 minutes without air. All the constituents in air are not harmful. Depending upon the chemical nature of the pollutants some may be harmful when present in air in small concentrations and others only if they are present in high concentrations. The duration of exposure of the body to polluted air is also important factor. The state of health of the receptor along with his age group are prime factors.

Susceptibility to the effects of air pollution is great among children and elderly persons. Also those with chronic diseases of the lungs or hearts are at great risk.

27.9.1 The Various Health Effects

The various health effects are listed below:

(1) Eye, nose, throat irritation

(2) H_2S, NH_3 cause odour nuisance

(3) Increase of asthmatic troubles

(4) CO combines with haemoglobin and creates problems to those suffering from cardiovascular and pulmonary diseases

(5) Particulate matter, dust etc. create respiratory problems

(6) Lead and its some of the compounds enter lungs and is poisonous.

27.9.2 Effect of Specific Pollutants

(i) **Sulphur di-oxide.** It is an irritant gas which affects the mucous membranes when inhaled. It causes severe bronchospasms at relatively low levels of concentration.

(ii) **Carbon Monoxide.** This gas has strong affinity to combine with haemoglobin of the blood and forms carboxyhaemoglobin (COHb). This reduces the O_2 carrying capacity of haemoglobin to the body tissues. CO affects the central nervous system, may cause heart attacks.

(iii) **Oxides of Nitrogen.** Out of seven oxides of nitrogen known to exist in the ambient air, only two namely NO and NO_2 are thought to affect human health.

Nitrogen dioxide causes occupational disease e.g., people working in a manufacturing unit of nitric acid, exposures of farmers to high nitrate fertilisation, electric arc welding and in mining where nitrogen compounds are used as explosives. The eye and nasal irritation is observed after exposure to about 15 ppm of NO_2 and pulmonary discomfort after brief exposure to 25 ppm of NO_2.

(iv) Hydrogen Sulphide. It is a foul smelling gas. Exposure to this gas for short periods causes fatigue of the sense of smell.

(v) Ozone. This gas has an irritant action in the respiratory treat reaching much deeper into the lungs than the oxides of sulphur.

(vi) Lead. The main source of lead in urban atmosphere is the automobile. It causes gastro-intestinal damage, liver and kidney damage, abnormalities in fertility and pregnancy.

More or less, for animals also they cause the same diseases as for human beings.

27.9.3 Pollutants Affecting Plants

They are as under:

(i) SO_2
(ii) O_3
(iii) Nitrogen oxides (NO, NO_2, etc.)
(iv) Smog
(v) NH_3
(vi) H_2S
(vii) Herbicides (Weed killers)
(viii) Ethylene

The above pollutants interfere with plant growth and the phenomenon of photosynthesis. Smog, dust etc. reduce the amount of light reaching the leaf and also clog the stomata, may reduce CO_2 intake to one extent thus interfere with photosynthesis.

Effects of pollutants on the plants are:

(i) SO_2, causes Interveinal chlorotic bleaching of the leave.
(ii) Ozone suppresses the growth of leaves.
(iii) NO_x suppresses growth and bleaching of leaves etc.

27.10 POLLUTION MEASURING EQUIPMENTS AND CONTEMPORARY POLLUTION NORMS

I. For Petrol Engines (2-stroke and 4-stroke engines) – Idling CO-meter

Keeping the engine running at idle conditions, when the probe is inserted into the exhaust gases i.e., in the silencer, it analyses and displays the constituents i.e., it shows the % of CO and HC ppm directly. Note that these are harmful emissions and hence these are to be kept under controlled limits.

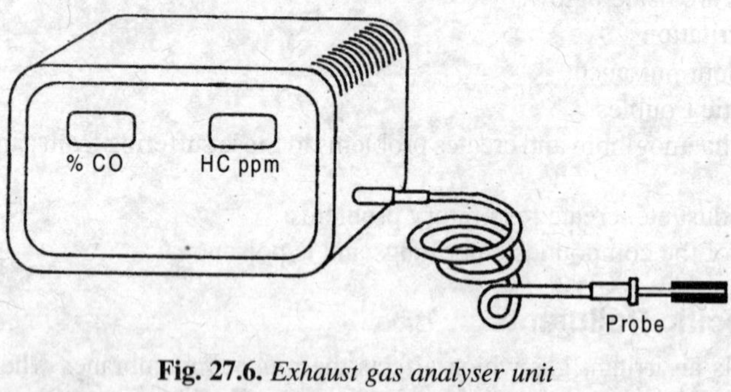

Fig. 27.6. *Exhaust gas analyser unit*

Pollution Under Control (PUC) contemporary norms are as follows:

"For petrol vehicles vide rule 115 of Central Motor Vehicle (CMV) Rules 1989 emission norms are as follows:

(i) Idling CO emission limit for all four wheeler petrol driven vehicle shall not exceed 3% by volume.
(ii) For 2 and 3 wheeled petrol driven vehicle shall not exceed 4.5% by volume.

II. For Diesel Engines—Hartridge Smoke-meter (Exhaust Gas Analyser)

Figure 27.7 shows the exhaust gas analyser, when the probe is inserted into the exhaust gases, it analyses and displays smoke density directly.

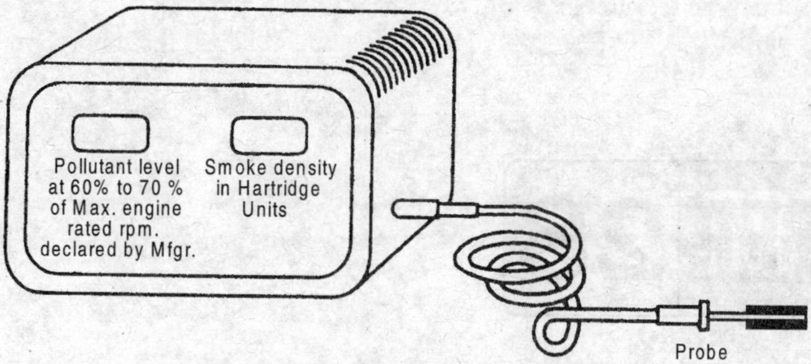

Fig. 27.7. *Exhaust gas analyser unit*

Pollution Under Control contemporary norms are as follows:

"For diesel vehicles vide rule 115 of CMV Rules 1989 is as follows:

Smoke density on full load at 60% to 70% of Maximum engine rated RPM declared by manufacturer : 75 Hartridge Units OR Free acceleration: 65 Hartridge Units."

27.11 PROPOSED EMISSION NORMS FOR THE YEAR 2000, EURO— I, II, III AND IV

I. Petrol Driven Passenger Car

	India 2000/Euro I	*Euro II*	*Euro III*	*Euro IV*
CO (g/km)	2.72	2.2	1.3	1.0
HC (g/km)	–	–	0.2	0.1
NO_x (g/km)	–		0.15	0.08
HC + NO_x	0.97	0.57		

II. Diesel Passenger Cars

	India 2000	*Euro I*	*Euro II*	*Euro III*
CO (g/km)	2.72 to 6.9	2.2	1.3	1.0
HC (g/km) + NO_x (g/km)	0.97 to 1.7	0.97	0.7	0.56
PM (g/km) (particulate matter)	0.14 to 0.25	0.14 to 0.19	0.08	0.05

III. Diesel Light Duty Vehicles below 3.5 Tonnes Capacity

	India 2000	*Euro I*	*Euro II*
CO (g/km)	2.75 to 6.9	2.72 to 6.9	1.5
HC (g/km) + NO_x (g/km)	0.97 to 1.7	0.97 to 1.7	0.7 to 1.3
PM (g/km)	0.14 to 0.25	0.14 to 0.7	–

IV. Diesel Heavy Duty Vehicles above 3.5 Tonnes Capacity

	India 2000	*Euro I*	*Euro II*	*Euro III*
CO (g/kwh)	4.5	4.5	4.0	2.1
HC (g/kwh)	1.1	1.1	1.1	0.66
NO_x (g/kwh)	8.0	8.0	7.0	5.0
PM (g/kwh)	0.36 – 0.61	0.36 to 0.61	0.15	0.1

■■■

CHAPTER 28

Fuel Supply Systems for SI and CI Engines

28.1 INTRODUCTION

In SI engines, highly volatile fuel like petrol, alcohol, benzol are used. In this case air-fuel mixture is produced outside the engine cylinder. This air-fuel mixture is supplied to the engine cylinder during suction through the engine intake manifold. Then this air-fuel mixture is compressed by the upward moving piston and just before the end of compression, the mixture is ignited by means of spark, produced by spark plug.

Carburettor is a device, which atomises and vapourises the fuel and mixes it with air in varying proportions so as to suit the changing conditions (or changing loads) of SI engines.

The process of automisation-vapourisation and mixing with air is known as **carburretion**. The air-fuel mixture so produced is called **Combustible mixture**.

This combustible mixture (air-fuel mixture) is carried to the engine cylinder through intake manifold and quantity of air-fuel mixture is controlled by **throttle valve**. Now it is essential to know the correct meaning of following terms.

1. **Vaporisation.** Vaporisation means change of phase from liquid phase to vapour phase.
2. **Automisation.** It is the mechanical breaking up of the liquid fuel into small particles i.e., when the liquid fuel is sprayed into air, the fuel particles absorb heat of air and get vaporized instantly.

Figure 28.1 shows the basic fuel supply system of a Petrol engine or Induction system.

28.2 FACTORS EFFECTING CARBURETTORS

Following factors affect carburettion process:

1. **Engine speed.** Directly affects the time available for preparation of mixture by the carburettor. When the engine speed increases, velocity of air must be increased by providing venturi in the carburettor.
2. **Temperature of air at the inlet.** When the temperature of atmosphere air increases, vaporisation increases, and decreases the density of air. Ultimately it results in decrease of efficiency and output power.
3. **The vaporisation characteristics of fuel.** It is always desirable to use more volatile fuels, otherwise it is required to heat the fuels in the intake manifold to produce vaporisation. However this is expensive and reduce power output.

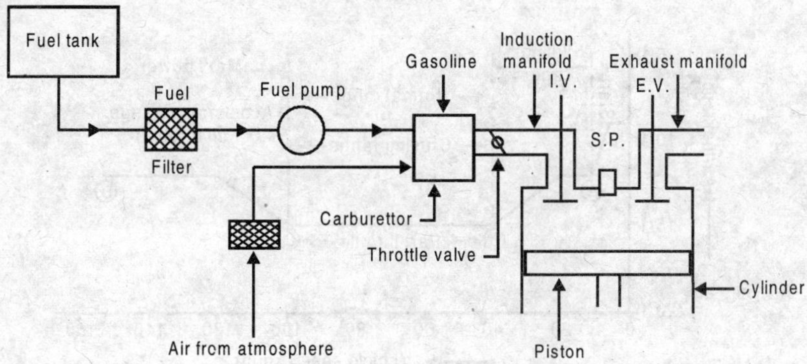

Fig. 28.1. *Induction systems of SI engine*

4. Design of intake manifold. In case of multicylinder engines, proper design of intake manifold is a must to ensure proper distribution of fuel to each cylinder.

28.3 TYPES OF AIR-FUEL MIXTURES

1. Stoichiometric Air-Fuel Mixture or Chemically Correct Air-Fuel Mixture

(a) This mixture contains just sufficient air for the complete combustion of fuel.

(b) Note after combustion all the fuel is burnt and all the air is consumed.

(c) For 1 kg of Octane (C_8H_{18}) fuel, 15.14 kg of air is needed for complete combustion.

Therefore, 15.14 : 1 is called chemically correct air-fuel ratio

2. Rich Mix. More fuel-less air

Therefore, 14, 13, 12, 10 or less : 1 is called **Rich Mix.**

3. Lean Mix. Less fuel-more air

Therefore, 17, 18, 20, 22, or more : 1 is called **Lean Mix.**

The range of homogenous mixture which can be ignited in SI engines:

7:1 Air-fuel by mass on Rich Side and @ 20:1 on Lean Side.

28.4 MIXTURE REQUIREMENTS

1. For Power and Economy. It is important to know the value of air-fuel at which max power and minimum fuel consumption is obtained.

Figure 28.2 shows the graph of power output and bsfc against air-fuel ratio.

Maximum power is obtained at a considerably rich mixture with air-fuel ratio around 12:1. (This mixture is the best power mixture)

The minimum brake specific fuel consumption (bsfc) is obtained at a slightly leaner mixture with air-fuel ratio around 16:1. (This mixture is the best economy mixture).

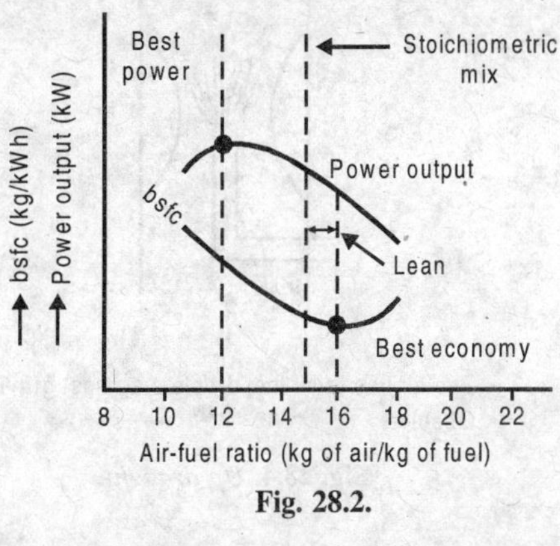

Fig. 28.2.

2. Mix Requirements for Automotive Engines. Figure 28.3 shows air-fuel ratio at different engine speeds.

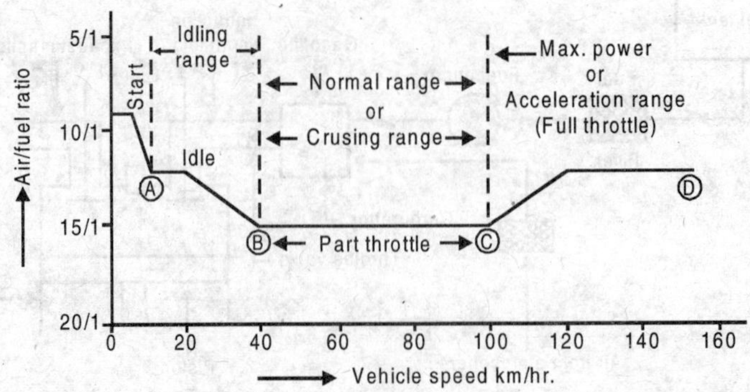

Fig. 28.3

- ❐ For the start mixture required is very rich, because the engine is cold-fuel does not vaporise properly.

Idling range (A–B). There is no external load on the engine—has to produce frictional horse power (fhp).

- ❐ Since the engine is not running hot a rich mixture is required. For idling air-fuel required is @12:1.

Normal or crushing range (B–C). Throttle valve is gradually opened and load increases above 20%.

- ❐ Fuel economy is prime consideration.
- ❐ Air-fuel ratio required @16 or 17:1.

Maximum Power or Acceleration range (C–D). For max. power rich mixture is required

- ❐ Nearly full throttle.
- ❐ Air-fuel ratio required 12 to 14:1.

28.5 CARBURETTOR TYPES

1. Open Choke Carburettors. Venturi is of fixed dimensions. Metering is affected by varying the pressure drop across it. Example, Solex, Zenith, Carter carburettors. (Venturi-fixed dimensions; pressure drop varied).

2. Constant Vacuum Carburettors. Area of air passage is varied, while pressure drop is kept constant. Example, S.U. carburettor.

(Venturi dimensions varied; pressure drop constant.)

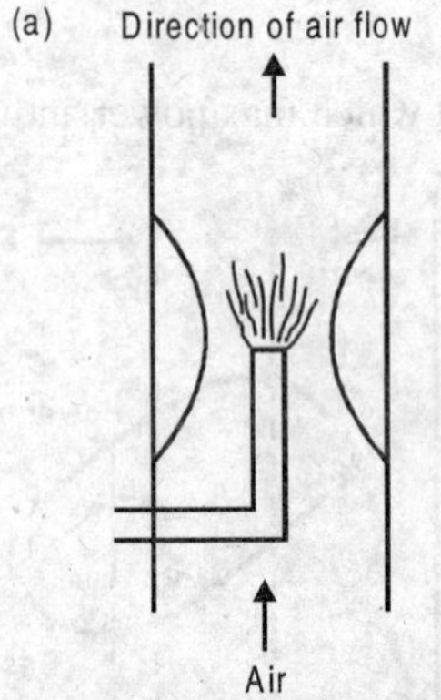

- The air has to lift fuel droplets against gravity
- Obsolete

Fig. 28.4. *Up draught*

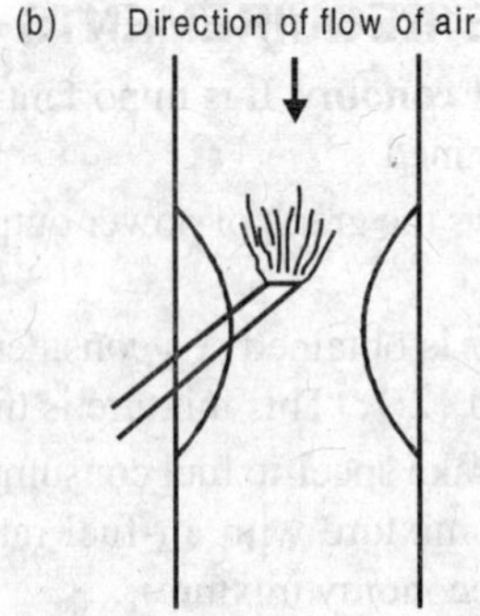

- Flow of air
- Mix flow is assisted by gravity

Fig. 28.5. *Down draught*

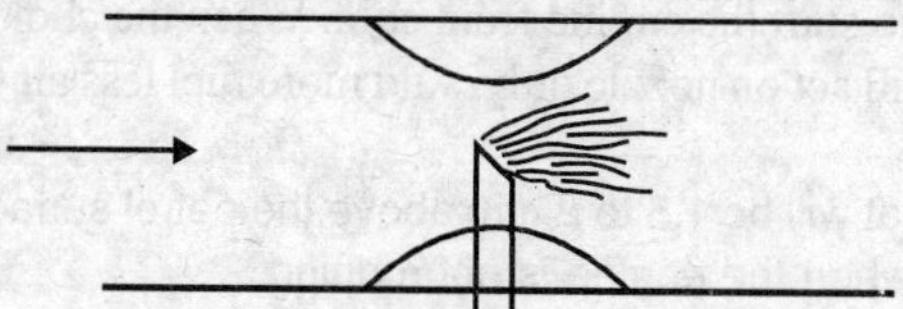

- Direction of flow of air or mix is horizontal
- Used where under bonnet space is limited

Fig. 28.6. *Horizontal draught*

28.6 A SIMPLE CARBURETTOR

Figure 28.7 shows a simple carburettor. It mainly consists of

(a) A float chamber with hollow float, needle valve, air bleed hole.
(b) A cylindrical pipe called air-horn. Narrowest c/s of which is called venturi.
(c) A throttle valve.
(d) A choke valve.
(e) A main petrol jet.

Float with needle valve maintains constant level of petrol in the float chamber. As the petrol level goes down—float move down—and needle valve opens—then petrol flows into the float chamber—level rises and comes to predecided level.

Through the air-bleed hole atmospheric pressure will be acting on the petrol surface.

During suction stroke, filtered air is drawn into the cylinder. This air passes through the **Venturi**. Venturi has a varying c/s, which first decreases to a minimum and then increases. Its narrowest c/s is called as the **throat**. As the air flows through the venturi, its velocity increases and pressure falls to a minimum at throat. The pressure differences between throat and float chamber exists and is called as **Intake Depression** or (Carburettor depression). Because of this intake depression, petrol will be sprayed through the nozzle into the throat. It is to be noted that the float chamber is connected to the venturi through nozzle as shown.

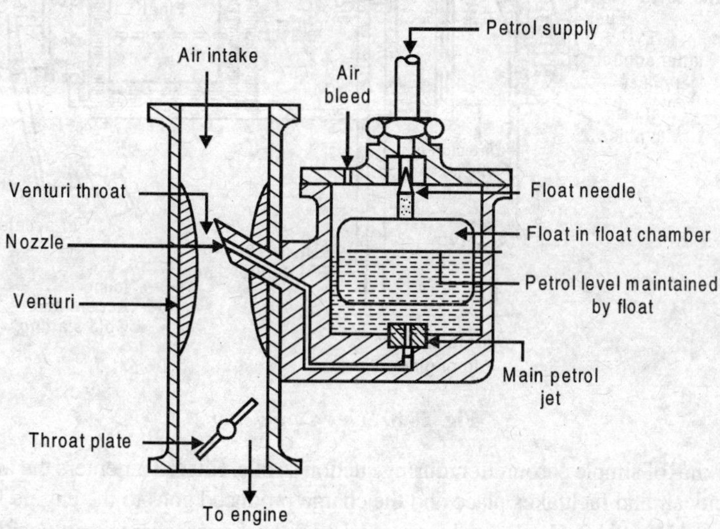

Fig. 28.7

When the petrol is sprayed in the high velocity air, petrol particles will absorb heat of air gets vaporised instantly. Then the combustible charge is produced and this charge admitted to engine cylinder. The amount of mixture is controlled by the **throttle valve**, which can be opened (or operated) by means of acceleration lever (or pedal) depending upon power requirements.

A choke valve is also provided, to start the engine from cold. When the choke lever is operated, it almost closes the air horn. The suction pressure will act on nozzle only. And more fuel less air (i.e., rich mixture) will go to engine and engine starts.

Note. Nozzle opening in the throat will be 1.5 to 2 mm above the petrol surface level. This is provided in order to avoid evaporation losses of petrol when the engine is not running.

Limitations of Simple Carburettor

1. It provides lean mixture at the starting warming up, idling range. Actually rich mixture is required.
2. At crusing it provides increasingly rich mixture.
3. At acceleration it provides lean mixture.

28.7 SOLEX CARBURETTOR

This carburettor has all the four basic circuits viz: Normal running; cold starting and warming up; Idling and slow running; Acceleration. It was used in Fiat cars, Ambassador cars, Gazal cars, Willys Jeeps etc. It is the basic carburetter—it is important to learn this carburetter so as to understand the working of any other newly developed carburettors.

1. Normal Running: Figure 28.8 shows the schematic diagram of Solex Carburettor. It consists of a float chamber. On the top of float needle valve is provided. Float chamber is connected to the air horn through main jet and emulsion tube. Emulsion tube has air jets, jets and orifices as shown and is provided in the venturi of air horn.

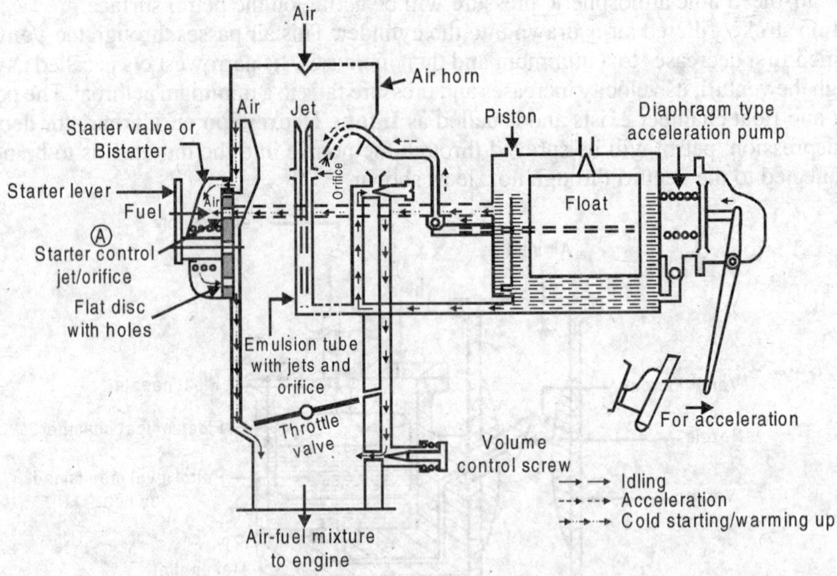

Fig. 28.8. *Solex Carburettor*

Similar to the working of simple carburettor, during suction stroke, filtered air enters the air horn and in the venturi mixing of high velocity air and fuel takes place and the charge produced goes to the engine for normal running.

2. Cold Starting and Warming Up. For cold starting—turn the starter lever to start position. When the starter flat disc with holes rotates and orifices (A) comes in contact with the dotted passage. Because of the engine suction which acts through the passage shown by (→. .→) and air comes through the passage—mixing of air and fuel takes place—the charge goes to the engine cylinder and engine starts—then simultaneously acceleration is given to warm-up the engine. (It is to be noted that some operations are performed simultaneously).

3. Idling and Slow Running. Once the engine starts and warm up—then for idle running the throttle valve will be almost closed (Not completely)—engine suction is applied through the passage shown by arrows (→)—Petrol from the main jet is sucked and air will enter from top at (14)— mixing of air and fuel takes place and —charge goes to engine and engine will be just running at no load conditions.

4. Acceleration Circuit. For acceleration— i.e., when more power is required—then acceleration lever/foot pedal in pressed as shown. Because of lever and fulcrum—diaphragm of acceleration pump is pressed towards LHS direction and extra petrol is supplied through acceleration jet—through injector into venturi. This fuel will be in addition to the rich mixture which is created because of more opening of throttle valve. It is also to be noted that throttle valve is also connected through the links to acceleration pedal.

28.8 CARTER CARBURETTOR

It is an American carburettor, used in jeeps.

It has three-venturies 3, 5, and 6.

At low speed-petrol is sprayed by nozzle 4, in the venturi 3.

Then the mixture flows through venturies 5 and 6, where homogenous mixture produced goes to engine.

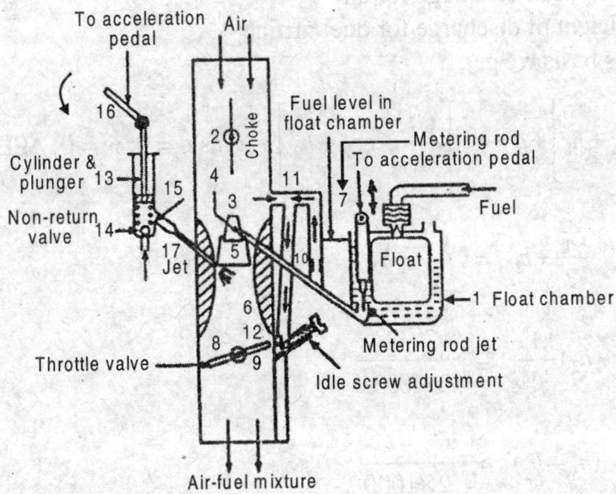

Fig. 28.9. *Downdraught Carter carburettor*

1. Metering. It has mechanical metering system. The metering rod (7) has 2–3 steps. At full throttle the rod 7 moves up.

2. Starting. When choke is operated, more petrol-less air (rich mixture) goes to engine, and engine starts.

3. Idle and Low Speed. Throttle valve almost closed, air from air horn, petrol from idle jet (10) produce mixture goes to engine.

4. Acceleration Circuit. Acceleration pump provided to supply extra fuel.

28.9 CARBURETTOR CALCULATIONS

P_1 – Pressure of air at sec. 1–1 in N/m^2

A_1 – Air flow area at sec. 1–1 in m^2

C_1 – Velocity of air at sec. 1–1 m/sec

v_1 – Sp. vol air sec. 1–1 m^3/kg

T_1 – Absolute temperature of air sec. 1–1K

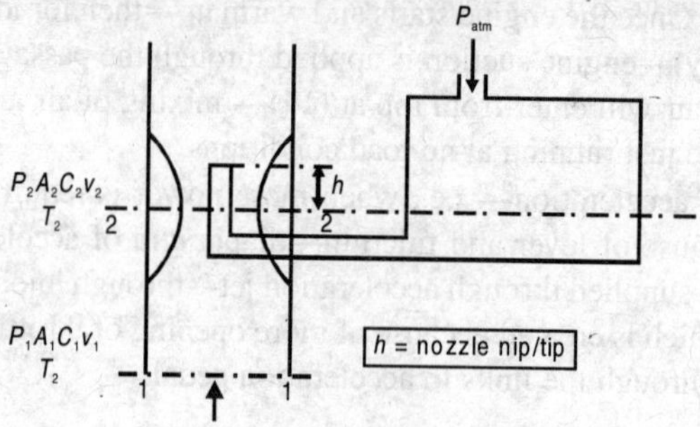

Fig, 28.10

Similarly $P_2 A_2 C_2 v_2$ and T_2 at sec. 2–2 respectively.

C_f – Velocity of fuel at nozzle tip in m/sec.

ρ_f – Density of fuel kg/m³

ρ_a – Density of air kg/m³

Cd_a – Co-efficient of discharge for air

Cd_f – Co-efficient of discharge for duel orifice

Applying SFEE on mass basis we get,

$$\left[\frac{C_1^2}{2} + gz_1 + h_1\right] + q = \left[\frac{C_2^2}{2} + gz_2 + h_2\right] + w \text{ as } q = 0;\ w = 0;\ \Delta PE = 0$$

$$\frac{C_1^2}{2} + h_1 = \frac{C_2^2}{2} + h_2$$

$$(h_1 - h_2)\frac{\text{kJ}^*}{\text{kg}} = \frac{C_2^2 - C_1^2}{2\times1000^{**}}$$

i.e.
$$(h_1 - h_2) = \frac{C_2^2 - C_1^2}{2\times1000}$$

$$C_p(T_1 - T_2) = \frac{C_2^2 - C_1^2}{2\times1000} \quad C_1 \text{ is neglible compared to } C_2$$

$$\therefore \sqrt{2\times1000\times C_p\times(T_1 - T_2)} = C_2$$

$$C_2 = \sqrt{2\times1000\times C_p\times T_1\left(1 - \frac{T_2}{T_1}\right)}$$

$$C_2 = \sqrt{2\times1000\times C_p\times T_1\left(1 - \left(\frac{P_2}{P_1}\right)^{\frac{\gamma-1}{\gamma}}\right)}$$

Note: * Exact — Compressibility considered; ** Approx. —Compressibility negelected

2. Exact air-fuel ratio (*when compressibility is considered*)

$$\frac{\dot{m}_a}{\dot{m}_f} = \frac{Cd_a}{Cd_f}.\frac{A_2}{A_f}.\frac{P_1}{R\sqrt{T_1}} \times \frac{\sqrt{2C_p\left\{\left(\frac{P_2}{P_1}\right)^{2/r} - \left(\frac{P_2}{P_1}\right)^{\frac{\gamma+1}{\gamma}}\right\}}}{\sqrt{2.\rho_f.\left(\Delta P - \underset{\text{nip height}}{h}.g.\rho_f\right)}}$$

where A_f = Area of fuel orifice

h = Nipple height (tip height above fuel surface)

3. Air-Fuel Ratio (Neglecting Compressibility)

Approximate method

$$\frac{\dot{m}_a}{\dot{m}_f} = \frac{Cd_a}{Cd_f}.\frac{A_2}{A_f}.\frac{\sqrt{2.\rho_a.\Delta P}}{\sqrt{2.\rho_f.\left(\Delta P - h.g.\rho_f\right)}} \quad (1)$$

Note. $Cd_a \simeq 0.8$ to 0.85

If *h is neglected* (*Nozzle tip*)

$$\frac{\dot{m}_a}{\dot{m}_f} = \frac{Cd_a\ .\ A_2}{Cdf\ .\ Af} \times \sqrt{\frac{\rho_a}{\rho_f}} \quad (2)$$

Example. *A simple carburettor has to supply 5 kg of air/min. The air is at a pressure of 1.013 bar and at a temperature of 27°C. Calculate the throat diameter of the choke tube for the actual air flow velocity 91 m/sec. Take velocity co-efficient to be 0.8. Assume isentropic flow and assume the flow to be compressible i.e., using exact method.*

Solution

$$\dot{m}_a = \frac{5}{60} \text{ kg/sec}$$

$$P_1 = 1.013 \times 10^5 \text{ N/m}^2$$

$$T_1 = 27 + 273 = 300 \text{ K}$$

$$d_2 = ?$$

$$C_2 \text{ actual} = 91 \text{m/sec.}$$

$$\text{Velo} - \text{Co-eff.} = 0.8$$

We know that velocity,

$$C_2 = \sqrt{2 \times 1000 \times C_P \times T_1\left\{1 - \left(\frac{P_2}{P_1}\right)^{\frac{\gamma-1}{\gamma}}\right\}}$$

$$91 = 0.8\sqrt{2 \times 1000 \times 1.005 \times 300\left\{1 - \left(\frac{P_2}{1.013_1}\right)^{\frac{1.4-1}{1.4}}\right\}}$$

$$\Rightarrow \quad P_2 = 0.9389\,\text{bar} = \text{Pressure at throat}$$

We also know that,

$$P_1 v_1 = R.T_1$$

$$\Rightarrow \quad \frac{P_1}{RT_1} = \frac{1}{v_1} = \rho_1$$

$$\therefore \quad \rho_1 = \frac{P_1}{RT_1} = \frac{1.013 \times 100}{0.287 \times 300} = 1.1729 \text{ kg/m}^3$$

Since $PV^{\gamma} = C$ for isentropic flow

$$\Rightarrow \frac{P}{\rho^{\gamma}} = \text{Constant}$$

$$\frac{P_1}{\rho_1^{\gamma}} = \frac{P_2}{\rho_2^{\gamma}}$$

$$\frac{\rho_2}{\rho_1} = \left(\frac{P_2}{P_1}\right)^{\frac{1}{\gamma}}$$

$$\rho_2 = \rho_1 \times \left(\frac{P_2}{P_1}\right)^{\frac{1}{\gamma}}$$

$$= 1.1729 \times \left(\frac{0.9389}{1.013}\right)^{\frac{1}{1.4}}$$

$$\rho_2 = 1.1109 \text{ kg/m}^3$$

Now to find throat area, using continuity equation

$$\dot{m}_{air} = \rho_1, A_1, C_1 = \rho_2, A_2, C_2$$

$$\frac{5}{60} = 1.1109 \times A_2 \times 91$$

$$\Rightarrow A_2 = 8.243 \times 10^{-4} \quad m^2 = 8.243 \text{ cm}^2$$

$$\mathbf{d_2 = 3.2397 \text{ cm} = \text{Throat diameter}}$$

28.10 FUEL SUPPLY SYSTEM FOR CI ENGINES

28.10.1 Introduction

The fuel supply system of a diesel engine can be called as the heart of the engine, since the engine performance directly depends upon the proper functioning of this system—which must supply, meter, inject and atomize the fuel.

Fuel injection systems are manufactured with great accuracy, hence they are costlier.

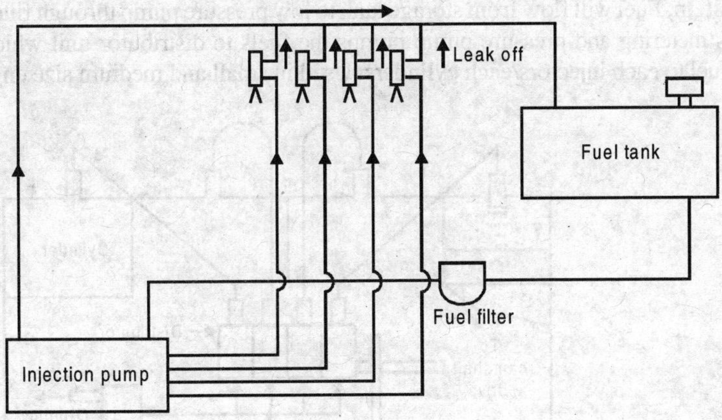

Fig. 28.11

Fuel will flow either because of gravity or fuel feed pump, which is provided to supply fuel through the filter to the injection pump. Which pumps the fuel to the injectors which are provided in the cylinder heads.

The fuel injection systems are of 2 types.

1. Air Injection System. In this case fuel is injected under the pressure of air. For supplying high pressure air multistage air compressors are required, which are very much costly and hence this system is not in use.

2. Solid Injection System. In this case diesel fuel is directly injected by fuel pump (Bosch Pump).

Further these are of 3 types of solid injection systems.

A. Individual Pump System. (Figures 28.12 – 24.14 are self explanatory).

As shown fuel will flow from the storage tank to filters to low pressure pumps. This low pressure pump pumps the fuel to 4 separate metering and pressure pumps.

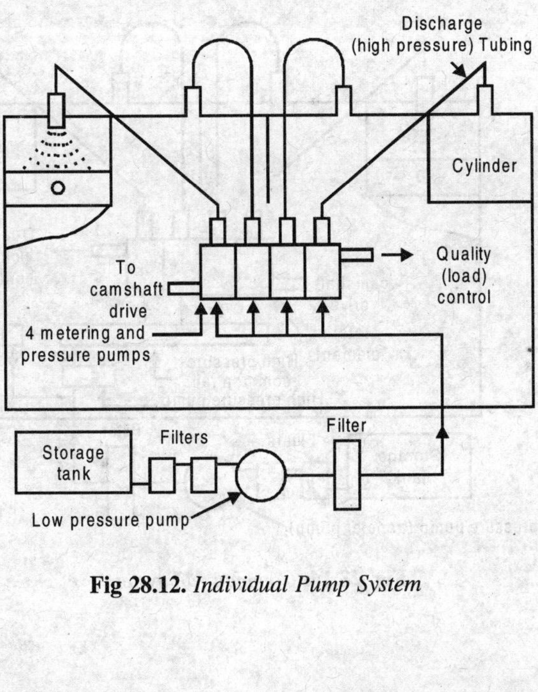

Fig 28.12. *Individual Pump System*

These separate metering and pressure pumps will pump the fuel to individual injectors which are provided in the cylinder heads. These are used in large slow speed engines.

B. Distributor System. Fuel will flow from storage tank to low pressure pump through filters, then to metering and pressure pumps. This metering and pressure pump pumps the fuels to distributor unit which distributes and sends required quantity of fuel to each injectors/each cylinders. Used in small and medium size engines.

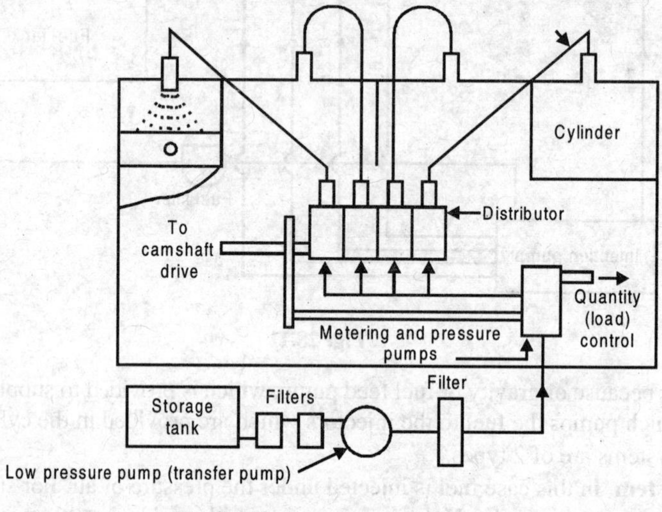

Fig. 28.13. *Distributor system*

C. Common Rail System. In this case fuel flows from storage tank to low pressure pump through filters. Low pressure pump, pumps the fuel to high pressure pump, which pumps the fuel to high pressure pump, which pumps the fuel to common rail. Thus high pressure fuel is collected in common rail and from here, through the metering devices required quantity of fuel goes to injectors/cylinders. Generally Cummins and multicylinder engines use this system.

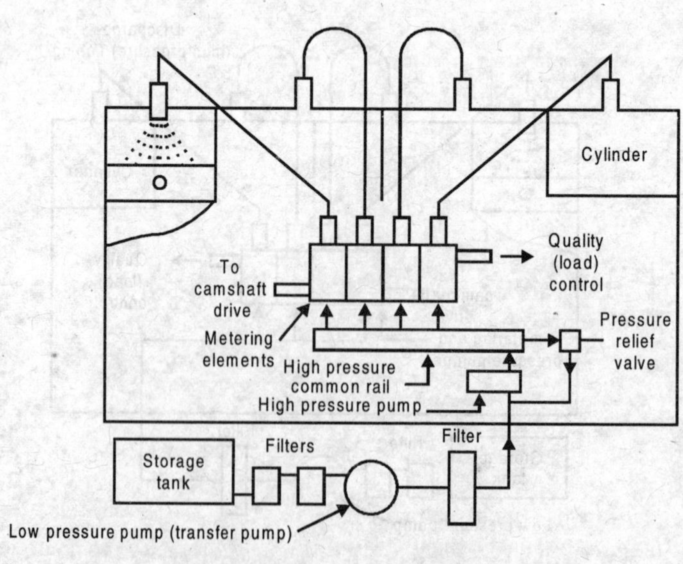

Fig. 28.14. *Common rail system*

28.11 FUEL PUMP AND INJECTOR

Rack (1) is connected to the accelerator pedal or governor, which moves in and out when the accelerator pedal is operated.

Rack is in contact with the Quadrant gear (2) (a part of gear), which has cylindrical bottom part (Skirting cylinder). This skirting cylinder has cross slot. In this cross slot, cross bottom part of the plunger (3) is held. As the rack moves in and out—quadrant gear rotates—in turn plunger with helical groove moves in the cylinder (4).

The cylinder has inlet and overflow ports. This fuel pump and injector operates under primed conditions.

Valve (5) is resting on the valve seat by the spring (6).

Injector and pump are connected by the delivery pipe (7).

In the injector (8) is the nozzle body, (9) is the nozzle valve and (10) is valve cap nut, (11) is the spindle-held in position by spring (12).

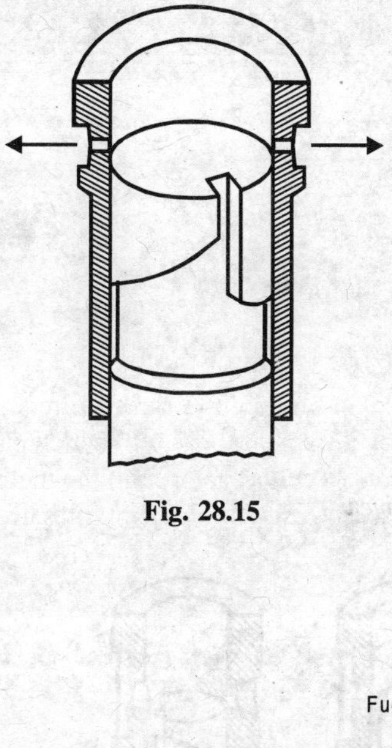

Fig. 28.15

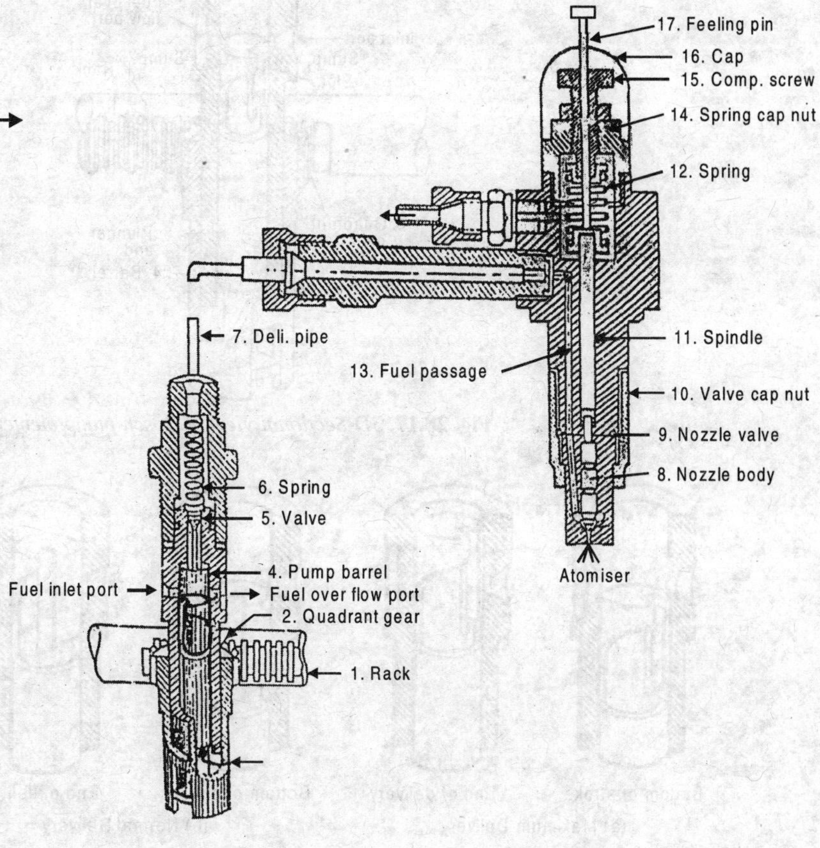

Fig. 28.16. *Fuel pump and injector*

It is to be noted that the plunger has up and down reciprocating motion—which is obtained by the camshaft below it and has rotary motion because of rack. When the rack moves in and out depending upon the power requirement. Quadrant gear moves—in turn plunger rotates—plunger has helical groove—so height of groove with respect to ports varies—so amount of fuel to be injected will vary. During the upward motion of plunger—once the ports are closed—valve is lifted from its seat because of fuel pressure and fuel flows through the delivery pipe—through the fuel passage (13) to nozzle valve (9). Because of fuel pressure—nozzle valve (9) is lifted against the compression of spring (12) and fuel is injected till the edge of helical groove comes in contact with port—when pressures of fuel is released and injection stops.

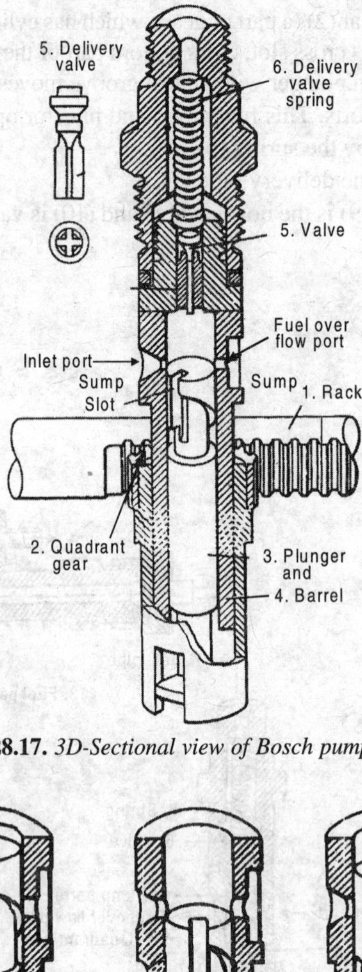

Fig. 28.17. *3D-Sectional view of Bosch pump element*

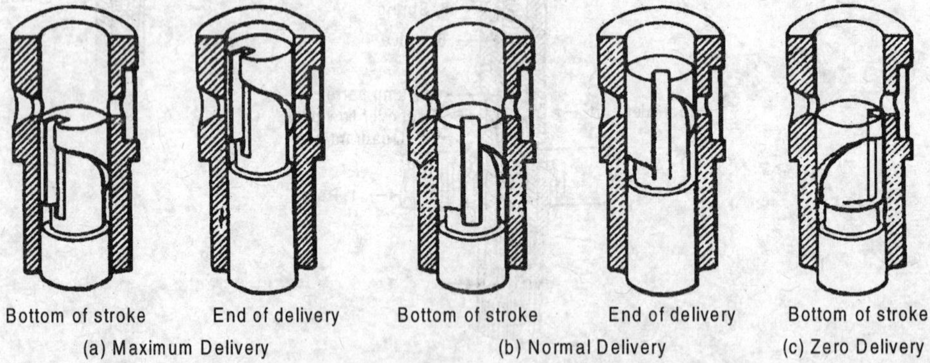

Fig. 28.18. *Position of the helix for various load conditions*

28.12 TYPES OF FUEL INJECTORS

1. Air Blast Injectors. These are used in air injection systems. Nowadays air injection systems are not in use as they require multistage compressors. And hence these injectors are no more used.

2. Mechanically Operated Injectors. These injectors are operated by a mechanism similar to that used to operate. IC engine valves i.e., it uses camshaft, push rods, rocker arms, etc. Cam operates the plunger.

3. Automatic Fuel Injector. All automobile CI engines use these Automatic Fuel Injectors. They comprise needle valve, which is lifted up by fuel pressure. This fuel pressure is created by fuel pump.

28.13 TYPES OF NOZZLES

Following types of nozzles are normally used with diesel engines.

1. Single hole type
2. Multiple hole type
3. Pintle type

1. Single Hole Type. At the centre of nozzle body a hole of 0.2 mm dia is provided.

Spray cone ∠ is @ 15°.

Used in open combustion chambers.

High pressure is required to get same velocity. No good mixing with air. It has a tendency to dribble.

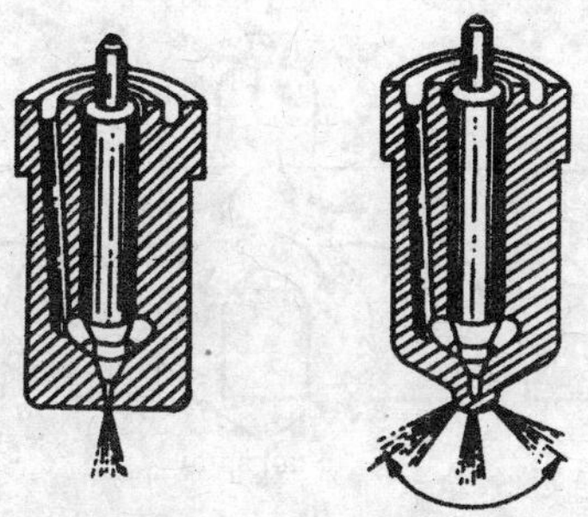

(a) Single and Multiple hole type

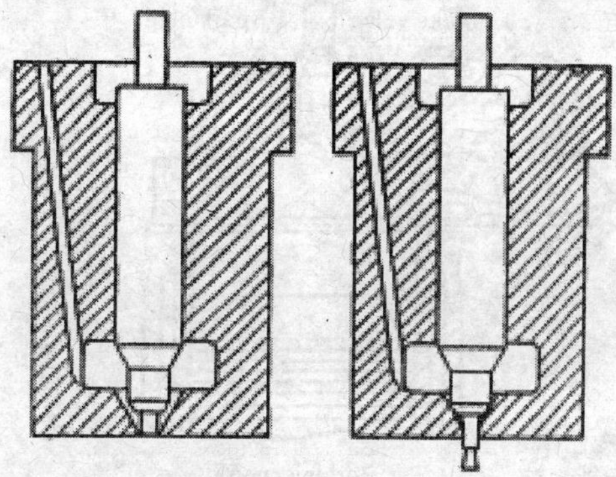

(b) Pintle type

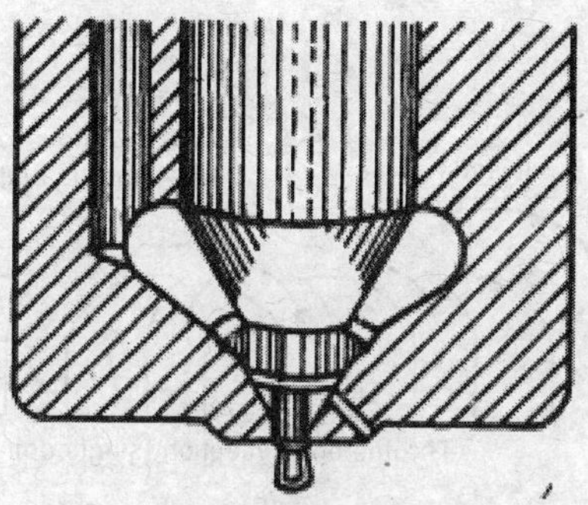

(c) Pintaux nozzle

Fig. 28.19

2. Multiple Hole Type. Proper mixing with air 4 to 18 holes. Size of the holes will be 0.35 to 1.5 mm. Ref. Fig. 28.19 (a).

3. Pintle Type. To avoid weak injection and dribbling, the spindle is provided with projection called **Pintle**. It protrudes through the mouth of nozzle body. It may be cylindrical or conical in shape. Dribbling is avoided.

Used in precombustion chambers, air cells, swirl chambers. Ref. Fig. 28.19 (b)

4. Pintaux. Good for cold starting. It is a development of pinter nozzle. It has an auxiliary hole in the nozzle body Ref. Fig. 28.19 (c). It results in good cold starting.

Disadvantage. Side hole may be choked–better filter needed.

28.14 ELECTRONIC FUEL INJECTION (EFI) SYSTEM

Electronics is introduced in automobiles in 1965. About 30–40% of cost of vehicles is for electronic items. Max. power and best economy are attained by using electronics and computers in automobiles.

EFI systems was various sensors to sense various parameters like temperature, pressure of gases, position of throttle valve, air flow rate etc.

Sensors feed this data to Electronic Control Unit (ECU)—which is basically a computer.

This ECU—processes the data and operates injectors and other devices to have maximum power, with best economy, and low emissions.

28.15 MULTIPOINT FUEL INJECTION SYSTEM (MPFI)

Multipoint Fuel Injection system is used to supply air-fuel mixture of proper strength and in required quantity to each cylinder of a multicylinder engine, at all engine speeds loads.

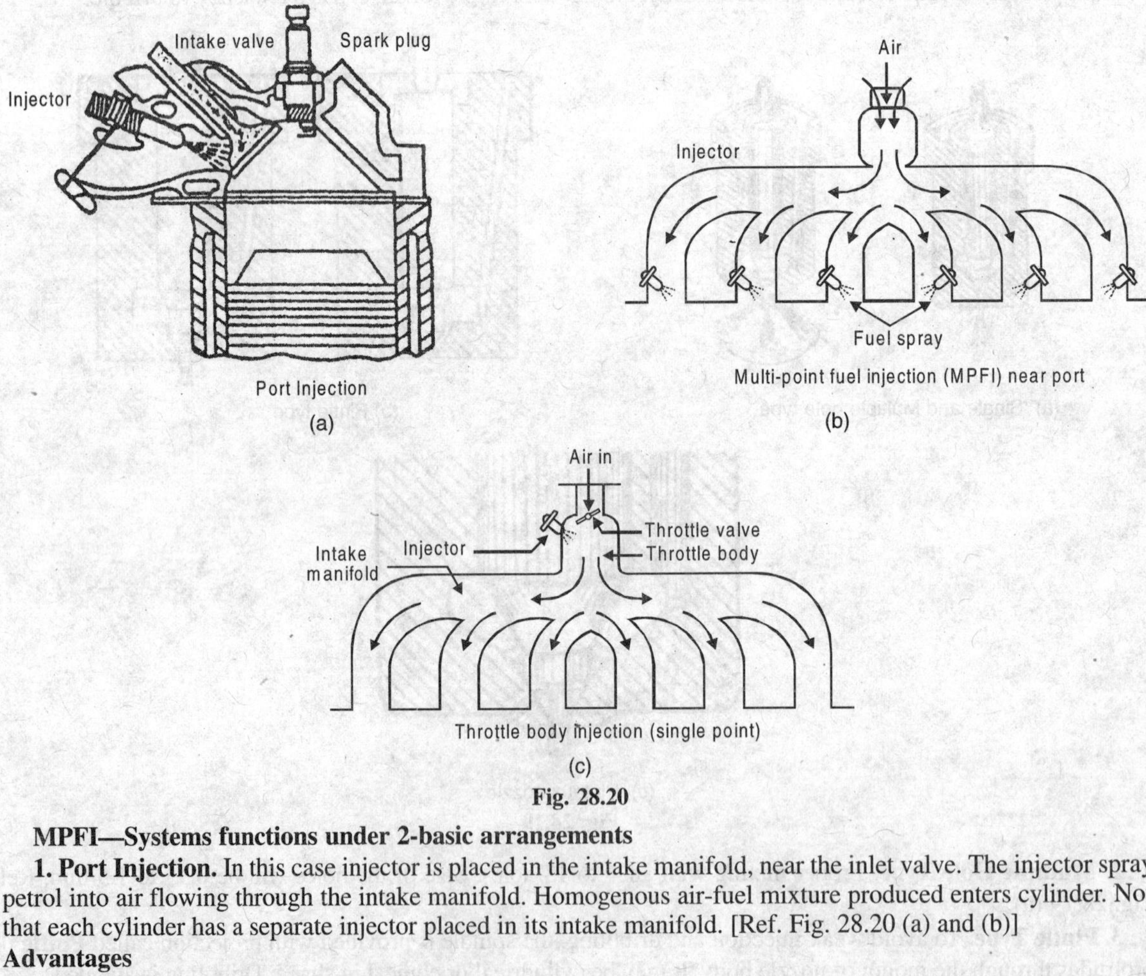

Port Injection
(a)

Multi-point fuel injection (MPFI) near port
(b)

Throttle body injection (single point)
(c)

Fig. 28.20

MPFI—Systems functions under 2-basic arrangements

1. Port Injection. In this case injector is placed in the intake manifold, near the inlet valve. The injector sprays petrol into air flowing through the intake manifold. Homogenous air-fuel mixture produced enters cylinder. Note that each cylinder has a separate injector placed in its intake manifold. [Ref. Fig. 28.20 (a) and (b)]

Advantages

1. Uniform fuel distribution
2. Increase in power output
3. More precise control of air-fuel ratio.

2. Throttle Body Injection. In this case injector is provided at a single point in the throttle body. Throttle valve controls the amount of air entering intake manifold. [Ref. Fig. 28.20 (c)]

28.16 QUANTITY OF FUEL AND SIZE OF NOZZLE ORIFICE

1. Velocity of fuel $C_f = Cd_f \cdot \sqrt{\frac{2\Delta P}{\rho_f}}$

$$C_f = Cd_f \cdot \sqrt{\frac{2(P_i - P_c)}{\rho_f}}$$

where
Cd_f = Co-eff. of discharge for fuel orifice
P_i = Fuel pressure at the inlet to injector N/m²
P_c = Cylinder pressure N/m²
ΔP = Change of pressure
ρ_f = Density of fuel

2. Also C_f in terms of h

$C_f = Cd_f \cdot \sqrt{2gh}$ where h = Pressure different between injection pressure in m of fuel

3. Volume of fuel injected/sec, Q

$$Q = \left(\frac{\pi}{4} d^2 \cdot n\right) \times C_f \times \left[\frac{\theta}{360} \times \frac{60}{N}\right] \times \frac{N_i}{60}$$

where
n = No. of orifices
N = Speed in rpm
$N_i = \frac{N}{2}$ for 4-stroke
N_i = N for 2-stroke
θ = Duration of fuel inbjection in crank angle degree
N_i = No. of injections/min.

4. Fuel consumed/cycle $= \frac{\text{bsfc} \times \text{Power}}{\text{Cycles/hr}}$

Cycles/hr $= \frac{N}{2} \times 60$ (4-stroke engine)
$= N \times 60$ (2-stroke engine)

5. Time of injection $= \left(\frac{Q}{360} \times \frac{60}{N}\right)$ sec

Example *In a solid fuel injection system find the velocity of fuel injection, when the difference in fuel pressure and cylinder pressure is 90 bar. Assuming S = 0.905 and Cd for orfifice = 0.86.*

Solution

Data
$\Delta P = 90 \times 10^5$ N/m²
$S = 0.905$
$\rho = 0.905 \times 10^3$ kg/m³
$Cd_f = 0.86$

We know that,

$$C_f = Cd_f \times \sqrt{\frac{2\Delta P}{\rho_f}} = 0.86 \times \sqrt{\frac{2 \times 90 \times 105}{905}}$$

$$\mathbf{C_f = 121.285\ m/sec}$$

■■■

CHAPTER 29

Ignition Systems of SI Engines

29.1 INTRODUCTION

We know that in case of Internal Combustion (IC) engines, combustion of air and fuel takes place inside the engine cylinder and the products of combustion expand to produce reciprocating motion of the piston. This reciprocating motion of the piston is in turn converted into rotary motion of the crank shaft through connecting rod and crank.

This rotary motion of the crank shaft is in turn used to drive the generators for generating power.

We also know that there are 4-cycles of operations viz., suction; compression; power generation and exhaust.

These operations are performed either during the 2-strokes of piston or during 4-strokes of the piston and accordingly they are called as 2-stroke cycle engines and 4-stroke cycle engines.

In case of petrol engines during suction operation, charge of air and petrol fuel will be taken in. During compression this charge is compressed by the upward moving piston. And just before the end of compression, the charge of air and petrol fuel will be ignited by means of the spark produced by means of for spark plug. And the ignition system does the function of producing the spark in case of spark ignition engines.

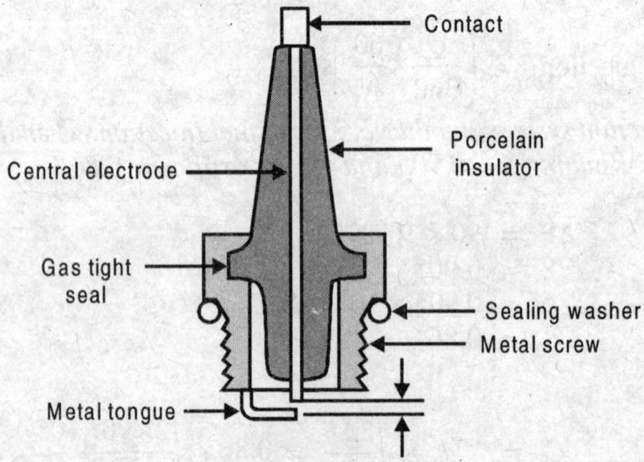

Fig. 29.1. *Spark plug*

Figure 29.1 shows a typical spark plug used with petrol engines. It mainly consists of a central electrode and metal tongue. Central electrode is covered by means of porcelain insulating material. Through the metal screw the spark plug is fitted in the cylinder head plug. When the high tension voltage of the order of 30000 volts is applied across the spark electrodes, current jumps from one electrode to another producing a spark.

Whereas in case of diesel (Compression Ignition—CI) engines only air is taken in during suction operation and in compressed during compression operation and just before the end of compression, when diesel fuel is injected it gets ignited due to heat of compression of air.

Once the charge is ignited, combustion starts and products of combustion expand i.e., they force the piston to move downwards i.e. they produce power and after producing the power the gases are exhausted during exhaust operation.

29.2 IGNITION SYSTEM TYPES

Basically Convectional Ignition systems are of 2 types:

1. Battery or Coil Ignition System
2. Magneto Ignition System

Both these conventional, ignition systems work on mutual electromagnetic induction principle.

Battery ignition system was generally used in 4-wheelers, but nowadays it is more commonly used in 2–wheelers also (i.e., Button start, 2-wheelers like Pulsar, Kinetic Honda; Honda-Activa, Scooty. Fiero etc.). In this case 6 V or 12 V battery will supply necessary current in the primary winding.

Magneto ignition system is mainly used in 2-wheelers, kick start engines. (Example, Bajaj Scooters, Boxer, Victor, Splendor, Passion etc.).

In this case magneto will produce and supply current to the primary winding. So in magneto ignition system magneto replaces the battery.

(I) Battery or Coil Ignition System

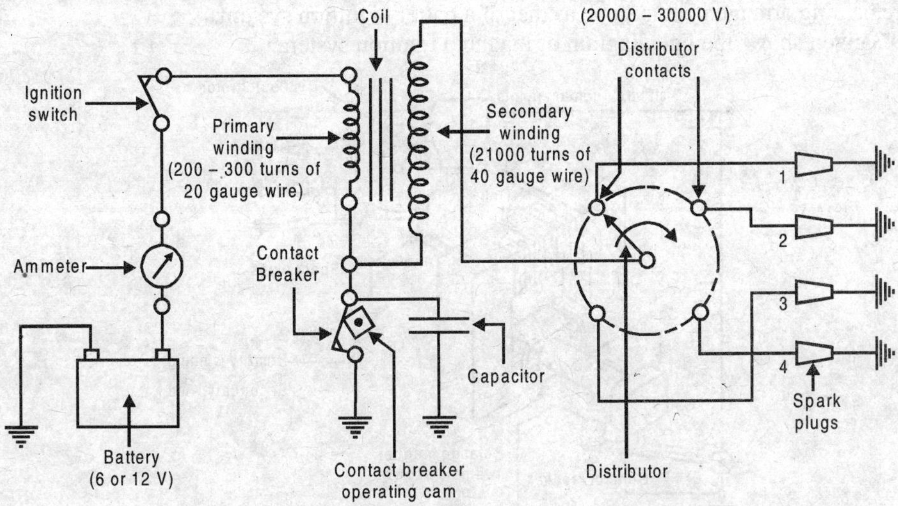

Fig. 29.2. *Schematic diagram of Coil/Battery Ignition System*

Figure 29.2 shows line diagram of battery ignition system for a 4-cylinder petrol engine. It mainly consists of a 6 or 12 volt battery, ammeter, ignition switch, auto-transformer (step up transformer), contact breaker, capacitor, distributor rotor, distributor contact points, spark plugs, etc.

Note that the Fig. 29.2 shows the ignition system for 4-cylinder petrol engine, here there are 4-spark plugs and contact breaker cam has 4-corners. (If it is for 6-cylinder engine it will have 6-spark plugs and contact breaker cam will be a perfect hexagon).

The ignition system is divided into 2-circuits.

(i) Primary circuit. It consists of 6 or 12 V battery, ammeter, ignition switch, primary winding—it has 200–300 turns of 20 SWG (Sharps Wire Gauge) gauge wire, contact breaker, capacitor.

(ii) Secondary circuit. It consists of secondary winding. Secondary winding consists of about 21000 turns of 40 (SWG) gauge wire. Bottom end of which is connected to bottom end of primary and top end of secondary winding is connected to centre of distributor rotor. Distributor rotor rotates and makes contacts with contact points and are connected to spark plugs which are fitted in cylinder heads (engine earth).

(iii) Working. When the ignition switch is closed and engine in cranked, as soon as the contact breaker closes, a low voltage current will flow through the primary winding. It is also to be noted that the contact beaker cam opens and closes the circuit 4-times (for 4 cylinders) in one revolution. When the contact breaker opens the contact, the magnetic field begins to collapse. Because of this collapsing magnetic field, current will be induced in the secondary winding. And because of more turns (@ 21000 turns) of secondary, voltage goes upto 28000–30000 volts.

This high voltage current is brought to centre of the distributor rotor. Distributor rotor rotates and supplies this high voltage current to proper stark plug depending upon the engine firing order. When the high voltage current jumps the spark plug gap, it produces the spark and the charge is ignited-combustion starts-products of combustion expand and produce power.

Note:

(1) The Function of the capacitor is to reduce arcing at the contact breaker (CB) points. Also when the CB opens the magnetic field in the primary winding begins to collapse. When the magnetic field is collapsing capacitor gets fully charged and then it starts discharging and helps in building up of voltage in secondary winding.

(2) Contact breaker cam and distributor rotor are mounted on the same shaft.

In 2-stroke cycle engines these are motored at the same engine speed. And in 4-stroke cycle engines they are motored at half the engine speed.

II. Magneto Ignition System

In this case magneto will produce and supply the required current to the primary winding. In this case as shown, we can have rotating magneto with fixed coil or rotating coil with fixed magneto for producing and supplying current to primary, remaining arrangement is same as that of a battery ignition system.

Figure 29.3 given shows the line diagram of magneto ignition system.

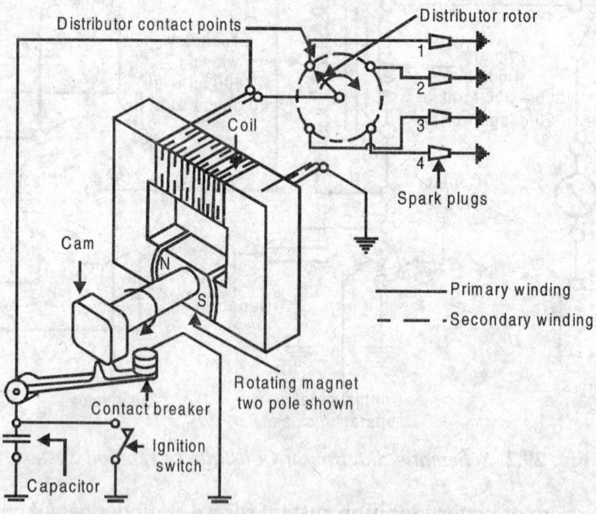

Fig. 29.3. *Schematic diagram of Magneto Ignition System*

29.3 DRAWBACKS (DISADVANTAGES) OF CONVENTIONAL IGNITION SYSTEMS

Following are the drawbacks of conventional ignition systems.

(i) Because of arcing, pitting of contact breaker point and which will lead to regular maintenance problems.

(ii) Poor starting: After few thousands of kilometres of running, the timing becomes inaccurate, which results into poor starting (Starting trouble).

(iii) At very high engine speed, performance is poor because of inertia effects of the moving parts in the system.

(iv) Sometimes it is not possible to produce spark properly in fouled spark plugs.

In order to overcome these drawbacks Electronic Ignition system is used.

29.4 ADVANTAGES OF ELECTRONIC IGNITION SYSTEM

Following are the advantages of electronic ignition system

1. Moving parts are absent—so no maintenance
2. Contact breaker points are absent—so no arcing
3. Spark plug life increases by 50% and they can be used for about 60000 km without any problem
4. Better combustion in combustion chamber, about 90–95% of air fuel mixture is burnt compared with 70–75% with conventional ignition system
5. More power output
6. More fuel efficiency

29.5 TYPES OF ELECTRONIC IGNITION SYSTEM

Electronc Ignition System are as follows:

1. Capacitance Discharge Ignition system
2. Transistorised system
3. Piezo-electric Ignition system
4. The Texaco Ignition system

1. Capacitance Discharge Ignition System

Figure 29.4 illustrated the Capacitance Discharge Ignition System.

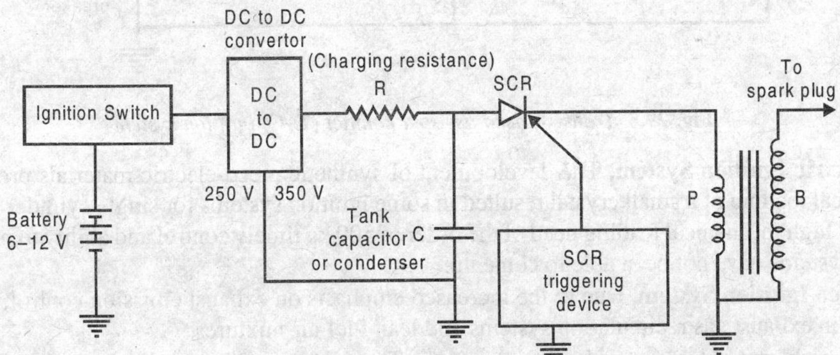

Fig. 29.4. *Capacitance Discharge Ignition System*

It mainly consists of 6 – 12 V battery, ignition switch, DC to DC convertor, charging resistance, tank capacitor, Silicon Controlled Rectifier (SCR), SCR-triggering device, step up transformer, spark plugs.

A 6 – 12 volt battery is connected to DC to DC converter i.e., power circuit through the ignition switch, which is designed to give or increase the voltage to 250–350 volts. This high voltage is used to charge the tank capacitor (or condenser) to this voltage through the charging resistance. The charging resistance is also so designed that it controls the required current in the SCR.

Depending upon the engine firing order, whenever the SCR triggering device, sends a pulse, then the current flowing through the primary winding is stopped. And the magnetic field begins to collapse. This collapsing magnetic field will induce or step up high voltage current in the secondary, which while jumping the spark plug gap produces the spark, and the charge of air fuel mixture is ignited.

2. Transistorised Assisted Contact (TAC) Ignition System: Figure 29.5 shows the TAC system.

Advantages

1. The low breaker-current ensures longer life.
2. The smaller gap and lighter point assembly increase dwell time, minimise contact bouncing and improve repeatability of secondary voltage.
3. The low primary inductance reduces primary inductance reduces primary current drop-off at high speeds.

Disadvantages

1. As in the conventional system, mechanical breaker points are necessary for timing the spark.
2. The cost of the ignition system is increased.
3. The voltage rise-time at the spark plug is about the same as before.

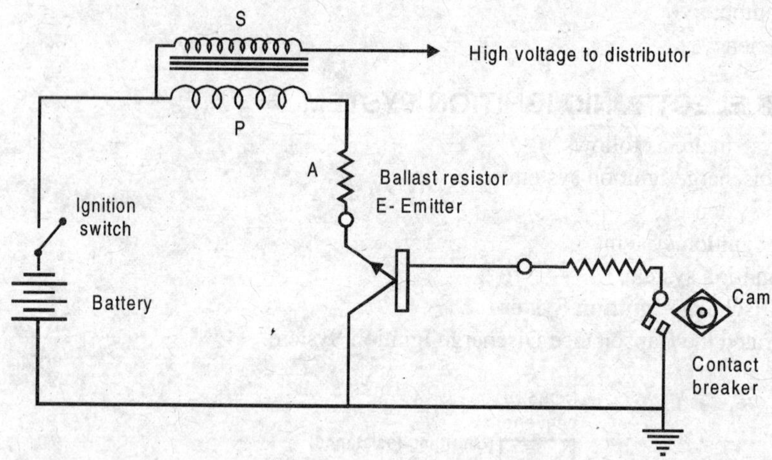

Fig. 29.5. *Transistorised assisted contact (TAC) ignition system*

3. Piezo-electric Ignition System. The development of synthetic piezo-electric materials producing about 22 kV by mechanical loading of a small crystal resulted in some ignition systems for single cylinder engines. But due to difficulties of high mechanical loading need of the order of 500 kg timely control and ability to produce sufficient voltage, these systems have not been able to come up.

4. The Texaco Ignition System. Due to the increased emphasis on exhaust emission control, there has been a sudden interest in exhaust gas recirculation systems and lean fuel-air mixtures.

To avoid the problems of burning of lean mixtures, the Texaco Ignition system has been developed. It provides a spark of controlled duration which means that the spark duration in crank angle degrees can be made constant at all engine speeds. It is a AC system. This system consists of three basic units : a power unit, a control unit and a distributor sensor.

This system can give stable ignition up to A/F ratios as high as 24 : 1.

CHAPTER 30

Cooling Systems in IC Engines

30.1 INTRODUCTION

We know that in case of Internal Combustion engines, combustion of air and fuel takes place inside the engine cylinder and hot gases are generated. The temperature of gases will be around 2300 –2500°C. This is a very high temperature and may result into burning of oil film between the moving parts and may result into seizing or welding of the same. So, this temperature must be reduced to about 150 – 200°C at which the engine will work most efficiently. Too much cooling is also not desirable since it reduces the thermal efficiency. So, the object of cooling system is to keep the engine running at its most efficient operating temperature.

It is to be noted that the engine is quite inefficient when it is cold and hence the cooling system is designed in such a way that it prevents cooling when the engine is warming up and till it attains to maximum efficient operating temperature, then it starts cooling.

It is also to be noted that,

(i) About 20 – 25% of total heat generated is used for producing brake power (useful work).

(ii) Cooling system is designed to remove 30 – 35% of total heat.

(iii) Remaining heat is lost in friction and carried away by exhaust gases.

30.2 METHODS OF COOLING

There are mainly 2-types of cooling systems:

(i) Air cooled system

(ii) Water cooled system

(i) Air Cooled System. Air cooled system is generally used in small engines say upto 15–20 kW and in aeroplane engines.

In this system fins or extended surfaces are provided on the cylinder walls, cylinder head etc. Heat generated due to combustion in the engine cylinder will be conducted to the fins and when the air flows over the fins, heat will be dissipated to air.

The amount of heat dissipated to air depends upon:

(i) amount of air flowing through the fins

(ii) fin surface area

(iii) thermal conductivity of metal used for fins.

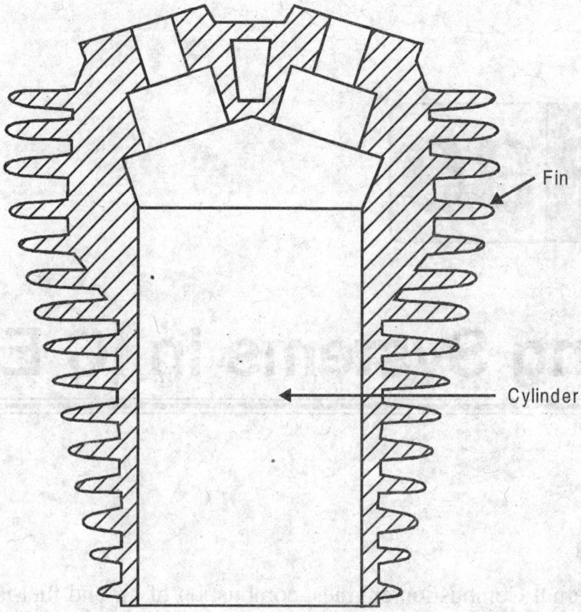

Fig. 30.1. *Cylinder with fins*

Advantages of Air Cooled System

Following are the advantages of air cooled system:

(a) Radiator/pump are absent hence the system is light.

(b) In case of water cooling system there are leakages, but in this case there are no leakages.

(c) Coolant and antifreeze solutions are not required.

(d) This system can be used in cold climates, where if water is used it may freeze.

Disadvantages

(a) Comparatively it is less efficient.

(b) It is used in aeroplanes and motorcycle engines where the engines are exposed to air directly.

30.3 WATER COOLING SYSTEM

In this method, cooling water jackets are provided around the cylinder, cylinder head, valve seats etc. The water when circulated through the jackets, it absorbs heat of combustion. This hot water will then be cooling in the radiator partially by a fan and partially by the flow developed by the forward motion of the vehicle. The cooled water is again recirulated through the water jackets.

Types of Water Cooling System

There are two types of water cooling system:

(i) Thermosiphon System. In this system the circulation of water is due to difference in temperature (i.e., difference in densities) of water. So in this system pump is not required but water is circulated because of density difference only.

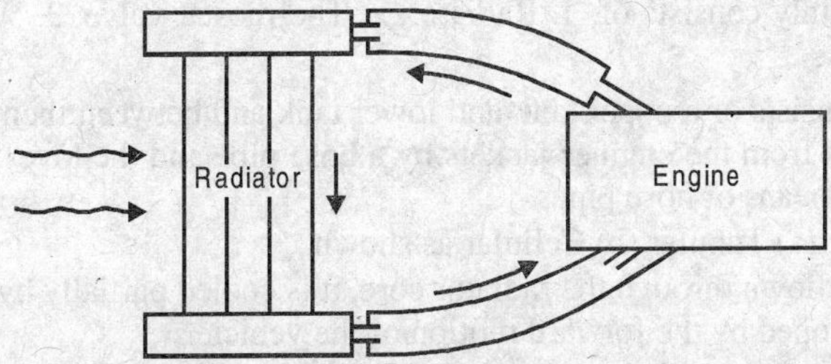

Fig. 30.2. *Thermosiphon system of cooling*

(ii) Pump Circulation System. In this system circulation of water is obtained by a pump. This pump is driven by means of engine output shaft through V-belts.

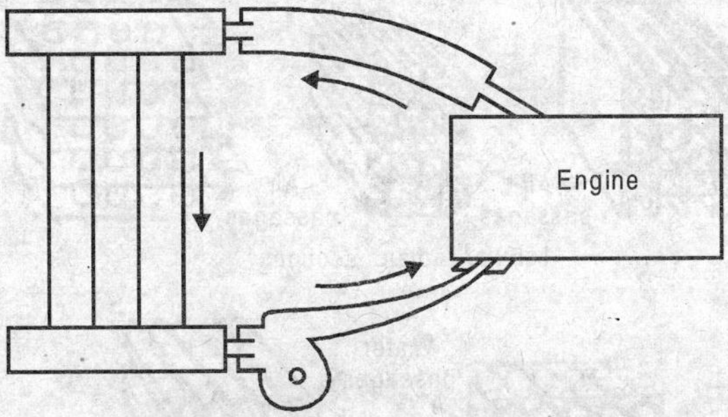

Fig. 30.3. *Pump circulation system*

30.4 COMPONENTS OF WATER COOLING SYSTEM

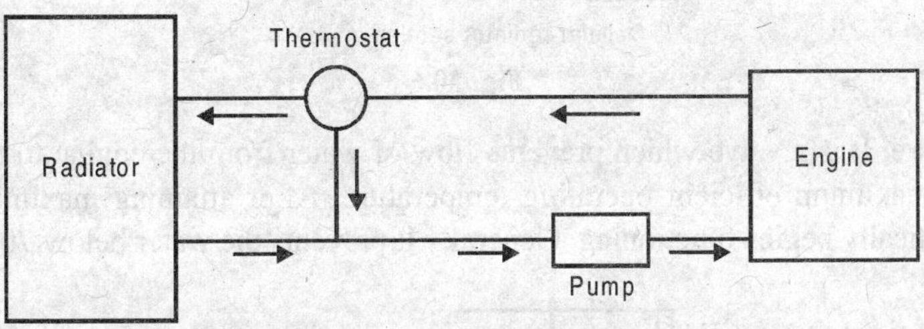

Fig. 30.4. *Water cooling system using thermostat valve*

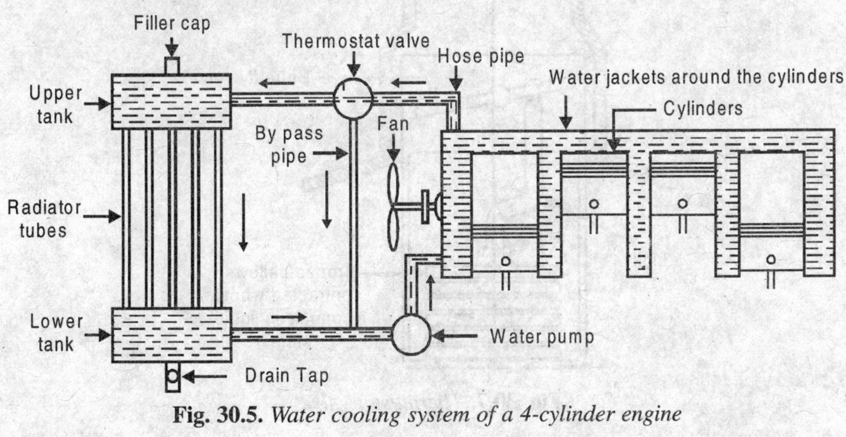

Fig. 30.5. *Water cooling system of a 4-cylinder engine*

Water Cooling System mainly consists of: 1. Radiator, 2. Thermostat valve, 3. Water pump, 4. Fan, 5. Water Jackets, 6. Antifreeze mixtures.

(1) Radiator. It mainly consists of a upper tank and lower tank and between them is a core. The upper tank is connected to the water outlets from the engines jackets by a hose pipe and the lover tank is connect to the jacket inlet through water pump by means of hose pipes.

There are 2-types of cores: (a) Tubular (b) Cellular as shown.

When the water is flowing down through the radiator core, it is cooled partially by the fan which blows air and partially by the air flow developed by the forward motion of the vehicle.

As shown through water passages and air passages, water and air will be flowing for cooling purpose.

It is to be noted that radiators are generally made out of copper and brass and their joints are made by soldering.

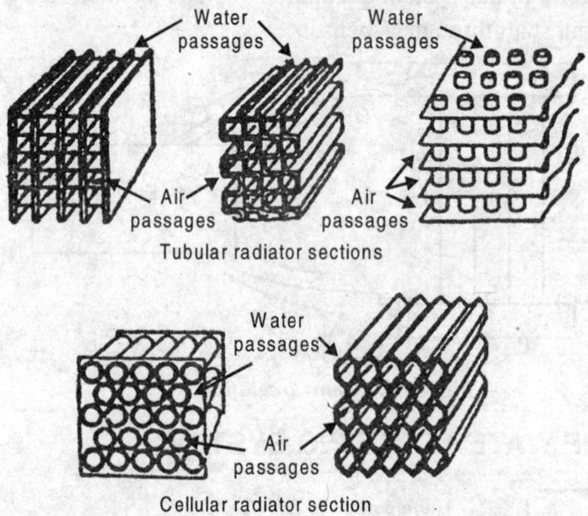

Fig. 30.6

(2) Thermostat Valve. It is a valve which prevents flow of water from the engine to radiator, so that engine readily reaches to its maximum efficient operating temperature. After attaining maximum efficient operating temperature, it automatically begins functioning. Generally it prevents the water below 70°C.

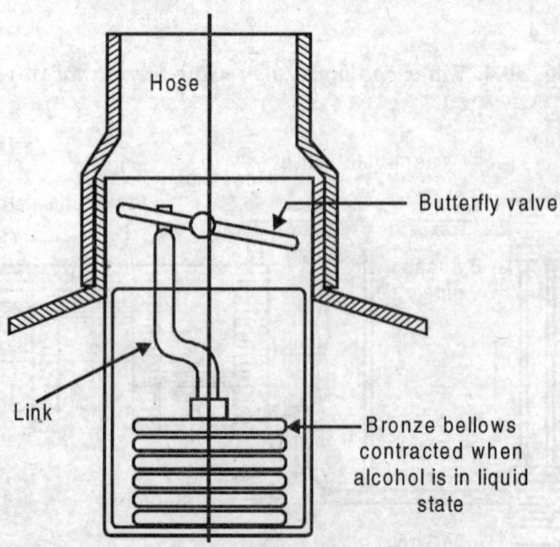

Fig. 30.7. *Thermostat valve*

Figure 30.7 shows the Bellow type thermostat valve which is generally used. It contains a bronze bellow containing liquid alcohol. Bellow is connected to the butterfly valve disc through the link.

When the temperature of water increases, the liquid alcohol evaporates and the bellow expands and in turn opens the butterfly valve, and allows hot water to the radiator, where it is cooled.

(3) Water Pump. It is used to pump the circulating water. Impeller type pump will be mounted at the front end. Pump consists of an impeller mounted on a shaft and enclosed in the pump casing. The pump casing has inlet and outlet openings.

The pump is driven by means of engine output shaft only through belts. When it is driven water will be pumped.

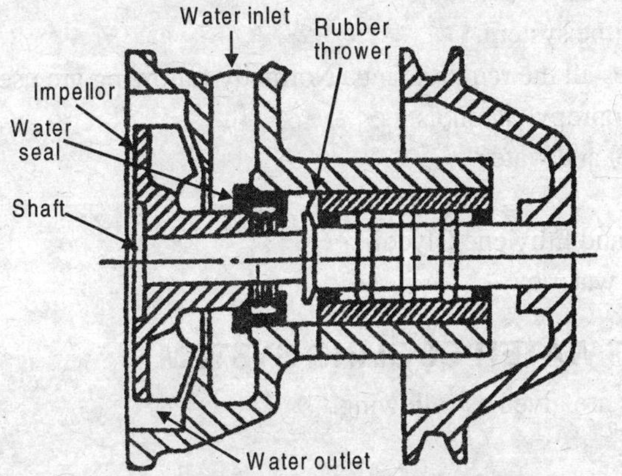

Fig. 30.8. *Water pump*

(4) Fan. It is driven by the engine output shaft through same belt that drives the pump. It is provided behind the radiator and it blows air over the radiator for cooling purpose.

(5) Water Jackets. Cooling water jackets are provided around the cylinder, cylinder head, valve seats and any hot parts which are to be cooled. Heat generated in the engine cylinder, conducted through the cylinder walls to the jackets. The water flowing through the jackets absorbs this heat and gets hot. This hot water will then be cooled in the radiator (Ref. Fig. 30.9).

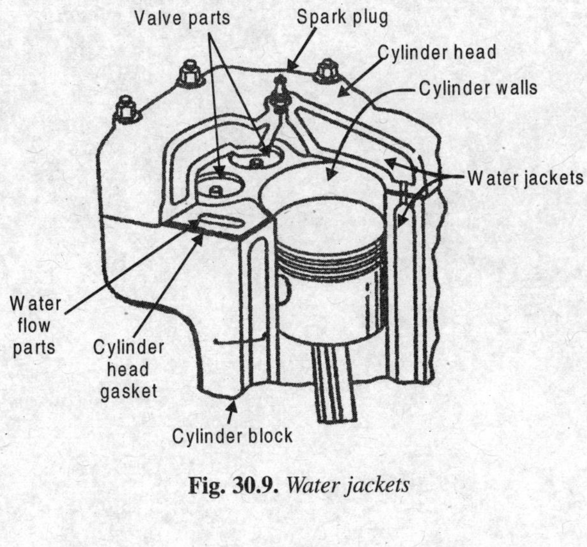

Fig. 30.9. *Water jackets*

(6) Antifreeze Mixtures. In Western countries if the water used in the radiator freezes because of cold climates, then ice formed has more volume and produces cracks in the cylinder blocks, pipes, radiator. So, to prevent freezing antifreeze mixtures or solutions are added in the cooling water.

The ideal antifreeze solutions should have the following properties.

1. It should dissolve in water easily.
2. It should not evaporate.
3. It should not deposit any foreign matter in cooling system.
4. It should not have any harmful effect on any part of cooling system.
5. It should be cheap and easily available.
6. It should not corrode the system.

No single antifreeze satisfies all the requirements. Normally following are used as antifreeze solutions.

(a) Methyl, ethyl and isopropyl alcohols
(b) A solution of alcohol and water
(c) Ethylene Glycol
(d) A solution of water and Ethylene Glycol
(e) Glycerin along with water etc.

30.5 ADVANTAGES OF WATER COOLING SYSTEM

Advantages and disadvantages are given as following.

Advantages

1. Uniform cooling of cylinder, cylinder head and valves.
2. Specific fuel consumption of engine improves by using water cooling system.
3. If we employ water cooling system, then engine need not be provided at the front end of moving vehicle.
4. Engine is less noisy as compared with air cooled engines, as it has water for damping noise.

Disadvantages

(a) It depends upon the supply of water.
(b) The water pump which circulates water absorbs considerable power.
(c) If the water cooling system fails then it will result in severe damage to engine.
(d) The water cooling system is costlier as it has more number of parts. Also it requires more maintenance and care for its parts.

CHAPTER 31

Lubrication Systems in IC Engines

31.1 INTRODUCTION

Lubrication of moving parts by means of oil, grease etc. is essential for proper functioning and to reduce friction between the moving parts of automobiles.

31.2 PURPOSE AND FUNCTIONS OF LUBRICATION

Following are the purposes and functions of lubrication:

(i) To make the moving parts function properly by reducing friction
(ii) By reducing friction wear of moving parts will be minimised
(iii) Lubricating oil will take up the shocks/vibrations by small extent
(iv) Corrosion of parts will be prevented
(v) To reduce noise
(vi) Because of lubricating oil film between piston rings and cylinder walls, it does not allow any leakage of gases across the rings.

31.3 LUBRICATING SYSTEM

The functions of lubrication system is to provide a sufficient quantity of filtered oil to all moving parts. There are 3 basic types of lubrication systems.

(i) Petroil or Mist System of Lubrication. In this system, lubricating oil will be added to petrol tank only i.e., it is mixed in the petrol in a definite proportion (i.e., generally 30–60 ml/litre e.g., Bajaj scooters, M-80, Luna, Pride etc.). Lubricating oil particles will go to engine cylinder and various parts for lubrication along with air fuel mixture. The various parts to be lubricated are as under:

Parts of IC Engines which requires lubrication are:

(1) Cylinder	(2) Piston
(3) Piston rings	(4) Gudgeon pin
(5) Crank pins	(6) Main bearings
(7) Cam and cam shaft	(8) Timing gears
(9) Rocker arms	(10) Valves etc.

(ii) Wet Sump System. These are used in relatively small engines. The bottom of the crankcase contains oil pan (sump) which acts as oil reservoir and oil cooler. Oil supplied to different parts by 2 different methods.

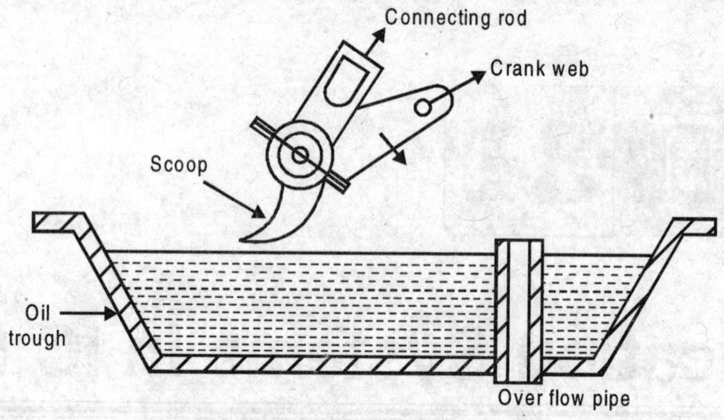

Fig. 31.1. *Splash system of lubrication*

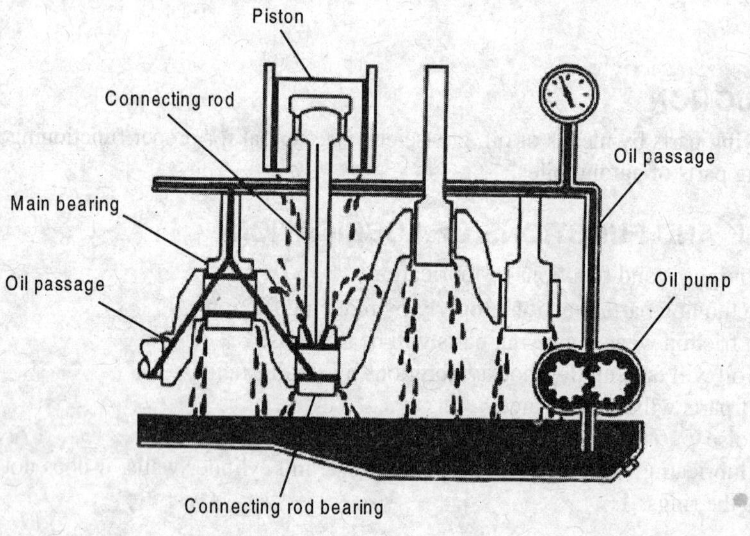

Fig 31.2. *Schematic diagram of wet sump lubrication system (Pressure force feed)*

(a) Splash Lubrication. In this type, the oil from the sump is pumped to the oil trough which is located below the crank pins. The extended end of connecting rod known as dipper strikes the oil surface and splashes oil over various parts of engine. The oil returns to sump due to gravity.

(b) Pressure Lubrication. This system mainly consists of a gear pump, oil strainer, oil gallery made up of copper tubing. In this system oil flows to different parts through the drilled passages in the crank shaft and connecting rod.

Oil pump draws filtered oil through the oil strainer and pumps it to the main oil gallery. From where it goes to the main bearings. After lubricating main bearings, lubricating oil will go to crank pin bearings through the drilled holes. And from the crank pin bearings, oil will flow to piston pin bearings through the oil hole drilled through connecting rod web. Small amount of oil is also sprayed on the cylinder walls to lubricate between cylinder walls and piston rings.

Cam shaft, valves mechanisms are lubricated through the separate lines.

(iii) Dry Sump System. In this system the oil falling from various parts is removed by a scavenging pump. Oil passes through a filter and goes to the supply tank, oil is again pumped from the supply tank and supplied to various engine parts. Thus oil is prevented from accumulating in this base of engine.

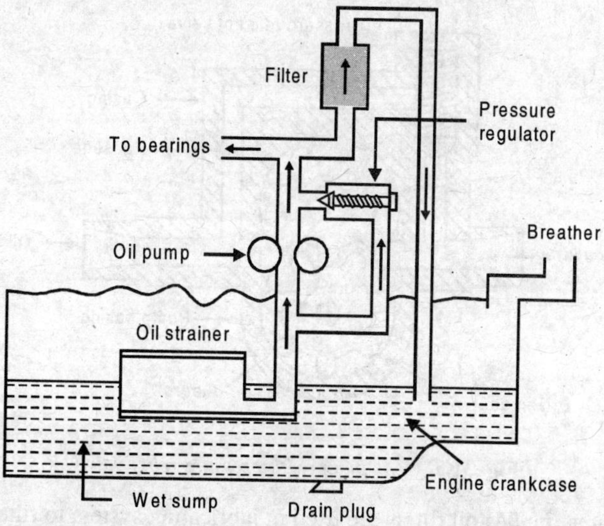

Fig. 31.3. *Schematic diagram of wet sump lubricating system*

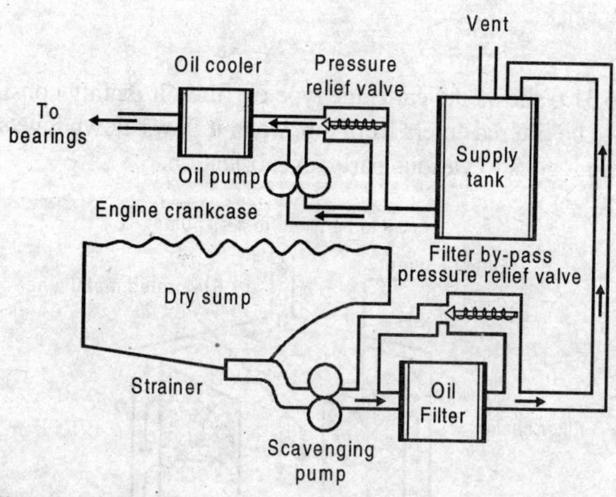

Fig. 31.4. *Schematic diagram of a dry sump lubricating system*

This uses only forced (pressure) types feed system. A filter and strainer are used in the oil circuit to remove the foregin materials and impurities. In large, IC engines oil is cooled with the help of cold water to maintain its viscosity in the required range and to prevent it from lubrication.

31.4 IMPORTANT PARTS OF LUBRICATION SYSTEM

Following are the important parts of a lubrication system:

(a) Oil sump. It contains lubricating oil and it is the bottom most part of crank case. It is made out of casting.

(b) Oil pump. Oil pump is a pump which pumps lubricating oil to various parts to be lubricated. Generally oil pump will be a gear type pump. It consists of two inter meshing gears in a pump casing. When one of the gears is rotated, other gear will also rotate and the oil will be trapped in the pockets between the pump casing and gear teeth and oil will be pumped. As shown in Fig. 31.5 it also consists of pressure relief valve to release pressures.

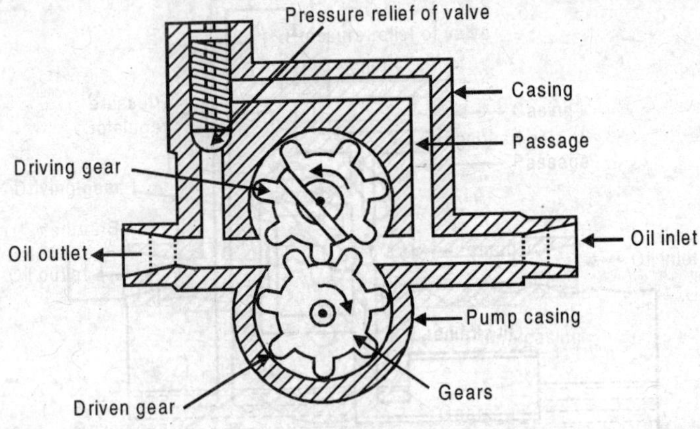

Fig. 31.5. *Gear pump with pressure relief valve*

(c) Oil filters. As the name implies oil filters are used in lubricating system to filter out the dirt particles. Various types of oil filters are as under:

(i) Cartridge type
(ii) Edge type
(iii) Centrifugal type

(i) Cartridge type. Figure 31.6 shows the cartridge type oil filter. It mainly consists of filter element enclosed in a casing as shown. The oil to be filtered enters from top, when it flows down it gets filtered and clean oil will go to main oil gallery, from where it goes to various parts for lubrication.

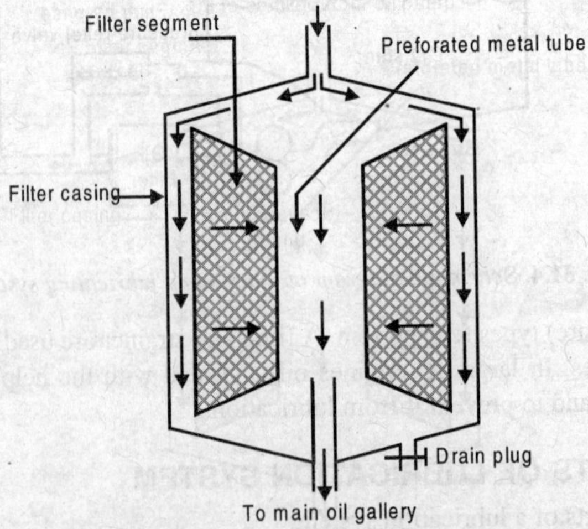

Fig. 31.6. *Cartridge type oil filter*

Filter cartridge element is cleaned oftenly or replaced if required during servicing.

(ii) Edge type oil filter. Figure 31.7 shows an edge type oil filter. It mainly consists of a central spindle to which a number of discs are attached as shown. Also there is a square rod having alternate discs attached. The clearance between the discs on central spindle and square rod is very very small. When the impure oil enters from oil inlet and flows through the discs to oil outlet, impurities are collected in the clearance between the discs and clean oil comes out from oil outlet.

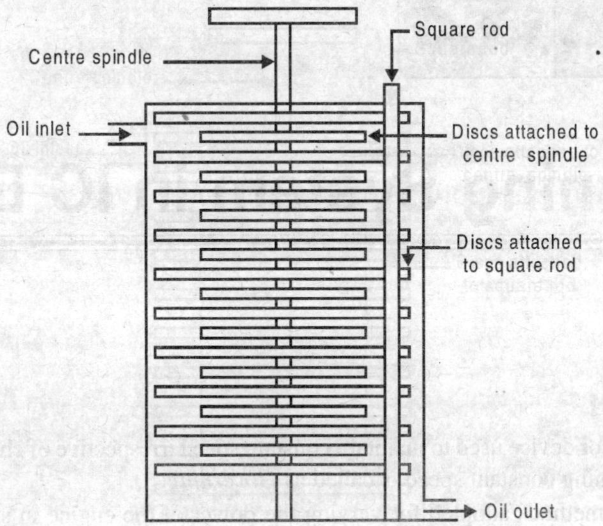

Fig. 31.7. *Edge type oil filter*

(iii) Centrifugal type of filter. Figure 31.8 shows the centrifugal type oil filter. It mainly consists of a central hallows spindle, rotor casing, oil tubes, and stationary casing as shown.

Impure oil enters from oil inlet and flows to oil tubes through the spindle holes. From the oil tubes oil jets are issued under pressure. Oil jets when issued it impinges on the walls of stationary casing and rotor starts rotating, pure oil gets separated and comes out from oil outlet.

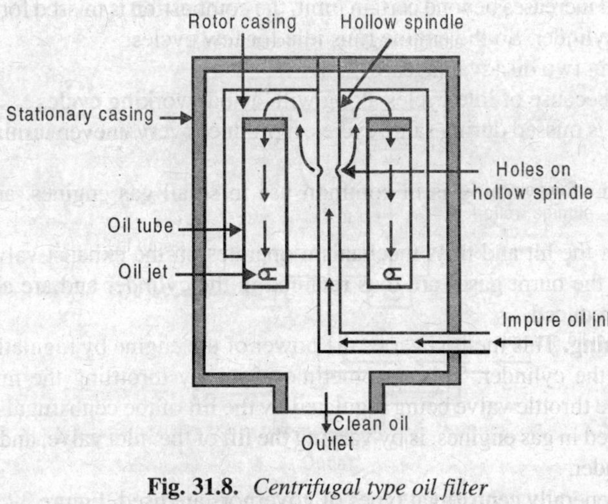

Fig. 31.8. *Centrifugal type oil filter*

(d) Oil strainers. It consists of a wire mesh. It prevents any impurities from entering the oil pump. ■■■

CHAPTER 32

Governing System in IC Engines

32.1 INTRODUCTION

Governor is the mechanism or device used to maintain constant speed irrespective of changes of load on the engine. And the method of maintaining constant speed is called as *Governing*.

There are three distinct methods adopted for varying the power of the engine in sympathy with the action of governor, these methods will again differ in the mechanical means adopted in regulating the power. In all the cases the ordinary centrifugal governor is used for controlling the mechanical devices.

32.2 METHODS OF GOVERNING

The following is the usual classification adopted for the various methods of governing.

(a) Hit and Miss Governing. In this type of governing, the action of the governor causes the fuel inlet valve to remain closed during a cycle so that no fuel is admitted. The engine will thus perform an idle cycle. Or in other words, when the engine speed increases beyond certain limit, the combustion is missed for a few cycles by preventing the charge to enter into the cylinder. So the engine runs idle for few cycles.

It results into the following two disadvantages.

1. Efficiency decreases because of idle cycles along with actual working cycles.
2. Since the combustion is missed during same cycles, it produces very uneven turning movement for the crank shaft.

The hit and miss method of governing is in common use in small gas engines, and is actually quantitative governing.

In small, light oil engines the hit and miss mechanism operates on the exhaust valve and prevents it opening when the speed is too high; the burnt gases are thus retained in the cylinder and are alternately compressed and expanded until the speed is reduced.

(b) Quantitative Governing. This method varies the power of the engine by regulating the quantity of mixture (fuel and air) which enters the cylinder. This is sometimes done by throttling the mixture before it enters the cylinder, the movement of the throttle valve being regulated by the lift of the centrifugal governor. Another method of quantitative governing, used in gas engines, is by varying the lift of the inlet valve, and thus reducing the quantity of mixture entering the cylinder.

So, as discussed above, generally centrifugal types of governors are used. Figure 32.1 shows a centrifugal type of governor used with petrol engine.

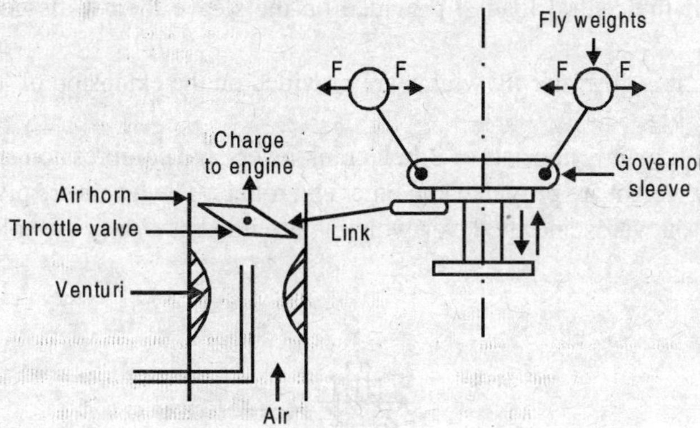

Fig. 32.1. *Centrifugal governor*

Cylindrical part is the air horn and the narrowest cross-section of which is the venturi. Throttle valve is provided to regulate the amount of charge (air + fuel mixture) entering the engine cylinder. Throttle valve is connected to the Governor as shown.

Now when the load on the engine increase, the speed of the engine decreases and because of reduced speed, fly weights move inwards (shown by dotted arrows) and the sleeve moves downwards. Then its opens the throttle valve more and hence more charge enters the engine cylinder to develop more power and to bring engine speed constant.

And when the load on the engine decreases—speed increases—fly weights move outwards (shown by arrow)—close the throttle valve accordingly—reduce the supply of charge. Hence the engine develops required less power and maintain constant speed.

(c) Qualitative Governing. In this method of governing the strength of the mixture is altered by the action of the centrifugal governor, thus varying the quality of the mixture. This is done by regulating the amount of fuel entering the cylinder, air supply remaining constant. This method of governing is used in all heavy oil engines using pure air compression.

The amount of fuel entering the cylinder is varied by the action of the governor. It may vary the stroke of the oil pump or it may by pass part of the fuel back to the oil tank or it may delay the closing of the suction valve of the fuel pump. All of these methods will cause a variation in the point of cut off.

32.3 IMPORTANT TYPES OF CENTRIFUGAL GOVERNORS USED WITH THE ENGINES

Following are the important centrifugal governors.

(i) Watt Governer (Simple Conical Governer). It is named after Watt, when the load decreases-speed increases fly weights move outwards as shown by arrows and lift the sleeve. On the other hand when the load increase speed decreases, weight move inwards (shown by dotted arrows) and sleeve moves downwards. (Ref. Fig. 32.2)

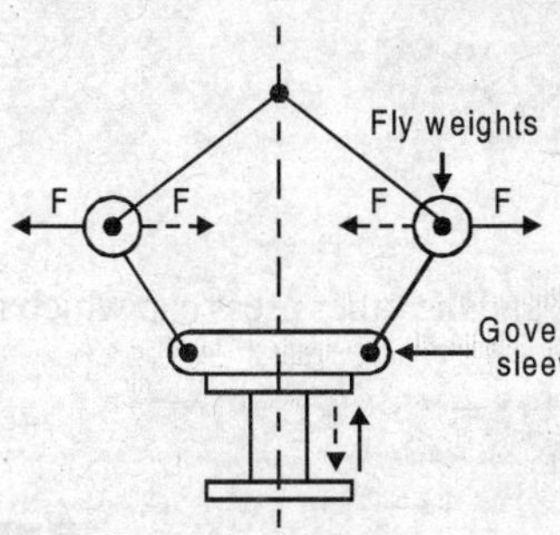

Fig. 32.2

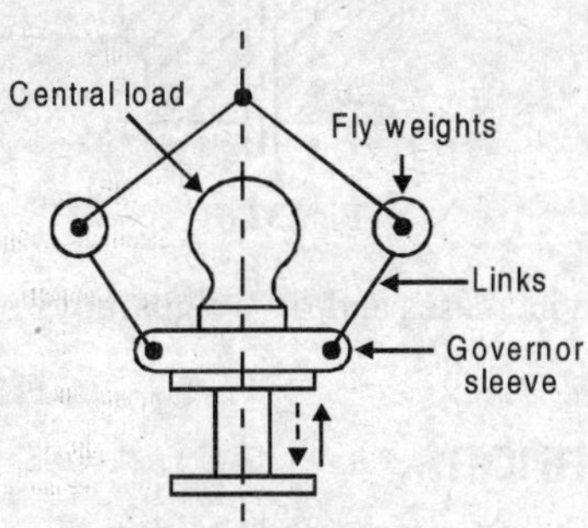

Fig. 32.3

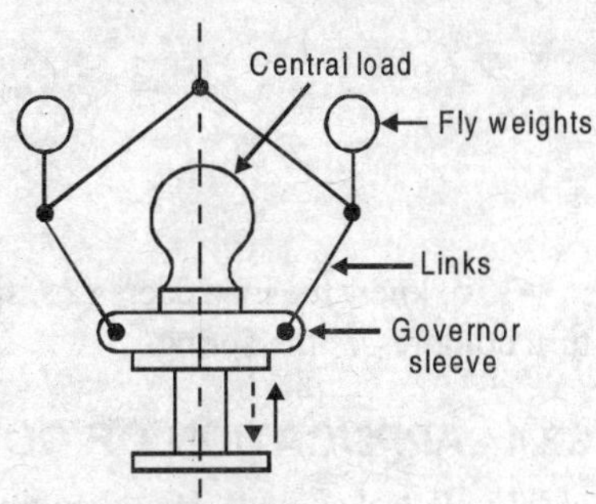

Fig. 32.4

(ii) Porter governor: If the central load is provided on the sleeve then watts governor is called its Porter Governor. (Ref. Fig. 32.3)

(iii) Proell governor. In this governer fly wights are provided on the extension of lower links as shown and it contains central load. (Ref. Fig. 32.4)

(iv) Hartnell governor. It mainly consists of 2-bell crank levers and compression spring as shown. To the one end of bell crank levers, fly weight are provided and other end rollers. The levers are pivoted at *O*. When the speed increases fly weight move outwards and the lever which is provided at *O* lifts the rollers and sleeve against the compression of spring.

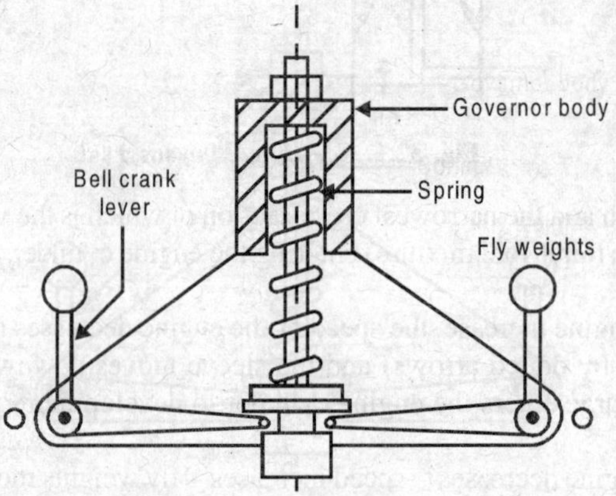

Fig. 32.5

(v) Spring Controlled Governor of Gravity Type. Figure 32.6 shows spring controlled gravity governor. It mainly consists of 2-bell crank levers. They are provided with rollers at one end and fly weights at the other. Rollers will be resting over the cap which forms top end of shaft. Under the cap spring is provided as shown.

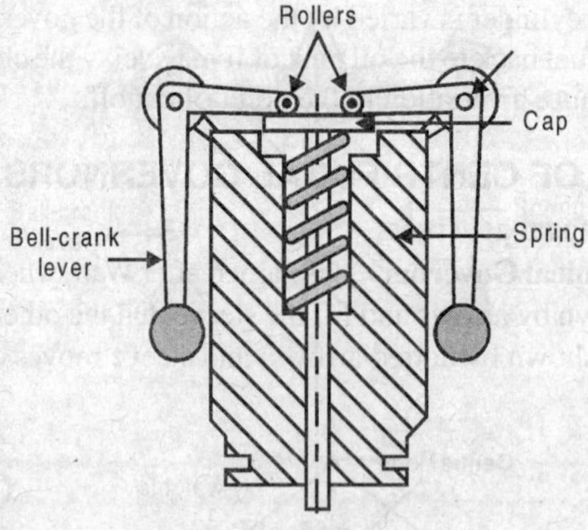

Fig. 32.6

As we know the load decreases, speed increases and fly weight move outwards and the other press cap which in turn compresses the spring.

32.4 APPLICATION OF GOVERNORS

Used in Petrol engines, Diesel engines, Steam engines and Steam turbines etc.

CHAPTER 33

Supercharging and Turbocharging

33.1 INTRODUCTION

Basically supercharging means supplying high pressure air to the engine. When the high pressure air is supplied to the engine, the mass flow rate of fuel also increases and consequently it results in more power output. Hence the efficiency of the engine also increases.

So, the output power developed by the engine mainly depends upon:

(i) Amount of air supply
(ii) Amount of air used in combustion for generating power
(iii) Thermal efficiency

The amount of air supplied to the engine may be increased by increasing the speed of engine or by supplying high pressure air. The process of supplying high pressure air is known as **Supercharging**. And equipment used for supplying high pressure air is known as **Supercharger**.

Note that **Turbocharging** is a particular case of supercharging in which compressor or high pressure air supplying equipment in driven by the exhaust gas driven turbine. This will be clearly understood while studying the methods of supercharging.

33.2 NEED FOR SUPERCHARGING OR OBJECTS OF SUPERCHARGING

Objects of Supercharging are discussed as follows:

(1) Supercharging is essentially required in order to increase the power of the given engine.
(2) Supercharging is essential in order to increase the output power of the given engine, for given weight. It is to be noted that the weight factor is important in aeroplane engines.
(3) We know that, as we go higher and higher, the air density decreases, so the engines get less air/oxygen and the output of the engines decreases and this decrease in power will be about 10% in one kilometre.

33.3 SUPERCHARGING OF PETROL (SI) ENGINES

In case of petrol (SI) engines supercharging is used only when the power output is the main consideration. So it is used in the following engines

(a) Aeroplane engines
(b) Racer car engines.

As we have studied, by using supercharger, we will be supplying high pressure air so:

(1) it increases air supply
(2) it increases the volumetric efficiency of the engine
(3) pressure of air at the inlet to the engine increases
(4) as the pressure of air increases, temperature of air at the inlet also increases.

Thus when the high pressure and high temperature air is supplied to the engine, it decreases the ignition delay and this leads into preignition and more detonation problems.

During supercharging more amount of air is supplied and consequently more amount of fuel is supplied and this results in more fuel consumption. But power developed will be more than the naturally aspirated engines. (Note that the naturally aspirated engines are those engines which take air directly from the atmosphere and if the high pressure air is supplied to the engines, then they are known as supercharged engines).

In order to summarise, supercharging of petrol (SI) engines is not generally used because of more fuel consumption, but is used when more power is to be developed. This is also more clear from the graph. (Fig. 33.1).

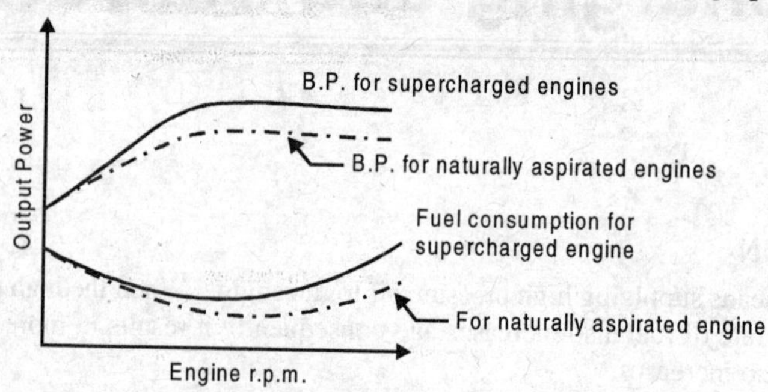

Fig. 33.1

33.4 SUPERCHARGING OF DIESEL (CI) ENGINES

In case of petrol engines supercharging results in more specific fuel consumption. But whereas in case of diesel engines when high pressure, high temperature air is supplied to the engine, it results in better combustion. Also any existing combustion problem will be sorted out because of higher temperatures. Also it decreases ignition delay. Because of this improved combustion any low quality of diesel fuel can also be used in the supercharged engines.

Fuel consumption of supercharged diesel engines is comparatively less than the naturally aspirated engines because of better combustion.

Supercharged diesel engines are generally used because of their better combustion and improved mechanical efficiency.

33.5 SUPERCHARGING EFFECTS ON THE ENGINE PERFORMANCE

Following are the effects on the performance of the engine because of supercharging:

(1) As we have studied, in supercharged engines more amount of high pressure and high temperature air is supplied, which results in increased volumetric efficiency and more output power.
(2) Because of more pressures, scavenging (i.e., removing the burnt gases out of the engine cylinder) will be improved.
(3) By supercharging mechanical efficiency increases.
(4) For petrol engines supercharging results in better mixing of air and fuel. Note that as more amount of air is supplied, more amount of fuel is also required to be supplied. Hence specific fuel consumption will be more when compared with naturally aspirated engines.
(5) For Diesel engines by supercharging there is improvement in combustion and hence increase in mechanical efficiency.

33.6 TYPES OF SUPERCHARGERS

Following are the important types of superchargers which are generally used for supercharging.

1. Reciprocating Compressors. Reciprocating air compressor takes air from the atmosphere during the suction stroke and it will be delivered to the engine after increasing the pressure. These type of reciprocating air compressors are very rarely used because of their heavy weight and more bulk. However they are used with some stationary engines.

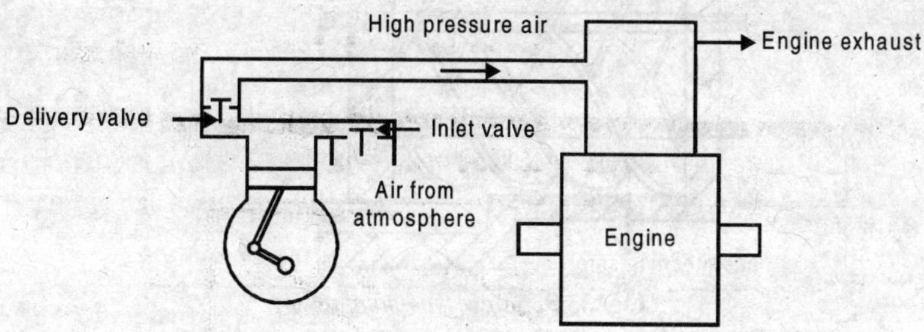

Fig. 33.2

2. Vane Blower. Figure 33.3 shows a vane blower. It consists of a rotating disc with radial slots into which vanes can slide. The disc will be rotating eccentrically in the casing. When the disc is rotated, the vanes will move outwards and press against the casing due to centrifugal action. And the air entering from the atmosphere will be trapped between the vanes and casing and will be supplied to the engine after increasing the pressure.

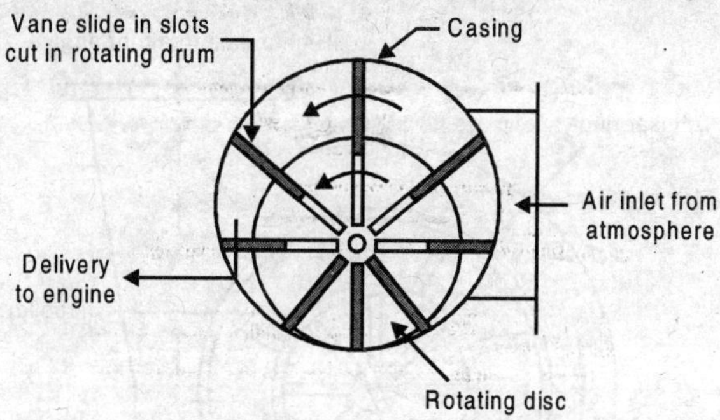

Fig. 33.3. *Vane type compressor*

Because of vanes the supply of air is not continuous and vanes produce noise. Hence these are not used nowadays.

3. Roots Blower. Figure 33.4 shows the Roots Blower. It essentially consists of two intermeshing lobes rotating in the casing. Air from the atmosphere enters at the inlet and the pressure of air increases because of squeezing action of lobes rotating in the casing. And the high pressure air is supplied from the delivery end.

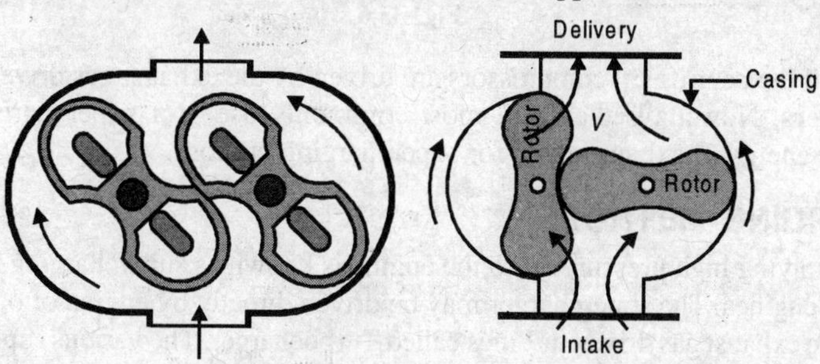

Fig. 33.4. *Root blower*

4. Screw Compressor. Screw type compressors is shown in Fig. 33.5. It has helical screws that match and mesh with the helix of the screw. The air is taken in from one side. The entrapped air progressively passes through the narrow passage ways formed by the locks and cover.

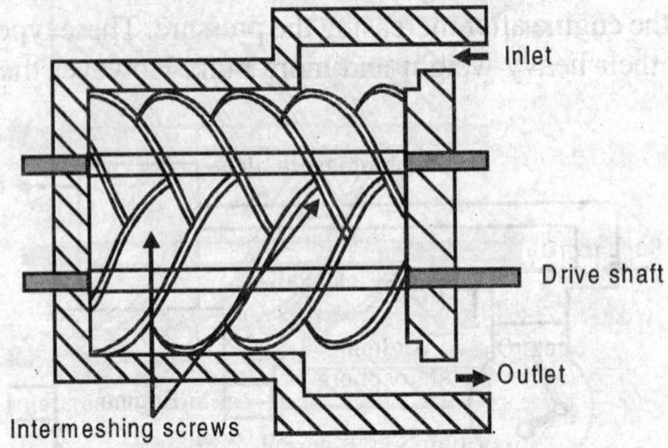

Fig. 33.5. *Screw type compressor*

When the entrapped air reaches the wall it gets compressed and is taken out of the outlet valve. This process happens in a fraction of a second because the screw is rotated at a very high speed.

(5) Centrifugal Compressors. It consists of an impeller rotating in the casing as shown. In this case air enters axially at the centre or at the hub of impeller and will be turned through right angles because of blades of impeller. And in doing so it imparts high velocity to the air because of centrifugal action. The pressure of this high velocity air is increased as it flows through the diffuser and this high pressure air is supplied to the engine.

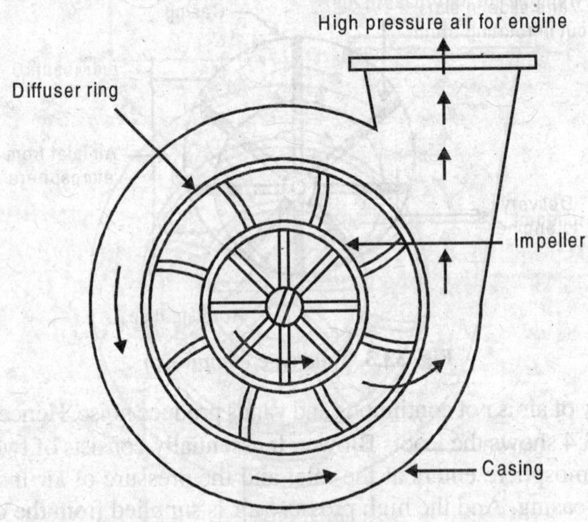

Fig. 33.6

As we have studied, if the centrifugal compressors are driven by the exhaust gas driven turbines, then they are known as **Turbochargers**. Nowadays these are most commonly used for supercharging the engines. These turbochargers utilise the energy of exhaust gases for supercharging purpose.

33.7 SUPERCHARGING METHODS

As we have studied, supplying high pressure air to the engine is known as supercharging. And the supercharger is used to supercharge the engines. The supercharger may be driven directly by engine of output shaft through some gears or it is driven by the exhaust gas driven turbines called Turbocharger. The various supercharging arrangements generally used are as under.

(i) As shown in Fig. 33.7 the compressor is driven by the engine output shaft through some gears. The compressor supplies high pressure air to the engine. Finally the gases are exhausted as shown. Note that a part of total output is used for driving the compressor and net output power increases because of supercharging.

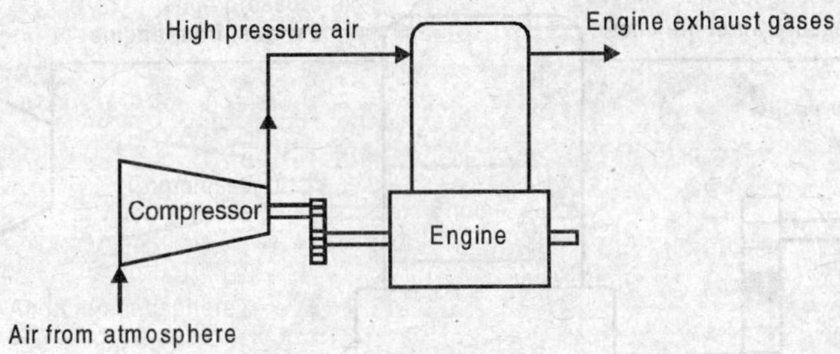

Fig. 33.7

(ii) As shown in Fig. 33.8 (a) exhaust gases from the engine drive the turbine which in turn drives the compressor. In Fig. 33.8 (b) turbine and compressor are coupled as shown. The compressor takes the air from atmosphere and after increasing the pressure it will be supplied to the engine. And this type of arrangement is nowadays generally used and is called as **Turbocharging.**

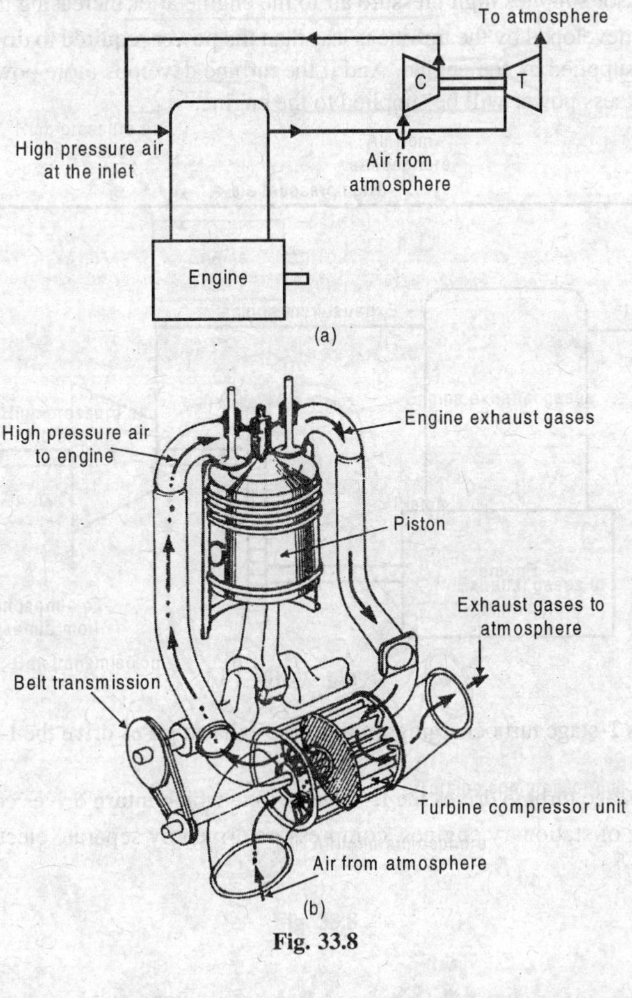

Fig. 33.8

(iii) As shown in Fig. 33.9 the complete power developed by the engine is supplied to drive the compressor. The compressor takes the air from atmosphere and after increasing the pressure it will be supplied to the engine. Finally exhaust gases from the engine drive the turbine to give power output.

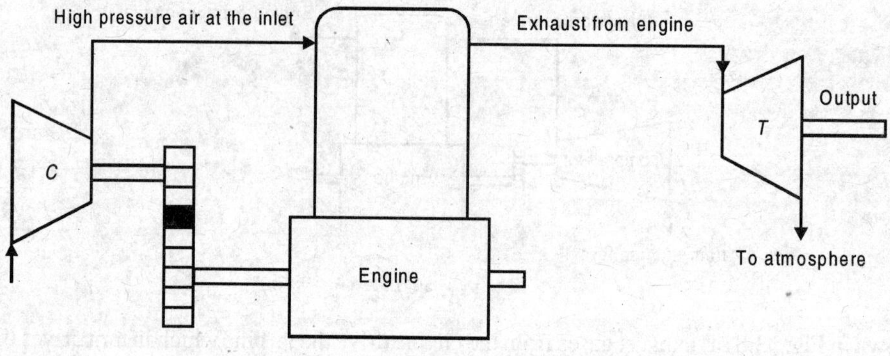

Fig. 33.9

(iv) As shown in Fig. 33.10 in this case exhaust from the engine drives the turbine and the turbine drives the compressor. The compressor supplies high pressure air to the engine after increasing the pressure.

Note that if the power developed by the turbine is less than the power required to drive the compressor, then the necessary power will be supplied by the engine. And if the turbine develops more power than what is required to drive compressor, then excess power will be supplied to the engine.

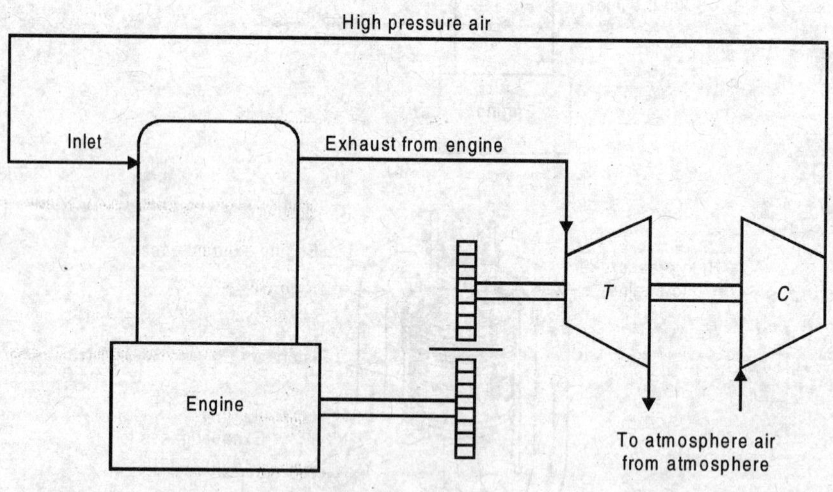

Fig. 33.10

(v) Figure 33.11 shows 2-stage turbocharging. In this case exhaust gases drive the I-stage turbine which in turn drives the compressor-I.

Then the exhaust of I-stage turbine drives the II-stage turbine which inturn drives compressor II.

(vi) For supercharging of stationary engines, compressors driven by separate electrical motors may be used.

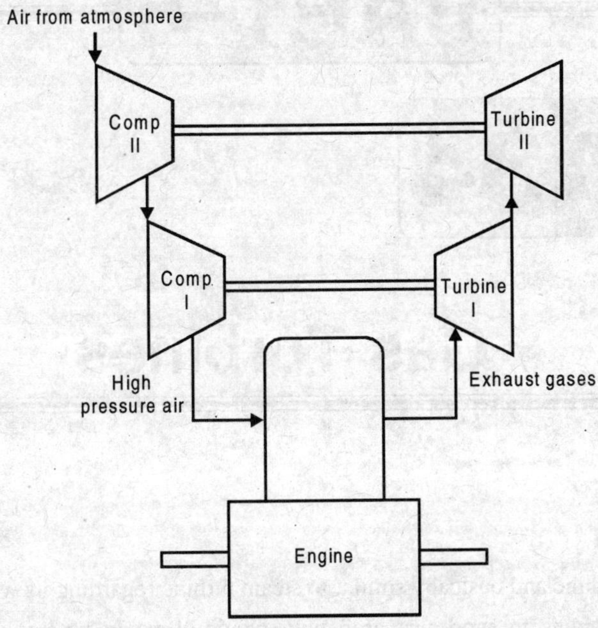

Fig. 33.11

33.8 LIMITATIONS OF SUPERCHARGING FOR SI AND CI ENGINES

The following are the limitations of supercharging for SI and CI engines.

1. For petrol (SI) engines the main limitation of supercharging is knocking. As we know during supercharging pressure and temperature of air and fuel increases, which reduces ignition delay and results in engine knocking problems.
2. For Diesel (CI) engines the main limitation of supercharging is because of thermal loading. In this case the temperatures of piston and cylinder is very-very high and because of this it results in scuffing of piston rings and wearing of cylinder liners. Also because of higher pressures loads on the bearings will increase.
3. In general for supercharged engines because of high pressure and high temperature of products of combustion, the engine components are to be designed to take up (sustain) higher pressures and temperatures i.e., the engine components are to be made of better materials, so that they can withstand high temperatures encountered in supercharged engines.
4. If the air pressure is to be increased, power input to the supercharger is to be increased. Note that the net power increases because of supercharging.

■■■

PART–6

CHAPTER 34

Gas Turbines

34.1 INTRODUCTION

A gas turbine is a turbo-machine and basically similar to steam turbine regarding its working principle.

The first turbine to produce useful work was probably a windmill, where no compression and no combustion exist. The gas turbine as we know today include a compression process and a heat addition or combustion process. Joule and Braylon independently proposed the cycle that is the ideal prototype of the actual unit. In 1906, a unit that produced net-power had been built.

There were two main reasons or obstacles to be overcome as revealed by thermodynamic analysis. To produce and deliver a sizeable amount of power, it is noted that—

- the temperature at the beginning of expansion must be high and the lack of materials to withstand these high temperatures.
- the compressor and turbine must operate at a high efficiency.
- for high efficiency, the lack of knowledge of aerodynamic and thermodynamics of flow mechanism.

Metallurgical developments are raising the highest permissible temperatures of about 800ºC to 1100ºC.

A better knowledge of aerodynamics has been responsible for improving the efficiency of both the centrifugal and axial compressors.

The gas turbines are also driven by the exhaust of internal combustion engines. Such turbines are used for supercharging IC Engines.

34.2 CLASSIFICATION OF GAS TURBINES

Basically, gas turbines are classified according to three main events. These are—

1. Combustion process
2. Path of working substance
3. Action of combustion gases in turbine.

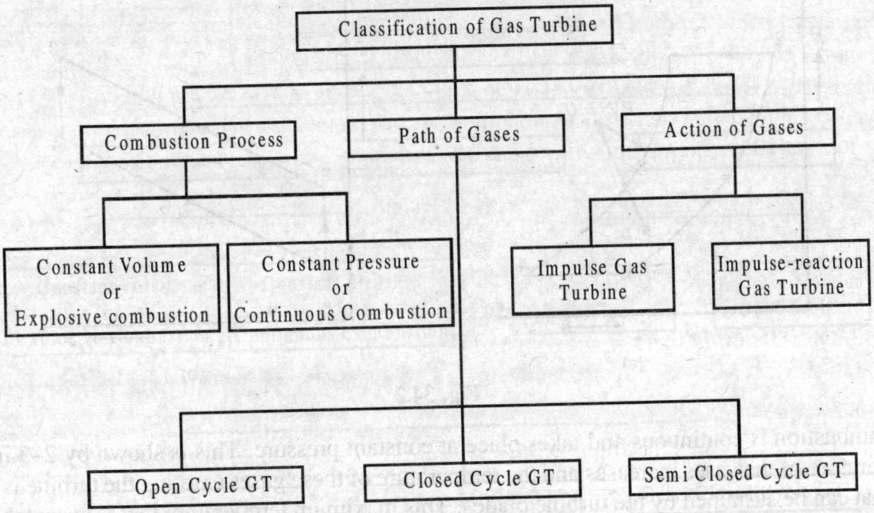

34.3 WORKING OF SIMPLE, CONSTANT PRESSURE, OPEN-CYCLE GAS TURBINE

The schematic diagram of a simple, Constant Pressure Cycle is shown in Fig. 34.1 as given below.

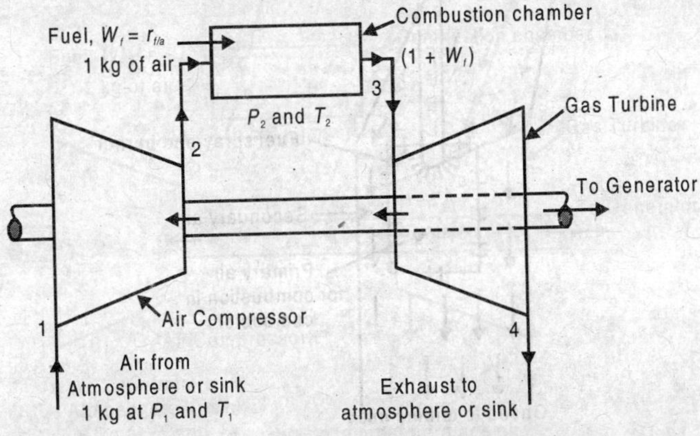

Fig. 34.1

This gas turbine works on Joule or Brayton cycle. Constant Pressure Cycle is Joule cycle. Generally, standard cycle for gas turbine working on constant pressure cycle is Joule cycle and here mass of fuel is generally neglected.

In Brayton cycle analysis, mass of fuel is taken into consideration.

Many times, Joules cycle and Brayton cycle are taken as identical.

Air continuously enters the compressor from atmosphere at P_1, v_1, and T_1. This state is shown on diagram by 1. The compressor may be of axial or centrifugal type. Air is compressed isentropically to state 2 given by P_2, v_2, T_2.

This process of compression is shown by 1–2 on *P–V* and *T–S* diagrams, as shown in Fig. 34.2 (a) and (b).

After compression, air enters the combustion chamber in the open-cycle. Some of this air goes around the outside of the combustor proper and some supplies oxygen for burning the fuel which is continuously injected as shown in Fig. 34.3. Both these streams mix together so that the temperature of the gases formed is within the limits required for the gas turbine inlet.

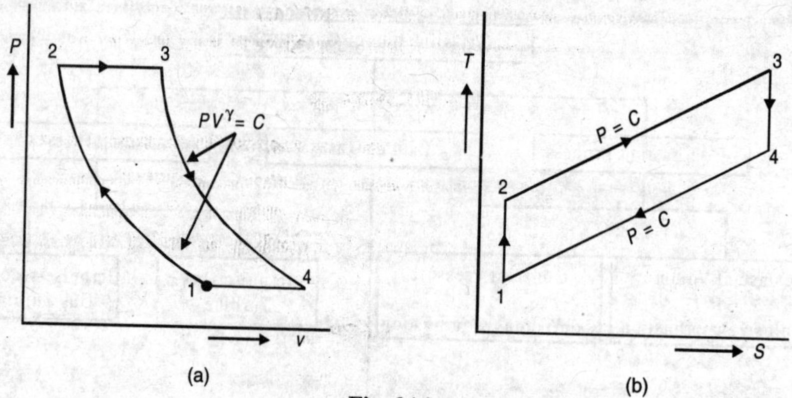

Fig. 34.2

Here the combustion is continuous and takes place at constant pressure. This is shown by 2–3 in Fig. 34.2 (a) and (b). Temperature of the gases increases and the temperature of these gases entering the turbine is the maximum temperature that can be sustained by the turbine blades. This maximum temperature that is acceptable to the blade material is known as *ceiling temperature.*

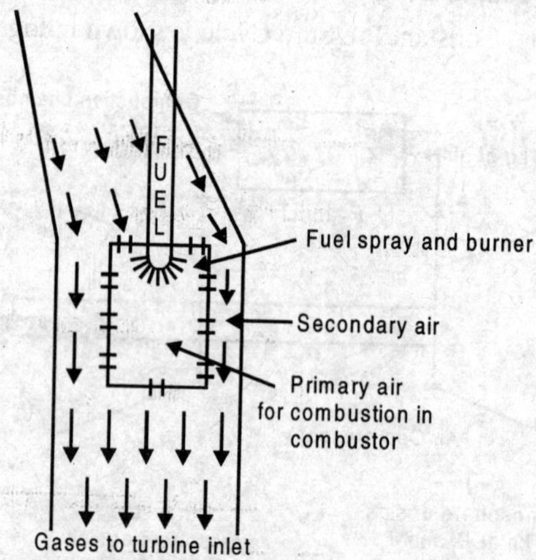

Fig. 34.3. *Combustion chamber*

Gas at the state 3 expands isentropically through the turbine and this expansion is represented by 3–4 as shown in Fig. 34.2. This expansion is complete expansion upto atmospheric pressure as against an expansion in an IC Engine and is an incomplete expansion.

After the expansion, gases will be exhausted—in open cycle—to the atmosphere in condition 4. During this process, the heat is rejected to the atmosphere. This rejection of heat takes place at constant pressure. This process is represented by 4–1 on *P–v* and *T–S* diagrams of Fig. 34.2. Thus the cycle is complete.

You will note that all the processes are continuous. Thus we say that

1–2 is isentropic compression.

2–3 is constant pressure heat addition

3–4 is isentropic expansion and

4–1 is constant pressure heat rejection.

Generally, turbine is coupled to compressor and generator. Turbine supplies power required to drive the compressor and the remaining power to generator or to the aircraft (jet or propeller or both).

It is observed that 60–70 % of the power developed by the turbine is absorbed by the compressor and the remaining power is used to generate power or power required to operate the aircraft.

34.4 ANALYSIS OF THE SIMPLE JOULE OR BRAYTON CYCLE

To analyse the simple cycle, we have to assume certain facts. These assumptions are

(a) Compression and expansion are isentropic.

(b) Heat addition process in the combustion chamber and the heat rejection process are assumed to take place at constant pressure.

(c) There is no pressure loss in the combustion chamber, connecting pipes, and heat exchangers.

(d) The working fluid is assumed to be perfect or ideal gas and that the composition of the fluid does not change during its passage through the cycle.

(e) The flow of fluid is steady flow and the mass flow is constant throughout the cycle is constant.

(f) Other losses during the operation of the machine/turbine are absent. (Frictional loses at the bearings etc. are assumed to be absent).

(g) The velocities of the gas flowing between the two components remain constant.

With these assumptions we will analyse the Joule/Brayton cycle. Consider 1 kg of fluid.

Work of compressor is given by

$$W_C = \frac{P_1v_1 - P_2v_2}{\gamma-1}$$

$$= \frac{R}{\gamma-1}[T_1 - T_2]$$

$$= C_V[T_1 - T_2] \text{ and is –ve.}$$

$$= C_V(T_2 - T_1) \text{ Positive quantity.}$$

Similarly, work of expansion in turbine is given by

$$W_t = \frac{P_3V_3 - P_4V_4}{\gamma-1}$$

$$= \frac{R}{\gamma-1}[T_3 - T_4] \text{ and is positive.}$$

Flow work during 2–3 is given by

$$W_{f_{2\text{-}3}} = (P_3v_3 - P_2v_2) = R[T_3 - T_2]$$

and during 4–1,

$$W_{f_{4\text{-}1}} = (P_4v_4 - P_1v_1) = R[T_4 - T_1]$$

∴ The net work done/kg

$$W_n = C_V(T_3 - T_4) - C_V(T_2 - T_1) + R(T_3 - T_2) - R(T_4 - T_1)$$

$$= C_V[T_3 - T_2 - T_4 + T_1] + R[T_3 - T_2 - T_4 + T_1]$$

$$= (C_V + R)(T_3 - T_2 - T_4 + T_1)$$

$$= C_P(T_3 - T_2 - T_4 + T_1)$$

$$= C_P(T_3 - T_2) - C_p(T_4 - T_1)$$

$$= Q_A - Q_R \quad \text{...(1)}$$

where Q_A = Heat added/kg

Q_R = Heat rejected/kg

OR

Work of a open cycle or flow cycle 1–2–3–4 is given by the area of P–v diagram or area 1–2–3–4–1.

$$\therefore \quad W_n = \text{Area } 1\text{–}2\text{–}3\text{–}4\text{–}1$$

$$= \begin{bmatrix} \text{Area behind the} \\ \text{curve } 3-4 \end{bmatrix} - \begin{bmatrix} \text{Area behind the} \\ \text{curve } 2-1 \end{bmatrix}$$

$$= \frac{\gamma}{\gamma-1}[P_3v_3 - P_4v_4] - \frac{\gamma}{\gamma-1}[P_2v_2 - P_1v_1]$$

$$= \frac{\gamma R}{\gamma-1}[T_3 - T_4] - \frac{\gamma R}{\gamma-1}[T_2 - T_1]$$

$$= C_P(T_3 - T_4) - C_p(T_2 - T_1)$$

$$= C_P[(T_3 - T_2) - (T_4 - T_1)]$$

$$= C_P(T_3 - T_2) - C_p(T_4 - T_1) \text{ as before}$$

$$= Q_A - Q_R \quad \text{...(2)}$$

$$= (h_3 - h_4) - (h_2 - h_1) \quad \text{...(3)}$$

= Turbine work – Compressor work

$$= W_t - W_c$$

∴ Net work of the cycle/kg

W_n = Work of turbine – Work of compressor

$$= W_t - W_c \quad \text{...(4)}$$

Heat added during the cycle/kg of fluid or air is given by

$$Q_A = C_P(T_3 - T_2) \quad \text{...(5)}$$

∴ Thermal efficiency of Joule cycle or Brayton cycle is given by

$$\eta_{th} = \frac{\text{Net work of cycle}}{\text{Heat supplied}}$$

$$= \frac{W_t - W_c}{Q_A} = \frac{Q_A - Q_R}{Q_A}$$

$$= 1 - \frac{Q_R}{Q_A}$$

$$= 1 - \frac{C_P(T_4 - T_1)}{C_P(T_3 - T_2)}$$

$$\boldsymbol{\eta_{th} = 1 - \frac{T_4 - T_1}{T_3 - T_2}} \qquad \text{...(6)}$$

This efficiency is given in terms of temperatures.
Referring to *P–v* diagram of Fig. 34.2, we have

$$\frac{T_2}{T_1} = \left(\frac{P_2}{P_1}\right)^{\frac{\gamma-1}{\gamma}}$$

and
$$\frac{T_3}{T_4} = \left(\frac{P_3}{P_4}\right)^{\frac{\gamma-1}{\gamma}}$$

But
$$\frac{P_2}{P_1} = \frac{P_3}{P_4} = \text{Pressure ratio}$$

$$= r_p$$

$$\therefore \quad \frac{T_2}{T_1} = (r_p)^{\frac{\gamma-1}{\gamma}} \text{ and } \frac{T_3}{T_4} = (r_p)^{\frac{\gamma-1}{\gamma}}$$

$$\therefore \quad T_2 = T_1 - r_p^{\frac{\gamma-1}{\gamma}} \qquad \text{...(7)}$$

$$T_3 = T_4 \cdot r_p^{\frac{\gamma-1}{\gamma}} \qquad \text{...(8)}$$

Substituting these values in Eq. 6 we get

$$\eta_{th} = 1 - \frac{T_4 - T_1}{T_4 \cdot r_p^{\frac{\gamma-1}{\gamma}} - T_1 \cdot r_p^{\frac{\gamma-1}{\gamma}}}$$

$$= 1 - \frac{T_4 - T_1}{r_p^{\frac{\gamma-1}{\gamma}} (T_4 - T_1)}$$

$$= 1 - \frac{1}{r_p^{\frac{\gamma-1}{\gamma}}}$$

$$\boldsymbol{\eta_{th} = 1 - \frac{1}{r_p^{\frac{\gamma-1}{\gamma}}}} \qquad \text{...(9)}$$

It is observed from Eq. (9) that the thermal efficiency of Joule or Brayton cycle depends only on r_p, the pressure ratio. As the pressure ratio increases, thermal efficiency of the Joule cycle increases and this variation is shown by a curve given in Fig. 34.4.

At $r_p = 1$, $\eta_{th} = 0$

Ceiling temperature limits the pressure ratio.

We have, from Eq. 2

$$W_n = C_p [T_3 - T_4 - T_2 + T_1]$$

and $$\frac{T_2}{T_1} = \frac{T_3}{T_4} = r_p^{\frac{\gamma-1}{\gamma}}$$

$\therefore$ $$T_4 = \frac{T_3 \times T_1}{T_2}$$

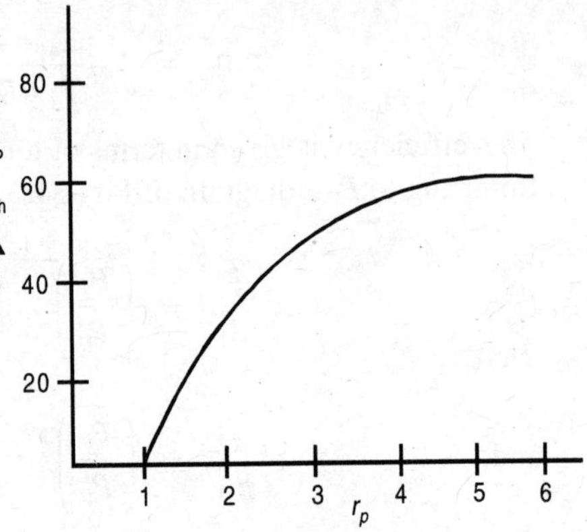

Fig. 34.4

Now work is a function of T_2 that is work depends on T_2 only as

T_1 = Initial temperature and is constant.

T_3 = Ceiling temperature and is constant.

$\therefore$ $$\text{Work } W_n = C_p\left(T_3 - \frac{T_1 \times T_3}{T_2} - T_2 + T_1\right) \quad \ldots(10)$$

This work is maximum when $\frac{dW_n}{dT_2} = 0$

$\therefore$ Differentiating Eq. 10 and equating it to zero, we get

$$\frac{dW_n}{dT_2} = 0 = \frac{T_1 \times T_3}{T_2^2} - 1$$

or $$T_2^2 = T_1 \times T_3$$

$$\mathbf{T_2 = \sqrt{T_1\, T_3}} \quad \ldots(11)$$

34.5 WORK RATIO

To know the extent to which the work developed by the turbine is required or supplied to run the compressor, another ratio called *Work Ratio* is introduced in the study. Work ratio is defined as the ratio of the net work $(W_t - W_c)$ to the turbine work developed (W_t). It is denoted by *WR*.

$\therefore$ $$WR = \frac{\text{Net work}}{\text{Turbine work}}$$

$$= \frac{W_t - W_c}{W_t} = 1 - \frac{W_c}{W_t}$$

$$= 1 - \frac{C_p (T_2 - T_1)}{C_p (T_3 - T_4)}$$

$$= 1 - \frac{T_1\left(\frac{T_2}{T_1} - 1\right)}{T_3\left(1 - \frac{T_4}{T_3}\right)} \quad \ldots(12)$$

But
$$\frac{T_2}{T_1} = r_p^{\frac{\gamma-1}{\gamma}}$$

$$\frac{T_3}{T_4} = r_p^{\frac{\gamma-1}{\gamma}}$$

Substituting these ratios in the $WR = 1 - \dfrac{T_1\left(r_p^{\frac{\gamma-1}{\gamma}} - 1\right)}{T_3\left(1 - \dfrac{1}{r_p^{\frac{\gamma-1}{\gamma}}}\right)}$

$$= 1 - \frac{T_1}{T_3}\left[\frac{r_p^{\frac{\gamma-1}{\gamma}}\left(r_p^{\frac{\gamma-1}{\gamma}} - 1\right)}{\left(r_p^{\frac{\gamma-1}{\gamma}} - 1\right)}\right]$$

$$\mathbf{WR = 1 - \frac{T_1}{T_3} \times r_p^{\frac{\gamma-1}{\gamma}}} \qquad \text{...(13)}$$

From this relation we note that, for the given temperatures T_1 and T_3, WR is maximum when $r_p^{\frac{\gamma-1}{\gamma}}$ is minimum.

Similarly, for the given pressure ratio r_p, again WR is maximum when $\dfrac{T_1}{T_3}$ is minimum. i.e., T_1 is minimum and T_3 is maximum.

Maximum T_3 is according to metallurgical considerations and varies between 800°C to 1000°C.

Similarly T_1 will be depending on the atmospheric conditions and may be taken as varying between 15°C and 30°C.

34.6 CONCEPT OF MAXIMUM PRESSURE RATIO AND OPTIMUM PRESSURE RATIO

We are having, now, two quantities that will be considered. One is the net work (W_{net}) and the work ratio (WR).

$$\text{Net work} = W_t - W_c$$

$$= C_{p_g}(T_3 - T_4) - C_{p_g}(T_2 - T_1)$$

$$= C_{p_g} T_3\left(1 - \frac{T_4}{T_3}\right) - C_{p_g} T_1\left(\frac{T_2}{T_1} - 1\right)$$

$$= C_{p_g} T_3\left(1 - \frac{1}{r_p^{\frac{\gamma-1}{\gamma}}}\right) - C_{p_g} T_1\left(r_p^{\frac{\gamma-1}{\gamma}} - 1\right)$$

$$= C_{p_g}\left[T_3\left(\frac{r_p^{\frac{\gamma-1}{\gamma}}-1}{r_p^{\frac{\gamma-1}{\gamma}}}\right)-C_{p_g}\,T_1\left(r_p^{\frac{\gamma-1}{\gamma}}-1\right)\right]$$

Therefore, net work is a function of r_p only for constant T_3 (ceiling temperature) and T_1 (atmospheric temperature). Similarly, work ratio which is given by

$$WR = 1-\frac{T_1}{T_3}\cdot r_p^{\frac{\gamma-1}{\gamma}}$$

is also a function r_p — the pressure ratio and temperature ratio.

Thus we note that the efficiency of a theoretical cycle is a function of pressure ratio while the performance of the cycle is a function of both the pressure ratio and temperature ratio.

Consider now the different cycles with different r_p and the same maximum cycle temperature and the same initial temperature and pressure.

Consider a new-cycle 1– 2"– 3"– 4"– 1(Fig. 34.5) where the pressure is increased such that the temperature is very near the ceiling temperature T_3. In this case, heat added is less. This heat addition will be zero in case the temperature after compression is same as ceiling temperature. Thus in the limit when $T_{2"} = T_{3"} = T_{max}$, no heat being added in the combustion chamber and the cycle plots as a single vertical line and no network is available ($W_t = W_c$). The maximum pressure ratio can be obtained equating net output equal to zero.

i.e. $$W_{net} = W_t - W_c = 0$$

$\therefore$ $$0 = C_p(T_3-T_4)-(T_2-T_1)C_p$$

$\therefore$ $$C_pT_3\left(1-\frac{1}{r_p^{\frac{\gamma-1}{\gamma}}}\right) = C_p\left(r_p^{\frac{\gamma-1}{\gamma}}-1\right)T_1$$

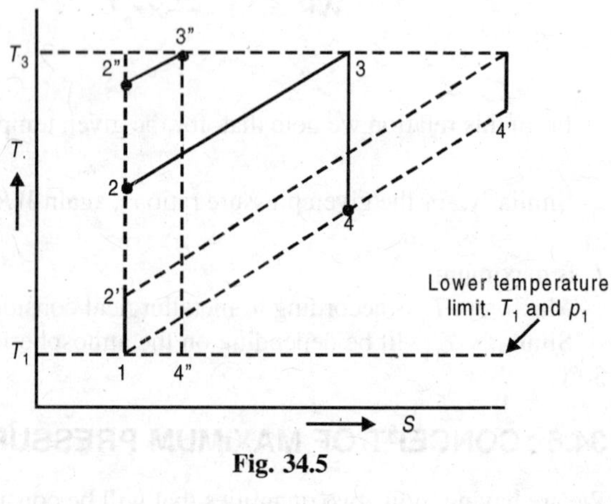

Fig. 34.5

or $$\frac{T_3}{T_1} = \frac{\left(r_p^{\frac{\gamma-1}{\gamma}}-1\right)\cdot r_p^{\frac{\gamma-1}{\gamma}}}{r_p^{\frac{\gamma-1}{\gamma}}-1}$$

$$= r_p^{\frac{\gamma-1}{\gamma}}$$

This pressure ratio is maximum pressure ratio written as $(r_p)_{max}$ and is given by

$$(\mathbf{r_p})_{max} = \left(\frac{\mathbf{T_3}}{\mathbf{T_1}}\right)^{\frac{\gamma}{\gamma-1}} \quad \text{...(14)}$$

We will now consider another case where the lowest possible pressure is considered. Consider another cycle 1–2'–3'–4'–1 where some low pressure ratio is considered, and the thermal efficiency will be very poor. In this cycle, it will be noticed that the heat added in the combustion chamber is very large as compared with the work obtained. Also average temperature at which heat is added and rejected is very small. In the limit, the pressure ratio will be one (1) and the heat addition and rejection lines are same on T–S diagram and net work is gain zero.

Between the extreme values of pressure ratios of $r_p = 1$ and $r_p = (r_p)_{max}$, there lies an optimum value for which the specific output, the net output per unit mass flow is maximum. Thus taking T_1 and T_3 as constant temperatures, differentiate expression for W_{net} with respect to r_p and equating it to zero, we get an optimum pressure ratio $(r_p)_{opt}$.

Consider, therefore, a cycle 1–2–3–4–1, Fig. 34.5.

$$\text{Net work/kg} = W_n = W_t - W_c$$

$$= C_p(T_3 - T_4) - C_p(T_2 - T_1)$$

$$= C_p T_3 \left(1 - \frac{1}{r_p^{\frac{\gamma-1}{\gamma}}}\right) - C_p T_1 \left(r_p^{\frac{\gamma-1}{\gamma}} - 1\right)$$

Differentiating this equation w.r.t. r_p and equating it to zero, we get

$$O = \frac{dW_n}{dr_p} = C_p T_3 \left[\frac{\frac{\gamma-1}{\gamma} \cdot r_p^{\frac{\gamma-1}{\gamma}-1}}{r_p^{\frac{2(\gamma-1)}{\gamma}}}\right] - C_p T_1 \left[\frac{\gamma-1}{\gamma} \cdot r_p^{\frac{\gamma-1}{\gamma}-1}\right]$$

$$\therefore \quad \frac{T_3}{T_1} = \frac{r_p^{-\frac{1}{\gamma}} \cdot r_p^{\frac{2(\gamma-1)}{\gamma}}}{r_p^{-\frac{1}{\gamma}}} = r_p^{\frac{2(\gamma-1)}{\gamma}} = (r_p)_{opt}^{\frac{2(\gamma-1)}{\gamma}}$$

$$\therefore \quad (r_p)_{opt} = \left(\frac{T_3}{T_1}\right)^{\frac{\gamma}{2(\gamma-1)}}$$

$$\text{But} \quad \frac{T_3}{T_1} = (r_p)_{max}^{\frac{\gamma-1}{\gamma}}$$

$$\therefore \quad (r_p)_{opt} = (r_p)_{max}^{\frac{\gamma-1}{\gamma} \times \frac{\gamma}{2(\gamma-1)}} = (r_p)_{max}^{\frac{1}{2}}$$

$$\therefore \quad \mathbf{(r_p)_{opt}} = \sqrt{\mathbf{(r_p)_{max}}} \qquad \ldots(15)$$

Figure 34.6 shows the variation of work ratio and efficiency w.r.t. r_p — the pressure ratio.

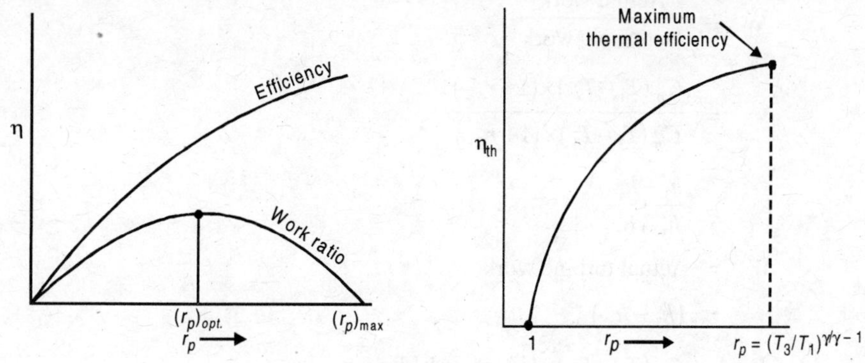

Fig. 34.6

34.7 ACTUAL GAS-TURBINE CYCLE WITH FLUID FRICTION

As you know, the efficiency of an actual cycle is the actual net work delivered, divided by the actual energy chargeable against the cycle. However, since three or more different pieces of equipment are needed in order for the cycle to operate, and since the efficiency of each is significant, it is customary to study each separately. As steady flow machines with $\Delta K = 0$, the works of the fluid are adiabatic flow are

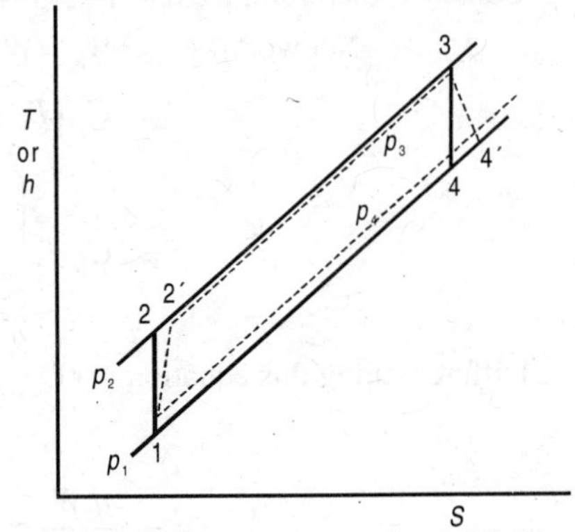

Fig. 34.7. *Brayton cycle with fluid frcition*

I. Compressor

$$\text{Ideal work} = W_c = h_1 - h_2$$

$$= C_p (T_1 - T_2)$$

$$= -\Delta h_s \text{ kJ/kg air}$$

But actual work is more than this ideal or isentropic compression work because of the fluid friction. Compressor isentropic efficiency takes into cosideration this fluid friction and is given by –

$$\eta_c = \frac{\text{Isentropic work/kg}}{\text{Actual work of compressor}}$$

and is most commonly used. 1–2' is the actual compression process and the actual work is given by

$$\text{Actual work} = W_{c'} = h_1 - h_{2'}$$

$$= C_p (T_1 - T_{2'})$$

$$= -\Delta h_{\text{act}} \text{ kJ/kg}$$

$$\therefore \quad \eta_c = \frac{C_p (T_2 - T_1)}{C_p (T_{2'} - 1)} = \frac{T_2 - T_1}{T_{2'} - 1}$$

II. Turbine

$$\text{Ideal work} = W_t = (h_3 - h_4)(1 + r_{f/a})$$

where r_{fla} = Fuel to air ratio

$$\therefore \quad W_t = (1 + r_{f/a}) C_p (T_3 - T_4)$$

Again, actual work is less than this isentropic work because of the fluid friction. Turbine isentropic efficiency takes into consideration this fluid friction and is given by

$$\eta_t = \frac{\text{Actual work}}{\text{Isentropic work}}$$

$$= \frac{C_p (T_3 - T_{4'}) \times (1 + r_{fla})}{C_p (T_3 - T_4) \times (1 + r_{fla})}$$

$$= \frac{h_3 - h_{4'}}{h_3 - h_4}$$

$$\therefore \quad W_{t'} = \text{Actual turbine work}$$

$$= (h_3 - h_{4'})$$

$$= C_p (T_3 - T_{4'}) \times (1 + r_{f/a}) \text{ kJ/kg air}$$

$$= -\left(1+r_{f/a}\right)\Delta h_{t'}$$

Note. In making an *air standard cycle analysis*, we generally ignore the amount of fuel ($r_{f/a} = 0$). With the current temperature ceilings, the maximum value of $r_{f/a}$ will be small, of the order of 0.02 or less.

$$\therefore \quad \text{Net actual work} = (W_n)_{\text{actual}}$$

$$= W_{t'} - W_{c'}$$

$$= C_p(T_3 - T_{4'}) - C_p(T_{2'} - T_1). \text{ Neglecting } r_{f/a}$$

$$= C_p\, \eta_t\, T_3\left(1 - \frac{T_4}{T_3}\right) - \frac{C_p\, T_1}{\eta_c}\left(\frac{T_2}{T_1} - 1\right)$$

But
$$\frac{T_2}{T_1} = (r_p)^{\frac{\gamma-1}{\gamma}} = \frac{T_3}{T_4} \text{ neglecting pressure drop.}$$

$$\therefore \quad \text{Net actual work/kg} = C_p\, \eta_t\, T_3\left[1 - \frac{1}{r_p^{\frac{\gamma-1}{\gamma}}}\right] - \frac{C_p\, T_1}{\eta_c}\left(r_p^{\frac{\gamma-1}{\gamma}} - 1\right)$$

$$\text{Also, heat supplied/kg} = C_p(T_3 - T_{2'})$$

$$\approx C_p(T_3 - T_2)$$

This approximation is only for simplification of the problem without any appreciable error.

Substitute x for $(r_p)^{\frac{\gamma-1}{\gamma}}$ and $\dfrac{T_3}{T_1} = y$

$\therefore$ Heat supplied/kg air

$$= Q_A = C_p(T_3 - T_2)$$

$$= C_p(y \cdot T_1 - T_1 \cdot x) - C_p T_1 (y - x)$$

Therefore, the actual thermal efficiency of the cycle is given by

$$(\eta_{\text{th}})_{\text{act.}} = \frac{\text{Actual net work}}{\text{Heat supplied}}$$

$$= \frac{\eta_t\, T_3\left(1 - \dfrac{1}{x}\right) - \dfrac{T_1}{\eta_c}(x-1)}{T_1(y-x)}$$

$$= \frac{\eta_t\, T_1\, y\left(1 - \dfrac{1}{x}\right) - \dfrac{T_1}{\eta_c}(x-1)}{T_1(y-x)}$$

$$= \frac{\eta_t\, y\left(1 - \dfrac{1}{x}\right) - \dfrac{(x-1)}{\eta_c}}{(y-x)}$$

For maximum thermal efficiency, we have

$$\frac{d\eta_{\text{th}}}{dx} = 0 = \frac{\left\{(y-x)\left[\eta_t\, y\times\frac{1}{x^2}-\frac{1}{\eta_c}\right]+\left[\eta_t\, y\left(1-\frac{1}{x}\right)-\frac{x-1}{\eta_c}\right]\right\}}{(y-x)^2}$$

$$= \left[\frac{y(y-x)}{x^2}\eta_t - \frac{y-x}{\eta_c}\right]+\left[\frac{\eta_t y(x-1)}{x}-\frac{x-1}{\eta_c}\right]$$

$$= \left[\frac{\eta_c\,\eta_t\,y(y-x)-x^2(y-x)}{\eta_c\,x^2}\right]+\left[\frac{\eta_t\,\eta_c\,y(x-1)-x(x-1)}{x\,\eta_c}\right]$$

$$= (y-x)\left[\frac{\eta_c\,\eta_t\,y-x^2}{x^2\,\eta_c}\right]+(x-1)\left[\frac{\eta_t\,\eta_c\,y-x}{\eta_c\,x_2}\right]$$

$$= \frac{(y-x)\left(\eta_c\,\eta_t\,y-x^2\right)+x(x-1)(\eta_t\,\eta_c\,y-x)}{\eta_c\,x^2}$$

$$= (y-x)\left(\eta_c\,\eta_t\,y-x^2\right)+x(x-1)(\eta_t\,\eta_c\,y-x)$$

$$= y^2\eta_c\eta_t - x^2y - xy\,\eta_c\eta_t + x^3 + \left(x^2-x\right)\left(\eta_c\eta_t y - x\right)$$

$$= y^2\eta_c\eta_t - x^2y - xy\eta_c\eta_t + x^3 + x^2y\eta_c\eta_t - x^3 - xy\eta_c\eta_t + x^2$$

$$= x^2y(\eta_c\eta_t-1) - 2xy\eta_c\eta_t + y^2\eta_c\eta_t + x^2$$

$$= x^2\left[y(\eta_c\eta_t-1)+1\right] - 2xy\eta_c\eta_t + y^2\eta_c\eta_t$$

$$= \left[y(\eta_c\eta_t-1)+1\right] - \frac{2y}{x}\eta_c\eta_t + \frac{y^2}{x^2}\eta_c\eta_t$$

$$\text{O} = \frac{1}{x^2}\left(y^2\eta_c\eta_t\right) - \frac{1}{x}(2y\eta_c\eta_t) + \left[y(\eta_c\eta_t-1)+1\right]$$

This is a quadratic in x.

$$\therefore \quad \frac{1}{x} = \frac{2y\eta_c\eta_t \pm \sqrt{4y^2\eta_c^2\eta_t^2 - 4\left(y^2\eta_c\eta_t\right)\left[y(\eta_c\eta_t-1)+1\right]}}{2y^2\eta_c\eta_t}$$

Taking + ve root only, we have

$$\frac{1}{x} = \frac{1+\sqrt{(1-y)+\dfrac{y-1}{\eta_c\eta_t}}}{y}$$

$$\therefore \quad x = \frac{y}{1+\sqrt{(1-y)+\dfrac{y-1}{\eta_c\eta_t}}}$$

$$= \frac{y}{1+\sqrt{(y-1)\left[\frac{1}{\eta_c\eta_t}-1\right]}}$$

Substituting the values of y we get

$$x = (r_p)_{opt} = \left\{ \frac{\frac{T_3}{T_1}}{1+\sqrt{\left(\frac{T_3}{T_1}-1\right)\left[\frac{1}{\eta_c\eta_t}-1\right]}} \right\}^{\frac{\gamma}{\gamma-1}}$$

Following the same procedure for the condition for actual optimum pressure ratio for specific output, we get

$$(r_p)_{opt} = \left((r_p)_{opt}\frac{T_3}{T_1}\right)^{\frac{\gamma}{2(\gamma-1)}}$$

34.8 AIR RATE

Generally we use symbol m_f to designate the mass of fuel used. Its units are defined by the context, usually either m_f kg fuel/kg air or m_f kg fuel/unit of work say kg fuel/kWh. In this case it is generally called the *Specific Fuel Consumption*.

Similarly, symbol m_a is used to designate the mass of air used. Its units are defined by the context, usually either m_a kg air/kg fuel or m_a kg air/unit of work say kg air/kWh.

∴ Air rate of an engine or turbine is given by the ratio.

$$\dot{m}_a \text{ (kg/kWh)} = \frac{3600 \text{ (kJ/hour)}}{\text{Work done/kg air}}$$

$$\dot{m}_a = \text{Air rate} = \frac{3600}{\text{Work done / kg air}}$$

34.9 METHODS USED TO IMPROVE GAS TURBINE PERFORMANCE

Although the performance of the gas turbine is not especially attractive compared with the efficiencies possible in Diesel and Petrol engine power plants, a simple gas turbine has advantages in weight, size and vibration compared to the engine and in size and cost compared to small steam plant. It is also superior to both in quantity of water used, for the simple gas turbine plant uses almost no cooling water.

Even if the components used in the gas turbine power plants are improved in design, the efficiency and the specific output of the simple gas turbine cycle are quite low. The efficiency handicap is surmountable, at the expense of adding complexity to the gas turbine plant. The principle refinements which accomplish this are : (a) regeneration (b) intercooling and (c) reheating.

We write for thermal efficiency of the Brayton cycle as

$$\eta_{th} = \frac{Q_A - Q_R}{Q_A}$$

where Q_A = Heat added or supplied/kg

Q_R = Heat rejected/kg

Also $$\eta_{th} = \frac{\text{Net work}}{\text{Heat supplied}}$$

$$= \frac{W_t - W_c}{Q_A}$$

where W_t = Gas turbine work kJ/kg

W_c = Compressor work kJ/kg.

(*a*) Q_A— the heat added or supplied can be decreased by the process called regenerative heating of gas before entering the combustion chamber.

(*b*) Compressor work W_c can be decreased when multistage compressors with intercooling are used.

(*c*) Turbine workoutput can be increased by using a multistage turbine with reheating of the gas in between the two stages.

Thus the modifications in the simple gas turbine cycle required for improving the performance of the cycle are

(a) Regeneration

(b) Intercooling and

(c) Reheating

We will now discuss these modifications.

(a) Regeneration

We observe that the temperature of the turbine exhaust at (4), Fig 34.8, is higher than the temperature at the end of compression at 2, and, therefore, we may think of applying Ericsson's notion of regeneration. In this event, the exhaust at 4 gives up heat to the air at 2. This transfer of heat is called *Regenerative heating* and the heat exchanger used for this is called *Regenerator*. This results in cooling of final exhaust gases and thus a reduction in heat rejection takes place. Figure 34.9 shows the diagrammatic layout with regeneration.

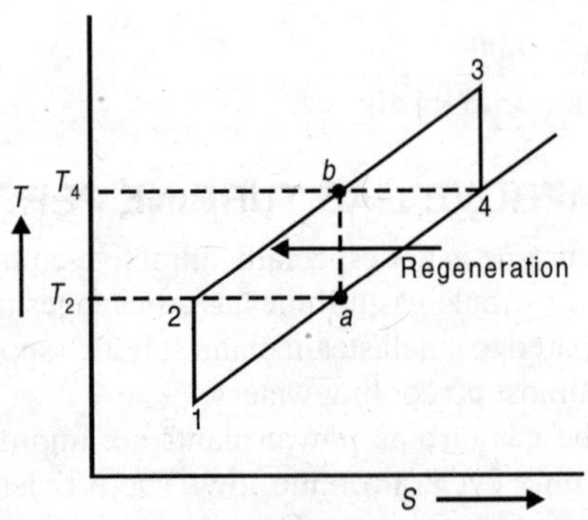

Fig. 34.8

Theoretically, if the heat exchanger were large enough and flow were slow enough, the air from the compressor could be heated reversibly to temperature 4 at state *b*, while the exhaust cools to temperature 2 at state *a*. Some of the formerly discharged heat $h_4 - h_a$ is exchanged within the system and the heat to the sink is now only $h_a - h_1$. Moreover, it is necessary to add only the heat equal to $h_3 - h_b$ instead of $h_3 - h_2$ as before. Consequently less fuel is needed and this additional piece of equipment should materially increase the efficiency of the ideal cycle. From the figure we find the thermal efficiency for a constant mass of 1 kg as

$$\eta_{th} = \frac{W_n}{Q_A} = \frac{\Sigma Q}{Q_A} = \frac{W_t - W_c}{Q_A}$$

$$= \frac{(h_3 - h_4) - (h_2 - h_1)}{h_3 - h_b}$$

$$= \frac{(h_3 - h_b) - (h_a - h_1)}{h_3 - h_b}$$

$$= \frac{(h_3 - h_4) - (h_2 - h_1)}{(h_3 - h_4)}$$

$$= 1 - \frac{h_2 - h_1}{h_3 - h_4}$$

$$= 1 - \frac{T_2 - T_1}{T_3 - T_4}$$

$$= 1 - \frac{T_1\left(\frac{T_2}{T_1} - 1\right)}{T_3\left(1 - \frac{T_4}{T_3}\right)}$$

$$= 1 - \frac{T_1}{T_3}\left[r_p^{\frac{\gamma-1}{\gamma}}\right]$$

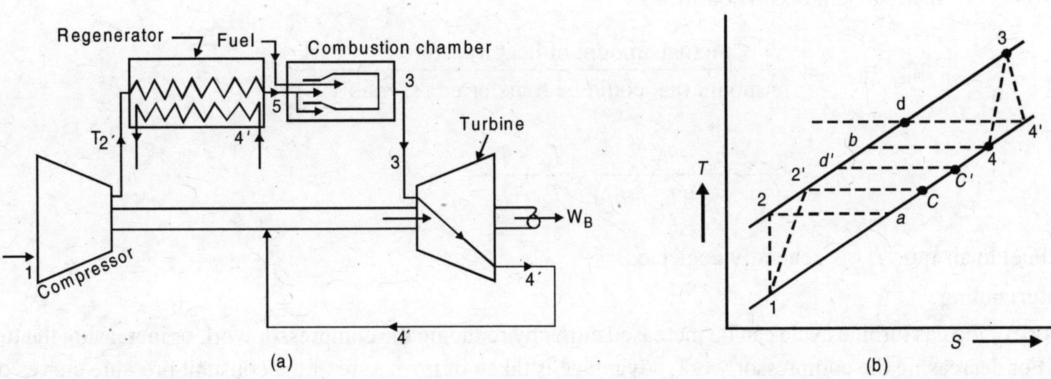

Fig. 34.9

With a fixed initial temperature T_1 the above equation shows that with a regenerator, the thermal efficiency increase as T_3 increases and the thermal efficiency decreases as the pressure ratio increases.

Without the regenerator, the cycle efficiency increases as the pressure ratio increases.

Effectness of Regenerator

Figure 34.10 is the T–S diagram for imperfect regeneration with fluid friction in turbine.

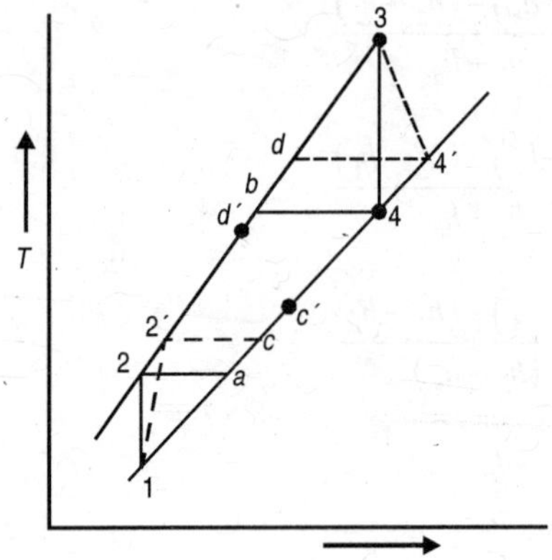

Fig. 34.10. *Imperfect regenration with fluid friction in machines*

1–2 Isentropic compression of air in compressor.

1–2' Actual compression.

3–4 Isentropic expansion of gases in turbine.

3–4' Actual expansion of gases in turbine.

If regeneration is perfect then

d is the state of air before combustion chamber

c is the state of gases before the gases are exhausted

and in this case heat given by gases from turbine is equal to the heat given by gases from turbine is equal to the heat gained by air in the regenerator. But the actual state of air will be *d'* because the regenerator will not be perfect. Consequently the regenerator will not be 100 % efficient. The performance of the regenerator is given by the *Effectiveness* of the regenerator. It is defined as

$$\eta_{reg.} = \varepsilon = \frac{\text{Actual amount of heat transferred}}{\text{Amount that could be transferred reversibly}}$$

$$\therefore \quad \frac{h_{d'} - h_{2'}}{(1+r_{f/a})(h_{4'} - h_{c})} \simeq \frac{T_{d'} - T_{2'}}{T_{4'} - T_{2'}}$$

Fuel to air ratio $r_{f/a}$ is generally neglected,

(b) Intercooling

Net work of the gas turbine cycle can be increased either by reducing the compressor work or increasing the turbine work. For decreasing the compressor work, advantage is taken of the nature of the constant pressure curves of h–s and T–S diagrams. The vertical distance between any two constant pressure lines goes on decreasing to the left and goes on increasing to the right.

If the compression is achieved in two or more stages, the air delivered by the 1st stage of the compressor, is cooled, on its way to the next stage. This cooling of air in between the two stages is called *intercooling*. When the air is cooled to the temperature of air entering any stage, intercooling is called *perfect intercooling*. Refer Fig. 34.11.

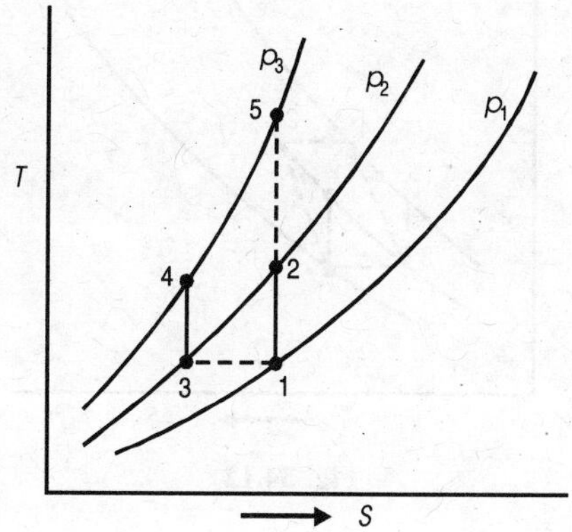

Fig. 34.11

1–2 Isentropic compression in first stage
2–3 Intercooling between the stages
3–4 Isentropic compression in the second stage
1–5 Isentropic compression without intercooling

From Fig. 34.11, vertical distance between 3–4 is less than the vertical distance 2–5 and therefore,

$$[(1-2)+(3-4)] < (1-5)$$

$\therefore$ The compression work is reduced while the turbine work remains same when other data remains same.

$$\therefore \qquad \text{Net work} = W_t - W_{c_i}$$

$$= \text{Constant} - \text{reduces } W_c$$

$$> W_t - W_c$$

where $W_c = h_5 - h_1$

W_{c_i} = Compression work with intercooling.

When the intercooling is perfect and when the intermediate pressure is the geometric mean $\left(p_2 = \sqrt{p_1 \times p_3}\right)$ then the compression work is minimum.

Usually water-cooled surface coolers are employed. Low and high pressure stages must necessarily separately encased.

Figure 34.12 shows intercooling and fluid friction during compression. The diagram is self-explanatory the same working substance flows through the different components of the plant again and again, receiving heat and rejecting heat at the approximate events in the cycle.

If the entire flow comes from the atmosphere and is returned to the atmosphere, the turbine is said to work on the *Open Cycle*.

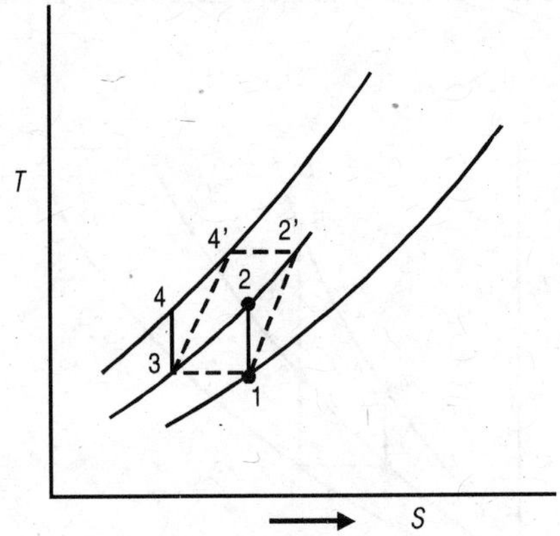

Fig. 34.12

The basic difference between open and closed cycle gas of turbines is in the method of heating the air after compression. In case of an open cycle gas turbine, the fuel is burned in the air itself to increase its temperature i.e., fuel is mixing with air and then the products of combustion are passed on to the atmosphere through turbine. For the next cycle, a fresh supply of air is taken in the compressor and the processes are repeated.

On the other hand, in a closed cycle gas turbine, the same air or working substance is circulated over and over again. The working substance is heated in a heat exchanger where a separate hot gas is obtained by burning the fuel in the supply of additional air in a combustion chamber. The heat exchanger is of the shell and tube type so that working substance does not come into contact with the products of combustion.

Most of the gas turbines in use are open cycle plants. But the recent developments made the closed-cycle plants work producing more than 1500 kW and having gas turbine inlet gas temperature of 800°C and having a thermal efficiency of the order of 30 %.

Figure 34.13 shows schematically the simplest closed cycle components with the flow of the working substance and energy.

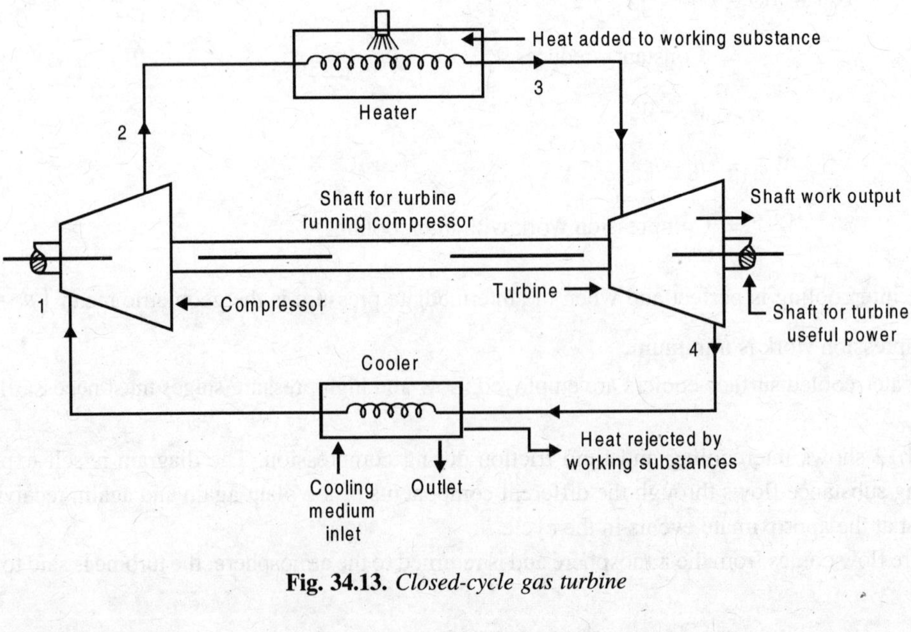

Fig. 34.13. *Closed-cycle gas turbine*

At state 1, cold gas enters a compressor, where shaftwork is to be done on the compressor to increase the pressure and temperature. The gas leaves the compressor at the state 2. This gas enters the heater (heat exchanger) where heat is supplied at constant pressure. The temperature of the gas increases and the gas leaves the heater at state 3. This gas from the heater enters the gas turbine where it is expanded to the lower pressure so that the temperature is reduced, shaft work is produced. Part of this shaft work is utilized to run the compressor and the rest is supplied to the load or the useful power or net-power.

(c) Reheating and Reheat Cycle

Another variation in the simple Brayton cycle is the use of two turbines instead of one, in which case, one turbine is used to drive the compressor and the other produces the net work output. The shafts of the turbines may collinear or as shown in Fig. 34.14. In between these two turbine, we may or may not arrange to burn more fuel by installing another combustor as shown.

Reheating is the increase of temperature of partially expanded gas by burning more fuel, in a device called *Reheater*. Figure 34.14 shows reheating and regeneration.

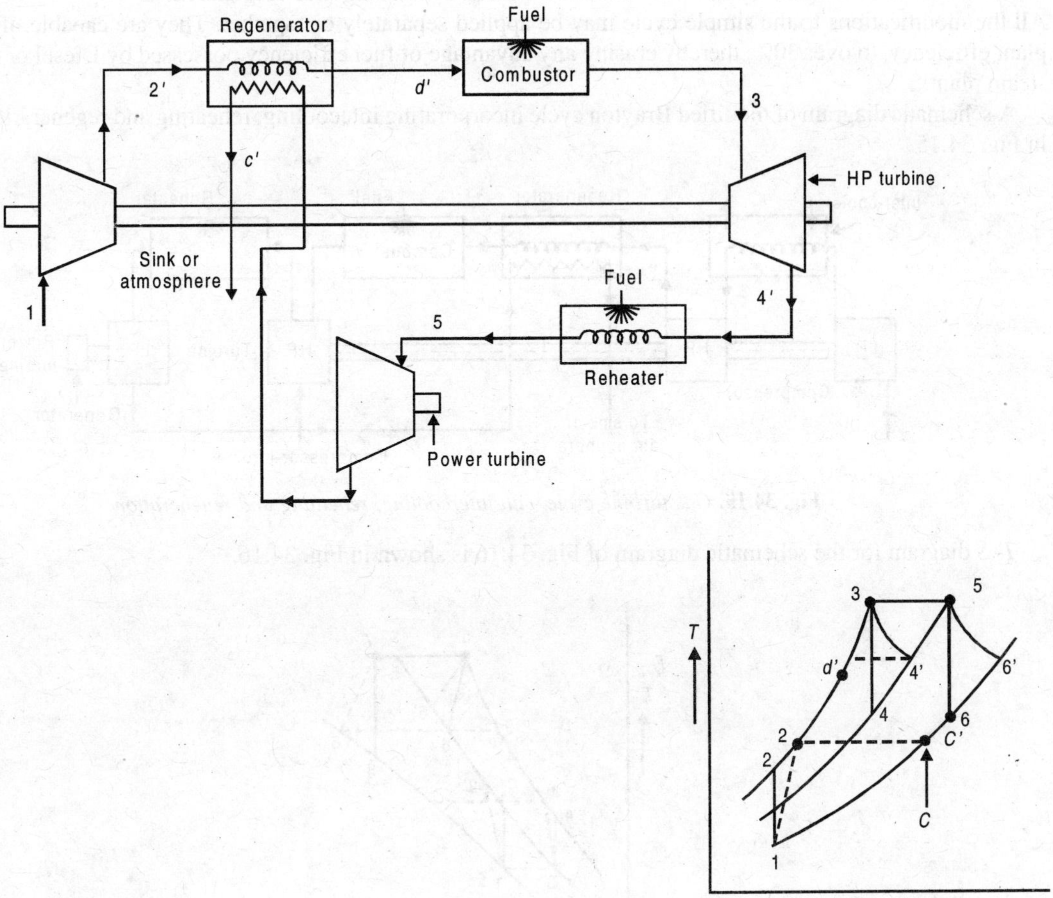

Fig. 34.14. *Reheat cycle*

$$W_{c'} = \text{Compression work}$$
$$= h_{2'} - h_1$$
$$W_{HP} = (1+r_{f/a})(h_3 - h_{4'})$$
$$\text{Net work output} = \text{Output of Power turbine}$$
$$= (1+r_{f/a}+m_{fr})(h_5 - h_{6'})$$

where $r_{f/a}$ = mass of fuel/kg air in combustor

m_{fr} = mass of fuel in reheater

Thus we can find the thermal efficiency of the cycle with regeneration and reheating.

$$\eta_{th} = \frac{(1+r_{f/a}+m_{fr})(h_5 - h_{6'})}{(h_3 - h_{a'}) + m_{fr} \times \text{C.V.}}$$

(d) Gas Turbine (Brayton cycle) with all Intercooling, Reheating and Regeneration

All the modifications to the simple cycle may be applied separately or together. They are capable of raising the plant efficiency to over 30%, thereby erasing any advantage of fuel efficiency possessed by Diesel or condensing steam plant.

A schematic diagram of modified Brayton cycle incorporating intecooling, reheating and regeneration is shown in Fig. 34.15.

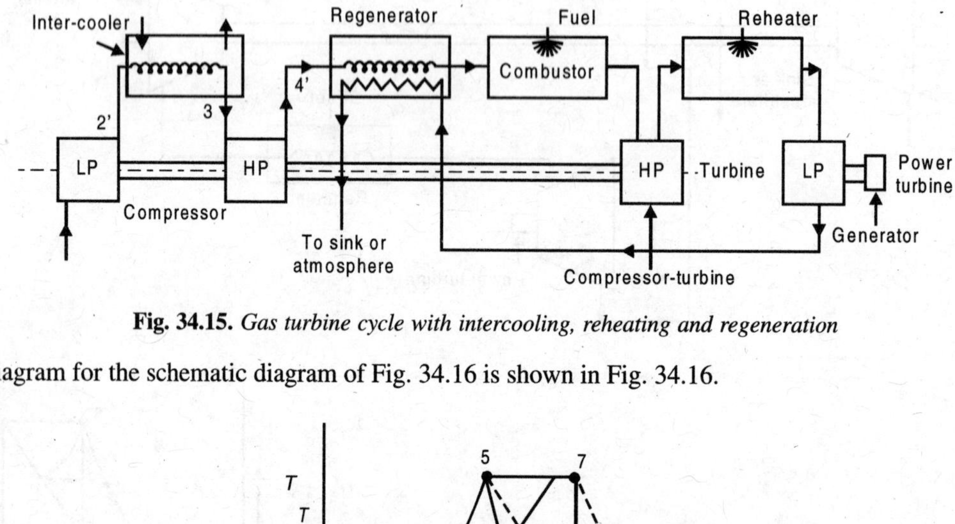

Fig. 34.15. *Gas turbine cycle with intercooling, reheating and regeneration*

T–S diagram for the schematic diagram of Fig. 34.16 is shown in Fig. 34.16.

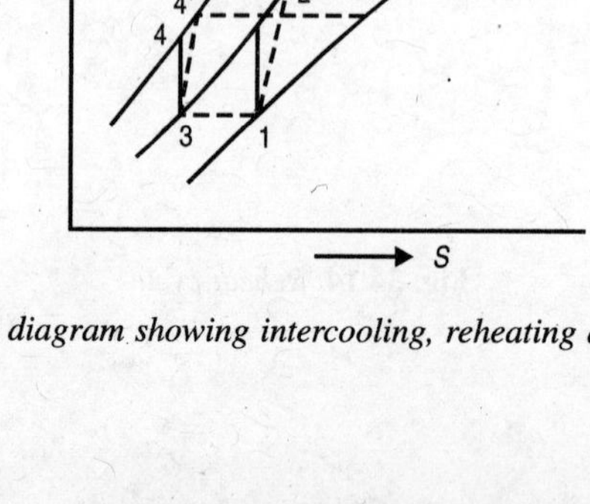

Fig. 34.16. *T-s diagram showing intercooling, reheating and regeneration*

Analysis

$$\text{Compressor work} = W_c = (h_{2'} - h_1) + (h_{4'} - h_3)$$

$$\text{Turbine work} = W_t = (h_5 - h_{6'}) + h_7 - h_{8'}$$

Effectiveness of regenerator

$$= \frac{T_9 - T_{4'}}{T_{8'} - T_{4'}}$$

from which T_9 is calculated.

$$\therefore \quad \text{Heat added} = C_{pa}(T_5 - T_9) = Q_A$$

$$\therefore \quad \eta_{th} = \text{Thermal efficiency}$$

$$= \frac{\text{Net work}}{\text{Heat added}}$$

$$= \frac{W_t - W_c}{Q_A}$$

(e) Effect of modifications on the Performance of Brayton Cycle–Gas Turbine

Various graphs showing the effect of intercooling, reheating and regeneration on the cycle efficiency are given in Fig. 34.17.

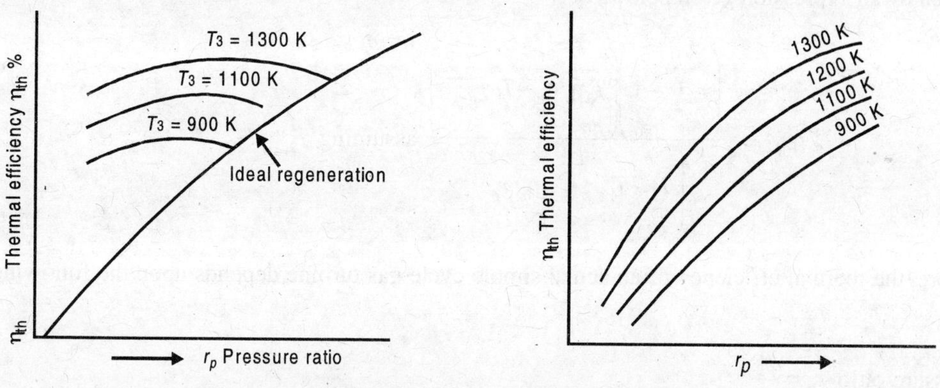

Fig. 34.17. *Regenerative cycle*

These graphs gives the nature of the curves only.

A—Simple cycle.

B—Simple cycle with regeneration.

C–-Simple cycle with intercooling and regeneration.

D—Simple Open cycle with regenerator, reheater and inter-cooling.

Combine effect of intercooling, reheating and regeneration on gas turbine cycle on thermal efficiency and specific output for different maximum temperatures are shown in Fig. 34.18 (a) and (b).

(f) Water Injection

This is one of the methods of improving the performance of the gas turbine is to inject water into the working air at the entrance to the compressor. This way the compressed air is cooled by absorbing, from air, the latent heat of vaporisation of water. By injecting of water, the total mass flow of the working medium is increased by the mass of the injected water and hence the power output of the cycle is increased, the work ratio is also increased in addition to the lowring of the air rate.

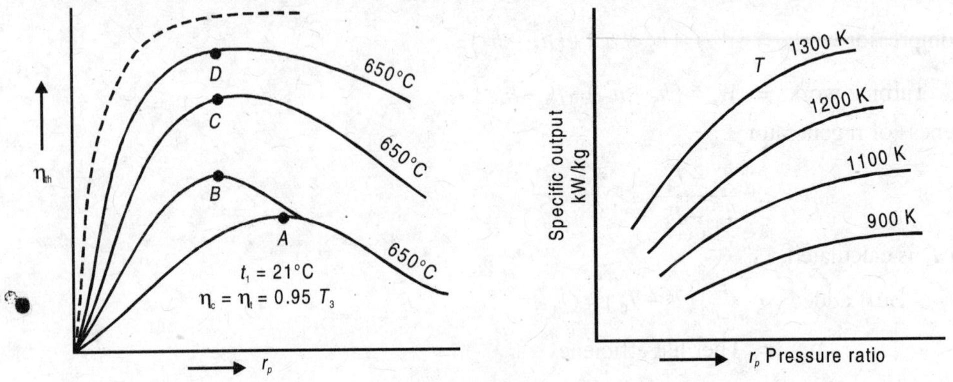

Fig. 34.18

Water injection system is commonly used as a power book for take off and emergencies on the jet propelled air-craft. It is possible for marine or land based gas turbines to use water injection for long duration or continuous water injection. The water to be injected should be pure, otherwise, it may cause corrosion or deposits on the blades.

34.10 EFFECT OF OPERATING VARIABLES ON THERMAL EFFICIENCY

The thermal efficiency of the Gas-turbine plant taking into consideration the compressor and turbine efficiencies, will be given by an expression given below

$$\eta_{th} = \frac{\left(r_p^{\frac{\gamma-1}{\gamma}} - 1\right)\left(T_3 \eta_c \eta_t - T_1\, r_p^{\frac{\gamma-1}{\gamma}}\right)}{\left(T_3 - T_1\, r_p^{\frac{\gamma-1}{\gamma}}\right) \eta_c\, r_p^{\frac{\gamma-1}{\gamma}}} \quad \text{assuming } T_2 \simeq T_{2'}$$

Therefore, the thermal efficiency of an actual simple cycle gas turbine depends upon the following operating variables

(a) Pressure ratio $r_p = \dfrac{P_2}{P_1}$

(b) Turbine isentropic efficiency η_t

(c) Compressor isentropic efficiency η_c

(d) Turbine inlet temperature T_3

(e) Compressor inlet temperature T_1

(f) Specific fuel consumption

I. Effect of Variation of Turbine Inlet Temperature and Pressure Ratio

Figure 34.19 shows the variation of turbine inlet temperature T_3 and pressure ratio $r_p = \dfrac{P_2}{P_1}$.

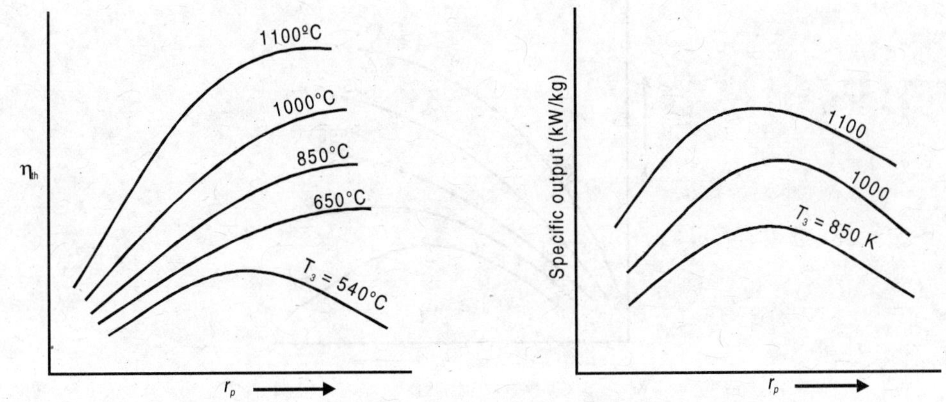

Fig. 34.19. *variation of thermal efficiency with pressure ratio*

As the turbine inlet temperature T_3 increases, the thermal efficiency increases with other variables being held constant. For each value of turbine inlet temperature there is an optimum pressure ratio for maximum thermal efficiency. For lower value of turbine inlet temperature the thermal efficiency first increases and after reaching a maximum value it decreases rapidly. For higher values of turbine inlet temperature the peaks of the curves are flatter giving a greater range of optimum value of pressure ratio. The turbine inlet temperature is limited by turbine blading material. Similarly specific output increases first and then decreases as the pressure ratio increases.

II. Effect of Variation of Turbine and Compressor Efficiency

Figure 34.20 shows the effect of compressor and turbine efficiencies. From the equation for thermal efficiency of the Brayton Cycle it is very clear that the thermal efficiency is very much sensitive to the variations of turbine and

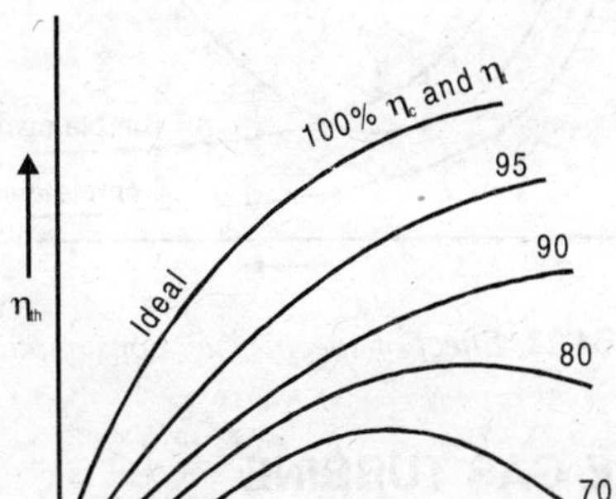

Fig. 34.20. *Effect of compressor and Turbine efficiencies*

compressor efficiencies. As the turbine and compressor efficiencies increase, the thermal efficiency of the cycle increases. There is an optimum pressure ratio for each set of component efficiency.

III. Effect of Variation of Compressor Inlet Temperature

Figure 34.21 shows the effect of compressor inlet temperature. If the air inlet temperature is lowered, the thermal efficiency increases keeping the same pressure ratio. Peaks are at a higher pressure ratio and have a flatter curve giving optimum value of a pressure ratio at a greater range.

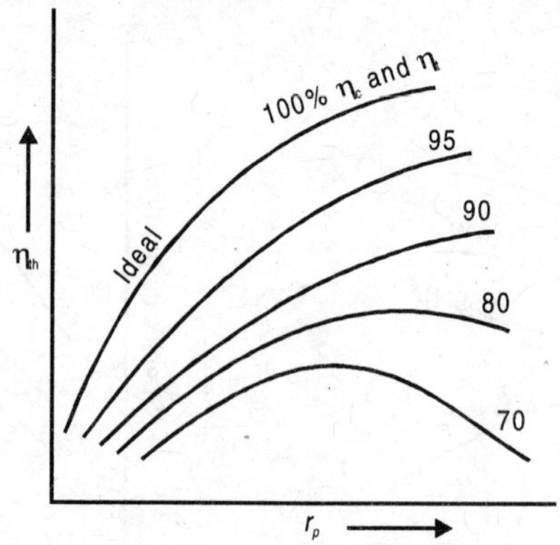

Fig. 34.21. *Effect variation of compressor inlet temperature*

IV. Effect of Specific Fuel Consumption of Thermal Efficiency

Figure 34.22 shows for a certain turbine inlet temperature T_3, the effect of regenerative, simple and complete cycle on the specific fuel consumption. It suggest that for all values of pressure ratios the fuel consumption is minimum in complete cycle i.e., cycle with regeneration, reheating and intercooling.

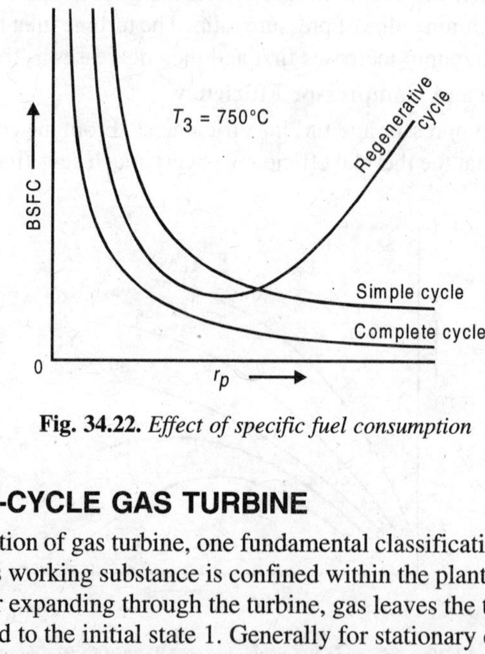

Fig. 34.22. *Effect of specific fuel consumption*

34.11 SIMPLE CLOSED-CYCLE GAS TURBINE

As per the methods of classification of gas turbine, one fundamental classification of gas turbine cycles is the path of the working substance. If this working substance is confined within the plant, the gas turbine is said to work on a *closed cycle*. In this case, after expanding through the turbine, gas leaves the turbine at the state 4 and passes on to the cooler where gas is cooled to the initial state 1. Generally for stationary closed-cycle gas turbine plant will have water as the cooling medium in the cooler. Gas from the cooler is again taken to the compressor and the cycle is repeated.

The *T–S* diagram for this cycle is shown in Fig. 34.23 and is same as that for open cycle.

In ideal cycle, compression and expansion are isentropic processes while heat addition and heat rejection is at constant pressure. All types of loses are assumed to be absent and hence pressure drops when passing through the heat exchangers are absent.

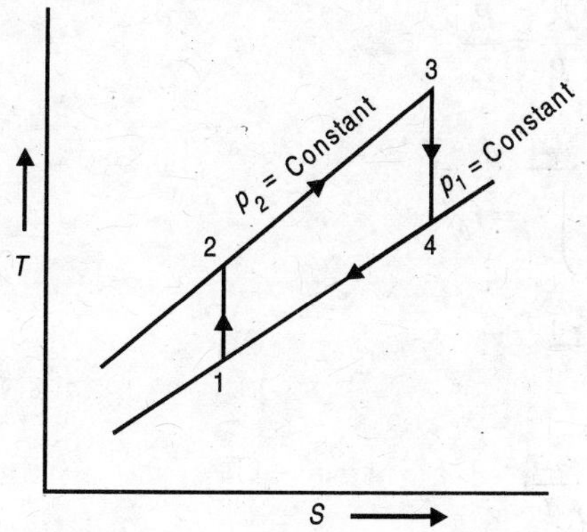

Fig. 34.23. *T–S diagram for closed cycle gas turbine*

Simple closed-cycle gas turbines can work with modifications like intercooling, reheating and regeneration and the numerical analysis is exactly same as that for open cycle gas turbines.

These closed-cycle gas turbine plants are highly suitable for nuclear and high output power stations. The efficiency of such plant is more than 30 %.

Compressor is a steady flow machine and hence the work done on the working substance is given by

$$W_c = (h_2 - h_1) \text{ kJ/kg}$$

$$= C_p (T_2 - T_1) \text{ kJ/kg}$$

Similarly, work done by the working substance in turbine is given by

$$W_t = h_3 - h_4 \text{ kJ/kg}$$

$$= C_p (T_3 - T_4) \text{ kJ/kg}$$

Also, heat supplied/kg $= h_3 - h_2$ kJ/kg

$$= C_p (T_3 - T_2) \text{ kJ/kg}$$

∴ Net work supplied/kg $= W_t - W_c$

$$= C_p (T_3 - T_4) - C_p (T_2 - T_1)$$

∴ Thermal efficiency of the cycle

$$\eta_{th} = \frac{\text{Net work done/kg}}{\text{Heat supplied/kg}}$$

$$= \frac{C_p (T_3 - T_4) - C_p (T_2 - T_1)}{C_p (T_3 - T_2)}$$

$$= \frac{(T_3 - T_2) - (T_4 - T_1)}{(T_3 - T_2)}$$

$$= 1 - \frac{T_4 - T_1}{T_3 - T_2}$$

and, pressure ratio $= r_p = \frac{P_2}{P_1} = \frac{P_3}{P_4}$

and $\frac{T_2}{T_1} = \left(\frac{p_2}{p_1}\right)^{\frac{\gamma-1}{\gamma}} = r_p^{\frac{\gamma-1}{\gamma}}$

$\frac{T_3}{T_4} = (r_p)^{\frac{\gamma-1}{\gamma}}$

$\therefore \quad T_2 = T_1 \times r_p^{\frac{\gamma-1}{\gamma}}$

$T_3 = T_4 \times r_p^{\frac{\gamma-1}{\gamma}}$

$$\therefore \quad \eta_{th} = 1 - \frac{T_4 - T_1}{T_4 \times r_p^{\frac{\gamma-1}{\gamma}} - T_1 \cdot r_p^{\frac{\gamma-1}{\gamma}}}$$

$$= 1 - \frac{T_4 - T_1}{(T_4 - T_1) r_p^{\frac{\gamma-1}{\gamma}}}$$

$$= 1 - \frac{1}{r_p^{\frac{\gamma-1}{\gamma}}}$$

as for open cycle.

Also, the work ratio which is defined as the ratio of the network obtained by the plant to the turbine work.

$$\therefore \quad \text{Work ratio } r_w = \frac{\text{Net work}}{\text{Turbine work}}$$

$$= \frac{W_t - W_c}{W_t}$$

$$= 1 - \frac{W_c}{W_t}$$

$$= 1 - \frac{C_p (T_2 - T_1)}{C_p (T_3 - T_4)}$$

$$= 1 - \frac{T_2 - T_1}{T_3 - T_4}$$

$$= 1 - \frac{T_1\left(\frac{T_2}{T_1} - 1\right)}{T_4\left(\frac{T_3}{T_4} - 1\right)}$$

$$= 1 - \frac{T_1}{T_4}\left[\frac{(r_p)^{\frac{\gamma-1}{\gamma}} - 1}{(r_p)^{\frac{\gamma-1}{\gamma}} - 1}\right]$$

$$= 1 - \frac{T_1}{T_4}$$

$$= 1 - \frac{T_1}{T_3} \cdot r_p^{\frac{\gamma-1}{\gamma}}$$

As before, we observe from this relation and conclude that the cycle efficiency is a function of (r_p) pressure ratio only while the work ratio is a function of pressure ratio and the temperature ratio of the cycle.

34.12 ADVANTAGES AND DISADVANTAGES OF CLOSED CYCLE GAS TURBINE

(a) Advantages

1. The working of this cycle is not dependent on the atmospheric pressure and therefore any pressure can be used so that the specific output of the plant can be increased. By using high pressure, the sizes of the components are reduced.
2. We can use any gas with favourable properties. Helium and helium-carbon dioxide mixture gives high efficiencies and smaller components. It is highly suitable for nuclear plants.
3. When helium is used, then other alternative materials can be used as no oxidation occurs with these inert gases.
4. The turbine blades are not fouled by the products of combustion.
5. Regulation of the closed cycle gas turbine is simpler.
6. Low thermal stresses exist because of the constant temperatures at all loads.
7. As the working substance or fluid does not come in contact with the fuel or products of combustion, very low quality fuel can be used. Any kind or type of the fuel – solid, liquid, and gases and even nuclear can be used.

(b) Disadvantages

1. Strong heat exchanger will be required if the high pressures are utilized at the inlet of the compressor.
2. Poor combustion efficiency and poor heat transfer results because of the indirect type of heat exchanger.
3. Because air heater and coolers are used the cost of the plant increases. Also the total plant becomes heavy. Due to all these factors, the cost of the plant is more than that of open type.
4. A coolant is required for precooler and is a disadvantage as compared to an open cycle.

34.13 SEMI-CLOSED CYCLE

As in an internal combustion engine, there is a combination of constant volume heat addition and constant pressure heat addition process, in gas turbine practice, there is a combination of open-cycle gas turbine and closed cycle gas turbine. In this case, the cycle is called as *Semi-Closed cycle*. In semi-closed cycle, there are two turbines one of which is a closed-cycle gas turbine and the other is an open-cycle gas turbine. Closed cycle gas turbine drives the compressor while an open cycle gas turbine is driving a power generator. Figure 34.24 shows the schematic diagram of a *semi-closed cycle* gas turbine plant.

The different components of the semi-closed cycle gas turbine shown in Fig. 34.24 are as follows

C_1 — Low Pressure compressor (LP)
C_2 — High Pressure compressor (HP)
T_C — Compressor turbine
CC — Combustion chamber
G — Generator
PT — Power turbine
m_2 — Air to LP compressor (from atmosphere)
m_1 — Air/gas from TC admitted to HP compressor after precoder.

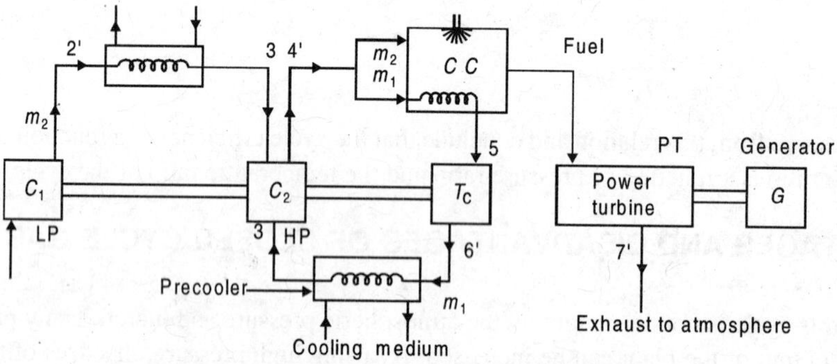

Fig. 34.24. *Schematic of semi-closed cycle gas turbine*

In the semi-closed cycle gas turbine, some part of the total fluid, is confined to the plant and another part of the fluid is taken from the atmosphere. With reference to the Fig. 34.24, m_1 is the mass of the fluid confined to the closed cycle while m_2 is the fluid following an open cycle. The closed-cycle part is a high pressure system and hence the parts are smaller than those for open-cycle parts.

It is observed that the closed-cycle plant gives a better part load performance as compared to open-cycle system. Similarly the air-heater is smaller than that of closed cycle plant.

Figure 34.25 shows the T–S diagram for the schematic of the semi-closed cycle gas turbine.

Different processes shown on the T–S diagram are as follows:

1–2 Isentropic compression in LP stage.
1–2' Actual compression in LP stage.
2'–3 Intercooling in intercooler.
3–4 Isentropic compression in HP stage.
3–4' Actual compression in HP stage.
4'–5 Heating of air in combustion chamber or heater
5–6 Isentropic expansion in compressor turbine.
5–6' Actual expansion in compressor turbine

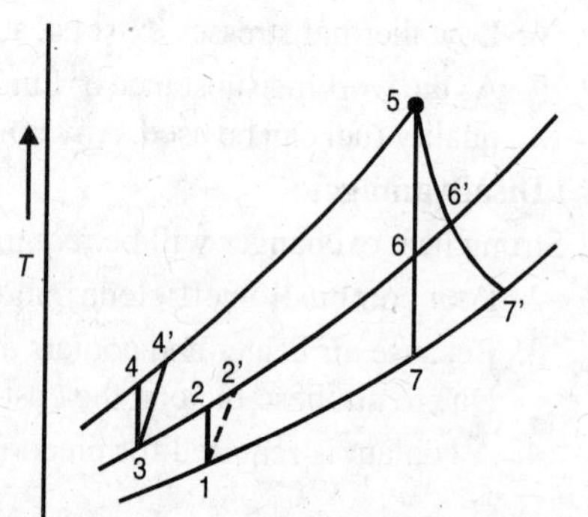

Fig. 34.25

5–7 Isentropic expansion in Power turbine

5–7' Actual expansion of gases in Power turbine.

An illustrative example given below shows the calculations for semi-closed cycle gas turbine plant.

Illustrative Example : (Semi-Closed Cycle)

Following is the data given for a semi-closed cycle gas turbine unit. The plant producers 2000 kW.

1. LP Compressor – Inlet 1 bar and 17°C
   Outlet 2 bar
2. Perfect intercooling
   HP compressor outlet 4 bar.
3. Compressor turbine inlet temperature 800°C
4. Power turbine inlet at 4 bar and 850°C.

Determine the following

(i) Temperature of the power turbine exhaust.

(ii) The ratio of mass flow of air in both circuits.

(iii) The overall efficiency.

(iv) Power required to drive the compressor.

Assume:

Efficiency of heat exchanger = 75 %

$\eta_c = 0.84, \quad \eta_t = 0.86$

Efficiency of combustion = 0.98

Mech. efficiency of power turbine = 0.99

$C_{pa} = 1, \ \gamma \text{ for air} = 1.4, \ C_{pg} = 1.125, \ \gamma = 1.34 \text{ for gases}$

Solution

Figure 34.26 (a) shows the schematic diagram and (b) shows the *T–S* diagram for the same schematic.

Given:

$$P_1 = P_9 = 1 \text{ bar}, \quad T_1 = 290 \text{ K}$$

$$P_2 = P_7 = 2 \text{ bar}, \quad T_6 = 800 + 273 = 1073 \text{ K}$$

$$P_4 = P_8 = 4 \text{ bar}, \quad T_8 = 850 + 273 = 1123 \text{ K}$$

(i) The Temperature of the Power Turbine Exhaust

For the power turbine, we have

$$\frac{T_8}{T_9} = \left(\frac{P_8}{P_9}\right)^{\frac{\gamma-1}{\gamma}} = \left(\frac{4}{1}\right)^{\frac{1.34-1}{1.34}}$$

$$= 4^{0.2537} = \mathbf{1.422}$$

$$\therefore \quad T_9 = \frac{1123}{1.422} = \mathbf{789.73 \text{ K}}$$

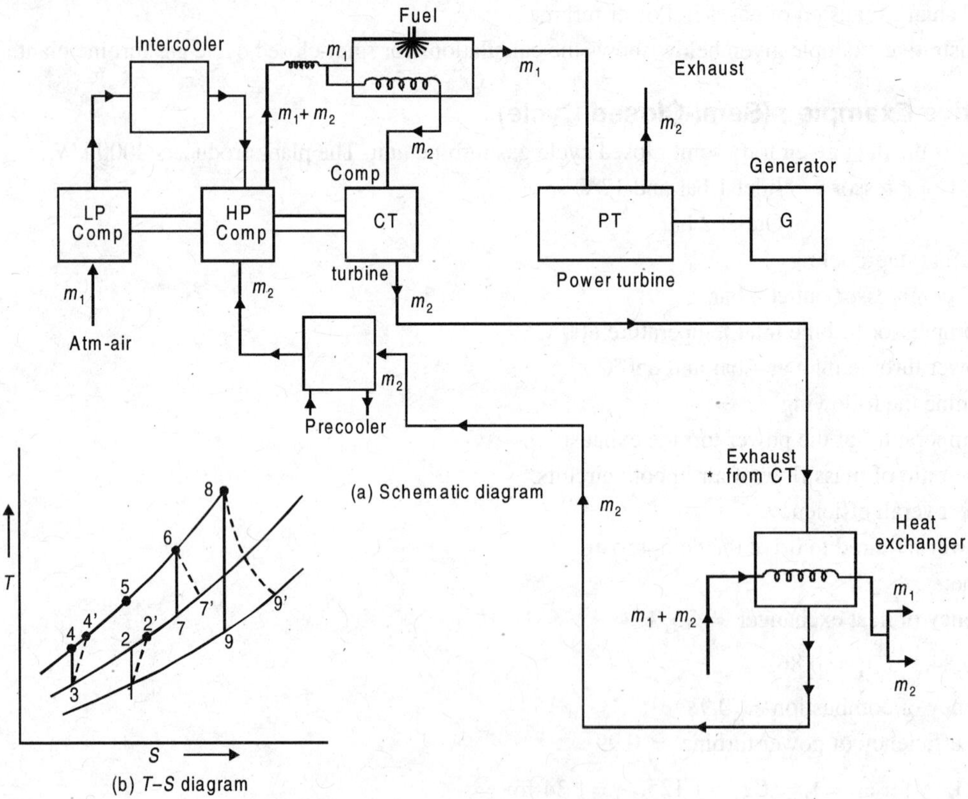

Fig. 34.26. *Semi-closed cycle gas turbine*

For turbine,
$$\eta_t = \frac{T_8 - T_{9'}}{T_8 - T_9}$$

$$0.86 = \frac{1123 - T_{9'}}{1123 - 789.73} = \frac{1123 - T_{9'}}{333.27}$$

$$\therefore \quad T_{9'} = 1123 - 333.27 \times 0.86$$
$$= \mathbf{834.4\ K}$$
$$= \mathbf{563.4\ °C}$$
= Temperature of the power turbine exhaust

(ii) The Ratio of Mass Flow in both Circuits

For compressor,

$$\frac{T_2}{T_1} = \left(\frac{P_2}{P_1}\right)^{\frac{\gamma-1}{\gamma}}$$

where P_2 = Intermediate pressure

$$= \sqrt{P_1 P_3} \text{ for perfect intercooling}$$

$$= \sqrt{1 \times 4} = 2 \text{ bar}$$

$$\therefore \quad \frac{T_2}{T_1} = (2)^{\frac{1.4-1}{1.4}} = 2^{0.2857}$$

$$= \mathbf{1.219}$$

$$\therefore \quad T_2 = 290 \times 1.219 = \mathbf{353.5\ K}$$

Now $$\eta_c = \frac{T_2 - T_1}{T_{2'} - T_1}$$

$$\therefore \quad 0.84 = \frac{353.5 - 290}{T_{2'} - 290} = \frac{63.5}{T_{2'} - 290}$$

$$\therefore \quad T_{2'} = 290 + \frac{63.5}{0.84} = 290 + 75.6$$

$$= \mathbf{365.6\ \ K}$$

For perfect intercooling, the pressure ratio for the second stage is again 2 only.

$$\therefore \quad T_4 = T_2 = 353.5 \text{ K}$$

$$T_{4'} = T_{2'} = 365.6 \text{ K}$$

Consider now the compressor turbine. We have

$$\frac{T_6}{T_7} = \left(\frac{P_6}{P_7}\right)^{\frac{\gamma-1}{\gamma}} = 2^{0.2857}$$

$$= 1.219$$

$$\therefore \quad T_7 = \frac{1073}{1.219} = \mathbf{880.23\ K}$$

and $$\eta_t = \text{turbine efficiency}$$

$$= \frac{T_6 - T_{7'}}{T_6 - T_7} = \frac{1073 - T_{7'}}{1073 - 880.23}$$

$$= \frac{1073 - T_{7'}}{192.77}$$

$$\therefore \quad T_{7'} = 1073 - 192.77 \times 0.86$$

$$= 1073 - 165.75$$

$$= \mathbf{907.25\ \ K}$$

Again consider the power turbine.
We have

$$\text{Work done/kg} = C_{pg}(T_8 - T_{9'})$$

$$= 1.125(1123 - 836.4)$$

$$= \mathbf{322.425\ \ kJ/kg}$$

$$\text{Generator output} = 2000 \text{ kW}$$

and $$\eta_{mech} = 0.99$$

$$\therefore \text{ Power turbine output} = \frac{2000}{0.99} \text{ kJ/sec}$$

$$= m_1 \times 322.425$$

$$\therefore \quad m_1 = \text{mass of air through power turbine}$$

$$= \frac{2000}{0.99} \times \frac{1}{322.425}$$

$$\therefore \quad m_1 = \mathbf{6.266 \ kg/sec}$$

Again we will consider compressor.

m_1 is the mass of air that is flowing through the LP stage of the turbine.

∴ Work required for LP compressor

$$W_{LP} = C_{pa}(T_{2'} - T_1)$$

$$= 1 \times (365.6 - 290)$$

$$= \mathbf{75.6 \ kJ/kg}$$

$$= 75.6 \times m_1 = 75.6 \times 6.266 \text{ kJ/sec}$$

$$= \mathbf{473.71 \ (kJ/kg = kW)}$$

For the second stage, power required is given by

$$W_{HP} = (m_1 + m_2) \times \text{work/kg}$$

$$= (m_1 + m_2) \times 75.6$$

$$= (6.266 + m_2) \times 75.6$$

$$= 473.71 + 75.6\, m_2$$

Here m_2 = mass of air required by the compressor turbine (kg/sec)

Power produced by the compressor turbine

$$W_{CT} = m_2 (T_6 - T_{7'})$$

$$= m_2 (1073 - 907.25)$$

$$= m_2 \times 165.75$$

$$\therefore \quad W_{C} = W_{LP} + W_{HP}$$

$$m_2 \times 165.75 = (473.71) + (473.71 + 75.6\, m_2)$$

$$\therefore \quad m_2 = \frac{947.42}{90.15}$$

$$= \mathbf{10.51 \ kg/sec}$$

∴ Required ratio of flow of air

$$= \frac{m_2}{m_1} = \frac{10.51}{6.266}$$

$$= \mathbf{1.677}$$

(iii) Overall Efficiency

For heat exchanger (regenerator) we have

$$\varepsilon = \frac{T_5 - T_{4'}}{T_{7'} - T_{4'}}$$

$$\therefore \quad 0.75 = \frac{T_5 - 365.6}{907.25 - 365.6} = \frac{T_5 - 365.6}{541.65}$$

$$\therefore \quad T_5 = 365.6 + 0.75 \times 541.65$$

$$= 365.6 + 406.24$$

$$= \mathbf{771.84\ K}$$

$$\therefore \quad \text{Heat supplied/kg} = Q_A$$

$$= m_1 C_{pg}(T_8 - T_5) + m_2 C_{pg}(T_6 - T_5)$$

$$= \frac{[6.266 \times 1.25(1123 - 771.84) + 10.51 \times 1.125(1073 - 771.84)]}{0.98}$$

$$= \mathbf{5755.71\ kJ/sec}$$

∴ Overall efficiency is given by

$$\eta_{ove.} = \frac{\text{Net output}}{\text{Heat supplied}}$$

$$= \frac{2000}{5755.71}$$

$$= 0.3475$$

$$= \mathbf{34.75\ \%}$$

***(iv)* Power Required to Drive the Compressor**

$$\begin{bmatrix}\text{Power required to} \\ \text{drive compressor}\end{bmatrix} = \begin{bmatrix}\text{Power produced by} \\ \text{compressor turbine}\end{bmatrix}$$

$$= m_2(T_6 - T_{7'})$$

$$= 10.51 \times [1073 - 907.25]$$

$$= 10.51 \times 165.75$$

$$= \mathbf{1742\ (kJ/sec = kW)}$$

34.14 ADVANTAGES AND DISADVANTAGES OF GAS TURBINE OVER STEAM TURBINE

Advantages of gas turbines are

1. No feed-water supply system is required.
2. No need of condensing plant
3. No boiler is required.
4. There are less components to be designed.
5. Maintenance cost is low.
6. It is quick to start and hence there are no banking loss. This is very much required for aviation, locomotive

and stand-by or peak load power plants.

7. It has lower specific weight and size. For aviation this is important.
8. Pressures are less.
9. This is an independent system except closed cycle gas-turbine.
10. Useful for mobile vehicles.
11. Costwise it is good (Lower cost).

Disadvantages of gas turbines are

1. The part load efficiency is lower.
2. The efficiency is lower.
3. The power developed is less than steam turbine because the compressor consumes a large power from the power developed by the turbine.
4. For marine application, it is not suitable because air rate is high and the air is containing salts which is harmful to compressors and turbines.

34.15 ADVANTAGES AND DISADVANTAGES OF GAS TURBINE OVER IC ENGINES

The advantages are

1. Well balanced because reciprocating masses are not present.
2. Better mechanical efficiency. This is because there are no sliding parts where friction power is present. Only bearing loss is present.
3. Greater capacity.
4. Specific weight is lower and hence the size is smaller for the same capacity.
5. Cheaper fuel can be used. Even solid fuels like pulverized coal can be used.
6. It requires less lubrication.
7. It has higher operating speed.
8. Smokeless exhaust.
9. Comparatively it works silently because of continuous exhaust.
10. Can be used in jet propulsion system because of the continuous exhaust.
11. Modifications in the Joules or Brayton simple cycle are possible, namely, reheating, regenerative and intercooling arrangements can be added to increase output and efficiency.
12. Lower inlet temperature of air which increases the efficiency of the system.

The disadvantages are

1. It has a poor part load efficiency.
2. High air rate, and open cycle gas turbine is not suitable for marine application.
3. It is very much sensitive to compressor and turbine efficiencies.

SOLVED EXAMPLES

Example 34.1. *The air enters the compressor of an open cycle constant pressure gas turbine at 1 bar, 20°C. The pressure of air at the end of compression is 4 bar. The isentropic efficiency of compressor and turbine are 80% and 85% respectively. The air to fuel ratio is 90. If the air low rate is 3 kg/s, find*

(i) Power developed, (ii) Thermal efficiency.

(Given: C_p of air = 1.0 kJ/kg-K, γ = 1.4 for air, Heat value of fuel = 41800 kJ/kg)

Solution

The gas turbine cycle is shown on *T–S* diagram given below.

Given : $P_1 = 1 \text{ bar} = P_4$

$$T_1 = 20 + 273 = 293 \text{ K}$$
$$P_2 = 4 \text{ bar} = P_3$$

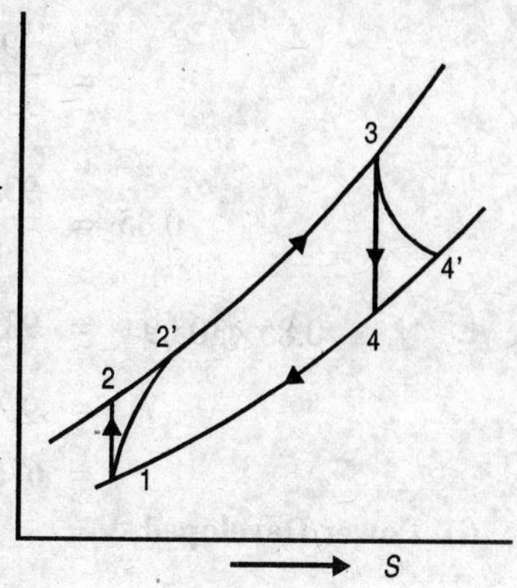

Fig. Ex. 34.1

$$\frac{\text{Air}}{\text{Fuel}} = \frac{90}{1}$$

$$\eta_c = 80\,\%, \eta_t = 85\,\%$$

$$\text{C. V. of fuel} = 41800 \text{ kJ/kg}$$

First we will calculate all the temperatures at the salient points.

$$\frac{T_2}{T_1} = \left(\frac{P_2}{P_1}\right)^{\frac{\gamma-1}{\gamma}} = \left(\frac{4}{1}\right)^{\frac{1.4-1}{1.4}}$$

$$= (4)^{0.2857} = 1.486$$

$$\therefore \quad T_2 = 293 \times 1.486 = \mathbf{435.4\ K}$$

Again, $$\eta_c = \frac{T_2 - T_1}{T_{2'} - T_1}$$

$$\therefore \quad 0.8 = \frac{435.4 - 293}{T_{2'} - 293} = \frac{142.4}{T_{2'} - 293}$$

$$\therefore \quad T_{2'} = 293 + \frac{142.4}{0.8}$$

$$= 293 + 178 = \mathbf{471\ K}$$

$$\text{Fuel supplied/kg air} = \frac{1}{90} \text{ kg}$$

$$\therefore \quad \text{Heat supplied/kg air} = \frac{1}{90} \times 41800 = \mathbf{464.44\ kJ/kg}$$

$$\therefore \quad 464.44 = 1 \times 1 \times (T_3 - T_{2'})$$

$$\therefore \quad T_3 = 464.44 + 471 = \mathbf{935.44\ K}$$

$$\therefore \quad T_4 = \text{Temperature after expansion}$$

$$= \frac{T_3}{1.486}$$

$$= \frac{935.44}{1.486}$$

$$= \mathbf{629.5\ K}$$

Again $$\eta_t = \frac{T_3 - T_{4'}}{T_3 - T_4}$$

$$= \frac{935.44 - T_{4'}}{935.44 - 629.5}$$

$$\therefore \quad 0.85 = \frac{935.44 - T_{4'}}{305.94}$$

$$\therefore \quad 0.85 \times 305.94 = 935.44 - T_{4'}$$

$$\therefore \quad T_{4'} = 935.44 - 260.05$$

$$= \mathbf{675.39 \ K}$$

(i) Power Developed

$$\text{Compressor work/sec} = mC_p\,(T_{2'} - T_1)$$

$$= 3 \times 1 \times (471 - 293)$$

$$= 3 \times 178$$

$$= 534 \ \text{kJ/sec}$$

$$= \mathbf{534 \ kW}$$

$$\text{Turbine work/sec} = m_g C_p\,(T_3 - T_4)$$

$$= \left(3 + \frac{3}{90}\right) \times 1 \times (935.44 - 675.39)$$

$$= \frac{91}{30} \times 1 \times 260.05$$

$$= \mathbf{788.82 \ kJ/s = kW}$$

$$\therefore \text{Net power developed} = W_t - W_c$$

$$= 788.82 - 534$$

$$= \mathbf{254.82 \ kW}$$

(ii) Thermal Efficiency of Cycle

$$\text{Thermal efficiency} = \frac{\text{Work developed}}{\text{Heat supplied}}$$

$$= \frac{254.82}{464.44} = \mathbf{0.5486 = 54.86\ \%}$$

Example 34.2. *(a) Describe with neat sketch and T–S diagram an open cycle and closed cycle gas turbine. List the advantages of closed cycle gas turbine over open cycle gas turbine.*

(b) In a gas turbine plant air compressor takes in air at a temperature of 15°C and compresses it to 4 times the initial pressure with an isentropic efficiency of 82%. The air is then passed through a regenerator before reaching the combustion chamber. The effectiveness of the regenerator is 78%. The maximum temperature of the cycle is 600°C and the isentropic efficiency of the turbine is 70%. Find the thermal efficiency of the plant. Take R = 0.287 kJ/kg-K.

Solution

Given :

$$T_1 = 15+273 = 288 \text{ K}$$

$$T_3 = 600+273 = 873 \text{ K}$$

$$r_p = 4,\ \eta_c = 82\ \%$$

$$\eta_t = 70\ \%,\ \eta_R = 78\ \%$$

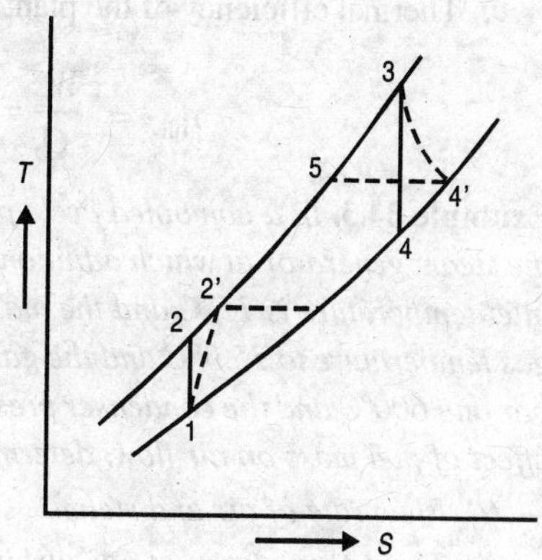

Fig. Ex. 34.2

$$\frac{T_2}{T_1} = \frac{T_3}{T_4} = \left(r_p\right)^{(\gamma-1)/\gamma}$$

$$= (4)^{0.2857} = 1.486$$

$$\therefore \quad T_2 = 1.486 \times 288 = \mathbf{427.96\ K}$$

$$T_4 = \frac{873}{1.486} = \mathbf{587.48\ K}$$

For compressor, $$\eta_c = \frac{T_2 - T_1}{T_{2'} - T_1} = \frac{427.96-288}{T_{2'}-288}$$

$$\therefore \quad T_{2'} = 288 + \frac{139.96}{0.82} = 288 + 170.68 = \mathbf{458.68\ K}$$

Similarly for turbine, $$\eta_t = \frac{T_3 - T_{4'}}{T_3 - T_4} = \frac{873 - T_{4'}}{873 - 587.48}$$

$$0.7 \times 285.52 = 199.86 = 873 - T_{4'}$$

$$\therefore \quad T_{4'} = 873 - 199.86 = \mathbf{673.14\ K}$$

$$\therefore \quad \text{Net work/kg} = W_{tur} - W_{com} = 1.005 \times (199.86 - 170.68)$$

$$= 1.005 \times 29.18$$

$$= \mathbf{29.33\ \ kJ/kg}$$

For generator, $$\eta_R = \frac{T_5 - T_{2'}}{T_{4'} - T_{2'}}$$

$$\therefore \quad 0.78 = \frac{T_5 - 458.68}{673.14 - 458.68}$$

$$\therefore \quad T_5 = 458.68 + 0.78 \times 214.46$$

$$= 458.68 + 167.28$$

$$= \mathbf{625.96\ \ K}$$

∴ Heat added in combustion chamber/kg

$$= C_p\left(T_3 - T_5\right) = 1.005 \times (873 - 625.96)$$

$$= 1.005 \times 247.04$$

$$= \mathbf{248.27\ kJ/kg}$$

∴ Thermal efficiency of the plant is given by

$$\eta_{th} = \frac{W_n}{Q} = \frac{29.33}{248.27} = \mathbf{0.1182\ or\ 11.82\ \%}$$

Example 34.3. *In a combined cycle power plant, the exhaust gas from the open cycle gas turbine is the supply to the steam generator at which additional fuel is burnt in the gas. The pressure ratio for the gas turbine is 7.5, the air inlet temperature is 15°C and the maximum cycle temperature is 750°C. Combustion of additional fuel raises the gas temperature to 75.0°C and the gas leaves steam generator at 100°C. The steam is supplied to the turbine at 50 bar and 600°C and the condenser pressure is 0.1 bar. The total power output of the plant is 200 MW. Neglecting the effect of fuel mass on air flow, determine*

(i) Flow rate of air and steam.

(ii) The power output of gas turbine.

(iii) Power output of steam turbine.

(iv) Efficiency of combined plant.

(v) Air fuel ratio.

The heating value of fuel = 43300 kJ/kg

$C_{pg} = 1.11$ *kJ/kg-K and* $\gamma_g = 1.33$, $C_{pa} = 1.005$ *kJ/kg-K and* $\gamma_a = 1.4$

Solution

(i) Gas turbine

Given : $T_1 = 288$ K

$$\frac{P_2}{P_1} = 7.5, T_3 = 750 + 273 = 1023 \text{ K}$$

$$\frac{T_2}{T_1} = \left(\frac{P_2}{P_1}\right)^{\frac{\gamma-1}{\gamma}} = (7.5)^{2/7}$$

$$= (7.5)^{0.2857} = 1.7783$$

$$\frac{P_2}{P_1} = 7.5$$

$$\therefore \quad \frac{T_2}{T_1} = (7.5)^{0.2857} = 1.7783$$

$$\therefore \quad T_2 = 1.7783 \times 288 = \mathbf{512.15\ K}$$

$$\therefore \quad T_4 = \frac{1023}{1.649} = 620.38 \text{ K}$$

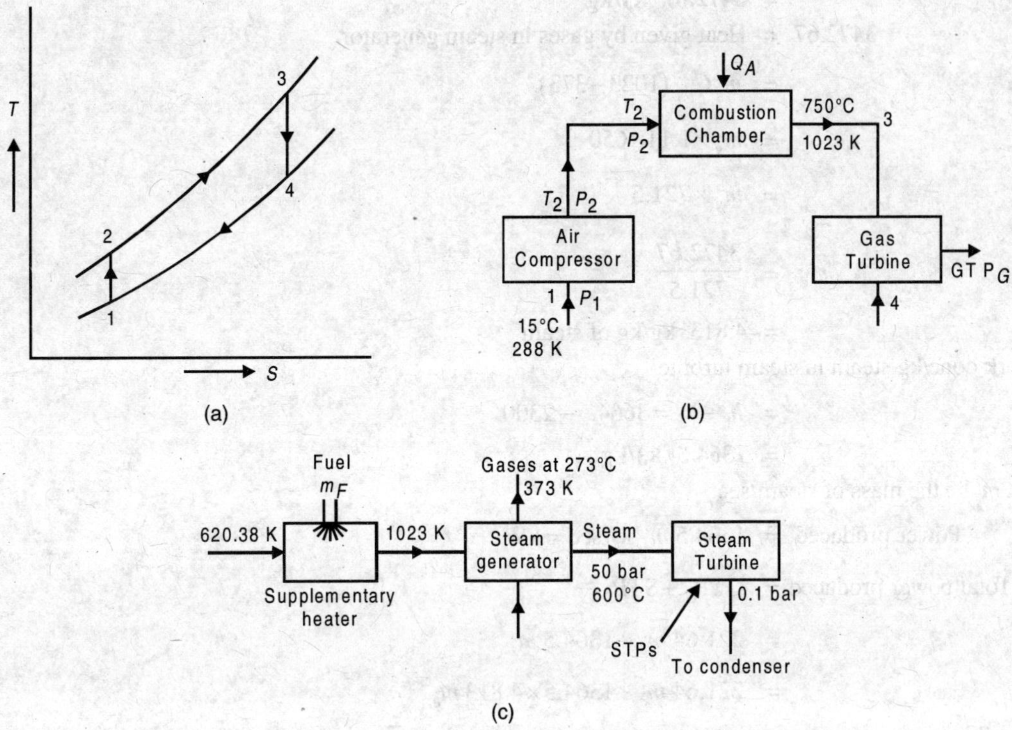

Fig. Ex. 34.3

$$\text{Compressor work/kg} = C_{pa}(T_2 - T_1)$$
$$= 1.005 \times (512.15 - 288)$$
$$= \mathbf{225.27 \ kJ/kg}$$

$$\therefore \quad \text{Turbine work/kg} = C_{pg}(T_3 - T_4)$$
$$= 1.11 \times (1023 - 620.38)$$
$$= 446.91 \ \text{kJ/kg}$$

$$\therefore \quad \text{Net work/kg} = 446.91 - 225.27$$
$$= \mathbf{221.91 \ kJ/kg}$$

If m_a is the mass of air admitted/sec, therefore work done/sec by Gas Turbine plant is

$$\text{GTP}_G = m_a \times 221.64 \text{ kJ/sec} = \text{kW} \qquad \text{...(i)}$$

(ii) Steam Turbine Unit

For steam turbine, $h_1 = 3664.5$ kJ/kg

$h_2 = 2300.0$ kJ/kg

$h_{f2} = 191.83$ kJ/kg

Heat required to generate 1 kg steam

$$= h_1 - h_{f2} = 3664.5 - 191.83$$

$$= \mathbf{3472.67\ kJ/kg}$$

$$\therefore \quad 3472.67 = \text{Heat given by gases in steam generator}$$

$$= m_g C_{pg}(1023-373)$$

$$= m_g \times 1.11 \times 650$$

$$= m_g \times 721.5$$

$$\therefore \quad m_g = \frac{3472.67}{721.5}$$

$$= 4.813 \text{ kg/kg of steam}$$

Work done/kg steam in steam turbine

$$= h_1 - h_2 = 3664.5 - 2300$$

$$= \mathbf{1364.5\ kJ/kg}$$

Let m_s be the mass of steam/sec.

$$\therefore \quad \text{Power produced} = 1364.5\, m_s \text{ kJ/sec} = \text{kW}$$

$$\therefore \text{Total power produced} = \text{GTP}_G + \text{STP}_S$$

$$= 221.64\, m_a + 1364.5\, m_s$$

$$= 221.64\, m_a + 1364.5 \times 4.813\, m_a$$

$$\therefore \quad 200 \times 1000 = 221.64 \times 4.813\, m_s + 1364.5\, m_s$$

$$= 1066.75\, m_s + 1364.5\, m_s$$

$$= 2431.25\, m_s$$

$$\therefore \quad m_s = \text{Mass of steam/sec}$$

$$= \frac{200 \times 1000}{2431.25}$$

$$= \mathbf{82.26\ kg/sec}$$

$$m_a = \text{Mass of air/sec}$$

$$= \mathbf{395.93\ kg/sec}$$

(ii) Power Output of Gas Turbine Unit

$$= \frac{395.93 \times 221.64}{1000}$$

$$= \mathbf{87.754\ MW}$$

(iii) Power Output of Steam Turbine Unit

$$= \frac{82.26 \times 1364.5}{1000}$$

$$= \mathbf{112.246\ MW}$$

(iv) Efficiency of Combined Plant

Heat supplied in combustion chamber of gas turbine/sec

$$= m_a C_{pa}(T_3 - T_2)$$

$$= 395.93 \times 1.005 \times (1023 - 512.15)$$

$$= 395.93 \times 1.005 \times 510.85$$

$$= \mathbf{203272.14 \ kJ/sec}$$

$$\mathbf{m_{f1}} = \mathbf{4.6945 \ kg/sec}$$

Heat supplied in supplementary combustion chamber/sec

$$= m_g C_{pg}(T_2 - T_1)$$

$$= 395.93 \times 1.11 \times (1023 - 620.38)$$

$$= 395.93 \times 1.11 \times 402.62$$

$$= \mathbf{176944.36 \ kJ/sec}$$

$$= \mathbf{4.0865 \ kg/sec}$$

$\therefore$ Total heat supplied/sec $= 203272.14 + 176944.36$

$$= \mathbf{380216.50 \ kJ/sec}$$

$\therefore$ Efficiency of combined plant

$$= \frac{200 \times 1000}{380.2165 \times 1000} = 0.5263$$

$$= \mathbf{52.63 \ \%}$$

(v) Air : Fuel Ratio

Total fuel supplied/ sec $= 4.6945 + 4.0865$

$$= \mathbf{8.7810 \ kg/s}$$

Air supplied/sec $= 395.93$ kg/s

$$\therefore \quad \text{Air to fuel ratio} = \frac{395.93}{8.781} = \mathbf{45.1 \ kg}$$

Example 34.4. *(a) Bring out the difference between the closed cycle and open cycle gas turbine power plants.*

(b) A gas turbine power plant consists of a two stage air compressor with intercooling and a single stage turbine with a regenerator. Air enters the compressor at 1 bar, 20°C. The maximum temperature of the cycle is limited to 900°C and the maximum pressure ratio is 6. The effectiveness of the regenerator is 0.7. The air flow rate is 210 kg/s and the calorific value of the fuel is 40.8 MJ/kg. The $(\eta_{isen})_{comp} = 0.82$ *and* $(\eta_{isen})_{turbine} = 0.92$. *Combustion efficiency = 95%,* $\eta_{mech} = 96\%$, *generator efficiency = 95%.*

Assuming perfect intercooling, and neglecting and heat losses, estimate:

(i) Air: Fuel ratio

(ii) The cycle efficiency

(iii) The power supplied by the plant

(iv) The specific fuel consumption of the plant and

(v) The fuel consumed per hour.

Take for air: $C_p = 1.005$ *kJ/kg-K,* $C_v = 0.7178$ *kJ/kg-K.*

For gases: $C_p = 1.08$ *kJ/kg-K,* $\gamma = 1.33$.

Draw schematic of the plant.

Solution

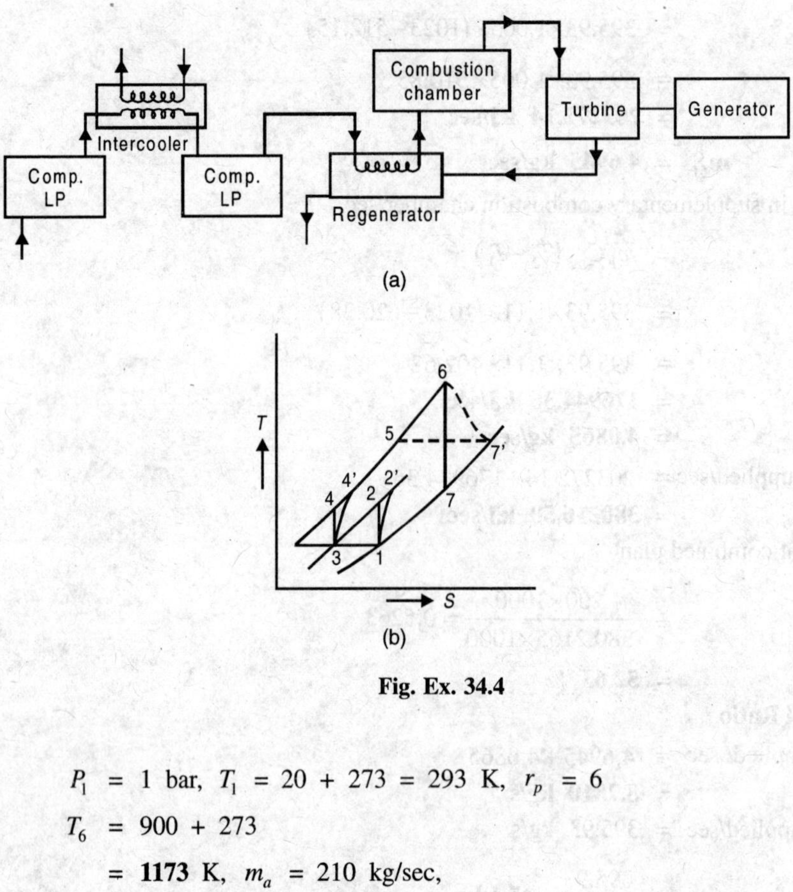

Fig. Ex. 34.4

Given :

$$P_1 = 1 \text{ bar}, \ T_1 = 20 + 273 = 293 \text{ K}, \ r_p = 6$$

$$T_6 = 900 + 273$$

$$= \mathbf{1173} \text{ K}, \ m_a = 210 \text{ kg/sec},$$

$$\eta_c = 0.82, \ \eta_t = 0.92, \ \eta_{comb} = 0.95,$$

$$\eta_{mech} = 96\,\%, \ \eta_{reg} = 0.95$$

$$\text{C. V.} = 40.8 \text{ MJ/kg}$$

$$= \frac{40.8 \times 10^6}{1000} = 40800 \text{ kJ/kg}$$

Compressor : $$\frac{P_2}{P_1} = \frac{P_4}{P_3} = \sqrt{6} = 2.45$$

$$\gamma_{air} = \frac{1.005}{0.7178} = \mathbf{1.4}$$

$$\therefore \quad \frac{T_2}{T_1} = (2.45)^{\frac{\gamma-1}{\gamma}} = (2.45)^{0.2857} = 1.292 = \frac{T_4}{T_3}$$

$$\therefore \quad T_2 = 293 \times 1.292 = 378.5 \text{ K} = T_4$$

$$\eta_c = 0.82 = \frac{T_2 - T_1}{T_{2'} - T_1} = \frac{378.5 - 293}{T_{2'} - 293} = \frac{85.5}{T_{2'} - 293}$$

$$\therefore \quad T_{2'} = T_{4'} = \frac{85.5}{0.82} + 293 = 104.27 + 293 = \mathbf{397.27\ K}$$

Turbine :

$$\frac{T_6}{T_7} = (6)^{1.66-1/1.33} = (6)^{0.2481} = 1.5598$$

$$\therefore \quad T_7 = \frac{1173}{1.5598} = \mathbf{752.02\ K}$$

$$\therefore \quad \eta_t = 0.92 = \frac{T_6 - T_{7'}}{T_6 - T_7} = \frac{1173 - T_{7'}}{1173 - 752.02}$$

$$420.98 \times 0.92 = 387.3 = 1173 - T_{7'}$$

$$\therefore \quad T_{7'} = 1173 - 387.3 = \mathbf{785.7\ K}$$

Generator :

$$\eta_R = \frac{\text{Actual temperature rise}}{\text{Maximum temperature rise}}$$

$$= \frac{T_5 - T_{4'}}{T_{7'} - T_{4'}}$$

where T_5 = Temperature of air before combustion chamber.

$$\therefore \quad 0.95 = \frac{T_5 - 397.27}{785.7 - 397.27}$$

$$= \frac{T_5 - 397.27}{388.43}$$

$$\therefore \quad T_5 = 397.27 + 0.95 \times 388.43$$

$$= 397.27 + 369$$

$$= 766.27 \text{ K}$$

Combustion chamber :

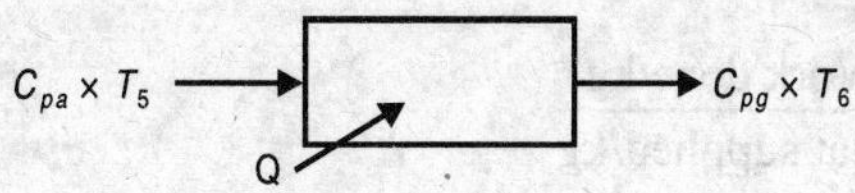

Fig. Ex. 34.4 (c)

Actual heat supplied/kg air (received by air)

$$= C_{pg}\, T_6 - C_{pa}\, T_5 = 1.08 \times 1173 - 1.005 \times 766.27$$

Neglecting the mass of fuel,

$$Q = 1.08 \times 1173 - 1.005 \times 766.27$$

$$= 1266.84 - 770.1$$

$$= \textbf{496.74 kJ/kg of air}$$

$$\therefore \text{Mass of fuel/kg of air} = \frac{496.74}{43300 \times 0.95} = \frac{0.0115}{0.95} = \mathbf{0.0121}$$

Let x be the mass of the fuel supplied/kg air.

$$C_{pa} \times T_5 + x \times \text{C.V.} \times \eta_{\text{comb}} = C_{pg}(1+x)T_6$$

$$\therefore 1.005 \times 766.27 + x \times 43300 \times 0.95 = 1.08 \times (1+x) \times 1173$$

$$\therefore \quad 770.1 + 41135\,x = 1266.84 + 1266.84\,x$$

$$\therefore (41135 - 1266.84)x = 1266.84 - 770.1$$

$$\therefore \quad x = \frac{496.74}{39868.16} = \mathbf{0.0125}$$

(i) Air : Fuel Ratio: Neglecting mass of fuel, we have,

$$\text{Air : Fuel ratio} = \frac{1}{1.0121} = \textbf{82.64 kg/kg fuel}$$

If fuel is considered, this ratio will be

$$\text{Air : Fuel ratio} = \frac{1}{0.0125}$$

$$= \textbf{78.125 kg of air/kg fuel}$$

(ii) Cycle Efficiency

$$\text{Compressor work/kg} = 2C_{pa}(T_{2'} - T_1)$$

$$= 2 \times 1.005 \times (397.27 - 293)$$

$$= \textbf{209.583 kJ/kg air}$$

$$\text{Turbine work/kg} = C_{pg}(T_6 - T_{7'})$$

$$= 1.08 \times (1173 - 785.7)$$

$$= 1.08 \times 387.3$$

$$= \textbf{418.284 kJ/kg}$$

$$\therefore \quad \text{Cycle efficiency} = \frac{\text{Work done/kg}}{\text{Heat supplied/kg}}$$

$$= \frac{W_t - W_c}{Q}$$

$$= \frac{423.83 - 209.583}{496.4} = \frac{214.247}{496.4}$$

$$= \mathbf{0.4316} = \mathbf{43.16\ \%}$$

(iii) Power Supplied by Plant

$$\text{Mass of air/sec} = 210 \text{ kg/s}$$

$$\therefore \quad \text{Work done/s} = 210 \times 214.247 \times 0.96 \times 0.95$$

$$= 44992 \times 0.96 \times 0.95$$

$$= \mathbf{41032 \text{ kW}}$$

(iv) Specific Fuel Consumption

$$\text{BSFC} = \frac{\text{Mass of fuel/hour}}{\text{BP}}$$

$$= \frac{0.01325 \times 210 \times 3600}{41032}$$

$$= \mathbf{0.36 \text{ kg/kWH}}$$

(v) Fuel Consumption/Hour

$$\text{Fuel consumption/hour} = 0.01325 \times 210 \times 3600$$

$$= \mathbf{14787 \text{ kg/hour}}$$

Example 34.5. *(a) Explain with a neat sketch a closed cycle gas turbine power plant.*

(b) A gas turbine set having a heat exchanger gave the following data at design speed.

Isentropic efficiency of compressor = 80%

Isentropic efficiency of turbine = 85%

Pressure ratio = 5 : 1

Mass flow of air = 20 kg/hr

Effectiveness of heat exchanger = 0.8

Maximum cycle temperature = 725°C

Ambient pressure and temperature = 1.013 bar, 25°C

Determine

(i) Power output,

(ii) Specific fuel consumption,

(iii) Thermal efficiency.

Take C_p = 1.05 kJ/kg-K, γ = 1.4. Heating value of fuel = 43000 kJ/kg

Solution

(i) Power Output

$$T_1 = 25 + 273 = 298 \text{ K}$$

$$P_1 = 1.013 \text{ bar}$$

$$\frac{P_2}{P_1} = 5, \; m_a = 20 \text{ kg/hr}$$

$$T_3 = 725 + 273 = 1098 \text{ K}$$

$$\eta_c = 0.8, \; \eta_t = 0.85, \; \varepsilon = 0.8,$$

$$\frac{T_2}{T_1} = (5)^{0.256} = 1.5098$$

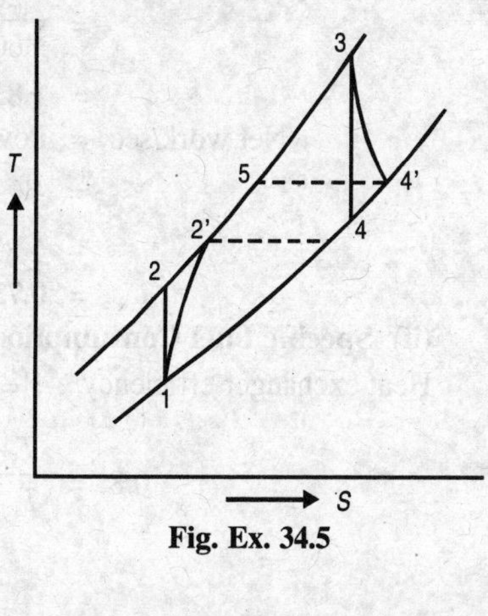

Fig. Ex. 34.5

$$\therefore \quad T_2 = 298 \times 1.5098 = \mathbf{450\ K}$$

$$\eta_c = 0.8 = \frac{T_2 - T_1}{T_{2'} - T_1}$$

$$\therefore \quad T_{2'} - T_1 = \frac{450 - 298}{0.8} = \frac{152}{0.8} = 190$$

$$\therefore \quad T_{2'} = 298 + 190 = \mathbf{488\ K}$$

Again,
$$\frac{T_3}{T_4} = 1.5098$$

$$\therefore \quad T_4 = \frac{1098}{1.5098} = \mathbf{727.25\ K}$$

and
$$\eta_t = 0.85 = \frac{T_3 - T_{4'}}{T_3 - T_4}$$

$$\therefore \quad T_3 - T_{4'} = 0.85 \times (T_3 - T_4)$$

$$\therefore \quad T_{4'} = T_3 - 0.85 \times (T_3 - T_4)$$

$$= 1098 - 0.85 \times (1098 - 727.25)$$

$$= 1098 - 0.85 \times 370.75 = 1098 - 315.14 = \mathbf{782.86\ K}$$

Compressor work,
$$W = mC_p (T_{2'} - T_1)$$

$$= \frac{20}{3600} \times 1.05 \times (488 - 298)$$

$$= \frac{20 \times 1.05}{3600} \times 190 = 1.108 \text{ kJ/sec} = 199.5 \text{ kJ/kg}$$

Turbine work/sec
$$= mC_p (T_3 - T_{4'})$$

$$= \frac{20}{3600} \times 1.05 \times 315.14$$

$$= \mathbf{1.838\ kJ/s\ =\ 330.9\ kJ/kg}$$

Net work/sec = Power output

$$= W_t - W_c$$

$$= 1.838 - 1.108$$

$$= \mathbf{0.730\ \ kJ/s(=kW)}$$

(ii) Specific Fuel Consumption

Heat exchanger efficiency, $\varepsilon = 0.8$

$$\therefore \quad 0.8 = \frac{T_5 - T_{2'}}{T_{4'} - T_{2'}}$$

$$\therefore \quad T_5 = T_{2'} + 0.8(T_{4'} - T_{2'})$$

$$= 488 + 0.8 \times (782.86 - 488)$$

$$= 488 + 0.8 \times 294.86$$

$$= 488 + 235.89$$

$$= \mathbf{723.89\ K}$$

$$\therefore \text{ Heat supplied/hour} = m_a C_p (T_3 - T_5)$$

$$= 20 \times 1.05 \times (1098 - 723.89)$$

$$= 20 \times 1.05 \times 374.11$$

$$= \mathbf{7856.31\ kJ/hr = 392.82\ kJ/kg\ air}$$

$$\text{Fuel supplied/hour} = \frac{7856.31}{43000}$$

$$= \mathbf{0.1827\ kg/hr}$$

$$\therefore \text{Specific fuel consumption} = \frac{0.1827}{0.73} = \mathbf{0.25\ kg/kWH}$$

(iii) Thermal Efficiency

$$\text{Thermal efficiency} = \frac{\text{Work done/kg air}}{\text{Heat supplied/kg air}}$$

$$= \frac{131.4}{392.82} = \mathbf{0.3345 = 33.45\ \%}$$

Example 34.6. *In a gas turbine plant air at 10°C and 1 bar is compressed to 4 bar with compression efficiency of 80%. The air is heated in the regenerator and the combustion chamber till its temperature is raised to 700°C and during the process the pressure falls by 0.14 bar. The air is then expanded in the turbine and then passes to a regenerator which has 75% effectiveness and causes a pressure drop of 0.14 bar. If the isentropic efficiency of the turbine is 85%, determine the thermal efficiency of the plant.*

Solution

The cycle is shown on *T–S* diagram (Fig. Ex. 34.6) 700ºC, 4 – 0.14 = 3.86 bar

$$P_1 = 1 \text{ bar}, \; P_2 = 4 \text{ bar}$$

$$T_1 = 283 \text{ K}$$

$$\frac{T_2}{T_1} = \left(\frac{P_2}{P_1}\right)^{0.2857}$$

$$= (4)^{0.2857} = \mathbf{1.486}$$

$$\therefore \quad T_2 = 283 \times 1.486 = \mathbf{420.53\ K}$$

$$\eta_c = \frac{T_2 - T_1}{T_{2'} - T_1}$$

$$\therefore \quad 0.8 = \frac{420.53 - 283}{T_{2'} - 283} = \frac{137.53}{T_{2'} - 283}$$

$$\therefore \quad T_{2'} = 283 + \frac{137.53}{0.8} = 283 + 171.91 = \mathbf{454.91\ K}$$

$$P_3 = 4 - 0.14 = 3.86 \text{ bar}$$

$$T_3 = 700 + 273 = 973 \text{ K}$$

$$\therefore \quad \frac{T_3}{T_4} = \left(\frac{P_3}{P_4}\right)^{\frac{n-1}{n}} = \left(\frac{3.86}{1}\right)^{0.2857} = \mathbf{1.471}$$

$$\therefore \quad T_4 = \frac{973}{1.471} = \mathbf{661.45\ K}$$

$$\eta_t = 0.85 = \frac{T_3 - T_{4'}}{T_3 - T_4} = \frac{973 - T_{4'}}{973 - 661.45}$$

$$= \frac{973 - T_{4'}}{311.55}$$

$$\therefore \quad T_{4'} = -(0.85 \times 311.55) + 973$$

$$= 973 - 264.82$$

$$= 708.18 \text{ K}$$

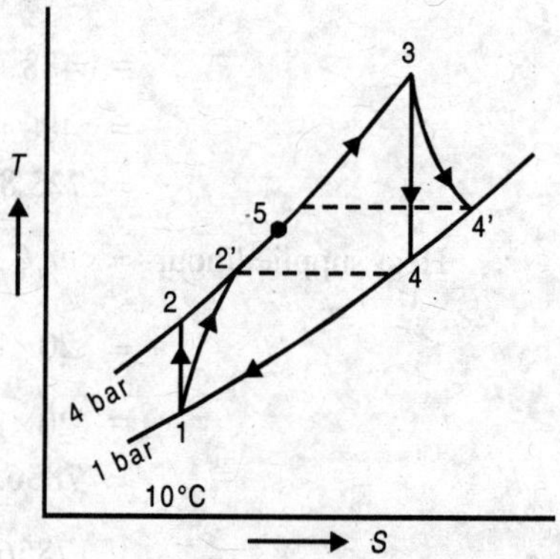

Fig. Ex. 34.6

$$\text{Regenerator effectiveness} = \frac{\text{Actual temperature rise}}{\text{Theoretical temperature rise}}$$

$$0.75 = \frac{T_5 - T_{2'}}{T_{4'} - T_{2'}}$$

$$= \frac{T_5 - 454.91}{708.18 - 454.91}$$

$$= \frac{T_5 - 454.91}{253.27}$$

$$T_5 = (0.75 \times 253.27) + 454.91$$

$$= 189.95 + 454.91$$

$$= \mathbf{644.86\ \ K}$$

$$\text{Work done/kg} = C_{pg}(T_3 - T_{4'}) - C_{pa}(T_{2'} - T_1)$$
$$= 1.1 \times (973 - 708.18) - 1.005 \times (454.91 - 283)$$
$$= 1.1 \times 264.82 - 1.005 \times 171.91$$
$$= 291.3 - 172.77 = \mathbf{118.53\ kJ/kg}$$

$$\text{Heat supplied/kg} = C_{pg}(973 - 644.86)$$
$$= 1.1 \times 328.14$$
$$= 360.95 \text{ kJ/kg}$$

$$\therefore \text{Theoretical efficiency} = \frac{118.53}{360.95} = \mathbf{0.3284 = 32.84\ \%}$$

Example 34.7. *(a) Write short note on closed cycle gas turbine plant.*

(b) In a gas turbine plant air compressor takes in air at a temperature of 15°C and compresses it to 4 times the initial pressure with an isentropic efficiency of 82%. The air is then passed through a regenerator before reaching the combustion chamber. The effectiveness of the regenerator is 78%. The maximum temperature in the cycle is 600°C and the isentropic efficiency of turbine is 70%. Find the thermal efficiency of the plant.

Take R = 0.287 kJ/kg-K.

Solution

(b)
$$\frac{P_2}{P_1} = 4 = \frac{P_3}{P_4}$$
$$\eta_c = 82\ \%,\ \eta_t = 70\ \%$$
$$\eta_R = 78\ \%,\ T_3 = 873\ \text{K}$$
$$T_1 = 15 + 273 = 288\ \text{K}$$
$$\therefore \frac{T_2}{T_1} = \left(\frac{P_2}{P_1}\right)^{\frac{\gamma-1}{\gamma}} = (4)^{\frac{0.4}{1.4}} = 1.4866$$
$$\therefore \mathbf{T_2 = 1.4866 \times 288 = 428.136\ K}$$
$$\therefore \eta_c = \frac{T_2 - T_1}{T_{2'} - T_1}$$
$$\therefore 0.82 = \frac{428.136 - 288}{T_{2'} - 288} = \frac{140.136}{T_{2'} - 288}$$
$$\therefore T_{2'} = \frac{140.136}{0.82} + 288 = 170.9 + 288 = \mathbf{458.9\ K}$$

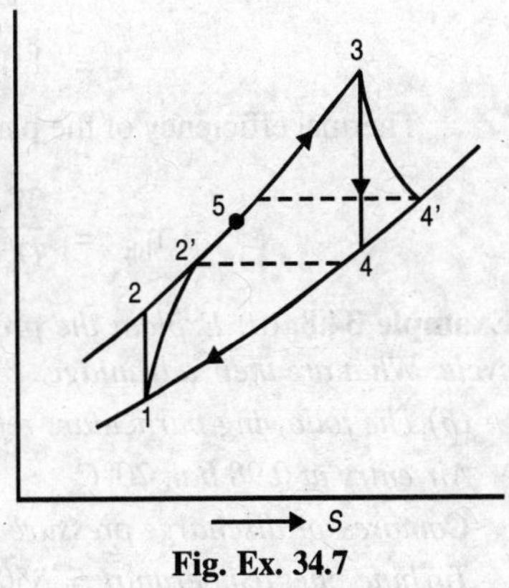

Fig. Ex. 34.7

Similarly,
$$\frac{T_3}{T_4} = \left(\frac{P_3}{P_4}\right)^{\frac{\gamma-1}{\gamma}} = 1.4866$$

$$\therefore \quad T_4 = \frac{873}{1.4866} = \mathbf{587.25\ K}$$

$$\therefore \quad \eta_t = \frac{T_3 - T_{4'}}{T_3 - T_4}$$

$$\therefore \quad 0.7 \times (873 - 587.25) = 873 - T_{4'}$$

$$\therefore \quad T_{4'} = 873 - 200.025 = \mathbf{672.975}$$

$$\therefore \quad \text{Net work/kg} = C_{pg}(T_3 - T_{4'}) - C_{pa}(T_{2'} - T_1)$$

$$= 1.01 \times (873 - 672.975) - 1.005 \times (458.9 - 288)$$

$$= 202.025 - 171.754$$

$$= \mathbf{30.271\ kJ/kg}$$

Regenerator:

$$\eta_r = \frac{T_5 - T_{2'}}{T_{4'} - T_{2'}}$$

$$\therefore \quad 0.78 = \frac{T_5 - 458.9}{672.925 - 458.9}$$

$$= \frac{T_5 - 458.9}{214.025}$$

$$\therefore \quad 214.025 \times 0.78 = 166.94 = T_5 - 458.9$$

$$\therefore \quad T_5 = 166.94 + 458.9 = \mathbf{625.84\ K}$$

$$\therefore \quad Q_A = \text{Heat added/kg}$$

$$= C_{pg}(T_3 - T_5) = 1.01 \times (873 - 625.84)$$

$$= 1.01 \times 247.16 = \mathbf{249.63\ kJ/kg}$$

$\therefore$ Thermal efficiency of the plant,

$$\eta_{th} = \frac{W}{Q_A} = \frac{30.271}{249.63} = \mathbf{0.1213\ or\ 12.13\ \%}$$

Example 34.8. *(a) Explain the processes of multistaging, intercooling and reheating as applied to gas turbine cycle. What are their advantages ?*

(b) The following particulars refer to a gas turbine cycle.

Air entry at 0.98 bar, 20°C

Compressor discharge pressure = 4.2 bar

Turbine inlet temperature = 850°C

Efficiency of compressor and turbine = 0.8 each

Calculate thermal efficiency of actual and theoretical cycles.

Solution

(b)

$$T_1 = 293 \text{ K}$$

$$P_1 = 0.98 \text{ bar}$$

$$P_2 = 4.2 \text{ bar}$$

$$T_3 = 850 + 273 = \mathbf{1123 \ K}$$

$$\eta_c = \eta_t = 0.8$$

$$\frac{T_2}{T_1} = \left(\frac{4.2}{0.98}\right)^{\frac{\gamma-1}{\gamma}} = (1.2857)^{0.286} = 1.5162$$

$\therefore$ $$T_2 = 293 \times 1.5162 = \mathbf{444.25 \ K}$$

Similarly, $$T_4 = \frac{1123}{1.5162} = \mathbf{740.67 \ K}$$

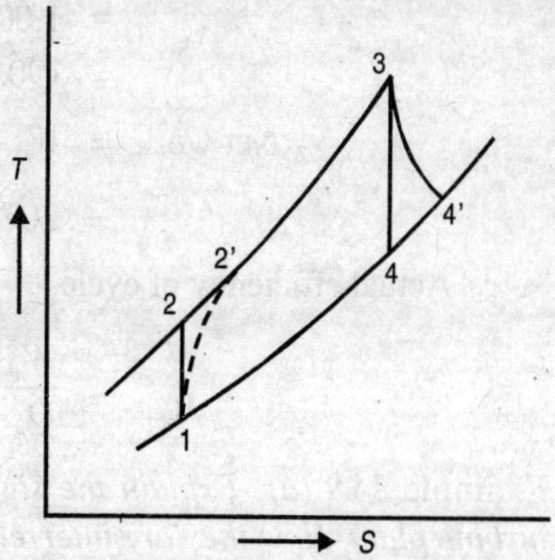

Fig. Ex. 34.8

Theoretical efficiency,

$$\eta_{\text{theo}} = 1 - \frac{T_4 - T_1}{T_3 - T_2} = 1 - \frac{740 - 293}{1123 - 444.25}$$

$$\eta_{\text{theo}} = 1 - \frac{447.67}{678.75} = 1 - 0.6596$$

$$= \mathbf{0.3404 \ or \ 34.04 \ \%}$$

$$\eta_c = \frac{T_2 - T_1}{T_{2'} - T_1}$$

$\therefore$ $$0.8 = \frac{444.25 - 293}{T_{2'} - 293}$$

$\therefore$ $$T_{2'} = 293 + \frac{444.25 - 293}{0.8}$$

$\therefore$ $$T_{2'} = 293 + \frac{151.25}{0.8} = \mathbf{482.1 \ K}$$

$\therefore$ $$Q_{A'} = T_3 - T_{2'} = 1123 - 482.1 = \mathbf{640.9 \ kJ/kg}$$

$$\eta_t = \frac{T_3 - T_{4'}}{T_3 - T_4}$$

$$0.8 = \frac{1123 - T_{4'}}{1123 - 740.67}$$

$$382.33 \times 0.8 = 1123 - T_{4'}$$

$$T_{4'} = 1123-305.864 = 817.136 \text{ K}$$

$$W_{t'} = 1.005\times(1123-817.136)$$

$$= \mathbf{307.393\ kJ/kg}$$

$$W_{c'} = C_p(T_{2'}-T_1) = (482.1-293)\times 1.005$$

$$= 1.005\times 189.1$$

$$= 190.045 \text{ kJ/kg}$$

$\therefore$ Net work $= W_{t'} - W_{c'}$

$$= 307.393-190.045 = \mathbf{117.348\ kJ/kg}$$

$\therefore$ Actual efficiency of cycle,

$$\eta_{act} = \frac{W_{n'}}{Q_{A'}} = \frac{117.348}{640.9} = \mathbf{0.183 = 18.3\ \%}$$

Example 34.9 *(a) Explain the significance of optimum pressure ratio and maximum pressure ratio for a gas turbine plant. How they are interrelated ?*

(b) A gas turbine plant operates between l bar and 9 bar with minimum and maximum cycle temperature of 25°C and 1250°C respectively. The two stage compressor has perfect intercooling and ideal intermediate pressure. Gases are reheated to 1250°C after the expansion in the H.P. turbine and are expanded in the L.P. turbine such that output power from each turbine is same. The isentropic efficiencies of the compressor and the turbine is 0.83. Assume ideal regeneration. Calculate: (i) Efficiency of the cycle, (ii) Power output for a mass flow rate of 5 kg/s, (iii) Work ratio.

Sketch the plant schematic and the cycle on T–S diagram.

Solution

(b) Schematic diagram of plant is given in Fig. Ex. 34.9

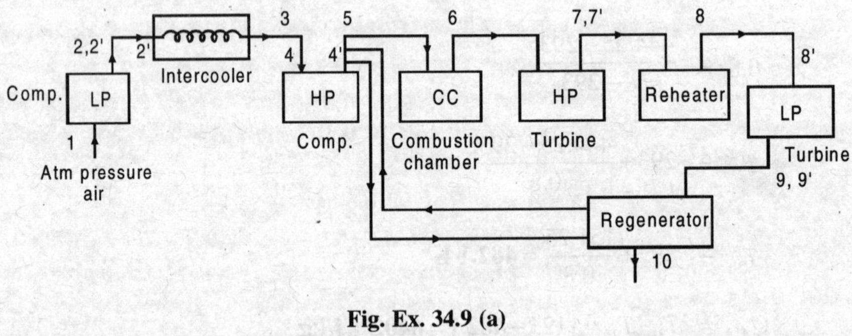

Fig. Ex. 34.9 (a)

Cycle of *T–S* Diagram

Intercooling is perfect.

$$\therefore \quad P_2 = \sqrt{P_1 P_3} = 3 \text{ bar}$$

$$\frac{T_2}{T_1} = (3)^{\gamma-1/\gamma} = (3)^{0.286} = 1.3692$$

$$\therefore \quad T_2 = 298\times 1.3692 = \mathbf{408\ K}$$

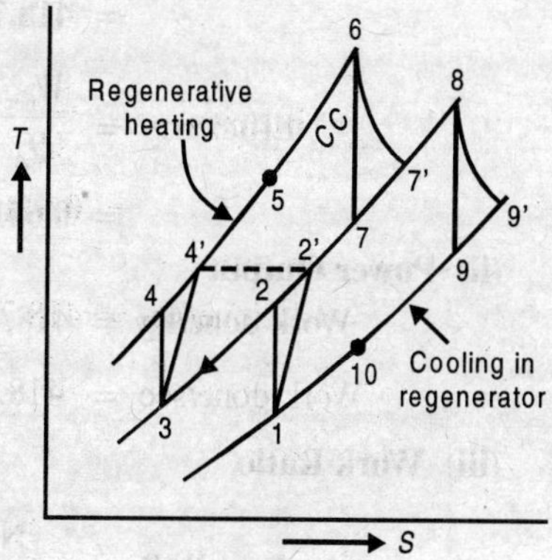

Fig. Ex. 34.9 (b)

$$\eta = \frac{T_2 - T_1}{T_{2'} - T_1}$$

$$\therefore \quad 0.83 = \frac{408.298}{T_{2'} - 298} = \frac{110}{T_{2'} - 298}$$

$$\therefore \quad T_{2'} = \frac{110}{0.83} + 298 = 132.53 + 298 = \mathbf{430.53\ K}$$

Also, $T_{4'} = \mathbf{430.53\ K}$

$$\text{Compressor work} = 2C_p\left(T_{2'} - T_1\right)$$
$$= 2 \times 1.005 \times (430.53 - 298.00)$$
$$= \mathbf{266.38\ kJ/kg}$$

Work in turbine is same and as T_6 and T_8 are same, the intermediate pressure is 3 bar only.

$$\therefore \quad \frac{T_6}{T_7} = (3)^{0.286} = 1.3692$$

$$\therefore \quad T_7 = \frac{1250 + 273}{1.3692} = \mathbf{1112.33\ K} = T_9$$

Again, $$\eta_t = \frac{T_6 - T_{7'}}{T_6 - T_7}$$

$$\therefore \quad 0.83 = \frac{1523 - T_{7'}}{1523 - 1112.33}$$

$$\therefore \quad T_{7'} = T_{9'} = -(410.67 \times 0.83) + 1523$$
$$= -340 - 86 + 1523 = \mathbf{1182.14\ K}$$

$$\therefore \quad T_5 = \mathbf{1182.14\ K}$$

(i) Cycle Efficiency

$$\text{Heat supplied/kg} = (h_6 - h_5) + (h_8 - h_{7'})$$
$$= C_{pa}(T_6 - T_5) + C_{pa}(T_8 - T_{7'})$$
$$= 1.005 \times [(1523 - 1182.14) + (1523 - 1182.4)]$$
$$= 2 \times 1.005 \times 340.86 = \mathbf{685.13\ kJ/kg}$$

$$\text{Turbine work/kg} = 2C_p(T_6 - T_{7'})$$
$$= 2 \times 1.005 \times (1523 - 1182.14) = \mathbf{685.13}$$

$$\therefore \quad \text{Net work/kg} = W_t - W_c = 685.13 - 266.38$$

$$= \mathbf{418.75\ kJ/kg}$$

$$\therefore \quad \text{Efficiency} = \frac{W_{net}}{Q_A} = \frac{418.75}{685.13}$$

$$= \mathbf{0.6112 = 61.12\ \%}$$

(ii) Power Output

$$\text{Work done/kg} = 418.75\ \text{kJ/kg}$$

$$\therefore \quad \text{Work done/sec} = 418.75 \times 5 = \mathbf{2093.75\ (kJ/sec = kW)}$$

(iii) Work Ratio

$$\text{W R} = \frac{\text{Net work}}{\text{Turbine work}}$$

$$= \frac{418.75}{685.13} = \mathbf{0.6112 = 61.12\ \%}$$

Example 34.10. *(a) Prove that for a gas turbine*

$$(r_p)_{opt} = (r_{p\ max})^{1/2} \text{ with usual notations.}$$

(b) A gas turbine plant compresses air with a pressure ratio of 6. The compression begins at 15°C. The maximum temperature in the cycle is restricted to 750°C. The expansion is carried out in two stages with an overall pressure ratio 6. Air is reheated to 750ºC after its expansion in the first stage. A heat exchanger of effectiveness 0.75 provided in the plant. The isentropic efficiencies of compressor and turbine are 0.8 and 0.85 respectively.

Determine

(i) thermal efficiency,

(ii) work ratio,

(iii) work done/kg of air.

Sketch the cycle on T–S diagram.

Solution

T–S diagram for the example is given in Fig. Ex. 34.10

$$\frac{P_2}{P_1} = 6$$

$$T_1 = 288\ \text{K}$$

$$T_3 = 750 + 273 = 1023\ \text{K}$$

$$T_5 = 750 + 273 = 1023\ \text{K}$$

$$\varepsilon = 0.75$$

$$\eta_c = 0.8, \ \eta_t = 0.85$$

$$\frac{T_2}{T_1} = (6)^{0.286} = 1.67$$

$$\therefore \quad T_2 = 288 \times 1.67 = 480.78\ \text{K}$$

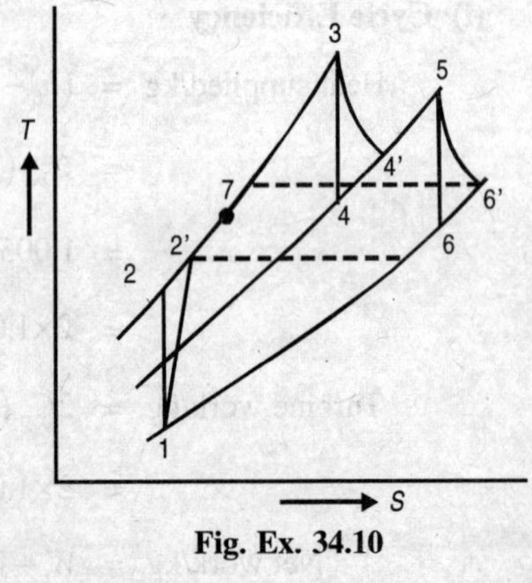

Fig. Ex. 34.10

$$\eta = \frac{T_2 - T_1}{T_{2'} - T_1}$$

$$\therefore \quad 0.8 = \frac{480.78 - 288}{T_{2'} - 288}$$

$$\therefore \quad T_{2'} = 288 + \frac{192.78}{0.8} = 528.975 \text{ K}$$

$$\therefore \text{ Compressor work/kg} = C_p\left(T_{2'} - T_1\right)$$

$$= 1.005 \times (528.975 - 288)$$

$$= \mathbf{242.18 \ kJ}$$

Turbine: Compressor turbine.

$$\text{Work done/kg} = C_p\left(T_3 - T_{4'}\right)$$

$$\therefore \quad 242.18 = 1.005 \times (1023 - T_{4'})$$

$$\therefore \quad 1023 - T_{4'} = 240.975$$

$$\therefore \quad T_{4'} = 1023 - 240.975 = 782.025 \text{ K}$$

$$\therefore \quad 0.85 = \frac{T_3 - T_{4'}}{T_3 - T_4}$$

$$T_3 - T_4 = \frac{240.975}{0.85}$$

$$\therefore \quad T_3 - T_4 = 283.5$$

$$\therefore \quad T_4 = 1023 - 283.5 = \mathbf{739.5 \ K}$$

$$\therefore \quad \left(\frac{P_3}{P_4}\right)^{\frac{\gamma-1}{\gamma}} = \frac{1023}{739.5} = 1.3834$$

$$\therefore \quad \frac{P_3}{P_4} = (1.3835)^{\frac{\gamma}{\gamma-1}} = (1.3835)^{\frac{1.4}{0.4}} = (1.3835)^{3.5} = 3.114$$

$$\frac{P_3}{P_6} = \frac{P_3}{P_4} \times \frac{P_4}{P_6}$$

$$\therefore \quad \frac{P_4}{P_6} = \frac{6}{3.114} = 1.927$$

$$\therefore \quad \frac{T_5}{T_6} = \left(\frac{P_5}{P_6}\right)^{0.286} = (1.927)^{0.286} = 1.2063$$

$$\therefore \quad T_6 = \frac{1023}{1.2063} = \mathbf{848.05\ K}$$

$$\therefore \quad 0.85 = \frac{T_5 - T_{6'}}{T_5 - T_6}$$

$$\therefore \quad T_5 - T_{6'} = 0.85 \times (1023 - 848.05) = 0.85 \times 174.95 = 148.71$$

$$\therefore \quad T_{6'} = 1023 - 148.71 = 874.29$$

$$\therefore \quad \text{Turbine power/kg} = C_p (T_5 - T_{6'})$$

$$= 1.005 \times 148.71 = \mathbf{149.45\ kJ/kg}$$

$$\text{Heat supplied/kg} = C_p (T_3 - T_7) + C_p (T_5 - T_{4'})$$

$$= 1.005 \times (1023 - T_7 + 1023 - 782.025)$$

$$= 1.005 \times (2046 - 782.025 - T_7)$$

$$= 1.005 \times (1263.975 - T_7)$$

$$\text{Heat exchanger effectiveness} = \frac{\text{Actual rise in temperature}}{\text{Theoretical temperature rise}}$$

$$\therefore \quad 0.75 = \frac{T_7 - T_{2'}}{T_{6'} - T_{2'}} = \frac{T_7 - 528.975}{874.29 - 528.975}$$

$$0.75 \times 345.315 = 258.99 = T_7 - 528.975$$

$$\text{or} \quad T_7 = 258.99 + 528.975 = 787.965$$

$$\therefore \quad Q_A = 1.005 \times (1263.975 - 787.965)$$

$$= 1.005 \times 476.010$$

$$= \mathbf{478.39\ kJ/kg}$$

$$\therefore \quad \text{Thermal efficiency} = \frac{W}{Q_A} = \frac{149.45}{478.39}$$

$$= \mathbf{0.3124 \text{ or } 31.24\ \%}$$

$$\text{Work ratio} = \frac{W_t - W_c}{W_t}$$

$$= \frac{149.45}{149.45 + 242.18}$$

$$= \frac{149.45}{391.63} = \mathbf{0.3816 \text{ or } 38.16\ \%}$$

THEORY QUESTIONS

34.1. What is a Gas turbine? In what respect it is different from IC Engines?

34.2. What are the assumptions made while analysing a Joule or Brayton cycle?

34.3. Derive the condition for maximum efficiency and maximum network from Gas turbine.

34.4. What is work ratio? In what way maximum pressure ratio is different than optimum pressure ratio? Derive the relation between them.

34.5. What are the methods used for improving the performance of a simple gas turbine plant? Explain them briefly.

34.6. How will you compare gas turbine with IC engines and steam turbine.

34.7. Explain with neat sketch the working of a semi-closed cycle gas-turbine. What are the advantages and disadvantages of semi-closed cycle gas turbine?

PROBLEMS FOR PRACTICE

34.1. The following data refer to a gas turbine plant.

Compressor inlet pressure	=	1 bar
Compressor inlet temperature	=	37 ºC
Compressor efficiency	=	75 %
Pressure ratio	=	4
Turbine inlet temperature	=	1113 K
Turbine efficiency	=	83 %
Calorific value of fuel	=	44 MJ/kg
Air flow rate	=	75 kg/min.

Find :

(a) Air-fuel ratio

(b) Net Power output

(c) Overall efficiency of plant

Assume : $C_{pa} = 0.996$, $C_{va} = 0.71$,
$C_{pg} = 1.046$, $C_{vg} = 0.744$

[*Ans.* (a) 114.5, (b) 35.18 kW, (c) 7.32 %]

34.2. The following refers to a closed-cycle gas turbine plant using helium as working fluid and incorporation two-stage compression with intercooling and two stage expansion with reheating, temperature at entry to each compression stage is 270ºC; pressure at entry to each compression stage is 270ºC, pressure at entry to first stage compression stage and exit from the second turbine stage is 1 bar; first compression stage pressure ratio is 6 and isentropic efficiency is 85% temperature at inlet to each expansion stage is 1150 ºC and isentropic efficiency of each expansion stage is 0.9; reheat pressure is 6 bar; for helium adiabatic index is 1.24 and cycle $\gamma_p = 10.05$. Calculate the cycle thermal efficiency. [*Ans.* 22.31%]

34.3. In a gas turbine plant the air pressure and temperature before compression are 1 bar and 28ºC. Pressure ratio = 3.5 and the temperature of gas before expansion is 700°C. If the isentropic efficiencies of the compressor and turbine are 80 % and 85 % respectively, calculate

(a) The efficiency of the cycle

(b) The turbine exhaust temperature

(c) % change in efficiency if the regenerator of 0.7 effectiveness is added to the plant.

[*Ans.* (a) 17.14 %, (b) 724.157 K, (c) 56.12 %]

34.4. A gas turbine operates on Brayton cycle. The temperature range is 1000 K and 288 K. Find pressure ratio for maximum power output. Also determine thermal efficiency, work ratio, specific power output. Take $C_p = 1.005$ kJ/kg.K and $\gamma = 1.4$. Expansion and compression is isentropic.

[*Ans.* $(r_p)_{opt.} = 8.831$, $\eta_{th} = 46.34$ %, $WR = 46.34$ %, sp. output = 215.83]

34.5. An open circuit, continuous combustion, constant pressure gas turbine takes in air at 1.01 bar and 15°C. The air is compressed through a pressure ratio of 6. The air then passes at constant pressure through a heat exchanger, the effectiveness of which is 65 %. From the heat exchanger air passes at constant pressure through a combustion chamber in which the temperature is raised to 870°C. From the combustion chamber the air passes through a gas turbine in which it is expanded to a pressure of 1.01 bar and it then passes through the heat exchanger to exhaust. The isentropic efficiency of the compressor is 85 % while that of turbine is 80 %. Neglect the mass of fuel and take air mass flow rate as 4 kg/sec. Take γ = 1.4 and C_p = 1.005 kJ/kg.K.

Determine :

(a) The net power output of the plant.

(b) The exhaust temperature from the heat exchanger.

(c) The thermal efficiency of the plant.

(d) The thermal efficiency of the plant if there was no heat exchanger.

[**Ans.** (a) 612 kW, (b) 600 K, (c) 33.5 % (d) 24.1 %]

IMPORTANT POINTS TO BE REMEMBERED

1. *Thermal efficiency of ideal Brayton cycle*

$$\eta_{th} = 1 - \frac{1}{r_p^{\gamma-1/\gamma}}$$

2. *For maximum work*

$$T_2 = \sqrt{T_1 T_3}$$

3. *Work ratio (WR)*

$$WR = 1 - \frac{\pi}{T_3} \times r_p^{\gamma-1/\gamma}$$

4. *Maximum pressure ratio* $(r_p)_{max}$

$$(r_p)_{max} = \left(\frac{T_3}{T_1}\right)^{\gamma/\gamma-1}$$

5. *Optimum pressure ratio* $(r_p)_{opt.}$

$$(r_p)_{opt} = \sqrt{(r_p)_{max}}$$

6. *Air Rate*

$$\dot{m}_a = \frac{3600}{\text{Work done/kg}} \text{ kg/hour/kW}$$

7. *Thermal efficiency (actual cycle)*

$$\eta_{th} = \frac{W_t - W_c}{Q_A}$$

8. *Regenerator efficiency* (η_{Reg})

$$\eta_{Reg} = \frac{\text{Actual heat transferred}}{\text{Reversibly heat transferred}}$$

9. *Intercooling*

For perfect intercooling, $P_2 = \sqrt{P_1 P_3}$

10. *Reheat cycle*

For maximum turbine work

$$R_{P_L} = R_{P_H} = \sqrt{r_p}$$

CHAPTER 35

Jet Propulsion and Rocket Propulsion

JET PROPULSION

35.1 HISTORY

Since the beginning of aviation in the early 1900s, the speed and flight endurance of powered aircraft have been determined by the power and efficiency of their propulsion systems—piston engines at first and now powerful jet engines.

Jet engines are the most widely used form of propulsion on commercial and military aircraft. They generate power when a mixture of compressed air and fuel is ignited and the resulting hot gases escape through an exhaust nozzle. The reaction of the hot expanding gases exiting the exhaust nozzle produces thrust in the opposite direction and moves the aircraft forward.

Most jet engines today fall into three categories — turbojet, turbofan and turboprop. Turbojets incorporate a turbine-driven compressor to pull air into the engine and compress it before fuel is injected into the combustion chamber and ignited. Turbofan engines are turbojets in which additional power is generated by compressor blades that extend beyond the exterior of the main engine casing. Turboprop engines are turbojets that have propellers which provide extra thrust.

Today's jet engines have succeeded in flying faster than speed of sound, making best use of sophisticated intake systems that minimize drag even at hypersonic speeds, efficient fuel control systems, newly invented materials that can sustain very high temperatures, auxiliary power units that ensure starting in worst conditions and vectored thrust nozzle that can direct thrust in required direction.

Here, we focus on modern jet engines which are used for military and civil purposes, their major components, and systems. Type comparison helps to identify the best engine for a particular application and set of situations. It also focuses on advanced designs like Ramjet, Scramjet, etc.

35.2 HISTORICAL EVOLUTION

Jet engines can be dated back to the first century AD, when Hero of Alexandria invented the aeolipile. This used steam power directed through two jet nozzles so as to cause a sphere to spin rapidly on its axis. So far as is known, it was never used for supplying mechanical power, and the potential practical applications of Hero's invention of the jet engine were not recognised. It was simply considered a curiosity.

Jet propulsion only literally and figuratively took off with the invention of the rocket by the Chinese in the 11th century. Rocket exhaust was initially used in a modest way for fireworks but gradually progressed to propel some quite fearsome weaponry; and there the technology stalled for hundreds of years.

The problem was that rockets are simply too inefficient to be useful for general aviation. Instead, by the 1930s, the piston engine in its many different forms (rotary and static radial, air-cooled and liquid-cooled inline) was the only type of powerplant available to aircraft designers. This was acceptable as long as only low performance aircraft were required, and indeed all that were available.

However, engineers were beginning to realise conceptually that the piston engine was self-limiting in terms of the maximum performance which could be attained; the limit was essentially one of the propeller efficiency. This seemed to peak as blade tips approached the speed of sound. If engine, and thus aircraft, performance were ever to increase beyond such a barrier, a way would have to be found to radically improve the design of the piston engine, or a wholly new type of powerplant would have to be developed. This was the motivation behind the development of the gas turbine engine, commonly called a "jet" engine, which would become almost as revolutionary to aviation as the Wright brothers' first flight.

The earliest attempts at jet engines were hybrid designs in which an external power source supplied the compression. In this system (called a thermojet by Secondo Campini) the air is first compressed by a fan driven by a conventional piston engine, then it is mixed with fuel and burned for jet thrust. The examples of this type of design were the Henri Coandâ's Coandâ-1910 aircraft, and the much later Campini Caproni CC.2, and the Japanese Tsu-11 engine intended to power Ohka kamikaze planes towards the end of World War II. None were entirely successful and the CC.2 ended up being slower than the same design with a traditional engine and propeller combination.

The key to a practical jet engine was the gas turbine, used to extract energy to drive the compressor from the engine itself. The gas turbine was not an idea developed in the 1930s: the patent for a stationary turbine was granted to John Barber in England in 1791. The first gas turbine to successfully run self-sustaining was built in 1903 by Norwegian engineer Ægidius Elling. The first patents for jet propulsion were issued in 1917. Limitations in design and practical engineering and metallurgy prevented such engines reaching manufacture. The main problems were safety, reliability, weight and, especially, sustained operation.

35.3 JET ENGINES

35.3.1 Introduction

A jet engine is an engine that discharges a fast moving jet of fluid to generate thrust in accordance with Newton's third law of motion. This broad definition of jet engines includes turbojets, turbofans, rockets and ramjets and water jets, but in common usage, the term generally refers to a **gas turbine Brayton cycle engine** used to produce a jet of high speed exhaust gases for special propulsive purposes in Illustration 1.

The principle of jet propulsion is based on Newton's second and third laws of motion. Momentum is imported to a mass of fluid in such a manner that the reaction of the imparted momentum gives a propulsive force. This is done by expanding the gas which is at high pressure and temperature through a nozzle so that the gas with markedly increased velocity in the form of jet comes out into the atmosphere and its reaction in the opposite direction gives the propulsive force. The open cycle gas turbine is most suited for jet propulsion. If the working fluid is expanded in the turbine such that the power developed is only sufficient to drive the compressor and accessories and the rest expansion is achieved in a nozzle which is placed just after the turbine. This unit will serve as a jet propulsion system. The gas from the turbine while passing through the nozzle will be accelerated and will come out in the form of jet with a tremendous velocity. The reaction of this jet propels the unit in the forward direction (i.e., opposite direction of the jet).

Illustration 1. *Jet Engine*

The jet propulsion engines are classified as follows:

1. Atmospheric jet engines.
2. Rocket engines.

1. Atmospheric jet engines require oxygen from atmospheric air for combustion of the fuel. As a result their performance depends to a great extent on the forward speed of the engine and on the atmospheric pressure and temperature.

2. The rocket engine carries its own oxidizer for the combustion of the fuel and is, therefore, independent of atmospheric air. It requires maximum specific fuel consumption among all types of propulsion systems.

Atmospheric jet engines are further classified as follows.

(a) Steady Combustion Systems (Continuous Air Flow)

(i) Turbo jet

(ii) Turbo jet with after burner (also known as turboram jet, turbojet with tail-pipe burning and turbojet with reheater).

(iii) Turboprop (also known as prop-jet)

(iv) Ram jet (also known as athodyds and Lorin tube)

(b) Intermittent Combustion System-Intermittent air flow.

(i) Pulse jet (also known as aeropulse, resojet, schmidt tube and intermittent jet).

The turbojet, the turbojet with after burner and turbo-prop are all modified forms of simple open-cycle gas turbine. The ram jet and pulse jet are athodyds (aero-thermodynamic ducts) i.e., straight duct-type of jet engines without compressor and turbine wheels.

Rocket engines are further classified as: (i) Liquid-propellent and (ii) solid propellent.

Rocket propulsion should be regarded as a source of power for achieving objectives obtainable by other methods. Some of the rocket applications are

(a) artillery barrage rockets
(b) anti tank 'bazcoka' rockets
(c) all types of guided missiles
(d) aircraft launched rockets
(e) jet assisted take-off for air planes
(f) engines for long range, high speed guided missiles and pilot less aircraft
(g) transonic airplanes such as D-558.

35.4 THE TURBOJET

The turbojet engine consists of a diffuser at the entrance which slows down the entrance air and thereby compresses it, called the raming effect; a simple open gas turbine cycle and an exit nozzle which expands the gas and converts the thermal energy of the exit gas into kinetic energy which is emitted out in the form of a jet. The increased velocity of air thereby produces thrust. A schematic arrangement of a turbo-jet engine is shown in Fig. 35.1. A corresponding *T–S* diagram is shown. The basic cycle for the turbojet engine is the joule or Brayton cycle. In process 0 to 1, the air entering from the atmosphere is diffused isentropically from velocity V_a down to zero i.e., $V_1 = 0$. This means that the diffuser has an efficiency of 100 percent. This is termed as **ram-compression**. Process 0–1' is the actual compression (ram). Processs 1′–2 shows the isentropic compression of air and 1'–2' shows the actual compression of air. 2′–3 shows the heat addition at constant pressure $p_2 = p_3$. Process 3–4 shows isentropic expansion of gas in the turbine and 3–4′ shows the actual expansion in the turbine. Process 4'–5 shows the isentropic expansion of the gas in the nozzle and 4'–5' shows the actual expansion of gas in the nozzle. Consider 1 kg of working fluid flowing through the system.

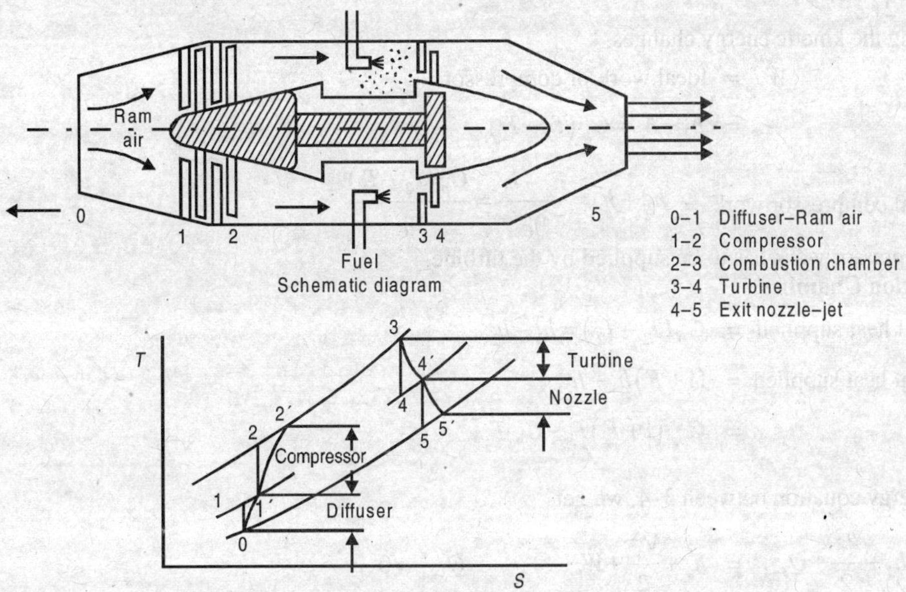

Fig. 35.1

Diffuser: Energy equation between 0 and 1 gives

$$\frac{V_a^2}{2} + h_0 + Q_{01} = \frac{V_1^2}{2} + h_1 + W_{01}$$

V_a = Velocity of air entering from atmosphere.

In ideal diffuser, $V_1 = 0,\ Q_{01} = 0,\ W_{01} = 0$

$$\therefore \quad \frac{V_a^2}{2} + h_0 = h_1$$

or

$$C_p T_1 = C_p T_o + \frac{V_a^2}{2}$$

$$T_1 = T_o + \frac{V_a^2}{2C_p} \quad \text{...(1)}$$

Process 0–1′ shows the actual process in the diffuser.

$$\text{Diffuser efficiency} = \eta_d = \frac{T_1 - T_0}{T' - T_0} = \frac{h_1 - h_0}{h_{1'} - h_0}$$

$$\therefore \quad h_{1'} = h_0 + \frac{V_a^2}{2\,\eta_d}$$

$$\therefore \quad T_{1'} = T_0 + \frac{V_a^2}{2\,\eta_d \cdot C_p} \qquad \ldots(2)$$

Compressor

Using energy equation,

$$h_1 + \frac{V_1^2}{2} + Q_{1-2} + W_c = h_2 + \frac{V_2^2}{2}$$

Neglecting the kinetic energy changes,

$$W_c = \text{Ideal work of compressor}$$

$$= h_2 - h_1 = C_p (T_2 - T_1)$$

$$\text{The actual compressor work} = h_{2'} - h_1 = \frac{h_2 - h_1}{\eta_C} = \frac{C_p (T_2 - T_1)}{\eta_C}$$

This compressor work has to be supplied by the turbine.

Combustion Chamber

$$\text{The ideal heat supplied} = C_p (T_3 - T_2) = h_3 - h_2$$

$$\text{The actual heat supplied} = (1+F) h_3 - h_{2'}$$

$$= C_{pg} (1+F) T_3 - C_{pa} T_{2'}$$

Turbine

Using energy equation between 3–4, we get

$$h_3 + \frac{V_3^2}{2} + Q_{3-4} = h_4 + \frac{V_4^2}{2} + W_t \qquad Q_{3-4} = 0$$

then

$$W_t = \text{turbine work}$$

$$= (h_3 - h_4) + \frac{V_3^2 - V_4^2}{2}$$

$$= h_3 - h_4 \qquad \text{neglecting } \Delta\text{KE}$$

$$= C_p (T_3 - T_4)$$

$$\text{Actual turbine work} = h_3 - h_{4'} = C_p (T_3 - T_{4'})$$

$$= C_p (T_3 - T_4)\eta_t$$

For simplification, for turbo-jet engine,

$$W_c = W_t$$

or

$$C_p (T_3 - T_{4'}) = C_p (T_3 - T_4)\eta_t = \frac{(T_2 - T_1) C_p}{\eta_c}$$

or $$T_{4'} = T_3 - (T_3 - T_4)\eta_t = T_3 - \frac{T_2 - T_1}{\eta_c}$$

Jet-nozzle

Using energy equation between states 4 and 5,

$$h_4 + \frac{V_4^2}{2} = h_5 + \frac{V_5^2}{2}$$

In actual case

$$h_{4'} + \frac{V_{4'}^2}{2} = h_{5'} + \frac{V_{5'}^2}{2}$$

$$\therefore \quad \frac{V_{5'}^2}{2} = h_{4'} - h_{5'} \qquad \text{as } V_4 << V_5$$

$$\therefore \quad V_{5'} = \sqrt{2C_p(T_{4'} - T_{5'})}$$

$$= \sqrt{2C_p\eta_n(T_{4'} - T_5)} \qquad ...(3)$$

$$\therefore \quad \text{Thermal efficiency} = \frac{(h_3 - h_{5'}) - (h_{2'} - h_0)}{h_3 - h_{2'}}$$

$$= \frac{(T_3 - T_{5'}) - (T_{2'} - T_0)}{T_3 - T_{2'}}$$

35.5 MAJOR COMPONENTS

The major components of a jet engine are similar across the major different types of engines, although not all engine types have all components. The major parts include.

35.5.1 Air Inlets

35.5.1.1 Subsonic Inlets

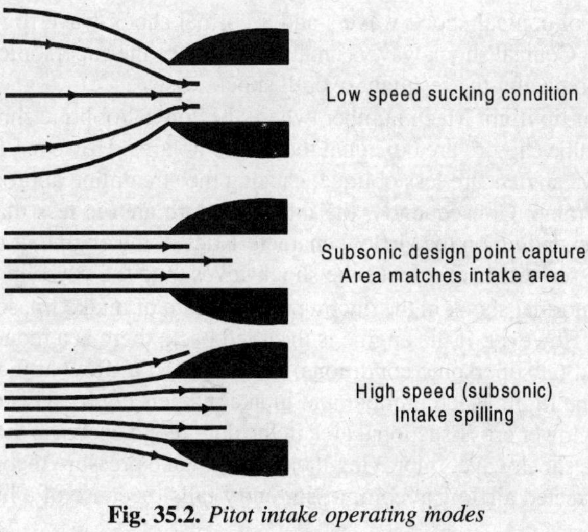

Fig. 35.2. *Pitot intake operating modes*

Pitot intakes are the dominant type for subsonic applications. A subsonic pitot inlet is little more than a tube with an aerodynamic fairing around it.

At zero airspeed (i.e., rest), air approaches the intake from a multitude of directions: from directly ahead, radially, or even from behind the plane of the intake lip.

At low airspeeds, the stream tube approaching the lip is larger in cross-section than the lip flow area, whereas at the intake design flight Mach number the two flow areas are equal. At high flight speeds the stream tube is smaller, with excess air spilling over the lip.

Beginning around 0.85 Mach, shock waves can occur as the air accelerates through the intake throat.

Careful radiusing of the lip region is required to optimise intake pressure recovery (and distortion) throughout the flight envelope.

35.5.1.2 Supersonic Inlets

Supersonic intakes exploit shock waves to decelerate the airflow to a subsonic condition at compressor entry.

There are basically two forms of shock waves:

1. Normal shock waves lie perpendicular to the direction of the flow. These form sharp fronts and shock the flow to subsonic speeds. Microscopically the air molecules smash into the subsonic crowd of molecules like alpha rays. Normal shock waves tend to cause a large drop in stagnation pressure. Basically, the higher the supersonic entry Mach number to a normal shock wave, the lower the subsonic exit Mach number and the stronger the shock (i.e., the greater the loss in stagnation pressure across the shock wave).

2. Conical (3-dimensional) and oblique shock waves (2D) are angled rearwards, like the bow wave on a ship or boat, and radiate from a flow disturbance such as a cone or a ramp. For a given inlet Mach number, they are weaker than the equivalent normal shock wave and, although the flow slows down, it remains supersonic throughout. Conical and oblique shock waves turn the flow, which continues in the new direction, until another flow disturbance is encountered downstream.

A sharp-lipped version of the pitot intake, described above for subsonic applications, performs quite well at moderate supersonic flight speeds. A detached normal shock wave forms just ahead of the intake lip and 'shocks' the flow down to a subsonic velocity. However, as flight speed increases, the shock wave becomes stronger, causing a larger percentage decrease in stagnation pressure (i.e., poorer pressure recovery). An early US supersonic fighter, the F-100 Super Sabre, used such an intake.

An unswept lip generate a shock wave, which is reflected multiple times in the inlet. The more reflections before the flow gets subsonic, the better pressure recovery.

More advanced supersonic intakes, excluding pitots,

(a) exploit a combination of conical shock wave/s and a normal shock wave to improve pressure recovery at high supersonic flight speeds. Conical shock wave/s are used to reduce the supersonic Mach number at entry to the normal shock wave, thereby reducing the resultant overall shock losses.

(b) have a design shock-on-lip flight Mach number, where the conical/oblique shock wave/s intercept the cowl lip, thus enabling the streamtube capture area to equal the intake lip area. However, below the shock-on-lip flight Mach number, the shock wave angle/s are less oblique, causing the streamline approaching the lip to be deflected by the presence of the cone/ramp. Consequently, the intake capture area is less than the intake lip area, which reduces the intake airflow. Depending on the airflow characteristics of the engine, it may be desirable to lower the ramp angle or move the cone rearwards to refocus the shockwaves onto the cowl lip to maximise intake airflow.

(c) are designed to have a normal shock in the ducting downstream of intake lip, so that the flow at compressor/fan entry is always subsonic. However, if the engine is throttled back, there is a reduction in the corrected airflow of the LP compressor/fan, but (at supersonic conditions) the corrected airflow at the intake lip remains constant, because it is determined by the flight Mach number and intake incidence/yaw. This discontinuity is overcome by the normal shock moving to a lower cross-sectional area in the ducting, to decrease the Mach number at entry to the shockwave. This weakens the shockwave, improving the overall intake pressure recovery. So, the absolute airflow stays constant, whilst the corrected airflow at compressor entry falls (because of a higher entry pressure). Excess intake airflow may also be dumped overboard or into the exhaust system, to prevent the conical/oblique shock waves being disturbed by the normal shock being forced too far forward by engine throttling.

35.5.1.3 Inlet Cone

35.5.1.3.1 Purpose

The main purpose of an inlet cone is to slow down the flow of air from supersonic flight speed to a subsonic speed, before it enters the engine. At supersonic flight speeds a conical shock wave, sloping rearwards, forms at the apex of the cone. Air passing through the conical shock wave (and subsequent reflections) slows down to a low supersonic speed. The air then passes through a strong normal shock wave, within the diffuser passage, and exits at a subsonic velocity. The resulting intake system is more efficient (in terms of pressure recovery) than the much simpler pitot intake.

Illustration 2. *Inlet Cone*

35.5.1.3.2 Shape

The inlet cone is shaped so that the shock wave that forms on its apex is directed to the lip of the intake; this allows the engine to operate properly in supersonic flight. As speed increases, the shock wave becomes increasingly more oblique. As a result, some inlet cones are designed to move axially to maintain the shock-on-lip and allow efficient operation over a wider range of flight speeds.

35.5.1.3.3 Opeartion

At subsonic flight speeds, the conical inlet operates much like pitot intake. However, as the vehicle goes supersonic a conical shock wave appears, emanating from the cone apex.

Conical (and oblique) shock waves are akin to the bow wave on a ship. As the flight Mach number increases, the conical shock wave becomes more oblique and eventually impinges on the intake lip.

With increasing mach number the cone is not moved out of the inlet as it would make sense for the external compression, but inwards, because the cone sits in the cowl like a plug: By moving it inwards the path between the cone and the inlet narrows as it is needed for higher speed. The compression occurring in this path is called "internal compression" (opposed to the "external compression" on the cone) and can be subject to stall.

Care must be taken to prevent the normal shock wave, which lies behind the throat of the diffuser, coming forward through the throat, replacing the oblique shocks and reducing stagnation pressure and leading to excessive inlet temperatures burning the compressor.

- The cone is moved in and out of the pitot intake controlled by a computer with temperature and pressure sensors in the intake to match the throat to the mach number.
- With a ramjet, this occurs if excessive fuel is injected into the combustor, raising internal pressure too far. However, with a turbojet or turbofan, the problem arises when the engine is throttled back, causing a mismatch between intake airflow and engine mass flow. A trapdoor is needed to dump excess flow overboard. Or a variable geometry for the stators of the fan is used.

35.5.2 Compressors

35.5.2.1 Axial Flow Compressor

Axial flow compressor uses a series of fan-like rotating rotor blades to progressively compress the gasflow. Stationary stator vanes, located downstream of each rotor, redirect the flow onto the next set of rotor blades. The area of the gas passage diminishes through the compressor to maintain a roughly constant axial Mach number. Axial-flow compressors are normally used in high flow applications, such as medium to large gas turbine engines. They are almost always multi-staged.

35.5.2.2 Centrifugal Compressor

It uses a vaned rotating disk or impeller in a shaped housing to force the gas to the rim of the impeller, increasing the velocity of the gas. A diffuser (divergent duct) section converts the velocity energy to pressure energy. With multiple staging, they can achieve extremely high output pressures greater than 60 MPa.

Axial compressors rely on spinning blades that have aerofoil sections, similar to aeroplane wings. As with aeroplane wings in some conditions the blades can stall. If this happens, the airflow around the stalled compressor can reverse direction violently. Each design of a compressor has an associated operating map of airflow versus rotational speed for characteristics peculiar to that type.

35.5.2.3 Anti Stall Systems for Compressors

At a given throttle condition, the compressor operates somewhere along the steady state running line. Unfortunately, this operating line is displaced during transients. Many compressors are fitted with anti-stall systems in the form of bleed bands or variable geometry stators to decrease the likelihood of surge. Another method is to split the compressor into two or more units, operating on separate concentric shafts.

Another design consideration is the average stage loading. This can be kept at a sensible level either by increasing the number of compression stages (more weight/cost) or the mean blade speed (more blade/disc stress).

Although large flow compressors are usually all-axial, the rear stages on smaller units are too small to be robust. Consequently, these stages are often replaced by a single centrifugal unit. Very small flow compressors often employ two centrifugal compressors, connected in series. Although in isolation centrifugal compressors are capable of running at quite high pressure ratios (e.g. 10 : 1), impeller stress considerations limit the pressure ratio that can be employed in high overall pressure ratio engine cycles.

35.5.3 Combustor/Flame Holder/Combustion Chamber

A flame holder is a component of a jet engine designed to help maintain continual combustion.

All continuous-combustion jet engines require a flame holder. A flame holder creates a low-speed eddy in the engine to prevent the flame from being blown out.

The simplest design, often used in amateur projects, is the can-type flame holder, which consists of a can covered in small holes. Much more effective is the H-gutter flame holder, which is shaped like a letter H with a curve facing and opposing the flow of air. Even more effective, however, is the V-gutter flame holder, which is shaped like a V with the point in the direction facing the flow of air. Some studies have suggested that adding a small amount of base bleed to a V-gutter helps reduce drag without reducing effectiveness.

35.5.4 Turbine

The turbine acts like a windmill, extracting energy from the hot gases leaving the combustor. This energy is used to drive the compressor through the shaft, or bypass fans, or props, or even (for a gas turbine-powered helicopter) converted entirely to rotational energy for use elsewhere. Relatively cool air, bled from the compressor, may be used to cool the turbine blades and vanes, to prevent them from melting.

Because a turbine expands from high to low pressure, there is no such thing as turbine surge or stall. The turbine needs fewer stages than the compressor, mainly because the higher inlet temperature reduces the delta T/T (and thereby the pressure ratio) of the expansion process. The blades have more curvature and the gas stream velocities are higher.

Designers must, however, prevent the turbine blades and vanes from melting in a very high temperature and stress environment. Consequently bleed air extracted from the compression system is often used to cool the turbine blades/vanes internally. Other solutions are improved materials and/or special insulating coatings. The discs must be specially shaped to withstand the huge stresses imposed by the rotating blades. They take the form of impulse, reaction, or combination impulse-reaction shapes. Improved materials help to keep disc weight down.

35.5.5 Turbopumps

Turbopumps are used to raise the fuel pressure above the pressure in the combustion chamber so that it can be injected. Turbopumps are very commonly used with rockets, but ramjets also have been known to use them. The turbopump is usually driven by a gas turbine.

35.5.5.1 Centrifugal Turbopump

Most turbopumps are centrifugal type—the fluid enters the pump near the axis and the rotor accelerates the fluid circumferentially and compresses it against the rim, generating high pressures, and if the outlet backpressure is not too high, high flow rates.

35.5.5.2 Axial Turbopump

In this case the axle has essentially propellers attached to the shaft and the fluid is forced by these parallel with the main axis of the pump. Generally, axial pumps tend to give much lower pressures than centrifugal pumps. However they are still useful —axial pumps are commonly used as 'inducers' for centrifugal pumps; these raise the inlet pressure enough to prevent excessive cavitation from occurring within the centrifugal portion of the pump.

35.5.6 Nozzles

The primary object of a nozzle is to expand the exhaust stream to atmospheric pressure, thereby producing a high velocity jet, relative to the vehicle. If the fully expanded jet velocity exceeds the flight velocity, there will be a forward thrust on the airframe.

35.5.6.1 Convergent Nozzles

Simple convergent nozzles are used on many jet engines. If the nozzle pressure ratio is above the critical value (about 1.8:1) a convergent nozzle will choke, resulting in some of the expansion to atmospheric pressure taking place downstream of the throat (i.e., smallest flow area), in the jet wake. Increasing the nozzle pressure ratio further will not increase the throat Mach number beyond unity. Downstream (i.e., external to the nozzle) the flow is free to expand to supersonic velocities.

35.5.6.2 Convergent-divergent Nozzle (De Laval Nozzle)

At high nozzle pressure ratios, much of the expansion will take place downstream of a convergent nozzle, which is somewhat inefficient. Consequently, some jet engines incorporate a convergent-divergent nozzle, to allow most of the expansion to take place within the nozzle to maximise thrust. However, unlike the con-di nozzle used on a conventional rocket motor, when such a device is used on a jet engine it has to be a complex variable geometry device, to cope with the wide variation in nozzle pressure ratio encountered in flight and engine throttling. This further increases the weight and cost of such an installation.

35.5.6.3 Ejector Nozzle

The simpler of the two is the *ejector nozzle*, which creates an effective nozzle through a secondary airflow and springloaded petals. At subsonic speeds, the airflow constricts the exhaust to a convergent shape. As the aircraft speeds up, the two nozzles dilate, which allows the exhaust to form a convergent-divergent shape, speeding the exhaust gasses past Mach 1. More complex engines can actually use a tertiary airflow to reduce exit area at very low speeds. Advantages of the ejector nozzle are relative simplicity and reliability. Disadvantages are average performance (compared to the other nozzle type) and relatively high drag due to the secondary airflow. Notable aircraft to have utilized this type of nozzle include the SR-71, Concorde, F-111, and Saab Viggen.

35.5.6.4 Iris Vectored Thrust Nozzle

For higher performance, it is necessary to use an *Iris nozzle*. This type uses overlapping "petals" which mechanically adjusts the petals with hydraulics. Although more complex than the ejector nozzle, it has significantly higher performance and smoother airflow. As such, it is employed primarily on high-performance fighters such as the F-14, F-15, F-16, though is also used in high-speed bombers such as the B-1B. Some modem iris nozzle additionally have the ability to change the angle of the thrust.

Illustration 3. *Iris Vectored thrust nozzle*

Rocket motors also employ convergent-divergent nozzles, but these are usually of fixed geometry, to minimise weight. Because of the much higher nozzle pressure ratios experienced, rocket motor con-di nozzles have a much greater area ratio (exit/throat) than those fitted to jet engines.

35.5.7 Afterburner (Thrust Booster or Augmentation)/Tail Bunching

Many military combat engines incorporate an afterburner (or reheat) in the engine exhaust system. Its purpose is to provide a temporary increase in thrust for situations such as take-off, or in military aircraft, combat or supersonic flight. This is achieved by injecting additional fuel into the jet pipe downstream of the turbine. This fuel is ignited by the hot exhaust gasses and adds greatly to the thrust of the engine. The advantage of afterburning is significantly increased thrust, the disadvantage of afterburning is its very high fuel consumption and inefficiency but this is acceptable for the short periods in which reheat is usually used.

Jet engines are referred to as operating wet when reheat is being used, and dry when the engine is used without afterburner.

A jet engine afterburner is an extended exhaust section containing extra fuel injectors, and since the jet engine upstream (i.e., before the turbine) will use little of the oxygen it ingests, the afterburner is, at its simplest, a type of ramjet. When the afterburner is turned on, fuel is injected, which ignites readily, owing to the relatively high temperature of the incoming gases. The resulting combustion process increases the afterburner exit (nozzle entry) temperature significantly, resulting in a steep increase in engine net thrust. As well as an increase in afterburner exit stagnation temperature, there is also an increase in nozzle mass flow (i.e., afterburner entry mass flow plus the effective afterburner fuel flow), but a decrease in afterburner exit stagnation pressure owing to a fundamental loss due to heating plus friction and turbulence losses.

The resulting increase in afterburner exit volume flow is accommodated by increasing the throat area of the propulsion nozzle. Otherwise, the upstream turbomachinery rematches (probably causing fan surge in a turbofan application).

35.5.7.1 Limitations

Due to their high fuel consumption, afterburners are not used for extended periods. They are only used when it is important to have as much thrust as possible. This includes takeoffs from short runways (as on an aircraft carrier) and air combat situations.

35.5.7.2 Efficiency

Since the exhaust gas already has reduced oxygen due to previous combustion, and since the fuel is not burning in a highly compressed air column, the afterburner is generally inefficient compared with the main combustor. Afterburner efficiency also declines significantly if, as is usually the case, the tailpipe pressure decreases with increasing altitude.

Illustration 4. *A statically mounted Pratt and Whitney J58 engine with full after burner on disposing of the last of the SR-71 fuel prior to program termination*

Afterburners do produce markedly enhanced thrust as well as (typically) a very large, impressive flame at the back of the engine. This exhaust flame may show *shock-diamonds*, which are caused by shock waves being formed due to slight differences between ambient pressure and the exhaust pressure. These imbalances cause oscillations in the exhaust jet diameter over distance and cause the visible banding where the pressure and temperature is highest.

35.6 TURBOJET ENGINE

Turbojets are the simplest and oldest kind of general purpose jet engine. Two different engineers, Frank Whittle in Britain and Hans von Ohain in Germany, developed the concept independently during the late 1930s. On 27 August 1939 the Heinkel He 178 became the world's first aircraft to fly under turbojet power, thus becoming the first practical jet plane.

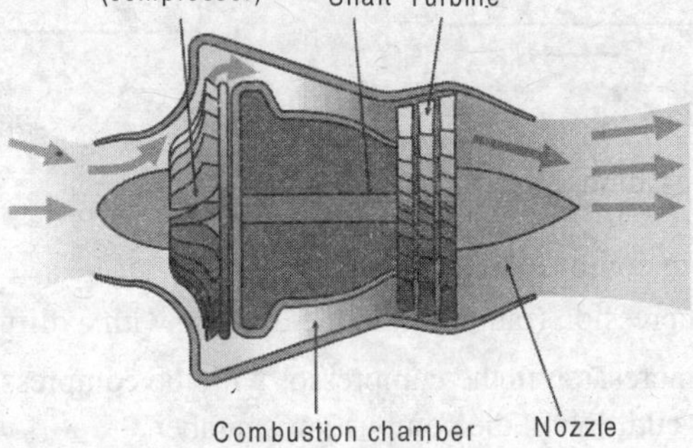

Fig. 35.3. *Schematic diagram showing the operation of a centrifugal flow turbojet engine. The compressor is driven via the turbine stage and throws air outwards, requiring it to be redirected parallel to the axis of thrust*

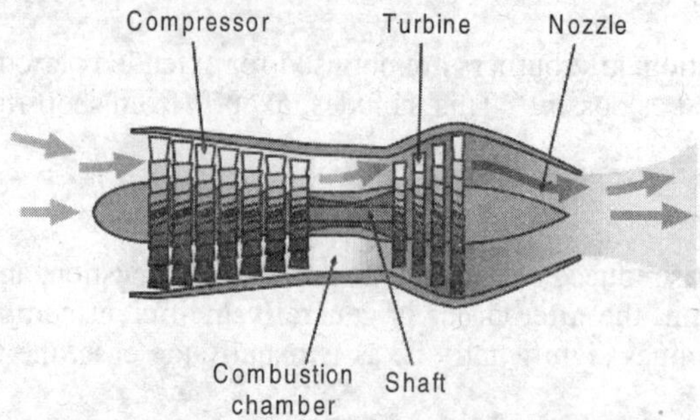

Fig. 35.4. *Schematic diagram showing the operation of an axial flow turbojet engine. Here, the compressor is again driven by the turbine, but the air flow remains parallel to the axis of thrust*

A turbojet engine is used primarily to propel aircraft. Air is drawn into the rotating compressor via the intake and is compressed to a higher pressure before entering the combustion chamber. Fuel is mixed with the compressed air and ignited by flame in the eddy of a flame holder. This combustion process significantly raises the temperature of the gas. Hot combustion products leaving the combustor expand through the turbine, where power is extracted to drive the compressor. Although this expansion process reduces the turbine exit gas temperature and pressure, both parameters are usually still well above ambient conditions. The gas stream exiting the turbine expands to ambient pressure via the propelling nozzle, producing a high velocity jet in the exhaust plume. If the momentum of the exhaust stream exceeds the momentum of the intake stream, the impulse is positive, thus, there is a net forward thrust upon the airframe.

35.6.1 Turbojet Engine–Thermodynamic Cycle

Figure 35.5 shows the *P–V* diagram of turbojet engine and Fig. 35.6 is the *T–S* diagram of the turbojet engine.

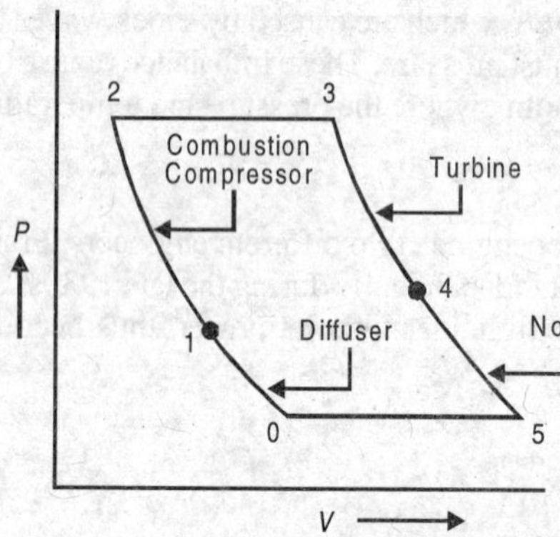

Fig. 35.5

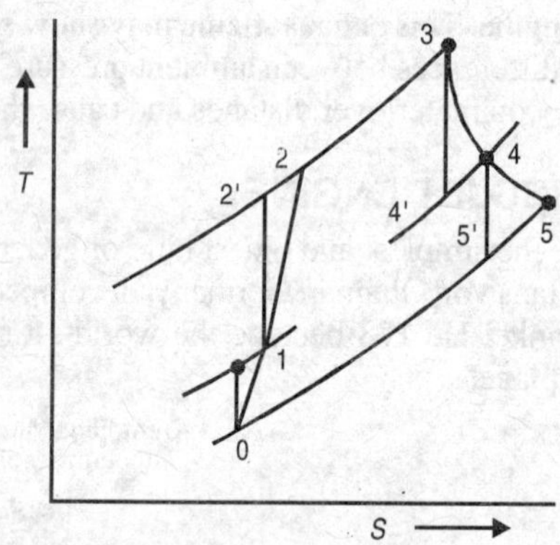

Fig. 35.6

While analysing the cycle we assume that

1. Specific heat is constant.
2. There is no pressure drop in combustion chamber.

0 – 1 is the actual compression (ramming) in the diffuser with a diffuser efficiency of η_d.

1 – 2 is the actual compression in the compressor with the compression efficiency η_c.

2 – 3 Heating or heat addition in the combustion chamber.

3 – 4 Actual expansion of gases in gas-turbine with an isentropic efficiency of η_t.

4 – 5 Actual expansion of gases through jet-nozzle with an nozzle efficiency of η_n.

After solving this *T–S* diagram, we get,

$$\text{Exit Velocity} = \left\{2\eta_n \cdot C_p \cdot T_{o_4}\left[1-\left(\frac{P_5}{P_{o_4}}\right)^{\frac{\gamma-1}{\gamma}}\right]\right\}^{1/2} \quad ....(1)$$

Let m_a = mass of air inducted in kg/sec

m_f = mass of fuel injected in kg sec

$\frac{F}{A}$ = Fuel to air ratio

$V_p = V_a$ = Velocity of incoming air at diffuser inlet

$\therefore$ Initial momentum of air = $m_a \times V_a$

and final or momentum at exit

$$= (m_a + m_f)V_e$$

$\therefore$ Force = Thrust developed

$$T = (m_a + m_f)V_e - m_a \times V_a$$

$$= m_a\left[\left(1+\frac{m_f}{m_a}\right)V_e - V_a\right] \quad ...(2)$$

Note that velocity of plane is equal to the velocity of incoming air and sometimes the inlet velocity is also denoted by V_p.

$$\therefore \quad T = m_a\left[\left(1+\frac{m_f}{m_a}\right)V_e - V_a\right]$$

Here, $(m_a + m_f)V_e$ is known as gross thrust and $m_a V_a$ is known as inlet drag.

If we neglect the fuel per kg of air or fuel to air ratio is very very small and many times it is neglected.

$$\therefore \quad \text{Thrust} = m_a V_e - m_a V_a$$

$$= m_a (V_e - V_a)$$

$$= m_a \cdot V_a\left[\frac{V_e}{V_a} - 1\right]$$

Here $\frac{V_a}{V_e}$ = Effective speed ratio = α

$$\therefore \quad \text{Thrust } T = m_a V_a\left[\frac{1}{\alpha} - 1\right]$$

$$\mathbf{T = m_a\ V_a\left[\frac{1}{\alpha} - 1\right]} \quad ...(3)$$

The thrust developed by the engine should overcome the drag on the air-craft and in doing so develops power, which is called *thrust power* and is given by

$$\begin{aligned} P_T &= \text{Thrust power} \\ &= \text{Thrust} \times \text{Velocity at inlet} \\ &= T \times V_a \\ &= m_a V_a [V_e - V_a] \end{aligned}$$

$$\mathbf{P_T = m_a\ V_a [V_e - V_a]} \qquad \text{...(4)}$$

Specific thrust is the thrust produced kg of air flow. It is also sometimes called as *specific impulse*.

$$\therefore \qquad \text{Specific thrust} = \frac{T}{m_a}$$

$$= \frac{m_a (V_e - V_a)}{m_a}$$

$$= V_e - V_a \qquad \text{...(5)}$$

$$= V_e \left(1 - \frac{V_a}{V_e}\right)$$

$$= V_e (1 - \alpha) \qquad \text{...(6)}$$

Specific thrust is a criterion of the size of the engine required to produce a given total thrust.

From the equation for the thrust (2a) we observe that if the effective jet velocity V_e is kept constant, then the specific thrust decreases with increase in plane velocity $V_a = V_p$ or flight speed. Specific thrust will be maximum when $\alpha = 0$ i.e., $V_a = 0$ or flight velocity is zero.

35.7 THERMAL EFFICIENCY OF A TURBOJET ENGINE

Thermal efficiency is defined as the ratio of propulsive power developed at the exhaust nozzle to the heat supplied by the fuel. This ratio indicates the degree of utilisation of fuel in accelerating the fluid flow.

∴ Thermal efficiency is given by

$$\eta_{th} = \frac{\text{Propulsive power}}{\text{Heat supplied}}$$

Here, the propulsive power is given by the net increase in energy of the working fluid between inlet and exit.

∴ Propulsive power P is given by

$$P = \frac{1}{2}\left[(m_a + m_f)V_e^2 - m_a V_a^2\right]$$

$$= \frac{1}{2}\left[m_a V_e^2 - m_a V_a^2\right] \quad \text{if } m_f \text{ is neglected}$$

$$= \frac{m_a}{2}\left(V_e^2 - V_a^2\right) \qquad \text{...(7)}$$

$$\therefore \qquad \eta_{th} = \frac{P}{\text{Heat supplied}}$$

$$= \frac{\frac{m_a V_e^2}{2}\left[1-\left(\frac{V_i}{V_e}\right)^2\right]}{m_f \times \text{C.V.}}$$

$$= \frac{\frac{m_a V_e^2}{2}\left[1-\alpha^2\right]}{m_f \times \text{C.V.}}$$

$$= \frac{\frac{V_e^2}{2}\left(1-\alpha^2\right)}{\frac{m_f}{m_a} \times \text{C.V.}}$$

$$= \frac{V_e^2\left(1-\alpha^2\right)}{2 \cdot \frac{F}{A} \times \text{C.V.}}$$

$$= \frac{V_e^2\left(1-\alpha^2\right)}{2 \cdot f \cdot \text{C.V.}} \quad \text{where } f = \frac{m_f}{m_a}$$

$$\therefore \qquad \boldsymbol{\eta_{th} = \frac{V_e^2\left(1-\alpha^2\right)}{2f \times C.V.}} \qquad \ldots(8)$$

35.8 PROPULSIVE EFFICIENCY

Propulsive efficiency is defined as the ratio of useful propulsive power or thrust power to the sum of thrust power and unused kinetic energy of the jet. It is generally denoted by η_p. Here the kinetic energy of the jet is relative to the earth viz. $\frac{m(V_e - V_a)^2}{2}$.

$$\therefore \qquad \eta_p = \frac{m_a (V_e - V_a) V_a}{m_a\left[V_a (V_e - V_a) + \frac{(V_e - V_a)^2}{2}\right]}$$

$$= \frac{V_a (V_e - V_a)}{(V_e - V_a)\left[V_a + \frac{(V_e - V_a)}{2}\right]}$$

$$= \frac{V_a}{V_a + \frac{V_e - V_a}{2}}$$

$$= \frac{2V_a}{2V_a + V_e - V_a}$$

$$= \frac{2V_a}{V_a + V_e}$$

$$= \frac{2}{1 + \frac{V_e}{V_a}}$$

$$= \frac{2V_a}{V_e\left(1 + \frac{V_a}{V_e}\right)}$$

$$= \frac{2\alpha}{(1+\alpha)} \qquad \ldots(9)$$

For maximum propulsive efficiency, we differentiate Eq. (9) with respect to α and equating to zero.

$$\frac{d\eta_p}{d\alpha} = 0$$

or when $\alpha = 1$

$$\eta_p = 1 \quad \text{i.e., } 100\ \%$$

$\therefore \quad V_e = V_a$

and in this case no thrust is available or produced and thus no useful work output can be obtained.

∴ To get the thrust power, $V_e > V_a$ or $V_e > V_p$

∴ From Eq. (4) we have

$$\text{Thrust power} = TP = m_a V_a (V_e - V_a)$$

and $\alpha = \frac{V_a}{V_e}$

$\therefore \quad V_a = V_e \cdot \propto$

$$TP = m_a V_e \alpha (V_e - V_a)$$

$$= m_a V_e^2 \alpha (1 - \alpha)$$

Differentiating this w.r.t. α, and equating it to zero, we get the condition for maximum thrust power.

$$\therefore \quad \frac{d(TP)}{d\alpha} = 0 = m_a V_e^2 - m_a V_e^2 \times 2\alpha$$

$$\therefore \quad 0 = m_a V_e^2 [1 - 2\alpha]$$

or $\alpha = \frac{1}{2} = \mathbf{0.5} \qquad \ldots(10)$

∴ For this condition, the propulsive efficiency is given by

$$\eta_p = \frac{2\alpha}{1+\alpha}$$

$$= \frac{2\times 0.5}{1+0.5} = \frac{1}{1.5}$$

$$= 0.667 \text{ or } \mathbf{66.7\ \%}$$

35.9 OVERALL EFFICIENCY OF PROPULSIVE SYSTEM

Overall efficiency of the propulsive system gives the performance of the propulsive efficiency. It indicates the extent to which the system utilizers the energy supplied. Also, the overall efficiency is defined as the ratio of the rate of doing useful propulsive work and the rate at which energy is supplied to the system. It is indicated by η_o.

∴ $$\eta_o = \text{Overall efficiency}$$

$$= \frac{\text{Useful propulsive work}}{\text{Energy supplied}}$$

$$= \frac{\text{System output}}{\text{Energy input}} \times \frac{\text{Transmission output}}{\text{System output}} \times \frac{\text{Useful propulsive work}}{\text{Transmission output}}$$

$$= \eta_{th} \times \eta_{tr} \times \eta_p$$

$$\boldsymbol{\eta_o = \eta_{th} \times \eta_{tr} \times \eta_p} \qquad \text{...(11)}$$

35.10 THRUST SPECIFIC FUEL CONSUMPTION (TSFC)

Similar to engines and turbines, the specific fuel consumption in case of jet propulsive system is on the basis of thrust produced.

Hence, the thrust specific fuel consumption (TSFC) is defined as the amount of fuel required to produced unit thrust per hour. Thrust take here is Newton. Therefore,

$$\text{TSFC} = \frac{\text{Mass of fuel/hour}}{\text{Thrust}}$$

$$= \frac{m_f \times 3600}{F} \text{ kg/Nh} \qquad \text{...(12)}$$

Here m_f = Mass of fuel/sec

and F = Thrust in Newtons

Again, $$\text{TSFC} = \frac{m_f \times 3600}{F} = \frac{\frac{m_f}{m_a} \times 3600}{\frac{F}{m_a}}$$

Here $$\frac{m_f}{m_a} = \frac{\text{Fuel}}{\text{Air}} \text{ Ratio} = f$$

∴ $$\text{TSFC} = \frac{3600 f}{\frac{F}{m_a}} \qquad \text{...(13)}$$

Sometimes, specific fuel consumption is based on thrust power. Then, Thrust Power Specific Fuel Consumption (TPSFC) is defined as the ratio of fuel consumed per hour per unit thrust power.

$$\therefore \qquad \text{TPSFC} = \frac{m_f \times 3600}{TP} \quad \frac{\text{kg of fuel}}{\text{thrust kWH}} \qquad \ldots(14)$$

The nature of the variation curves for thrust, thrust power, TSFC and TPSFC against air or flight speed are shown in Fig. 35.7.

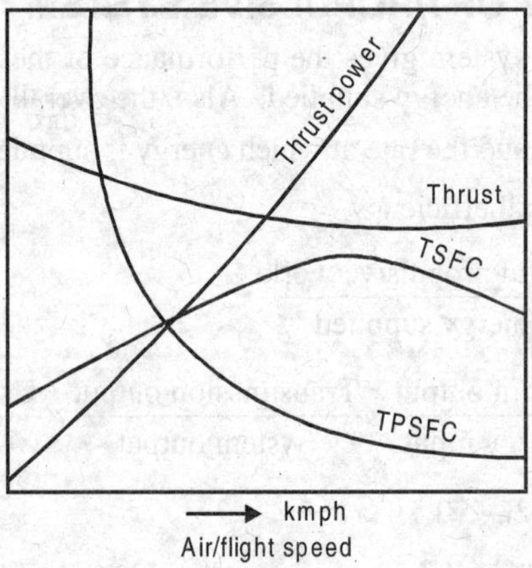

Fig. 35.7. *Performance curves for jet engine*

The performance of the jet propulsion system is affected mainly by two parameters as

1. Forward speed of the air-craft and
2. Altitude of the air-craft.

35.11 CYCLE IMPROVEMENTS

Increasing the overall pressure ratio of the compression system raises the combustor entry temperature. Therefore, at a fixed fuel flow and airflow, there is an increase in turbine inlet temperature. Although the higher temperature rise across the compression system, implies a larger temperature drop over the turbine system, the nozzle temperature is unaffected, because the same amount of heat is being added to the system. There is, however, a rise in nozzle pressure, because overall pressure ratio increases faster than the turbine expansion ratio. Consequently, net thrust increases, whilst specific fuel consumption (fuel flow/net thrust) decreases.

Thus turbojets can be made more fuel efficient by raising overall pressure ratio and turbine inlet temperature in unison. However, better turbine materials and/or improved vane/blade cooling are required to cope with increases in both turbine inlet temperature and compressor delivery temperature. Increasing the latter requires better compressor materials.

35.12 ADVANTAGES AND DISADVANTAGES OF JET PROPULSION OVER THE OTHER SYSTEM

Following are the advantages of jet propulsion

1. **Low-specific weight.** The specific weight of the jet propulsion is one-fourth to one-half of the reciprocating engine.
2. **No unbalance force.** There are no reciprocating parts and so jet propulsion is free from unbalanced forces. Greater reliability is thus achieved.

3 **Small frontal area.** The frontal area of jet propulsion is less than one-fourth the frontal area of the reciprocating engines which decreases drag greatly and hence makes available greater power, particularly at high loads. This also reduces the air cooling problem.

4. **No restriction in power output.** Compared to reciprocating engine the jet propulsion with greatly increased power output can be built because the power is not limited by detonation. The unit can work/operate over a large range of mixture strength.

5. **High-speed.** The speed of jet propulsion is not limited by propeller. High speed can be obtained.

6. **Neither lubrication nor radiators.** Jet propulsion requires neither internal lubrication nor radiators as reciprocating engines require.

7. At high speed greater than 900 kmph and at an altitude greater than 10,000 metres, the efficiency of the jet is much higher than that of a propeller.

8. Combustion and delivery of power is continuous, whilst peak and fluctuating pressures do not occur.

9. There is no ship stream loss, the drag is reduced, and warm compressed air is available for cabin heating.

10. The unit permits of a better position of the pilot whilst the absence of a propeller permits a smaller under carriage.

Disadvantages

The disadvantages from which jet propulsion suffers are

1. Particularly at low pressure, the thermal efficiency is lower. At low altitude and speed upto 150 m/sec/540 kmph, the fuel consumption is 2 to 3 times that of a reciprocating engine.
2. The plant is very noisy, materials costly and life short.
3. The compression-pressure ratio is not constant as in the reciprocator, but varies approximately with the square of the speed.
4. There are certain difficulties which are encountered in the running of the propulsive unit.

35.13 TURBOFAN

A turbofan is a type of airplane engine, similar to a turbojet. It essentially consists of a ducted fan with a smaller diameter turbojet engine mounted behind it that powers the fan. Part of the airstream from the ducted fan passes through the turbojet where it is burnt to power the fan, but the majority of the flow bypasses it, and produces most of the thrust.

All of the jet-engines that power currently-manufactured commercial jet aircraft are turbofans. They are mainly used commercially because they are highly efficient and relatively quiet in operation.

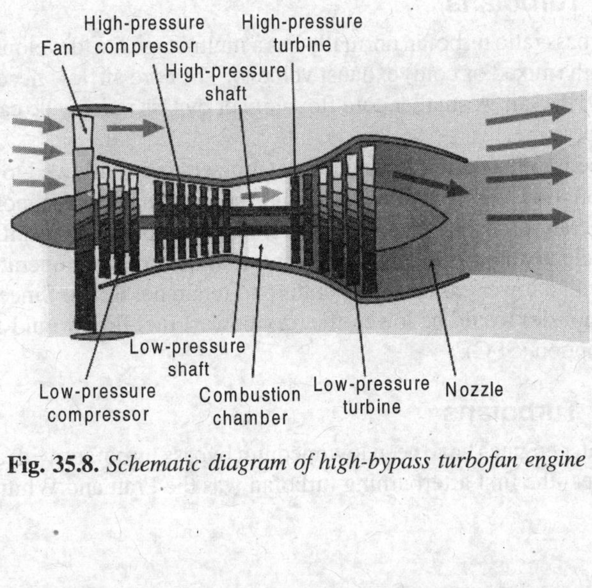

Fig. 35.8. *Schematic diagram of high-bypass turbofan engine*

After World War II, 2-spool (shaft) turbojets were developed to make it easier to throttle-back compression systems with a high design overall pressure ratio (i.e., combustor inlet pressure/intake delivery pressure). Adopting the 2-spool arrangement enables the compression system to be split into two, with a Low Pressure (LP) Compressor supercharging a High Pressure (HP) Compressor. Each compressor is mounted on a separate co-axial shaft, driven by its own turbine. Otherwise a 2-spool turbojet is much like a single spool engine.

Modern turbofans evolved from the 2-spool axial-flow turbojet engine, essentially by increasing the relative size of the Low Pressure (LP) Compressor to the point where some of the air exiting the unit actually bypasses the core (or gas generator) stream, passing through the main combustor. This bypass air either expands through a separate propelling nozzle, or is mixed with the hot gases leaving the Low Pressure (LP) Turbine, before expanding through a Mixed Stream Propelling Nozzle. Owing to a lower jet velocity, a modern civil turbofan is quieter than the equivalent turbojet. Turbofans also have a better thermal efficiency. In a turbofan, the LP Compressor is often called a fan. Civil turbofans usually have a single fan stage, whereas most military turbofans have multi-stage fans.

35.13.1 Bypass Ratio

It is a parameter often used for classifying turbofans, defined as ratio of bypassed airflow to combustor airflow.

The noise of any type of jet engine is strongly related to the velocity of the exhaust gases. High bypass ratio (i.e., low specific thrust) turbofans are relatively quiet compared to turbojets and low bypass ratio (i.e., high specific thrust) turbofans. A low specific thrust engine has a low jet velocity by definition, as the following approximate equation for net thrust implies:

$$F_n = \dot{m}.(V_{jfe} - V_a)$$

where

$\dot{m}$ = intake mass flow

V_{jfe} = fully expanded jet velocity (in the exhaust plume)

V_a = aircraft flight velocity

Rearranging the above equation, specific thrust is given by:

$$\frac{F_n}{\dot{m}} = (V_{jfe} - V_a)$$

So for zero flight velocity, specific thrust is directly proportional to jet velocity. Relatively speaking, low specific thrust engines are large in diameter to accommodate the high airflow required for a given thrust.

35.13.2 Low Bypass Turbofans

A high specific thrust/low bypass ratio turbofan normally has a multi-stage fan, developing a relatively high pressure ratio and, thus, yielding a high (mixed or cold) exhaust velocity. The core airflow needs to be large enough to give sufficient core power to drive the fan. A smaller core flow/higher bypass ratio cycle can be achieved by raising the (HP) turbine rotor inlet temperature.

A bypass flow can only be introduced if the turbine inlet temperature is allowed to increase, to compensate for a correspondingly smaller core flow. Improvements in turbine cooling/material technology would facilitate the use of a higher turbine inlet temperature, despite increases in cooling air temperature, resulting from a probable increase in overall pressure ratio. Efficiently done, the resulting turbofan would probably operate at a higher nozzle pressure ratio than the turbojet, but with a lower exhaust temperature to retain net thrust. Since the temperature rise across the whole engine (intake to nozzle) would be lower, the (dry power) fuel flow would also be reduced, resulting in a better specific fuel consumption (SFC).

35.13.3 Afterburning Turbofans

Since the 1970s, most jet fighter engines have been low/medium bypass turbofans with a mixed exhaust, afterburner and variable area final nozzle—the first afterburning turbofan was the Pratt and Whitney TF30.

Unlike the main combustor, where the integrity of the downstream turbine blades must be preserved, an afterburner can operate at the ideal maximum temperature (i.e., about 2100 K). Now, at a fixed total applied fuel: air ratio, the total fuel flow for a given fan airflow will be the same, regardless of the dry specific thrust of the engine. However, a high specific thrust turbofan will, by definition, have a higher nozzle pressure ratio, resulting in a higher afterburning net thrust and, therefore, a lower afterburning specific fuel consumption. However, high specific thrust engines have a high dry SFC. The situation is reversed for a medium specific thrust afterburning turbofan: i.e., poor afterburning SFC/good dry SFC. The former engine is suitable for a combat aircraft which must remain in afterburning combat for a fairly long period, but only has to fight fairly close to the airfield (i.e., cross border skirmishes). The latter engine is better for an aircraft that has to fly some distance, or loiter for a long time, before going into combat. However, the pilot can only afford to stay in afterburning for a short period, before his/her fuel reserves become dangerously low.

Modern low-bypass military turbofans include the Pratt and Whitney F119, the Eurojet EJ200 and the General Electric F110, all of which feature a mixed exhaust, afterburner and variable area propelling nozzle. Non-afterburning engines include the Rolls–Royce/Turbomeca Adour (afterburning in the SEPECAT Jaguar) and the unmixed, vectored thrust, Rolls–Royce Pegasus.

35.13.4 High-bypass Turbofans

The low specific thrust/high bypass ratio turbofans used in today's civil jetliners (and some military transport aircraft) evolved from the high specific thrust/low bypass ratio turbofans used in such aircraft back in the 60's.

Low specific thrust is achieved by replacing the multi-stage fan with a single stage unit. Unlike some military engines, modern civil turbofans do not have any stationary inlet guide vanes in front of the fan rotor. The fan is scaled to achieve the desired net thrust.

The core (or gas generator) of the engine must generate sufficient Core Power to at least drive the fan at its design flow and pressure ratio. Through improvements in turbine cooling/material technology, a higher (HP) turbine rotor inlet temperature can be used, thus facilitating a smaller (and lighter) core and (potentially) improving the core thermal efficiency. Reducing the core mass flow tends to increase the load on the LP turbine, so this unit may require additional stages to reduce the average stage loading and to maintain LP turbine efficiency. Reducing core flow also increases bypass ratio (5 : 1, or more, is now common).

Further improvements in core thermal efficiency can be achieved by raising the overall pressure ratio of the core. Improved blade aerodynamics reduces the number of extra compressor stages required. With multiple compressors (i.e., LPC, IPC, HPC) dramatic increases in overall pressure ratio have became possible. Variable geometry (i.e., stators) enable high pressure ratio compressors to work surge-free at all throttle settings.

The first high-bypass turbofan engine was the General Electric TF39, built to power the Lockheed C-5 Galaxy military transport aircraft. The civil General Electric CF6 engine used a derived design. Other high-bypass turbofans are the Pratt and Whitney JT9D, the three-shaft Rolls-Royce RB211 and the CFM International CFM56. More recent large high-bypass turbofans include the Pratt and Whitney PW4000, the three-shaft Rolls–Royce Trent, the General Electric GE90, and the General Electric GEnx.

The significantly higher thrust provided by high-bypass turbofan engines also made civil wide-body aircraft practical and economical. In addition to the vastly increased thrust, these engines are also generally quieter. This is not so much due to the higher bypass ratio, as to the use of low pressure ratio, single stage, fans, which significantly reduce specific thrust and, thereby, jet velocity. The combination of a higher overall pressure ratio and turbine inlet temperature improves thermal efficiency. This, together with a lower specific thrust (better propulsive efficiency), leads to a lower specific fuel consumption.

For reasons of fuel economy, and also of reduced noise, almost all of today's jet airliners are powered by high-bypass turbofans. Although modern military aircraft tend to use low bypass ratio turbofans, military transport aircraft (e.g. C-17) mainly use high bypass ratio turbofans (or turboprops) for fuel efficiency.

35.13.5 Turbofan Configurations

Turbofan engines come in a variety of engine configurations.

35.13.5.1 Single Shaft Turbofan

This is probably the simplest configuration, comprising a fan and high pressure compressor driven by a single turbine unit, all on the same shaft. Despite the simplicity of the turbomachinery configuration, it requires a variable area mixer to facilitate part-throttle operation.

35.13.5.2 Aft Fan Turbofan

One of the earliest turbofans was a derivative of the General Electric J79 turbojet, known as the CJ805, which featured an integrated aft fan/low pressure (LP) turbine unit located in the turbojet exhaust jetpipe. Hot gas from the turbojet turbine exhaust expanded through the LP turbine, the fan blades being a radial extension of the turbine blades. One of the problems with the Aft Fan configuration is hot gas leakage from the LP turbine to the fan.

35.13.5.3 Basic Two Spool

Many turbofans have the Basic Two Spool configuration where both the fan and LP turbine are mounted on a second (LP) shaft, running concentrically with the HP spool. At the smaller thrust sizes, instead of all-axial blading, the HP compressor configuration may be axial-centrifugal, double-centrifugal or even diagonal/centrifugal.

35.13.5.4 Boosted Two Spool

Higher overall pressure ratios are achieved by either raising the HP compressor pressure ratio or adding an Intermediate Pressure (IP) Compressor between the fan and HP compressor, to supercharge or boost the latter unit.

35.13.5.5 Three Spool

The IP compressor is mounted on a separate (IP) shaft, running concentrically with the LP and HP shafts, and is driven by a separate Intermediate Pressure (IP) Turbine.

35.13.5.6 Geared Fan

As bypass ratio increases, the mean radius ratio of the fan and LP turbine increases. Consequently, if the fan is to rotate at its optimum blade speed the LP turbine blading will run slow, so additional LPT stages will be required, to extract sufficient energy to drive the fan. Introducing a reduction gearbox, with a suitable gear ratio, between the LP shaft and the fan, enables both the fan and LP turbine to operate at their optimum speeds. This is not a popular solution, since high power gearboxes tend to be unreliable.

The advantage of turbofan or bypass engine is that, at lower jet velocities, it gives better propulsive efficiency and fuel economy compared with a turbojet engine for a given thrust, as less kinetic energy is wasted to atmosphere. This applies particularly at aircraft speeds below sonic velocity and for long range aircraft. The bypass air helps for fire protection by the cool air in the bypass which surrounds the hot parts of the engine. Such engines are mostly used on jumbo-jets for civil aviation.

35.14 TURBOPROP

A Turboprop engine is a type of gas turbine engine which uses most of its power to drive a propeller. The propeller of a turboprop is very similar to that used by piston or reciprocating engines, but turboprops usually use a constant velocity propeller. Turboprop engines are generally used on small or slow subsonic aircraft, but some aircraft outfitted with turboprops have cruising speeds.

A turboprop consists of an intake, compressor, combustor, turbine and a propelling nozzle. Air is drawn into the intake and compressed by the compressor. Fuel is then added to the compressed air in the combustor. The hot combustion gases expand through the turbine, to provide power to the turbine. Part of this power goes to the compressor, and the rest, through the reduction gearing, to the propeller. Further expansion of the gases occurs in the propelling nozzle, where the gasses are expelled, providing approximately 25% of the thrust.

A turboprop engine is similar to a turbojet, but has additional fan blades in the turbine stage to recover more power from the engine to turn the propeller.

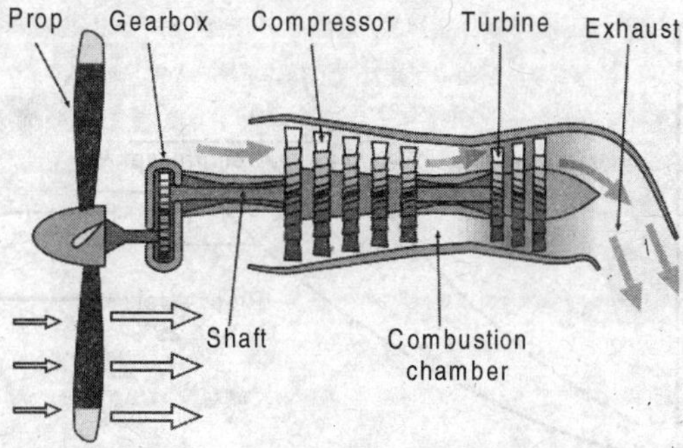

Fig. 35.9

In a turboprop much of the jet thrust is sacrificed in favour of shaftpower, which is obtained by extracting additional power (to that necessary to drive the compressor) from the turbine expansion process. While the power turbine may be integral with the gas generator section, many turboprops today feature a Free Power Turbine, on a separate coaxial shaft. This enables the propeller to rotate freely, independent of compressor speed. Owing to the additional expansion in the turbine system, the residual energy in the jet is fairly low (<10% of total thrust, including that of the propeller).

Because the propeller is very much larger in diameter than the power turbine, the tip speed of the propeller can become supersonic. Consequently, to prevent this, a speed reduction gearbox is inserted between the power turbine and propeller shafts. The gearbox is part of the engine, whereas in a turboshaft the (helicopter) rotor reduction gearbox is remote from the engine.

Residual thrust on a turboshaft is avoided by further expansion in the turbine system and/or truncating and turning the exhaust through 90 degrees, to produce two opposing jets.

Apart from the above, there is very little difference between a turboprop and a turboshaft.

Turboprops are very efficient at modest flight speeds (below 724 km/h or 450 mph), because the jet velocity of the propeller (and exhaust) is relatively low. Consequently, small commuter aircraft and military transports tend to feature turboprop engines. Although turboprops are used in some General Aviation applications, their high price deters more widespread acceptance except for very high performance STOL applications. One of the most common turboprop military aircraft is the C-130 Hercules transport.

While most modern turbojet and turbofan engines use axial-flow compressors, turboprop engines usually contain at least one stage of centrifugal compression, because of the small size of the engines.

Propellers lose efficiency as aircraft speed increases, which is why turboprops are not used on higher-speed aircraft.

35.15 PROP-JET TURBINE PLANT

A combined propeller and jet drive has been developed for aircraft propulsion. In this device a gas turbine drives a propeller and the exhaust gases are used as a propulsion jet, such a combination is termed a prop-jet engine. It is being developed for heavy bombers and civil airliners for which it provides a satisfactory method of propulsion, because the propeller has a high efficiency at the lower altitudes whilst the efficiency of the jet propulsion increases at the higher altitudes.

During a flight test on a Theseus (Th-21) prop-jet engine installed in a plane, 1500 kW, was developed in the propeller shaft, together with a 227 kg jet thrust at sea level and a speed of 480 kmph. The fuel consumption at this speed was found to be 0.352 kg/kW, the fuel used in the test being aviation kerosene. The equivalent brake power is defined as the propeller shaft power plus jet-power divided by the propulsion efficiency which is about 80 %.

The Armstrong Siddeley prop-jet engine 'Python' had a shaft power of 3000 kW plus a thrust of 450 kg. It will be noticed that in both of these prop-jet engines the propulsion power obtained from the jet is 25 % of the propeller shaft power.

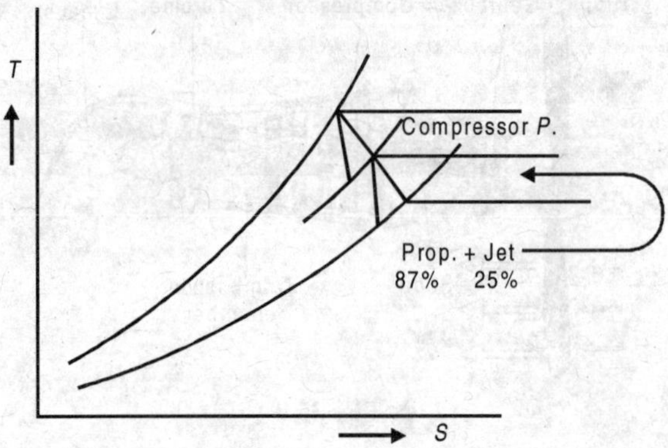

Fig. 35.10

Advantages of Turbo-prop-jet Engine

1. The turbo prop-jet, has good take off and climb characteristics in the lower subsonic speed range due to the large thrust and power available.
2. The turbo prop-jet engine combines the advantages of turbojet and turbo-prop engines i.e., low specific weight, small frontal area, simplicity, low vibrations and merits of propeller engine like high power for take off and high propulsive efficiency at low speeds.
3. The overall efficiency of the turbo-prop i.e., the fuel economy is superior to the turbo-jet engine at speeds below 800 kmph.
4. The turbo prop-jet engine has high thrust per m^2 of frontal area compared with reciprocating engine.
5. This has less drag losses.

Disadvantages

1. The addition of the reduction gear and propeller adds weight to the prop-jet engine.
2. At the altitude increases, the thrust of turbo-prop decreases.

35.16 OTHER DESIGNS

35.16.1 Thermojet

A thermojet is a rudimentary type of jet engine. At its heart is an ordinary piston engine, but instead of this driving a propeller, it drives a compressor. The compressed air is channelled into a combustion chamber, where fuel is injected and ignited. The high temperatures generated by the combustion cause the gases in the chamber to expand and escape at high pressure from the exhaust, creating a reactive force that drives the engine.

Illustration 5. *The Campini Carroni CC.2 Thermo-jet powered aircraft*

Thermojet engines provide greater thrust than a propeller mounted on a piston engine; this has been demonstrated in a number of different aircrafts.

35.16.2 Rocket

It carries all the propellants on board and does not require air intake. Oxidizers are provided on board for combustion. It is used as space shuttle vehicle.

These are suitable for high altitudes, even in vacuum. Efficient at very high speeds. Disadvantages include risk of carrying oxidizers, extreme thermal stresses of combustion chamber, not reusable, extraordinarily noise.

35.16.3 Pulsejet

Air is compressed and combusted intermittently instead of continuously. Some designs use valves. Very simple design, commonly used on model aircraft. Noisy, inefficient (low compression ratio), works poorly on a large scale, valves on valved designs wear out quickly.

Pulse jet is very similar to ram jet in construction except that in addition to diffuser at intake, combustion chamber and exhaust nozzle, it has mechanically operated flapper valves which can allow or stop air flow in the combustion chamber. Thus pulse jet is an intermittent flow, compressorless type of device with minimum number of moving parts. Pulse jet was the power plant of German V-1 bomb popularly known as 'Buzz bomb' or flying Bomb first used in World War II in 1939-45. Figure 35.11 shows that schematic diagram of the pulse jet engine.

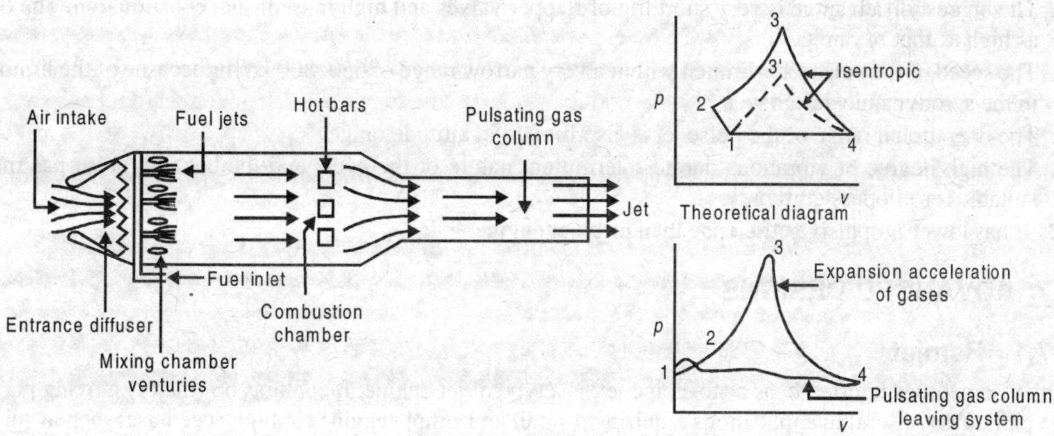

Fig. 35.11

The operation of the pulse-jet is as follows: During starting compressed air is forced into the inlet which opens the spring loaded flapper valves; the air enters combustion chamber into which fuel is injected and burnt with the help of a spark plug. Combustion occurs with a sudden explosion (process 2–3) i.e., combustion is at constant volume instead of constant pressure as in other propulsive devices. The pulse-jet cycle is more near to Otto-cycle. Ram action can also be used to increase the pressure of the cycle. Due to increase in pressure in combustion chamber, the flapper-valves close stopping the air suuply and at the same time the gases are forced out of the nozzle with a very high velocity. Because of this high velocity, the pressure continues to fall since the vacuum is created in the combustion chamber. This allows the flapper valves to open and freash air enters the combustion chamber. Some air also reenters from the tailpipe because the pressure in the tailpipe is also very low, and the next cycle starts. This increase in pressure due to back flow helps in better combustion as it will be hot by the time reaches combustion chamber and also increases the thrust due to additional filling. The frequency of cycles depends upon the duct shape and working temperature. In V-1 rocket it was about 40 cycles per second which corresponds to about 2400 rpm of a two-stroke reciprocating engine.

Once the cycle is started neither the initial flow of compressed air nor the spark-plug is necessary as the inlet diffusion will produce sufficient compression and any residual flame or even a hop-spot in the combustion chamber will ignite the fuel. Though the air flow is intermittent, the fuel is continuously supplied.

The pulse jet has low thermal efficiency and limited speed range. Speed is limited by (1) design of difuser at high speeds and (2) the flapper valves, the only mechanical part in the pulse jet, also have certain natural frequency and if resonance with the cycle frequency occurs and the valves may remain open and no compression will take place. Also as the speed increases it is difficult for the air to flow back. This reduces the total compression pressure as well as the mass flow air which results in incomplete combustion and lower thrust. The reduction thrust and efficiency is quite sharp as the speed increases. At subsonic speeds it might not to operate as the the speed is not sufficient to raise the air pressure to combustion pressure.

Advantages

1. This is very simple device only next to ramjet and is light in weight. It requires very small and occasional maintenance.
2. Unlike ramjet, it has static thrust because of the compressed air starting; the static thrust is even more than the crusie thrust.
3. It can run on almost any type of liquid fuels without any much effect on the performance. It can also operate on gaseous fuel with a little modifications.
4. Pulse jet is relatively cheap.

Disadvantges

1. The biggest disadvantge is very short life of flapper valves and high rates of fuel consumption. The SFC is as high as that of ramjet.
2. The speed of the pulse jet is limited within a very narrow range of 650–800 km/hr becasue of the limitations in the aerodynamic design.
3. The operational range of the pulse jet is also limited in altitude range.
4. The high degree of vibrations due to intermittent nature of the cycle and the buzzing noise has made it suitable for pilotless craft only.
5. It has lower propulsive efficiency than turbojet engine.

35.17 ADVANCED DESIGNS

35.17.1 Ramjet

A ramjet, sometimes referred to as a stovepipe jet, is a type of jet engine. It contains no (major) moving parts and can be particularly useful in applications requiring a small and simple engine for high speed use; such as missiles. They have also been used successfully, though not efficiently, as tipjets on helicopter rotors.

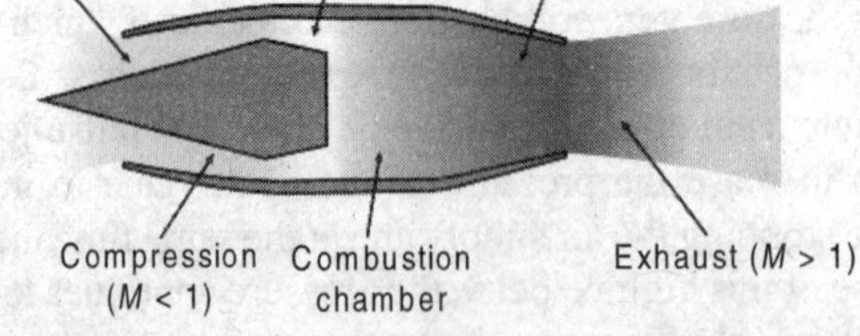

Fig, 35.12. *Schematic diagram showing simple ramjet operation, with Mach numbers of flow shown*

35.17.1.1 Design

In its simplest form a turbojet consists of an air intake, compressor, combustor, turbine and nozzle. In a ramjet, owing to the high flight speed, the ram compression is sufficient to dispense with the need for a compressor and a turbine to drive it. So a ramjet is virtually a 'flying stovepipe', a very simple device comprising of an air intake, a combustor, and a nozzle. Normally the only moving parts are those within the turbopump, which pumps the fuel to the combustor, in a liquid fuel ramjet. Solid fuel ramjets are even simpler.

Ramjets try to exploit the very high total pressure within the streamtube approaching the air intake lip. A reasonably efficient intake will recover much of the freestream stagnation pressure, to support the combustion and expansion processes. Most ramjets operate at supersonic flight speeds and use one or more conical (or oblique) shock waves, terminated by a strong normal shock, to decelerate the airflow to a subsonic velocity at intake exit. Further diffusion is then required to get the air velocity down to level suitable for the combustor.

Since there is no downstream turbine, a ramjet combustor can safely operate at stoichiometric fuel: air ratios, which implies a combustor exit stagnation temperature of the order of 2400 K for kerosene. Normally the combustor must be capable of operating over a wide range of throttle settings, for a range of flight speeds/altitudes. Usually a sheltered pilot region enables combustion to continue when the vehicle intake undergoes high yaw/pitch, during turns. Other flame stabilisation techniques make use of flame holders, which vary in design from combustor cans to simple flat plates, to shelter the flame and improve fuel mixing. Overfuelling the combustor can cause the normal shock within a supersonic intake system to be pushed forward beyond the intake lip, resulting in a substantial drop in engine airflow and net thrust.

Because nozzle pressure ratios are relatively high, ramjet engines are normally fitted with a convergent/divergent propelling nozzle. Given sufficient initial flight velocity, a ramjet will be self-sustaining. Indeed, unless the vehicle drag is extremely high, the engine/airframe combination will tend to accelerate to higher and higher flight speeds, substantially increasing the air intake temperature. As this could have a detrimental effect on the integrity of the engine and/or airframe, the fuel control system must reduce engine fuel flow to stabilize the flight Mach number and, thereby, air intake temperature to sensible levels.

35.17.1.2 Applications

As a ramjet contains no (major) moving parts, it is lighter than a turbojet and can be particularly useful in applications requiring a small and simple engine for high speed use; such as missiles. They have also been used successfully, though not efficiently, as tipjets on helicopter rotors.

35.17.1.3 Flight Speeds

Ramjets generally give little or no thrust below about half the speed of sound, and they are inefficient (less than 600 seconds due to low compression ratios) until the airspeed exceeds 1000 km/h (600 mph). Even above the minimum speed a wide flight envelope (range of flight conditions), such as low to high speeds and low to high altitudes, can force significant design compromises, and they tend to work best optimised for one designed speed and altitude (point designs). However, ramjets generally outperform gas turbine based jet engine designs at supersonic speeds (Mach 2–4). Although inefficient at the slower speeds they are more fuel-efficient than rockets over their entire useful working range.

35.17.2 Scramjet

35.17.2.1 Design

A scramjet (supersonic combustion ramjet) is a variation of a ramjet with the key difference being that the flow in the combustor is supersonic. At higher speeds it is necessary to combust supersonically to maximize the efficiency of the combustion process. Like a ramjet, a scramjet essentially consists of a constricted tube through which inlet air is compressed by the high speed of the vehicle, fuel is combusted, and then the exhaust jet leaves at higher speed than the inlet air. Also like a ramjet, there are few or no moving parts. A scramjet requires supersonic airflow through the engine, thus, similar to a ramjet, scramjets have a minimum functional speed. This speed is uncertain due to the low number of working scramjets, relative youth of the field, and the largely classified nature of research using complete scramjet engines. However it is likely to be at least Mach 5 for a pure scramjet, with higher Mach numbers 7–9 more likely. Thus scramjets require acceleration to hypersonic speed via other means.

The first working scramjet was developed by a team from the University of Queensland, Australia led by Professor Allan Paull.

35.17.2.2 Working

A scramjet is a type of engine which is designed to operate at the high speeds normally associated with rocket propulsion. It differs from a classic rocket by using air collected from the atmosphere to burn its fuel, as opposed to an oxidizer carried with the vehicle. Normal jet engines and ramjet engines also use air collected from the atmosphere in this way. The problem is that collecting air from the atmosphere causes drag, which increases quickly as the speed increases. Also, at high speed, the air collected becomes so hot that the fuel doesn't burn properly any more.

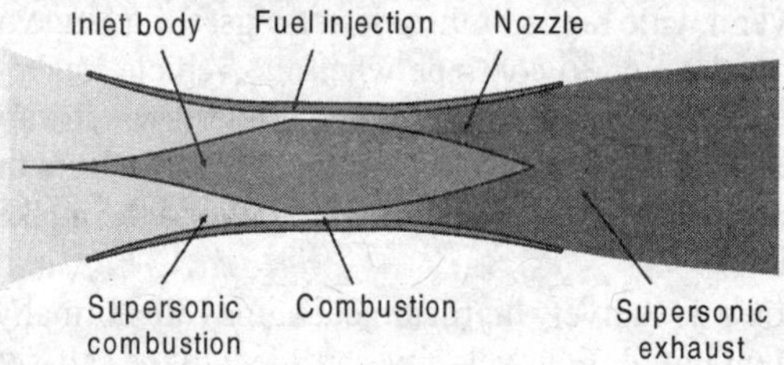

Fig. 35.13. *Diagram illustrating the principle of scramjet operation*

The scramjet is a proposed solution to both of these problems, by modifications of the ramjet design. The main change is that the blockage inside the engine is reduced, so that the air is not slowed down as much. This means that the air is cooler, so that the fuel can burn properly. Unfortunately the higher speed of the air means that the fuel has to mix and burn in a very short time, which is difficult to achieve.

To keep the combustion of the fuel going at the same rate, the pressure and temperature in the engine need to be kept constant. Unfortunately, the blockages which were removed from the ramjet were useful to control the air in the engine, and so the scramjet is forced to fly at a particular speed for each altitude. This is called a "constant dynamic pressure path" because the wind that the scramjet feels in its face is constant, making the scramjet fly faster at higher altitude and slower at lower altitude.

The inside of a very simple scramjet would look like two kitchen funnels attached by their small ends. The first funnel is the intake, and the air is pushed through, becoming compressed and hot. In the small section, where the two funnels join, fuel is added, and the combustion makes the gas become even hotter and more compressed. Finally, the second funnel is a nozzle, like the nozzle of a rocket, and thrust is produced.

Note that most artists' impressions of scramjet-powered vehicle designs depict waveriders where the underside of the vehicle forms the intake and nozzle of the engine. This means that the intake and nozzle of the engine are asymmetric and contribute directly to the lift of the aircraft. A waverider is the required form for a hypersonic lifting body.

35.17.2.3 Applications

Seeing its potential, organisations around the world are researching scramjet technology. Scramjets will likely propel missiles first, since that application requires only cruise operation instead of net thrust production. Much of the money for the current research comes from governmental defence research contracts.

35.17.3 Other Advanced Designs

35.17.3.1 J-58 Combined Ramjet/Turbojet

The SR-71's Pratt and Whitney J58 engines were rather unusual. They could convert in flight from being largely a turbojet to being largely a compressor-assisted ramjet. At high speeds (above Mach 2.4), the engine used variable geometry vanes to direct excess air through 6 bypass pipes from downstream of the fourth compressor stage into the afterburner. 80% of the SR-71's thrust at high speed was generated in this way, giving much higher thrust, improving specific impulse by 10–15%, and permitting continuous operation at Mach 3.2. The name coined for this configuration is turbo-ramjet.

35.17.3.2 Pre-cooled Turbojets

Engines that may need to operate at low hypersonic speeds could theoretically have much higher performance if a heat exchanger is used to cool the incoming air. The low temperature allows lighter materials to be used and combustors to inject more fuel (ordinarily, fuel flow must be reduced at high speed to prevent the turbines from melting, but doing so greatly reduces thrust—precooling the air avoids this.)

This idea leads to plausible designs like SABRE, that might permit single-stage-to-orbit, and ATREX, that might permit jet engines to be used as boosters for space vehicles.

35.17.3.3 Nuclear-powered Ramjets

Project Pluto was a nuclear-powered ramjet, intended for use in a cruise missile. Rather than combusting fuel as in regular jet engines, air was heated using a high-temperature, unshielded nuclear reactor. This raised the specific impulse of the engine by stupendous amounts, and the ramjet was predicted to be able to fly for months at supersonic speeds (Mach 3 at tree-top height). However, there was no obvious way to stop it once it had taken off, which is a great disadvantage. Unfortunately, because the reactor was unshielded, it was dangerous to be in or around the flight path of the vehicle (although the exhaust itself was not radioactive).

35.18 APPLICATION OF VARIOUS PROPULSIVE ENGINES

(a) Turbo-prop. Turbo-prop jet engines are used for medium and long range transport and bomber aircraft. They fly with subsonic speed.

(b) Turbojet. Turbojets are mainly used in military as fighters, bombers and for transport applications. Generally they fly with supersonic speeds, *e.g.* MIGS, Mirage, Knat, Jaugar etc. The only turbojet used for civil aviation is concord.

(c) Turbofan. Used mainly for civil aviation. Flies at subsonic speed.

(d) Ramjet. Ramjets are used for missiles, as pilotless aircrafts. They fly at supersonic speed.

(e) Pulsejet. They are used for missiles, applications and fly at supersonic speed.

35.19 COMPARISON AND EVALUATION OF TURBOPROP, TURBOJET AND TURBOFAN ENGINES

By converting the shaft power of the turboprop into thrust and the fuel consumption per power into fuel consumption per unit thrust, a comparison between turbojet, turboprop and turbofan can be made.

Assuming that the engines are equivalent as to compression ratio and temperature and that the engines are installed in equal-sized aircraft best suited to the particular type of the engine being used. Figure 35.14 shows the various engines compared as to thrust and thrust specific fuel consumption versus air speed.

Turbojet – Characteristics and Uses

1. Low thrust at low forward speed.
2. Relatively high TFSC at low altitudes and air speeds. This disadvantage decreases as altitude and air speed increases.
3. Long take off road required.
4. Small frontal area results in ground clearance problems.
5. Lightest specific weight.
6. Ability to take advantage of high ram-pressure ratio.

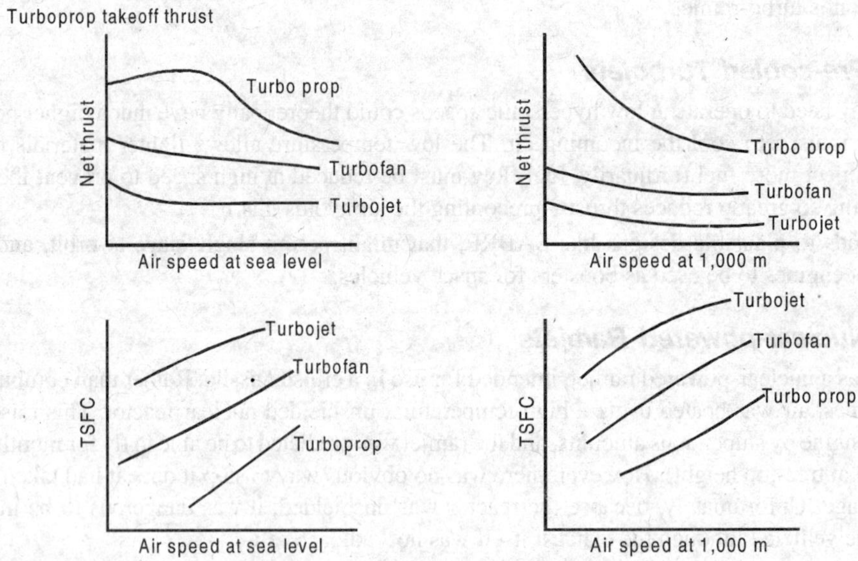

Fig. 35.14

Turbo-prop—Characteristics and Uses

1. High propulsive efficiency at low speeds, which falls off rapidly as airspeed increases. This results in shorter rolls. This engine is able to develop very high thrust at low air speeds because the propeller can accelerate large quantities of air at zero forward velocity of the airplane.
2. More complicated and heavier than turbojet.
3. Lowest TFSC.
4. Large frontal area of propeller and engine combination necessitates longer landing gears for low-wing airplanes but does not necessarily increase parasitic drag.
5. Efficient reverse thrust possible.

These characteristics show that turboprop engines are superior for lifting heavy loads off short and medium runways. They are limited in speeds to 800 km/hr approximately. Since propeller efficiency falls off rapidly with increase in speeds and altitude.

Turbo-Fan—Characteristics and Uses

1. Increased thrust at forward speeds similar to a turbo prop results in a relatively short take off. But unlike the turbo prop the turbo fan thrust is not penalised with increasing air speeds upto approximately. Mach-1 with current fan designs.
2. Weight falls between turbojet and turbo prop.
3. Ground clearances are less than turbo prop, but not as good as turbojet.
4. TSFC and specific weight fall between turbo prop and turbojet.
5. Considerable noise reduction over the turbojet. This results in the reduction of acoustic fatigue in surrounding aircraft parts and it is less objectionable to the people on ground.
6. Superior to turbojet 'hot-day' performance.

From the above characteristics it is learnt that the turbofan engine is suitable for long range, relatively high speed flight and has a definite place in the prolific gas turbine family.

Type Comparison

Comparative suitability for (left to right) turboshaft, low bypass and turbojet to fly at 10 km altitude in various speeds. Horizontal axis—speed, m/s. Vertical axis carries only logical meaning.

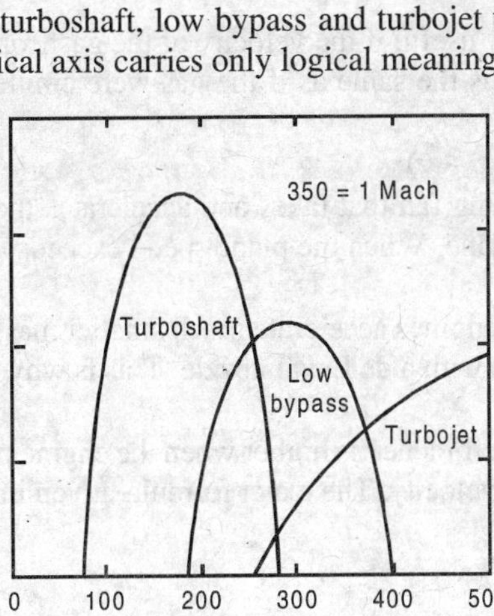

Fig. 35.15

Efficiency as a function of speed of different Jet types. Although efficiency plummets with speed, greater distances are covered, it turns out that efficiency per unit distance (per km or mile) is roughly independent of speed for Jet engines as a group; however airframes become inefficient at supersonic speeds.

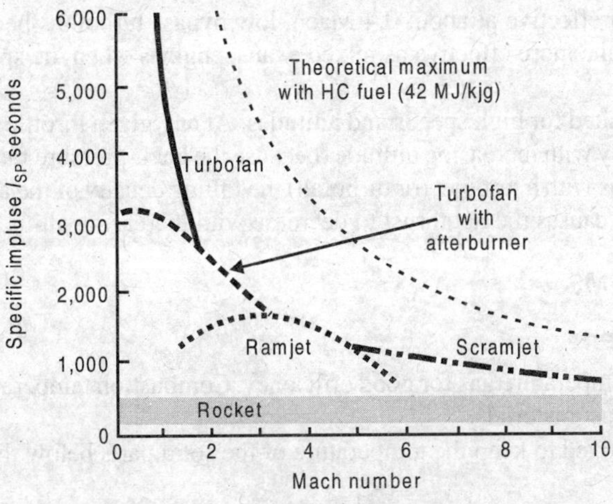

Fig. 35.16. *Propulsion performance.*

The motion impulse of the engine is equal to the air mass, multiplied by the speed that the engine emits this mass:

$$I = mc$$

where m is the air mass per second and c is the exhaust speed. In other words, the plane will fly faster if the engine emits the air mass with a higher speed or if it emits more air per second with the same speed. However, when the plane flies with certain velocity v, the air moves towards it, creating the opposing ram drag at the air intake = mv

Most types of jet engine have an air intake, which provides the bulk of the gas exiting the exhaust. Conventional rocket motors, however, do not have an air intake, the oxidizer and fuel both being carried within the airframe. Therefore, rocket motors do not have ram drag; the gross thrust of the nozzle is the net thrust of the engine.

Consequently, the thrust characteristics of a rocket motor are completely different from that of an air breathing jet engine.

The air breathing engine is only useful if the velocity of the gas from the engine, c, is greater than the airplane velocity, v. The net engine thrust is the same as if the gas were emitted with the velocity $c - v$. So the pushing moment is actually equal to

$$S = m\,(c - v)$$

The turboprop has a wide rotating fan that takes and accelerates the large mass of air but only till the limited speed of any propeller driven airplane. When the plane speed exceeds this limit, propellers no longer provide any thrust ($c - v < 0$).

The turbojets and other similar engines accelerate much smaller mass of the air and burned fuel, but they emit it at the much higher speeds possible with a de Laval nozzle. This is why they are suitable for supersonic and higher speeds.

From the other side, the energy efficiency is higher when the engine pushes as large as possible mass of air at the speed, comparable to the airplane velocity. The exact formula, given in the literature, is

$$\eta = \frac{2}{1+\frac{c}{v}}$$

The low bypass turbofans have the mixed exhaust of the two air flows, running at different speeds (c_1 and c_2). The pushing moment of such engine is

$$S = m_1\,(c_1 - v) + m_2\,(c_2 - v)$$

where m_1 and m_2 are the air masses, being blown from the both exhausts. Such engines are effective at lower speeds, than the pure jets, but at higher speeds than the turboshafts and propellers in general. For instance, at the 10 km attitude, turboshafts are most effective at about 0.4 Mach, low bypass turbofans become more effective at about 0.75 Mach and true jets become more effective as mixed exaust engines when the speed approaches 1 Mach—the speed of sound.

Rocket engines are best suited for high speeds and altitudes. At any given throttle, the thrust and efficiency of a rocket motor improves slightly with increasing altitude (because the back-pressure falls thus increasing net thrust at the nozzle exit plane), whereas with a turbojet (or turbofan) the falling density of the air entering the intake (and the hot gases leaving the nozzle) causes the net thrust to decrease with increasing altitude.

35.20 ENGINE SYSTEMS

35.20.1 Cooling System

All jet engines require high temperature gas for good efficiency. Combustion temperatures can be as high as 3500°, above the melting point of most materials.

Cooling systems are employed to keep the temperature of the solid parts below the failure temperature.

35.20.1.1 Air Systems

A complex air system is built into most turbine based jet engines, primarily to cool the turbine blades, vanes and discs.

Air, bled from the compressor exit, passes around combustor and is injected into the rim of the rotating turbine disc. The cooling air then passes through complex passages within the turbine blades. After removing heat from the blade material, the air (now fairly hot) is vented, via cooling holes, into the main gas stream. Cooling air for the turbine vanes undergoes a similar process.

Small quantities of compressor bleed air are also used to cool the shaft, turbine shrouds, etc. Some air is also used to keep the temperature of the combustion chamber walls below critical. This is done using primary and secondary airholes which allow a thin layer of air to cover the inner walls of the chamber preventing excessive heating.

Exit temperature is dependent on the turbine upper temperature limit depending on the material. Reducing the temperature will also prevent thermal fatigue and hence failure. Accesories may also need their own cooling systems using air from the compressor or outside air.

Air from compressor stages is also used for heating of the fan, airframe anti-icing and for cabin heat. Which stage is bled from depends on the atmospheric conditions at that altitude.

35.20.1.2 Cooling System for Rocket Engines

Rocket engines have extreme cooling requirements, due to the simultaneous combination of both high pressures and high temperatures typically found in the combustion chamber.

Rocket engines often use liquid coolant, typically the propellant is passed around the hot parts of the engine as the fuel is a good conductor of heat. The amount of heat is controlled by temperature difference, heat transfer coefficient and the velocity of the internal flow in the chamber or the nozzle. The fuel is then fed into a special gas generator or injected into the main combustion chamber. This is called *Regenerative cooling*. Other techniques include *Radiative cooling*.

35.20.2 Fuel System

Apart from providing fuel to the engine, the fuel system is also used to control propeller speeds, compressor airflow and cool lubrication oil. Fuel is usually introduced by an atomized spray, the amount of which is controlled automatically depending on the rate of airflow.

So the sequence of events for increasing thrust is, the throtttle opens and fuel spray pressure is increased, increasing the amount of fuel being burned. This means that exhaust gases are hotter and so are ejected at higher acceleration, which means they exert higher forces and therefore increase the engine thrust directly. It also increases the energy extracted by the turbine which drives the compressor even faster and so there is an increase in air flowing into the engine as well.

Obviously, it is the rate of the *mass* of the airflow that matters since it is the change in momentum (mass × velocity) that produces the force. However, density varies with altitude and hence inflow of mass will also vary with altitude, temperature etc. which means that throttle values will vary according to all these parameters without changing them manually.

This is why fuel flow is controlled automatically. Usually there are 2 systems, one to control the pressure and the other to control the flow. The inputs are usually from pressure and temperature probes from the intake and at various points through the engine. Also throttle inputs, engine speed etc. are required. These affect the high pressure fuel pump.

35.20.2.1 Fuel Control Unit (FCU)

This element is something like a mechanical computer. It determines the output of the fuel pump by a system of valves which can change the pressure used to cause the pump stroke, thereby varying the amount of flow.

Take the possibility of increased altitude where there will be reduced air intake pressure. In this case, the chamber within the FCU will expand which causes the spill valve to bleed more fuel. This causes the pump to deliver less fuel until the opposing chamber pressure is equivalent to the air pressure and the spill valve goes back to its position.

And when the throttle is opened, it releases i.e., lessens the pressure which lets the throttle valve fall. The pressure is transmitted (because of a back-pressure valve i.e., no air gaps in fuel flow) which closes the FCU spill valves (as they are commonly called) which then increases the pressure and causes a higher flow rate.

The engine speed governor is used to prevent the engine from over-speeding. It has the capability of disregarding the FCU control. It does this by use of a diaphragm which senses the engine speed in terms of the centrifugal pressure caused by the rotating rotor of the pump. At a critical value, this diaphragm causes another spill valve to open and bleed away the fuel flow.

At initial acceleration, more fuel is required and the unit is adapted to allow more fuel to flow by opening other ports at a particular throttle position. Changes in pressure of outside air i.e., altitude, speed of aircraft etc. are sensed by an air capsule.

35.20.2.2 Fuel Pump

Fuel pumps are used to raise the fuel pressure above the pressure in the combustion chamber so that the fuel can be injected. Fuel pumps are usually driven by the main shaft, via gearing.

Turbopumps are very commonly used with liquid-fuelled rockets and rely on the expansion of an onboard gas through a turbine.

Ramjet turbopumps use ram air expanding through a turbine.

35.20.3 Engine Starting Systems

The fuel system as explained above, is one of the two systems required for starting the engine. The other is the actual ignition of the air/fuel mixture in the chamber. Usually, an *Auxiliary power unit* is used to start the engines. It has a starter motor which has a high torque transmitted to the compressor unit. When the optimum speed is reached, i.e., the flow of gas through the turbine is sufficient, the turbines take over. There are a number of different starting methods such as *electric, hydraulic, pneumatic* etc.

The *electric starter* works with gears and clutch plate linking the motor and the engine, The clutch is used to disengage when optimum speed is achieved. This is usually done automatically. The electric supply is used to start the motor as well as for ignition. The voltage is usually built up slowly as starter gains speed.

Some military aircraft need to be started quicker than the electric method permits and hence they use other methods such as a *turbine starter*. This is an impulse turbine impacted by burning gases from a cartridge. It is geared to rotate the engine and also connected to an automatic disconnect system. The cartidge is set alight electrically and used to turn the turbine.

Another turbine starter system is almost exactly like a little engine. Again the turbine is connected to the engine via gears. However, the turbine is turned by burning gases—usually the fuel is iso-propyl-nitrate stored in a tank and sprayed into a combustion chamber. Again, it is ignited with a spark plug. Everything is electrically controlled, such as speed etc.

Most Commercial aircrafts and large Military Transport airplanes usually use what is called an *auxiliary power unit* or *APU*. It is normally a small gas turbine. Thus, one could say that using such an APU is using a small jet engine to start a larger one. High pressure air from the compressor section of the APU is bled off through a system of pipes to the engines where it is directed into the starting system. This "bleed air" is directed into a mechanism to start the engine turning and begin pulling in air. When the rotating speed of the engine is sufficient to pull in enough air to support combustion, fuel is introduced and ignited. Once the engine ignites and reaches idle speed, the bleed air is shut off.

The APUs on aircraft such as the Boeing 737 and Airbus A320 can be seen at the extreme rear of the aircraft. This is the typical location for an APU on most commercial airliners. The APUs also provide enough power to keep the cabin lights, pressure and other systems on while the engines are off. The valves used to control the airflow are usually electrically controlled. They automatically close at a pre-determined speed.

Usually an APU is started by its own electric starter motor which is switched off at the proper speed automatically. When the main engine starts up and reaches the right conditions, this auxiliary unit is then switched off and disengages slowly.

35.20.4 Ignition

Usually there are 2 igniter plugs in different positions in the combustion system. A high voltage spark is used to ignite the gases. The voltage is stored up from a low voltage supply provided by the starter system. It builds up to the right value and is then released as a high energy spark. Depending on various conditions, the igniter continues to provide sparks to prevent combustion from failing if the flame inside goes out.

Of course, in the event that the flame does go out, there must be provision to relight. There is a limit of altitude and air speed at which an engine can obtain a satisfactory relight.

35.20.5 Lubrication System

A lubrication system serves to ensure lubrication of the bearings and to maintain sufficiently cool temperatures, mostly by eliminating friction.

The lubrication system as a whole should be able to prevent foreign material from entering the plane, and reaching the bearings, gears, and other moving parts. The lubricant must be able to flow easily at relatively low temperatures and not disintegrate or break down at very high temperatures.

Usually the lubrication system has subsystems that deal individually with the pressure of an engine, scavenging, and a breather.

The pressure system components are an oil tank and de-aerator, main oil pump, main oil filter/filter bypass valve, pressure regulating valve (PRV), oil cooler/by pass valve and tubing/jets.

Usually the flow is from the tank to the pump inlet and PRV, pumped to main oil filter or it is bypass valve and oil cooler, then through some more filters to jets in the bearings.

Using the PRV method of control, means that the pressure of the feed oil must be below a critical value (usually controlled by other valves which can leak out excess oil back to tank if it exceeds the critical value). The valve opens at a certain pressure and oil is kept moving at a constant rate into the bearing chamber.

If the engine speed increases, the pressure within the bearing chamber also increases, which means the pressure difference between the lubricant feed and the chamber reduces which could reduce slow rate of oil when it is needed even more. As a result, some PRVs can adjust their spring force values using this pressure change in the bearing chamber proportionally to keep the lubricant flow constant.

ROCKET PROPULSION

35.21 INTRODUCTION

Rockets are becoming important as an aircraft power plant. The primary use of rockets is in military applications, but they offer the promise for long range, high speed transport aircraft and also as a power plant for space travel. Efforts of scientists made Neil Armstrong to land on Moon in 1969.

A rocket is similar to other jet power plants in many ways as has been discussed in previous chapter. The major difference between rocket engine power plant and other jet propulsion systems is that it carries the entire propellant (fuel and oxidizer) with it, while the other jet propulsion systems depend on atmospheric air. i.e., Air breathing jet propulsion systems utilize oxygen from the atmosphere (surroundings), whereas rocket engine utilizes oxidizer which is carried in its tanks. This fundamental distinction between the two power plants produces different performance capabilities, which are summarised in the Table 35.1.

Table 35.1

Air-Breathing Engine	*Rocket Engine*
There is limitation of altitude.	There is no altitude limitation. Space travel is possible.
Thrust decreases with altitude (due to low air pressure).	There is slight improvement in thrust with altitude.
Engines ram increases with flight speed.	Engine has no ram, hence thrust is constant with speed.
Flight speed is always less than jet velocity.	Flight speed is not limited. It can be greater than jet velocity.
Efficiency is higher than rocket engine. Duration of flight is more.	Efficiency is low except at extremely high flight speeds. Duration of flight is low.

35.22 CLASSIFICATION OF ROCKETS

The classification of rockets is based on the form of energy and momentum utilised for producing the thrust. Thus rockets are classified into three basic types.

(i) Chemical rocket engines.
(ii) Nuclear rocket engines.
(iii) Electric propulsion rockets.

35.22.1 Chemical Rockets

The oldest types of rockets are chemical rockets, in which two or more chemicals mix together, producing a chemical reaction. The reaction produces hot gases that are forced out of a nozzle, pushing the rocket in the other direction.

There are two basic types of chemical rocket engines, based on the fuel used.

(a) Liquid propellant
(b) Solid propellant

(a) Liquid Propellant Rocket Engines. These engines may be divided into various categories. These include cryogenic, monopropellant, storable bi-propellant system and hybrid system. Hybrid system is a combination of liquid and solid propellant systems. A liquid bi-propellant system has greater potential and is discussed in detail.

Figure 35.17 shows the main elements of a liquid bi-propellant system. It consists of an oxidizer tank, fuel tank, control valve lines, and the rocket motor, which is the heart of the rocket engine. The basic rocket motor consists of an injection plate *I*, combustion chamber *C*, and discharge nozzle *N*. The injection plate receives liquid oxidizer and fuel, which after mixing produce a chemical reaction in the combustion chamber. The very high pressure and temperature gases produced in the combustion chamber, expand in the nozzle and produce a high supersonic exit velocity (1500 – 3000 m/s). The net thrust produced is the product of exit velocity and mass flow rate of gases.

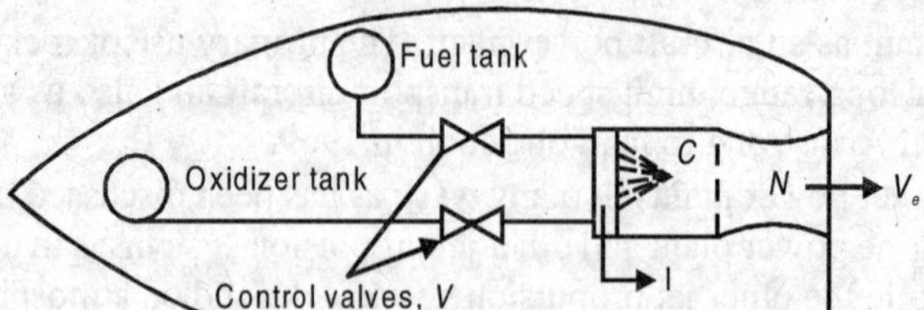

Fig. 35.17. *Liquid propelled rocket*

An ignition system uses hypergolic combinations for igniting the mixture. Hypergolic fuels commonly used are hydrazine, and monomethyl hydrazine. They ignite spontaneously on contact. A cooling system is also necessary to keep the walls of the motor from melting because the temperature during reaction exceeds 2700°C.

The pressure at which the propellants have to be supplied from the storage tank should be higher than the pressure in the combustion chamber. There are two methods which are used for transporting liquid propellants from their storage tanks to the rocket motor. They are:

(i) Gas pressurisation system and (ii) Pump pressurisation system.

(i) Gas pressurisation system. In this system, an inert gas (nitrogen) is used. This gas is stored at high pressure and is supplied through pressure regulator valves to force the liquid propellants through the lines, control valves, injector plate and into the combustion chamber. The gas is supplied at a pressure of about 150 bar as shown in Fig. 35.18. The mixture ratio supplied is 3 to 5 (oxidizer to fuel flow). The system is first energized by opening the system energizing valve and then the rocket motor is turned on by opening of bi-propellant valve. The motor can be stopped and restarted at will by closing or opening of the bi-propellant valve.

This system is simple. Its disadvantage is heavy weight of the storage tanks because of the high pressures at which they are to be stored. This limitation makes gas pressurisation system applicable only to short duration operations because lesser quantity of fuel can be stored due to weight considerations.

(ii) Pump pressurisation system. In this system liquid oxidizer and fuel are stored at low pressure so that tanks are light in weight and forced into rocket motor at high pressure by fuel and oxidizer pumps. The power driving the pumps is supplied by gas turbine which is supplied with steam and oxygen obtained by decomposing hydrogen peroxide by a catalyst.

Because of the use of third liquid (H_2O_2), gas turbine, pumps and additional lines that are necessary, pump pressurisation system is more complex than gas pressurisation system. The design of pumps is the biggest problem because any leakage of liquids will lead to explosion. Liquid oxidizers are generally acids, liquid oxygen, concentrated hydrogen peroxide (H_2O_2), nitrogen tetra-oxide (N_2O_4) etc. and hence special pump materials are required for the oxidizer pump and impeller.

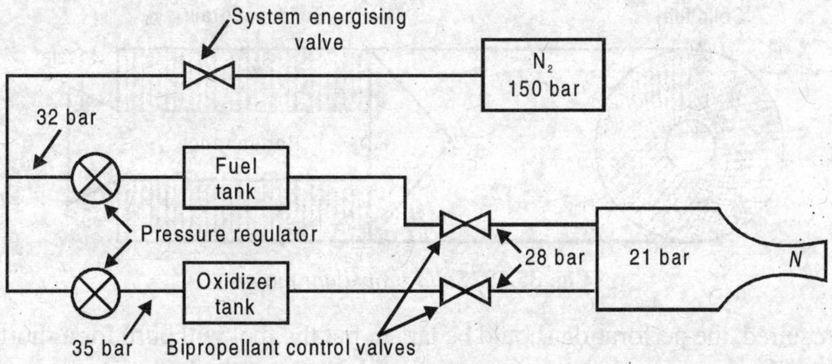

Fig. 35.18. *Gas pressurisation system*

Rocket Propellants. The desirable properties of rocket propellants are:

1. It should produce a high combustion chamber temperature. This means the calorific value of the propellant should be high.
2. The molecular weight of the products of combustion should be low, which will cause high jet velocity and specific thrust.
3. Easy to store and handle.
4. It should be readily ignitable.
5. It should not react with motor system, tanks, piping, valves and ignition nozzles.
6. It should have high density to reduce the overall size and weight of the system.

Requirements (1) and (2) are contradictory and hence a balance should be reached.

An ideal propellant is not yet available. For example, if liquid hydrogen is used as a fuel, then the jet velocities are very high but the size of the fuel tank is large due to the low specific weight of hydrogen. Also, very low temperatures are required for the storage of liquid hydrogen.

The liquid propellants used are given in the Table 35.2.

Table 35.2

Oxidizer	*Fuel*
Oxygen	Gasoline, Hydrogen Hydrazine, (N_2H_4), Ethyl alcohol Methane, Kerosene
Nitric acid	Ethyl alcohol Aniline Furfural alcohol

Oxidizer	*Fuel*
Hydrogen peroxide	Ethyl alcohol Methyl alcohol Hydrazine (Methyl and dimethyl) Hydrazine

(b) Solid Propellant. Solid propellant motors are the simplest of all rocket designs. They consist of a casing, usually steel, filled with a mixture of solid compounds (fuel and oxidizer) which burn at a rapid rate, expelling hot gases through nozzle to produce thrust. When ignited, a solid propellant burns from the centre towards the sides of the casing. The shape of the centre channel determines the rate and pattern of burning, thus providing a means to control thrust. Unlike liquid propellant engines, solid propellant motors cannot be shut down. Once they are ignited, they will burn till all the propellant is exhausted.

Solid propellants are of two types: Homogeneous and composite. Both are dense, stable at ordinary temperatures and easily storable.

Working of Solid Propellants. Figure 35.19 shows a solid propellant system. By changing the shape and size of perforation, we can control the rate and duration of burning and thus control thrust.

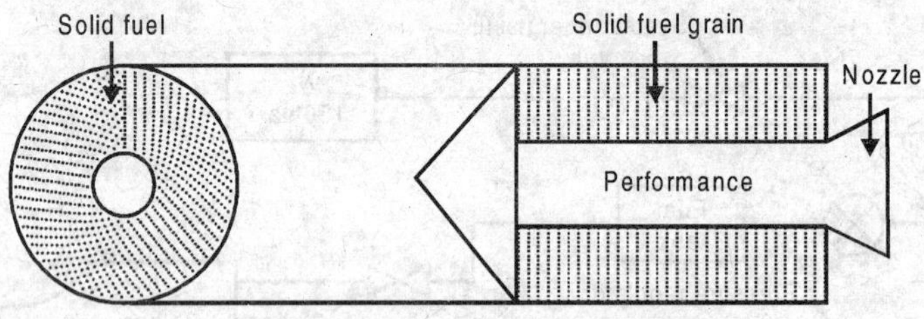

Fig. 35.19. *Solid propellant motor*

If more thrust is required, the perforation should be larger, but the fuel will burn for a short time and vice versa. The burning period and the thrust depends on the type of perforation in the fuel.

Advantages and disadvantages of solid propellants are as following:

Advantages

1. They are stable and easily storable.
2. They do not require turbopumps and complex propellant feeding devices.

Disadvantages

1. The solid propellant motor cannot be shut down. Once ignited, fuel burns till the end.
2. The propellant is temperature sensitive.

35.22.2 Nuclear Rocket Engine

Work on nuclear rocket engine began in 1956 and further development was done by NASA (National Aeronautics and Space Administration, USA) for space flight applications, for manned and unmanned space exploration.

Nuclear thermal rocket is very simple concept. The elements of la nuclear rocket engine are shown in Fig. 35.20. It consists of a propellant (H_2) tank, a pump (P), valve (V), nuclear fission reactor (R) and discharge nozzle (N).

Hydrogen runs through the reactor (gets heated) and into the nozzle. Hydrogen gives the best exhaust velocity. But the exhaust velocity is limited or fixed by the melting point of the reactor. The reactor elements have to be durable, since erosion will contaminate the exhaust with fissionable materials. The temperature in the reactor may go upto 3000°C and hence cooling of nozzle is required to prevent it from melting.

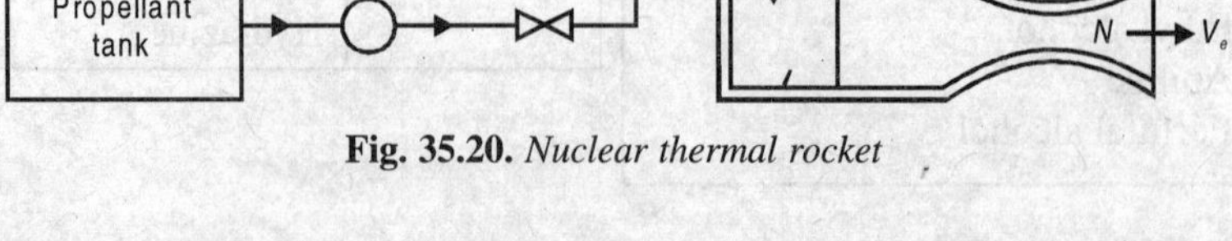

Fig. 35.20. *Nuclear thermal rocket*

35.22.3 Electric Propulsion Engines

The basic requirement for space propulsion is the generation of high exhaust velocities so that the propellant consumption is low. Chemical rocket engines produce low exhaust velocities meaning requiring more fuel. Hence electric propulsion engines are used which produce 4 to 100 times more exhaust velocities. There are three main types of electric rocket propulsion engines.

(a) Arc plasma rocket
(b) Ion rocket
(c) Magneto plasma rocket.

The energy supplied to these is from a separate propulsion device.

(a) Arc Plasma Rocket Engine. It is one of the simplest types of electrical propulsion systems. The schematic is shown in Fig. 35.21.

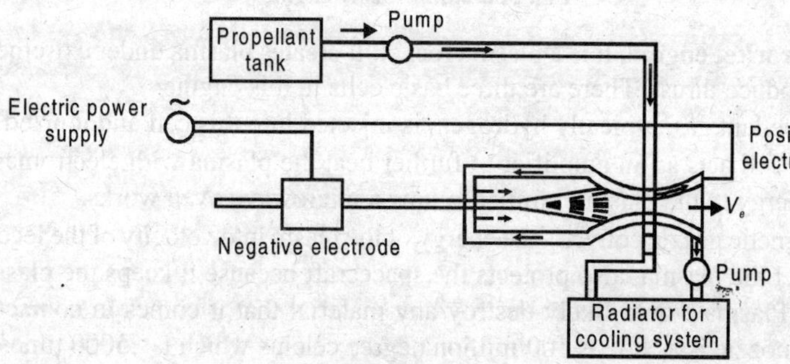

Fig. 35.21. *Arc plasma rocket engine*

The major components of this system are:

(i) Propellant tank
(ii) Thrust chamber
(iii) Electric supply system
(iv) Cooling system, and
(v) Pumps.

The thrust chamber contains two electrodes. The electrode within the chamber is the anode and the nozzle wall serves as cathode. The propellant gets heated to an electrically neutral plasma in passing through the arc formed between the two electrodes. Thrust results from the expansion of heated plasma through the nozzle. A velocity of 4000 – 15000 m/s is produced. The propellant can be used to cool the chamber regeneratively before passing into the chamber.

The arc plasma or electro-thermal rocket is similar to the chemical rocket, the main difference being that electrical rather than chemical energy is used for heating the propellant.

(b) Ion Rocket Engine. The major components of ion or electrostatic rocket engine are

(i) Propellant tank and feed mechanism
(ii) Thrust chamber
(iii) Electric power supply.

The thrust chamber incorporates a vaporising chamber, an ionisation chamber, an accelerating grid and an electron emitter. In the vaporization chamber, the propellant is heated and vaporised from where it goes to the ionisation chamber where electrons are stripped to ionize it. The ions are then accelerated electrostatically by electrodes to produce thrust. For examples, potassium seeded argon is ionized and ions are accelerated. Though it has admirably high exhaust velocity, there are theoretical limits that ensure all ion drives have low thrust.

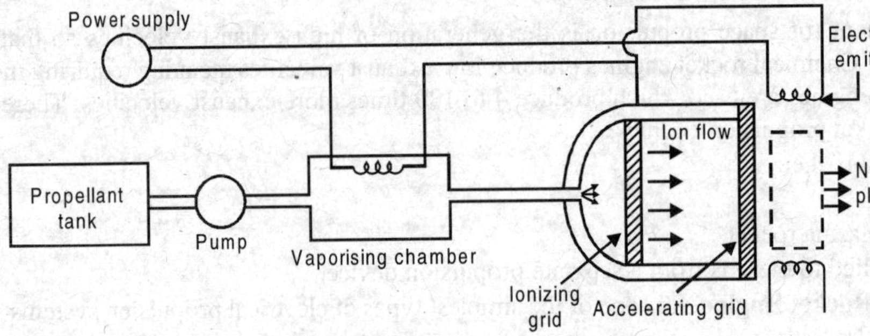

Fig. 35.22. *Ion rocket engine*

(c) Magneto plasma rocket engine. It is a plasma rocket. It creates plasma under extremely hot conditions and then expels plasma to produce thrust. There are three basic cells in this engine.

(i) *Forward cell.* Propellant gas, typically hydrogen, is injected into this cell and ionized to create plasma.

(ii) *Central cell.* This cell acts as an amplifier to further heat the plasma with electromagnetic energy. Radio-waves are used to add energy to the plasma, similar to how a microwave oven works.

(iii) After cells. A magnetic nozzle converts the energy of the plasm into velocity of the jet exhaust. The magnetic field, that is used to expel the plasma, also protects the spacecraft because it keeps the plasma from touching the shell of the space-craft. Plasma would likely destroy any material that it comes in contact. The temperature of plasma exiting from the nozzle is as hot as 100 million degree celcius which is 25000 times hotter than the gases expelled from the space shuttle.

35.23 SOLAR ROCKET PROPULSION

It is a form of space-craft propulsion that makes use of solar power to directly heat reaction mass and hence does not require electrical generator. A solar thermal rocket only has to carry a means of capturing solar energy, such as concentrators and mirrors. The heated propellant is fed through a conventional rocket nozzle to produce thrust. The engine thrust is directly related to the surface area of the solar collector and to the local intensity of solar radiation. But still it is in the developmental stage.

Most proposed designs for solar thermal rockets use hydrogen as propellant due to its low molecular weight which gives excellent specific impulse of 900 seconds (9 kNs/kg).

In the shorter term, solar thermal propulsion has been proposed as a good candidate for use in reusable inter-orbital tugs, as it is a high efficiency, low thrust system that can be refueled with relative ease.

35.24 STAGING OF ROCKETS

Staging of 'rockets is done to improve the payload capability of vehicles with a high change in velocity (ΔV) requirement, such as launch vehicles or interplanetary spacecraft. In a multi-stage rocket, the propellant is stored in smaller, separate tanks rather than in a larger single tank as in a single stage rocket. As each tank is discarded when empty, energy expended to accelerate the empty tanks, so a higher total ΔV is obtained. For convenience, separate tanks are usually bundled with their own engines, with each discardable unit called a "stage".

The first stage is at the bottom and is usually the largest, the second stage is above it and is usually the next largest, etc. In the typical case of linear staging, when the first stage's motor fire, the entire rocket is propelled upwards. When the first stage's motor runs out of fuel, the first stage is detached from the rest of the rocket (usually with some kind of explosive charge) and falls away. This leaves a smaller rocket, with the second stage on the bottom, which then fires. This process is repeated until the final stage's motor burns out.

Advantages of Multl-stage Rockets

1. The space and structure of each stage are useless and only add weight to the vehicle, which slows down its future acceleration. By dropping the stages which are no longer useful, the rocket lightens itself.
2. The weight of the future stages is able to provide more acceleration, than if the earlier stages were still attached.
3. Less total fuel is required to reach given velocity or altitude.

Disadvantages of Multi-stage Rockets

1. Multi-staging requires the vehicle to lift motors which not required until later, hence rocket is more complex and difficult to build.
2. Cost of launching is very high.

35.25 APPLICATIONS OF ROCKET ENGINES

Rocket engines are used for launching of space-craft and satellites into space. The rockets that have been used are Atlas–Centaur (1962), Titan-II (1964), Saturn-V (1967), Space shuttle (1981), Delta-II (1989), etc.

To launch satellites, the velocity of rocket should be greater than the escape velocity (11 m/s) for which multi-stage rockets should be used. Some other applications are

1. Jet assisted take-off.
2. Lethal weapons.
3. Long range artillery.
4. Research.

SOLVED EXAMPLES

Example 35.1. *A jet propelled engine having two jets and working on turbojet system has a velocity of 210 m/s, when flying at an altitude of 12000 m, the density of air at this altitude being 0.172 kg/m³. The resistance or drag of the plane is 6670.8 N and propulsion efficiency of the jet is 50%. Calculate*

(i) Absolute velocity of the jet.
(ii) Quantity of air compressed/minute.
(iii) Diameter of the jet.
(iv) Net output of gas turbine plant.
(v) TSFC.
(vi) Air-fuel ratio.

Take $\eta_{overall} = 18\%$, $CV = 4.895 \times 10^4$ *kJ/kg.*

Solution

Given : $V_i = 210$ m/s, CV $= 4.895 \times 10^4$ kJ/kg , $\rho_{air} = 0.172$ kg/m^2. Thrust, $F = 6670.8$ N.

(i) Propulsion efficiency, $\eta_p = \dfrac{2\alpha}{1+\alpha}$; $\alpha = \dfrac{V_i}{V_e}$

$$= \frac{2V_i}{V_e + V_i}$$

$$\therefore \quad 0.5 = \frac{2 \times 210}{V_e + 210}$$

$$\therefore \quad V_e = 630 \text{ m/s}$$

But absolute velocity of jet, $V = V_e - V_i$

$$= 630 - 210$$

$$= \mathbf{420\ m/s}$$

(ii) Propulsive thrust, $F = \dot{m}_i (V_e - V_i)$

$$6670.8 = \dot{m}_i (420)$$

$$\therefore \quad \dot{m}_i = 15.88 \text{ kg/s}$$

$$\rho_{air} = 0.172 \text{ kg/m}^3$$

$$\therefore \text{ Volume of air compressed/min} = \frac{15.88}{0.172} \times 60 = \mathbf{5539.8\ m^3/min}$$

(iii) Let a_e be the area of one jet. Neglecting mass of fuel,

$$\text{Volume compressed/min} = 2(a_e V_e) \times 60$$

$$\therefore \quad a_e = \frac{5539.8}{2 \times 60 \times 630} = 0.0733 \text{ m}^2$$

But $$a_e = \frac{\pi}{4} d_e^2 = 0.0733$$

$$\therefore \quad d_e = 0.305 \text{ m}$$

or $$d_e = 30.5 \text{ cm}$$

(iv) $$\eta_p = \frac{\text{Thrust power}}{\text{Net output of turbine}}$$

$$\therefore \text{ Net output of turbine plane} = \frac{6670.8 \times 210}{1000 \times 0.5} = \mathbf{2801.7\ kW}$$

(v) Let $\dot{m}_f$ be fuel consumed/hour.

$$\eta_{overall} = \frac{\text{Propulsive work}}{\text{Heat supplied (by fuel)}}$$

$$\therefore \quad 0.18 = \frac{2801.7 \times 3600}{\dot{m}_f \times 48950}$$

$$\therefore \quad \dot{m}_f = 1144.7 \text{ kg/h}$$

$$\text{TSFC} = \frac{\dot{m}_f}{\text{Thrust}} = \frac{1144.7}{6670.8 \times V_i}$$

$$= \frac{1144.7}{\frac{6670.8}{1000} \times 210} = \mathbf{0.817\ kg/kWh}$$

(vi) Air-fuel ratio, $$\text{A/S} = \frac{\text{Mass of air}}{\text{Mass of fuel}}$$

$$= \frac{15.88 \times 60 \times 60}{1144.7} = \mathbf{49.94}$$

Example 35.2. *A German Jumo turbojet plant uses petrol having a calorific value of 4.184 × 10⁴ kJ/kg. The fuel consumption is 0.15 kg/Nh of thrust when the thrust is 8829 N. The air-craft velocity is 425 m/s. The mass of air passing through the compressor is 19.8 kg/s. Calculate the air-fuel ratio and overall efficiency.*

Solution

Given : CV = 41840 kJ/kg, TSFC = 0.156 kg/Nh, F = 8829 N,

V_i = 425 m/s, $\dot{m}_i$ = 19.8 kg/s

To find : η_o, A/F

$$\dot{m}_f = \frac{0.15}{3600} \times 8829 = 0.3678 \text{ kg/s}$$

$$A/F = \frac{\text{Mass flow rate of air}}{\text{Mass flow rate of fuel}}$$

$$= \frac{19.8}{0.3678} = \mathbf{53.8:1}$$

Thrust power, $T_p = F \times V_i$

$$= 8829 \times 425 = 3752.3 \text{ kW}$$

Heat input, $Q_i = \dot{m}_f \times \text{CV}$

$$= 0.3678 \times 41840$$
$$= 15388.7 \text{ kW}$$

Overall efficiency, $\eta_o = \frac{T_p}{Q_i} \times 100$

$$= \frac{3752.3}{15388.7} \times 100$$
$$= \mathbf{24.4\ \%}$$

Example 35.3. *A turbojet engine consumes air at the rate of 60 kg/s when flying at a speed of 950 km/h. Calculate*

(i) Exit velocity of jet when the enthalpy change for the nozzle is 210 kJ/kg.
(ii) Fuel flow rate in kg/s if A/F = 65.
(iii) TSFC.
(iv) Thermal efficiency of the plant when combustion efficiency is 93% land CV of fuel is 42500 kJ/kg.
(v) Propulsive power.
(vi) propulsive efficiency.
(vii) Overall efficiency.

Solution

Given : $\dot{m}_i$ = 60 kg/s, Δh = 210 kJ/kg, A/F = 65 : 1, V_i = 950 km/h,

η_{comb} = 0.93, CV = 42500 kJ/kg

$$V_i = \frac{950 \times 1000}{3600} = 263.9 \text{ m/s}$$

(i) Exit velocity of jet, $V_e = \sqrt{2\,\Delta h \times 1000}$

$$= \sqrt{2 \times 210 \times 1000}$$
$$= \mathbf{648.1 \text{ m/s}}$$

(ii) Fuel consumption, $$\dot{m}_f = \frac{\text{Air consumption}}{(A/F)}$$

$$= \frac{60}{65} = \mathbf{0.923\ kg/s}$$

(iii) Thrust produced or propulsive thrust,

$$F = m_i (V_e - V_i)$$

$$= 60 \times (648.1 - 263.9)$$

$$= 23052\ \text{N}$$

Thrust specific fuel consumption,

$$\text{TSFC} = \frac{\dot{m}_f}{T_p}$$

$$= \frac{0.923}{23052}$$

$$= \mathbf{4 \times 10^{-5}\ kg/N\ thrust/s}$$

(iv) $$\eta_{th} = \frac{\text{Work output}}{\text{Heat supplied}}$$

$$= \frac{\text{Gain of K.E./kg}}{\dot{m}_f/\text{kg of air} \times CV \times \eta_{comb} \times 1000}$$

$$= \frac{\dfrac{(648.1)^2 - (263.9)^2}{2}}{\dfrac{1}{65} \times 42500 \times 0.93 \times 1000} = \mathbf{28.81\ \%}$$

(v) $$\text{Propulsive power} = m_i \left(\frac{V_e^2 - V_i^2}{2} \right) \times \frac{1}{1000}$$

$$= \frac{60 \times \left[(648.1)^2 - (263.9)^2 \right]}{2 \times 1000}$$

$$= \mathbf{10511.7\ kW}$$

(vi) Propulsive efficiency, $$\eta_p = \frac{2V_e}{(V_i + V_e)}$$

$$= \frac{2\alpha}{(1+\alpha)} \quad \text{where } \alpha = \frac{V_i}{V_e} = 0.4072$$

$$= \frac{2 \times 0.4072}{(1 + 0.4072)} = \mathbf{57.87\ \%}$$

(vii) Overall efficiency,

$$\eta_o = \frac{(V_e - V_i)V_i}{10^3 \times (F/A) \times CV \times \eta_{comb}}$$

$$= \frac{(648.1 - 263.9) \times 263.9}{10^3 \times (1/65) \times 42500 \times 0.93}$$

$$= \mathbf{15.9\ \%}$$

Example 35.4. *A turboprop air-craft is flying at 600 km/h at an altitude where the ambient conditions are 0.45 bar and –15 °C. Compressor pressure ratio is 9 : 1. Maximum gas temperature is 1200 K. The intake duct efficiency is 90% and the total head isentropic efficiency of compressor and turbine are 88% and 92% respectively. Calculate specific power output in kW/kg of air/s and thermal efficiency of the unit taking mechanical efficiency of transmission as 98% and neglecting losses other than specified. Assume that exhaust gases leave the air-craft at 600 km/h relative to the air-craft. Take specific heat of exhaust gases,* C_{pg} *= 1.15 kJ/kg-K.*

Solution

Given :

$$V_i = 600 \times \frac{1000}{3600}$$

$$= 166.7 \text{ m/s}$$

$$V_e = 600 \text{ km/h}$$

$$= 166.7 \text{ m/s}$$

As it is a turboprop engine, thrust = 0.

$(\eta_{is})_c = 0.88,\ (\eta_{is})_t = 0.92,\ \eta_m = 0.98$

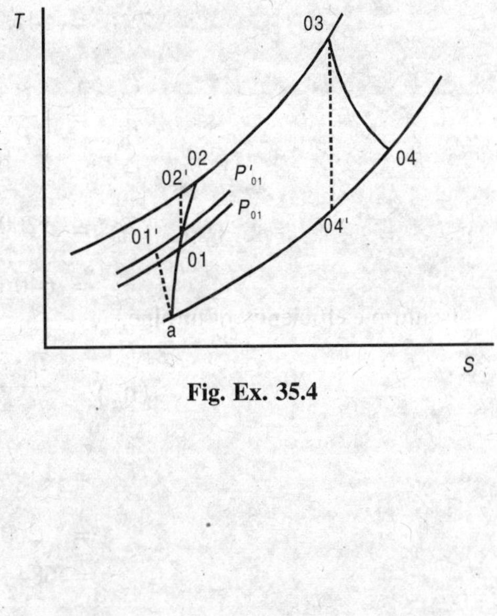

Fig. Ex. 35.4

Because

$$T'_{o_1} = T_a + \frac{V_i^2}{2C_p};\ C_p \text{ is in J/kg-K}$$

$$= 258 + \frac{(166.7)^2}{2 \times 1005}$$

$$= 271.8 \text{ K}$$

$$P'_{o_1} = P_a \left(\frac{T'_{o_1}}{T_a}\right)^{\frac{\gamma}{\gamma-1}}$$

$$= 0.45 \times \left(\frac{271.8}{258}\right)^{\frac{1.4}{0.4}}$$

$$= 0.54 \text{ bar}$$

$$\eta_{duct} = \frac{P_{o_1} - P_a}{P'_{o_1} - P_a}$$

$$P_{o_1} = 0.9 \times \left(P'_{o_1} - P_a\right) + P_a$$

$$= 0.9 \times (0.54 - 0.45) + 0.45$$

$$= 0.531 \text{ bar}$$

$$\frac{P_{o_2}}{P_{o_1}} = 9$$

$$\therefore \quad P_{o_2} = 9 \times 0.531$$
$$= 4.779 \text{ bar}$$

$$\frac{T'_{o_2}}{T_{o_1}} = \left(\frac{P_{o_2}}{P_{o_1}}\right)^{\frac{\gamma-1}{\gamma}}; \quad T_{o_1} = T'_{o_1}$$

$$\therefore \quad T'_{o_2} = 271.8 \times \left(\frac{9}{8}\right)^{0.286}$$
$$= 509.5 \text{ K}$$

But $$(\eta_{is})_c = \frac{T'_{o_2} - T_{o_1}}{T_{o_2} - T_{o_1}}$$

$$\therefore \quad T_{o_2} = \frac{T'_{o_2} - T_{o_1}}{(\eta_{is})_c} + T_{o_1}$$

$$= \frac{509.5 - 271.8}{0.88} + 271.8$$
$$= 541.9 \text{ K}$$

$$\frac{T'_{o_4}}{T_{o_3}} = \left(\frac{P_{o_4}}{P_{o_3}}\right)^{\frac{\gamma-1}{\gamma}}$$

$$\therefore \quad T'_{o_4} = 1200 \times \left(\frac{1}{9}\right)^{0.286}$$
$$= 640.1 \text{ K}$$

Isentropic efficiency of turbine,

$$(\eta_{is})_t = \frac{T_{o_3} - T_{o_4}}{T_{o_3} - T'_{o_4}}$$

$$T_{o_4} = T_{o_3} - 0.92\left(T_{o_3} - T'_{o_4}\right)$$

$$= 1200 - 0.92(1200 - 640.1)$$
$$= 684.9 \text{ K}$$

Net work, $$W_N = \text{Turbine work} - \frac{\text{Compressor work (Theoretical)}}{\eta_m}$$

$$= C_{pg}\left(T_{o_3} - T_{o_4}\right) - C_{pa}\left(\frac{T_{o_2} - T_{o_1}}{\eta_m}\right)$$

$$= 1.15 \times (1200 - 684.9) - 1.005 \times \left(\frac{541.9 - 271.8}{0.98}\right)$$

$$= 592.4 - 277$$

$$W_N = 315.4 \text{ kJ/kg}$$

Thermal efficiency, $$\eta_{th} = \frac{\text{Work output}}{\text{Heat supplied}} = \frac{W_N}{C_{pg}(T_{o_3} - T_{o_2})}$$

$$= \frac{315.4}{1.15 \times (1200 - 541.9)}$$

$\therefore$ $$\eta_{th} = 41.67 \ \%$$

Example 35.5. *A turbojet engine travels at 216 m/s in air at 0.8 bar and –7.5°C. Air first enters the diffuser in which air is brought to rest relative to the unit and it is then compressed through a pressure ratio of 5.8 and fed to a turbine at 1130° C. The gases then expand through turbine and nozzle to the atmospheric pressure of 0.78 bar. The efficiencies of diffuser, nozzle and compressor are 90%, 92% and 90% respectively. Pressure drop in the combustion chamber is 0.17 bar. Take turbine efficiency as 85%, calculate: Taking*

(i) Air-fuel ratio

(ii) Specific thrust of the unit

(iii) Total thrust, if the inlet cross-section of the diffuser is 0.14 m².

Take CV of fuel as 44000 kJ/kg.

Solution

Given : $V_i = 216$ m/s, $P_a = 0.8$ bar, $T_a = -7.5°\text{C} = 265.5$ K,

$\eta_c = 0.9$, $\eta_d = 0.90$, $\eta_n = 0.92$, $\eta_t = 0.85$, $T_{o_3} = 1130°\text{C}$

Pressure drop in combustion chamber = 0.17 bar

$$T_{o_1} = T_a + \frac{V_i^2}{2C_p}$$

$$= 265.5 + \frac{(216)^2}{2 \times 1005}$$

$$= 288.7 \text{ K}$$

and diffuser efficiency, $$\eta_d = \frac{T'_{o_1} - T_a}{T_{o_1} - T_a}$$

$$T_{o_1} = \frac{T'_{o_1} - T_a}{\eta_d} + T_a$$

$\therefore$ $$T_{o_1} = 291.3 \text{ K}$$

$$\frac{T'_{o_1}}{T_a} = \left(\frac{P_{o_1}}{p_a}\right)^{\frac{\gamma-1}{\gamma}}$$

$\therefore$ $$P_{o_1} = \left(\frac{T'_{o_1}}{T_a}\right)^{\frac{\gamma}{\gamma-1}} \times P_a = \left(\frac{288.7}{265.5}\right)^{\frac{1.4}{0.4}} \times 0.8$$

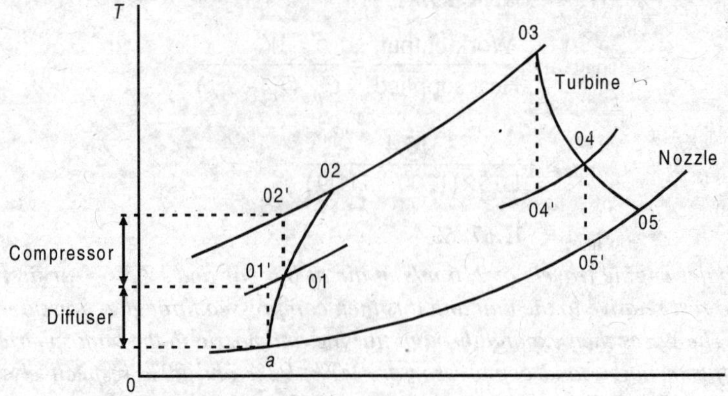

Fig. Ex. 35.5

$$\therefore \quad P_{o_1} = 1.073 \text{ bar}$$

$$\left(\frac{P_{o_2}}{P_{o_1}}\right)^{\frac{\gamma-1}{\gamma}} = \left(\frac{T'_{o_2}}{T_{o_1}}\right)$$

$$\therefore \quad T'_{o_2} = T_{o_1}(5.8)^{\frac{0.4}{1.4}}$$

$$= 291.3 \times (5.8)^{0.286} = 481.6\text{K}$$

But compressor efficiency, $$\eta_c = \frac{T'_{o_2} - T_{o_1}}{T_{o_2} - T_{o_1}}$$

$$\therefore \quad T_{o_2} = \frac{T'_{o_2} - T_{o_1}}{\eta_c} + T_{o_1}$$

$$= \frac{481.6 - 291.3}{0.9} + 291.3$$

$$= 502.7 \text{ K}$$

$$T_{o_3} = 1130\,°\text{C} = 1403 \text{ K}$$

As specific heat of air and gas are not given, we will assume $C_{pa} = C_{pg} = 1.005$ kJ/kg-K

Heat supplied, $$Q = (\dot{m}_i + \dot{m}_f)C_p\, T_{o_3} - \dot{m}_i\, C_p T_{o_2}$$

But $$Q = \dot{m}_f \times \text{Calorific value of fuel}$$

$$\therefore \quad 44000 \times \dot{m}_f = (\dot{m}_i + \dot{m}_f) \times 1.005 \times 1403 - \dot{m}_i \times 1.005 \times 502.7$$

$$\therefore \quad \dot{m}_i \times 1.005 \times (1403 - 502.7) = \dot{m}_f\,(44000 - 1.005 \times 1403)$$

$$\therefore \quad \frac{\dot{m}_i}{\dot{m}_f} = 47.3$$

(i) ∴ Air-fuel ratio is 47.3.

(ii) $$P_{o_3} = P_{o_2} - 0.17$$

But $$P_{o_2} = 5.8\,P_{o_1} = 5.8 \times 1.073 = 6.22 \text{ bar}$$

∴ $$P_{o_3} = 6.22 - 0.17 = 6.05 \text{ bar}$$

Assuming the power required to drive the compressor is developed by the turbine only, then

$$C_p\left(T_{o_2} - T_{o_1}\right) = C_p\left(T_{o_3} - T_{o_4}\right)$$

∴ $$T_{o_4} = -\left(T_{o_2} - T_{o_1}\right) + T_{o_3}$$

$$= 1403 - (502.7 - 291.3)$$

∴ $$T_{o_4} = 1191.6 \text{ K}$$

Turbine efficiency, $$\eta_t = \frac{T_{o_3} - T_{o_4}}{T_{o_3} - T'_{o_4}}$$

∴ $$T'_{o_4} = -\frac{T_{o_3} - T_{o_4}}{\eta_t} + T_{o_3}$$

$$= 1403 - \frac{1403 - 1191.6}{0.85}$$

∴ $$T'_{o_4} = 1154.3 \text{ K}$$

But $$\left(\frac{T_{o_3}}{T'_{o_4}}\right) = \left(\frac{P_{o_3}}{P_{o_4}}\right)^{\frac{\gamma-1}{\gamma}}$$

∴ $$P_{o_4} = P_{o_3}\left(\frac{T'_{o_4}}{T_{o_3}}\right)^{\frac{\gamma}{\gamma-1}}$$

$$= 6.05 \times \left(\frac{1154.3}{1403}\right)^{3.5}$$

$$P_{o_4} = 3.06 \text{ bar}$$

and $$\frac{T_{o_4}}{T'_{o_5}} = \left(\frac{P_{o_4}}{P_{o_5}}\right)^{\frac{\gamma-1}{\gamma}}$$

∴ $$T'_{o_5} = \frac{T_{o_4}}{\left(\frac{P_{o_4}}{P_{o_5}}\right)^{\frac{\gamma-1}{\gamma}}}$$

$$= \frac{1191.6}{\left(\dfrac{3.06}{0.78}\right)^{0.286}}$$

$$\therefore \quad T_{o_5}' = 806 \text{ K}$$

Nozzle efficiency,

$$\eta_n = \frac{T_{o_4} - T_{o_5}}{T_{o_4} - T_{o_5}'}$$

$$\therefore \quad T_{o_5} = T_{o_4} - \frac{T_{o_4} - T_{o_5}'}{\eta_n}$$

$$= 1191.6 - \frac{1191.6 - 806}{0.92}$$

$$\therefore \quad T_{o_5} = 772.5 \text{ K}$$

Velocity of jet at exit of nozzle,

$$V_e = \sqrt{2C_p\left(T_{o_4} - T_{o_5}\right)}$$

$$= \sqrt{2 \times 1050 \times (1191.6 - 772.5)}$$

$$\therefore \quad V_e = 917.8 \text{ m/s}$$

$$\text{Specific thrust} = \left(1 + \dot{m}_f\right)V_e$$

$$= \left(1 + \frac{1}{47.3}\right) \times 917.8$$

$$= \mathbf{937.2 \text{ N/kg of air/s}}$$

(iii) Volume of air flowing through diffuser,

$$V_a = \text{Area} \times \text{Velocity}$$

$$= 0.14 \times 216$$

$$\therefore \quad V_a = 30.24 \text{ m}^3\text{/s}$$

Mass flow rate of air,

$$\dot{m}_a = \frac{P_a V_a}{RT_a}$$

$$= \frac{0.8 \times 30.24 \times 10^2}{0.287 \times 265.5}$$

$$\therefore \quad \dot{m}_a = 31.75 \text{ kg/s}$$

$$\therefore \quad \text{Total thrust} = \text{Specific thrust} \times \dot{m}_a$$

$$= 937.2 \times 31.75$$

$$= \mathbf{29756.1 \text{ N}}$$

■■■

THEORY QUESTIONS

35.1. Compare air breathing engines and rocket engines.

35.2. How rocket engines are classified?

35.3. Write a note on 'rocket propellants'.

35.4. Explain 'solar rocket propulsion'.

35.5. What is staging of rockets? What are the advantages and disadvantages of staging?

35.6. Give applications of rocket engines.

PROBLEM FOR PRACTICE

35.1. What is meant by thrust ? Derive the thrust equation for a general propulsion system.

35.2. Explain clearly the various efficiencies associated with a propulsion device.

35.3. What do you mean by thrust augmentation ? Explain how it is effected.

35.4. What is meant by jet propulsion ? What are the basic differences between jet propulsion cycle and shaft power cycle ?

35.5. A naval air-craft is powered by a turbojet engine, with provision for flap blowing. When landing at 55 m/s, 15% of the compressor delivery is bled off for flap blowing and it can be assumed to be discharged perpendicular to the direction of flight. If a propelling nozzle of area 0.13 m^2 is used, calculate the net thrust during landing. Given that the engine operating conditions are as follows:

Compressor pressure ratio	= 9.0
Compressor isentropic efficiency	= 0.82
Turbine inlet temperature	= 1275 K
Turbine isentropic efficiency	= 0.87
Combustion pressure loss	= 0.45 bar
Nozzle isentropic efficiency	= 0.95
Mechanical efficiency	= 0.98
Ambient conditions	= 1 bar, 288 K

The ram pressure and temperature rise can be regarded as negligible.

[*Ans.* 20794 N]

35.6. A turbojet engine flying at a speed of 960 km/hr consumes air at the rate of 54.5 kg/s. Calculate:

(i) Exit velocity of jet when the enthalpy change for the nozzle is 200 kJ/kg and velocity coefficient is 0.97.

(ii) Fuel flow rate in kg/s when air-fuel ratio is 75 : 1.

(iii) Thrust specific fuel consumption.

(iv) Thermal efficiency of the plant when the combustion efficiency is 93% and calorific value of the fuel is 45000 kJ/kg.

(v) Propulsive power.

(vi) Propulsive efficiency and

(vii) Overall efficiency.

[*Ans.* 613.5 m/s, 0.727 kg/s, 4.3×10^{-5} kg/N/s, 25.4%, 8318 kW, 60.6%,16.6%]

35.7. A turbo-jet has a speed of 750 km/hr while flying at an altitude of 10000 m. The propulsive efficiency of the jet is 50% and overall efficiency of the turbine plant is 16%. The density of air at 10000 m altitude is 0.173 kg/m^3. The drag on the plant is 6250 N. The calorific value of the fuel is 48000 kJ/kg. Calculate

(i) Absolute velocity of the jet

(ii) Volume of air compressed/min

(iii) Diameter of the jet

(iv) Power output in kW and

(v) Air-fuel ratio.

[*Ans.* 417.3 m/s, 5194 m^3/min, 415 mm, 2500 kW, 46%]

IMPORTANT POINTS TO BE REMEMBERED

1. Thrust (T)

 (a) Specific thrust $T = (V_e - V_a)$ N/kg with m_f = Neglected

$$= [V_e (m_f + M_a) - V_a m] \quad m_f = \text{kg/kg air}$$

 (b) *Thrust power* (*TP*)

$$TP = V_a [V_e (m_f + m_a) - V_a m_a]$$

$$= V_a m_a [V_e - V_a] \text{ when } m_f \text{ neglected}$$

2. *Propulsive power* (*PP*)

$$PP = \left(\frac{V_e^2}{2} - \frac{V_a^2}{2}\right) \quad \text{for } [m_a = 1, \quad m_f \text{ neglected}]$$

$$= \left[(1 + m_f)\frac{V_e^2}{2} - \frac{V_a^2}{2}\right] \quad m_f \text{ considered}$$

3. *Thermal efficiency*

$$\eta_{th} = \frac{PP}{m_f \times \text{cal. val.}}$$

$$= \frac{1}{2}\frac{(V_e^2 - V_a^2)}{m_f \times \text{cal. val.}} \quad m_f \text{ is for 1 kg air}$$

$$= \frac{\frac{1}{2}\left[(1 + m_f)V_e^2 - V_a^2\right]}{m_f \times \text{C.V.}}$$

4. *Propulsive efficiency*

$$PE = \eta_P = \frac{\text{Thrust power}}{PP}$$

$$= \frac{V_a (V_e - V_a)}{\frac{1}{2}(V_e^2 - V_a^2)} \quad m_f \text{ neglected}$$

$$= \frac{2}{V_e + V_a}$$

This is maximum when $V_e = V_a$.
Then the system does not work

For maximum thrust power,

$$\frac{V_a}{V_e} = 0.5$$

and $$PP = 66.7\%$$

5. *Overall efficiency*(η_o)

$$\eta_o = \eta_{th} \times \eta_P$$

If transmission efficiency is taken into consideration then

$$\eta_o = \eta_{th} \times \eta_{tr} \times \eta_P$$

6. *Thrust specific fuel consumption*

$$TSFC = \frac{3600F}{F/m_a} \qquad f = \frac{\text{Fuel}}{\text{Air}} \text{ ratio}$$

$$F = \text{Thrust}$$

PART–7

CHAPTER 36

Refrigeration

36.1 INTRODUCTION

The term *refrigeration* is used to denote the maintenance of a system or body at a lower temperature than of its *surroundings*. The system maintained at the lower temperature is known as *refrigerated system*.

A machine that produces temperature less than atmospheric temperature is called a *refrigerator*.

Generally any reversible heat engine will act as refrigerator if it is run in the reversed direction by means of external power.

36.2 METHODS OF REFRIGERATION

There are different methods of refrigeration. They are

1. Refrigeration by non-cyclic methods
   (a) Refrigeration by using ice or snow
   (b) Refrigeration by evaporation
2. Refrigeration by cyclic methods or by means of mechanical or artificial systems as
   (a) by using non-condensable gases such as air, CO_2 and SO_2
   (b) by using condensable vapours
      (i) vapour compression refrigeration system
      (ii) vapour absorption refrigeration system
3. Other methods of refrigeration.
   (a) Magnetic refrigeration system
   (b) Vortex tube system
   (c) Thermo-electric refrigeration system
   (d) Steam-jet refrigeration system
   (e) Ultra-sound refrigeration.

36.2.1 Natural Methods

(a) By using ice or snow. The natural methods include the utilization of ice or snow produced naturally in cold climate. Ice melts at 0°C, so when it is placed in a space or system warmer than 0°C heat flows into ice and the space is cooled. The ice in turn melts into water by absorbing its latent heat at the rate of 335 kJ/kg. The river water, if it is quite cold, may also be used for cooling certain systems, but the removal of heat will not be as fast and satisfactory as in the case when ice or snow is used.

The natural methods of refrigeration were actually used in early days. With the present developments in science and technology and the subsequent rise in the standard of leaving the refrigeration requirements have become so large that the natural methods have become inadequate and therefore obsolete.

When ice is to be used for refrigeration, winter ice can be stored by packing it in straw and dried weeds, for use in the summer months.

(b) Evaporative cooling process or Adiabatic cooling process. In the old days, it was observed by the people that evaporation of water naturally causes cooling effect and hence they used to put their wine or other liquids into porus earthenware containers placed in the open atmosphere such as roof.

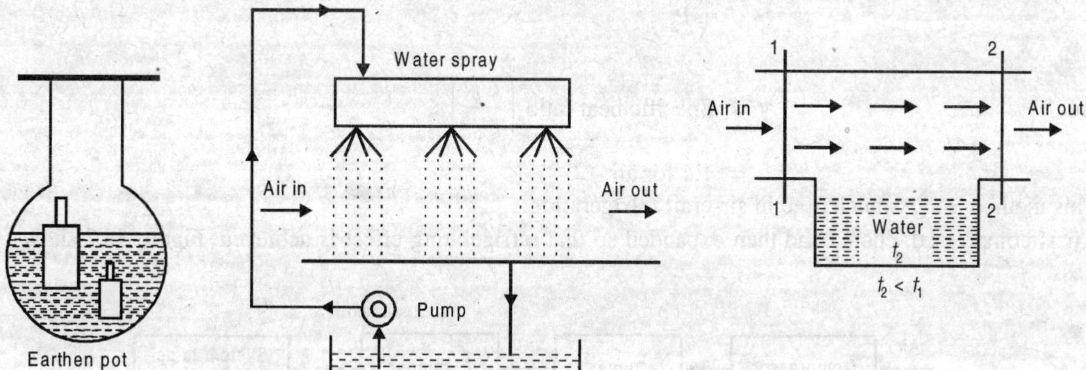

Fig. 36.1. *Examples of evaporative cooling*

We have every day experience that water kept in an earthen pot becomes cooled because of the evaporation of water absorbing latent heat of evaporation from water in the pot and coming out through the pores. This effect is experienced very well in summer.

Air flowing over the surface of water whose temperature is less than the incoming air, then the air is cooled because of the evaporation during which latent heat is taken from air and water sump.

Same effect is obtained when air is flowing across the spray of water as shown in Fig. 36.1. The cooling is taking place without exchanging the heat with the surrounding and this type of the cooling is called as *Adiabatic cooling*.

Evaporation of sweat on our body thereby producing cooling effect is the best example of evaporative cooling.

36.2.2 Mechanical Refrigeration

Artificial or *mechanical* methods of producing *cold* have been used since 1850. Methods of mechanical refrigeration include the use of non-condensable gases and condensable gases or vapours. When non-condensable gases are used, refrigeration can be achieved by two methods such as

(a) reversible or isentropic expansion of gases against the restraining force such as a piston-cylinder mechanism.

(b) irreversible adiabatic expansion of gases from the high pressure to a low pressure through a very minute orifice or similar restriction.

(a) Isentropic expansion of gases. If we expand high-pressure air or gas in a piston-cylinder mechanism, then the temperature of gas decreases depending on the extend of expansion and the temperature of gas becomes less than atmosphere. This cold gas can be used to cool the object or the system.

We know that if,

T_1 = Initial temperature of gas (K) before expansion

T_2 = Final temperature of gas (K) after expansion
P_1 = Pressure of gas before expansion (bar)
P_2 = Pressure of gas after expansion (bar)

then,

$$\frac{T_1}{T_2} = \left(\frac{P_1}{P_2}\right)^{\frac{\gamma-1}{\gamma}}$$

and

$$T_2 = \frac{T_1}{\left(\frac{P_1}{P_2}\right)^{\frac{\gamma-1}{\gamma}}}$$

where γ = Specific heat ratio $\left(\frac{C_P}{C_V}\right)$

= 1.4 for air

This method is very much used in air-craft refrigeration.

Air is compressed, cooled and then expanded so that refrigerating effect is achieved. Figure 36.2 shows this process.

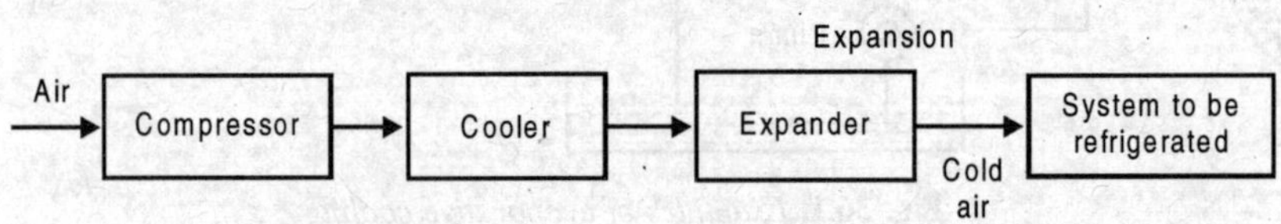

Fig. 36.2. *Refrigeration by expansion of air*

Advantage of such a metbod is that the gas (air) is available free but the major disadvantage will be the lubrication difficulties for the piston cylinder mechanism.

(b) Refrigeration by irreversible adiabatic expansion of gases (Throttling) or Joule–Thomson Effect. Throttling is an irreversible process in which a fluid, flowing across a restriction, undergoes a drop in total pressure. Such a process occurs in the flow through a porous plug, a partially closed valve, or a small orifice. Joule and Thomson performed the basic throttling experiments in the period 1852–62, and their experiments clarified the process and led to use of throttling as a method for determining certain properties of gaseous substances.

The apparatus used in the Joule–Thomson experiment is shown in Fig. 36.3.

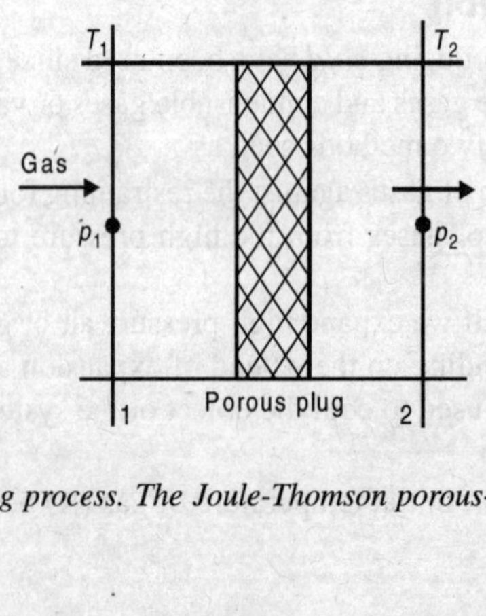

Fig. 36.3. *Throttling process. The Joule-Thomson porous-plug experiment*

A steady stream of gas flows through a porous plug contained in a horizontal tube. This system is open, is thermally insulated ($Q = 0$) and does not exchange work with its environment ($W = 0$). At sections 1 and 2 both the temperature and the pressure are measured. If the kinetic energy does not change significantly as the fluid passes through the porous plug, the steady flow energy equation reduces to

$$h_1 = h_2 \qquad \ldots(1)$$

Hence, in an adiabatic throttling process the enthalpy remains constant. When a series of Joule–Thomson experiments is performed at the same initial temperature t_1 and pressure p_1 but with different downstream pressure, it is found that the temperature t_2 changes. The results from these experiments can be plotted as a constant enthalpy curve on *T–P* plane. If several different rates of flow are established at each condition of t_1 and p_1, a series of isenthalpic curves is obtained. These are shown in Fig. 36.4. The maximum point on each curve is called the *inversion point* and the locus of the inversion points is called the inversion curve. The slope of an isenthalpic curve is called the Joule–Thomson coefficient μ_h and is expressed by

$$\mu_h = \left(\frac{\partial T}{\partial p}\right)_h \qquad \ldots(2)$$

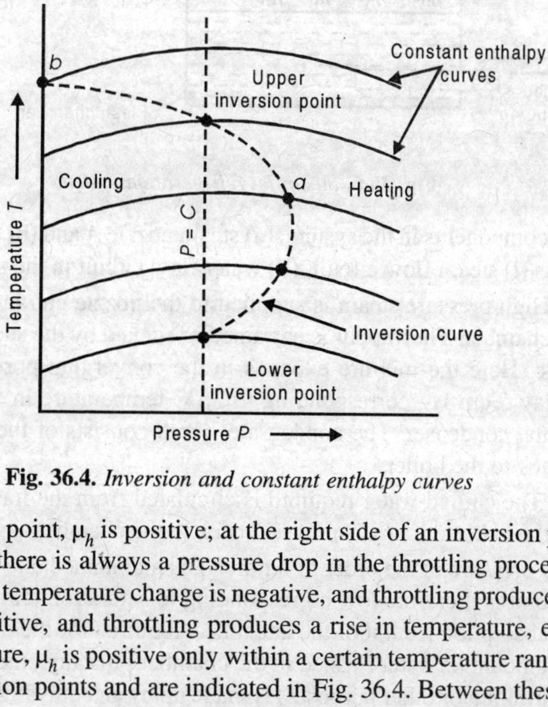

Fig. 36.4. *Inversion and constant enthalpy curves*

At the left side of an inversion point, μ_h is positive; at the right side of an inversion point, μ_h is negative; at the inversion point μ_h is zero. Since there is always a pressure drop in the throttling process, ΔP is always negative. Therefore when μ_h is positive, the temperature change is negative, and throttling produces cooling. Similarly, when μ_h is negative, the change is positive, and throttling produces a rise in temperature, even though the process is adiabatic. At any particular pressure, μ_h is positive only within a certain temperature range. These temperatures are called the upper and lower inversion points and are indicated in Fig. 36.4. Between these two temperatures, throttling causes a drop in the temperature of the gas; outside of this temperature range, throttling results in a temperature rise. Above a certain pressure, throttling can cause only a cooling effect. This is point *a* in Fig. 36.4. Similarly, above a certain temperature throttling can cause only a heating effect. This is point *b* in Fig. 36.4.

In case of a perfect gas, enthalpy is a function of temperature alone, and therefore the temperature of the gas remains constant so that $\mu_h = 0$. In case of a real gas the temperature generally does change with pressure and μ_h is a function of both temperature and pressure.

36.2.3 Steam-Jet Refrigeration

This system is one of the methods employed by chemical agent machines for achieving mechanical refrigeration. Evidently water is used here as refrigerant.

Steam–Jet system is primarily used in installations where chilled water is required or the systems like cold storage where the temperatures required are more than 5°C as water starts changing its volume at 4°C and starts freezing at 0°C. The principle of flash cooling is employed in achieving the objective. A typical refrigeration system using steam-jet is shown diagrammatically in Fig. 36.5.

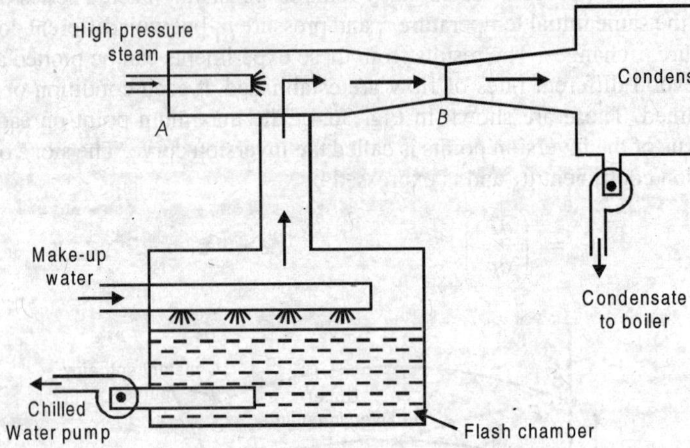

Fig. 36.5. *Steam-jet refrigeration system*

There are two important components in the system, (a) steam nozzle *A* and (b) the flash chamber. Also, there are two distinct circuits of fluids (i) steam flow circuit, (ii) water flow circuit in the system.

(i) Steam flow circuit. High pressure steam is supplied to the nozzle entrance at *A*. Here it takes with it any vapour flashed in the flash chamber. The vapour is entrained or sucked by the suction created by the high velocity of steam flow in the nozzle. Here the mixture expands in the converging portion upto the throat *B*. It is then compressed to a pressure say 5 cm Hg corresponding to 36°C temperature in the diverging diffuser. The compressed fluid passes on to the condenser. The condensate which consists of the flashed vapour and steam from supply source is pumped back to the boiler.

(ii) Water flow circuit. The chilled water required is circulated from the flash chamber, by a pump on to the point of consumption, and the balance if any is returned back to the chamber as spray.

In the *flash chamber* low pressure of the order of 6 mm Hg is maintained due to the suction of vapour by steam flow at nozzle. The flash point is therefore low corresponding to saturation temperature at that pressure say 5°C. The water is chilled due to absorption from it of the latent heat required for the evaporation of a certain fraction of water during flashing. The amount of water flashed in the chamber plus any other quantity of chilled water drawn from the chamber (say for drinking or other purposes) is made up by an equivalent amount of fresh supply through the spray.

It will be noted that for obtaining water at 5°C, with a condensing temperature say 35°C, the operating pressures are 6 mm Hg and 5 mm Hg respectively, which are extremely low values. The compression ratio of approximately 8 also approaches the limit of efficiency operation. Note, if temperatures below 0°C are required, antifreez should be added to water.

36.2.4 Refrigeration by using Liquid-Gases

We know that a liquid can be vaporized at any desired temperature by changing its pressure. Further, heat is required to be added to the liquid during vaporization when the liquid phase changes to the gaseous phase. Accordingly, a vapourizing liquid can be used to produce refrigeration at any temperature. For instance, at a pressure of about 1 atm. ammonia may be made to boil at –33°C or at a pressure of about 5 atm. Freon -22 boils at 0°C.

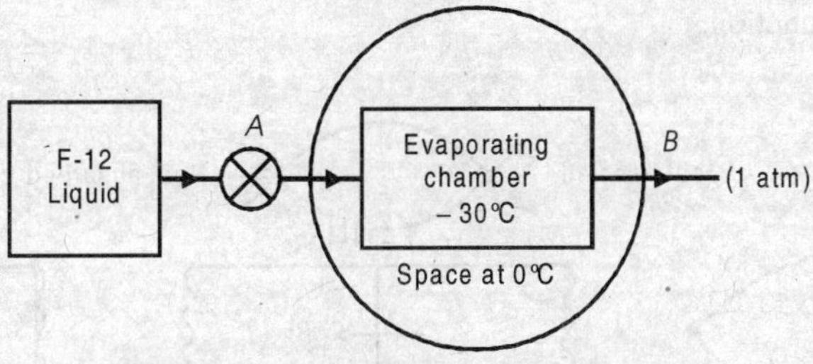

Fig. 36.6. *Refrigeration by using liquid-gas*

Hence in an arrangement in which a system containing some refrigerant in liquid form at a certain pressure is exposed to another system at temperature greater than refrigerating temperature.

A simple scheme of such an arrangement is shown in Fig. 36.5. Liquid Freon F-12 is supplied to the evaporating chamber through a valve at a pressure of about 1 atm. Since the boiling temperature of F-12 at this pressure is – 30°C, heat flows from the surround space at 0°C and makes the F-12 boil. Thus the space will be cooled as long as there is a supply of liquid F-12 to the evaporating chamber.

The arrangement of Fig. 36.6, however, suffers from two important drawbacks.

(a) The cost of replacing Freon-12 (an expensive refrigerant) will be prohibitive as the evaporated vapour leaks out to the atmosphere.

(b) If the system were to use some other refrigerant like ammonia, it may become a hazard to life due to its discharge to atmosphere since it is a highly toxic and irritating fluid.

To eliminate these drawbacks, it will be necessary to let the refrigerant fluid work in a *closed loop* and be used again and again.

36.2.5 Thermo-Electric Refrigeration

Thermoelectric effects refer to phenomenon involving the interchange of heat and electrical energy. One familiar example is the heating effect due to the flow of an electric current in a resistor. The electric power dissipation depends on the current flow I and the potential V according to the equation $P = VI$ or $P = VI = I^2R$. This type of energy conversion is irreversible, although some thermoelectric effects are reversible.

There are three thermoelectric effects. They are: (i) The Seebeck effect, (ii) Peltier effect, and (iii) Thomson effect.

(i) The Seebeck Effect

In 1833 Seebeck held the junctions of two dissimilar electrical conductors at different temperatures and noted that an open-circuit voltage was generated that was proportional to the difference in temperature between the two junctions. A familiar example which illustrates the Seebeck effect is the ***thermocouple***. The heat supplied at one of the junctions is converted into electrical energy, which is then measured.

(ii) The Peltier Effect

In the Peltier effect, which is the inverse of the Seebeck effect, an electric current is sent through a junction of two dissimilar materials A and B, and this results in either a heating or cooling of the junction. The rate of heating or cooling is proportional to the current flow, and is given by the equation

$$\dot{Q} = \pi_{AB} I$$

where the coefficient π_{AB} is a function of temperature called the ***Peltier coefficient***. The subscript AB indicates that the Peltier coefficient also depends on the two materials A and B. A typical unit of π_{AB} is the volt. If current flows in the direction of the potential drop characteristic of the junction, then the temperature of the junction decreases, and a refrigerating effect occurs in the space surrounding the junction. If current flows in the direction of the junction's potential rise, the temperature of the junction increases and heat must be transferred from the junction to maintain constant temperature at the junction.

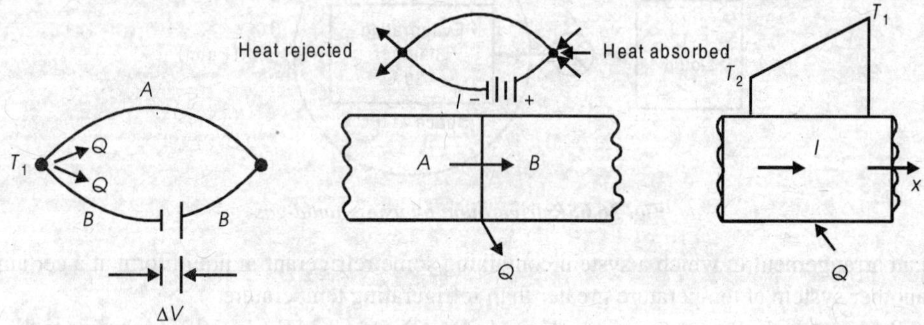

Fig. 36.7 *Thermoelectric effects*

(*iii*) The Thomson Effect

When an electric current flows through a homogeneous material, heat is exchanged between the environment and the material, and the rate of transfer of heat depends on both the quantity of current that flows through the material and the temperature gradient that exists in the material. The Thomson coefficient τ is defined as

$$\tau_A = \frac{d\dot{Q}/dx}{I\,(dT/dx)}$$

where $d\dot{Q}/dx$ is the rate of heat interaction per unit length of the conductor and dT/dx is the temperature gradient along the length of the conductor. Like the Seebeck and Peltier effects, the Thomson effect is reversible; on the other hand, it involves only a single substance.

The Seebeck, Peltier and Thomson effects occur in both metals and semiconductors, but *these effects are more pronounced in semiconductors*. The three effects are shown in Fig. 36.7.

If Seebeck would probably have tried reversing the direction of flow of electrons in the thermoelectric circuit (by externally applying a potential difference in the reverse direction), he would have found a refrigerating effect. But this very important effect (refrigeration) was discovered by Mr. Jean Charles Athanase Peltier, in 1834. He observed during his experiments that when a small current was passed through the junction of two dissimilar wires, the junction was cooled as shown in Fig. 36.7.

Peltier effect is the basis for *thermoelectric refrigeration*. A practical thermoelectric refrigeration circuit using semiconductor materials is shown in Fig. 36.8.

Thermoelectric refrigerators cannot compete with the vapour compression refrigeration systems because of its low coefficient of performance, small refrigerating effect. They are preferred in some specific applications because of their small size, simplicity, quietness and reliability.

Mention may be made here of certain non-conventional refrigeration system which are used for specific applications only and will not be discussed here.

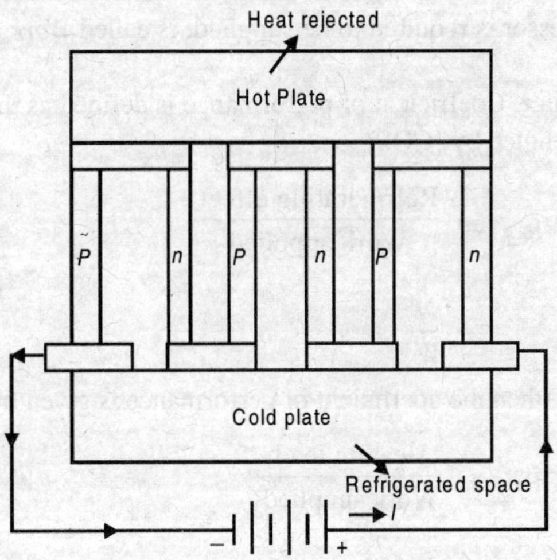

Fig. 36.8. *Thermoelectric refrigerator*

Non-conventional refrigeration systems are,

(i) Magnetic refrigeration system

(ii) Vortex-tube system

(iii) Pulse jet refrigeration system

(iv) Ultrasound refrigeration system

36.3 AIR REFRIGERATION SYSTEM

The virtues of air refrigeration, including cheapness, availability and safety of the refrigerant used made it quite appealing to the early inventors of refrigeration machines. Air refrigeration dates back to the middle of nineteenth century, but it was not until 1877 that a reliable unit was developed. It was used mostly for cold-storage of meat on ships. However, the equipment was bulky and inefficient. The bulkiness was due to the reciprocating compressor and expanders and low speed of rotation. Rapid developments of the more efficient and less bulky vapour compression systems made air-refrigeration rather short-lived.

Recent developments in the rotory compressors and gas turbines have renewed interest in air refrigeration systems. Rotory machines caused great reduction in weight per unit of refrigeration.

36.3.1 Terms and Definitions used in Refrigeration

For further studies of refrigeration following terms and definitions will be of importance.

(i) Refrigeration. Refrigeration is defined as the process (cyclic or non-cyclic) by means of which temperatures of the particular body or space, less than that of surroundings or atmosphere, are produced.

(ii) Refrigerants. The working substance used in the refrigeration system is called Refrigerants. It may be air, gas or chemical compounds.

(iii) Mechanical Refrigeration. When mechanical components such as compressors and electric motor are used for refrigeration, the system is called as *Mechanical refrigeration system.*

(iv) Refrigerating Effect. The amount of heat extracted from a body or space for cooling purpose is called *Refrigeration Effect*. It is denoted by *N*.

(v) Heating Load. The amount of heat energy supplied to the body or space, for heating purposes, by a heat pump is called *Heating load* for the system.

(vi) Work or Power Supplied. To run the system, energy in the form of Heat energy directly or to the electric motor to drive the compressor is required to be supplied, is called *Work or Power Supplied.* Generally it is denoted by letter *W*.

(vii) Coefficient of Performance. Coefficient of performance is defined as the ratio of the refrigerating effect to the work supplied. It is denoted by $[\text{COP}]_{\text{Ref.}}$.

$$\therefore \quad (\text{COP})_{\text{Ref.}} = \frac{\text{Refrigerating effect}}{\text{Work supplied}}$$

$$= \frac{N}{W}$$

When heat pump is used, then the coefficient of performance is given by

$$(\text{COP})_H = \frac{\text{Heating load}}{\text{Work supplied}}$$

$$= \frac{N+W}{W}$$

$$= 1+\frac{N}{W}$$

$$= 1 + (\text{COP})_{\text{Ref.}}$$

COP is, therefore, the extent to which the work or energy supplied is utilized in producing the desired effect. The properties of the refrigerant should be such that it will give highest COP.

(viii) Unit of Refrigeration. (*Rating for refrigeration*) Unit of refrigeration is Tonne Refrigeration shortly written as TR.

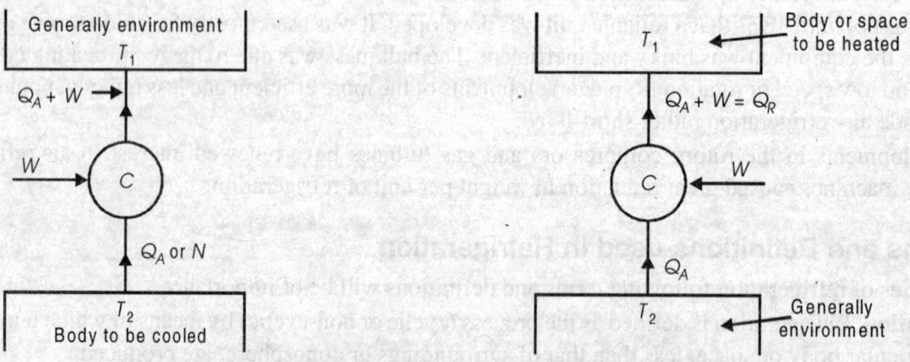

Fig. 36.9. *Schematics for refrigeration and heat pump*

One TR is defined as the heat extracted from 1 tonne water at 0°C to produce ice at 0°C in one day. One tonne in refrigeration practice is the smaller tonne and is corresponding to 2000 lbs (pounds). Latent heat of ice is taken as 335 kJ.

$$\therefore \quad 1 \text{ TR} = \frac{2000}{2.216} \times \frac{335}{24 \times 3600}$$

$$\simeq 3.516 \text{ (kJ/sec = kW)}$$

$$\simeq \mathbf{3.52\ kW}$$

(ix) Subcooling or Undercooling. In case of vapour compression cycle, vapours after compression are passed through a condenser where the heat is rejected by the refrigerant to the cooling medium. If the temperature of the condensed liquid refrigerant coming out of the condenser is less than the saturation temperature of the refrigerant liquid at the pressure of compressor outlet, the liquid is said to be *undercooled* or *subcooled*.

(x) EER (Energy Efficiency Ratio) and SEER (Seasonal Energy Efficiency Ratio). We have defined the above term $(COP)_R$ for the refrigerating machine as

$$(COP)_R = \frac{\text{Refrigerating effect}}{\text{Energy supplied}}$$

Here the units for numerator and denominator are same so that COP is a pure number.

The other way to expressing the performance of the machine is the termed Energy Efficiency Ratio (EER). At present this term is used in United States of America and is defined as the ratio of the amount of heat energy removed per watt-hour of electrical energy consumed. It is noted here that the two terms in the numerator and denominator are same as COP but their units are different.

$$\therefore \qquad \text{EER} = \frac{\text{Cooling capacity in kJ/hour}}{\text{Power input in Watt}}$$

$$= \frac{\text{kJ}}{\text{Wh}}$$

$$\mathbf{EER} = \frac{\mathbf{kJ}}{\mathbf{Wh}}$$

The anticipated performance of a refrigerating device, over an average season is sometimes called the *Seasonal Energy Efficiency Ratio* (SEER) and it is also expressed in kJ/Wh

$$\therefore \qquad \mathbf{SEER} = \frac{\mathbf{kJ}}{\mathbf{Wh}}.$$

36.3.2 Refrigeration Load

As the temperature of the body or space is less than that of the surroundings, it is but natural that heat will flow into the space by the combined effect of conduction, convection and radiation. In addition to this heat flow, there is always some air leakage and ventilation and hence additional heat removal is required. Products when placed fresh, then the temperature of the products or bodies is more than the surroundings. And, therefore, to maintain the temperature constant in the refrigerator, all heat coming in, should be removed. This total heat to be removed is called the *refrigeration load*. This is the actual *refrigerating effect*.

Hence the heat sources in a refrigerator are as follows:

1. Heat transmission/transfer
2. Product heat
3. Air leakage and ventilation
4. Miscellaneous sources

36.3.3 Heat Pump

A heat pump is a thermodynamic system operating in a cycle which removes heat from a low temperature body and delivers heat to a high temperature body. In accomplishing this, the heat pump receives external energy in the form of work. The system or working substance undergoes a series of processes which constitute the heat pump cycle.

The heat pump may be used as a refrigerator where the primary function is the transfer of heat from a low temperature system. In a heating system, utilising a heat pump, the primary function is the transfer of heat to a high temperature system.

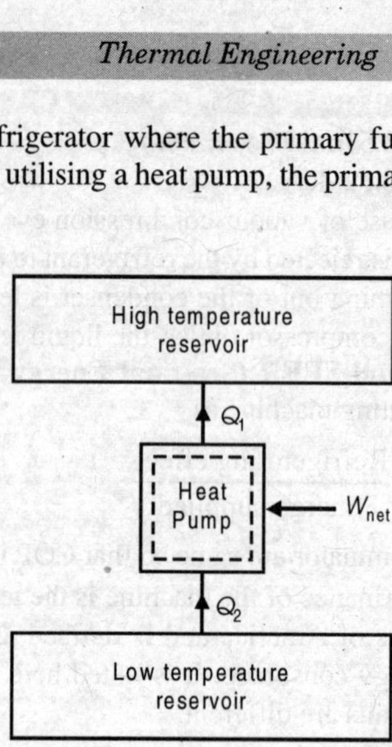

Fig. 36.10. *Principle of heat pump*

An index of performance of a heat pump is the coefficient of performance, COP.

Referring to Fig. 36.10, let

Q_1 = Heat rejected by system

Q_2 = Heat transferred to system

W_{Net} = Net work required by system

Applying the first law of thermodynamics to the cycle gives

$$\oint dQ + \oint \partial W = 0$$

or

$$Q_1 - Q_2 = W_{Net}$$

∴ COP for a heat pump is

$$(\text{COP})_{\text{Heat pump}} = \frac{Q_1}{Q_1 - Q_2}$$

$$= \frac{Q_2 + W_{Net}}{W_{Net}}$$

$$= 1 + \frac{Q_2}{W_{Net}}$$

$$= 1 + (\text{COP})$$

36.3.4 Carnot Cycle for Refrigeration

A refrigeration system removes heat from a low-temperature region and transfers heat to a high-temperature region.

The Carnot cycle can serve as the model of the ideal refrigeration cycle. Considering all refrigerators operating between the same two thermal reservoirs, the maximum coefficient of performance is attained by the Carnot refrigeration cycle.

Figure 36.11 shows the Carnot refrigeration cycle on *T–S* and *P–V* diagram. The working substance is a condensable vapour or it may be any gas or air (non-condensable). Starting at state 1 and following the arrows, the four processes of the cycle are

1–2 reversible adiabatic compression
2–3 reversible isothermal heat rejection
3–4 reversible adiabatic expansion
4–1 reversible isothermal heat absorption

During the heat rejection process (2–3), condensation takes place at a constant temperature T_h and the area *a*–3–2–*b*–*a* in the *T–S* diagram represents the heat removed Q_h. Similarly, during the heat absorption process 4–1, evaporation takes place as heat is absorbed from the surrounding space at a temperature T_c, and the area *a*–4–1–*b*–*a* in the *T–S* diagram represents the refrigeration effect Q_c. The coefficient of performance of the Carnot refrigeration cycle is given by

$$(\text{COP})_{\text{Ref.}} = \frac{Q_c}{Q_h - Q_c}$$

$$= \frac{T_c}{T_h - T_c} \quad \ldots(3)$$

It is not practical, however, to operate a refrigeration unit according to the Carnot cycle requirements. The main difficulty lies in the isothermal heat interaction processes. These processes proceed at a very low rate so that very large heat transfer surfaces are needed. Furthermore, the presence of the liquid phase of the working fluid in the compressor creates problems, for it causes severe erosion of the compressor. This difficulty is avoided in the Rankine refrigeration cycle. When air or gas are used for Carnot cycle, Coefficient of Performance is given by

$$(\text{COP})_{\text{Ref.}} = \frac{T_c}{T_h - T_c}$$

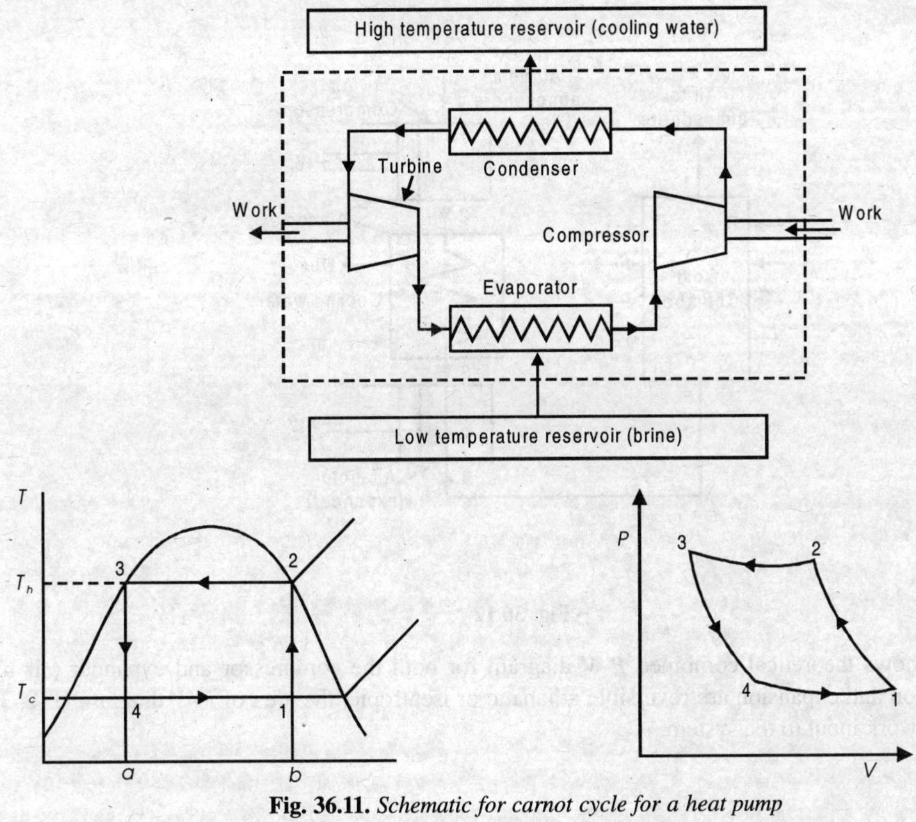

Fig. 36.11. *Schematic for carnot cycle for a heat pump*

The ideal gas refrigeration cycle operates on a *reversed Brayton* or *Joule cycle*. This reversed Joule cycle is called *Bell-Coleman Cycle*.

36.3.5 Practical Cycle for Air Refrigeration–Bell-Coleman Cycle

Modification of the ideal reversed Carnot cycle so as to make it practicable has resulted in Bell-Coleman Cycle.

The distinguishing features of this system as compared to vapour system is that the *refrigerant air* remains in the gaseous phase throughout the cycle and does not undergo liquification and evaporation as a vapour refrigeration system.

Air refrigeration cycle may be further classified as (i) Open cycle and (ii) Closed or dense air cycle.

The isothermal operations are replaced by constant pressure operations.

(i) Open Cycle. Figure 36.12 shows a flow diagram of the open air system. The name is applied because air is discharged into cold chamber at atmospheric pressure and allowed to come in direct contact with the cold body. The compressor draws the air from the cold chamber, compresses it, and delivers it to the cooler. As air flows through the cooler, heat is removed from it causing a great decrease in its volume. In the ideal case, the pressure remains constant during cooling; in the actual case there is slight pressure drop. The high pressure air is then allowed to expand and made to do work; this work will be done at the expense of its enthalpy (total heat) and there will be a big drop in its temperature. This expansion of air is carried out in the expansion cylinder (air motor). The low temperature air leaving the expansion cylinder then enters the cold chamber and abstracts heat from the bodies kept in the chamber and in turn lowers temperature in the cold chambers.

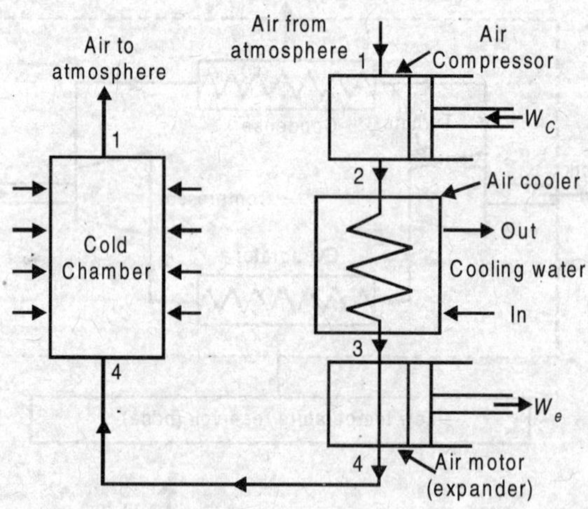

Fig. 36.12

Figure 36.13 shows theoretical combined *P–V* diagram for both the compressor and expander (air motor). Ideally, compression and expansion are reversible adiabatic or isentropic, the area of *P–V* diagram (1–2–3–4–1) represents the net work input to the system.

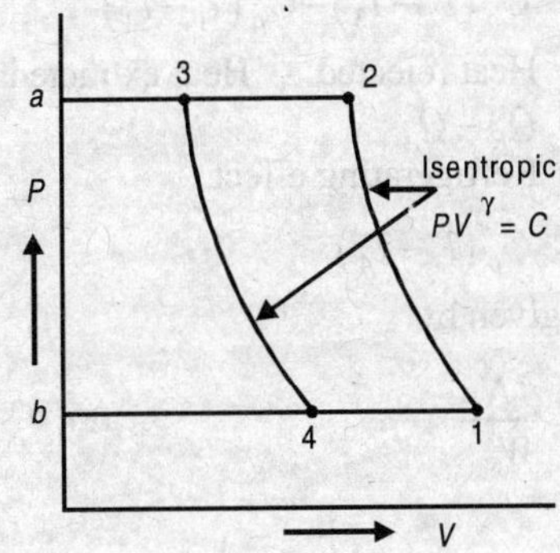

Fig. 36.13. *Combined P–V diagram for Bell-Coleman*

Bell-Coleman cycle is also shown on *T–S* diagram of Fig. 36.14.

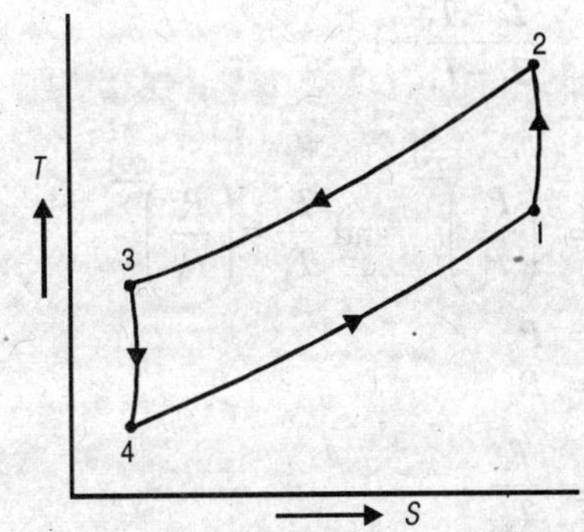

Fig. 36.14. *Bell–Coleman cycle on T–S diagram*

1–2 Isentropic compression
2–3 Constant pressure heat-rejection
3–4 Isentropic expansion in expander
4–1 Constant pressure heat-addition
Let T_1, T_2, T_3, T_4 be the absolute temperatures at the points 1, 2, 3 and 4 respectively.
From the *P–V* diagram, we get

$$\begin{aligned}
\text{Work done} &= \text{Area 1–2–3–4–1} \\
&= \text{Area (1–2–}a\text{–}b\text{–1)} - \text{Area (4–3–}a\text{–}b\text{–4)} \\
&= \frac{(P_1V_1 - P_2V_2)\gamma}{\gamma-1} - \frac{(P_3V_3 - P_4V_4)\gamma}{\gamma-1} \\
&= \frac{\gamma R}{\gamma-1}[T_2 - T_1] - \frac{\gamma R}{\gamma-1}[T_3 - T_4] \\
&= \frac{\gamma R}{\gamma-1}[T_2 - T_1 - T_3 + T_4] \\
&= C_p\left[(T_2 - T_3) - (T_1 - T_4)\right]
\end{aligned}$$

$$= C_p(T_2 - T_3) - C_p(T_1 - T_4)$$

$$= \text{Heat rejected} - \text{Heat extracted}$$

$$= Q_R - Q_A$$

$$\text{Heat extracted} = \text{Refrigerating effect}$$

$$= C_p(T_1 - T_4)$$

∴ Coefficient of performance is given by

$$\text{COP} = \frac{N}{W_{\text{Net}}}$$

$$= \frac{C_p(T_1 - T_4)}{C_p(T_2 - T_3) - C_p(T_1 - T_4)}$$

$$= \frac{1}{\frac{T_2 - T_3}{T_1 - T_4} - 1}$$

But $\frac{T_2}{T_1} = \left(\frac{P_2}{P_1}\right)^{\frac{\gamma-1}{\gamma}}$ and $\frac{T_3}{T_4} = \left(\frac{P_3}{P_4}\right)^{\frac{\gamma-1}{\gamma}}$

But $P_2 = P_3$

and $P_1 = P_4$

∴ $\frac{T_2}{T_1} = \frac{T_3}{T_4}$

∴

$$\text{COP} = \frac{1}{\frac{T_1(r_p)^{\frac{\gamma-1}{\gamma}} - T_4 \cdot r_p^{\frac{\gamma-1}{\gamma}}}{T_1 - T_4} - 1}$$

$$= \frac{1}{\frac{r_p^{\frac{\gamma-1}{\gamma}}(T_1 - T_4)}{T_1 - T_4} - 1}$$

$$= \frac{1}{r_p^{\frac{\gamma-1}{\gamma}} - 1}$$

$$= \frac{1}{\frac{T_3}{T_4} - 1}$$

$$= \frac{T_4}{T_3 - T_4}$$

$$\mathbf{COP} = \frac{\mathbf{T_4}}{\mathbf{T_3 - T_4}} \quad \ldots(5)$$

If the expansion and compression are not isentropic, then expansion and compression may be considered as polytropic and $PV^n = C$ is the law of compression and expansion.

$$\therefore \quad \text{Work done/kg} = W_{\text{Net}} = \frac{n(P_2V_2 - P_1V_1)}{n-1} - \frac{n}{n-1}(P_3V_3 - P_4V_4)$$

$$= \frac{nR}{n-1}[T_2 - T_1] - \frac{nR}{n-1}[T_3 - T_4]$$

Note here that, $W_{\text{Net}} \neq Q_R - Q_A$

but heat during compression and expansion are to be taken into consideration.

(ii) Closed or Dense Air Cycle. The basic air standard cycle is shown in Fig. 36.15. Air is contained in the system and is compressed from B by a reciprocating or rotory *compressor*. The compressed air at C is passed on to the *air cooler* where it cools down at constant pressure, to stare D.

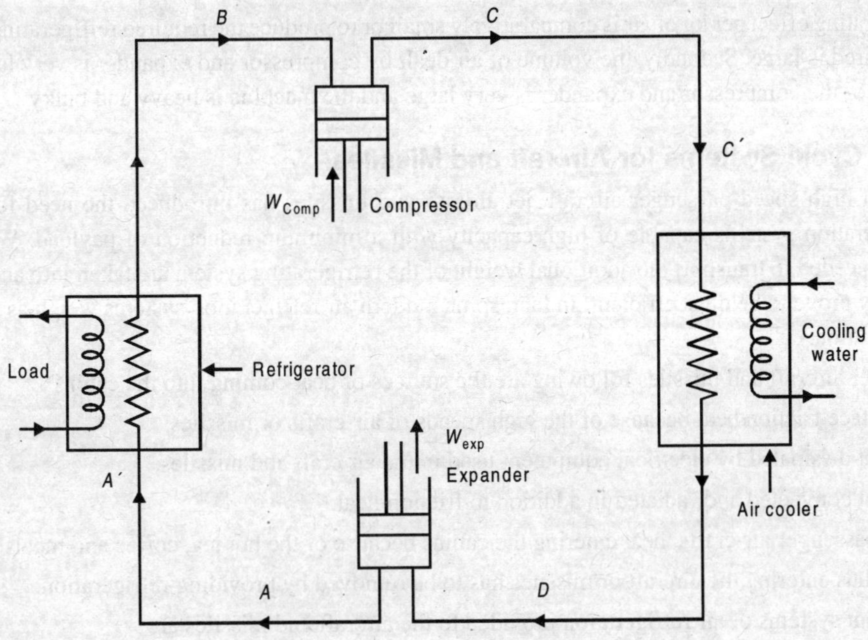

Fig. 36.15. *Dense or closed air cycle*

The high pressure gas at D is then passed on to an expansion cylinder. Here the air expands upto A, delivers work and gets cooled. The work delivered may be fed back to the compressor by mounting the compressor and the expander on the same shaft. The cooler air at A is passed through the refrigerator. The heat picked up by the air in the refrigerator causes its state to change to point B and the cycle repeats.

The cycle discussed above is known as *closed or dense air system.* If the cool air is used as such for air conditioning or other purposes and is not taken back to the compressor, then the fresh charge of air is drawn by the compressor and the system is called *Open Air System* as discussed in the previous cycle or system.

The open air system is inherently limited to operation at atmospheric pressure only whereas in the dense air system the pressure after expander may be even lower than atmospheric, hence lower temperatures may be attained.

The *P–V* and *T–S* diagrams will be same as for open cycle and the numerical analysis of the cycle or system is also same.

36.3.6 Advantages and Disadvantages of Air Refrigeration

The main advantage of air refrigeration is that the working substance or air refrigerant is available always in the atmosphere. Further it is free or no cost is involved. The open air cycle system is simple. Thus we can say that

1. Availability of air is there
2. Air is free
3. System is very simple

The disadvantages are

1. Freezing of the moisture in the air used in the system. This is particularly for open system. A deposit of snow is liable to choke up the valves and ducts carrying air.
2. Coefficient of performance (COP) of the sytem is very low.
3. Refrigerating effect per kg of air is comparatively small or to produce the required refrigerating effect mass of air required is large. Secondly, the volume of air dealt by compressor and expander is very large. Therefore, the size of the compressor and expander is very large and the machine is heavy and bulky.

36.3.7 Air Cycle Systems for Aircraft and Missiles

The advent of high speed passenger aircraft, jet aircraft and missiles has introduced the need for compact and simple refrigeration systems capable of high capacity with a minimum reduction of payload. When the power requirements needed to transport the additional weight of the refrigerating system are taken into account, air cycle system usually prove to be most efficient. In fact, some modern air refrigeration systems weigh as low as 2 kg per ton of refrigeration.

In passenger aircraft and missiles following are the sources of heat coming into the cabins

(i) Surface friction heat because of the high speeds of air-crafts or missiles.

(ii) Heat dissipated by electrical equipment used in the air craft and missiles.

(iii) Heat convected and radiated in addition to friction heat.

(iv) In passenger air-crafts, heat entering the cabins because of the hot tea, coffee and meals served.

The heat, thus entering the aircraft or missiles has to be removed by providing refrigeration.

The different systems of air refrigeration provided to the aircraft and missiles are

(i) Simple air cycle-evaporating cooling system.

(ii) Bootstrap air-cycle cooling system.

(iii) Reduced ambient air-cycle system.

The schematic diagrams of these systems are given below without elaborating much the working of these systems.

(i) Simple air Cycle-Evaporating Cooling System

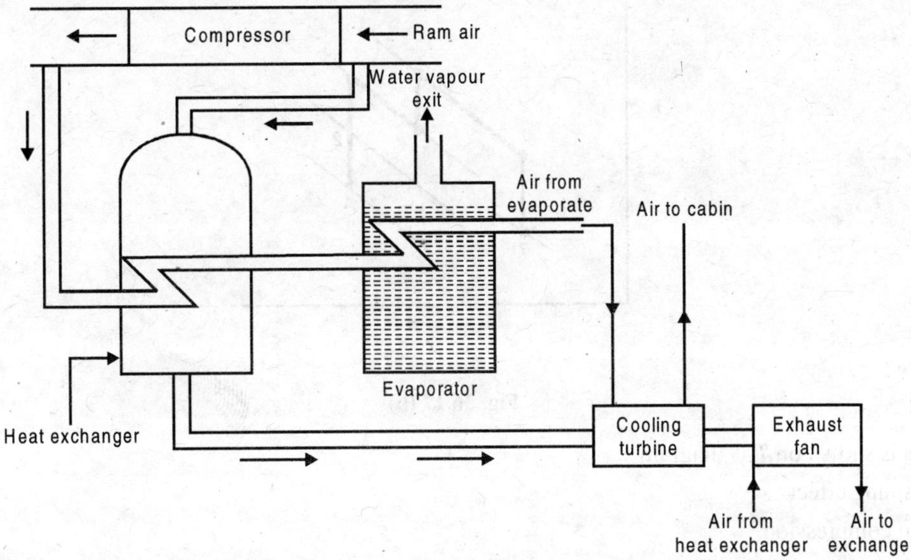

Fig. 36.16. *Simple air cycle—Evaporating cooling system*

Ram air is at lower temperature than compressed air and hence compressed air is cooled by ram air in heat exchanger. Ram air is inducted by a cooling fan. This fan is driven by the cooling turbine as shown in Fig. 36.16. Compressed air leaves the heat exchanger and is passed through the evaporator where water is evaporated and the heat required is taken from the air and is cooled. This air is expanded through a cooling turbine and this air is supplied to the cabin.

(ii) Bootstrap Air-Cycle Cooling System. This system is shown in Fig. 36.17 (a) and (b). The cycle is shown on *T–S* diagram.

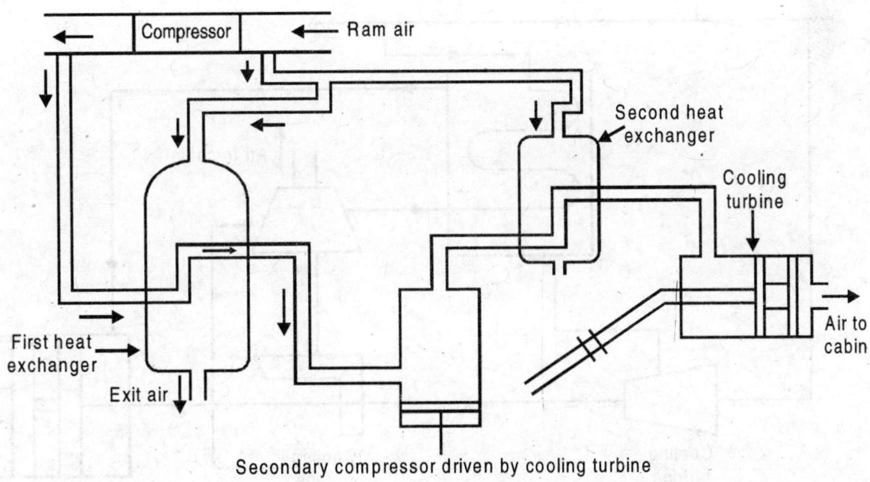

Fig. 36.17 (a) *Bootstrap air cooling cycle*

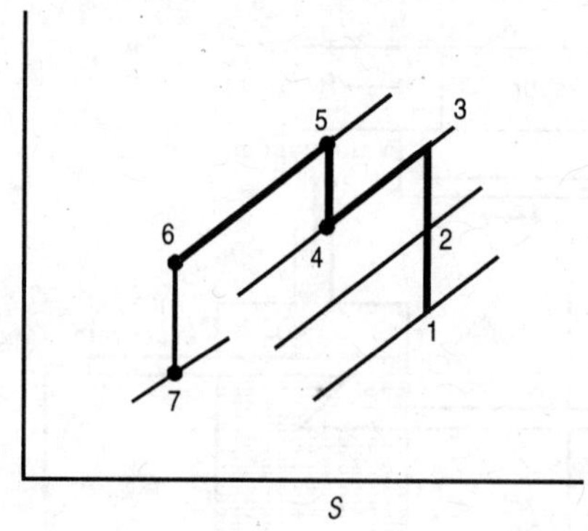

Fig. 36.17 (b)

The cycle is shown on *T–S* diagram

1–2 Ramming effect.

2–3 Main compression.

3–4 Cooling in first *H-E*.

4–5 Compression in secondary compressor.

5–6 Cooling second *HE*.

6–7 Expansion in cooling turbine.

(iii) Reduced Ambient Air-Cycle Cooling System. This system is diagrammatically shown in Fig. 36.18 (a) and *T–S* diagram of Fig. 36.18 (b).

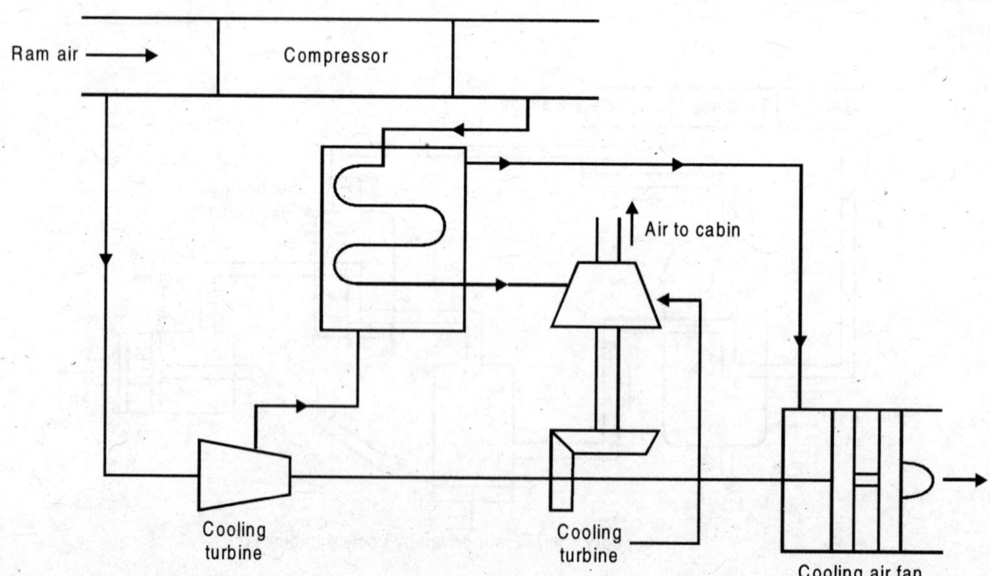

(a)

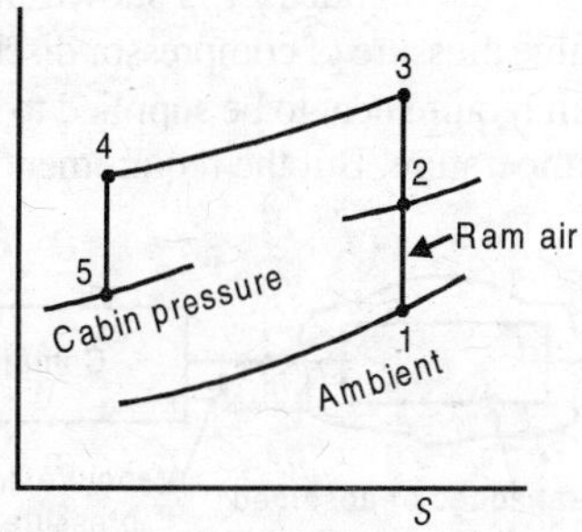

Fig. 36.18. *Reduced ambient air cycle*

1–2 Ramming effect.
2–3 Compression in main compressor.
3–4 Cooling in heat exchanger.
4–5 Expansion in the cooling turbine.

36.4 SIMPLE VAPOUR COMPRESSION REFRIGERATION SYSTEM

A refrigerator may be defined as a machine for producing cold. Refrigerators are used for the manufacture of ice and for the cooling of storage chambers in which perishable food is stored. Theoretically, any reversible heat engine would act as a refrigerator when run in a reversed direction by means of external power; the engine then becomes a heat pump, which pumps heat from a cold body and delivers it to a hot body. The cold body usually consists of a tank of brine which is kept at a low temperature by the refrigerator; the cold brine is circulated through pipes, around the storage chamber which is to be cooled. The hot body is a supply of water which abstracts the heat from the working substance of the refrigerator. The net action of the refrigerator is to pump heat from the brine and to deliver this heat to the cooling water. By doing this it is transferring heat from a cold body to a hot body; this operation, according to the *Second Law of Thermodynamics*, can only be performed by the aid of external work. Hence, it is necessary to supply power from an external source in order to drive the refrigerator.

We have seen in Sec. 35.2.4 that there must be a closed loop for the refrigerant (condensing) so that no fresh supply of the refrigerant is required and thus avoiding the hazard to surroundings because of the refrigerant coming out of the system. In this arrangement, heat addition to the refrigerant i.e. evaporation and heat rejection by condensation of the refrigerant take place.

Depending upon the equipment employed the systems of refrigeration may be broadly classified into three types :

(a) Vapour compression system.
(b) Vapour absorption system.
(c) Steam-jet refrigeration system.

36.4.1 Vapour Compression System

For a closed cycle of refrigeration employing the condensable refrigerant vapour, the following processes are required.

(i) Compression of the vapour, thereby increasing pressure.
(ii) Condensing these vapours and rejecting heating to the cooling medium (usually water or atmospheric air).
(iii) Expanding the condensed liquid refrigerant thereby lowering the pressure and corresponding saturation temperature.
(iv) Evaporating the liquid refrigerant thereby absorbing heat from the body or space to be cooled or refrigerated.

It is due to this requirements of compression that the system is called *Vapour Compression System* and the cycle of operation is called *Vapour Compression Cycle of Refrigeration*.

This cycle, incorporating the compressor and condenser is shown in Fig. 36.19. Here the liquid at state *D*, the discharge of the condenser, is still at the same pressure of compressor discharge. If it were allowed to boil as it is, it will do so at the saturation temperature and will require heat to be supplied to it. Heat could possibly flow only from some temperature higher than this saturation temperature. But the requirement is to achieve a low temperature.

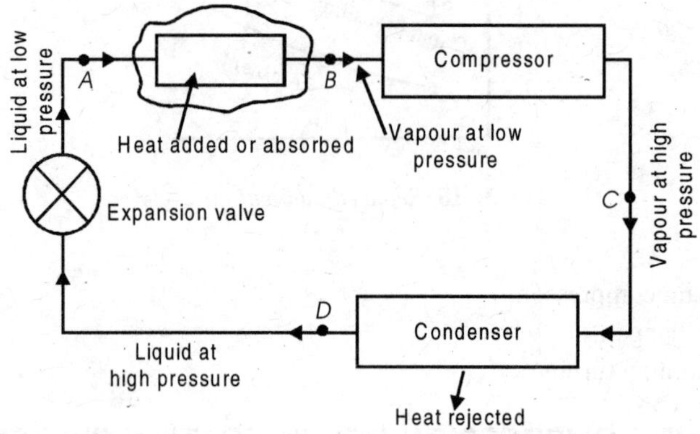

Fig. 36.19. *Schematic of vapour compression cycle*

In order to produce the temperature less than the temperature of the body, say –10ºC, the liquid at state *D* should be throttled down to the pressure which is the saturation pressure corresponding to –10ºC. For this purpose an expansion device is provided at point between condenser and evaporator. The state of the liquid at *A*, is low pressure liquid having the saturation temperature of –10ºC. Hence, the system once charged with Freon or Ammonia will continuously work, where the refrigerant will liquify and evaporate alternatively, producing refrigeration effect at evaporator and heat rejection at condenser. Neither the cylinder of Freon nor any fresh supply of the same is required.

A simple vapour compression system of refrigeration, consists of the equipments as follows.

(a) Compressor

(b) Condenser

(c) Expansion valve

(d) Evaporator

The schematic diagram of the arrangement is shown in Fig. 36.20. The low temperature, low pressure vapour at state *B* is compressed by the *compressor*. This vapour is condensed into high pressure liquid at state *D* in the *condenser* and then passes through the *expansion valve*. Here the liquid is throttled down to a lower pressure and passes on to the evaporator, where it absorbs heat from the surroundings or the circulating fluid (being refrigerated) and vaporises to low pressure vapour of state *B*. The cycle then repeats. The exchange of energy is as follows.

(a) Compressor requires work *W*. The work energy is supplied by the system from the surroundings.

(b) During condensation heat Q_1, equivalent of latent heat of condensation etc. is lost. This heat energy flows out *from the system to the surroundings.*

(c) During evaporation heat Q_2, equivalent of latent heat of vaporisation is absorbed by the refrigerant. This heat is again an inflow of energy *into the system.*

(d) There is *no exchange of energy* during the throttling process through the expansion valve, as this process occurs at constant enthalpy.

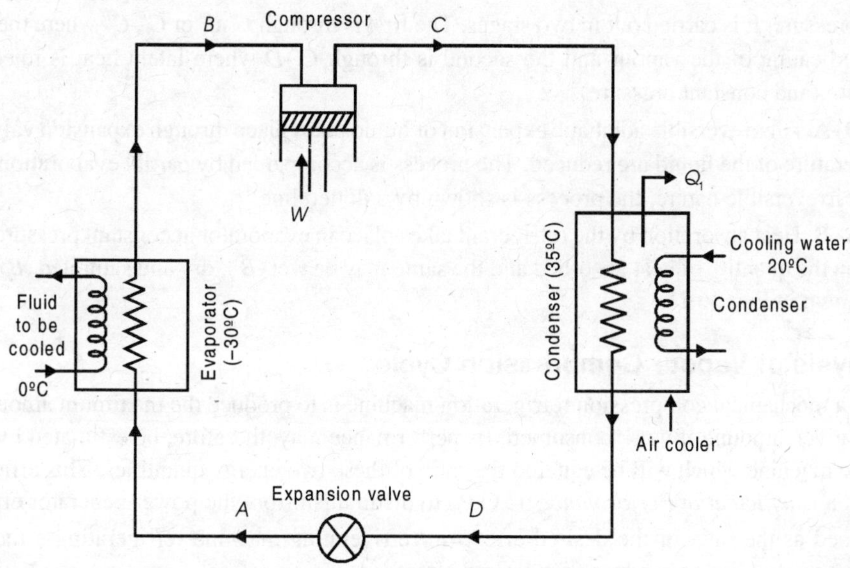

Fig. 36.20

Hence, by making an energy balance we get,

$$Q_1 = Q_2 + W$$

Figure 35.21 shows on *T–S* diagram for the equipment arrangement of Fig. 36.20. The cycle works between the temperatures T_1 and T_2 representing the condenser and evaporator temperature respectively.

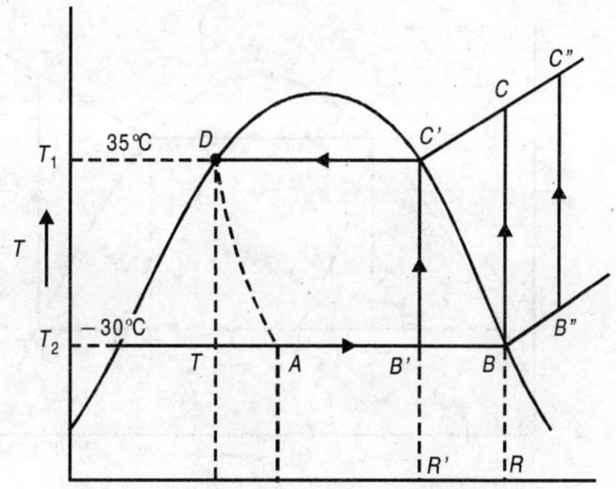

Fig. 36.21

The various processes of the cycle *A–B–C–D* are as given below:

(i) Process B–C. Isentropic compression of the vapour from state is *B* to *C*. If the vapour state, *B* or superheated, (*B*"), the compression is called *Dry Compression*. If initial state is wet, (*B*'), the compression is called *wet compression* as represented by *B'–C'*.

(ii) Process C–D. Heat is rejected by the refrigerant, in condenser, to the cooling medium (water ar air), at constant pressure. It is carried out in two stages. The first is through *C–C'* or *C'–C"* where the heat is rejected by desuperheating of the vapour and the second is through *C'–D* where latent heat is rejected at constant temperature (and constant pressure).

(iii) Process D–A. An irreversible adiabatic expansion of liquid takes place through expansion valve. The pressure and temperature of the liquid are reduced. The process is accompanied by partial evaporation of some liquid. Due to its irreversible nature, the process is shown by a dotted line.

(iv) Process A–B. Heat absorption by the refrigerant takes place in evaporator at constant pressure. The final state depends on the quantity of heat absorbed and the same may be wet (*B'*), dry and saturated (*B*), or superheated (*B"*) as shown in Fig. 36.21.

36.4.2 Analysis of Vapour Compression Cycle

The purpose of a mechanical compression refrigeration machine is to produce the maximum amount of refrigerating effect for a given amount of work consumed. Its performance may, therefore, be estimated by calculating the efficiency of the machine which will be equal to the ratio of these two energy quantities. This efficiency in refrigeration is called a *Coefficient of Performance* (C.O.P.) to distinguish from the power generator efficiency.

COP is defined as the ratio of the desired effect (in refrigerating machine refrigerating effect is the desired effect) to the energy supplied to produce that desired effect.

$$\therefore \qquad \text{COP} = \frac{\text{Refrigerating effect}}{\text{Energy supplied}}$$

Work of compression/kg is given by $W = h_2 - h_1$ and is supplied. (Fig. 36.22)

Similarly, the refrigerating effect is the heat absorbed by the refrigerant in the evaporator at constant temperature and pressure and is equal to the change in enhalpy during the process.

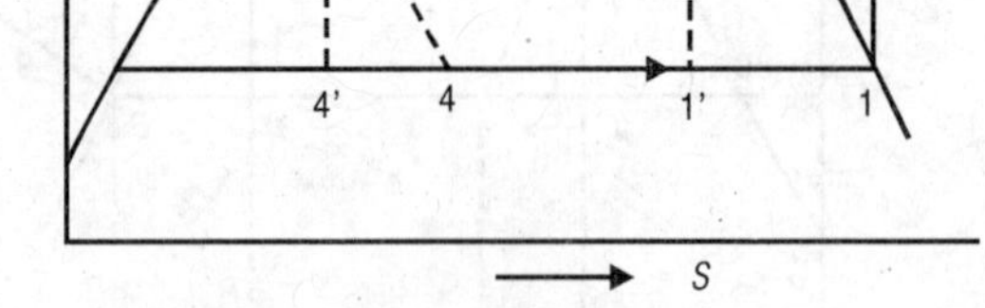

Fig. 36.22

$$\therefore \qquad \text{Refrigerating effect/kg} = h_1 - h_4 \text{ kJ/kg}$$

Letters *A*, *B*, *C*, *D* are replaced by 4,1, 2 and 3 and shown *T–S* diagram of Fig. 36.22.

$$\therefore \qquad \text{COP} = \frac{h_1 - h_4}{h_2 - h_1}$$

But $\qquad h_3 = h_4$ (isenthalpic process)

$$\therefore \quad \text{COP} = \frac{h_1 - h_3}{h_2 - h_1}$$

Instead of throttle valve or expansion valve, if an expander (piston-cylinder arrangement) is used, the expansion is taken as isentropic as compression. This isentropic expansion is shown on *T–S* diagram of Fig. 36.22 by a vertical line 3–4'. Then the expansion work is to be taken into consideration.

$$\text{Expansion work/kg} = h_3 - h_{4'}$$

and for cycle analysis, it is supplied to compressor.

∴ Net work supplied from outside/kg

$$W = \text{Work of compression} - \text{Work of expansion}$$

$$= (h_2 - h_1) - (h_3 - h_{4'})$$

Also, refrigerating effect/kg is given by

$$N = h_1 - h_{4'}$$

and

$$\text{COP} = \frac{N}{W} = \frac{h_1 - h_{4'}}{(h_2 - h_1) - (h_3 - h_{4'})}$$

$$\therefore \quad \mathbf{(COP)_{Ref.}} = \frac{\mathbf{h_1 - h_4}}{\mathbf{h_2 - h_1}} \quad \textbf{with expansion valve}$$

$$\mathbf{(COP)_{Ref.}} = \frac{\mathbf{h_1 - h_{4'}}}{\mathbf{(h_2 - h_1) - (h_3 - h_{4'})}} \quad \textbf{with expansion cylinder}$$

If we know the total capacity of the refrigerating system say *x TR* then, mass flow rate of refrigerant can be determined as

Mass flow rate of refrigerant (kg/sec)

$$= \frac{3.516\,TR}{\text{Ref. effect/kg}} = \frac{3.516x}{\text{Ref. effect/kg}}$$

36.4.3 Pressure-Enthalpy (Total heat) Chart

The pressure-total heat, or pressure-enthalpy chart is probably the most convenient chart for refrigeration calculations; this is the chart recommended by the refrigeration sub-committee of IME. Such a chart giving common features is shown in Fig. 36.23. Horizontal lines represent constant pressure lines, vertical lines are constant enthalpy lines, while lines of constant temperature and of constant entropy are also plotted and are shown in Fig. 36.23.

For different refrigerants, different *P–h* charts are available.

(a) Constant pressure process is shown by the horizontal line.
(b) Constant enthalpy lines represent throttling or isenthalpic process.
(c) Constant entropy line represents isentropic or adiabatic (reversible) process.
(d) Constant temperature lines represent isothermal processes.

36.4.4 Simple Vapour Compression Cycle on P-h Chart

A simple vapour compression cycle is shown by 1–2–3–4–1 on *P–h* chart of Fig. 36.24.

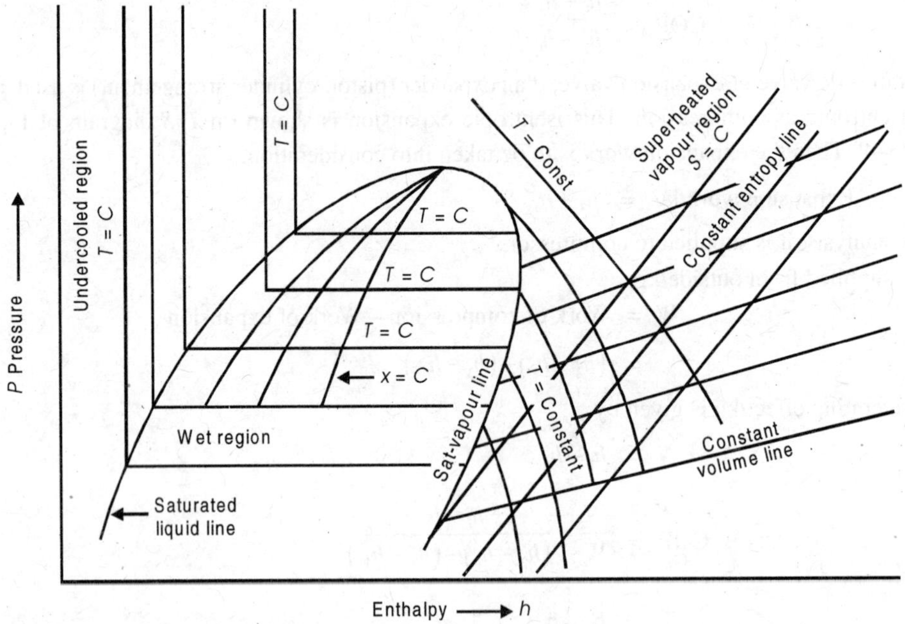

Fig. 36.23. *P-h chart*

1–2 Isentropic compression in compressor.
2–3 Constant pressure cooling (Heat rejection).
3–4 Isenthalpic expansion through expansion valve.
4–1 Constant pressure heat absorption.

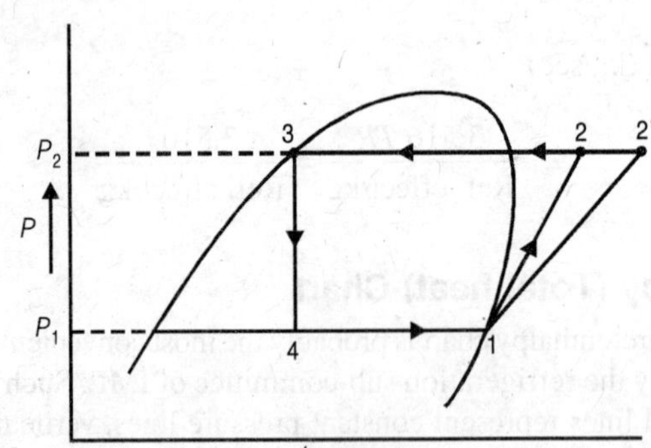

Fig. 36.24

(a) Mass of Refrigerant in Circulation

$$m_r = \frac{\text{Total refrigerating effect/sec}}{\text{Refrigerating effect/kg}}$$

$$= \frac{3.516 \times \text{TR}}{\text{Ref. effect/kg}}$$

$$= \frac{3.156(\text{TR})}{\text{Ref. effect/kg}} \text{ kg/sec}$$

(b) Piston Displacement

Let the specific volume of the vapour at 1 i.e., at the suction to the compressor, be V_1 and let the total volumetric efficiency of the compressor be $\eta_{vol.}$, then the piston displacement per min. $= \frac{V_1}{\eta_{Vol.}} \times m_r$ where m_r is the mass of refrigerant per minute.

(c) Power Required

(i) If the compression is isentropic, then work of compression $= h_2 - h_1$ kJ/kg

$$\text{Hence the power required} = \frac{m_r(h_2 - h_1)}{60} \text{ kJ/sec or kW}$$

(ii) If the actual compression is not isentropic, but it is according to any other law (actual) as shown by line 1–2', then work of compression is given by $(h_{2'} - h_1)$ kJ/kg .

$$\therefore \quad \text{Actual power required} = \frac{m_r(h_{2'} - h_1)}{60} \text{ kW}$$

(d) Heat Rejected in Condenser

Heat rejection process is 2–3 and is at constant pressure.

$$\therefore \quad \text{Heat rejected/kg} = h_2 - h_3 \text{ kJ/kg}$$

$$\text{or} \quad \text{Heat rejected/min} = m_r(h_2 - h_3) \text{ kJ/min}$$

$$\text{or} \quad \text{Heat rejected/sec} = \frac{m_r(h_2 - h_3)}{60} \text{ kW}$$

36.4.5 Actual Vapour Compression Refrigeration Cycle

In the preceding sections we have discussed the theoretical vapour compression cycle. The cycle as applied in practice, however, differs considerably from the theoretical cycle.

An actual vapour compression cycle is shown in Fig. 36.25 on *T–S* diagram. A few of the important points of deviations of this cycle as compared to the ideal theoretical cycle are explained below.

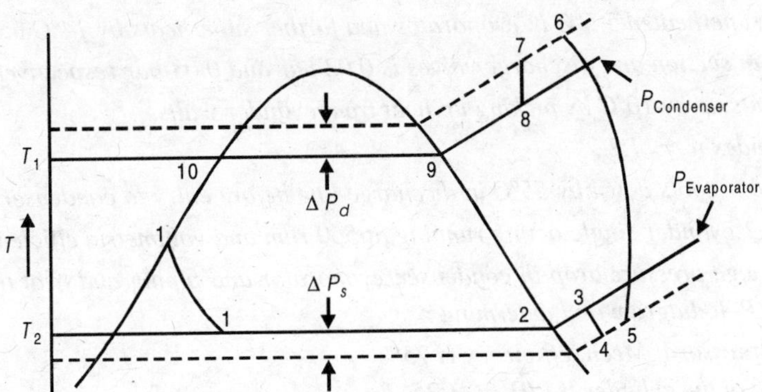

Fig. 36.25 *Actual vapour compression cycle*

(i) Process 1–2–3. This process represents the passage of refrigerant through the evaporator coil at $P_{evaporator}$ and T_2 represents the entry of vapour to compressor in superheated condition. Superheating upto this point may be either due to larger removal of heat from the evaporator; (this is an advantage) or due to heat picked up by the vapour in suction piping which is a disadvantage.

(ii) Process 3–4–5–6–7–8. This process represents the passage of vapour through the compressor, and shows several departures from the theoretical process.

(a) Both the suctions and discharge valves of the compressor are actuated by pressure difference. Hence the actual suction pressure p_s in the cylinder is lower than the evaporator pressure (p_{evap}). Similarly the actual discharge pressure (p_d) will be higher than the condenser pressure (p_{cond}). There is thus *throttling effect* in both the suction and discharge valves. Pressure 3–4 and 7–8 represent this throttling effect through Δp_s and Δp_d respectively.

(b) As soon as the cold vapour enters the cylinder it gets heated by coming in contact with the hot cylinder walls. Hence the temperature of the vapour increases. This heating is represented by process 4–5. Similarly there is a cooling effect at the discharge which is represented by process 6–7.

(c) States 5 and 6, therefore, represent the initial and final condition of vapour during actual compression. The compression may neither be isentropic nor polytropic. It may be assumed that the heat absorbed by the gas during the first part of compression is equal to the heat rejected during the latter part. The actual compression process is as shown by process 5–6 in Fig. 36.25.

(iii) Process 8–9–10–11. This process represents the passage of refrigerant through the condenser at pressure (p_{Cond}). Process 8–9 is actually the removal of superheat, 9–10 is the removal of latent heat and 10–11 represents the subcooling of refrigerant liquid.

(iv) Process 11. This process represents the throttling of subcooled liquid from state 11 so as to result in state 1 at the entry to the evaporator, after expansion.

The following example illustrates the method of considering most of the departures of actual cycle from the theoretical cycle discussed above.

Illustrative Example

An ammonia compression machine is required to cater for a load of 100 tons. The cooling water temperature requires the condenser to work at 35°C and the brine temperature requires evaporator to work at –30°C. The other data is given below :

(i) *Temperature at entry to expansion valve 30°C.*

(ii) *The vapour is superheated 5.5°C in evaporator and further superheats by 14°C in suction line.*

(iii) *Pressure drop in suction and discharge valves is 0.02 bar and 0.03 bar respectively.*

(iv) *The vapour heats up by 10°C by picking up heat from cylinder walls.*

(v) *Compression index n = 1.2.*

(vi) *Discharge vapour cools down by 55°C in discharge line before entry to condenser.*

(vii) *Compressor is 2-cylinder single acting running at 560 rpm and volumetric efficiency is 0.57.*

Ignore the losses e.g., pressure drop in condenser, evaporator and piping and heat transfer in liquid line etc. Draw the cycle on a P–h diagram and determine :

(a) *B.P. of the compressor if Mech. efficiency is 75%*

(b) *Bore and stroke of the cylinder if L/D = 1.25*

(c) *Condenser water required for a temperature rise of 6°C.*

Solution The various state points of the cycle shown in Fig. 36.26 may be obtained with the help of *P-h* chart as given below :

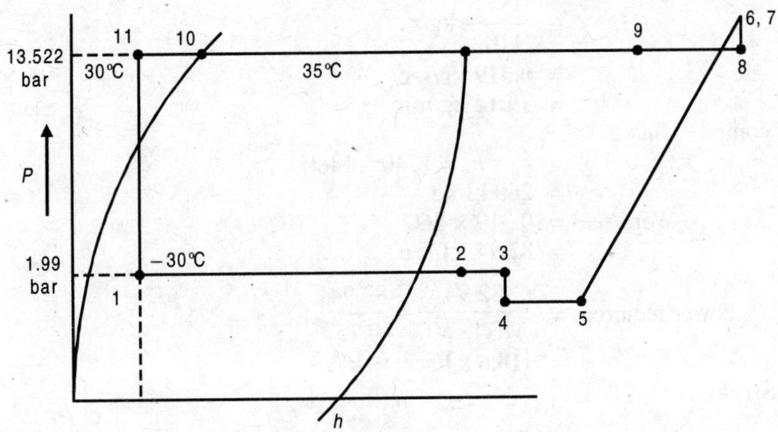

Fig. 36.26

$$P_1 = P_2 = P_3 = 1.199 \text{ bar}$$

$$t_2 = -30 + 5.5 = -24.5°C, \qquad h_2 = 1440 \text{ kJ/kg}$$

$$t_3 = -24.5 + 14 = -10.5°C, \qquad h_3 = 1456 \text{ kJ/kg}$$

and $v_3 = 1.04 \text{ m}^3/\text{kg}$

$$P_4 = P_5 = 1.199 - 0.02 = 1.179 \text{ bar}, \qquad h_3 = h_4 = 1456 \text{ kJ/kg}$$

$$t_5 = -10.5 + 10 = 0.5°C, \qquad h_5 = 1480 \text{ kJ/kg}$$

$$V_5 = 1.3 \text{ m}^3/\text{kg}$$

$$P_8 = P_9 = P_{10} = P_{11} = 13.522 \text{ bar}$$

$$P_6 = 13.522 + 0.03 = 13.552 \text{ bar}$$

For compression, $P_5\, v_5^{1.2} = P_6\, v_6^{1.2}$

$$\therefore \quad \left(\frac{V_5}{V_6}\right)^{1.2} = \frac{P_6}{P_5} = \frac{13.552}{1.179} = 11.494$$

$$\therefore \quad \frac{V_5}{V_6} = (11.494)^{\frac{1}{1.2}} = (11.494)^{0.8333} = 7.651$$

$$\therefore \quad V_6 = \frac{1.3}{7.651} = 0.1699 \text{ m}^3/\text{kg}$$

and $h_6 = h_8 = 1740$ kJ/kg

$$t_8 = 138°C$$

$$\therefore \quad t_9 = 138 - 55 = 83°C$$

$$\therefore \quad h_9 = 1630 \text{ kJ/kg}$$

$$h_{11} = 338 \text{ kJ/kg}, \; t_{11} = 30°C$$

$$= h_1$$

Now we can find the required quantities.

(a) The mass of NH_3 in circulation :

m_r = Mass of refrigerant kg/sec

$$= \frac{\text{TR} \times 3.516}{h_2 - h_1} = \frac{3.516 \times 100}{1440 - 338}$$

$$= \frac{351.6}{1102} \text{ kg/s}$$

$= \mathbf{0.319\ kg/sec}$

$= \mathbf{19.14\ kg/min}$

Work of compression/kg

$$W = h_6 - h_5 = 1740 - 1480$$

$= 260$ kJ/kg

$\therefore$ Work/sec $= 0.319 \times 260$

$= \mathbf{82.94\ kJ/sec}$

$$\therefore \quad \text{Power required} = \frac{82.94}{\text{Mech. eff.}} = \frac{82.94}{0.75}$$

$= \mathbf{110.6\ kJ/sec = kW}$

(b) Bore and Stroke :

$$\text{Total displacement/min} = \frac{m_r \cdot V_3}{\eta_{Vol.}}$$

$$= \frac{19.14 \times 1.04}{0.57}$$

$= \mathbf{34.922\ m^3/min}$

$$= 2 \times \frac{\pi d^2}{4} \times L \times N$$

$$= \frac{2 \times \pi \times d^2 \times 1.25d}{4} \times 560$$

$= \mathbf{1100\ d^3}$

$$\therefore \quad d^3 = \frac{34.922}{1100} = 0.03174$$

$\therefore$ $d = 0.3166$ m $=$ **31.66 cm**

$\therefore$ $L = 1.25 \times 31.66 =$ **39.58 cm**

(c) Heat Rejected in Condenser and Water Required :

$$\text{Heat rejected/min} = m_r (h_g - h_{11})$$

$= 19.14\ (1630 - 338)$

$= 19.14 \times 1292$

$= \mathbf{24729\ kJ/min}$

$$\therefore \quad \text{Water required/min} = \frac{24729}{6 \times 4.187}$$

$= \mathbf{984.35\ kg/min}$

36.4.6 Under Cooling or Subcooling

Undercooling or subcooling of the liquid takes place when the liquid is cooled below the saturation temperature corresponding to condenser pressure, before admitted to the throttle valve or expansion valve. Undercooling of the liquid is generally along the constant pressure line. This is represented by line 3–3' both on *T–S* and *P–h* diagrams of Fig. 36.27. Many times this undercooling is shown wrongly along the liquid line of *T–S* diagram (3–3").

Undercooling is brought about by circulating greater quantity of cooling water through the condenser.

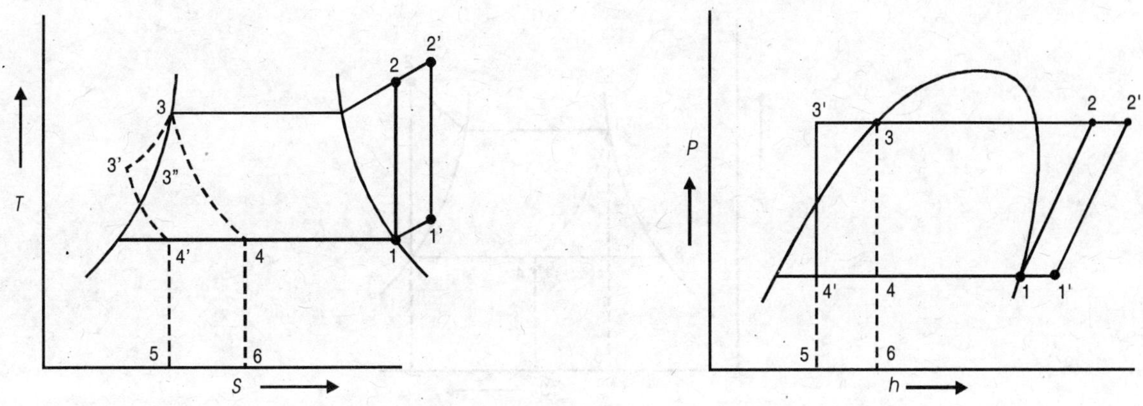

Fig. 36.27

Many times undercooling of liquid refrigerant coming out of the condenser is brought about by the vapours coming from the evaporator thus vapours are superheated to some extent. This arrangement is shown in Fig. 36.28.

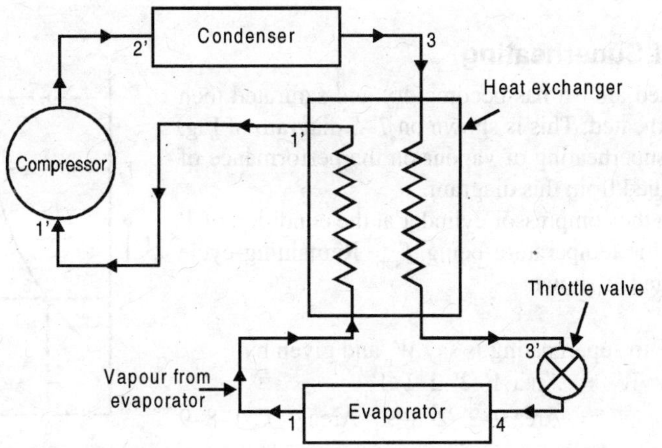

Fig. 36.28. *Schematic showing arrangement for undercooling liquid*

36.4.7 Effect of Undercooling

The coefficient of performance COP may be increased by what is known as precooling or undercooling. It will be noticed from the diagrams of Fig. 36.27 that the net refrigerating effect has been increased by the area 4'–4–6–5'–4'. The work done in compressed is not increased in that proportion and hence the coefficient of performance is increased by undercooling the liquid.

36.4.8 Effect of an Expansion Cylinder

If, instead of expanding the high-pressure liquid at 3 (coming from the condenser) through an expansion valve, it is expanded in an expansion cylinder or expander, thus driving the piston, work may be obtained which could be utilized in helping to drive the compression cylinder or compressor. By this method less external power would be required to drive or run the refrigerator. The work saved by the expansion cylinder is represented by the area 3–3'–4'. The high pressure liquid at 3 is expanded isentropically to 4' and the wet vapour is then forced through the brine tank or refrigerator. The net refrigerating effect is now increased by the area 4'–4–5–6–4' and the work required to drive the compression is less and hence the coefficient of performance is increased to a large extent.

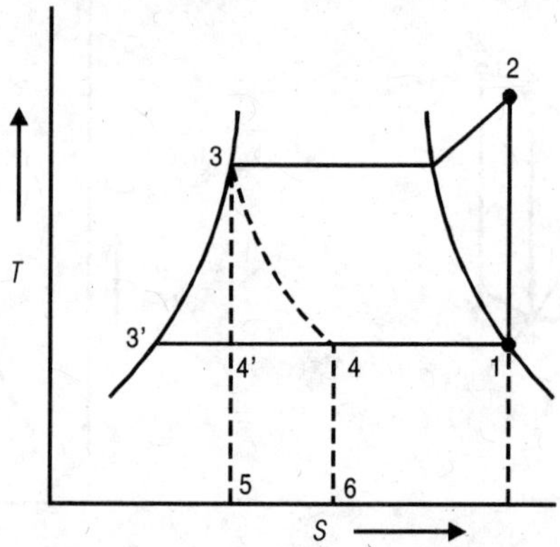

Fig. 36.29. *Effect of expansion cylinder*

The expansion cylinder is not used in practice, as the work saved is insufficient to overcome the friction of the necessary mechanism.

36.4.9 The Effect of Superheating

If the vapour is compressed after it has become dry and saturated then the vapour becomes superheated. This is shown on *T–S* diagram of Fig. 36.30. The effect of this superheating of vapour on the performance of the cycle COP can be studied from this diagram.

The vapour is drawn in the compressor cylinder at the condition of 1' and is compressed to 2', the temperature being T_{sup}. Remaining cycle processes remain unchanged as shown.

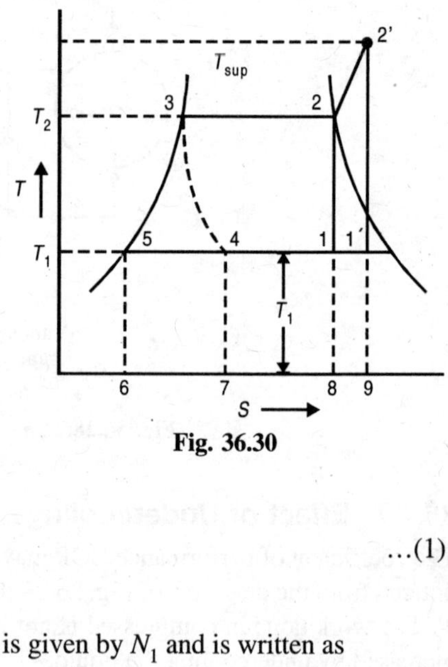

Fig. 36.30

Referring to Fig. 36.30,

Additional work done due to superheating is say W_1 and given by

$$W_1 = \text{Area } 1'\text{–}2'\text{–}2\text{–}1\text{–}1'$$

$$= \text{Area } 9\text{–}2'\text{–}2\text{–}8\text{–}9 - \text{Area } 9\text{–}1'\text{–}1\text{–}8\text{–}9$$

$$= \text{Superheat} - T_1 \times C_{pv} \log_e \frac{T_{sup}}{T_2}$$

$$= C_{PV}\left(T_{sup} - T_2\right) - T_1 \times C_{PV} \log_e \frac{T_{sup}}{T_2} \qquad \ldots(1)$$

where C_{pv} is the specific heat of the superheated vapour.

Similarly, addition heat abstracted from brine due to superheating is given by N_1 and is written as

$$N_1 = \text{Addition refrigerating effect}$$

$$= \text{Area } 9\text{–}1'\text{–}1\text{–}8\text{–}9$$

$$= T_1 \times C_{pv} \log_e \frac{T_{sup}}{T_2} \qquad \ldots(2)$$

Now, if W = Work without superheating

= Area 1–2–3–4–1

and N = Refrigerating effect

= Heat abstracted without superheating

Then, COP = Coefficient of performance with superheating

$$= \frac{N + N_1}{W + W_1}$$

From the diagram it is noticed that the ratio of $\frac{W_1}{W}$ is much greater than the ratio of $\frac{N_1}{N}$. The effect of this that the coefficient of performace is lowered or decreased by superheating.

In practice it is not usual to superheat.

36.4.10 Coefficient of Performance when T_1, T_2, Specific Heat of Liquid and Latent Heat at Higher Temperature Given

Let Fig. 36.31 represents the *T–S* diagram of the cycle and in this case let the vapour have a dryness fraction *x* at the end of compression.

From the diagram of Fig. 36.31, we have

$$x = \frac{3\text{–}2}{3\text{–}11}$$

and 3–11 = Change in entropy during evaporation

$$= \frac{h_{fg_2}}{T_2}$$

$$\therefore \quad 3\text{–}2 = \frac{x \cdot h_{fg_2}}{T_2}$$

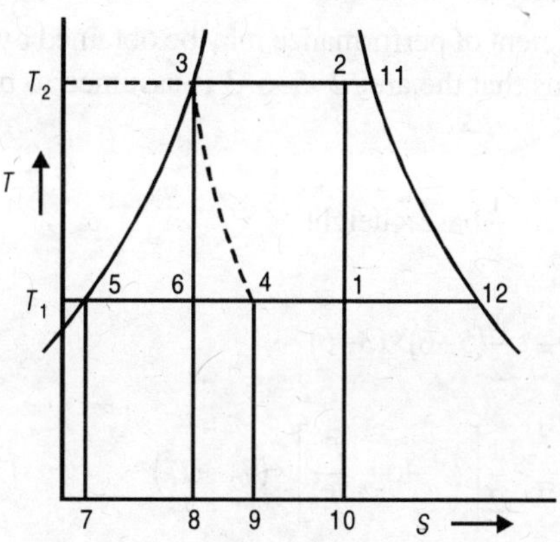

Fig. 36.31

Also, area 3–5–6–3 = Liquid heat at T_2 above that of T_1.

$$= C_{pi}\,(T_2 - T_1)$$

Similarly, 5–6 = Entropy change during liquid heating

$$= C_{p_i} \log_e \frac{T_2}{T_1}$$

Also, as 3–4 is an isenthalpic process,

Area 3–5–6–3 = Area 4–6–8–9–4

From these equation we get

$$C_{p_l}(T_2 - T_1) - C_{p_l}\log_e\frac{T_2}{T_1} = (6-4)\times T_1$$

from which (6–4) can be calculated.

Then for the performance of coefficient,

$$
\begin{aligned}
W &= \text{Work done} \\
&= \text{Area 1–2–3–4–1} \\
&= \text{Area 3–5–6–3 + Area 1–2–3–6–1} \\
&= T_1\times(6-4) + x\cdot h_{fg_2}\times(T_2 - T_1)
\end{aligned}
$$

and

$$
\begin{aligned}
N &= \text{Refrigerating effect} \\
&= \text{Area 1–4–9–10–1} \\
&= \{(2-3)-(6-4)\}\times T_1 \\
&= \left\{\frac{x\cdot h_{fg_2}}{T_2} - (6-4)\right\}\times T_1
\end{aligned}
$$

and $\text{COP} = \dfrac{N}{W}$ can be calculated.

36.4.11 Approximate Solution for Coefficient of Performance

An approximate solution for the coefficient of performance may be obtained by assuming the liquid line (3–5) to be a straight line. (Fig. 36.30). This means that the area 3–5–6–3 is assumed to be a triangle.

Then triangle 3–5–6–3

$$
\begin{aligned}
&= \frac{1}{2}\text{base}\times\text{height} \\
&= \frac{1}{2}(5-6)\times(3-6) \\
&= \frac{1}{2}\left[C_{p_l}\log_e\frac{T_2}{T_1}\right]\times(T_2 - T_1)
\end{aligned}
$$

$$
\begin{aligned}
\text{Also triangle 3–5–6–3} &= \text{rectangle 4–6–8–9} \\
&= T_1\times(4-6)
\end{aligned}
$$

From these two equations,

$$(4-6)\times T_1 = \frac{1}{2}\left[C_{p_l}\log_e\frac{T_2}{T_1}\right]\times(T_2 - T_1)$$

$$\therefore \quad (4-6) = \frac{T_2 - T_1}{T_1}\times\frac{1}{2}C_{p_l}\log_e\frac{T_2}{T_1}$$

By using this value of (4–6), we can calculate work done W and refrigerating effect N by using the value of (4–6).

Note here that for CO_2 as a refrigerant liquid line 3–5 cannot be regarded as a straight line.

36.4.12 Limitations of Air Refrigeration System

After noting down the working of Bell–Coleman air refrigeration systems, we observe certain limitations in using air-refrigeration systems. These limitations may be enlisted as follows.

1. The power required to get unit of refrigeration is comparatively large and hence the performance is very poor. It is observed that approximately 4 kW are required to produce 1 TR. For vapour compression refrigeration system, power requirement is approximately 1 kW only for every TR capacity.
   Therefore, the air refrigeration system is to be used when other systems can not be used.
2. When the air refrigeration plant is to be stationary, *then atmosphere should not be dusty*. In spite of the dusty atmosphere, if air refrigeration system is to be used, then dust-separators must be provided, otherwise wear and tear of the components will result.
3. Moisture contained in the air is a big problem as moisture condenses and even freezes choking the valve and creating problems for the running of this air system.
4. To have the required capacity of the air refrigerating system, the quantity of air required is very large and the volume of air to be handled is very large as compared to vapour compression system. In this case the size and weight of the compressor is very large and the compressor is heavy and bulky.

Thus when the air refrigerating system is absolutely necessary, then and then only this system is used.

36.5 VAPOUR ABSORPTION REFRIGERATION SYSTEM

We now come to the discussion of the vapour absorption refrigeration system. The power utilized in vapour compression system is high grade energy i.e., *electric power* for running the compressor motor. In absorption refrigeration system, however, heat is directly utilized as source of energy. Of course, it may be preferable to utilize heat as such as it avoids undergoing through the various energy transformations required in the generation of electrical energy from heat energy.

It may be recalled that in the compression system the vapour was compressed by undergoing a great change in volume during the compression process. Accordingly the major part of the power was consumed in the process. If means were available for rising this pressure of the refrigerant without appreciably altering its volume, the work requirements will be enormously reduced (by about 95% or so). This may possibly be done by dissolving the refrigerant in some absorbent and supplying the heat to the solution for compression purposes. The absorption cycle achieves this objective by placing the refrigerant in solution before the so called compression process and by removing from the solution immediately after the process. The absorption of the vapour is governed by **Raoult's law.**

The basic difference between vapour compression and vapour absorption cycles will thus be to replace the compressor of the vapour compression cycle by a set of equipment which fulfils the objective discussed above. The other important element i.e, condenser, expansion device and evaporator will exist in both systems.

36.5.1 Simple Vapour Absorption Cycle

Figure 36.32 illustrates the simplest scheme of equipment required for the replacement of the compressor.

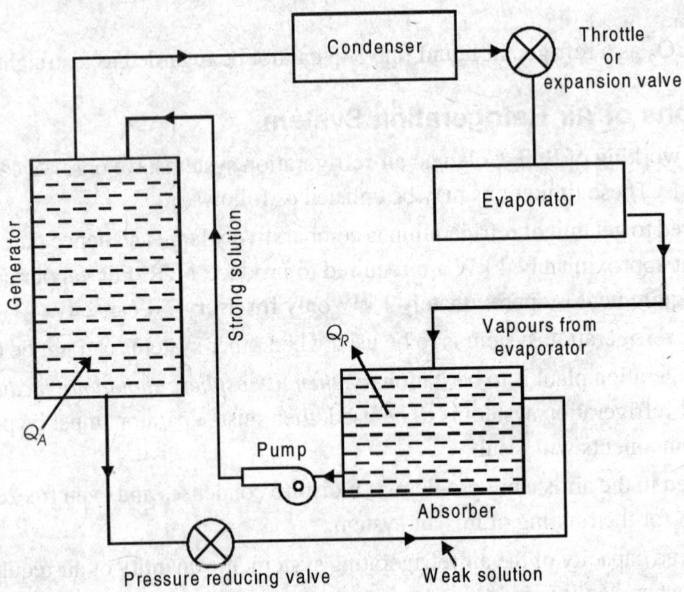

Fig. 36.32. *Equipment replacing the compressor (Simple vapour absorption system of refrigeration)*

Q_A = Heat added or supplied

Q_R = Heat rejected to coolant.

Figure 36.32 equipment replacing the compressor (Simple vapour absorption system of refrigeration).

The most commonly used fluids in the absorption system are ***water*** as absorbent and ***ammonia*** as refrigerant. The vapour from the evaporator is allowed to be mixed and *absorbed* in the absorber. The heat of absorption generated in the process is rejected from the absorber to the circulating cold water in a heat exchanger dipped in the solution contained in the absorber. The strong acqua-ammonia solution from the absorber is pumped upto the condenser pressure and fed to the generator which is the main energy consuming element of the system. Heat is supplied to the generator. The boiling point of refrigerant NH_3, is lower than that of the absorbing liquid H_2O, hence the vapours leaving the generator are predominantly those of refrigerant. These vapours then pass on to the condenser. The liquid refrigerant from the condenser, then, passes through an expansion valve or throttle valve to the evaporator where it absorbs heat from the substances or bodies to be refrigerated. Liquid refrigerant is then evaporated and the vapours enter the absorber completing the cycle.

The weak acqua-ammonia solution in the generator left due to separation of refrigerant vapour is drained back to the absorber for repeating the cycle.

The weak acqua-ammonia solution leaving the generator is at high pressure and the pressure in the absorber is the evaporator pressure which is less than the generator or condenser pressure, and hence a pressure reducing valve is provided in the weak solution line to the absorber.

The energy requirements of the system are at the generator and at the pump as compared to those at compressor in the vapour compression system. Since the volume of liquid handled by the pump is too small, the power required here is almost negligible as compared to that by the generator.

36.5.2 Practical Absorption Refrigeration Cycle

The replacement of the compressor by the simple arrangement of Fig. 36.33 is not very economical in practice. In order to make improvements certain additional auxiliary items are provided in the system. They include, ***analyzer***, a ***rectifier***, and two ***heat exchangers***. The practical absorption cycles as developed after incorporating these auxiliaries is shown in Fig. 36.33.

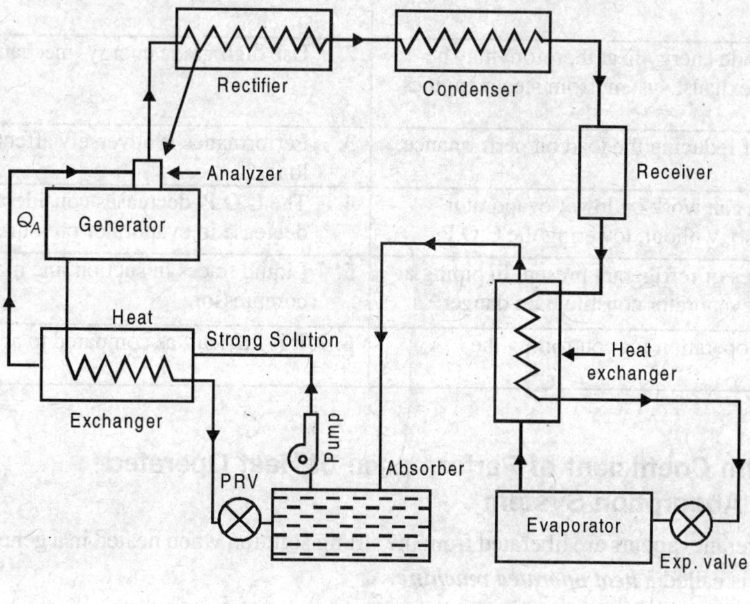

Fig. 36.33. *Practical absorption system*

(a) Analyzer. The ammonia vapours leaving the generator may contain certain moisture, and therefore it should be freed from any trace of water vapour before passing on to the condenser and then to the expansion valve, otherwise the water vapour is likely to freeze in the small valve passage and choke the flow.

The function of the analyzer is to remove the moisture as far as possible. It is an open types of cooler and forms an integral part of the generator, mounted on its top. Both the strong acqua-ammonia solution from the absorber and the condensate removed in rectifier are introduced from the top and flow downwards. The hot rising vapour of ammonia therefore comes in contact with the same and gets cooled. Thus most of the water vapour is condensed and drips back into the generator. This helps in salvaging a certain portion of heat in outgoing vapour which would otherwise have been rejected out through the condenser.

(b) Rectifier. It is a closed type of cooler and is actually a miniature condenser where any traces of water vapour left in the ammonia vapour, are *removed* by condensation .The cooling is achieved by circulating water as is done in an ordinary condenser. The condensed acqua is drained back to the generator through the analyzer.

(c) Heat Exchangers. Two heat exchangers are provided to internally exchange heat from the higher temperature fluid to the lower temperature fluid so that one is cooled and the other is heated.

One heat exchanger is provided between liquid receiver and evaporator so that the liquid is *subcooled* and vapour is heated up. Another heat exchanger is located between generator and absorber so that the strong acqua is *heated up* before going on to the analyzer and weak acqua is cooled before entering the absorber.

36.5.3 Comparison between Vapour Compression and Absorption System

Some of the important points of difference between the vapour absorption are listed below.

Absorption	*Compression System*
1. Moving parts are only in the pump, which is a small element of the system. Hence quiet in operation.	1. Moving parts are in the compressor. More wear and tear and noisy operation.
2. Use low grade energy heat therefore may be worked on exhaust system from steam engines etc.	2. Use high grade energy–mechanical work.
3. No effect of reducing the load on performance.	3. Performance is adversely affected at partial loads.
4. The system can work on lower evaporator pressure also, without, lowering the C.O.P.	4. The C.O.P. decreases considerably with decrease in evaporator pressure.
5. Liquid traces of refrigerant present in piping at the exit of evaporator constitute no danger.	5. Liquid traces in suction line may damage the compressor.
6. Automatic operation for controlling the capacity.	6. It is difficult as compared to absorption unit.

36.5.4 Maximum Coefficient of Performance of Heat Operated Vapour Absorption System

We know that refrigerant vapours are liberated from the strong solution when heated in a generator. This type of the machine or system is called a ***heat operated machine***.

Let T_h be the temperature at which heat is supplied to the strong solution.

We will think of a system shown in Fig. 36.34.

T_k is the temperature at which heat is rejected to atmosphere; T_o is the temperature of the body to be refrigerated.

Thermal efficiency of the engine is given by

$$\eta_{th} = \frac{\text{Work}}{\text{Heat supplied}}$$

$$= \frac{\text{Work}}{Q_h}$$

$$\text{Work of the engine} = \eta_{th} \times Q_h$$

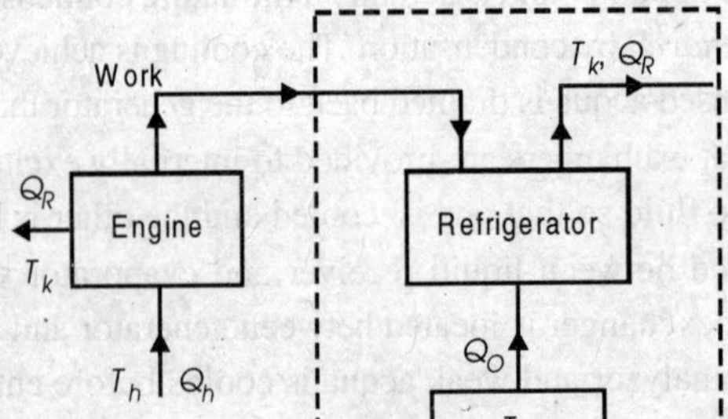

Fig. 36.34

This work, it is assumed, is used for refrigeration system.

For refrigeration system, T_o is the temperature at which heat is given to refrigerant (actually this is the heat abstracted from the body to be refrigerated) and the heat is rejected the atmosphere at T_k temperature.

To get the maximum COP, we have to consider the Carnot reversed engine.

$$(\text{COP})_{\text{carnot}} = \frac{T_o}{T_k - T_o} = \frac{Q_o}{Q_R - Q_o}$$

Similarly, we consider the engine as a Carnot engine and

$$(\eta_{\text{th}})_{\max} = \frac{T_h - T_k}{T_h} = \frac{\text{Work}}{Q_h}$$

$$Q_o = (\text{COP})_{\text{carnot}} \times \text{Work}$$

and
$$Q_h = \frac{\text{Work}}{(\eta_{\text{th}})_{\text{carnot}}}$$

∴ COP of the cycle

$$= \frac{Q_o}{Q_h} = \frac{\text{Work} \times \eta_{\text{carnot}}}{\text{Work}/\eta_{\text{th}}}$$

$$= \eta_{\text{th}} \times \eta_{\text{carnot}}$$

This is maximum when η_{th} is maximum which is the Carnot thermal efficiency

$$\mathbf{(COP)_{max}} = (\eta)_{\textbf{carnot}} \times \mathbf{(COP)_{carnot}}$$

Note here that this maximum C.O.P. is not to be compared with COP of vapour compression system which is $\frac{Q_o}{W}$.

$$\therefore \quad \mathbf{(COP)_{max}} = \frac{\mathbf{T_h - T_k}}{\mathbf{T_h}} \times \frac{\mathbf{T_o}}{\mathbf{T_k - T_o}}$$

In this case

T_h = Temperature of the source should be as high as possible (source temperature)

T_k = Sink temperature which should be as possible.

and T_o = Refrigerator temperature should be as possible.

36.5.5 Single-Effect and Double-Effect Cycles

In vapour absorption refrigeration system, the temperature of the heat source play an important role. The heat given by the source may be used in one single stage or in more than one stage. Accordingly the system is called as *single stage* or *single effect cycle* of operation and if the heat is supplied in two stages then the system is called *two stage* or *double effect cycle* of operation.

It is found that when the temperature of the heat source is upto 105°C, the heat is utilized in a single stage and the cycle is a *single effect cycle.* When the temperature of heat source is greater than 105°C, then generally the heat is utilized in two stages and the system becomes a *double-effect cycle*.

1. **Single Effect Cycle.** A simple and practical absorption system using ammonia as refrigerant and water as absorbent described in the previous articles is an example of single-effect cycle system for vapour absorption refrigeration system.

2. Double Effect Cycle. The schematic sketch of double-effect system is shown in Fig. 36.35. Vapours of refrigerants are generated in two stages as shown. For this purpose two heat exchangers are used. Figure 36.35 for double-effect system is self explanatory.

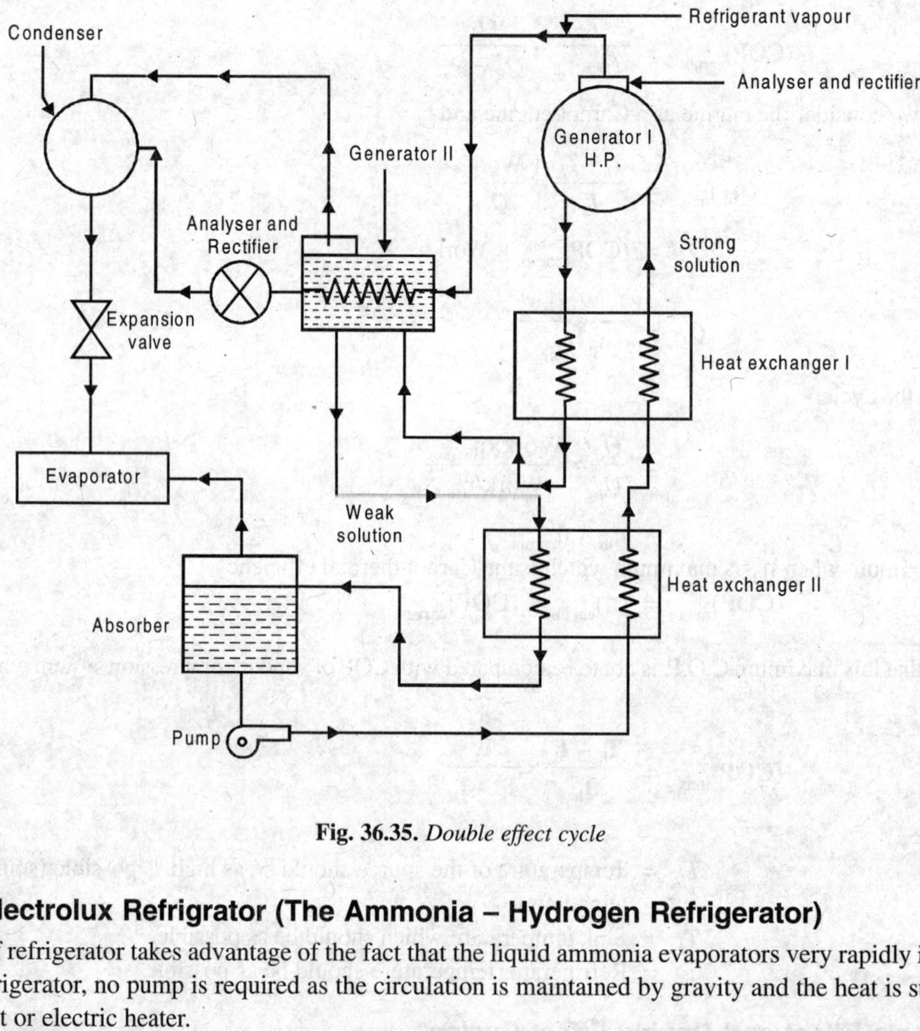

Fig. 36.35. *Double effect cycle*

36.5.6 Electrolux Refrigrator (The Ammonia – Hydrogen Refrigerator)

This type of refrigerator takes advantage of the fact that the liquid ammonia evaporators very rapidly in hydrogen. For this refrigerator, no pump is required as the circulation is maintained by gravity and the heat is supplied by a small gas jet or electric heater.

Figure 36.36 shows a diagrammatic view of Electrolux refrigerating system. This system consist of an absorber *A* which contains a strong solution of ammonia dissolved in distilled water. When a gas burner *B* is lighted the circulation of the system commences because of the warming of ammonia solution in the boiler or heater *D*. The strong ammonia solution now flows from *A* through the intercharger *C*, where it is heated by the hot work solution returning from *D*. It then passes around the heater *D*, which causes the ammonia vapour to be driven out of the solution. Now the released ammonia vapour passes through a rectifier, after which it enters the condenser *E*. The object of the rectifier is to drain off any particles of water coming with the ammonia vapours. The hot weak solution left behind *D* drains back into the absorber *A*. In doing so it passes through the intercharger *C*, where it gives its heat to the strong solution flowing from *A*.

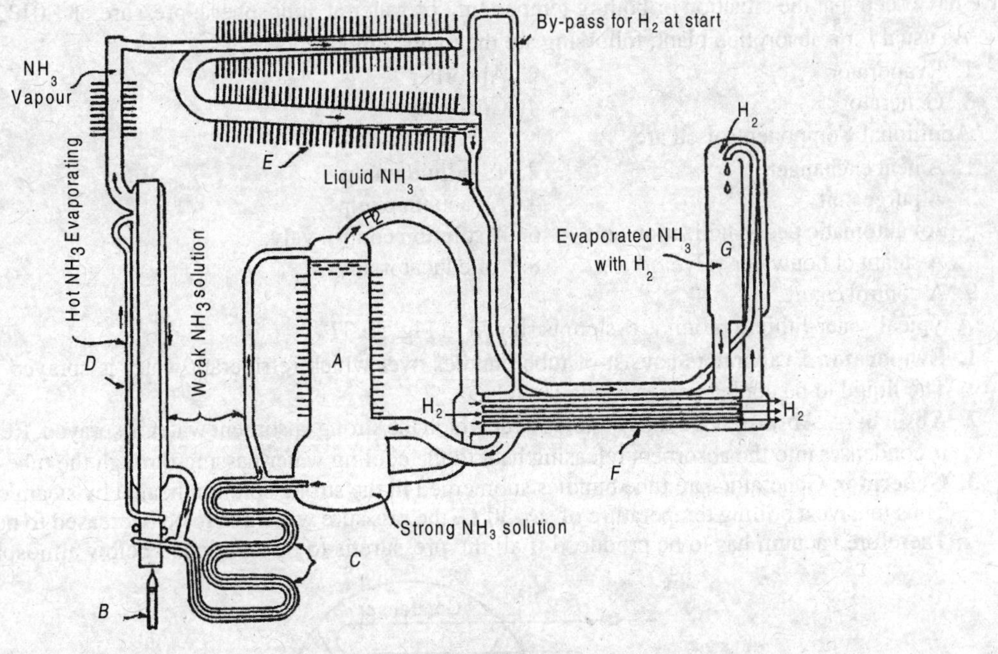

Fig. 36.36. *Electrolux refrigerator*

After condensation in the condenser, ammonia vapours forms the liquid and this ammonia liquid flows by gravity into the evaporator *G*. This is full of hydrogen into which the liquid ammonia evaporates. Liquid ammonia absorbs latent heat producing intense cold around *G*, which is situated in the food cabinet or any space which is to be refrigerated and thus the cabinet or space is cooled and the objects place in the cabinet or space are cooled. The mixture of hydrogen and ammonia vapour now flows through the gas heat exchanger *F*, where it cools the fresh supply of hydrogen flowing into *G*. From *F* the NH_3 ammonia vapour and hydrogen gas flows into the absorber *A*, where the hydrogen rises to the top and flows back to the evaporator *G*. Thus, the cycle is complete.

The main advantage of this Electrolux refrigerator is that there is no moving parts like compressor, pump or fan and therefore there is no noise and no machinery to give mechanical troubles.

The coefficient of performance of this refrigerator is given by the equation

$$\text{COP} = \frac{\text{Heat absorbed by evaporator } G}{\text{Heat supplied by } B}$$

Total pressure in the system is about 15 bar. H_2 gas is present in the evaporator at about 11–12 bar pressure. The quantity of H_2 is so proportioned that the practical pressure of NH_3 within evaporator will be only a smaller portion of the total pressure for the pressure of NH_3 vapours in the evaporator will only 4–3 bar. This low pressure causes NH_3 to boil or evaporator at low temperatures. Here the Dalton's law of partial pressure is applied.

Other substances that can be used in vapour absorption system are

(i) Water as the refrigerant and lithium bromide salt solution as absorbant.

(ii) Water as the refrigerant and lithium chloride brine as the absorbant.

(iii) Methylene chloride as refrigerant and dimethyl ether of tatraethylene glycol as absorbant.

36.5.7 Lithium Bromide-water Absorption Refrigeration System

We have seen that the situation or boiling temperature of water at atmospheric pressure of 1.01325 bar is 100ºC.

As usual for a absorption plant, following are the main components

1. Evaporator
2. Absorber
3. Generator
4. Condenser

Additional Components used are

1. A heat exchanger
2. Two fluid pump
3. A purge unit
4. A vacuum pump
5. An automatic decrystallization unit
6. A solution control valve
7. A steam or hot water valve
8. An educator and
9. A control centre

A typical water-lithium bromide system is shown in Fig. 36.37.

1. **Evaporator.** Evaporators consist of tube bundles over which refrigerant water is sprayed and evaporated. The liquid to be cooled passes inside the tube.
2. **Absorbers.** Absorbers are tube bundles over which the strong absorbent water is sprayed. Refrigerant vapour is condenser into the absorbent releasing heat to the cooling water passing through the tubes.
3. **Generator.** Generators are tube bundles submerged in the strong solution heated by steam or hot water. Thus to have a boiling temperature of say 30ºC, the pressure will have to be decreased to nearly 0–042 bar. Therefore,vacuum has to be produced if all the pressure is to be mentioned below atmospheric pressure.

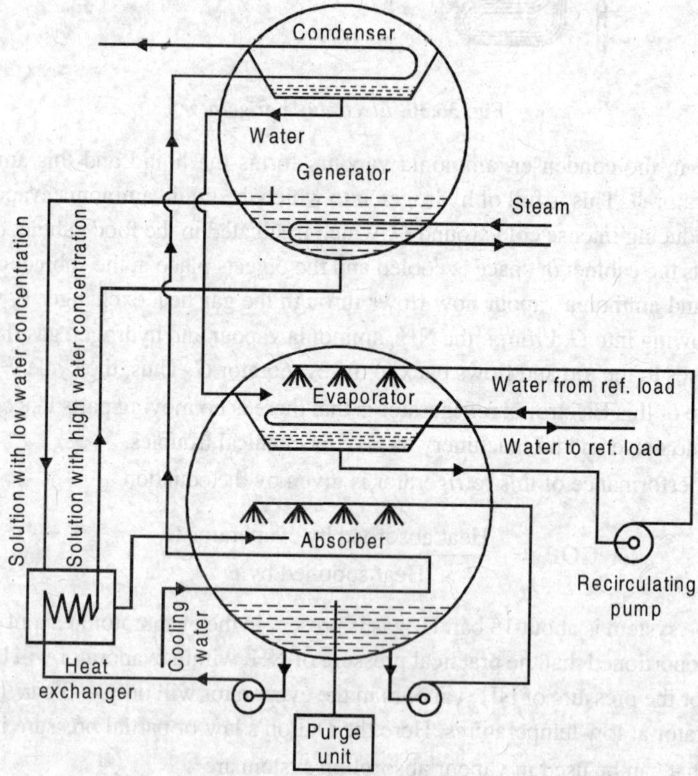

Fig. 36.37

When water is to be used as a refrigerant, the boiling or evaporating temperature cannot be reduced below 5ºC. Therefore, if the refrigerator or evaporation temperature is above 5ºC, then water can be used as a refrigerant. Then the absorbent for water must have a great affinity for water. The substances which can be used as absorbent are common salt (sodium chloride), calcium chloride, lithium bromide etc.

According to the tests made on these different salts, *lithium bromide* was found to possess the best overall characteristics required for an absorbent to be used in large capacity absorption chillers.

Lithium bromide is available in crystals and these crystals are dissolved in water and the mixture is called *Lithium Bromide Solution* which is used in the absorption chiller. The amounts of lithium bromide and water in a solution are measured by weights and not by volume 40% concentration means that 40 % of the total weight is lithium bromide.

4. **Condensers.** Condensers are tube bundles located over which vapours are present coming from the generators and the cooling water flows through the tubes. This cooling water first passes through the absorber.
5. **Heat Exchangers.** These are of steel shelf and tube type and tube type heat exchangers.
6. **Purge Unit.** This is used to remove non-condensable gases. These gases are present in small quantities and will rise the total pressure in the absorber sufficiently and the evaporator pressure changes thus changing the evaporating temperature. A very small pressure rise can cause appreciable change in the evaporating temperatures.
7. **Expansion Device.** Normal mechanical expansion valves are not used in absorption units. The flow of refrigerant liquid to evaporator is controlled by an orifice or other fixed restriction between the condenser and the evaporator (just like capillary tube in small Household refrigerator).
8. **Pumps.** Pumps which are used for recirculation are shown in Fig. 36.37.

36.6 REFRIGERANTS

A refrigerant is a working substance which is used in refrigerating machine. It is a working fluid through which heat is transferred from the cold body to the hot body. Formerly the number of useful refrigerants was very limited. Air, ammonia (NH_3), carbon dioxide (CO_2), sulphur dioxide (SO_2) were the only four refrigerants which has survived the adolescent stage. Somewhat later methyl chloride (CH_3Cl) came in general use for domestic and commercial units but since the development of the fluorinated hydrocarbons, a great variety of refrigerants has been put at the disposal of the refrigerating engineer. They are largely used in modern plants. The most common Freon refrigerants in use are

Freon-11 ($CHCl_3$), Freon-12 (CF_2Cl_2)
Freon-13 (CF_3Cl), Freon-21 ($CHFCl_2$)
Freon-22 (CHF_2Cl), Freon-113 ($C_2F_3Cl_3$)
and Freon-114 ($C_2F_4Cl_2$)

Curiously enough the use of air as a refrigerant is both older and more recent than the other vapour refrigerants.

36.6.1 Desirable Properties of Refrigrant

Desirable refrigerant possess such chemical, physical and thermodynamic properties which permit their practical application in refrigerating machines in an efficient way. No refrigerant may be ideal for all jobs. The best suited refrigerant may be selected on the basis on its characteristics and the type of service under consideration. The best suited refrigerant may be selected on the basis of its characteristics and the type of service under consideration. The referigerator selected for service should possess as many qualities as practicable out of the following:

1. The condensing pressure should not be excessive.
2. The boiling temperature should be low.
3. The critical temperature should be high.
4. The latent heat of vaporisation should be high.
5. The specific heat of liquid should be low.
6. The specific volume of vapour should be low.
7. It should be non-corrosive to metals and inert to oils.
8. The refrigerant should be chemically stable.
9. The refrigerant should be non-inflammable and non-explosive.

10. It should be non-toxic and harmless to human beings or living cells.
11. It should be easily detectable for leaks.
12. It should be easily available and have low cost.
13. It should have satisfactory heat transfer and viscosity coefficient.

36.6.2 Classification of Refrigerants

As refrigeration is a process of heat removal, for this purpose a medium for heat transfer is necessary. Such a medium of heat transfer, which absorbs heat by evaporating at low temperature and gives up heat by condensing at high temperature and pressure is termed as *refrigerant.* In the broadest sense the term refrigerant is also applied to the cooling medium like cold water, brine solution and antifreezes. Thus the refrigerant may be classified as ***Primary*** refrigerants and ***Secondary*** refrigerants.

Primary refrigerants as commonly interpreted include only those working medium which actually pass through the cycle of evaporation, compression, condensation and liquefaction e.g., NH_3, CO_2, SO_2, Freon group etc.

Secondary refrigerants include those working medium which transfer heat from a remote point to the evaporator of the refrigeration system. Cooling medium such as ice and solid CO_2 are therefore not primary refrigerants but secondary refrigerants. Desirable properties of secondary refrigerants mainly are its specific gravity, freezing point, thermal conductivity, viscosity and specific heat.

Advantages of the secondary refrigerants over the direct cooling system are,

1. Easy to handle and easy to control its flow.
2. Temperatures can be controlled by controlling the flow.
3. A toxic refrigerant, such as NH_3, will be away from the refrigeration load.
4. Lengths of the primary refrigerants pipe or tubing can be eliminated thereby pressure drop and leakages of the primary refrigerants are eliminated to a large extent.

Nowadays number of refrigerants are commercially available and these are classified into several groups as under:

I. Halocarbon Compounds. This group is a chemical compound formed from the base methane CH_4 and ethane C_2H_6 and include refrigerants which contain one or more of the three halogens i.e., Chlorine, Fluorine and Bromide. These refrigerants are:

F-11 Trichloromonefuoro methane CCl_3F
F-12 Dichlorodifluoro methane CCl_2F_2
F-22 Monochlorodifuloro methane $CHClF_2$
F-114 Dichlorotetrafluoro ethane $CClF_2CClF_2$

II. Azeotropes. This group consist of mixtures of refrigerants which do not separate into their compounds with pressure or temperature changes and have fixed thermodynamic properties. e.g., F-500(F-12-.73.8%+ F-152, 26.2%).

III. Hydrocarbons. This group contain organic compounds, and many hydrocarbons are used as refrigerant in industrial installations. They are highly flammable but possess satisfactory thermodynamic properties.

Examples are: CH_4 (methane), C_2H_6 (ethane) etc.

IV. Inorganic Compound. These refrigerants are,

711 Ammonia NH_3
718 Water H_2O
729 Air
744 Carbon dioxide CO_2
764 Sulphur dioxides SO_2

V. Oxygen Compounds. This group contains oxygen element e.g., ethyl-ether $C_2H_5OC_2H_6$

VI. Nitrogens Compounds. This group contains nitrogen element e.g., methyl-amine CH_3NH_2.

VII. Unsaturated Organic Compounds. This group is mainly hydrocarbon group of base ethylene C_2H_4 and propylene C_3H_6. There are thus number of refrigerants commercially available and are sold under different trade names.

36.6.3 Properties of Important Refrigerants

Although a refrigerant has not been found yet which has all the desirable properties, Freon-12 carbon dioxide and ammonia are commonly used for very low temperature work because of their low boiling points. But for the units used in household refrigerants, there is a wide range of available refrigerants each of which in comparison with others has certain desirable qualities.

The properties of some of the common refrigerants in use are given below:

1. Ammonia (NH_3). This primary refrigerant has long been used as refrigerant in large ice-making and cold storage plants. At atmospheric pressure it boils at –33°C and is therefore suitable for low temperature work. It is much less expensive than some other refrigerants. It is well suited for one or two stage reciprocating compressors. It is regarded as the most useful refrigerants because of its high latent heat, moderate working pressures, high critical temperature and small sensible heat compared with the latent heat. It requires low volume of gas to be circulated per ton of refrigeration. Its heat transfer properties are excellent. It is very soluble in water forming ammonium hydroxide. When heat is applied to ammonia hydroxide, ammonia is released in vapour form. It is used in operating on absorption system of refrigeration. It is not miscible with oil. Ammonia is almost never used in comfort cooling for public places.

It is toxic and has extremely offensive and pungent (irritaing) odour. Leaks can be easily detected on account of its pungent odour. It attacks many non-ferrous metals like copper, zinc, brass etc. in presence of humidity. It has no corrosive effect on iron and steel. The exact location of leak is shown by the dense white fumes produced by burning sulphur taper or by use of test papers.

The density of ammonia is smaller than that of air and lubricating oil.

2. Sulphur-dioxide (SO_2). It is one of the refrigerants commonly used in household refrigerators because of the very low working pressure. Its boiling point at atmospheric pressure is –10°C. It has a fairly large latent heat and high critical temperature. It is non-inflammable and non-explosive. It is highly toxic. SO_2 when used in a hermatically sealed compressor within domestic refrigerating units, does not present a serious problem even though it is toxic. It has a very pungent and suffocating odour. Leakage is easily detected on account of its pungent odour and the white smoke which forms when strong ammonia water (28% solution) is brought in the presence of gas.

It is very corrosive when in contact with moisture and therefore extreme care should be taken to prevent even the slightest bit of moisture entering to system to avoid the formation of sulphurous (H_2SO_3) acid which is corrosive in effect.

3. Carbon Dioxide (CO_2). Formerly this refrigerants was prepared for marine work and in dairies. The heat extracted per unit volume is the highest known for many refrigerants, but the high working pressure (75 ata), the low critical temperature (31°C) the low COP and the excessive power required per ton of refrigeration has restricted its use. Now its use as a refrigerant in the proper sense of the word is on the decline but it is increasingly used as dry ice. It is used for reciprocating compressor only. It is non-toxic but will suffocate in high concentration. Leaks of CO_2 gas are difficult to locate for the reason that the gas is colourless and odourless. It is non-inflammable. BP at atm-pressure = – 78.5°C. It has no corrosive effect on copper, copper alloys, iron or steel. It has no effect on oil and greases. If is miscible with oil. Due to high working pressure, the compressors, condenser and pipe connecting them must be designed for greater strength.

Dry ice is very well suited for cooling of railway refrigerator vans, lorries etc. as well as for ice-cream during distribution and sales etc. It is readily available and its cost is low.

4. Methyl Chloride (CH_3Cl). Methyl chloride is suitable for small refrigerators and domestic units. It has low working pressure, fairly high latent heat and high critical temperature. Its boiling point is –23.7°C at atmospheric pressure. It is inflammable but does not been reality. It is colourless and has sweet odour. Its odour resembles that of chloroform and hence leaks can be easily detected. It is somewhat toxic. Large concentration of vapour in air may produce unconciousness or even death but when well diluted with air it has little effect on human body.

In the presence of moisture it is very corrosive to zinc magnetism and aluminium and it produces solvent action on rubber It does affect copper, iron or steel-methyl chloride dissolves practically all types of oil and glycerine.

When the fluorinated hydrocarbons gained foothold as refrigerants methyl chloride was abandoned in most countries and in some the use is now restricted by the authorities.

5. Fluorinated Hydrocarbons. These refrigerants are halogen derivaties of saturated hydrocarbon i.e., of methane, ethane etc. Fluorine and chlorine are substituted for one or more of the hydrogen atoms and investigations have shown that an increasing number of fluorine atoms reduces health hazard and flammability and increases stability.

The different types in use cover a large range of normal boiling points and molecular weights. Thus some are not suitable for reciprocating and some for centrifugal compressors and some are well suited for low temperature work and some for heat pumps.

They are all more or less miscible with oils. They are all safe as long as they do not come into contact with red hot surfaces or open fire. In these cases they decompose forming poisonous products, viz. hydrochloric acid (HCl), Hydrofluric acid (HF) and phosgene (CCL_2O). They are, however, inflammable.

Like most other organic compounds their heat transfer properties are not so good as those of water and ammonia.

In order to avoid confusion an international agreement has been proposed to use letter F (Freon).

F-11 (Monofluotrichloro Methane). This refrigerant is low pressure type, well suitable for centrifugal compressors. High molecular weight. Reduces the number of stages for a given pressure ratio. F-11 is the least stable refrigerant of Freon group.

F-12 (Difluorodichloro Methane). The development of the fluorinated refrigerants with the manufacture of F-12 which rapidly gained a widespread application in household units and commercial plants. BP at atm =29.8ºC.

It is well suited for reciprocating and rotary compressors. However, at normal temperatures the swept volume necessary for a certain output is about 70% larger than when ammonia is used.

F-12 attacks magnesium and zinc, but copper and its alloys can be safely used so that pipe lines are easily laid. F-12 attacks natural rubber.

The discharge pressure are lower for F-12 than for ammonia at the same temperatures. Mass of F-12 is much more for a given output than ammonia plants.

F-21 (Monofluorodichloro Methane). This is a low pressure refrigerant. In certain countries it has been applied in household refrigerators. With rotory compressors and in absorption units. It has been proposed as a refrigerant in plants working with centrifugal compressors.

The properties of F-21 are similar to those of other fluorinated refrigerants.

F-22 (Difluromonochloro Methane).

BP at atm. pressure = – 40.77ºC
Critical temperature = – 96.2ºC
Corrosive action = Nil

Swept volume is necessary very much less than that for F-12.

F-22 has been increasingly used in recent years as a substitute for F-12. The range of application goes from small commercial units to medium size industrial units. It is mostly used for reciprocating compressors. With regard to miscibility with oil F-22 presents the peculiarity that is fully miscible with oil at the temperatures in the high side of the plant, whereas it is only partially miscible at usual evaporating temperatures. For this an effective oil separator should always be used.

Water (H_2O). Water freezes at 0º C at atmospheric pressure and therefore the applications requiring operating temperatures above this point, water has been used as a working fluid. The specific volume of water vapour at atmospheric temperatures is very great and hence reciprocating compressors can not be used. But centrifugal compressors are used with success.

36.6.4 Refrigerant Drying Agents

To remove moisture from the refrigerating systems the refrigerant drying agents are frequently used. These agents commonly used are

1. Silica gel (silicon dioxide)
2. Activated alumina (Aluminium oxide)
3. Anhydrous calcium sulphate

First two are available in granular form and are of absorption type. The last is available in the granular form and cast sticks and is of absorption type.

36.6.5 CFCs and Non-CFCs

From the properties of the halocarbon compounds containing one or more of the three halogens viz. Chlorine, Fluorine and Bromine, reduces health hazards and flammability. It is also observed that the stability increases as the number of fluorine atoms increases. That way these refrigerants are more or less ideal refrigerants with few exceptional qualities. The referigerants which do not contain hydrogen atoms, containing only carbon, fluorine and chlorine are known as CFCs (carbon-fluorine compounds) and are very much detrimental for the ozone layer round the earth.

If these compounds contain hydrogen atoms in addition to carbon, fluorine and chlorine, they are called as HCFs. Thus we have CFCs as CFCs— (1) F-12, (2) F-11 and many others.

Similarly HFCs F-22, 9($CHCF_2$) F-123 ($CHCl_2$ CF_3) and many others.

Refrigerants which do not contain any halogens i.e., chlorine, fluorine and bromine are termed as **non-CFCs**.

36.7 APPLICATIONS OF REFRIGERATION

Mechanical refrigeration serves human beings throughout their life. It has found tremendous applications in manufacture of ice, preservation of food stuff and in air conditioning. An analysis of the various applications of refrigeration indicates that they may be classified into three general groups:

1. The industrial process.
2. Preservation of perishable goods.
3. Providing comfortable environments.

Mechanical refrigeration provides aid in oil refining and rubber manufacturing, in processing food stuff and beverages and in treatment of metals or concrete for dams. It has helped in saving large quantities of food from being spoiled, by the development of frozen foods industry where the perishable food can be stored at low temperatures for a considerable amount of time without loosing its nutritive value or flavour. Further it has made the tropical regions more comfortable to live in by summer air conditioning of buildings. There are just a few examples wherein the science of refrigeration has helped us.

Specifically speaking the various applications of refrigeration may be listed as follows:

I. (a) Processing of foods products.
(b) Processing of farm crops.
(c) Processing of textiles, printing work, photographic materials etc.
(d) Cooling of concrete for dams.
(e) Treatment of air for blast furnaces.
(f) Processing of tobacco, petroleum and other chemical products.

II. (a) Manufacture of ice.
(b) Freezing or chilling, storage and transportation of food stuff including beverages, meat, poultry products, dairy products, fish, fruits, vegetables, fruit juices etc.

III. (a) Industrial air conditioning
(b) Comfort air conditioning of hospitals, residences, hotels, restaurant, theaters, offices etc.

SOLVED EXAMPLES

(Bell–Coleman Cycle)

Example 36.1 *In a Bell–Coleman cycle, air is drawn into the cylinder of a compressor at a pressure of 1 bar and –5°C. It is compressed adiabatically to 5 bar at which pressure it is cooled to 15°C. It is then expanded in the expansion cylinder to atmospheric pressure and discharged into the refrigerator chamber. Law of expansion is $PV^{1.2}$ = Constant.*

Find the work done on the air and the COP of the refrigeration plant and also the capacity of the plant in TR for a mass of flow of 10 kg/min.

Solution

The cycle for this problem is given in Fig. Ex. 36.1

For diagram shown we have,

$$P_1 = 1 \text{ bar}, T_1 = 273 - 5 = 268 \text{ K}, P_4 = 1 \text{ bar}$$
$$P_2 = 5 \text{ bar}, T_3 = 273 + 15 = 288 \text{ K}, P_3 = 5 \text{ bar}$$

We require to calculate all temperatures at the salient points.

We have
$$\frac{T_2}{T_1} = \left(\frac{P_2}{P_1}\right)^{\frac{\gamma-1}{\gamma}} = (5)^{\frac{0.4}{1.4}}$$
$$= 5^{0.286}$$
$$= \mathbf{1.5845}$$
$$T_2 = 268 \times 1.5845$$
$$= \mathbf{424.65\ K}$$

Similarly,

$$\frac{T_3}{T_4} = \left(\frac{P_3}{P_4}\right)^{\frac{n-1}{n}} = 5^{\frac{1.2-1}{1.2}}$$
$$= (5)^{\frac{1}{6}} = \mathbf{1.3084}$$
$$T_4 = \frac{T_3}{1.3084} = \frac{288}{1.3084}$$
$$= \mathbf{220.12\ K}$$

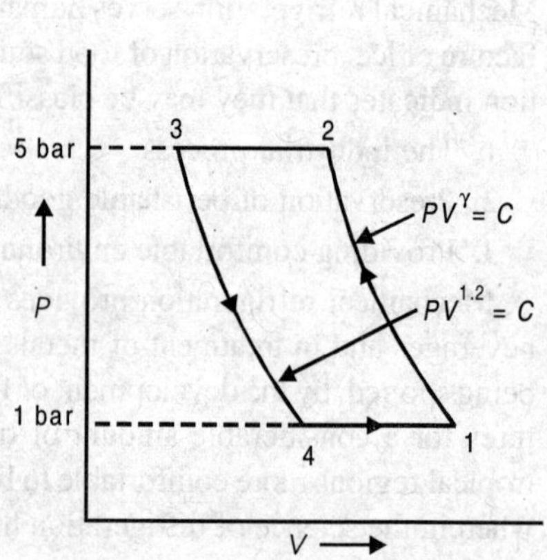

Fig. Ex. 36.1

$\therefore$ Work done/kg = Area of P–V diagram

$$= \frac{\gamma}{\gamma-1}[P_2V_2 - P_1V_1] - \frac{n}{n-1}[P_3V_3 - P_4V_4]$$
$$= \frac{1.4}{0.4}\cdot R[T_2 - T_1] - \frac{1.2}{0.2}\cdot R[T_3 - T_4]$$
$$= 3.5 \times 0.287\ (424.65 - 268) - 6 \times 0.287\ (288 - 220.12)$$
$$= 157.355 - 116.89$$
$$\mathbf{W} = \mathbf{40.465\ kJ/kg}$$

Refrigerating effect/kg = $C_p\,(T_1 - T_4)$

$$N = 1.005\,(268 - 220.12)$$
$$= \mathbf{48.12\ kJ/kg}$$

$$\therefore \quad \text{COP} = \frac{N}{W}$$
$$= \frac{48.12}{40.465}$$
$$= \mathbf{1.19}$$

Total refrigerating effect/sec

$$= \text{(Mass flow of air/sec)} \times \text{(Refrigerating effect/kg)}$$
$$= \frac{10}{60} \times 48.12$$
$$= \mathbf{8.02\ kJ/s}$$

or
$$= \mathbf{8.02\ kW}$$

$\therefore$ Capacity of the plant in TR

$$= \frac{\text{Total refrigerating effect/sec}}{3.516}$$
$$= \frac{8.02}{3.516}$$
$$= \mathbf{2.281\ TR}$$

Example 36.2 *In a refrigerator working on Bell–Coleman cycle, air is drawn into the compressor from the cold chamber at a pressure of 0.95 bar and temperature of –5°C. After compression (isentropic) to 5 bar, air is cooled at constant pressure to temperature of 15°C. The polytropic expansion is according to $PV^{1.3} = C$ and air is expanded to 1 bar to the chamber. Calculate (a) COP, (b) TR if m_a = 1.5 kg/min.*

Solution

The process taking place in a refrigerator are shown on *P–V* diagram of Fig. Ex. 36.2.

We have
$$P_1 = 0.95 \text{ bar},\ P_4 = 1 \text{ bar}$$
$$T_1 = -5 + 273 = 268 \text{ K}$$
$$T_3 = 15 + 273 = 288 \text{ K}$$
$$P_2 = P_3 = 5 \text{ bar}$$

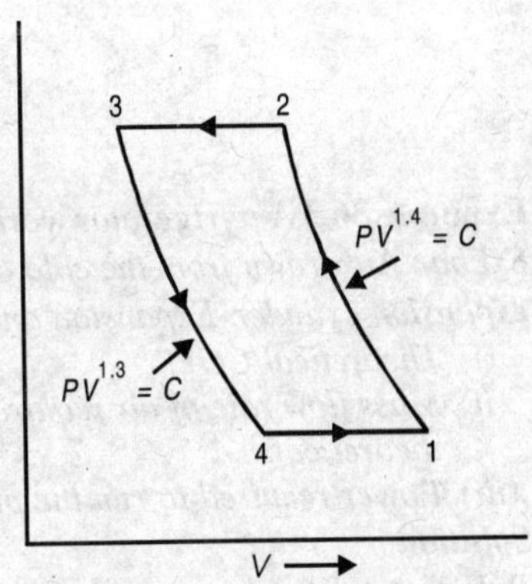

Fig. Ex. 36.2

We have

$$\frac{T_2}{T_1} = \left(\frac{P_2}{P_1}\right)^{\frac{\gamma-1}{\gamma}}$$
$$= \left(\frac{5}{0.95}\right)^{0.286}$$
$$= \mathbf{1.608}$$

$$\therefore \quad T_2 = 1.608 \times 268 = \mathbf{431\ K}$$

Similarly,
$$\frac{T_3}{T_4} = \left(\frac{P_3}{P_4}\right)^{\frac{n-1}{n}} = \left(\frac{5}{1}\right)^{\frac{1.3-1}{1.3}}$$
$$= (5)^{\frac{0.3}{1.3}} = \mathbf{1.45}$$

$$\therefore \quad T_4 = \frac{T_3}{1.45} = \frac{288}{1.45}$$
$$= \mathbf{198.62\ K}$$

(a) COP

Refrigerating effect/kg

$$N = C_p\,(T_1 - T_4)$$
$$= 1.005\,(268 - 198.48)$$
$$= \mathbf{69.87\ kJ/kg}$$

$$\text{Work done/kg} = C_{pa}\,(T_2 - T_1) - \frac{n}{n-1}\cdot R(T_3 - T_4)$$

$$W = 1.005\times(431-268) - \frac{1.3\times0.287}{0.3}\times(288-198.62)$$
$$= 1.005 \times 163 - 1.213 \times 89.38$$
$$= 163.81 - 108.45$$
$$= 55.36\ \text{kJ/kg}$$

$$\therefore \quad \text{COP} = \frac{N}{W} = \frac{69.87}{55.36}$$
$$= \mathbf{1.2621}$$

(b) Capacity in TR

$$\text{Refrigerating effect/sec} = \frac{\text{Mass of air}}{\text{sec}} \times \frac{\text{Refrigerating effect}}{\text{kg}}$$

$$= \frac{1.5}{60}\times 69.87$$
$$= \mathbf{1.7467\ \ kJ/sec}$$
$$= \frac{1.7467}{3.516}\ \text{TR}$$
$$= \mathbf{0.497}$$

Example 36.3 *A refrigerator working on Bell–Coleman cycle operates between the pressure limts of 1.05 bar and 8.5 bar. Air drawn from the cold chamber at 10°C is compressed and then it is cooled to 30°C before entering the expansion cylinder. Expansion and compression follows the law $PV^{1.35}$ = C. Determine*

(i) *Theoretical COP*

(ii) *Mass flow rate of air required in kg/min for a refrigerating capacity of 10 TR if the actual COP is 0.5 of the theoretical.*

(iii) *Power required to run the plant.*

Solution

P–V diagram for the example data is shown in Fig. Ex. 36.3.

For calculating the temperatures of the salient points we have,

$$\frac{T_2}{T_1} = \left(\frac{P_2}{P_1}\right)^{\frac{n-1}{n}} = \left(\frac{P_3}{P_4}\right)^{\frac{n-1}{n}} = \frac{T_3}{T_4}$$

$$\therefore \quad \frac{T_2}{T_1} = \left(\frac{8.5}{1.05}\right)^{\frac{0.35}{1.35}} = \mathbf{1.7199}$$

$$\therefore \quad T_2 = 283 \times 1.7199 = \mathbf{486.73\ K}$$

$$T_4 = \frac{T_3}{1.7199} = \frac{303}{1.7199} = \mathbf{176.2\ K}$$

Now we calculate the required facts.

(i) Theoretical COP

$$\text{Work done/kg} = \text{Area of } P\text{–}V \text{ diagram}$$

$$= \frac{n}{n-1}(P_2V_2 - P_1V_1) - \frac{n}{n-1}(P_3V_3 - P_4V_4)$$

$$= \frac{1.35}{0.35}\cdot R(T_2 - T_1) - \frac{1.35}{0.35}\cdot R(T_3 - T_4)$$

$$= \frac{1.35}{0.35}\times 0.287[486.73 - 283 - 303 + 107.334]$$

$$= \mathbf{85.16\ kJ/kg}$$

Refrigerating effect/kg

$$N = C_{pa}(T_1 - T_4) = 1.005(283 - 176.2)$$

$$= \mathbf{107.334\ kJ/kg}$$

$$\therefore \quad \text{COP} = \frac{N}{W} = \frac{107.334}{85.16}$$

$$= \mathbf{1.26}$$

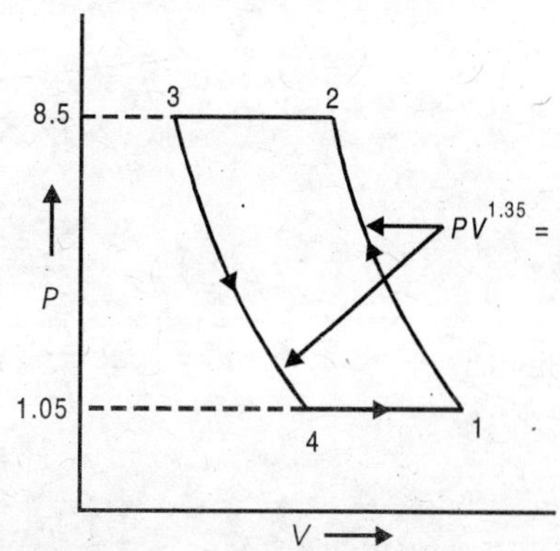

Fig. Ex. 36.3

(ii) Air Flow Rate

Actual refrigerating effect = 10 TR

= 10 × 3.516

= **35.16 kW** = kJ/sec

Refrigerating effect/kg = 107.334 kJ/kg

$$\therefore \quad \text{Actual mass flow rate} = \frac{35.16}{107.334} \text{ kg/sec}$$

∴ Required mass flow rate in kg/min

$$= \frac{35.16 \times 60}{107.334}$$

$$= \mathbf{19.65\ kg/min}$$

(iii) Power Required

$$\text{Actual COP} = 0.5 \times 1.26 = 0.63$$

$$= \frac{\text{Actual refrigerating effect}}{\text{Actual work}}$$

$$= \frac{35.16}{\text{Actual work}}$$

$$\therefore \quad \text{Actual work} = \frac{35.16}{0.63}$$

$$= 55.81 \text{ kJ/sec}$$

$$= \mathbf{55.81\ kW}$$

Example 36.4 *A refrigeration unit working on Bell–Coleman cycle operates between the pressure limits of 1 bar and 7 bar and maintains a refrigeration temperature of –13°C. Heat from the air compressed according to $PV^{1.3}$ = Constant, is removed such that it gets cooled to 37°C before expansion. Index of expansion is 1.35. Calculate:*

(i) Theoretical COP

(ii) Mass flow rate of air required to manufacture of ice, at 0°C from water at 30°C, at the rate of 5 tonnes per day.

(iii) Power required to run the plant.

Take C_{pa} = 1 kJ/kg-K

γ = 1.4

Latent heat of fusion of ice at 0°C = 336 kJ/kg

Specific heat of water = 4.187 kJ/kg-K

Solution

The *P–V* diagram is shown in Fig. Ex. 36.4. We have,

$$T_1 = -13 + 273 = \mathbf{260\ K}$$

$$T_3 = 307 + 273 = \mathbf{310\ K}$$

$$P_1 = P_4 = 1 \text{ bar}$$

$$P_2 = P_3 = 7 \text{ bar}$$

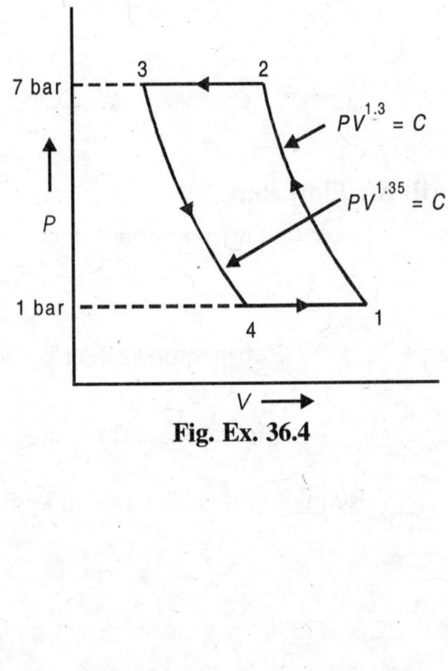

Fig. Ex. 36.4

We calculate all temperatures

$$\frac{T_2}{T_1} = \left(\frac{P_2}{P_1}\right)^{\frac{n-1}{n}} = \left(\frac{7}{1}\right)^{\frac{1.3-1}{1.3}} = 1.567$$

$$\therefore \quad T_2 = 360 \times 1.567$$

$$= \mathbf{407.42\ K}$$

Similarly,

$$\frac{T_3}{T_4} = \left(\frac{P_3}{P_4}\right)^{\frac{n-1}{n}} = \left(\frac{7}{1}\right)^{\frac{1.35-1}{1.35}}$$

$$= \mathbf{1.6563}$$

$$\therefore \quad T_4 = \frac{311}{1.6563}$$

$$= \mathbf{187.76\ K}$$

(i) Theoretical COP

$$\text{COP} = \frac{\text{Refrigerating effect}}{\text{Work supplied}}$$

$$\text{Refrigerating effect/kg} = C_{pa}(T_1 - T_4)$$

$$N = 1 \times (260 - 187.76)$$

$$= \mathbf{72.24\ kJ/kg}$$

$$\text{Work done/kg} = W$$

$$= \frac{1.3}{0.3} \cdot R[T_2 - T_1] - \frac{1.35}{0.35} \cdot R[T_3 - T_4]$$

$$= \frac{1.3\times0.287}{0.3}[407.42-260]-\frac{1.35\times0.287}{0.35}[310-187.76]$$

$= 183.34 - 136.43$

$= \mathbf{46.91\ kJ/kg}$

$\therefore$ $\text{COP} = \dfrac{72.24}{46.91}$

$= \mathbf{1.54}$

(ii) Air flow Rate

Actual refrigerating effect/min

= Actual heat extracted from water to produce ice at 0ºC/min

= Sensible heat + Latent heat

$= m_w C_{pw} (t_w - 0) + m_w \times \text{Latent heat}$

$$= \frac{5\times1000}{24\times60}\{4.187\times30+336\}$$

$$= \frac{5000}{1440}\times461.61$$

$= \mathbf{1598.75\ kJ/min}$

$\therefore$ Mass flow rate of air

$$= \frac{1598.75}{72.24}$$

$= \mathbf{22.131\ kg/min}$

(iii) Power Required

Work done per kg $= 46.91$ kJ/kg

Mass flow rate/sec $= \dfrac{22.131}{60}$ kg/sec

$\therefore$ Work done/sec $= 46.91\times\dfrac{22.131}{60}$ kJ/sec

$\therefore$ Power required $= \mathbf{17.3\ kW}$

Example 36.5 *A refrigerating machine of 8 TR capacity works on ideal Bell–Coleman cycle. The pressure of air entering the compressor is 1 bar and corresponding temperature is 20ºC. The highest pressure reached in the cycle is 6 bar. The air after compression gets cooled to 40ºC before entering the expansion cylinder. Assuming expansion and compression to be adiabatic with γ = 1.4, calculate*

(i) *COP of the cycle*
(ii) *Mass flow rate of refrigerant*
(iii) *Power required*
(iv) *Piston displacement of compressor*
(v) *Bore and stroke of compressor cylinder if compressor used is double acting and runs at 300 rpm (Given: Bore = Stroke)*

Solution

The cycle processes are shown on *P–V* diagram of Fig. Ex. 36.5.

Given : $P_1 = P_4 = 1$ bar

$$P_2 = P_3 = 6 \text{ bar}$$
$$T_3 = 40 + 273 = 311 \text{ K}$$
$$T_1 = 20 + 273 = 293 \text{ K}$$
$$\text{Capacity} = 8 \text{ TR}$$

For temperatures T_2 and T_4, we have

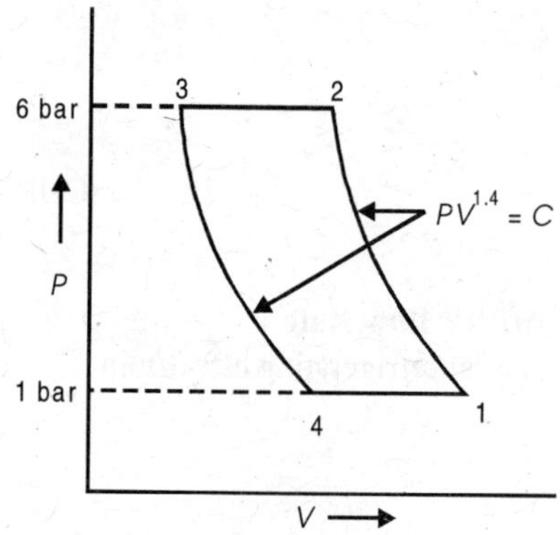

Fig. Ex. 36.5

$$\frac{T_2}{T_1} = \frac{T_3}{T_4} = \left(\frac{P_2}{P_1}\right)^{\frac{\gamma-1}{\gamma}} = (6)^{\frac{0.4}{1.4}}$$
$$= \mathbf{1.6694}$$

$\therefore$ $$T_2 = 293 \times 1.6694 = \mathbf{489.12\ K}$$

and $$T_4 = \frac{311}{1.6694} = \mathbf{187.5\ K}$$

(i) COP of Cycle

$$\text{Work done/kg} = Q_{2-3} - Q_{1-4}$$
$$= C_{pa}(T_2 - T_3) - C_{pa}(T_1 - T_4)$$
$$= 1.005\,(489.12 - 311 - 293 + 187.5)$$
$$= \mathbf{70.95\ kJ/kg}$$

Note: As compression and expansion are adiabatic, $W = Q_R - Q_A$.

$$\text{Refrigerating effect/kg} = C_{pa}(T_1 - T_4)$$
$$= 1.005\,(293 - 187.5)$$
$$= \mathbf{106.03\ kJ/kg}$$

$\therefore$ COP of the cycle is given by

$$\text{COP} = \frac{N}{W} = \frac{106.03}{70.95}$$
$$= \mathbf{1.494}$$

(ii) Mass Flow Rate of Refrigerant

$$\text{Capacity of machine} = 8 \text{ TR} = 8 \times 3.516$$
$$= 28.128 \text{ kW}$$
$$= \mathbf{28.128\ kJ/sec}$$

But refrigerating effect/kg = 106.03 kJ/kg

$\therefore$ Mass flow rate of refrigerant

$$= \frac{28.128}{106.03} = \mathbf{0.265\ kg/s}$$
$$= \mathbf{15.91\ kg/min}$$

(iii) Power Required

$$\text{Work done/sec} = \text{Mass flow rate/sec} \times \text{Work/kg}$$
$$= 0.265 \times 70.95$$
$$= 18.8 \text{ kJ/sec}$$
$$= \mathbf{18.8\ kW}$$

(iv) Piston Displacement of Compressor. Neglecting clearance – V_1 is the piston displacement/sec

= Volume of air in flow/sec

$$= \frac{mRT_1}{P_1}$$

$$= \frac{0.265 \times 287 \times 293}{1 \times 10^5}$$

$$= \mathbf{0.22284\ m^3/sec}$$

(v) Bore and Stroke of Compressor Cylinder. As comperssor is double acting,

$$V_1 = 2 \times \frac{\pi d^2}{4} \times L \times \frac{\text{RPM}}{60}$$

$$\therefore \quad 0.22284 = 2 \times \frac{\pi d^3}{4} \times \frac{300}{60}$$

$$\therefore \quad d^3 = 0.0284$$

$$d = L = 0.305 \text{ m}$$

$$= \mathbf{30.5\ cm}$$

Example 36.6. *An air refrigerator operates between pressure limits of 1 bar and 8 bar on the reversed Joule cycle. Air enters the compressor at 6.7°C and enters the expander turbine at 26.7°C. The compression and expansion follows the law $PV^{1.35} = C$.*

Determine : (i) COP, (ii) Specific power required in kW/TR if the flow rate of air is 6 kg/min.

Take C_{pa} = 1.005 and C_v = 0.718 kJ/kg-K

Solution

Given :

$$P_1 = P_4 = 1 \text{ bar}$$

$$P_2 = P_3 = 8 \text{ bar}$$

$$T_1 = 273 + 6.7 = 279.7 \text{ K}$$

$$T_3 = 273 + 26.7 = 299.7 \text{ K}$$

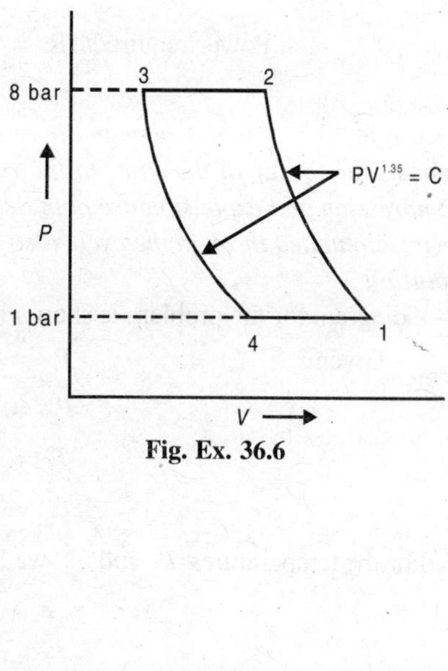

Fig. Ex. 36.6

We have

$$\frac{T_2}{T_1} = \frac{T_3}{T_4} = \left(\frac{P_2}{P_1}\right)^{\frac{\gamma-1}{\gamma}} = (8)^{\frac{1.35-1}{1.35}}$$

$$= \mathbf{1.7146}$$

$$\therefore \quad T_2 = 279.7 \times 1.7146 = \mathbf{479.6\ K}$$

$$T_4 = \frac{299.7}{1.7146} = \mathbf{174.79\ K}$$

(i) COP

$$\text{Refrigerant effect/kg} = C_{pa}(T_1 - T_4)$$

$$= 1.005(279.7 - 174.79)$$

$$= \mathbf{105.43\ kJ/kg}$$

$$\text{Work done/kg} = \frac{n}{n-1}[P_2V_2 - P_1V_1] - \frac{n}{n-1}[P_3V_3 - P_4V_4]$$

$$= \frac{1.35}{0.35} \cdot R[T_2 - T_1 - T_3 + T_4]$$

$$= \frac{1.35 \times 287}{0.35}[479.6 - 279.7 - 299.7 + 174.79]$$

$$= \mathbf{83.7\ kJ/kg}$$

$$\therefore \quad COP = \frac{\text{Refrigerating effect}}{\text{Work}}$$

$$= \frac{105.43}{83}$$

$$= \mathbf{1.27}$$

(ii) Specific Power Required (kW/TR)

$$\text{Work done kJ/sec} = \frac{\text{Mass}}{\text{sec}} \times \frac{\text{Work}}{\text{kg}}$$

$$= \frac{6}{60} \times 83$$

$$= 8.3\ \text{kJ/sec}$$

$$= \mathbf{8.3\ kW}$$

$$\text{Refrigerating effect kJ/sec} = \frac{6}{60} \times 105.43$$

$$= \mathbf{10.543\ kJ/s}$$

$$\therefore \quad \text{TR capacity} = \frac{10.543}{3.516}$$

$$= \mathbf{2.9986\ TR}$$

$$\therefore \quad \text{Power required/TR} = \frac{8.3}{2.9986}$$

$$= \mathbf{2.7678\ kW/TR}$$

Example 36.7. *A Bell–Coleman air refrigeration cycle working between the pressure limits of 1 and 8 bar. The temperature of air at the entry to the compressor is 20°C and at the entry to the expander is 30°C. The laws of compression and expansion are both according to the law of $PV^{1.3}$ = Constant. Find the refrigerating effect of kg or air circulated through the cycle also find the COP. Take C_{pa} = 1.05 kJ/kg/K.*

Solution

P–V diagram for the problem is shown in Fig. Ex. 36.7

Given :

$$P_1 = P_4 = 1\ \text{bar}$$

$$P_2 = P_3 = 8\ \text{bar}$$

$$T_1 = 20 + 273 = 293\ \text{K}$$

$$T_3 = 30 + 273 = 303\ \text{K}$$

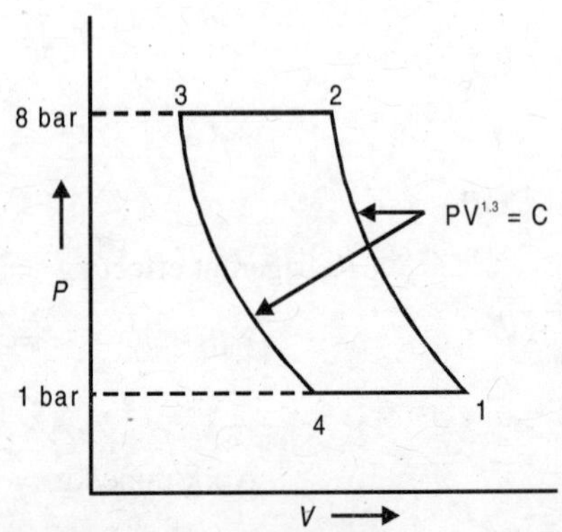

Fig. Ex. 36.7

For the temperatures T_2 and T_4 we have,

$$\frac{T_2}{T_1} = \frac{T_3}{T_4} = \left(\frac{P_2}{P_1}\right)^{\frac{n-1}{n}}$$

$$= (8)^{\frac{1.3-1}{1.3}} = \mathbf{1.616}$$

$\therefore \qquad T_2$ = Temperature after compression

$= 293 \times 1.616$

$= \mathbf{473.5\ K}$

and $\qquad T_4$ = Temperature after expansion

$$= \frac{303}{1.616} = \mathbf{187.5\ K}$$

$\therefore$ Refrigerating effect/kg air

$$= C_{pa}(T_1 - T_4)$$

$$= 1.05 \times [293 - 187.5]$$

$$= \mathbf{110.775\ kJ/kg}$$

To find COP of the cycle, we have to calculate work/kg air

$$\text{Work done/kg} = \frac{n}{n-1} \cdot R[T_2 - T_3 - T_1 + T_4]$$

because n for the expansion and compression are is same.

$$\therefore \qquad W = \frac{1.3}{0.3} \times 0.287[473.5 - 303 - 293 + 187.5]$$

$$= \mathbf{80.84\ kJ/kg}$$

$$\therefore \qquad \text{COP} = \frac{N}{W} = \frac{110.775}{80.84}$$

$$= \mathbf{1.378}$$

Example 36.8. *A 20 TR air refrigeration plant working on the Bell–Coleman cycle draws air from the cold chamber at 1 bar and –10°C compressed it to 4 bar. The air is then cooled to 40°C at constant pressure before it enters the expander. Assume compression and expansion index as 1.4. Calculate : (i) COP (ii) mass flow rate of air circulated per minute (iii) piston diameter for the expander and compressor both running at 500 rpm and are single acting having a stroke of 20 cm (iv) power required. Neglect clearance.*

Solution

Given :

$$P_1 = P_4 = 1 \text{ bar}$$

$$P_2 = P_3 = 4 \text{ bar}$$

$$T_1 = -10 + 273 = 263 \text{ K}$$

$$T_3 = 40 + 273 = 313 \text{ K}$$

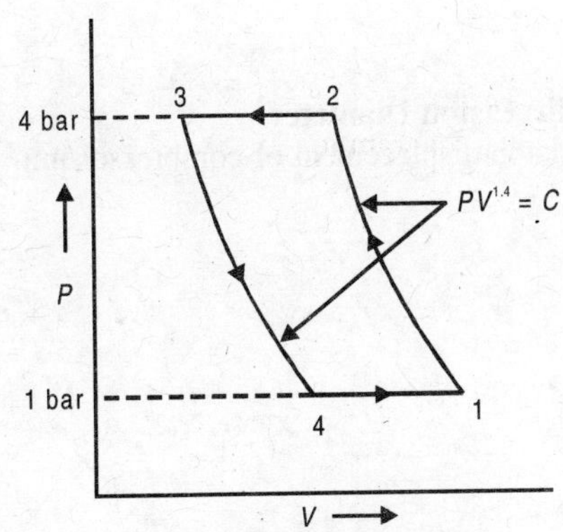

Fig. Ex. 36.8

First calculate T_2 and T_4

$$\therefore \qquad \frac{T_2}{T_1} = \left(\frac{P_2}{P_1}\right)^{\frac{\gamma-1}{\gamma}} = (4)^{\frac{0.4}{1.4}}$$

$$= \mathbf{1.4866}$$

$\therefore \qquad T_2$ = Temperature after compression

$= 263 \times 1.4866$

$= \mathbf{390.97\ K}$

Similarly,

$$\frac{T_3}{T_4} = \left(\frac{P_3}{P_4}\right)^{\frac{\gamma-1}{\gamma}} = (4)^{\frac{0.4}{1.4}} = \mathbf{1.4866}$$

$$\therefore \quad T_4 = \frac{313}{1.4866} = \mathbf{210.55\ K}$$

(i) C.O.P.

$$\text{COP} = \frac{N}{W}$$

where

$$N = \text{Refrigerating effect/kg}$$
$$= C_{pa}(T_1 - T_4)$$
$$= 1.005(263 - 210.55)$$
$$= \mathbf{52.71\ kJ/kg}$$
$$W = \text{Work done/kg}$$
$$= \frac{\gamma}{\gamma-1} \cdot R[T_2 - T_3 - T_1 + T_4]$$
$$= C_{pa}[T_2 - T_3 - T_1 + T_4]$$
$$= 1.005[390.97 - 313 - 263 + 210.55]$$
$$= \mathbf{25.66\ kJ/kg}$$

$$\therefore \quad \text{COP} = \frac{52.71}{25.66}$$
$$= \mathbf{2.054}$$

(ii) Mass Flow Rate of Air kg/min

$$\text{Capacity of plant} = 20\ \text{TR}$$
$$= 20 \times 3.516$$
$$= \mathbf{70.32\ kW = kJ/sec}$$

∴ Mass flow rate of air kg /min

$$= \frac{\text{Ref. eff./min}}{\text{Ref. eff./kg}}$$
$$= \frac{70.32 \times 60}{52.71}$$
$$= \mathbf{80.06\ kg/min}$$

(iii) Piston Diameter

Piston displacement of compressor/min

$$= V_1/\text{min}$$
$$= \frac{mRT_1}{P_1}$$
$$= \frac{80.06 \times 287 \times 263}{1 \times 10^5}$$
$$= 60.385\ \text{m}^3/\text{min}$$
$$= \frac{\pi d^2}{4} \times L \times N$$

$$= \frac{\pi d^2}{4} \times 0.2 \times 500$$

$$\therefore \quad d_c^2 = \frac{60.385 \times 4}{\pi \times 0.2 \times 500}$$

$$= 0.7692$$

$\therefore \quad d_c$ = Diameter of compressor piston

$$= 0.877 \text{ m}$$

$$= \mathbf{87.7 \text{ cm}}$$

Similarly, V_4 = Displacement of expander piston

$$= \frac{mRT_4}{P_4}$$

$$= \frac{80.06 \times 287 \times 210.55}{1 \times 10^5}$$

$$= \mathbf{48.252 \text{ m}^3/\text{min}}$$

$$= \frac{\pi}{4} d_e^2 \cdot LN$$

$$\therefore \quad d_e^2 = \frac{48.252 \times 4}{\pi \times 0.2 \times 500}$$

$$= 0.6141$$

$\therefore \quad d_e$ = Diameter of expander piston

$$= 0.7836 \text{ m} = \mathbf{78.36 \text{ cm}}$$

(iv) Power Required

Work done/kg = **25.66 kJ/kg**

Mass flow rate = 80.06 kg/min

$$\therefore \quad \text{Power required} = \frac{25.66 \times 80.06}{60} \text{ kJ/s}$$

$$= \mathbf{34.24 \text{ kW}}$$

Example 36.9. *A dense air machine operates on reverse Brayton cycle and is required for a capacity of 10 TR. The cooler pressure is 4.2 bar and refrigerator pressure is 1.4 bar. The air is cooled in the cooler at a temperature of 50°C and temperature of air at inlet to compressor is –20°C. Determine for the ideal cycle:*

(i) C.O.P.

(ii) Mass of air circulated per min.

(iii) Theoretic piston displacement of compressor and expander.

(iv) Net power per tonne of refrigeration.

Solution

Theoretical cycle is shown on *P–V* diagram of Fig. Ex. 36.9.

Given :

$$P_1 = P_4 = 1.4 \text{ bar}$$

$$P_2 = P_3 = 4.2 \text{ bar}$$

$$T_1 = -20 + 273 = 253 \text{ K}$$

$$T_3 = 50 + 273 = 323 \text{ K}$$

$$\therefore \quad \frac{T_2}{T_1} = \left(\frac{P_2}{P_1}\right)^{\frac{\gamma-1}{\gamma}} = \left(\frac{4.2}{1.4}\right)^{\frac{0.4}{1.4}} = (3)^{\frac{2}{7}}$$

$$= \mathbf{1.3692}$$

$$\therefore \quad T_2 = 253 \times 1.3692 = \mathbf{346.4\ K}$$

Similarly, $$T_4 = \frac{T_3}{1.3692} = \frac{323}{1.3692} = \mathbf{235.9\ K}$$

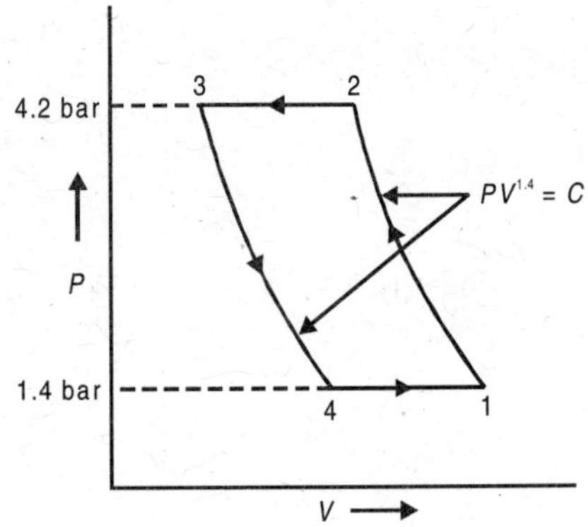

Fig. Ex. 36.9

(i) COP

$$\text{Refrigerating effect/kg} = C_{pa}(T_1 - T_4)$$

$$= 1.005(253 - 235.9)$$

$$N = \mathbf{17.1855\ kJ/kg}$$

Work done/kg

$$W = C_{pa}[T_2 - T_3 - T_1 + T_4]$$

$$= 1.005[346.4 - 323 - 253 + 235.9]$$

$$= \mathbf{6.3315\ kJ/kg}$$

$$\therefore \quad \text{COP} = \frac{N}{W} = \frac{17.1855}{6.3315}$$

$$= \mathbf{2.714}$$

(ii) Mass of Air Circulated

$$\text{Required capacity} = 10\ \text{TR}$$

$$= 10 \times 3.516\ \text{kW}$$

$$= \mathbf{35.16\ kJ/sec}$$

$$\text{Refrigerating effect/kg} = 17.1855\ \text{kJ/kg}$$

$$\therefore \quad \text{Mass of air/sec} = \frac{35.16}{17.1855}\ \text{kg/sec}$$

$$\therefore \quad \text{Air/min} = \frac{35.16 \times 60}{17.1855}$$

$$= \mathbf{122.75\ kg/min}$$

(iii) Piston Displacement of Compressor and Expander

$$\text{Piston displacement for compressor} = V_1 = \frac{mRT_1}{P_1}$$

$$= \frac{122.75 \times 287 \times 253}{1.4 \times 10^5}$$

$$= \mathbf{63.66\ m^3/min}$$

Similarly piston displacement of expander

$$= V_4 = \frac{mRT_4}{P_4}$$

$$= \frac{122.75 \times 287 \times 235.9}{1.4 \times 10^5}$$

$$= \mathbf{59.36\ m^3/min}$$

(iv) Net Power Required: (kW/TR)

$$\text{Work done/kg} = 6.3315 \text{ kJ/kg}$$

$$\text{Mass of air/sec} = \frac{122.75}{60} \text{ kg/s}$$

$$\therefore \quad \text{Work done/sec} = \frac{6.3315 \times 122.75}{60}$$

$$= \mathbf{12.953\ kJ/sec = kW}$$

$$\therefore \quad \text{Power required/TR} = \frac{12.953}{10}$$

$$= \mathbf{1.2953\ kW/TR}$$

Example 36.10. *Air is drawn from the cold chamber at 1 bar and 16°C. It is compressed to 5.2 bar and then cooled to 41°C before passing through an expander. Assuming both compression and expansion to be isentropic, calculate*

(i) COP

(ii) Mass flow rate of air in kg/s

(iii) Bore of compressor and expansion cylinders if the stroke length is 200 mm, speed is 240 rpm and the units are single acting.

Take – $C_{pa} = 1.005$ *kJ/kg/K,* $\gamma = 1.4$*, capacity of the machine = 5 TR*

Solution

P–V diagram for the cycle is shown in Fig. Ex. 36.10.

Given :

$$P_1 = P_4 = 1 \text{ bar}$$

$$P_2 = P_3 = 5.2 \text{ bar}$$

$$T_1 = 16 + 173 = 289 \text{ K}$$

$$T_3 = 41 + 273 = 314 \text{ K}$$

$$\text{TR capacity} = 5$$

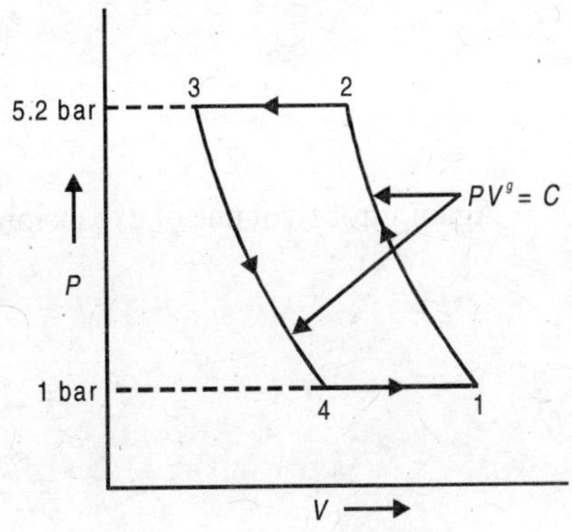

Fig. Ex. 36.10

First we will calculate for T_2 and T_4.

$$\frac{T_2}{T_1} = \left(\frac{P_2}{P_1}\right)^{\frac{\gamma-1}{\gamma}} = (5.2)^{\frac{2}{7}} = 1.602$$

$$\therefore \quad T_2 = 289 \times 1.602 = \mathbf{462.87\ K}$$

and

$$T_4 = \frac{314}{1.602} = \mathbf{196\ K}$$

(i) COP

$$\text{Refrigerating effect/kg} = C_{pa}(T_1 - T_4)$$

$$N = 1.005(289 - 196)$$

$$= \mathbf{93.465\ kJ/kg}$$

$$\text{Work done/kg } W = \frac{\gamma}{\gamma-1}\cdot R[T_2 - T_3 - T_1 + T_4]$$

$$= 1.005[462.87 - 314 - 289 + 196]$$

$$= \mathbf{56.15\ kJ/kg}$$

$$\therefore \quad \text{COP} = \frac{N}{W} = \frac{93.465}{56.15}$$

$$= \mathbf{1.665}$$

(ii) Mass Flow Rate

$$\text{TR capacity} = 5$$

$$\therefore \quad \text{Refrigerating effect/sec} = 5 \times 3.516 = \mathbf{17.580\ kJ/sec}$$

$$\text{Refrigerating effect/kg} = 93.465\ \text{kJ/kg}$$

$$\therefore \quad \text{Air flow/sec} = \frac{17.58}{93.465}$$

$$= \mathbf{0.1881\ kg/s}$$

(iii) Bore of Compressor and Expander

$$\text{Stroke Volume of Compressor} = V_1$$

$$= \frac{mRT_1}{P_1} = \frac{0.1881 \times 287 \times 289}{1 \times 10^5}$$

$$= \mathbf{0.156\ m^3/sec}$$

$$= \frac{\pi d_c^2}{4} \times \frac{LN}{60} = \frac{\pi d_c^2}{4} \times \frac{200}{1000} \times \frac{240}{60}$$

$$\therefore \quad d_c^2 = \frac{0.156 \times 4}{\pi} \times \frac{5}{4} = \frac{0.156 \times 5}{\pi}$$

$$= \mathbf{0.2483}$$

$$\therefore \quad d_c = \text{Bore of compressor cylinder}$$

$$= 0.4983\ \text{m}$$

$$= \mathbf{49.83\ cm}$$

Again, stroke volume of expansion cylinder

$$V_4 = \frac{mRT_4}{P_4} = \frac{0.1881 \times 287 \times 196}{1 \times 10^5}$$

$$= 0.1058\ \text{m}^3/\text{sec}$$

$$= \frac{\pi d_e^2}{4} \times L \times \frac{N}{60} = \frac{\pi d_e^2}{4} \times 0.2 \times \frac{240}{60}$$

$$\therefore \quad d_e^2 = \frac{0.1058 \times 4}{\pi} \times \frac{1}{0.2 \times 4}$$

$$= \frac{0.1058 \times 5}{\pi}$$

$$= 0.1682$$

$$\therefore \quad d_e = \text{Bore of expander cylinder} = 0.41\ \text{m}$$

$$= \mathbf{41\ cm}$$

Example 36.11 *A vapour compression refrigeration plant operates between evaporation and condensation temperatures of –10°C and 45°C respectively. The refrigerant is dry and saturated vapour at the entry to the compressor. It is discharged at 102°C from the compressor. The bore and stroke of the compressor are 80 mm each. It runs at 720 rpm with a volumetric efficiency of 80%. The liquid refrigerant enters the expansion valve at 35°C. Determine*

(1) COP

(2) Mass low rate of refrigerant

(3) Capacity of the plant in TR

Take specific heat of liquid refrigerant =1.62 kJ/kg-K. You may use the properties of refrigerant tabulated below.

Sat. temperature	V_g	h_f kJ/kg	h_g kJ/kg	S_f kJ/kg-K	S_g kJ/kg-K
–10°C	0.233	45.4	460.7	0.183	1.762
45°C	0.046	133.0	488.6	0.485	1.587

Solution

(i) COP

The cycle of operation is shown on *T–S* diagram in Fig. Ex. 36.11.

In Fig. Ex. 36.11, 1–2 is an isentropic process

$$\therefore \quad S_1 = S_2$$

$$\therefore \quad S_{g_1} = S_{g_2} + C_{p_v} \log_e \frac{T_2}{T_{\text{sat}}2}$$

$$\therefore \quad 1.762 = 1.587 + C_{p_v} \log_e \frac{102+273}{45+273}$$

$$\therefore \quad 1.762 - 1.587 = C_{p_v} \log_e \frac{375}{318} = \mathbf{0.175}$$

$$\therefore \quad C_{p_v} = \frac{0.175}{\log_e 1.1792}$$

$$= \mathbf{1.061 \ kJ/kg\text{-}K}$$

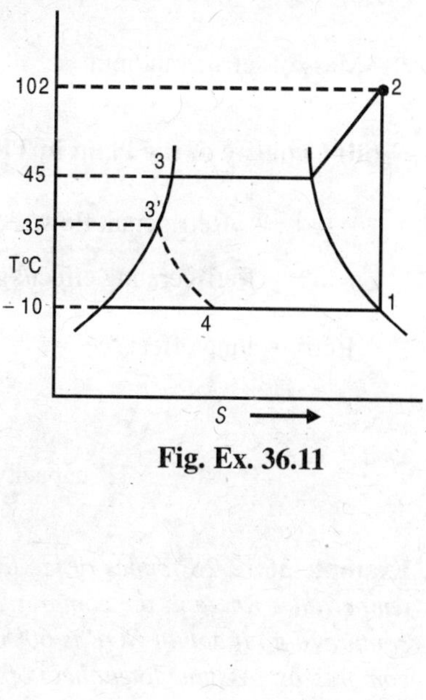

Fig. Ex. 36.11

Again, Work done/kg $= h_2 - h_1$

where $h_1 = 460.7$ kJ/kg

$$h_2 = h_{g_2} + C_{p_v}(T_2 - T_{\text{sat}\,2})$$

$$= 488.6 + 1.061\,(102 - 45)$$

$$= \mathbf{549.08 \ kJ/kg}$$

$$\therefore \quad \text{Work done/kg} = 549.08 - 460.7$$

$$= \mathbf{88.38 \ kJ/kg}$$

Also, refrigerating effect/kg

$$N = h_1 - h_4 = h_1 - h_{3'}$$

where $h_{3'} = h_3 - C_{p_a}(45-35)$

$$= 133 - 1.62 \times 10 = 133 - 16.2$$

$$= \mathbf{116.8 \ kJ/kg}$$

$\therefore \quad N = 460.7 - 116.8$

$= 343.9$ kJ/kg

$\therefore \quad \text{COP} = \dfrac{N}{W} = \dfrac{343.9}{88.38}$

$= \mathbf{3.89}$

(ii) Mass Flow Rate of Refrigerant

Theoretical volume taken in the compressor/min

= Displacement volume of compressor/min

$= \dfrac{\pi d^2}{4} \times L \times N$

$= \dfrac{\pi}{4} \times \left(\dfrac{80}{1000}\right)^2 \times \dfrac{80}{1000} \times 720 = \mathbf{0.2895\ m^3/min}$

$= \dfrac{\pi}{4} \times \left(\dfrac{80}{1000}\right)^2 \times \dfrac{80}{1000} \times 720 = \mathbf{0.2895\ m^3/min}$

$\therefore$ Actual volume of refrigerant vapour/min

= Theoretical volume × Volumetric efficiency

$= 0.2895 \times 0.8$

$= \mathbf{0.2316\ m^3/min}$

But specific volume of vapour $= \mathbf{0.233\ m^2/kg}$

$\therefore$ Mass of refrigerant/min $= \dfrac{0.2316}{0.233}$

$= \mathbf{0.994\ kg/min}$

(iii) Capacity of the Plant in TR

Refrigerant flow/sec $= \dfrac{0.9941}{60}$ kg/s

Refrigerating effect/kg $= 343.9$ kJ/kg

$\therefore$ Refrigerating effect/sec $= \dfrac{343.9 \times 0.9941}{60}$

$= \mathbf{5.698\ kJ/sec = kW}$

$\therefore$ TR capacity $= \dfrac{5.698}{3.576}$

$= \mathbf{1.62\ TR}$

Example 36.12 *28 tonnes of ice from and at 0°C is produced per day in an ammonia refrigeration plant. The temperature range in the compressor is from 25°C to –15°C. The refrigerant is dry and saturated at the end of compression. If actual COP is 60% of the theoretical COP, calculate the power supplied or required to drive the compressor. Assume latent heat of ice =335kJ/kg. Use properties of refrigerant given below:*

Temperature °C	h_f *kJ/kg*	h_g *kJ/kg*	S_f *kJ/kg-K*	S_g *kJ/kg-K*
+25	+100.04	1319.22	+0.3473	4.4852
–15	–54.56	13.04.99	–2.1338	5.0585

Solution

T–S diagram for the cycle is shown in Fig. Ex. 36.12

(i) Theoretical C.O.P.

$$\text{Theoretical C.O.P} = \frac{h_1 - h_4}{h_2 - h_1} = \frac{h_1 - h_3}{h_2 - h_1}$$

First we have to find all the enthalpies.

$$h_2 = \text{Enthalpy of vapour after compression}$$
$$= \mathbf{1319.22\ kJ/kg}$$

3–4 is isenthalpic process

$$\therefore \quad h_3 = h_4$$

or $$h_{f_3} = h_4 = \mathbf{100.04\ kJ/kg}$$

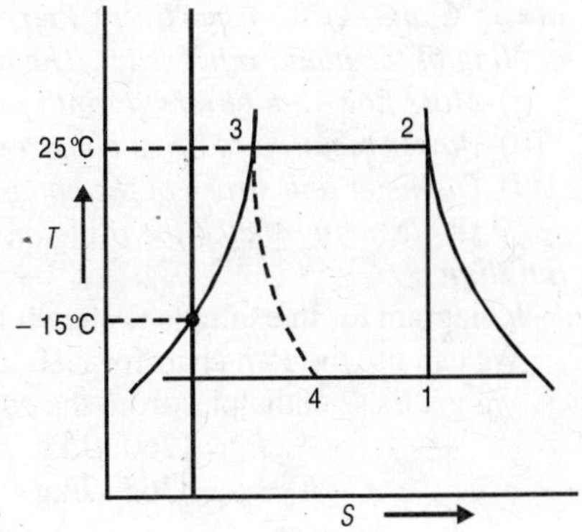

Fig. Ex. 36.12

To find h_1,we will require to know the quality of vapour at compressor inlet. For this – $S_2 = S_1$

$$\therefore \quad 4.4852 = S_{f_1} + x_1 S_{fg_1}$$
$$= -2.1338 + x_1\left(S_{g_1} - S_{f_1}\right)$$
$$= -2.1338 + x_1\ (5.0585 + 2.1338)$$
$$\therefore \quad x_1 = \frac{4.4852 + 2.1338}{7.1923} = \frac{6.6190}{7.1923}$$
$$= \mathbf{0.9203}$$
$$\therefore \quad h_1 = h_{f_1} + x_1 h_{fg_1}$$
$$= -54.56 + 0.9203\ (1304.99 + 54.56)$$
$$= -54.56 + 0.9203 \times 1359.55$$
$$= \mathbf{1196.62\ kJ/kg}$$

∴ Theoretical COP

$$\frac{h_1 - h_3}{h_2 - h_1} = \frac{1196.62 - 100.04}{1319.22 - 1196.62}$$
$$= \mathbf{8.9444}$$
$$\therefore \quad \text{Actual COP} = 0.6 \times 8.9444$$
$$= \mathbf{5.3664}$$

Again actual refrigerating effect /sec

$$= m_w\left[C_{p_w}\ (t_w - 0) + \text{Latent heat}\right]$$
$$= \frac{28 \times 1000}{24 \times 3600}[0 + 335]$$
$$= \mathbf{108.565\ kJ/kg}$$
$$\therefore \quad \text{Actual COP} = \frac{\text{Actual referigerating effect/sec}}{\text{Power required}}$$
$$= \frac{108.56(\text{kW})}{\text{Power}(\text{kW})}$$
$$\therefore \quad \text{Power (kW)} = \frac{108.565}{5.3664}$$
$$= \mathbf{20.23\ kW}$$

Example 36.13 *An NH_3 refrigeration plant has a capacity of 20 TR. The condensation and evaporation temperatures are 35°C and –20°C respectively. Refrigerant is dry and saturated at the entry to the compressor. There is no under cooling of the liquid refrigerant. The actual COP is 0.7 times the theoretical COP, determine,*

(i) *Mass flow rate of refrigerant*
(ii) *Power required to drive compressor*
(iii) *Diameter and stroke of the compressor running at 4 RPS and L= D. Its volumetric efficiency is 80% and is single acting sketch the cycle on P–h diagram.*

Solution

P–h diagram for th example is shown in Fig. Ex. 36.13.

We can use the *P–h* chart for NH_3 and find the enthalpies from the chart.

We get these enthalpies from the chart.

$$h_1 = 1260 \text{ kJ/kg} \qquad h_2 = 1488 \text{ kJ/kg}$$
$$h_3 = h_4 = 365 \text{ kJ/kg}$$

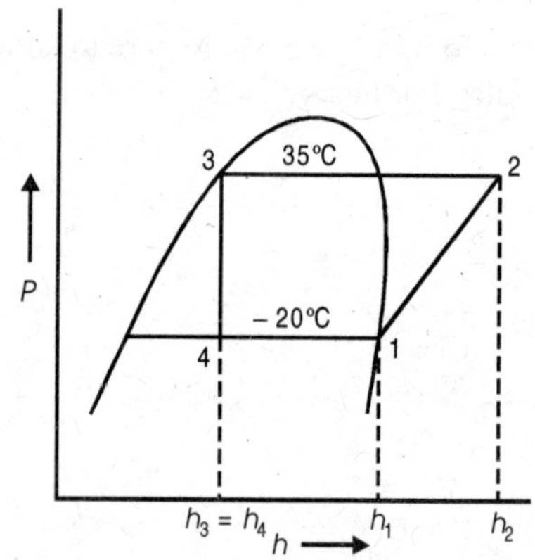

Fig. Ex. 36.13

$$\therefore \quad \text{C.O.P.} = \frac{h_1 - h_4}{h_2 - h_1} = \frac{1260 - 365}{1488 - 1260}$$
$$= \frac{895}{228}$$
$$= \mathbf{3.925}$$

But actual C. O. P. $= 0.7 \times 3.925$
$= \mathbf{2.7475}$

Actual refrigerating effect $= 20$ TR
$= 20 \times 3.516$ kJ/sec
$= \mathbf{70.32\ kJ/sec}$

But refrigerating effect/kg $= 895$ kJ/kg

$$\therefore \quad \text{Mass of refrigerant kg/s} = \frac{70.32}{895}$$
$$= \mathbf{0.0787\ kg/s}$$

$$\text{Now we have, Actual COP} = \frac{\text{Actual ref. effe. (kW)}}{\text{Actual power (kW)}}$$

$$\therefore \quad 2.7475 = \frac{70.32 \text{ (kW)}}{\text{Actual power (kW)}}$$

$$\therefore \quad \text{Actual power} = \frac{70.32}{2.7475}$$
$$= \mathbf{25.6\ kW}$$

Power required to drive the compressor
$= \mathbf{25.6\ kW}$

(iii) Diameter and Stroke of Compressor

Actual volume of refrigerant at the compressor inlet/kg = 0.54 m^3/kg

$\therefore$ Volume/sec $= 0.0787 \times 0.54$
$= \mathbf{0.0425\ m^3/sec}$

$\therefore$ Theoretical Swept volume/sec

$$= \frac{0.0425}{0.8}$$
$$= \mathbf{0.05312\ m^3/sec}$$

$$\text{But swept volume/sec} = \frac{\pi d^2}{4} \times L \times \frac{N}{60}$$

$$\text{But } D = L$$

$$\therefore \quad 0.05312 = \frac{\pi D^3}{4} \times \frac{N}{60}$$

$$= \frac{\pi D^3}{4} \times 4 = \pi D^3$$

$$\therefore \quad D^3 = \frac{0.05312}{\pi} = 0.0169$$

$$\therefore \quad D = \text{Diameter of cyclinder}$$

$$= (0.0169)^{1/3}$$

$$= 0.2565 \text{ m}$$

$$= \mathbf{25.65\ cm}$$

$$= \text{Stroke of piston}$$

Example 36.14 *An NH_3 vapour compressor cycle refrigerator operates between an evaporation and condensation temperature of – 10°C and 35°C respectively. The refrigerant is dry and saturated at entry to the compressor and is discharged at 102°C after compression. The single acting compressor has a base 80 mm and stroke 80 mm. It runs at 480 rpm with volumetric efficiency of 80%. Assuming liquid subcooling of 10°C, calculate:*

(i) COP
(ii) Mass flow of refrigerant
(iii) Capacity of the plant in TR
Sketch cycle on P–h chart

Solution

The *P–h* diagram is shown in Fig. Ex. 36.14.

Given
$$t_2 = 102°\text{C}$$
$$t_3 = 35–10$$
$$= 25°\text{C}$$
$$t_1 = 10°\text{C}$$
$$\eta_{vol} = 80\%$$
$$D = L = 80 \text{ mm, rpm} = 480$$
$$t_a = 35°\text{C undercooling by } 10°\text{C}$$

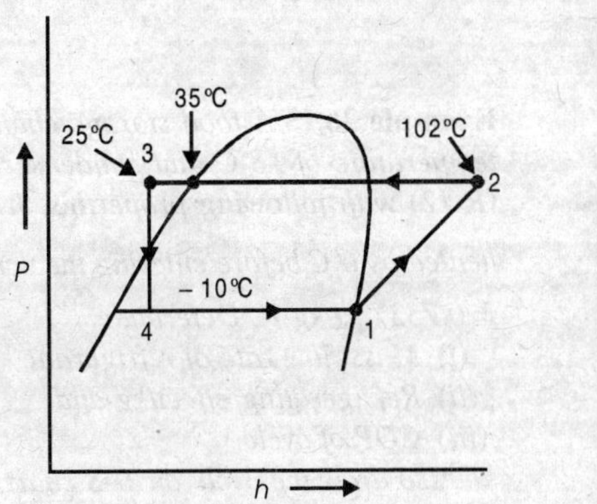

Fig. Ex. 36.14

From *P–h* chart for NH_3, we get the following enthalpies

$$h_1 = 1450 \text{ kJ/kg}, h_2 = 1680 \text{ kJ/kg}$$
$$h_3 = h_4 = 325 \text{ kJ/kg and } v_3 = 0.4175 \text{ m}^3$$

Now we calculate the required quantities.

(i) COP

$$\text{Refrigerating effect/kg} = h_1 - h_3 = 1450 - 325$$
$$= \mathbf{1125\ kJ/kg}$$
$$\text{Work supplied/kg} = h_1 - h_2 = 1680 - 1450$$
$$= \mathbf{230\ kJ/kg}$$

$$\therefore \quad \text{COP} = \frac{\text{Ref. effect/kg}}{\text{Work supplied/kg}}$$

$$= \frac{1125}{230} = \mathbf{4.891}$$

(ii) Mass Flow Rate of Refrigerant

$$\text{Stroke volume/min.} = \text{Theoretical volume/min.}$$

$$= \frac{\pi D^2}{4} \times L \times N$$

$$= \frac{\pi}{4} \times \left(\frac{80}{1000}\right)^3 \times 480$$

∴ Actual volume of refrigerant taken in/min

$$= \frac{\pi}{4} \times \left(\frac{80}{1000}\right)^3 \times \frac{480}{1} \times 0.8$$

$$= 0.618 \text{ m}^3\text{/min}$$

$$\text{Specific volume} = 0.4175 \text{ m}^3\text{/kg}$$

$$\text{Mass of refrigerant/min} = \frac{\text{Volume/min}}{\text{Specific volume}}$$

$$= \frac{0.618}{0.4175}$$

$$= \mathbf{1.48 \text{ kg/min.}}$$

(iii) Capacity of Plant in TR

$$\text{Refrigerating effect/sec} = \frac{1.48}{60} \times 1125 \text{ kJ/s}$$

and $$1 \text{ TR} = 3.516 \text{ kJ/sec}$$

∴ TR capacity of the plant

$$= \frac{1.48 \times 1125}{60 \times 3.516}$$

$$= \mathbf{7.8925 \text{ TR}}$$

Example 36.15 *A food storage chamber requires a refrigeration system of 12 TR capacity when an evaporator temperature of –8ºC and condenser temperature of 30ºC are maintained. The system uses F-12 as refrigerant (R-12) with following properties. It is subcooled by 5ºC before entering throttle valve and vapour is supplied heated by 6ºC before entering the compressor. The specific heats of refrigerant are C_{P_V} = 1.235 and C_P (liquid) = 0.7327 kJ/kg-K. Determine:*

(i) Mass flow rate of refrigerant
(ii) Refrigerating effect/kg and
(iii) COP of cycle

Also draw the cycle on T–S chart assuming isentropic compressor properties of R-12.

Temperature ºC	*Specific enthalpy*		*Specific entropy*	
	Liquid	*Vapour*	*Liquid*	*Vapour*
– 8	*411.3*	*569.8*	*4.1618*	*4.7564*
30	*447.9*	*586.5*	*4.2874*	*4.7438*

Solution

The *T–S* diagram for the cycle is shown in Fig. Ex. 36.15

Here 3–3' is under cooling

First find the enthalpies at the salient points.

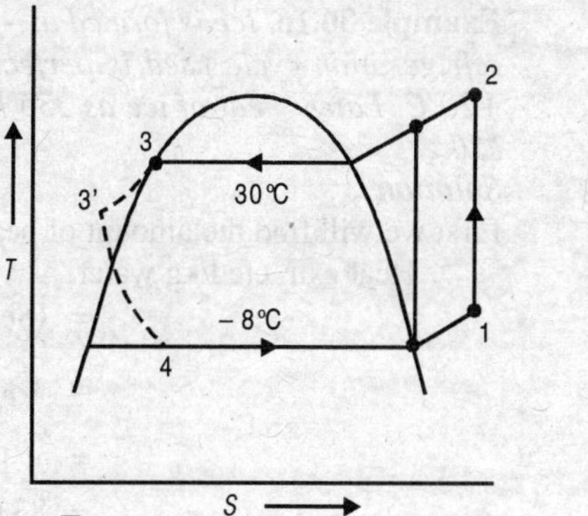

Fig. Ex. 36.15

$$h_1 = h_{g_1} + C_{p_v} \times (\text{degrees of superheat})$$
$$= 569.8 + 1.235 \times 6$$
$$= 569.8 + 7.310$$
$$= \mathbf{577.11\ kJ/kg}$$

For isentropic compression

$$S_1 = S_2$$

$$\therefore\ 4.7564 + C_{P_V} \log_e \frac{273-2}{265} = 4.7564 + 1.235 \log_e \times \log_e \frac{273}{265}$$
$$= 4.784 \text{ kJ/kg-K}$$

$$S_2 = S_{g_2} + C_{P_V} \log_e \frac{T_2}{303} = 4.7438 + 1.235 \log_e \frac{T_2}{303}$$

$$S_1 = S_2,\ T_2 = \mathbf{306.11\ K}$$

$$\therefore\ h_2 = h_{g_2} + C_{P_V}\left(t_{\text{sup}} - t_{\text{sat}}\right)$$
$$= 586.5 + 1.235\ (306.11 - 303)$$
$$= \mathbf{590.34\ kJ/kg}$$

Also
$$h_3 = h_3 - c_{p_v} \times \text{degrees of undercooling}$$
$$= 447.9 - 0.7327 \times 5$$
$$= \mathbf{444.24\ kJ/kg}$$

(i) Refrigerator Effect/kg

$$\text{Ref. effect/kg} = h_1 - h_4 = h_1 - h_{3'}$$
$$= 577.11 - 444.24$$
$$= \mathbf{132.87\ kJ/kg}$$

(ii) Mass Flow of Refrigerant

Actual refrigerator effect = 12 TR

= 12 × 3 – 516 kW = kJ/sec

∴ Mass flow rate of refrigerant kg/sec

$$= \frac{12 \times 3.516}{132.87}$$
$$= \mathbf{0.3175\ kg/s}$$
$$= \mathbf{19.05\ kg/min}$$

(iii) COP

$$\text{C. O. P.} = \frac{\text{Ref. effect}}{\text{Work supplied}}$$
$$= \frac{h_1 - h_{3'}}{h_2 - h_1}$$
$$= \frac{132.87}{590.34 - 577.11}$$
$$= \frac{132.87}{13.23}$$
$$= \mathbf{10.04}$$

Example 36.16. *Ice is formed at –10°C from water at 20°C. Find the mass of ice formed in kg/kWh. Assume the refrigeration cycle used is perfect reversed Carnot cycle working between temperature limits of –10°C and +20°C. Latent heat of ice as 335 kJ/kg and specific heat of ice as 2.09 kJ/kg-K. Specific heat of water = 4.18 kJ/kg.*

Solution

First we will find the amount of heat that is to be extracted from water to form the ice.

∴ Heat extracted/kg water

$$= \text{(Sensible heat of water)} + \text{Latent heat} + \text{Sensible heat of ice}$$

$$= C_{P_w} t_w + \text{Latent heat of ice} + C_{ice}\left(t_{ice} - 0\right)$$

$$= 4.18 \times 20 + 335 + 2.09 \times 10$$

$$= 83.6 + 335 + 20.9 = \mathbf{439.5\ kJ/kg}$$

For reversed Carnot cycle,

$$T_1 = \text{Lower temperature} = 10 + 273 = \mathbf{263\ K}$$

$$T_2 = \text{Higher temperature} = 20 + 273 = \mathbf{293\ K}$$

$$\therefore \quad \text{C. O. P.} = \frac{T_1}{T_1 - T_2} = \frac{263}{293 - 263}$$

$$= \frac{263}{30} = \mathbf{8.67}$$

$$\text{C. O. P.} = \frac{\text{Refrigerating effect}}{\text{Work supplied}}$$

$$= \frac{\text{Refrigerating effect}}{3600\ \text{(kWh)}}$$

$$\text{Refrigerating effect/kWh} = 8.67 \times 3600$$

$$= \mathbf{31562\ kJ}$$

To produce 1 kg ice, refrigerating effect required is **439.5 kJ**

Total ice produced/kWh

$$= \frac{\text{Refrigerating effect/kWh}}{\text{Ref. effect/kg}}$$

$$= \frac{31562}{439.5}$$

$$= \mathbf{71.8\ kg}$$

Example 36.17. *An ammonia refrigerator works between the temperature limits of –15°C amd + 30°C. The compressor has diameter 15 cm, stroke 20 cm, twin cylinder single acting type with volumetric efficiency of 80%. It runs 300 rpm. The refrigerant is subcooled to 21°C before expansion. Assuming no suction superheating, calculate:*

(i) Mass flow rate of refrigerant circulated/min.
(ii) Capacity of the point in TR and
(iii) COP
(iv) Power required

Temperature °C	V_{fl} m^3/kg	V_s m^3/kg	h_f *KJ/kg*	h_g *KJ/kg*	S_f	S_g
30°	1.68×10^{-3}	0.111	560.39	1705.91	4.675	8.454
–15	1.518×10^{-3}	0.509	345.43	1661.4	3.93	9.015
21	–	–	517.18	–	–	–

Solution

Figure Ex. 36.17 shows the cycle on *T-S* diagram. Assumption is that vapours are dry and saturated after compression.

$$\therefore \quad h_2 = 1705.91 \text{ kJ/kg}$$

$$S_2 = 8.454 \text{ kJ/kg-K}$$

$$h_{3^1} = 517.18 \text{ kJ/kg}$$

For isentropic compression,

$$S_1 = S_2$$

$$\therefore \quad 8.454 = S_{f_1} + x_1 S_{f_g}$$

$$= 3.93 + x_1(9.015 - 3.93)$$

$$= 3.93 + x_1 \times 5.085$$

$$\therefore \quad x_1 = \frac{8.454 - 3.93}{5.085}$$

$$= \frac{4.524}{5.085} = \mathbf{0.8897}$$

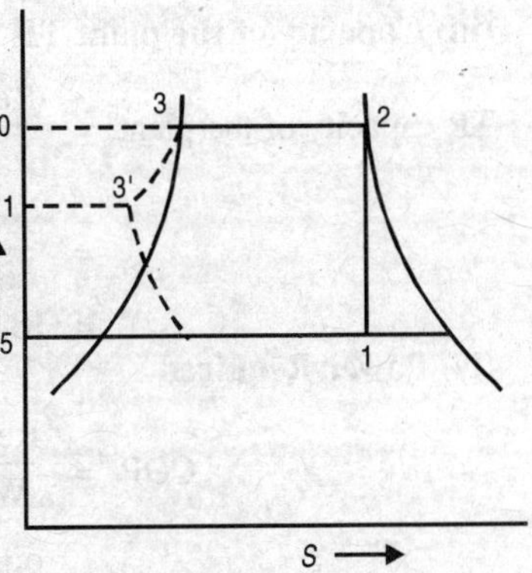

Fig. Ex. 36.17

$$h_1 = h_{f_1} + x_1 h_{fg_1}$$

$$= 345.43 + 0.8897\ (1661.4 - 345.43)$$

$$= \mathbf{1516.23\ kJ/kg}$$

(i) $\quad$ **COP** $= \mathbf{1516.23\ kJ/kg}$

$$\therefore \quad \text{Work done/kg} = h_2 - h_1 = 1705.91 - 1516.23$$

$$= \mathbf{189.68\ kJ/kg}$$

$$\text{Also} \quad \text{Ref. Effect/kg} = h_1 - h_4 = 1516.23 - 518$$

$$= \mathbf{998.23\ kJ/kg}$$

$$\text{COP} = \frac{998.23}{189.68}$$

$$= \mathbf{5.263}$$

(ii) Mass Flow of Refrigerant/min

First we can find the volume of refrigerant deal with by compressor inlet

$$V = \frac{\pi d^2}{4} \times L \times N = \frac{\pi}{4} \times \left(\frac{15}{100}\right)^2 \times \left(\frac{20}{100}\right) \times 300 \text{ m}^3\text{/min}$$

$\therefore$ Actual volume of refrigerant/min/cylinder

$$= V \times \eta_{vol_2} \text{ m}^3\text{/cylinder/min}$$

$$= \frac{\pi}{4} \times \left(\frac{15}{100}\right)^2 \times \left(\frac{20}{100}\right) \times 300 \times 0.85$$

$$= 06 \times 0.85 \text{ m}^3/\text{min}$$

$\therefore$ Total volume of refrigerant for two cylinders

$$= 2 \times 1.06 \times 0.85$$

$$= \mathbf{1.8025\ m^3/min}$$

Volume of vapour at compressor inlet/kg

$$= x_1 \times V_{s_1}$$

$$= 0.89 \times 0.509$$

$\therefore$ Mass flow rate of refrigerant/min.

$$= \frac{1.8025}{0.89 \times 0.509}$$

$$= \mathbf{3.98\ kg/min}$$

(iii) Capacity of the plant TR

$$\text{TR capacity of the plant} = \frac{\text{Ref. effect/sec}}{3.516}$$

$$= \frac{998.23 \times 3.98}{60 \times 3.516}$$

$$= \mathbf{18.83\ TR}$$

(iv) Power Required

$$\text{COP} = \frac{\text{Ref. effect/sec}}{\text{Word done/s}} = \frac{998.23 \times 3.98}{60 \times \text{Work supplied/sec}}$$

$$\therefore \quad \text{Work supplied/sec} = \frac{998.23 \times 3.98}{60 \times 5.263}$$

$$= \mathbf{12.58\ kJ/sec}$$

$$= \mathbf{12.58\ kW}$$

Example 36.18. *In an ammonia refrigerating machine, the temperature in the refrigerator is –10°C and after compression 30°C. In the cooler, the vapour is condensed at 30°C and then passes through an expansion valve into refrigerator. Estimate the coefficient of performance when the vapour at the end of compression is 90% dry. Take the latent heat of ammonia at 30°C as 1142 kJ/kg and specific heat of liquid at 407 kJ/kg-K.*

What will be the percentage improvement in the coefficient of performance if the expansion cylinder is used instead of an expansion valve?

Solution

Figure Ex. 36.18 is the *T–S* diagram for the cycle.

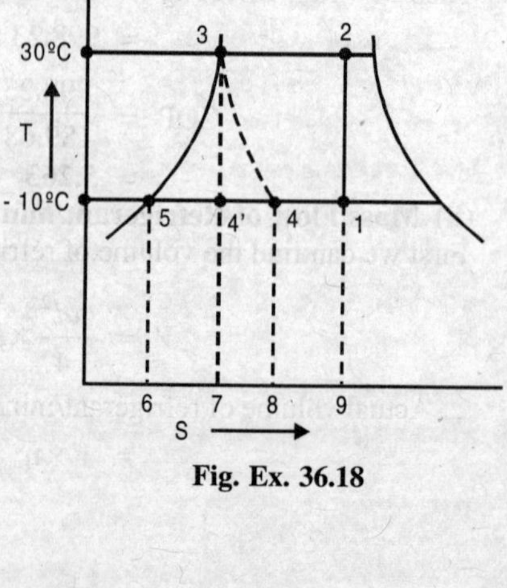

Fig. Ex. 36.18

Given

$$T_1 = -10 + 273 = 263 \text{ K}$$

$$T_2 = 30 + 273 = 303 \text{ K}$$

$$h_{fg_2} = 1142 \text{ kJ}$$

$$C_{p_l} = 4.7 \text{ kJ/kg-K}$$

$$x_1 = 0.9$$

Other properties of ammonia are not known. We will calculate the required COP by considering the area of this *T–S* diagram.

A. When Expansion valve is used

$$
\begin{aligned}
W = \text{Work done/kg} &= h_2 - h_1 \\
&= \text{Area behind } 1-2 \\
&= \text{Area } 1-2-3-5-1 \\
&= \text{Area } 1-2-3-4'-1 + \text{Area } 3-5-4'-3 \\
&= \left[(1-2)\times(2-3)\right]+\left[(3-5-6-7-3)-(4'-5-6-7-4')\right] \\
&= \left[(T_2-T_1)\times\left(\frac{x_2\times h_{fg_2}}{T_{sat}2}\right)\right]+\left[C_{p_r}(T_3-T_5)-T_1\times C_{p_r}\log_e\frac{T_3}{T_5}\right] \\
&= \left[(30+10)\times\left(\frac{0.9\times1142}{303}\right)\right]+\left[4.7(30+10)-263\times4.7\times\log_e\frac{303}{268}\right] \\
&= 135.68+\left[4.7\times40-175\right] \\
&= 135.68 + (188 - 175) \\
&= 135.68 + 13 \\
&= \mathbf{148.68\ kJ/kg}
\end{aligned}
$$

Refrigerating effect/kg

$$
\begin{aligned}
N &= h_1 - h_4 \\
&= \text{Area } 1\ 4\ 8\ 9\ 1 \\
&= \text{Area } 1-4'-7-9-1 - \text{Area } 4-4'-7-8-4 \\
&= T_1\times(1-4')-\text{Area}3-4'-5-3 \\
&= T_1\times(2-3)-\text{Area}3-4'-5-3 \\
&= 263\times3.392-13 \\
&= 892.1 - 13 \\
&= \mathbf{879.1\ kJ/kg}
\end{aligned}
$$

$$
\begin{aligned}
\text{COP} &= \frac{\text{Ref. effect}}{\text{Work done}}=\frac{N}{W} \\
&= \frac{879.1}{148.68} \\
&= \mathbf{5.913}
\end{aligned}
$$

For isentropic process 3 – 4,

$$h_3 = h_4$$

$\therefore$ Area $-3-5-6-7-3$ = Area $4-5-6-8-4$

$$\therefore\ \text{Area}(3-5-4'-3)+\text{Area}(4'-5-6-7-4') = \text{Area}(4-4'-7-8-4)+\text{Area}(4'-5-6-7-4')$$

or $\quad \text{Area}(3-5-4'-3) = \text{Area}(4-4'-7-8-4)$

B. When expansion cylinder is used instead of expansion valve

$$
\begin{aligned}
W = \text{Work done/kg} &= W_c - W_e \\
&= \text{Work of compression} - \text{Work of expansion}
\end{aligned}
$$

$$= (h_2 - h_1) - (h_3 - h_{4'})$$
$$= \text{Area } 1-2-3-4'-1$$
$$= \text{Area}\left[(1-2)\times(2-3)\right]$$
$$= (T_1 - T_2)\times\frac{xh_{fg_2}}{T_2}$$
$$= 40\times 3.392$$
$$= \mathbf{135.68\ kJ/kg}$$

Also, refrigerating effect is given by

$$N = h_1 - h_4$$
$$= \text{Area } (1-4'-7-9-1)$$
$$= \text{Area}\left[(1-9)\times(1-4')\right]$$
$$= \text{Area}\left[(1-9)\times(2-3)\right]$$
$$= T_1\times 3.392$$
$$= 263\times 3.392$$
$$= \mathbf{892.1\ kJ/kg}$$

$$\therefore \quad \text{COP} = \frac{N}{W}$$
$$= \frac{892.1}{135.68}$$
$$= \mathbf{6.575}$$

$\therefore$ Percentage of improved of COP

$$= \frac{6.575-5.913}{5.913}\times 100$$
$$= \mathbf{11.2\%}$$

Example 36.19. *A refrigeration plant is used for the production of ice. Saturation temperatures to condensor pressure and evaporator pressures are 25°C and – 5°C respectively. Refrigerant is 0.8 dry at the entry to the condenser. The quantity of refrigerant circulated through the cycle is 1 kg/sec. Find the ice formed in kg per hour assuming the relative efficiency of the cycle as 50%. The ice is formed from the water at 10°C and the ice is formed at – 5°C.*

Assume the water C_p = 4.187 kJ/kg-K

For ice C_p = 2.05 kJ/kg-K

Latent heat of fusion of ice = 335 kJ/kg

List the following properties of refrigerant

Temperature °C	h_f *kJ/kg*	h_{fg} *kJ/kg*	S_f *kJ/kg-K*
+ 25	81.5	121.2	0.2515
– 5	– 7.5	246.0	– 0.0419

Solution

The *T–S* diagram for the cycle is shown in Fig. Ex. 36.19

First find the heat to be extracted from water to produce 1 kg ice.

$$\therefore \quad N = \text{Sensible heat (water)} + \text{Latent heat} + \text{Sensible heat (ice)}$$

$$= C_{p_w}(t_w - 0) + \text{Latent heat} + C_{p_i}(t_i - 0)$$

$$= 4.187 \times 10 + 335 + 2.05 \times 5$$

$$= 41.87 + 335 + 10.25$$

$$= \mathbf{387.12\ kJ/kg}$$

For isentropic compression 1 – 2,

$$S_1 = S_2$$

$$S_1 = S_{f_1} + x_1 S_{fg1}$$

$$= -0.0419 + x_1 \times \frac{246}{268}$$

$$S_2 = S_{f_2} + x_2 S_{f_g}$$

$$= 0.2515 + 0.8 \times \frac{121}{298}$$

$$= 0.2575 + 0.3248$$

$$= 0.5763$$

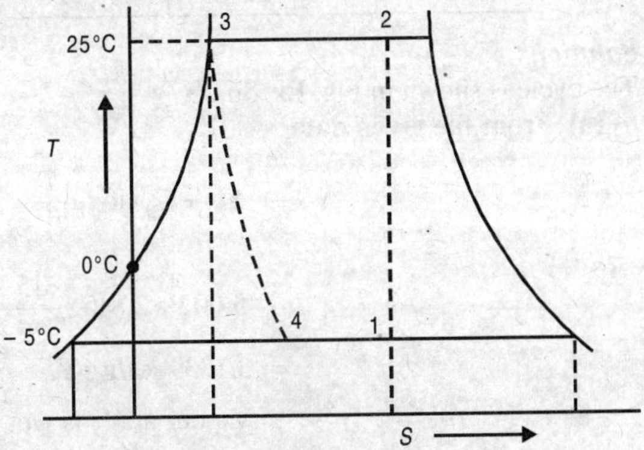

Fig. Ex. 36.19

$$\therefore \quad x_1 = \frac{0.5763 + 0.0419}{1} \times \frac{268}{246}$$

$$= 0.6182 \times \frac{268}{246}$$

$$= \mathbf{0.6735}$$

$$h_1 = h_{f_1} + x_1 h_{f_{g1}}$$

$$= -7.5 + 0.6735 \times (246 + 7.5)$$

$$= -7.5 + 170.73$$

$$= \mathbf{163.23\ kJ/kg}$$

Refrigeration effect/kg/sec

$$= h_1 - h_4 = h_1 - h_3$$

$$= 163.23 - 81.5$$

$$= \mathbf{81.73\ kJ/kg}$$

$\therefore$ Actual refrigerating effect/kg $= 81.73 \times 0.5 = 40.865$ kJ/kg

$\therefore$ Mass of ice formed/sec

$$= \frac{81.73 \times 0.5}{387.12}$$

$$= 0.1055 \text{ kg/sec}$$

$$= \mathbf{380\ kg/hour}$$

Example 36.20. *An ammonia refrigerator plant works between the pressure limits of 11 bar and 2.5 bar. The temperature of ammonia entering and leaving the condenser are 45°C and 15°C respectively. The quality of ice produced at –5°C from water at 22.2°C is 360 kg/hour. Assume that the vapour is compressed adiabetically and expansion valve is used. Find:*

(a) The condition of ammonia at entry to the compressor
(b) COP
(c) The quantity of ammonia flowing per minute

Specific heat of ice is 2.1 kJ/kg-K, latent heat of ice is 335 kJ/kg and sp. heat of ammonia vapour at constant pressure is 2.8 kJ/kg-K.

Pressure	t_{sat} °C	*Enthalpy kJ/kg*		*Entropy kJ/kg-K*	
		h_f	h_g	S_f	S_g
11	25	316	1473	1.1774	5.002
	15	255	1464	0.9722	5.1475
2.5	– 14	118	1432	0.4765	5.5327

Solution

The cycle is shown in Fig. Ex. 36.20

(a) From the given data

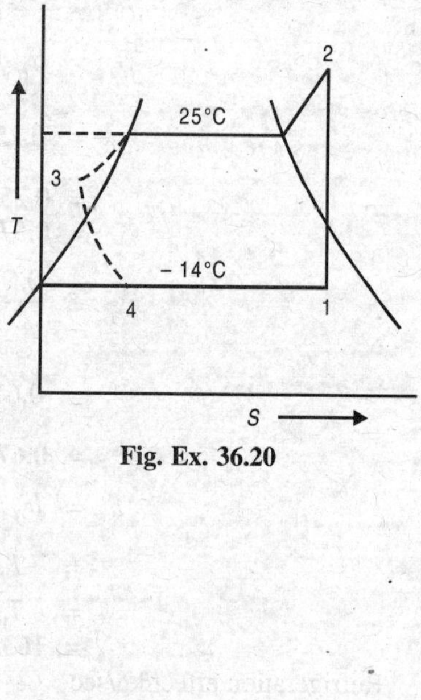

Fig. Ex. 36.20

$$S_2 = S_{g2} + C_{pv} \log_e \frac{T_2}{T_{sat_2}}$$

$$= 5.002 + 2.8 \log_e \frac{45+273}{25+273}$$

$$= \mathbf{5.1839\ kJ/kg\text{-}K}$$

$S_1 \le S_{g_1} \le S_2$ Vapour at (1) is wet

$$\therefore \quad 5.1839 = S_{f_1} + x_1 \left(S_{g_1} - S_{f_1}\right)$$

$$= 0.4765 + x_1\,(5.5327 - 0.4765)$$

$$x_1 = \frac{5.1839 - 0.4765}{5.0562}$$

$$= \mathbf{0.931}$$

∴ The condition of ammonia at entry to the compressor

$$\mathbf{x_1 = 0.931}$$

(b) COP

$$h_1 = h_{f_1} + x_1 h_{f_{g1}}$$

$$= 118 + 0.931 \times 1314 = \mathbf{1341.34\ kJ/kg}$$

$$h_2 = h_{g_2} + C_{p_v}\left(T_2 - T_{sat\,2}\right)$$

$$= 1473 + 2.8\,(45 - 25)$$

$$= \mathbf{1529\ kJ/kg}$$

and $h_3 = h_4 = 255$ kJ/kg

Refrigerating effect/kg $= h_1 - h_4 = 1341.334 - 255$

$= \mathbf{1086.334\ kJ/kg}$

Work done/kg $= h_2 - h_1$

$= 1529 - 1341.334$

$= \mathbf{187.664\ kJ/kg}$

$$\therefore \quad \text{COP} = \frac{\text{Ref. effect}}{\text{Work done}}$$

$$= \frac{1086.334}{187.664}$$

$$= \mathbf{5.789}$$

(c) Quantity of ammonia flowing

Required refrigerating effect/min

$$= \frac{360}{60}\left[C_{p_w} \times t_w + h_{f_g} + C_{p_i} t_i\right]$$

$$= 60[4.187 \times 22.2 + 335 + 201 \times 5.5]$$

$$= \mathbf{2631.6\ kJ/min}$$

∴ Mass of ammonia flowing/min

$$= \frac{2631.6}{1086.334}$$

$$= \mathbf{2.422\ kg/min}$$

Example 36.21. *In a vapour absorption refrigeration system, heating in generator cooling in condenser and refrigeration in evaporator takes place at 100°C, 20°C and –10°C respectively. Find maximum COP of the system.*

Solution

Maximum COP of the cycle is given by

$$(\text{COP})_{\max} = \frac{T_h - T_k}{T_n} \times \frac{T_o}{T_k - T_o}$$

where T_h = Temperature of heat source = 100°C

T_k = Temperature of condenser = 20°C

T_o = Temperature of refrigeration in evaporator = –10°C

$$\therefore \quad (\text{COP})_{\max} = \frac{(100+273)-(20+273)}{(100+273)} \times \frac{(-10+273)}{(+20+273)-(-10+273)}$$

$$= \frac{80}{(273+100)} \times \frac{263}{293-263}$$

$$= \frac{80 \times 263}{373 \times 30}$$

$$= \mathbf{1.88}$$

Example 36.22. *In an acqua ammonia vapour absorption plant, heat is supplied to generator by condensing steam at 2 bar and 0.9 dry. The evaporator is to be maintained at 5°C. Assuming the ambient temperature as 503°C. Calculate maximum possible COP condensate leaves the generator at 30°C.*

If actual COP is 70% of the maximum COP, calculate the mass of steam required per day for a plant capacity of 20 TR.

Take saturation temperature at 2 bar as 120°C and enthalpy of evaporation as 2200 kJ/kg.

Solution

The maxmimum COP is given by

$$(\text{COP})_{\max} = \frac{T_h - T_a}{T_h} \times \frac{T_o}{T_a - T_R}$$

$$= \frac{(120+273)-(30+273)}{393} \times \frac{278}{(30+273)-278}$$

$$= \frac{90}{393} \times \frac{278}{25} = \mathbf{2.517}$$

$$\text{Actual COP} = 0.7 \times 2.517$$
$$= \mathbf{1.7619}$$
$$= \frac{\text{Refrigerating effect}}{\text{Energy supplied}}$$
$$= \frac{20 \times 3.576\ (\text{kJ/kg})}{\text{Heat supplied (kJ/sec)}}$$

$$\therefore \quad \text{Heat supplied/sec} = \frac{20 \times 3.516}{1.7619}$$
$$= \mathbf{39.91\ kJ/kg}$$

$$\therefore \quad 39.91 = \frac{\text{Mass of steam}}{\text{Sec}} \times \left[h_f + xh_{fg} - h_{\text{cond.}}\right]$$
$$= \frac{m_s}{\text{sec}}[504.7 + 0.9 \times 2200 - 30 \times 4.187]$$
$$= \frac{m_s}{\text{sec}}[504.7 + 1980 - 125.61]$$
$$= (379.09 + 1980)\ m_s$$

$$\therefore \quad \frac{\text{Mass of steam}}{\text{sec}} = \frac{39.91}{2359.09} = \mathbf{0.01692\ kg/sec}$$

$$\therefore \quad \text{Mass of steam/day} = 0.01692 \times 3600 \times 24$$
$$= \mathbf{1461.67\ kg/day}$$

Example 36.23. *An aqua-ammonia absorption cycle has a generator temperature 120°C, an evaporator temperature – 20°C. The ambient temperature is 30°C. Estimate maximum possible COP. Also find the capacity in TR per kW of heat supplied.*

Solution

$$\text{Max. Possible COP} = \eta_{th} \times \eta_{Rc}$$
$$= \frac{T_g - T_a}{T_g} \times \frac{T_e}{T_a - T_c}$$

where T_g = Temperature of generator

T_a = Temperature of ambient

T_e = Temperature of evaporator

$$\therefore \quad (\text{COP})_{\max} = \frac{(120+273)-(273+30)}{(120+273)} \times \frac{-20+273}{303-253}$$

$$= \frac{90}{393} \times \frac{253}{50} = \mathbf{1.159}$$

Let refrigerating effect/sec = X kJ/s

Work supplied/sec = Y (kJ/s = kW)

$$= \frac{\text{Refrigerating effect (kJ/s)}X}{\text{COP}}$$

$$\therefore \text{ Capacity} \frac{\text{TR}}{\text{kW of heat}} = \frac{X/3.516}{X/\text{COP}}$$

$$= \frac{\text{COP}}{3.516}$$

$$= \frac{1.159}{3.516} = \mathbf{0.3296}$$

$$\simeq \mathbf{0.33}$$

Example 36.24. *A vapour absorption cycle has generator temperature 120ºC, evaporator temperature –10ºC and the ambient temperature 30ºC. Estimate the maximum possible COP. The actual COP is 0.5 of the maximum COP. If the capacity of the plant is 100 TR, calculate the fuel consumption of the plant. The calorific value of the fuel is 40 MJ/kg.*

Solution

Maximum possible COP is given by

$$(\text{COP})_{\text{max}} = \eta_{\text{th}} \times \eta_{Rc}$$

$$= \frac{T_g - T_a}{T_g} \times \frac{T_e}{T_a - T_c}$$

$$= \frac{(120+273)-(30+273)}{(120+273)} \times \frac{(-10+273)}{(30+273)-263}$$

$$= \frac{90}{393} \times \frac{263}{40} = \mathbf{1.506}$$

$$\text{Actual COP} = 0.5 \times 1.506$$

$$= \mathbf{0.753}$$

$$= \frac{\text{Refrigerating effect}}{\text{Heat supplied}}$$

$$= \frac{100 \times 3.516}{\text{Heat supplied}}$$

$$\text{Heat supplied/sec} = \frac{100 \times 3.516}{0.753}$$

$$= \mathbf{466.93 \text{ kJ/sec}}$$

$$\therefore \text{ Fuel consumption/hour} = \frac{466.93 \times 3600}{40,000}$$

$$= \mathbf{42.02 \text{ kg/hour}}$$

Example 36.25. *In an NH_3 absorption refrigerator, heat is supplied to NH_3 generator by steam condensing at 2.5 bar and 0.9 dry. The refrigerator cabinet is to be maintained at –5ºC. The ambient temperature is 35ºC. Estimate the maximum COP possible COP for this generator.*

If actual COP is 75% of the maximum COP, find the quantity of steam required in kg/hour for a refrigerator capacity of 25 TR.

Solution

For a heat operated refrigerator system, the Maximum COP is given by

$$(\text{COP})_{\max} = \eta_{th} \times \eta_{Rc}$$

where $\eta_{th} = \dfrac{T_{\max} - T_{\min}}{T_{\max}}$

$T_{\max}$ = Saturated temperature of the steam at 2.5 bar
= 127.42 °C
= **400.42 K**

$T_{\min}$ = Ambient temperature
= 35 + 273
= **308 K**

$\therefore \quad \eta_{th} = \dfrac{400.42 - 308}{400.42} = \mathbf{0.2308}$

Similarly, $\eta_{Rc} = \dfrac{T_R}{T_a - T_R}$

where T_R = Temperature of refrigerator
= – 5 + 273
= **268 K**

T_a = Ambient temperature
= 35 + 273
= **308 K**

$\therefore \quad \eta_{Rc} = \dfrac{268}{308 - 268} = \dfrac{268}{40} = \mathbf{6.7}$

$\therefore \quad (\text{COP})_{\max} = 0.2308 \times 607$
= **1.5464**

$\therefore$ Actual COP = 0.75 × 1.5464
= **1.16**

$$\therefore \quad 1.16 = \frac{\text{Refrigerating effect}}{\text{Heat supplied}}$$

$$= \frac{25 \times 3.516}{Q_A}$$

$$\therefore \quad Q_A = \frac{25 \times 3.516}{1.16} = \mathbf{75.776\ kJ/sec}$$

Heat supplied per kg of steam

$$= h_1 - h_{W_2} = \left(h_{f_1} + x_1 h_{f_{g1}}\right) - h_{\text{con}}$$

$$= 535.25 + 0.9 \times 2181.1 - 146.6$$

= **2351.64 kJ/kg**

$$\text{Mass of steam/sec} = \frac{75.776}{2351.64} = \mathbf{0.0322\ kg/sec}$$

= **116 kg/hr**

■■■

THEORY QUESTIONS

36.1. What is refrigeration ? Enumerate the applications of refrigeration.
36.2. What is the unit of refrigeration?
36.3. What do you understand under cooling or subcooling?
36.4. What is heat-pump?
36.5. Distinguish clearly between ideal refrigeration cycle and the Carnot reversed cycle.
36.6. Why do you require refrigeration in a aeroplane flying in a sub-zero temperature atmosphere at an altitude?
36.7. What is a regenerative system used in air refrigeration? Explain with neat sketch.
36.8. What are the limitations of air refrigeration system?
36.9. Explain the working of a simple vapour compression cycle machine. Derive the relation between work required, refrigerating effect and COP.
36.10. Draw a *P–h* chart and show on that the simple theoretical vapour compression cycle and actual cycle. Show the cycle with undercooling on *P–h* chart.
36.11. Explain with the neat sketch (*T–S* diagram) the effect of superheat, undercooling and condenser pressure on the performance of the working of vapour compression cycle.
36.12. What is the principle on which the vapour absorption cycle works?
36.13. Explain with the neat sketch a simple vapour absorption cycle working.
36.14. What modifications are required in the simple cycle to become a practical cycle.
36.15. What are the combinations of refrigerant and absorbent?
36.16. Explain with the neat sketch the working of Electrolux refrigerating system.
36.17. Compare vapour compression and vapour absorption cycles.
36.18. Explain with neat sketch the Lithium-bromide-water absorption system.
36.19. What is the single effect and double effect refrigeration systems. Explain.
36.20. Derive the relation for the maximum C.O.P. of a Heat operated vapour absorption cycle.

PROBLEMS FOR PRACTICE

36.1. An air refrigeration plant working on Carnot cycle between the temperatures limits of 34°C and –1°C has a capacity of 10 tons. Determine the power input to the plant. [*Ans.* 4.52 kW]

36.2. An air refrigeration plant working on Carnot cycle produces refrigeration capacity equivalent of production of tons of ice in 24 hours at 0°C from water at 20°C. The compressor temperature limits are 30°C and –15°C. Determine the power input to the plant. The latent heat of ice is 335 kJ/kg. [*Ans.* 8.45 kW]

36.3. In a refrigerator plant working on the Bell–Coleman cycle the air is drawn in from the cold chamber at –5°C and 1.03 bar. The same air is compressed adiabatically to 5.25 bar before being cooled in the heat exchanger to 15°C. After the exchanger the air is expanded adiabatically and reversibly to cold changer state. Assume C_p = 1.004 kJ/kg-K and C_v = 0.717 kJ/kg/K for air. Determine

(i) work done on the cycle per kg of air.
(ii) the refrigeration effect produced in cold chamber.
(iii) C.O.P.

[*Ans.* (i) 51.85 kJ/kg, (ii) 87.51 kJ/kg air, (iii) 1.69]

36.4. In a Bell–Coleman cycle, air is drawn from cold chamber at a pressure of 1.03 bar and –10°C, compressed isentropically to 0.18 bar. Cooled to 25°C. It is expanded with the law of $PV^{1.3}$ =Constant. C_p = 1.004 kJ/kg/K, C_v = 1.717 kJ/kg-K. Determine

(a) Work input/kg of air
(b) Refrigeration produced in cold chamber
(c) COP

[*Ans.* (a) 51.03 kJ/kg, (b) 66.16 kJ/kg, (c) 1.3]

36.5. Following data is related to a dense air refrigeration cycle. Capacity = 1 TR, pressure and temperature at compressor suction = 5 bar and –18°C, pressure and temperature at the inlet of expander = 20 bar and 16°C. Determine

(a) COP
(b) Heat removed in cooler
(c) Piston displacement of compressor.

[*Ans.* (a) 2.06, (b) 5.22 kJ/sec, (c) 0.32 m³/min]

36.6. A simple saturated ammonia vapour compression plant operates between 40ºC and –20ºC. Calculate
(a) Refrigerating effect/kg
(b) Work input/kg
(c) Heat rejected from condenser/kg
(d) C.O.P
(e) Mass flow rate in kg/sec per TR

Compare the above mentioned parameters for dry and saturated vapours at the suction with altering other conditions. Take C_{pv} = 3.52 kJ/kg-K.

Temperature ºC	Pressure bar	h_f kJ/kg	h_g kJ/kg	S_f kJ/kg -K	S_g kJ/kg -K
– 14	1.7	108.7	1437.6	0.65	5.9
+ 40	15.6	391.1	1489.9	1.65	5.15

[*Ans.* (a) 856.66 (b) 242.12 kJ/kg (c) 1098.8 (d) 3.54 (e) 0.0041]
[(a) 1046.5 (b) 313.93 (c) 1360.43 (d) 3.33 (e) 0.00336]

36.7. A F-12 refrigeration system operates between temperature limits of –6ºC and 36ºC. The vapour is 95 % dry at the beginning of isentropic compression and the liquid leaving the condenser is at 30ºC. Assume actual COP as 50% of theoretical COP, calculate the amount of ice produced in kg/kWh at 0ºC from water at 30ºC. Latent heat of ice = 335 kJ/kg. Specific heat of refrigerant in liquid stage is 1.24 kJ/kg. K.

Temperature ºC	h_f kJ/kg	h_g kJ/kg	S_f kJ/kg -K	S_g kJ/kg -K
– 6	30.53	184.94	0.1217	0.6996
+ 36	70.55	201.80	0.2591	0.6836

[*Ans.* 21.657 kg/kWh]

36.8. Compare the COP of an ammonia refrigeration cycle which uses wet compression with one which uses dry compression and operates between –20ºC and 30ºC. Assume compression to be isentropic and no subcooling of liquid refrigerant. In wet compression, vapour leaving the compressor is dry and saturated whereas in dry compression, vapour entering the compressor is dry and saturated.

Sat. Temperature ºC	h_f kJ/kg	h_{fg} kJ/kg	h_g kJ/kg	S_f kJ/kg -K	S_g kJ/kg -K
– 20	89.8	1330	1420	0.3684	5.6244
+ 30	323.1	1145.8	1468	1.2037	4.9842

[*Ans.* Wet compression COP = 1.025, Dry compression COP = 4.113]

36.9. 30,000 kg of ice per day is produced from and at 0ºC in an ammonia ice plant working on vapour compression cycle. The temperature range of the cycle is –15ºC to 25ºC. The vapour at the end of compression dry saturated. The relative COP of the system is 65 %. Find the power required to drive the plant. Latent heat of ice = 335 kJ/kg.

[*Ans.* 5.303 kW]

36.10. A refrigeration system using NH_3 works on standard vapour compression cycle. The condensing temperature is 30ºC. Investigate the performance of the system based on

(a) C O P

(b) Mass of refrigerant handled by compressor.

(c) Volume of refrigerant handled by compressor.

(d) Temperature of refrigerant at compressor delivery.

The system operates at temperature in the evaporator of –10ºC and 16ºC respectively.

Capacity of the system is 10 TR.

Sat. Temperature °C	*Sp. Volume* V_g m^3/kg	*Sp. Enthalpy kJ/kg*			*Specific entropy kJ/kg-K*	
		h_f	h_{fg}	h_g	S_f	S_g
–10	0.4189	112.4	1314.2	1426.6	0.4572	5.5789
+ 30	0.1107	323.1	1145.8	1468.9	1.2037	4.9842

[*Ans.* (a) 4.125, (b) 1.697 kg/min, (c) 0.7811 m³/min, (d) 116.2ºC]

IMPORTANT POINTS TO BE REMEMBERED

1. One tonne refrigeration $TR = 3.56 \text{ kW} \simeq 3.52 \text{ kW}$

2. $$(\text{COP})_{\text{Ref}} = \frac{N}{W} = \frac{\text{Refrigeration effect}}{\text{Work done or supplied}}$$

$$(\text{COP})_H = 1 + (\text{COP})_{\text{ref}}$$

3. *Bell–Coleman cycle*

$$\text{COP} = \frac{T_4}{T_3 - T_4}$$

4. *Vapour compression cycle*

$$(\text{COP})_{\text{ref}} = \frac{h_1 - h_4}{h_2 - h_1} \quad \text{with expansion valve}$$

$$= \frac{h_1 - h_{4'}}{(h_2 - h_1) - (h_8 - h_{4'})} \quad \text{with expansion cylinder}$$

Table 36.1: Properties of Refrigeration R-116

Saturation temp. in °C (t)	Saturation pressure in bar (p)	Specific volume in m^3/kg		Specific enthalpy in kJ/kg			Specific entropy in kJ/kg K	
		Liquid (v_f)	Vapour (v_g)	Liquid (h_f)	Vapour (h_g)	Latent (h_{fg})	Liquid (s_f)	Vapour (s_g)
–40	0.05096	0.000617	2.7625	0.00	203.48	203.48	0.0000	0.8730
–38	0.05942	0.000619	2.5360	1.65	202.83	204.48	0.0068	0.8701
–36	6.06787	0.000621	2.3094	3.30	202.18	205.48	0.0137	0.8671
–34	0.07633	0.000623	2.0827	4.95	201.53	206.48	0.0206	0.8642
–32	0.08478	0.000624	1.8562	6.60	200.88	207.48	0.0274	0.8613
–30	9.09323	0.000626	1.6296	8.25	200.23	208.48	0.0343	0.8583
–28	0.10345	0.000627	1.4448	9.90	199.58	209.48	0.0410	0.8554
–26	0.11586	0.000629	1.3123	11.55	198.94	210.49	0.0477	0.8533
– 24	0.12828	0.000631	1.1798	13.20	198.29	211.49	0.0544	0.8508
–22	0.14292	0.000633	1.0722	14.85	197.65	212.50	0.0611	0.8483
– 20	0.15869	0.000634	0.9770	16.51	196.99	213.50	0.0678	0.8462
– 18	0.17414	0.000636	0.8793	18.19	196.33	214.52	0.0741	0.8441
– 16	0.19280	0.000638	0.8045	19.82	195.71	215.53	0.0808	0.8420
– 14	0.21260	0.000639	0.7345	21.49	195.05	216.54	0.0875	0.8403
– 12	0.23390	0.000641	0.6713	23.17	194.38	217.55	0.0938	0.8387
10	0.25848	0.000643	0.6191	24.83	193.73	218.56	0.1009	0.8366
– 8	0.28306	0.000645	0.5671	26.50	193.07	219.57	0.1063	0.8349
– 6	0.30280	0.000647	0.5194	28.17	192.42	220.59	0.1126	0.8332
– 4	0.33967	0.000649	0.4801	29.85	191.76	221.61	0.1189	0.8315
– 2	0.36983	0.000651	0.4409	31.54	191.09	222.63	0.1252	0.8303
0	0.40360	0.000653	0.4069	33.22	190.42	223.64	0.1315	0.8296
1	0.42190	0.000653	0.3920	34.04	190.08	224.14	0.1344	0.8280
2	0.44021	0.000654	0.3770	34.90	189.75	224.65	0.1374	0.8274
3	0.45852	0.000655	0.3620	35.74	189.41	225.15	0.1405	0.8267
4	0.47683	0.000656	0.3471	36.59	189.06	225.65	0.1436	0.8261
5	0.49720	0.000657	0.3340	37.43	188.73	226.16	0.1465	0.8254
6	0.51917	0.000658	0.3225	38.28	188.38	226.66	0.1498	0.8248
7	0.54118	0.000659	0.3109	39.13	188.04	227.17	0.1526	0.8242
8	0.56317	0.000660	0.2994	39.98	187.70	227.68	0.1562	0.8236
9	0.58517	0.000661	0.2878	40.84	187.35	228.18	0.1586	0.8229
10	0.60717	0.000662	0.2763	41.68	187.01	228.69	0.1616	0.8223
11	0.63320	0.000663	0.2673	42.53	186.67	229.20	0.1646	0.8219
12	0.65921	0.000664	0.2584	43.39	186.32	229.71	0.1675	0.8215
13	0.68523	0.000665	0.2495	44.24	185.97	230.21	0.1706	0.8210

(*Contd.*)

Contd. Table 36.1

Saturation temp. in °C (t)	Saturation pressure in bar (p)	Specific volume in m^3/kg		Specific enthalpy in kJ/kg			Specific entropy in kJ/kg K	
		Liquid (v_f)	Vapour (v_g)	Liquid (h_f)	Vapour (h_g)	Latent (h_{fg})	Liquid (s_f)	Vapour (s_g)
14	0.71124	0.000666	0.2405	45.10	185.62	230.72	0.1738	0.8206
15	0.73727	0.000667	0.2316	45.95	185.27	231.22	0.1767	0.8200
16	0.76552	0.000669	0.2235	46.81	184.92	231.73	0.1796	0.8194
17	0.79655	0.000670	0.2163	47.67	184.56	232.23	0.1825	0.8170
18	0.82759	0.000671	0 2092	48.54	184.19	232.73	0.1855	0.8186
19	0.85862	0.000672	0.2021	49.40	183.84	233.24	0.1885	0.8180
20	0.88966	0.000673	0.1950	50.26	183.48	233.74	0.1913	0.8173
21	0.92070	0.000674	0.1878	51.12	183.12	234.24	0.1943	0.8169
22	0.95625	0.000675	0.1821	51.99	182.75	234.74	0.1972	0.8165
23	0.99237	0.000676	0.1764	52.85	182.39	235.24	0.2003	0.8163
24	1.02496	0.000677	0.1708	53.72	182.03	235.75	0.2031	0.8160
25	1.06117	0.000678	0.1651	54.59	181.66	236.25	0.2061	0.8157
26	1.10074	0.000679	0.1595	55.46	181.29	236.75	0.2089	0.8152
27	1.13880	0.000680	0.1542	56.32	180.93	237.25	0.2120	0.8150
28	1.17917	0.000681	0.1495	57.20	180.56	237.76	0.2148	0.8148
29	1.22273	0.000682	0.1452	58.07	180.18	238.25	0.2177	0.8144
30	1.26070	0.000684	0.1401	58.94	179.82	238.76	0.2206	0.8140
31	1.30665	0.000685	0.1362	59.82	179.43	239.25	0.2234	0.8135
32	1.34821	0.000686	0.1316	60.70	179.05	239.75	0.2261	0.8131
33	1.39568	0.000687	0.1278	61.57	178.67	240.24	0.2290	0.8128
34	1.44414	0.000689	0.1242	62.45	178.28	240.73	0.2320	0.8126
35	1.49276	0.000690	0.1205	63.32	177.91	241.23	0.2349	0.8123
36	1.54130	0.000691	0.1168	64.20	177.52	241.72	0.2378	0.8119
37	1.58983	0.000692	0.1131	65.07	177.15	242.22	0.2406	0.8116
38	1.64000	0.000693	0.1096	65.95	176.77	242.72	0.2433	0.8114
39	1.69571	0.000694	0.1066	66.84	176.66	243.20	0.2462	0.8112
40	1.75145	0.000696	0.1037	67.74	175.95	243.69	0.2490	0.8110
41	1.80718	0.000697	0.1007	68.63	175.55	244.18	0.2518	0.8108
42	1.86292	0.000698	0.0977	69.52	175.15	244.67	0.2546	0.8106
43	1.91866	0.000699	0.0947	70.41	174.75	245.16	0.2574	0.8104
44	1.97952	0.000700	0.0921	71.31	175.34	245.65	0.2600	0.8102
45	2.04300	0.000701	0.0896	72.20	173.94	246.14	0.2029	0.8100
46	2.10641	0.000702	0.0871	73.08	173.55	243.63	0.2658	0.8098
47	2.16980	0.000704	0.0847	73.98	173.14	247.12	0.2686	0.8096
48	2.23310	0.000705	0.0823	74.87	172.73	247.60	0.2713	0.8094
49	2.29670	0.000706	0.0798	75.76	172.31	248.07	0.2741	0.8092

Table 36.2: Properties of Refrigeration R-12

Saturation temp. in °C (t)	Saturation pressure in bar (p)	Specific volume in m³/kg		Specific enthalpy in kJ/kg			Specific entropy in kJ/kg K	
		Liquid (v_f)	Vapour (v_g)	Liquid (h_f)	Vapour (h_g)	Latent (h_{fg})	Liquid (s_f)	Vapour (s_g)
–100	0.01185	0.000600	10.1951	– 51.84	193.84	142.00	– 0.2567	0.8628
– 95	0.01864	0.000604	6.6231	– 47.56	191.78	144.22	– 0.2323	0.8442
– 90	0.02843	0.000608	4.4206	– 43.28	189.74	146.46	– 0.2086	0.8273
– 85	0.04254	0.000613	3.0531	– 39.00	187.73	148.73	– 0.1856	0.8122
– 80	0.06200	0.000617	2.1519	– 34.72	185.74	151.02	– 0.1631	0.7985
– 75	0.08826	0.000622	1.5462	– 30.42	183.74	153.32	– 0.1412	0.7861
–70	0.12298	0.000627	1.1314	– 26.12	181.75	155.63	– 0.1198	0.7665
– 65	0.16807	0.000632	0.8421	– 21.81	179.77	157.96	– 0.0988	0.7648
– 60	0.22665	0.000637	0.6401	– 17.48	177.77	160.29	– 0.0783	0.7558
– 55	0.30052	0.000643	0.4930	– 13.14	175.76	162.62	– 0.0581	0.7475
– 50	0.39237	0.000648	0.3845	– 8.78	173.73	164.95	– 0.0394	0.7401
– 45	0.50512	0.000654	0.3035	– 4.39	171.66	167.27	– 0.0190	0.7334
– 40	0.64190	0.000660	0.2422	0.00	169.60	169.60	0.0000	0.7274
– 38	0.70460	0.000663	0.2221	1.76	168.76	170.52	0.0075	0.7251
– 36	0.77196	0.000665	0.2040	3.53	167.91	171.44	0.0149	0.7230
– 34	0.84421	0.000667	0:1877	5.31	167.05	172.36	0.0224	0.7209
– 32	0.92776	0.000670	0.1729	7.08	166.20	173.28	0.0298	0.7190
– 30	1.00441	0.000673	0.1596	8.86	165.34	174.20	0.0371	0.7171
– 28	1.09311	0.000676	0.1475	10.64	164.47	175.11	0.0444	0.7153
– 26	1.18778	0.000678	0.1364	12.43	163.59	176.02	0.0516	0.7135
– 24	1.28858	0.000681	0.1265	14.22	162.71	176.93	0.0588	0.7118
– 22	1.39581	0.000683	0.1173	16.02	161.81	177:83	0.0660	0.7102
– 20	1.50972	0.000686	0.1090	17.82	160.91	178.73	0.0731	0.7087
– 18	1.63104	0.000689	0.1014	19.62	160.01	179.63	0.0801	0.7073
– 16	1.75963	0.000692	0.0944	21.43	159.10	180.53	0.0871	0.7059
– 14	1.89575	0.000695	0.0880	23.23	159.19	181.42	0.0941	0.7045
– 12	2.04605	0.000698	0.0821	25.05	157.26	182.31	0.1010	0.7032
– 10	2.19172	0.000701	0.0767	26.87	156.32	183.19	0.1080	0.7019
– 8	2.35272	0.000704	0.0717	28.70	155.36	184.06	0.1148	0.7007
– 6	2.52244	0.000707	0.0672	30.53	154.41	184.94	0.1217	0.6996
– 4	2.70116	0.000710	0.0630	32.37	153.43	185.80	0.1285	0.6986
– 2	2.88921	0.000713	0.0591	34.20	152.47	186.67	0.1352	0.6975
0	3.08690	0.000717	0.0555	36.05	151.48	187.53	0.1420	0.6965
1	3.18974	0.000719	0.0538	36.98	150.97	187.95	0.1453	0.6961

(Contd.)

Contd. Table 36.2

Saturation temp. in °C (t)	Saturation pressure in bar (p)	Specific volume in m³/kg		Specific enthalpy in kJ/kg			Specific entropy in kJ/kg K	
		Liquid (v_f)	Vapour (v_g)	Liquid (h_f)	Vapour (h_g)	Latent (h_{fg})	Liquid (s_f)	Vapour (s_g)
2	3.29513	0.000720	0.0521	37.90	150.48	188.38	0.1489	0.6956
3	3.40310	0.000721	0.0505	38.83	149.98	188.81	0.1521	0.6951
4	3.51367	0.000723	0.0490	39.76	149.47	189.23	0.1553	0.6947
5	3.62690	0.000725	0.0475	40.69	148.96	189.65	0.1568	0.6943
6	3.74280	0.000727	0.0461	190.07	41.62	148.45	0.1620	0.6938
7	3.86141	0.000729	0.0447	190.49	42.56	147.93	0.1653	0.6933
8	3.98283	0.000730	0.0434	190.91	43.50	147.41	0.1686	0.6929
9	4.10702	0.000732	0.0422	191.32	44.43	146.89	0.1719	0.6925
10	4.23407	0.000734	0.0410	191.74	45.37	146.37	0.1752	0.6921
11	4.36442	0.000736	0.0398	192.15	46.31	145.84	0.1784	0.6917
12	4.49763	0.000738	0.0386	192.56	47.26	145.30	0.1817	0.6913
13	4.63386	0.000739	0.0375	192.97	48.20	144.77	0.1850	0.6909
14	4.77312	0.000741	0.0365	193.38	49.15	144.23	0.1883	0.6906
15	4.91545	0.000743	0.0355	193.79	50.10	143.69	0.1915	0.6902
16	5.06087	0.000745	0.0345	194.19	51.05	143.14	0.1948	0.6898
17	5.20942	0.000747	0.0335	194.59	52.00	142.59	0.1981	0.6894
18	5.36117	0.000749	0.0326	194.99	52.95	142.04	0.2013	0.6891
19	5.51614	0.000751	0.0317	195.38	53.91	141.47	0.2046	0.6888
20	5.67441	0.000753	0.0308	195.78	54.87	140.91	0.2078	0.6884
21	5.83635	0.000756	0.0300	196.17	55.83	140.34	0.2110	0.6881
22	6.00171	0.000758	0.0292	196.56	56.79	139.77	0.2143	0.6878
23	6.17050	0.000759	0.0284	196.96	57.75	139.21	0.2174	0.6875
24	6.34269	0.000761	0.0276	197.34	58.73	138.61	0.2207	0.6872
25	6.51840	0.000764	0.0269	197.73	59.70	138.03	0.2239	0.6868
26	6.69765	0.000766	0.0262	198.11	60.67	137.44	0.2271	0.6865
27	6.88048	0.000768	0.0255	198.50	61.65	136.85	0.2303	0.6862
28	7.06704	0.000770	0.0248	198.87	62.63	136.24	0.2335	0.6859
29	7.25738	0.000772	0.0241	199.25	63.61	135.64	0.2368	0.6856
30	7.45103	0.000775	0.0235	199.62	64.59	135.03	0.2400	0.6853
31	7.64903	0.000777	0.0230	199.99	65.58	134.41	0.2431	0.6850
32	7.85089	0.000779	0.0225	200.36	66.57	133.79	0.2463	0.6847
33	8.05662	0.000782	0.0218	200.73	67.56	133.17	0.2495	0.6845
34	8.26621	0.000784	0.0212	201.09	68.56	132.53	0.2527	0.6842
35	8.48000	0.000786	0.0207	201.45	69.56	131.89	0.2559	0.6839

Contd. Table 36.2

Saturation temp. in °C (t)	Saturation pressure in bar (p)	Specific volume in m^3/kg		Specific enthalpy in kJ/kg			Specific entropy in kJ/kg K	
		Liquid (v_f)	Vapour (v_g)	Liquid (h_f)	Vapour (h_g)	Latent (h_{fg})	Liquid (s_f)	Vapour (s_g)
36	8.69766	0.000789	0.0202	201.80	70.55	131.25	0.2591	0.6836
37	8.91904	0.000792	0.0196	202.16	71.55	130.61	0.2623	0.6833
38	9.14483	0.000794	0.0191	202.51	72.56	129.95	0.2654	0.6830
39	9.37497	0.000796	0.0186	202.86	73.57	129.29	0.2685	0.6828
40	9.60897	0.000799	0.0182	203.20	74.59	128.61	0.2718	0.6825
41	9.84793	0.000802	0.0177	203.54	75.61	127.93	0.2750	0.6822
42	10.09131	0.000804	0.0173	203.87	76.62	127.25	0.2782	0.6820
43	10.33862	0.000807	0.0168	204.21	77.65	126.56	0.2814	0.6817
44	10.59021	0.000810	0.0164	204.55	78.68	125.87	0.2846	0.6814
45	10.84655	0.000813	0.0160	204.87	79.71	125.16	0.2878	0.6811
46	11.10758	0.000815	0.0156	205.19	80.75	124.44	0.2909	0.6808
47	11.37304	0.000818	0.0153	205.51	81.79	123.72	0.2941	0.6805
48	11.64290	0.000821	0.0149	205.83	82.83	123.00	0.2973	0.6802
49	11.91724	0.000824	0.0146	206.14	83.88	122.26	0.3005	0.6800
50	12.19655	0.000827	0.0142	206.45	84.94	121.51	0.3037	0.6797

Chart I

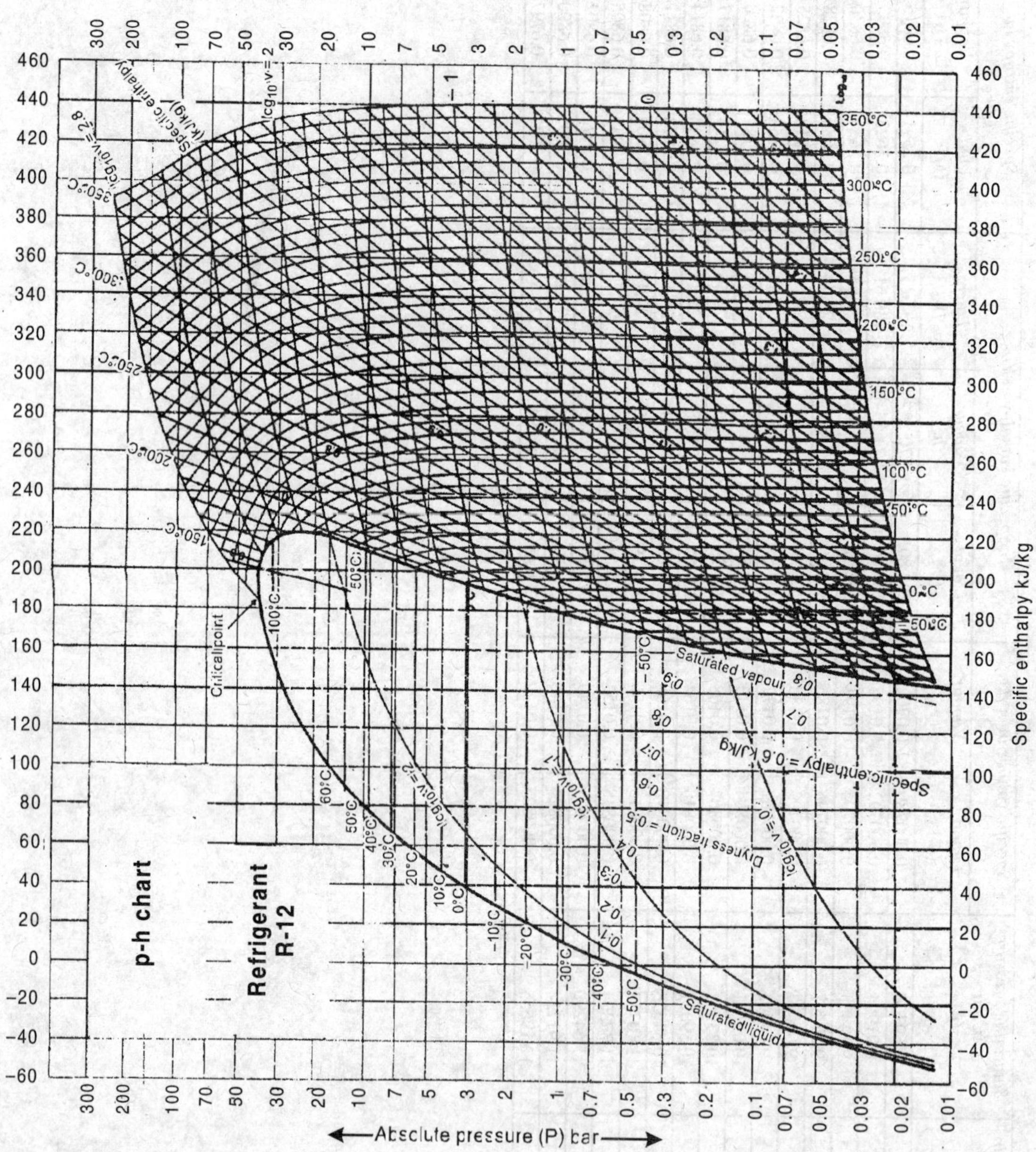
p-h chart
Refrigerant R-12
Specific enthalpy kJ/kg
Absolute pressure (P) bar
Saturated vapour
Saturated liquid
Critical point
Dryness fraction = 0.5
Specific enthalpy = 0.6 kJ/kg
Specific enthalpy (kJ/kg)

(C 1)

Table 36.3: Properties of Refrigeration R-22

Saturation temp. in °C	Saturation pressure in bar	Specific volume in m^3/kg		Specific enthalpy in kJ/kg			Specific entropy in kJ/kg K	
		Liquid (v_f)	Vapour (v_g)	Liquid (h_f)	Vapour (h_g)	Latent (h_{fg})	Liquid (s_f)	Vapour (s_g)
100	0.02009	0.000643	8.3412	– 63.45	267.18	203.73	–0.3144	1.2293
– 95	0.03150	0.000647	5.4344	– 58.14	264.33	206.19	–0.2843	1.2004
– 90	0.04792	0.000652	3.6381	– 52.87	261.51	208.64	–0.2550	1.1736
– 85	0.07731	0.000656	2.5204	– 47.61	258.72	211.11	–0.2269	1.1489
– 80	0.10393	0.000661	1.7816	– 42.40	256.00	213.60	–0.1989	1.1267
– 75	0.14759	0.000666	1.2842	– 37.17	253.28	216.11	–0.1721	1.1066
– 70	0.20517	0.000672	0.9420	– 31.93	250.55	218.62	–0.1461	1.0874
– 65	0.27965	0.000677	0.7037	– 26.68	247.84	221.16	–0.1206	1.0702
– 60	0.37448	0.000683	0.5351	– 21.42	245.09	223.67	–0.0959	1.0543
– 55	0.49621	0.000689	0.4131	– 16.13	242.31	226.18	–0.0712	1.0396
– 50	0.64758	0.000696	0.3229	– 10.81	239.50	228.69	–0.0473	1.0262
– 45	0.83241	0.000702	0.2556	– 5.40	236.60	231.20	–0.0234	1.0137
– 40	1.05586	0.000709	0.2049	0.00	233.67	233.67	0.0000	1.0024
– 38	1.15852	0.000712	0.1882	2.20	232.45	234.65	0.0096	0.9982
– 36	1.26910	0.000715	0.1728	4.40	231.25	235.65	0.0188	0.9940
– 34	1.38731	0.000719	0.1590	6.58	230.02	236.60	0.0280	0.9902
– 32	1.51324	0.000721	0.1465	8.83	228.77	'237.60.	0.0373	0.9860
– 30	1.64690	0.000724	0.1353	11.05	227.50	238.55	0.0464	0.9283
– 28	1.78938	0.000727	0.1253	13.29	226.23	239.52	0.0557	0.9785
– 26	1.94069	0.000731	0.1161	15.58	224.90	240.48	0.0645	0.9747
– 24	2.10207	0.000734	0.1077	17.77	223.64	241.41	0.0733	0.9710
– 22	2.27448	0.000738	0.1000	19.99	222.34	242.33	0.0821	0.9676
– 20	2.45793	0.000741	0.0930	22.21	221.04	243.25	0.0908	0.9638
– 18	2.65310	0.000744	0.0865	24.47	219.70	244.17	0.0996	0.9605
– 16	2.85903	0.000748	0.0806	26.72	218.36	245.08	0.1080	0.9576
– 14	3.07876	0.000752	0.0752	28.94	217.02	245.96	0.1164	0.9538
– 12	3.31172	0.000756	0.0701	31.16	215.68	246.84	0.1248	0.9508
– 10	3.55793	0.000759	0.0655	33.40	214.32	247.72	0.1336	0.9479
– 8	3.81321	0.000763	0.0612	35.66	212.94	248.60	0.1419	0.9454
– 6	4.09172	0.000767	0.0573	37.92	211.54	249.46	0.1503	0.9425
– 4	4.37972	0.000771	0.0536	40.19	210.11	250.30	0.1591	0.9396
– 2	4.68317	0.000775	0.0503	42.52	208.62	251.14	0.1675	0.9370

(*Contd.*)

Contd. Table 36.3

Saturation temp. in °C	Saturation pressure in bar	Specific volume in m^3/kg		Specific enthalpy in kJ/kg			Specific entropy in kJ/kg K	
		Liquid (v_f)	Vapour (v_g)	Liquid (h_f)	Vapour (h_g)	Latent (h_{fg})	Liquid (s_f)	Vapour (s_g)
0	5.00207	0.000779	0.0471	44.94	207.03	251.97	0.1763	0.9345
1	5.16841	0.000781	0.0457	46.16	206.21	252.37	0.1809	0.9333
2	5.33917	0.000783	0.0443	47.38	205.39	252.77	0.1855	0.9320
3	5.51434	0.000785	0.0429	48.64	204.53	253.17	0.1901	0.9303
4	5.69352	0.000787	0.0416	49.91	203.65	253.56	0.1943	0.9291
5	5.87621	0.000790	0.0403	51.16	202.79	253.95	0.1989	0.9278
6	6.06276	0.000792	0.0391	52.41	201.92	254.33	0.2035	0.9266
7	6.25393	0.000794	0.0379	53.68	201.03	254.71	0.2077	0.9253
8	6.44993	0.000797	0.0368	54.94	200.14	255.08	0.2123	0.9241
9	6.65034	0.000799	0.0357	56.22	199.22	255.44	0.2169	0.9228
10	6.85517	0.000801	0.0346	57.52	198.29	255.81	0.2211	0.9216
11	7.06621	0.000803	0.0336	58.80	197.35	256.15	0.2257	0.9203
12	7.27724	0.000805	0.0326	60.07	196.41	256.48	0.2303	0.9190
13	7.49793	0.000808	0.0316	61.38,	195.45	256.83	0.2345	0.9178
14	7.72552	0.000810	0.0307	62.72	194.45	257.17	0.2391	0.9165
15	7.95517	0.000813	0.0298	64.02	193.48	257.50	0.2437	0.9152
16	8.18758	0.000815	0.0289	65.32	192.48	257.81	0.2483	0.9140
17	8.42552	0.000817	0.0281	66.63	191.50	258.13	0.2529	0.9127
18	8.63241	0.000820	0.0273	67.95	190.49	258.44	0.2571	0.9115
19	8.91793	0.000822	0.0266	69.27	189.46	258.73	0.2617	0.9102
20	9.17241	0.000825	0.0258	70.59	188.41	259.00	0.2663	0.9089
21	9.43310	0.000827	0.0251	71.93	187.36	259.29	0.2709	0.9077
22	9.69931	0.000830	0.0244	73.31	186.27	259.58	0.2755	0.9065
23	9.97103	0.000833	0.0237	74.66	185.19	259.85	0.2801	0.9052
24	10.24828	0.000835	0.0230	76.04	184.07	260.11	0.2847	0.9039
25	10.53103	0.000838	0.0224	77.39	182.99	260.38	0.2889	0.9027
26	10.81931	0.000841	0.0218	78.79	181.85	260.64	0.2935	0.9014
27	11.11310	0.000844	0.0212	80.16	180.73	260.89	0.2981	0.9002
28	11.41241	0.000846	0.0206	81.54	179.58	261.12	0.3023	0.8[illegible]
29	11.69034	0.000849	0.0200	82.96	178.41	261.37	0.3069	0.8977
30	12.03448	0.000852	0.0194	84.38	177.22	261.60	0.3115	0.8964
31	12.35103	0.000855	0.0189	85.77	176.04	261.81	0.3161	0.8948
32	12.67310	0.000858	0.0184	87.17	174.85	262.02	0.3207	0.8935
33	13.00070	0.000861	0.0179	88.57	173.65	262.22	0.3249	0.8922

(*Contd.*)

Contd. Table 36.3

Saturation temp. in °C	Saturation pressure in bar	Specific volume in m^3/kg		Specific enthalpy in kJ/kg			Specific entropy in kJ/kg K	
		Liquid (v_f)	Vapour (v_g)	Liquid (h_f)	Vapour (h_g)	Latent (h_{fg})	Liquid (s_f)	Vapour (s_g)
34	13.33793	0.000864	0.0174	90.00	172.40	262.40	0.3295	0.8909
35	13.68276	0.000867	0.0169	91.43	171.15	262.58	0.3337	0.8893
36	14.03034	0.000870	0.0165	92.85	169.89	262.74	0.3383	0.8880
37	14.38207	0.000874	0.0161	94.24	168.64	262.88	0.3429	0.8868
38	14.74345	0.000877	0.0156	95.63	167.37	263.00	0.3471	0.8851
39	15.11103	0.000881	0.0152	97.03	166.10	263.13	0.3517	0.8838
40	15.48965	0.000884	0.0148	98.44	164.77	263.21	0.3563	0.8822
41	15.86827	0.000887	0.0144	99.82	163.47	263.29	0.3601	0.8809
42	16.25793	0.000891	0.0140	101.24	162.14	263.38	0.3651	0.8797
43	16.65380	0.000894	0.0137	102.68	160.80	263.48	0.3693	0.8780
44	17.05517	0.000898	0.0133	104.12	159.46	263.58	0.3739	0.8767
45	17.46552	0.000902	0.0130	105.58	158.09	263.67	0.3785	0.8755
46	17.88414	0.000906	0.0126	107.04	156.72	263.76	0.3827	0.8738
47	18.30827	0.000910	0.0123	108.51	155.36	263.87	0.3871	0.8723
48	18.73793	0.000914	0.0119	109.98	153.95	263.93	0.3915	0.8709
49	19.12414	0.000918	0.0116	111.42	152.57	263.99	0.3959	0.8694
50	19.61380	0.000922	0.0113	112.86	151.19	264.05	0.4003	0.8680

Chart II

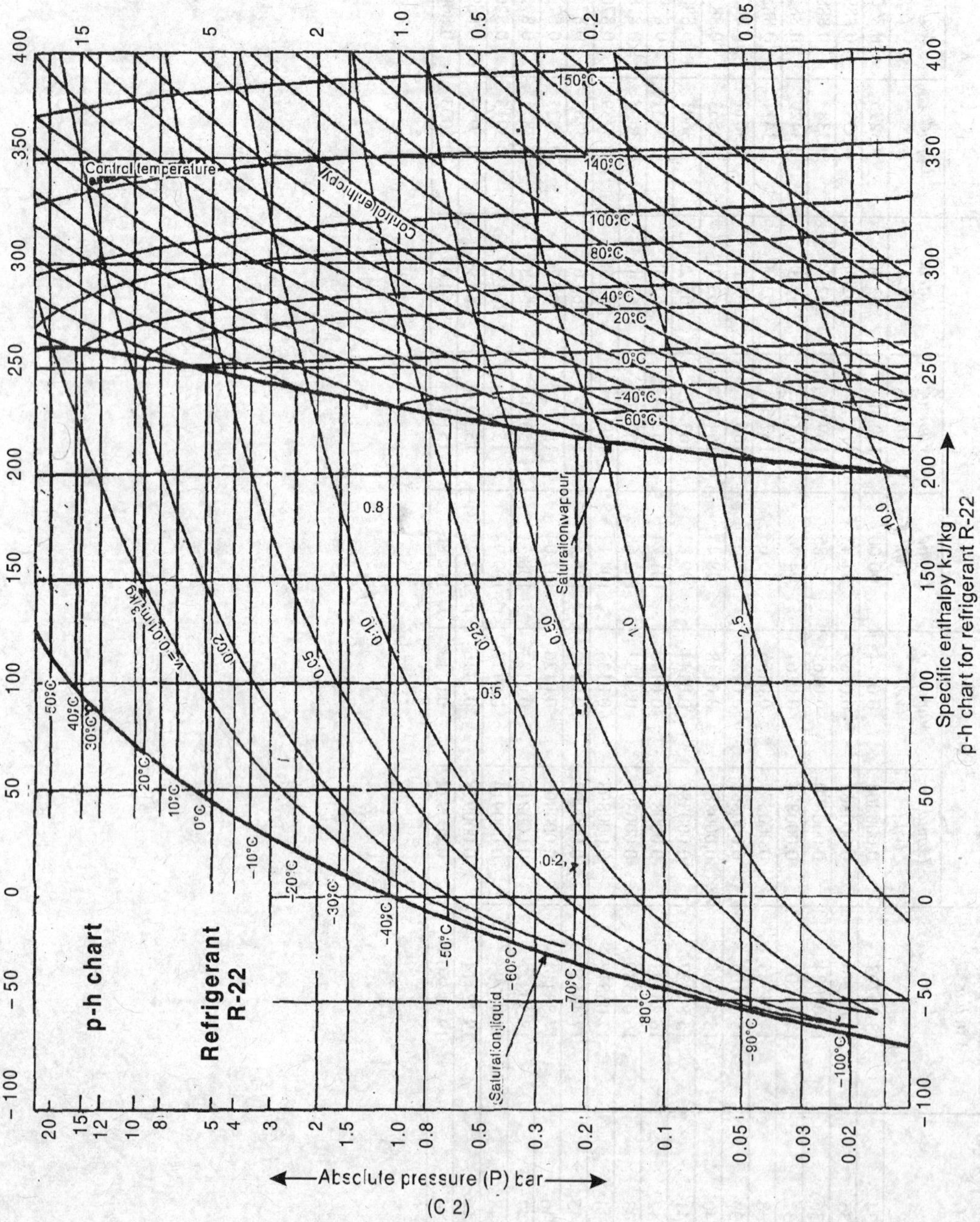
p-h chart
Refrigerant R-22
Control temperature
Control enthalpy
Saturation vapour
Saturation liquid
150°C
140°C
100°C
80°C
40°C
20°C
0°C
-40°C
-60°C
50°C
40°C
30°C
20°C
10°C
0°C
-10°C
-20°C
-30°C
-40°C
-50°C
-60°C
-70°C
-80°C
-90°C
-100°C
v = 0.01 m³/kg
0.02
0.05
0.10
0.25
0.50
1.0
2.5
10.0
0.8
0.5
0.2
Absolute pressure (P) bar
(C 2)
Specific enthalpy kJ/kg
p-h chart for refrigerant R-22

Table 36.4: Properties of Refrigerant R-134a

Temp. T °C	Pressure P, bar	Volume V_g m^3/kg	Enthalpy, kJ/kg		Entropy, kJ/kg K		Sp Heat, kJ/kg K	
			Liquid h_f	Vapour h_g	Liquid s_f	Vapour s_g	Liquid C_{pf}	Vapour C_{pg}
–60	0.1594	1.07700	123.96	361.51	0.6871	1.8016	1.220	0.685
–50	0.2948	0.60560	136.21	367.83	0.7432	1.7°?2	1.229	0.712
–40	0.5122	0.36095	148.57	374.16	0.7973	1.7649	1.243	0.740
– 30	0.8436	0.22596	161.10	380.45	0.8498	1.7519	1.260	0.771
– 28	0.9268	0.20682	163.62	381.70	0.8601	1.7497	1.264	0.778
–26	1.0164	0.18961	166.16	382.94	0.8704	1.7476	1.268	0.785
–24	1.1127	0.17410	168.70	384.19	0.8806	1.7455	1.273	0.791
– 22	1.2160	0.16010	171.26	385.43	0.8908	1.7436	1.277	0.798
–20	1.3268	0.14744	173.82	386.66	0.9009	1.7417	1.282	0.805
– 18	1.4454	0.13597	176.39	387.89	0.9110	1.7399	1.286	0.812
– 16	1.5721	0.12556	178.97	389.11	0.9211	1.7383	1.291	0.820
– 14	1.7074	0.11610	181.56	390.33	0.9311	1.7367	1.296	0.827
– 12	1.8516	0.10749	184.16	391.55	0.9410	1.7351	1.301	0.835
– 10	2.0052	0.09963	186.78	392.75	0.9509	1.7337	1.306	0.842
– 8	2.1684	0.09246	189.40	393.95	0.9608	1.7323	1.312	0.850
– 6	2.3418	0.08591	192.03	395.15	0.9707	1.7310	1.317	0.858
– 4	2.5257	0.07991	194.68	396.33	0.9805	1.7297	1.323	0.866
– 2	2.7206	0.07440	197.33	397.51	0.9903	1.7285	1.329	0.875
0	2.9269	0.06935	200.00	398.68	1.0000	1.7274	1.335	0.883
2	3.1450	0.06470	202.68	399.84	1.0097	1.7263	1.341	0.892
4	3.3755	0.06042	205.37	401.00	1.0194	1.7252	1.347	0.901
6	3.6186	0.05648	208.08	402.14	1.0291	1.7242	1.353	0.910
8	3.8749	0.05284	210.80	403.27	1.0387	1.7233	1.360	0.920
10	4.1449	0.04948	213.53	404.40	1.0483	1.7224	1.367	0.930
12	4.4289	0.04636	216.27	405.51	1.0579	1.7215	1.374	0.939
14	4.7276	0.04348	219.03	406.61	1.0674	1.7207	1.381	0.950
16	5.0413	0.04081	221.80	407.70	1.0770	1.7199	1.388	0.960
18	5.3706	0.03833	224.59	408.78	1.0865	1.7191	1.396	0.971
20	5.7159	0.03603	227.40	409.84	1.0960	1.7183	1.404	0.982
22	6.0777	0.03388	230.31	410.89	1.1055	1.7176	1.412	0.994
24	6.4566	0.03189	233.05	411.93	1.1149	1.7169	1.420	1.006
26	6.8531	0.03003	235.90	412.95	1.1244	1.7162	1.429	1.018

(Contd.)

Contd. Table 36.4

Temp. T °C	Pressure P, bar	Volume V_g m^3/kg	Enthalpy, kJ/kg		Entropy, kJ/kg K		Sp Heat, kJ/kg K	
			Liquid h_f	Vapour h_g	Liquid s_f	Vapour s_g	Liquid C_{pf}	Vapour C_{pg}
28	7.2676	0.02829	238.77	413.95	1.1338	1.7155	1.438	1.031
30	7.7008	0.02667	241.65	414.94	1.1432	1.7149	1.447	1.044
32	8.1530	0.02516	244.55	415.90	1.1527	1.7142	1.457	1.058
34	8.6250	0.02374	247.47	416.85	1.1621	1.7135	1.467	1.073
36	9.1172	0.02241	250.41	417.78	1.1715	1.7129	1.478	1.088
38	9.6301	0.02116	253.37	418.69	1.1809	1.7122	1.489	1.104
40	10.165	0.01999	256.35	419.58	1.1903	1.7115	1.500	1.120
42	10.721	0.01890	259.35	420.44	1.1997	1.7108	1.513	1.138
44	11.300	0.01786	262.38	421.28	1.2091	1.7101	1.525	1.156
46	11.901	0.01689	265.42	422.09	1.2185	1.7094	1.539	1.175
48	12.527	0.01598	268.49	422.88	1.2279	1.7086	1.553	1.196
50	13.177	0.01511	271.59	423.63	1.2373	1.7078	1.569	1.218
52	13.852	0.01430	274.71	424.35	1.2468	1.7070	1.585	1.241
54	14.553	0.01353	277.86	425.03	1.2562	1.7061	1.602	1.266
56	15.280	0.01280	281.04	425.68	1.2657	1.7051	1.621	1.293
58	16.033	0.01212	284.25	426.29	1.2752	1.7041	1.641	1.322
60	16.815	0.01146	287.49	426.86	1.2847.	1.7031	1.663	1.354
62	17.625	0.01085	290.77	427.37	1:2943	1.7019	1.686	1.388
64	18.464	0.01026	294.08	427.84	1.3039	1.7007	1.712	1.426
66	19.344	0.00970	297.44	428.25	1.3136	1.6993	1.740	1.468
68	20.234	0.00917	300.84	428.61	1.3234	1.6979	1.772	1.515
70	21.165	0.00867	304.29	428.89	1.3332	1.6963	1.806	1.567
72	22.130	0.00818	307.79	429.10	1.3430	1.6945	1.846	1.626
74	23.127	0.00772	311.34	429.23	1.3530	1.6926	1.890	1.693
76	24.159	0.00728	314.96	429.27	1.3631	1.6905	1.941	1.770
78	25.227	0.00686	318.65	429.20	1.3733	1.6881	2.000	1.861
80	26.331	0.00646	322.41	429.02	1.3837	1.6855	2.069	1.967

Chart III

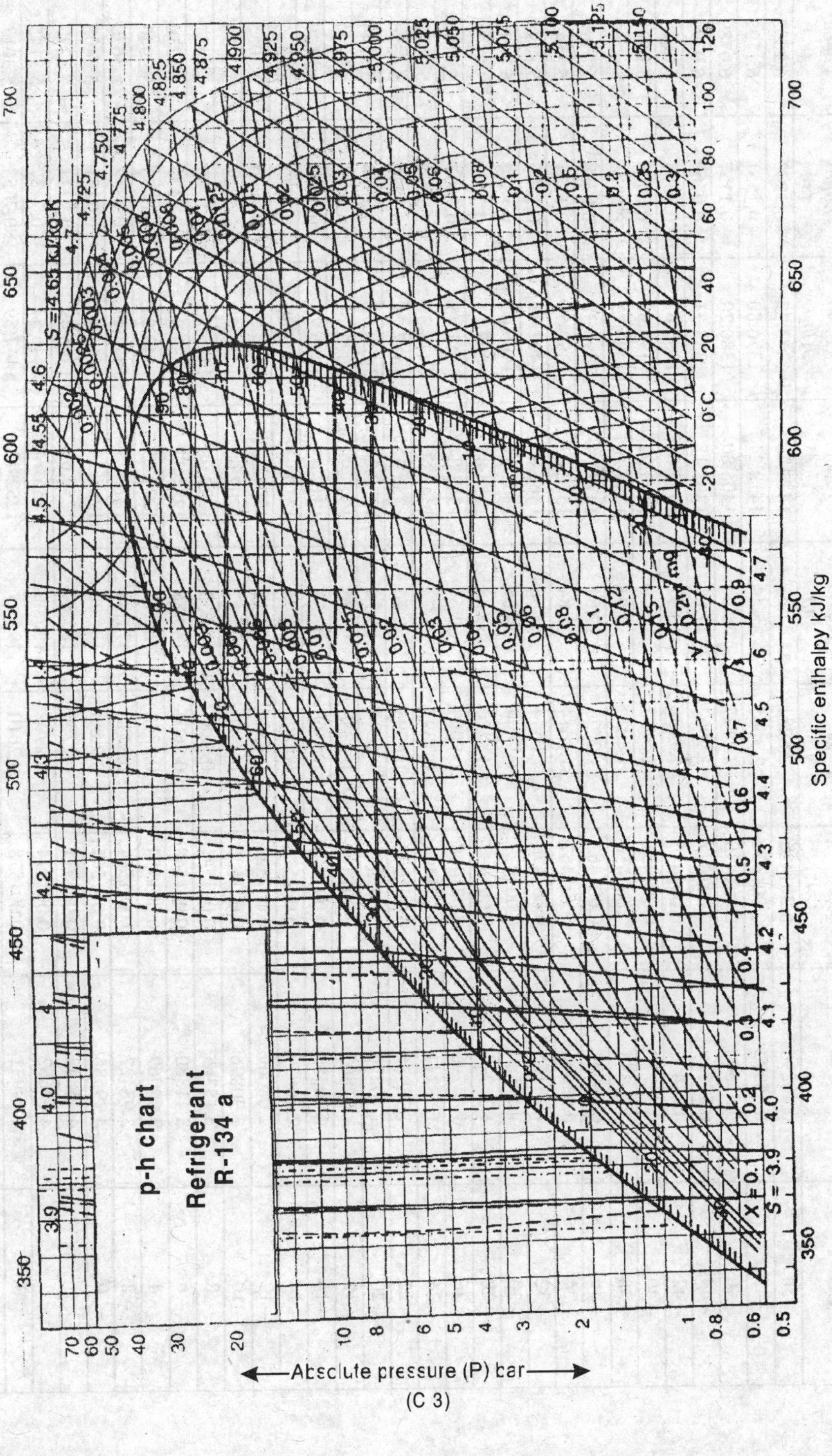
p-h chart
Refrigerant
R-134 a
Specific enthalpy kJ/kg
Absolute pressure (P) bar

(C 3)

Table 36.5: Properties of Ammonia

Saturation temp. in °C (t)	Saturation pressure in bar (p)	Specific volume in m^3/kg		Specific enthalpy in kJ/kg			Specific entropy in kJ/kg K	
		Liquid (v_f)	Vapour (v_g)	Liquid (h_f)	Vapour (h_g)	Latent (h_{fg})	Liquid (s_f)	Vapour (s_g)
– 50	0.40896	0.001426	2.6281	– 44.43	1417.70	1373.27	– 0.1943	6.1603
– 48	0.45972	0.001431	2.3565	– 35.44	1412.24	1376.80	– 0.1551	6.1192
– 46	0.51600	0.001436	2.1177	– 26.60	1406.80	1380.20	– 0.1157	6.0789
– 44	0.57710	0.001441	1.9062	– 17.81	1401.12	1383.31	– 0.0769	6.0394
– 42	0.64455	0.001446	1.7196	– 8.97	1395.64	1386.67	– 0.0384	6.0008
– 40	0.71793	0.001451	1.5537	0.00	1390.02	1390.02	0.0000	5.9631
– 38	0.79384	0.001456	1.4077	8.97	1384.16	1393.13	0.0381	5.9262
– 36	0.88607	0.001462	1.2775	17.81	1378.54	1396.35	0.0758	5.8900
– 34	0.98096	0.01467	1.1614	26.84	1372.67	1399.51	0.1131	5.8545
– 32	1.08165	0.001472	1.0574	35.68	1366.80	1402.48	0.1506	5.8198
– 30	1.19586	0.001477	0.9644	44.66	1360.94	1405.60	0.1876	5.7856
– 28	1.31724	0.001483	0.8820	53.68	1354.85	1408.53	0.2242	5.7521
– 26	1.44790	0.001488	0.8069	62.61	1348.84	1411.45	0.2607	5.7195
– 24	1.58841	0.001494	0.7397	71.73	1342.66	1414.39	0.2970	5.6872
– 22	1.73986	0.001500	0.6793	80.76	1336.52	1417.28	0.3330	5.6556
– 20	1.90276	0.001505	0.6244	89.78	1330.24	1420.02	0.3684	5.6244
– 18	2.07807	0.001511	0.5750	98.76	1323.96	1422.72	0.4043	5.5939
– 16	2.26551	0.001517	0.5303	107.83	1317.45	1425.28	0.4397	5.5639
– 14	2.46634	0.001523	0.4896	116.95	1310.93	1427.88	0.4747	5.5356
– 12	2.68069	0.001529	0.4526	126.16	1304.38	1430.54	0.5096	5.5055
– 10	2.90896	0.001536	0.4189	135.37	1297.68	1433.05	0.5443	5.4770
– 8	3.15365	0.001541	0.3884	144.35	1290.98	1435.33	0.5789	5.4487
– 6	3.41380	0.001548	0.3604	153.56	1284.37	1437.97	0.6139	5.4210
– 4	3.69069	0.001554	0.3348	162.77	1277.25	1449.02	0.6473	5.3940
– 2	3.98427	0.00 1561	0.3113	171.98	1270.19	1442.17	0.6812	5.3670
0	4.29586	0.001567	0.2.898	181.20	1263.25	1444.45	0.7151	5.3405
1	4.45848	0.001571	0.2798	185.80	1259.69	1445.49	0.7321	5.3277
2	4.62662	0.001574	0.2702	190.40	1256.14	1446.54	0.7487	5.3145
3	4.79931	0.001578	0.2610	195.17	1252.42	1447.59	0.7653	5.3017
4	4.97682	0.001582	0.2521	199.85	1248.78	1448.63	0.7818	5.2888
5	5.15862	0.001585	0.2436	204.46	1245.10	1449.59	0.7989	5.2765
6	5.34745	0.001589	0.2354	209.06	1241.43	1450,40	0.8154	5.2638

(Contd.)

Contd. Table 36.5

Saturation temp. in °C (t)	Saturation pressure in bar (p)	Specific volume in m^3/kg		Specific enthalpy in kJ/kg			Specific entropy in kJ/kg K	
		Liquid (v_f)	Vapour (v_g)	Liquid (h_f)	Vapour (h_g)	Latent (h_{fg})	Liquid (s_f)	Vapour (s_g)
8	5.73820	0.001595	0.2201	218.50	1234.04	1452.54	0.8487	5.2389
9	5.94186	0.0015 98	0.2128	223.11	1230.28	1453.39	0.8652	5.2266
10	6.15103	0.001603	0.2060	227.72	1226.50	1454.22	0.8814	5.2141
11	6.36641	0.001607	0.1992	232.53	1222.77	1455.30	0.8979	5.2020
12	6.58731	0.001610	0.1928	237.15	1218.96	1456.11	0.9142	5.1900
13	6.81420	0.001614	0.1866	241.92	1215.04	1456.96	0.9307	5.1780
14	7.04717	0.001617	0.1807	246.60	1211.20	1457.80	0.9470	5.1659
15	7.28276	0.001621	0.1751	251.44	1207.19	1458.63	0.9634	5.1542
16	7.53104	0.001624	0.1696	256.14	1203.33	1459.47	0.9794	5.1421
17	7.78138	0.001628	0.1643	259.88	1200.36	1460.24	0.9956	5.1302
18	8.03862	0.001633	0.1592	265.54	1195.38	1460.92	1.0118	5.1186
19	8.30551	0.001637	0.1543	270.35	1191.40	1461.75	1.0280	5.1073
20	8.57241	0.001641	0.1496	275.16	1187.44	1462.60	1.0442	5.0956
21	8.85172	0.001645	0.1450	279.77	1183.44	1463.21	1.0604	5.0843
22	9.13655	0.001648	0.1407	284.56	1179.28	1463.84	1.0763	5.0729
23	9.42690	0.001652	0.1365	289.37	1175.26	1464.63	1.0924	5.0616
24	9.72690	0.001656	0.1324	294.19	1171.14	1465.33	1.1083	5.0503
25	10.02760	0.001661	0.1284	298.90	1166.94	1465.84	1.1242	5.0391
26	10.34069	0.001665	0.1246	303.82	1162.77	1466.59	1.1402	5.0279
27	10.66137	0.001669	0.1210	308.63	1158.59	1467.22	1.1563	5.0170
28	10.99172	0.001673	0.1174	313.45	1154.40	1467.85	1.1721	5.0061
29	11.32758	0.001678	0.1140	318.26	1150.19	1468.45	1.1879	4.9951
30	11.66896	0.001682	0.1107	323.08	1145.79	1468.87	1.2037	4.9842
31	12.01655	0.001686	0.1075	327.89	1141.61	1469.50	1.2195	4.9733
32	12.37517	0.001691	0.1045	332.71	1137.23	1469.94	1.2350	4.9624
33	12.74482	0.001695	0.1015	337.52	1132.84	1470.36	1.2508	4.9517
34	13.12137	0.001700	0.0987	342.48	1128.44	1470.92	1.2664	4.9409
35	13.50345	0.001704	0.0960	347.50	1123.93	1471.43	1.2821	4.9302
36	13.89379	0.001709	0.0932	352.29	1119.41	1471.70	1.2978	4.9196
37	14.29517	0.001713	0.0907	357.25	1114.94	1472.19	1.3135	4.9091
38	14.70620	0.001718	0.0882	362.12	1110.33	1472.45	1.3290	4.8985
39	15.12276	0.001723	0.0857	367.11	1105.76	1472.87	1.3445	4.8885
40	15.54483	0.001727	0.0834	371.93	1101.37	1473.30	1.3600	4.8774

(*Contd.*)

Contd. Table 36.5

Saturation temp. in °C (t)	Saturation pressure in bar (p)	Specific volume in m^3/kg		Specific enthalpy in kJ/kg			Specific entropy in kJ/kg K	
		Liquid (v_f)	Vapour (v_g)	Liquid (h_f)	Vapour (h_g)	Latent (h_{fg})	Liquid (s_f)	Vapour (s_g)
41	15.98551	0.001732	.0.0811	376.95	1096.55	1473.50	1.3754.	4.8669
42	16.43172	0.001738	0.0789	381.79	1091.91	1473.70	1.3908	4.8563
43	16.88344	0.001742	0.0768	386.76	1087.15	1473.91	1.4065	4.8461
44	17.34482	0.001747	0.0747	391.79	1082.33	1474.12	1.4221	4.8357
45	17.81724	0.001752	0.0727	396.81	1077.52	1474.33	1.4374	4.8251
46	18.30070	0.001757	0.0708	401.84	1072.70	1474.54	1.4528	4.8147
47	18.79518	0.001763	0.0689	406.86	1067.82	1474.68	1.4683	4.8045
48	19.29931	0.001768	0.0670	411.89	1062.87	1474.76	1.4835	4.7938
49	19.80965	0:001773	0.0652	416.91	1057.93	1474.84	1.4990	4.7834
50	20.33103	0.001779	0.0636	421.94	1052.98	1474.92	1.5148	4.7732

Chart IV

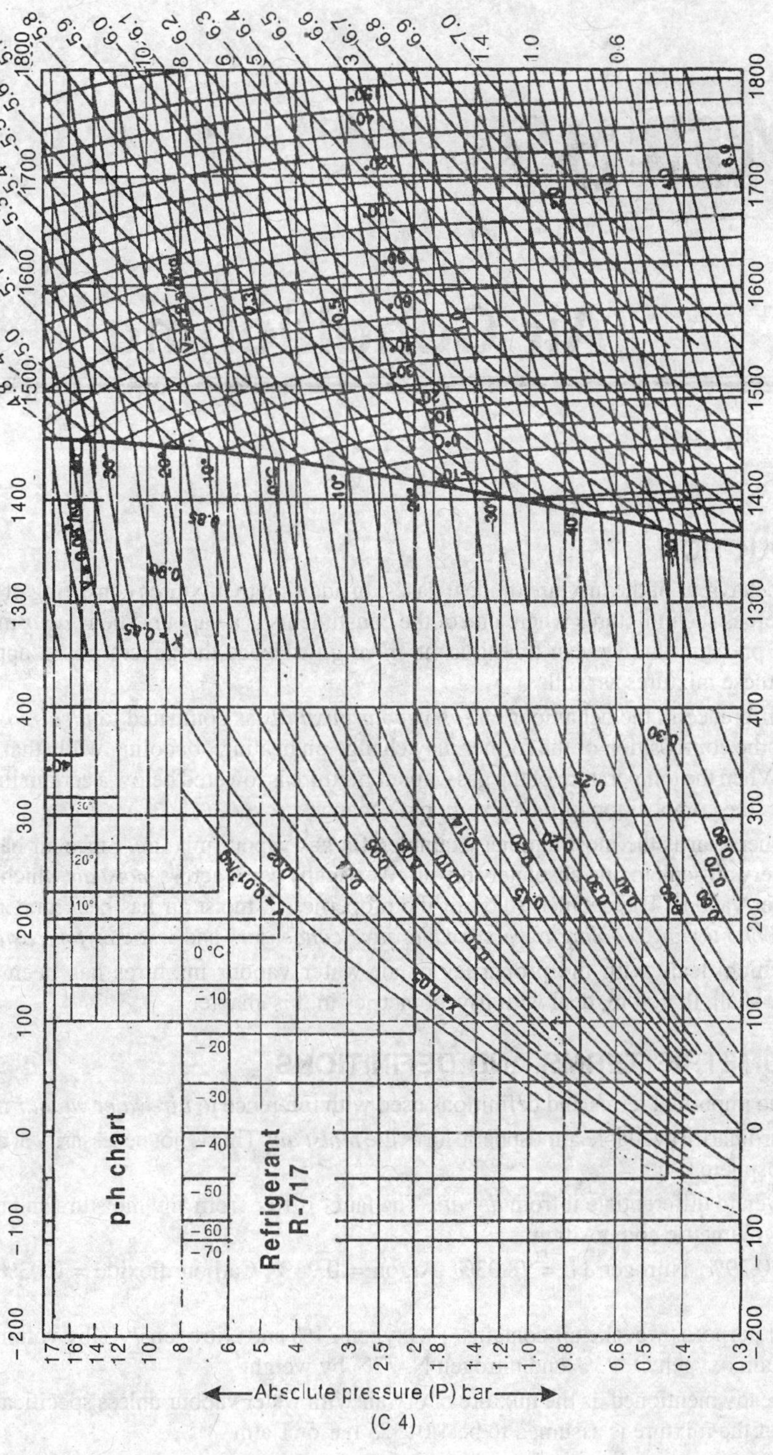
p-h chart
Refrigerant R-717
Absolute pressure (P) bar
Specific enthalpy kJ/kg

(C 4)

CHAPTER 37

Air Conditioning

37.1 INTRODUCTION

We have seen the behaviour of the mixture of ideal gases. In addition to mixtures consisting of ideal gases, we may frequently come across such mixtures where one of the constituents is a vapour. The *vapour* in these mixtures is at such a low partial pressure that we can treat it as an *ideal gas*. Hence the general rules, applicable to ideal gas mixtures apply to these mixtures as well.

A significant difference in the behaviour of *gas-vapour mixtures* as compared to *ideal-gas mixtures*, however, lies in the fact that the composition of the former may change on heating or cooling while that of the later does not change normally. When the temperature of the gas-vapour mixture is lowered below a certain limit, it may condense or solidify and thus the composition of the mixture may change.

Normal atmospheric air is the most familiar example of a gas-vapour mixture. Since air has a great affinity for water vapour it is very difficult to find absolutely dry air. We usually come across *moist air* which is actually a mixture of dry air and water vapour. The detailed study of the properties of moist air has been important due to its wide application in *air-conditioning*, *cooling towers*, *evaporative condensers* and *weather forecasting* etc.

The science which deals with the properties of air-water vapour mixtures has been given the name of *psychometrics*. We shall, therefore, deal with psychometrics in this chapter.

37.2 PSYCHOMETRIC TERMS AND DEFINITIONS

Following are some important terms and definitions used with reference to *air-water vapour* mixtures.

1. Moist-Air. Ordinarily available atmospheric air is the *moist air*. This is not necessarily at a particular pressure. It is a natural atmosphere.

The name is given to differentiate it from *dry air*. The latter is free from any moisture and technically speaking has the following volumetric composition.

Oxygen O_2 = 20.29%, Nitrogen N_2 = 78.03%, Argon = 0.94%, Carbon dioxide = 0.03% plus traces of rare gases.

For all practical purposes we take air contains — Oxygen 21% and Nitrogen 79% by volume. Similarly, dry air is assumed to contain oxygen O_2 23% and nitrogen N_2 77% by weight.

Moist air, as already mentioned, is the mixture of dry air with water vapour unless specifically stated otherwise, the total pressure of the mixture is assumed to be 1.01325 bar or 1 atm.

The relative amount of water vapour in the atmosphere or the moist air is very small, say about 1 to 2%. Hence the partial pressure of the water vapour in the atmosphere is also very small. At ordinary temperature, therefore, the moisture in atmosphere will be present in the *superheated state.*

2. Dry Bulb Temperature (DBT). The temperature of a mixture of gases as recorded or indicated by an ordinary accurate thermometer is called the *dry bulb temperature*, abbreviated as DBT or t_{db} or $t_{d.}$

3. Dew Point Temperature (DPT). If a sample of moist air is cooled at constant pressure then the temperature of both the dry air and water vapour part are reduced simultaneously. Dry air being a gas mixture remain unchanged in composition. The water vapour, however, may change its phase and therefore cause a change in the composition of the mixture. Figure 37.1 represents a *T–S* diagram of steam (for that purpose moisture vapour in air), show the process of cooling of water-vapour part of the mixture.

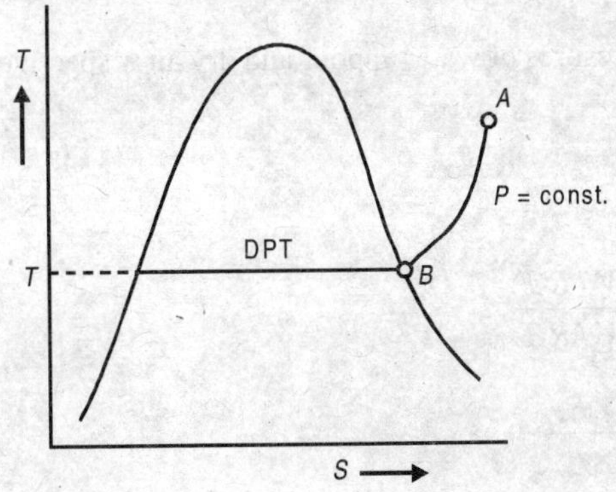

Fig. 37.1. *T–S diagram of water vapour part*

Point *A* represents the original state of water vapour in the mixture. This vapour is in a superheated state. The pressure at point *A* equals the partial pressure of the vapour and the temperature equals the temperature of the mixture. Now if we cool the mixture at constant pressure, the water vapour will be cooled at constant partial pressure as long as the composition of the mixture does not change.

Curve *AB* represents the cooling at constant partial pressure until the vapour reaches the saturated state at point *B*. Any further cooling will result in the *condensation* of vapour i.e., release of moisture. The temperature at *B* is called the *dew-point-temperature.* This temperature, therefore, corresponds to the *saturation temperature* of steam at the *partial pressure* of water vapour in the original mixture.

Dew Point Temperature of the given air is that temperature at which the given moisture or water vapour in air saturates the air. At this temperature, the moisture is on the verge of condensing if the air temperature is reduced below this temperature. Dew Point Temperature is denoted by DPT or t_{dpt}.

4. Saturated Air. It is the mixture of dry air and saturated water vapour at the dry-air temperature. In other words, moist air in which the partial pressure of water vapour equals the saturation pressure of steam at the temperature of air is known as *Saturated Air*. State *B* in Fig. 37.1 indicates the water vapour part of the saturated air at temperature *T*. It may be noted that that it is not the air which is saturated but it is actually water vapour is a saturated vapour at its partial pressure.

Air is saturated with water vapour means that the vapour holding capacity of air at its temperature is complete and maximum and at that temperature it cannot hold any more water vapour.

5. Unsaturated Air. Moist air at a particular temperature which is not saturated is known as *unsaturated air*. It has a capacity to hold or to absorb still more water vapour. Water vapour in an unsaturated air is in a superheated condition.

6. Specific Humidity or Humidity Ratio (W_H). Specific humidity is the ratio of the mass of water vapour m_v to the mass of the associated dry air in a sample of moist air. It is denoted by *W*.

$$\therefore \quad W_H = \frac{\text{Mass of water vapour}}{\text{Mass of air (dry)}}$$

$$= \frac{m_v}{m_a}$$

where $$m_v = \frac{p_v V}{R_v T}$$

and $$m_a = \frac{p_a V}{R_a T}$$

Here p_v and p_a are the partial pressures of water vapour and dry air respectively in the sample of moist air at the temperature T and volume V.

$$\therefore \quad W_H = \frac{m_v}{m_a}$$

$$= \frac{p_v V / R_v T}{p_a V / R_a T}$$

$$= \frac{p_v \times R_a}{p_a \times R_v}$$

$$= \frac{p_v}{p_a} \times \frac{287}{8314.3/14}$$

$$= \frac{0.622 p_v}{p_a}$$

But, if p = total pressure of air = barometric pressure

$$= p_v + p_a$$

$$\therefore \quad p_a = p - p_v$$

$$\therefore \quad W_H = \frac{0.622 p_v}{p - p_v} \qquad \text{...(1)}$$

7. Degree of Saturation (μ). Degree of saturation is defined as the ratio of the actual mass of water vapour present in the moist air to the maximum mass of water vapour in the air at the same dry bulb temperature. It is denoted by μ.

With maximum water vapour in the air, the air becomes saturated and the water vapour is m_s (mass of water vapour at saturation condition).

$$\therefore \quad \mu = \text{Degree of saturation}$$

$$= \frac{m_v}{m_s} \text{ or } = \frac{W_v}{W_s} \qquad \text{...(2)}$$

$$= \frac{0.622\ p_v/(p - p_v)}{0.622\ p_s/p - p_s}$$

$$= \frac{p_v}{p_s} \times \frac{p - p_s}{p - p_v} \qquad \ldots(3)$$

8. Relative Humidity. Water vapour contained in the air is called *humidity*. The *Relative Humidity* is defined as the ratio of the actual partial pressure of water vapour in the moist air to the saturation pressure p_s of the vapour at the same dry-bulb temperature of air. It is denoted by symbol ϕ or RH

$$\text{RH} = \phi = \frac{m_v}{m_s}$$

$$= \frac{p_v V / R_v T}{p_s V / R_v T}$$

$$= \frac{p_v}{p_s}$$

$$\phi = \frac{\mathbf{p_v}}{\mathbf{p_s}} \qquad \ldots(4)$$

$$\therefore \qquad \mu = \frac{p_v}{p_s} \times \frac{p - p_s}{p - p_v} = \frac{p_v}{p_s} \times \frac{p\left[1 - \left(p_s/p\right)\right]}{p\left[1 - \left(p_v/p\right)\right]}$$

$$= \phi . \frac{1 - \frac{p_s}{p}}{1 - \frac{p_s \phi}{p}}$$

$$\text{or} \qquad \phi = \frac{\mu}{1 - (1 - \mu)(p_s/p)} \qquad \ldots(5)$$

Equation (5) gives a relationship between the relative humidity and degree of saturation.

It may be pointed out that there is no convenient method to measure ϕ, μ or W_H. They must therefore, be calculated by suitable means. All the equations noted above require the values of p_s and p_v. The values of p_s may be obtained from the steam tables. There are two methods by which p_v can be calculated.

1. p_s is the saturation pressure at the dry bulb temperature of air. This is found from steam tables.

If we get or know the dew point temperature of air, then the saturation pressure corresponding to the dew-point temperature is the vapour pressure p_v. From Fig. 37.1, *B* corresponds to the dew point temperature and the pressure at T_B is the partial pressure of water vapour at *A*.

The accurate measurement of DPT, however, is difficult.

2. The second method involves the determination of wet bulb temperature which is discussed below.

9. Wet-Bulb Temperature (WBT) or t_{wb} or t_w. The temperature recorded by a thermometer whose bulb is covered by a wet fibrous cloth is known as the *wet-bulb temperature*. It is denoted either by WBT or t_{wb} or t_w. Frequently a wet bulb thermometer and a dry bulb thermometer are mounted together, and the combined instrument is known as *psychometer*.

Figure 37.2 shows a sling psychometer consisting of two thermometers side by side on a holder. A handle is provided so that the whole device may be whirled through the air. The vigorous rotation of the psychometer helps to get quick and dependable readings on the wet bulb thermometer.

When the surrounding moist air comes in contact with the wet bulb, vapourisation of water takes place on the bulb till the surrounding air becomes saturated. Since heat is required for vapourisation the temperature of both the air and that of water on the bulb decreases until equilibrium is reached. The equilibrium temperature is known as the *wet-bulb temperature*, abbreviated as WBT or t_{wb} or t_e.

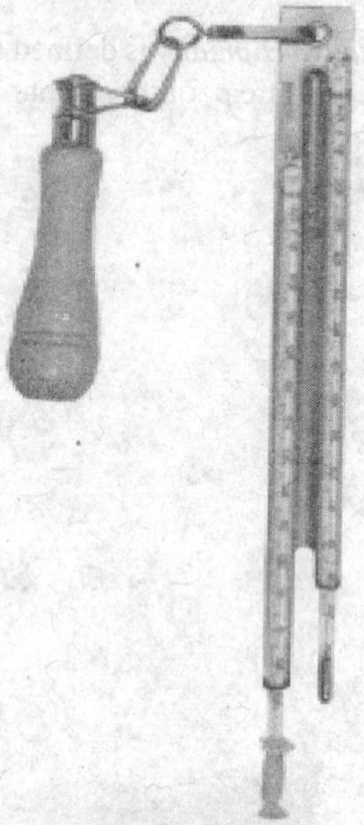

Fig. 37.2. *Sling Psychrometer*

Evidently, the less the relative humidity of the ambient air the greater will be its affinity to absorb more moisture, hence greater will be the evaporation of water from the bulb surface and therefore the greater will be its wet-bulb depression $(t - t')$ below dry bulb temperature value. If the ambient air is *saturated*, there will be no evapouration of the bulb and hence dry bulb and wet bulb temperatures will be the same.

$$\therefore \quad \text{Wet bulb depression} = t_d - t_w = t - t'$$

$$\phi = 1, \quad t_d = t_w = t'$$

or

$$\text{for Wet bulb depression} = 0,$$

$$\phi = 1$$

It may be noted that the wet bulb temperature is an indirect measure of the moisture content of the air. Due to various inaccuracies in the measurement (which is not carried on under ideal conditions), the temperature t_w can not be considered as a thermodynamic property. Therefore, no expression can be found analytically to obtain the value of p_v in terms of $t_w = t'$.

However, since the measurement of wet bulb temperature t_w is easy, certain empirical relations exist which are used to obtain the value of partial pressure p_v from the value of t_w. some of the empirical equations involving p_v and t_w are,

(i) Modified Apjohn Equation (as converted to metric system)

$$p_v = (p_s)_{t_w} - \frac{1.8p(t_d - t_w)}{2700} \qquad \text{...(6)}$$

(ii) Modified Ferret Equation (as converted to metric system).

$$p_v = (p_s)_{t_w} - 0.00066\, p(t_d - t_w)\left(1 + \frac{1.8t_d}{1571}\right) \qquad \text{...(7)}$$

(iii) The Carrier Equation (as converted to metric system)

$$p_v = (p_s)_{t_w} - \frac{(p - (p_s)_{t_w})(t_d - t_w)1.8}{2800 - 1.3(1.8t_d + 32)} \qquad \text{...(8)}$$

where

p_v = partial pressure of water vapour

$(p_s)_{t_w}$ = pressure of saturated vapour at t_w

P = barometric pressure

t_d = dry bulb temperature °C

t_w = wet bulb temperature °C

The pressures P, $(P_s)_{tw}$ and P_v may be in any consistent units.

In addition to these, two more commonly used empirical equations to find p_v, may be considered.

$$(iv)\quad p_v = (p_s)_{t_w} - \frac{\left[p - (p_s)_{t_w}\right](t_d - t_w)}{1527.4 - 1.3t_w} \qquad \text{...(9)}$$

(v) $$p_v = (p_s)_{t_w} - \frac{\left[p-(p_s)_{t_w}\right](t_d - t_w)}{1544 - 1.44t_w} \qquad \ldots(10)$$

10. Thermodynamic Wet Bulb Temperature or Temperature of Adiabatic Saturation. If the saturation process as discussed above for the measurement of wet bulb temperature is carried on under ideal conditions (no radiation and other heat losses), then the temperature so recorded will be a thermodynamic property and is known as thermodynamic wet bulb temperature or temperature of adiabatic saturation. It is denoted as t^*. The concept of this temperature is given below.

A mixture of dry air and water vapour at temperature t_a flows through an insulated duct, Fig. 37.3, pass over a surface of water. It is assumed that (i) the flow is steady (ii) there is no work done and (iii) the process is adiabatic.

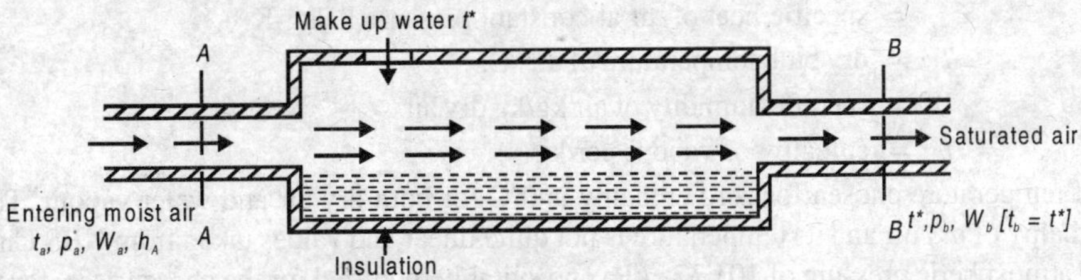

Fig. 37.3. *Adiabatic saturation process*

Since the air entering the duct is unsaturated, as it passes over the water surface some of the water is evaporated. Both air and water are cooled due to evapouration. Eventually, the water attains a *steady state* temperature when the heat energy transferred to water from air is exactly equal to the heat energy required to vapourise the water. When this state of thermal equilibrium is reached, the latent heat of vapourisation water is supplied completely by the air and no further drop in water temperature occurs. This *steady state temperature* of water is known as the adiabatic saturation temperature t^* of entering air. In order to maintain the thermal equilibrium the make up water must be supplied at the *adiabatic saturation temperature* t^*.

If the insulated duct is sufficiently long, the air leaving the duct will be saturated air and its temperature will equal to water temperature t^*.

A simple energy balance for the adiabatic process described above will show that, for a given initial state of air and for complete saturation of air at exit, the *exit air temperature* t^* can have but one value. Thus t^* is a property of the air-vapour mixture.

Note that the state of air continuously changes from the inlet to the exit of the duct. But for all these different states of air, the temperature of adiabatic saturation is the same because the steady-state water temperature has the same value all along the duct. Thus the above process occurs at a constant temperature of adiabatic saturation.

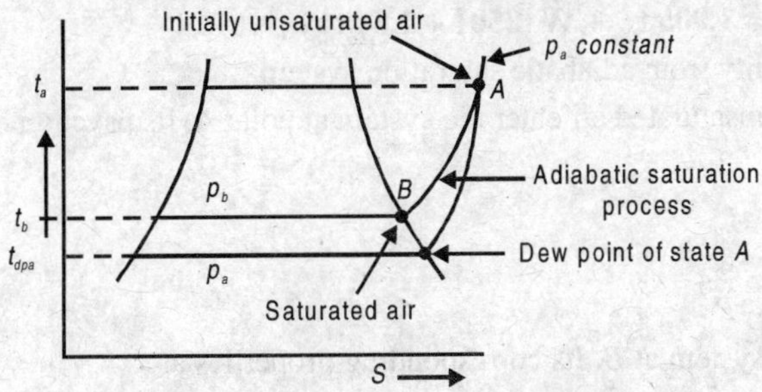

Fig. 37.4. *Adiabatic saturation process*

The process of adiabatic saturation may be traced on a T–S diagram by curve AB in Fig. 37.4. The partial pressure of water vapour increases during the process from p_a to p_b while the total pre-pressure p of the mixture remains the same.

The vapour initially at DBT t_a is cooled adiabatically to DBT t_b, which is equal to the wet bulb temperature t^*.

11. Enthalpy of Moist Air. The *enthalpy* of a mixture of dry air and water vapour is equal to the sum of the enthalpy of (i) 1 kg of dry air and (ii) the associated water vapour.

$$\therefore \quad h = h_a + Wh_v \qquad \ldots(11)$$

$$= C_{pa} t_d + wh_v$$

where

h = enthalpy of atmospheric air kJ/kg dry air

C_{pa} = specific heat of air at constant pressure kJ/kg-K

T_d = dry bulb temperature of air °C.

W = specific humidity of air kg/kg dry air

h_v = enthalpy of vapour per kg

The value of temperature chosen for zero enthalpy is 0°C for both dry air and water vapour. The relationship between the enthalpy of dry air and its temperature is not quite linear and values taken from NBS Circular No. 564, for the standard atmospheric pressure of 101.325 kPa and suitably modified for the chosen zero, for the basis of the tables of properties of humid air. An approximate equation for the enthalpy of dry air over the range of 0 °C to 60 °C is

$$h_a = 1.007\,t - 0.026 \qquad \ldots(12)$$

and for lower temperature down to – 10 °C, the approximate equation is

$$h_a = 1.005\,t \qquad \ldots(13)$$

The enthalpy of vapour over ice, however, is based on the information in the ASHRÁE Handbook of Fundamentals (1967).

For the purpose of approximate calculations, without recourse to the psychrometric Tables, we may assume that, in the range of 0°C to 60°C, the vapour is generated from water at 0°C and that the specific heat of superheated steam is constant. The following equation can then be used for the enthalpy of water vapour

$$h_v = 2501 + 1.84\,t \qquad \ldots(13a)$$

These equations can be combined to give us an approximate expression for the enthalpy of humid air at a barometric pressure of 101.325 kPa.

$$\therefore \quad h = h_a + wh_v$$

$$= (1.007\,t - 0.026) + W(2501 + 1.84t)$$

$$= 1.005\,t_{db} + W(2501 + 1.84\,t_{db})$$

$$\mathbf{h = 1.005\,t_{db} + W\,(2501 + 1.84\,t_{db})} \qquad \ldots(14)$$

We can find specific humidity from adiabatic saturation system.

Referring to Fig. 37.3, let unsaturated air enter the system at point A. Its psychometric properties are,

Enthalpy = h_a

DBT = t_a

Specific humidity = w_a

Total pressure = P

Let saturated air leave the system at B. Its corresponding properties are

Enthalpy = $h_B = h_{sb}$

DBT = WBT = $t^* = t_{wb}$

Specific humidity $= w_{bs}$

Total pressure $= P$

Water at t_{wb} which is supplied, has liquid enthalpy h_{cb}

System is adiabatic.

$\therefore \quad h_{sb} = h_a + (w_{bs} - w_a) h_{lb}$

$$\therefore \quad h_a - w_a \cdot h_{lb} = h_{sb} - w_{bs}\, h_{lb} \qquad \ldots(15)$$

But $h_a = (h_{\text{dry air}})_a + w_a$ (enthalpy of superheated vapour at t_a per kg)

$= h_{\text{dry air}} + w_a h_{vb}$

h_{sb} = enthalpy of 1 kg air at $t_{ws} + w_{bs}$ (enthalpy of saturated vapour at t_{wb} per kg of vapour)

$= (h_{\text{dry air}})_b + w_{bs}\, h_{vb}$

Knowing $t_{db} = t_{wb}$, and substituting these values in Eq. 15, we get,

$$\mathbf{w_a} = \frac{\mathbf{W_{bs}L_{bs} - 1.005(t_{da} - t_{db})}}{\mathbf{h_{va} - h_{wb}}} \qquad \ldots(16)$$

By applying fundamental procedures of statistical mechanics, Goff and Gratch calculated accurate thermodynamic properties of moist air for standard sea level conditions (total pressure 760 mm Hg).

Calculations for specific volume, specific enthalpy and specific entropy of *unsaturated moist air* are approximately given closely by the following relations

$$\left.\begin{array}{ll} \text{Volume} \rightarrow & v = v_a + \mu\,(v_{as} - v_a) \\ \text{Enthalpy} \rightarrow & h = h_a + \mu\,(h_{as} - h_a) \\ \text{Entropy} \rightarrow & s = s_a + \mu\,(s_{as} - s_a) \end{array}\right] \text{Refer table on next page 1404} \qquad \ldots(17)$$

37.3 PSYCHROMETRIC TABLE

Psychrometric properties of air at standard Barometric Pressure of 760 mm Hg or 1.01325 bar are tabulated for temperatures from – 10°C to 60°C.

The Psychrometric Tables are given on next page.

37.4 PSYCHROMETRIC CHART

In the preceding sections we considered various thermodynamic properties of moist air, and the equations relating them, with the help of which we accurately obtain the solution of problems concerning moist air. However, calculations in some cases is tedious and time consuming. A graphical plot of the properties of moist air may reduce the calculations as the various processes may directly be shown on such a graph and the required results obtained. Such a monograph is called a *psychrometric chart*. It is shown in Fig. 37.5.

If the pressure and two other independent properties of moist air are known, the thermodynamic state of moist air is uniquely fixed. A particular chart is therefore constructed for a specified value of barometric pressure usually 76 mm Hg. The choice of coordinates is arbitrary. However, most of the charts employ DBT and specific humidity as basic coordinates.

In 1923 Mollier introduced a chart using enthalpy and specific enthalpy as oblique co-ordinates, but the inclination of enthalpy lines was so chosen that specific humidity lines were horizontal. His chart, therefore, had the appearance similar to that of Fig. 37.5. Equation (1) indicates that the vapour pressure is approximately a linear function of the specific humidity; hence pressure scale will be an auxiliary ordinate scale parallel to the specific humidity scale.

Psychometric Properties of Air at Standard Barometric Pressure of 76 mm. Hg or 1.01325 bar

Properties of water and steam						Properties of dry air at 760 mm of Hg		Properties of mixture of dry air and saturated steam i.e. (1+ w) kg at 760 mm Hg		
Tem-pera-ture °C	Saturated pressure of water and steam p_s		Enthalpy of saturated water h_f or h_w	Enthalpy of saturated steam h_g	Specific volume of saturated steam V_g	Specific volume V_a	Enthalpy h_a	Volume of mixture (1 + w) kg of dry air V_{as}	Enthalpy of mixture (1 + w) kg per kg of dry air h_{as}	Specific humidity per kg of dry air W_s
°C	mm Hg	Bar	kJ/kg	kJ/kg	m^3/kg	m^3/kg	kJ/kg	m^3/kg	kJ/kg	kg/kg
0	4.580	0.006108	0.00	2501.6	206.31	0.7734	0.000	0.7781	9.473	0.003789
1	4.924	0.006566	4.17	2503.4	192.61	0.7762	1.006	0.7813	11.204	0.004076
2	5.291	0.007055	8.39	2505.2	179.92	0.7791	2.012	0.7845	12.982	0.004381
3	5.681	0.007575	12.60	12507.1	168.17	0.7819	3.018	'0.7878	14.811	0.004707
4	6.097	0.008129	16.80	2508.9	157.27	0.7848	4.024	0.7911	16.696	0.005054
5	6.539	0.008718	21.01	2510.7	147.16	6.7876	5.030	0.7944	18.639	0.005424
6	7.010	0.009345	25.21	2512.6	137.79	0.7904	6.036	0.7978	20.644	0.005818
7	7.510	0.010012	29.41	2514.4	129.06	0.7933	7.041	0.8012	22.713	0.006237
8	8.041	0.010720	33.60	2516.2	120.97	0.7961	8.047	0.8046	24.852	0.006683
9	8.605	0.011472	37.80	2518.1	113.44	0.7990	9.053	0.8081	27.064	0.007157
10	9.204	0.012270	41.99	12519.9	106.43	0.8018	10.059	0.8116	29.352	0.007661
11	9.839	0.013116	46.19	2521.7	99.909	0.8046	11.065	0.8152	31.724	0.008197
12	10.512	0.014014	50.38	2523.6	93.835	0.8075	12:071	0:8188	34.179	0.008766
13	11.225	0.014965	54.57	2525.4	88.176	0.8103	13.077	0.8225	36.726	0.009370
14	11.982	0.015973	58.75	2527.2	82.900	0.8132	14.084	0.8262	39.370	0.010012
15	12.781	0.017039	62.94	2529.1	77.978	0.8160	15.090	0.8300	42.113	0.010692
16	13.628	0.018168	67.13	2530.9	73.384	0.8188	16.096	0.8338	44.963	0.011413
17	14.524	0.019362	71.31	2532.7	69095	0.8217	17,103	0.8377	47.926	0.012178
18	15.470	0.020624	75.50	2534.5	65.087	0.8245	18.108	0.8414	51.008	0.012989
19	16.470	0.021957	79.68	2536.4	61.341	0.8274	19.114	0.8457	54.216	0.013848
20	17.53	0.023366	83.86	2538.2	57.838	0.8302	20.121	0.8498	57.555	0.014758
21	18.64'	0.024853	88.04	2540.0	54.561	0.8330	21.127	0.8540	61.035	0.015721
22	19.82	0.026422	92.23	2541.8	51.492	0.8359	22.133	0.8583	64.660	0.016741
23	21.06	0.028076	96.41	2543.6	48.619	0.8387	23.140	0.8627	68.440	0.017821

Contd.

Contd.

	Properties of water and steam					Properties of dry air at 760 mm of Hg		Properties of mixture of dry air and saturated steam i.e. (1+ w) kg at 760 mm Hg		
Tem-pera-ture °C	Saturated pressure of water and steam p_s		Enthalpy of saturated water h_f or h_w	Enthalpy of saturated steam h_g	Specific volume of saturated steam V_g	Specific volume V_a	Enthalpy h_a	Volume of mixture (1 + w) kg of dry air V_{as}	Enthalpy of mixture (1 + w) kg per kg of dry air h_{as}	Specific humidity per kg of dry air W_s
°C	mm Hg	Bar	kJ/kg	kJ/kg	m^3/kg	m^3/kg	kJ/kg	m^3/kg	kJ/kg	kg/kg
24	22.37	0.029821	100.59	2545.5	45.926	0.8416	24.146	0.8671	72.385	0.018963
25	23.75	0.031660	104.77	2547.3	43.402	0.8444	25.153	0.8717	76.500	0.020170
26	25.50	0.033597	108.95	2549.1	41.034	0.8472	26.159	0.8764	80.798	0.021448
27	26.73	0.035636	113.13	2550.9	38.813	0.8501	27.165	0.8811	85.285	0.022798
28	28.34	0.037782	117.31	2552.7	36.728	0.8529	28.172	0.8860	89.976	0.024226
29	30.03	0.040040	121.48	2554.5	34.769	0.8558	29.179	0.8910	94.878	0.025735
30	31.81	0.042415	125.66	2556.4	32.929	0.8586	30.185	0.8962	100.006	0.027329
31	33.69	0.044911	129,84	2558.2	31.199	0.8614	31.192	0.9015	105.369	0.029014
32	35.65	0.047534	134.02	2560.0	29.572	0.8643	32.198	0.9069	110.979	0.030793
33	37.72	0.050288	138.20	2561.8	28.042	0.8671	33.205	0.9125	116.857	0.032674
34	39.89	0.053180	142.38	2563.6	26.601	0.8700	34.212	0.9183	123.011	0.034660
35	42.17	0.056216	146.56	2565.4	25.245	0.8728	35.219	0.9242	129.455	0.036756
36	44.55	0.059400	150.74	2567.2	23.967	0.8756	36.226	0.9303	136.209	0.038971
37	47.06	0.062739	154.92	2569.0	22.763	0.8785	3 7.233	0.9366	143.290	0.041309
38	49.69	0.066240	159.09	2570.8	21.627	0.8813	38.239	0.9431	150.713	0.043778
39	52.44	0.069908	163.27	2572.6	20557	0.8842	39.246	0.9498	158.504	0.046386
40	55.31	0.73750	167.45	2574.4	19.546	0.8870	40.253	0.9568	166.683	0.049141
41	58.33	0.677773	171.63	2576.2	18592	0.8898	41.261	0.9640	175.265	0.052049
42	61.49	0.081986	175.81	2577.9	17.692	0.8927	42.268	0.9714	184.275	0.055119
43	64.80	0.086391	179.99	2579.7	16.841	0.8955	43.275	0.9792	193.749	0.058365
44	68.25	0.091001	184.17	2581.5	16.036	0.8983	44.282	0.9872	203.699	0.061791

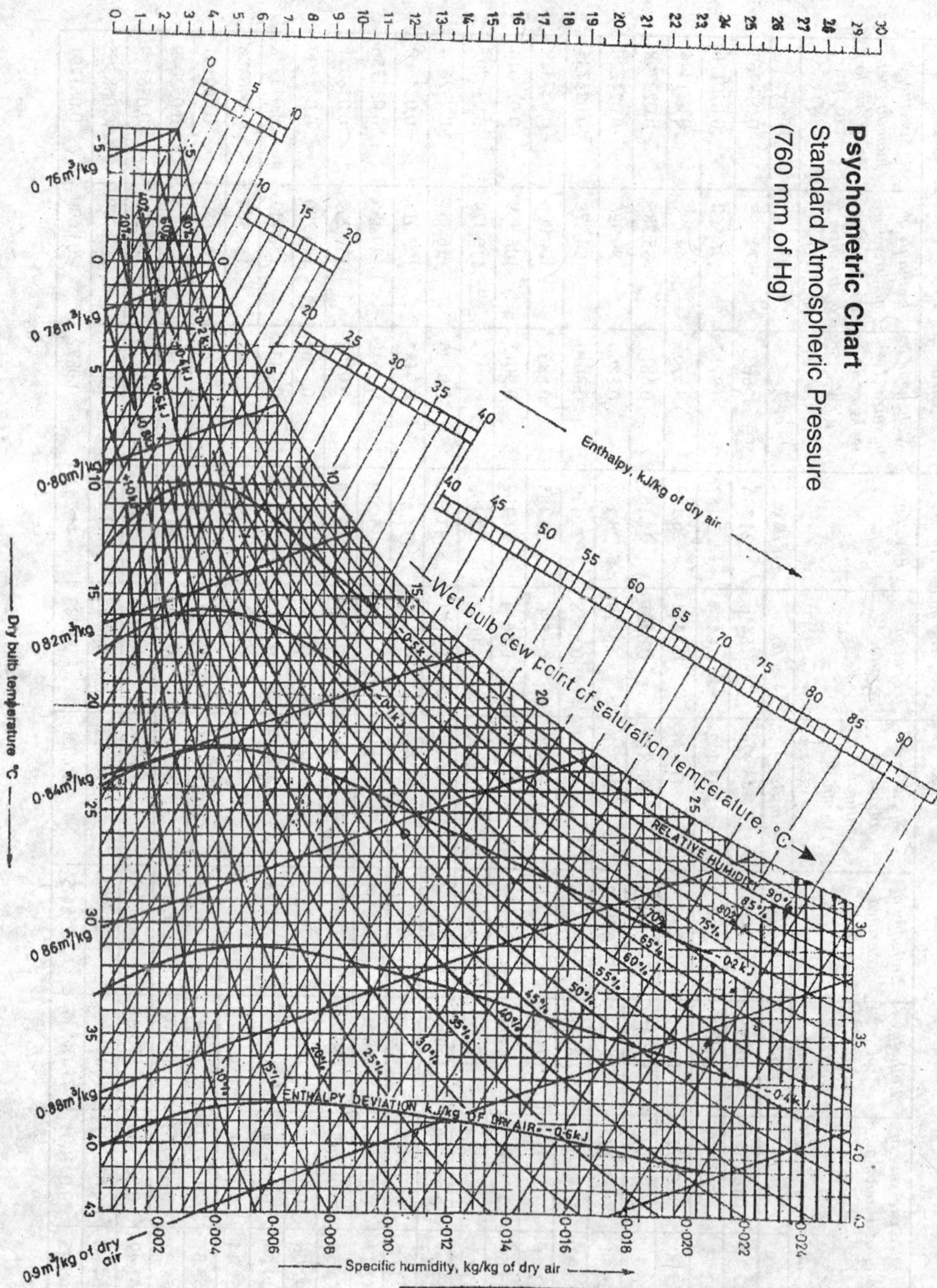

Fig. 37.5. *Psychometric chart.*

The mass basis of the chart is 1 kg. dry air plus associated water vapour. The various lines drawn on the psychrometric chart are discussed now with reference to Fig. 37.5.

(a) Constant Dry Bulb Temperature Lines. All vertical lines perpendicular to the horizontal d.b.t. scale are constant temperature lines. (–5°C to 43°C). Fig. 37.5 (a).

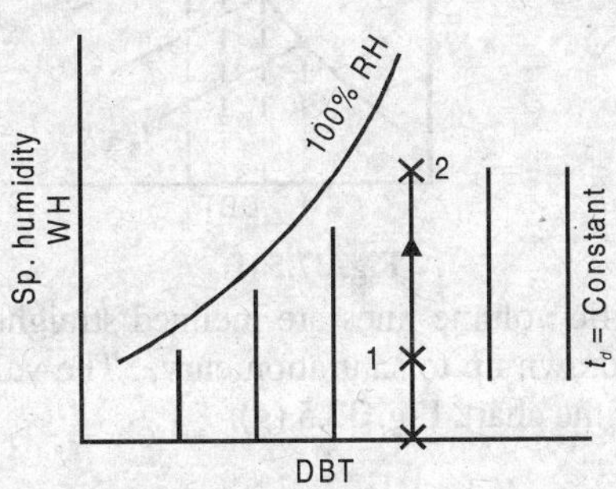

Fig. 37.5 (a)

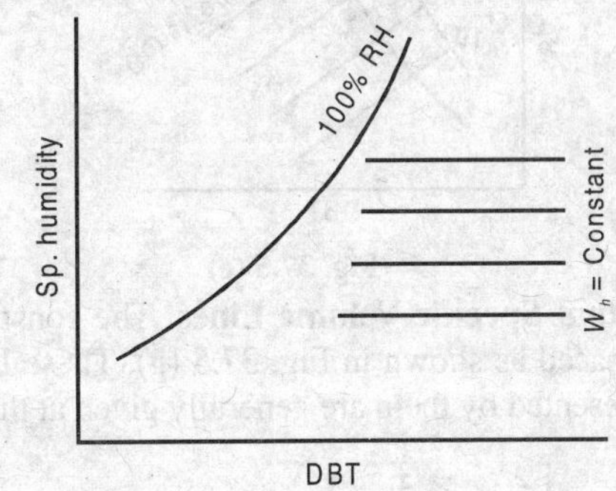

Fig. 37.5 (b)

(b) Constant Specific Humidity Lines. All horizontal lines perpendicular to the vertical specific humidity scale are constant specific humidity lines. $(kg)_{vapour}$ per $(kg)_{dry\ air}$ [0 to 0.026]. Fig. 37.5(b).

(c) Saturation Curve. The extreme left hand curve is the saturation curve which represents state of saturated air at different values of d.b.t. (DBT). It is a100% relative humidity line or a curve. Fig. 37.5 (c).

(d) Constant R.H. Lines. Suppose we want to construct a constant RH line of $\phi = 50\%$. We can select dry bulb temperatures t_1, t_2, t_3 etc. and find out the corresponding saturation pressures p_{s1}, p_{s2}, p_{s3} etc. from the Steam Tables. Since $p_v = \phi . p_s = 0.5\, p_s$, the values of p_{v1}, p_{v2}, p_{v3} etc. are readily calculated. Since $w = 0.622\, p_v$, we can find out w_1, w_2, w_3 etc. for $\phi = 50\%$. The $P - P_v$ curve passing through (t_1, w_1), (t_2, w_2), (t_3, w_3) is the required $\phi = 50\%$ line or curve. Fig. 37.5 (d).

(e) Constant Enthalpy Lines. The enthalpy lines are inclined straight lines and uniformly spaced as shown in Fig. 37.5 (e). These lines are parallel to wet bulb temperature lines and are drawn up to saturation curve. Most of these lines coincide with bulb temperature lines. Fig. 37.5 (e).

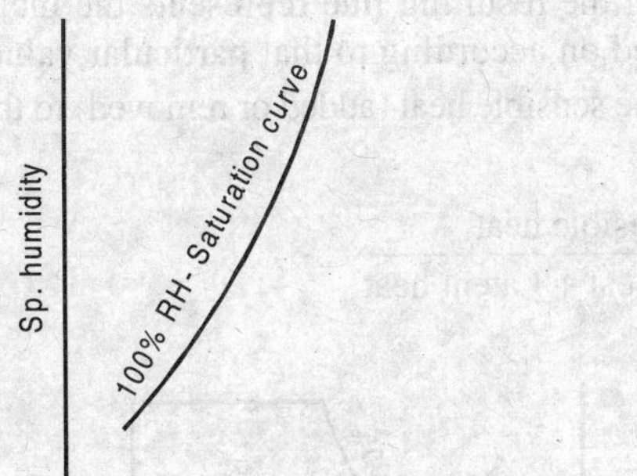

Fig. 37.5 (c)

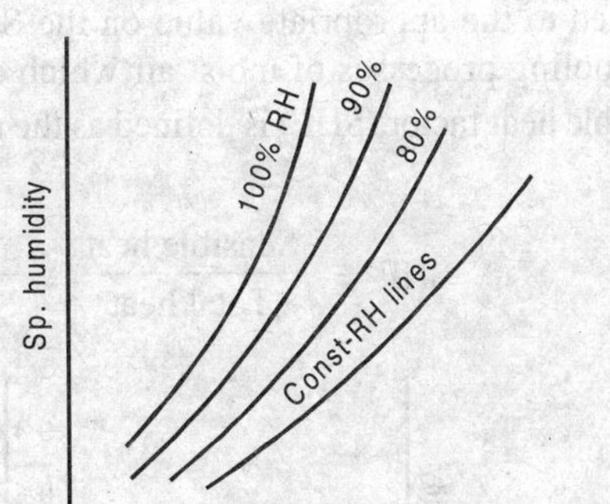

Fig. 37.5 (d)

(f) Constant Wet Bulb Temperature Lines. The wet-bulb temperature lines are inclined lines and non-uniformly spaced as shown in Fig. 37.5 (f). At any point on the saturation curve, the dry bulb and wet-bulb temperature are equal. The values of wet bulb temperature are generally given along the saturation curve of the chart Fig. 37.5 (f).

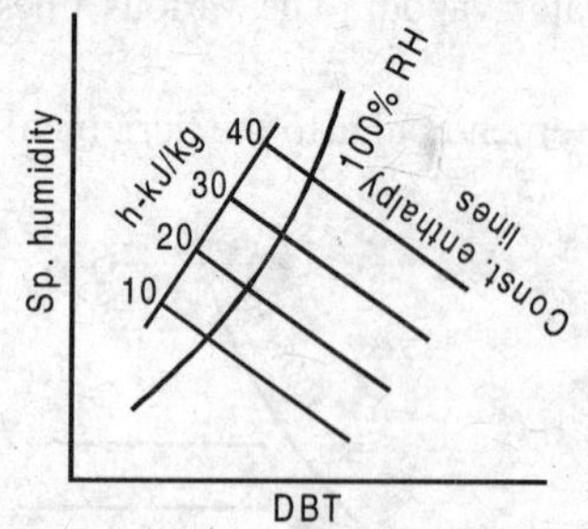

Fig. 37.5 (e)

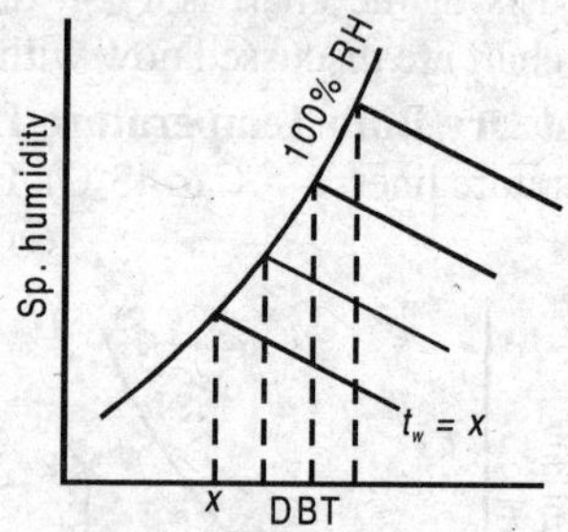

Fig. 37.5 (f)

(g) Constant Specific Volume Lines. The constant specific volume lines are inclined straight lines and are uniformly spaced as shown in Fig. 37.5 (g). These lines are drawn up to saturation curve. The value of specific volume represented by them are generally given at the base of the chart. Fig. 37.5 (g).

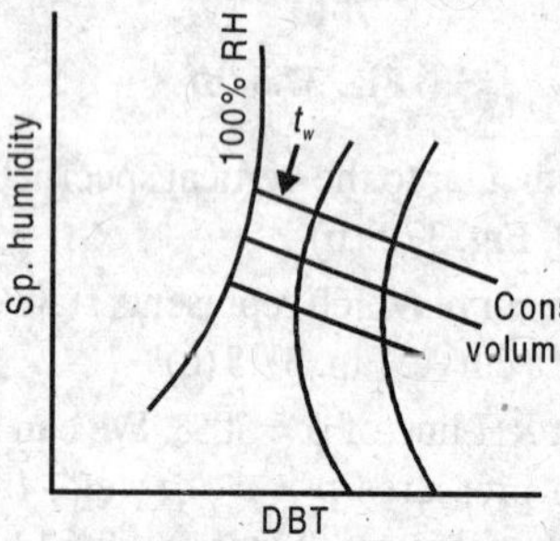

Fig. 37.5 (g)

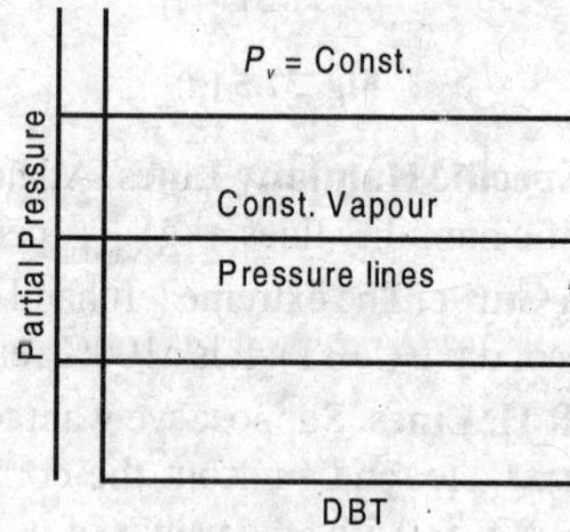

Fig. 37.5 (h)

(h) Vapour Pressure Lines. The vapour pressure lines are horizontal and uniformly spaced. Generally, the vapour pressure lines are not drawn in the main chart. But a scale showing vapour pressure in mm Hg is given on the extreme left side of the chart shown in Fig. 37.5 (h).

(i) Sensible Heat Factor (SHF) Scale. The scale on the extreme right is the sensible heat factor, SHF, scale which is drawn with reference to the point shown as a dark circle at 50% RH line. (near 27° DBT). When this point is joined to the appropriate value on the SHF scale, the resulting line represents the inclination of all heating or cooling processes of moist air which are carried on according to that particular value of SHF.

The sensible heat factor (SHF) is defined as the ratio of the sensible heat (added or removed) to the total heat in the process.

$$\therefore \quad \text{SHF} = \frac{\text{Sensible heat}}{\text{Total heat}} = \frac{\text{Sensible heat}}{\text{Sensible heat + Latent heat}}$$

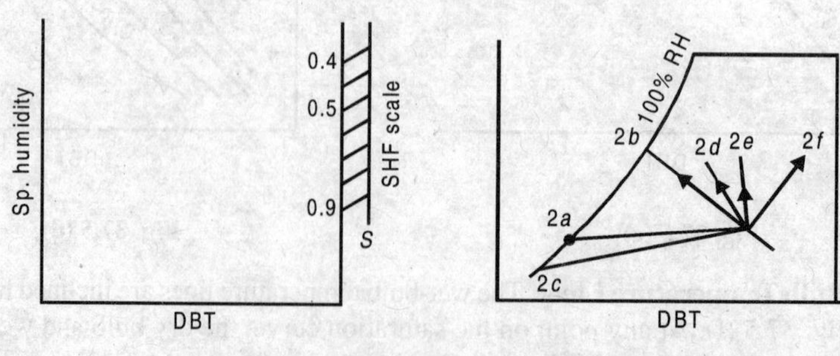

Fig. 37.5 (i)

37.5 PSYCHROMETRIC PROCESSES

The common *Psychrometric processes* which are usually encountered in practice and which may be easily solved with the help of a psychrometric chart are given below and are indicated in Fig. 37.6.

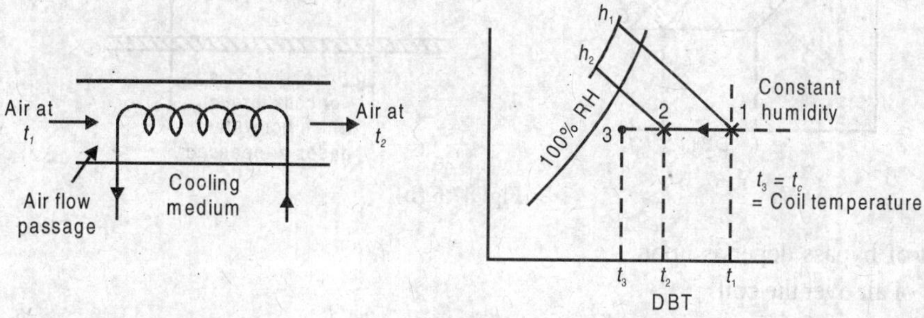

Fig. 37.6(a)

1. Simple heat transfer process:

(a) Sensible cooling process

(b) Sensible heating process.

2. Simple Humidification process at same DBT.

3. Simple dehumidification process at same DBT.

4. Heat transfer processes with change in humidity.

(a) Cooling and dehumidification

(b) Heating and humidification

(c) Heating and dehumidification.

(d) Cooling and humidification.

5. Adiabatic mixing of two samples of air. These processes will now be explained individually.

1. (a) Sensible Cooling. This process involves cooling of air at constant moisture content. Process 1–2 is a horizontal line from right to left. Evidently, such cooling of air has to be carried on by bringing it into contact with a body or a surface temperature of which is not below the dew point temperature of air at 1.

$$\text{Heat removed} = h_1 - h_2 \text{ kJ/kg of dry air.}$$

Generally all the air passing through the air duct or flow passage does come in contact with the coil surface and the temperature of the air coming out of the passage is not the temperature of the coil surface. With reference to Fig. 37.6 (a), let t_c be the temperature of the coil surface. Some of the air bypasses the coil surface and this bypassed air at t_1 mixes with the other part of air at temperature t_c and the final temperature of the mixture is, say, t_2.

The ratio of the bypassed air to the total air is called the bypass factor (BF).

$$\therefore \quad BF = \frac{\text{Bypassed air}}{\text{Total air}}$$

$$= \frac{t_2 - t_c}{t_1 - t_c} \text{ (In terms of temperature ratio)} \quad \ldots(18)$$

Given the bypass factor and the coil surface temperature, actual temperature of air can be found by using the equation.

1. (b) Sensible Heating Process. This process involves of heating of air without the change of humidity or humidity content.

Heating process line is a straight line from left to right Fig. 37.6 (b)

Heat supplied/kg of air = $h_2 - h_1$ kJ/kg dry air.

For this heating process,

$$\text{Bypass factor (BF)} = \frac{t_c - t_2}{t_c - t_1} \quad \ldots(19)$$

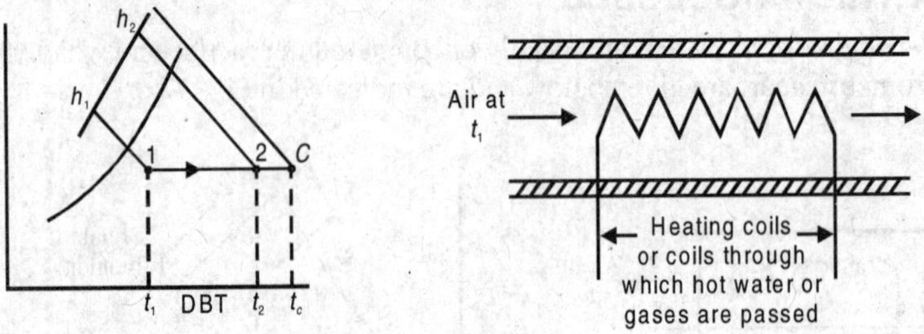

Fig. 37.6 (b)

The amount of bypass depends upon

(a) Velocity of air over the coil.

(b) Number of rows of coil pipes in the direction of flow.

(c) In case of a coil with fins attached, the bypass factor depends on the number of fins per unit length on the coil.

2. Simple Humidification Process at the same DBT. Adding moisture at constant dry bulb temperature can be represented by a vertical line on a psychrometric chart.

Pure humidifying process is possible only when heat of vaporisation is supplied from a source other that air without affecting the dry bulb temperature. This happens when water at the dry-bulb temperature of the entering air is sprayed in the chamber whose temperature is maintained at the air DB temperature.

$$\text{Heat supplied} = h_2 - h_1 \text{ kJ/kg dry air} = (w_2 - w_1) \times \text{Latent heat}$$

$$\text{Moisture added} = w_2 - w_1 \text{ kg/kg dry air.}$$

3. Simple Dehumidification Process at same DBT. Pure dehumidification occurs without the change of dry-bulb temperature of air, when air is passed over chemical absorbing moisture and heat evolved , when moisture is absorbed, is removed immediately by the surrounded walls. These walls are maintained at the dry-bulb temperature of air. The chemical is also maintained at the dry-bulb temperature of air. The process of pure dehumidification is shown in Fig. 37.6(d).

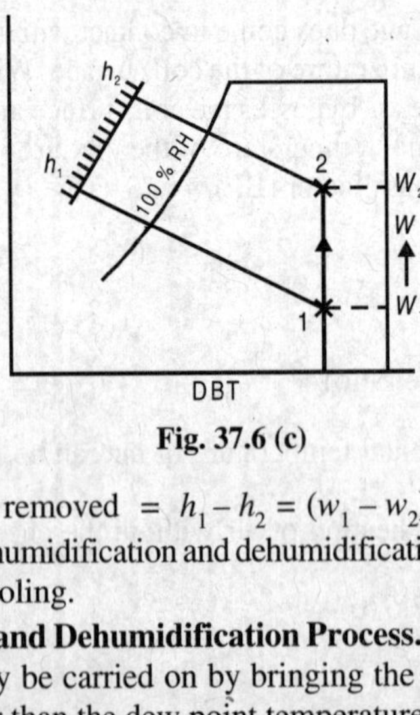

Fig. 37.6 (c)

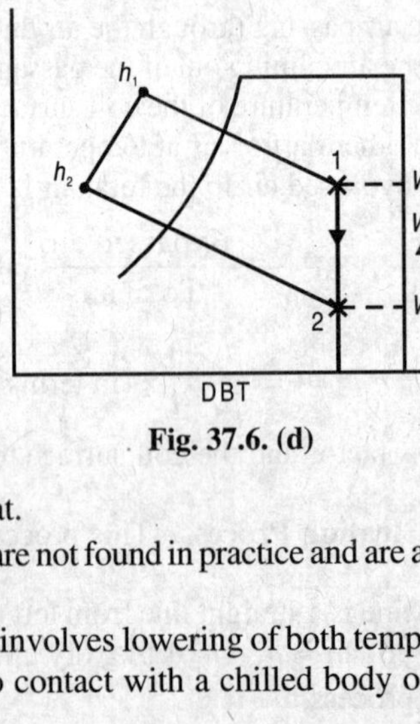

Fig. 37.6. (d)

$$\text{Heat removed} = h_1 - h_2 = (w_1 - w_2) \times \text{Latent heat.}$$

Pure or simple humidification and dehumidification processes are not found in practice and are always accompanied by heating and cooling.

4. (a) Cooling and Dehumidification Process. This process involves lowering of both temperature and specific humidity, and may be carried on by bringing the moist air into contact with a chilled body or surface having its temperature lower than the dew point temperature of the air.

The cooling and dehumidification process of moist air may be carried on by any one of the methods given below:

1. By causing the air to flow over a *chilled surface*, below dew point temperature, which may be cooled by circulating chilled water or by directly evaporating the refrigerant of a refrigeration system, within the coil.
2. By causing the air to flow through an air washer spraying chilled water below dew point temperature.

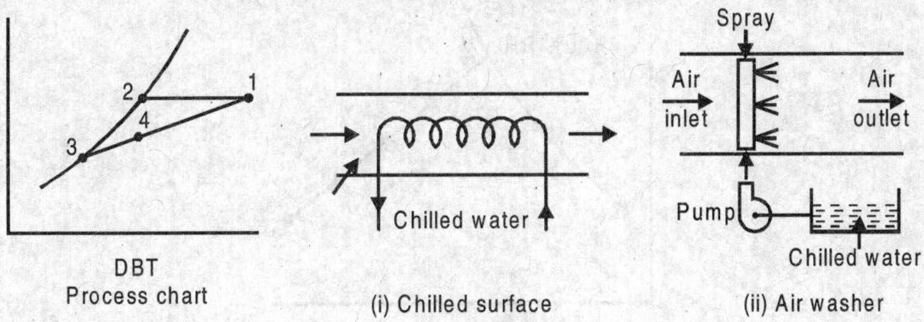

Fig. 37.6 (e)

These processes have practical applications in summer air conditioning. Since the actual process is difficult to trace on the chart, only end states are shown distinctly and joined by dotted curve. But possible processes can be shown as – (1–2) is the sensible cooling up to the dew point temperature and then cooling of air below dew point temperature as shown by (2–3). t_3 is the coil surface temperature and if the coil is having a bypass factor then the actual state of air may be given by point (4). t_3 is the apparatus dew point temperature ADP and the bypass factor is given by

$$BF = \frac{t_4 - t_3}{t_1 - t_3} = \frac{t_4 - ADP}{t_1 - ADP}$$

4. (b) Heating and Humidification Process. This process is the reverse of that discussed in 4 (a) above. In this process the specific humidity and the dry bulb temperature of air are increased.

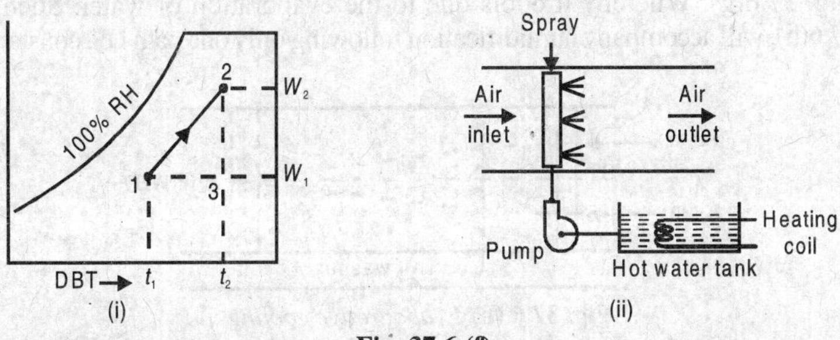

Fig. 37.6 (f)

This process is carried on by blowing the air through an air washer spraying hot water. The process has practical application in winter air conditioning. During this process (i) specific humidity, dry bulb temperature, wet bulb temperature, dew point temperature and enthalpy increase and (ii) RH may increase or decrease.

The final state 2 of the air may also be achieved by a combination of sensible heating 1–3 and then by sensible humidification process 3–2.

$$SHF = \frac{h_3 - h_1}{(h_3 - h_1) + (h_2 - h_3)} = \frac{h_3 - h_1}{h_2 - h_1}$$

Heating and humidification can also be achieved by injecting into air, the steam. This process is generally used in textile industry where high humidity is to be maintained. Rise in dry bulb temperature is generally very littie. Instead of hot water, steam will be injected.

4. (c) Heating and Dehumidification Process. There are certain chemical substances like *silica gel* and *activated alumina* which have an affinity for moisture. If air is passed over such chemicals it may be dehumidified. Usually in doing so the moisture is condensed and gives up its latent heat, raising the dry bulb temperature of air.

This process is reverse of adiabatic saturation process.

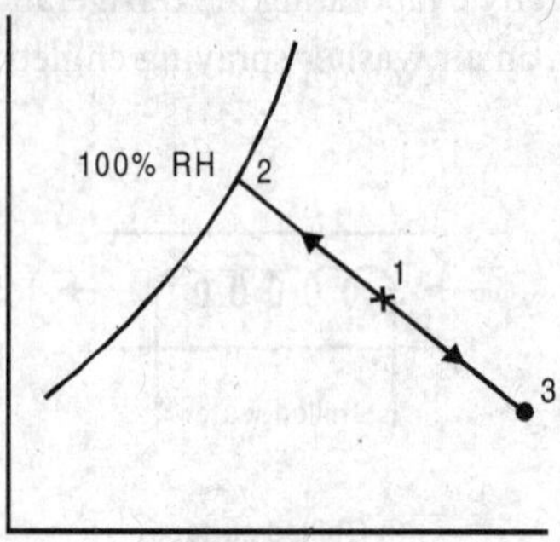

Fig. 37.6 (g)

This is shown in Fig. 37.6(g). In figure, 1–2 adiabatic saturation process 1–3 adiabatic dehumidification process. The path followed during this (adiabatic dehumidification) process is along the constant wet bulb temperature line or constant enthalpy line.

Actually there are two types of chemicals used for chemical dehumification. The are

1. **Absorbents.** They include water solutions or brines of calcium chloride, lithium chloride, lithium bromide and ethylene glycol. Generally these are used in absorption refrigeration systems.
2. **Adsorbents.** These are the substances in solid state and while absorbing moisture from air, they do not change chemically and physically. These are silica gel and activate alumina (porous amorphous form of aluminium dioxide).

4. (d) Cooling and Humidification (adiabatic humidification). This process is explained in Sec. 37.2.9 and is a constant wet bulb temperature process. Air passing through a spray chamber with recirculated water follows the process line 1–2 in Fig. 37.6(g). Whereby it cools due to the evaporation of water. Such cooling of air shown schematically in Fig. 37.6(h), will accompany humidification following only one path i.e., constant wet-bulb temperature line.

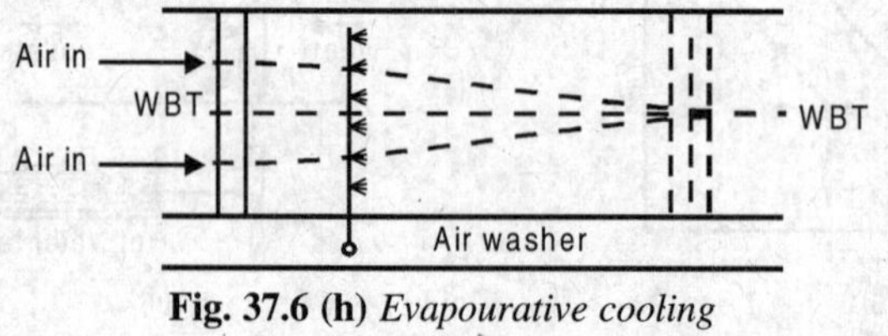

Fig. 37.6 (h) *Evapourative cooling*

This principle is made use of in *desert coolers* for cooling buildings in hot and dry climate so as to make them comfortable. Since the cooling cannot be achieved below the wet bulb temperature of the outside or incoming air, at which temperature the leaving air will be saturated, this technique of cooling has only limited use. Hence the system becomes ineffective in monsoon season.

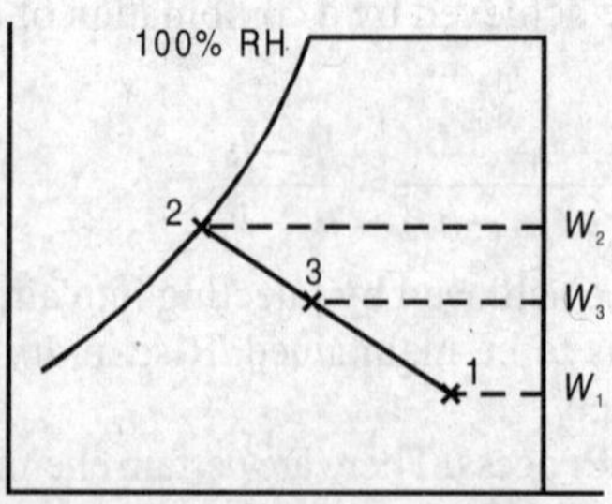

Fig. 37.6 (i)

Perfect humidification of air is not possible in actual practice due to inefficiency of spray chamber. Therefore, the final condition of air at outlet is represented by state 3 on the line 1–2 as shown in Fig. 37.6(i). The effectiveness or the humidifying efficiency of the spray chamber is given by

$$\eta_H = \frac{\text{Actual drop in DBT}}{\text{Ideal drop in DBT}} = \frac{t_1 - t_3}{t_1 - t_2}$$

$$= \frac{\text{Actual increase in humidity}}{\text{Ideal increase in humidity}} = \frac{w_3 - w_1}{w_2 - w_1}$$

5. Adiabatic Mixing of Air Streams. Very often air having different conditions are mixed in the air conditioning system, such as the mixing of fresh air and return or recirculated air before it enters the process plant. The psychrometric chart can be used to find the resulting condition of the mixture.

Figure 37.6(j) shows the schematic diagram showing two streams of air mixing together.

Let the two streams of masses m_1 and m_2, humidity rations w_1 and w_2 and enthalpies h_1 and h_2 respectively, mix together without any heat exchange with surroundings and also exchange of moisture to give a stream of mass $m_3 = m_1 + m_2$, humidity $w_3 = w_1 + w_2$ and enthalpy $h_3 = h_1 + h_2$.

$$\therefore \quad m_3 = m_1 + m_2$$

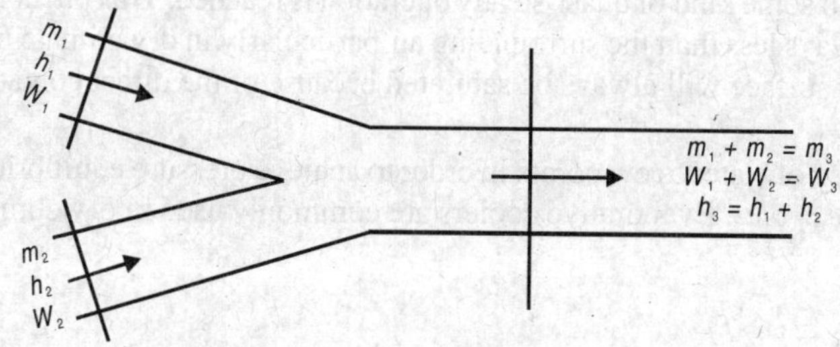

Fig. 37.6 (j) (a)

$$m_3 w_3 = m_1 w_1 + m_2 w_2$$

$$m_3 h_3 = m_1 h_1 + m_2 h_2$$

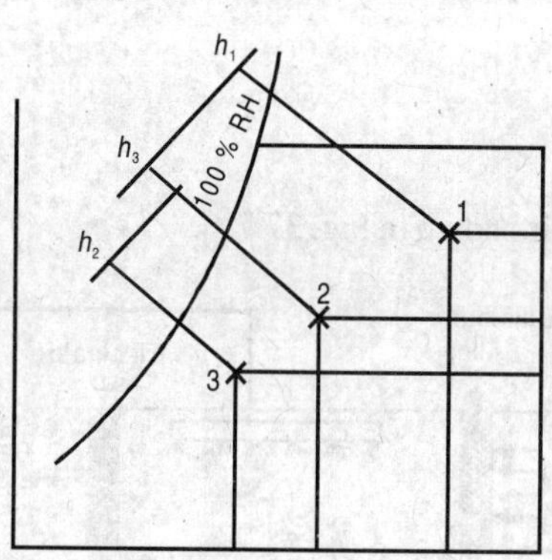

Fig. 37.6 (j) (b)

Eliminating m_3 and rearranging the above equations, we get

$$\frac{m_1}{m_2} = \frac{w_3 - w_2}{w_1 - w_3} = \frac{h_3 - h_2}{h_1 - h_3}$$

The final condition of the mixture can be found with the help of chart as follows.

First determine the percentage of air in each stream, in the mixture. Let in stream one, this is $x\%$. Therefore, $(100 - x)$ is the percentage of air mass in other stream.

Then locate the two points 1 and 2 from the given conditions of two streams. Join 1 and 2. Next step is very important. Multiply the dry bulb temperature of each air by its percentage in the mixture. Let this be $x.t_1$. This is the contribution of stream (1) DB degrees to the mixture. Similarly for second stream $(100 - x)\, t_2$ will be DB degrees contributing in the mixture.

$\therefore \quad x\, t_1 + (100 - x)\, t_2 = 100 \times t_3$

$\therefore \quad t_3$ can be found.

Draw a vertical at t_3 to meet the line 1–2 in 3. Then all other properties can be found from the chart.

37.7 EVAPOURATIVE COOLING

We know that liquid will evaporate if there is a difference between the vapour pressures in the air and the saturation pressure of liquid (say water) at the water temperature. The larger the difference between the vapour pressure and saturation pressure, the rate of evaporations larger or higher. The evaporation of water will have a cooling effect on water and thus reduce its temperature. Then the saturation pressure will reduce and the rate of evapouration will reduce. This will continue till some kind of quasi-steady operation is reached. This phenomenon, thus, explains why the water temperature is always less than the surrounding air particularly in dry climate like summer. It is therefore said that the air at the water surface will always be saturated because of the direct contact with water, and thus the vapour pressure.

Then, the natural tendency of water to evaporate in order to achieve pressure equilibrium, forms the basis for the operation of the *evaporative cooler.* Evaporative coolers are commonly used in dry climates (tropical regions) and provide effective cooling.

37.8 AIR WASHERS

Air washer is a device that may be used for

(a) Cooling and dehumidification
(b) Cooling and humidification
(c) Heating and humidification
(d) Humidification
(e) Adiabatic saturation
(f) Cooling.

The schematic diagram for air washer is shown in Fig. 37.7.

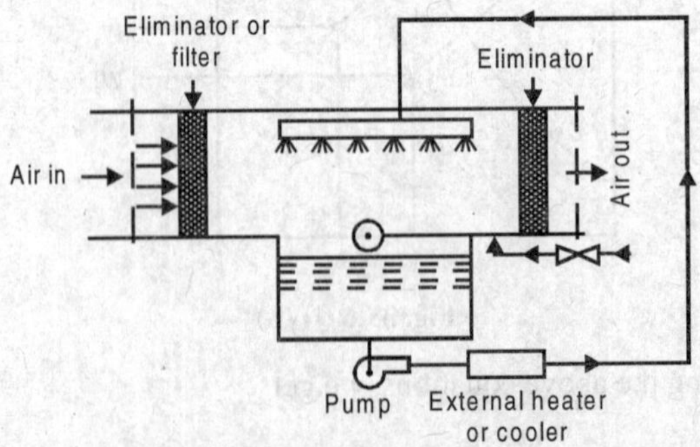

Fig. 37.7. *Air washer*

Following are the processes that can be carried out with the help of washer shown in Fig. 37.7 and they are shown on psychrometric chart of Fig. 37.8.

1–2a Sensible cooling. Water temperature is equal to dew-point temperature of air.

1–2b Adiabatic saturation. It takes place along the wet bulb temperature line. Water is at WBT

1–2c Cooling and dehumidification. Temperature of water is below the dew point temperature of air.

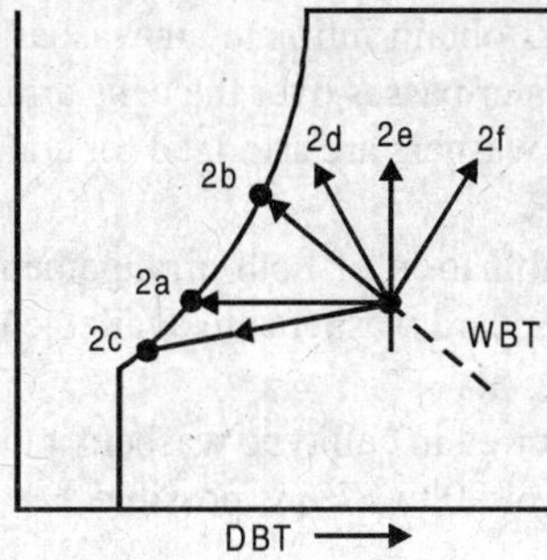

Fig. 37.8. *Various processes on psychrometric chart.*

1–2d Cooling and humidification. When wet bulb temperature is less than water spray temperature and this spray water temperature is less than the dry bulb temperature of air, this cooling and humidification process takes place. The water is required to be heated externally.

1–2e Humidification. Water spray temperature is equal to the dry bulb temperature of air. The water externally heated. (Latent heat of evaporation)

1–2f Heating and Humidification. If the spray water is heated and maintained at temperature higher than the dry bulb temperature of air entering the washer. In this case, the dry bulb temperature of air leaving the washer is more than the inlet temperature of air. Wet bulb is higher than the initial WBT and higher humidity ratio than the entering air.

In comparison with the cooling coil, the air washer has a number of disadvantages:

(a) It is more bulky

(b) Cost wise it is more expensive as far as maintenance is concerned.

(c) Water is required in greater quantity.

(d) Corrosion is a greater risk.

(e) Rusting, deposition of scales in the chilling plant is also a great problem. Because of rusting and deposition of scales, heat trouser efficiency is greatly affected and cost of refrigeration plant is increased.

The air washers allows humidification of air and therefore we can think of steam injection for humidification and it is safe.

There are three different types of air washers. They are:

(i) Spray type air washer.

(ii) High velocity sprays type air washer.

(iii) Cell-type air washers.

1. Spray Type Air Washer. There is no standardisation for air washers. The essential requirements in the air washer operations are

(a) uniform distribution of the air across the spray chamber.

(b) an adequate amount of spray water broken up into fine droplets.

(c) good spray distribution across the air stream.

(d) sufficient length of travel through the spray and the wetted surfaces.

(e) the elimination of free moisture from the outlet air.

Refer Fig. 37.7 for the simple air-washer used in practice. A pumps recirculates water at a rate greater than the evapouration rate. Washers are commonly available from 0.8 to 118 m^3\sec. capacity. This type of washer is used primarily as an evaporative cooler or humidifier.

2. High-Velocity Spray Type Air Washer. Air velocity range is 6 to 9 m/s –12 m/s. Maximum capacity up to 70 m^3/s.

These are generally rectangular in cross-section.

3. Cell-Type Air Washer. These washers obtain intimate air-washer contact by passing the air through cells packed with glass, metal or fiber screens. Water passes over the cells arranged in tires. Behind the cells are blade-type or glass-mat eliminates. Most cell-type washers are arranged for concurrent air and water flow. They are also constructed for special duty with counter.

Current flow characteristics or in a combination of both arrangements, cell washers come to many sizes of insulated or uninsulated construction and standard washers are available up to 10 cells high by 12 cells wide with a capacity of up to 100 m^3/s.

Atomisation of the spray water is not required in cell type washers, but good water distribution over the face of the cell essential. A saturation effectiveness of 90 to 97% is possible.

37.9 COMFORT – PHYSIOLOGICAL CONSIDERATIONS

Man and many other living creatures may be compared in some respects with an automatically controlled furnace. Food consisting mainly of carbon, hydrogen, oxygen, nitrogen and certain minerals air taken periodically by a man and likewise coal containing the more or less the same elements is fed to the boiler. Air is continually drawn in and the oxygen combines with carbon and hydrogen in coal and food producing CO_2 which is exhaled by man and leaves with the gases in boiler. O_2 combines with hydrogen to form water vapour which is also exhausted. Water leaves the man body through the sweat glands and evaporates from the skin surface. Human body has a marvelous control system that maintains our body temperature at about 37°C (or 98.6°F).

Heat evolved from this oxidation process is termed *metabolism* similar to combustion process in the furnace.

The amount of heat generated and dissipated by the human body varies considerably with *activity* and *age* as well as *size and gender*. The human body has a complex regulating system acting to maintain the deep body temperature of about 37% (98.6°F) regardless of the environmental conditions. A normal, healthy person generally feels most comfortable when the environment is maintained at conditions where the body can easily maintain a thermal balance with the surroundings. ASHRAE specifies conditions in which 80% or more of the occupants will find the environment thermally acceptable.

The environmental factors that affect a person's thermal balance and therefore which influence thermal comfort are

(a) the dry bulb temperature

(b) the humidity of the surrounding air

(c) the relative velocity, of the surrounding air

(d) The temperature of any surface that can directly view any part of the body and thus exchange radiation.

In addition, the personal variables that influence thermal comfort are *activity* and *clothing*.

The basic mechanisms that the body uses to control body temperatures are metabolism, blood circulation near the surface of the skin (cutaneous blood circulation), respiration and sweating. *Metabolism* determines the rate at which energy is converted from chemical to thermal form within the body, and *blood circulation* controls the rate at which the thermal energy is carried to the surface of the skin. In *respiration*, air is taken in at ambient conditions but leaves saturated with moisture and very near the body temperature. *Sweating* has a significant effect on the rate at which energy can be carried away from the skin by heat and mass transfer.

Heat is exchanged between the body and its environment by four modes:

1. Evapouration (E)

2. Radiation (R)
3. Convection (C)
4. Positive or negative heat (S).

Storage of that would cause the deep tissue temperature to rise or fall. For a normally clothed, healthy human being in a comfortable condition or environment and engaged in a non-strenuous activity, $S=0$ and the thermoregulatory system of the body is able to modify the losses by radition and convection to maintain a stable, satisfactory temperature. Evaporative losses occur in three ways: by exhalation of saturated water vapour from the lungs, by a continual normal process of insensible perspiration, and by an energy mechanism of sweating.

Insensible perspiration results from the body fluids oozing through skin under osmatic pressure and forming microscopic droplets on the surface which, because of their small size, evapourate virtually instantaneously, not being felt or seen and hence termed insensible.

If the body temperature tends to rise, the thermoregulatory system increases the evaporative loss by operating sweat-glands selectively and flooding strategic surfaces. In extreme cases the body is entirely covered with sweat that must evaporate on the skin to give a cooling effect. If the sweat rolls off or is absorbed by clothing, the cooling effect will be reduced.

Then we write the thermal energy balance in the equation form as:

$$M - W = E + R + C + S$$

where M = Thermal energy generated by metabolism
W = Work or energy spend in doing the activity

37.10 HUMAN REQUIREMENTS OF COMFORT

For a common person, the following conditions are desirable

1. Dry bulb temperature should be between 22°C and 26°C
2. The relative humidity should be between 20% and 60%
3. The average velocity of air in the room should preferably not to exceed 0.15 m/s
4. The carbon dioxide content should not exceed about 0.1%
5. The temperature difference between the feet and the head level should be as small as possible, normally not exceeding 1.5°C and should never exceed 3°C.

Not all these variables are directly amenable to regulation and no air-conditioning system is able to achieve control over all of them. The two most important are dry bulb temperature and air velocity.

37.11 HUMAN COMFORT AND EFFECTIVE TEMPERATURE

It is now logical to ask the desired temperature–humidity relationship is. People react in widely different ways to atmospheric conditions around them. A comfortable atmospheric condition for one person may be chilly to another or too warm for the third person. Some people may like the moist climate of the sea coast; others may prefer the dry weather. Dry bulb temperature and relative humidity are, therefore, not the indication for the comfort for the individual. Air movement is another factor that decides the comfort conditions for the human beings.

The ASHRAE has conducted research over many years involving hundreds of persons in different localities to find the combination of the three factors dry bulb temperature, relative humidity and air movements so that the people are comfortable. From these results of tests, a concept of effective temperature is developed.

Effective temperature is a measure of comfort which involves the combined effects of dry bulb temperature, relative humidity and air movement.

It was found that for a given air velocity, there are number of different combinations of dry bulb temperature, relative humidity which would give the same feeling of comfort to a high percentage of the people. *All these combinations were said to have the same effective temperature.*

Effective temperature scale is not present as is the case of dry-bulb temperature.

Yaglau and others using a survey of the responses of the people to a relatively short term exposure in different environments found the scale of *Effective Temperature*, defined as the temperature of still, and saturated air that gives a feeling of comfort similar to that of another combination of three relevant environmental variables.

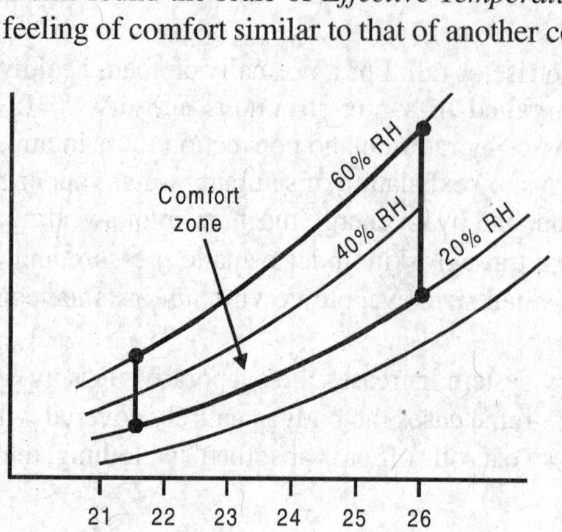

Fig. 37.9. *Normal Comfort Zone*

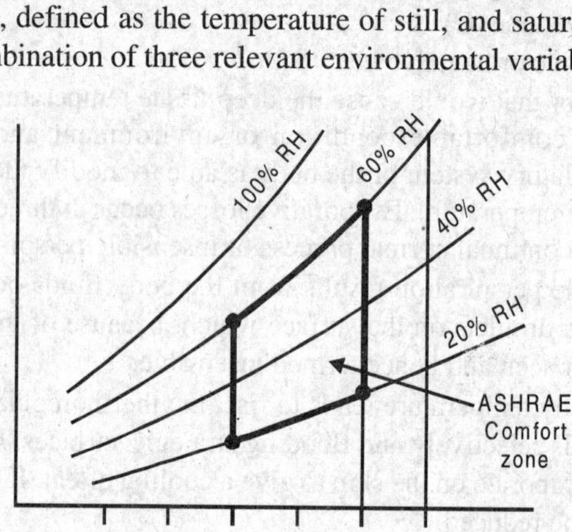

Fig. 37.10. *ASHRAE Comfort Zone*

Figure 37.9 shown normal comfort zone where DBT varies between 22°C, and 26°C, relative efficiency varies between 20% and 60% while air velocity varies from 0.1 to 0.15 m/s.

Figure 37.10 shows the new comfort zone as recommended by ASHRAE comfort standard. The comfort zone lies within about 23°C and 25°C and between 20% RH and 60% RH.

Thus we conclude that *effective temperature is the index of sensations of warmth produced* by the combined effect of air temperature, humidity and air motion.

Effective temperature cannot be measured directly by dry-bulb thermometer but from the effective temperature chart. Other forms of comfort zone and comfort charts are shown in Figs 37.11 and 37.12.

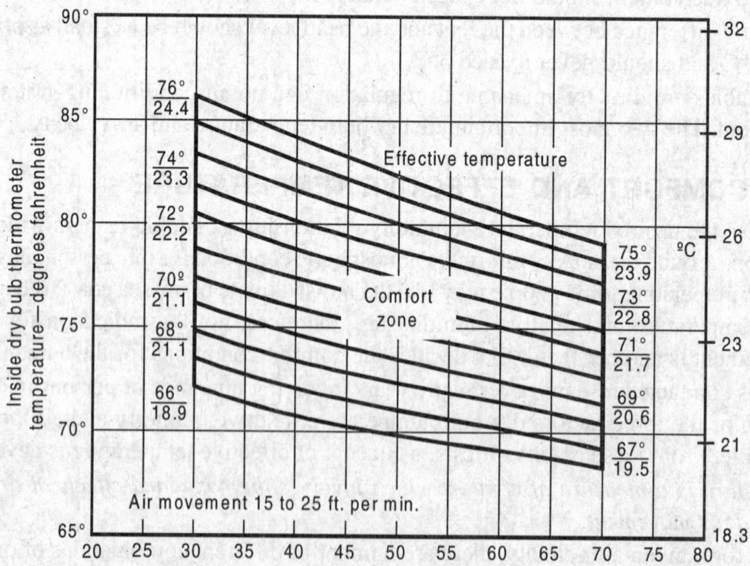

Fig. 37.11. *Comfort zone and chart*

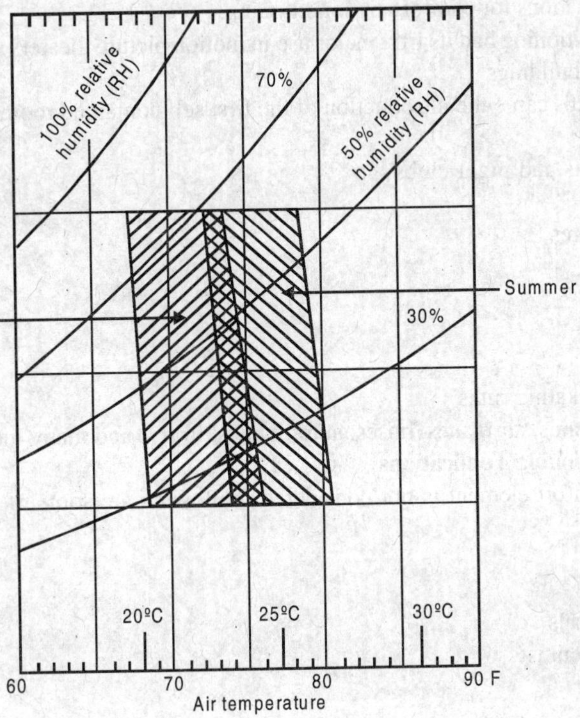

Fig. 37.12. *Other form of comfort-zone chart*

37.12 AIR CONDITIONING SYSTEM CLASSIFICATION

Air conditioning systems can be classified in a number of ways such as

I. The cooling or heating fluid that is used.

Under this category, there are three possible groups in regard to the fluid used:

(*a*) *All air systems:* These systems use only air for cooling and heating.

(*b*) *All water (hydronic) systems:* These systems use only water for cooling and heating.

(*c*) *Air-water combination systems:* These systems use both water and air for cooling and heating.

II. Unitary and Central Systems

(*a*) *A unitary system* uses packaged equipment. This is most, if not all, of the system components (fan, coils, refrigeration equipment) are furnished as an assembled package from the manufacturer.

(*b*) *A central or built system* is one whose components are furnished separately and installed and assembled by the contractor.

III. Single-Zone or Multiple Zone Systems

Out of these three ways of classification, here we will study about the unitary and central systems.

37.13 APPLICATIONS OF AIR CONDITIONING

Air conditioning is basically divided in two types:

1. Comfort air conditioning
2. Industrial air conditioning

We will enumerate the applications accordingly.

I. Comfort Air Conditioning

Following are the applications for comfort air conditioning.

1. Comfort air conditioning had its first major use in motion picture theaters (early 1920s).
2. Multiform office buildings
3. At the end of 1920s came the introduction of the first self contained room air conditioners
4. Apartments
5. Hotels, restaurants and night clubs
6. Hospitals
7. Departmental stores
8. Drug stores
9. Dress shops
10. Grocery shops
11. Banks
12. Dance halls and skating rinks
13. Transportation: trains, air planes, buses, automobiles, trollys and many more applications can be listed.

II. Industrial Air Conditioning Applications

In this case, human comfort element is not considered. Following are some of the applications of industrial air conditioning:

1. Textile industry
2. Printing industry
3. Drugs and chemicals
4. Libraries and museums
5. Metal working
6. Laboratories and many more.

37.14 UNITARY AND CENTRAL AIR CONDITIONING SYSTEMS

We know that the air conditioning is the preparation and distribution of air as per the requirements of the human beings — comfort air conditioning — and different industrial processes and preservation of certain raw materials — Industrial air conditioning. The requirements of both the applications are different.

There are various systems that are used for the preparation of air as required. These systems are mainly classified as to equipment arrangement.

1. Unitary system, which includes
   (a) Room air conditioner
   (b) Unitary packaged air and water cooled systems
   (c) Unitary system with combination of heating and cooling
   (d) Packaged heat-pump unit
   (e) Unitary split system for cooling.
2. Central air conditioning system, which includes
   (a) Chilled water system
   (b) Chilled water air washers
   (c) Centrifugal chillers system
   (d) DX system.

1. Unitary Air Conditioning Systems

If we make a survey of the market of air conditioning equipments or systems being sold, we will see that there is a continually increasing use of unitary systems.

These systems make use of air conditioners which are factory assembled or packaged. Many of the current unitary or packaged units are designed to be located outside the rooms or space they serve.

(a) Room Air Conditioner (Window Air Conditioner). A room air conditioner is an encased assembly designed as a unit primarily for mounting in a window or through a wall. These units are made to deliver cool or warm conditioned air to the room, generally without ducts. This unit includes the main source of refrigeration, dehumidification and means of circulating and cleaning air and also may include means of heating.

The basic function of any air conditioning plant is to provide comfort by cooling or heating, dehumidifying or humidifying, filtering or cleaning and recirculating the space air. All these functions are present in a room air conditioner or window air conditioner. It may provide ventilation by introducing outside air into the room and/or exhausting the room air to the outside. Room temperature can be controlled by providing a thermostatic setting in the window air conditioner. The conditioner may provide heating by heat pump operation, electric resistance elements or by a combination of both.

Figure 37.13 shows the schematic diagram of a typical room air conditioner or window air conditioner. Working or operation of this air conditioner is illutrated in Fig. 37.13 (b)

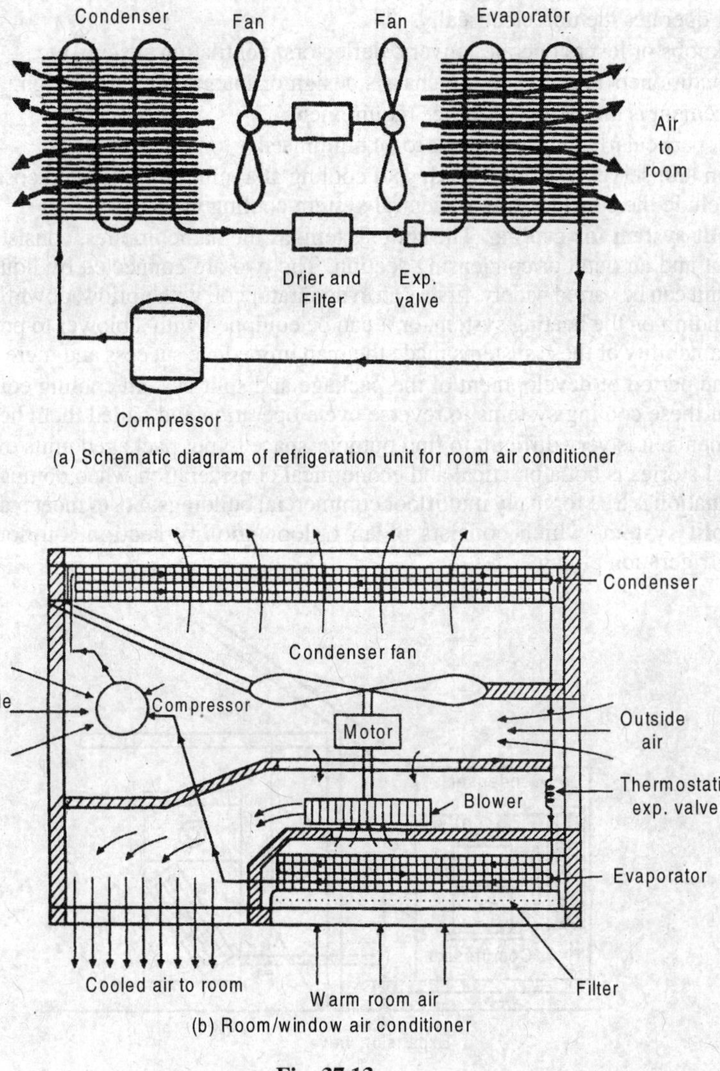

(a) Schematic diagram of refrigeration unit for room air conditioner

(b) Room/window air conditioner

Fig. 37.13

Warm room air (for recirculation) passes over the cooling or evaporator coil and in the process, gives up its sensible and latent heat also (in case of dehumidification is required). This conditioned air along with the fresh air (ventilation air) is then recirculated in the room by a fan or blower.

The heat from the warm air-vaporises the cold liquid refrigerant flowing through the evaporator. The vapour then carries heat to the compressor, which compresses the vapour and increase its temperature to a value higher than temperature of the outdoor air. In the condenser, the hot refrigerant vapour liquefies and gives up the heat from the room air to the outdoor air. The high pressure liquid refrigerant then passes through a restrictor—thermostatic expansion valve or capillary tube—which reduces its pressure and temperature. The cold and low pressure liquid refrigerant then re-enters the evaporator to repeat this refrigeration cycle.

The cooling capacities or commercially available window conditioners range from 0.75 to 10 kW.

Most room air conditioners are designed for bringing in outside air, exhausting room air or both. Controls usually permit these features to function independently. Shortly, we say that

(a) temperature is controlled by an adjustable built-in thermostat.
(b) one control operates the unit electrically.
(c) additional knobs or levers operate louvers, deflectors, ventilation system, etc.
(d) some air conditioners are of slide-out chassis design or integral chassis design.
(e) filter replacement is the essential service requirement.
(f) sound level, particularly in bed-room, is to be minimised.

Lastly, the room unit serves a definite need; spot cooling at a minimum installed cost and mobility of location, both of which preclude the consideration of central system cooling units.

(b) Unitary split-system for cooling. The split system, as the name implies, consists of two parts—an indoor cooling component and an outdoor condensing section. The two are connected by liquid and suction refrigerant lines. The indoor unit can be varied widely, first as an evapourator coil with upflow, downflow or a horizontal air flow arrangement depending on the heating system, or it can be equipped with a blower to provide its own air handling capability. The adaptability of these systems made them an immediate success and there is fast growth in their use.

During the same period of development of the package and split-system cooling equipment, the industry also converted many of these cooling systems to reverse cycle operation and called them heat pumps.

In many apartments, it is very difficult to find outdoor space to put packaged units except on the roof: running ducts down several stories is not a practical and economical consideration when compared to running refrigerant lines. The same situation is true for many multifloor commercial buildings. So, to meet many variations, the industry developed the split system, which consists of an indoor cooling section, outdoor condensing unit, and interconnecting refrigeration piping.

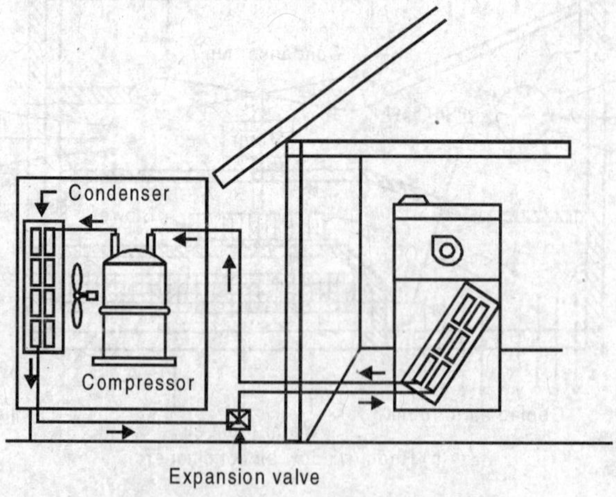

Fig. 37.14. *Split system*

Figure 37.14 shows the schematic diagram of a unitary split-system for cooling and is self-explanatory. The same unit can be used as a heat pump by providing a reversing valve.

Upto 7.5 tonnes, there is no distinct rule to classify a piece of equipment as commercial or residential, for there is a wide degree of application in both markets, using the same products. However, above 7.5 tonnes, the application becomes distinctly commercial, and the product designs are different in component standards. Commercial split-systems range from 7.5 tonnes to 100 tonnes and above.

2. Central Air Conditioning System

The name central station system or equipment is also commonly called as applied system or applied machinery. Central station equipment is associated with installations where the cooling plant is located in the basement or in a penthouse on the roof of multistorey buildings. It serves air handling equipment and air distribution systems throughout the building. Although size is not necessarily the crossover point between unitary and central station, it is usually acknowledged that central station equipment starts at 25 to 50 tonnes and extends upward to the multi-thousand-tonne systems.

Another distinguishing difference is that central station systems use the medium of liquid—mostly water— to transfer heating and cooling to a space air terminal, while unitary systems are based on distributing conditioned air directly to the conditioned space. Unitary equipment makes use of factory packaged, balanced and tested equipment which requires a minimum of onsite labour and. material to be operational. Central station systems are made up of separate components such as chillers, air-handlers, water towers, controls, etc. which can become quite complex in terms of on-site installations and labour and related trades and crafts. Central station equipment is closely associated with the plan and spee engineered systems put on by engineering firms. Equipment is then selected and/or built in order to compliance with these specifications.

Conventional Central Station System

Conventional Central Station System is shown in Fig 37.15. The major components are

1. Water chiller 2. Boiler 3. Air handling unit 4. Water cooling tower 5. Control system

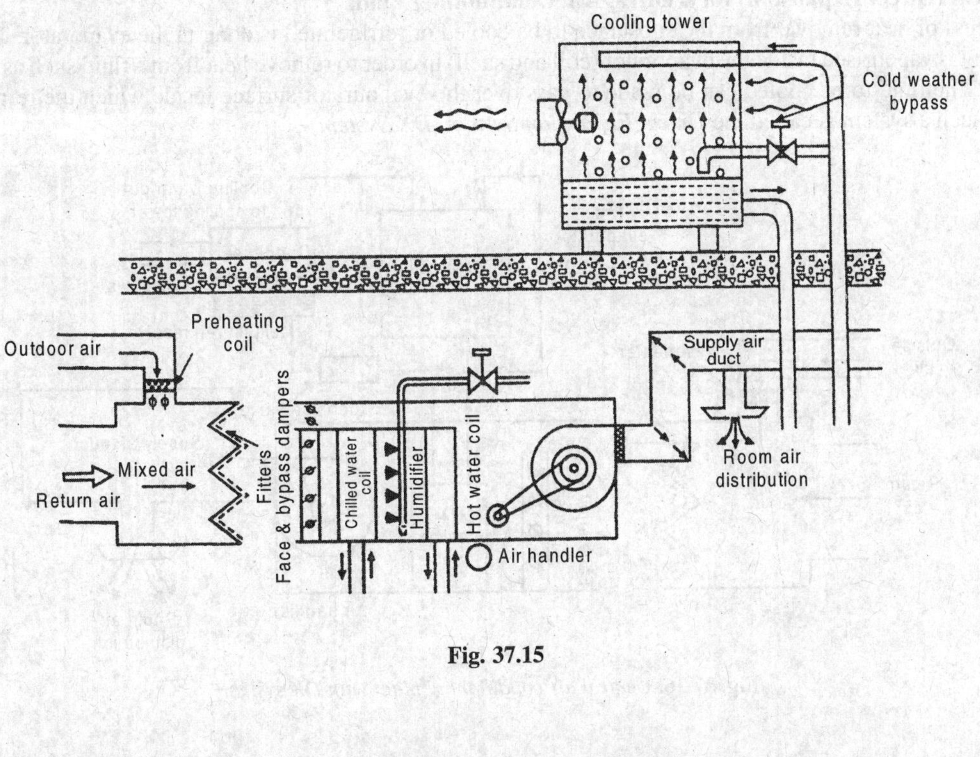

Fig. 37.15

The water chiller will produce 4°C – 7°C cold water, and by means of a pump, circulate it to the cold water coil in the air handler. Water off the coil will generally return at a 6°C rise.

Similarly, in winter, the boiler will produce hot water at 80°C to 90°C and pump it to the hot water coil. Note that it is possible to have the boiler and chiller operating at the same time, because in large buildings, there may be need for cooling and heating in different zones.

Condenser water off the chiller (36 – 37°C) is pumped to cooling tower spray nozzles where it is cooled within the tower to around 30°C and then is returned to the condenser.

The air handling unit or units, depending on the number of floors or zones, generally contain:

(a) Chilled water coils
(b) Main hot water coils (can be steam)
(c) Humidifier
(d) Filters
(e) Dampers for mixing return air and outside air
(f) Blower and motor.

A pre-heat coil is frequently required where large amounts of outside air at or below 0°C are needed.

The face and by-pass dampers either permit all the air to go through the humidifiers and the heating and cooling coils or allow some of it to be by-passed, depending on the particular situation. All air to be conditioned is always filtered and cleaned.

Water Chilling Equipments may be

1. Packaged chillers – Lower capacity.
2. Centrifugal chillers (Hermetic) – Very large capacities upto 1300 tonnes.
3. Screw compressor chillers.
4. Absorption chillers.

Cooling coils in central air conditioning station may use either chilled water circulating through the coils or there may be chilled water evaporators (chilled water sprays) or air-washers. In this case, chilled water works as a secondary refrigerant.

DX system (Direct Expansion) for Central Air Conditioning Plant

The process of heat removal from the substance to be cooled or refrigerated is done in the evaporator. The liquid refrigerant is vapourised inside the evaporator (coil and shell) in order to remove heat from a fluid such as air, water or brine. The fluid to be cooled can be made to pass over the evaporator surface inside which the refrigerant is boiling, such a system is called the *Direct Expansion System* (*DX System*).

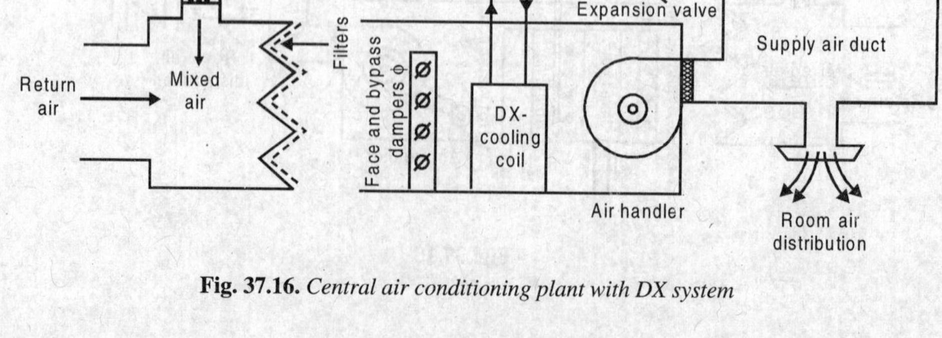

Fig. 37.16. *Central air conditioning plant with DX system*

In certain cases, such as in large air conditioning systems or in industrial processing, water or brine is chilled in the evaporator and the chilled fluid is circulated through copper or steel coils over which the air is passed. Such a system is called the *indirect system*. The coil, generally called cooling coil, acts as heat exchangers.

In cases, where condensing unit and evaporator are comparatively near to each other. Direct expansion (DX) system can be used.

Figure 37.16 shows a central air conditioning plant where DX system is used.

37.15 OTHER AIR CONDITIONING SYSTEMS

Air Conditioning Systems in addition to the equipment arrangement—are classified according to the seasons in which the systems are used. They are discussed as follows.

(a) Summer air conditioning system for hot and dry weather. In this plant, the process carried out will be cooling with humidification by using adiabatic humidifier. The schematic diagram is shown in Fig. 37.17. Places like Nagpur and North India use this type of the system.

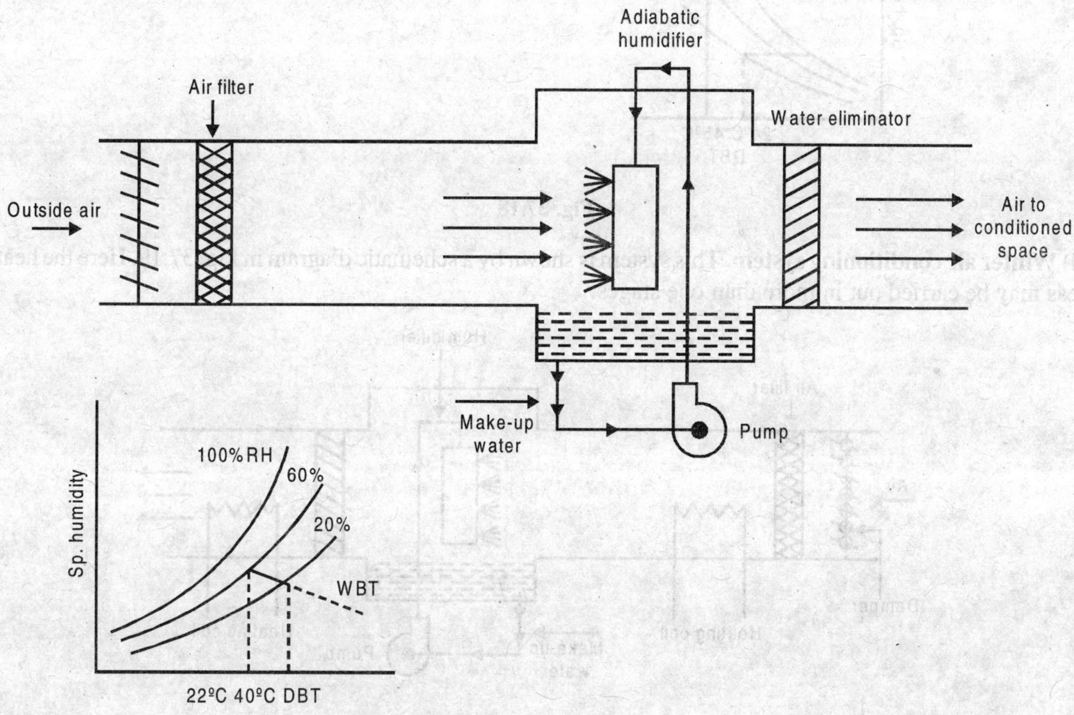

Fig. 37.17. *Air conditioning plant for hot and dry weather*

(b) Summer air-conditioning system for hot and wet weather. In this plant, the process carried out will be cooling with dehumidification. The schematic diagram is shown in Fig. 37.18. Places like Mumbai, Chennai use this type of system because along with the high temperature, humidity is also more and hence 'cooling and dehumidification' process is carried out.

(c) Winter air conditioning system. Winter air conditioning is altogether different than the summer air conditioning. For comfort, air has to be heated. By heating air, relative humidity decreases. To increase relative humidity, we have to humidify the air.

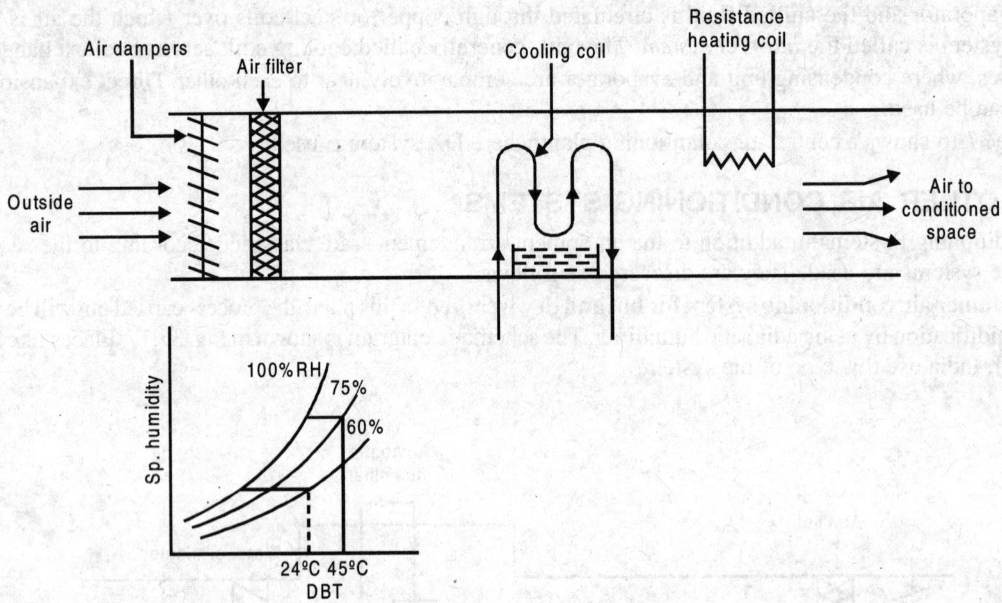

Fig. 37.18

(d) Winter air conditioning system. This system is shown by a schematic diagram in Fig. 37.19. Here the heating process may be carried out in more than one stages.

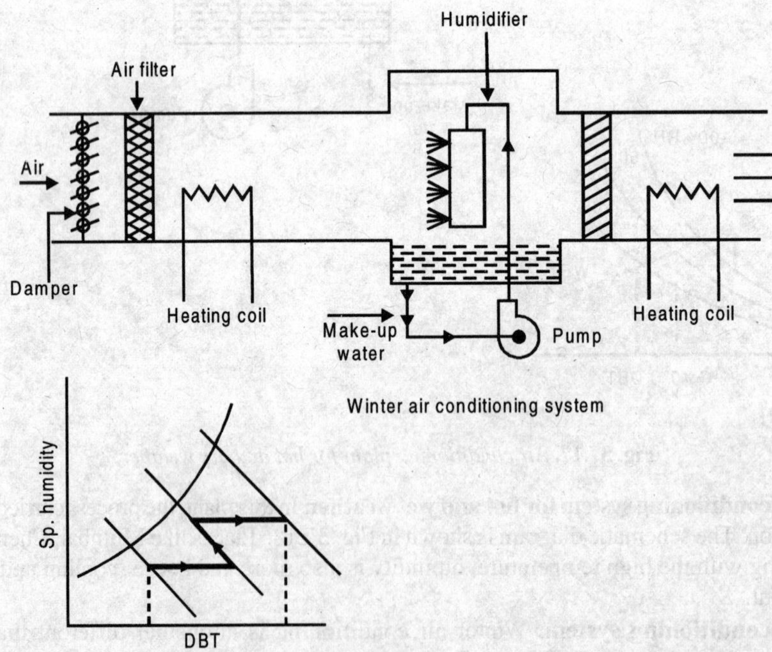

Fig. 37.19

■■■

SOLVED EXAMPLES

Example 37.1. *Atmospheric air at 35°C DBT and 60% RH is conditioned to 22°C DBT and 55% RH. This is done first by cooling and dehumidifying and then by heating. If the quantity of air flow is 60 m³/min, find:*

(i) Weight of water drained

(ii) Capacity of cooling coil

(iii) Capacity of heating coil.

Solution

Initial condition of air, then process and final conditions of air are represented on psychrometric chart as shown in Fig. Ex. 37.1.

1. Weight of Water Drained

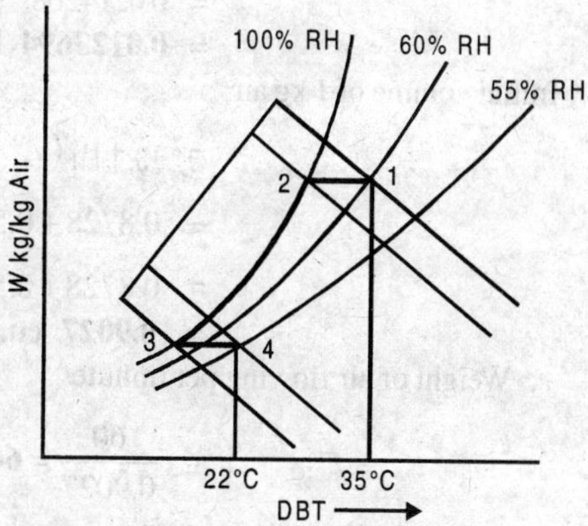

Fig. Ex. 37.1

$$p_{s_1} = 42.17 \text{ mm of Hg}$$

$$\text{RH} = \frac{p_{v_1}}{p_{s_1}}$$

$$\therefore \quad p_{v_1} = 42.17 \times 0.6$$
$$= 25.302 \text{ mm of Hg}$$

$$\therefore \quad w_1 = \frac{0.622 p_{v_1}}{(p - p_{v_1})} = \frac{0.622 \times 25.302}{760 - 25.302}$$

$$= \frac{0.622 \times 25.302}{734.698}$$

$$= \mathbf{0.0214208 \text{ kg/kg of dry air}}$$

and $$w_{s_1} = \mathbf{0.03663 \text{ kg/kg of dry air}}$$

$$\therefore \quad \mu_1 = \frac{w_1}{w_{s_1}}$$

$$= \frac{0.02142}{0.03663}$$

$$= \mathbf{0.5848}$$

Similarly, at the final state,

$$p_{s_4} = 19.82 \text{ mm of Hg}$$

$$\text{RH}_4 = \frac{p_{v_4}}{p_{s_4}}$$

$$\therefore \quad p_{v_4} = 0.55 \times 19.82$$
$$= \mathbf{10.901 \text{ mm of Hg}}$$

$$\therefore \quad w_4 = \frac{0.622 \times 10.901}{760 - 10.901} = \frac{0.622 \times 10.901}{749.099}$$

$$= \mathbf{0.00905 \text{ kg/kg of dry air}}$$

$$w_{s_4} = 0.01668 \text{ kg/kg of dry air}$$

$$\therefore \quad \mu_4 = \frac{w_4}{w_{s_4}}$$

$$= \frac{0.00905}{0.01668}$$

$$= \mathbf{0.5426}$$

∴ Weight of water drained/kg of dry air

$$= w_1 - w_4$$

$$= 0.0214208 - 0.0090514$$

$$= \mathbf{0.0123694 \ kg/kg \ of \ dry \ air}$$

Initial volume of 1 kg air

$$= v_a + \mu_1 (v_{as} - v_a)$$

$$= 0.8728 + 0.5848 \times (0.9239 - 0.8728)$$

$$= 0.8728 + 0.5848 \times 0.0511$$

$$= \mathbf{0.9027 \ cu.m/kg}$$

∴ Weight of air flowing per minute

$$= \frac{60}{0.9027} = \mathbf{66.47 \ kg/min.}$$

∴ Weight of water drained/min.

$$= 0.0123694 \times 66.47$$

$$= \mathbf{0.8222 \ kg/min.}$$

2. Capacity of Cooling Coil

Enthalpy of 1 kg at the initial condition

$$h_1 = h_{a_1} + \mu_1 (h_{as_1} - h_{a_1})$$

$$= 35.25 + 0.5848 \times (129.17 - 35.25)$$

$$= 35.25 + 54.92$$

$$= \mathbf{90.17 \ kJ/kg \ of \ dry \ air}$$

Dew point temperature of air at 4 is approximately 12.5°C

$$\therefore \quad h_3 = 35.445 \text{ kJ/kg of dry air}$$

and

$$h_4 = h_{a_4} + \mu_4 (64.522 - 22.149)$$

$$= 22.149 + 0.5426 \times (64.522 - 22.149)$$

$$= 22.149 + 22.992$$

$$= \mathbf{45.141 \ kJ/kg \ of \ dry \ air}$$

∴ Capacity of cooling air

$$= h_1 - h_3$$

$$= 90.17 - 35.445$$

$= 54.725$ kJ/kg

$= 54.725 \times 66.47$ kJ/min

$= 3637.57$ kJ/min

$= \mathbf{60.63\ kW}$

3. Capacity of Heating Coil

Capacity of Heating Coil $= h_4 - h_3$

$= 45.141 - 35.445$

$= 9.696$ kJ/kg of dry air

$= 9.696 \times 66.47$ kJ/min

$= \mathbf{644.493\ kJ/min}$

$= \mathbf{10.74\ kJ/sec\ or\ kW}$

Example 37.2. *(a) Obtain all psychrometric properties of moist air having 35°C DBT and 14°C DPT, (b) 200 c mm air at 30°C DBT and 90% RH is mixed with 300 cm of air at 10° C DBT and 100% RH. Determine the final condition of air after mixing. Also find the moisture separated if any.*

Solution

(a) DBT is 35°C $\therefore p_s = 42.17$ mm of Hg

DPT is 14°C $\therefore p_v =$ vapour pressure

$w =$ specific humidity at 14°C DPT

$= \mathbf{0.010012}$ **kg/kg of dry air.**

$$= \frac{0.622\, p_v}{760 - p_v} = \frac{0.622}{\dfrac{760}{p_v} - 1}$$

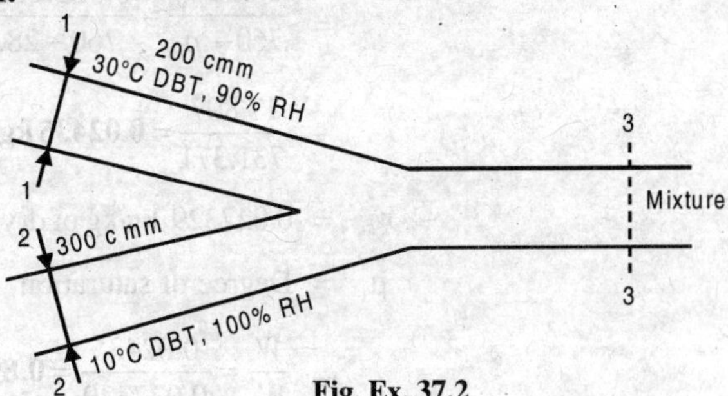

Fig. Ex. 37.2

$$\therefore \quad \frac{760}{p_v} = 1 + \frac{0.622}{0.010012}$$

$$= 1 + 62.125 = 63.125$$

$$\therefore \quad p_v = \frac{760}{63.125} = \mathbf{12.04\ mm\ of\ Hg}$$

$\therefore \quad \phi =$ Relative humidity

$$= \frac{p_v}{p_s} = \frac{12.04}{42.17}$$

$= \mathbf{0.2855}$ **or 28.55 %**

$\mu =$ Degree of saturation

$$= \frac{w}{w_s} = \frac{0.010012}{0.036756}$$

$= \mathbf{0.2724}$ **or 27.24 %**

$h =$ Enthalpy of 1 kg air

$$= h_a + \mu (h_{as} - h_a)$$

$$= 35.219 + 0.2724 \times (129.455 - 35.219)$$

$$= 35.219 + 0.2724 \times 94.236$$
$$= \mathbf{60.89 \text{ kJ/kg of dry air}}$$
$$t_{wb} = \text{Wet bulb temperature}$$
$$= \mathbf{20.9583^\circ C}$$
$$V = \text{Volume of 1 kg dry air}$$
$$= v_a + \mu(V_{as} - V_a)$$
$$= 0.8728 + 0.2724 \times (0.9242 - 0.8728)$$
$$= 0.8728 + 0.2724 \times 0.0514$$
$$= \mathbf{0.8868 \text{ cu-m/kg of dry air}}$$

Stream 1:

$$p_{s_1} = 31.81 \text{ mm of Hg}$$
$$\therefore \quad \phi_1 = \frac{p_v}{p_s} = \frac{p_v}{31.81} = 0.9$$
$$\therefore \quad p_{v_1} = 0.9 \times 31.81 = \mathbf{28.629 \text{ mm of Hg}}$$
$$\therefore \quad w_1 = \text{Specific humidity}$$
$$= \frac{0.622 p_v}{760 - p_v} = \frac{0.622 \times 28.629}{760 - 28.629}$$
$$= \frac{17.807}{731.371} = \mathbf{0.02435 \text{ kg/kg of dry air}}$$
$$w_{s_1} = 0.027329 \text{ kg/kg of dry air}$$
$$\therefore \quad \mu_1 = \text{Degree of saturation}$$
$$= \frac{W}{W_s} = \frac{0.02435}{0.027329} = \mathbf{0.891}$$
$$\therefore \quad h_1 = h_{a_1} + \mu_1 \left(h_{as_1} - h_{a_1}\right)$$
$$= 30.185 + 0.891(100.006 - 30.185)$$
$$= 30.185 + 0.891 \times 69.821$$
$$= \mathbf{92.396 \text{ kJ/kg of dry air}}$$
$$V_1 = V_{a_1} + \mu_1 \left(V_{as_1} - V_{a_1}\right)$$
$$= 0.8586 + 0.891(0.8962 - 0.8586)$$
$$= 0.8586 + 0.891 \times 0.0376$$
$$= \mathbf{0.8921 \text{ cu-m/kg of dry air}}$$
$$\therefore \quad m_1 = \text{Air flow/min}$$
$$= \frac{200}{0.8921} = \mathbf{224.19 \text{ kg/min}}$$

Stream 2:

$$p_s = p_v = 9.204 \text{ mm of Hg}$$

$$w_{s_2} = w_2 = 0.007661 \text{ kg/kg of dry air}$$

$$\phi_2 = \mu_2 = \mathbf{100\ \%}$$

$$h_2 = 29.352 \text{ kJ/kg of dry air}$$

$$V_2 = 0.8116 \text{ cu-m/kg}$$

$$\therefore \quad m_2 = \text{Mass flow rate}$$

$$= \frac{300}{0.8116} = \mathbf{369.64\ kg/min.}$$

Stream 3:
We can write

$$m_1 w_1 + m_2 w_2 = m_3 w_3$$

$$224.19 \times 0.02435 + 369.64 \times 0.007661 = \left[(224.19 + 369.64) \times w_3\right]$$

$$\therefore \quad w_3 = \frac{5.459 + 2.8318}{593.83}$$

$$= \mathbf{0.01396\ kg/kg\ of\ dry\ air}$$

Similarly,

$$m_1 h_1 + m_2 h_2 = m_3 h_3$$

$$224.19 \times 92.396 + 369.64 \times 29.352 = 593.83\, h_3$$

$$\therefore \quad h_3 = \frac{20714.259 + 10849.673}{593.83}$$

$$= \mathbf{53.153\ kJ/kg\ of\ dry\ air}$$

Referring the Tables we get wet bulb temperature as **18.699°C.**
Again enthalpy of air is given by

$$h_3 = 53.153 = t_{d_3} + w_3\left[2501 + 1.82\, t_{d_3}\right] \text{ assuming } C_{pa} = 1.$$

$$= t_{d_3} + 0.01396 \times 2501 + 0.01396 \times 1.82\, t_{d_3}$$

$$= t_{d_3} + 34.92 + 0.02541\, t_{d_3}$$

$$\therefore \quad t_{d_3} = \text{Dry bulb temperature of the mixture}$$

$$= \frac{53.153 - 34.92}{1.02541} = \frac{18.233}{1.02541} = \mathbf{17.78°C}$$

Example 37.3 : *(a) Define and explain the following terms used in air conditioning. (1) Psychrometry (2) DPT, (3) WBT, (4) RH.*

(b) What is Comfort Air Conditioning? Recommend suitable psychrometric processes for following atmospheric conditions. Show the processes on handmade psychrometric chart.

(1) 35°C DBT, 65% RH,

(2) 40°C DBT 10% RH.

(c) Atmospheric air at 30°C DBT and 60% RH is passed over a cooling coil with bypass factor of 0.15 at the rate of 200 m³/min. The coil surface temperature is 14°C. Find supply air DBT, cooling coil capacity in TR and amount of moisture separated per hour.

Solution

(b) Assume comfort conditioning requirements as—
DBT = 22°C; RH = 40%

1. 1 is atmospheric condition of air. 2 is required condition of air. Processes recommended: 1–a–sensible cooling. a–b: conndensation or separation of moisture and b–2 is sensible heating.

2. 1 is atmospheric condition.
2 is required condition.
To achieve 2, we proceed as—
1–a sensible cooling
a–b moisture removal or condensation.

b–2 sensible heating or use a cooling coil having surface temperature t_c having a bypass factor BF = $\dfrac{2-c}{1-c}$.

(c) Figure Ex. 37.3 shows the process.

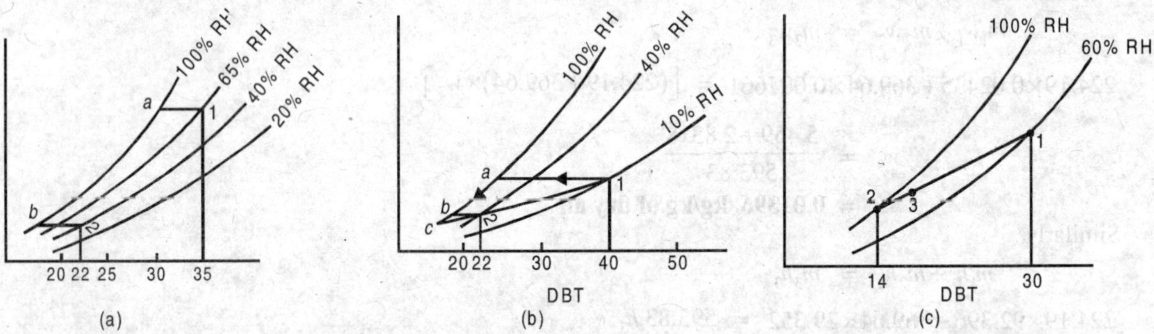

Fig. Ex. 37.3

1 is the condition of atmospheric air. 2 is the condition if the air is passed over the cooling coil having 100% efficiency or zero bypass factor.

1–2 is measured from chart. This is $2\frac{3''}{8}$ or $\frac{19''}{8}$ or 6 cm approximately. Cooling coil has BF = 0.15. Then point 3 is located on the chart and from chart DBT of the supply air is found to be 16.4°C approximately.

Specific volume of atmospheric air is 0.88 cu-m/kg (from chart).

$$\therefore \quad \text{Mass of air flowing} = m_a = \frac{200}{0.88} = \mathbf{227.273\ kg/min.}$$

$$\text{From chart, } h_1 = 71 \text{ kJ/kg; } W_{H_1} = 0.016 \text{ kg/kg of dry air}$$

$$h_3 = 44 \text{ kJ/kg; } W_{H_3} = 0.0108 \text{ kg/kg of dry air}$$

$$\therefore \quad \text{Moisture separated/hour} = \left(W_{H_1} - W_{H_3}\right) \times \text{Mass of air/hr}$$

$$= (0.016 - 0.0108) \times 227.273 \times 60$$

$$= 0.0052 \times 227.273 \times 60$$

$$= \mathbf{70.91\ kg/hour}$$

$$\text{Also heat removed from air/sec} = (h_1 - h_3) \times \text{Mass of air/sec}$$

$$= (71 - 44) \times \frac{227.273}{60}$$

$$= \frac{27 \times 227.273}{60}$$

$= $ **102.273 (kJ/sec = kW)**

$= $ Capacity of the cooling air.

Example 37.4. *(a) 100 m³/min. of air at 15°C DBT and 80% RH is heated until its temperature is 25°C. Find the following*

(i) Heat added to the air in kJ/min.

(ii) RH of the heated air.

Assume air pressure as 100 kPa.

(b) Define the following:

(i) Effective temperature.

(ii) Adiabatic saturation temperature.

(c) Compare central air conditioning system with unit air conditioning system.

Solution

(a) From chart, $h_1 = 36.5$ kJ/kg of dry air

Heating is sensible heating i.e., humidity ratio is to be same. Therefore, this heating process will be shown as a horizontal line on psychrometric chart.

$$h_2 = 47 \text{ kJ/kg of dry air}$$

$\therefore$ Heat added per kg of dry air $= h_2 - h_1$

$= 47 - 36.5$

$= 10.5$ kJ/kg of dry air

Specific volume of air $= v_1 = 0.8282$ cu-m/kg

$\therefore$ Mass of air $= \frac{100}{0.8282} = \mathbf{120.744\ kg/min.}$

$\therefore$ Heat added per minute $= 120.744 \times 10.5$

$= $ **1267.81 kJ/min**

From chart, RH of heated air $=$ **43%**

Example 37.5. *50 m³ of air at 38°C and 60% RH is cooled to 32°C maintaining its specific humidity constant. Without using psychrometric chart, determine*

(i) Relative humidity of cooled air. (ii) Heat removed from air.

Use the following formula to calculate specific enthalpy of moist air.

h = 1.005 (DBT) + w [2500 + 1.88 (DBT – DPT)]

where DBT = Dry bulb temperature and DPT = Dew point temperature.

Solution

(i) RH of cooled air : Given : 50 m³ of air at DBT = 38°C and RH = 60 %.

$$(\text{DBT})_2 = 32°\text{C}$$

Initial air (before cooling):

$$p_s = 49.69 \text{ mm Hg}$$

$$\phi = \text{RH} = 60\ \%$$

$$\therefore \quad \phi = 0.6 = \frac{p_v}{p_s} = \frac{p_v}{49.69}$$

$$\therefore \quad p_v = 0.6 \times 49.69 = 29.814 \text{ mm of Hg}$$

$$\therefore \qquad w = \frac{0.622 p_v}{p - p_v} = \frac{0.622 \times 29.814}{760 - 29.816} = \frac{0.622 \times 29.814}{730.184}$$

$$= 0.0254 \text{ kg/kg of air}$$

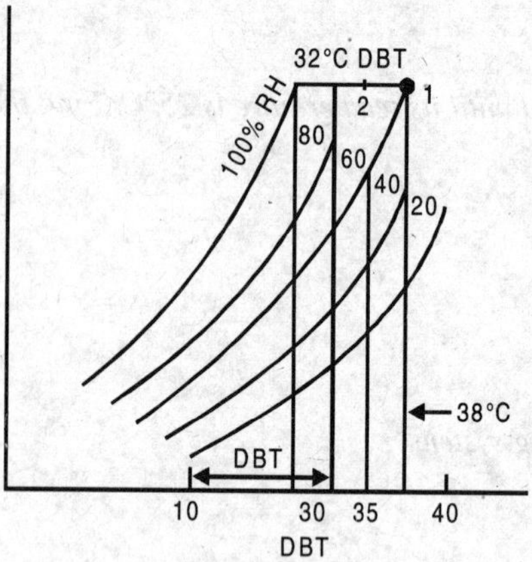

Fig. Ex. 37.5

Dew point temperature of air is that temperature at which the

$$w = 0.0254 \text{ kg/kg of dry air, saturates air.}$$

From table,

$$\text{DPT} = 29 - 0.222$$

$$= \mathbf{28.778\ °C}$$

After sensible cooling, temperature is 32ºC p_s at 32ºC

$$= 35.65 \text{ mm of Hg.}$$

$$p_v = 29.814 \text{ mm of Hg}$$

$$\therefore \qquad \text{RH at } 32°\text{C} = \frac{29.814}{35.65} = \mathbf{83.63\%}$$

(ii) Heat removed from air :

h_1 = Specific enthalpy before cooling

$$= 1.005 \times 38 + 0.0254 \times [2500 + 1.88(38 - 28.778)]$$

$$= 38.19 + 0.0254 \times [2500 + 1.88 \times 9.222]$$

$$= 38.19 + 0.0254 \times [2500 + 17.34]$$

$$= 38.19 + 63.94$$

$$= \mathbf{102.13 \text{ kJ/kg of dry air}}$$

h_2 = Specific enthalpy after cooling (32º DBT)

$$= 1.005 \times 32 + 0.0254 \times [2500 + 1.88(32 - 28.778)]$$

$$= 32.16 + 0.0254 \times [2500 + 6.057]$$

$$= 32.16 + 63.654$$

$$= \mathbf{95.814 \text{ kJ/kg}}$$

∴ Heat removed/kg of dry air

$$= 102.13 - 95.814$$

$$= \mathbf{6.316 \text{ kJ/kg of dry air}}$$

Initially, specific volume of air

$$v_a = \frac{m_a R_a T_a}{p_a} = \frac{1 \times 287 \times 311}{\frac{730.184}{760} \times 1.01325 \times 10^5} = \frac{287 \times 311}{0.9735 \times 10^5}$$

$$= \mathbf{0.9169 \text{ m}^3\text{/kg}}$$

$$\therefore \qquad \text{Mass of air} = \frac{50}{0.9169} = 54.53 \text{ kg}$$

$$\therefore \qquad \text{Total heat removed} = 54.53 \times 6.316$$

$$= \mathbf{344.42 \text{ kJ}}$$

Example 37.6. *Air at 10°C DBT and 90% RH is to be heated and humidified to 35°C DBT and 22.5°C WBT. The air is preheated sensibly before passing to the air washer in which water is circulated. The relative humidity of air coming out of the air washer is 90%. Find:*

(1) The temperature to which the air should be preheated.
(2) The total heating required.
(3) Make up water required in the air washer.
(4) The humidifying efficiency of the air washer.
(Use of psychrometric chart is expected while solving the problem).

Solution

1–2 Sensible heating $h_1 = 27.2,\ h_2 = 42$ kJ/kg

2–3 Sensible humidifying $h_3 = 66.5$ kJ/kg

3–4 Heating and dehumidifying $h_4 = 66.5$ kJ/kg

$$W_1 = W_2 = 0.00712,\ W_3 = 0.0168,\ W_1 = 0.0122$$

Referring to the diagram,

(1) The temperature to which the air should be preheated = T_2 = **24°C**

(2) The total heating required

$$= h_3 - h_1 = 66.5 - 27.2$$
$$= \mathbf{39.3\ kJ/kg}$$

(3) Make up water in the air washer

$$= w_3 - w_2$$
$$= 0.01680 - 0.00712$$
$$= \mathbf{0.00968\ kg/kg\ of\ dry\ air}$$

(4) Humidifying efficiency of the washer

$$\eta_H = \frac{w_3 - w_2}{w_5 - w_2}$$
$$= \frac{0.01680 - 0.00712}{0.01840 - 0.00712}$$
$$= \frac{0.00968}{0.01128}$$
$$= 0.8581$$
$$= \mathbf{85.81\ \%}$$

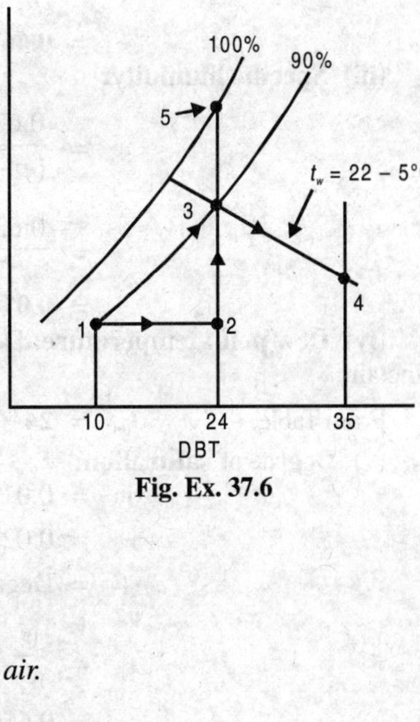

Fig. Ex. 37.6

Example 37.7. *(a) Derive the expression to calculate the enthalpy of moist air.*

(b) Define the terms
(1) Evapourative cooling
(2) Comfort chart.

(c) In a laboratory test, a sling psychrometer recorded 30°C DBT and 25°C WBT.

Calculate
(1) Vapour pressure
(2) Relative humidity
(3) Specific humidity
(4) DPT (Dew Point Temperature)
(5) Degree of saturation.

Solution

(c)

$$t_d = 30°\text{C}, \; t_w = 25°\text{C}$$

$$p_{sd} = 31.81, \; p_{sw} = 23.75 \text{ mm of Hg}$$

(i) Vapour pressure:

$$p_v = p_{sw} - \frac{(p - p_{sw})(t_d - t_w)}{1527.4 - 1.3\,t_w}$$

$$= 23.75 - \frac{(760 - 23.75) \times (30 - 25)}{1527.4 - 1.3 \times 25}$$

$$= 23.75 - \frac{736.25 \times 5}{(1527.4 - 32.5)}$$

$$= 23.75 - 2.4625$$

$$= \mathbf{21.2875 \text{ mm of Hg}}$$

(ii) Relative humidity: Partial presure of vapour at saturation at 30ºC is $p_s = 31.81$

$$\therefore \quad \text{Relative humidity} = \frac{p_v}{p_s} = \frac{21.2875}{31.81}$$

$$= \mathbf{0.6692} \text{ or } \mathbf{66.92\ \%}$$

(iii) Specific humidity:

$$w = \frac{0.622\,p_v}{(p - p_v)} = \frac{0.622 \times 21.2875}{760 - 21.2875}$$

$$= \frac{0.622 \times 21.2875}{738.7125}$$

$$= \mathbf{0.0179 \text{ kg/kg of dry air}}$$

(iv) Dew point temperature: Dew point temperature is the temperature at which the given moisture saturates the air.

From Table, $t_{dp} = \mathbf{24\ °C}$

(v) Degree of saturation:

$$w = 0.0179$$

$$w_s = 0.027329$$

$$\therefore \quad \mu = \text{Degree of saturation}$$

$$= \frac{w}{w_s} = \frac{0.0179}{0.027329}$$

$$= \mathbf{0.655} \text{ or } \mathbf{65.5\ \%}$$

Example 37.8. *(a) Define and discuss the significance of following terms:*

(1) Dew point temperature

(2) Specific humidity.

(3) Degree of saturation.

(4) Relative humidity.

(b) For a hall to be air conditioned,

Outdoor conditions: DBT = 40°C, WBT = 20°C

Required conditions: DBT = 20°C, RH = 60%

Seating capacity of hall = 1500.

Amount of outdoor air supplied = 0.3 m³/min/person.

If required conditions are achieved first by adiabatic humidification and then by cooling, estimate

(a) Capacity of cooling coils in tonnes (TR).

(b) Capacity of humidifier in kg/hr.

Solution

From Psychrometic chart,

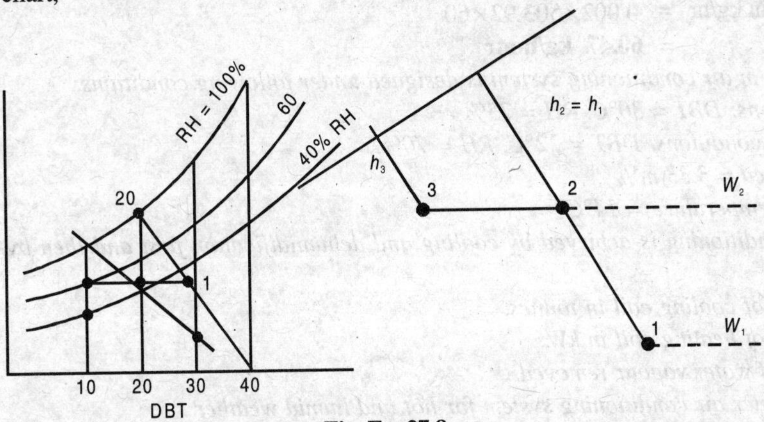

Fig. Ex. 37.8

$$h_1 = h_2 = 57.5 \text{ kJ/kg of dry air}$$

$$h_3 = 42.5 \text{ kJ/kg of dry air}$$

$$w_1 = 0.0066 \text{ kg/kg of dry air}, \quad w_2 = 0.0086 \text{ kg/kg of dry air}$$

1–2 – Adiabatic humidification process
This is along the WBT line.

2–3 – Sensible cooling.

Specific volume/kg of dry air at inlet = **0.893 cu-m/kg**

$$\therefore \quad \text{Mass of air supplied/min} = \frac{1500 \times 0.3}{0.893} \frac{\text{m}^3/\text{min}}{\text{m}^3/\text{kg}}$$

$$= \mathbf{503.92 \ kg/min}$$

(a) Capacity of cooling coil in TR

$$\text{Capacity of cooling coil/sec} = m_a (h_2 - h_3)$$

$$= \frac{503.92}{60} \times (57.5 - 42.5)$$

$$= \frac{503.92 \times 15}{60}$$

$$= 125.98 \text{ (kJ/sec = kW)}$$

$$1 \text{ TR} = \mathbf{3.516 \ kW}$$

$$\therefore \quad \text{Capacity in TR} = \frac{125.98}{3.516}$$

$$= \mathbf{35.83 \ TR}$$

(b) Capacity of humidifier in kg/hr

Capacity of humidifier/kg of dry air supplied

$$= w_2 - w_1$$

$$= 0.0086 - 0.0066$$

$$= 0.002 \text{ kg/kg of dry air}$$

$\therefore$ Capacity in kg/hr $= 0.002 \times 503.92 \times 60$

$= \mathbf{60.47 \text{ kg/hour}}$

Example 37.9. *(a) An air conditioning system is designed under following conditions:*

Outdoor conditions: DBT = 30°C, RH = 75%

Required indoor conditions: DBT = 22°C, RH = 70%.

Free air circulated = 3.33 m³/s

Coil dew point temperature = 14°C

The required conditioning is achieved by cooling and dehumidification first and then by heating.

Find :

(i) Capacity of cooling coil in tonnes.

(ii) Capacity of heating coil in kW.

(iii) Amount of water vapour removed.

(b) Explain summer air conditioning system for hot and humid weather.

(c) Discuss the process of chemical dehumidification.

Solution

(a) The process is shown in Psychrometric chart.

1–2 = Sensible cooling

2–3 = Dehumidification

3–4 = Sensible heating to 22°C and 70 %

From chart, $w_1 = w_2 = 0.0202$ kg/kg of air

$w_3 = w_4 = 0.0114$ kg/kg of air

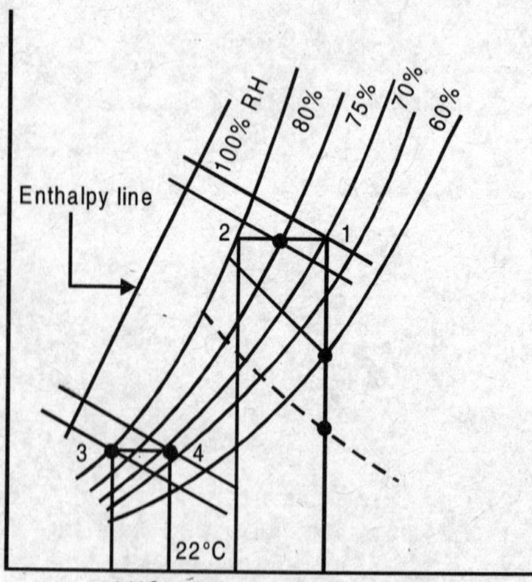

Fig. Ex. 37.9

$h_1 = 81.5, h_2 = 76.5, h_3 = 45.5$

$h_4 = 51$ kJ/kg of dry air

Specific volume at 1 $= \dfrac{8}{28} \times 0.02 + 0.88$

$= 0.88 + 0.00571$

$= 0.88571 \text{ m}^3\text{/kg}$

Free air is assumed to be at inlet conditions :

$\therefore$ Mass of air circulated/sec $= \dfrac{3.33}{0.88571}$

$= \mathbf{3.76 \text{ kg/s}}$

(a) Capacity of cooling coil in tonnes:

Capacity $= (h_1 - h_3) m_a = 3.76 \times (81.5 - 45.5)$

$= 3.76 \times 36$

$= 135.36$ kJ/s = kW

$= \dfrac{135.36}{3.516}$ TR

$= \mathbf{38.5 \text{ TR}}$

(b) Capacity of heating coil in kW

$$\text{Heating coil capacity} = m_a(h_4 - h_3)$$
$$= 3.76\times(51-45.5)$$
$$= 3.76\times5.5$$
$$= \mathbf{20.68\ kJ/s = kW}$$

(c) Amount of water removed/min

$$\text{Amount of water removed/min} = (m_a\times60)(w_1 - w_4)$$
$$= (3.76\times60)\times(0.0202-0.0114)$$
$$= 3.76\times60\times0.0088$$
$$= \mathbf{1.9853\ kg/min}$$

Example 37.10. *(a) Discuss the thermal balance mechanism of a human body. List the factors influencing human comfort. Explain the significance of effective temperature.*

(b) The atmospheric air at 25°C DBT and 12°C WBT flows at the rate of 100 cmm through a duct. Dry saturated steam at 100°C is injected into the air stream at the rate of 72 kg/hour. Calculate the specific humidity and enthalpy of air leaving the duct. Also find DBT, WBT and RH of air. Show the process on psychrometric chart.

(c) 200 cmm of saturated air at 10°C DBT and 100% RH is mixed adiabatically with 200 cmm of air at 30°C and 95% RH. Determine the final DBT, RH of air after mixing. Also find moisture separated, if any, during the process. Sketch the process on Psychrometric chart.

Solution

(b)

From chart :

$$w_1 = 0.003346$$
$$h_1 = 34.179\ \text{kJ/kg}$$
$$v_1 = 0.84901\ \text{m}^3/\text{kg}$$
$$\therefore\quad m_1 = \frac{100}{0.84901} = 117.78\ \text{kg/min}$$

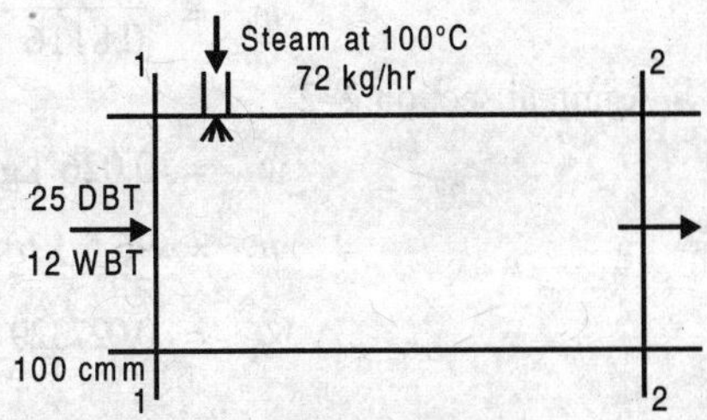

Fig. Ex. 37.10 (a)

Enthalpy of steam at 100°C = 2676 kJ/kg

$$= \frac{2676}{60}\times\frac{72}{1} = 3211.2\ \text{kJ/min}$$

$$m_1w_1 + w_2 = m_3w_3$$
$$117.78\times0.003346+1.2 = (117.78+1.2)w_3$$
$$0.3941+1.2 = 1.5941$$
$$w_3 = \frac{1.5941}{118.98} = \mathbf{0.0134\ kg/kg\ of\ dry\ air}$$
$$\therefore\quad \text{Specific humidity} = \mathbf{0.0134\ \ kg/kg\ of\ dry\ air}$$
$$m_1h_1 + m_2h_2 = (m_1+m_2)h_3$$
$$117.78\times34.179+1.2\times2676 = (117.78+1.2)h_3$$
$$4025.6+2676\times1.2 = 118.98\,h_3$$
$$4025.6+3211.3 = 7236.8 = 118.98\,h_3$$

$$h_3 = \frac{7236.8}{118.98} = \mathbf{60.82\ kJ/kg}$$

$$= 1.005\, t_d + w(2501 + 1.82\, t_d)$$

$$= 1.005\, t_d - 0.0134 \times (2501 + 1.82 \times t_d)$$

$$= 1.005\, t_d + 33.513 + 0.02439\, t_d$$

$$= 1.02939\, t_d + 33.513$$

$$60.820 - 33.513 = 27.307 = 1.02939\, t_d$$

$$t_d = \frac{27.307}{1.02939}$$

$$= \mathbf{26.53\ ^\circ C}$$

From Table, WBT = **21 °C**

RH = **61.5 %**

(c)

From Chart and Tables :

$$w_1 = 0.007661 \text{ kg/kg of dry air}$$

$$h_1 = 29.348 \text{ kJ/kgof dry air}$$

$$v_1 = 0.8116 \text{ cu-m/kg of dry air}$$

∴ Weight of air flowing through section 1–1

$$m_1 = \frac{200}{0.8116} = 246.43 \text{ kg/min}$$

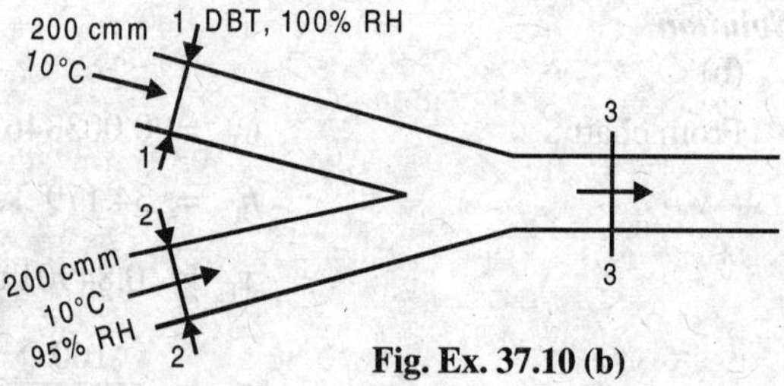

Fig. Ex. 37.10 (b)

Again at section 2–2 :

$$w_2 = 0.026 \text{ kg/kg of dry air}$$

$$h_2 = 96.5 \text{ kJ/kg of dry air}$$

$$w_{s_2} = 0.027329$$

$$\therefore \quad \mu = \text{Degree of saturation} = \frac{0.026}{0.02729} = 0.9527$$

$$\therefore \quad v_2 = v_a + \mu(v_{as} - v_a) \text{ approx.}$$

$$= 0.8586 + 0.9527[0.8962 - 0.8586]$$

$$= 0.8586 + 0.9527 \times 0.0376 = 0.8586 + 0.03582$$

$$= 0.89442 \text{ cu-m/kg}$$

∴ Weight of air flowing through section 2–2,

$$m_2 = \frac{200}{0.89442} = \mathbf{223.6\ kg/min}$$

$$\therefore \quad m_1 w_1 + m_2 w_2 = m_3 w_3$$

$$\therefore 246.43 \times 0.007661 + 223.6 \times 0.026 = (246.43 + 223.6) + w_3$$

$$\therefore \quad 7.7015 = 470.03 \times w_3$$

$$w_3 = \frac{7.7015}{470.03} = \mathbf{0.01638\ kg/kg\ of\ dry\ air}$$

Similarly, $m_1 h_1 + m_2 h_2 = m_3 h_3$

$$246.43 \times 29.348 + 223.6 \times 96.5 = (246.43 + 223.6)\, h_3$$

$$7232.23 + 21577.4 = 470.03\, h_3$$

$$\therefore \quad h_3 = \frac{28809.63}{470.03} = 61.2932 \text{ kg/kg of dry air}$$

We have, $h_3 = C_p\, t_{db} + w_3 (2501 + 1.82\, t_{db})$

$$= 1.005\, t_{db} + 0.01638 \times (2501 + 1.82\, t_{db})$$

$$61.2932 = 1.005\, t_{db} + 40.97 + 0.0298\, t_{db}$$

$$= (1.005 + 0.0298) t_{db} + 40.97$$

$$= 1.0348\, t_{db} + 40.97$$

$$\therefore \quad t_{db_3} = \frac{61.2932 - 40.97}{1.0348} = \frac{20.3232}{1.0348}$$

$$= \mathbf{19.64\ ^\circ C}$$

Assume for calculation purpose, $t_{db} = \text{DBT} = 20°\text{C}$

$$\therefore \quad w_{s_3} = 0.014758$$

$$p_{s_3} = 17.53 \text{ mm of Hg}$$

$$w_3 = \frac{0.622\, p_{v_3}}{p - p_{v_3}}$$

$$\therefore \quad 0.01638 = \frac{0.622\, p_{v_3}}{760 - p_{v_3}}$$

$$\therefore \quad \frac{p_{v_3}}{760 - p_{v_3}} = \frac{0.01638}{0.622} = 0.02633$$

$$p_{v_3} = 20.01 - 0.02633\, p_{v_3}$$

$$1.02633\, p_{v_3} = 20.01$$

$$p_{v_3} = \frac{20.01}{1.02633} = \mathbf{19.5\ mm\ of\ Hg}$$

All these calculations show that air at the particular temperature cannot have specific humidity greater than saturated air at that temperature.

∴ In practice, 0.01638 – 0.014758 = 0.001622 kg/kg of dry air is condensed.

∴ The air after mixing is 100% saturated and the relative humidity is 100%.

THEORY QUESTIONS

37.1. Explain the difference between comfort and industrial air conditioning.

37.2. Define the term 'Human Comfort'. What factors are involved in Human comfort air conditioning? Explain.

37.3. What do you understand by effective temperature? In what way it is connected with comfort air conditioning.

37.4. In what way the human body compensates and reacts automatically to the warm of hot environment? Explain.

37.5. Explain how does the human body attempt to compensate for a cool environment which tends to lower the internal temperature.

37.6. Draw a 'hand-drawn' psychrometric chart and show on it the comfort zone for both winter and summer air conditioning.

37.7. What is an air conditioning system? In what way it is different from refrigeration system?

37.8. Classify the air conditioning systems.

37.9. Distinguish clearly between unitary and central air conditioning systems.

37.10. What is the difference between winter and summer air conditioning systems? What different equipments are required?

PROBLEMS FOR PRACTICE

37.1. What do you understand by the term "Air Conditioning"? How will you classify the different air conditioning systems?

37.2. Distinguish clearly between comfort air conditioning and industrial air conditioning.

37.3. What are the applications of air conditioning?

37.4. In what way an industrial air conditioning is applied in practice? Give examples.

37.5. What are the factors on which comfort air conditioning depends? Explain.

37.6. How the human body responds to the following different environmental conditions?
(1) Dry hot weather
(2) Humid and hot weather and
(3) Cold weather.

37.7. Define the following terms as applied in air conditioning language:
(a) Dry bulb temperature (DBT)
(b) Wet bulb temperature (WBT)
(c) Dew point temperature (DPT)
(d) Humidity, specific humidity and absolute humidity
(e) Relative humidity
(f) Degree of saturation
(g) Saturated air (h) Enthalpy of air.

37.8. What do you understand by the terms "Adiabatic saturation temperature" and "Thermodynamic Wet Bulb Temperature"?

37.9. Are the dry bulb temperature and dew point temperature are same? When they are same?

37.10. When dry bulb temperature, wet bulb temperature and dew point temperature are numerically same?

37.11. Define and explain the terms "Effective temperature" and "Comfort zone".

37.12. Will "Cooling of air" mean air conditioning? Explain.

37.13. What is evaporative cooling? Under what conditions it will take place?

37.14. What is best example of evaporative cooling in air conditioning systems?

37.15. What are psychrometric tables and psychrometric charts?

37.16. How will you construct psychrometric chart?

37.17. Enumerate the different psychrometric processes that are taking place.

37.18. Explain with neat sketch/diagram each of these processes and show them on the hand-drawn psychrometric chart.

37.19. For the given state of air, how will you find dew-point temperature with the help of psychrometric chart and psychrometric table?

37.20. What is a by-pass factor? Explain and define. Where this factor is considered?

37.21. Distinguish clearly the central air conditioning system and unitary air conditioning system.

37.22. Explain with a neat sketch the working of "Window Air Conditioner". Can this be called as unitary air conditioning system?

37.23. What do you understand by 'Split Unit' air conditioner? What are the advantages and disadvantages of this unit?

37.24. What is DX type central air conditioning system? Explain.

37.25. What is an air washer? Explain the different types of air washers.

37.26. Where chilled water air washer will be used? Explain.

37.27. Explain the 'recirculation of air' in air conditioning system. Is there any advantage for this recirculation?

37.28. Find the relative humidity and specific humidity for air at 30°C and having dew point temperature of 15°C.

(*Ans.* RH = 40.169%, w_H = 0.010783 kg /kg of dry air)

37.29. Air is supplied to a room at 17°C and 60% RH. If the barometer reads 1.01325 bar, calculate the specific humidity and dew point of the air.

(*Ans.* w_H = 0.007213 kg/kg of dry air, DPT = 9.18°C)

37.30. One kg of air at 40°C DBT and 50% RH is mixed with 2 kg of air at 20°C DBT and 20° DPT. Calculate temperature and specific humidity of the mixture. (*Ans.* T_3 = 270 C, w_3 = 0.0179)

37.31. Outside air at 760 mm of Hg pressure has its DBT and WBTs 300 °C and 200° C respectively. Determine its specific humidity, relative humidity, vapour density, DPT and enthalpy of air.

(*Ans.* w = 0.0 1048, f = 39.56%, p_a = 0.01199 kg/m³ ; DPT = 14.75°C, h = 56.9 kJ/kg of dry air)

37.32. Outside air flowing at the rate of 2 kg/sec at 30°C DBT, 21°C WBT mixes adiabatically with return air stream flowing at the rate of 3 kg/sec at 21°C DBT and 50% RH. Determine the DBT, WBT, w and h of the mixed stream.

(*Ans.* h = 49.5 kJ/kg; 0.0092; DBT = 24.6°C; WBT = 17.6°C)

37.33. Air at 30°C DBT and 60% RH is passed through a cooling coil at the rate of 200 kg/min. The surface temperature of the coil is 14°C and by-pass factor is 0.12. Find the DBT, cooling coil capacity in TR, moisture condensed.

(*Ans.* 15.95°C; TR = 52.3; 1.08 kg/min)

37.34. 45 m³ /min of air at 20°C and 75% RH is heated till its DBT becomes 30°C. Calculate the mass flow rate of air, RH and heat added per hour.

(*Ans.* 53.3 kg/min; ϕ = 41%; Q = 33.42 kJ/hour)

37.35. A small room requires 6 m³/min of conditioned air at 20°C DBT and 60% RH, while the surrounding conditions in winter are 12°C DBT and 10°C WBT. Conditioned air is obtained by first heating the air-heating coil having a by-pass factor 0.4 and then by adiabatic humidifying. Determine the surface temperature of coil and the capacity of the humidifier.

(*Ans.* ADP = 32.75°C; 0.76 kg/hr)

37.36. 50 cu-m/min. of air at 15°C DBT and 80% RH is heated until its temperature becomes 22°C. Find,

(1) Heat added per hour

(2) RH of the heated air.

Assume barometer reads 760 mm of Hg.

(*Ans.* (1) 26126.9 kJ/hr; (2) RH = 51.7%)

37.37. Outside air at 12°C and 75% RH is to be conditioned to a temperature of 22°C DBT and 60% RH. Amount of air supplied is 100 m³/min. The required condition is achieved first by heating and then by adiabatic humidifying. Find the following

(1) Quantity of steam required per hour. Steam at 2 bar and 0.96 dry.

(2) Quantity of water required per hour in the humidifier. You can use chart.

(*Ans.* (1) 156.4 kg/hr, (2) 251 cg/hr)

37.38. Atmospheric air – 760 mm of Hg—at 40°C DBT and 60% RH is passed through the conditioning system wherein 10 gm of water per kg of dry air are removed. Find:

(a) Relative humidity

(b) Wet bulb temperature and

(c) Dew point temperature.

(*Ans.* (a) 69%, (b) 25.2°C, (c) 23.6°C)

37.39. Outside air – 760 mm of Hg—at 30°C DBT and 25°C WBT is heated to 40°C. If the air supply is 150 m³ /min, find the amount of heat added and RH, WBT of the air.

(*Ans.* 1762.73 kJ/min; 36.8%; 27.75 °C)

37.40. Cold air at 10° C and at 60% RH is required for an industrial processes at the rate of 100 cu-m/ hr. When the outside conditions are 41°C DBT and 25°C WBT, the required conditions are achieved first by cooling and dehumidifying and then by heating. Determine

(a) Heat removed from the air per minute,

(b) Relative humidity and wet bulb temperature.

IMPORTANT POINTS TO BE REMEMBERED

1. *Specific humidity or Humidity ratio* (W_H)

$$W_H = \frac{0.622P_V}{P - P_V}$$

2. *Degree of saturation* (μ)

$$\mu = \frac{W_V}{W_S}$$

3. *Relative humidity* ϕ

$$\phi = \frac{P_V}{P_S}$$

4. *Enthalpy of air* (h_a)

$$h_a = 1.007t - 0.026$$
$$= 1.005\ t_d$$

5. *Enthalpy of vapour* h_v

$$h_v = 2401 + 1.84\ t_d$$

6. *Enthalpy of moist air* (h)

$$h = h_a + h_v$$
$$= 1.005\ t_a + W_H\,(2501 + 1.84\ t_d)$$

7. *Approximate properties*

$$\text{Volume} = V = V_a + \mu\,(V_{as} - V_a)$$
$$\text{Enthalpy} = h = h_a + \mu\,(h_{as} - h_a)$$
$$\text{Entropy} = S = S_a + \mu\,(S_{as} - S_a)$$

8. *Vapour pressure* (p_v)

$$p_v = (P_s)_{tw} - \frac{(P - P_s)_{tw}(t_d - t_w)1.8}{2800 - 1.3(1.8t_d + 32)}$$

This is carrier equation

Also

$$P_v = (P_s)_w - \frac{[P - (P_s)_w][t_d - t_w]}{1527.4 - 1.3t_w}$$

$$= (P_s)_{tw} - \frac{[P - (P_s)_w][t_d - t_w]}{1544 - 1.44t_w}$$

9. *By pass factor BF* $= \dfrac{\text{Bypassed air}}{\text{Total air}}$

10. *Sensible heat factor :* SHF

$$\text{SHF} = \frac{\text{Sensible heat}}{\text{Total heat}}$$

CHAPTER 38

Conventional and Non-conventional Energy Sources

38.1 INTRODUCTION

Since ages man has been trying to harness energy to get more power which can be used to do work. Energy appears in many forms but it has ability to produce dynamic and vital effect. Energy is always associated with physical substances but it is not a substance itself. Energy manifests itself by excited state of the matter which gets capacity to do work.

38.2 SOURCES OF ENERGY

All energy eventually has common origin from the sun. Energy, especially electric energy is important for day-to-day living and the industry. Energy from various sources—like fossil fuels, coal, water etc. is converted into electrical energy and then distributed wherever needed.

Thermal

1. About 65% of power production is through thermal.
2. Production cost is less.
3. Stock of coal is limited and quality of coal is also poor.
4. Plants also have less production capacity.

Gas

1. Cost of gas is increasing in other countries, so costlier projects.
2. Countries supplying gas, have increased the cost of gas in recent days.
3. Supply of gas is also not continuous.

Hydro

1. Because of environmental and rehabilitation problems, projects are getting delayed.
2. Time required for construction and starting up is about 1.5 times than the schedule.

Nuclear

1. Less than 3% of total power share
2. Continuous supply of Uranium is necessary.
3. Thorium based energy plants are cheaper, but because of technical problems it requires more time.

38.2.1 Power Generation Scenario in the Following Countries

1. India

(a) Out of total energy requirement 3.5% is through Nuclear, upto 2050 MW it has to be increased to 2.5%.
(b) In India less capacity about 220 MW plants are about 17.
(c) Uranium deficity is the main cause.
(d) To develop technology it requires time.
(e) Tremendous stocks of Thorium, but technological problems, otherwise use can produce energy for a very long period of time.

2. America

(a) About 21% of total power requirements is provided through Nuclear energy source.
(b) In 32 states nearly 100 Nuclear power plants.

3. France

(a) About 77% of total power requirement is produced through Nuclear energy.
(b) For power generation, there are 19 plants in the country.

4. China

(a) More share is through coal and hydro.
(b) 10 Nuclear power plants are active.
(c) On 10 work is under progress and 31 other are sanctioned.

Challenges

(a) About half of our population resides in dark.
(b) Supplying electricity to all parts of country, it requires more production of electricity.
(c) Coal is depleting continuously stocks are available for about 100 more years.
(d) Energy generation technology is to be developed, otherwise it will effect economy of our country.

38.3 IMPORTANCE OF ELECTRIC ENERGY

Per capita consumption of electricity is an index of the progress and industrialisation of a country. It is convenient to convert energy from various sources into electric energy for ease of transmission. For example, the potential energy of stored water in a hydroelectric power plant is used to run water turbines, which run electric generators that produce electricity. The electricity so generated is distributed through transmission lines up to our homes and to the factories. The electric power is reconverted into mechanical energy to run fans, compressors, and machines or for cooling/heating or lighting.

In our country demand for electricity is growing rapidly—roughly at about 8 – 10% per year. Present per capita consumption is about 500 units (kWh) as compared to approximately 36 times of this in Canada. Power generation by thermal power plants is about (83%), by hydro-power stations (14%) and nuclear plants (3%).

38.4 CLASSIFICATION OF SOURCES OF ENERGY

Various sources of energy can be classified into as mentioned below:

(a) Conventional/Non-renewable. These sources of energy get consumed and are not renewable.

(b) Non-conventional/Renewable/Perennial. These sources are ever lasting and get renewed.

A broad classification of the sources of energy on various basis is given in the Table 38.1.

38.5 ADVANTAGES/DISADVANTAGES OF NON-CONVENTIONAL AND CONVENTIONAL ENERGY SOURCES

Advantages

1. Such sources are everlasting, renewable, are available in abundance in nature and virtually come free like solar energy, wind energy or energy from water.
2. They are nonpolluting.

Disadvantages

1. The energy is not concentrated in such sources. Energy conversion into electric energy is relatively costly.
2. The energy like solar energy or energy from wind is weather and time dependent.

(I) Energy - Basis - Conventional / Non-conventional

- Conventional
  - (1) Fossil fuels (Coal, Coke)
    (2) Vegetation (Wood, alcohol)
  - Animal waste (Gobar)
  - Water or hydraulic energy
  - Nuclear energy (Fission of nuclear fuels like U^{235}, Pu^{239})
- Non-conventional
  - Solar energy
  - Wind energy
  - Ocean wave energy
  - Ocean thermal energy
  - Geothermal or Terrestrial energy
  - Tidal energy

(II) Energy - Basis - Renewable / Non-renewable

- Renewable or Regenerative
  1. Vegetation
  2. Hydraulic energy
  3. Solar energy
  4. Wind energy
  5. Tidal energy
  6. Animal waste
  7. Ocean thermal energy
- Non-renewable or Non-regenerative
  1. Fossil fuels
  2. Nuclear fuels
  3. Geothermal energy

(III) Energy - Basis - Capital / Celestial energy

- Capital energy (Existing in or on earth)
  1. Fossil fuels
  2. Nuclear fuels
  3. Geothermal energy
  4. Vegetation
  5. Animal
- Incoming or celestial energy (Reaching earth from outer space)
  1. Solar energy
  2. Wind energy
  3. Tidal energy

ENERGY CONVERSION – NON-CONVENTIONAL/ RENEWABLE SOURCES

38.6 SOLAR ENERGY

In our country sunshine is available throughout the year and can be used for heating. The solar energy however has low energy density and therefore cannot be used directly for power generation. Solar energy received on earth rarely exceeds 1 kW/m^2 and the total radiation over a day at best is about 7 kWh/m^2. These are low values for conversion. The solar energy can be concentrated by using parabolic mirror collectors. The solar energy can be collected by flat plate or parabolic mirrors depending on the temperature requirements.

38.6.1 Flat Plate Heaters

(a) These are used for water or space heating, and drying etc.

(b) **Parabolic mirrors** used in Vapour engines, process heating, cooking.

(c) **Parabolic mirror arrays (For high temperature > 200° C).** These are used for steam generation.

38.6.2 Flat Plate Solar Heaters

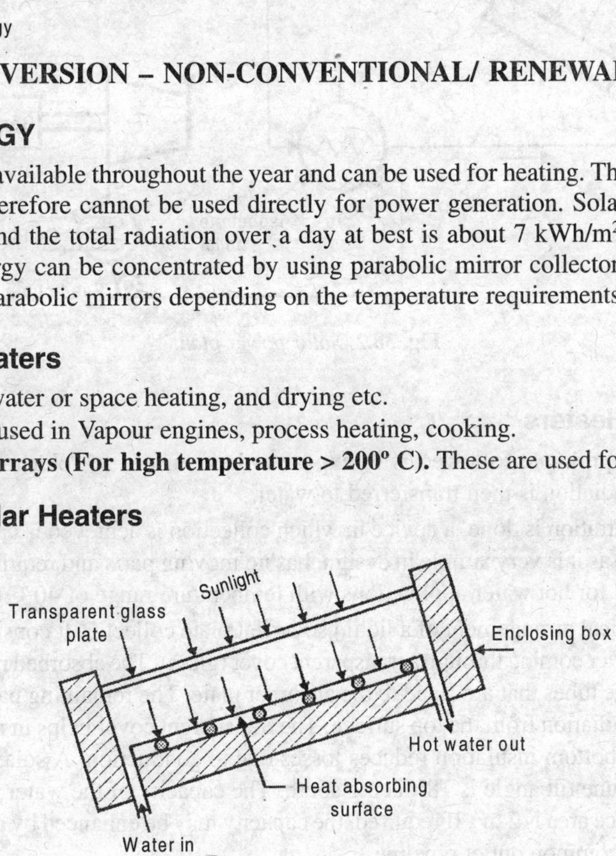

Fig. 38.1. *Flat plate collector*

Flat plate water heater can give hot water up to 70–80° C. Schematic diagram is given below. Water is heated usually by passing through tubes which are exposed to the sun's heat rays.

38.6.3 Application

(a) Hot water can be used to vaporize Isobutene in a boiler. The vapours run a turbine that is coupled to an electric generator to produce electricity. Schematic diagram of a solar power plant using Isobutene or Freon as working substance is shown in Fig. 38.2.

(b) Parabolic concentrating type mirror collectors are used to intensify the solar energy that can be used to heat water to produce steam at low pressure of 2–2.5 bar or to heat air. Steam and hot air are used to run turbines that are coupled to electric generators.

38.6.4 Solar Power Plant

From flat plate collector we get hot water. The temperature of water is 70–80°C maximum. This hot water is then passed through Isobutane boiler. Water gives its heat to Butane and Butane vapour is generated. This vapour will be then made to drive the butane turbine for generating the power (shown in Fig. 38.2).

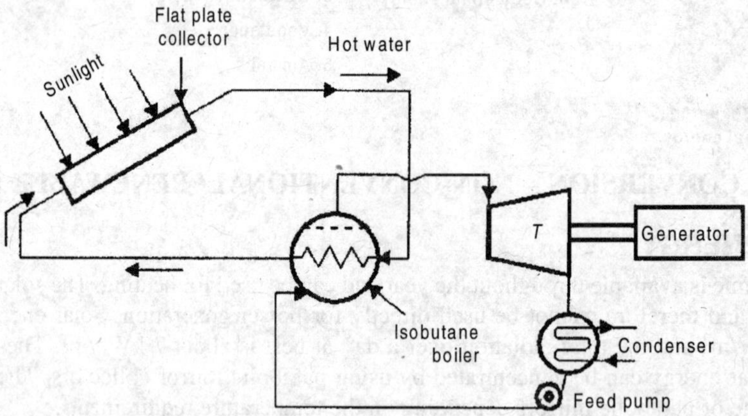

Fig. 38.2. *Solar power plant*

38.6.5 Solar Water Heaters

In Solar water heater, the principle followed is to expose a dark surface to solar radiation so that radiation is absorbed. This absorbed radiation is then transferred to water.

When no optical concentration is done, a device in which collection is achieved is called a flat plate collector. It is the most common device as it is very simple in design, has no moving parts and requires negligible maintenance. It is nowadays widely used for hot water applications with temperature range of 40°C to 80°C.

Figure 38.3 shows a typical arrangement of a liquid solar flat plate collector. It consists of an absorber plate on which solar radiations fall after coming through a transparent cover (glass). The absorbed radiation is partly transferred to water flowing through the tubes that are fixed to the absorber plate. The remaining part of absorbed radiation are lost by convection and re-radiation from the top surface. The transparent cover helps in reducing convection and re-radiation losses while, the bottom insulation reduces losses due to conduction. A solar (water heater) collector is usually held tilted and at Pune tilt angle is 18° facing south. The capacity of the water drum is usually 100 to 200 litres and the collector surface area is 2 m^2. If required, the capacity may be enhanced by arranging several individual collectors in row having a common outlet pipeline.

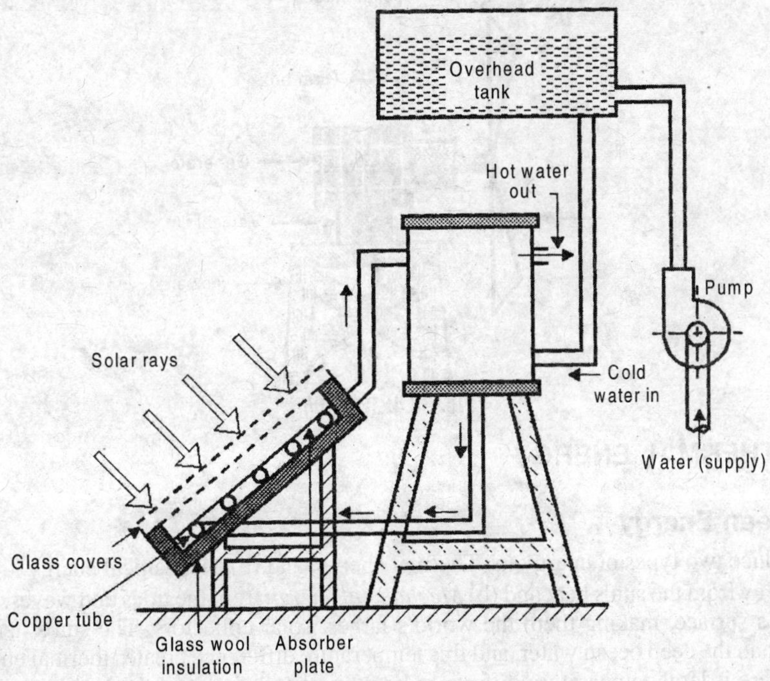

Fig. 38.3. *Solar water heater*

Apart from solar water heater, other applications of solar power are:

1. Solar dryers
2. Solar cookers
3. Solar photovoltaic converters etc.

Advantages of Solar Energy

1. This energy is freely available and in abundance
2. No environmental pollution is caused by solar energy
3. Equipments require negligible maintenance
4. This can be stored in dry batteries (as electricity) or in oil pipes (as heat energy).

Disadvantages

1. Initial investment cost in solar equipment is high.
2. Preservation of solar energy in the form of either heat or electricity is difficult.

Due to the above mentioned advantages, solar operated equipments are nowadays very popular in our country and the future is bright.

38.7 WIND ENERGY

(a) Wind speeds are reasonably high throughout the year. It can be used to produce electricity from wind mills. India has commissioned wind energy projects to produce 1340 MW and is presently ranked fifth in harnessing this source of energy after USA, Germany and Denmark. Maharasthra state leads in wind power development.

(b) The windmill consists of a propeller driven generator installed on a tower. The propeller is connected to the generator through gears. The propeller is kept in the direction of the wind and it drives a generator which produces electricity. Block diagram of the wind mill is given below.

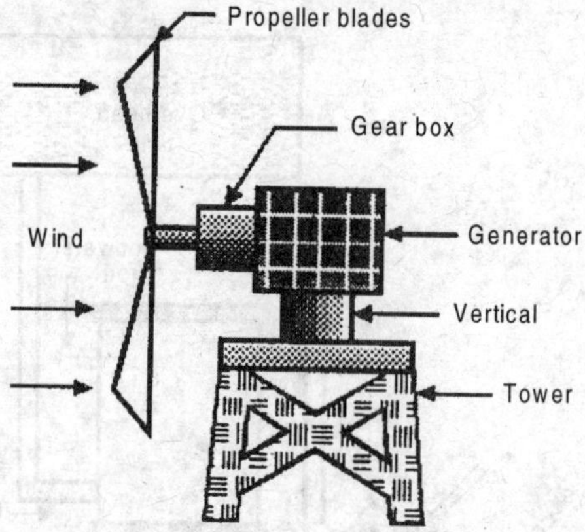

Fig. 38.4. *Wind mill*

38.8 OCEAN THERMAL ENERGY

38.8.1 The Ocean Energy

The ocean can produce two types of energy (a) Thermal energy, and (b) Mechanical energy.

(a) *Thermal energy* from the sun's heat and (b) *Mechanical energy* from the tides and waves. Oceans cover more than 70% of Earth's surface, making them the world's largest solar collectors. The sun's heat warms the water surface a lot more than the deep ocean water, and this temperature difference creates thermal energy. Due to this, in tropical climates there is large temperature difference between the surface and the lower levels. The temperature difference may be up to 20ºC within a 200–300 m depth.

38.8.2 Ocean Thermal Energy Conversion (OTEC)

Ocean thermal energy is used for many applications, including electricity generation. There are three types of electricity conversion systems: closed-cycle, open-cycle, and hybrid.

(a) Closed-cycle systems use the ocean's warm surface water to vaporize a *working fluid* like ammonia, propane or R-22 which have low boiling points. The vapour expands and runs turbine coupled to a generator to produce electricity. Vapour after expansion is condensed using cold water from depth by pumps. However a lot of energy is used to run the pump. (Ref. Fig. 38.5)

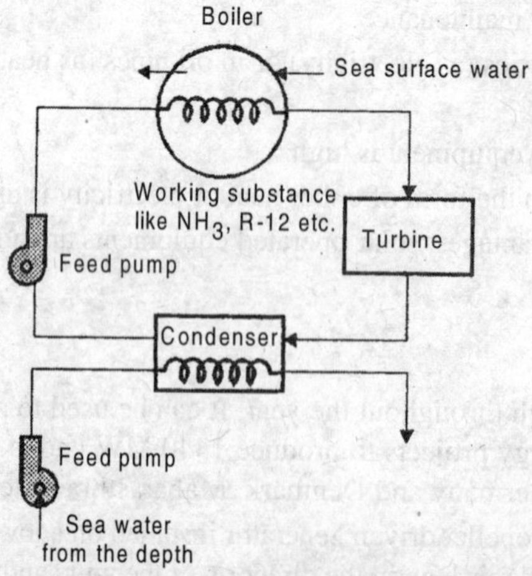

Fig. 38.5. *Closed sytem ocean thermal energy plant*

(b) In open-cycle systems sea water is boiled to generate steam at low pressures. The steam in turn operates a turbine/generator. Hybrid systems combine both closed-cycle and open-cycle systems.

38.9 GEOTHERMAL ENERGY

The earth, having been originally thrown off from the sun according to some theories, still has a molten core. Evidence for this lies in the volcanic action that takes place in many regions of the earth's surface. At some places steam, geysers and hot water exudes from the earth.

The geothermal steam has been used for many years in Italy for power generation and space heating. Recently projects are in operation in Mexico, Newzealand and California. Figure 38.6 shows the schematic layout of plant. Steam is tapped from drilled wells and passes through a drum to separate out moisture. From here the dry steam enters turbines which are connected to electric generators.

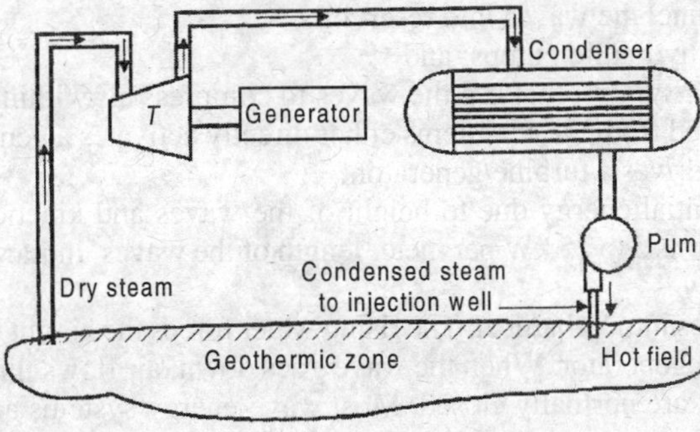

Fig. 38.6. *Geothermal energy plant*

38.10 TIDAL AND WAVE ENERGY

There are two ways to tap the ocean for the mechanical work (a) using tidal energy or (b) making use of the energy in the waves. Ocean mechanical energy is quite different from ocean thermal energy. Even though the sun affects all ocean activity, tides are driven primarily by the gravitational pull of the moon, and waves are driven primarily by the winds. As a result, tides and waves are intermittent sources of energy, while ocean thermal energy is fairly constant. Also, unlike thermal energy, the electricity conversion of both tidal and wave energy usually involves mechanical devices.

(a) Tidal Energy

(i) Tides are generated through combination of forces exerted by sun, moon and augmented by rotation of the earth. Local effects such as funneling, shelving and reflection increase the tides. An increase of about 5 m at low tide and high tide is needed. There are only a few places where this: tide change occurs around the earth Gulf of Cambay and Kutch in our country have lot of potential of the tidal energy.

(ii) Sea water can be trapped in reservoirs behind dams on the sea shores at the time of high tides. When the tide drops, the water behind the dam is let out just like in a regular hydroelectric power plant. For tidal energy to work well we need large increases in tides.

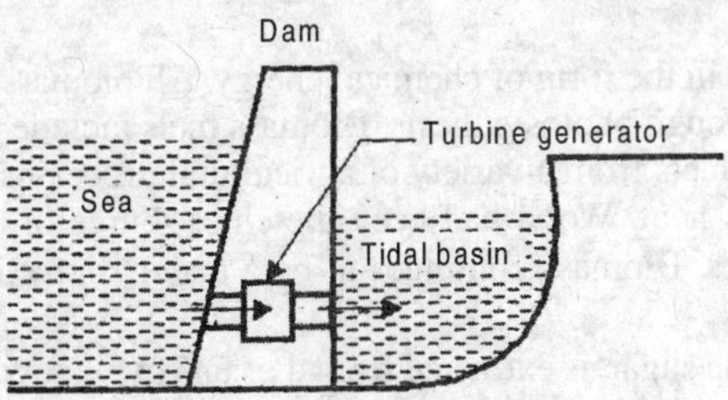

Fig. 38.7. *Filling the tidal basin during high side*

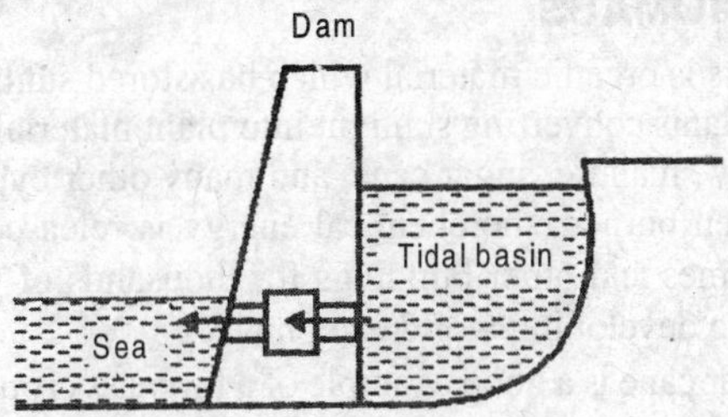

Fig. 38.8. *Power generation during low tide*

As shown in Fig. 38.7 during high tide conditions water from sea flows to the tidal basin via turbine. When the water flows through the turbine, it drives the turbine, which in turn drives the generator for generating power.

During low tide conditions as shwon in Fig. 38.8 water from tidal basin flows to the sea. As water flows through turbine, it will rotate the turbine, which in turn drives the generator to generator power.

Advantages	*Disadvantages*
1. It is pollution free	1. It can work where tides are quite high
2. Its cost of running is very low	2. It is liable to corrosion
3. It does not depend on rainfall	3. Its operation is intermittent

(b) Wave Energy

(i) For wave energy conversion, there are three basic systems:

- *Channel systems* that funnel the waves into reservoirs
- *Float systems* that drive hydraulic pumps; and
- *Oscillating water column systems* that use the waves to compress air within a container.

The mechanical power created from these systems either directly activates a generator or transfers to a working fluid, water, or air, which then drives a turbine/generator.

(ii) Ocean waves have potential energy due to height of the waves and kinetic energies due to wave length. Electricity generation possibility is 5 to 15 kW per meter length of the waves. India with coastline of some 6000 km can exploit this source of energy.

(iii) In Fig. 38.9 the wave rises into a chamber. The rising water forces the air out of the chamber. The moving air spins a turbine which can turn a generator. When the wave goes down, air flows through the turbine and back into the chamber through doors that are normally closed. Most wave-energy systems are very small. But, they can be used to power a warning buoy in seas or a small light house.

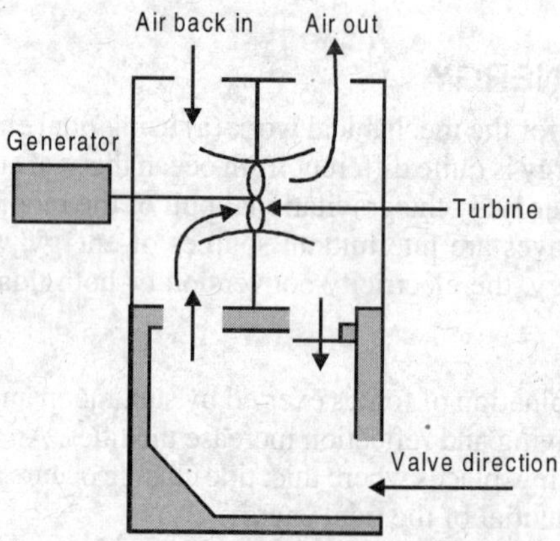

Fig. 38.9. *Wave energy conversion*

38.11 BIOMASS

(a) Biomass is organic material which has stored sunlight in the form of chemical energy. All biomass is produced by green plants converting sunlight into plant material through photosynthesis. Biomass fuels include wood, wood waste, straw, manure, sugar cane, and many other byproducts from a variety of agricultural processes.

(b) When burned, the chemical energy is released as heat. Wood had been the chief source of cooking and heating homes and other buildings for thousands of years. Biomass continues to be a major source of energy in much of the developing world even now.

(c) Sugar cane is a good example of a biomass crop. The sugar, is extracted from the cane by removing the juice; the remainder of the plant called bagasse produces heat when burnt.

(d) **The Biomass Resource.** Main sources of biomass energy are: municipal and industrial wastes, agricultural crop residues and energy plantations. Industrial biomass wastes are primarily produced by the forest products, industry, and animal waste. Many different types of biomass can be grown for energy production. Crops that have been used for energy include: sugarcane, corn, husk, grains, grass, seaweed and many others.

(e) **Conversion Technology**

(i) The simplest and most common method of obtaining energy from biomass is direct combustion. The heat of combustion is used to provide space or process heat, water heating or for producing electricity through the use of a steam turbine. In the developing world, many types of biomass such as dung and agricultural wastes are burned for cooking and heating.

(ii) Biochemical conversion of biomass is completed through alcoholic fermentation to produce liquid fuels and anaerobic digestion or fermentation, resulting in biogas. Fermentation of crops such as sugarcane produces ethanol for use in internal combustion engines has been common. Ethanol has been blended with gasoline for use in automobiles. Similarly in our country now blended biodiesel is being used by the railways.

(iii) Anaerobic digestion of biomass has been practised for almost a century, and is very popular in many developing countries such as China and India. The organic fraction of almost any form of biomass, including sewage sludge, animal wastes and industrial effluents, can be broken down through anaerobic digestion into methane and carbon dioxide. This "biogas" is a reasonably clean burning fuel which can be used for cooking, heating or electrical generation.

(iv) Wood and many other similar types of biomass which contain lignin and cellulose-agricultural wastes, cotton waste, wood wastes, peanut hulls etc. can be converted through thermochemical processes into solid, liquid or gaseous fuels. During the process biomass is heated in the absence of air and breaks down into a complex mixture of liquids, gases, and a residual char. If wood is used as the feedstock, the residual char is what is commonly known as charcoal.

38.12 BIO-GAS PLANT

In India a very large amount of cow dung is produced every year. Out of this only 50% of this is used for cooking purpose.

In the bio-gas plant, bio-gas is produced as a result of fermentation and decomposition of the dung.

Other than cow dung the material used in the production of bio-gas are corn husk, green leaves, dried leaves, vegetable waste and even human excreta.

Construction and Working. Figure 38.10 shows schematic diagram of bio-gas plant.

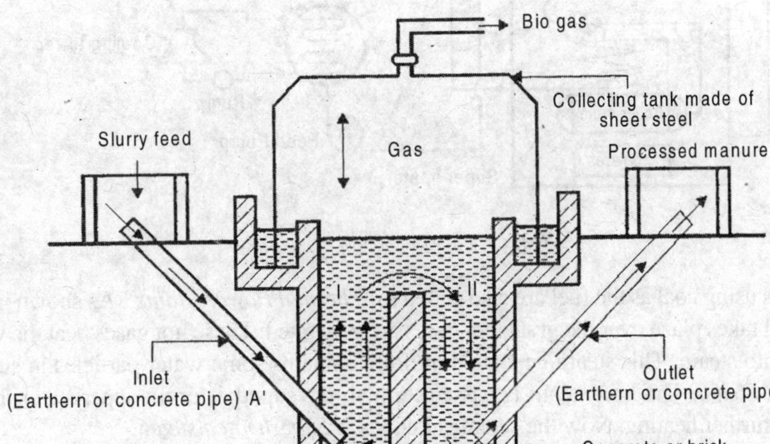

Fig. 38.10. *Bio-gas plant*

It mainly consists of a very large concrete or brick work chamber 'C' which is divided into two separate compartments by a partition wall.

On the top of the chamber 'C', a gas collecting tank 'D' is provided. Also it is arranged with inlet pipe 'A' and outlet pipe 'B' as shown.

Dung is mixed with sufficient amount of water so as to get dung slurry and this is then fed to the bottom of the chamber 'C' through the inlet pipe 'A'.

The fermentation process begins in the I–compartment. And as the atmospheric temperature increases, the decomposition of organic materials increases and the slurry rises up and over flows to the II–compartment, where further decomposition continues.

The gases produced during decomposition, bubble up through the slurry and get collected in the tank 'D'. The gas collected can be taken out through the valve.

From the II–compartment thick slurry will be thrown out through the outlet pipe 'B'. The calorific value of bio-gas is about 25 kJ/litre.

38.13 SIMPLE STEAM POWER PLANT (OR THERMAL POWER PLANT)

Figure 38.11 shows a simple steam power plant, it mainly consists of a boiler, steam turbine, condenser, feed pump etc.

1. Boiler. Boiler is a closed vessel in which water is heated and converted into steam, i.e., it is a steam generator. Boilers will be either oil fired or coal fired. In case of oil fired boilers, oil burners are provided, whereas in case of coal or wood fired boilers grate is provided. (Grate is the platform on which combustion of solid fuel takes place).

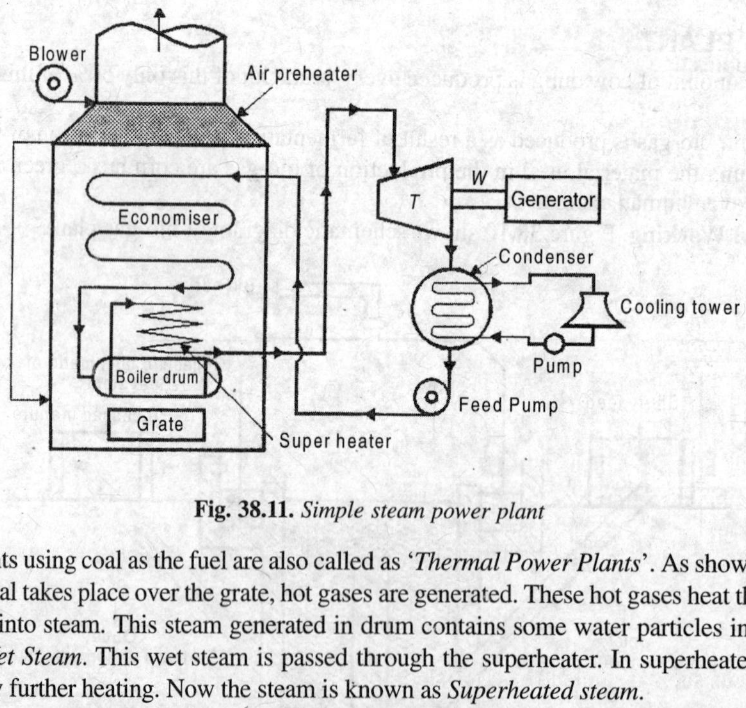

Fig. 38.11. *Simple steam power plant*

Steam power plants using coal as the fuel are also called as *'Thermal Power Plants'*. As shown in Fig. 38.11 when the combustion of coal takes place over the grate, hot gases are generated. These hot gases heat the water in the boiler drum and convert it into steam. This steam generated in drum contains some water particles in suspension, so thus steam is called as *Wet Steam.* This wet steam is passed through the superheater. In superheater the temperature of steam is increased by further heating. Now the steam is known as *Superheated steam.*

2. Turbine. This superheated steam is then allowed to expand into the steam turbine. Turbine consists of an impeller or a cylindrical hub on which blades are fixed. This turbine is mounted on the shaft. When the steam expands on the blades it rotates the turbine impeller, which in turn drives the generator to generate electrical power.

3. Condenser. The steam after generating power is exhausted into the condenser and gets condensed into water (this water is also called as condensate). In the condenser, it comes in contact with the cold surface of water tubes.

4. Pump. The condensate is then pumped by means of the feed pump into the boiler drum through the economiser.

When the water passes through the economiser, it recovers some of the heat of the flue gases and becomes hot. Then this hot water will enter the boiler drum. In the boiler drum thus hot water is again heated and converted into steam and the cycle repeats.

Thus by using economiser some of the heat of the flue gases will be recovered, which otherwise would have been lost to the atmosphere. Thus by using economiser, the efficiency of the plant will increase.

Advantages of Steam Power Plant

1. Fuel used is cheaper.
2. Less space is required in comparison with that for hydro-electric plants.
3. As these plants can be set up near the industry, transmission costs are reduced.
4. Cheaper in production cost and initial cost compared with diesel power stations.
5. Can be located very conveniently near the load centre.
6. They can respond to rapidly changing loads with difficulty.

Disadvantages of Steam Power Plant

1. The cost of plant increases with increase in temperature and pressure.
2. Maintenance and operating costs are high.
3. Long time required for errection and putting into action.
4. A large quantity of water is required.
5. Greater difficulty experienced in coal handling.
6. Presence of troubles due to smoke and heat in the plant.

38.14 NUCLEAR POWER PLANT

Nuclear reactions produce energy either by fusion or by fission reaction. Fusion means union of lighter nuclei, such as two nuclei of Hydrogen, after fusion produce one nucleus of Helium and one neutron and energy. Atomic Fission is exactly opposite of fusion.

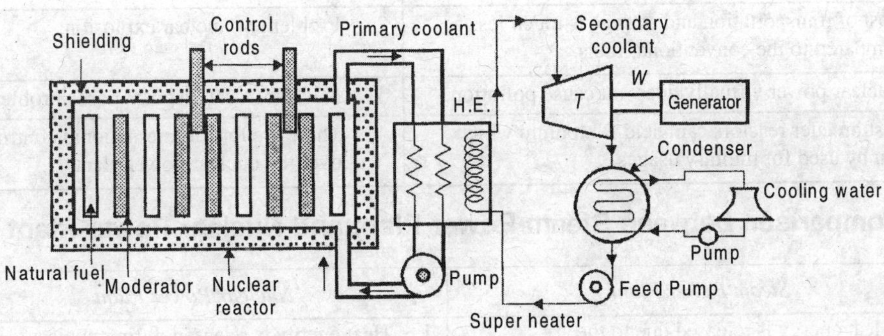

Fig. 38.12. *Schematic of nuclear plant*

Here a heavy nucleus such as nucleus of U^{235} is bombarded by means of high energy neutrons, then it splits into number of fragments more or less of equal mass. The combined mass of fissioned products is slightly less than the parent nucleus. This disappearing mass is converted into energy according to $E = mC^2$ where C is velocity of light. This process is associated with the release of tremendous amount of heat energy.

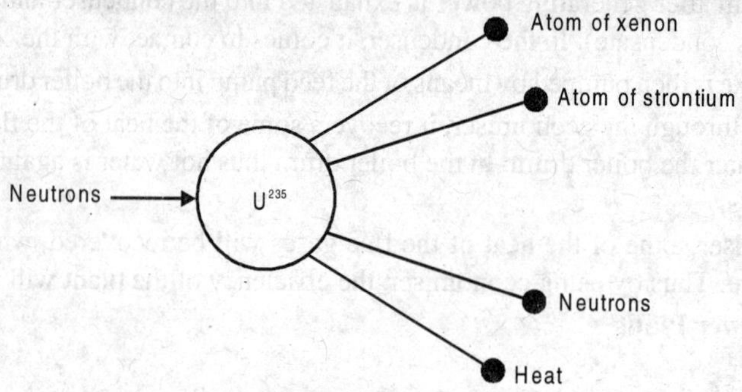

Fig. 38.13

The heat produced by the nuclear disintegration is taken up by the primary coolant (which may be Heavy water, Molten Sodium or an Organic liquid) and the primary coolant becomes hot. The heat of the primary coolant is transferred to the secondary coolant (i.e., water) in the heat exchanger. The water gets converted into steam. This steam then drives the turbine, which in turn drives the generator for generating power.

Note:

(i) **Moderators** are made of Graphite, Boron, Cadmium. These are used to slow down the fast neutrons to the right speed which is required for splitting up of U^{235}.

(ii) **Control rods** are made up of Boron or Cadmium. These are used to absorb excess neutrons released during fission process.

(iii) **Shielding** is provided to absorb harmful radiations like β- and γ-rays produced during fission reaction.

38.14.1 Advantages and Disadvantages of Nuclear Power Plants

Advantages	*Disadvantages*
1. Nuclear fuel has very high energy density i.e., very small quantity of nuclear fuel can produce very large amount of energy.	1. Problem of radiation hazard.
2. Cost of transportation and storage is much less as compared to the conventional fuels.	2. Problem of nuclear explosion.
3. Nuclear power virtually does not cause pollution.	3. Nuclear waste disposal poses problem.
4. Fast breeder reactors can yield Plutonium which can by used for military usages.	4. There are lots of international controls over the nuclear power plants.

38.14.2 Comparison between Steam Power Plant and Nuclear Power Plant

Steam Power Plant	*Nuclear Power Plant*
1. Heat energy is released due to the combustion of fuel.	1. Heat energy is released during nuclear reaction.
2. Concrete shielding is not required.	2. Shielding is required to absorb harmful radiations.
3. The number of workman required for the operation of steam power plant are more.	3. The number of workman required for the operation of nuclear power plant is much less than a steam power plant. This reduces the cost of operation.

38.15 HYDRO ELECTRIC POWER PLANT

Generation of power by hydro-electric stations is the utilisation of the part of hydrological cycle. Potential energy of rain falling on earth's surface, with respect to ocean is converted into mechanical energy by using water turbines. In order to generate energy by thus method economically, ample quantity of water at sufficient potential (head) must be available. Moreover, the past history of the place must be known in order to estimate the minimum and maximum quantity of water that can be made available for the purpose of power generation. As the availability of water depends on the natural phenomenon of rain, the maximum capacity of such plants is usually fixed on the basis of minimum quantity of water available. Usually storage reservoirs are constructed with such plant in order to store water during peak periods and to utilise the same during the off-peak periods. The civil engineering work involved in case of hydroelectric power station is considerable as it involves the construction of a dam and a large catchment area.

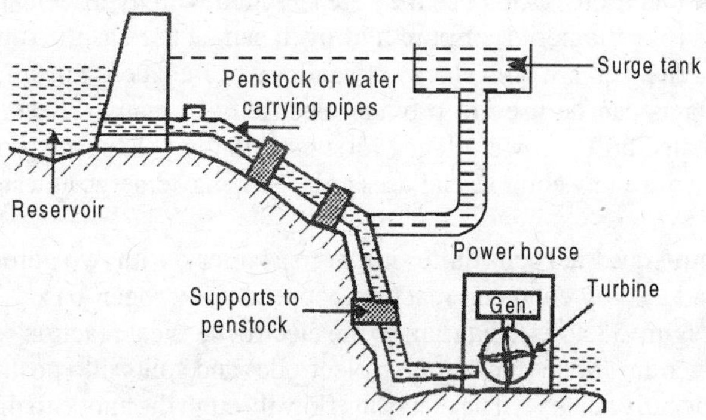

Fig. 38.14. *Hydro electeric power plant*

General arrangement of storage type Hydro-electric Power Plants is shown in Fig. 38.14 below and are self-explanatory.

(a) The water stored in dams have potential energy which is used to run water turbines which in turn are connected to coupled to elect generators which produce electricity.

(b) Water of river stored in the reservoir which is between narrow valley is closed by a high rise dam.

(c) **Penstock.** The water is brought to the turbines by large size pipes known as penstock.

(d) The penstocks are held in position by anchor blocks.

(e) **Surge Tank.** To avoid damage to the penstock due to water hammer i.e., sudden variation in pressure of water surge tanks are provided as shown.

(f) **Water turbine.** It is a device that converts kinetic energy in water into mechanical energy (rotation).

(g) **Generation.** It is electrical device that converts mechanical energy (rotation) into the electrical power.

Advantages of Hydraulic Power Plant

(a) Operating cost low.

(b) No fuel.

(c) Useful life 40–50 years as against 20–25 years for thermal plant.

(d) Fuel handling problem and ash problem absent.

(e) Cost of land is less (Remote place).

(f) Quick response.

(g) Effective change with time.

(h) No standby loss.

(i) Labour/operator less.

(j) Can be used for irrigation and flood control.

Disadvantages

(a) Capital cost much more.
(b) Depends on rain (natural).
(c) Site is remote.
(d) Long period for selection of plant.

38.16 FUEL CELL

(a) Fuel cells are electrochemical devices which directly convert hydrogen, or hydrogen-rich fuels into electricity without combustion. Thus process is much more efficient than traditional thermal power plants, converting up to 80% of the chemical energy in the fuel into electricity as compared to a maximum of 30–40% for conventional power plants. Although their structure is somewhat like that of a battery, fuel cells never need recharging or replacing and can consistently produce electricity as long as they are supplied with hydrogen and oxygen. Fossil fuels (coal, oil and natural gas), biomass (plant material) or pure hydrogen can be used as the source of fuel. If pure hydrogen is used, the emissions from a fuel cell are only electricity and water. Fuel cells are small and modular in nature and therefore fuel cell power plants can be used to provide electricity in many different applications, from electric vehicles to large, grid-connected utility power plants. First used in the U.S. space program in the 1950s, fuel cells are a developing technology with a few commercial uses today, but may emerge as a significant source of electricity in the near future.

(b) Fuel Cell Electrochemistry. Fuel cells quite similar to a battery, with two porous electrodes separated by an electrolyte. Electricity is produced by chemical reaction between a hydrogen-based fuel and an oxidant—usually oxygen inside the fuel cell. Figure 38.15 is a diagram of the electrochemical reaction which takes place inside a fuel cell. Hydrogen (H_2) flows over the anode (the negative electrode) and splits into positively charged hydrogen ions, and electrons which carry a negative charge. The electrons flow through the anode to the external circuit, performing useful work (this is the electric current generated) while the hydrogen ions pass through the anode and into the electrolyte, moving towards the cathode (the positive electrode). The electrons eventually return to the cathode which is supplied with oxygen (O_2). At this point the electrons, hydrogen ions and oxygen react to form water (H_2O) and heat. In stationary fuel cell power plants thus heat can be captured and used for process heat in industries or space heating (co-generation). As long as the fuel cell is supplied with hydrogen and oxygen, this electrical production can continue indefinitely.

(c) Electrolyte. The general design of most fuel cells is similar except for the electrolyte. Several different substances have been used as the electrolyte in fuel cells, each with their own advantages and disadvantages. Main types of fuel cells based on the electrolytes are:

(i) alkaline fuel cells
(ii) solid polymer fuel cells (also known as proton exchange membrane fuel cells)
(iii) phosphoric acid fuel cells
(iv) molten carbonate fuel cells and
(v) solid oxide fuel cells. Alkaline and solid polymer fuel cells operate at lower temperatures (50–260° C) and are mainly designed for use in transportation applications, while the other three operate at higher temperatures (up to 1000° C for solid oxide fuel cells and are being developed for use in cogeneration and large central power plants.

(d) Sources of Fuel and Oxidant. Fuel cells operate at maximum efficiency, when operating on pure hydrogen and pure oxygen. Pure oxygen is very expensive, and thus air is used as the source of oxygen in most applications except where the extra cost can be justified, as in the space program. Pure hydrogen is expensive, and difficult to transport and store, therefore, like pure oxygen it is only used in special cases. Gaseous mixtures of hydrogen (H_2) and carbon dioxide (CO_2) which can be created by the 'processing' of fossil fuels or biomass are used instead of hydrogen in most commercial uses of fuel cells. The most economical sources of the necessary H_2/CO_2 fuel mixture have been found to be gaseous hydrocarbons such as natural gas and propane. Light hydrocarbon liquids such as naphtha and methanol (from biomass), heavier hydrocarbon liquids such as fuel oil, and coal.

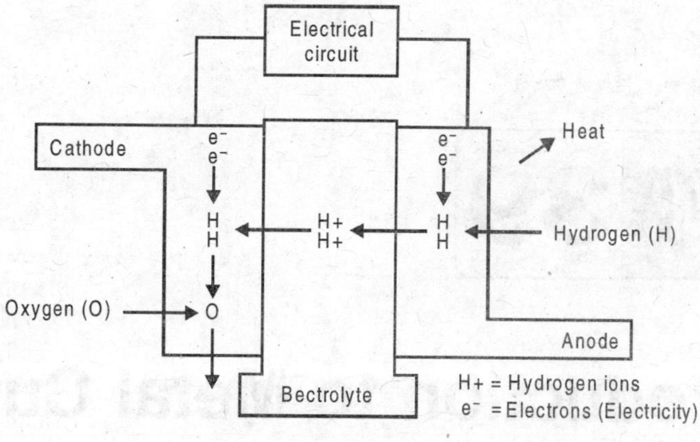

Fig. 38.15. *Fuel cell*

(e) Fuel Cell Power Plants. The fuel cell power section contains 'stacks' of one or more fuel cells. The individual fuel cells are small in size and produce between 0.5 and 0.9 volts of electricity. Therefore stacks of large number of fuel cells would be needed for power generation must have a large number of individual fuel cells. This eliminates the problems associated with large conventional power plants which can take up to ten years to build and have a fixed capacity. The fuel cells use air and hydrogen rich fuel and produce water, heat and direct current (DC) electricity.

The power conditioner section of a fuel cell power plant may also have an 'inventor' which converts the electricity to alternating current.

Uses of Fuel Cells. Fuel cells appear to be an important technology for the future as they.

THEORY QUESTIONS

38.1. Explain various engineering applications of solar energy.

38.2. What are the merits and demerits of Renewable Energy sources. Compare renewable and non-renewable energy sources.

38.3. What are the types of energy sources? State various forms of renewable and non-renewable energy sources.

38.4. Explain with the help of block diagram working of a hydroelectric power plant. What are the advantages of hydroelectric power plant?

38.5. Draw a neat sketch of steam power plant and explain the function of each component.

38.6. With the help of a neat labelled diagram explain the working of Ocean Thermal Power plant.

38.7. Draw the schematic diagram of steam power plant and explain functions of each component therein.

38.8. Draw a neat sketch of a steam power plant. State the functions of various elements of it.

38.9. State the function of condenser and cooling tower in a steam power plant.

38.10. Draw a neat sketch of steam power plant and state the functions of each component.

38.11. Write two important disadvantages of steam power plant.

38.12. Describe the working of steam power plant using Nuclear fuel.

38.13. With the help of neat sketches, explain the working of Nuclear power plant.

38.14. State the function of the following parts of a nuclear power plant.

38.15. (i) Nuclear reactor; (ii) Moderator; (iii) Control rod.

38.16. State the advantages and disadvantages of nuclear power plants.

CHAPTER 39

Introduction to Metal Cutting Processes

39.1 MACHINE TOOL

A machine tool is a power driven device or apparatus to perform certain metal removing operations and produce a desired form on the surface being machined. A cylindrical or flat surfaces are generally required for the product.

Machine tools may be classified as standard or special purpose machine tools. A standard machine tool is that machine which can be used for a variety of work. As against this, a special purpose machine tool is required to perform specific work and cannot be used for any other purpose. Its application is a limited one.

A centre lathe is a very good example of a standard machine tool with the help of which variety of work can be performed, which include

(i) turning between centers

(ii) turning, facing, drilling and boring

In this chapter we will study the following machines:

1. Centre lathe
2. Drilling machine
3. Grinding machine
4. Power saw
5. NC and CNC machines (Introduction only)

39.2 CENTRE LATHE

The lathe is the father of all machine tools. It is a universal machine tool which can be used—with certain attachment—for many processes. For its development to the form in which we now know it, we owe much to Henry Maudlsey, who developed the sliding carriage, and in 1800 built a screw cutting lathe on which he turned screws having from 6 to 40 threads per centimeter and which were the best screws that had been made up to that time.

About 1830 Maudlsey constructed a lathe, with 3 metre face plate, which was used to bore large cylinders and turn flywheels.

In its operation the lathe holds a piece of material between two rigid supports called centers or by some other device such as a chuck or a face plate. The spindle carrying the work is rotated whilst a cutting tool supported in a tool post is caused to travel in a certain direction, depending upon the form of the surface required. If the tool moves parallel to the axis of rotation of the work a cylindrical surface is produced Fig. 39.1(a) whilst if it moves perpendicular to this axis it produces a flat surface as at Fig. 39.1(b)

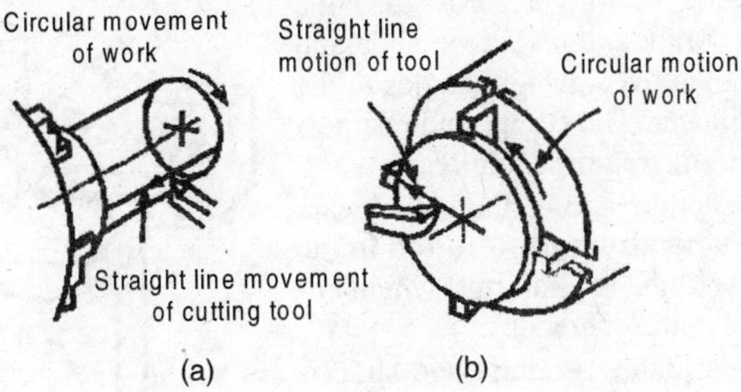

Fig. 39.1. *Surface formation of the lathe*

Figure 39.2 shows a simple centre-lathe or general purpose lathe showing its main components. The components are as follows.

(i) Bed. The bed of the lathe is made of a good quality gray iron casting or alloy and is supported on legs or a bench. It has accurately machined hardened and ground flat or inverted Vee-shaped inner and outer surfaces to guide the carriage head stock and tail-stock and to ensure accurate alignment.

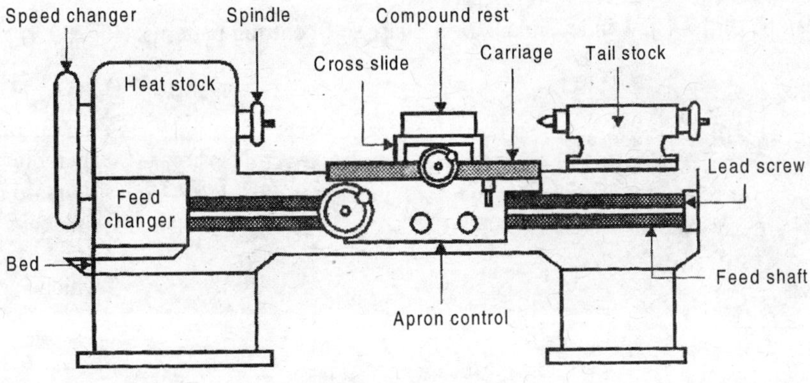

Fig. 39.2. *Centre lathe*

(ii) Head Stock. It is the powered end of a lathe which is securely fixed on the inner ways of the bed. It is always on the left side of the operator. It houses the speed changing gears and the driving spindle. Normally a built in electric motor supplies power to the spindle. The centre of the spindle is hollow so that long bars can be put through and be held by chucks for machining. It is made of cast iron.

(iii) Tail Stock. It is supported on the inner ways of the bed and is located on the right side of the operator. The main purpose of the tail-stock is to support the free end of the work piece when it is machined between centers. It is also used to hold tools when operations such as drilling, reaming, tapping etc. are performed on a lathe. (Ref. Fig. 39.3).

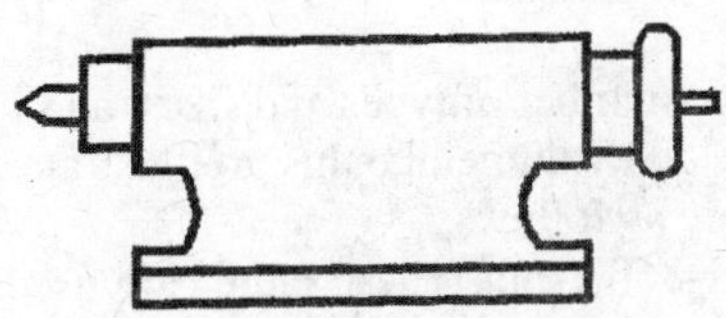

Fig. 39.3. *Tail stock*

(iv) Carriage. It consists of three main parts—a saddle, a cross slide and apron. It is used for moving the cutting tool along the lathe bed. (Fig. 39.4)

The saddle, an H-shaped casting, is bridged across the lathe bed and carries a cross-slide, a compound rest, a top slide and a tool post.The cross slide can be moved at right angles to the lathe bed either by hand or by power. The compound rest permits the swivelling of the tool to the required angle.

The apron is fastened to the saddle and contains the gears and clutches for transmitting motion from the feed rod to the carriage, and also contains the split nut (or half nut) which engages with the lead screw while cutting threads.

The compound slide. On most lathes a compound slide is interposed between the Tool post and the cross-slide. This slide may be swivelled to any angle and its use enables the tool to be moved in directions other than those permitted by the carriage and cross-slide. The compound slide is useful for turning and boring short tapers, chamfers and other jobs requiring an angular movement of the tool.

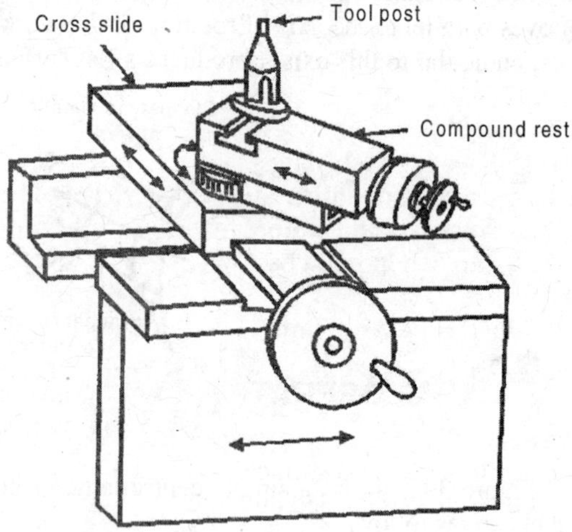

Fig. 39.4. *Lathe carriage*

(v) The tool post is located on the top of the compound-rest. The purpose of tool post is to hold the tool and to enable it to be adjusted to a convenient working position.

(vi) Chucks may be of the 4-jaw independent, or a 3-jaw self-centring patterns. (Fig. 39.5)

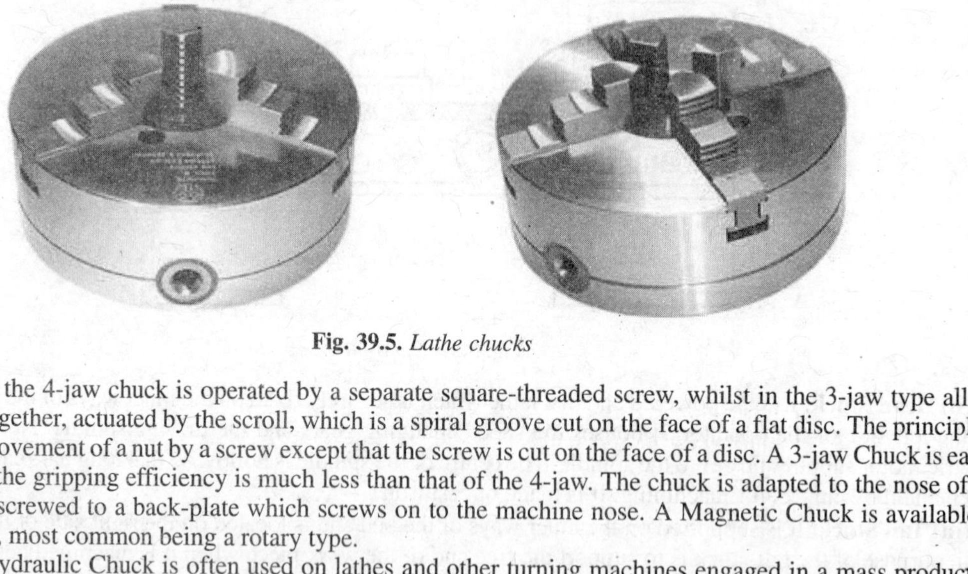

Fig. 39.5. *Lathe chucks*

Each jaw of the 4-jaw chuck is operated by a separate square-threaded screw, whilst in the 3-jaw type all the jaws close in together, actuated by the scroll, which is a spiral groove cut on the face of a flat disc. The principle is similar to the movement of a nut by a screw except that the screw is cut on the face of a disc. A 3-jaw Chuck is easier to operate, but the gripping efficiency is much less than that of the 4-jaw. The chuck is adapted to the nose of the lathe by being screwed to a back-plate which screws on to the machine nose. A Magnetic Chuck is available in several designs, most common being a rotary type.

An Air or Hydraulic Chuck is often used on lathes and other turning machines engaged in a mass production work. It is quick acting and grips the work strongly. The chuck can be worked by compressed air or by a fluid.

39.3 TYPES OF LATHES

Lathes are manufactured in such a large number of types and sizes, and for such variety of purposes that it is quite impossible to attempt within the scope of a volume like this, to illustrate completely the different classes of lathes. Some of the different types of lathes are as follows:

1. Speed lathe. Normal lathe 1200–3600 rpm. These lathes are used for polishing, metal spinning and wood turning Hand-operated tool.

2. A bench lathe. A small lathe which can be mounted on a work bench. All operation carried on centre lathe can be there with this lathe.

3. A precision lathe. It is a bench-type lathe capable of giving very accurate work.

4. A centre lathe. It is a general purpose lathe.

5. A tool room or tool maker's lathe. Similar in appearance to centre lathe, but is very accurate. Used for precision work on tools, jigs, dies etc.

6. A production lathe. Simple lathe for medium run production work. No lead screw, no compound rest.

7. A single spindle automatic lathe. It is used to produce components in large quantities.

8. A gap bed lathe. It has a bed with a removable section adjacent to the head-stock so that a large work can be swung.

9. Special purpose lathe. It has such characteristics that they are meant for only certain types of work.

39.4 LATHE OPERATIONS

The operations that can be performed on a lathe are known as turning operations. The word turning usually refers to machining external surfaces while machining internal surfaces is termed boring.

The operations carried out on a lathe can be grouped into the following categories:

(1) Plain turning between the centres
(2) Facing in chucks
(3) Taper turning
(4) Screw cutting
(5) Eccentric turning
(6) Drilling and boring
(7) Knurling
(8) Form turning

39.5 DRILLING MACHINES

Drilling is a manufacturing process by means of which cylindrical holes are produced in the work piece. The cutting tool used for this purpose is called a Drill.

Drills are of two types:

(1) Flat drill and (2) Twist drill.

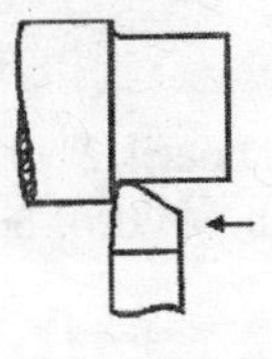

Fig. 39.6. *Plain or rough turning*

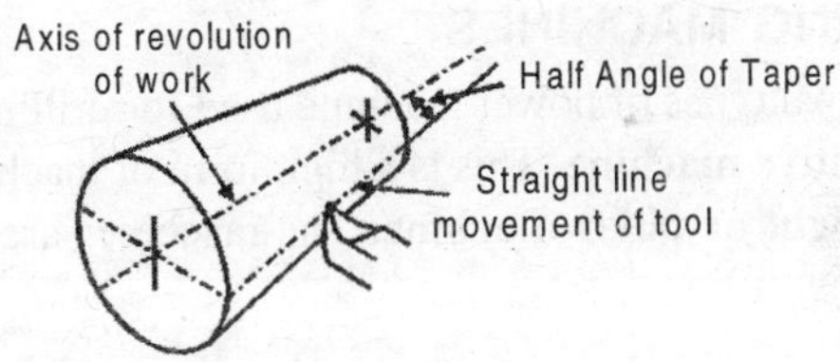

Fig. 39.7. *Taper turning*

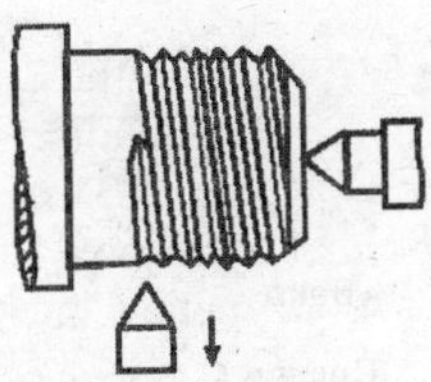

Fig. 39.8. *Thread cutting*

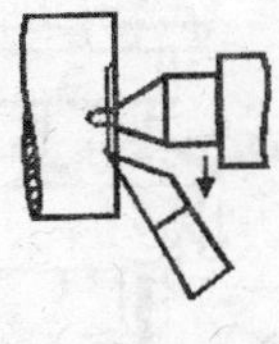

Fig. 39.9. *Facing*

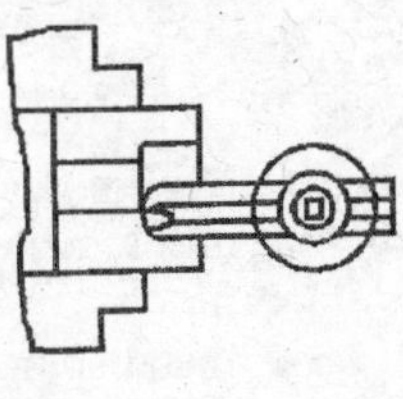

Fig. 39.10. *Drilling and boring*

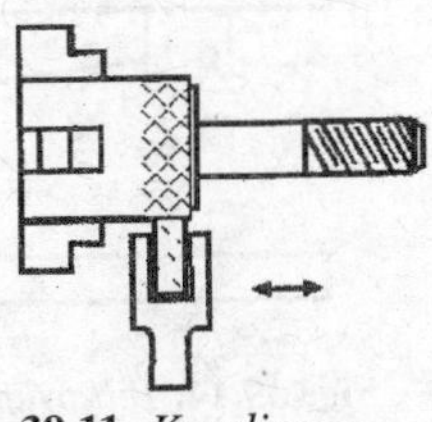

Fig. 39.11. *Knurling*

(1) Flat Drill. The flat drill, which is the forerunner of the twist drill is shown in Fig. 39.12 This drill is easily forged and ground to any required size, but the results it gives are not comparable with those obtained from a twist drill; this is mainly because the twist-drill point is backed up and kept true by the body of the drill following behind, while the point of the flat drill has no such influence to guide it.

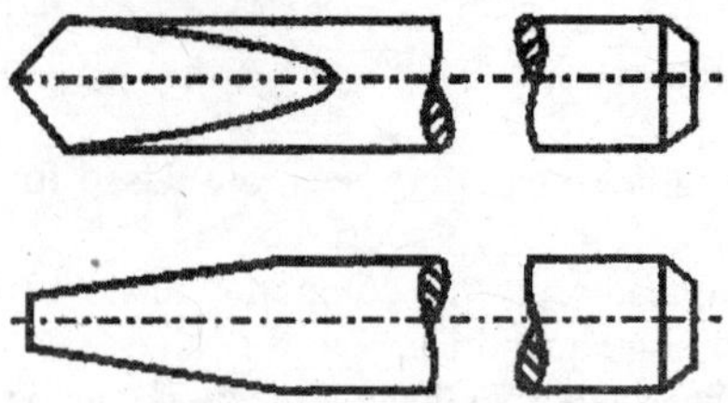

Fig. 39.12

The principal present day use of the flat drill is for boring very long holes, in which process a cutter in the form of a flat drill is carried at the end of a long bar. The arrangement gives the best results when the work rotates, the drill and bar being held stationary and fed into the hole (as in lathe). Generally the bar which carries the flat cutter has tube sunk into it for conveying cutting lubricant to flush away the swarf and avoid choking up of the drill.

(2) Twist Drill. The twist drill is the most commonly used variety of drill, and twist drills are specified according to the end by which they are held, called the Shank. Drills may be taper shank, parallel shank or jobber's drills and are shown in Fig. 39.13.

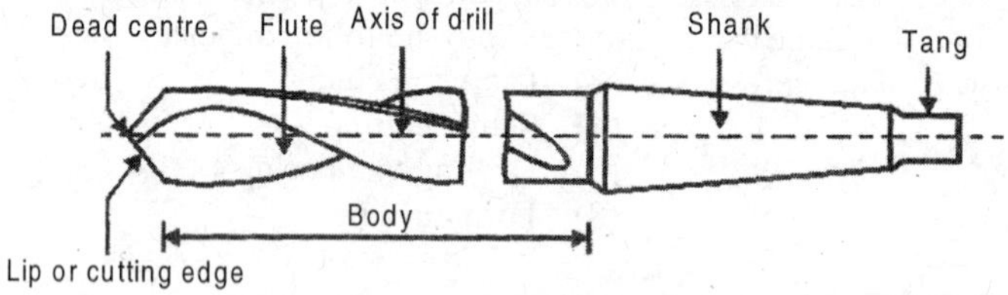

Fig. 39.13. *Twist drills*

The drills are held in the machine by special chucks. Drills may be made of either carbon steel or high speed steel.

39.6 DRILLING MACHINES

The three principal types of power machine used for drilling are briefed as below.

(a) The sensitive machine. This is a high form of machine used for holes upto about 1/2" (12 mm) diameter.

(b) The upright or pillar machine. The machines are in sizes upto that which can drill holes up to 2–3 in. diameter (50–75 mm).

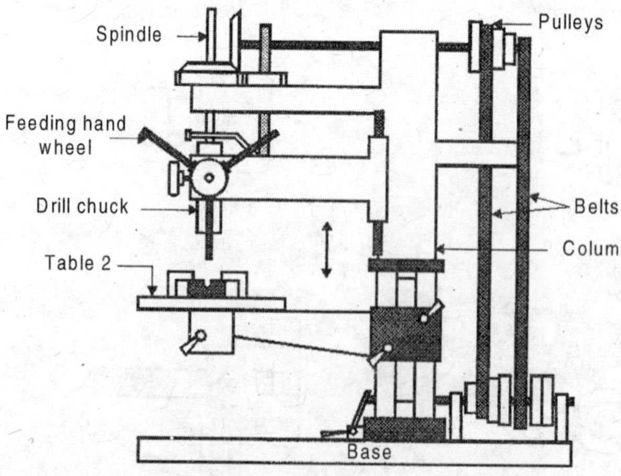

Fig. 39.14. *Pillar/upright drilling machine*

A diagram of a medium upright or pillar drilling machine shown in Fig. 39.14. The main components of this drilling machine are:

(i) A column fixed at the base which is resting on floor.

(ii) A table with job/work fixing arrangement and which can be moved up and down.

(iii) Drill chuck and feeding mechanism.

(iv) A set up belt and pulleys for changing the sped of drill.

(v) Base frame on which machine is mounted.

Working. Work piece is held on the table by means of fixtures. Table can be moved so that drilling can be possible by rotating the hand wheel connected for feeding the drill.

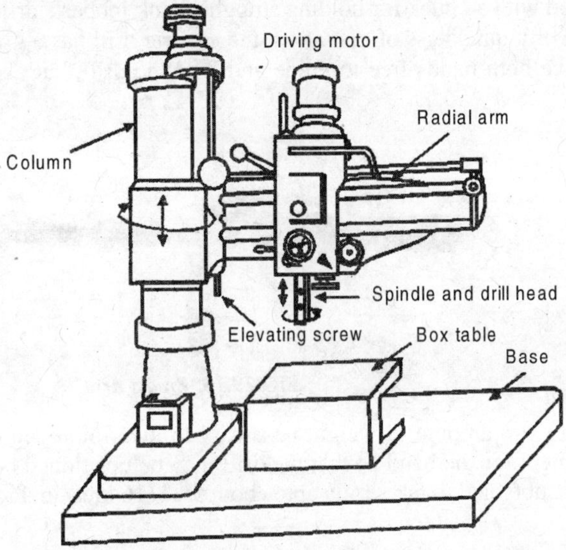

(a) *Schematic diagram of a radial drilling machine*

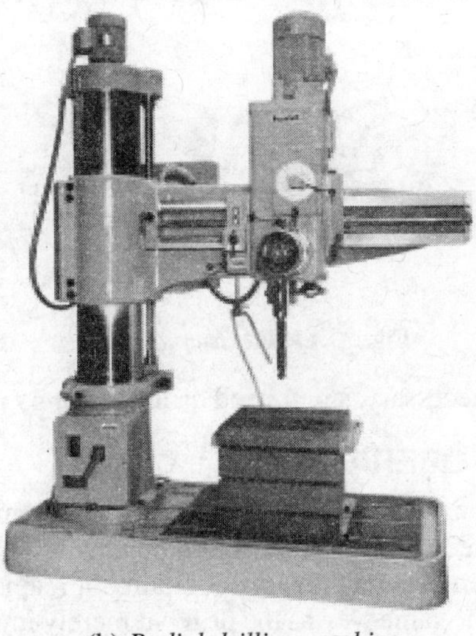

(b) *Radial drilling machine*

Fig. 39.15

The speed of the drill can be changed from 470–1440 rpm. This variation in speed can be achieved by providing an electrical pole changing motor.

(c) The radial drilling machine. In this machine the drill head is on an arm and may be swung or moved over the area of the machine table or base.

Radial drilling machine may be of the sensitive or heavy types.

(d) Hand-drilling machine. The fitter will have occasion to use various hand-drilling appliances for holes where it is not possible to bring the work up to a machine. The breast-drill 39.16 may be used for holes upto about 1/2 inch. (12 mm) diameter. Pressure is applied to the drill by pressing the body on the shaped end plate, whilst the drill is rotated by the handle.

The hand-drill for drills upto 1/4" (6 mm) is similar but smaller and has a handle instead of a breast plate. Both breast and hand drills are fitted with a chuck for holding straight shank jobber's drills.

(e) Electric hand drills. For a quickness of operation, the electric drill has a great advantage over the breast drill, for not only does it leave both hands free to guide and feed the drill, but also rotates the drill with more

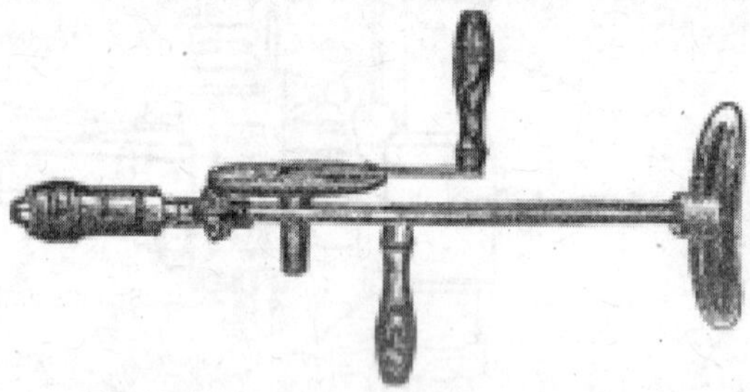

Fig. 39.16. *Breast drill*

uniformity and possesses that extra amount of weight necessary to give improved balance and manipulation. A switch is incorporated convenient for the hand so that the drill may be positioned before it is started-up. Electric drills of the breast type may be obtained to take drills upto about 5/8" (16 mm) in diameter, and an example of one is shown in Fig. 39.17.

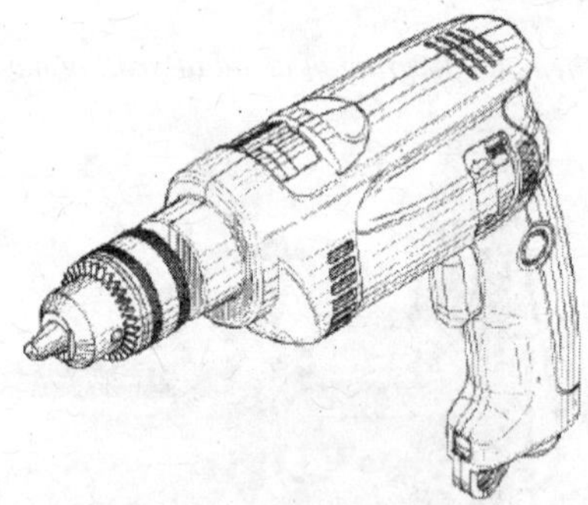

Fig. 39.17. *Electric hand drill*

Where extremely high speeds are necessary, small hand drills driven by a compressed air motor may be used.

39.7 DIFFERENT DRILLING OPERATIONS

Other than drilling, following operations can be performed on the drilling machine:

1. Reaming. A drill cannot be relied upon to produce a hole having sufficiently good qualities of finish and accuracy for many purposes, and when accurate holes are required a reamer is a must for finishing to size. The reamer does not originate the hole in the same way as the drill but merely imparts to the previously drilled hole the necessary smoothness, parallelism, roundness and accuracy in size. Reamers may be made of cast steel, case-hardened mild steel. Reamers are shown in Fig. 39.18.

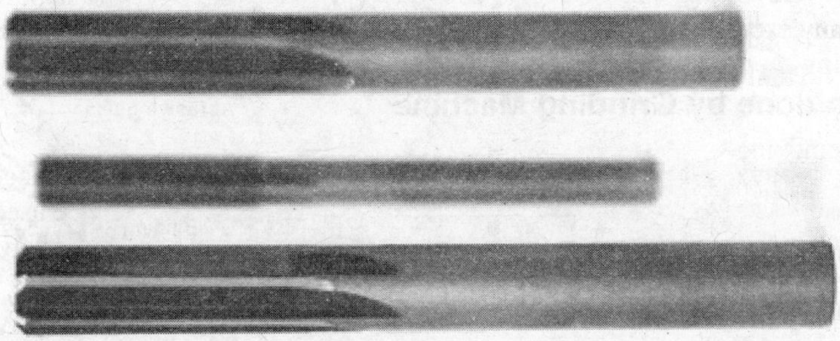

Fig. 39.18. *Reamers*

2. Countersinking. Countersunk head screws and wood screw require a 90° chamfer cut round their hole as a seating for the under side of the head. This is cut by means of a countersinking cutter as shown.

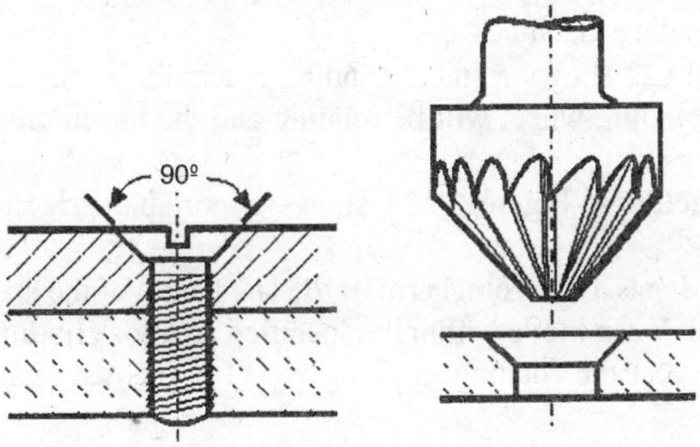

Fig. 39.19. *Countersinking*

3. Counter boring. The preparation of holes for certain purposes involves increasing the diameter of the hole for a certain distance down. This is called Counterboring, and is done with a cutter of the type shown.

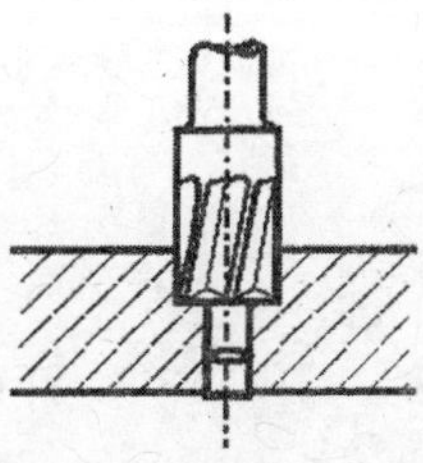

Fig. 39.20. *Counter boring*

39.8 GRINDING MACHINE

Introduction. Grinding is a process of metal cutting or metal removal, to obtain better finish by a rotating abrasive wheel called grinding wheel.

It removes very little quantity of metal and we get a very good surface finish.

39.8.1 Applications

1. To sharpen the cutting tools
2. To grind the gears
3. For thread grinding

4. To obtain better surface finish
5. For removing any excess material to get the required dimensions in very close tolerances.

39.8.2 Operation done by Grinding Machine

1. Excess metal removing
2. Finishing
3. Buffering

39.8.3 A Typical Grinding Wheel

A typical grinding wheel is shown in Fig. 39.21.

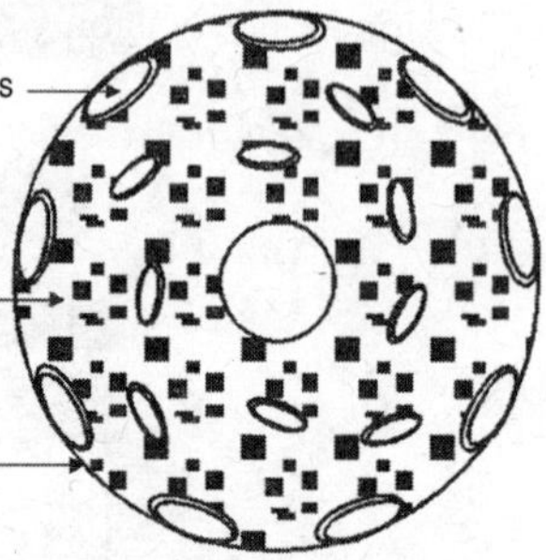

Fig. 39.21

39.8.4 Types of Grinding

(A) Rough Grinding. When the surface finish and accuracy are not of prime importance, then this method is used. In this the job is pressed against the rotating grinding wheel (or vice-versa), so as to remove excess metal.

Following are the Rough Grinding machines.

(1) Stand Grinder. Figure 39.22 shows a stand grinding machine. In this case on the cast stand motor and grinding wheel are mounted. Grinding wheel will be rotating and the job in pressed as shown to remove excess metal.

(2) Portable Grinding Machines. Figure 39.23 shows a portable grinding machine (The figure is self explanatory).

(3) Swing Frame Grinders. It has a horizontal arm *AOB*. The length of the arm will be around 2–3 m. *O* in the centre of gravity where it is used. It can move within the specified area for grinding purpose. For grinding we have to apply the force at end *B*, so that the rotating grinding wheel is pressed against the job and excess metal is removed.

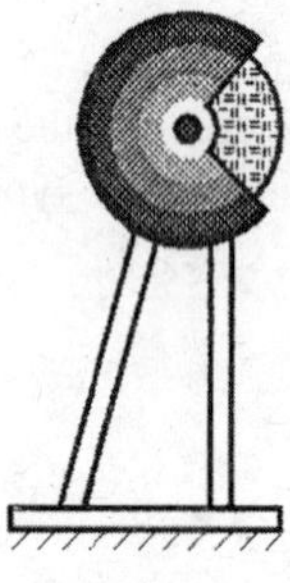

Fig. 39.22. *Stand grinder machine*

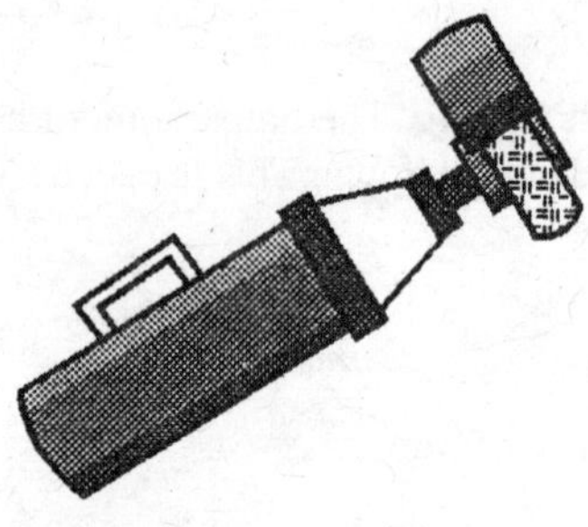

Fig. 39.23. *Portable grinder*

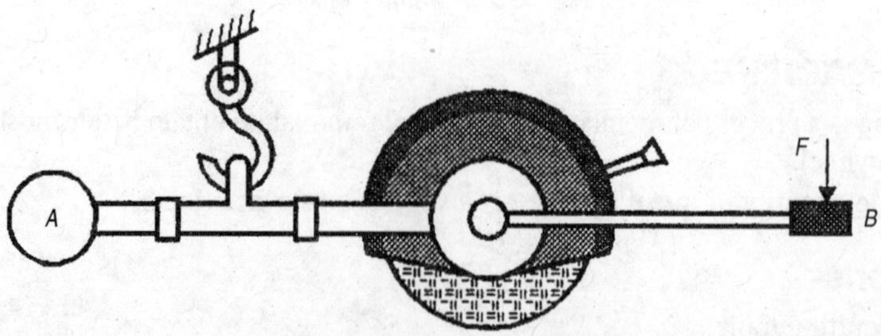

Fig. 39.24. *Swing frame grinder*

(B) Precision or Fine Grinding. Whenever surface finish and accuracy are of prime importance then this method is used to obtain finished products.

Classification Precision Grinding according to the type of surface:

(i) External cylindrical grinding

(ii) Internal cylindrical grinding

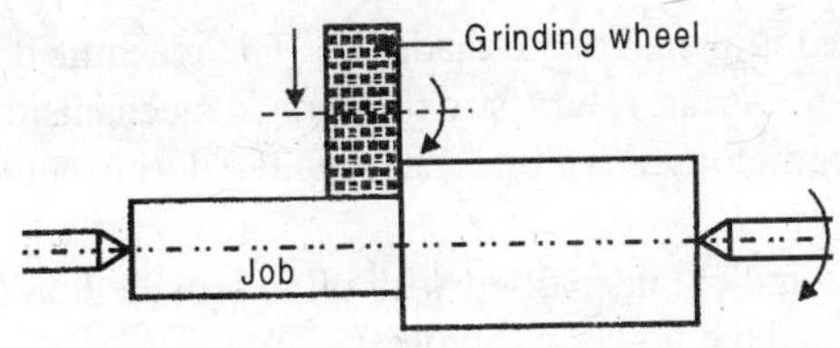

Fig. 39.25. *External cylindrical grinding*

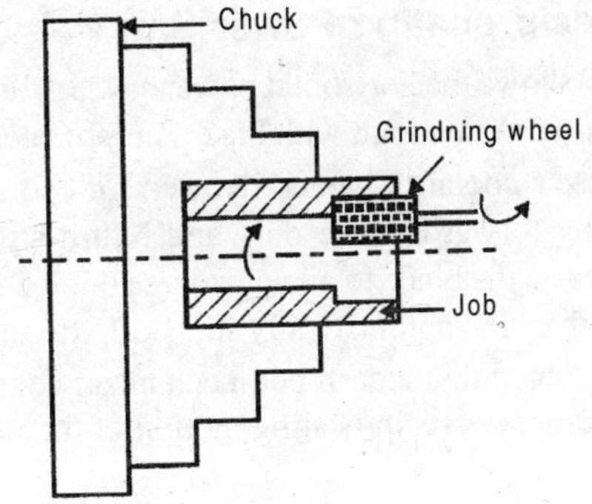

Fig. 39.26. *Internal cylindrical grinding*

(iii) External taper griding

(iv) Surface griding

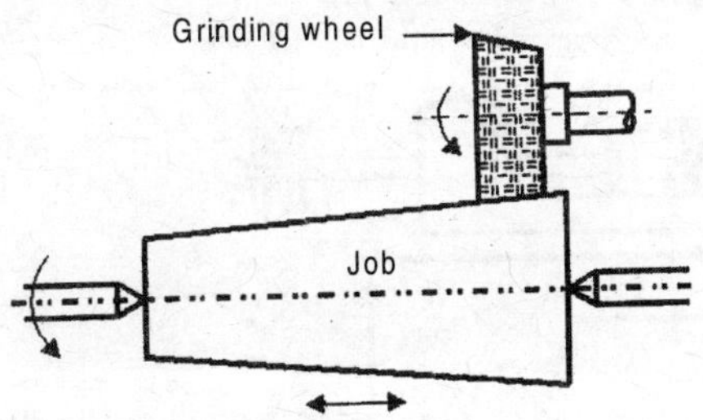

Fig. 39.27. *External taper grinding*

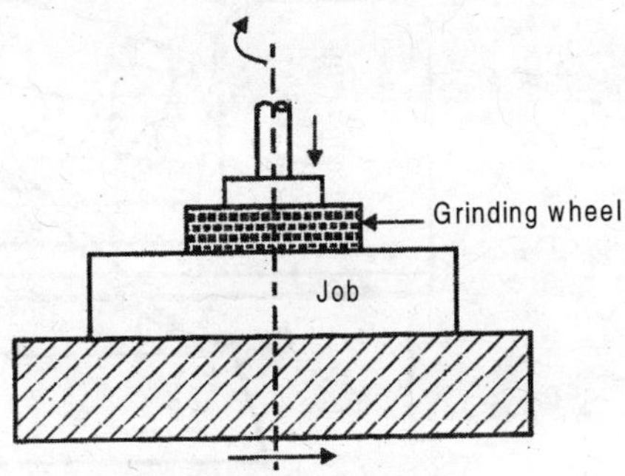

Fig. 39.28. *Surface grinding*

(v) Face grinding

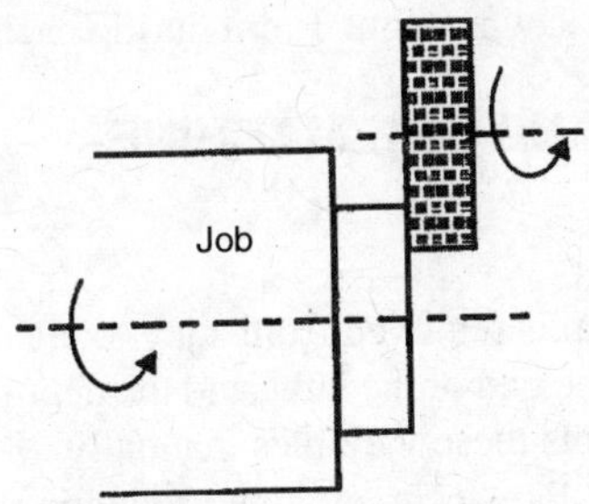

Fig. 39.29. *Face grinding*

39.9 POWER SAW

It is a machine, which is used to cut the metal bars or work pieces to the required lengths.

39.9.1 Types of Power Saw

(i) Reciprocating Power Saw.
(ii) Circular Power Saw.
(iii) Band Power Saw.

39.9.2 Reciprocating Power Saw

Figure 39.30 shows the horizontal power saw. In this case the work piece will be held by means of the clamps in the correct position. As the ram with blade reciprocates the job will be cut.

In this case a coolant is generally used for cooling purpose since it increases the blade life. The machine does cutting during forward stroke only and return stroke will be idle. During return stroke a crank mechanism in provided to raise the blade to avoid damage of blade. In this case continuous attention is not required during cutting because the machine automatically shuts off after cutting.

It is to be noted that a dash pot mechanism consisting of piston and cylinder filled with oil is provided on the machine. It produces reciprocating motion of the ram and blade when the power is supplied.

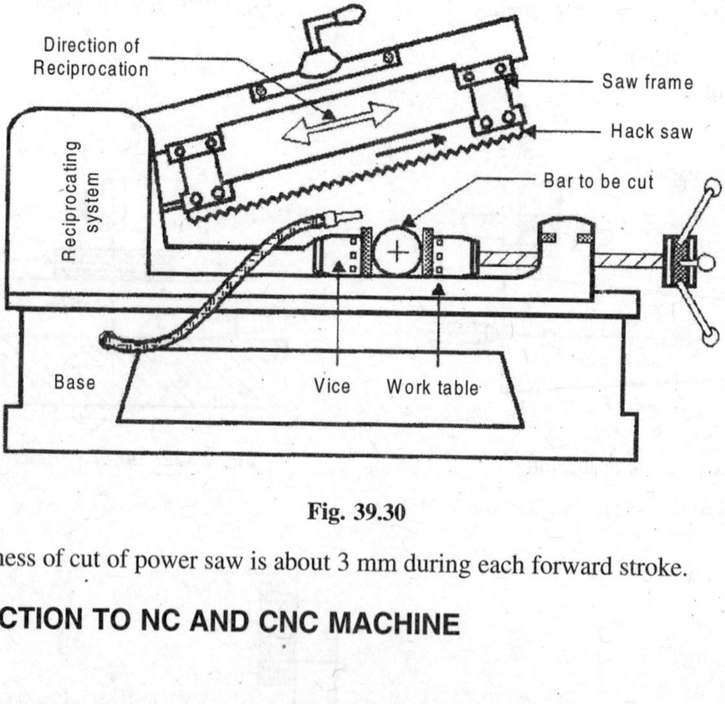

Fig. 39.30

The average thickness of cut of power saw is about 3 mm during each forward stroke.

39.10 INTRODUCTION TO NC AND CNC MACHINE

Introduction

In machine tools say a lathe machine-metal is removed from a job to create required shape-using cutting tools. For this purpose travel of the tool both along the bed of the lathe and the depth of the cut, speed etc. are to be controlled. This is done by skilled worker who controls these variables manually.

Manual operations have its limitations like accuracy, low production and high rejections. In *numerically controlled* machines most of the operations can be done automatically. A machine tool which operates automatically or semi-automatically as per coded instructions given to it is called numerically controlled machine (NC machine). In NC machines servos motors replace human operators in positioning of the work piece and in positioning/operation of the cutting tools.

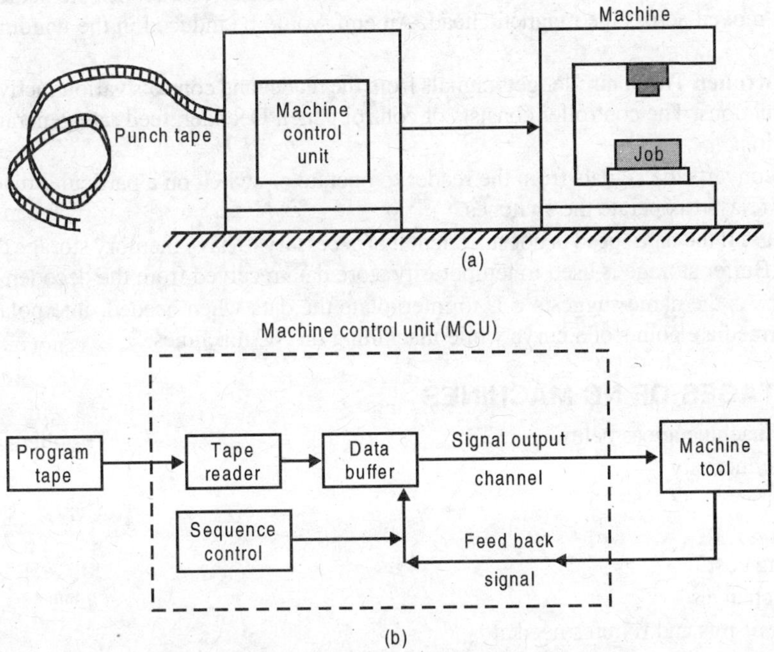

Fig. 39.31. *Block diagram of NC machine*

For the NC machines first programme to do the required operation has to be written and punched on the tapes. The reader reads the tapes and sends required electrical signals to the controller to execute the job. Block diagram of an NC machine is given below.

39.11 COMPONENTS OF NC MACHINE

Main components of an NC machine are as follows:

(a) Programme of Instruction. (*i*) It is part programme which contains coded instructions to execute the operation in a cycle of operation. Numbers, letters and symbols logically organized to direct a machine tool to the required job are called NC programme. The instructions are in the form of numerals, letters and symbols.

(*ii*) Programme Media. Punched tapes are most widely used media for programming in NC machines. Tapes are made of paper, paper-aluminum sandwich or plastic. Paper tapes are inexpensive but do not last long. The programme instructions can also be given on the magnetic tapes. Punched tapes are preferred to the magnetic tapes due to—

Low cost of tape
Low cost of punching machine and the corresponding reader
Ease of detecting error or damage on a punched tape

(b) Machine Controller Unit. The controller unit consists of hardware and electronics to read and interpret the programme of instructions and convert it into mechanical inputs to the machine tool. Typical elements of a controller are as below:

(*i*) Programme Reader. The tape reader and winds the punched tapes. Depending on the tape used the programme readers are—

Punched tape reader. The readers can be electromechanical, photoelectrical or pneumatic type. Depending whether there is hole on the tape electrical signals are generated in all these readers. For example in the photovoltaic reader when light passes through the perforated tape and falls on the photocell. This is turn would give the programmed signal to operate the machine.

Magnetic tape reader. It consists of a magnetic head which reads as well encodes the instructions. For reading propose the tape is moved across the magnetic head. An emf would be induced in the winding which is used for controlling.

(c) Machine controller. The controller gets signals from the reader and controls various activities of the machine to execute the instructions. The controller consists of control panel. Decoder, feed rate generator, interpolator and auxiliary rate function.

(d) Decoder. It converts the signals from the reader to operate or switch on a particular machine tool function. The decoders have relays to operate the switches.

(e) Buffer storage. If the tape speed is not sufficient it is necessary to have memory storage device for execution of the programme. Buffer storage is used to temporarily store data received from the decoder.

(f) Interpolator. As the name suggests it is to interpolate the data when needed. Interpolation is required for computing the intermediate points of a curve while machining on NC machines.

39.12 ADVANTAGES OF NC MACHINES

Advantages of NC machine are as below.

1. Increased productivity
2. Fewer rejections
3. Job accuracy
4. Lower tooling cost
5. Easy design changes
6. Less number of jigs and fixtures needed

39.13 DISADVANTAGES OF NC MACHINES

1. Higher initial investment
2. Higher maintenance cost
3. Need for having programmer

39.14 USES OF NC MACHINES

It is possible to have NC and CNC machines from simple single spindle drilling to complex machining centers like:

(a) Metal cutting operations. Turning, drilling, boring, grinding, milling etc.

(b) Production job. Where geometry is complex and need many operations, works where high accuracy is needed.

39.15 ILLUSTRATIVE EXAMPLE OF NC PROGRAMMING

To operate the NC machine tool as programmed apart from the coded instructions—electronic/electrical and mechanical interfacing is required to move and control the cross slide, tool post, control speeds etc. As way of illustration a simple turning job where 2 mm material is required to be removed from a shaft of 50 mm length, then

Step (a) Movement of the tool post through 2 mm along-Z-axis

Step (b) Move of the tool by 50 mm in-X-direction

Step (c) Move the tool post by 2 mm in + Z-axis

Step (d) Move the tool post by 50 mm along + X-axis

39.16 CNC MACHINES

Computer Numerically Controlled machines (CNC machines) are extensively used these days for precision machining. There is hardly a facet of manufacturing that is not in some way touched by what these innovative machine tools can do. It incorporates a dedicated computer or microprocessor. In the CNC machines a computer receives information from a tape directly or from a computer storage and controls the machine.

In the NC machines the physical components are connected through hardware whereas in the CNC the physical components are soft wired. There is lot of flexibility in the CNC machines as compared to the NC machines. The block diagram of a CNC and DNC machine along with that of NC machine is given below.

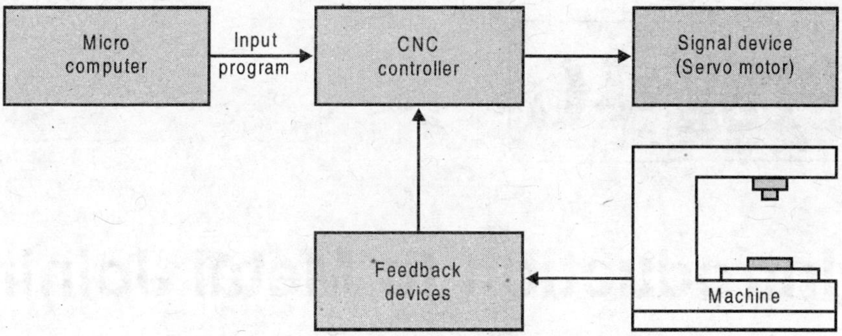

Fig. 39.32. *Block diagram of CNC machines*

THEORY QUESTIONS

39.1. Explain centre lathe and its components.

39.2. Explain pillar drilling machine with a sketch.

39.3. Explain different types of grinders.

39.4. Explain power saw.

39.5. Explain in detail NC/CNC machine in detail.

CHAPTER 40

Introduction to Metal Joining Process

40.1 WELDING

In olden days, in order to join two pieces of metals, rivetting was done. But nowadays for joining two pieces of metals welding is done.

So, welding is a process of joining two pieces of metals by heating them.

Nowadays welding is used in almost all the industries since it is the easiest and rapid method of joining the metals.

Nowadays methods have also been developed to weld dissimilar metals. One important feature of welding is that the strength of the welded joint is more than 100% when compared to the metals.

40.2 WELDING APPLICATIONS

It is used in joining metals in following fields:

(a) In fabrication of tanks, vessels, boilers.
(b) In automobile industry.
(c) In structural work, i.e., to fabricate trusses, frames.
(d) In pipe line fabrication.
(e) In ship building, motor building i.e., in general, in all the types of joining metals, welding is used.

40.3 REQUIREMENTS OF GOOD WELDED JOINTS

The following requirements of good welding joints are:

(i) For good welded joints, the surfaces should be clean. The cleaning can be done by using wire brush.
(ii) Proper edge preparation should be done as discussed below otherwise the joint may not have required strength. The edges must be prepared depending upon the thickness.

Various types of edge preparations for butt welds

The square butt weld is used for the sheets having thickness between 1–5 mm.

Single V or Single U butt welds are used for the plates having thickness between 5–15 mm.

Double V or Double U butt welds are used for welding plates above 15 mm thickness.

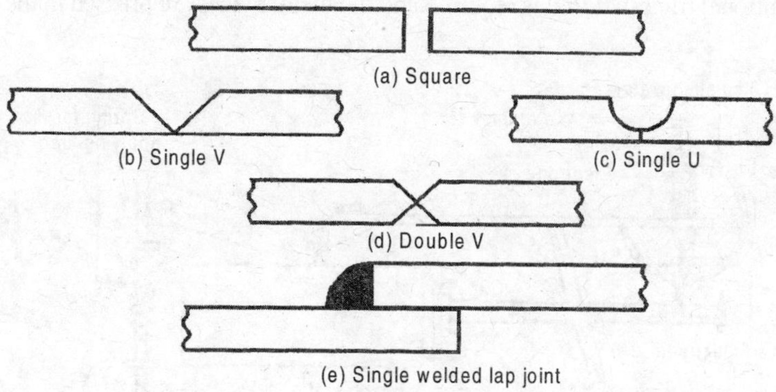

Fig. 40.1

40.4 WELDING CLASSIFICATION

Welding is broadly classified into two types:

(i) Plastic Welding. In this case the two pieces of metals to be joined are heated to plastic state and welding is completed by forcing them. In this case no filler material is added.

(ii) Fusion Welding. In this case the two pieces of metal are heated and brought to molten state and filler material is added and allowed to solidify.

Plastic and fusion welding processes are further classified as under.

(I) Plastic Welding (Under pressure without filler material)

(a) Forge Welding
(Heat is created by Blacksmith's fire)

(b) Resistance welding
(Heat is created by electric current)

Lap welding
Butt welding
Spot welding
Projection welding
Seam welding

(c) Thermit welding with pressure
(Heat is created by chemical reaction)

(II) Fusion Welding (No pressure but filler material is added)

(a) Gas welding
(Heat created by gas)

Oxy – Acetylene welding
Air – Acetylene welding

(b) Arc welding
(Heat is created by electric arc)

Metal arc
Submerged arc
Tungsten Inert gas welding
Gas metal arc welding.

(c) Thermit welding without pressure.

Now we will study the welding processes in detail.

(I) (A) Forge Welding

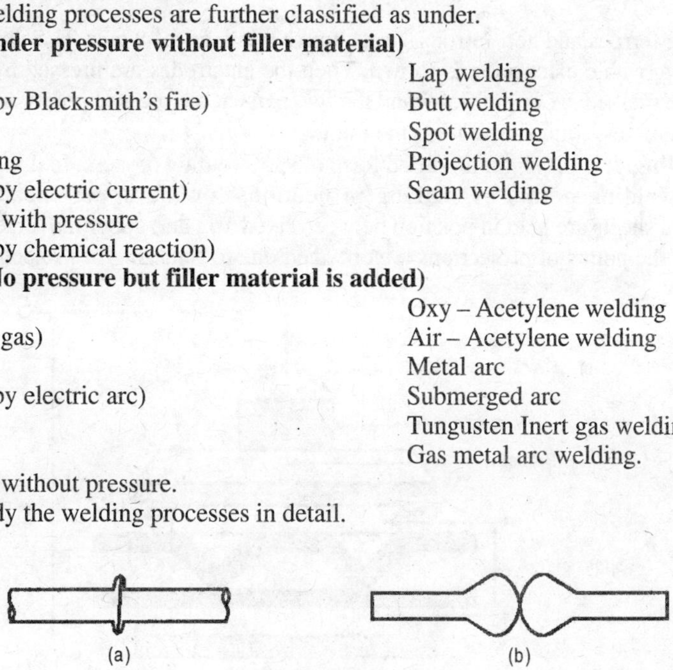

Fig. 40.2

It is the oldest method of joining two pieces of metals. In this case heat is created by blacksmith's fire and the two pieces of metals to be joined are heated and brought to plastic state. Then they are superimposed and hammered together to form the joint.

(I) (B) Resistance Welding. In this case a heavy electric current is passed through the pieces of metals to be joined. Then because of electric resistance, the metals are heated to plastic state and by applying force, welding can be completed.

In this case, no additional filler material is required and the metal pieces are pressed in the two copper electrodes.

(i) Spot welding

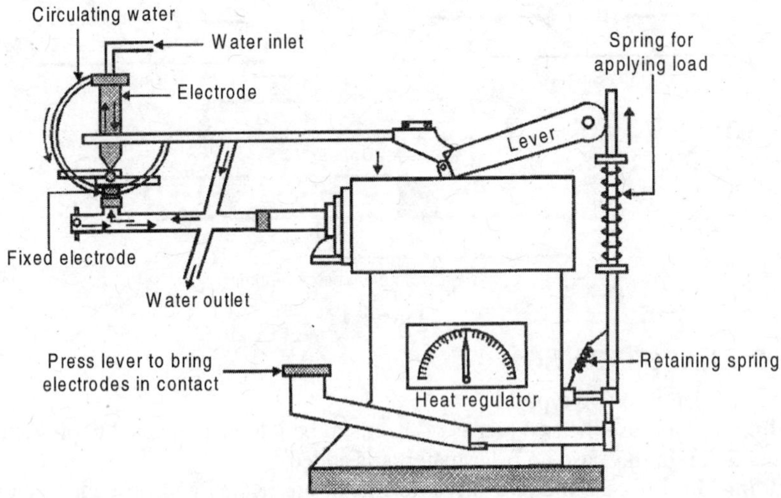

Fig. 40.3. *Spot welding machine*

It is used for welding ferrous and non-ferrous sheets upto a thickness of 8 mm. The sheets to be welded are held between the fixed and movable electrodes as shown. Then the electrodes are pressed by pressing the foot lever. When the electrodes are pressed, a current flows and the two parts at the pressed points are brought to plastic state. Then, this plastic metal mixes, solidifies to form the joint.

(ii) Projection Welding. It is the slight modified form of spot welding process. In this case current and pressure are localised at the welding points by making projections for the upper sheet of metal as shown in Fig. 40.9. The two metal sheets are held in position between fixed arm and upper movable arm. Then the current is passed and welds at all the points of projections are obtained due to flattering of projections.

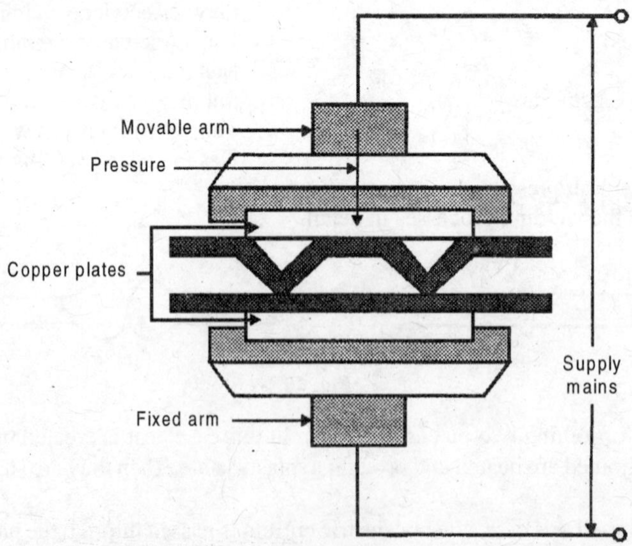

Fig. 40.4

(iii) Seam Welding

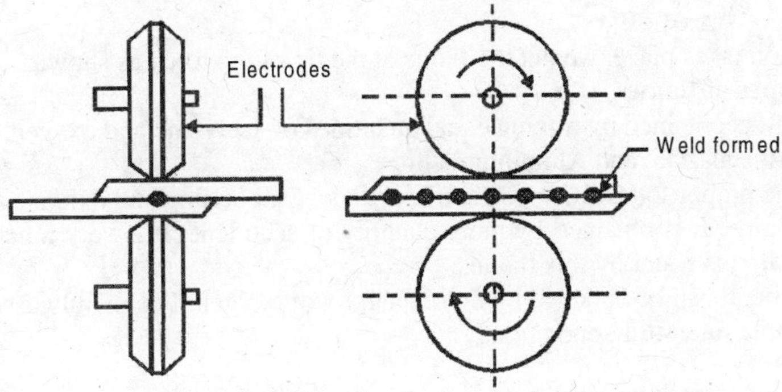

Fig. 40.5. *Seam welding*

For producing continuous welds between two over lapping sheets this seam welding is used. In this case the two sheets to be seam welded are held as shown between the electrodes. When the electrode wheels rotate continuous weld will be produced.

(II) (a) Gas Welding. In gas welding heat energy required to heat the surfaces is obtained by the combustion of mixture of two gases. Generally oxygen and acetylene are used for gas welding purpose. These gases are mixed in proper proportions in the welding torch. This welding torch is provided with two regulators to regulate the gases.

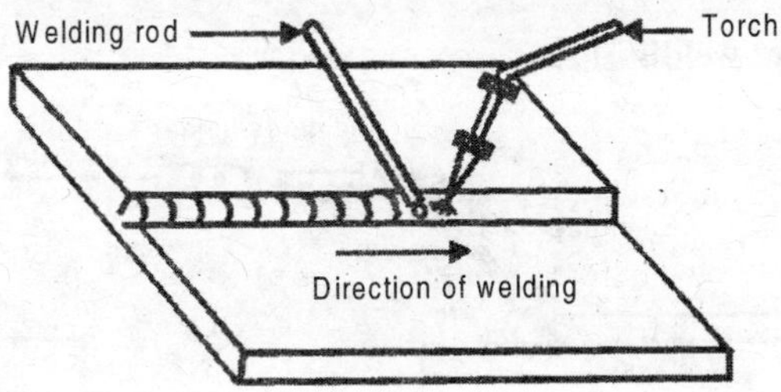

Fig. 40.6

The flame produced due to the combustion is used to heat the metal surfaces to plastic state and welding is completed by adding the filler material as shown.

Oxyacetylene Welding

As the name implies in this case the two gases used are oxygen and acetylene. It is to be noted that oxygen and acetylene are commercially available in cylinders. If required acetylene can be produced by the chemical reaction between calcium carbide and water as under

$$CaC_2 + 2\,H_2O \rightarrow \underset{\text{Acetylene}}{C_2H_2} + Ca\,(OH)_2$$

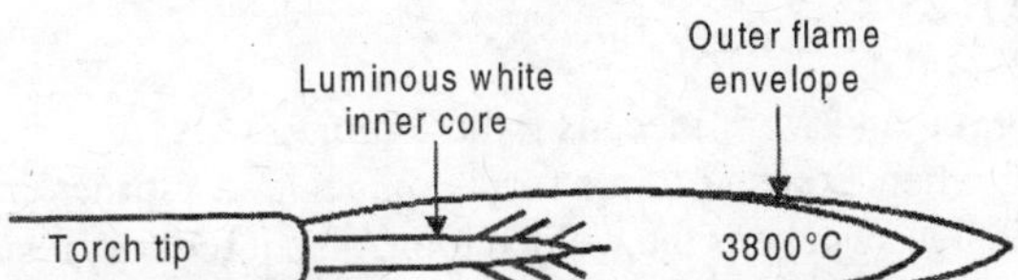

Fig. 40.7

And the chemical reaction for the combustion of acetylene is,

$$2\,C_2H_2 + 5O_2 \rightarrow 4\,CO_2 + 2\,H_2O$$

(Acetylene)

When the combustion takes place, we get the flame at the tip of the torch as shown.

Mainly there are 3 types of flames

(i) Neutral Flame. It is obtained by mixing equal quantities of acetylene and oxygen. It is used for welding all the metals like ferrous metals, Cu, and Aluminium alloys.

When we change the proportions of Oxygen and acetylene. Carburising and oxidising flames can be obtained.

(ii) Carburising Flame. It is obtained by more quantity of acetylene. It is very much suitable for welding of steel as the rate of welding is faster by this flame.

(iii) Oxidising Flame. It can be obtained by more quantity of oxygen. It is mainly used for welding of brass. It is also very much suitable for cutting operations.

Advantages

(1) Gas welding is more suitable for thin sheets.
(2) The equipment is portable, so suitable for out door repair works.
(3) By changing the nozzle in the torch, the torch can be used for gas cutting.

Limitations

(1) It is a slow process compared to arc welding.
(2) Gases used in the gas welding are more costly.

Note that we will be studying welding torch, and other welding equipments while studying gas cutting.

Air acetylene welding. As the name implies in this case air and acetylene are used as the gases to produce the flame. In this case temperatures produced are low compared to other gas welding processes. This method is generally used for Lead welding.

II (b) (i) Arc Welding (Metal arc welding)

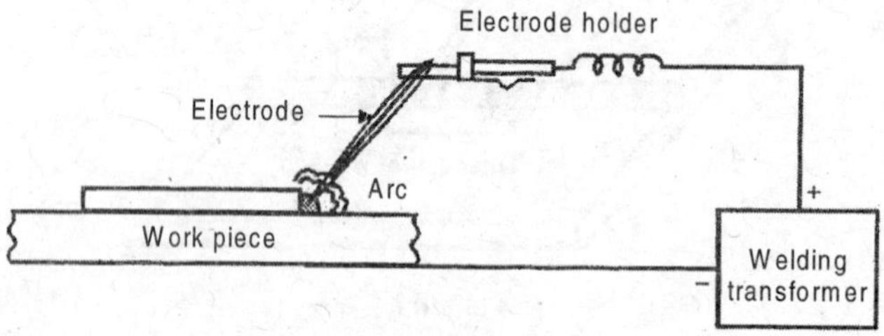

Fig. 40.8

Figure 40.8 shows an arc welding circuit. It mainly consists welding rectifier, electrode (or welding rod) holder, welding rod, and work piece.

When the welding rod touches the work piece, an arc is produced and tremendous amount of heat is liberated. The temperature of the arc is about 3600° C. This heat energy is utilised for melting work piece and welding rod. So, a small pool of molten metal is formed. This molten metal is agitated by the action of arc and the metal is perfectly mixed and after cooling it produces a sound joint.

Note

(i) Arc initiation voltage is 60–100 V.
(ii) Arc maintenance voltage is 25–45 V.
(iii) Power source may be a.c. or d.c.
(iv) The gap between welding rod and work piece is to be 3 mm.
(v) Molten metal in the pool when exposed to air forms oxides and nitrides in the steel, when it reacts with oxygen and nitrogen in air. This weakens the welded joint, also it reduces resistance to corrosion. So in order to avoid this, the welding rods are provided with flux coating. This flux after welding forms a slag covering on the welded joint and prevents from oxidation. This slag has to be chipped off afterwards.

Application. This arc welding is most commonly used in the fabrication of tanks, vessels, trusses, frames, boilers, automobile chassis and body buildings etc.

II (b) (ii) Submerged Arc Welding (SAW). It is the improved arc welding process, which is used for the production of butt welds of thick steel plates.

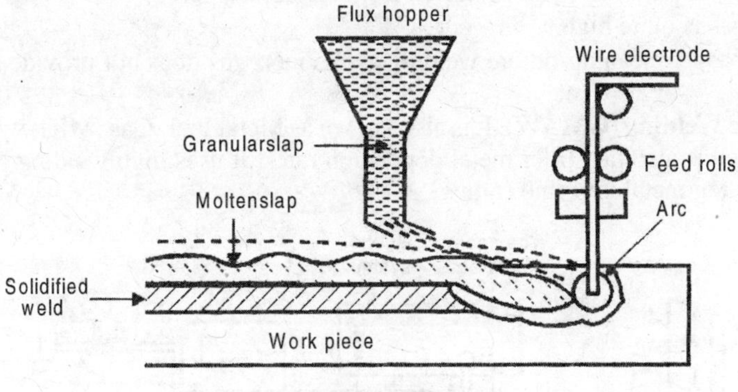

Fig. 40.9

In this case the arc produced is submerged (covered) in the flux, hence the name submerged arc welding.

As shown in Fig. 40.9 the two plates to be welded are placed in position on a back up plate. Through the flux hopper, flux is supplied and wire feeder feeds the bare wire continuously.

In this case arc produced is submerged in the flux. Because of heat of arc the wire and parent metals melt and form a pool of molten metal. Thus molten pool produces a welded joint after cooling.

II (b) (iii) Gas Tungsten Arc Welding (GTAW). It is also known as Tungsten Inert Gas welding (TIG).

It is a fast process, produces clean welds and it can weld metals considered to be impossible to weld. TIG–welding utilizes a non-consumable electrode in a special holder, a separate filler material and an inert gas i.e., argon gas cylinder, power supply source as shown in Fig. 40.10.

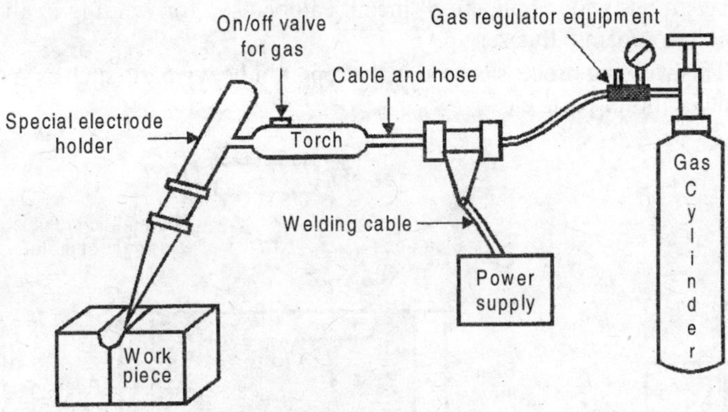

Fig. 40.10. *TIG welding*

When the tungsten electrode strikes the work piece, an arc will be produced. Around the arc inert gas shielding is formed as the gas is coming out from the torch. Because of heat of arc, work piece and filler material melt and form a molten pool. This molten pool after cooling forms a sound welded joint in the shielding of inert gas.

Advantages of TIG Welding

1. Welded joints are stronger more ductile and corrosion resistant than the other weld made by other methods.
2. The welding of non-ferrous metals is simplified as no flax is required.
3. Dissimilar metals can also be welded easily.
4. Due to presence of inert gas, there is less smoke.

5. The arc is transparent due to shielding inert gas, so the welder gas can clearly observe the weld as it is being made.

Limitations of TIG Welding

1. The process is relatively slow in operation.
2. The back side of weld joint has to be protected from the atmosphere.
3. The cost of inert gas is quite high.
4. Al joints require proper cleaning before welding as the inert gas does not provide any cleaning or fluxing action.

II (b) (iv) Gas Metal Arc Welding (GMAW). It is also known as Metal Inert Gas (MIG) welding. In MIG welding we get clean and good welds also fast filler metal deposition rates. It uses high welding current which is used to break the globules of molten metal into fine spray.

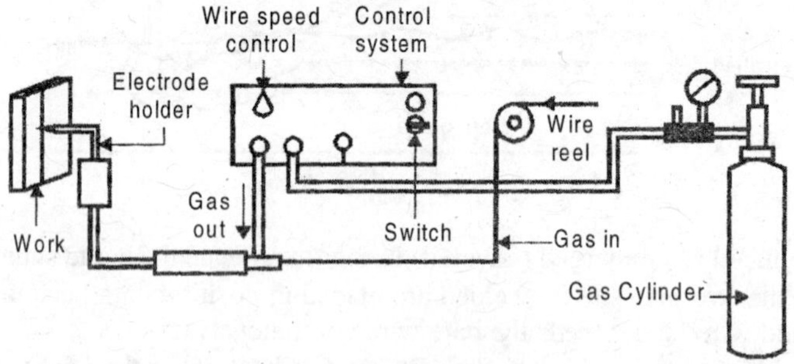

Fig. 40.11. *MIG welding set up*

Figure 40.11, shows schematic diagram of MIG welding. MIG uses a consumable electrode. It is supplied through an electrode holder into the arc. Then at the same speed electrode is melted and deposited into the weld. A small motor with adjustable speed will be used to remove the wire from the reel and fed into the arc.

Generally CO_2 or Argon are used as shielding gases. Mainly the process was developed for welding Aluminium and Titanium. But nowadays it has wide application since it can be used for welding in all the positions; and less skilled operators are required to operate this set up.

II (c) Thermit Welding. This welding process is used in the repair of heavy parts such as spokes of driving wheels, broken connecting rods, other motor parts etc.

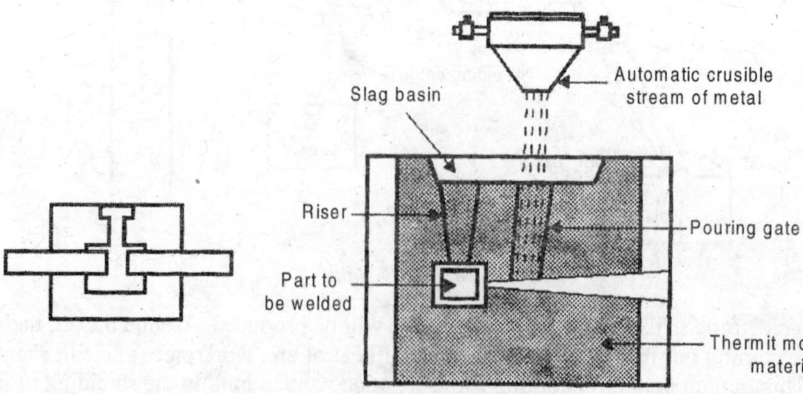

Fig. 40.12. *Thermit welding*

Thermit welding is based on the following exothermic chemical reaction.

$$8\,Al + 3\,Fe_3O_4 \longrightarrow 9\,Fe + 4Al_2O_3$$

So, oxygen from iron oxide unites with aluminium and forms aluminum oxide or slag and a superheated thermit steel will be formed. The temperature of thermit will be around 2700°C which is almost double when compared to the melting point of steel.

In thermit welding process, the parts which are to be welded are placed in the mould as shown. Then the superheated thermit steel is poured from the refractory lined crucible as shown. Due to the heat of thermit steel, the surfaces of metals to be welded are brought to plastic state and if required some mechanical pressure is applied to complete the weld.

40.5 BRAZING

Brazing is a process of joining two pieces of similar or dissimilar metals. In this case as non-ferrous metal or alloy in the molten form is supplied in between the metals to be joined and allowed to solidify.

The melting point of filler material is about 425°C but which is lower than the melting point of parent metals. During brazing no forging action is present and also the parent metal parts do not melt.

The various brazing metals and alloys are, Copper, Brass, Bronze, Silver alloys, Aluminium alloys etc.

Types of Brazing Processes

Based on heating source the various brazing processes are—

(i) Torch brazing
(ii) Furnace brazing
(iii) Resistance brazing etc.

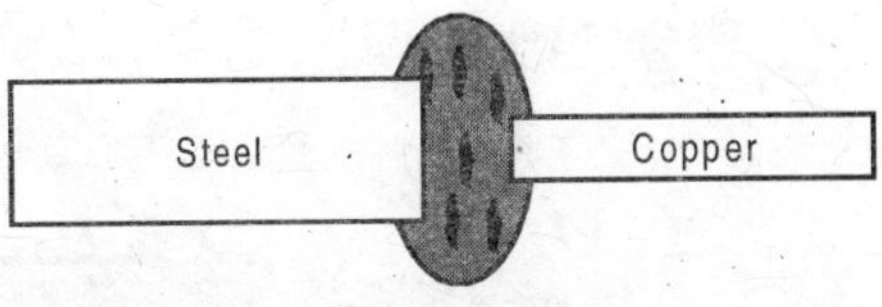

Fig. 40.13

Advantages

(i) Dissimilar metals can be joined
(ii) It is a quick process
(iii) It is used for joining pipe fittings, carbide tips on the tools etc.

Limitations of Brazing

1. Brazing cannot be performed on hardened steel.
2. Aluminum brazing needs a special technique and is very expensive.

40.6 SOLDERING

Soldering is a process of joining two pieces of metals. In this case a molten fusible metal called solder is supplied and allowed to solidify.

The melting point of solder is below 420° C. Generally Lead and Tin alloys are used as solder materials. Some of the solders are:

(i) Soft Solder. Lead 37% and tin 63%.
(ii) Medium Solder. Lead 50% and tin 50%.
(iii) Electricians Solder. Lead 58 % and tin 42 %.
(iv) Plumbers Solder. Lead 70% tin 30%.

As one might have seen with the radio repairers that, when the current is supplied to the solder gun, the front end of it becomes hot. This hot end when brought in contact with solder material, it melts. This molten material is used for joining wires in radios, taperecorders etc. The molten metal, when cools, produces solder joints.

Types of Soldering Processes

(i) Soft Soldering. It is used in sheet metal work and it has less strength.
(ii) Hard Soldering. It is employed when stronger joint is required.

Based on the method of heating source, the soldering processes may be, resistance soldering, torch soldering, furnace soldering etc.

Application. It is mainly used in joining wires in radios, taperecorders, printed circuit boards etc.

Advantages

1. It produces liquid and gas-tight joints quickly.
2. The cost of the joint is quite low.
3. The temperature of soldering is low.

4. The equipment used for soldering is simple, cheap and easy to handle.
5. It is the most economical method.

Drawbacks

The joints are not strong enough to withstand the constant jerks and pulls, which is the main drawback of the process.

40.7 GAS CUTTING PROCESS AND EQUIPMENTS OR (OXYACETYLENE GAS CUTTING PROCESS)

As we have studied in gas welding, heat energy required to heat the surfaces is obtained by the combustion of oxygen and acetylene. These gases are mixed in proper proportions in the torch which is provided with regulators. When the gases are burnt a flame is produced.

The flame so produced is used to heat the steel to red hot colour. Then a jet of pure oxygen is blown on the red hot surface. The steel is burnt i.e., iron oxide (slag) is produced, which under pressure falls down and the steel is cut.

The equipments used in Gas welding and gas cutting—

1. **Gas torch or welding torch.** In this gas torch proper mixing of oxygen and acetylene takes place. The torch is provided with regulators to regulate the mixture of gases. When the combustion takes place flame will be produced at the tip.

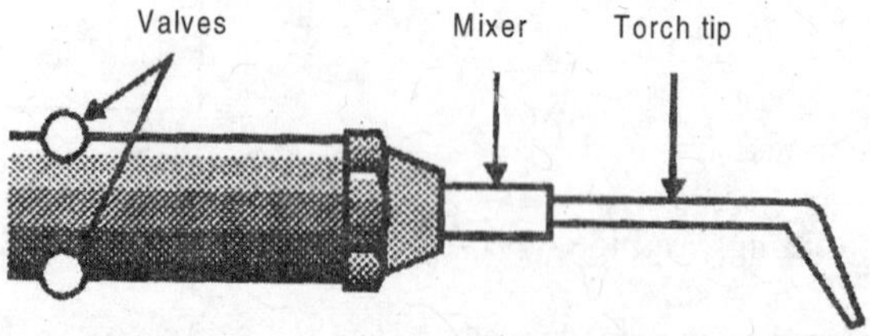

Fig. 40.14

2. **Torch tip.** It is that part of the torch through which gases pass just before their ignition and burning.
3. **Pressure regulator.** The function of a pressure regulator is to reduce the cylinder pressure to the required working pressure and also to produce steady flow of gas.
4. **Hose and Hose fittings.** Hose pipes are used to convey gases from the cylinder to the torch They should be strong and durable.
5. **Goggles.** Goggles fitted with coloured lenses are provided to protect the eyes of the operator from harmful heat.
6. **Gloves.** These are used to protect the hands of the operator.
7. **Spark lighters.** These are used for lighting the torch.
8. **Oxygen and Acetylene cylinders.** These are commercially available in the market or acetylene gas can be prepared as discussed earlier.

THEORY QUESTIONS

40.1. What are kinds of joints, that are normally employed for welding purposes? Give their sketches.

40.2. How the welding is classified? Explain them briefly.

40.3. Discuss the method of electric resistance welding. What are its advantages and disadvantages?

40.4. What do you understand by gas welding?

40.5. Explain with the help of a neat sketch, the principles of carbon arc and metal arc weldings.

40.6. Describe: (1) TIG Welding, (2) MIG Welding.

40.7. State the advantages of soldering and brazing

40.8. Describe the various processes used in cutting of metals.

40.9. Write a short note on Gas cutting tools.

CHAPTER 41

Introduction to Sheet Metal Working

41.1 SHEET METAL

(a) Generally sheet metals have thickness less than 5 mm and are produced by cold or hot rolling processes. The sheet metals can be in various materials like—

(i) **Base metals** – black iron, mild steel, aluminum, copper, lead
(ii) **Coated metals** – like Galvanized iron, (G,I), tin sheets
(iii) **Alloys** – Stainless steel, brass etc.

(b) Sheet metals are available in rectangular sizes varying from 1 to 1.5 m in width and 2 to 3 m in length. The sheets are also available in rolls. The thickness of the sheets is given in gauge numbers. These are given in the table at the end of the chapter.

(c) Most of the sheet metal in use is in mild steel and comes in the form of sheets or strips. The steel may be annealed, soft, half hard, extra spring hard etc. The soft steels are used for deep drawing while harder varieties can be used for blanking.

41.2 USES OF SHEET METALS

Sheet metal parts are used extensively for items such as,

(a) Household utensils, covers, pipes, funnels, boxes, bends etc. for day-to-day requirements.
(b) Automobiles-bicycle, cars, scooters, motorcycles and trucks parts
(c) Aircraft
(d) Railway locomotives and coaches
(e) Farm and construction equipment
(f) Household appliances
(g) Office furniture
(h) Computers and office equipment
(i) Packaging like containers, boxes
(j) Corrugated sheets for roofs

41.3 ADVANTAGES OF SHEET METAL PARTS

The components made out of sheet metals have the following advantages:

(a) High strength

(b) Good dimensional accuracy
(c) Good surface finish
(d) Relatively low cost of manufacture
(e) Economical mass production operations

41.4 METAL WORKING

The required items or components can be made from the sheet metals by hand tools or by press work. But most of the sheet metal work is done on the presses.

41.5 PRESS WORK

It is method of forming sheet metals and plates by cold working into required shapes by applying large force through press tools. The presses may be manually or power operated.

(a) Hand operated press. It is small press which is operated by hand. As the arm of the press is rotated, the ram moves up or down and the stored energy is transferred in doing work on the sheet metal. Hand presses are used for small jobs. (Fig. 41.1)

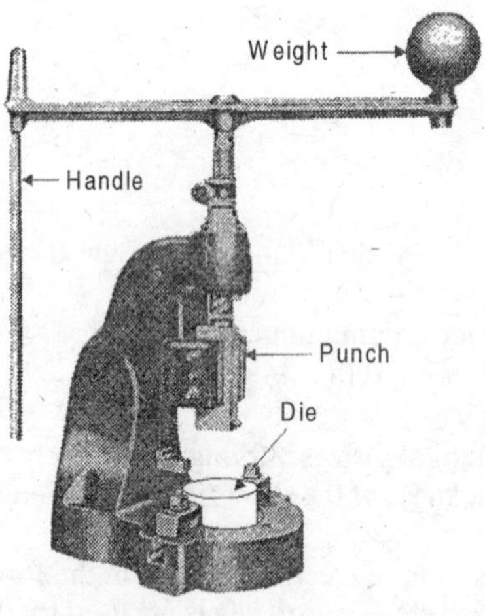

Fig. 41.1. *Hand operated fly press*

(b) Power Press. For large jobs and where high production is required hand operated presses are not suitable and power presses are used. There can be many types of power presses on the basis of power used—electric motor driven, pneumatic or hydraulic press etc. or on the type of jobs to be done like-shearing press, coining or punching press etc.

(c) Die and Punch. The sheet metal work on the press would require die and punch depending on the type of job. Die and punch for making a circular disc is shown in the Fig. 41.2. The die and punch are made of materials harder than that of the sheet metals.

(d) Hand tools. For sheet metal working following hand tools are used—

(i) *Measuring tools*—Like steel rule, folding rules, flexible, push-pull rule, vernier caliper, micrometer, thickness gauge etc. are used to measure the parts during manufacture.
(ii) *Straight Edge*—It is used to scribe long straight lines.
(iii) *Scriber*—It is a long thin rod with pointed end and handle at the other side. It is used to mark lines on the sheet metal.

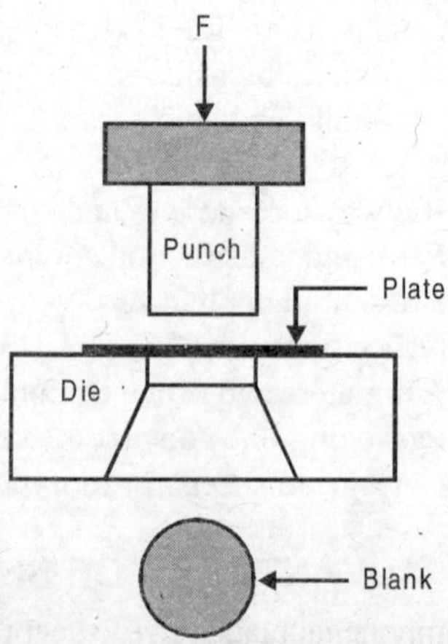

Fig. 41.2. *Die and punch for disc manufacture*

(iv) *Divider*—Used to draw small circles.
(v) *Trammel Compass*—It is used to draw large circles on sheets.
(vi) *Punch*—To locate centre or to make mar.
(vii) *Hammer*—Used for forming.
(viii) *Mallets*—These are soft hammers with rubber, raw hide, soft metals etc.
(ix) Miscellaneous tools like chisels, scissors etc.

41.6 SHEET METAL WORKING

Most of the sheet metal work is done on presses where a die and punch or other formed tools are required. In press work large force is applied on thin sheet metals to give the required shape or to cut it in to the desired shape. Press work is highly economical method of manufacturing. Various manufacturing methods on sheet metals can be classified as follows:

(a) Cutting and Shearing. Here the sheets are cut by using various tools. Blanking, piercing, perforation, notching etc. are various types of shearing operations.

(b) Bending Operation. The sheet metals are straining around a straight axis.

(c) Forming. Flanging and tube forming are forming operations to give shape to the sheets.

(d) Drawing Operations. Cupping, drawing, deep drawing are all drawing operations in which sheet metals are given cup or shell type shapes.

(e) Reducing Operations. Necking type of operations.

(f) Squeezing. For example coining operation is done in closed dies applying large force.

TYPICAL SHEET METAL WORKING

41.7 DRAWING OPERATION

(a) It is a forming process by pushing a punch against a flat sheet and forcing it into a die to take the required shape. Or in other words drawing is the operation of producing thin walled, hollow shaped parts from sheet metal. Both the die and the punch have shape of the part to be manufactured as shown in Fig. 41.3. Classifications of drawing operations could be cupping, redrawing and deep drawing. Components like cups, shells, household utensils etc. can be made by drawing.

(b) In the drawing operation soft material are used since the material is required to undergo large permanent deformation to acquire the intended shapes in the presses. Drawing operations could be for cup making or deep drawing.

(c) For making cup shapes on the press first a circular blank of proper size is taken. The die is usually open from the bottom side and has an ejector system placed below a pad for the removal of the component after pressing it into shape. The pressure pad ensures smooth flow of the blank over the die/punch.

(d) The ratio of the diameter of the blank (D) and that of the cup (d) is called draw ratio (D/d). If draw ratio is greater than about 1.8 then the cup cannot be drawn in one draw. Such items may require redrawing in several steps.

(e) In drawing operations lubricants are extensively used.

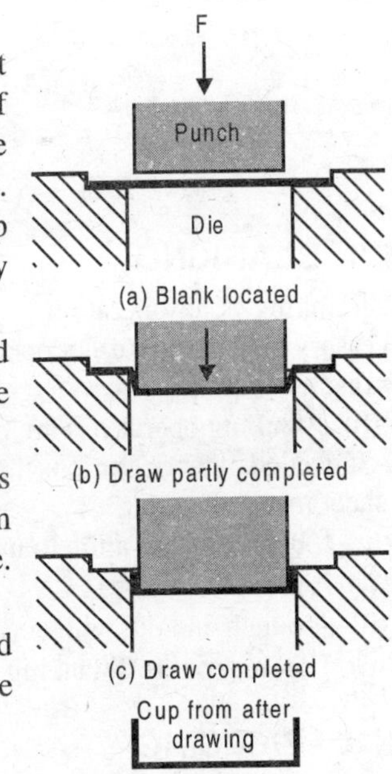

Fig. 41.3. *Drawing operation*

41.8 BENDING

It is a forming operation performed on a press to bend a sheet metal or a strip through required angle. Figure showing bending operation is shown in Fig 41.4. Salient features of bending are,

(a) It requires a press and tooling to do the bending operation.
(b) The sheet metal strip is permanently deformed to the required shape.
(c) When the sheet metal is pressed for bending in the die the outer portion is stretched while inner sides (towards the punch) are compressed.
(d) The radius of the corner should not be small i.e., the angle at the bent should not be too sharp—otherwise material can crack at the bend.
(e) After removal of the bent component from the press there is usually spring back up to about 4 degrees depending on the material hardness. Hard material have greater tendency to spring back while softer materials have less tendency to recover shape.

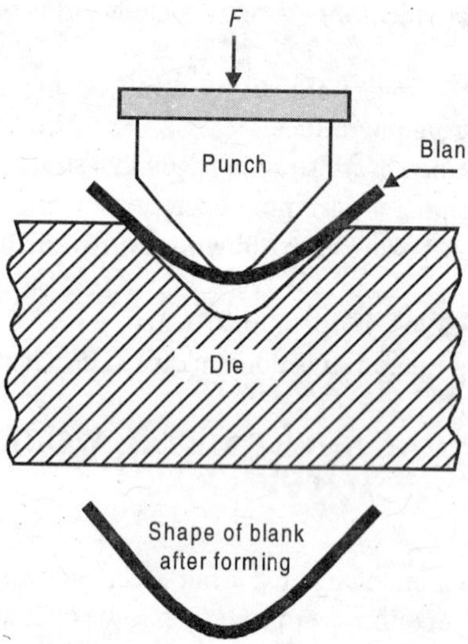

Fig. 41.4. *Bending operation*

41.9 BLANKING

(a) Blanking is an operation of cutting an object of given shape from sheet metal strip. It may or may not be necessary to perform further operations on the blank. Piece detached from the metal strip is called blank. Examples are discs, washers.

(b) Blanking operations are extensively used in sheet metal working to make components like washers, discs etc. If needed other operations may be combined with the blanking operation. Blanking is usually the first operation in sheet metal.

(c) A blanking operation to make disc from a steel strip would require—

(i) A press
(ii) A punch and die with clearance at the lower side so that the disc can fall down freely. (Ref. Fig. 41.2)
(iii) The material for blanking operation is usually hard.

41.10 PIERCING

It is a distinct process of making a hole in sheet metals. A pierced hole is shown in Fig. 41.3. It is characterized by:

(a) The punch is sharp and pointed tool which is able to penetrate or pierce through the sheet metal.
(b) There is no scrap from the hole.
(c) The hole has rough flanges around the hole.

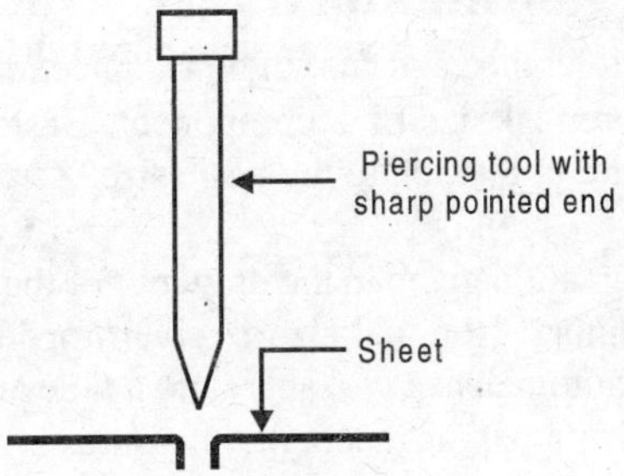

Fig. 41.5. *Pierced hole*

41.11 NOTCHING

Notching is removing small quantity of material from the edges of a sheet metal part. Notching may be done to avoid overlapping of material after bending at the seams. Proper notching would give the sheet metal joints a better fit. Various types of notching is shown in Fig. 41.6.

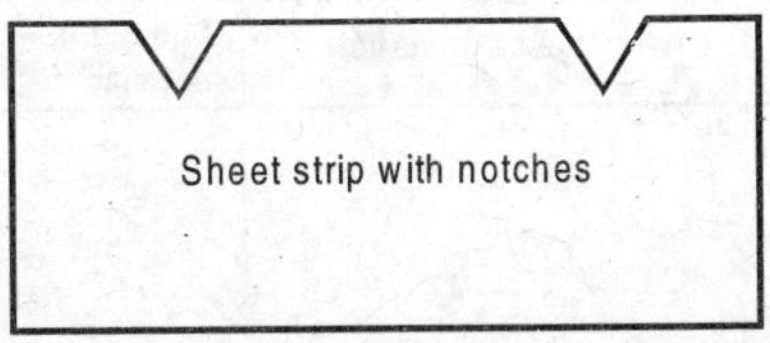

Fig. 41.6. *Notches*

LIMITS, TOLERANCE AND FITS

41.12 INTERCHANGEABILITY

(a) Modern industry has been developed on the basis of interchangeable manufacturing. By interchangeability it is meant that the parts manufactured anywhere can be assembled without rework. The production of any component requires interaction between man, machine and material. Defects or variations in any of these would result in variation in the dimensions of component.

(b) The variations may arise due to—

(i) Difference in the skills of the machine operator

(ii) Machine tool may have inherent inaccuracies.

(iii) The cutting tools may have variations and get worn out.

(iv) The materials may have defects.

(c) Cost of producing the components to exact dimensions would be very high and rather impossible. However it is possible to manufacture them with such variations which would not affect the working of the assembly. A component is said to have interchangeability if with the recognized deviations the component is able to be assembled at random with the mating part and also has required tightness or looseness after assembly. If the parts are not interchangeable selective assembly would be needed which is costly and time consuming. Advantages of interchangeability are as follows:

(i) It is possible to manufacture components on mass scale at various places.

(ii) It is also possible to replace worn out parts.

41.13 ELEMENTS OF INTERCHANGEABILITY

This leads us to the concept of limits; tolerance and fits. In this connection knowledge of following is essential.

(a) Basic Size. It is the exact theoretical size of a component. Basic size is chosen on the basis of design considerations. Limits and variations are in relation to the basic size. Zero line coincides with the basic size. (Fig. 41.7).

(b) Nominal Size. It is the size designation given in the drawing for the sake of convenience and identification.. It is usually rounded off to whole number. Nominal size is given for identifying a component and it does not represent accurate measurement. The nominal size of a shaft and a hole are same.

(c) Actual Size. It is the dimension of a part as it actually measures is actual size. Actual size should be within the specified limits.

(d) Zero Line. The limits and fits are represented graphically in relation to the zero line which represents the basic size.

(e) Limit of Size. It is the two extreme limits of sizes between which the actual size may lie.

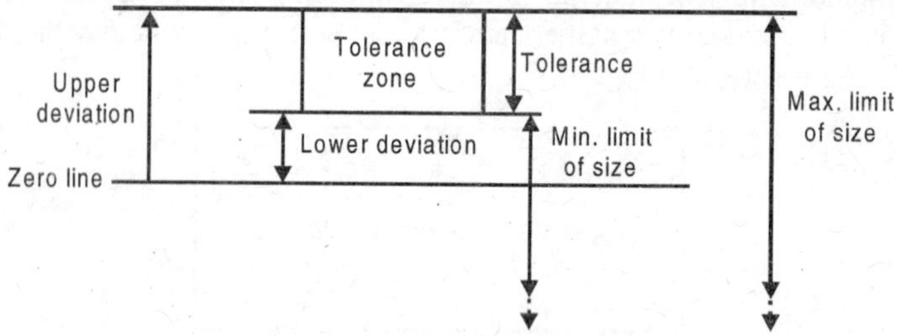

Fig. 41.7. *Basic and other sizes*

41.14 LIMITS

(a) As mentioned earlier it is impossible to manufacture large number of pieces to an exact dimension and there will always be some difference in size. Then limits of acceptable dimensions are found so that the component would perform satisfactorily within these limits. The component can differ from the proper size by the small amount and still be able to be used.

(b) For example if a component has basic size of 25.00 mm and it is found that it would perform satisfactorily if the size is between 25.02 mm on the higher side and 24.97 mm on the lower side.

Then upper deviation is: The maximum size minus the basic size (in this case 25.02 – 25.00 = 0.002).

Lower Deviation – Minimum deviation = Basic Size – Lowest size = (in this case 25.00 – 24.97 = 0.003)

(c) Maximum Limit or Upper Limit = Max. dimension which can be allowed = Basic size + Upper deviation = 25.02.

(d) Minimum Limit or Lower Limit Minimum dimension of component which can be allowed = Basic size – Lower deviation = 24.97.

41.15 TOLERANCE

(a) It is the total amount by which a dimension is allowed to vary. The Tolerance is the total amount by which the size of the component can differ from the Nominal Size. The difference between the maximum and the minimum limits of sizes is tolerance. It is equal to the algebraic difference between the upper and lower deviation.

(b) *Types of tolerances.* A Tolerance is said to be Bilateral if it is spread over both sides of the Nominal Size. If it is on side it is called unilateral tolerance.

(c) *Example of Unilateral tolerance—*

(i) If basic size is 20.00 mm and component can be accepted up to 20.01 min. It can be written as 20.00 + 0.01 where +.01 represents unilateral tolerance.

(ii) If the dimension is given as 16 – .02, it implies that the component has unilateral tolerance of –.02 and the component can vary between 16.00 – 02 = 15.98 to 16 mm.

(d) *Examples of bilateral tolerances.* Here the tolerance are spread on both sides of the zero line. For example 20.01 would mean that component can lie between 19.99 to 20.01 mm.

(e) *Tolerance zone.* The zone between the two limits of sizes as represented graphically in Fig. 41.7.

41.16 ALLOWANCE

(a) Allowance is basically the gap between components that work together. Allowance between parts that are assembled is very important. For example, the axle of a car has to be supported in a bearing otherwise it will fall to the ground. If there was no gap between the axle and the bearing then there would be a lot of friction and it would be difficult to get the car to move. If there was too much of a gap then the axle would be jumping around in the bearing. It is important to get the Allowance between the axle and the bearing, correct so that the axle rotates smoothly and easily. Allowance is intentional difference between a shaft and hole dimensions for any type of fit.

(b) The allowance is the intended difference in the sizes of mating parts. This allowance may be: positive which means there is intended clearance between parts; negative for intentional interference: or "zero allowance" if the two parts are intended to be the "same size". This last case is common to selective assembly.

41.17 FITS

Tolerance is applied to a single component. When two such components are assembled and are required to work together as a unit—the assembly characteristics is known as fit between mating parts. The relationship between parts when one is inserted in to another with degree of tightness or looseness is called fit. Generally speaking following are most common fits:

(i) Clearance Fit. When there is relative motion between the shaft and the hole it is called clearance fit. In clearance fit there is positive clearance between the maximum shaft size and the minimum whole size.

(ii) Interference Fit. When there is no relative movement between the shaft and the hole it is called interference fit. Here the shaft size is bigger than the hole.

(iii) Transitional Fit. When there is neither interference nor clearance fit–then it is called transitional fit.

41.18 HOLE AND SHAFT BASED FIT SYSTEM

Fits can be based on the basic hole size or basic shaft size.

	Basic Hole System		*Basic Shaft*
1.	Design size of the hole is taken as the basic size for both the hole and the shaft.	1.	The design size of the shaft is taken as basic size for both the hole and the shaft.
2.	It is most common system for limit dimensions. In this system the design size of the hole is taken to be equivalent to the basic size for the pair. This means that the lower limit of the hole dimension is equal to design size. The basic hole system is more frequently used since most hole generating devices are of fixed size like drills, reamers etc.	2.	When designing using bought out items with fixed outer diameters like bearings, bushings, etc. a basic shaft System may be used. Taking the basic size of a common reference in both following relationships exist.
3.	The position of the tolerance zone of the shaft may be above or below the zero line. Depending on the tolerance zone it can give various types of fits—interference, clearance	3.	The position of the tolerance zone of the shaft may be above or within the tolerance zone of the hole.
4.	Capital letters H is used for the hole. Shaft have different letters to indicate the position of tolerance zone for desired fit.	4.	Small letter *h* is used for the shaft and the holes have different letters.

Gauge Thickness Conversion Table

Sheet Metal Thickness		
Imperial Gauge	*Imperial in mm*	*Metric Sheet mm*
10	3.25	3.0
12	2.64	2.5
14	2.03	2.0
16	1.63	1.5
18	1.22	1.2
20	0.91	0.9
22	0.71	0.7
24	0.56	0.6
26	0.45	0.5

THEORY QUESTIONS

41.1. What is a sheet metal? Give different metals that can be classified under sheet metal.

41.2. Enumerate the uses of sheet metals.

41.3. What advantages are obtained from the parts which are formed from sheet metal.

41.4. Explain the working with a neat sketch of a hand operated flypress.

41.5. Give the procedures for sheet metal working like Bending, Blanking etc.

41.6. Write a note on limit, fits and tolerances.

CHAPTER 42

Machine Devices

DRIVES

42.1 INDIVIDUAL AND GROUP DRIVE SYSTEM

The arrangement used to transfer the power from a prime mover to the machine is known as '*driven system*'. Thus this system connects the output of the prime mover to the input of machine.

Individual Drive

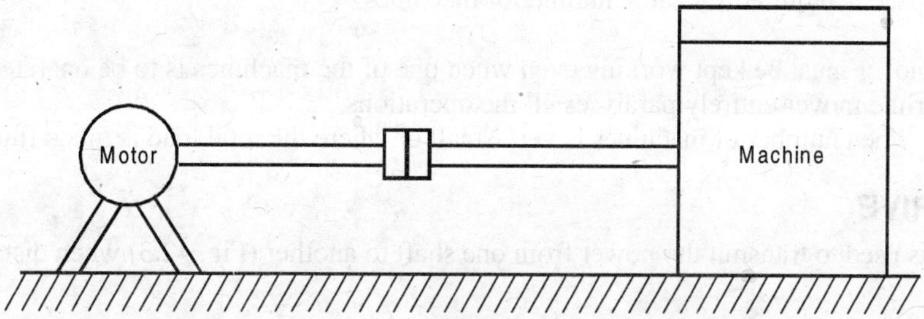

Fig. 42.1

When output from one prime mover is used to drive one and only one machine, the arrangement is known as *individual drive e.g.* electric motor driving a compressor or a pump.

Group Drive

When the output from one prime mover is used to drive number of machines, the arrangement is known as *group drive*. e.g., a car engine is used to drive the car through a gear box. Simultaneously it also operates the cooling water pump of radiator, cooling fan, compressor of air conditioner etc.

This type of arrangement is generally used in machine shops as shown in Fig. 42.2. The prime mover is first connected to the line shaft with belt and pulley arrangement. The line shaft is supported by bearings and runs over all the machines which are to be driven. On the line shaft, for each machine two pulleys are mounted. Out of those, one is known as loose pulley. It can run freely over the shaft while the other is known as *fast pulley* which is keyed to the line shaft and rotates along with it. Thus using this arrangement the machine can be started or stopped just by shifting the belt on 'fast' or 'loose' pulley respectively.

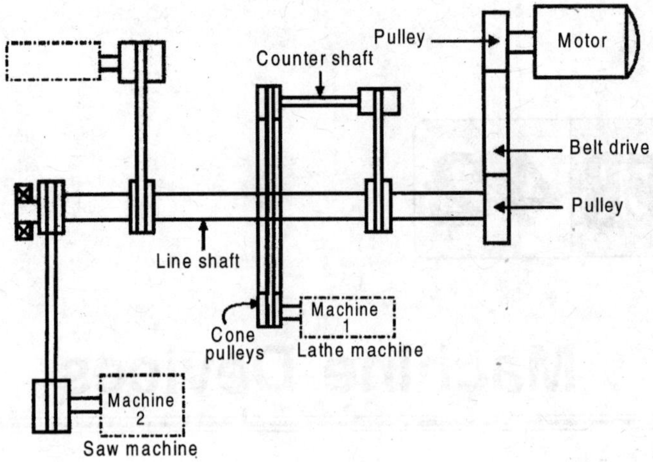

Fig. 42.2

42.1.1 Advantages and Disadvantages of Group and Individual Drives

Advantages

1. Only one prime mover generating sufficient power to drive all machines. Hence the initial cost is less.
2. Number of machines, types of machines can be increased without arranging for additional prime mover.
3. Prime mover can be located out of the machine shop, giving pollution free, noiseless working place.
4. Power losses in prime movers, friction are less giving high overall efficiency.
5. Suitable where number of machines are required to operate simultaneously and the total load remaining more or less constant.
6. Less floor space is required for same number of machines.

Disadvantages

1. The prime mover must be kept working even when one of the machines is to be operated.
2. Failure of prime mover entirely paralyses all the operations.
3. Not suitable when number of machines is very small or where the total load is much fluctuating.

42.2 BELT DRIVE

This arrangement is used to transmit the power from one shaft to another (Fig. 42.3) when distance between them is large.

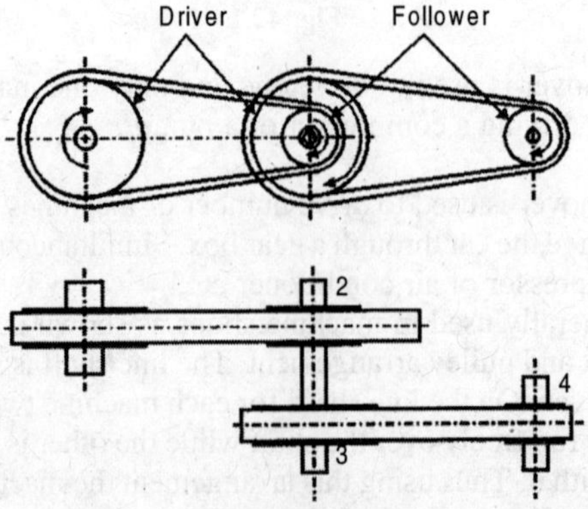

Fig. 42.3

The main components of this drive are discussed as follows:

1. Belt. These are made of material like canvas, leather, rubber etc. These materials give high friction coefficient between them and pulleys. Some times belts are reinforced with steel wires to improve the strength. They are manufactured with different cross sections and accordingly are known as *Flat Belt*. (with rectangular c/s) and '*V*' belt (trapezoidal c/s). The low initial cost, no maintenance are major *advantages*, while non positive transmission and low transmission efficiency (due to slipping of belt) are major disadvantages of belt drives.

2. Pulley. It is a circular disc with a flat surface or a groove. It is mounted on shaft with help of a key. Thus it rotates with the shaft. The belt is made to pass over it. Due to the friction between belt and pulley surface, the power is transmitted.

42.2.1 Comparison of Flat and V-belt

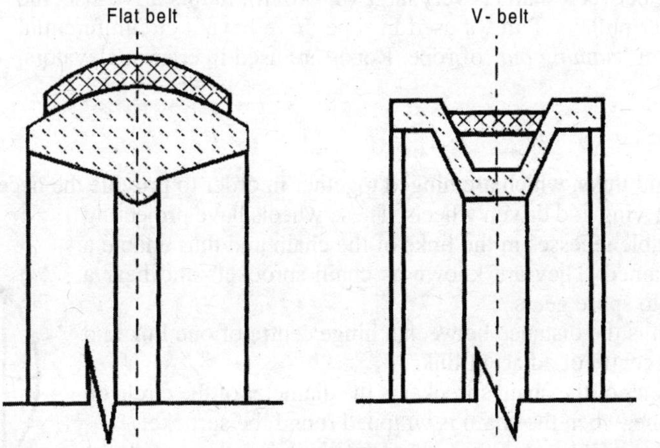

Fig. 42.4

	Flat Belt		*V-Belt*
1.	Cross section is rectangular.	1.	Cross section is trapezoidal.
2.	Made of canvas or leather.	2.	Made of molded rubber, reinforced with steel wires, covered with canvas sheath.
3.	Can be used for distant shafts (distance up to 5 m).	3.	Are used for relatively closer shaft (upto 1 m).
4.	Can be used for non-parallel shafts.	4.	Normally used for parallel shaft.
5.	Open as well as crossed belting is possible.	5.	Only open belting is possible.
6.	Have a joint.	6.	These are endless.
7.	High slippage over pulleys hence low efficiency.	7.	Due to wedging action, low slippage, hence high efficiency.
8.	Can be repaired.	8.	No repair possible.
9.	No standardisation of length.	9.	These are available in only standard lengths.

42.2.2 Advantages of V-belt over Flat Belt

Following are the few advantages of V-belt over flat belts:

1. Transmission efficiency of V-belt is more than flat belts as it is also in contact from sides.
2. V-belts are comparatively silent during operation.
3. In case of V-belts the contact area is more when compared with flat belts as they are at contact from sides.

bushes and the bushes turn freely on the pins. All the contact surfaces are hardened so as to resist wear and are lubricated so as to reduce friction.

Figure 42.7 (a) shows a simple roller chain, consisting of one strand only, but duplex and triplex chains, consisting of two or three strands, may be build as shown in Fig. 42.7(b), each pin passing right through the bushes in the two or three strands.

(b) The Inverted Tooth or Salient Chain. The construction of this type of chain is shown in Fig. 42.9. It is built up from a series of flat plates, each of which has two projections or teeth. The outer faces of the teeth are ground to give an included angle of 60° and they bear against the working faces of the sprocket teeth. The inner faces of the link teeth take no part in the drive and are so shaped as to clear the sprocket teeth. The required width of chain is built up from a number of these plates, arranged alternately and connected together by hardened steel pins which pass through hardened steel bushes inserted in the ends of the links. The pins are riveted over the outside plates. The chains may be prevented from sliding axially across the face of the sprocket teeth by outside guide plates without teeth.

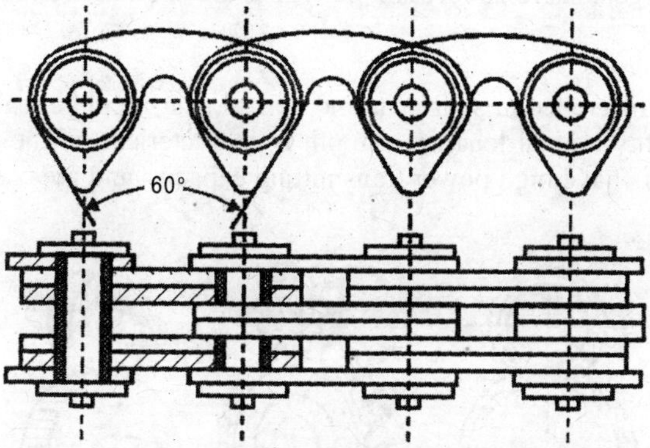

Fig. 42.8

When the chain is new, the position which it takes up on the sprocket is shown in Fig. 42.9.

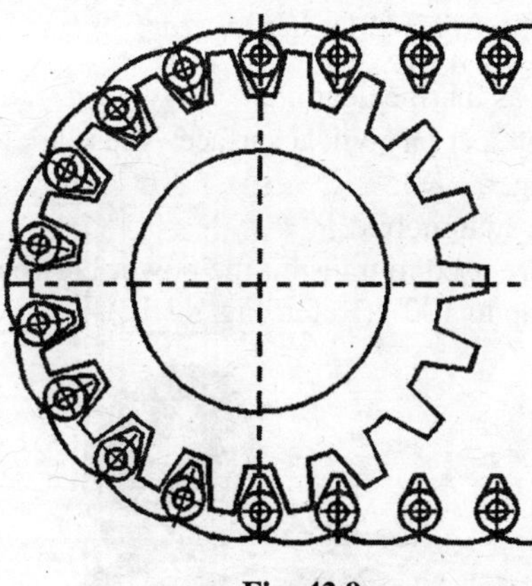

Fig. 42.9

The number of teeth on the smaller sprocket of a chain drive should preferably be not less than 17 and the speed ratio should not exceed 6 or 7 to 1.

42.5 GEAR DRIVES

Following types of gears are widely used in practice.

1. Spur gear. In this type, the teeth are cut on cylindrical surface of hub. The teeth in this gear are parallel to the axis of the shaft. These are used to transmit power between two parallel shafts. (Refer Fig. 42.10).

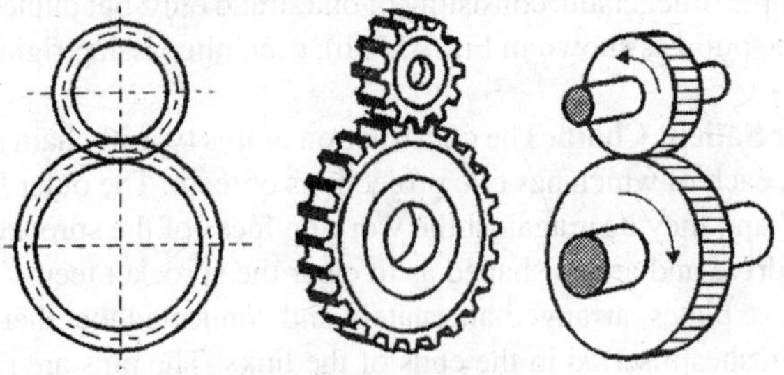

Fig. 42.10. *Spur gear*

The manufacturing of these gears is relatively simple. These are used where the gear ratio (reduction in speed) is upto 5.

Application. In small gear boxes.

2. Helical Gear. The only difference in spur gear and helical gear is, in helical gear the teeth are cut making certain angle with axis of shaft. Gradual loading of tooth is characteristic of the gear, which results in smooth, noiseless operation. These gears have high power transmitting capacity and are used upto gear ratio of 10. (Refer Fig. 42.11).

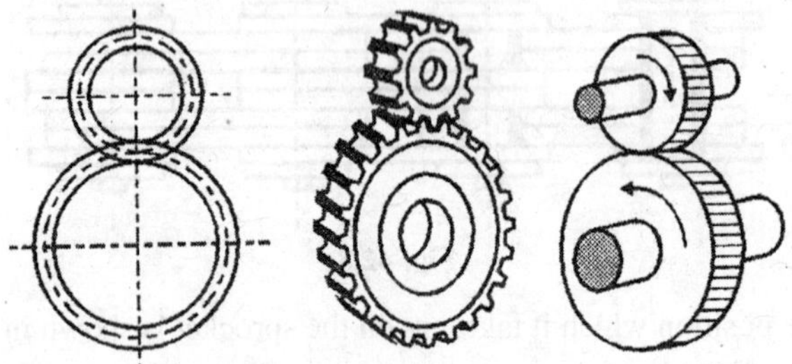

Fig. 42.11. *Helical gear*

Application. In special duty gear box as marine gear box.

3. Bevel Gear. In this type, the teeth are cut on conical surface. These are used to transmit power between two shafts having intersecting axes. (Refer Fig. 42.12).

Application. Differential of 4 wheeler automobile.

4. Worm and Worm Wheel. These are used to transmit the power between to non-intersecting shafts (scew). This gearing gives very high gear ratio (up to 100). (Refer. Fig. 42.12).

Fig. 42.12. *Bevel gear*

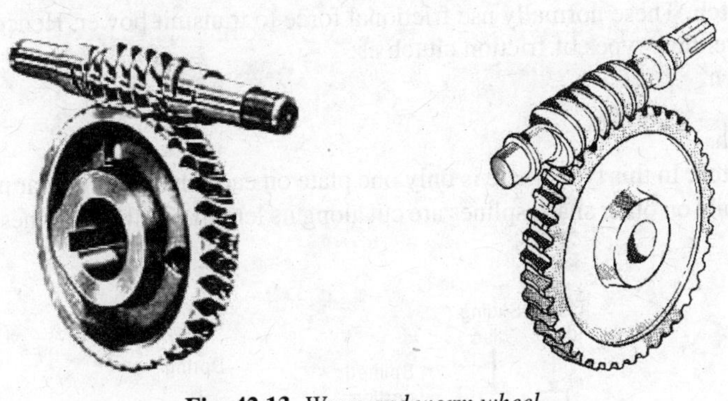

Fig. 42.13. *Worm and worm wheel*

Application: These are used in Lathe machine, feed shaft.

5. Rack and Pinion. The toothed flat strip is known as *rack* and small circular toothed wheel is known as *pinion*. This arrangement is used to convert rotary motion into linear motion.

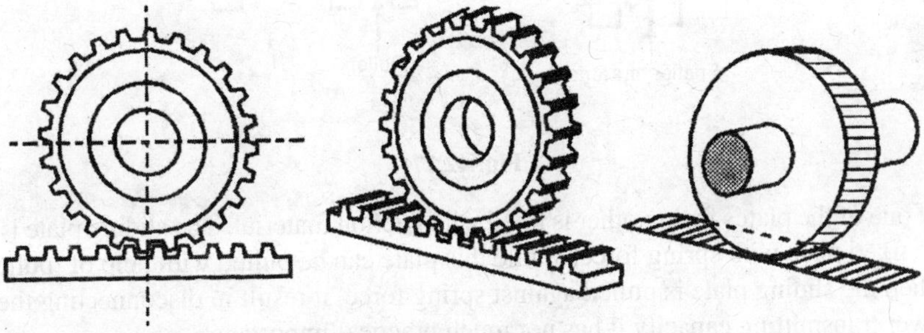

Fig. 42.14. *Rack and pinion*

Application. Plastic hand molding machine Fig. 42.14.

42.6 CLUTCHES

Clutch is a device used to connect or disconnect two coaxial shafts. Clutches have very wide application in automobiles.

With help of clutch we can bring the vehicle to rest, without stopping, the engine, just by disconnecting the input and output shafts. It also enables change of gear without creating a shock load on engine.

42.6.1 Types of Clutches

Clutches are classified in two types viz,

1. Positive Clutch. This type of clutch when engaged, connects two shafts rigidly.

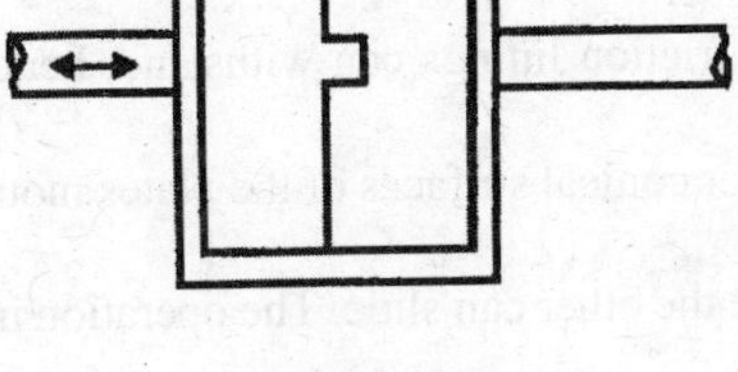

Fig. 42.15

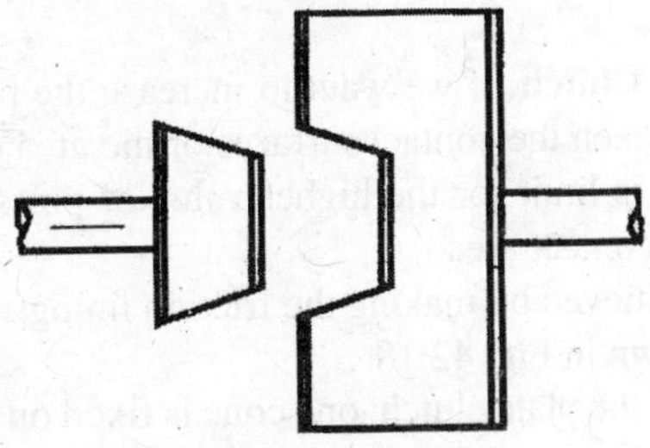

Fig. 42.16

2. Non Positive Clutch. These normally use frictional force to transmit power. Hence these are also known as *friction clutch*. There are three types of friction clutch as:

(a) Single plate clutch
(b) Cone clutch
(c) Centrifugal clutch.

(a) Single Plate Clutch. In this type, there is only one plate on each shaft. A circular plate is fixed to the end of one shaft with a key. While on other shaft, splines are cut along its length. On these splines another circular disk can slide freely.

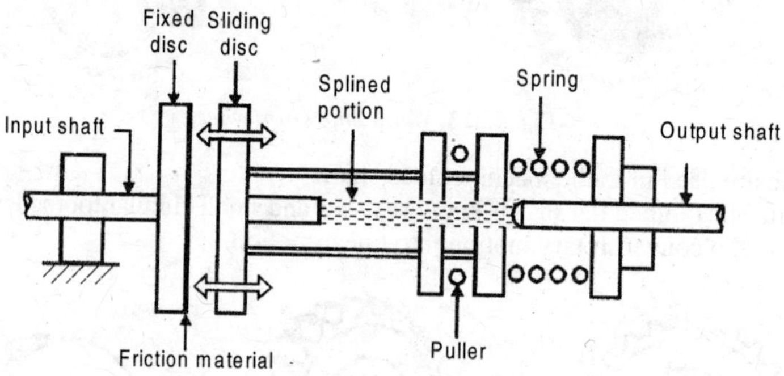

Fig. 42.17

The surfaces of one of the plates facing other is lined with friction material. The sliding plate is forced to remain in contact with the fixed plate with spring force. While this plate can be pulled with help of 'puller' which is fixed to '*neck*'. Thus when the sliding plate is pulled against spring force, it result in disconnecting the two shafts.

Due to low power transmitting capacity it has not much practical importance.

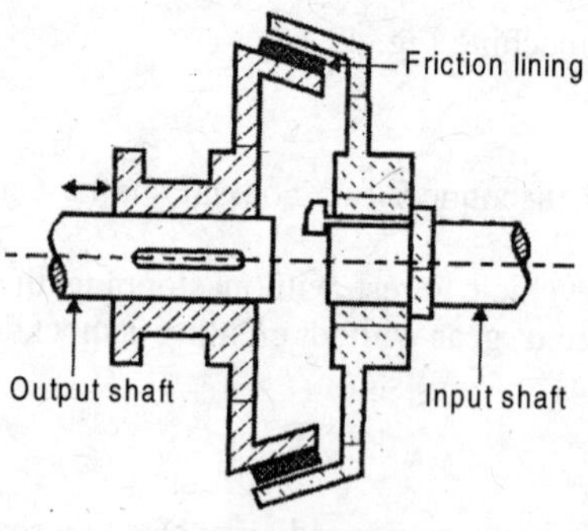

Fig. 42.18

(b) Cone Clutch. If we want to increase the power transmitting capacity of a clutch, we have to increase the pressure between the contact surfaces or the area of contact surface.

But there is limit for the higher value of pressure which the friction linings can withstand, hence one has to increase the contact area.

This is achieved by making the friction linings on inner or outer conical surfaces of the plates mounted on each shaft, as shown in Fig. 42.18.

Similar to the plate clutch, one cone is fixed on one shaft while the other can slide. The operation in same as the plate clutch.

(c) Centrifugal Clutch. This gets engaged due to the centrifugal force. Thus it operates automatically.

In this the shoes are held in the position as shown in Fig. 42.19 with springs, which are fixed to the shaft. When the shaft starts rotating due to centrifugal force the shoes start moving in radially outward direction and at a specific speed they make contact with the dish from inside. This dish is fixed to the output shaft. This type of clutch is widely used in small automobiles (below 50 CC).

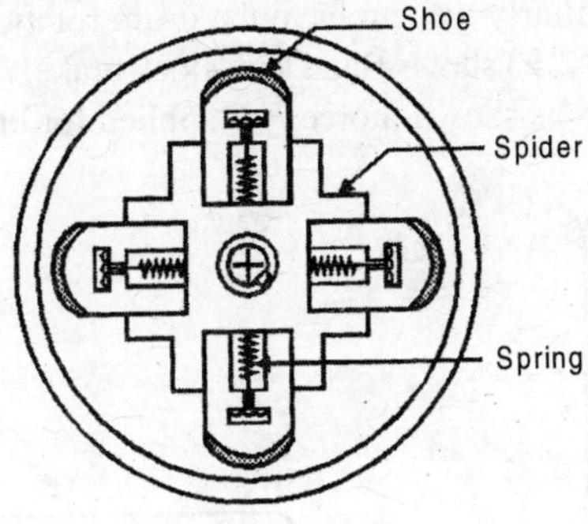

Fig. 42.19

42.7 BRAKES

We all are very much conversant with the term brakes. We apply brakes in bicycles, motor vehicles to retard the motion and to stop the vehicle. So we can define brake as a device or appliance which is used to retard or stop the vehicle.

42.7.1 Types of Brakes

Brakes are broadly classified into 3-types.

1. Electric brakes: e.g., Eddy current brakes.
2. Hydraulic brakes.
3. Mechanical Brakes

Mechanical brakes are further classified as:

(a) Single block or shoe brakes
(b) Double block or Double shoe brake
(c) Band brake
(d) Band and Block brake
(e) Internal Expanding brake.

(a) Single block or shoe brakes. Figure 42.20 shows a single block or shoe brake. *A* is the rotating wheel or drum to which a block or shoe. *B* is pressed so as to retard its speed. *OC* is the lever on which force *F* is applied to retard and stop the wheel.

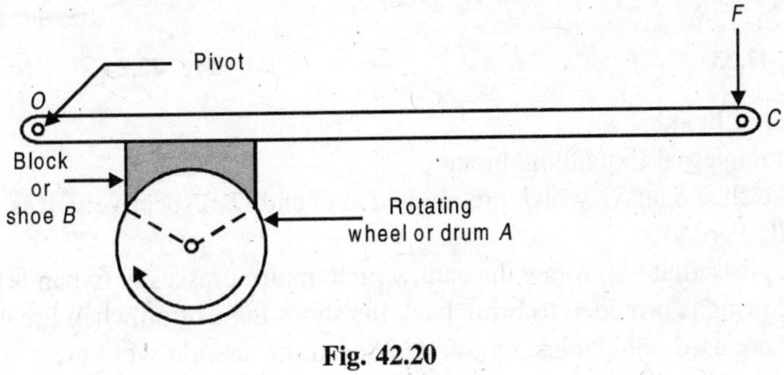

Fig. 42.20

(b) Double block or Double shoe brake. Figure 42.21 shows a double block or double shoe brake. A is the rotating wheel or drum to which two blocks B are pressed so as to retard its speed. OC and $O'c'$ are the two levers on which forces F and F' are applied to retard and stop the wheel.

(c) Band brake. Figure 42.22 shows a Band brake. A is the rotating wheel or drum to which a band generally belt/rope/flexible band (B) is pressed so as to retard its speed. OC is the lever on which force F is applied. T_1 and T_2 are the tensions in rope or belt. Similarly we can have the drum rotation in anti-clockwise direction also.

(d) Band and Block brake. Figure 42.23 shows Band and Block brake. As the name implies it consists both belt or flexible band and the wooden blocks as shown. Force F is applied for braking.

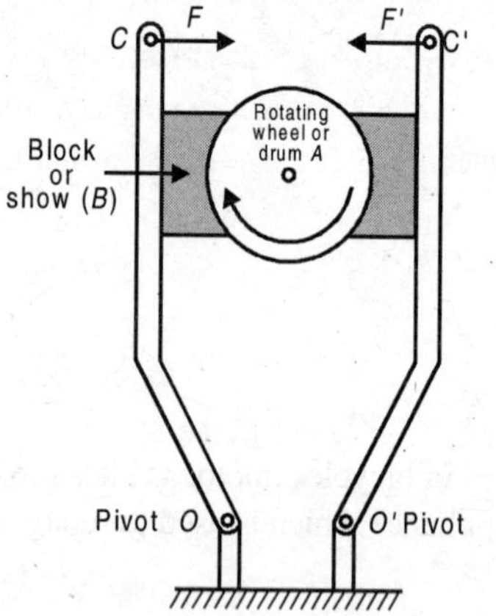

Fig. 42.21

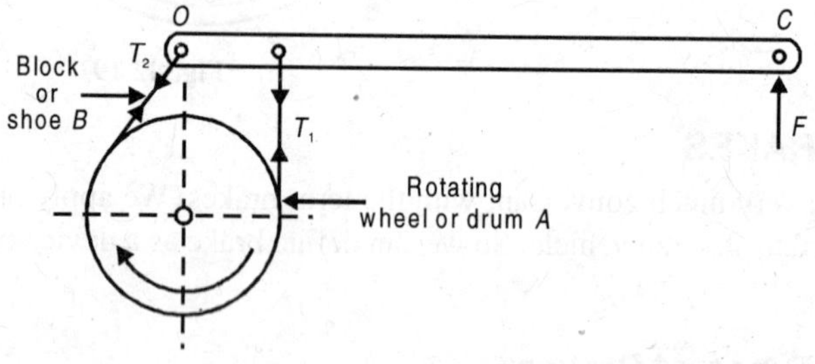

Fig. 42.22

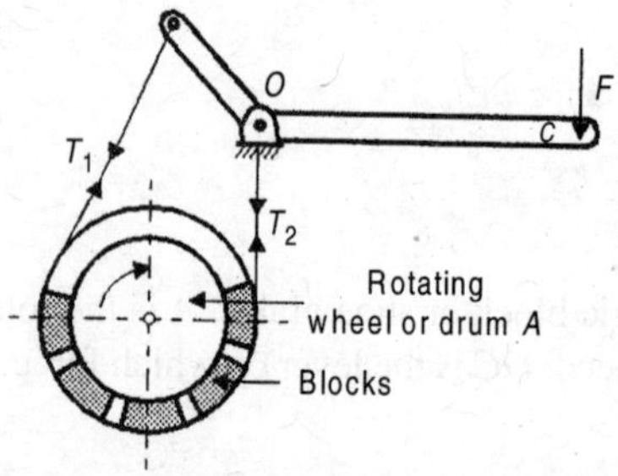

Fig. 42.23

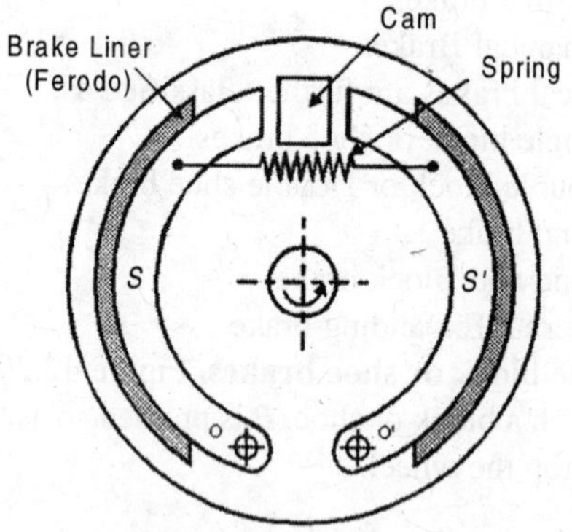

Fig. 42.24

(e) Internal Expanding brake

Figure 42.24 shows an internal Expanding brake.

It consist of two brake shoe S ans S' which are pivoted at O and O' respectively. It is to be noted that the shoes liner material is generally Ferodo.

When break lever is pressed, it will rotate the cam, which in turn presses or expands the shoes S and S' against the drum for breaking. Spring is provided to bring back the shoes in position when break is released.

Application. Breaks are used in bicycles, motor vehicles, tram cars, railways etc.

4. V-belts are endless and they generally used in the power transmission in the industry to run pumps, compressors.
5. The life of V-belts is more when compared with flat belts.

42.2.3 Application of Flat Belt and V-belts

1. Flat belts are generally used in floor mills and Lathe machines to transmit power.
2. V-belts are used to transmit power in industries i.e., to run compressors, pumps, blowers etc.
3. In automobiles V-belt is used to drive the fan to blow air over the radiator etc.

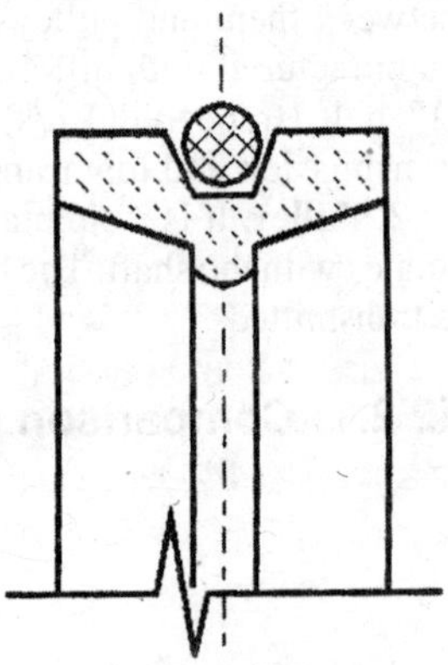

Fig. 42.5

42.3 ROPES

Rope is a bundle of steel wires twisted in a special way. These are used to transmit power when the distance between shafts is very large (8 – 10 m). In this drive also, the rope is made to pass over pulleys. Pulleys used in rope drive have a circumferential groove in order to prevent '*running out*' of rope. Ropes are used in cranes, elevators, etc.

42.4 CHAIN DRIVE

A chain is built up of rigid links, which are hinged together in order to provide the necessary flexibility for wrapping action around the driving and driven wheels. These wheels have projecting teeth, which fit into suitable recesses in the links of the chain and thus enable a positive drive to be obtained. They are known as chain sprockets and bear a superficial resemblance to spure gears.

The pitch of the chain is the distance between a hinge centre of one link and the corresponding hinge centre of adjacent link.

The pitch circle diameter of the chain sprocket is the diameter of the circle on which the hinge centres lie, when the chain is wrapped round the sprocket.

Referring to Fig. 42.6 it will be seen that, since the chain links are rigid, the pitch becomes a chord, not an arc of the pitch circle.

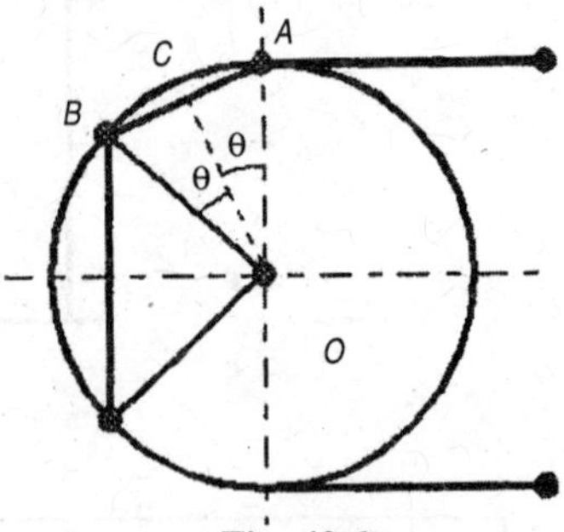

Fig. 42.6

Types of Chain

There are two types of chains in common use for transmitting power, namely (a) The roller chain and (b) The inverted tooth or salient chain.

(a) The Roller Chain. The construction of this type of chain is shown in Fig. 42.7.

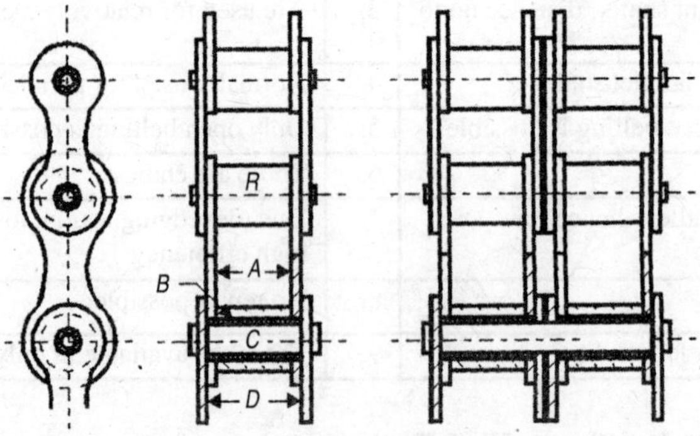

Fig. 42.7

The inner plates *A* are held together by steel bushes *B*, through which pass the pins *C* riveted to the outer links *D*. The roller *R* surrounds each bush *B* and the teeth of the sprockets bear on the rollers. The rollers turn freely on the

THEORY QUESTIONS

42.1. Explain individual and group drive system with sketch.

42.2. Compare the advantages and disadvantages of group drives over individual drives.

42.3. Compare Group drive and Individual drive.

42.4. Write a short note on 'Belt drives'.

42.5. Draw with neat sketches different types of belts commonly used in workshops.

42.6. State the function of the following mechanical element : V-belt

42.7. State the advantages of V-belt over Flat Belt.

42.8. Enumerate various types of belts with their applications.

42.9. Write a short note on rope drive.

42.10. What are different types of gears? Mention their applications.

42.11. Draw the sketches of the following types of gears and state their applications.

1. Spur gear
2. Helical gear
3. Bevel gear
4. Worm and worm wheel

42.12. Explain with diagrams which types of gears are used in the following situations:

1. Parallel shafts
2. Intersecting shafts
3. Non-parallel and non-intersecting shafts.

42.13. Compare the belt drive, chain drive and gear drive on the basis of following: layout/design, applications, advantages and limitations. You may present your answer in tabular form.

42.14. What is a clutch, what are its different types?

42.15. Explain the difference between a coupling and a clutch.

42.16. Write a short note on single plate clutch, cone clutch and centrifugal clutch.

42.17. Draw a neat sketch of single plate clutch and describe its working.

42.18. Draw the sketches of the Helical gear and state their applications.

42.19. Explain chain drive in detail.

42.20. Explain what is brake? What are its types? Explain in detail.

CHAPTER 43

Machine Elements

43.1 POWER TRANSMISSION SHAFTS

Shaft is a machine element, which is used to transmit power from one member to another member. When a tangential force is acting on the shaft, a turning moment is set up. This makes the shaft to rotate for transmitting power from one shaft to another. Pulleys, gears etc. are mounted on the shafts through keys.

Generally the cross-section of the shafts will be circular (i.e., cylindrical in length) also they are solid throughout their length. However square or rectangular shafts are also available depending upon specific applications.

43.2 TYPES OF SHAFTS

They are broadly classified into two types:

(i) Power transmission shafts. As the name implies, they transmit power from one member to another member.

(ii) Machine shafts. As the name suggests, they are the part of machine only. For example crank shaft.

43.3 SHAFT MATERIAL

Generally for regular shafts carbon steels are used. However for higher strengths Ni, Ni–Cr steels are used.

43.4 APPLICATION OF SHAFTS

As discussed earlier whenever power is to be transmitted from one member to another member, shafts are used.

Figure 43.1 given on next page shows a flywheel or gear mounted shaft. Shaft is supported in the bearing at *A* and *B*.

43.5 AXLE

A non-rotating member similar to that of a shaft is called as an axle. It simply supports a rotating body.

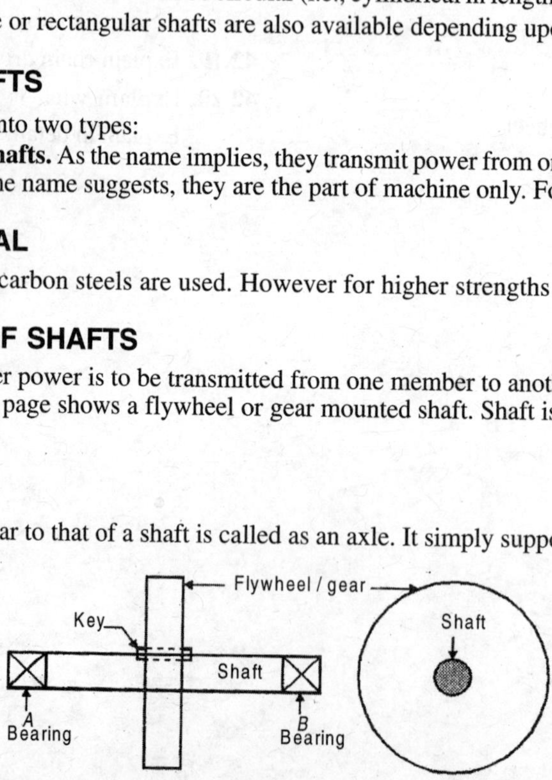

Fig. 43.1. *Flywheel mounted on a shaft*

43.6 APPLICATION

One might have seen the pulley mounted on the axle on the wells, where pulley is rotating, when the rope is pulled and the axle is stationary.

Key is a member used to prevent relative rotational motion between shaft and flange or pulley. It is a piece of metal, rectangular in cross-section. The slots on outer surface of shaft and inner surface of the bore are cut along the length. When these slots are matched they give a cross section same as that of the key. The key is then inserted in this hole which prevents the rotational motion of flange or pulley on the shaft.

43.7 KEYS

Various types of keys are used in practice as shown in Fig. 43.2. Key is metallic piece inserted mating shaft and hole to prevent their relative motion. Pulleys, gears, fly wheels etc. are secured to the shaft by means of keys.

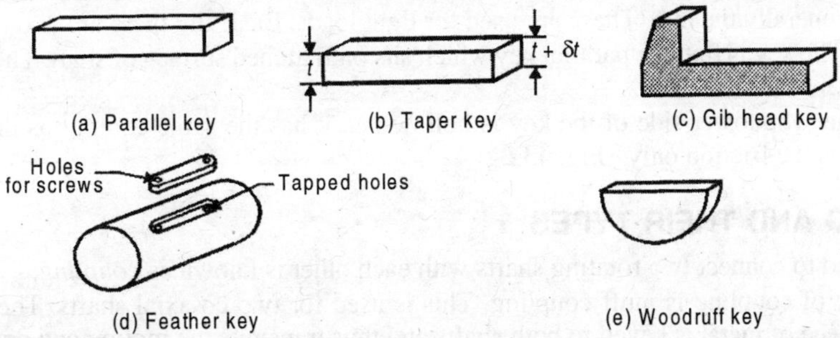

Fig. 43.2

Key way. Keys are inserted in the key ways cut in the mating parts i.e., shaft and hub (or gears, pulleys, fly wheels etc.).

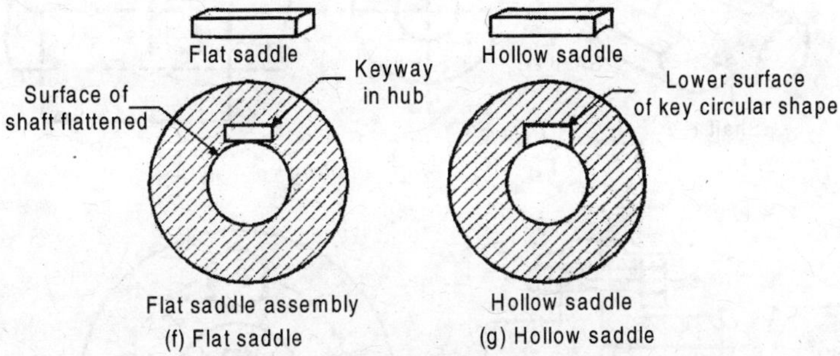

Fig. 43.2

Types of Keys

Main types of keys are :

(a) Sunk Keys. Here the keys are inserted in the key ways such that they are half in the shaft and remaining half in the hub portion. Various types under the sunk key category are:

(i) Parallel key. The key has uniform rectangular or square cross section throughout its entire length. Such keys are used for securing pulleys and gears.

(ii) Taper key. This is similar to the parallel sunk key excepting that it has taper in its thickness along the length. The taper may be 1 in 100. The width is uniform and only the thickness has the taper as shown. If the key has thickness, width and length of say 10, 20 and 100 mm at one end its thickness at the other end will be 9 mm while all other dimensions will remain same if it has a taper of 1 in 100. Due to taper insertion and removal is easy and it also prevents relative circular and axial moment.

(iii) Gib Head key. It is taper sunk key with head at one end. The head makes it easy to remove the key. Such keys are used where the shaft and hub are required to remove frequently.

(iv) Feather key. It is a special type of parallel sunk key and is secured to the shaft by screws as shown. The key allows transmission of turning moment as well as axial force. It is generally used in drilling machines and clutches.

(v) Woodruff key. It has semicircular cross section which partly sits in similar key way in the shaft portion. It can be used to assembled to any shaft or hub with taper.

(vi) Round key. It has a circular cross section and fits into the circular key way in the shaft and hub. There are chances of slip under heavy loads. Therefore such keys are used for light loads.

(b) Saddle keys. Saddle keys are attached to the hub portion and just sit over the shaft. There is key way only in one mating portion generally the hub. These are used for light loads. Basically there are two types:

(i) Flat saddle. The key is flat like parallel key which sits on flattened surface of shaft. There is key way in the hub. (Fig. 43.2f)

(ii) Hollow saddle. The lower side of the key is hollow i.e., it has the same contour as that of the shaft. The holding force is the friction only. (Fig. 43.2g)

43.8 COUPLING AND THEIR TYPES

The arrangement used to connect two rotating shafts with each other is known as *coupling*.

The simplest type of coupling is muff coupling. This is used for two co-axial shafts. The 'muff', which is a hollow cylindrical piece of metal is keyed to both shafts and thus transmits the motion and power.

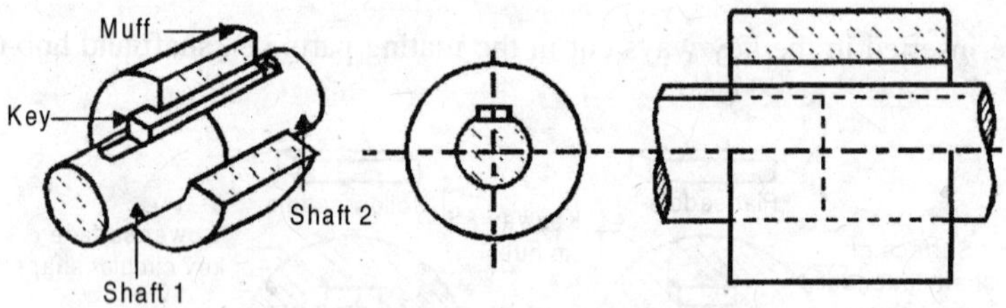

Fig. 43.3. *Muff coupling*

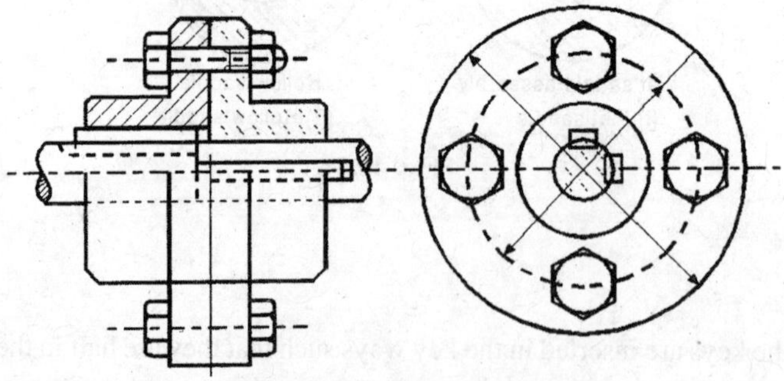

Fig. 43.4. *Flanged coupling*

The muff may be manufactured in one piece or in two halves. In the latter case, coupling is known as split muff coupling.

43.9 FLANGED COUPLING

Flange couplings are used to connect two co-axial shafts.

Flange is a circular disc with a bore at centre. A keyway is cut on the bore along its length. Bore diameter is slightly greater than shaft diameter. Number of holes are drilled on the flange. One flange each is keyed to the shaft, while the flanges are connected to each other with bolts.

The types are:

(1) Unprotected flange coupling – (Ref. Fig. 43.4)

(2) Protected type flange coupling. A projection is provided on the outer surface of each flange. This projection prevents any part of nut, bolt when broken, from moving away, because of centrifugal force and hence prevents the hazards.

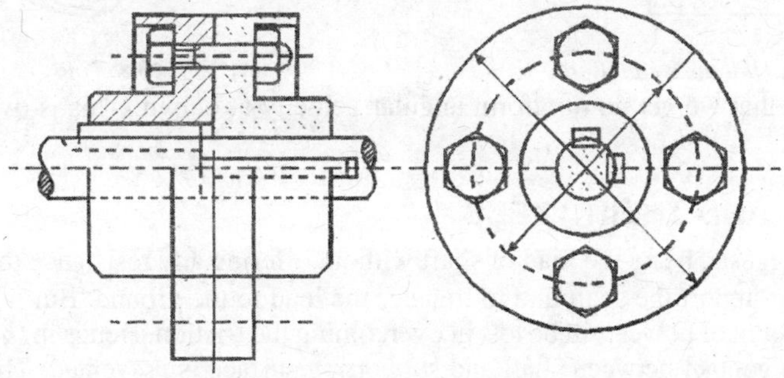

Fig. 43.5. *Protected Flanged Coupling*

(3) Flexible coupling

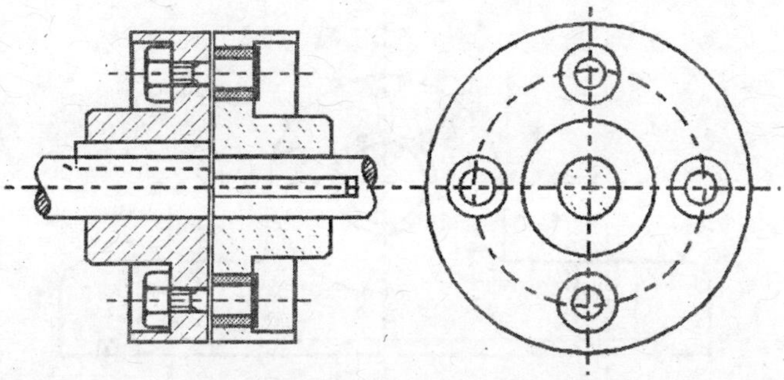

Fig. 43.6. *Protected flanged coupling*

If the load is fluctuating simple flange coupling will transfer these fluctuation to the prime mover. To avoid this, rubber bushes are put in the holes of one of the shaft. These bushes act as shock absorbers.

43.10 OLDHAM'S COUPLING

This is used to connect two parallel shafts with small distance between them. (Ref. Fig. 43.7)

43.11 UNIVERSAL COUPLING OR HOOKE'S JOINT

This type of coupling is used to connect two shafts having interesting axes. This type is widely used in transmitting power from crank shaft to the rear wheel axle in heavy automobiles.

In this type a fork is fixed to end of each shaft. Two forks are connected together with a cross head and pins.

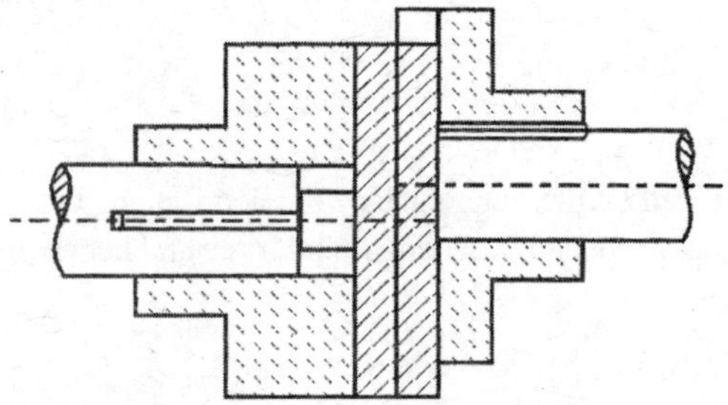

Fig. 43.7. *Oldham's coupling*

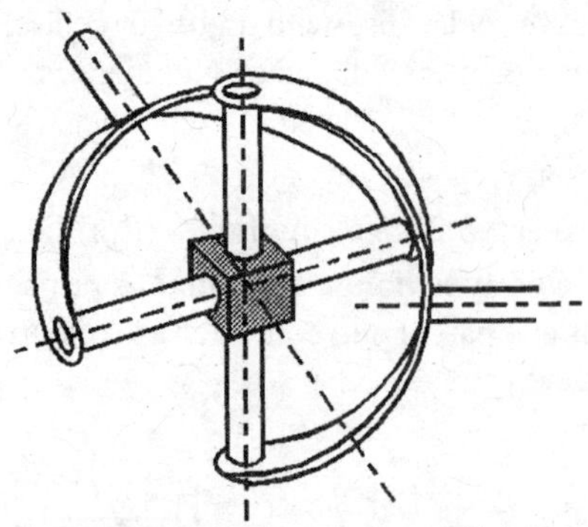

Fig. 43.8. *Universal coupling or hooke's joint*

The disadvantage is that we get non-uniform angular velocity at output. This is overcome by using two couplings in line.

43.12 BEARINGS AND THEIR TYPES

Bearing, as its name suggests, bears the load of shaft without offering any resistance to its rotational motion.

Bearings are used to support the shaft and to transfer the load to the ground. But if the shaft is supported on a rough surface, large amount of power will be lost in overcoming the friction. Hence in '*bearing*' by means of proper '*lubrication*', the direct contact between shaft and supporting surface is prevented. Thus the frictional losses are reduced by very large extent.

Normally two types of bearings are used.

(i) Journal Bearing. In this type the Journal bearing or the '*plummer block*' is lined with special material to reduce the friction. In this, the shaft theoretically makes line contact. Hence this type has high load bearing capacity.

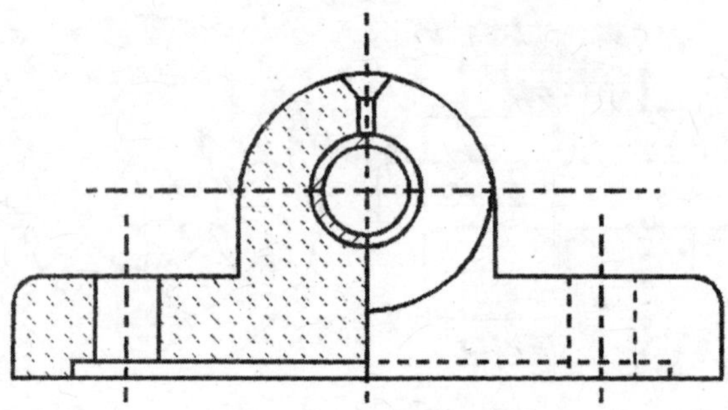

Fig. 43.9. *Journal bearing*

Journal Bearings Applications

Used for shafts for general purpose machinery with medium loads and medium speeds. For example, compressor shafts, overhang transmission shafts, shafts used with group drives in floor mills.

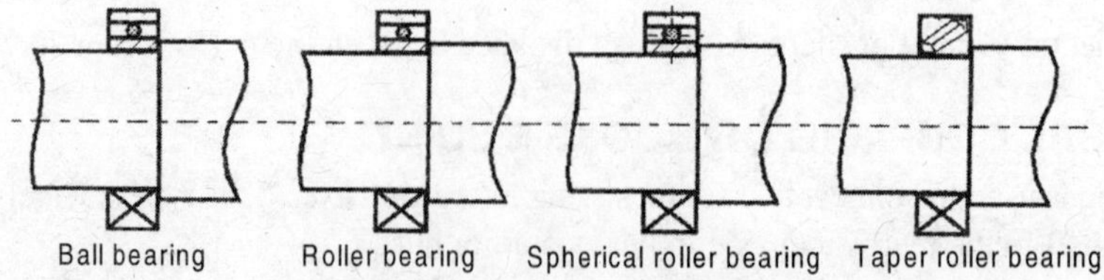

Fig. 43.10. *Rolling contact bearing*

(ii) Rolling Contact Bearings (Fig. 43.10). In this type two concentric rings, known as races are used with balls or rollers in the annular space between them. The inner race is fixed to the shaft while outer race is held stationary in plummer block. The balls or rollers transfer the shaft load to outer race. This type is known as rolling contact bearing. This gives large reduction in frictional losses, but have relatively low load carrying capacity. Hence these are also known as antifriction bearings.

Rolling Contact Bearing Applications

These are used in automobile wheels, or high speed machinery. We can design and select appropriate bearing for a given axial load, radial load and an expected life of bearing. These bearings are safe, as they make a squeaky noise before failing.

Thrust Bearings

There are two types of thrust bearings viz: foot step and collar thrust.

Foot step bearing is used to take the axial load (acting along its length) of the shaft.

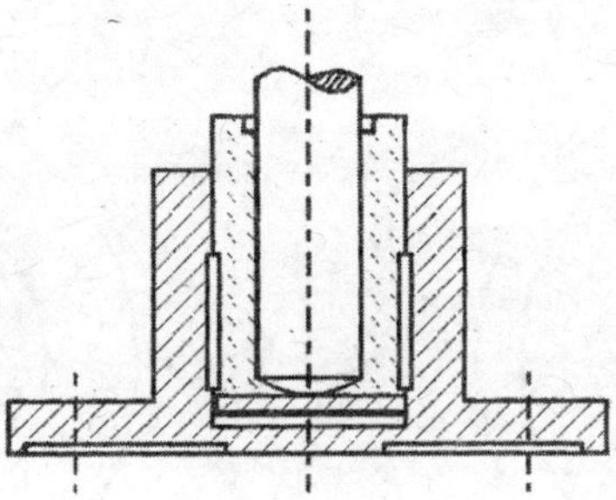

Fig. 43.11

It comprises of three parts. Two parts are identical circular discs having a groove along their diameters on one face. These discs are mounted on the shafts with keys. The third part is central disc. It is also circular disc of same size but it has two projections along its diameter but at right angle to each other on its either side. When these parts are assembled as shown in Fig. 43.11, transfer the rotating motion of one shaft to the other by means of sliding of the projection on central disc in the grooves in the end discs.

43.13 FLYWHEEL

Flywheel stores energy when more energy is being produced and gives back the same energy when less energy is being produced.

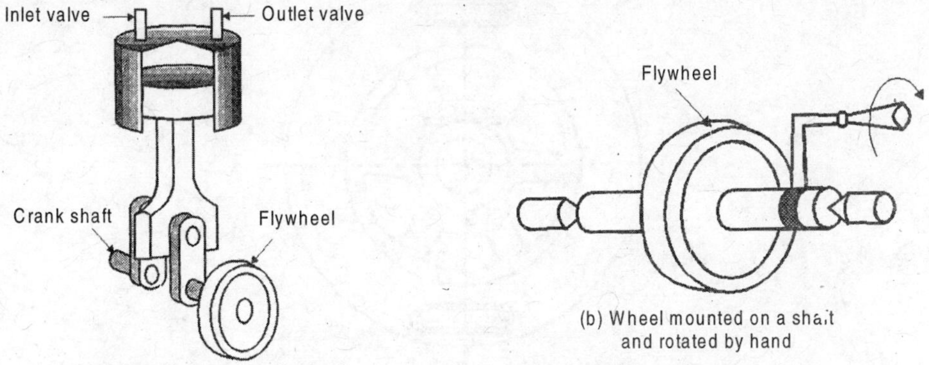

(a) 4-stroke engine with flywheel

(b) Wheel mounted on a shaft and rotated by hand

Fig. 43.12

Or in the other words, "it is an energy reservoir, it stores energy when more energy is produced and gives back when less energy is produced, so that the crank shaft has uniform speed".

If no fly wheel is used then the working of the engine will be jerky.

As we know in case of 4-stroke cycle engines there are four strokes suction, compression, power generation and exhaust. And power is produced only during the power stroke and no power is produced during suction, compression and exhaust strokes.

The energy produced during the power stroke will be stored in the flywheel in the form of inertia and is given back during other strokes the crack shaft so as to have the uniform speed.

As shown in Fig. 43.12(b) a wheel is mounted on the shaft and which is being rotated by hand. If the hand is removed wheel will continue rotating for some time depending upon energy imparted to it.

43.14 FLYWHEEL CONSTRUCTION AND TYPES

Flywheels are made out of castings. Depending upon the diameter and speed they are classed into 3-types.

(i) Smaller flywheel with web and without arms for sizes upto 550–650 mm diameter.

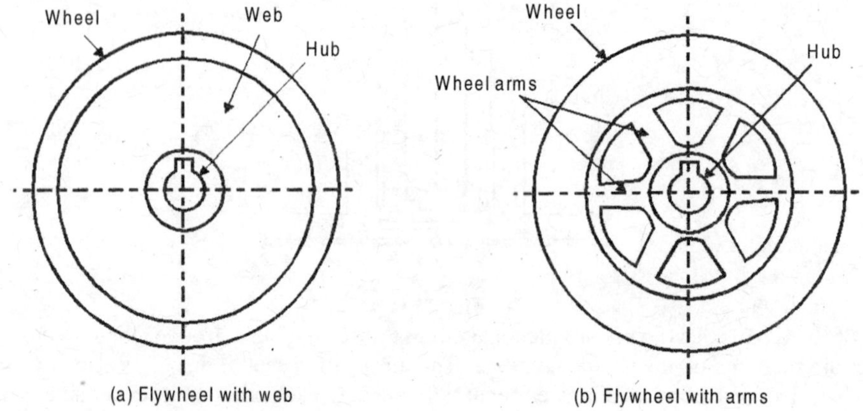

Fig. 43.13

(ii) Medium sized flywheel for sizes upto 2.25 m to 2.50 m dia. and arms provided as shown.

(iii) For sizes above 2.5 m diameter and for higher speeds split type of CI fly wheels are used as shown.

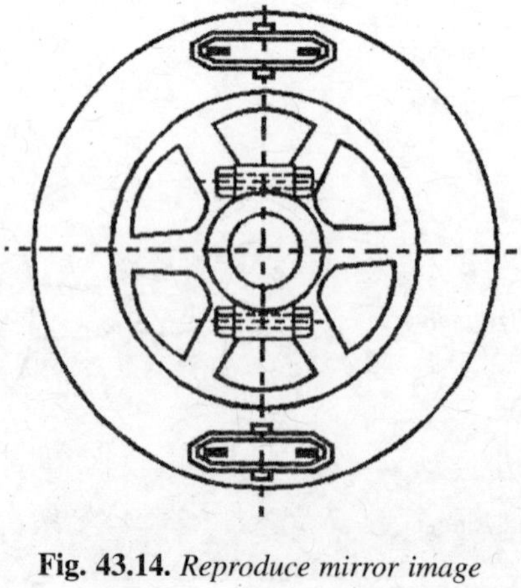

Fig. 43.14. *Reproduce mirror image*

THEORY QUESTIONS

43.1. What is a 'key'. Draw sketches of different types of keys.

43.2. What is a 'coupling'? Draw a sketch of Muff Coupling and Flanged Coupling.

43.3. Explain the difference between a Clutch and Coupling.

43.4. Draw a sketch of flanged coupling.

43.5. Sketch different types of flanged coupling showing their features.

43.6. Draw the sketches of the Helical gear and state their applications.

43.7. Explain with the help of a suitable diagram the construction, working and applications of a bush pin type flexible coupling.

43.8. Compare rigid and flexible coupling.

43.9. What are the different types of shaft couplings ?

43.10. With a neat sketch explain universal coupling.

43.11. What is function of bearing? What are its types.

43.12. Draw a neat sketch of a ball bearing and name its parts. What are its advantages.

43.13. What is a thrust bearing? Draw sketch of its types.

43.14. Explain the different types of bearings used to support shafts.

43.15. Draw neat sketches of:

(a) Bush bearing

(b) Thrust bearing

(c) Sunk and Woodruff key

43.16. State the advantages of rolling contact bearings over sliding contact bearings.

43.17. State the function of the following mechanical element: Bush Bearing.

43.18. Compare bush bearing and antifriction bearings.

43.19. State where Thrustbearing is used.

43.20. State the advantages of ball and roller bearings over plain journal bearings.

43.21. What is a 'key'. Draw sketches of different types of keys.

43.22. What is a 'coupling'? Draw a sketch of Muff Coupling and Flanged Coupling.

43.23. Explain the difference between a Clutch and Coupling.

43.24. Draw a sketch of flanged coupling.

43.25. Sketch different types of flanged coupling showing their features.

43.26. Explain with the help of a suitable diagram the construction, working and applications of a bush pin type flexible coupling.

43.27. Compare rigid and flexible coupling.

43.28. What are the different types of shaft couplings?

43.29. With a neat sketch explain universal coupling?

43.30. What is function of bearing? What are its types?

43.31. Draw a neat sketch of a ball bearing and name its parts. What are its advantages?

43.32. What is a thrust bearing? Draw sketch of its types.

43.33. Explain the different types of bearings used to support shafts.

43.34. Draw neat sketches of:

(a) Bush bearing

(b) Thrust bearing

(c) Sunk and Woodruff key

43.35. State the advantages of rolling contact bearings over sliding contact bearings.

43.36. State the function of the following mechanical element: Bush Bearing.

43.37. Compare bush bearing and antifriction bearings.

43.38. State where Thrustbearing is used.

43.39. State the advantages of ball and roller bearings over plain journal bearings.

CHAPTER 44

Introduction to Mechatronics

In Focus

Introduction, Definition of Mechatronics, System of Mechatronics, Measurement Systems and Control Systems — Open Loop Control System and Close Loop Control System (with simple block diagrams) their advantages, disadvantages. Introduction to microprocessor based controller.

44.1 INTRODUCTION AND DEFINITION OF MECHATRONICS

In recent years, the application of digital microelectronics computers in design and manufacturing sectors has considerably improved productivity and quality of many mechanical products. Automation and control methods, adopting integrated embedded technology has become relevant to industries machinories and consumer products. Automation and control represents a broad topic with application in many disciplines, including production, industrial and manufacturing engineering, process control, robotics, instrumentation, home automation and many others based upon sophistication, flexibility and state-of-the art technology. Mechatronics, an enabling discipline has already emerged to cater to the need for sophistication and flexibility and in fact has become a preferred choice for the current generation of real time automation and control solutions for traditional mechanical systems.

The term 'Mechatronics' was introduced in the late 1960's by Japan Yasakawa Electric Company and was derived from the synergistic combination of mechanical and electrical/electronic technologies.

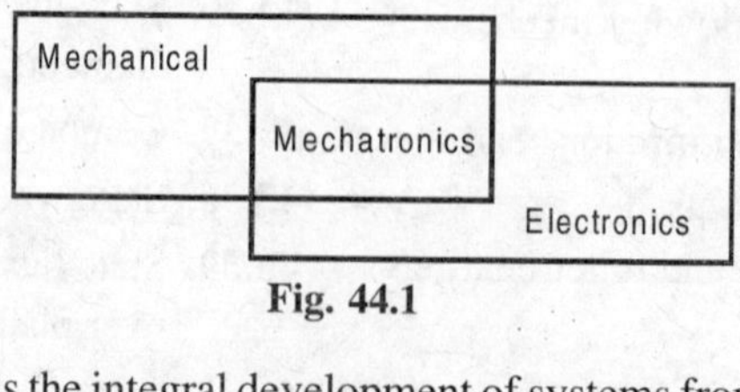

Fig. 44.1

The central focus of Mechatronics is the integral development of systems from technical components ("Mecha") which are to be intelligently controlled ("tronic"). Thus a system composed of mechanical and electrical parts, overlaid with sensors which record information, microprocessors which interpret process and analyze the information and assemblies which then react upon this information, this then becomes a complete mechatronics system.

Mechatronics is the term used for the integration of mechanical and electronic engineering disciplines to achieve the objective of developing efficient systems controlled intelligently.

44.2 DEFINITION OF MECHATRONICS

A definition of mechatronics has been adopted within the EEC as; Mechatronics is the synergetic combination of precision mechanical engineering, electrical control and systems required in the design of products and processes.

The interdisciplinary field of engineering dealing with the design of products whose function relies on the synergistic integration of mechanical and electronics components coordinated by control architecture.

44.3 SYSTEM OF MECHATRONICS

Mechatronics involves, what are termed systems. A system can be thought of as a box which has an input and an output and where we are not concerned with what goes on inside the box but only the relationship between the output and the input. Thus, for example, a motor may be thought of as a system which has as its input electric power and as output the rotation of a shaft.

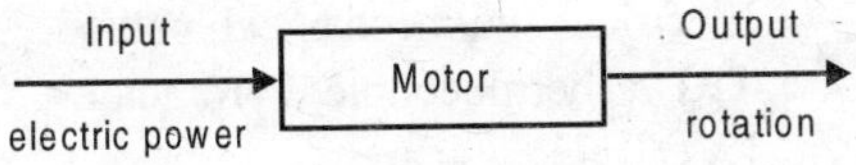

Fig. 44.2

A measurement system can be thought of as a black box which is used for making measurements. It has as its input the quantity being measured and its output the value of that quantity. For example, a temperature measurement system i.e., a thermometer has an input of temperature and an output of a number on a scale.

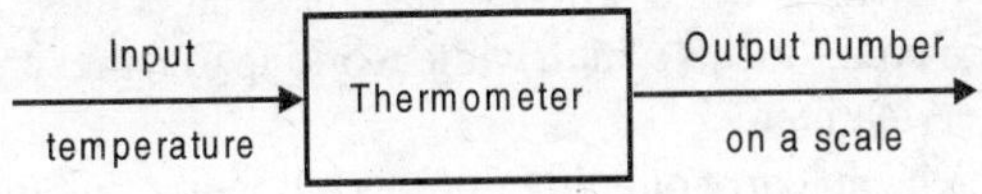

Fig. 44.3. *An example of a measurement system*

A control system can be thought of as a black box which is used to control its output to some particular value or particular sequence of values. For example, a domestic central heating control system has as its input the temperature required in the house and as its output the house at that temperature i.e.. you set the required temperature on the thermostat or controller and the heating furnace adjusts itself to pump water through radiators and so produce the required temperature in the house. Fig. 44.4.

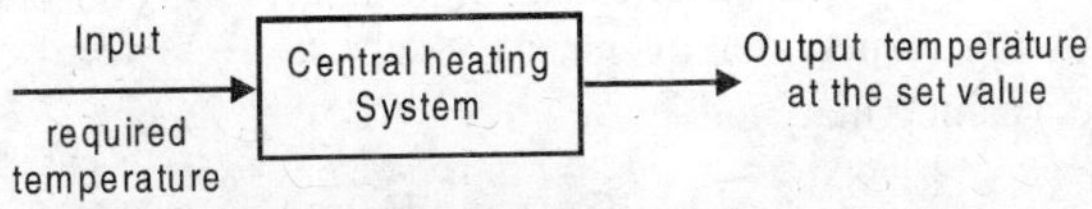

Fig. 44.4. *An example of a control system*

44.4 MEASUREMENT SYSTEMS

Measurement system is an essential part or a foundation of major mechatronics system. Measurement system consists of basically three parts namely sensors or transducer, signal processor and display system.

Consider an example of temperature measurement. Sensor used for temperature is Thermocouple and the input to the sensor is a temperature and it converts temperature to a voltage.

The signal processor is used to give the output from the thermocouple as a e.m.f., so an amplifier is used to amplify the voltage and a convertor is used to convert the analog signal to a digital signal.

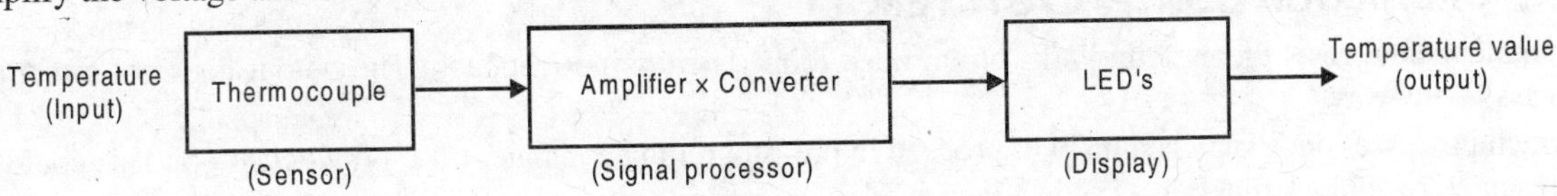

Fig. 44.5

The output from the signal processor is displayed on the LED's on a screen as digital value of a temperature as shown above.

44.5 DIFFERENCE BETWEEN TRANSDUCER AND SENSOR

The term transducer and sensor have been used synonymously although the concepts are different. Transducers are the physical element which is a part of a sensor. In fact, transducer is an essential element of a sensor. A sensor is merely a sophisticated transducer in the sense that it contains some signal conditioning circuits capable of amplifying and refining the weak and raw signal that is available at the output of the transducer. Some of the commonly used signal conditioning circuits are amplifiers, filters, Analog to Digital Converter (ADC) etc. Example of measurement system are:

Flow measurement	—	Rotameter, anemometer etc.
Pressure measurement	—	(Mcleodgauge)
Level measurement	—	Electrical and mechanical method
Temperature measurement	—	TRD, Thermocouple, Pyrometer
Displacement measurement	—	LVDT etc.

44.6 CONTROL SYSTEMS

The word control is very familiar to all of us. But still we could see what it actually means during our current study. Control means achieving certain desired results or completing certain task in a desired manner.

System means a group of devices which while working together can achieve certain function. Thus 'control system' refers to a system or group of components which while working together can achieve certain desired results or complete certain tasks in a desired manner.

A control system is a system which can carry out either of the following functions.

1. For particular input(s) the system output can be controlled to some desired particular value.
2. If certain conditions are satisfied, it can give a particular sequence of events or event as output, corresponding to given input(s).

An example of the above mentioned first type of control system is the household central heating system. In this case the controlled variable is the temperature in the house and the control system is designed to operate a furnace so that the required temperature is attained.

An example of second type of above mentioned control system is the household washing machine. Here the controller determines the following sequence of events:

1. Pumping water into the washing machine
2. Heating pumped water
3. Spinning water for a good wash of clothes
4. Emptying the washing machine after the wash etc.

The function of the controller in this case is to control the sequence of operations determining when particular events in the sequence are to occur as a result of certain conditions being met *e.g.* switching off the heater for the water when the desired right temperature is attained and then switching on next event.

The basic two types of control systems that are used are as follows:

1. Open-loop control system
2. Closed-loop control system.

44.7 OPEN-LOOP CONTROL SYSTEM

As the name suggests the system works on an open loop. There is no feedback. The output does not affect the process in any way.

Such processes once switched on, will proceed in pre-programmed manner till it is switched off. This could be represented by a block diagram as shown in Fig. 44.

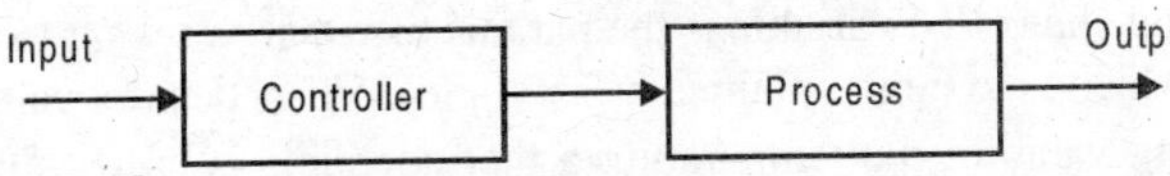

Fig. 44.6

We could see that the output does not affect the process in any way. For example, the washing machine belongs to this type of control. In the washing machine all the processes and operation are controlled by a Microcontroller.

We give the input in the form of a detergent and the clothes. Also the settings are given depending upon the type of cloth. Thereafter, we cannot have any control over the washing.

All process required for the washing will proceed as per the setting we have selected. The microcontroller will ensure that all the processes proceed in the correct sequence, correct speed and for the correct time. The output is the quality of washed cloth. This we will come to know only when we take out the clothes after all process are over.

This type of control or open loop control can be used only when the operation is predictable. That is, for a given input condition, the output conditions are predictable and known.

There is not going to be any frequent variation in the output which needs monitoring or in other words the output need not be controlled very accurately.

The washing machine e.g., is one which satisfies these conditions. In this, the quality of the washed clothes is the output. From experience, we know that if we carry out the processes of agitating the clothes for certain time and do the rinsing and washing certain acceptable quality will be achieved for the clothes.

Also depending upon the type of clothing or the condition of clothing, the parameter can be changed. Thus, the output is predetermined.

44.8 CLOSED-LOOP SYSTEM

This is also known as a feedback control system in figure. If we compare it with the open loop-system, we could see that the loop is closed by a providing a feedback of the output through a measuring element.

- If the output changes, the process also changes. That means the process is dependent on the output. This is the main difference between an open-loop system and a closed-loop system.
- Functioning of a closed-loop system could be understood very clearly by taking an example of temperature control system. The temperature of a space is to be maintained at 75°C and is not allowed to go up or down.
- If it goes up or down immediately some action is taken. If the temperature goes up a cooler is switched on so that temperature comes down back to 75°C. If it goes down a heater is switched on so that the temperature comes up do 75°C again.
- This means that decision should be taken to switch on or off the cooler or heater. This decision may be taken manually and the switching on or off is also done manually then such systems are called manual control system.
- But if the decision is given automatically, it is called automatic control system. We shall concentrate our discussion in respect of automatic control system only.
- To achieve this control we need to know the present value of the temperature at any instant. We should compare this value with the desired value. Depending upon the difference between two we take the decision whether to switch on the cooler or the heater. The whole process could be understood with help of a block diagram in Fig. 44.7.

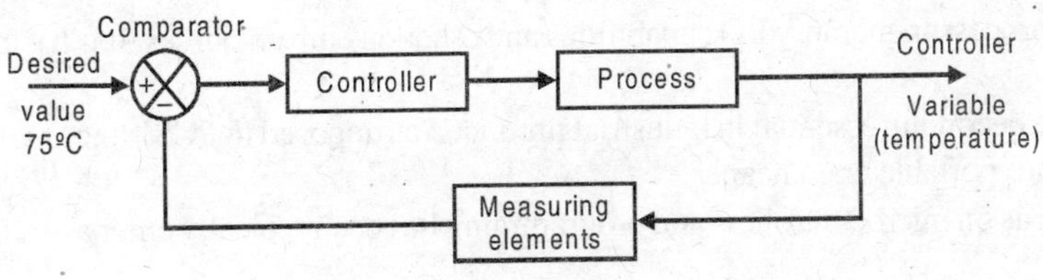

Fig. 44.7

- In the block diagram the controlled variable is the parameter which we want to control. In this case it is the temperature of room which is to be maintained at 75°C. So 75°C is desired value.
- To measure the existing value of the temperature a thermocouple is used as the measuring element. The thermocouple will constantly keeps on measuring the temperature at all time. The measured value of temperature will be then compared with the desired value of 75°C.
- These are three possibilities of this value
  (a) Temperature is exactly equal to 75°C
  (b) Temperature is more than 75°C
  (c) Temperature is less than 75°C

Based on which is the case, the decision will be taken by controller as under:

1. If the temperature is exactly equal to 75°C then, no action required. Everything will remain as it is.
2. If the temperature is more than 75°C, that means the temperature needs to be brought down. So that controller will send a signal to switch on the cooler. The cooler will start reducing the temperature is still above 75°C, the cooler will continue to remain. After some time, let us say that the temperature becomes equal to 75°C, the action stops now and the cooler is switched off.
3. If temperature is less than 75°C, the controller will send a signal to switch on the heater. Once the heater is on the temperature will start rising. Again the thermocouple will switched off and no action will be taken by the controller.

44.9 ADVANTAGES AND DISADVANTAGES OF CLOSED-LOOP SYSTEM AND OPEN-LOOP SYSTEM

Advantages of Closed-loop System

Since the output is constantly measured and compared with the desired value. The output can be very accurately maintained at desired value. Hence the effect of disturbances could be taken care of.

Disadvantages of Closed-loop System

- The major problem with closed loop system is the cost. This is mainly due to the more number of component. Also the power consumption will be more in closed-loop system.
- Yet another drawback is the problem of stability. A faculty design of a closed-loop system can lead to instability.

Advantages of Open-loop System

1. Open-loop system are simple in construction and maintenance.
2. Also it is less expensive and has less number of components
3. No stability problem in open-loop system.

Drawback of Open-loop System

Open-loop system are not suitable where frequent disturbances takes place which change the output frequently.

44.10 MICROCONTROLLERS

Microcontroller, which is a single IC containing specialised circuits and functions that are applicable to mechatronics system design.

It contains a microprocessor, memory I/O capabilities and other on chip resources. It is basically a microcomputer on single IC.

Microcontroller are becoming essential in industrial products (as opposed to PCS) since they provide functionally at lower costs. They are portable and cheaper.

This intense emphasis on microcontrollers and programming has been typically emphasised in most mechatronics system.

Authors References

A Compact High Intensity Combustion System. The Oil Engine and Gas Turbine, 1955.
Air Compressor Hand Book, Caterpillar Tractor Co. 1969.

- **Ashrae**
  Handbook of Fundamentals, 1986 (SI Units).
  Book of Car
- **Cengel**
  Introduction to Thermodynamics & Heat Transfer.
- **Chruch E.F.**
  Steam Turbines
- **E.F. Obert**
  I.C. Engines, International Text Book Co.
- **E.H. Lewitt**
  Thermodynamics Applied to Heat Engines, Sir Isaac Pitman & Sons Ltd.
- **Faires/ Simmong**
  Thermodynamics.
- **Hill P.G. & Peterson C.R.**
  Mechanics and Thermodynamics of Propulsion, Addison Weslay, New York 1965.
- **Holman & White**
  Heat Transfer.
- **Holman J.P.**
  Thermodynamics.
- **J.F. Lee & F.W. Sears**
  Thermodynamics, Addison Weslay Publishing Co.

- **James L. Threlkeld**
  Thermal Environment Engineering.
- **Jeenings B.H. & Roger W.C.**
  Gas Turbines Analysis & Practice.
- **Jordon & Priester**
  Refrigeration & Air Conditioning.
- **Judge A.W.**
  Modern Gas Turbine, Chaepman & Hall Publishing.
- **K.K. Ramalingam**
  I.C. Engines.
- **K.M. Gupta**
  Automobile Engineering Vol. 1 & 2.
- **Kearton**
  Steam Turbine Theory and Practice.
- **Keenan J.G.**
  Elemental Theory of Gas Turbines & Jet Propulsion.
- **Kreith & M.S. Bohn**
  Principles of Heat, 1986 Edition.
- **Longton N.H.**
  Rocket Propulsion, University of London, Press 1970.
- **Maxwell W.H. Holt**
  Thrust into Space, Rinehart & Winston Inc. New York, 1965.
- **R.C Patel & V.C. Bhavsar**
  Steam Turbine Plants.
- **Rayner Joel**
  Basic Engineering Thermodynamics.
- **S.V. Sukhatme**
  Heat Transfer, Orient Longman Publishers, 1993.
- **Turcotte D.L.**
  Space Propulsion, Blaisdell New York 1965.
- **W.P. Stoecker**
  Refrigeration & Air Conditioning.
- **Yahya S.M.**
  Turbomachines.

Index

D

E

F

G

H

I

J

K

L

M

N

O

P

Q

R

S

T

U

V

W

Z